THE LIPID HANDBOOK

THE LIPID HANDBOOK

SECOND EDITION

FRANK D. GUNSTONE
University of St Andrews, Scotland
JOHN L. HARWOOD
University of Wales, Cardiff, Wales
FRED B. PADLEY
Unilever Research, Sharnbrook, England

with sections contributed
by an international team
of lipid researchers

CHAPMAN & HALL
Electronic Publishing Division
London · Glasgow · Weinheim · New York · Tokyo · Melbourne · Madras

Published by Chapman & Hall, 2–6 Boundary Row, London SE1 8HN

Chapman & Hall, 2–6 Boundary Row, London SE1 8HN, UK

Blackie Academic & Professional, Wester Cleddens Road, Bishopbriggs, Glasgow G64 2NZ, UK

Chapman & Hall GmbH, Pappelallee 3, 69469 Weinheim, Germany

Chapman & Hall USA, One Penn Plaza, 41st Floor, New York NY 10119, USA

Chapman & Hall Japan, ITP-Japan, Kyowa Building, 3F, 2-2-1 Hirakawacho, Chiyoda-ku, Tokyo 102, Japan

Chapman & Hall Australia, Thomas Nelson Australia, 102 Dodds Street, South Melbourne, Victoria 3205, Australia

Chapman & Hall India, R. Seshadri, 32 Second Main Road, CIT East, Madras 600 035, India

First edition 1986

Second edition 1994

Reprinted 1995

Typeset in Great Britain by EJS Chemical Composition, Midsomer Norton, Bath, Avon

Printed in Great Britain at the University Press, Cambridge

ISBN 0 412 43320 6

A catalogue record for this book is available from the British Library

Library of Congress Catalog Card Number: 94-70866

∞ Printed on acid-free text paper, manufactured in accordance with ANSI/NISO Z39.48-1992 (Permanence of Paper)

CONTENTS

CONTRIBUTORS

G. Biernoth
Unilever Research Laboratory, Vlaardingen, The Netherlands

D. Chapman
Department of Biochemistry and Chemistry, Royal Free Hospital School of Medicine, London, UK

A. Cryer
Department of Biochemistry, University of Wales, Cardiff, UK

N.G.J. Davidson
Unilever Research, Port Sunlight, Wirral, Merseyside, UK

P. Dodds
Department of Biochemistry, Wye College (University of London), Ashford, Kent, UK

F.M. Goñi
Department of Biochemistry, Faculty of Science, University of Pais Vasco, Bilbao, Spain

F.D. Gunstone
School of Chemistry, University of St Andrews, Fife, UK

M.I. Gurr
Maypole Scientific Services, Maypole, St Mary's, Isles of Scilly, UK

J.L. Harwood
Department of Biochemistry, University of Wales, Cardiff, UK

P.W. Hendrikse
Unilever Research Laboratory, Vlaardingen, The Netherlands

M. Kates
Department of Biochemistry, University of Ottawa, Ottawa, Canada

N. Krog
Grindsted Products A/S, Braband, Denmark

K. Larsson
Chemical Center, University of Lund, Lund, Sweden

F.B. Padley
Unilever Research Laboratory, Sharnbrook, Bedford, UK

C. Poot
Unilever Research Laboratory, Vlaardingen, The Netherlands

P.J. Quinn
Department of Biochemistry and Biophysics, Kings College, University of London, London, UK

F.V.K. Young
Vernon Young Consultant Ltd, Formby, Liverpool, UK

PREFACE

The first edition of *The Lipid Handbook* was drafted during 1984 and 1985, and published in 1986. It was well received by reviewers and readers, and frequently appears as a cited reference. This suggests that it has been widely purchased and used, and this is borne out by the publication records. Late in 1991, aware of new developments in several aspects of lipid technology and lipid biochemistry, we agreed with our publisher that we should prepare a second edition of the *Handbook*, and writing was undertaken and completed during 1992 and 1993.

We have retained the 12 chapters of the first edition, but some new sections have been added. Among these is completely new material on: minor oils and fats, and 'gourmet oils'; data arising from plant breeding to produce oils with radically different fatty acid composition; algal and viral lipids; biological membranes (by P.J. Quinn); paint; important changes from dispersive to interferometric spectrometers that have revolutionized infrared spectroscopy; ^{13}C nuclear magnetic resonance spectroscopy; herbicides that affect acyl lipid metabolism; and xenobiotic lipids (by P. Dodds). Other sections have been extensively reorganized and rewritten, such as section 6.3 on other analytical procedures (i.e. non-standard methods). This has been prepared for us by M. Kates. The remainder of the *Handbook* has been extended, thoroughly revised and updated, and we expect this volume to be a worthy successor to the first edition. Typical of the greater amount of information is the increase in the number of tables in Chapter 3 devoted to the occurrence and characteristics of oils and fats.

We are grateful to our co-authors from the first edition who have helped us again, and to those who have newly joined the writing team. Their names and addresses are listed with ours in the List of Contributors.

The Dictionary Section of the *Handbook* has also grown in size. This material has been selected from the *Dictionary of Organic Compounds* (5th edition) and its 10 Supplements. With the growth of the chemical database, this section now contains a larger number of entries and also many updates derived from the Supplements, and is a unique feature of the book. It contains a lot of useful information and often provides quick access to the original literature. We are grateful to Dr Hilary Bradley at Chapman & Hall for making the Dictionary Section material available as part of the *Handbook*, and for checking the data included.

F.D Gunstone
J.L. Harwood
F.B. Padley
1994

PREFACE TO THE FIRST EDITION

For the purpose of this *Handbook*, lipids are taken to include long-chain fatty acids, closely related compounds such as the alcohols and the natural derivatives of these substances. Sterols are included only to a minor degree. Even with this restriction we are concerned with a wide range of important compounds of interest to chemists, biochemists, physicists, physiologists, technologists and clinicians. These materials are used in foods and to a lesser extent as oleochemicals. They are important to those engaged in their production and distribution, and to those who trade in them whether physically or in the futures market.

We have tried to produce a reference work which will be of value to those who are interested in lipids. Chapters 1 and 2 deal with the chemical structure of the fatty acids and the lipids derived from them. The long Chapter 3 covers the occurrence, composition and characteristics of a wide range of lipids with emphasis on those vegetable and animal fats of commercial importance. After a short chapter on separation and isolation procedures appropriate mainly in the laboratory, Chapter 5 is devoted to the processing of fats and oils at the commercial level. It covers the production of edible oils from natural sources, their modification by hydrogenation, interesterification or fractionation, and the production of margarine and butter, soap, other surfactants and emulsifiers, and paints and varnishes. Chapter 6 is devoted to analytical methods – always important for those who work with lipids. The major techniques are presented in sufficient detail to permit them to be undertaken. Chapter 7 describes synthetic procedures used mainly in the laboratory. Chapters 8 and 9 cover the increasingly important topic of the physical properties of lipids: Chapter 8 deals with crystalline form and thermal properties and Chapter 9 with spectroscopic properties. In Chapter 10 the chemical reactions of lipids are described. These are concerned mainly with olefinic centres in unsaturated acids or with the carboxyl group, usually in ester form. Chapter 11 deals with the metabolism of the fatty acids themselves and of various types of lipids and Chapter 12 covers the more important medical and agricultural aspects of lipids.

Unable ourselves to cover adequately this range of topics, we have secured the help of a number of colleagues who have written about areas of lipid science in which they are especially competent. We acknowledge their valuable assistance without which this publication would have been much poorer. Their names are listed with ours in the List of Contributors.

The Dictionary Section of the *Handbook* consists of entries selected from the *Dictionary of Organic Compounds* (5th Edition) and its first two Supplements along with appropriate indexes. We believe that this will be a valuable and much-used part of the Handbook and we thank the editors and publishers of the *Dictionary* for allowing us to use it in this way. The *Dictionary* itself with Supplements currently contains about 60 000 entries and there may be items in the *Dictionary* of interest to lipid scientists which we have overlooked.

We thank Dr John Buckingham of Chapman and Hall for his help in planning this work and in converting our typescript to this *Handbook*.

F.D. Gunstone
J.L. Harwood
F.B. Padley
February 1986

1

FATTY ACID STRUCTURE

F.D. GUNSTONE

1.1 Saturated acids

Most of the natural saturated (alkanoic) acids have unbranched structures with an even number of carbon atoms. Acids from C_2 to C_{30} (or longer) have been reported, but the most common and the most important fall in the range C_{12}–C_{22}. These may be designated by trivial or systematic names or by a

Table 1.1 Names and selected physical properties of the C_1–C_{30} alkanoic acids (Adapted from Gunstone, 1967, and Eissler and Hagemann, 1979)

Chain length	Systematic name	Trivial name	Acid M.p. (°C)	Acid B.p.[a] (°C)	Methyl ester M.p. (°C)	Methyl ester B.p.[a] (°C)	d_4^{20} [b]	ΔH_f (kJ/mol)	ΔH_s [c] (kJ/mol)
1	Methanoic	Formic	8.4	101	–	32	–	–	–
2	Ethanoic	Acetic	16.6	118	–	57	1.0492	–	–
3	Propanoic	Propionic	−20.8	141	–	80	0.9942	–	–
4	Butanoic	Butyric	−5.3	164	–	103	0.9583	–	–
5	Pentanoic	Valeric	−34.5	186	−80.7	127	0.9394	14.1	–
6	Hexanoic	Caproic	−3.2	206	−69.6	151	0.9276	16.1	–
7	Hepanoic	Enanthic	−7.5	223	−55.7	174	0.9181	15.0	–
8	Octanoic	Caprylic	16.5	240	−36.7	195	0.9105	21.4	–
9	Nonanoic	Pelargonic	12.5	256	−34.3	214	0.9052	20.3	–
10	Decanoic	Capric	31.6	271	−12.8	228	0.8858 (40)	28.0	118.8 (25)
11	Undecanoic	–	29.3	284	−11.3	250	0.8505 (80)	25.1	121.3 (25)
12	Dodecanoic	Lauric	44.8	130[1]	5.1	262	0.8477 (80)	36.6	132.6 (25)
13	Tridecanoic	–	41.8	140[1]	5.8	–	–	–	–
14	Tetradecanoic	Myristic	54.4	149[1]	19.1	114[1]	0.8439 (80)	44.8	141.0 (39)
15	Pentadecanoic	–	52.5	158[1]	19.1	127[1]	–	–	–
16	Hexadecanoic	Palmitic	62.9	167[1]	30.7	136[1]	0.8414 (80)	54.4	153.6 (57–60)
17	Heptadecanoic	Margaric	61.3	175[1]	29.7	148[1]	0.8355 (91)	51.0	–
18	Octadecanoic	Stearic	70.1	184[1]	37.8	156[1]	0.8390 (80)	63.2	166.1 (58 60)
19	Nonadecanoic	–	69.4	–	38.5	191[1]	–	–	–
20	Icosanoic	Arachidic	76.1	204[1]	46.4	188[2]	–	–	–
21	Henicosanoic	–	75.2	–	–	207[4]	–	–	–
22	Docosanoic	Behenic	80.0	–	51.8	206[2]	–	–	–
23	Tricosanoic	–	79.6	–	53.9	–	–	–	–
24	Tetracosanoic	Lignoceric	84.2	–	57.4	222[2]	–	–	–
25	Pentacosanoic	–	83.5	–	59.5	–	–	–	–
26	Hexacosanoic	Cerotic	87.8	–	63.5	237[2]	–	–	–
27	Heptacosanoic	Carboceric	87.6	–	64.6	–	–	–	–
28	Octacosanoic	Montanic	90.9	–	67.5	–	–	–	–
29	Nonacosanoic	–	90.4	–	68.8	–	–	–	–
30	Triacontanoic	Melissic	93.6	–	71.5	–	–	–	–

[a] B.p. at 760 mmHg or at 1, 2 or 4 mmHg as indicated by superscript.
[b] Measured at 20 °C unless otherwise indicated.
[c] Temperature of observation in parentheses.

numerical symbol such as 16:0, which indicates the unbranched C_{16} acid with no unsaturation. These names are given in Table 1.1, along with a range of physical properties: melting point (m.p.), boiling point (b.p.), density at 20 °C compared with water at 4 °C (d_4^{20}), heat of formation (ΔH_f) and heat of sublimation (ΔH_s).

Many physical properties associated with the solid state show alternation. For example, the melting points of acids with an even number of carbon atoms lie on a smooth curve, which falls above a similar curve for acids with an odd number of carbon atoms. Thus the melting points rise and fall (alternate) with increasing chain length. Some other physical properties of these saturated acids and their derivatives are discussed in section 8.5 (solid-state behaviour, melting point, heat of fusion, vapour pressure, boiling point, heat of vaporization, density, specific volume, viscosity, refractive index, solubility, surface tension, dielectric constant and other electrical properties. Some recent papers cover critical properties (Fisher, 1989), heats of combustion (Freedman and Bagby, 1989) and viscosity and density (Liew *et al.*, 1991).

The natural occurrence of these acids is discussed in sections 3.2.1 to 3.2.3. Where there is an appropriate source, these compounds can be isolated in a pure state. They may also be obtained from natural sources by hydrogenation of a more readily available olefinic acid, e.g. docosanoic acid (22:0) from erucic acid (22:1). Otherwise they are usually obtained from a readily available homologue. Chain extension by one, two, four, five or six carbon atoms is fairly easily achieved.

References and further reading

Eissler, R.L. and Hagemann, J.W. (1979) in *Fatty Acids* (ed. E.H. Pryde), American Oil Chemists' Society, Champaign, p. 173.

Fisher, C.H. (1989) *J. Am. Oil Chem. Soc.*, **66**, 1158.

Freedman, B. and Bagby, M.O. (1989) *J. Am. Oil Chem. Soc.*, **66**, 1601.

Gunstone, F.D. (1967) *An Introduction to the Chemistry and Biochemistry of Fatty Acids and Their Glycerides*, Chapman & Hall, London, p. 54.

Liew, K.Y., Seng, C.E. and Lau, E.K. (1991) *J. Am. Oil Chem. Soc.*, **68**, 488.

1.2 Monoene acids

Over a hundred naturally occurring monoene acids have been identified, but most of these are very rare compounds. Table 1.2 indicates the structure of over 30 monoene acids ranging in chain length from C_{10} to C_{30}. The more important of these are discussed below. In general, these acids have an even number of carbon atoms, and the double bond has *cis* or *Z* configuration and is most likely to start on the ninth carbon atom from the carboxyl group (Δ9) or from the methyl group ($n-9$).

The 9*c* isomer is the most widely occurring hexadecenoic acid. It is a minor component (less than 1 or 2%) in most animal and vegetable fats, is rather more significant in many fish oils (around 10%) and is a major component in a few seed oils (e.g. macadamia oil, *ca* 22%).

Oleic acid (**1**) may be formulated as shown. It is the most widely distributed and among the most extensively produced of all fatty acids and is the prototype of all the monoene members. Biosynthetically, it serves as the precursor of longer-chain $n-9$ monoene acids (Table 1.2) and of a family of $n-9$ polyene acids (section 1.3). Olive oil is the most readily available oil with a high content of oleic acid (72 ± 3%). Other useful sources include a range of nuts (filbert, 77 ± 3%; almond, 68 ± 3%; pistachio, 67 ± 5%; pecan, 60 ± 7%; macadamia, 57 ± 2%; and cashew, 56 ± 3%) and a valuable new palm from the Amazon region, *Jessenia bataua* (78 ± 3%). Some of these oils are available as specialty (exotic, gourmet) oils. High-oleic-acid varieties of sunflower and

$CH_3[CH_2]_7CH{=}CH[CH_2]_7COOH$

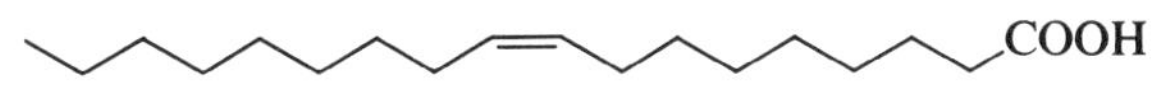

(**1**) Oleic acid (9*c*- or 9*Z*- or ($n-9$)-18:1)

safflower oils containing 75–80% of oleic acid are also available.

Elaidic acid (9*t*-18:1) is easily made from oleic acid (section 10.9) but is very rare among natural acids. Vaccenic acid (11*t*-18:1), present in small quantities in depot and milk fats of ruminant animals, is probably produced from dietary linoleic acid by biohydrogenation occurring in the rumen. The *cis* isomer is a major bacterial fatty acid (section 3.14). It also occurs as a minor component in many seed oils, and as a major component in a few rare examples.

Petroselinic acid (6*c*-18:1), the only other octadecenoic acid of significance, is present as a major component (often above 50%) in many species of Umbelliferae, Araliaceae and Garryaceae. Petroselinic acid-containing oils are being examined as potential new crops (section 3.27).

Among C_{20} monoene acids the 9*c* and 11*c* acids are frequently present in fish oils. Rapeseed oils contain the 11*c* and 13*c* isomers, and the 11*c* acid is a major component of a few seed oils.

The most important C_{22} monoene acid is undoubtedly erucic acid (13*c*). It occurs, generally at a high level, in seed oils of the Cruciferae (rape, mustard, crambe, nasturtium). This acid is used for the preparation of erucamide, which is employed extensively as a slip agent in polyethylene film.

Acids with more than 22 carbon atoms have been reviewed by Řezanka (1989).

Table 1.2 Some naturally occurring monoene acids

10:1	4*c*
12:1	4*c*
14:1	4*c*, 9*c*
16:1	3*t*, 4*c*, 5*t*, 6*c*, 6*t*, 9*c* [a], 11*c*
18:1	3*t*, 5*c*, 5*t*, 6*c* [b], 6*t*, 9*c* [c], 10*c*, 11*c* [d], 11*t* [e], 13*c*
20:1	5*c*, 9*c* [f], 11*c*, 13*c*, 15*c*
22:1	5*c*, 11*c* [g], 13*c* [h], 15*c*
24:1	15*c* [i]
26:1	9*c*, 17*c* [j]
28:1	9*c*, 19*c* [k]
30:1	21*c*

Trivial names: [a] palmitoleic; [b] petroselinic; [c] oleic; [d] *cis*-vaccenic, asclepic; [e] vaccenic; [f] gadoleic; [g] cetoleic; [h] erucic; [i] selacholeic, nervonic; [j] ximenic; [k] lumequic.

References and further reading

Badami, R.C. and Patil, K.B. (1982) *Prog. Lipid Res.*, **19**, 119.

Pohl, P. and Wagner, H. (1972) *Fette Seifen Anstrichm.*, **74**, 424.

Pryde, E.H. (1979) in *Fatty Acids* (ed. E.H. Pryde), American Oil Chemists' Society, Champaign, p. 1.

Řezanka, T. (1989) *Prog. Lipid Res.*, **28**, 147.

Smith, C.R. (1971) *Prog. Chem. Fats Other Lipids*, **11**, 137.

Smith, C.R. (1979) in *Fatty Acids* (ed. E.H. Pryde), American Oil Chemists' Society, Champaign, p. 29.

1.3 Methylene-interrupted and other non-conjugated polyene acids

Natural fatty acids with more than one double bond can be divided into a number of categories. The most important, in terms of both range of structure and extent of occurrence, are those where unsaturation is entirely methylene-interrupted and of *cis* (*Z*) configuration. Linoleic acid and arachidonic acid (discussed later in this section) are two important examples of this class. Another group of acids have all or part of their unsaturation in a conjugated system with *cis* (*Z*) and/or *trans* (*E*) double bonds (section 1.4). A third group – frequently described as non-methylene-interrupted dienes (NMID) (though more-unsaturated members are also known) – have double bonds that are not entirely in a methylene-interrupted arrangement. The 5,9,12-18:3 acid present in tall oil (section 3.3.36) is an example, and such acids will be discussed later in this section.

Table 1.3 Natural polyene acids (methylene-interrupted) arranged by families[a]

$n-1$ family derived from 9,12,15-16:3	16:3, 18:3
$n-2$ family derived from 9,12,15-17:3	15:3, 17:3, 17:4, 20:4
$n-3$ family derived from 9,12,15-18:3	15:2, 15:3, 15:4, 16:3, 16:4, 18:3[b], 18:4, 18:5, 20:2, 20:3, 20:4, 20:5, 21:5, 22:3, 22:5, 22:6, 24:3, 24:4, 24:5, 24:6, 26:5, 26:6, 28:7, 30:5
$n-4$ family derived from 9,12-16:2	16:2, 16:3, 18:2, 18:3
$n-5$ family derived from 9,12-17:2	15:2, 17:2, 17:3, 19:2, 19:4, 20:3, 20:4, 21:4, 21:5
$n-6$ family derived from 9,12-18:2	15:2, 16:2, 18:2[c], 18:3[d], 20:2, 20:3, 20:4[e], 22:2, 22:3, 22:4[f], 22:5, 24:2, 24:4, 25:2, 26:2, 30:4
$n-7$ family derived from 9-16:1	16:2, 17:2, 18:3, 19:2
$n-8$ family derived from 9-17:1	15:2, 16:2, 17:2, 18:2, 19:2
$n-9$ family derived from 9-18:1	17:2, 18:2, 20:2, 20:3, 22:3, 22:4
$n-11$ family	19:2
$n-12$ family	20:2

[a] Most acids, but not necessarily all, in each family group are derived from the C_{16}, C_{17} or C_{18} acid indicated.
Trivial names: [b] α-linolenic acid; [c] linoleic acid; [d] γ-linolenic acid; [e] arachidonic acid; [f] adrenic acid.

Oleic acid family ($n-9$)

$18{:}1 \xrightarrow{D} 18{:}2 \xrightarrow{E} 20{:}2 \xrightarrow{D} 20{:}3^{a}$

Linoleic acid family ($n-6$)

$18{:}2 \xrightarrow{D} 18{:}3 \xrightarrow{E} 20{:}3^{b} \xrightarrow{D} 20{:}4^{c} \xrightarrow{E} 22{:}4 \xrightarrow{D} 22{:}5$

Linolenic acid family ($n-3$)

$18{:}3 \xrightarrow{D} 18{:}4 \xrightarrow{E} 20{:}4 \xrightarrow{D} 20{:}5^{d} \xrightarrow{E} 22{:}5 \xrightarrow{D} 22{:}6^{e}$

Fig. 1.1 The $n-9$, $n-6$ and $n-3$ families of polyene acids. D = desaturation, E = elongation. [a] Mead's acid; [b] dihomo-γ-linolenic acid, source of PG_1 series; [c] arachidonic acid, source of PG_2 series; [d] eicosapentaenoic acid (EPA), source of PG_3 series; [e] docosahexaenoic acid (DHA).

Methylene-interrupted polyenes can be represented by symbols such as ($n-6$)-18:2 for linoleic acid and ($n-3$)-18:3 for linolenic acid. These symbols indicate straight-chain C_{18} acids with systems of two and three methylene-interrupted *cis* double bonds, which start on the sixth and third carbon atom respectively from the methyl end (formerly the terms ω6 and ω3 were used). The polyene acids of this type (at least 80 have been reported) are divided into groups according to the nature of the endgroup $H_3C[CH_2]_mCH{=}CH—$. The most important families are the $n-3$, $n-6$ and $n-9$ acids based on linolenic, linoleic and oleic acid respectively, though other less common families also exist (see Table 1.3). The acids within any single family are biosynthetically related, being interconverted by enzyme processes of desaturation, chain elongation and chain shortening. This is illustrated for $n-9$ acids (oleic family), $n-6$ acids (linoleic family) and $n-3$ acids (linolenic family) in Fig. 1.1.

Linoleic acid (**2**) is comparable with oleic acid in the extent of its occurrence and is the prototype of the polyene acids. It occurs in almost every vegetable fat and attains major proportions in many seed oils, including several of commercial importance (section 3.2.5). It is also present in animal lipids and fish oils at a much lower level. The natural acid is almost

$CH_3[CH_2]_4CH{=}CHCH_2CH{=}CH[CH_2]_7COOH$

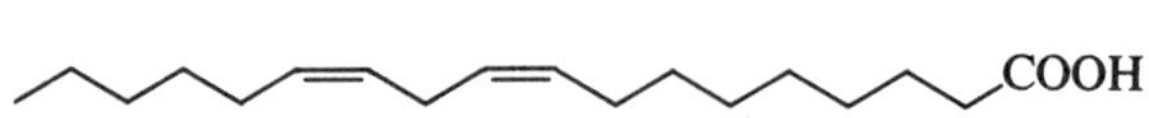

(**2**) Linoleic acid (9*c*,12*c*- or 9*Z*,12*Z*- or ($n-6$)-18:2)

$CH_3[CH_2]_4CH{=}CHCH_2CH{=}CHCH_2CH{=}CHCH_2CH{=}CH[CH_2]_3COOH$

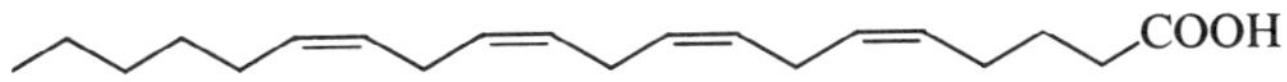

(**3**) Arachidonic acid (5*c*,8*c*,11*c*,14*c*- or 5*Z*,8*Z*,11*Z*,14*Z*- or ($n-6$)-20:4)

always the *cis*,*cis* isomer, but the 9*c*,12*t* (*Dimorphotheca sinuata*, 1.3%; *Crepis rubra*, 3.0%; Morris and Marshall, 1966) and the 9*t*,12*t* isomers (*Chilopsis linearis*, 15%; Chisholm and Hopkins, 1963) have been recognized in seed oils of the plants indicated.

γ-Linolenic acid, ($n-6$)-18:3, is the first intermediate in the bioconversion of linoleic to arachidonic acid. Therapeutic claims have been made for oils containing this acid in the treatment of atopic eczema, arthritis, multiple sclerosis, premenstrual tension, etc. A trace constituent of some animal phospholipids and milk fats, it is more readily available from seed oils of evening primrose (8–10%), borage (20–25%) and blackcurrant (15–18%).

Arachidonic acid (**3**), of considerable importance as a precursor of the prostaglandins, thromboxanes and leukotrienes (section 11.1.7), is present at low levels in many fish oils and at higher levels in animal fats (mainly phospholipids). It is usually isolated from liver or egg lipids. Though rare in the plant kingdom, it has been identified in some mosses and ferns and is a major component of marine algae.

α-Linolenic acid, ($n-3$)-18:3, is mainly of plant origin. It occurs in linseed oil (~60%), in several other highly unsaturated oils (section 3.2.6) and is the major fatty acid of leaves (section 3.11). The other two important $n-3$ acids are (e)icosapentaenoic acid (EPA, 20:5) and docosahexaenoic acid (DHA, 22:6), which (sometimes accompanied by stearidonic acid, 18:4) are major components of fish oils and marine algae. Fish oils containing 30% of EPA and DHA combined are sold as sources of these important $n-3$ acids. The C_{20} acid is a precursor of some prostaglandins and thromboxanes, and the C_{22} acid attains very high proportions in some animal lipids such as those in sperm and in parts of the eye.

Also identified are about 30 polyene acids in which unsaturation is not completely methylene-interrupted. These acids are probably derived from the more common monoene and polyene acids by insertion of an additional double bond, most often at Δ5 (though sometimes Δ2 or Δ3). Subsequent chain extension may shift these double bonds to the Δ7 or Δ9 positions. Such acids occur in some seed oils, in some micro-organisms and in some marine lipids (especially sponges). This last source, for example, has furnished 5,9,19-26:3, 5,9,19-28:3 and 5,9,23-30:3, which may be produced from hexadec-9-enoic, octadec-9-enoic and hexadec-9-enoic acids respectively (Litchfield *et al.*, 1980).

The C_{18}–C_{22} acids of plant or marine origin listed in Table 1.4 are probably related mainly to oleic acid (the dienes), linoleic acid (trienes) and linolenic acid (tetraenes). Tall oil (section 3.3.36) is a convenient source of several of these acids.

The micro-organism *Mycobacterium phlei* contains an unusual group of phleic acids of general structure $H_3C[CH_2]_n[CH{=}CHCH_2CH_2]_mCOOH$ ($n = 12$ or 14, $m = 4$–6) with 4,8,12,16,20-36:5 as the major component (Asselineau and Montrozier, 1976).

Reviews have been published by Smith (1971, 1979), Pohl and Wagner (1972), Pryde (1979) and Ackman and Cunnane (1992).

Table 1.4 Natural non-methylene-interrupted polyene acids (the double bonds are known to have or expected to have *cis* configuration unless otherwise indicated)

	Dienes	Trienes	Tetraenes
C_{18}		2*t*,9,12[a]	
		3*t*,9,12[b]	3*t*,9,12,15[c]
		5*t*,9,12[d–f]	
	5,9[g–j]	5,9,12[d,g–j,k]	5,9,12,15[g,i]
	5,11[l,m]	5,11,14[l]	
C_{20}	5,11[l,m]	5,11,14[g–j,l–n]	5,11,14,17[g,l,m]
	5,13[o]		
	7,11[o]	7,11,14[i]	
	7,13[o]		
C_{22}	5,11[o]		
	5,13[p]		
	7,13[o]		
	7,15[o]		
	7,17[o]		
	9,13[o]		
	9,15[o]		

Sources of acid (seed oils unless otherwise indicated): [a] pollen; [b] *Calea urticaefolia*; [c] *Tecoma stans*; [d] *Teucrium depressum*; [e] *Aquilegia vulgaris*; [f] *Thalictrum polycarpum*; [g] conifer leaf lipids; [h] tall oil; [i] Norwegian spruce; [j] *Larix leptlolepis*; [k] *Xeranthemum anuum*; [l] *Ginkgo biloba*; [m] *Ephadra campylopoda*; [n] *Podocarpus nagi*; [o] marine origin; [p] *Limnanthes douglasii*.

References and further reading

Ackman, R.G. and Cunnane, S.C. (1992) *Adv. Appl. Lipid Res.*, **1**, 161 (ed. F.B. Padley), JAI Press, London, NY.

Asselineau, C.P. and Montrozier, H.L. (1976) *Eur. J. Biochem.*, **63**, 509.
Chisholm, M.J. and Hopkins, C.Y. (1963) *Can. J. Chem.*, **41**, 1888.
Litchfield, C., Tyszkiewicz, J. and Dato, V. (1980) *Lipids*, **15**, 200.
Morris, L.J. and Marshall, M.O. (1966) *Chem. Ind.*, 1493.
Pohl, P. and Wagner, H. (1972) *Fette Seifen Anstrichm.*, **74**, 424.
Pryde, E.H. (1979) in *Fatty Acids* (ed. E.H. Pryde), American Oil Chemists' Society, Champaign, p. 1.
Smith, C.R. (1971) *Prog. Chem. Fats Other Lipids*, **11**, 137.
Smith, C.R. (1979) in *Fatty Acids* (ed. E.H. Pryde), American Oil Chemists' Society, Champaign, p. 29.

1.4 Conjugated polyene acids

Acids with conjugated unsaturation that also contain acetylenic bonds (section 1.5) or oxygen functions (sections 1.8 and 1.9) are discussed later and are excluded from this section.

Conjugated polyene acids do not occur to any significant extent in animal lipids and assume importance in only a few seed oils, of which only tung oil (section 3.3.39) is commercially available. The most common conjugated polyene acids are octadecatrienoic acids, of which seven isomers occur naturally. It was long thought that only one such acid was present in any single source, but recent ^{13}C nuclear magnetic resonance (NMR) and high-performance liquid chromatography (HPLC) studies have shown this to be not so (Takugi and Itabashi, 1981). Acids with conjugated unsaturation differ from their non-conjugated isomers (section 1.3) in their characteristic ultraviolet (UV) spectra (section 9.7), their higher melting points and the frequent presence of *trans* (*E*) unsaturation. The subject has been reviewed by Hopkins and Chisholm (1968), Hopkins (1972) and Pohl and Wagner (1972).

Stillingia oil (*Sapium sebiferum* seed oil, section 3.3.34) is unusual in that some of its glycerides (**4**) contain four acyl groups, including two conventional C_{18} acids, a C_8 allenic hydroxy acid (section 1.5) and deca-2*t*,4*c*-dienoic (stillingic) acid (Sprecher *et al.*, 1965).

The decatrienoic acids isolated from the latex of *Euphorbia pulcherrima* include the 2*t*,4*t*,6*c*; 2*c*,4*t*,6*t*; 3*t*,5*t*,7*c* and 3*c*,5*t*,7*t* isomers (Warnaar, 1977). Two unusual pentaenoic acids (3,5,7,9,11-12:5 and 3,5,7,9,11-14:5) have been identified in the body fat of the coccid *Llaveia axin* at around the 10% level (Cason *et al.*, 1971).

Most of the conjugated C_{18} acids are trienes, but a diene (10*t*,12*t* ~12%) is present in *Chilopsis linearis* seed oil along with some other less common acids. Conjugated octadecadienoic acids result from dehydration of ricinoleic and ricinelaidic acids (section 1.8) and from alkali isomerization of linoleic acid. Claims have been made for the formation of the 8*t*,10*t*, the four 9,11 and the four 10,12 dienes by one or both of these methods.

Seven conjugated octadecatrienoic acids have been recognized as natural compounds, including three 8,10,12-trienes and four 9,11,13-trienes. Detailed configurations, trivial names and some indication of occurrence are given in Table 1.5. They are all believed to arise from linoleic acid (9,12-18:2) by processes involving dehydrogenation and double-bond migration (Crombie and Holloway, 1985). Because of its occurrence in tung oil, α-eleostearic acid is the most readily available of these acids, but attempts are being made to grow calendula as a new crop (section 3.3.43e). This contains the 8*t*,10*t*,12*c*-18:3 acid. Each of them is readily isomerized to the all-*trans* form. Parinaric acid is a conjugated tetraenoic acid.

Reviews have been published by Smith (1971, 1979), Pohl and Wagner (1972), Hopkins (1972), Pryde (1979) and Badami and Patil (1982).

Table 1.5 Natural octadecatrienoic and octadecatetraenoic acids

Configuration	Trivial name	Typical source (seed oils)
Trienes		
8*c*,10*t*,12*c*	Jacaric	*Jacaranda mimosifolia*
8*t*,10*t*,12*c*	Calendic	*Calendula officinalis*
8*t*,10*t*,12*t*	–	*C. officinalis*
9*t*,11*t*,13*c*	Catalpic	*Catalpa ovata*
9*c*,11*t*,13*t*[a]	α-Eleostearic	*Aleurites fordii* (tung oil)
9*c*,11*t*,13*c*	Punicic	*Punica granatum*
9*t*,11*t*,13*t*	β-Eleostearic	*A. fordii*
Tetraenes		
9*c*,11*t*,13*t*,15*c*[b]	α-Parinaric	*Impatiens balsamina*
9*t*,11*t*,13*t*, 15*t*	β-Parinaric	

[a] Occurs also as the 18-hydroxy (kamlolenic acid, *Mallotus philippinensis*) and 4-oxo (licanic acid, *Licania rigida*) derivatives.
[b] Occurs also as the 4-oxo derivative (*Chrysobalanus icaco*).

References and further reading

Badami, R.C. and Patil, K.B. (1982) *Prog. Lipid Res.*, **19**, 119.

$$
\begin{array}{l}
CH_2OCO[CH_2]_7CH{=}CHCH_2CH{=}CHCH_2CH{=}CHCH_2CH_3 \\
| \\
CHOCO[CH_2]_7CH{=}CHCH_2CH{=}CH[CH_2]_4CH_3 \\
| \\
CH_2OCO[CH_2]_3CH{=}C{=}CHCH_2OCOCH{=}CHCH{=}CH[CH_2]_4CH_3
\end{array}
$$

(**4**) A glyceride of stillingia oil

Cason, J., Davis, R. and Sheehan, M.H. (1971) *J. Org. Chem.*, **36**, 2621.
Crombie, L. and Holloway, S.J. (1985) *J. Chem. Soc. Perkin Trans. I*, 2425.
Hopkins, C.Y. (1972) *Top. Lipid Chem.*, **3**, 37.
Hopkins, C.Y. and Chisholm, M.J. (1968) *J. Am. Oil Chem. Soc.*, **45**, 176.
Pohl, P. and Wagner, H. (1972) *Fette Seifen Anstrichm.*, **74**, 541.
Pryde, E.H. (1979) in *Fatty Acids* (ed. E.H. Pryde), American Oil Chemists' Society, Champaign, p. 1.
Smith, C.R. (1971) *Prog. Chem. Fats Other Lipids*, **11**, 137.
Smith, C.R. (1979) in *Fatty Acids* (ed. E.H. Pryde), American Oil Chemists' Society, Champaign, p. 29.
Sprecher, H., Maier, R., Barber, M. and Holman, R.T. (1965) *Biochemistry*, **4**, 1856.
Takagi, T. and Itabashi, Y. (1981) *Lipids*, **16**, 546.
Warnaar, F. (1977) *Lipids*, **12**, 707.

1.5 Allenic and acetylenic acids

Allenic acids, containing the —CH=C=CH— system, occur only rarely in lipids, but the known examples are of interest because the glycerides are optically active by virtue of the allene function. Allenic acids are isomeric with acetylenic compounds and each can be rearranged to the other.

Stillingia oil (*Sapium sebiferum* seed oil, see also sections 1.4, 3.3.6 and 3.3.34) contains unusual glycerides having four acyl groups. A C_8 hydroxy allenic acid and a C_{10} conjugated dienoic acid are attached to each other and to one of the primary hydroxyl groups in glycerol through the allenic acid (see (**4**), section 1.4 for structure). The C_8 and C_{10} combination may arise through rearrangement of a C_{18} acid. The natural allene probably has *R* configuration.

A C_{14} allenic acid (2,4,5-14:3) (**5**), also believed to be the *R*(−) form, occurs as a sex pheromone in the male dried bean beetle.

$$H_3C[CH_2]_7CH{=}C{=}CHCH\overset{t}{=}CHCOOH$$

(**5**)

Two *R*(−) C_{18} allenic acids (5,6-18:2 (**6**) and (5,6,16-18:3 (**7**)) have been identified in seed oils of several members of the Labiatae (mint family) (Hagemann *et al.*, 1967).

$$H_3C[CH_2]_{10}CH{=}C{=}CH[CH_2]_3COOH$$

(**6**) Laballenic acid

$$H_3CCH\overset{t}{=}CH[CH_2]_8CH{=}C{=}CH[CH_2]_3COOH$$

(**7**) Lamenallenic acid

The acetylenic acids are conveniently divided into those with only one triple bond and those with more than one. In the former group of acids one or more double bonds may also be present and unsaturation frequently appears in the conventional methylene-interrupted pattern.

The most important monoacetylenic acids are tariric (6*a*-18:1) and crepenynic (9*c*,12*a*-18:2), though other examples have also been recognized. Tariric acid (**8**) has long been known as a component of seed oils of the *Picramnia* species. Its isomer, stearolic acid (**9**), though easily made in the laboratory from oleic acid, occurs only rarely as a minor component in Santalaceae seed oils where more unsaturated mono- and polyacetylenic compounds are also present (see below). Two rare acids related to stearolic acid are 8*a*-17:1 and 9*a*,17*e*-18:2.

$$H_3C[CH_2]_{10}C{\equiv}C[CH_2]_4COOH$$

(**8**) Tariric acid

$$H_3C[CH_2]_7C{\equiv}C[CH_2]_7COOH$$

(**9**) Stearolic acid

Crepenynic acid (9*c*,12*a*-18:2) (**10**) is of special interest because it is recognized as an important biosynthetic intermediate between the more conventional C_{18} fatty acids and a large number of polyacetylenes of varying chain length but most commonly C_9, C_{13} or C_{17} (Lam and Hansen, 1990). This acid, first recognized in *Crepis foetida* seed oil (35–60%), occurs in other Crepis species, in *Saussurea candicans* seed oil (~33%) and in seed oils of the *Afzelia* species, where it is sometimes accompanied by a more unsaturated compound (dehydrocrepenynic acid) (**11**).

$$H_3C[CH_2]_4C{\equiv}CCH_2CH\overset{c}{=}CH[CH_2]_7COOH$$

(**10**) Crepenynic acid

$$H_3C[CH_2]_2CH\overset{c}{=}CHC{\equiv}CCH_2CH\overset{c}{=}CH[CH_2]_7COOH$$

(**11**) Dehydrocrepenynic acid

Three acetylenic acids present in the lipids of mosses (6*a*,9*c*,12*c*-18:3, 6*a*,9*c*,12*c*,15*c*-18:4 and 8*a*,11*c*,14*c*-20:3) are clearly related to the corresponding polyene acids.

Sterculynic acid (**12**) from *Sterculia alata* seed oil is unusual in that it contains an acetylenic group and a cyclopropene unit. It is possible that cyclopropenes are made in nature from acetylenic acids.

Most natural acetylenic acids are C_{18} compounds with conjugated unsaturation involving one or more acetylenic groups along with olefinic unsaturation, which may be *c* or *t*. Unsaturation usually starts at C-9 and appears between that carbon atom and the methyl end of the chain. Most of these acids occur together in seed oils of the Olacaceae, Santalaceae, Asteraceae, or Simurubiaceae. This group of acids is characterized by three additional features: the frequent appearance of a double bond at Δ17, the presence of a hydroxyl group at C-8 and the presence of C_{17} homologues of the C_{18} acids (see Table 1.6).

The seed oil of *Helichrysum bracteatum* contains

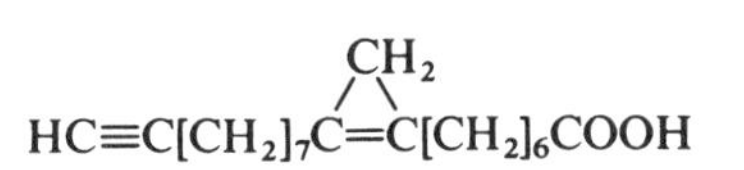

(**12**) Sterculynic acid

$$H_3C[CH_2]_4C{\equiv}CCH\overset{t}{=}CHCH(OH)[CH_2]_7COOH$$

(**13**) Helenynolic acid

$BrCH\overset{c}{=}CHCBr{=}CH[CH_2]_4CH\overset{t}{=}CHC{\equiv}C[CH_2]_3COOH$

(14)

$H_3C[CH_2]_4CH\overset{c}{=}CHCH_2CH\overset{c}{=}CHCH{=}CHC\overset{c}{\equiv}C[CH_2]_3COOH$

(15)

$H_2C{=}CHCH(OH)[C{\equiv}C]_2CH_2CH\overset{c}{=}CH[CH_2]_7COOH$

(16)

Table 1.6 The structure of natural acetylenic acids with conjugated unsaturation

	C_{18} acids					C_{17} acids[a]
Acetylenic acid	9*a*	9*a*,11*t*[b]	9*a*,11*a*	9*a*,11*a*,13*c*[f]	9*a*,11*a*,13*a*,15*e*	8*a*,10*t*[g]
Δ17*e* derivative	√	√	√[d]	√	√	√
8-OH derivative	–	√[c]	√	√	–	√
Δ17*e*,8-OH derivative	–	√	√[e]	–	–	√

Also recognized: 9*a*,11*c*-18:2, 9*a*,11*a*-18:2, 9*a*,11*t*,13*t*-18:3, 9*a*,11*a*,13*t*-18:3[h], 8-oxo-9*a*,11*a*- and 10*t*,12*a*-18:2, 8-oxo-9*a*,11*a*,17*e*- and 10*a*,12*a*,17*e*-18:3.

[a] Additional double bond at 16*e*, additional hydroxyl at C-7.

[b] ximenynic, santalbic; [c] ximenynolic; [d] isanic, erythrogenic; [e] isanolic; [f] bolekic; [g] pyrulic; [h] exocarpic.

The symbol *e* represents an ethylenic bond that does not display *cis–trans* isomerism.

an acetylenic acid (**13**) of a different kind in that the double bond at Δ10 appears before the triple bond at Δ12.

Yet other acetylenic acids include a C_{16} dibromo compound (**14**) from a marine sponge, a C_{18} acid (**15**) from red marine alga, which is reported to be highly toxic to fish, and a hydroxy C_{18} acid (**16**) from *Dendropanax trifidus* leaves.

This topic has been reviewed by Smith (1971), Hopkins (1972) and Badami and Patil (1982). References are also available in sections 3.2.11 and 3.2.15.

References and further reading

Badami, R.C. and Patil, K.B. (1982) *Prog. Lipid Res.*, **19**, 119.

Hagemann, J.M., Earle, F.R., Wolff, I.A. and Barclay, A.S. (1967) *Lipids*, **2**, 371.

Hopkins, C.Y. (1972) *Top. Lipid Chem.*, **3**, 37.

Lam, J. and Hansen, L. (1990) in *Methods in Plant Biochemistry* (eds J.L. Harwood and J.R. Bowyer), Academic Press, London, p. 159.

Smith, C.R. (1971) *Prog. Chem. Fats Other Lipids*, **11**, 137.

1.6 Branched-chain acids

Although branched-chain acids occur widely in Nature, they are usually present as minor components and only rarely do they account for a significant amount of the total fatty acids (e.g. in bacteria, section 3.14). The most common type of branching is a single methyl group, but sometimes there are longer branches or more than one branching methyl. Acids with a single methyl group most commonly have this at the $n-2$ (iso acids), $n-3$ (anteiso acids), 2 or 3 positions, though other positions are possible. Acids with more than one methyl group generally have them attached to even carbon atoms. There is a biosynthetic explanation for most of these generalizations concerning structure.

Iso and anteiso acids occur widely in animal fats and waxes (where they are sometimes major components) and in the lipids of some micro-organisms. The total number of carbon atoms of the iso and anteiso acids are usually even and odd respectively, and the anteiso acids are generally dextrorotatory with absolute configuration L(S). They are thought to result from a modification of the normal *de novo* pathway in which the usual starter acetate molecule is replaced by 2-methylpropanoate or 2-methylbutanoate:

$$H_3C\overset{CH_3}{\overset{|}{C}}HCOOH \xrightarrow{\text{Chain extension}} H_3C\overset{CH_3}{\overset{|}{C}}H[CH_2]_{2n}COOH \quad \text{(even)}$$

$$H_3CCH_2\overset{CH_3}{\overset{|}{C}}HCOOH \xrightarrow{\text{Chain extension}} H_3CCH_2\overset{CH_3}{\overset{|}{C}}H[CH_2]_{2n}COOH \quad \text{(odd)}$$

Branched-chain acids of this kind have been found in the depot, milk and blood lipids of animals, including humans. They are, for example, significant components of the acids (and alcohols) present in human skin lipids.

A mid-chain branch occurs in tuberculostearic acid (10D-methylstearic) present in the lipids of the human tubercle bacillus and in other bacterial lipids.

Ruminants produce branched-chain acids, particularly on a barley-rich diet, and more than 180 C_{10}–C_{18} acids with one to four branched methyl groups have been identified in the subcutaneous adipose tissue of lambs. The methyl group is always attached to an even carbon atom. These acids are probably formed by a modification of the *de novo* pathway in which one or more malonate units (from acetate) are replaced by methylmalonate (from propionate) (Smith *et al.*, 1979).

Similar polymethyl branched acids are dominant in the uropygial waxes produced by the preen gland of many birds. For example, the wax produced by the flamingo is made up of acids having two (40%) or three (60%) methyl groups and alcohols that, though mainly unbranched (87%), contain some monomethyl compounds.

Polymethyl branched acids isolated from tubercle bacilli and other bacterial lipids contain longer-chain acids of type (**17**) and (**18**). The C_{27} trimethyl acid also occurs as its 3-hydroxy derivative and in monoene (Δ2) and diene (Δ2,15) forms.

Other branched-chain acids of bacterial origin and of relatively high molecular weight are the mycolic and related acids (section 3.14). These are α-branched β-hydroxy acids of general structure RCH(OH)CH(R′)COOH probably resulting from condensation of acid RCOOH with acid $R'CH_2COOH$. Group R′ is usually a saturated alkyl chain (C_8–C_{24}); the group R is longer (up to 80 carbon atoms) and may be saturated or unsaturated and contain one or more cyclopropane rings and/or oxygenated functions (hydroxy, methoxy, oxo, epoxy).

Branched-chain acids (in addition to phytol-based acids (this section) and furanoid acids (section 1.9)) have been identified in a number of aquatic sources. These include monoene acids such as 7-methylhexadec-6- and -7-enoic acids and 5-methyl-tetradec-4-enoic acid. Some marine animals have a deposit of lipid in the head (melon), an organ that has been implicated in the echo-locating abilities of these animals. These lipids are rich in isovaleric (3-methylbutanoic) acid and iso acids of more conventional chain length, but there is no evidence that their unusual properties are associated with the branching nature of the lipid fatty acids.

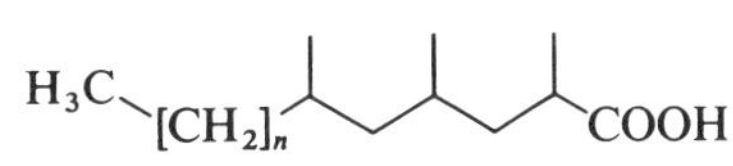

$n = 15, 17$ or 19

(**17**)

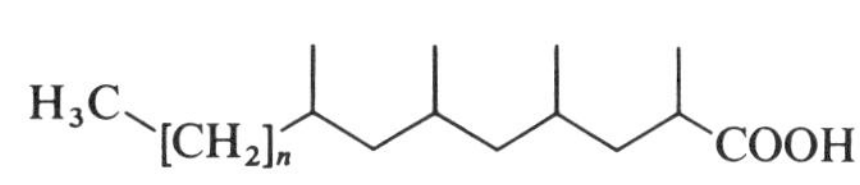

$n = 17, 19$ or 21

(**18**)

A further group of branched-chain acids differ from those already discussed in that they are produced from isoprenoid precursors and not by the normal *de novo* pathway or some simple modification of it. These acids are based on the diterpene phytol, derived from chlorophyll, and the more common acids are phytanic (C_{20}) (**19**), pristanic (C_{19}) (**20**) and a C_{16} homologue (**21**). These molecules contain chiral centres and the natural forms are not always the same enantiomer. These acids occur widely, but at low levels, in the lipids of land and aquatic animals, and have been identified in ancient sediments. In animals they occur in tissue, milk and blood lipids and are present as trace components of butterfat. The tissue lipids of humans suffering from the rare neurological disorder Refsum's disease contain up to 50% of phytanic acid because of an inability to metabolize this unusual branched-chain acid (section 12.6).

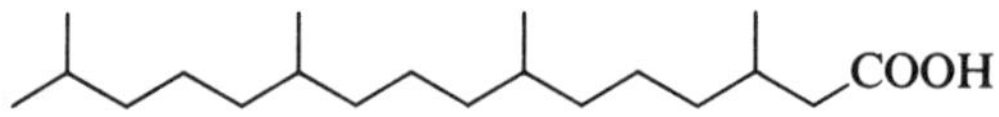

(**19**) 3,7,11,15-tetramethylhexadecanoic acid (phytanic)

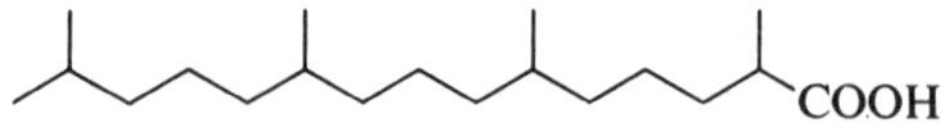

(**20**) 2,6,10,14-tetramethylpentadecanoic acid (pristanic)

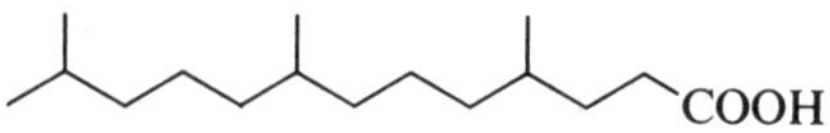

(**21**) 4,8,12-trimethyltridecanoic acid

References and further reading

Abrahamssohn, S., Stallberg-Stenhagen, S. and Stenhagen, E. (1963) *Prog. Chem. Fats Other Lipids*, **7**, 1.
Kates, M. (1975) *Adv. Lipid Res.*, **2**, 17.
Kolattukudy, P.E. (ed.) (1976) *Chemistry and Biochemistry of Natural Waxes*, Elsevier, Amsterdam.
Lederer, E. (1964) *Angew. Chem. Int. Edn*, **3**, 393.
Lough, A.K. (1975) *Prog. Chem. Fats Other Lipids*, **14**, 1.
Polgar, N. (1971) *Top. Lipid Chem.*, **2**, 207.
Smith, A., Calder, A.G., Lough, A.K. and Duncan, W.R.H. (1979) *Lipids*, **14**, 953.
Smith, C.R. (1970) *Top. Lipid Chem.*, **1**, 277.

1.7 Cyclic acids

Some natural fatty acids contain a carbocyclic unit, which may have three (cyclopropane and cyclopropene acids), five (prostaglandins – see section 11.1.7 – and cyclopentene acids) or six carbon atoms (cyclohexane acids). Aspects of this topic have reviewed (Christie, 1970; Lie Ken Jie, 1979; Badami and Patil, 1982). Other cyclic acids occur in tall oil (section 3.3.36).

Cyclopropane fatty acids occur frequently in bacterial membrane phospholipids. Also, they generally accompany the cyclopropene acids in seed oils (see following paragraph). Though other chain lengths have been reported, the most common cyclopropane acids are C_{17} (**22**) and C_{19} (lactobacillic acid) (**23**) compounds. They are probably formed from appropriate olefinic acids (9*c*-16:1 and 11*c*-18:1), which are widely distributed in bacterial lipids. The cyclopropane acids have *cis* configuration, but it is not clear whether they are individual enantiomers or racemic mixtures.

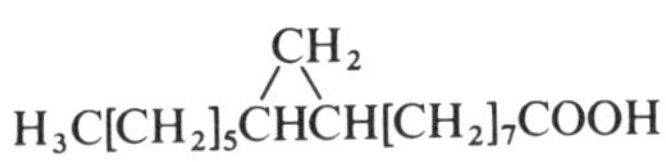

(**22**)

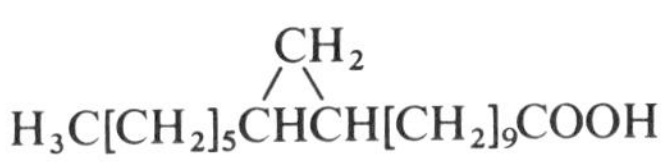

(**23**) Lactobacillic

Four cyclopropene acids (**24**)–(**27**) have been identified in seed oil triacylglycerols. They are found mainly among Malvales, especially Sterculiaceae, Malvalaceae, Bombaceae and Tiliaceae. The recognition of 2-hydroxysterculic acid as a natural product is interesting because it might be an intermediate in the α-oxidation of sterculic acid to malvalic acid. Further details of the occurrence of cyclopropene acids are given in section 3.2.14.

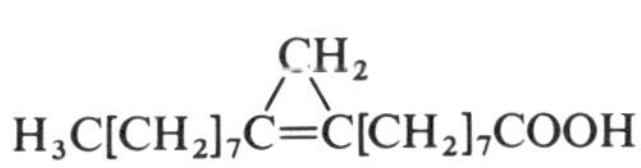

(**24**) Sterculic

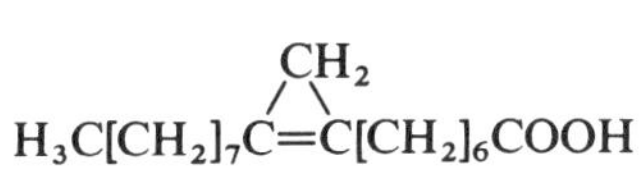

(**25**) Malvalic

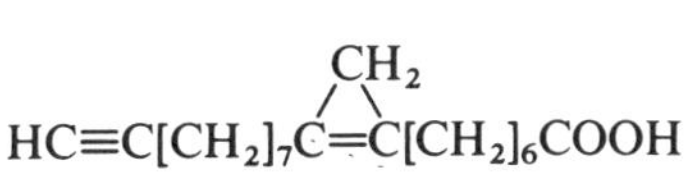

(**26**) Sterculynic

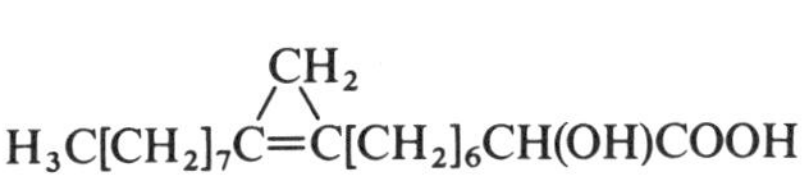

(**27**) 2-Hydroxysterculic

The cyclopropene acids inhibit the biodesaturation of stearic acid to oleic acid, and this leads to several adverse biochemical effects including the production of pink whites in eggs. Accordingly there was some concern when it was demonstrated that cottonseed oil (section 3.3.10) and the residual meal might contain up to 1% of cyclopropene acids. However, such acids are largely deactivated or removed from the oil by hydrogenation or during deodorization at 230–235 °C.

Mycolic acids are complex acids of high molecular weight occurring in a wide range of micro-organisms (section 3.14). They are 2-alkyl-3-hydroxy acids with the structure shown in (**28**) and may contain up to 80 carbon atoms. Variations arise in the R group, and there are many examples of acids with one or more cyclopropane units. Branched methyl groups and/or an oxygenated function may also be present (Polgar, 1971).

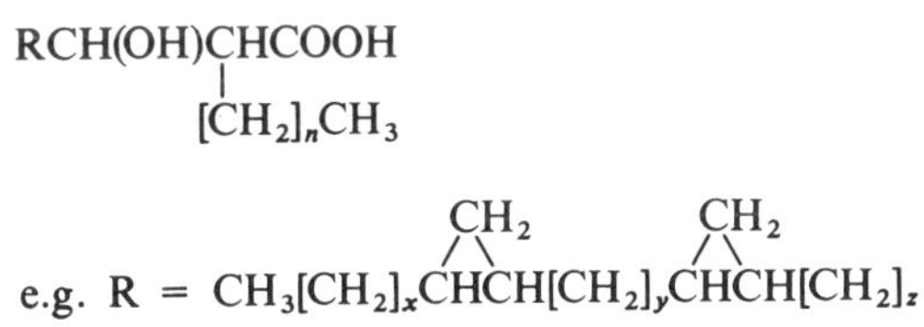

$x = 19, y = 14, z = 11, 13, 15,$ or 17

(**28**)

Table 1.7 Natural cyclopentenyl fatty acids (**29**) and (**30**)

Trivial name	Number of carbon atoms	*n*	*x*	*y*	Configuration
Aleprolic	6	0	–	–	*R*(+)
Alepramic	8	2	–	–	*S*(+)
Aleprestic	10	4	–	–	*R*(+)
Aleprylic	12	6	–	–	*R*(+)
Alepric	14	8	–	–	*R*(+)
Hydnocarpic	16	10	–	–	*R*(+)
Chaulmoogric	18	12	–	–	*R*(+)
Hormelic	20	14	–	–	*R*(+)
–	16	–	2	6	–
Manaoic	16	–	4	4	–
–	16	–	7	1	–
–	18	–	2	8	–
Gorlic	18	–	4	6	*R*(+)
–	18	–	7	3	–
Oncobic	20	–	6	6	–
–	20	–	7	5	–

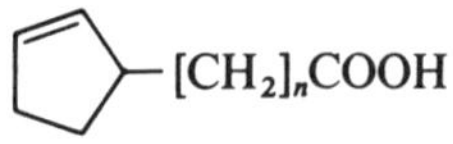

(29)

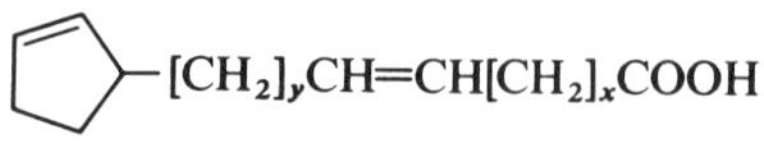

(30)

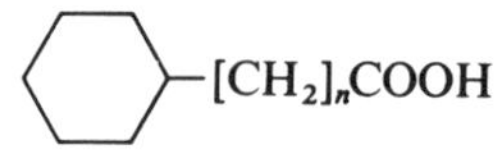

n = 10 or 12

(31)

Seed fats of the Flacourtiaceae are unique in containing cyclopentenyl acids with one (**29**) or two (**30**) double bonds (Table 1.7). All of them contain a chiral centre and are optically active. They vary in chain length from C_6 to C_{20} and have been briefly reviewed by Badami and Patil (1982). Interest in these unusual seed oils arises from their former use in the treatment of leprosy, though there is little evidence to relate their therapeutic value (if any) to these unsaturated cyclic fatty acids.

C_{17} and C_{19} cyclohexylalkanoic acids (**31**) have been identified. Both have been recognized in acido-thermophilic bacteria in hot springs and the C_{17} acid has also been detected in rumen bacteria (~3%), sheep perenephric fat (0.05%) and butter (0.01%).

References and further reading

Badami, R.C. and Patil, K.B. (1982) *Prog. Lipid Res.*, **19**, 119.
Christie, W.W. (1970) *Top. Lipid Res.*, **1**, 1.
Lie Ken Jie, M.S.F. (1979) *Chem. Phys. Lipids*, **24**, 407.
Pohl, P. and Wagner, H. (1972) *Fette Seifen Anstrichm.*, **74**, 541.
Polgar, N. (1971) *Top. Lipid Res.*, **2**, 207.
Smith, C.R. (1970) *Prog. Chem. Fats Other Lipids*, **11**, 137.

1.8 Hydroxy acids

Natural acids with a functional group other than the carboxyl group and one or more types of unsaturation are relatively unusual, but among such acids hydroxy and epoxy (section 1.9) compounds are the most common.

α-Hydroxy or 2-hydroxy acids occur in sphingolipids, as components of wool wax and skin lipids, and in a few unusual seed oils. 2-Hydroxylinolenic acid, reported in the seed oils of *Thymus vulgaris* (13%) and *Salva nilotica*, is accompanied by linolenic acid and its C_{17} homologue (8,11,14-17:3). The hydroxy acid may be an intermediate in the bioconversion of the C_{18} to the C_{17} acid. *Salva nilotica* seed oil also contains 2-hydroxyoleic and 2-hydroxylinoleic acid along with the corresponding (non-hydroxy) 17:1 and 17:2 acids. The hydroxy acids probably have D(*R*) configuration. See also sections 1.5 and 3.2.14 for discussion of 2-hydroxysterculic acid.

β-Hydroxy or 3-hydroxy acids occur in some bacterial lipids and the very long and branched-chain mycolic acids (sections 1.7 and 3.14) contain this structural feature.

Natural monohydroxy acids, which are either saturated or contain non-conjugated unsaturation, are listed in Table 1.8. Ricinoleic acid (12-hydroxyoleic) is the best known, being the major acid in castor oil. Isoricinoleic acid is an isomer with the double bond and hydroxyl group interchanged. Densipolic acid is an unsaturated analogue of ricinoleic acid. Lesquerolic and auricolic acids are the C_{20} homologues of ricinoleic and densipolic acid respectively. All these C_{18} and C_{20} hydroxy acids are β- and/or γ-hydroxyalkenes.

Some *Lesquerella* species are being developed as potential new crops (section 3.43g).

Mid-chain hydroxy acids with conjugated unsaturation (*cis* and/or *trans* olefinic and/or acetylenic) are listed in Table 1.9. With the exception of two C_{17} acids, these are C_{18} acids, and apart from kamlolenic acid (18-hydroxyeleostearic) they are unsaturated derivatives of 8-, 9- or 13-hydroxystearic acids. Those based on 8-hydroxystearic acid differ from the 9- and 13-hydroxy acids. The former group contain acetylenic unsaturation, sometimes have end-unsaturation (Δ17) and co-occur with the corresponding non-hydroxy acetylenic acids. They are found in the seed oils of *Ximenia caffra*, *Onguekoa gore* (isano oil), *Acanthosyris spinescens* and *Pyrularia pubera*. The unsaturated 9- and 13-hydroxy acids resemble in some respects the hydroperoxy acids produced by oxidation of linoleic or other methylene-interrupted polyene acids, and the hydroxy acid usually co-occurs with the unsaturated acid from which it might be formed. Some seed oils contain a 9-hydroxy acid only (*Calendula officinalis*, *Dimorphotheca sinuata*) or the 13-hydroxy acid only (*Coriara nepalensis*, *Monnina emarginata*), whilst others contain acids of both types (*Xeranthemum*

Table 1.8 Mid-chain hydroxy acids without conjugated unsaturation

Chain length and unsaturation	Position of hydroxyl group	Trivial name	Source
10:1	7	–	Convolvulacea species
14:0	6	Butolic	Shellac
14:0	11	–	Convolvulacea species
16:0	8	–	*Lycopodium complanatum*
16:0	11	–	Convolvulacea species
9*c*-16:1	12	–	*Lesquerella densipila*
18:0	9	–	*Strophanthus* species
18:0	10	–	Oroju (sulphur olive oil)
9*c*-18:1	12D(*R*)	Ricinoleic	*Ricinus communis* (castor oil)[a]
12*c*-18:1	9D(*S*)	Isoricinoleic	*Strophanthus* species[b]
9*c*,15*c*-18:2	12D(*R*)	Densipolic	*L. densipila*
11*c*-20:1	14D(*R*)	Lesquerolic	*L. densipila*[c]
11*c*,17*c*-20:2	14	Auricolic	*Lesquerella auriculata*
22:0	13L(*S*)	–	*Candida* glycolipids

[a] Also ergot oil, *Hiptage benghalensis*, *Linum mucronatum*.
[b] Also *Wrightia* spp., *Nerium* spp., *Holleherna antidysentrica*.
[c] Also *Heliophilia amplexicaulis*.

Table 1.9 Mid-chain C_{18} hydroxy acids with conjugated unsaturation[a]

8-OH acids		9-OH acids		13-OH acids	
9*a*,11*t*[b]	Ximenynolic	10*t*,12*c*	Dimorphecolic[c]	9*c*,11*t*	Coriolic[c,d]
9*a*,11*t*17*e*[b]	–	10*t*,12*t*	Dimorphecolic[c]	3*t*,9*c*, 11*t*	–
9*a*,11*a*	–	10*t*,12*a*	Helenynolic[c]		
9*a*,11*a*,17*e*	Isanolic	3*t*,10*t*,12*c*	–		
9*a*,11*a*,13*c*	–				
10*a*,12*a*	–				
10*a*,12*a*,17*e*	–				

[a] Also 18-OH 9*c*,11*t*,13*t*-18:3, kamlolenic acid (Table 1.5) and 8-OH 5*e*,6*e*-18:2 (section 1.5).
[b] Also occur as C_{17} homologues: 7-OH 8*a*,10*t* and 7-OH 8*a*,10*t*,16*e*.
[c] These acids have D(*R*) configuration.
[d] Also occur as the C_{16} (11-OH, 7*c*,9*t*-16:2) and C_{14} (9-OH 5*c*,7*t*-14:2) homologues.

annuum, *Tragopogon porrifolius*, *Tagetes erecta*, *Stenachaenium macrocephalum*).

Yet other hydroxy acids have their hydroxyl group at or near the methyl end of the chain. ω-Hydroxy acids (see also polyhydroxy acids later in this section) are frequently present in waxes (section 2.10) and in (polymerized) cutins and suberins (section 2.11). Yeasts frequently generate ω2-hydroxy acids as glycosides within their extracellular lipids. (*S*)-17-L-hydroxystearic and (*S*)-17-L-hydroxyoleic acids are the best-known examples.

Because of their hydroxyl group, these acids do not occur only as simple triacylglycerols, but sometimes as glycerides with more than three acyl groups through acylation of one or more hydroxyl groups or as lactones. There is no evidence that the glycerides of castor oil contain anything other than conventional triacylglycerols, but ergot oil (containing ricinoleic acid), lesquerella oils (several hydroxy acids), *Monnina emarginata* seed oil (coriolic acid) and kamala oil (kamlolenic acid) contain tetra- and penta-acylglycerols.

Table 1.10 Polyhydroxy acids

Chain length	Position of hydroxyl group and unsaturation (if any)
14:0	3,11
16:0	7,16; 8,16; 9,16; 10,15; 10,16; 15,16; 2,15,16 (ustilic); 9,10,16 (aleuritic)
18:0	9,10; 9,14; 9,10,14; 9,10,18; 9,10,12,13; 9,10,12,13,18
18:1	(9*t*)5,8,12; (10*t*)9,12,13; (11*t*)9,10,13; (12*c*)9,10
18:2	(10*t*15*c*)9,12,13
20:0	11,12
22:0	13,14; 8,9,13; 8,9,13-oxo
24:0	15,16
24:1	(9*t*)11,13 (axillarenic)

The following long-chain hydroxy acids have been found to occur as their lactones: 3,5-dihydroxydecanoic (as the 5-olide), 11-hydroxydodec-8-enoic, 5-hydroxydodeca-2,6-dienoic, 6-acetoxy-5-hydroxyhexadecanoic, 13-hydroxyoctadeca-9,11-dienoic and 12,13-dihydroxyoctadec-9-enoic acid (as the 13-olide).

Unsaturated oils often contain small amounts of hydroxy and/or epoxy acids after prolonged storage, probably as a result of enzymic and/or non-enzymic oxidation (sections 10.2 and 10.3).

Polyhydroxy acids with two to five hydroxyl groups are mainly (but not entirely) saturated: they range in chain length from C_{14} to C_{24}, though the C_{16} and C_{18} acids are the most common (see Table 1.10). They are present in cutins and suberins (and other sources), often in polymeric form. The hydroxyl groups are most likely to be found in the ω position and in *vic* positions in the chain where a double bond previously existed.

Gravier-Pelletier *et al.* (1990) have reviewed the synthesis of hydroxy fatty acids.

References and further reading

Badami, R.C. and Patil, K.B. (1982) *Prog. Lipid Res.*, **19**, 119.

Gravier-Pelletier, C., Dumas, J., LeMerrer, Y. and Depezay, J.C. (1990) *Prog. Lipid Res.*, **29**, 229.

Gunstone, F.D. (1980) in *Fats and Oils: Chemistry and Technology* (eds R.J. Hamilton and A. Bhati), Applied Science, London, p. 47.

Kishimoto, Y. and Radin, N.S. (1963) *J. Lipid Res.*, **4**, 139.

Kolattukudy, P.E. (1976) in *Chemistry and Biochemistry of Natural Waxes*, Elsevier, Amsterdam.

Pohl, P. and Wagner, H. (1972) *Fette Seifen Anstrichm.*, **74**, 541.

Smith, C.R. (1970) *Top. Lipid Chem.*, **1**, 316.

Smith, C.R. (1971) *Prog. Chem. Fats Other Lipids*, **2**, 157.

Smith, C.R. (1979) in *Fatty Acids* (ed. E.H. Pryde), American Oil Chemists' Society, Champaign, p. 29.

1.9 Epoxy and furanoid acids

Epoxy acids are present in a number of seed oils, where they occur as triacylglycerols, and in cutins, where they occur as polymers of hydroxy acids. They also develop in some seeds on prolonged storage. Acids with a furan unit (another type of cyclic ether) have been identified in plants, in fish oils and in blood and urine. Synthetic epoxy acids are discussed in section 10.4.1.

Natural epoxy acids occurring in seed oils are listed in Table 1.11. Apart from one *trans*-epoxide, all are *cis* isomers, and apart from one C_{20} acid (alchornoic), all are C_{18} compounds closely related to unsaturated acids usually present in the same source. Vernolic acid was the first natural epoxy acid to be correctly identified and it is still the most widely occurring acid of this type (section 3.2.9). It generally occurs in the (+) form but is present as the (−) enantiomer in some seed oils of the Malvaceae.

On prolonged storage, even at 0–5 °C, unsaturated acids in seeds are oxidized to hydroxy and epoxy acids. The regiospecific and stereospecific nature of these reactions suggests that they are under enzymic control. For example, sunflower seeds generate 9,10-epoxystearic acid and 9,10-epoxyoctadec-12-enoic acid, both in optically active form.

Cutins – the hard outer layer in many fruit skins – are polymers of C_{16} and C_{18} hydroxy acids, among which some of the C_{18} acids also contain a *cis* epoxy group (section 2.11). These are intermediates in the biosynthesis of 9,10,18-trihydroxystearic acid from oleic and 9,10,12,13,18-pentahydroxystearic acid from linoleic and include compounds (**32**)–(**34**).

Epoxide groups are frequently present in the complex long-chain acids present in bacterial lipids (section 3.14).

Table 1.11 Epoxy C_{18} acids[a] in seed lipids

9,10-epoxy	12,13-epoxy	15,16-epoxy
18:0[a,b]		
12*c*-18:1[c]	9*c*-18:1[d,e]	
12*a*-18:1		
12*c*,15*c*-18:2	6*c*,9*c*-18:2	9*c*,12*c*-18:2
3*t*,12*c*-18:2	6*t*,9*c*-18:2	

[a] All these epoxy acids have *cis* configuration except 9,10-epoxy stearic acid, which exists in *cis* and *trans* forms.
[b] Absolute configuration of *cis* enantiomers (+)-9L,10L (9*R*,10*S*), (−)-9D,10D(9*S*,10*R*).
[c] Coronaric acid, 9L,10L(9*R*,10*S*).
[d] Vernolic acid; (+)form 12D,13D(12*S*,13*R*), (−)form 12L,13L-(12*R*,13*S*).
[e] Also occurs as the C_{20} homologue (alchornoic acid, 14,15-epoxyicos-11-enoic acid).

X = $HOCH_2$ or OHC or HOOC

(32)

(33)

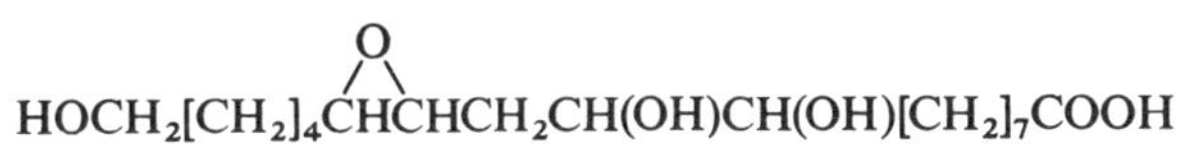

(34)

Natural furan-containing acids are unusual compounds with the general structure (**35**). The first acid of this type was identified as a C_{18} compound ($R^1 = R^2 = H$, $m = 6$, $n = 7$) in the seed oil of *Exocarpus cupressiformis*. A second group of eight or more acids have one or two methyl groups ($R^1 = CH_3$, $R^2 = H$ or CH_3) and belong to two families with $m = 2$ or 4 ($n = 6$, 8 or 10). These unusual acids occur widely but at low levels in fish lipids, where they are present as cholesterol esters or triacylglycerols but not as phospholipids. Their concentration increases markedly during prolonged fasting. One member of this group ($R^1 = CH_3$, $R^2 = H$, $m = 4$, $n = 8$) is the major component (~90%) of the triacylglycerols in rubber latex. There is evidence that furanoid acids are not produced *de novo* but accumulate in rat tissue having been detected in rat food. They are present, at low levels, in vegetables, leaves and fruit. They have also been shown to be present in soybean oil and are responsible for some of the flavour components associated with this oil (Guth and Grosch, 1991).

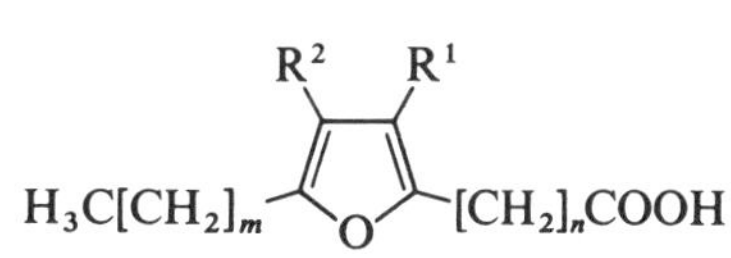

(35)

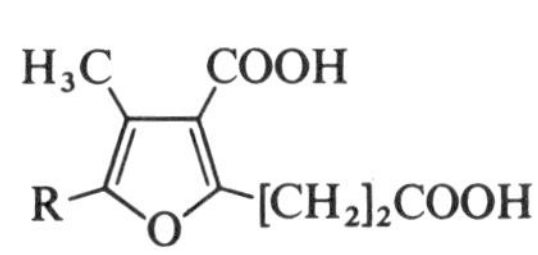

R = $CH_3[CH_2]_4$, $CH_3[CH_2]_2$, $CH_3CH_2CH(OH)$, CH_3CH_2CO

(36)

Some short-chain dibasic furanoid acids, which might be metabolites of the long-chain acids, have been isolated from urine and detected also in blood. They have been designated urofuranic acids (**36**).

References and further reading

Badami, R.C. and Patil, K.B. (1982) *Prog. Lipid Res.*, **19**, 119.

Gunstone, F.D. (1980) in *Fats and Oils: Chemistry and Technology* (eds R.J. Hamilton and A. Bhati), Applied Science, London, p. 53.

Guth, H. and Grosch, W. (1991) *Fat Sci. Technol.*, **93**, 249.

Pohl, P. and Wagner, H. (1972) *Fette Seifen Anstrichm.*, **74**, 541.

Smith, C.R. (1970) *Top. Lipid Chem.*, **1**, 277.

Smith, C.R. (1971) *Prog. Chem. Fats Other Lipids*, **11**, 155.

Smith, C.R. (1979) in *Fatty Acids* (ed. E.H. Pryde), American Oil Chemists' Society, Champaign, p. 29.

1.10 *Trans* acids

Ruminant depot and milk fats are characterized by their low level of unsaturation and by the presence of small amounts of *trans* acids resulting from rumen biohydrogenation (section 10.1.3). The most common acid appears to be vaccenic (11*t*-18:1) but other *trans* acids are also present. For example, butterfat contains 10 octadecenoic acids (7*t* through 16*t*) with the Δ11 isomer accounting for 72% of this group. *Trans* acids have been identified in human milk fat (2–4%), presumably arising from dietary *trans* acids (section 3.4).

Some marine sources of fat contain 6*t*-16:1 at levels up to 5%, sometimes accompanied by higher homologues (8*t*-18:1 and 10*t*-20:1) and by the 7-methyl analogue.

Trans acids also occur in some plant lipids. Acids with conjugated unsaturation – whether olefinic or acetylenic – frequently contain one or more *trans* double bonds (Table 1.5). Other plant lipids contain acids with a *trans* double bond at position 3, 5 or 6 sometimes superimposed on a more familiar pattern of *cis* unsaturation, e.g. 3*t*,9*c*,12*c*- and 5*t*,9*c*,12*c*-18:3 (see also Table 1.4).

Trans acids are easily made from the more readily available *cis* acids by stereomutation (section 10.9) and there is evidence that this occurs inadvertently during processing at high temperatures. Samples of linolenic acid (9*c*,12*c*,15*c*-18:3) have been shown to contain the 9*c*,12*c*,15*t*, 9*t*,12*c*,15*c*, 9*t*,12*c*,15*t* and 9*c*,12*t*,15*c* isomers. Important reactions such as oxidation (section 10.2) and catalytic hydrogenation (section 10.1) are accompanied by double-bond movement and/or stereomutation, so that partially hydrogenated fats may contain appreciable amounts of *trans* isomers.

It has been estimated that the daily intake of *trans* acids (in Western Germany) is 4.5–6.4 g per day. About 35–45% of this comes from ruminant fats (biohydrogenation) and the balance from partially hydrogenated fat. Some of the *trans* acids are deposited in the tissues, which may contain about 2% (1.0–4.3%) of these acids. Another study concludes that such acids are harmless if accompanied by an adequate dietary intake of linoleic acid.

References and further reading

Chardigny, J.M., Sébédio, J.L., Juanéda, P., Vatèle, J.M. and Grandgirard, A. (1993) *Nutrition Research*, **13**, 1105.

Ackman, R.G., Hooper, S.N. and Hooper, D.L. (1974) *J. Am. Oil. Chem. Soc.*, **51**, 42.

Heckers, H., Melcher, F.W. and Dittmar, K. (1979) *Fette Seifen Anstrichm.*, **81**, 217.

Juillet, M.Th. (1978) *Fette Seifen Anstrichm.*, **80**, 445.

Picciano, M.F. and Perkins, E.G. (1976) *Lipids*, **12**, 407.

Pohl, P. and Wagner, H. (1972) *Fette Seifen Anstrichm.*, **74**, 541.

Smith, C.R. (1971) *Prog. Chem. Fats Other Lipids*, **11**, 137.

2

LIPID STRUCTURE

J.L. HARWOOD

2.1 Acylglycerols (glycerides)

Acylglycerols are esters of glycerol (propane-1,2,3-triol) and fatty acids. Partial glycerides are important intermediates in metabolism, and triacylglycerols are the major constituents of natural fats and oils.

In order to designate the stereochemistry of glycerol-containing components, the carbon atoms of glycerol are numbered stereospecifically. When the glycerol molecule is drawn in a Fischer projection with the secondary hydroxyl group to the left of the central prochiral carbon atom, then the carbons are numbered 1, 2 and 3 from top to bottom. Molecules that are stereospecifically numbered in this fashion have the prefix '*sn*' immediately preceding the term 'glycerol' in the name of the compound to distinguish them from compounds that are numbered in a conventional fashion. The prefix '*rac*' in front of the full name shows that the compound is an equal mixture of both antipodes, whereas '*x*' is used if the configuration is unknown or unspecified.

Any glycerolipid will be chiral when the substituents at the *sn*-1 and *sn*-3 positions are different. Mirror-image molecules or enantiomers possess opposite but equal optical rotations. However, if both substituents are long-chain acyl groups, then the optical rotation will be extremely small (Myher, 1978).

2.1.1 Monoacylglycerols (monoglycerides)

These are fatty acid monoesters of glycerol and exist in two isomeric forms (**1**) and (**2**).

2.1.2 Diacylglycerols (diglycerides)

These are fatty acid diesters of glycerol and occur in two isomeric forms (**3**) and (**4**).

2.1.3 Triacylglycerols (triglycerides)

These are fatty acid triesters of glycerol. The fatty acids may be all different, two different, or all alike (see (**5**)).

2.1.4 Glycerides containing four or more acyl groups

When triacylglycerols contain a hydroxy fatty acid, then the hydroxyl group can be esterified with further fatty acids. Examples of these glycerides have been found in plants (Hitchcock, 1975) with the occurrence of tetra-, penta- and hexa-acid glycerides. The tetra-acid triacylglycerol (**6**) from *Sapium sebiferum* oil (Sprecher *et al.*, 1965) is an example.

References and further reading

Hitchcock, C. (1975) in *Recent Advances in the Chemistry and Biochemistry of Plant Lipids* (eds T. Galliard and E.I. Mercer), Academic Press, New York, p. 1.

Myher, J.J. (1978) in *Handbook of Lipid Research*, vol. 1 (ed. A. Kuksis), Plenum, New York, p. 123.

Sprecher, H.W., Maier, R., Barber, M. and Holman, R.T. (1965) *Biochemistry*, **4**, 1856.

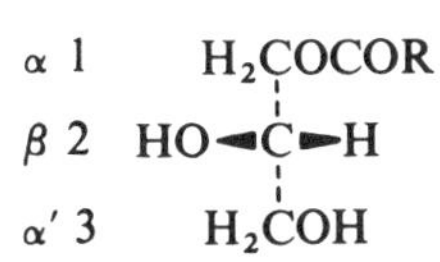

(**1**) 1-Monoacyl-*sn*-glycerol (α isomer)

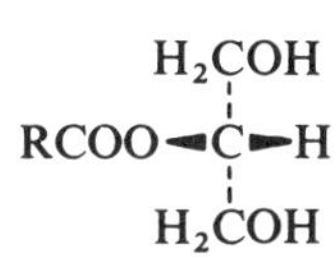

(**2**) 2-Monoacyl-*sn*-glycerol (β isomer)

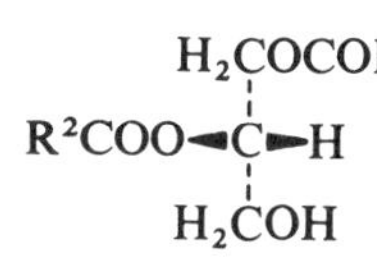

(**3**) 1,2-Diacyl-*sn*-glycerol (or α,β isomer)

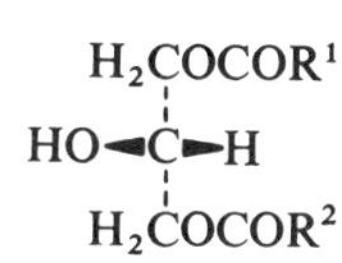

(**4**) 1,3-Diacyl-*sn*-glycerol (or α,α′ isomer)

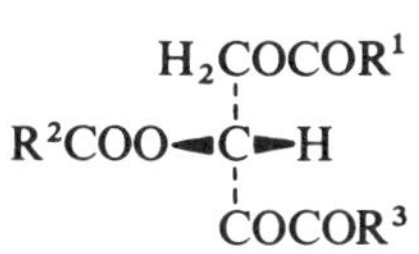

(**5**) Triacyl-*sn*-glycerol

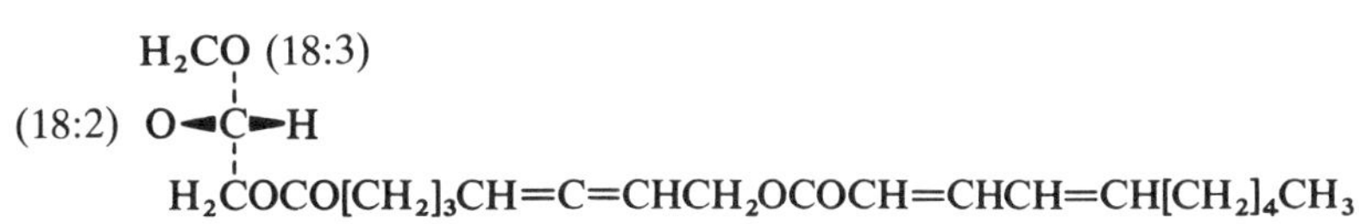

(**6**)

2.2 Glycerol ethers

Glycerol ethers can be divided into two general types: alkyl ethers and alk-1-enyl ethers (or plasmalogens).

2.2.1 Alkyl ethers

(a) Monoalkyl ethers (**7**) *and* (**8**)

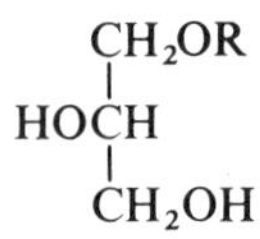
(**7**) *sn*-1 or α isomer

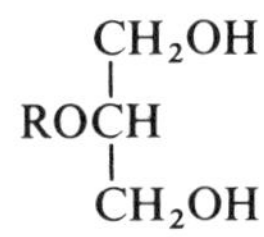
(**8**) *sn*-2 or β isomer

The α-isomer is the more common and is usually found to have the D or *sn*-1 configuration. Examples include batyl alcohol (*sn*-1-*O*-octadecylglycerol) and chimyl alcohol (*sn*-1-*O*-hexadecyl alcohol). *sn*-3-*O*-Tetramethylhexadecylglycerol is a reported example of an α-isomer with the L configuration. β-Monophytanylglycerol (2-*O*-tetramethylhexadecylglycerol) has been reported and is an example of a β-monoalkyl ether.

(b) Dialkyl ethers (**9**) *and* (**10**)

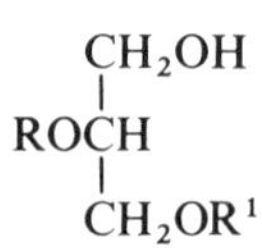
(**9**) *sn*-2,3 or α,β isomer

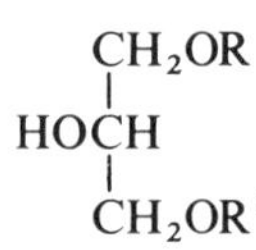
(**10**) *sn*-1,3 or α,α' isomer

Although diethers have been synthesized in the 1,2 and 1,3 forms, the only naturally occurring diether is *sn*-2,3-di-*O*-tetramethylhexadecylglycerol (diphytanylglycerol).

(c) Trialkyl ethers (**11**)

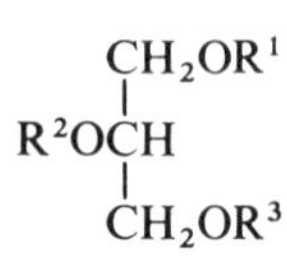
(**11**)

These compounds have not yet been reported in Nature but have been synthesized with various combinations of saturated or unsaturated chains. They have been proposed as non-fattening dietary lipids (Mangold and Paltauf, 1983).

(d) Acylated alkyl ethers (**12**) *and* (**13**)

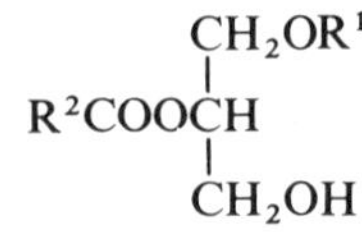
(**12**) *sn*-1-Alkyl-2-acylglycerol

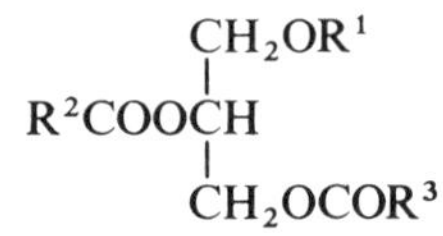
(**13**) *sn*-1-Alkyl-2,3-diacylglycerol

Both the monoacyl and diacyl derivatives are found in Nature.

2.2.2 Alk-1-enyl ethers

As with the alkyl ethers of glycerol, the monoalk-1-enyl ethers predominate in Nature with small amounts of the dialk-1-enyl ethers but no trialk-1-enyl ethers so far detected.

(a) Monoalk-1-enyl ethers (**14**) *and* (**15**)

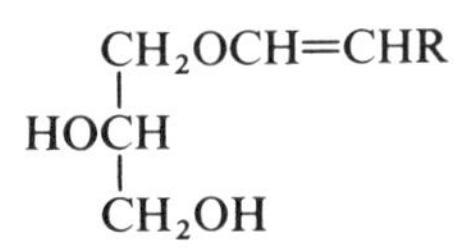
(**14**) *sn*-1 or α isomer

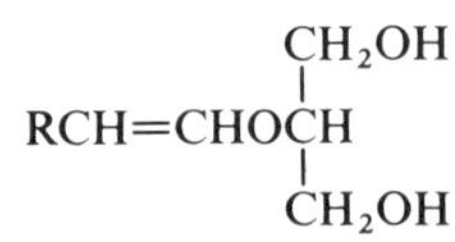
(**15**) *sn*-2 or β isomer

Only the *sn*-1 isomer occurs naturally. Chains of 16 and 18 carbons are usual.

(b) Dialk-1-enyl ethers (**16**) *and* (**17**)

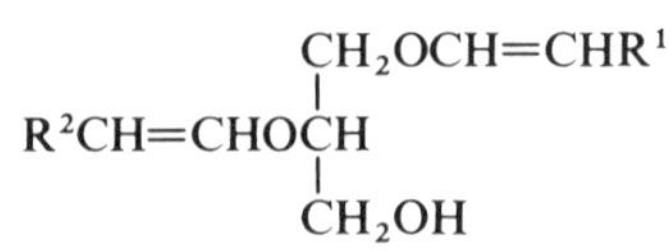
(**16**) *sn*-1,2 or α,β isomer

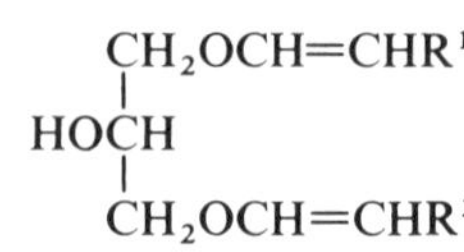
(**17**) *sn*-1,3 or α,α' isomer

(c) Acylated alk-1-enyl ethers (neutral plasmalogens) (**18**) *and* (**19**)

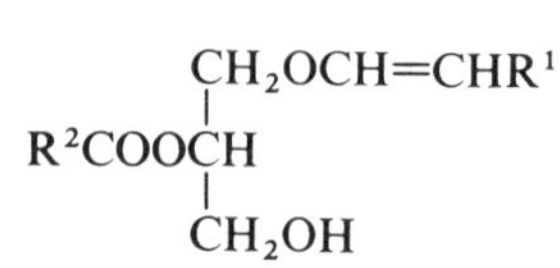
(**18**) *sn*-1-Alk-1′-enyl-2-acylglycerol

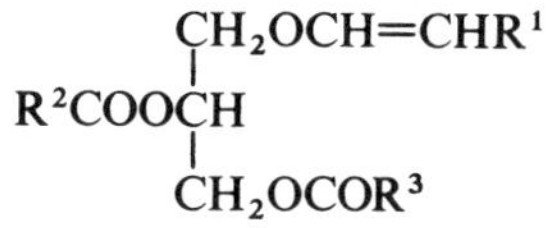

(**19**) *sn*-1-Alk-1′-enyl-2,3-diacylglycerol

Both the monoacyl and diacyl derivatives have been reported.

2.2.3 Ether lipids of Archaebacteria

The lipids of Archaebacteria are clearly distinguished from those of Eubacteria in consisting almost entirely of derivatives of a dialkylglycerol, diphytanylglycerol diether structure (**20**).

Variants of this structure including terpenyl-, glyco- and sulphoglyco- derivatives have been reported and identified. These lipids are found in all three classes of Archaebacteria – the extreme halophiles, the methanogens and the thermoacidophiles (see Kates, 1990).

2.2.4 Betaine lipids

Two betaine lipids, diacylglycerol-*O*-(*N*,*N*,*N*-trimethyl)homoserine (DGTS (**21**)) and diacylglycerol-*O*-(hydroxymethyl)-(*N*,*N*,*N*-trimethyl)-*β*-alanine (DGTA (**22**)), have been reported from a variety of lower plants and algae, often as major components (see Eichenberger, 1990).

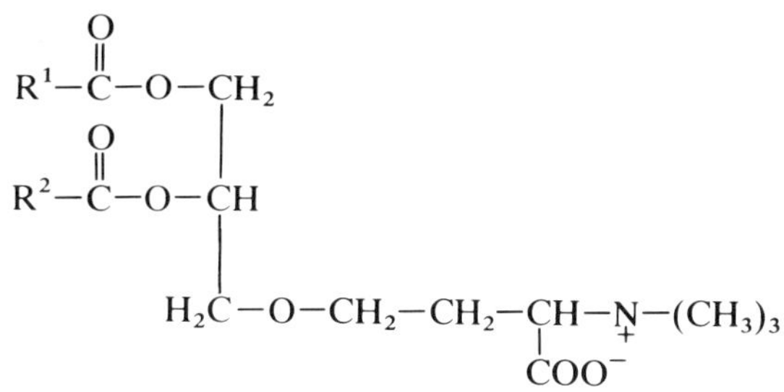

(**21**) DGTS

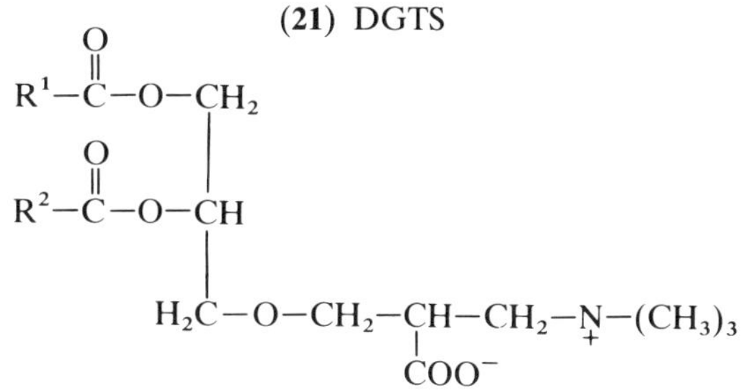

(**22**) DGTA

References and further reading

Eichenberger, W. (1990) in *Plant Lipid Biochemistry, Structure and Utilization* (eds P.J. Quinn and J.L. Harwood), Portland Press, London, p. 9.

Kates, M. (1972) in *The Ether Bond in Lipids* (ed. F. Snyder), Academic Press, New York, p. 351.

Kates, M. (1990) in *Handbook of Lipid Research*, vol. 6 (ed. M. Kates), Plenum, New York, p. 1.

Mangold, H.K. and Paltauf, H. (eds) (1983) *Ether Lipids*, Academic Press, New York.

Snyder, F. (1969) *Prog. Chem. Fats Other Lipids*, **10**, 287.

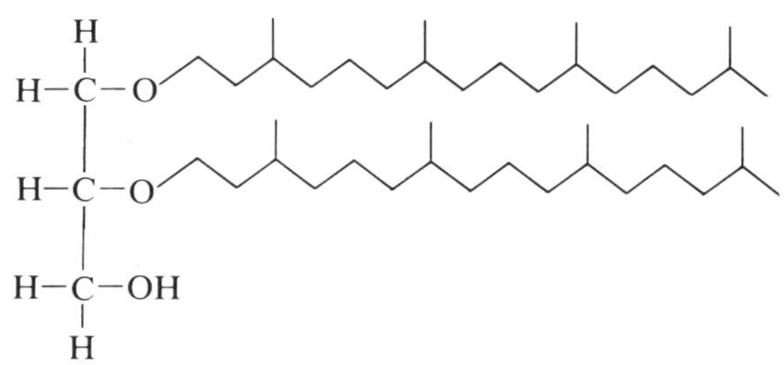

(**20**) 2,3-Di-*O*-phytanyl-*sn*-glycerol

2.3 Phospholipids

Phospholipids are divided into two main classes depending on whether they contain a glycerol or a sphingosyl backbone.

2.3.1 Glycerophospholipids

The compounds are named after and contain struc-

Table 2.1 Structure and distribution of important phosphoglycerides of general formula

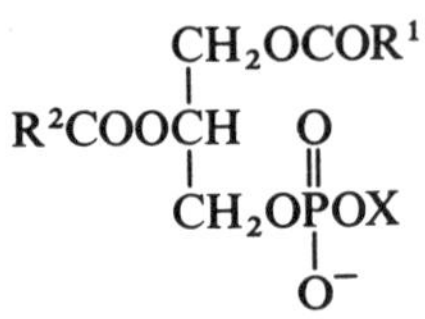

Substituent (X)	Phospholipid	Remarks
H	Phosphatidic acid (PtdOH)	Negatively charged lipid. Important metabolic intermediate, only occurring in trace amounts
$—CH_2CH(\overset{+}{N}H_3)COO^-$	Phosphatidylserine (PtdSer)	Negatively charged lipid. Serine is the L isomer. Widespread but minor lipid. *N*-Acylated derivatives have been found
$—CH_2CH_2\overset{+}{N}H_3$	Phosphatidylethanolamine (PtdEtn)	Widespread and major lipid. The partly methylated derivatives (phosphatidyl-*N*-monomethyl-ethanolamine, phosphatidyl-*N*-dimethyl-ethanolamine) are found in small amounts in many organisms. They are metabolic intermediates in the conversion of phosphatidyl-ethanolamine to phosphatidylcholine (see section 11.2.1). *N*-Acylated derivatives of phosphatidylethanolamine (and lyso-phosphatidylethanolamine) are found in small amounts in many tissues. In some tissues they may be significant components (Schmid *et al.*, 1990)
$—CH_2CH_2\overset{+}{N}(CH_3)_3$	Phosphatidylcholine (PtdCho)	Has a net neutral charge. The major animal phospholipid. Found in large amounts in plants and in small quantities in some bacteria
$—CH_2CH(OH)CH_2OH$	Phosphatidylglycerol (PtdGro)	Negatively charged lipid. Headgroup glycerol has *sn*-1 configuration. The major phospholipid in photosynthetic tissues and many bacteria. Some bacteria contain *O*-aminoacyl groups (lysine, ornithine, arginine or alanine) attached to position 3 of the base glycerol. Bisphosphatidic acid, the fully acylated analogue of PG, has been found in some plant tissues
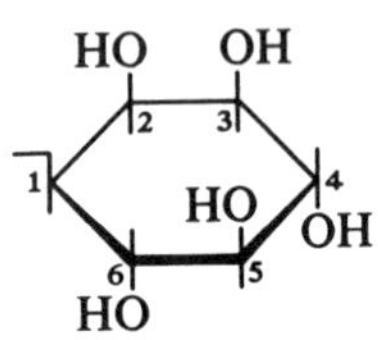	Phosphatidylinositol (PtdIns)	Negatively charged lipid. Inositol is the *myo* isomer. Widespread and usually minor lipid. The higher inositides, diphosphoinositide (1-(3-*sn*-phosphatidyl)-L-*myo*inositol 4-phosphate) and triphosphoinositide (1-(3-*sn*-phosphatidyl)-L-*myo*inositol 4,5-bisphosphate) have been found in most eukaryotes, including animals, plants and yeast
Phosphatidylglycerol	Diphosphatidylglycerol (DPG) (cardiolipin)	Negatively charged lipid. Common in bacteria. Localized in the inner mitochondrial membrane of eukaryotes

Abbreviations: R^1 and R^2 are long-chain alkyl groups.

tures that are based on phosphatidic acid (3-*sn*-phosphatidic acid). The X moiety attached to the phosphate includes nitrogenous bases (amino alcohols) or polyols. In Table 2.1 the major phosphoglycerides are listed. Some of the less common compounds are indicated in the remarks with the table.

2.3.2 Sphingophospholipids

Lipids that contain sphingosine (*trans*-D-*erythro*-1,3-dihydroxy-2-amino-4-octadecene) or a related amino alcohol are known as sphingolipids. The commonest phospholipid in this class is the phosphorylcholine ester of an *N*-acylsphingosine (or ceramide) that is more commonly called sphingomyelin (**23**).

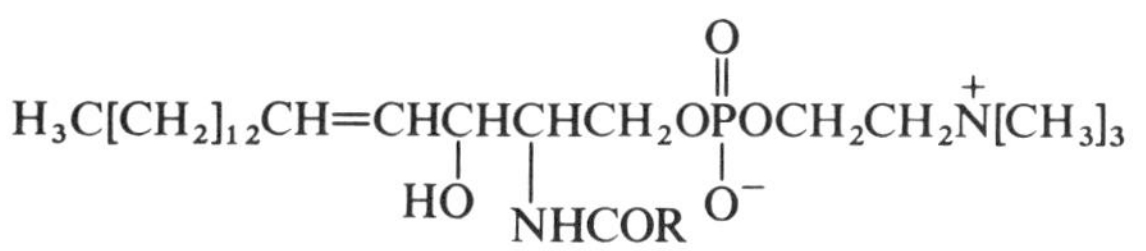

(**23**) *N*-Acyl-*trans*-4-sphingenine-1-phosphorylcholine (sphingomyelin containing sphingosine as the sphingosyl moiety)

Although sphingomyelin is a major lipid of certain membranes in animal (particularly nervous) tissues, it is of minor importance in plants and probably absent from bacteria. Even in animals, the non-phosphorus sphingolipids (section 2.4) are more widely distributed.

Glycolipids based on phytosphingosine that contain both inositol and phosphate were identified first by Carter and coworkers (Carter *et al.*, 1969). Their general structure is shown in Fig. 2.1.

Both phytosphingosine- and dehydrosphingosine-based variants have been reported. However, all contain inositol linked via a phosphodiester linkage to the ceramide and via a glycosidic bond to a chain of sugar residues of variable composition (Hetherington and Drobak, 1992). The components used to be called 'phytoglycolipids' but Laine and Hsieh (1987) suggested that the term 'glycophosphosphingolipid' be used to distinguish them from other plant glycosphingolipids.

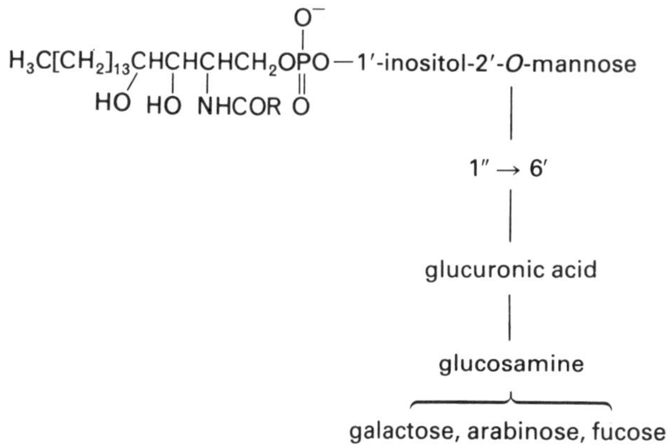

Figure 2.1 Phytoglycolipid.

A number of glycophosphosphingolipids have been reported (Laine and Hsieh, 1987; Stults *et al.*, 1989). Two structures are shown in (**24**) and (**25**); see Hetherington and Drobak (1992) for more details.

The analysis of plant glycophosphosphingolipids has been thoroughly discussed by Laine and Hsieh (1987).

2.3.3 Variants on the diacyl structure

These are summarized in Table 2.2.

(a) Monoacyl derivatives

Lyso derivatives (monoacylglycerophospholipids) are found in small amounts in most tissue, but their presence in large amounts is usually indicative of lipid degradation either before or during extraction. An exception is the presence of lysophosphatidylcholine and lysophosphatidylethanolamine as the major phospholipids of cereal grains.

(b) Plasmalogens

These are the monoacyl monoalk-1-enyl ether forms of phospholipids. Choline and ethanolamine

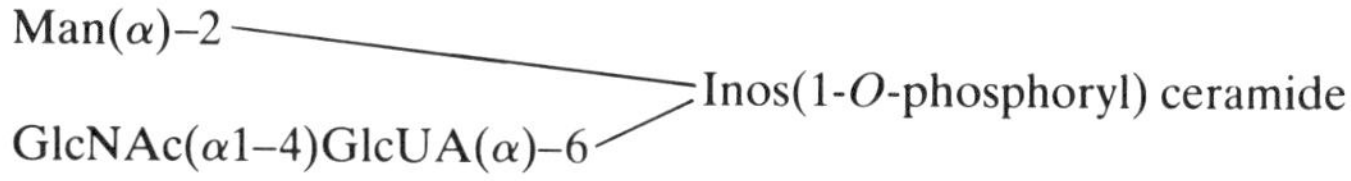

(**24**), found in corn

GlcNAc(α1–4)GlcUA(α1–2)Inos(1-*O*-phosphoryl) ceramide

(**25**), found in tobacco

GlcUA = glucuronic acid

Table 2.2 Variations in phosphoglyceride structures

Structure[a]	Structure form	Common phosphoglyceride[a]
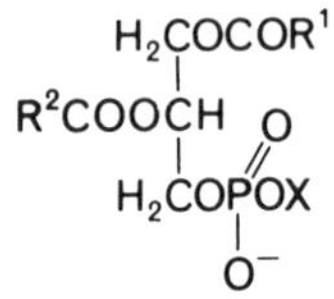	Diacyl ester form	All
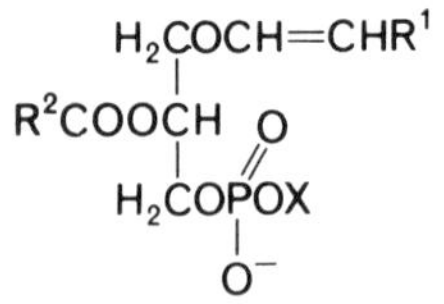	Monoacyl monoalk-1-enyl ether form (plasmalogen)	PS, PE, PC
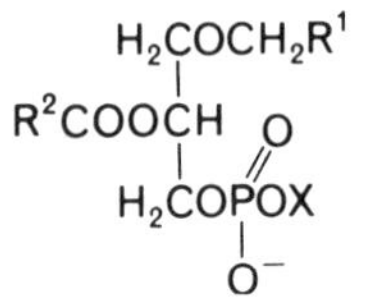	Monoacyl monoether form	PE, PC
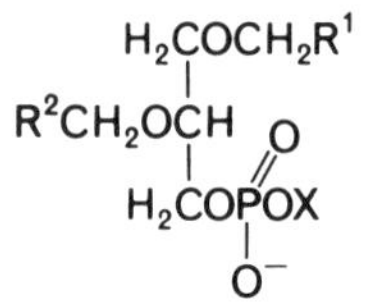	Diether form	PG, PGP
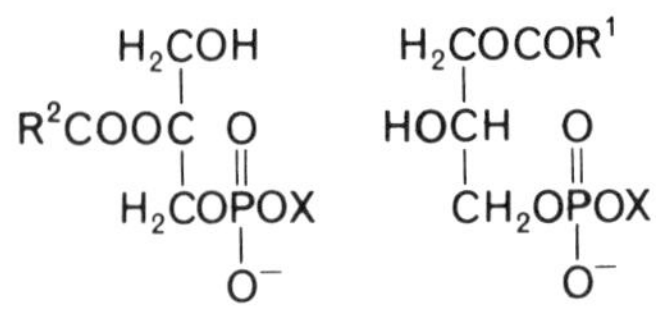 β isomer α isomer	Monoacyl (lyso) form	PS, PE, PC
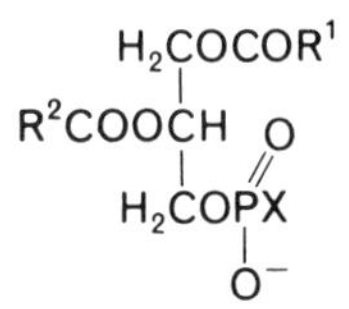	Phosphono form	PE

[a] For abbreviations see Table 2.1. R^1 and R^2 represent saturated or unsaturated alkyl chains. X represents a nitrogenous base or polyol residue. PGP is phosphatidylglycerol phosphate.

plasmalogens are the most common forms, although serine plasmalogen has also been found. The percentage of the plasmalogen form of a given phospholipid may be quite high (e.g. in mammalian brain) and is usually underestimated because most thin-layer chromatographic systems fail to resolve the plasmalogen from the diacyl form of a given lipid class. They are present in most animal tissues but are rare in plants.

Plasmanic acid and plasmenic acid represent the alkyl and alk-1-enyl analogues of phosphatidic acid, respectively. Thus, the choline plasmalogen (1-alkyl-2-acyl-*sn*-glycero-3-phosphocholine) is termed plasmanylcholine, while a 1-alk-1-enyl derivative would be plasmenylcholine. There are a number of reports of phospholipids in bovine heart and spermatozoa that have *O*-alkyl groups at both *sn*-1 and *sn*-2 positions. Halophilic bacteria contain large amounts of dialkylglycerolipids but of the opposite stereochemical configuration (*sn*-2,3 or D series; see Snyder, 1991).

Platelet-activating factor (1-alkyl-2-acetyl-*sn*-glycero-3-phosphocholine) (**26**) is a biologically active phospholipid of great current interest.

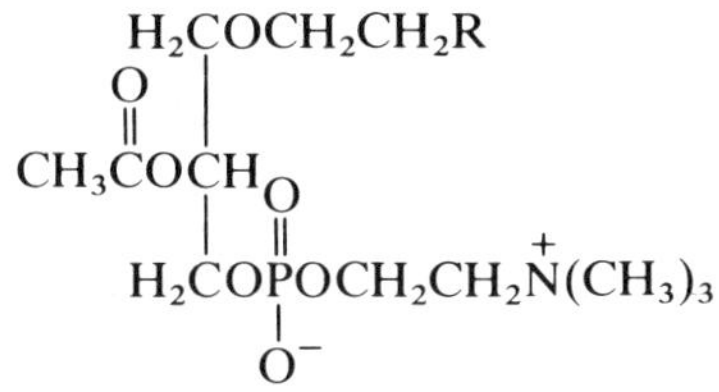

(**26**) 1-Alkyl-2-acetyl-*sn*-glycero-3-phosphocholine (platelet-activating factor, PAF)

(c) *Ether derivatives*

Monoacyl monoether and diether forms have been detected in very small amounts in various tissues (Mangold and Paltauf, 1983).

2.3.4 Phosphonolipids

Phosphono analogues of various phospholipids have been reported. These include phosphono analogues of sphingosylphosphatides, which have been found in various invertebrates, and of phosphatidylethanolamine (Fig. 2.2). Phosphonolipids are major constituents in three phyla and are synthesized by phytoplankton, the base of the food chains of the ocean (Kittredge and Roberts, 1969). The following phosphonic acids have been found in Nature: 2-aminoethylphosphonic acid, 2-methylaminoethylphosphonic acid, 2-dimethylaminoethylphosphonic acid, 2-trimethylaminoethylphosphinic acid and 2-amino-3-phosphonopropionic acid. The first of these phosphonic acids is the main component of the phosphonolipids. Phosphonolipids have been reviewed by Hori and Nozawa (1982).

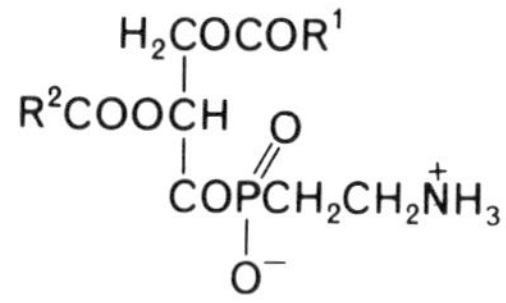

Phosphono form of phosphatidylethanolamine

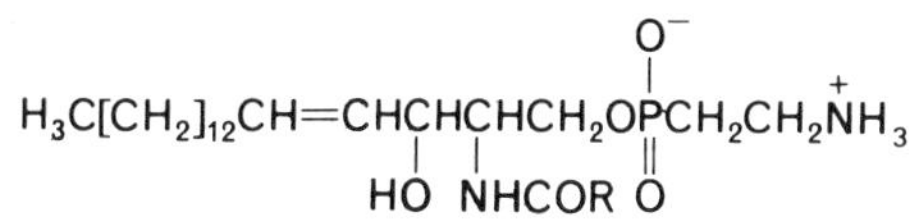

Ceramide aminoethyl phosphonate (ceramide ciliatine)

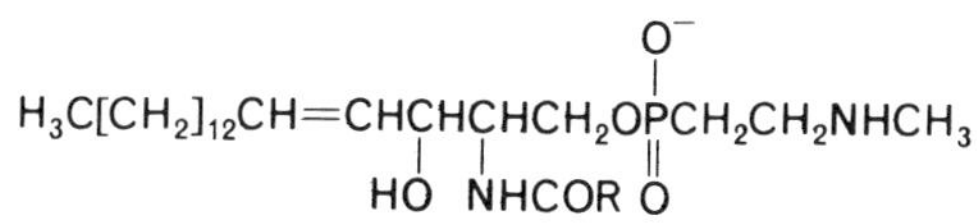

Ceramide *N*-methylaminoethyl phosphonate

Figure 2.2 Phosphonolipids.

References and further reading

Ansell, G.B., Dawson, R.M.C. and Hawthorne, J.N. (eds) (1973) *Form and Function of Phospholipids*, Elsevier, Amsterdam.

Carter, H.E., Strobach, D.R. and Hawthorne, J.N. (1969) *Biochemistry*, **8**, 383.

Hetherington, A.M. and Drobak, B.K. (1992) *Prog. Lipid Res.*, **31**, 53.

Hori, T. and Nozawa, Y. (1982) in *Phospholipids* (eds J.N. Hawthorne, R.M.C. Dawson and G.B. Ansell), Elsevier, Amsterdam, p. 95.

Kanfer, J.N. and Hakomori, S.-I. (1983) *Sphingolipid Biochemistry*, Plenum, New York.

Kates, M. (1972) *Techniques in Lipidology*, Elsevier, Amsterdam.

Kittredge, J.S. and Roberts, E. (1969) *Science*, **164**, 37.

Laine, R.A. and Hsieh, T.C.-Y. (1987) *Methods Enzymol.*, **138**, 186.

Mangold, H.K. and Paltauf, H. (eds) (1983) *Ether Lipids*, Academic Press, New York.

Ratledge, C. and Wilkinson, S.G. (eds) (1988) *Microbial Lipids*, vol. 1, Academic Press, London.

Schmid, H.H.O., Schmid, P.C. and Watarajan, V. (1990) *Prog. Lipid Res.*, **29**, 1.

Snyder, F. (1991) in *Biochemistry of Lipids, Lipoproteins and Membranes* (eds D.E. Vance and J. Vance), Elsevier, Amsterdam, p. 241.

Stults, C.L.M., Sweeley, C.C. and Macher, B.A. (1989) *Methods Enzymol.*, **179**, 167.

2.4 Glycosphingolipids

A large number of different glycosphingolipids are known, which differ in the nature and number of their glycosyl residues. A general reference is that of Kanfer and Hakomori (1983). Insect glycolipids are dealt with specifically by Wiegandt (1992) while a useful introductory article is that by Sweeley (1991).

2.4.1 Ceramides and glycosylceramides

Attachment of a fatty acid to the amino group of sphingosine or other related amino alcohol gives rise

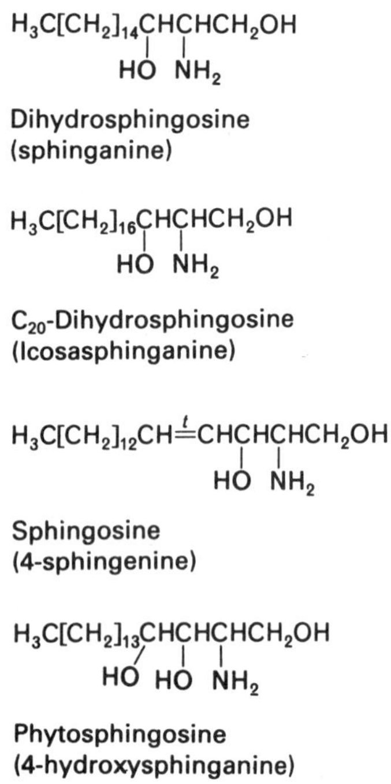

Figure 2.3 Amino alcohols found in sphingolipids.

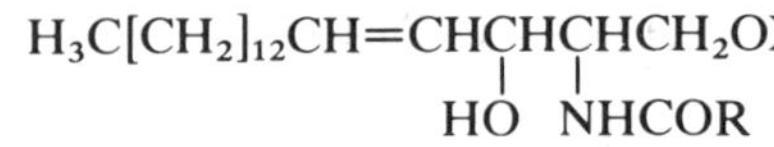

(**27**) Cerebrosides and neutral ceramides

to a ceramide. The most commonly found sphingosyl alcohols are shown in Fig. 2.3. Attachment of glucose or of galactose by *O*-ester linkage to the primary alcohol of the sphingosyl moiety yields a ceramide hexoside (or cerebroside) (see (**27**)). X = H for simple ceramides; X = a monosaccharide (usually galactose) for cerebrosides in animals and often glucose in plants. The linkage is 1-*O*-β. X = two to six sugar units for ceramide polyhexosides where the first sugar residue is glucose. The sphingosine base can vary. It is usually phytosphingosine in plant sphingolipids. The fatty acids in the R group are frequently 2-hydroxy compounds for the galactosylceramides. In animals the galactocerebroside is the most common. Confusingly, further attachment of hexosides to glucocerebroside yields the neutral ceramides (see (**27**)). These are usually written by a short-hand nomenclature, e.g. Glc-Gal-Gal-Glc-Cer would be glucosyl(1$\rightarrow$4)galactosyl(1$\rightarrow$4)galactosyl-(1$\rightarrow$4)glucosyl(1$\rightarrow$1)ceramide.

The nomenclature of simple glycosphingolipids as recommended by the IUPAC–IUB Commission on Biochemical Nomenclature is shown in Table 2.3. Further classification can be based on shared partial oligosaccharide sequences as shown in Table 2.4 and as discussed by Sweeley (1991).

Some of the neutral ceramides have immunochemical properties, such as the so-called 'Forsmann antigen' (CerGlc(4 $\leftarrow$ 1)Gal(4 $\leftarrow$ 1)Gal(3 $\leftarrow$ 1)-α-*N*-acetylgalactosamine). In general, each mammalian organ has a characteristic pattern of neutral

Table 2.3 Nomenclature for the classification of glycosphingolipids (from Sweeley, 1991)

Root name	Abbreviation	Partial structure[a] (IV III II I)
Ganglio	Gg	Galβ1–3GalNAcβ1–4Galβ1–4Glc–Cer
Lacto	Lc	Galβ1–3GlcNAcβ1–3Galβ1–4Glc–Cer
Neolacto	nLc	Galβ1–4GlcNAcβ1–3Galβ1–4Glc–Cer
Globo	Gb	GalNAcβ1–3Galα1–4Galβ1–4Glc–Cer
Isoglobo	iGb	GalNAcβ1–3Galα1–3Galβ1–4Glc–Cer
Mollu	Mu	GalNAcβ1–2Manα1–3Manβ1–4Glc–Cer
Arthro	At	GlcNAcβ1–4GlcNAcβ1–3Manβ1–4Glc–Cer

[a] Roman numerals define sugar positions in the root structure.

Table 2.4 Names and abbreviations of simple glycosphingolipids (Taken from IUPAC–IUB Commission on Biochemical Nomenclature, 1977)

Structure	Trivial name of oligosaccharide[a]	Symbol[b]
Gal(α1–4)Gal(β1–4)GlcCer	Globotriaose	$GbOse_3$
GalNAc(β1–3)Gal(α1–4)Gal(β1–4)GlcCer	Globotetraose	$GbOse_4$
Gal(α1–3)Gal(β1–4)GlcCer	Isoglobotriaose	$iGbOse_3$
GalNAc(β1–3)Gal(α1–3)Gal(β1–4)GlcCer	Isoglobotetraose	$iGbOse_4$
Gal(β1–4)Gal(β1–4)GlcCer	Mucotriaose	$McOse_3$
Gal(β1–3)Gal(β1–4)Gal(β1–4)GlcCer	Mucotetraose	$McOse_4$
GlcNAc(β1–3)Gal(β1–4)GlcCer	Lactotriaose	$LcOse_3$
Gal(β1–3)GlcNAc(β1–3)Gal(β1–4)GlcCer	Lactotetraose	$LcOse_4$
Gal(β1–4)GlcNAc(β1–3)Gal(β1–4)GlcCer	Neolactotetraose	$nLcOse_4$
GalNAc(β1–4)Gal(β1–4)GlcCer	Gangliotriaose	$GgOse_3$
Gal(β1–3)GalNAc(β1–4)Gal(β1–4)GlcCer	Gangliotetraose	$GgOse_4$
Gal(α1–4)GalCer	Galabiose	$GaOse_2$
Gal(1–4)Gal(α1–4)GalCer	Galatriaose	$GaOse_3$
GalNAc(1–3)Gal(1–4)Gal(α1–4)GalCer	*N*-Acetylgalactosaminylgalatriaose	GalNAc1–3$GaOse_3$

[a] Name of glycolipid is formed by converting ending '-ose' to '-osyl', followed by 'ceramide', without space, e.g. globotriaosylceramide.
[b] Should be followed by Cer for the glycolipid, without space, e.g.. $McOse_3Cer$, Mc_4Cer.
Abbreviations: GlcNAc, *N*-acetylglucosamine; Gal, galactose; Glc, glucose; Cer, ceramide.

ceramides, with kidney, lung, spleen and blood containing quite large amounts.

In micro-organisms, glucose or galactose are the usual carbohydrate residues, although glucuronic acid has been reported in the ceramides of *Pseudomonas paucimobilis*. Inositol and mannose residues may also be attached to sphingosine bases in fungi (Ratledge and Wilkinson, 1988). However, sphingolipids are not quantitatively important in most micro-organisms. A number of acylated sphingosines have been reported, where the hydroxyl as well as the amino group of the base can be reacted. For example, both triacetyl and tetraacetyl derivatives have been described in *Hansenula ciferri* (Brennan *et al.*, 1974).

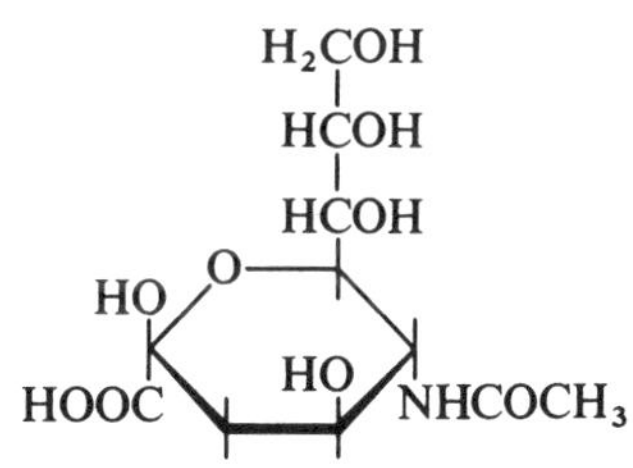

(**28**) Sialic acid

2.4.2 Gangliosides

These are glycosphingolipids that contain a ceramide linked to a glucosylgalactosylsialic acid (β-D-*N*-acetylneuraminic acid) moiety. Other monosaccharide and sialic acid residues may also be present. The structure of sialic acid is shown in (**28**).

Gangliosides have been reported with structures shown in Table 2.5.

2.4.3 Cerebrosides

These are glycosides of *N*-acyl long-chain bases (ceramides). Galactose and glucose are the monosaccharides commonly found. The structures of two representative molecules – one from mammalian brain and the other from a higher plant – are shown in (**29**) and (**30**).

Cerebrosides frequently contain large amounts of 2-hydroxy fatty acids, which are linked through the

Brain cerebrosides

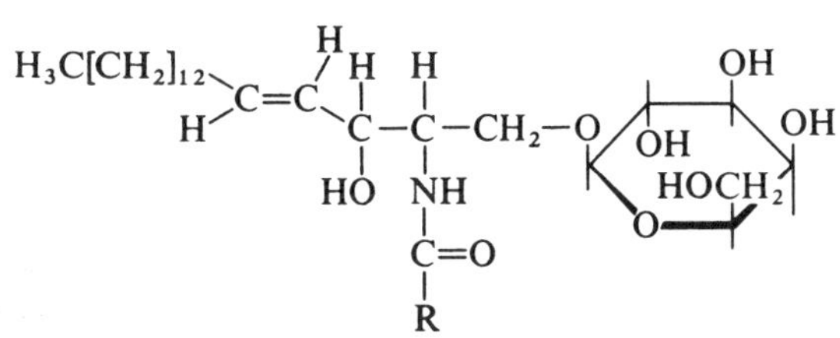

(**29**) *N*-Acylsphingosine-1-β-D-galactopyranoside

where R—CO =

lignoceryl ($C_{24}H_{47}O$)	in kerasin
cerebronyl (2-hydroxylignoceryl)	in phrenosin
nervonyl ($C_{24}H_{45}O$)	in nervone
2-hydroxynervonyl	in oxynervone

Plant cerebrosides

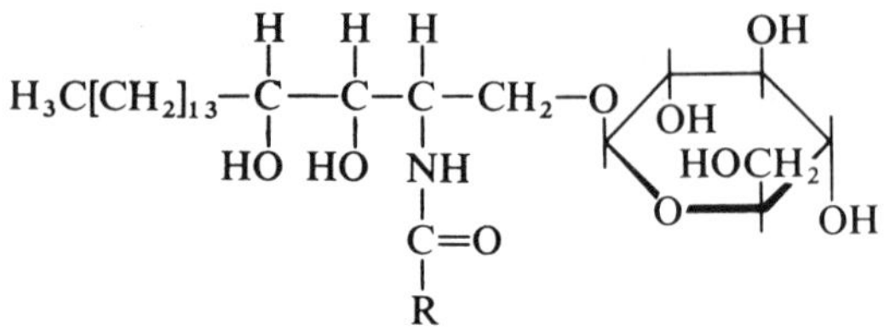

(30) *N*-(α-Hydroxy)acylphytosphingosine-1-*O*-β-glucopyranoside
where R—CO =
2-hydroxypalmitic, 2-hydroxystearic, 2-hydroxybehenic and 2-hydroxylignoceric acid chains
In addition to phytosphingosine, other long-chain bases found in plant cerebrosides are dihydrosphingosine and dehydrophytospingosine

amino group of the sphingosine base. In addition, odd-chain-length acids are also found. Typical fatty acids might be behenic (22:0), lignoceric (24:0) as in kerasin, nervonic (24:1) as in nervone, cerebronic (α-OH 24:0) as in phrenosin, and 2-hydroxynervonic (α-OH 24:1) as in oxynervone. The sugar composition of mammalian cerebrosides depends on the tissue source, brain cerebroside containing mainly galactose while that of blood contains mainly glucose. The myelin sheath of nerves contains particularly large amounts of cerebrosides, while they are also present in substantial quantities in lung and kidney.

Table 2.5 Abbreviated representation of gangliosides (From IUPAC–IUB Commission on Biochemical Nomenclature, 1977)

	Lipid[a]	Designation according to Wiegandt[b]	Svennerholm[c]
1	I^3NeuAc–GalCer	G_{Gal}1NeuAc	–
2	II^3NeuAc–LacCer	G_{Lac}1NeuAc	G_{M3}
3	II^3NeuGc–LacCer	G_{Lac}1NeuNG1	–
4	$II^3(NeuAc)_2$LacCer	G_{Lac}2NeuAc	G_{D3}
5	II^3NeuAc/NeuGc–LacCer	G_{Lac}2NeuAc/NeuNG1	–
6	II^3NeuGc–LacCer	G_{Lac}2NeuNG1	–
7	II^3NeuAc–GgOse$_3$Cer	G_{Gtri}1NeuAc	G_{M2}
8	II^3NeuAc–GgOse$_4$Cer	G_{Gtet}1NeuAc	G_{M1}
9	IV^3NeuAc–nLcOse$_4$Cer	G_{Lntet}1aNeuAc	$G_{M1\text{–}GlcNAc}$
10	IV^6NeuAc–nLcOse$_4$Cer	G_{Lntet}1bNeuAc	–
11	IV^2Fuc, II^3NeuAc–GgOse$_4$Cer	G_{Gfpt}1NeuAc	–
12	IV^3NeuAc–nLcOse$_4$Cer	–	–
13	$II^3(NeuAc)_2$–GgOse$_4$Cer	G_{Gtet}2bNeuAc	G_{D1b}
14	IV^3NeuAc, II^3NeuAc–GgOse$_4$Cer	G_{Gtet}2aNeuAc	G_{D1a}
15	$II^3(NeuAc)_3$–GgOse$_4$Cer	G_{Gtet}3bNeuAc	–
16	IV^3NeuAc, $II^3(NeuAc)_2$–GgOse$_4$Cer	G_{Gtet}3aNeuAc	G_{T1}
17	IV^3NeuAc, $II^3(NeuAc)_3$–GgOse$_4$Cer	G_{Gtet}4bNeuAc	–
18	$IV^3(NeuAc)_2$, $II^3(NeuAc)_3$–GgOse$_4$Cer	G_{Gtet}5NeuAc	–
19	IV^3NeuAc, II^3NeuAc–GgOse$_5$Cer	G_{Gpt}2aNeuAc	–

[a] To indicate linkage points and anomeric form: Fuc should be written (←1αFuc); NeuAc should be written (←2αNeuAc); $(NeuAc)_2$ should be written $(←2αNeuAc8)_2$; etc. If these features are assumed or defined, the short form used in this column is more convenient for use in texts and tables.
[b] The subscripts to G (for ganglioside), from lipid 7 onwards have the meanings: Gtri = gangliotriose, Gtet = gangliotetraose, Lntet = lactoisotetraose, Gpt = gangliopentaose, Gfpt = gangliofucopentaose (Wiegandt, 1973).
[c] G = ganglioside, M = monosialo, D = disialo, T = trisialo. Arabic numerals indicate sequence of migration in thin-layer chromatograms (Svennerholm, 1963).

Notes on composition and occurrence (see Gurr and Harwood, 1991)

Gangliosides appear to be confined to the animal kingdom. In man, cattle and horse, the main ganglioside outside the brain is G_{Lact}1NeuAc; *N*-glycolylneuraminic acid is chief sialic acid in erythrocyte and spleen ganglioside of horse and cattle.

Physical properties: Insoluble in non-polar solvents; form micelles in aqueous solution.

Major bases: C_{18} and C_{20} sphingosines. Minor amounts of dihydro analogues.

Fatty acids: Large amounts of 18:0 (86–95% in brain).

Occurrence: Mainly in grey matter of brain but also in spleen, erythrocytes, liver and kidney. Modern analytical techniques have shown them to be present in a much wider range of tissues than previously realized. Main gangliosides of human brain are G_{GNT}1,2a,2b,3a.

2.4.4 Sulphatides

Sulphatides are 3-sulphate esters of galactosylcerebrosides (**31**). Their distribution in mammalian tissues is generally similar to that of cerebrosides and their fatty acid and base compositions are also alike.

$H_3C[CH_2]_{14}CHCHCH_2O$—β-D-galactopyranosyl-3-sulphate
(with HO on the first CH and NHCOR on the second CH)

HO NHCOR

(**31**)

References and further reading

Brennan, P.J., Griffin, P.F.S., Losel, D.M. and Tyrrell, D. (1974) *Prog. Chem. Fats Other Lipids*, **14**, 49.

Gurr, M.I. and Harwood, J.L. (1991) *Lipid Biochemistry*, 4th edn, Chapman & Hall, London.

IUPAC–IUB Commission on Biochemical Nomenclature (1977) *Eur. J. Biochem.*, **79**, 11.

Kanfer, J.N. and Hakomori, S.-I. (1983) *Sphingolipid Biochemistry*, Plenum, New York.

Kates, M. (1986) *Techniques in Lipidology*, 2nd edn, Elsevier, Amsterdam.

Ratledge, C. and Wilkinson, S.G. (eds) (1988) *Microbial Lipids*, vol. 1, Academic Press, London.

Svennerholm, L. (1963) *J. Neurochem.*, **10**, 613.

Sweeley, C.C. (1991) in *Biochemistry of Lipids, Lipoproteins and Membranes* (eds D.E. Vance and J. Vance), Elsevier, Amsterdam, p. 327.

Wiegandt, H. (1973) *Hoppe-Seyler's Z. Physiol. Chem.*, **354**, 1049.

Wiegandt, H. (1992) *Biochim. Biophys. Acta.*, **1123**, 117.

2.5 Glycosylglycerides

These compounds, which are especially important in the photosynthetic membranes of Cyanobacteria, algae and higher plants, contain one to four sugars linked glycosidically to diacylglycerol. The two galactosylglycerides (**32**) and (**33**) contain large amounts of polyunsaturated fatty acids, e.g. >90% α-linolenic acid in chloroplastic diacylgalactosylglycerol. In diacyldigalactosylglycerol (digalactosyldiglyceride, often abbreviated as DGDG) the linkage between the galactose residues is 1→6. Although the additional linkages in diacyltrigalactosylglycerol and diacyltetragalactosylglycerol have not been fully characterized, they are also probably 1→6. These higher homologues are usually only found in small quantities and may be the result of stress to the plant. In bacteria and algae, a large number of different sugar combinations have been reported, some of those for bacteria being shown in Table 2.6. Structures of some 12 eubacterial monoglycosyldiacylglycerols, nine diglycosyldiacylglycerols, six triglycosyldiacylglycerols and three polyglycosyldiacylglycerols are given by Kates (1990). Their distribution in different bacteria is also reported.

In addition to diacylgalactosylglycerol and diacyldigalactosylglycerol, plant, algal and cyanobacterial photosynthetic membranes also contain large amounts of diacylsulphoquinovosylglycerol (**34**). This compound, also known trivially as the plant sulpholipid, is the only lipid with a sulphonic acid linkage that has been reported so far. It contains more saturated fatty acids (mainly palmitic) than the plant galactosylglycerides.

A number of gluco- and galactoglycerolipids have been isolated in small quantities from animal tissues. Their structures are given in Murray and Narasimhan

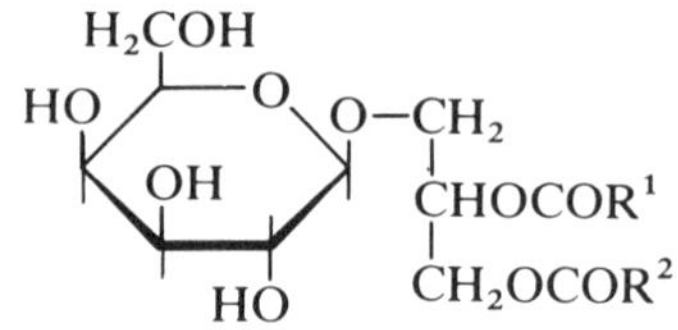

(**32**) 1,2-diacyl-[β-D-galactopyranosyl-(1′ → 3)]-*sn*-glycerol
(monogalactosyldiacylglycerol, MGDG)

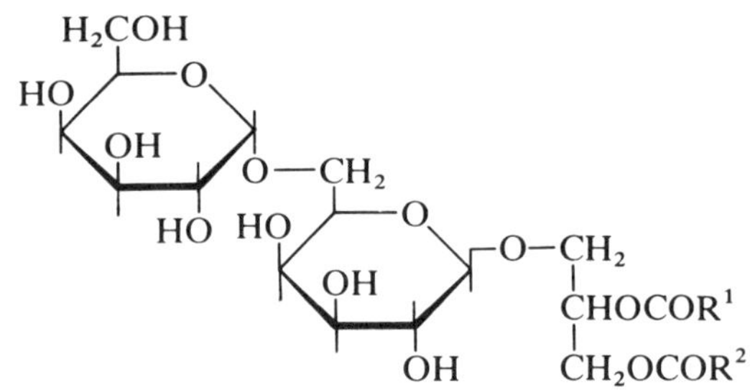

(**33**) 1,2-Diacyl-[α-D-galactopyranosyl-(1′ → 6′)-β-D-galactopyranosyl-(1′→3)]-*sn*-glycerol
(digalactosyldiacylglycerol, DGDG)

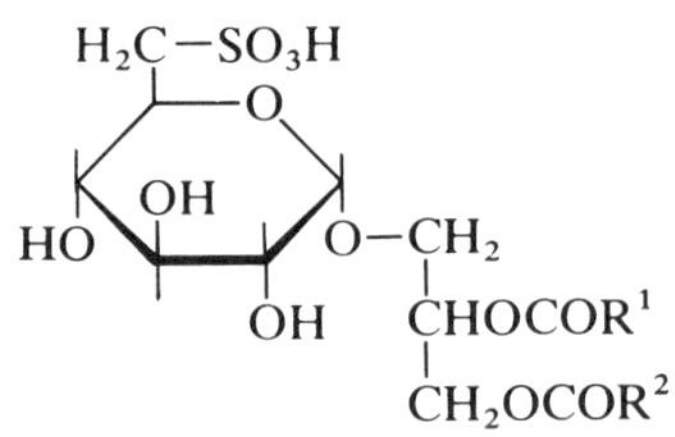

(**34**) 1,2-Diacyl-[6-sulpho-α-D-quinovopyranosyl-(1′ → 3)]-*sn*-glycerol
(plant sulpholipid, sulphoquinovosyldiacylglycerol, SQDG)

D-quinovose is 6-deoxy-D-glucose
(note the carbon–sulphur bond)

Table 2.6 Some bacterial *sn*-3-*O*-glycosyldiacylglycerols (Adapted from Kates, 1972)

Glyceride	Structure of glycoside moiety	Occurrence
Monoglucosyldiacylglycerol	α-D-Glucopyranoside	*Pneumococcus*, *Mycoplasma*
Diglucosyldiacylglycerol	β-D-Glucopyranosyl(1→6)-*O*-β-D-glucopyranoside	*Staphylococcus*
Diglucosyldiacylglycerol	α-D-Glucopyranosyl(1→2)-*O*-β-D-glucopyranoside	*Mycoplasma*, *Streptococcus*
Dimannosyldiacylglycerol	α-D-Mannopyranosyl(1→3)-*O*-D-mannopyranoside	*Microccus lysodeikticus*
Galactofuranosyldiacylglycerol	β-D-Galactofuranoside	*Mycoplasma*, *Bacteroides*
Galactosylglucosyldiacylglycerol	α-D-Galactopyranosyl(1→2)-*O*-α-D-glucopyranoside	*Lactobacillus*
Glucosylgalactosylglucosyldiacylglycerol	α-D-Glucopyranosyl(1→6)-*O*-α-D-galactopyranoyl(1→2)-*O*-α-D-glucopyranoside	*Lactobacillus*

(1990). The majority of galactoglycerolipids contain a single galactose residue, which is linked in a β-glycosidic link between the C-1 of galactose and the C-3 of glycerol. The glucoglycerolipids constitute a large number of compounds with up to eight glucose residues linked $\alpha(1 \rightarrow 6)$. Alkylacyl and diacyl lipids as well as sulphated forms have been reported (Slomiany *et al.*, 1987).

References and further reading

Harwood, J.L. (1980*a*) in *Biochemistry of Plants*, vol. 4 (eds P.K. Stumpf and E.E. Conn), Academic Press, New York, p. 1.

Harwood, J.L. (1980*b*) in *Biochemistry of Plants*, vol. 4 (eds P.K. Stumpf and E.E. Conn), Academic Press, New York, p. 301.

Kates, M. (1972) *Techniques in Lipidology*, Elsevier, Amsterdam.

Kates, M. (1990) in *Handbook of Lipid Research*, vol. 6 (ed. M. Kates), Plenum, New York, p. 1.

Murray, R.K. and Narasimhan, R. (1990) in *Handbook of Lipid Research*, vol. 6 (ed. M. Kates), Plenum, New York, p. 321.

Slomiany, B.L., Murty, V.L.N., Liau, Y.N. and Slomiany, A. (1987) *Prog. Lipid Res.*, **26**, 29.

2.6 Sulphur-containing lipids

In addition to cerebroside sulphates (section 2.4) and diacylsulphoquinovosylglycerol and sulphated gluco- or galactoglycerolipids (section 2.5), various other sulphur-containing lipids have been reported. These include alkyl sulphates (Mayers *et al.*, 1969) in micro-organisms (**35**).

$$H_3C[CH_2]_7CH(OSO_3^-)[CH_2]_{12}CH_2OSO_3^-$$

(**35**) (1,14*S*)-Docosanediol-1,14-disulphate

Chlorosulpholipids are found in some fungi (e.g. caldariomycin ($ClCH_2COCH_2CH_2COOH$)) and certain algae. *Ochromonas danica* contains particularly high amounts of chlorosulpholipids, where they represent almost half of the total membrane lipids. A whole family of compounds are found with two sulphate ester functions and from one to six chlorines (Haines, 1973) as shown in Table 2.7. Some other structures (e.g. trichloro derivatives) have been reported but not characterized (Haines, 1973).

An unusual glycolipid sulphate ester (**36**) has been reported in extremely halophilic bacteria by Kates and co-workers (cf. Kates, 1972), and a glycolipid sulphate (2,3,6,6′-tetraacetyl-α,α'-trehalose-2′-sulphate) in *Mycobacterium tuberculosis* (Goren, 1970).

For a recent review of sulphated glycolipids in Archaebacteria see Kates (1990), and for mycobacterial sulphoglycolipids see Goren (1990). Sulphated glycolipids and sterols are minor components of animal tissues (see Murray and Narasimhan, 1990).

A number of novel taurine-containing lipids have been isolated from the ciliated protozoan, *Tetrahymena*. These have the structures shown in (**37**).

Table 2.7 Structures of the sulpholipids of *Ochromonas danica*

Derivative	Position of substitution in 1,14-dicosane-disulphate	Position of substitution in 1,15-tetracosane-disulphate
Monochloro-	13	14
Dichloro-	11,15	
Pentachloro-	3,3,11,13,16	2,12,14,16,17
Hexachloro-	3,3,11,13,15,16	2,2,12,14,16,17

^-O_3SO—3-Gal(1→6)Man(1→2)Glc(1→1′)—OCH_2–$ROCH$–$ROCH_2$

R = $C_{20}H_{41}$ (phytanyl)

(**36**) Glycolipid sulphate ester

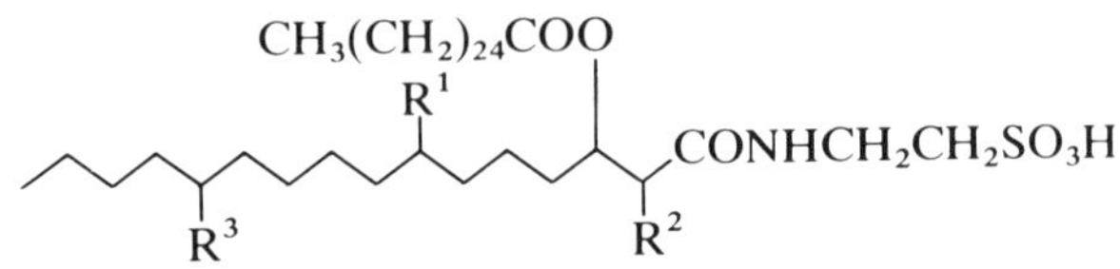

(**37**) Taurine-containing lipids

Lipid	R^1	R^2	R^3
Taurolipid A	OH	H	H
Taurolipid B	OH	OH	H
Taurolipid C	OH	OH	OH
7-Acyltaurolipid A	$CH_3(CH_2)_{14}COO$	H	H

An alkaline-stable taurine-containing lipid, lipotaurine (2-(7,13-dihydroxy-2-*trans*-octadecenoyl-amino)ethanesulphonic acid) was also detected and probably plays a role as a metabolic intermediate. Other related compounds, such as 2-(octadecanoyl-amino)ethanesulphonic acid, were also identified. The isolation, characterization and biochemistry of the taurolipids have been reviewed recently (Kaya, 1992).

References and further reading

Goren, M.B. (1970) *Biochim. Biophys. Acta*, **210**, 127.

Goren, M.B. (1990) in *Handbook of Lipid Research*, vol. 6 (ed. M. Kates), Plenum, New York, p. 363.

Haines, T.H. (1973) in *The Biochemistry of Lipids* (ed. T.W. Goodwin), MTP International Reviews of Science, Biochemistry Series 1, vol. 4, Butterworths, London, p. 197.

Kates, M. (1972) *Techniques in Lipidology*, Elsevier, Amsterdam.

Kates, M. (1990) in *Handbook of Lipid Research*, vol. 6 (ed. M. Kates), Plenum, New York, p. 1.

Kaya, K. (1992) *Prog. Lipid Res.*, **31**, 87.

Lederer, E. (1967) *Chem. Phys. Lipids*, **1**, 294.

Mayers, G.M., Pousada, M. and Harries, T.H. (1969) *Biochemistry*, **8**, 2981.

Murray, R.K. and Narasimhan, R. (1990) in *Handbook of Lipid Research*, vol. 6 (ed. M. Kates), Plenum, New York, p. 321.

2.7 Diol lipids

Only recently has the presence of diol lipids, such as diacylpropane-1,3-diol (**38**), been confirmed for a wide variety of tissues. This is probably because techniques for the elucidation of their structures are a recent development. Small quantities of such diols are found in mammalian and fish liver, mammalian adipose tissue, egg yolk, corn seed and yeast.

Diesters of butane-1,3-diol and butane-1,4-diol are produced by various yeasts (Ratledge and Wilkinson, 1988) and mixed acyl and alk-1-enyl derivatives of these and other simple diols (e.g. ethylene glycol) have been reported (Batrakov *et al.*, 1974). Extracellular acyl esters of arabinitol, xylitol or mannitol have been reported (Stodola *et al.*, 1967) and an acylated diol phospholipid has been isolated from the yeast *Lipomyces starkeyi*, after growth on propane-1,2-diol (Suzuki and Hasegawa, 1974).

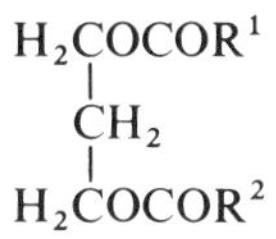

(**38**) Diacylpropane-1,3-diol

References and further reading

Batrakov, S.G., Panosyan, A.G., Konova, I.C. and Bergelson, L.D. (1974) *Biochim. Biophys. Acta*, **337**, 29.

Ratledge, C. and Wilkinson, S.G. (1988) in *Microbial Lipids*, vol. 1 (eds C. Ratledge and S.C. Wilkinson), Academic Press, London, p. 2.

Stodola, F.H., Deinema, M.H. and Spencer, J.F.T. (1967) *Bacterial Rev.*, **31**, 194.

Suzuki, T. and Hasegawa, K. (1974) *Agric. Biol. Chem.*, **38**, 613.

2.8 Other esters

A wide variety of other lipid esters have been reported. Wax esters are a typical example. Although the term 'wax' should, strictly speaking, only be used for esters of long-chain fatty acids with long-chain primary alcohols, common usage, unfortunately, often equates 'wax' with the entire mixture of lipids of which the true waxes are but a part. Ester waxes are found in animals and plants where they form part of the water-repellent surface coating (i.e. skin surface of animals and the leaf cuticle (cf. section 2.10)). The general formula for a simple wax is shown in (**34**). The preen glands of birds, in addition, contain esters of normal alcohols with mono- or multi-branched fatty acids (Odham, 1967).

$$H_3C[CH_2]_xCOO[CH_2]_yCH_3$$

where x and y are greater than 10

(**39**) A simple wax

Complex waxes are compounds where either the fatty acid or the alcohol component or both has (have) a complex structure. For example, the waxes of *Mycobacterium* spp. are diesters of phthiocerols (C_{33}–C_{35} branched-chain diols) with mycocerosic acids (C_{29}–C_{32} branched-chain acids) (Asselineau, 1966).

Animal skin-surface lipids have two types of diester waxes. In the first, a hydroxy acid has its hydroxyl group esterified to a normal fatty acid and its carboxyl group to a fatty alcohol. The second wax type consists of an alkane α,β-diol in which both hydroxyls are esterified with fatty acids (Nicolaides *et al.*, 1970).

Esters are found for most of the commonly occurring sterols (**40**). The fatty acid constituents usually reflect those of the acylglycerols from the same source.

Likewise, esters are found for vitamin alcohols such as vitamin A, the D vitamins and vitamin E. Examples are shown in (**41**) to (**43**).

Carotenoid esters have been reported from a few plant sources (Hitchcock and Nichols, 1971). Flower pigments, for example, are known to contain saturated fatty acid esters of carotenoid alcohols. These compounds are also found in green algae. Where dihydroxy alcohols are involved, then both substitutents are usually esterified, for example (**44**).

In addition, acyl esters of terpenoid alcohols have been reported. For example, Dunphy and Allcock (1971) showed that 30–60% of the total monoterpenoid alcohol content of rose petals occurred as acyl esters with geranyl stearate (**45**) predominating.

An important example of a carbohydrate ester is the so-called cord factor from *Mycobacteria* spp. This

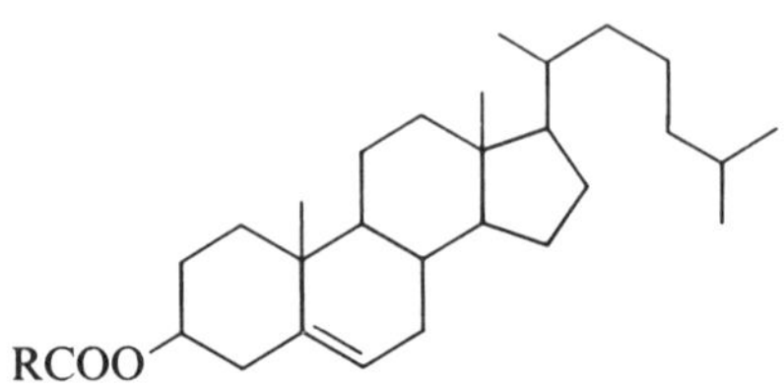

(**40**) Cholesterol ester

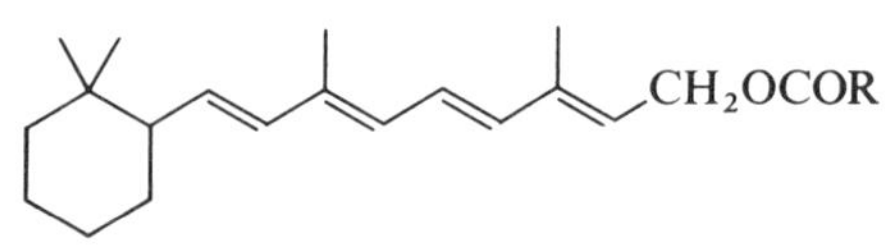

(**41**) Acyl retinol

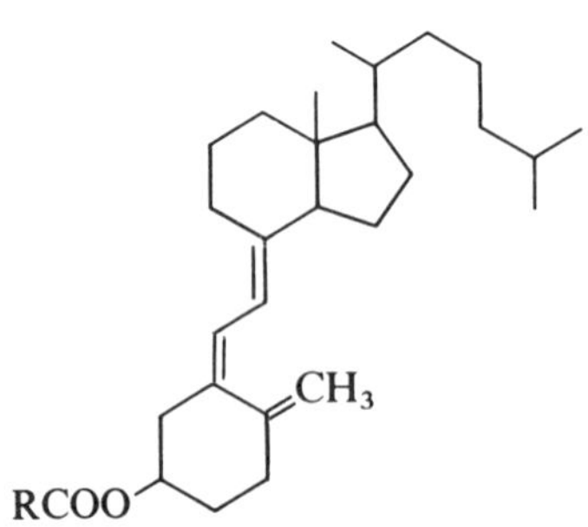

(**42**) Acyl cholecalciferol

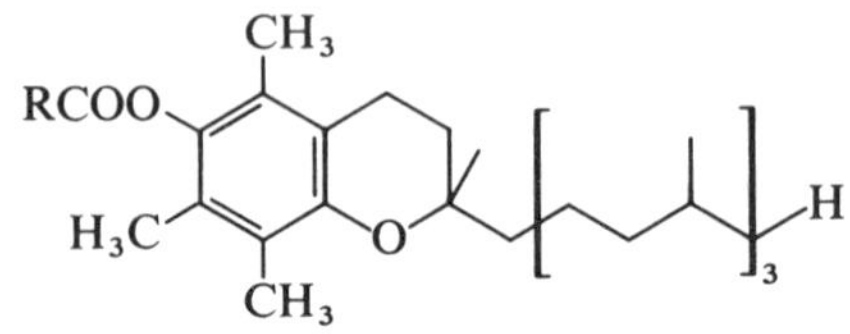

(**43**) Acyl α-tocopherol

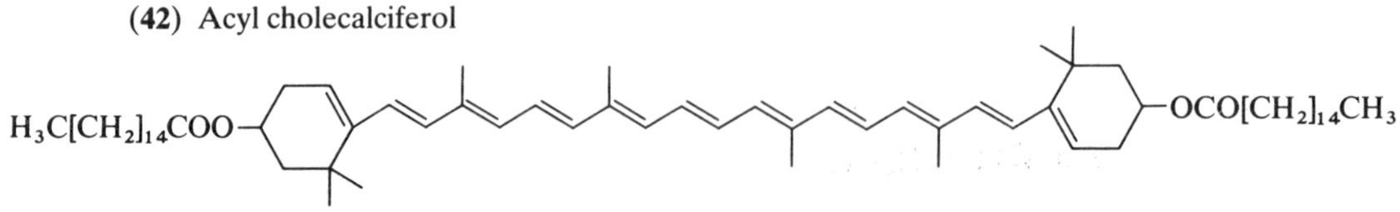

(**44**) Luteol dipalmitate

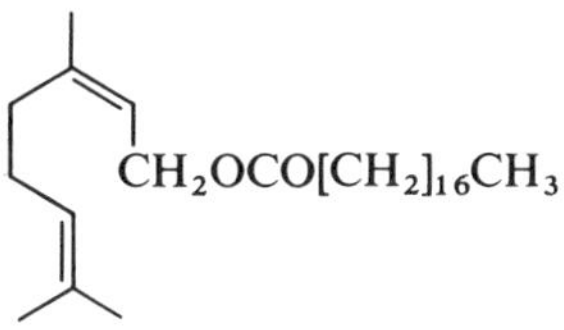

(**45**) Geranyl stearate

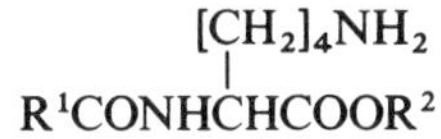

R^1COOH = normal, branched and β-hydroxy-branched acids
R^2OH = long-chain polyalcohols

(**46**) Siolipin A

contains an ester of the disaccharide, trehalose, with two molecules of a complex acid, mycolic acid. The latter is a general term embracing a whole series of fatty acids containing 60–90 carbons. They are hydroxy fatty acids that differ in their degree of unsaturation and chain branching (section 1.8). In the example given in Fig. 2.4 the mycolic acid is the 60-carbon compound found in *Mycobacterium smegmatis*.

Various other esters have been reported in different bacteria. For example, propionibacteria contain diacyl *myo*-inositol mannosides in which the mannose is glycosidically linked to position 2 of *myo*-inositol. The 1 and 6 positions of inositol are esterified with fatty acids. Other bacteria and yeasts contain esters of glucose and certain other sugars (Lederer, 1967; Weete, 1980).

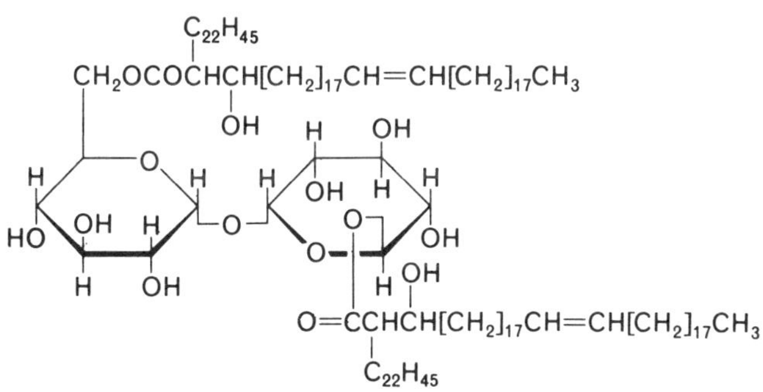

Figure 2.4 Cord factor from *Mycobacterium smegmatis*.

Not only are simple carbohydrate esters found in Nature, but fatty acyl derivatives of amino acids have also been reported. These include serratamic acid (*N*-(D-3-hydroxydecanoyl)-L-serine), siolipin A (**46**) and siolipin B. The latter is the ornithine analogue of siolipin A.

The peptidolipids that occur in Mycobacteria and *Nocardia* spp. are *N*-acyl oligopeptides. They often occur as glycoside derivatives (Kates, 1972).

References and further reading

Asselineau, J. (1966) *The Bacterial Lipids*, Hermann, Paris.

Dunphy, P.J. and Allcock, C. (1971) quoted by Hitchcock and Nichols (1971).

Hitchcock, C. and Nichols, B.W. (1971) *Plant Lipid Biochemistry*, Academic Press, London.

Kates, M. (1972) *Techniques in Lipidology*, Elsevier, Amsterdam.

Lederer, E. (1967) *Chem. Phys. Lipids,* **1**, 294.

Nicolaides, F., Fu, H.C. and Ansari, M.N.A. (1970) *Lipids*, **5**, 299.

Odham, G. (1967) *Fette Seifen Anstrichm.*, **69**, 164.

Weete, J.D. (1980) *Lipid Biochemistry of Fungi and Other Organisms*, Plenum, New York.

2.9 Glycosides

Several types of glycosides can be identified – those of hydroxy fatty acids, aromatic glycols and sterols.

Glycolipids of some micro-organisms, particularly yeasts, are extracellular products. Often the lipids contain a mono- or disaccharide glycosidically linked to a hydroxy acid. Examples would be a rhamnolipid from *Pseudomonas aeruginosa*, a sophorolipid from *Candida bombicola* (see Fig. 2.5) and cellobiolipids from *Ustilago maydis* (see Ratledge and Wilkinson, 1988).

Mycosides from Mycobacteria are glycosides of methylated sugars with a long-chain highly branched hydroxylated hydrocarbon terminated by a phenol group. The hydroxy groups of the long-chain glycol are esterified with mycocerosic and palmitic acids (Kates, 1972).

Acylated steryl glycosides of plants usually contain D-glucose attached via a β-glycosidic linkage to the 3-position of sterols such as sitosterol, stigmasterol, campesterol and cholesterol (see Fig. 2.6). The 6-position of the glucose is esterified with fatty acids such as palmitic, stearic, oleic, linoleic and linolenic (Mudd and Garcia, 1975). Although glucose is the major esterified sugar, galactose, mannose, xylose and gentiobiose have been found in isolated cases and an α-glycosidic link also reported (Harwood, 1980).

Nitrogen-fixing cyanobacteria produce heterocyst cells containing characteristic glycolipids (see Murata and Nishida, 1987). The chemical structures of major components of *Anabaena cylindrica* are shown in (**47**) and (**48**). Glycerol ester glycolipids (**49**) and (**50**) are also present (Murata and Nishida, 1987). The heterocyst lipids of *Nodularia harveyana* (a marine cyanobacterium) have been purified recently and fully characterized as 1-(*O*-α-D-glucopyranosyl)-3*R*,25*R*-hexacosanediol, 1-(*O*-α-D-glucopyranosyl)-3*S*,25*R*-hexacosanediol and 1-(*O*-α-D-glucopyrano-syl)-3-keto-25*R*-hexacosanol (Soriente *et al.*, 1992).

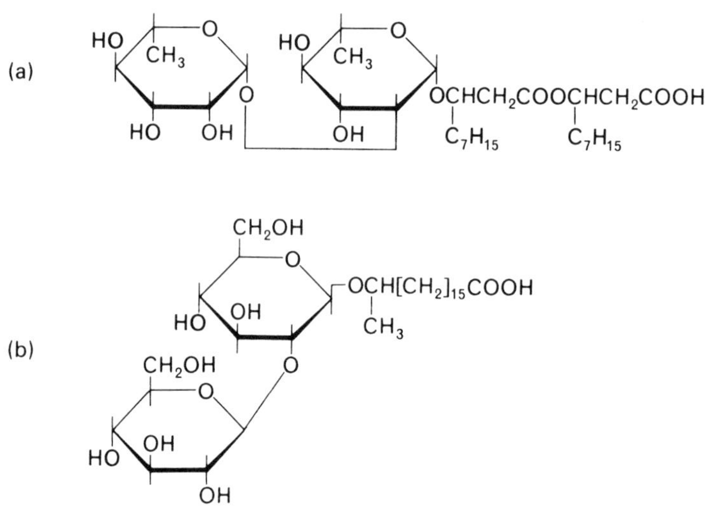

Figure 2.5 (a) Rhamnolipid of *Pseudomonas aeruginosa*; (b) sophorolipid (yeasts).

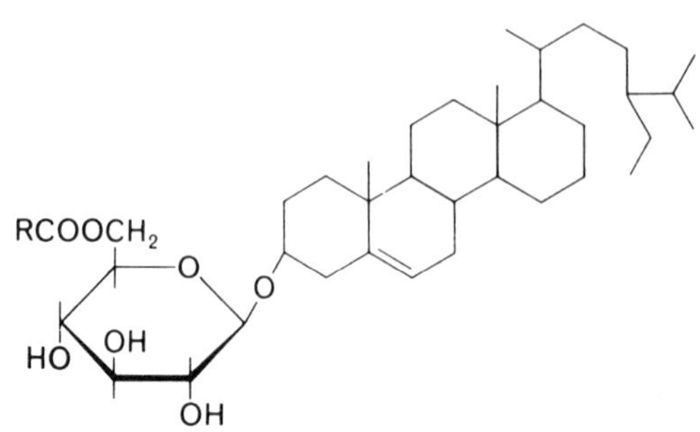

Figure 2.6 Acylated steryl glucoside (ASG).

Glycosidic glycolipids

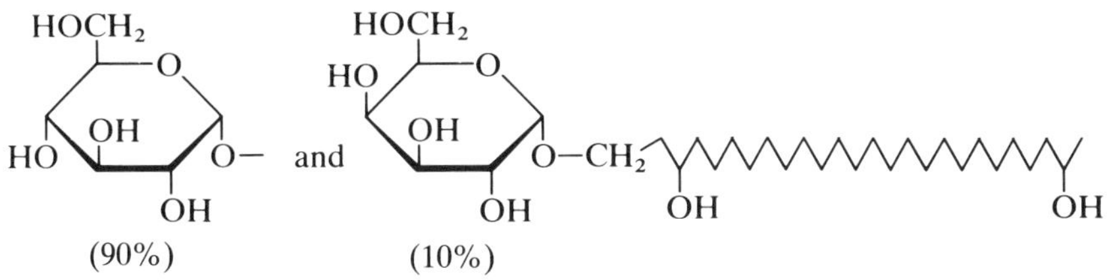

(**47**) 3,25-Dihydroxyhexacosanyl α-D-glycopyranoside

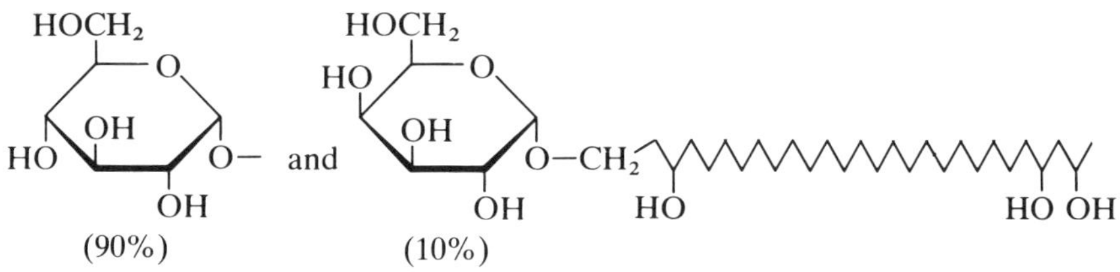

(**48**) 3,25,27-Trihydroxyoctacosanyl α-D-glycopyranoside

Glycosyl ester glycolipids

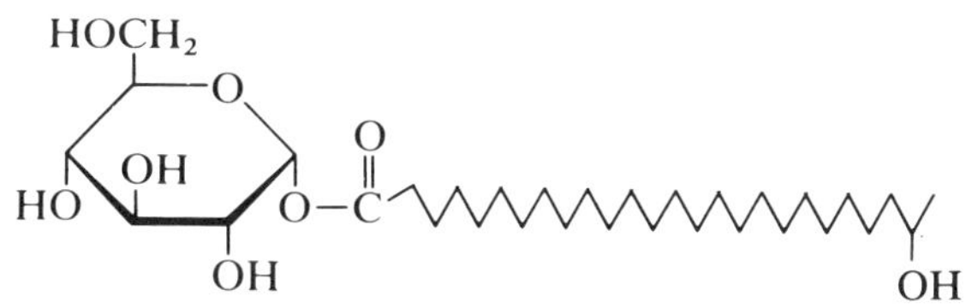

(**49**) α-D-Glucopyranosyl 25-hydroxyhexacosanate

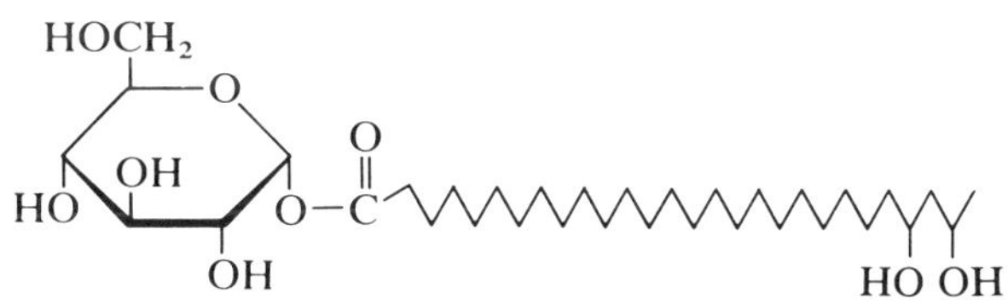

(**50**) α-D-Glucopyranosyl 25,27-dihydroxyoctacosanate

References and further reading

Harwood, J.L. (1980) in *Biochemistry of Plants*, Vol. 4 (eds P.K. Stumpf and E.E. Conn), Academic Press, New York, p. 1.

Kates, M. (1972) *Techniques in Lipidology*, Elsevier, Amsterdam.

Mudd, J.B. and Garcia, R.E. (1975) in *Recent Advances in the Chemistry and Biochemistry of Plant Lipids* (eds T. Galliard and E.I. Mercer), Academic Press, New York, p. 161.

Murata, N. and Nishida, I. (1987) in *The Biochemistry of Plants*, vol. 9 (eds P.K. Stumpf and E.E. Conn), Academic Press, New York, p. 315.

Ratledge, C. and Wilkinson, S.G. (eds) (1988) *Microbial Lipids*, vol. 1, Academic Press, London.

Soriente, A., Sodano, G., Gambacorta, A. and Trincone, A. (1992) *Tetrahedron*, **48**, 5375–84.

2.10 Waxes

The components of plant cuticular waxes have been reviewed by Kolattukudy (1980, 1987) and Walton (1990). Major components include hydrocarbons, very long-chain fatty acids, alcohols and monoesters. These and minor components are shown in Table 2.8.

The wax associated with suberin has also been examined, but only recently. Hydrocarbons, wax esters, very long-chain fatty acids, alcohols and terpenes have all been found. These are all typical components of cuticular wax, but certain differences have been noted. The hydrocarbons in suberin have a broader chain-length distribution with a predominance of shorter carbon chains and more even-numbered carbon chains than cuticular wax. Suberin-associated wax also contains a high proportion of free fatty acids. Free and esterified alkan-2-ols are also present (Kolattukudy, 1980).

Table 2.8 Cuticular wax components (Adapted from Kolattukudy, 1980, where further details can be found)

Compound	General structure	Notes
Hydrocarbons		
n-Alkanes	$H_3C[CH_2]_nCH_3$	Common; usually C_{29}, C_{31}
Isoalkanes	$[CH_3]_2CHR$	Not as common as n-alkanes
Anteisoalkanes	$H_3CCH_2CH(CH_3)R$	
Internally branched alkanes	$R^1CH(CH_3)R^2$	
Cyclic alkanes	$C_6H_{11}R$	Very small amounts
Alkenes	$R^1CH{=}CHR^2$	Sometimes important
Terpenoid hydrocarbons, e.g. farnesene, pristane		
Aromatic hydrocarbons		Anthracene or phenanthrene type with an attached alkane chain
Ketones	R^1COR^2	Not as common as alkanes
Ketols	$R^1CHOH[CH_2]_nCOR^2$	α-, β- and δ-ketols, all rare
Secondary alcohols	$R^1CH(OH)R^2$	About as common as ketones
β-Diketones	$R^1COCH_2COR^2$	Usually minor, sometimes major
Monoesters	R^1COOR^2	Common. Contain even-chain saturated acid and alcohol components
Phenolic esters		Rare
Diesters		Rare
Polyesters		Only important in gymnosperms
Primary alcohols	RCH_2OH	Common; even chains predominate
Aldehydes	RCHO	Usually minor. Not as common as alcohols
Acids		
Alkanoic acids	RCOOH	Very common. Even-chain saturated usually
Dicarboxylic acids	$HOOC[CH_2]_nCOOH$	Uncommon
γ-Lactones	$RCHCH_2CH_2CO$ (ring closed through –O–)	Rare. δ-Lactones also rare
ω-Hydroxy acids	$HOCH_2[CH_2]_nCOOH$	Not common
Diols		
α,ω-Diols	$HOCH_2[CH_2]_nCH_2OH$	Not very common
α,β-Diols	$HOCH_2CH(OH)[CH_2]_nCH_3$	Rare
Terpenes		Many types found. Not as common as other wax components
Flavones		Occur occasionally

References and further reading

Kolattukudy, P.E. (ed.) (1976) *Chemistry and Biochemistry of Natural Waxes*, Elsevier, New York.

Kolattukudy, P.E. (1980) in *Biochemistry of Plants*, vol. 4 (eds P.K. Stumpf and E.E. Conn), Academic Press, New York, p. 571.

Kolattukudy, P.E. (1987) in *Biochemistry of Plants*, vol. 9 (eds P.K. Stumpf and E.E. Conn), Academic Press, New York, p. 291.

Walton, T.J. (1990) in *Methods in Plant Biochemistry*, vol. 4 (eds J.L. Harwood and J.R. Boyer), Academic Press, London, p. 105.

2.11 Cutin and suberin

In many organisms the outer envelope or covering consists of polymers of carbohydrate or amino acids. In plants, however, the covering (cuticle) consists of a hydroxy fatty acid polymer called cutin. The underground parts and healed wound surfaces of plants are covered with an analogous material, suberin. Both cutin and suberin are embedded in or associated with a complex mixture of lipids, which is termed wax (section 2.10). The structure and composition of cutin and suberin are reviewed by Kolattukudy (1980, 1987) and by Walton (1990).

Cutin contains C_{16} and C_{18} families of acids. The former predominates in rapidly growing plants, while both are present in the thicker cuticle of slower-growing plants. The C_{16} family is based on palmitic acid, while the C_{18} family is based on oleic acid (Table 2.9) (see also Kolattukudy, 1980).

In the cutin structure, a polyester intramolecular structure exists where crosslinking is mainly influenced by the availability of secondary hydroxyl groups. Thus cutins that contain large amounts of epoxy, oxo and ω-hydroxy monomers must be predominantly linear (Table 2.10) (Deas and Holloway, 1977). Esterification appears to occur chiefly through the primary hydroxy groups of the monomers. A significant portion (up to 40%) of the monomers are also crosslinked through secondary hydroxyl groups (Tables 2.10 and 2.11).

Table 2.9 The major components of cutin, the cutin acids (For further details see Kolattukudy, 1977, and Harwood, 1980)

C_{16} family	C_{18} family[a]
$H_3C[CH_2]_{14}COOH$	$H_3C[CH_2]_7CH{=}CH[CH_2]_7COOH$
$HOCH_2[CH_2]_{14}COOH$	$HOCH_2[CH_2]_7CH{=}CH[CH_2]_7COOH$
$HOCH_2[CH_2]_xCHOH[CH_2]_yCOOH$ ($x + y = 13; y = 5$–8)	$HOCH_2[CH_2]_7\underset{O}{\underbrace{CH{-}CH}}[CH_2]_7COOH$
	$HOCH_2[CH_2]_7CHOHCHOH[CH_2]_7COOH$

[a] Δ12 unsaturated analogues also occur.

Table 2.10 Typical cutin monomers and their ability to form polyesters (Adapted from Deas and Holloway, 1977)

Cutin monomers (acids) capable of crosslinking a polyester polymer	
$HO[CH_2]_6CHOH[CH_2]_8COOH$[a]	10,16-dihydroxyhexadecanoic acid (and other positional isomers)
$HOOC[CH_2]_5CHOH[CH_2]_8COOH$	7-hydroxyhexadecanedioic acid (and other positional isomers)
$OHC[CH_2]_6CHOH[CH_2]_7COOH$[a]	9-hydroxy-16-oxohexadecanoic acid (and other positional isomers)
$HO[CH_2]_8CHOHCHOH[CH_2]_7COOH$[a]	9,10,18-trihydroxyoctadecanoic acid (and its Δ12 analogue)
$HO[CH_2]_5CHOHCHOHCH_2CHOHCHOH[CH_2]_7COOH$	9,10,12,13,18-pentahydroxyoctadecanoic acid
Cutin monomers (acids) capable of forming only a linear polyester	
Monobasic	
α,ω-Dibasic	
ω-Hydroxymonobasic namely	16:0, 18:0, 18:1 (9), 18:2 (9,12)
$HO[CH_2]_6CO[CH_2]_8COOH$[a]	16-hydroxy-10-oxohexadecanoic acid (and other positional isomers)
$HO[CH_2]_8\overset{O}{\overbrace{CH{-}CH}}[CH_2]_7COOH$[a]	9,10-epoxy-18-hydroxyoctadecanoic acid (and its Δ12 analogue)

[a] Major components of cutin.

Table 2.11 Polymeric form of dihydroxyhexadecanoic acid and related C_{18} acids in four plant cutins (Adapted from Deas and Holloway, 1977)

Polymeric form	Total monomers (%)			
	Tomato fruit	Rose-hip fruit	Blackcurrant leaf	Lemon fruit
Cutin—O[CH_2]$_{15}$COO—Cutin	5	7	18	14
Cutin—OOC[CH_2]$_5$CHOH[CH_2]$_8$COO—Cutin	1	6	–	2
Cutin—OOC[CH_2]$_5$CH(O—Cutin)[CH_2]$_8$COO—Cutin	4	5	5	1
HO[CH_2]$_6$CHOH[CH_2]$_8$COO—Cutin	2	5	3	2
Cutin—O[CH_2]$_6$CHOH[CH_2]$_8$COO—Cutin	48	50	38	25
Cutin—O[CH_2]$_6$CH(O—Cutin)[CH_2]$_8$COO—Cutin	36	24	30	5
HO[CH_2]$_6$CH(O—Cutin)[CH_2]$_8$COO—Cutin	3	3	4	1
Cutin—O—[CH_2]$_6$CO[CH_2]$_8$COO—Cutin	2	1	2	51

Table 2.12 Compositional differences between cutin and suberin (Kolattukudy, 1975)

Monomer	Cutin	Suberin
Dicarboxylic acids	Minor	Major
In-chain substituted acids	Major	Minor
Phenolics	Low	High
Very long-chain (C_{20}–C_{26}) acids	Rare and minor	Common and substantial
Very long-chain alcohols	Rare and minor	Common and substantial

Considerable diversity is evident when cutins from different sources are compared in detail. It is of interest, though, that the cutin composition of delicate tissues such as spinach leaves is essentially similar to that of much more substantial membranes. The wax part of the epidermal layer (section 2.10) is also usually similar between species but with differences in detail (Harwood, 1980; Kolattukudy, 1980).

The major aliphatic components of suberins are ω-hydroxy acids and dicarboxylic acids. Octadec-9-enedioic acid is the usual dicarboxylic acid, and 18-hydroxyoleic the major hydroxy fatty acid. The proportion of very long-chain fatty acids ($>C_{20}$) is usually much greater in the ω-hydroxy acid fraction than in the dicarboxylic acid fractions. Among the α,ω-diols, fatty alcohols and fatty acids, which are often found as significant components of suberin, long chains are common. Kolattukudy has suggested some basic rules for the classification of hydroxy acid phytopolymers as cutin or suberin (Table 2.12). However, these rules must be regarded only as a guide, since the examination of individual plant species has provided exceptions. Indeed, it should be noted that, apart from species or varietal differences, environmental conditions can cause large changes in surface lipids. Thus, light, temperature and age have all been found to affect leaf cuticular components (Harwood, 1980).

References and further reading

Deas, A.H.B. and Holloway, P.J. (1977) in *Lipids and Lipid Polymers in Higher Plants* (eds M. Tevini and H.K. Lichtenthaler), Springer-Verlag, Berlin, p. 293.

Harwood, J.L. (1980) in *Biochemistry of Plants*, vol. 4 (eds P.K. Stumpf and E.E. Conn), Academic Press, New York, p. 1.

Kolattukudy, P.E. (1975) in *Recent Advances in the Chemistry and Biochemistry of Plant Lipids* (eds T. Galliard and E.I. Mercer), Academic Press, New York, p. 203.

Kolattukudy, P.E. (1977) in *Lipids and Lipid Polymers in Higher Plants* (eds M. Tevini and H.K. Lichtenthaler), Springer-Verlag, Berlin, p. 271.

Kolattukudy, P.E. (1980) in *Biochemistry of Plants*, vol. 4 (eds P.K. Stumpf and E.E. Conn), Academic Press, New York, p. 571.

Kolattukudy, P.E. (1987) in *Biochemistry of Plants*, vol. 9 (eds P.K. Stumpf and E.E. Conn), Academic Press, New York, p. 291.

Walton, T.J. (1990) in *Methods in Plant Biochemistry*, vol. 4 (eds J.L. Harwood and J.R. Boyer), Academic Press, London, p. 105.

2.12 Bacterial-wall lipids

A full discussion of the varied lipid structures found in bacterial cell walls is beyond the scope of this book, but the reader will find detailed accounts in Rogers *et al.* (1980) and Goldfine (1982). Recent detailed updates on the biochemistry and distribution of the cell-wall lipids of Mycobacteria and other actinomycetes and of Gram-negative bacteria will be found in Brennan (1988) and Wilkinson (1988), respectively.

Gram-negative bacteria have a cell envelope containing two membranes, with the outer membrane having lipopolysaccharide in its outer leaflet. Lipopolysaccharide is complex and consists of four parts. On the outside is the O-antigen, which is a poly-

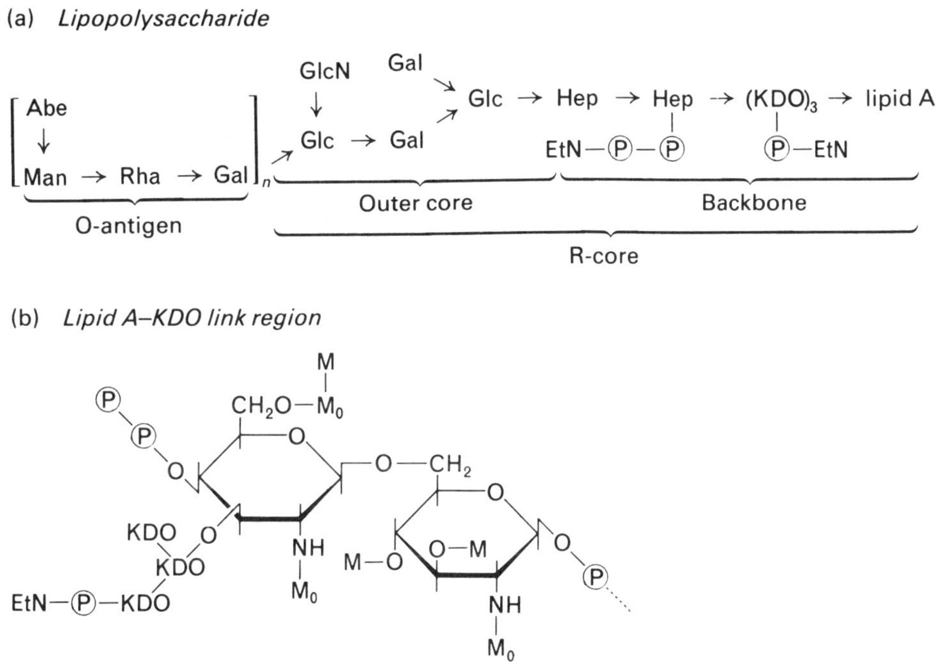

Figure 2.7 Generalized structures of lipopolysaccharide and lipid A. *Abbreviations*: Abe, abequose; Man, mannose; Rha, rhamnose; Gal, galactose; Glc, glucose; Hep, heptose; KDO, 3-deoxy-D-manno-octulosonic acid; Ⓟ, phosphate; EtN, ethanolamine; M, myristate; M_0, β-hydroxymyristate. Reproduced with permission from Harwood and Russell (1984).

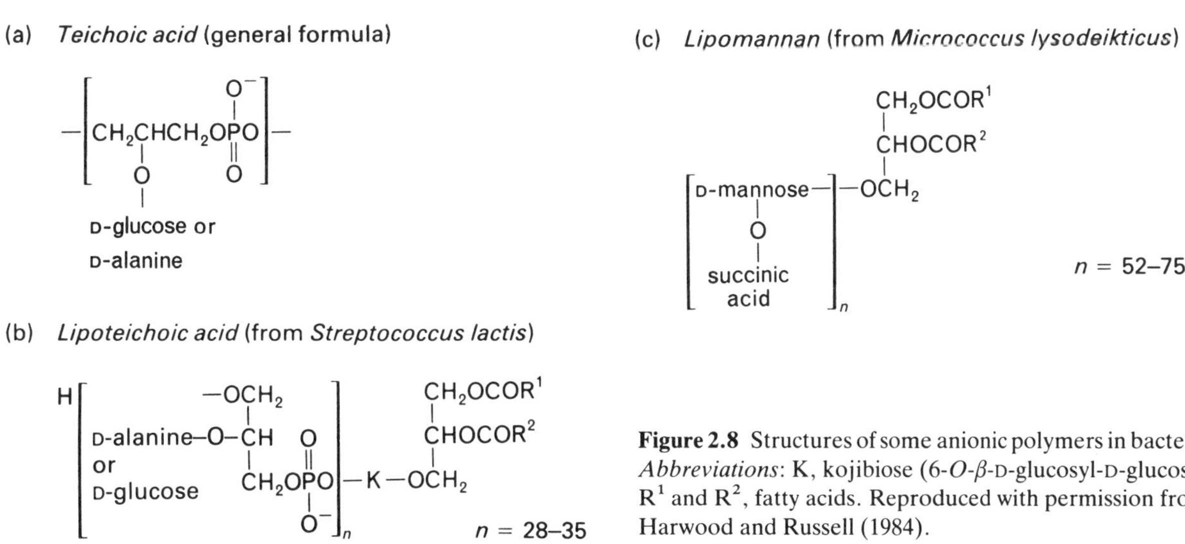

Figure 2.8 Structures of some anionic polymers in bacteria. *Abbreviations*: K, kojibiose (6-*O*-β-D-glucosyl-D-glucose); R^1 and R^2, fatty acids. Reproduced with permission from Harwood and Russell (1984).

saccharide of variable structure. This is attached to a core polysaccharide, which is in two parts, an outer core and a backbone. The backbone is connected to a glycolipid, called lipid A, through a short 'link' usually composed of 3-deoxy-D-manno-octulosonic acid (KDO). These structures are shown in Fig. 2.7.

The cell walls and membranes of most Gram-positive bacteria contain a series of highly anionic polymers. Quantitatively one of the most important of these is teichoic acid, which can be covalently linked to a glycolipid to give a lipoteichoic acid (Fig. 2.8). An alternative type of anionic polymer, which is found in Gram-positive bacteria such as *Micrococcus lysodeikticus*, is succinylated lipomannan (Fig. 2.8). Like teichoic acid, the lipomannan is embedded in the membrane by linkage to a diacylglycerol moiety.

References and further reading

Brennan, P.J. (1988) in *Microbial Lipids*, vol. 1 (eds C. Ratledge and S.G. Wilkinson), Academic Press, London, p. 203.

Goldfine, H. (1982) *Curr. Top. Membr. Transp.*, **17**, 1.

Harwood, J.L. and Russell, N.J. (1984) *Lipids in Plants and Microbes*, Allen and Unwin, Hemel Hempstead.

Rogers, H.J., Perkins, H.R. and Ward, J.B. (1980) *Microbial Cell Walls and Membranes*, Chapman & Hall, London.

Wilkinson, S.G. (1988) in *Microbial Lipids*, vol. 1 (eds C. Ratledge and S.G. Wilkinson), Academic Press, London, p. 299.

3

OCCURRENCE AND CHARACTERISTICS OF OILS AND FATS*

F.B. PADLEY, F.D. GUNSTONE and J.L. HARWOOD

3.1 Introduction

The characterization of oils and fats has been the subject of academic study for over 200 years. The commercial developments in the nineteenth and twentieth centuries placed a greater emphasis on the need for more precise methods of analysis. In the latter half of this century, an additional stimulus has been the growth of biochemistry and the desire to understand the mechanisms leading to the wide variety of natural products and their function.

The source of information concerned with characteristics of oils and fats is therefore very wide. Moreover, if one examines great reference works such as Hilditch and Williams (1964), it is immediately apparent that the available information is voluminous. The contents of this chapter are therefore necessarily restricted, and we do not pretend to offer an alternative to more detailed texts.

The main elements of this chapter are concerned first with describing the types of naturally occurring fatty acids and their origins. The next part describes the commercially important oils and fats. Certain minor oils and fats have been included that may be of current interest either on nutritional grounds or because they are seen to have some commercial potential. The final sections deal with the occurrence of other important lipids.

In compiling the data on commercial oils and fats, the following observations have been made:

The commercial importance of an oil cannot be directly correlated with the volume of published literature. Certain oils have been relatively neglected and modern analytical data are limited, e.g. castor oil, linseed oil, cottonseed oil.

Certain oils have been widely researched because of strong academic and/or commercial interest, e.g. cocoa butter.

Fatty acid compositional data are easily accessible for the majority of common oils and fats. The Codex ranges have been included in this review to indicate the wide range of compositions that can be associated with a single named oil. The use of these compositional ranges as a means of characterizing a fat, for example in a Court of Law, is, however, questionable. In normal commercial trading, the fatty acid range is generally much narrower within a given trading period.

Systematic triglyceride compositional data can now be obtained with the development of high-performance liquid and high-temperature gas–liquid chromatographic techniques.

Significant changes in the pattern of commercial trade are continuing. For example, although soybean still remains the major oil seed crop, there has been a virtually exponential growth of palm oil in Malaysia and the Far East. Exports of palm oil from Africa have, however, fallen to low levels. Rapeseed is increasingly grown in Europe. In the United Kingdom, for example, more rapeseed oil is now consumed than soybean oil.

The continuing interest in minor oil seed crops has led to the commercialization of evening primrose oil and Chinese vegetable tallow. There is also some commercial interest in crops such as *Cuphea*, *Crambe* and certain species of *Vernonia*.

Probably, the developments of greatest interest lie in the application of genetic engineering and tissue culture to already well commercialized crops in order to optimize yield and characteristics.

Information on world production of oils and fats is summarized in Tables 3.171–3.173, but it is of interest to note some significant predictions for the future. The editor of *Oil World* (*Oil World*, 1983 and 1991) reviewed the production of oils and fats for the 25 years 1958–82 and made informed predictions for the next 25 years (1983–2007). His major conclusions are summarized in Table 3.1. The most significant conclusion is the greatly increased production of

* The authors gratefully acknowledge the assistance of J.P. Brun and R.I. Howker, Unilever Research Laboratory, Colworth House, Sharnbrook, Bedford, MK44 1LQ.

Table 3.1 World production of oils and fats (Mt) in the quinquennia 1978–82 and 1986–90 and predicted for 2003–07 (annual averages). Market share is given in parentheses (From *Oil World*, 1983, and *Oil World Annual*, 1991)

	1978–82	1986–90	2003–07
Soya	12.4 (22)	15.3 (20)	23.0 (22)
Palm	4.5 (8)	9.1 (12)	23.0 (22)
Cotton		3.6 (5)	
Groundnut	10.7 (19)	3.7 (5)	15.7 (15)
Sunflower		7.3 (10)	
Lard			
Tallow	11.2 (20)	11.9 (16)	15.7 (15)
Greases			
Rape	3.9 (7)	7.5 (10)	8.4 (8)
Olive	7.3 (13)	1.9 (2)	8.4 (8)
Butter		6.4 (8)	
Coconut	3.4 (6)	3.1 (4)	7.3 (7)
Palmkernel		1.2 (2)	
Others	2.8 (5)	4.6[a] (6)	3.0 (3)
Total	56.2	75.6	104.5

[a] Including sesame (0.7), corn (1.3), marine (1.5), linseed (0.7) and castor (0.4).

palm oil from a quinquennial average of 4.50 Mt (million tonnes) in 1978–82 to 23 Mt in 2003–07. This leads to an increase in the market share from 8 to 22%, bringing it equal to soybean oil. Soya is expected to maintain its market share at 22%, whilst rape, palmkernel and coconut show modest increases in market share. Total oil and fat production is predicted to rise by 86% from 56.2 to 104.5 Mt per annum. Others have argued that the rise of palm oil production will not be quite so large (Arndt, 1984). Figures for the quinquennium 1986–90 are also shown in Table 3.1, and have been taken from *Oil World Annual* (1991). See also Tables 3.171 and 3.172.

References and further reading

Arndt, S. (1984) *J. Am. Oil Chem. Soc.*, **61**, 1810.

Hilditch, T.P. and Williams, P.N. (1964) *The Chemical Constitution of Natural Fats*, 4th Edn, Chapman & Hall, London.

Oil World (1983) (ed. S. Mielke), *The Past 25 Years and the Prospect for the Next 25*, ISTA Mielke, Hamburg.

Oil World Annual (1991) (ed. T. Mielke), ISTA Mielke, Hamburg.

3.2 Classification of oils and fats according to their major or unusual fatty acids*

3.2.1 Short-chain acids

The lower saturated acids (C_4–C_{10}) occur in milk fats (section 3.4.1) and in a few seed fats. Cow milk fat contains butanoic acid (butyric, C_4, 10 mol%, 4 wt%) as well as smaller amounts of C_6, C_8, C_{10} and C_{12} acids. Sheep and goat milk fats contain the same range of acids, with decanoic as the most significant (up to 10 mol%). Octanoic, decanoic or dodecanoic acids are major components of seed oils from the elm (10:0, 60%), from *Zelkova serrata* (10:0, >70%) and from the *Cuphea* genus (section 3.3.12), which contains high levels of octanoic (~70%), decanoic (~85%) or dodecanoic (~60%) in different species.

Lauric (12:0) and myristic acid (14:0) occur extensively in seed fats of the Lauraceae and Myristiceae respectively, a fact that is reflected in the trivial names of these acids. Lauric acid, for example, is the dominant acid in cinnamon oil (80–90%), babassu fat (40–55%, section 3.3.2), coconut oil (41–56%, section 3.3.8) and palmkernel oil (41–55%, section 3.3.25). Myristic acid occurs in nutmeg butter (60–70%) and ucuhuba oil (~70%, section 3.3.40) and at lower levels in coconut (13–23%) and palmkernel oil (14–20%).

Because they are so easily available, coconut and palmkernel oil are the most convenient source of the C_8, C_{10}, C_{12} and C_{14} acids, which can be isolated in a pure state by fractional distillation.

3.2.2 Palmitic and stearic acids

Palmitic acid is the most widely occurring saturated acid, being present in practically every fat examined. It is present in fish oils (10–30%), in the milk and depot fats of land animals (up to 30%) and in vegetable fats. Useful sources of this acid include cottonseed oil (17–31%, section 3.3.10), palm oil (32–59%, section 3.3.24), Chinese vegetable tallow (60–70%, section 3.3.6) and animal fats from pig, sheep and cattle (20–37%, section 3.4.2).

Despite popular impression, stearic acid (18:0) is less common than palmitic. It is present in most vegetable fats, though a significant component in only a few, such as the vegetable butters: cocoa butter (30–36%, section 3.3.7), illipe or Borneo tallow (46%, section 3.3.17) and shea butter (44%, section 3.3.32). Present in most animal fats, it is a major component in the tallow of ruminant animals (5–40%, section 3.4.2).

3.2.3 Long-chain acids

Saturated acids of molecular weight greater than that of stearic acid are major components in only a few uncommon seed oils. Arachidic (20:0), behenic (22:0) and lignoceric acid (24:0) are minor components of groundnut oil (total 5–8%, section 3.3.15) and this is probably the most convenient source of these acids. They are nevertheless present in larger amounts in the less well known seed oils of the Leguminosae and Sapindaceae. For example, rambutan tallow (*Nephelium lappaceum*, *ca* 35%) and kusum (*Schleichera trifuga*, 20–30%) are rich in arachidic acid. Behenic acid is a major component in *Lophira alata* (15–30%) and *L. procera* (20%) seed fats. Erucic acid from high-erucic rape provides on hydrogenation a ready source of behenic acid. Saturated acids with 24 to 30 carbon atoms are present as ester waxes in a number of commercially available waxes (section 3.6). See Hilditch and Williams (1964) for further information.

3.2.4 Oleic acid

Oleic acid, the most widely distributed of all the natural fatty acids, is present in practically all lipids and is a significant component in sources such as those listed below. Many of these are commercially available and are discussed in more detail in section 3.3. The least familiar is *Jessenia bataua*, which is a native Amazonian palm. Its oil is considered as a potential replacement for olive oil. The octadecenoic acids of plant origin are mainly oleic acid, though the 11*c* isomer is frequently present at low levels. Those of animal origin may contain other isomers.

Animal and vegetable fats rich in oleic acid include: filbert nuts (73–80%), *J. bataua* (70–76%), olive (section 3.3.23), almond (65–70%), pistachio (62–73%), pecan (53–67%), macadamia (50–59%), cashew (53–61%), zero-erucic rape (section 3.3.27), groundnut (section 3.3.15), shea nut (section 3.3.32), lard (section 3.4.2), palm (section 3.3.24), beef tallow (section 3.4.2), mutton fat (section 3.4.2), cocoa butter (section 3.3.7), illipe (section 3.3.17), high-oleic sunflower (section 3.3.35) and high-oleic safflower (section 3.3.29).

* Further information on these acids is given in Chapter 1.

Euphorbia lathyris (caper spurge) is a Mediterranean annual containing about 50% of oil in its seed and 80–90% of oleic acid in its oil. The seed shatters and the plant contains a co-carcinogenic milky sap, but attempts are being made to get round these problems.

3.2.5 Linoleic acid

Many vegetable oils available on a commercial scale are rich in linoleic acid and are employed extensively for dietary purposes in a refined state or after some modification, such as partial hydrogenation, which removes some of the linoleic acid, or inter-esterification, which does not. Some of these oils also contain linolenic acid at a lower level and are discussed in greater detail in several subsections of section 3.3. Oils with a significant proportion of linolenic acid are used mainly for industrial purposes (section 3.2.6). Oils rich in linoleic acid include safflower (section 3.3.29), sunflower (section 3.3.35), black walnut (~58%), corn (section 3.3.9), English walnut (~57%), soybean (section 3.3.33), cottonseed (section 3.3.10), pine nut (~46%), sesame (section 3.3.31) and groundnut (section 3.3.15).

3.2.6 Linolenic acid (α and γ)

The most readily available source of α-linolenic acid is linseed oil (section 3.3.19), which normally contains 45–60% of this acid. Other oils, richer in linolenic acid, but less readily available, include chia, camphor, black sesame and perilla (all in the range 60–70%). Candlenut (25–35%) and rubber (20–25%) contain useful quantities of linolenic acid in their seed oils along with appreciable proportions of linoleic acid (35–50 and 35–40% respectively). All of these can be used as drying oils, as can many of the linoleic-rich oils (section 5.5).

Linolenic acid is the major fatty acid of plant leaves, stems and roots and is also a significant component of other photosynthetic organisms. See Hilditch and Williams (1964) for more details.

γ-Linolenic acid is present in evening primrose (section 3.3.13), borage and blackcurrant seed oils, and these oils are now readily available.

3.2.7 Petroselinic acid

Petroselinic acid (16*c*-18:1) is an isomer of oleic acid that occurs widely in seed oils of the order Umbelliferae (35–85%), usually along with oleic acid and other saturated and unsaturated acids. The co-occurrence of petroselinic and oleic acids complicates the analysis of these seed oils, and the two isomers are usually distinguished after ozonolysis. ^{13}C NMR spectroscopy can also be used for this purpose (section 9.3). Kleiman and Spencer (1982) analysed over 200 petroselinic acid-containing oils and designated 14 that were especially rich sources of this acid. These contained 32–50% of oil and the oil had 52–87% of petroselinic acid. There is interest in developing coriander as a new crop.

3.2.8 Erucic acid

Although erucic acid (13*c*-22:1) is present in seed oils of most Cruciferae, reaching a level of 80% in nasturtium seed oil, the most important erucic-containing oils are those from rape, mustard and *Crambe abyssinica* (sections 3.3.27, 3.3.21 and 3.3.11). The last of these is a potential new crop with 55–60% of erucic acid. Mustard seed (22–50%) is widely cultivated in Asia, but high-erucic rapeseed oil is the most important. High-erucic acid rape oils contain up to 50% of erucic acid along with 7–10% of the C_{20} homologue. Even higher levels of erucic acid are being achieved through plant breeding.

Most of the present interest is in zero-erucic rape oils (section 3.3.27). This is one of the few oils produced on a commercial scale in cooler climates and is being cultivated increasingly.

3.2.9 Epoxy acids

The structure of natural epoxy acids (section 1.9) and their synthesis by oxidation of appropriate unsaturated acids (section 10.4) are described elsewhere. Reference should also be made to the sections dealing with sal (section 3.3.30) and with *Vernonia* oil (section 3.3.41).

The best-known and most widely occurring of the natural epoxy acids is vernolic acid (12,13-epoxyoleic), which is present in the seed lipids of several different plant families. Rich sources of the acid include the seed oils of *Vernonia anthelmintica* (72%), in which it was first discovered, *V. galamensis* (73–78%), Stokes aster (65–79%), *Cephalocroton cordofanus* (62%), *Euphorbia lagascae* (57–62%), *Erlangea tomentosa* (52%), *Crepis aurea* (52–54%) and *Crepis biennis* (68%). Crepis oils are of interest in that some contain vernolic acid, others contain crepenynic acid and yet others contain both these acids, even though there is no accepted biosynthetic relationship between the two. Most samples of vernolic acid are the (+) isomer but seed oils of the Malvaceae contain the (−) isomer.

Coronaric acid (9,10-epoxyoctadec-12-enoic), a positional isomer of vernolic acid, is the second most common epoxy acid, and sometimes the two acids co-occur. It is present, for example, in the seed oils of *Chrysanthemum coronarium* (14%), *Helichrysum bracteatum* (14%, including some 9,10-epoxyoctadec-12-ynoic acid) and *Xeranthemum annuum* (8%, along with vernolic acid and 9,10-epoxystearic acid).

9,10-Epoxystearic acid occurs as its *cis* (both (+) and (−) forms) and *trans* isomers. The latter is present in oroju (sulphur olive oil), whilst sal fat contains the (+)-*cis*-epoxy acid.

3.2.10 Hydroxy acids

Hydroxy acids (section 1.8) do not occur very extensively, though they are present in sphingolipids, stillingia oil (section 3.3.34), waxes (section 3.6) and in fungal (section 3.13) and bacterial lipids (section 3.14). The best-known hydroxy acid is ricinoleic acid, which is the major component of the readily available castor oil (section 3.3.4). The structures of the natural hydroxy acids have been discussed in section 1.8, where there is also an account of the occurrence of the less common members.

The presence of the hydroxyl group sometimes leads to their natural occurrence in unusual forms, so that in addition to conventional triacylglycerols they may be present as glycerides with more than three acyl groups, as ester waxes, as polymers, or as lactones.

3.2.11 Acetylenic acids

Acetylenic acids occur in a number of rare seed oils and in some mosses. No oil containing acetylenic acids is available commercially.

Tariric acid (6*a*-18:1) occurs in the seed oils of several *Picramnia* species, reaching 95% in *P. sow* (Steger and van Loon, 1933). Crepenynic acid (9*c*,12*a*-18:2) has been recognized in several species of *Crepis* (0–65%) (Earle *et al.*, 1966), in *Saussurea candicans* (33%) (Smith, 1974) and in several *Afzelia* species (17–36%) (Gunstone *et al.*, 1972). Helenynolic acid (9-OH 10*t*,12*e*-18:2) is present in *Helichrysum bracteatum* (70%) (Powell *et al.*, 1965). The more complex acetylenic acids listed in Table 1.5 generally co-occur and several have been found in *Ximenia caffra* (Ligthelm and Schwartz, 1950), *Santalum album* (Gunstone and Russell, 1955), *S. acuminatum* (Bu'lock and Smith, 1963), *Onguekoa gore* (Gunstone and Sealy, 1963; Morris, 1963), *Acanthosyris spinescens* (Powell *et al.*, 1966) and *Pyrularia pubera* (Hopkins and Chisholm, 1968).

3.2.12 Conjugated diene and triene acids

The most convenient source of conjugated diene acids is via the dehydration of castor oil (section 3.3.4) by heating (~250 °C) in the presence of an acidic catalyst. The product is complex and the dienes produced from ricinoleate are both conjugated (mainly Δ9,11) and non-conjugated (mainly Δ9,12) in the proportion 1:3 or 1:4. Several configurational isomers are produced, and double-bond migration complicates the product further. Alkaline isomerization of linoleic-rich oils also produces conjugated dienes (Δ9,11 and Δ10,12), but these processes have not proved to be of commercial value.

Conjugated octadecatrienoic acids are available in tung oil (α-eleostearic acid, 70–80%, section 3.3.39), in oiticia oil (licanic acid, (~80%, section 3.3.22) and in the seeds of some cucurbitaceae (punicic acid, (~20%). The occurrence of other isomers has been listed in Table 1.4.

Attempts are being made to develop calendula as a new crop (section 3.43).

3.2.13 Branched-chain acids

Branched-chain acids (for structures, see section 1.6) occur widely but mainly at low levels in animal fats, some marine oils and in the lipids of micro-organisms. They are very rare in plant lipids. They may be encountered in butter fats (section 3.4.1), tallows (section 3.4.2), bacterial lipids (section 3.14) and waxes (section 3.6). In this last group they may be a significant component of the total fatty acids. This is particularly true of skin lipids (including wool wax and vernix caseosa – the wax covering the newborn human) and of the uropygial waxes of birds. Branched-chain acids are major components of the lipids of Gram-positive bacteria (section 3.14). Wool wax is a convenient source of some acids of this class.

3.2.14 Cyclopropene acids

Cyclopropene acids such as sterculic and malvalic acids (section 1.7), frequently accompanied by their dihydro derivatives (cyclopropane acids), occur widely in plant families of the order Malvales (Sterculiaceae, Malvalaceae, Bombaceae and other families) but are not confined to these. In addition to the seeds, the cyclic acids are found in leaves, stems and roots and are present mainly as glycerides. The occurrence of these compounds has been reviewed by Christie (1970) and Badami and Patil (1982). Particularly rich sources of these acids include: malvalic acid in *Gnetum gnomen* (39%), *Pterospermum acerifolium* (32%), *Adansonia za* (28%) and *Eriolaena hookeriana* (26%); sterculic acid in *Sterculia foetida* (65%), *Bombacopsis glabra* (35%) and *Bombax oleaginum* (30%); and dihydrosterculic acid in *Litchi sinensis* (41%). 2-Hydroxysterculic acid has been recognized in *Pachira insignis* (~20%) and *Bombacopsis glabra* (~10%) and sterculynic acid in *Sterculia alata* (8%).

3.2.15 Allenic acids

Stillingia oil (section 3.3.34) contains a C_8 allenic acid. C_{18} allenic acids have been identified in seed oils of the Labiatae, especially in the subfamily Stachyoideae (Hagemann *et al.*, 1967). Laballenic acid (5*e*,6*e*-18:2) was first identified in *Leonotis nepetaefolia* (~15%) (Bagby *et al.*, 1965) and lamenallenic acid (5*e*,6*e*,16*t*-18:3) in *Lamium purpureum* (~14%) (Mikolajczak *et al.*, 1967).

References and further reading

Badami, R.C. and Patil, K.B. (1982) *Prog. Lipid Res.*, **19**, 119.

Bagby, M.O., Smith, C.R. and Wolff, I.A. (1965) *J. Org. Chem.*, **30**, 4227.

Bu'lock, J.D. and Smith, G.N. (1963) *Phytochemistry*, **2**, 289.

Christie, W.W. (1970) *Top. Lipid Res.*, **1**, 1.

Earle, F.R. (1970) *J. Am. Oil Chem. Soc.*, **47**, 510.

Earle, F.R., Barclay, A.S. and Wolff, I.A. (1966) *Lipids*, **1**, 325.

Gunstone, F.D. and Russell, W.C. (1955) *J. Chem. Soc.*, 3782.

Gunstone, F.D. and Sealy, A.J. (1963). *J. Chem. Soc.*, 5772.

Gunstone, F.D., Steward, S.R., Cornelius, J.A. and Hammonds, T.W. (1972) *J. Sci. Food Agric.*, **23**, 53.

Hagemann, J.M., Earle, F.R., Wolff, I.A. and Barclay, A.S. (1967) *Lipids*, **2**, 371.

Hilditch, T.P. and Williams, P.N. (1964) *The Chemical Constitution of Natural Fats*, 4th edn, Chapman & Hall, London.

Hopkins, C.Y. (1972) *Top. Lipid Chem.*, **3**, 37.

Hopkins, C.Y. and Chisholm, M.J. (1968) *J. Am. Oil Chem. Soc.*, **45**, 176.

Kleiman, R. and Spencer, G.F. (1982) *J. Am. Oil Chem. Soc.*, **59**, 29.

Kolattukudy, P.E. (ed.) (1976) *Chemistry and Biochemistry of Natural Waxes*, Elsevier, Amsterdam.

Krewson, C.F. (1968) *J. Am. Oil Chem. Soc.*, **45**, 250.

Lessman, K.J. and Anderson, W.P. (1981) *New Sources of Fats and Oils* (eds E.H. Pryde, L.H. Princen and K.D. Mukherjee), American Oil Chemists' Society, Champaign, p. 223.

Ligthelm, S.P. and Schwartz, H.M. (1950) *J. Am. Chem. Soc.*, **72**, 1868.

Mikolajczak, K.L., Rogers, M.F., Smith, C.R. and Wolff, I.A. (1967) *Biochem. J.*, **105**, 1245.

Morris, L.J. (1963) *J. Chem. Soc.*, 5779.

Morrison, W.R. (1970) *Top. Lipid Chem.*, **1**, 51.

Powell, R.G., Smith, C.R. and Wolff, I.A. (1965) *J. Am. Oil Chem. Soc.*, **42**, 165.

Powell, R.G., Smith, C.R., Glass, C.A. and Wolff, I.A. (1966) *J. Org. Chem.*, **31**, 528.

Smith, C.R. (1974) *Lipids*, **9**, 640.

Steger, A. and van Loon, J. (1993) *Rec. Trav. Chim.*, **52**, 593.

Wolf, R.B., Graham, S.A. and Kleiman, R. (1983) *J. Am. Oil Chem. Soc.*, **60**, 103.

3.3 Major vegetable fats

3.3.1 Aceituno oil
[*Simarouba glauca*; Quassia]

There is no reliable information available on world production. Other information is summarized in Tables 3.2 to 3.4.

The oil is obtained from the seeds of the 'Central American' or Salvador plant, which grows in Central and South America (Eckey, 1954). The tree is widely used to mark boundaries. The species was imported into India from Brazil in 1966 and large-scale plantations have been established (Satpathi, 1984; Rath, 1987). The fruit, similar to an olive, has a sweet edible pulp and contains seeds of which 30% is kernel and 70% shell.

Table 3.2 Typical fatty acid composition (wt%) of aceituno oil

14:0	16:0	18:0	20:0	16:1	18:1	18:2	18:3
tr[a]	11.5	27.5	1.5	tr	57.5	2.0	tr

[a] Throughout these tables tr indicates 'trace'.

Table 3.3 Characteristics of aceituno oil: typical values (Lognay *et al.*, 1981)

Relative density	Not available
Refractive index, n_D^{40} [a]	1.4596
Saponification value	192, 198 [b]
Iodine value	58, 51 [b]
Unsaponifiable matter (%)	0.4
Oil content of dry kernel (%)	55–65

[a] Refractive index of sodium D line at 40 °C.
[b] Two typical values.

Table 3.4 Typical triglyceride composition of aceituno oil (Lognay *et al.*, 1981)

S_3 [a]	SOS	SSO	SOO	SL	SLO	OOO	Others
1	29	3	42	1	2	15	7

[a] S, saturated; O, oleic; L, linoleic.

Table 3.5 World production of babassu oil (1000 tonnes)

1970/71	1975/76	1980/81	1983/84
105	170	–	130

The semi-solid fat, isolated by expelling and extraction, is used directly as a component of vegetable shortenings. It can be converted to a vegetable butter by either partial hydrogenation or fractionation. The meal is toxic to livestock (Bhatnagar *et al.*, 1983).

3.3.2 Babassu oil
[*Orbignya martiana*, *O. oleifera*; Palmaceae]

Information is summarized in Tables 3.5 to 3.8. The plant is one of a group of Central and South American palms of which babassu and tucum oils are of greatest economic importance. The problems facing larger-scale production of babassu and its effect on small farmers has been discussed (May *et al.*, 1985; Reufels, 1989). Babassu fruits are relatively large (20 cm long, 10 cm diameter) and have a particularly hard, thick shell. Special machines have been designed to crack the nut to obtain the four to eight kernels it contains.

Table 3.6 Fatty acid composition (wt%) of babassu oil

8:0	10:0	12:0	14:0	16:0	18:0	18:1	18:2
6	4	45	17	9	3	13	3

Table 3.7 Characteristics of babassu oil: values from draft Codex (Codex, 1979 and 1993)

Relative density, d_{20}^{25} [a]	0.914–0.917
Refractive index, n_D^{40}	1.448–1.451
Saponification value	245–256
Iodine value	10–18
Unsaponifiable matter (g/kg)	Not more than 12
Reichert value	4.5–6.5
Polenske value	8–10
GLC[b] ranges of fatty acid composition (%)	
8:0	2.6–7.3
10:0	1.2–7.6
12:0	40–55
14:0	11–27
16:0	5.2–11
18:0	1.8–7.4
18:1	9.0–20
18:2	1.4–6.6

[a] Density at 25 °C compared with water at 20 °C.
[b] Gas–liquid chromatography.

Table 3.8 Fatty acid composition of South American palmkernel oils

	8:0	10:0	12:0	14:0	16:0	18:0	18:1	18:2
Babassu	6	4	45	17	9	3	13	3
Cohune	8	7	46	16	9	3	10	1
Murumuru	1	1	42	37	5	2	11	1
Ouricuri [a]	10	8	46	9	8	2	13	2
Tucum	1	4	49	22	6	2	13	3

[a] Also 6:0, 2%, and 20:0, tr.

The oil provides an alternative to coconut and palmkernel oils and is used both as an edible oil and in soap-making (Arkcoll, 1988).

The compositions of some other South American palmkernel oils, cohune (*Attalea cohune*), murumuru (*Astrocaryum murumuru*), ouricuri (*Syagrus coronata*) and tucum (*Astrocaryum vulgare*), are compared with babassu in Table 3.8 (see also Table 3.91).

The lactones in babassu have been characterized (Van der Ven and de Jong, 1970; Hulsker *et al.*, 1989).

3.3.3 Buffalo gourd oil [*Cucurbita foetidissima*; Cucurbitaceae]

World production is not commercialized. Other information is summarized in Tables 3.9 and 3.10. Buffalo gourd has been studied as a potential commercial source of vegetable oil for many years. The gourds grow on vine-like plants, which come to maturity in their second year of planting. The plant is of interest because it grows on land too dry to support other crops. It is indigenous to both arid and temperate regions, and grows in many parts of the USA, northern Mexico, Lebanon and India.

The crop has not been commercialized but increasing interest is being shown in the USA. Experimental plantings in Texas yielded between 300 and 900 kg of seed per acre.

The oil is high in linoleic acid and contains 2–3% conjugated 18:2 (Bemis *et al.*, 1978; Vasconcellos *et al.*, 1981; Vasconcellos and Berry, 1982). It has been fed to rats and to chickens without ill effects (Vasconcellos *et al.*, 1981; Khoury *et al.*, 1982). Laboratory-refined oil gave a peroxide value of 100 after 4.9 h under active oxygen method (AOM) stability test conditions (Vasconcellos *et al.*, 1981).

The botany, applications and food quality have been reviewed (Scheerens and Berry, 1986; Wright, 1988; Thompson, 1990).

Table 3.9 Fatty acid composition (wt%) of buffalo gourd oil

16:0	18:0	18:1	18:2
9	2	25	64 [a]

[a] Of which 2.3% is conjugated.

Table 3.10 Characteristics of buffalo gourd oil: reported values

Relative density	Not available
Refractive index, 25 °C	1.471
Saponification value	201
Iodine value	144
Unsaponifiable matter (g/kg)	0.94
Oil content of seed (%)	28.5

3.3.4 Candlenut oil (lumbang oil) [*Aleurites moluccana*, *A. trisperma*; Geraniales]

World production is not significant. Composition is given in Table 3.11. The seeds come from a tropical tree widely distributed in Australia, Sri Lanka, Hawaii, Java and Malaysia. The oil gets its name from the custom of stringing the nuts on bamboo and using them for lighting. The oil is not appreciably commercialized, but there has been a significant interest in the oil over a number of years because of its high unsaturation.

The kernels contain 57–69% oil similar to linseed oil but of slightly lower iodine value. The triacylglycerol composition is given in Table 3.159 (see also Afaque *et al.*, 1987).

Table 3.11 Fatty acid composition (wt%) of candlenut oil

16:0	18:0	18:1	18:2	18:3
5.5	6.7	10.5	48.8	28.5

3.3.5 Castor oil
[*Ricinus communis*; Euphorbiaceae]

Summary information is given in Tables 3.12 to 3.15. This is the only commercially important oil that contains a hydroxy fatty acid, 12-hydroxyoctadecenoic acid (ricinoleic), as its major component (sections 1.8 and 3.2.10). Its history has been traced back to the Egyptians, who used it in embalming.

Table 3.12 Castor oil production, exports and disappearance (1000 tonnes)

	1982	1986	1990
Production			
India	112.0	109.4	178.3
Brazil	84.9	131.7	83.9
China	44.3	51.0	77.5
West Germany	10.8	16.3	16.5
Thailand	14.3	13.9	17.7
Italy	4.5	4.5	3.6
Ecuador	3.5	2.6	2.2
Exports			
India	59.1	48.0	122.0
Brazil	54.7	98.7	45.0
China	10.1	14.3	15.3
West Germany	3.4	4.8	5.7
Thailand	14.7	8.4	4.2
Ecuador	3.8	2.0	1.6
USA	0.4	1.5	
Netherlands	1.3	1.1	0.9
France	0.3	0.5	0.5
UK	0.3	0.2	0.2
Italy	0.1	1.2	0.1
Belgium	0.1	0.1	0.1
Sweden	0.1		
Disappearance			
Soviet Union	62.7	60.1	62.1
China	34.3	38.8	55.7
India	47.9	57.8	54.1
France	37.9	46.8	53.7
Brazil	31.5	27.2	31.9
USA	30.5	33.5	29.9
West Germany	21.5	20.5	24.0
Japan	17.2	19.0	19.2
Thailand	5.5	16.5	
Italy	5.4	7.2	
UK	8.3	6.8	
Poland	2.2	5.0	4.7
Romania	2.6	3.4	4.1

Source: *Oil World Annual*

Table 3.13 Fatty acid composition (wt%) of castor oil

16:0	18:0[a]	18:1	18:2	18:3	18:1(OH)[b]
1	1	3	4	tr	90

[a] Dihydroxystearic acid.
[b] Ricinoleic acid.

The castor plant is normally cultivated as an annual, but it can be grown as a perennial herb to a height of 8 m. The development of shatter-resistant varieties will aid the mechanical harvesting of the seed (8–15 mm long depending on variety). Brazil, India, the former USSR and USA are the main producers.

The relationship between cultivation site and time of sowing on fatty acid composition has been studied (Conte *et al.*, 1984). Site influenced the content of ricinoleic acid.

The oil is isolated by pressing and solvent extraction. The first pressing yields No. 1 oil (USA), a virtually colourless oil (free fatty acid 2%) suitable for medicinal use. The residual meal contains three toxins: ricinine, a mildly toxic alkaloid; ricin, an extremely poisonous protein; and a heat-stable allergen, which forms approximately 12% of the cake. The seed also contains a very active lipase.

The properties of castor oil are dictated by the triacylglycerols of the hydroxy acid (ricinoleic), which is strongly dextrorotatory. The oil is more viscous than normal oils. It has limited solubility in petroleum solvents but is soluble in alcohol and other polar solvents. The oil is, however, sufficiently soluble in hot hexane for this to be used in seed extraction.

The determination of the fatty acid composition of castor oil by gas–liquid chromatography (GLC) normally relies on first converting the hydroxy acid to a less polar derivative, e.g. acetate or trimethylsilyl ether. Great care must therefore be exercised in carrying out the derivatization and analysis to obtain meaningful results (Sreenivasan, 1969; Wood *et al.*, 1965). In addition to the major acids, an optically active 9,10-dihydroxystearic acid is present (*ca* 1%) (King, 1942).

An analysis of 36 castor bean varieties gave a range of 83 to 90% ricinoleic acid (Ramos *et al.*, 1984). A similar range for ricinoleic acid content (87–90%) was found in 17 varieties of castor seed grown in India (Lakshminarayana *et al.*, 1984).

The triacylglycerols of castor oil consist mainly of triricinolein and other triacylglycerols containing two ricinoleic acyl groups (Table 3.15).

The composition of the triterpene alcohols, 4-methylsterols and sterols are given in Tables 3.163–3.165.

The tocopherol content of a castor oil as determined by thin-layer chromatography (TLC) was α-tocopherol 28 mg/kg, β-tocopherol 29 mg/kg, γ-toco-

Table 3.14 Characteristics of castor oil

	Typical No. 1 oil	American Oil Chemists' Society (1946)
Relative density, d_{25}^{25}	0.957–0.961	0.945–0.965
Refractive index, 25 °C	–	1.473–1.477
Saponification value	179–185	176–187
Iodine value	82–88	81–91
Unsaponifiable matter (%)	–	1.0
Hydroxyl value	160–168	–
Acetyl value	–	144–150
Acid value	2–3	–
Colour (Gardner)	1–2	–
Viscosity, 25 °C (St)	6.5–8.5	–
Oil content of seed (%)	35–55 (typical 44)	–

Table 3.15 Triacylglycerol composition of castor oil (Achaya and Craig, 1964)

RRR[a]	RRO	RRSt	RRL	Others
68	7.5	5	8.5	11

[a] R, ricinoleic acid; O, oleic, St, stearic; L, linoleic.

pherol 111 mg/kg and δ-tocopherol 310 mg/kg. A range of 400–500 mg/kg total tocopherol has been reported (Muller-Mulot, 1976).

Unmodified castor oil is not used in edible products apart from its limited medicinal use. It is mainly used as a component of paints and varnishes after dehydration (dehydrated castor oil, DCO), where it is noted for its non-yellowing characteristics. Castor oil is used as a lubricant and is also added directly to alkyd and urethane systems as a plasticizer. The condensation of castor oil at high temperature yields polyglycerol polyricinoleate, which has unusual surface-active properties. When used in chocolate it reduces the yield value to virtually zero (Bamford *et al.*, 1970). The chemistry and uses of castor oil have been reviewed (*INFORM*, 1991).

Castor oil is converted to numerous derivatives typified by the following reactions.

1. Pyrolysis of methyl esters at 450–500 °C to give methyl 10-undecenoate. The acid is an intermediate in the manufacture of perfumes and polymers. Derivatives are also used in the manufacture of nylon-11, an engineering plastic.
2. Pyrolytic decomposition in the presence of alkali at 250–275 °C yields sebacic acid, 2-octanol and hydrogen. Sebacic acid (decanedioic acid) is condensed with hexamethylenediamine in the manufacture of nylon-6,10.
3. Esterification of the hydroxyl group with sulphuric acid gives 'turkey red oil', which has excellent wetting and emulsification characteristics. Sulphated oils are used in the textile industry to assist fibre wetting and as a dyeing agent (Kremers, 1971).
4. Hydrogenation to give hydrogenated castor oil, which is used in solvent-thinned paints to introduce thixotropy. Hydroxystearic acid (Naughton, 1974) is reported to gel hydrocarbon oils (Tachibana and Kambara, 1968). Salts of castor oil acids, e.g. lithium, are used in solid greases.
5. Dehydration by heating (230–280 °C) under vacuum with a suitable catalyst (e.g. mineral acids or acid clays) converts the hydroxyoleic acid to a mixture of geometric and positional isomers of octadecadienoic acid, both conjugated and non-conjugated (Achaya, 1971; Body and Shorland, 1965; Von Mikusch, 1955).

3.3.6 Chinese vegetable tallow and stillingia oil [*Sapium sebiferum*, *Stillingia sebifera*; Euphorbiaceae]

There are no current data on production available from China. Other information is given in Tables 3.16–3.18. Between 1951 and 1956 an average of 800 tonnes per year of stillingia oil was imported into the UK at between £70 and £126/tonne.

The Chinese tallow tree, a deciduous tree, has been cultivated in China for centuries, partly because the leaves yield a black dye used for dyeing silk. The tree grows to a height of 7–10 m under wet subtropical conditions and is found in China, the USA and Pakistan. Recent studies suggest that with proper harvesting techniques it could provide an oil yield per hectare similar to the oil palm (Morgan and Shultz, 1981; Glumac and Cowles, 1990).

Chinese vegetable tallow (Hopper and Nesbitt, 1937; Gupta and Meara, 1950; Jeffrey and Padley, 1991) is obtained from the outer seed coating (*ca* 50% of fat) by hexane extraction prior to crushing. Stillingia oil, a drying oil, is obtained from the seed kernel by solvent extraction after crushing. The tallow is of potential interest as an edible oil, for example in confectionery (Itagaki *et al.*, 1989; Qing, 1989). It is reported to be widely used in China both for edible use and soap-making. The edibility of the tallow will depend, amongst others, on the ability to produce a material free of stillingia oil (Jeffrey and Padley, 1991). The kernel oil has been used in paints and varnishes.

Chinese vegetable tallow consists primarily of trisaturated triacylglycerols (25%) and 2-oleo-disaturated triacylglycerols (70%), of which the major components are tripalmitin and 2-oleodipalmitin respectively (Wang *et al.*, 1988; Jeffrey and Padley, 1991).

Stillingia oil, the kernel oil, is unusual in that it contains significant amounts of deca-2,4-dienoic acid (8–10%) and the optically active allenic acid 8-hydroxy-octa-2,3-dienoic acid (8–10%) (Sprecher *et al.*, 1965; Christie, 1969). These short-chain acids are present as a C_{18} estolide, which can be separated by chromatography (Payne-Wahl and Kleiman, 1983; Jeffrey and Padley, 1991) (sections 1.4, 1.5, 1.8, 3.2.10 and 3.2.15).

Table 3.16 Fatty acid composition (wt%) of Chinese vegetable tallow

14:0	16:0	18:0	18:1	18:2
4	62	6	27	1

3.3.7 Cocoa butter
[*Theobroma cacao*; Sterculia]

Data are summarized in Tables 3.19–3.27. The commercial exploitation of cacao or cocoa beans was probably first practised by the Aztecs. The Spanish introduced the bean from Mexico into Europe in the sixteenth century, where it was consumed as a drink. Chocolate was not developed until the nineteenth century (Minnifie, 1980).

The cocoa plant is an evergreen tree growing 5–10 m in height but normally 6 m. It originated in tropical rainforests and grows within 20° of the equator (Martinson *et al.*, Alvim, 1989). The flowers and fruit grow from eyes or cushions located directly on either the main trunk or heavier branches of the tree. Flowers and fruit are born continually but harvesting reaches a maximum at certain times of the year. The fruit is a large pod approximately 15–20 cm long and 7 cm in diameter. The pod contains between 25 and 50 seeds or beans embedded in a soft, sweet, edible pulp.

The normal practice is to allow the beans with the adhering pulp to ferment after placing in boxes or heaps. Fermentation liquefies the pulp, which can be drained away. The flavour of the beans is also partly developed at this stage. The beans are then dried to 6–8% moisture. The full flavour of the beans is developed by roasting, which introduces a range of flavour compounds including pyrazine and flavanoids

Table 3.17 Fatty acid composition (wt%) of stillingia oil

16:0	18:0	ω-OH 8:2	conj. 10:2	18:1	18:2	18:3
5	2	5	5	13	23	47

Table 3.18 Characteristics of Chinese vegetable tallow (Hopper and Nesbitt, 1937) and stillingia oil (Bolley and McCormack, 1950)

	Vegetable tallow	Stillingia oil
Relative density	Not available	Not available
Refractive index, n_D	1.455–1.457 (40 °C)	1.482–1.484 (25 °C)
Saponification value	200–218	202–212
Iodine value	16–29	169–191
Unsaponifiable matter (%)	0.5–1.3	0.5–3
Oil content of seed (%)	20–23	10–17

(Marian, 1967; Rohan, 1969; Van der Waal *et al.*, 1968; Flament *et al.*, 1968; Mohr *et al.*, 1987).

Cocoa butter, a pale yellow solid at 20 °C, is obtained by hydraulic pressing of the kernels or 'nibs' of cleaned, roasted and decorticated beans. The expelled butter (free fatty acids *ca* 1%) has a distinctive flavour and is used directly in confections. Lower-grade cocoa butter of higher free fatty acid content is normally refined and deodorized. Certain butters are also deodorized to control the flavour level in the final product. Cocoa butter is relatively very stable to oxidation. The cocoa bean and cocoa butter are used mainly as ingredients of chocolate (Minnifie, 1980), which is a finely ground mixture of roasted cocoa nibs, sugar, additional cocoa butter and also milk solids in the case of milk chocolate. Cocoa powder, the by-product from cocoa butter expelling, is used in a variety of confections and as the major ingredient of chocolate drinks. Cocoa powder contains up to 3% of theobromine and also caffeine.

The variation in fatty acid composition of cocoa butter is given in Table 3.22. Typical glyceride compositions of some different types of cocoa butter are given in Tables 3.23, 3.24 and 3.158. The distribution of the major glycerides varies, but typical figures are POP 16%, POSt 46%, StOSt 33% and others 5%, expressed as a percentage of total 2-oleo-disaturated triacylglycerols, SOS (P, palmitic; O, oleic; St, stearic; S, saturated). This is only an approximation. Factors regulating the synthesis of triacylglycerols in the maturing cocoa bean have been reviewed by Harwood (1991).

When the composition of the cocoa butter glycerides is expressed in terms of carbon number, as determined by GLC, a straight-line relationship between the C_{50} and C_{54} triglycerides is observed, and this can be used to detect adulterated cocoa butters (Padley and Timms, 1980; Fincke, 1980; Young, 1984; Chaveron and Verdoia, 1985).

The major triglycerides are reported to be racemic (Sampugna and Jensen, 1969). Changes in compo-

Table 3.19 World production (1000 tonnes) of cocoa beans

1970	1975	1980	1982
1541	1542	1557	1531

Table 3.20 Fatty acid composition (wt%) of cocoa butter

14:0	16:0	18:0	20:0	22:0	16:1	18:1	18:2	18:3
0.1	26.0	34.4	1.0	0.2	0.3	34.8	3.0	0.2

Table 3.21 Characteristics of cocoa butter, from draft Codex (Codex, 1979)

Relative density, d_{15}^{99}	0.856–0.864
Refractive index, 40 °C	1.453–1.458
Saponification value	190–200
Iodine value	33–40
Unsaponifiable matter (%)	0.2–0.6
Oil content of whole bean (%)	40–50

Table 3.22 Variation in the fatty acid composition (mol%) of cocoa butter[a] (Fincke, 1976)

14:0	0.02–0.16	16:1	0.1–0.4
16:0	23.6–30.5	18:1	33.2–38.2
17:0	0.1–0.2	18:2	2.2–4.8
18:0	30.2–36.5	18:3/20:1	0.3
20:0	0.7–1.4		
22:0	0.2		

[a] Data on 57 samples of authentic cocoa butters.

Table 3.23 Typical glyceride compositions of cocoa butters of different origin[a] (Kattenberg, 1981)

	S_3[b]	SOS	SSO	S_2L	SO_2	SOL etc.
Ivory Coast	1.6	77.2	0.5	7.5	8.4	4.8
Ghana	1.4	76.8	0.4	6.9	8.4	6.1
Cameroon	1.3	75.2	0.5	7.1	11.0	4.9
Brazil	1.0	63.7	0.5	8.9	17.9	8.0

[a] Composition subject to natural variation.
[b] S, saturated; O, oleic; L, linoleic.

Table 3.24 Triacylglycerol composition of a cocoa butters from different countries of origin (Chaiseri and Dimick, 1989)

	Number of samples	Triacylglycerol (%)								
		PLP	POO	PLSt	POP	StOO	StLSt	POSt	StOSt	StOA
Bolivia	1	1.1	3.3	3.5	22.6	4.0	2.1	40.4	22.8	0.5
Brazil	6	0.9	3.9	3.7	17.9	6.7	3.2	37.1	26.0	0.04
Colombia	2	1.1	3.3	3.6	20.4	4.4	2.3	39.4	25.0	0.6
Ecuador	3	1.2	3.0	3.2	19.2	5.4	2.3	38.4	26.9	0.4
Peru	1	1.5	4.3	3.9	18.3	7.4	3.7	35.8	24.6	0.4
Venezuela	1	0.9	1.0	3.1	20.4	2.8	1.9	40.4	28.8	0.8
Costa Rica	1	1.0	2.6	3.5	17.8	5.5	3.0	38.7	27.4	0.4
Dominican Republic	4	1.1	3.3	3.2	18.4	6.1	2.7	38.2	26.5	0.6
Guatemala	1	1.0	2.3	3.4	19.3	4.9	2.2	39.0	27.5	0.4
Mexico	1	1.1	2.4	3.5	19.1	4.1	3.0	38.8	27.8	0.6
Panama	1	1.0	1.5	3.0	19.1	3.1	2.7	41.4	27.3	0.8
Cameroon	2	1.0	3.0	3.4	17.9	5.8	2.5	38.3	27.7	0.5
Gabon	1	0.9	3.7	3.5	17.5	7.3	3.0	37.1	26.5	0.4
Ghana	3	1.2	2.2	3.4	17.8	4.9	2.2	39.0	27.5	0.4
Ivory Coast	9	1.0	1.9	3.0	19.0	3.9	2.5	39.6	28.5	0.6
Nigeria	2	1.0	2.3	3.6	17.9	5.2	3.0	38.8	27.8	0.5
Indonesia	2	1.1	1.6	3.0	19.9	3.6	1.7	40.6	28.1	0.5
Malaysia	20	0.7	1.2	2.8	18.4	2.9	2.2	40.0	31.1	0.8
Solomon Islands	1	1.0	0.9	3.0	19.3	2.8	2.0	40.7	29.5	0.7

[a] P, palmitic (16:0); St, stearic (18:0); A, arachidic (20:0); O, oleic (18:1); L, linoleic (18:2).

Table 3.25 Tocopherols in cocoa butter (mg/kg) (Muller-Mulot, 1976)

	Cocoa butter A Non-deodorized	Cocoa butter B Deodorized
α-Tocopherol	10.9	18.5
β-Tocopherol	tr	3.9
γ-Tocopherol	170	196
δ-Tocopherol	17	14
α-Tocopherol	1.8	–

Table 3.26 Polymorphic behaviour of cocoa butter

Willie and Lutton (1966)	Larsson (1966)	M.p. (°C)
Form I	β'_2	16–18
Form II	α	21–22
Form III	Mixed	25.5
Form IV	β'_1	27–29
Form V	β_2	34–35
Form VI	β_1	36

Table 3.27 Solid fat content of typical West African cocoa butter

Temperature (°C)	20	25	30	32.5	35
Solid fat (%)	76	69	32	8	0

sition during maturation have been studied, and an increase in temperature produces a more saturated cocoa butter (Lehrian *et al.*, 1980). The biosynthesis of lipids in the cocoa bean has been studied by Harwood (1991).

Numerous minor components including phospholipids (0.05–0.13%), tocopherols (200 mg/kg), flavours and hydrocarbons have been determined. The diglyceride content of cocoa butter (1.5–2.5%) has been compared with the diglyceride level and composition of other confectionery fats (Fincke, 1982); see also Table 3.160.

The compositions of the sterols, 4-methylsterols and triterpene alcohols are given in Tables 3.163–3.165.

The tocopherols present in non-deodorized and deodorized cocoa butters are given in Table 3.25. These, together with phenolic and other unidentified antioxidants, ensure that cocoa butter is very stable to autoxidation.

The melting and crystallization behaviour of cocoa butter is complicated because of the polymorphic behaviour of the major glycerides (Table 3.26). The crystal structure of symmetrical mono-unsaturated glycerides has been described (de Jong *et al.*, 1991; Hagemann and Rothfus, 1992). A more detailed study of the crystal structures of the B_2 and B_1 forms (forms V and VI) of POP, POSt and StOSt has been described by Sato and co-workers (Yano *et al.*, 1994).

Typical solid fat contents of cocoa butters, determined by pulsed nuclear magnetic resonance (NMR), are given in Table 3.27.

In the chocolate-manufacturing process, the molten chocolate mass is passed through tempering units to nucleate and partially crystallize the fats so that when the bulk of fat solidifies it crystallizes in the stable form (Chapman *et al.*, 1971; Jewel and Bradford, 1981; Jeffery, 1991). A micro-scale study of continuous tempering has demonstrated the interrelationship between temperature, throughput and degree of temper (Cebula *et al.*, 1991). Incorrect processing and high-temperature storage can result in the cocoa butter recrystallizing and appearing as bloom on the chocolate surface (Duck, 1967; Kleinert, 1961, 1970; Wille and Lutton, 1966). The addition of the mono-unsaturated dibehenoyl triglyceride (BOB) delays the onset of bloom in chocolate stored at elevated temperatures (Hachiya *et al.*, 1989; Koyano *et al.*, 1990).

The system is further complicated by the addition of milk fat, which softens and extends the crystallization rate of cocoa butter. The phase behaviour of cocoa butter/milk fat has been determined (Chapman *et al.*, 1971; Timms, 1980, 1984; Sato, 1987; Schlichter *et al.*, 1988).

NMR imaging has been applied to chocolate confectionery and was able to distinguish different states of the fat. Water migration could also be detected by the method (Duce *et al.*, 1990).

Cocoa butter fractions have been developed to allow a more flexible use of this valuable component of chocolate (Weyland, 1992).

The relatively high price of cocoa butter has resulted in the development of a number of lower-priced confectionery fats (Best *et al.*, 1956; Gordon *et al.*, 1979; Thomas and Paulicka, 1976; Soeters, 1970). Cocoa butter equivalents (CBEs) are produced by blending palm oil mid-fraction, a source of 2-oleodipalmitin, with illipe butter and/or shea stearin, sources of 2-oleopalmitostearin and 2-oleodistearin (Best *et al.*, 1956).

A new technology, enzymic inter-esterification, is being applied to manufacture 2-oleo-1,3-disaturated triglycerides (Macrae and Coleman, 1976; Eigtved, 1992).

3.3.8 Coconut oil
[*Cocos nucifera*; Palmae]

Summary data are shown in Tables 3.28 to 3.32. The coconut palm (Child, 1964; Piggott, 1964; Langstraad, 1976) grows along the shores of most tropical and subtropical lands. The major growing areas lie between 20° north and south of the equator. Its history is obscure; the west coast of Central America and an Indian Ocean location have both been proposed as the origin. The tree grows to a height of 24 m, but these are increasingly being replaced by dwarf varieties, *ca* 12 m in height, which begin to yield four years after planting. The main producer is the Philippines.

The prospects for improving coconut palm propagation by tissue culture techniques have been reviewed (Jones, 1989; Sethuraj *et al.*, 1989).

The nuts are harvested either by allowing them to fall naturally, or by picking manually or mechanically. The fruit is ovoid, three-angled, 15–30 cm in diameter. It has a thick, fibrous outer husk (exocarp), which covers the hard, bony shell or endocarp. Adhering inside the shell is the testa with a thick endosperm layer. The shell is split open and the dried endosperm layer yields the copra of commerce. The coconut cavity contains a watery fluid known as coconut water.

Copra is the source of coconut oil and is available in two forms, cup copra and ball copra, depending on the method of drying: in the first case by sun, oven, kiln or smoke drying the isolated endosperm; in the second by drying in the shell and subsequently removing the dried endosperm. Smoke drying has the added advantage that it reduces microbial contamination but can introduce polycyclic aromatic hydrocarbons. Polycyclic aromatic hydrocarbons (PAHs) can be found in a range of crude oils including fish, soybean, sunflower, rape, palm, palmkernel and cottonseed oils. Apart from coconut oil and cottonseed oil from flame-delinted seed, the other oils were relatively free of PAHs. PAHs in coconut are largely eliminated during refining (Sagredos *et al.*, 1988). The yield of copra varies widely (300–3000 kg/ha) depending on the method of cultivation. Copra contains 65–68% oil and 4–7% moisture.

The microbiology of coconut in relation to plant growth, productivity and coconut products has been reviewed (Thomas *et al.*, 1987).

The oil is normally recovered by a two-stage process of expelling followed by solvent extraction. The free fatty acid content of the crude oil is generally higher than for other commercial edible oils and often exceeds 5%. The higher grades have a free fatty acid content of *ca* 3%. Good-quality crude oils are rela-

Table 3.28 World production of coconut oil (1000 tonnes)

1970/71	1975/76	1980/81	1982/83	1989/90
2375	3115	3150	2949	3187

Table 3.29 Fatty acid composition (wt%) of coconut oil

6:0	8:0	10:0	12:0	14:0	16:0	18:0	20:0	18:1	18:2
0.5	7.8	6.7	47.6	18.1	8.8	2.6	0.1	6.2	1.6

Table 3.30 Characteristics of coconut oil from draft Codex (Codex, 1979 and 1993)

Relative density, d_{20}^{40}	0.908–0.921
Refractive index, n_D^{40}	1.448–1.450
Saponification value	248–265
Iodine value	6–11
Unsaponifiable matter (g/kg)	Not more than 15
Reichert value	6–8.5
Polenske value	13–18
GLC ranges in fatty acid composition (%)	
6:0	0–0.6
8:0	4.6–9.4
10:0	5.5–7.8
12:0	45.1–50.3
14:0	16.8–20.6
16:0	7.7–10.2
18:0	2.3–3.5
18:1	5.4–8.1
18:2	1.0–2.1
18:3	0–0.2
20:0	0–0.2
20:1	0–0.2

tively low in pigments and almost free of phosphatides.

The fatty acid composition, as with other oils, shows a natural variation (Tables 3.30 and 3.32). The oil is characterized by the presence of short-chain acids (6:0 to 12:0), and by its low level of unsaturated acids.

The oil is used for industrial purposes, notably in detergents either as soap or as derivatives of the fatty alcohols obtained by hydrogenation (Berger and Ong, 1985). The use of coconut esters as a liquid fuel has been proposed (Banzon, 1980). It is also widely used as an edible oil, where its sharp, cooling, melting characteristics are appreciated. It is used in ice cream, ice cream couvertures, margarine, confectionery and filled milk. Blends of coconut oil with other long-chain edible oils foam badly when used for frying (Kim *et al.*, 1988; Lee and Timms, 1988). Soapy off-flavour may develop in food owing to the hydrolytic action of lipases releasing the short-chain fatty acids (Fritsch *et al.*, 1971). Ketonic rancidity can also develop as a consequence of fungal conversion of short-chain acids to methyl ketones one carbon less than the parent acid (Kinderlerer and Hatton, 1991).

Coconut and palmkernel (section 3.3.25) oils bear a superficial resemblance and have similar applications. Coconut oil is generally more stable because it is more saturated than palmkernel oil. Both oils can be fractionated to give sharp-melting stearins useful in confectionery products (Rossell, 1985; Timms, 1985). The melting point of these oils is raised, coconut only slightly, by hydrogenation.

Typical melting characteristics of coconut and palmkernel oil are compared in Table 3.31. The oils normally form eutectic compositions with other fats, leading to a significant lowering of solid fat content. This effect on melting behaviour can be particularly disadvantageous when formulating chocolate substitutes because of the incompatibility of lauric stearins with cocoa butter (Gordon *et al.*, 1979).

The triglyceride compositions of coconut and palmkernel oils are compared in Table 3.92 (Bezard, 1971; Bezard *et al.*, 1971). The 2 position analyses for coconut and palmkernel oils are compared in Table 3.162. The sterols, 4-methylsterols and triterpene alcohols are given in Tables 3.163–3.166 (Fedeli *et al.*, 1966; Gracian and Martel, 1969; Itoh *et al.*, 1973a, b).

The oil contains C_6–C_{14} δ-lactones (~150 ppm total), the major ones being δ-C_8 (46 ppm), δ-C_{10} (50 ppm) and δ-C_{12} (20 ppm) (Allen, 1965; Van der Ven and de Jong, 1970). The α-, β- and γ-tocopherols do not occur at any significant level in coconut oil. The major tocopherols are δ-tocopherol (2–3 ppm) and α-tocotrienol (20 ppm) (Muller-Mulot, 1976) (see also Tables 3.167–3.170).

Table 3.31 Melting characteristics of coconut and palmkernel oils[a]

	N_{10}	N_{15}	N_{20}	N_{25}	N_{30}
Coconut oil	81	64	35	0.5	0
Palmkernel oil	71	59	41	17	0

[a] N_{20} signifies solid fat content measured by NMR at 20 °C.

3.3.9 Corn oil (maize oil)
[*Zea mays* L.; Graminae, grass family]

A summary of relevant information is given in Tables 3.33–3.37. Corn is grown as an annual crop mainly for fodder. A relatively small proportion is grown for human consumption and industrial processing. The total kernel or grain contains approximately 5% oil

Table 3.32 Distribution of fatty acid compositions with sample origin for coconut oil (Rossell *et al.*, 1985)

Origin (no. of samples)	Fatty acid composition (wt%)										
	6:0	8:0	10:0	12:0	14:0	16:0	18:0	18:1	18:2	20:0	20:1
Philippines (11)	0.4–0.6	7.1–8.3	6.2–6.8	46.2–48.7	18.0–19.2	8.3–9.5	2.3–3.2	5.6–7.1	1.3–2.1	0.1–0.2	tr–0.1
Papua New Guinea (4)	0.4–0.6	6.9–7.5	6.4–6.8	47.1–50.3	17.8–18.1	8.3–9.3	2.3–3.0	5.8–6.3	1.4–1.6	0.1	tr
Vanuatu (5)	0.5–0.6	7.3–9.4	6.6–7.8	47.1–48.4	16.8–17.9	7.7–9.1	2.3–2.9	5.4–6.5	1.4–2.0	tr–0.2	tr–0.2
North Sulawesi (1)	0.5	7.3	6.3	45.9	18.1	9.7	2.7	7.4	1.9	0.1	tr
Sri Lanka (2)[a]	0.4–0.5	7.3–7.4	5.9–6.2	49.3–52.6	18.6–19.6	7.0–7.9	2.9	4.5–5.3	0.6–0.9	0.1	tr
Overall range (21)	0.4–0.6	6.9–9.4	6.2–7.8	45.9–50.3	16.8–19.2	7.7–9.7	2.3–3.2	5.4–7.4	1.3–2.1	tr–0.2	tr–0.2
Mean (21 samples)	0.5	7.8	6.7	47.5	18.1	8.8	2.6	6.2	1.6	0.1	tr

tr = trace (unquantified amount of less than 0.05%). Traces of 18:3 were found in all origins ranging up to 0.2% in one Vanuatu sample.

[a] These results relate to desiccated coconut and were not used in calculation of ranges and means. All other oil samples were extracted from samples of commercial copra.

of which 80% is concentrated in the germ. Corn oil is obtained from the germ after it has been separated from the grain by either wet or dry milling. The majority of corn oil is obtained by wet milling, a process used in the production of corn starch, sugar and syrup (Leibovitz and Ruckenstein, 1983).

The characteristics of the oil and its contribution to the American diet are reviewed (Dupont *et al.*, 1990).

The crude oil, dark reddish amber in colour, is isolated from the germ by a combination of expelling and solvent extraction. The crude oil normally contains less than 3% free fatty acid and 1–3% phospholipids. These are removed during conventional refining. Traces of wax (0.05%) can be removed by winterization.

The refined oil is highly regarded as an edible oil because of its good oxidative and flavour stability despite its high degree of unsaturation. Corn oil behaves in a similar way to sunflower oil when used in deep fat frying (Huang *et al.*, 1981). The oil is used as a salad and frying oil, and as a component of margarine, in some cases after partial hydrogenation.

There is considerable genetic variability, in terms of both yield and fatty acid composition. Percentage oil is positively correlated with oleic acid and both are negatively correlated with linoleic acid (Kabul, 1985).

The triglycerides of corn oil follow the normal distribution anticipated for a vegetable oil (Table 3.159). The stereospecific distribution of the fatty acids (de la Roche *et al.*, 1971) is influenced by the overall fatty acid composition. The triglyceride carbon-number distribution is given in Table 3.160. Reverse-phase HPLC was used to characterize the triglycerides in corn oil and 20 glycerides identified (Grossberger and Rothschild, 1989).

Corn oil contains approximately 1% of sterols, which is unusually high for a vegetable oil. The composition of the triterpene alcohols, methylsterols and sterols are given in Tables 3.163–3.166.

Table 3.33 World production of corn oil (1000 tonnes)

1970/71	1975/76	1980/81	1981/82	1982/83
325	440	665	680	725

Table 3.35 Characteristics of corn oil, from draft Codex (Codex, 1979 and 1993)

Relative density, d_{20}^{20}	0.917–0.925
Refractive index, n_D^{40}	1.465–1.468
Saponification value	187–195
Iodine value	103–128
Unsaponifiable matter (g/kg)	Not more than 28
GLC ranges in fatty acid composition (%)	
<12:0	0–0.3
14:0	0–0.3
16:0	8.6–16.5
16:1	0–0.4
18:0	1.0–3.3
18:1	20.0–42.2
18:2	39.4–62.5
18:3	0.5–1.5
20:0	0.3–0.6
20:1	0.2–0.4
20:2	0–0.1
22:0	0–0.5
22:1	0–0.1
24:0	0–0.4

Table 3.36 Quality characteristics of corn oil (Codex, 1979)

Colour	Characteristic of the designated product
Odour and taste	Characteristic of the designated product and free from foreign and rancid odour and taste
Acid value	
Virgin oil	Not more than 4 mg KOH/g oil
Non-virgin oil	Not more than 0.6 mg KOH/g oil
Peroxide value	Not more than 10 milliequivalents peroxide oxygen/kg oil

Table 3.37 Tocopherols in corn oil (mg/kg) (Muller-Mulot, 1976)

α-T	β-T	γ-T[a]	δ-T	α-T_3	β-T_3	γ-T_3
134	18	412	39	–	–	–

The non-polar fractions from corn, soybean and sunflower seed oils have been compared (Trost, 1989). In corn oil this is mainly steryl and triterpenyl esters.

Table 3.34 Fatty acid composition (wt%) of corn oil

12:0	14:0	16:0	18:0	20:0	18:1	18:2	18:3	20:1
0.1	0.2	13.0	2.5	0.5	30.5	52.0	1.0	0.2

The main tocopherol in corn oil is γ-tocopherol. Typical values are α-tocopherol 150 mg/kg, β-tocopherol 10 mg/kg, γ-tocopherol 400–900 mg/kg and δ-tocopherol 40 mg/kg. More recent analyses of refined corn oil are given in Table 3.37 (Muller-Mulot, 1976). In a separate study (Carpenter, 1979) no β-tocopherol was found. The oil also contains significant levels of ubiquinone (Dupont *et al.*, 1990).

Ferulic acid has been detected as an ester with dihydro-β-sitosterol (Tamurat *et al.*, 1958).

Adulteration has been detected using carbon isotope ratio measurement (Rossell, 1991).

3.3.10 Cottonseed oil
[*Gossypium hirsutum* (American), *G. barbadense* (Egyptian), *G. herbaceum* (Asiatic); Malvaceae]

Data for cottonseed oil are given in Tables 3.38–3.42. Cotton has been grown since antiquity. There are numerous varieties and species, ranging in size from herbs 25 cm high to some the size of a small tree. There are two main groups: the American or New World and the Asiatic or Old World cottons. These are genetically distinct and do not form natural fertile hybrids. Most cultivated varieties are grown as annual crops (Bailey, 1948; Cornelius and Raymond, 1967; Hyer, 1966).

Cottonseed oil is a by-product arising from the cultivation of cotton. The cotton is attached to the seed and up to 40 seeds occur in a single cotton boll. The cotton is removed from the seed by ginning the shorter fibres or linter still attached (10–15% of whole seed). Bald seed types have *ca* 2% of linter. The seed yields approximately 16% oil, 45% meal, 9% linters (depending on type) and 26% hull (Kromer, 1977).

Oil is extracted either by pressing or increasingly by solvent extraction. The crude oil has a strong characteristic flavour and odour and has a dark reddish brown colour. The free fatty acid content of the oil depends on the conditions of harvesting and moisture content of the seed: good-quality oil has free fatty acid of 0.5–1%. Normally it is increasingly difficult to bleach the oil satisfactorily, as the free fatty acid of the crude oil increases. However, there is no direct relation between the quality of the crude oil and the stability and flavour of the fully refined oil.

The oil is mainly used for edible purposes, in the manufacture of margarine and shortenings, as a frying oil and as a salad oil.

One feature distinguishing cottonseed oil from other commercial oils is its content of C_{20}, C_{22} and C_{24} saturated fatty acids. The presence of cyclopropenoid acids (0.1–1.3%) (Bailey *et al.*, 1966; Pandey and Suri, 1982) is of even greater importance. These acids are responsible for giving a positive Halphen test (Halphen, 1897), which has been used to characterize cottonseed oil for almost 100 years. The major cyclopropenoid acids are sterculic (C_{19}) (**1**) and malvalic (C_{18}) (**2**) (Shenstone and Vickery, 1959, 1961) (sections 1.7 and 3.2.14).

Adverse biological effects have been observed following the incorporation of sterculic acid in animal

Table 3.38 World production of cottonseed oil (1000 tonnes)

1970/71	1975/76	1980/81	1981/82	1982/83	1989/90
2450	2433	3035	3320	3090	3588

Table 3.40 Characteristics of cottonseed oil, from draft Codex (Codex, 1979 and 1993)

Relative density, d_{20}^{20}	0.918–0.926
Refractive index, n_D^{40}	1.458–1.466
Saponification value	189–198
Iodine value	99–119
Unsaponifiable matter (%)	Not more than 1.5
Halphen test	Positive [a]
GLC ranges in fatty acid composition (%)	
<12.0	0–0.2
14:0	0.6–1.0
16:0	21.4–26.4
16:1	0–1.2
18:0	2.1–3.3
18:1	14.7–21.7
18:2	46.7–58.2
18:3	0–0.4
20:0	0.2–0.5
20:1	0–0.1
22:0	0–0.6
22:1	0–0.3
22:2	0–0.1
24:0	0–0.1

[a] Heated and hydrogenated oils may not give a positive test.

Table 3.39 Fatty acid composition (wt%) of *Gossypium hirsutum*

14:0	16:0	18:0	20:0	16:1	18:1	18:2	18:3
0.8	24	2.6	0.3	0.8	19	52.5	tr

Table 3.41 Phospholipids of glandless cottonseed oil (Cherry, 1983)

Phospholipid	Composition (% of total phosphorus)
Uncharacterized, polar	4.1
Phosphatidylinositol	13.4
Phosphatidylserine	2.4
Phosphatidylcholine	23.2
Phosphatidylethanolamine	13.5
Phosphatidylglycerol	7.6
Phosphatidic acid	8.8
Lysophosphatidylcholine	2.6
Lysophosphatidylserine	–
Lysophosphatidylethanolamine	–
Unknown (sum of six TLC spots)	25.3

Table 3.42 Tocopherols in cottonseed oil (mg/kg) (Muller-Mulot, 1976)

α-T	β-T	γ-T	δ-T
402	1.5	572	7.5

$$H_3C[CH_2]_7C\overset{CH_2}{=}C[CH_2]_7COOH$$

(**1**) Sterculic acid

$$H_3C[CH_2]_7C\overset{CH_2}{=}C[CH_2]_6COOH$$

(**2**) Malvalic acid

diets (Phelps *et al.*, 1965; Kircher, 1964). Pink coloration of eggs is caused when sterculic acid (25 mg/day) is fed to laying hens. The cyclopropenoid group is destroyed either by hydrogenation (Zarins *et al.*, 1982) or by heating the oil in the presence of free fatty acid (Vickery, 1980; Fisher and Schuller, 1981; Hopkins *et al.*, 1973).

Methods of determining cyclopropene acids continue to be developed (Fisher and Schuller, 1981; Vickery, 1980). Methyl malvalate and methyl sterculate have been separated by spinning band distillation (Pawlowski *et al.*, 1981).

The presence of cyclopropene acids in 40 seed oils including a number of the Malvales is reported (Vickery, 1980).

The triglyceride composition of cottonseed oil follows the 2-random, 1,3-random theory within the limits of current experimental accuracy (Tables 3.159 and 3.160).

The disaturated triglycerides partially crystallize as the oil temperature is lowered to 5–10 °C. Cottonseed

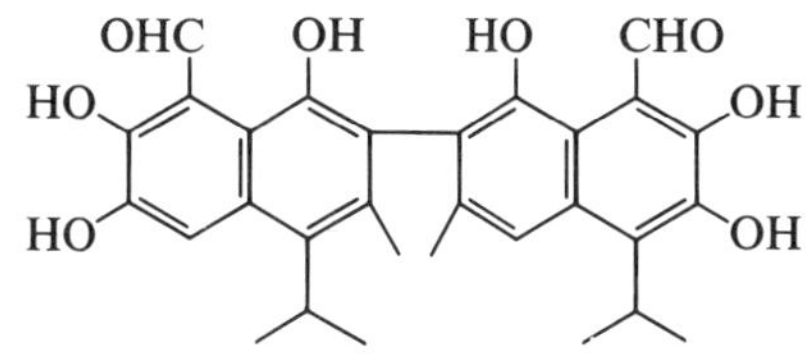

(**3**) Gossypol (aldehyde tautomer)

oil is for this reason dry-fractionated or winterized to produce oleins, which remain liquid at lower temperatures and can be used as salad oils.

Crude cottonseed oil contains numerous minor components (approximately 2%). These include gossypol (**3**) (a yellow pigment), phospholipids, sterols, carbohydrates, pigments and hydrocarbons, mainly squalene (0.004–0.012%).

Gossypol is a 2,2′-di-(1-naphthol) derivative. It is converted to the purple pigment gossypurpurin in seeds stored for prolonged periods. Gossypol occurs in the pigment glands of the seed and is toxic to certain animals (Abou-Donia, 1976). It has received recent attention as a possible natural contraceptive (Kobayashi and Yuyama, 1988). The pigment remains mainly in the meal after oil extraction, but any material extracted into the oil is removed during alkali refining and subsequent heat treatment. Non-gossypol pigments remaining in the oil confer on it a yellow–amber colour. A glandless variety of cottonseed containing no gossypol has been developed and is increasingly replacing older varieties (Kromer, 1977; Cherry, 1983). Analytical methods for determining gossypol (based on HPLC) have been developed (Abou-Donia *et al.*, 1981).

The composition of some of the other minor components of significance in the oil, sterols, 4-methylsterols and triterpene alcohols, are given in Tables 3.163–3.166. The phospholipids in glandless cottonseed oil are given in Table 3.41.

The tocopherols in crude cottonseed oil (Table 3.42), as with other oils, are partially removed during refining. This, together with the removal of gossypol and related phenolic compounds with antioxidant activity, is possibly the reason for cottonseed oil's relatively poor oxidative stability compared to groundnut oil, but this is probably only part of the explanation. Refined cottonseed oils still contain 74–94 mg/kg of tocopherols.

3.3.11 Crambe oil

[*Crambe abyssinica*, *C. hispanica*; Cruciferae]

Crambe oil is produced in limited quantities in Eastern Europe and is at the early stages of commer-

Table 3.43 Fatty acid composition (wt%) of crambe oil (Downey, 1971)

	14:0	16:0	18:0	20:0	22:0	16:1	16:3	18:1	18:2	18:3	20:1	22:1	24:1
C. abyssinica	0.1	1.7	1.0	1.3	2.7	0.3	1.0	16.7	7.8	6.9	2.9	55.7	2.9
C. hispanica	0.1	3.7	0.5	tr	1.6	0.4	0.5	20.1	10.7	5.5	4.6	52.4	–

Table 3.44 Characteristics of crambe oil (Lessman and Anderson, 1981)

Relative density, d_{25}^{25}	0.908
Refractive index, n_D^{20}	1.472
Saponification value	–
Iodine value	93
Unsaponifiable matter (%)	0.5
Oil content of seed, dehulled (%)	33–54

cial development in the USA. A two-year field study placed crambe in the moderately sensitive salt-tolerance category (Francois and Kleiman, 1990). Seed has been processed successfully in commercial oil seed facilities in the USA (Carlson *et al.*, 1985). Summary data are shown in Tables 3.43 and 3.44.

Crambe is an annual herb adapted as a spring crop in wheat-growing areas (Brun and Matchett, 1963; Cornelius and Simmons, 1969; Earle *et al.*, 1966; Lessman and Anderson, 1981; Mustakas *et al.*, 1965). It produces a great number of seeds borne singly at or near the terminus of branches. Yields of unhulled seed vary from 255 to 2000 g/ha depending on climatic conditions, etc.

The conversion of commercial rape (see section 3.3.27) to low-erucic-acid varieties has favoured the exploitation of crambe as a source of erucic acid, this despite the presence of undesirable thioglucosides (*epi*-progoitrin) and a bitter alkaloid (sinapin). Treatment with ammonia combined with heat treatment removes these materials and increases the nutritive value of the meal (Kirk *et al.*, 1966; Tookey *et al.*, 1980).

Erucic acid (**4**) is converted by ozonolysis (section 10.4) to a mixture of brassylic (tridecanedioic) (**5**) and pelargonic (nonanoic) (**6**) acids, which are separable by distillation. Crambe oil can also used as a lubricant in steel casting.

$$H_3C[CH_2]_7CH{=}CH[CH_2]_{11}COOH$$

(**4**)

$$\downarrow \text{Ozone | oxygen}$$

$$H_3C[CH_2]_7COOH + HOOC[CH_2]_{11}COOH$$

(**6**) (**5**)

3.3.12 *Cuphea* species oil [Lythraceae]

This oil is currently at the experimental stage. Interest in the genus arises because these annual species offer the possibility of producing medium-chain-length glycerides based on capric (10:0) or lauric (12:0) acid. Some species are rich in caprylic (8:0) or myristic (14:0) acid. The fatty acid patterns of *Cuphea* oils are very diverse, as is indicated in Table 3.45.

All *Cuphea* species exhibit wild-plant characteristics: dormancy; non-uniform germination; indeterminate flowering and seed maturation over a period of up to 8 weeks; extreme dehiscence and the presence of viscid glandular hairs on stem, leaves, flowers and fruits. Commercial exploitation of this crop is being studied, particularly in Germany and the USA (Wolf *et al.*, 1983a; Princen and Rothfus, 1984; Thompson, 1984; Hirsinger, 1985; Thompson and Kleiman, 1988). Various species are being studied and include *C. viscosissima* and *C. lanceolata* as sources of capric acid, C_{10} (Knapp *et al.*, 1991; Webb and Knapp, 1991; Bafor *et al.*, 1990). The development of *Cuphea* has been reviewed (Röbbelin, 1988; Graham, 1989).

3.3.13 Evening primrose oil [*Oenothera biennis*, *O. larmarkiana*; Onagraceae] and other sources of γ-linolenic acid

World production is approximately 100–150 tonnes per annum (1984). Oil obtained from *Oenothera biennis* or *O. larmarkiana* is designated as evening primrose oil. Interest in this oil has developed in recent years because of its content of γ-linolenic acid (Δ6,9,12-18:3) (Hudson, 1984). It is claimed that the dietary intake of this highly unsaturated oil is beneficial in combating certain diseases, e.g. multiple sclerosis, and is useful as a health supplement (Lovell *et al.*, 1981; Horrobin, 1982, 1992; Horrobin and Manku, 1989; Kies, 1989; Gunstone, 1992). The seeds are grown in the USA, Europe, Israel and New Zealand. Oil content of seed is 22–28%. Alternative

Table 3.45 Fatty acid composition of *Cuphea* seed oils expressed as percentages of total fatty acid composition (Graham, 1989)

Species[a]	8:0	10:0	12:0	14:0	16:0	18:0	18:1	18:2	18:3	20:0
Subg. *Cuphea*										
Sect. *Archocuphea* (3/1)										
mimuloides	–	–	25.0	10.0	20.0	3.0	10.0	30.0[b]	nr	nr
Sect. *Cuphea* (15/6)										
utriculosa	–	0.2	0.8	0.5	25.4	3.5	26.4	30.7	nr	nr
salicifolia$_2$	–	–	–	0.4	20.7	3.0	30.8	43.3	nr	nr
decandra	–	1.3	0.6	0.3	29.6	2.4	20.6	44.7	nr	nr
lindmaniana	0.1	0.5	0.6	0.6	17.3	1.6	18.2	55.3	–	4.1
racemosa	–	–	0.1	0.2	15.3	1.7	17.4	58.8	nr	nr
fruticosa	–	tr[c]	tr	0.1	16.8	0.4	12.8	67.2	–	2.0
Subg. *Bracteolatae*										
Sect. *Heteranthus* (10/3)										
setosa	–	1.1	62.2	26.0	5.3	–	1.3	1.4	nr	nr
tetrapetala	–	0.4	32.4	51.0	7.0	0.9	1.5	6.4	0.4	0.1
epilobiifolia	tr	0.3	31.8	55.3	5.2	0.8	1.2	4.9	0.1	0.1
Sect. *Melicyathium* (1/0)										
Sect. *Brachyandra* (22/5)										
elliptica	–	1.3	48.4	25.8	7.7	1.2	7.9	6.8	nr	nr
micrantha	–	2.4	53.2	17.8	6.3	0.8	6.8	12.9	nr	nr
carthagenesis$_5$	0.4	9.0	65.8	10.5	2.9	0.2	4.8	5.5	nr	nr
parsonsia	–	7.9	73.9	4.4	0.9	–	1.3	6.0	nr	nr
calophylla$_4$	0.1	11.1	76.9	4.1	1.9	0.1	2.1	3.1	0.3	nr
Sect. *Euandra* (74/11)										
linariodes	tr	0.2	3.2	3.1	17.7	2.1	11.6	62.1	nr	nr
linifolia	tr	tr	0.4	3.1	17.9	1.9	13.7	62.5	0.5	–
strigulosa										
ssp. *nitens*	0.9	18.1	13.8	37.3	10.3	1.3	7.3	11.0	–	–
ssp. *opaca*	tr	tr	33.7	45.2	7.0	0.8	5.2	7.1	0.8	0.2
thymoides	tr	3.6	55.7	8.2	7.3	0.9	7.0	14.5	0.2	–
sclerophylla	tr	tr	59.7	27.6	5.3	0.2	2.1	5.1	tr	–
diosmifolia	tr	tr	64.0	31.3	1.8	0.4	1.5	1.0	–	–
glutinosa$_3$	0.5	17.6	64.8	3.9	2.3	0.4	3.0	7.1	0.2	–
ingrata	0.2	12.5	68.8	8.0	1.9	–	2.9	4.0	nr	–
pseudovac$_2$	0.3	6.7	75.9	6.6	3.2	0.9	3.6	3.4	1.4	–
hyssopifolia	0.2	7.3	78.5	4.8	1.5	0.2	2.4	3.8	nr	nr
polymorpha	tr	7.4	80.1	3.6	2.0	0.1	2.5	4.3	–	–
Sect. *Amazoniana* (17/0)										
Sect. *Trispermum* (17/1)										
sessilifolia	1.1	–	10.2	36.6	19.1	9.9	23.1	–	–	–
Sect. *Pseudocircaea* (5/1)										
lutescens$_2$	0.5	0.9	70.9	22.8	2.2	0.1	1.5	1.1	0.2	–
Sect. *Heterodon* (28/26)										
purpurascens	2.2	–	0.9	3.1	19.0	2.2	17.2	17.6	36.1	–
ferrisiae[d]	0.7	–	0.8	35.2	12.0	2.7	20.0	22.5	–	–
ferrisiae	1.2	82.2	1.9	1.0	3.2	tr	2.7	4.4	6.4	–
lutea$_2$	0.4	29.4	37.7	11.1	4.1	0.7	9.2	7.0	nr	nr
wrightii$_9$	–	29.4	53.9	5.1	2.3	0.4	3.1	4.6	nr	nr
glossostoma	–	17.4	59.5	9.2	3.1	0.5	4.6	5.5	nr	nr
viscosa	0.2	0.3	59.9	31.4	2.1	0.2	1.9	4.0	–	–
trochilus	–	0.1	61.8	24.7	3.0	0.2	2.8	7.1	0.2	0.1
laminuligena$_2$	–	17.1	62.6	9.5	2.8	0.2	2.0	3.5	nr	nr

Table 3.45 continued

Species[a]	8:0	10:0	12:0	14:0	16:0	18:0	18:1	18:2	18:3	20:0
tolucana$_3$	–	23.0	63.3	4.5	1.8	–	1.9	5.0	nr	nr
palustris	19.7	1.4	2.0	63.7	6.7	–	3.0	2.9	nr	nr
lobophora	–	10.9	66.2	13.1	2.4	–	2.9	3.9	nr	nr
vesiculigera	tr	0.4	71.2	18.0	2.6	0.2	2.4	5.0	–	0.1
calcarata	–	64.3	6.3	6.9	6.5	0.9	7.7	6.9	nr	nr
viscosissima	9.1	75.5	3.0	1.3	3.1	0.3	1.9	4.7	0.5	0.3
procumbens$_3$	0.4	80.1	2.7	1.5	4.5	0.3	4.4	4.4	nr	nr
angustifolia$_2$	–	80.4	2.6	2.9	3.4	0.3	6.0	4.7	nr	nr
lophostoma	0.9	81.1	2.4	4.1	2.9	0.5	3.4	2.8	nr	nr
humifusa$_2$	1.2	82.0	2.9	1.6	3.1	0.4	5.4	5.1	nr	nr
lanceolata	0.6	83.2	2.1	2.0	3.4	–	3.4	4.6	nr	nr
caesariata	1.1	85.9	1.6	0.6	2.2	–	3.9	3.9	nr	nr
inflata	0.7	86.4	1.6	0.4	2.1	0.5	3.6	4.2	nr	nr
leptopoda$_3$	1.1	87.3	1.5	–	2.3	0.3	2.9	3.9	nr	nr
paucipetala$_2$	1.2	87.4	2.0	0.8	1.9	–	1.8	4.0	nr	nr
crassiflora$_2$	1.4	87.2	1.7	0.4	2.3	0.6	3.5	2.7	nr	nr
llavea$_2$	1.5	87.5	1.6	0.7	1.9	0.4	3.8	2.2	nr	nr
koehneana	0.1	91.6	1.5	0.6	1.3	0.3	1.1	3.1	0.2	0.1
Sect. *Melvilla* (39/9)										
flavovirens	tr	20.5	9.5	5.3	15.2	2.2	21.6	22.9	1.1	0.4
heterophylla$_3$	2.6	32.8	47.7	5.5	2.0	0.5	3.0	5.6	nr	nr
melvilla$_2$	0.2	1.6	49.0	13.9	7.5	0.8	10.0	15.3	0.6	–
jorullensis	0.1	32.0	53.1	4.1	1.5	0.6	2.7	5.0	–	0.1
retroscabra	3.1	24.7	54.9	4.2	2.0	–	3.8	6.5	nr	nr
micropetala	0.2	26.4	25.2	24.2	6.5	1.7	4.3	11.8	nr	nr
quaternata	0.1	62.9	8.2	15.6	4.1	0.5	2.2	5.1	1.4	tr
caeciliae	–	70.9	11.1	2.6	3.9	0.6	4.1	4.8	nr	nr
ignea	0.9	87.1	1.2	0.6	3.1	0.2	2.9	4.1	nr	nr
Sect. *Leptocalyx* (7/4)										
aequipetala	24.6	1.3	1.8	56.0	6.6	–	4.9	3.6	nr	nr
appendiculata$_4$	–	4.4	75.5	9.1	2.7	0.1	2.6	4.1	–	–
calaminthifolia	tr	43.7	tr	4.1	13.0	1.5	12.5	25.2	–	–
bustamanta	25.8	62.5	–	–	2.1	–	3.2	5.6	nr	nr
Sect. *Diploptychia* (20/6)										
spectabilis	0.6	0.2	1.2	2.7	15.7	1.2	26.0	22.2	30.8	–
nitidula	1.6	73.9	2.4	3.9	4.3	–	5.5	7.6	nr	nr
pinetorum	48.1	40.3	0.4	0.4	2.6	0.4	3.0	4.1	nr	nr
hookeriana$_3$	50.2	25.4	3.6	1.0	7.1	0.7	4.3	4.6	nr	nr
painteri	65.0	24.0	0.2	0.4	2.8	0.4	3.3	3.9	nr	nr
cyanea	67.8	24.4	–	–	1.4	–	2.4	3.1	nr	nr

[a] Each section is followed in parentheses by number of species in the section/number of species sampled. When more than one collection was sampled, the mean is presented and number of sampled collections indicated as a subscript. In a few instances, names or number of species in a section may differ from those in the original publications owing to subsequent taxonomic changes.

[b] The dominant fatty acid is in italic.

[c] nr = not recorded; tr = trace $<0.1\%$.

[d] Two collections of this species had inexplicably different patterns, both recorded here.

sources of γ-linolenic acid have been reviewed (Wolf *et al.*, 1983b) (Table 3.48).

Specific concentrations of γ-linolenic acid are achieved by urea fractionation (Traitler *et al.*, 1986). The seed oils from various *Oenothera* species have been characterized. The highest amount of γ-linolenic acid was found in *O. acerviphilla nova* (16%) and *O. paradoxa* (14%) (Lamer-Zarawska *et al.*, 1989).

Borage (starflower) (Whipkey *et al.*, 1988) is also an attractive source of γ-linolenic acid (GLA) apart from the fact that high losses are incurred during harvesting due to seed shattering. Probably a more attractive alternative source is from blackcurrent (Traitler *et al.*, 1988). The seed is a by-product of blackcurrent juice production; the extracted oil contains up to 20% of GLA.

3.3.14 Grapeseed oil
[*Vitis vinifera* and others; Vitaceae]

Data on world production are not readily available. Other information is summarized in Tables 3.46 and 3.47. Grapeseed oil is a by-product of the wine industry. The oil content of the seed varies: sweet white grape seeds approximately 20% oil, certain black grapes as low as 6% oil. The seeds are isolated from the wine pomace, freed of adhering pulp and dried. The seeds are then pressed and solvent-extracted. The best grades of oil are used for edible purposes (Curz Madueno, 1985; Rohne, 1984; Lanzini *et al.*, 1986); other grades are used for soap-making and in paint.

Table 3.46 Characteristics of edible grapeseed oil, from draft Codex (Codex, 1979 and 1993)

Relative density, d_{20}^{20}	0.923–0.926
Refractive index, n_D^{25}	1.473–1.477
Saponification value	188–194
Iodine value	130–138
Unsaponifiable matter (g/kg)	Not more than 20
Erythrodiol content (% of total sterols)	Not less than 2
GLC ranges in fatty acid composition (%)	
12:0	<0.5
14:0	<0.3
16:0	5.5–11
16:1	<1.2
18:0	3.0–6.0
18:1	12–28
18:2	58–78
18:3	<1.0
20:0	<1.0
22:0	<0.3
24:0	<0.1

3.3.15 Groundnut oil (peanut oil, arachis monkey nut oil, huile d'arachide)
[*Arachis hypogaea*; Leguminosae]

Data on groundnut oil are summarized in Tables 3.49–3.53. The oil is of major commercial importance (Woodroof, 1966, 1983; Ayres, 1983). The plant grows as an annual in tropical and subtropical climates. Some varieties are erect, others prostrate. The fruit is a legume, which is borne under the ground after the flowers have been fertilized in the normal manner above ground level. Several perennial species grow in South America, where it is believed to have originated. The plant was carried to Africa by Portuguese traders, to the Phillippines by Magellan in 1519 and hence to various parts of Asia. Groundnut is a major crop in Africa, India, China and the USA.

A critical review of the progress in cytogenetics and genetics of *Arachis* has been written (Wynne, 1989).

The mechanical properties and composition of the peanut are influenced by harvest date (Kim and Hung, 1991). The colour of the mesocarp can be used to group peanuts with similar shear-compression strength.

The groundnut or peanut is made up of the outer shell (or hull), the seed coat (or testa) and the kernel. The hull is separated either by hand or mechanically and the oil extracted by expelling and subsequent solvent extraction. The oil is pale yellow and has the flavour and odour of peanuts. A number of alternative oil extraction procedures have been proposed (Ayres, 1983). Crude oil can have a free fatty acid content as low as 0.3%, but most commercial oil is in the range 0.5–1.5%. Water activity (a_w) of the crude oil influences the rate of peroxide formation. Peroxide formation decreased when $a_w > 0.67$ (Gopalakrishna and Prabhakar, 1983).

Table 3.47 Fatty acid composition (wt%) of grapeseed oil (Fedeli, 1983)

	14:0	16:0	18:0	20:0	16:1	18:1	18:2	18:3
Typical	–	7	3	0.1	tr	22	67	0.9
Range	0–0.2	6–8	3–6	0–0.5	0–1.0	12–25	60–76	0–0.5

Table 3.48 Sources of γ-linolenic acid (Wolf *et al.*, 1983b)

Species	Total γ-linolenic acid (% of seed)	Oil (wt%)	Fatty acids (area % by GC)									
			Δ6,9,12-18:3	16:0	16:1	18:0	18:1	18:2	18:3[a]	18:4	20:0	20:1
Onagraceae												
Oenothera agrillicolla Mackenz.	1.7	25.1[a]	6.7	6.2	0.2	2.5	13.9	69.4	0.1		0.2	0.4
O. biennis L.	2.5	24.6	10.0	7.0	tr[b]	2.0	9.0	72.0				
O. brevipes Gray	0	18.9[a]		8.7	0.2	2.4	7.1	78.2	0.1		0.2	0.1
O. cardiophylla Torr.	0	35.6		11.2	0.1	2.2	6.4	77.7	0.2		0.3	0.1
O. clavaeformis Torr.	0	34.6		9.4	0.1	1.5	4.6	84.1	0.1		0.1	
O. depressa[c] Greene	1.7	26.3	6.3	7.3	0.1	2.0	8.2	69.8	0.2		0.3	0.2
O. drummondii Hook	0.1	29.1	0.5	10.8	0.1	2.0	10.4	76.2				
O. elata Hbk.	1.6	24.4[a]	6.7	5.9	0.1	2.5	10.4	73.3	0.4		0.2	0.2
O. grandiflora Ait.	0.3	3.7[a]	9.3	7.7	0.1	1.7	9.0	67.4	0.2		0.3	0.1
O. hookeri Torr. & Gray	1.7	24.1[a]	7.0	6.3	0.1	2.4	9.3	74.0	0.2		0.4	0.1
O. laciniata Hill	0.7	18.5[a]	3.9	7.1	0.1	2.4	7.6	77.4	0.1		0.2	0.1
O. lamarckiana Ser.	2.3	28.3	8.2	9.2	0.1	2.1	12.3	62.2	0.2		0.4	0.2
O. leptocarpa Greene	0	28.5		7.9	tr	2.4	6.5	82.8	0.4		tr	
O. missouriensis Sims	0	20.8		6.5	0.3	3.0	16.3	71.5	0.2		0.6	0.1
O. odorata Jacq.	0.5	28.8	1.7	10.8	0.1	3.7	10.9	70.5	0.1		0.3	0.2
O. parviflora L.	1.9	26.2[a]	7.1	5.4	0.1	2.7	11.9	70.9	0.3		0.6	0.4
O. rhombipetala Nutt.	1.9	30.7	6.1	7.6	0.1	2.5	8.2	75.0	0.1		0.2	0.1
O. rosea Ait.	0	31.9		10.2	0.1	2.8	7.7	77.9	0.4		0.3	0.2
O. serrulata Nutt.	0.3	22.0	1.2	10.2	0.2	3.7	11.1	68.3	0.2		0.5	0.1
O. stricta Steud.	0.6	30.7	1.9	9.8	0.1	2.8	8.7	75.8	0.3		0.3	0.2
O. strigosa (Rydb.) Mackenz. & Bush	2.0	29.0[a]	7.0	5.8	0.1	2.6	10.6	72.6	0.2		0.4	0.2
O. tetragona Roth	1.8	30.8	5.9	6.8	0.1	2.3	6.6	73.9	0.3		0.2	tr

Boraginaceae												
Adelocaryum coelestinum [d] (Lindl.) Brand	2.7	21.5	12.4	11.0	0.2	3.1	31.2	14.1	4.7	3.7	0.7	4.6
Alkanna froedinii Rech.	4.6	47.0 [a]	9.9	7.4	0.1	2.7	20.6	26.3	27.1	5.9		
Alkanna orientalis (L.) Boiss	2.8	22.9	12.4	5.8	0.1	2.6	15.3	26.1	31.7	4.9	tr	1.1
Amsinckia intermedia F. & M.	2.3	27.7	8.2	11.5		3.8	31.7	13.3	17.4	9.5	0.3	2.9
Amsinckia lunaris Macbr.	2.4	26.5	8.9	11.2		3.4	38.5	13.4	12.2	5.8	0.8	3.7
Anchusa strigosa Labill.	1.6	22.2	7.4	10.0	tr	2.1	38.0	38.0	tr			2.2
Brunnera orientalis (Schenk.) I. M. Jtn.	4.2	27.2	15.4	9.6	0.1	2.8	28.9	27.1	8.8	2.5	0.1	2.6
Cryptantha grayi (Macbr.) Jtn.	1.8	29.2	6.2	7.9	0.2	3.4	15.9	21.4	33.8	9.6	0.2	0.2
Cynoglossum nervosum [e] Benth. Exc. B. Clarke	1.9	24.2	7.8	10.1	0.4	1.9	32.7	21.7	7.0	3.1	1.6	4.7
Echium glomeratum Poir	1.1	17.1	6.6	8.0		2.3	13.2	15.3	44.6	9.1	0.9	
Hackelia floribunda [f] (Lehm.) Johnston	1.6	25.2	6.4	8.5	0.3	2.8	33.1	15.5	13.8	6.9	0.5	5.9
Nonnea macrosperma [g] Boiss & Heldr.	5.1	38.6	13.1	7.0	0.1	2.2	24.7	34.4	9.0			
Pectocarya platycarpa Munz. & Jonst.	2.3	15.0	15.2	9.5	0.1	2.9	19.0	17.9	19.9	12.0	0.1	1.5
Trichodesma zeylanicum R. Br. (Burm. F.)	1.4	31.9	4.3	9.5	0.2	6.2	28.6	20.5	24.3	5.0	0.6	0.8
Scrophulariaceae												
Scrophularia canina L.	1.4	31.1	4.5	8.4	0.1	2.8	28.1	55.9	tr		tr	0.2
S. grayana Maxim.	1.6	43.4	3.8	6.4	0.1	1.8	16.3	71.0	0.5		tr	0.1
S. koraiensis Nakai	1.5	39.9	3.7	7.2	0.3	2.8	14.8	70.5	0.1		0.3	
S. lanceolata Pursh.	2.1	26.3	8.0	5.8	0.1	2.1	16.6	66.0	0.5		0.4	tr
S. marilandica L.	3.6	37.9	9.6	7.2	0.6	2.5	14.9	62.6	0.6		0.9	0.2
S. michoniana Coss & Kral.	1.0	28.1	3.5	9.9	tr	2.6	28.1	55.8				0.1
Saxifragaceae												
Ribes alpinum L.	1.7	18.7	8.9	5.6	0.2	1.4	18.1	39.0	22.0	4.4	tr	0.1
R. inebrians Lindl.	0.6	17.7 [a]	3.4	5.4	0.1	2.0	17.6	40.4	26.4	2.4		0.8
R. montigenum McClatchie	0.8	22.0 [a]	3.7	5.3	0.1	1.5	22.3	32.5	28.9	3.3		0.5
R. orientale Desf.	0.2	11.7	1.9	6.8	0.2	1.4	18.1	49.1	19.6	0.9	tr	0.2

[a] Oil percentage calculated on fresh weight basis.
[b] tr = trace <0.1%.
[c] Unknown fatty acids comprise 4.8% of the seed oil.
[d] Also 2.2% 22:0 and 7.4% 22:1.
[e] Also 0.7% 22:0 and 5.3% 22:1.
[f] Also 4.3% 22:1 and 1.5% 24:1.
[g] Unknown fatty acids comprise 8.8% of the seed oil.

The crude oil does not present any problems in refining and has excellent oxidative stability. The refined or crude oils (in the Orient) are used mainly for edible purposes, e.g. margarine, shortenings, frying oil and salad oil. The frying performance of groundnut and cottonseed oils was compared (du Plessis *et al.*, 1981). Similar properties were observed, with groundnut giving better-flavoured products. The oil has a cloud point in the range 0–5 °C.

Work in the USA is aimed at increasing the oil content of peanuts both for edible purposes and as a diesel fuel (Hammons, 1981).

The compositions of oils from four major types of peanut are given in Table 3.54 together with ranges for the major sterols (Sheppard and Rudolf, 1991).

The presence of C_{20}, C_{22} and C_{24} fatty acids distinguishes the oil from other major edible vegetable oils (Table 3.50). The oil contains only traces of linolenic acid (*ca* 0.05%). The overall fatty acid composition, low linolenic acid content and the presence of tocopherols and possibly other antioxidants provides a relatively stable liquid oil.

The stereospecific distribution of the fatty acids (Sanders, 1980; Treadwell and Young, 1982) shows that the greatest concentration of 16:0 and 18:0 is at the *sn*-1 position and that the 20, 22 and 24 acids occur mainly at *sn*-3. The entire amount of 20:0 was found at the *sn*-3 position. See also Table 3.161. The effect of season and location has been studied (Sanders, 1982). As the seed matures, the overall level of 18:1 increases (Pattee *et al.*, 1969; Sanders, 1980).

A survey of the fatty acid compositions of over 50 different *Arachis* species is reported (Stalker *et al.*, 1989).

The triglyceride composition is summarized in Tables 3.159 and 3.160. Unsaponifiable components occur at levels of 0.2–1% of the oil and include hydrocarbons, aliphatic alcohols, triterpene alcohols (Table 3.165), sterols (Tables 3.163–3.166 and 3.52) and tocopherols (Table 3.53). β-Carotene and lutein are the major pigments, their level decreasing in the

Table 3.49 World production of groundnut oil (1000 tonnes)

1970/71	1975/76	1980/81	1981/82	1982/83	1989/90
3230	3565	2605	3265	2955	3787

Table 3.51 Characteristics of groundnut oil, from draft Codex (Codex, 1979 and 1993)

Relative density, d_{20}^{20}	0.914–0.917
Refractive index, n_D^{40}	1.460–1.465
Saponification value	187–196
Iodine value	80–106
Unsaponifiable matter (%)	1.0 or less
Oil content of seed (%) [a]	Range 38–50 [a]
GLC ranges of fatty acid composition (%)	
12:0	0–0.1
14:0	0–0.1
16:0	8.3–14.0
16:1	0–0.2
17:0	nd
17:1	nd
18:0	1.9–4.4
18:1	36.4–67.1
18:2	14.0–43.0
18:3	0–0.1
20:0	1.1–1.7
20:1	0.7–1.7
22:0	2.1–4.4
22:1	0–0.3
24:0	1.1–2.2
24:1	0–0.3

[a] Additional information, not in the draft Codex. nd, not detected.

Table 3.52 Sterol composition of groundnut oil [a] (Rossell *et al.*, 1983)

Sterol	Range (%)	Mean (%)
Cholesterol	0.5–3.8	1.5
Brassicasterol	0.0–0.4	0.0
Campesterol	11.4–19.8	17.0
Stigmasterol	4.8–13.3	8.7
β-Sitosterol	47.6–64.8	58.5
Δ^5-Avenasterol	8.3–19.0	12.3
Δ^7-Stigmasterol	0.0–5.2	2.1
Δ^7-Avenasterol	0.0–6.6	1.2
Total sterol (mg/kg)	901–2850	

Table 3.50 Fatty acid composition (wt%) of groundnut oil

16:0	18:0	20:0	22:0	24:0	18:1	18:2	18:3	20:1	22:1
12.5	2.5	1.2	2.5	1.3	37.0	41.0	0.3	0.7	1.0

Table 3.53 The tocopherol content of groundnut oil (ppm)

Reference	α-T	β-T	γ-T[a]	δ-T	ΣT_3[a]
Muller-Mulot (1976)	169	5.4	144	13	–
Rossell *et al.* (1983)	49–304[b]	1–41	99–389	–	–
Rossell *et al.* (1983)	178[c]	9.8[c]	220[c]		

[a] Tocotrienols are absent.
[b] Based on 43 samples.
[c] Mean of 43 samples.

oil as the seed matures. Changes in the major lipid classes during maturation have been studied (Pattee *et al.*, 1969; Sanders, 1980). Oleic acid increases as maturation progresses.

Aflatoxin can occur as a contaminant in crude oil, but this is removed during alkali refining and bleaching. Because this toxin is very prevalent in mouldy seeds, it is worth noting that seeds infected with 5500 ppb aflatoxin yielded a crude oil with 812 ppb aflatoxin. This was progressively removed by alkali refining (10–14 ppb in neutral oil) and bleaching (<1 ppb in oil) (Parker and Melnick, 1966). Electronic sorting can also be used to discard mouldy beans (Ayres, 1983).

The phospholipids of groundnut oil (0.3–0.4%) are composed of phosphatidylethanolamine (7.6%), phosphatidylcholine (22.6%), phosphatidylinositol (16.7%), phosphatidic acid (2.6%), phytoglycolipid (38.3%) and unidentified phospholipids (12.2%) (Singleton and Pattee, 1981; Wagner and Wolff, 1964).

3.3.16 Hemp seed oil (marijuana) [*Cannabis sativa*; Cannabinaceae]

There are no recent data available on world production, but an estimated 55 000 tonnes were produced in 1970; it is produced in the Far East, the former USSR, Poland, the former Yugoslavia and Turkey. Summary information is shown in Tables 3.55 and 3.56.

Hemp is an annual herbaceous plant, which is cultivated for its seed, its fibre and for narcotics. Substantial quantities are grown in southern Europe, the former USSR, China, Japan and Chile. The oil is obtained mainly by expelling; lighter-coloured oil is obtained by cold pressing. The seed oil is used for similar purposes to linseed oil, which it resembles. The oil is widely used for edible purposes in Asia. The presence of 6% of 12,13-epoxyoleic acid has been reported (Latif and El Daly, 1973).

3.3.17 Illipe butter (Borneo tallow, tenkawang) [*Shorea stenoptera*; Dipterocarpaceae] (see also Mowrah butter)

World production is confined to Indonesia, in small tonnages and very erratic: 7000 tonnes were reported in 1980/82. Other data are shown in Tables 3.57 and 3.58.

Illipe is indigenous to Borneo and is a large tree growing in the tropical rainforests. The trees bear large winged seeds, which are collected by natives (Raymond and Ward, 1956).

The availability of this uncultivated crop is unpredictable and only every 2–6 years is a crop collected that is of any commercial value. The harvested seed is subjected to a variety of treatments prior to export. These vary from kiln or sun drying to water soaking (to remove saponins) prior to drying. Good-quality oil is obtained from kiln- or sun-dried nuts that have not been water-soaked and provided that the drying is done reasonably quickly and in a controlled way. Soaking in water produces a seed with a higher oil content, but the oil has a high free fatty acid content and is of poor quality.

The seeds contain a pale yellow oil (45–50%), which can be obtained by expelling and subsequent solvent extraction. Commercial oils have a relatively high free fatty acid content (5–20%) depending on the age of the seed and the harvesting methods. The residual meal has a low nutritional value when used in animal feed. The analysis of 95 contractual samples showed oil contents of 45.6–59.7% with free fatty acid levels of 6.2–33.4% (Kershaw, 1987).

The refined oil of good quality melts sharply between 30 and 35 °C and is used almost solely as a partial replacer of cocoa butter in chocolate. The triglyceride composition is given in Table 3.158. Many of the minor components of illipe butter have been characterized, particularly the sterols and triterpene alcohols (Itoh *et al.*, 1974) (Tables 3.163–3.166.

Numerous fats available from uncultivated crops have also been considered as sources of cocoa butter-

Table 3.54 Mean concentrations [a] of fatty acids and sterols by groundnut type and groups of types with similar fatty acid composition (Sheppard and Rudolf, 1991)

Groundnut	Fatty acid [b] (g/100 g of groundnut)							Sterols (mg/100 g)		
	16:0	18:0	18:1	18:2	20:0	20:1	22:0	Campesterol	Stigmasterol	β-Sitosterol
Runner (R)	4.6 ± 0.49	1.5 ± 0.12	23.1 ± 2.1	12.4 ± 0.66	0.5 ± 0.58	1.0 ± 0.19	0.5 ± 0.75	26.3 ± 5.6	27.2 ± 5.5	135.5 ± 31.7
Virginia (V)	5.2 ± 0.25	1.5 ± 0.12	23.5 ± 0.34	13.7 ± 1.1	0.7 ± 0.43	0.9 ± 0.55	0.3 ± 0.37	31.1 ± 6.1	32.8 ± 3.4	179.2 ± 14.5
Valencia (VA)	5.0 ± 0.16	1.5 ± 0.12	17.2 ± 0.50	17.6 ± 1.1	0.3 ± 0.49	1.2 ± 0.0	0.2 ± 0.08	55.2 ± 4.8	25.6 ± 5.8	290.6 ± 40.2
Spanish (S)	5.6 ± 0.19	1.7 ± 0.12	18.0 ± 1.7	18.3 ± 0.96	0.5 ± 0.50	1.1 ± 0.10	0.3 ± 0.19	31.3 ± 6.7	32.4 ± 2.5	170.8 ± 35.8
R + V	4.9 ± 0.46	1.5 ± 0.11	23.3 ± 1.4	13.0 ± 1.1	0.5 ± 0.45	0.9 ± 0.38	0.4 ± 0.56	28.2 ± 5.8	30.0 ± 5.2	157.6 ± 32.5
VA + S	5.3 ± 0.34	1.6 ± 0.15	17.6 ± 1.2	17.9 ± 1.0	0.4 ± 0.47	1.1 ± 0.10	0.2 ± 0.13	41.5 ± 13.9	29.5 ± 6.3	222.1 ± 72.6
All types	5.1 ± 0.44	1.6 ± 0.14	17.7 ± 6.4	15.5 ± 2.7	0.5 ± 0.47	1.0 ± 0.29	0.3 ± 0.45	34.4 ± 12.1	29.8 ± 5.0	187.7 ± 62.5

[a] Mean ± standard deviation, based on 4, 4, 4, 4, 8, 8 and 16 values, respectively, for the seven groundnut types.
[b] Trace levels were arbitrarily assigned a value of 0.1 when mean concentrations were calculated.

Table 3.55 Fatty acid composition of hemp seed oil (Archer-Daniels-Midland, 1961)

16:0	18:0	18:1	18:2	18:3
6	2	12	55	25

Table 3.56 Characteristics of hemp seed oil (typical values)

Relative density, d_{25}^{25}	0.923–0.925
Refractive index, n_D^{45}	1.470–1.473
Saponification value	190–193
Iodine value	140–175
Unsaponifiable matter (%)	1
Oil content of seed (%)	30–35

Table 3.57 Fatty acid composition (wt%) of illipe butter

14:0	16:0	18:0	20:0	16:1	18:1	18:2
0.2	17.5	46.0	0.1	0.2	35.0	1.0

Table 3.58 Characteristics of illipe butter

	Typical	Range
Relative density, d_{15}^{100}	–	0.852–0.850
Refractive index, n_D^{65}	1.4487	–
Saponification value	194	189–200
Iodine value	34	29–38
Unsaponifiable matter (%)	0.9	0.4–2.0
Oil content of seed (%)	45–50 [a]	–

[a] Depends on pretreatment.

Table 3.59 Natural fats providing a source of StOSt, POSt and POP (Bracco, 1973; Lakshminarayana *et al.*, 1983)

Family	Botanical name (species)	Common name of oil	Analytical values			Fatty acid composition						Triglyceride composition		
			Iodine value	Saponification value	M.p. (°C)	14:0	16:0	18:0	18:1	18:2	20:0	Saturated	Monoene	Polyene
Dipterocarpeae	*Shorea robusta*	Sal oil	34–36	181–189	30–35	–	5	44	42	3	6	2	64	34
	S. stenoptera	Illipe butter	31–33	190–196	34–38	–	19–20	40–42	35–38	1	–	4	79	17
	Vateria indica	Malabar tallow	35–38	185–190	34–36	1	12	41	45	1	–	2	74	24
Euphorbiaceae	*Sapium sebiferum*	Chinese vegetable tallow	27–33	199–204	38–44	5	60	2	33	–	–	26	56	18
Gnetaceae	*Gnetum scandens*		–	199	–	–	14	56	27	3	–	–	–	–
Guttifera	*Garcinia indica*	Kokum butter	33–35	189–195	38–42	1	4	53	40	2	–	2	74	24
	G. morella	Gamboge	38–46	189–195	30–41	1	7	42	43	6	1	3	45	50
	Platonia insignis	Bacury	39–44	191–201	54–58	1	55	7	33	4	–	24	56	25
Sapotaceae	*Dumoria heckeli*	Maccore	53	192	24–26	–	3	38	58	1	–	1	10	89
	Madhuca butry	Phulwara butter	45	194	40–42	–	60	4	36	4	–	8	69	23
	Plaguium oleosum	Siak nuts	38	187	35–39	–	6	54	40	–	–	2	77	21
	Butyrosperum parkii	Karite-shea	51–60	180–198	32–34	–	6	40	50	4	–	–	52	48
Sterculiaceae	*Theobroma grandiflorum* S.	Cupu assu	37	195	32–34	1	8	34	41	2	9 [a]	–	–	–
	Theobroma bicolor	Lupu	38–45	188–189	30–32	–	–	–	–	–	–	–	–	–
	Mangifera indica	Mango kernel					3–18	24–57	34–56	1–13	tr–4			

[a] Also 22:0 2%.

like triglycerides. Some of these are summarized in Table 3.59.

3.3.18 Kapok seed oil
[*Ceiba pentandra (Eriodendron anfractuosum)*, *Bombax malarbaricum* (Indian kapok); Bombacaceae]

World production: data are not readily available; 22 000 tonnes were reported in 1978. Other information is given in Tables 3.60 and 3.61.

A number of tropical trees in the *Bombax* family, of which *C. pentandra* is the most commercialized, bear seed pods that contain fibre that finds use as a stuffing material. The seeds, a by-product of kapok fibre production, contain an oil resembling cottonseed oil (Earle and Jones, 1962). The oil is obtained by crushing followed by solvent extraction. In the past it has been refined to produce an 'edible' oil. This should not be permitted because of possible toxic effects arising from the presence of cyclopropene acids (13%). The oil gives a positive Halphen test. It does not, however, contain any gossypol, and can be distinguished from cottonseed oil by the Besson test (Mehlenbacher, 1936). Characterization of the triterpene alcohols offers a more definitive test. The sterols, 4-methylsterols and triterpene alcohols of kapok oil are given in Tables 3.163–3.166. The oil is used for lubrication and in soap-making.

3.3.19 Linseed oil
[*Linum usitatissimum*; Linaceae]

The relevant data are summarized in Tables 3.62 to 3.65. Linseed flax or common flax is an annual, 25–100 cm in height. There are two distinct strains of flax: linseed flax gives high yields of seed (oil content 35–44%) and linen flax is grown for its fibre. It is a subtropical or warm-temperature crop, and is harvested either manually or by combined harvester. It is grown principally in Argentina, India, the former USSR, the USA and Canada.

The small (5 mm long), oval, flat seeds are usually pressed then solvent-extracted, yielding crude linseed oil, which is dark amber in colour. The colour of the oil is improved by refining and bleaching. This also removes phosphatides. Waxes are removed from the refined oil by winterization.

The oil is a drying oil used principally as a component of paints, varnishes, linoleum and printing inks. It has been used as an edible oil, preferably after partial hydrogenation.

The triglycerides of linseed reflect its high linolenate composition. A wide variation in linolenate content has been noted (Table 3.65). This is related both to variety and to climatic conditions during maturation: the colder the climate, the higher is the iodine value. Early work (Dillman and Hopper, 1943) has also shown that moisture deficiency during growth tends to reduce the iodine value. There is a

Table 3.60 Characteristics of kapok seed oil (American Oil Chemists' Society, 1941)

Relative density, d_{15}^{15}	0.920–0.933
Refractive index, n_D^{25}	1.466–1.472
Saponification value	189–197
Iodine value	86–110
Unsaponifiable matter (%)	0.5–1.8
Oil content of seed (%)	22–25

Table 3.62 World production of linseed oil (1000 tonnes)

1980/81	1981/82	1982/83	1989/90
660	665	860	645

Table 3.63 Fatty acid composition (wt%) of linseed oil

16:0	18:0	16:1	18:1	18:2	18:3
6.1	3.2	0.1	16.6	14.2	59.8

Table 3.61 Fatty acid composition[a] (wt%) of oil from *E. anfractuosum* (Blanchini *et al.*, 1981)

16:0	18:0	18:1	18:2	M	MH_2	S	SH_2	Others
22	3	21	37	8	1	4	1	3

[a] This oil contains malvinic (M), dihydromalvinic (MH_2), sterculic (S) and dihydrosterculic (SH_2) acids.

Table 3.64 Characteristics of raw linseed oil (ASTM specification D234-48)

Relative density, $d_{15.5}^{15.5}$	0.931–0.936
Refractive index, n_D^{20}	1.400–1.4835 [a]
Saponification value	189–195
Iodine value, minimum	177
Unsaponifiable matter, maximum (%)	1.50
Acid number	4.0
Loss on heating at 105–110 °C, maximum (%)	0.3
Appearance	Clear and transparent at 65 °C
Colour	Not darker than a freshly prepared solution of 1.0 g of $K_2Cr_2O_7$ in 100 ml of pure H_2SO_4 (rel. dens. 1.84)
Foots, max. (%)	
heated oil	1.0
chilled oil	4.0

[a] Not part of the ASTM specification, a British Standard.

Table 3.65 Fatty acid composition (wt%) of linseed oil (*L. usitatissimum*)

Fatty acid	Variety					
	Krupnosemganny [a]		Abyssinian yellow [b]	Cawpore [b]	Low-linolenate [c]	
	1961	1962			Glenelg	Croxton
14:0	tr		tr	tr	–	–
16:0	6.8	6.3	7.0	7.0	8	8
16:1	tr	tr	tr	tr	–	–
18:0	5.8	4.1	4.0	4.0	5	3
18:1	23.5	14.7	15.0	39.0	21	18
18:2	15.0	18.8	18.0	15.0	64	69
18:3	48.9	56.1	56.0	35.0	2	2

[a] Grown in Russia; 1962 crop grown from 1961 seeds (Vereshchagin and Novitskaya, 1965).
[b] Data from Yermanos (1966).
[c] Green (1988).

close correlation between refractive index and iodine value (Zeleny and Coleman, 1936) (see also Von Schuster and Marquard, 1974).

The major triglycerides of linseed are given in Table 3.159. The sterols, methylsterols and triterpene alcohols of linseed are given in Tables 3.163–3.166. The technical application of linseed in paints is described in Chapter 5.

Flaxseed and flaxseed cake contain the cyanogenic glucoside linamarin, which, when hydrolysed by the endogenous enzyme linase, yields hydrogen cyanide. This can be prevented by denaturing the enzyme by heat.

Plant breeding via genetic modification has produced a low-linolenate variety of flax, which is more suitable for edible purposes (Table 3.65) (Green, 1988, *INFORM*, 1990).

3.3.20 Mowrah butter (bassia, illipe) [*Madhuca latifolia* (*M. indica*, *Bassia latifolia*), *M. longifolia* (*B. longifolia*); Sapotaceae] confused with Borneo tallow; see Illipe butter

World production was reported as 37 000 tonnes in 1977. Fatty acid data and characteristics are shown in Tables 3.66 and 3.67. The seeds are obtained from tropical trees growing in India. *M. longifolia* is an evergreen found in southern India. *M. latifolia* is deciduous and grows in northern India. Although they are botanically distinct, the trade does not distinguish between them. Confusion can occur because the seeds of *Shorea stenoptera* are also called illipe (Kaufmann and Thieme, 1954).

Table 3.66 Fatty acid composition (wt%) of mowrah butter (Tropical Products Institute, 1971)

	16:0	18:0	18:1	18:2
M. latifolia	24	19	43	14
M. longifolia	28	14	49	9

Table 3.67 Characteristics of mowrah butter

Relative density, d_{15}^{100}	0.856–0.870
Refractive index, n_D^{40}	1.458–1.462
Saponification value	188–200
Iodine value	53–70
Unsaponifiable matter (%)	1–3
Oil content of seed (%)	55–60

Commercial quantities of mowrah butter are obtained by expelling. The oil is used for soap-making and to a lesser extent as an edible oil. The meal contains a poisonous saponin, mowrin, which has been used to kill worms.

The oil is of significant commercial importance in India and is used for both edible and industrial purposes. The oil, either hydrogenated or not, has been proposed as a confectionery fat. Many crystallizations from solvent are apparently required to obtain a suitable fraction with the right melting properties.

A wide variation in the composition of the oil has been noted (Sen Gupta *et al.*, 1982): 15–32% palmitic, 16–26% stearic, 32–45% oleic, 14–18% linoleic and traces of 14:0 and 20:0.

Other minor oils produced from seeds of trees in India include neem, karanja and kusum. Typical fatty acid compositions of these oils are given in Table 3.68, and the triglycerides in mowrah are given in Table 3.69. For a detailed review, see Bringi (1987).

3.3.21 Mustard seed oil

[*Brassica alba* (white mustard), *B. hirta* (yellow m.), *B. nigra* (black m.), *B. juncea* (brown m.), *B. carinata* Braun (Ethiopian m.); Cruciferae]
see also Rapeseed oil

World production: there are no comprehensive data available; approximately 100 000 tonnes were traded in 1980. Other data are shown in Tables 3.70 and 3.71.

Mustard seed is normally associated in the West with table mustard, its taste arising from sulphur-based essential oils. *Brassica juncea* (brown mustard) is, however, grown for its oil in India, Pakistan and Bangladesh. Other species of Cruciferae grown for oil in southern Asia are *B. tournefortii* Gouan and rocket (*Eruca sativa* Hill).

Mustard is grown as both a summer and a winter

Table 3.68 Fatty acid composition of minor seed oils of India (wt%)

	14:0	16:0	18:0	20:0	22:0	18:1	18:2	18:3	Others
Karanja (*Pongamia glabra*)	0.3	10.7	7.2	1.0	6.7	47.9	23.6	–	1.6
Kusum *Schleichera trijuga*	1	7	6	26	–	42	4	tr	14[a]
Neem (*Melia azadirachta*)	2	14	18	2	–	55	9	–	–

[a] Includes 12% 20:1.

Table 3.69 The triglyceride composition[a] of mowrah butter (Bringi, 1987)

S_3	S_2U	SU_2	U_3	POP	POSt	StOSt	POO	StOO
1–2	42–56	36–44	6–12	7–12	12–16	4–7	8–12	7–10

[a] S, saturated; U, unsaturated; P, palmitic (16:0); O, oleic (18:1); St, stearic (18:0).

Table 3.70 Fatty acid composition (wt%) of *Brassica alba*

16:0	18:0	20:0	16:1	18:1	18:2	18:3	20:1	22:1	Others
2.6	0.8	0.7	0.1	23.2	8.9	10.4	8.1	43.1	2.1

Table 3.71 Characteristics of mustard seed oil, from draft Codex (Codex, 1970b, see Table 3.157)

Relative density, d_{20}^{20}	0.910–0.921
Refractive index, n_D^{40}	1.461–1.469
Saponification value	170–184
Iodine value	92–125
Unsaponifiable matter (g/kg)	15
Allylisothiocyanate (g/kg) (method specified)	Not more than 4

crop. The oil content of the seed varies (24–40%). The yield is approximately 1000 kg/ha of seed. The oil, in some cases unrefined, is used for edible purposes in a number of countries. There is increasing interest in the crop as a source of erucic acid for the chemical industry (Knowles *et al.*, 1981). At the present time there is no edible oil market for mustard oils in the USA. Rapeseed oil (low-erucic acid) has recently been permitted as an edible oil.

The relationship between the six brassica species has been described by Downey *et al.* (1975). Commercially grown high-erucic acid rape and mustard seed oil are similar in composition and are characterized by their high erucic acid content (McPherson, 1978; Tsunoda *et al.*, 1980).

The glyceride composition of mustard oil has not been reported but presumably is similar to that of high-erucic acid rape (Table 3.100).

As with rape, the oil contains large amounts of brassicasterol, a 'unique' component used for identifying the presence of Cruciferae oils (Matsumoto *et al.*, 1982). Minor amounts of brassicasterol have, however, been reported in other vegetable oils.

3.3.22 Oiticica oil
[*Licania rigida*; Rosaceae]

Comprehensive data on world production are not readily available; 14 000 tonnes reported in 1981. Brazil exported 3000 tonnes in 1981. Other information is summarized in Tables 3.72 and 3.73.

The oil comes from the kernels of the nuts of the oiticica tree, which grows in Brazil. The tree is large and spreading, similar to the mango tree. The seeds, contained in a tough fibrous shell (2.5–5 cm long), are readily split to provide the long reddish kernel. The seed must be decorticated, as the outer husk contains significant amounts of chlorophyll, which can transfer to the oil. The Mexican oiticica tree, *Licania arborea*, yields an oil of similar composition and is called Mexican oiticica or cacahuanache oil.

The oil is extracted by either expelling or solvent extraction or a combination of both. The crude oil is

Table 3.72 Fatty acid composition (wt%) of oiticica oil

16:0	18:0	18:1	oxo 18:3	OH 18
7	5	6	78	4

Table 3.73 Characteristics of oiticica oil (typical values)

Relative density, d_{20}^{40}	0.970
Refractive index, n_D^{25}	1.5140
Saponification value	190
Iodine value	148
Unsaponifiable matter (%)	0.5
Oil content of seed (%)	62

freed from mucilage at 120 °C with 5% water and filtered hot. The extracted oil isomerizes and gradually solidifies on standing at room temperature. This is prevented by heating the oil to 210–220 °C for 30 min.

The major fatty acid in oiticica, Mexican oiticica and poyoak (*Parinarium sherbroense*) oils is α-licanic acid, a conjugated trienoic C_{18} fatty acid with a keto function at C_4 (4-keto 9*c*,11*t*,13*t*-18:3) (Morrell and Davis, 1936). Hydroxy fatty acids (4%) have also been detected by GLC. Prior to GLC α-eleostearic acid had been reported (sections 1.4 and 3.2.12).

The oil superficially resembles tung oil, but oiticica dries more slowly and has poorer water- and chemical-resistance properties. The oil is used in paints and enamels and in the linoleum, oil cloth and printing ink industries.

Oiticica oil triglycerides have been separated by silica column chromatography into four fractions: a monoketo fraction (27.3%), diketo fraction (32.1%), a more polar triglyceride fraction (12.9%) and others (15.9%). The level of licanic acid was reported to be lower than that previously reported, i.e. 46% instead of 70–80% (Evans *et al.*, 1967).

The neutral unsaponifiables of Brazilian oiticica oil have also been characterized (Fedeli *et al.*, 1966).

3.3.23 Olive oil (aceituna, oliva, olivier)
[*Olea europea*; Oleaceae]

Data are summarized in Tables 3.74–3.79. Today, olive oil is the major edible vegetable oil of the Mediterranean countries (including Spain, Greece and Italy) and Portugal (Lunde, 1972; Tous, 1968). Spanish missionaries brought the olive to California in 1850 and Mediterranean immigrants brought it to South America. For recent reviews, see Kiritsakis

Table 3.74 World production of olive oil (1000 tonnes)

1970/71	1975/76	1980/81	1981/82	1982/83	1989	1990
1605	1760	1980	1580	2160	1911	1768

Table 3.75 Fatty acid composition of olive oil (wt%), two examples

16:0	18:0	20:0	16:1	18:1	18:2	18:3	20:1
10.3	2.3	0.4	0.7	78.1	7.3	0.6	0.3
18.3	2.5	0.5	2.4	56.6	18.9	0.8	–

Table 3.76 Characteristics[a] of olive oil, from draft Codex (Codex, 1979 and 1987)

Relative density, d_{20}^{20}	0.910–0.916
Refractive index, n_D^{20}	1.4677–1.4705
Saponification value	184–196
Iodine value	75–94
Unsaponifiable matter (%)	1.5
Bellier index	Not more than 17
Tests for specific oils (see recommended standard)	Negative
GLC ranges of fatty acid composition (%)	
14:0	<0.05
16:0	7.5–20
16:1	0.3–3.5
18:0	0.5–5.0
18:1	55–83
18:2	3.5–21
18:3	<1.5
20:0	<0.8
20:1	tr
22:0	<0.2
24:0	<1.0

[a] The composition and quality factors of olive oil vary widely from country to country. The physical and chemical characteristics are determined for each olive crop year and published in each producing country's *National Olive Oil Index File*.

and Markakis (1987) and Kiritsakis and Min (1988). The oil is obtained from the fruit of the evergreen olive tree. Trees 100–500 years old still bear fruit. The olive fruit is a drupe consisting, when ripe, of an oily pulp enclosing an ellipsoidal stone containing a small kernel. Prior to harvesting, the oil content increases (to 15–40%) and the olive progressively turns black. The kernels contain oil in smaller proportions (1% of total) than that found in the pulp, but both oils are similar in composition.

The kernels are not separated during normal methods of oil expelling or pressing. The first pressing yields virgin olive oil, which is separated from aqueous material by prolonged settling. This oil provides the finest quality of oil for export and is used in its unrefined state.

The Codex defines virgin olive oil as 'the oil obtained from the fruit of the olive tree by mechanical or other physical means under conditions, particularly thermal, which do not lead to alterations of the oil. Virgin olive oil is an oil which is suitable for consumption in its natural state.' The International Olive Oil Council (IOOC) has defined three grades of virgin olive oil (extra, fine and ordinary) based on flavour, free fatty acid content, peroxide value and ultraviolet (UV) absorbance coefficient criteria. Virgin olive oil not for consumption as it is, i.e. that

Table 3.77 IOOC quality standards of olive oil for international trade

Virgin olive oils	Flavour	Acidity (%)	Peroxide value	Specific UV absorbance	
				K_{270}	ΔK
Extra	Absolutely perfect	1.0	20	0.2	0.010
Fine	Absolutely perfect	1.5	20	0.2	0.010
Ordinary	Good	3.3	20	0.25	0.010

Table 3.78 Fatty acid composition (%) of olive oil, from different origins

Acid	Libya [a]	Iran [b]	Italy [c]	Turkey [c]	Portugal [c]
16:0	17.8	10.0	7–17	11.1	12.1
17:0	0.1	tr	–	–	–
18:0	2.2	2.6	1.5–4.0	2.5	2.5
20:0	0.4	0.2	0–0.7	0.4	0.1
22:0	–	–	0–0.3	0.1	0.1
24:0	–	–	0–0.5	0.4	0.4
16:1	2.5	0.8	0.3–3.0	0.7	1.5
17:1	0.1	tr	–	–	–
18:1	43.7	78.2	63–83	74.1	75.3
18:2	32.3	7.9	5–13.5	9.4	6.1
18:3	0.8	0.2	0–1.5	0.2	1.0
20:1	0.2	0.1	0–0.5	0.2	–

[a] Rana and Ahmed (1981).
[b] Paganuzzi (1979).
[c] Fedeli (1983).

Table 3.79 Minor components in olive oil (Tous, 1968)

Component	Level
Hydrocarbons (total)	0.125–0.750%
Squalene	0.125–0.700%
Sterols (total)	0.125–0.250%
Triterpene alcohols (total)	500 ppm
Chlorophyll	0.6–2.2 ppm
Carotenoids as:	
β-carotene (total)	0.6–9.5 ppm
tocopherols (total)	175–200 ppm

requires refining, is defined as virgin olive oil lampante.

The residual pulp is subjected to a second pressing with hot water. The oil is usually refined and blended with virgin olive oil. The pomace can be pressed a third time and the residue extracted with solvent (olive residue oil, ORO).

Inedible 'sulphur olive oil' or 'olive foots' are the names given to very low grades of olive oil obtained by extraction. The name 'sulphur' originates from the use of carbon disulphide, which is now rarely used as an extraction solvent. Higher grades of extracted oil are refined and used for edible purposes.

The oil is used almost solely for edible purposes, e.g. cooking oil, salad oil, but inedible grades are used for soap-making, in wool-combing and in cosmetics.

In a number of countries legal and commercial pressures have encouraged olive oil adulteration. Similarly, low-grade olive oils are 'blended' off with groundnut, cottonseed and soybean oil, etc. The practice of adulterating olive oil was highlighted in the tragic events in Spain in 1981 when it was reported that rapeseed oil denatured with aniline was used as an adulterant (Pastana and Munoz, 1982; Toxic Oil Syndrome References Archive, 1990).

Olive oil is unusual because it contains such a high proportion of oleic acid, 65–85%. The compositions of oils from various origins are given in Table 3.78. The triglyceride composition of a typical olive oil is given in Tables 3.158 and 3.160, and the positional distribution of fatty acids in Table 3.161. See also the paper by Santinelli *et al.* (1992). Numerous minor components in olive oil have been characterized (Table 3.79).

Detailed analyses of the sterols, 4-methylsterols and triterpene alcohols are given in Tables 3.163–3.166. A detailed analysis of the minor components of Iranian olive oil has been reported (Paganuzzi, 1979).

Refined olive oil was found to contain (Muller-Mulot, 1976) α-tocopherol (93 mg/kg) and γ-tocopherol (7 mg/kg). Other tocopherols and tocotrienols were not detected (see also Carpenter, 1979). The level of polyphenols has been related to oxidative stability (Gutfinger, 1981). The oil also contains the highest amount of squalene amongst vegetable oils (Table 3.80).

The phospholipid content varies from 40 to 135 ppm (Vitagliano and Turri, 1958; Vitagliano, 1960). Phosphatidylcholine and phosphatidylethanolamine are present in the kernel (Vioque and Maza, 1973).

The flavour components are reported (Fedeli, 1977) to consist of hydrocarbons, aliphatic hydroxy compounds, aldehydes, ketones, ethers, furan derivatives, thiophene derivatives, esters and terpene alcohols.

Table 3.80 Squalene content of some vegetable oils (Kiritsakis and Min, 1988)

Oil	Squalene (mg/100 g)
Olive	136–708
Corn	19–36
Cottonseed	4–12
Groundnut	13–49
Soybean	7–17
Sunflower	8–19
Tea seed	8–16

Detection of the adulteration of olive oil is a special topic in itself. Numerous specific tests have been proposed (Tous, 1968). The polyunsaturated fraction isolated by silver-ion chromatography has been used to detect adulteration with linoleate-rich oils (Kapoulas and Passaloglou-Emmanouilidou, 1981).

The flavour chemistry has been reviewed (Kiritsakis and Min, 1988).

3.3.24 Palm oil [*Elaeis guineensis* (African oil palm), *E. oleifera* (American oil palm); Palmae]

Summary data are shown in Tables 3.81–3.86. The oil palm *E. guineensis* is the source of two major edible oils: palm oil from the mesocarp, and palmkernel oil from the kernel. The composition and properties of palm oil have been widely described (Cornelius, 1977; Hartley, 1977; Jacobsberg, 1975; *Palm Oil*, 1977; Raymond, 1961).

There is strong evidence that the oil palm originated in Africa. Palm oil trade between Africa and Europe began in the eighteenth century. The introduction of *E. guineensis* to the East Indies occurred in 1848. The progeny of these palms were transferred to Deli in Sumatra and gave rise to the Deli palms, which now provide the basis for the palm oil industry in the Far East (Hartley, 1977).

E. guineensis now exists in Africa, South-East Asia and America, and of all the oil-bearing plants it is the highest yielding (5–10 tonnes/ha in plantations).

Table 3.81 World production of palm oil from *E. guineensis* (1000 tonnes)[a]

1970/71	1975/76	1980/81	1981/82	1982/83	1989/90
1745	2845	4625	5370	5340	10 917

[a] In 1982 Malaysia produced 3512, Nigeria 838, Indonesia 950 and the Ivory Coast 525 (× 1000 tonnes).

Table 3.82 Fatty acid compositions (wt%) for whole palm oil and fractions (Rossell *et al.*, 1985)

Origin (number of samples)	12:0	14:0	16:0	16:1	18:0	18:1	18:2	18:3	20:0
Malaysia (21)	nd–0.1	0.9–1.1	43.1–45.3	0.1–0.3	4.0–4.8	38.4–40.8	9.4–11.1	0.1–0.4	0.1–0.4
Mainland Malaysia (11)	nd–0.1	0.9–1.0	43.1–45.0	0.1–0.3	4.0–4.8	39.2–40.8	9.4–10.8	0.1–0.4	0.1–0.4
Sabah (8)	nd–0.1	0.9–1.1	43.2–45.3	0.1–0.3	4.3–4.7	38.4–40.1	10.1–11.1	0.2–0.4	0.1–0.4
Sarawak (2)	nd[a]	0.9–1.0	43.5–45.0	0.2–0.3	4.5–4.8	38.5–39.4	10.4–10.5	0.2–0.3	0.2–0.4
Ivory Coast (8)	0.1–0.2	0.8–1.0	43.4–45.2	0.1–0.2	4.9–5.5	37.1–39.9	9.6–10.9	0.2–0.4	0.2–0.4
Sumatra (4)	nd–0.2	1.1–1.3	44.6–46.3	0.1–0.2	4.1–4.6	36.7–38.6	10.1–11.5	0.2–0.3	0.2–0.4
Papua New Guinea (3)	nd–0.1	0.9–1.0	43.4–45.4	tr–0.1	4.4–4.6	37.5–39.4	11.0–11.1	0.3[a]	0.2–0.4
Solomon Islands (4)	nd–0.1	1.0–1.1	44.4–44.7	tr–0.1	4.4–4.8	37.6–38.9	10.3–11.0	0.3–0.4	0.4[a]
New Britain (4)	0.1[a]	1.0–1.3	43.5–44.1	tr–0.2	4.6–5.0	37.4–38.3	11.8–11.9	0.2–0.3	0.4[a]
Nigeria (1)	0.2	1.0	45.4	0.1	4.6	37.7	10.6	0.2	0.3
Overall range (45)	nd–0.2	0.8–1.3	43.1–46.3	tr–0.3	4.0–5.5	36.7–40.8	9.4–11.9	0.1–0.4	0.1–0.4
Mean	0.1	1.0	44.3	0.15	4.6	38.7	10.5	0.3	0.3
Stearins (8) (mainland Malaysia)	0.1–0.2	1.0–1.3	46.5–68.9	tr–0.2	4.4–5.5	19.9–38.4	4.1–9.3	0.1–0.2	0.1–0.3
Oleins (5) (Malaysia)	0.1–0.2	0.9–1.0	39.5–40.8	tr–0.2	3.9–4.4	42.7–43.9	10.6–11.4	nd–0.4	0.1–0.3

[a] All samples had the same value within experimental error. nd = not detected. tr = trace (unquantified level of less than 0.05%).

The fruit of the oil palm grows in bunches, which weigh 4–20 kg and contain 200–2000 individual fruits varying in size and shape. Three main fruits have been characterized, *dura*, *tenera* and *pisifera*. These forms are distinguished by both size of the kernel and thickness of shell. *Tenera* is a hybrid of *dura* and *pisifera*. The variability of fatty acid composition within bunches has been determined (Rajanaidu and Tan, 1983).

The nature and type of cultivated palm is likely to change dramatically in the coming years as genetic manipulation and cloning techniques are applied to optimize palm characteristics (Corley *et al.*, 1981; Jones, 1974, 1984; Paranjothy, 1989).

Table 3.83 Characteristics of palm oil, from draft Codex (Codex, 1979 and 1993)

Relative density, d_{20}^{50}	0.891–0.899
Refractive index, n_D^{50}	1.449–1.455
Saponification value	190–209
Iodine value	50–55
Unsaponifiable matter (g/kg)	Not more than 12
GLC ranges in fatty acid composition (%)	
12:0	0–0.4
14:0	0.5–2.0
16:0	40.1–47.5
16:1	0–0.6
18:0	3.5–6.0
18:1	36–44
18:2	6.5–12
18:3	0–0.5
20:0	0–1.0

Table 3.84 Carotenoids of palm oil (%)[a]

Compounds	Percentage of total carotenoids
α-Carotene	29–36
β-Carotene	55–62
γ-Carotene	3–4
Xanthophylls	2–3
Lycopene	2–4

[a] Range in palm oil, 500–700 mg/kg.

Table 3.85 Tocopherols in palm oil (mg/kg)[a] (Rossell *et al.*, 1983; Gapor *et al.*, 1981)

Tocol	Range	Mean
α-Tocopherol	3–185	85
β-Tocopherol	tr	0
γ-Tocopherol	4–36	18
α-Tocotrienol	4–336	124
γ-Tocotrienol	42–710	318
δ-Tocotrienol	tr–148	72
Total	98–1330	617

[a] From 46 samples.

Table 3.86 Sterol composition of palm oil[a] (Rossell *et al.*, 1983)

Sterol	Range[b] (%)	Mean[b] (%)
Cholesterol	2.2–6.7	4.1
Brassicasterol	0.0	0.0
Campesterol	19.8–29.1	22.8
Stigmasterol	8.3–13.0	11.3
β-Sitosterol	50.2–62.1	57.5
Δ^5-Avenasterol	0.0–2.8	1.5
Δ^7-Stigmasterol	0.0–2.8	1.0
Δ^7-Avenasterol	0.0–4.0	1.5

[a] From 15 samples.
[b] Total sterol 326–627 mg/kg.

The American oil palm *Elaeis oleifera* (also called corozo or *E. melanococca*) has a procumbent habit. The oil is relatively unsaturated, which has encouraged its hybridization with *E. guineensis* to produce experimental palms. The native Amazonian palm, *Jessenia bataua*, has a fatty acid composition similar to olive oil (Balick, 1981).

Crude palm oil is obtained from the mesocarp (Berger, 1983) either by centrifugation or by hydraulic pressing of the fruit after it has been 'digested'. Digestion involves mashing and sterilizing the stripped fruit in a steam-jacketed vessel at 95 °C. Unless the damaged fruit is quickly sterilized, fungal lipases rapidly hydrolyse the oil (Tombs and Stubbs, 1982). In West Africa many hand-operated screw and hydraulic presses are in operation.

The crude oil is deep red in colour owing to the presence of carotenoids (500–700 ppm) and is sometimes referred to in the trade as either hard or soft depending on whether the free fatty acid content is high or low. The taste of oil with a high free fatty acid content is preferred in many local parts of West Africa. Oils from plantations with good bleachability and oxidative stability are marketed under the name Zaire Special Prime Bleach (SPB) from West Africa and Special Quality (SQ) from Malaysia. These oils have low free fatty acid (2–3%) and iron contents (4–5 ppm). A 'standard bleaching test' has been devised for these oils. The anisidine test and UV absorbance at 233 and 269 μm are also considered to be good guides to oxidative stability. The high level of carotenoids mitigates against easy bleachability during refining, but refined oils of low residual colour can be obtained when starting with SPB or SQ oils.

The oil is bleached after neutralization using

bleaching earth or by heat bleaching at temperatures of 240–270 °C. In this latter case fatty acid stripping, bleaching and deodorization can be carried out in a single step. Partial hydrogenation also bleaches the oil.

The processing of palm oils in the countries of origin has increased immensely over the last 20 years to the extent that virtually no crude palm oil is now exported from Malaysia. Dry fractionation, approximately 2M tonnes/annum capacity in Malaysia 1994, is carried out on a large scale (Symposium, 1985). Under certain circumstances, it has been commercially attractive to reblend palm fractions to produce simulated palm oils or 'cocktails'. These cocktails can be distinguished in many cases from crude palm oil by determining their triglyceride composition and carotenoid and tocopherol contents.

Palm oil is widely used in food products (Berger, 1987). The nutritional aspects of palm oil have received increasing attention with suggestions that there may be beneficial effects, but current evidence is difficult to interpret unambiguously (Cottrell, 1991; Elson, 1992).

Palm fractions containing high levels (*ca* 70%) of SOS, produced by solvent fractionation (Bernardini, 1978; Berger, 1977), are used as components of cocoa butter equivalents (Best *et al.*, 1956; Thomas and Paulicka, 1976; Gordon *et al.*, 1979; Willems and Padley, 1985).

The characteristics of palm oils from various sources, including Malaysian oils and their fractions are given in Table 3.82. See also data on fatty acid, triglyceride compositions and solid fat contents of palm oil products (Siew *et al.*, 1992). The triglyceride composition of commercial palm oil (Table 3.158) is interesting in that it contains not only high levels of the symmetrical triglyceride SOS (30–36%) but also significant amounts of the unsymmetrical disaturated glyceride SSO (*ca* 7%) and trisaturated triglycerides, mainly tripalmitin. Unlike other vegetable oils, the 2 position of palm oil triglycerides contains approximately 10% of saturated acid.

The minor components of palm oil include carotenoids (Table 3.84), tocopherols (Table 3.85), sterols (Tables 3.86 and 3.163), triterpene alcohols (Table 3.165) and methylsterols (Table 3.164). Average figures for the minor components in plantation oils are given in Tables 3.84–3.86.

The major phospholipids of *E. guineensis* palm oil expressed as a mole percentage of the total phospholipid are phosphatidylcholines (36%), phosphatidylethanolamines (24%), phosphatidylinositols (22%), phosphatidylglycerols (9%) and others (9%). The total phospholipid content in crude palm oil depends on the method of extraction. Values ranging from 20 to 2000 ppm are reported; the lower values are for normal commercially extracted Malaysian oils. The high values are for laboratory solvent-extracted oils (Goh *et al.*, 1982).

The lipids and minor components present in palm leaflets from fronds of different ages have been determined (Gapor *et al.*, 1991). On a dry basis the leaflets contained on average 0.20–0.4% α-tocopherol, 0.05–0.22% carotenoids, 0.006–0.2% squalene for young and old leaflets respectively. Chlorophyll, 0.48–0.62%, reaches a maximum (0.62%) at an intermediate stage.

3.3.25 Palmkernel oil
[*Elaeis guineensis*; Palmae]

Relevant data are shown in Tables 3.87 to 3.92. The background to the oil palm is given in the preceding section (Cornelius, 1977; Hartley, 1977). After the palm oil has been recovered from the fruit, the residual cake from the hydraulic or screw presses is further separated to recover the nuts from which the palmkernels are obtained. The first step is to separate the nuts from the fibre in a stream of air. The nuts are dried, cracked and the kernel separated from the shells either by flotation or by hydrocyclones. Traditional hand methods are still used in certain

Table 3.87 World production of palmkernel oil (1000 tonnes)

1970/71	1975/76	1980/81	1981/82	1982/83	1985/86	1989/90
490	510	585	680	715	992	1421

Table 3.88 Fatty acid composition (wt%) of palmkernel oil

6:0	8:0	10:0	12:0	14:0	16:0	18:0	20:0	18:1	18:2
0.5	2.5	4	49	16	9	2	1	14	2

Table 3.89 Characteristics of palmkernel oil, from draft Codex (Codex, 1979 and 1993)

Relative density, d_{20}^{40}	0.899–0.914
Refractive index, n_D^{40}	1.448–1.452
Saponification value	230–254
Iodine value	13–23
Unsaponifiable matter (g/kg)	Not more than 10
Reichert value	4–7
Polenske value	8–12
GLC ranges of fatty acid composition (%)	
6:0	0–0.8
8:0	2.4–6.2
10:0	2.6–5.0
12:0	41–55
14:0	14–18
16:0	6.5–10
18:0	1.3–3.0
18:1	12–19
18:2	1.0–3.5
Σ18:3–24:1	0–1.0

parts of Africa. Mechanical cracking can lead to a greater proportion of broken kernels.

The kernels are dried to a moisture content of approximately 7% after separation. A rapid reduction in moisture content is important to ensure that the oil is not hydrolysed by yeasts or by fungal attack. Reduction in moisture content also inhibits the production of toxic fungal metabolites such as *Aspergillus flavus*.

Mature kernels, which contain 46–57% oil (moisture-free), vary widely in size and shape. Those from *tenera*, a hybrid of *dura* and *pisifera*, are smaller with a thinner shell than the *dura* (Hartley, 1977). The kernels mature earlier than the mesocarp oil.

The kernel oil is obtained by expelling, usually followed by solvent extraction. The oil superficially resembles coconut oil in chemical and physical properties. The free fatty acid of the crude oil varies from 0.5 to 10%, plantation oils normally having the lowest values. The refined oil keeps well but has a tendency to form soapy rancid off-flavours under conditions in which microbial or lipolytic activity has not been controlled. The off-flavours are from two sources, first the short-chain acids released by hydrolysis of the fat and secondly from methyl ketones, one carbon atom less than the parent acid, generated by fungal action (*Penicillium crustosum*) on the short-chain acids (Kinderlerer and Hatton, 1991).

A major use is in soaps because lauric acid imparts excellent solubility and lathering properties. Palmkernel acids are also used to produce short-chain alcohols and their derivatives for detergents. Edible uses include margarine, ice cream and biscuit cream fillings (Timms, 1986). The melting behaviours of palmkernel and coconut oils are compared in Table 3.32.

The oil foams badly when used in mixtures with other oils for frying. The oil can be fractionated and the high-melting point stearin fraction used as a chocolate substitute fat (Rossell, 1985; Timms, 1985, 1986). The melting point of the oil and its fractions can be further raised by hydrogenation. The effect is greater for palmkernel than coconut oil because the latter contains only low levels of unsaturated acids.

The high lauric acid content in palmkernel oil is typical of numerous species of palm (Tables 3.90 and 3.91). Kernel oils of intermediate composition have been obtained from hybrids of *E. oleifera* and *E. guineensis*. In the longer term we may therefore encounter commercial palmkernel oils of very different character to those of today (Tan and Berger, 1982).

Analyses of Brazilian palmkernel oils showed a similar range of fatty acid compositions as reported in Table 3.88 (Tavares *et al.*, 1990).

The complex fatty acid composition of palmkernel oil is reflected in its triglyceride composition. Gas–liquid chromatography is a convenient method of characterizing the triglycerides of both palmkernel and coconut oils (Bezard, 1971; Bezard *et al.*, 1971; Kuksis and McCarthy, 1962) according to carbon number. These two oils are compared in Table 3.92. The distribution of the major acids follows slightly different patterns in the two oils. Lauric acid is enriched at the 2 position in coconut oil (47% overall, 79% at the 2 position) but only slightly in palmkernel oil (54% overall, 59% at the 2 position). Conversely oleic acid is enriched at the 2 position (9% overall, 15% 2 position) in palmkernel oil but in coconut oil there is no appreciable difference (4.7% and 3.9% respectively) (Table 3.162).

The sterols, 4-methylsterols and triterpene alcohols are reported in Tables 3.163–3.166. Free and bound lactones were detected in Asiatic palmkernel oils (Van der Ven and de Jong, 1970). Data on tocopherols are given in Tables 3.167 and 3.170 but for phospholipids are not readily available.

3.3.26 Poppy seed oil
[*Papaver somniferium*; Papavaraceae]

World production: comprehensive data are not readily available; 12000 tonnes were reported in 1981/82, excluding the former USSR. Other data are summarized in Tables 3.93 and 3.94.

The poppy grows in various climates, but is cultivated for oil as an annual crop mainly in warm temperate zones, e.g. the former USSR, South-East

Table 3.90 Distribution of fatty acid compositions with sample origin for palmkernel oil (Rossell *et al.*, 1985)

Origin (number of samples)	Fatty acid composition (wt%)											
	6:0	8:0	10:0	12:0	14:0	16:0	18:0	18:1	18:2	18:3	20	20:1
Nigeria (7)	0.1–0.2	3.2–3.5	3.4–3.8	46.6–48.5	16.5–16.9	8.5–9.3	2.7–2.9	14.1–15.7	1.4–2.2	tr–0.1	0.1–0.3	tr–0.5
Sierra Leone (2)	0.2–0.4	3.1–3.4	3.4	45.6–45.7	16.7–16.8	8.9	2.7	16.0–16.7	2.2–2.6	tr–0.2	0.1	0.1
Ivory Coast (7)	0.2–0.3	2.6–3.4	3.0–3.5	46.4–47.6	16.0–16.9	8.5–9.0	2.3–2.8	15.3–16.7	1.9–2.5	tr–0.2	0.1	tr–0.1
Guinea Bissau (4)	0.1–0.2	2.5–3.0	2.8–3.2	43.6–46.8	16.9–17.2	9.0–10.0	2.3–2.9	16.4–18.5	1.4–1.9	tr	0.2	nd–0.1
Malaysia (8)	0.2–0.8	3.0–4.4	3.3–4.0	46.2–49.6	15.9–16.8	7.9–8.6	1.9–2.2	13.7–16.1	1.9–2.9	tr–0.2	0.1–0.3	0.1–0.4
Honduras (2)	0.2	3.4–3.6	3.6	48.5–49.8	15.3–16.1	7.8–8.4	2.0–2.4	13.9–15.6	1.9–2.5	0.1–0.7	0.1	0.1
New Britain (3)	0.2–0.3	3.0–3.8	3.4–3.7	48.0–49.3	15.7–16.0	7.8–8.1	2.2–2.3	14.7–15.0	2.8–3.3	tr–0.1	0.1	0.1–0.3
Solomon Islands (1)	0.2	3.3	3.3	49.2	16.1	8.6	2.2	14.2	2.7	tr	0.1	0.1
Sumatra (5)	nd–0.5	3.3–4.7	3.5–4.5	47.3–50.2	15.6–15.9	7.3–8.1	1.9–2.1	13.8–15.0	2.4–2.7	tr–0.1	0.1	0.1
Cameroons (6)	0.2–0.5	2.8–3.8	3.0–3.8	46.2–51.4	16.0–17.1	7.2–9.4	2.3–2.6	11.9–16.0	2.3–3.2	tr–0.3	0.1–0.2	tr–0.2
Costa Rica (1)	0.2	3.4	3.5	48.8	16.0	8.3	2.8	14.4	2.4	tr	0.1	0.1
Togo (4)	0.2–0.3	3.0–3.5	3.2–3.7	46.2–47.8	16.3–16.8	8.6–9.4	2.7–3.0	14.8–16.5	1.9–2.3	tr–0.2	0.1–0.2	0.1
Liberia (1)	0.5	3.5	3.5	46.6	16.4	8.5	2.7	15.7	2.4	tr	0.1	0.1
Papua New Guinea (1)	0.1	3.4	3.5	47.9	16.3	8.3	2.1	15.5	2.7	tr	0.1	0.1
Sarawak (1)	0.2	3.0	3.3	46.9	15.7	8.9	2.3	16.0	3.2	0.1	0.2	0.2
Sabah (1)	0.2	3.3	3.4	47.7	15.8	8.9	2.2	15.2	2.9	tr	0.2	0.2
Overall range (54)	nd–0.8	2.5–4.7	2.8–4.5	43.6–51.4	15.3–17.2	7.2–10.0	1.9–3.0	11.9–18.5	1.4–3.3	tr–0.7	0.1–0.3	nd–0.5
Mean	0.3	3.3	3.5	47.5	16.4	8.5	2.4	15.3	2.4	0.1	0.1	0.1

tr = trace (unquantified level of less than 0.05%). nd = Not detected.

Table 3.91 Fatty acid composition of palmkernel triglycerides (Litchfield, 1970)

	6:0	8:0	10:0	12:0	14:0	16:0	18:0	18:1	18:2
Cocoideae									
Astrocaryum murumuru (murumuru)	–	2.3	1.9	55.6	27.3	4.8	1.8	4.4	1.9
Astrocaryum vulgare (tucum)	–	1.3	4.4	48.9	21.6	6.4	1.7	13.2	2.5
Cocos nucifera	0.6	10.1	6.8	47.7	18.8	7.9	2.0	4.7	1.4
Elaeis guineensis	0.4	5.9	4.8	53.9	16.2	6.6	1.7	8.9	1.6
Orbignya oleifera	0.5	8.3	7.3	49.5	13.9	6.8	2.7	9.9	1.1
Elaeis oleifera	–	1.0	0.8	29.5	25.9	10.1	1.8	26.4	4.5
Coryphoideae									
Chamaerops humilis	–	3.2	3.7	26.5	9.6	9.6	2.0	31.4	14.0
Livistona chinensis	–	0.5	0.6	22.2	9.3	13.8	3.0	28.6	22.0
Sabal palmetto	–	0.5	0.4	21.7	10.2	9.1	1.4	31.2	25.5
Trachycarpus fortunei	–	0.2	0.2	14.6	12.4	9.5	2.3	28.3	32.5
Washingtonia robusta	–	0.8	0.5	17.7	12.7	9.7	1.9	37.5	19.2
Phoenicoideae									
Phoenix reclinata	–	0.6	0.6	29.2	13.5	8.9	1.4	34.3	11.5
Arecpodeae									
Archontophoenix cunninghamiana	–	0.2	0.3	23.9	14.2	12.6	2.7	15.1	31.0
Howeia fosteriana	–	1.5	1.9	45.8	12.2	5.9	2.7	21.1	8.9

Table 3.92 Triglyceride composition (%) of coconut and palmkernel oils (Bezard, 1971; Bezard *et al.*, 1971)

Carbon no.	Mol% in oil		Major glyceride type [a]	Mol% in oil	
	Coconut	Palmkernel		Coconut	Palmkernel
28	0.9	–	6,10,12	0.3	–
			8,8,12	0.4	–
30	4.2	1.3	6,12,12	1.6	0.6
			8,10,12	1.8	0.3
32	15.8	8.6	6,12,14	1.1	0.9
			8,12,12	11.9	6.4
34	19.0	9.8	8,12,14	9.1	3.6
			10,12,12	5.7	4.7
36	20.3	25.7	12,12,12	10.6	19.8
38	17.2	16.8	12,12,14	10.8	14.1
40	9.6	9.5	12,12,16	3.8	3.5
			12,14,14	3.1	3.6
42	6.4	7.8	8,16,18	0.4	0.9
			12,12,18	2.8	3.3
			12,12,16	2.7	2.7
44	3.2	5.8	8,18,18	0.3	1.0
			12,14,18	1.6	3.5
			12,16,16	0.7	0.8
46	1.5	4.0	12,16,18	1.0	2.7
			14,14,18	0.2	0.6
48	1.0	4.4	12,18,18	0.4	2.9
			14,16,18	0.4	1.2
50	0.5	2.2	14,18,18	0.1	1.2
			16,16,18	0.3	0.9
52	0.2	2.2	16,18,18	0.1	1.9
54	–	1.8	18,18,18	–	1.3

[a] 6,10,12 signifies all glycerides with acyl chains of 6, 10 and 12 carbons.

Table 3.93 Fatty acid composition (wt%) of poppy seed oil

16:0	18:0	18:1	18:2	18:3
10	2	11	72	5

Table 3.94 Characteristics of poppy seed oil, reported values (Awe, 1937; Ninomiya, 1940)

Density (60 °C)	0.898
Refractive index, n_D^{60}	1.4604
Saponification value	197.5
Iodine value	133.4
Unsaponifiable matter (%)	0.4–1.2
Oil content of seed (%)	36–50

Asia, China, India, Turkey and the former Czechoslovakia. The main cultivated variety is the opium poppy, grown for both oil and opium.

Opium does not occur in the seed but is obtained from the juice that exudes from the unripe capsules when they are manually scarred. Morphine, the active principle, is also extracted more cheaply from the straw in the former USSR after harvesting.

The oil is extracted from the seed by expelling and solvent extraction. Cold-pressed oil, light yellow in colour, is used for edible purposes without refining. The oil is a semi-drying oil and is also used in artists' paints, in white paint in conjunction with boiled poppy seed oil and in soap after hydrogenation. Poppy seeds are used for decorating baked goods and as bird seed (Anon, 1985).

The fatty acid composition of the principle cultivars and hybrids of the opium poppy grown in Uttar Pradesh, India, gave a mean linoleic acid content of 51.6% with a range of 41–68% (Singh *et al.*, 1990).

The glyceride composition has been determined by silver-ion TLC and agrees well with the data predicted by lipolysis. The 2 position is occupied essentially by unsaturated acids. The sterols, 4-methylsterols and triterpene alcohols are described in Tables 3.163–3.166.

3.3.27 Rapeseed oil (colza oil) [*Brassica napus*, *B. rapa*, *B. campestris*, turnip, mustard; Cruciferae)

Relevant data are given in Tables 3.95–3.101. Over the past two decades, the nature of commercial rapeseed and rapeseed oil has changed dramatically (Andersson, 1981). Rape, mustard (*B. alba*, *B. nigra*) and crambe (*Crambe abyssinica*) seed oils all belong to the same family, the Cruciferae. In the past, they have been grouped together because they all contained significant quantities of erucic acid. Some commercial rapeseed oils are now very different because of their low erucic acid content. The classification of rapeseed is given in Table 3.98.

Brassica napus and *B. campestris* both occur in winter and spring/summer forms. The first is sown in autumn and requires a vernalization period of some weeks below 5 °C. The spring variety (45–75 cm high) requires a relatively short growth period and is used particularly in Canada because of the very harsh winters. The winter (biennial) varieties (up to 140 cm high) give the highest yields, 2.7 t/ha (tonnes/hectare) in Europe, compared to 0.9–1.1 t/ha in Canada for the summer variety.

Plant breeders have developed strains that are not only low in erucic acid but also more disease-resistant, less prone to shattering (casting of seeds from the pod at maturity) and contain less glucosinolates (Rakow and McGregor, 1973; Downey *et al.*, 1969; Craig *et al.*, 1973; Grami and Stefansson, 1977).

Rape seeds are small, round (0.1–0.2 mm diameter) and can be yellow, brown or black. The oil content usually lies within the range of 40–60%. The

Table 3.95 World production of rapeseed oil (1000 tonnes)

1970/71	1975/76	1980/81	1981/82	1982/83	1985/86	1989/90
2220	2650	3675	4060	5000	6426	8025

Table 3.96 Fatty acid composition of rapeseed oil (two examples) (wt%)

	16:0	18:0	20:0	22:0	18:1	18:2	18:3	20:1	22:1
High-erucic-acid rape	3	1	1	tr	16	14	10	6	49
Low-erucic-acid rape	4	2	tr	tr	56	26	10	2	tr

Table 3.97 Characteristics of rapeseed oil[a] (Codex, 1979 and 1993)

	Low-erucic rape	Rape
Relative density, d_{20}^{20}	0.914–0.917	0.910–0.920
Refractive index, n_D^{40}	1.465–1.467	1.465–1.469
Saponification value	188–193	168–187
Iodine value	110–126	94–120
Crismer value	67–70	71–85
Unsaponifiable matter (g/kg)	Not more than 20	Not more than 20
Brassicasterol	Not less than 5% of total sterols	
Erucic acid	Not more than 5 mol% of the component fatty acids	
GLC ranges of fatty acid composition (%)[b]		
<14:0	nd	<0.1
14:0	<0.2	<0.2
16:0	3.3–6.0	1.5–6.0
16:1	0.1–0.6	0–3.0
18:0	1.1–2.5	0.5–3.1
18:1	52–66.9	8–60
18:2	16.1–24.8	11–23
18:3	6.4–14.1	5–13
20:0	0.2–0.8	0–3.0
20:1	0.1–3.4	3–15
20:2	0.0–0.1	0–1.0
22:0	0.0–0.5	0–2.0
22:1	0.0–2.0	5–60
22:2	0.0–0.1	<2.0
24:0	0.0–0.2	<2.0
24:1	0.0–0.4	<3.0

[a] Draft Codex standards for both 'edible low-erucuc-acid rape' and 'rape' have been proposed (Codex, 1993); see Table 3.157.
[b] 17:0 and 17:1 both 0.0–0.3.

oil can be obtained by either expelling or solvent extraction or a combination of the two. The various processing steps to achieve good-quality edible oil have been reviewed (Ohlson, 1992). The crude oil is dark amber in colour and contains a high level of phosphatides (3.5%), pigments, especially chlorophyll from unripe seeds, and sulphur compounds (Crawford, 1982; Mag, 1983).

Table 3.98 Classification of rapeseed (Appelqvist and Ohlsen, 1972)

Botanical name	Common name	Synonyms	Grown in
Brassica napus subsp. *oleifera*			
f. *biennis*	Winter rape	Oil rape, rapeseed, Swedish rape	Europe, Chile, China
f. *annua*	Summer rape	Oilseed rape	Canada
Brassica campestris subsp. *oleifera*			
f. *biennis*	Winter turnip rape	Rapeseed	Sweden
f. *annua*	Summer turnip rape	Oil turnip	Canada
f. *annua* var. *chinensis*	Summer turnip rape	Chinese mustard	
f. *annua* var. *pekinensis*	Summer turnip rape	Celery cabbage, Chinese kale	
f. *annua* var. *dichotoris*	Summer turnip rape	Toria	India, Pakistan
f. *annua* var. *trilocularis*	Summer turnip rape	Sarson	India, Pakistan
Brassica juncea	Brown mustard	Leaf mustard, Indian mustard	India, Pakistan

Table 3.99 Fatty acids in old and new varieties of rape (wt%) (Kroeze, 1982; Wong *et al.*, 1990; Knutson *et al.*, 1992; Voelker and Davies, 1992)

	Brassica napus (Kroeze)				*Brassica campestris* (Kroeze)		High-oleic (Wong)	High-stearic (Knutson)	High-lauric (Voelker)
	Winter (French)		Summer (Canada)		Summer (Canada)				
	Old major	New major	Old target	New tower	Old span	New candle			
12:0	–	–	–	–	–	–	–	–	29.3
14:0	–	–	–	–	–	–	–	–	3.5
16:0	3.5	4.5	3.0	3.9	4.0	3.8	4	4	3.5
18:0	1.2	1.5	0.1	0.3	0.1	0.2	2	22	1.3
18:1	14.2	60.5	19.5	65.0	58.0	53.5	79	40	41.0
20:1	10.9	0.9	10.0	1.2	3.8	1.4	2	–	–
22:1	46.9	0.2	39.5	1.0	3.7	1.0	–	–	–
18:2	13.8	21.6	16.0	19.0	20.0	23.5	7	18	11.3
18:3	9.1	10.3	11.0	8.6	14.0	14.0	5	13	10.1
Others	0.4	0.5	0.9	1.0	0.4	0.6	1	3	–

Table 3.100 Triglyceride composition (mol%) of high-erucic-acid rapeseed oil[a] (Adapted from Jurriens, 1968)

One double bond		Five double bond	
SOS	0.2	ErLL	4.5
SSEr	0.2	OLeEr	0.8
		ErOLe	2.2
Two double bonds		ErLeEr	16.3
SOEr	3.9	Others	1.1
Others	1.6		
		Six double bonds	
Three double bonds		ErLLe	3.0
OOEr	0.9	ErLeL	3.0
ErOEr	18.1	Others	0.8
ErErEr	0.9		
SLEr	5.2	Seven double bonds	
Others	0.9	LLLe	0.3
		ErLeLe	2.0
Four double bonds		Others	0.6
OLEr	1.2		
ErOL	3.4		
ErLEr	24.3		
SLeEr	3.4		
Others	1.2		

[a] Er, erucic acid; S, saturated; L, linoleic; O, oleic; Le, linolenic.

The fully refined oil is bland, pale yellow and free from waxes and essentially free of phosphorus and sulphur.

Until the early 1970s high-erucic-acid rape had been used freely in a number of countries for edible purposes. Biological tests on animals have highlighted potential dangers to the heart in humans. As a consequence, national legislation and FAO/WHO (Food and Agriculture Organization/World Health Organization) recommendations have all led to the use of only low-erucic-acid varieties of rape for edible purposes in numerous countries. In the EEC and Canada only oils with <5% erucic acid are permitted for edible use. The oil is used as a salad oil, for frying and in margarine and shortenings (Ackman, 1990; Ames, 1989; Kramer *et al.*, 1983).

The breeding of low-erucic-acid rape varieties means that the composition of rapeseed oil can vary enormously. Various names have been adopted commercially in an attempt to avoid confusion. 'Canbra' and 'Lear' were used by the Canadians to describe oils with <3.5% erucic acid. The tradename 'Canola' (Sebedio and Ackman, 1981; Vaisey-Genser and Eskin, 1982) has been registered in Canada to describe 'double-zero' varieties of rape, low in glucosinolates (26.5 μmol/g), which conform to the Codex draft standard of ≤5% erucic acid. 'Sinola' oil in Germany contains <2% erucic acid. Typical changes in composition are given in Table 3.99. A reduction in erucic acid (22:1) is accompanied by a major increase in oleic (18:1). This is further complicated by the development of high-oleic (Wong *et al.*, 1990), high-stearate (Knutson *et al.*, 1992) and high laurate varieties (Voelker and Davis, 1992).

The high-erucic-acid rapeseed oil has been used industrially, for example as a lubricant (erucamide) and as a source of brassylic acid (C_{13} dicarboxylic acid, tridecanedioic acid) and pelargonic acid (nonanoic acid) via ozonolysis. Low-erucic-acid rape is also used as a lubricating and cutting oil and, after hydrogenation, in soap.

Rapeseed meal contains glucosinolates, which are hydrolysed by an indigenous enzyme myrosinase, yielding toxic products such as isothiocyanates. This can be prevented by inactivating the enzymes by heat prior to seed extraction (Kramer *et al.*, 1983; Dietz and Harris, 1990). Plant breeding is also significantly reducing the levels of glucosinolates in the meal, to as low as 7 μmol/g (Crawford, 1982; Mag, 1983; Ackman, 1990; Unger, 1990; Lange and Linow, 1987).

Analysis of minor fatty acids (Sebedio and Ackman, 1981) of Canola oil showed that erucic acid, $(n-9)$-22:1, is accompanied by low levels of the $n-7$ isomer (*ca* 1% of total 22:1) (see Table 3.102).

The triglyceride compositions of high-erucic-acid varieties are given in Tables 3.100 and 3.160. Erucic

Table 3.101 Typical values for minor components in Canola oil (Mag, 1983; Teasdale, 1975; Pritchard, 1983)

Component	Crude water degummed	Alkali refined	Bleached and deodorized
Free fatty acid (%)	0.1–1	0.05	0.05
Phosphorus (ppm)	150–250	0–5	5
Sulphur (ppm)	3–15	2–7	
Chlorophyll (ppm)	5–25	1–25	No visual green
Soap (ppm)	–	0–50	tr
Moisture (%)	0.05	0.05	–
Fe (ppm)	1–3	0.1	0.1
Cu (ppm)	–	0.1	0.01

acid is enriched at the 1,3 positions in the glyceride (Table 3.161). Positional distribution of the acids is given in Table 3.102.

The major pigments of rapeseed oil are the chlorophylls a and b (Table 3.101) (Canola Council of Canada, 1981; Niewiadomski *et al.*, 1965) and their decomposition products pheophytin a and b.

The crude degummed oil contains 10–300 ppm of phosphorus depending on the type of degumming (acid or water) operation used (Table 3.101).

The sterols, 4-methylsterols and triterpene alcohols in rapeseed are given in Tables 3.163–3.166. Rapeseed oils contain brassicasterol and little or no stigmasterol (Matsumoto *et al.*, 1982).

The major phospholipids in low-erucic-acid varieties of rape are phosphatidylcholine (48%), phosphatidylinositol (19.5%) and phosphatidylethanolamine (8.9%). The major fatty acids in the phospholipids are palmitic, stearic and oleic acids (Sosulki *et al.*, 1981; Lawsson *et al.*, 1978).

The following tocopherols (mg/kg) in high-erucic-acid rapeseed oil were determined (Slover, 1971; Whittle and Pennock, 1967): α-tocopherol (18–24), γ-tocopherol (38–42), δ-tocopherol (1.1–1.2) and total tocopherol (67) (see Tables 3.167–3.170).

The effect of reducing the content of erucic acid on the crystallization behaviour of hydrogenated rapeseed oil is to slow down the crystallization rate and for the products to have a greater tendency to form large crystals in the β-polymorphic form (Persmark and Bengtsseon, 1976). Margarines developed a grainy texture when manufactured solely from rapeseed and hydrogenated rapeseed oil. The phase behaviour of unhydrogenated rapeseed oils depends on their erucic acid content (Riijner, 1970).

Table 3.102 Stereospecific analysis of rapeseed triacyl-*sn*-glycerols (Takagi and Ando, 1991)

Acyl group	Composition (mol%)			
	Total	*sn*-1	*sn*-2	*sn*3
14:0	0.1	0.1	0.2	0.1
16:0	3.9	5.6	1.6	4.5
$(n-9)$-16:1	0.1	nd[a]	0.1	0.6
$(n-7)$-16:1	0.3	0.3	0.3	0.2
$(n-8)$-17:1	0.1	nd	0.3	nd
18:0	1.3	1.9	0.4	1.7
$(n-9)$-18:1	17.2	13.7	32.4	5.1
$(n-7)$-18:1	1.4	2.8	1.0	0.5
$(n-6)$-18:2	16.1	1.2	39.3	7.4
$(n-3)$-18:3	8.7	3.3	21.3	1.3
20:0	0.8	0.6	0.3	1.5
$(n-9)$-20:1	8.9	15.5	1.5	10.0
$(n-7)$-20:1	1.5	1.9	0.4	2.0
$(n-6)$-20:2	0.5	1.2	0.1	0.2
22:0	0.7	nd	nd	2.1
$(n-9)$-22:1	37.4	50.2	0.8	61.3
$(n-7)$-22:1	0.6	0.9	nd	1.0
$(n-6)$-22:2	0.5	0.9	nd	0.7

[a] ND = not detected.

3.3.28 Rice bran oil
[*Oryza sativa*; Graminae]

World production: there are no comprehensive data available; 370 000 tonnes were reported in 1981. Other information is summarized in Tables 3.103 and 3.104.

Rice is the main cereal crop of Asia and is the basis of the diet of half the world's population (Gaydou and Raonizafinimanana, 1980; Rao and Murthi, 1961; Grist, 1961; Terstage, 1972; Vaccarino, 1977).

Rice grain has an outer siliceous hull, which is removed to give 'brown' rice. Brown rice is further milled to remove the bran and the germ, leaving the endosperm whole (white or polished rice).

The bran, 6–8% of the rice grain, contains about 15–20% oil, 0.4–1.5% wax, 5–8% protein, 40–50% soluble carbohydrate and 5–8% fibre. The rice-bran-oil industry has been hampered by various technical and logistic problems. The bulk of rice bran oil is wasted in a number of countries. The oil is quickly hydrolysed if it remains in the bran. Good-quality oil is therefore only obtained by extracting the oil immediately after milling (Saunders, 1985; Fellers, 1991). The increase in free fatty acid can also be reduced by heating and drying the bran prior to extraction. Hexane and isopropanol have been used as extraction solvents. Rice sugar is a by-product obtained when bran is extracted with isopropanol.

Good-quality oil has excellent oxidative stability arising from a number of naturally occurring anti-

Table 3.103 Fatty acid composition (wt%) of rice bran oil from *O. sativa* var. *japonica* from Malagasy, Madagascar (Gaydou and Raonizafinimanana, 1980)

14:0	16:0	18:0	20:0	16:1	18:1	18:2	18:3	20:1
0.4	16.4	1.8	0.4	0.1	42.3	37.1	1.3	0.2

Table 3.104 Characteristics of rice bran oil

Relative density, d_{25}^{25}	0.916–0.921
Refractive index, 25 °C	1.470–1.473
Saponification value	181–189
Iodine value	99–108
Unsaponifiable matter (%)	3–5
Oil content of bran (%)	15–20

oxidants including tocopherols and esters of ferulic acid. Rice bran lipids include glycolipids, phospholipids, ceramides and waxes (Nasirullah and Nagaraja, 1987). The sterols, 4-methylsterols and triterpene alcohols are given in Tables 3.163–3.166. The fatty acid and sterol compositions of six Malagasy varieties have been reported (Gaydou and Raonizafinimanana, 1980).

3.3.29 Safflower oil (safflowerseed, carmanthus, kurdee)

[*Carthamus tinctorius*; Compositae]

Data are summarized in Tables 3.105–3.108. The plant (false saffron, dyers' saffron) has been grown since antiquity as an oil seed crop and for its flowers, from which the dye (carthamin and safflower yellow) is produced. The plant is a spiny erect herb resembling the thistle (30–100 cm high). There are two distinct varieties, one very spiny, the other moderately spiny or spineless. The seeds resemble sunflower seeds but are approximately half the size; oil content is 36–43% when dehulled. The hull can form a large proportion of the seed (35–65%). The higher-oil-containing commercial seeds are low in hull content. Early commercial data are scanty.

The oil is obtained either by expelling or by solvent extraction (Knowles, 1955; US Department of Agriculture, 1965) and oil is refined in the conventional way. The oil is normally associated with very high levels of linoleic acid, but a commercial variety of high-oleic-acid safflower was introduced into the USA in 1968 (Fuller, 1966; Fuller *et al.*, 1967; Knowles, 1965; Kuemmel, 1964; Knowles, 1975).

The high-linoleic safflower is used as an edible oil and as a component of paints, giving fast-drying non-yellowing properties (no linolenic acid). High-oleic-acid safflower is used as a salad oil, as a stable cooking oil and in certain pharmaceutical and industrial applications.

The fatty acid compositions of both commercial and experimental oils are summarized in Table 3.108.

Statistical analysis of a large number of analyses has shown a strong negative correlation between oleic and linoleic acid (Yermanos *et al.*, 1967). The oil content of the seed is controlled by several genes, none of which is dominant.

The major glycerides of safflower (high-linoleic) are trilinolein, oleodilinolein and monosaturated dilinolein (Table 3.159). The minor components in

Table 3.105 World production of safflower oil (1000 tonnes)

1970/71	1975/76	1980/81	1981/82	1982/83
170	330	265	260	245

Table 3.107 Characteristics of high-linoleic safflower oil, from draft Codex (Codex, 1969, 1979 and 1987)

Relative density, d_{20}^{20}	0.912–0.927
Refractive index, n_D^{40}	1.467–1.470
Saponification value	186–198
Iodine value	135–150
Unsaponifiable matter (g/kg)	Not more than 15
GLC ranges of fatty acid composition (wt%)	
14:0	0–0.2
16:0	5.3–8.0
16:1	0–0.2
18:0	1.9–2.9
18:1	8.4–21.3
18:2	67.8–83.2
18:3	0–0.1
20:0	0.2–0.4
20:1	0.1–0.3
22:0	0.2–0.8
22:1	0–1.8[a]
24:0	0–0.2
24:1	0–0.2

[a] One sample only. Thirty samples had ≤0.4.

Table 3.106 Fatty acid composition (wt%) of safflower oil

	14:0	16:0	18:0	20:0	22:0	16:1	18:1	18:2
High-oleic[a]	0.1	5.7	2.3	0.7	0.3	–	73.6	15.8
High-linoleic[b]	0.1	6.5	2.9	0.4	0.2	0.1	13.8	75.3

[a] Others 1.5%.
[b] Others 0.7%.

Table 3.108 Fatty acid composition (wt%) of safflower oils

Fatty acid	Commercial oils		Experimental oils	
	High-linoleic (Kuemmel, 1964)	(Jurriens, 1968)[a]	High-linoleic (Yermanos *et al.*, 1967)	High-oleic (Craig, 1953)
12:0	0.1	0.2		
14:0	0.1	0.1		
16:0	6.5	6.9	7.6	4–8
18:0	2.9	3.1	2.1	4–8
20:0	0.4	0.1		
22:0	0.2	–		
24:0	0.1	–		
16:1	0.1	0.1		
18:1	13.8	12.9	33.1	74–79
18:2	75.3	76.5	57.2	11–19
20:1	$0.3(n-8)$	0.1		
22:1	$0.1(n-8)$	–		
24:1	$0.1(n-8)$	–		

[a] Triglyceride composition is shown in Table 3.159.

Table 3.109 Tocopherols in safflower oil (mg/kg)

α-T	β-T	γ-T[a]	δ-T	α-T_3	β-T_3	γ-T_3
223	7	33	3.9	–	–	–

[a] β-T_3 is included in the γ-T values.

safflower include sterols, methylsterols and triterpene alcohols (Tables 3.163–3.166). The most recent data show that the main tocopherol in safflower oil is α-tocopherol (Tables 3.109, 3.169 and 3.170).

3.3.30 Sal fat
[*Shorea robusta*; Dipterocarpaceae]

World production of sal fat was 35 000 tonnes reported in 1978. Other data are given in Tables 3.110–3.112.

Sal fat was commercially unknown until the mid-1970s (Bringi, 1987). The seed is the fruit of large subtropical trees growing in the natural forests of northern India. The tree is used mainly for its timber. Deforestation has to some extent been halted with the growing commercialization of seed collection (May–July). The winged seed consists of 60% kernel, 12–20% shell and 20–30% wings. The oil content of the kernel is 14–18%.

The oil is extracted with hexane to give a deep-green semi-solid fat. The quality of the oil depends upon a number of factors, but mainly on whether the seed is harvested before or after the monsoon and on the storage conditions of the seed. The oil is difficult to bleach by conventional refining methods; partial hydrogenation or hydrobleaching has been used.

The oil is used both as an edible oil, in the formulation of cocoa butter equivalents (Yella Reddy and Prabhakar, 1989) and also for soap-making. Certain grades of oil contain epoxy or dihydroxyoctadecanoic acids (Bringi *et al.*, 1972; Bringi, 1987; Yella Reddy and Prabhakar, 1987, 1989). The reason for the occurrence of these components is not fully understood and is probably related to either indigenous lipoxygenase or fungal enzymes modifying the unsaturated acids. Methods of determining these components by TLC and GLC have been reported (Bringi *et al.*, 1972; Belvadi *et al.*, 1979).

The fatty acid compositions of oils extracted from different Indian States in 1979 are given in Table 3.111 (Chakrabarty, 1979). A reported triglyceride composition is given in Table 3.112.

Table 3.110 Fatty acid composition (wt%) of sal fat (Bringi *et al.*, 1972)

14:0	16:0	18:0	20:0	22:0	18:1	18:2	18:3
tr	5.2	42.8	7.0	tr	40.4	3.8	0.8

Table 3.111 Fatty acid composition of sal oils extracted from sal seeds from various States of India (Chakrabarty, 1979)

State	Fatty acid (wt%)						Epoxy acid[a]
	16:0	18:0	18:1	18:2	18:3	20:0	
Bihar	4.6–7.3	32.6–49.3	36.9–48.2	0.6–3.1	0.5–1.2	4.1–8.5	0.6–7.0
Orissa	4.6–6.7	39.6–49.0	34.7–44.5	1.2–2.7	0.5–0.9	3.6–8.7	0.6–5.4
Madhya Pradesh	3.6–10.5	38.4–48.1	34.0–44.5	1.1–3.1	0.3–1.2	3.1–10.0	0.8–5.2
Uttar Pradesh	4.6–7.0	35.2–48.6	36.5–45.2	1.2–2.6	0.5–0.9	4.4–8.8	0.4–2.1
West Bengal	4.1–4.9	42.8–48.9	32.9–38.8	1.5–3.1	0.5–1.2	7.3–7.8	0.3–6.8

[a] As epoxystearic (chemical method).

Table 3.112 Triglyceride composition of a sal fat (Bringi *et al.*, 1972)

Triglyceride[a]	Content (%)
SSS	2
SSU	1
SUS	55
SEpS	9
SUU	26
SEpU	3
USU	1
UUU	3
UEpU	1

[a] S, total saturated acids; U, total unsaturated acids; Ep, *cis*-9,10-epoxystearic acid; SUS, etc., represents the total of all triglycerides of each type.

The minor components, i.e. sterols and triterpene alcohols, are given in Tables 3.163–3.166.

A number of investigations have demonstrated the wholesomeness of the oil (Gandhi *et al.*, 1976; Sundaravalli *et al.*, 1978).

3.3.31 Sesame oil
[*Sesamum indicum*, *S. orientale* is subsp.; Pedaliaceae]

Data are summarized in Tables 3.113–3.115. The plant is an annual herb 0.3–2 m high, of which there are many varieties, including shattering and non-shattering types. It grows in tropical, subtropical and warm-temperate regions.

The main producing countries are China (40%), India (30%), Sudan and Mexico. The seed contains 40–60% oil (Yermanos *et al.*, 1972), which is obtained by either expelling or solvent extraction. Cold-pressed oil is used directly for edible purposes. The colour of the crude oil depends on the type of seed and the extraction method used. Light-coloured oils are obtained by conventional refining methods. High-quality oils can be used in place of olive oil and also in margarine. The oil has an unusually high oxidative stability, until recently attributed to the presence of sesamol (see below). Lower grades are used in the perfumery industry, as a carrier fat for pharmaceuticals and insecticides and for soap-making. The seeds are used in confectionery and bakery products.

The mean fatty acid composition of 721 different samples of sesame seed obtained on a worldwide basis was palmitic 9.5%, stearic 4.4%, oleic 39.6% and linoleic 46.0%, with a range of 32.7–53.9% for oleic acid and 39.3–59.0% for linoleic (Yermanos *et al.*, 1972). There is a significant negative correlation between oleic and linoleic acid content. A more limited range has been observed by others (El Tinay *et al.*, 1976). Wide variations in compositions are observed within countries (Budowski and Markley, 1951).

The triglycerides of a sesame oil are given in Tables 3.159 and 3.160. The fatty acids follow the normal distribution pattern for vegetable oils.

Sesame oil contains two unusual components, sesamin (0.5–1.1%) and sesamolin (0.3–0.6%) (Budowski and Markley, 1951; Beroza and Kinman, 1955; Tashiro *et al.*, 1991). Both are based on a 2,7-dioxabicyclo[3.3.0]octane nucleus. Concentrates of these have been used for their synergistic effect with pyrethrum as an insecticide (Lyon, 1972), but this use has been superseded by synthetic alternatives. Only trace amounts of free sesamol are detected in the oil.

The sterols, 4-methylsterols and triterpene alcohols are given in Tables 3.163–3.166.

Table 3.113 World production of sesame oil (1000 tonnes)

1970/71	1975/76	1980/81	1981/82	1982/83
755	645	615	730	630

Table 3.114 Fatty acid composition (wt%) of sesame oil

16:0	18:0	20:0	22:0	16:1	18:1	18:2	18:3	20:1
9.2	5.8	0.5	tr	0.1	38.2	45.0	0.6	0.6

Table 3.115 Characteristics of sesame oil, from draft Codex (Codex, 1979) (See also Table 3.157)

Relative density, d_{20}^{20}	0.915–0.923
Refractive index, n_D^{40}	1.465–1.469
Saponification value	187–195
Iodine value	104–120
Unsaponifiable matter (g/kg)	Not more than 20
Modified Villavecchia test or sesame oil test (Baudouin)	Positive

The unusually high oxidative stability of the oil has until recently been attributed to the presence of small amounts of sesamol, which is usually present. Sesamol is released from sesamolin by hydrogenation, acid or acid bleaching earths or certain conditions of processing or storage (Budowski and Markley 1951; Budowski, 1964). Sesamolin, sesamin and sesamol have been separated by HPLC (Kikugawa *et al.*, 1983). Studies on the oxidative stability of the oil to which has been added controlled amounts of sesamol showed that the high oxidative stability could not be ascribed to sesamol but to another unknown powerful antioxidant (Kikugawa *et al.*, 1983).

3.3.32 Shea nut oil (shea butter, bambuk butter, karite)
[*Butyrospermum parkii*; Sapotaceae]

World production: the approximate trading volume in 1979 was 35 000 tonnes. Other data are shown in Tables 3.116 and 3.117.

The tree grows in the drier parts of equatorial central Africa, roughly in a belt from the equator to 15°N latitude (Adomako, 1985). The trees bear fruit 10–15 years after planting and reach maturity after 20–30 years. The fruit, 5 cm long, is shaped like a plum and has an edible yellow-green outer pulp enclosing the nut. The nuts are normally dried by sun and fire after the pulp has rotted away. The oil content of the nut is 45–55%. The fruit is available from June to September (Cobley, 1956; Ruyssen, 1957).

The oil is extracted locally, but inefficiently, by grinding the nut to a paste, boiling in water and skimming off the oil. It is extracted commercially by a combination of expelling and extraction. Commercial samples of shea nuts were shown to be very heterogeneous regarding moisture and oil content (Kershaw and Hardwick, 1981).

The oil is used in baking, in other edible applications and also in soap-making (locally). When fractionated from the solvent it yields a stearin, rich in 2-oleodistearin, used in producing cocoa butter equivalents (Soeters, 1970).

The triglyceride composition of shea nut oil is given in Table 3.158.

The oil contains a polyisoprenoid hydrocarbon. This is a gum-like material (3–4%), which can interfere with oil processing (Wolf, 1978; Paquot *et al.*, 1952).

Other minor components (Badifu, 1989) characterized include: tocopherols (90 mg/kg), triterpenol alcohols and cinnamic acid esters, sterols and 4-methylsterols. These are characteristic of the oil (Derbesy and Richert, 1979; Peers, 1977; Itoh *et al.*, 1974). The major triterpene alcohols are α- and β-amyrin, butyrospermol and lupeol. The major sterols are α-spinasterol, previously confused with β-sitosterol and Δ^7-stigmasterol (Derbesy and Richert, 1979; Itoh *et al.*, 1974) (Tables 3.163–3.166).

The selective extraction of shea components using supercritical CO_2 has been described (Turpin *et al.*, 1992).

Table 3.116 Fatty acid composition (wt%) of shea nut oil (Sawadogo and Bezard, 1982)

16:0	18:0	20:0	18:1	18:2
3.3	44.3	1.3	45.6	5.5

Table 3.117 Characteristics of shea nut oil, from draft Codex (Codex, 1970b)

Relative density, d_{15}^{100}	0.859–0.869
Refractive index, n_D^{40}	1.463–1.467
Saponification value	178–190
Iodine value [a]	53–65
Unsaponifiable matter (%)	2–11

[a] The oil contains 3–4% of polyisoprenoid gums with iodine value ~350.

3.3.33 Soybean oil
[*Sofa max*, *Glycine max*; Leguminosae]

Data relevant to soybean oil are summarized in Tables 3.118–3.121.

Soybean is one of the oldest crops cultivated by humans and was grown in China as far back as 2338 BC. Subsequently, it has been grown in Asia, Europe and America. The commercial production of soybean (*Glycine max*) started to expand significantly in the early 1940s, particularly in the USA, and it is now the major world oil seed crop. It is grown in over 50 countries, notably the USA, China, Brazil and Argentina. Papers (Soybean, 1981) and a book (Erikson *et al.*, 1980) review most aspects of the soybean and soybean oil. The chemical, biochemical and nutritional aspects have been reviewed (Lundberg, 1972).

The breeding of different strains of soybean relies in part on germplasm banks in India, Japan, South Korea and the USA (USDA, Soybean germplasm collection).

Soybeans (oil content 18–22%) can be safely stored at 12% moisture without any noticeable increase in free fatty acids. Long-term storage leads to slow but perceptible changes in oil composition, notably the formation of oxygenated fatty acids by the action of lipoxygenase (Barger, 1981; List and Erickson, 1980; Sen Gupta, 1972).

Crude soybean oil can be obtained from the bean by pressing but more normally by hexane extraction. Good-quality oil is light amber in colour with a free fatty acid content of 0.5%. Oil from damaged beans is darker, higher in free fatty acids and modified lipids and is more difficult to bleach. The residual meal is a major source of animal feed and is being increasingly used in human nutrition (Slavin, 1991).

The crude oil is 'degummed' by the controlled addition of water (2%), which precipitates the hydratable phosphatide (1.5–2.5%) to give a valuable by-product. The fully refined oil is used in margarine, salad and frying oils. The oil has a tendency to oxidative flavour reversion, which is largely associated with the linolenic acid in the oil. Its stability for edible purposes is enhanced by partial hydrogenation. Hydrogenation is also applied to soybean oil and other liquid oils to raise their melting range so that they can be used in products that require a certain level of solid fat.

Table 3.120 Characteristics of soybean oil, from draft Codex (Codex, 1979 and 1993)

Relative density, d_{20}^{20}	0.919–0.925
Refractive index, n_D^{40}	1.466–1.470
Saponification value	189–195
Iodine value	120–143
Unsaponifiable matter (g/kg)	Not more than 15
GLC ranges of fatty acid composition (%)	
<14	0–0.1
14:0	0–0.2
16:0	8–13.3
16:1	0–0.2
18:0	2.4–5.4
18:1	17.7–26.1
18:2	49.8–57.1
18:3	5.5–9.5
20:0	0.1–0.6
20:1	0–0.3
20:2	0–0.1
22:0	0.3–0.7
22:1	0–0.3
24:0	0–0.4

(a) Fatty acid and triglyceride composition

Soybean oil shows a wide variation in composition (Tables 3.120 and 3.121). Strong interrelationships

Table 3.118 World production of soybean oil (1000 tonnes)[a]

1970/71	1975/76	1980/81	1981/82	1982/83	1985/86	1989/90
7700	11 000	13 275	14 015	15 395	14 326	16 123

[a] In the period 1980/81, the following amounts of *soybean* were grown (1000 tonnes): USA 48 772; Brazil 15 500; Republic of China 7880; Argentina 3500; Europe 656; former USSR 525.

Table 3.119 Fatty acid composition (wt%) of normal and high-stearate soybean oils (Graef *et al.*, 1985; Knauf, 1987)

	16:0	18:0	20:0	16:1	18:1	18:2	18:3	20:1
Normal	11	4	1	0.5	22	53	7.5	1
High-stearate	8	30	–	–	21	35	6	–

Table 3.121 Fatty acid composition for soybean oil according to origin[a] (Rossell, 1991)

Origin (number of samples)	Fatty acids compositions (wt% methyl esters)										
	16:0	16:1	18:0	18:1	18:2	18:3	20:0	20:1	22:0	22:1	24:0
Argentina (6)	10.7–11.9	tr–0.2	3.7–4.3	17.7–19.1	55.0–56.8	7.9–9.5	0.2–0.5	0.2[b]	0.5–0.6	tr–0.2	nd–0.2
Brazil (6)	10.7–11.8	tr–0.2	3.7–4.2	21.9–23.9	53.1–54.7	5.9–7.3	0.4–0.5	0.2–0.3	0.5–0.7	tr–0.3	tr–0.2
Paraguay (4)	11.3–12.2	tr–0.2	3.7–3.9	22.5–22.7	54.0–54.4	5.6–6.4	0.3–0.5	0.2[b]	0.5–0.6	0.2–0.3	0.2[b]
Uruguay (2)	11.4–11.7	tr–0.1	4.2[b]	19.8–20.4	53.7–54.6	8.3–8.8	0.4[b]	0.2[b]	0.4–0.5	tr–0.1	0.2[b]
USA (8)	10.4–11.9	0.1–0.2	3.9–4.8	22.3–25.5	50.5–53.6	5.5–8.1	0.3–0.6	0.2–0.3	0.3–0.6	nd–0.1	nd–0.4
Canada (3)	10.8–10.9	0.1[b]	3.6–4.0	21.9–24.7	51.2–52.9	8.4–9.3	0.3[b]	0.2[b]	0.3–0.4	nd–tr	0.1[b]
S. Africa (3)	9.9–10.7	0.1–0.2	4.0–5.4	18.8–21.9	53.2–55.8	7.9–9.1	0.4–0.5	0.2[b]	0.4–0.5	tr[b]	0.1–0.2
Australia (2)	11.1–11.7	tr[b]	3.4–3.8	21.7–21.9	54.2–54.8	7.5–8.0	0.2–0.3	tr–0.1	0.4[b]	nd–0.1	0.1[b]
India (3)	9.7–11.7	tr–0.2	3.5–4.3	22.4–26.1	49.8–54.7	6.8–7.1	0.3–0.4	0.2–0.3	0.4–0.5	0.1–0.2	0.1–0.2
France (1)	10.9	tr	3.0	19.8	57.1	8.4	0.1	0.1	0.3	nd	0.1
China (1)	13.3	tr	4.0	20.4	51.9	9.1	0.5	0.3	0.4	nd	nd
Overall range (39)	9.7–13.3	tr–0.2	3.0–5.4	17.7–26.1	49.8–57.1	5.5–9.5	0.1–0.6	tr–0.3	0.3–0.7	nd–0.3	nd–0.4
Mean (39)	11.2	0.1	4.0	21.8	53.9	7.6	0.4	0.2	0.5	0.1	0.1
Coefficient of variation (%) (39)	6.0	68.4	9.9	9.3	2.9	15.7	26.1	25.6	19.2	124.6	61.1
Codex range[c]	7–14	0–0.5	3.0–5.5	18–26	50–57	5.5–10	0–0.6	0–0.5	0–0.5	–	0–0.5

[a] tr = trace (unquantified level of less than 0.05%); nd = not detected.
[b] Where single values are given and more than one sample was analysed, all samples had the same value within experimental error.
[c] Codex Standard 20–1981, as revised February 1987.

exist between the fatty acids. Linolenic and linoleic acids are directly correlated; linolenic and oleic acids are negatively correlated. Saturated acids range from 11 to 26% and are virtually independent of the iodine value (Howell and Collins, 1957; Singh, 1976; White *et al.*, 1961).

The most interesting developments relate to the breeding of varieties that have very different compositions to those associated with conventional soybean oil (Hammond and Fehr, 1984; Graef *et al.*, 1985; Knauf, 1987).

Epoxy fatty acids occur at levels between 0.08 and 1.04% in crude oils depending to a large extent on the quality of the beans (Sen Gupta, 1972; Spencer, 1973).

As with most vegetable oils, the saturated acids hardly occur at the 2 position of the triglyceride. The glyceride composition calculated by the 1,3-random, 2-random distribution agrees well with experimentally determined values. Typical figures are given in Table 3.159. The stereospecific distribution of fatty acids in soya does not show strong differences in the composition of the 1 and 3 positions (Tables 3.122 and 3.161).

(b) Minor components

Perhaps the most important from the refining point of view are the phospholipids (400–700 ppm phosphorus in crude oil). Hydratable phosphatides are removed by degumming with water, but the oil still contains varying amounts of residual phosphatides depending on the quality of the beans. The non-hydratable phospholipids complex with metal ions, notably iron and copper. These are normally removed by treatment with dilute phosphoric acid. Failure to remove phosphatides adversely influences the colour of the oil and hinders hydrogenation.

Soybean phospholipids (Chapman, 1980) contain phosphatidic acid (5%), phosphatidylinositol (20%), phosphatidylethanolamine (23%), phosphatidylcholine (39%) and unidentified components (13%). The phosphatides in crude and degummed soybean oil are given in Table 3.123 (Racicot and Handel, 1983).

Typical tocopherol, sterol, triterpene alcohol, chlorophyll and hydrocarbon levels in soybean oil are given in Table 3.124. Many of these compounds are at least partially removed during refining, bleaching and deodorization. More detailed analyses of the sterols, 4-methylsterols and triterpene alcohols are given in Tables 3.163–3.166, and for tocopherols in Tables 3.167–3.170.

The partial hydrogenation of soybean oil, apart from saturating a proportion of the double bonds, results in the isomerization and migration of double bonds along the fatty acid chain (Dutton and Emken, 1979; Coenen, 1976). The composition of partially hydrogenated soya oil appears in Table 3.125 (D'Alonzo *et al.*, 1981). A US Select Committee on GRAS substances concluded that there was no evidence to demonstrate or suggest that the current levels of partially hardened soya oil used in the USA should be reduced in food (Food and Drug Administration, 1976; Applewhite, 1981). However, recent

Table 3.122 Comparison of stereospecific distributions of acyl groups in soybean oil triacyl-*sn*-glycerols obtained by HPLC and enzymatic methods (Takagi and Ando, 1991)

Acyl group	Composition (mol%)					
	HPLC method			Enzyme method		
	sn-1	*sn*-2	*sn*-3	*sn*-1	*sn*-2	*sn*-3
14:0	0.1	0.1	0.1	0.3	0.3	−0.2
16:0	16.7	2.2	16.1	16.9	1.4	16.6
(*n* − 7)-16:1	0.1	0.1	0.1	0.1	0.1	0.0
17:0	0.1	0.0	0.1	0.1	0.0	0.0
18:0	5.4	0.3	4.6	5.2	0.3	4.7
(*n* − 9)-18:1	22.2	23.0	22.9	20.3	22.6	25.1
(*n* − 7)-18:1	2.1	0.4	1.7	2.0	0.4	1.8
(*n* − 6)-18:2	46.4	68.1	47.0	47.4	69.2	45.2
(*n* − 6)-18:3	6.4	5.7	7.0	7.5	5.7	5.9
20:0	0.5	nd[a]	0.4	0.2	nd	0.7

[a] nd, not detected.

Table 3.123 Individual phosphorus compounds of crude and degummed soybean oil (Racicot and Handel, 1983)

Compound[a]	Phosphorus content[b]			
	Crude		Degummed	
	(ppm)	(% of total P)	(ppm)	(% of total P)
LPC	10.5 ± 4.1	2.4	1.9 ± 1.2	4.4
PC	143.7 ± 11.4	32.7[e]	4.5 ± 2.6	10.5[e]
LPE	6.3 ± 2.4	1.4	2.2 ± 1.5[c]	5.1
PI	67.2 ± 15.3	14.3	4.5 ± 3.8[d]	10.5[d]
PA	73.7 ± 53.0	16.8	4.3 ± 3.0[d]	10.1[d]
PE	109.1 ± 25.3	24.8	8.5 ± 4.2[d]	19.9[d]
Unknown 1	6.0 ± 4.2	1.4[e]	8.9 ± 0.5	20.8[e]
Unknown 2	23.0 ± 10.7	5.2[e]	8.0 ± 5.2	18.7[e]

[a] LPC, lysophosphatidylcholine; PC, phosphatidylcholine; LPE, lysophosphatidylethanolamine; PI, phosphatidylinositol; PA, phosphatidic acid; PE, phosphatidylethanolamine.
[b] Average of four companies unless otherwise noted.
[c] Average of three companies.
[d] Average of eight samples from one company.
[e] Significantly different at the 1% level.

comparisons between *trans* acids and saturated acids may influence future policy.

Soybean oil is widely used in the chemicals industry (Erickson *et al.*, 1980). In the drying-oil industry it is used mainly in alkyd resins. The skipped methylene unsaturated acids can be conjugated by the action of catalysts, e.g. nickel, sulphur dioxide or iodine (Johnson, 1979), to produce faster-drying coatings.

Table 3.124 Some minor components in soybean oil

Tocopherols (total)	530–1500 mg/kg
α-Tocopherol	90–120 mg/kg
γ-Tocopherol	740–1020 mg/kg
δ-Tocopherol	240–300 mg/kg
Chlorophyll	1500 μg/kg
Hydrocarbons	300 mg/kg (squalene 150 mg/kg)
Sterols	1000–3500 mg/kg
Triterpene alcohols	600 mg/kg

Epoxidation is carried out with performic acid, peracetic acid or hydrogen peroxide in the presence of a strong acid or ion-exchange resin. The epoxidized oil is used as both a stabilizer and a plasticizer (Erickson *et al.*, 1980).

Dimerized fatty acids are used in adhesives, coatings, lubricants, printing inks and iron corrosion inhibitors. They are produced by condensing two unsaturated fatty acids in the presence of a catalyst, e.g. clay (Johnson, 1979; Pryde and Cowan, 1972; Berman and Loeb, 1975). They are further converted to polyamides for applications in adhesives, coatings and inks.

Table 3.125 Fatty acid composition (wt%) of a partially hydrogenated soybean oil (D'Alonzo *et al.*, 1981)

							Total
Saturated	12:0	14:0	16:0	18:0	20:0		
	0.2	0.2	9.2	4.2	tr		
Monoenes (18:1)	$n-10$	$n-9$	$n-8$	$n-7$	$n-6$	$n-5$	
cis	0.6	21.9	0.9	3.6	12.0	–	39.8
trans	0.8	1.7	3.9	3.2	1.7	0.7	11.6
Dienes							
cis,cis							29.4
cis,trans							3.0
trans,trans							0.0
Trienes							3.5

The main aspects of soybean oil chemistry and processing have been reviewed (Erickson *et al.*, 1980; Soybean, 1981).

3.3.34 Stillingia oil
[*Sapium sebiferum*; Euphorbiaceae]

This is dealt with alongside Chinese vegetable tallow in section 3.3.6).

3.3.35 Sunflower oil
[*Helianthus annus*; Compositae]

Data are summarized in Tables 3.126–3.130. The sunflower is native to the Americas, but was introduced into Europe in the sixteenth century (Putt, 1978). It is now widely exploited (Table 3.129) as an annual field crop taking four months to mature, 70 days for dwarf varieties.

There are numerous varieties of sunflower, which are identifiable in part on the basis of height but also on the basis of oil composition, i.e. linoleic acid rich and oleic acid rich. The flower heads are harvested mechanically, ideally when the moisture content has dropped to approximately 9–10%. Mechanical drying is frequently necessary. Yields of 1300–2000 kg/ha are obtained. Oil content of seed: 20–32% old strains, 40% new strains.

The seed has a hard woody pericarp, often white with black streaks, which is removed using a decorticator. The oil is extracted by cold and hot pressing followed by solvent extraction. Most of the oil is used for edible purposes, the best quality as table or cooking oil and in margarine. For certain applications, e.g. table oil, the oil is winterized to remove waxes. Sunflower seeds are used as bird seed and are also roasted and salted for confectionery.

Table 3.128 Characteristics of sunflower oil, from draft Codex (Codex, 1979 and 1993)

Relative density, d_{20}^{20}	0.918–0.923
Refractive index, n_D^{40}	1.467–1.469
Saponification value	188–194
Iodine value	110–143
Unsaponifiable matter (g/kg)	Not more than 15
GLC ranges of fatty acid composition (wt%)	
12:0	0–0.1
14:0	0–0.2
16:0	5.6–7.6
16:1	0–0.3
18:0	2.7–6.5
18:1	14–39.4
18:2	48.3–74.0
18:3	0–0.2
20:0	0.2–0.4
20:1	0–0.2
22:0	0.5–1.3
22:1	0–0.2
22:2	0–0.3
24:0	0.2–0.3

Table 3.129 Major sunflower-producing countries, 1981/82

Country	Tonnage of seeds (× 1000)
Former USSR	5 150
USA	2 120
Argentina	1 450
PR China	1 050
Rumania	740
Turkey	575
Hungary	520
S. Africa	495
France	380
Others	1 820
Total	14 300

Table 3.126 World production of sunflower oil (1000 tonnes)

1970/71	1975/76	1980/81	1981/82	1982/83	1985/86	1989/90
3335	3575	4695	5250	5910	6894	7767

Table 3.127 Fatty acid composition (wt%) of sunflower oil (Earle *et al.*, 1968; Knauf, 1987)

	14:0	16:0	18:0	20:0	22:0	24:0	16:1	18:1	18:2	18:3	20:1
Russian	0.2	6.0	5.6	0.3	0.8	0.1	0.1	17.8	68.7	0.2	0.1
American	0.1	5.5	4.7	0.3	0.9	0.2	0.1	19.5	68.5	0.1	0.1
High-oleate	0.1	3.6	4.9	0.4	1.2	0.3	0.1	80.6	8.4	0.1	0.3

Table 3.130 Component glycerides (mol%) of maturing sunflower seed oil (Gunstone and Padley, 1967)

Glyceride[a]	38-day			55-day			90-day		
	Ag[b]	L[b]	T[b]	Ag	L	T	Ag	L	T
222	21	20	20	41	44	45	37	41	41
221	22	27	27	24	28	28	25	26	25
220	20	16	16	19	14	13	20	18	18
211	7	12	12	8	6	6	7	5	5
210	9	14	15	5	6	6	9	7	8
111	5	2	2	0	0	0	0	0	0
200	5	4	3	2	1	1	1	2	2
110	6	3	3	1	1	1	1	1	1
100	5	2	2	0	0	0	0	0	0

[a] These figures indicate the number of double bonds in the three acyl chains. Each glyceride category includes all positional isomers.
[b] Ag, results from silver-ion chromatography; L, results from lipolysis calculated according to Coleman; T, results from component acids calculated according to Gunstone.

Low grades are used for industrial purposes: in soap (after hydrogenation), in blends with linseed oil, etc., in paints and varnishes and in alkyd resins. Paints based solely on sunflower oil have a good resistance to yellowing because of the low linolenic acid content. The use of the oil as a diesel fuel is also being explored (Zimmerman, 1981).

The composition of sunflower oil varies as a consequence of climate and type (Kinman, 1972; Putt, 1978; Von Schuster *et al.*, 1980; Sunflower Oil Today, 1987). The temperature during maturation correlates inversely with linoleic acid content, an observation made with many types of oil seed. The oil is normally renowned for its high linoleic acid content, but there is increasing interest in high-oleic varieties, which are already being grown commercially in the former USSR and the USA (Ficke, 1983; Knauf, 1987).

The triglyceride composition of normal sunflower oil is given in Table 3.159 and of maturing sunflower seed oil (Gunstone and Padley, 1967) in Table 3.130. Carbon-number data are given in Table 3.160.

cis-9,10-Epoxy- and 9,10-dihydroxystearic and -octadecenoic (12*c*) acids have been detected in sunflower oil extracted from seed stored for prolonged periods (Mikolajczak *et al.*, 1968). Only trace amounts of oxygenated acids were found in certain high-oil Russian varieties.

The major phospholipids (0.5–1.2% of crude sunflower oil) are phosphatidic acid (2.2% of total phospholipid), phosphatidylinositol (27.9%), phosphatidylethanolamine (21.2%) and phosphatidylcholine (48.7%) (Chapman, 1980).

The sterols, 4-methylsterols and triterpene alcohols of sunflower oil are given in Tables 3.163–3.166.

The presence of hydrocarbons (0.025–0.031%), sterols (0.542–0.584%) and waxes (0.008–0.044%) has been reported (Morrison and Robertson, 1975). The waxes are frequently removed by winterization to obtain oils with improved cloud points (Rivarola *et al.*, 1985).

α-Tocopherol (608 mg/kg), β-tocopherol (17 mg/kg) and γ-tocopherol (11 mg/kg) are present in sunflower oil (Muller-Mulot, 1976) (Table 3.167).

3.3.36 Tall oil

World production: comprehensive data are not readily available. Production capacity is currently estimated to be 1490 000 tonnes. Data are shown in Tables 3.131 and 3.132. Tall oil is a by-product of the pulping of pine wood using the sulphate or Kraft process.

Crude tall oil is a mixture of fatty acids (45%), rosin acids (42%) and unsaponifiables (triterpenes 13%). The characterization of the fatty and resin acids has been described by a number of workers including

Table 3.131 Fatty acid composition (wt%) of tall oil

16:0	18:0	20:0	18:1	18:2	18:3
5	3	2	46	41	3

Table 3.132 ASTM specifications for tall oil fatty acids

Properties	Type I		Type II		Type III	
	Min.	Max.	Min.	Max.	Min.	Max.
Iodine value	125	135	–	–	–	–
Acid value	197	–	192	–	190	–
Unsaponifiables (wt%)	–	1.0	–	2.0	–	10
Resin acids (wt%)	–	1.0	–	2.0	–	10
Fatty acids (wt%)	98	–	96	–	90	–
Colour (Gardner)	–	4	–	5	–	10

Holmbom (1977, 1978). The resin acids include abietic, dehydroabietic, palustric, pimaric and isopimaric acids.

Most of the crude tall oil is fractionally distilled to separate the resin and fatty acids, which are present in approximately equal amounts. The specifications for different grades of tall oil fatty acids are given in Table 3.132.

A food-grade oleic acid has been produced commercially from tall oil. The main uses of tall oil fatty acids are as intermediates in the production of oleochemicals and as components of alkyd resins. The absence of linolenic acid conveys good colour retention and non-yellowing properties for protective coatings.

3.3.37 Tea seed oil
[*Thea sasanqua*, *Camellia sasanqua*; Theaceae]

Comprehensive information is not readily available on world production; 100 000 tonnes were recorded in 1979, excluding India and parts of China. Other data are given in Tables 3.133 and 3.134.

Tea seed oil is obtained from the woody evergreen shrubs or small trees of *Thea sasanqua* growing in tropical and subtropical climates (oil content of seed 56–70%). The tea leaves used as a beverage normally come from *Thea sinensis* (*Camellia sinensis*), but this is not used to any extent as a source of tea seed oil (oil content of seeds 16–20%). The fatty acids in tea leaf have been determined (Bhuyan and Mahanta, 1989).

Table 3.134 Characteristics of tea seed oil

Relative density	–
Refractive index, n_D^{25}	1.467–1.469
Saponification value	193–196
Iodine value	83–89
Unsaponifiable matter (%)	~0.9
Oil content of seed (%)	56–70

Japanese tea seed oil, tsubaki, comes from *Thea japonica*. The oil is obtained either by pressing or by a combination of expelling and solvent extraction. The fully refined oil is virtually colourless and is very similar to olive oil in composition (Owuor *et al.*, 1985).

The oil is mainly used for edible purposes and only the very low grade oils are used for soap or other industrial purposes. Commercial data are very scarce.

The meal is reported to contain saponins; it has astringent properties and cannot be used in animal feeds.

The presence of tea seed oil in olive oil is 'detected' by the Fitelson test.

The sterols (0.6%) are mainly β-sitosterol, campesterol and stigmasterol. The triterpenes (0.04%) include cycloartenol, cycloartanol, 24-methylenecycloartenol and probably α-amyrin (Fedeli *et al.*, 1966).

Table 3.133 Fatty acid composition (wt%) of tea seed oil (Sengupta *et al.*, 1976; Cloughley, 1983)

	14:0	16:0	18:0	18:1	18:2	20:1	22:1
Camellia sasanqua	–	16.1	0.4	59.0	22.3	–	–
Camellia sinensis[a]	0.2	27.5	5.5	45.5	19.5	0.9	0.9

[a] The 'tea plant'.

3.3.38 Tobacco seed oil [*Nicotiana tabacum* and *N. rustica*; Solanaceae]

World production: data are not readily available. Other information is summarized in Tables 3.135–3.137.

Tobacco (*Wealth of India*, 1966) is grown in both temperate and tropical zones primarily for its leaf, which forms the basis of the tobacco industry. The seed heads are in many cases removed from the plant before they reach maturity as a method of controlling the quality of the tobacco. In many parts of India and to a lesser extent in Greece and Bulgaria, however, the seed is allowed to ripen and is harvested. Relatively small amounts of oil are produced in the USA and Turkey.

The oil is obtained by pressing or solvent extraction. The refined oil is used for edible purposes. It is also widely used as a drying oil because of its non-yellowing characteristics and as a component of alkyd resins. Both the seed and the oil are free of nicotine. The meal is rich in protein and is used in cattle feed. The triglyceride composition determined by silver-nitrate TLC and lipase hydrolysis is detailed in Table 3.138.

Detailed analyses of tobacco seed meal, fatty acid, sterol and triglyceride composition of the oil have been made (Frega *et al.*, 1991). Sterol composition is given in Table 3.137.

Table 3.135 Fatty acid composition (wt%) of *N. rustica* (Sengupta and Mazumbler, 1976)

16:0	18:0	20:0	18:1	18:2	18:3
8.4	3.8	0.1	15.7	71.2	0.8

Table 3.136 Analytical characteristics of *N. rustica* (AOCS values)

Relative density, d_{15}^{15}	0.923–0.925
Refractive index, 25 °C	1.474–1.483
Saponification value	186–197
Iodine value	129–197
Unsaponifiable matter (%)	Not over 1.5
Oil content of seed (%)	30–43

Table 3.137 Triglycerides (wt%) of tobacco seed oil (all isomers)

002[a]	012	022	211	221	222
3	10	20	5	24	38

[a] 0, 1, 2 indicate number of double bonds in fatty acid chains, e.g. 2 = linoleic (18:2).

Table 3.138 Composition[a] of the sterol fraction of tobacco seed oils (as trimethylsilyl (TMS) derivatives), compared to those of other oil-bearing seeds (Frega *et al.*, 1991)

RRT[b]	Identification[c]	Kentucky 104	Bright Italia	Bright V	Safflower	Grape seed	Tomato
0.62	Unknown	tr	tr	0.4	–	–	–
0.67	Cholesterol	8.6	7.3	9.8	tr	0.3	20.6
0.69	Cholestanol	0.4	0.2	tr	tr	tr	1.0
0.74	Brassicasterol	1.1	0.1	1.0	–	–	1.1
0.77	24-Methylencholesterol	0.5	0.9	0.4	1.3	–	0.9
0.78	Unknown	tr	0.5	0.7	0.1	–	–
0.83	Campesterol	13.0	13.4	13.6	13.6	9.5	4.8
0.84	Campestanol	0.3	0.3	tr	–	–	–
0.88	Stigmasterol	9.3	10.2	9.6	5.5	11.3	9.6
0.90	Stigmastenol	0.6	0.6	0.6	–	–	–
0.92	Δ^7-Campestanol	1.5	0.9	1.9	4.2	–	–
0.98	Clerosterol	0.7	0.8	1.3	–	–	0.5
1.00	β-Sitosterol	39.5	43.8	39.5	48.9	75.0	49.8
1.03	Δ^5-Avenasterol	22.7	19.3	19.2	4.0	1.7	10.2
1.06	Unknown	tr	0.2	0.3	–	–	–
1.08	$\Delta^{3,2}$-Stigmastadienol	1.0	0.8	0.8	–	–	–
1.11	Δ^7-Stigmastenol	0.1	0.1	0.2	18.5	1.4	1.0
1.16	Δ^7-Avenasterol	0.7	0.6	0.7	2.1	0.8	0.5

[a] Calculated on the basis of the areas found in high-resolution gas chromatography (HRGC), expressed as a percentage.
[b] Relative retention time, referred to β-sitosterol(TMS) = 1.00.
[c] Tentative identification on the basis of data reported in literature.

3.3.39 Tung oil (China wood oil) [*Aleurites fordii* Hemsl., *A. montana* Willd.; Euphorbiaceae]

Data relating to tung oil are summarized in Tables 3.139–3.141. *A. fordii* is a small tree (up to 12 m high) and *A. montana* a larger (20 m high) deciduous tree. *A. fordii* probably originated in western China and grows under warm-temperate and subtropical conditions. *A. montana* is more suited to a tropical climate. The oils from the two species are virtually identical and trade under the same name.

The fruit is about the size of a large walnut, with a fibrous outer hull inside which are usually four or five seeds. In China the fruit is reported to be harvested unripe, whereas in the USA the ripe fruit is collected after being allowed to fall naturally. The fruit is normally dried immediately. After dehulling and decortication, the kernels are ground, heated to 80 °C and either pressed or solvent extracted. The yield of air-dried seed varies from 1000 to 3000 kg/ha depending on the age of the tree and type of stock (Foster, 1963).

The meal is unsuitable as an animal feed because of the presence of a toxic protein (Mann, 1954).

Table 3.139 World production of tung oil (1000 tonnes)

1970/71	1975/76	1980/81	1981/82	1982/83
120	110	105	95	100

Table 3.140 Fatty acid composition (wt%) of tung oil (Gunstone *et al.*, 1965)

16:0	18:0	18:1	18:2	18:3 conj.
3.1	2.1	11.2	14.6	69

Table 3.141 Characteristics of tung oil

Relative density, d_{15}^{15}	0.938–0.945
Refractive index, n_D^{25}	1.5165–1.5200
Saponification value	189–195
Iodine value	160–175
Unsaponifiable matter (%)	0.5
Oil content of seed (% of dry seed)	16–25

Tung oil contains high levels of α-eleostearic acid (*cis*-9-*trans*-11-*trans*-13-octadecatrienoic) and is therefore inedible. It is used mainly as a component of quick-drying enamels and varnishes. Tung oil is normally liquid down to 7 °C but traces of catalysts such as sulphur, iodine, selenium or light will isomerize α-eleostearic acid to β-eleostearic acid (the all-*trans* isomer), thus raising the melting point (sections 1.4 and 3.2.12). The oil thickens and gels when heated and thus forms the basis for the Browne and Worstall tests.

The gelation time for *A. montana* is longer than that of *A. fordii*. Surprisingly, the oil is 'permanently' prevented from elaidinizing by heating to 200–225 °C for a short time. No explanation has been offered.

The contamination of edible oils with tung oil has been detected using a variety of colorimetric tests (Kitamara *et al.*, 1984).

Tung oil is reported to contain small amounts of 9,14-dihydroxyoctadeca-10,12-dienoic acid (Davies *et al.*, 1950). The estimation of unsaturation by the Wijs iodine method is not quantitative as a result of the relative inactivity of the conjugated system.

The drying of tung oil is reported to require less oxygen (5% absorbed) than the drying of non-conjugated oils such as linseed (12% oxygen adsorbed).

3.3.40 Ucuhuba [*Virola surinamensis*, *V. sebiteia*; Myristacaceae]

Data on world production are not readily available. Other data are given in Tables 3.142 and 3.143.

Both varieties are trees of medium height, the former growing along the swampy coast of northern Brazil and the Guianas, the latter growing inland near the Amazon and its tributaries. The fruits have a spherical hull enclosing a kernel the size of a hazelnut.

The oil is obtained by expelling. Earlier work (Steger and van Loon, 1935) showed that the commercial oil was not completely soluble in petrol. An ether-soluble 'resin fraction' (6.5%) has been isolated from the seed after extraction with petrol. This fraction is apparently largely removed by alkali refining owing to its acidic nature.

Table 3.142 Fatty acid composition (wt%) of ucuhuba (Culp *et al.*, 1965)

10:0	12:0	14:0	16:0	18:0	18:1	18:2
0.5	13	68.5	7	1	8	2

Table 3.143 Characteristics of ucuhuba

	Crude fat	Resin-free fat
Relative density, d_{15}^{100}	0.8715	
Refractive index, n_D^{25}	1.4525	1.4502
Saponification value	221	229
Iodine value	17.0	12.3
Unsaponifiable matter (%)	1–3	
Acid value	26.5	20.7
Oil content of seed (%)	65–76[a]	

[a] Including resins.

Table 3.144 Fatty acid composition (wt%) of vernonia oil

	14:0	16:0	18:0	18:1	18:2	18:3	V[a]	Others
V. anthelmintica	tr	2	1	2	9	0.5	79.5	6
V. roxburghii	tr	3.0	3.5	4.5	13.5	tr	75.2	0.3
V. galamensis	tr	~3	3–4	4–6	13–14	tr	72–78	tr

[a] Vernolic (*cis*-12,13-epoxyoleic) acid.

The major triglycerides found (Culp *et al.*, 1965) were laurodimyristin (31%), trimyristin (43%), lauromyristopalmitin (10%) and lauromyristoolein (12%).

3.3.41 Vernonia oil
[*Vernonia anthelmintica*, Indian iron weed, purple fleabane; Compositae]

World production of this oil is non-commercialized. Other data are shown in Tables 3.144 and 3.145.

The extracted oil is viscous and green. The oil is of interest because of the unusually high level of *cis*-12,13-epoxy-*cis*-9-octadecanoic acid (vernolic acid), the first of this type to be discovered (Gunstone, 1954). Epoxyoleic acids have also been identified in okra (*Hibiscus esculentis*) and in several species of Malvaceae. Oils of this type have been viewed with fresh interest as a natural source of epoxy fatty acids for industrial use (see sections 1.9 and 3.2.9).

Table 3.145 Characteristics of vernonia oil

Relative density, d_{25}^{25}	0.905–0.973
Refractive index, n_D^{25}	1.4742–1.4860
Saponification value	165–210
Iodine value	103.7–107.5
Unsaponifiable matter (%)	6.7
Optical rotation, 25 °C	+2.5
$[\alpha]_d^{25}$	+18.4
Oil content of seed (%)	24

Other species of *Vernonia* have been studied (Osman and Ahmad, 1981; Carlson *et al.*, 1981), and *V. roxburghii* (75.2% epoxy content) and *V. galamensis* (72–78% vernolic acid) have been proposed as being more commercially attractive than *V. anthelmintica* as sources of epoxy acids. *V. galamensis* seed oil has been studied in some detail (Carlson *et al.*, 1981). High enzymic activity was noted in many seeds and the enzyme has to be inactivated before the oil is extracted. Encouraging results were obtained when the oil was evaluated as an epoxy coating.

Detailed compositions of fatty acid and sterols have been determined (Ologunde *et al.*, 1992).

3.3.42 Wheat germ oil
[*Triticum aestivum*; Graminae]

World production: commercial information on oil is not readily available. Other information is shown in Tables 3.146 to 3.148.

Wheat is a monocotyledon. The grain consists of a starchy endosperm surrounded by a protective layer of bran. The germ is located at the proximal end of the grain. The bran and germ are separated from the endosperm in the production of white flour. The germ is isolated by suitable mechanical processing and is stabilized against rancidity by wet or dry heat treatment to inactivate lipases and lipoxygenases.

Wheat germ typically contains 10% fat (Al-Saleh *et al.*, 1986). Commercial oil is produced by expelling

Table 3.146 Fatty acid composition (wt%) of wheat germ oil (Barnes and Taylor, 1980)

	16:0	18:0	18:1	18:2	18:3
Pressure expelled (1)	13.7	1.5	21.8	57.9	5.1
Pressure expelled (2)	12.3	2.0	19.3	61.2	5.2
Solvent extracted	7.1	4.1	22.7	66.1	tr

Table 3.147 Characteristics of wheat germ oil

Relative density, d_{25}^{25}	0.900–0.933
Refractive index, n_D^{20}	1.470–1.480
Saponification value	184–185
Iodine value	120–130
Unsaponifiable matter (%)	2–6
Free fatty acid (%)	2.6–6.0

under pressure. Solvent extraction is reported to produce oils with higher iodine values and lower free fatty acid. The oil resembles corn oil and there is increasing interest in its use as a health food because of its high content of vitamin E (Kahlon, 1989).

The lipids of wheat germ and wheat germ oil have been reviewed (Barnes, 1982). The sterols, methylsterols and triterpene alcohols are reported in Tables 3.163–3.166. The oil contains unusually high levels of tocopherols (Table 3.148).

Recent work shows that the tocotrienols are not present in the germ. Pericarp/aleurone fractions contain predominantly β-T_3 and α-T_3. The endosperm contains β-T_3 (Barnes and Taylor, 1980).

Other minor components include squalene and a range of n-, iso- and anteiso-hydrocarbons. Ferulic and vanillic acids have been extracted from wheat germ, and esters of ferulic acid with dihydro-γ-sitosterol have been detected in the oil.

Both wheat germ and corn oil contain higher levels of gramisterol than other oils. These two oils may be distinguished from each other by means of the higher β-amyrin content in wheat germ oil (Kornfeldt and Croon, 1981).

Detailed compositions of the phospholipids have not been published.

3.3.43 Unclassified oils of interest and new oil seed crops

It is widely accepted that in many parts of the developed world there is a surplus of most traditional crops and consequently there is a search for new crops and for break crops. Unless a new vegetable oil has some specific property (like the oils containing γ-linolenic acid), it will have to compete with existing supplies of oils available at bulk prices. This exerts several constraints:

1. The new crops should be easily cultivated, harvested, processed and marketed. Even when these requirements are met, additional costs may result from the need to have separate and distinct harvesting, processing and marketing facilities.
2. The new crop must quickly become available in good and reliable yields, at appropriate volumes, and at acceptable prices.
3. The demand for and interest in new crops may come more from the oleochemical industry than from the food industry, though, traditionally, the oleochemical industry has used lower-grade and cheaper oils than the food industry.
4. The demand must be market-led (or at least market-maintained after the first few years), and there is an interest in new crops that produce high concentrations of a single acid such as lauric, oleic, erucic, petroselinic, or acids with hydroxy or epoxy groups.
5. Whilst agronomists must help to produce oils meeting these requirements, chemists and technologists must assist in the substitution of existing oils by new oils and in the development of new uses for new oils.

Table 3.148 Tocopherols in wheat germ oil (mg/kg) (Barnes, 1982)

		α-T	β-T	γ-T	δ-T	α-T_3	β-T_3
Laboratory extracted	(1)	2238	803	44	–	–	80
	(2)	1088	292	25	–	63	323
Commercial	(1)	1845	739	213	–	–	–
	(2)	984	386	681	105	–	–

New crops or existing crops with changed fatty acid composition can be obtained by domestication of wild plants (e.g. *Cuphea*) or by the modification of existing crops either by the traditional methods of plant breeders (e.g. Canola, low-erucic rape, or Linola, low-linolenic linseed; see sections 3.3.27 and 3.3.19, respectively) or by the application of gene splitting techniques (e.g. production of a lauric oil from rapeseed).

(a) *Lauric oils*

Glycerides rich in lauric and other medium-chain fatty acids are used extensively both in the food industry and in the production of surfactants, usually in the form of derivatives of the C_{12} alcohol. Whilst the availability of palmkernel oil should rise along with increasing supplies of palm oil, the supply and price of coconut oil has been sufficiently uneven in recent years to arouse interest in other ways of providing lauric acid. Three approaches have been tried: the domestication of *Cuphea* species (section 3.3.12); the production of petroselinic acid-containing oils, which can be oxidized to adipic and lauric acids (see section 3.3.43(c) below); and the application of gene-splitting techniques to rape to make it produce an oil rich in medium-chain acids.

(b) *Oleic oils*

Oleic acid is a very important source material for the oleochemical industry. It is usually obtained from tall oil (about 45%, section 3.3.36), palm oil (about 40%, section 3.3.24), or tallow (about 50%, section 3.4.2). These are cheap sources, but the oleic acid level is not very high and the proportion of oleic acid has to be increased by some fractionation process. Richer sources of oleic acid include rapeseed oil (56%, section 3.3.27), macadamia oil (56%, and 22% of 16:1), almond oil (61%), high-oleic safflower oil (74%, section 3.3.29), olive oil (78%, section 3.3.23), *Jessenia* palm (78%), high-oleic sunflower oil (sun oil, 82%, section 3.3.35) and *Euphorbia lathyris* seed oil (84%). Of these, almond and rape also contain significant amounts of linoleic acid (25–30%), which is undesirable in a source of oleic acid for non-food use.

E. lathyris (caper spurge) is a Mediterranean annual containing about 50% oil in its seed and 80–90% of oleic acid in the oil. At present it suffers from a number of deficiencies, especially seed shattering and the presence of a co-carcinogenic milky sap, but it is hoped to get rid of these problems by plant breeding.

Attempts are also being made to raise the level of oleic acid in rapeseed oil to around 80%.

(c) *Oils containing petroselinic acid*

Seed oils of the Umbelliferae are characterized by the presence of petroselinic acid, often at high levels. Oil content is frequently in the range of 30–60% with petroselinic acid at the 50–90% level. Many plants from this family are already cultivated for other purposes, so it should be possible to develop strains optimized for oil yield and petroselinic acid level. This acid melts at 33 °C in contrast to its common isomer (oleic acid, 12 °C). It is also less soluble in organic solvents and the two acids can be separated by crystallization. It can be oxidized to adipic and lauric acids and may have useful properties for cosmetic and pharmaceutical applications. Attempts are being made in Europe to develop coriander (80% petroselinic acid) into a useful oil crop.

(d) *Oils containing higher monoene acids* (C_{20}–C_{24})

Erucic acid, available in industrial rapeseed oil (high-erucic rapeseed, HEAR), is used extensively in the form of its amide as an antiblock and slip-promoting agent for polyolefins. It can also serve as a source of tridecanedioic acid for the production of nylon-13,13. Attempts are being made to raise the level of erucic acid to about 70%. Other potential sources of erucic acid include *Crambe* (section 3.3.12), honesty (containing 22:1 43% and 24:1 25%), nasturtium (20:1 25% and 22:1 65%) and mustard (section 3.3.21).

Another source of higher monoene compounds is jojoba oil, which has already been developed as a marketable but low-level commodity. It is produced by a drought-tolerant plant that withstands desert heat. This takes 3–5 years to harvest first, 10–17 years to full yield and has a lifespan of around 100 years. It is grown in south-west USA and Mexico mainly, but also in Latin America, Israel, South Africa and Australia. Yields are reported to be 1 tonne of oil/acre.

Jojoba oil is not a typical triacylglycerol but a mixture of wax esters containing mainly 20:1 and 22:1 acids and alcohols. It is thus a mixture of C_{40}, C_{42} and C_{44} esters with two isolated double bonds. The oil is a replacement for sperm whale oil, which is now prohibited in most countries because the sperm whale is an endangered species. At present it is a high-priced oil used mainly in cosmetics, but it has good lubricating properties and could be used extensively for that purpose if available in sufficient quantity at an appropriate price. The oil is fairly pure as extracted, has a light colour and, since the double bonds are well separated, it is resistant to oxidation.

The oil can be chemically modified by reaction (hydrogenation, stereomutation, epoxidation, sulphochlorination) of the double bonds in both the acid and alcohol portions of the esters.

Meadowfoam oil (*Limnanthes alba*) is another oil rich in C_{20} and C_{22} acids. It comes from a herbaceous annual grown in the Pacific coastal regions of North America. More than 95% of its component acids are C_{20} or C_{22} and include 5*c*-20:1 (63%), 5*c*-22:1 (12%), 13*c*-22:1 (3%) and 5*c*,13*c*-22:2 (18%). The double bond is mainly $\Delta 5$, which is an unusual position. The oil remains clear at refrigerator temperature and is fairly stable to oxidation. Potential uses of the oil include conversion to a liquid wax, hydrogenation to a solid ester, use of the isopropyl ester as a cosmetic, use of its glycerol (mono)ester and propylene glycol esters as surface-active compounds, vulcanization to a factice (i.e. a rubber substitute generated by reaction with sulphur), and conversion to amide and dimer acid.

(e) Oils containing C_{18} polyene acids

New interest in linseed oil and in a new low-linolenic variety of this oil (Linola) is discussed in section 3.3.19.

A potential new crop is calendula oil (marigold), which contains about 55% of calendic acid (8*t*,10*t*,12*c*-18:3). This oil with its conjugated triene acid will be a drying oil like tung oil but is being used for cosmetic purposes.

There is a small but growing and thriving industry based on oils containing γ-linolenic acid (6*c*,9*c*,12*c*-18:3). This $n-6$ acid is an isomer of α-linolenic acid and an intermediate in the important bioconversion of linoleic acid to arachidonic. Claims have been made that oils containing this acid are beneficial in the treatment of diseases such as multiple sclerosis, arthritis, atopic eczema and premenstrual tension (section 11.1). This acid is present in a number of seed oils, but commercial exploitation is restricted to evening primrose oil (8–12%, section 3.3.13), borage (19–25%) and blackcurrant (15–19%). γ-Linolenic acid also occurs in a number of microbiological lipids, and products have been developed in Japan and in Britain (oil of javanicus, about 18%) (Gunstone, 1992; Horrobin, 1992).

(f) Oils containing epoxy acids

The use of epoxidized soybean oil as a stabilizer–plasticizer for PVC provides a basis for the interest in natural epoxy acids, of which the best-known is vernolic acid (12,13-epoxyoleic, section 3.2.41). The acid occurs at high levels in the seed oils of *Vernonia galamensis* (73–78%), *Euphorbia lagascae*(62%) and *Stokes aster* (65–79%), and attempts are being made to make the first two into commercial crops. In addition to the possibility of epoxidizing them further, they can also be used as a source of C_9 and C_{12} dibasic acids.

(g) Oils containing hydroxy acids

Lesquerella fondleri produces 25% of oil in its seed and this contains 55% of lesquerolic acid (14-OH 11*c*-20:1), the bishomologue of the ricinoleic acid present in castor oil (section 3.3.5). These two hydroxy acids should have similar physical properties but they will give different products on pyrolysis (undecenoic and tridecenoic acids from the C_{18} and C_{20} acids, respectively) and on alkali fusion (sebacic and dodecanedioic acids).

Dimorphotheca oil contains dimorphecolic acid (9-OH 10,12-18:2). This hydroxy conjugated diene acid can easily be dehydrated to a conjugated triene.

(h) Other unusual acids

Still other examples are known of seed oils rich in a single acid of unusual structure. For example, *Picramnia* species contain the acetylenic acid tariric (6*a*-18:2) in their seed oils up to 90% level, and *Santalum album* (the sandalwood tree) contains over 90% of santalbic (ximenynic) acid (9*a*,11*t*-18:2) in its seed oil. There is no report that these are being developed for commercial purposes.

(i) Specialty (exotic, gourmet) oils

Specialty oils (also described as exotic or gourmet oils) are available in Europe and in the USA. They include low-volume oils such as almond, apricot, avocado, grapeseed (section 3.3.14), hazelnut, macadamia, mustard (section 3.3.21), pecan, passionflower, pistachio, pumpkin, sesame (section 3.3.31), walnut and wheat germ (section 3.3.42) in addition to some of the bulk oils such as coconut, corn, cottonseed, olive, groundnut, palmkernel, rape, safflower, soya and sunflower (Hancock and Houghton, 1990).

These are sold as food oils, in some cases on account of their distinctive flavour, and are also used in toiletries, cosmetics and pharmaceuticals. Careful sourcing of seeds or nuts is necessary to obtain products of the highest quality. The oil must then be extracted under appropriate conditions (cold-pressed, hot-pressed, solvent-extracted) and refined (if desired) using stainless-steel equipment and nitrogen blanketing. In order to achieve low or zero levels of pesticides, it may be necessary to grow the seeds or nuts under organic farming conditions.

The fatty acid compositions of these newer oils are given in Tables 3.150–3.153. These data include the compositions of a range of tree nuts, fruit pulp, fruit seeds and kernels, many of which have not been commercialized as a source of oil.

Some of these oils are also available after chromatographic purification carried out on a multi-kilogram scale. This removes impurities more and less polar than the triacylglycerols to give products that are >99% triacylglycerols.

(j) Crops of potential interest from India

A significant research effort has been devoted in India to the study and development of indigenous oil seed crops. These include sal, mowrah, mango, kokum, neem, pisa and a host of others. A number of these crops have been commercialized to varying degrees depending on ease of collecting the seed and the ability to convert the extracted oil into a suitable material for either soap-making or edible purposes. A detailed description of the major oils of interest has been written (Bringi, 1987).

The compositions of some typical Indian oils are given in Tables 3.149 and 3.154 (mango) and Table 3.155 (others).

(k) Oats and lupin

The relevance of oats in the diet has been highlighted. The fatty acid compositions of 12 strains are given in Table 3.156 (Sahasrabudhe, 1979).

Various varieties of lupin are being considered as a source of protein. The seeds contain approx. 10% oil. See Table 3.156 for fatty acid composition of *Lupinus polyphyllus* (Anizewski, 1993; Feldheim, 1991).

Table 3.149 Fatty acid composition in triglyceride (TG) fraction of mango fat (mol%)[a] (Ali *et al.*, 1985)

Variety	16:0	18:0	18:1	18:2	18:3	20:0
Brindaboni	9.2	39.1	41.1	7.6	0.7	2.3
Fazli	8.4	39.4	42.5	7.3	0.7	1.7
Kalabau	8.1	40.2	41.5	7.3	0.6	2.2
Kanchamitha	9.1	38.2	43.0	6.3	0.9	2.5
Kuipahari	10.0	38.3	41.3	7.1	0.9	2.5
Lakhanvoge	9.7	38.3	41.7	6.9	0.7	2.6
Lengra	8.5	39.3	42.4	6.8	0.7	2.2
Mohanvoge	9.0	38.6	41.4	7.6	0.9	2.4
Misrakanta	7.9	39.2	43.8	6.0	1.0	2.0
Ranipasand	7.9	39.9	42.0	7.4	0.6	2.2

[a] Mean results of three experiments.

Table 3.150 Fatty acid composition (%)[a] of fruit pulp fatty acids (Kamel and Kakuda, 1992)

Common name	Scientific name	Lipid (%)	14:0	16:0	16:1	18:0	18:1	18:2	18:3	Sat.	Unsat.
Apple[b]	*Malus sylvestris*	0.4	1.2	26.7	0.6	3.9	7.9	48.9	10.1	32.4	67.5
Apricot	*Prunus armeniaes*	0.4	–	8.8	–	1.1	62.0	28.1	–	9.9	90.1
Avocado	*Persea americana*	6.5–25.5	–	9.0	1.8	0.5	77.7	8.5	0.9	9.5	88.9
Banana	*Musa paradisiaca*	0.2	–	41.9	2.1	3.8	14.4	16.1	21.7	45.7	54.3
Blueberry	*Vaccinium corymbosum*	0.43–0.55	2.1	7.38	–	2.8	24.0	38.7	27.1	12.2	89.8
Cherry (sour)	*Prunus cerasus*	0.3	0.8	20.0	0.4	6.7	33.8	19.2	18.3	27.5	71.7
Cherry (sweet)	*Prunus avium*	1.0	0.7	19.8	0.4	6.8	33.8	19.2	18.5	27.3	71.9
Chironji fruit	*Buchanania lanzan*	–	–	34.4	–	2.0	62.3	1.2	0.2	36.4	63.7
Currant	*Ribes nigrum*	0.4	–	7.4	0.4	2.6	20.7	39.5	26.6	10.0	87.2
Durian	*Durio zibethinus*	4–5	0.5	39.8	8.5	0.8	45.8	1.8	2.7	41.1	58.8
Fig	*Citrus ficus carica*	0.3	0.8	17.0	–	4.4	24.4	53.3	–	22.2	77.7
Gooseberry	*Ribes* spp.	0.6	–	4.7	–	3.2	12.6	66.7	11.3	7.9	90.6
Grapefruit	*Citrus paradisi*	0.1	–	23.5	2.0	1.9	23.5	37.3	9.8	25.4	72.6
Grape	*Vitis vinifera*	0.6	3.4	42.5	–	5.8	6.0	34.1	10.2	51.7	50.3
Guava	*Psidium guajava*	0.6	2.5	30	0.6	3.3	10.8	37.9	14.8	35.8	64.1
Lemon	*Citrus limon*	0.3	0.7	25.2	0.7	1.4	7.2	45.3	18.7	27.3	71.9
Loquats[c]	*Eriobotrya japonica*	0.2	0.7	23.0	–	2.8	5.7	55.4	9.3	27.2	70.4
Mango	*Mangifera indica*	0.1–0.4	2.4	23.6	27.4	1.3	26.4	1.8	16.7	27.3	72.3
Nabk[d]	*Zizyphus jujuba*	4	–	11.3	1.1	6.7	53.8	20.6	–	19.9	79.1
Olive[e]	*Olea europea*	35–70	–	12.0	0.4	4.0	72.0	10.0	0.5	16.4	83.0
Orange	*Citrus sinensis*	0.3	–	20.6	4.7	–	31.7	28.6	11.1	20.6	76.1
Papaya[f]	*Carica papaya*	0.14	6.3	28.6	17.8	1.8	16.1	5.3	22.3	37.6	61.5
Peach	*Prunus persica*	0.9	–	10.1	1.1	1.1	38.0	49.4	1.1	11.2	89.6
Pear[g]	*Pyrus communis*	0.4	–	8.5	1.0	1.5	40.5	46.5	0.5	10.0	89.0
Pineapple	*Ananas comosus*	0.43	–	8.4	1.3	4.8	19.9	37.2	27.4	13.2	85.8
Plum	*Prunus* spp.	0.62	–	7.0	0.84	1.5	67.9	22.7	–	8.5	91.4
Quince[h]	*Cydonia oblonga*	0.1	–	7.29	–	2.1	37.5	51.1	–	9.4	89.6
Raspberry	*Rubus* spp.	0.55	–	3.4	–	0.8	12.7	54.0	27.3	4.2	94.0
Strawberry	*Fragaria xananassa*	0.37	–	5.4	0.4	1.5	19.8	41.9	30.2	6.9	92.3
Tangerine	*Citrus reticulata*	0.2	1.1	21.5	4.4	1.1	32.3	29.0	10.7	23.7	76.4
Tamarind	*Tamarindus indica*	0.6	1.3	32.8	–	11.7	35.3	11.5	–	45.8	46.8

[a] For references, see original publication.
[b] 0.6% 12:0; [c] 0.7% 12:0; [d] 1.9% 20:0, 1.1% 20:1, 1.4% 20:2, 1.1% 20:4; [e] 0.4% 20:0, 0.2% 20:1; [f] 0.9% 12:0; [g] 0.5% 20:1; [h] 1.0% 20:1.

Table 3.151 Fatty acid composition of oils from tree nuts[a] (Beuchat and Worthington, 1978)

Common name	Botanical name	Oil (%)	Fatty acid (wt%)											
			16:0	16:1	18:0	18:1	18:2	18:3	20:0	20:1	22:0	22:1	24:0	Others
Almond	*Prunus amygdalus*	52.7	6.7	0.5	1.2	66.3	22.3	tr	tr	tr	tr	–	–	3.0
Butternut	*Juglans cinerea*	60.0	1.6	tr	0.8	19.0	61.9	16.0	–	–	–	–	–	0.7
Brazil	*Berthrolletia myrtaceae*	68.7	14.1	0.3	8.6	29.0	46.6	tr	0.3	tr	tr	–	–	1.1
Cashew	*Anacardium occidentale*	47.7	10.2	0.4	8.5	60.9	18.3	–	0.8	tr	tr	–	–	0.9
Chestnut	*Castanea mollisima*	2.5	14.6	0.7	1.1	54.0	24.9	2.7	–	1.0	0.2	–	–	0.8
Filbert/hazelnut	*Corylus avellana*	62.3	4.7	0.2	1.6	76.4	16.3	0.1	0.1	0.2	–	–	–	0.4
Hickory	*Caryaovata ovata*	70.4	8.8	0.5	2.3	52.0	33.5	1.7	0.2	tr	–	–	–	1.0
Macadamia	*Macadamia tetraphylla*[b]	73.2	8.3	21.8	2.1	56.4	2.8	–	2.4	3.1	0.8	0.3	0.5	1.3
Pecan	*Carya illioensis*	70.3	5.7	0.1	2.2	66.9	22.1	1.1	0.2	0.4	–	–	–	1.3
Pignolia	*Pinus* sp.	48.1	5.8	0.2	3.8	38.1	46.4	0.8	0.7	0.9	–	–	–	3.3
Pistachio	*Pistachia vera*	57.0	8.6	0.7	2.3	68.8	17.8	0.3	0.3	0.6	tr	–	–	0.6
Walnut, black	*Juglans nigra*	59.0	3.1	tr	2.6	29.1	58.3	4.9	tr	tr	tr	–	tr	2.0
Walnut, English	*Juglans regia*	67.4	7.3	0.2	2.3	19.1	57.4	13.1	tr	–	–	–	–	0.6

[a] All fats contained tr (trace) of 17:0.
[b] 0.1% 17:1.

Table 3.152 Fatty acid content (%) of fruit seeds and kernels[a] (Kamel and Kakuda, 1992)

Common name	Scientific name	Lipid (%)	8:0	10:0	12:0	14:0	16:0	16:1	18:0	18:1	18:2	18:3	20:0	22:0	24:0	Sat.	Unsat.
Apple seed	*Malus sylvestris*	25.7	–	–	–	–	9.0	–	2.3	41.3	45.6	0.7	1.1	–	–	12.4	87.6
Apricot kernel	*Prunus armeniaea*	49.50	–	–	–	–	4.8	0.42	0.23	69.2	25.4	–	–	–	–	5.0	95.0
Avocado seed	*Persea americana*	1–1.5	–	–	–	0.8	22	3.2	0.8	25	41.5	4.5	tr	tr	–	23.6	74.2
Cantaloupe seed	*Cucumis melo*	36[m]	–	–	–	34	27.4	–	6.4	12.9	17.7	1.6	–	–	–	67.8	32.2
		34[n]	–	–	–	–	26.7	–	6.7	16.7	46.6	3.3	–	–	–	33.4	66.6
Cherry kernel	*Prunes avium*	30–38	–	–	–	–	7.7	–	2.4	60.8	30.8	tr	–	–	–	10.0	91.6
Coffee seed[b]	*Coffea arabica*	11.8	–	–	–	2.0	28.8	–	4.5	18.9	44.0	–	1.6	–	–	36.9	62.9
Currant seed (black)[c]	*Ribes nigrum*	30.5	–	–	–	–	6.4	–	1.3	10.3	46.5	13.6	–	–	–	7.7	92.3
Currant seed (red)[d]	*Ribes rubrum*	25.2	–	–	–	–	4.5	–	1.5	15	42	29	–	–	–	6.0	95
Date seed	*Phoenix dactylifera*	10.4	–	2.8	17.3	11.4	10.3	0.17	2.8	44.2	8.4	0.6	0.6	0.3	0.2	45.8	53.4
Durian seed[e]	*Durio zibethinus*	0.5	–	–	–	0.12	12.2	1.1	1.4	8.4	6.5	11.3	–	1.2	–	14.9	81.5
Egusi seed	*Colocynthis citrullus*	52	–	–	–	0.02	10.4	0.14	9.8	15.9	62.8	0.41	–	–	–	22.4	77.6
Fig seed	*Ficus carica*	30	–	–	–	–	7.2	–	2.6	14.9	30.6	44.7	–	–	–	9.8	90.2
Gooseberry seed[f]	*Ribes uva crispa*	18.3	–	–	–	–	7	–	1	17	40	19.5	–	–	–	8.0	92
Grape seed	*Vitis vinifera*	13–18.4	–	–	–	0.05	7.0	–	3.9	15.6	72.2	0.24	–	–	–	10.9	88.1
Grapefruit seed	*Citrus paradisi*	41.4	8.4	–	–	–	42.6	–	0.83	12.2	34.5	1.5	–	–	–	51.8	48.2
Guava seed	*Psidium guajava* L.	10.1	–	–	–	1.2	8.9	–	4.8	53.9	29.2	1.1	–	–	–	14.9	84.2
Jack fruit[g]	*Artocarpus integrifolia*	6.1	–	–	–	3.3	30.2	–	3.3	6.4	40.2	9.4	–	–	–	38.8	63.2
Lime seed	*Citrus aurantifolia*	42.7	0.8	–	0.99	–	19.1	0.28	0.84	17.1	18.0	42.3	–	–	–	21	78.5
Mandarin seed[h]	*Citrus reticulata*	40.2	3.1	4.5	4.4	4.9	25.7	2.9	11.1	23.9	19.2	–	–	–	–	53.7	46.0
Mango kernel	*Mangifera indica*	6–12	–	–	–	–	9.2	–	39.1	41.1	7.6	0.7	2.3	–	–	50.6	49.4
Melon seed[i]	*Citrullus vulgaris*	23.1	–	–	–	0.55	12.2	0.13	11.2	11.1	54.7	0.18	–	–	–	23.9	66.1
Melon seed	*Colocynthis vulgaris*	–	–	–	–	1.5	20	0.3	7.1	21.6	24.5	15.9	–	–	–	28.6	67.4
Orange seed[j]	*Citrus sinensis*	45.5	–	5.4	–	–	28.8	0.65	2.6	23.7	30.9	6.5	–	–	–	36.8	61.7
Papaya kernel	*Carica papaya*	50.1	–	–	–	0.2	17.2	1.3	3.6	77.3	0.4	–	–	–	–	21.0	79.0
Passion fruit	*Passiflora edulis* f.	20	–	–	–	–	8	–	2.2	12.6	77.2	–	–	–	–	10.2	89.8
Peach kernel	*Prunus persica*	45.1–54.5	–	–	–	0.88	13.4	0.16	6.4	63.8	15.4	tr	–	–	–	20.7	79.3
Prune kernel	*Prunus domestica*	28.6	–	–	–	–	8.0	–	1.7	61.2	29.1	–	–	–	–	9.7	90.3
Pumpkin seed	*Cucurbita mixta*	36	–	–	–	tr	16.4	–	7.7	33.8	42.0	–	–	–	–	24.7	75.8
Soursop seed	*Annona muricata*	24	–	–	–	0.34	17.0	–	5.8	61.0	12.1	–	–	–	–	23.1	73.1
Tamarind seed (India)	*Tamarindus indica*	7.4	–	–	tr	tr	14.8	–	5.9	27.0	7.5	5.6	4.5	12.2	22.3	59.7	40.1
Tamarind seed (Madagascar)[k]	*Tamarindus indica* L.	6–8	–	–	–	0.2	17.4	0.3	6.7	19.5	42.4	0.1	2.8	4.0	5.5	36.8	63.2
Tomato seed[l]	*Lysopersicon esculentum*	22.4	–	–	–	0.4	14.4	0.5	3.9	21.7	54.8	3.2	0.5	–	–	19.8	80.2
	Calodendram capense	60	–	–	–	–	23.6	–	4.5	33.7	35.6	1.4	–	–	–	29.1	70.7

[a] For references, see original publication.
[b] 0.2% behenic, 4.5% 18:4.
[c] 18.2% γ-18:3.
[d] 4% γ-18:3.
[e] 15.7% malvalic acid, 38.5% sterculic acid.
[f] 11% γ-18:3.
[g] 7.2% ricinoleic acid.
[h] 2.4% 15:0, 3.2% 15:1.
[i] 5% 21:0.
[j] 0.43% 15:1.
[k] 0.9% 20:1.
[l] 0.6% 17:0.
[m] Var. Malti Bahtim, 3.7% 18:4.
[n] Var. King Bahtim.

Table 3.153 Fatty acid composition (%) of fruit nuts[a] (Kamel and Kakuda, 1992)

Common name	Scientific name	Lipid (%)	14:0	16:0	16:1	18:0	18:1	18:2	18:3	20:0	20:1	Sat.	Unsat.
Acorn	*Quercus* spp.	23.8	–	12.5	–	1.1	66.2	20.1	–	–	–	13.6	86.3
Almond	*Prunus amygdalus*	50–65	0.1	12.2	1.5	0.1	44.5	40.5	0.6	–	–	12.8	87.2
Beechnut	*Sylvatica fagus* spp.	50	0.13	7.6	0.6	2.5	40.3	39.1	3.6	–	5.9	10.2	89.5
Black walnut	*Juglan nigra*	56–59	–	2.8	–	2.8	32.7	57.6	3.9	–	–	5.6	94.2
Brazil	*Berthrolletia myrtaceae*	68.7	–	14.1	–	0.3	8.6	29.0	46.6	0.3	–	22.7	76.2
Butternut	*Juglans cinerea*	60	–	1.6	–	0.8	19.0	61.9	16.0	–	–	2.4	96.9
Cashew	*Anacardium occidentale*	47.7	–	10.4	0.6	8.4	62.6	18.1	–	–	–	18.7	81.2
China chestnuts[b]	*Sterculia monosperma*	3.1	0.15	23.4	1.3	2.5	24.9	18.2	–	–	–	26.1	74.4
English walnut	*Juglan regia*	67.4	–	7.3	0.2	2.3	19.1	57.4	13.1	–	–	9.6	89.8
European chestnuts	*Castanea mollisima*	2.5	–	14.6	0.7	1.1	54.0	24.9	2.7	–	–	15.7	82.3
Faveleira nut[c]	*Cnidoscolus phyllacanthus*	32	–	17.4	–	9.4	15.1	55.4	1.0	0.4	0.2	27.6	71.7
Filbert	*Corylus avellana*	62.3	–	4.7	0.2	1.6	76.4	16.3	0.15	0.1	–	6.4	93.0
Ginko	*Ginko biloba*	1.68	0.4	19.1	5.2	1.0	33.9	38.3	1.4	–	0.7	20.5	79.5
Hickory	*Caryaovata ovata*	70.4	–	8.8	0.5	2.3	52.0	33.5	1.7	0.2	–	11.3	87.7
Macadamia	*Macadamia tetraphylla*	73.2–78	0.7	10.1	20.1	4.3	57.4	2.5	–	2.8	1.9	17.8	81.5
Mafura	*Trichilia emetica*	55–56	–	38.6	–	2.2	48.0	10.2	1.0	–	–	40.8	59.2
Pecan	*Carya illioensis*	70.3	–	5.7	0.11	2.2	66.9	22.1	1.1	0.21	0.4	8.1	90.6
Pine nut	*Pinus pinea*	50.7	–	8.0	0.45	3.6	39.2	45.2	1.4	–	2.1	11.6	88.3
Pistachio	*Pistachia vera*	54–57	–	8.6	0.7	2.3	68.8	17.8	0.3	0.3	0.6	11.2	87.7

[a] For references, see original paper.
[b] 3.2% 18:3 + 20:0.
[c] 0.4% 20:4.

Table 3.154 Physical and chemical characteristics of mango fats[a] (Ali *et al.*, 1985)

Variety	Moisture in fresh kernel (%)	Fat content (%)	Iodine value	Melting point (°C)	Saponification value	Unsaponifiable matter (%)	Peroxide value
Brindaboni	62.0	9.3	51.2	27–28	195	1.7	0.1
Fazli	68.0	7.1	55.0	38–39	190	1.8	0.5
Kalabau	50.0	9.0	44.0	27–28	192	1.8	0.1
Kanchamitha	50.0	7.5	50.4	26–27	194	1.8	0.1
Kuipahari	61.0	8.0	50.3	44–45	193	1.7	0.5
Lakhanovoge	52.6	10.0	46.8	28–29	191	1.5	0.1
Lengra	48.5	9.7	46.0	46–47	195	1.5	0.5
Mohanvoge	50.0	7.8	55.0	25–26	188	1.9	0.2
Misrakanta	48.5	8.9	50.6	25–26	191	1.7	0.1
Ranipasand	56.0	9.4	51.2	34–35	197	1.9	0.1

[a] Mean results of three experiments.

Table 3.155 Fatty acid compositions of some Indian oils (Bringi, 1987)

Name	Source	16:0	18:0	20:0	22:0	18:1	18:2	18:3	Others
Dhupa	*Vateria indica*	9	47	1	–	42	1	–	–
Karanja	*Pongamia glabra*	10	7	4	5	49	19	–	6
Kusum[a]	*Schleichera trijuga*	8	2	21	2	43	5	–	19
Mango	*Mangifera indica*	10	38	1	tr	45	5	1	–
Mowrah	*Madhuca indica*	25	22	–	–	36	17	–	–
Neem	*Azadirachta indica juss*	19	21	2	–	42	16	–	–
		8:0	10:0	12:0	14:0	16:0	18:0	18:1	
Pilu	*Salvadora oleoides persica*	2	4	45	32	10	5	2	
Pisa	*Actinodaphne hooker*	–	2	92	3	1	–	2	

[a] Chemistry very complex; cyanolipids form a significant part of the oil.

Table 3.156 Moisture, protein, lipid and fatty acid composition of 12 strains of oats (*Avena sativa*) (Sahasrabudhe, 1979) and Lupin (*Lupinus polyphyllus*) (Aniszewski, 1993)

Strain	Moisture (%)	Protein[a] (%)	Lipid[b] (%)	Fatty acids (%)						
				14:0	16:0	18:0	18:1	18:2	18:3	Other[c]
Oats										
Exeter	8.4	13.2	4.2	1.0	25.8	2.9	25.8	40.6	3.7	0.2
Garry	7.7	20.4	5.4	0.6	22.4	2.6	30.0	41.0	3.3	0.1
Hinoat	7.4	24.4	5.5	3.1	19.6	2.5	32.8	36.9	2.8	2.2
OA-290-5	7.8	18.8	5.5	3.6	23.1	3.0	39.8	38.0	2.5	nd[d]
Gemini	7.4	16.4	5.9	1.1	21.3	2.0	32.0	40.3	3.2	0.1
Random	8.0	17.9	6.3	0.5	21.0	2.2	36.1	37.4	2.4	0.3
Terra	7.7	16.2	6.5	4.9	19.7	2.4	35.6	32.3	2.0	3.1
Elgin	7.8	19.3	6.8	1.0	23.0	3.9	30.8	37.6	3.6	tr[e]
Dal	7.8	19.1	7.9	0.5	23.6	2.8	33.7	37.7	1.7	tr
CI-3387	7.7	14.1	11.0	2.7	16.5	1.6	40.2	35.5	2.2	1.3
Lodi	7.8	19.9	11.3	1.4	14.9	2.4	41.0	36.0	2.1	2.2
CI-4492	8.0	14.2	11.8	4.9	15.6	2.7	41.3	31.3	1.7	2.4
Lupin										
L. polyphyllus		43.0	9.1	0.1	14.3	1.2	33.4	39.0	11.3	0.7

[a] Percentage N × 5.7.
[b] Lipid extracted by hexane.
[c] Includes fatty acids C_{20} and C_{22}.
[d] Not detected.
[e] Indicates less than 0.1%.

3.3.44 Summary tables relating to vegetable oils and fats

The data given in this section have been compiled on the basis that in most cases the work has been done by a single research group. This gives the data some element of consistency. In some instances there may be more up-to-date analyses that can support the information supplied here. Triglyceride, sterol and tocopherol analyses present their own particular problems, in the first case because of the varied techniques used to quantify these components and in the second because of the natural variability that is known to occur.

The summary tables that are given here are as follows:

Codex ranges of fatty acid compositions of fats and oils (3.157);

Triglycerides of palmitic/stearic/oleic-rich fats (3.158);

Triglycerides of seed oils rich in linolenic and/or linoleic acid (3.159);

Free fatty acids, mono-, di- and triglycerides of fats according to carbon number (3.160);

Stereospecific positional distribution of fatty acids in triglycerides of seed oils (3.161);

Vegetable oils – fatty acid composition of triglycerides and of fatty acids at the 2 position (3.162);

Sterol composition of some commonly occurring vegetable oils (3.163);

4-methylsterols of some commonly occurring vegetable oils (3.164);

Triterpene alcohols of some commonly occurring vegetable oils (3.165);

Sterols in vegetable oils (3.166);

Ranges of tocopherol and tocotrienol levels in vegetable oils (3.167);

Vitamin E activity of tocopherols/ tocotrienols in vegetable oils (3.168);

Tocopherols determined in vegetable oils (3.169);

Tocopherol content and vitamin E activity of oils and fats (3.170);

World production of oils and fats (3.171);

Major oils and fats, summary of world supply and demand (3.172);

Oil-producing nations (3.173);

Fatty acid composition and lipid content of edible vegetables, cruciferae (3.174) and Asteraceae and Apiaceae (3.175). See also tables in section 3.3.43 (unclassified oils of interest and new oil seed crops).

Table 3.157 GLC fatty acid ranges in Codex standards for individual fats and oils (revised in 1993 at 14th Session CCFO, Al inorm 95/17) (per cent of total fatty acids). Ranges refer to typical commercial samples of *bona fide* fats and oils

Fatty acid	Arachis	Babassu	Coconut	Cottonseed	Grapeseed	Maize	Mustard seed	Olive oil	Palm	Palm kernel
C6:0	nd[c]	nd	0.0–0.6	nd	nd	nd	0.0–0.5*[d] (C6:0–C12:0)	nd	ns	0.0–0.8
C8:0	nd	2.6–7.3	4.6–9.4	nd	nd	nd		nd	ns	2.4–6.2
C10:0	nd	1.2–7.6	5.5–7.8	nd	nd	nd		nd	ns	2.6–5.0
C12:0	0.0–0.1	40.0–55.0	45.1–50.3	0.0–0.2	0.0–0.5	0.0–0.3		nd	0.0–0.4	41.0–55.0
C14:0	0.0–0.1	11.0–27.0	16.8–20.6	0.6–1.0	0.0–0.3	0.0–0.3	0.0–1.0	<0.05	0.5–2.0	14.0–18.0
C16:0	8.3–14.0	5.2–11.0	7.7–10.2	21.4–26.4	5.5–11	8.6–16.5	0.5–4.5	7.5–20	40.1–47.5	6.5–10.0
C16:1	0.0–0.2	nd	nd	0.0–1.2	0.0–1.2	0.0–0.4	0.0–0.5	0.3–3.5	0.0–0.6	ns
C17:0	nd	nd	nd	nd	nd	nd	nd	nd	ns	ns
C17:1	nd	nd	nd	nd	nd	nd	nd	nd	ns	ns
C18:0	1.9–4.4	1.8–7.4	2.3–3.5	2.1–3.3	3.0–6.0	1.0–3.3	0.5–2.0	0.5–5.0	3.5–6.0	1.3–3.0
C18:1	36.4–67.1	9.0–20.0	5.4–8.1	14.7–21.7	12–28	20.0–42.2	8.0–23	55–83	36.0–44.0	12.0–19.0
C18:2	14.0–43.0	1.4–6.6	1.0–2.1	46.7–58.2	58–78	39.4–62.5	10–24	3.5–21	6.5–12.0	1.0–3.5
C18:3	0.0–0.1	nd	0.0–0.2	0.0–0.4	0.0–1.0	0.5–1.5	6.0–18	<1.5	0.0–0.5	
C20:0	1.1–1.7	nd	0.0–0.2	0.2–0.5	0.0–1.0	0.3–0.6	0.0–1.5	<0.8	0.0–1.0	
C20:1	0.7–1.7	nd	0.0–0.2	0.0–0.1	nd	0.2–0.4	5.0–13	tr	ns	0.0–1.0% (C20:1–C24:1)
C20:2	nd	nd	nd	0.0–0.1	nd	0.0–0.1	0.0–1.0	nd	ns	
C22:0	2.1–4.4	nd	nd	0.0–0.6	0.0–0.3	0.0–0.5	0.2–2.5	<0.2	ns	
C22:1	0.0–0.3	nd	nd	0.0–0.3	nd	0.0–0.1	22–50	nd	ns	
C22:2	nd	nd	nd	0.0–0.1	nd	nd	0.0–1.0	nd	ns	
C24:0	1.1–2.2	nd	nd	0.0–0.1	0.0–0.1	0.0–0.4	0.0–0.5	<1.0	ns	
C24:1	0.0–0.3	nd	nd	nd	nd	nd	0.5–2.5	nd	ns	

[a] Revised 13th Session CCFO, Alinorm 87/17; 14:1 <0.2; 15:0, 15:iso, 22:0, each <0.1; 20:4 <1.0.

[b] Revised 13th Session CCFO, Alinorm 87/17; 14:iso <0.3; 14:1, 0.5–1.5; 15:0, 0.5–1.0; 15:iso + anteiso <0.5; 16:2 <1.0; 16:iso <0.5; 17:iso and anteiso <1.5; 20:4 <0.5.

[c] nd, not detected; ns, not specified.

[d] Total value for the fatty acids indicated.

[e] 1.8% 22:1 found in only one out of 31 samples. 25 samples ≤0.2%.

Palm stearin	Rapeseed	Rapeseed (low erucic acid)	Safflower seed	Sesame seed	Soya bean	Sunflower seed	Lard[a] and rendered pork fat	Premier jus[b] and edible tallow
nd	0.1	nd	nd	nd	nd	nd	<0.5	<2.5
nd		nd	nd	nd	nd	nd		
nd		nd	nd	nd	nd	nd		
0.1–0.4		nd	nd	nd	0.0–0.1	0.0–0.1		
1.1–1.8	0.2	0.0–0.2	0.0–0.2	0.0–0.1	0.0–0.2	0.0–0.2	0.5–2.5	1.4–7.8
48.4–73.8	1.5–6.0	3.3–6.0	5.3–8.0	7.9–10.2	8.0–13.3	5.6–7.6	20–32	17–37
0.05–0.2	0.0–3.0	0.1–0.6	0.0–0.2	0.1–0.2	0.0–0.2	0.0–0.3	1.7–5	0.7–8.8
nd	nd	0.0–0.3	nd	0.0–0.2	nd	nd	<0.5	0.5–2.0
nd	nd	0.0–0.3	nd	0.0–0.1	nd	nd	<0.5	<1.0
3.9–5.6	0.5–3.1	1.1–2.5	1.9–2.9	4.8–6.1	2.4–5.4	2.7–6.5	5–24	6.0–40
15.6–36.0	8–60	52.0–66.9	8.4–21.3	35.9–42.3	17.7–26.1	14.0–39.4	35–62	26–50
3.2–9.8	11–23	16.1–24.8	67.8–83.2	41.5–47.9	49.8–57.1	48.3–74.0	3–16	0.5–5.0
0.1–0.6	5–13	6.4–14.1	0.0–0.1	0.3–0.4	5.5–9.5	0.0–0.2	<1.5	<2.5
0.3–0.6	0.0–3.0	0.2–0.8	0.2–0.4	0.3–0.6	0.1–0.6	0.2–0.4	<1.0	<0.5
nd	3–15	0.1–3.4	0.1–0.3	0.0–0.3	0.0–0.3	0.0–0.2	<1.0	<0.5
nd	0.0–1.0	0.0–0.1	nd	nd	0.0–0.1	nd	<1.0	
nd	0.0–2.0	0.0–0.5	0.2–0.8	0.0–0.3	0.3–0.7	0.5–1.3	<0.1	
nd	5–60	0.0–2.0	0.0–1.8[c]	nd	0.0–0.3	0.0–0.2		
nd	0.0–2.0	0.0–0.1	nd	nd	nd	0.0–0.3		
nd	0.0–2.0	0.0–0.2	0.0–0.2	0.0–0.3	0.0–0.4	0.2–0.3		
nd	0.0–3.0	0.0–0.4	0.0–0.2	nd	nd	nd		

Table 3.158 Triglycerides (mol%) of palmitic/stearic/oleic-rich fats (Jurriens, 1968)

Triglyceride[a]	Cocoa butter	Illipe fat	Palm oil		Shea butter	Olive oil
			Congo	Sumatra		
No double bonds						
PPP	0.1	0.1	4.3	6.1	–	–
PPSt	0.5	0.4	1.1	0.9	0.3	–
PStP	–	0.1	0.4	0.3	–	–
StPSt	0.8	0.5			1.2	–
PStSt	0.3	0.8	0.3	0.3	0.4	–
StStSt	0.4	1.2			1.8	–
Others	0.1	–	0.5	0.9	–	0.1
One double bond						
MOP	0.9	–	1.2	1.3	–	
MOSt	1.3	–	–	–	–	
POP	12.0	6.6	24.1	25.9	0.5	4.3
POSt	34.8	34.3	7.0	3.1	6.4	
StOSt	25.2	44.5	0.5	–	29.6	
PPO	0.6	–	3.6	6.0	0.1	
StPO	0.9	0.1	0.5	0.3	1.1	
PStO	0.8	0.1		0.5	0.2	0.2
StStO	1.1	0.2	1.2	–	1.7	
Others	0.7	–		0.8	–	
Two double bonds						
POO	3.7	2.2	18.9	18.9	3.2	26.0
StOO	4.9	6.0	2.8	2.6	29.8	
OPO	–	–	1.0	1.2	0.3	0.3
PLP	1.2	0.1	7.8	6.8	–	
StLP	3.1	0.3	2.3	1.9	0.8	0.5
Others	3.1	0.4	2.2	3.6	4.6	
Three double bonds						
OOO	0.5	0.4	2.7	3.2	7.4	39.2
POL	0.4	0.4	4.0	2.6	0.3	4.5
StOL	0.4	1.1	0.6	–	2.2	
MLO	–	–	0.3	–	–	–
PLO	0.3	–	4.5	4.3	0.4	3.0
StLO	0.3	0.1	0.6	0.5	4.0	
Others	1.6	0.1	7.1	7.8	3.9	0.6
More than three double bonds						
LOO	–	–	–	–	–	13.7
OLO	–	–	0.5	0.2	–	4.6
Others	–	–	–	–	–	3.0

[a] P, palmitic; St, stearic; M, myristic; O, oleic; L, linoleic.

Table 3.159 Triglycerides of seed oils rich in linolenic acid and/or linoleic acid

Component glycerides[a]		Seed oils (mol%)									
		Candlenut[b]	Corn[c]	Cottonseed[c]	Groundnut[d]	Linseed[b]	Safflower[c]	Sesame[d]	Soya[b]	Sunflower[c]	Wild rose[b]
LeLeLe	(333)	2	–	–	–	24	–	–	–	–	5
LeLeL	(332)	10	–	–	–	14	–	–	1	–	17
LeLeO	(331)	6	–	–	–	15	–	–	–	–	6
LeLeS	(330)	3	–	–	–	12	–	–	1	–	4
LeLL	(322)	13	–	–	–	3	–	–	7	–	21
LeLO	(321)	14	–	–	–	5	–	–	5	–	10
LeLS	(320)	8	–	–	–	5	–	–	4	–	4
LeOO	(311)	5	–	–	–	5	–	–	2	–	1
LeOS	(310)	4	–	–	–	7	–	–	3	–	1
LeSS	(300)	–	–	–	–	1	–	–	2	–	–
LLL	(222)	6	15	11	4	1	47	10	15	14	13
LLO	(221)	8	21	9	20	2	19	20	16	39	8
LLS	(220)	6	17	27	12	–	18	9	13	14	4
LOO	(211)	5	14	5	17	1	5	22	8	19	2
LOS	(210)	5	17	20	22	3	7	17	12	11	2
LSS	(200)	1	5	17	6	–	2	3	4	1	1
OOO	(111)	1	6	2	6	1	1	8	2	–	1
OOS	(110)	2	4	3	10	2	1	8	5	1	–
OSS	(100)	1	1	6	3	–	–	3	–	1	–
SSS	(000)	–	–	–	–	–	–	–	–	–	–

[a] All isomers: linolenic (Le, 3); linoleic (L, 2); oleic (O, 1); saturated (S, 0).
[b] Gunstone and Padley (1965).
[c] Gunstone and Qureshi (1965).
[d] Jurriens (1968).

Table 3.160 Free fatty acids and mono-, di- and triglycerides (wt%) of fats according to carbon number[a] (D'Alonzo *et al.*, 1982)

	Cocoa butter	Corn oil	Cottonseed oil	Lard	Olive oil	Palm oil	Peanut oil	Rapeseed oil (high-erucic)	Tallow	Safflower oil	Sesame oil	Soybean oil	Sunflower oil
Fatty acid	0.6	–	–	0.2	0.1	0.1	–	0.1	2.3	–	0.1	–	0.1
Monoglyceride	0.2	–	–	–	0.2	–	–	0.1	–	–	–	–	–
Diglyceride	2.2	2.8	3.1	1.3	5.5	5.8	2.2	0.8	3.8	2.1	2.6	1.0	2.0
Triglyceride	96.0	95.8	95.0	97.9	93.3	93.1	93.3	96.8	89.6	96.0	95.0	97.9	95.6
Other	0.9	1.4	1.9	0.6	0.9	1.0	4.5	2.2	4.3	1.9	2.3	1.1	2.3
Fatty acid													
$C_{16}A$	0.2	–	–	0.1	–	–	–	–	1.0	–	–	–	–
$C_{18}A$	0.4	–	–	0.1	0.1	0.1	–	0.1	1.3	–	0.1	–	0.1
Monoglyceride													
$C_{16}M$	–	–	–	–	–	–	–	0.1	–	–	–	–	–
$C_{18}M$	0.1	–	–	–	0.2	–	–	–	–	–	–	–	–
$C_{20}M$	–	–	–	–	–	–	–	–	–	–	–	–	–
$C_{22}M$	–	–	–	–	–	–	–	–	–	–	–	–	–
Diglyceride													
$C_{30}D$	–	–	–	–	–	–	–	–	–	–	–	–	–
$C_{32}D$	0.1	–	0.1	0.1	0.1	0.6	–	–	0.3	–	–	–	–
$C_{34}D$	0.8	0.7	1.0	0.6	1.4	3.0	0.5	0.1	1.6	0.3	0.5	0.1	0.2
$C_{36}D$	1.3	2.1	2.0	0.6	4.0	2.2	1.7	0.7	1.9	1.8	2.1	0.9	1.8
$C_{38}D$	–	–	–	–	–	–	–	–	–	–	–	–	–
$C_{40}D$	–	–	–	–	–	–	–	–	–	–	–	–	–
Triglyceride													
$C_{46}T$	–	0.2	0.1	0.4	0.1	0.6	0.1	0.7	1.1	0.2	0.5	0.2	–
$C_{48}T$	0.3	1.2	1.7	1.9	0.3	6.6	0.1	0.9	5.2	0.3	0.1	0.1	0.6
$C_{50}T$	16.9	3.0	14.4	13.4	6.2	35.0	3.0	1.5	16.4	1.6	2.7	2.2	2.0
$C_{52}T$	43.2	25.2	39.7	59.1	33.2	37.6	22.2	4.5	43.4	16.8	21.3	24.2	16.9
$C_{54}T$	32.7	63.4	35.7	20.7	51.1	12.3	52.4	11.8	21.3	73.9	66.6	68.2	72.0
$C_{56}T$	2.3	2.5	2.8	2.4	2.1	1.0	9.3	13.3	2.2	2.5	3.2	2.2	2.6
$C_{58}T$	0.6	0.3	0.6	–	0.3	0.1	6.2	17.4	–	0.7	0.6	0.8	1.5
$C_{60}T$	–	–	–	–	–	–	–	19.0	–	–	–	–	–
$C_{62}T$	–	–	–	–	–	–	–	26.4	–	–	–	–	–
$C_{64}T$	–	–	–	–	–	–	–	1.3	–	–	–	–	–

[a] Determination by capillary gas chromatography.

Table 3.161 Positional distribution of fatty acids in triglycerides of seed oils (Brockerhoff and Hoyle, 1963; Brockerhoff and Yurkowski, 1966; Brockerhoff *et al.*, 1966)

Seed oils	Positions	Fatty acids (mol%)										
		16:0	16:1	18:0	18:1	18:2	18:3	20:0	20:1	22:0	22:1	24:0
Cocoa butter	1	34.0	0.6	50.4	12.3	1.3	–	1.0	–	–	–	–
	2	1.7	0.2	2.1	87.4	8.6	–	–	–	–	–	–
	3	36.5	0.3	52.8	8.6	0.4	–	2.3	–	–	–	–
Linseed	1	10.1	0.2	5.6	15.3	15.6	53.2	–	–	–	–	–
	2	1.6	0.1	0.7	16.3	21.3	59.8	–	–	–	–	–
	3	6.0	0.3	4.0	17.0	13.2	59.4	–	–	–	–	–
Maize	1	17.9	0.3	3.2	27.5	49.8	1.2	–	–	–	–	–
	2	2.3	0.1	0.2	26.5	70.3	0.7	–	–	–	–	–
	3	13.5	0.1	2.8	30.6	51.6	1.0	–	–	–	–	–
Olive	1	13.1	0.9	2.6	71.8	9.8	0.6	–	–	–	–	–
	2	1.4	0.7	–	82.9	14.0	0.8	–	–	–	–	–
	3	16.9	0.8	4.2	73.9	5.1	1.3	–	–	–	–	–
Peanut	1	13.6	0.3	4.6	59.2	18.5	–	0.7	1.1[a]	1.3	–	0.7
	2	1.6	0.1	0.3	58.5	38.6	–	–	0.3[a]	0.2	–	0.5
	3	11.0	0.3	5.1	57.3	10.0	–	4.0	2.7[a]	5.7	–	2.8
Rapeseed	1	4.1	0.3	2.2	23.1	11.1	6.4	–	16.4	1.4	34.9	–
	2	0.6	0.2	–	37.3	36.1	20.3	–	2.0	–	3.6	–
	3	4.3	0.3	3.0	16.6	4.0	2.6	–	17.3	1.2	51.0	–
Soybean	1	13.8	–	5.9	22.9	48.4	9.1	–	–	–	–	–
	2	0.9	–	0.3	21.5	69.7	7.1	–	–	–	–	–
	3	13.1	–	5.6	28.0	45.2	8.4	–	–	–	–	–

[a] Together with 18:3.

Table 3.162 Vegetable oils – fatty acid composition (mol%) of whole triglycerides and of fatty acids at the 2 position (Adapted from Mattson and Volpenheim, 1963)

Oil		6:0	8:0	10:0	12:0	14:0	16:0	18:0	18:1	18:2	18:3
Acorn	Triglyceride	–	–	–	–	–	9	1	70	20	–
	2 Position	–	–	–	–	–	1	0	68	31	–
Almond	Triglyceride	–	–	–	–	–	7	2	70	21	–
	2 Position	–	–	–	–	–	1	0	65	34	–
Brazil nut	Triglyceride	–	–	–	–	–	17	12	33	38	–
	2 Position	–	–	–	–	–	1	1	40	58	–
Cashew nut	Triglyceride	–	–	–	–	–	13	9	60	18	–
	2 Position	–	–	–	–	–	2	0	56	42	–
Chinese veg. tallow	Triglyceride	–	–	–	–	–	71	1	28	–	–
	2 Position	–	–	–	–	–	14	1	85	–	–
Coconut	Triglyceride	0.6	9.4	7.0	50.0	17.4	7.6	2.0	4.7	1.3	–
	2 Position	–	0.3	3.0	80.4	8.6	1.6	0.6	3.5	1.5	–
Corn	Triglyceride	–	–	–	–	–	12	2	29	57	–
	2 Position	–	–	–	–	–	2	0	26	72	–
Cotton seed (gland-free)	Triglyceride	–	–	–	–	–	27	3	17	53	–
	2 Position	–	–	–	–	–	2	0	22	76	–
Cotton seed	Triglyceride	–	–	–	–	–	26	2	17	55	–
	2 Position	–	–	–	–	–	2	1	20	77	–

(*Continued*)

Table 3.162 *Continued*

Oil							16:0	18:0	18:1	18:2	18:3
Cucumber seed	Triglyceride	–	–	–	–	–	16	5	7	72	–
	2 Position	–	–	–	–	–	2	1	4	93	–
Dunkwa	Triglyceride	–	–	–	–	–	2	53	45	–	–
Allanblackia	2 Position	–	–	–	–	–	2	5	93	–	–
Filbert nut	Triglyceride	–	–	–	–	–	7	2	83	8	–
	2 Position	–	–	–	–	–	1	0	87	12	–
Kokum butter	Triglyceride	–	–	–	–	–	2	61	37	–	–
	2 Position	–	–	–	–	–	2	2	96	–	–
Linseed	Triglyceride	–	–	–	–	–	6	4	22	15	53
	2 Position	–	–	–	–	–	1	0	27	23	49
Olive	Triglyceride	–	–	–	–	–	12	2	78	8	–
	2 Position	–	–	–	–	–	1	0	89	10	–
Onion seed	Triglyceride	–	–	–	–	–	8	2	26	64	–
	2 Position	–	–	–	–	–	1	0	25	74	–
Palm fruit	Triglyceride	–	–	–	–	–	45	6	40	9	–
	2 Position	–	–	–	–	–	11	2	65	22	–
Palmkernel	Triglyceride	0.4	5.7	4.7	52.8	16.2	7.6	2.2	9.2	1.2	–
	2 Position	–	2.7	2.5	62.7	15.8	2.9	0.3	11.8	1.3	–
Pecan	Triglyceride	–	–	–	–	–	9	3	58	30	–
	2 Position	–	–	–	–	–	1	0	57	42	–
Poppy seed	Triglyceride	–	–	–	–	–	10	2	11	77	–
	2 Position	–	–	–	–	–	1	0	9	90	–
Rice bran	Triglyceride	–	–	–	–	–	18	1	41	38	2
	2 Position	–	–	–	–	–	2	0	43	53	2
Safflower seed	Triglyceride	–	–	–	–	–	8	3	14	75	–
	2 Position	–	–	–	–	–	1	0	13	86	–
Sesame seed	Triglyceride	–	–	–	–	–	10	6	40	44	–
	2 Position	–	–	–	–	–	1	1	43	55	–
Soybean	Triglyceride	–	–	–	–	–	12	4	25	51	8
	2 Position	–	–	–	–	–	1	0	22	69	8
Spinach seed	Triglyceride	–	–	–	–	–	12	1	24	61	2
	2 Position	–	–	–	–	–	2	1	21	74	2
Squash seed	Triglyceride	–	–	–	–	–	17	7	16	60	–
	2 Position	–	–	–	–	–	1	1	17	81	–
Sunflower seed	Triglyceride	–	–	–	–	–	6	3	27	64	–
	2 Position	–	–	–	–	–	1	0	23	76	–
Tomato seed	Triglyceride	–	–	–	–	–	11	5	21	60	3
	2 Position	–	–	–	–	–	1	0	22	74	3
Walnut	Triglyceride	–	–	–	–	–	8	3	15	62	12
	2 Position	–	–	–	–	–	1	0	17	71	11
Wheat flour	Triglyceride	–	–	–	–	–	4	4	27	58	7
	2 Position	–	–	–	–	–	2	0	22	70	6
Wheat germ	Triglyceride	–	–	–	–	–	20	–	18	55	7
	2 Position	–	–	–	–	–	1	–	18	77	6

Table 3.163 Sterol composition (mg/kg) of some commonly occurring vegetable oils (Adapted from Itoh *et al.*, 1973a, b, 1974)

Vegetable oil	Cholesterol	Brassicasterol	Campesterol	Stigmasterol	β-Sitosterol	Δ^5-Avenasterol	Δ^7-Stigmastenol	Δ^7-Avenasterol	Unknown
Castor oil	–	–	285	627	1254	599	57	28	–
Cocoa butter	59	–	266	769	1746	88	29	–	–
Coconut	23	–	18	296	1322	319	136	–	–
Coffee seed	–	–	3488	3672	9914	1101	183	–	–
Corn oil	–	–	2691	702	7722	468	117	–	–
Cottonseed	–	–	170	42	3961	85	–	–	–
Illipe butter	–	–	739	323	3234	227	46	–	–
Kapok	–	–	256	57	2451	57	28	–	–
Linseed	42	–	1218	378	1932	546	84	–	–
Olive (France)	–	–	28	14	1310	29	58	–	–
Palm	26	–	358	204	1894	51	25	–	–
Palmkernel	40	–	118	145	924	79	13	–	–
Peanut	–	–	360	2160	1536	192	72	24	–
Poppy	–	–	605	83	1870	55	55	–	83
Rapeseed	–	612	1530	–	3549	122	306	–	–
Rice bran	–	–	5056	2709	8849	903	180	361	–
Safflower	–	–	452	313	1809	35	696	104	69
Safflower (oleic-rich)	–	–	576	384	1996	38	576	192	77
Sal	–	–	1469	918	3427	306	–	–	–
Sesame	–	–	1170	616	3819	431	123	–	–
Shea butters	–	–	–	–	–	–	944	281	132
Soybean	–	–	720	720	1908	108	108	36	–
Sunflower	–	–	313	313	2352	156	588	156	39
Wheat germ	–	–	5702	–	17336	1555	777	518	–

Table 3.164 4-Methylsterols (mg/kg) of some commonly occurring vegetable oils (Adapted from Itoh *et al.*, 1973a, b, 1974)

Oil	Relative retention time [a]													
	0.64	0.71	0.79	0.86	0.91	0.95 [b]	0.99	1.13 [c]	1.29	1.36	1.46	1.52 [d]	1.62	Others
Castor	–	–	–	4	–	70	16	54	4	10	–	34	–	8
Cocoa butter	–	8	–	19	–	25	16	19	1	24	–	46	–	–
Coconut	–	–	–	–	–	19	–	61	–	35	–	38	6	–
Coffee seed	–	–	–	57	–	1503	–	2485	–	173	–	1502	–	58
Corn	1	–	–	–	1	39	–	44	1	1	1	39	–	1
Cottonseed	–	–	–	–	–	113	33	76	8	12	–	172	–	4
Illipe	–	–	–	–	–	132	–	546	–	–	18	26	–	158
Kapok	–	–	–	5	–	32	10	45	7	27	–	110	10	3
Linseed	–	–	–	83	–	220	9	1078	5	4	–	58	–	–
Olive (France)	–	–	–	–	–	18	–	38	3	25	5	57	5	1
Palm	–	–	–	–	–	61	10	241	7	3	–	32	–	4
Palmkernel	–	4	–	4	–	52	–	80	12	40	40	126	8	40
Peanut	4	–	–	–	–	40	11	44	1	14	3	38	–	2
Poppy	–	–	–	6	–	96	–	108	–	18	–	–	–	72
Rapeseed	–	2	75	29	–	78	–	59	–	–	2	18	–	3
Rice	–	–	–	–	–	294	–	252	1638	–	336	–	1680	–
Safflower	–	–	–	–	–	96	29	55	12	19	–	28	–	–
Safflower (oleic-rich)	–	–	–	–	–	129	33	75	6	12	6	30	–	9
Sal	–	–	–	12	–	60	–	192	6	6	12	60	–	252
Sesame	–	–	–	–	–	812	568	1421	–	–	–	1258	–	–
Shea butter	–	–	–	–	–	224	–	41	51	31	–	–	–	673
Soybean	–	–	–	–	–	53	13	158	6	66	–	349	–	13
Sunflower	–	–	–	–	–	336	–	224	22	145	–	347	45	–
Wheat germ	–	–	–	–	–	70	34	475	–	46	–	533	–	–

[a] Retention time relative to β-sitosterol.
[b] Obtusifoliol.
[c] Gramisterol.
[d] Citrostadienol.

Table 3.165 Triterpene alcohols (mg/kg) of some commonly occurring vegetable oils (Adapted from Itoh *et al.*, 1973a, b, 1974)

Oil	Relative retention time [a]											
	0.73	1.02 [b]	1.06	1.13 [c]	1.17 [d]	1.23 [e]	1.27 [f]	1.32 [g]	1.38 [h]	1.50	1.65	1.68 [i]	Others
Castor	–	–	9	–	203	144	–	27	32	6	–	5	27
Cocoa butter	26	10	–	–	–	411	–	–	57	16	–	–	–
Coconut	–	14	20	–	–	503	–	–	122	–	20	–	–
Coffee seed	44	88	177	–	–	1635	–	–	2387	–	–	44	44
Corn	–	3	–	5	49	–	–	–	69	1	–	3	1
Cottonseed	–	–	–	19	96	–	–	187	158	–	–	–	19
Illipe	–	11	–	330	–	–	231	–	385	–	–	–	143
Kapok	124	19	–	10	–	266	–	–	437	29	–	57	10
Linseed	–	15	–	123	–	1063	–	–	339	–	–	–	–
Olive (France)	–	–	29	216	101	130	–	302	475	–	–	144	43
Palm	6.4	6.4	–	–	–	192	–	–	109	–	–	–	–
Palmkernel	–	–	–	29	65	295	209	94	29	–	–	–	–
Peanut	14	7	–	25	–	119	–	–	166	–	–	29	–
Poppy	–	4	–	63	–	67	–	(60[j])	126	–	–	–	32
Rapeseed	–	16	38	–	–	324	–	–	157	–	–	–	5
Rice bran	823	1058	–	–	–	4822	–	–	4939	–	–	–	–
Safflower	–	7	–	72.6	–	337	119	–	66	–	26	20	13
Safflower (oleic-rich)	6	–	–	84	–	360	78	–	36	–	12	24	–
Sal	–	–	–	624	–	1416	–	192	–	–	–	–	168
Sesame	–	36	–	36	–	619	–	–	1074	55	–	–	–
Shea butter	–	–	–	3060	9945	–	17595	6120	–	–	–	383	1148
Soybean	8	–	42	126	139	168	–	235	76	–	17	–	34
Sunflower	7	–	–	49	–	294	–	119	161	–	–	63	7
Wheat germ	–	67	–	403	–	381	179	–	986	45	–	45	–

[a] Retention time relative to β-sitosterol (30 min) taken as 1.0.
[b] Cycloartanol.
[c] β-Amyrin.
[d] Butyrospermol.
[e] Cycloartenol.
[f] α-Amyrin.
[g] Lupeol.
[h] 24-Methylenecycloartenol.
[i] Cyclobranol.
[j] Cycloaudenol.

Table 3.166 Ranges (means) of sterol compositions[a] in vegetable oils (percentage of sterol fraction; totals in mg/kg) determined on an OV-17 column in GLC stage of analysis (Rossell and Pritchard, 1990)[b]

Sterol	Oils analysed								
	Palmkernel	Coconut	Cottonseed	Soybean	Maize	Groundnut	Palm	Sunflower seed	Rapeseed
Cholesterol	1.0–3.7 (1.7)	0.6–3.0 (1.7)	0.7–2.3 (1.0)	0.6–1.4 (0.9)	0.2–0.6 (0.4)	0.6–3.8 (1.5)	2.7–4.9 (3.5)	0.2–1.3 (0.5)	0.4–1.3 (0.7)
Brassicasterol	nd–0.3 (0.1)	nd–0.9 (0.5)	0.1–0.9 (0.3)	nd–0.3 (0.1)	nd–0.2 (0.05)	nd–0.2 (0.0)	nd (nd)	nd–0.2 (<1)	5.0–13.0 (9.6)
Campesterol	8.4–12.7 (10.0)	7.5–10.2 (8.7)	7.2–8.4 (7.9)	15.8–24.2 (19.5)	18.6–24.1 (21.3)	12.3–19.8 (17.0)	20.6–24.2 (23.0)	7.4–12.9 (9.5)	18.2–38.6 (34.1)
Stigmasterol	12.3–16.1 (13.7)	11.4–13.7 (12.5)	1.2–1.8 (1.4)	15.9–19.1 (17.5)	4.3–7.7 (5.5)	5.4–13.3 (8.7)	11.4–11.8 (11.7)	8.6–10.8 (9.4)	nd–0.7 (tr)
Sitosterol	62.6–70.4 (67.0)	42.0–52.7 (46.7)	80.8–85.1 (83.2)	51.7–57.6 (54.3)	54.8–66.6 (63.4)	48.0–64.7 (58.5)	56.7–58.4 (57.7)	56.2–62.8 (59.9)	45.1–57.9 (49.9)
Δ^5-Avenasterol	4.0–9.0 (6.2)	20.4–35.7 (26.6)	1.9–3.8 (2.6)	1.9–3.7 (2.4)	4.2–8.2 (5.7)	8.3–18.8 (12.3)	2.1–2.7 (2.5)	1.9–6.9 (3.4)	nd–6.6 (4.5)
Δ^7-Stigmastenol	nd–2.1 (0.6)	ns–3.0 (2.4)	0.7–1.4 (1.0)	1.4–5.2 (2.9)	1.0–4.2 (1.8)	nd–5.2 (2.1)	0.4–0.8 (0.7)	7.0–13.4 (10.4)	nd–1.3 (0.5)
Δ^7-Avenasterol	nd–1.4 (0.1)	0.6–3.0 (1.1)	1.4–3.3 (0.7)	1.0–4.6 (0.6)	0.7–2.7 (0.5)	nd–5.5 (1.2)	0.4–2.5 (1.0)	3.1–6.5 (4.8)	nd–0.8 (0.4)
Others	nd–2.7 (0.7)	nd–3.6 (1.1)	nd–1.5 (0.7)	nd–1.8 (0.6)	nd–2.4 (0.5)	nd–1.4		nd–5.3 (2.0)	nd–4.2 (0.5)
Total (mg/kg)	792–1187 (1025)	470–1100 (807)	2690–5915 (4490)	1837–4089 (3199)	7931–22 137 (13 776)	901–2854 (1575)	389–481 (446)	2437–4545 (3387)	4824–11 276 (7516)

[a] nd = not detected; ns = not separated; tr = trace.
[b] For additional data on sterols in a wide range of oils see Homberg and Bielefeld (1989).

Table 3.167 Ranges (means) of tocopherol and tocotrienol levels (mg/kg) in vegetable oils[a] (Rossell and Pritchard, 1990)

Tocol	Relative retention times	Oils analysed									
		Palm kernel[b,c]	Coconut[b]	Cottonseed	Soybean	Maize	Groundnut	Palm	Sunflower seed	High-erucic rapeseed	Low-erucic rapeseed
αT	1.0	nd–44	nd–17	136–543 (338)	9–352 (99.5)	23–573 (282)	49–304 (178)	4–185 (89)	403–855 (670)	39–305	100–320 (202)
βT	1.5	nd–248	nd–11	nd–29 (16.9)	nd–36 (7.7)	nd–356 (54)	0–41 (8.8)	–	9–45 (27)	24–158	16–140 (65)
γT	1.8	nd–257	nd–14	158–594 (429)	409–2397 (1021)	268–2468 (1034)	99–389 (213)	6–36 (18)	nd–34 (11)	230–500	287–753 (490)
δT	2.7	–	nd–2	nd–17 (3.3)	154–932 (421)	23–75 (54)	3–22 (7.6)	–	nd–7 (0.6)	5–14	4–22 (9)
$αT_3$	1.1	nd–tr	nd–5	–	–	nd–239 (49)	–	4–336 (128)	–	–	–
$βT_3$	1.7	–	–	–	–	nd–52 (8)	–	–	–	–	–
$γT_3$	1.95	nd–60	nd–1	–	–	nd–450 (161)	–	42–710 (323)	–	–	–
$δT_3$	3.2	–	–	–	–	nd–20 (6)	–	tr–148 (72)	–	–	–
Total		nd–257[b]	tr–31[b]	410–1169 (788)	575–3320 (1549)	331–3402 (1647)	176–696 (407.4)	98–1327 (630)	447–900 (709)	312–928	424–1054 (766)

[a] Vitamin E activities are shown in Table 3.168.
[b] Means are negligible.
[c] High values may be due to migration of some palm oil into the palm kernels before separation.

Table 3.168 Vitamin E activity of tocopherols and tocotrienols[a] found in vegetable oils (Sheppard and Pennington, 1993)

Tocopherol	Activity as α-TE[b] (mg/mg compound)
RRR-α-tocopherol[c]	1.0
RRR-β-tocopherol	0.5
RRR-γ-tocopherol	0.1
RRR-δ-tocopherol	0.03
RRR-α-tocotrienol	0.3
RRR-β-tocotrienol	0.05
Synthetic α-tocopherol acetate	0.74

[a] The activities of γ-tocotrienol and δ-tocotrienol are unknown.
[b] α-Tocopherol equivalents.
[c] Naturally occurring α-tocopherol. See Sheppard and Pennington for a definition of modern nomenclature.

Table 3.169 Tocopherols determined in vegetable oils (mg/kg) (Compiled from Carpenter, 1979; Muller-Mulot, 1976; Ong, 1993)

Oil	α-T	β-T	γ-T	δ-T	α-T_3	γ-T_3	δ-T_3
Castor	28	29	111	310	–	–	–
Cocoa butter	11	–	170	17	2	–	–
Coconut[a]	–	–	–	2–4	20	–	–
Corn[a]	134	18	412	39	–	–	–
Cottonseed	573	40	317	10	–	–	–
Groundnut	169	5	144	13	–	–	–
Linseed	–	–	–	–	–	–	–
Mustard	75	–	494	31	–	–	–
Olive	93	–	7	–	–	–	–
Palm, from fibres[b]	1662	–	–	–	456	485	142
Palm[b]	150	–	–	–	117	297	80
Palm	279	–	61	–	274	398	69
Rape[a]	70	16	178	7	–	–	–
Rice bran[b]	324	18	53	–	236	349	–
Safflower	477	–	44	10	–	–	–
Sesame[a]	12	6	244	32	–	–	–
Soybean[a]	116	34	737	275	–	–	–
Sunflower[a]	608	17	11	–	–	–	–
Wheat germ	1179	398	493	118	tr	–	–

[a] Refined oils.
[b] Ong (1993).

Table 3.170 Tocopherol content and vitamin E activity of oils and fats (mg/kg) (Adapted from Sheppard and Pennington, 1993)

Fats and oil	Tocopherols				Tocotrienols		Vitamin E activity[a] as α-TE (mg)
	α (mg)	β (mg)	γ (mg)	δ (mg)	α (mg)	β (mg)	
Vegetable							
Canola	210.0	1.0	42.0	0.4	0.4	–[b]	215.0
Coconut[c]	5.0	–	–	6.0	5.0	0.1	7.0
Corn	112.0	50.0	602.0	18.9	–	–	198.0
Cottonseed	389.0	–	387.0	–	–	–	428.0
Olive	119.0	–	7.0	–	–	–	120.0
Palm[d]	256.0	–	316.0	70.0	146.0	3.2	335.0
Palmkernel	62.0	–	–	–	–	–	62.0
Peanut	130.0	–	214.0	21.0	–	–	152.0
Safflower	342.0	–	71.0	–	–	–	349.0
Sesame	136.0	–	290.0	–	–	–	165.0
Soybean[e]	75.0	15.0	797.0	266.0	2.0	0.1	171.0
Sunflower	487.0	–	51.0	8.0	–	–	492.0
Walnut	563.0	–	595.0	450.0	–	–	636.0
Wheat germ	1330.0	710.0	260.0	271.0	26.0	18.1	1736.0
Fish							
Cod liver	220.0	–	–	–	–	–	220.0
Herring	92.0	–	–	–	–	–	92.0
Menhaden	75.0	–	–	–	–	–	75.0
Animal							
Lard	12.0	–	7.0	–	7.0	–	15.0
Tallow	27.0	–	–	–	–	–	27.0

[a] α-Tocopherol equivalents.
[b] Dashes denote no value or trace.
[c] Contains 19.0 mg γ-tocotrienol.
[d] Contains 286.0 mg γ-tocotrienol.
[e] Contains 0.3 mg δ-tocotrienol.

Table 3.171 World production of oils and fats ($\times 10^6$ tonnes), January to December each year (*Oil World Annual*, 1991)

	1986	1987	1988	1989	1990	Average
Soya	14.45	15.56	15.42	14.99	16.05	15.29
Cotton	3.60	3.30	3.76	3.83	3.68	3.63
Groundnut	3.30	3.44	3.67	4.27	3.93	3.72
Sunflower	6.89	6.84	7.49	7.50	7.88	7.32
Rape	6.50	7.40	7.69	7.86	8.18	7.53
Sesame	0.67	0.66	0.66	0.66	0.66	0.66
Corn	1.23	1.24	1.29	1.27	1.37	1.28
Olive	1.83	1.80	2.04	1.67	1.92	1.85
Palm	7.74	8.04	8.86	10.37	10.64	9.13
Palmkernel	1.02	1.04	1.12	1.29	1.40	1.17
Coconut	3.41	3.14	2.74	2.69	3.22	3.04
Butter (as fat)	6.58	6.24	6.23	6.42	6.57	6.41
Lard	5.15	5.24	5.46	5.38	5.51	5.35
Fish oil	1.66	1.44	1.50	1.63	1.36	1.52
Linseed	0.78	0.81	0.71	0.64	0.65	0.72
Castor	0.39	0.37	0.34	0.38	0.43	0.38
Tallow/grease	6.44	6.44	6.76	6.63	6.67	6.59
Total	71.64	73.00	75.74	77.48	80.12	75.59

Table 3.172 Seventeen major oils and fats: summary of world supply-and-demand balances (1000 tonnes) (Data provided by *Oil World*)

Oil	Oct.–Sept. 1990/91	Oct.–Sept. 1989/90	Oct.–Sept. 1988/89	Oct.–Sept. 1987/88	Oct.–Sept. 1986/87
Soybean					
Opening stocks	2010	1993	2542	2196	1785
Production	16179	16050	14697	15509	15430
Imports	3812	4099	3770	3911	3783
Exports	3887	3948	3886	3813	3880
Disappeared	16310	16184	15131	15260	14922
Ending stocks	1804	2010	1993	2542	2190
Cotton					
Opening stocks	244	312	306	258	369
Production	3896	3669	3874	3677	3220
Imports	350	341	316	340	264
Exports	357	317	330	350	249
Disappeared	3863	3762	3854	3620	3346
Ending stocks	270	244	312	306	258
Groundnut					
Opening stocks	305	403	392	367	401
Production	3648	3787	4275	3561	3351
Imports	319	327	376	395	352
Exports	323	335	373	388	346
Disappeared	3688	3877	4267	3543	3390
Ending stocks	262	305	403	392	367
Sunflower					
Opening stocks	1023	944	922	767	867
Production	7722	7783	7397	7507	6858
Imports	2378	2508	2278	2310	1927
Exports	2360	2469	2282	2365	1956
Disappeared	7921	7743	7371	7297	6930
Ending stocks	842	1023	944	922	767

(*Continued*)

Table 3.172 *Continued*

Oil	Oct.–Sept. 1990/91	Oct.–Sept. 1989/90	Oct.–Sept. 1988/89	Oct.–Sept. 1987/88	Oct.–Sept. 1986/87
Rapeseed					
Opening stocks	759	683	912	734	805
Production	8231	7923	7684	7722	7250
Imports	1896	2080	1783	1821	1624
Exports	1899	1948	1912	1867	1641
Disappeared	8272	7979	7784	7499	7303
Ending stocks	715	759	683	912	734
Sesame					
Opening stocks	45	47	50	49	47
Production	653	623	659	667	661
Imports	22	20	17	16	12
Exports	21	20	17	15	13
Disappeared	654	624	662	667	659
Ending stocks	45	45	47	50	49
Corn					
Opening stocks	115	92	109	111	125
Production	1346	1331	1264	1262	1240
Imports	450	404	490	390	354
Exports	444	421	417	402	342
Disappeared	1357	1291	1293	1253	1266
Ending stocks	110	115	92	109	111
Olive					
Opening stocks	837	770	1076	763	960
Production	1777	1948	1559	2198	1670
Imports	545	589	569	519	732
Exports	543	592	563	515	732
Disappeared	1867	1878	1870	1890	1867
Ending stocks	749	837	770	1076	763
Palm					
Opening stocks	2223	2328	2015	1691	1640
Production	10760	10849	9718	8575	7918
Imports	8182	8076	7019	6749	6069
Exports	7957	8264	7120	6463	6146
Disappeared	11347	10766	9304	8538	7790
Ending stocks	1860	2223	2328	2015	1691
Palmkernel					
Opening stocks	266	182	201	195	184
Production	1427	1419	1236	1126	1030
Imports	978	899	799	772	643
Exports	968	915	837	744	663
Disappeared	1469	1320	1217	1148	1000
Ending stocks	234	266	182	201	195
Coconut					
Opening stocks	528	388	573	613	519
Production	3073	3175	2052	2929	3229
Imports	1467	1622	1007	1432	1445
Exports	1434	1596	1223	1449	1467
Disappeared	3148	2972	2872	3011	3113
Ending stocks	485	528	388	573	613

(*Continued*)

Table 3.172 *Continued*

Oil	Oct.–Sept. 1990/91	Oct.–Sept. 1989/90	Oct.–Sept. 1988/89	Oct.–Sept. 1987/88	Oct.–Sept. 1986/87
Butter, as fat					
Opening stocks	1350	1295	1483	2057	2257
Production	6569	6555	6339	6199	6295
Imports	1172	1062	1249	1640	1372
Exports	1175	1054	1261	1821	1367
Disappeared	6601	6508	6514	6792	6501
Ending stocks	1315	1350	1295	1483	2057
Lard					
Opening stocks	323	320	351	345	322
Production	5576	5457	5409	5409	5221
Imports	518	518	540	534	504
Exports	518	519	531	518	508
Disappeared	5568	5453	5450	5420	5193
Ending stocks	331	323	320	351	345
Fish oil					
Opening stocks	359	530	300	451	457
Production	1439	1442	1717	1402	1488
Imports	758	812	1062	734	878
Exports	736	841	971	822	839
Disappeared	1479	1584	1579	1463	1534
Ending stocks	341	359	530	300	451
Linseed					
Opening stocks	102	111	137	154	100
Production	675	648	637	738	830
Imports	228	219	224	273	297
Exports	230	224	200	273	305
Disappeared	662	651	688	756	768
Ending stocks	112	102	111	137	154
Castor					
Opening stocks	77	71	91	139	134
Production	482	401	366	331	385
Imports	241	211	182	168	193
Exports	249	201	188	159	197
Disappeared	447	404	380	388	376
Ending stocks	105	77	71	91	139
Tallow and grease					
Opening stocks	522	547	537	505	511
Production	6743	6657	6644	6743	6356
Imports	2650	2738	2835	2655	2560
Exports	2700	2708	2770	2743	2543
Disappeared	6689	6712	6700	6624	6379
Ending stocks	525	522	547	537	505
GRAND TOTAL					
Opening stocks	11087	11016	11997	11394	11481
Production	80195	79717	76127	75556	72433
Imports	25966	26435	24709	24731	33000
Exports	25800	26373	24880	24507	23193
Disappeared	81343	79708	76938	75168	72336
Ending stocks	10106	11087	11016	11997	11394

Table 3.173 Top oil-producing nations in 1991–92 (1000 tonnes); 1989–90 total in parentheses (Data from *Oil World*)

Soybean oil			Palm oil			Rapeseed/Canola oil		
USA	6325	(5898)	Malaysia	6560	(6095)	China	2235	(1831)
Brazil	2620	(2974)	Indonesia	3000	(2413)	India	1889	(1331)
Argentina	1288	(1080)	Nigeria	630	(580)	Germany	1180	(1042)
Japan	653	(680)	Colombia	292	(226)	Japan	803	(763)
Netherlands	617	(609)	Ivory Coast	280	(270)	Canada	647	(488)
Germany	482	(477)	Thailand	270	(226)	UK	486	(420)
China	458	(497)	Papua New Guinea	206	(138)	France	443	(412)
Taiwan	425	(407)	Ecuador	147	(120)	Belgium	311	(263)
Spain	382	(418)	Cameroon	107	(108)	Poland	258	(316)
India	344	(246)	Zaire	105	(101)	Netherlands	152	(140)
Sunflower seed oil			**Cottonseed oil**			**Peanut oil**		
Argentina	1320	(1458)	China	1112	(801)	India	1580	(1850)
CIS	1952	(2199)	CIS	662	(708)	China	1331	(1199)
France	557	(422)	USA	623	(438)	Nigeria	167	(123)
Spain	408	(398)	Pakistan	362	(300)	Senegal	152	(199)
USA	364	(217)	India	314	(371)	USA	120	(75)
Hungary	298	(287)	Brazil	186	(154)	Myanmar (Burma)	114	(101)
China	283	(219)	Turkey	132	(127)	Argentina	91	(27)
India	283	(182)	Argentina	75	(62)	Mali	72	(24)
Turkey	276	(407)	Egypt	66	(67)	Sudan	72	(77)
Romania	206	(223)	Australia	58	(42)			
			Paraguay	58	(54)			

Table 3.174 Fatty acid composition and lipid content of edible vegetables[a]: Cruciferae (Nasirulla *et al.*, 1984)

Fatty acid	Chinese cabbage (*Brassica chinensis* L.)	White cabbage (*Brassica oleracea* L. var. *capitata* L. f. *alba*)	Savoy (*Brassica oleracea* L. var. *sabauda* L.)	Kale (*Brassica oleracea* L. convar. *acephala* (DC.) Alef. var. *sabellica* L.)
10:0	tr	0.1	0.1	tr
12:0	0.1	0.2	0.1	tr
14:0	0.4	0.3	0.3	0.2
15:0	0.3	0.4	0.3	0.3
16:0	0.1	0.2	0.2	0.2
16:0	16.1	14.5	22.1	13.0
16:1	0.8	0.9	0.5	0.4
16:2	0.2	0.1	0.1	–
(*n* – 3)-16:3	3.0	1.1	0.9	6.6
17:0	1.3	0.4	2.2	4.4
17:0	0.2	0.2	0.2	0.3
18:0	1.9	1.1	1.2	1.2
(*n* – 12)-18:1	–	–	–	–
(*n* – 9)-18:1	1.3	1.3	0.7	0.7
(*n* – 7)-18:1	5.6	5.4	5.7	2.3
(*n* – 6)-18:2	7.5	14.8	15.7	18.5
(*n* – 3)-18:3	59.5	57.0	46.4	50.0
20:0	0.6	0.3	0.8	0.3
20:1	0.3	0.3	0.3	0.3
(*n* – 6)-20:2	–	0.2	0.3	0.1
(*n* – 3)-20:2	–	0.1	–	0.2
22:0	0.3	0.4	0.7	0.2
(*n* – 9)-22:1	–	tr	–	0.3
24:0	0.4	0.2	0.5	0.2
24:1	0.3	0.6	0.8	0.3
Plant total lipids (% of mass)	0.21	0.12	0.21	0.15

[a] Composition of total lipid extracted from edible parts of the vegetable, i.e. roots, stems, leaves, flowers, as appropriate.

Red cabbage (*Brassica oleracea* L. var. *capitata* L. f. *rubra*)	Brussels sprouts (*Brassica oleracea* L. var. *gemmifera* DC.)	Cauliflower (*Brassica oleracea* L. convar. *botrytis* var. *botrytis* L.)	Kohlrabi (*Brassica oleracea* L. convar. *acephala* var. *gongylodes L.)*	Swede (*Brassica napus* L. var. *napobrassica* L.)
–	0.1	–	0.1	0.1
0.1	0.2	0.1	0.2	0.4
0.2	0.5	0.2	0.3	0.4
0.5	0.4	0.4	0.4	0.6
tr	tr	tr	0.1	0.3
18.6	14.5	17.4	20.8	14.5
0.6	0.5	1.7	1.6	2.3
0.1	–	0.4	0.1	0.2
0.4	4.1	0.5	0.5	0.3
0.3	1.8	0.6	0.3	0.4
0.5	0.2	0.3	0.3	0.4
1.9	0.9	1.7	1.9	0.7
–	–	–	–	–
0.5	0.6	0.8	4.2	0.6
2.9	3.1	11.9	6.0	6.8
29.9	11.4	12.8	17.9	20.7
39.8	58.7	48.9	41.9	50.8
1.0	1.6	0.5	1.3	–
0.6	0.4	0.4	0.3	0
tr	0.2	–	–	–
0.1	0.2	0.2	–	–
0.5	0.1	0.2	0.4	–
–	–	0.3	–	–
1.0	0.2	0.3	1.1	0.2
0.5	0.3	0.4	0.3	0.1
0.15	0.18	0.28	0.12	0.17

Table 3.175 Fatty acid composition and lipid content of edible vegetables[a]: Asteraceae and Apiaceae (Nasirulla *et al.*, 1984)

Fatty acid	Cabbage lettuce (*Lactuca sativa* L.)	Parsley (*Petroselinum crispum* (Mill.) Nym. subsp. *crispum*)	Black salsify (*Scorzonera hispanica* L.)
10:0	tr	tr	tr
12:0	0.1	0.1	tr
14:0	0.4	0.3	0.3
15:0	0.2	0.2	0.7
16:0	0.1	tr	–
16:0	13.9	12.8	34.4
16:1	1.4	1.7	2.9
16:2	–	–	0.8
$(n-3)$-16:3	0.1	11.8	–
17:0	0.1	0.3	–
17:0	0.2	0.2	0.3
17:1	–	–	0.1
18:0	0.7	0.6	1.1
$(n-12)$-18:1	–	tr	–
$(n-9)$-18:1	0.6	0.6	0.5
$(n-7)$-18:1	0.3	0.2	1.4
$(n-6)$-18:2	23.0	25.6	44.4
$(n-3)$-18:3	54.6	41.7	10.0
20:0	0.4	0.6	0.9
20:1	0.6	0.3	tr
$(n-6)$-20:2	0.2	0.3	0.1
$(n-3)$-20:3	0.2	0.1	–
22:0	0.9	1.1	0.7
$(n-9)$-22:1	0.1	–	tr
24:0	1.3	1.4	0.6
24:1	0.6	0.1	0.8
Plant total lipids (% of mass)	0.11	0.33	0.38

[a] Composition of total lipid extracted from edible parts of the vegetable, i.e. roots, stems, leaves, flowers, as appropriate.

Carrot (*Daucus carota* L. subsp. *sativus* (Hoffm.) Arcang.)	Turnip-rooted parsley (*Petroselinum crispum* (Mill.) Nym. subsp. *tuberosum* Soo)	Florence fennel (*Foeniculum vulgare* Mill. var. *azoricum* (Mill.) Thell.)
–	–	0.1
0.1	0.1	0.2
0.2	0.1	0.2
0.5	0.5	0.9
–	–	–
20.2	19.6	20.6
tr	0.3	0.2
–	–	–
–	–	–
0.2	0.2	–
0.5	0.3	0.4
0.2	0.3	0.1
1.3	0.4	1.4
–	0.7	–
2.1	4.0	2.2
0.5	1.5	0.3
65.5	61.8	62.8
6.0	8.5	5.9
0.5	0.3	0.8
0.4	0.6	0.4
0.2	–	0.3
–	–	–
1.0	0.7	2.1
–	–	–
0.5	tr	1.1
–	–	–
0.25	0.60	0.15

References and further reading

Abou-Donia, M.B. (1976) *Resid. Rev.*, **61**, 125.

Abou-Donia, S.A., Lasker, J.M. and Abou-Donia, M.B. (1981) *J. Chromatogr.*, **206**, 606.

Achaya, K.T. (1971) *J. Am. Oil Chem. Soc.*, **48**, 758.

Achaya, K.T. and Craig, B.M. (1964) *J. Am. Oil Chem. Soc.*, **41**, 783.

Ackman, R.G. (1990) in *Canola Rapeseed* (ed. F. Shahidi), Van Nostrand Reinhold, New York, p. 81.

Adomako, D. (1985) *Tech. Bull. Cocoa Res. Inst. Ghana*, **11**, 12.

Afaque, S., Rauf, A., Khan, M. and Osman, S.M. (1987) *J. Oil Technol. Assoc. India*, **19**, 78.

Ali, M.A., Gafur, M.A., Rahman, M.S. and Ahmed, G.M. (1985) *J. Am. Oil Chem. Soc.*, **62**, 520.

Allen, R.R. (1965) *Chem. Ind.*, 1560.

Al-Saleh, A., Marion, D. and Gallant, D.J. (1986) *Food Microstruct.*, **5**, 131.

Alvim, R. (1989) *Agrotropica*, **1**, 89.

American Oil Chemists' Society (1941) *Oil Soap*, **18**, 194.

American Oil Chemists' Society (1946) *Official and Tentative Methods*, AOCS, Chicago, p. 1, table 1.

Ames, I.A. (1989) *J. Am. Coll. Nutr.*, **8**, 360.

Andersson, G. (1981) *Fette Seifen Anstrichm*, **83**, 1.

Anizewski, T. (1993) *J. Sci. Food Agric.*, **61**, 409.

Anon. (1985) *Zucker Suesswaren*, **38**, 19.

Appelqvist, L.A. (1966) *Acta Agric. Scand. Suppl.*, **16**, 215.

Appelqvist, L.A. and Ohlsen, R. (1972) *Rapeseed*, Elsevier, Amsterdam, p. 50.

Applewhite, T.H. (1981) *J. Am. Oil Chem. Soc.*, **58**, 260.

Archer-Daniels-Midland (1961) *Composition and Constants of Natural Fats and Oils by Gas–Liquid Chromatography*, Chart, Archer-Daniels-Midland, Minneapolis, 1961.

Arkcoll, D. (1988) *Econ. Bot.*, **42**, 195.

Awe, W. (1937) *Dtsch. Apoth. Ztg.*, **52**, 750.

Ayres, J.L. (1983) *J. Am. Oil Chem. Soc.*, **60**, 357.

Badifu, G.I.O. (1989) *J. Food Compos. Anal.*, **2**, 238.

Bafor, M., Stobart, A.K. and Stymne, S. (1990) *J. Am. Oil Chem. Soc.*, **67**, 217.

Bailey, A.E. (1948) *Cottonseed and Cottonseed Products*, Interscience, New York.

Bailey, A.V., Harris, J.A. and Skau, E.L. (1966) *J. Am. Oil Chem. Soc.*, **43**, 107.

Balick, M.J. (1981) in *New Sources of Fats and Oils* (eds E.H. Pryde, L.H. Princen and K.D. Muckerjee), American Oil Chemists' Society, Champaign, p. 141.

Bamford, H.L., Gardiner, K.J., Howat, J.R. and Thomson, A.F. (1970) *Confect. Prod.*, 359.

Banzon, J.A. (1980) *Phil. J. Coco. Stud.*, **5**, 47.

Barger, W.M. (1981) *J. Am. Oil Chem. Soc.*, **58**, 154.

Barnes, P.J. (1982) *Fette Seifen Anstrichm.*, **84**, 256.

Barnes, P.J. and Taylor, P.W. (1980) *J. Sci. Food Agric.*, **31**, 997.

Belvadi, V.K., Kamat, P.L., Mehta, D.T., Bhaskar, A. and Bringi, N.V. (1979) *J. Oil Tech. Assoc. India*, **11**, 3.

Bemis, W.P., Berry, J.W., Weber, C.W. and Whitaker, T.W. (1978) *Hort. Sci.*, **13**, 235.

Berger, K.G. (1977) *Oil Palm News*, **22**, 10.

Berger, K.G. (1983) *J. Am. Oil Chem. Soc.*, **60**, 206.

Berger, K.G. (1987) *Chem. Ind.*, 846.

Berger, K.G. and Ong, S.H. (1985) *Oleagineux*, **40**, 613.

Berman, L.U. and Loeb, M.L. (1975) in *Dimer Acids* (ed. E.C. Leonard), Humko Sheffield Chemical, Memphis, p. 9.

Bernardini, M. (1978) *Oleagineux*, **33**, 297.

Beroza, M. and Kinman, M.L. (1955) *J. Am. Oil Chem. Soc.*, **32**, 348.

Best, R.L., Crossley, A., Paul, S., Pardun, H. and Soeters, J. (1956) Br. Pat. 827172.

Beuchat, L.R. and Worthington, R.E. (1978) *J. Food Technol.*, **13**, 355.

Bezard, J.A. (1971) *Lipids*, **6**, 630.

Bezard, J.A., Begaut, M. and Clement, G. (1971) *J. Am. Oil Chem. Soc.*, **48**, 134.

Bhatnagar, S., Polonsky, J., Prange, T. and Pascard, C. (1983) *Tetrahedron Lett.*, **25**, 299.

Bhuyan, L.P. and Mahanta, P.K. (1989) *J. Sci. Food Agric.*, **46**, 325.

Blanchini, J.P., Ralaimarivo, A. and Gaydou, E.M. (1981) *Anal. Chem.*, **53**, 2194.

Body, D.R. and Shorland, F.B. (1965) *J. Am. Oil Chem. Soc.*, **42**, 5.

Bolley, D.S. and McCormack, R.H. (1950) *J. Am. Oil Chem. Soc.*, **27**, 84.

Bracco, U. (1973) *Agosto*, **2**, 145.

Bringi, N.V. (1987) *Non Traditional Oilseeds and Oils of India*, Oxford and IBH, New Delhi.

Bringi, N.V., Padley, F.B. and Timms, R.E. (1972) *Chem. Ind.*, 805.

Brockerhoff, H. and Hoyle, R.J. (1963) *Arch. Biochem. Biophys.*, **102**, 452.

Brockerhoff, H. and Yurkowski, M. (1966) *J. Lipid Res.*, **7**, 62.

Brockerhoff, H., Hoyle, R.J. and Wolmark, N. (1966) *Biochim. Biophys. Acta*, **116**, 67.

Bruckner, G. (1992) in *Fatty Acids in Foods and Their Health Implications* (ed. C.K. Chow), Marcel Dekker, New York.

Brun, J.H. and Matchett, J.R. (1963) *J. Am. Oil Chem. Soc.*, **40**, 1.

Budowski, P. (1964) *J. Am. Oil Chem. Soc.*, **41**, 280.

Budowski, P. and Markley, K.S. (1951) *Chem. Rev.*, **48**, 125.

Canola Council of Canada (1981) *Canola, A New Oilseed for Canada*, Winnipeg.

Carlson, K.D., Schneider, W.J., Chang, S.P. and Princen, L.H. (1981) in *New Sources of Oils and Fats* (eds E.H. Pryde, L.H. Princen and K.D. Mukherjee), American Oil Chemists' Society, Champaign, p. 297.

Carlson, K.D., Baker, E.C. and Mustakas, G.C. (1985) *J. Am. Oil Chem. Soc.*, **62**, 897.

Carpenter, A.P. (1979) *J. Am. Oil Chem. Soc.*, **56**, 668.

Cebula, D., Dilley, K.M. and Smith, K.W. (1991) *Manuf. Confect.*, May, 131.

Chaiseri, S. and Dimick, P.S. (1989) *J. Am. Oil Chem. Soc.*, **66**, 1772.

Chakrabarty, M.M. (1979) Private communication.

Chapman, G.M., Akehurst, E.E. and Wright, W.B. (1971) *J. Am. Oil Chem. Soc.*, **48**, 824.

Chapman, G.W. (1980) *J. Am. Oil Chem. Soc.*, **59**, 299.
Chaveron, H. and Verdoia, C. (1985) *Ann. Fals. Exp. Chim. Toxicol.*, **77**, 401.
Cherry, J.P. (1983) *J. Am. Oil Chem. Soc.*, **60**, 312A.
Child, R. (1964) *Coconuts*, Longmans Green, London.
Christie, W.W. (1969) *Biochim. Biophys. Acta*, **187**, 1.
Cloughley, J.B. (1983) *Trop. Agric. (Trinidad)*, **60**, 139.
Cobley, L.S. (1956) *An Introduction to the Botany of Tropical Crops*, Longmans Green, London, p. 128.
Codex (1969) Codex Alimentarius Commission, Recommended Standard, CAC/RS 27–1969.
Codex (1970a) Codex Alimentarius Commission, Document CAC/RS 33–1970: Recommended International Standard for Olive Oil, Virgin and Refined and for Refined Olive Oil Residue.
Codex (1970b) Codex Alimentarius Commission Report CAC/RS 34–1970.
Codex (1979) Codex Alimentarius Commission Report, 13th Session, 3–14 December (Report of the 10th Session of the Codex Committee on Fats and Oils, London, 4–8 December 1978), Alinorm 79/17.
Codex (1983) Codex Alimentarius Commission Report, (14th Session) (Report of the 12th Session of the Codex Committee on Fats and Oils, April 1982), Alinorm 83/17.
Codex (1983) Codex Alimentarius Commission Report (Report of the 14th Session of the Codex Committee on Fats and Oils, September 1993, London), Alinorm 95/17.
Coenen, J.W.E. (1976) *J. Am. Oil Chem. Soc.*, **53**, 382.
Conte, L.S., Moschetta, M. and Capella, P. (1984) *Riv. Ital. Sost. Grasse*, **61**, 563.
Corley, R.H.V. (1981) *J. Perak Planters' Assoc.*, 35.
Corley, R.H.V., Wong, C.Y., Wooi, K.C. and Jones, L.H. (1981) in *Oil Palm in Agriculture in the Eighties*, Incorporated Society of Planters, Kuala Lumpur.
Cornelius, J.A. (1977) *Prog. Chem. Fats Other Lipids*, **15**, 5.
Cornelius, J.A. and Raymond, W.D. (1967) *Trop. Sci.*, **9**, 75.
Cornelius, J.A. and Simmons, E.A. (1969) *Trop. Sci.*, **11**, 17.
Cottrell, R.C. (1991) *Am. J. Clin. Nutr.*, **53**, 989S.
Craig, B.M. (1953) *J. Am. Oil Chem. Soc.*, **31**, 202.
Craig, B.M., Mallard, T.M., Wight, R.E., Irvine, G.M. and Reynolds, J.R. (1973) *J. Am. Oil Chem. Soc.*, **50**, 395.
Crawford, R.V. (1982) *Food Proc. Ind.*, 24.
Culp, T.W., Harlow, F.D., Litchfield, C. and Reiser, R. (1965) *J. Am. Oil Chem. Soc.*, **42**, 974.
Curz Madueno, E. (1985) *Aliment. Equipos. Technol.*, **4**, 61.
D'Alonzo, R.P., Kozarek, W.H. and Wharton, H.W. (1981) *J. Am. Oil Chem. Soc.*, **58**, 215.
D'Alonzo, R.P., Kozarek, W.J. and Wade, R.L. (1982) *J. Am. Oil Chem. Soc.*, **59**, 292.
Davies, S.B., Conroy, E.A. and Shakespeare, N. (1950) *J. Am. Chem. Soc.*, **72**, 124.
de Jong, S., van Soest, T.C. and van Schick, M.A. (1991) *J. Am. Oil Chem. Soc.*, **68**, 371.
de la Roche, I.A., Weber, E.J. and Alexander, D.E. (1971) *Lipids*, **6**, 531.
Derbesy, M. and Richert, M.T. (1979) *Oleagineux*, **34**, 405.
Dietz, H. and Harris, R.V. (1990) *Food Control*, **1**, 84.
Dillman, A.C. and Hopper, P.H. (1943) *US Dept of Agriculture, Technical Bulletin*, No. 844.
Downey, R.K. (1971) *J. Am. Oil Chem. Soc.*, **48**, 718.
Downey, R.K., Craig, B.M. and Youngs, C.G. (1969) *J. Am. Oil Chem. Soc.*, **46**, 121.
Downey, R.K., Stringham, G.R. and MacGregor, D.I. (1975) in *Oil Seed and Pulse Crops in Western Canada* (ed. J.T. Harapiak), Western Co-operative Fertilisers Ltd, Calgary, p. 157.
Duce, S.L., Carpenter, T.A. and Hall, L.D. (1990) *Lebensm. Wiss. Technol.*, **23**, 545.
Duck, W. (1967) *Gordian*, **67**, 28.
du Plessis, L.M., Van Twisk, P., van Niekerk, P.J. and Steyn, M. (1981) *J. Am. Oil Chem. Soc.*, **58**, 575.
Dupont, J., White, P.J., Carpenter, M.P., Schaefer, E.J., Meydani, S.N., Elson, C.E., Woods, M. and Gorback, S.L. (1990) *J. Am. Coll. Nutr.*, **9**, 438.
Dutton, H. and Emken, E.A. (1979) *Geometrical and Positional Fatty Acid Isomers*, American Oil Chemists' Society, Champaign.
Earle, F.R. and Jones, Q. (1962) *Econ. Bot.*, **16**, 221.
Earle, F.R., Peters, J.E. and Wolff, I.A. (1966) *J. Am. Oil Chem. Soc.*, **43**, 330.
Earle, F.R., Vanetten, C.H., Clark, T.F. and Wolff, I.A. (1968) *J. Am. Oil Chem. Soc.*, **45**, 876.
Eckey, E.W. (1954) *Vegetable Fats and Oils*, Reinhold, New York, p. 556.
Eigtved, P. (1992) *Adv. Appl. Lipid Res.*, **1**, 1. (Ed. F.B. Padley), JAI Press, London, NY.
Elson, C.E. (1992) *Crit. Rev. Food Sci. Nutr.*, **32**, 79.
El Tinay, A.H., Khattab, A.H. and Khidir, M.O. (1976) *J. Am. Oil Chem. Soc.*, **53**, 648.
Erickson, D.R., Pryde, E.H., Brekke, O.L., Mounts, T.L. and Falb, R.A. (eds) (1980) *Handbook of Soy Oil Processing and Utilisation*, American Soybean Association and American Oil Chemists' Society, Champaign.
Evans, C.D., McConnell, D.G., Hoffman, R.L. and Peters, H. (1967) *J. Am. Oil Chem. Soc.*, **44**, 281.
Fedeli, E. (1977) *Prog. Chem. Fats Other Lipids*, **15**, 57.
Fedeli, E. (1983) *J. Am. Oil Chem. Soc.*, **60**, 356A.
Fedeli, E., Lanzani, A., Capella, P. and Jacini, G. (1966) *J. Am. Oil Chem. Soc.*, **43**, 254.
Feldheim, W. (1991) *Proc. 6th International Lupin Conference*, Temuco-Pucon, Chile, p. 124.
Fellers, D.A. (1991) *Food Rev. Int.*, **7**, 445.
Ficke, G.N. (1983) *J. Am. Oil Chem. Soc.*, **60**, 1252.
Fincke, A. (1976) *Gordian*, **76**, 324.
Fincke, A. (1980) *Dtsch. Lebens. Rundsch.*, **76**, 162.
Fincke, A. (1982) *Gordian*, **82**, 246.
Fisher, G.S. and Schuller, W.M. (1981) *J. Am. Oil Chem. Soc.*, **58**, 943.
Flament, I., Willhelm, B. and Stoll, M. (1968) *Helv. Chim. Acta*, **50**, 2223.
Food and Drug Administration (1976) *Evaluation of the Health Aspects of Hydrogenated Soybean Oil as a Food Ingredient*, Prepared for FDA by Federation of American Societies for Experimental Biology.

Foster, L.J. (1963) *World Crops*, **15**, 220.
Francois, L.E. and Kleiman, R. (1990) *Agron. J.*, **82**, 1110.
Frega, N., Bocci, F., Conte, L.S. and Teta, F. (1991) *J. Am. Oil Chem. Soc.*, **68**, 29.
Fritsch, C.W., Weiss, V.E. and Anderson, R.H. (1971) *J. Am. Oil Chem. Soc.*, **48**, 392.
Froyd, D.E. (1966) *Polyamide Resins*, 2nd edn, Reinhold, New York.
Fuller, G. (1966) *J. Am. Oil Chem. Soc.*, **43**, 477.
Fuller, G., Diamond, M.J. and Applewhite, T.H. (1967) *J. Am. Oil Chem. Soc.*, **44**, 264.
Gandhi, V.M., Cherian, K.M. and Mulky, M.J. (1976) *J. Oil Tech. Assoc. India*, **8**, 65.
Gapor, A.B., Top, M.D., Berger, K.G., Hashimoto, T., Kato, A., Tanabe, K., Mamuro, H. and Yamoaka, M. (1981) *Int. Conf. on Palm Oil Product Technology in the Eighties*, June 1981, Kuala Lumpur.
Gapor, A.B., Akio, K. and Ong, A.S.H. (1991) *Yukagaku*, **40**, 554.
Gardner, H.A. (1943) *Natl Paint Varnish Lacquer Assoc. Circ.* No. 654, p. 28.
Gaydou, E.M. and Raonizafinimanana, R. (1980) *J. Am. Oil Chem. Soc.*, **57**, 141.
Glumac, E. and Cowles, J. (1990) *Advances in New Crops, Proc. 1st Natl Symp. on New Crops*, Indianapolis, October 1988 (eds J. Janick and J.E. Simon), Timber Press, Portland, OR, p. 268.
Goh, S.H., Khor, H.T. and Gee, P.T. (1982) *J. Am. Oil Chem. Soc.*, **59**, 296.
Gopalakrishna, A.G. and Prabhakar, J.V. (1983) *J. Am. Oil Chem. Soc.*, **60**, 968.
Gordon, M.H., Padley, F.B. and Timms, R.E. (1979) *Fette Seifen Anstrichm.*, **81**, 116.
Gracian, J. and Martel, J. (1969) *Grasas Aceites*, **20**, 231.
Graef, G.L., Fehr, W.R. and Hammond, E.G. (1985) *Crop Sci.*, **25**, 1076.
Graham, S.A. (1989) *Crit. Rev. Food Sci. Nutr.*, **28**, 139.
Graham, S.A., Hirsinger, F. and Robbelen, G. (1981) *Am. J. Bot.*, **68**, 908.
Grami, B. and Stefansson, B.R. (1977) *Can. J. Plant Sci.*, **57**, 625.
Green, A.G. (1988) *Proc. World Conf. on Biotechnology for the Fats and Oils Industry, 1987* (ed. T.H. Applewhite), American Oil Chemists' Society, Champaign.
Grist, D.H. (1961) *World Crops*, **13**, 154.
Grossberger, T. and Rothschild, E. (1989) *LC–GC Int.*, **2**, 44, 47–8.
Gunstone, F.D. (1954) *J. Chem. Soc.*, 1611.
Gunstone, F.D. (1987) *Palm Oil*, Wiley, Chichester.
Gunstone, F.D. (1992) *Prog. Lipid Res.*, **31**, 145.
Gunstone, F.D. and Padley, F.B. (1965) *J. Am. Oil Chem. Soc.*, **42**, 957.
Gunstone, F.D. and Padley, F.B. (1967) *Chem. Phys. Lipids*, **1**, 429.
Gunstone, F.D. and Qureshi, M.I. (1965) *J. Am. Oil Chem. Soc.*, **42**, 961.
Gunstone, F.D., Hamilton, R.J., Padley, F.B. and Qureshi, M.I. (1965) *J. Am. Oil Chem. Soc.*, **42**, 965.
Gupta, S.S. and Meara, M.L. (1950) *J. Chem. Soc.*, 1337.
Gutfinger, T. (1981) *J. Am. Oil Chem. Soc.*, **58**, 966.
Hachiya, I., Koyano, T. and Sato, K. (1989) *J. Am. Oil Chem. Soc.*, **66**, 1757.
Hageman, J.W. and Rothfus, J.A. (1992) *J. Am. Oil Chem. Soc.*, **69**, 429.
Halphen, G. (1897) *J. Pharm.*, **6**, 390.
Hammond, E.G. and Fehr, W.R. (1984) *J. Am. Oil Chem. Soc.*, **61**, 1713.
Hammons, R.O. (1981) *Proc. Am. Peanut Res. Educ. Assoc.*, **13**, 12.
Hancock, J. and Houghton, C. (1990) *Lipid Technol.*, **2**, 90.
Hartley, C.W.S. (1977) *The Oil Palm*, Longmans, London.
Harwood, J.L. (1991) *Chem. Ind.*, 753.
Hess, D.C. (1977) *Cereal Foods World*, **22**, 98.
Hilditch, R.P. and Williams, P.N. (1964) *The Chemical Constitution of Natural Fats*, Chapman and Hall, London.
Hirsinger, F. (1985) *J. Am. Oil Chem. Soc.*, **62**, 76.
Holmbom, B. (1977) *J. Am. Oil Chem. Soc.*, **54**, 289.
Holmbom, B. (1978) *J. Am. Oil Chem. Soc.*, **55**, 876.
Homberg, V.E. and Bielefeld, B. (1989) *Fat Sci. Technol.*, **91**, 23.
Hopkins, D.T., Dahlgren, R.R., Davis, D., Munson, A.W. and Dupuy, H.P. (1973) *J. Am. Oil Chem. Soc.*, **50**, 381.
Hopper, T.H. and Nesbitt, L.L. (1937) *Oil Soap*, **14**, 34.
Horrobin, D.F. (ed.) (1982) *Clinical Uses of Essential Fatty Acids*, Eden, Montreal.
Horrobin, D.F. (1992) *Prog. Lipid Res.*, **31**, 163.
Horrobin, D.F. and Manku, M.S. (1989) *Prost. Leuk. Ess. Fatty Acids*, **37**, 255.
Howell, R.W. and Collins, R.I. (1957) *Agron. J.*, **49**, 593.
Huang, A.S., Hsieh, O.A., Huang, C.L. and Chang, S.S. (1981) *J. Am. Oil Chem. Soc.*, **58**, 997.
Hudson, B.J.F. (1984) *J. Am. Oil Chem. Soc.*, **61**, 540.
Hulsker, F.H., de Jong, K. and Turksma, H. (1989) Eur. Pat. Appl. EP 0321056A.
Hyer, A. (1966) Breeding for glandless cotton in California, *Cotton Gin Oil Mill Press*, **67**, 10, 14.
INFORM (1990) **1**, 1, 934, 944.
INFORM (1991) **2**, 692.
Itagaki, K., Yanagihara, T., Maruzeni, S. and Yasuda, N. (1989) Eur. Pat. Appl. EP 304493A.
Itoh, T., Tamura, T. and Matsumoto, T. (1973a) *J. Am. Oil Chem. Soc.*, **50**, 122.
Itoh, T., Tamura, T. and Matsumoto, T. (1973b) *J. Am. Oil Chem. Soc.*, **50**, 300.
Itoh, T., Tamura, T. and Matsumoto, T. (1974) *Oleagineux*, **29**, 253.
Jacobsberg, B. (1975) *Oleagineux*, **30**, 271.
Jeffery, M.S. (1991) *Manuf. Confect.*, **71**, 76.
Jeffrey, B.S.J. (1991) *J. Am. Oil Chem. Soc.*, **68**, 289.
Jeffrey, B.S.J. and Padley, F.B. (1991) *J. Am. Oil Chem. Soc.*, **68**, 123.
Jewell, G.G. and Bradford, L. (1981) *Manuf. Confect.*, 26.
Johnson, R.W. (1979) in *Fatty Acids* (ed. E.H. Pryde), American Oil Chemists' Society, Champaign, chap. 17.
Johnson, R.W. and Pryde, E.H. (1979) in *Fatty Acids* (ed. E.H. Pryde), American Oil Chemists' Society, Champaign, chap. 16.

Jones, L.H. (1974) *Oil Palm News*, **17**, 1.
Jones, L.H. (1984) *J. Am. Oil Chem. Soc.*, **61, 1717.**
Jones, L.H. (1989) *Biotech. Gen. Eng. Rev.*, **7**, 281.
Jurriens, G. (1968) in *Analysis and Characterisation of Oils and Fats and Fat Production* (ed. H.A. Boekenoogen), vol. 2, Interscience, New York.
Kabul, P. (1985) *Diss. Abs. Int. B*, **45**, 1973.
Kahlon, T.S. (1989) *Cereal Foods World*, **34**, 872.
Kamel, B.S. and Kakadu, Y. (1992) in *Fatty Acids in Foods and Their Health Implications* (ed. C.K. Chow), Marcel Dekker, New York.
Kapoulas, V.M. and Passaloglou-Emmanouilidou, S. (1981) *J. Am. Oil Chem. Soc.*, **58**, 694.
Kattenberg, H.P. (1981) *Manuf. Confect.*, 32.
Kaufmann, H.P. and Thieme, J.G. (1954) *Fette Seifen Anstrichm.*, **56**, 1001.
Kershaw, S.J. (1987) *Int. J. Food Sci. Technol.*, **22**, 67.
Kershaw, S.J. and Hardwick, E. (1981) *J. Am. Oil Chem. Soc.*, **58**, 706.
Khoury, N.N., Dagher, S. and Sawaya, W. (1982) *J. Food Technol.*, **17**, 19.
Kies, C. (1989) *Cereal Foods World*, **34**, 1016.
Kikugawa, K., Arai, M. and Kurechi, T. (1983) *J. Am. Oil Chem. Soc.*, **60**, 1528.
Kim, M.A., Matoba, T. and Hasegawa, K. (1988) *Agric. Biol. Chem. (Japan)*, **52**, 693.
Kim, N.K. and Hung, Y.C. (1991) *J. Food Sci.*, **56**, 1378.
Kinderlerer, J.L. and Hatton, P.V. (1991) *J. Appl. Bacteriol.*, **70**, 502.
King, G. (1942) *J. Chem. Soc.*, 387.
Kinman, M.L. (1972) *J. Am. Oil Chem. Soc.*, **49**, 36.
Kircher, H.W. (1964) *J. Am. Oil Chem. Soc.*, **41**, 4.
Kiritsakis, A.K. and Markakis, P. (1987) *Adv. Food Res.*, **31**, 453.
Kiritsakis, A.K. and Min, D.B. (1988) in *Flavour Chemistry of Lipids Foods* (eds D.B. Min and T.H. Smouse), American Oil Chemists' Society, Champaign, p. 196.
Kirk, L.D., Mustakas, G.C. and Griffin, E.L. (1966) *J. Am. Oil Chem. Soc.*, **43**, 550.
Kitamara, K. *et al.* (1984) *Yukagaku*, **33**, 800.
Kleinert, R.J. (1961) *Int. Choc. Rev.*, **16**, 201.
Kleinert, R.J. (1970) *Int. Choc. Rev.*, **25**, 162.
Knapp, S.J., Tagliani, L.A. and Roath, W.W. (1991) *J. Am. Oil Chem. Soc.*, **68**, 515.
Knauf, B.V. (1987) *TIB Tech.*, **5**, 40.
Knowles, P.F. (1955) *Econ. Bot.*, **9**, 273.
Knowles, P.F. (1965) *Calif. Agric.*, **19**, 15.
Knowles, P.F. (1975) *J. Am. Oil Chem. Soc.*, **52**, 374.
Knowles, P.F., Kearney, T.E. and Cohen, D.B. (1981) in *New Sources of Fats and Oils* (ed. E.H. Pryde, L.H. Princen and K.D. Mukherjee), American Oil Chemists' Society, Champaign, p. 255.
Knutson, D.S., Thompson, G.A., Radke, S.E., Johnson, W.B., Knauf, V.C. and Kridl, J.C. (1992) *Proc. Natl Acad. Sci. USA*, **89**, 2624.
Kobayashi, H. and Yuyama, A. (1988) *Nippon Jui. Zasshi*, **41**, 594.
Kornfeldt, A. and Croon, L.B. (1981) *Lipids*, **16**, 306.
Koyano, T., Hachiya, I. and Sato, K. (1990) *Food Struct.*, **9**, 231.
Kramer, J.K.G., Sauer, F.D. and Pigden, W.J. (eds) (1983) *High and Low Erucic Acid Rapeseed Oils*, Academic Press, London, Animal Research Centre, Ottawa.
Kremers, F.J. (1971) *J. Am. Oil Chem. Soc.*, **48**, 314.
Kroeze, H. (1982) Private communication.
Kromer, G.W. (1977) in *Glanders Cotton: Its Significance, Status and Prospects, Proc. Conf.*, Dallas, Texas, Agricultural Research Service, US Department of Agriculture.
Kuemmel, D.F. (1964) *J. Am. Oil Chem. Soc.*, **41**, 667.
Kuksis, A. and McCarthy, M.J. (1962) *Can. J. Biochem. Physiol.*, **40**, 679.
Lakshminarayana, G., Chandrasekhara, R.T. and Ramalingaswahy, P.A. (1983) *J. Am. Oil Chem. Soc.*, **60**, 88.
Lakshminarayana, G., Paulose, M.M. and Kumari, B.N. (1984) *J. Am. Oil Chem. Soc.*, **61**, 1871.
Lamer-Zarawska, E., Hojden, B. and Szymczak, J. (1989) *Herba Pol.*, **35**, 165.
Lange, R. and Linow, F. (1987) *Lebensmittelindustrie*, **34**, 113.
Langstraad, A. (1976) *J. Am. Oil Chem. Soc.*, **53**, 241.
Lanzini, A., Bondiolo, P., Gasparoli, A., Mariani, C., Fedeli, E., Ponzetti, A. and Peralisi, G. (1986) *Riv. Ital. Sost. Grasse*, **63**, 325.
Larsson, K. (1966) *Acta Chem. Scand.*, **20**, 2255.
Latif, A. and El Daly, M.A. (1973) *Agric. Res. Rev.*, **51**, 123.
Lawsson, I., Rydhag, L. and Wilton, I. (1978) *Proc. 5th Int. Rapeseed Conf.*, vol. 2, p. 187.
Lee, L.S. and Timms, R.E. (1988) *ASEAN Food J.*, **3**, 11.
Lehrian, D.W., Keeney, P.G. and Butler, D.R. (1980) *J. Am. Oil Chem. Soc.*, **57**, 61, 66.
Leibovitz, Z. and Ruckenstein, C. (1983) *J. Am. Oil Chem. Soc.*, **60**, 347A.
Lessman, K.J. and Anderson, W.P. (1981) *New Sources of Fats and Oils* (eds E.H. Pryde, L.H. Princen and K.D. Mukherjee), American Oil Chemists' Society, Champaign, p. 223.
List, G.R. and Erickson, D.R. (1980) in *Handbook of Soya Oil Processing and Utilisation* (eds D.R. Erickson, E.H. Pryde, O.L. Brekke, T.L. Mounts and R.A. Falb), American Soybean Association and American Oil Chemists' Society, Champaign.
Litchfield, C. (1970) *Chem. Phys. Lipids*, **4**, 96.
Lognay, G., Wathelet, J.-P. and Severin, M. (1981) *Rev. Fr. Corps Gras*, **28**, 67.
Lovell, C.R., Burton, J.L. and Horrobin, D.F. (1981) *Lancet*, **i** (8214), 278.
Lundberg, W.O. (1972) *Fette Seifen Anstrichm.*, **74**, 557.
Lunde, G. (1972) *J. Am. Oil Chem. Soc.*, **49**, 44.
Lyon, C.K. (1972) *J. Am. Oil Chem. Soc.*, **49**, 245.
Macrae, A. and Coleman, M.H. (1976) Br. Pat. 1577933.
Macrae, A.R. (1983) *J. Am. Oil Chem. Soc.*, **60**, 291.
Mag, T.K. (1983) *J. Am. Oil Chem. Soc.*, **60**, 380.
Mann, G.E. (1954) *Agric. Food Chem.*, **2** (5), 258.
Marian, J.P. (1967) *Helv. Chim. Acta*, **50**, 1509.
Martinson, V.A., Adomako, D. and Manu, M. (1987) *Cocoa Growers Bull.*, **38**, 39.

Matsumoto, T., Shimizu, N., Itoh, T., Iida, T. and Nishioka, A. (1982) *J. Am. Oil Chem. Soc.*, **59**, 521.
Mattson, F.H. and Volpenheim, R.A. (1963) *J. Lipid Res.*, **4**, 392.
May, P.H., Anderson, A.B., Balik, M.J. and Frazao, J.M.F. (1985) *Econ. Bot.*, **39**, 113.
McKinney, R.S. and Jamieson, G.C. (1936) *Oil Soap*, **13**, 10.
McMichael, S.C. (1959) *Agron. J.*, **51**, 630.
McPherson, A. (1978) *Rapeseed and Mustard, a Selected Bibliography*, Rapeseed Association of Canada, Winnipeg.
Mehlenbacher, V.C. (1936) *Oil Soap*, **13**, 136.
Mikolajczak, K.L., Freidinger, R.M., Smith, C.R. and Wolff, I.A. (1968) *Lipids*, **3**, 489.
Minnifie, B.W. (1980) *Chocolate, Cocoa and Confectionery: Science and Technology*, 2nd edn, Avi, Westport, CT.
Mohr, E., Wichmann, G. and Roche, G. (1987) *Gordian*, **87**, 77.
Morgan, R.P. and Shultz, E.B. (1981) *Chem. Eng. News*, **59**, 69.
Morrell, R.S. and Davis, W.R. (1936) *J. Chem. Soc.*, 1481.
Morrison, W.H. and Robertson, J.A. (1975) *J. Am. Oil Chem. Soc.*, **54**, 148.
Muller-Mulot, W. (1976) *J. Am. Oil Chem. Soc.*, **53**, 732.
Mustakas, G.C., Kapas, G. and Robinson, N. (1965) *J. Am. Oil Chem. Soc.*, **42**, 550A.
Nasirulla, B., Werner, G. and Seher, A. (1984) *Fette Seifen Anstrichm.*, **86**, 264.
Nasirullah, B. and Nagaraja, K.V. (1987) *J. Oil Tech. Assoc. India*, **19**, 2.
Naughton, F.C. (1974) *J. Am. Oil Chem. Soc.*, **51**, 65.
Niewiadomski, H., Bratkowska, I. and Mossakowska, E. (1965) *J. Am. Oil Chem. Soc.*, **42**, 731.
Ninomiya, M. (1940) *Rep. Inst. Sci. Manchoukuo*, **4**, 69.
Ohlson, R. (1992) *J. Am. Oil Chem. Soc.*, **69**, 195.
Oil World Annual (1991) (ed. T. Mielke), ISTA Mielke, Hamburg.
O'Keefe, S.F., Wiley, V.A. and Knauf, D.A. (1993) *J. Am. Oil Chem. Soc.*, **70**, 489.
Ologunde, M.O., *et al.* (1992) *J. Sci. Food Agric.*, **58**, 221.
Ong, A.S.H. (1993) in *Vitamin E in Health and Disease* (eds L. Packer and J. Fuchs), Marcel Dekker, New York.
Osman, S.M. and Ahmad, F. (1981) in *New Sources of Oil and Fats* (eds E.H. Pryde, L.H. Princen and K.D. Mukherjee), American Oil Chemists' Society, Champaign, p. 109.
Owuor, P.O., Chavanji, A.M. and Munavu, R.M. (1985) *Tea*, **6**, 23.
Padley, F.B. and Timms, R.E. (1980) *J. Am. Oil Chem. Soc.*, **57**, 286.
Paganuzzi, V. (1979) *J. Am. Oil Chem. Soc.*, **56**, 925.
Palm Oil (1977) *International Developments in Palm Oil, Proc. Malaysian Int. Symp.*, International Society of Planters, Kuala Lumpur.
Pandey, S.N. and Suri, L.K. (1982) *J. Am. Oil Chem. Soc.*, **59**, 99.
Paquot, C., Argoudn, S. and Querolle, M. (1952) *Oleagineux*, **7**, 397.
Paranjothy, K. (1989) *Elaeis*, **1**, 119.
Parker, W.A. and Melnick, D. (1966) *J. Am. Oil Chem. Soc.*, **43**, 635.
Passwater, R.A. (1981) *Evening Primrose Oil*, Keats, New Canaan, CT, USA.
Pastana, A. and Munoz, E. (1982) *Nature*, **298**, 608.
Pattee, H.E., Purcell, A.E. and Jones, E.B. (1969) *J. Am. Oil Chem. Soc.*, **46**, 629.
Pawlowski, N.E., Loveland, P.M. and Sinnhuber, R.O. (1981) *J. Am. Oil Chem. Soc.*, **58**, 692.
Payne-Wahl, K. and Kleiman, R. (1983) *J. Am. Oil Chem. Soc.*, **60**, 1011.
Peers, K.E. (1977) *J. Sci. Food Agric.*, **28**, 1000.
Persmark, W. and Bengtsseon, L. (1976) *Riv. Ital. Sost. Grasse*, **53**, 307.
Phelps, R.A., Shenstone, F.S., Kemmerer, A.R. and Evans, R.J. (1965) *Poult. Sci.*, **44**, 358.
Piggott, C.J. (1964) *Coconut Growing*, Oxford University Press, London.
Princen, L.H. and Rothfus, J.A. (1984) *J. Am. Oil Chem. Soc.*, **61**, 281.
Pritchard, J.R. (1983) *J. Am. Oil Chem. Soc.*, **60**, 332.
Pryde, E.H. and Cowan, J.C. (1972) in *Condensation Monomers* (eds J.K. Stille and T.W. Campbell), Wiley, New York, p. 91.
Putt, E.D. (1978) in *Sunflower Science and Technology*, Agronomy 19, American Society of Agronomy, Madison, WI.
Qing, X.B. (1989) *J. Am. Oil Chem. Soc.*, **66**, 479.
Racicot, L.D. and Handel, A.P. (1983). *J. Am. Oil Chem. Soc.*, **60**, 1098.
Rajanaidu, N. and Tan, B.K. (1983) *Oleagineux*, **38**, 581.
Rakow, G. and McGregor, D.I. (1973) *J. Am. Oil Chem. Soc.*, **50**, 400.
Ramos, L.C.D.S., Tango, J.S., Savi, A. and Leal, N.R. (1984) *J. Am. Oil Chem. Soc.*, **61**, 1841.
Rana, M.S. and Ahmed, A.A. (1981) *J. Am. Oil Chem. Soc.*, **58**, 630.
Rao, S.D.T. and Murthi, K.S. (1961) *Indian Oilseeds*, **5**, 121.
Rath, T. (1987) *Indian Forester*, **113**, 73.
Raymond, W.D. (1961) *Trop. Sci.*, **3**, 69.
Raymond, W.D. and Ward, J.B. (1956) *Colon. Plant Anim. Prod.*, **6**, 243.
Reufels, J. (1989) *Entwickl. Landlicher Raum*, **23**, 28.
Riijner, U. (1970) *J. Am. Oil Chem. Soc.*, **47**, 129.
Rivarola, G., Anon, M.C. and Calvelo, A. (1985) *J. Am. Oil Chem. Soc.*, **62**, 1508.
Röbbelin, G. (1988) *Proc. World Conf. on Biotechnology for the Fats and Oils Industry, 1987* (ed. T.H. Applewhite), American Oil Chemists' Society, Champaign, pp. 78–86.
Rohan, T.A. (1969) *Gordian*, **69**, 443.
Rohne, G.W. (1984) *Fette Seifen Anstrichm.*, **86**, 172.
Rose, W.G. and Jamieson, G.S. (1943) *Oil Soap*, **20**, 227.
Rossell, J.B. (1985) *J. Am. Oil Chem. Soc.*, **62**, 385.
Rossell, J.B. (1991) *Fat Sci. Technol.*, **93**, 526.
Rossell, J.B. and Pritchard, J.L.R. (eds) (1990) *Analysis of Oilseeds, Fats and Fatty Foods*, Elsevier Applied Science, London, pp. 315, 317.

Rossell, J.B., King, B. and Downes, M.J. (1983) *J. Am. Oil Chem. Soc.*, 285A.
Rossell, J.B., King, B. and Downes, M.J. (1985) *J. Am. Oil Chem. Soc.*, **62**, 221.
Ruyssen, B. (1957) *Agron. Trop.*, **12**, 137, 273.
Sagredos, A.N., Roy, D.S. and Thomas, A. (1988) *Fett. Wiss. Technol.*, **90**, 76.
Sahasrabudhe, M.R. (1979) *J. Am. Oil Chem. Soc.*, **56**, 80.
Sampugna, J. and Jensen, R.G. (1969) *Lipids*, **4**, 444.
Sanders, T.H. (1980) *J. Am. Oil Chem. Soc.*, **57**, 12.
Sanders, T.H. (1982) *J. Am. Oil Chem. Soc.*, **59**, 346.
Santinelli, F., Damiani, P. and Christie, W.W. (1992) *J. Am. Oil Chem. Soc.*, **69**, 552.
Sato, K. (1987) *Food Microstruct.*, **6**, 151.
Satpathi, P. (1984) *Indian Farming*, **34**, 22.
Saunders, R.M. (1985) *Food Rev. Int.*, **1**, 465.
Sawadogo, K. and Bezard, J. (1982) *Oleagineux*, **37**, 69.
Scheerens, J.C. and Berry, J.W. (1986) *Cereal Foods World*, **31**, 183, 186, 190.
Schlichter, A.J., Sarig, S. and Garti, N. (1988) *J. Am. Oil Chem. Soc.*, **65**, 1140.
Sebedio, J.L. and Ackman, R.G. (1981) *J. Am. Oil Chem. Soc.*, **58**, 972.
Sen Gupta, A.K. (1972) *Chem. Ind.*, 257.
Sengupta, A. and Mazumbler, U.K. (1976) *J. Am. Oil Chem. Soc.*, **53**, 680.
Sengupta, C., Sengupta, A. and Ghosh, A. (1976) *J. Sci. Food Agric.*, **27**, 1115.
Sen Gupta, S., Chakrabarty, M.M. and Bhattacharyya, D.K. (1982) *Fette Seifen Anstrichm.*, **84**, 226.
Sethuraj, M.R., Rao, G.G. and Asokan, M.P. (1989) *J. Plant. Crops*, **16**, 9.
Shenstone, F.S. and Vickery, J.R. (1959) *Poult. Sci.*, **38**, 1005.
Shenstone, F.S. and Vickery, J.R. (1961) *Nature*, **190**, 168.
Sheppard, A.J. and Pennington, J.A.T. (1993) in *Vitamin E in Health and Disease* (eds L. Packer and J. Fuchs), Marcel Dekker, New York.
Sheppard, A.J. and Rudolf, T.S. (1991) *Peanut Sci.*, **18**, 51.
Siew, W.L., Tang, T.S., Flingoh, C.H.O.H., Chong, C.L. and Tau, Y.A. (1992) *Elaeis*, **5**, 38.
Singh, B.B. (1976) PhD Thesis, University of Illinois, Urbana.
Singh, S.P., Khanna, K.R., Dixit, B.S. and Srivastava, S.N. (1990) *Indian J. Agric. Sci.*, **60**, 358.
Singleton, J.A. and Pattee, H.E. (1981) *J. Am. Oil Chem. Soc.*, **58**, 873.
Slavin, J. (1991) *J. Am. Diet. Assoc.*, **91**, 816.
Sleeter, R.T. (1981) *J. Am. Oil Chem. Soc.*, **58**, 239.
Slover, H.T. (1971) *Lipids*, **6**, 291.
Soeters, C.J. (1970) *Fette Seifen Anstrichm.*, **72**, 711.
Sosulki, F., Zadernowski, R. and Babachowski, K. (1981) *J. Am. Oil Chem. Soc.*, **58**, 561.
Soybean (1981) *Proc. World Conf. on Soya Processing and Utilisation*; *J. Am. Oil Chem. Soc.*, **58**, 121.
Spencer, G.F. (1973) *Chem. Phys. Lipids*, **10**, 191.
Sprecher, H.W., Maier, R., Barber, M. and Holman, R.T. (1965) *Biochemistry*, **4**, 1856.
Sreenivasan, R. (1969) *J. Am. Oil Chem. Soc.*, **46**, 459.
Stalker, H.T., Young, C.T. and Jones, T.M. (1989) *Oleagineux*, **44**, 419.
Steger, A. and van Loon, J. (1935) *Rec. Trav. Chim.*, **54**, 149.
Sundaravalli, O.E., Guttikar, H.A., Desai, B.L.M., Kantha, S. and Shiurpalekar, I. (1978) *J. Oil Tech. Assoc. India*, **10**, 147.
Sunflower Oil Today (1987) *Rev. Fr. Corps Gras*, **34**, 183.
Symposium (1985) World Conference, Processing of Palm, Palmkernel and Coconut Oils, Proceedings. *J. Am. Oil Chem. Soc.*, **62**, (2), 215.
Tachibana, T. and Kambara, H. (1968) *J. Colloid Int. Sci.*, **28**, 173.
Takagi, T. and Ando, Y. (1991) *Lipids*, **26**, 542.
Tamurat, T., Sakaedani, N. and Matsumoto, T. (1958) *Nippon Kayaku Zasshi*, **79**, 1011.
Tan, B.K. and Berger, K.G. (1982) *J. Sci. Food Agric.*, **33**, 204.
Tan, B.K. and Oh, F.C.H. (n.d.) *PORIM Technology Series*, No. 4, Kuala Lumpur.
Tashiro, T., Fukuda, Y. and Osawa, T. (1991) *Nippon Sak. Gakk. Kiji*, **60**, 116.
Tavares, M., Badolato, E.S.G., Carvalho, J.B.D. and Aued, S. (1990) *Rev. Inst. Adolfo Lutz*, **50**, 307.
Teasdale, B.F. (1975) in *Oilseed and Pulse Crops in Western Canada* (ed. J.T. Harapuik), Modern Press, Saskatoon.
Terstage, R.G. (1972) *Oil Mill Gaset.*, **77**, 9.
Thomas, E.A. and Paulicka, F.R. (1976) *Chem. Ind.*, 774.
Thomas, G.V., Shautaram, M.V. and Jayasankar, N.P. (1987) *J. Plant. Crops*, **15**, 72.
Thompson, A.E. (1984) *Hort. Sci.*, **19**, 352.
Thompson, A.E. (1990) in *Advances in New Crops, Proc. 1st Natl Symp. on New Crops*, Indianapolis, October 1988 (eds J. Janick and J.E. Simon), Timber Press, Portland, OR.
Thompson, G.V. and Kleiman, R. (1988) *J. Am. Oil Chem. Soc.*, **65**, 139.
Timms, R.E. (1980) *Lebensm. Wiss. Technol.*, **13**, 61.
Timms, R.E. (1984) *Prog. Lipid Res.*, **23**, 1.
Timms, R.E. (1985) *J. Am. Oil Chem. Soc.*, **62**, 241.
Timms, R.E. (1986) *Fette Seifen Anstrichm.*, **88**, 294.
Tombs, M.P. and Stubbs, M.J. (1982) *J. Sci. Food Agric.*, **33**, 892.
Tookey, H.L., Van Etten, C.H. and Daxenbichler, M.E. (1980) in *Toxic Constituents of Plant Foodstuffs*, 2nd edn (ed. I.E. Liener), Academic Press, New York, p. 103.
Tous, J.G. (1968) in *Analysis and Characterisation of Oils, Fats and Fat Products* (ed. H.A. Boekenoogen), Interscience, London.
Toxic Oil Syndrome References Archive (1990) Fondo de Investigaciones Sanitarias, Ministerio de Sanidadyconsumo, Madrid.
Traitler, H., Wille, H.J. and Studer, A. (1986) *Fette Seifen Anstrichm.*, **88**, 378.
Traitler, H., Wille, H.J. and Studer, A. (1988) *J. Am. Oil Chem. Soc.*, **65**, 755.
Treadwell, K. and Young, C.T. (1982) *Oleagineux*, **37**, 477.
Tropical Products Institute (1971) *Oils and Oil Seeds*, No. 1 in the series *Crop and Product Digests*, HMSO, London.

Trost, V.W. (1989) *J. Am. Oil Chem. Soc.*, **66**, 325.
Tsunoda, S., Hinata, K. and Gomez-Compo, C. (eds) (1980) *Brassica Crops and Wild Alliles*, Japan Scientific Society Press, Tokyo.
Turpin, P.E., Coxon, D.T. and Padley, F.B. (1992) *Fat Sci. Technol.*, **5**, 179.
Unger, E.H. (1990) in Canola Rapeseed (ed. F. Shahidi), Van Nostrand Reinhold, New York, p. 235.
US Department of Agriculture (1965) *Safflower Culture in the West Central Plains*, Agric. Res. Serv., Agric. Inform. Bull. No. 300.
Vaccarino, C. (1977) *Riv. Ital. Sost. Grasse*, **54**, 183.
Vaisey-Genser, M. and Eskin, N.M.A. (1982) *Canola*, Canola Council of Canada, Winnipeg.
Van der Ven, B. and de Jong, K. (1970) *J. Am. Oil Chem. Soc.*, **47**, 299.
Van der Waal, B., Sipma, B.G., Ketenes, D.K. and Semper, A. Th. (1968) *Rec. Trav. Chim.*, **87**, 238.
Vasconcellos, J.A. and Berry, J.W. (1982) *J. Am. Oil Chem. Soc.*, **59**, 79.
Vasconcellos, J.A., Bemis, W.P., Berry, J.W. and Weber, C.W. (1981) in *New Sources of Fats and Oils* (eds E.H. Pryde, L.H. Princen and K.D. Mukherjee), American Oil Chemists' Society, Champaign.
Vereshchagin, A.G. and Novitskaya, G.V. (1965) *J. Am. Oil Chem. Soc.*, **42**, 970.
Vickery, J.R. (1980) *J. Am. Oil Chem. Soc.*, **57**, 87.
Vioque, E. and Maza, M.P. (1973) *Grasas Aceites*, **24**, 323.
Vitagliano, M. (1960) *Riv. Ital. Sost. Grasse*, **37**, 136.
Vitagliano, M. and Turri, E. (1958) *Olearia*, **12**, 145.
Voelker, T.A. and Davies, H.M. (1992) Int. Pat. Appl. WO 92/20236.
Von Mikusch, J.W. (1955) *Paint Manuf.*, **25**, 386.
Von Schuster, W. and Marquard, R. (1974) *Fette Seifen Anstrichm.*, **74**, 207.
Von Schuster, W., Kubler, I. and Marquard, R. (1980) *Fette Seifen Anstrichm.*, **82**, 443.
Wagner, H. and Wolff, P. (1964) *Fette Seifen Anstrichm.*, **66**, 425.
Wang, J.P., Yu, F.L., Li, J.M. and Shan, X.Q. (1988) *Acta Bot. Sin.*, **30**, 619.
Wealth of India – Raw Materials (1966) Publ. Inform. Directorate, CSIR, New Delhi, vol. VII, p. 23.
Webb, D.M. and Knapp, S.J. (1991) *Crop Sci.*, **31**, 621.
Weyland, M. (1992) *Manuf. Confect.*, **72**, 53.
Whipkey, A., Simon, J.E. and Janick, J. (1988) *J. Am. Oil Chem. Soc.*, **65**, 979.
White, M.B., Quackenbush, F.W. and Probost, A.H. (1961) *J. Am. Oil Chem. Soc.*, **38**, 113.
Whittle, K.J. and Pennock, J.F. (1967) *Analyst*, **92**, 423.
Wille, R.L. and Lutton, E.S. (1966) *J. Am. Oil Chem. Soc.*, **43**, 491.
Willems, M.G.A. and Padley, F.B. (1985) *J. Am. Oil Chem. Soc.*, **62**, 454.
Wolf, J.P. (1978) *Manuel d'Analyse des Corps Gras*, Azoulay, Paris.
Wolf, R.B., Graham, S.A. and Kleiman, R. (1983a) *J. Am. Oil Chem. Soc.*, **60**, 103.
Wolf, R.B., Kleiman, R. and England, R.E. (1983b) *J. Am. Oil Chem. Soc.*, **60**, 1858.
Wong, R.S., Beversdorf, W.D., James, R., Grant, I. and Patel, J.D. (1990) Int. Pat. Appl. WO 90/10380.
Wood, R.D., Raju, P.K. and Reiser, R. (1965) *J. Am. Oil Chem. Soc.*, **42**, 81.
Woodroof, J.G. (1966) *Peanuts, Production, Processing, Products*, Avi, Westport, CT.
Woodroof, J.G. (1983) *Peanuts*, Avi, Westport, CT.
Wright, N.G. (1988) in *Arid Lands: Today and Tomorrow* (eds E.E. Whitehead, C.F. Hutchinson, B.N. Timmerman and R.G. Varady), Belhaven, London, p. 1341.
Wynne, J.C. (1989) *Crit. Rev. Plant Sci.*, **8**, 189.
Yano, J., Veno, S., Sato, K., Arishima, T., Sagi, N., Kaneko, F. and Kobayashi, M. (1994) *J. Phys. Chem.*, to be published.
Yella Reddy, S. and Prabhakar, J.V. (1987) *J. Am. Oil Chem. Soc.*, **64**, 97.
Yella Reddy, S. and Prabhakar, J.V. (1989) *J. Am. Oil Chem. Soc.*, **66**, 805.
Yermanos, D.M. (1966) *J. Am. Oil Chem. Soc.*, **43**, 546.
Yermanos, D.M., Hemstreet, S. and Garber, M.J. (1967) *Crop. Sci.*, **7**, 417.
Yermanos, D.M., Hemstreet, S., Saleeb, W. and Huszar, C.K. (1972) *J. Am. Oil Chem. Soc.*, **49**, 20.
Young, C.C. (1984) *J. Am. Oil Chem. Soc.*, **61**, 576.
Zarins, Z.M., White, J.L., Willich, R.K. and Feuge, R.O. (1982) *J. Am. Oil Chem. Soc.*, **59**, 511.
Zeleny, L. and Coleman, D.A. (1936) *Oil Soap*, **13**, 253.
Zimmerman, D.C. (1981) in *New Sources of Oils and Fats* (eds E.H. Pryde, L.H. Princen and K.D. Mukherjee), American Oil Chemists' Society, Champaign, p. 25.

3.4 Animal fats: milk and depot fats

3.4.1 Butterfat (bovine milk fat)

Data on world production and characteristics of butterfat are summarized in Tables 3.176 and 3.177.

Butterfat from cow milk is produced in huge quantities throughout the world. The lactating cow can produce up to 30 kg of milk/day. The butterfat content of cow milk is 2.6–6% and in butter >80%. Butter and milk sold to the consumer normally contain a specific minimum level of butterfat defined by law (see section 5.7). The fat crystal structure in cream and butter is described in detail based on electron microscope studies (Precht, 1988).

Surplus butter is stored for extended periods at −30 °C. Alternatively, the butterfat is isolated by melting and separating the butter. In this case the butterfat is refined to ensure that the fat is dry and free of metal pro-oxidants.

Butter is used almost solely as an edible product either as a spread, in baking or in frying. In India, butter obtained from either the cow or the water buffalo is converted into ghee by first dehydration and then allowing the fat to crystallize into coarse agglomerates. The properties of cream and butter powders have been reviewed (Prasad and Gupta, 1988) and also the composition of milk fat (Jensen *et al.*, 1991; Precht, 1992).

The biochemical pathways of milk fat synthesis are well established and have been reviewed (Dimick *et al.*, 1970; Storry, 1980; Smith, 1980; Patton and Jenson, 1976; Moore and Christie, 1979). Similarly, lipid digestion in the ruminant has received intensive research (Christie, 1979; Dawson and Kemp, 1970; Garton, 1974; Harrison and Leat, 1975; Moore and Christie, 1979). The complexity of milk fat fatty acids arises from a number of factors including (a) the synthesis of short-chain acids (C_4–C_{10}) within the mammary gland and (b) micro-organisms in the rumen that fully or partially hydrogenate C_{18} fatty acids in the diet, thereby introducing geometric and positional mono-unsaturated acid isomers and also branched-chain fatty acids into the total pool of fatty acids.

The composition of the lipids in whole milk and in the milk fat globule membrane are given in Table 3.178.

Butterfat is the most complex natural fat on the basis of its fatty acid (Table 3.179; Jensen *et al.*, 1962) and triglyceride (Table 3.183) compositions. There are approximately 15 major acids, but over 500 acids have been characterized. The composition of butterfat is further complicated because it contains numerous isomers of the monoenoic, dienoic and branched-chain acids (Table 3.180) (Jensen *et al.*, 1962; Hay and Morrison, 1970; Deman and Deman, 1983; Van der Ven and de Jong, 1967). Such complex mixtures are probably best characterized by combinations of HPLC, capillary GLC and mass spectrometry. The higher-molecular-weight fatty acids have been characterized in some detail (Table 3.181) (Iverson *et al.*, 1965).

The seasonal variation in milk fat composition is demonstrated in Table 3.179, winter milk fat being more saturated than summer milk fat. The *trans* fatty acid content is not frequently recorded. Literature values vary between 2 and 11%, with higher values in the summer. The determined values are very method-dependent and values in the range 4–8% are probably realistic (Deman and Deman, 1983). The main *trans* mono-unsaturated acid is *trans*-vaccenic acid, $(n-7)$-18:1 (Hay and Morrison, 1970; Lund and Jensen, 1983).

The positional distribution of the major fatty acids in butterfat (Kuksis *et al.*, 1973; Pitas *et al.*, 1967) has been determined using Brockerhoff's method (Bracco *et al.*, 1972; Brockerhoff, 1965; Brunner, 1969). Butyric (4:0) and caproic (6:0) acids occur only at the *sn*-3 position. Other acids are also distributed preferentially, e.g. 18:0 at the *sn*-1 position. This type of distribution is common to virtually all mammalian milk fats examined except the echidna (Parodi, 1982). The positional distributions of a number of mammalian milk fats (cow, goat, pig, sheep, rat and human) are compared in Table 3.182. The triglyceride carbon-number distribution of 76 Australian butterfats (Table 3.183) has been studied (Guyot, 1977; Timms, 1980b), and the relationship below has been observed for triglycerides with carbon numbers 40, 42 and 44:

$$R = 14.197C_{40} - 36.396C_{42} + 32.364C_{44}$$

Here C_{40}, C_{42} and C_{44} are the weight percentages of the triglycerides with these carbon numbers. The term R, calculated using experimental data, has a standard deviation of 0.71. A knowledge of the triglyceride carbon numbers and fatty acid composition permits the analyst to detect adulteration of butterfat (Kuksis and McKarthy, 1964; Timms, 1980a). Statistically derived formulae based on the carbon-number analysis of 755 different milk fats allow the detection and quantitation of adulteration by many foreign fats (Precht, 1992). In this latter study, the data for 755 samples of milk fat have been collated (Table 3.184).

Table 3.176 World butter production, including butter equivalent of butter oil (1000 tonnes) (International Dairy Federation)

	January–December		Forecast	Change (%)	
	1989	1990	1991	1990/89	1991/90
Belgium	59.2	55.1	na [a]	−6.9	na
Denmark	92.3	93.3	82.0	+1.1	−12.1
France	518.0	547.0	518.0	+5.6	−5.3
Germany	667.0	644.0	580.0	−3.4	−9.9
of which W. Germany	398.0	393.0	na	−1.3	na
Irish Republic	144.7	152.3	143.0	+5.3	−6.1
Italy	84.0	83.0	80.0	−1.2	−3.6
Netherlands	213.0	214.0	215.0	+0.5	+0.5
United Kingdom	131.0	138.0	120.0	+5.3	−13.0
Portugal	12.0	15.0	na	+25.0	na
Spain	30.6	46.0	46.0	+50.3	nc [a]
Total EC '12' excl. E. Germany	1715.0	1745.0	1715.0	+1.7	−1.7
Austria	34.0	35.0	33.0	+2.9	−5.7
Finland	55.4	55.8	47.8	+0.7	−14.3
Norway	27.8	29.1	29.0	+4.7	−0.3
Sweden	68.0	74.0	64.0	+8.8	−13.5
Switzerland	38.5	37.0	na	−3.9	na
Iceland	1.4	1.7	1.6	+21.4	−5.9
Malta	0.0	0.0	0.0	nc	nc
Total other Western Europe	225.1	232.6	na	+3.3	na
Canada	97.4	100.0	94.0	+2.7	−6.0
USA	587.6	590.7	na	+0.5	na
Total North America	685.0	690.7	na	+0.8	na
Australia	101.4	104.0	107.0	+2.6	+2.9
New Zealand	264.4	189.7	185.0	−28.3	−2.5
Japan	81.0	75.0	na	−7.4	na
Total Pacific	446.8	368.7	na	−17.5	na
Bulgaria	22.0	23.5	16.0	+6.8	−31.9
Former Czechoslovakia	156.0	159.0	na	+1.9	na
Hungary	34.2	39.0	35.0	+14.0	−10.3
Former USSR	1737.0	1728.5	na	−0.5	na
Other Eastern Europe	374.5	380.9	na	+1.7	na
Total Eastern Europe	2323.7	2330.9	na	+0.3	na
Argentina	38.9	39.9	42.0	+2.6	+5.3
Brazil	58.5	59.8	61.1	+2.2	+2.2
China	2.2	2.5	3.0	+16.3	+20.0
India	210.0	225.0	na	+7.1	na
Philippines	0.0	0.0	0.0	nc	nc
South Africa	15.8	20.9	17.3	+32.3	−17.0
Zimbabwe	2.0	1.6	1.7	−20.0	+6.3
Other countries	896.3	927.2	na	+3.4	na
Total world	6888.2	6895.7	na	+0.1	na

[a] na, Not available; nc, no change.

Table 3.177 Characteristics of butterfat (typical values)

Relative density, d_{20}^{60} [a]	0.887
Refractive index	1.4465
Saponification value	210–250
Iodine value	37 (28–43)
Unsaponifiable matter (%)	0.4
Reichert–Neissl value	22–34
Polenske value	2–4
Kirschner value	20–26
Per cent *trans* as elaidate	4–8

[a] At 60 °C relative to water at 20 °C.

Table 3.178 Composition of the milk fat globule membrane and whole milk lipids

Lipid component	Membrane lipids (wt%)[a]	Whole milk lipids (wt%)[b]
Carotenoids	0.45	} tr
Squalene	0.61	
Cholesterol esters	0.79	tr
Triglycerides	53.4	97–98
Free fatty acids	6.3[c]	0.10–0.44
Cholesterol	5.2	0.22–0.41
Diglycerides	8.1	0.28–0.59
Monoglycerides	4.7	0.016–0.038
Phospholipids	20.4	0.2–1.0

[a] Dimick *et al.* (1970).
[b] Dimick *et al.* (1970), Morrison (1970).
[c] Contained some triglycerides.

Table 3.179 Fatty acid composition (wt%) of milk fats – variation with season (Jensen *et al.*, 1962)

Fatty acid[a]	June[b]	December[b]	Average[c]
4:0	4.22	3.51	3.57
6:0	2.53	2.24	2.22
8:0	2.34	1.07	1.17
9:0	0.05	0.05	0.03
10:0	2.24	2.57	2.76
10:1	0.32	–	–
11:0	0.34	0.29	0.33
12:0	2.40	2.77	2.81
13:0 (12:1)	0.29	0.29	0.33
14 (br)	0.23	0.14	0.17
14:0	9.01	10.58	10.06
14:1 (15br)	1.54	1.61	1.63
15:0	1.29	1.11	1.09
16 (br)	0.42	0.39	0.38
16:0	22.05	25.98	24.97
16:1 (17br)	2.29	2.98	2.55
17:0	0.69	1.08	0.91
18:0 (br)	0.31	0.40	0.38
18:0	14.27	11.58	12.07
18:1	30.41	24.75	27.09
18:1 (br)	0.24	1.56	1.26
18:2	1.23	2.75	2.39
18:3 (20:0)	2.61	2.30	2.06

[a] Not including isomers – identification by relative retention time; br = branched.
[b] Average of four to eight samples, June and December.
[c] Average of 108 samples, June–June.

Table 3.180 Isomers of monoenoic acids (wt%) in butterfat (Hay and Morrison, 1970)

Position of double bond	*cis* isomer				*trans* isomer	
	14:1	16:1	17:1	18:1	16:1	18:1
5	1.0	tr	–	–	2.2	–
6	0.8	1.3	3.4	–	7.8	1.0
7	1.0	5.6	2.1	–	6.7	0.8
8	0.6	tr	20.1	1.7	5.0	3.2
9	96.6	88.7	71.5	95.8	32.6	10.3
10	–	tr	tr	tr	1.7	10.6
11	–	2.6	2.9	2.5	10.6	36.1
12	–	tr	tr	–	12.9	4.1
13	–	–	–	–	10.6	10.6
14	–	–	–	–	–	9.0
15	–	–	–	–	–	6.8
16	–	–	–	–	–	7.5

Table 3.181 Composition of higher butter oil fatty acids [a] (per cent of total fatty acid methyl esters) (Iverson *et al.*, 1965)

Saturated		Branched		Monoenes		Dienes		Polyenes		Multibranched	
Acid [b]	(wt%)	Acid [b]	(wt%)	Acid [b]	(wt%)	Acid [b]	(wt%)	Acid [b]	(wt%)	Acid [b]	(wt%)
8:0	0.69	12:0i	0.01	10:1	0.48	14:2	0.04	18:3	1.03	16:0br3	0.01
10:0	1.88	13:0i	trace	12:1	0.05	16:2	0.02	18:4	0.10	17:0br3	0.01
11:0	0.12	14:0i	0.03	13:1	0.003	18:2	2.22	20:3	0.05	18:0br3	0.16
12:0	2.96	14:0a	0.23	14:1	0.75	20:2	0.12	20:4	0.07	19:0br4	0.02
13:0	0.10	15:0i	0.14	15:1	0.02	22:2	0.14	20:5	0.02	20:0br4	0.14
14:0	11.2	16:0i	0.2	16:1	1.84	24:2	0.02	22:3	0.03	21:0br4	0.02
15:0	1.52	17:0i	0.36	17:1	0.2	26:2	0.0004	22:4	0.04	22:0br4	0.02
16:0	27.8	17:0a	0.23	18:1	30.3			22:5	0.02	23:0br4	0.01
17:0	0.71	18:0i	0.02	19:1	0.14					24:0br4	0.10
18:0	12.1	19:0br	0.01	20:1	0.52					25:0br4	0.10
19:0	0.05	20:0br	0.01	21:1	0.01					26:0br3	0.01
20:0	0.02	21:0br	0.01	22:1	0.02					26:0br4	0.04
21:0	0.06	22:0br	0.02	23:1	0.05					27:0br4	0.04
22:0	0.04	23:0br	0.01	24:1	0.0008					28:0br3	0.02
23:0	0.01	24:0br	0.02	25:1	0.0008					28:0br4	0.12
24:0	0.02	25:0br	0.004	26:1	0.0008					28:0br5	0.01
25:0	0.02	26:0br	0.004								
26:0	0.02										
27:0	0.0004										
28:0	0.0004										

[a] Acids with less than eight C atoms are not included.
[b] First number indicates number of carbon atoms; second number the number of double bonds; i = iso; a = anteiso; br = iso and/or anteiso; number after br indicates number of methyl branches for multibranched acids.

Table 3.182 Positional distribution of fatty acids in triglycerides of mammalian milk fats

Species	Position	Fatty acid composition (mol%)																
		4:0	6:0	8:0	10:0	12:0	14:0	14:1	15:0	15:1	16:0	16:1	17:0	17:1	18:0	18:1	18:2	18:3	Other
Cow [a,c]	TG	8.8	5.0	2.5	5.3	5.2	13.8	1.7	1.0	0.2	28.0	2.3	0.6	0.2	8.3	14.1	1.1	0.9	1.0
	sn-1	–	–	0.2	1.4	3.4	12.7	1.4	1.5	0.3	41.2	2.7	1.0	0.3	15.2	16.2	1.2	0.7	0.6
	sn-2	–	0.1	2.9	7.8	8.1	23.9	2.4	1.8	0.4	36.7	3.3	0.5	0.2	3.0	7.2	0.6	0.3	0.8
	sn-3	26.4	14.8	4.4	6.7	4.0	4.9	1.2	0.3	–	6.1	1.0	0.2	0.2	6.7	19.0	1.6	1.7	1.4
Goat [b,d]	TG	5.1	4.4	2.6	7.8	3.8	3.6	0.2	2.0	–	26.0	1.8	0.8	–	9.9	20.6	2.7	–	2.7
	sn-1	–	–	1.7	3.5	4.2	8.5	–	1.8	–	43.4	3.1	1.3	–	14.9	15.9	0.4	–	1.3
	sn-2	–	–	1.3	7.4	4.7	20.4	0.8	2.8	–	33.4	2.0	0.5	–	6.5	15.7	2.5	–	2.0
	sn-3	15.0	12.1	5.1	11.9	1.2	2.5	1.0	1.4	–	3.4	1.5	1.7	–	6.9	28.6	4.2	–	3.5
Pig [c]	TG	–	–	–	–	0.1	2.4	nd	0.2	nd	26.9	4.0	0.6	nd	6.8	30.3	27.3	1.4	–
	sn-1	–	–	–	–	0.1	1.8	nd	0.3	nd	26.4	3.5	0.8	nd	11.2	32.7	21.8	1.4	–
	sn-2	–	–	–	–	0.2	5.1	nd	0.5	nd	48.1	5.1	0.5	nd	1.8	15.5	21.9	1.2	–
	sn-3	–	–	–	–	0.1	0.4	nd	0.2	nd	6.2	3.4	0.5	nd	7.5	42.8	38.0	1.6	–
Sheep [b,d]	TG	4.4	3.6	2.4	5.5	3.5	9.8	0.6	2.8	–	21.2	1.7	1.0	–	14.0	21.8	4.4	2.6	0.7
	sn-1	–	–	0.7	1.5	1.6	8.6	0.2	2.7	–	38.2	2.1	1.6	–	18.9	18.6	2.9	2.4	tr
	sn-2	–	–	2.0	5.5	4.9	17.9	0.9	4.0	–	23.6	1.6	0.9	–	12.6	19.6	4.3	1.8	0.4
	sn-3	11.8	11.2	4.6	10.5	3.5	5.0	1.1	1.5	–	2.6	1.3	0.8	–	8.7	26.3	5.9	4.3	0.9
Human [e]	TG	–	–	–	0.5	3.0	5.8	0.7	0.7	–	26.6	5.3	0.8	–	6.7	36.4	11.1	0.9	1.5
	sn-1	–	–	–	0.2	1.3	3.2	0.3	0.4	–	16.1	3.6	0.9	–	15.0	46.1	11.0	0.4	1.5
	sn-2	–	–	–	0.2	2.1	7.3	0.2	1.0	–	58.2	4.7	0.8	–	3.3	12.7	7.3	0.6	1.6
	sn-3	–	–	–	1.1	5.6	6.9	1.6	0.7	–	5.5	7.6	0.7	–	1.8	50.4	15.0	1.8	1.4

[a] The 14:1, 15:1, 16:1 and 17:1 acids include 15:0br, 16:0br, 17:0br and 18:0br, respectively; 18:3 includes 18:2*c*, *t* conj.
[b] Calculated assuming long- and short-chain glyceride fractions present in equal amounts.
[c] Parodi (1982).
[d] Marai *et al.* (1969).
[e] Breckenbridge *et al.* (1969).

Table 3.183 Variation in the triglyceride analyses (wt%) of 76 samples of Australian milk fat determined by triglyceride GLC (Timms, 1980a)

Carbon no.	Minimum	Maximum	Mean	Standard deviation
26 [a]	0.39	0.76	0.58	0.10
28	0.26	0.74	0.54	0.15
30	0.56	1.35	1.03	0.25
32	1.44	2.94	2.33	0.46
34	4.15	6.29	5.45	0.58
36	9.37	11.28	10.52	0.54
38	11.88	14.36	13.13	0.60
40	8.35	12.04	10.52	1.07
42	4.95	7.27	6.31	0.66
44	4.05	6.44	5.57	0.54
46	4.88	7.19	6.38	0.58
48	6.80	10.63	8.44	1.06
50	10.13	15.08	11.74	1.53
52	8.88	15.34	11.29	1.69
54	4.72	8.07	5.96	0.97
56	0.10	0.41	0.22	0.08

[a] Including cholesterol.

Table 3.184 Variation in the triacylglycerol composition of 755 milk fat samples from the winter and summer feeding period (Precht, 1992)

Acyl group carbons	Mean value (wt%)	Standard deviation	Range	
			Min.	Max.
24	0.04	0.02	0.00	0.11
26	0.22	0.04	0.14	0.34
28	0.57	0.08	0.40	0.82
30	1.13	0.19	0.76	1.63
32	2.56	0.42	1.58	3.76
34	5.95	0.84	3.86	7.90
36	10.76	1.07	8.21	13.81
38	12.46	0.45	10.00	13.99
40	9.87	0.33	8.60	10.82
42	6.87	0.64	5.14	8.65
44	6.47	0.84	4.17	8.89
46	7.32	0.72	5.24	9.71
48	9.12	0.39	7.91	10.65
50	11.33	1.05	8.18	13.36
52	10.03	2.14	5.13	15.49
54	4.99	1.74	1.54	11.09

There is increasing commercial interest in the fractionation of butterfat to obtain a wider range of applications, e.g. soft 'butters' from the olein. The characteristics of the three major glyceride fractions obtained by solvent fractionation have been characterized (Timms, 1980b).

The influence of dietary fat on milk fat composition is complex (Storry, 1981) and results vary considerably (Banks *et al.*, 1976; Bitman *et al.*, 1973; Storry, 1981). Net transfer of acids from the diet into milk are reported for the following acids: <12:0, low transfer; 12:0, 27%; 16:0, 29–90%; 18 (total), 19–65%; 20–22 (total), low transfer. Some results are given in Table 3.185. The C_{18} polyunsaturated acids in the diet are hydrogenated in the rumen prior to adsorption.

Soft butters have been produced by feeding encapsulated polyunsaturated oils in order to overcome the problem of biohydrogenation (Barbano and Sherbon, 1980; Kuksis and McKarthy, 1964; Scott *et al.*, 1972a, b; Storry *et al.*, 1974, 1980). Using encapsulation techniques the 18:2 content of butterfat can be increased from 2–3% to 35%, but problems of oxidative stability arise (Edmunson *et al.*, 1974). The effect of feeding protected lipids on the composition of the milk fat globule membrane has been reported (Smith *et al.*, 1977) (Table 3.186).

The flavour components in butter and butterfat are complex and include γ- and δ-lactones (Boldingh and Taylor, 1962), methyl ketones (Forss, 1972; Langeer and Day, 1964), aldehydes (Badings, 1970) short-chain fatty acids (Forss, 1972; Langeer and Day, 1964), sulphides (Patton *et al.*, 1956; Day *et al.*, 1957; Badings *et al.*, 1975), alcohols (Forss, 1972; Langeer and Day, 1964), proteins, sugars and secondary reaction products (Teranishi *et al.*, 1981).

The flavour threshold of many of these compounds is at the parts per million (ppm) level and below, and depends on how they are distributed in the oil and water phases. The flavour thresholds for some fatty acids are given in Table 3.187 (Forss, 1972); for ketones, see Langeer and Day (1964).

The γ- and δ-hydroxy fatty acids have a special significance in relation to butter flavour. These acids are esterified almost solely at the 1 position of the triglycerides (Kinsella *et al.*, 1967) and are relatively unstable to heat (Boldingh and Taylor, 1962; Patton, 1965; Dimick *et al.*, 1966). The free hydroxy acids generated quickly form lactones, which contribute to butter flavour. Milk contains only a trace of lactones, but commercial samples of butter contain up to 30 ppm and on heating this increases to 93 ppm (Ellis and Wong, 1975), suggesting that most of the available lactone is present initially as bound or unbound hydroxy acid. The enhanced level of lactones is largely responsible for the characteristic flavours of the heated butter. The lactones of milk and butterfat have a low, positive, optical rotation (Van der Ven and de Jong, 1970), as does milk fat itself. This latter effect has been related to the occurrence of asymmetric glycerides (Anderson *et al.*, 1970).

Other minor components in butterfat include

Table 3.185 Effects of increasing amounts of dietary fats and oils on the composition of the major fatty acids in milk (g/100 g) (from Banks *et al.*, 1976)

Fatty acid	Red palm oil (%)				Tallow (%)					Coconut oil (%)					Soya oil (%)			
	0	2	4	8	0	2	4	7	10	0	2	4	7	10	0	5	7	10
4:0	3.3	3.3	3.3	3.8	3.1	2.9	3.4	3.5	3.9	3.3	4.0	3.8	3.3	3.3				
6:0	2.5	2.5	2.4	2.4	2.9	2.7	2.6	2.5	2.1	2.2	2.1	2.1	1.9	2.1				
8:0	1.7	1.7	1.6	1.5	1.8	1.8	1.6	1.4	1.1	1.3	1.3	1.3	1.0	1.2				
10:0	4.2	4.0	3.1	2.9	4.7	4.5	3.7	2.9	2.1	3.9	3.7	3.6	3.0	3.2	26.4	19.2	16.7	14.6
12:0	5.8	5.5	4.4	3.5	5.6	5.2	4.1	3.3	2.2	5.9	7.1	9.0	11.9	13.2				
14:0	15.3	14.4	12.5	11.1	15.3	14.8	13.0	11.8	9.3	15.4	16.2	17.1	17.9	19.3				
14:1	1.9	2.0	1.6	1.3	2.7	2.4	2.2	2.0	1.4	2.2	1.9	2.1	2.4	2.4				
16:0	39.4	36.3	35.2	36.8	36.5	32.6	31.8	30.1	29.5	41.7	41.6	38.7	34.8	31.2	43.4	30.0	27.7	26.2
16:1	2.7	3.0	3.0	2.8	3.3	2.8	2.9	3.1	3.4	3.3	3.2	3.0	3.3	2.9				
18:0	3.5	5.0	7.0	6.6	4.4	6.4	7.5	9.0	10.1	2.7	2.8	2.7	2.9	2.9	7.7	15.1	16.0	18.8
18:1	14.4	17.7	20.7	23.6	14.9	16.5	19.7	23.6	28.1	12.6	11.6	12.2	12.4	13.8	19.2	31.2	35.1	35.7
18:2	1.1	0.9	0.9	0.9	1.5	1.9	1.9	2.0	1.4	–	0.5	0.6	0.5	0.4	3.3	4.5	4.5	4.7

Table 3.186 Effects of protected fats and oils on the composition of the major fatty acids in milk (g/100 g) (Storry *et al.*, 1974, 1980; Wood *et al.*, 1975)

Fatty acid	Coconut oil		Tallow		Sunflower–soya oil	
	Unprotected	Protected	Unprotected	Protected	Unprotected	Protected
4:0	2.6	2.4	3.5	2.8	4.2	3.9
6:0	2.3	2.0	2.5	1.2	2.9	2.0
8:0	1.5	1.2	1.6	0.6	1.5	0.8
10:0	4.0	3.5	4.1	1.1	3.7	1.4
12:0	5.5	16.6	4.6	1.2	4.4	1.3
14:0	14.3	18.3	12.2	5.2	12.3	4.0
14:1	3.2	2.9	1.5	1.0	3.1	0.5
16:0	41.2	30.7	29.8	28.6	31.1	10.7
16:1	3.9	2.7	2.3	2.8	3.5	0.3
18:0	3.3	4.1	8.2	13.3	5.3	13.3
18:1	11.8	11.5	20.4	33.9	19.5	30.2
18:2	0.7	0.6	2.1	2.1	1.8	28.0
18:3	–	–	0.2	0.3	1.2	1.6

Table 3.187 Flavour threshold of short-chain fatty acids (ppm)

Acid	Water	Vegetable oil
Butyric (4:0)	6.8	0.6
Caproic (6:0)	5.4	2.5
Caprylic (8:0)	5.8	350
Capric (10:0)	3.5	200

Table 3.188 Phospholipid composition of bovine milk [a]

Phopholipid	Content (mol%)
Phosphatidylcholine	34.5
Phosphatidylethanolamine	31.8
Phosphatidylserine	3.1
Phosphatidylinositol	4.7
Sphingomyelin	25.2
Lysophosphatidylcholine	tr
Lysophosphatidylethanolamine	tr
Total choline phospholipids	59.7
Plasmalogens	3
Diphosphatidylglycerol	tr
Ceramides	tr
Cerebrosides	tr

[a] Adapted from Morrison (1970).

phospholipids (Table 3.188) (Boatman *et al.*, 1969; Jensen, 1973; Kayser and Patton, 1970; Nutter and Privett, 1967). Cholesterol is the major sterol (90%), present partly (10%) as cholesterol ester (Brewington *et al.*, 1970; Keenan and Patton, 1970). The remaining sterols include lanosterol and traces of dihydrolanosterol and β-sitosterol.

Vitamin A and carotene are important nutritional factors. Their concentration depends in part on the dietary intake and is normally in the range 6–12 μg/g for vitamin A and 2–10 μg/g for carotene, being lowest in winter and highest in summer (Baxter and Robson, 1942; Hickman, 1943; Jenness and Palmer, 1945). The level is normally determined by the amount of carotene in the feed but is further increased by incorporating vitamin A in the feed. Vitamin D is also present in butterfat in small amounts.

Apart from cow milk, a wide variety of mammalian milks are consumed. There is no comprehensive review of this area. The composition and properties of goat milk have been described and some typical data are given in Tables 3.189 and 3.190 (Juarez and Ramos, 1986). Information on camel milk is given in Tables 3.191 and 3.192 (Morton, 1984). See also Table 3.200 for sow's milk.

Table 3.189 Composition (%) of goat milk (Juarez and Ramos, 1986)

Country [a]	Total solids	Fat	Protein	Casein	Lactose	Ash
Australia	11.65	3.40	3.22	–	–	–
Bulgaria	12.85	4.04	3.62	2.77	4.42	0.78
Egypt	13.98–15.83	5.22–6.79	–	–	–	–
England	–	2.75–6.43	–	2.14–3.18	3.91–4.81	–
France	–	3.38	3.09	2.33	–	–
Greece	11.13–14.80	3.00–5.63	3.14–3.77	2.41–3.05	4.33–4.76	0.73–0.80
	13.65	4.90	3.67	–	4.41	0.85
India	13.15–14.25	4.09–4.97	–	–	–	–
	13.26	4.70	3.78	3.04	3.95	0.81
Israel	10.9–13.5	3.1–4.9	2.4–36	–	–	–
Italy	12.43	2.99	3.33	–	–	0.95
	13.34–16.25	4.5–5.4	3.45–5.02	2.52–3.97	–	–
	11.05–11.79	3.19–3.50	2.78–3.08	2.07–2.33	–	0.77–0.83
Malawi	16.3	6.7	2.2	–	6.3	1.1
Nigeria	15.7	4.6	4.7	–	5.3	–
Saudi Arabia	9.81–13.00	2.29–3.90	2.65–4.38	–	3.71–4.55	0.63–0.87
Spain	13.19–15.94	4.44–6.94	3.45–3.73	2.67–2.89	–	0.81–0.90
Switzerland	11.27	3.30	2.96	–	4.46	–
Turkey	11.33–13.39	3.45–4.63	2.89–3.48	–	4.01–4.52	0.82–0.85

[a] Values correspond to different breeds, different lactation stages, different conditions of hydration, different regions and breeds, and a.m. and p.m. milking.

Table 3.190 Fatty acid composition (%) of goat milk reported by different authors[a] and compared with those of cow milk (Juarez and Ramos, 1986) and camel milk fat (Hilditch and Williams, 1964)

Acid	Goat					Cow	Camel (India)
	1979	1979	1981	1984	1986	1979	1933
4:0	3.3–4.8	1.0–4.9	1.81	3.0	1.8–2.8	2.5–6.2	2
6:0	1.7–3.0	1.5–4.3	2.03	2.0	2.2–3.4	1.5–3.8	0.9
8:0	1.5–3.6	2.0–5.2	2.68	2.0	2.4–3.9	1.0–1.9	0.6
10:0	6.4–11.1	7.1–16.2	8.45	6.1	8.8–13.4	2.1–4.0	1.4
12:0	2.5–5.0	3.3–9.8	5.21	2.9	3.8–5.5	2.3–4.7	4.6
14:0	8.5–11.2	6.9–15.4	10.52	9.5	8.5–11.6	8.5–12.8	7.3
14:1	0.02–0.2	0.1–1.1	0.96	0.4	0.5–0.8	0.6–1.5	–
16:0	25.1–38.4	16.7–39.4	24.33	28.6	23.3–32.1	24.0–33.3	29.3
16:1	0.7–1.7	0.7–3.5	2.58	2.5	1.0–2.0	1.3–2.8	–
18:0	5.9–14.9	4.4–17.3	9.49	10.3	4.3–11.2	6.2–13.6	11.1
18:1	15.6–28.2	13.5–33.3	23.96	26.3	16.2–26.6	19.7–31.2	38.8
18:2	1.8–4.0	0.5–4.7	1.68	2.6	1.2–2.5	1.3–5.2	3.8

[a] For references, see original publication.

Table 3.191 Camel milk – yields from various countries of *C. dromedarius* [a] (Morton, 1984)

Country	Average body weight (kg)	Daily yield range (kg)	Lactation yield range (kg)	Lactation length range (month)	Daily yields as percentage of body weight range	Calculated 305 days lactation (kg)
Egypt	–	3.5–4.5	1600–2000	–	–	1078–1373
Libya	380–400	8.3–10	2700–4000	9–16	2.2–2.6	2532–3050
Somalia	350	5	1950	13	1.4	1525
Ethiopia	360	5–13	1872–2592	12–18	1.4–3.6	1525–3965
Pakistan						
heavy	450	15–35	5475–12 775	12	3.3–7.8	4575–10 675
desert	350	8–10	2920–3650	12	2.3–2.9	2440–3050
China, dromedary	–	7.5	3300	16–17	–	2288

[a] Most reports are based on less than 10 complete records.

Table 3.192 Average composition (%) of milk from different livestock species (Morton, 1984)

Species and source [a]	Water	Total solids	Fat	Protein	Lactose	Ash
Camel (one-humped)						
Knoess	85.6	14.3	5.5	4.5	3.4	0.9
Camel (one-humped)						
Dahl and Hjort	87.0	13.0	3.3	3.9	5.2	0.7
Camel (one-humped)						
Mukasa-Mugerwa (pooled average)	86.6	13.4	4.3	4.0	4.2	0.8
Cattle						
Ensminger	86.2	13.8	4.4	3.8	4.9	0.7
Goat						
Ensminger	87.1	12.9	4.1	3.7	4.2	0.8
Sheep						
Ensminger	82.0	18.0	6.4	5.6	4.7	0.9

[a] See original publication for sources.

3.4.2 Animal fats, lard and animal tallows

Some relevant data are summarized in Tables 3.193 to 3.197.

Animal fat is obtained commercially from animal tissues by rendering. In wet rendering, the tissues are heated with water, generally under pressure. Cooling ruptures the cells and allows the fat to rise to the surface, where it can be drained off. High-grade edible fats such as beef oleo stock are produced by this method. This process has been largely superseded by dry rendering, in which the tissues are heated by steam in the absence of added water. Liquid fat and protein solids are removed after cooling and separated by centrifugation or pressing. There are many variants on these processes, including solvent extraction instead of mechanical separations (Bates, 1968). The wet process is generally milder than the steam process and yields a milder-flavoured product (*INFORM*, 1992).

Overall production of animal tallows is predicted to decrease as the world consumption of meat products declines together with the drive for lower fat in meat. World production of beef tallow and grease is given in Table 3.194.

Lard produced by wet rendering is known as 'prime steam lard'. Neats foot oil is a low-melting inedible fat rendered from the feet of cattle. The quality of the rendered fat depends on the raw materials, but, under well controlled conditions, fat with a free fatty acid content of 0.3% is obtained and no subsequent refining is required for most applications.

Antioxidants are normally added to the rendered fats to prevent the development of oxidative

Table 3.193 World production (1000 tonnes) of lard and animal tallow

	1970/71	1975/76	1980/81	1981/82	1982/83	1989/90
Lard	4365	3895	4815	4720	4720	5344
Animal tallow	5270	5695	6365	6225	6210	6559

Table 3.194 World production of tallow and grease (1000 Mt) (*INFORM*, 1992)

Year	USA	Australia	Canada	New Zealand	Others	Total
1982	3237	340	215	138	2344	6274
1983	3351	298	263	138	2343	6393
1984	3309	315	213	129	2460	6426
1985	3279	334	206	144	2568	6531
1986	3080	333	201	137	2620	6371
1987	3111	360	198	145	2660	6474
1988	3250	362	210	149	2627	6598
1989	3182	401	207	159	2619	6568
1990	3145	450	205	123	2747	6670

Table 3.195 Selected fatty acid compositions (wt%, ignoring minor acids) of animal fats

	12:0	14:0	16:0	18:0	20:0	14:1	16:1	18:1	18:2	18:3
Lard	tr	2.0	27.1	11.0	tr	tr	4.0	44.4	11.4	tr
Beef	tr	2.5	27.0	7.4	tr	3.0	10.8	47.5	1.7	1.1
Mutton	0.6	5.6	27.0	31.7	tr	–	1.6	31.7	1.6	0.2
Goat	3.3	2.1	25.5	28.1	2.4	–	tr	38.4	tr	tr
Red deer [a]	0.3	5.7	30.9	35.7	–	0.4	2.0	15.4	1.1	–
Reindeer	–	1.2	23.0	30.0	–	–	1.6	43.0	0.8	0.4
Horse	0.1	2.0	26.7	6.2	–	–	2.8	32.8	12.8	16.6
Kangaroo [a]	tr	3.5	22.0	15.0	–	–	–	45.0	3.0	7.0

[a] Contains significant levels of 15:0 and 17:0 fatty acids (Garton and Duncan, 1971; Payne, 1971); data not normalized to 100%.

Table 3.196 Range of fatty acid composition [a] (wt%) of lard and two animal tallows

	Lard (Spencer *et al.*, 1976)	Beef tallow (Spencer *et al.*, 1976)	Mutton tallow (Screenivasan, 1968)
12:0	0.5	0.1	1.0
14:0	0.5–2.5	1.4–6.3	5.5
14:1	0.2	0.5–1.5	–
15:0	0.1	0.5–1.0	0.8
15:0iso	0.1	1.5	0.1
16:0	20–32	20–37	25.8
16:0iso	0.1	0.5	–
16:1	1.7–5.0	0.7–8.8	1.5
16:2	–	1.0	0.6
17:0	0.5	0.5–2.0	1.4
17:1	0.5	1.0	0.2
18:0	5.0–24	6–40	30.5
18:1	35–62	26–50	30.0
18:2	3–16	0.5–5.0	1.4
18:3	1.5	2.5	0.2
20:0	1.0	0.5	0.2
20:1	1.0	0.5	0.1
20:2	1.0	–	–
20:4	1.0	0.5	–
22:0	0.0	–	–

[a] Not fully comprehensive, e.g. *trans* acids not determined – see later.

Table 3.197 Typical characteristics of lard and beef tallow (*Oil and Soap*, 1943)

	Lard	Beef tallow
Relative density, $d^{99}_{15.5}$	0.858–0.864	0.860–0.870
Refractive index, 40 °C (Zeiss)	45–52	46–49
Saponification value	195–202	193–202
Iodine value	46–70	35–48
Unsaponifiable (%)	1.0	0.8
Titre	36–42	40–46

rancidity. The effect of tocopherols (Parkhurst *et al.*, 1968), BHA (butylated hydroxyanisole), BHT (butylated hydroxytoluene) and metal chelators on oxidative stability has been described.

(a) Lard

The major producers of lard and tallow are the USA, the former USSR and China. The composition of lard varies widely and depends on feed and origin, including animal type and location of fat on the animal. Fats from various sites on the pig have been characterized (von Brixius and Treiber, 1974; Christie and Moore, 1970; Chacko and Perkins, 1965; Coleman, 1961; Warmbier, 1974) (Table 3.198). This table lists the major fatty acids, but many more have been identified (Table 3.199). Animal fats are noted for the presence of odd-numbered, branched-chain and *trans* fatty acids (Iverson *et al.*, 1965; Magidman *et al.*, 1963).

The positional distribution of the major acids in fats from different tissues of the pig are given in Table 3.200 (Christie and Moore, 1970). The adipose tissue fats show similar fatty acid distributions, with palmitic and myristic in the 2 position and the remaining saturated acids in the 1 position.

The triglyceride composition of commercial lard is given in Table 3.201. Results for lard from different anatomical positions are given in Table 3.202 (Chacko and Perkins, 1965).

Cholesterol is the major sterol in lard (3000–4000 mg/kg). Tocopherols are present at low levels in animal fats (Bunnell, 1957). α-Tocopherol

Table 3.198 Fatty acid composition (%) of fat from different pig tissues (Chacko and Perkins, 1965)

	Anatomical position							
	Back fat layer				Perinephric	Back top	Back bottom	Flank
	Rump 1 cm depth		Shoulder					
	Bottom	Top	Upper 1 cm	Lower 1 cm				
14:0	1.4	2.1	1.7	1.9	1.7	2.0	2.2	2.0
16:0	31.6	29.2	24.4	25.9	30.6	27.1	29.2	24.3
16:1	2.9	3.3	2.8	2.8	2.9	4.0	3.5	5.2
18:0	13.4	18.2	11.9	15.0	11.3	11.0	12.7	16.9
18:1	38.8	31.2	46.5	39.8	41.8	44.4	41.5	38.0
18:2	11.9	16.0	10.5	14.5	11.2	11.4	11.0	13.6
18:3	–	–	1.2	–	–	–	–	–

Table 3.199 Fatty acids of lard [a] (wt%) determined by GLC of urea fractions (Iverson *et al.*, 1965)

Saturated		Branched [b]		Monoenes		Dienes		Polyenes		Saturated multibranched [b]	
Acid	(wt%)	Acid [c]	(wt%)	Acid	(wt%)	Acid	(wt%)	Acid	(wt%)	Acid [c]	(wt%)
12:0	0.08	23:0br	0.008	12:1	0.02	18:2	11:2	18:3	1.30	20:0br4	0.02
13:0	0.01	24:0br	0.004	14:1	0.03	20:2	0.54	20:3	0.08	22:0br4	0.02
14:0	1.27	25:0br	0.02	15:1	0.01	22:2	0.02	20:4	0.13	24:0br4	0.03
15:0	0.02	26:0br	0.02	16:1	2.82			20:5	0.04	26:0br4	0.03
16:0	24.0			17:1	0.22			22:3	0.02	28:0br3	0.09
17:0	0.29			18:1	43.5			22:4	0.05	28:0br4	0.20
18:0	13.0			19:1	0.06			22:5	0.02		
19:0	0.03			20:1	0.34						
20:0	0.34			21:1	0.004						
21:0	0.004			22:1	0.02						
22:0	0.04			24:1	0.0001						
23:0	0.02										
24:0	0.02										

[a] Fatty acids less than C_{12} not determined.
[b] Tentatively identified in appropriate urea fractions by semi-logarithmic plots of GLC retention times.
[c] br = branched (iso or anteiso position not determined); number after br indicates number of methyl branches.

Table 3.200 Positional distribution of fatty acids in triglycerides isolated from pig tissues [a] (Christie and Moore, 1970)

	Position	14:0	16:0	16:1	18:0	18:1	18:2	18:3	20:1	20:3 ($n-9$)	20:3 ($n-6$)	20:4	20:5	22:5	22:6
Milk	1	2.4	21.8	6.6	6.9	49.6	11.3	1.4	–	–	–	–	–	–	–
	2	6.8	57.6	11.2	1.1	13.9	8.4	1.0	–	–	–	–	–	–	–
	3	3.7	15.4	10.4	5.5	51.7	11.5	1.8	–	–	–	–	–	–	–
Inner back fat	1	0.7	9.8	1.7	38.8	42.7	6.3	–	–	–	–	–	–	–	–
	2	3.5	72.1	3.7	3.8	14.0	2.9	–	–	–	–	–	–	–	–
	3	0.6	5.4	2.1	11.3	65.4	15.2	–	–	–	–	–	–	–	–
Outer back fat	1	0.9	9.5	2.4	29.5	51.3	6.4	–	–	–	–	–	–	–	–
	2	4.1	72.3	4.8	2.1	13.4	3.3	–	–	–	–	–	–	–	–
	3	−0.2	0.4	1.5	7.4	72.7	18.2	–	–	–	–	–	–	–	–
Perinephric fat	1	2.5	11.7	2.3	35.2	43.7	4.6	–	–	–	–	–	–	–	–
	2	4.8	75.4	4.1	2.8	10.2	2.7	–	–	–	–	–	–	–	–
	3	−1.6	5.0	2.3	12.1	68.5	13.7	–	–	–	–	–	–	–	–
Adipose tissue adhering to stomach	1	1.5	15.1	2.4	37.9	38.8	4.3	–	–	–	–	–	–	–	–
	2	4.8	76.1	4.1	4.0	9.9	1.1	–	–	–	–	–	–	–	–
	3	0.6	12.0	1.6	19.0	58.4	8.4	–	–	–	–	–	–	–	–
Mesenteric fat	1	1.8	15.6	2.6	37.6	39.4	3.0	–	–	–	–	–	–	–	–
	2	4.5	78.4	3.9	3.0	8.9	1.3	–	–	–	–	–	–	–	–
	3	−0.3	2.3	1.6	19.1	66.6	10.7	–	–	–	–	–	–	–	–
Heart	1	1.5	12.7	3.1	26.9	50.3	5.5	–	–	–	–	–	–	–	–
	2	3.1	64.8	4.6	5.0	19.0	3.5	–	–	–	–	–	–	–	–
	3	−1.3	−0.1	0.4	15.8	73.8	11.4	–	–	–	–	–	–	–	–
Adrenal	1	1.4	11.9	1.9	40.2	40.3	4.3	–	–	–	–	–	–	–	–
	2	4.0	70.5	4.4	4.5	13.4	3.2	–	–	–	–	–	–	–	–
	3	0.3	8.2	1.2	13.5	63.6	13.8	–	–	–	–	–	–	–	–
Liver	1	1.3	41.6	2.9	20.2	20.7	7.4	0.7	1.1	0.8	0.7	1.4	0.2	0.5	0.5
	2	0.9	16.6	4.0	2.2	40.6	24.6	1.2	0.3	0.6	1.5	3.3	0.4	2.8	0.9
	3	1.1	18.0	0.9	7.9	41.3	19.3	0.8	0.4	0.1	0.4	5.8	1.5	2.1	0.4
Blood	1	2.8	29.5	3.3	22.0	33.2	6.2	0.6	0.7	0.8	–	0.9	–	–	–
	2	2.2	36.0	4.3	4.8	37.0	13.8	1.9	–	–	–	0.9	–	–	–
	3	0.4	17.9	2.6	6.5	54.3	15.1	0.5	0.8	1.9	–	0.9	–	–	–
Kidney	1	1.4	14.2	1.6	38.3	37.9	4.7	–	1.9	–	–	–	–	–	–
	2	3.8	71.4	3.7	3.8	13.4	3.9	–	–	–	–	–	–	–	–
	3	0.2	6.5	2.8	12.2	59.7	14.2	1.5	1.1	0.3	0.9	0.6	–	–	–

[a] Fatty acid composition at the 3 position is calculated from the appropriate measurements. Small errors can similarly lead to negative numbers. $n-9$ indicates ninth carbon from terminal methyl.

Table 3.201 Triglyceride composition [a] (mol%) of lard (Jurriens, 1968)

No double bonds		Two double bonds	
PMSt	0.4	POO	5.2
StMSt	0.4	StOO	6.1
PPP	0.5	OMO	1.6
PPSt	2.0	OPO	18.4
StPSt	2.0	OStO	1.2
PStSt	0.4	PPL	1.8
StStSt	0.4	StPL	2.1
Other	0.5	Other	1.5
One double bond		Three double bonds	
POP	0.6	OOO	11.7
POSt	1.9	StLO	0.6
StOSt	1.5	OML	0.6
PMO	0.4	OPL	7.2
StMO	0.7	OStL	1.2
MPO	0.8	Other	0.5
PPO	7.9		
StPO	12.8	Four or more double bonds	
PStO	0.9	OLL	1.4
StStO	1.6	OLO	1.5
Other	0.6	LPL	0.5
		Other	0.5

[a] P, palmitic; M, myristic; St, stearic; O, oleic; L, linoleic.

Table 3.202 Glyceride composition of pig fats [a] (Chacko and Perkins, 1965)

	SSS	SUS	SSU	USU	UUS	UUU
Bottom rear back	8.0	3.4	31.4	30.8	13.3	13.0
Kidney	7.7	8.9	37.3	44.7	4.3	5.1
Top rear back sirloin	4.9	1.1	38.4	46.6	8.6	10.5
Flank fat	5.9	2.3	29.5	36.3	11.6	14.3
Back fat layer (shoulder), upper 1 cm	2.6	0.7	23.0	51.4	6.8	15.2
Back fat layer (shoulder), lower 1 cm	7.7	7.7	22.6	18.2	23.9	18.3
Back top [b]	4.5	2.1	26.2	37.5	12.2	17.9
Back bottom [b]	5.8	1.6	30.8	41.2	8.8	11.7

[a] Calculations based on lipolysis data; no allowance made for minor acids. S, saturated; U, unsaturated.
[b] These results were obtained from samples from a different animal.

(1.2 mg/kg), α-tocotrienol (0.7 mg/kg) and γ-tocopherol (0.7 mg/kg) were found in lard (Glover *et al.*, 1969).

(b) Beef tallow

Edible beef tallow (*premier jus*) is obtained by rendering approved edible stock and frequently does not require further refining (free fatty acid 0.3%). The USDA exerts very close control on the processing and transport of edible tallow. Recent specifications for different grades of tallow are given in Tables 3.203 and 3.204. The fatty acid compositions of beef tallow are given in Table 3.196.

Changes in composition by virtue of the location on the animal are given in Table 3.205. Apart from animal type and feed, the climate also influences fat composition: the hotter the climate, the lower the iodine value (Achaya and Banerjee, 1964).

As with lard, beef tallow contains significant levels of branched and odd-chain acids and also *trans* acids (Hoffman and Meijboom, 1969). Tabulated data presented here do not reflect the complexity of the fatty acids present in tallow. *Trans*-9, -10- and -11-octadecenoic acids (5% total) have been reported.

Table 3.203 Standard grades: specifications and quality tolerances for tallows and greases as established by the American Fats and Oils Association [a]

Grades	Specifications				
	Titre, min.	FFA, max.	FAC [b], max.	R and B [c], max.	MIU [d], max.
Top white tallow	41.0	2	5	0.5	1
Extra fancy tallow	42.0	2	5	None	1
All beef packer tallow	42.0	2	None	0.5	1
Industrial extra fancy tallow	41.0	3	5	None	1
Fancy tallow	40.5	4	7	None	1
Bleachable fancy tallow	40.5	4	None	1.5	1
Prime tallow	40.5	6	13–11B	None	1
Special tallow	40.5	10	19–11C	None	1
No. 1 tallow	40.5	15	33	None	2
No. 2 tallow	40.0	35	None	None	2
Intermediate special tallow	39.0	10	21	None	1
'A' tallow	39.0	15	39	None	2
Choice white grease	36.0	4	13–11B	None	1
Yellow grease	36.0	15	37	None	2
Feed grade fats	Must be within the pesticide and industrial chemical tolerances set by FDA [e], EPA [f] and the California Department of Food and Agriculture				

[a] Effective 1 January 1983.
[b] Fat Analysis Committee of the American Oil Chemists' Society colour standards, range 1–45 (45 darkest).
[c] Red and blue.
[d] Moisture, insoluble impurities and unsaponifiables.
[e] Food and Drug Administration.
[f] Environmental Protection Society.

Similarly *trans* isomers occur in the dienoic series of acids, e.g. *trans*(*cis*)-11-, *trans*(*cis*)-15-octadecadienoic acid.

The triglyceride composition of tallow is very complex, but the results of silver-ion TLC are given in Table 3.206 (Jurriens, 1968). In contrast to most vegetable fats, beef tallow contains significant amounts of trisaturated triglycerides, unsymmetrical triglycerides and elaidoglycerides. The positional distribution of fatty acids determined by stereospecific analysis is given in Table 3.207 (Christie *et al.*, 1991).

The composition of beef tallow is modified by diet. Dietary fat can be encapsulated in protein hardened

Table 3.204 The Society of British Soap Makers specifications for tallows and greases (based on the methods of test contained in BS 3919 – 1976)

Grade	Free fatty acids, max. (mol%)	Crude oil colour, red	Bleached oil colour, red	Moisture and dirt (wt%)	Unsaponifiable matter, max. (wt%)	Titre (°C)	Iodine value, max.	Max. content of plastics and polymers [a] (mg/kg)
Tallow 1	3.0	–	Max. 0.5, 5¼ inch cell	0.5	0.5	Min. 40.0	57	200
Tallow 2	5.0	–	Max. 1.0, 5¼ inch cell	1.0	1.0	Min. 40.0	57	200
Tallow 3	8.0	–	Max. 3.0, 5¼ inch cell	1.0	1.0	Min. 40.0	57	200
Tallow 4	12.0	–	Max. 4.0, 1 inch cell	1.0	1.5	Min. 40.0	60	200
Tallow 5	15.0	–	Max. 12, 1 inch cell	1.0	1.5	Min. 40.0	60	200
Tallow 6	20.0	–	No limit	1.0	2.0	Min. 40.0	60	200
Grease	20.0	No limit	No limit	2.0	2.0	36.0–40.0	63	200

[a] Calculated as polyethylene.

Table 3.205 Fatty acid composition (%) of beef fats at different locations on the animal (Chacko and Perkins, 1965)

	Anatomical position				
	Back fat layer		Perinephric	Visceral pericardial	Ribs
	Rump (1 cm depth)				
	Bottom	Top			
14:0	6.9	5.2	5.2	2.1	4.4
14:1	1.5	2.5	0.6	0.7	1.2
16:0	33.1	33.7	35.3	26.9	32.5
16:1	5.4	6.0	2.5	5.0	4.0
18:0	14.0	9.5	20.3	34.2	15.7
18:1	39.1	42.3	36.2	29.0	39.9
18:2	–	0.9	–	2.2	1.5

with formaldehyde in order to protect the fat from the microflora in the rumen (see section 3.4). Unsaturated fats fed in this way produce more unsaturated depot fats and milk fat (Garrett *et al.*, 1976).

Cholesterol is the major sterol in animal fats including lard (0.37–0.42%), beef tallow (0.08–0.14%) and mutton tallow (0.23–0.31%).

(c) *Other tallows and animal fats*

The compositions of many animal tallows are superficially very similar to that of beef. Mutton tallow is normally slightly more saturated and therefore harder than beef tallow. Goat tallow is even harder. Some typical fatty acid compositions are given in Table 3.195. As with beef tallow, many of these fats

Table 3.206 Triglyceride composition (mol%) of beef tallow [a,b] (Jurriens, 1968)

No double bonds		One *trans* double bond		Two *cis* double bonds	
PMP	0.5	PESt	0.5	MOO	0.6
PMSt	1.0	PPE	0.8	POO	12.1
StMSt	0.5	StPE	0.7	StOO	5.9
MPP	0.5	PStE	0.6	OMO	0.5
MPSt	0.5	StStE	0.6	OPO	2.1
PPP	1.7	Others	1.9	OStO	0.5
PPSt	3.6			PLP	0.5
StPSt	1.8	One *cis* double bond		*trans* glycerides	2.3
MStSt	0.5	MOP	0.5	Others	1.3
PStP	1.6	MOSt	0.5		
PStSt	3.2	POP	5.8	Tree *cis* double bonds	
StStSt	1.6	POSt	10.5	OOO	4.9
Others	1.4	StOSt	4.8	POL	0.8
		PMO	1.5	PLO	0.7
		StMO	1.4	OPL	0.6
		PPO	3.9	*trans* glycerides	1.6
		StPO	3.6	Others	1.8
		PStO	1.6		
		StStO	1.5		
		trans glycerides	4.5		
		Others	0.2		

[a] E, elaidic.
[b] Others, sum of glycerides present at <0.5%.

Table 3.207 Fatty acid positional disribution in tallow (Christie *et al.*, 1991)

Fatty acid	TG	*sn*-1	*sn*-2	*sn*-3
14:0	3.4	0.2	4.3	2.9
15:0	0.6	0.4	0.7	0.5
16:0	29.5	45.7	21.0	25.5
16:1	1.9	0.4	2.4	1.3
16br	1.0	1.6	1.3	0.0
17:0	1.2	1.5	0.7	1.7
18:0	26.0	33.1	12.5	31.2
18:1	33.1	16.6	52.0	33.7
18:1*t*	1.8	0.2	1.4	2.7
18:2	1.5	0.4	3.7	0.4

Table 3.208 Positional distribution of fatty acids in triglycerides of depot fats of land animals (Brockerhoff *et al.*, 1966)

	Position	Fatty acids (mol%)						
		14:0	16:0	16:1	18:0	18:1	18:2	18:3
Cat	1	3	27	4	20	36	5	1
	2	5	20	5	5	49	11	1
	3	2	20	3	14	46	8	1
Dog	1	2	26	4	22	36	5	1
	2	9	27	7	4	35	12	1
	3	1	16	4	14	57	7	1
Horse	1	3	39	7	6	27	5	11
	2	7	9	10	1	29	17	25
	3	3	30	6	7	37	5	11
Man	1	4	39	5	10	33	3	1
	2	11	10	11	2	50	9	1
	3	1	25	4	9	51	5	1
Ox	1	4	41	6	17	20	4	1
	2	9	17	6	9	41	5	1
	3	1	22	6	24	37	5	1
Pig	1	2	16	3	21	44	12	1
	2	4	59	4	3	17	8	1
	3	1	2	3	10	65	24	1
Rabbit	1	3	34	9	6	25	14	2
	2	6	25	12	1	26	23	5
	3	1	24	7	3	35	22	5
Rat	1	2	32	5	9	32	15	2
	2	1	10	4	1	37	45	2
	3	2	27	5	7	37	17	2

Table 3.209 Positional distribution of fatty acids (mol%) in triglycerides of depot fats of birds and reptiles (Brockerhoff and Hoyle, 1963; Brockerhoff *et al.*, 1966)

Fat	Position	Fatty acids (mol%)											
		12:0	14:0	16:0	16:1	18:0	18:1	18:2	18:3	20:1	22:1	20:5	22:6
Chicken	1	–	2	25	12	6	33	14	2	–	–	–	–
	2	–	1	15	7	4	43	23	3	–	–	–	–
	3	–	1	14	12	6	35	14	3	–	–	–	–
Duck	1	–	1	26	7	8	43	14	2	–	–	–	–
	2	–	1	11	4	4	59	20	2	–	–	–	–
	3	–	1	27	7	6	45	15	1	–	–	–	–
Turkey	1	–	1	30	7	11	30	17	2	–	–	–	–
	2	–	2	13	5	4	34	39	2	–	–	–	–
	3	–	1	27	8	8	33	19	2	–	–	–	–
Herring gull	1	–	2	22	4	13	41	7	–	7	3	1	1
	2	–	3	15	3	9	48	11	–	6	4	1	1
	3	–	2	17	5	7	46	9	–	7	5	1	1
Cormorant	1	–	3	34	7	14	21	1	–	3	3	5	3
	2	–	2	21	10	6	37	2	–	4	4	4	3
	3	–	2	21	11	10	30	1	–	4	5	7	3
Frog	1	–	3	26	10	4	27	12	9	–	–	–	–
	2	–	2	17	8	1	30	25	13	–	–	–	–
	3	–	2	18	8	3	35	16	9	–	–	–	–
Turtle	1	7	18	24	7	12	1	1	–	5	5	1	1
	2	2	10	4	13	2	45	1	–	3	3	2	5
	3	9	14	10	8	5	19	1	–	11	6	4	2

contain significant amounts of *trans* and branched-chain fatty acids.

The positional distributions of fatty acids in the triglycerides of the depot fats of some land animals are given in Table 3.208. The positional distributions of fatty acids in bird and reptile fats are given in Table 3.209.

References and further reading

Achaya, K.T. and Banerjee, B.N. (1964) *Curr. Sci.*, **15**, 22.

Anderson, B.A., Sutton, C.A. and Pallansch, M.J. (1970) *J. Am. Oil Chem. Soc.*, **47**, 15.

Badings, H.T. (1970) *Neth. Milk Dairy J.*, **24**, 147.

Badings, H.T., Maarse, H., Kleipool, R.J.C., Tas, A.C. and Neeter, R. (1975) *Proc. Int. Symp. on Aroma Research*, Zeist (eds H. Maarse and P.S. Groenen), Centre for Agricultural Publishing and Documentation, Wageningen, Netherlands, p. 63.

Banks, W., Clapperton, J.L. and Ferrie, M.E. (1976) *J. Dairy Res.*, **43**, 219.

Barbano, D.M. and Sherbon, J.W. (1980) *J. Dairy Sci.*, **63**, 731.

Bates, R.W. (1968) *J. Am. Oil Chem. Soc.*, **45**, 420A.

Baxter, J.G. and Robson, C.D. (1942) *J. Am. Chem. Soc.*, **64**, 2411.

Berger, S.E., Krolewski, C.V. and Wizemann, L.C. (1970) *J. Am. Oil Chem. Soc.*, **47**, 168.

Bitman, J.L., Dryden, L.P., Goering, H.K., Wren, T.R., Yoncoskie, R.A. and Edmondson, L.F. (1973) *J. Am. Oil Chem. Soc.*, **50**, 93.

Boatman, V.E., Patton, S. and Parsons, J.G. (1969) *J. Dairy Sci.*, **52**, 256.

Boldingh, J. and Taylor, R.J. (1962) *Nature*, **194**, 909.

Bracco, U., Hidalgo, J. and Bohren, H. (1972) *J. Dairy Sci.*, **55**, 165.

Breckenbridge, W.C., Marai, L. and Kuksis, A. (1969) *Can. J. Biochem.*, **47**, 761.

Brewington, C.R., Caress, E.A. and Schwartz, D.P. (1970) *J. Lipid Res.*, **11**, 355.

Brockerhoff, H. (1965) *J. Lipid Res.*, **6**, 10.

Brockerhoff, H. and Hoyle, R.J. (1963) *Arch. Biochem. Biophys.*, **102**, 452.

Brockerhoff, H., Hoyle, R.J. and Wolmark, N. (1966) *Biochim. Biophys. Acta*, **116**, 67.

Brunner, J.R. (1969) in *Structural and Functional Aspects of Lipoproteins in Living Systems* (eds E. Tria and A.M. Scanu), Academic Press, London, p. 545.

Bunnell, R.H. (1957) *Poult. Sci.*, **36**, 413.

Chacko, G.K. and Perkins, E.G. (1965) *J. Am. Oil Chem. Soc.*, **42**, 1121.

Christie, W.W. and Moore, J.H. (1970) *Biochim. Biophys. Acta*, **210**, 46.

Christie, W.W. (1979) *Prog. Lipid Res.*, **17**, 245.

Christie, W.W., *et al.* (1991) *J. Am. Oil Chem. Soc.*, **68**, 695.

Coleman, M.H. (1961) *J. Am. Oil Chem. Soc.*, **38**, 685.

Dawson, R.M.C. and Kemp, P. (1970) in *Physiology of Digestion and Metabolism of the Ruminant* (ed. A.T. Phillipson), Oriel, Newcastle upon Tyne.

Day, E.A., Forss, D.A. and Patton, S. (1957) *J. Dairy Sci.*, **40**, 932.

Deman, L. and Deman, J.M. (1983) *J. Am. Oil Chem. Soc.*, **60**, 1095.

Dimick, P.S., Walker, N.J. and Kinsella, J.E. (1966) *Cereal Sci.*, **11**, 479.

Dimick, P.S., McCarthy, R.D. and Patton, S. (1970) in *Physiology of Digestion and Metabolism of the Ruminant* (ed. A.T. Phillipson), Oriel, Newcastle upon Tyne.

Edmunson, L.F., Youkoskie, R.A., Rainey, N.H., Douglas, F.W. Jr and Bitman, J. (1974) *J. Am. Oil Chem. Soc.*, **51**, 72.

Ellis, R. and Wong, N.P. (1975) *J. Am. Oil Chem. Soc.*, **52**, 252.

Forss, D.A. (1972) *Prog. Chem. Fats Other Lipids*, **13**, 177.

Garrett, W.N., Yang, Y.T., Dunckley, W.L. and Smith, L.M. (1976) *J. Anim. Sci.*, **42**, 1522.

Garton, G.A. (1974) *Biochem. Soc. Trans.*, **2**, 1200.

Garton, G.A. and Duncan, W.R.H. (1971) *J. Sci. Food Agric.*, **22**, 29.

Glover, H., Lehmann, J. and Valis, R. (1969) *J. Am. Oil Chem. Soc.*, **46**, 417.

Guyot, A.L. (1977) *Bull. Rech. Agron. Gembloux*, **12**, 299.

Harrison, F.A. and Leat, W.M.F. (1975) *Proc. Nutr. Soc.*, **34**, 203.

Hay, J.D. and Morrison, W.R. (1970) *Biochim. Biophys. Acta*, **202**, 237.

Hickman, K.C.D. (1943) *Annu. Rev. Biochem.*, **12**, 353.

Hilditch, T.P. and Sleightholme, J.J. (1931) *Biochem. J.*, **25**, 507.

Hilditch, T.P. and Williams, P.N. (1964) *The Chemical Constitution of Natural Fats*, 4th edn, Chapman & Hall.

Hoffman, G. and Meijboom, P.W. (1969) *J. Am. Oil Chem. Soc.*, **46**, 620.

INFORM (1992) *INFORM*, **3**, 152.

International Dairy Federation (n.d.) *Bull. Int. Dairy Fed.*, No. 266.

Iverson, J.L., Eisner, J. and Firestone, D. (1965) *J. Am. Oil Chem. Soc.*, **42**, 1063.

Jenness, R. and Palmer, L.S. (1945) *J. Dairy Sci.*, **28**, 473.

Jensen, R.G. (1973) *J. Am. Oil Chem. Soc.*, **50**, 186.

Jensen, R.G., Gander, G.W. and Sampugna, J. (1962) *J. Dairy Sci.*, **45**, 329.

Jensen, R.G., Ferris, A.M. and Lammi-Keefe, C.J. (1991) *J. Dairy Sci.*, **74**, 3228.

Juarez, M. and Ramos, M. (1986) *Bull Int. Dairy Fed.*, No. 202, 54.

Jurriens, G. (1968) in *Analysis and Characterisation of Oils and Fats and Fat Products*, vol. 2 (ed. H.A. Boekenoogen), Interscience, New York.

Kayser, S.G. and Patton, S. (1970) *Biochem. Biophys. Res. Commun.*, **41**, 1572.

Keenan, T.W. and Patton, S. (1970) *Lipids*, **5**, 42.

Kinsella, J.E., Patton, S. and Dimick, P.S. (1967) *J. Am. Oil Chem. Soc.*, **44**, 202.

Kuksis, A. and McKarthy, M.J. (1964) *J. Am. Oil Chem. Soc.*, **41**, 17.

Kuksis, A., Marai, L. and Myher, J.J. (1973) *J. Am. Oil Chem. Soc.*, **50**, 193.

Lange, W. (1950) *J. Am. Oil Chem. Soc.*, **27**, 414.
Langeer, J.E. and Day, E.A. (1964) *J. Dairy Sci.*, **47**, 1291.
Lund, P. and Jensen, F. (1983) *Milchenwissenschaft*, **38**, 193.
Magidman, P., Herb, S.F., Luddy, F.E. and Riemenschneider, R.W. (1963) *J. Am. Oil Chem. Soc.*, **40**, 86.
Marai, L., Breckenbridge, W.C. and Kuksis, A. (1969) *Lipids*, **4**, 562.
Moore, J.H. and Christie, W.W. (1979) *Prog. Lipid Res.*, **17**, 347.
Morrison, W.R. (1970) *Top. Lipid Chem.*, **1**, 51.
Morton, R.H. (1984) *J. Dairy Sci.*, **67**, 1548.
Nutter, L.J. and Privett, O.S. (1967) *J. Dairy Sci.*, **50**, 298.
Oil and Soap (1943) *Oil Soap*, **20**, 163.
Parkhurst, R.H., Skinner, W.A. and Sturm, P.A. (1968) *J. Am. Oil Chem. Soc.*, **45**, 641.
Parodi, P.W. (1982) *Lipids*, **17**, 437.
Patton, S. (1965) *Food Quality*, **77**, 165.
Patton, S. and Jensen, R.E. (1976) in *Biomedical Aspects of Lactation*, Pergamon, Oxford.
Patton, S., Forss, D.A. and Day, E.A. (1956) *J. Dairy Sci.*, **39**, 1469.
Payne, E. (1971) *J. Sci. Food Agric.*, **22**, 520.
Pitas, R.E., Sampugna, J. and Jensen, R.G. (1967) *J. Dairy Sci.*, **50**, 1322.
Prasad, S. and Gupta, S.K. (1988) *Agric. Rev.*, **9**, 181.
Precht, D. (1988) *Surfactant Sci. Ser.*, **31**, 305.
Precht, D. (1992) *Z. Lebensm. Unters. Forsch.*, **194**, 107.
Scott, T.W., Cook, L.J., Ferguson, K.A., McDonald, I.W., Buchanon, R.A. and Hills, G. (1972a) *Austr. J. Sci.*, **32**, 291.
Scott, T.W., Cook, L.J. and Mills, S.C. (1972b) *J. Am. Oil Chem. Soc.*, **48**, 358.
Screenivasan, R. (1968) *J. Am. Oil Chem. Soc.*, **45**, 259.
Smith, L.M., Bianco, D.H. and Dunckley, W.L. (1977) *J. Am. Oil Chem. Soc.*, **54**, 132.
Smith, S. (1980) *J. Dairy Sci.*, **63**, 337.
Spencer, G.F., Herb, S.F. and Gorminsky, P.J. (1976) *J. Am. Oil Chem. Soc.*, **53**, 95.
Storry, J.E. (1980) *Bull. Int. Dairy Fed.*, No. 125, 88.
Storry, J.E. (1981) in *Recent Advances in Animal Nutrition* (ed. W. Haresign), Butterworths, London, p. 88.
Storry, J.E., Hall, A.J. and Johnson, V.W. (1971) *J. Dairy Res.*, **38**, 73.
Storry, J.E., Brumby, P.E., Hall, A.J. and Johnson, V.W. (1974) *J. Dairy Sci.*, **57**, 61.
Storry, J.E., Brumby, P.E., Tuckley, B., Welch, V.A., Stead, D. and Fulford, R.J. (1980) *J. Agric. Sci. Camb.*, **94**, 503.
Teranishi, R., Flath, R.A. and Sugisawa, J. (eds) (1981) *Flavour Research, Recent Advances*, Marcel Dekker, New York, p. 361.
Timms, R.E. (1980a) *J. Dairy Sci.*, **47**, 295.
Timms, R.E. (1980b) *Lebensm. Wiss. Technol.*, **13**, 61.
Van der Ven, B. and de Jong, K. (1967) *Fette Seifen Anstrichm.*, **67**, 279.
Van der Ven, B. and de Jong, K. (1970) *J. Am. Oil Chem. Soc.*, **47**, 299.
von Brixius, L. and Treiber, H. (1974) *Fette Seifen Anstrichm.*, **76**, 83.
Warmbier, M. (1974) *Fette Seifen Anstrichm.*, **76**, 86.
Wood, T.S., Wood, F.W., Murphy, M.F. and Dunckley, W.L. (1975) *J. Dairy Sci.*, **58**, 839.

3.5 Marine oils: fish and whale oils

Data relating to marine oil and fat production are shown in Tables 3.210 and 3.211.

Marine oils form a significant proportion (2–3%) of the world's available edible oils. Future trends are difficult to predict, but, unless we learn to fish the sea in a controlled way, this natural resource will almost certainly decline. An additional danger arises from pollution. In isolated cases this has reached a point where the fish have become inedible, e.g. mercury poisoning in Japan (Wood, 1976).

Marine oils of commercial importance are obtained from 'oily fish', e.g. herring, pilchard, menhaden and anchovy. The fish livers of cod, halibut and shark contain approximately 50% oil and are an important source of vitamins A and D.

Further interest in fish oils has been stimulated because of the association of the long-chain highly unsaturated acids, 20:5 and 22:6, with the alleviation of a wide variety of diseases (Dyerberg *et al.*, 1978; Barlow, 1980; Ackman and Cunnane, 1992). Marine oils are, however, partially hydrogenated in order to obtain fats of adequate oxidative stability for conventional food application in margarine and shortenings. Industrial uses include soap, paints, varnishes, printing inks, leather treatment, water repellents and plasticizers.

3.5.1. Fish oils

Approximately one-third of the world fish catch (FAO, 1981), mainly small bony and oily fish, is used specifically for the production of fish meal and oil (Tables 3.212 and 3.213). Meal and oil are also obtained from offal and unwanted fish. Simple fat–water–protein relationships are, however, unlikely (Dabrowsky, 1978).

Table 3.210 World production (1000 tonnes) of oils and fats of aquatic animal origin (FAO, 1992)

1977	1980	1988	1989	1990
1080	1080	1565	1639	1398

Table 3.211 Production of fish oils and fats (1000 tonnes) by major production countries (FAO, 1992)

	1980	1989	1990
Capelin oil (Iceland)	80.1	43	64
Herring oil (Canada, Iceland)	7.5	9.8	10.5
Menhaden oil (USA)	132	99	124
Pilchard oil (Japan)	133	407	381
Anchoveta oil (Peru and Panama)	78	336	200

Almost all industrial fish are processed by cooking, pressing or centrifugation and subsequent drying of the fish meal for use as animal feed. Cooking is essential because raw fish loses very little water or oil even when exposed to very high pressures. The water and oil mixture isolated during straining and pressing is screened and the oil separated by centrifugation. The oil is polished by washing with hot water followed by centrifugation. The oil should be stored dry and free from solids (*INFORM*, 1992).

The proportion of major lipid classes in muscle or body tissue of five fish are given in Table 3.214. Lean muscle fish generally have fatty liver, e.g. cod (Jangaard *et al.*, 1967).

The compositions of virtually all marine oils can be described by reference to 8–10 major fatty acids, ignoring isomers. Numerous minor acids are also present. A wide variation in composition is observed within species. This can be due both to seasonal variation and to geographical location (Table 3.215) (Stansby, 1981; Gruger, n.d.). Both these factors are secondary to the all-important nature of the dietary intake. The composition of fish lipids has been reviewed by Ackman (1980) and Ackman and Cunnane (1992).

The major saturated acids are palmitic (16:0) and myristic (14:0). It has been suggested that the higher the iodine value (IV) of the oil, the higher the total saturated acid, i.e. low-IV oils are usually rich in monoenes, high-IV oils are rich in saturated and polyenoic acids. Not all the analytical data support this statement, however.

The monoethylenic fatty acids are composed mainly of palmitoleic acid $(n-7)$-16:1, oleic acid $(n-9)$-18:1 and its isomer *cis*-vaccenic acid $(n-7)$-18:1. These are often accompanied by large amounts of $(n-9)$-20:1 and $(n-9)$-22:1.

The polyunsaturated fatty acids are the most characteristic feature of marine oils. The acids with five and six double bonds originate in unicellular phytoplankton and seaweeds, which form part of the food chain. Linoleic and linolenic acids commonly found in vegetable oils occur to no more than 1 or 2% in fish oils. The major polyenoic acids are usually 20:5 and 22:6 (occasionally 18:4) of the $n-3$ family. The occurrence of $(n-3)$-22:6 in marine organisms is apparently related to the vertebrate nervous system.

Table 3.212 World production of fish oil and fish meal (millions of tonnes) (sources: USDA, *Oil World*)

		1985/86	1986/87	1987/88	1988/89	1989/90 (est.)	1990/91 (est.)
Fish oil							
World production		1.36	1.29	1.39	1.48	1.54	1.58
of which	Japan	0.49	0.42	0.42	0.42	0.41	0.41
	Peru	0.21	0.15	0.15	0.15	0.28	0.30
	USA	0.15	0.14	0.11	0.12	0.10	0.09
	Chile	0.20	0.17	0.18	0.18	0.19	0.20
Fish meal							
World production		5.97	5.99	6.04	6.91	6.50	6.60
of which	Japan	1.13	1.01	1.02	0.95	0.86	0.86
	Chile	1.17	1.07	1.06	1.30	1.04	1.06
	Peru	1.10	1.18	0.88	1.41	1.10	1.19
	Former USSR	0.73	0.72	0.70	0.77	0.79	0.79

Table 3.213 Utilization of world fish catch (1000 tonnes) From FAO *Yearbook of Fisheries Statistics 1990*, FAO 1992)

Utilization (live weight)	1980	1990
For human consumption	50 541	70 212
Marketing fresh	14 800	22 093
Freezing	15 690	24 261
Curing	9 850	11 158
Canning	10 200	12 700
For other purposes	21 650	27 034
Reduction [a]	20 650	25 534
Miscellaneous purposes	1 000	1 500
Total world catch	72 190	97 246

[a] For conversion to fish meal and oil.

This is also true for mammalian brain lipids (Crawford, 1974; Lenarchal, 1978). The eyeball of several fish species is known to contain largely tridocosahexaenoin (Nicol *et al.*, 1972).

The fatty acids of the copepod (small crustacean, a major fish feed) triglycerides and wax esters are compared with the compositions of commercial capelin, mackerel, Atlantic herring, Pacific herring and anchovy in Table 3.216 (Ackman, 1980).

Data on Californian sardine oil, also known as pilchard oil, from the fish *Sardinops caerulea*, are limited. Japanese sardine oil is obtained from *Clupanodon melanostica* (Table 3.217) (Ackman and Sipos, 1964; Itoh and Fukuzumi, 1963).

The fatty acid compositions of a number of fresh-

Table 3.214 Distribution of polar and non-polar lipids in cod, hake, trout, pilchard and wrasse (Adapted from Ackman, 1980)

Name	Cod [a]	Hake [b]	Trout [c]	Pilchard [d]	Wrasse [e]
Species	*Gadus morrhua*	*Merluccius capensis*	*Salmo iridis*	*Sardinia ocellata*	*T. duperreyi*
Sample	White muscle	Muscle	Muscle	Whole	Muscle
Total lipid (g/100 g)	0.59	1.55	5.3	5.0	0.5–1.0
Polar lipid (g/100 g)	0.52	0.46	0.63	0.91	0.5
PC (%)	69	63	67	53	59
PE (%)	19	21	21	25	13
Sphingomyelin (%)	–	4	2	6	12
PS (%)	5	3	4	2	10
Neutral lipid (g/100 g)	0.07	1.0	4.68	3.9	–
Triglyceride in neutral lipid (%)	35	100	94	100	100
Sterol in neutral lipid (%)	50	–	1.1	–	–

[a] Addison *et al*, (1968).
[b] de Koning (1966).
[c] Gray and Macfarlane (1961).
[d] de Koning and McMullan (1966).
[e] Patton (1975).

Table 3.215 Variation in composition of menhaden and herring oils showing seasonal and geographical variation (Adapted from Stansby, 1981)

Fatty acid	Herring (*Clupea harengus*)		Menhaden (*Brevoortia tyrannus*)	
	Alaska [a]	Nova Scotia [b]	Overall variation	Variation [c]
14:0	5.6–7.7	4.6–8.4	6.7–16.3	8.7–10.6
16:0	11.8–18.6	10.1–15.0	19.6–24.0	20.1–21.2
16:1	6.2–8.0	7.0–12.0	11.2–17.9	11.2–13.9
18:0	1.1–2.0	0.7–2.1	2.4–3.4	3.0–3.4
18:1	11.7–25.2	9.3–21.4	10.7–23.4	10.8–12.1
18:2	0.1–0.6	0.6–2.9	0.9–1.7	1.0–1.7
18:3	None	0.3–1.1	0.4–3.7	1.3–3.7
18:4	1.1–2.8	1.1–2.5	0.8–3.6	1.9–3.6
20:1	7.3–19.1	11.0–19.9	1.1–2.7	1.1–2.0
20:4	0.3–0.8	0.4–1.2	0.6–2.3	1.7–2.3
20:5	11.4–15.2	3.9–8.8	10.2–14.1	11.8–14.1
22:1	6.9–15.2	14.8–30.6	0.2–1.0	0.2–0.7
22:5	0.3–1.0	0.5–1.3	1.1–2.5	1.9–2.5
22:6	4.8–7.8	2.0–6.2	3.3–10.6	5.6–10.6
24:1	0.6–1.3	0.2–0.9	–	–
24:5	–	0.2–0.5	–	–

[a] Fish caught in 1964 and 1965.
[b] Fish caught in 1966.
[c] Without seasonal variation as a factor.

water fish fillets are given in Table 3.218 (Kinsella *et al.*, 1977). The data suggest that freshwater fish contain higher levels of the $n-6$ polyene acids than do marine fish.

An extensive tabulation by Gooch *et al.* of the compositions of finfish fatty acids is given in Table 3.219 (Gooch *et al.*, 1987).

The triglyceride composition of fish oils is complex, and detailed studies are not available. Work on fatty acid distribution suggests that polyunsaturated acids $(n-3)$-20:5 and $(n-3)$-22:6 occupy the 2 position, saturated and mono-unsaturated the 1 position and mono-unsaturated the 3 position (Brockerhoff *et al.*, 1968; Litchfield, 1972; Malins and Wekell, 1970). In marine mammals, however, the polyunsaturated acids are at the 1,3 positions (Table 3.220). Fish oil triglycerides have been separated by high-temperature gas–liquid chromatography (Litchfield, 1972; Litchfield *et al.*, 1978). Typical physical and chemical properties of the major commercial oils are given in Table 3.221. (Progress in the Chemistry of Fats and Other Lipids, 1958; Bailey *et al.*, 1952; Oil and Soap, 1942; Stansby, 1967).

Partially hydrogenated fish oils have been characterized (Ackman, 1974). The complete elimination of $(n-3)$-20:5 (eicosapentaenoic) fatty acid from Peruvian anchoveta oils was not achieved until the iodine value was reduced from 199 to 108 (see Table 3.222). The $(n-3)$-20:5 acid gives rise to 2-*trans*-4-*cis*-7-*cis*-decatrienal, which has a strong fishy flavour (Meijboom and Stroink, 1972).

Because of the increasing relevance given to the need for polyunsaturated acids in the diet, an increasing amount of research into methods of stabilizing unhydrogenated fish oil has led to new developments. The volatile components responsible for odour and flavour have been characterized (Table 3.223) (Hsieh *et al.*, 1989). Methods of stabilizing unhydrogenated fish oils have also been developed (Freeman *et al.*, 1987; Food Manufacture, 1988). Various practical aspects of fish oil characterization, stabilization of refined oils and methods of enriching $n-3$ fatty acids and their physiological relevance have been reviewed (Ackman, 1988; Ackman and Cunnane, 1992).

Cholesterol is the major (95–99%) sterol in fish oils (Feeley *et al.*, 1972; Kritchevsky *et al.*, 1967). Minor levels of plant sterols have been identified in mackerel, rainbow trout, smelt, sardine and chimera (Takagi *et al.*, 1979). Other fish lipids that have been characterized include wax esters (Nevenzel, 1970; Sargent, 1976b), glyceryl ethers (Myher, 1978), plasmalogens (Kauer *et al.*, 1978), phospholipids (Table 3.214; Addison *et al.*, 1968; de Koning, 1966; de Koning and McMullan, 1966; Gray and Macfarlane, 1961; Patton, 1975) and hydrocarbons. Wax esters fall into two main classes, those rich in 16:0 and those rich in 22:1 (Nevenzel, 1970; Sargent,

1976b). The total chain length is in the range C_{32} to C_{36}; the species rich in 16:0 alcohol will tend to be esterified to 16:0, 16:1 and 18:1 fatty acids. The 'castor oil' fish *Ruvettus pretiosus* contains high levels of wax esters (Nevenzel, 1970), which offers an explanation why the fish, when eaten, has a purgative action. Herring does not accumulate wax esters, capelin has a consistent proportion in its lipids and mackerel shows some variability (Ackman, 1980).

Ether lipids occur extensively in the fish liver oils (Sargent, 1976a) and as major components of the muscle lipids of the elasmobranches (Mori *et al.*, 1978). The general structure of ether lipids (**7**) and plasmalogens (**8**) found in marine lipids are given here (see also section 2.2). The compositions of some fish liver oils are given in Table 3.224. The plasmalogens occur only in minor amounts in fish lipids.

Table 3.216 Fatty acids of copepod triglycerides and wax esters and of commercial fish oils (Adapted from Ackman, 1980)

Fatty acid	Copepod		Total acids				
	Triglyceride	Wax esters	Capelin	Mackerel	Atlantic herring	Pacific herring	Pacific anchovy
12:0	0.32	0.62	0.16	0.11	0.10	0.20	–
Iso-14:0	0.07	0.25	0.06	0.03	0.03	0.03	–
14:0	19.84	36.42	7.85	7.81	8.77	6.81	8.26
Iso-15:0	0.31	0.59	0.14	0.20	0.22	0.01	0.39
Anteiso-15:0	0.12	0.19	0.04	0.07	0.09	0.09	–
15:0	1.05	0.97	0.28	0.41	0.34	0.34	0.64
Iso-16:0	0.19	0.06	0.03	0.12	0.03	0.03	–
16:0	28.98	11.15	8.81	15.93	14.84	22.74	19.46
Iso-17:0	0.06	0.05	0.03	0.15	0.03	0.01	–
Anteiso-17:0	0.07	0.05	0.03	0.05	0.01	0.07	–
17:0	0.83	0.42	0.17	0.46	0.33	1.29	1.11
Iso-18:0	nd	nd	nd	0.07	0.04	nd	–
18:0	1.04	0.35	0.72	1.73	0.97	2.72	3.28
19:0	0.09	0.16	0.06	0.19	0.10	0.01	0.37
20:0	0.51	0.38	0.08	0.11	0.15	0.05	0.28
22:0	0.20	nd	nd	0.01	nd	nd	0.07
24:0	0.09	nd	0.17	0.03	0.01	nd	0.34
Total saturated	53.8	52.7	18.6	27.5	26.1	34.4	34.20
14:1	0.07	nd	0.06	0.11	0.07	0.22	–
15:1	nd	nd	nd	nd	0.03	nd	–
$(n-9)$-16:1	0.05	0.35	0.03	0.29	nd	0.07	0.57
$(n-7)$-16:1	8.89	12.33	15.42	8.20	7.22	7.53	8.21
$(n-5)$-16:1	0.73	1.04	0.73	0.54	0.52	0.45	0.32
17:1	0.13	0.12	nd	0.15	0.04	nd	0.7
$(n-9)$-18:1	3.14	3.46	4.40	8.61	12.27	29.72	12.9
$(n-7)$-18:1	0.91	0.56	3.43	3.78	3.66	4.98	2.55
$(n-5)$-18:1	0.41	0.15	0.62	0.54	0.64	0.41	–
19:1	nd	nd	nd	0.04	0.07	nd	–
$(n-11)$-20:1	0.33	0.36	1.20	0.24	0.50	1.01	0.15
$(n-9)$-20:1	4.12	4.37	14.53	10.59	14.37	4.40	1.29
$(n-7)$-20:1	0.34	0.55	1.84	1.13	0.94	0.37	1.29
$(n-5)$-20:1	0.02	0.01	0.23	0.09	0.19	nd	–
$(n-11,13)$-22:1	5.16	4.59	17.45	12.74	20.92	3.92	1.4
$(n-9)$-22:1	0.34	0.65	1.70	1.00	1.36	0.87	0.16
$(n-7)$-22:1	0.11	0.11	0.42	0.19	0.33	0.12	0.02
24:1	0.49	0.24	0.59	0.69	0.52	0.47	–
Total monoethylenic	25.2	28.9	62.7	48.9	63.7	54.5	29.56

Table 3.216 *Continued*

Fatty acid	Copepod		Total acids				
	Triglyceride	Wax esters	Capelin	Mackerel	Atlantic herring	Pacific herring	Pacific anchovy
(*n*−6)-16:2	nd	0.20	0.84	nd	nd	0.75	–
(*n*−4)-16:2	0.09	0.52	0.06	0.45	0.23	0.27	0.47
(*n*−3)-16:3	0.26	0.01	0.82	0.52	0.39	0.44	–
(*n*−3)-16:4	0.32	1.35	nd	0.44	0.39	0.01	–
(*n*−1)-16:4	nd	nd	1.47	nd	0.03	0.42	1.39
(*n*−6)-18:2	0.97	0.83	0.78	1.28	0.78	0.67	0.93
(*n*−6)-20:2	0.04	0.04	0.05	0.24	0.09	0.14	0.08
(*n*−6)-18:3	0.16	0.14	0.01	0.07	0.04	0.07	0.12
(*n*−3)-18:3	1.08	1.00	0.20	0.99	0.39	0.18	0.41
(*n*−3)-18:4	3.23	5.63	1.36	2.47	0.93	0.67	1.78
(*n*−6)-20:4	6.39	0.19	0.14	0.36	0.24	nd	0.53
(*n*−3)-20:4	0.38	0.62	0.28	0.50	0.22	0.29	0.39
(*n*−3)-20:5	8.38	5.81	9.35	7.64	2.85	5.31	18.24
(*n*−3)-21:4	0.21	0.24	nd	0.41	0.13	0.09	0.60
(*n*−6)-22:5	0.09	0.01	0.03	0.01	0.03	nd	–
(*n*−3)-22:5	0.60	0.23	0.60	0.57	0.37	0.22	1.19
(*n*−3)-22:6	4.90	0.64	2.70	7.66	2.70	1.53	10.88
Total polyethylenic	21.0	17.5	18.7	23.6	10.2	11.1	36.24

At one time dogfish (*Squalus acanthus*) were considered as a commercial source of glyceryl ethers (Bailey *et al.*, 1952).

Table 3.217 Composition of Californian and Japanese sardine oils (Ackman and Sipos, 1964; Itoh and Fukuzumi, 1963)

Fatty acid	Californian sardine/pilchard (*Sardinops sagax*, *S. caerulea*)	Japanese sardine (*Clupanodon melanostica*)
14:0	7.6	6.6
15:0	0.6	–
16:0	16.2	15.5
16:1	9.2	9.5
17:0	0.7	–
18:0	3.5	3.7
18:1	11.4	17.3
(*n*−6)-18:2	1.3	2.5
(*n*−3)-18:3	0.9	1.3
(*n*−3)-18:4	2.0	2.9
(*n*−9)-20.1	3.2	8.1
(*n*−6)-20:4	1.6	2.5
(*n*−3)-20:5	16.9	9.6
(*n*−9)-22:1	3.6	7.8
(*n*−3)-22:5	2.5	2.8
(*n*−3)-22:6	12.9	8.5
Others	5.9	1.4

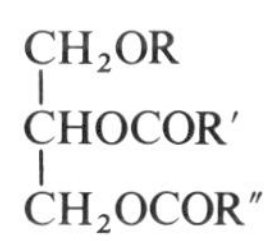

(**7**) Ether lipids

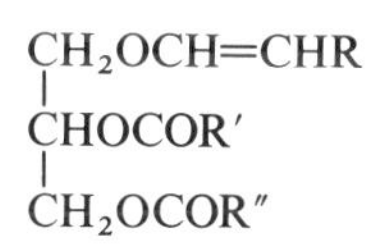

(**8**) Plasmalogen

The phospholipids in marine organisms are comparable to those in land animals (Silk and de Koning, 1964), e.g. phosphatidylcholine, phosphatidylethanolamine, cardiolipin, phosphatidylinositol, etc., were found in South African pilchard.

The fatty acid distribution at positions 1 and 2 have been determined (Brockerhoff *et al.*, 1963; Menzel and Olcott, 1964) for a number of species. The 20:5 and 22:6 acids were preferentially esterified at the 2 position.

Hydrocarbons occur in varying amounts in marine organisms. For example, the liver lipids of some sharks are almost entirely hydrocarbons (Heller *et al.*, 1957; Lambertsen and Holman, 1963), but some other marine organisms contain only trace amounts. Squalene is the predominant hydrocarbon in marine life, but small amounts of pristane and the unsaturated norphytoene are usually present (Heller *et al.*, 1957; Christiansen and Sorensen, 1951; Avigan and Blumer, 1968; Blumer and Thomas, 1965).

Twenty-one straight-chain and eight branched-

Table 3.218 The fatty acid composition (wt%) of total lipids from the fillets of several species of freshwater fish [a] (Adapted from Kinsella *et al.*, 1977)

Fish species	14:0	16:0	(*n*−7)-16:1	18:0	(*n*−9)-18:1	(*n*−6)-18:2	(*n*−3)-18:3	(*n*−3)-18:4
Bass, white (5)	2.6 ± 0.2	17.6 ± 0.8	11.4 ± 0.8	3.0 ± 0.5	29.5 ± 1.5	2.6 ± 0.4	3.3 ± 0.3	1.2 ± 0.2
Bulhead, brown (7)	2.4 ± 0.2	18.5 ± 1.6	13.8 ± 2.0	2.8 ± 0.7	25.7 ± 4.1	4.4 ± 1.5	5.4 ± 1.4	–
Burbot (6)	–	20.0 ± 0.8	3.8 ± 1.0	6.3 ± 0.6	15.9 ± 0.6	1.1 ± 0.2	tr	–
Crappie, black (6)	3.1 ± 0.8	20.3 ± 0.9	11.3 ± 2.7	3.3 ± 0.7	19.6 ± 5.1	3.1 ± 0.3	3.0 ± 0.7	1.1 ± 0.5
Drum, freshwater (6)	2.2 ± 0.4	19.5 ± 2.1	16.6 ± 3.0	3.3 ± 0.5	26.4 ± 5.9	3.1 ± 1.0	2.5 ± 2.0	–
Perch, yellow (10)	2.0 ± 0.4	20.3 ± 1.2	7.9 ± 1.8	4.7 ± 0.7	9.1 ± 1.4	1.6 ± 0.3	1.7 ± 0.6	1.3 ± 0.5
Pike, walleys (7)	1.7 ± 0.2	18.9 ± 1.0	9.4 ± 1.7	3.3 ± 0.4	18.8 ± 3.0	2.5 ± 0.4	1.3 ± 0.2	–
Salmon (1)	2.9	10.7	5.0	3.6	24.5	5.2	5.3	1.5
Smelt (6)	4.6 ± 0.2	13.8 ± 0.3	9.0 ± 0.4	1.3 ± 0.1	17.5 ± 1.2	3.6 ± 0.3	4.5 ± 0.2	1.7 ± 0.2
Sucker, white (5)	2.5 ± 0.7	15.3 ± 1.0	18.8 ± 2.4	2.2 ± 0.6	14.3 ± 2.1	2.7 ± 0.7	2.3 ± 0.6	1.9 ± 0.3
Sunfish, pumpkinseed (8)	2.3 ± 0.7	18.8 ± 1.3	7.9 ± 2.5	5.3 ± 0.7	13.2 ± 3.0	2.9 ± 0.6	1.9 ± 0.8	–
Trout, brook (8)	3.7 ± 0.2	17.9 ± 1.2	11.2 ± 2.4	4.0 ± 0.4	21.2 ± 3.5	5.5 ± 0.6	6.0 ± 0.6	2.8 ± 0.6

Fish species	(*n*−9)-20:1	(*n*−6)-20:4	(*n*−3)-20:4	(*n*−3)-20:5	(*n*−6)-22:4	(*n*−6)-22:5	(*n*−3)-22:5	(*n*−6)-22:6
Bass, white (5)	1.7	4.3 ± 0.3	–	7.1 ± 0.5	–	1.1 ± 0.1	1.4 ± 0.1	10.6 ± 0.8
Bulhead, brown (7)	1.1	4.9 ± 0.9	–	7.2 ± 1.9	–	–	2.4 ± 0.9	7.0 ± 1.2
Burbot (6)	–	15.8 ± 2.0	–	12.0 ± 1.5	1.1 ± 0.1	1.5 ± 0.3	2.8 ± 0.3	17.1 ± 2.7
Crappie, black (6)	–	5.6 ± 1.8	–	4.8 ± 1.6	–	1.4 ± 0.3	4.6 ± 0.3	14.7 ± 4.6
Drum, freshwater (6)	1.2 ± 0.4	4.9 ± 2.9	–	5.1 ± 1.4	–	1.1 ± 0.9	2.2 ± 0.6	6.9 ± 4.2
Perch, yellow (10)	–	7.7 ± 1.1	–	11.5 ± 0.9	1.2 ± 0.4	1.7 ± 0.4	2.5 ± 0.4	26.4 ± 2.0
Pike, walleys (7)	–	5.6 ± 1.3	–	8.2 ± 0.8	–	1.8 ± 0.3	1.8 ± 0.3	21.6 ± 2.7
Salmon (1)	1.0	5.3	2.3	4.5	2.2	2.3	5.0	17.0
Smelt (6)	–	3.5 ± 0.3	–	13.3 ± 1.0	–	1.1 ± 0.1	–	22.5 ± 1.8
Sucker, white (5)	1.2 ± 0.2	4.3 ± 0.4	–	10.3 ± 2.8	–	–	3.4 ± 0.3	14.9 ± 1.3
Sunfish, pumpkinseed (8)	1.0 ± 0.2	14.9 ± 2.9	–	7.1 ± 1.6	1.4 ± 0.2	3.0 ± 0.5	3.3 ± 0.5	13.7 ± 2.5
Trout, brook (8)	–	4.3 ± 0.4	1.0 ± 0.3	7.1 ± 0.9	–	–	1.0 ± 0.2	9.3 ± 2.8

[a] The number of fish analysed is given in parentheses. Fatty acids amounting to less than 1% are not included. Sucker contained 1.7% 17:1; and the 20:4 contains less than 1% 22:1.

Table 3.219 Proximate chemical composition and fatty acid contents of 37 finfish species of the southeastern USA [a] (Ackman and Cunnane, 1992)

Species [c]	Black sea bass	Bluefish	Gulf butterfish	Channel catfish [b]	Atlantic croaker	Dolphin	Southern flounder	Goosefish (monkfish)	Gag grouper	Yellowedge grouper	White grunt	Harvestfish	Speckled hind
No. of samples	2	4	4	3	4	3	2	2	3	2	3	3	3
Mean length (cm)	34.80	40.10	16.00	36.30	25.30	68.40	41.40	nd	70.40	74.55	33.20	18.40	47.60
Mean weight (kg)	0.61	0.93	0.10	0.55	0.20	2.83	1.01	nd	4.70	4.71	0.81	0.21	2.59
Proximate composition (g/100 g)													
Protein	18.0	20.0	20.0	16.2	18.0	20.2	17.4	16.2	21.0	19.3	20.5	18.9	21.1
Moisture	81.3	77.5	77.5	82.1	77.3	79.3	82.4	84.5	76.8	79.2	80.9	78.0	75.5
Fat	0.7	2.4	3.9	2.4	2.9	0.8	0.7	0.6	2.2	1.3	0.7	2.4	4.7
Ash	1.4	1.1	1.2	1.0	1.0	1.2	0.9	0.9	1.3	1.1	1.1	1.2	1.1
Fatty acid (g/100 g)													
14:0	0.005	0.064	0.137	0.045	0.046	0.005	0.009	0.003	0.060	0.020	0.007	0.057	0.149
15:0	–	–	–	–	–	–	–	–	–	–	–	–	–
16:0	0.101	0.491	0.820	0.377	0.667	0.115	0.098	0.065	0.482	0.188	0.100	0.516	1.305
18:0	0.043	0.154	0.237	0.102	0.171	0.079	0.027	0.027	0.152	0.064	0.051	0.186	0.329
$(n-7)$-16:1	0.018	0.166	0.127	0.119	0.323	0.010	0.023	0.010	0.152	0.039	0.017	0.048	0.441
$(n-9)$-18:1	0.060	0.346	0.632	0.489	0.364	0.056	0.043	0.049	0.317	0.096	0.066	0.447	0.726
$(n-7)$-18:1	0.016	0.083	0.116	0.098	0.082	0.014	0.014	0.011	0.062	0.032	0.017	0.054	0.111
$(n-9)$-20:1	0.004	0.035	0.102	0.030	0.022	0.002	0.003	0.003	0.020	0.036	0.003	0.047	0.066
$(n-11/13)$-22:1	0.000	0.006	0.002	0.001	0.006	0.000	0.001	0.001	0.004	0.027	0.002	0.010	0.002
$(n-9)$-22:1	0.003	0.002	0.035	0.001	0.003	0.000	0.000	0.002	0.003	0.006	0.000	0.014	0.013
$(n-6)$-18:2	0.004	0.039	0.031	0.053	0.011	0.005	0.009	0.005	0.016	0.010	0.004	0.013	0.025
$(n-6)$-20:4	0.024	0.033	0.072	0.069	0.054	0.022	0.013	0.015	0.051	0.033	0.048	0.056	0.051
$(n-6)$-22:4	0.006	0.008	0.027	0.009	0.021	0.000	0.002	0.002	0.014	0.009	0.011	0.019	0.016
$(n-3)$-20:5	0.027	0.083	0.127	0.076	0.125	0.022	0.018	0.025	0.065	0.036	0.033	0.053	0.087
$(n-6)$-22:5	0.010	0.011	0.039	0.029	0.014	0.016	0.006	0.005	0.023	0.017	0.010	0.032	0.032
$(n-3)$-22:5	0.012	0.050	0.100	0.053	0.057	0.009	0.014	0.007	0.039	0.031	0.014	0.046	0.065
$(n-3)$-22:6	0.121	0.174	0.380	0.094	0.125	0.196	0.095	0.090	0.276	0.209	0.088	0.235	0.403

Table 3.219 continued

Species [c]	Crevalle jack	Southern kingfish	Ladyfish	Chub mackerel	King mackerel	Striped mullet	Red porgy	Silver rag	Blue runner	Rough scad	Spotted seatrout	American shad
No. of samples	4	2	3	5	2	5	2	4	2	7	2	4
Mean length (cm)	46.60	29.20	41.30	23.70	94.80	37.00	35.30	16.00	26.30	21.70	39.00	40.10
Mean weight (kg)	4.95	0.30	1.04	0.19	5.53	0.75	0.91	0.05	0.41	0.13	0.69	1.12
Proximate composition (g/100 g)												
Protein	21.1	18.7	22.7	21.8	21.7	21.0	22.0	21.1	21.3	19.8	19.4	19.1
Moisture	74.3	77.3	73.0	72.0	76.6	74.4	77.1	72.5	75.3	76.8	79.4	65.4
Fat	3.9	4.5	4.4	5.5	1.7	5.1	0.9	5.2	1.6	2.1	2.7	14.6
Ash	1.3	1.1	1.2	1.5	1.4	1.0	1.5	1.4	1.3	1.4	1.1	1.3
Fatty acid (g/100 g)												
14:0	0.067	0.096	0.074	0.132	0.026	0.285	0.006	0.156	0.023	0.068	0.072	0.588
15:0	–	–	–	–	–	0.220	–	–	–	–	–	–
16:0	0.895	1.245	1.164	1.131	0.389	1.143	0.149	1.002	0.316	0.417	0.608	1.687
18:0	0.324	0.198	0.310	0.389	0.157	0.118	0.049	0.369	0.150	0.144	0.142	0.296
(n−7)-16:1	0.230	0.695	0.310	0.167	0.050	0.726	0.016	0.133	0.034	0.095	0.324	0.341
(n−9)-18:1	0.649	0.833	1.064	1.031	0.268	0.255	0.066	0.853	0.167	0.320	0.378	1.013
(n−7)-18:1	0.135	0.114	0.112	0.190	0.050	0.125	0.015	0.153	0.037	0.051	0.086	0.307
(n−9)-20:1	0.021	0.024	0.034	0.098	0.018	0.009	0.007	0.155	0.015	0.015	0.014	2.666
(n−11/13)-22:1	0.004	0.008	0.002	0.007	0.002	0.002	0.002	0.030	0.002	0.004	0.001	2.663
(n−9)-22:1	0.004	0.004	0.002	0.027	0.007	0.002	0.001	0.023	0.001	0.004	0.001	0.168
(n−6)-18:2	0.085	0.016	0.026	0.063	0.014	0.060	0.004	0.044	0.009	0.022	0.020	0.177
(n−6)-20:4	0.095	0.065	0.088	0.068	0.036	0.131	0.030	0.121	0.037	0.026	0.050	0.029
(n−6)-22:4	0.023	0.025	0.026	0.012	0.008	0.016	0.011	0.054	0.008	0.004	0.020	0.017
(n−3)-20:5	0.113	0.067	0.083	0.238	0.045	0.355	0.021	0.165	0.058	0.099	0.106	0.506
(n−6)-22:5	0.029	0.018	0.024	0.045	0.017	0.007	0.015	0.062	0.017	0.020	0.021	0.015
(n−3)-22:5	0.061	0.055	0.068	0.074	0.019	0.177	0.020	0.125	0.035	0.037	0.041	0.203
(n−3)-22:6	0.258	0.175	0.251	0.720	0.131	0.136	0.193	0.546	0.308	0.317	0.175	0.887

Table 3.219 continued

Species [c]	Atlantic sharpnose shark	Lemon shark	Scalloped hammerhead shark	Tiger shark	Sheepshead	Red snapper	Vermilion snapper	Spot	Tilefish	Blueline tilefish	Gray triggerfish	Weakfish
No. of samples	3	2	2	2	3	2	2	2	2	2	2	2
Mean length (cm)	53.90	183.00	229.00	nd	49.30	44.70	48.20	21.60	62.40	63.10	36.30	72.40
Mean weight (kg)	6.10	37.05	106.60	104.33	3.15	1.54	1.57	0.20	2.74	3.21	1.11	3.56
Proximate composition (g/100g)												
Protein	23.7	19.7	23.1	19.2	21.4	20.4	21.3	18.2	17.6	19.9	20.7	19.3
Moisture	75.1	79.5	76.0	80.1	77.9	77.8	78.0	71.4	81.1	78.3	79.1	78.1
Fat	0.8	0.6	0.7	0.7	1.6	1.4	0.6	8.0	1.0	2.5	0.8	2.1
Ash	1.4	1.3	1.4	1.2	1.2	1.2	1.2	1.0	1.3	1.3	1.2	1.1
Fatty acid (g/100g)												
14:0	0.001	0.002	0.001	0.002	0.034	0.029	0.009	0.240	0.018	0.067	0.002	0.052
15:0	–	–	–	–	–	–	–	–	–	–	–	–
16:0	0.100	0.055	0.064	0.048	0.310	0.276	0.088	1.987	0.139	0.419	0.087	0.378
18:0	0.085	0.044	0.064	0.073	0.086	0.107	0.036	0.492	0.040	0.129	0.071	0.089
(*n*−7)-16:1	0.005	0.004	0.006	0.009	0.087	0.049	0.009	0.718	0.032	0.102	0.006	0.134
(*n*−9)-18:1	0.038	0.037	0.036	0.089	0.284	0.199	0.045	1.400	0.129	0.406	0.065	0.277
(*n*−7)-18:1	0.023	0.013	0.023	0.020	0.043	0.031	0.007	0.244	0.029	0.077	0.020	0.042
(*n*−9)-20:1	0.003	0.002	0.003	0.003	0.011	0.015	0.003	0.085	0.021	0.076	0.002	0.079
(*n*−11/13)-22:1	0.001	0.002	0.000	0.002	0.001	0.006	0.001	0.010	0.003	0.040	0.001	0.095
(*n*−9)-22:1	0.000	0.000	0.001	0.000	0.003	0.002	0.002	0.011	0.006	0.022	0.000	0.008
(*n*−6)-18:2	0.002	0.001	0.003	0.006	0.017	0.008	0.004	0.030	0.005	0.019	0.003	0.021
(*n*−6)-20:4	0.047	0.025	0.030	0.045	0.063	0.032	0.014	0.093	0.018	0.042	0.061	0.020
(*n*−6)-22:4	0.035	0.026	0.030	0.013	0.015	0.008	0.003	0.060	0.007	0.021	0.006	0.004
(*n*−3)-20:5	0.013	0.006	0.011	0.007	0.051	0.040	0.014	0.347	0.022	0.060	0.024	0.066
(*n*−6)-22:5	0.019	0.011	0.010	0.007	0.011	0.015	0.012	0.031	0.009	0.021	0.009	0.009
(*n*−3)-22:5	0.020	0.010	0.021	0.011	0.037	0.022	0.006	0.153	0.026	0.075	0.015	0.025
(*n*−3)-22:6	0.112	0.049	0.075	0.052	0.082	0.216	0.104	0.293	0.147	0.284	0.155	0.233

Table 3.219 continued

[a] See Gooch *et al.* (1987) for original source, where data are expressed as fatty acid composition.
[b] Wild catfish, not cultured.
[c] Scientific names of fish listed

Species (common name)	Scientific name	Species (common name)	Scientific name
Black sea bass	*Centropristis striata*	Red porgy	*Pagrus pagrus*
Bluefish	*Pomatomus saltatrix*	Silver rag	*Ariomma bondi*
Gulf butterfish	*Peprilus burti*	Blue runner	*Caranx crysos*
Channel catfish	*Ictalurus punctatus*	Rough scad	*Trachurus lathami*
Atlantic croaker	*Micropogonias undulatus*	Spotted seatrout	*Cynoscion nebulosus*
Dolphin	*Coryphaena hippurus*	American shad	*Alosa sapidissima*
Southern flounder	*Paralichthys lethostigma*	Atlantic sharpnose shark	*Rhizoprionodon terraenovae*
Goosefish (monkfish)	*Lophius americanus*	Lemon shark	*Negaprion brevirostris*
Gag grouper	*Mycteroperca microlepis*	Scalloped hammerhead shark	*Sphyrna lewini*
Yellowedge grouper	*Epinephelus flavolimbatus*	Tiger shark	*Galeocerdo cuvieri*
White grunt	*Haemulon plumieri*	Sheepshead	*Archosargus probatocephalus*
Harvestfish	*Peprilus alepidotus*	Red snapper	*Lutjanus campechanus*
Speckled hind	*Epinephelus drummondhayi*	Vermilion snapper	*Rhomboplites aurorubens*
Crevalle jack	*Caranx hippos*	Spot	*Leiostomus xanthurus*
Southern kingfish (whiting)	*Menticirrhus americanus*	Tilefish	*Lopholatilus chamaeleonticeps*
Ladyfish	*Elops saurus*	Blueline tilefish	*Caulolatilus microps*
Chub mackerel	*Scomber japonicus*	Gray triggerfish	*Balistes capriscus*
King mackerel	*Scomberomorus cavalla*	Weakfish	*Cynoscion regalis*
Striped mullet	*Mugil cephalus*		

chain hydrocarbons (0.1% of total lipid) have been detected in herring lipids, of which pristane and squalene were the major ones (Heller *et al.*, 1957; Gershbein and Singh, 1969). Pristane accumulates in zooplanktonic copepods (*Calanus*) and they probably provide the major source of this hydrocarbon in higher marine life (Christiansen and Sorensen, 1951).

Tocopherols occur in fish lipids (Ackman, 1974).

Table 3.220 Fatty acid distribution in aquatic animal depot fats (Brockerhoff *et al.*, 1968)

	Position	Fatty acids (mol%)											
		14:0	16:0	16:1	18:0	18:1	18:2	20:1	22:1	20:5	22:5	22:6	Others
Fish, freshwater													
Burbot	1	3	16	28	6	38	1	11	1	2	1	1	–
	2	5	17	31	4	14	2	1	1	10	1	4	–
	3	1	3	17	2	50	4	2	1	9	3	2	–
Sheepshead	1	3	15	25	4	29	5	5	2	2	1	1	–
	2	4	28	22	1	11	5	5	2	4	5	6	–
	3	2	8	20	1	38	6	6	4	8	1	1	–
Trout A	1	2	13	8	7	24	6	11	9	4	2	3	–
	2	3	6	14	1	35	11	7	2	4	2	9	–
	3	4	13	8	8	25	5	12	9	6	2	1	–
Trout B	1	3	24	10	14	37	5	1	2	1	1	1	–
	2	1	3	17	1	45	12	3	4	1	3	7	–
	3	2	19	11	12	36	7	6	5	1	1	1	–
Goldfish	1	2	17	8	7	32	16	7	–	–	–	–	–
	2	4	32	8	3	20	22	4	–	–	–	–	–
	3	1	9	7	4	38	24	7	–	–	–	–	–

Table 3.220 continued

	Position	Fatty acids (mol%)											
		14:0	16:0	16:1	18:0	18:1	18:2	20:1	22:1	20:5	22:5	22:6	Others
Fish, marine													
Herring	1	6	12	13	1	16	3	25	14	3	1	1	–
	2	10	17	10	1	10	3	6	5	18	3	3	–
	3	4	7	5	1	8	1	20	50	4	1	1	–
Mackerel A	1	6	15	11	3	21	2	8	18	5	1	2	–
	2	10	21	6	1	9	1	5	5	12	3	0	–
	3	2	5	4	2	21	2	19	24	10	1	5	–
Mackerel B	1	6	15	12	3	19	2	16	14	4	1	1	–
	2	11	30	8	1	9	1	7	5	10	2	1	–
	3	4	7	10	2	27	2	9	25	6	1	4	–
Skate	1	2	19	12	5	30	1	12	8	4	1	5	–
	2	3	15	7	1	9	1	8	5	6	7	7	–
	3	1	6	6	1	28	2	19	11	11	2	1	–
Cod	1	2	12	11	5	35	1	14	12	3	1	3	–
	2	5	14	8	1	9	1	8	8	11	3	2	–
	3	2	6	9	1	25	2	18	11	11	1	2	–
Invertebrates													
Squid	1	4	28	8	21	4	6	3	12	1	4	–	–
	2	2	2	4	1	7	1	7	5	28	2	38	–
	3	4	12	9	2	26	1	17	13	5	1	10	–
Periwinkle	1	2	13	6	4	27	5	18	9	6	1	4	–
	2	9	17	9	1	12	5	13	3	16	2	7	–
	3	3	4	10	1	21	5	16	9	8	2	9	–
Lobster A	1	3	13	10	3	22	2	11	7	8	1	3	–
	2	4	12	7	1	17	2	10	2	13	1	15	–
	3	4	13	10	3	25	2	12	8	8	1	5	–
Lobster B	1	2	31	10	4	29	1	7	5	5	1	1	–
	2	2	8	5	1	19	2	11	6	20	4	17	–
	3	2	8	9	3	42	2	10	7	11	1	2	–
Scallop	1	3	19	5	6	19	3	7 [a]	8 [b]	22	1	5	–
	2	6	5	5	1	9	2	2 [a]	10 [b]	33	1	29	–
	3	10	12	10	1	10	2	5 [a]	14 [b]	22	1	8	–
Mammals													
Harbour seal	1	4	11	15	1	29	1	18	8	3	2	3	–
	2	11	13	30	1	30	3	3	1	1	1	1	–
	3	1	4	14	1	26	1	16	7	8	6	10	–
Harp seal	1	1	7	9	1	27	1	17	4	6	4	15	–
	2	6	9	27	2	36	5	4	1	2	1	3	–
	3	1	5	11	1	20	2	7	1	12	11	26	–
													18:4 20:4
Sei whale	1	3	13	3	4	14	1	33	10	3	1	6	1 5
	2	12	6	12	1	29	5	10	2	5	1	3	4 6
	3	4	6	2	2	7	1	28	16	6	3	16	1 3

[a] Unknown.
[b] 18:4.

Table 3.221 Analytical characteristics of commercial fish and whale oils

Characteristics	Whale	Sardine	Menhaden	Herring
Relative density, d_{25}^{25}	0.910–0.920	0.914–0.921	–	–
Refractive index, n_D^{25}	1.470–1.477	1.4785–1.4802	–	1.473–1.478
Saponification value	185–202	188–199	192–199	180–192
Iodine value	110–135	170–188	150–165	115–160
Unsaponifiable matter (%)	2.0	0.1–1.25	1.0	0.5–1.7

Table 3.222 Change in the fatty acid composition in menhaden and herring oils during hydrogenation, as percentage of oil (*INFORM*, 1992)

(a) Menhaden oil

Iodine value	159	150	140	130	120	96	90	85
Saturates	41.6	40.7	41.7	42.1	43.1	41.8	43.2	43.8
Monoenes	24.0	25.1	25.1	26.3	29.0	33.7	33.1	34.2
Dienes	39.0	4.2	5.1	7.7	8.0	10.5	10.5	13.2
Trienes	4.2	4.8	6.1	8.9	8.7	10.7	10.9	8.3
Tetraenes	5.6	5.4	5.7	6.5	6.0	3.0	2.1	0.4
Pentaenes + hexaenes	21.2	19.8	15.3	8.8	4.9	0.1	tr	tr

(b) Herring oil

Iodine value	119	107	101	88	79
Saturates	26.6	26.0	26.7	25.7	26.6
Monoenes	55.7	56.4	60.1	62.1	59.9
Dienes	2.8	3.3	4.3	8.3	11.5
Trienes	1.8	1.7	4.4	4.1	2.7
Tetraenes	3.3	2.5	1.3	0.1	0.0
Pentaenes + hexaenes	10.3	9.7	3.3	0.2	0.0

Table 3.223 Volatile components in crude menhaden oil as determined by dynamic head space/gas chromatography/mass spectrometry (DHS/GC/MS) (From Hsieh *et al.*, 1989)

Compound	Retention index	Odour characteristics	Ion *m/z*	Conc. (ppb)
Octane [a]	800		43	20
Nonane [a]	900		43	2
Pentanal [b]	979		58	360
Decane [a]	1000		57	90
2-Methyl-5-ethylfuran [c]	1014		110	10
1-Penten-3-one [a]	1022		84	240
(*E*)-But-2-enal [a,e]	1040	Painty	70	130
Hexanal [a]	1085	Cut grass, green	56	1380
Undecane [a]	1100		57	130
(*Z*)-Pent-2-enal [b,e]	1110		84	70
(*E*)-Pent-2-enal [a,e]	1130	Greasy green, musty	84	930
Heptan-3-one [a]	1162	Sickly sweet, cooling	57	8530
Heptanal [a]	1186	Waxy green, grassy	70	410
Dodecane [a]	1200		57	280
(*E*)-Hex-2-enal [a,e]	1218	Sharp green, oily	69	540
2-Pentylfuran [b]	1233		138	10
1,3,5-Trimethylbenzene [a]	1242	Pesticide-like	120	20
1,2,4-Trimethylbenzene [a]	1279	Pesticide-like	120	40

Table 3.223 *Continued*

Compound	Retention index	Odour characteristics	Ion *m/z*	Conc. (ppb)
Octanal [a]	1290	Citrus, fatty, orange	57	240
Tridecane [a]	1300		57	370
(*E*)-Hept-2-enal [a,e]	1323	Sharp green, greasy	83	140
1,2,3-Trimethylbenzene [a]	1332	Pesticide-like	120	20
Nonan-2-one [a]	1390	Musty with citrus topnote	58	620
Nonanal [a]	1395	Fatty floral	57	500
Tetradecane [a]	1400		81	230
(*E,E*)-Hexa-2,4-dienal [c,e]	1402		81	100
(*E*)-Oct-2-enal [a,e]	1429	Musty, waxy floral	70	280
Acetic acid [a]	1471	Irritating, vinegar-like	60	1380
Hepta-2,4-dienal [c,d,f]	1467	Vegetable green	81	1880
Decanal [a]	1488	Sweet, green fruity, fatty with citrus topnote	57	640
(*E,E*)-Hepta-2,4-dienal [a,e]	1493		81	2180
Pentadecane [a]	1500		57	520
(*Z,Z*)-Octa-3,5-dien-2-one [c,e]	1520		124	130
Benzaldehyde [a]	1521	Cherry, almond, sweet fruity	105	60
Nonenal [a]	1535	Fatty, waxy, musty	70	200
Propanoic acid [a]	1547	Astringent, acidic	74	2410
Octa-2,4-dienal [a,d,f]	1563		81	10
(*E,E*)-Octa-3,5-dien-2-one [c,e]	1570		124	40
Isobutanoic acid [c]	1579	Sweaty, dirty socks	73	100
(*E,Z*)-Nona-2,6-dienal [a,e]	1587		70	240
(*E,E*)-Octa-2,4-dienal [c,e]	1590		81	40
Hexadecane [a]	1600		57	10
Butanoic acid [a]	1636	Dirty socks	60	8110
Decenal [c]	1645		55	410
Heptadecane [a]	1700		85	80
Pentanoic acid [a]	1747	Parmesan cheese, dirty socks	60	330
5-Ethyl-2(5*H*)-furanone [c]	1755		112	20
Deca-2,4-dienal [a,d]	1764	Oxidized oil	81	150
Deca-2,4-dienal [a,d]	1806	Oxidized oil	81	250
Hexanoic acid [a]	1850	Sweaty, dirty socks	60	810
Nonatrienal [c,d,f]		Oxidized oil	79	60
Nonatrienal [c,d,f]		Oxidized oil	79	130
Decatrienal [c,d,f]		Oxidized fish oil	79	70
Decatrienal [c,d,f]		Oxidized fish oil	79	8
Phenol [a]	2014	Medicinal, disinfectant	94	110

[a] Identified by GC retention index and MS (standard and literature).
[b] Identified by GC retention index and MS (literature).
[c] Identified by MS (literature).
[d] Configuration of geometric isomers not determined.
[e] Prefix (*E*) denotes a *trans* isomer and prefix (*Z*) denotes a *cis* isomer.
[f] Calculation of MS response factors based on *trans,trans*-hepta-2,4-dienal.

The level of α-tocopherol shows an inverse relationship with total fish lipids, i.e. 500–600 μg/g in lean fish and 100–200 μg/g lipid in fish with high lipid levels. Typical levels in fatty fish are 200–300 μg/g lipid.

Fish oils are also of nutritional importance, because they contain high levels of vitamins A and D. Sardine, herring and menhaden oils are reported to contain, respectively, 500 and 100 USP units of vitamins A and D per gram. Fish liver oils are especially rich in these vitamins, e.g. 100 000–300 000 USP units of vitamin A per gram in halibut and soupfin shark liver oils (Springer and French, 1944; Reed *et al.*, 1944). Probably the most important commercially is cod liver oil (30–70% oil in liver), which contains 800–1500 IU vitamin A and 20–300 IU vitamin D per gram (Bailey *et al.*, 1952; Stansby, 1967). Note that IU and USP units are equivalent (Pharmacopeia, 1970).

Table 3.224 Quantitative analysis of some fish liver oils [a] (From Mangold and Mukherjee, 1975)

Lipid class	Amount (%)			
	Ratfish	Dogfish	Soupfin shark	Basking shark
Ionic and other polar lipids	2.4 ± 0.5	8.9 ± 0.8	4.6 ± 0.4	4.9 ± 0.6
Sterols	1.7 ± 0.2	–	–	–
Triacylglycerols	24.2 ± 0.7	53.6 ± 0.9	92.5 ± 0.7	42.7 ± 2.2
Alkyldiacylglycerols	66.0 ± 0.8	37.5 ± 0.4	–	–
Alk-1-enyldiacylglycerols	5.8 ± 0.6	–	–	–
Sterol esters and wax esters	–	–	1.5 ± 0.2	–
Hydrocarbons	–	–	1.4 ± 0.1	47.9 ± 2.5

[a] Average of four determinations by tubular thin-layer chromatography.

Table 3.225 Fatty acids of cod liver oil (*G. morrhua*) compared with other cod lipids (Addison *et al.*, 1968)

Fatty acid	Flesh		Liver	
	Phosphatidyl-ethanolamine	Phosphatidyl-choline	Triglyceride	Total liver triglyceride
14:0	0.5	1.4	4.4	4.9
16:0	8.8	23.9	14.1	12.4
18:0	4.4	2.6	2.6	1.8
$(n-7)$-16:1	1.7	4.2	12.0	11.6
$(n-9)$-18:1	14.4	12.6	22.4	22.6
$(n-9)$-20:1	4.0	1.7	12.1	7.6
$(n-11)$-22:1	0.3	0.4	11.4	5.2
$(n-6)$-18:2	0.8	0.9	0.9	1.4
$(n-3)$-18:3	0.3	0.2	0.2	1.3
$(n-3)$-18:4	0.3	0.3	1.2	1.9
$(n-6)$-20:4	1.6	2.4	0.5	0.5
$(n-3)$-20:5	10.7	19.4	7.3	12.6
$(n-4)$-20:5	1.9	1.6	0.4	1.0
$(n-3)$-22:6	40.5	27.3	7.4	10.6
Total	90.2	98.9	96.9	95.4

The compositions of cod flesh and cod liver oil lipids are given in Table 3.225 (Addison *et al.*, 1968).

3.5.2 Whale oils

Data on whale oils are summarized in Tables 3.226 to 3.228.

Partially hydrogenated whale oil has been widely used as an edible oil and also in soap and for a range of industrial uses. Whale stocks have become progressively depleted, and increasing pressure for conservation has been largely successful (Tables 3.226 and 3.229).

Table 3.226 World production (1000 tonnes) of marine mammals (FAO, 1992)

	1980	1989	1990
Whale	7.7	0.27	0.06
Sperm oil	3.1	–	–
Seal oil	1.0	0.38	0.36
Total marine mammals	11.9	0.65	0.42

Table 3.227 Fatty acid composition of whale oils

	14:0	16:0	18:0	16:1	18:1	20:1	22:1	($n-6$)-18:2	($n-3$)-18:3	($n-3$)-18:4	($n-6$)-20:4	($n-3$)-20:5	($n-3$)-22:5	($n-3$)-22:6
Atlantic														
Finwhale, Arctic [a]	5.0	12.0	1.7	12.0	30.6	12.1	9.7	1.3	0.5	0.7	0.3	3.7	1.6	4.1
Finwhale, Antarctic [a]	12.0	16.1	1.2	11.7	24.3	2.4	1.2	4.0	1.0	1.1	0.5	5.6	3.7	4.5
Pacific beaked (blubber) [c]	5.5	14.0	1.3	21.7	28.1	18.0	9.9	–	–	–	–	–		(others 0.5)
Sperm, Antarctic [d]	8.3	11.9	1.1	18.0	32.8	11.9	3.8	0.1	–	–	–	–		(others 0.4)
Sperm, Arctic [d]	7.4	10.4	1.0	15.9	27.7	16.6	8.6	0.7	–	–	–	–		(others 0.6)

[a] Ackman *et al.* (1965). [b] Ackman and Eaton (1966). [c] Litchfield *et al.* (1978). [d] Mori *et al.* (1964).

Table 3.228 Analytical characteristics of whale oils (Sonntag, 1979)

Species	Relative density, 15 °C	Iodine value	Saponification value
Sei	0.9196–0.9229	136.4–162.5	186.9–193.1
Fin	0.9137–0.9236	107.4–155.8	190.3–196.5
Blue	0.9140–0.9407	112.0–131.0	183.0–198.0
Humpback	0.9234–0.9154	120.3–159.4	183.5–190.1
Grey (Calif.)	0.9290	147.0–167.0	191.0–193.0
Bottlenose	0.876–0.885	79.7–88.7	121.5–135.9
Sperm	0.844–0.881	70.4–96.4	120.0–130.3

Table 3.229 Whale population estimates (rounded up to the third significant figure of the upper confidence limit) (Gambell, 1994)

Population	Year	Size	95% Confidence interval
Minke whales			
Southern hemisphere	1982/83–1988/89	760 000	510 000–1140 000
North Atlantic (excluding Canadian east coast area)	1987–89	118 000	84 000–154 000
North Pacific (North west Pacific and Okhotsk Sea)	1989–90	25 000	12 800–48 600
Fin whales			
(North Atlantic)	1969–89	47 300	27 700–82 000
Gray whales			
(Eastern North Pacific)	1987–88	21 000	19 800–22 500
Bowhead whales			
(Bering–Chukchi–Beaufort Seas)	1988	7 500	6400–9 200
Humpback whales			
(Western North Atlantic)	1979–86	5 500	8 120–2 890
Pilot whales			
(Central and eastern North Atlantic)	1989	780 000	440 000–1370 000

Because of the considerable scientific uncertainty over the numbers of whales of different species and in different geographical stocks, the International Whaling Commission decided in 1989 that it would be better not to give whale population figures except for those species/stocks which have been assessed in detail, and for which there is statistical certainty with respect to the numbers.

See also Allen, 1980.

Table 3.230 Composition of component acids and alcohols of waxes in sperm whale oil (*Physeter catodon*) (Nevenzel, 1970)

Homologue	Alcohol	Acid
10:0	–	0.9
12:0	0.2	5.5
12:1	–	0.6
14:0	9.9	10.7
14:1	0.7	7.4
15:0	1.5	0.2
16:0	27.2	4.5
16:1	10.1	21.4
17:0	1.0	0.2
17:1	–	1.1
18:0	2.8	1.3
18:1	38.6	35.0
20:1	6.9	9.2

Commercial whaling is now limited to certain species and is controlled by the International Whaling Commission. The use of whale-derived products is banned in a number of countries.

The unique properties of sperm whale oil (Table 3.230) are related to its high content of wax esters (76%) and relatively low triglyceride content (23%) (Mori *et al.*, 1964). There are no readily available comprehensive data on the fatty acid composition of whale oils. The wax esters, triglycerides and diacylglyceryl ethers of the jaw, melon and blubber fats of the Pacific beaked whale (*Berardius bairdii*) have been described (Ackman *et al.*, 1965). The positional distribution of the fatty acids in the triglycerides of the sei whale are given in Table 3.220. Alternatives to sperm whale oil have been proposed utilizing, for example, jojoba oil and derivatives of meadowfoam or crambe seed oils (Hamilton *et al.*, 1972; Jolliff, 1981).

References and further reading

Ackman, R.G. (1974) *Fishery Products* (ed. E. Kreuzer), Fishing News Books, Farnham, p. 112.

Ackman, R.G. (1980) in *Advances in Fish Science and Technology* (ed. J.J. Connell), Fishing News Books, Farnham, p. 86.

Ackman, R.G. (1988) *Chem. Ind.*, 142.

Ackman, R.G. and Cunnane, S.C. (1992) *Adv. Appl. Lipid Res.*, **1**, 161, (ed. F.B. Padley), JAI Press, London, NY.

Ackman, R.G. and Eaton, C.A. (1966) *Can. J. Biochem.*, **44**, 1561.

Ackman, R.G. and Sipos, J.C. (1964) *J. Fish. Res. Bd Can.*, **21**, 841.

Ackman, R.G., Eaton, C.A. and Jangaard, P.M. (1965) *Can. J. Biochem.*, **43**, 1513.

Addison, R.F., Ackman, R.G. and Hingley, J. (1968) *J. Fish Res. Bd Can.*, **25**, 2083.

Allen, K.R. (1980) *Conservation and Management of Whales*, University of Washington Press, Seattle.

Avigan, J. and Blumer, M.J. (1968) *J. Lipid Res.*, **9**, 350.

Bailey, B.E., Carter, N.M. and Swain, L.A. (1952) *Fish. Res. Bd Can. Bull.*, No. 89, p. 413.

Barlow, S.M. (ed.) (1980) *Nutritional Evaluation of Long Chain Fatty Acids in Fish Oil*, Academic Press, New York.

Blumer, M. and Thomas, D.W. (1965) *Science*, **148**, 370.

Brockerhoff, H., Ackman, R.G. and Hoyle, R.J. (1963) *Arch. Biochem. Biophys.*, **100**, 9.

Brockerhoff, H., Hoyle, R.J., Hwang, P.C. and Litchfield, C. (1968) *Lipids*, **3**, 24.

Christiansen, P.K. and Sorensen, N.H. (1951) *Acta Chem. Scand.*, **5**, 751.

Crawford, M.A. (1974) *Riv. Ital. Sost. Grasse*, **51**, 302.

Dabrowsky, K. (1978) *Experientia*, **34**, 1263.

de Koning, A.J. (1966) *J. Sci. Food Agric.*, **17**, 112.

de Koning, A.J. and McMullan, K.B. (1966) *J. Sci. Food Agric.*, **17**, 385.

Dyerberg, J., Bang, H.O., Stofferson, E., Moncada, S. and Vane, J.R. (1978) *Lancet*, 15 July, 117.

FAO (1992) *Yearbook of Fisheries Statistics 1990*, vol. 71, FAO.

Feeley, R.M., Criner, P.E. and Watt, B.K. (1972) *J. Am. Diet. Assoc.*, **61**, 134.

Food Manufacture (1988) *Food Manuf.*, October, p. 75.

Gambell, R. (1994) Secretary to International Whaling Commission, personal communication.

Freeman, I.P., van Lokeren, G.-J., Padley, F.B. and Polman, R.G. (1987) Br. Pat. 8718523.

Gershbein, L.J. and Singh, E.J. (1969) *J. Am. Oil Chem. Soc.*, **46**, 554.

Gooch, J.A., Hale, M.B., Brown, T. Jr., Bonnet, J.C., Brand, C.G. and Regier, L.W. (1987), NOAA (Natl. Oceanic Atmos. Adm.) Tech. Rep. NMFS (Natl. Mar. Fish Ser.) pp. 1–23.

Gray, G.M. and Macfarlane, M.G. (1961) *Biochem. J.*, **81**, 480.

Gruger, E.M., Jr (n.d.) in *Uses of Industrial Fish Oils in Industrial Fishery Technology*, 2nd edn (ed. M.E. Stansby), Krieger, New York.

Hamilton, R.J., Long, M. and Raie, N.Y. (1972) *J. Am. Oil Chem. Soc.*, **49**, 307.

Heller, J.H., Heller, M.S., Springer, S. and Clark, E. (1957) *Nature*, **197**, 919.

Hsieh, T.C.Y., Williams, S.S., Vejaphan, W. and Meyers, S.P. (1989) *J. Am. Oil Chem. Soc.*, **66**, 114.

INFORM (1992) *INFORM*, **3**, 988.

Itoh, S. and Fukuzumi, K. (1963) *J. Japan. Oil Chem. Soc.*, **12**, 278.

Jangaard, P.M., Brockerhoff, H., Burgher, R.D. and Hoyle, R.J. (1967) *J. Fish Res. Bd Can.*, **24**, 607.

Jolliff, G.D. (1981) in *New Sources of Oils and Fats* (eds E.H. Pryde, L.H. Princen and K.D. Mukherjee), American Oil Chemists' Society, Champaign.

Juarez, M. and Ramos, M. (1986) *Bull. Int. Dairy Fed.*, No. 202, 54.
Kauer, H., Grosse-Oetringhaus, S. and Hudalla, N. (1978) *J. Chromatogr.*, **154**, 93.
Kinsella, J.H., Shimp, J.L. and Weihrauch, J. (1977) *J. Am. Oil Chem. Soc.*, **54**, 424.
Kritchevsky, D., Tepper, S.A., Ditulla, N.W. and Holmes, W.L. (1967) *J. Food Sci.*, **32**, 64.
Lambertsen, G. and Holman, R.T. (1963) *Acta Chem. Scand.*, **17**, 281.
Lenarchal, P. (1978) *Rev. Fr. Corps Gras*, **25**, 303.
Litchfield, C. (1972) *Analysis of Triglycerides*, Academic Press, New York, p. 104.
Litchfield, C., Greenberg, A.J., Ackman, R.G. and Eaton, G.A. (1978) *Lipids*, **13**, 860.
Malins, D. and Wekell, J.C. (1970) *Prog. Chem. Fats Other Lipids*, **10**, 337.
Mangold, H.K. and Mukherjee, K.D. (1975) *J. Chromatogr. Sci.*, **13**, 398.
Meijboom, P.W. and Stroink, J.B.A. (1972) *J. Am. Oil Chem. Soc.*, **49**, 555.
Menzel, D.B. and Olcott, M.S. (1964) *Biochim. Biophys. Acta*, **84**, 133.
Mori, M., Iwakari, Y., Ozawa, A. and Shibata, S. (1964) *Bull. Japan. Soc. Sci. Fish.*, **30**, 161.
Mori, M., Yasuda, S. and Nishimuro, S. (1978) *Bull. Japan. Soc. Sci. Fish.*, **44**, 363.
Myher, J.J. (1978) *Handbook of Lipid Research: Fatty Acids and Glycerides* (ed. A. Kuksis), vol. 1, Plenum, New York, p. 123.
Nevenzel, J.C. (1970) *Lipids*, **5**, 308.
Nicol, J.A.C., Arnott, H.J., Mizuno, G.R., Ellison, E.C. and Chipault, J.R. (1972) *Lipids*, **7**, 171.
Oil and Soap (1942) *Oil Soap*, **19**, 140.
Patton, J.S. (1975) *Compar. Biochem. Physiol.*, **52B**, 105.
Perlstein, T., Eisner, A. and Schmeltz, I. (1974) *J. Am. Oil Chem. Soc.*, **51**, 335.
Pharmacopeia (1970) *Pharmacopeia of the United States*, 18th revision, Bethesda, MD.
Progress in the Chemistry of Fats and Other Lipids (1958) *Prog. Chem. Fats Other Lipids*, **5**, chaps 5, 6, 7.
Reed, G., Wise, E.C. and Frundt, R.J.L. (1944) *Ind. Eng. Chem., Anal. Edn*, **16**, 509.
Sargent, J.R. (1976a) in *Biochemical and Biophysical Perspectives in Marine Biology* (eds D.C. Malins and J.R. Sargent), vol. 3, Academic Press, London, p. 149.
Sargent, J.R. (1976b) *Spectrum*, **143**, 8.
Silk, M.H. and de Koning, A.J. (1964) *J. Am. Oil Chem. Soc.*, **41**, 619.
Sonntag, N.V. (1979) in *Bailey's Industrial Oil and Fat Products*, 4th edn (ed. D. Swern), Wiley, New York, p. 444.
Springer, S. and French, P.M. (1944) *Ind. Eng. Chem.*, **36**, 190.
Stansby, M.E. (1967) *Fish Oils, Their Chemistry, Technology, Stability, Nutritional Properties and Uses*, Avi, Westport, CT.
Stansby, M.E. (1981) *J. Am. Oil Chem. Soc.*, **58**, 13.
Takagi, T., Sakai, A., Hayashi, K. and Itaboshi, Y. (1979) *Lipids*, **14**, 5.
Wood, J.M. (1976) in *Biochemical and Biophysical Perspectives in Marine Biology* (ed. D.C. Malins), vol. 3, Academic Press, London.

3.6 Waxes

3.6.1 General aspects

Strictly speaking, the word 'wax' should refer to an ester of a monohydric long-chain alcohol with a long-chain acid. However, generally, the term 'wax' is used in the broad sense of the word to include surface lipids, which have properties that are broadly similar to honeycomb material. The types of compounds that can be included under this broad definition are listed in Table 3.231.

Surface waxes are exposed to the environment and therefore are chemically rather stable. Thus, there is an absence of functional groups, which might be susceptible to attack by atmospheric agents. Furthermore, the very long carbon chains of most wax components reduces their volatility. In addition, many of the compounds present in surface waxes are rather stable metabolically and are not readily susceptible to microbial degradation (Kolattukudy, 1976).

Certain general structural features of natural waxes have been described by Kolattukudy (1976), and these are summarized in Table 3.232. However, it must also be stressed that the structure and composition of surface waxes vary considerably from organism to organism. Thus, with regard to Table 3.232, the longer aliphatic chains are more abundant in plant waxes than in animal surface waxes, whereas bird waxes may contain appreciable amounts of chains of less than 16 carbons. With regard to branching, methyl branches are the most common, but in birds ethyl and propyl branches are found. Although polyunsaturated carbon chains are nearly always absent from surface waxes, in insects substantial proportions of di-unsaturated hydrocarbons have been found. In this case, autoxidation may be reduced by the simultaneous presence of cuticular phenolics. So far as the general composition of surface waxes is concerned, very-long chain hydrocarbons are common in insects and plants but rare in animals. Higher plant waxes contain the most complex mixture of components, while insects and birds have the simplest.

The biosynthesis of individual types of waxes has been thoroughly described for various classes of organisms in Kolattukudy (1976). So far as degradation is concerned, natural waxes often possess sufficient resistance to breakdown to appear in sewage sludge or fossil remains. However, although they have much slower turnovers than internal lipids, waxes are degraded by micro-organisms and do not (fortunately!) accumulate in excessive amounts in the environment.

Surface waxes serve a number of functions usually associated with protection. In plants and insects they prevent desiccation, and in birds they serve to waterproof feathers. While a few components present in surface lipids can prevent growth of pathogens, the total surface wax layer certainly functions to prevent microbial entry into the organism. Some surface lipids serve as chemical communicants, such as the hydrocarbon sex attractants and kairomone of insects. Although internal waxes are infrequently found in Nature, where they do occur they act as energy storage (e.g. in jojoba seeds or marine organisms). For a fuller review of all aspects of natural waxes, refer to Kolattukudy (1976).

3.6.2 Mammalian and bird waxes

The greatest amounts and variety of waxes in terrestrial mammals are associated with the skin. The epidermal cells play a role in synthesizing surface lipid, but the major input is provided by the sebaceous glands. These are connected via a duct into a hair canal, from which sebaceous lipid travels to the surface over a period of about a week.

Some examples of mammalian surface lipid composition can be seen in Table 3.233. It will be seen clearly that few generalizations can be made. Thus, the contents of sterols, sterol esters and wax esters are extremely variable, while squalene, non-esterified fatty acids and triacylglycerols, which are major components of human surface lipids, are minor or absent from other animals. For further details of human surface lipids see section 12.4, and for different animals refer to Downing (1976).

In birds, waxes are produced by the uropygial gland situated on the rump at the bottom of the tail feathers. The wax is then transferred to the feathers during the act of preening. Among uropygial secretions, monoester waxes predominate. In the fatty acid moieties, methyl branches are common, while the alcohol part usually contains a homologous series of unbranched and monomethyl-branched chains. Additionally, the uropygial gland wax composition varies significantly from order to order (cf. Table 3.234). Extensive details of compositions are given in Jacob (1976), and the biosynthesis of bird waxes is discussed by Buckner and Kolattukudy (1976).

Table 3.231 Classes of chemicals found in what are generally classified as natural waxes (From Kolattukudy, 1976)

Class of compound	Occurrence
1. Hydrocarbons	
(a) n-Alkanes	Bacteria, fungi, algae, higher plants, insects and higher animals including mammals [a]
(b) Branched alkanes	Bacteria, fungi, algae, higher plants, insects and higher animals including mammals
(c) Olefins	Bacteria, algae, higher plants (minor), insects and higher animals (minor)
(d) Cyclic alkanes	Plants
(e) Isoprenoid hydrocarbons (other than squalene)	Marine organisms and certain birds
2. Ketones	
(a) Monoketones	Higher plants, bacteria
(b) β-Diketones	Higher plants
3. Secondary alcohols	Higher plants, insects, bacteria (minor)
4. Alkanediols	
(a) Alkane-1,2-diols (diesters)	Birds, mammals, plants (minor)
(b) Alkane-2,3-diols (diesters)	Birds
(c) Alkane-α,ω-diols	Plants
5. Acids	
(a) Alkanoic and alkenoic acids	Every living organism [b]
(b) α- or ω-Hydroxy acids	Mammals (as diesters), higher plants
6. Wax esters	
(a) Primary alcohol esters	Bacteria, fungi, algae, higher plants, insects and higher animals including mammals
(b) Secondary alcohol esters	Insects
(c) Diester waxes	
Alkanediol diesters	Birds, mammals, insects
Diesters of hydroxy acids	Mammals, insects
(d) Triesters	
Triacylglycerols	Plants, animals
Triesters af alkane-1,2-diol, ω-hydroxy acid and fatty acid	Mammals [c]
Esters of hydroxymalonic acid, fatty acid and alcohol	Birds
Triesters of hydroxy acids, fatty acid and fatty alcohol	Insects [d]
Triesters of fatty acid, hydroxy acid and diol	Insects
(e) Polyesters of hydroxy fatty acids	Higher plants [e]
7. Primary alcohols	Bacteria, algae, higher plants, insects and higher animals including mammals
8. Aldehydes	Higher plants
9. Terpenoids	Plants and animals [f]

Note: In bacteria, fungi and algae, there is no evidence of the occurrence of surface waxes, but waxy materials have been isolated from them. In order to indicate the widespread occurrence of such compounds, these organisms are included in this table.

[a] The source of the hydrocarbons has not been established; proof of synthesis within the mammalian tissues is not available except in wool wax.

[b] Free fatty acids (very long) occur in the surface of waxes of mainly higher plants and humans.

[c] Thus far, it has been identified only in the rhino mutant mouse.

[d] Only beeswax that has been subjected to careful analysis has revealed the presence of the diesters and triesters indicated.

[e] Insoluble biopolyester called 'cutin' is the structural component of plant cuticle.

[f] Pentacyclic triterpenes are major components of certain plant waxes, while sterols occur in most higher animals.

3.6.3 Insect waxes

The surface of insects is covered by a waxy layer whose primary function is to prevent desiccation. Hydrocarbons comprise a majority of the cuticular lipids, while wax esters, sterol esters, alcohols and non-esterified fatty acids are also common components. In addition to preventing desiccation, insect waxes also function to prevent abrasion, to act as a barrier against microbial penetration and to reduce the absorption of toxic environmental chemicals (including insecticides), and in some cases some components may act in chemical communication (Jackson and Blomquist, 1976).

Some examples of the compositions of different insect waxes are given in Table 3.235. It will be seen that the overall composition varies considerably between species, but with hydrocarbons usually representing the major component. It will also be noticed that there are large differences between life stages of a given insect. This is particularly so for the stonefly, where the aquatic naiad has a much higher percentage of triacylglycerols.

The predominant n-alkanes of insects are odd chain lengths in the range C_{21}–C_{33}. Branched alkanes and alkenes are also usually odd-numbered. In primary alcohol wax esters, even-chain fatty acids and alcohols usually predominate, but secondary alcohol wax esters are unusual components. Triacylglycerols, when significant, usually have C_{16} and C_{18} saturated and unsaturated acyl groups. Sterols are often found in small amounts, with cholesterol the most common compound. Primary alcohols and non-esterified fatty acids are both usually even-chain compounds.

Table 3.232 General structural features of natural waxes

Chain length	Very long chains (up to C_{62}) are common
Branching	Branched carbon chains common, with methyl branches frequent
Unsaturation	Polyunsaturated chains nearly always absent; double bonds, when present, at different position from those of internal lipids
Functional types	Saturated hydrocarbons, olefins, wax esters, aldehydes, ketones, primary and secondary alcohols and terpenoids can be present; the bulk of the surface lipid is distinctly different from the major internal lipids of the same organism

Table 3.233 Composition (wt%) of unhydrolysed mammalian surface lipids (From Downing, 1976)

Species	Squalene	Sterols	Sterol esters	Wax esters	Diesters	Glyceryl ethers	Triacylglycerols	Free fatty acids	Free fatty acids
Human adult	12	2	3	25	–	–	41	16	–
Human newborn	9	9	33	12	7	–	26	0.5	–
Sheep	–	12	46	10	21	–	–	–	11
Rat	1	6	27	17	21	8	–	1	–
Mouse	–	13	10	5	65	–	6	–	–

Table 3.234 Composition (%) of the uropygial gland secretions of different birds (Data from Jacob, 1976)

Chain type	Heron (*Aredea cincrea*)			Kestrel (*Falco tinnunculus*)		Oyster-catcher (*Haematopus ostralegus*)		Eagle owl (*Bubo bubo*)	
	Triacylglycerol fatty acids	Wax fatty acids	Wax alcohols	Wax acids	Wax alcohols	Wax acids	Wax alcohols	Wax acids	Wax alcohols
Unbranched	98	100	100	1	32	1	50	3	23
Monomethyl	–	–	–	37	57	63	48	2[a]	48
Dimethyl	–	–	–	45	10	35	–	26	29
Trimethyl	–	–	–	16	–	–	–	3	–

[a] This species contains 42% monoethyl fatty acids.

Table 3.235 Major classes of surface lipids from nine insects (From Jackson and Blomquist, 1976)

	Lipid content (wt%)								
	Big stonefly (*Pteronarcys californica*)		Mormon cricket (*Anabrus simplex*)	Cockroaches (*Periplaneta australasiae*, *P. brunnea*, *P. fulginosa*)	Blowfly (*Lucilia cuprina*)		Eri silkworm (*Cynthia ricini*)	Mealworm (*Tenebrio molitor*)	Migratory grasshopper (*Melanoplus sanguinipes*)
	Naiad	Adult			Pupa	Adult			
Hydrocarbon	3	12	48–58	91	33	62	5	10	60
Wax esters	1	4	9–11	–	–	–	16	13	28
Triacylglycerols	78	7	–	7	25	16	–	–	1
Free fatty acids	12	49	15–18	2	16	7	tr	5	6
Aliphatic alcohols	–	–	2–3	–	–	–	70	58	2
Sterols	1	18	–	1	2	–	tr	1	1
Others	6	10	15–18	–	15	24	8	13	–

The biosynthesis of the individual insect wax components is discussed by Jackson and Blomquist (1976).

3.6.4 Plant waxes

Epicuticular wax of plants lies on and is interspersed with the insoluble part of the leaf cuticle known as cutin (section 2.11). The major components are indicated in Table 3.236. Thus, n-alkanes, monoesters, polyesters of ω-hydroxy acids, primary alcohols, acids, secondary alcohols and ketones, and β-diketones can be generally regarded as widespread and major constituents (Tulloch, 1976) (cf. Table 3.237).

The chain lengths of the hydrocarbons and primary alcohols of some frequently analysed waxes are shown in Table 3.238. Hydrocarbons of C_{27}, C_{29} and

Table 3.236 Consituents of plant waxes

Hydrocarbons	n-Alkanes probably present in all species but often in minor amounts, Waxes of many Crassulaceae contain 40–60% alkanes. Major alkanes odd-numbered in C_{21}–C_{37} range, with C_{31} most usual. Branched and unsaturated alkanes sometimes found
Monoesters	Widespread but often minor component. Acid and alcohol components of esters are usually long-chains and even-numbered
Diesters	Rarely present
Hydroxy esters	Rare
Polyesters of ω-hydroxy acids	Common components of gymnosperm waxes
Free primary alcohols	Represent 4–50% of total wax. Usually straight-chain components with even carbons. C_{22}–C_{34} usual range
Aldehydes	Minor components of most waxes (exception is sugarcane wax)
Free acids	Minor component (<10%) of most waxes. Straight chains with even carbon numbers (C_{14}–C_{36}) are usual
Secondary alcohols, ketones	Present in many waxes and may form up to 30% of total wax. C_{29} and C_{31} most common. Substitution is either in the centre of the chain or slightly towards one end
β-Diketones, hydroxy and oxo β-diketones	Major components. β-Diketones represent up to 70% of waxes of some *Eucalyptus* species and 10–50% of that of cereals. C_{29}, C_{31} and C_{33} usual with substitutions at 12,14; 14,16; and 16,18 the most common
Free diols	Rarely present and only in small amounts
Glycerides	Rare. *Rhus* wax contains significant amounts
Triterpenes	Sometimes major constituents of the wax of fruits

Table 3.237 Composition of waxes of some Gramineae (Data taken from Tulloch, 1976)

Species	Major components (% of total)				
	Hydro-carbon	Esters	Alcohols	β-Diketones	β-Hydroxy-diketones
Agropyron cristatum	15	20	48	–	–
Avena sativa	5	10	45	5	3
Elymus cinereus	8	9	8	34	21
Festuca ovina	5	5	7	70	1
Lolium perenne	5	5	70	–	–
Secale cereale	5	10	15	40	13
Stipa tenacissima	20	10	10	–	–
Triticum aestivum	10	15	20	20	10

Table 3.238 Composition of hydrocarbons and free alcohols from some plant waxes (Data taken from Tulloch, 1976)

Chain length	Grape leaf		Apple fruit		Pea leaf		Sugarcane stem	
	H [a]	A [a]	H	A	H	A	H	A
22	–	1	–	2	–	–	–	–
24	4	7	–	6	–	2	1	–
25	10	–	1	–	–	–	7	–
26	6	21	–	33	–	58	5	14
27	17	–	13	3	–	1	56	–
28	8	42	1	32	–	39	3	73
29	30	1	81	5	1	–	13	–
30	3	16	1	13	–	–	2	7
31	12	–	1	–	98	–	4	–
32	–	4	–	–	–	–	–	6
33	–	–	–	–	–	–	2	–

[a] Abbreviations: H = hydrocarbon, A = alcohols.

C_{31} are the most common, while the free alcohols are usually of C_{26}, C_{28} and C_{30}. Free acids have chain lengths that are generally similar to those of the alcohols from the same species, although the waxes of apple and rose have free acids that contain a much higher proportion of C_{16}–C_{22} components than the free alcohols.

Jojoba accumulates a storage wax and is commercially important (sections 3.3.4(*d*) and 3.5.2).

Tulloch (1976) has described taxonomic and

Table 3.239 Occurrence of wax esters in marine animals [a]

Phylum	Order/family	Wax esters (% lipid)
Coelenterata	Acinaria (sea anenomes)	15–40
	Madreporaria (scleractinian corals)	92
	Zoantharia (zoanthid corals)	30
Mollusca	Cranchiidae (squid)	32
Arthropoda	Amphipoda	12–80
Copepoda	Actideidae	11–44
	Calanidae	21–92
	Eucalanidae	11–69
	Euchaetidae	31–82
	Heterorhabdidae	67–69
	Lucicutidae	52–63
	Metridiidae	41–76
Decapoda	incl. Euphausia (krill)	15–69
Chordata	Elasmobranchii (e.g. frill sharks)	58
	Beryciformes	90
	Clupeiformes (e.g. bristlemouths)	16–85
	Crossopterygii (e.g. hatchet fishes)	10–97
	Gadiformes (e.g. cod)	25–60
	Myctophiformes (e.g. lantern fishes)	12–90
	Perciformes (e.g. dolphins, mullets)	18–92
	Zeiformes	76
	Odontoceti (e.g. dolphins, sperm whale)	30–80

[a] For full informatiom refer to Sargent *et al.* (1976).

physical aspects of plant waxes, while Kolattukudy *et al.* (1976) and Kolattukudy (1980) have detailed their biosynthesis and degradation.

A recent review by Walton (1990) is a useful source of information, particularly concerning methods for the analysis of cuticular wax (as well as cutin and suberin). The book edited by Linskens and Jackson (1991) also contains information on plant wax analysis.

3.6.5 Marine waxes

The phytoplankton contain little or no wax esters and these compounds are primarily an animal product; exceptions are the marine cryotomonad, *Chroomonas salina*, which makes large amounts of wax ester under certain growth conditions (Henderson and Sargent, 1989), and the dinoflagellate, *Peridinium foliaceum* (Lee and Patton, 1989). To date, seven phyla have been found to contain species where wax esters are major lipid components (>10% of the total lipid). Some examples of families where wax analyses have been made are shown in Table 3.239. In addition to those animal phyla listed in Table 3.239, the Ctenophora, Chaetognatha and Annelida contain species in which wax esters are major components, i.e. >10% total lipid (Lee and Patton, 1989). Copepods, which are the dominant zooplankton in most areas of the ocean, contain wax esters in an oil sac. The size of the sac changes with the amount of food digested and is presumed to have a reserve energy function (Sargent *et al.*, 1976; Bauermeister and Sargent, 1979). Fish also often contain significant amounts of wax esters, and it has been calculated that much of the organic matter produced in the euphotic zone will be transformed to wax esters (Sargent *et al.*, 1976).

In contrast, although hydrocarbons occur in all marine organisms, they generally account for 1% or less of the total lipid. Notable exceptions are the liver oils from some sharks, which may contain over 30% of their lipid as squalene.

The chain lengths of some examples of marine wax esters are shown in Table 3.240. Most fish species, marine mammals and deep-water zooplankton have C_{32}, C_{34}, C_{36} and C_{38} as major components, while upper-water zooplankton (e.g. *Calanus hyperboreas*) often have C_{42} as an additional significant component. When the fatty acid and alcohol components from wax esters are examined, some differences in chain length are obvious (Table 3.241). Thus, 16:0 and 18:1 chains are common for alcohols, while the acids are mainly 16:1 and 18:1, with 22:6 often being a significant component.

Most marine organisms contain odd-chain-length n-alkanes in the range C_{13}–C_{33}. The predominant hydrocarbons of most marine algae are polyunsaturated (particularly all-*cis*-3,6,9,12,15,18-henicosahexaene). Branched-chain hydrocarbons occur in some marine bacteria and cyanobacteria. Two terpenoid hydrocarbons with a wide distribution are squalene and pristane. Squalene is a major component of the liver fat of certain sharks and the eulachon. Pristane is the major hydrocarbon in copepods and other zooplankton.

Table 3.240 Wax ester chain-length composition (wt%) of some marine animals (Data taken from Sargent *et al.*, 1976)

Carbon number	*Gaussia princeps* (copepod; whole animal)	*Cyclothone acclinidene* (bristlemouth; whole animal)	*Gnathophausia ingens* (mysid shrimp; egg)	*Lampanyctus ritteri* (lantern fish; muscle)	*Mugil cephalus* (mullet; roe)	*Physeter catodon* (sperm whale; head oil)
26	–	–	–	–	–	2
28	–	1	–	–	–	6
30	1	6	tr	1	5	11
31	1	–	1	–	6	1
32	12	9	10	10	24	19
33	5	2	7	–	10	3
34	47	18	22	72	19	26
35	tr	3	9	–	5	3
36	15	13	15	16	12	20
37	2	3	–	–	4	1
38	11	17	25	1	12	8
39	1	3	–	–	1	–
40	3	17	8	–	2	–
42	2	8	–	–	–	–

Table 3.241 Fatty acid and alcohol compositions (wt%) from wax esters of marine animals (Data from Sargent *et al.*, 1976)

Fatty acid	*Gaussia princeps* (copepod; whole animal)		*Cyclothone acclinidene* (bristlemouth; whole animal)		*Gnathophausia ingens* (mysid shrimp; egg)		*Lampanyctus ritteri* (lantern fish; muscle)		*Mugil cephalus* (mullet; roe)		*Physeter catodon* (sperm whale; head oil)	
	Alcohol	Acid	Alcohol	Acid	Alcohol	Acid	Alcohol	Acid	Alcohol	Acid	Alcohol	Acid
12:0	–	–	–	2	–	–	–	–	–	–	tr	6
14:0	8	tr	3	2	8	tr	5	tr	10	1	10	11
14:1	tr	–	–	–	–	–	tr	–	–	–	1	7
15:0	tr	tr	–	tr	–	–	1	tr	6	tr	2	tr
16:0	55	1	22	3	53	1	82	2	54	4	27	4
16:1	tr	9	tr	12	7	12	1	4	14	23	10	21
16:2	–	tr	–	3	–	–	–	1	–	5	–	–
16:3	–	–	–	–	–	–	–	–	–	2	–	–
18:0	1	tr	1	1	6	tr	tr	1	6	1	3	1
18:1	12	71	9	42	12	51	7	72	5	13	39	35
18:2	1	2	1	2	–	1	–	3	–	4	–	–
18:3	–	1	6	–	–	1	tr	tr	–	3	–	–
18:4	–	–	–	–	–	–	–	–	–	3	–	–
20:1	5	2	27	6	1	2	tr	14	–	–	7	9
20:4	–	tr	–	4	–	1	–	–	–	4	–	–
20:5	–	6	–	6	–	7	–	tr	–	9	–	–
22:1	12	–	14	–	–	–	tr	2	–	–	–	–
22:5	–	–	–	–	–	–	–	–	–	5	–	–
22:6	–	2	–	13	–	12	–	tr	–	6	–	–
24:1	3	–	4	–	–	–	–	1	–	–	–	–

The biosynthesis and degradation of marine wax esters and hydrocarbons has been reviewed by Sargent *et al.* (1976). So far as functions are concerned, the wax esters may serve as a source of metabolic energy or of metabolic water, for buoyancy, as bio-sonar (e.g. in the head regions of whales and porpoises) or for thermal insulation. Further details can be found in Sargent *et al.* (1976) and in Lee and Patton (1989). The biochemistry of marine copepods is covered by Sargent and Henderson (1986) and Sargent and Falk-Petersen (1988).

References and further reading

Bauermeister, A. and Sargent, J.R. (1979) *Trends Biochem. Sci.*, **4**, 209.

Buckner, J.S. and Kolattukudy, P.E. (1976) in *Chemistry and Biochemistry of Natural Waxes* (ed. P.E. Kolattukudy), Elsevier, Amsterdam, p. 147.

Downing, D.T. (1976) in *Chemistry and Biochemistry of Natural Waxes* (ed. P.E. Kolattukudy), Elsevier, Amsterdam, p. 17.

Henderson, R.J. and Sargent, J.R. (1989) *Phytochemistry*, **28**, 1355.

Jackson, L.L. and Blomquist, G.J. (1976) in *Chemistry and Biochemistry of Natural Waxes* (ed. P.E. Kolattukudy), Elsevier, Amsterdam, p. 201.

Jacob, J. (1976) in *Chemistry and Biochemistry of Natural Waxes* (ed. P.E. Kolattukudy), Elsevier, Amsterdam, p. 93.

Kolattukudy, P.E. (ed.) (1976) *Chemistry and Biochemistry of Natural Waxes*, Elsevier, Amsterdam, p. 1.

Kolattukudy, P.E. (1980) in *Biochemistry of Plants* (eds P.K. Stumpf and E.E. Conn), vol. 4, Academic Press, New York, p. 571.

Kolattukudy, P.E., Croteau, R. and Buckner, J.S. (1976) in *Chemistry and Biochemistry of Natural Waxes* (ed. P.E. Kolattukudy), Elsevier, Amsterdam, p. 289.

Lee, R.F. and Patton, J.S. (1989) in *Marine Biogenic Lipids, Fats and Oils* (ed. R.G. Ackman), CRC Press, Boca Raton, FL, p. 73.

Linskens, H.F. and Jackson, J.F. (eds) (1991) *Modern Methods of Plant Analysis*, vol. 12, Springer, Berlin.

Sargent, J.R. and Falk-Petersen, S. (1988) *Hydrobiologia*, **167/168**, 101.

Sargent, J.R. and Henderson, R.J. (1986) in *The Biological Chemistry of Marine Copepods* (eds E.D.S. Corner and S.C.M. O'Hara), Clarendon, Oxford, p. 59.

Sargent, J.R., Lee, R.F. and Nevenzel, J.C. (1976) in *Chemistry and Biochemistry of Natural Waxes* (ed. P.E. Kolattukudy), Elsevier, Amsterdam, p. 49.

Tulloch, A.P. (1976) in *Chemistry and Biochemistry of Natural Waxes* (ed. P.E. Kolattukudy), Elsevier, Amsterdam, p. 235.

Walton, T.J. (1990) in *Methods in Plant Biochemistry* (eds J.L. Harwood and J.R. Boyer), vol. 4, Academic Press, London, p. 105.

3.7 Egg lipids

Egg yolks are a well known source of lipid and are frequently used for the isolation of phosphatidylcholine (Wells and Hanahan, 1969). Between 32 and 35% (depending on species) of the egg's weight is contributed by the yolk, of which about one-third is lipid. A typical composition – that of hen yolk lipids – is shown in Table 3.242. It will be seen that triacylglycerol and phosphatidylcholine are the major constituents (Long, 1961). The lipids appear to be largely complexed to proteins *in vivo*.

Hen's egg yolk has been resolved into four major fractions using ultracentrifugation procedures (Cook and Martin, 1969). These are the low-density fraction (70%), a water-soluble fraction (8%), the low-density fraction of the granule (4%) and the phosvitin–lipovitellin fraction (18%) of total yolk solids. The low-density fraction can be split further into two subfractions (see Kuksis, 1992) and is rich in lipid, particularly triacylglycerol and phospholipid (Table 3.243). The phosvitin–lipovitellin fraction contains α-lipovitellin (41%), β-lipovitellin (38%) and phosvitin (21%). The protein content of these fractions is high (Table 3.243). Aspects of hepatic synthesis and oocyte uptake of precursors have been recently reviewed (Kuksis, 1992).

There is much interest in dietary cholesterol, and eggs are recognized as being a major source of this sterol (see section 12.1). The cholesterol of eggs is predominantly unesterified and represents about 5% of the total lipid (0.3 g per hen's egg).

The fatty acid composition of yolk lipids has been estimated. The major fatty acids are palmitic and oleic, but their enrichments in triacylglycerols and phospholipids differ. Triacylglycerols contain about 53% oleate, 25% palmitate, 9% linoleate and 8% stearate as major acids. Phospholipids contain about 43% oleate, 32% palmitate, 8% linoleate and 4% stearate as major acids (Long, 1961). It should be borne in mind that the fatty acid composition of yolk lipids is very dependent on the hen's diet. In contrast, dietary fat has little effect on yolk total lipid or cholesterol content (Kuksis, 1992).

There is now much information on the detailed molecular species composition of major classes of yolk lipids. For triacylglycerol (Table 3.244) the saturates are minor components and three species

Table 3.242 The lipid composition of hen's egg yolk (Data from Long, 1961)

Lipid class	Amount (g)	Percentage lipid (wt/wt)
Neutral lipids	3.71	61.7
Sterols (cholesterol)	0.29	4.8
Phospholipids	2.01	
Phosphatidylcholine	1.47	24.5
Lysophosphatidylcholine	0.12	0.8
Sphingomyelin	0.05	0.1
Phosphatidylethanolamine	0.30	5.0
Lysophosphatidylethanolamine	0.04	0.7
Inositol phospholipids	0.01	tr
Plasmalogen	0.02	tr
Aminoacylphospholipid	tr	tr

Table 3.243 Composition and properties of egg yolk lipoproteins (From Johnson and Davenport, 1971)

Lipoprotein	Molecular weight	Sedimentation		Protein (%)	Phospholipid (%)	Cholesterol		Triacylglycerol (%)
		S_f^0	S_w^0			Free (%)	Ester (%)	
β-Lipovitellin	4×10^5	–	11	78	12	0.9	0.1	9
Low-density	2×10^6 to 10^7	15	–	18	22	3.5	0.2	58

Table 3.244 Major molecular species of egg yolk triacylglycerols (from Kuksis, 1992)

Molecular species	Fraction (mol%)
Saturates (0.7%)	
16:0–16:0–16:0	18
16:0–18:0–16:0	47
16:0–18:0–18:0	29
Monoenes (16.4%)	
16:0–18:1–16:0	24
16:0–18:1–18:0	43
18:0–18:1–16:0	14
Dienes (38.5%)	
16:0–18:1–18:1	74
Trienes (28.4%)	
16:0–18:2–18:1	51
18:1–18:2–16:0	10
18:1–18:1–18:1	19
Polyenes (16.0%)	
16:1–18:2–18:1	24
16:0–18:2–18:2	18
18:1–18:2–18:1	44

dominate, the monoene, triene and polyene fractions. By contrast, the dienes are mainly the palmitoyl–dioleoyl species (Table 3.244). As mentioned above the exact balance of species can be changed significantly by the hen's diet. However, the final fatty acid composition of hen's yolk does not reflect the dietary percentage directly, owing to rapid metabolism and limits to the extent of incorporation for individual acids (see Anderson *et al.*, 1989; Kuksis, 1992; Watkins, 1991).

The major phospholipid components of yolk are phosphatidylcholine and phosphatidylethanolamine. These differ significantly in their fatty acid contents and molecular species distributions (Table 3.245).

Table 3.245 Major molecular species (mol%) of egg yolk phosphatidylcholine and phosphatidylethanolamine (from Kuksis, 1992)

Molecular species	Phosphatidyl-choline	Phosphatidyl-ethanolamine
16:0–18:1	38.2	11.5
18:0–18:1	9.3	10.0
16:0–18:2	21.8	6.2
18:0–18:2	11.2	12.0
18:0–20:4	3.4	28.5
Others	16.1	31.8

Both are rich in polyunsaturated (and essential) fatty acids, but phosphatidylethanolamine is notable by having major proportions of arachidonate and docosahexaenoate. Further details, including analytical results for other acyl lipids, are given in Kuksis (1992).

References and further reading

Anderson, G.J., Connor, W.E., Corliss, J.D. and Lin, D.S. (1989) *J. Lipid Res.*, **30**, 433.

Cook, W.H. and Martin, W.G. (1969) in *Structural and Functional Aspects of Lipoproteins in Living Systems* (eds E. Tria and A.M. Scanu), Academic Press, New York, p. 579.

Johnson, A.R. and Davenport, J.B. (1971) *Biochemistry and Methodology of Lipids*, Wiley-Interscience, New York.

Kuksis, A. (1992) *Biochim. Biophys. Acta*, **1124**, 205.

Long, C. (ed.) (1961) *Biochemists' Handbook*, E. & F.N. Spon, London.

McIndoe, W.M. (1959) *Biochem. J.*, **73**, 45.

Watkins, B.A. (1991) *J. Nutr.*, **121**, 1475.

Wells, M.A. and Hanahan, D.J. (1969) in *Methods in Enzymology* (ed. J.M. Lowenstein), vol. 14, Academic Press, New York, p. 178.

3.8 Milk lipids*

The lipid content of mammalian milk varies from 16 g/litre (horse) to 105 g/litre (deer). Human milk contains about 38 g of fat per litre (Long, 1961). When the fat content is expressed in terms of the total calories available, then most mammals, including human, have a rather similar value, with about half of the total calories being provided by lipid.

The principal lipids of milk are triacylglycerols, which may represent up to 98% of the total lipids (Table 3.246). Human milk contains significant quantities of phospholipids, sterols, sterol esters, unesterified fatty acids and monoacylglycerols. In contrast, the proportion of phospholipids in bovine milk is low (Table 3.246).

The major phospholipids of milk from mammals examined so far are phosphatidylethanolamine, phosphatidylcholine and sphingomyelin. These each comprise about 30% of the total phospholipids (Morrison, 1970).

The fatty acid composition of different milk fats is shown in Table 3.247. It will be seen clearly that palmitic and oleic acids are the main components. Other acids usually present as significant components are capric, lauric, myristic, palmitoleic and linoleic acids. Although medium-chain acids (C_8–C_{12}) usually represent 8–15% of the total acids of milk triacylglycerols, in rabbit they may comprise up to 65% and in elephant about 100% of the total. The chain-termination mechanism by which medium-chain fatty acids are released from fatty acid synthetase is of some interest to biochemists (see section 11.1.1; Dils and Parker, 1982). When present, the medium-chain fatty acids tend to be concentrated at the *sn*-3 position (Gurr and Harwood, 1991).

The fatty acid compositions of phosphatidylcholines and sphingomyelins from several mammals are shown in Table 3.248. In contrast to triacylglycerols, the fatty acids of these phospholipids are mainly long- and very long-chain. Palmitic, stearic, oleic and, sometimes, linoleic acids are major components of the phosphatidylcholine fraction. In contrast, sphingomyelin has much smaller amounts of unsaturated components. Palmitic, stearic, arachidic, behenic and lignoceric acids are major components. In sheep milk sphingomyelin, tricosanoic acid is a major constituent, while in the other species shown nervonic acid is the major unsaturated fatty acid (Table 3.248).

A summary description of the formation of the milk fat globule and its movement into the lumen of the mammary gland has been given (Gurr and Harwood, 1991, and references therein). At birth, quite large changes in lipid metabolism occur. Whereas the fetus relied extensively on glucose, the sole source of nutrition for the newborn is milk, from which about 50% of energy comes from fat. The enzymes for fatty acid synthesis are suppressed and the baby's metabolism becomes geared to using fat directly from the diet. Human milk contains quite a high proportion of polyunsaturated fatty acids, and this may be a response to the needs of the still-developing nervous tissues. There are, however, quite large differences in milk fat composition depending on the mother's diet (Table 3.249). Despite the low content of polyunsaturated fatty acids in cow milk, there is no evidence that babies given mainly formulae based on cow milk develop less well than those primarily breast-fed. Nevertheless, manufacturers have made many modifications to the fat in baby formulae to mimic more closely human milk. Some of these formulae affect body composition (Gurr and Harwood, 1991).

Many of the compounds responsible for the flavour of milk and milk products are derived from milk lipids. Lactones and methyl ketones are derived from specific minor triacylglycerols. Other flavour volatiles have also been described (cf. Morrison, 1970). In small amounts these various compounds give rise to desirable flavours, but if present in larger quantities they may produce off-flavours.

A comprehensive review of milk lipids is given by Morrison (1970) and a recent discussion of dietary implications of milk fats is that of Gurr (1989). Useful recent references giving information on the composition and structure of milk lipids are Jensen (1988), Jensen and Clark (1988), Jensen *et al.* (1990, 1991) and Christie (1994).

* See also section 3.4.1.

Table 3.246 Composition of lipids in bovine, human and rat milks (Adapted from Morrison, 1970)

	Bovine	Human	Rat
Sterol esters	tr	1.4	1.5
Free esters	0.4	7.5	2.5
Triacylglycerols	95.5	81.0	83.3
Diacylglycerols	3.0	2.7	3.5
Monoacylglycerols	0.2		1.9
Unesterified acids	0.4	2.8	6.2
Phosphoglycerides	0.5	4.6	0.5

Table 3.247 Fatty acid composition (wt%) of the milk fats of various species (Data taken from Morrison, 1970)

Species	6:0	8:0	10:0	12:0	14:0	16:0	18:0	16:1	18:1	18:2	18:3
Red kangaroo	–	–	–	–	1.0	22.4	6.7	5.1	55.2	6.3	2.7
Rhesus macaque	–	5.4	7.0	2.4	1.9	20.9	4.7	4.7	30.8	10.7	2.6
Human	–	–	1.3	3.1	5.1	20.2	5.9	5.7	46.4	13.0	1.4
Domestic rabbit	–	22.4	20.1	2.9	1.7	14.2	3.8	2.0	13.6	14.0	4.4
Bat (*T. brasillensis*)	0.3–0.6	3.1–5.8	10.9–16.6	10.5–11.4	10.1–14.1	29.1–31.8	2.4–4.0	1.8–2.1	14.9–15.3	5.8–9.6	–
Rat	–	–	tr	tr	3.0	23.8	5.6	1.9	49.8	14.3	–
Fin whale	–	–	–	0.1	5.5	22.9	3.9	0.5	24.7	1.1	0.7
Dog	–	–	tr–1	tr–1.0	4.0	27.3	4.4	6.4	41.8	12.6	1.9
Black bear	–	–	tr–0.3	tr–0.8	3.4	26.2	tr–5.9	4.4	49.8	11.0	1.4
Ferret	–	–	–	0.4	4.3	23.2	tr	4.9	56.6	7.6	2.1
Cat	–	–	1.0	1.9	7.2	24.6	10.4	5.1	37.4	8.5	2.0
Grey seal	–	–	–	tr	3.2	17.9	1.8	15.9	30.6	0.6	tr
Horse	tr	1.8	5.1	6.2	5.7	23.8	tr–2.3	7.8	20.9	14.9	12.6
Pig	–	–	0–0.7	tr–0.5	4.0	32.9	3.5	11.3	35.2	11.9	0–0.7
Llama	–	0.4	0.5	0.4	6.0	36.8	18.9	7.3	24.1	2.0	1.6
Reindeer	1.1	0.4	1.0	1.3	10.7	29.0	11.6	–	15.0	3.7	–
Goat	2.9	2.7	8.4	3.3	10.3	24.6	12.5	2.2	28.5	2.2	–
Sheep	2.8	2.7	9.0	5.4	11.8	25.4	9.0	3.4	20.0	2.1	1.4

Table 3.248 Fatty acid composition (mol%) of milk phosphatidylcholines (PC) and sphingolipids (Sph) from various species (From Morrison, 1970)

Fatty acid (major only)	Sheep		Pig		Ass		Human	
	PC	Sph	PC	Sph	PC	Sph	PC	Sph
14:0	4.6	2.5	1.8	0.4	6.1	3.9	4.5	2.0
16:0	38.2	22.5	39.9	15.1	52.2	28.0	33.7	12.8
16:1	0.5	1.0	6.3	0.3	3.1	3.4	1.7	0.6
18:0	10.6	8.1	10.3	6.9	6.5	4.7	23.1	11.8
18:1	26.6	6.2	21.8	0.5	10.9	2.9	14.0	1.0
18:2	4.3	0.5	15.9	–	14.6	–	15.6	0.3
18:3	2.8	–	1.5	–	6.0	–	1.3	–
20:0	–	0.5	–	10.5	–	7.0	–	8.9
20:1	–	0.7	–	0.6	–	1.3	–	0.5
20:3	0.3	–	0.3	–	0.2	–	2.1	–
20:4	0.3	–	1.3	–	0.4	–	3.3	–
22:0	–	7.5	–	17.0	–	13.7	–	19.5
23:0	–	27.2	–	3.5	–	13.5	–	4.0
24:0	–	17.0	–	20.2	–	14.6	–	19.5
24:1	–	2.0	–	22.0	–	16.1	–	15.4

Table 3.249 Influence of diet on the fatty acids of human milk (From Gurr and Harwood, 1991)

	Energy[a] as fat (%)	Fatty acids (g/100 g total fatty acids)							
		12:0	14:0	16:0	18:0	16:1	18:1	18:2	Others
Subject A	7	17	13	18	6	1	40	1	4
Subject B	51	6	7	24	7	2	40	11	3

Total energy intakes were 11.7 MJ (subject A) and 10.5 MJ (subject B).

References and further reading

Christie, W.W. (1994) in *Advanced Dairy Chemistry* (ed. P.F. Fox), Part 2, Elsevier, London, in press.

Dils, R.R. and Parker, D.S. (1982) in *Biochemical Development of the Fetus and Neonate* (ed. C.T. Jones), Elsevier, Amsterdam, p. 573.

Gurr, M.I. (1989) in *Fats for the Future* (ed. R.C. Cambie), Ellis Horwood, Chichester, p. 41.

Gurr, M.I. and Harwood, J.L. (1991) *Lipid Biochemistry*, 4th edn, Chapman & Hall, London.

Jensen, R.G. (1988) *Lipids of Human Milk*, CRC Press, Boca Raton, FL.

Jensen, R.G. and Clark, R.W. (1988) in *Fundamentals of Dairy Chemistry* (eds W.P. Wong, R. Jenness, M. Keaney and E.H. Marth), 3rd edn, Van Nostrand Reinhold, New York, pp. 171–213.

Jensen, R.G., Ferris, A.M., Lammi-Keefe, C.J. and Henderson, R.A. (1990) *J. Dairy Sci.*, **73**, 223.

Jensen, R.G., Ferris, A.M. and Lammi-Keefe, C.J. (1991) *J. Dairy Sci.*, **74**, 3228.

Long, C. (ed.) (1961) *Biochemists' Handbook*, E. & F.N. Spon, London.

Morrison, W.R. (1970) in *Topics in Lipid Chemistry* (ed. F.D. Gunstone), vol. 1, Logos, London, p. 51.

3.9 Liver and other tissue lipids

With the exception of lean skeletal muscle, all major organs of the body contain appreciable lipid. The lipid is present partly as storage lipid (triacylglycerol), in, for example, liver, but mainly as phospholipid (and cholesterol) in the tissue's membranous structures. A simple distribution is shown in Table 3.250.

Mammalian liver is generally quite rich in lipid. Fat represents about 50% of the total calories available for metabolic purposes, with a range of 28% (horse) to 69% (rat) (Girard and Ferre, 1982). In terms of lipid content/kg fresh weight, then figures for five species range from about 60 g lipid/kg fresh weight for human, ox and rat through 91 for chicken to 116 g lipid/kg fresh weight of cat liver.

The high content of myelinated nerves in brain tissue means that phospholipid and sterol contents will be relatively high. In other tissues, sterols are much less important (Table 3.250).

Table 3.250 Lipid content of different mammalian tissues, as g/kg fresh weight (Data from Long, 1961)

	Total lipid	Phospho-lipid	Total sterol
Brain (man)	104	31–78	26–44
(rat)	97	45 (46)[b]	15
Muscle (lean)	0.6	– (9)	–
Gastric mucosa (tripe)	20–30	–	–
Intestine	65	–	–
Kidney (rat)	96[a]	24–27 (33)	<0.1
Liver (rat)	60	30	2.5

[a] Total fatty acids.
[b] Values in parentheses are from data in White (1973).

When the total fatty acids of liver are examined, it is found that land animals contain 30–40% of the total acids as saturated (mainly palmitic and stearic). In contrast, aquatic animals contain a much higher proportion of unsaturated acids, with a saturated total of 10–20%. Very long-chain unsaturated fatty acids are typical, with 12–30% C_{20} unsaturated and 8–18% C_{22} unsaturated acids in aquatic animals (Long, 1961).

A thorough analysis of the lipid composition of normal human liver has been made by Kwiterovich *et al.* (1970). They found that the contents of total cholesterol and total phospholipid were 3.9 ± 0.8 and 25.1 ± 2.7 mg/g wet weight, respectively. As expected, triacylglycerol content was much more variable and was 19.4 ± 15.9 mg/g wet weight. Human liver glycolipids were analysed, with dihexosides being the major neutral ceramides and GM_3 (hematoside) the major ganglioside. The phospholipid compositions of several mammalian species are shown in Table 3.251. It will be seen clearly that phosphatidylcholine and phosphatidylethanolamine are the major components.

When different sheep or ox tissues are compared for their phospholipid composition (Table 3.252), although phosphatidylcholine and phosphatidylethanolamine are always major components, their plasmalogen derivatives are much more variable. Heart contains large amounts of the latter. Phosphatidylinositol, phosphatidylserine and sphingomyelin are always significant components. Diphosphatidylglycerol, which is localized in mitochondria, is found in appreciable quantities in those tissues, such as heart muscle, which contain large numbers of these organelles.

Table 3.251 Phospholipid composition (%) of livers from several mammalian species (from White, 1973)

	Human	Rat	Sheep	Bovine	Mouse
Phosphatidylcholine[a]	43.6	50.8	40.5	55.7	45.6
Phosphatidylethanolamine[a]	27.9	25.2	31.5	13.0	25.1
Phosphatidylinositol	8.6	7.2	8.2	7.9	7.7
Phosphatidylserine	3.1	3.2	2.3	4.2	3.9
Phosphatidylglycerol	–	–	5.3	–	0.1
Diphosphatidylglycerol	3.7	4.8	1.1	4.1	4.7
Sphingomyelin	4.6	4.2	5.0	5.8	4.7
Others	8.5	5.6	6.1	9.3	8.2

[a] Includes plasmalogen derivatives when analysed.

Table 3.252 Phospholipid composition of various sheep or ox tissues (from White, 1973)

	Sheep				Ox		
	Brain	Heart	Liver	Kidney	Spleen	Lung	Skeletal muscle
Phosphatidylcholine	37.3	25.5	39.7	35.8	40.5	39.5	42.5
Choline plasmalogen	0.9	20.4	0.8	3.2	3.0	0.3	–
Phosphatidylethanolamine	7.7	12.4	27.9	10.6	25.0	21.2	26.6
Ethanolamine plasmalogen	16.5	11.0	3.6	7.4	9.8	0.2	–
Phosphatidylinositol	2.1	4.7	8.2	4.6	1.6	3.3	5.6
Phosphatidylserine	9.2	2.2	2.3	3.2	2.1	9.4	4.1
Sphingomyelin	12.8	5.5	5.0	12.1	12.5	16.1	4.5
Phosphatidylglycerol	–	–	5.3	–	–	2.0	0.3
Diphosphatidylglycerol	2.0	5.8	1.1	5.9	1.5	1.0	8.9

Table 3.253 Influence of diet on the lipid composition of different rat tissue samples. Results as g/kg fresh weight tissue; 15-week feeding regime (Data from Fasman, 1975)

Sample		Diet				
		Fat-free	Hydrogenated soybean fat	Corn oil	Milk fat	Beef tallow
Liver	Neutrals[a]	35 ± 1	46 ± 2	45 ± 2	46 ± 1	48 ± 2
	CE	5 ± 1	5 ± tr	5 ± 2	5 ± 1	8 ± 1
	PL	26 ± 1	35 ± 1	33 ± 1	31 ± 1	32 ± 1
Heart	Neutrals	59 ± 1	63 ± 2	49 ± 2	41 ± 1	42 ± 2
	CE	2 ± tr	5 ± 1	4 ± 1	4 ± 1	3 ± tr
	PL	22 ± 1	18 ± 1	33 ± 1	26 ± 1	22 ± 1

[a] Neutrals = primarily triacylglycerols but also cholesterol and non-esterified fatty acids; CE = cholesterol esters; PL = phosphoglycerides.

Not surprisingly, dietary factors can play a significant role in influencing the lipid composition of different tissues, particularly those where lipid storage is significant. Aspects of diet are dealt with in section 12.1. In Table 3.253 an example of data showing alterations in lipid composition during a feeding regime in rat is given. Of note is that structural lipid content is also altered in many cases quite significantly. As discussed by Gurr (1992), it is very important to bear in mind that different animals react quite differently and that animal models may not always be appropriate with regard to, for example, human obesity or heart disease.

References and further reading

Fasman, G.D. (ed.) (1975) *CRC Handbook of Biochemistry and Molecular Biology*, 3rd edn, CRC Press, Cleveland, OH.

Girard, J. and Ferre, P. (1982) in *The Biochemical Development of the Fetus and Neonate* (ed. C.T. Jones), Elsevier, Amsterdam, p. 517.

Gurr, M.I. (1992) *Prog. Lipid Res.*, **31**, 195.

Kwiterovich, P.O., Sloan, H.R. and Frederickson, D.S. (1970) *J. Lipid Res.*, **11**, 322.

Long, C. (ed.) (1961) *The Biochemists' Handbook*, E. & F.N. Spon, London.

White, D.A. (1973) in *Form and Function of Phospholipids* (eds G.B. Ansell, R.M.C. Dawson and J.N. Hawthorne), Elsevier, Amsterdam, p. 441.

3.10 Cereal lipids

There has been increasing interest in cereal lipids because of their importance in food technology (e.g. in wheat flour) or as potential sources of significant amounts of dietary polyunsaturated fatty acids (Price and Parsons, 1974). Morrison and co-workers have conducted a systematic study of the distribution of acyl lipids in different parts of wheat and maize seed. In addition, there has been considerable interest in the rather unusual content of lipids in starch (Morrison, 1981).

The lipid content of total cereal seeds varies considerably. Furthermore, the various anatomical parts of seeds have very different lipid contents. Thus, in wheat seeds the germ contains 25–30% lipid while the endosperm has only 1% lipid. Considerable amounts of wheat germ lipid are lost during milling (Galliard and Barnes, 1980). Cereal germ oils are also a very good source of tocopherols.

Maize has been examined in detail (Tan and Morrison, 1979b). Pericarp lipids seem to be the remnants of those of spherosomes (i.e. triacylglycerol, non-esterified fatty acids, sterol esters, diacylglycerol). Maize germ has a very high lipid content (~45%), which is mostly triacylglycerol with small amounts of phosphoglycerides. Indirect analysis of the aleurone layer by difference (Tan and Morrison, 1979a) showed that it contained most of the endosperm triacylglycerol.

Wheat showed many similarities to maize except that the germ of wheat was much smaller and there was less lipid degradation in the developing wheat endosperm. Pericarp lipids were mostly triacylglycerol, diacylglycerol and sterol esters (as in maize pericarp) (Hargin and Morrison, 1980). The aleurone and germ contain similar amounts and proportions of lipids. In comparison to maize, more phospholipid (14–18% total lipid) is found. Endosperm non-starch lipids are mainly triacylglycerol, which is in spherosome structures. Glycosylglycerides are present, presumably from the amyloplast membranes (Hargin *et al.*, 1980).

Sorghum lipids resemble maize and there is considerable similarity between wheat, rye, triticale and (to a lesser extent) oat lipids. Rice lipids resemble those of barley or oats except that there is sometimes very little glycolipid or phospholipid in milled rice (Choudhury and Juliano, 1980a, b).

For representative analyses of maize and wheat

Table 3.254 Lipid composition of fractions from wheat and maize seeds

Lipid class	H51 maize[a] (μg/kernel)		LG11 maize[b] (μg/mg seed)		Wheat[c]	
	Germ	Endosperm	Germ	Non-starch endosperm	Germ	Non-starch endosperm
SE			818	49	8–10	4–17
TG	7479 (SE–FFA)	1008 (SE–FFA)	36 953	344	196–244	46–70
DG			1244	33	0–13	10–18
FFA			305	133	3–6	8–21
DGDG	24	320	436	12	0–7[e] (DGDG + DGMG)	32–74
DGMG	5[d]	296[d]	172	12		4–24
N-AcylPE	8	5	110	3	tr	8–50
PE	51	20	194	1	4–7	1–9[f]
PC	252	154	748	3	20–28	5–20
PI	96	14	247	3	6–9	n.d.

[a] Weber (1979).
[b] Tan and Morrison (1979a, b).
[c] Hargin and Morrison (1980).
[d] Includes SQDG; a number of other minor lipids were quantified (see references).
[e] Includes other glycosylglycerides.
[f] Includes PG.

Abbreviations: SE = sterol ester; TG = triacylglycerol; DG = diacylglycerol; FFA = non-esterified fatty acid; DGDG = digalactosyldiacylglycerol; DGMG = digalactosylmonoacylglycerol; *N*-acylPE = *N*-acylphosphatidylethanolamine; PC = phosphatidylcholine; PI = phosphatidylinositol; SQDG = diacylsulphoquinovosylglycerol; PG = phosphatidylglycerol.

Table 3.255 Lipids in cereal starches (mg/100 g) (Data from Morrison, 1983)

Lipid	Wheat A-granules	Maize	Rice
SE, TG, DG	1–2	3	2–12
FFA	15–54	379	221–355
MG, ASG*	1–2	9	19–31
MGMG, DGMG*	4–12	13	20–40
LPG	30–41	7	38–48
LPE	41–75	20	86–112
LPC	451–734	262	453–513
Others	27–62	9	–

Abbreviations: see Table 3.254; MG = monoacylglycerol; ASG = acylated sterol glycoside; LPG = monoacylphosphatidylglycerol; LPE = monoacylphosphatidylethanolamine; LPC = monoacylphosphatidylcholine; MGMG = monogalactosylmonoacylglycerol.
* Probably from contaminating non-starch lipids.

lipids see Table 3.254 and refer to Morrison (1978) for reviews.

Cereal starches contain lipids that are very difficult to extract. The analyses in Table 3.255 are for internal starch lipids, although the presence of surface and non-starch lipids can modify the properties of starch significantly (Morrison, 1981). The internal lipids are almost exclusively monoacyl lipids (86–94%). The presence of significant quantities of non-esterified fatty acids and other neutral lipids (Becker and Acker, 1976) may indicate that the starches that were analysed were not pure and the neutral lipids arose from non-starch or surface lipids (Morrison, 1983). In contrast, maize and rice starch internal lipids have more non-esterified fatty acid than lysophospholipid (Morrison and Milligan, 1982). Precautions to be taken for the accurate estimation of starch lipids have been discussed (Morrison, 1988) and methods for the use of alcohol/water mixtures in extracting internal lipids described (Morrison and Coventry, 1985).

The fatty acids of cereal lipids are generally linoleate, palmitate, oleate, α-linolenate and stearate in order of abundance. Considerable variations in the fatty acid compositions of bulk lipids have been achieved by breeding in barley, maize and oats. The fatty acid content and distribution in cereals has been reviewed (cf. Barnes, 1983). If the acyl composition of individual lipids is analysed, then 200–500 mg of starch are needed per analysis (Morrison and Karkalas, 1990).

Considerable interest has been shown in the study of lipid metabolism during the germination of cereal seeds, and this subject has been reviewed (Tomos and Laidman, 1979). Moreover, the activity of endogenous lipases causes considerable spoilage of wholewheat flours. This is due to bran-localized lipases and a germ-localized lipoxygenase (Galliard, 1986).

References and further reading

Barnes, P.J. (ed.) (1983) *Lipids in Cereal Technology*, Academic Press, London.

Becker, G. and Acker, L. (1976) *Die Lipide der Getreidestarken. Handbuch der Stärke in Einzeldarstellung* (ed. M. Ulmann), vol. VI-5, Paul Parey, Berlin.

Choudhury, N.H. and Juliano, B.O. (1980a) *Phytochemistry*, **19**, 1063.

Choudhury, N.H. and Juliano, B.O. (1980b) *Phytochemistry*, **19**, 1385.

Galliard, T. (1986) *J. Cereal Sci.*, **4**, 33.

Galliard, T. and Barnes, P.J. (1980) in *Biogenesis and Function of Plant Lipids* (eds P. Mazliak, P. Benveniste, C. Costes and R. Douce), Elsevier, Amsterdam, p. 191.

Hargin, K.D. and Morrison, W.R. (1980) *J. Sci. Food Agric.*, **31**, 877.

Hargin, K.D., Morrison, W.R. and Fulcher, R.G. (1980) *Cereal Chem.*, **37**, 320.

Morrison, W.R. (1978) *Adv. Cereal Sci. Technol.*, **2**, 221.

Morrison, W.R. (1981) *Stärke*, **33**, 408.

Morrison, W.R. (1983) in *Lipids in Cereal Technology* (ed. P.J. Barnes), Academic Press, London, p. 11.

Morrison, W.R. (1988) *J. Cereal Sci.*, **8**, 1.

Morrison, W.R. and Coventry, A.M. (1985) *Stärke*, **37**, 83.

Morrison, W.R. and Karkalas, J. (1990) in *Methods in Plant Biochemistry* (ed. P.M. Day and J.B. Harborne), vol. 2, Academic Press, London, p. 323.

Morrison, W.R. and Milligan, T.P. (1982) in *Maize: Recent Advances in Chemistry and Technology* (ed. G.E. Inglett), Academic Press, New York, p. 1.

Price, P.B. and Parsons, J.G. (1974) *Lipids*, **9**, 560.

Tan, S.L. and Morrison, W.R. (1979a) *J. Am. Oil Chem. Soc.*, **56**, 531.

Tan, S.L. and Morrison, W.R. (1979b) *J. Am. Oil Chem. Soc.*, **56**, 759.

Tomos, A.D. and Laidman, D.L. (1979) in *Recent Advances in the Biochemistry of Cereals* (eds D.L. Laidman and R.G. Wyn Jones), Academic Press, London, p. 119.

Weber, E.J. (1979) *J. Am. Oil Chem. Soc.*, **56**, 637.

3.11 Leaf lipids

Both the lipid and the fatty acid compositions of plant leaves show a remarkable consistency between species. Furthermore, these compositions are only slightly altered by external factors and developmental period (Hitchcock and Nichols, 1971). The only small exception to these comments is the division of plants into the so-called '16:3-plants' (those containing hexadecatrienoic acid mainly esterified in diacylgalactosylglycerol) and '18:3-plants' (which contain little, if any, hexadecatrienoic acid) (Heinz, 1977). These differences are also reflected in similar variations in lipid metabolism in 16:3-plants and 18:3-plants (Heinz and Roughan, 1982).

Up to 7% of the dry weight of leaves is lipid, with the galactosylglycerides being major constituents (Table 3.256). The plant sulpholipid (diacylsulphoquinovosylglycerol) is also a significant component, representing about 5% of the total acyl lipids. The remainder of the lipid consists principally of phospholipid, with phosphatidylcholine, phosphatidylglycerol and phosphatidylethanolamine as major components. Other phospholipids found in minor amounts are phosphatidylinositol and diphosphatidylglycerol, while phosphatidylserine, phosphatidic acid and phytoglycolipid are usually found in trace amounts. Other minor lipids include triacylglycerol, diacylglycerol and glucocerebroside (Hitchcock and Nichols, 1971). The waxy cuticle also contains typical lipid constituents (see section 2.10), while carotenoids, chlorophylls and plastoquinone are major lipids of the photosynthetic membranes (Table 3.257).

Table 3.256 The acyl lipid composition of plant leaves (Reproduced from Harwood, 1980, with permission)

Plant	Lipids (% of total acyl lipids)							
	MGDG	DGDG	SQDG	PC	PG	PE	PI	Others
Clover	46	28	4	7	6	5	1	3
Sugarbeet	26	14	9	24	9	12	5	1
Maize	42	31	5	6	7	3	1	2
Barley	43	26	5	11	6	4	1	4
Rye grass	39	29	4	10	7	5	2	4
Broad bean	38	30	6	7	6	4	2	7

Abbreviations: MGDG = diacylgalactosylglycerol; DGDG = diacyldigalactosylglycerol; SQDG = diacylsulphoquinovosylglycerol; PC = phosphatidylcholine; PG = phosphatidylglycerol; PE = phosphatidylethanolamine; PI = phosphatidylinositol.

Table 3.257 Pigment composition of spinach chloroplast membranes

	Thylakoids[a]		Envelope (μg/mg of protein)[c]
	Moles of lipid/100 mol of chlorophyll[b]	μg/mg of protein[c]	
Chlorophyll a	70	–	–
Chlorophyll b	30	–	–
Carotenoids			
β-Carotene	6	–	1.42
Violaxanthin	3	4.52	6.50
Lutein + zeaxanthin	10	6.05	2.50
Antheraxanthin	–	–	0.79
Neoxanthin	3	–	0.40

[a] Typical thylakoid membranes are about 50% lipid, of which 25% is pigment.
[b] Lichtenthaler and Park (1963).
[c] Douce and Joyard (1979).

The overall distribution of lipids in leaves reflects the high content of chloroplasts in the tissue. The thylakoid membranes of the latter have a rather simple acyl lipid composition (cf. Harwood, 1980), which resembles that of cyanobacterial membranes. Plants grown in the dark have pale yellow leaves, which contain etioplasts rather than chloroplasts. The lipid composition of such etioplasts is different from normal green leaves in that glycosylglycerides are less important and there are changes in the fatty acid composition (Table 3.258). These differences are also seen if the white and green parts of variegated leaves are compared (Table 3.258).

The major fatty acids of leaves are palmitic, linoleic and α-linolenic. The main monoenoic acids are oleic and palmitoleic, with consistent amounts of *trans*-Δ^3-hexadecenoic acid in green leaves. This unusual fatty acid is located at the 2 position of phosphatidylglycerol (Harwood, 1980). Other fatty acids are only present in minor amounts, with the exception of 16:3-plants such as spinach or tobacco, which contain all-*cis*-7,10,13-hexadecatrienoic acid (Table 3.258).

As mentioned above, fatty acids of unusual structure that characterize seed oils from a variety of plants are seldom found in the lipids of the corresponding leaf tissue. However, there are two exceptions to this rule. First, some members of the family Malvaceae accumulate cyclopropenoid fatty acids in their leaves as well as their seeds (Shenstone and Vickery, 1961). Secondly, the γ-linolenic and octadecatetraenoic acids that characterize the seed oils of the Boraginaceae are also found esterified to the leaf lipids (Jamieson and Reid, 1968, 1969). Stymne *et al.* (1987) have shown that both γ-linolenic and octadecatetraenoic acids are present at the *sn*-2 position of monogalactosyldiacylglycerol, where octadecatetraenoic acid was the major component (74%) in *Borago officinalis*.

In view of the fact that leaf lipids are dominated by those of the chloroplasts, it is worth emphasizing certain features of the latter's thylakoids. In contrast to mammalian and most bacterial membranes, phospholipids are minor components. Phosphatidylglycerol is the only phosphoglyceride of importance. Over 75% of the acyl lipid content is glycosylglyceride (Table 3.259). In addition, the fatty acid composition of chloroplast acyl lipids is unusually rich in polyunsaturates. α-Linolenic acid, the main fatty

Table 3.258 Fatty acid composition of leaf tissue (From Harwood, 1980)

Plant	Fatty acids (% of total acids)							
	16:0	9*c*-16:1	3*t*-16:1	16:3	18:0	18:1	18:2	18:3
Maize	8	1	3	0	2	7	8	66
Barley	13	3	3	0	2	6	6	64
Broad bean	12	1	4	0	2	7	14	56
Broad bean (etiolated)	16	2	0	0	4	4	29	41
Chlorophytum (green part)	12	1	2	0	4	3	24	48
Chlorophytum (white part)	19	1	0	0	5	4	39	24
Pea	12	1	2	0	1	2	25	53
Spinach[a]	13	–	3	5	tr	7	16	56

[a] Data from Debuch (1961).

Table 3.259 The lipid composition of chloroplast membranes

Plant	Fraction	Percentage of total lipids					
		MGDG	DGDG	SQDG	PG	PC	PI
Spinach[a]	Envelope	22	32	5	8	27	1
	Thylakoid	51	26	7	9	3	1
Wheat[b]	Envelope	22	43	11	10	14	–
	Thylakoid	45	36	8	10	2	–

Abbreviations: see Table 3.256.
[a] Douce *et al.* (1973).
[b] Bahl *et al.* (1976).

Table 3.260 The fatty acid composition of chloroplast lipids

Plant	Lipid	Fatty acid composition (% of total fatty acids)						
		16:0	16:1[a]	16:3	18:0	18:1	18:2	18:3
Spinach	MGDG	tr	–	25	–	1	2	72
	DGDG	3	–	5	–	2	2	87
	SQDG	29	1	–	1	7	26	36
	PG	11	32	–	–	2	4	47
Barley	MGDG	3	1	–	1	1	3	91
	DGDG	9	2	–	1	3	7	78
	SQDG	32	3	–	1	2	5	55
	PG	18	27	–	3	2	11	38

Abbreviations: see Table 3.256.
[a] 16:1 = *cis*-9-hexadecenoic acid in MGDG, DGDG and SQDG and *trans*-3-hexadecenoic acid in PG.
18:1 = oleic acid; 18:2 = linoleic acid; 18:3 = α-linolenic acid.

acid, is particularly enriched in the two galactosylglycerides. Phosphatidylglycerol also contains high amounts of α-linolenic acid, but also has large amounts of palmitic and *trans*-Δ^3-hexadecenoic acids, while diacylsulphoquinovosylglycerol also contains palmitic acid as a major component (Table 3.260).

For recent references to work on plant lipids, see Quinn and Harwood (1990), and for methodological discussion, refer to Harwood and Boyer (1990). Research making use of mutants of *Arabidopsis* (and other species) that have changed lipid compositions is covered by a comprehensive review (Ohlrogge *et al.*, 1991).

References and further reading

Bahl, J., Francke, B. and Monegar, R. (1976) *Planta*, **129**, 193.

Debuch, H. (1961) *Z. Naturforsch.*, **16B**, 561.

Douce, R. and Joyard, J. (1979) *Adv. Bot. Res.*, **7**, 1.

Douce, R., Holz, R.B. and Benson, A.A. (1973) *J. Biol. Chem.*, **248**, 7215.

Harwood, J.L. (1980) in *Biochemistry of Plants* (eds P.K. Stumpf and E.E. Conn), vol. 4, Academic Press, New York, p. 1.

Harwood, J.L. and Boyer, J.R. (eds) (1990) *Methods in Plant Biochemistry*, vol. 4, Academic Press, London.

Heinz, E. (1977) in *Lipids and Lipid Polymers in Higher Plants* (eds M. Tevini and H.K. Lichtenthaler), Springer-Verlag, Berlin, p. 102.

Heinz, E. and Roughan, P.G. (1982) in *Biochemistry and Metabolism of Plant Lipids* (eds J.F.G.M. Wintermans and P.J.C. Kuiper), Elsevier, Amsterdam, p. 169.

Hitchcock, C. and Nichols, B.W. (1971) *Plant Lipid Biochemistry*, Academic Press, London.

Jamieson, G.R. and Reid, E.H. (1968) *J. Sci. Food Agric.*, **19**, 628.

Jamieson, G.R. and Reid, E.H. (1969) *Phytochemistry*, **8**, 1489.

Lichtenthaler, H.K. and Park, R.B. (1963) *Nature*, **198**, 1070.

Ohlrogge, J.B., Browse, J. and Somerville, C.R. (1991) *Biochim. Biophys. Acta*, **1082**, 1.

Quinn, P.J. and Harwood, J.L. (eds) (1990) *Plant Lipid Biochemistry, Structure and Utilization*, Portland Press, London.

Shenstone, F.S. and Vickery, J.R. (1961) *Nature*, **190**, 168.

Stymne, S., Griffiths, G. and Stobart, A.K. (1987) in *The Metabolism, Structure and Function of Plant Lipids* (eds P.K. Stumpf, J.B. Mudd and W.D. Nes), Plenum, New York, p. 405.

3.12 Algal lipids

There is increasing commercial interest in the exploitation of algae (both freshwater and marine) as sources of foodstuffs and other important natural products. One subject of particular importance is that of polyunsaturated fatty acids – since many algae are rather rich in these components.

Algae are an extremely diverse group of organisms. It is unsurprising, therefore, that their lipids are varied. Most attention has been paid to algal fatty acid content. Freshwater algae contain similar fatty acids to terrestrial plants (see section 3.11 for leaf composition). However, the proportions of these acids vary considerably and, in general, freshwater algae contain higher proportions of C_{16} fatty acids and less of C_{18} components than do plant leaves. Representative fatty acid compositions of some freshwater and salt-tolerant algae are shown in Table 3.261. It will be noted that, whereas C_{20} fatty acids are absent (or only found in trace amounts) in freshwater species, in those from high-salt environments, such as the Dead Sea, arachidonate and $(n-3)$ eicosapentaenoate are major components (Pohl and Zurheide, 1979; Harwood and Jones, 1989).

Marine algae contain a bewildering array of major fatty acids. Some representative examples of compositions for phytoplankton and macroalgae are shown in Table 3.261. The major saturated fatty acid is invariably palmitate and, in contrast to higher plants, palmitoleate is the major monoene. C_{18} fatty acids are much less abundant than in leaves, and the C_{20} polyunsaturated acids (particularly arachidonate and eicosapentaenoate) are very important. In some individual lipid classes, these latter acids can account for a very high proportion of the total acyl groups (see Harwood and Jones, 1989). Interestingly, in view of its postulated role in granal stacking (see Bolton *et al.*, 1978), *trans*-Δ^3-hexadecenoate is found in phosphatidylglycerol of algae that have a quite different chloroplast morphology to higher plants, e.g. brown and red algae and diatoms (see Harwood and Jones, 1989).

Cyanobacteria have a much simpler lipid and fatty acid composition than eukaryotic algae. Murata and Nishida (1987) have divided cyanobacteria into four groups based on the original classification of Kenyon *et al.* (1972). The fatty acid compositions of examples from each of the four groups are shown in Table 3.262. Group 1 cyanobacteria only synthesize saturated and mono-unsaturated acids. The other three groups all synthesize the polyunsaturated linoleate and, in addition, accumulate α-linolenate, γ-linolenate and $(n-3)$ octadecatetraenoate, respectively.

Cyanobacteria also differ from eukaryotic algae in having a very simple lipid class composition. Only the typical 'chloroplast' lipids (monogalactosyldiacylglycerol, digalactosyldiacylglycerol, sulphoquinovosyldiacylglycerol and phosphatidylglycerol) are found, with, sometimes, minor amounts of monoglucosyldiacylglycerol (Table 3.263). In nitrogen-fixing cyanobacteria, unusual glycolipids have been reported. These are most easily studied in the heterocysts of the filamentous heterocystous strains. Chemical structures for the main glycolipid components have been reported (Bryce *et al.*, 1972) and minor components described (see Murata and Nishida, 1987).

For eukaryotic algae, the glycosylglycerides are major components and all the usual phosphoglycerides are found. Some representative examples of different compositions are shown in Table 3.264.

Certain classes of algae may accumulate significant amounts of more unusual complex lipids. DGTS (diacylglycerol-*O*-(*N*,*N*,*N*-trimethyl)homoserine) is an unusual compound found in Chlorophyta but first reported in *Ochromonas danica* (see Harwood and Jones, 1989). Its β-alanine derivative diacylglycerol-*O*-hydroxymethyl-(*N*,*N*,*N*-trimethyl)-β-alanine (DGTA) has been identified in Phaeophyta (see Eichenberger, 1990). Many freshwater algae contain chlorosulpholipids (Mercer and Davies, 1979). These unusual lipids have been reviewed (Haines, 1973a, b). Certain diatoms have been found to contain other novel sulphur-containing lipids (see Kates, 1987).

The metabolism of algal lipids has been reviewed by Harwood and Jones (1989). Some aspects of the use of algae as a commercial source of lipids, particularly hydrocarbons, are covered by Scragg and Leathers (1988). *Botrycoccus braunii* and *Dunaliella salina* produce hydrocarbons at levels of 15–75%, while the red alga *Porphyridium cruentum* has been proposed as a possible source of arachidonic acid. The importance of such essential fatty acids when microalgae are used as feedstocks in aquaculture is reviewed by Volkman (1989), who also covers some more general aspects of lipid analysis and content of algae.

Table 3.261 Total fatty acid compositions (% total) of some algae (Harwood and Jones, 1989)

	16:0	16:1	16:2	16:3	16:4	18:1	18:2	18:3	18:4	20:4	20:5	22:6
Marine phytoplankton												
Monochrysis lutheri (Chrysophyceae)	13	22	5	7	1	3	1	tr	2	1	18	7
Olisthodiscus spp. (Xanthophyceae)	14	10	2	2	1	4	4	6	18	2	19	2
Lauderia borealis (Bacillariophyceae)	12	21	3	12	1	2	1	tr	–	1	3	–
Amphidinium carterae (Dinophyceae)	24	1	1	tr	–	5	1	2	15	–	14	25
Dunaliella salina (Chlorophyceae)	41	15	tr	–	–	11	8	19	–	–	–	–
Hemiselmis brunescens (Cryptophyceae)	13	3	3	tr	tr	2	tr	9	30	tr	14	–
Marine macroalgae												
Fucus vesiculosus (Phaeophyceae)	21	2	tr	–	tr	26	10	7	4	15	8	–
Chondrus crispus (Rhodophyceae)	34	6	tr	–	–	9	1	1	4	18	22	–
Ulva lactuca (Chlorophyceae)	18	2	tr	1	18	9	2	17	24	1	2	tr
Freshwater species												
Scenedesmus obliquus	35	2	tr	tr	15	9	6	30	2	–	–	–
Chlorella vulgaris	26	8	7	2	–	2	34	20	–	–	–	–
Chlamydomonas reinhardtii	20	4	1	4	22	7	6	30	3	–	–	–
Salt-tolerant species												
Ankistrodesmus spp.	13	3	1	1	14	25	2	29	2	–	1	–
Isochrysis spp.	12	6	–	–	15	4	6	17	–	–	2	13
Nannochloris spp.	9	20	7	9	–	4	1	1	–	–	27	–

Table 3.262 Representative examples of the fatty acyl compositions of cyanobacteria[a]

Group	Organism	Fatty acid composition (% total)						
		16:0	16:1	18:1	18:2	α-18:3	γ-18:3	18:4
1	*Anacystis nidulans*	46	46	3	0	0	0	0
2	*Anabaena variabilis*	32	22	11	17	16	0	0
3	*Synechocystis 6714*	28	4	5	17	0	31	0
4	*Tolypothrix tenuis*	22	3	16	15	6	13	11

[a] Data taken from Murata and Nishida (1987), where references will be found.

Table 3.263 The acyl lipid compositions of cyanobacteria and *Prochloron* spp.[a]

Organism	Lipid (% total)				
	MGDG	MGlcDG	DGDG	SQDG	PG
Anabaena variabilis	54	1	17	11	17
Anacystis nidulans	57	n.m.	11	11	21
Prochloron spp.	55	3	11	26	5

[a] Data taken from Murata and Nishida (1987), where the original references can be found.
n.m. = not measured. MGlcDG, monoglucosyldiacylglycerol.

Table 3.264 The acyl lipid compositions of some algae (From Harwood and Jones, 1989)

Lipid (%)	*Chattonella antiqua*	*Dunaliella parva*	*Acetabularia mediterranea*	*Chlamydomonas reinhardtii*	*Chondrus crispus*	*Fucus vesiculosus*
PC	5	9	0.1		30.1	4.2
PE	3	n.d.	1.2	5	1.5	6.2
PI	1	2	0.6	2	tr	2.9
PG	3	6	3	10	7.7	2.2
DPG	–	–	–	–	1.9	5.4
MGDG	29	21	37	47	16.9	15.0
DGDG	18	11	20	16	14.8	11.3
SQDG	29	7	20	7	15.5	22.0
Other glycolipids	–	1	–	–	–	–
DGTS	6	15	20	16	n.d.	n.d.
DGTA	–	–	–	–	–	24.1
Neutral lipids	–	15	–	–	3.4	4.9
Free fatty acids	–	13	–	–	2.2	–

Abbreviations: PC, phosphatidylcholine; PE, phosphatidylethanolamine; PI, phosphatidylinositol; PG, phosphatidylglycerol; DPG, diphosphatidylglycerol; MGDG, monogalactosyldiacylglycerol; DGDG, digalactosyldiacylglycerol; SQDG, sulphoquinovosyldiacylglycerol; DGTS, diacylglycerol trimethylhomoserine ether lipid; DGTA, diacylglycerol hydroxymethyltrimethyl-β-alanine lipid.

References and further reading

Bolton, P., Wharfe, J. and Harwood, J.L. (1978) *Biochem. J.*, **174**, 67.

Bryce, T.A., Welti, D., Walsby, A.E. and Nichols, B.W. (1972) *Phytochemistry*, **11**, 295.

Eichenberger, W. (1990) in *Plant Lipid Biochemistry, Structure and Utilization* (eds P.J. Quinn and J.L. Harwood), Portland Press, London, p. 9.

Haines, T. (1973a) *Annu. Rev. Microbiol.*, **27**, 403.

Haines, T. (1973b) in *Lipids and Biomembranes of Eukaryotic Microorganisms* (ed. J.A. Erwin), Academic Press, New York, p. 197.

Harwood, J.L. and Jones, A.L. (1989) *Adv. Bot. Res.*, **16**, 1.

Kates, M. (1987) in *The Metabolism, Structure and Function of Plant Lipids* (eds P.K. Stumpf, J.B. Mudd and W.D. Nes), Plenum, New York, p. 613.

Kenyon, C.N., Rippka, R. and Stanier, R.Y. (1972) *Arch. Microbiol.*, **83**, 216.

Mercer, E.I. and Davies, C.L. (1979) *Phytochemistry*, **18**, 457.

Murata, N. and Nishida, I. (1987) in *The Biochemistry of Plants* (eds P.K. Stumpf and E.E. Conn), vol. 9, Academic Press, New York, p. 315.

Pohl, P. and Zurheide, F. (1979) in *Marine Algae in Pharmaceutical Science* (eds H.A. Hoppe, T. Levring and Y. Tanaka), Walter de Gruyter, Berlin, p. 473.

Scragg, A.H. and Leathers, R.A. (1988) in *Single Cell Oil* (ed. R.A. Moreton), Longman, Harlow, p. 71.

Volkman, J.K. (1989) in *Fats for the Future* (ed. R.C. Cambie), Ellis Horwood, Chichester, p. 263.

3.13 Fungal lipids

The lipid content of many fungi has been reported, and has proved to be highly variable depending on not only the individual species but also the growth conditions. In addition, the lipids of mycelium and yeast cells, reproductive structures as well as those of cells in different stages of development all vary. In general, phospholipids, sterols (and their esters), triacylglycerols and fatty acids are major components (Tables 3.265 and 3.266). For a thorough survey of fungal lipids, the reader is referred to two books by Weete (1974, 1980) and to reviews covering yeast lipids (Rattray, 1988) and other fungi (Losel, 1988).

3.13.1 Overall lipid content

Vegetative hyphae contain 1–56% of their dry weight as lipid, depending on species, developmental stage and growth conditions. Although certain species, such as the yeast *Saccharomyces cerevisiae* (ATCC 7754), have a very high lipid content, fungi generally average about 17% lipid. Several genera seem to be capable of producing high amounts of lipids. These include *Claviceps*, *Penicillium*, *Aspergillus*, *Mucor*, *Fusarium* and *Phycomyces* (Woodbine, 1959). Thus, lipids can represent up to 64% of the dry weight of *Penicillium spinulosum* (see Losel, 1988). The fluctuating proportions of total cellular lipid present in membranes correspond closely to the amount of these structures as seen in electron micrographs. The relative proportions of these and of storage lipid also vary considerably with growth conditions and development stage (Losel, 1988). In comparison, yeasts usually average 5–15% of their dry weight. 'Fat' yeasts produce 30% or more lipid and, apart from *Saccharomyces* spp. mentioned above, other 'fat' yeasts include some *Rhodotorula* and *Lipomyces* spp. (Hunter and Rose, 1971; Rattray *et al.*, 1975).

The lipid content of fungal spores varies considerably with species and the conditions of spore formation. Values between 1 and 45% of spore dry weight have been reported. When individual genera are examined, there are still considerable variations in total lipid between species. For example, the lipids of *Ustilago* spp. vary between 4 and 22%. Spores from mesophilic species of *Mucor* have a lower lipid content than those of thermophilic or thermotolerant species (Sumner and Morgan, 1969). In some species, such as the rust fungi, the endogenous lipid is used as a source of energy during germination. However, not all the lipid is present as globules that are easily available for mobilization; sometimes most of the lipid is phospholipid and carotenoids (as in *Neurospora crassa*), and these compounds are little utilized during germination.

3.13.2 Cell-wall lipids

Most of the fungal cell wall is composed of polysaccharides (80–90%), with the remainder consisting mainly of lipid and protein. There seems to be no correlation between cell-wall lipid amounts and the genus examined. Most of the cell-wall lipid is 'bound' and relatively difficult to extract. In *S. cerevisiae* neutral lipids and phosphoglycerides are major

Table 3.265 Lipid composition of some yeasts (Data from Rattray, 1988)

Species	Lipid content (% dry wt)	Percentage composition				
		TG	FA	Sterol	SE	PL
Debaryomyces hanseni	7	27	4	6	1	60
Hansenula anomala	13	77	tr	3	15	7
Lipomyces starkeyi	16	60	22	3	1	9
Saccharomyces cerevisiae (agar plate)	7	7	4	4	22	62
Saccharomyces cerevisiae (batch culture)	9	40	6	–	20	30
Yarrowia lipolytica (*Candida lipolytica*)	36	52	5		4	9
Candida utilis	11	55	1	2	1	38
Rhodotorula rubra	15	63	6	2	1	25

Abbreviations: TG = triacylglycerols; FA = non-esterified fatty acids; SE = sterol esters; PL = phospholipids; tr = trace.

Table 3.266 Lipid composition of some fungi (Data from Losel, 1988)

Class	Species	Lipid content (% dry wt)	NL	Polar	PL	TG	Sterols	SE
Dictyosteliomycetes	*Dictyostelium discoideum*	12	21	79	–	–	8	–
Chrytridiomycetes	*Blastocladiella emersonii*	8–11	12–47	25–82	35–69	8–28	5–13	18[a]
Oomycetes	*Achyla americana* (oospores)	33	98	2	0.4	89	0.3	0.8
	Pythium ultimum	4–18	55–91	9–45	–	40–58	–	–
Zygomycetes	*Mucor* spp.	3–15	55–65	35–45	20–35	–	4–16	5–10
	Rhizopus arrhizus	3–15	35–68	32–65	–	2–18	2–18	1–12
Ascomycetes	*Aspergillus niger*	4	31	69	41	5	19	4
	Neurospora crassa	9–19	–	–	93–98	<1	–	1–4
Hymenomycetes	*Agaricus bisporus*	5–6	68–74	22–26	–	12–13	2.3	2[a]
Urediniomycetes	*Puccinia graminis*	5–18	64–92	8–36	–	–	1–7	–

Abbreviations: NL = neutral lipids; Polar = polar lipids; PL = phospholipids; TG = triacylglycerols; SE = sterol esters.
[a] Only one sample analysed.

Table 3.267 Unsaturation indices in fungi grown at different temperatures

Class	Fungus	Growth temperature (°C)	Unsaturation (Δ/mol fatty acid)	Reference
Phycomycetes	*Mucor mucedo*	25	1.16–1.47	Sumner and Morgan (1969)
	M. miehei	25	0.95	Sumner and Morgan (1969)
		48	0.78	Sumner and Morgan (1969)
	M. strictus	10	1.13–1.26	Sumner *et al.* (1969)
		20	0.98–1.24	Sumner *et al.* (1969)
	Rhizopus sp. III	36	0.76–0.84	Sumner *et al.* (1969)
		48	0.60–0.86	Sumner and Morgan (1969)
Ascomycetes	*Candida lipolytica*	25	1.00	Kates and Baxter (1962)
	C. scottii	10	1.72	Kates asnd Baxter (1962)
	Chaetomium thermophile	45	0.65	Mumma *et al.* (1970)
	C. globosum	25	0.96	Mumma *et al.* (1970)
	Stilbella thermophile	45	0.56	Mumma *et al.* (1970)
	Stilbella sp.	25	1.47	Mumma *et al.* (1970)

components, while in *Candida albicans* sterol esters, sterols, fatty acids, triacylglycerols and phospholipids are all contained. Thus, in contrast to *Saccharomyces* spp., the *Candida* spp. contain mainly sterols and sterol esters (40–60% total). In fungi the nature of the wall lipid is undefined, although glycolipids have been noted in some cases. Hydrocarbons and sporopollenin, a carotenoid polymer highly resistant to decomposition, may occur on the surface of spores (see Losel, 1988).

3.13.3 Changes in lipid content with growth conditions

There are several major factors involved in determining the influence of the environment on fungal lipids. These include temperature, pH, carbon source, inorganic nutrients, aeration and growth factors. A particularly comprehensive account of factors influencing fatty acid composition in yeasts is given by Rattray (1988).

The response of fungi to growth temperatures varies with species. Many yeasts and yeast-like fungi contain more lipid when they are grown at temperatures below their optimum (Hunter and Rose, 1971). However, in other fungi (e.g. *Rhodotorula gracilis*) there are opposite changes. Fungi can be classified into mesophilic, psychrophilic, thermophilic and thermo-tolerant species depending on their optimal growth temperatures. There do not appear to be any consistent differences in lipid content between these four groups. However, there is an increase in unsaturation of fatty acids of mesophilic fungi when compared to thermophilic fungi (Table 3.267) and this is consistent with a general increase with unsaturation at lower temperatures (Table 3.267).

The most important carbon source for fungal growth is glucose, while the best disaccharide is maltose. These and other sugars are easily converted to lipid. However, there are significant differences in the amount of lipid synthesis from different sugars. For example, in *Penicillium chrysogenum*, sucrose > fructose > glucose as sources for fat production. In addition, the actual concentration of the sugar has been shown to be important in many cases (cf. Weete, 1980; Rattray *et al.*, 1975). Often, optimal lipid synthesis occurs at around 40% sucrose or glucose in the growth medium, but individual fungi vary considerably. With the increasing use of microorganisms for the utilization of inexpensive industrial waste products, many fungi have been studied with regard to the efficiency of their conversion of carbon substrates to lipid. These include, for example, fungi that can degrade alkanes (cf. Gerasimova *et al.*, 1975).

There are two aspects to fungal requirements for inorganic nutrients. First, there are, clearly, absolute requirements for certain elements. Secondly, the levels of individual nutrients may alter lipid metabolism. For example, phosphorus levels may alter the phospholipid composition of *Candida utilis*, and increasing sodium chloride (from 0 to 10%) in the medium increases the lipid content (from 0.3 to 6.3%) in *Candida* spp. The most important nutritional parameter for fungal lipid synthesis is the ratio of carbon to nitrogen. High fat accumulation is associated with low protein synthesis and is, therefore, favoured by high ratios of carbon/nitrogen. Indeed, the highest accumulation of fats by many fungi is obtained in nitrogen-deficient media. In general, lipid accumulation is favoured more when nitrogen is supplied from an inorganic source, since organic nitrogen tends to stimulate growth and protein synthesis.

The effect of pH on fungal lipid accumulation is not clear. For some species, increasing medium pH increases lipid synthesis, but for others there is no effect (Weete, 1980). On the other hand, the degree of aeration is important because of the requirement for oxygen in sterol and unsaturated fatty acid synthesis. In addition, the degree of aeration has profound effects on fungal growth and differentiation. For example, in *Mucor rouxii*, growth under aerobic conditions results in filamentous growth, while anaerobic conditions cause the cells to be yeast-like. Under the latter conditions there are relatively high concentrations of short- to medium-chain (C_8–C_{14}) fatty acids and hardly any unsaturation.

Although many fungi are able to grow successfully on media containing a carbon source and mineral salts, some also require vitamins or certain amino acids. In general, vitamin deficiencies result in a reduction of lipid content. However, certain yeasts, when grown on inositol-deficient media, accumulate large amounts of triacylglycerols (cf. Weete, 1980).

3.13.4 Fungal fatty acids

The fatty acid compositions of fungi have been extensively reviewed (Shaw, 1966; Wassef, 1977; Weete, 1980; Rattray, 1988; Losel, 1988). Fungi produce saturated and unsaturated aliphatic acids of 10–24 carbons in chain length. Palmitic is the major saturated acid, oleic the main monoenoic acid and linoleic the most common polyunsaturated acid.

In Phycomycetes the presence of γ-linolenic acid is notable since the higher fungi (Ascomycetes, Fungi Imperfecti and Basidiomycetes) produce only α-linolenate. Some higher homologues in the $n-6$ series, such as 20:3 and 20:4, may also be found. The

Table 3.268 Fatty acid compositions of some fungi (Data taken from Weete, 1980)

Species	Fatty acids (% total)								
	12:0	14:0	16:0	16:1	18:0	18:1	18:2	18:3	20:4
Phycomycetes									
Blastocladiella emersonii	–	1	13	3	3	39	17	11	16
Entomophthora muscae	–	6	14	17	3	38	5	2	14
Mucor strictus	–	6	22	2	9	35	12	14	–
Phycomyces blakesleeanus	–	tr	11	2	21	30	35	2	–
Rhizophylyctis rosea (aquatic)	–	tr	14	–	6	74	3	3	–
Ascomycetes									
Candida albicans	–	1	12	8	7	36	25	10	–
Saccharomyces cerevisiae	–	3	18	52	2	20	–	–	–
Torulopsis candida	–	tr	28	4	9	43	12	3	–
Fungi Imperfecti (Deuteromycetes)									
Aspergillus niger	–	tr	16	1	7	21	38	16	–
Penicillium chrysogenum	–	3	13	–	12	19	43	6	–
Homobasidiomycetes									
Calvatia gigantea	12	3	9	1	tr	9	62	tr	1[a]
Lactarius vellerens	–	tr	7	tr	54	22	14	2	–
Polyporus hirsutus	1	1	20	7	2	22	50	2	tr[a]
Heterobasidiomycetes									
Ustilago scitaminca	1	2	20	3	7	30	32	2	–

[a] 20:0.

Phycomycetes also tend to produce more C_{20} and C_{22} acids than the higher fungi, and several of their aquatic species have quite distinct fatty acid patterns. A few examples of fatty acid patterns are given in Table 3.268. Because of the recent interest in γ-linolenic acid for human nutrition, commercial processes have been developed by using moulds of the Mucorales for producing significant amounts of γ-linolenate-enriched oil (see Ratledge, 1989a).

Further discussions of this aspect and the use of *Mortierella* species for γ-linolenic and eicosapentaenoic acids is given by Ratledge (1989b). The industrial production of lipids by yeasts is reviewed by Davies and Holdsworth (1992), with physiological aspects of lipid accumulation summarized by Moreton (1988) and other topics, including economics, by Ratledge (1988).

So far as the Ascomycetes are concerned, there seem to be no differences in fatty acid composition that distinguish the subgroups ascosporogenous (true yeasts) from asporogenous (yeast-like) yeasts. *Saccharomyces* and *Candida* are the most-studied examples of these two groups. *Saccharomyces* spp. contain rather high levels of 16:1 and with few exceptions do not synthesize polyunsaturated fatty acids. In general, 16:1 > 18:1 > 16:0 and these acids account for >90% of the total. The relative amounts of acids in *Candida utilis* vary with strain and culture conditions but 16:0, 18:1 and 18:2 are principal fatty acids as for most asporogenous yeasts.

Several yeasts or yeast-like fungi secrete lipids. These extracellular lipids include hydroxy-, acetoxy- or long-chain fatty acids. For example, *Saccharomyces melanga* produces 3-D-hydroxypalmitic acid. Extracellular glycolipids called sophorosides are also produced by some yeast-like fungi, e.g. *Torulopsis magnaliae*.

In contrast to the Ascomycetes, Fungi Imperfecti do not usually contain large quantities of 16:1. Their major fatty acids are 16:0, 18:1 and 18:2, with significant amounts of α-18:3.

The basidiomycete fungi are divided into Homobasidiomycetes (saprophytic, e.g. mushrooms, puffballs) and the Heterobasidiomycetes (parasitic, e.g. the rust and the smut fungi). The Homobasidiomycetes have 16:0, 18:1 and 18:2 as their main fatty acids (Table 3.268), with linoleate usually predominating. In some mushrooms, hydroxy fatty acids are significant and may be the major acids (Prostenik *et al.*, 1978). The rust fungi contain *cis*-9,10-epoxyoctadecanoic acid, which is used to confirm them in culture.

3.13.5 Glycerol-based lipids in fungi

Triacylglycerols are the most abundant type of lipid in many fungi. However, their quantities vary considerably according to species, stage of development and growth conditions (see previous sections and Tables 3.269 and 3.270). The alkyl and alkenyl ethers are not common constituents of yeasts or mycelial fungi. Certain unusual neutral lipids, such as diesters or 1-alkenyl esters of dihydric alcohols (e.g. ethanediol, butane-1,4-diol), may be present in a few species (Weete, 1980).

A comparison of the fatty acid content of the principal storage lipid, triacylglycerol, with that of a major membrane lipid, phosphatidylcholine, for several yeasts is made in Table 3.271. Although the same fatty acids are usually present in both lipids, their proportions vary considerably. As a generalization, the membrane lipid is much more unsaturated than the storage lipid.

The glycerophospholipids of fungi (Griffin *et al.*, 1970), yeasts (Hunter and Rose, 1971; Rattray *et al.*, 1975) and various organisms (Kates and Wassef, 1970) have been reviewed. The glycerophospholipid content of fungi averages 41% of the total lipids and is higher in the log phase than in the stationary phase of growth. Sample compositions are shown in Tables 3.272 and 3.273. It will be seen that the major lipids are phosphatidylcholine and phosphatidylethanolamine, with phosphatidylinositol often a major

Table 3.269 Lipid composition of some yeasts (Rattray, 1988)

Species	Lipid content (% dry wt)	Percentage composition				
		TG	FA	Sterol	SE	PL
Debaryomyces hanseni	7	27	4	6	1	60
Hansenula anomala	13	77	tr	3	15	7
Lipomyces starkeyi	16	60	22	3	1	9
Sacchromyces cerevisiae (agar plate)	7	7	4	4	22	62
Sacchromyces cerevisiae (batch culture)	9	40	6	–	20	30
Yarrowia lipolytica (Candida lipolytica)	36	52	5	4		9
Candida utilis	11	55	1	2	1	38
Rhodotorula rubra	15	63	6	2	1	25

TG: triacylglycerols; FA = nonesterified fatty acids; SE = sterol esters; PL = phospholipids; tr = trace.

Table 3.270 Lipid composition of some fungi (data from Losel, 1988)

Class	Species	Lipid content (% dry wt)	NL	Polar	PL	TG	Sterols	SE
Dictyosteliomycetes	*Dictyostelium discoideum*	12	21	79	–	–	8	–
Chytridiomycetes	*Blastocladiella emersonii*	8–11	12–47	25–82	35–69	8–28	5–13	18*
Oomycetes	*Achyla americana* (oospores)	33	98	2	0.4	89	0.3	0.8
	Pythium ultimum	4–18	55–91	9–45	–	40–58	–	–
Zygomycetes	*Mucor* spp.	3–15	55–65	35–45	20–35	–	4–16	5–10
	Rhizopus arrhizus	3–15	35–68	32–65	–	2–18	2–18	1–12
Ascomycotina	*Aspergillus niger*	4	31	69	41	5	19	4
	Neurospora crassa	9–19	–	–	93–98	<1	–	1–4
Hymenomycetes	*Agaricus bisporus*	5–6	68–74	22–26	–	12–13	2–3	2*
Urediniomycetes	*Puccinia graminis*	5–18	64–92	8–36	–	–	1–7	–

NL = neutral lipids; Polar = polar lipids; PL = phospholipids; TG = triacylglycerols; SE = sterol esters.
* Only one sample analysed.

Table 3.271 Comparison of the fatty acid compositions of the neutral lipid or triacylglycerol fraction with phosphatidylcholine from different yeasts (From Rattray, 1988)

Organism	Lipid	Fatty acid (%)					
		14:0	16:0	16:1	18:1	18:2	18:3
Candida albicans	NL	4	52	–	33	6	3
	PC	1	15	25	47	9	–
Lipomyces starkeyi	NL	–	34	7	51	4	–
	PC	–	9	16	52	20	–
Saccharomyces cerevisiae	TG	3	16	42	27	–	–
	PC	tr	6	58	35	–	–
Trichosporon pullulans	TG	tr	15	tr	57	24	1
	PC	–	6	tr	8	83	3

Abbreviations: NL = neutral lipid; TG = triacylglycerol; PC = phosphatidylcholine; tr = trace.

Table 3.272 Phospholipid composition of some yeasts (From Rattray, 1988)

Organism	Total (% lipid)	Composition (%)					
		PS	PI	PC	PE	DPG	PA
Ascosporogenous yeasts							
Hansenula polymorpha	31	17	22	24	26	–	11
Lipomyces starkeyi	16	11	12	43	25	5	1
Saccharomyces cerevisiae	32	5	12	20	56	7	–
Schizosaccharomyces pombe	43	12	14	50	13	6	–
Asporogenous and basidiosporogenous yeasts							
Candida albicans	15	12 (PS + PI)		40	25	5	–
Candida utilis	30	10	12	17	48	15	–
Rhodotorula rubra	25	16	7	16	29	12	6
Trichosporan cutaneum	40	12	11	29	15	6	7

Abbreviations: PS = phosphatidylserine; PI = phosphatidylinositol; PC = phosphatidylcholine; PE = phosphatidylethanolamine; DPG = diphosphatidylglycerol; PA = phosphatidic acid.

Table 3.273 Phospholipid composition of some fungi (From Losel, 1988)

Organism	Composition (%)					
	PS	PI	PC	PE	DPG	PA
Dictyostelium discoideum						
spores	–	13	29	44	3	7
early culmination	–	11	11	41	4	7
Phytophora parasitica	1	7	39	20	3	3
Choanephora cucurbitarum [a]	3	3	28	39	8	7
Glomus mosseae (mycelium) [a]	17	–	16	34	14	–
Aspergillus niger	5	–	51	29	2	–
Neurospora crassa	8	6	45	36	2	3
Cephalosporum lalciforme	9	2	50	15	6	–
Uromyces phaseoli	8	6	54	27	3	–

Abbreviations: PS = phosphatidylserine; PI = phosphatidylinositol; PC = phosphatidylcholine; PE = phosphatidylethanolamine; DPG = diphosphatidylglycerol; PA = phosphatidic acid.

Contain significant amounts of phosphatidylglycerol (13% *C. cucurbitarum*, 19% *G. mosseae*).

component. Unlike many other organisms, fungi often contain phosphatidylserine in significant amounts (usually 5–10%). Some detail is known of the subcellular distribution of different phospholipids. Most information is known for isolated mitochondria. In addition, the effects of growth temperature, morphogenesis and reproduction and of genetic or nutritional factors on phospholipid class composition have been reported (see Losel, 1988). On the other hand, there is a wealth of detail of the phospholipids in individual membranes from yeasts such as *S. cerevisiae* (see Rattray, 1988).

3.13.6 Other fungal lipids

Sphingolipids such as cerebrosides and ceramides are minor but widely distributed constituents of fungi. Certain species contain characteristic sphingosine-containing lipids, such as the inositol-containing phosphorylceramides of *S. cerevisiae* (cf. Weete, 1980). Other minor lipids include hydrocarbons, while sterols (and sometimes carotenoids) are often major components. A full description of such lipids is outside the remit of this section and, for further details, the reader is directed to Weete (1980), Brennan *et al.* (1974), Goodwin (1973) and Losel (1988).

When grown aerobically, most fungi can synthesize sterols. Exceptions often occur with pathogens of plants and animals, particularly the Oomycetes. Fungal sterols and their biosynthesis have been well reviewed (Elliott, 1977; Goodwin, 1973; Wassef, 1977; Weete, 1980). The exact pattern of sterol accumulation depends very much on the species. For example, desmosterol is found in many oomycetes but seldom in other fungi (Losel, 1988). As with acyl lipids, the amounts of sterols vary with environmental factors. Sterol esters often appear to function as a store for membrane sterol precursors. In yeasts, lanosterol, zymosterol, dehydroergosterol and ergosterol are common (Rattray, 1988).

Carotenoids have been found in 60% of all fungi examined (see reviews by Goodwin, 1973; Weete, 1980). The major component is β-carotene. Xanthophylls appear to be absent from lower fungi but have a wide, if sporadic, distribution in higher fungi. In many of the Discomycetes, xanthophylls are the major carotenoids. A vast range of terpenoids have been reported also (see Losel, 1988).

References and further reading

Brennan, P.J., Griffin, P.F.S., Losel, D.M. and Tyrrell, D. (1974) *Prog. Chem. Fats Other Lipids*, **14**, 51.

Davies, R.J. and Holdsworth, J.E. (1992) *Adv. Appl. Lipid Res.*, **1**, 120.

Elliott, C.G. (1977) *Adv. Microb. Physiol.*, **15**, 121.

Gerasimova, N.M., Le Zui Lin and Bekhtereva, M.N. (1975) *Microbiology*, **44**, 408.

Goodwin, T.W. (1973) in *Lipids and Biomembranes of Eukaryotic Microorganisms* (ed. J.A. Erwin), Academic Press, New York, p. 1.

Griffin, P.F.S., Brennan, P.J. and Losel, D.M. (1970) *Biochem. J.*, **119**, 11P.

Hunter, K. and Rose, A.H. (1971) in *The Yeasts* (ed. A.H. Rose and J.S. Harrison), vol. 2, Academic Press, New York, p. 211.

Kates, M. and Baxter, R.M. (1962) *Can. J. Physiol.*, **40**, 1213.

Kates, M. and Wassef, M.K. (1970) *Annu. Rev. Biochem.*, **39**, 323.

Losel, D.M. (1988) in *Microbial Lipids* (eds C. Ratledge and S.G. Wilkinson), vol. 1, Academic Press, London, p. 699.

Moreton, R.S. (1988) in *Single Cell Oil* (ed. R.S. Moreton), Longman, Harlow, p. 1.

Mumma, R.D., Fercus, C.L. and Sekwa, R.D. (1970) *Lipids*, **5**, 100.

Prostenik, M., Burcar, I., Castek, A., Coscovic, C., Golem, J., Jandric, Z., Kijaic, K. and Ondrusek, V. (1978) *Chem. Phys. Lipids*, **22**, 97.

Ratledge, C. (1988) in *Single Cell Oil* (ed. R.S. Moreton), Longman, Harlow, p. 33.

Ratledge, C. (1989a) *Microbial Lipids* (eds C. Ratledge and S.G. Wilkinson), vol. 2, Academic Press, London, p.567.

Ratledge, C. (1989b) in *Fats for the Future* (ed. R.C. Cambie), Ellis Horwood, Chichester, p. 153.

Rattray, J.B.M. (1988) in *Microbial Lipids* (eds C. Ratledge and S.G. Wilkinson), vol. 1, Academic Press, London, p.555.

Rattray, J.B.M., Schibeci, A. and Kidby, D.K. (1975) *Bacteriol. Rev.*, **39**, 197.

Shaw, R. (1966) *Adv. Lipid Res.*, **4**, 107.

Sumner, J.L. and Morgan, E.D. (1969) *J. Gen. Microbiol.*, **59**, 215.

Sumner, J.L., Morgan, E.D. and Evans, H.C. (1969) *Can. J. Microbiol.*, **15**, 515.

Wassef, M.K. (1977) *Adv. Lipid Res.*, **15**, 159.

Weete, J.D. (1974) *Fungal Lipid Biochemistry*, Plenum, New York.

Weete, J.D. (1980) *Lipid Biochemistry of Fungi and Other Organisms*, Plenum, New York.

Woodbine, M. (1959) *Prog. Ind. Microbiol.*, **1**, 181.

3.14 Bacterial lipids

Any discussion of the occurrence of lipids in bacteria is complicated by the wide variety of such compounds that are found, often only in a few species. Nevertheless, some broad generalizations can be made, and these will be highlighted in the following section – though it must always be remembered that there is an exception to every rule. For a more complete coverage of bacterial lipids, the reader is referred to Razin and Rottem (1982), Finnerty (1978), Cronan (1978), Raetz (1978), Shaw (1975), Goldfine (1972) and Ratledge and Wilkinson (1988, 1989).

3.14.1 Fatty acid distribution

Fatty acids in the C_{12}–C_{20} chain-length range account for the majority of bacterial fatty acids, which are usually saturated or mono-unsaturated. Bacteria generally do not contain polyunsaturated fatty acids, and some isolated reports of their occurrence are probably due to contamination from the growth medium. Notable exceptions are some cyanobacteria, which contain linoleate and linolenate (section 3.12), gliding bacteria, which have large amounts of arachidonate in their acyl lipids (Fautz *et al.*, 1979), and a range of eubacteria found in marine environments (e.g. Intriago and Floodgate, 1991; Matsui *et al.*, 1991).

Besides the ubiquitous even-chain saturated and unsaturated fatty acids, bacteria characteristically contain odd-chain and branched fatty acids as well as 3-hydroxy- and cyclopropane derivatives. These fatty acids are present in lipopolysaccharide, cell-wall lipoprotein and lipoteichoic acid, as well as membrane glycerolipids (Table 3.274). As a generalization, the Gram-positive bacteria contain higher proportions of branched-chain acids than do Gram-negative bacteria. In those Gram-positive bacteria where branched chains are prevalent, there is usually a corresponding decrease in the proportions of unsaturated acids (O'Leary and Wilkinson, 1988).

Table 3.274 Major fatty acids in bacteria

Saturated	Tetradecanoic acid Hexadecanoic acid Octadecanoic acid
Unsaturated	Octadecenoic acid (11*c*)
Branched chain	13-Methyltetradecanoic acid 14-Methylhexadecanoic acid
Substituted	3-Hydroxytetradecanoic acid *cis*-9,10-Methylene-hexadecanoic acid *cis*-11,12-methyleneocta-decanoic acid

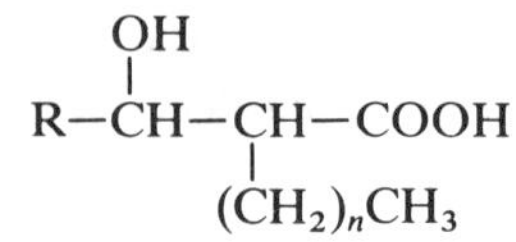

(9)

n = 5–23 (odd numbers only)

R may be up to C_{59} and include other functional groups

Exceptionally long-chain, branched fatty acids termed mycolic acids and of general formula (**9**) are found in members of the *Mycobacterium–Nocardia–Corynebacterium* group, *Rhodococcus* and the 'aurantiaca' toxin (Brennan, 1988) (see also section 2.12).

3.14.2 Glycerol-based acyl lipids

In contrast to animals and plants, bacteria seldom accumulate triacylglycerols or other 'fats'. The monoacyl and diacyl derivatives are important metabolic intermediates and will be present in small but detectable amounts. Thus the lipids of bacteria are essentially associated with one or other of the cell membranes, with the exception of storage polymers (notably poly(3-hydroxybutyrate) (PHB)) or hydrocarbon inclusions (e.g. from hydrocarbon-grown *Acinetobacter* strains). PHB occurs in many prokaryotes – both Gram-positive and Gram-negative. Its biosynthesis is reviewed by Schweizer (1989) and some commercial applications are discussed by Ratledge (1989). The structure of PHB is shown in (**10**). Economically, the production of PHB is much more expensive than that of cheap chemically produced polypropylenes and polyethylenes. However, its biodegradable nature is a big advantage in medical applications, and it is likely to become more important in the future.

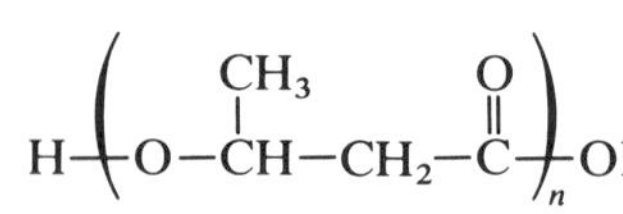

(10)

where n = 600–25 000

Poly(3-hydroxyalkanoate)s have also been reported to accumulate in inclusion bodies in different bacteria such as *Pseudomonas* spp. (de Smet *et al.*, 1983). In contrast to the inclusion bodies for PHB and poly(3-hydroxyalkanoate)s, where the surrounding membrane is an intracytoplasmic one, the hydrocarbons of *Acinetobacter* spp. are contained by a monolayer limiting membrane with a lipid composition different from that of other cellular membranes (Scott and Finnerty, 1976).

Glycerophospholipids are usually the major constituents of bacterial membranes. These lipids are usually enriched with saturated fatty acids at the *sn*-1 position and with unsaturated acids at the *sn*-2 position. In cyanobacteria, the distribution is governed by chain length rather than unsaturation, with *sn*-1 C_{18} and *sn*-2 C_{16} being the rule (Zepke *et al.*, 1978). The cyanobacteria are also exceptional in that they synthesize (and accumulate) only one phospholipid, phosphatidylglycerol.

With the exception of the Actinomycetes (e.g. mycobacteria), phosphatidylglycerol is present in all species of bacteria. It is frequently the major phospholipid. The *sn*-3 hydroxyl of the glycerol headgroup may be esterified with an amino acid (usually alanine, ornithine or lysine) to form an aminoacylphosphatidylglycerol (Goldfine, 1982). These derivatives are more common in Gram-positive species. The *sn*-2 hydroxyl of the glycerol headgroup may be glycosidically linked to glucosamine or the *sn*-3 hydroxyl group can be acylated. However, these substitutions are not very common.

Cardiolipin (or diphosphatidylglycerol) usually occurs along with phosphatidylglycerol. The proportions of these two lipids can be easily changed by growth conditions. The presence of cardiolipin as a major constituent in bacteria is in striking contrast to eukaryotes, where it is only present in the inner mitochondrial membrane.

The major glycerophospholipid of Gram-negative bacteria is phosphatidylethanolamine. It is also a major component of some Gram-positive species such as *Bacillus*. In contrast, phosphatidylcholine is seldom a major lipid in bacteria. In a few genera the ethanolamine headgroup may be partly methylated (giving the mono- and dimethyl derivatives).

Some examples of phospholipid composition in Gram-positive and Gram-negative bacteria are shown in Table 3.275 and general features of the lipid composition of Gram-negative bacteria are shown in Table 3.276.

Phosphatidylinositol is uncommon in bacteria and is found in a few Gram-positive species only. In Actinomycetes and a few other bacteria, mannosides of phosphatidylinositol may be present. Other glycerophospholipids such as phosphatidic acid and phosphatidylserine have a widespread occurrence in bacteria but only in small amounts. Both of these lipids play an important role as metabolic intermediates (Finnerty, 1978; Raetz, 1978).

Table 3.275 Examples of phospholipid compositions of bacteria[a]

	Major phospholipids (% total)[b]		
	PE	PG	DPG
Gram-positives			
Micrococcus spp.	0	26–88	10–67
Planococcus spp.	8–14	23–38	44–45
Staphylococcus spp.	0	52–90	1–10
Streptococcus spp.	0	8–38	8–62
Gram-negatives			
Escherichia coli	74	19	3
Klebsiella pneumoniae	38–82	5–13	6–14
Vibrio spp.	23–81	10–54	tr–2
Haemophilus spp.	78–85	15–18	0–3
Pseudomonas spp.	40–87	9–76	0–35
Acetobacter spp.	25–31	20–31	4–13
Neisseria spp.	51–77	19–22	2–20

[a] For further information see O'Leary and Wilkinson (1988) and Wilkinson (1988).
[b] Others include 36–38% lysyl PG in *Staphylococcus* spp., 5–16 phosphoglycolipids in *Streptococcus* spp., and 32–44% phosphatidylcholine in *Acetobacter* spp.

Instead of a phosphate unit at the *sn*-3 position, a carbohydrate moiety may be present. Such glycosylglycerides have quite a widespread but usually minor occurrence in bacteria. The cyanobacteria have already been mentioned as having only one phospholipid. They have a very simple lipid pattern, with the remaining compounds being three glycosylglycerides: diacylgalactosylglycerol, diacyldigalactosylglycerol and diacylsulphoquinovosylglycerol. Thus, in contrast to other bacteria and like the photosynthetic membranes of plants and algae, these organisms contain glycosylglycerides as their major membrane lipids.

Glycosylglycerides are more important in Gram-positive than Gram-negative bacteria. Except in green photosynthetic bacteria and *Treponema* spp., where they are the most common lipids, the monoglycosyldiacylglycerols are less common than the diglycosyl lipids. The disaccharide moiety of the latter is usually either two glucose, galactose or mannose sugars linked $\alpha(1\rightarrow 2)$ or $\beta(1\rightarrow 6)$ (section 2.5 and Table 2.6).

The glycosyldiacylglycerols also give rise to some unusual phosphoglycolipids and constitute the hydrophobic moiety of lipoteichoic acids, and the lipo-

Table 3.276 Summary of the lipid content of Gram-negative bacteria[a]

Usual features	
Main fatty acids	16:0, 16:1 (Δ9), 18:1 (Δ11), 3-OH 14:0
Main phospholipids	PE with variable proportions of PG, DPG
Minor/rare acyl lipids	Hydrocarbons, acylglycerols, waxes
Absent	Steroids
Unusual compositions	
Fatty acids	
Mainly straight-chain odd-carbon	e.g. *Desulphobulbus* spp.
Mainly branched-chain	e.g. *Pseudomonas putrefaciens*
Polyenoic acids significant	e.g. *Vibrio marinus*
Phospholipids	
PE absent	e.g. *Pseudomonas vesiculans*
PC a major component	e.g. *Agrobacterium* spp.
Plasmenyl lipids significant	e.g. *Butyrobvibrio* spp.
Plasmanyl lipids significant	e.g. *Stigmatella* spp.
Other polar lipids	
Glycolipids significant	e.g. Some *Pseudomonas* spp.
Ornithine amide lipids present	e.g. *Acetobacter* spp.
Sphingolipids present	e.g. *Bacteriodes* spp.
Sulphonolipids present	e.g. *Flavobacterium* spp.

[a] For more details refer to Wilkinson (1988).
Abbreviations: PE, phosphatidylethanolamine; PG, phosphatidylglycerol; DPG, diphosphatidylglycerol; PC, phosphatidylcholine.

conjugates of most Gram-positive bacteria (section 2.12). A distinctive feature of the streptococci is the variety and abundance of phosphoglycolipids linking the glycolipids with the lipoteichoic acids. Typically the phosphoglycolipids are derived from α-kojibiosyldiacylglycerol by substitution at one or both of the 6 positions on glucose by a (i) 3-*sn*-phosphatidyl group, (ii) *sn*-glycerol-1-phospho group or (iii) fatty acyl group (see O'Leary and Wilkinson, 1988). In mycobacteria phosphatidylinositol mannosides may act as membranous anchors with lipoarabinomannins as the functionally important biopolymers, while glycosyldiacylglycerols, phosphatidylglucosyldiacylglycerols and glycerophosphoglycosyldiacylglycerols act as membranous anchors with the lipoteichoic acids as the corresponding extensions (Brennan, 1988). In a few species (e.g. *Mycobacterium tuberculosis*) both anchors may coexist.

Some phosphoglycolipids are also present in certain Gram-positive organisms. Phosphatidylglycolipids are found in streptococci, *sn*-glycero-3-phosphoglycolipids in mycoplasmas and the *sn*-glycero-1-phosphophosphatidylglycolipids in several Gram-positive genera.

3.14.3 Other glycolipids

Acylated sugars such as acylglucoses of mycobacterial 'cord factor', a diacylated glucose attached to D-glyceric acid in the cell envelope of *Nocardia caviae* (Pommier and Michel, 1981) and the rhamnolipid of *Pseudomonas aeruginosa* are specific examples of sparsely occurring glycolipids. Other examples are the mycosides of mycobacteria. These are glycosides of *p*-substituted phenols with normal and branched-chain fatty acids esterified to the oligosaccharide. The glycosides are involved in the pathogenicity of *Mycobacteria*.

In comparison to other bacteria, *Mycobacteria* (and the related *Nocardia* and *Corynebacteria*) contain large amounts of lipid, which is mainly in the cell wall.

The mycobacterial cell wall contains three main components. First, there is a skeleton composed of arabinogalactan mycolate covalently linked through a phosphodiester bond to peptidoglycan. There are also peptides that are removed by proteinolysis and free lipids that are easily extracted by solvents.

The skeleton consists of a branched polymer of D-arabinose and D-galactose in a 5/2 ratio, where about every tenth arabinose contains a mycolic acid esterified to the 5′-hydroxyl. These mycolic acids are C_{60}–C_{90} fatty acids that are 2-branched and 3-hydroxylated. They may also contain cyclopropane rings, methyl branches and methoxy groups. In *Nocardia* similar arabinogalactan mycolates are present, known as nocardomycolic acids (40–60 carbons with a C_{14} or C_{16} branch). In *Corynebacterium* spp., corynomycolic acids (28–40 carbons with a C_{14} branch) are found (Minnikin, 1982).

Some 25–30% of the weight of mycobacterial cell walls consists of extractable lipids. These are a mixture of cord factors, mycosides, sulpholipids and wax D. Cord factors are a mixture of trehalose 6,6′-mycolates. The constituent mycolic acids are of the same composition as those found in wax D and are characteristic of the organism from which they are isolated (Fig. 3.1). The sulpholipids consist of trehalose, which is sulphated at the 2 position and acylated at several positions on both sugar residues. The acyl groups are a mixture of palmitic acid and very long-chain (31–46 carbons) methyl-branched acids called phthioceranic acids (Minnikin, 1982). The outermost layer of the mycobacterial cell wall contains mycosides, which are either phenolic glycolipids (mycosides A and B) or glycolipid peptides (mycoside C). Their complex structure is shown in Fig. 3.2. For a comprehensive review of the lipids of mycobacteria and other actinomycetes, refer to Brennan (1988).

Gram-negative bacteria contain lipopolysaccharide in the outer membrane of their cell envelope. The lipopolysaccharide is a complex polymer with four parts. On the outside is a polysaccharide of very variable structure, which is then attached to a core polysaccharide, which itself is divided into an outer core and a backbone. The outer cores are variable in composition but the link that connects the backbone to lipid A is usually composed of 3-deoxy-D-manno-octulosonic acid (KDO). The presence of KDO (Fig. 3.3) is often used as a marker for lipopolysaccharide or outer membrane.

Lipid A is composed of a disaccharide of glucosamines. The amino groups are substituted with 3-hydroxymyristate, while the hydroxyl groups contain saturated (12–16 carbons) acids and 3-myristoxymyristate. Phosphate and KDO groups are also substituted (Fig. 2.7). The fatty acids are sometimes referred to as 'bound fatty acids' while lipopoly-

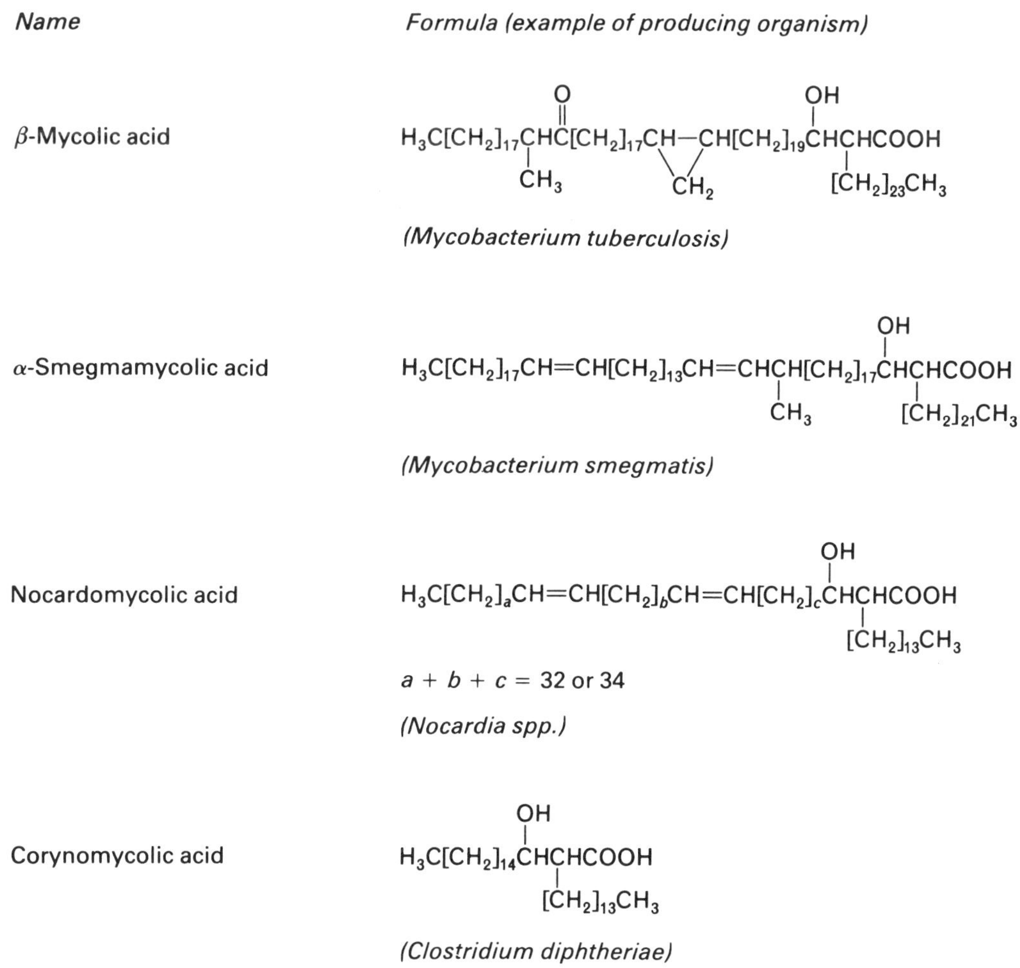

Figure 3.1 Some major mycolic acids found in mycobacteria and related actinomycetes. Reproduced with permission from Harwood and Russell (1984).

(a) *Trehalose mycolate*

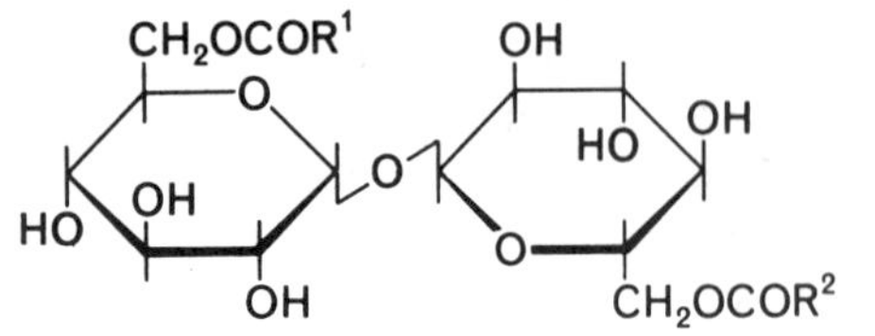

R^1OH and R^2OH are mycolic acids

(b) *Sulpholipid*

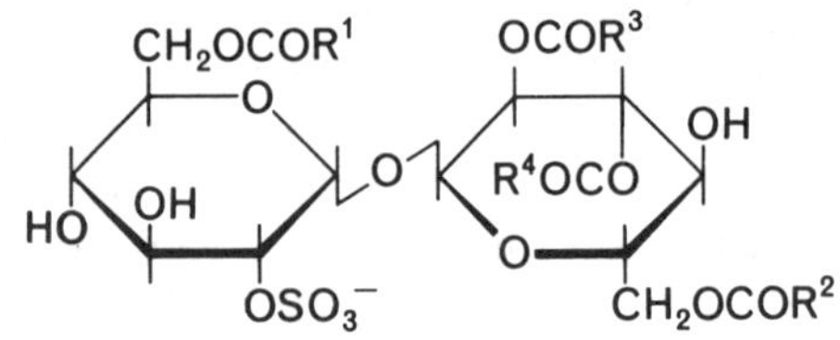

R^1OH–R^4OH are palmitic acid or very-long-chain branched fatty acids

(c) *Mycosides A and B*

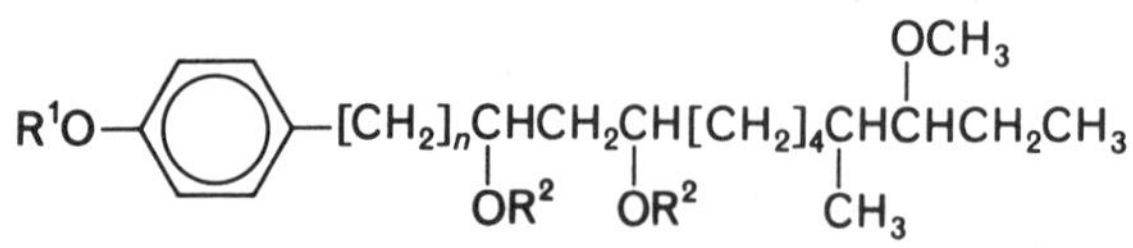

R^1OH is a mono- or trisaccharide

R^2OH is a C_{12}–C_{18} saturated fatty acyl or a mycocerosic acyl, e.g. $H_3C[CH_2]_{21}\left(\underset{CH_3}{CH}CH_2\right)_2\underset{CH_3}{CH}CO$

(d) *Mycoside C*

RO-Phe-alloThr-Ala-alaninol-sugar
|
O-sugar

ROH is a long-chain (C_{28}–C_{34}) methoxy fatty acid e.g. $H_3C[CH_2]_{22}CH{=}CH\underset{OCH_3}{C}HCH_2COOH$

Phe, phenylalanine; alloThr, allo-threonine; Ala, alanine

Figure 3.2 Structures of mycobacterial lipids. Reproduced with permission from Harwood and Russell (1984).

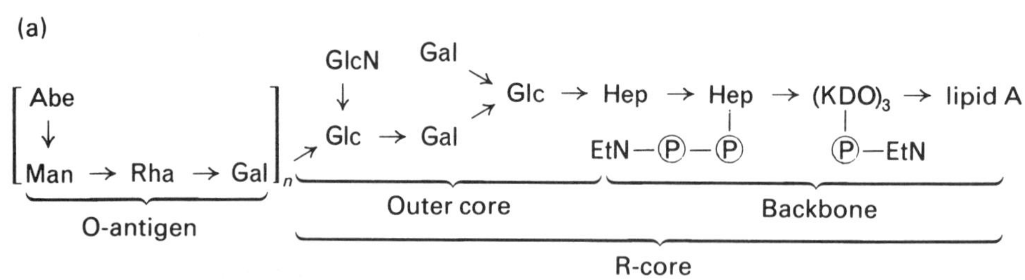

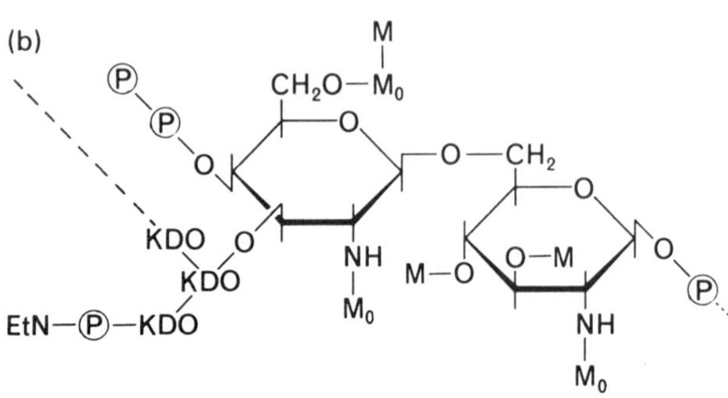

Figure 3.3 Generalized structures of (a) lipopolysaccharide and (b) lipid A showing the KDO link region. Abe, abequose; Man, mannose; Rha, rhamnose; Gal, galactose; Glc, glucose; Hep, heptose; KDO, deoxy-D-manno-octulosonic acid; Ⓟ, phosphate; EtN, ethanolamine; M, myristate; M_0, β-hydroxymyristate; --- represents the connection to the rest of the molecule (see (a)). Reproduced with permission from Harwood and Russell (1984).

saccharide is 'bound lipid'. This is because common lipid extraction procedures do not solubilize them and they are only released after hot phenol/water treatment. Lipopolysaccharide not only contains hydroxy fatty acids but also lacks unsaturated or cyclopropane fatty acids, which are characteristic of other lipids in Gram-negative bacteria (Goldfine, 1982).

Gram-positive bacteria contain teichoic acid and other highly anionic polymers in their cell walls and membranes. Some of the membrane teichoic acids are covalently linked to glycolipids such as diacylglycosylglycerol to give a lipoteichoic acid (Fig. 2.8). The amounts of esterification and glycosylation and the type of sugars or glycolipid vary amongst bacteria.

3.14.4 Other lipid types

Several other types of lipids have been reported in bacteria. Anaerobic bacteria, including those of the rumen, contain large quantities of plasmalogens with both ether and ester linkages. In contrast, plasmalogens are found in very few aerobic or facultative anaerobes.

Archaebacteria contain two types of phytanylglycerol ethers that lack ester bonds (Tornabene *et al.*, 1979; Bayley and Morton, 1978). Extreme halophiles contain diglycosyldiphytanylglycerol diether analogues of phosphatidylglycerol, phosphatidylglycerophosphate and phosphatidylglycerol sulphate. Acidophilic thermophiles contain diethers and tetraethers of diglyceroldiphytanyldiglycerol. In methanogenic bacteria, both types of ether are found.

Other lipid types that have been found in bacteria include wax esters, hydrocarbons, sphingolipids, steroids and terpenoids. These occur sporadically, however, and are seldom major components. An obvious exception is the presence of carotenoids and the ornithine-containing lipid of purple photosynthetic bacteria. Further details of minor lipid components of bacteria can be found in Razin and Rottem (1982), Shaw (1975) and Goldfine (1972).

3.14.5 Archaebacteria and other specialized bacteria

For further information on these bacteria refer to Smith (1988) and De Rosa and Gambacorta (1988).

Archaebacteria have distinctive features that allow them to be classified within a separate kingdom from other prokaryotes. These features include the presence of large amounts of ether-linked lipids (section 2.2). They often occupy extreme environments and include methanogens, the extreme halophiles and most of the thermoacidophiles.

Neutral lipids are a significant fraction of the total, with isoprenoid hydrocarbons making up 35–95% of the apolar lipids. The complex lipids of archaebacteria are unique in that they are based upon alkyl ethers of glycerol or some other polyol. The diether has the structure 2,3-di-*O*-phytanyl-*sn*-glycerol. The major phospholipid of halophiles is the diethyl analogue of phosphatidylglycerophosphate (i.e. 2,3-di-*O*-phytanyl-*sn*-glycerol-1-phosphoryl-3′-*sn*-glycerol-1′-phosphate). Various unidentified polar lipids are present in *Thermoplasma*, of which the major compound is a phosphoglycolipid.

Photosynthetic prokaryotes include those which carry out anoxygenic photosynthesis (green and purple bacteria) and those which utilize oxygenic photosynthesis (cyanobacteria and prochlorophyta). Of course, these organisms contain photosynthetic as well as cytoplasmic membranes – both with a distinct lipid composition. The photosynthetic pigments (chlorophylls, carotenoids, etc.) are confined to the photosynthetic membranes. Glycosyldiacylglycerols are major components in green bacteria and cyanobacteria but are generally absent from purple bacteria, apart from small amounts of sulphoquinovosyldiacylglycerol. The purple bacteria have phosphatidylglycerol as their major phospholipid but also possess significant amounts of phosphatidylethanolamine. Phosphatidylglycerol is also the major phospholipid in green bacteria and is the only significant phospholipid of cyanobacteria. An ornithine lipid is a significant component of purple bacteria.

The mollicutes (e.g. *Acholeplasma laidlawii*) are unique among prokaryotes in lacking an outer cell wall. They are thus very useful organisms for biochemical experiments and have been extensively studied. Glycolipids are usually major components with three types (glycosylacylglycerols, polyterpenol glycosides, acylated sugars) occurring. About 50% of the lipids of mollicutes are polar, mainly phosphoglycerides. Phosphatidylglycerol and diphosphatidylglycerol are always found and aminoacyl derivatives of the former are present in some species. Phosphatidylethanolamine is only found in *Ureaplasma*. Unique phosphoglycolipids containing glyceraldehyde and phosphotriester groupings occur in some *Acholeplasma* spp. (see also section 2.3).

References and further reading

Bayley, S.T. and Morton, R.A. (1978) *CRC Crit. Rev. Microbiol.*, **6**, 151.

Brennan, P.J. (1988) *Microbial Lipids* (eds C. Ratledge and S.G. Wilkinson), vol. 1, Academic Press, London, p. 203.

Cronan, J.E. (1978) *Annu. Rev. Biochem.*, **47**, 163.

De Rosa, M. and Gambacorta, A. (1988) *Prog. Lipid Res.*, **27**, 153.

de Smet, M.J., Eggink, G., Witholt, B., Kingma, J. and Wynberg, H. (1983) *J. Bacteriol.*, **154**, 870.

Fautz, E., Rosenfelder, G. and Grotjahn, L. (1979) *J. Bacteriol.*, **140**, 852.

Finnerty, W.R. (1978) *Adv. Microb. Physiol.*, **18**, 177.

Goldfine, H. (1972) *Adv. Microb. Physiol.*, **8**, 1.

Goldfine, H. (1982) *Curr. Top. Membr. Transp.*, **17**, 1.

Harwood, J.L. and Russell, N.J. (1984) *Lipids in Plants and Microbes*, Allen and Unwin, Hemel Hempstead.

Intriago, P. and Floodgate, G.D. (1991) *J. Gen. Microbiol.*, **137**, 1503.

Matsui, Y., Susuki, S., Susuki, T. and Takama, K. (1991) *Lett. Appl. Microbiol.*, **12**, 51.

Minnikin, D.E. (ed.) (1982) *Lipids: Complex Lipids, Their Chemistry, Biosynthesis and Roles in the Biology of the Mycobacteria*, vol. 1, *Physiology, Identification and Classification*, Academic Press, New York.

O'Leary, W.W. and Wilkinson, S.G. (1988) *Microbial Lipids* (eds C. Ratledge and S.G. Wilkinson), vol. 1, Academic Press, London, p.117.

Pommier, M.T. and Michel, G. (1981) *Eur. J. Biochem.*, **118**, 329.

Raetz, C.R.H. (1978) *Microbiol. Rev.*, **42**, 614.

Ratledge, C. (1989) *Microbial Lipids* (eds C. Ratledge and S.G. Wilkinson), vol. 2, Academic Press, London, p. 567.

Ratledge, C. and Wilkinson, S.G. (eds) (1988) *Microbial Lipids*, vol. 1, Academic Press, London.

Ratledge, C. and Wilkinson, S.G. (eds) (1989) *Microbial Lipids*, vol. 2, Academic Press, London.

Razin, S. and Rottem, S. (eds) (1982) *Curr. Top. Membr. Transp.*, **17**, 1–383.

Schweizer, E. (1989) *Microbial Lipids* (eds C. Ratledge and S.G. Wilkinson), vol. 2, Academic Press, London, p. 3.

Scott, C.C.L. and Finnerty, W.R. (1976) *J. Bacteriol.*, **127**, 481.

Shaw, N. (1975) *Adv. Microb. Physiol.*, **12**, 141.

Smith, P.F. (1988) *Microbial Lipids* (eds C. Ratledge and S.G. Wilkinson), vol. 1, Academic Press, London, p. 489.

Tornabene, T.G., Langworthy, T.A., Holzer, G. and Oro, J. (1979) *J. Mol. Evol.*, **13**, 73.

Wilkinson, S.G. (1988) *Microbial Lipids* (eds C. Ratledge and S.G. Wilkinson), vol. 1, Academic Press, London, p. 299.

Zepke, H.D., Heinz, E., Radunz, A., Linscheid, M. and Pesch, R. (1978) *Arch. Microbiol.*, **119**, 157.

3.15 Lipids of viruses

Lipids are found in a wide range of both RNA and DNA viruses. Usually the lipids are present in an external membrane, the viral envelope, for those viruses which infect eukaryotic cells. For other viruses, such as bacteriophages, the lipid is in internal membranes. A comprehensive summary of viral lipids is available (Steiner and Steiner, 1988).

Phospholipids account for 50–75% of the total viral lipid in most cases. Examples of the phospholipid composition are shown in Table 3.277 for both RNA and DNA viruses. The major components are phosphatidylcholine, phosphatidylethanolamine, sphingomyelin and phosphatidylserine. Phosphatidylinositol is frequently absent in the RNA viruses. Percentages of the three zwitterionic phospholipids are somewhat variable (Table 3.277). As a generalization, high proportions of phosphatidylcholine are accompanied by relatively lower percentages of both phosphatidylethanolamine and sphingomyelin and vice versa. Sphingomyelin is much less important in the DNA viruses – only *Pseudorabies* virus having more than 4%.

So far as fatty acid composition is concerned, comparisons have usually been made between viral composition and the host cell. The viruses often contain lowered amounts of polyenoic acids and may accumulate large proportions of acids that are only trace components in the host. For example, the *Rubella* virus contains about 25% odd-carbon fatty acids (C_{15}, C_{17}, C_{19}) (Voiland and Bardeletti, 1980). A considerable proportion of the total fatty acids may be acylated to protein (see Steiner and Steiner, 1988). In the human immunodeficiency virus (HIV), the p17 protein of the interior surface is myristoylated (Tomaselli *et al.*, 1991).

Viruses that bud from the plasma membrane often have quite a high ratio of cholesterol to phospholipid. Thus, for the simian virus 5, the ratio is 0.84, and is about 0.75 for the host MDBK cells (Klenk and Choppin, 1969). For plant viruses, sterols such as stigmasterol, β-sitosterol and campesterol will be found.

For further details of other viral lipids and their metabolism and function, the reader is referred to Steiner and Steiner (1988).

References and further reading

Klenk, H.-D. and Choppin, P.W. (1969) *Virology*, **38**, 255.

Steiner, M. and Steiner, S. (1988) in *Microbial Lipids* (eds C. Ratledge and S.G. Wilkinson), vol. 1, Academic Press, London, p. 83.

Tomaselli, A.G., Howe, W.J., Sawyer, T.K. *et al.* (1991) *Chimicaoggi*, May, 6.

Voiland, A. and Bardeletti, G. (1980) *Arch. Virol.*, **64**, 319.

Table 3.277 Phospholipid composition of viruses (From Steiner and Steiner, 1988)

	Percentage of phospholipids				
	PC	PE	Sph	PS	PI
RNA viruses					
Paramyxoviridae, e.g. Sendai, measles	8–53	10–41	12–30	2.17	0–11
Rhabdoviridae, e.g. rabies	16–38	20–34	16–31	7–20	0–10
Orthomyxoviridae, e.g. influenza	10–39	12–45	16–23	7–22	0–8
Togaviridae, e.g. rubella	21–49	19–35	7–29	9–21	0–9
Retroviridae, e.g. Rose's sarcoma	16–30	26–42	22–33	12–17	0–4
DNA viruses					
Herpetoviridae, e.g. herpes simplex	51–57	21–25	4–21	4–13	1–13
Poxviridae, e.g. vaccinia	34–49	12–14	2–4	0–5	12–19

Abbreviations: PC = phosphatidylcholine; PE = phosphatidylethanolamine; Sph = sphingomyelin; PS = phosphatidylserine; PI = phosphatidylinositol.

4

SEPARATION AND ISOLATION PROCEDURES

F.D. GUNSTONE, M. KATES and J.L. HARWOOD

This chapter describes the separation and isolation procedures used by lipid chemists and biochemists to concentrate and purify individual components from natural mixtures. Brief reference is also made to industrial applications of these procedures, which are also discussed in section 5.5.3.

4.1 Crystallization

Crystallization at ordinary, zero and sub-zero temperatures is still a useful method of purifying fatty acids and their derivatives. It is, for example, important in the purification of synthetic glycerides and of polyacetylenic acids before their partial hydrogenation (section 10.1). It is often used as one of several purification procedures in the isolation of novel acids prior to their identification and of common acids from natural sources (section 4.8).

Saturated acids are usually solid at ambient temperature and can be crystallized at temperatures down to 0 °C. Crystals are separated from the mother liquor by any conventional filtration procedure.

Unsaturated acids with lower melting points and higher solubilities must be crystallized at lower temperatures (0 to −90 °C), and filtration must be carried out at an appropriate low temperature. The solution should be reduced to the crystallizing temperature slowly (2–4 h) and maintained thereat for 4–24 h. This produces larger crystals, which are easier to filter.

Crystallization is a mild procedure especially suitable for the polyene acids, which are so easily oxidized or otherwise modified at elevated temperatures. Separation of saturated acids from monoene acids by crystallization is good; separation of polyene acids from one another is less satisfactory. Useful solvents include methanol, ether, petroleum ether and acetone, usually at a dilution of 5–10 ml of solvent per gram of acid. Repeated crystallization may give an improved product, but the isolation of pure compounds (>99%) usually involves the application of more than one separation procedure (section 4.8).

Two examples are now given. A concentrate of linoleic acid (750 g, 84% linoleic acid) crystallized from acetone at −75 °C gave a precipitate (575 g, 93%) and a mother liquor (145 g, 35%). A concentrate of oleic acid (518 g, 74% oleic acid) crystallized from petroleum ether at −40 °C gave a precipitate (335 g, 90%) and a mother liquor (168 g, 41%). Recrystallization of this precipitate several times eventually gave acid of 99% purity (216 g).

The crystallization of lead salts from ethanol to separate saturated from unsaturated acids and of lithium salts from acetone to separate polyene from non-polyene acids are classical procedures not much employed at the present time. However, in a recent report on the isolation of docosahexaenoic acid (DHA) from cod liver oil, the first step was the crystallization of lithium salts from acetone to concentrate the polyene acids in the mother liquor. DHA was isolated from the polyene concentrate by iodolactonization (Wright *et al.*, 1987; section 10.5).

Crystallization is applied on an industrial scale to natural triacylglycerol (triglyceride) and fatty acid mixtures (see section 5.5.3). Sometimes crystallization is from the melt (dry fractionation), though better results are obtained using a solvent (solvent fractionation). Resulting from this latter process are a crystalline fraction enriched in saturated acids and of higher melting point and a liquid fraction enriched in unsaturated acids and less likely to crystallize when stored. In the case of natural oils, the resulting glyceride mixtures are referred to as *stearins* and *oleins* respectively.

There are two major procedures using solvent and two without solvent. The solvents commonly employed are methanol (Emerson process) or acetone (Armour–Texaco process), and these have been employed to isolate 'stearic' (a mixture of palmitic and stearic acids with low iodine value) and oleic acids from tallow fatty acids, oleic and linoleic acids from tall oil, and stearic and isostearic acids from the monomer fraction remaining after dimerization. Attempts have also been made to use propane and methyl formate.

Crystallization without solvent occurs in the Tirtiaux process, which involves slow cooling and slow agitation, and in the hydrophilization process. In the latter, crystallization produces a slurry of solid and liquid material, which are separated after addition of an aqueous solution of a wetting agent (sodium laurylbenzenesulphonate) and of magnesium sulphate.

Separation processes applied on an industrialized scale for the production of fatty acid concentrates have been reviewed by Fritz (1988) and by Johnson (1990).

References and further reading

Brown, J.B. and Kolb, D.G. (1955) *Prog. Chem. Fats Other Lipids*, **3**, 57.

Fritz, E. (1988) *Fatty Acids in Industry* (eds R.W. Johnson and E. Fritz), Dekker, New York, p. 173.

Johnson, R.W. (1990) *World Conference on Oleochemicals into the 21st Century* (ed. T.H. Applewhite), American Oil Chemists' Society, Champaign, p. 189.

Sonntag, N.O.V. (1979) *Fatty Acids* (ed. E.H. Pryde), American Oil Chemists' Society, Champaign, p. 125.

Wright, S.W., Kuo, E.Y. and Corey, E.J. (1987) *J. Org. Chem.*, **52**, 4399.

4.2 Urea fractionation

Urea normally crystallizes in tetragonal form, but in the presence of certain aliphatic compounds it forms hexagonal prisms containing some of the aliphatic material. These prisms are built up from urea: six molecules form a unit cell (1.11 nm long and 0.82 nm in diameter) containing a channel in which an open-chain molecule may be held so long as it fulfils certain dimensional qualifications. It must not be too short or it will not be held within the channel, and it must not be too wide if it is to fit into the free space, variously estimated at 0.40–0.60 nm. Many straight-chain acids and their esters satisfy these conditions and thus readily form complexes (adducts, inclusion compounds) with urea.

Saturated acids form stable complexes more readily than do unsaturated acids, and oleic acid enters into these inclusion compounds more readily than do the polyunsaturated acids. In practice, either urea and mixed acids are dissolved in hot methanol or urea and methyl esters are dissolved in a hot mixture of methanol and ethanol. The solution is crystallized at room temperature or at 0 °C. The adduct and the mother liquor will furnish the acids or esters when mixed with water and extracted with ether or petroleum ether in the usual way.

The procedure is employed for two purposes. It separates straight-chain acids from branched-chain or cyclic compounds, with the former normally concentrating in the adduct and the latter normally in the mother liquor. It has thus been used to concentrate branched-chain and cyclic acids in natural mixtures, particularly when these are present only as trace constituents, and to isolate cyclic acids prepared from polyene acids. One report, for example, describes the isolation of cyclic acids produced from linseed oil by high-temperature reaction with alkali in 90–95% yield and 95% purity.

Ackman *et al.* (1988) and Ratnayake *et al.* (1988) have described optimum conditions for obtaining concentrates of $n-3$ acids (69–85%) from fish oil fatty acids by urea fractionation. These consist mainly of 18:4, 20:5 and 22:6. The separation has been scaled up to 40 kg batches.

The acids from blackcurrant seed oil (section 3.3.13), which contains α and γ-18:3 and 18:4 acids, have been fractionated with the help of the urea technique followed by HPLC. The urea mother liquor contains γ-18:3 (79.6%) and 18:4 (15.5%) and a reduced level of α-18:3 (2.6%). This separation has been carried out on the tonne scale. Pure γ-18:3 (94–96%) and 18:4 (>90%) are obtained by HPLC (section 4.4.4) (Traitler *et al.*, 1988; Willie and Traitler, 1988; Wille *et al.*, 1991).

References and further reading

Ackman, R.G., Ratnayake, W.M.N. and Olsson, B. (1988) *J. Am. Oil Chem. Soc.*, **65**, 136.

Ratnayake, W.M.N., Olsson, B., Matthews, D. and Ackman, R.G. (1988) *Fat Sci. Technol.*, **90**, 381.

Schlenk, H. (1954) *Prog. Chem. Fats Other Lipids*, **2**, 243.

Traitler, H., Wille, H.J. and Studer, A. (1988) *J. Am. Oil Chem. Soc.*, **65**, 755.

Wille, H.J. and Traitler, H. (1988) *Fat Sci. Technol.*, **90**, 476.

Wille, H.J., Traitler, H. and Lagarde, M. (1991) *Fat Sci. Technol.*, **93**, 363.

4.3 Distillation

Since the effectiveness of distillation depends on differences of boiling point, chain length is more important than degree of unsaturation. It is possible to separate the esters of C_{12}, C_{14}, C_{16}, C_{18} and C_{20} acids from each other, and distillation is commonly used for this purpose. On the other hand, stearate (18:0), oleate (18:1), linoleate (18:2) and linolenate (18:3) are not usefully separable by distillation (section 8.5.3).

The most widely used distillation procedure is fractional distillation of methyl esters under reduced pressure (0.1–1.0 mmHg). Even under these conditions, moderately high temperatures are required (bath temperatures around 200 °C); the more highly unsaturated acids, especially those with conjugated unsaturation or those longer than C_{18}, are prone to polymerization, cyclization and stereomutation of double bonds. Heated columns packed with glass helices or some form of metal packing are in common use despite the disadvantage of a significant hold-up and pressure drop through the column. Spinning band columns do not suffer from these disadvantages. More sophisticated systems are used on a commercial scale (Huibers and Fritz, 1988; Stage, 1990).

Distillation at still lower pressures has been used in the isolation of some highly unsaturated acids, and is particularly valuable in polymerization studies to separate monomeric, dimeric and polymeric material, and in the separation of monoacylglycerols from mono- and diacylglycerol mixtures (section 5.8.1).

Distillation of fatty acids is important in industry, where it is used primarily to separate saturated acids such as (i) 16:0 and 18:0 from hydrogenated tallow and seed oils, (ii) 18:0, 20:0, 22:0 and 24:0 from hydrogenated fish oils and high-erucic-acid rapeseed oil, and (iii) 8:0, 10:0, 12:0 and 14:0 from hydrogenated lauric oils (coconut, palmkernel, babassu). The temperatures range from 160 to 230 °C at pressures below 20 mmHg in specially constructed stills (up to 7 ft (2.1 m) in diameter), which run continuously. It has been claimed that it should be possible to separate stearic and oleic acids by distillation (Stage, 1984).

Table 4.1 lists the boiling points of the C_6 to C_{18} acids at a range of pressures.

References and further reading

Huibers, D.T.A. and Fritz, E. (1988) *Fatty Acids in Industry* (eds R.W. Johnson and E. Fritz), Dekker, New York, p. 85.

Poole, W.A. and Ralston, A.W. (1942) *Ind. Eng. Chem.*, **34**, 1104.

Sonntag, N.O.V. (1979) *Fatty Acids* (ed. E.H. Pryde), American Oil Chemists' Society, Champaign, p. 125.

Stage, H. (1984) *J. Am. Oil Chem. Soc.*, **61**, 204.

Stage, M.A.E. (1990) *World Conference on Oleochemicals into the 21st Century* (ed. T.H. Applewhite), American Oil Chemists' Society, Champaign, p. 142.

Table 4.1 Boiling points (°C) of some common saturated acids (Poole and Ralston, 1942)

Pressure (mmHg)	6:0	8:0	10:0	12:0	14:0	16:0	18:0
1	61.7	87.8	110.3	130.2	149.2	167.4	183.6
2	71.9	97.9	121.1	141.8	161.1	179.0	195.9
4	82.8	109.1	132.7	154.1	173.9	192.2	209.2
8	94.6	121.3	145.5	167.4	187.6	206.1	224.1
16	107.3	134.6	159.4	181.8	202.4	221.5	240.0
32	120.8	149.2	174.6	197.4	218.3	238.4	257.1
64	136.0	165.3	191.3	214.6	236.3	257.1	276.8
128	152.5	183.3	209.8	234.3	257.3	278.7	299.7
256	171.5	203.0	230.6	256.6	281.5	303.6	324.8
512	192.5	225.6	254.9	282.5	309.0	332.6 [a]	355.2 [a]
760	205.8	239.7	270.0	298.9	326.2 [a]	351.5 [a]	376.1 [a]

[a] Extrapolated values.

4.4 Liquid chromatography

Chromatography of lipids on a column of suitable material eluted with an appropriate solvent system (liquid chromatography, LC) is a versatile and useful procedure for fractionating lipid mixtures on a preparative scale or on a micro-scale. Three types of liquid chromatography are in general use: adsorption (solid–liquid) chromatography, partition (liquid–liquid) chromatography and ion-exchange chromatography. The interested reader is referred to reviews by Rouser *et al.* (1969, 1976), Carroll and Serdarevich (1967) and Carroll (1976) and to compendia by Marinetti (1967–69, 1976), Christie (1982) and Kates (1986).

4.4.1 Adsorption chromatography

In adsorption chromatography, compounds are bound directly to the solid adsorbent (as stationary phase) largely by polar (hydrogen bonding) and ionic forces and to a smaller extent by non-polar or van der Waals' forces. Separation of lipid mixtures thus takes place according to the relative polarities of the individual lipids, which are determined by the number and type of polar groups (e.g. OH, C=O, COOH, $—H_2PO_3^-$, HSO_3^-, etc.) in the molecules, but may also be influenced somewhat by the number and type of non-polar or hydrophobic groups. In general, elution of the column by increasingly polar solvents (e.g. hexane, chloroform, benzene, ethyl ether, acetone, methanol) resolves the lipid mixture into increasingly polar classes of compounds, in the following order: saturated hydrocarbons, unsaturated hydrocarbons, wax esters, steryl esters, long-chain aldehydes, triglycerides, long-chain alcohols, fatty acids (free), quinones, sterols, diglycerides, monoglycerides, glycosyl diglycerides, cerebrosides, sulpholipids, acidic glycerophosphatides, phosphatidylethanolamine, lysophosphatidylethanolamine, lecithin, sphingomyelin, lysolecithin (see reviews by Rouser *et al.*, 1976; Carroll and Serdarevich, 1967; Carroll, 1976; Morris, 1966; Malins, 1966; Stein and Slawson, 1966; Aitzetmüller, 1975; Kuksis, 1977; Lie Ken Jie, 1980).

Complete separation of the neutral lipid classes given above (see section 4.4.1(a)(i)) may be achieved by a single-column chromatography (Carroll and Serdarevich, 1967), but this cannot be achieved with phospholipids because of overlapping polarities. Further fractionation using preparative thin-layer chromatography (TLC) or another type of column chromatography is usually necessary for isolation of pure individual phospholipid classes (see section 4.4.1(a)(ii)). For a given lipid class, individual members differing in fatty acid chain length and degree of unsaturation are generally not resolved, although some partial separations on this basis have been achieved.

Several separations based on adsorption have been described in the patent literature and are reviewed by Johnson (1990). They include the separation of the following pairs of acids: lauric and myristic, palmitic and stearic, palmitic/stearic and oleic, oleic and linoleic, and fatty acids and rosin acids. Large-scale chromatography (multi-kilogram scale) is employed on a commercial basis to produce purified lipids. These may be vegetable oils (>99% triacylglycerols) freed from less-polar (hydrocarbons) and more-polar impurities (diacylglycerols, phospholipids). Also produced by this method are phospholipids of entirely one lipid class from different sources. Examples include phosphatidylcholines or phosphatidylethanolamines from rape, soya or egg. These products are in demand for cosmetic and pharmaceutical purposes (Herslof and Kaufmann, 1990).

(a) Column chromatography

Several kinds of adsorbents have been used in column chromatography of lipids, e.g. alumina, magnesium oxide, Florisil (magnesium silicate), acid-treated Florisil, or silicic acid. Of these, the most versatile and effective adsorbent is silicic acid. Acid-treated Florisil, which is essentially a coarse-mesh silicic acid, has similar chromatographic properties. Adsorption chromatography can be carried out in glass columns (macro-scale) or Pasteur pipettes (micro-scale) packed with the adsorbent, or on thin layers of the adsorbent (TLC) spread on glass plates, or on aluminium or plastic sheets, provided the latter are inert to the developing solvents. With columns, 60–100 g silica per gram of lipid is generally used, but loading will depend on the mixture of lipids used and the kind of adsorbent chosen.

A general fractionation procedure (see Table 4.2, step 1) may be used for preliminary separation of most lipid mixtures into three groups of lipids, i.e. neutral lipids, glycolipids and phosphatides, respectively, by elution of the silicic acid column with the solvent sequence chloroform → acetone → methanol (Rouser *et al.*, 1976; Kates, 1986). The composition

Table 4.2 Separation of lipids by column chromatography

Silicic acid (Christie, 1982)		Silicic acid (Gray, 1976)		Florisil (acid-washed) (Carroll, 1976; Kates, 1986)	
Solvent	Eluted lipid	Solvent	Eluted lipid	Solvent	Eluted lipid
Step 1 [a]					
Chloroform	Neutral lipids	C/M (49/1) T/B (1/4)	Neutral lipids Cholesterol	Chloroform C/M (95/5)	Neutral lipids Cerebrosides, glycosyldiacylglycerols
Acetone	Glycolipids + sulpholipids	T/DM/M/H_2O (10/6/4/1)	DPG, PG, PI, PE	C/M (9/1)	PS, PE, PG, ceramide polyhexosides
Methanol	Phospholipids	C/M/H_2O (4/16/1)	PC + sphingomyelin + lysoPC	C/M (1/1)	PI + PC
				Methanol	Sphingomyelin
Step 2 [b]					
C/M (95/5)	DPG + PA				
C/M (80/20)	PE + PS				
C/M (50/50)	PI + PC				
Methanol	Sphingomyelin + lysoPC				

[a] Chromatography of total lipids on a silicic acid column.
[b] Rechromatography of methanol fraction on another silicic acid column.
Abbreviations: Solvents — C, chloroform; M, methanol; T, tetrahydrofuran; B, benzene; DM, dimethoxymethane. Lipids — DPG, diphosphatidylglycerol; PG, phosphatidylglycerol; PI, phosphatidylinositol; PE, phosphatidylethanolamine; PC, phosphatidylcholine; PA, phosphatidic acid.

of each fraction is as follows:

1. *Chloroform eluates*. For lipids of most animal and plant tissues or microbial cells, this fraction contains the neutral lipids, i.e. hydrocarbons, carotenoids and chlorophyll pigments (in plants), sterols and sterol esters (absent from bacteria), glycerides, waxes, fatty alcohols, aldehydes and fatty acids.
2. *Acetone eluates*. With animal lipids, containing large proportions of glycolipids and little cardiolipin (e.g. brain lipids), this fraction consists of cerebrosides, sulphatides and ceramide polyhexosides. With lipids of plants (e.g. seeds, leaves, algae), the acetone eluate contains mono- and digalactosyldiglycerides, cerebrosides, steryl glycosides and sulpholipid (plus small amounts of cardiolipin and phosphatidic acid). With lipids of Gram-positive bacteria (e.g. streptococci, micrococci), the acetone eluates contain glycosyldiglycerides and traces of acidic phosphatides. When samples contain relatively high proportions of cardiolipin (e.g. heart, liver, *Bacillus cereus* lipids), a considerable amount of this acidic phosphatide appears in the acetone eluate.
3. *Methanol eluates*. With lipids from all sources, this fraction contains mostly phospholipids, with only traces of glycolipids.

Each of these fractions is then further fractionated chromatographically as follows: neutral lipids, by preparative TLC (section 4.4.1(b)) or silicic acid column chromatography (section 4.4.1(a)(i)); glycolipids and sulpholipids, by preparative TLC or chromatography on DEAE-cellulose or CM-cellulose (section 4.4.3); phospholipids, by preparative TLC or combination of different types of column chromatography (section 4.4.1(a)(ii)) (Kates, 1986).

For lipid mixtures rich in glycolipids, e.g. bacterial, leaf, or brain lipids, the elution sequence may be modified as follows (monitored by microslide TLC, section 4.4.1(b)): chloroform (non-polar lipids and pigments); chloroform–acetone (1/1 v/v) and acetone (diglycosyldiglycerides, cerebrosides, sterylglycosides, sulphatides and sulpholipids, and also green chlorophyll breakdown products); chloroform–methanol (90/10 v/v) (traces of glycolipids, phosphatidic acid and cardiolipin); chloroform–methanol (1/1 v/v) and methanol (cardiolipin, phosphatidylglycerol, phosphatidylethanolamine, lecithin and sphingomyelin) (Vorbeck and Marinetti, 1965; Carroll *et al.*, 1968).

(i) Separation of neutral lipids (Carroll and Serdarevich, 1967)

Separation of neutral lipids (particularly those of animal tissues) containing free fatty acids is most conveniently carried out on a column of Florisil (magnesium silicate) hydrated with 7% water, which has the advantage of allowing a high flow rate without loss of resolving power, and also of strongly retaining free fatty acids. However, Florisil causes isomerization of monoglycerides and of diglycerides by acyl migration, but this can be prevented by using Florisil impregnated with 10% (w/w) boric acid. Chromatography on silicic acid or acid-treated Florisil also avoids isomerization of diglycerides, but not of monoglycerides; again, impregnation of the adsorbent with 10% boric acid prevents isomerization of monoglycerides. However, for routine separation of neutral lipid classes, in which isolation of individual isomers of mono- and diglycerides is not desired, chromatography on these three adsorbents unimpregnated with boric acid will suffice.

Columns are prepared in hexane and eluted with the solvent sequences given in Table 4.3, which also lists the components eluted by the various solvents, as monitored by microslide TLC. Separation of each neutral lipid class into molecular species or individual compounds may be achieved by reversed-phase partition chromatography (see section 4.4.2(b)).

(ii) Separation of phospholipids (Carroll, 1963, 1976; Carroll *et al.*, 1968; Kates, 1986)

Silicic acid or acid-treated Florisil (coarse silicic acid) columns can be used for fractionation (partial) of the phospholipid fraction. Acid-treated Florisil columns have the advantage of rapid elution flow rates (compared to silicic acid columns) and are particularly useful for lipids rich in acidic phospholipids or sulpholipids. The column of acid-washed Florisil (12 g) may be loaded with up to 100 mg total lipids (liver, blood, brain, kidney, egg yolk, yeast or bacterial lipids) in chloroform (5 ml) and eluted with the solvent sequence given in Table 4.2.

The eluted fractions (monitored by microslide TLC) have the following components (see Table 4.2): fraction 1, chloroform (neutral lipids such as hydrocarbons, sterols and sterol esters, glycerides, fatty acids); fraction 2, chloroform–methanol (95/5 v/v) (washings); fraction 3, chloroform–methanol (90/10 v/v) (cerebrosides, glycosyldiglycerides,

Table 4.3 Column chromatography of neutral lipids on various adsorbents [a] (Carroll and Serderavich, 1967)

Adsorbent	Eluting solvent	Eluate volume (ml)	Components eluted
Florisil (7% hydrated)	Hexane	40	Hydrocarbons
	Hexane–ethyl ether (95/5)	90	Steryl esters, fatty acid methyl esters, wax esters
	Hexane–ethyl ether (85/15)	150	Triacylglycerols
	Hexane–ethyl ether (75/25)	100	Cholesterol
	Hexane–ethyl ether (50/50)	100	Diacylglycerols
	Ethyl ether–methanol (98/2)	135	Monoacylglycerols
	Ethyl ether–acetic acid (96/4)	50	Free fatty acids
Acid-treated Florisil	Hexane	45	Hydrocarbons
	Hexane–ethyl ether (99/1)	80	Squalene, steryl esters, fatty acid methyl esters, wax esters
	Hexane–ethyl ether (95/5)	125	Triacylglycerol, fatty acids
	Hexane–ethyl ether (85/15)	75	Alcohols
	Hexane–ethyl ether (82/18)	100	Cholesterol, diacylglycerols
	Hexane–ethyl ether (75/25)	60	Monoacylglycerol
Silicic acid	Hexane	45	Hydrocarbons, squalene
	Hexane–ethyl ether (99/1)	95	Steryl esters, fatty acid methyl esters, wax esters
	Hexane–ethyl ether (95/5)	60	Triacylglycerols
	Hexane–ethyl ether (92/8)	75	Fatty acids and alcohols
	Hexane–ethyl ether (85/15)	115	Cholesterol, diacylglycerols
	Ethyl ether	45	Monoacylglycerols

[a] *Column*: Hydrated Florisil 12 g, 1.2 × 15 cm (column volume, 5 ml). Acid-treated Florisil 12 g, 1.2 × 25 cm (column volume, 15 ml). Unisil silicic acid 12 g, 1.2 × 24 cm (column volume, 15 ml). Boric acid-impregnated adsorbents (see text) may be used when isomerization of mono- and diglycerides is to be avoided. Load: 120–180 mg of total neutral lipids. Flow rate: 2–3 ml/min.

sulpholipid, sulphatides, phosphatidic acid and cardiolipin); fraction 4, chloroform–methanol (3/1 v/v) (phosphatidylserine, phosphatidylethanolamine, ceramide polyhexosides, phosphatidylglycerol); fraction 5, chloroform–methanol (1/1 v/v) (phosphatidylinositol, lecithin); and fraction 6, methanol (sphingomyelin and lysolecithin).

Similar separations can be achieved on columns of silicic acid (Table 4.2), but elution of the columns is relatively slow. Use of alumina columns is restricted by the alkalinity of alumina, which tends to promote hydrolytic and other changes in the lipids. However, alumina can be an effective adsorbent for large-scale separation of phosphatidylcholine from phosphatidylethanolamine in lipid mixtures containing high proportions of these two phospholipids (e.g. in egg yolk). For egg yolk phospholipids, a weight ratio of alumina/phospholipids of 2.5/1 is used and the column is eluted with chloroform–methanol (9/1 v/v) to elute lecithin and with chloroform–methanol–water (2/5/2) to elute phosphatidylethanolamine (PE), sphingomyelin, lysophosphatides, etc. (Singleton *et al.*, 1965; Nielsen, 1980). Final purification of lecithin and PE is achieved by chromatography on silicic acid columns or by preparative TLC.

(iii) Separation of glycolipids

Glycolipids (with the exception of gangliosides) can be separated from neutral lipids and most phospholipids by column chromatography on silicic acid or acid-washed Florisil (Table 4.2). Chloroform will elute simple lipids, and glycolipids are eluted with acetone (Christie, 1982; Kates, 1986). Silicic acid or acid-washed Florisil columns are also often used (in combination with ion-exchange chromatography on DEAE- or TEAE-cellulose or preparative TLC) for the isolation of pure fractions (see sections 4.4.1(a) and 4.4.3). Neutral glycolipids are eluted in the order ceramide, cerebroside and glycosyldiacylglycerol, ceramide dihexoside, sphingomyelin, with increasing concentrations of methanol in chloroform (Table 4.2).

Examples of fractionation schemes for ion-exchange chromatography are given later in Table 4.6. The method is considerably improved if phospho-lipids have been previously removed by an acetone elution procedure (Table 4.2). Solvent systems that can be used for preparative TLC of glycolipids are many and varied. Useful examples are listed in Kates (1986, 1990), Christie (1982), Stahl (1969) and Kundu (1981b). It is usually necessary to use preparative TLC finally to purify a lipid that has been obtained by column chromatography. TLC is also useful to separate ceramides with normal fatty acids from those with hydroxy fatty acids (Rouser *et al.*, 1969).

A general method for the purification of the three common plant glycosylglycerides (galactosyldiacylglycerol, digalactosyldiacylglycerol and sulphoquinovosyldiacylglycerol) has been described by O'Brien and Benson (1964) (see Sastry and Kates, 1964; Kates, 1990).

A number of specific staining reagents can be used to detect glycolipids, apart from the standard general sprays for lipids (see Kates, 1986). Periodate–Schiff stain (Shaw, 1968) yields a blue spot with glycolipids (but also a mauve one with phosphatidylglycerol). The α-naphthol stain (Siakotos and Rouser, 1965) gives blue–purple spots with glycolipids. Other polar lipids give yellow spots, and cholesterol a grey–red spot. Gangliosides can be detected with a resorcinol stain (Svennerholm, 1957) and sphingolipids with a sodium hypochlorite–benzidine spray (see Kates, 1986). Unfortunately, benzidine is a potent carcinogen and its use is inadvisable, and is banned by law in the UK. Orcinol–sulphuric acid mixtures are widely used for the general detection of glycolipids (Svennerholm, 1956), and the choline moiety of sphingomyelin (and other quaternary ammonium lipids) can be detected by the Dragendorff spray (Wagner *et al.*, 1961). Non-destructive sprays for lipids include rhodamine 6G and primuline (Kundu, 1981b), with the latter being especially useful when further analysis of hexoses, sphingosine bases or fatty acids by GLC is to be undertaken.

Glycosphingolipids are relatively minor components of the total lipids of most tissues, except brain and nervous tissue, and must be separated from other lipids before analysis. Typical chromatographic methods are given above (section 4.4.1(a)(ii)). Alternatively, phosphoglycerides, neutral glycerides and other lipids having *O*-acyl-bound fatty acids can be degraded by mild alkaline hydrolysis, and the alkali-resistant sphingolipids (cerebrosides, sphingomyelin, etc.), which contain alkali-stable *N*-acyl groups, can be isolated by preparative TLC in chloroform–methanol–ammonia (65/25/5 v/v) or by column chromatography on silicic acid (see Kates, 1986; Christie, 1982).

For complete separation of glycosphingolipids from phospholipids, it is recommended that the former be acetylated with acetic anhydride–pyridine before Florisil column chromatography or preparative TLC. The isolated acetylated glycosphingolipids can then be deacetylated with sodium methoxide (Saito and Hakomori, 1971; Hakomori, 1983). Usually the nature of the hexose unit has little effect on chromatographic mobility, but ceramide glucoside can be separated from ceramide galactoside by TLC on silica gel plates impregnated with 5% sodium tetraborate (Young and Kanfer, 1965).

Gangliosides are so polar that special precautions have to be taken during extraction to ensure that they are recovered. For example, gangliosides are found in the aqueous-methanol fraction of the Folch *et al.* (1957) or Bligh and Dyer (1959) extraction procedures. They can be conveniently purified by Sephadex chromatography (Siakotos and Rouser, 1965) and obtained with other sphingolipids by Florisil chromatography of acetylated lipid extracts (Hakomori, 1983). Individual gangliosides can be separated by TLC using neutral or alkaline solvent systems (Christie, 1982). Acidic systems must be avoided, as they cause degradation.

(b) Thin-layer chromatography

TLC is one of the most effective and versatile techniques for separation of intact complex polar lipids (phospholipids, glycolipids, etc.) and their lipid moieties, as well as for neutral lipids (hydrocarbons, acylglycerols, fatty acids, sterols, etc.). TLC of lipids is usually carried out on silicic acid as adsorbent, but other adsorbents, such as alumina, ion-exchange celluloses, Sephadex, cellulose, etc., may also be used. As with column chromatography, TLC may be used for adsorption chromatography, ion-exchange chromatography and partition (liquid–liquid) chromatography (normal and reversed phase).

TLC has the advantages of simplicity and ease of manipulation, sensitivity and high resolving power. TLC has further advantages that make it the method of choice in most instances: rapidity, detectability of all organic compounds by charring with sulphuric acid, and preparative-scale operation (Christie, 1982; Kates, 1986).

As an illustration of the advantage of rapidity, a microslide TLC plate will take up to six spots and may be run in 4–5 min; thus the eluates of a silicic acid column may be monitored by microslide TLC as the column is running, giving an almost immediate indication of which fractions to combine and when to change the eluting solvent. Another advantage of rapidity in TLC is that techniques for increasing the resolving power of chromatography, such as two-dimensional development and one-dimensional multiple development, may be carried out in a relatively short time and do not involve any specialized equipment other than adsorbent-coated plates and covered glass tanks.

The procedures involved in thin-layer chromatography have been extensively reviewed (Mangold *et al.*, 1964; Nichols, 1964; Morris, 1966; Mahadevan, 1976; Renkonen and Varo, 1967; Renkonen and Luukkonen, 1976; Rouser *et al.*, 1976; Mangold, 1969; Stahl, 1969; Kates, 1986).

The following solvents are useful for general separation of the indicated lipid classes by one-dimensional TLC.

1. *Hydrocarbons* (normal, saturated and unsaturated, isoprenoids, carotenes) and *wax esters*: heptane–benzene (9/1 v/v); and benzene.
2. *Neutral lipids* (wax esters, sterol esters, fatty acid methyl esters, fatty aldehydes, ketones, quinones, tocopherols, triglycerides, sterol, long-chain alcohols, fatty acids, diglycerides, monoglycerides, glycerol monoethers): hexane or petroleum ether (b.p. 60–70 °C)–ethyl ether–acetic acid (90/10/1 v/v); Mangold and Malins, 1960); hexane or petroleum ether (b.p. 60–70 °C)–ethyl ether-acetic acid (80/20/1 or 70/30/1 v/v; Mangold, 1969); petroleum ether–ethyl ether–acetic acid (82/18/1 v/v; Schlierf and Wood, 1965); heptane–ether–methanol–acetic acid (90/20/2/3 v/v; Brown and Johnston, 1962); hexane–isopropyl ether–ethyl ether–acetone–acetic acid (85/12/1/4/1 v/v; Lepage, 1967); isopropyl ether–acetic acid (96/4 v/v) then petroleum ether–ethyl ether–acetic acid (90/10/1 v/v) in the same direction (Skipski and Barclay, 1969); ethyl ether–benzene–ethanol–acetic acid (40/50/2/0.2 v/v; Freeman and West, 1966); and heptane–isopropyl ether–acetic acid (60/40/4 v/v; Breckenridge and Kuksis, 1968). For *prostaglandins* and derivatives, on $AgNO_3$-impregnated plates: ethyl acetate–methanol–water (8/2/5 v/v) (Green and Samuelsson, 1964; Ramwell and Daniels, 1969; Daniels, 1976).
3. *Phospholipids* and *glycolipids*: chloroform–methanol–water (65/35/5 v/v or 65/25/4 v/v; Lepage, 1964); chloroform–methanol–acetic acid–water (25/15/4/2 v/v; Skipski *et al.*, 1964); chloroform–methanol–acetic acid–water (85/15/10/3.5 v/v; Nichols, 1963); and chloroform–methanol–conc. ammonium hydroxide (65/35/5 v/v; Rouser *et al.*, 1976).

For two-dimensional TLC of phospholipids and glycolipids, the following pairs of solvents are useful: first Nichols' (1964) system chloroform–methanol–7 N ammonium hydroxide (65/30/4 v/v) and then chloroform–methanol–acetic acid–water (170/25/25/6 v/v); first Lepage's (1967) system chloroform–methanol–water (65/15/2 v/v) and then chloroform–methanol–acetone–acetic acid–water (65/10/20/10/3 v/v); first Rouser *et al.*'s (1970, 1976) system chloroform–methanol–conc. ammonia (65/35/5 v/v) and then chloroform–acetone–methanol–acetic acid–water (10/4/2/2/1 v/v); and, finally, first Evans *et al.*'s (1982) system, for algal lipids, chloroform–methanol–acetone–diethylamine–water (120/35/

37.5/6/4.5 v/v) and then chloroform–methanol–28% ammonia (65/25/5 v/v).

TLC mobilities of neutral lipids and of polar lipids in the above-mentioned solvent systems are given in Tables 4.4 and 4.5 respectively.

In addition to the universal (destructive) procedure for detection of lipids on TLC plates by charring with sulphuric acid or sulphuric acid–dichromate, lipids may be detected by non-destructive, general stains that are useful in preparative TLC, such as: rhodamine 6G (0.01%) (viewed under UV light), iodine vapour and 2′,7′-dichlorofluorescein (viewed under UV light; only neutral lipids are detected). The following specific staining reagents are useful for detection and characterization of polar lipids: molybdate/sulphuric acid for phosphate groups (phospholipids); ninhydrin for primary amino groups (amino lipids); α-naphthol for sugar groups (glycolipids); and Dragendorff for choline (lecithin, lysolecithin and sphingomyelin), etc. (see Christie, 1982; Kates, 1986).

The adsorption properties of silica can be modified in a number of useful ways by impregnating the silica with an appropriate chemical. Silver nitrate is most commonly employed, but boric acid, oxalic acid and sodium arsenite have also been used (Christie, 1982). This latter group of compounds, particularly boric acid (see Kates, 1986), influences the interaction between silica and hydroxy (particularly polyhydroxy) compounds, so that it is possible to get improved separation of positional isomers (e.g. 1- and 2-acyl- or alkylglycerols) and even of diastereoisomers of polyhydroxy acids. Trimethyl borate is also reported to decrease acyl migration in mono-acyl- and diacylglycerols during chromatography (Pollack *et al.*, 1971).

Table 4.4 R_f values of neutral lipid classes in various solvent systems on TLC [a] (Kates, 1986)

Lipid class	R_f value × 100 in solvent [b]						
	1	2	3	4	5	6	7
Hydrocarbons (paraffins and olefins)	80	95	98	98	95	95	95
Squalene	40	95	98	98	90	–	–
Carotenes	20–30	–	–	–	–	–	–
Trialkylglycerol ethers	–	90	85	–	–	–	–
Sterol esters	–	90	94	–	85	80	90
Wax esters	–	90	88	–	90	–	–
Dialkylmonoacylglycerols	–	70	88	–	–	–	–
Alkenyldiacylglycerols	–	65	82	–	–	–	–
Alkyldiacylglycerols	–	55	78	–	–	–	–
Fatty acid methyl esters	8	65	77	90	75	–	–
Long-chain methyl ketones	–	–	63	–	–	–	–
Long-chain aldehydes	–	55	73	–	70	–	–
Aldehyde dimethyl acetals	–	–	73	–	–	–	–
Vitamin K	7	55	–	–	–	–	–
Triacylglycerols	–	35	60	90	70	60	82
Coenzyme Q	2	25	–	–	–	–	
α-Tocopherol	–	19	–	–	–	–	
Fatty acids	–	18	39	65	62	25	50
Long-chain alcohols	0	15	30	40	50	–	–
Sterols	0	10	19	30	38	38	30
1,2-*O*-Dialkylglycerol ethers	0	9	30	55	–	–	–
1,3-Diacylglycerols	0	8	21	–	46	50	40
1,2-Diacylglycerols	0	8	15	32	41	45	25
1-*O*-Monoalkylglycerol ether	0	0	3	–	14	–	–
Monoacylglycerols	0	0	2	7	10	15	5

[a] The R_f values on silica gel H listed will vary somewhat in absolute value depending on the temperature, humidity, type of chamber, etc. However, the relative order of mobilities will be as given in this table.

[b] *Solvent*: 1, heptane–benzene (9/1); 2, petroleum ether (b.p. 60–70 °C)–ethyl ether–acetic acid (90/10/1) (Mangold and Malins, 1960); 3, petroleum ether (60–70 °C)–ethyl ether–acetic acid (80/20/1) (Mangold, 1969); 4, petroleum ether (60–70 °C)–ethyl ether–acetic acid (70/30/1); 5, isopropyl ether–acetic acid (96/4) then solvent 2 in same direction (Skipski and Barclay, 1969); 6, ethyl ether–benzene–ethanol–acetic acid (40/50/2/0.2) then ethyl ether–hexane (6/94) (Freeman and West, 1966); and 7, heptane–isopropyl ether–acetic acid (60/40/4) (Breckenridge and Kuksis, 1968).

Table 4.5 R_f values of polar lipids on TLC in various solvents (Spanner, 1973; Kates, 1986, 1990)

Lipid component	R_f values × 100 in solvents [a]					
	1	2	3	4	5	6
Phospholipids						
Lysophosphatidylcholine	–	–	10	8	15	–
Phosphatidylinositol	23	14	47	11	42	13
Sphingomyelin	16	–	18	22	26	–
Phosphatidylglycerol	48	30	–	37	75	39
Phosphatidylethanolamine	62	35	83	41	65	52
Phosphatidylserine	15	6	55	5	44	–
Phosphatidyl-*N*-methylethanolamine	83	–	62	–	–	–
Phosphatidyl-*N*,*N*-dimethylethanolamine	52	–	66	–	–	–
Cardiolipin	71	41	–	38	85	72
Phosphatidylglycerophosphate	–	–	–	6 [b]	–	–
Lysophosphatidic acid	70	75	–	5	–	–
Phosphatidic acid	74	79	–	5	66	87
Glycolipids						
Sulphoquinovosyldiacylglycerol	42	22	–	46	–	32–37
Digalactosyldiacylglycerol	62	18	–	30	–	38
Cerebrosides	70–76	–	–	45–51	78–84	58
Cerebroside sulphate	36	–	–	18–20	66–70	–
Sterol glycoside	73	43	–	64	–	68
Esterified sterol glycoside	78	65	–	90	–	80
Monogalactosyldiacylglycerol	77	51	–	84	–	84

[a] *Solvents*: 1. chloroform–methanol–water (65/25/4) (Lepage, 1964; Rouser *et al.*, 1976); 2, diisobutyl ketone–acetic acid–water (40/25/5) (Lepage, 1964; Nichols, 1964); 3, chloroform–methanol–acetic acid–water (25/15/4/2) (Skipski *et al.*, 1964); 4, chloroform–methanol–28% ammonia (65/25/5); 5, chloroform–acetone–methanol–acetic acid–water (6/8/2/2/1); and 6, chloroform–acetone–methanol–acetic acid–water (10/4/2/2/1) (Rouser *et al.*, 1969, 1970, 1976).
[b] R_f in chloroform–methanol–ammonia (65/35/5) (Davidson and Stanacev, 1970).

Silver ions deposited on silica interact with the π electrons of olefinic compounds and retard their chromatographic mobilities. A mixture of silica and silver nitrate (10–30%) is used in a column or as a thin layer (for experimental details see Morris, 1966; Mangold, 1969; Christie, 1982; Kates, 1986). Long-chain esters are then separated mainly according to the number of double bonds they contain. In addition, *trans* compounds have greater R_f values than *cis* isomers, and there are minor differences depending on double-bond position along the chain and on their position with respect to other double bonds (Barve *et al.*, 1972; Gunstone and Jacobsberg, 1972): conjugated, methylene-interrupted and non-methylene-interrupted polyenes behave differently from each other.

Silver-ion (argentation) chromatography, particularly in the thin-layer mode, has been very valuable for isolating compounds with novel structure, and a typical example is described by Ratnayake and Ackman (1979). Lipids of a single class such as triacylglycerols can be separated on the basis of their degree of unsaturation (Litchfield, 1972). Intact phospholipids (e.g. phosphatidylcholine of yeast (Paradis and Kates, 1973)) can be examined in a similar way (Arvidson, 1965). However, more detailed analysis of phospholipids is obtainable after conversion to the diacylglycerol moiety by phospholipase C hydrolysis and acetylation with acetic anhydride or trifluoroacetic anhydride. Silver-ion chromatographic separations are used in molecular-species analytical investigations of acylglycerols and phospholipids (sections 6.3.3 and 6.3.4).

Interaction of double bonds with mercuric salts is also the basis of a separation process, but these addition compounds must be prepared prior to chromatographic separation and the alkene esters must be regenerated after separation (see Radin, 1965; Sebedio and Ackman, 1981; Christie, 1982).

Preparative TLC is usually carried out on relatively thick layers (0.75–1.0 mm) of silica gel H or P (no binder), but can also be done on analytical plates (0.25 mm thick). Silver nitrate (5–10%)–silica gel for separation of unsaturated lipids and boric acid–silica gel for separation of positional isomers of acyl- or alkylglycerols may also be used (Kates, 1986). Samples (20–35 mg per plate) are applied by a streaking device and plates are developed one-dimensionally in the solvent systems given earlier in this section (see Tables 4.4 and 4.5). Two-dimensional chromatography may also be used. The lipid bands are located by spraying with water or with 0.01% aqueous rhodamine 6G or by staining with iodine vapour, and eluted from the silica with chloroform–methanol–ether (1/1/1 v/v) for neutral lipids, or with chloroform–methanol–water (1/1/0.2 v/v) (see Kates, 1986).

4.4.2 Liquid–liquid partition chromatography

In liquid–liquid partition chromatography, compounds are partitioned between an immobilized liquid (stationary phase) retained by an inert support (e.g. celite, cellulose, Sephadex) and an immiscible liquid (mobile phase) that flows through the column. Separation is based on differences in partition coefficient of the components in a mixture. There are two operational modes of partition chromatography: (a) the normal mode, in which the stationary phase is the more-polar solvent while the mobile phase is the less-polar solvent; and (b) the reversed-phase mode, in which the stationary phase is non-polar while the mobile phase is polar.

(a) Normal partition chromatography

This mode of operation is covered by Carroll and Serdarevich (1967) and Rouser *et al.* (1976), to which the reader is referred. An example of normal partition chromatography is the use of Sephadex, a crosslinked hydrophilic dextran gel, as inert support (see review on Sephadex gels by Fischer (1970). This procedure is useful for preliminary fractionation of crude total lipids (e.g. bile) to separate non-lipid contaminants from lipids prior to further fractionation on silicic acid columns (section 4.4.1(a)); it is particularly useful for quantitative removal of water-soluble contaminants and for separation of gangliosides and bile salt conjugates from other lipids (Rouser *et al.*, 1976).

A column of Sephadex G-25 (coarse, beaded; 1 × 30 cm, prepared in methanol–water, 1/1) is washed with two cycles of the solvent sequence: (1) chloroform–methanol (19/1) saturated with water (ca. 5 ml/l); (2) chloroform–methanol (19/1) 850 ml, glacial acetic acid 170 ml, and water to saturation (ca. 25 ml); (3) chloroform–methanol (9/1) 580 ml, glacial acetic acid 170 ml, and water to saturation (ca. 42 ml); and (4) methanol–water (1/1). The column is loaded with 250 mg total crude lipids and eluted with the above solvent sequence.

Composition of fractions in the above procedure is as follows. Solvent (1): all neutral lipids (hydrocarbons, glycerides, sterols, waxes, etc.) and free fatty acids; all glycolipids (glycosyldiacylglycerols, cerebrosides, ceramide polyhexosides, sulphatides, plant sulpholipid); all phosphatides including lysophosphatides; unconjugated bile acids or salts (cholic, deoxycholic acids, etc.) and conjugated (glycine, taurine) bile acids; and some solvent-soluble non-lipids. Solvent (2): gangliosides; glycine-conjugated bile salts; uncharacterized degraded acidic phosphatides; urea and other solvent-soluble non-lipids. Solvent (3): traces of gangliosides; taurine-conjugated dihydroxycholanic (deoxy and chenodeoxy) bile salts; non-polar amino acids (phenylalanine, tyrosine, tryptophan, etc.); and uncharacterized non-lipids. Solvent (4): taurocholate; and water-soluble non-lipids (salts, amino acids, sugars, etc.). The neutral lipids, glycolipids and phosphatides eluted with solvent (1) are then fractionated as described in sections 4.4.1(a) and/or 4.4.1(b).

In another procedure for removal of non-lipid and aqueous contaminants, columns of Sephadex G-25 are pretreated successively with the upper and lower phases of a chloroform–methanol–water (8/4/3 v/v) mixture and then a crude lipid extract applied. Lipids are eluted with the lower phase, and gangliosides and non-lipid contaminants are present in the upper phase. The original procedure of Wells and Dittmer (1963) has been modified by Wuthier (1966) to give a simpler method more suitable for large numbers of samples.

Liquid–liquid partition chromatography on silicic acid impregnated with water has been used for separation of fatty acids (Vandenheuvel and Hayes, 1952) and triglycerides (Youngs, 1961).

(b) Reversed-phase partition chromatography

For more details on this mode, the reader should consult Carroll and Serdarevich (1967) and Kates (1986). Two examples of reversed-phase partition chromatography will be described: on Factice, a hydrophobic polymerized product of soybean oil, and on lipophilic Sephadex derivatives. Both of these reversed-phase procedures are useful for separation of individual wax esters, cholesteryl esters, mono-, di- and triacylglycerols and fatty acid methyl esters; see Carroll and Serdarevich (1967), Carroll (1976) and Christie (1982) for more extensive accounts of this subject.

Reversed-phase partition chromatography of

neutral lipids on Factice (Hirsch, 1963) is carried out with the solvent system acetone–2% water (v/v) and gives good separations of neutral lipid classes, eluted in the order most polar to least polar, e.g. 2-monoacylglycerols, 1,3-diacylglycerols, cholesterol, triacylglycerols and individual sterol esters. However, separations may be influenced by the nature of the fatty acyl substituents, shorter chains being eluted before longer chains, and unsaturated chains before saturated chains.

With 2% water in acetone, individual saturated cholesterol esters, from the acetate to the stearate, are eluted after the triglycerides in separate peaks, the elution time of each being a logarithmic function of the fatty acid chain length (as for GLC). Unsaturated sterol esters are also separable, appearing as a leading shoulder on the peak for the saturated ester with two carbons less, e.g. cholesterol oleate precedes the palmitate, and the linoleate precedes the myristate.

Elution with acetone–5% water (v/v) is used for resolution of triglyceride mixtures; acetone–12% water (v/v) for C_{14}–C_{26} fatty acid methyl esters; and acetone–15% water (v/v) for C_4–C_{14} fatty acid methyl esters. To remove interfering free fatty acids and phosphatides from the sample of neutral lipids, a chloroform solution of the sample should be passed through a small column of Florisil (Carroll, 1961).

With 5% water in acetone, saturated triglycerides (tributyrin to tristearin) are eluted in separate peaks of which the logarithms of emergence times are directly proportional to chain lengths. However, the presence of a *cis* double bond in the fatty acid chain reduces the emergence time to that of a chain with two carbons less, e.g. triolein and tripalmitin, trilinolein and trimyristin, and trilinolenin and trilaurin are inseparable pairs.

Elution of fatty acid methyl esters with 12–15% water in acetone also yields peaks of which the logarithms of the emergence times for saturated chains are proportional to chain length; precise equivalence between two-carbon shortening and one double bond also applies here, e.g. methyl palmitate and oleate are inseparable at all column temperatures (2–30 °C).

Reversed-phase chromatography can also be carried out on columns of lipophilic Sephadex, e.g. hydroxyalkoxypropyl Sephadex with C_{11}–C_{14} or C_{15}–C_{18} alkoxy groups (Lipidex 1000 or 5000; Packard-Becker) using the eluting solvents: (1) heptane–methanol (5/95 v/v) for sterols; (2) heptane–acetone–water (4/15/1 v/v) for cholesterol esters or triacylglycerols; (3) heptane–acetone–methanol (1/2/4 v/v) for wax esters; (4) heptane–chloroform–isopropanol–water (1/9/70/80 v/v) for 1-monoacylglycerols; and (5) ethylene chloride–methanol–water (25/95/5 v/v) for fatty acid methyl esters (Ellingboe *et al.*, 1970; Nyström and Sjövall, 1975).

For sterols differing by one methylene group in the side chain, the order of elution with solvent (1) is cholesterol (C_{27}), campesterol (C_{28}) and sitosterol (C_{29}). Cholesteryl esters with saturated C_{11}–C_{14} acyl chains are eluted with solvent (2), in order of increasing acyl chain length. Triglycerides are also eluted with solvent (2) in order of increasing number of acyl carbon atoms, but mono- and diacylglycerols require the more-polar solvent system (4).

As in reversed-phase chromatography on Factice, unsaturated acyl esters have shorter emergence times than corresponding saturated esters, a *cis* double bond reducing the emergence time to that of the corresponding saturated chain with two carbons less. Fatty acid methyl esters are eluted with solvent (5) in logarithmic order of increasing chain length. However, because of the equivalence in emergence time between a *cis* double bond and two-carbon shortening, oleate–palmitate, linoleate–myristate and linoleate–laurate are inseparable pairs of esters.

Saturated straight-chain wax esters with 28–36 carbon atoms are well separated in solvent (3), the logarithms of the emergence times being directly proportional to the total number of carbon atoms.

Reversed-phase partition chromatography can also be carried out on kieselguhr or Celite (Hyflo Super Cel) treated with dichlorodimethylsilane to make it lipophilic and thus easily impregnated with non-polar solvents. Impregnation of lipophilic Celite with liquid paraffin, silicone oil or castor oil can provide the stationary phase, with acetone–water, methanol–water or acetonitrile–water as mobile phase for separation of normal or hydroxy fatty acid methyl esters (C_{12}–C_{20}), and mono-, di- and triacylglycerols (see Carroll and Serdarevich, 1967).

4.4.3 Ion-exchange chromatography

Ion-exchange chromatography (Rouser *et al.*, 1969, 1976; Christie, 1982; Kates, 1986) is a useful technique for the initial separation of lipids from crude mixtures. Separation of lipids by ion-exchange chromatography is based primarily on ion exchange with the ionic groups of the polar lipids, but the presence of polar non-ionic groups (such as hydroxyl groups) also exerts an influence on the adsorption and elution of the lipid. Three general groups of lipids are readily separated: non-ionic, zwitterionic and acidic. Within each group, further separation may also be achieved on the basis of differences in polarity and/or acidity.

There are three ion-exchange materials that have proved to be very useful for chromatography of lipids:

diethylaminoethylcellulose (DEAE-cellulose), triethylaminoethylcellulose (TEAE-cellulose) and carboxymethylcellulose (CM-cellulose); cellulose-based ion exchangers are described in detail by Peterson (1970). DEAE-cellulose is used for general separation of lipid classes; and TEAE-cellulose is particularly useful for separation of lipids having carboxyl groups as the only ionic group (e.g. fatty acids, bile acids, gangliosides) and for separation of phosphatidylethanolamine from ceramide polyhexosides; for preparation of columns see Rouser *et al.* (1969, 1976).

Ion-exchange chromatography on DEAE- or TEAE-cellulose is very useful in combination with silicic acid chromatography (see sections 4.4.1(a) and 4.4.1(b)). Examples of fractionation schemes are given in Table 4.6. Non-polar lipids are eluted with solvents such as chloroform and polar non-ionic lipids and non-acidic lipids are partly resolved by increasing concentrations of methanol in chloroform. Weakly acidic lipids can then be eluted by inclusion of acetic acid in the solvent, while strongly acidic lipids require additions such as ammonia and salts (Table 4.6). Although DEAE-cellulose columns give very reproducible results with little variation between batches of packing material, care should be taken in the preparation of the column (Rouser *et al.*, 1964, 1969) and in not changing back from polar to non-polar solvents too quickly when the column packing may be disturbed with resultant poor resolutions. DEAE-cellulose in the borate form (instead of the usual acetate form) and TEAE-cellulose in the hydroxyl form are useful alternatives.

TEAE-cellulose has a much higher capacity for

Table 4.6 Ion-exchange chromatography of lipids on DEAE-cellulose and TEAE-cellulose columns

Column	Eluting solvent	Column volumes	Lipid components eluted
DEAE-cellulose [a]			
Animal lipids	Chloroform	8–10	Neutral lipids
	Chloroform–methanol (9/1)	9	Cerebrosides, phosphatidylcholine, lysophosphatidylcholine
	Chloroform–methanol (1/1)	9	Ceramide polyhexosides, phosphatidylethanolamine, lysophosphatidylethanolamine
	Methanol	10	Oxidation products [b], salts
	Glacial acetic acid	10	Phosphatidylserine
	Methanol	4	None
	Chloroform–methanol (4/1)–ammonia, salt	10	Cardiolipin, phosphatidic acid, phosphatidylglycerol, phosphatidylinositol, cerebroside sulphate
Plant lipids [a,d]	Chloroform	10	Neutral lipids
		10	Monogalactosyldiacylglycerol
	Chloroform–methanol (95/5)	10	Phosphatidylcholine, cerebroside, steryl glycoside
	Chloroform–methanol (9/1)	10	Digalactosyldiacylglycerol
	Chloroform–methanol (6/4)	10	Phosphatidylethanolamine
	Chloroform–methanol (4/1)–ammonia, salt [c]	10	Phosphatidylglycerol, phosphatidylinositol, plant sulpholipid
TEAE-cellulose [a]			
Animal lipids	Chloroform	5	Neutral lipids
	Chloroform–methanol (9/1)	8	Cerebrosides, phosphatidylcholine, sphingomyelin
	Chloroform–methanol (2/1)	8	Ceramide polyhexosides
	Chloroform–methanol–glacial acetic acid (20/10/0.3)	6	Phosphatidylethanolamine, phosphatidyl-*N*-methylethanolamine, phosphatidyl-*N*,*N*-dimethylethanolamine, ceramide aminoethylphosphonate (from invertebrates)
	Glacial acetic acid	6	Phosphatidylserine
	Chloroform–methanol (4/1)–ammonia, salt [c]	6	Acidic phospholipids, cerebroside sulphate, plant sulpholipid, free fatty acids, bile acids

[a] Rouser *et al.* (1969, 1976).
[b] Uncharacterized oxidation products derived from phosphatidylethanolamine.
[c] Chloroform–methanol (4/1) made 0.01 to 0.05 M with ammonium acetate or potassium acetate and diluted with freshly prepared 28% aqueous ammonia (20 m/l).
[d] Nichols and James (1964).

acidic lipids, and ceramide polyhexosides and phosphatidylethanolamines are well separated (Rouser *et al.*, 1976). The borate form of DEAE-cellulose can be used for quantitative separation of ceramide polyhexosides (Rouser *et al.*, 1969). Kundu (1981a) has reviewed the use of ion-exchange chromatography for the separation of sphingolipids. His discussion includes the use of DEAE-silica gel and DEAE-Sephadex for the isolation of ganglioside species.

CM-cellulose may be used for separation of phospholipid mixtures, particularly acidic phospholipids (Comfurius and Zwaal, 1977). The solvent elution sequence given in Table 4.7 shows that individual acidic phospholipids as well as phosphatidylcholine and phosphatidylethanolamine can be isolated in fairly pure form by essentially a gradient elution with increasing concentrations of methanol (3–50%) in chloroform. Lysophospholipids are also well separated from the corresponding diacyl forms. Separation of the unresolved fractions (sphingomyelin, glycosyldiacylglycerols, etc.) and final purification of the individual lipid fractions can be readily achieved by silicic acid preparative TLC (section 4.4.1(b)).

Because the chromatographic properties of acidic phospholipids depend on the salt form, it is often advantageous to prepare sodium or potassium salts in order to prevent multiple bands of the same lipids. This can be achieved by passing the lipid extract through a chelating resin (Carter and Weber, 1966). Alternatively, acidic lipids can be converted to the free acid form and neutralized with the appropriate alkali-metal hydroxide or with ammonia for the ammonium salt (see Kates, 1986).

Table 4.7 Column chromatographic separation of phospholipids on CM-cellulose [a]

Eluting solvent, chloroform–methanol (% methanol)	Lipid eluted
0 (three bed volumes)	Neutral lipids
3	Diphosphatidylglycerol
4	Phosphatidylcholine
5	Sphingomyelin, glycosyldiacylglycerol, phosphatidylmethanol or -ethanol
9	Phosphatidylethanolamine
12.5	Phosphatidic acid
15	Diglycosyldiacylglycerol
20	Phosphatidylglycerol
23	Lysophosphatidylcholine
30–35	Phosphatidylserine
35–50	Phosphatidylinositol

[a] Comfurius and Zwaal (1977).

4.4.4 High-performance liquid chromatography

HPLC is a highly efficient form of adsorption, partition, or ion-exchange liquid chromatography, which owes its efficiency to the use of very uniform, finely divided, microspherical (5–10 μm diameter) supports of controlled porosity and degree of hydration. The absorbent or support is tightly packed into a stainless-steel column (10–30 cm long, 2–4 mm diameter), necessitating the use of high-pressure pumps that will deliver an adequate and constant flow of solvent through the column. Elution of the column may be carried out either isocratically, with a solvent mixture of constant composition, or by gradient elution, in which the solvent composition may be varied linearly or in stepwise fashion with both binary and ternary solvent systems. The column eluates are monitored continuously by means of a flow-through detector, which should be insensitive to solvent flow rate, temperature or solvent composition. The eluates may be collected, by means of a fraction collector, for further analysis.

The most versatile and sensitive detector is the variable-wavelength UV monitor. Although lipids lack specific absorption peaks, they may be detected by 'end absorption' in the region 203–214 nm owing to the presence of double bonds in the fatty acyl groups, or to functional groups such as carbonyl, carboxyl, phosphate, amino and quaternary ammonium groups. However, chloroform–methanol mixtures must be avoided since they show strong absorption below 245 nm. This difficulty has been overcome by the use of solvents such as acetonitrile–methanol–water (Jungalwala *et al.*, 1976) or hexane–2-propanol–water (Geurts van Kessel *et al.*, 1977) mixtures, which are transparent in the UV region 200 nm or greater. Phospholipids and glycolipids may also be detected by specific UV absorption after suitable derivatization (e.g. benzoylation), in which case chloroform- and/or methanol-containing solvent mixtures can be used (see Evans and McCluer, 1972). A universal 'mass' (light-scattering) detector is now available for detection of non-derivatized lipids (see Christie, 1987).

Detectors such as refractive index or moving wire–flame ionization detectors are generally not sufficiently sensitive or suitable for gradient elution procedures. Radioactivity detectors can be used for labelled lipids, and fluorescence detectors can be used for fluorescent lipids such as carotenoids and vitamins A and E.

Useful reviews on HPLC have been written by Ackman (1980, 1981), Aitzetmüller (1975, 1982), Hammond (1981), Kates (1986) and Christie (1987).

(a) Elution procedures for neutral lipids

The reader should see Aitzetmüller (1975) for useful information. Neutral lipid classes such as hydrocarbons, wax esters, fatty acid methyl esters, mono-, di- and triacylglycerols, cholesterol esters, cholesterol, long-chain alcohols, etc., have been separated by adsorbent-mode (normal-phase) HPLC on a LiChrosorb SI60 (5 or 10 μm) column eluted with solvent gradients: carbon tetrachloride–isooctane (34/66 v/v) → chloroform–methanol–diisopropyl ether (34/36/30 v/v) (Aitzetmüller and Koch, 1978). Separation of homologous series of alkanes and alkenes, as well as aromatic hydrocarbons, has been carried out by reversed-phase HPLC on a Zorbax ODS-C_{18} (Du Pont) column with isocratic elution using methylene chloride–acetonitrile (1/4 v/v) or tetrahydrofuran–acetonitrile (1/9 v/v) (Parris, 1978).

Fatty acid methyl esters may be separated by reversed-phase HPLC using isocratic elution with acetonitrile or acetonitrile–methanol (9/1 v/v). Unsaturated fatty acid methyl esters such as oleate, linoleate and linolenate have been separated in up to 10 g quantities of pure esters (>99%), from fatty acid methyl esters derived from suitable natural oils, by HPLC on Prep.Pack-500/C_{18} reversed-phase columns eluted with methanol, but this can only be achieved after urea fractionation to reduce the level of saturated fatty acid esters to below 5% (Bascetta *et al.*, 1984). These workers separated oleic, linoleic and α- and γ-linolenates from appropriate concentrates, obtaining up to 10 g of these materials at a purity level greater than 99% in 20–30 min. A similar technique was used by Traitler *et al.* (1988), Wille and Traitler (1988) and Wille *et al.* (1991) to obtain methyl γ-linolenate (94–96%) and methyl octadecatetraenoate (>90%) from concentrates of these esters derived from blackcurrant seed oil by urea fractionation (section 4.2). Carlson *et al.* (1990) used a 500 g cartridge to obtain methyl lesquerolate (>98% pure) from 30 g batches of lesquerella esters. It was possible to obtain 200 g of this purified hydroxy ester in 8 h.

Separation of a homologous series of triacylglycerols (1,2-dipalmitoyl-3-acyl-*sn*-glycerols, acyl groups C_2–C_{16}) has been achieved by reversed-phase HPLC on an Altex Ultrasphere-C_{18} bonded column at room temperature with UV detection at 220 nm, using a solvent gradient starting with 2-propanol–acetonitrile–tetrahydrofuran–water (29/63/6/3 v/v) and changing linearly over a period of 30 min to a final composition of 55/40/4/0 (Kodali *et al.*, 1984). A homologous series of monacylglycerols (C_8–C_{22}) has also been separated using similar conditions but with a linear gradient of 2-propanol–acetonitrile–tetrahydrofuran–water (5/64.4/7.6/23 v/v) changing to a final composition of 70/26.8/3.2/0 over a period of 30 min (Kodali *et al.*, 1985). Natural triacylglycerol mixtures have been resolved by reversed-phase HPLC on a Supelcosil LC-18 (Supelco) column with a gradient of 30–90% propionitrile in acetonitrile at 30 °C; detection was by chemical ionization mass spectrometry (Myher *et al.*, 1984).

Enantiomeric monoacylglycerols have been separated as their 3,5-dinitrophenylurethane derivatives by HPLC on a chiral stationary phase, and chiral monoalkyl-, dialkyl- and diacylglycerols, as their diastereomeric 1-(1-naphthyl)ethylurethane derivatives, have been separated on an achiral stationary phase (see review by Takagi, 1990).

For separation of carotenoids and retinoids, see review by Taylor (1983). Dolichol esters may be separated by reversed-phase HPLC on a Biosil ODS-10 column eluted with 2-propanol–methanol–hexane–water (65/30/5/1 v/v) at 1 ml/min flow rate (Tavares *et al.*, 1977). Polyprenols and dolichols can be separated by reversed-phase HPLC on a C_{18} column by gradient elution with methanol (10–80%)–isopropanol at 45 °C (Ravi *et al.*, 1983).

(b) Elution procedures for polar lipids

The interested reader should find more details in Aitzetmüller (1982). Separation and isolation of individual phospholipid classes on a preparative as well as analytical scale have been achieved by several groups using normal-phase HPLC and various mixtures of hexane–2-propanol–water as eluting solvents (Geurts van Kessel *et al.*, 1977, 1981; Patton *et al.*, 1982; Blank and Snyder, 1983). Good separation of all phospholipid classes was achieved, except for some overlapping of sphingomyelin with phosphatidylcholine. Christie (1985) has developed a ternary gradient elution system for separation of major lipid classes (cholesteryl esters to lysolecithin) on a silica column using a 'mass' detector, and Hurst and Martin (1984) have described an isocratic elution system for separation of phospholipid classes. Molecular species separation of individual phospholipids (Patton *et al.*, 1982; Jungawala *et al.*, 1979) and of mono- and digalactosyldiacylglycerols (Lynch *et al.*, 1983) has also been achieved using reversed-phase HPLC. Neutral glycosylceramides have been separated as their fully benzoylated derivatives (Evans and McCluer, 1972; Gross and McCluer,

1980) or as their *O*-acetyl-*N*-*p*-nitrobenzoyl derivatives (Suzuki *et al.*, 1980) by HPLC in normal mode. Dolichyl phosphates (from 17 to 21 isoprene units) have been resolved by reversed-phase HPLC (Chaudhary *et al.*, 1982).

References and further reading

Ackman, R.G. (1980) *Fette Seifen Anstrichm.*, **82**, 351.
Ackman, R.G. (1981) *Chem. Ind.*, **17**, 715.
Aitzetmüller, K. (1975) *J. Chromatogr.*, **113**, 231.
Aitzetmüller, K. (1982) *Prog. Lipid Res.*, **21**, 171.
Aitzetmüller, K. and Koch, J. (1978) *J. Chromatogr.*, **145**, 195.
Arvidson, G.A.E. (1965) *J. Lipid Res.*, **6**, 574.
Barve, J.A., Gunstone, F.D., Jacobsberg, F.R. and Winlow, P. (1972) *Chem. Phys. Lipids*, **8**, 117.
Bascetta, E., Gunstone, F.D. and Scrimgeour, C.M. (1984) *Lipids*, **19**, 801.
Blank, M.L. and Snyder, F. (1983) *J. Chromatogr.*, **273**, 415.
Bligh, E.G. and Dyer, W.J. (1959) *Can. J. Biochem. Physiol.*, **37**, 911.
Breckenridge, W.C. and Kuksis, A. (1968) *Lipids*, **3**, 291.
Brown, J.L. and Johnston, J.M. (1962) *J. Lipid Res.*, **3**, 480.
Carlson, K.D., Chaudry, A., Peterson, R.E. and Bagby, M.O. (1990) *J. Am. Oil Chem. Soc.*, **67**, 495.
Carroll, K.K. (1961) *J. Lipid Res.*, **2**, 135.
Carroll, K.K. (1963) *J. Am. Oil Chem. Soc.*, **40**, 413.
Carroll, K.K. (1976) in *Lipid Chromatographic Analysis*, 2nd edn, vol. 1 (ed. G.V. Marinetti), Dekker, New York, p. 173.
Carroll, K.K. and Serdarevich, B. (1967) in *Lipid Chromatographic Analysis*, 1st edn, vol. 1 (ed. G.V. Marinetti), Dekker, New York, p. 205.
Carroll, K.K., Cutts, J.H. and Murray, F.D. (1968) *Can. J. Biochem.*, **46**, 899.
Carter, J.E. and Weber, E.J. (1966) *Lipids*, **1**, 16.
Chaudhary, N., Freeman, D.J. and Carroll, K.K. (1982) *Lipids*, **17**, 558.
Christie, W.W. (1982) *Lipid Analysis*, 2nd edn, Pergamon, Oxford, p. 93.
Christie, W.W. (1985) *J. Lipid Res.*, **26**, 507.
Christie, W.W. (1987) *HPLC of Lipids*, Pergamon, Oxford.
Comfurius, P. and Zwaal, R.F.A. (1977) *Biochim. Biophys. Acta*, **488**, 36.
Daniels, E.G. (1976) in *Lipid Chromatographic Analysis*, 2nd edn, vol. 2 (ed. G.V. Marinetti), Dekker, New York, p. 611.
Davidson, J.B. and Stanacev, N.Z. (1970) *Can. J. Biochem.*, **48**, 633.
Ellingboe, J., Nyström, E. and Sjövall, J. (1970) *J. Lipid Res.*, **11**, 266.
Evans, J.E. and McCluer, R.H. (1972) *Biochim. Biophys. Acta*, **270**, 565.
Evans, R.W., Kates, M., Ginzburg, M. and Ginzburg, B.-Z. (1982) *Biochim. Biophys. Acta*, **712**, 186.
Fischer, L. (1970) in *Laboratory Techniques in Biochemistry and Molecular Biology*, vol. 1, part II (eds T.S. Work and E. Work), North-Holland, Amsterdam, p. 151.
Folch, J.M., Lees, G.A. and Sloane-Stanley, G.H.S. (1957) *J. Biol. Chem.*, **226**, 497.
Freeman, C.P. and West, D. (1966) *J. Lipid Res.*, **7**, 324.
Geurts van Kessel, W.S.M., Hax, W.M.A., Demel, R.A. and DeGier, J. (1977) *Biochim. Biophys. Acta*, **486**, 524.
Geurts van Kessel, W.S.M., Tieman, M. and Demel, R.A. (1981) *Lipids*, **16**, 58.
Gray, G.M. (1976) in *Lipid Chromatographic Analysis*, 2nd edn, vol. 3 (ed. G.V. Marinetti), Dekker, New York, p. 897.
Green, K. and Samuelsson, B. (1964) *J. Lipid Res.*, **5**, 117.
Gross, S.K. and McCluer, R.H. (1980) *Anal. Biochem.*, **102**, 429.
Gunstone, F.D. and Jacobsberg, F.R. (1972) *Chem. Phys. Lipids*, **9**, 112.
Hakomori, S. (1983) in *Handbook of Lipid Research*, vol. 3, *Sphingolipid Biochemistry* (eds J. Kanfer and S. Hakomori), Plenum, New York, p. 1.
Hammond, E.W. (1981) *Chem. Ind.*, **17**, 710.
Herslof, B. and Kaufmann, P. (1990) *Lipid Technol.*, **2**, 100.
Hirsch, J. (1963) *J. Lipid Res.*, **4**, 1.
Hurst, W.J. and Martin, R.A. (1984) *J. Am. Oil Chem. Soc.*, **61**, 1462.
Johnson, R.W. (1990) *World Conference on Oleochemicals into the 21st Century* (ed. T.H. Applewhite), American Oil Chemists' Society, Champaign, p. 189.
Jungalwala, F.B., Evans, J.E. and McCluer, R.H. (1976) *Biochem. J.*, **155**, 55.
Jungawala, F.B., Hayssen, V. Pasquini, J.M. and McCluer, R.H. (1979) *J. Lipid Res.*, **20**, 579.
Kates, M. (1986) *Techniques of Lipidology*, 2nd rev. edn, Elsevier, Amsterdam, pp. 187–215, 232–78.
Kates, M. (1990) in *Handbook of Lipid Research*, vol. 6, *Glycolipids, Phosphoglycolipids and Sulfoglycolipids* (ed. M. Kates), Plenum, New York, pp. 1–122, 235–320.
Kodali, D.R., Atkinson, D., Redgrave, T.G. and Small, D.M. (1984) *J. Am. Oil Chem. Soc.*, **61**, 1078.
Kodali, D.R., Redgrave, T.G.D., Small, D.M. and Atkinson, D. (1985) *Biochemistry*, **24**, 519.
Kuksis, A. (1976) in *Lipid Chromatographic Analysis*, 2nd edn, vol. 1 (ed. G.V. Marinetti), Dekker, New York, p. 215.
Kuksis, A. (1977) *J. Chromatogr.*, **143**, 3.
Kuksis, A. (ed.) (1978a) *Handbook of Lipid Research*, vol. 1, *Fatty Acids and Glycerides*, Plenum, New York.
Kuksis, A. (1978b) in *Handbook of Lipid Research*, vol. 1, *Fatty Acids and Glycerides* (ed. A. Kuksis), Plenum, New York, pp. 1–76.
Kundu, S.K. (1981a) in *Methods in Enzymology*, vol. 72 (ed. J.M. Lowenstein), Academic Press, New York, p. 174.
Kundu, S.K. (1981b) in *Methods in Enzymology*, vol. 72 (ed. J.M. Lowenstein), Academic Press, New York, p. 185.
Lepage, M. (1964) *J. Lipid Res.*, **5**, 587.

Lepage, M. (1967) *Lipids*, **2**, 244.
Lie Ken Jie, M.S.F. (1980) *Adv. Chromatogr.*, **18**, 1.
Litchfield, C. (1972) *Analysis of Triglycerides*, Academic Press, New York, p. 50.
Lynch, D.V., Gundersen, R.E. and Thompsom, G.A., Jr (1983) *Plant Physiol.*, **72**, 903.
Mahadevan, V. (1976) in *Lipid Chromatographic Analysis*, 2nd edn, vol. 3 (ed. G.V. Marinetti), Dekker, New York, pp. 777–89.
Malins, D.C. (1966) *Prog. Chem. Fats Other Lipids*, **8**, 301.
Mangold, H.K. (1969) in *Thin Layer Chromatography* (ed. E. Stahl), Springer, New York, pp. 155–200, 363–421.
Mangold, H.K. and Malins, D.C. (1960) *J. Am. Oil Chem. Soc.*, **37**, 383.
Mangold, H.K., Schmid, H.H.O. and Stahl, E. (1964) *Methods Biochem. Anal.*, **12**, 393.
Marinetti, G.V. (ed.) (1967–69) *Lipid Chromatographic Analysis*, 1st edn, vols 1 and 2, Dekker, New York.
Marinetti, G.V. (ed.) (1976) *Lipid Chromatographic Analysis*, 2nd edn, vols 1, 2 and 3, Dekker, New York.
Morris, L.J. (1966) *J. Lipid Res.*, **7**, 717.
Myher, J.J., Kuksis, A., Marai, L. and Manganaro, F. (1984) *Biochim. Biophys. Acta*, **283**, 289.
Nichols, B.W. (1963) *Biochim. Biophys. Acta*, **70**, 417.
Nichols, B.W. (1964) *New Biochemical Separations*, Van Nostrand, Princeton, NJ, p. 321.
Nichols, B.W. and James, A.T. (1964) *Fette Seifen Anstrichm.*, **66**, 1003.
Nielsen, J.R. (1980) *Lipids*, **15**, 481.
Nyström, E. and Sjövall, J. (1975) in *Methods of Enzymology*, vol. 35B (ed. J.M. Lowenstein), Academic Press, New York, pp. 378–95.
O'Brien, J.S. and Benson, A.A. (1964) *J. Lipid Res.*, **5**, 432.
Paradis, M. and Kates, M. (1973) *Can. J. Biochem.*, **51**, 184.
Parris, N.A. (1978) *J. Chromatogr.*, **157**, 161.
Patton, G.M., Fasulo, J.M. and Robins, S.J. (1982) *J. Lipid Res.*, **23**, 190.
Peterson, E.A. (1970) in *Laboratory Techniques in Biochemistry and Molecular Biology*, vol. 2, part II (eds T.S. Work and E. Work), North-Holland, Amsterdam, p. 225.
Pollack, J.D., Clark, D.S. and Somerson, N.L. (1971) *J. Lipid Res.*, **12**, 563.
Radin, N.S. (1965) *J. Am. Oil Chem. Soc.*, **42**, 569.
Ramwell, P.W. and Daniels, P.W. (1969) in *Lipid Chromatographic Analysis*, 1st edn, vol. 2 (ed. G.V. Marinetti), Dekker, New York, p. 313.
Ratnayake, W.N. and Ackman, R.G. (1979) *Lipids*, **14**, 580.
Ravi, J., Rip, W. and Carroll, K.K. (1983) *Biochem. J.*, **283**, 1953.
Renkonen, O. and Luukkonen, A. (1976) in *Lipid Chromatographic Analysis*, 2nd edn, vol. 1 (ed. G.V. Marinetti), Dekker, New York, p. 1.
Renkonen, O. and Varo, P. (1967) in *Lipid Chromatographic Analysis*, 1st edn, vol. 1 (ed. G.V. Marinetti), Dekker, New York, p. 41.
Rouser, G., Galli, C., Lieber, E., Blank, M.L. and Privett, O.S. (1964) *J. Am. Oil Chem. Soc.*, **41**, 836.
Rouser, G., Kritchevsky, G., Yamamoto, A., Simon, G., Galli, C. and Bauman, A.J. (1969) in *Methods in Enzymology*, vol. 14 (ed. J.M. Lowenstein), Academic Press, New York, p. 272.
Rouser, G., Fleischer, S. and Yamamoto, A. (1970) *Lipids*, **5**, 494.
Rouser, G., Kritchevsky, G. and Yamamoto, A. (1976) in *Lipid Chromatography Analysis*, 2nd edn, vol. 3 (ed. G.V. Marinetti), Dekker, New York, p. 713.
Saito, T. and Hakomori, S. (1971) *J. Lipid Res.*, **12**, 257.
Sastry, P.S. and Kates, M. (1964) *Biochemistry*, **3**, 1271.
Schlierf, G. and Wood, P. (1965) *J. Lipid Res.*, **6**, 317.
Sebedio, J.L. and Ackman, R.G. (1981) *Lipids*, **16**, 461.
Shaw, N. (1968) *Biochim. Biophys. Acta*, **164**, 435.
Siakotos, A.N. and Rouser, G. (1965) *J. Am. Oil Chem. Soc.*, **42**, 913.
Singleton, W.S., Gray, M.S., Brown, M.L. and White, J.L. (1965) *J. Am. Oil Chem. Soc.*, **42**, 53.
Skipski, V.P. and Barclay, M. (1969) in *Methods in Enzymology*, vol. 14 (ed. J.M. Lowenstein), Academic Press, New York, p. 530.
Skipski, V.P., Petterson, R.F. and Barclay, M. (1964) *Biochem. J.*, **90**, 374.
Spanner, S. (1973) in *Form and Function of Phospholipids* (eds G.B. Ansell, R.M.C. Dawson and J.N. Hawthorne), Elsevier, Amsterdam, p. 43.
Stahl, E. (ed.) (1969) *Thin Layer Chromatography* (English edn), Springer, New York.
Stein, R.A. and Slawson, V. (1966) *Prog. Chem. Fats Other Lipids*, **8**, 375.
Suzuki, A., Kundu, S.K. and Marcus, D.M. (1980) *J. Lipid Res.*, **21**, 473.
Svennerholm, L. (1956) *J. Neurochem.*, **1**, 42.
Svennerholm, L. (1957) *Biochim. Biophys. Acta*, **24**, 604.
Takagi, T. (1990) *Prog. Lipid Res.*, **29**, 277.
Tavares, I.A., Johnson, N.J. and Hemming, F.W. (1977) *Biochem. Soc. Trans.*, **5**, 1771.
Taylor, R.F. (1983) *Adv. Chromatogr.*, **22**, 157.
Traitler, H., Wille, H.J. and Studer, A. (1988) *J. Am. Oil Chem. Soc.*, **65**, 755.
Vandenheuvel, F.A. and Hayes, E.R. (1952) *Anal. Chem.*, **24**, 960.
Vorbeck, M.L. and Marinetti, G.V. (1965) *J. Lipid Res.*, **6**, 3.
Wagner, H., Horhammer, L. and Wolff, P. (1961) *Biochem. Z.*, **334**, 175.
Wells, M.A. and Dittmer, J.C. (1963) *Biochemistry*, **2**, 1259.
Wille, H.J. and Traitler, H. (1988) *Fat Sci. Technol.*, **90**, 476.
Wille, H.J., Traitler, H. and Lagarde, M. (1991) *Fat Sci. Technol.*, **93**, 363.
Wuthier, R.E. (1966) *J. Lipid Res.*, **7**, 558.
Young, O.M. and Kanfer, J.N. (1965) *J. Chromatogr.*, **19**, 611.
Youngs, C.G. (1961) *J. Am. Oil Chem. Soc.*, **38**, 62.

4.5 Gas–liquid chromatography

Gas–liquid partition chromatography, a technique first introduced by James and Martin (1952), is the method of choice for rapid quantitative and qualitative analysis of volatile lipid components such as hydrocarbons, fatty acid esters, sterols, etc. The method involves partitioning of the components of the mixture in the vapour state between a mobile stable gas phase and a stationary non-volatile liquid phase dispersed on an inert support. The mode of operation of GLC is analogous to that of reversed-phase partition chromatography. Detailed accounts of the theoretical basis of GLC and of the design and operation of gas chromatographs are available in several treatises on GLC of lipids (e.g. James, 1960; Horning *et al.*, 1964; Ackman, 1969; Kates, 1986; Christie, 1989). The GLC apparatus consists of a column (inert support-coated or wall-coated with the stationary liquid phase) housed in a thermostated oven maintained at the required temperature (from 170–220 to 300 °C, stability ±0.1 °C), with a sample injection inlet at the top of the column and a detection device (hydrogen flame ionization detector) at the outlet. A flow of inert gas (N_2, He) is maintained at constant rate. The signal from the ionization detector is amplified and recorded as a function of time, and retention times and areas of the peaks corresponding to the emerging components may be printed out electronically.

There are three types of GLC columns in general use: packed columns, wide-bore support-coated open tubular (SCOT) and wall-coated open tubular (WCOT) capillary columns. Packed columns may be metal (stainless steel, aluminium) or preferably glass, with dimensions 2–4 mm internal diameter and 1–2 m long, and are usually coiled. They are packed with a suitable inert support material (usually diatomaceous earth, alkali- and acid-washed and silanized) coated with non-polar (1–3 to 10%) or polar (10–15%) non-volatile liquid stationary phases. Examples of non-polar stationary phases are high-molecular-weight hydrocarbons such as Apiezon L or M grease, and the silicones SE-30, SE-52 and OV-1 (gums). Some polar stationary phases are butanediol succinate polyester (BDS), ethyleneglycol adipate (EGA), ethyleneglycol succinate (EGS), diethyleneglycol succinate (DEGS) and alkylpolysiloxanes with nitrile groups (e.g. cyanopropyl silicones such as SP-2330), etc., in increasing order of polarity. With non-polar liquid phases, homologous series (C_8–C_{32}) of saturated, normal or branched-chain hydrocarbons, long-chain alcohols (and their methoxy or acetate derivatives), aldehydes (and their dimethyl acetal derivatives), ketones and fatty acid methyl esters are well resolved into their respective molecular species, the logarithm of their emergence times being directly proportional to the number of carbon atoms. Unfortunately, non-polar liquid phases do not clearly discriminate between saturated and unsaturated species of the same molecule, since the separations are based mainly on molecular-weight differences (see Table 4.8).

In contrast, polar stationary phases resolve mono-, di-, tri-, etc., unsaturated species very well from their saturated counterparts, because of the higher polarity of the unsaturated species and their longer retention times relative to those in the non-polar phases. This is clearly evident from the data in Table 4.8 on the functional-group separation factors for different stationary phases. Polar liquid phases are thus more versatile than non-polar liquid phases for analysis of natural mixtures containing unsaturated species, but the non-polar phases can be useful in helping to discriminate between unsaturated and branched-chain species (see Table 4.8).

Support-coated open tubular (SCOT) capillary columns are wide-bore (0.5 or 0.75 mm internal diameter, 10–30 m long) columns filled with a finely powdered inert solid support coated with the liquid phase. They have a higher load capacity, shorter analysis times and a longer working life than narrow-bore capillary columns, and have better resolving power than packed columns. Wall-coated open tubular (WCOT) capillary columns of both narrow and wide bore (0.25–0.75 mm internal diameter, 15–60 m long) are made of fused silica or borosilicate glass with the liquid phase chemically bonded directly to the capillary wall as thick (0.5–1 μm) or thin (0.2–0.4 μm) films. These columns are very durable and reproducible, and have the highest resolving power. They are commercially available in a wide variety of sizes and liquid phase coatings, including those mentioned above (Table 4.8). Separation of chiral fatty acid methyl esters, e.g. 2-hydroxyisovalerate, as (−)-methoxycarbonyl derivatives, has been achieved by GLC on a QF-1 or OV-210 (non-polar) column (see Takagi, 1990).

GLC is used primarily as an analytical tool, but can also be used preparatively by inserting a stream splitter between the end of the column and the detector, which allows a small portion of the effluent to enter the detector, while the remainder can be collected by means of a suitable collecting device. A portion of the split effluent may also be by-passed to a mass spectrometer or radioactivity detector for

Table 4.8 Functional-group separation factors for various stationary phases in GLC (Kates, 1986)

Separation factor	Non-polar phases		Polar phases		
	Apiezon L (10%) (197 °C)	OV-1 (3%) (180–210 °C)	BDS (10%) (178 °C)	EGA (10%) (180–185 °C)	SP-2330 (10%) (185–190 °C)
CH_2 factor (F) [a]	1.53	1.40–1.39	1.41	1.41	1.31
Iso branch [b]	0.85–0.88	0.89	0.86–0.87	–	–
Anteiso branch [c]	0.88–0.90	0.92	0.90–0.92	–	–
Cyclopropane ring [j]	0.91	–	1.17	–	–
Monoene [d]	0.85–0.87	0.90–0.93	1.10–1.14	1.12–1.15	1.20
Diene [e]	0.81	–	1.33	1.36	1.48
Triene [f]	0.80	–	1.63	1.78	1.72
Tetraene [g]	0.64	–	1.8	1.8	1.9
1-OH [h]	3.0	2.9	13.5	–	17.7
2-OH [i]	1.6	–	4.0	14.3	3.0
3-OH [i]	2.0	–	5.5	–	4.2
—CHO (free ald.) [j]	3.2	–	6.3	–	–
—$CH(OCH_3)_2$ (ald. DMA) [j]	4.2	–	6.7	7.9	–
CH_3—C(=O) (methyl ketone) [j]	2.1	2.0	6.2	8.5	9.8
—O—CO—CH_3 (acetate) [j]	4.6	–	10.6	–	–
—O—CH_3 (methoxy) [j]	1.5	–	2.0	12.5	–
—$COOCH_3$ (methyl ester) [j]	3.3	3.4	7.6	9.8	9.9

Retention time ratios for: [a] methyl stearate/methyl palmitate or methyl palmitate/methyl myristate; [b] methyl isomyristate/methyl myristate; [c] methyl 12-methyl tetradecanoate/methyl pentadecanoate; [d] methyl oleate/methyl stearate; [e] methyl linolenate/methyl stearate; [f] methyl α-linolenate/methyl stearate; [g] methyl arachidonate/methyl eicosanate; [h] 1-octadecanol/octadecane; [i] 2- or 3-hydroxymethyl palmitate/methyl palmitate (O'Brien and Rouser, 1964).

[j] *Retention times* of the compound relative to that of the corresponding hydrocarbon, e.g. octadecanal/octadecane; 2-heptadecanone/heptadecane; octadecyl acetate/octadecane; 1-methoxyoctadecane/octadecane; methyl octadecanoate/octadecane.

mass spectral analysis or radioactivity measurement, respectively (see Christie, 1982, 1989; Kuksis, 1976, 1978a, b; Kates, 1986).

References and further reading

Ackman, R.G. (1969) in *Methods in Enzymology*, vol. 14 (ed. J.M. Lowenstein), Academic Press, New York, p. 329.

Christie, W.W. (1982) *Lipid Analysis*, 2nd edn, Pergamon, Oxford, p. 93.

Christie, W.W. (1989) *Gas Chromatography of Lipids*, Oily Press, Ayr.

Horning, E.C., Karmen, A. and Sweeley, C.C. (1964) *Prog. Chem. Fats Other Lipids*, 7, 167.

James, A.T. (1960) *Methods Biochem. Anal.*, **8**, 1.

James, A.T. and Martin, A.J.P. (1952) *Biochem. J.*, **50**, 679.

Kates, M. (1986) *Techniques of Lipidology*, 2nd rev. edn, Elsevier, Amsterdam, pp. 187–215, 232–78.

Kuksis, A. (1976) in *Lipid Chromatographic Analysis*, 2nd edn, vol. 1 (ed. G.V. Marinetti), Dekker, New York, p. 215.

Kuksis, A. (ed.) (1978a) *Handbook of Lipid Research*, vol. 1, *Fatty Acids and Glycerides*, Plenum, New York.

Kuksis, A. (1978b) in *Handbook of Lipid Research*, vol. 1, *Fatty Acids and Glycerides* (ed. A. Kuksis), Plenum, New York, pp. 1–76.

O'Brien, J.S. and Rouser, G. (1964) *Anal. Biochem.*, **7**, 288.

Takagi, T. (1990) *Prog. Lipid Res.*, **29**, 277.

4.6 Enzymic procedures

Enzymic processes provide a new way of separating mixtures of fatty acids or their esters. Whilst only a few examples have been described so far, they are likely to become more common and more important. Under appropriate conditions, lipase can be used to promote hydrolysis or esterification. Such enzymes show limited specificity for fatty acid structure, in particular for acids of short chain length, which react quicker, and for acids with a double bond close to the carboxyl group, which react slower than the usual range of acids. This latter effect will be apparent in acids having unsaturation at Δ4 (e.g. docosahexaenoic acid (DHA) 22:6), Δ5 (e.g. eicosapentaenoic acid (EPA) 20:5) and Δ6 (e.g. γ-linolenic acid (GLA) 18:3). Two examples of this phenomenon are given here.

Langholz *et al.* (1989) used an immobilized enzyme from the fungus *Mucor miehei* to prepare concentrates of DHA. After enzyme-catalysed preparation of the methyl esters of acid containing 22% DHA, the residual free acids contained 48% of this Δ4 polyene acid. In the enzyme-catalysed reaction (acidolysis) between free acids and their methyl esters, the DHA content of the free acids was 72%.

Hills *et al.* (1990) exploited the reduced reactivity of GLA in lipase-mediated reactions to produce concentrates of this acid starting with evening primrose acids containing 9.5% of GLA. The acids were reacted with butanol in hexane solution at 30 °C in the presence of lipase from rape seedlings or of the commercial preparation Lipozyme (*Mucor miehei*). Under these conditions, the GLA remains as free acid whilst the remaining acids are converted to butyl esters. Typical results with Lipozyme are given in Table 4.9. After the recommended reaction time of 6 h the free acid fraction (7.2%) contains 84.7% GLA and butyl esters contain only 3.7% of this acid. This represents 64% of the original GLA in the concentrate and 36% in the butyl esters.

References and further reading

Hills, M.J., Kiewitt, I. and Mukherjee, K.D. (1990) *J. Am. Oil Chem. Soc.*, **67**, 561.

Langholz, P., Anderson, P., Forskov, T. and Schmidtsdorff, W. (1989) *J. Am. Oil Chem. Soc.*, **66**, 1120.

Table 4.9 Reaction of evening primrose oil (EPO) acids with butanol in presence of Lipozyme

Time (h)	0	1	2	4	6	8	20
Unesterified acid (%)	100	39	23	9.3	7.2	6.5	2.3
GLA in free acid (%)	9.5	26.1	43.6	82.8	84.7	81.5	56.0
butyl ester (%)	–	tr	tr	2.0	3.7	4.5	8.3

4.7 Other separation methods

Since many lipid molecules have approximately the same molecular weight, gel filtration has not been extensively used for separating lipid fractions. However, a number of useful methods are available for specific purposes. First, low-molecular weight compounds like fatty acids can be separated from large molecules such as sterols or triacylglycerols by using columns packed with the organic solvent-resistant medium Sephadex LH-20 (e.g. Calderon and Baumann, 1970). Secondly, columns of Sephadex G-25 have been used to separate gangliosides or specific bile acid components from other lipids (Rouser *et al.*, 1967). However, the most useful application of gel filtration columns to lipid isolation has been in the removal of non-lipid and aqueous contaminants (see section 4.4). Columns of Sephadex G-25 are pretreated successively with the upper and lower phases of a chloroform–methanol–water (8/4/3 v/v) mixture and then a crude lipid extract applied. Lipids are eluted with the lower phase, and gangliosides and non-lipid contaminants are present in the upper phase. The original procedure of Wells and Dittmer (1963) has been modified by Wuthier (1966) to give a simpler method more suitable for large numbers of samples.

Where stereoisomerism exists in a particular lipid molecule, then it may be desirable to utilize chiral chromatography to resolve the isomers. Various modifications of the technique have been developed to make use of existing technologies. These include ligand exchange, high-performance liquid, gas and thin-layer chromatography. A useful general text on chiral chromatography is that of Lough (1989), where discussion includes consideration of preferred methods and optimization. A more specialized review on the chromatographic resolution of chiral lipid derivatives is that by Takagi (1990).

The utilization of supercritical fluids (particularly CO_2) for the extraction and chromatography of lipids has been reviewed recently (Bartle and Clifford, 1992). This article deals with a number of actual and potential applications – such as in triacylglycerol separations – as well as a discussion of the comparative merits of supercritical chromatography with other chromatographic techniques in lipid analysis.

References and further reading

Bartle, K.D. and Clifford, A.A. (1992) *Adv. Appl. Lipid Res.*, **1**, 217.

Calderon, M. and Baumann, W.J. (1970) *J. Lipid Res.*, **11**, 167.

Lough, W.J. (ed.) (1989) *Chiral Liquid Chromatography*, Blackie, Glasgow.

Rouser, G., Kritchevsky, G. and Yamamoto, A. (1967) *Lipid Chromatographic Analysis*, vol. 1 (ed. G.V. Marinetti), Dekker, New York, p. 99.

Takagi, T. (1990) *Prog. Lipid Res.*, **29**, 277.

Wells, M.A. and Dittmer, J.C. (1963) *Biochemistry*, **2**, 1259.

Wuthier, R.E. (1966) *J. Lipid Res.*, **7**, 558.

4.8 Isolation of natural acids

This section provides an account of procedures by which fatty acids can be isolated from natural sources, and is confined to the isolation of the more common acids using one or more of the separation procedures described in the preceding sections. For reasons discussed elsewhere (section 7.1) some acids are best obtained by synthesis, but others can be more cheaply isolated from natural sources. Most of these acids are available from commercial suppliers, but some are very expensive, particularly if required in large quantities.

Some acids are conveniently made by a combination of isolation and synthesis. For example, dihomo-γ-linolenic acid, $(n-6)$-20:3, is best obtained from γ-linolenic acid, $(n-6)$-18:3, itself isolated from evening primrose oil, by chain extension by two carbon atoms (malonation).

It will be seen that in most of the examples discussed below, purification ($\geqslant$99%) requires the use of several processes. The selection of these depends on the separation being effected and on the scale of operation. Thus, whilst low-temperature crystallization and urea fractionation are easily employed on a large scale, chromatography, and especially silver-ion chromatography, is less convenient or appropriate for very large amounts of material (Gunstone *et al.*, 1976).

It is not advisable to distil polyunsaturated acids or esters if this can be avoided since they are liable to undergo double-bond migration, stereomutation, cyclization and dimerization under extreme thermal conditions. Where distillation is unavoidable, artefacts must subsequently be removed.

Preparative HPLC provides a quick way of producing pure acids or esters on a 1–10 g scale, though even here preliminary concentration or final purification by other procedures may be required (Bascetta *et al.*, 1984).

Starting material must be selected with care. It should be easily available, be rich in the desired acid and contain little or none of any other acid that can be separated from the desired acid only with great difficulty.

Saturated acids with an even number of carbon atoms per molecule are usually isolated by distillation of esters from an appropriate source. These include esters from lauric-rich oils (C_{10}, C_{12}, C_{14}), palm oil and cottonseed oil (C_{16}), hydrogenated animal fats and seed oils (C_{18}) and hydrogenated high-erucic-acid rape oil (C_{22}). These acids are commercially available in high purity at a reasonable cost.

Oleic acid or ester is best obtained from a vegetable source in which oleate is virtually the only monoene acid present. Olive oil is a useful starting material. Animal fats are less suitable since they are more likely to contain isomeric octadecenoic acids. The following procedure has been recommended: (a) urea fractionation to reduce saturated acids to below 1% – two treatments may be necessary; (b)crystallization from petroleum ether at −40 °C three or four times until oleic acid exceeds 99%; and (c) chromatography of acids or methyl esters to remove coloured impurities and oxidized products. Further purification by repeated anaerobic low-temperature crystallization from acetonitrile has been described (Arudi *et al.*, 1983).

Petroselinic acid (6*c*-18:1) can be isolated from parsley seed oil by methanolysis followed by crystallization from acetone at −30 °C and urea fractionation.

Erucic acid (13*c*-22:1) is available from high-erucic-acid rape oil by distillation and crystallization.

Linoleic acid or ester can be obtained from any linoleic-rich oil that has less than 1% of linolenic acid (section 3.2.5) by the following steps: (a) urea fractionation to reduce the combined content of saturated and oleic acids in the mother liquor to <5%; (b) repeated low-temperature crystallization from acetone at −75 °C until the precipitate contains >90% linoleic acid; (c) further urea fractionation to reduce the combined content of saturated and oleic acid to below 0.8%; and (d) low-temperature crystallization at −75 °C to give pure acid (>99%), which is finally eluted from a column of silica. The presence of linolenic acid at levels initially above 1% renders the purification more difficult. Further purification by repeated anaerobic low-temperature crystallization from acetonitrile has been described (Arudi *et al.*, 1983). Christie (1982) reports obtaining pure methyl linoleate from safflower oil by partitioning the mercuric acetate adducts between 5% aqueous methanol and hexane in separating funnels.

α-Linolenic acid or ester is usually prepared from linseed oil, but this is a tedious purification because of the difficulty of separating linoleic from linolenic acid. The proportion of linolenic acid (55–60%) in linseed oil is raised to 65–75% by low-temperature crystallization from acetone at −75 °C and then to 80–90% by urea fractionation. This material is further upgraded by careful elution from a column of silica.

γ-Linolenic acid can be isolated from the acids of evening primrose or borage oil by a combination of urea fractionation or high-performance liquid chromatography. This same acid and stearidonic acid (18:4) can be obtained from blackcurrant seed oil by similar procedures (Traitler *et al.*, 1988; Wille and Traitler, 1988; Wille *et al.*, 1991).

Arachidonic acid is obtained from pig liver lipid by distillation of the methyl esters after chromatographic removal of cholesterol, followed by urea fractionation and silver-ion chromatography.

Eicosapentaenoic acid, $(n-3)$-20:5, and docosahexaenoic acid, $(n-3)$-22:6, must be obtained from a suitable fish oil (menhaden oil, tuna oil), or fish oil preparations already enriched in these acids, by a combination of urea fractionation, low-temperature crystallization and chromatography. Traditional distillation of the methyl esters may also be necessary.

References and further reading

Arudi, R.L., Sutherland, M.W. and Bielski, B.H.J. (1983) *J. Lipid Res.*, **24**, 485.

Bascetta, E., Gunstone, F.D. and Scrimgeour, C.M. (1984) *Lipids*, **19**, 801.

Christie, W.W. (1982) *Lipid Analysis*, 2nd edn, Pergamon, Oxford, p. 92.

Gunstone, F.D., McLaughlin, J., Scrimgeour, C.M. and Watson, A.P. (1976) *J. Sci. Food Agric.*, **27**, 675.

Traitler, H., Wille, H.J. and Studer, A. (1988) *J. Am. Oil Chem. Soc.*, **65**, 755.

Wille, H.J. and Traitler, H. (1988) *Fat Sci. Technol.*, **90**, 476.

Wille, H.J., Traitler, H. and Lagarde, M. (1991) *Fat Sci. Technol.*, **93**, 363.

5

PROCESSING OF FATS AND OILS

F.V.K. YOUNG, C. POOT, E. BIERNOTH, N. KROG, N.G.J. DAVIDSON and F.D. GUNSTONE

5.1 Introduction

The edible-oil consumer requires that the marketed product should be of acceptable quality irrespective of whether the product is a virgin or crude oil, for example olive oil or premier jus tallow, or a fully refined oil.

The quality of the product is influenced by many factors. In the case of vegetable raw materials, these include seed or plant type, growth conditions, pests and micro-organisms, chemical and weed contaminants, harvesting, storage and shipment. Factors of importance for animal or marine raw materials are the part of the animal or fish used and its freshness. Thereafter, product quality of both vegetable and animal oils and fats is affected by the processes of extraction, handling of the crude oil, refining, modifying, processing, packaging and distribution. There have been cases of raw materials, whether seed or oil, being contaminated by a chemical or other foreign substance that has been transported previously in a ship's hold or tank or in a road tanker. This problem has been addressed in discussions between suppliers, transport companies and customers under the chairmanship of the Federation of Oils, Seeds and Fats Associations Ltd (FOSFA) and the United States National Institute of Oilseed Products (NIOP). The result of these discussions has been the compiling of and agreement on lists of acceptable and unacceptable cargoes prior to carrying edible oils (FOSFA International, 1992; NIOP, 1992).

The subjects to be examined in this chapter are the storage, handling and extraction of the raw materials, and the refining and modifying of the resultant oils and fats.

The aspects of importance in edible-oil products are eating quality, appearance, oxidative or shelf-life stability and consistency of quality. By eating quality it is meant that the product should not be harmful and that its flavour and odour should be acceptable to the consumer. These aspects must be achieved by the processor as efficiently as possible.

The impurities with which the processor has to contend are as follows:

1. Naturally occurring components of the oil/raw material, such as pigments, phosphatides, trace metals.
2. Contaminants resulting from agronomy and harvesting, such as trash or dockage, dirt and chemicals, toxic weeds (List and Spencer, 1979).
3. Contaminants resulting from frost damage or wet weather, such as high chlorophyll levels.
4. Breakdown products of the natural raw material as a result of enzymic or oxidative attack. Such products are free fatty acids, partial glycerides, non-hydratable phosphatides, sulphur compounds, carbonyl compounds, pigment degradation or condensation products, protein fragments.
5. Contaminants introduced by processing, such as trace metals, soaps.

The processor is protected to a large extent by contractual standards for grades of raw materials (Pritchard, 1983). In international trade about 80% of all oil seeds, oils and fats are handled on FOSFA contracts. Other organizations deal with standards for specific countries or raw materials, for example the NSPA (National Soybean Processors' Association) in the USA and the CGSB (Canadian Government Specifications Board), which is primarily responsible for rapeseed specifications. It has, however, been pointed out (Gustafson, 1978) that trading rules may not adequately consider what it means to oil processors when, for example, seeds have been damaged or oils have become oxidized.

Living animals and plants use enzymes as catalysts for the synthesis of oils and fats, proteins, carbohydrates, etc. Under adverse conditions such as frost, high moisture or mechanical damage, the orderly structure is broken down. The result is degradation of and reactions between the various cell chemicals, again catalysed by enzymes until they are deactivated either by heat or by chemical action. Deactivation of enzymes is generally accomplished at temperatures between 70 and 90 °C. Many of the breakdown products of these reactions are oil-soluble, whereas the original substrates were not, thus introducing impurities into the oil. On dead tissues or harvested seeds, the same conditions of moisture and warmth that favour the cell enzymic degradation process will also support the growth of bacteria and moulds, thus further accelerating the breakdown of the raw material and increasing the level of impurities in the oil or fat.

Three forms of enzymic action relating to soybeans were identified by Ong (1980). Lipolytic enzymes are responsible for the hydrolysis of triglycerides (triacylglycerols), thus increasing the free fatty acid content of the oil. Lipoxygenases in the presence of oxygen catalyse the formation of peroxides of linoleic or similar acid chains. The peroxides in turn break down to form secondary oxidation compounds. Phospholipases catalyse the hydrolysis of phosphatides. Phospholipase D is responsible for the production of phosphatidic acid and in the presence of calcium or magnesium the respective phosphatidates. The acid and its calcium and magnesium salts have been reported as being the principal constituents of non-hydratable phosphatides in vegetable oil. It is clear that these and similar reactions are also responsible for many of the impurities found in other oils.

References and further reading

FOSFA International (1992) *Oils Fats Int.*, **8** (1), 25.

Gustafson, E.H. (1978) *J. Am. Oil Chem. Soc.*, **55**, 751.

List, G.R. and Spencer, G.F. (1979) *J. Am. Oil Chem. Soc.*, **56**, 706.

NIOP (1992) *Oils Fats Int.*, **8** (2), 14.

Ong, T.L. (1980) Lecture: The impact of the preparation and extraction conditions of soybeans on the oil quality, *ISF/AOCS Conference*, New York.

Pritchard, J.R. (1983) *J. Am. Oil Chem. Soc.*, **60**, 322.

5.2 Storage and handling

Edible-oil raw materials can be divided into three groups with regard to storage and handling. Fatty material of animal or fish origin is processed as soon as possible after death because of the powerful enzyme systems present. Oil-bearing fruits such as oil palm and olive are similar to animal tissues in that their high moisture content is favourable to enzymic action and, in consequence, they must be processed promptly after harvesting so as to minimize degradation of the oil. The third group, oil-bearing seeds, is less liable to deterioration because of its lower moisture content and may be stored under suitable conditions for long periods. The subject of the handling and storage of oilseeds, oils, fats and meal has been examined in detail by Patterson (1989).

5.2.1 Animal and fish

Animal fat tissues are obtained from abattoirs and from the off-cuts of butchers' shops. The abattoir fatty material may be whole carcases, depot fat portions or viscera. Of these, the depot fat yields the highest-quality product and, because of heavy enzyme and haem contamination, the viscera produce the lowest quality, frequently high in free fatty acid, iron and oxidation product contents and both dark in colour and difficult to bleach. Enzymes also catalyse the breakdown of protein with the consequent introduction into the oil of sulphur compounds. The rendering plant is sited near the abattoir and is best run on a daily basis, that is with all raw material being cleared through the plant on the day of receipt. The above types of raw material are usually stored and processed separately. Storage areas are protected from the sun and are of hygienic construction; they are cleaned as they are cleared of raw material. A particular hazard of animal fats is the presence of traces of polythene in the oil, resulting from the use of polythene to protect the carcases after slaughter. The polythene, which is frequently in small pieces in the raw material, must be removed whenever possible.

Fish raw material is primarily the whole fish, although fish oil is also extracted from the waste material from fish markets and canning factories. The enzymic breakdown of fish tissue is even more rapid than from the viscera of animals, and has more pronounced effects on the oil and its downstream processing. Fish oils are highly unsaturated and are therefore more liable to oxidative breakdown. Virtually all fish oil for human food is hydrogenated and therefore it is particularly important that oxidation products and sulphur compounds, both of which are hydrogenation catalyst poisons, should be as low as possible. A number of volatile sulphur compounds from fish oils have been identified by Christensen *et al.* (1981).

The fish are landed daily from inshore fishing boats or stored in ice in the case of deep-sea boats. The fish oil/meal plants are either in mother ships or, more usually, located in the fishing ports. The same conditions of storage, hygiene and daily use apply as in the treatment of animal fats.

5.2.2 Palm fruit

The fruit of the oil palm grows in bunches, each bunch weighing up to 20 kg and containing of the order of 1500 fruits (Berger, 1983; Cornelius, 1983). Palm oil is obtained from the fibrous and pulpy mesocarp that surrounds the nut or kernel, which is enclosed by a woody shell. The kernel is the source of palmkernel oil. Most of the oil in the fruit is synthesized in the last two weeks of ripening and therefore good judgement is required to provide high yield without over-ripeness. Bruising of over-ripe fruit occurs when the bunches are harvested and when single fruits fall to the ground. Bruising ruptures the oil cell wall, permitting enzymic attack on the oil. The principal feature of such attack in the case of palm oil is the rapid rise in free fatty acid content of the oil to the order of 5%. The fruit must therefore be transported, usually by truck, as quickly as possible to the mill, where the first step in the extraction process is sterilization to deactivate enzymes and destroy microorganisms.

5.2.3 Oil seeds

It is stated above that oil seeds can be stored for long periods under the correct conditions. The principal factor involved in these conditions is the moisture content of the seed. It has been suggested (Raymond, 1961) that the maximum permissible figure for safe storage of oil seeds generally is 14% of moisture calculated on the non-oleaginous portion of the seed. Taking palmkernels as an example, with an oil content of 52% on a dry basis, this would amount to 7.25% moisture on the whole seed (Cornelius, 1966). However, according to Galloway (1935) even lower moisture contents are required to prevent the growth of moulds, and the safe limit is the moisture content

Table 5.1 Maximum acceptable moisture levels of seeds

Seed	Moisture (%)
Copra	7.0
Rapeseed	7.0
Palmkernel	8.0
Sunflower seed	9.5
Cottonseed	10.0
Soybeans	13.0

attained when the material is in equilibrium with an atmosphere of 70% relative humidity. From considerations such as these and from practical experience, the maximum acceptable safety levels for several seeds are shown in Table 5.1. The deterioration of the seed in storage is caused not only by the enzymes within the seed but also by the external effect of insects, yeasts, moulds and bacteria. The action of the latter is particularly rapid on split or damaged seeds.

The production of copra begins with the husking and splitting of the mature coconut, usually by hand (Cornelius, 1973). The opened wet coconut meat must be dried as soon as possible to prevent the attack of bacteria and fungi. The drying process is preferably carried out by indirect heat. 'Smoke-dried' copra, from direct driers in which flue gases are passed over the copra, sometimes has a burned odour and the colour of the extracted oil may be fixed. After drying, copra is transported in bags to the oil mill, which is now usually in the producing country. Because of the ease of hydrolysis of coconut triglycerides, particular care must be taken over moisture content during storage.

Palmkernels are separated from the press-cake from palm oil production. Drying in a hot air current separates the nuts from the fruit fibre. Before shelling, the nut is dried to make the shell brittle, and the kernel shrinks away from the shell. The nut is cracked and the shell separated from the kernel in a winnowing column followed by a hydrocyclone. The kernels are usually dried to 6–7% moisture content in silos incorporating the circulation of air at 50–55 °C and are then bagged for dispatch (Cornelius, 1983). The low-molecular-weight triglycerides of palmkernel oil are as susceptible to hydrolytic splitting as are those of coconut oil. The degradative reactions of lipolysis and oxidation are accompanied by the evolution of heat, which, if allowed to get out of control, can fix the colour of the oil, making it difficult to decolorize during refining. Temperature and moisture conditions must therefore be monitored carefully.

Moisture control is also the most important factor in the storage of the commoner oil seeds, soybean, rapeseed and sunflower seed (List *et al.*, 1992). The farmer aims to harvest the seed when it is ripe and dry. Unripe seed, in particular soybeans and rapeseed, is green owing to relatively high levels of chlorophyll. This is transferred to the oil during extraction, and is difficult to remove from the oil during refining (Daun, 1982; Thomas, 1982). Wet seed containing, in the case of soybeans, over 13% moisture and stored at over 10 °C may begin to germinate as well as to suffer the attack of fungi, insects and mites. Germination generates heat, which may, apart from causing damage to the oil and meal, cause fires by spontaneous combustion (Barger, 1981). Rapeseed contains glucosinolates, which are hydrolysed in the presence of thioglucoside glucohydrolase. The products are further decomposed to form oil-soluble sulphur compounds, which are catalyst poisons (Rutkowski *et al.*, 1982). Storage silos or bins therefore should have the capability of drying the seed. The seeds are transported by truck, rail and barge to industrial elevators for processing or export. Such transport and handling should minimize the separation of hulls from whole soybeans and the formation of split seeds. List *et al.* (1992) and Mounts *et al.* (1979, 1990) reported the rise in non-hydratable phosphatides in soybean oil that resulted from an increase in split beans during export handling.

References and further reading

Barger, W.M. (1981) *J. Am. Oil Chem. Soc.*, **58**, 154.

Berger, K.G. (1983) *J. Am. Oil Chem. Soc.*, **60**, 206.

Christensen, B.W., Kjaer, A. and Øgaard Madsen, J. (1981) *J. Am. Oil Chem. Soc.*, **58**, 1053.

Cornelius, J.A. (1966) *J. Sci. Food Agric.*, **17**, 57.

Cornelius, J.A. (1973) *Trop. Sci.*, **15**, 15.

Cornelius, J.A. (1983) *Processing of Oil Palm Fruit and its Products*, Tropical Development and Research Institute, London.

Daun, J.K. (1982) *J. Am. Oil Chem. Soc.*, **59**, 15.

Galloway, L.D. (1935) *J. Text. Inst.*, **26**, T123.

List, G.R., Mounts, T.L. and Lancer, A.C. (1992) *J. Am. Oil Chem. Soc.*, **69**, 443.

Mounts, T.L., List, G.R. and Heakin, A.J. (1979) *J. Am. Oil Chem. Soc.*, **56**, 883.

Mounts, T.L., Snyder, J.M., Hinsch, R.T., Bongers, A.J. and Class, A.R. (1990) *J. Am. Oil Chem. Soc.*, **67**, 743.

Patterson, H.B.W. (1989) *Handling and Storage of Oilseeds, Oils, Fats and Meal*, Elsevier Applied Science, London.

Raymond, W.D. (1961) *Soc. Chem. Ind. Monogr.*, No. 11, p. 177.

Rutkowski, A., Gwiazda, S. and Krygier, K. (1982) *J. Am. Oil Chem. Soc.*, **59**, 7.

Thomas, A. (1982) *J. Am. Oil Chem. Soc.*, **59**, 1.

5.3 Preparation and extraction

The preparation of oil-bearing raw materials prior to extraction is in all cases concerned first with the separation of extraneous matter and secondly with transforming the raw material into a state that will give a high yield of good-quality oil and meal.

The oil is extracted from the prepared raw material mainly by one of the following methods.

1. Rendering: cooking followed by mechanical separation; used for fish oil and animals fats.
2. Wet screw-pressing or expelling: used for olive and palm oils.
3. High-pressure screw-pressing: used for many oil seeds of high oil content, such as copra, palmkernels, groundnuts.
4. Pre-pressing followed by solvent extraction: very widely used for oil seeds such as copra, palmkernels, groundnuts, sunflower seed, rapeseed, cottonseed.
5. Solvent extraction used alone: used primarily for oil seeds of low oil content, such as soybeans and ricebran, but also for some highly permeable seeds of high oil content, e.g. palmkernels.

The choice of process depends, as stated, on the type of raw material but also on the desired plant capacity. Continuous solvent extraction plants are supplied for daily seed throughput of 100 to 3000 tonnes. The first three processes listed and batch solvent extraction can be used for less than 100 tonnes per day.

5.3.1 Rendering

Animal fat tissue is conveyed via an open screw or belt conveyor to a size-reducing hasher. During conveying an opportunity is provided for the safe removal of pieces of polythene and other foreign matter. After size reduction the fatty material may be subjected to batch or continuous wet or dry rendering, the continuous methods being the most widely used (Düpjohann and Hemfort, 1975).

In the wet process, direct steam is injected into the pipeline carrying the raw material so as to reach about 100 °C, and the mixture is intensely agitated before passing to a decanter centrifuge. The liquor from the decanter is separated and the oil polished in two further centrifuges.

In the dry process, the hashed raw material is subjected to indirect steam heat in a horizontal cylindrical cooker equipped with an agitator. On leaving the cooker the mixture is transferred via a strainer, in which liquor is drained off, to a screw-press. The oil–water mix from the press is mixed with the liquor from the strainer and passed through a decanter and a centrifuge in order to produce the crude oil. To remove fines from the oil it is preferable to add sufficient water to wet the fines and then remove them by centrifuging (CSIRO, 1978).

Fish oil is extracted by the continuous dry process outlined above (Young, 1982). In both the wet and the dry processes, the enzymes referred to in section 5.1 are deactivated at the temperatures employed.

5.3.2 Wet screw-pressing

The extraction of palm oil has been described by Bek-Nielsen (1974), Berger (1983) and Maycock (1987). Fresh oil palm fruit bunches, in mills handling more than 6 tonnes of bunches per hour, are loaded into cages and moved by rail into a horizontal sterilizer. The purposes of the sterilizer (Berger, 1983) are inactivation of enzymes, loosening of the fruit on the bunch, softening of the fruit, conditioning of the nuts and coagulation of proteins. These objectives are attained by the application of live steam at between 2.5 and 3.0 kg/cm^2 for 1 h. The temperature reaches 132 °C (Bek-Nielsen, 1974). The cages are removed from the sterilizer and emptied into the hopper feed of a rotating cylindrical drum in which the fruit is shaken from the bunches. The fruit and oil cells are then broken up in a digester, which is a vertical, cylindrical, steam-jacketed vessel fitted with beater arms. The digester is kept full and the temperature is maintained at 90–95 °C. The pulp from the digester is fed to a screw-press from which the liquor, with a composition of approximately 53% oil, 7% solids and 40% moisture, is run through a vibrating screen into settling tanks. Residence time in the tanks is about 2 h, after which the oil layer is centrifugally clarified and vacuum dried to 0.10–0.15% moisture content. The sludge from the settling tanks is centrifuged and the recovered oil returned to the settling tanks.

5.3.3 High-pressure screw-pressing

The preparation of vegetable oil seeds is varied according to the seed being processed. It consists of some or all of the following basic steps: cleaning, drying, dehulling or decorticating, size reduction, cooking and flaking (Buhr, 1990; Galloway, 1976).

In cleaning, the dirty seed is passed through a mechanical and/or pneumatic system in which the foreign matter is either screened out or blown away. Drying may be necessary if the seed moisture content is higher than is desirable for downstream processing. The purposes of decortication are to obtain a high-protein meal by reduction of fibre, to reduce wear in the screw-press, to increase the capacity of the screw-press and to reduce impurities in the oil. Reduction of the seed is used to break down or weaken the oil cell walls to the point where the oil is available to be expelled, and is achieved in hammer grinders or by using breaker rolls, sometimes followed by flaking rolls.

Cooking the seed serves the following purposes (Ward, 1976):

1. Completion of the breakdown of oil cells.
2. Lowering of the viscosity of the oil.
3. Coagulation of the protein in the meal.
4. Adjustment of the moisture level in the meal to the optimum level for pressing or extraction.
5. Sterilization of the seed.
6. Detoxifying undesirable seed constituents such as gossypol in cottonseed.
7. Fixing certain phosphatides in the cake, thus lowering subsequent refining losses.

The cooking process is normally carried out in 'stack cookers', which consist of a number of closed, cylindrical steel kettles mounted one on top of another, the number and size of which depend primarily on the unit capacity but also on seed specification. The kettles have both live and jacketed steam heating, a sweep-type agitator for each level with a common drive, level controls and exhaust piping on the lower levels for the removal of moisture. The seed enters at the top level and passes down, receiving its specified treatment at each stage before being fed directly to the screw-press. Cooking conditions are varied depending on the seed being processed, but residence times are of the order of 30–60 min and the maximum temperature lies between 90 and 115 °C. The final stage in the cooker is usually a drying operation to produce the desired moisture level in the seed for maximum oil yield from the screw-press. Examples are 3% moisture for cottonseed and 2% for copra.

An oil-seed screw-press has a horizontal main wormshaft, which carries the worm assembly. This assembly revolves within a 'barrel' or 'cage', which normally consists of axially placed 'lining bars' or 'barrel bars' contained within substantial frames. Most screw-presses have feed gear, by which seed is rammed into the press, and choke gear, used for adjusting the cake thickness. The press is so constructed that the material passing through is subjected to increasing pressure and oil is expelled through the slots between the cage lining bars. The detailed design and operation of screw-presses have been well described by Allen (1982), Bredeson (1983), Stainsby (1988) and Ward (1976). The oil leaving the press contains foots or fines of solid material. In the single high-pressure operation, these foots must be separated from the oil and returned to the cooker or screw-press so as to reduce losses in oil refining.

High-pressure expelling of oil seeds is a very efficient operation. Modern presses are capable of high oil yields such that only 3–6% of oil is left in the cake, depending on seed type. Depending upon commercial factors, therefore, the process can in some cases be compared favourably with solvent extraction.

5.3.4 Pre-pressing followed by solvent extraction

For pre-pressing prior to solvent extraction, the seed is prepared through the cleaning, drying, decorticating, size reduction and cooking stages in generally the same manner as for high-pressure pressing.

Sunflower seed may be taken as an example. The decorticating of sunflower seed is a borderline case, in that it is practised in some parts of the world, notably Eastern Europe, but not in others. The ideal level of hulls in meats for pre-pressing and solvent extraction is 7%, and this reduction, from about 25%, also results in notable reduction of wax content in the extracted oil, as the waxes are to be found mainly in the hulls. The principal difficulty in decorticating is the disposal of the hulls.

Size reduction or flaking is carried out on decorticated seed to produce a flake thickness of 0.3–0.4 mm. The cooking operation has a residence time of about 40 min with a final temperature of 95–105 °C. About 5–7% moisture in the cooked meats is generally sufficient for good pre-press operation.

In the pre-press operation, the pressures employed are about half of those used for high-pressure expelling, that is, about 2.0 kg/mm^2 compared to 4.5 kg/mm^2. The main function of pre-pressing is to obtain a cake of excellent permeability for the following solvent extraction stage. The principle of the pre-press is the same as that of the high-pressure expeller, but its capacity is considerably higher, e.g. 200 tonnes per day. In practice, sunflower seed cake after pre-pressing has an oil content of 16–20% and moisture content of 5–7%. The oil from the press is strained to remove coarse dirt, filtered and cooled for storage. It may be water-degummed (see section

5.4.3) alone or together with the solvent-extracted oil. The pre-press cake is broken into pieces of 5–8 mm in size and cooled to 60–65 °C for feeding to the solvent extraction plant (Ward, 1982, 1984).

There are basically two designs of solvent extraction equipment, the percolation type, in which the solid material forms a fixed bed through which the solvent percolates, and the immersion type, in which the pretreated seed or cake is dispersed in the solvent. Both designs employ the countercurrent technique. The percolation design is the more widely used. It is to be found in a number of forms of which the most common are the rotary types, and the vertical and horizontal chain and basket types. In all percolator designs the fresh solvent is pumped into the extractor before discharging the spent extracted flakes, which are allowed to drain before discharge. The miscella is pumped countercurrently to the flow of flakes, becoming richer as oil is extracted, and finally is pumped over a bed of fresh flakes for filtration before leaving the extractor as full miscella. The theory of solvent extraction and details of plant design are given by Christensen (1983), Lajara (1990), Lusas *et al.* (1990), Milligan (1976) and Norris (1982).

The efficiency of the solvent extraction process is such that oil levels in meal are generally below 1% and frequently about 0.5%. This degree of extraction can, however, be counterproductive for the refiner, because the final 1% of oil removed is very rich in phosphatides and other impurities, which will cause considerable losses in refining and result in a product of markedly poorer quality.

The miscella may contain 'fines', which are removed by filtration, usually through leaf filters. Solvent is recovered from the miscella, which may contain from 20 to 30% oil, by passing the miscella through multistage rising film evaporators with the final stage incorporating the injection of live steam. In some plants the live steam is used to hydrate phosphatides in the oil, which, on leaving the solvent recovery plant, is passed directly to the degumming centrifuges.

5.3.5 Direct solvent extraction

In the introduction to this section on extraction, cottonseed was given as an example of a raw material processed by the pre-press/solvent extraction method. Such is the case in countries where a wide variety of oil seeds is handled (French, 1988). Delinted and decorticated cottonseed contains about 30% oil, which is a borderline level between pre-pressing followed by solvent extraction and direct solvent extraction. Improvements in efficiency of the latter process have favoured its adoption for cottonseed in the USA, and this movement has been encouraged by the introduction of the extruder, also known as the 'expander' or 'enhancer' (Lusas *et al.*, 1990; Lusas and Watkins, 1988; Bredeson, 1983), in which the seed tissue and cell walls are heat-denatured, thereby enhancing the release of oil.

In the extruder, the conditioned and flaked seed is injected with live steam under screw treatment such that, before the extrusion die, the moisture is 18 to 25% (Lusas and Watkins, 1988) and the temperature is in excess of 100 °C. On leaving the die the compacted seed expands, releasing steam, and forming 'collets' of increased porosity and density. The collets must be cooled to below 60 °C and contain less than 9% moisture for hexane extraction. The increase in porosity improves percolation, which, together with the increased density, results in higher extraction plant capacities and efficiencies being obtainable (Lusas *et al.*, 1990). Apart from cottonseed, the extruder has proved of value in the processing of soybeans and also rice bran.

The preparation of soybeans when not using the extruder follows the series of steps outlined in section 5.3.3. The conventional process and modern developments incorporating fluid-bed driers (Florin and Bartesch, 1983) or microwave drying are described by Moore (1983). Soybeans are initially tempered for 7–10 days, during which time the temperature should not be higher than 24 °C and their moisture content should be between 10 and 12%. The tempering loosens the hull from the bean and allows the bean moisture to attain equilibrium. The beans are next cleaned and subsequently cracked to one-sixth to one-eighth size followed by dehulling. For high-protein meal, the hull content (initially 7%) is reduced to about 2.5–3%. The cracked beans are conditioned at 60–76 °C and 10–12% moisture and are flaked to 0.25–0.30 mm thickness while hot. The temperature in the extractor is kept constant mainly by means of the flake temperature entering the extractor (Serrato, 1981). For this purpose the flakes are maintained at about 55–65 °C depending upon the installation, and the extraction temperature lies between 40 and 60 °C. The oil is separated from the miscella and polished as described in section 5.3.4, and must pass a negative flashpoint test at 120 °C.

The conditioning of soybeans with particular reference to crude oil quality and the physical refining of soybean oil has been investigated by Kock (1983). He reported that deactivation of the lipoxygenases and phospholipases in soybeans prior to extraction results in a significant reduction in the non-hydratable phosphatide content in the extracted oil. In the 'Alcon' process that resulted from this work the beans are cracked, dehulled, heated and flaked before entering a three-stage conditioning treatment. In the first stage, the flakes are steam-heated to 95–110 °C and a moisture content of 15%. The second stage has a tempering function and duration of 15 min. Finally, the flakes are dried and cooled before entering the extractor.

5.3.6 Solvents

The most widely used solvent for oil-seed extraction is a light paraffinic petroleum fraction, consisting mainly of n-hexane and methylpentane (DGF, 1975–76), with a boiling-point range of 65–70 °C, and commonly referred to as 'hexane'. Johnson and Lusas (1983) give a typical purchase specification for hexane and discuss the choice of solvents in general. The solvent:flake ratio for hexane is 1:1, which compares favourably with most alternatives. Losses reported in the solvent extraction process are about 2 gallons per tonne of seed, approximately equal to 0.7% (Norris, 1982), and 2.0 litres per tonne of seed, 0.14% (Dahlen and Lindh, 1983); 78% of the latter figure emanated from the meal desolventizer–toaster (DT). The total loss was later reduced to 1.6 litres (0.11%) by modification to the DT.

Although in many ways an ideal solvent for the process, hexane has the major drawback of fire and explosion hazards. The problems posed in this respect by dust and hexane in oil mills are discussed in detail by Radandt (1982), and Sandvig (1983) has outlined safety precautions in solvent extraction plants. Because of the flammability problem and because of the potential shortage of petroleum products, other solvents are being examined. The use of liquefied gases and supercritical gases is reported by Mangold (1983) and Lusas *et al.* (1990), and a pilot-plant process for the extraction of soybeans using aqueous isopropylalcohol has been developed (Baker and Sullivan, 1983). The crude desolventized oil obtained from the latter method was of appreciably higher quality than that obtained using hexane. Lusas *et al.* (1990) reported on the work that had been carried out investigating alternative solvents to 1989, including methylene chloride, ethyl and isopropyl alcohols, mixed solvents and supercritical carbon dioxide.

5.3.7 Other methods of extraction

(a) Aqueous extraction

Cater *et al.* (1974) described a process for the aqueous extraction of comminuted seed. Unit operations involved are grinding the seed, extraction, solid/liquid separation, centrifugation, demulsification and drying of the products. Experimental work with the process had been largely confined to coconuts and groundnuts. An economic evaluation of the process applied to coconuts was reported by Hagenmaier *et al.* (1975). Suggested advantages compared with solvent extraction are lower capital investment, improved safety since no flammable solvent is involved, simplicity, high-quality products requiring little further processing and the capability to utilize certain chemicals to remove or inactivate undesirable substances. Disadvantages include a lower yield of extracted oil.

(b) Extraction of oil by enzymic hydrolysis

The extraction of oil from crushed soybeans and rapeseed on a laboratory scale by enzymic hydrolysis has been described by Fullbrook (1983). Experimental work is described using a number of different enzyme systems both with and without the presence of a solvent to extract the released oil. When working with rapeseed, 50% more oil was extracted in the presence of hexane. A yield of 90% of the Soxhlet-extractable oil was obtained from soybeans using 3% of enzymes from *Aspergillus niger*. Advantages claimed for the process are low energy and solvent usage and better than traditional quality of the protein and oil products. Similar conclusions were reached by Sosulski *et al.* (1988) working with carbohydrase enzymes and Canola cultivars.

References and further reading

Allen, R.R. (1982) *Bailey's Industrial Oil and Fat Products*, 4th edn, vol. 2 (ed. D. Swern), Wiley, New York.

Baker, E.C. and Sullivan, D.A. (1983) *J. Am. Oil Chem. Soc.*, **60**, 1271.

Bek-Nielsen, B. (1974) *Technical and Economic Aspects of the Palm Fruit Processing Industry*, United Nations, New York.

Berger, K.G. (1983) *J. Am. Oil Chem. Soc.*, **60**, 206.

Bredeson, D.K. (1983) *J. Am. Oil Chem. Soc.*, **60**, 211.

Buhr, N. (1990) *Edible Fats and Oils Processing, World Conf. Proc.*, Maastricht, American Oil Chemists' Society, Champaign, p. 43.

Cater, C.N., Khee, C.R., Hagenmaier, R.D. and Mattil, K.F. (1974) *J. Am. Oil Chem. Soc.*, **51**, 137.

Christensen, B.W., Kjaer, A. and Øgaard Madsen, J. (1981) *J. Am. Oil Chem. Soc.*, **58**, 1053.

Christensen, P.L. (1983) *J. Am. Oil Chem. Soc.*, **60**, 214.

CSIRO (1978) *Meat Research News Letter*, No. 78/4.

Dahlen, J.Å.H. and Lindh, L.A. (1983) *J. Am. Oil Chem. Soc.*, **60**, 2009.

DGF (1975–76) Gewinnung von Fetten und Oelen aus

pflanzlichen Rohstoffen durch Extraktion, *Fette Seifen Anstrichm.*
Düpjohann, J. and Hemfort, H. Jr (1975) *Fette Seifen Anstrichm.*, **77**, 81.
Florin, G. and Bartesch, H.R. (1983) *J. Am. Oil Chem. Soc.*, **60**, 193.
French, D.P. (1988) *J. Am. Oil Chem. Soc.*, **65**, 1409.
Fullbrook, P.D. (1983) *J. Am. Oil Chem. Soc.*, **60**, 476.
Galloway, J.P. (1976) *J. Am. Oil Chem. Soc.*, **53**, 271.
Hagenmaier, R.D., Cater, C.M. and Mattil, K.F. (1975) *J. Am. Oil Chem. Soc.*, **52**, 5.
Johnson, L.A. and Lusas, E.W. (1983) *J. Am. Oil Chem. Soc.*, **60**, 229.
Kock, M. (1983) *J. Am. Oil Chem. Soc.*, **60**, 198.
Lajara, J.R. (1990) *Edible Fats and Oils Processing, World Conf. Proc.*, Maastricht, American Oil Chemists' Society, Champaign, p. 49.
Lusas, E.W. and Watkins, L.R. (1988) *J. Am. Oil Chem. Soc.*, **65**, 1109.
Lusas, E.W., Watkins, L.R. and Rhee, K.C. (1990) *Edible Fats and Oils Processing, World Conf. Proc.*, Maastricht, American Oil Chemists' Society, Champaign, p. 56.
Mangold, H.K. (1983) *J. Am. Oil Chem. Soc.*, **60**, 226.
Maycock, J.H. (1987) *Palm Oil* (*Critical Reports on Applied Chemistry*, vol. 15), Society of Chemical Industry, and Wiley, Chichester, p. 29.
Milligan, E.D. (1976) *J. Am. Oil Chem. Soc.*, **53**, 286.
Moore, N.H. (1983) *J. Am. Oil Chem. Soc.*, **60**, 189.
Norris, F.A. (1982) *Bailey's Industrial Oil and Fat Products*, 4th edn, vol. 1, *Raw Materials Handling and Storage*, vol. 2, *Extraction of Fats and Oils* (ed. D. Swern), Wiley, New York.
Radandt, S. (1982) *Fette Seifen Anstrichm.*, **84**, 235.
Sandvig, H.J. (1983) *J. Am. Oil Chem. Soc.*, **60**, 243.
Serrato, A.G. (1981) *J. Am. Oil Chem. Soc.*, **58**, 157.
Sosulski, K., Sosulski, F.W. and Coxworth, E. (1988) *J. Am. Oil Chem. Soc.*, **65**, 357.
Stainsby, M.A. (1988) *J. Am. Oil Chem. Soc.*, **65**, 1413.
Ward, J. (1976) *J. Am. Oil Chem. Soc.*, **53**, 261.
Ward, J. (1982) Lecture: The pre-pressing of rapeseed and sunflower, *AOCS Meeting*, Toronto.
Ward, J. (1984) Private communication.
Young, F.V.K. (1982) *Nutritional Evaluation of Long-Chain Fatty Acids in Fish Oil*, Academic Press, London.

5.4 Production of edible oils

5.4.1 Introduction

The further processing of edible oils after extraction from the raw material is concerned with purification, or refining, and modification. The refining treatment is needed to remove, or reduce as far as is possible, those contaminants of the crude oil which will adversely affect the quality of the end-product and the efficient operation of the modification processes. Certain non-glyceride compounds in the crude oil, notably the tocopherols and tocotrienols as antioxidants, have a beneficial effect on product quality, and therefore the refining treatment should be designed to retain as much as possible of these compounds in the oil. The process conditions used should be such that the nutritional value of the oil should not be adversely affected, either in the loss of beneficial compounds or in the formation of undesirable artefacts.

The modification processes that will be described are hydrogenation, fractionation and inter-esterification. These processes are used to widen the applicability of oils and fats and also, in the case of hydrogenation, to make products of acceptable oxidative and flavour stability from unstable or relatively unstable raw materials.

Impurities that are known to have a deleterious effect on product quality or processing efficiency are moisture and dirt, including protein degradation products, free fatty acids, partial glycerides, phosphatides, oxidation products, pigments and compounds containing trace elements, for example, copper, iron, sulphur and the halogens. Carbohydrates, glucosides and pectins may adversely affect fractionation by increasing viscosity. The effect of other classes of natural impurities such as sterols and hydrocarbons depends on their structure. The saturated molecules are heat-stable and inert, but unsaturated examples have been reported as having a nauseous taste and smell (Marcelet, 1936a, b).

In addition to the natural components, crude oils have been shown to contain very small amounts, in the parts per billion range, of chlorinated pesticide residues. Gas chromatographic determinations have been carried out on samples of oil seeds, crude and processed oils and by-products from soybeans (Chaudry *et al.*, 1976, 1978; Hashemy-Tonkabony and Soleimani-Amiri, 1976) and on sunflower-seed and cottonseed oils (Hashemy-Tonkabony and Soleimani-Amiri, 1976).

Trading specifications for crude oils have been discussed by Pritchard (1983). Contracts, trading rules and specifications are issued by a number of national and international organizations, amongst which are FOSFA (Federation of Oils, Seeds and Fats Associations), NSPA (National Soybean Processors' Association), CGSB (Canadian Government Specifications Board) and the American Fats and Oils Association. Examples of the specifications for a few of the major oil types are set out in Table 5.2. The Food and Agriculture Organization/World Health Organization (FAO/WHO) Codex Alimentarius Commission have published *Recommended International Standards* for individual oils, and national standards bodies, for example the British Standards Institute, issue specifications.

As can be seen from Table 5.2, the amount of information in the contract specification varies considerably and, in fact, tells the refiner very little about the quality of the oil. In very few cases, for example, are values included for parameters having a bearing on the oxidative state of the oil. Claims or complaints on such matters must be based on the FAQ/GMQ clause. The refiner must carry out laboratory tests on the oil designed to give foreknowledge of problems to be expected during processing in the plant. Information on oxidation and trace impurity levels in commercial oils has been published for soybean oil (Ong, 1980b; Thomas, 1981), rapeseed oil (Thomas, 1982a, b; Kramer *et al.*, 1983), palm oil (Young, 1982), lauric acid oils (Young, 1983) and fish oil (Barlow and Stansby, 1982).

Two methods are in use for the refining of oils and fats. These are termed 'physical' and 'chemical', from the means by which the free fatty acids are removed from the oil. The fatty acids are distilled off in the physical process, and in the chemical process are neutralized using an alkaline reagent, thus forming soaps, which are removed from the oil by phase separation. The process steps that are involved in the two methods are depicted in Fig. 5.1, and the principal impurities that are removed from the oil in those steps are shown in Table 5.3.

5.4.2 Storage and handling of crude oils

The storage and handling of crude oils is of the utmost importance in obtaining a high yield of a high-quality refined product. Factors that must be considered in the design of a crude oil tank farm are: prevention of leakage, i.e. loss of oil; prevention of contamination between different oil types; minimization of free fatty

Table 5.2 Specifications of oil types (Data from Pritchard, 1983)

	Soybean oil	Rapeseed oil	Palm oil	Palmkernel oil	Coconut oil	Sunflower-seed oil
State	Crude degummed	Crude degummed	Crude	Crude	Crude	Crude
Contract	NSPA	CGSB internal	Malaysian Standard MS 719: 1981	FOSFA 54	FOSFA 54	American Fats and Oils, Rule 14, export
Source	USA	Canada	Malaysia	Malaysia	Philippines	USA
Quality	FAQ[a]	–	Standard	GMQ[b]	GMQ	FAQ
Free fatty acid (%)	0.75 (oleic), 1.25 max. with allowances	1.0 max. (oleic)	5.0 (palmitic)	5.0 max. (lauric) shipment, 5.5 max. arrival	3.0 max. (lauric) shipment, 4.0 max. arrival	2.0 max. (oleic)
Moisture and impurities (%)	0.3 max.	0.3 max.	Moist, 0.2, imp. 0.05	0.5 max.	1.0 max.	Moist. and volatiles 0.5 max., imp. 0.3 max.
Flashpoint	121 °C min.[c]	–	–	121 °C min.	121 °C min.	121 °C min.
Phosphorus (%)	0.02 0.025 max. with allowances	0.02	–	–	–	–
Unsaponifiable matter (%)	1.5 max.	–	–	–	–	1.3 max.
Erucic acid (%)	–	5.0 max.	–	–	–	–
RB[d] colour (red)	–	1.5 max.	–	–	–	Refined 2.5 red max.
Neutral oil (%)	–	98.5 min.	–	–	–	–
Carotene (mg/kg)	–	–	500–1000	–	–	–
Saponification value	–	–	–	–	–	188–194
Linolenic acid (%)	–	–	–	–	–	1.0 max.

[a] Fair average quality.
[b] Good merchantable quality.
[c] Quoted as 250 °F min.
[d] Refined bleached.

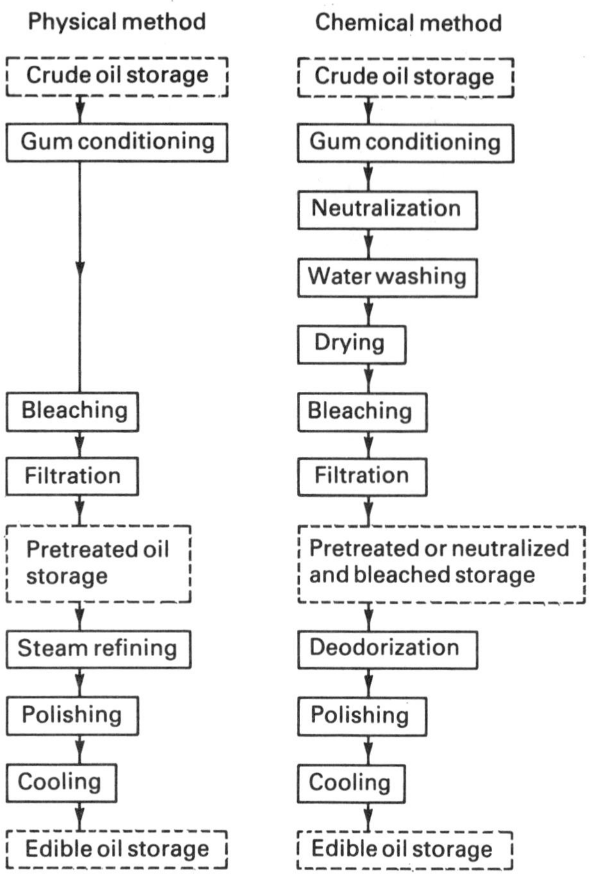

Fig. 5.1 Oil refining stages.

acid increase, of oxidation and of colour fixation; and unintentional fractionation.

Loss of oil in a tank farm occurs from flexible pipes, pumps, valves and the tanks themselves, and is caused by poor design, poor-quality plant items, poor maintenance and poor operation. Oil may also be lost by occlusion in the foot that settle to the bottom of the tank. Foots are a mixture of water, dirt and other oil-insoluble material, for example phosphatides, which are too thick to be pumped and must therefore be manually removed from the tank. In some cases, it is preferable for the overall refining efficiency to prevent such settlement by the use of a side-mounted agitator on the tank.

It is good practice, and in many countries a legal requirement, to prevent heavy loss of oil due to pipeline, valve or tank fracture by the construction of a bund wall around the tank farm.

Considerable disruption of refinery operation and loss of value and quality can be caused by inter-oil contamination. It is therefore desirable in multi-oil refineries to segregate the oils by class, for example, animal fats from vegetable oils, lauric acid oils from high-iodine-value liquid oils. Crude palm oil should be separated from other oils because of its dark colour. Soft fish oils must be handled in a separate installation because only minimal contamination is needed to produce a fishy smell and taste in another oil type that is used unhydrogenated.

Water in crude oils is responsible for the increase of free fatty acid content by hydrolysis particularly in the presence of dirt or lipolytic enzymes. High levels of moisture can cause hydration degumming, which results in increased neutral oil loss by occlusion in the precipitated gums. Moist air in storage tanks will condense on cool surfaces, thus, in mild steel tanks, producing rust, which will result in increased trace iron in the oil. The increased iron content impairs the bleachability of the oil and its oxidative stability. To reduce such risk, tanks can be insulated and the inside of the tank, particularly the roof, coated with an epoxy paint of suitable grade (Leong and Berger, 1982). The latter treatment is normally only applic-

Table 5.3 Refining-unit processes

Stage	Principal impurities reduced or removed
Degumming	Phospholipids, trace metals, pigments, carbohydrates, protein
Neutralization	Fatty acids, phospholipids, pigments, trace metals, sulphur, oil-insolubles, water-solubles
Washing	Soap
Drying	Water
Bleaching	Pigments, oxidation products, trace metals, sulphur, traces of soap
Filtration	Spent bleaching earth
Deodorization	Fatty acids, mono- and diglycerides, oxidation products, pigment decomposition products, pesticides
Physical refining	Fatty acids, mono- and diglycerides, oxidation products, pigment decomposition products, pesticides
Polishing	Last traces of oil-insolubles

able to new tanks owing to the difficulty of obtaining a clean surface in a used tank.

The oil should be stored cool to reduce the rates of oxidation and hydrolysis and minimize colour fixation. When heating the oil, low-pressure (less than 3 bar) steam or hot water should be used. Local overheating can be prevented by the use of side-mounted agitators. To reduce further the chance of oxidation, storage tanks should be designed to give a low surface-to-mass ratio of the oil, and, for general reduction of loss of quality, the size and number of tanks should be arranged so that they can be emptied and if necessary cleaned before re-use.

Where the climate is suitable, crude-oil storage tanks are sometimes used to obtain an initial de-waxing or winterization of the oil. However, accidental fractionation of oils having melting points above ambient temperature will lead to inconsistency of solid fat content in final products. To counter this problem, while oil is being drawn from the tank the oil should be held at a temperature 5–10 °C above its melting point. Tank insulation and mixing by the use of a side-mounted agitator are again to be advised.

The storage and handling of crude oils is discussed by Burkhalter (1976), Carr (1978) and Patterson (1989).

5.4.3 Pretreatment

The pretreatment of oils comprises the processes to which the oils are subjected prior to the removal of free fatty acids by either chemical or physical means. The purpose of the processing is either to recover phosphatides as the by-product lecithin or to remove from the oil impurities that will interfere with other refining or modification stages.

(a) Hydration degumming

The recovery of phosphatides is normally carried out in the extraction plant and is commonly known as degumming. The process consists of the precipitation of the phosphatides, or gums, by hydration. The mixture is agitated and separated, usually by centrifugation. The effects of varying process conditions on the efficiency of phosphatide removal and the quality of the lecithin product have been investigated by List *et al.* (1981). In this work on soybean oil, it was shown that the quantity of water added should be approximately equal to the phosphatide content in the oil. The mixing time depends on the type of plant used, e.g. batch or continuous. The temperature of reaction is usually between 60 and 75 °C; darkening of the lecithin product results from the use of temperatures over 60 °C. Figure 5.2 is a flow diagram of a continuous water-degumming plant in which, after separation of the aqueous phase, the oil is dried before storage. The use of a self-cleaning centrifuge is preferable for the phase separation because of the tendency for dirt in the oil to accumulate in the centrifuge bowl.

The phospholipid fraction of crude vegetable oils obtained from first-grade raw materials is composed mainly of phosphatidylcholine, phosphatidylinositol and phosphatidylethanolamine (Schneider, 1986). When oils of this quality are water-degummed, in excess of 90% of the phosphorus content of the crude oil is removed. As a result of poor conditions of harvesting, handling, storage and processing, enzymatic action takes place, resulting in the formation of the calcium and magnesium salts of phosphatidic acid and the ethanolamine and inositol compounds. These salts are hydratable, but more slowly than the original three compounds (Segers and

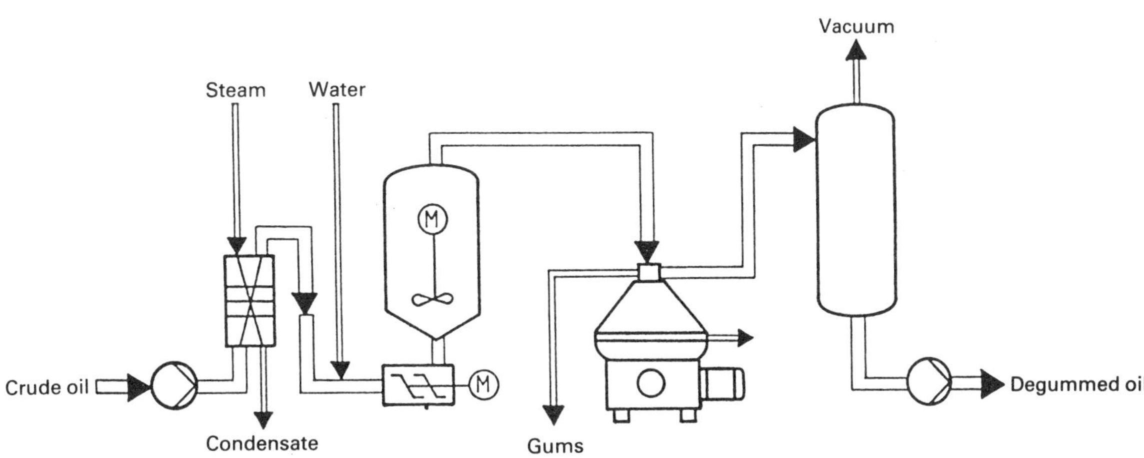

Fig. 5.2 Continuous Westfalia degumming plant (Courtesy of Westfalia Separator Ltd).

van de Sande, 1990). In consequence, the water-degumming process reduces the phosphorus content of average-quality oils to between 80 and 200 ppm, and in excess of 200 ppm in low-grade oils. The remaining phosphatides are emulsifying agents, and consequently, if not inactivated, are responsible for high losses and poor-quality products in subsequent refining. Phosphatides also have a poisoning effect on the catalysts used for hydrogenation at levels equivalent to 4 ppm and above of phosphorus.

With increasing interest in physical refining, much work was carried out on the removal of non-hydratable phosphatides and traces of iron from crude oils and, in particular, from seed oils such as soybean, sunflower-seed and rapeseed (List *et al.*, 1978a, b, c; Diosady *et al.*, 1982). The investigations proved that the decomposition of the calcium and magnesium salts of the phosphatides by using mineral or organic acids, e.g. phosphoric or citric acids, in the water-degumming process increased the hydratability of the phosphatides, and thereby reduced levels of phosphorus were obtained in the oil. Phosphoric acid should not be used in cases where the phosphatides are to be recovered as commercial lecithin, as such treatment tends to yield a dark-coloured lecithin due to charring. Rapeseed oil, which, when given a standard water-degumming treatment, will yield an oil containing 200 ppm phosphorus, will, when citric acid is used in the process, give an oil containing about 50 ppm phosphorus. The process has become known as acid degumming.

Segers and Ringers (1977) patented a development of the acid-degumming process. In this process, a strong, food acid such as citric acid is mixed with the crude oil at 70 °C for 5 to 15 min, cooled to 40 °C maximum, mixed with about 2% water, given a hydration time of 1–3 h and then quickly heated to 60–90 °C and centrifuged. Provided the heating is rapid enough, the phosphatide micelles (Sen Gupta, 1988), which when cold are in the aqueous phase, will not transfer to the oil phase. The process is known as 'super-degumming' (Segers, 1982) and produces an oil with a phosphorus level of maximum 30 ppm.

Kanamoto *et al.* (1981) reported that the degrees of hydration of phosphatidylethanolamine and phosphatidic acid are increased in the presence of phosphatidylcholine owing to the formation of mixed micelles. This association forms the basis of a method of oil purification by the addition to an oil of hydratable phosphatides. The method is the subject of US Patent 4162260.

Oils apart from soybean and rapeseed that are frequently water-degummed are corn oil and sunflower oil.

(b) Gum conditioning

Gum conditioning is the step in the chemical refining process that precedes the neutralization of the free fatty acids by alkali. Its purpose is primarily to convert the non-hydratable phosphatides into a form in which they can be removed from the oil during the soapstock separation without causing losses due to emulsification. The chemical most frequently used is phosphoric acid in concentrated (75–85%) form. Phosphoric acid decomposes the non-hydratable phosphatides, forming a dark-brown sludge. It also decomposes iron salts and has a noticeable effect on the reduction of chlorophyll in the oil. Because of the decomposition of calcium and magnesium salts, the soapstock is more easily separated, and consequently higher yields of refined oil are obtained and the quality of the refined oil is improved.

Gum conditioning is employed in batch, semi-continuous and continuous processing. The quantity of phosphoric acid used is based primarily on the phosphorus content of the crude or water-degummed oil, and lies usually between 0.05 and 0.2% of the oil weight. Larger amounts may be needed for oils containing over 200 ppm of phosphorus or for oils, e.g. rapeseed, containing high levels of chlorophyll, in which cases up to 0.5% has been used.

Although a number of different methods of alkali refining are used (section 5.4.4), the basic requirements of gum conditioning are the same in all cases. The acid must be intimately mixed with the oil so as to obtain contact between the reacting molecules. After the intense mixing it is good practice to allow a residence time of about 5 min with low or no agitation other than oil flow turbulence before addition of the alkali. The residence time permits agglomeration of the denatured phosphatide particles, which are then more easily separated as a semi-solid phase in the soapstock centrifuge. Because of the formation of this sludge, it is normal to install a self-cleaning centrifuge for the soapstock separation stage of a refining line to be used for oils of high phosphatide content.

(c) Pretreatment for physical refining

For more details on this topic, the interested reader should consult Young (1990). The advantages of physical compared to chemical refining are: improved product yield; in many cases reduced refining costs in both plant and processing; elimination of soapstock splitting; and reduction of effluent treatment costs.

The process of physical refining has been in use since the 1930s for crude oils containing high levels of free fatty acids, for example, lauric acid oils, palm oil and tallow. Its wider application has been influenced by economic and environmental factors and has been made feasible by improvements in technology and a considerable volume of research and development work.

The problem of applicability of the process is concerned with the types and quantities of impurities in the oil and the efficiency of present pretreatment methods to reduce them to acceptable levels at a cost that is competitive with chemical refining. Cottonseed oil cannot be physically refined because current pretreatment methods will not remove the gossypol class of pigments; in this context the new glandless varieties of cottonseed are of interest. Crude fish oils are not physically refined because of widely varying contents of sulphur, halogens, phosphatides, pigments and oxidation products. Phospholipid content and type is of importance in, for example, soybean and rapeseed oils. Ong (1980a), in recommendations for the quality of crude degummed soybean oil required for the chemical and physical refining processes, gave for physical refining upper limits of 1 ppm iron and 20 ppm phosphorus. Kramer *et al.* (1983) similarly recommend 50 ppm phosphorus as the upper limit for the starting oil for the physical refining of rapeseed oil, and the same figure has been quoted for crude degummed soybean oil. The levels of oxidation products, pro-oxidant trace metals and colours 'fixed' as a result of oxidation are also important in deciding whether or not an oil can be satisfactorily physically refined. The principal requirements of the pretreatment processes are that the phosphorus content should be reduced to a maximum of 4 ppm, iron preferably less than 0.2 ppm, oxidation levels be substantially reduced and non-heat-degradable pigments reduced to a level such that the colour after distillation will be satisfactory. The pretreatment process therefore consists of degumming and adsorption either combined or as separate steps.

In the simplest case, used with high-quality low-phosphatide oils, for example some lauric acid oils, some tallows and soybean oil from the Alcon process (Kock, 1983), the crude oil is bleached under vacuum with 1.0–1.5% of an activated bleaching earth at 90–110 °C for 15–30 min and filtered. In an extension of this process, the oil is first intimately contacted with phosphoric acid and/or citric acid solutions before the oil is dried and bleached. This process is used on poorer qualities of the above oils and on oils containing up to 50 ppm of phosphorus, for example, palm oil (Young, 1982) and crude degummed rapeseed oil (Kramer *et al.*, 1983).

The 'wet', or two-step, pretreatment process is used for seed oils, crude corn or sunflower-seed oils or water-degummed sunflower-seed, rapeseed and soybean oils. A typical process, in this case for corn oil, is described by Forster and Harper (1983). Crude corn oil is treated with 0.15% phosphoric acid at 60–80 °C for 20 min. After addition of water, the gums are separated by centrifuge and the oil is dried and bleached. List *et al.* (1978a, b, c) noted the beneficial effect on removal of iron from soybean oil by using phosphoric acid in combination with water-degumming and bleaching.

In the acid-degumming process, the calcium and magnesium salts of the phosphatides are decomposed, with consequent increase in the hydratability of the phosphatides. The rate of hydration of some of these molecules, notably phosphatidic acid and phosphatidylethanolamine, is still slow, so that the reaction is incomplete when the phases are separated. Nilsson-Johansson *et al.* (1988) and Dijkstra and van Opstal (1989) demonstrated that neutralization with caustic soda solution to pH 6.0 after the acid addition produces the sodium salts of the phosphatides, and that substantially the latter remain in the aqueous phase on separation by centrifuge, with consequent improved reduction of phosphorus in the oil. The process is also successful in removing iron to less than 0.2 ppm. The oil phase after the separation in the Dijkstra process contains some gums that are separated from the oil as a high-oil-content gum phase is a second centrifuge. The latter gum phase is recycled to the inlet of the first separator. The process is known as 'total degumming'.

The super-degumming and total-degumming processes, when carried out on crude or water-degummed oils of satisfactory quality, and when followed by treatment with activated bleaching earth accompanied in some cases by acid treatment, will produce a pretreated oil of acceptable quality for physical refining.

The introduction in the mid-1980s of synthetic amorphous silicas, which will adsorb soap, phosphatides and trace metals, has provided another means for the pretreatment of oils prior to physical refining. The affinity of these silicas for phosphatides is enhanced by the presence of soap in the oil. In the process, a water-degummed oil is treated with sufficient strong phosphoric acid to decompose the calcium and magnesium salts of the non-hydratable phosphatides. Caustic soda is then added to the reaction mixture to produce the desired level of soap. The mix is then dried to approximately 0.2% moisture content and agitated under atmospheric pressure with the silica for about 15 min before being bleached under vacuum with an activated bleaching earth and filtered (Welsh and Toeneboehn, 1991).

5.4.4 Neutralization, washing and drying

(a) General considerations

The objectives of the neutralization step are removal of the free fatty acids, and of the dirt and denatured phosphatides not previously separated, together with pigments and other impurities that are saponifiable. These objectives must be accomplished with minimum neutral oil losses, which occur due to emulsification, occlusion in the soapstock and saponification. The process variables include:

1. The type of alkali used, for example, sodium hydroxide (caustic soda), sodium carbonate (soda ash), ammonium hydroxide, or sodium silicate.
2. The strength of the alkali solution.
3. The excess used over the stoichiometrically required amount.
4. The temperature of addition.
5. The degree of agitation.
6. The oil flow rate through the plant.

The most widely used alkali is sodium hydroxide because of its thorough cleansing action on the oil, but sodium hydroxide when used in strong concentrations saponifies triglycerides. It has therefore been used in conjunction with the milder sodium carbonate or, for degumming purposes, with sodium silicate. Pardun (1979) has described two methods for the continuous refining of soybean oil using ammonia. The first treats water-degummed oil with aqueous ammonia followed by weak (0.5 M) caustic soda; and in the second, crude undegummed oil is treated with citric or formic acid solution followed, without gum separation, by neutralization with ammonia. The soapstock is separated and the ammonia boiled off, leaving, in the second case, a by-product of lecithin enriched with fatty acids. These processes avoid the environmental hazards of the splitting of soapstock with mineral acids. The oil from both processes is washed, dried, bleached and deodorized according to normal practice.

A strong caustic soda, for instance 20 °Bé (degrees Baumé), is effective for the removal of impurities, especially gums and pigments, but saponification losses are higher. The soap is also thicker than when using weak lye (6 °Bé), with a consequently higher loss due to oil occlusion in the soapstock.

On the other hand, emulsion losses are higher when using dilute solutions. Similarly, a large excess (30–50%) has a high purifying action but a greater tendency to saponification. Emulsions are more likely to be encountered with low excess lye usage (5–15%).

It should be noted that two ways are used to express the excess caustic soda. In American refineries, the excess is expressed in terms of dry sodium hydroxide on a percentage basis, calculated on the weight of oil (Norris, 1982; Farr, 1983). In most other refineries, the excess is stated as a percentage of the weight or volume of the caustic soda solution stoichiometrically equivalent to the free fatty acid content of the oil. The latter method is used in this chapter.

A long contact time, high mixing intensity and high temperature are responsible for high saponification losses, but a long contact time helps to remove pigments and high temperatures improve phase separation.

A double neutralization technique is often used for the treatment of poor-quality oils. In the first stage a strong, for example 20 °Bé, caustic soda is used and, after soap separation, the oil is given a second lye treatment using strengths between 6 and 20 °Bé, depending on the quality of the oil.

The modern fully continuous centrifugal refining line allows for many variations of throughput, mixing speed, contact time and temperature, so that the influences mentioned can be utilized to optimize quality and yield for a wide variety of feedstocks.

After separation of the soapstock from the oil, the oil contains about 2000 ppm of soap from a batch process using simple gravity separation, and generally between 500 and 1000 ppm following centrifugal separation. The soap is removed primarily by washing with water. The water for this purpose and for lye make-up should be soft; steam condensate is suitable and it should be very hot, always hotter than the oil and preferably over 90 °C. The quantity of water used is about 8% of oil weight per wash in batch refining, and between 10 and 20% of oil flow in the continuous process. To break emulsion formation in batch treatment, salt can be either added to the oil prior to the first wash or dissolved in the first wash water to make a brine solution of about 5% strength.

Where two washing stages are used in centrifugal refining, plant manufacturers offer a countercurrent washing system in which the spent wash water from the final centrifuge is used again to wash the oil in the first washing step.

The importance of thorough washing lies in the fact that soap is adsorbed on bleaching earth, thus reducing the surface area available for the removal of other impurities. Soap also has a catalytic action in the hydrolysis of triglycerides. The efficiency of water washing is to a large extent dependent on the removal of phosphatides with the soapstock, but with a well refined oil the soap content in the batch process will be about 100 ppm after three washes, and less than 50 ppm after two washing stages in a continuous plant. In order to eliminate soap from the washed oil, citric or phosphoric acid may be added to the final wash water or a citric acid solution can be added to the oil prior to drying. The oil after washing is normally at about 90 °C and is dried at an absolute pressure of 25–50 mmHg to a water content of less than 0.1%.

The soapstock and water washes are combined and 'split' using mineral acid, usually sulphuric acid. The oil phase, known as 'acid oil', is settled, and in some cases washed, to achieve the required specification and is sold as a by-product. The aqueous phase is run off to the effluent system.

The affinity of the amorphous silicas for soapstock has been used in the modified caustic refining process (Bogdanor and Welsh, 1987) to eliminate the need for at least one washing stage, with consequent reduction of power and effluent loading. After phosphoric acid and caustic soda additions, the soapstock is separated centrifugally. Provided that the soap content of the oil is less than 600 ppm, the oil can be treated with the silica without washing and carried on to the bleaching stage. Soap levels in excess of 600 ppm require a water wash to achieve the desired level.

(b) Plant

The batch neutralizer–bleacher has to a very large extent been superceded by semi-continuous and fully continuous refining lines because of the improved oil quality and yield, better process control and lower costs possible with the latter units. The batch plant has still, however, a very useful part to play. In large refineries it can be used for treating the inevitable small amounts of oil that accumulate and for refining oils that either are of very bad quality, for example tank bottoms, or are required in such low quantity that they would seriously interfere with the production output of the high-capacity continuous plant. Its advantage as a comparatively low-cost means of oil refining for the small-tonnage refiner is more obvious, particularly when, as is the case in many countries, the crude oil quality is very variable.

Two basic systems are in use for the continuous centrifugal refining of edible oils (Haraldsson, 1983; Hendrix, 1990). They differ in the contact time and temperature of initial contact between the oil and the caustic soda solution. In the USA, the lye is added to the oil at about 30 °C and the oil–lye mixing time is between 5 and 15 min. At the end of this time the mixture is heated to 75–80 °C to coagulate the soap and is then immediately separated by centrifuge. This is known as the 'long-mix' process. In Europe and many other parts of the world, the 'short-mix' process is used. A flow diagram of a continuous caustic soda refining plant is shown in Fig. 5.3. In this case the oil is heated to 90 °C, treated with phosphoric acid for gum conditioning and then mixed with lye for 1–15 s before separation. Re-refining, if required, washing and drying are similar for the two systems. The continuous refining line can be assembled to various degrees of complexity, which are dictated by the types of oil to be refined and the quality of those oils. The principal variations are the time and the means of contact with phosphoric acid in the gum conditioning step, and the number and type of centrifuges employed for separation. In its simplest form the line incorporates two solid-bowl disc centrifuges. The first is used for soapstock separation and, to aid the removal of soapstock from the bowl, is equipped for flushing the bowl periphery with water. The second centrifuge is used for water washing. This type of line may be used for the treatment of clean, low-phosphatide, high-quality oils that can be refined easily, for example good-quality soybean and palm oils. However, dirt in the crude oil and denatured phosphatide sludge from gum conditioning hamper efficient line operation by accumulating on the surface of solid bowls. It is therefore frequent practice for soapstock separation to use a split-bowl centrifuge with programmed periodic discharge of the dirt.

In most parts of the world the variability of oil quality or refining flexibility of oils refined require that the continuous line should contain three separation stages. This is the most commonly installed form of the line and is the one depicted in Fig. 5.3. For reasons stated above, the first centrifuge, which is used for soapstock separation, is of the split-bowl automatic-discharge type. The second stage is used either for a second alkali refine or for a first wash. The plant is therefore equipped at this stage for the addition to the oil of either lye or water, and the centrifuge has a solid bowl with the possibility of using bowl-flush water. The third separation stage is used only for water washing and therefore does not require the bowl-flush water system. After washing, as stated earlier, the oil may be mixed with citric acid solution to decompose remaining traces of soap before entering the drier where water is removed from the oil by cascading or being sprayed under vacuum.

Dirt in the crude oil and the sludge resulting from gum conditioning tend to stabilize emulsions formed after the addition of the neutralizing lye and may be carried with the soap to the soapstock splitting plant where again they stabilize emulsions and necessitate the use of additional acid. The self-cleaning centrifuge for soap separation helps to reduce these problems, but some refiners have found it beneficial on quality and yield grounds to install a self-cleaning centrifuge after gum conditioning and before lye

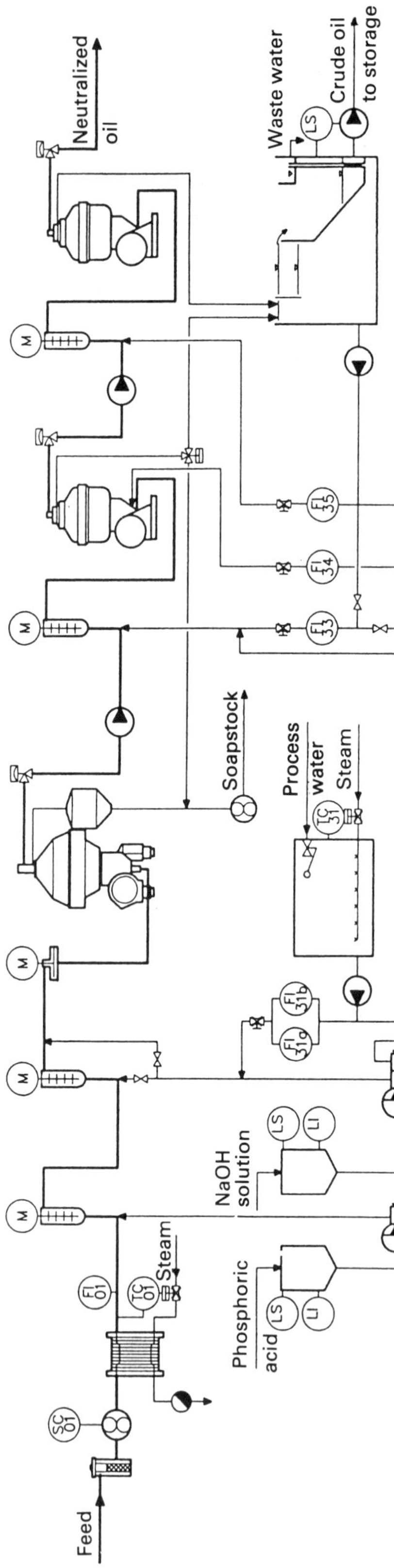

Fig. 5.3 Continuous three-stage caustic soda refining plant (Courtesy of Alfa Laval Engineering Ltd).

addition. In this case a solid-bowl centrifuge with bowl-flush arrangements is adequate for soap separation. This configuration would therefore consist of four separation stages: degumming (self-cleaner); first refine (solid bowl with bowl-flush); second refine or first wash (solid bowl with bowl-flush); and first or second wash (solid bowl).

Miscella refining (Cavanagh, 1990) is a special case of continuous caustic soda refining. The most favourable conditions for the process are those of the solvent extraction plant, where it is also desired to refine the oil and perhaps process it further by winterization. For this reason cottonseed is the oil most usually processed in this manner. The miscella from the extraction plant is evaporated to the desired oil:hexane proportion, approximately 65:35 in the case of cottonseed oil. The corrected miscella is then pretreated with phosphoric acid, neutralized and the soapstock centrifugally separated as for the normal 100% oil process. Soapstock separation is very efficient and therefore the miscella does not need to be washed. The solvent is evaporated from the refined oil and the oil is bleached and deodorized. The process is capable of giving higher yields of refined oil and improved colour of the bleached oil by comparison with non-miscella refined oils.

Alkali refining is also carried out in a semi-continuous manner, the Zenith process (Cavanagh, 1990). In this process small batches of oil are measured and treated with phosphoric acid in a pre-treatment unit. From this unit the oil is pumped continuously via a centrifuge, which removes dirt and denatured phosphatides, to one of two neutralizer vessels, which are used alternately. The oil enters the neutralizers, which contain a dilute (0.3 to 0.8 M) caustic soda solution, via a dispersing system and rises through the lye in streams of small droplets. In rising through the lye the free fatty acids are neutralized and the soap thus formed is dissolved in the very weak solution. The oil settles out on top of the lye before passing continuously to a citric acid dosing and bleaching unit, followed by filtration. The oil leaving the neutralizer contains less than 100 ppm of soap and therefore does not need to be washed. The traces of soap are decomposed by the citric acid before the oil is bleached. The removal of dirt before neutralization and the particularly mild neutralization conditions result in high yields being obtained in the refining of oils that are readily saponified, for example the lauric acid oils and palm oil.

5.4.5 Bleaching

(a) General considerations

For many years the bleaching process (Mag, 1990), which consists of contacting the oil with an adsorbent material, was considered purely from the standpoint of the removal of pigments from the oil. These pigments are primarily carotenoids, chlorophyll, gossypol and related compounds, and the products of degradation and condensation reactions that occur during the handling, storage and treatment of the raw materials and the extracted oils. It was later realized that activated adsorbents, in particular, are responsible for removing, at least partially, other impurities such as soap, trace metals, phosphatides and compounds containing sulphur. Oxidation levels are also reduced by breakdown of the hydroperoxide primary oxidation product on the adsorbent surface, followed by adsorption of the carbonyl compounds that are the secondary oxidation products. Owing to the bleaching action that occurs during modern high-temperature deodorization, it is these latter properties of the adsorbent which have become the more important.

Originally the bleaching process involved the use of a natural bentonite clay known as fuller's earth (Richardson, 1978). Later, activated bleaching earths or clays were introduced, made by washing a bentonite clay containing a high proportion of montmorillonite with either sulphuric or hydrochloric acid. The effect of the acid treatment in leaching out minerals from the natural clay is to reduce particle size (increase specific volume) and increase the volume of the micropores (Fahn, 1973). The changes in surface structure increase the adsorption activity but also increase the oil retention of the used earth. However, when bleaching to a specification, the increased adsorption power of the activated earth enables less earth to be used and consequently results in a reduced loss of oil by comparison with treatment of the oil using a natural earth.

Other adsorbents that are available to the refiner for the removal of impurities in the oil are activated carbon and amorphous silicas. The use of these materials is complementary to activated bleaching earth in most cases because of a degree of selectivity of action. The silicas, as stated in section 5.4.3(c), are noted for their affinity for soap, phosphatides and trace metals, but they have little or no effect on pigments, e.g. chlorophyll and carotene, or on oxidation products. It is therefore possible (Welsh *et al.*, 1990) to use the silica to remove soap, phosphatides and trace metals, and thereby make the whole of the bleaching earth activity available for the removal of pigments and oxidation products. For the best effect, the treatments must be sequential, but it is not necessary to filter the oil between the two steps. The modified bleaching process, as it is called, enables an overall reduction in quantity of adsorbent used, thus reducing oil loss and spent cake disposal.

Activated carbon was used extensively in the past, but the improvement of the activity of bleaching earths has rendered the use of carbon unnecessary except for specialist purposes. One such purpose is the removal from certain oils of traces of polyaromatic hydrocarbons (PAHs), which are not adsorbed by bleaching earths. Again, a sequential process is best in which the oil is first treated with the earth and then, after about 15 min, with the carbon.

The mechanism of the adsorption reaction is complicated and has been investigated (Norris, 1982). Brimberg (1981, 1982), in work on the kinetics of bleaching rapeseed and palm oils, has postulated that the pigments are particulate and are colloidally dispersed in the oil. The rate formula for the reaction,

$$\log(c/c_0) = -k\sqrt{t}$$

is characteristic of the flocculation of colloids, where t is the time from the addition of the clay, c is the concentration of the pigment at time t, c_0 is the concentration at time $t = 0$ and k is the rate constant. The evidence indicates initial destabilization of the colloidal suspension by deposition or adsorption on the earth surface of the surface-active lipids, which act as stabilizers, together with associated pigment particles. When sufficient stabilizer has been removed from the oil, the pigment particles coalesce and are in turn adsorbed until an equilibrium is reached.

Segers (1983) described the properties of activated bleaching earth as:

1. Adsorption capacity, depending on the surface area, pore size and pore volume.
2. Catalytic activity, dependent on the degree of acid treatment and temperature of bleaching.
3. Ion-exchange properties, dependent on acid activation for the replacement of aluminium ions by protons, which give the earth certain metal-binding properties.

The bleaching process must be carried out in the absence of oxygen. There are indications that the activated clay may act as a catalyst for oxidation if oxygen is present during the process. Without this catalytic possibility, oxidation would in any case occur in the presence of oxygen at the elevated temperature used. The process is therefore carried out under steam or a nitrogen blanket or, more usually on plant scale, under vacuum.

Whereas vacuum has a beneficial effect in the removal of oxygen, its effect in removing water from the system is more open to question. Activated bleaching earths contain about 8–12% of 'free' moisture, which is driven off at 120–130 °C under atmospheric pressure, and about 6–7% of 'structural' water, which is only removed at 450–650 °C. The oil after vacuum drying contains less than 0.1% moisture and therefore if 1% of earth is used for bleaching, the system will initially contain approximately 0.15–0.20% of available or free water. Brimberg (1981) in bleaching rapeseed oil increased the initial moisture content of the system from 0.056% to 0.477%, with consequent increase in both the rate of reaction and overall removal of carotene and chlorophyll. However, in later work on palm oil (Brimberg, 1982), the best results on carotene adsorption were obtained with 0.1% moisture in the system. Zschau (1982) has reported that, although the colour of palm oil after bleaching with earth and water was not as good as with activated earth alone, it was better after the subsequent deodorization step. Following experiments on the pretreatment of soybean oil for physical refining, Kock and Penk (1981) have reported that the best results in the removal of phosphatides and colour were obtained when equal proportions (1% each) of activated bleaching earth and water were used, and that the effect increased with increasing activity of the bleaching earth employed. They ascribed the effect to the leaching of the acid from the activated earth into the aqueous phase followed by attack of the acid on the non-hydratable phosphatides.

Apart from the pressure of the reaction, other process parameters of importance are the temperature and the earth–oil contact time. Temperatures between 80 and 180 °C have been advocated for the process. Adsorption effects are usually improved by the use of high temperatures, but in the treatment of low-quality oils the reverse can be true (Zschau, 1982). Because of damage to the oil caused by the ingress of air in industrial plant, the temperatures employed are normally between 90 and 120 °C.

From the kinetic studies reported by Brimberg (1981, 1982) the equilibrium stage of the adsorption of pigments is substantially reached after 4–40 min for activated earths, depending on the concentration and activity of the earth and the temperature of the process. Industrially, contact times of 10–30 min prior to filtration are employed, but in a patent Mag (1980) claims that efficient adsorption can be achieved in less than 3 min contact time. In the work on palm oil, Brimberg (1982) showed that degree of agitation was not important, but it is obvious that, in systems employing short contact times, the agitation should be such as to enable contact between the adsorbent and the impurities.

The quantity of activated earth required is dependent on the quality of the oil, the activity of the earth and the process conditions, and is normally between 0.2 and 2.0% of the weight of the oil. Apart from the cost of the earth and the oil retention on the spent earth, it is advisable to use as little earth as possible because of disposal problems and because of an increase in the level of conjugation of double bonds in the fatty acids of the oil and a reduction in the level of tocopherol in the oil, both of which tend to render the oil less stable to oxidation.

(b) Plant

The bleaching process is carried out in batch, semi-continuous and fully continuous plant. In the batch process the bleaching step is often carried out in the same vessel as the neutralization, washing and drying stages. The vessel is referred to as a neutralizer–bleacher. Alternatively, a separate vacuum bleaching vessel is used. The advantages of batch treatment are as described for batch neutralization in section 5.4.4(b). In multi-oil refineries, batch operation reduces the possibility of inter-oil contamination. The batch operation is not as efficient as the other two methods, involves long earth–oil contact time during the filtration step and, because the batch is usually open to the atmosphere during filtration, the oil is more prone to oxidation damage.

A continuous plant is shown in Fig. 5.4. The oil is pumped via heat recovery and steam heat exchangers to a mixer, where water, citric acid or other solution can be added before an adsorbent is dosed from the silo (603) into the mixer (635). The oil in the mixer is protected against oxidation by a blanket of steam emanating from the adsorbent. More than one of these combinations of silo and mixer can be included if it is required to use adsorbents sequentially. The oil–adsorbent mix is sucked into the bleacher (621/622) by the reduced pressure in that vessel. During its residence of about 15 to 30 min in the bleacher, the oil is agitated by live steam, thereby maintaining an atmosphere of steam in the process. At the end of the required bleaching time, the oil is filtered in one of the two tank filters, passed through a residue or guard filter and cooled before storage.

One of the main functions of the bleaching process is the reduction of the trace-metal content of the oil, in particular of iron and copper. Materials of construction are therefore of importance, and wide use is made of stainless steel in the construction of continuous and semi-continuous plants.

The method of addition of the earth to the oil is important. This can be effected in three ways: the earth is added as a powder to the main stream of the oil; the earth is slurried with a minor proportion of the feed oil and the slurry is then dosed into the main stream of the oil; or the slurry is made using bleached and filtered oil, which is then added to the main stream. Brimberg (1982) has indicated that in the first method of earth addition the bleaching effect is achieved more rapidly than when using a slurry.

The spent earth is removed from the oil by the use

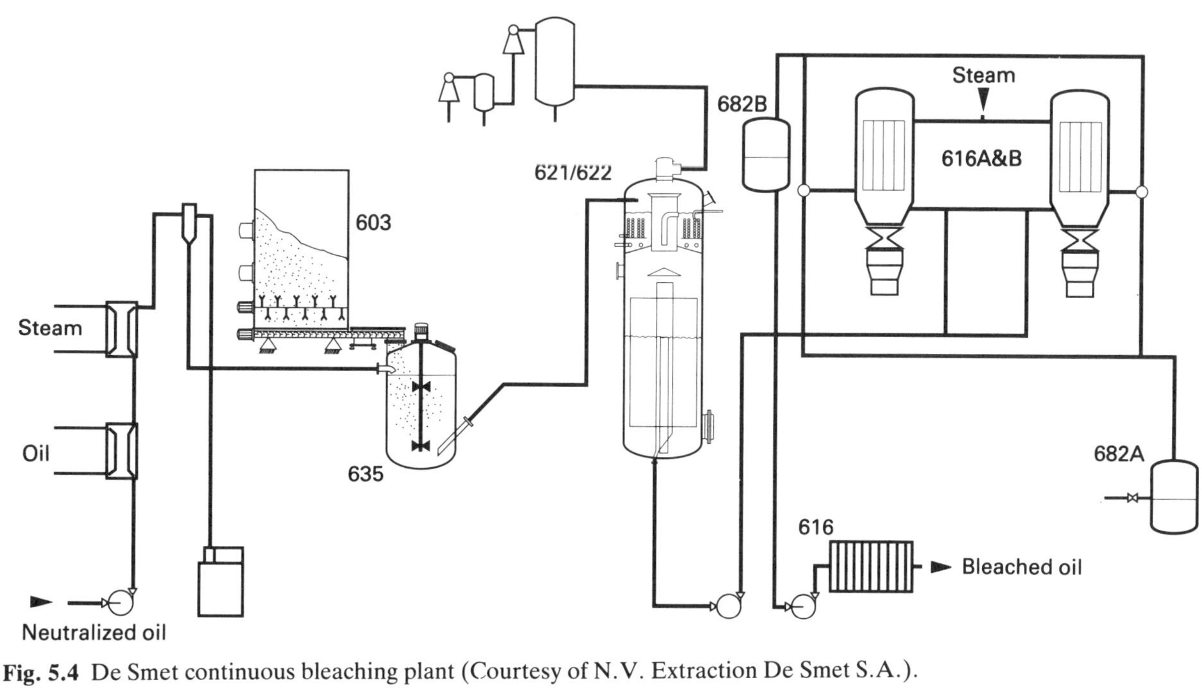

Fig. 5.4 De Smet continuous bleaching plant (Courtesy of N.V. Extraction De Smet S.A.).

of various types of filter. Examples are:

1. Plate and frame presses with open or closed discharge.
2. Chamber presses, sometimes fitted with diaphragms for squeezing the oil from the spent filter cake.
3. Vertical or horizontal tank filters with vertical leaves of stainless-steel mesh.
4. Vertical tank filters with horizontal leaves attached to a central shaft, which can be rotated for cake discharge.

All of these filters are preferably backed up by a polishing or 'guard' filter to prevent the accumulation of spent earth in the bleached-oil storage tanks. The oxidation products on such spent earth promote further oxidation of the stored oil. Similarly care must be taken to prevent contamination of the bulk of the filtered oil with low-quality oil obtained during the blowing, by nitrogen or steam, of the filter cake. Such blowing is carried out to reduce the loss of oil in the cake. The recovered oil should be returned to crude storage or be rebleached, depending on its quality. Oil retention in the blown cake is affected by the degree of activation of the earth and the type of filter and blowing medium used. It lies between 20 and 40% by weight of the cake. There is a risk of spontaneous combustion if the oil content of the cake from the filtration of highly polyunsaturated oils is reduced below 30%. Ong (1983) has discussed briefly the recovery of oil from spent bleaching earth.

An efficient adsorption action can be obtained by the so-called 'press effect'. In this operation a calculated amount of bleaching earth sufficient for a particular tonnage of oil is precoated on the filter. The rest of the oil is then passed through the precoated bed on the filter leaves. Although the contact time between oil and earth is very short, the relative concentration of earth to oil is very high compared to a conventional bleaching slurry. The press effect is useful as part of the modified bleaching process (Welsh *et al.*, 1990), as it enables the bleacher to be used for the silica, so that the phosphatides, soap and trace metals have been removed from the oil before it reaches the earth on the filter. The earth's efficiency, for e.g. chlorophyll removal, is thereby enhanced.

5.4.6 Deodorization and physical refining

Deodorization is the final major stage in the refining of edible oils (DGF, 1981–92). Its purpose is to produce a product of bland flavour and odour and of good shelf-life stability. To accomplish this, the process must remove from the oil free fatty acids and the aldehydes and ketones that are formed by oxidation of the unsaturated fatty acids and are principally responsible for the unacceptable taste and smell of the oil after bleaching (DGF, 1981–92, I). Also removed in part are pigments, mainly carotenoids by thermal breakdown, tocopherols and tocotrienols, which are naturally occurring antioxidants, and other unsaponifiable compounds such as hydrocarbons and alcohols. Some of these display pro-oxidant activity. Thomas (1982a, b) has reported on the almost complete removal under modern deodorization conditions from the oil of nitrosamines, traces of extraction solvent, organochlorine pesticides and volatile sulphur compounds.

The distillation stage of physical refining is a special case of deodorization. The objectives are the same, but the quantity of free fatty acids to be removed is greater, generally between 0.5 and 6%.

The ability of the deodorizing step to produce an oil of good quality is limited by the quality of the crude oil, the processing and storage treatment of the oil prior to deodorizing and the deodorizing plant itself.

The process does not effect a reduction in trace metals or phosphatide levels. If the oil has been neutralized with caustic soda, soap removal should be complete at the bleached and filtered stage, otherwise high losses will be encountered due to foaming in the deodorizer. The deodorizer will only reduce the total oxidation, or totox, value (totox value = 2 × peroxide value + anisidine value) by about 50%, and so air contact in prior refining equipment must be minimized, and during storage of the bleached oil the temperature should be held below 50 °C for soft oils and not more than 10 °C above the melting point for hard fats.

Bleaching by thermal degradation with distillation of the breakdown products is a feature of the modern process. Loncin *et al.* (1971) showed that the carotene content of palm oil is reduced after 40 min at 240 °C to 1% of the starting level. However, they also reported that a crude oil stored for 7 days at 75 °C in an oxygen atmosphere yielded a product of colour 3.3 red and 35 yellow (Lovibond 5¼ inch (~ 13.5 cm) cell), whereas the colour of the control product was 1.6 red and 18 yellow.

(a) Process theory and description

(i) Theory

Commercial deodorization is carried out by steam distillation under vacuum.

The mathematics for batch deodorization were developed by Bailey (1951) and modified by Szabo Sakardi (1958) by the introduction of an activity coefficient A to allow for the deviation of fatty acid solutions from the ideal, thus producing the following equation:

$$S = \frac{PO}{EP_v A} \ln\left(\frac{V_1}{V_2}\right) \quad (5.1)$$

or

$$\ln\left(\frac{V_1}{V_2}\right) = \frac{EP_v AS}{PO} = K\frac{P_v S}{PO} \quad (5.2)$$

where S = moles of stripping steam, O = moles of oil, P_v = vapour pressure of the free fatty acid, P = total system pressure, V_1 = initial number of moles of free fatty acid in the oil, V_2 = final number of moles of free fatty acid in the oil, E = vaporization efficiency, A = activity coefficient and $K = EA$ = experimental constant.

Gavin (1978) derived an equation for the operation of continuous deodorization:

$$\frac{V_1}{V_2} = 1 + K\frac{P_v}{P}\frac{S}{O} \quad (5.3)$$

where V_1, V_2 = moles of free fatty acid in the oil entering and leaving the deodorizer, S = moles of stripping steam per hour, O = moles of oil per hour, and K, P, P_v are the same as for batch deodorization (as equation (5.1)).

The vapourization efficiency is a measure of the steam's ability to become saturated with the free fatty acids and other volatile substances during its passage through the oil and, in modern continuous or semi-continuous plant with provision for intense mixing, approaches unity.

The activity coefficient is usually determined experimentally and is dependent on fatty acid type and concentration. It can also be calculated from an equation by Wilson (1964) as exemplified by Stage *et al.* (1979), who produced a curve showing that, for mixtures of stearic acid and tristearin containing up to 5% of the acid, the activity coefficient lies between 0.63 and 0.65.

Equations (5.2) and (5.3) govern the relationship between initial and final content of volatiles and the three main process variables, namely system pressure, steam-to-oil quantity ratio and temperature. The temperature is proportional to the vapour pressure P_v of the free fatty acid or other volatile component.

(ii) Deaeration

Bearing in mind the necessity for protecting the oil against oxidation at all stages and the increase in rate of oxidation with increasing temperature, it is advisable to remove air from the oil prior to heating. This is done either by pumping the pretreated oil to the first heating stage through a deaerating vessel connected to plant vacuum, or by cascading the oil as a thin film into the deodorizer, which is under vacuum.

(iii) Heating

Following deaeration the oil is heated to the required process temperature. The need for energy conservation has seen the introduction of heat-exchange procedures, which can, depending on the type of plant, save between 50 and 80% of heating costs.

The oil temperature after heat exchange will vary between 160 and 220 °C depending on the plant and process conditions used. Final heating to the required process temperature is effected by high-pressure steam or other heat-exchange medium such as thermal oil or the eutectic mixture of diphenyl and diphenyl oxide widely known as 'Dowtherm A'. When carried out in an internal section or tray of the deodorizer, sparging steam is used to improve efficiency of heat exchange, prevent burning of the product and initiate the deodorization process.

(iv) Deodorization/distillation

As stated earlier, the deodorization process is concerned with system pressure, temperature and steam/oil ratio.

It can be seen from equation (5.1) that, for fixed values of free fatty acid in feed and product (V_1 and V_2), halving the system pressure halves the requirement for stripping steam. However, because of energy costs involved in achieving high vacuum, most commercial plants operate at 2.5–5.0 Torr. The vacuum should be the same in the various stages of the process so as to reduce the chance of refluxing.

The temperature for deodorization and deacidification is dependent on the vapour pressure of the compounds to be removed. Thus for low-molecular-weight fatty acids as found in palmkernel oil, a temperature of 185 °C in batch equipment and 200 °C in continuous or semi-continuous plant will suffice for deodorization. Although, as stated earlier, carotene is substantially destroyed and removed at 240 °C, in practice temperatures of up to 270 °C are required to break down compounds formed by oxidation during storage. Temperatures should, however, be kept as low as possible, first to reduce losses and secondly to reduce the possibility of isomerization and other thermochemical reactions (Rossell *et al.*, 1981; Eder, 1982). It should also be remembered that it is in the interest of the refiner to minimize the loss from the oil of natural antioxidants, tocopherols and tocotrienols. There is a point at which the advantage to shelf-life of removal of aldehydes and ketones is cancelled out by the loss of these antioxidants.

The quality of stripping steam frequently does not receive the attention it merits. The boiler feed water used to produce the steam should be passed through an efficient deaerator and be chemically treated so that the steam contains no air, and boiler operation must not permit boiler water solids to become entrained in the steam. The steam should be dry to prevent losses due to foaming, but need not be superheated.

The quantity of steam used is dependent on system pressure and temperature and is normally between 1 and 4% of oil flow or quantity. The vaporization efficiencies of the most widely used plants employing mammoth pumps, bubble caps or perforated plates are similar.

It can be seen from equation (5.1) that the volume of steam required for deodorization is directly proportional to the system pressure and inversely proportional to the vapour pressure of the free fatty acid. Thus, a reduction in the former and an increase in the latter, brought about by increasing temperature, result in a reduction of time on temperature for a set steaming rate. This is correct for the simple reduction of fatty acid levels. However, oils vary in their content of pigments and oxidation products. Practical experience has shown that, whereas these products can be removed in the time required to reduce free fatty acid to the desired level from a good-quality feed oil, this is not so with oxidized oils. For such oils, an extended time at the selected temperature is required to allow thermal reactions to take place in which some of the oxidation products are further decomposed and the derivatives removed from the oil (Andersen, 1962; Brekke, 1980). If such oxidation products are not removed, the deodorized product will have a poorer taste and reduced oxidative stability. The limitations of this aspect of the deodorization process can be noted in the fact that the anisidine value, which is a measure of secondary oxidation products in the oil, is not reduced to zero. Commercial plants are currently designed for holding time on temperature of 30–120 min, but all are capable of extension.

(v) Cooling

In modern plant the first stage of cooling is achieved by heat exchange against the ingoing feed oil. Cooling is continued from a temperature in the region of 120–150 °C by heat exchange against water, either internally under vacuum with stripping steam or externally against feed oil and/or water. The temperature of the cooled oil is usually 60–65 °C. For stability reasons further cooling to 30 °C for liquid oils and 10 °C above the melting point for fats is recommended. Citric acid may be added at about 120 °C during the cooling of the oil for the chelation of trace metals. If desired, antioxidants may be added at the same stage.

Two factors should be remembered about the cooling stage. First, at 250 °C the oil is at its most sensitive as far as oxidation is concerned and therefore should be protected completely against air leaks. Secondly, the compounds responsible for off odours and flavours may remain in the oil after the main stripping stage of the deodorizer or develop during the initial part of the cooling. These compounds will be removed from the oil if the cooling is carried out under vacuum and with sparge steam, or if the oil is given a further steam and vacuum treatment following the initial cooling.

(vi) Process losses (Bailey, 1951)

Quantitatively the most important impurities to be removed during deodorization are the free fatty acids, which are present in the feed oil at about 0.1–0.15%. The theoretical loss is therefore of the order of 0.1%, but actual losses are about 0.5%. The most important source of these losses is the carry-over of neutral oil droplets entrained in the vapour leaving the deodorizer. The entrainment is caused by the velocity of the stripping steam.

Increasing temperature and reducing system pressure are also responsible for increased losses if the stripping steam flow is not reduced to compensate for the resultant increase in steam velocity.

(vii) Environmental factors (Gilbert and Tandy, 1979; Athanassiadis and Raes, 1986)

The entrained neutral oil is carried to the vacuum equipment and thence via the condenser water to the drain. To reduce contamination significantly, it is usual to place a vapour scrubber between the deodorizer and the vacuum equipment. These units make use of co- or countercurrent spraying of oil and demisting devices to condense and coalesce fatty material in the vapour. The level of fat in the condenser water is reduced from about 300 ppm to between 10 and 50 ppm depending on the molecular weight of the fatty acids.

To save water, the condenser water is normally recycled via cooling towers. The water in the cooling tower gives off the objectionable smelling compounds taken from the oil, and these are airborne to the neighbouring district. To overcome this problem, a number of refiners in city areas have installed a double-loop system in which pure water is circulated through one of two plate heat exchangers used alternately where the condenser water is cooled before being returned to the vacuum equipment. The pure water is cooled in the cooling tower, which thus remains clear of fat. The heat exchanger must be cleaned periodically with hot water or hot caustic soda solution to remove the fat that builds up on the plates.

(b) Plant

This topic is reviewed by Stage (1981), Gavin (1981) and DGF (1981–92, III), to whom the interested reader is referred.

(i) Batch units

The batch deodorizer is a very useful vessel for a refiner. It can be used for small parcels of oil, which, if introduced into the programme of a semi-continuous or particularly a continuous deodorizer, would reduce the latter's efficiency. Further, the chance of contamination can virtually be eliminated, which is of value in a plant processing oils whose special properties would be affected by very low levels of contamination.

The modern batch deodorizer has a capacity from 5 to 30 tonnes. It is still frequently constructed of mild steel, though stainless steel is preferred. It is heated by high-pressure steam or, in some cases, by Dowtherm or thermal oil, and its vaporization efficiency has been improved by the inclusion of mammoth pumps and effective head space insulation to prevent refluxing. A vapour scrubber can be installed to recover oil and fatty acids from the distillate vapour. The complete batch cycle time is between 4 and 8 h depending on the feedstock. The unit is not, however, suitable for physical refining and it is expensive to operate compared with continuous and semi-continuous plant.

(ii) Continuous and semi-continuous units

The history of deodorization is a continual search for increased efficiency of operation and improvement of product quality, which has resulted in the following designs.

1. Continuous: the oil flows continuously through the plant, receiving a part of the process treatment described above, for example heating, stripping, or cooling, in discrete sections of the plant.
2. Semi-continuous: relatively small batches of oil are collected and treated in the same manner; the oil flow from the plant is discontinuous.
3. Continuous/semi-continuous: the oil flows continuously through the deaerating and heating stages and is then converted to a batch system for the main deodorization or stripping stage; following this the oil flow becomes continuous again for cooling and discharge.

The continuous process has the advantage of simplicity and is particularly suitable for refineries operating on a few oil types, permitting long runs on individual oils. The consumption of utilities is constant throughout these runs and not subject to the peaks of demand that characterize a batch process, and therefore its energy efficiency with regard to heat exchange is greater.

The semi-continuous unit has the advantage of a defined dwell-time at the deodorization temperature rather than the probability spread of process times of even a well-baffled continuous plant. It is also suitable for frequent feedstock changes without suffering heavy production loss and can be operated so that contamination between feedstocks is very low.

The continuous/semi-continuous plants have been introduced to combine some of the advantages of both plants. They have a finite dwell-time at the stripping stage and have the same advantages as the continuous plant for heat exchange. The problem of feedstock change is the same as in the case of the continuous unit. This problem has of recent years been somewhat reduced owing to the practice by refineries of deodorizing blends rather than individual oils when, for example, supplying attached margarine factories. In this way the number of stock changes required and, by good programming, the effects of contamination, are reduced.

Whereas the semi-continuous unit is most logically designed as a vertical arrangement so that parcels of oil drop by gravity from one tray to the next, the continuous plant can just as effectively be designed using a horizontal still. Recent design modifications based on the older plant geometry have concentrated on heat conservation and methods of reducing size and therefore cost. Improved efficiency of steam distribution has resulted in the use of one deodorizing tray where previously two were used.

One of the problems of batch deodorizers is the chance of air leaks directly into the hot oil, for example through the drop valve at the bottom of the vessel. To overcome the problem, continuous and semi-continuous deodorizers were designed with individual small tanks or trays fabricated from stainless steel enveloped in mild-steel shells. By this method any leaking air did not come into contact with the hot oil. The principle is still valid today, but the cost of the double-shell vessel is high, particularly when used for physical refining, for which process, owing to the high volume of fatty acids in the distillate vapour, stainless steel must be used for the trays, shell and vapour take-off piping. To reduce cost, many deodorizers are produced with a single stainless-steel shell and designed to minimize the possibility of air–oil contact.

Plants are available with capacities ranging from 5000 lb/h (about 50 tonnes/day) to 60 000 lb/h (over 600 tonnes/day) and are automated to a very high degree. All the semi-continuous and continuous deodorizers are available, suitably modified, for physical refining.

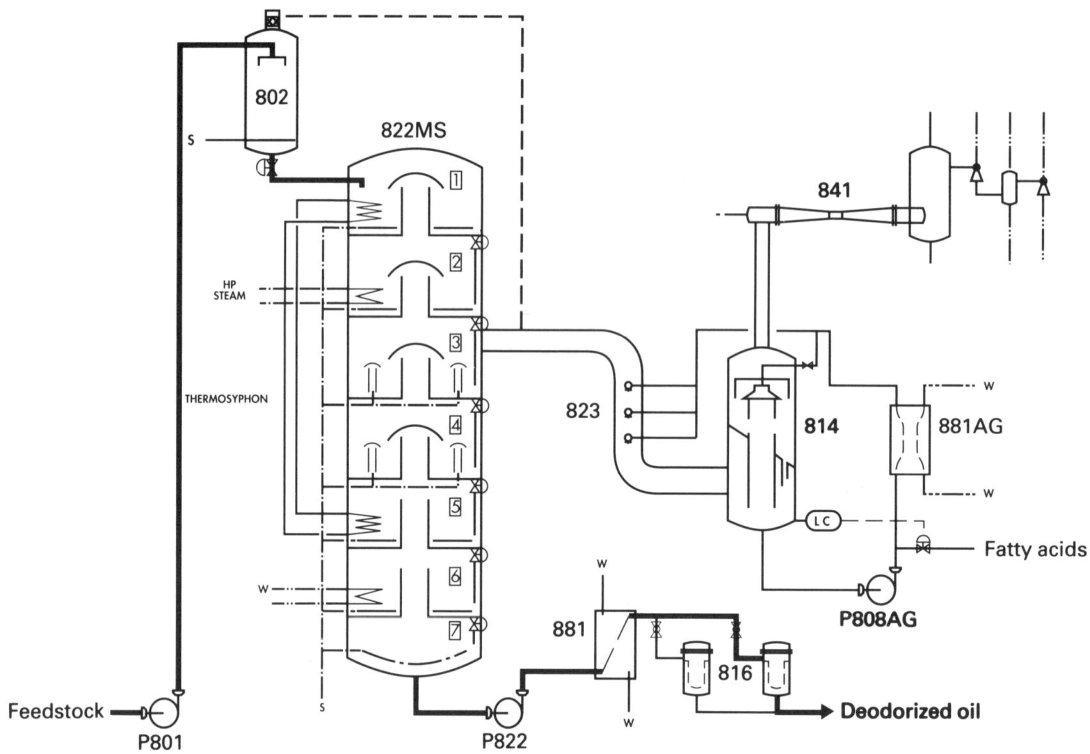

Fig. 5.5 De Smet 'Multistock' deodorizer (Courtesy of N.V. Extraction De Smet S.A.).

Figure 5.5 illustrates the single-shell semi-continuous deodorizer, which is also used for physical refining. Such a vessel with a capacity of 400 tonnes/day would have a batch size of 6.25 tonnes and a cycle time of 22½ min for deodorization and 25 min for physical refining. The operating pressure in the deodorizer could be 3 mbar using steam jets with a total condensation system (Hupe *et al.*, 1989) to minimize effluent problems. Live steam usage at that pressure is of the order of 1.2–1.3 kg/tonne oil. The oil is deaerated (802) before passing to the main vessel. In the first tray, the oil is heated with live steam agitation and by a water-filled heat recovery loop to about 170 °C. In tray 2, heating is continued by high-pressure steam to the temperature required, taken in this example to be 250 °C. The next two trays are the main distillation section where the majority of the live steam is used and, as a result, the temperature of the oil leaving tray 4 is about 245 °C. Tray 5 is for heat recovery, as also is tray 6, but in the latter case the heat is recovered in the form of hot process water. Citric acid is added in tray 6. The oil is finally cooled and passed through a polishing filter. The unit is fully automated for up to 10 stock changes per day.

5.4.7 Novel refining methods

Brief mention should be made of some novel methods that have been proposed for the refining of edible oils.

Austrian Patent 356228 is concerned with refining in miscella. Degumming is achieved by membrane ultrafiltration of the miscella, which is then refined by passing it through a silica or aluminium oxide column (see also Sen Gupta, 1975). If necessary, the oil can be bleached in miscella, after which the solvent is distilled off and the oil deodorized. UK Patent 2118568A also uses membrane ultrafiltration of a miscella. In this case a phosphatide solute and ammonia or other specified neutralizing chemical are added to the miscella before ultrafiltration. Membrane technology for edible oil refining is also discussed by Koseoglu (1991).

In UK Patent 2032789A supercritical gases are used to refine oils by separation of impurity fractions. The use of supercritical carbon dioxide is claimed in US Patent 4156688 in combination with adsorption of impurities on active carbon.

References and further reading

Andersen, A.J.C. (1962) *Refining of Oils and Fats*, 2nd edn, Pergamon, Oxford, p. 166.

Athanassiadis, A. and Raes, W. (1986) *Emerging Technologies, World Conf. Proc.*, Cannes, American Oil Chemists' Society, Champaign, p. 137.

Bailey, A.E. (1951) *Industrial Oil and Fat Products*, 2nd edn, Interscience, New York.
Barlow, S.M. and Stansby, M.E. (1982) *Nutritional Evaluation of Long-Chain Fatty Acids in Fish Oil*, Academic Press, London, p. 1.
Beal, R.E., Black, L.T., Griffin, E.L., Meng, J.C. and Farmer, G.S. (1973) *J. Am. Oil Chem. Soc.*, **50**, 260.
Bogdanor, J.M. and Welsh, W.A. (1987) Silica refining of edible oils, Presented at *78th AOCS Annual meeting*, New Orleans.
Brekke, O.L. (1980) *Handbook of Soy Oil Processing and Utilisation*, American Soybean Association, St Louis, MO, p. 166.
Brimberg, U.I. (1981) *Fette Seifen Anstrichm.*, **83**, 184.
Brimberg, U.I. (1982) *J. Am. Oil Chem. Soc.*, **59**, 74.
Burkhalter, J.P. (1976) *J. Am. Oil Chem. Soc.*, **53**, 332.
Carr, R.A. (1978) *J. Am. Oil Chem. Soc.*, **55**, 765.
Cavanagh, G.C. (1976) *J. Am. Oil Chem. Soc.*, **53**, 361.
Cavanagh, G.C. (1990) *Edible Fats and Oils Processing, World Conf. Proc.*, Maastricht, American Oil Chemists' Society, Champaign, p. 101.
Chaudry, M.M., Nelson, A.I. and Perkins, E.G. (1976) *J. Am. Oil Chem. Soc.*, **53**, 695.
Chaudry, M.M., Nelson, A.I. and Perkins, E.G. (1978) *J. Am. Oil Chem. Soc.*, **55**, 851.
DGF (1981–92) The deodorization of edible fats and oils. I – Introduction (1981) *Fette Seifen Anstrichm.*, **83**, 89; II – Theory (1988) *Fat Sci. Technol.*, **90**, 207; III – Deodorizers (1989) *Fat Sci. Technol.*, **91**, 7; IV – Ancillary equipment (1989) *Fat Sci. Technol.*, **91**, 253; V – Developments (1989) *Fat Sci. Technol.*, **91**, 452; VI – Quality control (1990) *Fat Sci. Technol.*, **92**, 115; VII – By- and waste products (1990) *Fat Sci. Technol.*, **92**, 459; VIII – Parallel reactions (1992) *Fat Sci. Technol.*, **94**, 1.
Dijkstra, A.J. and van Opstal, M. (1989) *J. Am. Oil Chem. Soc.*, **66**, 1002.
Diosady, L.L., Sleggs, P. and Kaji, T. (1982) *J. Am. Oil Chem. Soc.*, **59**, 313.
Eder, S.R. (1982) *Fette Seifen Anstrichm.*, **84**, 136.
Fahn, R. (1973) *Fette Seifen Anstrichm.*, **75**, 77.
Farr, W.E. (1983) *J. Am. Oil Chem. Soc.*, **60**, 571.
Forster, A. and Harper, A.J. (1983) *J. Am. Oil Chem. Soc.*, **60**, 265.
Gavin, A.M. (1978) *J. Am. Oil Chem. Soc.*, **55**, 783.
Gavin, A.M. (1981) *J. Am. Oil Chem. Soc.*, **58**, 175.
Gilbert, W.J. and Tandy, D.C. (1979) *J. Am. Oil Chem. Soc.*, **56**, 645A.
Grothues, D. (1981) *Proc. 2nd ASA Symp. on Soybean Processing*, Antwerp.
Haraldsson, G. (1983) *J. Am. Oil Chem. Soc.*, **60**, 251.
Hashemy-Tonkabony, S.E. and Soleimani-Amiri, M.J. (1976) *J. Am. Oil Chem. Soc.*, **53**, 753.
Hendrix, B. (1990) *Edible Fats and Oils Processing, World Conf. Proc.*, Maastricht, American Oil Chemists' Society, Champaign, p. 94.
Hupe, A., Gehring, H., Schlinke, W. and Binnewitt, W. (1989) *Fat Sci. Technol.*, **91**, 593.
Hvolby, A. (1971) *J. Am. Oil Chem. Soc.*, **48**, 503.
Kanamoto, R., Wada, Y., Miyajama, G. and Kito, M. (1981) *J. Am. Oil Chem. Soc.*, **58**, 1050.
Kock, M. (1983) *J. Am. Oil Chem. Soc.*, **60**, 198.
Kock, M. and Penk, G. (1981) *Proc. 2nd ASA Symp. on Soybean Processing*, Antwerp.
Koseoglu, S.S. (1991) *Oils Fats Int.*, **7** (5), 16.
Kramer, J.K.G., Sauer, F.D. and Pigden, W.J. (1983) *High and Low Erucic Acid and Rapeseed Oils*, Academic Press, Toronto.
Leibovitz, Z. and Ruckenstein, C. (1983) *Fette Seifen Anstrichm.*, **85** (1), 538.
Leong, W.L. and Berger, K.G. (1982) *PORIM Tech. Palm Oil Res. Inst. Malaysia*, No. 7.
List, G.R., Evans, C.D., Black, L.T. and Mounts, T.L. (1978a) *J. Am. Oil Chem. Soc.*, **55**, 275.
List, G.R., Mounts, T.L., Warner, K. and Heakin, A.J. (1978b) *J. Am. Oil Chem. Soc.*, **55**, 277.
List, G.R., Mounts, T.L. and Heakin, A.J. (1978c) *J. Am. Oil Chem. Soc.*, **55**, 280.
List, G.R., Avellaneda, J.M. and Mounts, T.L. (1981) *J. Am. Oil Chem. Soc.*, **58**, 892.
List, G.R., Mounts, T.L. and Lancer, A.C. (1992) *J. Am. Oil Chem. Soc.*, **69**, 443.
Loncin, M., Jacobsberg, B. and Evrard, G. (1971) *Rev. Fr. Corps Gras*, **18**, 69.
Mag, T.K. (1980) US Patent 4230630.
Mag, T.K. (1990) *Edible Fats and Oils Processing, World Conf. Proc.*, Maastricht, American Oil Chemists' Society, Champaign, p. 107.
Marcelet, H. (1936a) *C.R. Acad. Sci.*, **202**, 867, 1809.
Marcelet, H. (1936b) *J. Pharm. Chim.*, **24**, 213.
Mounts, T.L., List, G.R. and Heakin, A.J. (1979) *J. Am. Oil Chem. Soc.*, **56**, 883.
Nilsson-Johansson, L., Brimberg, U. and Haraldsson, G. (1988) *Fat Sci. Technol.*, **90**, 447.
Norris, F.A. (1982) *Bailey's Industrial Oil and Fat Products*, 4th edn, vol. 2 (ed. D. Swern), Wiley, New York, p. 269.
Ohlson, R. and Svensson, S. (1976) *J. Am. Oil Chem. Soc.*, **53**, 8.
Ong, T.L. (1980a) *Fette Seifen Anstrichm.*, **82**, 169.
Ong, T.L. (1980b) Lecture: The impact of the preparation and extraction conditions of soybeans on the oil quality, *IFS/AOCS Conf.*, New York.
Ong, T.L. (1983) *J. Am. Oil Chem. Soc.*, **60**, 314.
Pardun, H. (1979) *Fette Seifen Anstrichm.*, **81**, 297.
Patterson, H.B.W. (1989) *Handling and Storage of Oilseeds, Oils, Fats and Meal*, Elsevier Applied Science, London.
Pritchard, J.R. (1983) *J. Am. Oil Chem. Soc.*, **60**, 322.
Richardson, L.L. (1978) *J. Am. Oil Chem. Soc.*, **55**, 777.
Rossell, J.B., Kochar, S.P. and Jawad, I.M. (1981) *Proc. 2nd ASA Symp. on Soybean Processing*, Antwerp.
Schneider, M. (1986) *Emerging Technologies, World Conf. Proc.*, Cannes, American Oil Chemists' Society, Champaign, p. 160.
Segers, J.C. (1982) *Fette Seifen Anstrichm.*, **84** (1), 543.
Segers, J.C. (1983) *J. Am. Oil Chem. Soc.*, **60**, 262.
Segers, J.C. and Ringers, H.J. (1977) US Patent 4049686.
Segers, J.C. and van der Sande, R.L.K.M. (1990) *Edible Fats and Oils Processing, World Conf. Proc.*, Maastricht, American Oil Chemists' Society, Champaign, p. 88.
Sen Gupta, A.K. (1975) Br. Pat. 1564402.

Sen Gupta, A.K. (1988) *Fat Sci. Technol.*, **90**, 251.
Stage, H. (1981) *Proc. 2nd ASA Symp. on Soybean Processing*, Antwerp.
Stage, H., Kuhns, W. and Hammmer, H. (1979) UK Patent 2046606A.
Szabo Sakardi, D.S. (1958) *J. Am. Oil Chem. Soc.*, **35**, 472.
Thomas, A. (1981) *Proc. 2nd ASA Symp. on Soybean Processing*, Antwerp.
Thomas, A. (1982a) *J. Am. Oil Chem. Soc.*, **59**, 1.
Thomas, A. (1982b) *Fette Seifen Anstrichm.*, **84**, 133.
Welsh, W.A. and Toeneboehn, G.J. (1991) Silica refining, quality and environmental advantages, Presented at *80th AOCS Annual Meeting*, Chicago.
Welsh, W.A., Bogdanor, J.M. and Toeneboehn, G.J. (1990) *Edible Fats and Oils Processing, World Conf. Proc.*, Maastricht, American Oil Chemists' Society, Champaign, p. 189.
Wiedermann, L.H. (1981) *J. Am. Oil Chem. Soc.*, **58**, 159.
Wilson, G.M. (1964) *J. Am. Oil Chem. Soc.*, **41**, 127.
Young, F.V.K. (1981) *Proc. 2nd ASA Symp. on Soybean Processing*, Antwerp.
Young, F.V.K. (1982) PORIM *Tech. Palm Oil Res. Inst. Malaysia,* No. 1.
Young, F.V.K. (1983) *J. Am. Oil Chem. Soc.*, **60**, 374.
Young, F.V.K. (1990) *Edible Fats and Oils Processing, World Conf. Proc.*, Maastricht, American Oil Chemists' Society, Champaign, p. 124.
Zschau, W. (1982) *Fette Seifen Anstrichm.*, **84**, 498.

5.5 Modification processes

The refining processes are used to purify edible oils. The modification processes increase their usability. The latter processes consist of hydrogenation, interesterification and fractionation. They can be used alone or in combination with each other, and they can be used on single oils or fats or on blends. Examples of their use are as follows:

1. The hydrogenation of fish oils to provide an acceptable edible oil.
2. Improving by hydrogenation the stability to oxidation of soybean oil.
3. Improving by interesterification the baking performance of lard.
4. The production of dietary margarines using interesterification.
5. Fractionation of palm oil to yield a liquid fraction, or olein, which may be used as at least a partial substitute for the natural liquid oils.
6. To produce substitutes for expensive natural oils and fats, for example cocoa butter, by hydrogenation or fractionation alone or combined with interesterification.

In this manner the processes have a stabilizing influence on raw material prices.

5.5.1 Hydrogenation*

Most of the fatty acids of which the triglycerides of naturally occurring oils and fats are composed are unsaturated. The unsaturated acids contain from one to six double bonds. The number of double bonds in the chain of the fatty acids and the triglyceride composition are responsible for the physical state of the oil, i.e. liquid or solid at a given temperature, and for its susceptibility to oxidation. When hydrogen is added to a fatty acid double bond, it becomes saturated, with consequent increases in the oxidative stability and melting point of the oil of which it is part. The process, which is frequently referred to as 'hardening', can be stopped at any point up to complete saturation. The principal uses of the hydrogenation reaction have therefore been to increase oxidative stability and to produce high- or middle-melting-point fats (>40 °C and 30–40 °C respectively) for use in margarines, shortenings and speciality fats.

* See also section 10.1.

(a) Process theory and description

Hydrogenation is a catalytic reaction. The theory of the reaction has been described by, amongst others, Allen (1982). Bitner *et al.* (1973, 1977, 1980, 1982), Larsson (1983), Patterson (1983), Ucciani (1983), Beckmann (1983) and Ariaansz (1992).

A simplified reaction sequence for the hydrogenation of linolenic (Ln), linoleic (L) and oleic (O) acids is:

$$\text{linolenic} \xrightarrow{k_1} \text{linoleic} \xrightarrow{k_2} \text{oleic} \xrightarrow{k_3} \text{stearic}$$

where k_1, k_2 and k_3 are the reaction rate constants for the sequence.

The selectivity ratio for linolenic to linoleic acid is $S_{\mathrm{Ln}} = k_1/k_2$, and for linoleic to oleic acid $S_{\mathrm{L}} = k_2/k_3$. In a simple system the reaction proceeds in a stepwise manner because S_{Ln} is greater than S_{L}. However, because of the complexity of triglyceride systems, impurities in the oil and the effect of processing conditions, in reality all the steps of the sequence take place simultaneously and the selectivity of the reaction depends on the relative rates at which they occur. Reactions in which polyunsaturated acids are hydrogenated in preference to mono-unsaturated, i.e. high selectivity, are required for improving the oxidative stability of an oil.

During contact with the hydrogenation catalyst, isomerization takes place, in which some of the double bonds move to a new position or become transformed from the *cis* to the *trans* form. The *trans* fatty acids have higher melting points and thus contribute to the increase in melting point of the fat.

The hydrogenation system consists of the oil, a catalyst and hydrogen gas, and the reaction is directed by changes in conditions affecting mass transfer, e.g. of hydrogen to the catalyst surface and of oil to and from the surface. Process conditions and their effect on isomerization and selectivity are shown in Table 5.4.

It can be seen from Table 5.4 that, with the exception of catalyst activity, the changes in conditions have the same effect on selectivity and isomerization. As hydrogenation progresses and the melting point of the fat increases, the level of *trans* isomers reaches a maximum and then declines. This maximum occurs in the region of 'slip' (capillary rise) melting point of 35 °C. At this melting-point region using fresh active catalyst, the product will contain about 30% *trans* isomers compared with about 55% when a used catalyst is employed. These figures are

Table 5.4 Effects of changes in hydrogenation process conditions

Condition (increase in)	Effect on	
	Selectivity	Isomerization
Temperature	Up	Up
Pressure	Down	Down
Agitation	Down	Down
Catalyst concentration	Up	Up
Catalyst activity	Up	Down

obtained at customary hardening temperatures of 180–190 °C. When melting points of 40 °C and higher are desired, the choice of conditions with regard to selectivity versus isomerization is of progressively less importance.

Trienoic and higher polyunsaturated fatty acids such as are found in soybean, rapeseed and fish oils can, under extreme conditions, form cyclic compounds, which are biologically undesirable (Patterson, 1983). To prevent the formation of these cyclic compounds, a two-stage process has been worked out (Coenen *et al.*, 1967) and should be used in the hydrogenation of such oils. The process consists of using high hydrogen pressures, 3–4 bar, and low temperature, 150 °C maximum, for the first part of the reaction and thereafter conditions may be adjusted as desired.

(b) Reaction components

The oil normally used for hydrogenation is of caustic-refined and bleached quality. With the increase of physical refining, refiners with no caustic-refining facility must decide from the quality of the oil whether to use a pretreated or fully refined feedstock. Ong (1979) suggested the following quality for neutralized and bleached soybean oil for hydrogenation:

Free fatty acid (as oleic)	<0.05%
Soap	<25 mg/kg
Phosphatides (as P)	<2 mg/kg
Moisture	<0.05%
Peroxide value	<0.5 meq/kg
p-Anisidine value	<10

Ottesen (1975) demonstrated the effect on catalyst activity of a number of known catalyst poisons. His results showed that 1 ppm of sulphur poisons 0.004% of nickel, 1 ppm of phosphorus poisons 0.0007% of nickel, 1 ppm of bromine poisons 0.00125% of nickel, 1 ppm of nitrogen poisons 0.00144% of nickel and 1 part per thousand of oxy acids poisons 0.0046% of nickel. Drozdowski and Zajac (1980), when using phospholipids, allyl isothiocyanate, free fatty acids, sodium soaps and lipid oxidation products to study the effect of catalyst poisons on hydrogenation kinetics, noted that the poisoning effect was immediate, indicating immediate adsorption of the poison on the catalyst surface. Mørk and Norgard (1976) examined the poisoning effect of halogen-containing compounds found in fish oils and noted that the poisoning effect was in the order Cl < Br < I.

Many catalysts, both homogeneous and heterogeneous, have been used in the laboratory for the hydrogenation of edible oils. These and industrial catalysts were reviewed by the Deutsche Gesellschaft für Fettchemie (DGF, 1976) in a series of four articles on the hydrogenation process, and at the American Oil Chemists' Society's colloquium on hydrogenation in 1986 (Hastert, 1987). Reaction selectivity is required to improve oxidative stability but, for example, when commercial nickel catalysts are used for the light hydrogenation of soybean oil, there is a significant increase in stearic and *trans* acid content. On the other hand, the formation in the reaction of high levels of *trans* acids produces a fat that has a steep solid fat content (SFC) curve. Such fats may be used as cocoa butter substitutes. The object of the research work has therefore been to develop catalysts (a) with very high polyunsaturated acid selectivity accompanied by no isomerization and (b) with a combination of high selectivity and high isomerization capability. Copper catalysts (Rozendaal, 1983) very largely fulfil specification (a) but they are not as active as nickel catalysts; they are less resistant to poisons and, after their use, traces of copper must be rigorously excluded from the oil. To produce catalyst (b) use has been made of the increase in *trans* acid formation when a nickel catalyst is poisoned with sulphur (Beckmann, 1983). In this manner *trans* acid levels of about 70% can be obtained.

In industrial practice, although some copper catalysts have been used, it is customary to use dry or wet reduced nickel catalysts supported on a natural earth such as kieselguhr and suspended in a hydrogenated fat, usually hardened palm or soybean oil. The nickel content of the commercial catalyst is between 17 and 25% and a similar amount of earth is incorporated.

There are two methods for the use of catalyst, so-called single-use and multi-use. In the single-use system sufficient catalyst is used for one batch and the catalyst is discarded after filtration. In the multi-use system a larger quantity of catalyst is used and this is recovered after filtration for subsequent batches. As the catalyst loses its activity, a proportion of active (fresh) catalyst is added and a similar amount of the used catalyst discarded (Osinga, 1979). The advantages of the single-use system are simpler plant, simpler processing schedules and less variation in product characteristics. The advantages of the multi-use system are a fuller use of catalyst activity and, in multi-oil hardening plants, use of optimum activity, selectivity and *trans* acid formation where it is most needed. Generally simpler single-oil plants employ single-use and more complex operations employ the re-use system.

The purity of hydrogen is of great importance in the hydrogenation process. The standard of purity obtained in most cases is better than 99.5%, but gases containing only 98.5% hydrogen can be usable depending on the composition of the impurities in the gas. The impurities can be divided into two types, those which will act as catalyst poisons and others which, by accumulating in the headspace of a dead-end converter, can slow down and eventually stop hydrogenation by restricting the flow of hydrogen. In the first category are sulphur, chlorine and carbon monoxide, which are adsorbed on the catalyst surface, and oxygen and moisture, the action of which on the oil produces compounds acting as catalyst poisons. In the second category are the inert gases, e.g. nitrogen and argon, and methane, which is similarly inert in the reaction.

The two principal methods for the production of hydrogen are the electrolysis of water (Vandenborre, 1983) and the steam–hydrocarbon or hydrocarbon re-forming process (Allen, 1982). The purity of hydrogen by electrolysis of aqueous solutions of, for example, potassium hydroxide is minimum 99.8%. Impurities are traces of inert gases, oxygen and water. High-purity oxygen is a by-product of this process. The composition of the gas from the methanators of the steam–hydrocarbon process is CO_2 0.01%, CO 10 ppm, CH_4 0.45% and H_2 99.5%. In some plants this gas is given a final adsorption treatment to remove carbon monoxide, carbon dioxide, methane and nitrogen, thereby achieving a purity of 99.97%.

(c) Process control

Control of the process is a combination of the control of process parameters and of the degree of hydrogenation attained. The process parameters, temperature, hydrogen pressure, catalyst concentration and activity or type, are set according to the product required; agitation is usually of fixed speed. The degree of reaction is monitored by one or more of the following characteristics of the oil: refractive index (RI), iodine value (IV), melting point, congeal point, solid fat content (SFC) and, in certain cases, gas chromatography or differential scanning calorimetry (Huizenga, 1983). Of these values the SFC is usually the most important and is determined by nuclear magnetic resonance (NMR) spectrometry. Alternatively, the solid fat index (SFI) can be measured by dilatometry. These methods are too slow to be used during the hydrogenation reaction, and so use is made of the fact that, when a specific oil is hardened under the same conditions, the SFC can be approximately related to its RI and IV and also to the hydrogen usage. In practice, therefore, when using the dead-end batch converter described below, a pre-set flow meter is used to control the quantity of gas supplied to the oil. The process is stopped when the required amount of gas has been used and the batch is then checked in the laboratory and corrected, or is cooled and filtered and corrected by adjustment to the end-point of a subsequent batch.

(d) Plant

The hardening process is carried out predominantly in batch equipment. Some continuous plants are in use (Hastert, 1990), but they are suitable mainly for long runs of single melting points from vegetable oils for which extended catalyst life is attainable.

Two batch systems exist: gas recycling and dead-end. In the former the hydrogen gas is passed through the oil in the converter, out of the vessel, through cleansing and recompressing steps and back into the oil. In the dead-end plant the hydrogen does not leave the converter except when the latter is vented. Since 1960 the dead-end system has grown in popularity because plant and installation costs are lower and because the process is easier to control. The Buss closed-loop reactor is a special case of the dead-end design in which the oil and catalyst slurry are circulated via a heat exchanger for temperature control and through a venturi tube in which intimate mixing with hydrogen is obtained (Duveen and Leuteritz, 1982; Urosevic, 1986).

Dead-end converters are vertical, cylindrical vessels designed to harden 5–30 tonne batches. The vessel is built to withstand full vacuum and pressures up to 7 or 10 bar and is constructed from mild steel for the hardening of oils and from stainless steel for fatty acids. Heating and cooling are carried out by steam and water respectively, usually through the same coil. The coil size is determined by the cooling required to control the exothermic reaction. The IV reduction rate at the start of hydrogenation of good-quality high-IV oil with fresh catalyst is of the order of 3 IV units/minute. The heat of reaction is sufficient to raise the temperature of the oil by 1.6 °C per unit IV drop. Therefore the cooling effect must be capable of extracting approximately 5 °C per minute from the oil to maintain a stable temperature. The cooling water should be treated to prevent coil corrosion. Some hydrogenators have found it economically viable to install stainless-steel coils.

Agitator designs in dead-end converters are critical because of the need to suck or splash the hydrogen from the headspace of the vessel into the oil (Patterson, 1983; Hastert, 1987, 1990). The usual design consists of a shaft with two or three turbine-type impellers, the top one being sited just below the oil level, such that a vortex is produced that sucks the headspace gas into the oil. It is therefore important that the charge size should be reasonably accurately measured. Agitator shaft speed is usually between 75 and 150 rev/min.

During the reaction the temperature and pressure may be controlled either manually or automatically. The process lends itself to automation.

Figure 5.6 is a flow diagram of a dead-end batch hydrogenation plant. The plant is arranged (a) to take full advantage of the energy-saving possibilities resulting from the exothermic reaction, (b) to give high efficiency of utilization of the converter and (c) for automated operation.

When the end-point of the reaction is reached, the batch is cooled and filtered with initial recirculation through the filter so that the nickel content of the filtered oil is minimized.

The filters most widely used for this purpose are the closed discharge plate and frame filter press and the pressure leaf tank filter. If the oil is not to be used immediately after filtration, it should be cooled to 10–15 °C above its melting point for storage.

Hydrogen is a highly flammable gas that also forms explosive mixtures with oxygen. The safety precautions that should be taken in hydrogenation plants are described by Patterson (1983).

(e) Post-refining

During hydrogenation, an increase of the free fatty acid content of the oil occurs. The level reached depends on the quality of the oil and the time of hydrogenation, and is usually between 0.15 and 0.3%. Also the catalyst filtration leaves about 2–3 ppm of nickel in the oil, mainly in the form of soaps. Nickel is an autoxidation catalyst (Johansson, 1976), and therefore post-refining is needed to reduce the nickel content to less than 0.2 ppm and also to produce an oil having acceptable colour, odour and oxidative stability. To prevent unnecessary damage to the oil caused by the relatively high levels of free

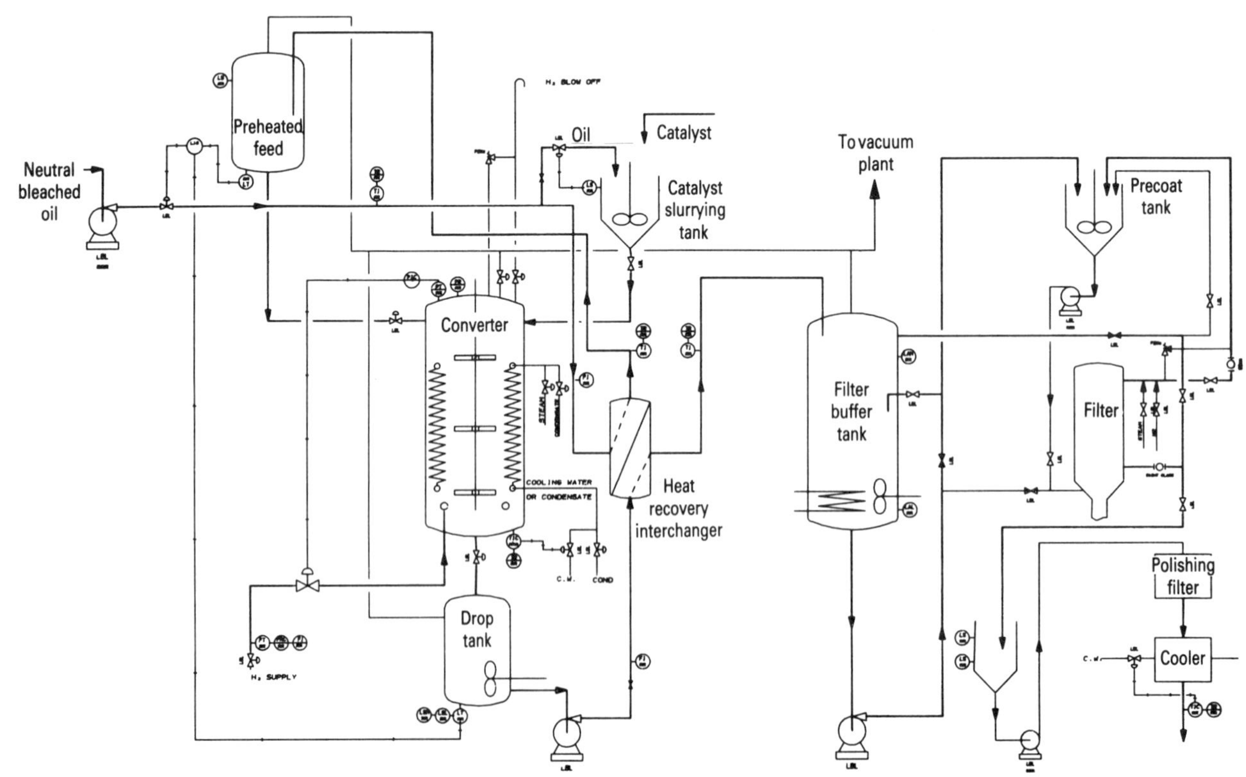

Fig. 5.6 Flow diagram of a dead-end batch hydrogenation plant (Courtesy of N.V. Extraction De Smet S.A.).

fatty acid and nickel, the oil should be post-refined as soon as possible after the catalyst filtration.

The treatment given to the oil is primarily dependent on the type of deodorizer installed in the plant. In plants using relatively low-temperature (180–200 °C) batch deodorizers, the oil is neutralized with caustic soda, washed, dried, bleached and filtered before deodorization. However, if semi-continuous or continuous deodorizers capable of treatment at up to 250 °C are available, then, because of the efficiency of such units in the distillation of fatty acids, neutralization and attendant washing is not required. In such cases it is usually sufficient to treat the oil with about 0.02% of citric acid in aqueous solution to split the nickel soaps and act as a chelating agent, followed by drying, bleaching and filtration. Processing and plant for these operations are generally as described in sections 5.4.4 and 5.4.5.

5.5.2 Interesterification

(a) Process description

Inter-esterification is the name given to the process by which the arrangement of fatty acids in a triglyceride molecule is changed. It incorporates the following specific cases (Sonntag, 1982):

1. Acidolysis, in which the triglyceride reacts with an organic acid, which is thereby introduced into the triglyceride molecule.
2. Alcoholysis, in which, as a result of reaction with an alcohol, partial glycerides and other esters are formed.
3. Transesterification, in which the positional rearrangement takes place first within individual triglycerides (interesterification) and secondly between all the triglycerides or other acyl esters present (ester interchange or, more commonly, interesterification). In the oils and fats industry, this last case has found wide application because it can result in beneficial changes in the physical characteristics of an oil or fat or blend of oils and fats.

The process involves the formation of an equilibrium mixture of the triglycerides formed by a random distribution of the fatty acids on the glycerol molecules. The process is therefore known as randomization or random interesterification.

The mathematics for the calculation of triglyceride composition following randomization of a fat system containing known amounts of individual fatty acids have been described by the DGF (1973).

If the equilibrium is disturbed by, for example, the crystallization from the reaction mixture of high-melting-point triglycerides, it is then referred to as 'directed' interesterification.

An extension of the process has been introduced (Macrae, 1983) by the use of microbial lipases as catalysts for the reaction. The effects of three types of enzyme are described: non-specific; 1,3 positional specific on the triglyceride; and fatty acid specific.

The lipase-catalysed reactions of glyceride hydrolysis and synthesis are reversible. Thus, if the water content of the reaction system is limited, the non-specific enzyme results in random interesterification, but the 1,3-specific enzyme and the fatty acid-specific enzyme can be used to introduce desirable fatty acids into the triglyceride molecule and to produce preferred positional fatty acid arrangements in the triglyceride molecule.

The conventional interesterification reaction can be instigated by the use of high temperatures, in the region of 250–300 °C, but reaction time is long and may result in decomposition and polymerization of the starting material. Normal practice is therefore to use alkali metals or alkali-metal alcoholates. Using such catalysts the reaction will take place at between 25 and 120 °C and randomization can be complete in as short a time as 5 min, although normally a reaction time of 30 min is allowed. Catalysts and conditions for the reaction have been reviewed by Sreenivasan (1978).

The alkali metals and their alcoholates react with water, fatty acids and peroxides. The quantities of catalyst that are inactivated by these substances are given in Table 5.5. The figures show the importance of using a dry and well-refined oil as the starting material for the reaction. With such an oil the quantity of catalyst used for the reaction is of the order of 0.1–0.2% alkali metal or 0.2–0.3% of methylate.

Rozendaal (1992) has discussed the mechanism and kinetics of the reaction. He states that the added catalyst is a precursor for the real interesterification catalyst, which is formed slowly during the initial stages of the reaction. The real catalyst is believed

Table 5.5 Inactivation of catalysts (Sreenivasan, 1978)

Substance type	Level	Catalyst inactivated (kg/1000 kg of oil)	
		Sodium	Sodium methylate
Water	0.01%	0.13	0.3
Fatty acid	Acid value = 0.1	0.04	0.1
Peroxide	Peroxide value = 1.0	0.023	0.054

to be the sodium derivative of a diacylglycerol. The formation of this compound is presented for the use of sodium ethylate and a sodium hydroxide/glycerol mix as examples of the precursor catalysts.

Methods employed for the detection of the end-point of the reaction have been reviewed by Sreenivasan (1978) and Joly and Lang (1978). They consist of:

1. Melting point (Sonntag, 1982), cloud point, cooling curve and differential scanning calorimetry.
2. Solid fat content or dilatometry.
3. Glyceride analysis by mass spectrometry, gas–liquid and thin-layer chromatography, and by pancreatic lipase hydrolysis.
4. Crystal form by microscopy or X-ray diffraction.
5. Plasticity or penetration.

The process consists of the steps of drying under vacuum, addition of the catalyst followed by a reaction time, killing the catalyst using water, acid or alkali, and post-refining by the normal practices of soap separation, washing, drying, bleaching and deodorization. In the special case of directed interesterification the reaction mixture is cooled to the required temperature after addition of the catalyst. When crystallization is complete the catalyst is killed and the oil may then be heated to melt and mix the new triglyceride arrangement, or the crystallized fraction may be separated by centrifugation. Batch, semi-continuous and continuous plants have been used for the reaction (DGF, 1973; Sreenivasan, 1978).

(b) Applications

Interesterification has found its widest applications in two areas:

1. The changing or stabilization of crystal habit.
2. Changing the melting and crystallization characteristics of single oils or oil blends.

Edible fats are polymorphic and the most important crystal forms in the production of margarines and shortenings are α, β' and β, of which the last two are the stable forms. The desired form for most margarines and shortenings is the β' structure. Individual oils and fats show tendencies towards either the β or the β' structure (Wiedermann, 1978). Lard crystallizes naturally in the β form because of the high proportion of palmitic acid in the 2 position of its disaturated glycerides. On randomizing, this proportion is reduced from 64 to 24% and, in consequence, the interesterified product crystallizes in the β' form with resultant improvement in its shortening performance (Weiss *et al.*, 1961; Sreenivasan, 1978). Similarly the crystal structures of margarines based on sunflower oil (Faur, 1977) and Canola oil (Teasdale, 1980) are stabilized in the β' form by random interesterification.

The fact that interesterification results in changes in the melting characteristics of fats has been used to produce dietary margarines high in polyunsaturated acids (Wieske, 1977) and with low to zero *trans* fatty acid content (List *et al.*, 1977; Kattenberg and Verburg, 1977) and confectionery fats with rapid mouth melt-down properties (Sreenivasan, 1978). By directed interesterification a shortening of improved plastic range is obtained from lard (Holman and Going, 1959) and, after fractionation, a salad oil can be produced from palm oil (Babin, 1974).

The uses cited by Coleman and Macrae (1980) for the lipase-catalysed reaction are examples of acidolysis and transesterification. Acidolysis is illustrated by the increase in 1(3)-palmitoyl-3(1)-stearoyl-2-olein (POSt) resulting from the reaction of palm oil mid-fraction with stearic acid in the presence of a 1,3-specific enzyme, and by the increase in the content of linoleic acid in olive oil when that oil is reacted with linoleic acid in the presence of an enzyme with fatty acid specificity. An example of transesterification is shown by the increase in POSt triglycerides when a 1,3-specific enzyme is used to catalyse the interesterification of shea butter and palm oil mid-fraction.

The *Mucor miehei* 1,3-specific enzyme is described by Hansen and Eigtved (1986), and its immobilized use for the production of cocoa butter equivalents is reported on by Padley and Macrae (1991). The latter authors list some of the enzymes available commercially at the time of their presentation. A more general review describing the use of enzymes in lipid modification has been published (Eigtved, 1992).

5.5.3 Fractionation

(a) General considerations

Oils and fats are mixtures of triglycerides that, because of their different fatty acid compositions, have melting points spanning from −50 to +80 °C, each oil having its own melting range. The melting range limits the use of a particular oil or fat. Hydrogenation is used to increase ‘hardness’; interesterification produces changes in characteristics by rearrangement; and fractionation is a thermomechanical process by which the raw material is separated into two or more portions, which widens the use of the oil. Thermomechanical separation processes include distillation and crystallization.

Distillation is commercially unsuitable for the fractionation of triglyceride mixtures because of their low vapour pressures and because of their relative instability at high temperatures. Separation can, however, be effected by crystallization.

'Winterization' is a special case of fractionation, deriving its name from the early practice of using the cold winter weather to separate a solid triglyceride fraction, or stearin, from cottonseed oil. The liquid fraction, which is known as the olein, thereafter when bottled did not suffer from unsightly deposits and did not cause emulsion separation when mayonnaise or salad dressings made from it were refrigerated. The process of dewaxing is similar to triglyceride fractionation but it is used to remove naturally occurring waxes found in, for example, sunflower, safflower, corn and sesame oils.

The crystallization process consists of three distinct stages (Taylor, 1976), which are supercooling of the melt, formation of crystal nuclei or nucleation and crystal growth.

The α, β' and β crystal forms of fats are of increasing stability in that order. Further, the rate of crystallization of the α form is greater than that of the β', which in turn is greater than the β polymorph. If supercooling is carried out too rapidly, crystallization of the α modification occurs, resulting in a mass of very small crystals.

To obtain good separation, the stearin crystals should be firm and of uniform spherical size, which is a condition found when they are mainly in the β' form. To obtain this state the first requirement is that crystal inhibitors such as phosphatides, carbohydrates, soap, mineral acids and monoglycerides should be absent from the oil (Deffense and Tirtiaux, 1982). Initially the oil is heated to a temperature higher than the slip melting point of the highest-melting triglyceride present, for example to 70–75 °C in the case of palm oil. Thereafter the oil is cooled slowly, typically over 4–12 h in the case of palm oil, and with careful control, to the separation temperature. During this time supercooling occurs, which results in nucleation and crystal growth (Coenen, 1974a, b; Berger, 1977). The phases are then separated.

(b) Processes

Because of the high viscosity of the crystalline slurry, separation of the olein from the stearin is difficult. Three methods are used to effect this separation.

(i) The dry process

Separation is effected by filtration in a diaphragm filter press or in a continuous-belt vacuum filter (Deffense and Tirtiaux, 1982).

(ii) The Lanza or Lipofrac process

Crystallization is carried out in a number of tanks programmed to produce a continuous supply of slurry, which is mixed with a wetting agent (sodium lauryl sulphate) and an electrolyte (magnesium sulphate). These solutions displace the oil from the surface of the stearin crystals so that, on centrifuging, the olein is discharged as the oil phase and the stearin forms part of the aqueous phase. The aqueous phase is heated to melt the stearin and is centrifuged to recover the stearin and the detergent/electrolyte solution, which is recycled. The oil fractions are washed to remove traces of the detergent (Haraldsson, 1978).

(iii) The solvent process

The oil is crystallized in miscella using mainly acetone or hexane as the solvent. In this way the viscosity is reduced, thereby improving crystallization and filterability (Bernardini, 1978).

The difference in efficiency between the three techniques lies in the degree of contamination of the stearin by the olein after separation. This was shown by Coenen (1974a, b) by the increasing yield of olein in the order: dry (vacuum filter process), Lanza process, solvent process. The introduction of the diaphragm filter in the dry process for palm oil increased the olein yield to about 78% (vacuum filter about 60%), which is close to the 80% achieved using the Lanza process.

Winterization of oils is carried out using the same processes as for fractionation but with conditions suitable for the raw material. Thus in the dry process for the winterization of cottonseed oil (Neumunz, 1978) the cooling cycle lasts for 48–60 h, during which time the oil temperature is reduced from 35 to 6 °C and the slurry is then held for a further 12 h at 6 °C before being filtered. The long crystallization time is needed to produce a large crystal. When the process is carried out in miscella, crystallization is effected in a very much shorter time and yields of the winterized oil are appreciably higher.

Dewaxing is frequently referred to as winterization. The process in which, for example, up to 0.4% of high-melting-point (76–77 °C) waxes are removed from sunflower oil is similar to those described. Crystallization is carried out to 6–8 °C, but in order to improve separation the temperature is raised to 15–18 °C before filtration or centrifugation (Kehse, 1979). A filter aid is always used to assist filtration in the dry process.

Dewaxing has been incorporated in the continuous caustic soda refining process. The oil is degummed using phosphoric acid, neutralized and the soapstock is separated. The oil is then cooled to crystallize the waxes and, after adjustment of the soap level (which acts as the wetting agent) to approximately 1000 ppm by the addition of a small quantity of caustic soda, the waxes are removed by centrifuging. The normal refining process is then continued by heating, washing and drying the oil. A single treatment of this type will remove about 90% of the waxes, which is usually insufficient for the oil clarity required. The final stage is therefore often a dry treatment after deodorization. This also guards against an increase in solids resulting from possible contamination with higher-melting-point oils and fats.

Quality testing of the products from fractionation is conducted according to the purposes for which they are intended. In many cases iodine values and melting points are sufficient in addition to the common general quality tests. The American Oil Chemists' Society's Cold Test is used for salad oils. Palm olein is assessed by its cloud point and sometimes by a 'cold stability' test in which the length of time for which the oil remains clear at 18–20 °C is measured. Other tests of importance are the solid fat content by NMR or dilatometry and differential scanning calorimetry. Deffense and Tirtiaux (1982) published HPLC chromatograms for the raw materials and products of the fractionation of palm oil and beef tallow.

(c) Applications

The fractionation process is used for three main purposes: extending the applicability of fats, the development of special properties and the improvement of quality. These three purposes are illustrated by the following examples.

(i) Palm oil
Single-stage fractionation yields an olein of cloud point less than 10 °C, which can be used as a substitute for soft oils, and a stearin of melting point 44–50 °C, which is used as a margarine or shortening hardstock (PORIM, 1981). Double solvent fractionation in which the olein from a first separation is refractionated produces a mid-fraction or second stearin that is used in the production of cocoa butter substitutes.

(ii) Palmkernel oil
Palmkernel stearin is a high-quality 'hard butter' or couverture fat. It was originally produced by the dry process but this method has now been partly replaced by solvent fractionation (Thomas and Paulicka, 1976; Willner *et al.*, 1990).

(iii) Butter oil
Fractionation of butter oil produces stearins and oleins, which by blending can be used to enhance baking properties or produce a soft butter spreadable from the refrigerator (Tirtiaux, 1990).

(iv) Soybean oil
A salad oil of high oxidative stability is produced in 80% yield when soybean oil is selectively hydrogenated to an iodine value of 100–105 and is then fractionated at 2 °C. If the hydrogenation is continued to IV 77 and the resultant oil subjected to double fractionation, the first stearin (yield 28%) is usable as a shortening hardstock, the mid-fraction (yield 33%) finds use as a hard butter and the second olein (yield 39%) may be used as a high-stability cooking oil (Tirtiaux and Deffense, 1984).

References and further reading

Allen, R.R. (1982) *Bailey's Industrial Oil and Fat Products*, 4th edn, vol. 2 (ed. D. Swern), Wiley, New York.

Ariaansz, R.F. (1992) Hydrogenation theory, Presented at *AOCS Conf.*, Budapest.

Babin, A. (1974) *Oleagineux*, **29**, 375.

Beckmann, H.J. (1983) *J. Am. Oil Chem. Soc.*, **60**, 282.

Berger, K.G. (1977) *Oil Palm News*, No. 22, Tropical Products Institute, London, p. 10.

Bernardini, M. (1978) *Oleagineux*, **33**, 297.

Bitner, E.D., Ernst, J.O. and Dutton, H.J. (1973) *J. Am. Oil Chem. Soc.*, **50**, 436.

Bitner, E.D., Snyder, J.M., Ernst, J.O. and Dutton, H.J. (1977) *Fette Seifen Anstrichm.*, **79**, 483.

Bitner, E.D., Mounts, T.L. and Dutton, H.J. (1980) *J. Am. Oil Chem. Soc.*, **57**, 209.

Bitner, E.D., Snyder, J.M., Mounts, T.L., Dutton, H.J. and Baker, G. (1982) *J. Am. Oil Chem. Soc.*, **59**, 286.

Coenen, J.W.E. (1974a) *Rev. Fr. Corps. Gras*, **21**, 403.

Coenen, J.W.E. (1974b) *Rev. Fr. Corps. Gras*, **21**, 343.

Coenen, J.W.E., Wieske, T.H., Cross, R.S. and Rinke, H. (1967) *J. Am. Oil Chem. Soc.*, **44**, 344.

Coleman, M.H. and Macrae, A.R. (1980) UK Patent 1577933.

Deffense, E. and Tirtiaux, A. (1982) Paper presented at *AOCS Conf.*, The Hague, Netherlands, October; Summary. *J. Am. Oil Chem. Soc.*, **60**, 473.

DGF (1973) *Fette Seifen Anstrichm.*, **75**, 467, 587, 663.

DGF (1976) *Fette Seifen Anstrichm.*, **78**, 385.

Drozdowski, B. and Zajac, M. (1980) *J. Am. Oil Chem. Soc.*, **57**, 149.

Duff, A.J. and von Klosterlein, C.B. (1979) *Fette Seifen Anstrichm.*, **81**, 257.

Duveen, R.F. and Leuteritz, G. (1982) *Fette Seifen Anstrichm.*, **84** (1), 511.

Eigtved, P. (1992) *Advances in Applied Lipid Research* (ed. F.B. Padley), JAI Press, London, p. 1.

Faur, L. (1977) *Rev. Fr. Corps. Gras*, **24**, 85.

Hansen, T.T. and Eigtved, P. (1986) *Emerging Technologies, World Conf. Proc.*, Cannes, American Oil Chemists' Society, Champaign, p. 365.
Haraldsson, G. (1978) Paper presented in Kuala Lumpur, Malaysia, 29 April.
Hastert, R.C. (1981) *J. Am. Oil Chem. Soc.*, **58**, 169.
Hastert, R.C. (ed.) (1987) *Hydrogenation: Proc. AOCS Colloquium*, American Oil Chemists' Society, Champaign, pp. 41, 47, 65, 109.
Hastert, R.C. (1990) *Edible Fats and Oils Processing, World Conf. Proc.*, Maastricht, American Oil Chemists' Society, Champaign, p. 142.
Holman, G.W. and Going, L.H. (1959) US Patent 2875066.
Huizenga, T. (1983) *Proc. 3rd ASA Symp. on Soybean Processing*, Antwerp.
Johansson, G. (1976) *J. Am. Oil Chem. Soc.*, **53**, 410.
Joly, F. and Lang, J.-P. (1978) *Rev. Fr. Corps. Gras*, **25**, 423.
Kattenberg, H.R. and Verburg, C.C. (1977) US Patent 4055679.
Kehse, W. (1979) *Fette Seifen Anstrichm.*, **81**, 463.
Larsson, R. (1983) *J. Am. Oil Chem. Soc.*, **60**, 198.
List, G.R., Emken, E.A., Kowleck, W.F., Simpson, T.D. and Dutton, H.J. (1977) *J. Am. Oil Chem. Soc.*, **54**, 408.
Macrae, A.R. (1983) *J. Am. Oil Chem. Soc.*, **60**, 291.
Mørk, P.C. and Norgard, D. (1976) *J. Am. Oil Chem. Soc.*, **53**, 506.
Neumunz, G.M. (1978) *J. Am. Oil Chem. Soc.*, **55**, 396A.
Ong, T.L. (1979) *Proc. Quality Control Seminar in Soybean Crushing Plants and Soybean Oil Processing Plants*, ASA, Brussels.
Osinga, Th.J. (1979) *Fette Seifen Anstrichm.*, **81**, 108.
Ottesen, I. (1975) Lipidforum lecture, Denmark.
Padley, F.B. and Macrae, A.R. (1991) Production of cocoa butter equivalents using enzymic interesterification. Presented at *Society of Chemical Industry Meeting*, Liverpool, May. (Personal communication.)
Patterson, H.B.W. (1983) *Hydrogenation of Fats and Oils*, Applied Science, London.
PORIM (1981) *Tech. Palm Oil Res. Malaysia*, No. 4.
Rozendaal, A. (1983) *Proc. 3rd ASA Symp. on Soybean Processing*, Antwerp.
Rozenaal, A. (1992) *INFORM*, **3**, 1232.
Sonntag, N.O.V. (1982) *Bailey's Industrial Oil and Fat Products*, 4th edn, vol. 2 (ed. D. Swern), Wiley, New York.
Sreenivasan, B. (1978) *J. Am. Oil Chem. Soc.*, **55**, 798.
Taylor, A.M. (1976) *Oleagineux*, **31**, 73.
Teasdale, B.F. (1980) Edible oil products from low erucic acid rapeseed oil (Canola oil), Paper presented at *ISF/AOCS Congress*, New York.
Thomas, A. and Paulicka, F.R. (1976) *Chem. Ind.*, 18 September, 774.
Tirtiaux, A. (1990) *Edible Fats and Oils Processing, World Conf. Proc.*, Maastricht, American Oil Chemists' Society, Champaign, p. 136.
Tirtiaux, A. and Deffense, E. (1984) Paper presented at *AOCS Meeting*, Dallas, April/May.
Ucciani, E. (1983) *Proc. 3rd ASA Symp. on Soybean Processing*, Antwerp.
Urosevic, D. (1986) *Emerging Technologies, World Conf. Proc.*, Cannes, American Oil Chemists' Society, Champaign, p. 128.
Vandenborre, H. (1983) *Proc. 3rd ASA Symp. on Soybean Processing*, Antwerp.
von Klösterlein, C.B. and Vas, H.J. (1983) *J. Am. Oil Chem. Soc.*, **60**, 443.
Weiss, T.J., Jacobson, G.A. and Wiedermann, L.H. (1961) *J. Am. Oil Chem. Soc.*, **38**, 396.
Wiedermann, L.H. (1978) *J. Am. Oil Chem. Soc.*, **55**, 823.
Wieske, T. (1977) US Patent 4045588.
Willner, T., Sitzmann, W. and Muench, E.-W. (1990) *Edible Fats and Oils Processing, World Conf. Proc.*, Maastricht, American Oil Chemists' Society, Champaign, p. 239.
Witte, N.H. (1980) *J. Am. Oil Chem. Soc.*, **57**, 854A.

5.6 Process and production control

Since 1970 considerable changes have taken place in the means by which processes are controlled. Pneumatic controllers and relay logic have been replaced by solid-state logic and programmable controllers. During the same time the cost of the hardware has fallen and the availability of appropriate software has increased while other process costs have risen. The inclusion of such equipment has therefore become standard for new plant.

Using the distributed control system a network of control can be built up in modular fashion. Individual process control units are applied to single plant units and communicate either with each other or with a central computer via a data highway. Via the same highway, links are also available for production management information, production planning, quality control involvement, whether on- or off-line, and costing operations (von Klösterlein and Vos, 1983).

Examples of process control that have been reported in the literature are as follows:

1. The operation of an export grain elevator (Witte, 1980).
2. The use of a mini-computer in the control of an extraction plant (Witte, 1980).
3. The control of a feedmill batching system (Witte, 1980).
4. The monitoring of tank farm stocks, charging and discharging of oils with safety interlocks (Duff and von Klösterlein, 1979, Pechey, 1982).
5. The proportioning of caustic soda and oil in continuous refining (Farr, 1983).
6. The continuous monitoring of losses across continuous caustic soda refining lines (Duff and von Klösterlein, 1979; Elliott, 1982; Farr, 1983).
7. The control of the running-off of soapstock from batch neutralization and the splitting of soapstock with sulphuric acid (Duff and von Klösterlein, 1979).
8. Proportioning bleaching earth addition to oil flow (Farr, 1983).
9. The operation of a semi-continuous bleaching plant (Pechey, 1982; Harper and Foster, 1984).
10. The automation of filtration and monitoring of filtration flow rate (Farr, 1983; Harper and Foster, 1984).
11. Hydrogenation and hydrogen control (Pechey, 1982; Harper and Foster, 1984).
12. The making up of oil blends (Duff and von Klösterlein, 1979).
13. The control of the deodorization process (Wocker, 1982; Harper and Foster, 1984).
14. Boiler efficiency and fuel consumption (Pechey, 1982).

Sensors capable of transmitting information on plant operating conditions, for example temperatures, pressures, weights and flows, are widely available, but an obstacle for the further development of automation in the edible oil industry is the shortage of on-line instrumentation for product quality parameters. Bitner and colleagues at the United States Department of Agriculture's North Regional Research Centre have reported their work on automated sampling and analysis. In brief these reports have been on the following subjects:

1. 'Computer monitoring of a mini-refinery', giving details of transducers for off-line measurement of colour, sodium content and free fatty acid concentration (Bitner *et al.*, 1973).
2. 'Development of an automated vegetable oil mini-refinery', in which the scope of automation has been extended to include bleaching, hydrogenation and deodorization and attendant test methods (Bitner *et al.*, 1977).
3. 'Monitoring of oil composition by an automated system for esterification–GC analysis', with details of the automated production of fatty acid methyl esters, which can be used for monitoring hydrogenation (Bitner *et al.*, 1980).
4. 'Analytical chain to monitor vegetable oil hydrogenation', in which the progress of the hydrogenation reaction is followed by on-line *trans*-acid measurement by infrared attenuated total reflectance and off-line measurements of refractive index and fatty acid composition by gas chromatography (Bitner *et al.*, 1982).

References and further reading

Bitner, E.D., Ernst, J.O. and Dutton, H.J. (1973) *J. Am. Oil Chem. Soc.*, **50**, 436.

Bitner, E.D., Snyder, J.M., Ernst, J.O. and Dutton, H.J. (1977) *Fette Seifen Anstrichm.*, **79**, 483.

Bitner, E.D., Mounts, T.L. and Dutton, H.J. (1980) *J. Am. Oil Chem. Soc.*, **57**, 209.

Bitner, E.D., Snyder, J.M., Mounts, T.L., Dutton, H.J. and Baker, G. (1982) *J. Am. Oil Chem. Soc.*, **59**, 286.

Duff, A.J. and von Klösterlein, C.B. (1979) *Fette Seifen Anstrichm.*, **81**, 257.

Elliott, P.F. (1982) Paper presented at *AOCS Meeting*, Toronto, Canada.

Farr, W.E. (1983) *J. Am. Oil Chem. Soc.*, **80**, 571.

Harper, A.J. and Foster, D.W. (1984) Paper presented at *AOCS Meeting*, Dallas, April/May.

Pechey, R. (1982) *Process Eng.*, April, p. 57.

von Klösterlein, C.B. and Vos, H.J. (1983) *J. Am. Oil Chem. Soc.*, **60**, 443.

Witte, N.H. (1980) *J. Am. Oil Chem. Soc.*, **57**, 854A.

Wocker, M. (1982) *Fette Seifen Anstrichm.*, **84**, 120.

5.7 Margarine and butter production

5.7.1 Introduction

The history of margarine begins with its invention by H. Mège-Mouriès in 1869 following an order from Emperor Napoleon III of France to produce a cheap and more stable butter substitute. He succeeded in producing such a product by a churn process also used for butter production, and started from cream prepared from water and/or milk and oleomargarine. He obtained this oleomargarine by fractionation of fresh beef tallow at 25–30 °C.

Both the margarine and its constituent oleomargarine (a shortening) were marketed in France on a modest scale (Van Alphen, 1964). But, although the cradle of margarine stood in France, production on an industrial scale got off to a good start in The Netherlands.

In the 1870s and 1880s, the process was introduced into many European countries and the USA, which greatly increased world production. In 1975, the world production of margarine was about 5.7×10^6 tonnes in relation to a total oil and fat production of 40×10^6 tonnes (Carrière, 1969). This total margarine volume was reached after a 10-fold increase between 1875 and 1925 and two 2-fold increases during the periods 1925–50 and 1950–60. In 1960, a volume of 4×10^6 tonnes was reached (Van Stuyvenberg, 1969). In 1973, world consumption was about 25% higher than that of butter, whereas the consumption of both products was almost equal in the European Community (EC) countries (Lesieur, 1976). In 1992 world margarine production is even expected to reach 9.2×10^6 tonnes.

This fast growth was made possible by a number of technical developments and innovations in the field of raw material utilization and processing and also by product diversification. Major developments were the hydrogenation process (see section 5.5.1) invented by Normann in Germany in 1902, the use of semi-continuous cooling drum/hopper/vacuum kneader systems since around 1930 (Gerstenberg complector, Schröder and Atlas Lardin), continuous votator processing equipment developed for margarine manufacture by Girdler USA in 1938 (see section 5.7.3(a)) and the launch of table margarine in plastic tubs for the first time in West Germany in 1955. Nowadays, a broad range of vegetable and animal fats is available for margarine production (see Chapter 3).

The major original function of margarine was, like that of butter, to provide a readily available source of energy. This function has not changed, but in the ongoing discussions on the health implications of fats, more emphasis is placed now on the role of essential fatty acids, which is accepted to be important to improve the balance between saturated and unsaturated fatty acids (WHO, 1982; Chapter 12).

Important non-nutritional functions of margarine are:

1. Improvement of palatability of bread and baked goods by lubrication.
2. Contribution to formation of an open structure of baked goods by trapping and retaining air in the batter during the mixing process.
3. Contribution to flavour release of foods.
4. Improvement of heat transfer during frying.

To obtain the plastic structure required for easy spreading on bread and handling in the bakery, the fat phase must consist of a solid (small fat crystals) and a liquid phase in which the crystals are homogeneously dispersed. The ratio between these phases and the crystalline character of the solid phase determine the product's consistency and firmness. Solid content and crystallinity both depend on fat blend composition and process conditions.

In the past, margarine was purely regarded as a substitute for butter and its physical properties were much like those of butter, which is plastic and spreadable at ambient temperature but very hard under cooler conditions. In the past three decades, developments shifted margarine more in the direction of tailor-made products, which differ significantly from classical margarine and butter as regards properties and fat composition. These new properties are in line with present-day market requirements, like spreadability from the refrigerator, a high content of polyunsaturated fatty acids and an optimal effect in, for instance, baked goods (Gander, 1976).

The fat phase of such tailor-made products varies in the first place as regards the solid/liquid ratio. Haighton (1976) showed that the product hardness is an exponentional function of the percentage of solid fat under constant processing conditions.

It requires much skill to find an optimal compromise between fat blend composition, processing conditions and desired production properties (see section 5.7.2(a)).

Margarine and butter are water-in-oil emulsions with a fat content of about 80%. Margarine legislation may vary slightly from country to country but, usually, a minimum fat level between 80 and 84% and a maximum water level between 16 and 20% are prescribed, the balance relating to permitted and/or prescribed ingredients (see sections 5.7.2(b)–(d)). Because the gross compositions of margarine and butter are very much alike and margarine fats are fully miscible with butterfat, it is very easy to manufacture products containing both margarine and butterfat. Until recently legislation in many European countries restricted severely the possibilities to market such blended products, in order to protect the dairy industry. Encouraged by more recent nutritional views, authorities are more prepared to admit such mixtures, and EC legislation is currently being drafted that will allow the sale of mixtures with 10 to 80% milkfat on total fat content. This development is an important step towards a uniform margarine legislation in Europe. Furthermore, the Codex Alimentarius Commission (1981) of the FAO/WHO has issued updated recommendations for margarine legislation.

A relatively recent development is the production of low-fat spreads in developed countries to meet demands for a reduced dietary intake. These spreads (fat content about 40%) were first introduced in the USA in 1964, and in Holland and the UK in 1969 and 1968 respectively. Legislation regarding this type of product is less strict with respect to mixing margarine and butterfat. In various countries, such mixtures are already on the market (Sweden, UK, France, Japan).

More recently, in anticipation of the new EC legislation, manufacturers in several Western European countries are producing spreads with fat levels between 80 and 41% or even with levels below 39% to meet the demands of reduced fat intake in the industrialized countries. In the USA this development has been under way for some time.

Production of spreads with up to 60% fat is, in fact, not much different from that of margarine (see sections 5.7.2 and 5.7.3). Production of low-calorie spreads and spreads with very low fat levels requires increased emphasis on the properties of the water phase. In this connection, control of the water droplet size is very important, and the use of thickeners (starch, gelatine, milk protein, etc.) may often be required (Moran, 1993). Butter production is different and is still based on the original churn process (for a discussion, see section 5.7.4) (Rajah and Moran, 1994).

5.7.2 Margarine raw materials and ingredients

(a) Oils and fats

Fats contain so many different triacylglycerols differing in melting point and degree of solubility in oil that the solid/liquid ratio in fats and in mixtures of fats and oils varies with the temperature of the product.

The extent to which the percentage of solids in various margarine blends and butter varies with temperature is shown in Fig. 5.7. Such curves are generally determined by dilatation or nuclear magnetic resonance techniques (see sections 6.2.2 and 6.2.4).

As already stated, the physical properties of margarine are controlled, for a significant part, by the solids content of the fat phase. This can also be demonstrated by the curves mentioned. Table and kitchen margarine as well as butter melt just below body temperature, which is an essential requirement for a good melt in the mouth. Table margarine is spreadable at a relatively low temperature, as is the highly polyunsaturated fatty acid-containing product. Tropical margarines must have sufficient body at high ambient temperatures and should, therefore, contain a large amount of relatively high-melting triacylglycerols. Consequently, the meltdown characteristics will be poor. In this context, puff-pastry margarine is referred to as an example of a very hard product. This is related to its ability to introduce a flaky structure in the cooked pastry. In general, bakery margarines are relatively harder than standard retail cooking margarines. The actual required solids levels depend on the desired effect

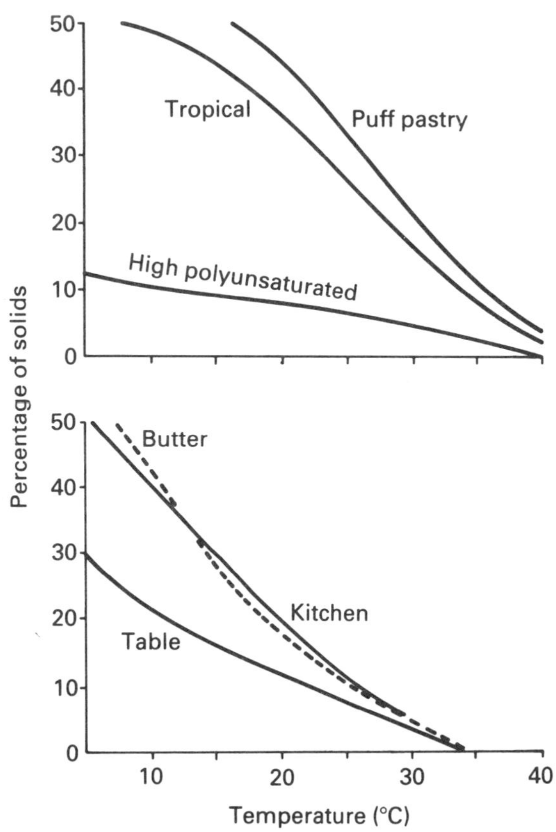

Fig. 5.7 Percentage of solids versus product temperature of the fat phase of some typical margarines and butter.

in the dough and the dough preparation procedures applied in the bakery. Products suitable for cake preparation, Danish pastry and cookies in order of increasing hardness should be mentioned. They are still softer than a puff-pastry product.

Taking into account the world market situation, prices and national legislation or regulations, margarine manufacturers have to make an optimal raw material choice enabling them to compose mixtures with the desired solids versus temperature curves and so obtain the product properties desired. Haighton (1976) showed how the solids curve of a blend can be calculated on the basis of individual raw material data. Calculations may not always result in an optimal match with the solids curve desired, because the crystallization of fats is a very complex process with too many different triacylglycerols playing a role. Practice may differ somewhat from theory so that further fine tuning will be required by adapting the calculated concentrations of the individual raw materials within narrow limits and by checking the blend and product properties ultimately obtained. In this way also the effects of crystal size distribution, crystal contacts and interlocking can be observed and corrected for, if necessary.

Examples of typical margarine blends are given by Andersen and Williams (1965) and Van Stuyvenberg (1969). For an optimal taste stability of the finished product, the raw materials should be refined and deodorized properly (see section 5.4).

(b) Water phase

The water phase of a margarine may be 16–20%, dependent on local legislation, and consists of water, cream, whole milk, skim milk (powder), whey (powder) and buttermilk in various ratios. Some countries may place restrictions on the use of certain components (section 5.7.1). In general, for cost and taste stability reasons, components not containing butterfat will be chosen.

The use of milk proteins is advantageous in relation to the palatability, flavour and taste stability of the margarine, while it also influences its frying properties (browning, sediment). There is a risk of bacterial infection, which can be minimized by pasteurizing the milk phase. Other measures taken to inhibit the growth of bacteria, moulds and yeasts are: a fine water distribution (see section 5.7.2(c)); addition of salt (to be effective, more than 13% NaCl in the water phase is desirable, which, in many countries, is not appreciated for taste reasons); a low pH (preferably <4.5); use of preservatives (like benzoic or sorbic acid or their sodium salts); and hygienic processing.

A low pH can be obtained by bacteriological souring of the milk phase, which also has a positive effect on the aroma of this phase. The pasteurized and cooled milk phase is then inoculated with 1–2% starter culture. Strains used are *Streptococcus lactis*, *S. cremoris* and *Betacoccus cremoris*. They produce lactic acid and minor amounts of aroma-producing compounds such as diacetyl (2,3-butanedione) and acetylmethylcarbinol (2-hydroxy-3-butanone). After a ripening period of about 8 h at 20–25 °C, the milk phase is cooled down to 5–8 °C before further processing.

Another way to lower the pH is the addition of an acidity regulator such as lactic or citric acid. Thickeners can be used to stabilize the water phase, particularly in finished products with 40% or less fat. The formation of loose moisture can thus be prevented, but at the risk of increased bacteriological vulnerability of the product. Not only is the water phase coarsened, but the concentration of nutrients for the micro-organisms is also increased. As a consequence additional precautions may have to be taken to prevent microbiological spoilage, during processing, packaging and distribution.

(c) Emulsifiers

In the first stage of margarine production (see section 5.7.3), the fat–water mixture is emulsified by vigorous agitation. This process is facilitated by adding 0.1–0.2% monoacylglycerols to the fat phase. During further processing, the emulsion is stabilized by the fat crystals formed. However, the monoacylglycerols also play a role because coalescence of water droplets and even water leakage might occur during the final working and kneading stages of the process without such emulsifiers being present.

For bacteriological reasons, the water droplets should be very fine (preferably all droplets $<5\ \mu m$, but larger droplets incidentally present should never be $>20\ \mu m$). On the other hand, they should not be too fine to present a greasy and dry sensation on the palate.

A compromise should be found by choosing a well balanced emulsifier system, taking the presence of milk protein and lecithin into account as well. The lecithin (0.3–0.4% on fat phase) is, in the first place, used to influence properties such as spattering, sedimentation and browning during frying.

For further information on emulsifiers, see section 5.8.

(d) Other ingredients

(i) Flavouring

Many substances identical to natural compounds are used to impart a butter-like flavour and taste to margarine. Most such flavour cocktails are composed on the basis of analytical data on butter. Diacetyl, fatty acids and ketones are the most important components.

(ii) Colouring agents and vitamins

Colouring agents generally applied in margarines are based on carotenes (e.g. red palm oil, natural anatto or synthetic β-carotene), and vitamins (A and D) are often added to increase the nutritive value of the product (a legal requirement in some countries). In this context, it should be noted that β-carotene is the provitamin of vitamin A.

(iii) Antioxidants

Rancidity, caused by chemical oxidation of fats, can be prevented or retarded by antioxidants, such as tocopherols, which are naturally present in some margarine oils. Well-known synthetic antioxidants are BHA (t-butylhydroxyanisole), BHT (di-t-butylhydroxytoluene), TBHQ (t-butylhydroquinone) and various alkyl gallates (Sherwin, 1976).

With the exception of some flavour components, all the ingredients mentioned are added to the fat phase. Addition may be subjected to legal restriction or obligation (vitamins); for example, see Van Stuyvenberg (1969).

5.7.3 Margarine processing

(a) Introduction

The basic margarine processing steps are: emulsification (see also section 5.7.2(c)); crystallization of the solid part of the fat phase; and plastification of the crystallized emulsion.

In the past (see section 5.7.1), these operations were carried out separately. In the multitube-cooler plants (votator process) almost exclusively used now, they are performed more or less simultaneously, although there are slight differences. Emulsification is concentrated in the first part and plastification in the last part of the line.

This continuous votator process was first used for margarine production by Girdler USA in 1938, introduced into the UK during the Second World War and in the rest of Europe in the 1950s. In some factories, the semi-continuous cooling drum/hopper/vacuum kneader process is still used, but almost exclusively for special-purpose margarines.

Votator processing has the following advantages compared with drum processing: reduced risk of bacteriological contamination and oxidative damage; more constant product quality; and lower production costs (labour, energy and equipment).

In the following paragraphs, the general principle of votator margarine processing (about 80% fat) is described. Significantly lower fat levels require adjustment of the described conditions to allow for the higher water levels and the possible use of thickeners.

(b) Multitube-cooler plant for wrapped margarine

Figure 5.8 shows a flowsheet of a modern plant. The various units and systems are discussed below.

(i) Dosing system

The fat phase and water phase are prepared in separate premix vessels and contain the fat-soluble and water-soluble ingredients respectively. They are then accurately dosed by means of a proportioning pump, after which the mass is pumped to an A-unit through a common tube where some mixing takes place. More sophisticated dosing systems can also proportion the ingredients separately. On the other hand, the whole composition can be prepared in one premix tank. Then, only a simple dosing pump is required.

(ii) A-unit

An A-unit is here a single scraped-surface tube cooler*. In Europe, it is generally externally cooled by liquid ammonia. The steel rotor is fitted with rows of scraper blades, pressed against the cooled inner surface of the tube by the high rotational speed (300–800 rev/min) of the rotor. High local pressures and shear induce a rapid nucleation, crystallization and fine mixing in 4–10 s. Usual dimensions of the unit are: internal diameter, 10–25 cm; annular space between rotor and inner wall, 5–10 mm; and length, 100–130 cm.

In general, a processing line consists of more than one A-unit. Modern packaging equipment can handle line capacities of 3–4 tonnes/h. In such cases, margarine with an average solids content is often processed with three A-units with a total cooling surface area of about 2.5 m^2.

Drawings of a Girdler A-unit and a crystallizer (discussed in the next section) have been given by Haighton (1976).

* In the open literature, an A-unit is often a combination of several scraped-surface tube coolers.

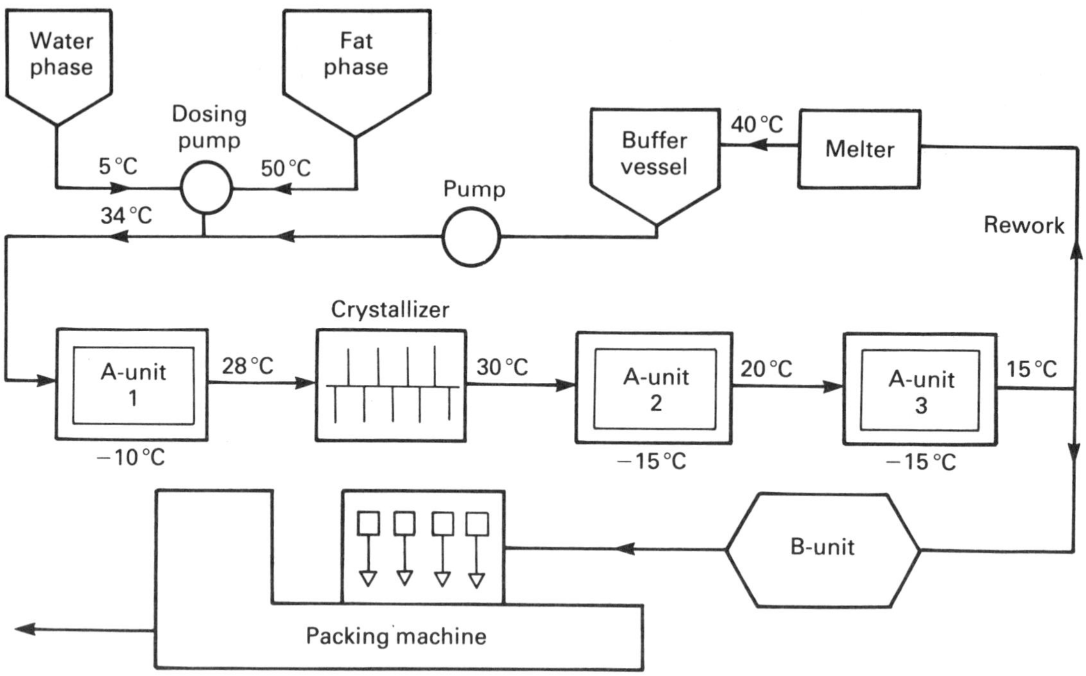

Fig. 5.8 Flow sheet of a modern multitube-cooler plant for wrapped margarine (votator line).

(iii) Crystallizer
A crystallizer is a cylinder fitted with pins on its inner wall and rotor. It ensures a sufficiently long residence time (some minutes) to create equilibrium conditions with respect to crystallization and keeps the crystallized mass pumpable for further processing. Usual parameters are: volume, 50–150 litres; and rotor speed, 100–200 rev/min.

(iv) B-unit
Most of the structure of the product is formed in a (tubular) B-unit. The residence time should be related to the rate of crystallization of the fat blend so that, at the end of the tube, crystallization is practically complete. In general, a residence time of some minutes is sufficient. To obtain a smooth, plastic and packable product, some working is usually carried out by placing a metal perforated plate or screen at the end of the unit. Without working there, a brittle and too hard structure may result.

(v) Rework
In the case of a breakdown of packaging equipment, it is not necessary to stop the votator because the processed material, which is ready for packing, is then passed from the inlet of the B-unit through a rework pipe to a buffer vessel. The rework margarine is heated in a melter to a temperature above its melting point to ensure complete melting of the fat crystals and kept at that temperature in the buffer vessel. By stopping fresh raw material intake and circulating rework, the votator can continue running under the adjusted conditions, thus avoiding the risk of blocking the line by the formation of too many solids. The production can immediately be resumed as soon as the packaging unit is in working order again.

(c) Process principles

A-units are very efficient coolers so that the fat crystallization speed cannot keep pace with their heat removal capacity. As a consequence, the emulsion leaves the first A-unit in a supercooled state. By placing the crystallizer after this A-unit, the higher-melting fat fraction can crystallize more or less separately, which may result in somewhat larger crystals, a softer product and better palatability. Further crystallization and supercooling take place simultaneously in the next A-units and crystallization is completed in the B-unit, followed by a final work step (sieve plate) at the end of the process line. Process conditions have to be chosen in such a way that the product is plastic and hard enough for packing and its amount of solid phase should be more or less in equilibrium with the ambient temperature. If this is not the case, there is a risk of post-crystallization in the pack, which may cause brittleness. For the same reason, a uniform temperature during storage and distribution has a beneficial effect.

In the first A-unit, emulsification is started (see section 5.7.2(c)) and continued in the second cooling stage and the B-unit with sieve plate. Very heavy working in the latter unit may cause droplet coagulation and even loose moisture formation.

The way of processing described so far is called 'straight-through with pre-crystallization'.

(d) Other processing routes

Another way to induce pre-crystallization is by recirculating part of the supercooled emulsion back into the fresh product flow. The effects are similar to those of a pre-crystallization by the process described.

Although they may have a negative effect on palatability, processes without pre-crystallization are sometimes applied, particularly when an extra-tough structure is required. The line then has some A-units and only one B-unit.

For tub products, a much softer filling consistency is required than for wrapping. Therefore, tub products are processed on lines with a (post-)crystallizer instead of a B-unit. The filling consistency can be adjusted by varying the rotational speed of the crystallizer. Pre-crystallization can be applied in the tub filling lines as well.

Detailed information on processing plants is given by Heesch (1963) and Andersen and Williams (1965). The processing of fats (shortenings) does not require different processing techniques.

5.7.4 Butter processing

(a) Introduction

Butter contains 80–82% milk fat and 18–20% aqueous phase. For processing reasons it is, therefore, tempting to start from a material of the same composition, e.g. an 80% cream or a mixture of milk fat and aqueous phase. Some processes that circumvent the buttermilk by-product problem are the Alfa-Laval (Gemmill, 1946), the Cherry Burrell (Cherry Burrell Corp., 1949), the Meleschin (Krupin, 1959) and the Creamery Package process (Creamery Package Manufacture Co., 1946). These processes are no longer used because of insufficient final product quality. At present, cream containing 30–45% fat is the raw material usually used for butter production.

Butter manufacture can be carried out as follows:

MILK (3–4% milk fat)
→ *Centrifugal separation* →
CREAM (30–45% milk fat) | SKIM MILK (0.05% milk fat)

CREAM →
Pasteurization
Physical ripening
Bacteriological ripening (optional)
Churning
Draining
→ BUTTER GRANULES | BUTTERMILK (0.2–0.6% milk fat)

BUTTER GRANULES →
Washing (optional)
Salting (optional)
Kneading
→ BUTTER (80–82% milk fat)

BUTTER → *Packaging* → FINAL PRODUCT

(b) Cream preparation and treatment

Cream is prepared by centrifugal separation of whole milk at 40–50 °C. To improve the keepability of the butter, the cream must be pasteurized (82 °C/20 s to 100 °C/1 s) to destroy any microbial contamination and to inactivate enzymes. After pasteurization, the cream is quickly cooled.

To reduce the churning time and the loss of fat in the buttermilk and to obtain a good butter consistency, the cream has to rest for several hours at reduced temperature (physical ripening) to allow the milk fat in the globules to crystallize properly.

To optimize this crystallization and finally obtain an oil/fat ratio optimal for the churning process, a defined temperature regime must be applied (see below). This regime depends strongly on the composition of the milk fat, which is subject to seasonal variations; the iodine value of winter milk fat is about 30 and that of summer milk fat about 40.

When lactic butter instead of sweet butter must be produced, not only physical ripening but also bacteriological ripening (souring) is necessary. These processes are carried out simultaneously. Souring is achieved by addition of 2–5% starter culture. During souring, lactose is partly converted into lactic acid. This process reduces the pH from 6.5 to 4.7 and forms flavour substances, mainly diacetyl, giving the butter a delicate flavour.

(c) Churning

During churning, phase inversion of the ripened cream and separation of buttermilk take place. Air is whipped into the cream, thus forming a foam with air–serum interface. The fat globules are concentrated at this interface and are mostly damaged due to mechanical and interfacial forces. Squeezed liquid milk fat spreads over the interfaces and promotes the agglomeration of fat globules and fat crystals of the damaged globules. This process is continued till pea-sized butter granules are formed and buttermilk is separated.

The churning process is carried out batchwise in rotating butter vats or continuously since the development of continuous butter-making machines (Fritz process; Pearce, 1948). In these machines, ripened cream is mechanically treated by rotating beaters.

(d) Draining and washing

The separated buttermilk is drained. Its milk fat content should be as low as possible, which is mainly influenced by the method of physical cream ripening. Sweet buttermilk contains more milk fat (0.3–0.6%) than sour buttermilk (0.2–0.3%). The butter granules are washed with cool tap water to reduce the buttermilk solids content. This step is omitted in cases where buttermilk drainage is sufficient. The process of buttermilk and wash water removal is controlled in such a way that a low residual water content (about 13%) is obtained, so that the final water content can be adjusted to 16% by controlled addition of flavour concentrates/cooking salt solution/water. Addition of sour flavour concentrates allows the manufacture of lactic butter without souring the cream (NIZO, 1975).

(e) Salting

For taste reasons, sweet as well as lactic butter is optionally salted. To this end, brine is added before kneading. For relatively high salt contents (e.g. 2%, in Scandinavian countries), salt is added in the continuous process as a dispersion in a saturated solution.

(f) Kneading

The butter granules together with the added solutions are kneaded by rotating the vat (batch process) or by worm conveyors (continuous process). In big continuous equipment, the kneader consists of two sections. The first removes excess water at a low rotational speed, whereas the second section finely disperses the serum, added solutions and some air at a higher rotation. Perforated plates and revolving knives promote this dispersion.

(g) Packaging

Butter is packed directly after manufacture or after setting in mini, consumer or bulk packs. Packaging after storage is done to obtain less firm products; this necessitates kneading (work softening) at the packaging stage.

(h) Firmness of butter

Apart from work softening, controlled crystallization of the milk fat in the fat globules of the cream is the most frequently applied method to reduce the firmness of butter, especially at lower (refrigerator) temperatures. Owing to the unique fatty acid pattern of the milk fat triacylglycerols, milk fat crystallizes in two fractions. By a separate crystallization of these fractions, mixed crystal formation should be prevented and the solids content be reduced. Therefore, winter cream, for instance, is kept for some hours at 8, 21 and 12 °C according to the so-called Alnarp method (Samulsson and Petterson, 1937). The precise temperature regime depends on the fat crystallization behaviour and the melting curve of each cream. The firmness of winter butter can be further reduced by adding frozen summer cream or vegetable oil when legally permitted (e.g. in Sweden).

Nowadays, butter-making is a fully automated, continuous process. The machines have a throughput of 7.5–10 tonnes/h and produce a product of constant quality and composition.

References and further reading

Andersen, A.J.C. and Williams, F.N. (1965) *Margarine*, Pergamon, Oxford.

Carrière, P. (1969) *Iterg-Journées d'Information sur la Margarine*, 21.

Cherry Burrell Corp. (1949) US Patent 2466896.

Codex Alimentarius Commission (1981) CDEX STAN 32-1981.

Creamery Package Manufacture Co. (1946) US Patent 2406819.
Gander, K.F. (1976) *J. Am. Oil Chem. Soc.*, **53**, 417.
Gemmill, A.V. (1946) *Food Ind.*, **18**, 841.
Haighton, A.J. (1976) *J. Am. Oil Chem. Soc.*, **53**, 397.
Heesch, A. (1963) *Fette Seifen Anstrichm.*, **65**, 125, 205.
Krupin, C.W. (1959) *Maschinen und Apparate der Molkerei und Milchindustrie*, VEB-Verlag Technik, Berlin, p. 239.
Lesieur, B. (1976) *J. Am. Oil Chem. Soc.*, **53**, 414.
Moran, D.P.J. (1993) *Porim Technol.*, **15**, Palm Oil Research Institute, Malaysia, Selangor, Malaysia.
NIZO (1975) DOS 2652558.6.
Pearce, J.A. (1948) *Can. Dairy Ice Cream J.*, **27**, 48.
Rajah, K.K. and Moran, D.P.J. (1994) *Fats in Food Products*, Blackie, Glasgow.
Samulsson, E. and Petterson, K.J. (1937) *Arsskrift för Alnarps laut brils-, mejeri- og trädgardinstitut.*
Sherwin, E.R. (1976) *J. Am. Oil Chem. Soc.*, **53**, 430.
Van Alphen, J. (1964) *Oleagineux*, **19**, 525.
Van Stuyvenberg, J.H. (1969) *Margarine*, University Press, Liverpool.
WHO (1982) Expert Committee, *Technical Report Series*, No. 678, Geneva.

5.8 Surfactants and emulsifiers

The consumption of industrially manufactured food products is increasing in modern society. Consumers are changing their way of purchasing food items from a daily to a weekly habit, and therefore a longer shelf-life of distributed foods is needed. To produce food products with the quality and shelf-life required by the consumer, the food industry utilizes surface-active lipids (surfactants, emulsifiers), which are chemical derivatives of natural edible fats and oils, or fatty acids.

The terms 'surfactants' and 'emulsifiers' are used in this section synonymously for amphiphilic lipids, having both hydrophilic and lipophilic properties that can be influenced by varying their chemical composition. They play an important role in many foods, not only in emulsions but also as complexing agents in starch-containing products or crystal modifiers in fats.

This section reviews food-grade surfactants, their chemical and physical characteristics, applications and functions in food products and legal aspects.

Surfactants are partial esters of fatty acids with chain lengths from C_{12} to C_{22} and polyvalent alcohols like glycerol, propylene glycol, sorbitol/sorbitan and sucrose, or organic hydroxy acids like lactic acid. Furthermore the partial esters, e.g. monoglycerides, may be esterified with organic acids such as acetic acid, lactic acid, diacetyltartaric acid, succinic acid or citric acid, forming a wide range of esters with big variations in properties and functions. The most commonly used types of emulsifiers are listed in Table 5.6.

Comprehensive reviews of food emulsifiers, their chemistry and applications are given by Schuster (1985). The world production of food emulsifiers is estimated to be approx. 200 000 tonnes (Als and Krog, 1990).

5.8.1 Monoglycerides (monoacylglycerols)

Monoglycerides of saturated and unsaturated fatty acids are by far the most commonly used food surfactants, accounting for over 70% of total world use. On an industrial scale monoglycerides are normally produced by interesterification of fats or oils with glycerol (also called glycerolysis; section 10.12), by reacting the components at 180–230 °C under alkaline catalysis. Synthetic procedures for monoglycerides are dealt with in Chapter 7. Production of monoglycerides by biotechnology is increasingly being studied. Modifications of fats and oils by lipases are reviewed by Kurashige *et al.* (1989), Schuch and Mukherjee (1989) and Yamaguchi and Mase (1991). Yamaguchi *et al.* (1986) describe a method to produce concentrated monoglycerides from fatty acids and glycerol in the presence of a lipase from *Penicillium cyclopium*. Holmberg and Osterberg (1988) describe production of monoglycerides (80% yield) by enzyme-catalysed hydrolysis of palm oil using a microemulsion (L2 phase) technique. Weiss (1990) obtained 95% yield of monoglycerides using selective conditions regarding lipase, temperature and water content. The monoglycerides were finally purified by thin-film distillation.

The composition of a glycerolysis product depends on the ratio of glycerol to fat and on the temperature, pressure and time of the reaction. Usually the objective is to obtain a product with maximum monoglyceride and minimum triglyceride. The acids and glycerol should be less than 1%, and remaining catalyst should be removed or neutralized. For example, a blend of fat and 30% glycerol reacted with an alkaline catalyst at about 225 °C for 1 h will give a product containing 58% monoglyceride, 36% diglyceride and 6% triglyceride. Most commercial mono-/diglyceride products contain 30–50% of 1-monoglyceride. With less glycerol the proportion of diglyceride increases, so that with about 8% glycerol the product will be 20% monoglyceride, 48% diglyceride and 32% triglyceride.

The fats and oils most commonly used in the manufacture of monoglycerides include refined lard or tallow or soybean, cottonseed, palm, palmkernel, sunflower or safflower oils, each of which may be used in hydrogenated or non-hydrogenated form. The fatty acid profile of a monoglyceride preparation may thus vary considerably, and this is reflected in the physical forms available (powder, block, or semi-liquid) and in the functional properties of the product.

Completely random distribution of acyl groups on the available hydroxyl groups is not achieved because of the differing rates of esterification of primary and secondary hydroxyl groups. The equilibrium constant favours esterification of primary over secondary hydroxyl groups, and is between 2 and 3 at 200 °C and between 6 and 10 at room temperature. Therefore monoglycerides undergo acyl migration from 2- to 1-monoglyceride on cooling (Brandner and Birkmeier, 1960).

Distilled monoglyceride, made by molecular distillation of mono-/diglyceride, has been commercially produced in the USA since 1950 and in Europe since

1964. The use of distilled product has increased with the desire to use purer and better-defined products and because of their enhanced behaviour.

To produce distilled monoglyceride, glycerolysis product with 40–60% 1-monoglyceride is submitted to high-vacuum thin-film molecular distillation using either a multi-stage centrifugal short-path distillation procedure or a one-stage wiped falling-film cylindrical type of still. The distilled product will contain about 95% monoglyceride along with diglyceride (about 3%), triglyceride (<1.0%) and free fatty acids (<1.0%). The undistilled residue of monoglyceride (20%), diglyceride (70%) and triglyceride (10%) can be returned to the glycerolysis reactor.

Mono-/diglyceride mixtures as well as distilled monoglycerides are polymorphic, like triglycerides. Mono-/diglycerides crystallize from the melt in an α form and on further cooling transform into a sub-α form at a temperature 20–25 °C below the melting point of the α form. The α form is metastable and will transform into a higher-melting stable β crystal form. The crystallography of monoglycerides is described in detail in section 8.7.

Monoglycerides are completely soluble in fats and oils and dispersible in water under certain conditions. Distilled monoglycerides form mesomorphic phases in water, depending on temperature, concentration, fatty acid chain length and degree of unsaturation.

Reviews of the phase behaviour in water of commercial distilled monoglycerides and some of their derivatives have been described by Krog (1990). Lipid–water interactions are dealt with in detail by Larsson in section 8.2.

5.8.2 Monoglyceride derivatives

Both mono-/diglyceride mixtures and distilled monoglycerides are used as raw materials for making either more lipophilic or more hydrophilic derivatives, which are utilized in the food industry for applications in which the monoglyceride itself does not fulfil the functional requirements.

Esters of monoglycerides with organic acids such as acetic acid, lactic acid, diacetyltartaric acid, succinic acid and citric acid are widely used as food emulsifiers. The esters are produced in standard batch reactors normally by reacting monoglycerides with the organic acid or its anhydride. A direct esterification of fatty acids, glycerol and the organic acid is also possible for some products.

(a) Acetoglycerides

Acetylated monoglycerides, also called acetoglycerides, can be made by acetylation of distilled monoglycerides with acetic anhydride (min. 98.5% pure). The reaction takes place at a temperature of 120 °C for about 1 h and acetic acid is removed by deodorization.

Acetylation may be partial or complete, depending on the ratio between acetic anhydride and monoglyceride. Normally products with 50%, 70% or 90% acetylation are used in the industry depending on application. It is also possible to make acetylated monoglycerides by inter-esterification of monoglycerides with triacetin.

Chemically pure acetoglycerides are polymorphic lipids. According to Martin and Lutton (1972) 1-aceto-3-stearin shows the following polymorphic transitions:

$$\text{sub } \alpha_1 \xrightarrow{3.5\,°\text{C}} \text{sub } \alpha_2 \xrightarrow{12.5\,°\text{C}} \alpha \xrightarrow{45\,°\text{C}} \beta' \xrightarrow{48.5\,°\text{C}} \text{melt}$$

Commercially available acetoglycerides always contain small amounts of other glycerides such as mono- and diglycerides, which stabilize the α crystal form obtained by cooling from the melt. In practice, therefore, acetoglycerides are considered to be stable in the α crystal form (α-tending).

The α crystalline state of acetoglycerides is remarkable because of its flexibility and film-forming properties. Feuge *et al.* (1952) have measured elongation of acetoglyceride samples, and found that they could be stretched eight times their original length (800%), whereas monoglyceride samples could only be stretched to a negligible extent (4%) before breakage occurred.

(b) Lactic acid esters

The manufacture of lactic acid esters of monoglycerides (lactylated monoglycerides) is normally based on a reaction between 88% pure lactic acid and monoglycerides.

Direct esterification of glycerol, lactic acid and fatty acids is also a possible method. The lactic acid ester usually contains 15–35% esterified lactic acid. Free lactic acid and its glycerol ester are normally removed by washing with water since such water-soluble compounds can give a bitter off-taste in the finished product where the emulsifiers are used.

The structural formula for the lactylated monoglycerides shown in Table 5.6 represents the main component in the product. However, lactic acid easily forms polymers (condensation products) and, therefore, components containing di-, tri-, etc., up to hexalactyl esters of monoglycerides or diglycerides are present as well. By reacting the components, glycerol–fatty acids–lactic acid, at a molar ratio of

Table 5.6 Commonly used types of emulsifiers

Surfactant type	Chemical structure of main component (n = 10–20)	EC No.	US/FDA 21 CFR[a]
Monoglycerides	$CH_2OCO[CH_2]_nCH_3$ $CHOH$ CH_2OH	E 471	§ 182.1505 GRAS[b]
Organic acid esters of monoglycerides:			
Acetic acid esters	$CH_2OCO[CH_2]_nCH_3$ $CHOH$ CH_2OCOCH_3	E 472a	§ 172.828
Lactic acid esters	$CH_2OCO[CH_2]_nCH_3$ $CHOH$ $CH_2OCOCHOHCH_3$	E 472b	§ 172.852
Succinic acid esters	$CH_2OCO[CH_2]_nCH_3$ CH_2OH $CH_2OCOCH_2CH_2COOH$	–	§ 172.830
Diacetyltartaric acid esters	$CH_2OCO[CH_2]_nCH_3$ $CHOH$ $CH_2OCOCH(OAc)CH(OAc)COOH$	E 472e	§ 182.1101 GRAS
Citric acid esters	$CH_2OCO[CH_2]_nCH_3$ $CHOH$ $CH_2OCOCH_2C(OH)CH_2COOH$ $COOH$	E 472c	–
Polyglycerol esters of fatty acids (x = 1–8)	$CH_2OCO[CH_2]_nCH_3$ $CHOH$ $CH_2[OCH_2CH(OH)CH_2]_xOCH_2CH(OH)CH_2OH$	E 475	§ 172.854
Propylene glycol esters	$CH_2OCO[CH_2]_nCH_3$ $CHOH$ CH_3	E 477	§ 172.856
Sorbitan esters of fatty acids	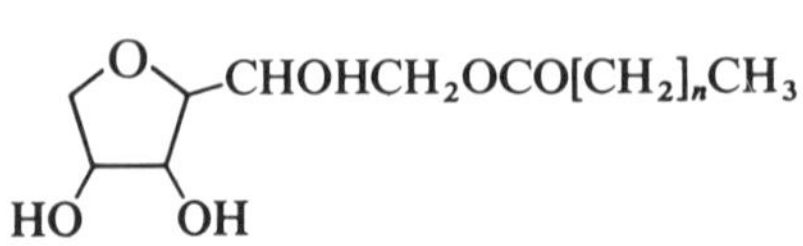	E 491 E 492	§ 172.842 –
Polyoxyethylene (20) sorbitan esters (polysorbate 60, 65, 80) ($w + x + y + z$ = 20)	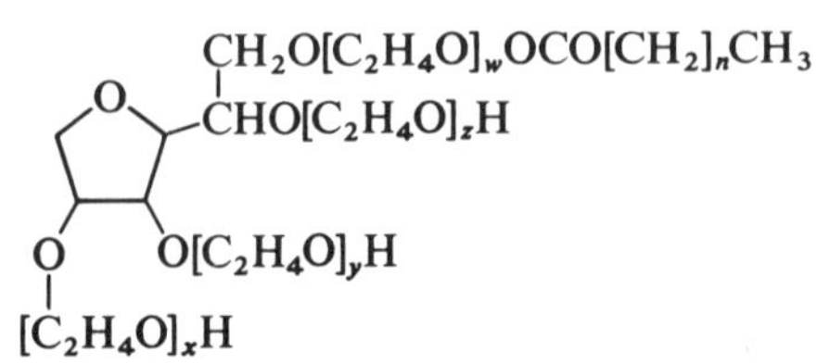	E 435 E 436 E 433	§ 172.836 § 172.838 § 172.840

Table 5.6 *Continued*

Surfactant type	Chemical structure of main component (n = 10–20)	EC no.	US/FDA 21 CFR [a]
Sodium stearoyl lactylate	CH_3 $CHOCO(CH_2)_{16}CH_3$ $COONa$	E 481	§ 172.846
Calcium stearoyl lactylate	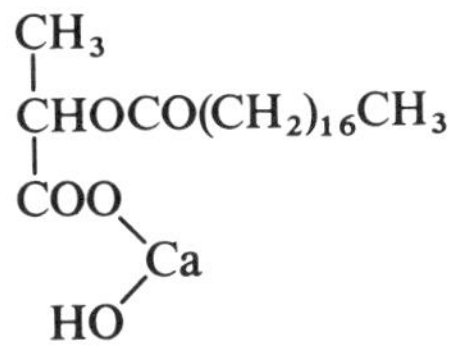	E 482	§ 172.844
Sucrose esters of fatty acids	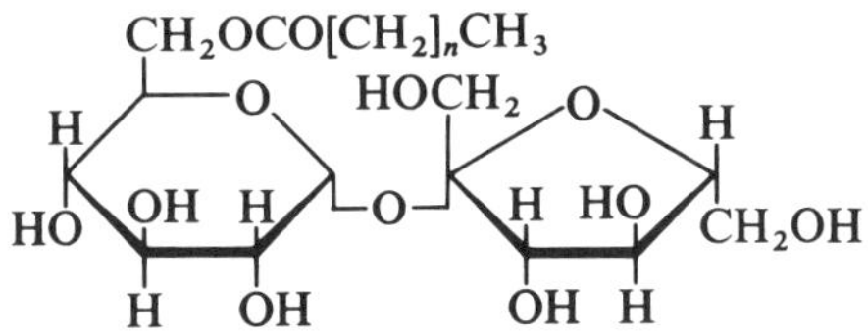	E 473	§ 172.859
Sucroglycerides		E 474	–

[a] Central Federal Register, Title 21.
[b] GRAS, generally recognized as safe.

1:1:1 the resulting product may contain a fair amount of glycerol–lactyl–fatty acid ester where the fatty acid is esterified to the OH group on the lactic acid molecule.

Lactylated monoglycerides are of the non-ionic lipophilic type of emulsifiers, completely soluble in oils and fats and only slightly water-dispersible. Like acetylated monoglycerides, they are less surface-active than the corresponding monoglycerides.

Lactylated monoglyceride is stable in the α crystal form obtained on cooling from the melt and is often described as an α-tending emulsifier, like aceto-glycerides and propylene glycol esters.

(c) Succinic acid esters

Succinic acid esters of monoglycerides (SMG) are anionic active emulsifiers made by reacting saturated C_{16}/C_{18} fatty acid monoglycerides (preferably distilled) with pure succinic anhydride at a ratio of approximately 4:1 by weight for about 1 h at 120 °C. The resulting succinic acid half ester of mono-glycerides has a free carboxyl group, which may be partially neutralized with alkali.

When cooled from the melt to about 55 °C, SMG shows a complex crystallization pattern. First it goes through a thermotropic liquid-crystalline state in a temperature range from 55 °C down to 48–50 °C. The liquid-crystalline state is characterized by having a long-range ordered structure caused by the polar group orientation, head to head, whereas the fatty acid hydrocarbon chains are still in a liquid-like state. X-ray diffraction of SMG at 50 °C shows long spacing of the order of 36–38 Å corresponding to a single chain length packing mode.

By cooling below about 48 °C an α-crystalline state is formed, which at room temperature transforms into a stable β form with a double chain length packing as normally found in monoglycerides.

Owing to the acidity of a free carboxyl group, the SMG molecule does not swell in water. When neutralized, SMG forms lamellar mesophases in water at a temperature above 45 °C. At higher water concentrations, dispersions of lamellar aggregates in equilibrium with water are formed.

(d) Diacetyltartaric acid esters

Diacetyltartaric acid esters of monoglycerides (DATEM) are produced by reacting acetylated tartaric anhydride with saturated or unsaturated monoglycerides under vacuum at a temperature of 120 °C for 2 h. Acetylated tartaric acid is made by reacting three parts of tartaric acid with seven parts of pure acetic anhydride at 110 °C for 30 min. Acetic acid is distilled off. The final step in the production of DATEM involves the reaction of diacetyltartaric anhydride with monoglycerides. Depending on the type of monoglycerides used as raw material, DATEM can be crystallized in an α-like crystal form (e.g. powder or flake form) or it can be semi-liquid.

According to the European Community (EC) regulations, DATEM may contain from 10 to 40% esterified tartaric acid, whereas in the USA the content of tartaric acid is limited to be within 17–20% by the Food Chemicals Codex.

In aqueous blends DATEM lowers the pH to about 2–3, owing to its free carboxyl group. If the pH is increased to about 4–5 by neutralization with alkali, DATEM forms lamellar mesophases, which are very stable over a wide temperature range (20–80 °C).

(e) Citric acid esters

Citric acid esters of monoglycerides (CMG) are manufactured by esterification of monoglycerides with citric acid, and the resulting product is often partially neutralized by alkali (Na salts).

Commercially available products are normally based on saturated monoglycerides with 60–90% monoglycerides, containing approximately 20% (min. 12%) esterified citric acid. The products are sold in powder or flake form.

CMG crystallizes as an α-like crystal form, which is stable. It forms a milky emulsion or dispersion in water, but owing to the complex composition it forms no ordered mesomorphic structures.

5.8.3 Polyglycerol esters of fatty acids

Polyglycerol esters (PGE) are made by esterifying condensed glycerol with edible fatty acids using an alkaline catalyst at 230 °C under vacuum. Interesterification of condensed glycerol with triglycerides is an alternative method.

The polyglycerol can have a polymerization degree from diglycerol to decaglycerol depending on the temperature (230–250 °C) and time (5–7 h) of the condensation process.

According to the EC regulations, the polyglycerol moiety shall predominantly be di-, tri- and tetraglycerol, and the product must not contain more than 10% of polyglycerols equal to or higher than heptaglycerol. In the USA it is permitted to use a polymerization degree up to decaglycerol.

PGE crystallizes from the melt in an α-like crystal form and does not show the polymorphism characteristic of mono-, di- or triglycerides. When dispersed in water above its melting point, it forms a hexagonal II mesophase, which on subsequent cooling forms a relatively stable α-crystalline gel.

5.8.4 Propylene glycol esters of fatty acids

Propylene glycol esters (PGME) can be manufactured by esterifying propylene glycol with edible fatty acids using an alkaline catalyst at about 200 °C under vacuum. After removal of excess propylene glycol, the reaction blend may contain approximately 55% propylene glycol monoester and 45% diester. An alternative method is interesterification of triglycerides and propylene glycol, yielding a reaction mixture containing the propylene glycol mono- and diesters together with monoglycerides and minor amounts of diglycerides as well as triglycerides.

The propylene glycol monoester can be concentrated by molecular distillation. Distilled PGME contains at least 90% monoesters, and is normally based on saturated C_{16}/C_{18} fatty acids.

PGME crystallizes from the melt in an α-like crystal form, and PGME with mixed fatty acid compositions are not polymorphic. Synthetically produced PGME based on pure fatty acids, however, does show polymorphism (Martin and Lutton, 1965), and racemic propylene glycol-1-monostearate shows four different polymorphic crystal forms, whereas optically active L(+)-propylene glycol-1-monostearate is monomorphic (Stauffer, 1967).

5.8.5 Sorbitan esters of fatty acids

Sorbitan is derived from sorbitol by dehydration and then esterified with fatty acids as described for PGME. Depending on the amount of fatty acids used for esterification, the equilibrium reaction mixture may contain primarily sorbitan monoesters or triesters. Sorbitan esters of palmitic or stearic acid crystallize from the melt in a stable α-like crystal form.

5.8.6 Polyoxyethylene (20) sorbitan esters

These very hydrophilic emulsifiers, also called polysorbates, are produced by reacting sorbitan mono-fatty acid esters with 5–20 mol of ethylene oxide, and the resulting products are very complex in composition and not very well defined in structure.

During the ethoxylation process, a transesterification takes place, resulting in a product where the main portion of the fatty acid is esterified to the terminal ethoxylene groups instead of being esterified to the sorbitan molecule.

Polysorbates are among the most hydrophilic emulsifiers used in foods. Polysorbates form hexagonal I mesophases when dissolved in high concentrations in water. In low concentrations they form micellar solutions, and the critical micelle concentration in water at 25 °C is of the order of 0.01–0.03 g/l (Becher, 1967).

5.8.7 Stearoyl lactylates (Na and Ca salts)

Esterification of stearic acid with lactic acid (minimum 88% pure) in the presence of sodium or calcium hydroxide (or carbonate) results in a mixture of stearoyl lactylates (sodium or calcium salts), fatty acid salts and free fatty acids. The main component is a stearoyl monolactyl ester, but esters containing di-, tri- and polylactic acids are also present.

Sodium stearoyl lactylate (SSL) is not completely neutralized when manufactured according to the EC or US/FDA regulations, which specify an acidity of not less than 60 mg of KOH per gram.

SSL crystallizes from the melt in an α-like crystal form, which may have a single or double chain length packing mode with corresponding variations in melting points from about 45 to 65 °C.

SSL is water-dispersible and forms hexagonal II mesophases at temperatures above 45 °C at pH 5.0. If neutralized to pH 6.5–7.0 in water, SSL forms lamellar mesophases, which can be diluted to form stable dispersions of lamellar units (Krog, 1990). Calcium stearoyl lactylate is less water-dispersible, but more soluble in oils and fats than the sodium salt.

5.8.8 Sucrose esters of fatty acids and sucroglycerides

Sucrose esters are manufactured by transesterification of sucrose with fatty acid methyl esters in a suitable solvent system where both components are soluble, such as dimethylformamide or dimethylsulphoxide (Benson, 1967). The sucrose esters are purified by distillation and crystallization, from e.g. ethyl acetate, in order to remove the solvent completely. For food use the content of dimethylformamide must be less than 1 mg/kg. An alternative method is to react the sucrose and fatty acid methyl esters in a microemulsion using 1,2-propanediol as solvent (Osipow and Rosenblatt, 1967).

The primary site of esterification of sucrose with fatty acid methyl esters is the sixth position on the glucose unit of the sucrose molecule. Commercial products contain, however, a mixture of sucrose mono-, di- and triesters.

Sucroglycerides are products made by the reaction of sucrose with triglycerides in e.g. dimethylformamide at about 140 °C in the presence of an alkaline catalyst. The sucroglycerides contain a mixture of 40–60% sucrose esters, the balance being mono-, di- and triglycerides.

Sucrose esters of palmitic and stearic acids, which are the most common types available, are solid at room temperature. They crystallize in a stable α-like crystal modification.

Sucrose esters with high HLB (hydrophilic–lipophilic balance) numbers are water-soluble and form lamellar mesophases in concentrated bulk solutions. In aqueous solutions and emulsions they have properties similar to polysorbates.

5.8.9 Food applications

The main applications of emulsifiers in the food industry are as processing aids in emulsions such as margarine, low-calorie spreads, dairy products, dressings, bakery products, chocolate and confectionery products.

The functional properties of a surfactant or emulsifier can be multiple, and vary from one application to another. As an example, monoglycerides are used (a) to stabilize water-in-oil emulsions (margarine, section 5.7), (b) to promote aeration of ice-cream emulsions due to a controlled destabilization of the emulsions, (c) to extend shelf-life of starch-based foods (wheat bread and cereal products) due to interactions with the amylose or amylopectin, or (d) to control viscosity of peanut butter, etc.

(a) Emulsions and foams

The surface-active properties of surfactants play a vital role for stabilization as well as destabilization of food emulsions. Many food emulsions contain proteins, which are adsorbed on emulsified oil droplets, forming interfacial films and stabilizing the emulsion droplets against coalescence (Dickinson and Stainsby, 1982). However, in whippable emulsions, such as ice cream, imitation dairy cream and toppings, a controlled destabilization due to desorption of interfacial protein films is necessary in order to obtain good whippability and foam structure (Buchheim and Dejmek, 1990), and monoglycerides are used widely in ice cream for that purpose.

Lipophilic emulsifiers, such as acetylated or lactylated monoglycerides and propylene glycol monostearate, are mainly used in non-dairy creams and toppings to facilitate aeration and improve foam texture.

Hydrophilic emulsifiers, such as SSL, DATEM, polyglycerol esters or polysorbates, etc., are used to stabilize salad dressings, beverage emulsions, flavour and vitamin emulsions, etc.

(b) Bakery and cereal products

There are three main functions of emulsifiers in bakery products: (a) dough strengthening of yeast-raised products (bread, rolls and buns), (b) aeration of cake emulsions, and (c) improvement of crumb softness and shelf-life.

The anionic-active emulsifiers DATEM, SSL, CSL, SMG together with polysorbates are primarily used as dough strengtheners to facilitate industrial production of yeast-raised doughs, in which they provide shock resistance, improved rheological properties of the dough, increased volume, and improved structure of the baked goods. The exact function of surfactants in doughs is not known in detail, but is probably similar to that of the native flour lipids (Pomeranz *et al.*, 1984), involving interactions between polar lipids and flour proteins, thereby forming the gluten network structure.

The emulsifiers used in cakes, cookies, etc., are acetylated or lactylated monoglycerides and propylene glycol monostearate, often in combinations with monoglycerides or polyglycerol esters.

Distilled, saturated monoglycerides are commonly used as starch-complexing agents in bakery products (bread) and other starch-based foods. The interaction with starch components is primarily with the amylose, which forms a water-insoluble complex with water-dispersible monoacyl lipids, such as monoglycerides (Krog, 1971). The starch-complexing effects of emulsifiers (monoglycerides) in wheat bread result in a softer crumb structure and prolonged shelf-life (anti-staling effect) of wheat bread and related products (Krog *et al.*, 1989).

Monoglycerides are often used in bakeries in the form of so-called hydrates, which consist of a suspension of 20–40% distilled monoglycerides in water. The production of hydrates involves formation of lamellar, liquid-crystalline phases at 50–70 °C as an intermediate stage before the mix is cooled and the monoglyceride crystallizes in its stable β crystal form. Phase diagrams of monoglycerides–water systems are shown in section 8.2.3. The advantage of using hydrates rather than monoglycerides direct in powder form is a finer particle size of the hydrate crystals, ensuring an optimum distribution in the dough.

In cereal products, pasta and snack foods, emulsifiers (monoglycerides, SSL) are used as extrusion aids and texturizing agents.

(c) Miscellaneous foods

Monoglycerides are used as starch-complexing agents in processed potato products. The ability of distilled, saturated monoglycerides to form aqueous dispersions at 60–65 °C is advantageous for the incorporation of emulsifiers during the potato processing.

Sorbitan tristearate is an effective β'-tending crystal modifier in confectionery fats (chocolate coatings) and margarine blends of β-crystal-forming fats. Crystal transformations leading to grainy texture of margarine or 'bloomed' surfaces of chocolate coatings can thus be eliminated. Polyglycerol esters are used in chocolate to reduce viscosity and provide better coating properties.

There are numerous other applications of emulsifiers in foods or confectionery products made on a commercial scale, and only some of the most important applications have been mentioned here.

5.8.10 Non-food applications

Lipid surfactants are used as emulsifiers, texturizing agents, lubricants and plasticizers in pharmaceuticals, cosmetics, textiles, plastics and oil drilling fluids, etc.

Polysorbates and sorbitan esters are used in pharmaceutical emulsions or dispersions containing vitamins or hormones. Lipophilic surfactants such as monoglycerides, propylene glycol ester or sorbitan esters are used as suppository bases. Monoglycerides and other lipophilic surfactants are typical compounds in cosmetic water-in-oil or oil-in-water creams, lotions and gels. Hydrophilic surfactants are used to solubilize perfume oils in various cosmetic oils and lotions.

The plastics industry uses surfactants as lubricants in processing of plastic products made from poly(vinyl chloride), polystyrene, etc., or anti-statics and anti-fogging agents. Monoglycerides are also used as fabric softeners in laundry detergents.

5.8.11 Legal aspects

Before a new type of food surfactant is permitted for use by the health authorities, it has to be tested in a number of toxicological studies including short-term and long-term feeding trials on several species together with studies on metabolism (WHO, 1974).

The FAO/WHO Joint Expert Committee on Food Additives has established acceptable daily intake (ADI) values for surfactants, and these values are used as a guidance by health authorities in the various countries where food laws exist.

Surfactants for which ADI values are not specified are often permitted for use in foods without limiting the use level provided that it does not exceed good manufacturing practice. For surfactants with limited ADI values the permitted maximum use level in various foods is calculated on the basis of the current eating habits in the individual countries of the world.

The EC Council Directive of 18 June 1974, 74/329/EC with later amendments, lists permitted surfactants, which are assigned an E number, and such surfactants may be authorized for use in food by all member states of the EC. A new international number system using the EC declaration numbers without the EC prefix is approved by FAO/WHO for use for declaration of emulsifiers in all countries.

In the USA surfactants are regulated by the US Food and Drug Administration, Washington, DC. Surfactants that are generally recognized as safe (GRAS) by US/FDA when used in accordance with good manufacturing practice can be used within practical and needed limits. Surfactants that are not GRAS substances are regulated in use level in specified food products according to the specifications given for each regulated surfactant in the Central Federal Register, Title 21.

Surfactants that have been approved for use in food by the EC or the US/FDA regulations should be regarded as safe food ingredients, when used as prescribed by the health authorities.

5.8.12 Analytical methods

Emulsifiers can be characterized by classical analytical methods for fat such as acid value, saponification value, or hydroxyl value depending on the nature of the emulsifier. See Chapter 6 for more details of these methods.

More detailed information about the composition of emulsifiers can be obtained by chromatographic methods. TLC is the most versatile chromatographic technique for emulsifiers. An emulsifier normally consists of a range of related components, which in TLC will show a characteristic pattern for almost every emulsifier type. More quantitative results can be obtained for some emulsifiers by gas chromatography, where the free hydroxyl groups are converted by a silylation reagent to the equivalent silyl ether to improve the separation. Mono- and diglycerides, lactic acid esters, acetylated monoglycerides and propylene glycol esters can be analysed by this GC technique (Brüschweiler and Dieffenbacher, 1991), whereas emulsifiers with free carboxylic groups are better analysed by HPLC. A new chromatographic technique that has shown interesting results is supercritical CO_2 chromatography (White and Houck, 1985). Especially for some of the higher-molecular-weight emulsifiers like polyglycerol esters, this technique has its advantages (Chester and Innis, 1986).

Size exclusion chromatography is also a technique that can be used for characterization of emulsifiers (Christopoulou and Perkins, 1986).

It has also been shown that quantitative information of the position-specific composition of mono-, di- and triglycerides can be obtained by ^{13}C NMR spectroscopy by complete identification of the glycerol carbon atoms and the C-1 and C-2 carbon atoms in the fatty acid molecule attached to each glycerol carbon atom (Gunstone, 1991).

References and further reading

Als, G. and Krog, N. (1990) *World Conf. on Oleochemicals Into the 21st Century* (ed. T.H. Applewhite), American Oil Chemists' Society, Champaign, p. 67.

Becher, P. (1967) *Nonionic Surfactants* (ed. M.J. Schick), Marcel Dekker, New York, p. 481.

Benson, F.R. (1967) *Nonionic Surfactants* (ed. M.J. Schick), Marcel Dekker, New York, p. 247.

Brandner, J.D. and Birkmeier, R.L. (1960) *J. Am. Oil Chem. Soc.*, **37**, 390.

Brüschweiler, H. and Dieffenbacher, A. (1991) *Pure Appl. Chem.*, **63**, 1153.

Buchheim, W. and Dejmek, P. (1990) *Food Emulsions* (eds K. Larsson and S.E. Friberg), Marcel Dekker, New York, p. 203.

Chester, T.L. and Innis, D.P. (1986) *J. High Resol. Chromatogr. Chromatogr. Commun.*, **9**, 178.

Christopoulou, C.N. and Perkins, E.G. (1986) *J. Am. Oil Chem. Soc.*, **63**, 679.

Dickinson, E. and Stainsby, G. (1982) *Colloids in Food*, Applied Science, London.

Feuge, R.O., Vicknoir, E.J. and Loogren, N.V. (1952) *J. Am. Oil Chem. Soc.*, **29**, 11.

Gunstone, F.D. (1991) *Chem. Phys. Lipids*, **58**, 219.

Holmberg, K. and Osterberg, E. (1988) *J. Am. Oil Chem. Soc.*, **65**, 1944.

Krog, N. (1971) *Stärke*, **23**, 206.

Krog, N. (1990) *Food Emulsions* (eds K. Larsson and S.E. Friberg), Marcel Dekker, New York, p. 127.

Krog, N., Olesen, S.K., Toernaes, H. and Joensson, T. (1989) *Cereal Foods World*, **34**, 281.

Kurashige, J., Matsuzaki, N. and Makabe, K. (1989) *J. Dispersion Sci. Technol.*, **10**, 531.

Martin, J.B. and Lutton, E.S. (1965) *J. Am. Oil Chem. Soc.*, **42**, 529.

Martin, J.B. and Lutton, E.S. (1972) *J. Am. Oil Chem. Soc.*, **49**, 683.

Osipow, L.I. and Rosenblatt, W. (1967) *J. Am. Oil Chem. Soc.*, **44**, 307.
Pomeranz, Y., El-Baya, A.W., Seibel, W. and Stephan, H. (1984) *Cereal Chem.*, **61**, 136.
Schuch, R. and Mukherjee, K.D. (1989) *Appl. Microbiol. Biotechnol.*, **30**, 332.
Schuster, G. (1985) *Emulgatoren für Lebensmittel*, Springer-Verlag, Berlin.
Stauffer, C.E. (1967) *J. Am. Oil Chem. Soc.*, **44**, 443.
Weiss, A. (1990) *Fett. Wiss. Technol.*, **92**, 392.
White, C.M. and Houck, R.K. (1985) *J. High Resol. Chromatogr. Chromatogr. Commun.*, **8**, 293.
WHO (1974) *Food Additives Series* No. 5, World Health Organization, Geneva, p. 7.
Yamaguchi, S. and Mase, T. (1991) *J. Ferment. Bioeng.*, **72**, 162.
Yamaguchi, S., Mase, T. and Asada, S. (1986) Eur. Pat. Appl. 191217 to Amano Pharmaceutical Co. Ltd, Japan.

5.9 Soap manufacture

5.9.1 Introduction

Soap is a substance of ancient origins, the manufacture of which has evolved from primitive beginnings into a sophisticated chemical process. It is mentioned in the Old Testament, 'For He is like the refiner's fire and like fuller's soap' (Malachi 3: 2), and extensively in writings of scholars and physicians whose centres of culture once bordered the Mediterranean (Levey, 1954). The practice of soap-making was substantially developed in Europe in the Middle Ages, notably at Savonna in Italy, Castille in Spain and Marseilles in France. Various technical innovations had a great influence on the development of soap processing, but it was the chemical discoveries of the late eighteenth century and early nineteenth century, particularly the introduction of the Le Blanc process (1804) for sodium carbonate production on a large scale, that allowed the industry to progress to its current state of development.

Soap is manufactured by a variety of techniques worldwide (Schulerud, 1963; Ryer, 1969). The basic unit operations may be conveniently grouped to describe the major facets of the most commonly used processes as shown in Fig. 5.9, i.e. soap-making, drying and finishing.

There is a further commercially important step associated with soap manufacture, namely glycerol recovery. Glycerol is a by-product of saponification or fat hydrolysis and represents a credit in the overall economics of soap production.

5.9.2 Raw materials

The principal raw materials for soap manufacture are natural oils and fats, and sodium hydroxide or, for specific applications, the more expensive potassium hydroxide. Mixtures of potassium and sodium hydroxide are used in some physically soft or translucent soap products, and ammonia is used in some speciality products.

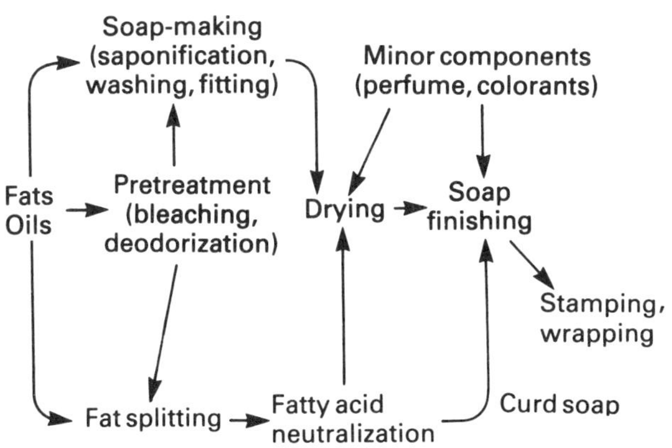

Fig. 5.9 Scheme of soap manufacture.

Common salt is also required in large quantities for glycerol extraction in the fat saponification process. Preservatives, colorants, superfatting agents and particularly perfumes, although present in relatively small amounts, represent a major proportion of raw material costs. For hard or laundry-grade soaps, some of the minor ingredients may be absent from the formulation and there may be partial substitution with inorganic fillers such as sodium silicate.

The most important triglycerides used for soap manufacture are tallow and coconut oil, although many of the natural oils and fats are or have been used (Davidsohn *et al.*, 1953). Those still frequently used include palm oil, palmkernel oil, babassu oil, soybean oil, grease, lard and marine oils.

The choice of a particular fat/oil blend is largely determined by economics within the overall constraint of producing a soap bar of the desired properties. These properties of resistance to oxidation, lather production, bar hardness, colour and odour are largely determined by the chain-length distribution and level of unsaturation of the fatty raw materials (Davidsohn *et al.*, 1953). Consequently, some constituent oils must be hardened or even dehydrated prior to their use for soap-making. For the most common fatty raw material charges, tallow and coconut oil, there is a range of blends from 85:15 to 50:50 that are frequently encountered in soap production. Soaps with chain-length distributions of between 12 and 18 carbon atoms have desirable features, particularly for lather generation, where the short chain length (laurate and myristate from coconut oil) and unsaturated soaps (oleate from tallow) dissolve quickly to give rapid lather generation, which is sustained by the longer-chain-length (palmitate and stearate) components (Ryer, 1969).

Empirical rules (Webb, 1927) are used to aid the selection of a fat blend. However, the final bar properties are also affected by the physical work process in soap finishing and the levels of the minor ingredients; therefore, within the broad limits described, there is scope for the soap manufacturer to accommodate a range of potential feedstocks.

For direct saponification routes to soap, once a fat blend has been selected then, particularly for toilet soap formulations, some method of raw material pretreatment is often applied. Pretreatment methods such as degumming or desliming particularly for soybean oil and alkali refining to upgrade low-quality tallows are used. However, the most commonly used pretreatment method is adsorption bleaching with diatomaceous earth and, less commonly, deodorization by steam stripping under vacuum (Davidsohn and Davidsohn, 1944). These processes are designed to improve odour and colour characteristics and to improve the stability of the product. The main components removed by pretreatment include coloured components, such as carotenoids and related oxidation products, polythene, which can give rise to black specks in soap, odorous components or their precursors such as isolinoleic acid (Von Barren and Klimmek, 1977) and finally free water, which can promote triglyceride hydrolysis on storage with a subsequent loss of glycerine.

In the production of soap from fatty acids, the distillation process normally associated with fat splitting is usually a sufficient pretreatment for the raw materials (Von Barren and Klimmek, 1977). Methods are available, however, to treat the fatty acids prior to distillation to prevent undesirable by-product formation, e.g. acid washing to remove trace metals. It is good practice to store fatty acids for the manufacture of high-quality soaps for as short a time as possible to prevent deterioration.

5.9.3 Soap phase equilibria

The phase behaviour of soaps is complex, but it is the application of the understanding of these relationships probably more than any other single factor that has allowed soap manufacturing to develop from an art to a science (McBain *et al.*, 1938; McBain and Lee, 1943; Vold, 1940, 1941; Maclennan, 1923). The binary phase diagram for sodium myristate–water is shown in Fig. 5.10.

Anhydrous and low-moisture-content soaps show a variety of distinct phases as determined by X-ray, optical and thermal techniques (Sanesi, 1980). Typically, a single pure anhydrous soap, such as sodium palmitate, passes through several successive phases when heated (Vold, 1940): curd fibre, a hard opaque, white solid, up to 117 °C; waxy-type phases between 117 and 208 °C; sub-waxy soap, slightly more translucent and less hard than curd fibre (117–135 °C); waxy soap, slightly softer than sub-waxy (135–172 °C); super-waxy soap, again slightly softer than waxy soap (172–208 °C); from 208 to 295 °C, neat soap phases, not to be confused with soap boilers' neat soap; sub-neat soap, a phase that has a marked increase in translucency and decrease in rigidity compared with the waxy phase (208–253 °C); neat soap, a liquid-crystalline, free-flowing, translucent phase (253–295 °C); above 295 °C, an isotropic solution, which is quite clear and of relatively low viscosity.

On the addition of water the anhydrous phases are replaced by the aqueous phases, which are of great practical significance in soap-making. Soap boilers' neat soap is a rather turbid, lamellar, liquid-crystalline phase, which flows readily (Tiddy, 1980). At higher temperatures super-neat soap forms. This is very similar to soap boilers' neat soap. Further water addition produces middle soap, which is again a liquid-crystalline phase, probably hexagonal (Luzzati and Husson, 1962), and unlike soap boilers' neat soap it appears as a viscous, plastic, translucent material. It is found in soap-making when the electrolyte concentration is allowed to fall and, once formed, this phase, also referred to as water lump, is difficult to remove.

Under the practical conditions that exist in most batch and continuous fat saponification soap-making routes, the ternary phase diagram at 90 °C (McBain and Langden, 1925), incorporating electrolyte content as the third axis, allows interpretation and therefore control of the changes that occur as a result of the processing steps involved. In addition, the pioneering work of J.H. Wigner provides a grounding in the fundamentals of soap manufacture (Wigner, 1940). A modified ternary phase diagram for a commercial soap formulation is shown in Fig. 5.10. This illustration shows the part of the phase diagram of most practical significance in soap-making. Commercial soap behaves essentially as a single soap over the range of conditions and compositions encountered in the soap-making process. The important single-phase regions are kettle wax, soap boilers' neat soap, middle soap, curd soap, nigre (isotropic solution of soap and electrolyte) and lye (electrolyte solution with minimal dissolved soap).

Two- and three-phase regions are combinations of these single phases. The ternary phase diagram can be used to illustrate features of the soap-making process. For example, soap at a point X in the two-phase region H of the diagram will separate along a tie line to give materials whose compositions are represented by points X_1 and X_2 respectively. Consequently a settling operation would yield soap boilers' neat soap with approximately 65% anhydrous soap, which is the raw material for drying to curd soap.

The solid curd phase region that forms on drying and cooling soap boilers' neat soap is also of practical significance, particularly for soap finishing. On phase diagrams, this region is normally shown as a heterogeneous mixture of solid soap phases and isotropic solution (Madelmont and Perron, 1976). In practice, there is a range of solid soap phases that enmesh a

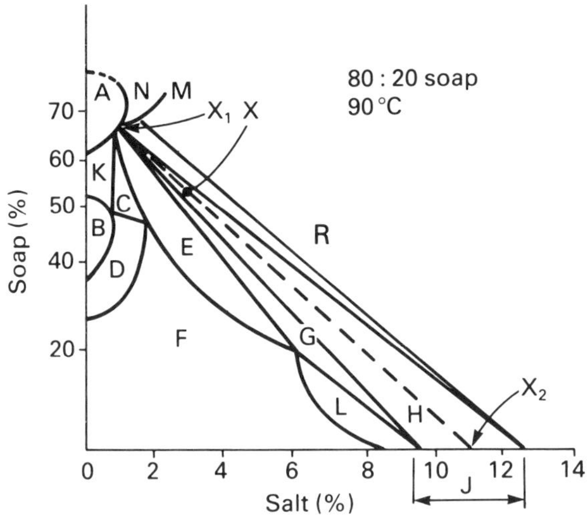

Key

Single-phase regions

A Neat soap
B Middle soap
F Nigre
J Brine (lye)
N Kettle wax

Two-phase regions

D Middle soap – nigre
E Neat soap – nigre
H Neat soap – brine (lye)
K Neat soap – middle soap
L Nigre – brine (lye)

Three-phase regions

C Neat soap – middle soap – nigre
G Neat soap – nigre – lye
M Neat soap – kettle wax – lye

Fig. 5.10 Soap/salt/water phase diagram.

liquid phase (Ferguson *et al.*, 1943). These phases are probably not in true equilibrium because of the reduced mobility of the species, but their approach to equilibrium can be enhanced by physical work (Buerger *et al.*, 1947).

A wide variety of solid soap phases have been postulated, but it is the four phases α, β, ω and δ that are of greatest interest in the soap-finishing processes of soap manufacture for their distinctive physical properties (Ferguson *et al.*, 1943). The α form is a slightly hydrated phase that has been observed in pure sodium palmitate and stearate soap. On dehydration α phase transforms to β phase. The β phase can exist in a translucent form, it lathers easily and may be distinguished from the ω phase by its tendency to swell and disintegrate in water. The ω phase is most readily formed by rapidly chilling neat soap without agitation, and occurs when low-moisture-content soaps are cooled from a molten condition. The ω phase is the predominant form in commercial soap formulations following the drying step in soap manufacture and is particularly favoured when there is a proportion of low-molecular-weight material, e.g. coconut oil, in the original fat charge. The formation of δ-phase soap is favoured by higher-molecular-weight materials, higher moisture contents and lower temperatures. It appears in high-moisture-content soaps on cooling and may be produced by physically working β-phase soap at low temperatures. Its effect in reducing the firmness of soap bars is quite marked, but its formation of lather is superior to that for ω phase.

The transformation from one soap phase to another is critically dependent on the temperature, moisture content and composition. Physical work aids the transformation, which if not applied results in the presence of mixed phases. Table 5.7 summarizes the main characteristics of the four important phases.

Finally, in addition to the consumer properties dictated by the composition and phase structure of soap, they are also affected by the orientation and compactness of the crystals imparted by the physical work processes. When soap is induced to flow, as during extrusion, the crystallites tend to align themselves so as to offer the least resistance (Mills, 1942). This orientation affects the penetration of water into the soap bar, which therefore influences the solubility and related features such as 'slush' in the finished product.

5.9.4 Soap-making

Soap-making, whether in batch or continuous processes, may generally be considered to comprise three key steps: saponification, soap washing or glycerol extraction, and fitting. The direct neutralization of fatty acids to produce soap differs in that the soap washing stage is not required.

The saponification reaction between triglycerides and sodium hydroxide solution is normally considered to proceed stepwise by the following reaction sequence (Mildwidsky, 1980):

$$\text{Triglyceride} \xrightarrow{\text{NaOH}} \text{soap} + \text{diglyceride}$$

$$\xrightarrow{\text{NaOH}} \text{soap} + \text{monoglyceride}$$

$$\xrightarrow{\text{NaOH}} \text{soap} + \text{glycerol}$$

The reaction is highly exothermic (250 kJ/mol of fat saponified). In practice mono- and diglycerides have not been detected in normal soap boiling, which suggests that, although they can be detected under certain specialized soap-making conditions, saponification of mono- and diglycerides must be faster than

Table 5.7 Properties of four phases of a typical soap (Ferguson *et al.*, 1943)

Property	α form	β form	ω form	δ form
Formation		Cooling of neat soap, then agitating the partially solidified mass	Very rapid chilling of neat soap without agitation, or slow cooling of low-moisture soaps	Cooling of high-water-content soaps, e.g. Na–P, or milling of β-phase soaps at low temperature
Form	Slightly hydrated	Translucent (degree of transparency)		
Stability/ temperature			Favoured by higher temperatures Increase in the order $\delta \rightarrow \beta \rightarrow \omega$	Favoured by low temperature
Effect of extrusions and mechanical working, e.g. milling, plodding		Increase extrusion increases β phase Hardens the soap, make it less opaque, tends to produce orientation and compactness		
Effect of moisture		Na–P exists in very wide range of hydrates between 2.4 and 95% water uptake	Has a much more limited range for Na–P	
Molecular weight		Favoured by high molecular weight	Favoured by low molecular weight	Favoured by high molecular weight
Lather of soap (at 21 °C, relative amount)		Good lathering ability	Poor lathering	Intermediate lathering ability
Soap rubbed off bar in use (%)[a]		2.4	0.5	1.7
Firmness (arbitrary units)		8.0[b]	7.2	3.0
Reaction to water when soaked		Can swell and disintegrate	No swelling or disintegration	Cracks with little swelling
Lattice spacing, *d/n* (Å)	2.45 and 3.65	2.75	2.85 and 3.55	2.95

[a] Measure of ease of lathering of soap bar.
[b] Very firm if produced under conditions tending to orientate and press together crystals. If crystals are small, and not joined, soap is much less firm.
Na–P, sodium phosphate.

that of triglycerides probably due to their greater solubility in aqueous solution (Davidsohn *et al.*, 1953). The rate of saponification depends on the triglyceride composition, the temperature, reactant concentration and the purely physical rate process associated with heterogeneous reactions. This latter factor is important in that fats and alkali are essentially immiscible and therefore have to be brought into intimate contact for the reaction to proceed.

When completely neutral fats or oils are saponified, the progress of the reaction is typically autocatalytic (Norris and McBain, 1922). At the start of reaction the components are immiscible and hence the reaction is slow, a feature commonly referred to as the induction period. The duration of this period may be reduced by increasing the surface area of the reactants, by homogenization or boiling, or by the addition of soap to catalyse the reaction.

Increasing the temperature increases the rate of reaction (Smith, 1932). The phases that are encountered during saponification can also dramatically affect the course of the reaction. For the commonly used batch kettle process, the ternary phase diagram shown in Fig. 5.10 gives a good guide to the conditions that should be observed to promote the most rapid saponification. It is important to avoid the middle soap region B of the phase diagram. Middle-phase soap is difficult to dissolve and is highly viscous. Limiting the amount of water and maintaining sufficient electrolyte in the system accompanied by sufficient agitation is usually adequate to ensure that this phase is not produced (Jones, 1958). It is generally considered that for most efficient saponification the electrolyte concentration should be such as to maintain the reacting mass in regions G (neat soap–nigre–lye) or E (neat soap–nigre) of the phase diagram. The electrolyte may be present as excess sodium hydroxide at the start of the reaction or as added sodium chloride or brine.

Soap washing, the next stage in a soap-making process, is an important feature in modern processes, although in some of the simplest batch processes glycerol is not extracted and remains with the soap (Davidsohn and Davidsohn, 1962). Apart from glycerol extraction, soap washing can also help to remove impurities from the soap, which are transferred into spent lye.

The most important feature of soap washing in batch and continuous systems is to ensure that soap is present in a two-phase region of the phase diagram H (neat soap–lye) or R (kettle wax–lye).

Soap in region H (neat soap–lye) is often considered to be the optimum for soap washing, as the phases separate quickly and the soap occludes the minimum amount of spent lye. In batch processes soap washing involves a simple mixing and settling operation, which is both time-consuming and wasteful in terms of energy and glycerol losses (Govan, 1946). The lyes that separate following saponification and lye addition normally contain between 30 and 45% of the available glycerol at concentrations between 6 and 8%. The further addition of lye accompanied by vigorous agitation and then settling allows up to 70% of the glycerol to be extracted in two washes. There is an economic optimum balance between the quantity of glycerol removed and its concentration. The same rule applies in continuous processes, where continuous countercurrent washing systems are employed. Dependent upon the equipment used and the ratio of washing brine to soap, over 99% of the glycerol may be extracted at concentrations of over 30% glycerol in spent lye.

Soap fitting and separation are the final operations normally encountered in soap-making. Soap from the fitting stage should have several desirable properties, namely, a high fatty matter content, a low dirt content, homogeneity, and a specified and controlled electrolyte content. In batch processes these requirements are achieved somewhat at the discretion of the skilled soap-maker, relying on such factors as taste, feel and the wetting characteristics of the soap on a heated trowel to judge the condition of the fitted soap. In modern processes, standard chemical analytical techniques (Hoover *et al.*, 1968) or automatic methods (Palmquist, 1961) are used to control the fitting operation.

In the batch process low-salt-content brine solution is normally added to washed soap to bring the soap into region E (neat soap–nigre) or region G (neat soap–nigre–lye) of the ternary phase diagram. The soap is left for several days to settle, as the low density difference of the phases present, and the relatively high viscosity of nigre compared with lye, makes gravity separation difficult. On separation the nigre component retains most of the dirt, colouring matter and other unwanted impurities. The separated soap may be transferred directly to storage prior to drying, but the nigre, which contains up to 30% soap, requires further treatment to avoid soap losses. In some cases, nigres are used to initiate the saponification of lower-quality soaps where purity considerations are less important.

In continuous processes where effective triglyceride pretreatment steps are employed, soap fitting may be considerably simplified. Dirt removal in this case is a lesser problem and therefore soap is 'fitted' in the two-phase region H (neat soap–lye) of the phase diagram. Sodium hydroxide solution is often used in order to modify the electrolyte content of the soap. The final separation is often accelerated by centrifugal as opposed to gravity separators, and separated lyes or nigres are recycled to the soap-washing stage. Any dirt present collects in the centrifuge bowl, which must therefore be cleaned periodically.

Most batch and continuous processes combine saponification washing and fitting in some form to comprise the total soap-making process.

(a) Batch processes

Cold process saponification is the simplest of the batch processes and is usually limited to small-scale manufacture (TPI, n.d.). Triglycerides and alkali are mixed together vigorously in a suitable container to form an intimate mix, which reacts slowly. Solid fats, such as tallow, are melted separately and added once the reaction between the liquid triglyceride component, usually coconut oil, and alkali has started so as to utilize the heat of reaction to prevent solidification. The whole of the mix, which becomes considerably more viscous as the reaction proceeds, is poured into frames where the reaction is completed during cooling and solidification. The soap thus produced, containing glycerol and all the impurities originally present in the fat, is finally broken out of the frames and cut into slabs or bars.

Semi-boiled saponification is similar to the cold process except that higher temperatures (70–80 °C) are used.

Full-boiled saponification or the kettle process utilizes all the concepts of saponification, washing and fitting required to produce high-quality soaps. There have been some developments particularly at the saponification stage of this process. For example, the Unilever method uses a type of cross-jet to emulsify the reactants and deliver them to a pan rather than simply boiling the reactants together (Owen, 1951). This procedure allows savings to be made in steam consumption.

Although batch processes, particularly the full-boiled process, are still widely used, the trend has been to move towards continuous processes, where the time taken to produce soap is shortened and the consistency of the final product is more even.

(b) Continuous processes

In the Monsavon process (Webb, 1958), fat at about 70 °C and alkali at a slight excess are metered to a colloid mill where they are homogenized and passed to a kettle to complete saponification. The kettle temperature is 95–100 °C. Glycerol is extracted continuously from soap in a six-section, counter-current washing column, a vertical mixer–settler arrangement, operated at 90 °C. Soap from the last section of the washing column is fitted with water, brine or sodium hydroxide solution, depending upon the required electrolyte specification, and passed to a settling pan. Nigre or lye from this separation stage is recycled to the washing column. The complete process takes about 24 h.

The Mazzoni SCN-N process is somewhat different in concept from other continuous processes in that the electrolyte content in the final product is not achieved by a conventional fitting operation involving the addition of electrolyte solution to washed soap followed by settling. The saponification reaction takes place between metered flows of fats and alkali in a multistage autoclave operated at 125 °C. Soap recycle from the last stage to the first is used to accelerate the reaction, while brine and recycled lye are also introduced to maintain the system in the required region of the phase diagram. The soap from the reactor is transferred to a cooler–mixer operating at 95 °C before washing to extract glycerol in a four-stage mixer–settler system. The wash liquor used in the final stage is 50% sodium hydroxide solution, and therefore after centrifugal separation the soap is 'fitted' by the addition of a metered flow of fatty acids contolled by a pH-sensitive electrode, which modifies the alkalinity of the soap and gives the required, specified electrolyte concentration.

In principle, the Sharples process (Jones, 1958; Scott, 1942) most closely follows the traditional full-boiled batch soap-making technique in that there are two saponification stages. In these two stages, fat passes countercurrently to the flow of spent lyes from the washing stages. By maintaining a balance between the free alkali in spent lye, fresh sodium hydroxide added and the percentage conversion of fats in the first-stage reactor, the spent lyes finally produced from the system are substantially neutral. Mixers and centrifugal separators are used for both completing saponification and washing. The soap/salt/water balance is adjusted in the final mixer to yield a neat soap–nigre or neat soap–lye separation in the final centrifuge depending upon the required electrolyte specification of neat soap. Dirt and other extraneous matter is sedimented in the centrifuges, which therefore need to be cleaned periodically. The residence time of fatty matter in the process is of the order of 2 h.

The De Laval method (Palmquist and Sullivan, 1959; Palmquist, 1939; Platt, 1970) for the continuous saponification of fats employs sophisticated process-control techniques for saponification, washing and fitting, based on the relationship between the viscosity of soap and the anhydrous soap content and electrolyte level. The fine adjustment of the addition of sodium hydroxide solution to the high-temperature (95–130 °C) agitated loop reactor is controlled by measuring the differential pressure across the pump in the recirculation loop. The reaction, which proceeds rapidly owing to the presence of soap, which acts as a catalyst, is completed in a holding column that provides residence time. The soap leaving the reactor is diluted with spent lye and is washed countercurrently, normally with brine, in a system comprising two centrifugal extractors and a post-washing centrifuge. Fitting with electrolyte solution is again controlled by monitoring pressure drops over a recirculation system, after which separation is by centrifuge. The minimum residence time for fatty matter in this process is claimed to be 15 min.

There are several soap-making processes that utilize the continuous neutralization of fatty acids. In the Mills process (McBride, 1947), distilled fatty acids at 75 °C are fed with a metered supply of sodium hydroxide solution to a high-speed mixer. Sufficient sodium chloride is added to the alkali so that the final soap composition is similar to that obtained by the triglyceride saponification route.

Mazzoni offer two fatty acid neutralization processes. The SC process uses sodium hydroxide to neutralize fatty acids, prior to drying, in a loop reactor fitted with an in-line dynamic homogenizer.

Additional sodium chloride is metered as a separate stream. The SCC process used sodium carbonate for the bulk of the reaction in substantially similar equipment to that used in the SC process. In this case sodium hydroxide is used to complete the reaction.

The Armour process (Ladyn, 1964) uses a De Laval column to neutralize fatty acids; while Alfa Laval have developed their own system, where again a De Laval column is used with concentrated (50%) sodium hydroxide to produce an 86% anhydrous soap, which therefore requires no drying step, a considerable process improvement.

5.9.5 Drying

When soap is produced from triglyceride saponification or fatty acid neutralization soap-making processes, it is obtained as soap boilers' neat soap at a moisture content of between 30 and 35%. The function of the drying stage in toilet soap manufacture is to reduce this water content of the soap to a carefully defined and controlled level, usually between 10 and 15% for toilet soap bars. Soap at this moisture content will generally produce, after subsequent processing, a bar of the required appearance and consumer properties. Preservatives and other minor ingredients are also normally added at the drying stage.

A great variety of driers and drying techniques have been used for laundry soap (McCutcheon, 1958). Specific methods such as frame cooling, drum cooling and atmospheric flash cooling are still widely used. By far the most important class of driers for toilet-grade soaps are the vacuum spray driers, of which the Mazzoni type are widely used (Mazzoni, 1959). During the drying process the lamellar, liquid-crystalline phase of soap boilers' neat soap is transferred into a mixture of solid soap phases with liquid-crystalline and liquid phases. For super-fatted soaps, the system will not be in a state of equilibrium and the phase structure of the soap will change slowly on standing or more rapidly when the dried soap is subjected to the various physical work processes of soap finishing.

5.9.6 Soap finishing

The finishing operation of soap manufacture is designed to fulfil several requirements. Minor ingredients such as colorants, perfume and titanium dioxide must be accurately dosed and mixed uniformly into the soap, grit particles that can form during drying must be removed, the soap must be converted from a particulate mass into a compact solid containing no occluded air, and the crystal structure of the soap may be modified to produce specific phase structures.

The first step in soap finishing involves mixing soap with minor ingredients in a batch mixer. A coarse blend of these materials is normally achieved using low-shear horizontal mixers. Mixing is followed by milling in which soap is physically worked between rolls, three or five, which rotate at different speeds and thereby supply a shearing action. The soap chips or flakes from the mill are compacted into a continuous bar suitable for cutting and stamping into bars by the plodding process. In practice, this stage also involves a process known as refining, where the soap chips are forced through wire meshes into a vacuum chest primarily for screening, physical mixing and for the removal of occluded air. The plodding process itself involves screw extrusion through a die to produce a soap billet. This is cut to a size suitable for stamping into bars. The final stage is wrapping and/or cartoning.

5.9.7 Glycerine refining

The glycerine-refining process associated with soap manufacture comprises several unit operations (Ziels, 1956; Tosh, 1958). Spent lyes derived from soap washing in the soap-making process can contain up to 30% glycerol, sodium hydroxide (1–2%), considerable quantities of sodium chloride (10–16%) and dissolved soap and unsaponifiable organic components derived from the oils and fats used for saponification, commonly referred to as MONG (material, organic, non-glycerol). The glycerol solution or sweet water derived from the fat-splitting route to fatty acids normally contains between 12 and 20% glycerol with low levels of MONG, fats and fatty acids. After the appropriate pretreatment, glycerine is concentrated by evaporating the water using multiple effect evaporators to obtain an 80% crude product. For soap lyes, salt is removed from the product by gravity or centrifugal sedimentation.

The penultimate step in glycerine refining is most commonly distillation under vacuum to produce the various glycerine grades: CP (chemically pure) glycerine, technical grade and dynamite grades (Dsouza, 1979). Batch and continuous stills are used for this duty. Modern processes tend to distil glycerine at low pressures (4–10 mmHg). This allows the distillation temperature to be reduced below 150 °C, which prevents glycerol losses due to polymerization. Ion-exchange processes have been used to eliminate the distillation step from some glycerine-refining processes. For spent lyes, salt still needs to be removed by evaporation, but for sweetwater treatment ion exchange can be applied directly and the product concentrated by simple evaporation. Cationic and anionic resins are used in series to remove ionizable substances; non-ionizable substances such as polyglycerols and glycols are not removed but their level is usually such that their presence does not significantly affect the final product quality.

The final step in glycerine refining is normally post-treatment. The quality of distilled glycerine in particular varies considerably with the quality of the feedstock, particularly MONG levels. Improvements to glycerine quality may be achieved by bleaching with activated carbon at 70 °C, ion exchange with cationic resins to remove nitrogenous impurities in particular, and deodorization by steam stripping at temperatures between 100 and 120 °C under vacuum.

References and further reading

Buerger, M.J., Smith, L.B. and Ryer, F.V. (1947) *J. Am. Oil Chem. Soc.*, **24**, 193.

Davidsohn, J. and Davidsohn, A. (1944) *Food Ind.*, **16**, Sept., 92.

Davidsohn, J. and Davidsohn, A. (1962) *Soap Chem. Spec.*, **38**, 229.

Davidsohn, J., Better, E.J. and Davidsohn, A. (1953) *Soap Manufacture*, vol. 1, Interscience, New York.

Dsouza, G.B. (1979) *J. Am. Oil Chem. Soc.*, **58**, 158A.

Ferguson, R.H. (1944) *Oil Soap*, **21**, 6.

Ferguson, R.H., Rosevear, F.B. and Stillman, R.C. (1943) *Ind. Eng. Chem.*, **35**, 1005.

Govan, W.J. (1946) *Oil Soap*, **23**, 229.

Hoover, W.E., Ginn, M.E. and Jungermann, E. (1968) *J. Am. Oil Chem. Soc.*, **45**, 169.

Jones, L.D. (1958) *J. Am. Oil Chem. Soc.*, **35**, 630.

Ladyn, H.W. (1964) *Chem. Eng.*, **71**, 106.

Levey, M. (1954) *J. Chem. Educ.*, **31**, 521.

Luzzati, V. and Husson, F. (1962) *J. Cell. Biol.*, **12**, 207.

Maclennan, K. (1923) *J. Soc. Chem. Ind.*, **42**, 393T.

Madelmont, C. and Perron, R. (1976) *Colloid Polym. Sci.*, **254**, 581.

Mazzoni, C. (1959) *Soap Chem. Spec.*, **3**, 145.

McBain, J.W. and Langden, G.M. (1925) *J. Am. Oil Chem. Soc.*, **127**, 852.

McBain, J.W. and Lee, W.W. (1943) *Oil Soap*, **20**, 17.

McBain, J.W., Vold, R.D. and Vold, M.J. (1938) *J. Am. Oil Chem. Soc.*, **60**, 1666.

McBride, G.W. (1947) *Chem. Eng.*, **54**, 94–7, 134–7.

McCutcheon, J.W. (1958) *J. Am. Oil Chem. Soc.*, **35**, 640.

Mildwidsky, B. (1980) *HAPPI (Household and Personal Products' Industry)*, 50.

Mills, V. (1942) US Patent 2295594.

Norris, M.H. and McBain, J.W. (1922) *J. Chem. Soc.*, **121**, 122, 1363.

Owen, R.V. (1951) US Patent 2566359.

Palmquist, F.T.E. (1939) US Patent 2727915.

Palmquist, F.T.E. (1961) *Soap Chem. Spec.*, Sept., 169.

Palmquist, F.T.E. and Sullivan, F.E. (1959) *J. Am. Oil Chem. Soc.*, **36**, 173.

Platt, G.R. (1970) *Soap Perfum. Cosmet.*, **43**, 233.

Ryer, F.V. (1969) *Encycl. Chem. Technol.*, **18**, 415.

Sanesi, M. (1980) *Thermodynamic and Transport Properties of Organic Salts*, Pergamon, Oxford.

Schulerud, A.L. (1963) *J. Am. Oil Chem. Soc.*, **40**, 609.

Scott, A.T. (1942) US Patent 2300749.

Smith, E.L. (1932) *J. Soc. Chem. Ind.*, **51**, 337.

Sonntag, N.O.V. (1981) *J. Am. Oil Chem. Soc.*, **58**, 155A.

Tiddy, G.J.T. (1980) *Phys. Rep.*, **57**, 1.

Tosh, A.K. (1958) *J. Am. Oil Chem. Soc.*, **35**, 615.

TPI (n.d.) *The Small Scale Manufacture of Soap: An Economic Study*, Tropical Products Institute, London.

Vold, R.D. (1940) *Soap*, **16**, 27.

Vold, R.D. (1941) *J. Am. Chem. Soc.*, **63**, 2915.

Von Barren, B. and Klimmek, H. (1977) *Seifen Öle Fette Wasche*, p. 325.

Webb, E.T. (1927) *Modern Soap and Glycerine Manufacture*, Davis Brothers, London.

Webb, E.T. (1958) *Soap Chem. Spec.*, **43**, 145.

Wigner, J.H. (1940) *Soap Manufacture, The Chemical Process*, Chemical Publishing, New York.

Ziels, N.W. (1956) *J. Am. Oil Chem. Soc.*, **33**, 556.

5.10 Paints and varnishes

5.10.1 Introduction

Although coatings of various types have been used from ancient times to decorate and protect materials, the modern paint industry can be said to have started with the production of paints from linseed oil. The special features of this oil are that it is a liquid in which white or coloured pigments can be readily dispersed. The resulting paint is easily applied in a thin film to a surface by brushing, and on exposure to the air the liquid oil is converted to a solid gel by oxidation, that is, the paint dries. Only linseed and a few other highly unsaturated oils possess this property of changing from liquid to solid within a fairly short time. They are known as drying oils. Of the pigments used, white lead has the property of easy dispersibility to give products of exceptional durability, once a measure of quality. Lead in little used now on account of its toxicity.

From the time of these original oil paints to the present, there have been several phases: first, the simple oils were replaced by heat-thickened oils; natural resins, followed by synthetic resins, were then blended into the oils; and, finally, the oils were chemically incorporated into oil-modified synthetic resins, notably the alkyd resins.

Varnishes are transparent coatings, essentially the paint binders without the pigment, but there is now only a limited market for such clear glossy finishes applied on site.

One result of the change from the simple oil to the alkyd resin as the main paint binder is that there is more flexibility in the choice of oils. The requirement that the oil should dry within a certain time is not so essential and semi-drying oils, such as soybean oil, are often preferred because they possess other properties.

The reader is referred to Turner (1980), Hamburg and Morgans (1979) and Sward (1972) for a more detailed treatment of paint chemistry and technology.

5.10.2 Oil paints

It was soon recognized that the property that determined how rapidly a film of oil dried was related to the degree of unsaturation of its fatty acids. This characteristic was commonly expressed in terms of the iodine value (section 6.2.8). Linseed oil was the predominant drying oil throughout the world, but other oils rich in linolenic acid, such as perilla oil, were equally suitable, when available economically.

Linseed oil was commonly used in paints after refining with alkali or acid. The acid-refined oil was popular for lead paints, as the refining process generated a proportion of free fatty acid, which assisted the dispersion of the pigments. Alternatively, the oil was used in the form of a boiled oil, in which it was lightly blown with air, and metallic driers incorporated, originally by heating (boiling). Metallic driers are compounds of polyvalent metals, particularly cobalt and lead, that catalyse the drying process, reducing it from a few days to a few hours.

Only simple milling or mixing equipment was required to make these paints, so decorators could make their own products.

5.10.3 Bodied oil paints

It was realized by the 1930s that, rather than use a thin natural oil, a better product could be obtained if the oil were first heat-thickened. Less oxidation of the film was then required for conversion to the solid gel stage. These heat-thickened oils were known as stand oils, and a temperature of about 300 °C was required for the thickening process.

White lead was not so suitable with the stand oils, but good results were obtained with zinc oxide. The zinc oxide stand oil paints were harder and glossier than the old paints and were called enamels, the term unfortunately creating confusion with the established vitreous enamels.

5.10.4 Varnish paints

The next stage of development was to incorporate a resin into the bodied oil to increase further the viscosity, which was adjusted with a thinner, usually white spirit, to facilitate application. The resins used were initially the hard natural resins such as Congo copal, which required pyrolysis (running) to render them miscible with the oil. These were followed by synthetic phenolic resins. It was at this stage that unpigmented products (varnishes), particularly copal varnishes, were widely used.

Also at this time the merits of the conjugated oils were recognized. These contain conjugated triene acids (section 3.2.12). Foremost of these was tung oil, originally from China. Compared to linseed oil, tung oil could be heat-bodied in less than half the time. Although it dried rapidly, the film of the raw oil was frosted and the oil could be used in paints or varnishes only in the bodied form. The water resistance of the products was outstanding.

The erratic nature of supplies from China and price fluctuations led to a search for alternative sources of conjugated drying oils. The only economic natural product was oiticica oil (section 3.3.22) from Brazil, but attempts were made to grow tung oil (section 3.3.39) in other parts of the world. The most useful synthetic substitute was dehydrated castor oil produced by thermal dehydration of castor oil (section 3.3.5).

5.10.5 Alkyd paints

Arriving now at current practice, the oil is chemically incorporated as a component of a polyester resin, known as an alkyd resin or sometimes as an oil-modified alkyd resin.

The components of the alkyd resin are a polyhydric alcohol, commonly glycerol and/or pentaerythritol, a polycarboxylic acid, usually phthalic acid, and the fatty acids of the oil.

There are two processes for making alkyd resins, the fatty acid process, in which the three components are heated together at around 200 °C, and the monoglyceride process, in which the oil and polyol are first heated together to form partial glycerides, which are then reacted with the polycarboxylic acid or its anhydride. In the latter case the glycerol of the oil forms part of the total polyol component.

Both drying and semi-drying oils can be used to make alkyd resins that dry by oxidation. Non-drying oils give alkyds suitable as plasticizers, particularly for nitrocellulose lacquers or stoving amino resin paints.

Linseed oil is widely used for the drying alkyds, but it has been largely replaced by semi-drying oils for white or pale shades. Alkyds from these oils give paints that yellow less on aging of the paint film. The commonest semi-drying oil is soybean oil (section 3.3.33) but tall oil fatty acids (section 3.3.36), a by-product of the paper industry, are sometimes used instead. Other oils, equally suitable when available, are safflower (section 3.3.29), sunflower (section 3.3.35) or tobacco seed (section 3.3.38). Alkyds can be made from the conjugated oils, particularly dehydrated castor oil. For the non-drying alkyds, coconut oil (section 3.3.8), palmkernel oil (section 3.3.25) and castor oil are preferred.

5.10.6 Modified alkyd paints

One trend in alkyd technology as applied to paints has been to increase still further the polymeric structure, thereby reducing the amount of oxidation required to obtain the gel structure on drying. This can be done by incorporating styrene (vinylbenzene) or (meth)-acrylic esters into the structure, to give products known as styrenated, or acrylated alkyds. These alkyds are used for speciality quick-drying paints.

A more important trend has been to replace part of the polycarboxylic acid with a polyisocyanate, usually toluene diisocyanate, thereby introducing a polyurethane structure into the alkyds. These products, which are known as urethane alkyds, form the basis of the most widely used gloss paints in the UK at the present time, although the straight alkyds are the more common in the USA and other parts of Europe.

A final modification, sometimes practised, is to introduce a measure of non-drip property into the paint. This is done by incorporating a small amount of a special polyamide into the alkyd, with which it is only partly compatible. These polyamides are derived from dimerized unsaturated fatty acids and diamines (section 10.10).

5.10.7 Water-borne paints

The use of water in place of organic solvents as a paint thinner has obvious attractions. The first important commercial product of this type was the oil-bound water paint, the vehicle being an emulsion of oil or varnish in an aqueous solution of gelatine. The preferred white pigment was lithopone (zinc sulphide/barium sulphate), and a high ratio of pigment to binder was used. This product was eventually replaced by the emulsion paint (known as latex paint in USA and dispersion paint on the continent of Europe). It used a synthetic vinyl polymer as a binder, mainly a copolymer of vinyl acetate with either vinyl versatate or butyl acrylate, with titanium dioxide as the white pigment. Attempts to use an alkyd emulsion based on oil as the vehicle were commercially unsuccessful. The droplet size of the dispersion in emulsion is normally about 0.2–1 μm.

5.10.8 Composition of paint oils

As indicated, the major oils required by the paint industry are:

1. Drying oils.
2. Semi-drying oils.
3. Conjugated drying oils.
4. Non-drying oils.

(a) Drying oils for fast-drying products

These are generally oils rich in non-conjugated triene acids, in particular α-linolenic acid. The most readily available and widely used drying oil is linseed oil, which contains 55–60% of α-linolenic acid and about 15% of linoleic acid (section 3.3.19). Other oils in this category include chia, camphor, black sesame and perilla (all in the range 60–70% of α-linolenic acid). Candlenut (25–35%) and rubber (20–25%) contain useful quantities of α-linolenic acid in their seed oils along with appreciable proportions of linoleic acid (35–50 and 35–40% respectively). None of these oils is readily available. Fish oils contain acids with 4–6 double bonds (non-conjugated) at levels up to about 30% along with significant levels of saturated and of monoene acids. Attempts to use fish oils such as menhaden, sardine, or pilchard have met with only limited success, probably because few of the glycerol esters contain more than one polyene chain (section 5.10.9).

(b) Semi-drying oils for slower-drying but less-yellowing products

These are oils with little or no α-linolenic acid but rich in linoleic acid, e.g. soybean (50–55% linoleic, 7% α-linolenic acid), sunflower (65–70% linoleic acid) and safflower oil (75% linoleic acid).

(c) Conjugated drying oils

These generally dry more quickly than the common drying oils and furnish products of high water resistance. Tung oil contains about 70% of α-eleostearic acid (9*c*,11*t*,13*t*-18:3) and oiticica oil has 75–80% of licanic acid (4-oxo 9*c*,11*t*,13*t*-18:3). Other oils with conjugated triene acids are even less available (section 3.2.12). Castor oil is characterized by the presence of ricinoleic acid (about 90%). This is a hydroxy acid that can be dehydrated to give a mixture of 18:2 isomers including acids with conjugated (9*c*,11*t*) and non-conjugated (9*c*,12*t*) unsaturation in the ratio 1:3 or 1:4. Dehydrated castor oil (DCO) is readily available.

(d) Non-drying oils for plasticizers

These are not highly unsaturated and include coconut oil (section 3.3.8), palmkernel oil (section 3.3.25) and castor oil (section 3.3.5). These properties result from oxidation and/or polymerization, and are related to the fatty acid composition of these oils and, in a more subtle manner not yet fully understood, to the composition of the triacylglycerols.

5.10.9 Modified glyceride oils

(a) Isomerized oils

Since oils with conjugated unsaturation dry quicker than non-conjugated oils with the same number of double bonds, attempts have been made to modify oils such as linseed or soybean. The process involves treatment with catalysts such as nickel, alkali, or alkali-metal alkoxides to shift double bonds so that they become conjugated. Products with conjugated diene (up to 50%) and triene (up to 15%) can be obtained. These behave more like dehydrated castor oil than tung oil. They are not used commercially at the present time.

(b) Segregated oils

The object of segregation is to separate natural oils into components some of which are more suitable for use as paint oils, i.e. to obtain fractions of higher iodine value enriched in polyene acids. Solvents used for extraction include furfural and liquid propane. Since glycerol esters usually contain two or three different fatty acids, the possibility of obtaining highly unsaturated fractions from natural oils is limited. Some success has been achieved with fish oils, and segregated fish oils have achieved limited commercial use in the past. It would be possible to raise the content of polyene acids in glycerol esters by acidolysis, for example by reaction of soybean oil and a linolenic acid concentrate. This approach does not seem to have been used.

(c) Blown oils

Blown oils result when an oil such as linseed is heated (75–120 °C) whilst air is being blown through it. In effect, the oxidative drying process required to form a paint film (section 5.10.11) is started at an elevated temperature before the paint ingredients are mixed.

(d) Boiled oils

Boiled oils – that have not been boiled – contain metal salts (cobalt, manganese) of fatty acids, which promote the drying process. These can be introduced in several ways: by blending oil and drier in the cold (but still designated boiled oil), by heating oil and metal oxides or salts at 95–120 °C whilst blowing air through, or by heating oils and metal oxides or salts at 230–270 °C in the absence of air.

(e) Heat-bodied oils, stand oils

Stand oils result when oils are heated at 200–260 °C in the absence of air or oxygen. This reaction leads to the production of dimers or trimers with molecular weights of almost 2000 and 3000 respectively. With tung oil a gel is obtained in 15 min at 276 °C. Dimerization results through Diels–Alder reaction between a conjugated diene and a second olefinic system:

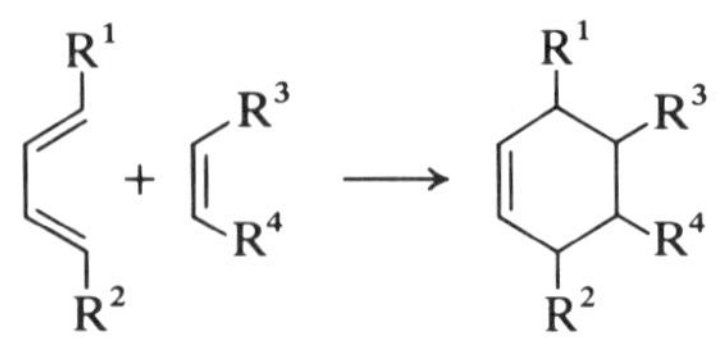

R^1–R^4 represent the remainder of the unsaturated acids

Non-conjugated polyenes react more slowly. Some molecules undergo double-bond migration to give conjugated dienes, which then take part in a Diels–Alder reaction, but most molecules react by radical addition to produce acyclic C–C dimers:

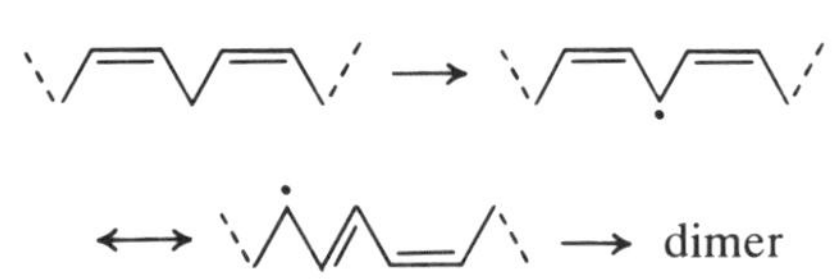

C–C bonded dimers are more stable than dimers linked through one or two oxygen atoms. These reactions occur with glycerol triesters (molecular weight about 1000), quickly giving compounds of higher molecular weight.

5.10.10 Autoxidation of drying oils

The process by which thin liquid films of oil are converted to a solid gel structure – the drying process – is fundamental to paint technology. The process relies essentially on autoxidation (section 10.2), which results eventually in crosslinked polymeric material.

The autoxidation reactions are promoted by catalysts, which are organic salts of multivalent heavy metals. The most powerful are cobalt salts (e.g. cobalt 2-ethylhexanoate or cobalt naphthenate) and these are usually accompanied by other salts (e.g. zirconium or calcium). The use of lead salts is being abandoned in many countries on toxicity grounds. A new group of 'through' driers being introduced are aluminium alkoxide derivatives.

Studies of autoxidation have concentrated on the primary stage of hydroperoxide formation and on the subsequent decomposition of such material to short-chain flavorous compounds. This is because of the importance of rancidity to food science and technology, where great efforts are made to inhibit this undesirable change. In paint technology, the situation is different. Autoxidation is to be promoted and there is interest in the nature of the secondary processes leading to the final polymeric products. Unfortunately these have been less extensively studied. Oxidative polymerization results from new carbon–carbon linkages via one or two oxygen atoms. These join acyl chains at olefinic or allylic centres. The union of appropriate radicals (ROO$^\bullet$, RO$^\bullet$, R$^\bullet$) produces dimers of the type ROR or ROOR, where units of molecular weight about 1000 are being joined. Reaction continues, producing compounds of still higher molecular weight, and internal cross-linking leads to compounds that are harder and more stable. Radicals once formed also react additively with double bonds, and this is an important polymerizing process at temperatures below about 60 °C:

$$\mathrm{ROO^\bullet + {-CH{=}CH-} \longrightarrow \underset{-CH-\dot{C}H-}{ROO} \longrightarrow polymer}$$

The intermediate radical can react with oxygen or with a second double bond to give a dimer radical capable of still further reaction. These changes are not well understood.

5.10.11 Future prospects for oil-based paints

It was prophesied in the 1960s that oil-based paints would disappear from the market in about 20 years in favour of petrochemical-based products. This has not happened, and the main protective paints are still based on fatty oils. In fact, there is now a trend back to raw materials from renewable resources. None of the new technologies, such as the use of powder coatings, water-borne paints, or radiation-curing paints, which would not have been based on oils, have proved as versatile as expected. Self-coloured materials, such as plastics or anodized aluminium, and adhesive plastics film also had their limitations. There is therefore every reason to assume that oil-based paints will be here in the year 2000. Since the general nature of the mechanisms of autoxidation and polymerization was realized, there have been few significant scientific papers and these subjects are not current research items. Sufficient is now known to ensure the most effective use of the fatty oils in commercial paint products.

References and further reading

Hamburg, H.R. and Morgans, W.M. (eds) (1979) *Hess's Paint Film Defects*, Chapman & Hall, London.

Rushman, D.F. (1953) *Paint Research Association Bulletin*, No. 59, p. 25.

Sward, G.G. (ed.) (1972) *Paint Testing Manual*, 13th edn, American Society for Testing and Materials.

Turner, G.P.A. (1980) *Introduction to Paint Chemistry and Principles of Paint Technology*, Chapman & Hall, London.

6
ANALYTICAL METHODS

P.W. HENDRIKSE, J.L. HARWOOD and M. KATES

6.1 Introduction

This chapter describes the analytical procedures currently used for the analysis of oils and fats. Irrespective of the quality of the analytical procedure, all experimental observations are subject to errors due to uncontrolled and often unknown influences, even if every effort is made to keep the experimental conditions constant.

A set of results has a statistical distribution, the shape of which is affected by these errors. The most commonly occurring distribution is the normal distribution. It is characterized by a bell-shaped (Gaussian) curve with a central value (μ) and a width that is expressed in terms of the standard deviation (σ). The central value μ is a combination of the true value and the systematic error or bias of the method. The standard deviation of a method is a measure of the magnitude of the random errors.

The objective of an analysis is, in fact, the estimation of an unknown true value. This requires knowledge of the bias and the scatter (σ) inherent in the test method. Therefore, these two aspects should be investigated for each analytical method used.

The bias can be estimated by analysing a sample of known composition. The difference between the average result $\bar{x}$ obtained from n such determinations and the known content can be considered as an estimate of the bias of the method. The standard deviation σ is estimated by the standard deviation s of the set of n results based on $n-1$ degrees of freedom. The smaller the standard deviation of a method, the more precise the method is said to be. A method can be considered to be accurate if it is both correct (small bias) and precise (small standard deviation). The equation for s is

$$s = \sqrt{\left(\frac{\Sigma\,(x-\bar{x})^2}{n-1}\right)} \qquad (6.1)$$

The value s^2 is known as the sample variance. The maximum variability occurs between results obtained by different laboratories and is referred to as reproducibility. For practical purposes (e.g. quality control), the within-laboratory variability (the repeatability) is relevant. It comprises not only the measuring error but also the contribution of the operator and the influence of environmental conditions such as temperature, reagents and apparatus.

6.2 Standard analytical methods

6.2.1 Oil/fat content determination by direct extraction

(a) *Definition*

Oil is all the substances extracted under the operating conditions specified in the test method. The oil content is expressed as a percentage by mass of the seeds as received.

(b) *Brief description of procedure*

In order to obtain a representative test portion, the laboratory sample is rasped or ground to a fine meal (maximum dimensions of the grains 2–3 mm). The sample should not be reduced to an impalpable powder, generate heat or lose too much moisture. Small seeds (e.g. linseed, rapeseed) can be sampled without prior mechanical grinding.

In the case of small seeds, the test portion should be ground, taking care not to leave any seeds intact. The ground sample is transferred to an extraction thimble, which is placed in a suitable extraction apparatus provided with 250 ml flasks and an electrical heating bath.

If the seeds are very moist (moisture and volatile matter contents above 10%), the filled thimble is left for some time in the oven at a temperature not higher than 80 °C, to reduce the moisture content to less than 10%. The seeds are extracted with gently boiling hexane (150 ml) in a weighed flask (A). After extraction for 4 h, the thimble is removed and the greater part of the solvent evaporated in a current of air.

The thimble is emptied and the contents ground as finely as possible. The mixture in the thimble is replaced (flask A is replaced) and extracted for 2 h (second extraction). After cooling, the thimble is removed, the solvent evaporated and grinding repeated. A third extraction with 150 ml hexane in flask B is carried out for another 2 h.

The greater part of the hexane from flasks A and B is removed by distillation and the last traces by heating the flasks for about 20 min in the oven at 103 ± 2 °C. After cooling, the flasks are weighed and heating is repeated under the same conditions for 10 min to constant weight.

(c) *Calculation*

Sum of the masses (mg) of the oil from flasks A and B $= W$

Mass (mg) of the test portion $= W_0$

$$\text{Oil content (\%) of the material as received} = \frac{100W}{W_0} \qquad (6.2)$$

(d) *Comment*

The method described here is comparable to the IUPAC method 1.122 (Paquot and Hautfenne, 1985). The American Oil Chemists' Society (AOCS, 1978) describes separate extraction methods for the various oil seeds and oil fruits (14 methods in all).

6.2.2 Oil/fat content determination by nuclear magnetic resonance

(a) *Definition*

Oil is all the organic substances contained in oleaginous seeds and fruits that are liquid at the measuring temperature (in principle 20 °C).

The method is applicable to oleaginous seeds and fruits giving a fully liquid oil at the measuring temperature (in principle 20 °C). Under the conditions of the test method, it is not applicable to seeds and fruits that do not give an entirely liquid oil at 20 °C (shea nut, palmkernel, cocoa, etc.).

(b) *Brief description of procedure*

An amount of seeds equal to 20 g is accurately weighed and dried in a metal dish at 103 ± 2 °C. The water content of the seed is determined, for instance by the IUPAC method 1.121 (Paquot and Hautfenne, 1985).

After an empty measuring tube is placed into the magnetic field, the continuous-wave low-resolution NMR spectrometer is set at zero. Amounts of 5, 10, 20 and 30 g of a crude oil of the same variety and origin as the seeds to be analysed are accurately weighed into four measuring tubes (matched tubes are used). The spectrometer response is measured. It gives four mean values of five readings for each tube. A straight line is drawn giving the values of the response as a function of the mass of oil in the tubes.

This straight line should pass through the origin. If not, the adjustment of the spectrometer should be checked.

The previously dehydrated seeds are quantitatively transferred into a measuring tube and the response is measured. It gives the mean value of five readings at the previously chosen integration time. The result is finally corrected for the response of the cake: an amount of fat-free and dehydrated cake is accurately weighed into a measuring tube, after which the response is measured as before.

The crude oil of the same variety and origin as the seeds to be analysed can be extracted by the method described in section 6.2.1. It should be prepared less than one month in advance and be protected against autoxidation. The cake from seeds of the same variety and origin as the seeds to be analysed can be freed from fat according to the method described in section 6.2.1 less than one month in advance and dried at 103 ± 2 °C.

(c) Calculation

Mass (mg) of the test portion before drying $= W_0$
Mass (mg) of the sample of x g of crude oil $= W_x$
Response of the NMR spectrometer for the test portion $= R$
Response of the NMR spectrometer for x g of crude oil $= R_x$
Apparent oil content in the seeds (mass %) $= a$

$$a = 100RW_x/R_xW_0 \tag{6.3}$$

Mass of the defatted and dehydrated cake (mg) $= W_c$
Response of the NMR spectrometer for W_c mg of defatted and dehydrated cake $= R_c$
Contribution of the cake (mass %) $= c$

$$c = 100R_xW_x/R_cW_c \tag{6.4}$$

Moisture content of the seeds (mass %) $= m$

$$\text{Oil content of the seed (mass \%)} = \frac{a - c(1 - 0.01\,m)}{1 - 0.01\,c} \tag{6.5}$$

(d) Comment

The method described here is identical to the IUPAC method 1.123 (Paquot and Hautfenne, 1985). The use of a low-resolution pulsed NMR spectrometer as applied in section 6.2.5 for the determination of the solid fat content is considered by Paquot (1979) as an auxiliary method. The dimensions and location of the measuring tubes allow only test portions of 250–300 mg, i.e. measurements with a small amount of seeds. Consequently, a large number of measurements are required to analyse at least 20 g of seeds. On the other hand, the measurements are very rapid.

The NMR method for the determination of the oil content of seeds is meant to be a quick, cheap and reliable routine method without the use of flammable extraction reagents. A drawback is that the oil content in the seeds has to be measured in dried seeds taking at least 1 h to dry.

A similar method has been presented to the EC Working Group for Oils and Fats and will, if accepted, be proposed as a routine method using extraction as the reference method.

6.2.3 Slip point

(a) Definition

The slip point of a fat is the temperature at which a column of fat, of specified length, starts rising in an open capillary tube under the influence of a hydrostatic pressure under the specified conditions of the test.

(b) Brief description of procedure

There are many methods, often yielding different values, for determining the 'melting point' or 'setting point' of an oil or fat. These melting or setting points are but little related to a real melting point, since fats have a melting range, which is also dependent on the pretreatment of the sample, i.e. on its state of crystallization (Boekenoogen, 1964).

The transition temperatures between crystal modifications and the rate of transition vary for different fats, and this means that it is impossible to draft a uniform standard method applicable to all fats. Any method is bound to compromise: the results obtained will only have a comparative value, and their reproducibility will only be accepted if the method is accurately standardized.

For scientific purposes, the determination of the 'melting point' or 'setting point' should be replaced by a dilatation measurement (section 6.2.4). For commercial reasons, however, the determination of a slip point has to be used. There are two methods for the determination of the slip point: a method containing a stabilization period of 16 h and a routine method yielding satisfactorily reproducible results in a short time.

(i) Method with stabilization step for fats not showing a substantial tendency to polymorphic behaviour

The sample is melted as quickly as possible at a temperature not higher than 10 °C above the point of complete fusion of the fat. The capillary tubes are inserted into the liquid sample so that 10 ± 2 mm long columns of fat are obtained.

The fat column is solidified by quickly cooling the tubes to 15 °C. The tubes are kept at 16 ± 1 °C for a minimum of 16 h. For most normal fats, 16 h is sufficient to obtain accurate and reproducible results, but in exceptional cases, a period of 24–30 h may be required. Two of the prepared capillary tubes are connected to a calibrated thermometer in such a way that the bottom of the tubes is at the same level as the bulb of the thermometer. Suspend the thermometer in the centre of a beaker containing 350 ml of water at 15 °C in such a way that the bottom of the tube is 30 mm below the surface of the water.

While agitating the water, the beaker is slowly heated so that the temperature rises at a rate of 1 °C/min, but the rate should be decreased to 0.5 °C/min when the melting point is approached. The temperature of the water when the fat column starts rising is noted as the slip point.

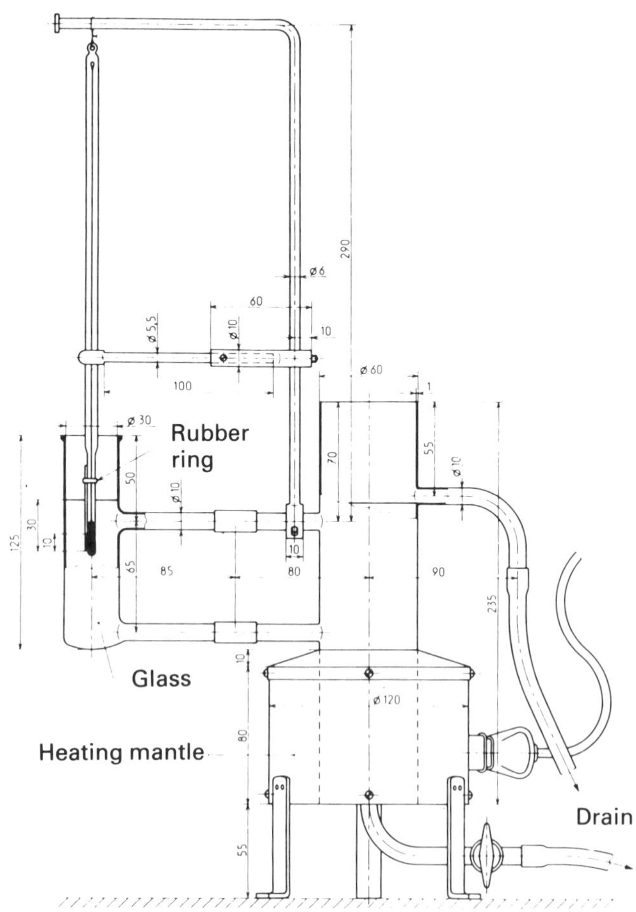

Fig. 6.1 Slip-point apparatus (dimensions in mm).

(ii) Routine method for fats not showing a substantial tendency to polymorphic behaviour

A special slip-point apparatus is used (Fig. 6.1), consisting of a glass water jacket fitted with inlet and outlet tubes for heated water in which the slip-point determination is carried out and a water heater capable of delivering a slow stream of water of gradually increasing temperature, 3–4 °C/min.

The preparation of the test samples is identical with that described before. The tubes are placed against a beaker filled with ice for a few seconds so that solidification of the fat sets in. The thermometer with attached tubes is placed for 5 min in a cooling bath kept at −10 to −12 °C.

The water heater is filled with water of such a temperature that the temperature is not higher than 10 °C when circulation through the water jacket is started.

The thermometer with the two capillary tubes is placed in the centre of the water jacket, so that the lower end of the capillary tubes is 30 mm below the surface of the water. The water is heated and the temperature at which the fat column starts rising is the slip point. If the slip points of the two capillary tubes show a difference of 0.5 °C, the determination should be repeated.

(c) Comment

The problems encountered in trying to draft a standard method for the 'melting point, or 'setting point' of oils and fats explain, in the authors' opinion, the omission of standard methods published by IUPAC. In the AOCS method Cc 1-25 (AOCS, 1978), the filled capillary tubes are placed overnight in a refrigerator at 4–10 °C. For the actual determination of the slip point, the starting temperature of the bath is adjusted to 8–10 °C below the expected slip point of the sample and heat is applied to the bath so as to increase the bath temperature at a rate of ca. 0.5 °C/min. It is obvious that a slip point determined in this way will deviate from that measured according to the method given here.

6.2.4 Dilatation

For more details of this technique, the interested reader is referred to Boekenoogen (1964).

(a) Definition

The dilatation of a fat is the isothermal expansion (due to the transition from the solid to the liquid state) of the fat previously solidified under well defined conditions. The dilatation indicates therefore the difference between the volume the fat would occupy at the measuring temperature if it were completely liquid (supercooled) and the actual volume of the fat at this temperature.

The dilatation affords an empirical indication of the amount of crystalline matter in the fat at the measuring temperature. The dilatation is expressed in millilitres per kilogram (ml/kg) of fat (μl/g) and also in μl/25 g of fat. In the latter case the value for the dilatation obtained by this test method must be multiplied by 25.

(b) Brief description of procedure

(i) Apparatus

The dilatometer should be made of borosilicate glass; shape and dimensions are given in Fig. 6.2. The bulb of the dilatometer should have a volume of 7.0 ± 0.5 ml and the calibrated capillary tube a volume of 900 μl. The zero mark on the capillary tube should be at the same level as the top of the neck of the dilatometer bulb. The stopper should be filled with sufficient lead shot to prevent the dilatometer from floating in the water bath.

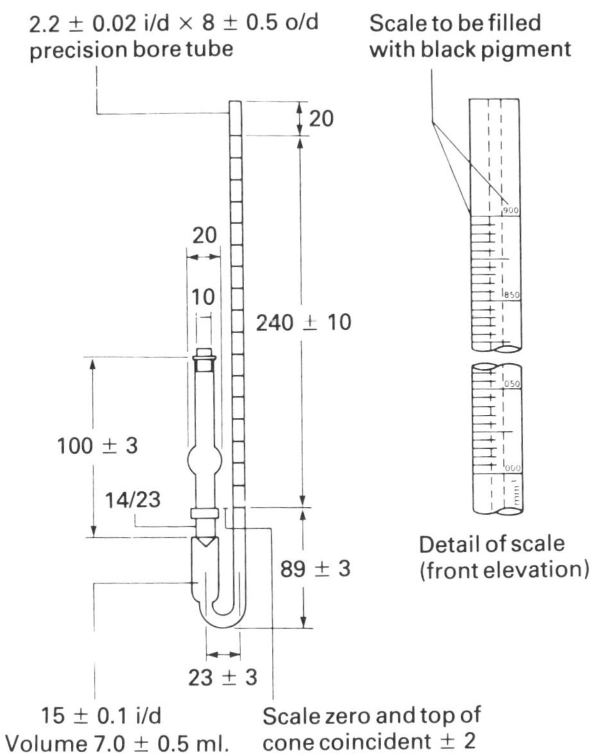

Fig. 6.2 Dilatometer (dimensions in mm).

The capillary should meet the following requirements:

1. The nominal volume of 200 μl should not deviate by more than 1% over the total length of the capillary.
2. The total volume of 900 μl should not deviate by more than 0.5%.

Constant-temperature water baths should be fitted with:

1. A thermostat capable of maintaining the water at any selected temperature between 10 and 60 °C.
2. A powerful stirrer.
3. A calibrated thermometer accurate to within 0.05 °C.

A bath at 0 °C (a cooling unit that can be maintained at 0 °C is recommended) must be filled with a liquid freezing below 0 °C.

(ii) General method not suitable for cocoa butter and cocoa-butter-like fats

Three dilatation procedures may be used:

1. Normal short-range – measuring temperatures of 20, 25, 30 and 35 °C and increases of 5 °C up to 60 °C if required.
2. Normal long-range – measuring temperatures of 0, 10, 15, 25, 30 and 35 °C and increases of 5 °C up to 60 °C, as required.
3. Rapid method – measuring temperatures of 20 and 30 °C and higher, if required.

Exactly 1.5 ml of sealing liquid (boiled water coloured with dye) is introduced into the dilatometer. After weighing the dilatometer to the nearest 10 mg, it is filled with the melted fat, stoppered and weighed again.

The dilatometer is immersed in the water bath maintained at 60 °C. After 30 min the level of the sealing liquid in the capillary is read. The dilatometer is removed from the 60 °C bath and immersed in the 0 °C bath for 90 min. The level of the meniscus in the capillary is noted if a D_0 value is required.

The dilatometer is removed from the 0 °C bath and immersed in a water bath maintained at the lowest temperature at which the dilatation is to be measured. The level of the meniscus in the capillary is read between 30 and 35 min after the water bath has reached the temperature required.

Readings of meniscus levels at successively higher temperatures are taken similarly. The reading at 60 °C should always be repeated. This reading should agree to within 2 μl with the value found at the start.

When reporting the results, the designation 'short-range', 'long-range' or 'rapid method' must be used since the number of temperature steps applied influences the dilatation values.

(iii) Method for cocoa butter and cocoa-butter-like fats

Unhardened fats mainly consisting of symmetrically disaturated 2-oleoyl-1,3-diacylglycerols display a characteristic polymorphism and must be stabilized in a particular way before the dilatations are measured. However, for cocoa-butter-like fat substitutes based on lauric acid derivatives, whether hydrogenated or not, the general method can be applied: stabilization, 40 h at 26 °C; measuring temperatures of 20, 25, 30, 35 and 40 and steps of 5 °C up to 60 °C, if required.

The dilatometer is placed in a 60 °C bath for 30 min and, after the level of the sealing liquid has been read, immersed in the 0 °C bath for 90 min. The dilatometer is then transferred to a water bath fitted with a stirrer and maintained at 26 ± 0.2 °C; it is kept there for 40 ± 0.5 h.

After the stabilization period, the dilatometer is again immersed in the 0 °C bath for 90 min. The level of the meniscus is noted if D_0 is required and the dilatometer placed in a water bath maintained at the lowest temperature at which the dilatation is to be measured.

In this case, however, the level of the meniscus in the capillary is read between 60 and 65 min instead of between 30 and 35 min. The designation 'cocoa butter' should be stated in the test report.

(c) Calculation

Mass (g) of fat in the dilatometer $= M$
Reading (μl) of the meniscus at temperature 60 °C $= A_{60}$
Reading (μl) of the meniscus at temperature T $= A_T$
Correction (μl) for the expansion of sealing liquid and glass between 60 °C and T, according to Table 6.1 $= W_T$
Expansion of oil (μl/g), i.e. the difference in volume of 1 g of melted fat at 60 °C and super-cooled to T, according to Table 6.2 $= V_T$
Dilatation at temperature T $= D_T$

$$D_T = [(A_{60} - A_T - W_T)/M] - V_T \qquad (6.6)$$

For specific information on calculations and tables, reference is made to the chapter on dilatation by Hannewyk *et al.* in Boekenoogen (1964).

(d) Comment

The method described here (Boekenoogen, 1964) is identical with IUPAC method 2.141 (Paquot and Hautfenne, 1985). The dilatometric method described in the AOCS method Cd 10-57 (AOCS, 1978) is based on the work of American investigators (Chapman *et al.*, 1960) and differs considerably from the method described here. The dilatation is expressed as the *melting* dilatation in ml/kg by

Table 6.1 Correction (W_T) for the expansion of the sealing liquid and the glass

Level of the meniscus[a] (μl)	Correction (W_T) at temperature T (°C)											
	0	10	15	20	25	30	35	40	45	50	55	60
0	24	22	21	20	18	17	15	12	10	7	3	0
100	22	21	20	18	17	15	13	11	9	6	3	0
200	20	19	18	17	16	14	12	10	8	6	3	0
300	18	17	16	15	14	13	11	9	7	5	3	0
400	16	15	15	14	13	12	10	9	7	5	2	0
500	14	14	13	12	11	10	9	8	6	4	2	0
600	12	12	12	11	10	9	8	7	5	4	2	0
700	11	10	10	9	9	8	7	6	5	3	2	0
800	9	9	8	8	7	6	6	5	4	3	1	0
900	7	7	7	6	6	5	5	4	3	2	1	0

[a] To find the correction W_T (μl), round off the observed level of the meniscus to multiples of 100 and read W_T at the corresponding temperature.

Table 6.2 Expansion of oil (V_T) at temperatures between 0 and 60 °C

Temperature T (°C)	V_T (μl/g)
0	50.6
10	42.4
15	38.3
20	34.1
25	29.9
30	25.7
35	21.5
40	17.2
45	13.0
50	8.6
55	4.3
60	0

assuming that a 100% solid fat has a melting dilatation of 100 ml/kg. The solid fat index (*SFI*) can be found as the ratio:

$$SFI = \frac{\text{dilatation (ml/kg)}}{100\,\text{mg/kg}} \times 100\%$$

This figure should not be equated to percentage solid fat measured for example by NMR.

6.2.5 Solid fat content determination by nuclear magnetic resonance

(a) Definition

The percentage of solids determined by pulse NMR may be defined as the ratio between the response from the hydrogen nuclei in the solid phase and that from all the hydrogen nuclei in the sample. The product of this ratio and 100 is termed the pulse NMR solids percentage. No correction is made for variations in the density of hydrogen nuclei in solid and liquid phases. The percentage of solids in a partly crystallized oil or fat mixture is measured after solidification under carefully defined conditions. In contrast to the conventional serial measurements of the dilatation method (described in section 6.2.4), by which the dilatation is measured at a number of temperatures, a quicker parallel procedure is employed for the determination of the pulse NMR solids percentage. The term 'parallel measurements' means that a separate tube is used for each measuring temperature. An advantage of this procedure is that all measurements are independent of each other. A considerable reduction in analysis time is achieved by applying the parallel procedure, while the accuracy and reliability of the measurements are comparable with those of serial measurements.

(b) Brief description of procedure

A suitable low-resolution pulse NMR spectrometer is the Brucker Minispec P20i. Briefly, the principle of the technique is as follows (Van den Enden *et al.*, 1978, 1982): the intensities of the relaxation signal caused by the nuclei of the protons of the triacylglycerols, which are solid at the measuring temperature, and that due to the nuclei of the protons of the solid and the liquid triacylglycerols at the same temperature are deduced from the response given by the pulse NMR spectrometer 10 and 70 μs, respectively, after the pulse.

Further apparatus is required: one constant-temperature bath for each measuring temperature, which must be capable of keeping the temperature of the liquid at any set temperature between 10 and 60 °C (±0.1 °C); a 0 °C bath; a cooling unit maintained at 0 °C and filled with a suitable cooling liquid (e.g. 20% aqueous ethylene glycol); and a water bath maintained at 80 °C.

All constant-temperature baths are fitted with metal blocks with holes for the sample tubes or metal racks (for stabilizing at 60 and 0 °C). The sample tubes have the following dimensions (mm): external diameter 10.0 ± 0.25; length, 150; wall thickness, 0.9 ± 0.25.

Special calibration samples consist of a plastic material immersed in liquid paraffin to give percentages of solids of approximately 0, 35 and 70. These must be standardized precisely before being used to calibrate the pulse NMR spectrometer. The pulse NMR spectrometer must first be checked by means of the calibration samples. If the values are found to deviate by more than ±0.3%, the instrument should be adjusted. The standard deviation of the pulse NMR spectrometer should always be smaller than 0.3% solids.

(i) Method for margarine and confectionery fats (Van den Enden *et al.*, 1982)

The fat is heated to 80 °C in a water bath and a sample tube is fitted for each measuring temperature. The tubes are filled to a height of at least 3 cm and should be clean and dry on the outside.

The sample tubes with the completely melted fat are placed in the 60 °C water bath and kept there for at least 5 min. After tempering at 60 °C, the tubes are placed in the 0 °C bath for 60 ± 2 min.

The sample tubes are removed from the 0 °C bath and placed in the blocks of the constant-temperature baths at the measuring temperatures desired.

After 30–35 min, the sample tube is quickly transferred to the sample holder of the pulse NMR spectrometer, ensuring that the tube touches the bottom of the holder; the solids content is read on the meter. The procedure is repeated till all samples have been measured.

Suitable measuring temperatures are 10, 15, 20, 25, 30, 35 and 40 °C and increases of 5 up to 60 °C, as required. However, unlike in the dilatation method, it is not necessary to measure at all these temperatures. In general, the temperatures chosen will correspond to those for the long- or short-range dilatation (section 6.2.4).

The requirement to measure the pulse NMR solid percentage after 30–35 min at the measuring temperature chosen determines the number of sample tubes that can be handled in one run (about 20). If more tubes have to be measured, a second run of 20 tubes can be started from the 60 °C bath about 10 min after the first.

(ii) Methods for cocoa butter and cocoa-butter-like fats needing extended stabilization

The mode of operation is essentially the same as described for non-stabilized fats, except for the following modifications. After the sample tubes have been kept at 60 °C for at least 5 min, they are placed in the 0 °C bath for 90 ± 5 min. The samples are then stabilized at 26 °C for 40 ± 0.5 h before being returned to the 0 °C bath for another 90 ± 5 min. Finally the sample tubes are placed for 60 min in the blocks of the constant-temperature baths at the measuring temperatures desired.

After 60–65 min the sample tube is transferred to the sample holder of the pulse NMR spectrometer. The solids content is then read.

(c) Calculation

Preferably, the pulse NMR spectrometer should be equipped with a computer performing automatically three successive measurements at 10 and 70 μs and the calculation of the mean solids content, which is read directly on the digital voltmeter. The measuring time is then equal to 6 s.

In the case of a solids content determination of cocoa butter and cocoa-butter-like fats, the computer has to be modified slightly to enable a 6 s interval between two successive measurements.

(d) Comment

In the IUPAC method 2.150 (Hautfenne, 1983; Paquot and Hautfenne, 1985), a standard is given describing methods for the determination of the solids content in fats by low-resonance NMR. There are three different methods:

(i) Methods for routine use without special thermal pretreatment applicable to fats that do not require stabilization of their crystalline form (for fats used in the margarine, shortening and confectionery fats industries).

Pulse NMR parallel measurements as well as continuous-wave NMR serial measurements can be used for the determination of the solids content.

(ii) Methods for routine use with special thermal pretreatment applicable to fats showing a high-degree polymorphism and able to crystallize in an unstable form (cocoa butter and cocoa-butter-like fats).

Pulse NMR parallel measurements as well as continuous-wave NMR serial measurements can be used for the determination of the solids content.

Methods (i) and (ii) using pulse NMR parallel measurements are identical to the methods described above (section 6.2.5).

(iii) Methods for specific applications. These are applicable to all types of fats, which are subjected to particular thermal pretreatment to achieve an absolute stable crystalline form in equilibrium with the liquid phase (for investigatory or development work). Serial measurements have to be used with pulse as well as continuous-wave NMR instruments.

The use of continuous-wave NMR in methods (i) and (ii) has the disadvantage that serial measurements have to be used, which will eliminate most of the gain in analysis time obtained with pulse NMR.

Method (iii) is very time-consuming as it comprises a special thermal pretreatment, which makes it only suitable for serial measurements: 2 h at 70 °C; 2 h at −15 °C; 1 h at the temperature of the flow point ±0.25 °C; 1 h at −15 °C; overnight at 5 °C below the first measuring temperature; 45 min at the first measuring temperature; and then NMR measurement.

AOCS has recently published AOCS Recommended Practice Cd 16-81 (AOCS, 1978) based on the use of the Praxis pulse NMR SFC 900.

Only the pulse NMR signal of the liquid fat is measured. The total liquid content is obtained by melting and measuring the sample at 60 °C. The pulse NMR signal at other temperatures of the partially crystallized fat is proportional to the total liquid measurement at 60 °C. An olive oil liquid reference is measured at all temperatures to compensate for instrument and sample differences.

The tempering procedure prescribed in the AOCS

Recommended Practice greatly deviates from that in the methods described in the IUPAC method 2.323 and the methods described in this handbook in section 6.2.5, namely: the sample is melted at 70 °C; 30 min tempering at 60 °C; 15 min equilibration at 26.7 °C; 15 min equilibration at 0 °C; 30 min tempering at 26.7 °C; 15 min cooling at 0 °C; followed by equilibration for 30 min at each of the measurement temperatures 10, 21.1, 26.7, 33.3 or 40.6 and 37.8 or 40.6 °C.

It is obvious that the solid fat indices obtained by the AOCS method will greatly deviate from those obtained by the methods used in Europe. The values obtained by the AOCS method are not directly comparable to those obtained by dilatometry (*SFI*), whereas the pulse NMR solid percentage according to the method described in section 6.2.5 can readily be converted into dilatation values determined according to the method given in section 6.2.4 (Van den Enden *et al.*, 1978).

6.2.6 *Trans* content determination

(a) *Definition*

The concentration in oils and fats of unsaturated components containing non-conjugated *trans* double bonds is the percentage of glycerol trielaidate present in a mixture of glycerol trielaidate and tristearate of which the absorbance at 970 cm^{-1} (10.3 nm) is identical with that of the oil or fat.

(b) *Brief description of procedure*

A self-recording double-beam infrared spectrophotometer suitable for quantitative measurements with a frequency range between 1100 and 900 cm^{-1} (wavelength between 9 and 11 nm) is required.

(i) Calibration

Into a series of 10 ml measuring flasks respectively 0 (reference), 25, 50, 100 and 150 mg of glycerol trielaidate is weighed accurately and each sample is made up to 200 mg with glycerol tristearate. Each of these mixtures is dissolved in carbon disulphide and made up to the mark at 20 °C. The analysis should be performed with solution, cells and spectrophotometer as near as possible at 20 °C.

The AOCS method Cd 14-61 (AOCS, 1978) is practically identical to the method described in this section. The method prescribes the use of elaidic acid, methyl elaidate or glycerol trielaidate as primary standards. Also secondary standards calibrated against the primary standards may be used. The AOCS method advises use of methyl esters if the *trans* content lies below 15%.

The absorbance (or transmission) of the four standard solutions is measured against the reference (glycerol tristearate) in the frequency range 1050–900 cm^{-1} (wavelength 9.5–11 nm). On each of the four spectral curves thus obtained, a baseline is drawn, i.e. the straight line between the maxima (or minima) at 1015 and 910 cm^{-1} (10.0 and 10.8 nm) (see Fig. 6.3).

If the spectrophotometer records the absorbance, the absorbances A_P at P and A_Q at Q are read. The absorbance A is then given by

$$A = A_Q - A_P \qquad (6.7)$$

If the spectrophotometer records the transmission, the distances QR and PR are measured. The absorbance A is then given by

$$A = \log(PR/QR) \qquad (6.8)$$

The calibration line is constructed by plotting the four values found for A against the corresponding amounts of glycerol trielaidate (mg).

(ii) Sample determination

Approximately 200 mg of the liquefied sample is accurately weighed into a 10 ml measuring flask, dissolved in carbon disulphide and made up to volume.

The absorption or transmission of the sample solution is measured against the reference solution

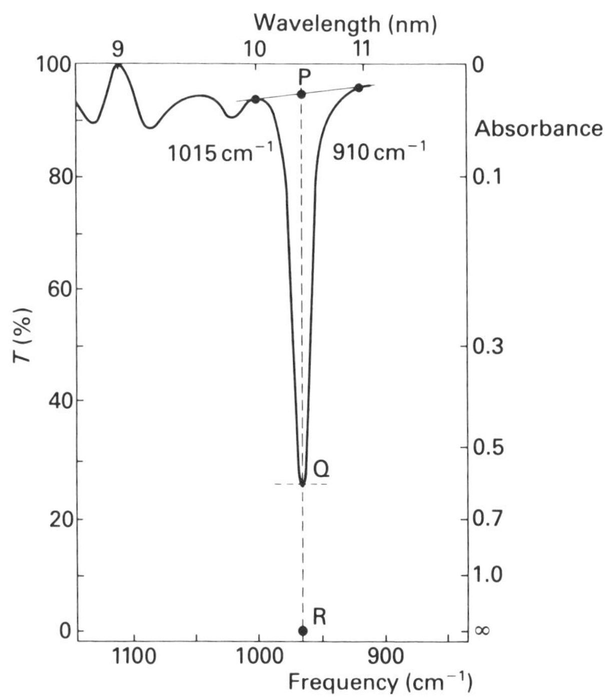

Fig. 6.3 *Trans* content determination.

(glycerol tristearate) in the 1050–900 cm^{-1} (9.5–11 nm) range, using the same instrument setting as in the calibration procedure. If isolated *trans* unsaturated components are present, there is a maximum (or minimum) at about 970 cm^{-1} (10.3 nm). A straight line is drawn between the absorption minima at around 1000 and 925 cm^{-1} (10.0 and 10.8 nm) (see Fig. 6.3).

(c) Calculation

When the absorbance is recorded, the absorbance at the peak maximum $A_S = A_Q - A_P$ is calculated; when the transmission is recorded, $A_S = \log(PR/QR)$ is calculated (see Fig. 6.3).

From the calibration line, the value B, i.e. the amount of glycerol trielaidate (mg) corresponding to the absorbance of the sample (A_S) is determined.

Mass (mg) of sample	$= W$
Trans content (%)	$= 100B/W$

(d) Comment

The accuracy of the determination depends on the effectiveness of compensation between the sample and reference cells. Ideally, there should be no absorption on either side of the *trans* peak. The determination becomes less accurate when interfering peaks occur close to the *trans* peak. The error becomes more significant at lower *trans* peak height. When the sample type and/or level of *trans* content is such that too great an interference by extraneous peaks is seen on either side of the *trans* peak, or the *trans* content is below 5%, the analysis should preferably be performed by the derived methyl ester method: glycerol trielaidate is replaced with methyl elaidate and glycerol tristearate with methyl stearate; a new calibration line should be constructed using these standard and reference materials; the sample is converted into its corresponding fatty acid methyl esters as described in section 6.2.19.

The IUPAC method 2.207 (Paquot and Hautfenne, 1985) is in fact identical to the method described in this section but follows the methyl ester method exclusively. The method described in this section has been issued by OICC (Office International du Cacao et du Chocolat) as an official method.

6.2.7 Titre

(a) Definition

The 'titre' is the value (°C) that denotes the solidification point of the water-insoluble fatty acids.

As the fatty acids change from the liquid state to the solid state, latent heat of fusion evolves. This causes a break in the cooling rate and an increase in temperature. The highest point then reached by a thermometer immersed in the fatty acids is recorded as the titre. It is, therefore, more accurate to define the titre as the highest temperature reached when fatty acids are crystallizing under controlled conditions.

(i) Apparatus (see Fig. 6.4)
The following are needed: a heating bath, e.g. a beaker of 2 litres (low model); a glass vessel, volume ca. 450 ml, height 190 mm, weighed down in the heating bath by lead shot; a wide test tube, length 100 mm, marked at a distance of 57 mm from the bottom to show the height to which the tube should be filled; a stirrer made of glass or stainless-steel wire, 2–3 mm in diameter with the lower end forming a loop of 19 mm diameter; and a laboratory thermometer and a precision thermometer with readings in 0.1 or 0.2 °C.

(b) Brief description of procedure

The melted fatty acids are poured into the test tube to the mark, i.e. 57 mm from the bottom; the acids should at least be 15 °C above the titre point. The

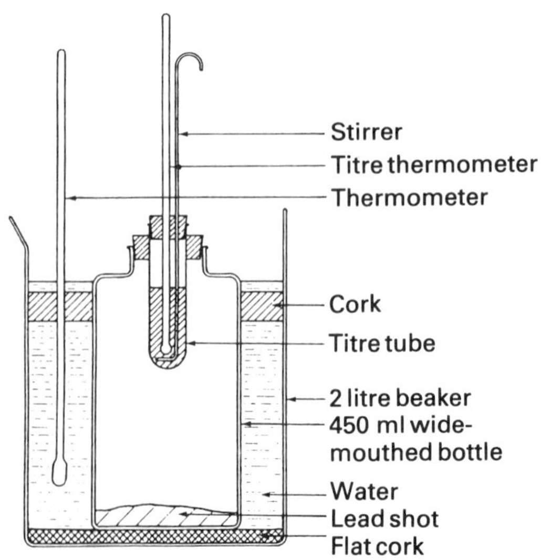

Fig. 6.4 Apparatus for determination of titre.

apparatus is then assembled as shown in Fig. 6.4 and the precision thermometer is inserted such that the bottom of the bulb is 45 mm below the surface of the fatty acids.

Before the temperature of the fatty acids drops to a value 10 °C above its expected titre point, stirring the fatty acids in a vertical manner (at a rate of 100 cycles per minute) is started. Stirring is continued until the temperature remains constant for 30 s. Stirring is then discontinued, the stirrer is removed from the fatty acids and the increase in temperature established. The highest temperature reached by the thermometer is recorded as the titre.

(c) Remark

With some samples the crystallization process occurs so rapidly that the stirrer cannot be removed from the sample. In such cases the observation may be made with the stirrer in the sample.

(d) Comment

The method described here deviates from the IUPAC method 2.121 (Paquot and Hautfenne, 1985). In the IUPAC method, the test tube is filled to a height of 57 mm and the initial temperature of the melted fatty acid is 25 °C above the expected titre point. Stirring is carried out by means of rapid circular movements with the thermometer when the temperature of the solidifying fatty acids appears to be stationary. During stirring, the temperature falls sharply. After replacing the thermometer in the centre of the test tube the temperature rises again. The maximum value is taken as the solidification point (titre).

The AOCS method Cc 12-59 (AOCS, 1978) is practically identical to the method described in this section.

6.2.8 Iodine value (Wijs method)

(a) Definition

The iodine value denotes the percentage by weight of halogen, calculated as *iodine* adsorbed under the conditions of the test.

(b) Brief description of procedure

The amount of sample to be weighed depends on the iodine value in the following way:

Iodine value expected	Weight to be taken for test (g)
0–5	3.0
5–20	1.0
20–60	0.34
60–80	0.25
80–130	0.15
>130	0.10

The amount of sample weighed is dissolved in 15 ml tetrachloromethane in a conical flask. From a burette, 25 ml of Wijs solution is added. (Note that Wijs solution is commercially available – ICl solution, 0.2 mol/l.) The flask is closed, the contents are mixed and allowed to stand in the dark at about 20 °C for: (i) 1 h, if it is certain that no polymerized or oxidized material is present; (ii) 2 h, if polymerized or oxidized materials may be present.

After standing, 20 ml potassium iodide 10% aqueous solution and approximately 150 ml water are added. Then the solution is titrated with a sodium thiosulphate solution, $C(Na_2S_2O_3) \simeq 0.05$ mol/l, using starch indicator towards the end of the titration. A blank determination with the same quantities of reagents is carried out at the same time and under the same conditions.

Studies are being carried out within IUPAC and AOCS to replace tetrachloromethane by a mixture of acetic acid and cyclohexane.

(c) Calculation

Mass (g) of fat weighed $= W$
Volume (ml) of thiosulphate solution used in test $= V_1$
Volume (ml) of thiosulphate solution used in blank $= V_2$
Atomic mass of iodine $= 126.9$
Titre of thiosulphate solution, $C(Na_2S_2O_3) \simeq 0.05$ mol/l $= T$

$$\text{Iodine value} = 12.69 \times 2T(V_2 - V_1)/W \quad (6.9)$$

(d) Comment

The method is practically the same as the IUPAC method 2.205 (Paquot and Hautfenne, 1985) and the AOCS method Cd 1-25 (AOCS, 1978).

6.2.9 Free fatty acid content

(a) Definition

The 'free fatty acid content' or 'acid content' is the percentage of free fatty acids present in the fat. The 'acid value' is the number of milligrams of potassium hydroxide required to neutralize 1 g of fat. The 'degree of acidity' is the number of millilitres of alkali hydroxide (1 mol/l) required to neutralize 100 g of fat.

(b) Brief description of procedure

Approximately 10 g of fat is weighed with an accuracy of 0.01 g and dissolved (if necessary by gentle heating) in 50 ml of a solvent consisting of equal volumes of 96% ethanol and diethyl ether neutralized shortly before use by titration with sodium hydroxide solution, $C(\mathrm{NaOH}) \simeq 0.1$ mol/l, using phenolphthalein as indicator. The solution of the fat is titrated while swirling with sodium hydroxide solution, $C(\mathrm{NaOH}) \simeq 0.1$ mol/l.

(c) Calculation

Mass (g) of fat weighed	$= W$
Volume (ml) of sodium hydroxide solution used	$= V$
Molar mass of *potassium hydroxide* (g/mol)	$= 56.1$
Titre of the sodium hydroxide solution, $C(\mathrm{NaOH}) \simeq 0.1$ mol/l	$= T$
Mean molar mass (g/mol) of the fatty acids	$= M$

Mostly, the value $M = 282$ g/mol (molar mass of oleic acid) is taken.

$$\text{Free fatty acid content (\%)} = MVT/10W \quad (6.10)$$

$$\text{Acid value} = 56.1VT/W \quad (6.11)$$

$$\text{Degree of acidity} = 100VT/W \quad (6.12)$$

(d) Comment

In cases where the mean molar mass of the fatty acids deviates greatly from that of the oleic acid, other factors for M giving a more realistic picture are recommended:

Nature of fat	Expressed as	Molar mass (g/mol)
Coconut, palmkernel and the like	Lauric acid	200
Palm oil	Palmitic acid	256
All other oils	Oleic acid	282

In the IUPAC method 2.201 (Paquot and Hautfenne, 1985), two methods are described for the determination of the free fatty acid content: 'acid value by the indicator method', a method comparable to the method described in this section, and 'acid value by the potentiometric method'. In the latter method, the free fatty acids are titrated potentiometrically in a non-aqueous medium with a solution of potassium hydroxide in isopropanol. This method is employed for the determination of free acidity in highly coloured oils and fats.

The AOCS method Ca 5a-40 (AOCS, 1978) is comparable to the method described in this section but provides a table to determine the quantities of sample to be used with various ranges of free fatty acid content.

6.2.10 Saponification value

(a) Definition

The saponification value, which is related to the molar mass of the fat, denotes the number of milligrams of potassium hydroxide required to saponify 1 g of fat, i.e. to neutralize the free fatty acids and the fatty acids combined as acylglycerols.

The 'saponification equivalent' is the amount of oil or fat saponified by 1 mol of potassium hydroxide, and is equal to 56 108 divided by the saponification value.

(b) Brief description of procedure

Approximately 4 g of fat is accurately weighed into a 250 ml conical flask provided with a reflux condenser. After addition of 50 ml of ethanolic potassium hydroxide, $C(\mathrm{KOH}) \simeq 0.5$ mol/l, the reflux condenser is connected to the flask. The mixture is heated and, as soon as the ethanol boils, the flask is occasionally shaken till the fat is completely dissolved. The solution is boiled under reflux for 30 min after the fat is completely dissolved. The hot soap solution is slowly titrated with hydrochloric acid, $C(\mathrm{HCl}) = 0.5$ mol/l, using phenolphthalein as indicator to determine the excess of potassium hydroxide. A blank determination using the same quantity of potassium hydroxide is carried out at the same time.

(c) Calculation

Mass (g) of fat weighed	$= W$
Volume (ml) of hydrochloric acid used in test	$= V_1$
Volume (ml) of hydrochloric acid used in blank	$= V_2$
Titre of hydrochloric acid, $C(\mathrm{HCl}) \simeq 0.5$ mol/l	$= T$

$$\text{Saponification value} = 56.1T(V_2 - V_1)/W \qquad (6.13)$$

(d) Comment

The relationship between the molar mass (M) of acylglycerols and the saponification value (SV) is given by: for pure monoacylglycerols, $M = 56\,108/SV$; for pure diacylglycerols, $M = 112\,216/SV$; for pure triacylglycerols, $M = 168\,324/SV$.

The IUPAC method 2.202 (Paquot and Hautfenne, 1985) describes two methods, the indicator method and the potentiometric method. The indicator method is to be preferred when it is applicable. This method deviates slightly from that described in this section in that 2 g of sample is used and refluxing takes 1 h.

The potentiometric method should be employed for the determination of the saponification value of highly coloured oils and fats. It is based on reflux boiling of the sample with isopropanolic potassium hydroxide solution and potentiometric titration of the excess of potassium hydroxide by hydrochloric acid in a non-aqueous medium.

The AOCS method Cd 3-25 (AOCS, 1978) is identical to the method described in this section.

6.2.11 Hydroxyl value

(a) Definition

The hydroxyl value is the number of milligrams of potassium hydroxide required to neutralize the amount of acetic acid capable of combining by acetylation with 1 g of oil or fat.

This method for the determination of the hydroxyl value does not differentiate between hydroxyl groups in the fatty acid chains and those present in mono- or diacylglycerols. Hence, partially hydrolysed fats may show a high hydroxyl value without containing hydroxy acids.

(b) Brief description of procedure

The amount of sample to be weighed and the amount of acetylating agent depend on the hydroxyl value in the following way:

Hydroxyl value expected	Weight to be taken for test (g)	Volume of acetylating agent to be added (ml)
10–100	2.0	5
100–150	1.5	5
150–200	1.0	5
200–250	0.75	5
250–300	1.2	10
300–330	1.0	10

To the amount of sample weighed into the acetylation flask, an accurately measured volume of acetylating agent is added through the air condenser. After mixing, the flask is placed in an oil bath at 95–100 °C. After 1 h the flask is cooled and 1 ml of distilled water added through the air condenser. After shaking, heating is continued for 10 min to convert anhydride into acid. After cooling again, 5 ml of 95% ethanol is added and the contents of the acetylation flask are titrated with ethanolic potassium hydroxide, $C(\mathrm{KOH}) \simeq 0.5$ mol/l, using phenolphthalein as indicator. A blank determination is carried out at the same time.

(c) Calculation

Mass (g) of fat weighed	$= W$
Volume (ml) of potassium hydroxide solution used in test	$= V_1$
Volume (ml) of potassium hydroxide solution used in blank	$= V_2$
Molar mass of potassium hydroxide (g/mol)	$= 56.1$
Titre of potassium hydroxide solution, $C(\mathrm{KOH}) \simeq 0.5$ mol/l	$= T$

$$\text{Hydroxyl value} = \frac{56.1T(V_2 - V_1)}{W} + \text{acid value}$$

For the acid value, see analytical method for free fatty acid content (section 6.2.9).

(d) Comment

The method described is comparable to IUPAC method 2.241 (Paquot and Hautfenne, 1985) as well as to the AOCS method Cd 13-60 (AOCS, 1978), although the latter uses different quantities of sample for the various ranges of expected hydroxyl value. The details of the procedure are also slightly different.

6.2.12 Unsaponifiable matter content

(a) Definition

The unsaponifiable matter is the whole quantity of substances present in the oil or fat that, after saponification by potassium hydroxide and extraction by a specified solvent, are not volatile under the conditions of the test.

The unsaponifiable matter includes lipids of natural origin such as sterols, higher aliphatic alcohols, pigments, hydrocarbons as well as any foreign organic matter non-volatile at 103 °C (e.g. mineral oils).

(b) Brief description of procedure

Approximately 5 g of fat is accurately weighed into a 250 ml round-bottomed flask fitted with a reflux condenser. About 50 ml of ethanolic potassium hydroxide, $C(KOH) \simeq 1$ mol/l is added and the mixture heated on a boiling water bath for 1 h.

The contents are transferred to a 500 ml separating funnel and the flask is first rinsed with 100 ml water and then with 100 ml ether. The water and ether are added to the contents of the separating funnel. The contents are shaken vigorously while still warm. After separation of the layers the ethanol–water layer is drawn off into the flask used for the saponification and the ether layer is transferred to a second 500 ml separating funnel containing 40 ml of water.

The ethanol–water layer is shaken twice, each time with 100 ml of ether. The three quantities of ether are collected in the second separating funnel.

The separating funnel with the ether solution and the 40 ml of water is gently rotated about its own axis without shaking (to prevent emulsion formation). When the layers have separated the water phase is drawn off. After washing the ether solution twice with 40 ml of water, it is shaken successively with 40 ml of potassium hydroxide solution, $C(KOH) \simeq 0.5$ mol/l, 40 ml of water, again 40 ml of potassium hydroxide solution and finally with 40 ml portions of water until free of hydroxide.

The ether solution is then transferred to a 500 ml conical flask, part of the ether is evaporated, after which the solution is transferred to a weighed 200 ml conical flask.

After addition of 6 ml of acetone, the volatile solvent is expelled quantitatively. The residue is dried to constant weight at 100 °C. The residue is dissolved in 20 ml of ethanol and titrated with ethanolic potassium hydroxide solution, $C(KOH) \simeq 0.1$ mol/l, using phenolphthalein as indicator.

(c) Calculation

Mass (g) of fat weighed $= M$
Mass (g) of the residue $= M_1$

$$\text{Unsaponifiable matter content (\%)} = \frac{100\, M_1}{M} \qquad (6.15)$$

The determination is carried out with the original fat. However, if it is desirable to express the content of unsaponifiable matter as a percentage of the pure fat, the value found in the determination is corrected taking into consideration the content of water and other volatile constituents and impurities as mentioned above. In this case, when stating the content of unsaponifiable matter, it must be recorded as 'calculated on the pure fat':

Content of water and volatile constituents (mass %) $= W$
Content of impurities (mass %) $= V$

$$\text{Unsaponifiable matter content calculated on the pure fat (\%)} = \frac{10\,000\, M_1}{(100 - W - V)\, M} \qquad (6.16)$$

(d) Comment

The IUPAC method 2.401 (Paquot and Hautfenne, 1985) describes two methods: the diethyl ether method and the hexane method. The first is in agreement with the method described in this section.

The hexane method differs from the hexane method described in the fifth edition of the *Standard Methods* (II.D.5.5) (IUPAC, 1966) by prescribing nine extractions instead of three in order to obtain experimental figures comparable to those obtained by the diethyl ether method.

The diethyl ether method, however, should be regarded as the reference method.

The AOCS method Ca 6a-40 (AOCS, 1978) is meant for normal animal and vegetable oils and fats. It excludes oils and fats such as marine oils containing excessive amounts of unsaponifiable matter. In these cases AOCS method Ca 6b-53 should be applied. The AOCS method Ca 6a-40 prescribes light petroleum ether as extractant. At least seven extractions have to be carried out.

6.2.13 Peroxide value

(a) Definition

The peroxide value of an oil or fat is the quantity of peroxide oxygen present in the sample, expressed in mmol active oxygen per 2 kg of the sample.

In the literature, the Lea value is sometimes mentioned: the Lea value is expressed in mmol peroxide oxygen per kg of fat. Numerically, the peroxide value is therefore twice as high as the Lea value.

(b) Brief description of procedure

Two methods are presented: the Wheeler method and the Sully method. Although the Wheeler method is preferred, in some countries the Sully method is legally required. The Sully method usually gives higher values.

(i) Wheeler method
The amount of sample to be weighed depends on the peroxide value in the following way:

Peroxide value expected (mmol active oxygen/2 kg)	Weight to be taken for test (g)
0–2	5
2–10	2
10–25	1
25–50	0.5
50–100	0.3

Liquid oils should be carefully stirred before weighing, taking care that no aeration of the oil occurs. Solid fats should not be melted, the sample being taken from the centre of the mass with a fat trier.

Approximately 0.3–5 g of sample, depending on the peroxide value expected, is weighed accurately; 25 ml of a solvent consisting of a mixture of 2 volumes of acetic acid and 1 volume of trichloromethane is added, after which the air above the liquid is displaced with carbon dioxide or nitrogen. After addition of 1 ml of a saturated potassium iodide solution, a homogeneous solution usually results. After standing for 1 min (or, if the mixture is not homogeneous after stirring for 1 min), 35 ml of water is added and the mixture if titrated with a sodium thiosulphate solution, $C(Na_2S_2O_3) \simeq 0.001$ mol/l (for peroxide values higher than 25, a thiosulphate solution of $C(Na_2S_2O_3) \simeq 0.01$ mol/l should be used), using starch solution as indicator. A blank determination is carried out at the same time.

(ii) Sully method (DGF method F-1 3b (68); DGF, 1950–75)
Approximately 1 g of sample is accurately weighed into a microbeaker made of Teflon. Into a round-bottomed flask – fitted with a condenser tube and an immersion condenser (Fig. 6.5) – 20 ml solvent (a mixture of equal volumes of acetic acid and trichloromethane) is introduced, after which the solvent is deaerated by heating it without boiling for about 2 min. A saturated potassium iodide solution is added and boiling is continued for another 2 min. Then the microbeaker with the sample is placed into the round-bottomed flask without interrupting boiling, which is continued for another 3–4 min.

After removal of the flame, 50 ml of water is added at once and the mixture is titrated with a sodium thiosulphate solution, $C(Na_2S_2O_3) \simeq 0.05$ mol/l (for peroxide values lower than 20, a thiosulphate solution of $C(Na_2S_2O_3) \simeq 0.005$ mol/l should be used), using starch solution as indicator. Simultaneously, a blank determination is carried out using all the reagents.

(c) Calculation

Mass (g) of oil or fat weighed $= W$
Volume (ml) of sodium thiosulphate solution used in test $= V_1$
Volume (ml) of sodium thiosulphate solution used in blank $= V_2$
Titre of sodium thiosulphate solution (mol/l) $= T$

$$\text{Peroxide value (mmol active oxygen/2 kg)} = \frac{1000\,T(V_1 - V_2)}{W} \qquad (6.17)$$

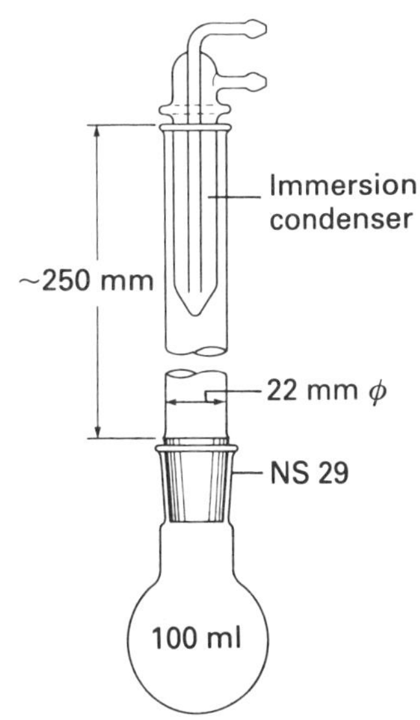

Fig. 6.5 Apparatus for the determination of the peroxide value.

As already stated above, the peroxide value is also expressed in other units: milliequivalents of active oxygen per kilogram, millimoles of active oxygen per kilogram or milligrams of active oxygen per kilogram. These units may be converted into mmol of active oxygen per 2 kg (Standard International Units) using the following factors:

Unit	Multiplication factor
meq/kg	1
mmol/kg (Lea value)	0.5
mg/kg	8

(d) Comment

The IUPAC method 2.501 (Paquot and Hautfenne, 1985) is comparable to that described in subsection (i) above, the Wheeler method. The AOCS method Cd 8-53 (AOCS, 1978) is also comparable to the method described in subsection (i), with the exception that 5.0 g of sample is used in all cases and the titration is carried out with sodium thiosulphate solution of $C(Na_2S_2O_3) \simeq 0.1$ mol/l. The Sully method, subsection (ii) above, is mostly used in Germany (DGF, 1950–75) and tends to give higher values compared to the Wheeler method.

6.2.14 Anisidine value

(a) Definition

The anisidine value is, by convention, 100 times the absorbance (measured in a 1 cm cell) of a solution resulting from the reaction of 1 g of oil or fat in 100 ml of solution with 'anisidine reagent'.

(b) Brief description of procedure

The anisidine reagent is a solution of 0.25% m/v of *colourless p*-anisidine in acetic acid (water content of the acetic acid should be less than 0.1%). The anisidine reagent should have an absorbance less than 0.200 when measured at 350 nm in a 1 cm cell.

It is essential that the oil or fat be perfectly clear and dry. An amount of 0.5–4.0 g of sample is weighed accurately and dissolved in 25 ml of optically pure hexane or isooctane. The absorbance of this solution is measured at 350 nm in a 1 cm cell, using the reference cell fitted with the solvent as a blank. Of the sample solution, 5 ml is pipetted into one test tube and 5 ml of the solvent into a second. To each tube, 1 ml of anisidine reagent is added. After exactly 10 min, the absorbance of the solution in the first test tube is measured at 350 nm using the solution of the second test tube as the blank in the reference cell.

(c) Calculation

Mass (g) of oil or fat weighed $= W$
Absorbance of the sample solution after reaction with the anisidine reagent $= A_s$
Absorbance of the fat solution $= A_b$

$$\text{Anisidine value} = 25\,(1.2A_s - A_b)/W \quad (6.18)$$

In the IUPAC method 2.504 (Paquot and Hautfenne, 1985) it is noted that *p*-anisidine in the presence of acetic acid reacts with aldehydic components in oils and fats. The intensity of the colour of the yellowish reaction product formed depends not only on the amount of aldehydic components present but also on their structure. It has been found that a double bond in the carbon chain conjugated with the carbonyl double bond increases the molar absorbance four to five times. This means that 2-alkenals, in particular, will substantially contribute to the value found.

The only difference between the method described here and the IUPAC method is that IUPAC indicates the anisidine value as *p*-anisidine value (*p-AV*).

6.2.15 Phosphorus estimation

(a) Principle

The oil is charred in the presence of magnesium carbonate and then ashed. The ash is dissolved in dilute hydrochloric acid and phosphorus is determined colorimetrically in the resulting solution by a molybdenum blue method.

(b) Brief description of procedure

An amount of 0.1–10 g of oil or fat, depending on the expected phosphorus content, is accurately weighed into a quartz crucible. For each gram of oil or fat, 50 mg of magnesium carbonate is added and mixed well with the sample. The mixture is heated on a flame to complete combustion and finally ignited until a white ash is obtained. The ash is dissolved in hydrochloric acid, $C(HCl) \simeq 8$ mol/l, and transferred into a 100 ml measuring flask. After neutralization, the solution is made up to the mark with distilled water.

Exactly 25 ml of the solution is pipetted into a second 100 ml measuring flask, and ca. 20 ml of water, 20 ml of reducing solution and 10 ml of sulphate–molybdate reagent are added subsequently. The reaction mixture is allowed to stand for 20 min in the dark, then 20 ml of sodium acetate is added and the mixture is made up to the mark with distilled water.

A 1 cm cell is filled with the blue-coloured solution and the absorbance at 720 nm measured against distilled water as a blank. From a calibration curve previously prepared with a series of dipotassium hydrogenphosphate solutions, the amount of phosphorus in the blue-coloured solution is found in milligrams of phosphorus per litre.

(c) Calculation

Mass (g) of oil or fat weighed $= W$
Amount of phosphorus (mg/l) read from the calibration curve $= P$

$$\text{Phosphorus content } (\%) = P/25W \qquad (6.19)$$

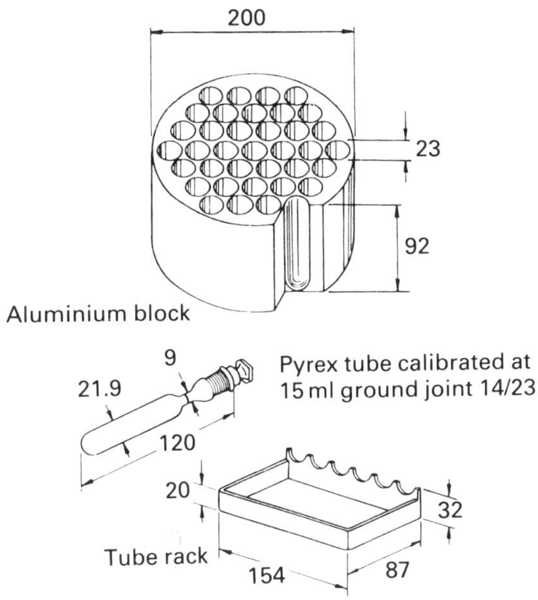

Fig. 6.6 Apparatus for the determination of phosphorus in oils (dimensions in mm).

(d) Comment

Various modifications of the colorimetric determination of phosphorus by the molybdenum blue method are available. The method described above has proved to be the most suitable.

An attractive modification of the method described above is the following. An amount of 200 mg of oil (10 droplets) and 20 mg of magnesium carbonate are weighed into calibrated 15 ml Pyrex tubes. The Pyrex tubes are placed in an aluminium block with holes (Fig. 6.6). The aluminium block is heated on an electrical hot plate at a maximum 350 °C till the oil is charred to a dry black mass. The tubes are then placed in a rack (Fig. 6.6) and ashed in a muffle furnace at ca. 550 °C to a white ash. After the ash is dissolved, the reaction with reduction solution and sulphate–molybdate reagent is carried out directly in the Pyrex tube. After making up to 15 ml, the absorption at 720 nm is also measured directly in the Pyrex tube.

In the IUPAC method 2.421 (Paquot and Hautfenne, 1985), the yellow phosphovanadomolybdic complex is used for the determination of the phosphorus content of oils and fats. The IUPAC method also differs on other points: the oil is ignited in the presence of magnesium oxide and the ash is dissolved in nitric acid.

The AOCS method Ca 12-55 (AOCS, 1978) is based on the molybdenum blue method but uses zinc oxide instead of magnesium carbonate in the ashing step. The method differs somewhat from that described in this section with respect to the details of the procedure.

An alternative method is the determination of the phosphorus content in oils and fats by atomic absorption spectrometry. In the recently published IUPAC method 2.423 (Hendrikse and Dieffenbacher, 1991; Dieffenbacher and Pocklington, 1991), graphite furnace atomic absorption spectrometry is used as a direct, rapid and accurate method for quality control in an oil mill.

6.2.16 Fatty acid composition by gas–liquid chromatography

(a) Principle

The oil or fat is converted into the corresponding fatty acid methyl esters, which, dissolved in heptane, are separated gas chromatographically on a packed column. From the diagram obtained, the fatty acid composition of the sample can be calculated.

(b) Brief description of procedure

(i) Preparation of fatty acid methyl esters
Approximately 150 mg of a dry fat sample is weighed into a 50 ml round-bottomed flask fitted with a reflux condenser. To the fat sample, 4 ml 0.5 M methanolic sodium hydroxide solution is added and the mixture is boiled under reflux for 5–10 min. Via the condenser,

5 ml of boron trifluoride solution (commercially available) is added and boiling is continued for another 2 min. Subsequently, 2–5 ml of heptane is added via the condenser to the boiling mixture, which is then boiled under reflux for 2 min.

After removal of the heating source, a sufficient amount of saturated sodium chloride solution is added to the mixture to bring the liquid level into the neck of the round-bottomed flask. About 1 ml of the heptane layer is transferred to a test tube. The solution of fatty acid methyl esters in heptane is ready for injection into the gas chromatographic column.

For the preparation of fatty acid methyl esters containing four or more carbon atoms from oils and fats, a different procedure can be used (the method is not suitable for acid oils and fats (acid value > 2)): approximately 1 g of dry sample is weighed into a test tube; the sample is dissolved in 10 ml of heptane and 0.5 ml of methanolic potassium hydroxide (2 mol/l) is added; the contents of the tube are mixed by inversion (20 s); the heptane layer is transferred to a test tube containing some granules of anhydrous sodium sulphate; the solution of methyl esters in heptane is ready for injection into a gas chromatographic column.

(ii) Gas chromatographic analysis

The procedure described is related to ordinary equipment used for GLC employing a packed column and a flame ionization detector. A column (length 1–3 m, internal diameter 2–4 mm) constructed of glass or stainless steel (material inert to the substances to be analysed) is used. The packing of the column is described in terms of a support material and a stationary phase with which it is impregnated. Support: acid-washed silanized diatomaceous earth, or other suitable inert support with a narrow range (25 μm) of grain size between the limits 125 and 200 μm, the average grain size being related to the internal diameter and length of the column. Stationary phase: polyester type of polar liquid (e.g. diethyleneglycol succinate, butanediol polysuccinate, ethylene glycol polyadipate) or any other suitable liquid (e.g. cyanosilicones); the stationary phase should amount to 5–20% of the packing; a non-polar stationary phase can be used for certain separations.

The sample for examination should be 0.1–2 μl of the heptane solution of fatty acid methyl esters prepared as described above. In the case of methyl esters not in solution, an approximately 10% solution in heptane is prepared and 0.1–1 μl of this solution is injected. If the search is for constituents present only in trace amounts, the sample size may be increased (up to 10-fold).

As a rule, the following operating conditions will give a column efficiency of at least 2000 theoretical plates for methyl stearate, its elution within about 15 min and a resolution of at least 1.25 for the separation of methyl stearate and methyl oleate:

Internal diameter of column (mm)	Carrier gas supply (ml/min)
2	15–25
3	20–40
4	40–60

Concentration of stationary phase (%)	Column temperature (°C)
5	175
10	180
15	185
20	185

Sometimes, it is possible to apply temperature programming. For example, if the sample contains the methyl esters of fatty acids below C_{12}, the sample should be injected at 100 °C, and the temperature raised at 4–8 °C/min to the optimum. In some cases, the two procedures can be combined: after the programmed heating, the elution is continued at a constant temperature.

Nowadays the use of capillary columns has increased to such an extent that various Standardization Institutes are working out standard methods for capillary GLC. Capillary GLC offers faster analysis of the fatty acid composition and results in better separations owing to the much higher number of theoretical plates in a capillary column.

A gas chromatograph suitable for capillary columns and equipped with a flame ionization detector can be used. The injector may be of the split or splitless type or a cold-on-column injector. Mostly, moderately polar phases will be suitable for the fatty acid analysis; only for special cases like the separation of *cis–trans* or positional isomers will a much more polar phase be needed. In many cases temperature programming is being applied. The calculation of the results is practically identical to that for GLC on packed columns.

(c) Calculation of results

Generally, the method of internal standardization is used for the determination of the fatty acid composition, i.e. it is assumed that all the components of the sample are represented on the chromatogram, so that the total of the areas under the peaks represents 100% of the constituents (total elution). If the equipment includes an integrator, the figures obtained therefrom are used. If not, the area under each peak is determined by triangulation.

If significant amounts of components below C_8 are absent, the content of a given constituent (expressed as percentage of methyl esters) is calculated by determining the percentage represented by the area of the corresponding peak relative to the sum of the areas of all the peaks.

Area of the peak corresponding to component i $= A_i$
Sum of the areas under all the peaks $= \Sigma A_i$

$$\text{Percentage of peak corresponding to component } i \text{ expressed as methyl ester (mass \%)} = \frac{100 A_i}{\Sigma A_i} \quad (6.20)$$

In certain cases, particularly in the presence of acids below C_8 or of acids with secondary groups, correction factors must be used to convert results calculated as percentage of peak areas into percentage mass of the components. The correction factors are determined with the help of a chromatogram derived from the analysis of a reference mixture of fatty acid methyl esters of known composition.

Mass (mg) of component i in the reference mixture $= m_i$
Total of masses (mg) of components in the reference mixture $= \Sigma m_i$

$$\text{Percentage of component } i \text{ (mass \%)} = \frac{100 m_i}{\Sigma m_i} \quad (6.21)$$

For the chromatogram of the reference mixture, the percentage (area/area) of component i is calculated as follows:

Area under the peak corresponding to component i $= B_i$
Sum of the areas under all the peaks $= \Sigma B_i$

$$\text{Percentage (area/area) of component } i = \frac{100 B_i}{\Sigma B_i} \quad (6.22)$$

Whence correction factor K_i:

$$K_i = m_i \Sigma B_i / B_i \Sigma m_i \quad (6.23)$$

Usually, the correction factors are given in relative factors related to the correction factor of palmitic acid K_{16}:

$$\text{Correction factor } K_i' = K_i / K_{16} \quad (6.24)$$

The content of each component in the samples is then given by:

$$\text{Content of component } i \text{ expressed as methyl ester (mass \%)} = \frac{100 K_i A_i}{\Sigma (K_i' A_i)} \quad (6.25)$$

In certain cases (notably the assay of C_4 and C_6 acids and the determination of acids when all the fatty acids are not eluted), an internal standard should be used and the correction factor for the internal standard be determined.

Mass (mg) of the internal standard $= m_s$
Mass (mg) of the test portion $= m$
Correction factor for the component standard $= K_s$
Correction factor for the corresponding i $= K_i$
Area of the peak corresponding to the internal standard $= A_s$
Area of the peak corresponding to component i $= A_i$

$$\text{Content of component } i \text{ expressed as methyl ester (mass \%)} = \frac{100 m_s K_i' A_i}{m K_s' A_s} \quad (6.26)$$

(d) Comment

The IUPAC method 2.302 (Paquot and Hautfenne, 1985) is identical to the method described above. IUPAC has a separate method 2.301 (Paquot and Hautfenne, 1985) for the preparation of the fatty acid methyl esters, which is also identical to the procedures described above. IUPAC refers to the procedure using methanolic potassium hydroxide as the 'special method'. IUPAC method 2.301 also contains alternative methods not involving the use of boron trifluoride.

The AOCS method Ce 1-62 (AOCS, 1978) is identical to the method described in this section. Also the method for the preparation of the fatty acid methyl esters is identical to the method described here.

6.2.17 Triacylglycerol analyses by pancreatic lipase hydrolysis

(a) Principle

The method is based on the partial and specific hydrolysis of triacylglycerols by pancreatic lipase, during which 2-monoacylglycerols are formed preferentially. The hydrolysis results in a mixture containing non-hydrolysed triacylglycerols, monoacylglycerols and free fatty acids. This mixture is separated by means of thin-layer chromatography, after which the monoacylglycerols are isolated. From these monoacylglycerols, the fatty acid methyl esters are prepared and analysed by gas–liquid chromatography.

(b) Brief description of procedure

(i) Preparation of the sample

If the sample has an acidity (determined according to section 6.2.9) below 1%, it is subjected to hydrolysis by pancreatic lipase without previous neutralization.

If the sample has an acidity between 1 and 3% inclusive, it is neutralized by means of alumina. A chromatographic column is filled with a suspension of 15 g of activated alumina in 50 ml of hexane. A solution of 5 g of sample in 25 ml of hexane is passed through the column and the whole of the effluent is collected in a 100 ml flask.

If the sample has an acidity above 3%, it is neutralized by shaking in a separating funnel of 500 ml: approximately 10 g of sample, 100 ml of hexane, 50 ml of 2-propanol, a few drops of phenolphthalein solution and an amount of sodium hydroxide solution (120 g/l) corresponding to 1.1 times the free acidity of the sample.

The hexane layer is separated and washed with 25 ml portions of a mixture of 2-propanol and water (1:1) till free of alkali. The solution is evaporated in a rotary evaporator at 30–40 °C. Subsequently the purification is carried out over alumina.

The effluent of the alumina is finally evaporated in a rotary evaporator at 30–40 °C.

(ii) Procedure

Approximately 0.1 g of the prepared sample is weighed into a centrifuge tube. If the sample is not liquid, it is melted at 60–65 °C; immediately thereafter, 20 mg of pancreatic lipase and subsequently 3 ml of trishydroxymethylaminomethane buffer (1 mol/l) acidified to pH 8 are added. After shaking, 0.5 ml of sodium cholate solution (1 g/l) and 0.2 ml of calcium chloride solution (220 g/l) are added. After cautious shaking, the tube is immediately placed in a water bath maintained at 40 ± 0.5 °C. After 1 min the tube with contents is shaken vigorously by means of an electric shaker for exactly 2 min.

Then the tube is cooled in running water, 1 ml of hydrochloric acid (6 mol/l) and 1 ml of diethyl ether are added and the mixture is homogenized by means of the electric shaker. After separation of the phases on standing, the organic layer is removed by means of a syringe, if necessary after centrifuging. The extract is applied to a chromatographic plate (0.25–0.50 mm, silica gel G, commercially available) in a thin uniform line by means of an applicator. The TLC plate is developed in a well saturated developing tank with a mixture of hexane, diethyl ether and formic acid (60/40/1.6, by volume) at about 20 °C up to about 1 cm from the top edge of the plate. The plate is dried in air and sprayed with a 2′,7′-dichlorofluorescein solution. The monoacylglycerol band is identified under ultraviolet light.

The monoacylglycerol band is removed with a spatula and transferred to a conical flask of 50 ml. The monoacylglycerols are dissolved in diethyl ether. The filtrate is collected in a round-bottomed flask of 25 ml for the preparation of the methyl esters (see section 6.2.16). The methyl esters thus obtained are examined by gas chromatography as described in section 6.2.16.

(c) Calculation

The fatty acid composition at the 2 position is calculated to one decimal place as given in section 6.2.16 (equation (6.20)). It is always also advisable to determine the total fatty acid composition of the same sample, since the comparison with that of the acids at the 2 position will help in the interpretation of the figures obtained.

(d) Comment

Except for some differences in the preparation of the sample, the developing solvent and the methylation of the isolated monoacylglycerols, this method is practically identical to IUPAC method 2.210 (Paquot and Hautfenne, 1985).

OICC (Office International du Cacao et du Chocolat) have issued a text for the determination of the composition of the fatty acids at the 2 position of the acylglycerols of oils and fats used in chocolate and confectionary products, which is very similar to the IUPAC method.

The ISO (International Organization for Standardization) commission ISO/TC 34/SC 11 'Animal and Vegetable Fats and Oils' has been involved for some years in drafting an ISO method that is also based on the IUPAC method.

6.2.18 Triacylglycerol analysis by thin-layer chromatography

(a) Principle

Prior to the analysis, a triacylglycerol internal standard, 1,3-diheptadecanoyl-2-6(*Z*)-octadecenoylglycerol (17:0, 18:1, 17:0), is added to the sample. The sample is then separated into its triacylglycerol components by silver nitrate TLC. The band for β-SOS (symmetrically disaturated 2-oleoyl-1,3-diacylglycerols) is isolated and the total β-SOS acylglycerols are converted to methyl esters, which are then analysed by GLC.

(b) Brief description of procedure

Commercially available TLC plates with 0.25 mm silica gel type 60 are cleaned by developing them in trichloromethane. After evaporation of the trichloromethane, the plate is immersed for 10 s in silver nitrate solution (20% m/v). After drying (0.5 h) in an oven at 50 °C followed by 0.5 h in an oven at 110 °C, the plate is stored in a nitrogen-filled desiccator in the dark.

To prevent oxidation of the triacylglycerols, the following procedure is carried out under nitrogen. Approximately 0.5 g of fat, accurately weighed into a 10 ml measuring flask, is dissolved in isooctane, made up to the mark and reweighed. Approximately 0.5 g of the internal standard (17:0, 18:1, 17:0 triacylglycerol) is accurately weighed into a 10 ml measuring flask, dissolved in isooctane, made up to the mark and the flask reweighed.

Exactly 1 ml of the fat solution is pipetted into a small stoppered tube (previously weighed) and reweighed. Then, exactly 1 ml of the internal standard solution is added by pipetting and the tube reweighed. From the masses of the solutions used and their concentrations, the mass of fat and the mass of internal standard in this standard mixture are calculated.

Approximately 4 mg of triacylglycerol as standard mixture is applied to the silver nitrate TLC plate as a narrow band by means of an applicator. The plate is developed in the developing solvent toluene/hexane (8:2, v/v) until the front is about half-way up the plate (about 20 min). The plate is then removed from the TLC developing tank and all the solvent evaporated. The plate is returned to the developing tank and developed further (about 1.25 h). Finally, the solvent is removed again.

The dry plate is sprayed with a dichlorofluorescein solution, after which the triacylglycerol bands are identified under 360 nm ultraviolet light.

The order of elution of the bands is: SSS (trisaturated acylglycerols), SES (symmetrically disaturated monoelaidoyl glycerols (if present)), SOS (symmetrically disaturated monooleoylglycerols), SSO (asymmetrically disaturated monooleoylglycerols), SOO (asymmetrical monosaturated dioleoylglycerols), SLinS (symmetrically disaturated monolinoleoylglycerols), more unsaturated triacylglycerols. Figure 6.7 shows a typical (diagrammatic) chromatogram.

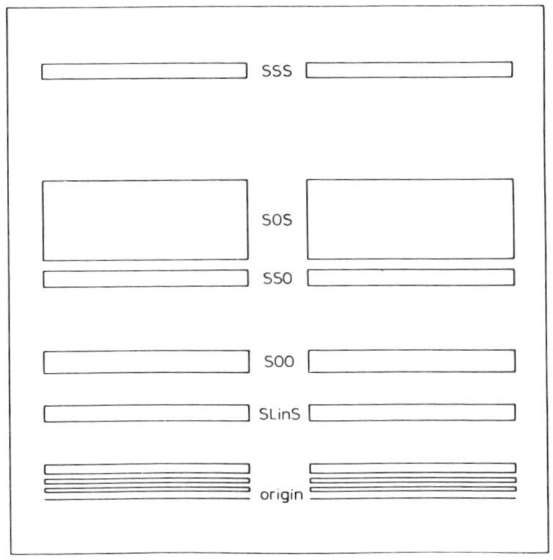

Fig. 6.7 Thin-layer chromatogram (diagrammatic) of a cocoa-butter equivalent fat on a silver nitrate-impregnated plate.

The positions of the upper and lower boundaries of the SOS band and the SSO bands are marked. The gap between the SOS and SSO bands is equally divided by a line and the silica between this line and the upper boundary of the SOS band is scraped off and collected in a 50 ml round-bottomed flask for methylation.

After the addition of 4 ml of methanolic sodium hydroxide (1.3 mol/l) to the silica gel scrapings in the 50 ml flask, the preparation of the methyl esters is carried out as described in section 6.2.16 with the exception of 10 ml of boron trifluoride being used instead of 5 ml.

About 2 μl of the solution of methyl esters in heptane thus obtained is analysed by GLC as described in section 6.2.16 under operating conditions that will give virtually complete separation of 16:0 from 17:0 and 18:0 from 18:1.

(c) Calculation

Mass of fat (mg) in standard mixture $= W_f$
Mass of internal standard (mg) in standard mixture $= W_i$
Area (%) of internal standard $= A_i$

$$\text{Disaturated monooleoylglycerol content (\%)} = 100\,W_i - \frac{100 - 1.521\,A_i}{W_f\,1.521\,A_i} \qquad (6.27)$$

Corrections for the GLC detector response should be made (see section 6.2.16, equation (6.23)).

(d) Comment

The molar ratio of saturated to unsaturated acids should be 2:1. This is a useful guide to confirm that the methyl ester preparation and GLC analysis proceeded satisfactorily and that a pure SOS band was isolated.

The proposed EC definition of a CBE (cocoa butter equivalent) fat includes a limit on the percentage of SOS-type triacylglycerols in the fat. To determine the percentage of SOS, a lipolysis method as described in section 6.2.17 followed by 1,3-random, 2-random calculation would be unsuitable except for whole natural fats.

The method described in this section has been published as a provisional method by OICC (Office International du Cocoa et du Chocolat).

6.2.19 Detection of adulteration in fats

(a) Principle

The following procedure is followed: saponification of the oil or fat, then separation of the unsaponifiable matter; isolation of the sterols from the unsaponifiable matter by TLC; analysis by GLC of the sterol fraction so isolated (or of the trimethylsilyl ethers prepared from the sterol fraction) and interpretation of the chromatograms.

(b) Brief description of procedure

The unsaponifiable matter from 5 g of oil or fat is extracted by the method in section 6.2.12 (diethyl ether method) but modified as follows: instead of drying the unsaponifiable matter at 103 ± 2 °C, it is dried at a temperature below 50 °C under partial vacuum.

The unsaponifiable matter is weighed and dissolved in 10 times its mass of trichloromethane (0.5–1.0 ml). Approximately 50 μl of this solution is applied to a thin-layer chromatographic plate (0.25 mm silica gel G, commercially available) in a continuous line, leaving a space of 2.5 cm from each side edge of the plate. Approximately 0.3 μl of a cholesterol solution in chloroform (100 g/l) is applied 1 cm from the right edge and 1 cm from the left edge of the plate.

The plate is developed in a developing tank till the solvent is 1 cm below the top edge. After the solvent is evaporated the dry plate is sprayed with a suitable indicator solution, e.g. a solution of Rhodamine 6G (0.5 g/l) or 2′,7′-dichlorofluorescein (0.5 g/l), and the plate is examined under ultraviolet light.

The position of the sterol fraction is recognized by comparison with the cholesterol reference and marked. The sterol fraction is removed with a spatula and transferred to a 25 ml conical flask fitted with a reflux condenser. After the addition of 5 ml of trichloromethane, the mixture is boiled under reflux for about 15 min. After cooling, the mixture is filtered, the silica removed from the filter and the extraction repeated four times. The filtrates are collected and the solvent is evaporated in a current of nitrogen. The residue is dissolved in a few millilitres of diethyl ether for analysis by GLC.

It is also possible to perform GLC on the trimethylsilyl ethers of the sterols. In that case, the residue is gently heated with a mixture of dimethylformamide, hexamethylsilazane and trimethylchlorosilane to prepare the trimethylsilyl ethers. The gas chromatographic method described provides a qualitative analysis of sterols in oils and fats. However, it can be employed for quantitative analysis within certain limitations: (a) the response of each sterol is a function of the type of detector used; (b) an accurate quantitative interpretation of the chromatogram is only possible if the individual sterols used as reference standards (e.g. cholesterol, stigmasterol, brassicasterol, β-sitosterol and campesterol) are pure; if this is not the case, the results must be expressed as percentage areas of the peaks.

(i) Qualitative analysis

The operating conditions are slightly different from those given in section 6.2.16. A glass column (length 180–200 cm and 2–4 cm diameter) is filled with diatomaceous earth, acid-washed, silanized and impregnated with 2–5% of a suitable stationary phase, e.g. polysiloxane, methyl silicone (e.g. SE 30) or 50% phenylmethyl silicone (e.g. OV 17). The column temperature should be 230–240 °C and the rate of the N_2 carrier gas 30–50 ml/min.

The resolution of the peaks of campesterol and stigmasterol (R) must be above 1.0. This resolution is given by the following formula:

$$R = 2(dR_s - dR_c)/(w_s + w_c) \qquad (6.28)$$

where dR is the retention distance (mm) and w is the width (mm) of the peak at its base.

Of the various pure reference sterols, solutions of 20 g/l are prepared in diethyl ether or in another suitable solvent. The relative retention distances (or times) of the various sterols compared with cholesterol are normally within the following ranges: cholesterol, 1.00; brassicasterol, 1.13–1.15;

campesterol, 1.32–1.34; stigmasterol, 1.44–1.46; β-sitosterol, 1.66–1.68; Δ^7-stigmasterol, 1.85–1.95.

A suitable amount of the solution of the sterol fraction, isolated from the test sample, is then injected into the GLC column and the sterols are identified by measuring the retention distances (or times) of the peaks on the chromatogram.

(ii) Quantitative analysis
The method of internal standardization is used, i.e. it is assumed that all the sterols present in the solution of the sterol fraction are represented by the chromatogram (total elution).

If pure sterols can be assayed, their correction factors K_i with reference to cholesterol ($K = 1$) are determined.

(c) Calculation

The content of sterol i is calculated as follows:

Area of the peak corresponding to sterol i $= A_i$
Correction factor for sterol i relative to cholesterol $= K_i$

$$\text{Content of sterol } i \text{ (mass \%)} = \frac{100\, A_i K_i}{\Sigma\,(A_i K_i)} \qquad (6.29)$$

(d) Comment

The method described is identical to the IUPAC method 2.403 (Paquot and Hautfenne, 1985). The AOCS method Ce 3-74 (AOCS, 1978) describes a method for the gas chromatographic determination of tocopherols and sterols in soya sludges and residues. In this method, the sample is saponified and the extracted unsaponifiable matter is reacted with butyric anhydride. The butyrate esters of the tocopherols and the sterols present in the unsaponifiable matter are subjected to gas chromatographic analysis and determined quantitatively. Relative retention times for the various tocopherol and sterol butyrates are given.

A drawback is that the method is restricted to soya sludges and residues.

(i) Detection of adulteration in fats
If a peak is observed in the gas chromatogram with the relative retention distance of β-sitosterol and a peak height of at least 2% of the maximum, this indicates the presence of vegetable fat in an animal fat. The conclusion can be supported by the presence of peaks of brassicasterol, campesterol and stigmasterol in the chromatogram. The total sterol content determined by the method of subsection (ii) above (quantitative analysis) has also to be taken into consideration when the conclusion is formulated.

If the objective is to detect the presence of animal fats in vegetable oils or fats, it is necessary to use other analytical tests in addition to the identification of the sterols, for example, a determination of the composition of the fatty acids by GLC (section 6.2.16): in practice, branched-chain acids and fatty acids with an odd number of carbon atoms are only present in significant amounts in animal fats and oils.

It is even possible to estimate that a mixture consists of *a*% fat A, *b*% fat B, etc., on the basis of the fatty acid composition according to section 6.2.16. Usually it is also in this case necessary to include other constants such as refractive index and *trans* content (see section 6.2.16). The melting behaviour can often give important indications (e.g. the dilatation method in section 6.2.4). Whether a mixture of mainly triacylglycerols is indeed involved can easily be checked with the help of TLC (section 6.2.18).

A worthwhile tool in estimating mixture compositions could be the publication concerning fatty acid, sterol and tocopherol profiles in vegetable and animal oils and fats issued by the Netherlands Standards Institute in 1989.

The publication is a translation of the Netherlands Code of Practice known as NPR 6305. The document is a result of gathering data on oils and fats obtained by Dutch laboratories in industry, trade and government. This Code of Practice makes it possible to adduce arguments in support of purity; however, without the force of law. It also guides the newcomer through the labyrinth of chromatographic data of fatty acid profiles, sterol compositions and tocopherol contents used for the identification of pure oils and fats, or for the estimation of the authenticity of the product. The document may also be applied to calculate compositional ratios of the oils and fats in blends within certain margins. The publication is available from the Netherlands Standards Institute as 'NPR 6305 English: 1989' entitled 'Vegetable and animal oils and fats. Interpretation of results obtained by chromatographic examination. Fatty acids, sterols and tocopherols profiles'. It can also be obtained from any national standards institution.

Finally, the analysis of the triacylglycerols by GLC (this section) can give information on adulteration by the determination of the composition of the fatty acids at the 2 position of the acylglycerols: to check whether oils (e.g. olive oil) or fats have been re-esterified; to check tallow on admixing with lard; to check, as already mentioned, cocoa butter equivalent fats against the internationally proposed requirements (see also Padley and Timms, 1980; Fincke, 1980).

References and further reading

AOCS (1978) *Official and Tentative Methods of the American Oil Chemists' Society*, 3rd edn, American Oil Chemists' Society, Champaign.

Boekenoogen, H.A. (1964) *Analysis and Characterisation of Oils, Fats and Fat Products*, vol. 1, Wiley, New York.

Chapman, D., Richards, R.E. and Yorke, R.W. (1960) *J. Am. Oil Chem. Soc.*, **37**, 243.

Deutsche Gesellschaft für Fettwissenschaft (1950–75) *DGF-Einheitsmethoden*.

Dieffenbacher, A. and Pocklington, W.D. (1991) *Standard Methods for the Analysis of Oils, Fats and Derivatives*, 1st supplement to the 7th edn, Blackwell Scientific, Oxford.

Fincke, A. (1980) *Deutsche Lebensmittel-Rundschau*, **76**, 162.

Hautfenne, A. (1983) *Pure Appl. Chem.*, **54**, 2766.

Hendrikse, P.W. and Dieffenbacher, A. (1991) *Pure Appl. Chem.*, **63**, 1191.

IUPAC (1966) *Standard Methods for the Analysis of Oils, Fats and Soaps*, 5th edn, Butterworth, London.

Padley, F.B. and Timms, R.E. (1980) *J. Am. Oil Chem. Soc.*, **57**, 286.

Paquot, C. (1979) *Pure Appl. Chem.*, **51**, 2505.

Paquot, C. and Hautfenne, A. (1985) *Standard Methods for the Analysis of Oils, Fats and Derivatives*, 7th edn, Blackwell Scientific, Oxford.

Van den Enden, J.C., Haighton, A.J., Vermaas, L.F. and Waddington, D. (1978) *Fette Seifen Anstrichm.*, **80**, 180.

Van den Enden, J.C., Rossell, J.B., Vermaas, L.F. and Waddington, D. (1982) *J. Am. Oil Chem. Soc.*, **59**, 433.

6.3 Other analytical procedures

6.3.1 Extraction of lipids from natural samples

It is best to extract lipids from natural samples as soon as possible to minimize the degradative changes that otherwise take place. When samples must be stored, they are usually frozen as quickly as possible at $-20\,°C$ or lower in an atmosphere of nitrogen. All-glass containers or glass containers with Teflon-lined caps are also advised. Obviously, if an analysis of a particular subcellular component is required, then subcellular fractionation must be undertaken before the freezing process, which disrupts the tissue.

Many tissues, especially those from plants, contain very active lipid-degrading enzymes, which are activated by organic solvents and are active at low temperatures. These enzymes are best inactivated before extraction of the tissue by, for example, brief exposure of the tissue to steam or boiling water or hot isopropanol (Harwood, 1980; Kates, 1986).

The actual method of extraction to be used depends on both the chemical nature of the sample and also the type of extract desired. For example, skin- or leaf-surface waxes and hydrocarbons may be extracted with non-polar solvents such as ethyl ether or chloroform (see Nicolaides *et al.*, 1970; von Wettstein-Knowles, 1979). Extraction of membrane-associated lipids, however, requires the use of solvent mixtures containing alcohols to disrupt the hydrogen bonding and ionic forces between the lipids and proteins (Klein and Kemp, 1977).

The most useful solvent mixtures are those containing chloroform and methanol. A typical procedure is that of Bligh and Dyer (1959) where a one-phase methanol system (chloroform–methanol–water, 1/2/0.8, by volume) rapidly extracts lipids from most tissues and the extract after filtration or centrifugation (Kates, 1986) is then diluted with 1 volume each of chloroform and methanol to yield a two-phase mixture: chloroform and methanol–water (10/9, by volume). The upper (aqueous methanol) phase contains water-soluble contaminants, while the lower chloroform phase contains the lipids. Water-soluble contaminants can also be removed from solutions of lipids in organic solvents by passage through Sephadex columns (Rouser *et al.*, 1967; Wuthier, 1966; see section 4.4.2(a)). After rotary evaporation of the solvent, any traces of water remaining with the lipid may be removed by azeotropic distillation with benzene or toluene.

The Bligh and Dyer method is satisfactory for most animal tissues, most simple and complex lipids being readily extracted by this procedure. Kates (1986) and Christie (1982) detail practical points. For quantitative extraction of the polyphosphoinositides (and other very polar lipids), it is advisable to use high concentrations of salts (Garbus *et al.*, 1963), or to add acid to the upper phase (Wells and Dittmer, 1965) or to use an acidified one-phase extraction solvent (chloroform–methanol–0.2 N HCl, 1/2/0.8, by volume) (Kates, 1986). The two phases can also be washed with fresh upper or lower phases, as appropriate, to ensure both quantitative extraction and removal of water-soluble contaminants. Note that highly polar lipids such as sulphated polyglycolipids, gangliosides, alkyl chlorosulphates, etc., may partition into the methanol–water phase; this should be checked by TLC of the methanol–water phase in appropriate solvents. The extraction of plant tissue poses particular problems in that endogenous lipid-degrading enzymes have to be inactivated. Standard procedures usually using boiling isopropanol are given in Hitchcock and Nichols (1971), Harwood (1980) and Kates (1986). Extraction of certain algae (Nichols and Appleby, 1969) and dry cereals (Colborne and Laidman, 1975; Fishwick and Wright, 1977; Kates, 1986) require special techniques. A method for lipid extraction from bacteria, including halophilic bacteria, is given in Kates (1986).

For successful recovery of non-contaminated lipid samples, special precautions should be taken. All solvents should be distilled before use (methods are given in Kates (1986)), plastic apparatus should not be used and conditions that might give rise to oxidation should be avoided. The latter is reduced by carrying out all procedures at room temperature or lower, adding an antioxidant such as 2,6-di-t-butyl-*p*-cresol (BHT), and using a nitrogen atmosphere when possible. Some lipids are also light-sensitive, and exposure to direct sunlight should be avoided.

Lipids should not be left for any length of time in the dry state but should be redissolved in a solvent such as chloroform–methanol (2/1, v/v). When storing lipids for any length of time, it is best to fill the container with as much solvent as possible, to add BHT or to seal under N_2, and to store at $-20\,°C$ or lower in the dark.

6.3.2 Structural determination of fatty acids

The definitive structural determination of fatty acids usually requires the isolation of adequate quantities of purified compounds (though techniques such as gas chromatography/mass spectrometry (GC/MS) require very little material). Methods for the preliminary fractionation and subsequent purification of fatty acids will be found in Kates (1986), Christie (1982) and Kuksis (1978) (see section 4.4.1).

Gas–liquid chromatography is widely used in structural studies of fatty acids. Retention data are usually recorded in the form of equivalent chain length (Miwa *et al.*, 1960; Christie, 1982, 1989; Kates, 1986) and, together with the group separation factors for the particular liquid phase used (see Table 4.8), deductions concerning fatty acid chain length and type (normal, branched, unsaturated, cyclic, etc.) can be made.

Usually fatty acids are chromatographed as their methyl esters, but the analyses of short-chain (C_2–C_6) and medium-chain (C_8–C_{12}) fatty acids pose particular problems. It is necessary to take special precautions in the analysis of their volatile methyl esters to avoid excessive losses during solvent evaporation or difficulty in quantification owing to the emergence of the fatty acid peaks too rapidly from conventional GLC columns. The less volatile short-chain free acids can be separated on a variety of liquid phases fortified with a non-volatile acid such as phosphoric acid (Cochrane, 1975; Christie, 1982, 1989), which reduces dimerization of the free acids. Relative retention times for saturated short-chain fatty acids on three stationary phases are given by Hrivnak *et al.* (1972). Separation of various saturated and unsaturated acids as alkyl or isoalkyl esters has been recorded by Ashes and Haken (1974a, b, 1975).

GLC of medium- and long-chain fatty acids as their methyl esters should be carried out on both polar (e.g. SP-2330, DEGS) and non-polar (e.g. SE-30, OV-1) columns (see section 4.5). Comparison of retention times relative to methyl palmitate and other standards will allow preliminary identification of normal saturated and unsaturated acids and of iso and anteiso acids. Hydroxyl groups can be recognized by GLC behaviour before and after conversion to an acetate, trifluoroacetate, methoxy or trimethylsilyl ether derivative (see Table 4.8 for group separation factors).

Saturated, branched, unsaturated and cyclopropane ester peaks can be distinguished by GLC of the sample before and after hydrogenation with Pt catalyst in methanol. Saturated, branched and cyclopropane fatty acid ester peaks remain unchanged after hydrogenation, but unsaturated acid peaks are now eluted at the same point as the corresponding saturated acids; subsequent treatment of the hydrogenated sample with bromine in ethyl ether results in the disappearance of cyclopropane fatty acid peaks. Alternatively, hydrogenation with Pt catalyst in glacial acetic acid causes opening of the cyclopropane ring to give the corresponding straight-chain ester and a mixture of the two possible branched-chain isomers (see Kates, 1986).

In addition, the presence of unsaturated groups (alkenoic or alkynoic) and of branched and cyclic groups is often indicated by GLC retention times, and the use of capillary or support-coated open tubular (SCOT) columns may even suggest the position of the unsaturated centre or functional group (for useful references and details of retention times, see Gunstone (1975) and Kuksis (1978)).

GLC combined with mass spectrometry (GC/MS) can be used for definitive identification of fatty acids present in natural mixtures. However, preliminary fractionation is advisable for complex mixtures containing different chain types (e.g. normal, branched, unsaturated, etc.). Unsaturated fatty acids are usually first separated by argentation TLC prior to analysis by GLC or GC/MS. This separates acids on the basis of their total number of ethylenic bonds, and, in addition, distinguishes *trans* from *cis* isomers (Emken and Dutton, 1974) (see section 4.4.1(b)). Branched and cyclic chains can be separated from normal (straight) chains by treating the mixture with a 5 Å molecular sieve, which retains the normal chains (see Kates, 1986).

Excellent reviews have been published on the use of GC/MS for identification of fatty acids and other lipids (Sun and Holman, 1968; McCloskey, 1969, 1970; Odham and Stenhagen-Stallberg, 1972; Klein and Kemp, 1977). Oxygenated functions are readily determined by mass spectrometry; hydroxy esters are best chromatographed as trimethylsilyl ethers and epoxy esters as hydroxymethoxy or methoxytrimethylsiloxy derivatives (Kleiman and Spencer, 1973). Deduction of the position of branched methyl and cyclic groups in otherwise unbranched acids is also fairly straightforward (McCloskey, 1970). Cyclopropane groups are generally unstable to electron bombardment but can be 'fixed' by preliminary chemical reactions such as hydrogenolysis, oxidation and reaction with methanolic BF_3. These reactions lead to products that can then be readily identified by mass spectrometry (Gunstone, 1975).

The location and stereochemistry of double (or triple) bonds cannot be directly determined by mass spectrometry. However, mass spectra of derivatized monoenoates, such as hydroxylated or epoxidized derivatives, can be easily interpreted (Gunstone, 1975). Some interpretation is also possible from the fragmentation patterns of underivatized monoenoates (McCloskey, 1969, 1970). In addition, good results have been obtained when the acid is derivatized to an amide, particularly a pyrrolidide (Andersson and Holman, 1974). A few empirical rules for interpreting fragmentation are given in

Klein and Kemp (1977) and in Kates (1986) (cf. section 9.6) and a typical fragmentation pattern for a saturated methyl ester (methyl stearate) is shown in Fig. 6.8.

Unsaturated fatty acids can be further studied by chemical methods. Estimation of the number of double bonds can be made by quantitative catalytic hydrogenation (Brown, 1967) or by determination of the iodine value (Kates, 1986). Partial hydrogenation of acetylenic bonds to *cis* double bonds can be performed with lead-poisoned palladium catalyst (Lie Ken Jie, 1975).

The most widely used method for determining double-bond positions is oxidative cleavage with ozone or potassium permanganate and identification of the cleavage products by GLC (section 10.4). Use of the latter reagent has been discussed by Roehm and Privett (1969) and by Downing and Green (1968) (see Kates, 1986).

Ozonolysis is widely employed and the reaction can be influenced by the choice of solvent. The ozonized product can be decomposed in various ways to give alcohols, acids or aldehydes – the choice depending on how easily the products can be identified (cf. Pryde and Cowan, 1971). Reductive ozonolysis gives unequivocal patterns for monoenoates and dienoates but not for acids with more than two double bonds. Therefore, in the latter case partial hydrogenation with hydrazine can be combined with ozonolysis. The mixture of mono-unsaturated isomers produced by hydrazine hydrogenation is separated into *cis* and *trans* derivatives by argentation TLC before cleavage by reductive ozonolysis (Roehm and Privett, 1969). Other methods are discussed by Gunstone (1975) (see Kates, 1986).

Fatty acids having double bonds close to the carboxyl end can be distinguished by using methods based on the intramolecular oxymercuration of alcohols derived from the acids. Double bonds that are 3*t*, 4*c*/*t* or 5*c*/*t* can be identified in this way (Gunstone and Inglis, 1973).

The presence of hydroxy groups is indicated by the behaviour of the parent hydroxy acid methyl ester on TLC (decreased R_f values) and GLC (increased relative retentions), and the position of the OH group in the chain is shown by GLC relative retention data (see Tables 4.4 and 4.8) and by mass spectrometry fragmentation patterns (cleavage α to the OH group). The hydroxyl function can also be oxidized to an oxo group, which can be identified by mass spectrometry of the ketone or its derivative (McCloskey, 1969).

Epoxides can easily be cleaved with periodic acid after ring opening to the diol, and the resulting aldehydes identified by GLC (see Table 4.8). Branched methyl groups can be located by subjecting the branched acid to vigorous oxidation with potassium permanganate or chromic acid (Gunstone, 1975); for example, iso- and anteiso-methyl esters yield acetone and 2-butanone, respectively, and these can be analysed by GLC on a Poropak column (Nicolaides and Fu, 1969).

Ultraviolet spectroscopy can be used to distinguish various chromophores. However, it is of limited use for most fatty acids because of low molar extinctions or overlapping absorptions for different groups. Sometimes the number of methylene-interrupted double bonds can be found by isomerization to conjugated isomers (Gunstone, 1975). On the other hand, infrared spectroscopy can be used to distinguish easily between *cis* and *trans* unsaturation (Kates, 1986). This also applies to polyenoates with various combinations of *cis* and *trans* double bonds (Gunstone, 1975). Wolff and Miwa (1965) also reported infrared absorption bands characteristic for hydroxyl, oxo, cyclopropene, epoxide, vinyl and con-

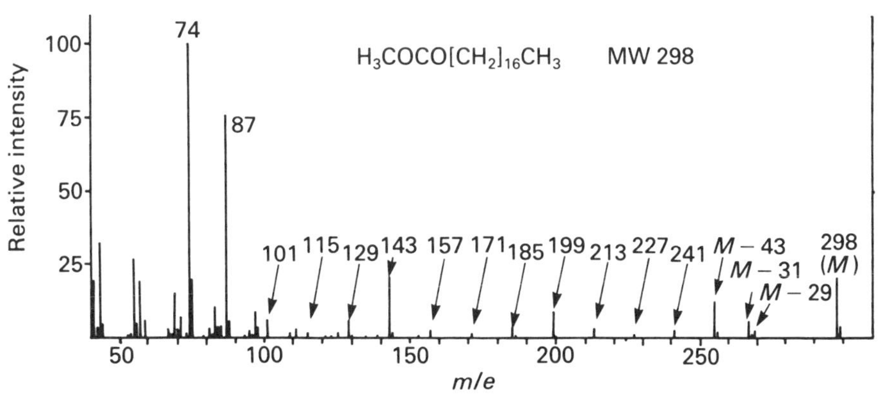

Fig. 6.8 Mass spectrum of methyl stearate. Reproduced with permission from McCloskey (1969).

jugated enyne groups (see Kates, 1986). Olefinic and acetylenic groups can be distinguished by Raman spectra (Andersson *et al.*, 1974).

NMR signals can be used to determine patterns of unsaturation, branching or substitution (Klein and Kemp, 1977; Kates, 1986). A wide variety of fatty acids have been investigated and peaks representing various special groups identified. Cyclopropane ring systems give very distinctive signals, and the use of derivatives of transition metals such as europium and praseodymium permits characteristic shifts in NMR signals to be noted. Useful references are Gunstone and Inglis (1971), Gunstone and Jacobsberg (1972), Klein and Kemp (1977), Kuksis (1978) and Kates (1986) (see section 9.4).

In summary, the most useful standard procedures for determining the structure of fatty acids are a combination of TLC and GLC separations together with mass spectrometry, infrared and NMR spectroscopy and chemical degradation. Unfortunately, many research workers publish results on inadequately characterized fatty acids, and it cannot be overemphasized that GLC relative retention times alone are insufficient to identify conclusively a given compound.

6.3.3 Analysis of acylglycerols

For a comprehensive discussion of this topic, the reader should consult Myher (1978). Before analysis of the three classes of acylglycerols, mono-, di- and triacylglycerols, each class should be separated from the neutral lipid fraction by silicic acid column chromatography (section 4.4.1(a)(i); Table 4.3) and finally purified by preparative TLC using the solvents given in Table 4.4 (solvents 3, 5 and 6). More efficient resolution of 1,2- and 1,3-diacylglycerols is achieved by preparative TLC on silica G impregnated with boric acid developed first in ethyl ether–petroleum ether (8/92) (b.p. 40–60 °C), then in ethyl ether–petroleum ether (40/60) (Brockerhoff, 1967); hexane–ethyl ether (1/1) may also be used (Christie and Moore, 1969). TLC separation may also be carried out on silica gel G with heptane–isopropyl ether–acetic acid (60/40/4) (Breckenridge and Kuksis, 1968). Separation of 1- and 2-monoacylglycerols is carried out on boric-acid-impregnated silica gel using chloroform–acetone (94/4) as developing solvent (Thomas *et al.*, 1965). Good separation without isomerization of the mono- or diglyceride isomers is achieved by these procedures. Mono-, di- and triacylglycerols may also be separated by preparative HPLC (see section 4.4.4).

The identity of isolated acylglycerol classes may now be checked by examining their characteristic infrared and NMR spectra; and by determination of their molecular ratios of fatty acids/ester groups/glycerol (3/3/1, 2/2/1, 1/1/1 for tri-, di- and mono-acylglycerols, respectively); see Kates (1986) for analytical procedures.

Each of the glyceride classes should also be characterized by GLC analysis of their total fatty acids (section 6.3.2). Separation of each acylglycerol class into molecular species differing in fatty acid chain length and degree of unsaturation may now be carried out by a combination of argentation TLC (section 4.4.1(b)) and GLC (section 4.5), or by reversed-phase HPLC (section 4.4.4), as discussed in section 6.3.2 for fatty acids.

Separation of triacylglycerols on the basis of degree of unsaturation can be achieved by TLC on silica gel G impregnated with 10–30% silver nitrate using the solvent systems reported by Barrett *et al.* (1963), Kuksis and Marai (1967) and Privett and Nutter (1967). With these solvents the saturated, monoenoic, dienoic, trienoic and tetraenoic glycerides are well separated from each other and from the more highly unsaturated species, which remain near the origin. The latter may, however, be resolved by use of more polar solvents such as mixtures of chloroform and methanol or chloroform–methanol–water for highly unsaturated glycerides (6–12 double bonds) (Renkonen and Varo, 1967). Separations of positional and configurational triglyceride isomers, as well as triglyceride isomers with different fatty acids and the same number of double bonds, are also obtained by argentation TLC (Kuksis, 1976; Renkonen and Varo, 1967). For silver nitrate phase HPLC see Jeffrey (1991).

Separation of diacylglycerols or monoacylglycerols based on number of double bonds per mole can be achieved by argentation TLC, after conversion to diglyceride acetates or monoglyceride diacetates, respectively, in the same way as triacylglycerols (Kuksis and Marai, 1967; Myher, 1978).

The saturated, monoenoic, dienoic, etc., subfractions of monoacyl- or diacylglycerol acetates and triacylglycerols eluted from the $AgNO_3$–SiO_2 TLC plates are then analysed for molecular species by GLC on non-polar liquid phases using the procedures developed by Kuksis (1976) (see also Myher, 1978; Kuksis *et al.*, 1984, 1985). In general, triacylglycerols are separated on the basis of the total number of carbons in their fatty acid chains, the number of double bonds having little influence on non-polar phases, except for triglycerides with six or more double bonds; separation of glycerides differing by only two carbons is readily achieved. Mono- and diacylglycerols are also resolved by GLC of their diacetates and monoacetates, respectively; 1- and 2-monoacylglycerols, and 1,3- and 1,2-diacylglycerols, as their respective acetates, are also separated. GLC analysis of free monoacylglycerols can also be carried out after conversion to allyl esters of the fatty acids (McInnes *et al.*, 1960).

Better resolution of diacylglycerol molecular species is achieved by GLC on polar capillary columns (SP-2330) using the tetramethylsilyl (TMS) or t-butyldimethylsilyl (t-BDMS) ether derivatives (Myher and Kuksis, 1982). This procedure is also applicable to alkenylacylglycerols (Myher and Kuksis, 1984).

It is well known that the fatty acids in a triglyceride are not randomly distributed among the three hydroxyl positions of the glycerol molecule, but are stereospecifically arranged in the molecule, i.e. hydroxyls on carbons 1 and 3 of the glycerol are not identical but can be distinguished enzymatically. The procedure for analysing fatty acids in each position on the glycerol molecule was devised by Brockerhoff (1965, 1967). When it is applied to the argentation TLC fractions or the triglyceride GLC peaks obtained as described above, it is possible to determine the complete stereospecific structure of triglyceride molecules in a natural fat, e.g. pig depot fat and corn oil (Christie and Moore, 1969), human milk fat (Breckenridge *et al.*, 1969) and bovine milk fat (Breckenridge and Kuksis, 1968; Brockerhoff *et al.*, 1966); see review by Breckenridge (1978).

The stereospecific analysis procedure (Brockerhoff, 1965, 1967), as modified by Christie and Moore (1969), consists essentially of partial degradation of the triacylglycerol by treatment with ethylmagnesium bromide to a mixture of 1,2-, 2,3- and 1,3-diglycerides; isolation of 1,2- and 2,3-diglycerides by TLC on boric acid–silica gel G and conversion to phosphatidylphenols by reaction with phenyl dichlorophosphate in pyridine; and finally cleavage of the fatty acids from position 2 by treatment with snake venom. Fatty acids in the 1 position are obtained by analysis of the lysophosphatide formed, and those in position 3 are calculated by difference; fatty acids in position 2 are also obtained by analysis of the monoglyceride formed by lipase action on the original glyceride mixture.

A detailed description of the procedure and of the errors and corrections involved is beyond the scope of this book; the reader should refer to the papers by Brockerhoff (1965, 1967) and Christie and Moore (1969) for the methodology involved, and to the reviews by Brockerhoff (1971), Breckenridge (1978), Christie (1982) and Takagi (1990).

6.3.4 Analysis of phospholipids

Once separation of a phospholipid mixture into individual phosphatides has been attained (see sections 4.4.1(a)(i) and 4.4.1(b)), then the compound(s) may be identified and quantified. Provisional identification is made by comparing R_f values with known standards and by differential colour spray reagents (cf. Kates, 1986; Christie, 1982; section 4.4.1(b)). Useful stains/sprays are those for phosphate (molybdate), primary amino groups (ninhydrin), choline (Dragendorff) and sphingolipid (amido or NH-group).

The tentative identification of the phospholipid should now be confirmed by analytical and spectral means. Elemental analysis (C, H, N and P) and estimation of particular groups, such as vicinal hydroxyl, glycerol and ester groups, can be undertaken (cf. Kates, 1986; Christie, 1982; Hawthorne and Ansell, 1982). More definitive evidence, particularly when confirming known compounds, is provided by alkaline or acid hydrolysis. Methods developed by Dawson (1967) use mild alkaline hydrolysis (methanolic 0.2 M NaOH at room temperature for 15 min) to deacylate phospholipids (and glycodiacylglycerols). The water-soluble deacylated products are then separated by two-dimensional chromatography on paper or cellulose TLC plates; phosphate esters, glycerol, inositol and other groups can be detected by spraying the paper chromatograms or cellulose plates with specific reagents (cf. Kates, 1986). Acid hydrolysis, which is much more vigorous, is generally performed with methanolic HCl (0.6, 1 or 2 M) because the acid can be conveniently removed by desiccation over potassium hydroxide before the hydrolysate is subjected to paper chromatography or cellulose TLC. Various solvent systems are available to separate the liberated glycerol, glycerophosphates, inositol phosphates and water-soluble bases, which can be identified and quantified (cf. Spanner, 1973; Christie, 1982; Kates, 1986).

Identification of the phospholipids can be confirmed by the use of a number of commercially available phospholipid-hydrolysing enzymes (phospholipases A_1, A_2, B, C, D). The stereospecificity of the enzymes also allows the stereochemistry of the lipid to be established. The action of such enzymes is detailed in Table 6.3, and incubation details are given in Lowenstein (1969), Kates (1986) and Waite (1987).

Quantification of phospholipids can be done in a number of ways. TLC plates can be charred and the amounts of individual (phospho)lipids estimated by densitometry (Rouser *et al.*, 1967). Skill and experience are required for this method to be of any use at all. Although it has the advantages of quickness and non-specificity, it does not generally give very reproducible results. A recent adaptation of this technique using the charred spots to quench radioactive scintillation counting (Shand and Noble, 1980) gives reproducible results but is not very sensitive. When

Table 6.3 Enzymic hydrolysis products of phospholipids (From Kates, 1986, with permission)

Lipid substrate	Enzyme	Lipid products	Water-soluble products
Phosphatidic acid (PA)	Phospholipase A_2	*sn*-1-Acyl-3-glycerophosphate (lyso-PA) + fatty acid	None
	Phospholipase B	Fatty acids	*sn*-3-Glycerophosphate (GP)
	Phosphatidic acid phosphatase	*sn*-1,2-Diacylglycerol	P_i
Phosphatidylcholine (PC)	Phospholipase A_2	*sn*-1-Acyl-3-glycerophosphoryl-choline (lyso-PC) + fatty acid	None
	Phospholipase B	Fatty acids	*sn*-3-Glycerophosphorylcholine (GPC)
	Phospholipase C	*sn*-1,2-Diacylglycerol	Phosphorylcholine
	Phospholipase D	*sn*-1,2-Diacyl-3-glycerophosphate	Choline
Phosphatidylethanolamine (PE)	Phospholipase A_2	*sn*-1-Acyl-3-glycerophosphoryl-ethanolamine (lyso-PE) + fatty acid	None
	Phospholipase B	Fatty acids	*sn*-3-Glycerophosphoryl-ethanolamine (GPE)
	Phospholipase C	*sn*-1,2-Diacylglycerol	Phosphorylethanolamine
	Phospholipase D	*sn*-1,2-Diacyl-3-glycerophosphate	Ethanolamine
Phosphatidylserine (PS)	Phospholipase A_2	*sn*-1-Acyl-3-glycerophosphoryl-serine (lyso-PS) + fatty acid	None
	Phospholipase B	Fatty acids	*sn*-3-Glycerophosphorylserine (GPS)
	Phospholipase D	*sn*-1,2-Diacyl-3-glycerophosphate	Serine
Phosphatidylglycerol (PG)	Phospholipase A_2	*sn*-1-Acyl-3-glycerylphosphoryl-glycerol (lyso-PG) + fatty acid	None
	Phospholipase B	Fatty acids	*sn*-3-Glycerophosphorylglycerol (GPG)
	Phospholipase C	*sn*-1,2-Diacylglycerol	*sn*-1-Glycerophosphate
	Phospholipase D	*sn*-1,2-Diacylglycerophosphate	Glycerol

phospholipids are separated on silicic-acid-coated rods they can be scanned by flame ionization detection (Iatroscan) (Ackman, 1981). This is a promising technique, but its use at present is limited to very simple lipid mixtures when components are well separated on the rods.

Phospholipids can be quantified more specifically by TLC separation of the mixture, elution of each separated spot (and blank spots) with chloroform–methanol (1/1) and phosphate analysis of the eluates, making appropriate corrections for the 'blank' values (Renkonen and Luukonen, 1976; Kates, 1986).

Phosphate analysis is usually carried out on perchloric acid digests (care!) of the lipid sample by the method of Bartlett (1959) (see Kates, 1986; section 6.2.15). Various new methods have been developed more recently, none of which has any distinct advantage over Bartlett's. However, for small samples a mixed acid method has been recommended (cf. Cook and Daughton, 1981), and this may be used for routine analysis.

In a more general procedure, acyl lipids separated on TLC plates may be quantified by scraping them from the plate, subjecting them to acid methanolysis to form fatty acid methyl esters (FAME) and quantifying the FAME by GLC using an internal standard; the fatty acid analysis of each lipid is thereby also determined. As a check, phospholipids and glycolipids may be estimated by phosphate analysis or sugar analysis, respectively, of the methanol–water phase of the methanolysate (Kates, 1986).

Inositol that has been released by strong acid hydrolysis is measured by bioassay or by GLC of its trimethylsilyl derivative (Spanner, 1973). Fatty acids can be measured and provisionally identified (section 6.3.2) following GLC of their methyl esters produced by acid- or base-catalysed transmethylation (Kates, 1986). Plasmalogens yield methyl esters of fatty acids and dimethyl acetals of the fatty aldehydes with acidic methanol. These can then be analysed by GLC, although the acetals are unstable on certain types of column (Stein and Slawson, 1966) and various methods for their analysis have been developed (cf. Spanner, 1973; Vishnubhatla *et al.*, 1988). Ethanolamine and serine bases can be determined as their dinitrofluorobenzene derivatives (Axelrod *et al.*, 1953) or trinitrobenzenesulphonic acid derivatives (Siakotos, 1967) after being released from phospholipids by acid hydrolysis. Choline can be estimated by spectrophotometry following its precipitation as a periodide derivative and redissolving the precipitate in 1,2-dichloroethane (Barak and Tuma, 1981).

Recent interest in the detailed metabolism of individual phospholipids has led to the development of methods for the separation and analysis of molecular species of phospholipids (Renkonen and Luukonen, 1976). Most of these are based on argentation TLC (section 4.4.1(b)). Phosphatidylcholine, phosphatidylethanolamine and phosphatidylinositol have been fractionated by these means (cf. Arvidson, 1965; Spanner, 1973). Although it is possible, by careful attention to technical details, to obtain good separations by these means (e.g. Harwood, 1976), a major snag is that the difference in binding between different species is small in relation to the polarity of the headgroup. For that reason it is usual to hydrolyse the phospholipid with phospholipase C and then separate the resultant diacylglycerols or their acetates by argentation TLC (Christie, 1982; Myher, 1978; see section 6.3.3). Careful attention must be paid to the choice of enzyme for this purpose, as phospholipase C preparations have high substrate specificities, and controls must always be included to ensure that hydrolysis is complete and recoveries quantitative. Argentation TLC usually fails to distinguish the position of fatty acid attachment (Morris, 1966), but the 1 and 2 positions of separated molecular species can be separately analysed by the use of appropriate phospholipase A enzymes (cf. Brockerhoff, 1975; Christie, 1982). A promising technique for the direct quantification of molecular species of phosphatides is the use of reversed-phase HPLC on phospholipid classes after prior separation by 'straight' HPLC (Patton *et al.*, 1982) (see Kates, 1986).

More efficient procedures employ GC/MS or HPLC(LC)/MS for analysis of the diacyl-, alkylacyl- and alk-1-enylacylglycerol moieties of phospholipids (released by phospholipase C hydrolysis; Table 6.3); either the acetate, TMS or t-BDMS derivative of the diradylglycerols can be used (see reviews by Kuksis *et al.*, 1984, 1985). For example, LC on Lipidex-5000 has been used to separate diacyl-, alkylacyl- and alk-1-enylacylglycerol molecular species derived from phosphatidylcholine (Curstedt, 1977), and HPLC on the acetate derivatives has been used for phosphatidylethanolamine species (Nakagawa and Horrocks, 1983). A potentially important variation of this approach is to carry out the molecular species analysis on the phospholipase-C-derived diacylglycerols after catalytic reduction with deuterium, conversion to the t-BDMS derivative and analysis by GC/MS (Dickens *et al.*, 1982).

Mass spectrometric analysis of intact, underivatized phospholipids can be achieved by fast atom bombardment mass spectrometry (FAB-MS), in positive- and or negative-ion mode (see Rinehart, 1982; Aberth *et al.*, 1982; May and Desiderio, 1983; Kates, 1986).

6.3.5 Analysis of glycolipids

Procedures for analysis and identification of the two major groups of glycolipids, glycosphingolipids and glycoglycerolipids, have been reviewed by Hakomori (1983) and Kates (1990), respectively.

Preliminary identification of a glycosyldiacylglycerol is made on the basis of its relative R_f value on TLC in several solvents (Table 4.5) and staining behaviour with specific reagents, such as the HIO_4–Schiff stain or α-naphthol stain for sugars (Kates, 1986). Analytical data at this point are desirable to establish molar ratios of sugar/glycerol/acyl ester groups/fatty acids.

To identify the sugar group(s) present, acid methanolysis followed by aqueous acid hydrolysis is carried out, and the water-soluble products (sugars, glycerol, etc.) identified by their mobilities on paper chromatography and their staining behaviour (Kates, 1986) or by GLC of their TMS ether derivatives (Sweeley and Vance, 1967). Mild alkaline deacylation yields the glycosylglycerol moiety, which is identified by paper chromatography (see Brundish *et al.*, 1965, 1966; Kates, 1986); the fatty acid components released as methyl esters are analysed by GLC (see sections 4.5 and 6.3.2). The glycosylglycerols are isolated from the aqueous alkaline hydrolysate (after treatment with anion–cation exchange resins) by preparative paper chromatography on Whatman 3MM paper in the solvent system pyridine–ethyl acetate–water (2/5/5, upper phase); bands corresponding to the separated glycerolglycosides are eluted with water, concentrated to dryness and crystallized from a suitable solvent (Sastry and Kates, 1964a; Carter *et al.*, 1956; Heinz, 1967). They may then be characterized by melting point, optical rotation and infrared spectroscopy (Heinz, 1967). The glycerolglycoside moiety of some glycosyldiglycerides (e.g. mono- and digalactosyldilinolenin) may also be obtained after hydrolysis with specific galactosyldiglyceride acyl hydrolases (galactolipases) from bean leaves (Sastry and Kates, 1964b, 1969) or the non-specific acyl hydrolase from potato tubers (Galliard, 1975).

To determine the linkages between the sugars, the glycosyldiacylglycerol or derived glycerolglycoside is permethylated in dimethylformamide with methyl iodide and silver oxide for 30–36 h at 30 °C (Heinz, 1967; Brundish *et al.*, 1967). Permethylation of glycosyldiacylglycerols, glycosyl diphytanylglycerol ethers or glycosyl dibiphytanyldiglycerol tetraethers is more conveniently carried out by heating under reflux in methyl iodide and silver oxide, followed by purification of the permethylated glycolipid by TLC in ethyl ether–petroleum ether (9/1) (Komaratat and Kates, 1975; Kushwaha *et al.*, 1981).

The permethylated glycolipid or glycosylglycerol is methanolysed in refluxing 2.5% methanolic HCl, and the methyl glycosides of the partially methylated sugar moieties are identified by GLC on Carbowax 6000 at 175 °C (Kates *et al.*, 1967) or on butanediol succinate polyester at 175 °C (Brundish *et al.*, 1967), or by GC/MS on neopentylglycol succinate polyester at 160 °C (or 140 °C for TMS ethers) (Matsubara and Hayashi, 1974); or they may be hydrolysed in 0.5 M H_2SO_4 to the free methylated sugars which are separated and identified by paper chromatography in butanol–ethanol–water (5/1/4, upper phase) (Heinz, 1967). More definitive analysis of the partially methylated sugars is achieved by GLC or GC/MS of the alditol acetate derivatives (prepared by reduction of the free sugars with borohydride followed by acetylation) on a column of 3% ethylene glycol succinate polyester–cyanoethylsilicone polymer (ECNSS-S) at 170 °C (Kates and Deroo, 1973) or 3% OV-225 at 175 °C (see Hakomori, 1983). The results of the methylation studies should be confirmed by quantitative periodate oxidation studies on the intact glycosyldiglyceride or the glycosylglycerol moiety (see Carter *et al.*, 1956; Heinz, 1967).

Configuration of the glycosidic linkages is determined by hydrolysis of the glycosylglycerols with specific α- and β-glycosidases and chromatographic identification of the free sugars liberated (Carter *et al.*, 1956; Brundish *et al.*, 1966; Hakomori, 1983). Anomeric configuration of the glycosidic linkages can also be determined by CrO_3 oxidation of the acetylated glycolipid or glycosylglycerol (Laine and Renkonen, 1975; see Hakomori, 1983). β-Linked hexopyranosides are oxidised and the remaining α-linked sugars are converted to alditol acetates and analysed by GLC on 3% OV-225 at 200 °C. The CrO_3 oxidation procedure is not applicable to furanosides since both α- and β-linked furanosides are oxidized (Smallbone and Kates, 1981; see Kates, 1990). Anomeric configuration of the glycosidic linkages can also be assigned by ^1H and ^{13}C NMR spectroscopy (Usui *et al.*, 1973; Ritchie *et al.*, 1975; Gorin and Mazurek, 1976; Heinz *et al.*, 1974; see Kates, 1990) and by molecular rotation data using Hudson's isorotation rule (Green, 1966; see Kates, 1990).

The order of linkages of the sugars in di- or triglycosyldiacylglycerols may also be determined by mild (partial) acid hydrolysis of the glycerolglycosides and identification of the free sugars, polyols or glycosides in the hydrolysate by paper chromatography (Brundish *et al.*, 1965, 1966). In di-, tri- or tetraglycosyl diphytanylglycerols the linkage order is determined by mild hydrolysis in $CHCl_3$–methanol–2.5% methanolic HCl (9/8/4 v/v) at room temperature, separation of the resulting mono-, di- or triglycosyl diethers by TLC and identification of their sugar moieties after strong acid hydrolysis. These results can be checked by identifying the sugar glycosides or free sugars released by the mild hydrolysis using TLC on Kieselguhr G–silica gel G in ethyl acetate–methanol–water (68/23/9, v/v) (Kates and Deroo, 1973). Final confirmation of the structure of the glycolipids should be obtained by mass spectrometry of the acetylated or TMS-derivatized glycolipid (see Heinz *et al.*, 1974; Karlsson, 1976; Hakomori, 1983).

Plant galactosyldiacylglycerols (Sastry and Kates, 1964, a, b; Heinz, 1967) and the plant sulpholipid (cf. Harwood, 1980) have been analysed by the above procedures and their structures determined (see Kates, 1990). Fractionation of molecular species of glycosyldiacylglycerols (Joyard *et al.*, 1980) and the plant sulpholipid (Heinz and Harwood, 1977) as well as their mass spectral analysis have been reported.

Glycosphingolipid classes (cerebrosides, ceramide polyhexosides and gangliosides), after isolation, should then be analysed quantitatively for sugars, long-chain base and fatty acid components. Acid methanolysis liberates the sugar (as methyl glycosides), fatty acid and long-chain base moieties but gives rise to some by-products from the long-chain bases. The method of Gaver and Sweeley (1965) using 1 N methanolic HCl (conc. HCl–water–methanol, 8.6/9.4/82, v/v) causes the least by-product formation. The sphingosine bases can be separated by chromatography on columns of silicic acid (Barenholz and Gatt, 1975) (Fig. 6.9), by TLC as their dinitrophenyl derivatives on borate–silica (Karlsson, 1970a) (Fig. 6.9) or on silver nitrate–silica gel (Christie, 1982), and by GLC on non-polar columns (Gaver and Sweeley, 1965; Hakomori, 1983). GLC of bases as the aldehyde dimethylacetal derivatives is described in Kates (1986). Phytosphingosine can be distinguished from sphingosine bases by oxidizing the intact lipid with periodic acid. Long-chain aldehydes are only formed from the phytosphingosine-containing lipids (see Kates, 1986).

The structures of long-chain bases can best be determined by mass spectrometry (Karlsson, 1970b). The dinitrophenyl derivatives with their free hydroxyl groups methylated give characteristic fragmentation patterns for determination of their molecular weights and the primary structure of the base. Double-bond positions can be located if they are first hydroxylated and derivatized to TMS (trimethylsilyl) ethers. Mass spectra can also be obtained (after GLC) of the *N*-acetyl, *O*-TMS ether derivatives of the base (Polito *et al.*, 1969). *Trans* double bonds can be identified by infrared absorption spectroscopy.

Analysis of the fatty acid moieties of sphingolipids is carried out by GLC after trans-esterification of the *N*-acyl groups with HCl–methanol. Hydroxy fatty acids are often major constituents, and appropriate GLC and other techniques must be used (section 6.3.2; see Kates, 1986).

The carbohydrate moieties obtained as the methyl glycosides can be analysed as such by GLC as the TMS derivatives (Sweeley and Walker, 1964;

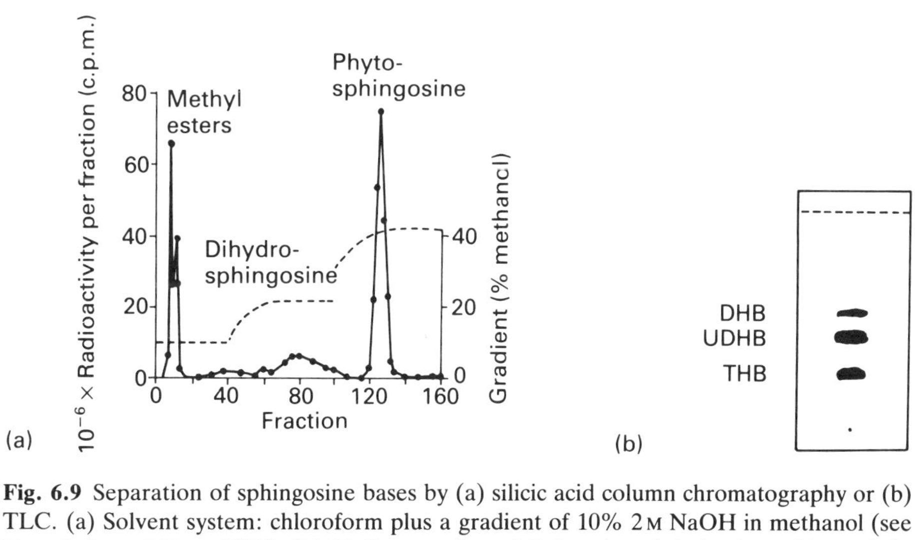

Fig. 6.9 Separation of sphingosine bases by (a) silicic acid column chromatography or (b) TLC. (a) Solvent system: chloroform plus a gradient of 10% 2 M NaOH in methanol (see Barenholz and Gatt, 1975). (b) TLC separation of dinitrophenyl derivatives of long-chain bases on layers of silica gel G containing 2% (w/w) boric acid. Solvent system: chloroform–hexane–methanol (50/50/14, by volume). Abbreviations: DHB, saturated dihydroxy bases; UDHB, Δ^4-*trans*-unsaturated dihydroxy bases; THB, trihydroxy bases. Reproduced with permission from (a) Barenholz and Gatt (1975) and (b) Christie (1982).

Hakomori, 1983; Christie, 1982; Sweeley and Vance, 1967), or as alditol acetate derivatives (Hakomori, 1983), which avoids the problem of multiple peaks from anomeric methyl glycosides in the TMS analysis (Fig. 6.10). Identification of the free sugars, obtained by aqueous acid hydrolysis of the methyl glycosides, by paper chromatography in pyridine–ethyl acetate–water also can be used (see Kates, 1986).

All the common sugars can be identified by these methods, though special care has to be taken when *N*-acetylgalactosamine is present (see Hakomori, 1983). Total hexose in glycolipids can be quantified readily by the colorimetric procedure of Dubois *et al.* (1956) (see Kates, 1986), in which the furfural derivatives formed by strong acids react with phenol to form coloured compounds, or by the procedure of Yamamoto and Rouser (1970), in which the furfurals are coupled with anthrone or α-naphthol.

The *N*-acetylneuraminic acid content of gangliosides can be estimated by a resorcinol reaction (Svennerholm, 1957) or by a thiobarbituric acid assay (Diringer, 1972). For other references, see Kates (1986).

Sugar sequences of the carbohydrate moieties are determined by permethylation analysis as described above for glycosyldiacylglycerols (see Hakomori and Watanabe, 1976; Hakomori, 1983; Kates, 1986, 1990). The anomeric configuration of the glycoside linkages can be established by CrO_3 oxidation, NMR spectroscopy and molecular rotation data (see Hakomori, 1983; Kates, 1986). The final structure should be confirmed by mass spectrometry (Karlsson, 1976; Hakomori, 1983).

The isolation, analysis and structure determination of gangliosides have been reviewed in detail by Hakomori (1983).

A variety of methods have been described for the separation of molecular species of sphingomyelin and glycosphingolipids. Silica gel TLC systems impregnated with sodium arsenite, sodium borate or silver nitrate have been used for this purpose, and cerebrosides have been separated into molecular species by GLC of their TMS ethers (see Hakomori, 1983; Christie, 1982). Most information is available on ceramide moieties of sphingolipids, which can be fractionated into individual molecular species by TLC of the acetylated derivatives on silica–silver nitrate plates (Samuelsson and Samuelsson, 1969) followed by GLC or GC/MS of the TLC fractions, or by direct GLC or GC/MS of the TMS ether derivatives; see Christie (1982) and Hakomori (1983) for details.

6.3.6 Structural determination of waxes, cutin and suberin

Identification procedures for wax esters must be adjusted to suit the particular mixture of wax esters, hydrocarbons, sterol esters, glycerides, etc., present in the neutral lipid fraction (Table 4.2) from different sources or extracted with non-polar solvents from animal or plant surface layers (see section 6.3.1). In general, the identification procedure involves a preliminary fractionation of the neutral lipids on a column of silicic acid (sections 4.4.1(a) and

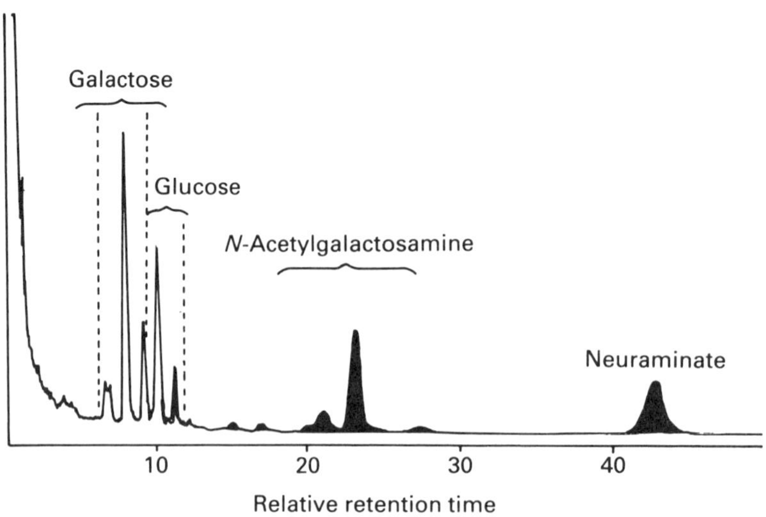

Fig. 6.10 GLC of TMS derivatives of carbohydrates derived from sphingolipids by acidic hydrolysis. Stationary phase 2.5% SE-30 at 160 °C. Reproduced with permission from Sweeley and Walker (1964). *Analytical Chemistry*, **36**, 1461. Copyright 1964 American Chemical Society.

4.4.1(a)(i); Table 4.3), in which the ester fraction is eluted with hexane–ether (99/1) or benzene (Kolattukudy, 1970b) or hexane–chloroform (9/1 then 4/1) (Tulloch, 1971); the pure wax monoester fraction is then isolated by preparative TLC in benzene ($R_f = 0.65$–0.70; Kolattukudy, 1970b; Tulloch, 1970), and the wax diesters by TLC in hexane–ether (95/5) ($R_f = 0.28$–0.36; Fu and Nicolaides, 1969). Wax monoesters can be separated from steryl esters by TLC on magnesium hydroxide–celite (1/1, w/w) in ethyl acetate–hexane (1/99, v/v) or by HPLC on magnesium hydroxide (Stewart and Downing, 1981). Identity of the wax ester class is then checked by infrared and NMR spectroscopy (Tulloch, 1971; Kates, 1986), and by examining the saponification products: monoesters give one mole each of long-chain alcohol and fatty acid; diesters give two moles of fatty acids and one mole of alkanediol, or one mole each of long-chain alcohol, hydroxy fatty acid and normal fatty acid.

At this stage it is advantageous to fractionate the wax ester fraction into individual species according to chain length, using preparative GLC under the following conditions: for monoesters (C_{40}–C_{50}), a column of 3% SE-30 at 335 °C and 40 ml/min flow rate (Kolattukudy, 1970b); for diesters (C_{40}–C_{60}), a column of 1.5% OV-101 on Chromosorb G with temperature programming from 300–400 °C in 16 min and helium flow rate of 35 ml/min. The columns are standardized for chain-length identification with synthetic esters of normal long-chain alcohols and acids covering the desired range. Alternatively, wax esters may be fractionated by preparative reverse-phase HPLC (see section 4.4.4).

Each individual wax ester peak is then saponified and the separated alcohol and fatty acid (converted to methyl esters) moieties are identified by GLC as described in section 4.5. When there is insufficient material, the total wax ester fraction is saponified and the total alcohol and acid moieties subjected to GLC analysis; the distribution and identity of individual wax ester species is calculated assuming a random combination of the alcohol and acid moieties (Kolattukudy, 1970b). These results should be checked by mass spectrometry of the unhydrolysed ester fraction, or by GLC and mass spectrometry, whereby each ester peak is identified as it emerges from the GLC. Complex branched acids, such as in preen gland waxes (Odham, 1967; Jacob, 1976), should also be identified by GLC and mass spectrometry.

The composition of wax esters of various natural sources has been extensively reviewed. The following articles should be consulted for further information (see Kolattukudy, 1976): animal waxes, Nicolaides *et al.* (1970); insect waxes, Tulloch (1970, 1971, 1976), Jackson and Baker (1970); plant waxes, Kolattukudy (1970a), von Wettstein-Knowles (1979), Walton (1991); and marine organisms, Nevenzel (1970), Holz (1981).

The surface layers of plants are covered in polymeric cutin or suberin (section 2.11), in addition to waxes (section 2.10), especially in the cuticle. There are two problems encountered in the structural determination of these compounds. First, cutin and suberin are polymers whose extraction from tissues and subsequent analysis pose all the problems usually associated with such materials. Secondly, waxes, cutin and suberin have variable compositions depending on environmental and developmental conditions and, especially, on the nature of the tissue being analysed.

Cuticular waxes are removed from surface layers of intact tissues by washing with organic solvents such as hexane or chloroform. Redistilled solvents should be used to avoid contamination with residual waxy material present in undistilled solvents (see Kates, 1986). Internal waxes are extracted from tissues following homogenization by the usual methods (section 6.3.1). Waxes have also been extracted from suberin-rich barks (Martin and Juniper, 1970).

For removal of the cuticle from plant tissues, it is necessary to disrupt the pectinaceous layer external to the epidermal cell wall. In addition, carbohydrates attached to the cuticle have to be removed. A combination of enzyme (pectinase, cellulase) and chemical (ammonium oxalate–oxalic acid, cuprammonium ion and $ZnCl_2$–HCl) treatments is usual (Matic, 1956; Eglinton and Hunneman, 1968; Walton and Kolattukudy, 1972). Waxes associated with the cuticle can be removed by extraction with organic solvents.

Suberin is tightly associated with the cell wall and cannot be removed in a pure form. The outer layer is usually removed physically and then treated with pectinase, cellulase and organic solvents to yield an enriched suberin-containing fraction. The crude suberin must then be subjected to chemical treatment in order to depolymerize it for chemical analysis (see references in Kolattukudy, 1980). Crude suberin preparations have also been isolated from endodermis (Robards *et al.*, 1976) and hypodermis–epidermis (Clarkson *et al.*, 1978), but the amounts obtained from these sources are usually too small for meaningful analyses to be performed.

Most of the linkages in cutin are intermonomeric ester bonds with, perhaps, a small number of peroxide bridges (Kolattukudy, 1980). Methods used to break the ester linkages of cutin are also applied to suberin analysis. Techniques that have been used include alkaline hydrolysis (Eglinton and Hunneman, 1968), trans-esterification with $NaOCH_3$–CH_3OH or BF_3–CH_3OH (Holloway and Deas, 1973;

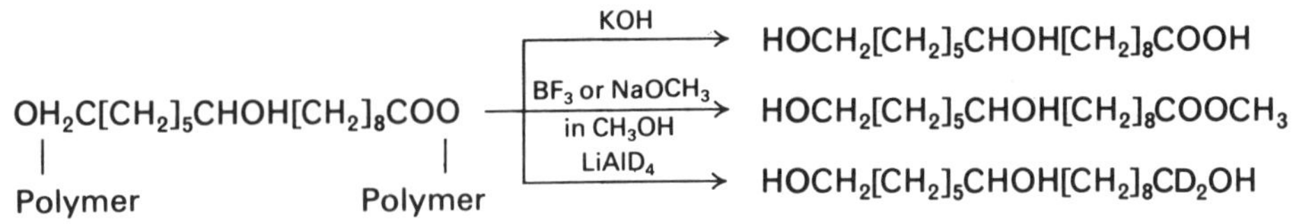

Fig. 6.11 Chemical depolymerization of biopolyesters. Reproduced with permission from Kolattukudy (1980).

Kolattukudy *et al.*, 1973) and hydrogenolysis with $LiAlH_4$ in tetrahydrofuran (Walton and Kolattukudy, 1972). The method of choice depends on the particular sample being analysed, but usually two methods yield sufficient structural information (Fig. 6.11).

The free hydroxy acids and phenolic compounds produced from suberin by the alkaline hydrolysis–depolymerization process can undergo secondary reactions. Some of these can be prevented by including a treatment with $LiAlH_4$ during depolymerization. However, this will not protect keto, aldehyde and epoxide groups. If $LiAlD_4$ is used, then the monomers become labelled and can be more easily analysed (Walton and Kolattukudy, 1972). The functional groups, such as the epoxides, can be studied following generation of esters with BF_3–CH_3OH (cf. Kolattukudy, 1980).

TLC is commonly used for the separation of different classes of wax components or for analysis of monomers from cutin and suberin depolymerization. A typical separation is shown in Fig. 6.12. By such methods, it is possible to separate hydrocarbons, wax esters, primary alcohols, secondary alcohols and β-diketones from plant waxes (von Wettstein-Knowles, 1979). Products of hydrogenolysis from cutin can be separated by TLC into alkan-1-ols, alkane-α,ω-diols, C_{16}–C_{18} triols and C_{18} tetrols

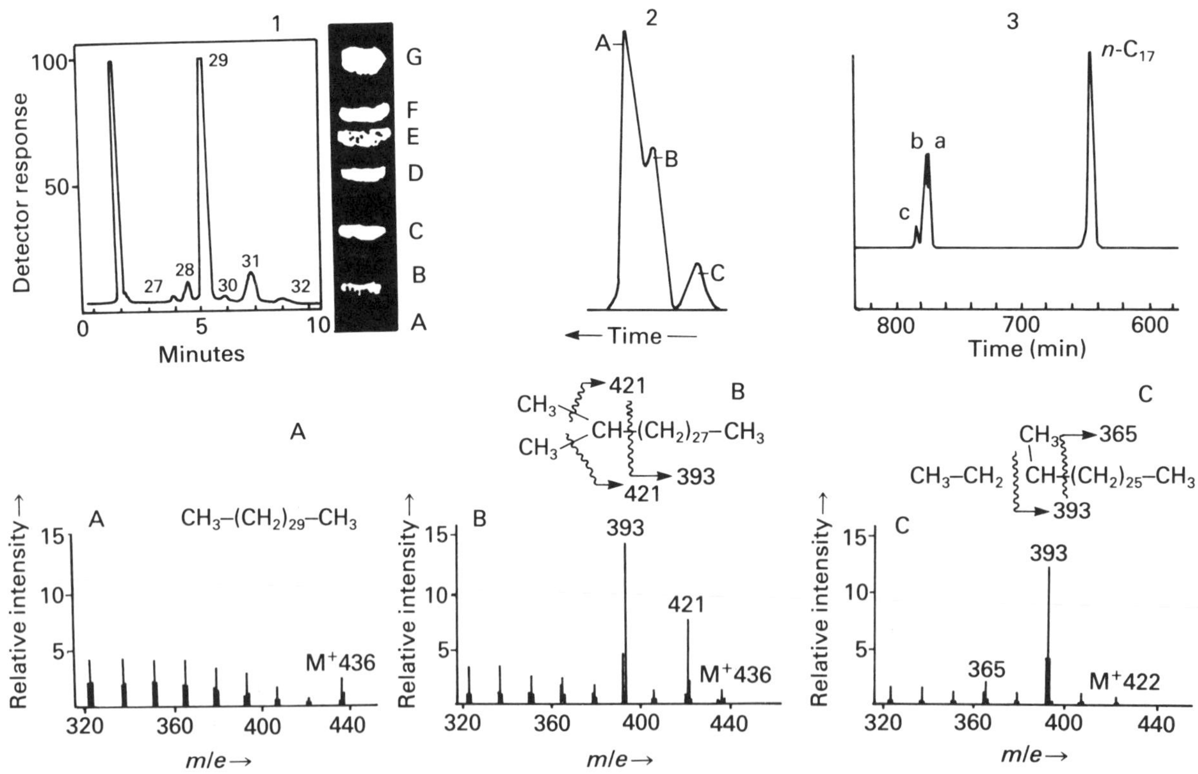

Fig. 6.12 Chromatography and mass spectrometry of plant waxes. (1) Thin-layer chromatogram of broccoli leaf wax and gas–liquid chromatogram of the hydrocarbon fraction isolated from this wax. (2) Partial gas–liquid chromatogram of the alkane fraction from the surface lipids of tobacco leaves. (3) Partial gas–liquid chromatogram of the hydrocarbons from *N. muscorum* illustrating resolution of (a) 8-, (b) 7- and (c) 6-monomethyl heptadecanes on a 750 ft capillary column. (A)–(C) Partial mass spectra taken at points indicated in (2) illustrating identification of normal and branched alkanes. Reproduced with permission from Kolattukudy (1980).

(Kolattukudy, 1980). Unsaturated components can be resolved by argentation TLC (Tulloch, 1976), and *threo* or *erythro* diastereoisomers can be separated by boric acid–silica gel TLC (Eglinton and Hunneman, 1968). Straight-chain compounds can be preferentially removed from branched compounds as their urea complexes (Kolattukudy, 1980).

The fractions obtained by TLC are then analysed by GLC on packed or capillary columns after suitable derivatization (e.g. acids as methyl esters). High temperatures are frequently necessary to analyse the very long chains found in most natural wax components, so that stable liquid phases must be employed, along with suitable volatile derivatives of the components. Fatty alcohols, diols and hydroxy acids are analysed as trimethylsilyl ethers and aldehydes as *O*-methyloximes (Tulloch, 1976; Kolattukudy, 1980).

When combined with mass spectrometry, GLC can confirm the identity of most of the closely related components of complex mixtures found in waxes, cutins or suberins. Examples of particular uses of this technique include the identification of branched fatty acids (Tulloch, 1976; Jackson and Blomquist, 1976), location of methyl branches in alkanes (Fig. 6.13), location of functional groups such as carbonyl and hydroxyl moieties on aliphatic chains following α-cleavage (Tulloch, 1976), and identification of wax esters (Kolattukudy, 1980).

Hydroxy acids are major aliphatic components of cutin and suberin, and these are readily identified by GLC/mass spectrometry. Major ions generated from the usual cutin components are listed in Kolattukudy (1977). The position of the hydroxyl group in the chain is easily seen because cleavage occurs on either side of the substituent (Fig. 6.13). The rather

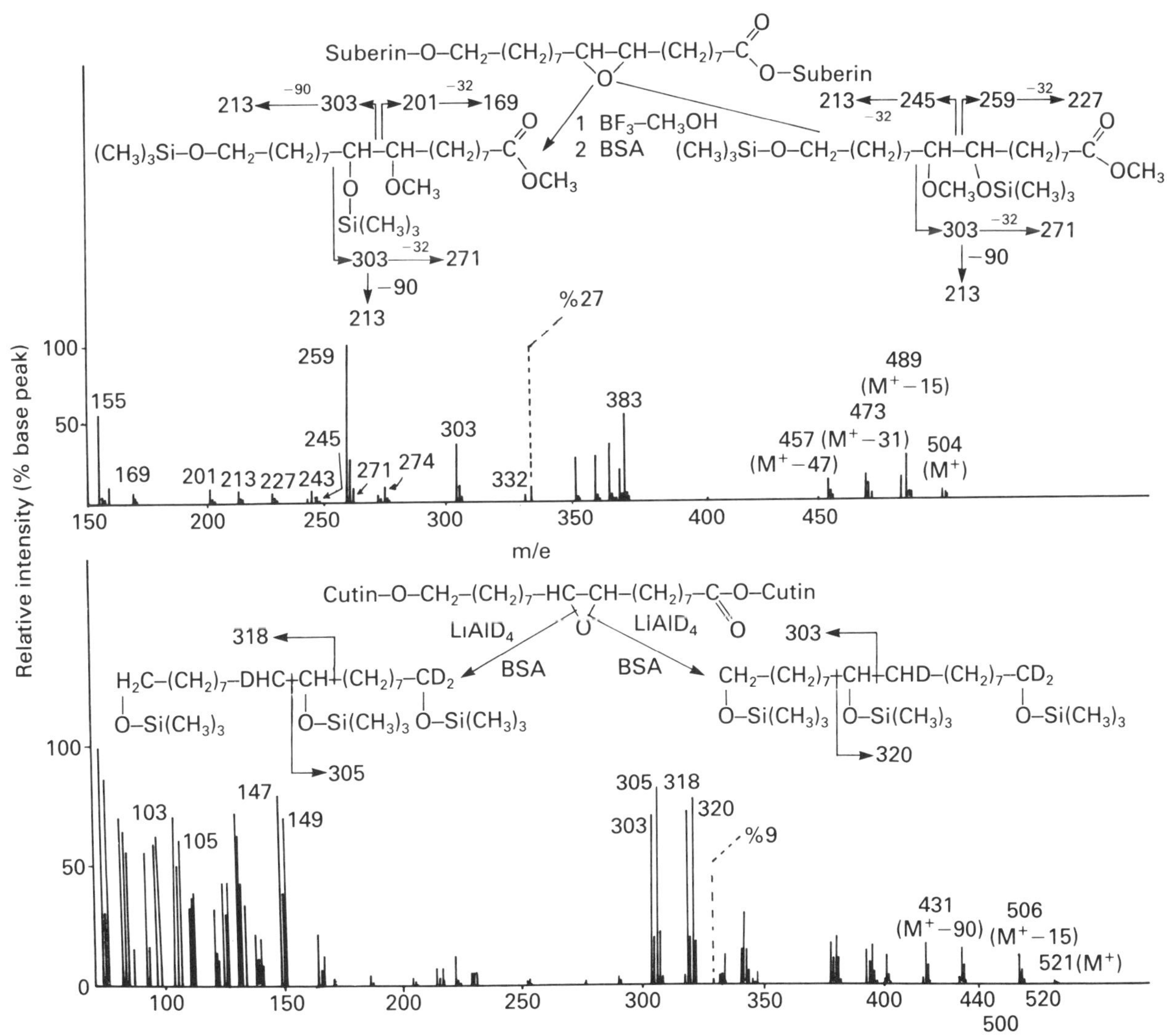

Fig. 6.13 Mass spectra of products derived from 18-hydroxy-9,10-epoxy C_{18} acid by cleavage of lipid-derived polymer with BF_3–CH_3OH or $LiAlD_4$. Reproduced with permission from Kolattukudy (1980).

simple phenolic compounds yielded by reductive depolymerization of cutin and suberin are also very suitable for mass spectrometric analysis (Riley and Kolattukudy, 1975).

Some of the cuticular lipids are also suitable for spectroscopic analysis. β-Diketones absorb at 273 nm so that ultraviolet spectroscopy can be used (Tulloch, 1976). The functional groups of various fractions obtained by TLC can be confirmed by infrared analysis. Aldehydes and β-diketones give useful signals when examined by NMR and this technique can be made considerably more sensitive by the application of ^{13}C-NMR for some wax components (Tulloch, 1976).

For further information on the analysis of waxes, cutins and suberins, the reader is referred to reviews by Martin and Juniper (1970), Pohl and Wagner (1972), Tulloch (1976), Kolattukudy (1976, 1977, 1980) and Walton (1991).

References and further reading

Aberth, W., Straub, K.M. and Burlingame, A.L. (1982) *Anal. Chem.*, **54**, 2029.

Ackman, R.G. (1981) in *Methods in Enzymology* vol. 72 (ed. J.M. Lowenstein), Academic Press, New York, p. 205.

Andersson, B. and Holman, R.T. (1974) *Lipids*, **9**, 185.

Andersson, B., Anderson, W.H., Chipault, J.R., Ellison, E.C., Fenton, S.W., Gellerman, J.L., Hawkins, J.M. and Schlenk, H. (1974) *Lipids*, **9**, 506.

Arvidson, G.A.E. (1965) *J. Lipid Res.*, **6**, 574.

Ashes, J.R. and Haken, J.K. (1974a) *J. Am. Oil Chem. Soc.*, **49**, 318.

Ashes, J.R. and Haken, J.K. (1974b) *J. Chromatogr.*, **101**, 103.

Ashes, J.R. and Haken, J.K. (1975) *J. Chromatogr.*, **111**, 171.

Axelrod, J., Reichenthal, J. and Brodie, B.B. (1953) *J. Biol. Chem.*, **204**, 903.

Barak, A.J. and Tuma, D.J. (1981) in *Methods in Enzymology*, vol. 72 (ed. J.M. Lowenstein), Academic Press, New York, p. 287.

Barenholz, Y. and Gatt, S. (1975) in *Methods in Enzymology*, vol. 35 (ed. J.M. Lowenstein), Academic Press, New York, p. 315.

Barrett, C.R., Dallas, M.S.J. and Padley, F.B. (1963) *J. Am. Oil Chem. Soc.*, **40**, 580.

Bartlett, G.R. (1959) *J. Biol. Chem.*, **234**, 466.

Bligh, E.G. and Dyer, W.J. (1959) *Can. J. Biochem. Physiol.*, **37**, 911.

Breckenridge, W.C. (1978) in *Handbook of Lipid Research*, vol. 1 (ed. A. Kuksis), Plenum, New York, p. 197.

Breckenridge, W.C. and Kuksis, A. (1968) *Lipids*, **3**, 291.

Breckenridge, W.C., Marai, L. and Kuksis, A. (1969) *Can. J. Biochem.*, **47**, 761.

Brockerhoff, H. (1965) *J. Lipid Res.*, **6**, 10.

Brockerhoff, H. (1967) *J. Lipid Res.*, **8**, 167.

Brockerhoff, H. (1971) *Lipids*, **6**, 942.

Brockerhoff, H. (1975) in *Methods in Enzymology*, vol. 35 (ed. J.M. Lowenstein), Academic Press, New York, p. 315.

Brockerhoff, H., Hoyle, R.S. and Wolmark, N. (1966) *Biochim. Biophys. Acta.*, **116**, 67.

Brown, C.A. (1967) *Anal. Chem.*, **39**, 1882.

Brundish, D.E., Shaw, D.N. and Baddiley, J. (1965) *Biochem. J.*, **97**, 158.

Brundish, D.E., Shaw, N. and Baddiley, J. (1966) *Biochem. J.*, **99**, 546.

Brundish, D.E., Shaw, N. and Baddiley, J. (1967) *Biochem. J.*, **105**, 885.

Carter, H.E., McCluer, R.H. and Slifer, E.D. (1956) *J. Am. Chem. Soc.*, **78**, 3735.

Christie, W.W. (1982) *Lipid Analysis*, 2nd edn, Pergamon, Oxford.

Christie, W.W. (1989) *Gas Chromatography of Lipids*, Oily Press, Ayr.

Christie, W.W. and Moore, J.H. (1969) *Biochim. Biophys. Acta*, **176**, 445.

Clarkson, D.T., Robards, A.W., Sanderson, J. and Peterson, C.A. (1978) *Can. J. Bot.*, **56**, 1526.

Cochrane, G.C. (1975) *J. Chromatogr. Sci.*, **13**, 440.

Colborne, A.J. and Laidman, D.L. (1975) *Phytochemistry*, **14**, 2639.

Cook, A.M. and Daughton, C.G. (1981) in *Methods in Enzymology*, vol. 72 (ed. J.M. Lowenstein), Academic Press, New York, p. 292.

Curstedt, T. (1977) *Biochim. Biophys. Acta*, **489**, 79.

Dawson, R.M.C. (1967) in *Lipid Chromatographic Analysis*, vol. 1 (ed. G.V. Marinetti), Marcel Dekker, New York, p. 163.

Dickens, B.F., Ramesha, C.S. and Thompson, G.A., Jr (1982) *Anal. Biochem.*, **127**, 37.

Diringer, H. (1972) *Hoppe-Seyler's Z. Physiol. Chem.*, **353**, 39.

Downing, D.T. and Green, R.S. (1968) *Lipids*, **3**, 96.

Dubois, M., Gilles, K.A., Hamilton, J.K., Rebers, P.A. and Smith, F. (1956) *Anal. Chem.*, **28**, 350.

Eglinton, G. and Hunneman, D.H. (1968) *Phytochemistry*, **7**, 313.

Emken, E.A. and Dutton, H.J. (1974) *Lipids*, **9**, 272.

Fishwick, H.J. and Wright, A.J. (1977) *Phytochemistry*, **16**, 1507.

Fu, H.C. and Nicolaides, N. (1969) *Lipids*, **4**, 170.

Galliard, T. (1975) in *Recent Advances in the Chemistry and Biochemistry of Plant Lipids* (eds T. Galliard and E.I. Mercer), Academic Press, New York, p. 319.

Garbus, J., de Luca, H.F., Loomans, M.E. and Strong, F.M. (1963) *J. Biol. Chem.*, **238**, 59.

Gaver, E.C. and Sweeley, C.C. (1965) *J. Am. Oil Chem. Soc.*, **42**, 294.

Gorin, P.A.J. and Mazurek, M. (1976) *Carbohydr. Res.*, **48**, 171.

Green, J.N. (1966) *Adv. Carbohydr. Chem.*, **21**, 95.

Gunstone, F.D. (1975) in *Recent Advances in the Chemistry and Biochemistry of Plant Lipids* (eds T. Galliard and E.I. Mercer), Academic Press, New York, p. 21.

Gunstone, F.D. (1990) in *Methods in Plant Biochemistry*,

Volume 4, Lipids, Membranes and Aspects of Photobiology (eds J.L. Harwood and J.R. Bower), Academic Press, New York, p. 1.
Gunstone, F.D. and Inglis, R.P. (1971) in *Topics in Lipid Chemistry*, vol. 2 (ed. F.D. Gunstone), Paul Elek, London, p. 287.
Gunstone, F.D. and Inglis, R.P. (1973) *Chem. Phys. Lipids*, **10**, 73, 89, 105.
Gunstone, F.D. and Jacobsberg, F.R. (1972) *Chem. Phys. Lipids*, **9**, 26.
Hakomori, S. (1983) in *Handbook of Lipid Research*, vol. 3 (eds S. Hakomori and J.N. Kanfer), Plenum, New York, p. 1.
Hakomori, S. and Watanabe, K. (1976) in *Glycolipid Methodology* (ed. L.A. Witting), American Oil Chemists' Society, Champaign, p. 13.
Harwood, J.L. (1976) *Phytochemistry*, **15**, 1459.
Harwood, J.L. (1980) in *Biochemistry of Plants*, vol. 4 (eds P.K. Stumpf and E.E. Conn), Academic Press, New York, p. 1.
Hawthorne, J.N. and Ansell, G.B. (eds (1982) *Phospholipids*, Elsevier, Amsterdam.
Heinz, E. (1967) *Biochim. Biophys. Acta*, **144**, 333.
Heinz, E. and Harwood, J.L. (1977) *Hoppe-Seyler's Z. Physiol. Chem.*, **358**, 897.
Heinz, E., Rullkotter, S. and Budzikiewicz, H. (1974) *Hoppe-Seyler's Z. Physiol. Chem.*, **355**, 612.
Hitchcock, C. and Nichols, B.W. (1971) *Plant Lipid Biochemistry*, Academic Press, New York.
Holloway, P.J. and Deas, A.H.B. (1973) *Phytochemistry*, **12**, 1721.
Holz, G.G., Jr (1981) in *Biochemistry and Physiology of Protozoa*, 2nd edn, vol. 4 (eds H. Levandowsky and S.H. Hutner), Academic Press, New York, p. 309.
Hrivnak, J., Sojak, L., Beska, E. and Janak, J. (1972) *J. Chromatogr.*, **68**, 55.
Jackson, L.L. and Baker, G.L. (1970) *Lipids*, **5**, 239.
Jackson, L.L. and Blomqvist, G.J. (1976) in *Chemistry and Biochemistry of Natural Waxes* (ed. P.E. Kolattukudy), American Elsevier, New York, p. 201.
Jacob, J. (1976) in *Chemistry and Biochemistry of Natural Waxes* (ed. P.E. Kolattukudy), Elsevier, Amsterdam, p. 94.
Jeffrey, B.S.J. (1991) *J. Am. Oil Chem. Soc.*, **68**, 289.
Joyard, J., Douce, R., Siebertz, H.P. and Heinz, E. (1980) *Eur. J. Biochem.*, **108**, 171.
Karlsson, K.A. (1970a) *Chem. Phys. Lipids*, **5**, 6.
Karlsson, K.A. (1970b) *Lipids*, **5**, 878.
Karlsson, K.A. (1976) in *Glycolipid Methodology* (ed. L.A. Witting), American Oil Chemists' Society, Champaign, p. 97.
Kates, M. (1986) *Techniques in Lipidology*, 2nd rev. edn, Elsevier, Amsterdam.
Kates, M. (1990) in *Handbook of Lipid Research*, vol. 6 (ed. M. Kates), Plenum, New York, pp. 1, 235.
Kates, M. and Deroo, P.W. (1973) *J. Lipid Res.*, **14**, 430.
Kates, M., Palameta, B., Perry, M.B. and Adams, G.A. (1967) *Biochim. Biophys. Acta*, **137**, 213.
Kleiman, R. and Spencer, G.F. (1973) *J. Am. Oil Chem. Soc.*, **50**, 31.
Klein, R.A. and Kemp, P. (1977) in *Methods in Membrane Biology*, vol. 8 (ed. E.D. Korn), Plenum, New York, p. 51.
Kolattukudy, P.E. (1970a) *Lipids*, **5**, 259.
Kolattukudy, P.E. (1970b) *Lipids*, **5**, 398.
Kolattukudy, P.E. (ed.) (1976) *Chemistry and Biochemistry of Plant Waxes*, American Elsevier, New York.
Kolattukudy, P.E. (1977) *Recent Adv. Phytochem.*, **11**, 185.
Kolattukudy, P.E. (1980) in *Biochemistry of Plants*, vol. 4 (eds P.K. Stumpf and E.E. Conn), Academic Press, New York, p. 571.
Kolattukudy, P.E., Walton, T.J. and Kushwaha, R.S.P. (1973) *Biochemistry*, **12**, 4488.
Komaratat, P. and Kates, M. (1975) *Biochim. Biophys. Acta*, **398**, 464.
Kuksis, A. (1976) in *Lipid Chromatographic Analysis* (ed. G.V. Marinetti), Dekker, New York, pp. 215–35.
Kuksis, A. (1978) in *Handbook of Lipid Research*, vol. 1 (ed. A. Kuksis), Plenum, New York, p. 1.
Kuksis, A. and Marai, L. (1967) *Lipids*, **2**, 217.
Kuksis, A., Myher, J.J. and Marai, L. (1984) *J. Am. Oil Chem. Soc.*, **61**, 1582.
Kuksis, A., Myher, J.J. and Marai, L. (1985) *J. Am. Oil Chem. Soc.*, **62**, 762.
Kushwaha, S.C., Kates, M., Sprott, G.D. and Smith, I.C.P. (1981) *Biochim. Biophys. Acta*, **664**, 156.
Laine, R.A. and Renkonen, O. (1975) *J. Lipid Res.*, **16**, 102.
Lie Ken Jie, M.S.F. (1975) *J. Chromatogr.*, **111**, 189.
Lowenstein, J.M. (ed.) (1969) *Methods in Enzymology*, vol. 14, *Lipids*, Academic Press, New York.
Martin, J.T. and Juniper, B.E. (1970) *The Cuticles of Plants*, St Martin's Press, New York.
Matic, H. (1956) *Biochem. J.*, **63**, 168.
Matsubara, T. and Hayashi, A. (1974) *Biomed. Mass Spectrom.*, **1**, 62.
May, H.E. and Desiderio, D.M. (1983) *J. Chem. Commun.*, 72.
McCloskey, J.A. (1969) in *Methods in Enzymology*, vol. 14 (ed. J.M. Lowenstein), Academic Press, New York, p. 382.
McCloskey, J.A. (1970) in *Topics in Lipid Chemistry*, vol. 1 (ed. F.D. Gunstone), Logos Press, London, p. 369.
McInnes, A.G., Tattrie, N.H. and Kates, M. (1960) *J. Am. Oil Chem. Soc.*, **37**, 7.
Miwa, T.K., Mikolajczak, K.L., Earle, F.R. and Wolff, I.A. (1960) *Anal. Chem.*, **32**, 1739.
Morris, L.J. (1966) *J. Lipid Res.*, **7**, 717.
Myher, J.J. (1978) in *Handbook of Lipid Research*, vol. 1, *Fatty Acids and Acylglycerols* (ed. A. Kuksis), Plenum, New York, p. 123.
Myher, J.J. and Kuksis, A. (1982) *Can. J. Biochem.*, **60**, 638.
Myher, J.J. and Kuksis, A. (1984) *Can. J. Biochem. Cell Biol.*, **62**, 352.
Nakagawa, Y. and Horrocks, L.A. (1983) *J. Lipid Res.*, **24**, 1268.
Nevenzel, J.C. (1970) *Lipids*, **5**, 308.
Nichols, B.W. and Appleby, R. (1969) *Phytochemistry*, **8**, 1907.

Nicolaides, N. and Fu, H.C. (1969) *Lipids*, **4**, 83.
Nicolaides, N., Fu, H.C. and Ansari, M.N.A. (1970) *Lipids*, **5**, 299.
Odham, G. (1967) *Fette Seifen Anstrichm.*, **69**, 164.
Odham, G. and Stenhagen-Stallberg, E. (1972) in *Biochemical Applications of Mass Spectrometry* (ed. G.R. Waller), Interscience, New York, p. 2111.
Patton, G.M., Fasulo, J.M. and Robins, S.J. (1982) *J. Lipid Res.*, **23**, 190.
Pohl, P. and Wagner, H. (1972) *Fette Seifen Anstrichm.*, **74**, 541.
Polito, A.J., Naworal, J. and Sweeley, C.C. (1969) *Biochemistry*, **8**, 1811.
Privett, O.S. and Nutter, L.J. (1967) *Lipids*, **2**, 149.
Pryde, E.H. and Cowan, J.C. (1971) in *Topics in Lipid Chemistry*, vol. 2 (ed. F.D. Gunstone), Logos Press, London, p. 1.
Renkonen, O. and Luukonen, A. (1976) in *Lipid Chromatographic Analysis*, 2nd edn, vol. 1 (ed. G.V. Marinetti), Marcel Dekker, New York, p. 1.
Renkonen, O. and Varo, P. (1967) in *Lipid Chromatographic Analysis*, vol. 1 (ed. G.V. Marinetti), Marcel Dekker, New York, p. 41.
Riley, R.G. and Kolattukudy, P.E. (1975) *Plant Physiol.*, **56**, 650.
Rinehart, K.L. (1982) *Science*, **218**, 254.
Ritchie, R.G.S., Cyr, N., Korsch, B., Koch, H.J. and Perlin, A.S. (1975) *Can. J. Chem.*, **53**, 1424.
Robards, A.W., Payne, H.L. and Gunning, B.E.S. (1976) *Cytobiologie*, **13**, 85.
Roehm, J.N. and Privett, O.S. (1969) *J. Lipid Res.*, **10**, 245.
Rouser, G., Kritchevsky, G. and Yamamoto, A. (1967) in *Lipid Chromatographic Analysis*, vol. 1 (ed. G.V. Marinetti), Marcel Dekker, New York, p. 99.
Samuelsson, B. and Samuelsson, K. (1969) *J. Lipid Res.*, **10**, 47.
Sastry, P.S. and Kates, M. (1964a) *Biochemistry*, **3**, 1271.
Sastry, P.S. and Kates, M. (1964b) *Biochemistry*, **3**, 1280.
Sastry, P.S. and Kates, M. (1969) in *Methods in Enzymology*, vol 14 (ed. J.M. Lowenstein), Academic Press, New York, p. 9.204.
Shand, J.H. and Noble, R.C. (1980) *Anal. Biochem.*, **101**, 427.
Siakotos, A.N. (1967) *Lipids*, **2**, 87.
Smallbone, B.W. and Kates, M. (1981) *Biochim. Biophys. Acta*, **665**, 551.
Spanner, S. (1973) in *Form and Function of Phospholipids* (eds G.B. Ansell, R.M.C. Dawson and J.N. Hawthorne), Elsevier, Amsterdam, p. 43.
Stein, R.A. and Slawson, V. (1966) *J. Chromatogr.*, **25**, 204.
Stewart, M.E. and Downing, D.T. (1981) *Lipids*, **16**, 355.
Sun, K.K. and Holman, R.T. (1968) *J. Am. Chem. Soc.*, **45**, 810.
Svennerholm, L. (1957) *Biochim. Biophys. Acta*, **24**, 604.
Sweeley, C.C. and Vance, D.E. (1967) in *Lipid Chromatographic Analysis*, vol. 1 (ed. G.V. Marinetti), Marcel Dekker, New York, p. 465.
Sweeley, C.C. and Walker, B. (1964) *Anal. Chem.*, **36**, 1461.
Takagi, T. (1990) *Prog. Lipid Res.*, **29**, 277.
Thomas, A.E., III, Scharoun, J.E. and Ralston, H. (1965) *J. Am. Oil Chem. Soc.*, **42**, 789.
Tulloch, A.P. (1970) *Lipids*, **5**, 247.
Tulloch, A.P. (1971) *Chem. Phys. Lipids*, **6**, 235.
Tulloch, A.P. (1976) in *Chemistry and Biochemistry of Natural Waxes* (ed. P.E. Kolattukudy), American Elsevier, New York, p. 235.
Usui, T.N., Yamaoka, N., Matsuda, K., Tuzimura, K., Sugiyama, H. and Seto, S. (1973) *J. Chem. Soc., Perkin Trans. I*, 2425.
Vishnubhatla, I. Kates, M. and Adams, G.A. (1988) *Lipids*, **23**, 609.
von Wettstein-Knowles, P. (1979) in *Advances in the Biochemistry and Physiology of Plant Lipids* (eds L.-A. Appelqvist and C. Liljenberg), Elsevier, Amsterdam, p. 1.
Waite, M. (1987) *Handbook of Lipid Research*, vol. 5, *The Phospholipases*, Plenum, New York.
Walton, J.J. (1991) in *Methods in Plant Biochemistry* (eds J.L. Harwood and J.R. Bowyer), vol. 4, Academic Press, London, p. 105.
Walton, T.J. and Kolattukudy, P.E. (1972) *Biochemistry*, **11**, 1885.
Wells, M.A. and Dittmer, J.C. (1965) *Biochemistry*, **4**, 2459.
Wolff, I.A. and Miwa, T.K. (1965) *J. Am. Oil Chem. Soc.*, **42**, 208.
Wuthier, R.E. (1966) *J. Lipid Res.*, **7**, 558.
Yamamoto, A. and Rouser, G. (1970) *Lipids*, **5**, 442.

7

SYNTHESIS

F.D. GUNSTONE and J.L. HARWOOD

7.1 Fatty acid synthesis via acetylenes

The laboratory synthesis of unsaturated fatty acids has been pursued extensively. Whilst a few acids are easily isolated from appropriate natural sources (section 4.8), it is necessary to resort to chemical synthesis when the acid occurs only in obscure sources or at low levels, or when an isotopically labelled sample is needed. Many synthetic procedures have been employed, but those based on the reactivity of acetylene (ethyne) and its derivatives are the most common. The topic has been reviewed by Kunau (1973) and by Sprecher (1977, 1979). Gravier-Pelletier *et al.* (1990), in their review on the synthesis of long-chain hydroxy acids, describe several routes that proceed through acetylenic intermediates. The Wittig reaction, which is also useful in this connection, and the synthesis of isotopically labelled compounds are covered in sections 7.2 and 7.3 respectively.

The usefulness of ethyne (**1**) and its derivatives depends on the ease with which such compounds can be alkylated (Normant and Alexakis, 1981) and subsequently reduced stereospecifically to *cis*-alkenes (**2**) (Marvell and Li, 1973). Reduction to a *trans* isomer is also possible, but not so simple.

Mono-unsaturated acids are prepared from three units, (**1**), (**3**) and (**4**). Ethyne may be condensed first with the alkyl halide to give $H_3C[CH_2]_xC{\equiv}CH$ or with the difunctional entity to give $HC{\equiv}C[CH_2]_yY$. The carboxyl group can be generated by oxidation of an alcohol or by reaction of halide with cyanide and then with acidic methanol to give a methyl ester. Condensation with acetylene or an ethynyl compound requires the sodium or lithium derivative and is usually conducted in liquid ammonia. Modification of this approach is necessary when the unsaturated centre is close to either end of the molecule.

$$\underset{(\mathbf{1})}{HC{\equiv}CH} \rightarrow RC{\equiv}CH \rightarrow RC{\equiv}CR' \rightarrow \underset{(\mathbf{2})}{RCH{=}CHR'}$$

$$\underset{\substack{(\mathbf{3})\\ X = \text{Br or I}}}{H_3C[CH_2]_xX} \qquad \underset{(\mathbf{1})}{HC{\equiv}CH} \qquad \underset{\substack{(\mathbf{4})\\ Y = \text{OH, Cl or COOH}}}{X[CH_2]_yY}$$

Selective reduction is best achieved with Lindlar's catalyst. This is a palladium catalyst (5%) on calcium carbonate to which some lead acetate has been added. Hydrogenation is carried out at ordinary temperature and pressure in ethyl acetate, petroleum ether, an alcohol of low molecular weight, or acetone. In all cases the presence of a little quinoline enhances selectivity. The reaction proceeds smoothly only with *pure* acetylenic acids, and it may be necessary to crystallize these several times prior to partial hydrogenation. Impurities present after reduction include saturated acids and compounds with *trans* unsaturation, and these are best removed by silver-ion chromatography.

Conversion of an acetylenic compound to a *trans* alkene is achieved with sodium or lithium and liquid ammonia, though this reduction is not usually as simple as that using Lindlar's catalyst. Alternatively the *cis*-alkenes can be converted to *trans*-alkenes by stereomutation (section 10.8) and the *trans* isomer isolated from the equilibrium mixture by crystallization and/or silver-ion chromatography.

These procedures have been extensively employed, and the examples in Schemes 7.1(i)–(iii) are taken from the work of Gunstone and Ismail (1967) and Barve and Gunstone (1971), who prepared all the octadecynoic acids and from them all the *cis*- and *trans* octadecenoic acids, and of Gilman and Holland (1974), who prepared all the C_{10}, C_{11}, C_{12}, C_{13} and C_{14} monoynoic acids. Pomonis and Hakk (1990) prepared long-chain monoene acids (C_{24} and C_{28}) from $CH_3(CH_2)_7C{\equiv}CLi$ and $Br(CH_2)_nCOOH$ ($n = 13$ or 17).

Linoleic acid was first synthesized via acetylenic intermediates in the early 1950s, but it was not until 1961 that a general procedure for obtaining the natural methylene-interrupted polyenes were outlined (Osbond *et al.*, 1961). This has been modified in useful ways by Kunau (1971) and, in the original or modified form, it has been used to prepare a large number of polyene acids (Sprecher, 1979). The necessary poly-ynoic acid is prepared first. This is a crystalline solid, which can be thoroughly purified by

Scheme 7.1(i) Octadec-8-enoic acid

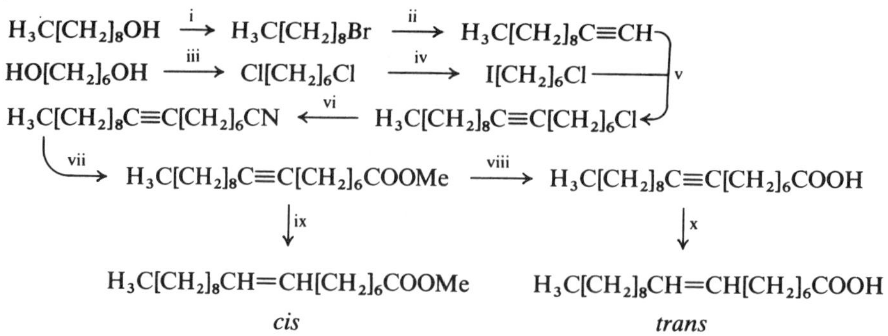

Reagents: i, HBr, H_2SO_4; ii, $NaC{\equiv}CH$, NH_3; iii, $SOCl_2$; iv, NaI, $COMe_2$; v, $NaNH_2$; vi, KCN, DMSO; vii, MeOH, H^+; viii, KOH, H_3O^+; ix, H_2, Pd; x, Li, NH_3

Scheme 7.1(ii) Octadec-7-enoic acid

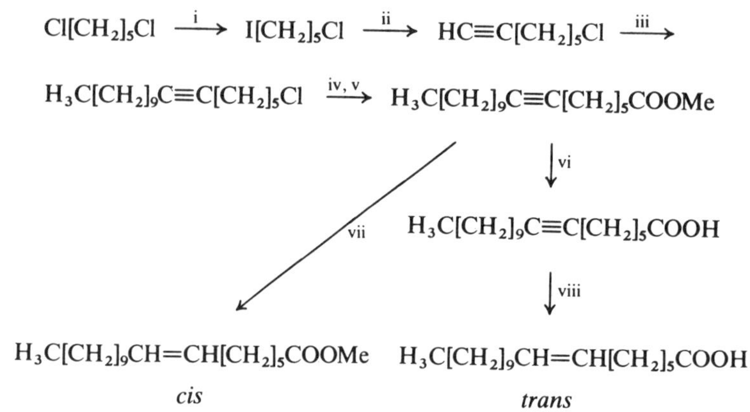

Reagents: i, NaI, $COMe_2$; ii, $NaC{\equiv}CH$, NH_3; iii, $NaNH_2$, NH_3, $H_3C[CH_2]_9Br$; iv, KCN, DMSO; v, MeOH, H^+; vi, KOH, H_3O^+; vii, H_2, Pd; viii, Li, NH_3

Scheme 7.1(iii) Dodec-5- and -6-ynoic acids

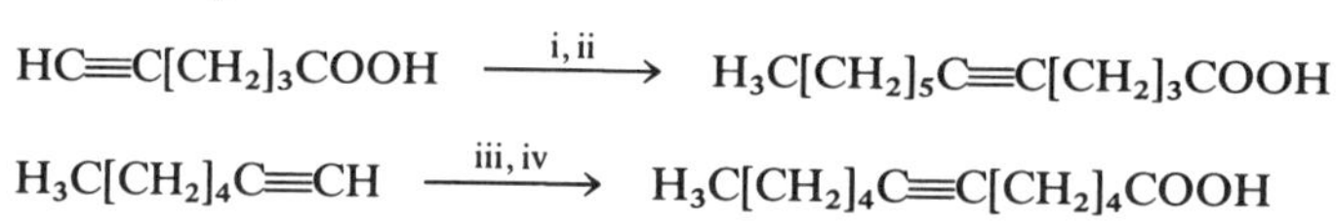

Reagents: i, BuLi, HMPA; ii, $H_3C[CH_2]_5Br$; iii, BuLi, HMPA; iv, $Br[CH_2]_4COOH$

crystallization, prior to partial hydrogenation with Lindlar catalyst.

Poly-ynoic acids are obtained from the fragments indicated in (**5**). The acidic fragment may be replaced by a chloride (X = Cl, converted to acid or ester via the nitrile), an alcohol (X = CH_2OH, oxidized to acid by chromic acid), or an acetal (X = $CH(OR)_2$, converted to acid via the aldehyde). The C_3 unit is usually propargyl alcohol or some derivative of it. The synthesis may be shortened (made more convergent) if the acid fragment is diynoic ($HC{\equiv}CCH_2C{\equiv}C[CH_2]_pCOOH$) and/or the propargyl unit is replaced by hexadiynol ($HC{\equiv}CCH_2C{\equiv}CCH_2OH$).

The essential condensation step involves a propargylic bromide (**6**) and the magnesium bromide derivative (**7**) of an ethynyl compound and occurs in

$$H_3C[CH_2]_mC{\equiv}CCH_2 \,\vdots\, [C{\equiv}CCH_2]_n \,\vdots\, C{\equiv}C[CH_2]_pX$$

(**5**) (X = COOH)

$$RC{\equiv}CCH_2Br + BrMgC{\equiv}CR' \rightarrow RC{\equiv}CCH_2C{\equiv}CR'$$

(**6**) (**7**)

Scheme 7.1(iv) Octadeca-5,8,11,14-tetraenoic acid (all-cis)

$$H_3C[CH_2]_2C{\equiv}CCH_2Br \xrightarrow{i,\,ii,\,iii} H_3C[CH_2]_2[C{\equiv}CCH_2]_3Br \xrightarrow{iv}$$

$$H_3C[CH_2]_2[C{\equiv}CCH_2]_4[CH_2]_2COOH \xrightarrow{v} H_3C[CH_2]_2[CH{=}CHCH_2]_4[CH_2]_2COOH$$

Reagents: i, $BrMgC{\equiv}CCH_2C{\equiv}CCH_2Otpy$, CuCl (tpy = tetrahydropyranyl); ii, $4\text{-}CH_3C_6H_4SO_3H$, CH_3OH; iii, PBr_3, C_5H_5N; iv, $BrMgC{\equiv}C[CH_2]_3COOMgBr$, CuCN; v, H_2, Pd

Scheme 7.1(v)

$$H_3C[CH_2]_xC{\equiv}CCH_2I + HC{\equiv}CCH_2OMe \xrightarrow{i} H_3C[CH_2]_xC{\equiv}CCH_2C{\equiv}CCH_2OMe \xrightarrow{ii}$$

$$H_3C[CH_2]_xC{\equiv}CCH_2C{\equiv}CCH_2I \xrightarrow{iii} H_3C[CH_2]_xC{\equiv}CCH_2C{\equiv}CCH_2C{\equiv}CCH_2C{\equiv}C[CH_2]_y\,CH_2Otpy$$

$$\xrightarrow{iv} H_3C[CH_2]_xC{\equiv}CCH_2C{\equiv}CCH_2C{\equiv}CCH_2C{\equiv}C[CH_2]_yCOOH$$

Reagents: i, EtMgBr; ii, CH_3COI, CH_2Cl_2, ZnI_2; iii, $BrMgC{\equiv}CCH_2C{\equiv}C[CH_2]_yCH_2Otpy$, CuCl; iv, H^+, MeOH, CrO_3

Scheme 7.1(vi)

$$CH_3[CH_2]_4CH{=}CHCH_2CH{=}CHCH_2C{\equiv}CH + BrCH_2C{\equiv}C[CH_2]_3Otpy \xrightarrow{i\text{–}iv} CH_3[CH_2]_4[CH{=}CHCH_2]_4[CH_2]_2Br$$

Reagents: i, EtMgBr, THF; ii, CuBr, THF; iii, P-2Ni, H_2, EtOH; iv, Ph_3P, CBr_4

the presence of Cu(I) ions. The propargylic bromide is made from the alcohol by reaction with phosphorous tribromide or from the methanesulphonate by reaction with magnesium bromide or iodide. C_3 or C_6 alcohols may be used as the bismagnesium bromide derivatives or as the magnesium bromide derivative of the tetrahydropyranyl ether. A number of these points are illustrated in the synthetic sequence in Scheme 7.1(iv) (Beerthius *et al.*, 1971).

Kunau (1971) modified Osbond's procedure to make it more suitable for acids with four to six triple bonds as in Scheme 7.1(v).

A more recent example covers the synthesis of arachidonic acid (Heitz *et al.*, 1989) and the related C_{19} bromide required for the preparation of labelled arachidonic acid (Scheme 7.1(vi)).

References and further reading

Barve, J.A. and Gunstone, F.D. (1971) *Chem. Phys. Lipids*, **7**, 311.

Beerthius, R.K., Nugteren, D.M., Pabon, H.J.J., Steenhoek, A. and van Dorp, D.A. (1971) *Rec. Trav. Chim. Pays Bas*, **90**, 943.

Gilman, N.W. and Holland, B.C. (1974) *Chem. Phys. Lipids*, **13**, 239.

Gravier-Pelletier, C., Dumas, J., LeMerrer, Y. and Depezey, J.C. (1990) *Prog. Lipid Res.*, **29**, 229.

Gunstone, F.D. and Ismail, I.A. (1967) *Chem. Phys. Lipids*, **1**, 209.

Heitz, M.P., Wagner, A., Mioskowski, C., Noel, J.P. and Beaucourt, J.P. (1989) *J. Org. Chem.*, **54**, 500.

Kunau, W.H. (1971) *Chem. Phys. Lipids*, **7**, 101, 108.

Kunau, W.H. (1973) *Chem. Phys. Lipids*, **11**, 254.

Marvell, E.N. and Li, T. (1973) *Synthesis*, 457.

Normant, J.F. and Alexakis, A. (1981) *Synthesis*, 841.

Osbond, J.M., Philpott, P.G. and Wickens, J.C. (1961) *J. Chem. Soc.*, 2779.

Pomonis, J.G. and Hakk, H. (1990) *Lipids*, **25**, 821.

Sprecher, H. (1977) *Polyunsaturated Fatty Acids* (eds W.H. Kunau and R.T. Holman), American Oil Chemists' Society, Champaign, p. 69.

Sprecher, H. (1979) *Prog. Chem. Fats Other Lipids*, **15**, 219.

7.2 Fatty acid synthesis by the Wittig reaction

The Wittig reaction is comparable in importance to the acetylenic route as a means of synthesizing olefinic fatty acids and has also been used extensively to prepare long-chain alcohols and their esters, which are important as insect pheromones.

The essential requirements for the Wittig reaction are an aldehyde (or ketone) and an alkyl triphenylphosphonium salt or its ylid (derived from an alkyl bromide), which interact to make an alkene (**8**). The group X is selected to give the necessary functionality at the end of the long chain.

$$\left.\begin{array}{l} RCHO + Ph_3P{=}CH[CH_2]_nX \\ RCH{=}PPh_3 + OHC[CH_2]_nX \end{array}\right\} \longrightarrow \underset{(\mathbf{8})}{RCH{=}CH[CH_2]_nX}$$

For effective use of this reaction, it is important to appreciate the factors that influence the configuration of the double bond, of which the most important is the structure of the ylid. Stabilized ylids with an α aromatic group, an α double bond, or an α electron-withdrawing group (such as an ester) give the *trans* alkene in >95% purity whilst non-stabilized ylids with electron-donating groups (such as alkyl) give *cis*-alkenes in >98% purity. If the reaction is conducted in a polar aprotic solvent (such as dimethylformamide, DMF), these results are unaffected by added salts (Lewis bases); but in non-polar solvents, the presence of iodides or bromides or certain amines increases the proportion of *trans* isomer.

Formation of the ylid requires a suitable base and, whilst several have been used (see examples in Schemes 7.2(i)–(vii)), particularly good results are claimed for sodium bistrimethylsilylamide at −78 °C.

The aldehydes required for the Wittig reaction have been prepared from unsaturated fatty acids or cycloalkenes by ozonolysis, from α,ω-diols by partial oxidation, and from α,ω-alkadienes by partial hydroboration followed by oxidation.

The topic has been reviewed by Bergelson and Shemyakin (1964) and by Bestmann and Vostrowsky (1979). Gravier-Pelletier *et al.* (1990) have reviewed synthetic routes to long-chain hydroxy acids, many of which involve the Wittig reaction. A number of examples of the use of this reaction are shown in Schemes 7.2(i)–(vii).

Viala and Santelli (1988) described the preparation of the phosphonium salt $Ph_3P^+[CH_2]_2CH[OPr^i]_2Br^-$ from acrolein (propenal) and found it to be very effective as a three-carbon homologating agent under Wittig conditions. They converted hexanal to methyl

Scheme 7.2(i) cis-Dodec-4-enoic acid (Bestmann and Vostrowsky, 1979)

$$H_3C[CH_2]_6CHO + Ph_3P{=}CH[CH_2]_2COOEt \xrightarrow{\text{i}} H_3C[CH_2]_6CH{=}CH[CH_2]_2COOH$$

Reagent: i, Wittig reaction, ester hydrolysis by KOH, H_3O^+

Scheme 7.2(ii) cis-Octadec-9-en-12-ynoic acid (crepenynic acid) (Bradshaw *et al.*, 1971)

$$H_3C[CH_2]_4C{\equiv}CH \xrightarrow{\text{i–iv}} H_3C[CH_2]_4C{\equiv}CCH_2CH{=}PPh_3 \xrightarrow{\text{v, vi}}$$

$$H_3C[CH_2]_4C{\equiv}CCH_2CH{=}CH[CH_2]_7COOH$$

Reagents: i, Li, NH_3, ethylene oxide; ii, PBr_3; iii, PPh_3, C_6H_6; iv, BuLi, Et_2O; v, $OHC[CH_2]_7COOMe$; vi, KOH (20 °C), H_3O^+

Scheme 7.2(iii) Octadeca-9-cis,11-trans,13-trans-trienoic acid (catalpic acid) (Bergelson and Shemyakin, 1964)

$$H_3C[CH_2]_3CH\overset{t}{=}CHCH\overset{t}{=}CHCHO + Ph_3P{=}CH[CH_2]_7COOEt \xrightarrow{\text{i, ii}}$$

$$H_3C[CH_2]_3CH\overset{t}{=}CHCH\overset{t}{=}CHCH\overset{c}{=}CH[CH_2]_7COOH$$

Reagents: i, DMF, I^-, ii, KOH, H_3O^+

Scheme 7.2(iv) Tetradeca-9-cis,11-trans-dienyl acetate (large-scale synthesis; Hall *et al.*, 1975)

$$HO[CH_2]_9OH \longrightarrow HO[CH_2]_9Otpy \longrightarrow Br[CH_2]_9Otpy \xrightarrow{i,ii}$$

$$Ph_3P{=}CH[CH_2]_8Otpy \xrightarrow{iii-v} H_3CCH_2CH{=}CHCH{=}CH[CH_2]_8Otpy$$

$$\longrightarrow H_3CCH_2CH{=}CHCH{=}CH[CH_2]_8OAc$$

Reagents: i, Ph_3P, CH_3CN, K_2CO_3; ii, dimsyl sodium, DMSO; iii, $H_3CCH_2CH\overset{t}{=}CHCHO$; iv, MeOH, H^+; v, Ac_2O, pyridine

Scheme 7.2(v) Pentadeca-11-trans,13-cis-dienyl acetate (Bestmann and Vostrowsky, 1979)

$$H_2C{=}CH[CH_2]_8CHO + Ph_3P{=}CHCHO \longrightarrow CH_2{=}CH[CH_2]_8CH\overset{t}{=}CHCHO$$

$$\xrightarrow{i} CH_2{=}CH[CH_2]_8CH{=}CHCH{=}CHCH_3 \xrightarrow{ii} AcO[CH_2]_{10}CH\overset{t}{=}CHCH\overset{c}{=}CHCH_3$$

Reagents: i, $H_3CCH{=}PPH_3$; ii, 9-BBN (9-borabicyclo[3.3.1]nonane); Ac_2O, pyridine

Scheme 7.2(vi) Hexadeca-cis-6,cis-11-dienal (Bestmann and Vostrowsky, 1979)

$$H_3C[CH_2]_3CH{=}CH[CH_2]_3CHO + Ph_3P{=}CH[CH_2]_5OAc \xrightarrow{i,ii} H_3C[CH_2]_3CH\overset{c}{=}CH[CH_2]_3CH\overset{c}{=}CH[CH_2]_4CHO$$

Reagents: i, KOH, H_3O^+; ii, $C_5H_5NHCrO_3Cl$

Scheme 7.2(vii) Triacontanol (Subramanian and Rastogi, 1991)

$$CH_3[CH_2]_{13}P^+Ph_3Br^- + OHC[CH_2]_7CH{=}CH[CH_2]_5CO_2Me$$

$$\xrightarrow{i} CH_3[CH_2]_{12}CH{=}CH[CH_2]_7CH{=}CH[CH_2]_5CO_2Me \xrightarrow{ii,iii} CH_3[CH_2]_{28}CH_2OH$$

Reagents: i, K_2CO_3, 1,4-dioxan, formamide (trace); ii, Pd/C; iii, $LiAlH_4$

Scheme 7.2(viii) Methyl arachidonate (Viala and Santelli, 1988)

$$CH_3[CH_2]_4CHO \xrightarrow{i,ii} CH_3[CH_2]_4CH{=}CHCH_2CHO \xrightarrow{i,ii} CH_3[CH_2]_4[CH{=}CHCH_2]_2CHO$$

$$\xrightarrow{i,ii} CH_3[CH_2]_4[CH{=}CHCH_2]_3CHO \xrightarrow{iii,iv} CH_3[CH_2]_4[CH{=}CHCH_2]_4[CH_2]_2COOMe$$

Reagents: i, $Ph_3P^+[CH_2]_2CH[OPr^i]_2Br^-$, $NaN(SiMe_3)_2$, THF, HMPA; ii, H_2O, TsOH; iii, $Ph_3P^+[CH_2]_4COOH\ Br^-$, $NaN(SiMe_3)_2$, THF, HMPA; iv, CH_2N_2

arachidonate in four steps with an overall yield of 58% (Scheme 7.2(viii)).

References and further reading

Bergelson, L.D. and Shemyakin, M.M. (1964) *Angew. Chem. Int. Edn*, **3**, 250.

Bestmann, H.J. and Vostrowsky, O. (1979) *Chem. Phys. Lipids*, **24**, 335.

Bradshaw, R.W., Day, A.C., Jones, E.R.H., Page, C.B., Thaller, V. and Vere Hodge, R.A. (1971) *J. Chem. Soc.*, C, 1156.

Gravier-Pelletier, C., Dumas, J., LeMerrer, Y. and Depezey, J.C. (1990) *Prog. Lipid Res.*, **29**, 229.

Hall, D.R., Beevor, P.S., Lester, R., Poppi, R.G. and Nesbitt, B.F. (1975) *Chem. Ind.*, 216.

Subramanian, G.B.V. and Rastogi, A. (1991) *Chem. Ind.*, 436.

Viala, J. and Santelli, M. (1988) *J. Org. Chem.*, **53**, 6121.

7.3 Isotopically labelled acids

Isotopically labelled fatty acids and the lipids derived from them are used in studies of reaction mechanism (e.g. hydrogenation), lipid biosynthesis and metabolism, and NMR spectra. Compounds containing radioactive isotopes (^3H, ^{14}C) are examined by liquid scintillation counting or radio gas chromatography. Non-radioactive labels are studied by mass spectrometry or by NMR spectroscopy.

Though some labelled acids are commercially available, it is frequently necessary to synthesize these compounds as required, sometimes by suitable adaptation of the synthetic procedures described in sections 7.1 and 7.2. This account is confined to acids labelled with one or more of the hydrogen (^2H, ^3H) or carbon (^{13}C, ^{14}C) isotopes.

The subject has been reviewed by Kunau (1973), Sprecher (1977, 1979), Emken (1978, 1979), Tulloch (1979) and Rakoff (1982).

7.3.1 Acids labelled with ^2H

Tulloch (1979) divided the methods of preparing ^2H fatty acids into five categories: exchange, reduction, reaction with small deuterated molecules, chemical synthesis with intermediates prepared in the preceding ways, and biological synthesis in the presence of ^2H_2O. He lists about 100 deuterated acids that have been prepared.

Hydrogen on carbon adjacent to an aldehyde or ketone group is replaced by deuterium from ^2H_2O in the presence of mildly acidic or basic catalysts:

$$RCH_2CHO \xrightarrow[\text{pyridine}]{^2H_2O} RC^2H_2CHO$$

Methylene groups next to a carboxyl group react similarly but require more vigorous conditions:

$$RCH_2COOH \xrightarrow[150\text{–}200\,°C]{^2HO^-} RC^2H_2COOH$$

Useful reduction processes include the partial reduction of alkynes ($^2H_2$, Lindlar catalyst), reduction of alkenes with tetradeuterohydrazine, and reduction of carbonyl compounds with sodium borodeuteride or lithium aluminium deuteride as in Scheme 7.3(i).

Catalytic hydrogenation with a heterogeneous catalyst is not a suitable procedure because reduction is accompanied by extensive exchange between hydrogen and deuterium atoms (section 10.1).

Small deuterated compounds such as $^2H_3CCOO^2H$ and C^2H_3I are easily incorporated into long-chain acids as in the anodic synthesis of deuterated stearate:

$$^2H_3CCOOH + HOOC[CH_2]_{16}COOMe \longrightarrow {}^2H_3C[CH_2]_{16}COOMe$$

Scheme 7.3(i)

$$RC{\equiv}CR' \xrightarrow[\text{Lindlar's catalyst}]{^2H_2} RC^2H{=}C^2HR'$$

$$RCH{=}CHR' \xrightarrow{N_2{}^2H_4} RCH^2HCH^2HR'$$

$$RCOOMe \xrightarrow{LiAl^2H_4} RC^2H_2OH \xrightarrow[NEt_3]{MsCl} RC^2H_2OMs \xrightarrow{\text{malonation}} RC^2H_2CH_2COOH$$

Once a ^2H-labelled compound has been made, it can be incorporated into a long-chain acid by chain extension (via the nitrile, by malonation, or by enamine synthesis) or by involvement in a Wittig or acetylenic synthesis as in the synthesis of methyl 11-dideuterooleate (Scheme 7.3(ii)).

The synthesis of several unsaturated acids containing deuterium has been reported recently. They include $n-6$ acids: 18:2 (several stereoisomers; Adlof *et al.*, 1991) and 18:3 (Rakoff, 1984, 1990); and $n-3$ acids: 18:3 (Rakoff, 1988), 18:4 (Rakoff, 1990), 20:3 and 20:4 (Rakoff, 1991). These compounds are prepared from appropriate starting materials via the acetylenic (section 7.1) or Wittig (section 7.2) routes.

Baenziger *et al.* (1990) used a biological system (the protozoan *Tetrahymena*) to convert several deuterated oleic acids to the corresponding 6,9-diene acids.

7.3.2 Acids labelled with ^3H

The methods described for ^2H labelling can also be applied to tritiation, in particular reduction of alkenes with $N_2{}^3H_4$, partial reduction of alkynes with

Scheme 7.3(ii)

$$H_3C[CH_2]_6C^2H_2CHO \xrightarrow{\text{i–iii}} H_3C[CH_2]_6C^2H_2CH{=}PPh_3$$

$$\xrightarrow{\text{iv}} H_3C[CH_2]_6C^2H_2CH{=}CH[CH_2]_7COOMe$$

Reagents: i, $LiAlH_4$; ii, PBr_3; iii, PPh_3, base; iv, $OHC[CH_2]_7COOMe$

$^3H_2$ and Lindlar's catalyst, and reduction of mesylates (or tosylates) with $LiAl^3H_4$. For example:

$$>CHOH \longrightarrow >CHOMs(Ts) \longrightarrow >CH^3H$$

7.3.3 Acids labelled with ^{14}C

When preparing fatty acids labelled with ^{14}C, it is usual to incorporate the label as late in the synthesis as possible and preferably in a reaction proceeding in high yield. $K^{14}CN$ and $^{14}CO_2$ (from labelled carbonate) are useful reagents, which can be used as shown in Scheme 7.3(iii).

Once prepared the 1-^{14}C-labelled acid can be chain-extended by any standard procedure (malonation, enamine synthesis, anodic synthesis).

The formyl ester $MeO_2C[CH_2]_7{}^{14}CHO$, prepared from [1-^{14}C]elaidate, has been used in Wittig reactions to give fatty acids labelled in a mid-position.

Fatty acids labelled non-specifically can be obtained from plants or photosynthetically active micro-organisms grown in a $^{14}CO_2$ atmosphere.

7.3.4 Acids labelled with ^{13}C

Acids labelled with ^{13}C have become more important with the development of ^{13}C NMR spectroscopy. Such acids can be prepared by adaptation of the methods used for ^{14}C labelling.

Scheme 7.3(iii)

$$RCl \xrightarrow{K^{14}CN} R^{14}CN \xrightarrow{MeOH,\ H^+} R^{14}COOMe$$

$$RBr \xrightarrow[^{14}CO_2,\ H_3O^+]{Mg,\ Et_2O;} R^{14}COOH$$

$$RCH_2OH \xrightarrow[NEt_3]{MsCl} RCH_2OMs \xrightarrow{K^{14}CN} RCH_2{}^{14}CN$$

$$\xrightarrow{MeOH,\ H^+} RCH_2{}^{14}COOMe$$

References and further reading

Adlof, R.O., Rakoff, H. and Emken, E.A. (1991) *J. Am. Oil Chem. Soc.*, **68**, 303.

Baenziger, J.E., Smith, I.C.P. and Hill, R.J. (1990) *Chem. Phys. Lipids*, **54**, 171.

Emken, E.A. (1978) *Handbook of Lipid Research*, vol. 1, *Fatty Acids and Glycerides* (ed. A. Kuksis), Plenum, New York, p. 77.

Emken, E.A. (1979) *Fatty Acids* (ed. E.H. Pryde), American Oil Chemists' Society, Champaign, p. 90.

Kunau, W.H. (1973) *Chem. Phys. Lipids*, **11**, 254.

Rakoff, H. (1982) *Prog. Lipid Res.*, **21**, 225.

Rakoff, H. (1984) *Chem. Phys. Lipids*, **35**, 117.

Rakoff, H. (1988) *Lipids*, **23**, 280.

Rakoff, H. (1990) *Lipids*, **25**, 130.

Rakoff, H. (1991) *Lipids*, **26**, 301.

Sprecher, H. (1977) *Polyunsaturated Fatty Acids* (eds W.H. Kunau and R.T. Holman), American Oil Chemists' Society, Champaign, p. 69.

Sprecher, H. (1979) *Prog. Chem. Fats Other Lipids*, **15**, 219.

Tulloch, A.P. (1979) *Chem. Phys. Lipids*, **24**, 391.

7.4 Acylglycerols

7.4.1 General methods

General procedures for the preparation of racemic (and enantiomeric) acylglycerols are described by Quinn *et al.* (1967), Jensen and Pitas (1976), Bhati *et al.* (1980) and Gunstone and Norris (1983). A number of key intermediates for preparation of enantiomeric acylglycerols are dealt with in section 7.4.4.

(a) Acylating agents

Acid chlorides or acid anhydrides (see also section 10.13) provide rapid and reliable acylation. The usual method of forming saturated acid chlorides is by reaction of the unesterified fatty acid with thionyl chloride (cf. Jensen and Pitas, 1976). Unsaturated fatty acid chlorides are best made with oxalyl chloride (Mattson and Volpenheim, 1962). Notes on precautions to be taken during the preparations are detailed in Buchnea (1978).

Fatty acid anhydrides can be prepared via mixed-acid anhydride intermediates using either trifluoroacetic anhydride (Bourne *et al.*, 1949) or acetyl chloride (Mattson *et al.*, 1964). After refluxing, the excess reagent is removed by distillation and the mixed-acid anhydrides converted to fatty acid anhydrides by heating under pressure.

Mazur *et al.* (1991) claim that the selectivity of acylation is improved when carried out in the presence of Lipozyme, a 1,3-specific enzyme.

Acylation with free acid in the presence of 1,1′-dicyclohexylcarbodiimide and 4-dimethylaminopyridine (DMAP) occurs at room temperature in tetrachloromethane solution and is complete in 2 h (Kodali, 1987).

(b) Selection of protective groups

Since the preparation of acylglycerols often involves the selective attack on a particular hydroxyl moiety, other hydroxyls must be blocked by suitable protective groups. In addition, since racemic derivatives of *sn*-glycerol cannot be resolved, at least one protective group must be used during the initial preparation of the enantiomeric glycerol. Eibl and Woolley (1986) have compared the benzyl, allyl and trityl groups.

(i) Benzyl group

This was first used as a blocking group for the synthesis of monoacid diacyl-*sn*-glycerols when the group was added by reacting sodium or potassium isopropylidene-*sn*-glyceroxide with benzyl chloride. 1-Benzyl- and 3-benzyl-*sn*-glycerols can be further protected by trityl groups (see below) so that mixed-acid diacyl- and triacyl-*sn*-glycerols can be made. Removal of the benzyl group requires catalytic hydrogenolysis, so its use is limited to preparations with saturated fatty acids, although, of course, unsaturated fatty acids can be introduced after the benzyl group has been removed (cf. Buchnea, 1978). 4-Methylbenzyl (Golding, 1988) is reported to be a better protecting group than benzyl. It is removed by reaction with 2,3-dichloro-5,6-dicyano-1,4-benzoquinone (DDQ).

(ii) Isopropylidene group

This is a widely used protective group for chemical reactions involving multiple hydroxyl functions. Its first use for the preparation of intermediates in acylglycerol synthesis was by Fischer and Baer (1937), who synthesized 1,2- and 2,3-isopropylidene-*sn*-glycerols. Such substances can then be acylated or blocked (e.g. by benzyl groups) in the remaining hydroxyl moiety. Care must be taken that no racemization or loss of the protective group takes place on storage (Buchnea, 1967a). The isopropylidene group is best removed by trimethyl borate in the presence of boric acid (Mattson and Volpenheim, 1962). Kodali (1987) claims improved results with dimethylboron bromide (Me_2BBr) at −50 °C.

(iii) Triphenylmethyl (trityl) group

Triphenylmethyl chloride reacts mainly with the primary hydroxyls of glycerol, and the trityl group has been extensively used for acylglycerol preparations (Verkade, 1953). This blocking group is removed by catalytic hydrogenolysis, with hydrogen chloride in diethyl ether or petroleum ether, or with a molar equivalent of boron trifluoride–methanol in dichloromethane at 22 °C for 15 min (Iwama and Foglia, 1988). Detritylation leads to the formation of by-products, which must be removed by crystallization or by silicic acid column chromatography. Therefore, the benzylidene protective group is more useful for many purposes (cf. Buchnea, 1978).

(iv) 2,2,2-Trichloroethoxycarbonyl group
This can be used during the preparation of unsaturated 1,2- and 2,3-diacyl-*sn*-glycerols. The 2,2,2-trichloroethoxycarbonyl group is introduced on to the free primary hydroxyl group as the carbonyl chloride. It is retained, without migration, when the isopropylidene group is removed and is, itself, removed with zinc in acetic acid at room temperature (Pfeiffer *et al.*, 1968).

(v) Other blocking groups
Various other blocking groups are discussed by Buchnea (1978) and Gunstone and Norris (1983) with respect to their use in acylglycerol synthesis. Moss *et al.* (1987) recommend the use of 2-iodobenzoyl as a protecting group, which can be removed by chlorination followed by mild basic hydrolysis with methanolic sodium carbonate or bicarbonate.

7.4.2 Racemic acylglycerols

(a) 1-Acylglycerols

Glycerol is reacted with acetone in the presence of *p*-toluenesulphonic acid to yield 1,2-isopropylidene glycerol, which can be acylated with a fatty acid under refluxing conditions in the presence of the acid catalyst (above). Acid catalyst and excess isopropylidene glycerol are removed with sodium acetate solution, and the extent of the acetonation and acylation reactions can be determined by quantification of the amount of water released (Jensen and Pitas, 1976). The isopropylidene ester can then be cleaved with boric acid and the 1-acylglycerol purified by crystallization. Purity can be checked by TLC using a solvent system where 1- and 2-monoacylglycerols are separated (cf. Thomas *et al.*, 1965). An alternative method of cleavage is the use of hydrogen chloride (Buchnea, 1971).

1-Monoacylglycerols can also be prepared using glycidol (section 7.4.4).

(b) 2-Acylglycerols

The starting material for these syntheses is 1,3-benzylideneglycerol. In this compound, only the free 2-hydroxyl is available for acylation. A typical method is that of Mattson and Volpenheim (1962).

Benzaldehyde is condensed with glycerol in the presence of the acid catalyst, *p*-toluenesulphonic acid. The product is crystallized, collected by filtration and further purified (cf. Jensen and Pitas, 1976). Acylation is carried out with an acid chloride (section 10.13) and, after removal of pyridine hydrochloride, the acyl-1,3-benzylideneglycerol is crystallized. Purity can be estimated by TLC (Jensen and Pitas, 1976). A critical stage now comes, where the blocking group has to be removed. Cleavage of the benzylidene acetal is carried out by treatment with boric acid and water. The presence of boric acid inhibits isomerization to the more stable 1-monoacylglycerol (the equilibrium ratio is 9:1 in favour of the 1 isomer).

If only small amounts of 2-acylglycerols are required, then it may be found more convenient to hydrolyse an appropriate triacylglycerol with pancreatic lipase. This enzyme is specific for the 1 and 3 positions and thus yields a 2-acylglycerol product. By keeping the incubation times short (e.g. 5 min), acyl migration is minimized. The 2-acylglycerol can then be purified by a boric acid TLC system (Jensen and Pitas, 1976).

(c) 1,3-Diacylglycerols

These can be synthesized by acylation of a 1-acylglycerol with an acid chloride in the presence of pyridine and chloroform. The completeness of the reaction can be followed by TLC. Generally 24 h at room temperature is more than sufficient. After crystallization from hexane or hexane–ethanol mixtures, 50–70% yields are obtained (cf. Jensen and Pitas, 1976).

(d) 1,2-Diacylglycerols

Several methods are available for synthesizing 1,2-diacylglycerols (Jensen, 1972). All procedures have to overcome the problems associated with the inherent instability of the isomer and its isomerization to 1,3-diacylglycerol. In addition, it is difficult to crystallize 1,2-diacylglycerols because of their high solubility in appropriate solvents. Again, if only small amounts are needed, then it is more convenient to use pancreatic lipase and to isolate 1,2-diacylglycerol by TLC (Jensen and Pitas, 1976). Also, passage of a 1,3 isomer through neutral alumina will result in conversion of about 50% to the 1,2 isomer, which can then be purified by TLC (Jensen *et al.*, 1966).

Of the different chemical methods, that of Gigg and Gigg (1967a) is quite useful. They prepared the diacid derivatives by removing the acyl moiety from the 1 position of a 1,2-diacyltetrahydropyranylglycerol. Pancreatic lipase was used for this purpose. Reacylation with a different fatty acid then yields a 1,2-diacylglycerol.

One of the most convenient methods uses the tetrahydropyranyl ether of glycerol. Two procedures can be used to generate this material. Dihydropyran reacts with allyl alcohol to form allyl tetrahydropyranyl ether, which can be oxidized with potassium permanganate to give tetrahydropyranylglycerol (Barry and Craig, 1955). Alternatively, glycerol carbonate can be converted to the tetrahydropyranyl ether and the carbonate removed with potassium hydroxide (Cunningham and Gigg, 1965).

The preparation of 1,2-dioleoylglycerol (Krabisch and Borgstrom, 1965; Turner *et al.*, 1968) and of 1,2-dipalmitoylglycerol (Jensen and Pitas, 1976) have been detailed. Acylation of tetrahydropyranylglycerol is performed with an acid chloride, the product crystallized and purity checked by TLC. The blocking group is then removed with hydrogen chloride under carefully controlled conditions. Alternatively, the blocking ether can be removed with boric acid. This has the advantage of reducing the tendency to form 1,3-diacylglycerols by isomerization (Gigg and Gigg, 1967a).

1,2-Diacylglycerols can also be prepared using glycidol (section 7.4.4).

(e) Triacylglycerols

(i) Monoacids

These are often commercially available or can be easily prepared by reaction of glycerol with an appropriate acid chloride. Purification can be effected by crystallization or by TLC.

(ii) Diacid

The general principle of the synthesis of diacid triacylglycerols is to react an appropriate monoacylglycerol or diacylglycerol with the desired acid chloride. Jensen and Pitas (1976) have detailed the formation of 1-palmitoyl-2,3-distearoylglycerol, 1-palmitoyl-2,3-dioleoylglycerol, 1-palmitoyl-2,3-dilinoleoylglycerol, 1-oleoyl-2,3-dipalmitoylglycerol and 2-palmitoyl-1,3-dioleoylglycerol.

For the acylation, additional amounts (50% molar excess) of pyridine and acid chloride are beneficial. The reaction takes from hours to days at room temperature. The triacylglycerol product can be crystallized with good purity in some cases, but in others it is necessary to use an alumina column in addition. Jensen and Pitas (1976) emphasize that, for the latter to be used efficiently, the triacylglycerol must be soluble in hexane–anhydrous diethyl ether (9:1) at room temperature. Many saturated triacylglycerols are not soluble in this mixed solvent and so have to be purified by crystallization alone. When the latter is used, then ethanol is included in the crystallization solvent so that excess acid chloride is converted to the ester, which remains in solution. When alumina columns are employed, care must be taken to eliminate all traces of alcohol, chloroform or water from the eluting solvents, since these can ruin the separation. Alumina columns can be run rapidly and recoveries are 80–90% (Jensen *et al.*, 1966).

Diacid triacylglycerols containing one acid in the 1 position and another in the 2 and 3 positions are prepared by acylation of the appropriate 1-monoacylglycerol. When the single acid is at the 2 position, then the 1,3-diacylglycerol is acylated.

The purity of the triacylglycerols is determined by TLC and the proper ratio of the acyl components can be established by GLC. Hydrolysis with pancreatic lipase will then confirm that the acyl groups are esterified to the correct positions.

(iii) Triacid

Jensen and Pitas (1976) have described the preparation of 1-palmitoyl-2-oleoyl-3-stearoylglycerol. The general principle is the same as for the synthesis of diacid triacylglycerols. In this case 1-palmitoyl-3-stearoylglycerol is reacted with oleoyl chloride. After crystallization and purification on an alumina column, the triacylglycerol was recovered at better than 99% purity. By an appropriate selection of 1,3-diacylglycerols it is possible to prepare triacid triacylglycerols with any desired combination of acids.

Triacid triacylglycerols can also be prepared using glycidol (section 7.4.4).

7.4.3 Enantiomeric acylglycerols

The question of the stereospecific description of the absolute configuration of glycerol derivatives has received considerable attention. The various methods are discussed by Buchnea (1978), who emphasizes the advantages of the stereospecific numbering (*sn*) system, which is now recommended by the IUPAC–IUB Commission. This system recognizes that the two primary carbonol groups of glycerol are not interchangeable in their reactions with chiral structures such as enzymes. If the secondary hydroxyl group is shown to the left of C-2 in a Fischer projection, then the carbon atom above is called C-1 and that below is termed C-3.

7.4.4 Preparation of important intermediates for the synthesis of enantiomeric acylglycerols

(a) Preparation and use of D- *and* L-*mannitols*

These two hexoses have proved extremely useful in the synthesis of enantiomeric *sn*-glycerol derivatives.

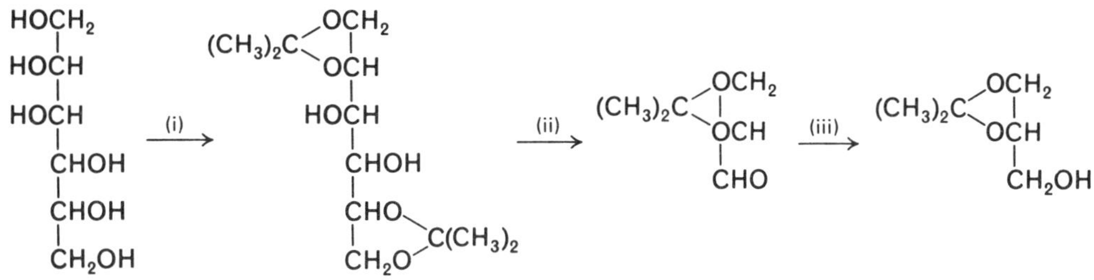

Fig. 7.1 Preparation of 1,2-isopropylidene-*sn*-glycerol from D-mannitol.

D-Mannitol occurs naturally and is available commercially. L-Mannitol, in contrast, has to be synthesized. There are several methods available. Baer and Fischer (1939) obtained L-mannonic lactone from L-arabinose and then reduced the former to L-mannitol. More conveniently, L-mannose is obtained from L-arabinose by the nitromethane route (Sowden, 1962) and L-mannose is then reduced to L-mannitol with sodium borohydride (Wright and Tove, 1967). L-Mannitol can also be made from quebrachitol via L-inositol (Buchnea, 1971).

The D- and L-mannitols are then converted to 1,2:5,6-diisopropylidene-D- and -L-mannitols. Zinc chloride powder is used as catalyst and is suspended in acetone. The excess of zinc chloride is removed before the solution is added to D- or L-mannitol. The 1,2:5,6-diisopropylidene-D- and -L-mannitols (Fig. 7.1) are then recovered and recrystallized (Baer, 1945; Buchnea, 1978). Yields of up to 70% can be obtained.

Baer (1945, 1952) described methods for the conversion of 1,2:5,6-diisopropylidene-D- and -L-mannitols into 1,2- and 2,3-isopropylidene-*sn*-glycerols. The procedure is shown in Fig. 7.1. L-Arabinose can also be used as a convenient source of 2,3-isopropylideneglycerol (Kodali, 1987).

A new method for obtaining 1,2- and 2,3-isopropylidene-*sn*-glycerols of the D and L series (Lok *et al.*, 1976) avoids the need to use D- and L-mannitols. Its principle is shown in Fig. 7.2. Deamination of the serines to glyceric acid is followed by esterification (without an intermediate isolation) to give substance C. This is acetonated to D, which is then reduced to yield isopropylidene-*sn*-glycerol. L-Serine gives the 2,3 derivative and D-serine yields the 1,2-isopropylidene-*sn*-glycerol. Although this is a more convenient procedure than that starting from mannitols, it is much more expensive.

1-Benzyl- and 3-benzyl-*sn*-glycerols can be formed from the appropriate 2,3 or 1,2-isopropylidene-*sn*-glycerol by reaction with benzyl chloride under alkaline conditions (Buchnea, 1978). The 3-benzyl-

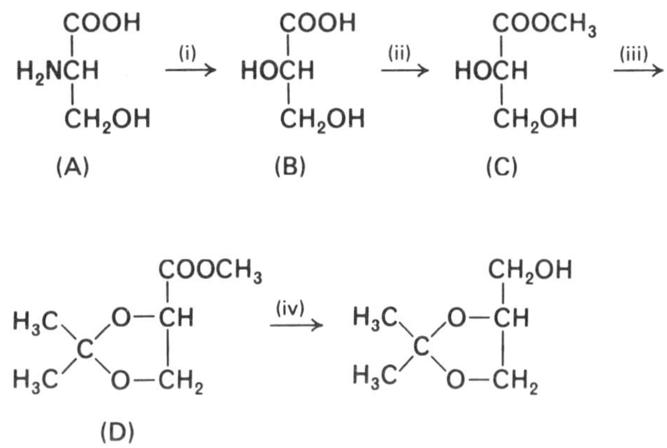

Fig. 7.2 Synthesis of 1,2- or 2,3-isopropylidene-*sn*-glycerol from D- and L-serine, respectively.

1,2-isopropylidene- or 1-benzyl-2,3-isopropylidene-*sn*-glycerol products are separated by vacuum distillation and the isopropylidene protective group removed by hydrolysis with 10% acetic acid. 3,4-Isopropylidene-D-mannitol is the equivalent of a glycerol molecule protected at the *sn*-3 position (Eibl and Woolley, 1988).

For preparation of enantiomeric alkyl- and alkenyl-*sn*-glycerols see Baumann and Mangold (1964), Palameta and Kates (1966), Buchnea (1978) and Gunstone and Norris (1983).

(b) *Glycidol*

Glycidol (2,3-epoxypropanol) is available in racemic and in *R* and *S* forms, and these are convenient starting points for the preparation of racemic and enantiomeric mono-, di- and triacylglycerols. Two examples are given in Schemes 7.4(i) and (ii).

Scheme 7.4(i) 1-Monoacyl and 1,2-diacylglycerols (Burgos *et al.*, 1987)

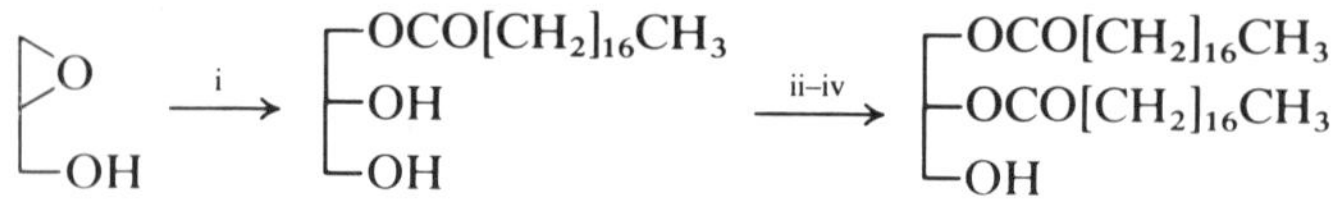

Reagents: i, stearic acid, $Ti(OPr^i)_4$; ii, Me_2Bu^tSiCl, imidazole; iii, $CH_3[CH_2]_{16}COCl$; iv, NBS, DMSO, H_2O, TMF

Scheme 7.4(ii) Triacylglycerols (Sonnet, 1991)

Reagents: i, stearic anhydride, LiBr, THF or benzene; ii, $C_{15}H_{31}COCl$, THF, HMPA, 50–55 °C, 5 h

(c) Configurational conversion and inversion of sn-*glycerol derivatives*

These methods are useful since they avoid the necessity of preparing L-mannitol for the synthesis of 2,3-isopropylidene-*sn*-glycerol and 1-benzyl-*sn*-glycerol.

The configurational conversion of 1,2-isopropylidene- into 2,3-isopropylidene-*sn*-glycerol was described by Baer and Fischer (1945b). Six steps are involved (cf. Fig. 7.3).

The inversion of 3-benzyl-*sn*-glycerol into 1-benzyl-*sn*-glycerol was reported by Lands and Zschocke (1965). Their method is shown in Fig. 7.4 and uses a 'Walden inversion'. The procedure has been theoretically discussed and constraints on individual steps described by Ingold (1973).

An alternative method for the synthesis of 1-benzyl-*sn*-glycerol from D-mannitol has been described (Gigg and Gigg, 1967b). Mannitol is converted into 1,3;2,5;4,6-*O*-trimethylene-D-mannitol, and this is converted via 2,5-*O*-methylene-D-mannitol and 1,6-dibenzoyl-3,4-benzylidene-D-mannitol into the benzylglycerol derivative. The benzylidene group is removed with dilute hydrogen chloride and 1,6-di-*O*-benzyl-2,5-*O*-methylene-D-mannitol is formed, which can be oxidatively cleaved and then reduced with lithium aluminium hydride to yield methylene-bis-2-*sn*-(benzyl-*sn*-glycerol). The latter is hydrolysed with acid to give 1-benzyl-*sn*-glycerol (cf. Buchnea, 1978, for more details).

7.4.5 Enantiomeric monoacylglycerols

1-Monoacyl- and 3-monoacyl-*sn*-glycerols were synthesized from L- and D-mannitol respectively by Baer and Fischer (1945a). The procedure is outlined in Fig. 7.5. The diisopropylidenemannitol and isopropylidene-*sn*-glycerol intermediates are formed as in sections 7.4.4(a) and 7.4.4(c). Unsaturated as well as saturated acid chlorides can be used. The protective groups were removed by acid hydrolysis in the original procedure. Monoacylglycerols with short-chain (C_8 or less) fatty acids were formed

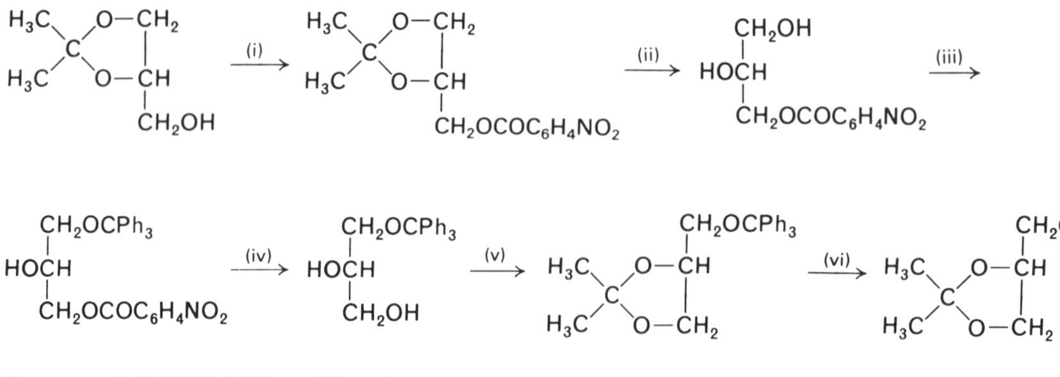

Reagents: i, $O_2NC_6H_4COCl$, pyridine; ii, HCl, 70 °C, iii, Ph_3CCl, pyridine; iv, 10 N NaOH; v, $CO(CH_3)_2$, $ZnCl_2$; vi, Pd, H_2, EtOH

Fig. 7.3 Conversion of 1,2-isopropylidene-*sn*-glycerol into 2,3-isopropylidene-*sn*-glycerol (Baer and Fischer, 1945b).

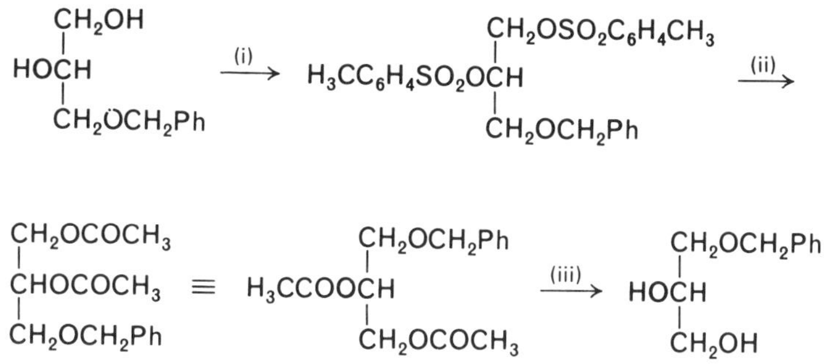

Reagents: i, $CH_3C_6H_4SO_2Cl$, pyridine; ii, CH_3COOK; iii, H_3O^+

Fig. 7.4 Inversion of 3-benzyl-*sn*-glycerol into 1-benzyl-*sn*-glycerol (Lands and Zschocke, 1965).

following hydrolysis with 10% acetic acid. Other monoacylglycerols were produced after hydrolysis with concentrated hydrochloric acid–diethyl ether (1:1). Both these methods cause some isomerization. An improved removal of blocking groups is to replace them with trimethyl borate groups, which can themselves be removed by hydrolysis in water at room temperature without acyl migration (cf. Hartman, 1959). Schlenk (1965a, b) used the latter method to synthesize a whole series of enantiomeric monoacylglycerols. The topic of acyl migration in monoacylglycerols has been carefully studied by a number of workers (cf. Buchnea, 1978).

Buchnea and Baer (1960) and Buchnea (1971) prepared saturated and monoenoic 1-monoacyl-*sn*-glycerols from 3-benzyl-*sn*-glycerol. The latter was tritylated in the *sn*-1 position and then acylated in the *sn*-2 position. When the trityl group is removed with hydrogen chloride, the acyl group migrates quantitatively to the *sn*-1 position. Since catalytic reduction was then used to remove the benzyl group, the method was unsuitable for polyenoic fatty acyl groups (mono-unsaturated fatty acids were protected by reversible bromination).

7.4.6 Enantiomeric 1,3-diacyl-*sn*-glycerols

These compounds have been synthesized by several routes. Lok *et al.* (1976) prepared asymmetrical 1,3-diacyl-*sn*-glycerols using D-glycidol and fatty acid in equimolar proportions. This mixture was then heated and agitated after the addition of tetraethylammonium bromide until all the glycidol ester had reacted. The details of the method have to be slightly changed with fatty acids of different chain lengths.

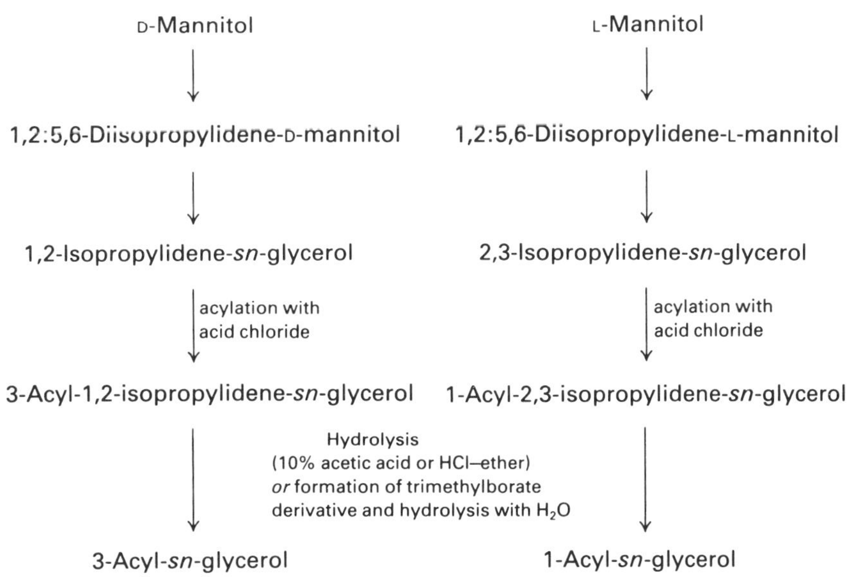

Fig. 7.5 Preparation of enantiomeric 1- and 3-monoacylglycerols. The scheme outlines the method of Baer and Fischer (1945a).

Schlenk (1965a, b) prepared a series of 1,3-diacyl-*sn*-glycerols using methods developed by Baer and Fischer (1945a).

Buchnea (1967a) isomerized 1,2- or 2,3-diacylglycerols with hydrogen chloride in diethyl ether solution at 0 °C to form enantiomeric 1,3-diacyl-*sn*-glycerols. Enantiomeric 1,3-diacylglycerols can be obtained from glycidol (section 7.4.4).

For the preparation of enantiomeric alkyldiacyl-*sn*-glycerols it is useful to have enantiomeric 1-alkyl-3-acyl- and 1-acyl-3-alkyl-*sn*-glycerols available. Methods for preparing these compounds have been discussed by Baumann (1972) and by Palameta and Kates (1966).

7.4.7 Enantiomeric 1,2- and 2,3-diacyl-*sn*-glycerols

These compounds have received considerable attention because of their importance in biosynthetic pathways and as precursors for the synthesis of enantiomeric phospholipids (cf. sections 7.5 and 11.2). There are several procedures, which involve blocking the *sn*-3-hydroxyl in 1,2-isopropylidene-*sn*-glycerol with various groups, followed by cleavage of the ketal, reacylation with specific fatty acids and then removal of the blocking group (Jensen, 1972).

Sowden and Fischer (1941) were the first to prepare optically active monoacid 1,2-diacyl-*sn*-glycerols. They blocked the free *sn*-3-hydroxyl group in the 1,2-isopropylidene-*sn*-glycerol with a benzyl group. The isopropylidene group was removed by acid hydrolysis and the free hydroxyls could then be acylated. Since the benzyl group was removed by catalytic hydrogenation, the method could only be used for saturated glycerides. When 1-benzyl-*sn*-glycerol was prepared as an intermediate, 2,3-diacyl-*sn*-glycerols could be synthesized.

Baer and Buchnea (1958) adapted the above procedure for the preparation of monoenoic-acid-containing glycerides. In their modification the *cis* double bonds were protected by bromine and regenerated when required with zinc dust in diethyl ether. Small amounts of the *trans* isomer also formed were removed by crystallization.

In a further modification, Buchnea and Baer (1960) published a method to permit preparation of mixed-acid saturated and saturated/mono-unsaturated 1,2- and 2,3-diacyl-*sn*-glycerols. The benzyl-*sn*-glycerol intermediates are tritylated to block the remaining primary hydroxyl. After acylation, the trityl group is removed under acid conditions when the acyl group migrates from the 2 position to the *sn*-1 or *sn*-3 position (depending on the benzyl ether selected). Acylation of the free 2 position is then followed by removal of the benzyl group by catalytic hydrogenation.

The benzyl-*sn*-glycerol was converted to 1,2-carbonate-3-benzyl-*sn*-glycerol with phosgene (Gigg and Gigg, 1967a). After removal of the benzyl group, reaction with dihydropyran was initiated followed by removal of the carbonate group with KOH to yield 3-tetrahydropyranyl-*sn*-glycerol. Acylation at the 1 and 2 positions was followed by pancreatic lipase treatment to remove the acyl group from the *sn*-1 position, which could then be re-acylated with a different acyl group. The protective tetrahydropyranyl group was removed with boric acid. If 1-benzyl-*sn*-glycerol was used, then a mixed-acid 2,3-diacyl-*sn*-glycerol could be prepared (Gigg and Gigg, 1967b).

Perhaps the most useful method for the synthesis of mixed-acid saturated and unsaturated 1,2- and 2,3-diacyl-*sn*-glycerols is that of Buchnea (1971, 1978). Carefully defined enantiomeric monoacyl-*sn*-glycerols are used. These are tritylated to produce monoacyltrityl-*sn*-glycerols, which can then be acylated in the free 2 position. The resultant 1,2- or 2,3-diacyltrityl-*sn*-glycerols are then purified by chromatography on silicic acid–boric acid columns, when the trityl group is removed. Details of yields and compounds prepared are given by Buchnea (1978).

1,2-Diacyl-*sn*-glycerols have also been made by acylation of 3-iododeoxy-*sn*-glycerols. Trityl protective groups were used and the iodo group was removed by replacing it with a phosphorylcholine moiety, which was hydrolysed by phospholipase C (cf. Haverkate and van Deenen, 1965). Stoffel and Pruss (1969) have used the procedure to make diacylglycerols containing highly unsaturated groups such as arachidonate. Buchnea (1978) has suggested that the method could be used to prepare 2,3-diacyl-*sn*-glycerols because *Bacillus cereus* phospholipase C is not stereospecific.

Buchnea (1978) has also given details of the preparation of 1-alkyl-2-acyl- and 1-alkenyl-2-acyl-*sn*-glycerols. 1,2-Dialkyl-*sn*-glycerols have also been synthesized by Kates *et al.* (1963) using alkyl bromides and 3-benzyl-*sn*-glycerol. Mixed dialkyl-*sn*-glycerol ethers were made by Baumann and Mangold (1966) by alkylation of 1-alkyl-3-trityl-*sn*-glycerol with alkyl methanesulphonates. In addition, optically active 2,3-dialkyl-*sn*-glycerols were synthesized from 3-alkyl- or 3-alkenyl-*sn*-glycerols (cf. Palameta and Kates, 1966).

7.4.8 Enantiomeric triacylglycerols

Diacid triacyl-*sn*-glycerols are made from appropriate enantiomeric monoacyl-*sn*-glycerols. The original method is that of Baer and Fischer (1939). For a more recent publication see Gronowitz *et al.* (1975). Diacid triacyl-*sn*-glycerols were also prepared by Sowden and Fischer (1941) using monoacid 1,2- and 2,3-diacyl-*sn*-glycerols.

Mixed-triacid triacyl-*sn*-glycerols have been made by Lok *et al.* (1976) using 1,3-diacyl-*sn*-glycerols. By the use of appropriate mixed-acid 1,2-, 2,3- and 1,3-diacyl-*sn*-glycerols, all varieties of stereospecifically substituted triacyl-*sn*-glycerols can be prepared (cf. Buchnea, 1978).

Enantiomeric triacylglycerols can also be obtained from glycidol (section 7.4.4).

7.4.9 General notes on synthetic procedures

Jensen and Pitas (1976) have provided useful general notes concerning the synthesis, storage and analysis of glycerides. These points are also discussed by Kates (1972) and Christie (1982). Special points of note include choice of glassware apparatus, sources of fatty acids, purification of solvents, storage of products and methods of crystallization.

General instructions for TLC are given in Kates (1972) and Christie (1982). Useful solvent systems for use with silica gel G plates are listed by Jensen and Pitas (1976). For assay of polar contaminants, the acylglycerols should be separated with petroleum ether (35–45 °C fraction)/diethyl ether/acetic acid (75/25/1, by volume). Non-polar contaminants can be separated using a petroleum ether (35–45 °C fraction)/diethyl ether (94/6, v/v) solvent. Compounds can be conveniently visualized with Bromothymol Blue spray followed by exposure to ammonia vapour (Mangold, 1961). Mono- and diacylglycerols may also have to be separated using more polar solvent mixtures. Two such mixtures are chloroform/acetone/acetic acid/methanol (151/50/1/4, by volume) and chloroform/acetone (24/1, v/v) using 95% silica gel G/5% boric acid plates; 1% or less of contaminating materials can be detected by TLC.

Fatty acid composition and quantification can be carried out most conveniently by GLC. Such compositions should be expressed as a mole percentage to allow accurate comparisons with molecules of different molecular weights. Again, the methods are covered well by Kates (1972) and Christie (1982).

Verification of the structure of the final glyceride is provided by a variety of chemical, enzymic and physical methods. Enzyme hydrolysis is particularly useful because of the stereospecific nature of the attack (see section 6.2.17).

Procedures for pancreatic lipase use are described by Jensen and Pitas (1976), and for lipoprotein, lingual and hepatic lipases by Paltauf *et al.* (1974). Stereospecific analysis of fatty acids in triacylglycerols has been fully discussed by Breckenridge (1978), and more recently by Christie (1992).

Various sensitive physical techniques can be used for demonstrating such features as the asymmetry of diacid triacylglycerols.

All natural triacylglycerols so far examined have been found to be optically active. The optical rotations measured with the sodium D line vary with the concentration of the diacylglycerol and the nature of the solvent. Morris (1965a, b) has demonstrated that a higher optical activity may be recorded by simple polarimetry for diacylglycerols than for triacylglycerols. Therefore, brief random digestion with pancreatic lipase is carried out, and the diacylglycerol products are isolated and converted to trimethylsilyl ethers. Each 1,2- and 2,3-diacyl-*sn*-glycerol mixture is then separated by TLC on the basis of the fatty acids in positions *sn*-1 and *sn*-3 and their optical rotations determined.

The rotatory dispersion and circular dichroism spectra of enantiomeric acylglycerols have been studied by Gronowitz *et al.* (1975). Some of the data from their experiments are discussed by Buchnea (1978). Again, as with polarimetry, there are distinct solvent effects – the highest rotations being obtained with benzene and chloroform.

High-frequency NMR has been used in the analysis of enantiomeric acylglycerols with the use of signal shift reagents, tris(3-heptafluorobutyryl-*d*-camphorato)europium(III) being particularly useful. The greatest separation of proton signals is obtained when there is a large difference between the fatty acids in the *sn*-1 and *sn*-3 positions. Bus *et al.* (1976) have described enantiomeric shift differences of proton signals from various acylglycerols in $CDCl_3$. Another physical technique, that of X-ray diffraction, has also given a limited amount of information about acylglycerols. For example, Schlenk (1965a, b) has developed a method that permits configurational types of acylglycerols to be distinguished using reference acylglycerols as standards.

High-resolution ^{13}C NMR spectroscopy also provides useful insights into the structure of glycerol esters (section 9.3.4).

References and further reading

Baer, E. (1945) *J. Am. Chem. Soc.*, **67**, 338.

Baer, E. (1952) in *Biochemical Preparations*, vol. 2 (ed. E.G. Ball), Wiley, New York, p. 31.

Baer, E. and Buchnea, D. (1958) *J. Biol. Chem.*, **230**, 447.

Baer, E. and Fischer, H.O.L. (1939) *J. Biol. Chem.*, **128**, 475.
Baer, E. and Fischer, H.O.L. (1945a) *J. Am. Chem. Soc.*, **67**, 2031.
Baer, E. and Fischer, H.O.L. (1945b) *J. Am. Chem. Soc.*, **67**, 944.
Barry, P.J. and Craig, B.M. (1955) *Can. J. Chem.*, **33**, 716.
Baumann, W.J. (1972) in *Ether Lipids* (ed. F. Snyder), Academic Press, New York, p. 51.
Baumann, W.J. and Mangold, H.K. (1964) *J. Org. Chem.*, **29**, 3055.
Baumann, W.J. and Mangold, H.K. (1966) *J. Org. Chem.*, **31**, 498.
Bhati, A., Hamilton, R.J. and Steven, D.A. (1980) in *Fats and Oils: Chemistry and Technology* (eds R.J. Hamilton and A. Bhati), Applied Science, London, p. 59.
Bourne, E.J., Stacey, M., Tatlow, T.C. and Tedder, J.M. (1949) *J. Chem. Soc.*, 2976.
Breckenridge, C.W. (1978) *Handbook of Lipid Research*, vol. 1 (ed. A. Kuksis), Plenum, New York, p. 197.
Buchnea, D. (1967a) *Chem. Phys. Lipids*, **1**, 113.
Buchnea, D. (1967b) *Chem. Phys. Lipids*, **1**, 177.
Buchnea, D. (1971) *Lipids*, **6**, 734.
Buchnea, D. (1978) in *Handbook of Lipid Research*, vol. 1 (ed. A. Kuksis), Plenum, New York, p. 233.
Buchnea, D. and Baer, E. (1960) *J. Lipid Res.*, **1**, 405.
Burgos, C.E., Ayer, D.E. and Johnson, R.A. (1987) *J. Org. Chem.*, **52**, 4973.
Bus, J., Lok, C.M. and Groenewengen, A. (1976) *Chem. Phys. Lipids*, **16**, 123.
Christie, W.W. (1982) *Lipid Analysis*, 2nd edn, Pergamon, Oxford.
Christie, W.W. (1992) in *Advances in Lipid Methodology*, (ed. W.W. Christie), Oily Press, Dundee, Scotland, p. 121.
Cunningham, J. and Gigg, R. (1965) *J. Chem. Soc.*, 1553.
Eibl, H. and Woolley, P. (1986) *Chem. Phys. Lipids*, **41**, 53.
Eibl, H. and Woolley, P. (1988) *Chem. Phys. Lipids*, **47**, 47.
Fischer, H.O.L. and Baer, E. (1937) *Naturwissenschaften*, **25** (3b), 588.
Gigg, J. and Gigg, R. (1967a) *J. Chem. Soc.*, 431.
Gigg, J. and Gigg, R. (1967b) *J. Chem. Soc.*, 1865.
Golding, B.T. (1988) *Chem. Ind.*, 617.
Gronowitz, S., Herslöf, B., Ohlson, R. and Toregard, B. (1975) *Chem. Phys. Lipids*, **14**, 174.
Gunstone, F.D. and Norris, F.A. (1983) in *Lipids in Foods*, Pergamon, Oxford, p. 85.
Hartman, L. (1959) *J. Chem. Soc.*, 4134.
Haverkate, F. and van Deenen, L.L.M. (1965) *Biochim. Biophys. Acta*, **106**, 78.
Ingold, C.K. (1973) quoted in Morrison, R.T. and Boyd, R.N. (1973) *Organic Chemistry*, 3rd edn, Allyn and Bacon, Boston, MA, p. 459.
Iwama, T. and Foglia, T.A. (1988) *J. Am. Oil Chem. Soc.*, **65**, 954.
Jensen, R.G. (1972) *Top. Lipid Chem.*, **3**, 1.
Jensen, R.G. and Pitas, R.E. (1976) in *Advances in Lipid Research* (eds R. Paoletti and D. Kritchevsky), Academic Press, New York, p. 213.
Jensen, R.G., Marks, T.L., Sampugna, J., Quinn, J.G. and Carpenter, D.L. (1966) *Lipids*, **1**, 451.
Karnovsky, M.L. and Wolff, D. (1960) in *Fifth International Conference on Biochemistry of Lipids* (ed. G. Popjak), Pergamon, New York.
Kates, M. (1972) *Techniques in Lipidology*, Elsevier, Amsterdam.
Kates, M., Chan, T.H. and Stanacev, N.Z. (1963) *Biochemistry*, **2**, 394.
Kodali, D.R. (1987) *J. Lipid Res.*, **28**, 464.
Krabisch, L. and Borgstrom, B. (1965) *J. Lipid Res.*, **6**, 156.
Lands, W.E.M. and Zschocke, A. (1965) *J. Lipid Res.*, **6**, 324.
Lands, W.E.M., Pieringer, R.A., Slakey, S.P.M. and Zschocke, A. (1966) *Lipids*, **1**, 444.
Lok, C.M., Ward, J.P. and van Dorp, D.A. (1976) *Chem. Phys. Lipids*, **16**, 115.
Mangold, H.K. (1961) *J. Am. Oil Chem. Soc.*, **38**, 708.
Mattson, F.H. and Volpenheim, R.A. (1962) *J. Lipid Res.*, **3**, 281.
Mattson, F.H., Volpenheim, R.A. and Martin, J.B. (1964) *J. Lipid Res.*, **5**, 374.
Mazur, A.W., Miller, G.D., Lee, S.S.C., Armstrong, M.P. and Wendel, J.D. (1991) *Chem. Phys. Lipids*, **60**, 189.
Morley, N.H., Kuksis, A. and Buchnea, D. (1974) *Lipids*, **9**, 481.
Morris, L.J. (1965a) *Biochem. Biophys. Res. Commun.*, **20**, 340.
Morris, L.J. (1965b) *Biochem. Biophys. Res. Commun.*, **18**, 495.
Moss, R.A., Scrimin, P., Bhattacharya, S. and Chatterjee, S. (1987) *Tetrahedron Lett.*, **28**, 5005.
Palameta, B. and Kates, M. (1966) *Biochemistry*, **5**, 618.
Paltauf, F., Esfandi, F. and Holasek, A. (1974) *FEBS Lett.*, **40**, 119.
Pfeiffer, F.R., Cohen, S.R., Williams, K.R. and Weisbach, A. (1968) *Tetrahedron Lett.*, **32**, 3549.
Quinn, J.G., Sampagna, J. and Jensen, R.G. (1967) *J. Am. Oil Chem. Soc.*, **44**, 439.
Schlenk, W. (1965a) *J. Am. Oil Chem. Soc.*, **42**, 945.
Schlenk, W. (1965b) *Angew. Chem.*, **77**, 161.
Sonnet, P.E. (1991) *Chem. Phys. Lipids*, **58**, 35.
Sowden, J.C. (1962) *Methods Carbohydr. Chem.*, **1**, 132.
Sowden, J.C. and Fischer, H.O.L. (1941) *J. Am. Chem. Soc.*, **63**, 3244.
Stoffel, W. and Pruss, H.D. (1969) *Hoppe-Seyler's Z. Physiol. Chem.*, **350**, 1385.
Tattrie, N.H., Bailey, R.A. and Kates, M. (1958) *Arch. Biochem. Biophys.*, **78**, 319.
Thomas, A.E., Scharoun, J.E. and Ralston, H. (1965) *J. Am. Oil Chem. Soc.*, **42**, 789.
Turner, D.L., Silver, M.J., Holburn, R.R. and Baczynski, E. (1968) *Lipids*, **3**, 228.
Verkade, P.E. (1953) *Chim. Ind. (Paris)*, **69**, 239.
Wright, W.R. and Tove, S.B. (1967) *Biochim. Biophys. Acta*, **137**, 54.

7.5 Glycerophospholipids

7.5.1 General comments

Two excellent reviews dealing with phospholipid synthesis are those of Rosenthal (1975) and Slotboom and Bonsen (1970). Other more recent references that deal with general aspects include a discussion of the effect of solvents and catalysts on acylation (Mangroo and Gerber, 1988), and the synthesis of phospholipids using a phosphotriester approach (Woolley and Eibl, 1988) and uniformly cleavable β-hydrogenated protecting groups (Lemman *et al.*, 1990). Stepanov and Slivets (1986) describe the formation of phosphoester bonds in phosphoglyceride synthesis and include discussion of various methods of phosphorylation used for phosphomono-, -di- and -triester formation. Particular attention was paid to a comparison of the chlorophosphate versus silver salt methods.

Several papers deal with the preparation of derivatives of a number of phospholipids. An example of the synthesis of spin-labelled phospholipids is that of Lai *et al.* (1988), while Reynolds *et al.* (1991) describe the preparation of synthetic phospholipase substrates with *p*-nitrophenol linked to the *sn*-2 moiety of the phospholipid. Yu and Dennis (1991) detail the preparation of chiral thiophospholipids. Isaacson *et al.* (1990) describe the synthesis of phospholipids with hydroxylated fatty acids at the *sn*-2 position, while Lebeau *et al.* (1988) have synthesized phospholipids linked to steroid hormone derivatives.

7.5.2 Phosphatidic acids

The first syntheses of phosphatidic acid used either 1,2-diacylglycerols or 1,2-diacyloxy-3-iodopropanes. Unfortunately, 1,2-diacylglycerols are inherently unstable, so the 1,2-diacyloxy-3-iodopropanes have been used more frequently and the phosphoric acid moiety introduced by means of a disubstituted silver phosphate containing protecting groups. A second method uses α-glycerophosphoric acid as starting material (Rosenthal, 1975).

(a) Method 1

In the synthesis of phosphatidic acids from 1,2-diacyloxy-3-iodopropanes, the first step is to synthesize isopropylidene-D-glycerol and convert that into 3-iododeoxy-*sn*-glycerol. The general procedure is outlined in Fig. 7.6 and is summarized by Rosenthal (1975). Specific references to individual steps are shown on the figure.

The 3-iododeoxy-*sn*-glycerol can then be acylated in a stepwise manner to give iododeoxy mixed-acid diacylglycerol with almost any desired combination of acids at the 1 and 2 positions. The procedure of de Haas and van Deenen (1961) is most often used, and is described in detail by Rosenthal (1975). The 1 position is acylated first using an acyl chloride, and any unwanted iododeoxydiacylglycerol removed during recrystallization. The 1-acyl-3-iododeoxyglycerol, which is light-sensitive, is then acylated using a second acyl chloride.

Silver di-t-butyl phosphate is used to introduce the phosphate moiety. It is synthesized from di-t-butyl phosphate via barium di-t-butyl phosphate as modified by de Haas and van Deenen (1961). The silver salt, generated using silver sulphate, is condensed with the diacyliododeoxyglycerol (Bonsen and de Haas, 1967). Labile t-butyl esters are then

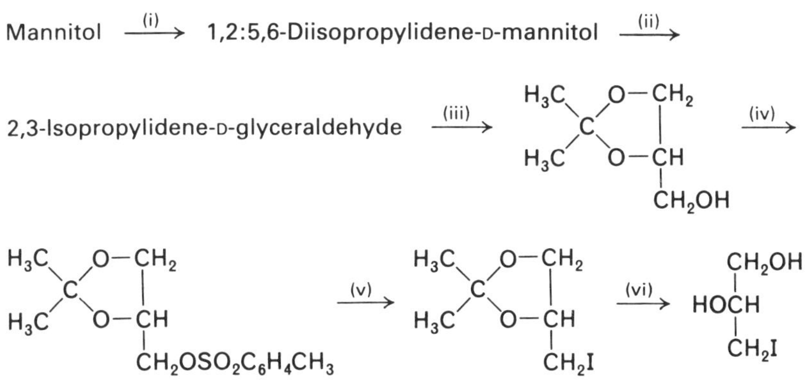

Reagents: i, $CO(CH_3)_2$, $ZnCl_2$; ii, $NaIO_4$; iii, $NaBH_4$; iv, $CH_3C_6H_4SO_2Cl$, pyridine; v, NaI; vi, H_3O^+

Fig. 7.6 Synthesis of 3-iododeoxy-*sn*-glycerol (Rosenthal, 1975).

removed and the phosphatidate product crystallized as the barium salt, from which the free acid or other salts can be generated.

(b) Method 2

In this procedure, α-glycerol phosphate is acylated directly using acyl anhydrides (Lapidot *et al.*, 1969). The acyl anhydrides are prepared with dicyclohexylcarbodiimide (Rosenthal, 1975). Optically active or racemic phosphatidic acids can be prepared and either saturated or unsaturated acyl residues added. The method depends on the ability of the fatty acid salt to suppress the formation of mixed phosphoric–carboxylic anhydrides with the consequent phosphorylation of adjacent hydroxy groups. The use of acyl anhydrides rather than acyl chlorides probably prevents the formation of glycerol chlorhydrin esters (Aneja and Chadha, 1971a).

Various other derivatives such as ether-containing and glycerol analogues of phosphatidic acid have been synthesized (Rosenthal, 1975). Methods for the synthesis of chiral vinylic phosphonolipid (Schwartz *et al.*, 1989a) and α-hydroxyphosphonolipid analogues of phosphatidic acid (Schwartz *et al.*, 1989b) have been described.

7.5.3 Phosphatidylcholines

The two general methods for the chemical synthesis of phosphatidylcholine (lecithin) involve the use of, respectively, phosphatidic acid or glycerylphosphorylcholine.

(a) Synthesis of phosphatidylcholine from phosphatidic acid

Readily available phosphatidic acids (see section 7.5.2) can be esterified with a suitable choline salt. Suitable condensing agents include carbodiimides, trichloroacetonitrile and sulphonyl chlorides (Rosenthal, 1975). Not only are these procedures relatively straightforward, but they are also suitable for the preparation of phosphatidylcholines radiolabelled in the choline moiety.

The method of Aneja and Chadha (1971b) utilizes triisopropylbenzenesulphonyl chloride and is claimed to give products of better optical purity than other procedures. In this technique the phosphatidic acid is mixed and dried with choline acetate (in twice molar excess). The sulphonyl chloride derivative in dry pyridine is then added, and, after reaction at 70 °C for 1 h, the procedure is continued at room temperature for 4 h. Excess choline acetate can then be filtered off and the product purified, after passage through a mixed-bed resin column, by chromatography on a neutral alumina column.

(b) Partial synthesis of phosphatidylcholine from sn-glycerol-3-phosphorylcholine

sn-Glycerol-3-phosphorylcholine is available from chemical suppliers, but can be prepared from an inexpensive source of phosphatidylcholine (e.g. egg yolk). To do this, selective alcoholysis with methanolic tetrabutylammonium hydroxide solution is employed (Brockerhoff and Yurkowski, 1965). *sn*-Glycero-3-phosphorylcholine can then be esterified by the reaction with the sodium salt of the desired fatty acid together with its anhydride at 70 °C for 48 h. The product is then purified by the use of silicic acid chromatography. Yields of 70% are typical (Cubero Robles and van den Berg, 1969). Such a method will yield phosphatidylcholines containing the same fatty acid at both positions. Should mixed-acid species be required, then these can be generated by action of a phospholipase followed by reacylation of the monoacyl product (Cubero Robles and van den Berg, 1969).

(c) Recent methods for preparing phosphatidylcholines

Recent advances in the chemical synthesis of phosphatidylcholine have been reported for (constrained) short-chain acyl derivatives (Barlow *et al.*, 1988; Roberts, 1991) and for mixed-acid species (Hermetter *et al.*, 1989).

(d) Other phosphatidylcholine derivatives

The synthesis of ether phosphatidylcholines (other than plasmalogens; see section 7.5.9) has been fully described by Rosenthal (1975). A large variety of other phosphate analogues have also been synthesized, and these are reviewed by Bonsen *et al.* (1972). A summary of the various compounds for which methods have been developed is made in Table 7.1.

For the preparation of mixed chain phosphatidylcholine analogues with bulky headgroups or with one acyl chain twice the length of the other, see Ali and Bittman (1989). Schwartz *et al.* (1989a) have described the synthesis of chiral vinylic phosphonolipid analogues, while Moschidis (1987) reported the preparation of 1-*O*-alkylethylene glycol phosphonic

Table 7.1 Phosphatidylcholine analogues (Bonsen *et al.*, 1972)

Type of analogue	Compound
Stereoisomers of normal PC	1-*sn*-Phosphatidylcholine
Structural isomer of normal isomer	*rac*-1,3-Diacylglycerol-2-phosphorylcholine
Modification of acyl chain at the 2 position of PC	2-Methyl derivatives of acyl chain 2,2-Dimethyl derivatives of acyl chain 3,3-Dimethyl derivatives of acyl chain *rac*-1,2-Dibenzoylglycero-3-phosphorylcholine
Modification of ester linkages at the 2 position of PC	*rac*-1-Acyl-2-acylamido-2-deoxyglycero-3-phosphorylcholine 1-Acyl-2-alkanesulphonyl-*sn*-glycero-3-phosphorylcholine *rac*-1-Acyl-2-alkylglycerol-3-phosphorylcholine *rac*-1-Acyl-2-alkyl-2-deoxyglycero-3-phosphorylcholine
Compounds without an acyl group at the 2 position of PC	1-Acyl-*sn*-glycero-3-phosphorylcholine 1-Acyl-2-deoxyglycero-3-phosphorylcholine
Modification of glycerol backbone	1-Acylglycol-2-phosphorylcholine *rac*-1,2-Diacylbutanetriol-4-phosphorylcholine 2,2-(Diacylhydroxymethyl)propanol-1-phosphorylcholine
Modification of the phosphate group	1,2-Diacyl-3-deoxy-*sn*-glycero-3-phosphorylcholine *rac*-1,2-Diacyl-3-deoxyglycero-3-sulphonic acid

acid analogues of phosphatidylcholine. Thiol ester and thiol ether analogues are described by Hendrickson and Hendrickson (1990), while the formation of phosphatidylcholine labelled with 8-(2-anthroyl)-octanoic acid, for use as a fluorescent probe, is detailed by Perochon and Tocanne (1991). General methods for fluorophore-labelled phospholipid substrates, including derivatives of phosphocholine, are given by Hendrickson (1991).

7.5.4 Phosphatidylethanolamines

Phosphatidylethanolamine is usually synthesized from phosphatidic acid or from iodohydrin diesters. The analogous route to that of phosphatidylcholine from glycerophosphorylcholine is not usually used because it is difficult to obtain the substrate in a sufficiently pure state. Care must be taken during the synthesis of phosphatidylethanolamine to protect the unsubstituted amino function.

(a) *Synthesis of phosphatidylethanolamine from phosphatidic acid*

The first problem encountered in this route is the blocking of the primary amino group by a reagent that can be removed easily and under conditions that are not accompanied by ester hydrolysis or fatty acyl modification. The trityl group can be used for this purpose. In a typical procedure (Aneja *et al.*, 1970) *N*-tritylethanolamine is condensed with a phosphatidic acid by means of 2,4,6-triisopropylbenzene-sulphonyl chloride. The reaction is fast and, after heating for 1 min at 60 °C to dissolve the phosphatidic acid, proceeds in 90% yield at room temperature in 1.5 h. The product is purified by triethylamine-impregnated silicic acid chromatography. Removal of the trityl group can be accomplished by hydrogenolysis (not appropriate for unsaturated acyl groups) or by very mild acid hydrolysis (Rosenthal, 1975).

Alternative condensing agents to sulphonyl chlorides include carbodiimides and trichloro-acetonitrile.

(b) *Synthesis of phosphatidylethanolamine from α-iododeoxyglycerol diesters*

This method is especially attractive for the synthesis of mixed-acid phosphatidylethanolamines because of the ease of preparation of specific diesters of α-iododeoxyglycerol. The phosphorylethanolamine function is introduced via the silver salt of a substituted derivative. As in the preparation of phosphatidylethanolamine from phosphatidic acid, it is essential that the protective groups can be easily removed. Silver t-butyl(*N*-t-butyloxycarbonyl-2-aminoethyl) phosphate is the most useful reagent because it allows the simultaneous removal of both phosphoester and amino-protecting groups under conditions that do not affect unsaturation or ester groups (Rosenthal, 1975). Yields of about 50% for both the preparation of the silver salt of the substituted phosphorylethanolamine and for the condensation with the iododeoxyglycerol can be expected.

(c) Synthesis of phosphatidylethanolamine from glycerylphosphorylethanolamine

As mentioned above, this method is not often used. The amino nitrogen is blocked with phthaloyl or trityl groups before acylation to yield a phosphatidylethanolamine with two identical fatty acids. Mixed-acid phosphatidylethanolamines can be synthesized by deacylation (phospholipase A)–reacylation (Slotboom and Bonsen, 1970).

(d) Synthesis of phosphatidylethanolamine from diacylglycerols

Two approaches have been made here: the phosphorylation of diacylglycerols with appropriate N-substituted derivatives of phosphorylethanolamine, or of isopropylideneglycerols with the N-substituted derivative followed by removal of the isopropylidene group and acylation (Slotboom and Bonsen, 1970).

(e) Recent methods

Hermetter *et al.* (1987) have described a simple method for the preparation of mixed-acid phosphatidylethanolamines using 1-*O*,*N*-ditritylglycerophosphoethanolamine. The preparation of phosphatidylethanolamines and their dimethyl derivatives is reported by Calderon and Yague (1986).

(f) Other phosphatidylethanolamine derivatives

The synthesis of ether phosphatidylethanolamines is analogous to that for the formation of the corresponding phosphatidylcholines (Rosenthal, 1975). The *N*-methyl and *N*,*N*-dimethyl derivatives have been synthesized from their corresponding phosphatidic acids (Aneja *et al.*, 1970). See also Calderon and Yague (1986) for the preparation of *N*,*N*-dimethylphosphatidylethanolamines. The formation of *N*-acylphosphatidylethanolamines and other analogues is detailed by Slotboom and Bonsen (1970) and that of ethanolamine plasmalogens in section 7.5.9. Synthesis of 1-palmitoyl-2-diphenylhexatrienylpropanoylphosphatidylethanolamine for use as a fluorescent probe is given by Beck *et al.* (1990), while dialkylphosphatidylethanolamines are detailed by Ruocco and Makriyannis (1989). Moschidis (1987) described the preparation of 1-*O*-alkylethylene glycol phosphonic acid analogues of phosphatidylethanolamine.

7.5.5 Phosphatidylserines

Three basic methods can be used, as now discussed.

(a) Synthesis of phosphatidylserine from phosphatidic acid

One simple procedure allows the synthesis of phosphatidylserine by condensation of phosphatidic acid with an amino- and carboxy-protected serine. *N*-Carbobenzoxyl-DL-serine benzyl ester was condensed with the phosphatidic acid in the presence of triisopropylbenzenesulphonyl chloride. The protecting groups were then removed by hydrogenation, which limits the method to saturated phosphatidylserines. However, the use of different protecting groups should allow this method to be used for unsaturated compounds. An alternative procedure involves the introduction of the phosphate and serine functions via a complex silver salt to glycerol iodohydrin diesters (de Haas *et al.*, 1964).

(b) Synthesis of phosphatidylserine from substituted glycerophosphorylserine

1,2-Isopropylideneglycerol is treated successively with phosphorus oxychloride and *N*-phthalylserine. Hydrolysis of the ketal and the phosphomonochloride groups then yields glycero-3-phosphoryl-*N*-phthalylserine as the barium salt after neutralization with barium hydroxide. Acylation with an excess of acid chloride then furnishes the desired phosphatidylserine. In addition, the acylation can be accomplished in a stepwise manner to give mixed-acid phosphatidylserines (Shvets *et al.*, 1967).

(c) Use of phosphotriester intermediate

Woolley and Eibl (1988) have described the preparation of enantiomerically pure phosphatidylserines. They used a procedure that is applicable to the formation of many phospholipids. Their method was to make an appropriate diacylglycerol (or dialkylglycerol), to react this with phosphorus oxychloride, then an alkyl donor and finally use methanolysis to create a phosphate triester. The nature of the alkyl donor determined the headgroup of the final phospholipid, while demethylation was used at a late stage. In the case of phosphatidylserine the yield was 95%.

(d) Compounds related to phosphatidylserine

Phosphatidylthreonine has been prepared by a method involving phosphorylation of isopropylideneglycerol with phosphorus oxychloride and a sulphonyl chloride-activated condensation (Moore and Szelke, 1970).

7.5.6 Phosphatidylglycerols

Phosphatidylglycerol can be synthesized using 2,3-dibenzyl-*sn*-glycerol as one of the starting materials (Saunders and Schwartz, 1966). The blocking groups must be removed by catalytic hydrogenolysis, so the method can only be used for saturated molecular species. Furthermore, the use of phosphorus oxychloride also has disadvantages (Slotboom and Bonsen, 1970).

Of the other reported syntheses of phosphatidylglycerol, the method of Bonsen *et al.* (1966) is the most widely used. It allows preparation of saturated or unsaturated mixed-acid and same-acid phosphatidylglycerols of any configuration. It should be borne in mind that stereochemistry is especially important for this phospholipid since in the natural compound the unacylated glycerol has the *sn*-1′ configuration. Therefore, 2,3-isopropylidene-*sn*-glycerol is a starting material (Fig. 7.7). The critical stage is the coupling of a silver salt and an iodo derivative. For various reasons (Slotboom and Bonsen, 1970), 1,2-diacyl-3-iododeoxy-*sn*-glycerol and the silver salt of *sn*-glycerol-1-phosphate, containing appropriate blocking groups, were used.

Dibenzylphosphoric acid and dicyclohexylcarbodiimide were used to generate tetrabenzyl phosphate. This could then be reacted with 2,3-isopropylideneglycerol in the presence of imidazole to yield 2,3-isopropylidene-*sn*-glycerol-1-dibenzyl phosphate. Debenzylation of the latter was carried out with barium iodide and the barium salt was converted to the silver salt, which was reacted with 3-iododeoxy-1-oleoyl-2-palmitoyl-*sn*-glycerol. The fully protected phosphatidylglycerol was debenzylated, followed by reaction with boric acid in trimethyl borate, which effected removal of the isopropylidene group (Rosenthal, 1975). Yields of 30% are obtained (Slotboom and Bonsen, 1970).

Rosenthal (1975) provides notes on certain parts of the preparative procedure. This includes preparation of dibenzylphosphoric acid and of 2,3-isopropylidene-*sn*-glycerol.

The preparation of enantiomerically pure phosphatidylglycerol is described by Woolley and Eibl (1988), while Calderon and Yague (1988) reported the preparation of phosphatidylglycerols from diacylglycerols.

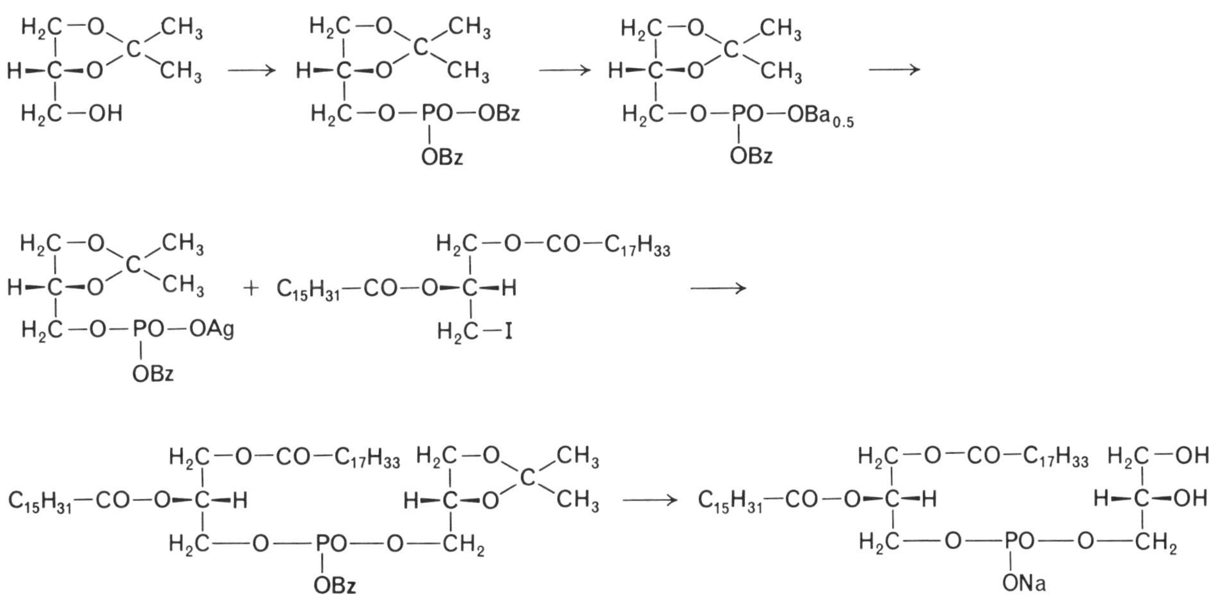

Fig. 7.7 Synthesis of a mixed-acid phosphatidylglycerol. Bz = benzyl. Scheme taken from Slotboom and Bonsen (1970) with permission.

Halophilic bacteria contain a diphytanyl analogue of phosphatidylglycerol with the opposite stereochemical configuration of the diphytanylglycerol moiety. Using diphytanylglycerol (from a natural source) as starting material, Joo and Kates (1969) prepared the iodo derivative. Condensation with the silver salt of di-*p*-nitrobenzyl phosphate was followed by debenzylation to give the silver salt of 2,3-diphytanyl-*sn*-glycerol-1-(mono-*p*-nitrobenzyl) phosphate. The latter was reacted with 1-benzyl-3-iododeoxy-2-t-butyl-*sn*-glycerol or its enantiomer, and the protecting groups were released by catalytic hydrogenolysis and then treatment with hydrogen chloride. Careful optical rotation measurements established that the natural compound was 1-*sn*-phosphatidyl-3′-*sn*-glycerol (i.e. both glycerol moieties had the opposite configurations to the glycerol moieties of the diacyl phospholipid from other bacteria or from plants). For a fuller discussion of diether structural analogues of phosphatidylglycerol, phosphatidylglycerophosphate and phosphatidylglycerol sulphate, see Kates (1978).

Methods have been reported for the synthesis of aminoacyl esters (Bonsen *et al.*, 1965) and glucosaminyl esters (Verheij *et al.*, 1971) of phosphatidylglycerol.

Phosphatidylglycerol phosphate was synthesized by Bonsen and de Haas (1967). 1,3-Diiodo-2-*O*-t-butylglycerol was reacted with the silver salt of 1,2-diacyl-*sn*-glycerol-3-(monobenzyl) phosphate to produce the mono condensation product. This was reacted with the silver salt of di-t-butyl phosphate, and then anionic debenzylation and treatment with hydrogen chloride yielded 3-*sn*-phosphatidyl-1′-*rac*-glycero-3′phosphate.

The diether analogue of phosphatidylglycerol phosphate, which occurs in halobacteria, was synthesized by Joo and Kates (1968, 1969).

7.5.7 Diphosphatidylglycerols (cardiolipins)

1,3-Diphosphatidylglycerol appears to be the structure of natural cardiolipin, but this has only been proven in a few cases (Slotboom and Bonsen, 1970). Some derivatives of diphosphatidylglycerol have also been isolated from various tissues. The lipid can be isolated in high yields from beef heart using chromatography (cf. Slotboom and Bonsen, 1970) or by recrystallization as the barium salt followed by chromatography on ammonia-treated silicic acid (Takahashi *et al.*, 1967).

Several methods of synthesizing diphosphatidylglycerol have been described. De Haas and van Deenen (1965) reacted 1 mol of 1,3-diiodo-2-*O*-t-butyloxypropane with 2 mol of the silver salt of 1-stearoyl-2-oleoyl-*sn*-glycerol-3-(monobenzyl) phosphate. Anionic debenzylation and then hydrogen chloride successively released the blocking groups (see Fig. 7.8; Rosenthal, 1975).

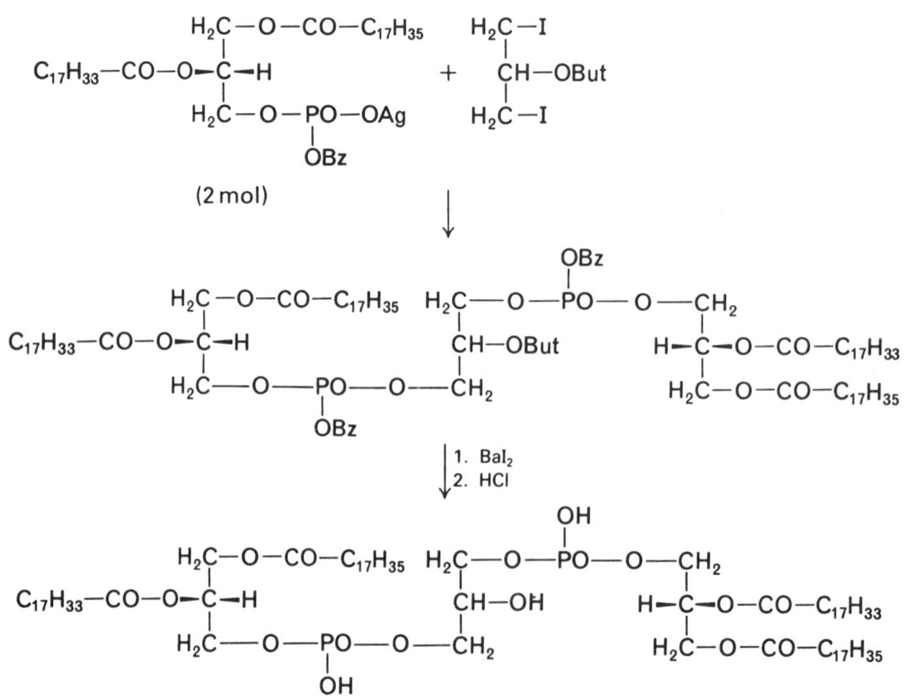

Fig. 7.8 Synthesis of mixed-acid unsaturated diphosphatidylglycerol by the method of de Haas and van Deenen (1965). Taken from Slotboom and Bonsen (1970) with permission.

Inoue and Nojima (1968) used a similar method for the synthesis of racemic diphosphatidylglycerol. These workers used a benzyl protecting group instead of the t-butyl group and, therefore, had to use catalytic hydrogenolysis for its removal. Consequently, their method can be used only for saturated molecular species.

Saunders and Schwartz (1966) used 1,2-distearoyl-*sn*-glycerol, phosphorus oxychloride and 2-benzylglycerol as starting materials. Because they also removed the benzyl group by catalytic hydrogenolysis, only saturated derivatives could be made.

Acyl derivatives of diphosphatidylglycerol have been prepared by de Haas and van Deenen (1963) and by Shvets *et al.* (1967). A series of diphosphatidylalkanediols have been synthesized by Inoue and Nojima (1968). Moschidis (1988) has described the preparation of phosphono analogues of diphosphatidylglycerol, while Fowler *et al.* (1988) reported the synthesis of various photoreactive analogues.

7.5.8 Phosphatidylinositols and other inositol lipids

The asymmetry of the inositol moiety poses a considerable problem for the synthetic chemist. A synthetic asymmetrically substituted inositol derivative must, therefore, be used, and this should contain an optically active inositol moiety (the natural isomer being *myo*-inositol).

Unsaturated phosphatidylinositol has been synthesized by Molotkovsky and Bergelson (1971) using an optically active glycerol derivative but a racemic *myo*-inositol moiety, which was added from an acetylated inositol phosphate benzyl ester. Klyashchitski *et al.* (1971) prepared a completely optically active dipalmitoylphosphatidylinositol. However, because they used benzyl and phenyl protecting groups, their method is not suitable for unsaturated phosphatidylinositols.

In the procedure of Molotkovsky and Bergelson (1971) *rac*-2,3,4,5,6-penta-*O*-acetyl-*myo*-inositol was first converted to *rac*-1-dibenzylphosphoryl-2,3,4,5,6-penta-*O*-acetyl-*myo*-inositol-1-benzyl phosphate, which in turn was made into a silver salt. The latter was reacted with 3-iododeoxy-1-lauroyl-2-oleoyl-*sn*-glycerol and the resultant triester was debenzylated with sodium iodide. The product was purified, reacted with hydrazine hydrate and the phosphatidylinositol finally purified by chromatography. Care was necessary to avoid deacylation of the product, and the method adopted seemed to avoid the danger of unprotected *myo*-inositol phosphotriesters undergoing positional migration. For details refer to Rosenthal (1975) and Bergelson (1980).

The synthesis of enantiomerically pure lyso- and alkylphosphatidylinositols is described by Filthath and Eibl (1992). A chromatogenic substrate for phosphatidylinositol-specific phospholipase C enzymes has been prepared by Shashidhar *et al.* (1991). A method for the synthesis of isopropylidene derivatives of phosphatidylinositol is given by Noda and Keenan (1990).

The higher inositides, diphosphoinositide (1-(3-*sn*-phosphatidyl)-L-*myo*-inositol 4-phosphate) and triphosphoinositide (1-(3-*sn*-phosphatidyl)-L-*myo*-inositol 3,4-bisphosphate) are isolated from natural sources. Great care must be taken at several stages in order to obtain reasonable yields of material (cf. Schacht, 1981).

Verheij *et al.* (1970) have synthesized phosphatidylglucose, and Lukyanov *et al.* (1965) have made a phosphatidylinositol aminoethyl phosphotriester isomer.

7.5.9 Plasmalogens

Plasmalogens are a group of phosphoglycerides that are phosphorylated derivatives of 1-(alk-*cis*-1-enyl)-2-acyl-*sn*-glycerol. The similarity in properties between plasmalogens and the corresponding 1,2-diacyl phospholipids has made isolation procedures difficult. Moreover, chemical synthesis is made difficult by the presence of both alkali-labile fatty acid ester bonds and a reactive, acid-labile *cis* vinyl ether bond.

The synthesis of 1-alkenylglycerols, 1-alkenyl-2,3-diacylglycerols (neutral plasmalogens) or 1-alkenyl-2-acyl-3-halodeoxyglycerols is detailed by Slotboom and Bonsen (1970) and Paltauf (1983). These syntheses are carried out in five ways: (a) methods involving debromination of cyclic glycerol-(1′-bromoalkyl) acetals; (b) syntheses involving partial reduction of a triple bond; (c) procedures involving thermal elimination of 1,2-diacylglycerol from bis(1,2-diacylglycerol) acetals; (d) methods involving elimination of *p*-toluenesulphonic acid or hydrogen iodide; and (e) syntheses involving elimination of hydrogen from 1-chloroethers.

Synthesis of a choline plasmalogen has been reported by Slotboom *et al.* (1967). An addition reaction between hexadec-1-enyl ethyl ether and *rac*-1,2-dipalmitoylglycerol was catalysed by *p*-toluenesulphonic acid to yield a mixed acetal (Fig. 7.9). Thermal elimination of ethanol in the presence of sulphanilic acid gave a neutral plasmalogen, which was modified so that the palmitic

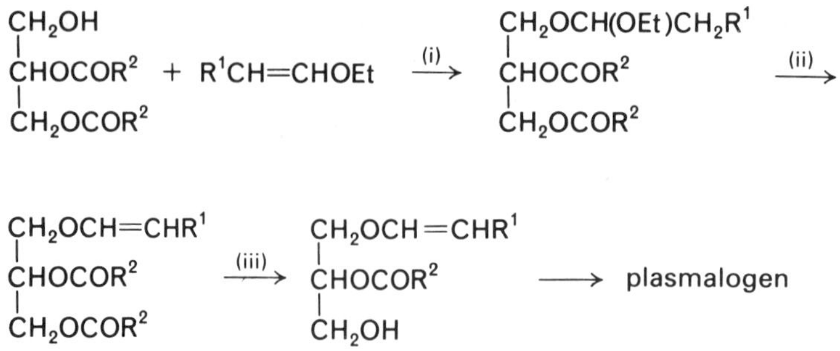

Reagents: i, p-$CH_3C_6H_4SO_3H$; ii, sulphanilic acid (thermal elimination); iii, pancreatic lipase.

Fig. 7.9 Synthesis of choline plasmalogen.

acid at position 2 was replaced with oleic acid. (This was because natural plasmalogens usually contain an unsaturated moiety at position 2). The modified neutral plasmalogen was hydrolysed with pancreatic lipase, which yielded 1-hexadec-1′-enyl-2-oleoylglycerol. (The advantages of using pancreatic lipase, which does not attack vinyl ether linkages, is fully discussed by Slotboom and Bonsen (1970).) The above derivative was then converted to the choline plasmalogen by the method of Hirt and Berchtold (1958).

Radioactive ethanolamine plasmalogen can be conveniently prepared and isolated from protozoan cell cultures (Achterberg *et al.*, 1986a). Radioactive ethanolamine plasmalogen can also be used to prepare dimethylethanolamine and choline analogues by means of a phospholipase D-catalysed reaction (Achterberg *et al.*, 1986b).

A partial synthesis scheme has been used for the formation of 1-(alk-*cis*-1′-enyl)-2-acyl-*sn*-glycero-3-phosphoric acid (Eibl and Lands, 1970).

A phosphotriester derivative of 1-dodec-1′-enyl-2-stearoylglycerol was made by Serebrennikova *et al.* (1969a), but this was not converted into a plasmalogen. However, Serebrennikova *et al.* (1969b) have used 1-alkenyl-2-acyl-3-bromodeoxyglycerol as a starting compound to form the *N,N*-dimethyl derivative of 1-hexadec-1′-enyl-2-stearoylglycero-3-phosphorylethanolamine. See also Rosenthal (1975) for more details of this method.

Chacko *et al.* (1967) have reported the preparation of a synthetic plasmalogen analogue – an acetal phosphatidylcholine. The synthesis of plasmalogens and a number of substituted or double-labelled derivatives has been summarized by Horrocks and Sharma (1982).

Dialkylphosphatidylcholines and -phosphatidylethanolamines can be made by a method detailed by Abdelmageed *et al.* (1989). General methods for the preparation of alkyl ether and vinyl ether substrates for phospholipases are given by Paltauf and Hermetter (1991). These authors also describe the preparation of the biologically active platelet-activating factor (PAF) (1-alkyl-2-acetyl-*sn*-glycero-3-phosphocholine), while Wang and Tai (1990) reported the synthesis of an aldehydic analogue of PAF for use in the preparation of antibodies.

7.5.10 Monoacylphosphoglycerides

Since 1- and 2-monoacylphosphoglycerides are important metabolic intermediates, there has been some interest in their chemical synthesis. In addition, specific phospholipases can be used to prepare either the 1- or 2-monoacyl compounds. Thus, pancreatic lipase attacks the 1 position of a phosphoglyceride to yield the 2-monoacyl product, while snake venom phospholipase removes only the 2-acyl moiety (cf. section 11.2.2). For the chemical methods a major problem is acyl migration. This is particularly rapid for the 2-monoacylphosphoglyceride, since the primary ester is more stable. Therefore, the primary hydroxyl must be protected throughout the synthesis and the blocking group only released at the final stage.

So far, methods are available for saturated monoacylphosphoglycerides. These are discussed by Slotboom and Bonsen (1970). For production of a 1-monoacyl compound, the 2′ position of glycerol can be protected with a benzyl ether group. Slotboom *et al.* (1963, 1967) prepared *rac*-1-stearoyl-lysophosphoglycerides starting from *rac*-1-stearoyl-2-benzyl-3-iododeoxyglycerol (cf. Rosenthal, 1975). Other methods are detailed in Slotboom and Bonsen (1970). In addition, Fujiwara *et al.* (1967) have reported that 80–90% of the 2-acyl ester of phosphatidylcholine can be cleaved with laurylamine or certain alkoxides.

Several methods for the synthesis of 2-monoacylphosphoglycerides are described in Slotboom and Bonsen (1970). These use benzyl or trityl groups to protect the free hydroxyl group and the blocking group is removed by hydrogenolysis. If this is carried out in neutral medium, then acyl migration is minimized. A representative method is that of de Haas and van Deenen (1965), who prepared four out of the six possible monoacylphosphatidylcholines.

Billimoria and Lewis (1968) used trityl groups for blocking, but showed that removing them by refluxing with 90% aqueous acetic acid for 10 min rather than by hydrogenolysis gave 100% of the 1-monoacyl product. Although this complete migration demonstrated the importance of neutral solutions in the final step, it does suggest that their method could be used for the preparation of unsaturated 1-monoacylphosphoglycerides (cf. Rosenthal, 1975).

Eibl *et al.* (1970) formed a D-α-monoacylphosphatidylcholine using a 1,2,5,6-D-mannitol tetrabenzyl ether intermediate. Arnold *et al.* (1967) described a method for the preparation of DL-α- and -β-ether monoacylphosphatidylcholines, while *O*-methylated and acetylated monoacylphosphatidylcholines were synthesized by Weltzien and Westphal (1967). Enantiomerically pure ester and other lysophospholipids were prepared by Eibl and Woolley (1988), while Filthath and Eibl (1992) reported the synthesis of lysophosphatidylinositol.

7.5.11 Phosphonolipids

These compounds occur naturally in a large number of organisms, particularly sea anemones, aquatic molluscs and ciliates (see section 2.3). Methods have been reported for the synthesis of: (1) phosphonolipids with the C–P bond between the phosphorus moiety and the polar headgroup, i.e. diacyl-*sn*-glycerol-3(*X*)-phosphonates; (2) phosphonolipids derived from 1,2-propanediol with a phosphonic acid attached to C-3, i.e. 1,2-diacyloxy- (or alkoxy-) propylphosphonic acid (or ester); (3) the phosphono analogue of sphingolipids, i.e. ceramide-(*X*)-phosphonate; and (4) phosphinic acid analogues with two C–P bonds, i.e. 1,2-diacyloxy- (or alkoxy-) propyl-(*X*)-phosphinate. These structures are indicated in Fig. 7.10. For details of the synthesis of such compounds, see Slotboom and Bonsen (1970) and Rosenthal (1975). Hori and Nozawa (1982) have reviewed phosphonolipids in detail.

A number of phosphonolipid analogues of phospholipids have been prepared. These include 1-*O*-alkylethylene glycol phosphonic acid analogues of phosphatidylcholine and phosphatidylethanolamine (Moschidis, 1987). a phosphono analogue of diphosphatidylglycerol (Moschidis, 1988), chiral vinylic analogues of phosphatidic acid and phosphatidylcholine (Schwartz *et al.*, 1989a) and α-hydroxyphosphonolipid analogues of phosphatidic acid (Schwartz *et al.*, 1989b).

References and further reading

Abdelmageed, O.H., Duclos, R.I., Griffin, R.G., Siminovitch, D.J., Ruocco, M.J. and Makriyannis, A. (1989) *Chem. Phys. Lipids*, **50**, 163.

Achterberg, V., Fricke, H. and Gercken, G. (1986a) *Chem. Phys. Lipids*, **41**, 137.

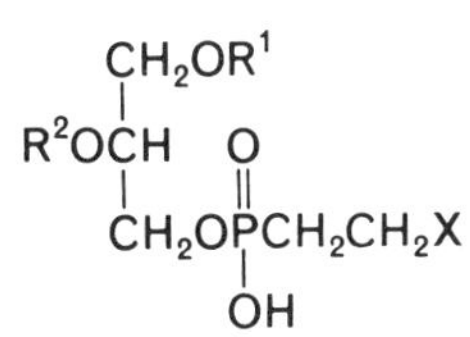

(1)

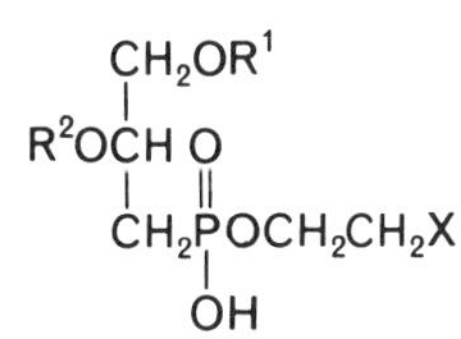

(2)

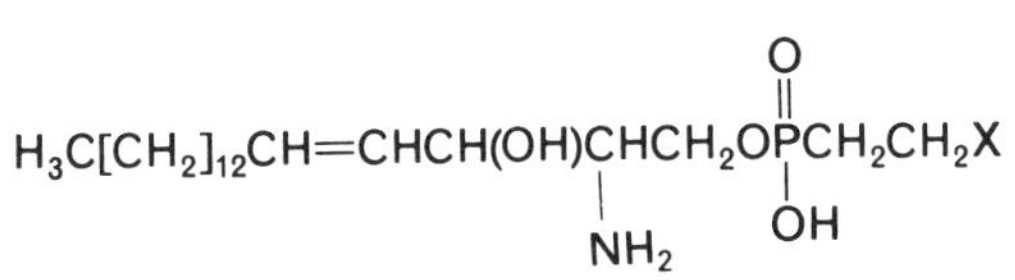

(3)

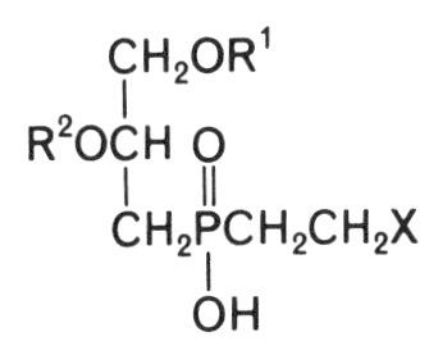

(4)

X represents a polar head-group in each case.
Note the two C–P bonds for compounds in group 4.

Fig. 7.10 General structures of type of phosphonolipids that have been synthesized.

Achterberg, V., Fricke, H. and Gercken, G. (1986b) *Chem. Phys. Lipids*, **41**, 349.
Ali, S. and Bittman, R. (1989) *Chem. Phys. Lipids*, **50**, 11.
Aneja, R. and Chadha, J.S. (1971a) *Biochim. Biophys. Acta*, **239**, 84.
Aneja, R. and Chadha, J.S. (1971b) *Biochim. Biophys. Acta*, **248**, 455.
Aneja, R., Chadha, J.S. and Davies, A.P. (1970) *Biochim. Biophys. Acta*, **218**, 102.
Arnold, D., Weltzien, H.U. and Westphal, O. (1967) *Ann. Chem.*, **40**, 230.
Barlow, P.N., Vidal, J.-C., Lister, M.D., Hancock, A.J. and Sigler, P.B. (1988) *Chem. Phys. Lipids*, **46**, 157.
Beck, A., Heissler, D. and Duportail, G. (1990) *Chem. Phys. Lipids*, **55**, 13.
Bergelson, L.D. (1980) *Lipid Biochemical Preparations*, Elsevier, Amsterdam.
Billimoria, J.D. and Lewis, K.O. (1968) *J. Chem. Soc.*, (C) 1404.
Bonsen, P.P.M. and de Haas, G.H. (1967) *Chem. Phys. Lipids*, **1**, 100.
Bonsen, P.P.M., de Haas, G.H. and van Deenen, L.L.M. (1965) *Biochim. Biophys. Acta*, **106**, 93.
Bonsen, P.P.M., de Haas, G.H. and van Deenen, L.L.M. (1966) *Chem. Phys. Lipids*, **1**, 33.
Bonsen, P.P.M., Burbach-Westerhuis, G.J., de Haas, G.H. and van Deenen, L.L.M. (1972) *Chem. Phys. Lipids*, **8**, 199.
Brockerhoff, H. and Yurkowski, M. (1965) *Can. J. Biochem.*, **43**, 1777.
Calderon, J. and Yague, P. (1986) *Chem. Phys. Lipids*, **41**, 147.
Calderon, J. and Yague, P. (1988) *Chem. Phys. Lipids*, **46**, 121.
Chacko, G.K., Schilling, K. and Perkins, E.G. (1967) *J. Org. Chem.*, **32**, 3698.
Cubero Robles, E. and van der Berg, D. (1969) *Biochim. Biophys. Acta*, **187**, 520.
de Haas, G.H. and van Deenen, L.L.M. (1961) *Rec. Trav. Chim.*, **80**, 951.
de Haas, G.H. and van Deenen, L.L.M. (1963) *Rec. Trav. Chim.*, **82**, 1163.
de Haas, G.H. and van Deenen, L.L.M. (1965) *Biochim. Biophys. Acta*, **106**, 315.
de Haas, G.H., van Zutphen, H., Bonsen, P.P.M. and van Deenen, L.L.M. (1964) *Rec. Trav. Chim.*, **83**, 99.
Eibl, H. and Lands, W.E.M. (1970) *Biochemistry*, **9**, 423.
Eibl, H. and Woolley, P. (1988) *Chem. Phys. Lipids*, **47**, 63.
Eibl, H., Westphal, O., van den Bosch, H. and van Deenen, L.L.M. (1970) *Liebig's Ann. Chem.*, **738**, 161.
Filthath, E. and Eibl, H. (1992) *Chem. Phys. Lipids*, **60**, 253.
Fowler, W.T., Lambeth, J.D. and Powell, G.L. (1988) *Chem. Phys. Lipids*, **47**, 261.
Fujiwara, Y., Masuoka, M. and Urakami, C. (1967) *Bull. Chem. Soc. Japan*, **40**, 230.
Hendrickson, H.S. (1991) *Methods in Enzymology*, vol. 197. Academic Press, New York, p. 90.
Hendrickson, H.S. and Hendrickson, E.K. (1990) *Chem. Phys. Lipids*, **53**, 115.
Hermetter, A., Stutz, H., Lohner, K. and Paltauf, F. (1987) *Chem. Phys. Lipids*, **43**, 69.
Hermetter, A., Stutz, H., Franzmair, R. and Paltauf, F. (1989) *Chem. Phys. Lipids*, **50**, 57.
Hirt, R. and Berchtold, R. (1958) *Pharmacol. Acta Helv.*, **33**, 349.
Hori, T. and Nozawa, Y. (1982) in *Phospholipids* (eds J.N. Hawthorne and G.B. Ansell), Elsevier, Amsterdam, p. 95.
Horrocks, L.A. and Sharma, M. (1982) in *Phospholipids* (eds J.N. Hawthorne and G.B. Ansell), Elsevier, Amsterdam, p. 51.
Inoue, K. and Nojima, S. (1968) *Chem. Pharm. Bull. (Tokyo)*, **16**, 76.
Isaacson, Y., Sherbourne, C.D., Gross, R.W. and Stenson, W.F. (1990) *Chem. Phys. Lipids*, **52**, 217.
Joo, C.N. and Kates, M. (1968) *Biochim. Biophys. Acta*, **152**, 800.
Joo, C.N. and Kates, M. (1969) *Biochim. Biophys. Acta*, **176**, 278.
Kates, M. (1978) *Prog. Chem. Fats Other Lipids*, **15**, 301.
Klyashchitski, B.A., Zhelvakova, E.G., Pimenova, V.V., Shvets, R.P., Evstigneeva, R.P. and Preobrazhenskii, N.A. (1971) *Zh. Obshch. Khim.*, **41**, 1386.
Lai, C., Zhang, T. and Joseph, J. (1988) *Anal. Biochem.*, **172**, 397.
Lapidot, Y., Barzilay, I. and Hajdu, J. (1969) *Chem. Phys. Lipids*, **3**, 125.
Lebeau, L., Mioskowksi, C. and Oudet, P. (1988) *Chem. Phys. Lipids*, **46**, 57.
Lemman, P., Buchweitz, K.M. and Stumpf, R. (1990) *Chem. Phys. Lipids*, **53**, 65.
Lukyanov, A.V., Lyutik, A.I., Schvets, V.I. and Preobrazhenskii, N.A. (1965) *Zh. Org. Khim.*, **36**, 1029.
Mangroo, D. and Gerber, G.E. (1988) *Chem. Phys. Lipids*, **48**, 99.
Molotkovsky, J.G. and Bergelson, L.D. (1971) *Tetrahedron Lett.*, 4791.
Moore, J.W. and Szelke, M. (1970) *Tetrahedron Lett.*, 4423.
Moschidis, M.C. (1987) *Chem. Phys. Lipids*, **43**, 39.
Moschidis, M.C. (1988) *Chem. Phys. Lipids*, **46**, 253.
Noda, N. and Keenan, R.W. (1990) *Chem. Phys. Lipids*, **53**, 53.
Paltauf, F. (1983) in *Ether Lipids* (eds H.K. Mangold and F. Paltauf), Academic Press, New York, p. 107.
Paltauf, F. and Hermetter, A. (1991) *Methods in Enzymology*, vol. 197, Academic Press, New York, p. 134.
Perochon, E. and Tocanne, J.-F. (1991) *Chem. Phys. Lipids*, **58**, 7.
Reynolds, L.H., Washburn, W.N., Deems, R.A. and Dennis, E.A. (1991) *Methods in Enzymology*, vol. 197, Academic Press, New York, p. 3.
Roberts, M.F. (1991) *Methods in Enzymology*, vol. 197, Academic Press, New York, p. 31.
Rosenthal, A.F. (1975) in *Methods in Enzymology*, vol. 35 (ed. J.M. Lowenstein), Academic Press, New York, p. 429.
Ruocco, M.J. and Makriyannis, A. (1989) *Chem. Phys. Lipids*, **50**, 163.

Saunders, R.M. and Schwartz, H.P. (1966) *J. Am. Chem. Soc.*, **88**, 3844.
Schacht, J. (1981) in *Methods in Enzymology*, vol. 72 (ed. J.M. Lowenstein), Academic Press, New York, p. 626.
Schwartz, P.W., Tropp, B.E. and Engel, E. (1989a) *Chem. Phys. Lipids*, **49**, 1.
Schwartz, P.W., Tropp, B.E. and Engel, E. (1989b) *Chem. Phys. Lipids*, **49**, 131.
Serebrennikova, G.A., Ovechkin, P.L., Vtorov, I.B. and Preobrazhenskii, N.A. (1969a) *Zh. Org. Khim.*, **5**, 546.
Serebrennikova, G.A., Titov, V.I. and Preobrazhenskii, N.A. (1969b) *Zh. Org. Khim.*, **5**, 550.
Shashidhar, M.S., Volwerk, J.J., Griffith, O.H. and Keana, J.F.W. (1991) *Chem. Phys. Lipids*, **60**, 101.
Shvets, V.I., Petrova, M.K., Kazenova, G.A. and Preobrazhenskii, N.A. (1967) *Zh. Obshch Khim.*, **37**, 1454.
Slotboom, A.J. and Bonsen, P.P.M. (1970) *Chem. Phys. Lipids*, **5**, 301.
Slotboom, A.J., de Haas, G.H. and van Deenen, L.L.M. (1963) *Rec. Trav. Chim.*, **82**, 469.
Slotboom, A.J., de Haas, G.H. and van Deenen, L.L.M. (1967) *Chem. Phys. Lipids*, **1**, 192.
Stepanov, A.E. and Slivets, V.I. (1986) *Chem. Phys. Lipids*, **41**, 1.
Takahashi, T., Hashimoto, T. and Hara, I. (1967) *Bull. Soc. Chim. Biol.*, **49**, 65.
Verheij, H.M., Smith, P.F., Bonsen, P.P.M. and van Deenen, L.L.M. (1970) *Biochim. Biophys. Acta*, **218**, 97.
Verheij, H.M., Bonsen, P.P.M. and van Deenen, L.L.M. (1971) *Chem. Phys. Lipids*, **6**, 46.
Wang, C.-J. and Tai, H.-H. (1990) *Chem. Phys. Lipids*, **55**, 265.
Weltzein, H.A. and Westphal, O. (1967) *Liebig's Ann. Chem.*, **709**, 240.
Woolley, P. and Eibl, H. (1988) *Chem. Phys. Lipids*, **47**, 55.
Yu, L. and Dennis, E.A. (1991) *Methods in Enzymology*, vol. 197, Academic Press, New York, p. 65.

Additional reference

Ceve, G. (ed.) (1993) *Phospholipids Handbook*, Marcel Dekker, New York.

7.6 Sphingolipids

Two general reviews of the chemistry and synthesis of sphingolipids are those of Hakomori (1983) and of Shapiro (1969). Table 7.2 indicates the general methods that have been used for the formation of sphingosines or glycosphingolipids. Specific references for the various methods are shown in the table and see text below.

Gigg and Gigg (1966a, b) have described two ways of synthesizing 4-D-hydroxysphinganine from D-galactosamine and 4-galactose. Gaver and Sweeley (1966) and Mendershausen and Sweeley (1969) prepared 3-oxo derivatives of sphinganine, *N*-acetyl-4-*trans*-sphingenine and *N*-carbobenzosphinganine. Stoffel and Sticht (1967) have described a simple method for the synthesis of 3-dehydrosphinganine hydrochloride. For the preparation of radiolabelled sphinganines and sphingenines refer to Stoffel and Sticht (1967), and to Stoffel *et al.* (1971) for the synthesis of radiolabelled sphingomyelin. For a summary of these syntheses and other chemical preparations of sphingolipids, see Shapiro (1969).

Barenholz and Gatt (1975) have described techniques for the separation of sphingosine bases by silicic acid chromatography. Bovine spinal cord lipids are used as starting material and hydrolysed in acid–methanol. Sphingosine and dihydrosphingosine can then be isolated from the hydrolysate by chromatography. For the isolation of phytosphingosine and dihydrosphingosine, a fraction obtained from the growth medium of the yeast *Hansenula ciferri* was used.

Ceramides are often prepared from sphingomyelins by phospholipase C digestion (cf. Bergelson, 1980). Hay and Gray (1969) prepared ceramides by such methods. These were then tritylated and the 3-*O*-benzoyl ester was formed. After detritylation, the 3-*O*-benzoylceramide was condensed with the appropriate acetobromo sugar. Flowers (1967) described a two-step method for lactosylceramide preparation. Individual species of ceramides can be made by *N*-acylation of DL-sphinganine or sphingenine with stearic acid or with DL-2-acetoxypalmitic acid in the presence of a mixed carbodiimide. The free hydroxy acid ceramide can then be generated by alkaline hydrolysis from the acetate and separation of diastereoisomers by preparative silica gel TLC (Hammarstrom, 1971).

The ceramides prepared by one of the above methods can be conveniently analysed by TLC and characterized by enzymic hydrolysis. Particular points to pay attention to during these syntheses are summarized by Bergelson (1980).

Galactosylceramide can be conveniently isolated from bovine brain tissue. If this organ is extracted successively with acetone, diethyl ether and then hot aqueous ethanol, sphingolipids are obtained in the final fraction. These precipitate on cooling and the residue is washed with acetone and dissolved in hot acetic acid. On cooling they can again be precipitated, dissolved in pyridine and passed through an aluminium oxide column to separate cerebrosides from sulphatides and gangliosides. After reprecipitating the cerebrosides, these are benzoylated and the benzoate separated by silicic acid chromatography into fractions containing non-hydroxy and hydroxy fatty acids. Finally, the benzoyl groups are hydrolysed with methanolic sodium methoxide (Bergelson, 1980).

Glucosylceramide can be synthesized by condensation of acetobromoglucose with 3-*O*-benzoylceramide. Benzoylceramide can be prepared by the method of Hay and Gray (1969) and is condensed with acetobromoglucose in the presence of mercuric cyanide catalyst (Bergelson, 1980). The benzoyl group is removed with methanolic sodium methoxide and the *O*-β-D-glycosyl(1$\rightarrow$1)ceramide purified by silicic acid column chromatography.

Fluharty *et al.* (1974) have reported the isolation of ^{35}S-labelled cerebroside sulphate. They injected the brains of young rats intracerebrally with [^{35}S]-sulphate and sacrificed the animals 3 days later. Total lipids were extracted and glycerolipids destroyed by alkaline hydrolysis. TEAE-cellulose column chromatography was then used to purify the [^{35}S]-sulphatide. Jatzkewitz and Nowoczek (1967) have synthesized D-galactose 3-sulphate and shown it to be identical to the galactose sulphate present in brain sulphatides.

Sulphatide can also be made by sulphation of 2,4,6-tri-*O*-acetylgalactopyranosylceramide followed by deacetylation (Ishizuku *et al.*, 1980). In addition, cerebroside 6-sulphate has been prepared by sulphation of cerebroside (Holmgren *et al.*, 1980).

Psychosine (1-*O*-galactosylsphingosine) was prepared by Radin (1974) starting from *O*-β-D-galactosyl(1$\rightarrow$1)ceramide. The latter was deacylated by refluxing with aqueous potassium hydroxide in butanol. The potassium hydroxide was removed by precipitation as potassium perchlorate, and fatty acids with hexane extraction. Finally, a chloroform/methanol/water mixture allows partition of excess perchloric acid into the upper phase, the chloroform phase containing the psychosine product (cf. Bergelson, 1980).

Table 7.2 Summary of methods for the chemical synthesis of sphingosines and glycosphingolipids (See Hakomori, 1983, for details)

Compound	Method
Sphingosines	
DL-Sphingosine	Reduction of DL-*erythro*-2-acetamido-3-*keto*-4-*trans*-octadecanoic acid
Synthesis and separation of D- and L-sphingosines	Resolution of DL-*erythro*-2-acetamido-3-hydroxy-4-*trans*-octadecanoate through L(+)-acetylmandeloyl ester
Differential synthesis of *trans*, *erythro* and *threo*; *cis*, *erythro* and *threo*	Condensation of 2-hexadecynal and 2-nitroethanol, followed by reduction
Dihydrosphingosine	
Dihydrosphingosine, DL-*threo*-, *erythro*- mixture	Oxamination of 3-*keto*-octadecanoate, followed by reduction with $LiAlH_4$
Dihydrosphingosine	Condensation of palmital and β-nitroethanol
D-*Erythro*-dihydrosphingosine	$LiAlH_4$ reaction of α-benzamido-β-hydroxyoctadecanoic acid methyl ester
D-*Erythro*- or D-*threo*-dihydrosphingosine	Ammonolysis of *trans* or *cis* epoxide
D-*Erythro*-dihydrosphingosine	Reduction of β-*N*-acetylamino-β-*keto*-octanoate by $LiAlH_4$ followed by treatment with methyl dichloroacetate and differential crystallization
^{14}C-labelled DL-dihydrosphingosine followed by racemic separation	Separation of DL-sphingosine into D and L diastereoisomers
Phytosphingosine	
D-*Ribo*-D-amino-1,3,4-trihydroxysphinganine	Glucosamine derivative (methyl-α-2-benzamido-2-deoxy-glucofuranoside) is converted to the oxazoline, the 2-allosamine derivative; compound is converted by oxidation to aldehyde, which is then coupled to olefin by Wittig reagent – triphenyltridecylphosphonium bromide; resulting product is reduced by catalytic hydrogenation
Cerebrosides	
(cerasine, phrenosine and glucocerebroside)	*N*-Fatty acyl-3-*O*-benzoylsphingosine condensed with acetobromo sugars by Koenig–Knor type reaction; $Hg(CN)_2$ as a catalyst
Lactosylceramide	
(cytolipin H)	*N*-Fatty acyl-3-*O*-benzoylsphingosine condensed with heptaacetyllactosyl bromide; $Hg(CN)_2$ as a catalyst
Gangliotriaosylceramide	
(Tay–Sachs' globoside)	Condensation of decaacetylgangliotriaosyl chloride and *N*-stearoyl-3-*O*-benzoylsphingosine
Sulphatide	
(3-*O*-sulphate galactosylceramide)	Reaction of 2,4,6-tri-*O*-acetyl-α-D-galactopyranosyl bromide with *N*-octadecanoyl-3-*O*-benzoyl-DL-dihydrosphingosine
Glycolipid analogues	
Gal-$\beta(1\rightarrow 6)$Gal-$\beta(1\rightarrow 1)$ceramide Gal-$\beta(1\rightarrow 3)$Gal-$\beta(1\rightarrow 1)$ceramide	Reaction analogous as for synthesis of lactosylceramide (see above)

Sphingomyelin can be obtained from bovine brain following acetone treatment, by ethanol and petroleum ether extraction. It is separated from cerebrosides by aluminium oxide column chromatography and phosphoglycerides were removed by mild alkaline hydrolysis. Silicic acid column chromatography allows final purification (Hanahan, 1961). DL-2-*N*-Stearoyldihydrosphingomyelin was synthesized by Zvonkova *et al.* (1974). They started with DL-*N*-stearoylsphinganine and tritylated the primary hydroxyl group on C-1. The hydroxyl on C-3 was then benzoylated, when the trityl group could be removed with hydrogen chloride. The unmasked hydroxyl group was then phosphorylated with β-chloroethylphosphodichloride to yield 1-*O*-(β-chloroethylphospho)-3-*O*-benzochloride. The latter was treated with trimethylamine to generate the quaternary ammonium of choline. Removal of the benzoyl group then yielded DL-2-*N*-stearoyl-dihydrosphingomyelin.

Stoffel (1975) has reported a method for the chemical synthesis of choline-labelled sphingomyelins. These compounds are demethylated to the corresponding ceramide-1-phosphoryl-*N*,*N*-dimethylethanolamines. The purified compounds are then quaternized in high yield with $[^{14}C]$- or $[^{13}C]$-methyl iodide to yield the respective sphingomyelins, which are labelled in their choline moieties. The technique allows the preparation of sphingomyelins of high specific radioactivity regardless of their degree of unsaturation.

The isolation of gangliosides from brain has been reviewed by Kanfer (1969) and by Bergelson (1980). Other specific references are Suzuki (1965), Svennerholm *et al.* (1972) and Laine *et al.* (1974), and general comments on ganglioside separations will be found in Kates (1986) and Christie (1982).

A typical isolation procedure makes use of the great hydrophilicity of gangliosides. Thus, a chloroform–methanol extract, when subjected to Folch partition, has most of the gangliosides in the upper phase. Phospholipid contaminants of this phase can be removed by alkaline hydrolysis, and the gangliosides precipitate as their barium salts. From these the free acids can be generated and then the gangliosides can be crystallized from solution. Other methods of isolating total gangliosides use gel filtration and DEAE-cellulose chromatography for purification (cf. Bergelson, 1980).

Some specific isolations are as follows. Haematoside (G_{M3}) can be obtained from rat liver by Folch partition, dialysis, Sephadex G-50 filtration and preparative thin-layer chromatography on silica gel G. Chloroform/methanol/2.5 M NH_4OH (60/35/8, by volume) is used as solvent, and the zones are detected with iodine vapour or, better, with a non-destructive spray such as primuline (Skipski, 1975). Haematoside is the least polar of the gangliosides and has a R_f of about 0.5. It can be eluted from the silica gel with chloroform/methanol/water (60/35/8, by volume).

Monosialyltetrahexosylceramide (G_{H1}) and disialyltetrahexosylceramide ($G_{D1\alpha}$) were isolated from bovine brain total gangliosides by silicic acid column chromatography (Svennerholm *et al.*, 1972). Ganglioside $G_{D1\alpha}$ has also been tritiated. The method involves catalytic addition of tritium gas to the unsaturated centres of the ganglioside isolated from beef brain. Because this results in a partial destruction of the ganglioside, the tritiated $G_{D1\alpha}$ has to be purified by Folch partition, dialysis and preparative TLC (Bergelson, 1980). Alternatively, purified beef brain ganglioside $G_{D1\alpha}$ can be oxidized selectively in its *N*-acetylneuraminic acid residue with sodium metaperiodate. The oxidized ganglioside is then dialysed and reduced with tritiated borohydride. After dialysis, the product is passed through a Dowex-50 column and freeze-dried (Bergelson, 1980).

Successful synthesis of ganglio-*N*-triosylceramide (Tay-Sachs' globoside) was based on the successful synthesis of ganglio-*N*-triose. After the trisaccharide had been made, it was converted to octaacetyl bromide and, subsequently, condensed with 3-*O*-benzoyl-*N*-stearoylsphingosine (see Hakomori, 1983). Analogous procedures can be used to synthesize various neutral ceramides (Flowers, 1966). Gangliosides with fluorescent or paramagnetic probes on the lipid moiety have also been synthesized chemically (Acquotti *et al.*, 1986).

The Tay-Sachs' ganglioside (G_{M2}) has been prepared, which is specifically labelled in either the *N*-acetylneuraminosyl or *N*-acetylgalactosaminyl portions (Tallman *et al.*, 1975). For the first type of labelling, G_{H1} is an intermediate and, for the second, G_{H3} is isolated from erythrocytes first and then converted to G_{M2}.

Two useful review articles covering the isolation and purification of sphingolipids are those of Rouser *et al.* (1969) and Kundu (1981). Rouser *et al.* (1969) cover the use of DEAE- and TEAE-cellulose column chromatography, while Kundu (1981) gives a thorough update of the TLC methods. The latter includes solvent systems, detection methods, analytical and preparative methods and radioautography (fluorography). Skipski (1975) has also discussed separation methods. He recommends the use of primuline as a non-destructive spray for sphingolipids because it is sensitive and does not interfere with the subsequent analysis of carbohydrate, fatty acids and sphingosines by GLC or with the colorimetric determination of sphingosines by the method of Yamamoto and Rouser (1970).

Various analogues of sphingolipids have been synthesized, particularly those in which the glycosidic linkages are different from natural compounds (see Hakomori, 1983).

References and further reading

Acquotti, D., Sonnino, S., Masserini, M., Casella, L., Fronza, G. and Tettamanti, G. (1986) *Chem. Phys. Lipids*, **40**, 71.

Barenholz, Y. and Gatt, S. (1975) in *Methods in Enzymology*, vol. 35 (ed. J.H. Lowenstein), Academic Press, New York, p. 529.

Bergelson, L.D. (1980) *Lipid Biochemical Preparations*, Elsevier, Amsterdam.
Christie, W.W. (1982) *Lipid Analysis*, 2nd edn, Pergamon, Oxford.
Flowers, H.M. (1966) *Carbohydr. Res.*, **2**, 371.
Flowers, H.M. (1967) *Carbohydr. Res.*, **4**, 42.
Fluharty, A.L., Davies, M.L., Kihara, H. and Kritchevsky, G. (1974) *Lipids*, **9**, 865.
Gaver, R.C. and Sweeley, C.C. (1966) *J. Am. Chem. Soc.*, **88**, 3643.
Gigg, J. and Gigg, R. (1966a) *J. Chem. Soc.*, 1876.
Gigg, J. and Gigg, R. (1966b) *J. Chem. Soc.*, 1878.
Hakomori, S.-I. (1983) in *Sphingolipid Biochemistry* (eds J.N. Kanfer and S.I. Hakomori), Plenum, New York, p. 1.
Hammarstrom, S. (1971) *J. Lipid Res.*, **12**, 760.
Hanahan, D.G. (1961) *Biochem. Prep.*, **8**, 121.
Hay, J.B. and Gray, G.M. (1969) *Chem. Phys. Lipids*, **3**, 59.
Holmgren, J., Svennerholm, L., Elving, H., Fredman, P. and Shannegard, O. (1980) *Proc. Natl. Acad. Sci. USA*, **77**, 1947.
Ishizuku, I., Ivamori, M., Ueno, K. and Yamakaura, T. (1980) *J. Biol. Chem.*, **253**, 898.
Jatzkewitz, H. and Nowoczek, G. (1967) *Chem. Ber.*, **100**, 1667.
Kanfer, J.W. (1969) *Methods in Enzymology*, vol. 14 (ed. J.M. Lowenstein), Academic Press, New York, p. 660.
Kates, M. (1986) *Techniques in Lipidology*, 2nd edn, Elsevier, Amsterdam.
Kundu, S.K. (1981) *Methods in Enzymology*, vol. 72 (ed. J.M. Lowenstein), Academic Press, New York, p. 185.
Laine, R.A., Stellner, K. and Hakomori, S.-I. (1974) *Methods Membr. Biol.*, **2**, 205.
Mendershausen, P.B. and Sweeley, C.C. (1969) *Biochemistry*, **8**, 2633.
Radin, N.S. (1974) *J. Lipid Res.*, **17**, 290.
Rouser, G., Kritchevsky, G., Yamamoto, A., Simon, G., Galli, C. and Bauman, A.J. (1969) *Methods in Enzymology*, vol. 14 (ed. J.M. Lowenstein), Academic Press, New York, p. 272.
Shapiro, D. (1969) *Chemistry of Sphingolipids*, Hermann, Paris.
Skipski, V.P. (1975) in *Methods in Enzymology*, vol. 35 (ed. J.H. Lowenstein), Academic Press, New York, p. 396.
Stoffel, W. (1975) in *Methods in Enzymology*, vol. 35 (ed. J.M. Lowenstein), Academic Press, New York, p. 533.
Stoffel, W. and Sticht, G. (1967) *Hoppe-Seyler's Z. Physiol. Chem.*, **348**, 1345.
Stoffel, W., Lekim, D. and Tschung, T.S. (1971) *Hoppe-Seyler's Z. Physiol. Chem.*, **352**, 1058.
Suzuki, K. (1965) *J. Neurochem.*, **12**, 629.
Svennerholm, L., Whistler, R.L. and Miller, J.N. (1972) in *Methods in Carbohydrate Chemistry*, vol. 6, Academic Press, New York, p. 464.
Tallman, J.F., Kolodny, E. and Brady, R.O. (1975) in *Methods in Enzymology*, vol. 35 (ed. J.D. Lowenstein), Academic Press, New York, p. 541.
Yamamoto, A. and Rouser, G. (1970) *Lipids*, **5**, 442.
Zvonkova, E.N., Mitzner, B.I., Bushnev, A.S., Elker, K.I., Soldatova, S.A. and Evstigneeva, R.P. (1974) *Khim Prir. Sojed*, 553.

7.7 Glycosylglycerides

Analytical separation and small-scale isolation of galactosylglycerides can be carried out conveniently by TLC on silica gel G (cf. Hitchcock and Nichols, 1971; Harwood, 1980). Sastry (1974) has discussed the various solvent systems in some detail, and the method of Khan and Williams (1977) is particularly useful. For larger-scale preparations, column chromatography by the method of O'Brien and Benson (1964) will be found to be good for the preparation of all three major plant glycosylglycerides. Yields of the two galactosylglycerides may be improved by the use of repeated chromatography on silica gel or on carbon–Celite columns (van der Veen *et al.*, 1967). A rapid method of batch elutions from silica gel has been reported by de Stefanis and Ponte (1969). Most of the fractions obtained by these methods are slightly contaminated but can be conveniently cleaned up by TLC or, in the case of phospholipid contaminants, by passage through Amberlite MB-3 (Sastry and Kates, 1964).

A useful reference for general chemical procedures to synthesize glycoglycerolipids is that of Gigg and Gigg (1990). They cover the synthesis of mono-, di- and triglycosyldiacylglycerols, glycosylalkylacylglycerols, glucosyldialkylglycerols and some phosphatidylglucosyldiacylglycerols.

Wehrli and Pomeranz (1969) prepared the chiral β-galactosyldiacylglycerol by a method shown in Fig. 7.11. Condensation of acetobromogalactose with the glycerol derivative, bis(1-acyl-2-methyleneglycerol), was followed by acid hydrolysis to give the acetylated galactosylmonoacylglycerol. Reaction with acid chloride then gave acetylated galactosyldiacylglycerol. Hydrazinolysis was used to remove acetyl groups and yield monogalactosyldiacylglycerol stereochemically identical to the naturally occurring plant lipid. The same route could be used to prepare galactosyldiacylglycerols with two different fatty acids (Gigg and Gigg, 1990).

Various β-glycosyldiacylglycerols have been prepared by an orthoacetate glycosidation procedure to diacylglycerols (see Shvets *et al.*, 1973). Glycosides of 2-deoxy-D-glucose with racemic diacylglycerols and glucosyl- or mannosyldiacylglycerols containing fluorescent labels in the lipid and sugar portions have been made (see Gigg and Gigg, 1990). A saturated β-galactosyldiacylglycerol was prepared via an intermediate formed by condensation of acetobromogalactose with 1,2-di-*O*-benzyl-*sn*-glycerol (Gent and Gigg, 1975). The latter compound was also used by Batrakov *et al.* (1976) for the preparation of β-galactosyl-, β-glucosyl- and α-mannosyldiacyl-

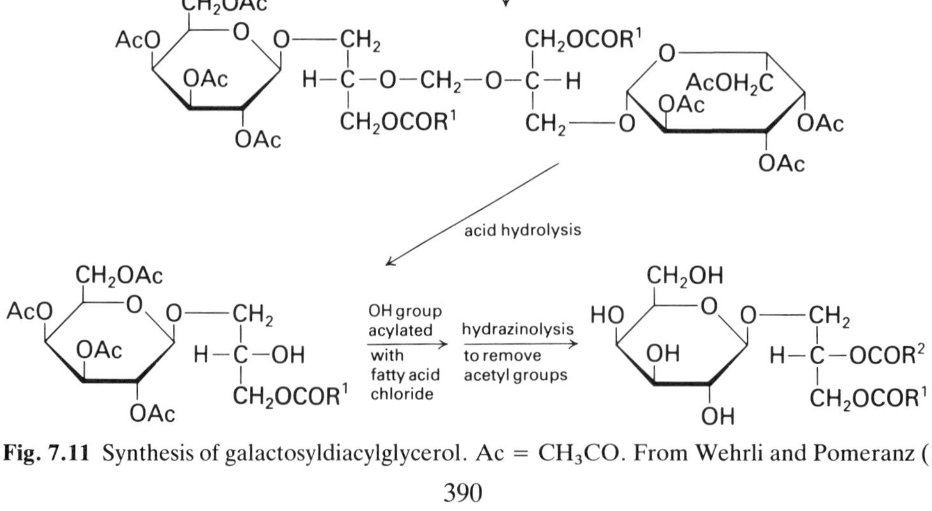

Fig. 7.11 Synthesis of galactosyldiacylglycerol. Ac = CH_3CO. From Wehrli and Pomeranz (1969).

glycerols. Improvements in their procedure resulting in high yields (70–75%) have been reported (Mannock *et al.*, 1987; van Boeckel *et al.*, 1985).

Seminolipids (sulphated galactosylacylalkylglycerols occurring in testes and brain of animals) (see Murray and Narasimhan, 1990), α-glucosyldiacylglycerols (occurring in bacteria) (see Kates, 1990b) and monoglycosyldialkylglycerols have all been synthesized chemically (see Gigg and Gigg, 1990, for more details).

Glycosylglycerols have been synthesized by several workers. Wickberg (1958) reported the synthesis of *O*-α-D-galactopyranosyl- and *O*-β-D-galactopyranosyl(1→1)-D-glycerols using a condensation reaction between tetra-*O*-acetyl-α-D-galactopyranosyl bromide and methylene-bis-2-*O*-(3-*O*-benzoyl-D-glycerol). He also prepared the corresponding L-glycerol galactosides. A second method was used by Charlson *et al.* (1957) for the formation of *O*-α-D-galactopyranosyl- and *O*-β-D-galactopyranosyl(1→2)glycerols. Brundish and Baddiley (1968) have reported the formation of *O*-α-D-glucopyranosyl- and *O*-β-D-glucopyranosyl(1→1)-D-glycerols and their anomeric L-glycerol galactosides. They have also synthesized four positional isomers of diglucosylglycerols. The major glycosylglyceride of *Pneumococcus*, *O*-α-D-galactopyranosyl(1→2)-*O*-α-D-glucopyranosyl(1→1)-D-glycerol, was synthesized by Brundish *et al.* (1967). Sulphoquinovosylglycerol has been synthesized by Miyano and Benson (1962).

Wehrli and Pomeranz (1969) reported the synthesis of 1,2-dipalmitoyl- and 1-palmitoyl-2-linoleoyl-3-*O*-(β-D-galactopyranosyl)-*sn*-glycerols. They acylated the primary hydroxyl groups of 2,5-methylene-D-mannitol and then cleaved the mannitol moiety between C-3 and C-4 with lead tetraacetate. The resultant aldehyde was reduced and galactose attached by the Koenigs–Knorr reaction. The acetal was then hydrolysed and the hydroxyl group acylated. Hydrazinolysis of the acetylated glycolipid completed the procedure. A simpler method, but which involved loss of optical activity at C-2 of the glycerol, was reported by Bashkatova *et al.* (1971a). They formed 1,2-dipalmitoyl-3-*O*-(β-D-glucopyranosyl)-*sn*-glycerol as well as the mannose equivalent. Their procedure has also been used to synthesize diglycosyldiacylglycerols (cf. Bashkatova *et al.*, 1971b; Shvets *et al.*, 1973).

A useful technique described by Heinz (1971) has been used to prepare different molecular species of monogalactosyl- and digalactosyldiacylglycerols. He used a semi-synthetic method starting from compounds isolated from natural sources. The

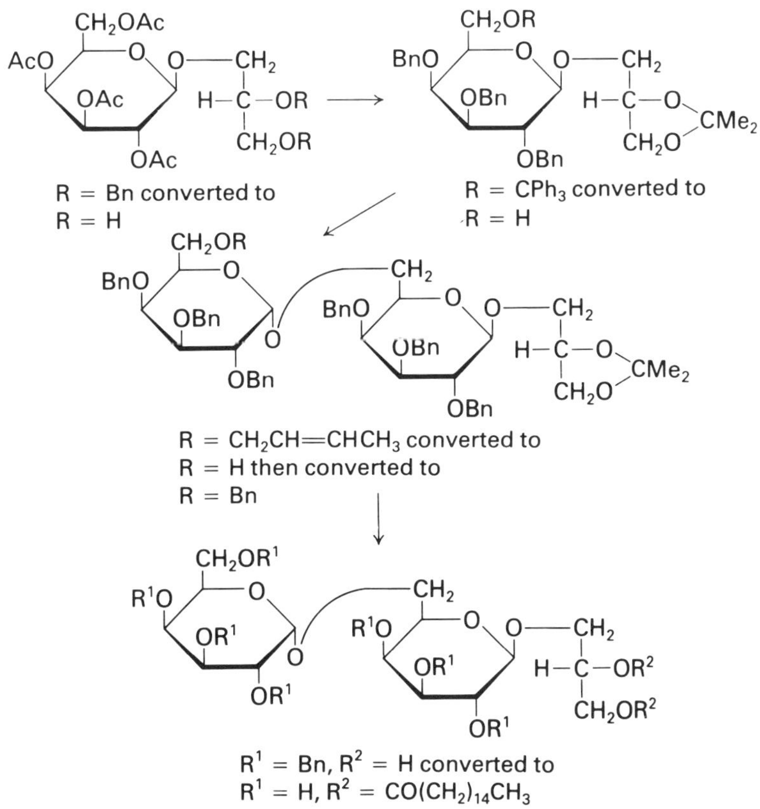

Fig. 7.12 Synthesis of digalactosyldiacylglycerol. Ac = CH_3CO, Bn = CH_2Ph. From Gent and Gigg (1975).

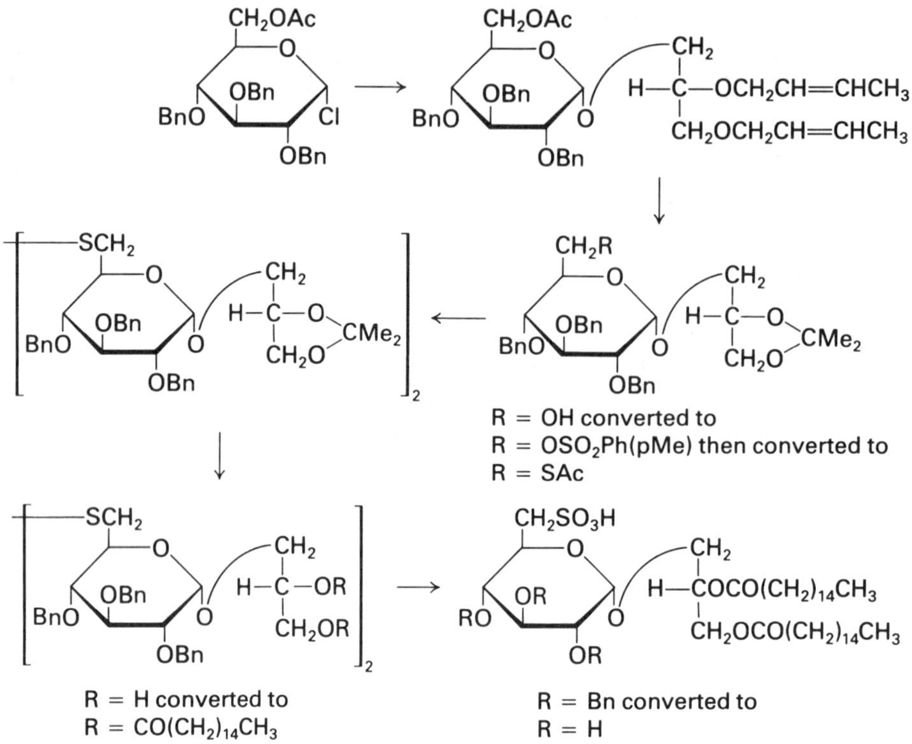

Fig. 7.13 Synthesis of sulphoquinovosyldiacylglycerol. Ac = CH_3CO, Bn = CH_2Ph. From Gigg and Gigg (1990).

hydroxyl hydrogens of the galactose residues of the substrates were substituted with *O*-(1-methoxyethyl) groups and the acyl groups were then removed with sodium methoxide. Acylation was then carried out with the desired fatty acyl chloride, and the protecting groups were removed with boric acid. Heisig and Heinz (1972) also used a similar method for preparing galactosylmonoacylglycerols.

The first synthetic diglycosyldiacylglycerol was prepared by Wehrli and Pomeranz (1969) using a condensation reaction between racemic isopropylideneglycerol and acetobromocellobiose. The isopropylidene group was removed by acid hydrolysis. Acylation with fatty acids and deacetylation with hydrazine yielded β-D-glucopyranosyl(1→4)-D-glucosyldiacylglycerol. Another diglucosyldiacylglycerol (derived from kojibiose, α-D-glucopyranosyl(1→2)-D-glucose) was prepared by van Boeckel and van Boom (1985). They used a route involving condensation of perbenzylated glucosylbromide with the α-glucosyldiacylglycerol derivative.

The important plant glycolipid, digalactosyldiacylglycerol (α-D-galactopyranosyl(1→6)-β-D-galactopyranosyl(1→3)-1,2-di-*O*-acylglycerol) was synthesized using a route beginning with the condensation of acetobromogalactose with 1,2-di-*O*-benzyl-*sn*-glycerol as shown in Fig. 7.12. The β-galactosyldiacylglycerol was also prepared. Other diglycosyldiacylglycerols, such as those important in bacteria and for serological activity, have been synthesized (see Gigg and Gigg, 1990, for more details).

The plant sulpholipid (sulphoquinovosyldiacylglycerol) was synthesized (Gigg, 1978; Gigg *et al.*, 1980) by a route shown in Fig. 7.13. A glucosyl chloride derivative was condensed with 1,2-di-*O*-(but-2-enyl)-*sn*-glycerol to give α-glucoside, which was converted to its acetonide. The toluene-β-sulphonate was converted to the thioacetate, which was then hydrolysed and oxidized (with iodine) to give the disulphide. The isopropylidene group was removed by acid hydrolysis and the alcohol product was acetylated with hexadecanoyl chloride. The resultant ester was oxidized with 3-chloroperbenzoic acid and subsequent debenzylation yielded sulphoquinovosyldiacylglycerol.

Methods for the chemical synthesis of 6-*sn*-phosphatidylglucosyldiacylglycerol (found in streptococci), diglycosyldialkylglycerols, triglycosyldiacylglycerols (occurring in plants, bacteria and gastric secretions) (see Kates, 1990a) and triglycosyl-*O*-phytanylglycerol (found in halobacteria) (see Kates, 1990b) are discussed by Gigg and Gigg (1990).

References and further reading

Bashkatova, A.I., Smirnova, G.V., Shvets, V.I. and Evstigneeva, R.P. (1971a) *Zh. Org. Khim.*, **7**, 1644.

Bashkatova, A.I., Shvets, V.I. and Evstigneeva, R.P. (1971b) *Zh. Org. Khim.*, **7**, 2627.

Batrakov, S.G., Ilina, E.F. and Panosyan, A.G. (1976) *Bull. Acad. Sci. USSR, Chem.*, 626.

Brundish, D.E. and Baddiley, J. (1968) *Carbohydr. Res.*, **8**, 308.

Brundish, D.E., Shaw, N. and Baddiley, J. (1967) *Biochem. J.*, **105**, 885.

Charlson, A.J., Gorin, P.A.J. and Perlin, A.S. (1957) *Can. J. Chem.*, **35**, 365.

de Stefanis, V.A. and Ponte, J.G. (1969) *Biochim. Biophys. Acta*, **176**, 198.

Gent, P.A. and Gigg, R. (1975) *J. Chem. Soc., Perkin Trans.*, 364.

Gigg, R. (1978) *ACS Symp. Ser.*, **77**, 44.

Gigg, J. and Gigg, R. (1990) in *Glycolipids, Phosphoglycolipids and Sulfoglycolipids* (ed. M. Kates), Plenum, New York, p. 489.

Gigg, R., Penglis, A.A.E. and Conant, R. (1980) *J. Chem. Soc., Perkin Trans.*, 2490.

Harwood, J.L. (1980) in *Biochemistry of Plants*, vol. 4 (eds P.K. Stumpf and E.E. Conn), Academic Press, New York, p. 1.

Heinz, E. (1971) *Biochim. Biophys. Acta*, **231**, 537.

Heisig, O.M.R.A. and Heinz, E. (1972) *Phytochemistry*, **11**, 815.

Hitchcock, C. and Nichols, B.W. (1971) *Plant Lipid Biochemistry*, Academic Press, London.

Kates, M. (ed.) (1990a) *Glycolipids, Phosphoglycolipids and Sulfoglycolipids*, Plenum, New York.

Kates, M. (1990b) in *Glycolipids, Phosphoglycolipids and Sulfoglycolipids* (ed. M. Kates), Plenum, New York, p. 1.

Khan, M. and Williams, J.P. (1977) *J. Chromatogr.*, **140**, 179.

Mannock, D.A., Lewis, R.N.A.H. and McElhaney, R.N. (1987) *Chem. Phys. Lipids*, **43**, 113.

Miyano, M. and Benson, A.A. (1962) *J. Am. Chem. Soc.*, **84**, 59.

Murray, R.K. and Narasimhan, R. (1990) in *Glycolipids, Phosphoglycolipids and Sulfoglycolipids* (ed. M. Kates) Plenum, New York, p. 321.

O'Brien, J.S. and Benson, A.A. (1964) *J. Lipid Res.*, **5**, 432.

Sastry, P.S. (1974) *Adv. Lipid Res.*, **12**, 251.

Sastry, P.S. and Kates, M. (1964) *Biochemistry*, **3**, 1271.

Shvets, V.I., Bashkatova, A.I. and Evstigneeva, R.P. (1973) *Chem. Phys. Lipids*, **10**, 267.

van Boeckel, C.A.A. and van Boom, J.H. (1985) *Tetrahedron*, **41**, 4567.

van Boeckel, C.A.A., Visser, G.M. and van Boom, J.H. (1985) *Tetrahedron*, **41**, 4557.

van der Veen, J., Hirota, K. and Olcott, H.S. (1967) *Lipids*, **2**, 406.

Wehrli, H.P. and Pomeranz, Y. (1969) *Chem. Phys. Lipids*, **3**, 357.

Wickberg, B. (1958) *Acta Chem. Scand.*, **12**, 1187.

7.8 Fatty acid esters

This section is devoted to the synthesis of fatty acid esters – a process carried out on the milligram and the tonne scale. Other matters bearing on this topic are covered in sections 7.4 (acylglycerols) and 10.12 (esterification, etc.). Esters can be made from acids (or more reactive acyl derivatives) or from existing esters by procedures in which alcohol (alcoholysis) or acid moieties (acidolysis) are exchanged.

Esters result from interaction of an acid and an alcohol, usually in the presence of a suitable catalyst or enzyme. If the acid is replaced by an acid chloride or anhydride, then a catalyst is not generally required. The catalysts include sulphuric acid, 4-toluenesulphonic acid, hydrogen chloride, boron trifluoride and 4-dimethylaminopyridine with dicyclohexylcarbodiimide. Methyl esters can also be produced from acids with diazomethane (Christie, 1989, 1993).

Alcoholysis is an important method of converting glycerol esters to other esters in the presence of a basic or enzymic catalyst. Acidic catalysts can also be used, but they are generally less efficient. Oils and fats are easily converted to methyl esters by reaction with methanolic sodium methoxide. So long as the glyceride does not contain free acid to neutralize the catalyst, reaction occurs quickly at room temperature or a little above. The equilibrium is displaced in favour of methyl ester through the presence of excess methanol, which serves as reagent and solvent.

Methanolysis is carried out on an industrial scale at 240 °C and 100 bar. Methyl esters are the favoured starting material for hydrogenolysis to produce alcohols (section 7.9). They can also serve as an alternative for diesel fuel. The so-called bio-diesel (or other similar name) is produced from a wide variety of oils and fats and is being tested or used in several countries, partly on environmental grounds. This is an interesting development, but the annual production of petroleum oils (3 billion tonnes) and vegetable oils (100 million tonnes) should not be overlooked.

Reaction of oils and fats with glycerol (glycerolysis) in the presence of an alkaline catalyst produces an equilibrium mixture of glycerol with mono-, di- and triacylglycerols. With an appropriate quantity of glycerol, the product is mainly mono- (40–60%) and diacylglycerol (50–30%).

By molecular distillation the content of monoacylglycerol is raised to about 95%. Both the mixture and the distilled product are important surface-active compounds widely used in the food industry. More specific products are made by further reactions of the mono- and diacylglycerols with acetic anhydride, lactic acid, diacetyltartaric anhydride or citric acid (section 5.8.2).

Alcoholysis can be promoted by lipases. Such reactions occur under milder conditions, can also occur with free acids, and allow for easy isolation of glycerol. Reaction is quicker with ethanol (especially 96%) than with methanol. Lipases obtained from *Pseudomonas*, *Candida* and *Mucor* have been examined (Mittelbach, 1990). When reactions occur under milder conditions, the products are usually purified more easily (with less waste) and the capital cost of plant is reduced.

There is growing interest in structured lipids. Some of these are designed to maintain the desired functional and sensory properties of conventional triacylglycerols but to reduce or eliminate absorption and so reduce caloric value.

Olestra is a sucrose derivative with 6–8 acyl groups obtained from common vegetable oils such as soybean, corn, cottonseed, sunflower or rapeseed. It is solid or semi-solid depending on the fatty acids present and is very suitable as a frying oil. It has high thermal stability, and any decomposition product is the same as that observed with triacylglycerols under the same thermal conditions. It can also be exchanged for fats in products such as ice cream, margarine, cheese and baked goods, and can be blended with vegetable oils. The material is non-toxic, non-carcinogenic, and is so poorly absorbed as to have zero caloric value. It has been claimed that, as part of a balanced diet, olestra is able to reduce total and saturated fat levels of food in which it is used without sacrificing functional purpose. Permission for its use has not yet been given (1993).

This mixture of acylated sucrose molecules with 6–8 acyl groups (usually 70–95% octaacyl) is prepared by reaction of high-grade sucrose and high-grade methyl esters in the presence of sodium or potassium soaps to improve solubility and an alkali carbonate as catalyst (Spinner, 1990).

Caprenin is a reduced-calorie fat (~5 kcal/g) with functional properties similar to cocoa butter and is used for soft candies and for confectionery coatings. It is a triacylglycerol containing caprylic (C_8), capric (C_{10}) and behenic (C_{22}) acids along with lower levels of the C_6, C_{20} and C_{24} acids (present in the sources of the major components). The major acids are present roughly in a 1:1:1 (molar ratio) or 1:1:2 (weight ratio). The glycerides are mainly C_{38}–C_{42} and consist of one unit of behenic acid and two units of the medium-chain acids. When liberated, the behenic acid is poorly absorbed and so caprenin produces fewer calories than the more common triacylglycerols.

In one method of producing this kind of product, a mixture of behenic glycerides (mono-, di- and tribehenin) is converted to triglycerols by esterification with capric acid, and this intermediate product is inter-esterified with medium-chain triacylglycerols containing an appropriate mixture of caprylic and capric esters under the influence of sodium hydroxide at 80 °C. The product is steam-stripped to remove medium-chain triacylglycerols and then subjected to molecular distillation followed by non-solvent fractionation. The product contains about 92% of mono-long-chain triacylglycerols, with C_{38} (~22%), C_{40} (~48%) and C_{42} (~24%) compounds as the major components.

Meffert (1984) has reviewed the technical uses of fatty acid esters.

References and further reading

Christie, W.W. (1989) *Gas Chromatography and Lipids*, Oily Press, Ayr, p. 64.

Christie, W.W. (1993) in *Advances in Lipid Methodology – Two* (ed. W.W. Christie), Oily Press, Dundee, pp. 69–111.

Meffert, A. (1984) *J. Am. Oil Chem. Soc.*, **61**, 255.

Mittelbach, M. (1990) *J. Am. Oil Chem. Soc.*, **67**, 168.

Spinner, J. (1990) *World Conference on Oleochemicals* (ed. T.H. Applewhite), American Oil Chemists' Society, Champaign, p. 231.

7.9 Alcohols

The hydrogenolysis of natural glycerol esters to long-chain alcohols is an important industrial process (about 1 million tonnes per annum expected to rise to about 1.5 million tonnes by 2000). The C_{12} alcohol is produced from lauric oils and the C_{16}/C_{18} alcohols from tallow, palm and palm stearin. Dodecanol is also produced from ethylene by the petrochemical industry, but production from natural fats provides more than half of the total and this share is rising.

The reaction occurs with glycerol esters, methyl esters and free acids, but the first of these is seldom used because of the loss of glycerol, which is an important part of the economics of the process. Production from the methyl esters is now most favoured. It is effected at 250–300 °C and 200–300 bar. Traditionally the catalyst has been copper chromite, but double bonds are then also reduced. With zinc chromite double bonds are unaffected and oleic esters furnish oleyl alcohol. Over-reduction produces hydrocarbon as a by-product. A rhenium–osmium catalyst has been described that promotes reaction under milder conditions (100–120 °C, 25–100 bar) (Yoshino *et al.*, 1990). In the laboratory, esters are conveniently reduced to alcohols with lithium aluminium hydride. Sodium borohydride will also effect this change, but only with appropriate solvents such as a mixture of methanol and t-butanol (Rao *et al.*, 1989).

The long-chain alcohols show very good environmental and toxicological properties. They are extensively used as surfactants, mainly as sulphate (**9**), ethoxylates (**10**), ether sulphates (**11**), esters and alkyl polyglucosides (**12**).

Dimeric (Guerbet) alcohols (**13**) are produced from the monomeric alcohol by reaction with alkali and can be oxidized to acids. Both the acids and the alcohols are branched-chain compounds. They remain liquid at room temperature and below, and are used as lubricants and in cosmetics.

RCH_2OH — Alcohol

RCH_2OSO_3H — (**9**) Sulphate

$RCH_2O[CH_2CH_2O]_nH$

(**10**) Ethoxylate

$RCH_2O[CH_2CH_2O]_nCH_2CH_2OSO_3H$

(**11**) Ether sulphate

$(\text{Glucose})_x\text{—}OCH_2R$

(**12**) Alkyl polyglucoside

The staff of Henkel (1982) have produced a book on fatty alcohols. Other aspects of production and use are reviewed by Egan *et al.* (1984), Houston (1984), Kreutzer (1984) and Voeste and Buchold (1984).

$$RCH_2CH_2OH \rightarrow RCH_2CH_2CH(R)CH_2OH$$
$$\rightarrow RCH_2CH_2CH(R)COOH$$

(**13**) Guerbet alcohol

References and further reading

Egan, R.E., Earl, G.W. and Ackerman, J. (1984) *J. Am. Oil Chem. Soc.*, **61**, 324.

Henkel (1982) *Fatty Alcohols: Raw Materials, Methods, Uses*, Henkel, Dusseldorf.

Houston, C.A. (1984) *J. Am. Oil Chem. Soc.*, **61**, 179.

Kreutzer, E.R. (1984) *J. Am. Oil Chem. Soc.*, **61**, 343.

Rao, R.Y., Pantalu, A.J. and Lakshminarayana, G. (1989) *J. Am. Oil Chem. Soc.*, **66**, 597.

Voeste, T. and Buchold, H. (1984) *J. Am. Oil Chem. Soc.*, **61**, 350.

Yoshino, K., Kajiwara, Y., Takaishi, N., Inamoto, Y. and Tsuji, J. (1990) *J. Am. Oil Chem. Soc.*, **67**, 21.

7.10 Dibasic acids

The industrially important dibasic acids are adipic (hexanedioic), azelaic (nonanedioic), sebacic (decanedioic), dodecanedioic, C_{19} and C_{21} dibasic acids, and dimer acids (C_{36}).

Azelaic (nonanedioic) acid is produced along with pelargonic (nonanoic) acid by ozonolysis of oleic acid. Oleic acid mixed with nonanoic acid is fed into an ozone absorber against a flow of oxygen gas containing about 2% of ozone. Ozonation occurs at 25–45 °C and external cooling is required. The product is decomposed and oxidized in another vessel at 100 °C (section 10.4.3). The product is distilled to remove nonanoic acid and other low-boiling acids, and azelaic acid is recovered from the residue by extraction with hot water, in which it is soluble. Brassylic (tridecanedioic) acid could be prepared in a similar way from erucic acid.

Other oxidizing agents convert oleic acid to give azelaic acid, but none is used industrially. The C_9 dibasic acid can also be made by dimerization of butadiene, carbonylation of the cyclooctadiene, and caustic fusion of the C_9 product.

Sebacic (decanedioic) acid is made on a commercial scale from ricinoleic acid or castor oil (section 3.3.4). Reaction with sodium hydroxide at 180–270 °C eventually furnishes octan-2-ol and sebacic acid. The reaction probably involves oxidation of secondary alcohol to ketone, double-bond migration and retro-aldol cleavage of the 12-oxooctadec-10-enoic acid. Sebacic acid has also been made by electrolysis of methyl hydrogen adipate (monomethyl hexanedioate).

Dodecanedoic acid is obtained from cyclododecane by oxidation first with air to give cyclic alcohol and ketone and then with nitric acid to the dibasic acid. The cyclic C_{12} hydrocarbon is made by trimerization of butadiene followed by hydrogenation.

Oleic acid and carbon monoxide interact under a variety of conditions to produce branched-chain dibasic acids (and other difunctional molecules) by hydroformylation (oxo reaction), hydrocarboxylation (Reppe reaction), or by the Koch reaction:

$$-CH{=}CH- \xrightarrow[\text{Co(CO)}_8]{\text{CO, H}_2} -\underset{\text{CHO}}{\underset{|}{\text{C}}}\text{HCH}_2-$$

$$-CH{=}CH- \xrightarrow[\text{Co(CO)}_8]{\text{CO, ROH}} -\underset{\text{COOR}}{\underset{|}{\text{C}}}\text{HCH}_2-$$

$$-CH{=}CH- \xrightarrow[\text{H}_2\text{SO}_4]{\text{CO, ROH}} -\underset{\text{COOR}}{\underset{|}{\text{C}}}\text{HCH}_2-$$

The formyl group (CHO) in the first reaction can be reduced to CH_2OH or oxidized to COOH. Hydrolysis of the ester group (COOR) in the second and third reactions leads to a C_{19} dibasic acid.

C_{21} diacid is prepared from tall oil fatty acids by reaction with iodine and acrylic (propenoic) acid at 220–250 °C. The product is obtained from linoleic acid, and unreacted oleic acid can be distilled out. The C_{21} dibasic acid can be further purified by thin-film or molecular distillation:

Linoleic acid $\xrightarrow{I_2}$ *t,t*-conjugated diene acids

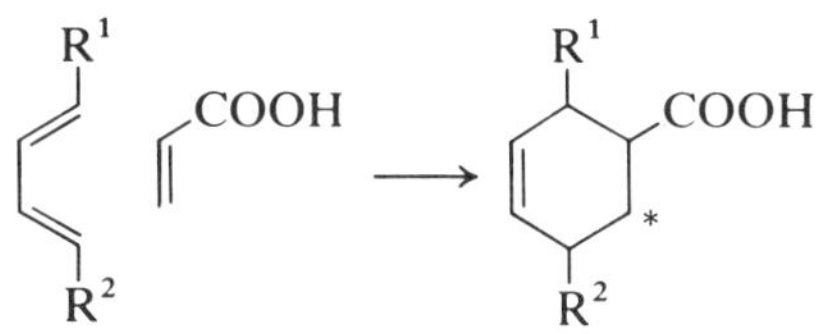

$R^1 = [CH_2]_7COOH$, $R^2 = H_3C(CH_2)_5$
or $R^1 = [CH_2]_8COOH$, $R^2 = H_3C[CH_2]_4$

* Alternative position for COOH group

Dimer acids (section 10.10) are made from tall oil fatty acids by heating with a clay catalyst and water at 220–280 °C under pressure. Remaining monomeric acids are distilled off.

The C_9, C_{10} and C_{12} dibasic acids, as their butyl, hexyl, 2-ethylhexyl and octyl esters, are used as plasticizers for PVC, as engine lubricants and as intermediates in the production of polyamides, polyesters and polyurethane adhesives, coatings, fibres, inks and resins. The C_{21} diacid is used primarily as a surfactant. By condensation with diamines, dimer acids are used extensively to furnish polyamide resins. Their major applications are printing inks, hot-melt adhesives and coatings. Reactive polyamides are used as curing agents for epoxy resins.

References and further reading

Johnson, R.W. (1984) *J. Am. Oil Chem. Soc.*, **61**, 241.
Johnson, R.W. (1988) in *Fatty Acids in Industry* (eds R.W. Johnson and E. Fritz), Marcel Dekker, New York, p. 327.

7.11 Nitrogen derivatives of long-chain acids

Compounds of this type are prepared on an industrial scale and used mainly as cationic surfactants to the extent of about 300 000 tonnes per annum. They include several types of nitrogen-containing compounds, and have one or more long alkyl chains derived mainly from tallow (C_{16}/C_{18}) or coconut oil (C_{12}).

7.11.1 Amides and diamides

Fatty acid amides are produced by reaction of ammonia with fatty acids (180–200 °C, 3–7 bar, 10–12 h) or esters (220 °C, 120 bar, 1 h). The amides of oleic acid and erucic acid are important compounds. They are used as antislip and antiblocking additives at the 0.1–0.5% level for polyethylene film and are permitted in food packaging material. They are also used in printing inks, lubricant additives and mould release agents. Diamides are made from diamines such as 1,2-diaminoethane:

$$H_2N[CH_2]_2NH_2 \xrightarrow{RCOOH} RCONH[CH_2]_2NHCOR$$

Bistline *et al.* (1991) have demonstrated that amides can be made by reaction of methyl or glycerol esters of C_8–C_{16} acids with primary amines such as butanamine at 20 °C in the presence of lipases like that from *Rhizomucor miehei*. Yields are better than 50%, but the reaction does not occur with free acid.

7.11.2 Primary amines

Monoamines (RCH_2NH_2) are made from fats via the acid and the nitrile:

$$\text{fat} \longrightarrow \underset{+\,\text{glycerol}}{RCOOH} \xrightarrow{NH_3} [RCOONH_4]$$

$$\longrightarrow [RCONH_2] \longrightarrow RCN \xrightarrow{H_2} RCH_2NH_2$$

Acids are converted to nitriles by reaction with ammonia at atmospheric pressure at 280–360 °C in the presence of catalysts such as bauxite, zinc oxide, or manganese or cobalt salts. Hydrogenation of the nitrile proceeds via the aldimine (RCH=NH) at 120–180 °C and 20–100 bar in the presence of nickel or cobalt catalysts.

7.11.3 Secondary amines

Under slightly different conditions, hydrogenation of nitrile furnishes a secondary amine ($(RCH_2)_2NH$). The reaction temperature is higher, pressure is somewhat lower, and the ammonia liberated in this reaction must be removed. Secondary amines can also be prepared directly from primary amines or from a mixture of alcohol and primary amine under similar reaction conditions. These compounds are used mainly to prepare quaternary ammonium compounds with two long alkyl chains, i.e. $R_2N^+(CH_3)_2X^-$.

7.11.4 Tertiary amines

Tertiary amines may contain three (R_3N), two (R_2NMe) or one ($RNMe_2$) long alkyl chain. They are used as corrosion inhibitors, fuel additives, flotation agents, fungicides, emulsifying agents, foaming agents and cosmetic ingredients and for the production of quaternary ammonium compounds. Compounds of the type R_3N are made by reduction of nitriles, of nitrile and amine, or of alcohol and amine or ammonia. Methyl groups are introduced when a primary or secondary amine is reacted with formic acid or formaldehyde. Tertiary amines (RCH_2NMe_2) can be made directly from the alcohol (RCH_2OH) and dimethylamine in the presence of thallium sulphate at 360 °C. Tertiary amines of another type result when primary or secondary amines react with ethylene oxide. They have the structure shown in (**14**) and (**15**).

$$RN\begin{cases}(CH_2CH_2O)_xH\\(CH_2CH_2O)_yH\end{cases} \qquad R_2N(CH_2CH_2O)_xH$$

(**14**) (**15**)

7.11.5 Quaternary ammonium compounds

Quaternary ammonium compounds (commonly called quats) make up about 40% of all cationic and amphoteric surfactants. They result from the reaction of primary, secondary or tertiary amine, a polyamine, or an oxyethylated amine with a short-chain alkylating agent such as methyl chloride, methyl sulphate, ethylene oxide, ethyl sulphate, or benzyl chloride. Some typical structures are shown in (**16**)–(**18**). R is often C_{16}/C_{18} and derived from hydrogenated tallow. These compounds are widely employed as fabric and textile softeners, for the manufacture of organo-modified clay for drilling mud, as bactericides and as cationic emulsifiers for asphalt.

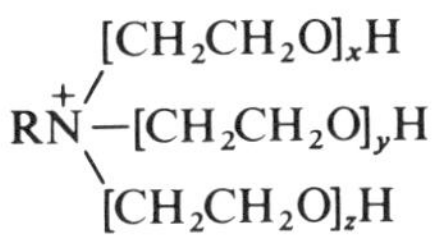
(16)

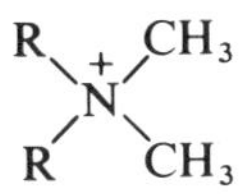
(17)

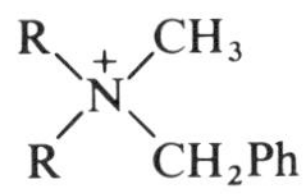
(18)

7.11.6 Polyamines

Polyamines of the type shown in (**19**)–(**21**) are made from amines (or alcohols) and acrylonitrile. These can undergo further reaction through ethoxylation or alkylation. The polyamines and their derivatives are employed as flotation agents in mining, as emulsifiers, as adhesive agents for asphalt, and as water-repellent agents.

$RNH[CH_2]_3NH_2$

(19)

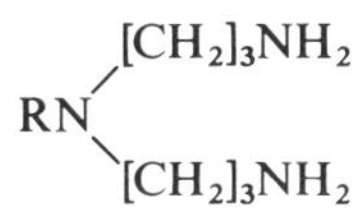
(20)

$RO[CH_2]_3NH_2$

(21)

7.11.7 Amine oxides

Amine oxides such as $RN^+(^-O)(CH_3)_2$, prepared from tertiary amines and hydrogen peroxide, are widely used in the cosmetics industry, and, as cleaners and detergents, can be incorporated into bar soaps (Smith *et al.*, 1991). They are non-toxic surfactants and are readily metabolized and non-mutagenic.

7.11.8 Amphoteric surfactants

Amphoteric surfactants contain a basic nitrogen group and a carboxylic (or sulphonic) acid group. Two possible routes to such compounds are given in the following equations:

$$RNH_2 + CH_2{=}CHCOOH \rightarrow R\overset{+}{N}H_2CH_2CH_2CO\overset{-}{O}$$

$$RN[CH_3]_2 + ClCH_2COOH \rightarrow R\overset{+}{N}[CH_3]_2CH_2CO\overset{-}{O}$$

These compounds are widely used in the cosmetics industry for hair and body care preparations.

References and further reading

Billenstein, S. and Blaschke, G. (1984) *J. Am. Oil Chem. Soc.*, **61**, 353.

Bistline, R.G., Jr, Bilyk, A. and Feairheller, S.H. (1991) *J. Am. Oil Chem. Soc.*, **68**, 95.

Reck, R.A. (1988) in *Fatty Acids in Industry* (eds R.W. Johnson and E. Fritz), Marcel Dekker, New York, p. 177.

Smith, K.R., Borland, J.E., Corona, R.J. and Saver, J.D. (1991) *J. Am. Oil Chem. Soc.*, **68**, 619.

8

PHYSICAL PROPERTIES: STRUCTURAL AND PHYSICAL CHARACTERISTICS*

K. LARSSON and P.J. QUINN

The most characteristic property of lipids from a physicochemical point of view is their amphiphilic behaviour, resulting from the occurrence of a water-repelling (hydrophobic) part and a water-attracting (hydrophilic) part of the molecule. Our environment is full of interfaces dominated by lipids – Nature's most surface-active compounds – from the envelope of every living cell to the surface of the sea. Numerous biological functions and technical applications are based on the amphiphilic properties of lipids. Important milestones in the understanding of the physical chemistry of lipids are determinations of molecular arrangement and phase behaviour in the solid state, in aqueous systems and in monomolecular surface films.

This chapter is focused on relations between structure and physical properties of lipids. After a presentation of general features in the physical chemistry of lipids, the different groups of common lipids are described.

8.1 Solid-state behaviour of lipids

Multiple-melting phenomena of fatty substances were observed during the nineteenth century, and Malkin (1930) was able to explain this behaviour as being due to *polymorphism*, i.e. the occurrence of alternative crystal structures in the solid state. Later the structural features of lipids in the crystalline state were seen in the first crystal structure determinations by Vand *et al.* (1949). Numerous crystal structure determinations can now be found in the literature, and some general features in the solid-state packing of lipid molecules are summarized below.

8.1.1 Molecular arrangement in layers

Simple lipids with one straight and saturated hydrocarbon chain linked to a polar headgroup, such as a fatty acid, will first be considered. The amphiphilic nature of the molecule is reflected in its solid-state structure. Thus the non-polar parts and the polar headgroups form separate regions. This makes possible a very effective interaction between the polar headgroups. A fragment of such a crystal is illustrated in Fig. 8.1. The molecules are arranged in parallel in layers with extended hydrocarbon chains. The polar groups form sheets, and the molecules in adjacent layers are oriented head-to-head. In this way an infinite lipid bilayer structure is formed, with a methyl endgroup gap in the middle and with surfaces consisting of the polar headgroups. This bilayer unit is the most important structural element in lipids.

If the forces between the polar headgroups are very weak, the molecules are arranged head-to-tail, as shown in Fig. 8.2. Simple esters, such as butyl stearate, form such structures. The alcohol chains shield the forces between the polar groups.

There are two main types of arrangements of molecules in the lipid bilayer type of structure. The molecules can be either vertical or tilted in relation to the endgroup chain, and both alternatives are shown for comparison in Fig. 8.3. The tilted type of structure gives increased lateral space for the polar headgroup, and it is thus possible to accommodate successively larger polar headgroups into the bilayer structure by a corresponding increase in the angle of tilt. As will be discussed further below, however, the chain tilt can only adopt certain discrete values.

In lipids with very large polar headgroups, an even higher angle of tilt is not enough to make sufficient

* Professor Larsson is grateful to Gunnel Lundh and Anita Sandberg for help in preparing his parts of this chapter and to Dr Krister Fontell for valuable discussions.

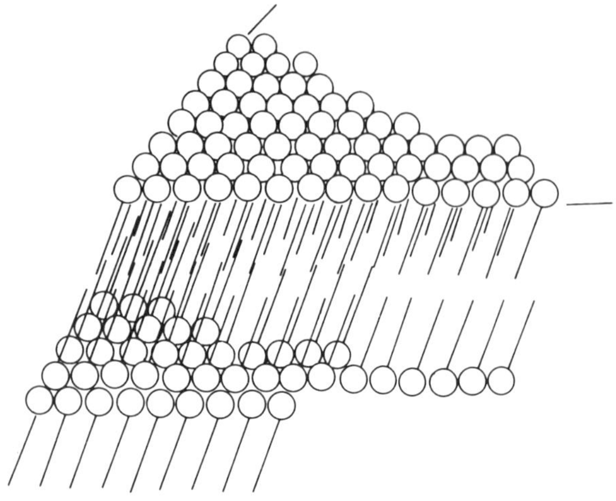

Fig. 8.1 Characteristic structure of lipids in the solid state. A fragment of the crystal is indicated showing the parallel arrangement of the chains and the packing of the polar heads in surface layers, so that bimolecular-unit layers are formed.

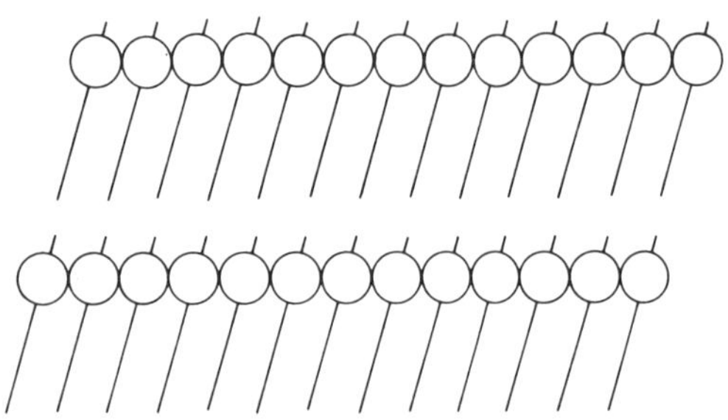

Fig. 8.2 Head-to-tail arrangement of lipids with weak forces between the polar headgroups. In esters of normal alcohols and fatty acids, the attraction between the polar groups is shielded enough to give such a monolayer type of structure.

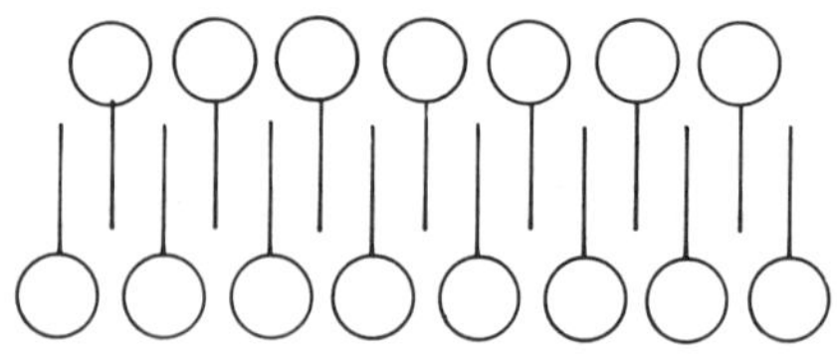

Fig. 8.4 Chain penetration structure, characteristic for lipids with large polar headgroups such as psychosine hydrochloride (Abrahamsson *et al.*, 1972).

space. An example of such a lipid is psychosine hydrochloride (Abrahamsson *et al.*, 1972), containing a sugar group in the polar region. This type of structure is shown in Fig. 8.4. The penetration of chains through the methyl endgroup gap increases the space available for the polar headgroup by a factor of 2, and the structural unit is a hydrocarbon-chain monolayer. Also this type of structure can have vertical chains, as in Fig. 8.4, or tilted chains.

In complex lipids with more than one chain per acyl group, and with different hydrocarbon chain structures, there are additional factors that will determine the arrangement in layers. One such factor is a chain-segregation mechanism, so that saturated chains are located in one layer and unsaturated chains in a separate layer. These special structural features will be described in connection with the different lipid classes below.

8.1.2 Chain packing

All lipids form crystal structures with the hydrocarbon chains arranged in layers. The chains are extended in an all-*trans* conformation, so that a planar zig-zag carbon chain is formed. The chains are close-packed with parallel chain axes, except for one crystal form of soaps that pack with crossed chain axes. Such a complex chain packing is due to the dominating influence of the ionic endgroups in relation to the effect of chain packing on the lattice energy.

It should also be mentioned that different chain-packing arrangements have been derived theoretically by Kitaigorodskii (1957) from the geometry of extended chains. The results are in good agreement with the experimentally determined chain-packing structures described below (Hagemann and Rothfus, 1992).

The packing arrangement of hydrocarbon chains can be described by the so-called *subcell*, which is

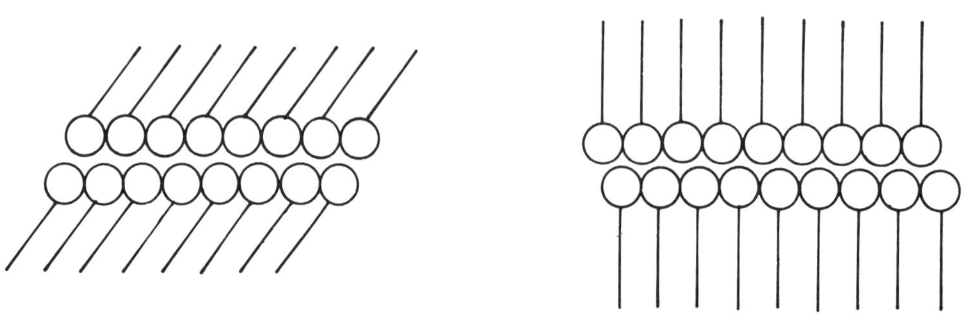

Fig. 8.3 Tilted and vertical long-chain molecules in the usual type of lipid bilayer structure.

the smallest crystallographic unit in a hydrocarbon-chain layer. By lateral repetition of the subcell, the whole structure of the chain region is obtained. The subcells have been described by a letter indicating the symmetry, for example T for triclinic, followed by a symbol indicating the geometry of the chain planes.

(a) *Simple chain-packing subcells with parallel chain planes*

The stable crystal form of triglycerides exhibits the triclinic chain packing T∥, and this subcell structure was first determined by Vand and Bell (1951). The subcell contains two CH_2 groups related by a centre of symmetry. The dimensions in the case of n-octadecane (Nyburg and Lüth, 1970) are: $a = 4.285$ Å, $b = 5.414$ Å, $c = 2.539$ Å, $\alpha = 80.99°$, $\beta = 112.2°$ and $\gamma = 121.78°$. The cross sectional area per chain is 18.23 Å^2, which demonstrates that this is a very effective close-packing arrangement.

There is also a monoclinic chain-packing subcell with all chain planes parallel, M∥, and it is shown together with the triclinic, T∥ in Fig. 8.5. This chain packing has been observed in 3-thiadodecanoic acid (Abrahamsson and Westerdahl, 1963), in monoglycerides (Larsson, 1964) and in 2-DL-hydroxytetradecanoic acid (Dahlén *et al.*, 1972). The dimensions of M∥ in the hydroxy fatty acids are: $a = 4.30$ Å, $b = 4.74$ Å, $c = 2.54$ Å and $\gamma = 110.9°$. This symmetry is $P2_1/m$, and the cross-sectional area per chain is 19.04 Å^2.

(b) *Simple chain-packing subcells with two orientations of chain planes*

The two orthorhombic chain-packing arrangements, where every second chain plane is perpendicular to the rest, are shown in Fig. 8.6. The most common one, O⊥, was first analysed in polyethylene (Bunn, 1939), and it is also found in crystal forms of n-alkanes and fatty acids. The dimensions in the case of polyethylene, where the chain endgroup disturbances are minimal, are: $a = 4.93$ Å, $b = 7.40$ Å, $c = 2.534$ Å, with a cross-sectional area per chain of 18.24 Å^2. The subcell contains four CH_2 groups and the space group is Pbnm. This orthorhombic chain packing and the triclinic packing (T∥) are the closest packing arrangements that can be achieved; these two packing arrangements are therefore the only ones occurring in stable forms of simple lipids. Whether T∥ or O⊥ occurs depends on the ideal tilt of the molecules, as discussed further below.

An alternative simple orthorhombic chain packing, O′⊥, which is closely related to O⊥, has been observed in three lipids, two of which are branched fatty acids (Abrahamsson, 1959). The space group is the same (Pbnm); the main difference is a kind of inversion of the a and b axes. In the unusual chain packing O′⊥, $a \simeq 7.4$ Å and $b \simeq 5.0$ Å. In 2-D-methyloctadecanoic acid (Abrahamsson, 1959), the dimensions are: $a = 7.43$ Å, $b = 5.01$ Å, $c = 2.50$ Å, with a cross-sectional area of 18.61 Å^2.

(c) *Complex chain-packing subcells with parallel chain planes*

Two orthorhombic chain-packing subcells with parallel chain planes, O∥ and O′∥, have been found. A branched fatty acid, 2-methyl-2-ethylicosanoic acid, exhibits O∥ (von Sydow, 1959), and the dimensions are: $a = 8.15$ Å, $b = 9.22$ Å, $c = 2.55$ Å. The cross-sectional area per chain is 18.77 Å^2, and the unit

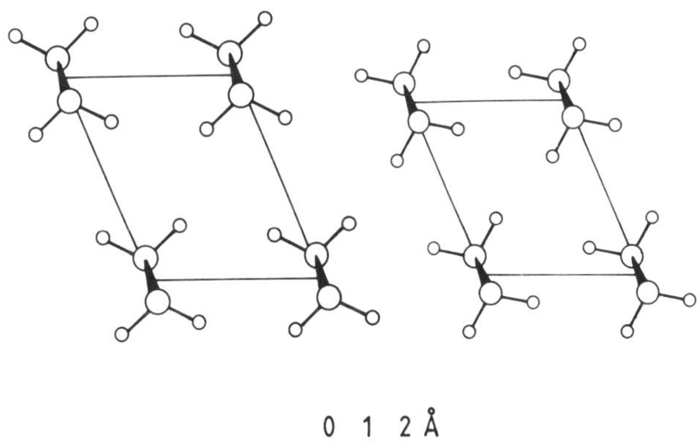

Fig. 8.5 The triclinic chain packing T∥ (left) and the monoclinic chain packing M∥ (right) viewed along the chain axes. The zig-zag units of the hydrocarbon chains are shown (carbon atoms illustrated by large spheres and hydrogen atoms by small spheres).

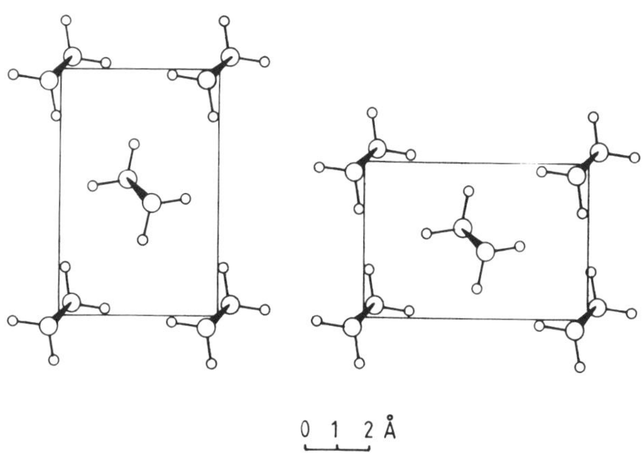

Fig. 8.6 The orthorhombic chain-packing arrangements O⊥ (left) and O′⊥ (right) (the same notations as in Fig. 8.5).

cell contains eight CH_2 groups arranged according to the space group $B22_12$. The other chain arrangement, O′‖, was found in the low-wetting crystal form of oleic acid (Abrahamsson and Ryderstedt-Nahringbauer, 1962), and the dimensions of the unit cell are $a = 7.93$ Å, $b = 4.74$ Å and $c = 2.53$ Å. The cross-sectional area per chain is 18.79 Å^2, and the space group is Pma2. These two orthorhombic chain-packing structures are shown in Fig. 8.7.

(d) Hybrid types of subcells

Abrahamsson *et al.* (1978) have described two hybrid subcells in complex lipids. Thus a twinned version of the O⊥ subcell, with dimensions a = 10.3 Å, $b = 7.5$ Å, $c = 2.54$ Å, was observed in the crystal structure of a cholesteryl ester (Abrahamsson and Dahlén, 1977), in 1,2-dilauroyl-DL-phosphatidylethanolamine (Hitchcock *et al.*, 1974) and also in the corresponding dipalmitoyl analogue (Dorset, 1976).

Another hybrid packing was observed in a cerebroside crystal structure (Pascher and Sundell, 1977) with dimensions a = 9.2 Å, b = 9.8 Å, c = 2.55 Å, and γ = 98°. This packing contains structural features of O⊥ and O′⊥ as well as of T‖, M‖ and O′‖. The hybrid subcells are shown in Fig. 8.8

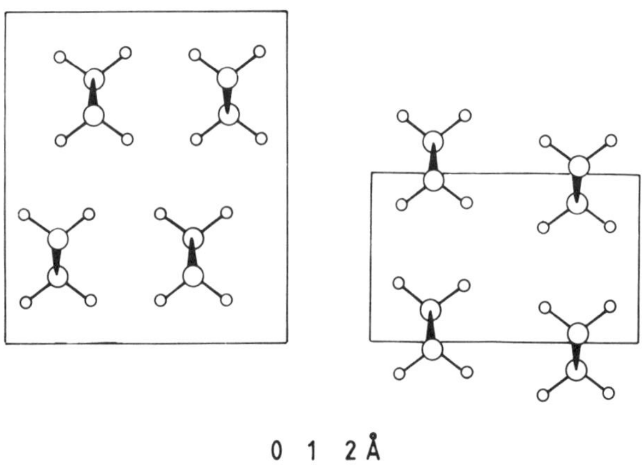

Fig. 8.7 The orthorhombic chain-packing subcells O‖ (left) and O′‖ (right) (the same notations as in Fig. 8.5).

(e) The hexagonal chain packing

Near the melting point of crystal forms with the orthorhombic chain packing O⊥, the dimensions of the subcell approach those of an orthohexagonal cell. Thus the a axis remains almost constant or is somewhat reduced in length, whereas the b axis increases towards 8.3 Å. Many lipids, like alkanes, n-alcohols, simple esters and glycerides, solidify from the melt in a crystal form with this hexagonal chain packing. When the temperature is lowered, there is a transition into a crystal form with the orthorhombic chain packing O⊥. An X-ray single-crystal analysis of this crystal form of n-nonadecane (Larsson, 1967) showed a hexagonal subcell with orthohexagonal axes $a = 4.79$ Å and $b = 8.30$ Å, and an intensity distribution that corresponds to complete rotational disorder of the chains. Also the physical properties of the crystals are in agreement with the rotational freedom of the molecules; the crystals are thus plastic from a rheological point of view. Owing to these properties, esters occurring in this form have been called waxes.

Intermediate pseudohexagonal states between the chain geometry of the orthorhombic subcell O⊥ and the hexagonal subcell have also been observed. The structure of sodium dodecyl sulphate (Sundell, 1977) is one example.

(f) Determination of chain packing from X-ray data

Each of the chain-packing subcells described above is characterized by a unique set of X-ray diffraction

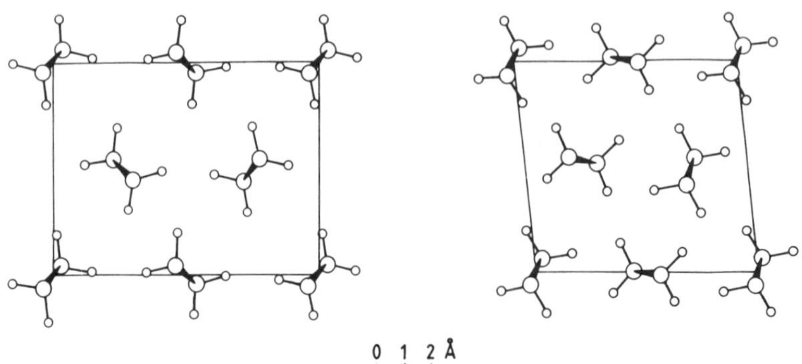

Fig. 8.8 The hybrid subcells HS1 (left) and HS2 (right) (the same notations as in Fig. 8.5).

lines in the wide-angle region between 3.5 and 4.5 Å, which correspond to the different crystallographic planes of the subcell. X-ray data of the most common subcells in lipids will be given here.

The hexagonal chain arrangement is easy to identify unambiguously, as it exhibits only one dominant diffraction line in the short-spacing region. This line is observed in the interval 4.15 to 4.20 Å (if this value is multiplied by $2/\sqrt{3}$, the hexagonal cell axis is obtained).

The triclinic subcell, T∥, gives a series of relatively weak diffraction lines with only one dominant line, which occurs near 4.6 Å.

The two orthorhombic chain-packing subcells, O⊥ and O′⊥, are characterized by two strong diffraction lines at 3.8 and 4.2 Å.

8.1.3 Polymorphism and nomenclature of lipid crystal forms

The existence of alternative crystal structures is a characteristic property for lipids. This is due to the possibilities for different chain-packing arrangements and for variations in the tilt of the chains within the layers as discussed above.

The chains can only adopt specific angles of tilt, which are characteristic for each chain-packing subcell. If we consider one particular chain packing, for example O⊥, it is natural to describe the orientation of the subcell in relation to the true unit cell by giving the subcell indices of the endgroup plane. Thus O⊥ (001) means that the subcell (*ab*) plane coincides with the endgroup plane, the chains being vertical in such a structure. The orientation O⊥ (101) means that the *ac* diagonal and the *b* axis of the subcell are parallel to the endgroup plane, and this corresponds to an angle of tilt of about 66°. Only three tilt angles of this chain packing have been observed in simple lipids: they are O⊥ (101), O⊥ (111) and O⊥ (021).

In a similar way there are just a few endgroup plane alternatives in each of the other chain-packing subcells. The occurrence of these chain tilt alternatives is due to the tendency of the methyl endgroups of one chain layer to lie in one plane. The angle of tilt is usually limited to angles above about 50°. Higher angles of chain tilt mean that there are direct contacts between methyl groups of one chain layer and methylene groups of the layer on the side of the endgroup gap, and this is energetically unfavourable.

Two types of polymorphism occur in lipids. When each form is thermodynamically stable in a definite range of temperature and pressure, the forms are *enantiotropic*. Thus one enantiotropic form changes to another at a certain transition temperature. If, on the other hand, one of two forms is thermodynamically unstable, the two forms are said to be *monotropic*. The phase transitions in lipids are usually of the monotropic type, for example in glycerides.

Furthermore, phase transitions can be of different orders. A first-order transition is characterized by a discontinuous entropy and enthalpy curve versus temperature at constant pressure, whereas a so-called continuous phase transition shows continuous entropy and enthalpy curves.

A nomenclature for lipid crystal forms originally proposed for glycerides (Larsson, 1966) was later applied also to other lipids and will be used in the present chapter. The nomenclature is based on the following three criteria:

1. A form that gives only one strong short-spacing line in the X-ray diffraction pattern near 4.15 Å is termed α.
2. A form that gives two short-spacing lines near 3.80 and 4.20 Å and also shows a doublet in the 720 cm^{-1} region of the infrared absorption spectrum is termed β'.
3. A form that does not satisfy criteria 1 or 2 is termed β.

When, on this basis, more than one crystal form of the same compound has the same symbol, they are then distinguished by subscripts (for example, β_1, β_2, etc.) in order of decreasing melting points.

Crystal forms of fatty acids and soaps have long been termed A, B, C, etc., in order of increasing angle of tilt. Both this earlier and the new nomenclature will be used in parallel for fatty acids.

References and further reading

Abrahamsson, S. (1959) *Ark. Kemi*, **14**, 65.

Abrahamsson, S. and Dahlén, B. (1977) *Chem. Phys. Lipids*, **20**, 43.

Abrahamsson, S. and Ryderstedt-Nahringbauer, I. (1962) *Acta Crystallogr.*, **15**, 1261.

Abrahamsson, S. and Westerdahl, A. (1963) *Acta Crystallogr.*, **16**, 404.

Abrahamsson, S., Pascher, I., Larsson, K. and Karlsson, K.-A. (1972) *Chem. Phys. Lipids*, **8**, 152.

Abrahamsson, S., Dahlén, B., Löfgren, H. and Pascher, I. (1978) *Prog. Chem. Fats Other Lipids*, **16**, 125.

Bunn, C.W. (1939) *Trans. Faraday Soc.*, **35**, 482.

Dahlén, B., Sundell, S. and Pascher, I. (1972) *Acta Crystallogr.*, **B28**, 2396.

Dorset, D.L. (1976) *Biochim. Biophys. Acta*, **424**, 396.

Hagemann, J.W. and Rothfus, J.A. (1992) *J. Am. Oil Chem. Soc.*, **69**, 429.

Hitchcock, P.B., Mason, R., Thomas, K.M. and Shipley, G.G. (1974) *Proc. Natl Acad. Sci. USA*, **71**, 3036.

Kitaigorodskii, A.I. (1957) *Organic Chemical Crystallography*, Consultants' Bureau, New York.
Larsson, K. (1964) *Ark. Kemi*, **23**, 35.
Larsson, K. (1966) *Acta Chem. Scand.*, **20**, 2255.
Larsson, K. (1967) *Nature*, **213**, 383.
Malkin, T. (1930) *J. Am. Chem. Soc.*, **52**, 3739.
Nyburg, S.C. and Lüth, H. (1970) *Acta Crystallogr.*, **B26**, 645.
Pascher, I. and Sundell, S. (1977) *Chem. Phys. Lipids*, **20**, 175.
Sundell, S. (1977) *Acta Chem. Scand.*, **A31**, 799.
Vand, V. and Bell, I.P. (1951) *Acta Crystallogr.*, **4**, 465.
Vand, V., Lomer, T.R. and Lang, A.R. (1949) *Acta Crystallogr.*, **2**, 214.
von Sydow, E. (1959) *Acta Chem. Scand.*, **12**, 777.

8.2 Lipid–water interaction and liquid-crystalline phases

For more than 2000 years liquid-crystalline phases of soaps have been used for cleaning. The physical structure of these phases was revealed only recently. This was mainly due to the opinion that soap molecules should be regarded as rigid rods. In an infrared absorption spectroscopy study of anhydrous soaps, Chapman (1958) concluded that a high-temperature phase transition was caused by melting of the chains. Later Luzzati *et al.* (1960) conclusively demonstrated the liquid nature of the hydrocarbon chains as a fundamental feature in the structure of liquid-crystalline phases. They were also able to determine the structures of the most common liquid-crystalline phases. A database on lipid phase transitions is now available from the National Institute of Standards and Technology, Gaithersburg, MD 20899, USA. The reader is also referred to section 9.3.5.

8.2.1 Lamellar and hexagonal liquid-crystalline phases

If a lipid crystal (see Fig. 8.1) is heated, it either melts directly or is transformed into a liquid-crystalline phase (alternatively called a mesomorphic phase). Such a liquid-crystalline phase obtained by heating is classified as thermotropic, and its formation can be explained in the following way. From temperature factors determined in crystal structure studies of lipids, it is known that the thermal vibrations of the hydrocarbon chain atoms increase strongly with temperature; the corresponding dimensional changes were discussed in the preceding section. In polar lipids, which are able to form liquid-crystalline phases, the bonds in the sheets formed by the polar headgroups are quite strong compared to the weak van der Waals' interaction between adjacent chains. It is therefore not surprising that a temper
can exist before melting where the c
'melted' although the polar groups are still associated in sheets, so that the overall structure consists of lipid bilayers with disordered chains. Such a structure with short-range disorder and long-range order in one dimension is called liquid-crystalline.

More commonly in lipids, however, a liquid-crystalline phase is formed in the presence of water, as shown in Fig. 8.9. Above a critical hydrocarbon-chain 'melting' temperature, water penetrates the polar region and a lamellar lipid–water structure is formed with water layers alternating with lipid bilayers.

Two other structures, which together with the lamellar phase are the most important liquid-crystalline phases in lipid–water systems, are shown in Fig. 8.9. It should be pointed out that the classification of lipids into polar and non-polar is best defined from their interaction with water. Lipids that do not give lipid–water phases are thus non-polar, whereas those forming aqueous phases are classified as polar lipids.

The two hexagonal structures shown in Fig. 8.9 are termed H_I and H_{II}. The H_I phase consists of infinite cylinders of lipid molecules with their polar groups exposed on the surface and with a core of hydrocarbon chains in a disordered liquid-like conformation. These cylinders are hexagonally arranged in a water medium. The H_{II} phase is the inverse of H_I, i.e. water forms cylinders and hydrocarbon chains form the continuous medium with the polar groups at the water–hydrocarbon chain interface.

The lamellar and hexagonal phases can be identified unambiguously from their X-ray diffraction patterns. The lamellar phase shows a series of reflections corresponding to one-dimensional periodicity. From the X-ray data it is possible to

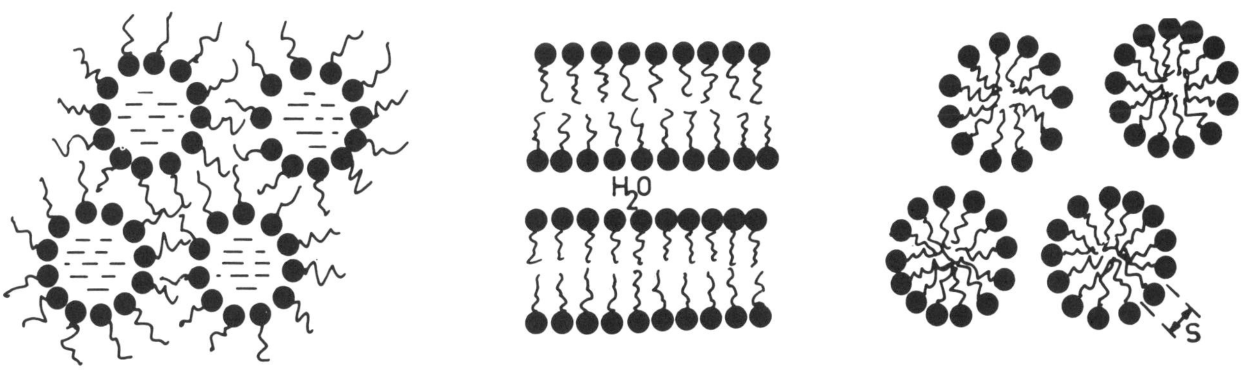

Fig. 8.9 Structure of the three most common liquid-crystalline phases. The lamellar L_α phase is shown in the middle, the hexagonal H_I phase to the right and the hexagonal H_{II} phase to the left.

determine the thickness of the lipid bilayer, the cross-section per polar headgroup and the water-layer thickness. In the same way the diameter of the cylinder in the H_I and H_{II} phases can be determined as well as the cross-sectional area per polar headgroup and the distance between adjacent cylinders. These procedures are described in the fundamental work by Luzzati *et al.* (1960), in which the most important structures of amphiphile–water phases were first revealed.

The liquid nature of the chains was derived from X-ray data. It was thus found that the thickness of the lipid bilayer decreases with temperature with a quite large coefficient of linear thermal expansion: about 10^{-3}/K. Such an effect is only consistent with a highly disordered chain conformation; furthermore, it was found that the X-ray scattering characteristics are the same as in liquid paraffins.

The lamellar liquid-crystalline phase has been termed L_α.

8.2.2 Cubic phases

Several cubic lipid–water phases have been observed. Usually they are formed in the temperature and existence range between H_{II} and L_α, L_α and H_I, or between H_I and the micellar solution. Their structure has been the subject of numerous studies, and the last word has certainly not yet been said.

A cubic phase in soap–water systems was described in the work by Luzzati *et al.* (1960) in which the lattice was then assumed to be face-centred. Cubic phases were also observed in other lipid systems (cf. Clunie *et al.*, 1965). A structure consisting of water spheres in a hydrocarbon-chain matrix was then proposed by Luzzati and Reiss-Husson (1966). Soon afterwards Luzzati and Spegt (1967) analysed the structure of the cubic phase of strontium myristate, which was reported to consist of two networks of rods formed by the polar groups embedded in a hydrocarbon-chain matrix. The proposed structure is shown in Fig. 8.10. The space group Ia3d was determined from 13 observed reflections, and calculated structure factors according to this model were in good agreement with the observed ones.

This rod–matrix structure was considered to be the general structure for cubic lipid–water phases (Luzzati *et al.*, 1968). Thus anhydrous strontium myristate, lecithin and galactolipids from maize chloroplasts with low water content were thought to have the structure shown in Fig. 8.10, with the polar groups arranged in two network systems of rods. The inverse structure, i.e. with hydrocarbon chains forming the rod systems and water forming the outer medium, was proposed to exist in aqueous systems of potassium soaps of lauric, myristic and palmitic acid and in aqueous systems of lauroyl- and palmitoyl-trimethylammonium bromide.

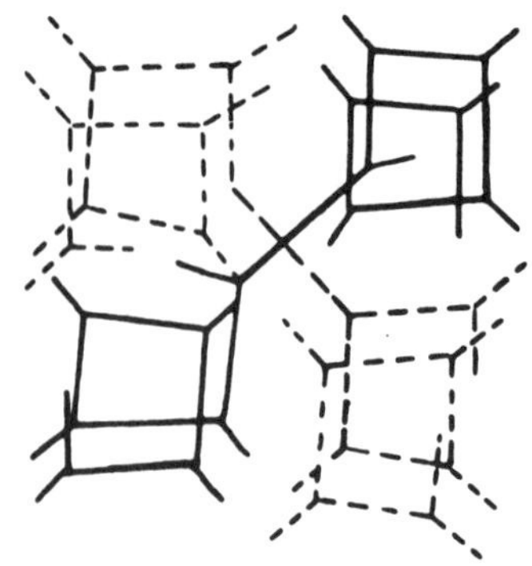

Fig. 8.10 Structure determined by Luzzati and Spegt (1967) in strontium myristate. The polar groups form two networks (indicated by full and broken lines respectively), and the disordered hydrocarbon chains fill the space in between.

Studies of the cubic monoglyceride–water phase indicated that the structure was closely related to that of the lamellar phase (Lindblom *et al.*, 1979). Thus heating the L_α phase results in a transition into the cubic phase. The $d(001)$-spacing of the L_α phase persists through the transition and goes over to cubic spacing. Furthermore it was concluded from NMR diffusion measurements that the local environment of the lipid molecules was similar to the lamellar phase and different from phases with molecules arranged in rod systems. Another conclusion from the NMR measurements was that the structure was both lipid- and water-continuous. The structure was proposed to consist of lipid bilayer units with hexagonal shape connected into a network according to space group Im3m. It was pointed out that the hexagonal units would be expected to be curved in order to avoid sharp edges. Later it was realized (Larsson *et al.*, 1980) that such a structure is identical to the primitive cubic alternative of an *infinite periodic minimal surface* (IPMS) according to Schwarz. The structure is shown in Fig. 8.11. The methyl endgroup gap of the bilayer is identical to the IPMS, and the infinite lipid bilayer thus formed separates two identical water channel systems. It should be pointed out that the space group of the surface is Pm3m, but the twofold symmetry of the lipid bilayer curved in the same way means that the structure should have space group Im3m and thus be body-centred as observed experimentally.

Later, however, Longley and McIntosh (1983) reported that monoolein–water forms a cubic phase with a primitive lattice. They gave convincing evidence for a structure consisting of a lipid bilayer forming an IPMS, as discussed above, which

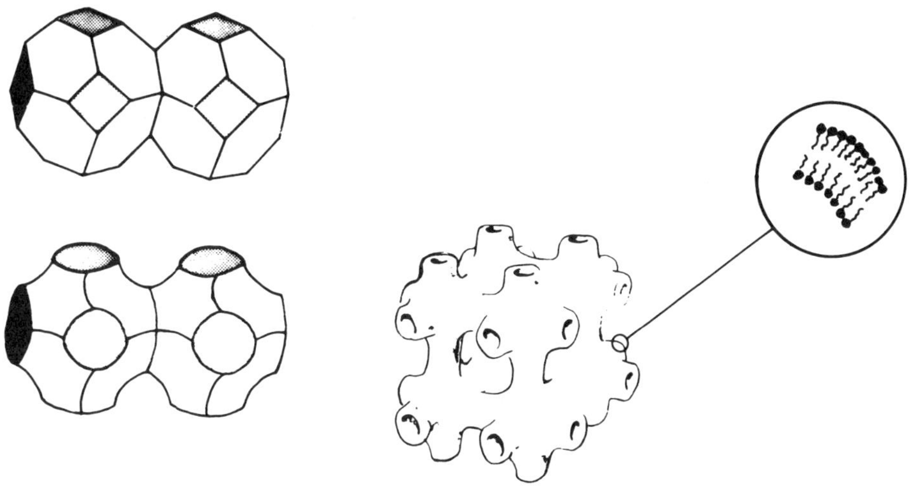

Fig. 8.11 One of the body-centred structures in lipid systems. Two structural elements are shown to the left (consisting of planar hexagons above and deformed to correspond to an infinite periodic minimum surface below). The tube network shown to the right separates two water channel systems.

separates two identical water channel systems with tetrahedral geometry (according to the diamond lattice). From these results it was concluded that monoolein in fact forms two cubic phases (Larsson, 1983), one with a body-centred lattice at low water content (where the L_α phase is formed by cooling the cubic phase), and the other with a primitive lattice at high water content.

The possibility that IPMS can occur in lipid systems as a structure of microemulsions or liquid-crystalline phases was first pointed out by Scriven (1976). The concept of IPMS and the possibilities of cubic phases consisting of lipid bilayers curved as an IPMS have attractive features that merit consideration. The lipid bilayer is the dominant structure type in lipid systems. In most cases there is an L_α phase on one side of the cubic phase in the phase diagram. The IPMS is intersection-free, so that the main difference compared to the infinite bilayer structure of the L_α phase is the curvature, which can allow for wedge-shaped molecules (see further discussions in section 8.2.8). The hydrostatic pressure is the same everywhere in an IPMS structure. All molecules can in fact have a similar close-packing environment, with lateral diffusion freedom similar to that of the L_α phase.

There are three fundamental IPMS structures with cubic symmetry. One is the body-centred lattice shown in Fig. 8.11 and another is the primitive diamond lattice proposed for monoolein at high water content (Longley and McIntosh, 1983). The third is the so-called gyroid (Shoen, 1970), which is also body-centred with space-group number 214. This IPMS separates two helical labyrinths with opposite handedness. The labyrinths or channels are centred on the cubic (100) directions and (111) directions. These directions are in fact identical with the backbone of the rod system proposed by Luzzati *et al.* (1968) (see Fig. 8.10). The symmetry of a bilayer structure, which is an IPMS, is higher than that of the minimum surface itself owing to the twofold symmetry relating molecules on both sides of the methyl endgroup gap. It is therefore possible to regard the structure of cubic Ia3d phases as a single lipid bilayer separating two helical water channel systems.

The body-centred lattice of the cubic phase of monoglycerides at low water content is consistent with the gyroid lattice, as well as with the structure shown in Fig. 8.11. Unfortunately available X-ray powder data cannot unambiguously exclude one of these two alternatives.

The concept of IPMS should be involved in connection with structural discussions on cubic lipid–water phases. These three fundamental cubic IPMS structures occur through the need to account for the wide variations in the shape of lipid molecules. Although the structures with the methyl endgroup gap located in the IPMS appear to be the most probable type, the reversed alternatives, i.e. bilayers with the gap between the polar headgroups forming the IPMS, might exist.

A new type of cubic structure has recently been reported by Luzzati *et al.* (1992). It consists of two types of reversed micelles embedded in a continuous hydrocarbon-chain matrix. This phase is formed in many systems of water-swelling lipids mixed with lipids that do not form aqueous phases. The proposed structure follows space group Fd3m.

Recent reviews on cubic lipid phases and their

biological significance have been reported by Lindblom and Rilfors (1989), Larsson (1989) and Seddon (1990).

Finally it should be mentioned that NMR diffusion studies indicate that another type of cubic phase exists with the amphiphiles forming micelles arranged in a cubic lattice surrounded by water. Such a structure seems to be due to strong electrostatic repulsion between the molecules. As it has mainly been observed in detergent-type systems, it is not considered further here. Among lipids, it occurs in aqueous systems of lysophosphatidylcholine (Arvidsson *et al.*, 1985).

8.2.3 Lipid–water phase diagrams

Lipid–water systems of a particular polar lipid of biological origin, for example lecithin, behave like two components with regard to the phase rule, although the lipid can be a mixture of hundreds of molecular species. Phase equilibria in binary lipid–water systems are illustrated by composition–temperature phase diagrams, as shown in Fig. 8.12. A homologous series of monoglycerides is used here as an example. At chain lengths C_8–C_{12}, only the lamellar L_α is formed. When the chain length is increased, there is also a cubic region with the two phases discussed in a previous section. Finally, at chain lengths above C_8, a phase termed L2 is formed when the temperature is increased; the structure of this phase is discussed in the next subsection.

There are two types of lipid–water phase diagrams. The first type, discussed above, is obtained from polar lipids, which are 'insoluble' in water (i.e. the solubility is quite small; monolaurin for example has a solubility of about 10^{-5} M). Figure 8.12 illustrates the principles of phase equilibria in this type of lipid–water system. The second type of binary system is obtained when the lipid is soluble as micelles in water. Examples of such lipids are fatty acid salts and lysolecithin. An aqueous soap system is illustrated in Fig. 8.13. When the lipid concentration in the micellar solution is increased, the spherical micelles are transformed into rod-shaped micelles. At still higher lipid concentrations the lipid cylinders are hexagonally

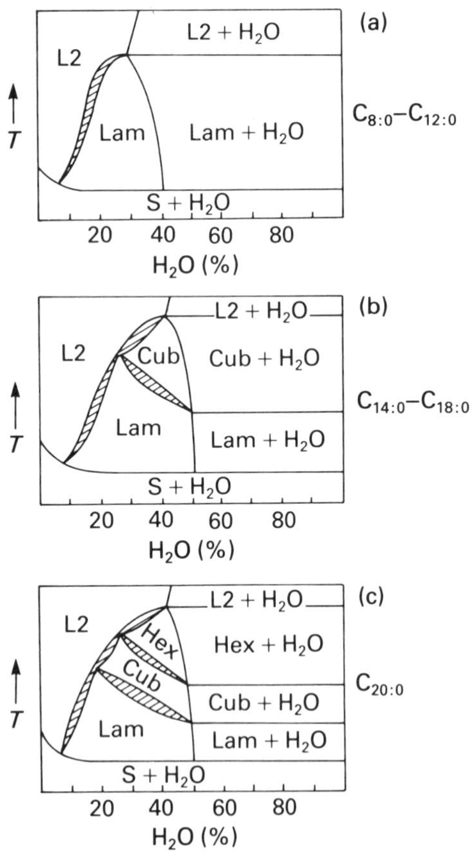

Fig. 8.12 Main features of monoglyceride–water phase diagrams at different hydrocarbon chain lengths (Lam, Cub and Hex correspond to regions of lamellar, cubic and hexagonal H_{II} phases respectively). Below the hydrocarbon-chain melting temperature, there is solid (S) in equilibrium with water. Small two-phase regions are shaded.

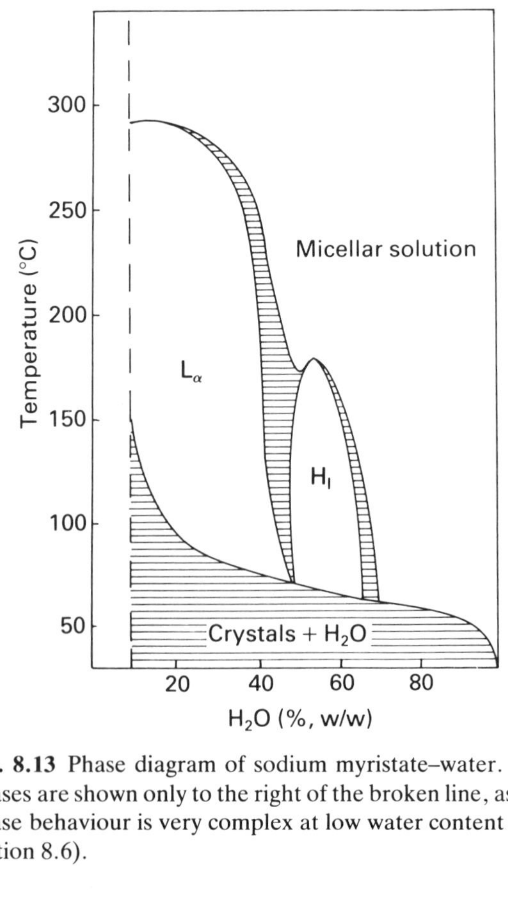

Fig. 8.13 Phase diagram of sodium myristate–water. The phases are shown only to the right of the broken line, as the phase behaviour is very complex at low water content (see section 8.6).

arranged and the liquid-crystalline phase H_I is formed. The lamellar liquid-crystalline phase L_α is usually formed in the region between H_I and the anhydrous lipid. Excellent reviews of the association behaviour of amphiphiles of this type have been published (Wennerström and Lindman, 1979; Lindman and Wennerström, 1980).

When a non-polar lipid is also introduced, the phase behaviour gets quite complex. An example of such a ternary lipid system (Lindström *et al.*, 1981) is shown in Fig. 8.14. The triglyceride component, as well as the monoglyceride component, contains numerous compounds with different chains and both behave as one component with respect to the phase rule. It is interesting to note that the transition $L_\alpha \rightarrow$ cubic $\rightarrow H_{II} \rightarrow$ L2, obtained by heating the binary monoglyceride system (Fig. 8.12), is obtained by solubilization of triglyceride molecules into the bilayer according to this phase diagram. This is due to an increased wedge shape of the average lipid molecule, which is further discussed in section 8.2.8.

Ternary systems consisting of lecithin–water–triglyceride oil, described by Rydhag and Wilton (1981), are dominated by a three-phase region where water, oil and the lamellar L_α phase exist in equilibrium.

According to a ^{13}C NMR study by Hamilton and Small (1981) of the solubilization of triolein in lecithin bilayers in excess water, the maximum solubility of triolein is about 2.8% (w/w), and beyond this limit the triolein forms a separate oil phase. The triglyceride is solubilized in the bilayer with the glycerol group at the surface. Recent phase diagrams involving triglyceride oil, water and phospholipids have been reported by Söderberg (1990).

8.2.4 L2 phases, L3 phases and microemulsions in aqueous lipid systems

Ekwall introduced the terms L1 for an ordinary micellar solution and L2 for the reversed type of liquid phase, i.e. water aggregates in a continuous hydrocarbon-chain medium (cf. Ekwall, 1975). The L2 phase is formed in binary lipid systems above the existence temperature of liquid-crystalline phases

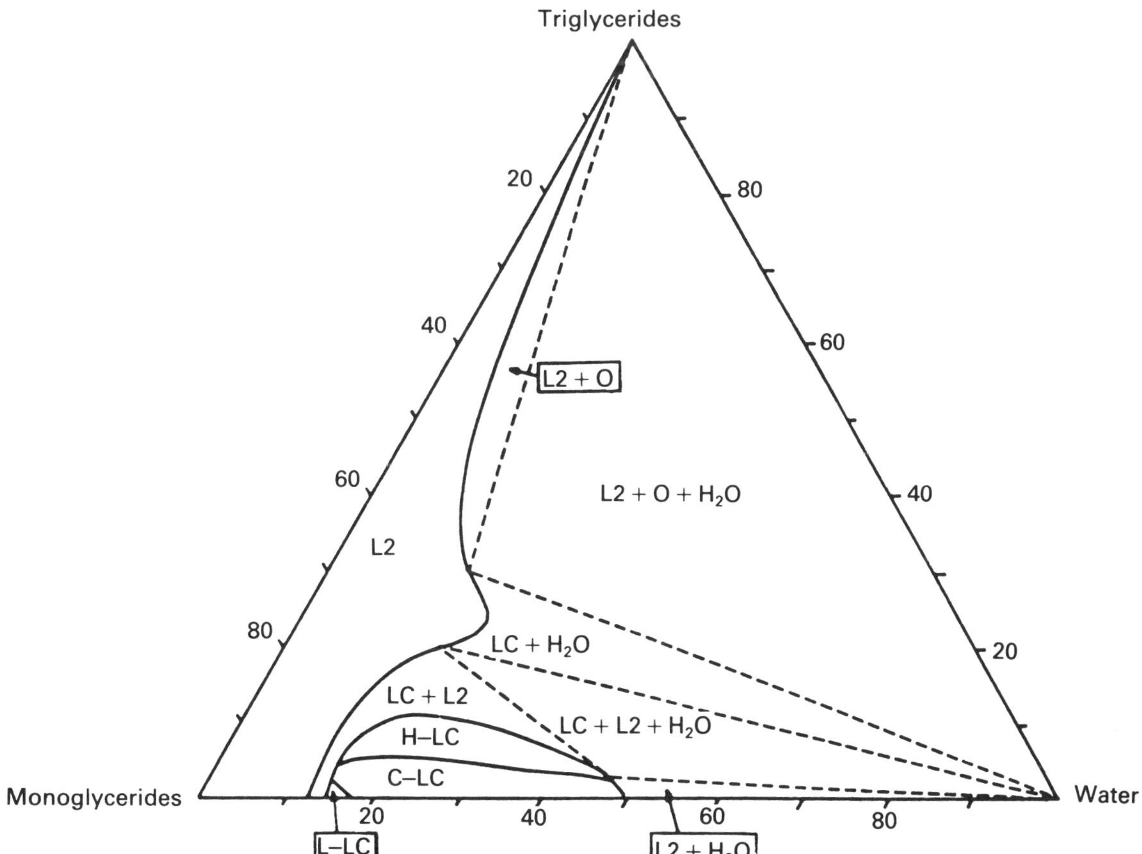

Fig. 8.14 Phase equilibria at 40 °C of the ternary system soybean oil–water–monoglycerides of sunflower oil (Lindström *et al.*, 1981). The liquid-crystalline (LC) phases are lamellar (L-LC), cubic (C-LC) and hexagonal (H-LC). (Axes in percentages).

and in ternary systems (polar lipid–non-polar lipid–water) by the disorder introduced when enough of the non-polar lipid is solubilized in the actual lipid–water phases (see Fig. 8.14). On the basis of an X-ray study of L2 phases in binary systems, Larsson (1979) proposed that the structure consists of water lamellae separated by lipid bilayers, as shown in Fig. 8.15. Thus on heating the corresponding liquid-crystalline phase it 'melts' to ordered domains of reduced size (a few hundred ångströms) according to the broadening of the X-ray diffraction lines, with the main structural features unchanged. This model of the structure has been confirmed by freeze–fracture–freeze–etch electron microscopy (Gulik-Krzywicki and Larsson, 1984), and it now seems generally accepted.

Criteria for the identification of microemulsions have been published (Lindman and Danielsson, 1981). According to this definition the L2 phase in aqueous systems of polar lipids and triglyceride oil (Fig. 8.14) is a microemulsion. An X-ray study of this system and of some other ternary systems (Fontell *et al.*, 1983) indicates that the general relation described above between the L2 phase and corresponding liquid-crystalline phase obtained by cooling is also true when the L2 phase is a microemulsion (see Fig. 8.16). Thus this microemulsion is regarded as a lipid bilayer type of structure (see also section 8.2.8). Diffusion studies by Lindman and co-workers (Brown *et al.*, 1983) of different microemulsions at high water content show that they behave both as hydrocarbon- and water-continuous systems (cf. Fabre *et al.*, 1981). If the other extreme is considered, i.e. at very low water content, close to the triglyceride corner in Fig. 8.14 the water aggregates surrounded by a monoglyceride monolayer must be expected to adopt a spherical shape.

Decyltriethyleneglycol (Tiddy, 1980) was found to exhibit a liquid phase between the lamellar phase and

Fig. 8.15 Proposed structure of the L2 phase in aqueous monoglyceride systems (Larsson, 1979). The black areas represent water and polar headgroup.

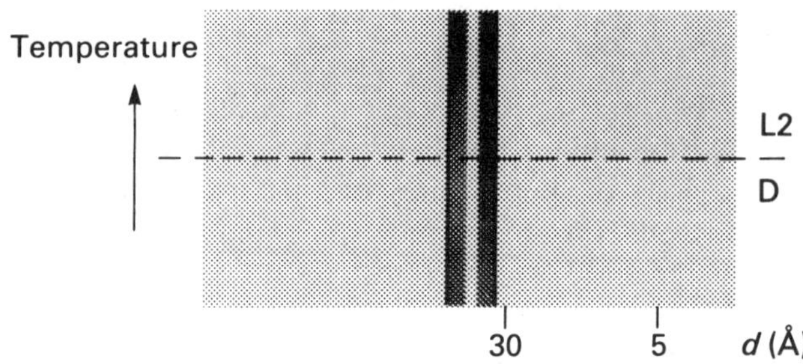

Fig. 8.16 An X-ray diffraction pattern of the transition between the lamellar liquid-crystalline phase (denoted D) and the L2 phase in the system triglyceride–water–monoglyceride (Fig. 8.14) of weight ratio 3:15:82 (Fontell *et al.*, 1983).

pure water in its aqueous system. This third liquid phase has been termed L3, and it has also been observed in non-ionic lipid–water systems, e.g. in glycerol ethers (Persson, 1984). Recent studies of this phase (also called sponge phase) indicate a structure related to the IPMS discussed in section 8.2.2 (Anderson and Wennerström, 1990).

Some microemulsions also seem to be related to the IPMS structure. If we imagine this zero-curvature surface with oil on one side and water on the other, with a monolayer of oriented amphiphilic molecules in between, such a disordered structure will be a liquid. If the volume ratio oil:water is slightly different from 1, the curvature can still be close to zero, with a similar structure.

8.2.5 The gel phase

The gel phase consists of crystalline lipid bilayers alternating with water layers. When the L_α phase is cooled through the hydrocarbon-chain crystallization temperature, a gel phase can be formed, which is usually metastable. There are even amphiphile–water systems that exhibit thermodynamically stable gel phases as the only type of lipid–water phase. One example is the tetradecylamine–water system (Larsson and Al-Mamun, 1974) shown in Fig. 8.17. Other lipids that only give gel phases in their aqueous systems are cholesterol sulphate and cholesterol phosphate (Abrahamsson *et al.*, 1977). The gel phase of tetradecylamine consists of bilayers with vertical chains in the orthorhombic chain packing. At low water content this structure swells to a water-layer thickness of 14 Å. At very high water concentrations, however, another gel phase with the same lipid bilayers but with a water layer several hundred ångströms thick is formed. The reason for this seems to be that the amount of water is enough to ionize a critical proportion of the amphiphile molecules. The

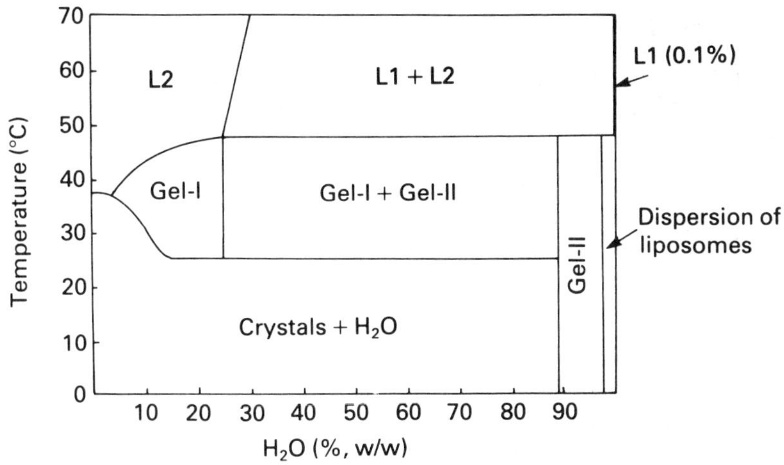

Fig. 8.17 Phase diagram of the binary system tetradecylamine–water (Larsson and Al-Mamun, 1974).

thick water layer obtained then is due to electrostatic repulsion. Similar effects were observed in the gel state formed by monoglycerides (Larsson and Krog, 1973).

In the case of monostearin, with a lipid-bilayer thickness of 50.5 Å, the gel swells to a water-layer thickness of 14 Å. When 1–2% (w/w) of a charged lipid is also present, however (for example sodium stearate), it is possible to achieve swelling to water-layer thickness values near the wavelength of light. Such gel phases can exhibit the colours of the visible spectrum.

A gel phase with a ripple-like periodicity in the bilayer, observed in synthetic lecithins, will be discussed in section 8.10.

8.2.6 Hydration force and hydrophobic force

The swelling of lamellar L_α phases and gel phases has been considered above and some effects of electrostatic repulsion have been discussed. The existence of an additional repulsive force that is very important in lipid–water systems was demonstrated by Rand, Parseigan and co-workers (cf. de Neveu *et al.*, 1976). This short-range force with exponential fall-off was called the hydration force. There is still no theory on this phenomenon that has been generally accepted.

Shear of a liquid-crystalline lipid phase that is in equilibrium with a water phase can result in exposure of hydrocarbon chains to water, which results in a strong tendency to fuse with other bilayers with exposed hydrocarbon chains. The formation of liposomes, discussed in the next subsection, is a consequence of such rupture and fusion behaviour. Experimental measurements of the force behind such adhesion, the hydrophobic force, have been reported (Israelachvili and Paskey, 1982). It was found to be a long-range force with the same range of existence as the van der Waals' interaction and shows an exponential fall-off with a decay length of 10 Å.

8.2.7 Liposomes

When water is added to the lamellar liquid-crystalline phase above its limit of swelling, so that there is a two-phase mixture of water and the liquid-crystalline phase, liposomes can be formed by minor shearing effects, such as stirring. A liposome consists of spherically concentric lipid bilayers alternating with water layers. The structure is similar to the lamellar liquid-crystalline phase, the difference being that the bilayers are curved into a closed aggregate. Liposomes can be several micrometres in diameter, and the minimal diameter of the innermost layer is a few hundred ångströms. A liposomal dispersion, once it has been formed, can be kinetically stable for years. It has been shown that the stability of such a dispersion follows the DVLO theory (Derjaguin–Verwey–Landau–Overbeck) for colloidal systems (Rydhag *et al.*, 1982). Unilamellar aggregates, so-called vesicles, can be formed under certain conditions, for example by ultrasonication, and they are frequently used as model systems for cell membranes.

The formation of particles consisting of concentric lipid bilayers was first described by Bangham (cf. Bangham and Horne, 1964) in very dilute phospholipid systems. Later it was shown that the existence range of this type of aqueous dispersion is the whole region of the phase diagram where water and the lamellar liquid-crystalline phase occur in equilibrium (Larsson, 1967). In dilute systems the liposomes

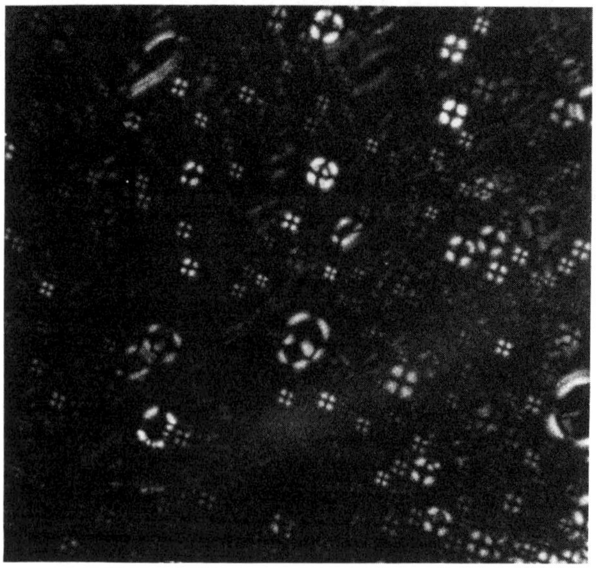

Fig. 8.18 Characteristic texture of a liposomal dispersion viewed in the polarizing microscope.

appear regular with perfect spherical shape. Liposomes are easily identified in the polarizing microscope where they show strong birefringence with an extinction cross parallel to the polarization planes (see Fig. 8.18).

Methods to prepare well defined large liposomes and vesicles have been reported (Deamer and Bangham, 1976; Servuss *et al.*, 1976).

8.2.8 Relations between shape of lipid molecules and structure of liquid-crystalline phases

Luzzati and co-workers (cf. Luzzati, 1968) observed that the cross-sectional area per polar group increases regularly with water content, even when phase boundaries are passed. There is an obvious tendency for formation of H_I phases in aqueous systems of lipids with a large polar headgroup compared to the cross-sectional area of the hydrocarbon-chain region. The tendency for formation of the H_{II} phase in the case of lipids with a bulky hydrocarbon-chain region compared to the size of the polar headgroup is also well known. A quantitative theory relating the wedge shape of the amphiphile molecule to the phase that is formed with water has been introduced by Israelachvili *et al.* (1976). Helfrich (1981) has presented a theory explaining the different amphiphile–water structures on the basis of the curvature elasticity of fluid monolayers and bilayers. Furthermore, a thermodynamic model for the description of the association behaviour of ionic amphiphiles in water has been introduced by Jönsson and Wennerström (1981), and their calculation of stability regions of the different phases in soap–water systems and in the aerosol OT–water system are in good agreement with the experimental observations. A few additional aspects of structural relationships between phases will be considered here.

Increase in temperature or water content results in increased molecular disorder, and this disorder corresponds to an increased wedge shape of the molecule, as the disorder within the lipid molecule increases towards the end of the hydrocarbon tail. The slope of the transition line L_α to cubic and of the transition cubic to H_{II} in Fig. 8.12 clearly illustrates these effects. If then the changes within the cubic and H_{II} regions are considered, a dilemma with regard to water swelling becomes obvious. Increased water content means that the lattice axes are increased, and this in turn results in a reduced curvature of the bilayer and less wedge shape for the molecules. Swelling of the lattice should thus be expected to result in strain if the composition range is broad. The transition from one cubic phase to another within the cubic region of the phase diagram might be due to the build-up of such strain.

The swelling behaviour of the H_{II} phase of monogalactosyldiglycerides, existing over a large composition range, is shown in Fig. 8.19. The swelling is

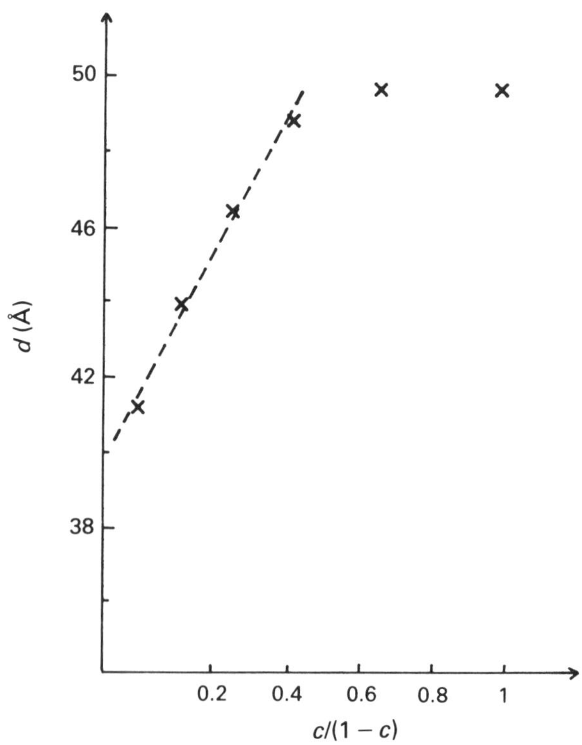

Fig. 8.19 Water swelling of the H_{II} phase of wheat monogalactosyldiglycerides at 20 °C. The crosses show the $d(11)$-spacing versus water weight ratio and the broken line shows the ideal behaviour in a lamellar system.

quite close to the ideal swelling of a lamellar phase. It has also been observed that in monoglyceride systems (Larsson *et al.*, 1980) one spacing of the body-centred cubic lattice persists through the transition into the H_{II} phase at different water contents. Both these observations are consistent with an H_{II} structure with a lipid bilayer separating water channels with hexagonal cross-sections, rather than cylindrical water channels in a hydrocarbon-chain matrix. The calculated dimensions at water swelling of such an H_{II} phase, consisting of a bilayer with constant thickness, are the same as in an L_α phase with constant lipid-bilayer thickness (Larsson *et al.*, 1980).

The relations between X-ray diffraction spacings of different liquid-crystalline phases discussed here also involve the L2 phase. These observations demonstrate the significance of the lipid bilayer as a common structural element of all phases in this type of lipid–water system.

8.2.9 Formation of non-aqueous liquid-crystalline phases

Water can be fully replaced by glycerol in different liquid-crystalline phases. A ternary phase diagram of monoglycerides–water–glycerol is shown in Fig. 8.20.

El-Nokaly *et al.* (1981) have shown that lecithin forms a lamellar liquid-crystalline phase with solvents other than water, such as ethylene glycol. The swelling of this phase is rather similar to the corresponding aqueous phase, with a somewhat higher limit of swelling. An analogue to the 'hydration force', discussed above, also exists in lamellar non-aqueous phases (Bergenståhl and Persson, 1985).

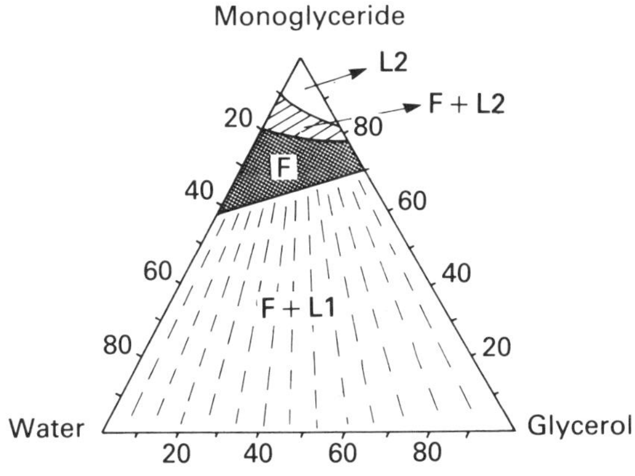

Fig. 8.20 Ternary phase diagram of the system sunflower oil monoglycerides–water–glycerol at 50 °C. F stands for the hexagonal phase H_{II} and L1 is the water–glycerol solution.

References and further reading

Abrahamsson, J., Abrahamsson, S., Hellqvist, B., Larsson, K., Pascher, I. and Sundell, S. (1977) *Chem. Phys. Lipids*, **19**, 213.
Anderson, D. and Wennerström, H. (1990) *J. Phys. Chem.*, **94**, 8683.
Arvidsson, G., Khan, A., Brentel, I., Lindblom, G. and Fontell, K. (1985) *Eur. J. Biochem.*, **152**, 753.
Bangham, A.D. and Horne, R.W. (1964) *J. Mol. Biol.*, **8**, 660.
Bergenståhl, B.A. and Persson, P.K.T. (1985) *Biophys. J.*, **47**, 743.
Brown, W., Johnson, R., Stilbs, P. and Lindman, B. (1983) *J. Phys. Chem.*, **87**, 4758.
Chapman, D. (1958) *J. Chem. Soc.*, 784.
Clunie, J.S., Corkill, J.M. and Goodman, J.F. (1965) *Proc. R. Soc.*, **A285**, 520.
Deamer, D. and Bangham, A.D. (1976) *Biochim. Biophys. Acta*, **443**, 629.
de Neveu, D.M., Rand, R.P., Ginger, D. and Parseigan, A. (1976) *Science*, **191**, 399.
Ekwall, P. (1975) in *Advances in Liquid Crystals*, vol. 1, Academic Press, New York, chap. 1.
El-Nokaly, M., Ford, L.D., Friberg, S.E. and Larsen, D.W. (1981) *J. Colloid Interface Sci.*, **84**, 228.
Fabre, H., Kamenka, N. and Lindman, B. (1981) *J. Phys. Chem.*, **85**, 3493.
Fontell, K., Hernqvist, L., Larsson, K. and Sjoblom, J. (1983) *J. Colloid Interface Sci.*, **93**, 453.
Gulik-Krzywicki, T. and Larsson, K. (1984) *Chem. Phys. Lipids*, **35**, 127.
Hamilton, J.A. and Small, D.M. (1981) *Proc. Natl Acad. Sci. USA*, **78**, 6878.
Helfrich, W. (1981) in *Physics of Defects* (ed. R. Balian), North-Holland, Amsterdam, p. 1.
Hernqvist, L., Herslöf, B., Larsson, K. and Podlaha, J. (1981) *J. Sci. Food Agric.*, **32**, 1197.
Israelachvili, J. and Paskey, R. (1982) *Nature*, **300**, 342.
Israelachvili, J., Mitchell, J. and Ninhan, B. (1976) *J. Chem. Soc., Faraday Trans. II*, **72**, 1525.
Jönsson, B. and Wennerström, H. (1981) *J. Colloid Interface Sci.*, **80**, 482.
Larsson, K. (1967) *Hoppe-Seyler's Z. Physiol. Chem.*, **56**, 173.
Larsson, K. (1979) *J. Colloid Interface Sci.*, **72**, 152.
Larsson, K. (1983) *Nature*, **304**, 664.
Larsson, K. (1986) *J. Colloid Interface Sci.*, **113**, 299.
Larsson, K. (1989) *J. Phys. Chem.*, **93**, 7304.
Larsson, K. and Al-Mamun, M.A. (1974) *Chem. Phys. Lipids*, **12**, 176.
Larsson, K. and Krog, N. (1973) *Chem. Phys. Lipids*, **10**, 177.
Larsson, K., Fontell, K. and Krog, N. (1980) *Chem. Phys. Lipids*, **27**, 321.
Lindblom, G. and Rilfors, L. (1989) *Biochim. Biophys. Acta*, **988**, 211.
Lindblom, G., Larsson, K., Johansson, L., Fontell, K. and Forsén, S. (1979) *J. Am. Chem. Soc.*, **101**, 5465.
Lindman, B. and Danielsson, I. (1981) *Colloids Surf.*, **3**, 391.

Lindman, B. and Wennerström, H. (1980) *Top. Curr. Chem.*, **87**, 1.
Lindström, M., Ljusberg-Wahren, H., Larsson, K. and Borgstrom, B. (1981) *Lipids*, **16**, 749.
Longley, W. and McIntosh, T.J. (1983) *Nature*, **303**, 612.
Luzzati, V. (1968) in *Biological Membranes* (ed. D. Chapman), Academic Press, New York, p. 71.
Luzzati, V. and Reiss-Husson, F. (1966) *Nature*, **210**, 1351.
Luzzati, V. and Spegt, P.A. (1967) *Nature*, **215**, 701.
Luzzati, V., Mustacchi, H., Skoulios, A. and Husson, F. (1960) *Acta Crystallogr.*, **13**, 660.
Luzzati, V., Tardieu, A., Gulik-Krzywicki, T., Rivas, E. and Riess-Husson, F. (1968) *Nature*, **220**, 485.
Luzzati, V., Vargas, R., Gulik, A., Mariai, P., Seddon, J.M. and Rivas, E. (1992) *Biochemistry*, **31**, 279.
Persson, P. (1984) Thesis, Royal Institute of Technology, Stockholm.
Rydhag, L. and Wilton, I. (1981) *J. Am. Oil Chem. Soc.*, **58**, 830.
Rydhag, L., Stenius, P. and Odberg, L. (1982) *J. Colloid Interface Sci.*, **86**, 274.
Scriven, L.E. (1976) *Nature*, **263**, 123.
Seddon, J.M. (1990) *Biochim. Biophys. Acta*, **1031**, 1.
Servuss, R.M., Harlich, W. and Helfrich, W. (1976) *Biochim. Biophys. Acta*, **436**, 900.
Shoen, A.H. (1970) *NASA Technical Note* TND-5541.
Söderberg, I. (1990) Thesis, University of Lund, Sweden.
Tiddy, G. (1980) *Phys. Rep.*, **57**, 1.
Wennerström, H. and Lindman, B. (1979) *Top. Curr. Chem.*, **87**, 1.

8.3 The liquid state of lipids

As discussed in the preceding section, the lipid molecules are disordered in the liquid-crystalline phases and the hydrocarbon chains give an X-ray scattering pattern 'characteristic for liquid paraffins' (Luzzati *et al.*, 1960). The structure in the hydrocarbon-chain regions of liquid-crystalline phases and in the liquid state of lipids in general will be considered further below.

8.3.1 Short-range disorder

The frequency of the X-irradiation is several orders of magnitude higher than that of molecular motions, which means that chains in the liquid state are observed in static positions. The significance of the 4.5 Å halo in the X-ray scattering of the lamellar phase will first be discussed here, as no geometrical analysis has been reported earlier as far as this author knows.

In the solid state the chains are in the all-*trans* conformation, and the degree of disorder of hydrocarbon chains can be defined by the relative amount of *gauche* conformation. This can be illustrated by the directions of the zig-zag units or the *c* axis of the paracrystalline subcell. In the solid state all *c* axes are linearly linked, whereas an increasing degree of disorder leads to successively larger deviations from a parallel linking of the subcell chain directions. A snapshot of the chain structure, as 'seen' by an X-ray beam, of such a lamellar phase with disordered chains is illustrated in Fig. 8.21. It is obvious for space-filling reasons that there must be a high degree of cooperativity in the chain motions. The average direction of the zig-zag units, or the *c* axes, in relation to the plane of the hydrocarbon-chain layer can be calculated from the thickness of the bilayer in the L_α phase in relation to the thickness of a crystalline bilayer with vertical chains. Using X-ray data from the monoglyceride L_α phase and the crystalline phases, a value of the average tilt of the *c* axes of 43° is obtained.

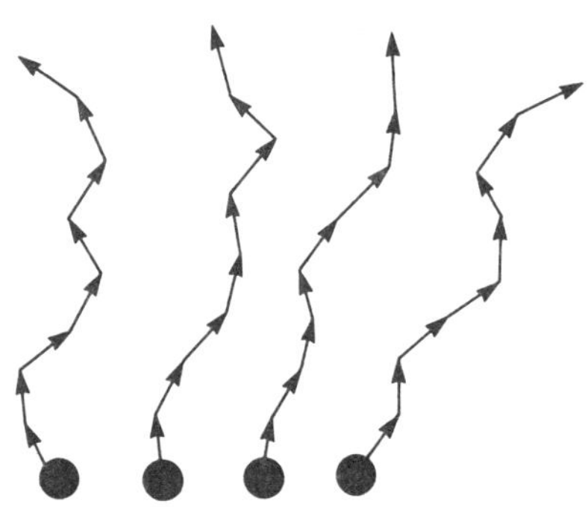

Fig. 8.21 Schematic illustration of C_{16} chains in the liquid state according to the c_s directions, i.e. the zig-zag units starting at the polar headgroup. The positions indicate the degree of *gauche* conformation and the need of cooperativity in the chain mobility.

The 4.5 Å spacing observed in the diffraction pattern of L_α phases and in the liquid state of lipids is due to the average lateral distance between CH_2 groups of adjacent chains. There are two alternatives of coordination number of lateral CH_2 groups to be considered: a fourfold or sixfold coordination. If there is tetragonal coordination the cross-sectional area per CH_2 group is 20.25 Å^2, and if the coordination should be hexagonal a cross-sectional area of 17.5 Å^2 is obtained. From the chain-packing subcells shown in section 8.1, it is known that the cross-sectional value of a CH_2 group in a close-packed arrangement is about 18.2 Å^2. The sixfold coordination can therefore be ruled out, as the cross-sectional value is too small. The difference in CH_2 cross-sectional values between a liquid with tetragonal coordination of CH_2 groups and a crystal with close-packed chain is consistent with the difference in density between the liquid and solid state of hydrocarbon-chain-dominated lipids.

Another support for a tetragonal coordination is the value of the cross-section per chain in the plane of the hydrocarbon-chain layer. Using the value of the cross-sectional area per CH_2 group of 20.25 Å^2 and the average direction of the zig-zag units in the L_α phase calculated above, a value of 30 Å^2 is obtained. This is in good agreement with observed values in L_α phases.

This result, indicating a tetragonal coordination in the L_α phase, can be generalized to the liquid state of lipids, according to the preferred arrangement in layers, which is treated in the next subsection.

It should finally be mentioned that the chain conformation in lipids has been examined by different spectroscopic techniques.

8.3.2 Structure at medium-range distances

There is a certain degree of molecular order over medium-range distances (a few hundred ångströms) in the liquid state of polar lipids. If the changes in the X-ray pattern during the melting of a monoglyceride are followed, it can be seen that a broad band is formed in the long-spacing region where the sharp

$d(0\,0\,1)$ diffraction line due to the crystalline state disappears. The integrated intensity of this band is almost as large as that of the $d(0\,0\,1)$ diffraction line and the position corresponds to a reduced spacing compared to $d(0\,0\,1)$. This has been taken as an indication of a bilayer type of structure similar to that discussed above in L2 phases (Larsson, 1979). The size of the lamellar units in the plane of the lamellae was estimated to be a few hundred ångströms from an X-ray line-broadening analysis.

The medium-range structure in the liquid state occurs in all lipids that form liquid-crystalline phases with water. Simple lipids, which do not form liquid-crystalline phases, give no X-ray patterns corresponding to medium-range order in the liquid state. Fatty acids, alcohols and esters are examples of lipids with no sign of order in the melt over the distances that give X-ray scattering in the low-angle region.

Triglycerides appear to have a certain degree of order in the melt intermediate in relation to the two types of lipids discussed above. The proposed structure based on the X-ray-scattering data is shown in Fig. 8.22 (Larsson, 1972). The occurrence of a bilayer structure in the melt is consistent with the observation that during fat crystallization the melt keeps a 'memory' of its structure before melting. Thus it tends to crystallize in the same form and this knowledge has been utilized for a long time in thermal processing before fat crystallization. Also ^{13}C NMR studies of tristearin have been reported that indicate a certain order in the melt (Callaghan and Jolley, 1977). There is a difference compared to the X-ray results, however, as the NMR data indicate that this order occurs only close to the melting point. There are also discontinuities in the viscosity of the melt close to the melting point (see section 8.9).

A recent neutron diffraction study of trilaurin in the liquid and solid states has been reported by Cebula *et al.* (1992). The diffraction data using selectively deuterated trilaurin were interpreted as evidence for a nematic type of structure, with a common molecular direction as the only order. A strong argument against such a structure, however, is the lack of molecular direction, which is a basic characteristic of the lipid hydrocarbon-chain conformation (Larsson, 1992).

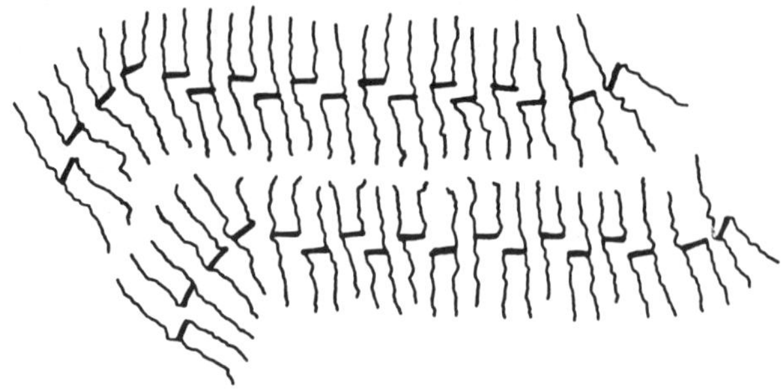

Fig. 8.22 Proposed structure of triglycerides in the liquid state (Larsson, 1972).

The reason for the order of triglyceride molecules in the melt is probably related to the complex shape of the molecules. It is not possible to pack in an effective way three chains that are allowed to have all directions.

References and further reading

Callaghan, P.T. and Jolley, K.W. (1977) *J. Chem. Phys.*, **67**, 4773.

Cebula, D.J., McClements, D.J., Povey, M.J.W. and Smith, P.R. (1992) *J. Am. Oil Chem. Soc.*, **69**, 130.

Larsson, K. (1972) *Fette Seifen Anstrichm.*, **74**, 136.

Larsson, K. (1979) *J. Colloid Interface Sci.*, **72**, 152.

Larsson, K. (1992) *J. Am. Oil Chem. Soc.*, **69**, 835.

Luzzati, V., Mustacchi, H., Skoulios, A. and Husson, F. (1960) *Acta Crystallogr.*, **13**, 660.

8.4 Lipid monolayers at the air–water interface

A well known property of lipids is their ability to form monomolecular films on a water surface. By orientation of the molecules, with the polar head-group in contact with water and the chain directly upwards, the surface energy is reduced. Langmuir introduced the surface-film balance as an experimental technique to investigate monolayers, and as long ago as 1917 he gave a correct molecular description of surface-film architecture (Langmuir, 1917). The surface balance consists of a trough filled with water, where a known amount of the lipid can be spread on the surface from an organic solvent. By a moving barrier the film can be compressed against another barrier, where the film pressure is measured. In this way it is possible to record the surface film pressure (Π) as a function of molecular area (A), and obtain pressure–area (Π–A) isotherms. Alternative devices to measure film pressure have since been developed, such as the Wilhelmy plate (cf. Gaines, 1966).

The spreading pressure of the lipid film tends to increase the surface, i.e. it works in a direction opposite that of surface tension (γ). Thus the reduction of the surface tension ($\Delta\gamma$) of water by a lipid is equivalent to the film pressure (Π).

It is remarkable that lipid molecules, which are spread on a water surface, remain at the surface for quite long times, even though the lipid is soluble to a certain extent in bulk water. Most lipids form such insoluble monolayers on water.

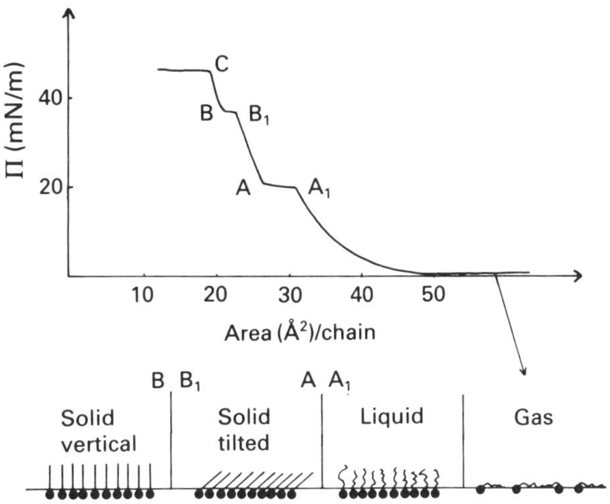

Fig. 8.23 Characteristic pressure–area isotherm of an insoluble lipid monolayer with proposed structures indicated below.

8.4.1 Structure of lipid monolayer phases

The most intensive period of monolayer research took place during the decade before the Second World War. Thus the basic interpretations of Π–A isotherms were done before the characteristic crystalline and liquid-crystalline structural features were known. Unfortunately, much of the later monolayer work has also been limited to the early concepts only, without the knowledge on three-dimensional lipid phases being fully utilized. Reports of molecular arrangement in monolayers are therefore often misleading and the nomenclature used is inadequate. A structural description of monolayer phases will first be given based on relations with crystalline and liquid-crystalline phases. A hypothetical pressure–area (Π–A) isotherm shown in Fig. 8.23 will be used and the nomenclature according to Harkins (1952) is applied.

When the molecules are far apart, they are considered to lie flat on the surface and distributed one by one in a two-dimensional *gas* phase (Gaines, 1966). An ideal gaseous monolayer should follow the two-dimensional gas equation:

$$\Pi A = kT$$

where k is the Boltzmann constant. When Π is given in mN/m and A in Å^2, the product ΠA should be 411.7 at 25 °C, which corresponds to about 1 mN/m at 400 Å^2 per molecule. Experimental work has demonstrated a reasonably good agreement at surface pressures below 0.1 mN/m, but there are various limitations in this treatment. One is that the interaction with water molecules in the subphase is ignored, and another is the area taken up by the film-forming molecules, which is not negligible as assumed in this equation (Gaines, 1966).

When the gaseous monolayer is compressed, a transition into the so-called *liquid-expanded* phase usually takes place. This phase has been the subject of many controversies. Most of this discussion, however, took place before the structure of the lamellar liquid-crystalline phase was known. It should be mentioned in this connection that Phillips *et al.* (1969) used such correlations in their interpretation of the monolayer structure of dipalmitoylphosphatidylcholine, which will be discussed further in section 8.10.

From data available on lipid–water L_α phases it is known that the cross-sectional area per chain is always near 30 Å^2, and when the chain disorder is exceptionally high (for example at high temperatures) the value can approach 40 Å^2. When the Π–A isotherm of a lipid is considered, which at the actual temperature forms an L_α phase in equilibrium with excess water, there is always a striking agreement between the hydrocarbon cross-section value of the L_α phase and that obtained from the monolayer formed in the presence of excess L_α phase, i.e. at the so-called equilibrium spreading pressure of the monolayer. This monolayer phase can be identified by expanding it to obtain the Π–A isotherm. Comparisons of monoglycerides, lecithins and galactolipids show this agreement in molecular area. The rheological properties of this monolayer phase are also consistent with a liquid type of disorder. It is therefore proposed that the *liquid-expanded* monolayer phase has the same hydrocarbon chain structure as a monolayer of an L_α phase.

The formation of the liquid-expanded phase at compression of a monolayer means that condensation from a gaseous phase to a coherent film in the liquid state takes place. Further compression can give phases with extended hydrocarbon chains tilted in relation to the surface. Such phases are called *liquid-condensed* (L_2) phases, and a consequence of this misleading term is that the formation from a liquid-expanded phase is described as a condensation.

There are numerous possible L_2 phases, as the angle of tilt of the chains can vary as well as the lateral packing of the chains. By further compression the smallest possible molecular area can be obtained when the extended zig-zag chains are vertical. When the vertical chains of the monolayer are hexagonally arranged, as in the α crystal form, this corresponds to the *super-liquid* (LS) phase. The name refers to the rheological properties of the film; dust can be observed to move as on a liquid surface. This is a result of the molecular mobility when the molecules are hexagonally arranged. This was discussed in connection with the α phase. At further compression, the LS phase can undergo a transition into a so-called *solid* (S) phase. This phase consists of close-packed chains, usually with the same packing as in the β' crystal form. The most common chain packing in β forms, the triclinic T$\|$, cannot occur in simple lipids with vertical chains, only in tilted chain structures. A summary of the commonly used nomenclature (Harkins, 1952) and the corresponding monolayer phase description is given in Table 8.1.

The evidence for the relations between L_2, LS and S phases and three-dimensional crystal forms is quite convincing. Lundqvist (1978) has reported extensive comparisons between condensed monolayer phases and polymorphic forms in the crystalline state of non-polar lipids, and a summary of this work, together with additional data on polar lipids, will be given below.

All monolayer phases with crystalline chains show linear Π–A curves, and there are always well defined plateau values in the Π–A isotherms at phase transitions involving tilted chains. As discussed in connection with hydrocarbon chain-packing alternatives (section 8.1), only certain discrete tilt values are possible. For example, in the case of the orthorhombic chain packing, one of the most common chain tilt alternatives occurring in simple lipids is about 66°. The corresponding hydrocarbon chain structure, described by the indices of the endgroup plane is O$\perp$ (101). Monolayer Π–A isotherms of ethyl stearate and of docosyl acetate show an L_2 phase, which has exactly this chain tilt, as calculated from the Π–A isotherms. Furthermore, these two esters crystallize in β' form with this hydrocarbon chain structure (O$\perp$ (101)).

It is possible to calculate the energy involved in the monolayer phase transition using a two-dimensional analogue of the Clausius–Clapeyron equation. The values obtained for the esters (Lundqvist, 1978) are in good agreement with those obtained (by differential scanning calorimetry) for the corresponding transitions between crystal forms. By recording Π–A isotherms at various temperatures, it is also possible

Table 8.1 A summary of structural features of lipid monolayer phases based on relations with crystalline and liquid-crystalline phases

Harkin's (1952) monolayer phase description	Relation between structure of monolayer phases and three-dimensional states
Gaseous (G)	Molecules distributed separately
Liquid-expanded (L_1)	Liquid chains, structure as in the lamellar liquid-crystalline phase
Intermediate (I)	Two-phase region, one with liquid chains, one with crystalline chains
Liquid-condensed (L_2)	Tilted molecules, crystalline chains
Super-liquid (LS)	Vertical molecules with chains in the hexagonal arrangement
Solid (S)	Vertical close-packed chains as in α and β' crystal forms

to construct a Π–*T* phase diagram for the monolayer, and the critical temperatures of the monolayer phase transitions obtained are in agreement with the transition temperatures between crystalline forms. Even alternation in transition temperatures between even and odd members with tilted chains can be seen from the monolayer Π–*T* diagrams (Lundqvist, 1978).

These close relationships between different monolayer phases and three-dimensional polymorphic forms occurring in non-polar lipids are considered here as conclusive evidence for the occurrence of the same structures. In the case of polar lipids, however, the interaction with water means that a relationship between a monolayer phase and a bulk phase can only be expected when the monolayer phases are compared with lipid–water phases, such as L_α and gel phases.

As an illustrative example the monolayer behaviour of glycerol 1-myristate (1-monomyristin) will be considered. With excess water monomyristin gives a lamellar liquid-crystalline phase at temperatures above about 42 °C and a metastable gel phase below this temperature. A Π–*A* isotherm of monomyristin is shown in Fig. 8.24. First a monolayer phase with liquid chains is obtained, referred to as form I, and at higher pressure a phase with crystalline chains, form II, is formed. At the first-order transition, represented by the plateau where the two forms coexist, the molecular area of form I is 28.5 Å^2, in good agreement with that of the L_α phase of monomyristin. The molecular area of form II is 22.5 Å^2, which is identical to the area in the gel phase of the aqueous system. Furthermore the monolayer Π–*T* phase diagram (see Fig. 8.24) shows that above 42 °C there is no monolayer phase with crystalline chains. The same relation between monolayer phases and the aqueous L_α and gel phases is also observed in other monoglycerides (Krog *et al.*, 1985).

Lipids that form liquid-crystalline phases with water give Π–*A* isotherms of two types. The first type is obtained above the chain melting temperature in the actual lipid–water system. Under these conditions there is only one monolayer phase, which exists up to a film pressure of about 40 mN/m. As discussed above, the structure is considered to be the same as in the lipid monolayer of the L_α phase. Below the chain transition temperature, the second type of Π–*A* isotherm is obtained. A phase with liquid chains is formed at low surface pressure, but at a certain film pressure it is transformed into a monolayer phase with crystalline chains. It is also possible that other phases with crystalline chains can be formed under further compression, as shown in section 8.10.

The equilibrium spreading pressure (ESP) of monolayers from polar lipids has the same value as the plateau pressure defining the transition between the form with liquid chains and that with crystalline chains. Above the chain melting temperature, the ESP value is the same as the collapse pressure of the monolayer phase with liquid chains.

Similar relations exist between the ESP and pressure values of monolayer phase transitions in non-polar lipids. Thus, if the lipid is above its melting point, the ESP is the same as the collapse pressure, whereas the ESP value obtained for crystals of the lipid coincides with the pressure of a monolayer phase transition below the collapse point.

In this connection it should be pointed out that a monolayer phase with liquid chains is always a stable phase, whereas monolayer phases with crystalline

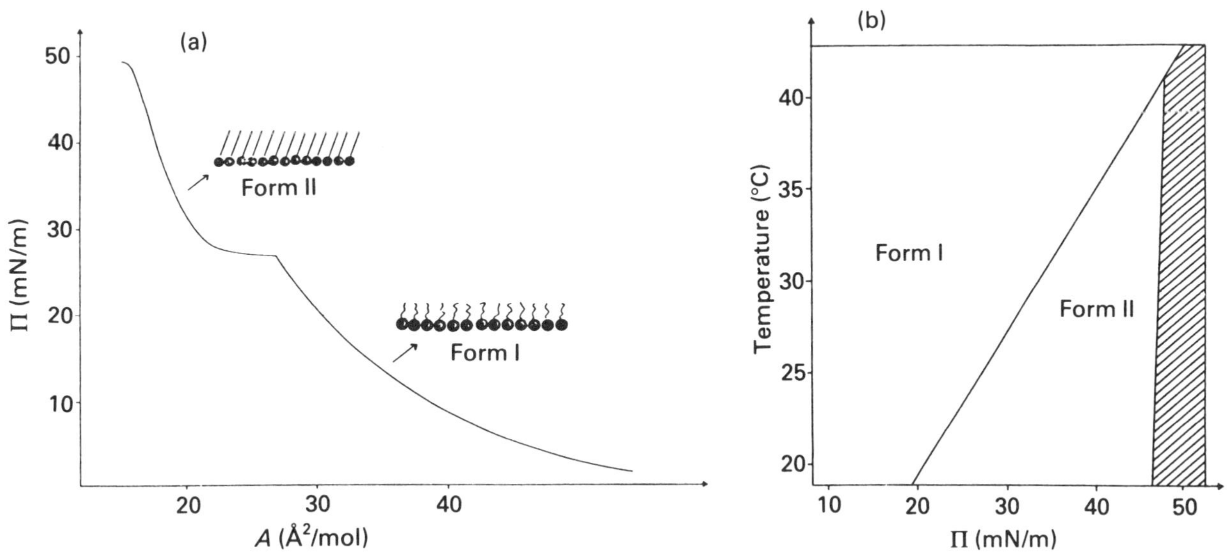

Fig. 8.24 Pressure–area isotherm of 1-monomyristin at 25 °C (a) with the proposed monolayer structures indicated, and the corresponding pressure–temperature phase diagram (b).

chains are often metastable. This behaviour is also in common with three-dimensional phases where those with crystalline chains can be metastable, contrary to those with liquid chains.

8.4.2 Lipid multilayers formed from monolayers

Langmuir and Blodgett (cf. Blodgett, 1935) found that monolayers of lipids on a water surface can be transferred to a solid, and lipid multilayers can be built up in this way. Thus a solid hydrophobic material, which is dipped through a monolayer on water, will acquire the monolayer with the hydrophilic surface oriented towards the water. Another monolayer is transferred on removing the object from the water to give a hydrophobic surface in air. Repeated dipping will thus give multilayers.

Lipid monolayers on a water surface can also under certain conditions be compressed to multilayers. In this way a kind of transitory state between a monolayer and a crystal can be obtained. Experience from various lipid systems (Larsson, 1973b) shows that the only monolayer phase that forms multilayers is the one with crystalline chains arranged according to hexagonal chain packing. The formation of a triple layer of triglycerides is shown in Fig. 8.25. Thus when a monolayer of tripalmitin at 40 °C is compressed beyond the collapse point, the pressure is constant and then suddenly it increases to an area corresponding to one-third of the cross-sectional area of the triglyceride molecule. The structure proposed is also shown in Fig. 8.25. The extended molecular conformation with pairs of molecules arranged head-to-head is the only structure type occurring in triglyceride crystals. The monolayer in contact with water, however, must have a conformation with all three chains pointing upwards in order to allow an interaction with water. When there are enough molecules above this monolayer to form a bilayer, it is natural that the ideal triglyceride conformation is obtained with an extended conformation of the molecules.

Multilayer formation can also be observed in many other lipids such as fatty acids, long-chain alcohols and esters of short-chain alcohols. Lundqvist (1971) has reported monolayer formation in fatty acid ethyl esters, where it is possible to observe from the Π–*A* isotherms how multifilms consisting of two, three, four, five and six molecular layers are formed. In this case the forces between the polar groups are so weak that the molecules can be arranged head-to-tail as observed in the crystal forms of ethyl stearate (see section 8.5).

The formation of multifilms is of interest with regard to the crystallization process. The fact that molecules that are squeezed out from a monolayer can form a crystalline bimolecular layer (provided that the chains are hexagonally arranged) shows that crystal forms with hexagonal chain packing can be obtained by two-dimensional nucleation.

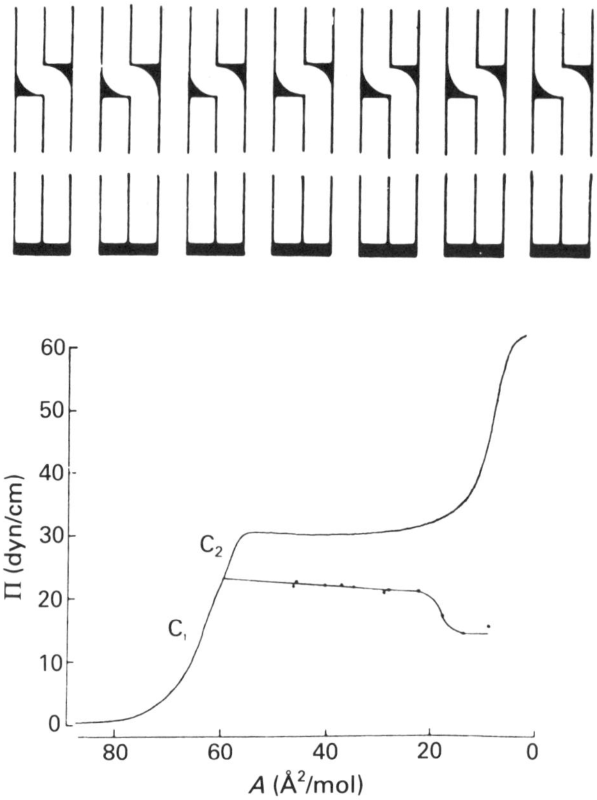

Fig. 8.25 Π–*A* isotherm of tripalmitin at 40.7 °C, where the filled circles correspond to equilibrium pressures. The proposed structure of the triple chain layer is also shown (Larsson, 1973a).

References and further reading

Blodgett, K.B. (1935) *J. Am. Chem. Soc.*, **57**, 1007.

Gaines, G.L., Jr (1966) in *Insoluble Monolayers at Liquid–Gas Interfaces*, Interscience, New York, p. 136.

Harkins, W.D. (1952) in *Physical Chemistry of Surface Films*, Reinhold, New York, p. 106.

Krog, N., Riisom, T. and Larsson, K. (1985) *Encyclopedia of Emulsion Technology*, vol. 2 (ed. P. Becher), Marcel Dekker, New York, chap. 4.

Langmuir, I. (1917) *J. Am. Chem. Soc.*, **39**, 1848.

Larsson, K. (1973a) *Chem. Phys. Lipids*, **10**, 177.

Larsson, K. (1973b) in *Surface and Colloid Science*, vol. 16 (ed. Matijevic), Wiley, New York, p. 261.

Lundqvist, M. (1971) *Chem. Scr.*, **1**, 5.

Lundqvist, M. (1978) *Prog. Chem. Fats Other Lipids*, **16**, 101.

Phillips, M.C., Williams, R.M. and Chapman, D. (1969) *Chem. Phys. Lipids*, **3**, 234.

8.5 Fatty acids and esters of monohydric alcohols

The physical properties of fatty acids and their simple esters will be discussed in considerable detail. Fatty acids are key components in most lipids, and the present section should provide a basis for the understanding of physical effects due to variations in fatty acid structure such as chain length, presence of double bonds, branches, etc.

8.5.1 Solid-state behaviour

(a) Polymorphism in normal-chain carboxylic acids

Fatty acids usually crystallize in three different crystal forms. The X-ray long spacing, related to the tilt of the molecules in the crystals, has been the basis for the nomenclature. Thus the three crystal forms occurring in fatty acids with an even number of carbon atoms have been termed A, B and C in order of increasing tilt of the molecules towards the end-group plane. The three common crystal forms observed in fatty acids with an odd number of carbon atoms are analogously termed A′, B′ and C′. Although there are structural relations between the pairs A/A′, B/B′ and C/C′, all are not similar enough to be classified as isomorphous or isostructural, as will be described below. In addition to these common six crystal forms, two other modifications have been observed, the forms A-super and B_1, which also are discussed below. Additional structures have been observed in branched and unsaturated acids.

The most complete investigation of the polymorphism of normal fatty acids in the interval C_{12}–C_{29} was reported by Stenhagen and von Sydow (1953). Their results can be summarized by the following schematic phase relationships:

$$A \rightarrow B \rightarrow C \rightleftharpoons \text{liq.}$$

All three crystal forms can be obtained from solvent, whereas only the C form is obtained from the melt. The irreversible character of the transitions A to B and B to C shows that the C form is the stable crystal form. It should be pointed out, however, that a transition in the solid state from the C form to the crystal form A-super has been observed in lauric acid (see subsection 8.5.1(b)).

The phase relations in odd acids can be described by:

$$A' \rightarrow B' \rightleftharpoons C' \rightarrow \text{liq.}$$
(with an arrow returning from liq. to B′)

The C′ form is only stable in a small interval close to the melting point, and the melt is supercooled so that the C′ form is not obtained on cooling. At other temperatures, the B′ form is the stable form.

Carboxylic acids with shorter chains (C_1–C_{10}) have been examined by Baun (1961). The X-ray data indicate that their structures are related, and the long-spacing values can be described by the relation $2.22n + 0.86$, where n is the number of carbon atoms in the chain. The complete crystal structures have been determined for formic acid (Holzberg *et al.*, 1953) and acetic acid (Jones and Templeton, 1958). The molecules are linked by hydrogen bonds, forming infinite chains, not dimers, which are most common in fatty acids.

The different crystal forms will be described below. In parallel with earlier nomenclature of fatty acid polymorphs, the classification defined in section 8.1 will also be used.

(b) The A and A′ forms (β forms)

The spontaneous transition in the solid state from form C of lauric acid to form A-super was described by Lomer (1955) and the crystal structure of this form was determined by von Sydow (1956a). The structure consists of lauric acid dimers. Later Lomer (1965) determined the crystal structure of a closely related form, which he called A_1. There are no reasons from available data to believe that any A form other than these two exists, and in the following text they will be called form A and form A-super. The crystal structure first determined is a superstructure with 12 molecules in the unit cell shown in Fig. 8.26. In one molecular layer three adjacent molecules have their carboxyl groups in the same directions, whereas the carboxyl groups of the next three molecules are directed in the opposite way. This structure is closely related to that of the A form as also shown in Fig. 8.26. The molecules form dimers with two centrosymmetrically related molecules in the triclinic unit cell. The molecules form dimers with the carboxyl groups of one dimer adjacent to methyl groups of the neighbouring dimer. Thus the unit structure is a monolayer and the structure of the A-super form can be regarded as a kind of transitory arrangement between the monolayer and the normal bilayer type of structure. Lomer (1965) found that the unit cell of the A form fits exactly with a three times larger cell of the A-super form, within the limits of experimental errors. The A-super structure is obtained from the

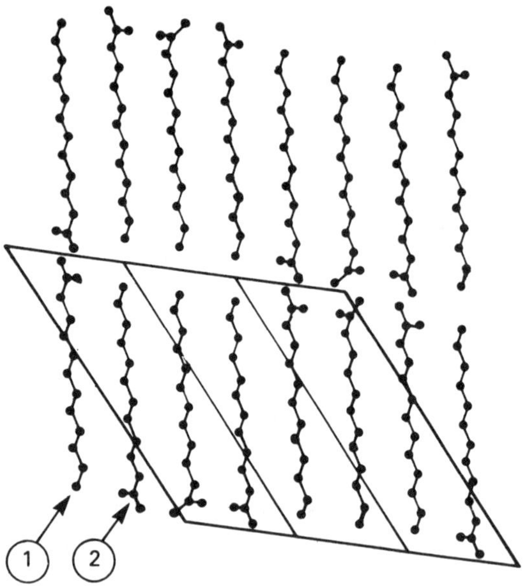

Fig. 8.26 Structure of the lauric acid crystal form A-super and the relations with the A form (after von Sydow, 1956a; Lomer, 1965). If the structural unit is equivalent to the two molecules denoted 1 and 2, the crystal structure of the A form is obtained.

structure of the A form by a translation of every third dimeric unit along the chain axis, a distance equal to the thickness of the monolayer.

The crystal structure of the A′ form of pentadecanoic acid (von Sydow, 1954) is quite similar to the A form and the form A-super. It has the same triclinic chain packing (thus these forms can be called β forms). The angle of tilt is between 66 and 71°, which is in agreement with that of the A form (67°). An interesting difference, however, is that the A′ form is the usual lipid bilayer structure.

(c) The B and B′ forms (β_2' forms)

The crystal structure of the B form of stearic acid (von Sydow, 1955) is shown in Fig. 8.27. The molecules are arranged in an orthorhombic chain-packing structure, in bilayers. The orientation of the chain-packing subcell in relation to the methyl end-group planes is O⊥ (101) (see section 8.1). A modified B form was observed in 11-bromoundecanoic acid (Larsson, 1963a), which was termed B_1. It is isomorphous with the B form according to X-ray diffraction data, but the OH out-of-plane vibration in the infrared absorption spectrum is different. A new form of stearic acid (Holland and Nielsen, 1963), described according to infrared absorption data, appears to be the same form as B_1 in 11-bromoundecanoic acid. The first crystal structure analysis of

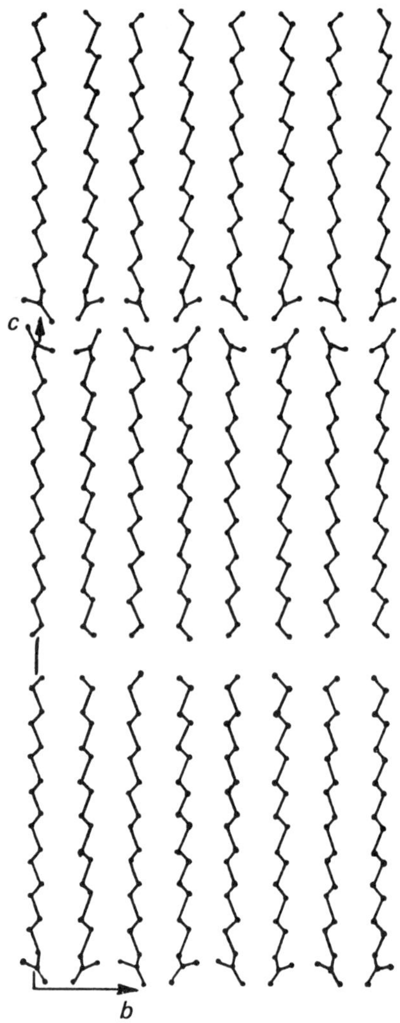

Fig. 8.27 The projection of the B form of stearic acid along the shortest axis (von Sydow, 1954).

the B form was determined in one projection only (von Sydow, 1955). A 'normal' three-dimensional arrangement of extended molecules described by Larsson and von Sydow (1966) was found to be wrong by Goto and Asada (1978), as evident from the structure in the projection on the *ac* plane shown in Fig. 8.28. The carbon atoms in 2 and 3 positions exhibit *gauche* conformation, giving the chains a bend.

The structure of the B′ form is shown in Fig. 8.29. It has the same chain packing as the B form but is somewhat more tilted. Thus the methyl endgroup plane in relation to the chain-packing subcell has indices O⊥ (111).

(d) The C and C′ forms (β_1' forms)

The crystal structure of the C form of lauric acid was the first fatty acid structure determined (Vand *et al.*,

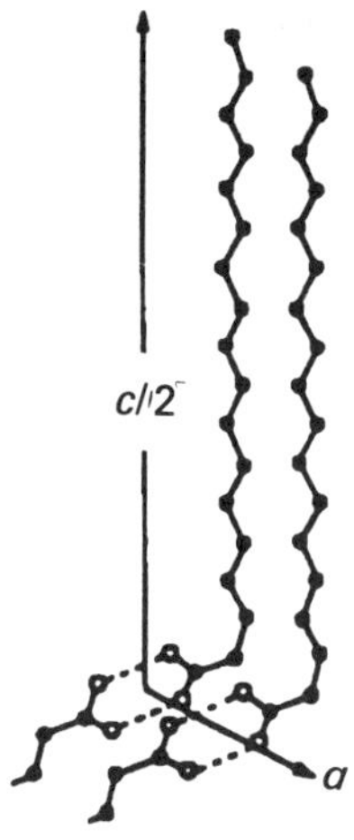

Fig. 8.28 Structure of the B form of stearic acid seen in the *ac* projection (after Goto and Asada, 1978).

1951). The structure is of the usual bilayer type with the chains arranged into the orthorhombic chain packing. The structure of the C′ form of undecanoic acid has also been determined (Fig. 8.30) (von Sydow, 1956b). It is strictly isostructural with that of the C form. The molecular arrangement in the bilayer is thus the same, with the endgroup plane related to the chain-packing subcell according to O⊥ (0 2 1).

The C′ form was found to have the carbon atoms located in a distorted way with regard to the planar zig-zag conformation. Also the dimensions of the orthorhombic subcell indicate an intermediate arrangement towards hexagonal chain packing. This is certainly a consequence of the existence of the C′ form in a narrow temperature range close to the melting point. Similar disorder effects in the C form close to the melting point were discussed in section 8.1.

(e) ω-Bromine- and ω-iodine-substituted fatty acids

The van der Waals' radius of a methyl group is similar enough to that of a bromine or iodine atom to make possible isomorphous replacement of terminating

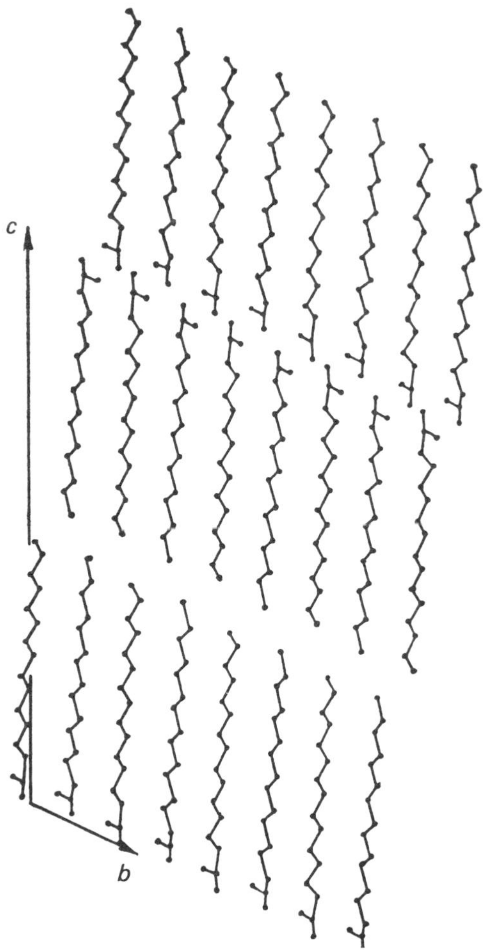

Fig. 8.29 The projection of the B′ form of pentadecanoic acid along the shortest unit-cell axis (von Sydow, 1956a).

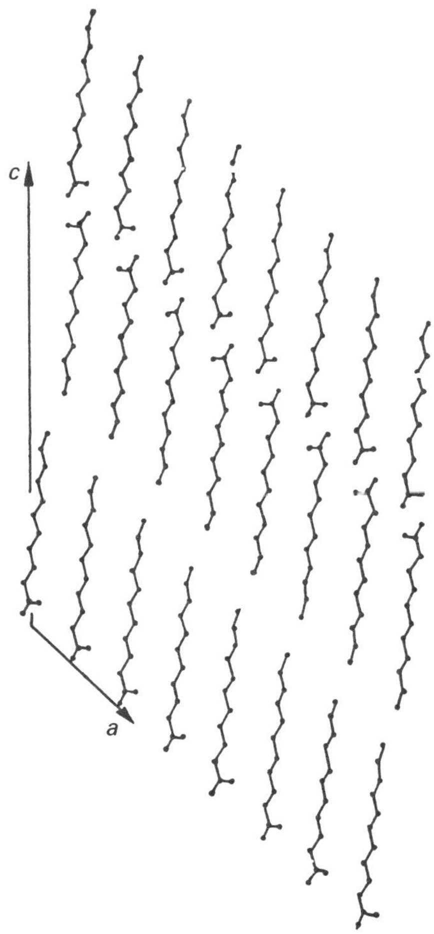

Fig. 8.30 The projection of the C′ form of undecanoic acid along the shortest unit-cell axis (von Sydow, 1956a).

methyl groups in lipids. This has been a useful feature for crystal structure determinations.

17-Bromoheptadecanoic acid and 17-iodoheptadecanoic acid exhibit B and C forms but no A form (Larsson, 1964). There is no transition B→C on heating, however, and the B form has the highest melting point (17-bromoheptadecanoic acid: B → liq. at 74.8 °C and C → liq. at 73.0 °C; 17-iodoheptadecanoic acid: B → liq. at 83.2 °C and C → liq. at 82.7 °C). The unit-cell dimensions of these substituted fatty acids are almost identical to those of stearic acid.

The more complex polymorphism of 11-bromoundecanoic acid (Larsson, 1963a) can be described by the scheme:

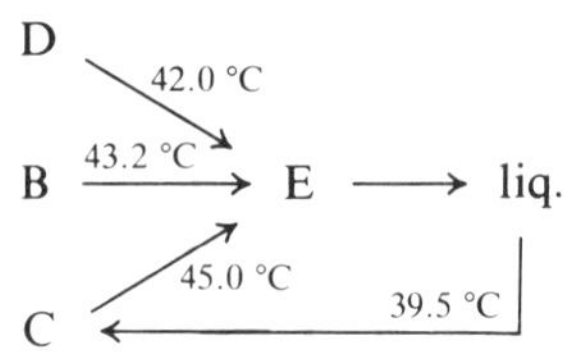

The crystal structure of the C form (Larsson, 1962) was determined for comparison with the C form of lauric acid. The only significant difference observed was a closer packing of the ω-bromine atoms over the endgroup gap than of methyl endgroups (a difference of 0.1 Å).

The crystal structure of the D form is shown in Fig. 8.31. The angle of tilt towards the endgroup plane is 41°. A remarkable feature is the alternation of the tilt of the molecules between adjacent bilayers (Larsson, 1963b).

The crystal structure of the E form is shown in Fig. 8.32. The lateral packing of the molecules head-to-tail is similar to that in the A form, although this form is more tilted (40°) (Larsson, 1963c). The chain-packing subcell in both the E and D forms is triclinic. They can therefore be termed β_1 and β_2, respectively.

(f) Branched saturated fatty acids

Methyl-branched fatty acids occur sometimes in Nature, for example in bacteria, and their solid-state behaviour has been investigated in detail (cf. Abrahamsson, 1959a). Two alternatives of accommodating a methyl branch into the close packing of the molecules have been found. One, called the τ form, is tilted so that the branch is accommodated in the methyl endgroup gap or carboxyl endgroup, and an example of this structure is shown in Fig. 8.33. The other alternative, called the ν form, makes space for the branch by change of the molecular direction at the branch, so that the molecule is V-shaped. An example of this structure alternative is shown in Fig. 8.34.

Racemic methyl-branched octadecanoic acids can exist in two forms, a high-melting form with the τ structure (Fig. 8.33), and a low-melting form with the ν structure (Fig. 8.34).

The optically active methyl-branched octadecanoic acids, although less studied, show similar structures. The 2-D-methyloctadecanoic acid has a high-melting τ form where the molecules are arranged head-to-tail

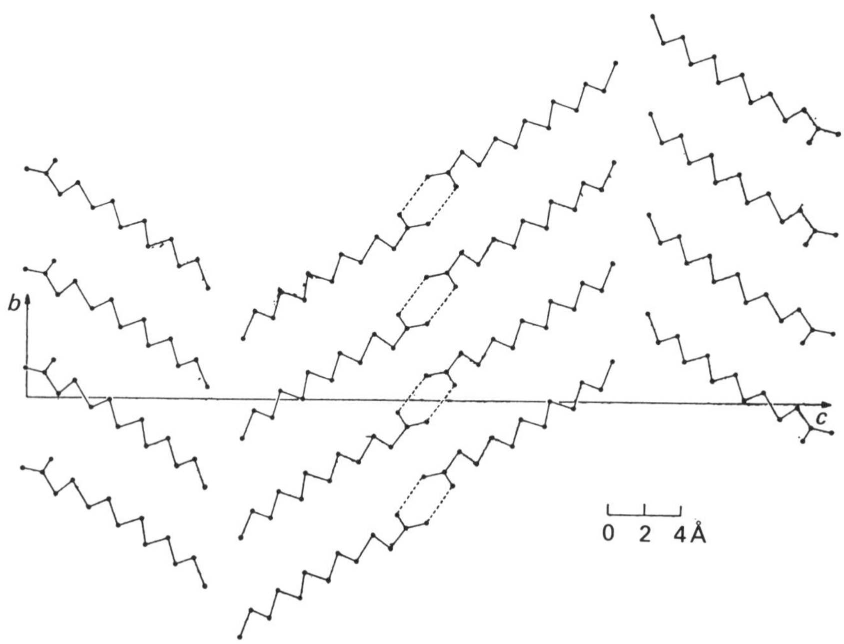

Fig. 8.31 The crystal structure of the D form of 11-bromoundecanoic acid projected along the *a* axis (5.63 Å) (Larsson, 1963b).

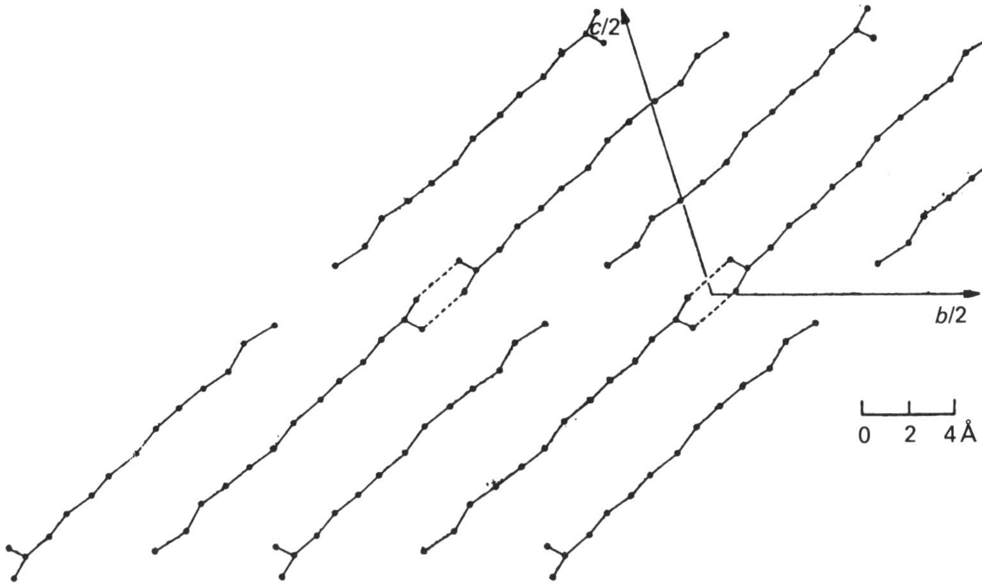

Fig. 8.32 The crystal structure of the E form of 11-bromoundecanoic acid projected along the shortest axis (Larsson, 1963c).

in monolayer units. The methyl branch is located in the gap between the endgroup planes and the hydrogen-bond system is of the infinite helical type (Abrahamsson, 1959c).

(g) Unsaturated fatty acids

The only complete crystal structure determination of an unsaturated lipid is that of the low-melting form of oleic acid (Abrahamsson and Ryderstedt-Nahringbauer, 1962), shown in Fig. 8.35. At the *cis* double bond there is a change in the direction of the chain tilt and the chain packing is of the unusual O′‖ type with hydrogen-bonded dimers arranged in bilayers.

Although there is no crystal structure determination available for any *trans*-unsaturated fatty acid, it is still possible to conclude that a *trans* double bond is accommodated into the chain packing without any change in the chain direction. This is possible from work by Lutton and Kolp (1949) on 18:1 fatty acids with a *trans* double bond in different positions. Thus they found that the molecules are arranged vertically in bilayers in certain members of

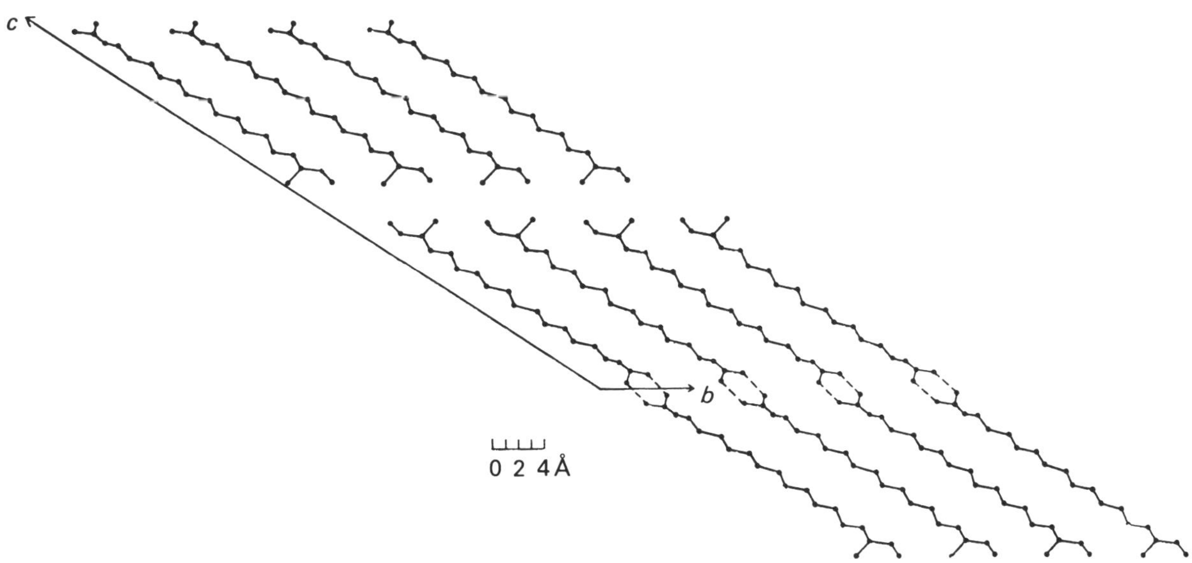

Fig. 8.33 Crystal structure of 16-DL-methyloctadecanoic acid viewed along the shortest unit-cell axis (Abrahamsson, 1958).

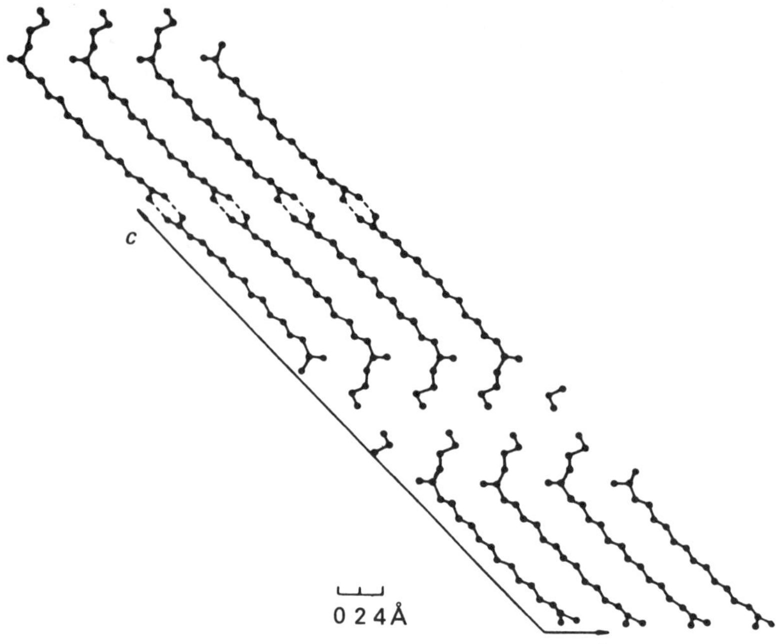

Fig. 8.34 Crystal structure of 14-DL-methyloctadecanoic acid viewed along the shortest unit-cell axis (Abrahamsson, 1959b).

this series, which means that the chain must be completely straight.

These data on *cis*- and *trans*-unsaturated double bonds combined with X-ray data from other lipids particularly from glycerides (see section 8.9), indicate that a *trans* double bond can be directly accommodated into any arrangement of extended chains. Ideal accommodation of a *cis* double bond, however, requires a change in the tilt of the chains.

Phase properties of binary mixtures of unsaturated fatty acids such as petroselenic acid–oleic acid, have been reported by Yoshimoto *et al.* (1991).

(h) Fatty acid esters of monohydric alcohols

Methyl esters of saturated fatty acids show one stable crystal form. It was originally termed β, but according to the nomenclature used here it is a β' form. The crystal structure of this form in methyl stearate (Aleby and von Sydow, 1960) is shown in Fig. 8.36. Although the forces between the polar groups must be very weak in an ester, the molecules are arranged head-to-head in bilayers. The chain packing is the orthorhombic O$\perp$, and the angle of tilt of the chains towards the endgroup plane is 62°. A second β' form (β'_2), which is closely related in structure according to the X-ray diffraction data, can be obtained from the melt or by rapid crystallization from solvents.

The methyl esters of fatty acids with an odd number of carbon atoms in the chain crystallize from the melt in an α form with a large stability interval. In this respect the odd esters are similar to normal paraffins where odd numbers give a stable α form over a broad temperature range.

Ethyl stearate separates from the melt as an α form with the characteristic transparent and plastic character due to the rotational disorder. At lower temperature this form is transformed into a β' form (originally termed β on the basis of its highest melting point). This β' form can also be obtained from solvents. The crystal structure has been determined (Aleby, 1968). In this ester, as well as in esters of longer alcohols, all molecules have the same direction, with the alcohol groups of one molecular layer facing the endgroups of the fatty acid chain in the adjacent layer (head-to-tail). Such an arrangement should be expected, as the polar forces between the carboxyl groups must be very weak at these long distances (the molecules have to be extended for space-filling reasons).

8.5.2 Melting point and heat of fusion

The melting of longer homologues of saturated fatty acids takes place as C$\rightarrow$liq. in even numbers and C'$\rightarrow$liq. in odd ones. The melting points are given in Table 8.2. There is a pronounced alternation between even and odd members; the even members show a melting curve at a somewhat higher temperature than the curve formed by odd acids. The alternation of

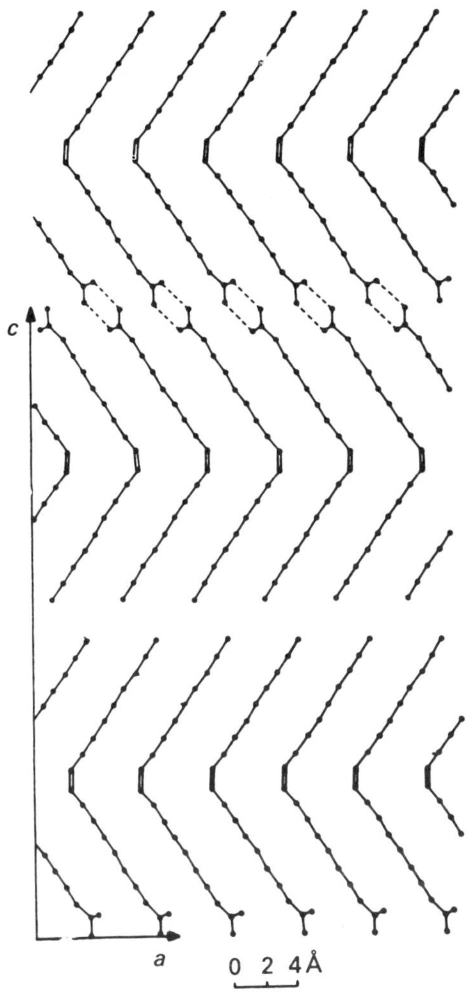

Fig. 8.35 Structure of the low-melting form of oleic acid as seen along the shortest unit-cell axis (Abrahamsson and Ryderstedt-Nahringbauer, 1962).

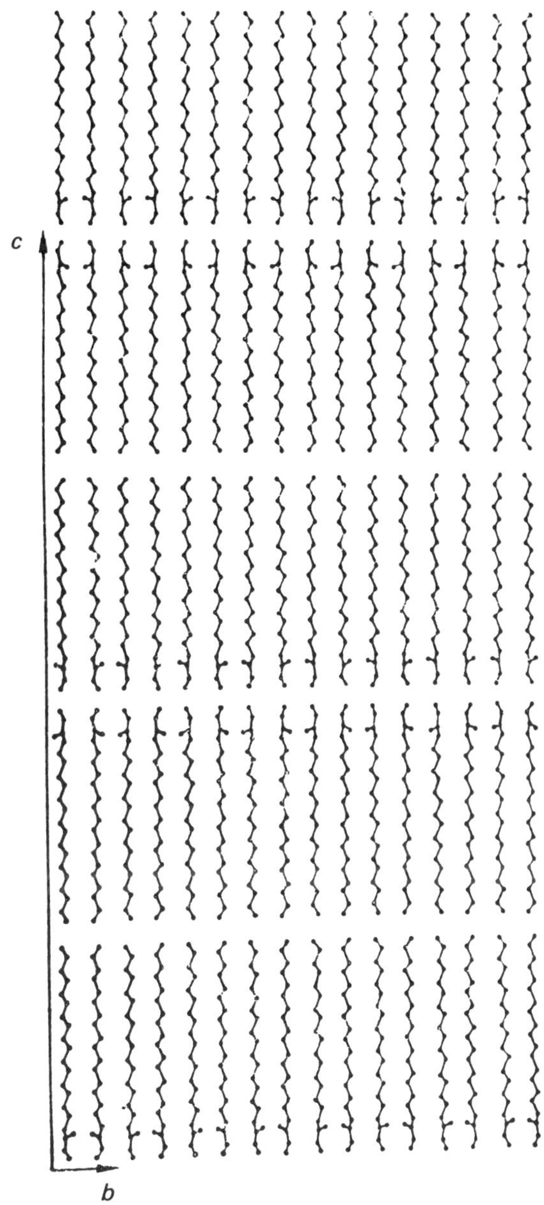

Fig. 8.36 A projection of the crystal structure of methyl stearate viewed along the shortest unit-cell axis (Aleby and von Sydow, 1960).

melting points can be explained by the methyl end-group structure (Larsson, 1966). The crystal structures of the forms C and C′ are isomorphous, which means that the difference is only due to packing differences over the methyl endgroup gap in the two forms. A denser packing is achieved when the last bond in the chain (CH_2–CH_3) is almost perpendicular to the endgroup plane, and this is the case in the C form of even fatty acids.

The melting-point curves show a convergence behaviour towards an asymptotic value of about 120–125 °C for infinitely long chains.

A few of the naturally occurring unsaturated fatty acids are also shown in Table 8.2. Depending upon the position of the double bond, there is a well known reduction in the melting point compared to the corresponding saturated fatty acids. This temperature reduction in a C_{18} chain with one double bond is only a few degrees when the double bond is near the carboxyl group (Δ2–Δ5), whereas it increases successively towards the methyl end to about 50 °C at the position Δ12. The reduction in melting point by a *trans* double bond is much smaller, but otherwise it follows the same behaviour as *cis* double bonds. A *trans* double bond in position Δ10–Δ12 causes a reduction in melting point of about 20 °C compared with the corresponding saturated chain (m.p. of stearic acid 69.6 °C and of elaidic acid 43.7 °C).

The heat of fusion of the even-membered saturated long-chain fatty acids involves the C form according to the polymorphic behaviour discussed above. The value in stearic acid (Singleton *et al.*, 1950), for

Table 8.2 Melting points of fatty acids according to Bailey (1950). Values given by Fasman (1975), when different, are shown afterwards within parentheses

Fatty acid		M.p. (°C)
Butanoic	4:0	−7.9
Pentanoic	5:0	−34.5
Hexanoic	6:0	−3.4
Heptanoic	7:0	−7.5 (−10.5)[a]
Octanoic	8:0	16.7
Nonanoic	9:0	12.5
Decanoic	10:0	31.6
Undecanoic	11:0	29.3
Dodecanoic	12:0	44.2
Tridecanoic	13:0	41.5
Tetradecanoic	14:0	54.4 (53.9)
Pentadecanoic	15:0	52.3
Hexadecanoic	16:0	62.9 (63.1)
Heptadecanoic	17:0	61.3
Octadecanoic	18:0	69.6
Nonadecanoic	19:0	68.65 (68.6)
Icosanoic	20:0	75.35 (76.5)
Henicosanoic	21:0	74.3
Docosanoic	22:0	79.95 (81.5)
Tricosanoic	23:0	79.1
Tetracosanoic	24:0	84.15 (86.0)
Pentacosanoic	25:0	83.5
Hexacosanoic	26:0	87.7 (88.5)
cis-unsaturated		
Palmitoleic	16:1	0.5
Oleic	18:1	16.25 and 13.36
Gadoleic	20:1	25
Erucic	22:1	33.4
Linoleic	18:2	−5.0
Linolenic	18:3	−11.0
Arachidonic	20:4	−49.5

[a] Solidification point.

example, is $\Delta H_f = 16.36$ kcal/mol. Above C_{10} in chain length the endgroup contribution to the heat of fusion remains constant, so there is a regular increase with the number of CH_2 groups according to the formula:

$$\Delta H_f = (n - 2)a + b$$

with a and b equal to 1.03 and −1.55 for even fatty acids, and 0.97 and −2.50 for odd fatty acids respectively (Bailey, 1950). The expression $(n-2)$, with n equal to the number of carbon atoms, is equal to the number of CH_2 groups in a saturated chain. A statistical-mechanical theory of melting has been reported by Aggarwal and Bhatnagar (1972), and calculated values of the heat of fusion are in fair agreement with experimental data.

Finally some thermal data for the different polymorphic forms of esters will be given.

Table 8.3 Melting point of the β' form of different esters (Malkin, 1936)

	Palmitate	Stearate
Methyl	–	38.8
Ethyl	23.5	30.5
n-Propyl	20.4	30.5
n-Butyl	16.9	27.5
n-Amyl	19.4	30
n-Octyl	22.5	31.8
n-Hexadecyl	51.6	56.6

1. *Methyl esters*: Two crystal forms with close melting points can be obtained from the melt. Their melting points are 28.95 and 29.20 °C in methyl palmitate, 37.37 and 37.80 °C in methyl stearate, and 51.65 and 51.84 °C in ethyl behenate (Francis and Piper, 1939).
2. *Ethyl esters*: The α form is obtained by cooling to 19.8, 30.75 and 40.5 °C for ethyl palmitate, ethyl stearate and ethyl icosanoate respectively. The α form is transformed into the stable crystal form, which (according to the nomenclature in section 8.1) is a β' form. The melting points of these esters are 25, 34 and 41 °C respectively (Malkin, 1933).
3. *Higher esters*: Melting points of a series of esters of palmitic and stearic acid are given in Table 8.3. The α form, obtained from the melt, is transformed to a β' form. It can be seen that there is a minimum in the melting points at the butyl ester of both series.

8.5.3 Vapour pressure, boiling point and heat of vaporization

Boiling points at various pressures of the most important fatty acids are given in Table 8.4 and of some esters in Table 8.5. Boiling points of C_1 to C_{18} fatty acids at atmospheric pressure are listed in Table 8.6.

The temperature dependence of the vapour pressure is related to the enthalpy of sublimation (ΔH_s) of the crystals by the Clapeyron equation:

$$dP/dT = \Delta H_s/T(V_g - V_c)$$

where V_g and V_c are the molal volume of the vapour and the crystal respectively. The enthalpy of sublimation can also be obtained from the enthalpy of fusion and the enthalpy of vaporization ($\Delta H_s = \Delta H_f + \Delta H_{vap}$) provided that these enthalpies are determined at the same temperature. The enthalpy of sublimation is very interesting from

Table 8.4 Boiling points of some important fatty acids at pressures between 1 and 760 mmHg (Jordan, 1954)

Pressure (mmHg)	Temperature (°C)				
	Dodecanoic	Tetradecanoic	Hexadecanoic	Octadecanoic	Oleic
1	121.0	142.0	153.6	173.7	176.5
5	150.6	174.1	188.1	209.0	208.5
10	166.0	190.8	205.8	225.0	223.0
20	183.6	207.6	223.8	243.4	240.0
40	201.4	223.5	244.4	263.3	257.2
60	212.7	237.2	256.0	275.5	269.8
100	227.5	250.5	271.5	291.0	286.0
200	249.8	272.3	298.7	316.5	309.8
400	273.8	294.6	329.0	343.0	334.7
760	299.2	318.0	353.8	370.0	360.0

Table 8.5 Boiling points of some fatty acid methyl esters at pressures between 1 and 20 mmHg (Jordan, 1954)

Pressure (mmHg)	Temperature (°C)		
	Palmitate	Stearate	Oleate
1	134.3	–	–
2	–	166	166.2
4	–	181	182
5	166.8	–	–
6	–	191	192
8	–	199	199.5
10	184.3	204	205.3
20	202.0	–	–

a theoretical point of view as it is equal to the lattice energy of the crystal structure (if the vapour phase behaves ideally). Selected values of heats of sublimation and heats of vaporization are given in Tables 8.7–8.9.

8.5.4 Density and specific volume

Dilatometric methods are frequently used in studies of phase transitions in lipids, as the dilatometric curves are very informative in that respect. Such a dilatometric curve is obtained by plotting specific volume versus temperature. It has been found that the ratio between specific heat and the coefficient of volume thermal expansion is nearly constant at different temperatures. This can be a useful relationship. If, for example, complete data on specific heats are available, a single determination of the coefficient of expansion can be used for estimation of the specific volume over a wide temperature range.

The density of the B form of stearic acid from flotation measurements was found to be 1.036 g/cm^3, and the calculated density from unit-cell dimensions was 1.041 ± 0.005 g/cm^3 (von Sydow, 1955). Calculation of the density changes of this B form according to unit-cell data between 0 °C and the transition point near 45 °C shows very small density differences, from 1.0444 to 1.0349 g/cm^3 (Degerman and von Sydow, 1959). The C form of stearic acid shows much stronger density variations with temperature. Thus between 0 °C and the melting point, near 70 °C, the calculated density changes from 1.0293 to 0.9995 g/cm^3 (Degerman and von Sydow, 1959). The C form has obviously a much more flexible structure, which can allow a higher degree of thermal mobility of the chains. The density for the A form, although the least-stable form, is close to those of the B and C forms. The density according to flotation for the A′ form of pentadecanoic acid was found to be 1.02 g/cm^3 and the calculated density from unit-cell dimensions was 1.034 ± 0.016 g/cm^3 (von Sydow, 1954).

Table 8.6 Boiling points of fatty acids[a] (Singleton, 1960)

Fatty acid	Boiling point (°C/mmHg)
Methanoic	100.5
Ethanoic	118.2
Propanoic	141.1
Butanoic	163.5
Pentanoic	187
Hexanoic	205.8
Heptanoic	223.0
Octanoic	239.7
Nonanoic	255.6
Decanoic	270
Undecanoic	284
Dodecanoic	225/100
Tridecanoic	236/100
Tetradecanoic	250/100
Pentadecanoic	202.5/10
Hexadecanoic	268/10
Heptadecanoic	220/10
Octadecanoic	213/5
cis-9-Octadecenoic (oleic)	234/15
trans-9-Octadecenoic (elaidic)	288/100
cis-9,*cis*-12-Octadecadienoic (linoleic)	202/1.4

[a] At 760 mmHg unless otherwise stated.

Table 8.7 Calculated values of heat of vaporization of some fatty acids at various temperatures (after Singleton, 1960)

Fatty acid	Temperature (°C)	ΔH_{vap} (cal/g)
Lauric	164	97
	243	77
Myristic	182	93
	209	87
	243	75
Palmitic	202	84
	244	76
Stearic	242	67

Table 8.8 Heat of vaporization of some short-chain esters (Svensson, 1980)

Ester	ΔH_{vap} (kJ/mol)
Methyl ethanoate	32.60 ± 0.06
Methyl propanoate	35.82 ± 0.09
Methyl butanoate	39.28 ± 0.22

Table 8.9 Heat of sublimation of fatty acids (Davies and Malpass, 1961)

Fatty acid	ΔH_s (kcal/mol)	Temperature range (K)
Decanoic acid	28.0	290–301
Dodecanoic acid	33.5	296–314
Tetradecanoic acid	33.4	312–325
Hexadecanoic acid	36.9	320–333
Octadecanoic acid	39.8	331–340
Icosanoic acid	47.7	337–346
Docosanoic acid	46.2	345–352

Table 8.10 Density of branched-chain fatty acids (Abrahamsson, 1956, 1958, 1959a)

Branched octadecanoic acid	Density (g/cm^3)	
	Observed	Calculated
2-D-Methyl	1.015	1.015 ± 0.010
2-DL-Methyl	0.998	1.001 ± 0.010
9-DL-Methyl	0.998	0.991 ± 0.010
14-DL-Methyl	0.94	0.95 ± 0.02
16-DL-Methyl	0.998	1.001 ± 0.010

The density for branched-chain fatty acids is considerably lower than that of the corresponding normal fatty acid. A methyl branch at different positions along a C_{18} chain gives small variations in density as shown in Table 8.10.

8.5.5 Viscosity

Dynamic viscosity measurements have been reported (Fernandez-Martin and Montes, 1976) for the saturated fatty acids C_{12}–C_{18} between 70 and 90 °C. A Newtonian behaviour was obtained at all shear rates with the values of η (cP) at 70 and 90 °C respectively:

C_{12}	4.415	2.982
C_{14}	5.984	3.906
C_{16}	7.682	4.885
C_{18}	9.583	6.294

Many binary, ternary and quaternary mixtures of these acids were also examined, and it was found that the viscosity of these mixtures falls on the same line as that of the components, provided that an average chain length is defined:

$$n = \Sigma\, x_i n_i$$

where x_i is the molar fraction and n_i the number of carbon atoms of the ith component. The lines of the experimental values were fitted by regression methods, and the regression parameters obtained for the data at 70 °C were

$$\log \eta = -0.602\,802 + 0.134\,844\bar{n} - 0.002\,591(\bar{n})^2$$

The data of the multicomponent mixtures were thus in agreement with the 'principle of congruence' introduced by Brönsted and Kofoed (1946).

Using a rotational–torsional surface viscometer, the surface shear viscosity of the C_{15}–C_{20} straight-chain fatty acids has been determined (Moo-Young *et al.*, 1981). The even fatty acids were found to show a surface Newtonian behaviour, with η_s values of 1.5, 2 and 3 mN s/m for the C_{16}, C_{18} and C_{20} members respectively. The odd acids, however, showed a surface pseudoplastic behaviour. Thus C_{17} gave decreasing surface shear viscosity up to a sharp rate of 8/s, and beyond that value the viscosity remained constant at about 0.6 mN s/m. These data indicate different monolayer structures of the even and odd members.

Viscosity data on fatty acids in the liquid state at various temperatures in solutions can be found in a review by Singleton (1960).

8.5.6 Refractive index

The ratio of the speed of electromagnetic radiation in vacuum to that in a particular medium is known as the absolute index of refraction of the medium, n:

$$n = c/v$$

According to Maxwell's relation,

$$n = \sqrt{K_\varepsilon}$$

where K_ε is the static dielectric constant. However, this relation is true only for simple gases. The interaction between the applied electric field and the molecules of the medium results in additional field components. The refractive index depends on the frequency of the radiation, which was known already by Isaac Newton (dispersion of white light by glass prisms).

Values of the refractive index for various fatty acids are given in Table 8.11.

Table 8.11 Refractive indices[a] of some fatty acids and their methyl esters (from Singleton, 1960)

Substance	n_D^t
	$n_D^{85.6}$
Palmitic acid	1.42545
Stearic acid	1.42830
Arachidic acid	1.43066
Behenic acid	1.43257
	n_D^{35}
Oleic acid	1.45442
Icosenoic acid	1.45574
Erucic acid	1.45674
	$n_D^{60.2}$
Methyl palmitate	1.42549
Methyl stearate	1.42897
Methyl arachidate	1.43165
Methyl behenate	1.43391

[a] Refractive index of sodium D line at temperature t (°C).

8.5.7 Solubility

Ideal solubilities, derived from the Clausius–Clapeyron equation, can be expressed by the equation:

$$2.3 \log X = \frac{\Delta H_f}{R}\left(\frac{1}{T_f} - \frac{1}{T}\right)$$

where X is the mole fraction of the solute, ΔH_f is the heat of fusion of the solute and T_f is the melting point of the solute (K).

The solubilities of fatty acids, although deviating from ideal behaviour, are known to vary regularly with molecular weight according to the formula:

$$\log S = c + kM$$

where S is the solubility in mol%; thus

$$S(\text{mol\%}) = \frac{\text{moles dissolved}}{\text{moles solvent} + \text{moles dissolved}} \times 100$$

Using the early data of fatty acid solubility in water from Ralston *et al.* (1942), the values $c = 0.0176$ and $k = -0.0173$ are obtained for the C_{10}–C_{18} saturated fatty acids. Even and odd fatty acids follow the same curve, i.e. there is no alternation. If the solubility of the same fatty acids in benzene is compared, however, they alternate (Brensch and Kirkali, 1968). Thus the C_{12}–C_{18} even members follow a line with $k = -0.0261$ and $c = 0.2696$, whereas the C_{13}–C_{17} odd members give values of $k = -0.0250$ and $c = 0.4761$. A probable explanation for this difference in alternation is the difference in methyl endgroup structure. Different endgroup structures in even and odd members mean that there is a difference in surface energy (see also section 8.5.2). A solvent like benzene has a surface energy close to that of the methyl endgroup surface of the crystals, whereas water has a far higher value. Thus the interfacial energy at the water–crystal surface interface is so high that the variations in surface energy between even and odd members can be neglected, contrary to the situation at the benzene–crystal surface interface.

The solubility in water will be considered first. Accurate values are difficult to determine, owing to strong variations with pH, and it is therefore necessary to take into account the amount of carbon dioxide dissolved in the water. Bell (1973) has published a critical analysis of the solubility in water. Instead of the molecular weight, given in the formula above, he has used the number of carbon atoms in the chain (n), i.e.

$$\log S = c + kn$$

The most reliable data for fatty acids were found to have the values $c = 2.32$ and $k = -0.60$. The corresponding values for normal alcohols are rather close, with $c = 2.3$ and $k = -0.58$. The scattered values of the fatty acid dissociation constants are also considered in the paper by Bell. It is concluded that at C_{10} and above this chain length, the dissociation constant is 1.20×10^{-5} mol/l.

There is one additional difficulty in the determination of water solubility of lipids, which often is neglected. The surface activity of lipid molecules means that monolayers are formed, and at very low solubilities the monolayer formed at the air–water interface should be allowed for (such a monolayer of a fatty acid corresponds to about 0.2 μg/cm^2 at a chain length near C_{18}).

The solubility of fatty acids in organic solvents has been treated in a series of papers by Skau and co-workers (Skau and Boucher, 1954; Skau and Bailey, 1959; Skau, 1970) and by Bailey *et al.* (1969), and the isopleth method was found to be useful in predicting solubility data. The solubility data for an ideal solution at constant pressure should follow the equation

$$\ln N = -\frac{\Delta H_f}{RT} + c$$

where N is the mole fraction of the compound with which the solution is saturated, T is the freezing point of the solution (K), ΔH_f is the heat of fusion (cal/mol) of the solute, R is the molar gas constant and c is a constant. If the same equation is written for another compound, then

$$\ln N' = -\frac{\Delta H'_f}{RT'} + c$$

The two equations can be combined at the same concentration in the same solvent ($N = N'$) to give

$$\frac{1}{T} = \frac{\Delta H'_f}{\Delta H_f T'} + c$$

For ideal solutions the $\log N$ versus $1000/T$ plots are straight lines, and $\Delta H'_f/\Delta H_f$ are constant over the whole concentration range. As deviations from ideal behaviour in homologous series are quite regular, plots of $1/T_a$ (analogue) will be linear against $1/T_r$ (reference) over small concentration ranges.

The procedure to use the isopleth reference method in order to predict solubility data for homologous and analogous compounds is this: an isopleth curve is first constructed for the reference compound by plotting $\log N$ against $1000/T$. Values of $1/T_r$ are taken out from the curve at N values where the solubility temperature for the analogous compound is known, and a $1/T_a$ versus $1/T_r$ plot is constructed (three points are generally sufficient). The $1/T_a$ value for any desired concentration can thus be determined by graphical interpolation (Bailey *et al.*, 1969).

An illustration of the use of the method is given in Table 8.12, containing solubility data for a few fatty acids that have been determined partly experimentally and partly by the isopleth reference method.

Solubilities of some fatty acid methyl esters in acetone are given in Table 8.13. The solubilities are rather close to those of the corresponding acids and the differences (the greater solubility of the esters) can be related to their lower melting points compared to the fatty acids.

A computer analysis of solubility data has been reported by Horvath (1973). Least-squares curve fitting of experimental data from homologous series of fatty acids and esters is given, based on the equation $x_2 = aM^b$, where x_2 is the mole fraction of the solute, M is its molecular weight and a and b are constants for a homologous series.

Knowledge of the solubility of binary mixtures of fatty acids is important as it provides a basis for fractionation. Such solubility data have been extensively discussed by Bailey (1950), and the use of isothermal phase diagrams is a useful way to represent the data. Such a ternary phase diagram is shown in Fig. 8.37. The point of highest solubility at any temperature is very close to the eutectic composition of the binary fatty acid systems (the broken line in the figure). The eutectic effect can also be described as a 'solubilizing' effect of one fatty acid in another.

8.5.8 Surface tension

The lowering of surface tension of lipids, which form insoluble monolayers, is identical to the spreading pressure of the monomolecular surface (see section 8.4), and some surface film data of normal fatty acids are given in Table 8.14.

Table 8.12 Solubility (wt%) of some fatty acids in acetone (Skau, 1970). Values in parentheses were determined experimentally and the other values were theoretically derived

Behenic		Stearic		Palmitic	
S(%)	T(°C)	S(%)	T(°C)	S(%)	T(°C)
95.9	77.2	(94.37)	(65.8)	(94.64)	(59.1)
89.8	74.5	(89.00)	(63.5)	(87.88)	(55.6)
79.6	70.5	(76.11)	(58.5)	(74.64)	(50.5)
59.4	65.4	(59.50)	(53.9)	(56.40)	(45.5)
39.5	61.1	(38.63)	(48.4)	(32.91)	(38.4)
33.8	60.2	(30.20)	(46.3)	(27.74)	(36.8)
(28.72)	(58.6)	(26.81)	(45.1)	(22.61)	(35.0)
19.6	55.9	(17.11)	(41.6)	(15.54)	(31.3)
10.7	51.2	(9.41)	(35.8)	(8.13)	(24.6, 26.7)
(5.09)	(46.2)	(6.41)	(32.6)	(4.26)	(18.4)
4.5	45.0	(3.28)	(27.2)	(3.44)	(16.3)
3.4	43.0	(1.54)	(19.1)	(2.59)	(13.5)
2.3	40.2	(0.82)	(14.3)	(1.90)	(10.0, 12.0)
1.04	(34.5)	(0.43)	(7.8)	(0.84)	(2.5, 5.5)
0.6	30.6	(0.10)	(−4.3)	(0.44)	(−3.6)

Table 8.13 Solubility (wt%) of methyl stearate and methyl palmitate in acetone (Bailey *et al.*, 1970). Notations as in Table 8.12

Methyl stearate		Methyl palmitate	
S(%)	T(°C)	S(%)	T(°C)
(85.82)	(37.4)	82.3	22.7
(47.37)	(31.9)	60.8	18.1
(10.83)	(24.5)	(56.12)	(17.5)
(2.77)	(18.6)	34.1	13.8
(0.40)	(6.2)	(20.11)	(10.9)
(0.06)	(−7.7)	10.7	6.9
		(4.53)	(0.7)
		0.5	−4.3
		0.25	−9.8
		0.1	−17.0

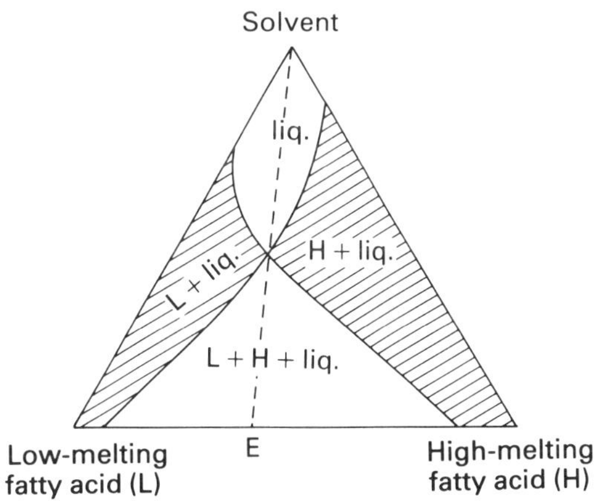

Fig. 8.37 Schematic illustration of the solubility of a binary fatty acid system in the form of a phase diagram (after Bailey, 1950). The broken line cuts the binary fatty acid system at the eutectic composition E.

Table 8.14 Monolayer data of palmitic acid and stearic acid (Sims and Zográfi, 1971)

Fatty acid	Compression rate (Å²/mol min)	Transition pressure (mN/m)	Collapse pressure (mN/m)
Palmitic acid	2.4	21.3	34.2
	1.2	21.9	32.0
	0.60	21.1	30.8
	0.24	20.9	26.5
Stearic acid	2.4	23.8	51.3
	1.2	23.8	49.8
	0.60	24.1	41.0
	0.24	23.1	34.6

There is a transition point at the Π–A isotherm from a monolayer with crystalline chain in a tilted arrangement to a monolayer with vertical (and crystalline) chains. The monolayer phase with vertical chains is unstable, which is not surprising as no stable crystal form exists with vertical chains. At variations in compression rate (Table 8.14) this transition pressure is constant, whereas the collapse pressure is reduced with decreasing compression rate. The equilibrium spreading pressure of these fatty acids is equal to the transition pressure in Table 8.14. As discussed in section 8.4 the reduction in surface tension, when crystalline fatty acids are present, is equal to the equilibrium spreading pressure.

The lowering of the surface tension of an aqueous solution of a fatty acid as a function of the natural logarithm of the concentration (g/kg) forms a sigmoidal curve. Hisey (1972) has analysed the data of the short-chain fatty acids, and gives the expression for the straight line of the surface tension reduction as

$$\Delta\gamma = aRT \log c + K$$

where c is the concentration (g solute/kg solution), aRT is the slope and K is the intercept. The value of these constants are given in Table 8.15. The intercept K (mN/m), obtained by the extrapolation of the linear part to the concentration 50% (w/w), represents a real value for the short chains, whereas for long chains this concentration lies in the two-phase region, acid + water. The molecular increment of the intercept is about 14 mN/m per CH_2 group for the homologous series shown in Table 8.15 and is independent of chain branching.

8.5.9 Dielectric constants and other electric properties

The dielectric constant K_ε (relative permittivity, $\varepsilon/\varepsilon_0$) of a medium is defined as the ratio of the field strength in vacuum to that in the medium for the same distribution of charge. It is usually measured as the ratio between the capacitance C of a condenser filled with the actual material and the capacitance C_0 of the empty condenser:

$$K_\varepsilon = \varepsilon/\varepsilon_0 = E_0/E = C/C_0$$

The relationship between the dielectric constant and the refractive index was discussed in section 8.5.6.

The Debye equation relates the dielectric constant to the polarizability (α) of the molecule, which is the dipole moment induced by unit electric field (1 esu):

$$\frac{K_\varepsilon - 1}{K_\varepsilon + 2} = \frac{4\pi N}{3}\left(\alpha + \frac{\mu}{3kT}\right)$$

Here N is the number of molecules per cubic centimetre, μ is the permanent electric dipole and k is the molecular gas constant.

Values of the dielectric constants for the liquid state of some fatty acids and esters are given in Table 8.16. Dielectric measurements are particularly useful for studies of molecular motion. Organic crystals

Table 8.15 Surface tension reduction of fatty acids ($aRT \log c + K$) according to Hisey (1972), where C is the concentration (g/kg)

Fatty acid	Temp. (°C)	aRT	K (mN/m)
Methanoic	30	5.943	21.45
Ethanoic	25	7.718	34.78
	30	7.752	33.97
	35	7.724	34.08
Propanoic	25	8.844	47.29
	35	8.721	46.16
Butanoic	25	10.58	64.26
	35	10.37	62.75
2-Methylpropanoic	25	10.16	62.48
	35	10.08	61.74
Pentanoic	25	10.59	78.74
	35	10.39	77.21
3-Methylbutanoic	25	10.39	74.88
	35	10.40	74.27
Heptanoic	20	11.40	107.4

Table 8.16 Some values of the dielectric constant of liquid fatty acids and esters (Landolt-Börnstein, 1959)

Fatty acid or ester	Temperature (°C)	K_ε
Palmitic acid	63	2.39
	69.7	2.24
	75	2.40
Stearic acid	74.5	2.22
Oleic acid	21.9	2.43
Linoleic acid	0	2.61
	20	2.71
	70	2.70
	120	2.60
Linolenic acid	−10	2.55
	20	2.76
	60	2.93
	100	3.01
Ethyl palmitate	20	3.20
	30	3.07
	104	2.71
	182	2.46
Ethyl oleate	28	3.18

without rotational disorder exhibit dielectric constants in the range 2–3, with a sharp increase at the melting point. If a crystal form with rotation of the molecules is formed from the melt, however, there is an increase in the dielectric constant related to the rotational freedom of the molecular dipoles in the applied field.

Studies of dielectric constants of simple esters that form α forms indicate that whole molecules rotate in the lattice (cf. Baker and Smyth, 1938). The behaviour of ethyl palmitate shown in Fig. 8.38 is characteristic of simple lipids in the α form. The peak obtained when the α form crystallizes was considered to be due to molecular rotation. A theory relating the degree of molecular rotation in lipids to the dielectric constant was later presented (Hoffman and Smyth, 1950).

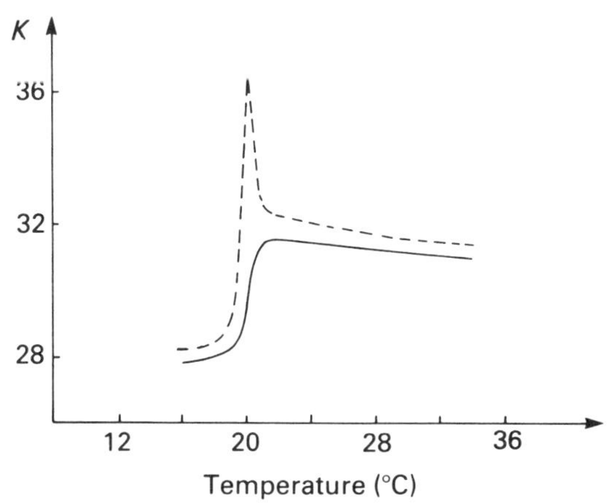

Fig. 8.38 Temperature dependence of the dielectric constant at 40 kHz of ethyl palmitate (after Baker and Smyth, 1938). The full curve shows heating and the broken curve cooling of the sample.

The dielectric properties of lipid–water phases have recently been analysed by Fontell and Lundström (1977). The measured dielectric constant of the lamellar liquid-crystalline phase is much lower than what should be expected for a lipid–water mixture. The difference was explained by local anisotropy.

The lateral electrical conductivity of the lamellar L_α phase has also been studied (Hamnerius *et al.*, 1978). The conductivity (σ) was obtained from plots against resistance (R):

$$\sigma = \left(\frac{\Delta(Rbs)}{\Delta L}\right)^{-1}$$

where b and s are the dimensions of contact between the electrodes and the oriented sample and L is the distance between the electrodes. The conductivity as a function of lamellar repeat distance was found to vary from 3.16 S/m at a distance of 37.5 Å to 0.07 S/m at a repeat distance of 152 Å.

Sjöblom *et al.* (1983) have studied the dielectric constant in the reversed micellar solution (the L2 phase) in the ternary system sodium octanoate–decanol–water, and some of their data are given in Fig. 8.39. It can be seen that changes in the dielectric constant are related to variations in association behaviour. Thus in the premicellar region there is a decrease in the dielectric constant, which can be explained by the predominant interaction between the alcohol and the water. In the region considered to consist of spherical water micelles there is a linear increase in the dielectric constant. When finally the region consisting of asymmetric micelles is reached, there is a sharp increase in the dielectric constant. An interesting analysis of the results is also presented, which makes possible the prediction of conditions in the palisade layer surrounding the aqueous core.

The paraffin hydrocarbons are highly insulating molecules. The electrical conductivity of about 10^{-22} S/m observed in liquid n-hexane has been attributed to the effect of cosmic rays (Gernant, 1950). Baker and Smyth (1938) have reported that the specific conductance of ethyl palmitate is below 10^{-11} S/m.

A remarkable phenomenon has been observed in simple esters, known as electret behaviour. Electrets are electrical analogues of magnets (cf. Wieseman and Feaster, 1957). Thus carnauba wax, which is a

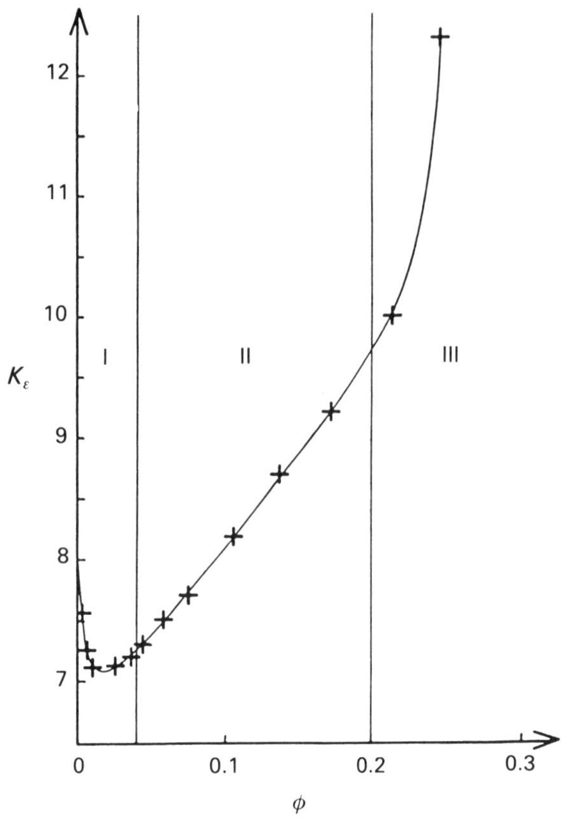

Fig. 8.39 Dielectric constant values as a function of the volume fraction of the aqueous micellar core in the L2 phase of the system sodium octanoate–decanol–water. The regions I–III correspond to the premicellar solution, the region of spherical micelles and the region with rod-shaped micelles respectively (Sjöblom *et al.*, 1983).

mixture of long-chain fatty acid esters of long-chain alcohols, when crystallized in an electric field, is positive at one end and negative at the other. Such an electret can keep its charge for years, and this phenomenon has been attributed to orientation of the dipoles in the bulk of the crystals.

Studies of charge injection into fatty acid multilayers and storage of charges in the lipid have been reported by Lundström and McQueen (1973) from capacitance measurements.

References and further reading

Abrahamsson, S. (1956) *Acta Crystallogr.*, **9**, 663.
Abrahamsson, S. (1958) *Acta Crystallogr.*, **11**, 270.
Abrahamsson, S. (1959a) *Ark. Kemi*, **14**, 65.
Abrahamsson, S. (1959b) *Acta Crystallogr.*, **12**, 206.
Abrahamsson, S. (1959c) *Acta Crystallogr.*, **12**, 304.
Abrahamsson, S. and Ryderstedt-Nahringbauer, I. (1962) *Acta Crystallogr.*, **15**, 1261.
Aggarwal, S.K. and Bhatnagar, H.L. (1972) *J. Indian Chem. Soc.*, **49**, 1293.
Aleby, S. (1968) *Acta Chem. Scand.*, **22**, 811.
Aleby, S. and von Sydow, E. (1960) *Acta Crystallogr.*, **13**, 487.
Bailey, A.E. (1950) *Melting and Solidifcation of Fats*, Interscience, New York.
Bailey, A.E., Harris, J.A. and Skau, E.I. (1969) *J. Am. Oil Chem. Soc.*, **46**, 583.
Bailey, A.E., Harris, J.A. and Skau, E.I. (1970) *J. Am. Oil Chem. Soc.*, **47**, 583.
Baker, W.O. and Smyth, C.P. (1938) *J. Am. Oil Chem. Soc.*, **60**, 1229.
Baun, W.L. (1961) *J. Phys. Chem.*, **65**, 2122.
Bell, G.H. (1973) *Chem. Phys. Lipids*, **10**, 1.
Brensch, F.L. and Kirkali, A. (1968) *Fette Seifen Anstrichm.*, **70**, 864.
Brönsted, J.N. and Kofoed, J. (1946) *K. Dan. Vidensk. Selsk. Mat.-Fys. Medd.*, **22**, 1.
Davies, M. and Malpass, V.E. (1961) *J. Chem. Soc.*, 1048.
Degerman, G. and von Sydow, E. (1959) *Acta Chem. Scand.*, **13**, 984.
de Jong, S., van Soest, T.C. and van Schaick, M.A. (1991) *J. Am. Oil Chem. Soc.*, **68**, 371.
Fasman, G.D. (ed.) (1975) *Handbook of Biochemistry and Molecular Biology: Lipids, Carbohydrates, Steroids*, 3rd edn, CRC Press, Cleveland, OH, p. 484.
Fernandez-Martin, F. and Montes, F. (1976) *J. Am. Oil Chem. Soc.*, **53**, 130.
Fontell, K. and Lundström, I. (1977) *J. Colloid Interface Sci.*, **59**, 360.
Francis, F. and Piper, S.H. (1939) *J. Am. Chem. Soc.*, **61**, 577.
Gernant, A. (1950) in *Physical Properties of Hydrocarbons*, vol. 1 (ed. A. Farkas), Academic Press, New York, p. 215.
Goto, M. and Asada, E. (1978) *Bull. Chem. Soc. Japan*, **51**, 2456.
Hamnerius, Y., Lundström, I., Paulsson, L.E., Fontell, K. and Wennerström, H. (1978) *Chem. Phys. Lipids*, **22**, 135.
Hisey, A. (1972) *J. Colloid Interface Sci.*, **10**, 133.
Hoffman, J.D. and Smyth, C.P. (1950) *J. Am. Chem. Soc.*, **72**, 171.
Holland, R.F. and Nielsen, J.R. (1963) *Acta Crystallogr.*, **16**, 902.
Holzberg, F., Post, B. and Fankuchen, I. (1953) *Acta Crystallogr.*, **6**, 127.
Horvath, A.L. (1973) *Chem. Eng. Sci.*, **28**, 299.
Jones, R.E. and Templeton, D.H. (1958) *Acta Crystallogr.*, **11**, 484.
Jordan, T.E. (1954) *Vapor Pressure of Organic Compounds*, Interscience, New York.
Landolt-Börnstein (1959) *Eigenschaften der Materie in ihren Aggregatzuständen*, Springer-Verlag, Göttingen, vol. 6, Tabelleneinrichtung 27450.
Larsson, K. (1962) *Acta Chem. Scand.*, **16**, 1751.
Larsson, K. (1963a) *Acta Chem. Scand.*, **17**, 215.
Larsson, K. (1963b) *Acta Chem. Scand.*, **17**, 199.
Larsson, K. (1963c) *Acta Chem. Scand.*, **17**, 221.
Larsson, K. (1964) *Ark. Kemi*, **23**, 35.
Larsson, K. (1966) *J. Am. Oil Chem. Soc.*, **43**, 559.

Larsson, K. and von Sydow, E. (1966) *Acta Chem. Scand.*, **20**, 1203.
Lomer, T.R. (1955) *Nature*, **176**, 653.
Lomer, T.R. (1965) *Acta Crystallogr.*, **16**, 984.
Lundström, I. and McQueen, D. (1973) *Chem. Phys. Lipids*, **10**, 181.
Lutton, E.S. and Kolp, D.G. (1949) *J. Am. Chem. Soc.*, **71**, 4113.
Malkin, T. (1933) *Trans. Faraday Soc.*, **29**, 977.
Malkin, T. (1936) *J. Chem. Soc.*, 103.
Moo-Young, M., Fulford, G.D. and Deysarkar, A.K. (1981) *Can. J. Chem. Eng.*, **59**, 366.
Ralston, A.W., Hoerr, C.W. and Sedwich, R.S. (1942) *J. Org. Chem.*, **7**, 546.
Sims, B. and Zográfi, G. (1971) *Chem. Phys. Lipids*, **5**, 109.
Singleton, W.S. (1960) *Fatty Acids* (ed. K.S. Markley), Interscience, New York, p. 499.
Singleton, W.S., Ward, T.L. and Dollear, F.G. (1950) *J. Am. Chem. Soc.*, **27**, 143.
Sjöblom, J., Jönsson, B., Nylander, C. and Lundström, I. (1983) *J. Colloid Interface Sci.*, **96**, 504 and **100**, 27.
Skau, E.L. (1970) *J. Am. Oil Chem. Soc.*, **47**, 233.
Skau, E.L. and Bailey, A.V. (1959) *J. Phys. Chem.*, **63**, 2047.
Skau, E.L. and Boucher, R.E. (1954) *J. Phys. Chem.*, **58**, 460.
Stenhagen, E. and von Sydow, E. (1953) *Ark. Kemi*, **6**, 309.
Svensson, C. (1980) Thesis, University of Lund, Sweden.
Vand, V., Morley, W.M. and Lomer, T.R. (1951) *Acta Crystallogr.*, **4**, 324.
von Sydow, E. (1954) *Acta Crystallogr.*, **7**, 529.
von Sydow, E. (1955) *Acta Crystallogr.*, **8**, 557.
von Sydow, E. (1956a) *Ark. Kemi*, **9**, 231.
von Sydow, E. (1956b) *Acta Chem. Scand.*, **10**, 1.
Wieseman, G.G. and Feaster, G.R. (1957) *J. Chem. Phys.*, **26**, 521.
Yoshimoto, N., Nakamura, T., Suzuki, M. and Sato, K. (1991) *J. Phys. Chem.*, **95**, 3384.

8.6 Soaps

Soaps are metal salts of fatty acids, and the alkali-metal salts are the most important members of this group of amphiphiles. Applications of soaps are related to their properties as association colloids and surface-active agents.

8.6.1 Polymorphism in anhydrous soaps

Alkali-metal salts of fatty acids have a very complex phase behaviour. There are different crystal forms between 100 °C (approximately) and the temperature where the lamellar liquid-crystalline phase ('neat') is formed. A series of phases exist that here are termed *intermediate* phases. All these phases represent a stepwise increase in the molecular disorder with temperature.

(a) Crystal forms

It is hard to obtain good single crystals of soaps for X-ray analysis. Potassium soaps appear to give the best-quality single crystals, and the most complete structural knowledge on soaps, therefore, is from the potassium salts.

Four crystal forms have been observed. These have been termed A, B, C and D in order of increasing angle of tilt of the chains. The crystal structure of the A form of potassium decanoate was first determined by Vand *et al.* (1949), and the original structure and a later revised version (Lewis and Lomer, 1969) are shown in Fig. 8.40. The molecules form the usual bilayer type of structure, and a remarkable feature is that the hydrocarbon chains have a crossed arrangement in the layers. This illustrates the dominating effect of the structure in the ionic region in determining the overall structure of soaps. The monoclinic space group is $P2_1/a$, and the dimensions of potassium soaps of different chain lengths are given in Table 8.17. The long spacings fit the curve $5.70 \pm 2.165n$ Å, where n is the number of carbon atoms in the chain.

At higher temperatures there is a transition to the C form, which has a closely related structure (crossed chain arrangement and the same space group). The difference is that the last four carbon atoms of the chains are highly disordered (Lomer, 1952). The dimensions are given in Table 8.18. The D form is a hydrate formed at low temperature.

The B form of soaps has a bilayer structure with parallel chains (Lomer, 1952). The chains are tilted 50–52°, and the dimensions of the triclinic unit cell are given in Table 8.18. The long spacings $d(001)$ fit the curve $6.80 + 1.95n$ Å, where n is the number of carbon atoms.

Sodium soaps have complex structures with large unit cells. The most complete single-crystal data have been reported by Segerman (1963). It is reported that the C_6–C_{18} sodium soaps form monoclinic crystals with space group $P2_1$ or $P2_1/m$, with 20 molecules in the unit cell and a bilayer structure with a crossed chain arrangement.

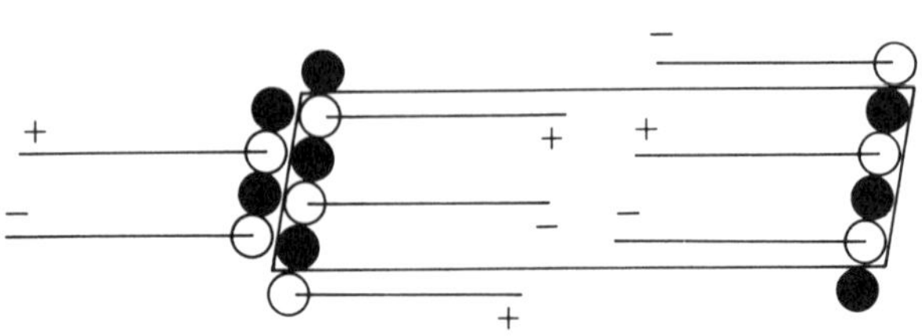

Fig. 8.40 Schematic illustration of the structure of the A form of potassium decanoate (Lewis and Lomer, 1969). The plus and minus signs indicate that the chains are directed up and down, respectively, in relation to the plane of the paper.

Table 8.17 Dimensions of the monoclinic unit cell of the A form (at 25 °C) and C form (at 75 °C) of potassium soaps (Lomer, 1952) of different chain length C_n

		a (Å)	b (Å)	c (Å)	β (deg)
Form A	C_6	8.00	5.74	18.84	91.9
	C_8	7.92	5.68	23.01	92.1
	C_{10}	8.08	5.68	27.58	93.5
	C_{12}	7.97	5.67	31.78	95.7
Form C	C_8	8.09	5.73	23.81	91.7
	C_{10}	8.12	5.68	28.07	91.8
	C_{12}	8.08	5.67	32.36	94.2
	C_{14}	8.08	5.69	36.94	92.1
	C_{16}	8.05	5.71	41.03	90.1
	C_{18}	8.07	5.67	45.46	93.2

Table 8.18 Dimensions of the triclinic unit cell of the B form of potassium soaps (Lomer, 1952)

	a (Å)	b (Å)	c (Å)	α (deg)	β (deg)	γ (deg)
C_{12}	4.14	5.60	30.04	92.4	92.0	94
C_{14}	4.15	5.61	34.07	91.6	91.5	93
C_{16}	4.17	5.61	38.27	92.7	91.3	93
C_{18}	4.16	5.57	42.11	91.5	91.6	94

Crystallization of molecular complexes of fatty acids and soaps (acid soaps) have also been reported, and unit-cell data are given in a recent review on organic salts (Fontell, 1978).

(b) Intermediate phases

When a crystalline phase of an anhydrous sodium soap is heated, it passes through a series of intermediate phases before the lamellar L_α phase is formed. These phases have phenomenologically been termed subwaxy, waxy, superwaxy and subneat. Skoulios and Luzzati (1961) have given a structural description of this stepwise disordering process based on an X-ray diffraction analysis. The temperature range for the intermediate phases is 100–255 °C for sodium laurate, with a regular increase of the lower limit to 117 °C and the upper limit to 257 °C for sodium stearate. Each intermediate phase shows an existence range of about 20 °C. The structure of these phases has been proposed to consist of infinite ribbons of bilayer units as shown in Fig. 8.41. The stepwise melting was explained as an effect caused by increasing hydrocarbon-chain disorder with persisting structure of the polar sheets, which are successively broken up into smaller ribbon units. The lateral size of the infinite ribbon units corresponds to some 10–20 laterally packed molecules.

8.6.2 Soap–water systems

The phase diagram shown in Fig. 8.13 is characteristic for a soap–water system. The structures occurring and the phase relations were determined in the classic work by Luzzati *et al.* (1960). Soaps of different chain lengths show an L_α phase (termed 'neat' in old soap literature) at low water content and at higher water content there is a large region where the H_I phase (termed 'middle' in earlier literature) exists. Between the H_I and the L_α phases, Luzzati *et al.* (1960) described some complex liquid-crystalline phases that with present knowledge on amphiphile–water systems probably can be better explained as two-phase mixtures also involving cubic phases. Structural parameters characteristic for the L_α and H_I phases are given in Fig. 8.42.

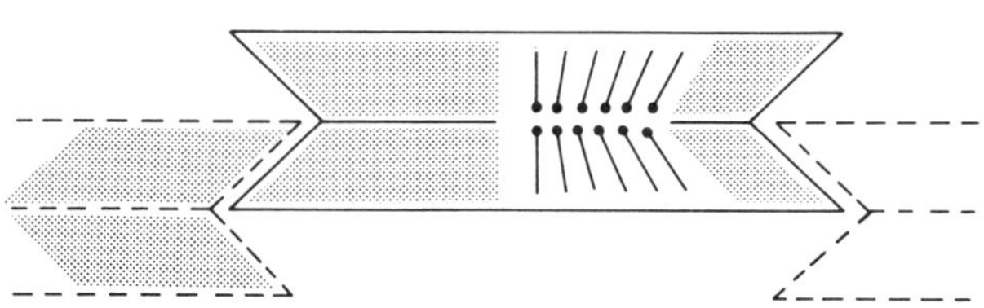

Fig. 8.41 Proposed structure of the ribbon-like units in the intermediate phases subwaxy, waxy, superwaxy and subneat of sodium soaps (Skoulios and Luzzati, 1961).

The structure in the micellar region of the phase diagrams of potassium soaps has been analysed by Reiss-Husson and Luzzati (1969) using X-ray methods. A common feature in soaps of saturated fatty acids is that spherical micelles exist at low concentrations, and at increased concentrations a transition into rod-like micelles occurs. Sodium oleate, however, was found to give rod-shaped micelles at all concentrations. The micellar association and phase behaviour have been reviewed by Wennerström and Lindman (1979) and Lindman and Wennerström (1980).

8.6.3 Viscosity

The viscosity of micellar soap solutions will first be discussed. Complete measurements of the viscosity of sodium octanoate solutions over the whole micellar composition range (0–40.5% (w/w)) have been reported by Ekwall and Holmberg (1965). In the premicellar solution the viscosity follows the relation:

$$\eta_{rel} = 1 + 0.005\sqrt{C} + 0.940C$$

where C is the concentration in moles per litre. Above the critical micellar concentration (CMC) ($C = 0.356$) and up to $C = 1.2$ the relation between viscosity and concentration follows the equation:

$$\log \eta_{rel} = 0.321C/(1 - 0.274C)$$

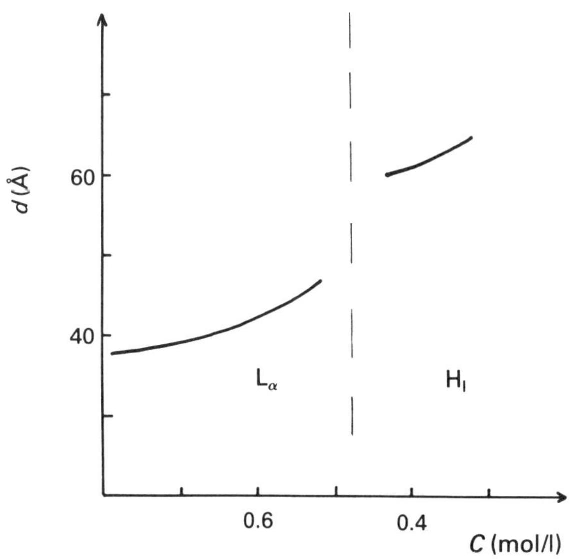

Fig. 8.42 Lattice parameters of the L_α phase and the H_I phase in the aqueous system of sodium stearate as a function of water content. The curve shows the lamellar repeat unit of the L_α phase and the distance between the centre of the cylinders in the H_I phase.

Table 8.19 Krafft points, given as the lowest temperature reported before crystallization at the CMC

	Decanoate	Dodecanoate	Tetradecanoate	Hexadecanoate	Octadecanoate
Sodium salt	20	17	21	45	50
Potassium salt	25	15	25	35	55

The viscosity data in this range indicate that the micelles are spherical in shape and, in the range $C = 1.2$ (second CMC) to $C = 1.8$ (third CMC), the data indicate that there is a transition from spherical to cylindrical micelles. In this range the viscosity increases with concentration according to:

$$\log(\eta_{sp}/C) = 0.74 + 0.95(C - 1.8)$$

Viscosity studies of the L2 solution of ternary systems involving soap/alcohol and water have been reported by Ekwall and Solyom (1967) and Sjöblom and Liljeström (1983). In the system sodium octanoate–pentanol–water, the viscosity values are quite small, with 3.4 cP for pure pentanol and values around 10 cP at the phase boundary towards the L_α phase. Much higher values are observed in the system sodium octanoate–decanol–water, where the L2 phase is formed towards a region of the hexagonal H_{II} phase. A strong increase in viscosity can be related to a concentration where the shape of the water micelles is considered to change from spherical to rod-like. Viscosities as high as 110 cP were recorded near the phase boundary towards the region of the H_{II} phase.

Interesting studies have been reported on the rheological properties of liquid-crystalline phases in soap systems. According to a theory of flow in concentrated colloidal systems (Bohlin, 1980), it is possible to relate the flow in ordered liquids, like liquid-crystalline phases, to the structure. This theory has been successfully applied to the liquid-crystalline phases of soaps and other amphiphiles, for example phases in the system cetyltrimethylammonium bromide (CTAB) or sodium octanoate–decanol–water (Bohlin and Fontell, 1978; Faucompre, 1982).

8.6.4 Krafft points

There is a critical temperature of amphiphilic compounds for formation of a micellar solution. Below this temperature the amphiphile forms crystals coexisting with water or a gel phase, and above this temperature the micellar solution is formed. This temperature varies with amphiphile concentration. The Krafft point has been defined as the knee in the melting curve of the amphiphile in the presence of water, which corresponds to a kind of triple point with coexistence of a micellar solution, a molecular solution and a solid phase (cf. Shinoda, 1981). This temperature is considerably lower than the hydrocarbon-chain melting point of liquid-crystalline phases, which are formed at much higher concentrations.

The Krafft temperature is a semi-quantitative concept, which cannot be determined, as the changes in physical properties at the Krafft point are continuous. Some temperatures for sodium and potassium soaps are given in Table 8.19.

References and further reading

Bohlin, L. (1980) *J. Colloid Interface Sci.*, **74**, 423.

Bohlin, L. and Fontell, K. (1978) *J. Colloid Interface Sci.*, **67**, 272.

Ekwall, P. and Holmberg, P. (1965) *Acta Chem. Scand.*, **19**, 455.

Ekwall, P. and Solyom, P. (1967) *Acta Chem. Scand.*, **21**, 1619.

Faucompre, B. (1982) Thesis, Université du Languedoc, Montpellier.

Fontell, K. (1978) in *Thermodynamics and Transport Properties of Organic Salts* (eds P. Franzosini and M. Sanesi), IUPAC Chem. Data Ser. 28, Pergamon, Oxford, p. 343.

Lewis, E.L.V. and Lomer, T.R. (1969) *Acta Crystallogr.*, **B25**, 702.

Lindman, B. and Wennerström, H. (1980) *Top. Curr. Chem.*, **87**, 1.

Lomer, T.R. (1952) *Acta Crystallogr.*, **5**, 11.

Luzzati, V., Mustacchi, H., Skoulios, A. and Husson, F. (1960) *Acta Crystallogr.*, **13**, 660.

Reiss-Husson, F. and Luzzati, V. (1969) *J. Phys. Chem.*, **68**, 3504.

Segerman, E. (1963) *Acta Crystallogr.*, **16**, part 13 suppl., A76.

Shinoda, K. (1981) *J. Phys. Chem.*, **85**, 3311.

Sjöblom, J. and Liljeström, M. (1983) *Colloid Polym. Sci.*, **261**, 251.

Skoulios, A.E. and Luzzati, V. (1961) *Acta Crystallogr.*, **14**, 278.

Vand, V., Lomer, T.R. and Lang, A.R. (1949) *Acta Crystallogr.*, **2**, 214.

Wennerström, H. and Lindman, B. (1979) *Top. Curr. Chem.*, **87**, 1.

8.7 Monoglycerides (monoacylglycerols)

8.7.1 Solid-state behaviour

(a) 2-Monoglycerides

2-Monoglycerides have a quite simple phase behaviour. When the melt is cooled, a thermotropic liquid-crystalline phase is first formed (at 47 °C in 2-monolaurin). A monotropic transition into the β form (at 40 °C in 2-monolaurin) then takes place. There is only one crystal form in the solid state, which is unusual in lipids. The crystal structure of 2-monolaurin (Larsson, 1964a) is shown in Fig. 8.43. The hydrogen-bond system, where only hydroxyl groups are involved, links the molecules both laterally and over the gap between the two halves of the bilayer. The chains are packed according to the triclinic chain packing T∥, and tilted 44° towards the endgroup plane.

The reason why a liquid-crystalline phase and no α form is obtained, when the melt is cooled, is probably related to the bulky polar headgroup. The cross-sectional area per molecule in the plane of the bilayer in the β-crystal form (Fig. 8.43) is in fact very close to that in the lamellar liquid-crystalline phase (see section 8.2).

(b) 1-Monoglycerides

1-Monoglycerides show a rather complex phase behaviour, contrary to the 2 isomer. A study of the racemic saturated monoglycerides (Larsson, 1964b) shows the following transition scheme:

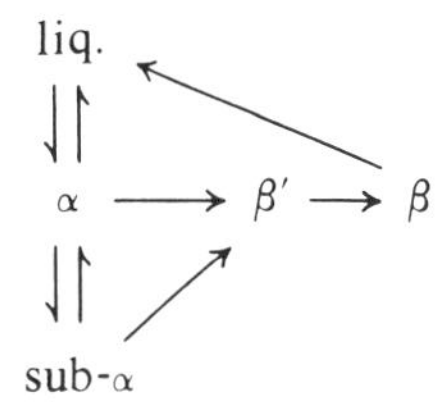

An α form separates from the melt, and is transformed reversibly into a sub-α form (a β form) at about 25 °C lower temperature. Alternatively the α form is transformed irreversibly into a β' form. There

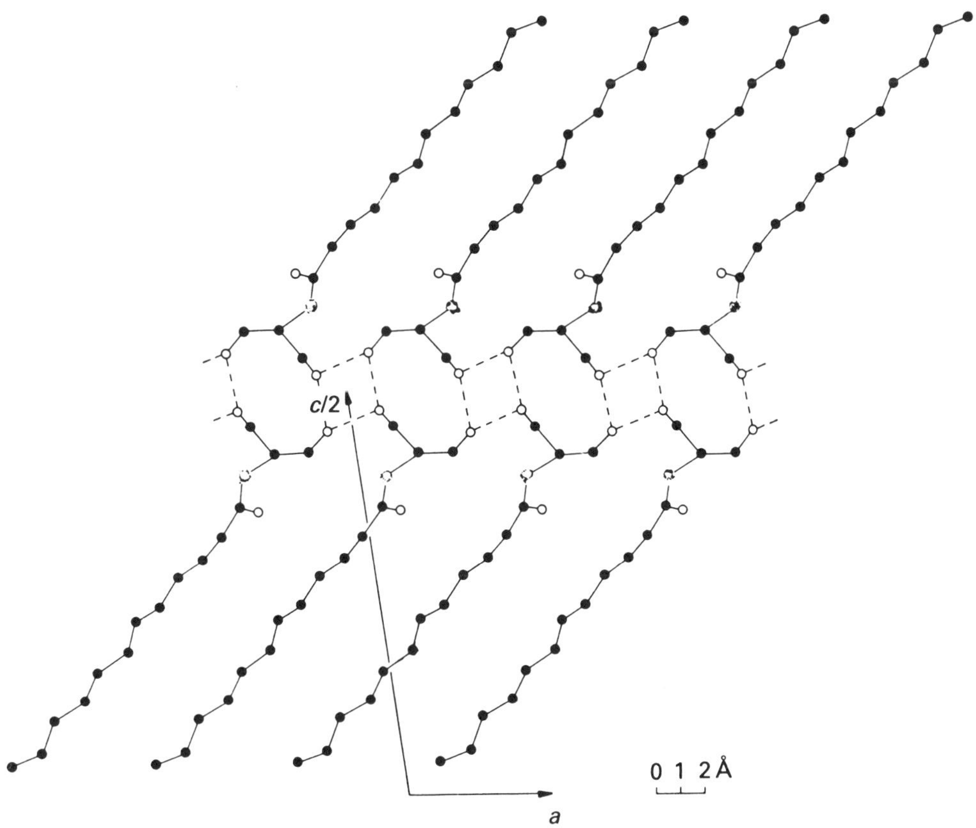

Fig. 8.43 Crystal structure of 2-monolaurin viewed along the shortest unit-cell axis (Larsson, 1964a). Hydrogen bonding is indicated by broken lines.

are furthermore two β forms, but they are so closely related in structure (the same bilayer units) that only one form is indicated in the scheme above. The β' and β forms can also be crystallized from solvents.

A remarkable feature in these polymorphic changes is that the bilayer thickness is constant in all crystal forms, which illustrates the dominant effect of the structure in the polar region. The chain packing in the β forms is the monoclinic M|| and these crystal forms are truly racemic. The β' form, however, is isomorphous with the β' form of optically active 1-monoglycerides, so obviously the two antipodes separate into different D-β' and L-β' crystals. This appears to be the main difference between the sub-α and the β' forms, which both have the same angle of tilt and the same chain packing (O⊥).

The crystal structure of the L-1-monoglyceride of 11-bromoundecanoic acid is shown in Fig. 8.44. As in the 2-monoglyceride β form, only the hydroxyl groups participate in the hydrogen-bond system, which seems to be a general feature in monoglyceride crystal forms.

8.7.2 Melting points and transition temperatures

The thermal transition points of the different polymorphic forms of saturated *1-monoglycerides* have been reported by Malkin (1954), and selected values are given in Table 8.20 (using the nomenclature defined in section 8.1). The values are in close agreement with the data reported by Lutton and Jackson (1948).

2-Monoglycerides have also been examined by Malkin (1954), and the melting and solidification points of saturated members are given in Table 8.21. It should be pointed out that, in the gap between the reported melting point and the solidification point, a thermotropic liquid-crystalline phase is formed as described above.

The melting points of the most stable forms of 1-monoolein and 1-monoelaidin, which are β forms, are 35.0 and 58.5 °C respectively. When 1-monoolein solidifies from the melt, a thermotropic liquid-crystalline phase with lamellar structure is formed as a metastable transition state.

8.7.3 Interaction between monoglycerides and water

Phase diagrams of monoglyceride–water systems have been reported by Lutton (1966) (C_{12}–C_{22}), Larsson (1967) (C_6–C_{10}) and Krog and Larsson (1968) (industrially distilled monoglycerides). The main features of the different monoglyceride–water systems were given in section 8.2 as an illustration of lipid–water phase equilibria (Fig. 8.12). Only additional comments to this earlier work will be given here.

The phase described by Lutton (1966) as a middle

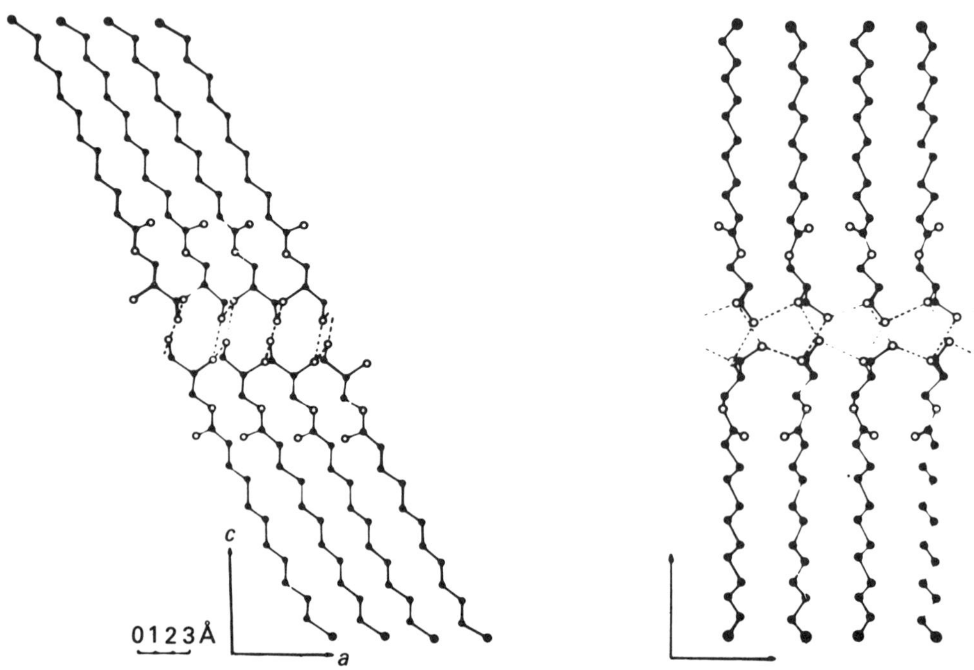

Fig. 8.44 Crystal structure of the L-1-monoglycerides of 11-bromoundecanoic acid (Larsson, 1966). Hydrogen bonding is indicated by broken lines.

Table 8.20 Melting point and $\alpha \rightleftharpoons$sub-$\alpha$ transition of saturated monoglycerides according to Malkin (1954)

Glycerol 1-alkanoate	Temperature (°C)			
	β	β'	α	$\alpha \rightleftharpoons$sub-$\alpha$
Decanoate	53	49	27	8
Undecanoate	56.5	52	36.5	3
Dodecanoate	63	59.5	44	15
Tridecanoate	65	61	50	9
Tetradecanoate	70.5	67.5	56	24
Pentadecanoate	72	69	62	17
Hexadecanoate	77	74	66.5	34
Heptadecanoate	77	74.5	70	28
Octadecanoate	81.5	79	74	47

Table 8.21 Melting and solidification points of 2-monoglycerides according to Malkin (1954)

Glycerol 2-alkanoate	M.p. (°C)	S.p. (°C)
Decanoate	40.2	34.0
Dodecanoate	51.0	47.5
Tetradecanoate	61.3	58.0
Hexadecanoate	69.0	65.5
Octadecanoate	75.2	70.0

phase, which is the term used in soap–water systems for the hexagonal phase H_I, is in fact the inverse structure, i.e. phase H_{II}. Furthermore, the phase diagrams given for monolaurin and monoerucin contain some errors. A revised phase diagram of monolaurin is given in Fig. 8.45; the revised phase diagram of monoerucin has been published earlier (Larsson *et al.*, 1978).

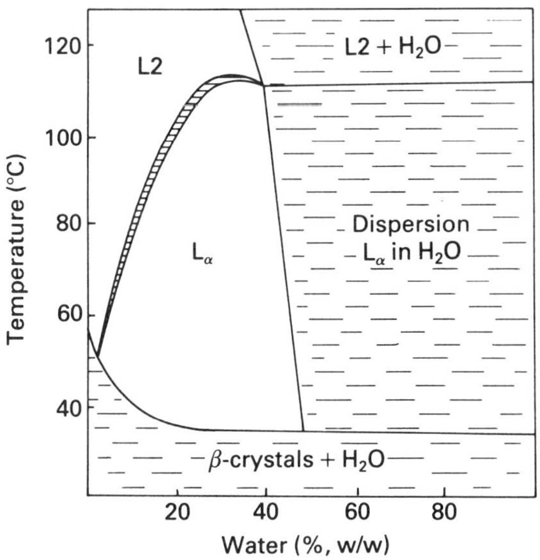

Fig. 8.45 Phase diagram of the binary system monolaurin–water.

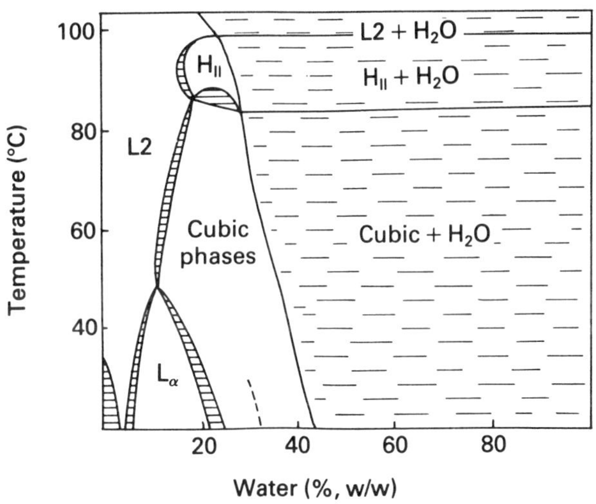

Fig. 8.46 Phase diagram according to Lutton (1966) and Krog *et al.* (1986) of monoolein–water. The broken line indicates the transition between the two cubic phases at low temperatures.

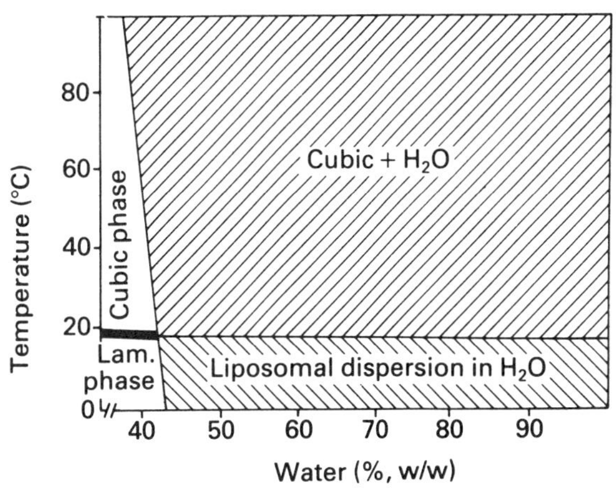

Fig. 8.47 Main features of the phase diagram 2-monoolein–water (Ljusberg-Wahren *et al.*, 1983).

The phase diagram of monoolein is given in Fig. 8.46, where the occurrence of the two cubic phases (see section 8.2) is indicated. It is also of interest to compare this phase diagram with that of 2-monoolein indicated in Fig. 8.47. All monoglycerides of the aqueous systems discussed above consist of an equilibrium mixture of about 90% (w/w) of the 1 isomer and 10% of the 2 isomer owing to the rapid acyl migration in monoglycerides. A comparison between the equilibrium mixture of monoolein (Fig. 8.46) and the phase boundaries in 2-monoolein (Fig. 8.47) shows pronounced differences. Thus the L_α phase in 2-monoglycerides can coexist with water. At about 19 °C it is transformed into a cubic phase. The differences in phase properties of the 1 and 2 isomers can be explained by differences in the wedge shape of the molecules discussed in section 8.2. Thus the polar headgroup requires a larger cross-sectional area than that of the 1 isomer.

References and further reading

Krog, N. and Larsson, K. (1968) *Chem. Phys. Lipids*, **2**, 129.

Krog, N.J., Riisom, T.H. and Larsson, K. (1985) in *Encyclopaedia of Emulsion Technology*, vol. 2, (ed. P. Becher), Marcel Dekker Inc., NY, Basle, pp. 321–65.

Larsson, K. (1964a) *Ark. Kemi*, **23**, 23.

Larsson, K. (1964b) *Ark. Kemi*, **23**, 35.

Larsson, K. (1966) *Acta Crystallogr.*, **21**, 267.

Larsson, K. (1967) *Z. Physiol. Chem.*, **56**, 173.

Larsson, K., Gabrielsson, K. and Lundberg, B. (1978) *J. Sci. Food Agric.*, **29**, 909.

Ljusberg-Wahren, H., Herslöf, M. and Larsson, K. (1983) *Chem. Phys. Lipids*, **33**, 211.

Lutton, E.S. (1950) *J. Am. Oil Chem. Soc.*, **27**, 276.

Lutton, E.S. (1966) *J. Am. Oil Chem. Soc.*, **42**, 1068.

Lutton, E.S. and Jackson, F.L. (1948) *J. Am. Oil Chem. Soc.*, **70**, 2445.

Malkin, T. (1954) *Prog. Chem. Fats Other Lipids*, **11**, 1.

8.8 Diglycerides (diacylglycerols)

8.8.1 Solid-state behaviour

The first diglyceride crystal structure to be determined contained heavy-atom-substituted fatty acids, the symmetric diglyceride of 3-thiododecanoic acid (Larsson, 1963). The structure is shown in Fig. 8.48. The most important feature is that the molecule is V-shaped, so that the polar groups are arranged in a sheet between the hydrocarbon-chain layers formed by the two acyl groups. The hydrogen-bond system links the molecules into infinite chains in one direction, which are parallel with the needle direction of the crystals. The same type of crystal structure was also found in the 1,3-diglyceride of 11-bromoundecanoic acid (Hyble and Dorset, 1971).

Another type of diglyceride crystal structure found in the chiral diglyceride, 2,3-dilauroyl-*sn*-glycerol (Pascher and Sundell, see Sundell, 1980), is shown in Fig. 8.49. In this structure the two hydrocarbon chains are adjacent and parallel in the same monolayer. Bilayer units are formed by the head-to-head orientation of molecules in adjacent monolayers. An interesting feature in this structure is that the conformation of the polar group (glycerol and carboxyl groups) is identical to that found in phospholipids (cf. section 8.10).

The polymorphism of racemic monoacid 1,2-diglycerides has been studied by Chapman and co-workers (cf. Chapman, 1965). A metastable α form separates from the melt, which then trans-

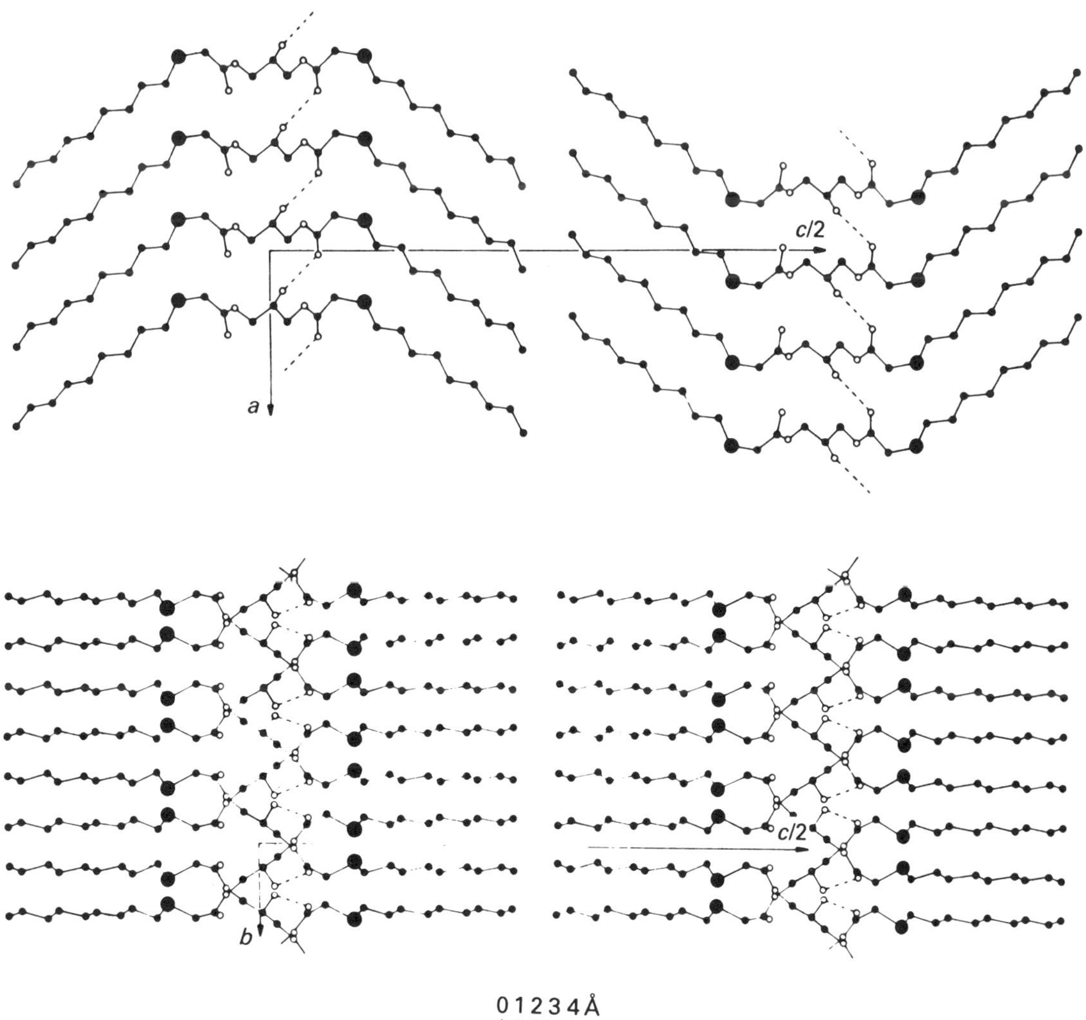

Fig. 8.48 Crystal structure of the 1,3-diglyceride of 3-thiododecanoic acid viewed along the shortest unit-cell axis (Larsson, 1963).

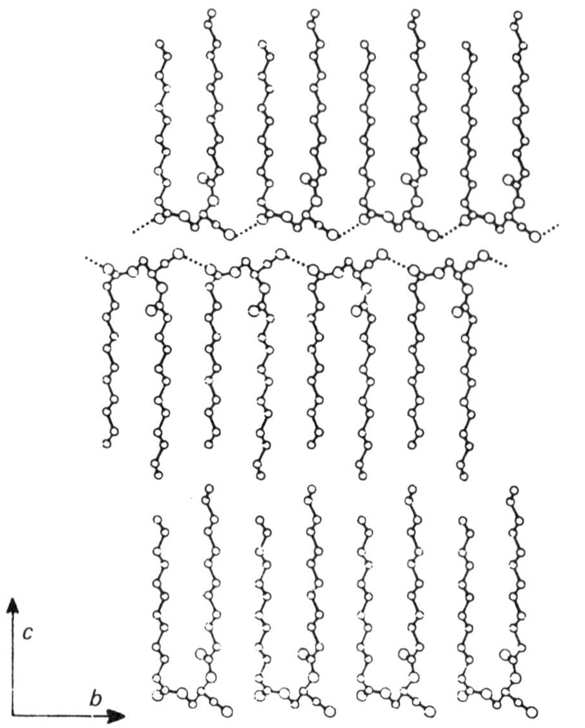

Fig. 8.49 Crystal structure of 2,3-dilauroyl-D-glycerol projected along the *a* axis (cf. Sundell, 1980).

forms into the stable β' form. This type of diglyceride is able to stabilize the β' form (of triglycerides) in fats (Hernqvist *et al.*, 1981).

Symmetrical saturated diglycerides crystallize in two β forms, β_1 and β_2, with melting points quite close (56.4–56.7 °C and 54.2–54.6 °C in dilaurin). A structure analysis of the β form of dilaurin shows that the structure is similar to that shown in Fig. 8.48, with V-shaped molecules forming a unit structure consisting of four hydrocarbon-chain layers (Larsson, 1964). The β_2 form is obtained from the melt, and it is remarkable that there is no evidence for the existence of an α form.

8.8.2 Melting points and transition temperatures

The most complete thermal analysis of diglycerides has been reported by Malkin (1954). Melting points of the polymorphic forms of symmetrical saturated diglycerides are given in Table 8.22. The forms earlier termed β' and β are in fact both β forms (Larsson, 1964).

Table 8.22 Melting points (°C) of symmetrical saturated diglycerides (after Malkin, 1954)

Glycerol 1,3-dialkanoate	β_1	β_2
Didecanoate	42	44.5
Didodecanoate	54	56.5
Ditetradecanoate	63	65.5
Dihexadecanoate	71.5	72.5
Dioctadecanoate	77	78

Asymmetrical monoacid diglycerides crystallize in an α form from the melt. The stable crystal form is a β' form. Melting points of some saturated monoacid 1,2-diglycerides are given in Table 8.23.

Table 8.23 Melting points (°C) of saturated monoacid 1,2-diglycerides (after Malkin, 1954)

Glycerol 1,2-dialkanoate	α form	β' form
Didodecanoate	20.0	39.0
Ditetradecanoate	37.5	54.0
Dihexadecanoate	50.0	63.5
Dioctadecanoate	59.5	71.0

The determination of thermal data for diglycerides is seriously complicated by the rapid isomerization due to acyl migration. A few hours in the liquid state is enough for a pure diglyceride to form an equilibrium mixture of about equal amounts of the 1,2 and 1,3 isomers.

References and further reading

Chapman, D. (1965) *The Structure of Lipids*, Methuen, London.

Hernqvist, L., Herslof, B., Larsson, K. and Podlaha, J. (1981) *J. Sci. Food Agric.*, **32**, 1197.

Hyble, A. and Dorset, D. (1971) *Acta Crystallogr.*, **B27**, 977.

Larsson, K. (1963) *Acta Crystallogr.*, **16**, 741.

Larsson, K. (1964) *Ark. Kemi*, **23**, 35.

Malkin, T. (1954) *Prog. Chem. Fats Other Lipids*, **11**, 1.

Sundell, S. (1980) Thesis, University of Göteborg, Sweden.

Shannon, R.J., Fenerty, J., Hamilton, R.J. and Padley, F.B. (1992) *J. Sci. Food Agric.*, **60**, 405.

8.9 Triglycerides (triacylglycerols)

8.9.1 Solid-state behaviour

The literature on triglyceride polymorphism was very confusing for a long time. The main reason was controversy between Malkin (1954) and Lutton (1950) on the existence of a vitreous state and the α form. The use of different nomenclature by these two groups led to a complete deadlock. This was broken by Chapman (1962, 1965), who introduced infrared spectroscopy. On the basis of observed spectral variations correlated with X-ray diffraction data, he was able to give a clear description of the polymorphic transitions. Shortly afterwards, Larsson (1964a) gave new X-ray diffraction results on the crystal forms and the transitions involved. In this connection it was also natural to suggest a nomenclature based on the crystal structures as described in section 8.1. Later, this study was extended to the odd series of monoacid triglycerides, which demonstrated the existence of a second β' form (Hernqvist and Larsson, 1982).

Crystallization of fats involves different solid phases with a successive improvement of the close packing of the molecules. The morphology of the various crystal forms determines the rheological properties of the fat. The order in the liquid state and

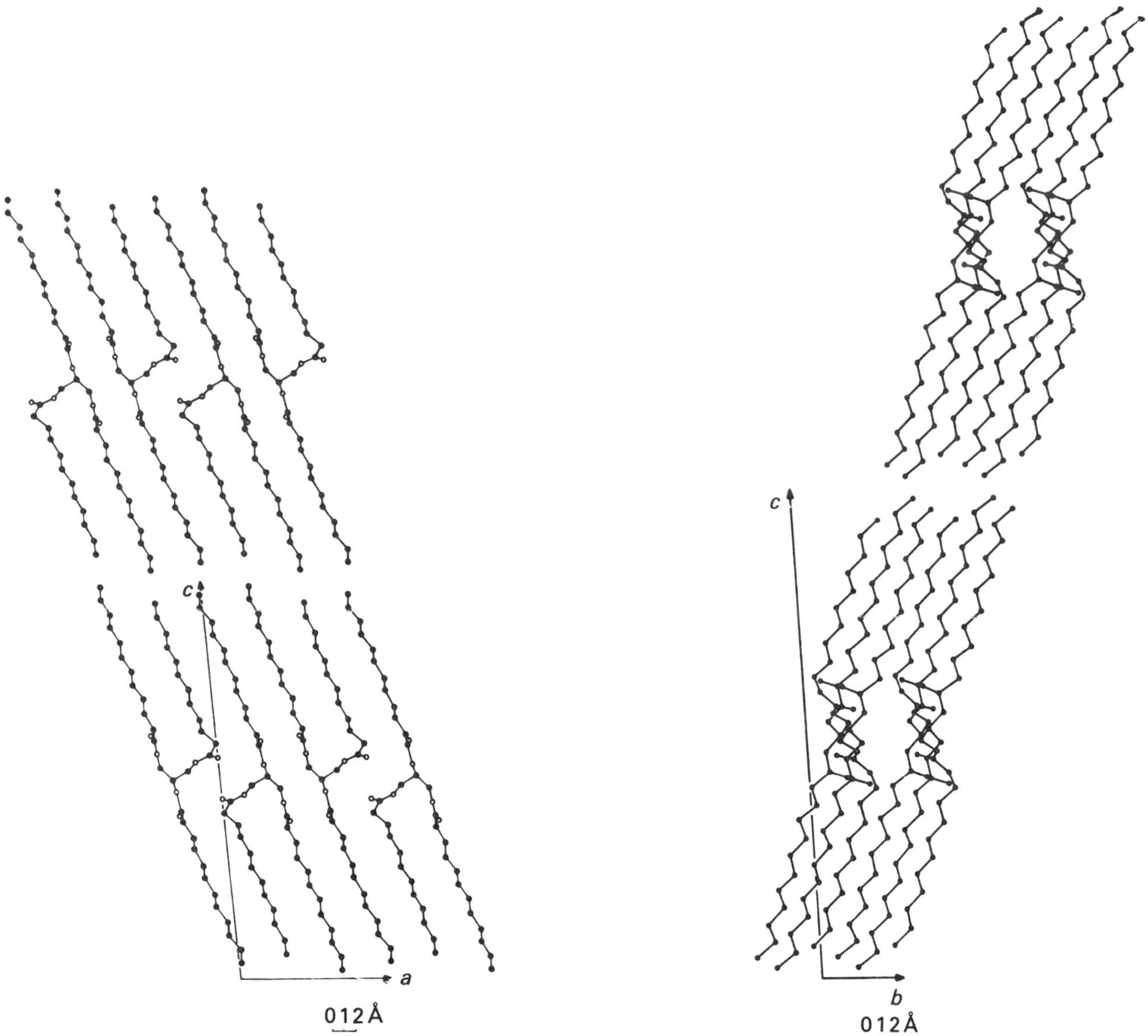

Fig. 8.50 Crystal structure of the β form of trilaurin (Larsson, 1964a).

the layer growth of the α form are important from a biological point of view and explain, for example, how fat-rich tissues such as seeds can be cooled without the cell rupturing because of the fat crystallizing.

(a) Crystal structures and phase transitions in saturated monoacid triglycerides

The crystal structure of the stable form of simple triglycerides, the β form, was determined independently for trioctanoin (Jensen and Mabis, 1963), trilaurin and the isomorphous triglyceride of 11-bromoundecanoic acid (Larsson, 1963). The crystal structure of trilaurin is shown in Fig. 8.50. The fatty acid group in 1 and 3 positions and the glycerol group form one extended chain, with the fatty acid chains in the 2 position aligned in parallel, giving the molecule a shape like a chair. If two such molecules are arranged middle-to-middle, a space-filling structure is obtained, which occurs in the other crystal forms.

The medium-range order occurring in the triglyceride melt was discussed in section 8.3, and a recent work on the structure of the other polymorphic forms (Hernqvist and Larsson, 1982) will be summarized here. This description is based on the assumption that the main structural features persist through the transitions with dimeric units of chair-shaped molecules.

The formation of the α form is the first step in the ordering process towards the most effective molecular packing. The chains are vertical in the bilayers and laterally arranged according to the hexagonal chain packing. To account for dimensional changes with temperature and changes in the Raman spectra, a structure shown in Fig. 8.51 was suggested. The methyl endgroup region forms a zone with disordered chains as in the lamellar liquid-crystalline phase. It should also be pointed out that the α form does not show supercooling, and in this respect it behaves like smectic (lamellar) liquid crystals (van den Tempel, 1979).

If the α form is cooled rapidly it can form a sub-α form reversibly at very low temperatures (about $-50\,°C$ in trimargarin (Chapman, 1965)). These extreme temperatures are seldom reached and the 'normal' phase change for a triglyceride α form is an irreversible transition into a β' form. This transition takes place within seconds when pure monoacid triglycerides are cooled.

Studies of the odd series of triglycerides have shown that there are in fact two β' forms. The β'_2 form of triundecanoin was studied by a single-crystal technique and, although the complete structure could not be determined, the main features could be derived as shown in Fig. 8.52. According to the geometry of the endgroup region, it is probable that the glycerol part has a structure similar to that of the

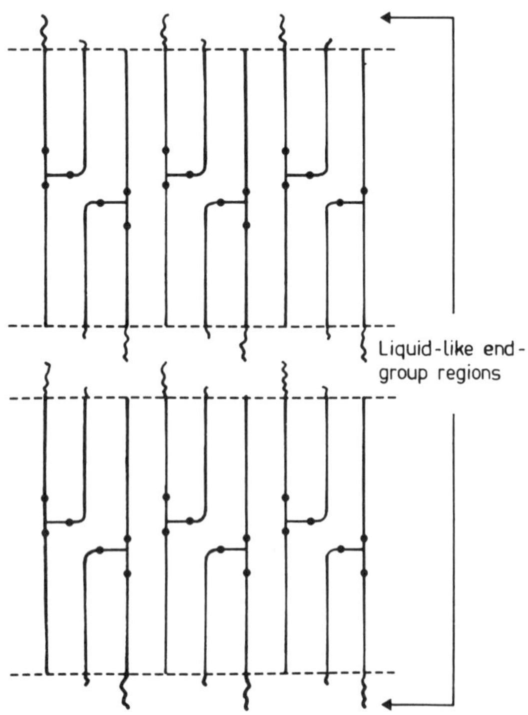

Fig. 8.51 Proposed structure in the α form of triglycerides (Hernqvist and Larsson, 1982).

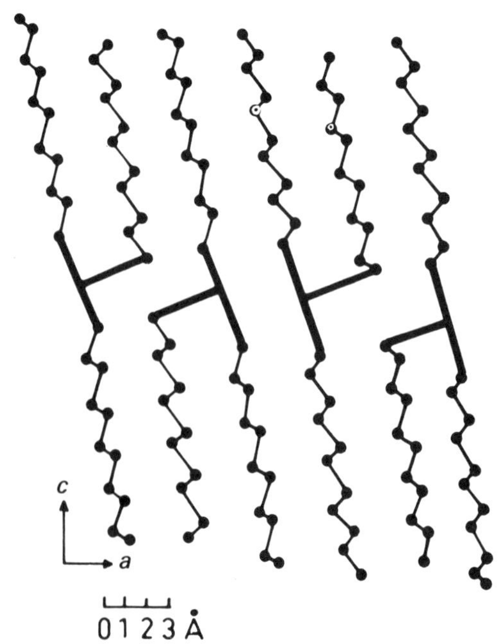

Fig. 8.52 Main structural features of the β'_1 form of triundecanoin projected along the shortest unit-cell axis (b).

β form. The chains are packed according to the orthorhombic chain packing, with the methyl endgroup planes defined by indices O⊥ (111). The relation between β_1' and β_2' is that the β_2' form has the methyl endgroup planes defined by O⊥ (101). The transition from the α form seems to take place in two steps. When the chains adopt the orthorhombic chain packing, a tilted structure in the *ac* plane (Fig. 8.52) is formed and the chain remains vertical in the other projection direction. The next step is the $\beta_2' \rightarrow \beta_1'$ transition when an additional tilt in the other direction is adopted. This chain tilt alternates between opposite directions in adjacent layers as shown in Fig. 8.53. It should be pointed out that, as the structure of the glycerol group region is not known, it is not possible to determine unambiguously where the chain tilt changes from the X-ray data. The proposed position of the alternation in Fig. 8.53 is based on the assumption, mentioned above, that the molecular conformation remains mainly unchanged through the transitions.

The nature of these polymorphic transitions has been frequently discussed, particularly whether true melting takes place or whether the transitions occur in the solid state. Continuous measurements of the diffraction pattern versus temperature show persisting short-spacing lines through the transition, which can be taken as an indication of some degree of order during the transitions, at least arrangement in a bilayer. Whether the β_2' structure is formed as a state of transitory existence in even triglycerides is another open question.

In conclusion, the phase transition in simple triglycerides can be described by the following scheme, where the β' can consist of two forms:

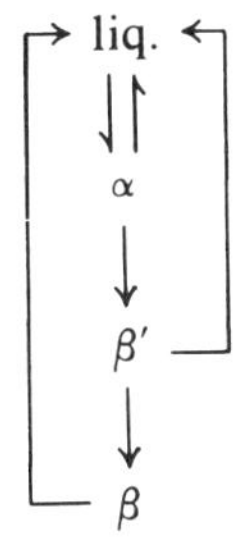

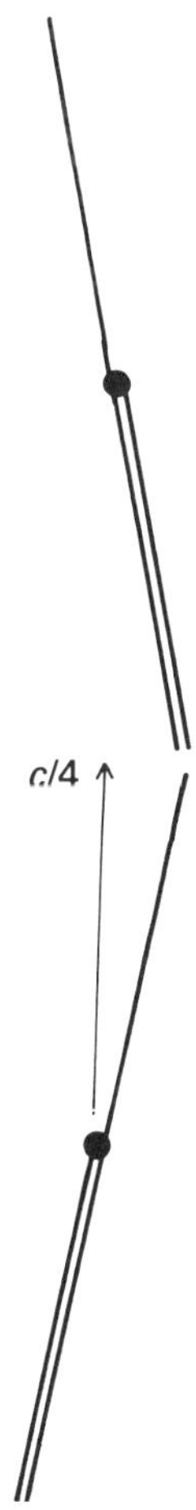

Fig. 8.53 Proposed position of the alternation in the tilt of the chains in the β_1' form projected along the *a* axis.

(b) *Fat polymorphism*

It is remarkable that the same polymorphic forms that are observed in saturated monoacid triglycerides also occur in triglycerides with different acyl chains as well as in natural fats, which are complex mixtures of a large number of triglycerides. Saturated acid triglycerides, such as 2-stearodipalmitin, show the same transitions $\alpha \rightarrow \beta_2' \rightarrow \beta_1'$ as odd monoacid triglycerides. Saturated triacid triglycerides and triglycerides containing unsaturated acyl groups have a more complex behaviour. There are obvious restrictions in chain lengths in the possibilities of packing different triglyceride molecules in the β structure owing to the requirement of tight stacking of the bilayers at the methyl endgroup planes. De Jong (1980) has done an extensive theoretical analysis of these molecular packing possibilities, including the β form, where the unit layer consists of three hydrocarbon-chain layers. An example of a triple-chain-length structure is shown in Fig. 8.54. This type of triglyceride, which dominates in cocoa butter, has an unsaturated chain in the 2 position. The unsaturated and saturated chains are best packed if they are located in separate layers, and this is possible by adopting the triple-chain-length structure. Similarly, if one chain is much shorter or much longer than the other two, the triple-chain-length structure makes possible the arrangement of chain layers with similar chain lengths. For more detailed discussions see Chapman, 1965; Timms, 1984; de Jong *et al.*, 1991; Hagemann and Rothfus, 1992 and Yano *et al.*, 1994.

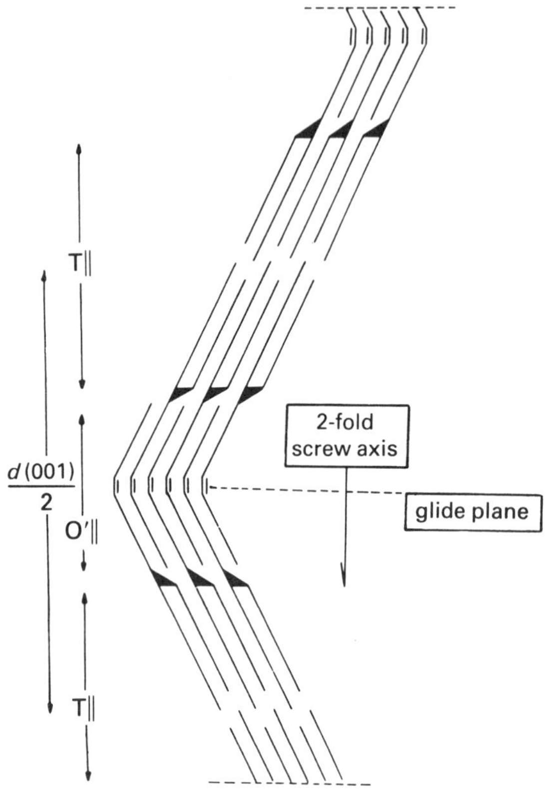

Fig. 8.54 Proposed structure of 2-oleodistearin (Larsson, 1972).

Gunstone (1964) has correlated the polymorphic forms with the long spacings for triglycerides and other long-chain compounds. All the saturated triglycerides examined were observed to fall into a few groups. Symmetrical and unsymmetrical isomers for example, were found to have almost identical long-spacing values. These relations can be explained from our present knowledge of crystal structures. The triclinic chain-packing subcell of the β form is oriented in the same way in relation to the endgroup plane independent of the acyl groups or of whether a bilayer or triple-chain-layer constitutes the structural unit. In the same way the β_1' form always has the orthorhombic subcell oriented according to $O\perp$ (111).

8.9.2 Melting points, heats of fusion and phase behaviour of mixtures

Owing to the industrial importance of melting properties of fats, many reports on thermal data of triglycerides can be found in the literature. Only those triglycerides for which we have a relatively good structural understanding of the solid state will be considered here. For rough estimations the melting point of a triglyceride is approximately equal to the average value of the melting points of its fatty acid components, and the most complete review of melting properties is that by Malkin (1954).

A probable explanation of Malkin's misinterpretation of the two polymorphic forms 'vitreous' and 'α' (which are α and β', respectively) is the observation that neither showed any alternation of melting points between even and odd members in the series of monoacid triglycerides. Earlier experience has shown that only structures with vertical hydrocarbon chains were non-alternating. The fact that the β' form is tilted and still non-alternating can be explained, however, from the methyl endgroup structure of the β' form, which is the same in even and odd members (Larsson, 1966).

Melting points of some even triglycerides are given in Table 8.24, and data on the two β' forms in odd-

Table 8.24 Selected values of the melting points (°C) of triacylglycerol (triglyceride) crystal forms (from Malkin, 1954) with modified nomenclature (section 8.2). The β' form of short-membered monoacyl triglycerides is directly transformed into the β form, and therefore there are no reliable melting-point data

Triacylglycerol	α	β_1'	β_2'	β
Tricaprin	−15.0	–	–	31.5
Trilaurin	15.0	–	–	46.4
Trimyristin	33.0	54.5	–	57.0
Tripalmitin	45.0	63.5	–	65.5
Tristearin	54.5	70.0	–	72.0
2-Laurodidecanoin	6	37.5	34	–
2-Myristodilaurin	24	48	44	–
2-Palmitodimyristin	38	58.5	55	–
2-Stearodipalmitin	49	68	65	–
Triolein	−32	−12	49	–
Trielaidin	15.5	37	42	–
Trierucin	6	17	30	–

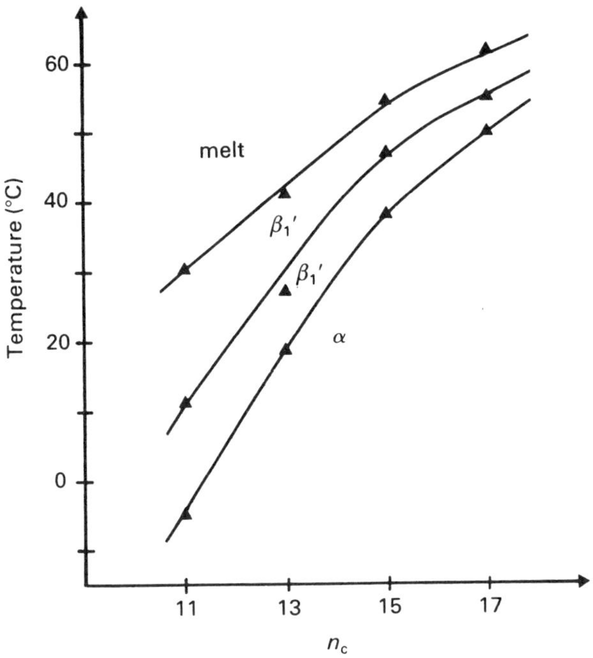

Fig. 8.55 Transition points of the α and the two β' forms in odd monoacid triglycerides (Hernqvist and Larsson, 1982).

membered monoacyl triglycerides are shown in Fig. 8.55.

In a critical review of the heat of fusion of triglycerides, Timms (1978) has summarized the most reliable data of numerous triglycerides. Some selected values are given in Table 8.25.

Values of the heat of fusion for saturated triglycerides were found to follow the linear relation:

$$\Delta H_f(\text{kcal/mol}) = 1.023(C_n - 7.79)$$

where C_n is the total number of carbon atoms in the three acyl groups (not the number of CH_2 groups). This equation can also be used for unsaturated acids with the following effective carbon numbers: oleic acid = 10.4, elaidic acid = 14.0, linoleic acid = 8.9 and erucic acid = 13.9. As an example, the experimental value of SOS is $\Delta H_f = 39.9$ kcal/mol. When the equation above is used (carbon number = 18 + 10.4 + 18), a calculated value of ΔH_f equal to 39.9 kcal/mol is obtained.

The equation given above corresponds to the β form, and Timms (1978) also found that values of heat of fusion of β' forms can be obtained from:

$$\Delta H_f(\beta') = 0.76\Delta H_f(\beta)$$

The melting points and values of heat of fusion of *mixed palmitic/stearic acid triglycerides* compared to tripalmitin and tristearin are given in Table 8.26. Knoester *et al.* (1972) examined the 15 binary systems that can be formed by these glycerides. Most of these diagrams are of the eutectic type with ideal miscibility in the melt and no or very low miscibility in the crystalline state. In systems with an asymmetric triglyceride, however, there is a considerable degree of miscibility in the solid state, and values of the mutual solubilities are given in Table 8.27. It was also found in this work that differences in polymorphic form have no effect on the mutual solubility.

Di- and triacid triglycerides with one or two unsaturated chains have more complex phase

Table 8.25 Best experimentally observed heats of fusion according to Timms (1978)

Glyceride[a]	ΔH_f (kcal/mol)
CCC	22.3
LLL	29.5
MMM	35.4
PPP	41.0
SSS	47.0
CLC	21.4
LLM	27.8
MML	29.3
PCP	29.4
SCS	34.6
PSM	36.2
PSP	39.7
PPS	40.4
SPS	40.9
PSS	41.4
POP	35.8
PPO	27.8
POS	35.0
PSO	26.5
SPO	30.1
SOS	39.9
SSO	30.0
POO	22.6
OPO	30.0
SOO	26.2
OOO	22.8

[a] C = capric acid, L = lauric acid, M = myristic acid, P = palmitic acid, S = stearic acid and O = oleic acid.

Table 8.26 Melting points and heats of fusion of triacylglycerols (triglycerides) containing palmitic (P) and stearic (S) acids (after Knoester *et al.*, 1972)

Triacylglycerol	Melting point (°C)	Heat of fusion ΔH_f (kJ/mol)
SSS	72.5	180
SSP	66.2	179
SPS	67.7	168
PSP	68.6	166
PPS	63.5	168
PPP	66.0	168

Table 8.27 Mutual solubilities of palmitic (P) and stearic acid (S) triglycerides (mole fraction) (after Knoester *et al.*, 1972)

Component 1	Component 2	Solubility of 2 in 1	Solubility of 1 in 2
SSS	SPS	0.10	0
SSS	PSP	0.05	0
SSS	PPS	0.30	0
SSS	PPP	0.10	0
SSP	PPP	0.10	0.10
SPS	PSP	0.05	0.05
SPS	PPS	0.20	0
SPS	PPP	0	0
PSP	PPS	0	0
PSP	PPP	0	0

behaviour. Cocoa butter, which is dominated by triglycerides with oleic acid in the 2 position and palmitic and stearic acids in the 1 and 3 positions, exhibits six crystal forms. Transitions into the most stable form cause so-called blooming of chocolate. An interesting study of the polymorphism and the changes induced by the addition of olive oil has been reported by Lovegren *et al.* (1976). The phase behaviour of fats and binary mixtures of triacylglycerols has been reviewed by Rossell (1967) and Timms (1984).

Studies of the binary systems SOS and SSO (Engström *et al.*, 1992) and of SOS and OSO have demonstrated the existence of molecular compounds due to chain segregation into layers of saturated chains separated from layers with unsaturated chains.

8.9.3 Specific heat

A curve of the specific heat (C) versus temperature of an organic substance starts with a zero value and a zero slope at 0 K, followed by a curve up to about 30 K where the slope is proportional to T^3. After an inflection point near 30 K, the curve is approximately linear up to the melting point, where there is a sharp increase in the slope.

Curves of the specific heat of tristearin (Charbonnet and Singleton, 1947) are shown in Fig. 8.56. It can be seen how the specific heat of the α form is rather close to that of the melt. This reflects the high degree of molecular mobility in the α form compared to that in the stable β form.

8.9.4 Density and specific volume

The significance of measurement of specific volume versus temperature (dilatometry) for studies of lipid phase transitions was discussed in section 8.5, and there are numerous dilatometric studies of fat systems.

The measured density of the β form of trilaurin was reported by Vand and Bell (1951) to be 1.04 g/cm^3, and the calculated density from the unit-cell volume according to X-ray data is equal to 1.027 ± 0.014 g/cm^3. As should be expected, the density is lower in the complex glyceride mixtures constituting natural fats. Singleton and Bailey (1945) have reported the specific volumes of various fats at −38.65 °C, and they obtained 0.9950 cm^3/g for lard, 0.9992 cm^3/g for olive oil and 1.0000–1.0015 cm^3/g for partially hydrogenated cotton-

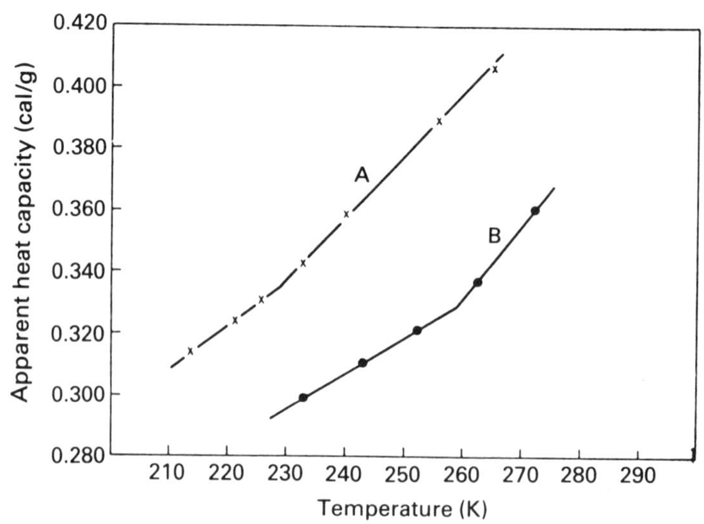

Fig. 8.56 The specific heat of the α form (curve A) and the β form (curve B) of tristearin (Charbonnet and Singleton, 1947).

seed oil. According to calorimetric studies, these fats were completely solid at 38.65 °C. They also estimated a thermal expansion of 0.000 30 $cm^3 g$ K.

The melting of triglycerides results in a decrease in density of about 10%. The melting dilation of a fully hardened cottonseed oil is 0.1100 cm^3/g (Singleton and Bailey, 1945) and a value of 0.090–0.095 cm^3/g has been estimated for cocoa butter.

An example of a dilatometric curve that illustrates the successive melting of butter fat is given in Fig. 8.57. Owing to the differences in dilation between the solid and liquid states, it is possible to estimate solid/liquid ratios at various temperatures from dilatometric curves.

Phipps (1962) has given an equation for the density of the liquid state of simple triglycerides that fits within 0.05% with the corresponding experimental measurements between the m.p. and 95 °C. The equation has the form:

$$\delta = A - Bt$$

where t is the temperature (°C). The values of the constants A and B respectively are:

trilaurin	0.9380	7.24×10^{-4}
trimyristin	0.9279	7.02×10^{-4}
tripalmitin	0.9222	6.84×10^{-4}
tristearin	0.9149	6.62×10^{-4}

8.9.5 Viscosity

An accurate study of the viscosity of simple triglycerides has been reported by Phipps (1962). An interesting result from this study was the observation of a discontinuity of supercooled melts (measurements were done from the β melting point down to the neighbourhood of the melting/solidification point of the α form). When the logarithm of the viscosity was plotted against the reciprocal of the absolute temperature, there were two linear regions. This difference in slope corresponds to an increase in the activation energy for viscous flow when supercooling occurs and was taken as an indication of enhanced molecular aggregation in the melt when it passes into the supercooled region. The observed behaviour fits well with the structural features of liquid triglycerides discussed in section 8.3.

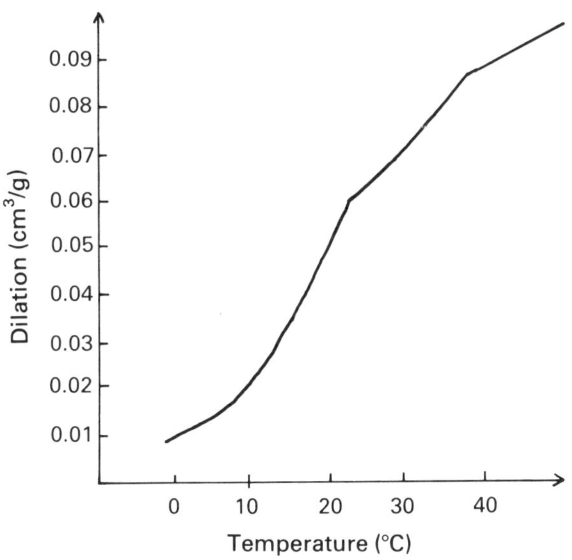

Fig. 8.57 Dilatometric curve for American butterfat (Bailey, 1950).

The experimental data were found to fit the equation:

$$\log_{10}(\log_{10}\eta) = C - Dt$$

with t = temperature (°C), within 0.5% between the melting point and 95 °C. The values of the constants C and D respectively are:

trilaurin	0.3414	5.05×10^{-3}
trimyristin	0.3649	4.68×10^{-3}
tripalmitin	0.3894	4.47×10^{-3}
tristearin	0.3972	4.19×10^{-3}

8.9.6 Refractive index

The physical significance of the refractive index was discussed in section 8.5. Values of the refractive index of the most common triglyceride oils are given in Table 8.28. Studies relating the refractive index of triglycerides to the fatty acid composition have been reported by Huygehebaert and Hendrickx (1970) and Kahn *et al.* (1974).

Table 8.28 Refractive index at 40 °C of some natural oils (after Fasman, 1975)

Triglyceride oil	Refractive index
Butter fat	1.4548
Castor oil	1.4770
Cocoa butter	1.4568
Coconut oil	1.4493
Cod-liver oil	1.4810
Corn oil	1.4735
Cottonseed oil	1.4782
Olive oil	1.4679
Palm oil	1.4578
Palmkernel oil	1.4569
Rapeseed oil	1.4706
Soybean oil	1.4729

8.9.7 Solubility

Solubility data of triglycerides in various solvents are important with regard to technical processes involving separation and fractionation of fats. Reliable data on simple triglycerides reported by Loskit (1928), Robinson *et al.* (1932) and Chen and Daubert (1945) have given extensive information on triglyceride solubilities in acetone, ethanol, ethyl ether and petroleum ether.

The solubility of a number of triglycerides in ethanol of various concentrations was reported by Arnold *et al.* (1963). Some of their results are shown in Fig. 8.58.

Solubilities of some series of mixed triglycerides are given in Table 8.29. The solubilities were determined at a temperature (25 °C) well below the melting points. A general tendency of increasing solubility with decreasing molecular weight can be seen in the polar as well as non-polar solvents. Supercritical carbon dioxide is increasingly used as a triglyceride solvent. Above 31 °C, which is the critical temperature, carbon dioxide cannot be compressed into a liquid. At high pressures, however, it has the density of a liquid but is still a gas, which can easily diffuse into fatty materials. It can thus absorb 2.5–2.7% of its weight of soybean oil.

8.9.8 Enantiomeric triglycerides

Although naturally occurring triglycerides are optically active, almost all studies of the physical chemistry of triglycerides have been done on the racemic compounds. An extensive study of the polymorphic behaviour of enantiomeric triglycerides has been reported (Ståhl *et al.*, 1976).

1,2-Dilauroyl-3-myristoyl-*sn*-glycerol and the enantiomer 1-myristoyl-2,3-dilauroyl-*sn*-glycerol crystallize from the melt in an α form, which is transformed into β'_2 and then into β'_1. The racemic mixture of these antipodes, however, shows the transitions $\alpha \rightarrow \beta' \rightarrow \beta$ under the same conditions. The antipodes and the racemic form of dilauroylstearoylglycerol behave analogously. 1,2-Dipalmitoyl-3-stearoyl-*sn*-glycerol and the enantiomer 1-stearoyl-2,3-dipalmitoyl-*sn*-glycerol show the transition $\alpha \rightarrow \beta'$ (the β' form is stable for days), whereas the racemic

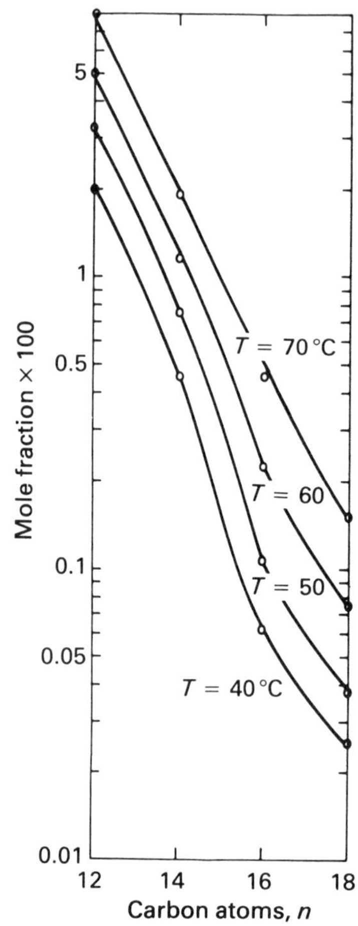

Fig. 8.58 Relation between the solubilities in 100% ethanol of trilaurin, trimyristin, tripalmitin and tristearin (Arnold *et al.*, 1963).

Table 8.29 Solubilities at 25 °C (g triglyceride/100 g solution) of some mixed triglycerides (triacylglycerols) (Singleton, 1960)

Glycerol triester	2-Alkanoate	Diethyl ether	Petroleum ether	Acetone	Ethanol
1-Stearate, 2-alkanoate, 3-palmitate	Myristate	10.97	7.59	0.18	0.03
	Laurate	16.49	9.49	0.31	0.03
	Caprate	22.87	10.60	0.59	0.03
1-Stearate, 2-alkanoate, 3-myristate	Palmitate	11.03	5.46	0.18	0.03
	Laurate	30.68	16.26	0.68	0.04
	Caprate	53.75	37.03	1.96	0.08
1-Stearate, 2-alkanoate, 3-laurate	Palmitate	72.63	58.88	1.47	0.06
	Myristate	112.59	81.44	2.53	0.07
	Caprate	192.13	179.56	13.49	0.39

mixture shows the transitions $\alpha \rightarrow \beta' \rightarrow \beta$. The antipodes and the racemic form of distearoyl-palmitoylglycerol show identical phase transitions: $\alpha \rightarrow \beta' \rightarrow \beta$.

There are clear differences in polymorphic behaviour between the antipodes and the racemic mixture, and this may have technical significance.

References and further reading

Arnold, L.K., Choudhury, B.B.R. and Guzman, A. (1963) *J. Am. Oil Chem. Soc.*, **4**, 33.

Bailey, A.E. (1950) *Melting and Solidification of Fats*, Interscience, New York.

Chapman, D. (1962) *Chem. Rev.*, **62**, 433.

Chapman, D. (1965) *The Structure of Lipids*, Methuen, London.

Charbonnet, G.H. and Singleton, W.S. (1947) *Oil Soap*, **24**, 140.

Chen, C. and Daubert, B.F. (1945) *J. Am. Chem. Soc.*, **67**, 1256.

de Jong, S. (1980) Thesis, University of Utrecht.

de Jong, S., van Soest, T.C. and van Schaik, M.A. (1991) *J. Am. Oil Chem. Soc.*, **68**, 371.

Engström, L., Larsson, K. and Lindman, B. (1992) US Pat. 5151272.

Fasman, G.D. (ed.) (1975) *Handbook of Biochemistry and Molecular Biology: Lipids, Carbohydrates, Steroids*, 3rd edn, CRC Press, Cleveland, OH, p. 484.

Gunstone, F.D. (1964) *Chem. Ind.*, 84.

Hagemann, J.W. and Rothfus, J.A. (1992) *J. Am. Oil Chem. Soc.*, **69**, 429.

Hernqvist, L. and Larsson, K. (1982) *Fette Seifen Anstrichm.*, **84**, 349.

Huygehebaert, A. and Hendrickx, H. (1970) *Milchwissenschaft*, **25**, 506.

Jensen, L.H. and Mabis, A.J. (1963) *Nature*, **197**, 681.

Kahn, A.R., Emery, D.A. and Singleton, J.A. (1974) *Crop Sci.*, **14**, 464.

Knoester, M., de Bruijne, P. and van den Tempel, M. (1972) *Chem. Phys. Lipids*, **9**, 309.

Koyano, T., Hachiya, L. and Sato, K. (1992) *J. Phys. Chem.*, **96**, 10514.

Larsson, K. (1963) *Proc. Chem. Soc.*, 87.

Larsson, K. (1964a) *Ark. Kemi*, **23**, 35.

Larsson, K. (1964b) *Ark. Kemi*, **23**, 23.

Larsson, K. (1964c) *Acta Chem. Scand.*, **18**, 272.

Larsson, K. (1966) *J. Am. Oil Chem. Soc.*, **43**, 559.

Larsson, K. (1972) *Fette Seifen Anstrichm.*, **74**, 136.

Loskit, K. (1928) *Hoppe Seyler's Z. Physiol. Chem.*, **134**, 135.

Lovegren, N.V., Gray, M.S. and Trenge, R.O. (1976) *J. Am. Oil Chem. Soc.*, **53**, 108.

Lutton, E.S. (1950) *J. Am. Oil Chem. Soc.*, **27**, 276.

Malkin, T. (1952) *Prog. Chem. Fats Other Lipids*, **1**, 1.

Malkin, T. (1954) *Prog. Chem. Fats Other Lipids*, **2**, 1.

Phipps, L.W. (1962) *Nature*, **193**, 541.

Robinson, H.E., Roche, J.N. and King, C.G. (1932) *J. Am. Oil Chem. Soc.*, **54**, 705.

Rossell, J.B. (1967) *Advances in Lipid Research*, vol. 5, Academic Press, London.

Singleton, W.S. (1960) *Fatty Acids* (ed. K.S. Markley), Interscience, New York, p. 499.

Singleton, W.S. and Bailey, A.E. (1945) *Oil Soap*, **22**, 295.

Ståhl, P.O., Persmark, U. and Herslöf, B. (1976) *13th World Congress of the International Society for Fat Research/Abstracts*, Imprimerie Louis-Jean, Marseille, p. 77.

Timms, R.E. (1978) *Chem. Phys. Lipids*, **21**, 113.

Timms, R.E. (1984) *Prog. Lipid Res.*, **23**, 1.

van den Tempel, M. (1979) *Physicochimie des Composés Amphiphiles*, Colloq. Nat. du CNRS 938, p. 261.

Vand, V. and Bell, I.P. (1951) *Acta Crystallogr.*, **4**, 465.

Yano, J., Veno, S., Sato, K., Arishishma, T., Sagi, N., Kaneko, F. and Kobayashi, M. (1994) *J. Phys. Chem.*, to be published.

8.10 Phospholipids

8.10.1 Crystalline phases

Hitchcock *et al.* (1974) reported the first crystal structure of a phospholipid, that of dilauroylphosphatidylethanolamine. Later the structure of dimyristoyllecithin was determined by Pearson and Pascher (1979). The structure of two monoacylphospholipids has also been determined. Hauser *et al.* (1980) reported the crystal structure of a lysophosphatidylcholine analogue, and Pascher and Sundell have determined the crystal structure of a lysophosphatidylethanolamine (see Sundell, 1980).

The molecular conformations in the bilayer structure of phosphatidylcholine are shown in Fig. 8.59. The two enantiomeric molecules are closely related in the structure of the headgroup region.

In his thesis Sundell (1980) has presented an interesting conformational analysis of all the known structures of phospholipids, and pronounced similarities in the region of the polar headgroups were found. The conformation of the glycerol group in relation to the C-2 to C-3 bond is identical in these lipids as well as in the diglyceride shown in Fig. 8.49 and in the monoglyceride shown in Fig. 8.44. Furthermore the rest of the polar group is nearly parallel to the layer surface, bringing the positive ammonium group close to the negative phosphate group.

A handbook on lipid bilayer structures, mainly on

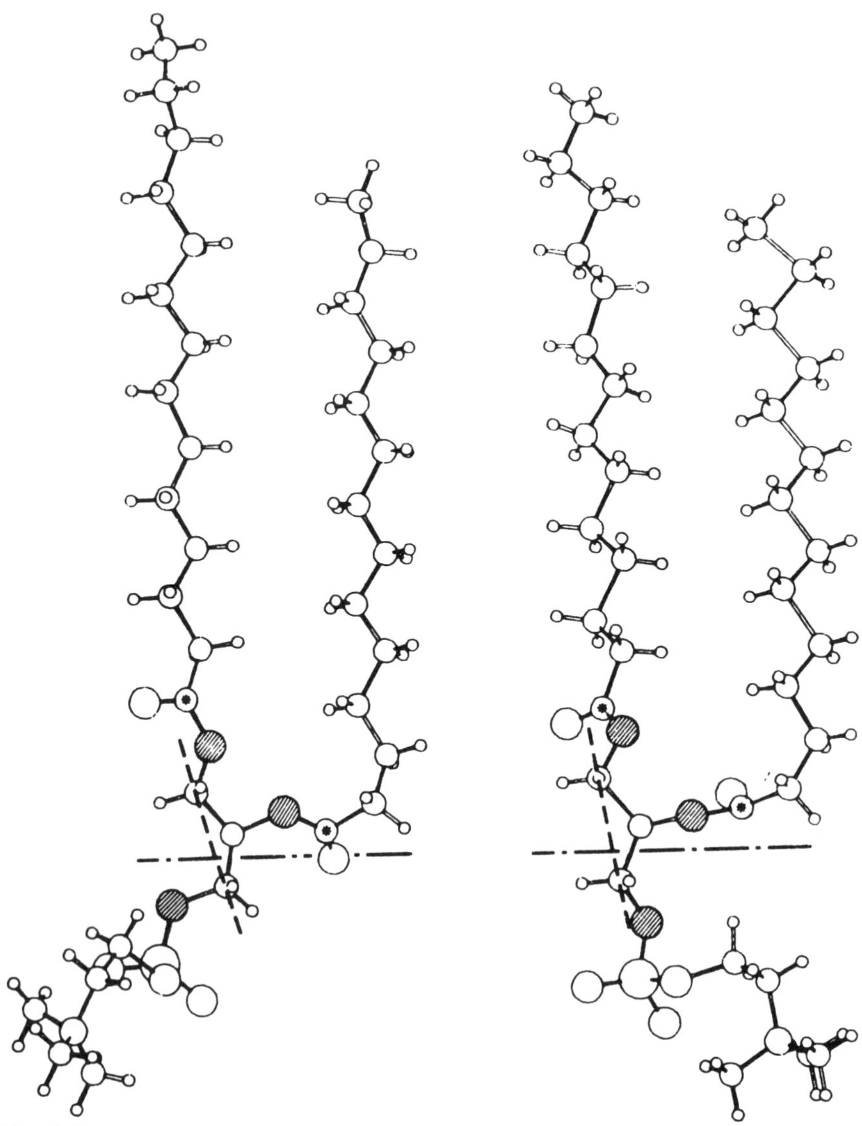

Fig. 8.59 Molecular conformation of the two enantiomeric molecules in the crystal structure of dimyristoylphosphatidylcholine (Pearson and Pascher, 1979).

phospholipids but also including galactolipids, has been published recently (*CRC Handbook of Lipid Bilayers*, 1990).

8.10.2 Phospholipid–water systems

The first structural description of aqueous phases of biological lipids was reported by Luzzati and Husson (1962). An extract of phospholipids from human brain was studied, and this was the first occasion when the hexagonal phase H_{II} was determined. Later the phase diagrams of the systems egg lecithin–water (Small, 1967) and dipalmitoyllecithin–water (Chapman *et al.*, 1967) were reported. An extensive study of monoacid lecithins with saturated chains in the range C_8–C_{18} was then reported (Tardieu *et al.*, 1973). The phase diagram of dimyristoyllecithin, together with the phases occurring in egg lecithin at low water content, is characteristic for all lecithin–water systems, and the corresponding phase diagrams are shown in Fig. 8.60. An NMR study of the dipalmitoyllecithin–water system has been reported by Ulmius *et al.* (1977). A database on phospholipid phase transitions was recently published by Caffrey *et al.* (1991).

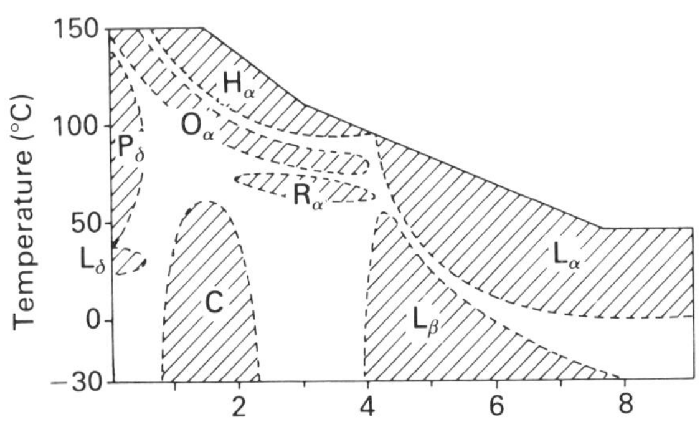

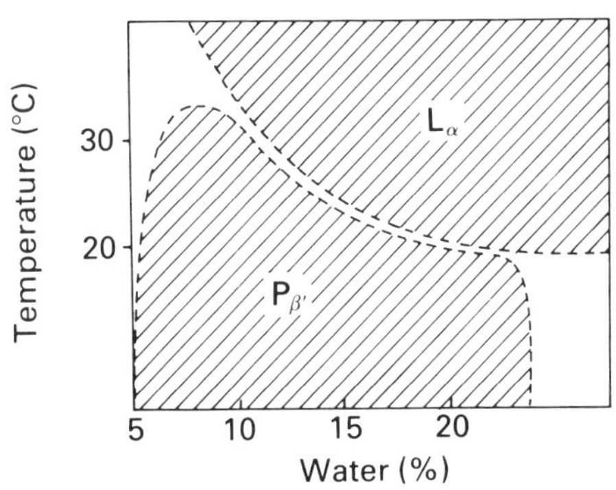

Fig. 8.60 Phase diagram of egg lecithin of low water content (a) and of the dimyristoyllecithin–water system (b). After Tardieu *et al.* (1973).

The L_α phase dominates in all lecithin–water systems. Tardieu *et al.* (1973) found a phase below the chain 'melting' transition in dimyristoyllecithin with two sets of long-spacing values, one corresponding to the bilayer thickness and the other considerably larger. The structure was proposed to be a rippled bilayer phase and the phase was called $P_{\beta'}$ (see Fig. 8.60). This phase was not observed, however, in dipalmitoyl- or distearoyllecithin. Rand *et al.* (1975) analysed dipalmitoyllecithin below the 'melting' of the chains at 42 °C and described the X-ray diffraction characteristics of a transition at 35 °C with the interpretation that the tilted chains in the bilayer below 35 °C become perpendicular at the transition. Later Janiak *et al.* (1976) reported that this so-called pretransition is a transition from the tilted bilayer structure of the rippled $P_{\beta'}$ phase described by Tardieu *et al.* (1973).

The packing rules of crystalline chains described in section 8.1, however, with only certain discrete values of the tilt of the chains, cannot be fulfilled in a rippled structure, where the tilt of the chains must vary continuously along the ripple. An alternative structure has therefore been proposed, which is consistent with experiences on chain close-packing geometry (Larsson, 1977). This structure consists of folded bilayers with alternation of units with tilted chains and units with perpendicular chains. The structure has been confirmed by later work (cf. Laggner and Kriechbaum, 1991).

8.10.3 Monolayers

An extensive analysis of the polymorphism in phospholipid monolayers has been reported by Albrecht *et al.* (1978), who established a phase diagram of dipalmitoyllecithin monolayers involving four phases. Two of these have the liquid type of disordered chains (liquid-expanded films). It was proposed that one has disordered chains oriented perpendicular to the water surface and the other has disordered chains in a tilted arrangement. Furthermore, two crystalline phases were described, one with tilted chains and the other with perpendicular chains.

The relation between the chain melting transition in the three-dimensional aqueous systems and in monolayers was also analysed in this work. It was then assumed that the heat of transition (calculated from the Clausius–Clapeyron equation) and the molecular area in the monolayer are the same as in the fully hydrated aqueous phases. A discrepancy in the transition temperature between the monolayer and bilayer phases obtained in this way was explained as an effect of the external pressure applied. Phase transitions in phospholipid monolayers were recently reported by Weis (1991) with direct observations of the fluid–solid coexistence region.

Monolayer phase diagrams of lecithin–phosphatidic acid and lecithin–cholesterol and cholesterol–phosphatidic acid were also reported by Albrecht *et al.* (1981). An interesting feature in these binary monolayer systems is the indication of a lecithin–phosphatidic acid 1:1 complex with crystalline chains.

References and further reading

Albrecht, O., Gruler, H. and Sackmann, E. (1978) *J. Physique*, **39**, 301.

Albrecht, O., Gruler, H. and Sackmann, E. (1981) *J. Colloid Interface Sci.*, **79**, 319.

Caffrey, M., Moynikan, D. and Hogan, J. (1991) *Chem. Phys. Lipids*, **57**, 275.

Chapman, D., Williams, R.M. and Ladbrooke, D.D. (1967) *Chem. Phys. Lipids*, **1**, 445.

CRC Handbook of Lipid Bilayers (1990) CRC Press, Boca Raton, FL.

Hauser, H., Pascher, I. and Sundell, S. (1980) *J. Mol. Biol.*, **137**, 249.

Hitchcock, P.B., Mason, R., Thomas, K.M. and Shipley, G.G. (1974) *Proc. Natl Acad. Sci. USA*, **71**, 3036.

Janiak, M.J., Small, D.M. and Shipley, G.G. (1976) *Biochemistry*, **15**, 4574.

Laggner, P. and Kriechbaum, M. (1991) *Chem. Phys. Lipids*, **57**, 121.

Larsson, K. (1977) *Chem. Phys. Lipids*, **20**, 225.

Luzzati, V. and Husson, F. (1962) *J. Cell Biol.*, **12**, 207.

Pearson, R.H. and Pascher, I. (1979) *Nature*, **281**, 499.

Rand, R.P., Chapman, D. and Larsson, K. (1975) *Biophys. J.*, **15**, 1117.

Small, D.M. (1967) *J. Lipid Res.*, **8**, 551.

Sundell, S. (1980) Thesis, University of Goteborg, Sweden.

Tardieu, A., Luzzati, V. and Reman, F.C. (1973) *J. Mol. Biol.*, **75**, 711.

Ulmius, J., Wennerström, H., Lindblom, G. and Arvidson, G. (1977) *Biochemistry*, **16**, 5742.

Weis, R.M. (1991) *Chem. Phys. Lipids*, **57**, 227.

8.11 Cholesterol and cholesteryl esters

Cholesterol and its interaction with phospholipids have been studied intensively in order to understand the role of cholesterol in biomembranes.

Crystals of cholesterol monohydrate occur in various pathological conditions (Bogren and Larsson, 1961). The crystal structure of cholesterol monohydrate has been determined by Craven (1976). Cholesterol sulphate has been found in marine organisms and can form aqueous bilayer phases of the gel type (Abrahamsson *et al.*, 1977).

Cholesterol exhibits interesting monolayer properties, which certainly are related to its function in membranes. A characteristic Π–A isotherm is shown in Fig. 8.61. The molecular cross-sectional area of the condensed monolayer is about 36 $Å^2$, close to that in crystals. The monolayer is in a disordered state, which can be observed if a powder is spread on the surface. The mobility is about the same as that of monolayers of long-chain amphiphiles with the hexagonal chain packing (LS phases, see section 8.4). The formation of a triple-layer film, illustrating the relation between the monolayer and a corresponding bilayer structure, is also evident from the Π–A isotherm.

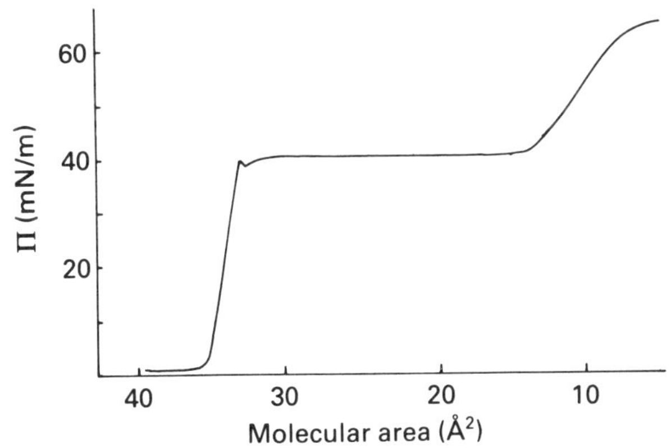

Fig. 8.61 Pressure–area isotherm of cholesterol at 13.6 °C at the air–water interface.

The first crystal structure determined of a cholesteryl ester was that of cholesteryl 17-bromoheptadecanoate (Abrahamsson and Dahlén, 1977), shown in Fig. 8.62. The cholesteryl skeletons are arranged in a bilayer separated from the hydrocarbon chains, which are penetrated by one another in pairs. Although the molecular cross-sectional area of the cholesterol part (36.8 $Å^2$) is almost exactly twice that of hydrocarbon chains in the different close-packing geometries, the arrangement of cholesterol and hydrocarbon chains in the same layer is still avoided. This is probably due to the length of the hydrocarbon chains compared to that of the cholesterol skeletons. In the crystal structure of cholesteryl laurate, however, the hydrocarbon chains and the steroid skeletons are arranged in the same layer (Dahlén, 1979), and this chain length fits well with the length of the cholesterol part of the molecule. The crystal structures of cholesteryl octanoate (Craven and Guernina, 1979a) and cholesteryl oleate (Craven and Guernina, 1979b) have also been determined.

The L_α phase of lipid–water systems is able to solubilize cholesterol. Usually one molecule of cholesterol requires at least two hydrocarbon chains, independent of the lipid type. Small (1970) has described the physical state of cholesteryl esters with saturated and unsaturated chains, as well as phase diagrams of cholesterol, triglycerides and cholesteryl esters at different temperatures.

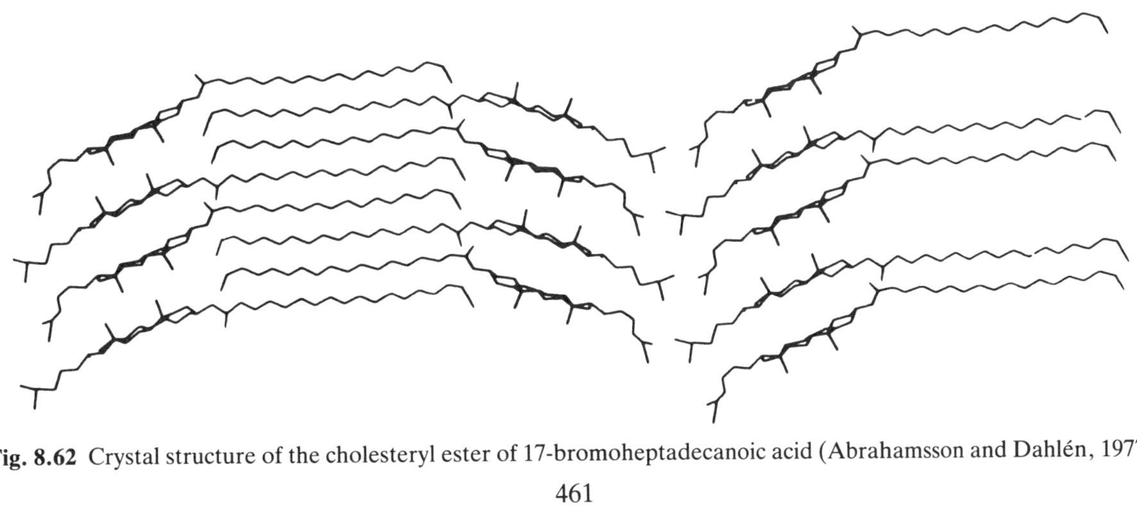

Fig. 8.62 Crystal structure of the cholesteryl ester of 17-bromoheptadecanoic acid (Abrahamsson and Dahlén, 1977).

References and further reading

Abrahamsson, J., Abrahamsson, S., Hellqvist, B., Larsson, K., Pascher, I. and Sundell, S. (1977) *Chem. Phys. Lipids*, **19**, 213.

Abrahamsson, S. and Dahlén, B. (1977) *Chem. Phys. Lipids*, **20**, 43.

Bogren, H. and Larsson, K. (1961) *Biochim. Biophys. Acta*, **75**, 66.

Craven, B.M. (1976) *Nature*, **260**, 727.

Craven, B.M. and Guernina, N.G. (1979a) *Chem. Phys. Lipids*, **24**, 157.

Craven, B.M. and Guernina, N.G. (1979b) *Chem. Phys. Lipids*, **24**, 91.

Dahlén, B. (1979) *Chem. Phys. Lipids*, **23**, 179.

Small, D.M. (1970) in *Surface Chemistry of Biological Systems*, Plenum, New York, p. 55.

8.12 Interaction between lipids and proteins

8.12.1 General

The basis for the interaction between lipids and proteins is related to their amphiphilic nature and is due to their influence on the water structure, the so-called hydrophobic effect (Tanford, 1980). In general terms, four alternative types of phases can occur in lipid–protein–water systems. An aqueous lipid solution can coexist with a protein in the same solution; alternatively a solution of molecular lipid–protein complexes can be formed. It is also possible that the lipid forms a liquid-crystalline phase (or an L2 phase) with water; such a phase can either solubilize a protein or coexist with a protein solution. The first two interaction alternatives in water solution have been thoroughly discussed (Tanford, 1980). When the lipid concentration is below the CMC there is no interaction besides the eventual association of a few lipid molecules to the protein at certain high-affinity binding sites. At lipid concentrations above the CMC there is a so-called mass cooperative binding of numerous lipid molecules to each protein molecule, and this leads to unfolding of the protein structure.

Liquid-crystalline lipid–water phases can, under certain conditions, incorporate proteins. On the basis of studies of liquid-crystalline phase in various lipid–protein–water systems (Gulik-Krzywicki *et al.*, 1969; Rand and SenGupta, 1972), it was concluded that a prerequisite for the formation of composite liquid-crystalline phases is that the lipid is charged (Rand, 1976). Recently, however, a system with a neutral lipid was examined that appeared to form a lipid–protein–water phase over a large composition range (Ericsson *et al.*, 1983). It was even possible to construct a ternary phase diagram, which is shown in Fig. 8.63. From determination of the denaturation temperature and enthalpy of the lipid–protein–water phase and from diffusion properties, it was concluded that lysozyme is distributed in the water channel systems of the cubic monoolein phase without association with the lipid bilayer. The same phase was observed with numerous other proteins, for example bovine serum albumin, α-lactalbumin, pepsin, myoglobin and glucose oxidase.

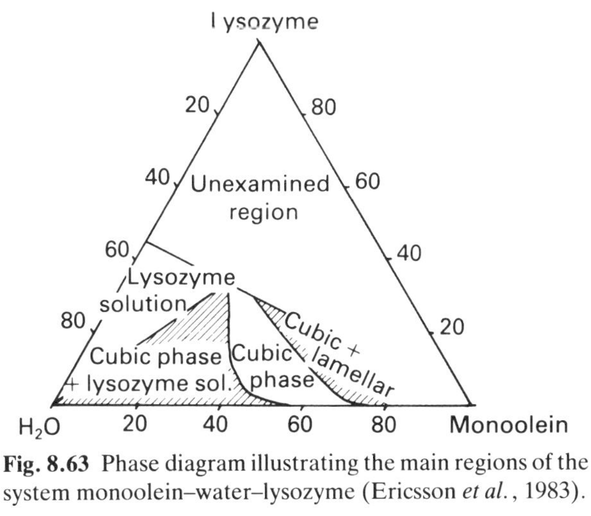

Fig. 8.63 Phase diagram illustrating the main regions of the system monoolein–water–lysozyme (Ericsson *et al.*, 1983).

A specific type of interaction between lipids and proteins is found in lipoproteins that transport triglycerides and cholesteryl esters in the plasma of mammals. The largest lipoproteins, chylomicrons, with a diameter between 800 and 5000 Å, and very-low-density lipoproteins (VLDL), with a diameter of 300–800 Å, resemble emulsion droplets with a core of non-polar lipid and a surface coat of phospholipids and proteins (cf. Brown *et al.*, 1981). A physical characterization of chylomicrons has been reported (Parks *et al.*, 1981). Most of the plasma cholesterol occurs in low-density lipoprotein (LDL), which is a particle with a diameter of 200 Å. The core consists of almost pure cholesteryl esters, and it has a surface coat of a phospholipid monolayer and four tetrahedrally arranged apoproteins (Gulik-Krzywicki *et al.*, 1979). The smallest particle, high-density lipoprotein (HDL), is a kind of molecular lipid–protein complex.

The structures of egg-yolk low-density lipoproteins have also been examined (Kamat *et al.*, 1972). These lipoproteins differ from plasma low-density lipoproteins mainly in the composition of the non-polar lipids. Thus there is only about 0.8% of the particle weight of cholesteryl esters, whereas there are about 60% triglycerides. It follows that the particles with a diameter of about 300 Å consist of a liquid core of neutral lipids covered by phospholipids and proteins.

The most important condensed phase, formed by specific lipid–protein interaction, is the cell membrane discussed in section 8.13.

8.12.2 Technical applications

There are numerous important technical applications of lipid–protein interaction, mainly in the fields of food technology and the pharmaceutical industry. Food emulsions such as margarine and ice cream have polar lipids and proteins at the fat–water interface.

Baking performance of cereals depends upon lipid–protein interaction. The properties of starch products can be monitored by formation of an amylose–lipid complex as a direct consequence of hydrophobic interaction.

Among pharmaceutical applications, the use of liposomes as carriers for drugs should be mentioned. An active component can be entrapped in liposomes or vesicles, giving a controlled-release system. By solubilizing proteins in the lipid bilayer that exposes a 'label' on the surface, it may even be possible to achieve drug targeting. Clinical experiences so far, however, appear to be very limited. The reason is that there have been numerous obstacles in producing and applying liposomes as a drug-carrier system; the most important one is that a liposomal dispersion is not thermodynamically stable.

References and further reading

Brown, M.S., Kovanen, P.T. and Goldstein, J.L. (1981) *Science*, **212**, 628.

Ericsson, B., Larsson, K. and Fontell, K. (1983) *Biochim. Biophys. Acta*, **729**, 23.

Gulik-Krzywicki, T., Shechter, E., Luzzati, V. and Faure, M. (1969) *Nature*, **223**, 1116.

Gulik-Krzywicki, T., Yates, M. and Aggerbeck, L.P. (1979) *J. Mol. Biol.*, **131**, 475.

Gulik-Krzywicki, T., Aggerbeck, L.P. and Larsson, K. (1984) *Surfactants in Solution* (eds B. Lindman and K.L. Mittal), Plenum, New York, p. 237.

Kamat, V.B., Lawrence, G.A., Barratt, M.D., Darke, A., Leslie, R.B., Shipley, G.G. and Stubs, J.M. (1972) *Chem. Phys. Lipids*, **9**, 1.

Parks, J.S., Atkinson, D., Small, D.M. and Rudel, L.L. (1981) *J. Biol. Chem.*, **256**, 12992.

Rand, R.P. (1976) *Food Emulsions* (ed. S. Friberg), Marcel Dekker, New York, p. 277.

Rand, R.P. and SenGupta, S. (1972) *Biochemistry*, **11**, 945.

Tanford, C. (1980) *The Hydrophobic Effect*, 2nd edn, Wiley, New York.

8.13 Structure of biological membranes

8.13.1 Introduction

A full discussion of lipid–protein interactions in cell membranes is beyond the scope of this book. Indeed, many books and even series have been devoted to membrane structure and function, and the reader is referred to Hoffman and Giebisch (1970–), Martonosi (1976), Manson (1971–), Jamieson and Robinson (1976) and Finean and Michell (1981).

The ability of lipid molecules to associate in an aqueous environment into infinite lipid bilayers is the basic mechanism behind the formation of biomembranes, and the ability to separate compartments of functional units by such membranes was an important step in the development of organized life. The significance of the ordered but fluid state of lipid molecules, particularly for the interaction with membrane proteins, was revealed by studies of lipid–water systems during the 1960s (see Chapman *et al.*, 1967).

Current ideas on the underlying structure of biological membranes are set out in the so-called fluid-mosaic model of membrane structure (Singer and Nicolson, 1972), in which the lipids are arranged in a bilayer configuration and form the matrix within which the various proteins are associated. With one or two notable exceptions, such as the purple membrane of halophilic bacteria, the lipid bilayer is thought to be in a fluid state such that the components are able to diffuse relative to one another subject to various constraints. It is remarkable that membranes form such a stable unified structure given the diversity of lipid classes and different types of proteins associated with each morphologically distinct membrane. Moreover, the biochemical mechanism(s) that is (are) responsible for creating and preserving this diversity, including that of the relative portions of the different lipids and proteins, is (are) presently unknown.

Many membrane functions can be directly related to the physical properties of the liquid-crystalline lipid bilayer. Variations in curvature and area per unit mass and diffusion properties can thus be explained by the liquid-crystal model. Membrane proteins contain hydrophobic segments in an α-helical conformation, which constitute the anchor in the hydrophobic core of the bilayer. The proteins contain short chains of sugar groups on the outer surface of the plasma membrane, and this surface zone is involved in recognition phenomena. A covalent link between the peptide and the oligosaccharide chain consists of the amino acid asparagine and *N*-acetylglucosamine (Struck and Lennarz, 1980). Other attachment sites are via the hydroxyl residues of serine and threonine. Glycolipids of the membrane contribute to the membrane surface zone in a similar way, for example, by providing blood group receptors and receptors for hormones, and toxins (Karlsson, 1978; Watenabe *et al.*, 1979).

Pearse and Bretscher (1981) have discussed the role of coated vesicles in membrane synthesis and function. Eukaryotic cells are able to take up macromolecules specifically by absorptive endocytosis. The macromolecules are usually transferred to lysosomes, where they may be degraded. The first stage of the process involves the binding of macromolecules to receptors that are localized in coated pits. The latter are indented sites on the plasma membrane, and the coated pit buds into the cytoplasm to form a coated vesicle in which lie the endocytosed macromolecules. The coated vesicle sheds its coat rapidly and the endocytic vesicles fuse with each other. This allows receptors to be returned to the plasma membrane while the contents are transferred to the lysosomes. In order to explain how lysosomal and plasma membranes remain different, it was suggested that the coated pits are able to accept certain macromolecules while excluding others. The accepted proteins enter the coated pit and were presumed to bind directly or indirectly to clathrin. Clathrin, a protein (of molecular mass 180 kDa) on the cytoplasmic face of coated pits, provides the polyhedron skeleton for the coated vesicles. Examples of the use of coated vesicles for mediated endocytosis are in the uptake of low-density lipoprotein from the blood, and in humans for the transport of immunoglobulins from the mother to the child. For other mammals such as the rat, the antibodies are selectively absorbed from the mother's milk by the intestinal epithelium. Coated vesicles also provide a mechanism for virus transport into cells.

Fusion is a fundamental process in membranes. The fusion process can be regarded as a local phase transition at adjacent lipid bilayers. Whereas lecithin forms the lamellar L_α structure in a wide pH range in the presence of various ions, other phospholipid constituents of the membrane favour other phases. Thus the presence of unsaturated phosphatidylethanolamines tends to give a hexagonal phase (H_{II} type) (Cullis and De Kruijff, 1979). Non-bilayer structures can be induced in membranes containing certain lipids by divalent cations. They were first described in aqueous dispersions of diphosphatidylglycerol and phosphatidylcholine in the presence of Ca^{2+} (Cullis *et al.*, 1978) and have been observed in lipid extracts from chloroplast thylakoid with high Mg^{2+} concentrations (Carter and Staehelin, 1980).

It has been known for a long time that monoolein and lysolecithin have strong fusogenetic effects on membranes. The cubic phase of monoolein has been discussed above; lysolecithin systems also exhibit cubic phases (Arvidsson *et al.*, 1985). It is therefore tempting to assume that fusion is induced by local occurrence in the membrane of the cubic structure, which is a bilayer in a three-dimensional arrangement. Cubic tube systems connecting planar lipid bilayers have in fact been observed (Harbich *et al.*, 1978). The occurrence of an isotropic phase has also been demonstrated in connection with monoolein-induced fusion (Tilcock and Fisher, 1982).

Although it is possible that the ability of major membrane lipids to adopt the H_{II} phase *in vitro* may be important for their function *in vivo* (Lugtenberg and van Alphen, 1983), it must be emphasized that such phases have never been unequivocally demonstrated for any biological membrane under physiological conditions (Gounaris, 1986). They can be induced by special conditions such as heat damage (Gounaris *et al.*, 1983). Even with membranes such as the chloroplast thylakoids, which contain large amounts of the H_{II} lipid monogalactosyldiacylglycerol, the membranes are present as bilayers. Presumably this is because of lipid–protein interactions (Gounaris, 1986).

So-called non-bilayer phases have had much attention during recent years in connection with biomembranes. The term refers to the H_{II} phase and cubic phases (which in fact *are* bilayer phases). Reviews on the significance of the cubic phase in membrane structure–function discussion have been reported by Lindblom and Rilfors (1989) and by Larsson (1989).

Some of the structural features of biological membranes have been revealed by application of different biophysical techniques. The properties of lipid–water systems using such techniques have been extensively reviewed earlier in this chapter, and the information contained therein forms a useful basis upon which their behaviour in biological membranes may be inferred. Some of the structural features of membrane proteins will also be reviewed, so that the factors that govern the interaction between the membrane proteins and lipids may be examined in an informed context.

8.13.2 Lipid arrangement in biological membranes

The principal lines of evidence for the arrangement of lipids in biological membranes are based on studies of certain classes of lipid that form lamellar structures in aqueous dispersions, and are supported by arguments along the lines of the amphipathic character of the molecules. This approach ignores the presence, often in very significant proportions, of lipids that do not form lamellar structures when dispersed in aqueous media at temperatures approximating to that of the growth temperature of the organism from which they are extracted. Indeed, it is commonly observed that dispersions of total polar lipid extracts of many membranes do not form pure lamellar phases but exhibit a mixture of different phases consisting of both lamellar and non-lamellar structures. The implications of these observations are either that the arrangement of the polar lipid in the membrane is, in part, lamellar and coexisting with non-lamellar structures, or that interaction with other membrane components, such as the protein, imposes a lamellar arrangement on those molecular species of lipid which would otherwise form a non-lamellar structure.

The methods used to determine the arrangement of lipids in biological membranes include diffraction and spectroscopic measurements as well as freeze–fracture electron microscopy and monomolecular-film area determinations. Some of this evidence will be examined in the following sections.

(a) Monomolecular-film area measurements

Amongst the earliest molecular models of the structure of biological membranes were those based on studies of the permeability of solutes of differing polarity. These indicated that the barrier consisted, at least in part, of a domain of hydrophobic character. Based on this evidence and the knowledge that lipids orient at an air–water interface, Gorter and Grendel (1925) formulated a model of the structure of membranes, which consisted of a bimolecular layer of lipids with the polar groups oriented on the outside of the hydrophobic components buried in the interior. Their experiment consisted of comparing the area occupied by the lipids extracted from a known area of membrane, in their case the human erythrocyte membrane. There were, however, a number of flaws in the experimental protocol: the solvent employed resulted in incomplete extraction of the lipid, but this was compensated to some extent by inaccuracy in measurement of membrane area. There was also uncertainty about the packing density of lipid molecules in the surface film at which the area should be measured; in fact the measurements were made on closely packed films near the collapse pressure. Under these conditions, it was found that the area occupied by the lipid at the air–water interface was about twice that of the area of membrane from which it was extracted. This led to the conclusion that the structure of cell membrane was composed of a lipid bilayer.

It is now known that the density of phospholipids in a bilayer configuration is somewhat less (equivalent to a monomolecular-film pressure of 30–35 mN/m) than the collapse pressure of a monomolecular film. Moreover, the Gorter and Grendel experiment has been repeated since by other workers to ensure complete lipid extraction and with accurate measurements of membrane area. The results almost invariably show that the area occupied by the lipids in a surface film is less than that required to form two layers over the entire membrane area. Table 8.30 shows the maximum proportion of membrane area that could be accounted for as lipid in a bilayer arrangement. It can be seen that, as the proportion of protein to lipid increases in a membrane, the amount of lipid that could be accounted for as lipid bilayer decreases.

It is important to recognize that no conclusions can be drawn about the arrangement of lipids in biological membranes using this approach. If, however, a bilayer configuration is *assumed* for the lipid, then the remaining area could be accounted for by intrinsic proteins. This is the assumption used in the approach adopted by Engelman (1969) in calculating the surface area occupied by lipids in the human erythrocyte membrane. Accordingly, the total volume of membrane occupied by the hydrocarbon chains of the phospholipids was calculated using an average chain length of 16.5 carbon atoms with 1.26 unsaturated bonds per chain. To this volume was added that occupied by the cholesterol, obtained from density

Table 8.30 Theoretical area of membrane consisting of lipids in bilayer arrangement calculated from the area occupied by total lipid extracts at the air–water interface

Membrane	Protein:lipid (wt/wt)	Lipid bilayer (%)
Plasma membrane		
Erythrocyte	1.50	67
Myelin	0.28	103
Acholeplasma laidlawii	1.78	62
Endoplasmic reticulum	0.90	83
Mitochondria		
Inner membrane	3.55	40
Outer membrane	1.22	72

measurements, assuming that the entire molecule resides in the hydrophobic region of the membrane. The value obtained was too large for a membrane consisting of a single layer of lipid molecules but insufficient for one covering the entire red-cell surface in a bilayer. It was suggested that the discrepancy could be accounted for by non-lipid components if it was assumed that the lipid formed a bilayer.

(b) Diffraction studies

The principal methods of diffraction analysis rely on examination of assemblies of molecules that are arranged in a regular repeating lattice. This is necessary because scattering occurs only weakly and amplification can be achieved most conveniently with a large ensemble, e.g. crystals. As we have noted above, membranes are said to be fluid, which implies that the molecules are in a state of disorder. The notable exception is the purple membrane, referred to above, which exists in a rigid, two-dimensional crystalline array. The purple membrane has proved to be amenable to detailed diffraction studies (Ceska and Henderson, 1990).

In these studies, the two-dimensional order in the membrane has been examined using electron diffraction methods, since the dose of X-rays required to detect a diffraction pattern from such a small assembly would cause radiation damage to the membrane. This problem is avoided in transmission electron microscopic methods, which have been modified to enable the detection of electron scattering intensities and their subsequent conversion into structural coordinates of the two-dimensional protein–lipid crystals. The intensity with which electrons are scattered varies according to the electrical potential within the structure, which, for practical purposes, is roughly proportional to electron density and, in turn, to atomic number. In this sense the electron diffraction method is analogous to X-ray diffraction, since the Fourier transform of the electron scattering yields the profile of atomic density through the structure.

Apart from the usual problems of phase determination, the advantage of the electron scattering method is that very low doses of electrons can be employed (less than one electron per unit cell) and, in order to reconstruct the image, information is combined from a large number of unit cells. This avoids radiation damage to the sample, which would otherwise destroy the structure. The strategy for constructing a three-dimensional image of a two-dimensional array of molecules is achieved by analysing specimens tilted with respect to the angle of incidence of the electron beam. This method relies on the fact that the Fourier transform of a transmission electron micrograph of a two-dimensional crystal is a central section to a three-dimensional Fourer transform of the crystal, that is, through a lattice of lines in reciprocal space perpendicular to the plane of the crystal. Thus the amplitudes and phases of Fourier terms along each lattice line can be obtained by combining the Fourier transforms from an appropriate number of different, tilted views of the two-dimensional array. As in the X-ray diffraction analysis, an inverse Fourier transform produces a three-dimensional map of the density distribution within the monomolecular lattice.

The picture that emerges from these studies is that of a protein that consists of seven rods, each about 4 nm long and 1 nm apart. The rods span the lipophilic portion of the membrane, three approximately vertical to the plane of the membrane and four tilted by 10° to 20° to the vertical. Each rod is an α-helix of the polypeptide chain, and, since the complete sequence of the protein is known, it has been possible to show that some of the chain connecting the seven rods is on one side of the membrane and some on the other. Most of the amino acid side chains of the rods are neutral. By contrast, those parts of the protein which are near or extend into the water on either side of the membrane contain 19 charged amino acid residues, and those parts which protrude into the water on the outside of the membrane contain six charged groups. Within the rods there are nine charged groups, five negative, four positive, which may be sufficiently close to neutralize each pair of charges in a predominantly non-aqueous environment. The length of the protein in contact with the lipid is approximately equal to the length of two molecules of lipid. The individual bacteriorhodopsin molecules are aggregated within the structure into trimers, which produce a three-fold axis of symmetry within the membrane. The lipids of the membrane occupy the interstitial spaces between the protein molecules, including the central region between the protein trimers. The lipids of the purple membrane are typical archaebacterial lipids containing phytanyl chains and a major sulphated triglycosylarchiol lipid dominating the composition, with minor amounts of archiol-based phospholipids reflecting the high salt concentrations of the environment in which these bacteria live.

The clear interpretation of the arrangement of the lipids in the purple membrane is that they are arranged in a bilayer configuration. This is not the arrangement adopted by the lipids when the proteins are removed. Studies of dispersions of total polar lipid extracts of a related species of halobacterium in growth medium (5 M NaCl) have been reported by Quinn *et al.* (1986). Freeze–fracture electron microscopy of these dispersions revealed a mixture of non-lamellar phases, and these are shown in Fig. 8.64. Dispersions of binary mixtures of the two major components, namely diphytanyl ether analogues of phosphatidylglycerol phosphate and glycolipid sulphate, are also shown in the figure, from which it can be seen that, as the proportion of phosphatidylglycerol phosphate to glycolipid sulphate in the mixture is increased, there is an increase in the proportion of the lipid structure in non-lamellar arrangement at the expense of structure that is purely lamellar. The heterogeneity of the overall structure is exemplified by the 2:1 mole ratio mixture (Fig. 8.64(c)), which shows evidence of lamellar, cubic and a variety of intermediate phases all in close juxtaposition. Spin label studies have indicated that the hydrocarbon chains of the lipids in the membrane were relatively immobilized, while the lipids in aqueous dispersions have hydrocarbon chains that are comparatively mobile even towards the ether linkage of the phytanyl chains to the glycerol (Esser and Lanyi, 1973). All these studies indicate that there is a strong interaction between the lipid and the protein in the membrane of the halophilic bacteria, which leads to a considerable modification of the phase behaviour of the lipids observed in isolated dispersions.

To obtain structural information from most biological membranes using X-ray or neutron diffraction methods, it is necessary to examine larger arrays of membranes that are suitably arranged in a regular array. Electron density profiles can be derived from phased X-ray scattering data after Fourier synthesis (Luzzati, 1968). Selective deuterium substitution for protons can provide valuable positional information within the profile using neutron diffraction methods. Electron density distributions through the unit cell obtained from membranes are usually compared with those obtained from Langmuir–Blodgett films of phospholipids deposited in bimolecular layers on a surface (Levine and Wilkins, 1971).

Electron density profiles have been constructed from X-ray diffraction data collected from biological membranes that are oriented naturally within the biological tissue, such as myelin and retinal rod outer segment disc membranes, or oriented by processes such as centrifugation. Studies of a variety of membranes at resolutions of about 1.1 nm have shown the structures to be generally asymmetric with respect to the distribution of electron density. The sarcoplasmic reticulum membrane has been examined in detail by both X-ray and neutron diffraction methods (Herbette *et al.*, 1985). Electron density profiles for water, lipid and protein have been constructed from these data, which show that the major protein of the

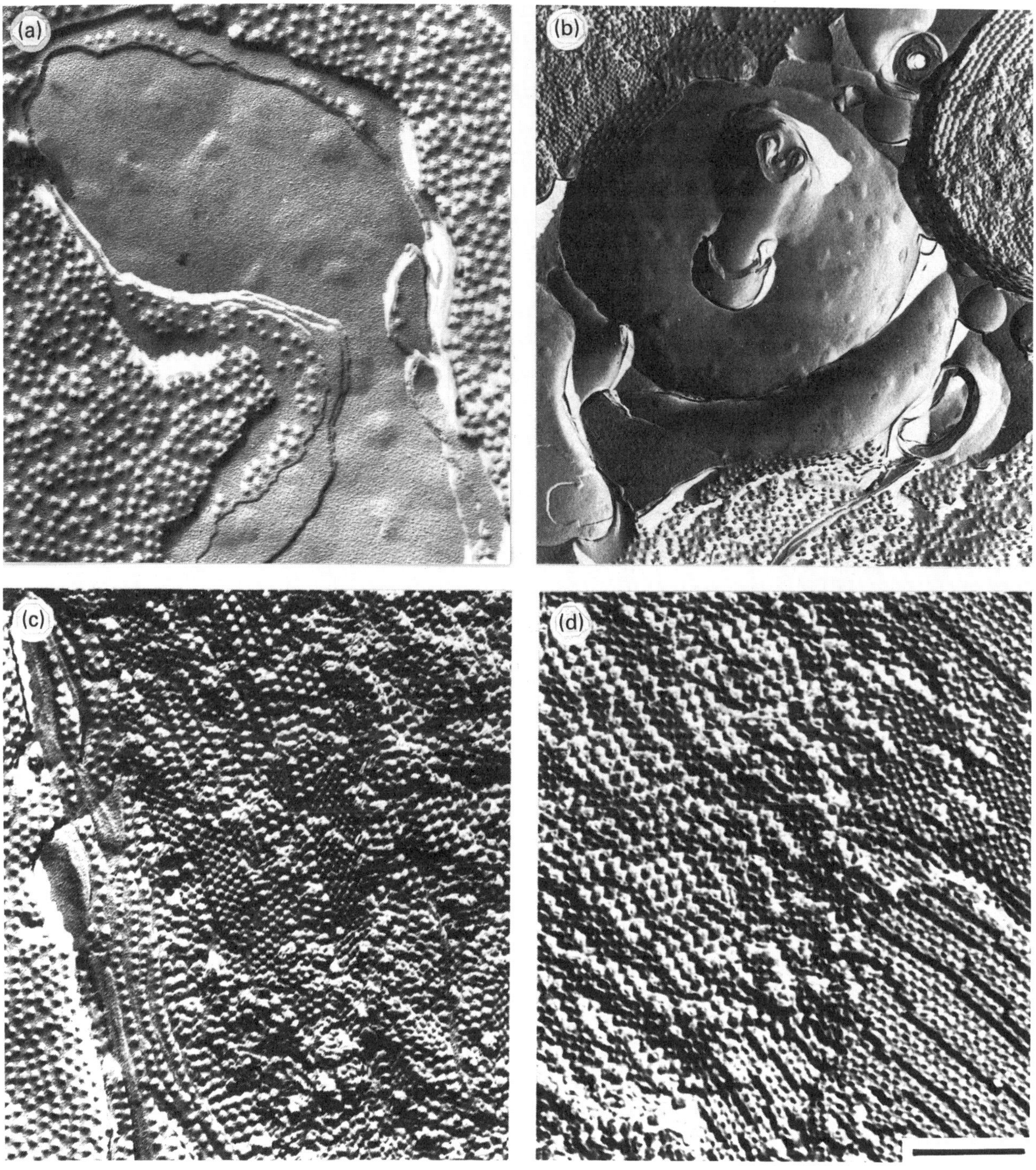

Fig. 8.64 (a) Electron micrograph of freeze–fracture replica prepared from a dispersion of an extract of total polar lipids of *Halobacterium cutirubrum* in 5 M NaCl thermally quenched from 20 °C. Panels (b)–(d) show corresponding micrographs of mixed dispersions of the two major polar lipids, phosphatidylglycerol phosphate and glycolipid sulphate, in mole ratios of 1:2 (b), 2:1 (c) and 3.5:1 (d) prepared as in (a). Bar represents 250 nm in (a), (b) and (d) and 500 nm in (c).

membrane, Ca^{2+} ATPase, spans the membrane and is in contact with water on both surfaces. A major portion of the protein protrudes substantially into the extravesicular aqueous domain. The extent of this protrusion depends on the functional configuration of the enzyme (Blasie *et al.*, 1990). By the use of deuterated lipids, it was possible to determine the arrangement of the lipid in the form of a bimolecular structure with the polar groups of the lipids residing at the aqueous interface on either surface of the structure and with the hydrocarbon chains extending into the interior. As with the protein, the membrane

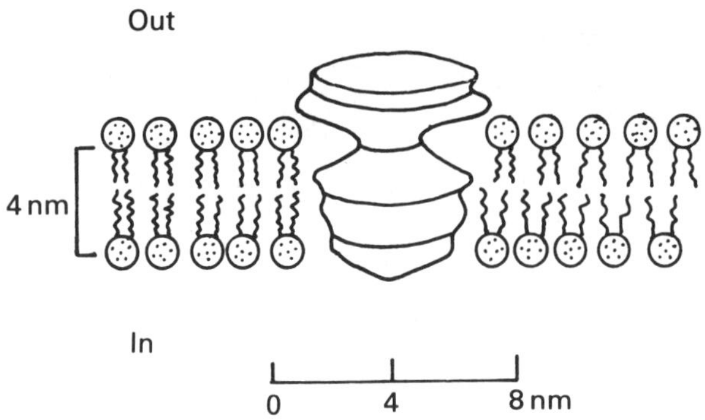

Fig. 8.65 A model of the sarcoplasmic reticulum membrane constructed from electron density distribution profiles derived from diffraction studies of oriented membranes. Redrawn from Herbette *et al.* (1985).

lipid and water profiles are asymmetric, consistent with the complementary asymmetry in the protein mass distribution within the lipid bilayer domain of the sarcoplasmic reticulum membrane profile structure. A model of the structure of the sarcoplasmic reticulum membrane constructed from the diffraction and other data is illustrated in Fig. 8.65.

(c) Freeze–fracture electron microscopy

The introduction of freeze–fracture electron microscopy (Moor *et al.*, 1961) soon highlighted discrepancies in the unit membrane hypothesis current at the time, since particles appeared in the fracture planes running longitudinally in the membrane. These were identified as enzyme complexes (Frey-Wyssling and Schwegler, 1965), which were supposedly rearranged during conventional fixation and staining required for thin-section electron microscopy. Several lines of evidence were used to establish that the fracture plane occurs through a central region of the membrane rather than along the membrane–water interface. One was the observation that when a layer of radioactively labelled fatty acid applied to a glass surface and sandwiched with another layer of unlabelled fatty acid was thermally quenched and prised apart, all the radioactivity remained on the original plate. This experiment is somewhat removed from the situation in biological membranes and brings into question the nature of the forces that anchor the fatty acids to the substrate surface. More convincing evidence was obtained from etching studies. These showed that, by allowing water to sublimate from the fracture surface before shadowing with heavy metal to form the replica, different membrane surfaces could be discerned as the water table was lowered. Failure of etching to alter the appearance of the initial fracture surface indicated that this surface was devoid of water and represented a hydrophobic domain within the membrane. Those surfaces of membrane originally in contact with the aqueous medium were exposed by the etching process.

Interpretation of features exposed in lateral fracture planes through biological membranes is still conjectural (Sjöstrand, 1983). Evidence that membrane-associated particles consist, at least in part, of intrinsic membrane proteins is supported by observations that removal of lipid by phospholipase digestion causes an increase in density of particles, suggesting that the surrounding smooth areas of fracture face represent lipid. Nevertheless, there are claims that vertical movement of proteins in the membrane matrix is possible, so that domains of intrinsic proteins that are normally present in the hydrophobic domain of the membrane do not appear as membrane-associated particles on the longitudinal fracture planes (Funk *et al.*, 1982).

What is more difficult to explain is that in complementary fracture faces there are no holes or pits corresponding to the particles present in the opposite surface. Plastic deformation of the lipid between the fracture and shadowing operations resulting in obliteration of the pits has been suggested as a possible explanation, but contemporary methods of freeze–fracture make this unlikely. This is illustrated by the electron micrograph shown in Fig. 8.66. The replica was prepared from a mixed aqueous dispersion of mono- and digalactosyldiacylglycerols (2:1, mole ratio) extracted from bean chloroplasts, and shows membrane-associated particles, believed to represent spherical inverted micelles sandwiched between leaflets of a lipid bilayer structure, and pits, representing the location of particles in the complementary fracture face. Clearly in this pure lipid system the pits have been faithfully preserved during preparation of the replica and little evidence for plastic deformation can be observed. It is also noteworthy that membrane-associated particles that consist entirely of non-lamellar lipid aggregates can be formed, and such particles would be indistinguishable from the putative protein complexes seen in replicas prepared from biological membranes.

It is clear that we do not yet have an unambiguous interpretation of all the features seen in freeze–fracture replicas of biological membranes. Nevertheless, taken together with other evidence, the current assignments of smooth regions to the internal cleavage along the central plane of lipid bilayer and membrane-associated particles as representing intrinsic membrane protein complexes are consistent with the fluid-mosaic model of membrane structure.

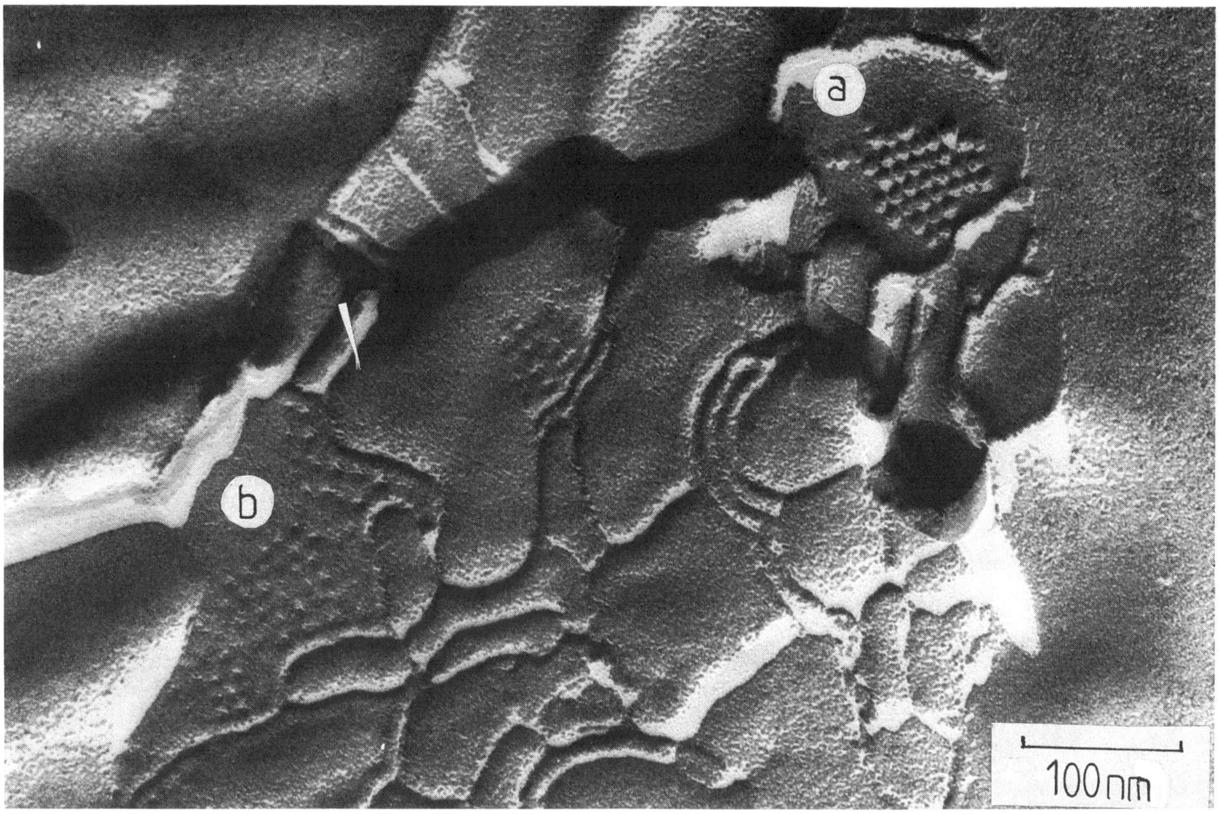

Fig. 8.66 An electron micrograph of a replica obtained from a freeze–fractured sample of a mixed aqueous dispersion (1:1 mole ratio) of mono- and digalactosyldiacylglycerols showing arrays of particles (a) and arrays of pits (b) within a lamellar structure. Reproduced from Sen *et al.* (1982).

(d) Nuclear magnetic resonance spectroscopic methods

The two most relevant nuclear magnetic resonance methods (see Chapter 9) for providing information about the conformation of lipids in biological membranes are ^{2}H and ^{31}P-NMR spectroscopy. In the case of ^{2}H-NMR, the average conformation of the lipid acyl chains and the polar headgroups can be deduced as well as an indication of specific and non-specific interactions between molecules in the structure. In addition, it is possible to obtain an order parameter representing a measure of the angular distribution of molecules about a preferred molecular orientation. In bilayer arrays, for example, there is effective axial symmetry about the normal to the bilayer surface, referred to as the director (for further details, see Stockton *et al.*, 1974; Seelig and Niederberger, 1974; McLaughlin *et al.*, 1975).

NMR methods exploiting quadrupolar splitting of deuterium nuclei have been widely used to study lipid polymorphism and the structure of lipids in biological membranes. The versatility of the method relies on the fact that it is possible to substitute ^{2}H and ^{1}H in all domains of the lipid without significantly perturbing the physical properties of the membrane. It is also possible to substitute deuterated water for ^1H_2O. When deuterated lipids are in an anisotropic arrangement, the deuterium quadrupolar resonance is split to an extent that depends on the size of the lipid structures and decreases as motional averaging tends to an isotropic type. It is possible to derive an order parameter from the extent of splitting of the deuterium resonances for particular residues of lipids in different polymorphic phases.

An example of the use of deuterium magnetic resonance methods to define the structure of lipids in a biological membrane is illustrated in Fig. 8.67. This shows a ^{2}H-NMR spectrum, recorded at 30 °C, of a fatty acid auxotroph of *Escherichia coli* grown on (^{2}H)-oleic acid, together with a spectrum of an aqueous dispersion of a total lipid extract of the cells recorded under the same conditions. A rather broad powder pattern is observed for the deuterons incorporated into the cell membranes, with two peaks separated by a $\Delta\nu_q$ of 6.7 kHz and a sharp isotropic signal, which can be assigned to a small amount of residual HOD in the sample (Killian *et al.*, 1992). The value of $\Delta\nu_q$ is related to the segmental order of the C–D bond (Seelig, 1977), and the large intrinsic line-

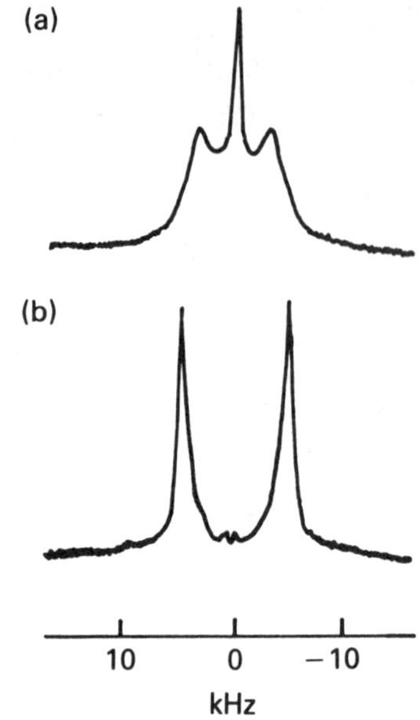

Fig. 8.67 ^{2}H NMR spectra recorded at 30 °C of (a) a suspension of a fatty acid auxotroph of *Escherichia coli* grown on (^{2}H)-oleic acid, and (b) an aqueous dispersion of a total lipid extract from the organism used to record spectrum (a). Data from Killian *et al.* (1992).

width is typical of biological membranes and reconstituted membrane systems (Bloom and Smith, 1985). The ^{2}H-NMR spectrum obtained from an aqueous dispersion of a total lipid extract of the cells is characterized by two sharp peaks, separated by 9.9 kHz, which is considerably greater than $\Delta\nu_q$ of the lipids in the membranes. The spectrum is typical of lipids in a liquid-crystalline bilayer configuration. The general conclusion from studies of this type is that the lipids in biological membranes are arranged in a bilayer with the motion of the lipids constrained by their interaction with other membrane components.

^{31}P-NMR signals from phospholipid assemblies can be used to characterize the arrangement of the constituent molecules and the particular phase that is formed. The information is obtained from the lineshapes of proton-decoupled spectra, which indicate chemical shielding anisotropy of the lipid phosphate moiety. Thus, different orientations of the phosphate segment yield characteristic resonances at different frequencies. In the case of lipids in a bilayer configuration, the ^{31}P-NMR spectrum is typical of a shielding tensor that is axially symmetric around a director axis consisting of a low-field shoulder and high-field peak. Oriented bilayers give rise to individual narrow resonances with an angular-dependent chemical shift (McLaughlin *et al.*, 1975; Seelig and Gally, 1976; Hemminga and Cullis, 1982). The interpretation of these features is that molecules with their long axis perpendicular to the magnetic field contribute to the high-field peak, and molecules with a parallel orientation correspond to resonances at a position in the shoulder region, and that the director axis coincides with the bilayer normal about which the phosphate segment rotates.

The rate of reorientation resulting from Brownian motion of lipid molecules in bilayers can also be determined from the lineshape (Burnell *et al.*, 1980). When the rate of reorientation of the phosphate segment is fast, the lineshape narrows until eventually, when motion is effectively isotropic on the timescale of the measurement (<10 μs), motional averaging leads to a narrow and symmetric resonance line. Motion of this type is experienced by phospholipid molecules in small, rapidly tumbling bilayer vesicles, micelles, inverted micelles and a variety of cubic phases. Although there is virtually no difference in the motion of the phosphate segment of phospholipid in bilayer and hexagonal-II arrangement, the macroscopic orientation of the lipid cylinders within the spectrometer magnetic field gives rise to additional averaging of the chemical-shift anisotropy that clearly distinguishes these two phases.

The use of ^{31}P-NMR methods to characterize phospholipid arrangement in biological membranes is illustrated in Fig. 8.68. This shows a ^{31}P-NMR

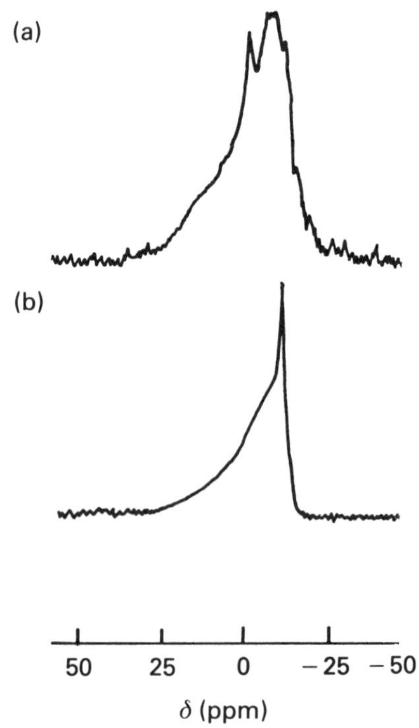

Fig. 8.68 ^{31}P NMR spectra of (a) a suspension of inner membrane vesicles isolated from a fatty acid auxotroph of *Escherichia coli* recorded at 30 °C, and (b) an aqueous dispersion of a total lipid extract from the organism used to obtain spectrum (a) and recorded at 37 °C.

spectrum recorded at 30 °C of a suspension of inner membrane vesicles isolated from an *Escherichia coli* fatty acid auxotroph. The spectrum is dominated by a single major component characterized by a low-field shoulder and a high-field peak. A minor isotropic component can be seen at zero chemical-shift anisotropy. The lineshape of the major component is typical of lipids arranged in a liquid-crystalline bilayer configuration in which the chemical-shift interaction is averaged by fast long-axis rotation of the lipids. There is a residual chemical-shift anisotropy of 29 ppm. The isotropic component is probably due to the presence of a small population of vesicles in which the lateral diffusion of the lipids and vesicle tumbling result in a complete motional averaging of the chemical-shift anisotropy and not to motion of the lipid alone within the membrane. A comparison with an aqueous dispersion of a total lipid extract of the cells is also shown in Fig. 8.68. This spectrum is characteristic of liquid-crystalline bilayers and indicates a similarity between the phase formed by the lipid in the intact membrane and in the isolated lipid dispersion. The chemical-shift anisotropy, however, is increased to 36 ppm. The dispersion, moreover, is not stable, and with time or at an elevated temperature a broad isotropic component appears in the signal superimposed on the bilayer signal, which gradually disappears altogether with a sharpening of the isotropic signal. This change in phase is likely to be due to the high proportion of non-bilayer-forming lipids in the membranes of this organism.

(e) Transmembrane lipid asymmetry

While the general consensus is that the lipid matrix of biological membranes consists of a fluid lipid bilayer, detailed studies have revealed that in some membranes the distribution of lipid species is not symmetrical. The most convincing evidence has been obtained from erythrocytes and platelets, in which sphingomyelin and most of the remaining choline phosphatides are oriented on the outer leaflet of the plasma membrane, whereas the aminophospholipids, mainly phosphatidylserine and phosphatidylethanolamine, are abundant on the inner leaflet (Op den Kamp, 1979). Similar studies of the distribution of phospholipids in the envelope of certain viruses have led to confirmation of the data obtained with blood cells (Allan and Quinn, 1989). The experimental evidence is generally obtained by examination of the products of phospholipase A_2 and sphingomyelinase C hydrolysis upon digestion of sealed vesicles or when exposed exclusively to the outer surface of the membrane.

Two mechanisms have been proposed to explain the creation and maintenance of membrane lipid asymmetry during the normal lifespan of the cell. It has been suggested that endofacial proteins of the membrane skeleton interact via electrostatic forces with aminophospholipids, thereby ensuring a preferred orientation of these lipids towards the cytoplasmic surface (Haest, 1982). *In vitro* binding studies have shown that selective interactions between phosphatidylserine and cytoskeletal proteins of the red cell take place, which provides some support for this idea. The second mechanism was proposed by Seigneuret and Deveaux (1984), who identified an aminophospholipid translocase activity of the plasma membrane based on the observation that introduction of spin-labelled aminophospholipids in the outer leaflet of the red-cell membrane is followed by a rapid transfer to the inner leaflet. This process leads to phospholipid asymmetry because translocation of phosphatidylcholine occurs on a much slower timescale and it cannot be detected in sphingomyelin (Deveaux, 1988). The involvement of a protein or group of proteins in the translocation mechanism is supported by the observation that the process is ATP-dependent, is inhibited by SH-oxidizing agents and is competitively inhibited by D isomers of phosphatidylserine and phosphatidylethanolamine. An integral membrane protein of molecular weight 34 000 in red blood cells has been implicated in the translocation mechanism (Connor and Schroit, 1990).

The asymmetric phospholipid distribution, at least in blood cells, has direct physiological significance. It has been shown that increased surface exposure of phosphatidylserine on the erythrocyte, for example, provides a signal for sequestration of the cell by the reticuloendothelial system (Schroit *et al.*, 1985). Furthermore, partial loss of the asymmetric orientation of phosphatidylserine may occur in sickled red cells, possibly due to the separation of the membrane from the underlying cytoskeleton when the cells become deformed (Davoust *et al.*, 1982; Sugihara *et al.*, 1992). In platelets, transbilayer asymmetry is rapidly lost upon activation by certain platelet agonists, and this may serve as an important function of platelets in the haemostatic process, since two sequential reactions of the coagulation cascade are markedly accelerated in the presence of a phospholipid surface containing phosphatidylserine (Rosing *et al.*, 1988).

8.13.3 Membrane lipid dynamics

With the notable exception of lipids in the purple membrane of *Halobacterium*, which adopt more or less a crystal structure, the lipid matrix of biological membranes is said to be fluid. The term 'fluidity' is not a precise quantitative parameter, but it implies that the molecules of the structure exhibit motion with respect to one another. The measurement of this motion is often performed by the use of membrane lipid analogues that incorporate paramagnetic, fluorescent or other probes, and it is important to recognize that it is the motion of the probe that is observed and not that of the native molecules. It is not unknown for probe molecules to exert a considerable perturbation on the motion and structure of the surrounding lipids or to behave in a manner distinctly different from the lipids they are designed to mimic.

Another problem that also has to be addressed is the need to introduce the probe into the membrane without disturbing the structure. The most common method is to incubate cells or membrane suspensions in a medium into which a small aliquot of lipid probe has been injected from a concentrated solution in organic solvent such as ethanol. Alternatively, the probe may be co-dispersed with unlabelled phospholipid or mild detergent. The strategy of choice often depends on whether the probe increases or decreases the hydropathy of the lipid analogue compared to the native lipids that comprise the membrane. Possibly the least-perturbing method is the use of phospholipid transfer proteins to perform the task, but this method relies on close physical resemblance between the analogue and the phospholipid for which the protein has affinity.

Nevertheless, with these restrictions in mind the general picture that has emerged is that lipid molecules are relatively mobile within the membrane, and constraints on this motion occur by interactions with the proteins and other membrane constituents.

(a) Motion of lipids in membranes

The two principal methods that have been exploited to examine lateral diffusion of lipids in the plane of biological membranes and from one leaflet of the membrane lipid bilayer to the other are spin-label (Marsh, 1988) and fluorescence (Lakowicz, 1983) probe methods. An example of the use of fluorescence probes has been reported by synthesis and use of Lucifer yellow derivatives of phospholipids and cholesterol (Nothnagel, 1989). Introduction of these labels into human erythrocyte ghost membranes and living melanoma cells was performed using organic solvent vectors. Lateral diffusion rates were compared with probe incorporated into aqueous dispersions of phospholipid using a spot fluorescence photobleaching recovery method. The method involves bleaching of probe molecules in a defined surface area of membrane using an intense pulse of laser light and monitoring recovery of fluorescence as unbleached probe molecules from the surrounding area of membrane diffuse into the bleached domain. It was found that Lucifer yellow derivatives of cholesterol and phosphatidylethanolamine diffused rapidly with a diffusion coefficient $>10^{-8}\,cm^2/s$ in lipid dispersions at temperatures greater than the gel to liquid-crystalline phase transition temperature, but the diffusion rate decreased dramatically for probes in gel-phase lipid ($<10^{-10}\,cm^2/s$). Intermediate values of diffusion coefficient were observed in erythrocyte ghost membranes and in the plasma membrane of live melanoma cells. In the latter case only approximately half the probe molecules were identified in the mobile fraction, indicating that motion of the probe is restricted by interaction with other membrane components. The order of diffusion coefficients recorded for these probes is similar to that of other fluorescent lipid probes in mammalian plasma membranes (Struck and Pagano, 1980; Packard *et al.*, 1984; Metcalf *et al.*, 1986).

Translational motion of lipid can also be determined from bimolecular collision frequencies that can be measured by perturbation of the signals derived from probe molecules. Such methods include fluorescence quenching or excimer formation for luminescent probes and spin–spin interactions between spin-labelled probes.

The diffusion of lipids does not take place through a homogeneous lipid bilayer matrix, as there is evidence from photoactivatable fluorescent probes for the non-homogeneous distribution of lipids within the plane of the membrane. Studies of lateral distribution of lipids in the bacterium *Micrococcus luteus* have been undertaken using a reversible photocross-linker, anthracene phospholipid analogue (de Bony *et al.*, 1989). The results clearly showed that the two major polar lipid components, phosphatidylglycerol and dimannosyldiacylglycerol, were not homogeneously distributed in the plane of the membrane. The lateral diffusion coefficient was found to be $(1–2) \times 10^{-9}\,cm^2/s$, in line with the above reports.

In addition to translational motion, membrane lipids undergo rotational motion about their long axes perpendicular to the plane of the membrane as well as motion within the molecule. The latter involves rotational motion of the polar group and *trans–gauche* isomerizations of the hydrocarbon chains. Rotational diffusion rates can be determined from decay of polarization of fluorescence of probe molecules excited by plane-polarized light. The lifetime of the excited fluorescent state is of the order of a few nanoseconds, which is appropriate to the rotational relaxations of lipids in biological membranes. Time-resolved or steady-state depolarization methods can be used in these measurements (Van der Meer, 1988). Information regarding molecular rotational motion can also be derived from motionally induced modulation of the angular anisotropy of the interaction in the magnetic resonance spectra. The timescale of the motional sensitivity depends on the frequency equivalent of the anisotropic interactions that are being modulated. For spin probes this lies in the nanosecond range, and for nuclear magnetic resonance it covers the time domain 1–100 ms.

(b) Lipid fluidity and motion of membrane proteins

An important consequence of the fluid nature of the membrane lipid matrix is that the intrinsic proteins are able to diffuse laterally in the plane of the membrane and rotate about an axis perpendicular to this plane. The mode of synthesis of membrane proteins and the way they interpolate into the lipid bilayer matrix ensures that they retain a unique orientation with respect to the membrane. The rate of rotational diffusion of membrane proteins can be determined by the time-resolved technique of flash photolysis (Cherry, 1978; Thomas, 1986) or the time-averaged method of saturation transfer electron paramagnetic resonance (Hemminga and de Jager, 1989; Beth and Robinson, 1989). Both methods provide information on the microsecond to millisecond timescale. Studies have been reported using triplet or spin-label probes attached to intrinsic membrane proteins, and the general conclusions are that the factor constraining such motion is the viscosity of the lipid domain in which the protein is embedded.

One protein that has been examined by both saturation transfer electron spin resonance and flash photolysis is the Ca^{2+}-ATPase of sarcoplasmic reticulum (Squier *et al.*, 1988; Birmachu and Thomas, 1990). The pump protein exists in the membrane in the form of monomers rotating independently about an axis perpendicular to the plane of the membrane in equilibrium with aggregates of more than one pumping unit. Increasing the viscosity of the aqueous medium in which the membrane vesicles were suspended had no effect on rotational relaxation rates of the protein, whereas decreasing the viscosity of the lipid with addition of diethyl ether markedly increased the rate of rotation of the protein. It was also reported that there was a tendency for the protein to aggregate as the viscosity of the lipid increases. Variation of membrane lipid viscosity with temperature (7.7 P at 4 °C cf. 2.6 P at 20 °C) was reflected in rotational rates of the Ca^{2+}-ATPase monomers, and calculation of the effective radius of rotating protein gave the same value (2.3 nm) at both temperatures. It follows that, in the absence of constraints on rotation of intrinsic membrane proteins by interaction with other proteins such as the cytoskeleton, the rotational motion about the axis perpendicular to the plane of the membrane is directly related to the viscosity of the lipid.

The fluorescence recovery after photobleaching method described above for the measurement of lateral diffusion of lipids can be applied equally to the measurement of protein diffusion. The bleached region of membrane into which unbleached molecules from the surrounding membrane diffuse need not be in the form of a spot, and a variety of bleaching patterns have been employed according to the particular aspect of the structure or dynamics of the membrane system examined. Periodic pattern bleaching (Koppel and Sheetz, 1983) measures dynamics over large areas of one or more regions. Spot or localized line bleaching (Koppel, 1979) yields spatially resolved information in which comparisons can be made between the properties of different lateral domains and different locations within the same region. Bleaching segments of membrane surface with a scanning slit provides information on inter-regional translocation (Koppel *et al.*, 1986). Typical lateral diffusion rates for proteins in biological membranes occur in the range of 10^{-10}–10^{-12} cm^2/s. This rate is some 10^3–10^5 times slower than values recorded for soluble proteins diffusing in aqueous media, and again reflects the effect of membrane lipid viscosity on lateral diffusion rates of membrane proteins.

(c) Lipid phase separations in membranes

Freeze–fracture and immunofluorescence studies have shown that, with the exception of certain specialized structures such as membrane junctional complexes, the distribution of protein complexes in the plane of the membrane tends to be random. When membranes are subjected to temperatures outside the normal range of growth temperatures, lateral phase separation of the lipid and protein components have been observed. Such phase separations appear to be driven largely by the phase behaviour of the lipid, because modification of membrane lipid structure influences the temperature at which membrane phase separations are observed.

An example of phase separations in biological membranes illustrating these effects has been reported by Furtado *et al.* (1979). They conducted freeze–fracture electron microscopic investigations of the blue-green alga, *Anacystis nidulans*, cultured at a temperature of 38 °C and thermally quenched for freeze–fracture at 35 °C compared with those equilibrated at 15 °C before thermal quenching. The results showed that, at the growth temperature, membrane-associated particles were randomly distributed in the plane of the membrane. Cooling the cells to 15 °C caused a lateral phase separation of the membrane-associated particles and the creation of large domains, both in the plasma membrane and in the photosynthetic membranes of these organisms, of smooth fracture face. This phase separation is illustrated in the electron micrograph shown in Fig. 8.69. The smooth regions are believed to represent gel-phase lipid domains from which the intrinsic membrane proteins have been excluded. In control experiments, where cells were grown at low temperatures and thermally quenched from 15 °C, a random distribution of membrane-associated particles was observed, and this correlates with changes in the lipid composition of the membrane. Thus the membrane lipids possess fatty acyl residues that are considerably less saturated than when the organism is grown at high temperature.

The question as to which lipids are located in the phase-separated domains and which lipids phase-separate together with the proteins has been investigated by differential scanning calorimetry, low-angle X-ray diffraction and freeze–fracture electron microscopy (Mannock *et al.*, 1985). The results of the calorimetric study are illustrated in Fig. 8.70. This shows that organisms cultured at 28 °C exhibit a broad endotherm in their heating scans and a mid-point temperature of 14 °C, which presumably corresponds with the transition of gel-phase lipid domains into a liquid-crystalline configuration. A similar endotherm is observed in cells cultured at 38 °C, although judging from the enthalphy of these transitions, the amount of gel-phase lipid separating out in cells cultured at 38 °C is greater than in the corresponding membranes of cells cultured at 28 °C. There is, however, a dramatic change in the mid-point temperatures of the endotherms observed in heating thermograms of total polar lipid extracts of

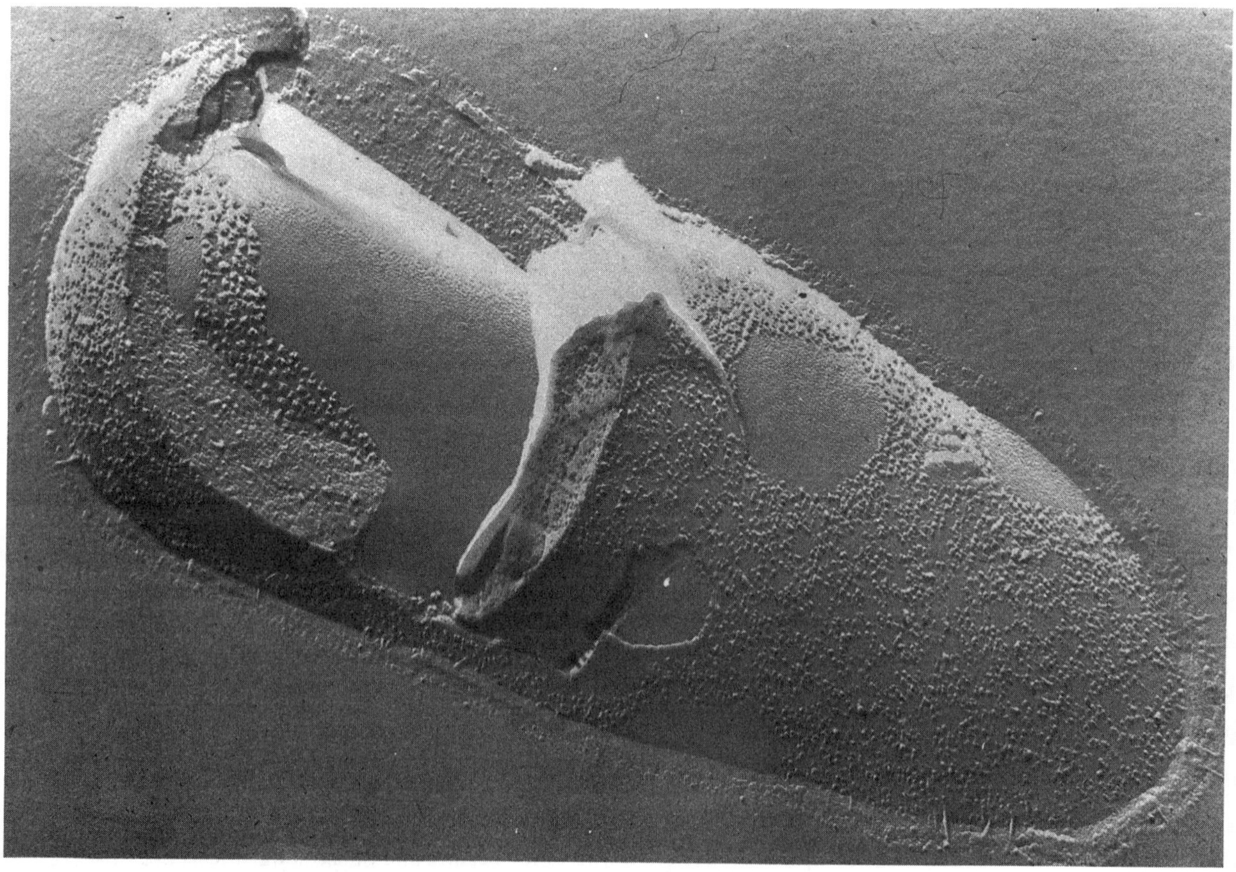

Fig. 8.69 Electron micrograph of a freeze–fracture replica prepared from *Anacystis* cells cultured at 38 °C and thermally quenched from 15 °C. The particle-free domains of the plasma membrane and photosynthetic membrane are believed to represent domains of gel-phase lipid from which the remaining fluid-phase lipid and protein have been excluded.

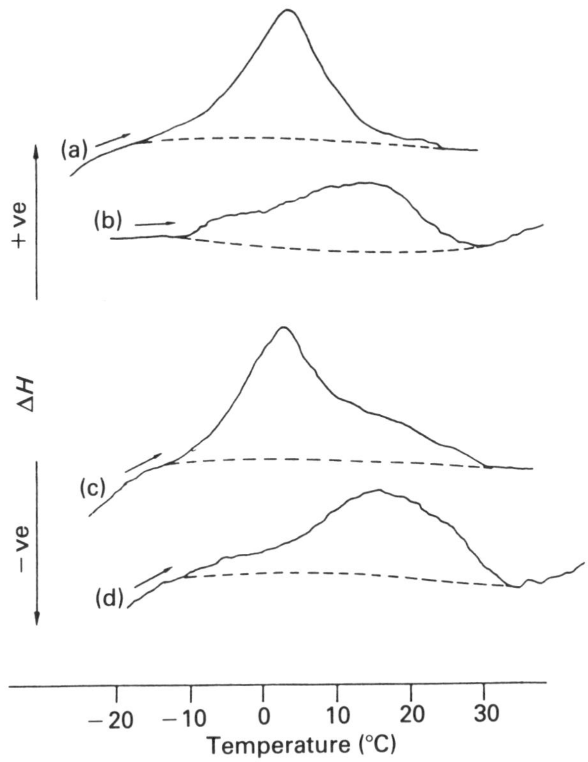

Fig. 8.70 Differential scanning calorimetric heating curves of total polar lipid extracts of *Anacystis nidulans* cultured at (a) 28 °C and (c) 38 °C. Thermograms of whole cells grown at (b) 28 °C and (d) 38 °C are also shown. Data from Mannock *et al.* (1985).

organisms cultured at 28 and 38 °C, which occur at 2 and 4 °C respectively. This observation clearly shows that it is the high-melting-point lipids that phase-separate from the membrane proteins in the intact biological membrane. The question remains as to which lipids have higher melting points and which lipids have low melting points. This has important implications with regard to the ability of the membrane to restore a random distribution of components after thermally induced phase separations. It is well known, for example, that the phase separations of the type observed in *Anacystis nidulans* result in irreversible changes and loss in viability of the cells, which are unable to repair their leaky membranes (Brand *et al.*, 1979).

It is clear that the lipid composition of *Anacystis* is relatively simple, and appreciable proportions of the membrane lipids will undergo phase transitions over a relatively narrow range of temperatures. With more complex mixtures, typical of many biological membranes, the transition endotherms observed on heating membranes previously cooled to low temperatures exhibit transitions that are invariably broad and extend over tens of degrees. Attempts have been made to resolve these broad transitions in erythrocyte membranes and lipid mixtures simulating the membrane lipid composition into components contributed to by each of the major lipid classes present in the membrane (Van Dijck *et al.*, 1976). The results of these experiments using human erythrocyte phospholipids have shown that the phase separation of the sphingolipids dominated the higher-temperature phase behaviour of the outer leaflet of the membrane, and the phosphatidylethanolamines appear to phase-separate initially in the inner leaflet upon cooling the membrane.

Other mechanisms that are responsible for the creation of particle-free domains in membranes have been described. Particle-free domains in the plasma membrane of rye protoplasts, for example, have been shown to be associated with freeze-induced dehydration injury to these cells (Steponkus *et al.*, 1990; Langis and Steponkus, 1991). It is thought that close apposition of the plasma membrane with the underlying subcellular membranes causes the exclusion of intrinsic membrane proteins. Irreversible injury to the cells is sustained when the phase-separated lipid forms a non-lamellar phase consisting of aggregates of tubular inverted micelles of lipid.

A lipid phase separation model has been proposed to account for the irreversible damage to membranes when cells are exposed to low temperatures, which provides a molecular explanation for the consequences of phase separation of components within the membrane (Quinn, 1985). The model takes into account the phase behaviour of membrane lipids and the likely events that occur during cooling and subsequent reheating of the membrane. Figure 8.71 is a schematic illustration of the molecular redistributions that probably take place during these processes. Thus, when membranes are cooled from the growth temperature, the first transition to which the membrane lipids will be subjected will be a hexagonal-II to liquid-crystalline lamellar phase transition of those lipids which tend to form non-bilayer phases at the growth temperature. It has been shown that intrinsic membrane proteins impose a lamellar configuration on these lipids at the growth temperature (Taraschi *et al.*, 1982) consistent with the role of these lipids in packaging of proteins and sealing them into the lipid bilayer matrix. When this constraint is no longer imposed, and in the absence of any specific interactions between the proteins and the non-lamellar lipids, it may be predicted that the lipids will diffuse to form a homogeneous fluid bilayer of lipids in the absence of any small component of lipid that is below the gel to liquid-crystalline phase transition temperature. Such lipids may form small domains of gel-phase lipid in the fluid membrane matrix.

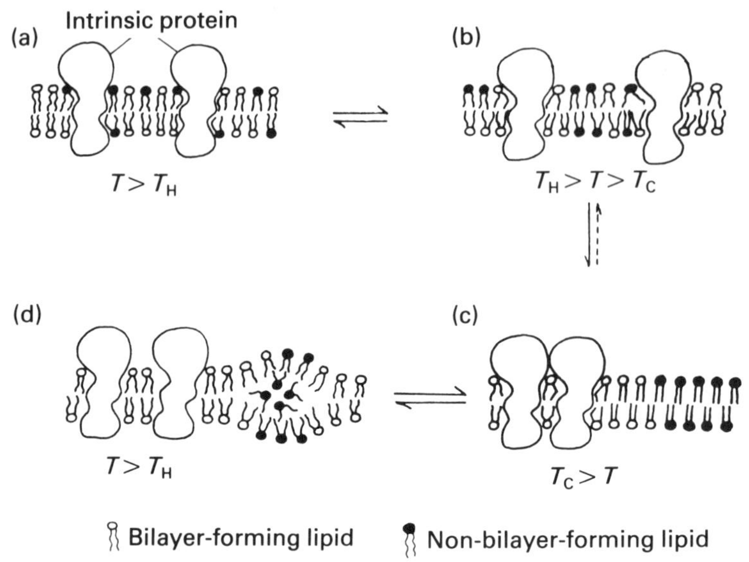

Fig. 8.71 Schematic representation of the phase separations in biological membranes associated with cooling from the growth temperature (a) to a temperature below the hexagonal-II to liquid-crystalline phase transition temperature of the non-bilayer-forming lipids (b) and subsequently below the gel to liquid-crystalline phase transition of all the membrane lipids (c). The effect of reheating to the growth temperature is shown in (d).

There is, as yet, no direct information on the extent to which bilayer- and non-bilayer-forming lipids may be heterogeneously distributed within membranes, but, where this is due to a tendency of the particular lipids to form non-bilayer structures, this constraint would be expected to be removed on cooling below the hexagonal-II to lamellar phase transition temperature. With the possible exception of membranes subjected to chilling damage, such as those described above for *Anacystis nidulans*, this type of phase separation is not likely to cause irreversible damage to membranes, and any heterogeneity may be restored on reheating to the growth temperature. Further cooling results in transitions from liquid-crystalline to gel phase beginning first with the higher-melting-point lipids, which, as we have seen, will segregate into domains of pure gel-phase lipid from which the intrinsic membrane proteins and low-melting-point lipids are excluded.

The central argument of the lipid phase separation model of irreversible damage to biological membranes is that the higher-melting point lipids present in gel-phase domains will be rich in the hexagonal-II-forming lipids. This assumption is based on the fact that, with equivalent hydrocarbon substituent, lipids that tend to form hexagonal-II structures have liquid-crystalline to gel-phase transitions at temperatures that are considerably higher than those corresponding to bilayer-forming lipids (see Quinn, 1985). Where phase separations between lipid classes of this type are created, the changes are not likely to be reversed on reheating to temperatures above the hexagonal-II to lamellar phase transition temperature. In practice, it would be improbable for there to be a single temperature where all the non-lamellar lipids would coexist together with the bilayer-forming lipids in the fluid phase, allowing the original distribution of the lipids and the proteins to be restored. Damage to the membrane in these circumstances would be expected to result when the membrane is reheated to temperatures where the domains of the phase-separated hexagonal-II-rich lipids tend to form non-bilayer structures. The creation of non-bilayer structures such as inverted lipid micelles, unless suppressed or dealt with by normal homoeostatic mechanisms operating within the membrane, would serve to destroy the permeability barrier properties of the membrane. If this breakdown of membrane barrier was of sufficient duration to permit loss of essential components or irreversible alterations in the intracellular compartmentation, loss of cell viability would result.

Phase separations of the constituents of biological membranes can also be driven by exposure of membranes to high temperatures. The structural changes in thylakoid membranes of higher plant chloroplasts subjected to thermal stress have been examined in detail by freeze–fracture electron microscopy by Gounaris *et al.* (1983). It was reported that a normal morphology of chloroplasts was preserved during brief exposure (5 min) of suspensions to temperatures up to 35 °C. Incubation at temperatures of between 35 and 45 °C caused complete destacking of the grana, and incubation at higher temperatures caused a phase separation of non-bilayer lipids into stable aggregates of cylindrical inverted micelles. Interpretation of the effects of temperatures greater than 45 °C is based on phase conditions that result in a release of the constraints imposed by interaction of the major non-bilayer-forming lipid of the membrane, monogalactosyldiacylglycerol, with other membrane components and its segregation into domains of non-bilayer lipid structure. Gross phase separations of this type require that the shift in thermal stability of the stacked membrane is relatively large because three-dimensional aggregates of lipid are not observed if the chloroplast membrane is destacked by manipulation of the ionic environment before heat treatment.

A careful analysis of the ultrastructural changes associated with temperature-induced destacking of thylakoid membranes has been carried out using electron microscopy (Gounaris *et al.*, 1984), where it was shown that normal granal stacks are progressively replaced by modified thylakoid attachment sites in which contact between opposing membranes becomes restricted to regions of focal contact. Changes in the distribution of particles observed in the exoplasmic face of stacked and unstacked regions and corresponding regions of the protoplasmic face due to heat treatment reveal a disappearance of the intramembranous particles in the exoplasmic profiles of the attachment sites formed during exposure to high temperature. The reverse occurs outside the stacked regions of the exoplasmic face, where the density of the particles increases significantly. Changes in particle size and regions of contact between membranes in profiles of exposed exoplasmic and protoplasmic fracture faces were interpreted as a dissociation of light-harvesting units of photosystem-II in which the antennae complexes cluster together, maintaining regions of membrane adhesion, whilst excluding the core complexes of photosystem-II and light-harvesting units of photosystem-I. Accordingly, the electrostatic interactions that maintain contact between the membranes are not perturbed by the brief exposure to high temperature, but a shift in the phase behaviour of the membrane lipids leads to dissociation of oligomeric complexes. This is consistent with functional changes observed by Schreiber and Armond (1978) in photosystem-II of chloroplasts subjected to heat stress.

It has been suggested that a shift in the phase of the non-bilayer lipid, monogalactosyldiacylglycerol, underlies these structural changes and that the functional role of this non-bilayer lipid may be to package the light-harvesting chlorophyll a/b–protein complexes together with the photosystem-II core protein complex into an efficient functional unit localized within the grana stack (Quinn and Williams, 1983). A similar functional role of non-bilayer lipids including phosphatidylethanolamines and monogalactosyldiacylglycerols has been proposed on the basis of reconstitution studies of Ca^{2+}-ATPase in which the efficiency of pumping is markedly improved by the presence of non-bilayer-forming lipids (Navarro *et al.*, 1984).

Lipid phase separations in biological membranes can also be induced by screening membrane surface charges or by removal of acidic lipids from the membrane. It has been reported, for example, that treatment of mitochondria with high concentrations of manganese (van Venetie and Verkleij, 1982) and chloroplasts with high concentrations of magnesium (Carter and Staehelin, 1980) causes phase separation of hexagonal-II lipid structures within these membranes. Similar types of phase separation have been observed when chloroplasts are exposed briefly to low pH (pH 4.5, 5 min) or subjected to digestion with phospholipase A_2 (Thomas *et al.*, 1985). Figure 8.72 shows an electron micrograph of a freeze–fracture replica prepared from a suspension of pea chloroplasts exposed to pH 4.5, where a large domain of non-lamellar lipid from which the protein appears to be excluded can be seen embedded in the stacked thylakoid region. It should be emphasized that phase separations such as those observed in this figure only appear in stacked chloroplasts. No non-bilayer lipid aggregates are found in chloroplasts treated with phospholipase A_2, for example, under conditions in which stacking is prevented. This suggests that the formation of such structures is dependent on the existence of two or more bilayers containing 'destabilized' non-bilayer lipids in close proximity to each other.

A similar conclusion has been drawn from studies of freeze-stressed protoplasts (Gordon-Kam and Steponkus, 1984). Failure to observe such aggregates does not necessarily imply the absence of any phase separation of non-bilayer lipids in unstacked membranes. It is quite possible that phase separations do occur but involve smaller, more transient, non-bilayer structures such as inverted micelles sandwiched within the bilayer. All these effects are interpreted as a removal of the influence of the acidic lipids in imposing a bilayer phase on non-bilayer-forming lipids of the biological membrane.

8.13.4 Function of lipids in membranes

Some of the functions of the lipid component of biological membranes have been mentioned in previous sections. These include the formation of the matrix in which the various functional membrane

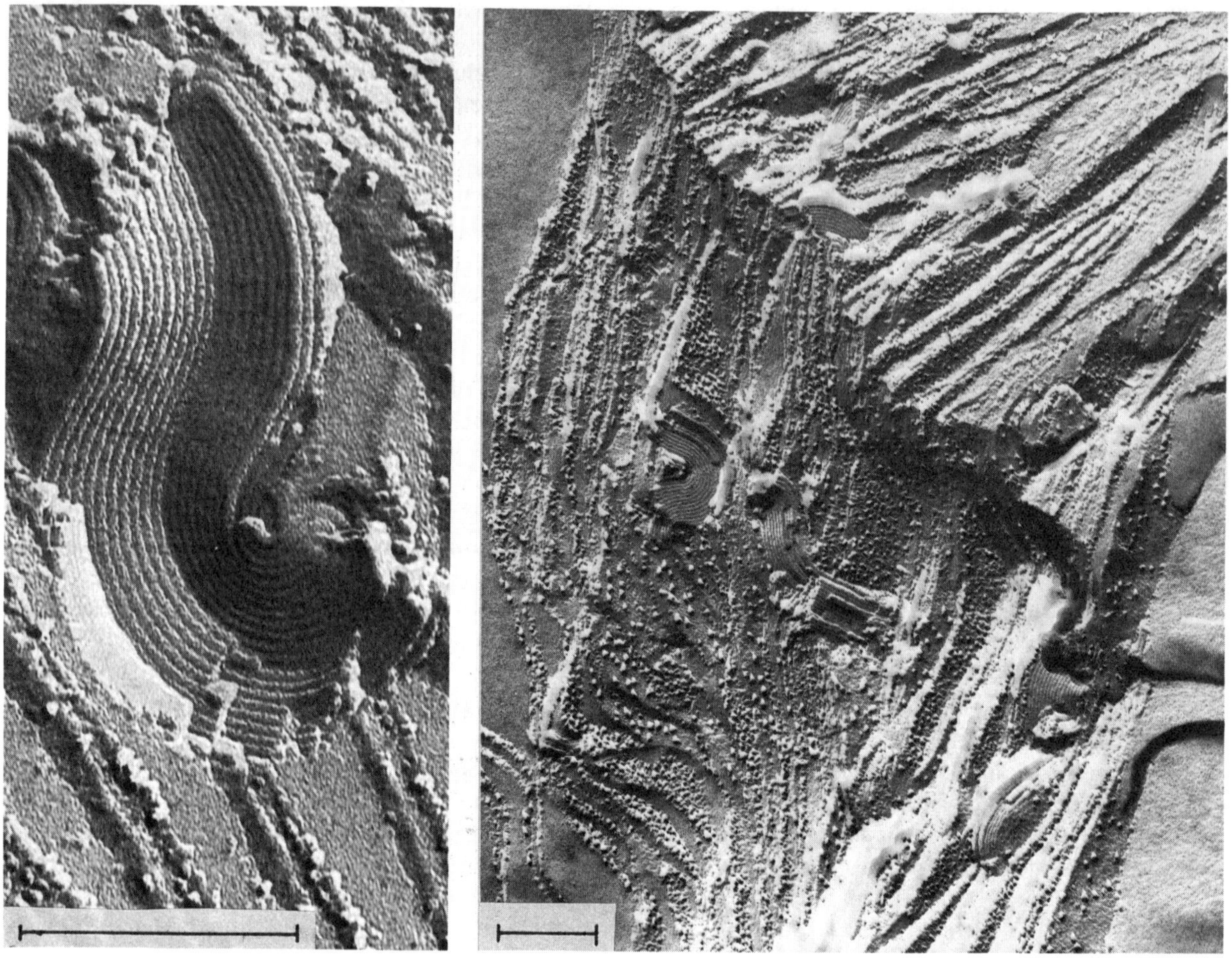

Fig. 8.72 Electron micrographs of a freeze–fracture replica prepared from pea chloroplasts exposed to a medium of pH 4.5 for 5 min before being restored to pH 7.6 prior to thermal quenching. From Thomas *et al.* (1985). Bars represent 200 nm.

proteins are supported, the creation of a passive barrier to the diffusion of solutes, and the packaging of intrinsic proteins into oligomeric complexes. Additional functions are involvement in membrane fusion processes, transmembrane signal transduction and as a reservoir of precursors of mediator agents such as platelet-activating factor, prostaglandins and related compounds (see Chapter 11).

(a) Membrane fusion

The fusion between membranes is required in secretion of products from cells and is associated with the biogenesis and differentiation of membranes during cell growth. All such fusions involve cytoplasmic membrane surfaces. Another type of fusion is that between cells or cells and enveloped viruses, where initial contact is established on the outer surface of the plasma membrane. The process of endocytosis, where plasma membrane is internalized within the cell, also falls into this category of membrane fusion.

The process of membrane fusion involving cytoplasmic surfaces may differ because the presence of carbohydrates, representing the so-called glycocalyx on the outer plasma membrane surfaces, impedes close apposition of the fusing membranes. Nevertheless, a perturbation of the bilayer arrangement of lipids is believed to be associated with all fusion events irrespective of the surfaces involved. Model systems consisting of liposomes have been extensively employed to simulate the fusion process. Using these methods, the sequence of fusion has been shown to begin with the approach of membrane surfaces, which establish close contact in the region where fusion takes place. This is associated with a partial dehydration of the surface and, in multicomponent systems, there may be phase separations in which certain membrane constituents are segregated into the domains of close contact. These constituents may be 'fusogenic molecules' including lipids that prefer non-lamellar configurations. Charged membrane constituents, on the other hand, may be excluded from such domains by electrostatic repulsion.

Studies combining assays of leakage of vesicle contents and mixing of contents of two vesicle populations with biophysical methods able to detect structural changes in the vesicle membrane have been undertaken to characterize the fusion process (Ellens *et al.*, 1989). It was found that fusion between the vesicles was correlated with the appearance of a small component of the membrane phospholipid that experiences isotropic motion in an overall transition from a bilayer membrane to a hexagonal phase in which all the contents of the original vesicles leak to the extravesicular medium. This isotropic component is believed to represent molecules participating in the intermembrane attachment at the fusion site.

The entry of certain viruses into cells involves fusion between the viral envelope and the plasma membrane of the target cell. Fusion in this case is known to be mediated by specific, membrane-spanning fusion proteins (Stegmann *et al.*, 1989). Of these fusion proteins, the influenza virus haemagglutinins have been most extensively characterized. Several haemagglutinins have been sequenced and the crystal structure of the major component of one haemagglutinin has been determined. The way in which haemagglutinin causes lipid bilayers to fuse is not yet understood despite a detailed knowledge of the conditions required to achieve fusion and the rearrangements in the haemagglutinin that accompany the fusion event. Patch clamp studies of fusion between erythrocyte ghosts and fusion-competent fibroblasts together with fluorescence microscopic assay of cytoplasmic mixing have been reported by Spruce *et al.* (1989). They showed that the earliest recorded event was the sudden opening of an aqueous pore connecting the cytoplasms of the two cells. In the initial stages a fluctuation in pore conductance was observed, which indicated that the pore was undergoing cycles of conductance and occlusion. The size of the pore was estimated to be about the same as proteinaceous channels formed by gap junction complexes between cells or that formed by the terminal components of complement factors. The pore eventually dilates over a period of tens of seconds: the constraints on expansion were thought to be due to the interaction of the proteins of the erythrocyte membrane with the underlying cytoskeleton.

(b) Receptor-linked phospholipid hydrolysis

A variety of mechanisms are known to be involved in the transduction of signals across cell membranes. These include the facilitated diffusion of solutes and ions through gated channels, modulation of adenylate cyclase activity, activation of protein kinases and activation of phospholipases to enhance turnover of specific membrane phospholipids. The hydrolysis of a relatively minor phospholipid, phosphatidylinositol-4,5-bisphosphate, is known to be directly linked to binding of a variety of hormones and neurotransmitters to receptors located in the plasma membrane (Berridge, 1987; Michell, 1988; see also section 11.2.1). Indeed, about half of all known agonists act in this way (see Table 8.31). The enzyme responsible is phospholipase C activated by a specific G-protein. The two hydrolytic products are a water-soluble inositol-1,4-5-trisphosphate (IP_3) and diacylglycerol, which remains associated with the membrane. Both products have second messenger functions. The structure of the phospholipid and cleavage products is illustrated in Fig. 8.73.

The second messenger function of IP_3 is to release Ca^{2+} from subcellular storage compartments such as the endoplasmic reticulum. It does so by binding with high affinity to a calcium channel, causing facilitated diffusion of Ca^{2+} down a concentration gradient maintained by a calmodulin-sensitive Ca^{2+}-ATPase. IP_3 concentrations are subsequently reduced by a stepwise dephosphorylation mediated by specific phosphatases, although removal of the first phosphate at position 5 by a phosphatase partly associated with the plasma membrane is sufficient to abolish its channel-opening role. An alternative pathway is the further phosphorylation of IP_3 catalysed by an ATP-dependent kinase, which forms inositol-1,3,4,5-tetrakisphosphate. It has been shown that this compound can facilitate diffusion of Ca^{2+} across the plasma membrane, resulting in a further increase in cytoplasmic calcium concentration. The metabolic transformations of inositol phosphates are illustrated in Fig. 8.74. It can be seen that eventually complete dephosphorylation of inositol phosphates provides *myo*-inositol, which can be resynthesized to inositol phospholipids.

The diacylglycerol product of inositol phospholipid hydrolysis by phospholipase C causes a change in the physical properties of the lipid domain of the plasma membrane, resulting in activation of protein kinase C. These changes in membrane lipid phase behaviour are also known to promote fusion between membranes. The action of diacylglycerol is switched off by further metabolism. One pathway is catalysed by a diacylglycerol kinase employing a substrate of ATP, yielding phosphatidic acid, which is a key phospholipid involved in phospholipid biosynthesis including resynthesis of inositol phospholipids. Phosphatidic acid also effectively reverses the change induced by

Table 8.31 Receptor-stimulated inositol phosphatide turnover agonists

Agonist/receptor	Responsive cell types
α-Adrenergic	Liver, muscle, central nervous system, fat cells
Muscarinic–cholinergic	Smooth muscle, central nervous system, heart, pancreatic islets, sympathetic ganglia, chromaffin cells, avian salt gland
Histaminergic	Central nervous system, astrocytoma cells, chromaffin cells
Serotoninergic	Central nervous system, platelets
Platelet-activating factor	Platelets, liver, exocrine glands
Glutaminergic	Oocytes
Purinergic/ATP	Liver, endothelial cells
Glucose	Pancreatic islets, yeast
Hypothalamic-releasing hormone	Pituitary
Intestinal peptide hormones	Pancreas
Peptide hormones:	
Vasopressin	Liver, vascular smooth muscle, sympathetic ganglia, platelets, fibroblasts
Angiotensin II	Liver, vascular smooth muscle, anterior pituitary, kidney
Bradykinin	Fibroblasts, carcinoma cells
Substance P	Smooth muscle, nerve cells, parotid
Growth factors	3T3 tissue culture cells
Thrombin/collagen	Platelets
Antigens	Mast cells, macrophages
Sperm	Oocytes
Light	Photoreceptors

diacylglycerol on membrane lipid phase behaviour. An alternative pathway is mediated by a lipolysis reaction in which arachidonic acid, present in high proportions at the C-2 position of the diacylglycerol, is released.

Many agonists that stimulate turnover of inositol phospholipids also promote enhanced turnover of the more ubiquitous choline phospholipids (Pelech and Vance, 1989). Phosphatidylcholine turnover cycles can also generate second messengers such as diacylglycerol and arachidonic acid. GTP-binding proteins have been implicated in ligand–receptor activation of phospholipases A_2, C and D, which catalyse the degradation of phospholipid. Many agonists also appear to stimulate pathways of phosphatidylcholine biosynthesis via an effect on CDP-choline synthesis.

(c) Lipid mediators, hormones and receptors

As noted above, essential fatty acids such as arachidonic acid are constituents of membrane lipids, where they serve as precursors for synthesis of a variety of lipid mediators (see section 11.1.7). Other lipid precursors that have been shown to be released as a consequence of stimulation by specific agonists include platelet-activating factor, lysophospholipids and diacylglycerols (Hanahan *et al.*, 1980; Vargaftig *et al.*, 1981; Prescott and Majerus, 1981; Mahadevapa and Hollub, 1983). Further metabolism of arachidonic acid released from complex membrane lipids by the action of phospholipase A_2 enzymes takes place in an integrated series of reactions to produce the eicosanoid group of compounds, including the prostaglandins, thromboxanes and leukotrienes. These compounds all have potent effects on white blood cells and are involved in inflammatory and hypersensitivity reactions.

Gangliosides displayed on the outer surface of the plasma membrane act as receptors for certain toxins including cholera toxin. The B-subunit of the toxin binds to the ganglioside, resulting in dissociation of the B-subunit from the A-subunit, which then locates in the membrane. Activation of the A-subunit takes

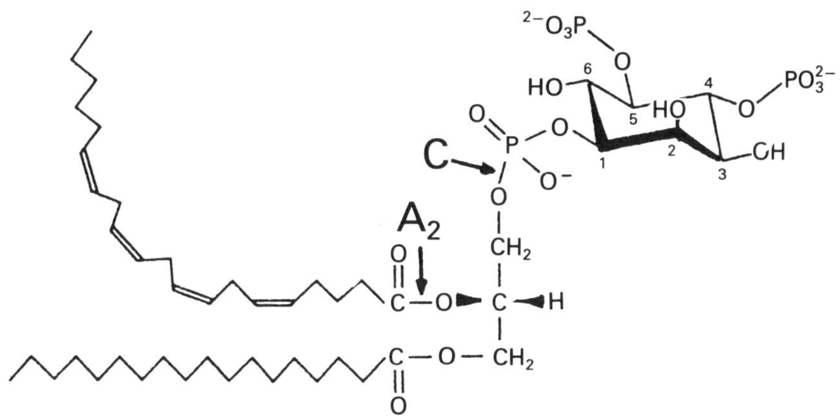

Fig. 8.73 Structure of phosphatidylinositol-4,5-bisphosphate, showing the sites of phospholipase A_2 and C hydrolysis. The structure shown is the 1-stearoyl, 2-arachidoyl molecular species.

place by formation of A_1 and A_2 components. The A_1 component catalyses the ADP-ribosylation by NAD of the GTP-bound α-subunit of G-protein, bringing about an irreversible dissociation from the β-complex and uncontrolled activation of adenylate cyclase. This leads to enhanced water and salt secretion into the gut and dehydration, which is the classical manifestation of infection.

Membrane lipids can act as targets for certain toxins. Many venoms, for example, contain phospholipases and agents that increase the susceptibility of membrane phospholipids to attack by these phospholipases. Sterols may also represent targets for toxins. These include thiol-activated cytolysins from streptococci (e.g. streptolysin O), antifungal polyenoic agents such as filipin, ampotericin, etc., and plant alkaloids like the saponins. It appears that in each case there is formation of toxin–sterol complexes,

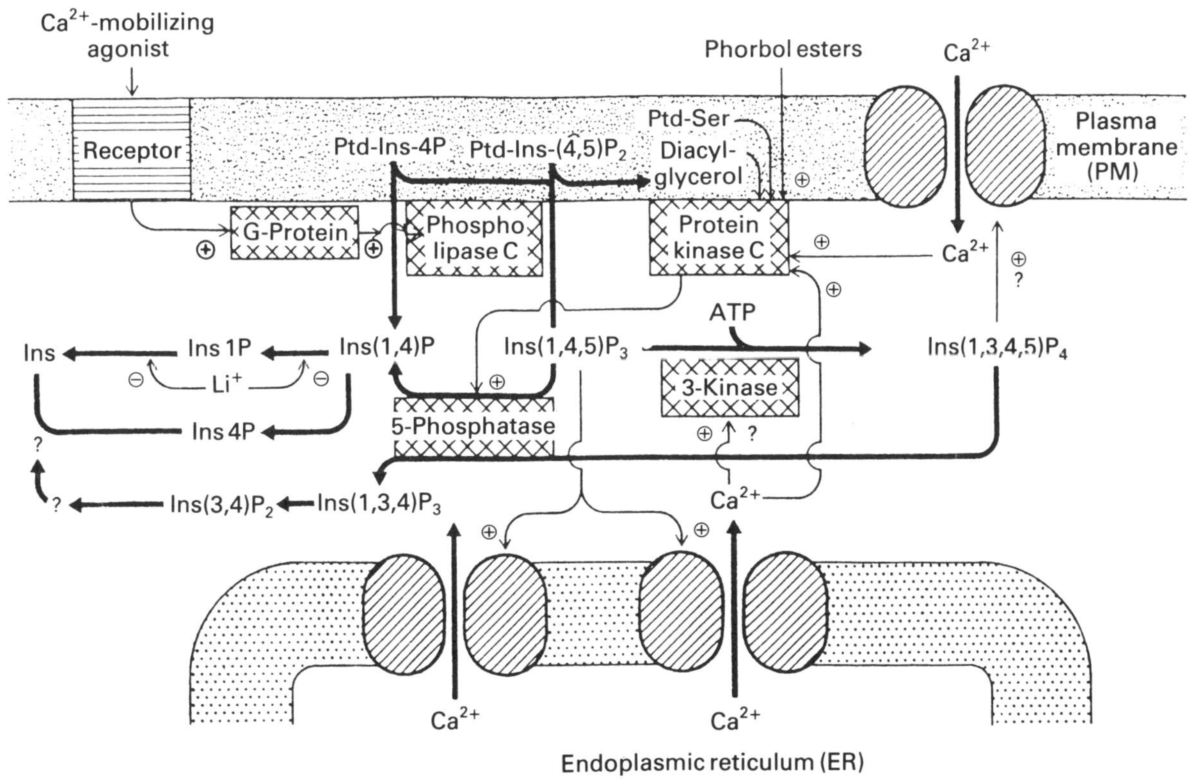

Fig. 8.74 Schematic diagram of cellular responses triggered by receptor-mediated hydrolysis of phosphatidylinositol-4,5-bisphosphate in the plasma membrane.

which create water-filled channels in the membrane, leading to a breakdown in permeability barrier properties.

Membrane lipids also act as cell surface antigens. The most notable of these are the blood group antigens. Gangliosides present in the erythrocyte membrane contain the determinants for ABO blood groups, although the specific oligosaccharide representing the determinant is also a component of a glycoprotein.

References and further reading

Allan, D. and Quinn, P. (1989) *Biochim. Biophys. Acta*, **987**, 199.

Arvidsson, G., Breutel, I., Khan, A., Lindblom, G. and Fontell, K. (1985) *Eur. J. Biochem.*, **152**, 753.

Berridge, M.J. (1987) *Annu. Rev. Biochem.*, **56**, 159.

Beth, A.H. and Robinson, B.H. (1989) *Biol. Magn. Reson.*, **8**, 179.

Birmachu, W. and Thomas, D.D. (1990) *Biochemistry*, **29**, 3904.

Blasie, J.K., Pascolini, D., Asturias, F., Herbette, L.G., Pierce, D. and Scarpa, A. (1990) *Biophys. J.*, **58**, 687.

Bloom, M. and Smith, I.C.P. (1985) in *Progress in Protein–Lipid Interactions*, vol. 1 (eds A. Watts and J.J.H.H.M. de Pont), Elsevier, Amsterdam, p. 61.

Brand, J.J., Kirchanski, S.J. and Ramirez-Mitchell, R. (1979) *Planta*, **145**, 63.

Burnell, E., van Alphen, L., Verkleij, A. and De Kruijff, B. (1980) *Biochim. Biophys. Acta*, **597**, 492.

Carter, D.P. and Staehelin, L.A. (1980) *Arch. Biochem. Biophys.*, **200**, 374.

Ceska, T.A. and Henderson, R. (1990) *J. Mol. Biol.*, **213**, 539.

Chapman, D., Williams, R.M. and Ladbroke, D.D. (1967) *Chem. Phys. Lipids*, **1**, 445.

Cherry, R.J. (1978) *Methods Enzymol.*, **54**, 47.

Connor, J. and Schroit, A.J. (1990) *Biochemistry*, **29**, 37.

Cullis, P.R. and De Kruijff, B. (1979) *Biochim. Biophys. Acta*, **559**, 399.

Cullis, P.R., Verkleij, A.J. and Ververgaert, P.H.J.T. (1978) *Biochim. Biophys. Acta*, **513**, 11.

Davoust, J., Deveaux, P.F. and Leger, L. (1982) *EMBO J.*, **1**, 1233.

de Bony, J., Lopez, A., Gilleron, M., Welby, M., Lanelle, G., Rousseau, B., Beaucourt, J.P. and Tocanne, J.F. (1989) *Biochemistry*, **28**, 3728.

Deveaux, P.F. (1988) *FEBS Lett.*, **234**, 8.

Ellens, H., Siegel, D., Alford, D., Yeagle, P., Boni, L., Lis, L.J., Quinn, P.J. and Bentz, J. (1989) *Biochemistry*, **28**, 3695.

Engelman, D.M. (1969) *Nature*, **223**, 1279.

Esser, A.F. and Lanyi, J.K. (1973) *Biochemistry*, **12**, 1933.

Finean, J.B. and Michell, R.H. (eds) (1981) *Membrane Structure*, Elsevier, Amsterdam.

Frey-Wyssling, A. and Schwegler, F. (1965) *J. Ultrastruct. Res.*, **13**, 543.

Funk, J., Wunderlich, F. and Kreutz, W. (1982) *J. Mol. Biol.*, **161**, 561.

Furtado, D., Williams, W.P., Brain, A.P.R. and Quinn, P.J. (1979) *Biochim. Biophys. Acta*, **555**, 352.

Gordon-Kam, W.J. and Steponkus, P.L. (1984) *Proc. Natl Acad. Sci. USA*, **81**, 6373.

Gorter, E. and Grendel, F. (1925) *J. Exp. Med.*, **41**, 439.

Gounaris, K., Brain, A.P.R., Quinn, P.J. and Williams, W.P. (1983) *FEBS Lett.*, **153**, 47.

Gounaris, K., Brain, A.P.R., Quinn, P.J. and Williams, W.P. (1984) *Biochim. Biophys. Acta*, **766**, 198.

Gounaris, K. (1986) in *Ion Interactions and Energy Transfer in Biomembranes, Proc. Int. Workshop* (eds G.C. Papageorgiou, J. Barber and S. Papa), Plenum, New York, p. 251.

Haest, C.W.H. (1982) *Biochim. Biophys. Acta*, **694**, 331.

Hanahan, D.J., Demopoulos, C.A., Liehr, J. and Prinkard, R.N. (1980) *J. Biol. Chem.*, **255**, 5514.

Harbich, W., Servuss, M. and Helfrich, W. (1978) *Z. Naturforsch.*, **33a**, 1013.

Hemminga, M.A. and Cullis, P.R. (1982) *J. Magn. Reson.*, **47**, 307.

Hemminga, M.A. and de Jager, P.A. (1989) *Biol. Magn. Reson.*, **8**, 131.

Herbette, L., DeFour, P., Fleischer, S., Pascolini, D., Scarba, A. and Blasie, J.K. (1985) *Biochim. Biophys. Acta*, **817**, 103.

Hoffman, J.F. and Giebisch, G. (eds) (1970–) *Current Topics in Membranes and Transport* (annual series), Academic Press, New York.

Jamieson, G.A. and Robinson, D.M. (eds) (1976) *Mammalian Cell Membranes* (several volumes), Butterworths, London.

Karlsson, K.-A. (1978) *Prog. Chem. Fats Other Lipids*, **16**, 207.

Killian, J.A., Fabrie, C.H.J.P., Baart, W., Morein, S. and De Kruijff, B. (1992) *Biochim. Biophys. Acta*, **1105**, 253.

Koppel, D.E. (1979) *Biophys. J.*, **28**, 281.

Koppel, D.E. and Sheetz, M.P. (1983) *Biophys. J.*, **43**, 175.

Koppel, D.E., Primakoff, P. and Myles, D.G. (1986) in *Applications of Fluorescence in Biomedical Sciences* (eds D.L. Taylor, A.S. Waggoner, R.F. Murphy, F. Lanin and R.R. Birge), Alan R. Liss, New York, p. 477.

Lakowicz, J.R. (1983) *Principles of Fluorescence Spectroscopy*, Plenum, New York.

Langis, R. and Steponkus, P.L. (1991) *Cryo-Lett.*, **12**, 107.

Larsson, K. (1989) *J. Phys. Chem.*, **93**, 7304.

Levine, Y.K. and Wilkins, M.H.F. (1971) *Nature, New Biol.*, **230**, 72.

Lindblom, G. and Rilfors, L. (1989) *Biochim. Biophys. Acta*, **988**, 211.

Lugtenberg, B. and van Alphen, L. (1983) *Biochim. Biophys. Acta*, **737**, 51.

Luzzati, V. (1968) in *Biological Membranes* (ed. D. Chapman), Academic Press, London, p. 71.

Mahadevapa, V.G. and Hollub, B.J. (1983) *J. Biol. Chem.*, **258**, 5337.

Mannock, D.A., Brain, A.P.R. and Williams, W.P. (1985) *Biochim. Biophys. Acta*, **821**, 153.

Manson, L.A. (1971–) *Biomembranes* (series), Plenum, New York.

Marsh, D. (1988) in *Physical Properties of Biological Membranes and Their Functional Implications*, vol. 4 (ed. C. Hidalgo), Plenum, New York, p. 123.

Martonosi, A. (ed.) (1976) *The Enzymes of Biological Membranes*, Plenum, New York.

McLaughlin, A.C., Cullis, P.R., Hemminga, M.A., Hoult, D.I., Radda, G.K., Ritchei, G.A., Seeley, P.J. and Richards, R.E. (1975) *FEBS Lett.*, **57**, 213.

Metcalf, T.N., Wang, J.L. and Schindler, M. (1986) *Proc. Natl Acad. Sci. USA*, **83**, 95.

Michell, R.H. (1988) *Br. Med. J.*, **295**, 1320.

Moor, H., Mühlethaler, K., Waldner, H. and Frey-Wyssling, A. (1961) *J. Biophys. Biochem. Cytol.*, **10**, 1.

Navarro, J., Tovio-Kunnuncan, M. and Racker, E. (1984) *Biochemistry*, **23**, 130.

Nothnagel, E.A. (1989) *Biochim. Biophys. Acta*, **980**, 209.

Op den Kamp, J.A.F. (1979) *Annu. Rev. Biochem.*, **48**, 47.

Packard, B.S., Saxton, M.J., Bissell, M.J. and Klein, M.P. (1984) *Proc. Natl Acad. Sci. USA*, **81**, 449.

Pearse, B.M.F. and Bretscher, M.S. (1981) *Annu. Rev. Biochem.*, **50**, 85.

Pelech, S. and Vance, D.E. (1989) *Trans. Biochem. Soc.*, **14**, 28.

Prescott, S.M. and Majerus, P.W. (1981) *J. Biol. Chem.*, **256**, 579.

Quinn, P.J. (1985) *Cryobiology*, **22**, 28.

Quinn, P.J. and Williams, W.P. (1983) *Biochim. Biophys. Acta*, **737**, 223.

Quinn, P.J., Brain, A.P.R., Stewart, L.C. and Kates, M. (1986) *Biochim. Biophys. Acta*, **863**, 213.

Rosing, J., Speijer, H. and Zwaal, R.F.A. (1988) *Biochemistry*, **27**, 8.

Schreiber, U. and Armond, P.A. (1978) *Biochim. Biophys. Acta*, **502**, 138.

Schroit, A.J., Madsen, J.W. and Tanaka, Y. (1985) *J. Biol. Chem.*, **260**, 5131.

Seelig, J. (1977) *Q. Rev. Biophys.*, **10**, 353.

Seelig, J. and Gally, H.U. (1976) *Biochemistry*, **15**, 5199.

Seelig, J. and Niederberger, W. (1974) *J. Am. Chem. Soc.*, **96**, 2069.

Seigneuret, M. and Deveaux, P.F. (1984) *Proc. Natl Acad. Sci. USA*, **87**, 3751.

Sen, A., Williams, W.P., Brain, A.P.R. and Quinn, P.J. (1982) *Biochim. Biophys. Acta*, **685**, 297.

Singer, S.J. and Nicolson, G.L. (1972) *Science*, **175**, 720.

Sjöstrand, F.S. (1983) in *Membrane Fluidity in Biology*, vol. 1 (ed. R.C. Aloia), Academic Press, London, p. 83.

Spruce, A.E., Iwata, A., White, I.M. and Almers, A. (1989) *Nature*, **342**, 555.

Squier, T.C., Hughes, S.E. and Thomas, D.D. (1988) *J. Biol. Chem.*, **263**, 9162.

Stegman, T., Doms, R. and Helenius, A. (1989) *Annu. Rev. Biophys. Chem.*, **18**, 187.

Steponkus, P.L., Lynch, D.V. and Uemura, M. (1990) *Phil. Trans. R. Soc. B*, **326**, 571.

Stockton, G.W., Polnaszek, C.F., Leitch, L.C., Tulloch, A.P. and Smith, I.C.P. (1974) *Biochem. Biophys. Res. Commun.*, **60**, 844.

Struck, D.K. and Lennarz, W.J. (1980) in *The Biochemistry of Glucoproteins and Proteoglucans* (ed. W.J. Lennarz), Plenum, New York, p. 1.

Struck, D.K. and Pagano, R.E. (1980) *J. Biol. Chem.*, **255**, 5404.

Sugihara, T., Sugihara, K. and Hebbel, R.P. (1992) *Biochim. Biophys. Acta*, **1103**, 303.

Taraschi, T.F., De Kruijff, B., Verkleij, A.J. and van Echfeld, C.J.A. (1982) *Biochim. Biophys. Acta*, **685**, 153.

Thomas, D.D. (1986) in *Techniques for the Analysis of Membrane Proteins* (eds R.J. Cherry and I. Ragan), Chapman & Hall, London, p. 377.

Thomas, P.G., Brain, A.P.R., Quinn, P.J. and Williams, W.P. (1985) *FEBS Lett.*, **183**, 161.

Tilcock, C.P.S. and Fisher, D. (1982) *Biochim. Biophys. Acta*, **685**, 340.

Van der Meer, B.W. (1988) *Subcell. Biochem.*, **13**, 1.

Van Dijck, P.W.M., Van Zoelen, E.J.J., Seldevijk, R., van Deenen, L.L.M. and de Gier, J. (1976) *Chem. Phys. Lipids*, **17**, 336.

Van Venetie, R. and Verkleij, A.J. (1982) *Biochim. Biophys. Acta*, **692**, 397.

Vargaftig, B.B., Chignard, M., Benveniste, J., Lefort, J. and Wal, F. (1981) *Ann. NY Acad. Sci.*, **370**, 119.

Watenabe, K., Powell, M.E. and Hakomori, S. (1979) *J. Biol. Chem.*, **254**, 8223.

9

PHYSICAL PROPERTIES: OPTICAL AND SPECTRAL CHARACTERISTICS*

D. CHAPMAN and F.M. GOÑI

Much of our present knowledge of lipids is based on their spectral properties. Some of the well established methods (ultraviolet absorption or X-ray diffraction) continue to give useful information, while a host of new ones, particularly in the fields of magnetic resonance and mass spectroscopy, provide us with a variety of previously unknown insights of lipid structure and dynamics; finally, the incorporation of new elements into traditional techniques (e.g. interferometric optics and computerization to infrared spectroscopy) add new dimensions to research and hence to our knowledge of the field. In this chapter we shall review a series of spectroscopic techniques as applied to lipid research. In each case, we shall briefly consider the physical foundations of the technique, later comment on a few essential points on instruments and experimentation, and then describe some important results, with an emphasis on the more recent ones. Reference will be made in each section to more specific reviews so that the interested reader may look for further details.

9.1 Infrared spectra

Infrared (IR) spectroscopy has been used since the early days of lipid chemistry (Barceló and Bellanato, 1953; Wheeler, 1954) and still finds frequent application in these studies (Fringeli and Günthard, 1981; Amey and Chapman, 1983).

Among the reasons for its popularity we shall mention its non-destructive and non-perturbing character. In addition, spectra can be obtained relatively rapidly, frequently with little or no sample preparation, and both qualitative and quantitative results can be obtained. IR spectroscopy is both sensitive and specific, requires relatively small amounts of sample (in the milligram range) and can respond to subtle changes in molecular environment, particularly polymorphic transitions. There are drawbacks, especially when infrared is compared with ultraviolet spectroscopy: IR absorptions are typically less intense than electronic spectra, thus higher concentrations of sample are required; also IR detectors are often less sensitive than visible or ultraviolet instruments; finally, water absorbs strongly in many of the IR regions where lipid absorption bands are found, although this is only a minor problem for lipids, these being hydrophobic substances. In general, the disadvantages just mentioned may be overcome by the use of the modern Fourier-transform instruments.

9.1.1 Basic principles

Infrared spectra arise from the interaction between matter and electromagnetic radiation of wavelength between about 2 and 50 μm (respectively 5000 and 200 cm^{-1}). The energy associated with such radiations corresponds to transitions between vibrational levels of the ground electronic state of a molecule. The absorbance is determined by Beer's law.

Classically, the interaction of light with matter is described as the induction of dipoles by the light electric field, E:

$$\mu_{\text{ind}} = \alpha E \qquad (9.1)$$

where α is the polarizability and μ_{ind} is the induced dipole moment. Since E is a function that fluctuates

* The authors are grateful to Dr J.L.R. Arrondo, Dr A. Alonso and Prof. F.D. Gunstone for their help in the preparation of this chapter. This work was supported in part by a grant from DGICYT (PB91-0441).

with time, μ_{ind} fluctuates also. The dipole induced by light on a system is called a transition dipole moment. The transition dipole for each pair of states is in a fixed orientation relative to the structure of the molecule; the intensity of a vibrational band will depend on the magnitude of the light-induced dipole. The atoms of a molecule can vibrate in a number of modes, and for absorption of energy to occur it is essential that there should be a change in the dipole moment of a molecule during the vibration. The number of normal modes of vibration for a non-linear molecule is equal to $3n - 6$ (where n = number of atoms in the molecule). Each normal mode of vibration can occur independently of the other modes, and the absorption band for each mode is known as its fundamental band or fundamental frequency. The fundamental frequencies are usually found in the 2–15 μm region. Most of the vibrational modes involve the very complicated motion of all the atoms, and the particular values of the frequencies are, therefore, highly characteristic for a particular molecule. In some cases the mode of vibration is highly localized in the molecule and involves the motions of atoms in a particular group, giving rise to a characteristic group frequency. Empirically it has been shown that these frequencies are almost independent of the structure of the remainder of the molecule, and hence afford a good diagnostic method for the presence of specific groupings (Chapman, 1965). In addition, the molecular environment can modify slightly the characteristic group frequency. These principles will be illustrated in detail in section 9.1.3.

9.1.2 Experimental information

Important changes in instrumentation, particularly the evolution from dispersive to interferometric spectrometers, are responsible for the recent upsurge of infrared spectroscopy in its application to lipids and other biomolecules (Mendelsohn, 1984; Griffiths and de Haseth, 1986). Early infrared work was performed on conventional dispersive instruments; however, several characteristics of these spectrometers prevented their use as routine instruments in the study of biological samples. The conventional apparatus consists of an infrared source whose radiation focuses onto the sample and is dispersed by some optical mechanism, usually a grating. Then the beam goes onto a slit that blocks all but a narrow range of frequencies from reaching the detector; therefore, resolution is dependent on the degree to which the slit is able to distinguish between closely spaced frequencies of light; the narrower the slit, the better the resolution. Hence, to obtain a complete spectrum, the angle of the grating had to be continuously changed with respect to the incident infrared beam; thereby, only one spectral resolution could be measured at a time. The result was a low-energy spectrum because (i) the necessarily short distance from sample to source limited the power of the latter in order to avoid sample heating, and (ii) only a fraction of the infrared beam coming out from the source reached the detector. Spectral acquisition was then very slow and dependent on several mechanical devices that made accuracy not very reliable and sample stability a problem.

Perkin–Elmer introduced in 1975 the first microprocessor-controlled commercial dispersion infrared spectrometer, and advantage was taken of this facility for lipid studies a few years later (Chapman *et al.*, 1980). However, the main problems arising from conventional dispersive spectrometry were only overcome by the use of interferometric methods. Interferometers replace the conventional monochromator and offer several advantages. The Michelson interferometer, still the most commonly used, is able to divide a beam of radiation into two paths and then recombine the two beams produced after a path difference has been introduced. It consists of two mirrors, one stationary (F) and one moving (M) placed at 90°, along with a beam-splitter (B) mounted at a 45° angle from the mirrors. The beam-splitter is a semi-reflecting film that reflects 50% of the incident beam and transmits the other 50%; it is usually made of Ge. When the collimated beam coming out from the source reaches the beam-splitter, half is reflected to the moving mirror and half is transmitted to the fixed mirror. After reflection of the beams, they come together again in the beam-splitter, where 50% is reflected and 50% transmitted, the reflected beam going through the sample before reaching the detector. A monochromatic light will give a cosine wave as the interference pattern as M moves along a fixed distance. Each light frequency will give its own interference pattern and the result will be the interferogram, a function of the path-length difference between the two mirrors. This function in the time domain is then Fourier-transformed into the frequency domain. The Fourier transform is calculated by means of the fast Fourier algorithm developed by Cooley and Tukey (1965), and, after apodization and phase correction, the output is a single-beam spectrum of the sample that is ratioed against a background spectrum (a single-beam spectrum collected with no sample present), giving rise to the intensity versus frequency spectrum.

The advantages of Fourier-transform infrared (FTIR) spectroscopy can be summarized as follows:

1. The detector receives simultaneously all the

frequencies, and a complete spectrum may be obtained in less than 1 s. The interferogram signal is termed 'multiplexed' and the consequence is a greater signal-to-noise ratio in the same spectral acquisition time (Felgett advantage), with the possibility of performing kinetic measurements. This effect is the same as in other Fourier-transform-based techniques, i.e. NMR.

2. The IR beam is incident fully on the sample without the need of a slit to define resolution or to limit the amount of energy reaching the detector. Thus, more energy is admitted to flow through the system, allowing greater sensitivity (Jacquinot advantage) and the use of special sampling techniques such as attenuated total reflection (ATR) or infrared microscopy.
3. FTIR instruments use an internal laser reference to facilitate digitization of data, and also for frequency calibration. Since the laser frequency is known accurately, any measured data point is calibrated by the system to better than 0.001 cm^{-1} (Connes advantage). However, peak positions of observed infrared bands are not measured with such accuracy, for practical reasons.
4. Additional advantages of FTIR spectroscopy include smaller sample heat-up since the sample is well removed from the source, no sample emission or stray light contributions, ease in spectral mathematical treatments, etc.

Fluid samples are examined by FTIR in special sample cells. These cells are limited by a pair of 'windows' made of various, more or less transparent, materials (Table 9.1). In the mid-infrared region there is no clear-cut best choice for window material: CaF_2 has the least reflection loss but is expensive; KBr is the cheapest, has almost as low a reflection loss as CaF_2 but, as with NaCl, it is soluble in water; and AgCl and AgBr both have low water solubility but suffer from high reflection losses. High-density polyethylene is the choice below 300 cm^{-1}.

Specialized techniques such as attenuated total reflection (ATR), polarized IR spectroscopy, etc., are also described by Amey and Chapman (1983). Applications of ATR to biological samples have been reviewed by Braiman and Rothschild (1988) and by Goormaghtigh and Ruysschaert (1990). Another technique, which is potentially useful for lipid studies, is the combination of gas chromatography (GC) or liquid chromatography (LC) with Fourier-transform infrared (FTIR) spectroscopy, so that the column effluents are analysed by an on-line spectrometer. GC/FTIR and LC/FTIR have been reviewed by the CECON Group (1990). The coupling of a micrscope with an infrared instrument is an attractive possibility in order to study lipids in their native environments, although its use is not yet widespread (Kodali *et al.*, 1991).

9.1.3 Characteristic absorption frequencies and spectral compilations

Infared absorption frequencies characteristic of lipid molecules are tabulated in several reviews (Chapman, 1965; Wallach *et al.*, 1979). Table 9.2,

Table 9.1 Properties of infrared window materials (From Amey and Chapman, 1983)

Material	Useful range (cm^{-1})	Reflection loss at 1000 cm^{-1} (%)	Solubility (g per 100 ml H_2O at 20 °C)	Relative cost
Sodium chloride (NaCl)	<4000–625	7.5	40	1.0
Potassium bromide (KBr)	<4000–400	8.5	70	1.3
Calcium fluoride (CaF_2)	<4000–1000	5.5	Almost insoluble	3.6
Irtran-2 (polycrystalline ZnS)	<4000–700	12	Insoluble	10
Silver chloride (AgCl)	<4000–450	19.5	Insoluble	7
Silver bromide (AgBr)	<4000–280	25	Almost insoluble	7
KRS-5 (TlBr/TlI)	<4000–250	28	0.1	10
High-density polyethylene	625–10	n.a.	Insoluble	0.6

Table 9.2 Major infrared absorption bands in spectra of various lipids in chloroform or carbon tetrachloride or as oils (From Kates, 1986)

Functional group	Absorption mode	Frequency (cm^{-1})	Intensity[a]
—CH_3	CH stretch	2962; 2872 ± 10	s
	CH bend { asymm.	1450 ± 20	m
	CH bend { symm.	1375 ± 5	m
—OCH_3	CH bend	1430	m
$N(CH_3)_3$	CH stretch (asymm.)	3040	w
	CH_3 deformation (asymm.)	1470	w
		960–975	w
$S(CH_3)_2$	CH stretch (asymm.)	3023	w
	CH_3 deformation (asymm.)	1435	w
—CH_2	CH stretch	2926; 2853 ± 10	s
	CH bend	1465 ± 10	m
	$[CH_2]_4$ skeletal	750–720	m
	—CH_2 in cyclopropane { CH stretch	3100–3000	m
	—CH_2 in cyclopropane { skeletal	1025–1000	m
—$C(CH_3)_2$ (isopropyl)	CH bend	1385; 1365 ± 5	m (db)
	skeletal	1170 ± 5; 1150 ± 5	m (db)
—$C(CH_3)_3$ (t-butyl)	CH bend	1395–1385 (m); 1365	s
	skeletal	1250–1200	s
—CH (tertiary)	CH stretch	2890 ± 10	w
	CH bend	1340 ± 20	w
C=C (aliphatic)	C=C stretch, non-conjugated	1680–1620	v
	C=C stretch, phenyl conjugated	1625 ± 20	v
	C=C stretch, C=O or C=C conjugated	1600 ± 20	v
	CH stretch	3040–3010	m
—CH=CH (*trans*)	CH out-of-plane deformation	970–960	s
	CH in-plane deformation	1310–1295	s–w
—CH=CH (*cis*)	CH out-of-plane deformation	*ca* 690	m
—CH=CH_2 (vinyl)	CH stretch	3095–3075	m
	CH out-of-plane deformation	995–985	s
	CH_2 out-of-plane deformation	915–905	s
	CH_2 in-plane deformation	1420–1410	s
R_1R_2C=CH_2	CH stretch	3095–3075	m
	CH_2 out-of-plane deformation	895–885	s
	CH_2 in-plane deformation	1420–1410	s
R_1R_2C=CHR_3	CH out-of-plane deformation	840–790	s
—$C(CH_3)$=CH— (isoprenoid)	CH out-of-plane deformation	835	s
C=C (aromatic)	=C—H stretch	3030	sharp
	=C—H out-of-plane:		
	5 H atoms	770–730; 710–690 (2 bands)	vs
	4 H atoms	770–735 (1 band)	vs
	3 H atoms	810–750	vs
	2 H atoms	860–800	vs
	C=C skeletal	1600; 1500	v

Table 9.2 *Continued*

Functional group	Absorption mode	Frequency (cm^{-1})	Intensity[a]
RC≡CR	C≡C stretch	2260–2190	m
RC≡CH	C≡C stretch	2140–2100	m
	C—H stretch	3300	m
—C—OH	O—H stretch; free OH	3650–3590 (sharp)	v
	O—H stretch; associated OH (hydrogen bonded)	3400–3200 (broad)	
—C—OH	C—O stretch	1150–1140 (tertiary)	s
		1120–1100 (secondary)	s
		1075–1010 (primary)	s
—C—F	alkyl, stretch	1400–1000	s
—C—Cl	alkyl, stretch	800–600	s
	aryl, stretch	1100–1050	s
—C—Br	alkyl, stretch	600–500	s
—C—I	alkyl, stretch	*ca* 500	v
—C—O—C—alkyl ether	C—O stretch	1150–1060	vs
—C—O—C aryl ether	C—O stretch	1270–1230	vs
$-\overset{H}{C}=\overset{H}{C}-O-C$ (*cis*-vinyl ether)	C—O stretch	1270–1230	s
	C=C stretch	1670	m
	CH deformation	732–730	m
—C—O—C—	C—C stretch	1250	s
	cis	830	m
	trans	890	m
Alkyl peroxides	C—O stretch	890–820	v–w
Acyl peroxides	C=O vibrations	1820–1810; 1800–1780	s
Aldehyde C=O	C=O vibrations:		
	saturated aldehyde	1740–1720	s
	α,β-unsaturated aldehyde	1705–1680	s
Ketone C=O	C=O stretch:		
	saturated ketone	1725–1705	s
	α,β-unsaturated ketone	1685–1665	s
Carboxyl, RCOOH	C=O stretch:		
	saturated fatty acid	1725–1700	s
	α,β-unsaturated acid	1715–1690	s
	C—O stretch	1320–1210	s
$R-COO^-$	C=O stretch (asymm.)	1610–1550	s
	C—O stretch (symm.)	1400–1300	s
Ester, RCOOR	C=O stretch:		
	saturated ester	1750–1730	s
	α,β-unsaturated ester	1730–1717	s
	β-keto ester (enol)	1650–1540	s
	C—O stretch:		
	formates	1200–1180	m
	acetates	1250–1230	m
	propionates and higher	1200–1150	m
Primary amide, $RCONH_2$	NH stretch (free NH)	3500; 3400	m
	NH stretch (bonded NH)	3350; 3180	m
	C=O stretch	1690–1650	s
	NH_2 deformation	1620–1590	s
	C—N stretch	1420–1400	v

Table 9.2 *Continued*

Functional group	Absorption mode	Frequency (cm^{-1})	Intensity[a]
Secondary amide, RCONHR	NH stretch (free NH)	3440–3400	m
	NH stretch (bonded NH)	3320–3140	m
	C=O stretch	1680–1630	s
	NH deformation	1550–1510	s
Tertiary amide, $RCONR_2$	C=O stretch	1670–1630	s
Anhydride, CO—O—CO	C=O stretch	1850–1800; 1790–1740	s
	C—O stretch	1170–1050	
Primary amine, —NH_2	NH stretch	3500–3300 (2 bands)	m
	NH deformation	1650–1590	s–m
Secondary amine, —NH	NH stretch	3500–3300 (1 band)	m
	NH deformation	1650–1550	vw
Phosphates			
R—O—PO—O^-	P=O stretch (free)	1300–1250	s
|	P=O stretch (bonded)	1250–1200	vs
O^-	P—O—C (aliphatic)	1050–1000	vs
	P—O—C (aromatic)	1240–1190	s
		995–850	s
	P—O^-	1110–1090	s
		2700–2560	w
R—O—$PO(OH)_2$ (free acid)	P—OH	2700–2560	w
	P=O stretch	1240–1180	m
	P—O—C	1070–1000	s
R—O—PO—$(OCH_3)_2$	P=O stretch	1275–1180	s
	P—O—CH_3	1190	w
		1040	s
Phosphonates			
R—O—PO—C	P=O stretch (bonded)	*ca* 1200	s
|	P—C stretch	750–650	v
O^-			
Sulphates			
R—O—SO_2—O^-	S=O stretch	1265–1200	s
		1080–1040	v
	C—O stretch	1040–990 (primary)	m
		990–930 (secondary)	v
	S—O stretch	840–760	w
R—O—SO_2—O—R′	S=O stretch	1440–1350	s
		1230–1150	v
	C—O stretch	1040–930	m
	S—O stretch	850–760	m
Sulphonates			
R—SO_2—O^-	S=O stretch	1220–1210	s
		1055–1050	v
	C—S stretch	800–600	w
R—SO_2—O—R′	S=O stretch	1420–1330	s
		1200–1145	v
	S—O stretch	900–860	m
	C—S stretch	800–600	w
	C—O stretch	1000	w
R—SO_2—OH	S=O stretch	1200; 1050	s
	S—O stretch	650	s

[a] *Abbreviations*: w, weak; m, moderate; s, strong; v, variable; vs, very strong; db, doublet.

taken from Kates (1986), lists some of the more frequently encountered absorption bands.

Although there is no compilation of spectra of lipid materials at present, several general collections are available and may be of interest. *Analytical Chemistry* publishes regular useful reviews with infared data of organic molecules (e.g. Nyquist *et al.*, 1990; Putzig *et al.*, 1992). The *Coblentz Society Desk Book of Infrared Spectra* (Craver, 1977) contains 860 high-quality spectra selected from their collection of 11 000 spectra. It also contains information on sample preparation, spectrometer operation, spectral interpretation and a bibliography of other sources of spectra. The *Sadtler Handbook of Infrared Spectra* (Simons, 1978) contains 3000 high-quality grating spectra from their collection of 59 000 spectra, with sections on spectral interpretation and a name index. More recent compilations include those by Lin-Vien *et al.* (1991) and Durig (1990). Infrared technology, i.e. detectors, sources, optical materials, etc., is covered in the books by Wolfe and Zissis (1978) and Lettington (1990). The third edition of the *Aldrich Library of Infrared Spectra* (Pouchert, 1981) is an excellent collection of 12 000 spectra organized conveniently according to chemical class. Both Aldrich and Sigma are publishing libraries of FTIR spectra (Pouchert 1985–89; Keller, 1986), with the additional advantage of the corresponding databases being also available on diskette. Reference spectra are essential to the infrared user. Seldom is one able to assign a structure exclusively from its infrared spectrum without use of the spectrum of some reference material. IUPAC has published their *Tables of Wavenumbers for the Calibration of Infrared Spectrometers* (Cole, 1977). Chemical Abstracts Services offers through its CA Selects (CAS) two abstract journals related to infrared spectroscopy. These are based on computer searches of the *Chemical Abstracts Services Information Base* and contain the complete abstracts. Note also that the subject indexes of *Chemical Abstracts* contain a heading 'infrared spectra', which lists compounds and classes of compounds whose spectra are found in published papers.

9.1.4 Infrared absorption of long-chain compounds

(a) Vibrations of methylene groups

The vibrations of a methylene group in the chain are shown in Fig. 9.1. The asymmetrical and symmetrical stretching and the deformation vibrations are the three modes due to the isolated methylene groups. When the methylene group forms part of a chain, rotational motions, called wagging, twisting and rocking, occur in relation to the chain. Some of these vibrations absorb at the same frequency for all methylene groups in the chain, while others have different frequencies for each methylene according to the phase differences of neighbouring oscillators. A measurable phase difference of neighbouring oscillators occurs when the vibration of one methylene group disturbs the vibration of the next methylene group.

The characteristic frequency of a vibrator (assuming linear elastic interaction according to Hooke's law) is given by:

$$\nu = \frac{1}{2\pi c}\left[\kappa\left(\frac{1}{m_1} + \frac{1}{m_2}\right)\right]^{1/2} = \frac{1}{2\pi c}\left(\frac{\kappa}{\mu}\right)^{1/2} \quad (9.2)$$

It is proportional to the strength of the spring (force constant κ of the bond in in dynes per centimetre) and reciprocally proportional to both masses in grams (ν is the wavenumber of frequency $1/\lambda$ in cm^{-1}, c is the velocity of light and μ is the reduced mass).

Coupling effects appear when two vibrators are joined by a common atom or a common bond, provided their individual frequencies are not too dissimilar and the vibrations belong to the same symmetry species. For example, the C—H stretching mode at $3000\,cm^{-1}$ and the C—C stretching mode at $1000\,cm^{-1}$ do not influence each other, but the two

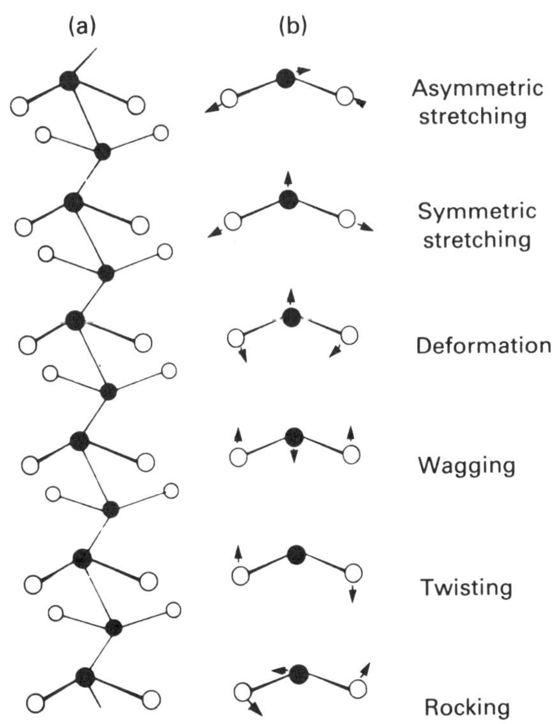

Fig. 9.1 (a) Section of a polymethylene chain; (b) vibrations of the methylene groups in the chain. From Fischmeister (1975).

C—H bonds in the methylene group do. Their asymmetric stretching mode absorbs at 2920 cm^{-1}, their symmetric mode at 2850 cm^{-1}.

Bending modes behave similarly. Strong coupling between equivalent modes occurs when both motions have one bond in common. If no common bond is involved, only weak coupling occurs and neighbouring groups absorb at similar frequencies. This is the case for the CH_2 deformation mode, where the two C—H bonds bend against each other without the need for a compensating motion by the carbon atom. In this case (Fig. 9.2) the next CH_2 group is hardly affected.

If, however, the C—H bending has to be compensated by a deformation of the C—C bond angle, the next CH_2 group is 'disturbed' by the first one and strong coupling results. This is the case for the wagging, twisting and rocking vibrations. Coupled vibrations can be 'in phase' or 'out of phase'. Snyder (1960) discusses this in terms of the symmetry property of the motion of two oscillators with respect to an inversion through the midpoint of the C—C bond. Out of phase implies that they can be brought to coincide by inversion.

(b) *Vibrational spectra of hydrocarbons*

A number of investigations have contributed to the understanding of the vibrational spectra of hydrocarbon-type chain molecules (see Fischmeister, 1975). The distributions of the methylene wagging, twisting, rocking and C—C stretching frequencies in the spectra for the homologous hydrocarbons were first established by the systematic analysis based on the symmetry properties and selection rules of the crystalline compounds by Brown *et al.* (1950, 1954) and by Tschamler (1954). In Fig. 9.3 a schematic representation of the vibrational spectra of the homologous n-paraffins (alkanes) is shown, divided into regions characteristic of the various CH_2 and skeletal vibrations.

(c) *Vibrational spectra of long-chain compounds with polar endgroups; band progression*

The distributions of the vibrational frequencies of the methylene chain were established in detail for the hydrocarbon spectra after they were initially observed and described for spectra of long-chain fatty acids and esters and dicarboxylic esters and keto esters. The band progression is especially prominent because the methylene wagging absorption is found to be enhanced in the presence of polar endgroups. As early as 1952, Jones demonstrated the dependence of the progression bands on the chain length, endgroups and substituents in the chain (Primas and Günthard, 1953). Figure 9.4 is an example of this.

Metal salts of saturated and unsaturated long-chain fatty acids have been studied (Kirby *et al.*, 1965) with the aim of determining the chain length from the band progressions. For the saturated salts, both wagging and rocking progressions indicated the length of the whole chain. In unsaturated compounds, the wagging progression revealed the position of the double bond for both *cis* and *trans* unsaturation, and so did the

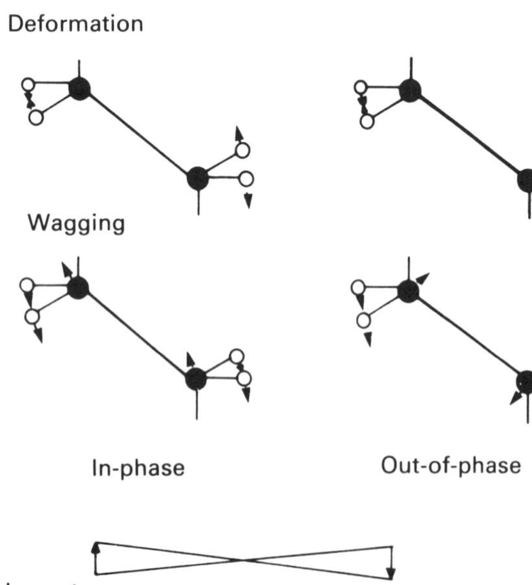

Fig. 9.2 In-phase and out-of-phase methylene deformation and wagging modes in the chain (Snyder, 1960).

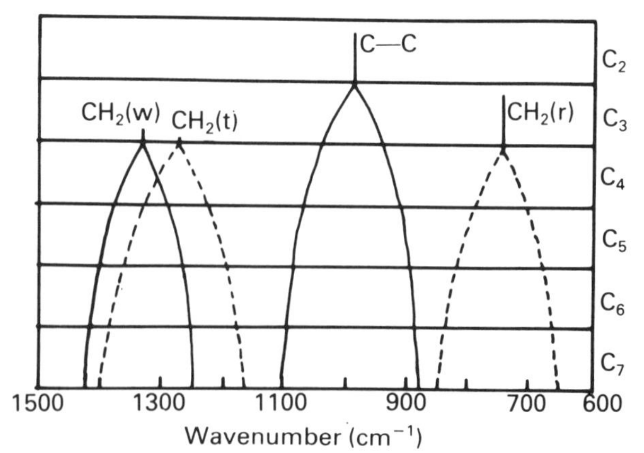

Fig. 9.3 A schematic representation of the vibration spectra of the n-paraffins divided into regions characteristic for the various CH_2 and skeletal vibrations. $CH_2(t) = CH_2$ twisting; $CH_2(w) = CH_2$ wagging; $CH_2(r) = CH_2$ rocking; C—C = skeletal stretching. From Fischmeister (1975.)

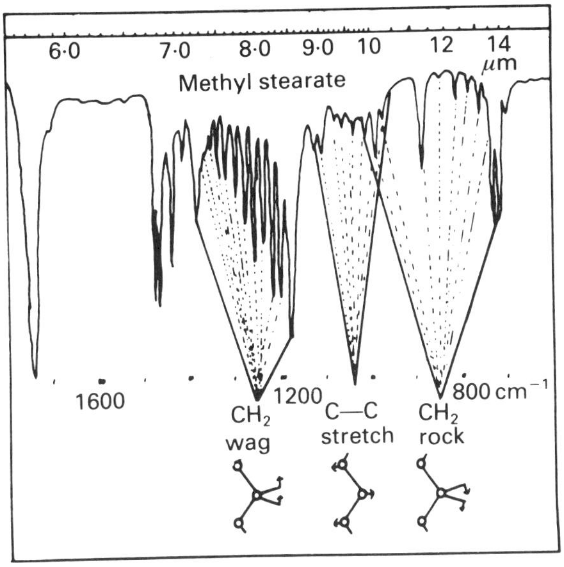

Fig. 9.4 Spectrum of methyl stearate from 1800 to $600\,cm^{-1}$. Bands are assigned to the different progressions. From Fischmeister (1975).

rocking progression in *cis* compounds. In the *trans* compounds, remarkably, the rocking progression indicated the whole chain length.

The progression bands were valuable not only for identification of homologues, but also for distinguishing different crystal forms of fatty acids. The intensity of the wagging progression bands is dependent on the crystal structure. Nujol spectra of the A′, B′ and C′ forms of two odd-numbered acids (chain length C_{15}, C_{17}) and of the A, B and C forms of one even-numbered acid (C_{16}) have been published (von Sydow, 1955). The spectral features of the different crystal forms were studied using polarized spectra and variable-temperature techniques (Susi and Smith, 1960; Chapman, 1962).

Chapman settled, by solid-state infrared spectroscopy, the controversy concerning the correlation of melting points of the polymorphic mono-, di- and triacylglycerols with structural models derived from X-ray powder data. The intensity of wagging bands and the single or double appearance of the $720\,cm^{-1}$ rocking band were found to be valuable criteria for the crystal forms. Meiklejohn's rule of band counting was used for chain-length determination.

A systematic infrared spectroscopic study of the mono-acid triacylglycerols of the homologous fatty acids C_{2-22} in the solid state at $-180\,°C$ (and as liquids) for the different polymorphic forms was performed (De Ruig, 1971). The wagging, rocking–twisting and C—C stretching chain absorptions were coordinated with the well established distributions of vibrational modes in hydrocarbon spectra.

(d) Polymorphism and phase transitions of long-chain molecules using IR spectroscopy

Among the long-chain molecules, polymorphism occurs quite frequently. Different crystal forms of one solid compound are usually stable in different temperature ranges, but there are also cases where two forms crystallize simultaneously from a solution. Each crystal form usually has its own spectrum, which means that a polymorphic substance can have as many spectra as it has crystal forms. When such substances are investigated, it is necessary to state the crystal form of the sample, or otherwise describe thoroughly the solvent and temperature of crystallization.

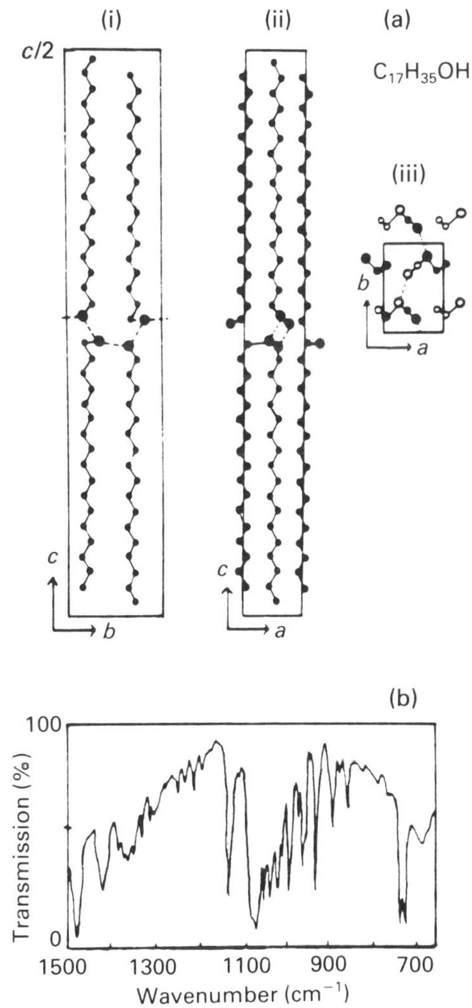

Fig. 9.5 (a) Molecular arrangement in the β-form crystal ($C_{17}H_{35}OH$) as viewed along the *a* axis (i), along the *b* axis (ii) and along the molecular axis (iii). Large circle represents the O atom. Full and open circles in (iii) represent the atoms belonging to the upper and lower layers, respectively. (b) Infrared spectrum and the β form of $C_{17}H_{35}OH$ (solid film). Adapted from Fischmeister (1975.)

The structural differences can influence the chain packing. The different packing modes are quite well known from X-ray crystallographic work, and their correlation with spectral features is so well established that it is frequently used to indicate chain packing for substances of unknown structure. Structural differences can also concern the polar group, which can be arranged in several orientations to give good packing. Such structural differences seem to be quite widespread, but too few structural details are available for correlation with the absorption spectra.

In studies of the polymorphic forms of acylglycerols Chapman (1955, 1956a, b, 1957a, b, 1958a, b, c) used the single or double appearance of the band at 720 cm^{-1} to infer the chain packing. A recent application of FTIR to distinguish between crystalline forms of tripalmitin may be found in Eads *et al.* (1992).

The polymorphism of long-chain alcohols has been related to the rotational isomerism of the alcoholic hydroxyl group (Tasumi *et al.*, 1964). In both crystal forms, β and γ, the chain packing is orthorhombic, but in β (Fig. 9.5) alternation of molecules occurs with the hydroxyl group in an in-plane *trans* conformation and with an out-of-plane *gauche* conformation giving a certain type of hydrogen-bond connection. In the γ form only *trans* hydroxyl orientation occurs, leading to a different hydrogen-bond system (Fig. 9.6(a)).

The spectra of the γ crystals show a regular simple band progression between 1380 and 1150 cm^{-1}, and weak bands in the region 950–850 cm^{-1} (Fig. 9.6(b)). The spectrum of the β-form crystals (Fig. 9.5), however, has a complex absorption between 1380 and 1150 cm^{-1} and strong bands between 950 and 850 cm^{-1}.

Phase transitions may be detected through changes in specific IR parameters. The more commonly used parameters are the frequency and the half-bandwidths of the individual vibrational modes. These parameters have the advantage of relating to specific phenomena at the molecular level and can be used to discriminate between different motions of the absorbing groups. Specific examples of phospholipid phase transitions as seen by IR spectroscopy will be shown in section 9.1.5.

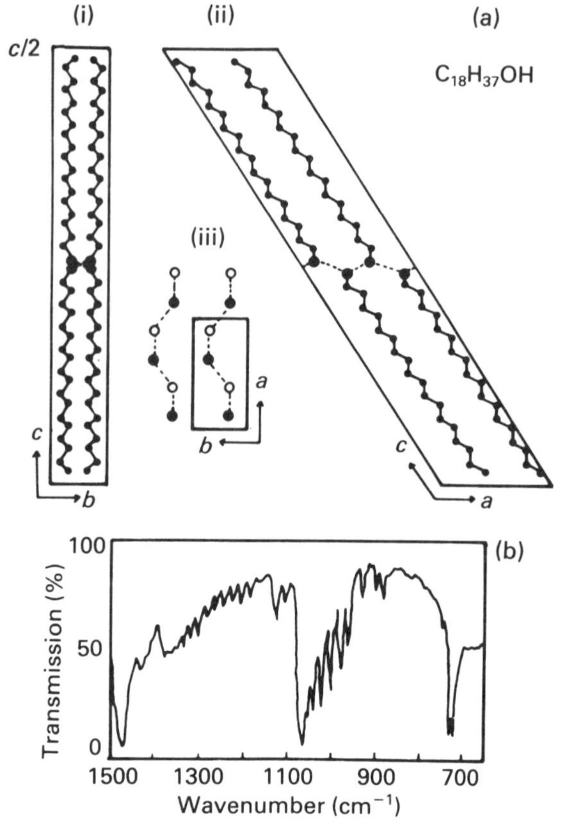

Fig. 9.6 (a) Molecular arrangement in the γ crystal ($C_{18}H_{37}OH$) as viewed along the *a* axis (i) and along the *b* axis (ii). Large circle represents the O atom. In (iii) the O atoms are projected perpendicular to the *ab* plane. Full and open circles represent the atoms belonging to the upper and lower layers, respectively. (b) Infrared spectrum for the γ form of $C_{18}H_{37}OH$ (solid film). From Fischmeister (1975).

9.1.5 Selected applications

(a) Fatty acids

One of the most common uses of IR spectroscopy in the analysis of fatty acids and fats is the determination of *trans* unsaturation. An interesting procedure for *trans* unsaturation using IR attentuated total reflectance (ATR) spectroscopy has been described (Sipos and Ackman, 1978). The samples do not have to be weighed or diluted. A special ATR cell is used to make readings at 10.3 μm. The system can be used to monitor a continuous-flow system, and the presence of catalyst does not affect the readings.

Fatty acid methyl esters have also been examined by gas chromatography/FTIR (Mantz, 1977). The chromatogram and selected spectra are shown in Fig. 9.7. The differences in retention times and the similarity of the spectra indicate a homologous series. More recently, Doumenq *et al.* (1990) have applied a similar procedure to the study of fatty acid methyl esters in marine environments.

(b) Phospholipids and membranes

Chapman *et al.* (1966, 1967) studied the spectra of films of pure phospholipids, including phosphatidyl-

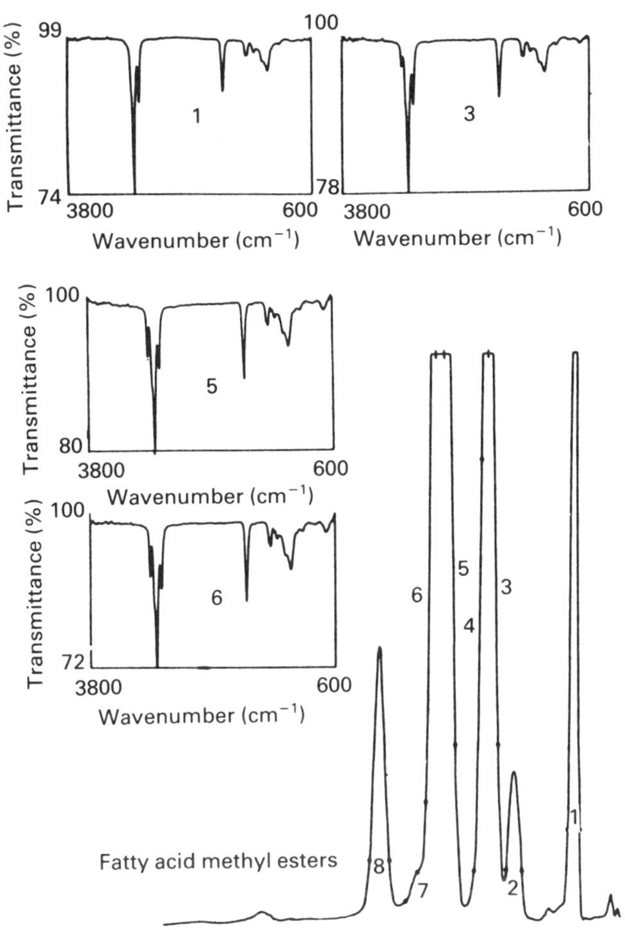

Fig. 9.7 Gas chromatogram and spectra obtained from GC/FTIR analysis of fatty acid esters (Mantz, 1977).

choline and phosphatidylethanolamine. Wallach *et al.* (1978) studied multi-bilayer films of pure phospholipids and phospholipid mixtures as a function of fatty acid composition, temperature and degree of hydration. The FTIR era of membrane lipid studies was inaugurated by Cameron *et al.* (1980). Table 9.3 summarizes the main band frequencies observed in these studies.

The infrared spectrum of membrane polar lipids can be separated into spectral regions that originate from the hydrophobic hydrocarbon tail, the interfacial region and the polar headgroup. Recent general reviews include those by Fringeli and Günthard (1981). Amey and Chapman (1983), Casal and Mantsch (1984), Lee and Chapman (1986) and Mantsch and McElhaney (1991). Infrared data from specific kinds of membrane lipids are also available, e.g. archaebacterial lipids (De Rosa and Gambacorta, 1988) or N-acylated glycerophospholipids (Schmid *et al.*, 1990).

(i) The acyl chain region

Bands arising from vibrational modes of the lipid acyl chains have been assigned previously by comparison with those of fatty acid esters and other polymethylene chain compounds. C—H stretching vibrations give rise to bands in the 3100–2800 cm^{-1} region. Asymmetric and symmetric CH_2 bands, at 2920 and 2853 cm^{-1} respectively, are the strongest ones in a phospholipid spectrum. Vibrational bands corresponding to terminal methyl residues are found around 2953 cm^{-1} (asymmetric stretching) and 2870 cm^{-1} (symmetric stretching), both appearing as shoulders of the stronger methylene bands. Olefinic group bands, =C—H, arising from unsaturated hydrophobic chains, are usually located around 3010 cm^{-1}. Methylene bending bands are located around 1470 cm^{-1} and can split into two components in motion-restricted chains (Dluhy *et al.*, 1985). The so-called 'wagging progression' is found between 1380 and 1180 cm^{-1} in the gel phase of saturated phospholipids.

Full or partial isotopic substitution of the acyl chains can be used to avoid interferences with other components, i.e. with proteins in lipid–protein interaction studies. In addition, selective deuteration can be used as an internal probe to study the characteristics of one specific methylene. Asymmetric and symmetric CD_2 vibrational bands are located around 2195 and 2090 cm^{-1} respectively, whereas CD_3 bands are found at 2212, 2169 and 2070 cm^{-1} (Castresana *et al.*, 1991). Selectively deuterated CD rocking modes have been used to quantify the number of *gauche* bonds in phospholipid acyl chains (Mendelsohn *et al.*, 1989).

Aqueous dispersions of phospholipids exhibit a thermotropic mesomorphism that results in changes in the infrared bands corresponding to methylene vibrations. The gel to liquid-crystal main phase transition of 1,2-dipalmitoyl-*sn*-glycero-3-phosphocholine (DPCC), a well defined phospholipid, produced changes in the methylene band parameters. The CH_2 stretching bands can be used to detect this phase transition: the gel phase contains mainly all-*trans* conformers; whereas, in the liquid-crystal phase, *gauche* conformers are introduced, producing changes in band position and bandwidth. Figure 9.8 shows the temperature dependence of the asymmetric CH_2 stretching band position and bandwidth of DPPC and DMPC, in the presence and absence of cholesterol.

The frequency values for the gel phase (2850 cm^{-1}) are indicative of the physical state of the lipid, but do not imply a quantification of the number of bonds in the *gauche* conformation. Therefore, methods have been developed recently to quantify the proportion of *gauche* rotamers. Mendelsohn *et al.* (1989) applied a method requiring the measurement of CD_2 rocking modes of specifically deuterated phospho-

Table 9.3 Infrared-active bands (cm^{-1}) of various phosphatidylcholines (PC) and phosphatidylethanolamines (PE)[a] (Wallach *et al.*, 1979)

Band	DPPC (dry)	DPPC (monohydrate)	DLPC (dry)	Egg PC (dry)	DPPE (dry)	DPPE (monohydrate)	Egg PE (dry)
$\nu_{symmetric}(CH_2)$	2848	2849	2848	2853	2848	2848	2848
$\nu_{asymmetric}(CH_2)$	2915	2918	2915	2920	2918	2915	2918
$\nu_{asymmetric}(CH_3)$	2957	2958	2953	2959	2958	2956	2958
$\delta(CH_2)$	1461	1466	1467	1469	1462	1469	1462
					1471		
$\nu(C{=}O)$	1724	1733	1734	1740	1739	1720	1740
	1734					1730	
$\nu(P{=}O)$	1254	1248	1252	1260	1222	1215	1224
					1241s		
Terminal CH_2 rock	722	721	720	720	715	720	720
					725		
$\nu(NH_3^+)$					2128	2108	2558
					2328	2328	2638
					2558	2548	2708
					2646		
					2688		
					2728	2728	
$\delta(NH_3^+)$					1550	1580	1559
					1625	1620	1638
$\nu(CH_2)$ of N^+CH_3	3028	3028	3028	3013			
$\nu(OH)$		3373					
$\delta(OH)$		1645					

[a] All films were prepared from solutions in chloroform/methanol (2:1, v/v), except for DPPE films, which were prepared from a chloroform solution. DPPC, dipalmitoylphosphatidylcholine; DLPC, dilauroylphosphatidylcholine; DPPE, dipalmitoylphosphatidylethanolamine; s, shoulder.

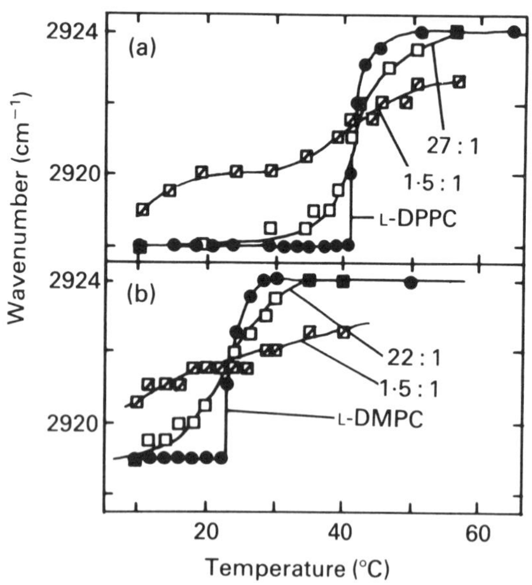

Fig. 9.8 Temperature dependence of the maximum wavenumber of the CH_2 asymmetric stretching vibrations in (i) dipalmitoylphosphatidylcholine (DPPC)–cholesterol and (ii) dimyristoylphosphatidylcholine (DMPC)–cholesterol at the molar ratios indicated. The temperature dependence for pure lipids (full circles) is also given. From Cortijo *et al.* (1982).

lipids, whereas Casal and McElhaney (1990) described a method based on the positions and intensities of isolated CH_2 wagging modes in disordered phases, and Senak *et al.* (1992) studied the relative intensities of the CH_2 wagging progression.

The approach of Mendelsohn *et al.* (1989) is based on the finding that the frequency, in the rocking mode, of a CD_2 substituted into a hydrocarbon chain is sensitive to conformation in the immediate neighbourhood of the CD_2 group (Maroncelli *et al.*, 1985). Thus, the local conformation of a C—C—C skeleton immediately surrounding a central CD_2 group partially determines the CD_2 rocking frequencies. Integrated band intensities can then provide a quantitative measure of the fraction of *gauche* rotamers at a particular chain position. DPPC deuterated at positions 4, 6 and 10 was studied at 48 °C, in the liquid-crystal phase, and a population percentage of 20.7 ± 4.2, 32.3 ± 2.3 and 19.7 ± 0.8 *gauche* rotamers was obtained for 4-d_4-DPPC, 6-d_4-DPPC and 10-d_4-DPPC, respectively. The *gauche* rotamers are not easily measured in the gel phase, because of the small number of *gauche* bonds contributing to band absorption. The gel phase of the DPPC molecules deuterated at positions 6 and 10 is highly ordered, but a population of *gauche* rotamers can be masured for the DPPC deuterated at position 4. The method implies the possibility of studying the effect of foreign molecules on the phospholipid order and motion. Cholesterol effects on conformational disorder in DPPC have been studied at various depths, i.e. positions 4, 6 and 12; an inhibition of *gauche* rotamer formation has been detected at positions 6 and 4, but the ability of the sterol to increase acyl chain order is much reduced at the 12 position. The gel phase is only slightly disordered by the sterol (Davies *et al.*, 1990).

The spectral region 1400–1300 cm^{-1} contains bands due to CH_2 wagging that are specific for the different types of *gauche* conformers, e.g. the *gauche–trans–gauche* sequences (or kinks) give a band at 1367 cm^{-1}, etc. The intensities of the different *gauche* bands yield a 'per chain' intensity, which is directly related to the concentration of the different types of conformational defects. The number of *gauche* bonds obtained is in good agreement with those estimated by the previous method (Casal and McElhaney, 1990). Also, the intensity of the CH_2 wagging progression has been used to monitor the all-*trans* conformational state of phospholipids and the effect of cholesterol (Senak *et al.*, 1992).

(ii) The interfacial region

The structure of the interfacial region of lipid assemblies can be studied by means of the band arising from the C═O stretching vibrational mode, located between 1750 and 1700 cm^{-1}. The midpoint of the band shows a 2 cm^{-1} shift to higher frequencies at the pretransition temperature in DPCC, whereas at the main transition temperature there is a shift of around 4 cm^{-1} to lower frequencies in DPPC (Casal and Mantsch, 1984) and PE (Castresana *et al.*, 1992a).

This stretching vibrational mode is composed of two bands, one centred near 1740 cm^{-1} and the other near 1727 cm^{-1}. It was postulated in previous studies (Levin *et al.*, 1982; Mushayakarara *et al.*, 1982) that the bands at 1727 and 1740 cm^{-1} could be assigned respectively to the *sn*-2 and *sn*-1 carbonyl groups. The assignment was reasonable, taking into account the conformational inequivalence of the *sn*-1 and *sn*-2 carbonyl groups in hydrated lipid bilayers (Hauser *et al.*, 1988), with the *sn*-2 carbonyl located in a more polar environment. However, if the carbonyl ^{12}C is changed into ^{13}C in either the *sn*-1 or the *sn*-2 position, it is shown (Blume *et al.*, 1988) that, in aqueous dispersions, the difference accountable by the conformational inequivalence is very small (less than 4 cm^{-1}) when compared with the experimental values (11–17 cm^{-1}). Moreover, it is also found that each of the two components present in the C═O stretching band contour of the unlabelled lipids is in fact the summation of comparable contributions from both the *sn*-1 and *sn*-2 carbonyl groups. Therefore, it is concluded that both carbonyl groups of hydrated phospholipids take part in hydrogen bonding to water, and that both C═O stretching component bands are a reflection of instantaneous subpopulations of hydrated and non-hydrated (or free and hydrogen-bonded) carbonyl groups. This interpretation has also been corroborated as a collateral result in a work undertaken to study the structures of the subgel phases of n-saturated diacylphosphatidylcholine bilayers with ^{13}C- and ^{2}H-labelled lipids (Lewis and McElhaney, 1992).

Hübner and Mantsch (1991) have studied by ATR-FTIR the orientation of individual *sn*-1 and *sn*-2 C═O double bonds and *sn*-1 and *sn*-2 CO—O single bonds in hydrated diacylphosphatidylcholines, using specifically ^{13}C═O labelled phosphatidylcholine multilayers. They have shown that both ester carbonyl double bonds are aligned in the membrane plane with a tilt angle greater than 60°; the two ester CO—O single-bond vibrations, located around 1175 cm^{-1}, are oriented differently, and the picture arising is one in which the glycerol moiety has to be oriented nearly perpendicular to the membrane plane, and a 90° bend must exist near the C-2 carbon of the *sn*-2 chain in order that both chains become parallel.

(iii) The phosphate group region

The phosphate moiety of the phospholipid headgroup exhibits strong bands corresponding to the PO_2 stretching vibrations in the 1300–1000 cm^{-1} region. These vibrations are sensitive to hydration, with a shift (of approximately 20 cm^{-1}) to lower wavenumbers upon hydration in phosphatidylcholines (Arrondo *et al.*, 1984; Grdadolnik *et al.*, 1991). The asymmetric vibration band is centred around 1222 cm^{-1}, and the symmetric one is centred around 1086 cm^{-1} in fully hydrated samples (Table 9.4). A shoulder near 1060 cm^{-1} on the PO_2 has been attributed to a R—O—P—O—R′ stretching mode, where both phosphate ester substituents are different (Arrondo *et al.*, 1984). Changes in the polar headgroup modify the positions of the bands attributed to the phosphate moiety in the spectral region 1300–1000 cm^{-1} (Goñi and Arrondo, 1986). These bands are also sensitive to changes in the microenvironment of the phosphate group in phospholipid arrays. Phosphatidylethanolamine (PE) undergoes gel–fluid and lamellar–hexagonal transitions that produce shifts to higher wavenumbers in the PO_2 stretching vibrations and show that the lamellar–hexagonal, but not the gel–fluid, transition is accompanied by a

Table 9.4 Maximum wavenumbers of the main bands appearing in the 1000–1300 cm^{-1} region of FTIR spectra of phosphate-containing molecules[a] (From Arrondo *et al.*, 1984)

Compound	Maximum wavenumbers (cm^{-1})	Assignment
Inorganic phosphate KBr disc	1046 (intense, broad)	PO_2^- stretch
Inorganic phosphate (buffer)	1077 (intense)	Symmetric PO_2^- stretch
	1160	Asymmetric PO_2^- stretch
L-α-Glycerophosphate	1085	Symmetric PO_2^- stretch
	1102 (shoulder)	Asymmetric PO_2^- stretch
Phosphorylcholine	1095	Symmetric PO_2^- stretch
	1117 (shoulder)	Asymmetric PO_2^- stretch
Glycerophosphorylcholine	1054 (shoulder)	R—O—P—O—R′
	1085	Symmetric PO_2^- stretch
	1217	Asymmetric PO_2^- stretch
Phosphatidic acid	1076	Symmetric PO_2^- stretch
	1181	Asymmetric PO_2^- stretch
DPPC KBr disc	1069 (shoulder)	R—O—P—O—R′
	1093	Symmetric PO_2^- stretch
	1245	Asymmetric PO_2^- stretch
DPPC (slurry)	1064 (shoulder)	R—O—P—O—R′
	1089	Symmetric PO_2^- stretch
	1168	
	1230	Asymmetric PO_2^- stretch
DPPC (buffer)	1060 (shoulder)	R—O—P—O—R′
	1086	Symmetric PO_2^- stretch
	1180 (shoulder)	
	1222	Asymmetric PO_2^- stretch

[a] All spectra were recorded in buffer except when otherwise stated.

weakening in the shell of hydrogen-bonded water (Castresana *et al.*, 1992a). Cations, such as Ca^{2+}, can bind to the phosphate groups of phosphatidylserine (PS), shifting the asymmetric stretching band towards higher wavenumbers, and showing that cation binding produces dehydration of the phosphate group but does not affect the interfacial region (Casal *et al.*, 1987a, b). Mg^{2+} binds in a similar way to DMPS, producing also dehydration; however, whereas in PS–Ca^{2+} complexes both P—O ester bonds have the *gauche*–antiplanar conformation, PS–Mg^{2+} complexes preserve the more usual *gauche–gauche* conformation (Casal *et al.*, 1989). Choi *et al.* (1991) have shown that the presence of cholesterol does not alter the nature of binding of Ca^{2+} ions to PS.

(iv) Lipid mixtures in bilayers

IR spectroscopy is ideally suited to examine the behaviour of molecules other than phospholipids, dispersed with them in bilayer arrays. Representative examples include studies of the effects of local anaesthetics on lipid bilayers (Trudeau *et al.*, 1981; Auger *et al.*, 1990), short-chain alcohols in DMPE (Centeno and O'Leary, 1990), or bilirubin in lipid bilayers under high pressure (Zakim and Wong, 1990). The effect of cholesterol on the location of organic molecules in lipid bilayers was examined by Muga and Casal (1990). Mixtures of isotopically labelled palmitic acid and DPPC have been studied by Villalain and Gómez-Fernández (1992). FTIR has also been applied to the detection of effects of monoglycerides (Nilsson *et al.*, 1991) and diglycerides (Ortiz *et al.*, 1992) on phospholipid bilayers. Finally, this technique has also shown its usefulness as a diagnostic tool of lipid phases in a study on the physical states of ubiquinone-10, either in pure form or incorporated into phospholipid bilayers (Castresana *et al.*, 1992b).

(c) Food lipids

The introduction of improved sample preparation and separation techniques has made IR a popular identification tool in food chemistry (Eskamani, 1975). One of the most useful applications of IR spectroscopy in food processing is the identification of *cis/trans* unsaturation. This subject has been partly dealt with in section 9.1.5(a). In the food industry, *cis/trans* determination may be helpful in the detection of fat adulteration. Bartlett and Chapman (1961) used IR spectroscopy to determine hydrogenated (saturated) fats in butter. Butter contains *cis/trans* conjugated unsaturation as well as isolated *trans* unsaturation, while hydrogenated fats contain only the latter. Therefore, addition of hydrogenated fats greatly increases the isolated *trans* double bands, which absorb at 967 cm^{-1}, but leaves the conjugated diene (948 and 925 cm^{-1}) essentially unchanged (Fig. 9.9). A GC/FTIR method has also been developed to determine *cis*- and *trans*-octadecanoic acids in margarine, made from partially hydrogenated vegetable oils (Ratnayake *et al.*, 1990).

Studies of autoxidation and rancidity in relation to off-flavour production in foods cannot be overemphasized. Rancidity development in food products such as milk, meat, dehydrated food, potato chips, fruits and vegetables is a prime concern of food chemists. The process of autoxidation and rancidity involves not only *cis* to *trans* isomerization but also production of many other compounds, such as ozonides, peroxides and hydroperoxides. IR spectroscopy has been used successfully to follow the autoxidation process qualitatively and quantitatively – see, e.g., a recent study on various isomeric ozonides from methyl oleate (Wu *et al.*, 1992).

Arnold and Hartung (1971) have successfully explored the utility of IR spectroscopy in the determination of degree of unsaturation of complex triacylglycerol mixtures found in food oils and fats. The ratio of absorbances of olefinic/saturated C═H stretching correlates (0.98) well with iodine number. The unsaturation in many food oils and fats such as corn, lard, milk, fat, peanut, soybean, vegetable,

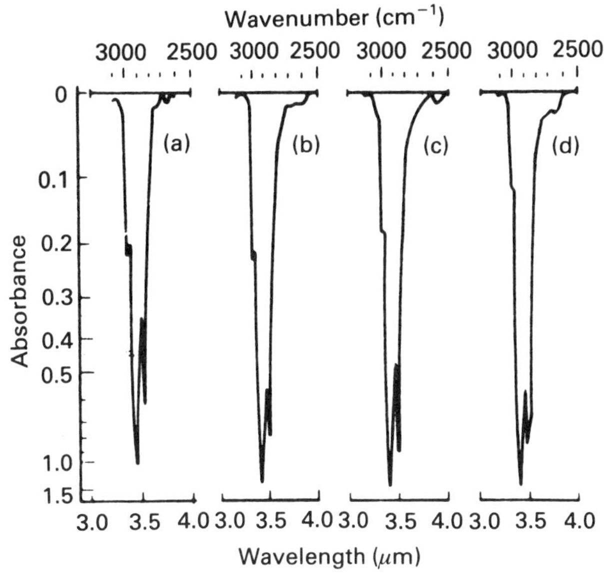

Fig. 9.10 IR absorption patterns in 3–4 μm region of: (a) safflower oil, iodine value 144; (b) soybean oil, iodine value 109; (c) lard, iodine value 73; (d) milk fat, iodine value 33. From Arnold and Hartung (1971).

cottonseed and many other oils can be determined by this technique. Some important IR absorption patterns for a few oils are illustrated in Fig. 9.10. The authors have developed a 'prediction equation' that enables one to calculate iodine value from IR data as shown below:

$$\text{Iodine value} = 495\,(\text{IR ratio}) - 54.1 \qquad (9.3)$$

$$\text{IR ratio} = \frac{\text{absorbance}_{3.3\mu m}}{\text{absorbance}_{3.5\mu m}} \qquad (9.4)$$

This IR technique should have potential in the development of IR instruments specially designed to monitor unsaturation in process control applications.

Recent IR studies in oils extracted from plant seeds include the characterization of cholesterol in such fluids (Yasaei *et al.*, 1989) and the analysis of fatty acid spectral assignments (Panford and De Man, 1990).

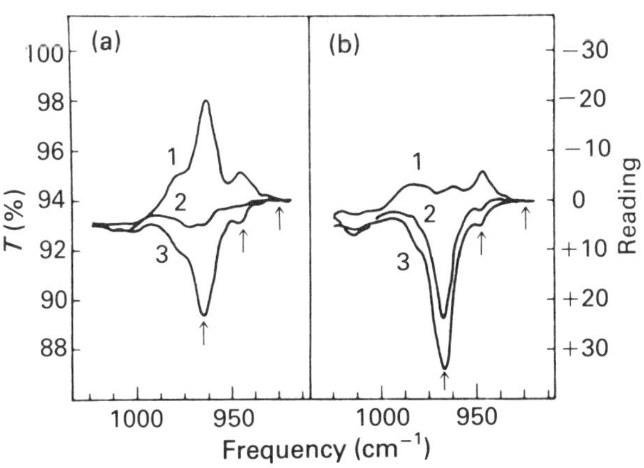

Fig. 9.9 Differential IR spectra: (a) pure butter; (b) butter containing 10% hydrogenated fat (Bartlett and Chapman, 1961).

(d) Haems

Haem groups and their derivatives (bilirubins, etc.) are not always included among the lipids, although they should be considered in that group because of their structure, metabolism and physical properties. The first established structure was that for protohaem, or haem B (Fig. 9.11), which is commonly isolated from blood (haemoglobin). More recently, the structure of haem A, isolated from beef heart, has been elucidated (Fig. 9.12). Another natural

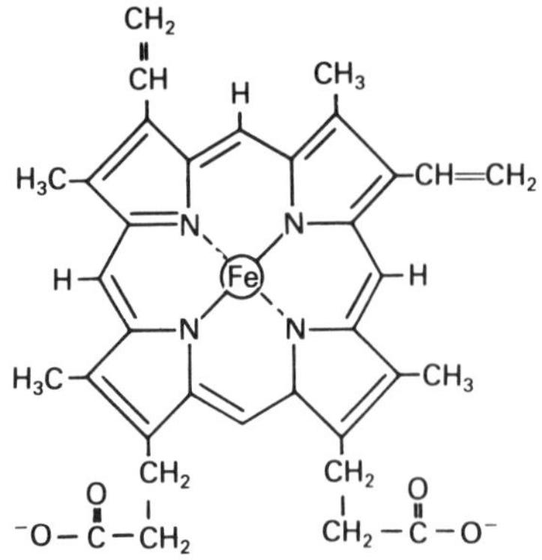

Fig. 9.11 The structure of haem B.

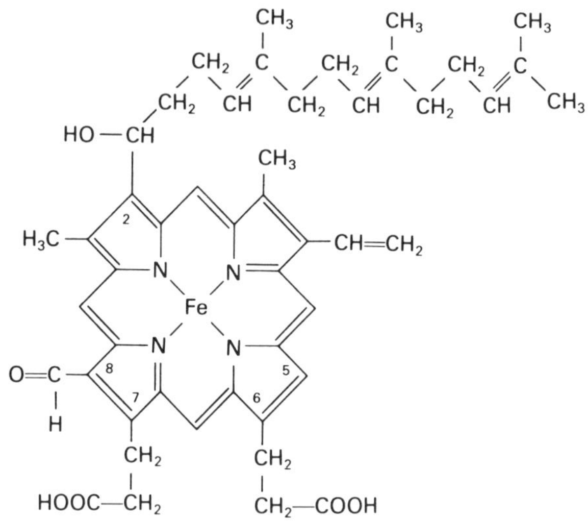

Fig. 9.12 The structure of haem A.

haem, haem S, is similar in structure to haem B except that the vinyl group at position 2 is replaced by a formyl.

A number of functional groups in haem molecules are conveniently detected by infrared spectra (Table 9.5). IR spectra may be obtained in KBr pellets with as little as 0.5 mg of haem or haemin in 200 mg of KBr. Solution spectra may also be recorded at concentrations 10 μM or less.

(e) Carotenoids

The conjugated polyene system typical of carotenoids gives rise to only very weak bands in IR spectroscopy. However, this technique has proved to be of value for detecting certain special structural features such as acetylenic, allenic, hydroxy and unreactive keto groups, e.g. like those in fucoxanthin and capsanthin; (for a comprehensive discussion of carotenoid chemistry, see Vetter *et al.* (1971). The structures of allenic and acetylenic carotenoids may be determined with the help of IR spectroscopy, because of the unusual peak at 1928 cm^{-1}, indicative of allenic groups, and the weak band at 2170 cm^{-1}, characteristic for acetylenic carotenoids (Vetter *et al.*, 1971).

Carotenoids with *trans*-disubstituted ethylenes show a strong band near 965 cm^{-1} due to C—H out-of-plane deformation vibrations. For all-*trans* carotenes this is a sharp singlet; *retro*-carotenoids as well as some carbonyl-substituted or *cis*-polyenes

Table 9.5 Infrared spectra data for haems and haemins in KBr

Derivative	ν (cm^{-1})				
	Ester carbonyl	Formyl	Vinyl(s)	Fe—O—Fe	Pyridine
μ-Oxobishaemin A monomethyl ether dimethyl ester[a]	1738	1666	1631	870	–
μ-Oxobishaemin B dimethyl ester[b]	1740	–	1628	880	–
	–	–	1614	–	–
μ-Oxobishaemin S/iso S[c] dimethyl ester[b]	1730	1663	1626	885	–
Dipyridine haem A monomethyl ether dimethyl ester	1738	1653	1630	–	1485
	–	–	–	–	755
	–	–	–	–	690 sh.[e]
Dipyridine haem B dimethyl ester	1733	–	1628	–	1485
	–	–	1617	–	761
	–	–	–	–	691 sh.
Dipyridine haem S dimethyl ester[d]	1734	1656	1630	–	1484
	–	–	–	–	754

[a] From Caughey *et al.* (1975).
[b] From O'Keeffe *et al.* (1975).
[c] A mixture of the 2,4 positional isomers.
[d] O'Keeffe (1974).
[e] sh. = shoulder.

show a doublet in this region. This double peak seems to be conclusive evidence for the *retro* system. Splitting of a peak into a doublet is often observed with central-*cis* carotenoids. The C—H out-of-plane deformation vibrations of *cis*-disubstituted double bonds give rise to a strong absorption at 780 cm^{-1}. A disubstituted *cis* double bond shows weak absorption near 830 cm^{-1} and sometimes a peak is also observed at 1380 cm^{-1} (Vetter *et al.*, 1971).

References and further reading

Amey, R.L. and Chapman, D. (1983) in *Biomembrane Structure and Function* (ed. D. Chapman), Macmillan, London, p. 199.

Arnold, R.G. and Hartung, T.E. (1971). *J. Food Sci.*, **36**, 166.

Arrondo, J.L.R., Goñi, F.M. and Macarulla, J.M. (1984) *Biochim. Biophys. Acta*, **794**, 165.

Auger, M., Smith, I., Mantsch, H. and Wong, P. (1990) *Biochemistry*, **29**, 2008.

Barceló, M.J. and Bellanato, J. (1953) *Anal. R.S. Esp. Fis. Quim. (Madrid)*, **49**, 557.

Bartlett, J.C. and Chapman, D.G. (1961) *J. Agric. Food Chem.*, **39**, 50.

Blume, A., Hübner, W. and Messner, G. (1988) *Biochemistry*, **27**, 8239.

Braiman, M.S. and Rothschild, K.J. (1988) *Annu. Rev. Biophys. Biophys. Chem.*, **17**, 541.

Brown, J.K., Sheppard, N. and Simpson, D.M. (1950) *Discuss. Faraday Soc.*, 261.

Brown, J.K., Sheppard, N. and Simpson, D.M. (1954) *Phil. Trans. R. Soc.*, **247**, 35.

Cameron, D.G., Casal, H.L. and Mantsch, H.H. (1980) *Biochemistry*, **19**, 3665.

Casal, H.L. and Mantsch, H.H. (1984) *Biochim. Biophys. Acta*, **779**, 381.

Casal, H.L. and McElhaney, R.N. (1990) *Biochemistry*, **29**, 5423.

Casal, H.L., Mantsch, H.H., Paltauf, E. and Hauser, H. (1987a) *Biochim. Biophys. Acta*, **919**, 275.

Casal, H.L., Martin, A., Mantsch, H.H., Paltauf, E. and Hauser, H. (1987b) *Biochemistry*, **26**, 4408.

Casal, H.L., Mantsch, H.H. and Hauser, H. (1989) *Biochim. Biophys. Acta*, **982**, 228.

Castresana, J., Valpuesta, J.M., Arrondo, J.L.R. and Goñi, F.M. (1991) *Biochim. Biophys. Acta*, **1065**, 29.

Castresana, J., Nieva, J.L., Rivas, E. and Alonso, A. (1992a) *Biochem. J.*, **282**, 467.

Castresana, J., Alonso, A., Arrondo, J.L.R. and Goñi, F.M. (1992b) *Eur. J. Biochem.*, **204**, 1125.

Caughey, W.S., Smythe, G.A., O'Keeffe, D.H., Maskasky, J.E. and Smith, M.L. (1975) *J. Biol. Chem.*, **250**, 7602.

CECON GROUP (1990) *Pract. Spectrosc.*, **10**, 1.

Centeno, J. and O'Leary, T. (1990) *Biochemistry*, **29**, 7289.

Chapman, D. (1955) *Nature*, **176**, 216.

Chapman, D. (1956a) *J. Chem. Soc.*, 55.

Chapman, D. (1956b) *J. Chem. Soc.*, 2522.

Chapman, D. (1957a) *J. Chem. Soc.*, 2715.

Chapman, D. (1957b) *J. Chem. Soc.*, 4489.

Chapman, D. (1958a) *J. Chem. Soc.*, 784.

Chapman, D. (1958b) *J. Chem. Soc.*, 3186.

Chapman, D. (1958c) *J. Chem. Soc.*, 4680.

Chapman, D. (1962) *Chem. Rev.*, **62**, 433.

Chapman, D. (1965) *The Structure of Lipids*, Methuen, London.

Chapman, D., Byrne, P. and Shipley, G.G. (1966) *Proc. R. Soc.*, **290**, 115.

Chapman, D., Williams, R.M. and Ladbrooke, B.D. (1967) *Chem. Phys. Lipids*, **1**, 445.

Chapman, D., Gómez-Fernández, J.C., Goñi, F.M. and Barnard, M. (1980) *J. Biochem. Biophys. Methods*, **2**, 315.

Choi, S., Ware, W., Lauterbach, S. and Philips, W. (1991) *Biochemistry*, **30**, 8563.

Cole, A.R.H. (1977) IUPAC Commission on Molecular Structure and Spectroscopy, *Tables of Wavenumbers for the Calibration of Infrared Spectrometers* (2nd edn), Pergamon, Oxford.

Cooley, J.W. and Tukey, J.W. (1965) *Math. Comput.*, **19**, 297.

Cortijo, M., Alonso, A., Gómez-Fernández, J.C. and Chapman, D. (1982) *J. Mol. Biol.*, **157**, 597.

Craver, C.D. (1977) *The Coblentz Society Desk Book of Infrared Spectra*, Coblentz Society, Kirkwood, MO.

Davies, M.A., Schuster, H.F., Brauner, J.W. and Mendelsohn, R. (1990) *Biochemistry*, **29**, 4368.

De Rosa, M. and Gambacorta, A. (1988) *Prog. Lipid Res.*, **27**, 153.

De Ruig, W.G. (1971) *Infrared Spectra of Monoacid Triglycerides* (Agric. Res. Rep. 759), Centre for Agricultural Publishing and Documentation, Wageningen, The Netherlands.

Dluhy, R.A., Chowdry, B.Z. and Cameron, D.G. (1985) *Biochim. Biophys. Acta*, **821**, 437.

Doumenq, P., Guillano, M., Bertrand, J.C. and Mille, G. (1990) *Appl. Spectrosc.*, **44**, 1355.

Durig, J.R. (ed.) (1990) *Vibrational Spectra and Structure*, **vol. 18,** *Applications of GT-IR*, Elsevier, Amsterdam.

Eads, T.M., Blaurock, A.E., Bryant, R.G., Roy, R.G. and Croasmun, W.R. (1992) *J. Am. Oil Chem. Soc.*, **69**, 1057.

Eskamani, A. (1975) *Pract. Spectrosc.*, 1, 623.

Fischmeister, I. (1975) *Prog. Chem. Fats Other Lipids*, **14**, 91.

Fringeli, U.P. and Günthard, H.H. (1981) in *Membrane Spectroscopy* (ed. E. Grell), Springer-Verlag, Berlin, p. 270.

Goñi, F.M. and Arrondo, J.L.R. (1986) *Faraday Discuss. Chem. Soc.*, **81**, 117.

Goormaghtigh, E. and Ruysschaert, J.M. (1990) in *Molecular Description of Biological Membranes by Computer Aided Analysis*, **vol. I** (ed. R. Brasseur), CRC Press, Boca Raton, FL, p. 285.

Grdadolnik, J., Kidric, J. and Hadzi, D. (1991) *Chem. Phys. Lipids*, **59**, 57.

Griffiths, P.R. and de Haseth, J.A. (1986) *Fourier*

Transform Infrared Spectrometry (*Chem. Anal. Serv. Monogr. Anal. Chem. Appl.* **Vol. 83**), Wiley, New York.
Hauser, H., Pascher, I. and Sundell, S. (1988) *Biochemistry,* **27**, 9166.
Hübner, W. and Mantsch, H.H. (1991) *Biophys. J.,* **59**, 1261.
Kates, M. (1986) *Techniques of Lipidology,* 2nd edn, North-Holland, Amsterdam.
Keller, R.J. (1986) *The Sigma Library of FTIR Spectra*, **vols 1 and 2**, Sigma Chemical Co., St Louis, MO.
Kirby, E.M., Evans-Vader, M.J. and Brown, M.A. (1965) *J. Am. Oil Chem. Soc.,* **42**, 437.
Kodali, D.R., Small, D.M., Powell, J. and Krishnan, K. (1991) *Appl. Spectrosc.,* **45**, 1310.
Lee, D.C. and Chapman, D. (1986) *Biosci. Rep.,* **6**, 235.
Lettington, A.H. (ed.) (1990) *Infrared Technology and Applications,* SPIE, London.
Levin, I.W., Mushayakarara, E. and Bittman, R. (1982) *J. Raman Spectrosc.,* **13**, 231.
Lewis, R.N.A.H. and McElhaney, R.N. (1992) *Biophys. J.,* **61**, 63.
Lin-Vien, D., Colthup, N.B., Fateley, W.G. and Grasselli, J.G. (1991) *The Handbook of Infrared and Raman Characteristic Frequencies of Organic Molecules*, Academic Press, Boston.
Mantsch, H.H. and McElhaney, R.N. (1991) *Chem. Phys. Lipids,* **57**, 213.
Mantz, A.S. (1977) *Ind. Res.,* **19**, 90.
Maroncelli, M., Strauss, H.L. and Snyder, R.G. (1985) *J. Chem. Phys.,* **89**, 4390.
Mendelsohn, R. (1984) in *Techniques in the Life Sciences,* **vol. BS110** (ed. L.-F. Tipson), Elsevier, Dublin.
Mendelsohn, R., Davies, M.A., Brauner, J.W., Schuster, H.F. and Dluhy, R.A. (1989) *Biochemistry,* **28**, 8934.
Muga, A. and Casal, H.L. (1990) *ACS Symp. Ser.,* **447**, 56.
Mushayakarara, E., Albon, N. and Levin, I.W. (1982) *Biochim. Biophys. Acta,* **686**, 153.
Nilsson, A., Holmgren, A. and Lindblom, G. (1991) *Biochemistry,* **30**, 2126.
Nyquist, R.A., Leugers, M.A., McKelvy, M.L., Papenfuss, R.R., Putzig, C.R. and Yurga, L. (1990) *Anal. Chem.,* **62**, 223R.
O'Keeffe, D.H. (1974) Ph.D. Dissertation, Arizona State University, Tempe.
O'Keeffe, D.H., Barlow, C.H., Smythe, G.A., Fuschsman, W.H., Moss, T.J., Lilienthal, H.R. and Caughey, W.S. (1975) *Bioinorg. Chem.,* **5**, 125.
Ortiz, A., Aranda, F.J., Villalain, J., San Martin, C., Micol, V. and Gómez-Fernández, J.C. (1992) *Chem. Phys. Lipids,* **62**, 215.
Panford, J. and De Man, J. (1990) *J. Am. Oil Chem. Soc.,* **67**, 473.
Pouchert, C.J. (1981) *The Aldrich Library of Infrared Spectra*, 3rd edn, Aldrich Chemical Co., Milwaukee.
Pouchert, C.J. (1985–89) *The Aldrich Library of FTIR Spectra*, **vols 1–3**, Aldrich Chemical Co., Milwaukee.
Primas, H. and Günthard, H.H. (1953) *Helv. Chim. Acta,* **36**, 1791.
Putzig, C.L., Leugers, M.A., McKelvy, M.L., Mitchell, G.E., Nyquist, R.A., Papenfuss, R.R. and Yurga, L. (1992) *Anal. Chem.,* **64**, 270R.
Ratnayake, W.M.N., Hollywood, R., O'Grady, E. and Beare-Rogers, J.L. (1990) *J. Am. Oil Chem. Soc.,* **67**, 804.
Schmid, H.H.O., Schmid, P.C. and Natarajan, V. (1990) *Prog. Lipid Res.,* **29**, 1.
Senak, L., Moore, D. and Mendelsohn, R. (1992) *J. Phys. Chem.,* **96**, 2749.
Simons, W.W. (1978) *The Sadtler Handbook of Infrared Spectra, Sadtler, Philadelphia.*
Sipos, J.C. and Ackman, R.G. (1978) *J. Chromatogr. Sci.,* **16**, 443.
Snyder, R.G. (1960) *J. Mol. Spectrosc.* **4**, 411.
Susi, H. and Smith, A.M. (1960) *J. Am. Oil Chem. Soc.,* **37**, 431.
Tasumi, M., Shimanouchi, T., Watanabe, A. and Goto, R. (1964) *Spectrochim. Acta,* **20**, 629.
Trudeau, G., Dumas, J.M., Dupuis, P., Guérin, M. and Sendorgy, C. (1981) *Top. Curr. Chem.,* **93**, 91.
Tschamler, H. (1954) *J. Chem. Phys., 22*, 1845.
Vetter, W., Englert, G., Rigassi, N. and Schwieter, U. (1971) in *Carotenoids* (ed. O. Isler), Birkhäuser, Basel, p. 189.
Villalain, J. and Gómez-Fernández, J.C. (1992) *Chem. Phys. Lipids,* **62**, 19.
von Sydow, E. (1955) *Acta Chem. Scand.,* **9**, 1119.
Wallach, D.F.H., Verma, S.P. and Fookson, J. (1978) *Biochim. Biophys. Acta,* **382**, 153.
Wallach, D.F.H., Verma, S.P. and Fookson, J. (1979) *Biochim. Biophys. Acta,* **559**, 153.
Wheeler, D.H. (1954) *Progress in the Chemistry of Lipids,* vol. 2, Pergamon, London.
Wolfe, W.L. and Zissis, G.I. (1978) *The Infrared Handbook,* Erim, Ann Arbor, MI.
Wu, M., Church, D.F., Mahier, T.J., Barker, S.A. and Pryor, W.A. (1992) *Lipids,* **27**, 129.
Yasaei, P., Sheppard, A., Rudolf, T., Shen, C. and Aldridge, M.J. (1989) *Micronutr. Anal.,* **5**, 245.
Zakim, D. and Wong, P.T.T. (1990) *Biochemistry,* **29**, 2003.

9.2 Raman spectra

Technical advances in spectrometers, detection systems and especially lasers as excitation sources increased the interest of lipid chemists in Raman spectroscopy in the 1970s. Today, although still useful for specific applications, this technique has been somewhat superseded by Fourier-transform infrared spectroscopy. General introductions to Raman spectroscopy are found in Tobin (1971), Anderson (1971) and Colthup *et al.* (1975). A review of biological applications has been published by Carey and Salares (1980). Reviews dealing specifically with lipids are those by Wallach *et al.* (1979) and Verma and Wallach (1983).

Sample handling and preparation is very easy, as in the case of IR spectroscopy. Commercial instruments belong to two main types, conventional scanning and optical multichannel analysers. The main difference between the two types is that, in scanning spectrometers, the scattered light is collected and analysed, usually at $1\,cm^{-1}$ intervals (channels); whereas in optical multichannel analysers, a $300–600\,cm^{-1}$ section of the Raman spectrum, projected onto the detector, is rapidly recorded in a computer memory without mechanical scanning; the latter procedure is obviously much faster. More technical details may be found in Verma and Wallach (1983).

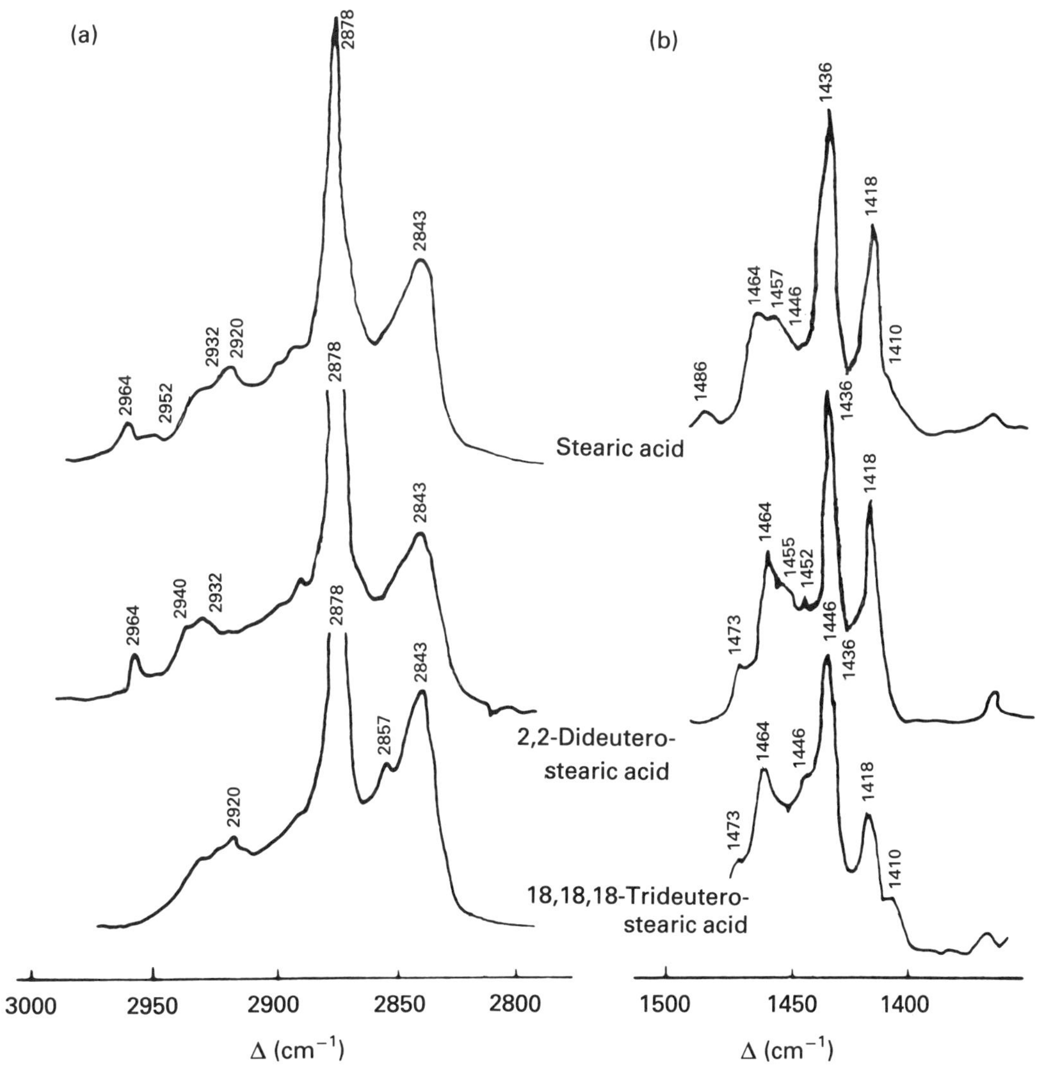

Fig. 9.13 Raman spectra of solid stearic acid. 2,2-dideuterostearic acid and 18,18,18-trideuterostearic acid in (a) the CH stretching region ($2800–3000\,cm^{-1}$), and (b) the HCH deformation range ($1400–1500\,cm^{-1}$). Recorded at 22 °C. From Verma and Wallach (1977).

Table 9.6 Characterization of selected lipids by Raman spectroscopy

System[a]	Parameters[a]	Spectral region (cm^{-1})	Comments	References
Phospholipids				
DMPC (M/U)	T_c	2800–3000 1064–1130	23 + 1 °C	Spiker and Levin (1976a)
DMPC (U)	T_c	2800–3000 1064–1300	Transition broadened	Spiker and Levin (1976a)
DMPC (M)	*trans*/*gauche* ratio	1064–1300	More *gauche* below T_c	Spiker and Levin (1976a)
DMPC–H_2O (gel)	*trans*/*gauche* ratio	1064–1300	Two *gauche* bonds/lipid below T_c and seven above	Yellin and Levin (1977b)
DMPC (M)	Cooperative units	2800–3000	400	Verma *et al.* (1980a)
DMPC (M)	*gauche* conformers	1130; 2930	*Gauche* bonds start to appear at −100 °C	Pink *et al.* (1980) Bunow and Levin (1977a)
DPPC (M)	T_c	2800–3000	41 °C	Gaber *et al.* (1978a)
DPPC (U)	T_c	1064–1130	T_c broadened	Spiker and Levin (1976a)
DPPC (suspension)	Chain order	Diff. spectr.[a] (2800–3000; 1064–1130)	Triclinic below 30 °C; loose hexagonal 30–40 °C	Gaber *et al.* (1978a)
DPPC (M)	Cooperative units	1064–1130	31	Yellin and Levin (1977a)
DPPC–H_2O (gel)	*trans*/*gauche* ratio	1064–1130	Two *gauche* bonds below T_c; 8–9 *gauche* bonds above	Yellin and Levin (1977b)
DSPC (M)	T_c	2800–3000 1064–1130	54 °C	Verma *et al.* (1980a)
DSPC (M)	Cooperative units	2800–3000	200	Verma *et al.* (1980a)
DSPC–H_2O (gel)	*trans*/*gauche* ratio	1064–1130	Six *gauche* bonds/molecule below T_c and 13–14 above	Yellin and Levin (1977b)
Egg PC	T_c	2800–3000 1064–1130	−6 °C	Verma and Wallach (1976b)
Sphingomyelin (dry)	T_c	1800–3000 1064–1130	85–87 °C	Mendelsohn *et al.* (1975)
Sphingomyelin (dry)	*trans*/*gauche* ratio	1064–1130	*Gauche* isomers below T_c	Mendelsohn *et al.* (1975)
PE (79% in water)	T_c	2800–3000	71 °C	Brown *et al.* (1973)
PS (water dispersion)	T_c	2800–3000 1064–1130	10 °C	Hark and Ho (1979)
DPPA + DMPC	T_c	1064–1130	20–24 °C (DMPC) and 57–59 °C (DPPA)	Lippert *et al.* (1980)

Phospholipids/additions				
PS + Mg^{2+}	T_c	2800–3000 1063–1130	Mg^{2+} shifts T_c from 10 to 20 °C	Hark and Ho (1979)
PS + Ca^{2+}	T_c	1063–1130	Ca^{2+} abolishes T_c	Hark and Ho (1980)
PS + DMPC + Ca^{2+}	*trans/gauche* ratio	1063–1130	More *trans* at/below T_c	
DPPC + I_2	*trans/gauche* ratio	1063–1130	I_2 raises *gauche* fraction	Loshchilova and Karvaly (1977)
Phospholipid/cholesterol				
DMPC/cholesterol (50 mol%)	T_c	2800–3000 1064–1130	Transition broadened	Spiker and Levin (1975)
DPPC or egg PC (50 mol%)	T_c	2800–3000 1064–1130	Transition widened by 160 °C	Spiker and Levin (1976a)
DMPC or DPPC/cholesterol (30 mol%)	1100–1600 cm^{-1} β-carotene as resonance Raman probe	Intensity of 1527 cm^{-1}	β-Carotene band reduces to a minimum (at 21 °C)	Verma and Wallach (1976b)
DMPC/cholesterol or DPPC/cholesterol	*trans/gauche* ratio	1064–1130	Cholesterol induces *gauche* conformation below and prevents *gauche* conformation above T_c	Lippert and Peticolas (1971) Spiker and Levin (1975)
DPPC/cholesterol frequencies	C=O stretch	1720–1740	Cholesterol does not form H bonds with C=O groups; induces non-equivalence of two ester linkages	Bush *et al.* (1980a, b) Umemura *et al.* (1980)
Egg PC/cholesterol	Chain conformation	Acoustic; lattice vibrations	Cholesterol impedes ordering of acyl chains	Verma and Wallach (1976a)
Egg PC/cholesterol	β-Carotene state; 1527 cm^{-1} resonance band	1400–1600	$\|I_{1527}/I_{1440}\|$ reaches minimum at cholesterol/lipid ratio of 0.9, indicating oxidation of carotenene	Verma and Wallach (1977)
DPPC/cholesterol/amphotericin B	Interaction of amphotericin B	1400–1600 cm^{-1} resonance band at 1556 cm^{-1}	Amphotericin B reacts specifically with cholesterol	Bunow and Levin (1977b)
Phospholipid/fatty acid				
DMPC, DPPC, DSPC, egg PC/oleic acid, linoleic acid, linolenic acid, linelaidic acid	T_c	2800–3000	*Cis*-unsaturated fatty acids split the transition; main T_c broadens and shifts to lower temperatures; linoleic acid most effective	Verma *et al.* (1980a)

[a] *Abbreviations*: DMPC, DPPC, DSPC, dimyristoyl-, dipalmitoyl-, distearoylphosphatidylcholine; DPPA, dipalmitoylphosphatidic acid; PE, phosphatidylethanolamine; PS, phosphatidylserine; (U) and (M), unilamellar and multilamellar liposomes; T_c, temperature of crystal–liquid crystal transition; Diff. spectr., difference spectroscopy.

The Raman spectra of lipids contain predominantly bands arising from vibrations of the hydrocarbon chains, with some contributions from those of the polar groups (Table 9.6). For the hydrocarbon chains, analysis of the Raman spectra is facilitated by the extensive vibrational studies on alkanes available in the literature (Snyder, 1967; Schaufele, 1968). The spectroscopic regions of interest in the Raman study of hydrocarbons are: (a) 2800–3000 cm^{-1}, the C—H stretching modes; (b) 1650–1700 and 1260–1310 cm^{-1}, C=C stretching and =C—H twisting modes; (c) 1400–1500 cm^{-1}, the HCH deformation range; (d) 1000–1200 cm^{-1}, the C—C stretching modes; (e) 10–400 cm^{-1}, the acoustic and lattice vibrations.

(a) The CH_2 and CH_3 residues of lipids yield prominent C—H stretching bands; bands due to symmetric and asymmetric CH_2 stretching can be

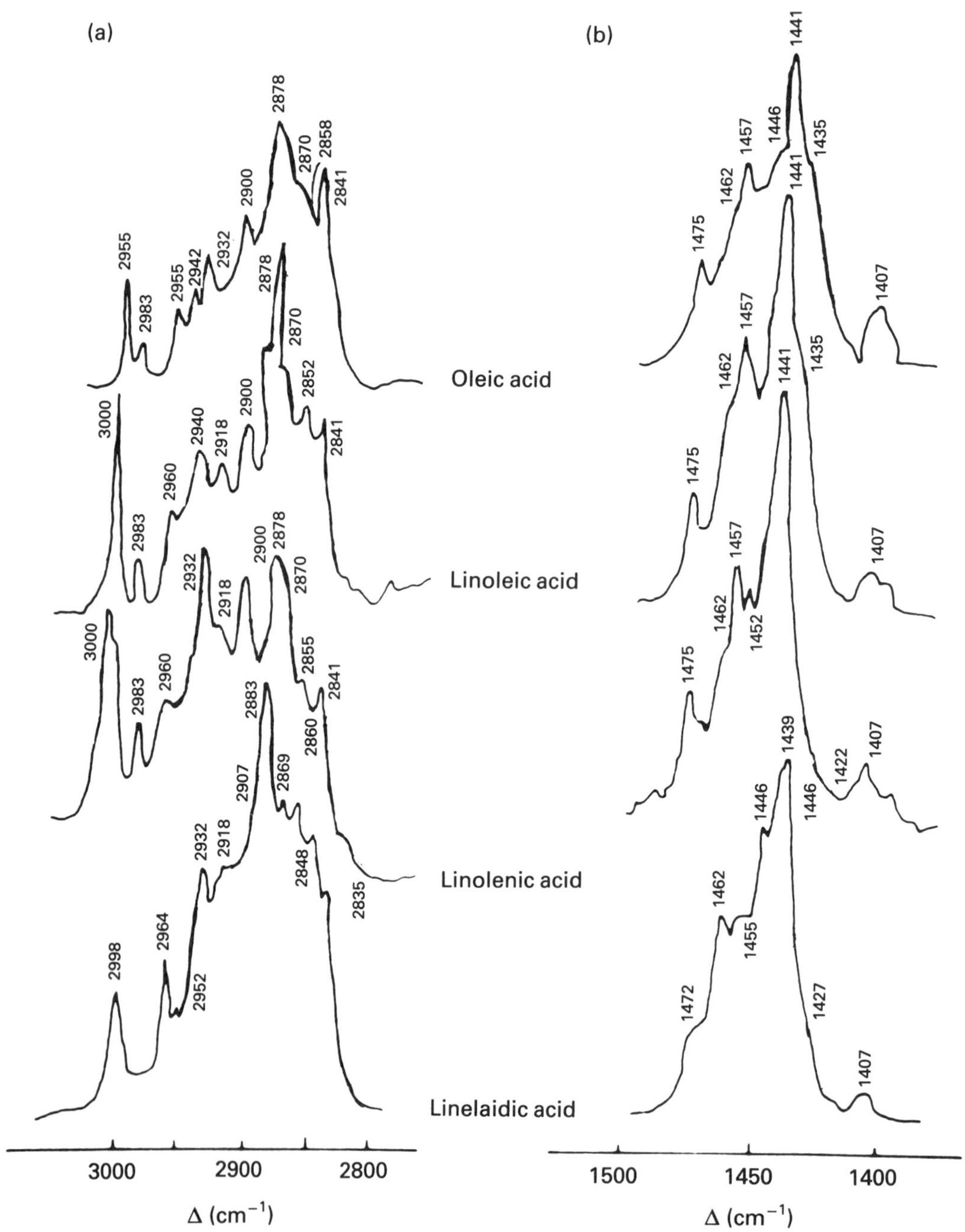

Fig. 9.14 Raman spectra of solid oleic, linoleic, linolenic and linelaidic acids in (a) the CH stretching region (2800–3100 cm^{-1}) and (b) the CHC deformation region (1440–1500 cm^{-1}). Recorded at $-40\,°C$. From Verma and Wallach (1977).

commonly differentiated, usually near 2850 and 2880 cm^{-1} respectively. Raman features due to methyl groups occur at 2932, 2952 and 2964 cm^{-1} (Verma and Wallach, 1977). Deuteration of stearic acid at specific positions reveals that CH_2 stretching vibrations of long methylenic segments are not representative of CH_2 residues adjacent to the CH_2 or to the carboxyl terminus (Fig. 9.13). The intensities and widths of CH_2 stretching bands depend on chain packing. Gaber and Peticolas (1977) have monitored thermotropic phase transitions of lipids through changes in the 2800–3000 cm^{-1} region of the Raman spectrum.

(b) *Cis* double bonds of unsaturated acyl chains give a peak at 1660 ± 2 cm^{-1} and *trans* double bonds at 1670 cm^{-1} (Davies *et al.*, 1975). Butler *et al.* (1979) have monitored unsaturation by use of the bands at 1303 and 1267 cm^{-1}, which are assigned to in-phase methylene twisting and =C—H in-plane deformation, respectively.

(c) Methylene bending modes are found in the 1400–1500 cm^{-1} region. Important changes are also seen in this region upon lipid melting. Thermally induced changes in the 2843–2850 cm^{-1} bands have been interpreted as reflecting reduced Fermi interaction, with an overtone of a symmetric methylene bending component contributing to the 1418 cm^{-1} feature (Gall *et al.*, 1972) (Fig. 9.14).

(d) Long-chain hydrocarbons yield three intense bands in the skeletal optical regions (C—C stretching, 1000–1200 cm^{-1}). The sharp features at 1064 and 1130 cm^{-1} are assigned to all-*trans* chain segments, and the broader third band, around 1090 cm^{-1}, is attributed to chains containing *gauche* conformations (Lippert and Peticolas, 1971; Pink *et al.*, 1980) (Fig. 9.15).

(e) The region below 600 cm^{-1} contains bands arising from accordion-like longitudinal vibrations of extended acyl chains. For example, solid lauric acid at −160 °C gives an acoustic mode at 205 cm^{-1} and torsional or lattice modes at 171 and 119 cm^{-1} (Mendelsohn *et al.*, 1975), while dilauroylphosphatidylethanolamine displays broader weaker features at 224 and 138 cm^{-1}, with shoulders at 218 and 148 cm^{-1}. The differences in band positions are related to chain packing.

Features from polar headgroup vibrations can also be identified, and detailed assignments have been made for the glycerophosphorylethanolamine (Akutsu and Hyogoku, 1975) and the glycerophosphorylcholine (Koyama *et al.*, 1977) headgroups. The latter study also showed that headgroup conformations are relatively insensitive to hydrocarbon chain melting.

The ability of Raman spectroscopy to evaluate the proportion of *trans* and *gauche* rotamers in hydrocarbon chains has been exploited in a large number of studies involving phospholipid bilayers and various model and biomembranes. These studies have been reviewed by Verma and Wallach (1983).

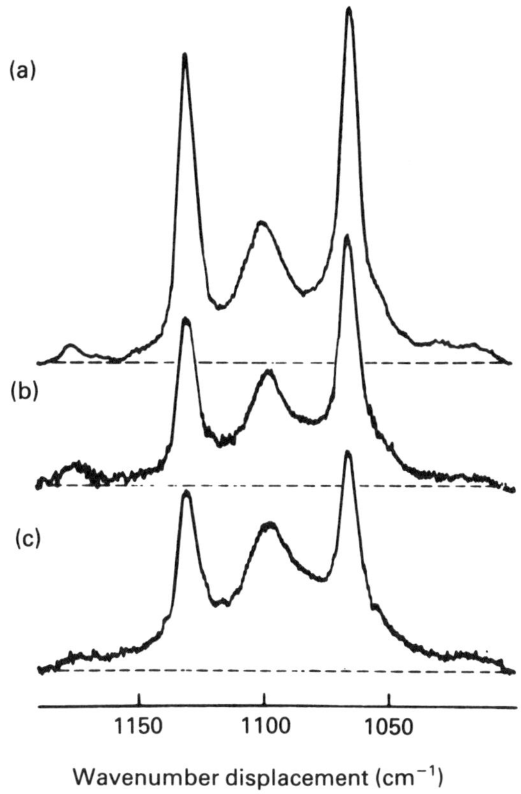

Fig. 9.15 Raman spectra of 1,2-dipalmitoylphosphatidylcholine in the (a) polycrystalline, (b) multilayer and (c) vesicle forms in the 1020–1170 cm^{-1} region utilizing 515.5 nm excitation. From Spiker and Levin (1976a).

More recently, interdigitated bilayer membranes have been examined by Raman spectroscopy (Slater and Huang, 1988). Also, Yang *et al.* (1991) have applied resonance Raman spectroscopy to study bilirubins and bilirubin–sphingomyelin liposomes to show that, in the mixtures, the bilirubin propionates form ion pairs with the quaternary ammonium ion of the choline moiety of sphingomyelin.

Fourier-transform Raman spectrometers have also become commercially available. The principles of FT-Raman have been reviewed by Schrader *et al.* (1990). Specific applications to the field of biomembranes have been covered by Levin and Lewis (1989). Also, an interesting application to the differentiation of steroid hormone isomers has been described (Butler *et al.*, 1990).

References and further reading

Akutsu, H. and Hyogoku, Y. (1975) *Chem. Phys. Lipids,* **14**, 113.

Anderson, A. (1971) *The Raman Effect*, vol. 1, Marcel Dekker, New York.

Brown, K.G., Peticolas, W.L. and Brown, E. (1973) *Biochem. Biophys. Res. Commun.,* **94**, 358.

Bunow, M.R. and Levin, I.W. (1977a) *Biochim. Biophys. Acta,* **487**, 388.

Bunow, M.R. and Levin, I.W. (1977b) *Biochim. Biophys. Acta,* **464**, 202.

Bush, S.F. and Levin, I.W. (1980b) *Chem. Phys. Lipids,* **27**, 101.

Bush, S.F., Adams, R.G. and Levin, I.W. (1980a) *Biochemistry,* **19**, 4429.

Butler, I., Dicaire, F., Cordier, C., Top, S., Vessieres, A., Jaoeun, G. and Guy, H. (1990) *J. Raman Spectrosc.,* **21**, 355.

Butler, M., Salem, N., Hoss, W. and Spoonhower, J. (1979) *Chem. Phys. Lipids,* **29**, 99.

Carey, P.R. and Salares, V.R. (1980) *Adv. Infrared Raman Spectrosc.* **7**, 1.

Colthup, N.B., Daly, L.H. and Wiberly, S.E. (1975) *Introduction to Infrared and Raman Spectroscopy.* Academic Press, New York.

Davies, J.E.D., Hodge, P., Gunstone, F.D. and Lie Ken Jie, M.S.F. (1975) *Chem. Phys. Lipids,* **15**, 48.

Gaber, B.P. and Peticolas, W.L. (1977) *Biochim. Biophys. Acta,* **465**, 260.

Gaber, B.P., Yager, P. and Peticolas, W.L. (1978a) *Biophys. J.,* **21**, 161.

Gaber, B.P., Yager, P. and Peticolas, W.L. (1978b) *Biophys. J.,* **24**, 677.

Gall, M.H., Hendra, P.J., Peacock, G.J., Cudby, M.E.A. and Willis, H.A. (1972) *Spectrochim. Acta,* **28A**, 1485.

Hark, S.K. and Ho, J.T. (1979) *Biochem. Biophys. Res. Commun.,* **91**, 655.

Hark, S.K. and Ho, J.T. (1980) *Biochim. Biophys Acta,* **601**, 54.

Koyama, Y., Toda, S. and Hyokoku, Y. (1977) *Chem. Phys. Lipids,* **19**, 74.

Levin, I. and Lewis, E. (1989) *Proc. SPIE,* **1145**, 99.

Lippert, J.L. and Peticolas, W.L. (1971) *Proc. Natl Acad. Sci. USA,* **68**, 1572.

Lippert, J.L., Lindsay, R.M. and Schultz, R. (1980) *Biochimm. Biophys. Acta,* **599**, 32.

Loshchilova, E. and Karvaly, B. (1977) *Chem. Phys. Lipids,* **19**, 159.

Mendelsohn, R., Sunder, S. and Bernstein, H.J. (1975) *Biochim. Biophys. Acta,* **413**, 329.

Pink, D.A., Green, T.J. and Chapman, D. (1980) *Biochemistry,* **19**, 349.

Schaufele, R.F. (1968) *J. Chem. Phys.,* **49**, 4168.

Schrader, B., Hoffmann, A., Simon, A., Podschadlowski, R. and Tischer, M. (1990) *J. Mol. Struct.,* **217**, 207.

Slater, J.L. and Huang, C.H. (1988) *Prog. Lipid Res.,* **27**, 325.

Snyder, R.G. (1967) *J. Chem. Phys.,* **47**, 1316.

Spiker, R.C. and Levin, I.W. (1975) *Biochim. Biophys. Acta,* **388**, 361.

Spiker, R.C. and Levin, I.W. (1976a) *Biochim. Biophys. Acta,* **433**, 457.

Spiker, R.C. and Levin, I.W. (1976b) *Biochim. Biophys. Acta,* **455**, 560.

Tobin, M.C. (1971) Laser Raman Spectroscopy. Wiley-Interscience, New York.

Umemura, J., Cameron, D.G. and Mantsch, H.H. (1980) *Biochim. Biophys. Acta,* **602**, 32.

Verma, S.P. and Wallach, D.F.H. (1976a) *Biochim. Biophys. Acta,* **416**, 616.

Verma, S.P. and Wallach, D.F.H. (1976b) *Biochim. Biophys. Acta,* **436,** 307.

Verma, S.P. and Wallach, D.F.H. (1977) *Biochim. Biophys. Acta,* **486**, 217.

Verma, S.P. and Wallach, D.F.H. (1983) in *Biomembrane Structure and Function* (ed. D. Chapman), Macmillan, London, p. 167.

Verma, S.P., Wallach, D.F.H. and Sakura, J.D. (1980a) *Biochemistry,* **19**, 574.

Verma, S.P., Schmidt-Ullrich, R. and Wallach, D.F.H. (1980b) *J. Receptor Res.,* **1**, 1.

Wallach, D.F.H., Verma, S.P. and Fookson, J. (1979) *Biochim. Biophys. Acta,* **382**, 542.

Yang, B., Morris, M.D., Xie, M. and Lightner, D.A. (1991) *Biochemistry,* **30**, 688.

Yellin, N. and Levin, I.W. (1977a) *Biochim. Biophys. Acta,* **468**, 490.

Yellin, N. and Levin, I.W. (1977b) *Biochemistry,* **16**, 642.

9.3 Nuclear magnetic resonance spectra

Nuclear magnetic resonance (NMR) spectroscopy is at present one of the most widely applied physical techniques in biology, and its potential applications increase day by day, as more sophisticated instrumentation becomes available and deeper theoretical knowledge is obtained. The phenomenon of NMR was discovered simultaneously by Purcell and his associates at Harvard University and by Bloch and co-workers at Stanford University, for which they were jointly awarded the Nobel prize in physics in 1952. In the lipid field there are two main types of NMR spectroscopy that are of interest: broad-line experiments, concerned mainly with the spectra obtained from samples in the solid state, or from oriented phases; and narrow-line, or high-resolution, experiments, carried out with samples in the liquid, solution or gas phases. Both types of NMR spectroscopy are extremely useful in the study of lipids. In addition, Fourier transform (FT) NMR has helped increase the sensitivity of the technique, and the so-called 'pulse method' of recording spectra has literally widened the prospect of NMR applications in the field of lipid research and industry. In particular, the technique is used routinely to measure the solid content of fats.

9.3.1 Theoretical considerations

(a) Interaction of nuclei with magnetic fields

NMR occurs as a result of the interaction of a static magnetic field H with a nucleus possessing a magnetic moment μ. Only particles with both an electric charge and a spin angular momentum L will have a magnetic moment. The magnitude of L is given by

$$L = \hbar[I(I + 1)]^{1/2} \tag{9.5}$$

where $\hbar$ is Planck's constant h divided by 2π and I is the spin quantum number of the nucleus; μ and L are related as follows:

$$\mu = \gamma L$$

γ being the magnetogyric ratio. From quantum-mechanical considerations, it is known that the value of γ is given by

$$\gamma = g_n(\beta_n/h) \tag{9.6}$$

where g_n is the nuclear g-factor and β_n is the nuclear magneton. The latter is a constant, with a value of 5.05×10^{-24} erg/G; γ is usually, but not always, positive. When interacting with an external magnetic field, the magnetic moment vector orients itself with respect to that field and experiences a movement of angular precession defined by the angular velocity vector $\omega = \gamma H$, where ω is called the Larmor frequency.

The energy of interaction of the nuclear dipole and the static field is given by the scalar product of vectors μ and H:

$$E = -\mu H \tag{9.7}$$

The minus sign means that the alignment with the lowest energy is with the magnetic moment parallel to the field direction, which is known as the z direction.

The energy E is quantized. Quantum-mechanical calculations constrain the values of the z component of μ that lies along the field to

$$\mu_z = m_I \gamma\hbar \tag{9.8}$$

where $m_1 = I, I - 1, \ldots, 0. \ldots, -(I - 1), -I$. For the common case of $I = 1/2$ (e.g.. ^1H, ^{13}C, ^{31}P), only two E values, or two alignments of μ, are allowed, so that

$$E = \pm 1/2\, \gamma\hbar H$$

The difference in energy between these two orientations is

$$\Delta E = \gamma\hbar H$$

The ratio of nuclei in the alignments depends on the Boltzmann factor $\exp(-\Delta E/kT)$, where k is Boltzmann's constant. At 300 K (about room temperature), for $H = 10\,000\ G$, the ratio of parallel/antiparallel protons is 1.000 007. Thus, nuclei are almost evenly distributed in the various orientations, with only a slight excess, 7 in 10^6 in the case of protons, in the parallel or low-energy orientation.

Therefore, one of the main difficulties of NMR is that ΔE is so small. From equations (9.7) and (9.8) it is clear that ΔE and sensitivity depend on γ. A list of nuclei, with their respective I, μ and ω values, and relative sensitivities is shown in Table 9.7. Only those nuclei which are able to give NMR signals have measurable magnetogyric ratios: these are nuclei with odd mass numbers (such as ^1H or ^{19}F) or those with even mass numbers and odd charge numbers (such as ^2H or ^{14}N).

(b) The NMR experiment

In this discussion we shall follow quite closely an analogy based on classical electrodynamics (Pople *et al.*, 1959; Becker, 1969). Let us consider a number

Table 9.7 Magnetic properties of selected nuclei (Data from Cantor and Schimmel, 1980)

Nucleus	Z	I	μ/β_n	γ (rad/G s)	$\omega/2\pi$ (MHz 10^4 G)	Natural abundance (%)	Relative sensitivity
^{1}H	1	½	4.84	27753	42.6	99.98	1.000
^{2}H	1	1	1.21	4107	6.5	0.016	0.0096
^{12}C	6	0	–	–	–	99.89	–
^{13}C	6	½	1.22	6728	10.7	1.11	0.016
^{14}N	7	1	0.57	1934	3.1	99.64	0.0010
^{16}O	8	0	–	–	–	99.76	–
^{19}F	9	½	4.55	24179	40.1	100	0.834
^{31}P	15	½	1.96	10840	17.2	100	0.066

of nuclei, each with its magnetic moment μ randomly oriented (Fig. 9.16). If a magnetic field H_z is imposed on the z axis, the nuclei will orient along and precess around this axis with a Larmor frequency ω. The vectorial sum of all the μ_z components is no longer zero, and is called magnetization M_z. Once the magnetic field is switched on, M_z increases with time from 0 to its equilibrium value $\bar{M}_z$.

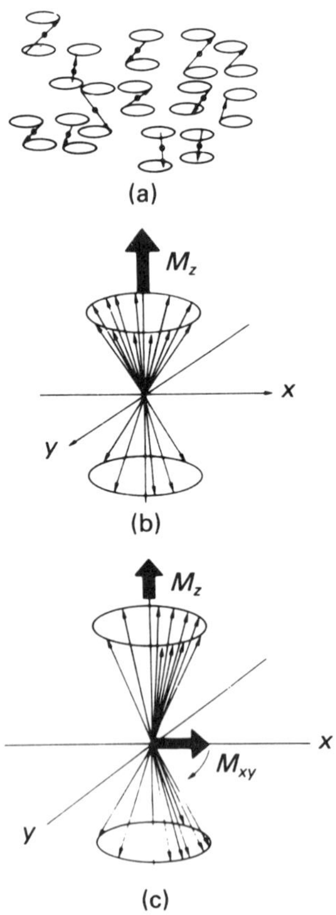

Fig. 9.16 An NMR experiment. Nuclei are represented by their magnetic moment μ. (a) Nuclei oriented at random. (b) Once the magnetic field H_z has been established, the nuclei become oriented, and the magnetization vector M_z appears. (c) In the presence of a second magnetic field H_{xy}, a new component M_{xy} appears and $M_z < \bar{M}_z$.

The nuclei are oriented only in relation to the field axis; therefore $M_x = M_y = 0$, and $M = M_z$. Let us imagine now that a second magnetic field H_{xy} is imposed on the xy plane, also rotating at the Larmor frequency ω. This second magnetic field may be the magnetic component of the electromagnetic radiation in an NMR experiment. As a consequence of H_{xy}, M will be tipped away from the z axis and give rise to a net M_{xy} component in the xy plane; under these conditions M_z will fall below its equilibrium value $\bar{M}_z$. The shift of M from the equilibrium, that is lowest energy, to its new orientation corresponds to an energy absorption. Note that energy absorption only occurs when H_{xy} oscillates at the Larmor frequency ω. As was stated in the previous subsection, for protons and other nuclei

$$\Delta E = \gamma \hbar H_z \tag{9.9}$$

but

$$\omega = \gamma H_z \tag{9.10}$$

and it also well known that $\omega = 2\pi\nu$; thus $\beta E = h\nu$. This means that H_{xy} must oscillate at a frequency such that its associated energy corresponds precisely to the quantized energy gap ΔE between the two allowed orientations. Thus,

$$h\nu = \Delta E = \gamma \hbar H_z \tag{9.11}$$

(c) *Relaxation: the Bloch formulation*

When the H_{xy} field is switched off, relaxation occurs: the longitudinal component of magnetization M_z goes back to its equilibrium value $\bar{M}_z$, whereas the transverse component M_{xy} vanishes. Bloch assumed that these were both first-order processes, so that

$$M_z = \bar{M}_z\,[1 - \exp(-t/T_1)] \tag{9.12}$$

and

$$M_{xy} = M_{xy}^0 \exp(-t/T_2) \tag{9.13}$$

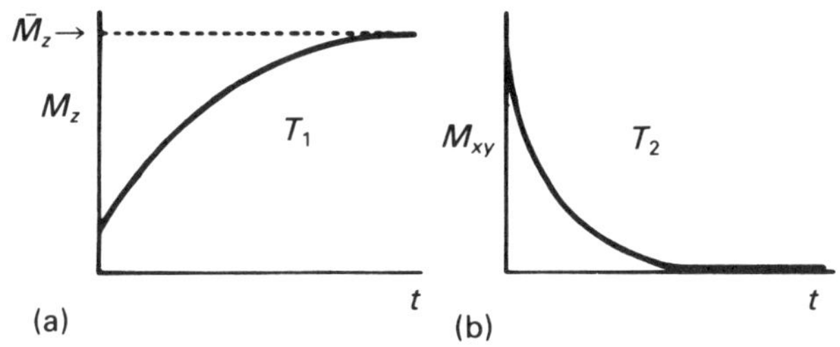

Fig. 9.17 Relaxation times: (a) longitudinal relaxation; (b) transverse relaxation.

where M^0_{xy} is the M_{xy} value before H_{xy} is turned off; t is the time after switching off H_{xy}; T_1 and T_2 are, respectively, the longitudinal and transverse relaxation times (Fig. 9.17).

Starting from this point, Bloch developed a series of useful equations relating each magnetization component with the various magnetic field components and the relaxation times. A simple and informative account of these calculations may be found in Cantor and Schimmel (1980). One of the most important results from this treatment may be expressed by the following equation:

$$g(\omega) = \frac{2T_2}{1 + T_2^2(\omega - \omega_1)^2} \qquad (9.14)$$

where $g(\omega)$ represents the intensity of the resonance signal at a frequency ω when the actual Larmor frequency of the resonating nuclei is ω_1. Equation (9.14) corresponds to the shape of the NMR spectral line; it is known as a Lorentzian lineshape. From equation (9.14), we can obtain useful information concerning the width of NMR lines. When $\omega = \omega_1$, in the centre of the peak, $g(\omega)$ is maximum

$$g(\omega)_{\max} = 2T_2 \qquad (9.15)$$

At half-height, $g(\omega) = \frac{1}{2}g\ (\omega)_{\max} = T_2$; this implies that the denominator of equation (9.14) equals 2. Thus

$$T_2^2(\omega - \omega_1)^2 = 1 \qquad (9.16)$$

$$1/T_2 = \omega - \omega_1 \qquad (9.17)$$

But $\omega - \omega_1$ is half the width of the resonance line, measured at its half-height. Therefore, if one takes into account that $\omega = 2\pi\nu$, then

$$\Delta\nu_{1/2} = 1/\pi T_2 \qquad (9.18)$$

(d) Free induction decay: Fourier transform

In FT-NMR, the radiofrequency irradiation is achieved by a short intense radiowave pulse, which is equivalent to the imposition of a second magnetic field H_x. This pulse will contain, among others, the Larmor frequencies of the nuclei under study, so that for a given nucleus there will be an H_x vector rotating at its own Larmor frequency. The pulses tilt M_2 a given angle (the nutation angle θ) from the original orientation, so that M_{xy} now has a finite value. A pulse producing a tilt of θ degrees is called a 'θ degree pulse'. The pulse lasts only a short time (1–200 μs), too short for relaxation processes to occur. After the pulse, the system is left in an unstable state. The rotation of M_{xy} around the z axis produces a radiation that is picked up by the radiofrequency coil in the receiver mode. This radiation is the FID (free induction decay), a sine wave of characteristic frequency decaying at a rate determined by T_2 of the particular nucleus as the M_{xy} component returns to zero. When various groups of nuclei are involved, the FID is actually a summation of as many sine waves.

If the amplitude of the FID is given by $g(t)$, it can be shown that the lineshape function $g(\omega)$ of the absorption spectrum is given by

$$g(\omega) = \text{const.} \int_0^\infty g(t)\cos(\omega t)\,\mathrm{d}t \qquad (9.19)$$

This is the mathematical basis for the interconversion of the FID or time-domain function $g(t)$, and the absorption spectrum or frequency-domain function $g(\omega)$. This interconversion is known as a Fourier transform (FT). Although mathematically $g(t)$ and $g(\omega)$ are easily interconvertible, in practice obtaining an FID takes a much shorter time. For instance, a single sweep of a spectrum of width 10^3 Hz, aiming at a 1 Hz resolution, takes about 10^3 s with a continuous-wave instrument; with the pulse method, an FID with the same degree of resolution is obtained in less than 1 s.

9.3.2 Experimental information

There are two basic ways to perform an NMR experiment: the continuous-wave and the pulse methods. The first one has been used in small instruments, and is described in many reviews (see, for instance, Chapman, 1965). We will concentrate our discussion on the pulse, or Fourier-transform, method.

(a) The spectrometer

The main elements of an NMR spectrometer are depicted in Fig. 9.18. There are four basic components: the magnet, the probe, the system of radiofrequency generation and detection, and the recording and computing system.

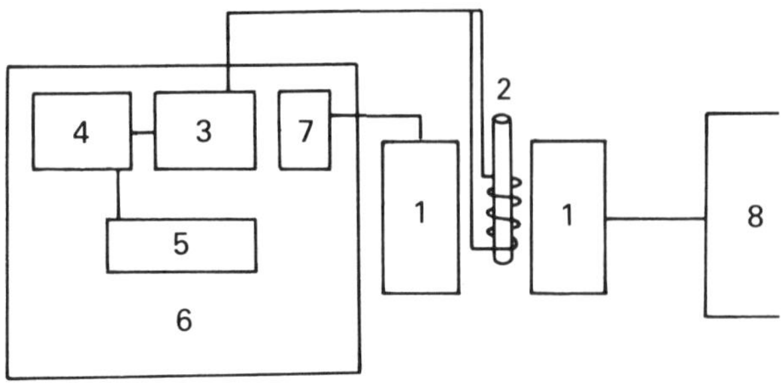

Fig. 9.18 A diagrammatic representation of the main components of an NMR spectrometer: (1) magnet; (2) probe (sample holder and radiofrequency coil); (3) radiofrequency amplifier and pulse generator; (4) computer; (5) recorder; (6) main console; (7) lock and shimming systems; and (8) magnet power source.

The magnet has to provide a magnetic field that is intense, homogeneous and invariant with time. An intense field is essential for sensitivity and resolution, as implied by equation (9.11). If the field is not homogeneous throughout the sample volume, identical nuclei in different positions are actually placed in different magnetic fields, and consequently resonate at slightly different frequencies, broadening the signal. Good resolution requires a field homogeneity better than 3 in 10^9, that is 0.6 Hz in 200 MHz. The best magnets cannot provide homogeneity better than 1 in 10^7; additional homogeneity is provided by auxiliary electric magnets (in the process known as shimming), as well as by spinning the sample. Field stability is important for the same reasons of homogeneity, since experiments may last for many hours. Invariance with time is impaired by variations in power supply, as well as by the movement of metal masses near the magnet. Stability is assured by a feedback system (known as a lock), which compares some strong resonance from the sample ^1H_2O or ^2H_2O with a reference radiofrequency. As soon as the field varies, the sample resonance is shifted, and the locking system applies a suitable correction. Only superconducting magnets provide the necessary specifications for biological NMR. A superconducting magnet consists simply of a solenoid immersed vertically in a Dewar vessel with liquid helium; once the current has been set going in the coil, the magnetic field remains very constant with time.

The probe contains the sample holder and the radiofrequency coil, as well as the sample spinning device, temperature control and part of the detection system. Commercial probes can hold samples from 5 to 25 mm in diameter, but, as we have mentioned, much larger sample holders are being built experimentally.

The radiofrequency has to be regulated as carefully as the magnetic field intensity. The radiation is generated by a crystal oscillator, amplified electronically and emitted through a coil placed in the probe. The same coil serves as a radiofrequency receiver, since irradiation and detection occur as events separated in time (time-sharing coils).

Finally, the signals received by the coil must be stored, summed and transformed by a computer into a spectrum. An oscilloscope and a potentiometric recorder complete this part of the instrument.

(b) The pulse method: Fourier transform

In conventional spectroscopy (absorption as well as NMR), irradiation is achieved by exposing the sample consecutively to electromagnetic waves of varying wavelengths until a given region of the spectrum has been scanned (continuous-wave method). In biological NMR, because of the complex biomolecules and small amounts of samples, a good signal-to-noise ratio can be obtained only by summation of many spectra. The continuous-wave method is too slow to produce in a reasonable time the thousands of spectra that are often required; instead, the pulse method is used (Farrar and Becker, 1971).

A pulse consists of a very short and intense discharge of radiofrequencies resulting from the superposition of a whole frequency spectrum. When the radiofrequency coil emits a pulse, all the resonances in the sample are excited simultaneously. After the pulse is switched off, the coil, now acting as a receiver, will receive a signal from the excited nuclei; the signal will decay as relaxation takes place. The signal, known as free induction decay (FID), is a sine wave; its frequency is related to the resonance frequency of the nuclei, and its decay rate is related to T_2.

The FID is a function of time (time domain) but, as we know, the spectra are represented as a function of field intensity or frequency (frequency domain). The time- and frequency-domain functions can be interconverted by a mathematical trick called the Fourier transform (FT).

(c) The sample

The amount of sample is very important in NMR spectroscopy, especially since the signal-to-noise ratio improves with the square root of the number of scans. But the sample requirements are not only quantitative. Among the qualitative factors, we should mention the following:

1. Paramagnetic ions, even in trace amounts, may produce measurable line broadening. Care must

be taken to avoid them in the samples. Complexing them with ethylenediamine tetraacetic acid (EDTA) is useful in preventing them from approaching the sample nuclei.

2. High-resolution NMR often requires the addition of D_2O for the locking system. In this case, the pH should be adjusted accordingly.
3. In some cases, it may be advisable, or necessary, to include a chemical-shift standard with the sample.
4. For long NMR runs, it may be necessary to add sodium azide or other antimicrobial agents to the sample.

(d) *Spectral characteristics*

Only resonances from a single class of nuclei (e.g. ^1H or ^{31}P or ^{19}F) can be recorded in a given spectrum. This is due to the fact that γ values vary considerably from one nucleus to another, and therefore the resonances to two classes of nuclei, for example ^1H or ^{13}C, occur at very different magnetic field intensities. Thus the first of the spectral characteristics, nuclear species, is fixed before the actual experiment is performed; we speak of ^1H (proton), ^{13}C, ^{19}F and so on, NMR spectra. When one of these spectra is examined, the various 'peaks' or signals may be characterized by their position in the spectrum of 'chemical shift', the width of the peak and the area under it. In addition, it is important to consider the possibility of multiple lines arising from a single nucleus.

The chemical shift δ indicates the position in the spectrum at which a signal occurs. The term 'shift' indicates that δ is measured with respect to a fixed zero. The latter corresponds to a predetermined NMR 'reference' signal, or is sometimes arbitrarily set at any convenient point in the spectrum. The chemical shift is usually expressed in parts per million (ppm) and may be defined as follows:

$$\delta \equiv \frac{\nu_{ref} - \nu_{sample}}{\text{reference}} \times 10^6 \qquad (9.20)$$

where ν_{ref} and ν_{sample} correspond to the resonance frequencies of the reference and sample nuclei, respectively. The shift δ is a dimensionless parameter and is therefore independent of the operating frequency of the instrument. In ^1H NMR most chemical shifts fall between 0 and 30 ppm; in ^{31}P spectra the corresponding values are between 0 and 100 ppm; whereas in ^{13}C spectra they fall in the region 0–200 ppm. Values are usually positive because the reference point is chosen in a high-field region; δ values increase towards the low-field region of the spectrum. Chemical shifts are very important, since the atoms in various functional groups have characteristic δ values, and thus NMR is a very efficient tool in the elucidation of the structure of small molecules. An elementary discussion on the origin of chemical shifts may be found in Goñi (1984).

The width of NMR lines is related to the phenomenon of relaxation. This concept has been defined. There are two main relaxation processes in NMR, namely, 'spin-lattice' (or longitudinal) relaxation and 'spin–spin' (or transverse) relaxation. Spin–lattice relaxation is the process by which the nuclei in the upper spin state lose energy externally to the crystal lattice (in a crystal) or, less effectively, to the surrounding liquid (in a solution). The speed at which spin–lattice relaxation occurs is characterized by the longitudinal relaxation time T_1.

In biological systems the conditions are such that relaxation depends mainly on dipole–dipole interactions; in addition, processes that increase spin–lattice relaxation also contribute to spin–spin relaxation. Therefore, T_2 relaxation is most important from our point of view. T_2 depends on molecular motion; this means that, as molecules move more rapidly in relation to each other, dipolar interactions become more difficult, because the nuclei can scarcely contact in the adequate orientation, and the relaxation rate is smaller. In summary, with a high molecular motion, T_2 is long and the line is narrow.

Given a number of conditions, the area under a peak in an NMR spectrum is proportional to the energy absorbed by each chemically distinct nucleus. It is thus an indication of the relative proportion of each nuclear population. Peak areas are especially useful in kinetic studies, where they can give a precise indication of the proportion of each chemical species present. For these quantitative studies, it is essential that no saturation occurs and that all the instrument conditions be kept constant.

Finally, a single nucleus may give rise to multiple lines: this is the so-called spin–spin splitting (Chapman, 1965; Knowles *et al.*, 1976; Goñi, 1984). Splitting occurs because each nucleus may be influenced by the various possible orientations of covalently linked neighbouring nuclei. This phenomenon may be very useful in the elucidation of the structure of small molecules; in some circumstances, however, splitting produces too complex spectra, and it must be eliminated. This is usually achieved by a double-resonance method in which a second electromagnetic radiation is applied precisely at the frequency where the nuclei producing the splitting are resonating. If the second radiation is sufficiently intense, the perturbing nuclei are rapidly saturated, so that they no longer influence their neighbours. This procedure is called spin decoupling, and is commonly used, for instance, in order to prevent perturbation by protons in ^{31}P NMR spectra.

Technical advance in NMR occur at a very fast pace. Good overviews of the subject are regularly presented in *Analytical Chemistry* (e.g. Haw, 1992).

9.3.3 High-resolution ^1H NMR

(a) Introduction

Lipids are commonly examined for high-resolution NMR in organic solution. Continuous-wave instruments require relatively concentrated solutions (35–50 mg/ml), while FT-NMR is less stringent in this respect, and lower concentrations (e.g. 1–10 mg/ml for a 200 MHz instrument) may be used. In principle, the solvent should not contain the nucleus under examination (e.g. CCl_4, CCl_3D or CS_2 for ^1H NMR). As an internal standard, tetramethylsilane (TMS) is often added to the solvent in a 5–10% concentration. A compilation of proton resonance signals that have diagnostic value for the major lipid classes is given in Table 9.8.

The ^1H spectrum of a straight-chain alkane $CH_3(CH_2)_nCH_3$ contains only two signals: a 6-H signal for the two end methyl groups (0.89 ppm) and a second signal (1.23 ppm) for the remaining hydrogen atoms. The spectrum becomes more complex – and more informative – when functional groups are incorporated into the alkane. The most important of these are acyl groups and unsaturated centres. Additional NMR signals arise from CH or CH_2 groups to which the unit is attached and also from nearby CH_2 and CH_3 groups under the influence of the functional group.

The effects of several functional groups are collected in Table 9.9. Since it is difficult to detect changes below 0.1 ppm, the influence of a functional group is seldom noticeable beyond the β position. With the exception of the methyl group, however, it is always significant in the α position. Small effects, not apparent on their own, may become so when a CH_2 or CH group is under the combined influence of two or three nearby functional groups. The effects of more than one functional group are generally additive, but they may not be so in a strict numerical sense. There are small, but not very useful, differences between monoene and methylene-interrupted polyene units. The figures in Table 9.9 can be used to calculate chemical shifts for a wide range of acids and esters.

(b) Methyl alkanoates

Instead of the two signals shown by an alkane, a methyl alkanoate (**1**) will usually show five signals: a, 0.90 (3H); b, 1.31; c, 1.58 (2H); d, 2.30 (2H); and e, 3.65 ppm (3H). The methyl ester has its own three-proton signal and its influence on the alkane chain is apparent in the α (C-2) and β (C-3) positions. The

Table 9.8 Proton (^1H) resonance signals for various structural groups (Taken from Kates, 1986; for a more detailed study see Frost and Gunstone, 1975)

Compound or group[a]	δ (ppm)
$(CH_3)_4Si$	0.00
—CH_2—, cyclopropane (*cis*)	0.2 (triplet)
—CH_2—, cyclopropene	0.6 (singlet)
CH_3—, terminal, in paraffin chain	0.88 (triplet)
CH_3, branched, saturated isoprenoid	0.85–0.90 (singlet or doublet)
$(CH_3)_2C$—, isopropyl methyls	1.2–1.3
$(\omega 1)CH_2$, saturated paraffin chain	1.2–1.3
—CH_2—, acyl C-3, saturated chains	1.58
—CH_2—, acyl C-4 to C-(ω3), saturated chains; (ω2)CH_2, saturated chains	1.2–1.3
$RS\underline{H}$, sulphydryl group	1.1–1.5[b]
$RN\underline{H}_2$, amino group	1.1–1.5(1.8)[b]
$R_2N\underline{H}$, imino group	0.4–1.6(2.2)[b]
R_3C—$\underline{H}$, saturated	1.4–1.7
$C\underline{H}_3$—C=C—, allylic methyl (e.g. squalene)	1.6–1.9 (doublet)
—$C\underline{H}_2$—C=C, allylic methylene	2.04 (doublet)
=C—$C\underline{H}_2$—C=, diallylic methylene	2.8 (triplet)

Table 9.8 *Continued*

Compound or group[a]	δ (ppm)
$-C\underline{H}_2-COOR$, acyl C-2	2.1–2.3 (triplet)
$-C\underline{H}_2-CO-$, α-methylene in ketone	2.2–2.5
$C\underline{H}_3COOR$, methyl in acetoxy group	1.9–2.6 (singlet)
$C\underline{H}_3-Ar$	2.1–2.5
$\underline{H}-C\equiv C-$, terminal acetylene, non-conjugated	2.5–2.7
CH_3-S-R, sulphide methyl	2.1
CH_3-N-R_2, amino methyl	2.1–3.0
$(CH_3)_2S^+$, sulphonium methyl	3.1–3.2 (singlet)
$(CH_3)_3N^+$, ammonium methyl	3.3–3.4 (singlet)
CH_3-O-, methoxy ether group, aliphatic	3.3–3.8 (singlet)
CH_3-O-, methyl ester group, aliphatic	3.6–3.8 (singlet)
$-C\underline{H}_2-N^+$	3.8 (singlet)
$-C\underline{H}_2-S^+$	3.9 (singlet)
$-C\underline{H}-OH$, *sn*-2 in glycerol	3.75 (multiplet)
$-C\underline{H}_2-OH$, *sn*-1 or *sn*-3 in glycerol	3.6 (doublet)
$-C\underline{H}_2-O-$, aliphatic saturated alcohol or ether	3.4–3.7
$-C\underline{H}_2-O-CO-R$, in *sn*-1 or *sn*-3 esterified glycerol	4.2–4.4
$-C\underline{H}-O-CO-R$, in *sn*-2 esterified glycerol	5.1–5.2 (quintet)
$-C\underline{H}_2-O-R$, in *sn*-1- or *sn*-3-*O*-alkylglycerol	3.5–3.6
$-CH-O-R$, in *sn*-2-*O*-alkylated glycerol	3.6–3.7
$-C\underline{H}-O-P$, in *O*-acylglycerol (*sn*-1 or *sn*-3)	3.9
$-C\underline{H}_2-O-P$, in *O*-alkylglycerol (*sn*-1 or *sn*-3)	3.9
$-C\underline{H}_2-O-P$, in choline or sulphocholine	4.3–4.4
$RO\underline{H}$, hydroxyl proton	3.0–5.3
$R-C\underline{H}=C\underline{H}-O$, vinyl ether	5.8 (*cis*)
	6.0 (*trans*)
$C\underline{H}_2=C$, terminal vinyl, non-conjugated	4.6–5.0
$\underline{H}-C=C-\underline{H}$, olefinic or cyclic, non-conjugated	5.1–5.9 (multiplet)
$R-\overset{\beta}{C}\underline{H}=\overset{\alpha}{C}\underline{H}-$ *cis*-Δ^2, in fatty acid chain	7.0 (β), 5.8 (α)
cis-Δ^3, in fatty acid chain	5.6
cis-Δ^4, in fatty acid chain	5.5
cis-Δ^5, in fatty acid chain	5.4
cis-Δ^6, in fatty acid chain	5.4
cis-Δ^9, in fatty acid chain	5.3
cis-Δ^{12}, in fatty acid chain	5.3
$\underline{H}-C=C(CH_3)-$, olefinic isoprenoid (e.g. squalene)	5.0–5.1
$C\underline{H}_2=C-$, terminal vinyl, conjugated	5.3–5.7 (6.2)
$\underline{H}-C=C-\underline{H}$, olefinic, conjugated (diene or triene)	5.8–6.5 (7.1)
$\underline{H}-N-CO-$, amide NH and CO	5.5–8.5
$Ar-H$, benzenoid protons and carbons	7.3–8.5
$RC\underline{H}O$, aldehyde proton and carbonyl:	
aliphatic, saturated	(9.5) 9.7–9.8
aliphatic, α,β-unsaturated	9.5–9.7
$RCOO\underline{H}$, carboxyl proton and carbon	10.5–12.0

[a] Assigned proton designated as $\underline{H}$; usual resonance range is as indicated, but values in parentheses apply to compounds that may absorb outside this range.
[b] The positions for the signals of these groups are concentration-dependent and are shifted to higher δ values in more dilute solutions.

Table 9.9 Deshielding effects of functional groups (ppm in CCl_4 solution) along an alkyl chain. Basic chemical shifts 0.885 ppm (CH_3) and 1.225 ppm (CH_2)

Functional group	α	β	γ	δ
COOH	1.04	0.36	0.10	0.06
$COOCH_3$	0.96	0.32	0.06	0.04
$COOC_2H_5$	0.94	0.32	0.06	0.03
CH_3	0.03	0.00	0.00	0.00
CH=CH (*cis*)	0.73[a]	0.07	0.02	0.02
CH=CH (*trans*)	0.69[b]	0.05	0.05	0.00
$CH{=}CHCH_2CH{=}CH$	0.77	0.08	0.03	0.03
$\overset{O}{\wedge}$ (*cis*)	0.17	0.19	0.10	0.06
HC—CH (*trans*)	0.17	0.15	0.08	0.04

[a] 0.69 for a CH_3 group.
[b] 0.74 for a CH_3 group.

$$\overset{a}{C}H_3(\overset{b}{C}H_2)_n\overset{c}{C}H_2\overset{d}{C}H_2COO\overset{e}{C}H_3$$

(**1**) Methyl alkanoate

integrals show the proportion of hydrogen atoms of each type and the splitting pattern is as expected: the OCH_3 signal is a singlet, the CH_3 and α-CH_2 are triplets, and the β-CH_2 is a quintet. The signal for the remaining methylene groups is complex and broad because it consists of a large number of overlapping signals.

(c) Monoene and polyene esters

In addition to the five signals in the spectrum of a saturated methyl ester, methyl oleate shows additional signals for two olefinic hydrogen atoms (5.35 ppm) and four allylic hydrogen atoms (2.05 ppm). Methylene-interrupted polyene esters also show a signal (2.77 ppm) for CH_2 adjacent to two *cis* double bonds as in the spectrum of methyl linoleate (**2**): a, 0.89 (3H); b, 1.31 (14H); c, 1.62 (2H); d, 2.05 (4H); e, 2.30 (2H); f, 2.77 (2H); g, 3.67 (3H), h, 5.35 ppm (4H). From the integrals of the spectrum of a methylene-interrupted polyene ester, it is easy to determine the number of double bonds, but it is only possible to locate the polyene unit if it is close to the $COOCH_3$ of the CH_3 group. This happens with the $n-3$ acids, where the final double bond is close enough to the methyl group to shift its signal from 0.89 to 0.98 ppm. The spectrum for a natural lipid containing $n-3$ and other acids will show two triplets for the CH_3 group: one for the $n-3$ acid(s) and another for the remaining acids.

When the double bond gets close to the ester function, the overlapping influence of the two functional groups gives rise to distinct signals. It is easy to distinguish Δ3 and Δ4 acids from those having Δ9 unsaturation, and careful inspection of good spectra shows minor differences for the Δ5–Δ7 esters (Frost and Gunstone, 1975). Monoene acids with unsaturation at these positions are rare – apart from petroselinic acid (Δ6) – but some important polyene acids have unsaturation close to the carboxyl group, e.g. γ-linolenic (6,9,12-18:3), arachidonic (5,8,11,14-20:4), eicosapentaenoic (EPA) (5,8,11,14,17-20:5) and docosahexaenoic (DHA) (4,7,10,13,16,19-22:6).

(d) Esters containing cyclopropane, cyclopropene, or oxygenated functions

Long-chain esters with cyclopropane and cyclopropene systems can be recognized by characteristic NMR signals. Cyclopropene compounds show a sharp two-proton singlet at 0.8 ppm but the cyclopropanes give more complex spectra. The *cis* isomer has a multiplet at −0.3 (1H) and a broad band at 0.6 (3H). The *trans* isomer shows a four-proton band between 0.6 and −0.1 ppm.

Epoxy esters have characteristic signals at 2.70 (*cis*) and 2.45 ppm (*trans*) and the effect of the epoxide function on neighbouring CH_2 groups is indicated in Table 9.9. Hydroxy esters have a one-proton signal at 3.4 for the C$\underline{H}$OH proton in carbon tetrachloride solution. In basic solvents this appears at 3.76 (pyridine) and 4.03 ppm (quinoline) and there are additional signals at 1.58 (in pyridine) and 1.77 ppm (in quinoline) for the CH_2 groups α and β to the CHOH group. These latter signals are not present when carbon tetrachloride is used as solvent.

(e) Acylglycerols and phospholipids

The discussion so far has been in terms of acids and simple esters. Consideration must now be given to the additional signals in acylglycerols and phospholipids. In triacylglycerols a one-proton signal (5.25 ppm) from the C$\underline{H}$OCOR unit overlaps with the signals from olefinic hydrogen atoms. The four protons from the two C$\underline{H_2}$OCOR units have signals at 4.12 and 4.28 ppm. The signals for PC and PE (**3**) are summarized as follows. PE, $R^1 = H$: a, 4.0–4.2; b, 5.2; c, 3.9; d, 4.0–4.2; e, 3.1; f, 7–8.5 ppm. PC, $R^1 = CH_3$: a, 4.13 and 4.40; b, 5.20; c, 3.94; d, 4.31; e, 3.79; f, 3.35 ppm.

$$\overset{a}{C}H_3(\overset{b}{C}H_2)_3\overset{d}{C}H_2\overset{h}{C}H{=}\overset{h}{C}H\overset{f}{C}H_2\overset{h}{C}H{=}\overset{h}{C}H\ \overset{d}{C}H_2(\overset{b}{C}H_2)_4\overset{c}{C}H_2\overset{e}{C}H_2COO\overset{g}{C}H_3$$

(**2**) Methyl linoleate

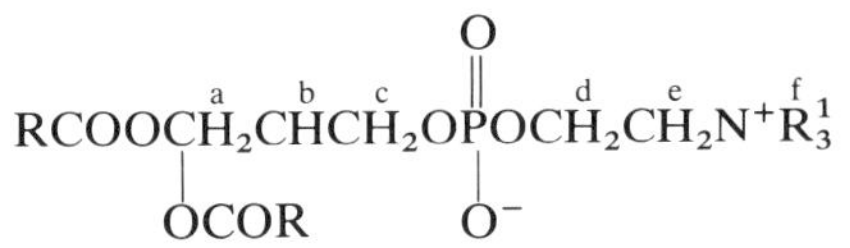

(3) PE/PC

(f) *Shift reagents*

Some metal complexes based on europium or praseodymium – known generally as lanthanide induced shift reagents (LIS) or chemically induced shift reagents (CSR) – interact with oxygen-containing groups to produce a marked effect on nearby CH_2 signals so that the neighbouring five or six CH_2 groups show separate signals. The shift from the normal position depends on the closeness of the CH_2 group to the oxygen-containing function and on the concentration of the shift reagent. Chiral shift reagents have been used to assist in the assignment of chemical shift to particular hydrogen atoms.

(g) *NMR and carotenoid structure*

NMR (especially ^{1}H NMR) has played a minor role in elucidating structures in carotenoid chemistry. In addition, this area has benefited enormously from the development of superconducting (220 MHz and bigger) spectrometers (Schwieter *et al.*, 1969). A detailed account of this rather specialized branch of lipid spectroscopy may be found in Vetter *et al.* (1971).

(h) *^{1}H NMR of porphyrins*

^{1}H NMR has also proved to be a useful technique in the characterization of haems and other substituted porphyrins (Smith and Caughey, 1976). As isolated the natural haemins are difficult to prepare as unique species of high purity, e.g. all molecules with the same ligand(s), oxidation state and form of carboxylic acid groups. A derivative that is easily prepared as a pure crystalline single species is the μ-oxobishaemin ester (Caughey *et al.*, 1975). ^{1}H NMR data for these esters are collected in Table 9.10.

The reduction of haemins (iron(III) porphyrins) to haems is of interest because the physical properties of the haems can be used to detect structure and because haems represent useful models for the reduced haemoproteins. Reduction is often carried out with stannous chloride in dry pyridine. ^{1}H NMR spectra provide an opportunity to detect directly the substitutents present in dipyridine haem dimethyl esters (Table 9.11).

9.3.4 High-resolution ^{13}C NMR

(a) *Introduction*

Natural-abundance carbon-13 NMR spectroscopy has become popular in the study of lipids after developments in instrumentation have successfully overcome the inherently small sensitivity of these experiments on non-enriched samples, which contain *ca* 1.1% ^{13}C. Present-day experimental technique takes advantage of stable spectrometers, wide-band proton decoupling and pulsed Fourier-transform (Rosenthal and Fendler, 1976). A considerable advantage of ^{13}C NMR spectroscopy is the large range of chemical shifts. Resonance frequencies for most organic molecules are spread over 200 ppm. Additionally, with complete proton decoupling, each magnetically non-equivalent carbon appears as a sharp line and individual lines are resolvable if separated by only 0.1 ppm. A review of high-resolution ^{13}C NMR spectroscopy of lipids has been published (Gunstone, 1993a).

Table 9.10 Proton magnetic resonance chemical shifts for μ-oxohaemin ester dimers in $CDCl_3$

μ-Oxo derivative	Ring substituent resonances						
	2, 4 or 8 —CHO	2 or 4 —C$\underline{H}$=CH$_2$	6, 7 α-CH$_2$—	2 or 4 —CH=C$\underline{H_2}$	1, 3, 5, (8) —CH$_3$	6, 7 β-CH$_2$—	ester —CH$_3$
Haemin A monomethyl ether	14.0	9.75	7.0	5.9	5.2	3.3	3.65
dimethyl ester[a]	–	–	6.3	–	–	3.1	–
Haemin B dimethyl ester[b]	–	9.6	6.19	5.9	5.18	2.81	3.54
Haemin S dimethyl ester[a]	13.32	8.97	5.44	5.95	4.75	2.46	3.57
Haemin iso-S dimethyl ester[a]	13.28	8.75	5.34	5.85	4.69	2.45	3.63
Haemin C tetramethyl ester[a]	–	–	6.39	–	5.36	2.96	3.64

[a] From Caughey *et al.* (1975).
[b] From O'Keeffe *et al.* (1975).

Table 9.11 Proton magnetic resonance chemical shifts for dipyridine haem dimethyl esters in pyridine-d_5

Haem	Ring substituent resonances[a]							
	2 or 8 —CHO	meso H	2 or 4 $C\underline{H}{=}CH_2$	2 or 4 $CH{=}C\underline{H_2}$	6, 7 α-CH_2—	1, 3, 5, (8) —CH_3	6, 7 β-CH_2—	6, 7 OCH_3
A	11.60	11.00 δ 10.39 α 10.15 β 9.98 γ	8.45	6.20	4.83–7 4.40–6	3.71–3 3.62–1 3.43–5	3.4–7 2.98–6	3.48–7 3.39–6
B		10.26 α 10.12 β 10.05 δ 9.94 γ	8.49	5.98 6.31	4.40	3.64 3.62 3.47(2)–5.8	3.40	–[b]
S	11.39	11.17 α 9.92 β 9.89 δ 9.80 γ	8.50	5.84 6.14	4.38 4.42	3.88(2) 3.78 3.53	3.34 3.38	3.48
iso-S	11.33	10.15 β 10.01 α 9.80 δ 9.74 γ		5.74 6.06	4.38 4.47	3.94–3 3.91–1 3.86–5 3.70–8	3.32	3.46
C		10.27(2) 9.98(2)			4.36	3.61(2)–1.3 3.46(2)–5.8		3.46

[a] From Caughey *et al.* (1975).
[b] Unesterified.

A ^{13}C NMR spectrum contains more signals from a ^{1}H spectrum because the effects of each functional group are larger and are effective through more atomic centres. Some typical and useful values are set out in Table 9.12, and comparison of these values with those listed for ^{1}H spectra in Table 9.8 confirms this claim. When a carbon atom comes under the influence of more than one functional group, the effects are approximately additive.

The ^{13}C NMR spectrum of a natural lipid contains 20–100 signals depending, in part, on the resolving power of the instrument in use. Some of these signals are clearly recognized and identified. The signals appear as single peaks in individual molecular species (usually synthetic compounds) but are apparent as clusters of several signals with natural mixtures and industrial products. This is because the carbon atom related to a particular signal occurs in several different environments (for example, in saturated, monoene and various polyene acids). The signals that are most easily recognized are listed in Table 9.13.

(b) Glycerol carbon atoms in glycerol esters and phospholipids

Triacylglycerols show signals at around 68.9 and 62.1 ppm for the β and α carbon atoms respectively. Some oils show additional small signals in this region, which can be related to low levels of diacylglycerols. Chemical shifts from the five types of glycerol esters are listed in Table 9.14 (Gunstone, 1991b). Phospholipids show three different signals for the glycerol carbon atoms and further signals for the nitrogen-containing unit (Table 9.14). Many of these signals appear as doublets. Changes in the detail of the glycerol unit have a small effect on the signals observed for the carbon atoms C-1 to C-3.

(c) ω1–ω3 signals

In any individual molecular species, such as tripalmitin, there are clear signals for the $\omega1$, $\omega2$ and $\omega3$ carbon atoms at around 14.1, 22.6 and 31.95 ppm respectively. These chemical shifts are, however, influenced by nearby double bonds, and it is easy to distinguish $n-3$, $n-6$ and $n-9$ (and saturated) acyl chains. Saturated and $n-9$ systems are more difficult to distinguish from each other, but may give two $\omega3$ signals after resolution enhancement.

Average chemical shifts for these carbon atoms based on a wide range of oils and fats are summarized in Table 9.13. Although the chemical shifts in

Table 9.12 Influence of functional groups (ppm in $CDCl_3$) on methylene (29.7 ppm) and methyl (14.1 ppm) signals

Functional group	Chemical shift						Chemical shift of functional group
	α	β	γ	δ	ε	ξ	
CH_3	−7.0	+2.3	−0.3				14.1
$COOCH_3$	+4.3	−4.7	−0.5	−0.4	−0.2	−0.1	174.1
CH═CH (*cis*)	−2.5	0	−0.4	−0.2		–	
CH═CH (*trans*)	+2.8	−0.1	−0.5	−0.2		–	
$(CH{=}CH)_3$ (*cis*)[a]	−1.8	0	−0.4	−0.1		–	
$(CH{=}CH)_3$ (*trans*)[a]	+3.1	0	−0.5	−0.2		–	
$\overset{O}{HC—CH}$ (*cis*)	−1.7	−2.9	−0.4				
(*trans*)	+2.6	−3.5	−0.2				
CH_2OH	+1.3	−3.9					62.9
CH_2OAc	−3.8	−0.5					64.7 (171.1, 20.9)
CH_2OOH	−1.9	−3.6					
CHOH	+7.8	−4.0	+0.1	−0.1	−0.1	−0.1	42.2
CHOAc	+4.4	−4.4	−0.2	−0.2	−0.1	−0.1	44.7
C═O	+13.1	−5.8	−0.4	−0.3	−0.2	−0.1	181.1
CHO	+13.4	−8.4					201.7
C≡C	−11.4	−1.2	−1.0	−0.6	−0.2	−0.1	
CH_2Cl	+3.0	−0.9					45.1
CH_2Br	+3.1	−1.0					
CH_2I	+3.9	+0.8					
$CH_2OCH_2CH_3$	−4.7	−0.5					70.8 (66.0, 15.3)

[a] These figures depend on the configuration of the nearest double bond in the conjugated triene system.
Example: The chemical shifts for the ω1–ω4 carbon atoms in a saturated acyl chain should be around 14.1, 22.7 (29.7 − 7.0), 32.0 (29.7 + 2.3) and 29.4 (29.7–0.3).

Table 9.13 Chemical shifts (ppm in $CDCl_3$) of easily identifiable signals present in glycerol ester spectra

	Usual values[a]	Other values often observed
C-1 α	173.2	173.05 (Δ6)[b], 172.92 (Δ5)[c], 172.40 (Δ4)[d]
β	172.8	172.66 (Δ6)[b], 172.54 (Δ5)[c], 172.03 (Δ4)[d]
C-2 β	34.24	34.11 (Δ6)[b], 33.50 (Δ5)[c], 34.00 (Δ4)[d]
α	34.07	33.94 (Δ6)[b], 33.31 (Δ5)[c], 33.86 (Δ4)[d]
C-3 β	24.90	24.54 (Δ6)[b], 24.70 (Δ5)[c], 22.62 (Δ4)[d]
α		24.51 (Δ6)[b], 24.60 (Δ5)[c]
ω1	14.13	14.09 ($n-6$), 14.29 ($n-3$), 14.09 (8:0), 14.12 (10:0)
ω2	22.60	22.72 ($n-6$), 20.57 ($n-3$), 22.63 (8:0), 22.69 (10:0)
ω3	31.95	31.55 ($n-6$), 31.83 ($n-7$), 31.69 (8:0), 31.90 (10:0)
Olefinic	128–132	See text
Allylic (single)	27.2	27.3 and 26.85 (Δ6)[b,e]
(double)	25.65 (L-11, Ln-11)[f] 25.55 (Ln-14)[f]	
Glycerol β	68.9	See also Table 9.14
α	62.1	
Phospholipids		See Table 9.14

[a] Observed with saturated glycerides and usually with acyl chains having Δ9 unsaturation.
[b] 6-18:1.
[c] ($n-6$)-20:4.
[d] DHA: ($n-3$)-22:6.
[e] Also (methyl esters) 22.81 (C-3, DHA), 26.57 (C-4, EPA), 26.88 (C-5, ($n-3$)-18:4) and 27.06 (C-6, ($n-3$)-22:5).
[f] *Notation*: L = linoleic, Ln = linolenic; L-11 means carbon atom number 11, from carboxyl end, in linoleic chain. This notation is also used later.

Table 9.14 Chemical shifts (ppm in $CDCl_3$) for glycerol carbon atoms

Glycerol ester	Glycerol carbon atoms				Others		
	2	1		3	CO	CN^+	$^+NMe_3$
1-Monoacyl[a]	70.29	[65.10		63.49][b]			
2-Monoacyl[a]	74.94		61.92				
1,2-Diacyl[a]	72.20	[62.18		61.50][b]			
1,3-Diacyl[a]	68.28		65.05				
Triacyl[a]	68.98		62.15				
PE	70.52 70.42	62.68		64.09 64.03	62.27	40.57	
PC	70.63 70.53	63.04		63.37 63.31	59.40 59.30	66.27 66.19	54.26

[a] *Acylglycerols*: These are average values and subject to some variation. Some of the values are fairly close together, but the signal at 65.10 ppm has an observed shift greater than that at 65.05 ppm, and the signals at 62.18, 62.15 and 61.92 ppm appear in that order.
[b] These assignments are probably correct but they may be interchanged.

different spectra may vary by ±0.02, the differences between signals in a single spectrum are more constant. The difference between $n-9$/saturated and $n-6$ signals for the $\omega1$, $\omega2$ and $\omega3$ carbon atoms is 0.04, 0.11 and 0.60. If the data are to be used qualitatively, the $\omega2$ and $\omega3$ signals will give better results than the $\omega1$ signal. In many oils containing saturated, oleic and linoleic acids, only one $\omega1$ signal is observed – especially when one acid is dominant – though this may be resolved by resolution enhancement. This problem is not apparent with the $\omega2$ and $\omega3$ signals. The presence of $n-3$ acids is easily indicated by the $\omega1$ and $\omega2$ signals.

A signal at around 31.83 ppm in the $\omega3$ cluster is indicative of $n-7$ acids such as 9-16:1 and/or 11-18:1.

The $\omega1$–$\omega3$ signals are also influenced by short acyl chains. This is apparent with the C_8 and C_{10} acids present in the lauric oils.

(d) C-1 to C-3 signals

For saturated acyl chains these carbons show signals for C-1 (173.2 and 172.8 ppm for α and β chains, difference 0.41 ppm), C-2 (34.2 and 34.05 ppm for β and α chains, difference 0.16 ppm) and C-3 (24.9 ppm, sometimes split on resolution enhancement, difference 0.025 ppm). There is only a small difference between these signals for saturated and the common unsaturated (Δ9) chains, but these can be seen after resolution enhancement. Larger changes are noted with acids having a double bond closer to these carbon atoms, e.g. Δ6 (petroselinic, γ-linolenic), Δ5 (columbinic, arachidonic acid, EPA) and Δ4 (DHA). Some of these values are given in Table 9.15 (Gunstone 1990a, 1991a).

Table 9.15 Chemical shifts (ppm) for carbon atoms C-1 to C-3[a] (From Gunstone, 1990a, 1991a)

	C-1	C-2	C-3
Saturated	173.26[b] 172.85	34.24 34.07	24.90
Oleic (9)	173.20[c] 172.83	34.20 34.04	24.87
Linoleic (9,12)	173.17[d] 172.77	34.19 34.02	24.85
Eicosenoic (11)	–[e]	–	–
Erucic (13)	173.23 172.82	34.21 34.04	24.88
Petroselinic (6)	173.05 172.66	34.11 33.95	24.54 24.51
γ-Linolenic (6,9,12)	173.04 172.64	34.07 33.92	24.51 24.49
$(n-6)$-20:4	172.92 172.54	33.50 33.31	24.70 24.60
DHA ($(n-3)$-22:6)	172.40 172.03	34.00 33.86	22.62

[a] The upper figure applies to α chains for C-1 and to β chains for C-2 and C-3.
[b–e] Wollenberg (1990) gives: b, 173.229; c, 173.216 and 172.794; d, 173.188 and 172.782; e, 173.216 ppm.

(e) Olefinic signals

Methyl oleate shows two olefinic signals at 129.8 and 130.0 ppm corresponding to carbon atoms 9 and 10 respectively. In a glycerol ester these values vary slightly depending on whether the oleic chain (O) is in the α or β position. The O-9 signal splits most easily into two (difference 0.026 ppm), but O-10 splits only by 0.015 ppm and is more difficult to observe. Linoleic (L) esters show four signals and α-linolenic (Ln) six – one for each olefinic carbon atom – but there may be some overlap, so that oleic and linoleic together only show five signals, and with α-linolenic acid also there are only nine or ten signals. The values quoted by Wollenberg (1990) are collected in Table 9.16. There is tendency for overlap between O-10 and L-9, Ln-9 and L-13, and between Ln-12 and Ln-13. The commonly observed values for these three acyl chains are: 131.87 (Ln-16), 130.19/130.14 (Ln-9, L-13), 130.00/129.99 (L-9, O-10), 129.73 (O-9), 128.26 (Ln-12 and Ln-13), 128.09 (L-10), 127.96 (L-12), 127.78 (Ln-10) and 127.17 (Ln-15). (These values differ slighly from those quoted in Table 9.16, but this variation is not unusual between measurements made at different times, in different laboratories and with different materials. Differences between related values are more constant and the sequential order is unchanged.)

Figures for more unsaturated acids and esters have been given by Gunstone *et al.* (1990b, 1991c), Aursand and Grasdalen (1992) and Aursand *et al.* (1993) (Table 9.17).

In a series of isomeric monoene acids or methyl esters, the two olefinic signals get closer together

Table 9.16 Chemical shifts[a] (ppm) for olefinic carbon atoms in oleic, linoleic and α-linolenic chains (Adapted from Wollenberg, 1990)

	Oleic (O), $(n-9)$-18:1	Linoleic (L), $(n-6)$-18:2	α-Linolenic (Ln), $(n-3)$-18:3
9	129.721, 129.696 (0.026)	130.002, 129.977[b] (0.025)	130.223, 130.198 (0.025)
10	130.022, 130.037 (0.015)	128.106, 128.124 (0.018)	127.794, 127.811 (0.017)
12	–	127.939, 127.927 (0.012)	128.262, 128.250 (0.012)
13	–	130.210, 130.218 (0.008)	128.311, 128.321 (0.010)
15	–	–	127.157, 127.150 (0.007)
16	–	–	131.946, 131.949 (0.003)

[a] Chemical shift for α chain, β chain (difference).
[b] Value corrected from the original paper by permission of the author.

Table 9.17 Chemical shifts (ppm) for olefinic carbon atoms in polyene acids[a]

$(n-3)$-18:4		$(n-6)$-20:4		$(n-3)$-20:5		$(n-3)$-22:6	
6	129.58	5	129.09	5	129.10	4	127.58
7	128.60	6	128.77	6	128.81	5	129.64
9	128.32[b]	8	128.27[b]	8	128.22[b]	7	128.01
10	128.18	9	128.12[b]	9	128.23[b]	8	128.35
12	127.98	11	127.88	11	128.14	10	128.28[b]
13	128.35[b]	12	128.60	12	128.32	11	128.31[b]
15	127.12	14	127.57	14	127.94	13	128.13
16	132.05	15	130.52	15	128.63	14	128.13
				17	127.08	16	127.92
				18	132.10	17	128.60
						19	127.08
						20	132.03

[a] $n-3$ acids: highest shift ω3, lowest shift ω4.
$n-6$ acids: highest shift ω6, lowest shift ω7.
[b] These assignments may be interchanged.

as unsaturation moves away from the acyl group (Gunstone *et al.*, 1977). This is illustrated among glycerol esters by the values for petroselinate (6*c*), oleate (9*c*), *cis*-vaccenate (11*c*) and erucate (13*c*) (Table 9.18).

Olefinic acids with *trans* unsaturation have their own distinct olefinic signals, which generally have a higher chemical shift than the *cis* isomers. These observations are of potential value in the study of partially hydrogenated fats, which contain positional and configurational isomers (section 10.1; Gunstone, 1991d). Data taken from an earlier paper (Gunstone *et al.*, 1977) for the *cis* and *trans* olefinic acids are summarized in Table 9.19, and a typical list of signals observed in a commercial hydrogenated oil is given in Table 9.20.

Table 9.18 Chemical shift (ppm) for olefinic carbon atoms in petroselenate, oleate, *cis*-vaccenate and erucate

	Chemical shift (ppm)		Difference
6*c*-18:1	130.55	128.95	1.60
9*c*-18:1	130.01	129.68	0.33
11*c*-18:1	129.92	129.84	0.08
13*c*-22:1	129.883	129.854	0.025

(*f*) *Allylic signals*

Allylic signals are of two major types: those adjacent to only one double bond (C-8 and C-11 in oleate, and C-8 and C-14 in linoleate) and those adjacent to two double bonds (C-11 in linoleate, and C-11 and C-14 in α-linolenate). The signal at about 27.2 ppm, corresponding to singly allylic systems, sometimes appears as two signals (difference about 0.05 ppm) because not all atoms of this type are identical. Doubly allylic carbon atoms give signals at 25.65 ppm (L-11 and Ln-11) and 25.55 ppm (Ln-14). Other values are observed when the allylic signal is also influenced by the carboxyl group, for example 26.85 (C-5) and 27.3 (C-8) in petroselinate, 22.1 (C-3, DHA) and 26.57 ppm (C-4, EPA), or by the methyl group as in α-linolenate (20.57 ppm, C-17).

Allylic groups next to *trans* unsaturation have higher chemical shifts, and attempts have been made to measure *cis* and *trans* unsaturation in partially hydrogenated oils from the intensities of appropriate signals (Gunstone, 1993b). Some useful values are summarized in Table 9.21.

Table 9.19 Olefinic signals[a] of the 18:1 acids (Δ5–Δ15, *cis* and *trans*)

trans	Chemical shift (ppm)[b]			*cis*	Chemical shift (ppm)[b]		
5	131.98	128.72	[3.26]	5	131.44	128.21	[3.23]
	(85)	(77)			(82)	(80)	
6	131.13	129.49	[1.64]	6	130.62	129.01	[1.61]
	(32)	(40)			(31)	(44)	
7	130.81	129.89	[0.92]	7	130.31	129.45	[0.86]
	(14)	(24)			(14)	(20)	
8	130.67	130.13	[0.54]	8	130.17	129.65	[0.52]
	(13)	(10)			(8)	(13)	
9	130.54	130.23	[0.31]	9	130.09	129.78	[0.31]
	(7)	(8)			(9)	(5)	
10	130.47	130.31	[0.16]	10	130.00	129.83	[0.17]
	(4)	(4)			(4)	(6)	
11	130.43	130.34	[0.09]	11	129.96	129.89	[0.07]
12	130.14			12	129.94		
13	130.39			13	129.90		
14	130.63	130.13	[0.50]	4	130.16	129.66	[0.50]
15	131.93	129.45	[2.48]	15	131.54	129.39	[2.15]

[a] See footnote to Table 9.20.
[b] Chemical shift (ppm) of olefinic carbon atoms:
[ ] differences between values (horizontal),
( ) differences between values (vertical).

Table 9.20 Olefinic signals observed in a commercial partially hydrogenated fats: assignment and chemical shift (ppm)[a]

5*t*	131.74	11*t*	130.20, 130.18	12*c*,13*c*	129.77, 129.75
15*t*	131.59	10*t*	130.15, 130.14	11*c*,7*c*	129.72, 129.70
15*c*	131.38	7*c*	130.12, 130.10	10*c*	129.68, 129.66
6*t*	130.95	9*t*	130.08, 130.06	9*c*	129.61, 129.58
?	130.79	14*c*	130.04	8*c*,14*c*	129.51, 129.48, 129.45
7*t*	130.64	8*c*, 14*c*	129.99	6*t*	129.32
?	130.58	8*t*,14*t*	129.96	7*c*,15*t*	129.28
8*t*,14*t*	130.48	9*c*	129.93, 129.90	15*c*	129.24
8*t*,14*t*,6*c*	130.45, 130.43, 130.42	10*c*	129.85		
9*t*	130.38	11*c*	129.82		
10*t*	130.33				
11*t*	130.30				
12*t*,13*t*	130.28, 130.25, 130.22				

[a] These chemical shifts differ somewhat from those in Table 9.19. The spectra were obtained in different laboratories. Nevertheless the assignments can be made on the basis of (i) results for oleic and elaidic glycerides run under the same conditions and (ii) the differences between the two olefinic resonances for a single isomer and between adjacent isomers (e.g. 9*t* and 10*t*), i.e. the horizontal and vertical differences listed in Table 9.19.

Table 9.21 Chemical shifts (ppm) for some allylic carbon atoms (typical values)

t,t	35.60	*c,c*	25.65, 25.55
t	32.60	*c*	27.24
t,c/c,t	30.44, 30.31		

9.3.5 Other high-resolution lipid spectra

(a) *High-resolution spectra of phospholipids in micelles*

Dennis and Owens (1973) showed that egg phosphatidylcholine in mixed micelles with Triton X-100 gives high-resolution ^{1}H NMR spectra in which the phospholipid gives rise to narrow lines and the Triton peaks, also narrow, are similar to those in pure Triton micelles (Fig. 9.19). High-resolution NMR spectra of unsonicated phospholipid preparations (multilamellar vesicles) result in low apparent intensities because of the large line broadening, whereas mixed micelles, or sonicated phospholipid micelles, show full intensities (Chapman and Salsbury, 1966). Proton T_1 and T_2^* (from linewidths) relaxation times have been determined for pure detergent micelles and mixed micelles of Triton and various lecithins (Dennis and Ribeiro, 1976). Some typical data for Triton X-100 micelles and mixed micelles with dimyristoylphosphatidylcholine are given in Table 9.22. It is apparent that the T_1/T_2^* ratios and frequency dependence of the hydrophobic t-butyl group (Fig. 9.19, peak a) are very similar in pure and mixed micelles, suggesting that in both cases the hydrophobic detergent moiety is freely mobile. T_1 measurements are compatible with the existence of *gauche–trans–gauche* isomerizations in the phospholipid hydrocarbon chains (McLaughlin *et al.*, 1973). Finally, the relatively long T_2^* values for the terminal methyl and choline methyl protons in the mixed micelles, as compared with those in pure phospholipid vesicles, have been interpreted as suggesting a less restricted environment in the micellar structure.

^{13}C NMR studies have complemented the above ^{1}H NMR results (Sears, 1975; Ribeiro and Dennis, 1976). The phospholipid in mixed micelles also exhibits similar T_1 values to the phospholipid in the unsonicated multilamellar vesicles, corresponding to similar fast segmental motions on the phospholipid fatty acid chains in all of these systems. However, the

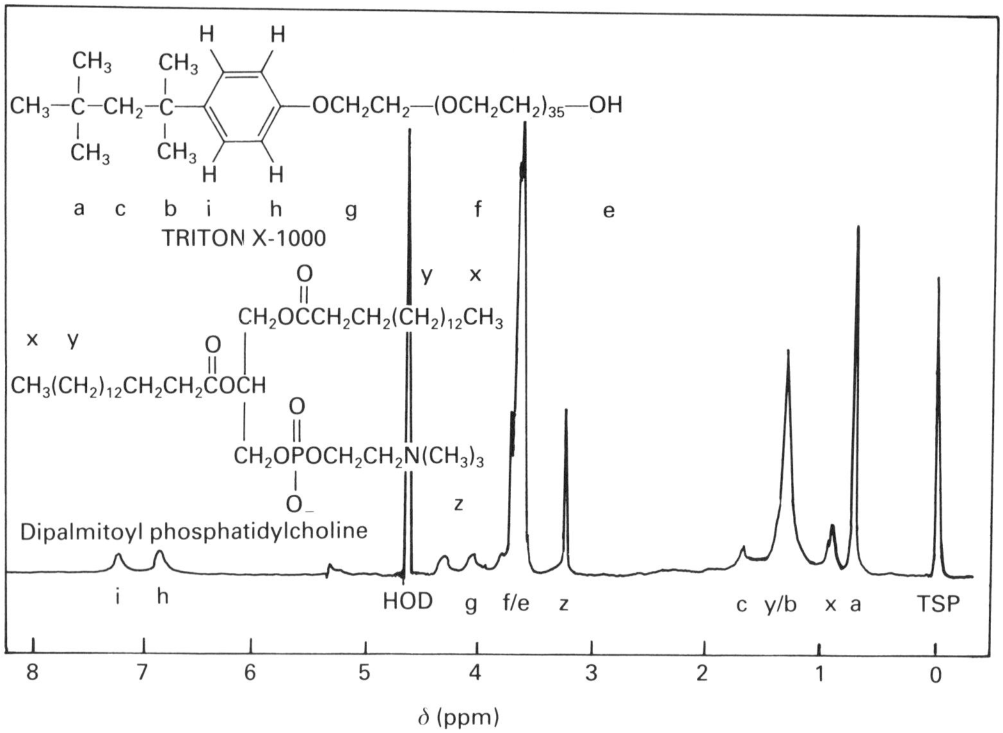

Fig. 9.19 The 200 MHz ^{1}H NMR spectrum recorded at 37 °C of a mixture of 200 mM Triton X-100 and 100 mM dipalmitoylphosphatidylcholine in D_2O and containing 3-trimethylsilylpropanoic acid. From Dennis and Ribeiro (1976).

phospholipid in mixed micelles very clearly exhibits narrower lines (longer T_2^* relaxation times) than in the bilayer systems.

More recently, ^{31}P NMR spectra of phospholipids in detergent micelles have been recorded (London and Freigenson, 1979; Roberts *et al.*, 1979; Jackson *et al.*, 1982; Urbaneja *et al.*, 1986). Such spectra can be characterized as high-resolution spectra. ^{31}P NMR spectra (with protein decoupling) of mixed micelles have the advantage of allowing resolution and quantification of individual classes of phospholipids in single peaks in the presence of detergents. For lysophospholipids (Plückthun and Dennis, 1982a, b) and short-chain phospholipids (Plückthun and Dennis, 1981), which form micelles in aqueous solution without detergent added, ^{31}P NMR can be used to identify individual isomers, as well as to determine critical micelle concentrations and aggregation states. Different phospholipids have been resolved in organic solvents and sonicated vesicles (Henderson *et al.*, 1974; Berden *et al.*, 1975), and phospholipids and lipoprotein structures have been studied and

Table 9.22 ^{1}H NMR spectra of phospholipids in micelles (Dennis and Ribeiro, 1976)

		Peak (see Fig. 9.19)					
		Micelles		Mixed micelles			
		a	b	a	x	y	z
100 MHz	T_1 (s)	0.23	0.098	0.23	0.50	0.36	0.43
	$\Delta\nu_{1/2}$ (Hz)	2.1	6.6	2.1	4.7	–	1.5
	T_2^* (s)	0.15	0.048	0.15	0.068	–	0.21
	T_1/T_2^*	1.5	2.0	1.5	7.3	–	1.9
220 MHz	T_1 (s)	0.27	0.12	0.26	0.68	0.53	0.51
	$\Delta\nu_{1/2}$ (Hz)	2.1	4.4	2.3	5.3	–	1.9
	T_2^* (s)	0.15	0.072	0.14	0.059	–	0.17
	T_1/T_2^*	1.9	1.7	1.9	11.4	–	3.0

found to give sharp resonances (Henderson *et al.*, 1974).

The chemical shifts of various phospholipids in organic solvent and Triton X-100 micelles are shown in Tables 9.23 and 9.24. A detailed discussion on ^{31}P NMR of phospholipids in micelles, including applications to membrane solubilization by detergents and phospholipase activity, may be found in the review by Dennis and Plückthun (1984).

(b) Miscellaneous examples

High-resolution spectra and spectral data from novel or less comon lipids have been published in recent years. Such is the case of ^{13}C data for archaebacterial isoprenoid glycerolipids (diethers and tetraethers), and ^{1}H and ^{13}C data for archaebacterial isoprenoid quinones (De Rosa and Gambacorta, 1988). ^{1}H NMR spectra of taurolipids and related fatty acids have been published (Kaya, 1992), as well as ^{1}H, ^{13}C and ^{31}P NMR spectra of N-acylated glycerophospholipids and their derivatives (Schmid *et al.*, 1990).

Proton magnetic resonance has been applied to the study of the aggregation of taurine-conjugated bile salts (Stevens *et al.*, 1992), while ^{13}C resonance has been used in the quantification of model digestive mixtures (Wang *et al.*, 1992). High-resolution ^{31}P NMR spectra of tissue phospholipids are used in cation-binding and other studies (Merchant and Glonek, 1992; Capuani *et al.*, 1992). Also, ^{1}H and ^{13}C NMR have found application in the characterization of oxidation products of methyl oleate (Wu *et al.*, 1992).

Of interest in the field of biomembranes, Oldfield *et al.* (1991) have shown the possibility of obtaining

Table 9.23 ^{31}P NMR chemical shifts of phospholipids in chloroform–methanol and analogues in aqueous solution (From Dennis and Plückthun, 1984)

Compound (20 mM)	Solvent[a]	Chemical shift[b] (ppm)
Egg PC	C–M	−0.9
N,N-Dimethyl-PE	C–M	0.3
N-Methyl-PE	C–M	0.0
PE	C–M	0.2
PS	C–M	0.0
Dipalmitoyl-PC	C–M	−0.8
Distearoyl-PC	C–M	−0.8
Dioleoyl-PC	C–M	−0.8
Distearoyl-PE	C–M	−0.1
Lyso-PC	C–M	−0.2
Lyso-PE[c]	C–M	0.2
Lyso-PS[d]	C–M	0.2
Plasmalogen PE	C–M	0.2
SM	C–M	0.0
PI	C–M	0.2
PA	C–M	2.8
PG	C–M	1.2
CL	–	–
Low-field band	C–M	1.0
High-field band	–	0.8
Phosphorycholine	Water	3.2
Phosphorylethanolamine	Water	3.7
Phosphorylserine	Water	3.7
Glycerophosphorylcholine	Water	−0.1
Glycerophosphorylethanolamine	Water	0.4
Glycerophosphorylserine	Water	0.1

[a] C–M is chloroform–methanol 2/1 (v/v); water contained 0.2 M EDTA, Na^{+} ion, pH = 7.0.
[b] Positive chemical shifts are shown in the direction of decreasing field strength.
[c] Concentration 7 mM.
[d] Concentration 5 mM.
Abbreviations: PC, phosphatidylcholine; PE, phosphatidylethanolamine; PS, phosphatidylserine; SM, sphingomyelin; PI, phosphatidylinositol; PA, phosphatidic acid; PG, phosphatidylglycerol; CL, cardiolipin; EDTA, ethylenediaminetetraacetate.

Table 9.24 ^{31}P NMR shifts of phospholipids and analogues in aqueous solution with Triton X-100[a,b] (From Dennis and Plückthun, 1984)

Compound	Triton X-100 (48 mM)	Chemical shift (ppm)
Dipalmitoyl-PC	+	−0.86
1-Palmitoyl-lyso-PC	−	−0.34
1-Palmitoyl-lyso-PC	+	−0.38
2-Palmitoyl-lyso-PC	−	−0.52
2-Palmitoyl-lyso-PC	+	−0.55
Dipalmitoyl-*β*-PC	+	−1.45
Palmitoyl-lyso-*β*-PC	−	−1.13
Dibutyryl-PC	−	−0.60
Dibutyryl-PC	+	−0.61
1-Butyryl-lyso-PC	−	−0.24
1-Butyryl-lyso-PC	+	−0.25
2-Butyryl-lyso-PC	−	−0.44
Dipalmitoyl-PE	+	−0.15
Palmitoyl-lyso-PE	+	0.26
Dipalmitoyl-*β*-PE	+	−0.71
Palmitoyl-lyso-*β*-PE	+	−0.46
Dipalmitoyl-*N*-methyl-PE	+	−0.30
Palmitoyl-lyso-*N*-methyl-PE	+	0.10
Dipalmitoyl-*N*-methyl-*β*-PE	+	−0.86
Palmitoyl-lyso-*N*-methyl-*β*-PE	+	−0.60
Dipalmitoyl-*N*,*N*-dimethyl-PE	+	−0.42
Palmitoyl-lyso-*N*,*N*-dimethyl-PE	+	−0.02
Dipalmitoyl-*N*,*N*-dimethyl-*β*-PE	+	−1.02
Palmitoyl-lyso-*N*,*N*-dimethyl-*β*-PE	+	−0.75
Glycero-3-phosphorylcholine	−	−0.08
Glycero-3-phosphorylcholine	+	−0.09
Glycero-2-phosphorylcholine	−	−0.79
Glycero-3-phosphate[c]	−	4.3
Glycero-2-phosphate[c]	−	3.9
Dodecylphosphorylcholine	+	−0.39
Bis(monoacylglycerol)phosphate	+	−0.70

[a] Positive chemical shifts are shown in the direction of decreasing field strength.
[b] $CaCl_2$ was generally present. When the compound readily dissolves in water in the absence of detergent, its chemical shift is also indicated. All compounds were analysed at pH 8.0.
[c] Contained 10 mM EDTA.

proton-coupled ^{13}C NMR spectra of lipid–water systems using the so-called 'magic angle' sample spinning (MAS) NMR technique. Spectra of cerebrosides (in the H_{II} hexagonal or L_α lamellar phase) and antidepressant drugs (desipramine, imipramine) in dimyristoylphosphatidylcholine (DMPC) provide information that is usually available only from isotopically enriched (e.g. ^2H-labelled) lipids.

9.3.6 Broad-line spectra

With an isolated nucleus, the absorption NMR spectrum would be a single, very sharp line corresponding to absorption at the magnetic field H_0 at the resonance condition. However, in a crystal, interactions from other nuclei modify this, so that the magnetic field at the nucleus is the sum of the applied magnetic field plus the field due to neighbouring nuclei. As a consequence, the different nuclei in the assembly will undergo resonance at widely different values of the applied magnetic field, and the resultant absorption line will be comparatively broad. Fats in the solid state, as well as phospholipids in bilayers, give rise to such broad, 'solid-state' spectra. The early studies in this field were reviewed by Chapman (1965); however, the advent of pulsed NMR methods has propitiated many interesting new investigations.

(a) *Applications to the oils and fats industry*

There are three main areas in which wide-line NMR can find application in the edible oils and fats field: as a replacement for dilatometry in the determination of solid fat contents; for determination of the fat in foods; and for determination of the oil content of oil seeds and meals. In addition, NMR is finding increasing application in the heavy oils and fats and petrochemicals industry (Waddington, 1981).

The most successful area of broad-line NMR application in industry has been the determination of solid-phase contents of a partially solidified fat, over a range of temperature, in order to characterize the melting behaviour of the fat (van den Enden *et al.*, 1978). Some manufacturers have developed inexpensive pulse NMR machines especially designed for these measurements, in which a meter reading gives directly the solid-phase contents of the sample. Alternatively, the specific liquid signal from the sample may be compared with that from a reference oil, using a standard spectrometer. These are called the direct and indirect procedures, respectively. Table 9.25 shows the results of Waddington (1981) applying both methods to a number of fats. See also van den Enden *et al.* (1982).

It is also possible to use broad-line NMR to determine the amount of fat in foods without solvent extraction. This technique has been applied, for instance, in the chocolate industry (Bruker, 1978). The method consists of raising the sample to a temperature such that all the fat in the sample is liquid, and then measuring the intensity of the liquid signal from the sample in an NMR spectrometer and comparing this with the signal from a known quantity of the same fat. In general, better than 1% absolute accuracy can easily be achieved.

Wide-line NMR spectroscopy has also been applied to the estimation of oil contents of oil seeds and meals without extraction procedures (Tiwari, 1977). Here, again, the basic principle involves the comparison of the specific NMR signal from the liquid oil in a seed sample with the specific signal from a known quantity of the same oil. A particular advantage of the NMR technique is that it can be used for the determination of the oil contents of single seeds, which may subsequently be used in seed breeding trials.

More basic studies, also in the field of edible fats, have been carried out in order to quantify molecular motions in solid triglycerides, using ^{2}H NMR quadrupole echo spectra of specifically deuterated tripalmitin (see below for more details on ^{2}H NMR); molecular motions and phase transitions of tripalmitin were conveniently determined in this way (Eads *et al.*, 1992). Other NMR studies of triglycerides have been reviewed by Eads (1991) and Pollard (1986).

Table 9.25 Some typical solid-phase content (percentage) results for 12 different fats (Waddington, 1981)

Fat	20 °C		25 °C		30 °C		35 °C	
	I[a]	D[a]	I	D	I	D	I	D
Lauric stearins	81.0	81.8	54.7	54.1	0.6	0.4	0.3	0.0
	84.3	84.5	69.8	69.0	34.1	33.4	0.0	0.2
	84.6	85.6	86.8	88.0	47.2	48.0	5.4	4.9
Hydrogenated lauric blends, etc	78.9	79.0	54.8	54.8	31.9	29.8	15.7	13.8
	80.2	79.8	57.0	56.8	26.7	25.5	3.2	2.5
	67.4	67.5	42.5	43.0	22.7	21.8	11.2	9.6
	79.4	79.6	51.9	52.1	21.1	19.6	5.9	4.3
	54.0	54.3	27.9	27.4	3.4	3.8	0.5	0.0
Fully hydrogenated palmkernel oil	86.1	86.2	62.7	62.7	31.3	29.6	11.6	10.3
Fully hydrogenated coconut oil	54.0	54.4	17.4	16.8	4.8	5.3	1.9	2.4
Partially hydrogenated liquid oils	86.0	88.3	71.2	72.4	35.2	34.3	1.4	0.3
	67.0	67.5	55.2	54.3	36.6	34.9	16.1	13.3

[a] I = indirect; D = direct.

(b) NMR spectroscopy of cell and model membranes

Solid-state NMR studies of biomembranes have been carried out with deuterium or phosphorus-31. ^{2}H NMR spectroscopy has been of great help in the study of the physical state of the hydrocarbon-chain region of phospholipid biolayers (Smith, 1979; Chan *et al.*, 1981). This requires the synthesis of specifically deuterated phospholipids. These lipids are then dispersed in water; proteins or other membrane components may be incorporated if required (Rice *et al.*, 1979). The deuterium signal can only be observed following a spin-echo pulse method, because of the rapidly decaying time-domain signals.

The ^{2}H NMR signal has a symmetric appearance (Fig. 9.20). The separation between the two maxima in hertz, $\Delta\nu_q$, is called the ^{2}H quadrupolar splitting and is considered to be an order parameter. (Order should not be confused with fluidity; order is measured by an equilibrium or a very long-time-average matrix, while the fluidity of the bilayer is reflected in the correlation times for the motions undergone by the acyl chains.) The ^{2}H quadrupolar splitting is the C–^{2}H vector angular fluctuation average over the timescale of the deuterium NMR experiment ($\simeq 10^{-5}$ s) (Seelig, 1977).

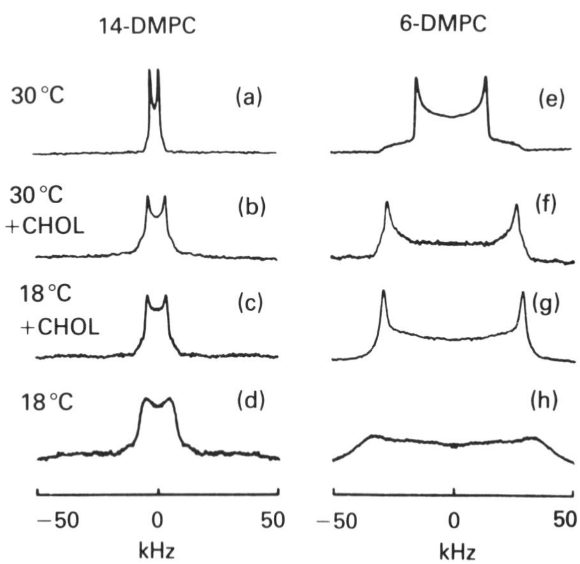

Fig. 9.20 A group of 34 MHz ^{2}H NMR spectra of dimyristoylphosphatidylcholine (DMPC) labelled in the C-14 methyl group ((a)–(d)) and the C-6 methylene group ((e)–(h)) of the 2-chain. Spectra (a) and (e) show the pure lipid above T_c and (g) shows how the addition of 50 mol% cholesterol (CHOL) affects the spectra. All spectra were obtained with the quadrupole echo technique. The quadrupole splitting $\Delta\nu_q$ is easily measured as the separation of two peaks. From Rice *et al.* (1979).

Figure 9.20 depicts the deuterium NMR spectra of dimyristoylphosphatidylcholine labelled at the terminal methyl group (spectra (a)–(b)) or at position 6′ (spectra (e)–(h)) of the 2-chain (Rice *et al.*, 1979). Spectra (a) and (e) show the pure lipid spectra in the liquid-crystalline state, whereas spectra (d) and (h) correspond to the respective samples in the gel state ($T_c = 23\,°C$). An increase in the order parameter upon cooling is apparent. The effect of cholesterol, at a concentration of 50 mol%, is seen in spectra (b), (c), (f) and (g). It is clear that cholesterol induces an intermediate-order state, increasing $\Delta\nu_q$ below T_c and decreasing it in the fluid state.

Intrinsic proteins can also modify the order of phospholipid hydrocarbon chains. When cytochrome oxidase is reconstituted into vesicles of dimyristoyl-phosphatidylcholine, deuterated at the fatty acid methyl end, it is seen that $\Delta\nu_q$ varies with the protein/lipid ratios. Experiments performed with the lipid in the liquid-crystalline (fluid) state show that lipid order decreases as the protein concentration in the bilayer increases (Kang *et al.*, 1979). This is attributed to the rough irregular protein surface in contact with the acyl chains.

A useful improvement in ^{2}H NMR spectral data treatment is the so-called de-Paking (Bloom *et al.*, 1981; Sternin *et al.*, 1983), which allows the resolution of the different methylene units in a full deuterated hydrocarbon chain, thus relaxing the requirement for specifically deuterated molecules (Fig. 9.21). De-Paked ^{2}H NMR spectra have been used to study the influence of membrane curvature in the lamellar to reverse-hexagonal phase transitions of PE (Thurmond *et al.*, 1990), or to measure the molecular areas of PC and PE (Thurmond *et al.*, 1991). Conventional ^{2}H NMR has recently been used in the study of gel–fluid transitions in binary phospholipid mixed bilayers (Sankaram and Thompson, 1992); the effective acyl chain lengths were computed from the spectral data, and used to estimate the effective bilayer thickness. Applications of ^{2}H NMR to the study of molecular motions and headgroup orientations in membrane lipids have been reviewed by Seddon (1990).

^{31}P NMR has been applied to membrane studies mainly for the detection of phase changes. Hydrated phospholipids may adopt a variety of conformations according to the nature of the lipid, the amount of water, and so forth. Two of these dispositions are depicted in Fig. 9.22, the so-called bilayer and hexagonal H_{II} arrangements; other phases such as cubic, micellar and inverted micellar are also possible. Classically, the various phases have been distinguished by means of X-ray diffraction. ^{31}P NMR has been shown to be useful in the diagnosis of lipid phases (Seelig, 1978; Cullis and de Kruijff,

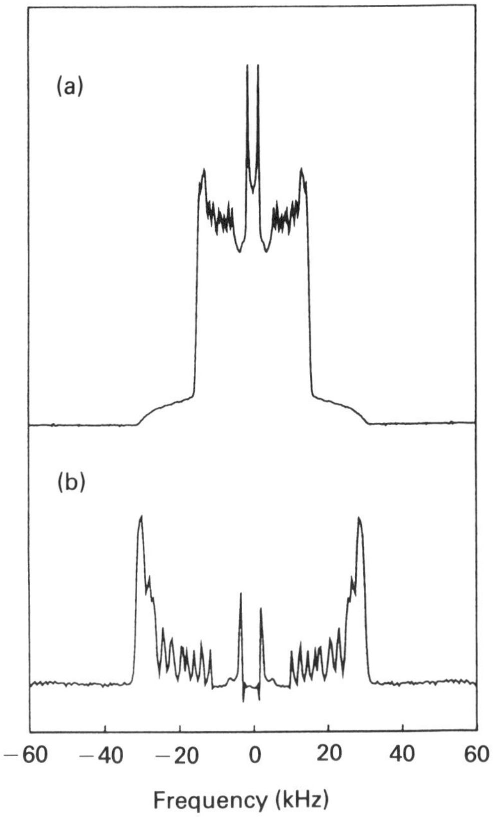

Fig. 9.21 Representative ^{2}H NMR spectra of multilamellar dispersions of DPPE (containing perdeuterated fatty acyl chains) at 69 °C in the liquid-crystalline L_α phase. (a) ^{2}H NMR powder-type (conventional) spectrum, corresponding to a random distribution of the lamellae. (b) De-Paked ^{2}H NMR spectrum, representing the $\theta = 0°$ orientation of the lamellar normal relative to the magnetic field. From Thurmond *et al.* (1991).

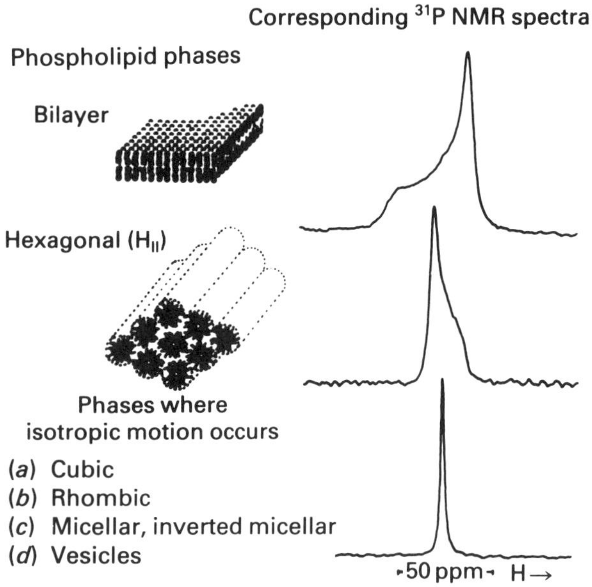

Fig. 9.22 Polymorphic phases available to hydrated liquid-crystalline phospholipids and the corresponding ^{31}P NMR spectra. The bilayer spectrum was obtained from aqueous dispersions of egg-yolk phosphatidylcholine; the hexagonal spectrum corresponds to soybean phosphatidyethanolamine; the 'isotropic motion' spectrum was obtained from a mixture of 15 mol% egg-yolk phosphatidylcholine and 85% soybean phosphatidylethanolamine. The spectra were obtained at 30 °C in the presence of broad-band proton decoupling. From Cullis and de Kruijff (1979).

1979); it is certainly a very convenient method, but an indirect one. Thus, phase identifications should ultimately rely on X-ray diffraction (Tilcock *et al.*, 1986; Seddon, 1990).

^{31}P NMR is based on the large chemical-shift anisotropy exhibited by lipid phosphorus. In the presence of proton decoupling, this anisotropy results in characteristically broad spectra whose shape and width depend on phospholipid motion; in turn, this is related to the lipid phase. In large bilayer structures (radius greater than 200 nm), as is the case for most biomembrane preparations, the only possibility of motion at the NMR timescale (10^{-5} s) is the rotation of the phospholipid molecules about their long axis. The result is a very broad spectrum with a low-field shoulder and a high-field peak (Fig. 9.22). In the hexagonal (H_{II}) phase, lipids experience additional motion averaging, because of the lateral diffusion around the small aqueous channels. As a result, ^{31}P NMR lineshapes occur with reversed asymmetry as

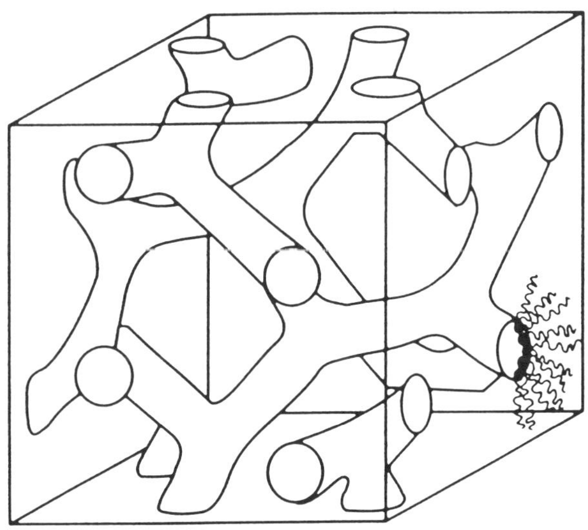

Fig. 9.23 Schematic representation of a cubic phase. In this particular cubic phase (Q^{230}), both lipid and water extend three-dimensionally, forming two infinite networks; the phase is said to be bicontinuous. Other cubic phases consist of an array of micelles, or inverted micelles; they are said to be discontinuous. From Mariani (1991).

compared to the bilayer spectra, and with about half their width. Finally, in the case of small sonicated liposomes, or in micellar configurations, effective isotropic motion occurs, and lateral diffusion succeeds in averaging all orientations, leading to a narrow symmetric ^{31}P NMR spectrum as shown in Fig. 9.22. A detailed review of the application of ^{31}P NMR to membrane studies may be found in Smith and Ekiel (1984).

In recent years, the interest in cubic phases has increased, partly as a result of their proposed implication in membrane fusion events (Gruner *et al.*, 1988). Cubic phases are isotropic, hence the narrow ^{31}P NMR spectral line; and example is shown in Fig. 9.23. NMR can assist X-ray diffraction in the study of cubic phases, mainly because the so-called NMR pulsed field gradient method may be used in determining the translational diffusion coefficients of both the lipid and water components (Eriksson *et al.*, 1987; Lindblom and Rilfors, 1989).

References and further reading

Aursand, M. and Grasdalen, H. (1992) *Chem. Phys. Lipids,* **62**, 239.

Aursand, M., Rainuzzo, J.R. and Grasdalen, H. (1993) *J. Am. Oil Chem. Soc.,* **70**, 971.

Becker, E.D. (1969) *High Resolution NMR*, Academic Press, New York.

Berden, J.A., Barker, R.W. and Radda, G.K. (1975) *Biochim. Biophys. Acta,* **375**, 186.

Bloom, M., Davis, J.H. and MacKay, A.L. (1981) *Chem. Phys. Lett.,* **80**, 198.

Bruker (1978) *Bruker Reports*, Bruker-Physik AG, Forschhein, Karlsruhe, Germany.

Cantor, C.R. and Schimmel, P.R. (1980) *Biophysical Chemistry, part II, Freeman, San Francisco.*

Capuani, G., Aureli, F., Miccheli, A., Di Cocco, M.E., Ramacci, M.T. and Delfini, M. (1992) *Lipids,* **27**, 389.

Caughey, W.S., Smythe, G.A., O'Keefe, D.H., Maskasky, J.E. and Smith, M.L. (1975) *J. Biol. Chem.,* **250**, 7602.

Chan, S.I., Bocian, D.F. and Petersen, N.O. (1981) in *Membrane Spectroscopy* (ed. E. Grell), Springer, Berlin, p. 1.

Chapman, D. (1965) *The Structure of Lipids*, Methuen, London.

Chapman, D. and Salsbury, N.J. (1966) *Trans. Faraday Soc.,* 2607.

Cullis, P.R. and de Kruijff, B. (1979) *Biochim. Biophys. Acta,* **559**, 399.

Dennis, E.A. and Owens, J.M. (1973) *J. Supramol. Struct.,* **1**, 65.

Dennis, E.A. and Plückthun, A. (1984) in *Phosphorus*-31 *NMR, Principles and Applications* (ed. D.G. Gorenstein), Academic Press, New York.

Dennis, E.A. and Ribeiro, A.A. (1976) *ACSD Symp. Ser.,* **34**, 453.

De Rosa, M. and Gambacorta, A. (1988) *Prog. Lipid Res.,* **27**, 153.

Eads, T.M., Blaurock, A.E., Bryant, R.G, Roy, D.J. and Croasmun, W.R. (1992) *J. Am. Oil Chem. Soc.,* **69**, 1057.

Eads, T.M. (1991) in *Analyses of Fats, Oils and Lipoproteins* (ed. E.G. Perkins), American Oil Chemists' Society, Champaign, pp. 409–57.

Eriksson, P.O., Lindblom, G. and Arvidson, G. (1987) *J. Phys. Chem.,* **91**, 846.

Farrar, T.C. and Becker, E.D. (1971) *Pulse and Fourier Transform NMR,* Academic Press, New York.

Frost, D.J. and Gunstone, F.D. (1975) *Chem. Phys. Lipids,* **15**, 53.

Goñi, F.M. (1984) in *Investigative Microtechniques in Medicine and Biology* (eds J. Chayen and L. Bitensky), Marcel Dekker, New York.

Gruner, S.M., Tate, M.W., Kirk, G.L., So, P.T.C., Turner, D.C., Keane, D.T., Tilcock, C.P.S. and Cullis, P.R. (1988) *Biochemistry,* **27**, 2853.

Gunstone, F.D. (1990a) *Chem. Phys. Lipids,* **56**, 201.

Gunstone, F.D. (1990b) *Chem. Phys. Lipids,* **56**, 227.

Gunstone, F.D. (1991a) *Chem. Phys. Lipids,* **58**, 159.

Gunstone, F.D. (1991b) *Chem. Phys. Lipids,* **58**, 219.

Gunstone, F.D. (1991c) *Chem. Phys. Lipids,* **59**, 83.

Gunstone, F.D. (1993a) in *Advances in Lipid Methodology – Two* (ed. W.W. Christie), Oily Press, Dundee, pp. 1–68.

Gunstone, F.D. (1993b) *J. Am. Oil Chem. Soc.,* **70**, 965.

Gunstone, F.D., Pollard, M.R., Scrimgeour, C.M. and Vedanayagam, H.S. (1977) *Chem. Phys. Lipids,* **18**, 115.

Haw, J.F. (1992) *Anal. Chem.,* **64**, 243R.

Henderson, T.O., Glorek, T. and Myers, T.C. (1974) *Biochemistry,* **13**, 623.

Jackson, M.L., Schmidt, C.F., Lichtenberg, D., Litman, D.J. and Albert, A.D. (1982) *Biochemistry,* **21**, 4576.

Kang, S.Y., Gutowki, H.S., Hsung, J.C., Jacobs, R., King, T.E., Rice, D. and Oldfield, E. (1979) *Biochemistry,* **18**, 3257.

Kates, M. (1986) *Techniques of Lipidology*, 2nd edn, North-Holland, Amsterdam.

Kaya, K. (1992) *Prog. Lipid Res.,* **31**, 87.

Knowles, P.F., Marsh, D. and Rattle, H.W.E. (1976) *Magnetic Resonance of Biomolecules,* Wiley, London.

Lindblom, G. and Rilfors, L. (1989) *Biochim. Biophys. Acta,* **988**, 221.

London, E. and Freigenson, G.W. (1979) *J. Lipid Res.,* **20**, 408.

Mariani, P. (1991) *Curr. Opin. Struct. Biol.,* **1**, 501.

McLaughlin, A.C., Podo, F. and Balié, J.K. (1973) *Biochim. Biophys. Acta,* **330**, 109.

Merchant, T.E. and Glonek, T. (1992) *Lipids,* **27**, 551.

O'Keefe, D.H., Barlow, C.H., Smythe, G.A., Fuschsman, W.H., Moss, T.J., Lilienthal, H.R. and Caughey, W.S. (1975) *Bioinorg. Chem.,* **5**, 125.

Oldfield, E., Adebodun, F., Chung, J., Montez, B., Park, K.D., Le, H. and Phillips, B. (1991) *Biochemistry,* **30**, 11025.

Plückthun, A. and Dennis, E.A. (1981) *J. Phys. Chem.,* **85**, 678.

Plückthun, A. and Dennis, E.A. (1982a) *Biochemistry,* **21**, 1743.

Plückthun, A. and Dennis, E.A. (1982b) *Biochemistry,* **21**, 1750.

Pollard, M. (1986) in *Analysis of Oils and Fats* (eds R.J. Hamilton and J.B. Rossell), Elsevier, New York.

Pople, J.A., Schneider, N.G. and Bernstein, H.J. (1959) *High Resolution NMR,* McGraw-Hill, New York.

Ribeiro, A.A. and Dennis, E.A. (1976) *J. Colloid Interface Sci.,* **55**, 94.

Rice, D.M., Meadows, M.D., Scheinman, A.O., Goñi, F.M., Gómez-Fernández, J.C., Moscarello, M.A., Chapman, D. and Oldfield, E. (1979) *Biochemistry,* **18**, 5893.

Roberts, M.F., Adamich, M., Robson, R.J. and Dennis, E.A. (1979) *Biochemistry,* **15**, 3301.

Rosenthal, S.N. and Fendler, J.H. (1976) *Adv. Phys. Org. Chem.,* **13**, 279.

Sankaram, M.B. and Thompson, T.E. (1992) *Biochemistry,* **31**, 8258.

Schmid, H.H.O., Schmid, P.C. and Natarajan, V. (1990) *Prog. Lipid Res.,* **29**, 1.

Schwieter, U., Englert, G., Rigassi, N. and Vetter, W. (1969) *Pure Appl. Chem.,* **20**, 365.

Sears, B. (1975) *J. Membr. Biol.,* **20**, 59.

Seddon, J.M. (1990) *Biochim. Biophys. Acta,* **1031**, 1.

Seelig, J. (1977) *Q. Rev. Biophys.,* **10**, 353.

Seelig, J. (1978) *Biochim. Biophys. Acta,* **515**, 105.

Smith, I.C.P. (1979) *Can. J. Biochem.,* **57**, 1.

Smith, I.C.P. and Ekiel, I.H. (1984) in *Phosphorus-31 NMR, Principles and Applications* (ed. D.G. Gorenstein), Academic Press, New York.

Smith, M.L. and Caughey, W.S. (1976) *Methods Enzymol.,* **58**, 421.

Sternin, E., Bloom, M. and MacKay, A.L. (1983) *J. Magn. Reson.,* **55**, 274.

Stevens, R.D., Ribeiro, A.A., Lack, L. and Killenbarg, P.G. (1992) *J. Lipid Res.,* **33**, 21–9.

Thurmond, R.L., Lindblom, G. and Brown, M.F. (1990) *Biochem. Biophys. Res. Commun.,* **173**, 1231.

Thurmond, R.L., Dodd, S.W. and Brown, M.F. (1991) *Biophys. J.,* **59**, 108.

Tilcock, C.P.S., Cullis, P.R. and Gruner, S.M. (1986) *Chem. Phys. Lipids,* **40**, 47.

Tiwari, P.N. (1977) *Proc. Symp. Basic Sci. Agric.,* Indian National Science Academy, New Delhi, p. 205.

Urbaneja, M.A., Arrondo, J.L.R., Alonso, A. and Goñi, F.M. (1986) in *Surfactants in Solution,* **vol. 5** (eds K. Mittal and P. Bothorel), Plenum, New York, p. 759.

van den Enden, J., Haighton, A., Van Putte, K., Vermaas, L. and Waddington, D. (1978) *Fette Seifen Anstrichm.,* **80**, 180.

van den Enden, J.C., Rossell, J.B., Vermaas, L.F. and Waddington, D. (1982) *J. Am. Oil Chem. Soc.,* **59**, 433.

Vetter, W., Englert, G., Rigassi, N. and Schweiter, U. (1971) in *Carotenoids* (ed. O. Isler), Birkhäuser, Basel, p. 189.

Waddington, D. (1981) *Fats and Oils: Chemistry and Technology,* Applied Science, London, p. 25.

Wang, D., Hadipour, N.L., Jerlin, E.A. and Stark, R.E. (1992) *J. Lipid Res.,* **33**, 431.

Wollenberg, K.F. (1990) *J. Am. Oil Chem. Soc.,* **67**, 487.

Wu, M., Church, D.F., Mahier, T.J., Barker, S.A. and Pryor, W.A. (1992) *Lipids,* **27**, 129.

9.4 Electron spin resonance spectra

Electron spin resonance (ESR) has found comparatively little use in the lipid field, especially because lipids do not usually give rise to ESR signals. However, the method is well suited for the study of free radicals, thus finding an application in lipid autoxidation. ESR has also been extensively applied to membrane studies, but this requires the addition of exogenous spin labels, and therefore risks perturbation of physiological processes by the probe. The advantages of this technique are high sensitivity to the signal generated by the probe, and a capacity to observe motion occurring over a large, biologically relevant timescale (10^{-7}–10^{-11} s for conventional ESR and 10^{-3}–10^{-7} s for saturation transfer methods).

Detailed reviews of ESR spectroscopy have been published. (Berliner, 1976, 1979; Knowles *et al.*, 1976; Schreier *et al.*, 1978; Marsh, 1981, 1990; Knowles and Marsh, 1991).

For those readers unfamiliar with the technique, the review by Marsh (1981) contains a good introduction to the theory and operation of ESR spectrometers. ESR spectra represent the absorption of microwaves as a function of a variable magnetic field. Owing to the electrical configuration of the spectrometer itself and the method used to detect the unabsorbed microwaves, the graphical display is not simply the absorption spectrum, but the first derivative of the absorption lineshape. Transitions between energy levels (spin states) of unpaired electrons are induced by the absorbed radiation and may be described by the equation:

$$h\nu = \beta g H \qquad (9.21)$$

Comparison of equation (9.21) with equations (9.9) and (9.6) reveals the resemblance of nuclear and electron resonance methods. In equation (9.21), β is the so-called Bohr magneton, and the electron g-value is a proportionality constant that is dependent upon the magnetic species and its environment. Only unpaired electrons give rise to ESR spectra, and this is why, except in the study of lipid autoxidation, lipids can only be studied through the use of spin labels, which are stable, usually nitroxide, radicals (Fig. 9.24).

Nitroxides yield three prominent spectral lines due to hyperfine interactions with the three ^{14}N nuclear spin states (Fig. 9.25). The appearance of the three-line nitroxide spectrum is determined in part by: (a) the polarity of the solvent surrounding the probe; (b) the type of motion exhibited by the probe; (c) the rate of motion; (d) the orientation of the probe

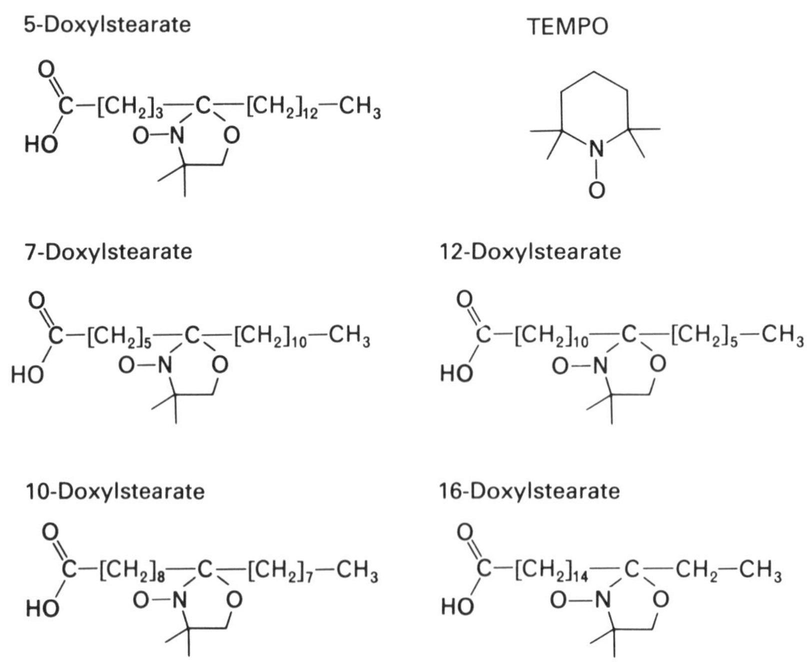

Fig. 9.24 Spin label structures. Shown are TEMPO (2,2,6,6-tetramethylpiperidine-1-oxyl) and the spin label derivatives of stearic acid with the spin label moiety at different positions.

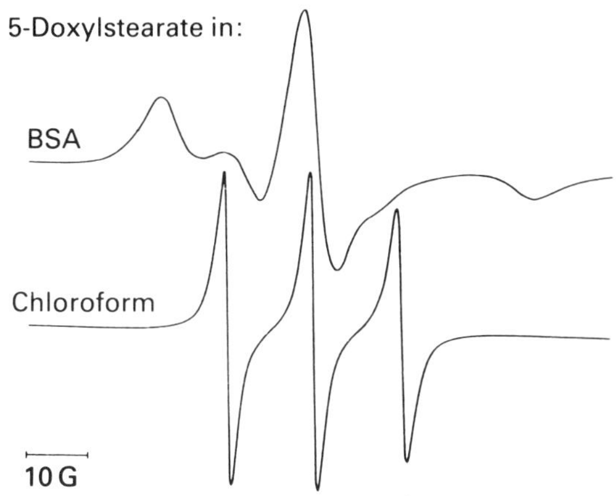

Fig. 9.25 Rapid and slow isotropic motion of 5-doxylstearate. Spectra of the spin-labelled fatty acid in chloroform solution and bound to bovine serum albumin are shown.

(actually the π orbital in which the unpaired electron resides) with respect to the applied magnetic field; and (e) the extent of probe–probe interactions.

Distances (in gauss, G) separating the hyperfine peaks are given by the isotropic hyperfine coupling constant, a'_N (Hubbell and McConnell, 1971). The polarity dependence of a'_N has been used to examine the membrane polarity profile in chromaffin granules (Fretten *et al.*, 1980) and the extent of water penetration into artificial and microsomal membranes (Griffith *et al.*, 1974). The stable free radical 2,2,6,6-tetramethylpiperidine-1-oxyl (TEMPO) has been utilized to detect thermotrophic phase changes evident as discontinuities of the TEMPO spectral parameter f versus temperature. In this manner, phase separations have been observed in binary and more complex lipid systems, and in prokaryotic and eukaryotic membranes (Kleeman and McConnell, 1974; Hatten *et al.*, 1978).

Upon insertion into a lipid bilayer, nitroxide fatty acid derivatives become aligned so that the carboxyl group becomes associated with the bilayer surface, and the terminal methyl group resides deep in the hydrophobic matrix (Hubbell and McConnell, 1969). Thus the spectra resulting from random distribution of nitroxide-containing membranes have an intermediate shape between the isotropic spectrum and the rigid lattice spectrum shown in Fig. 9.25. Separations between the hyperfine lines determine the experimental approximation of the order parameter, S. Increasing angular amplitudes of motion decrease the value of S, which is expressed as

$$S = \tfrac{1}{2}\,(<3\cos^2\theta> -1) \qquad (9.22)$$

where $<\theta>$ is the time-averaged angle formed by the long axis of the fatty acid spin label and the plane normal to the bilayer.

There are numerous examples of the application of ESR spectroscopy to the study of membrane dynamics. Among the most recent ones, they range from measurements of global fluidity, e.g. of the AIDS virus membrane (Aloia *et al.*, 1988), to more detailed studies of cholesterol dynamics in PC bilayers (Shin *et al.*, 1990), measurements of flexibility gradients, in parallel with ^{2}H NMR (Moser *et al.*, 1989), or studies on order and dynamics in oriented multi-bilayers (Korstanje *et al.*, 1989). The conformational heterogeneity of phospholipids in model membranes has been examined by ESR in parallel with frequency-domain fluorescence spectroscopy (Squier *et al.*, 1991). Spin-labelled lipids have been used to determine translational diffusion rates (King and Marsh, 1986; Cribier *et al.*, 1990) as well as transbilayer phospholipid motion (Bitbol and Devaux, 1988; Schreier *et al.*, 1992).

In addition, ESR has found wide application as a technique for detecting the perturbation produced on bilayer lipids by more or less 'foreign' molecules, e.g. anaesthetics (Firestone *et al.*, 1987), carotenoids (Subczynski *et al.*, 1992), metal ions (Mattai *et al.*, 1989) or retinoids (Wassall *et al.*, 1988). Studies of surface potential and polarity of phospholipid bilayers have benefited from the use of pH-sensitive spin labels (Khramtsov *et al.*, 1992). However, ESR is not particularly useful in the detection of non-lamellar lipid phases (Seddon, 1990).

Earlier applications of ESR to the study of lipid free radicals and autoxidation were reviewed by Chapman (1965). More recently, Brown and Wüthrich (1977) have shown an elegant example of a spin label study of lipid oxidation catalysed by haem proteins, and Bascetta *et al.* (1983, 1984) have used ESR to study the abstraction of hydrogen from saturated, olefinic and acetylenic fatty acids and esters.

References and further reading

Aloia, R.C., Jensen, F.C., Curtain, C.C., Mobley, P.W. and Gordon, L.M. (1988) *Proc. Natl Acad. Sci. USA*, **85**, 900.

Bascetta, E., Gunstone, F.D. and Walton, J.C. (1983) *J. Chem. Soc., Perkin Trans. II*, 603.

Bascetta, E., Gunstone, F.D. and Walton, J.C. (1984) *J. Chem. Soc., Perkin Trans. II*, 401.

Berliner, L.J. (1976, 1979) *Spin Labelling, Theory and Applications*, vols 1 and 2, Academic Press, New York.

Bitbol, M. and Devaux, P.F. (1988) *Proc. Natl Acad. Sci. USA*, **85**, 6783.

Brown, L.R. and Wüthrich, K. (1977) *Biochim. Biophys. Acta*, **464**, 356.

Chapman, D. (1965) *The Structure of Lipids*, Methuen, London.

Cribier, S., Morrot, G., Neumann, J.M. and Devaux, P.F. (1990) *Eur. Biophys. J.,* **18**, 33.

Firestone, L.L., Janoff, A.S. and Miller, K.W. (1987) *Biochim. Biophys. Acta,* **898**, 90.

Fretten, P., Morris, S.J., Walts, A. and Marsh, D. (1980) *Biochim. Biophys. Acta,* **598**, 247.

Griffith, O.H., Dehlinger, P.J. and Van, S.P. (1974) *J. Membr. Biol.,* **15**, 159.

Hatten, M.E., Scandella, C.J., Horwitz, A.F. and Burger, M.M. (1978) *J. Biol. Chem.,* **253**, 1972.

Hubbell, W.L. and McConnell, H.M. (1969) *Proc. Natl Acad. Sci. USA,* **64**, 20.

Hubbell, W.L. and McConnell, H.M. (1971) *J. Am. Chem. Soc.,* **93**, 314.

Khramtsov, V.V., Marsh, D., Weiner, L. and Reznikov, V.A. (1992) *Biochim. Biophys. Acta,* **1104**, 317.

King, M.D. and Marsh, D. (1986) *Biochim. Biophys. Acta,* **63**, 341.

Kleeman, W. and McConnell, H.M. (1974) *Biochim. Biophys. Acta,* **345**, 226.

Knowles, P.F. and Marsh, D. (1991) *Biochem. J.,* **274**, 625.

Knowles, P.F., Marsh, D. and Rattle, H.W.E. (1976) *Magnetic Resonance of Biomolecules,* Wiley, London.

Korstanje, L.J., Van Faasen, E.E. and Levine, Y.K. (1989) *Biochim. Biophys. Acta,* **980**, 225.

Marsh, D. (1981) in *Membrane Spectroscopy* (ed. E. Grell), Springer, Berlin.

Marsh, D. (1990) *FEBS Lett.,* **268**, 371.

Mattai, J., Hauser, H., Demel, R.A. and Shipley, G.G. (1989) *Biochemistry,* **28**, 2322.

Moser, M., Marsh, D., Meier, P., Wassmer, K.H. and Kothe, G. (1989) *Biophys. J.,* **55**, 111.

Schrier, S., Polnaszek, C.F. and Smith, I.C.P. (1978) *Biochim. Biophys. Acta,* **515**, 375.

Schreier, S.L., Zachowski, A., Herve, P., Kader, J.C. and Devaux, P.F. (1992) *Biochim. Biophys. Acta,* **1105**, 170.

Seddon, J.M. (1990) *Biochim. Biophys. Acta,* **1031**, 1.

Shin, Y.K., Moscicki, J.K. and Freed, J.H. (1990) *Biophys. J.,* **57**, 445.

Squier, T.C., Mahaney, J.E., Yin, J.J., Lai, C.S. and Lakowicz, J.R. (1991) *Biophys. J.,* **59**, 654.

Subczynski, W.K., Markowska, E., Gruszecki, W.I. and Sielewiesiuk, J. (1992) *Biochim. Biophys. Acta,* **1105**, 97.

Wassall, S.R., Phelps, T.M., Albrecht, M.R., Langsford, C.A. and Stillwell, W. (1988) *Biochim. Biophys. Acta,* **939**, 393.

9.5 X-ray diffraction

X-ray diffraction studies have been most valuable in the elucidation of the manner of arrangement of the lipid molecules, their multiple-melting phenomena, phase behaviour, and the characterization and identification of the structure of individual molecules. The early studies were reviewed by Chapman (1965) and Luzzati (1968). More recently, X-ray studies have been encouraged by the possibility of obtaining single crystals of phospholipids, as well as by the use of synchrotron radiation. Some of the newer results have been reviewed by Makowski and Li (1983), Ranck (1983), Caffrey (1989) and Mariani (1991).

9.5.1 Theoretical considerations

In this technique, X-irradiation is directed towards the electrons of atoms in a crystal. X-rays are electromagnetic waves with wavelengths of the order of 0.1 nm (1 Å), i.e. the same order as the interatomic distances. The particular feature of a crystal is that the atoms or molecules are arranged in such a way as to repeat a pattern in three dimensions. The distance at which the repetition occurs is termed the 'identity period'. In the three-dimensional structure three non-coplanar directions together with the identify periods describe the periodicity of the structure. The 'lattice' is the mathematical term expressing the mode of repetition producing the structure, and a set of identity periods along three non-coplanar directions is sufficient to define a lattice. A space lattice, defined by the choice of three coordinate axes, is divided into a set of identical parallelepipeds. The respective action of the lattice translational elements on one of these parallelepipeds generates the whole lattice. Such a parallelepiped is called the 'unit cell' of the lattice. By varying the choice of axes to describe the lattice, there is an infinite choice of unit cells.

In a three-dimensional infinite lattice, it can be shown that a plane passing through three non-collinear lattice points passes through an infinite number of lattice points. These planes are termed 'rational'. The action of lattice translations on a rational plane is to generate a family of parallel planes. The perpendicular distance d between successive planes is a particularly convenient single parameter. It is these interplanar spacings which are determined directly by X-ray powder photography. To describe these planes numerically and to express their angular relation to the axes of a unit cell of the lattice, Miller indices are used: $(h\,k\,l)$. A plane having these particular values makes intercepts on the axes proportional to a/h, b/k and c/l.

If we allow monochromatic X-rays to pass through a crystal, each atom scatters a very small fraction of the incident wave, and each becomes an emitter of secondary wavelets similar to the elements of the wavefront in the Huygens construction for the propagation of light waves. The scattered wavelets reinforce in a given direction only if they are in phase.

Thus, if parallel radiation incident at a glancing angle θ is reflected at the same angle from successive planes P_1 and P_2 that are distance d apart, the retardation between the two beams is $2d\sin\theta$. If $2d\sin\theta = n\lambda$, where n is an integer, there will be reinforcement for wavelength λ in the direction θ, whilst in all other directions there will be destructive interference. This equation is known as Bragg's law. For a specific wavelength, the directions in which diffraction is possible are determined by the interplanar spacings d of the rational planes. These are a property of the lattice and not of the distribution of the diffracting material within the unit cell.

The diffracted waves from different points within a unit cell differ in phase by an amount that is a function of their relative coordinates and the direction of diffraction. The amplitude of the total diffracted beams, therefore, depends upon the arrangement of the atoms within the unit cell. When the amplitudes are added, one obtains the structure factor F_{hkl}, which is related to the total number of atoms within the unit cell, the relative phase of the radiations scattered by each atom and the scattering power f of each atom:

$$F_{hkl} = \sum_{n=1}^{N} f_n \exp[2\pi i(hx_n + ky_n + lz_n)] \qquad (9.23)$$

A useful method for enabling a complete description of the X-ray diffraction properties of the crystal lattice is to use the reciprocal lattice. This is constructed by marking points distance $k(1/d)$ from the origin along the normal to the rational planes in real space (d is the spacing of the plane). If k is made equal to the wavelength of the radiation, the distance of a point from the origin is equal to $2\sin\theta$, and the direction of the reflected beam is obtained by simple geometrical construction. The set of reciprocal points form a true lattice in which each point corresponds to a reflecting plane.

More details on X-ray diffraction theory may be found in the excellent texts by Stout and Jensen (1968), Sherwood (1976), or Woolfson (1970).

9.5.2 Methodology

(a) Experimental information

X-rays are produced either by the bombardment of a metal target by electrons using an accelerating voltage of 30–50 kV or by a synchrotron source (Rosenbaum *et al.*, 1971). The most common source of X-rays is the K_α line of copper (Cu K_α), with a wavelength of 0.154 nm (1.54 Å). A point of line source of X-rays is defined and collimated by one of a variety of possible schemes using either pinholes or slits. A point-focus camera is essential when diffraction in more than one dimension is of interest, but a line focus allows the use of greater intensity when only one-dimensional information is to be obtained. Monochromation is achieved by a crystal monochromator or absorption filters. The X-ray beam can be focused by using a bent crystal monochromator or bent mirrors. X-rays will reflect from mirrors (glass optical flats usually coated with nickel) when they are incident at grazing angles below the critical angle for total external reflection (about 14 minutes of arc). Optimization of camera parameters for small-angle scattering has been discussed previously (for example, Harrison, 1969; Witz, 1969). Luzzati and co-workers recommend recording this spectra *in vacuo* with a Guinier-type camera (Mariani *et al.*, 1988).

The optimum size of a biological sample for X-ray diffraction is about 0.5–1.0 mm thick. A smaller sample will give rise to weaker scattering, while a larger sample will absorb too much of the scattered radiation. Samples can be kept hydrated by placing them in thin-walled glass or quartz capillaries or in a thin specimen chamber with mica, plastic or beryllium windows. Continuously rotating the sample cells during exposure helps to reduce spottiness.

X-rays are detected either by film or by a position-sensitive detector. For membrane systems that diffract in more than one dimension, film has the advantage of being able to record all detectable diffractions in a single exposure. The disadvantages of film include a fairly high background level and a limited dynamic range. The most intense reflections in a diffraction pattern from a membrane may be several orders of magnitude stronger than the weakest reflections. The only way the reflections with such dramatically different intensities can be simultaneously recorded with film is to stack three to six films, one behind the other, in a film pack. Each film absorbs about 70% of the radiation incident on it, and the subsequent films can then record stronger reflections without overexposure. Data on the films must be densitometered and scaled to one another.

The preparation of membrane specimens depends, of course, on the type of material being studied. The simplest form of specimen consists of a dispersion of phospholipid vesicles in buffer. For technical reasons, however, there are important advantages in using membrane stacks. Attempts have been made to stack membrane vesicles artificially by centrifugation and/or controlled dehydration. These attempts have had mixed success, since the structural integrity of membranes requires a high water activity, and in some cases even moderate drying can lead to the separation of membrane components. In particular, pure lipid phases readily separate out if the relative humidity is lowered too much. Another technique that has been used to produce membrane stacks consists of repeatedly dipping a substrate through a monolayer of purified membrane components spread at an air-water interface (Levine and Wilkins, 1971; McIntosh *et al.*, 1976, 1977). Oriented multilayers of pure lipids can also be formed simply by allowing a solution of the lipids in organic solvents to evaporate slowly under a gentle stream of humid nitrogen. Another kind of specimen is a crystal. By taking diffraction patterns with the crystal oriented in different ways relative to the X-ray beam, a three-dimensional set of intensities can be collected. Analysis of these diffraction patterns is the only proven method for obtaining electron density maps of macromolecules at atomic resolution.

(b) Structure determination in lipid mesophases

The characterization of lipid mesophases by diffraction is based first on symmetry (Luzzati, 1968; Seddon, 1990). Information on the long-range organization of the phase is contained in the low-angle region of the diffraction pattern. The long-range translational ordering of the lipid–water aggregates (bilayers, cylinders, micelles, etc.) onto one- (1D), two- (2D) or three-dimensional (3D) lattices gives rise to Bragg reflections whose reciprocal spacings ($S_{hkl} = 1/d_{hkl}$) are in characteristic ratios. For example:

Lamellar

$$S_l = l/d \tag{9.24a}$$

ratios: 1, 2, 3, 4, ...

Hexagonal

$$S_{hk} = 2(h^2 + k^2 - hk)^{1/2}/\sqrt{3}\,a \tag{9.24b}$$

ratios: 1, $\sqrt{3}$, 2, $\sqrt{7}$, 3, $\sqrt{12}$, $\sqrt{13}$, ...

Cubic

$$S_{hlk} = (h^2 + k^2 + l^2)^{1/2}/a \qquad (9.24c)$$

ratios: $1, \sqrt{2}, \sqrt{3}, 2, \sqrt{5}, \sqrt{6}, \sqrt{8}, 3, \ldots$

Here d_{hkl} is the spacing of the set of lattice planes $(h\,k\,l)$ characterized by the Miller indices h, k and l. A list of well established non-lamellar mesophases is given in Table 9.26.

It is essential in structural studies of lipid mesophases that the symmetry of the phase is correctly identified. However, this is not trivial in practice, since not all of the reflections mentioned above may be observed for a particular sample. In fact, only a few diffraction orders are detected, owing to the large thermal disorder inherent in liquid-crystalline phases. Thus, although X-ray diffraction is one of the least ambiguous methods for phase assignments in lipid systems, other techniques such as polarizing microscopy and NMR can be invaluable for confirming the phase identification (Lindblom and Rilfors, 1989; Seddon, 1990; Tate *et al.*, 1991).

Apart from characterizing a particular mesophase, the diffraction data (d-spacings, or positions of the Bragg peaks, and their intensities) can provide additional information. Positions can be used to estimate the lattice spacing and other phase parameters (lamellar thickness, diameter or hexagonal-phase cylinders, etc.), while the scattered intensities may be used to produce an electron density map of the structure. It should be noted that calculations of the lipid and water phase dimensions from d-spacings use the assumption that the water and lipid regions are distinct and separated by a sharp interface. Since, in practice, there is a partial interpenetration of both phases, some approximations are required (Luzzati and Husson, 1962; Seddon *et al.*, 1984).

Structural reconstruction from reflection intensities does not rely on a previous knowledge of the water contents. The procedure requires that an unknown phase angle, as well as the amplitude of each Bragg reflection, be determined. To this aim, the early trial-and-error approach (Luzzati and Spegt, 1967), which has nevertheless led to the determination of many structures, is being substituted by more direct approaches. Luzzati *et al.* (1988) have developed a 'pattern recognition' approach for phasing diffraction data, and have applied it to the inverted hexagonal and cubic structures. Alternatively, a phasing method has been proposed in which the diffraction pattern is changed through a systematic variation in the structure, in turn accomplished through changing the water contents (Stamatoff and Krimm, 1976; Turner and Gruner, 1990).

(c) *Time-resolved X-ray diffraction*

The recently developed method of time-resolved X-ray diffraction allows the study of the kinetics of phase transitions, including the detection of transition intermediates, that could provide

Table 9.26 Description of the known non-lamellar phases of lipids (From Seddon, 1990)

Type	Name	Space-group symbol	Space-group number	Type I example[a]	Type II example[a]
2D	M (oblique)	p2	2	SDS	K soaps (dry)
2D	P (centred rectangular)	cmm	9	SDS/decanol	PC
2D	H (hexagonal)	p6m	17	Lyso-PC	PE and PC
3D	O (body-centred orthorhombic)	Pmmm or P222	47 or 16	–	Na soaps (dry)
3D	T (body-centred) tetragonal	I422	97	SDS	Ca soaps (dry)
3D	R (rhombohedral)	R3m	166	SDS	PC
3D	Q (cubic)	$P4_332$	212	–	Cyt-c/MO
		Pm3n	223	Lyso-PC	–
		Pn3m	224	–	PE
		Fd3m	227	–	OA/MO
		Im3m	229	SDS	PE
		Ia3d	230	K soaps	PC and PE

[a] Except where specifically excluded, a water component (of unstated amount) is also present. MO, monooleoylglycerol; OA, oleic acid; SDS, sodium dodecyl sulphate; PC, phosphatidylcholine; PE, phosphatidylethanolamine; Cyt-c, cytochrome c.

information on the transition mechanism (Caffrey, 1985). The method combines the advantages of a focused, monochromatic, synchrotron-derived X-ray beam and a suitable real-time X-ray imaging device. The method is still in the development stage. An outline of the experimental set-up, early results and technical limitations can be found in Caffrey (1989).

9.5.3 Selected applications

The classical X-ray studies on soaps and neutral fats have been reviewed by Chapman (1965) and Luzzati (1968) among others. Here we shall concentrate on the results concerning polar lipids, and especially the recent studies on phospholipid crystals. Lipid crystals contain only a single species of lipid (there can be no heterogeneity of hydrocarbon chain length) and have a low water content.

(a) Phospholipid crystals

The crystal structure of 1,2-dilauroyl-DL-phosphatidylethanolamine (DLPE)–acetic acid has been studied by X-ray diffraction (Hitchcock *et al.*, 1974; Elder *et al.*, 1977) and the position and thermal parameters of all the non-hydrogen atoms have been determined. The phospholipids are arranged in bilayers in the crystal, with the hydrocarbon chains nearly perpendicular to the bilayer surface. The α fatty acid chain extends further into the bilayer than the β chain, as can be seen in the projection in Fig. 9.26. This is because the glycerol moiety extends nearly perpendicular to the bilayer surface. The β chain branches at right angles to the glycerol moiety

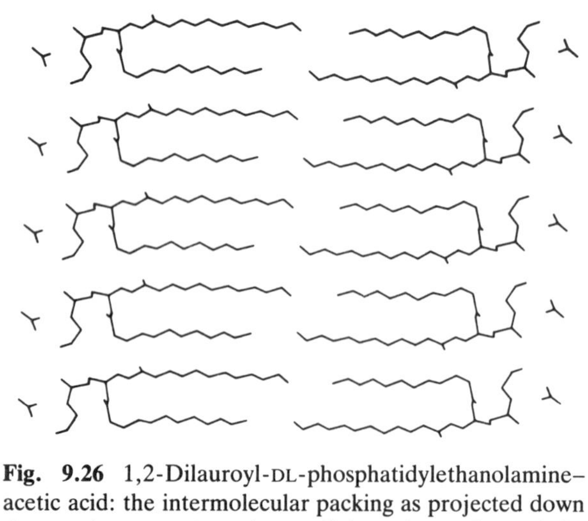

Fig. 9.26 1,2-Dilauroyl-DL-phosphatidylethanolamine–acetic acid: the intermolecular packing as projected down the *c* axis, approximately parallel to the surface of the bilayer in the crystal. From Elder *et al.* (1977).

and extends for several ångströms almost parallel to the bilayer surface. The arrangment of the glycerol group has also been observed in crystals of 1,2-dimyristoyl-*sn*-glycerol-3-phosphorylcholine (DMPC) (Pearson and Pascher, 1979) and cerebroside (Pascher and Sundell, 1977). In 2,3-dilauroyl-D-glycerol, the glycerol group is oriented almost parallel to the bilayer plane (Pascher *et al.*, 1981a). Apparently, the attachment of the polar headgroup alters the orientation of the glycerol moiety so that it is nearly perpendicular to the membrane surface.

In most lipid crystals the hydrocarbon chains appear to be almost hexagonally close-packed, as can be seen in the projection of the DLPE structure perpendicular to the bilayer surface shown in Fig. 9.27. The packing cannot be exactly hexagonal because at high resolution the hydrocarbon chains do not closely approximate circular cylinders. The disorder in the hydrocarbon chains, as indicated by the isotropic temperature factors of the carbon atoms, increases with distance from the polar headgroups in the crystal (Elder *et al.*, 1977). In the centre of the bilayer, where the ends of the α chains from lipid molecules on opposite sides of the bilayer interdigitate, the disorder is very large. In this region the interatomic distances between carbon atoms on different chains are greater than the sum of the expected van der Waals' radii.

The region of the polar headgroup is much more highly ordered. In most lipid crystals these regions are stabilized by a continuous network of hydrogen bonds. In the crystal structure of dimyristoylphosphatidylethanolamine (DMPE) (Elder *et al.*, 1977), the phosphatidylethanolamine dipoles are organized into antiparallel strands. In lysophosphatidylethanolamine (Pascher *et al.*, 1981b) the dipoles are oriented parallel to one another in double strands. A continuous network of hydrogen bonds stabilizes both of these arrangements. When water molecules are present, they also are seen to hydrogen-bond to the lipid polar headgroups (Pearson and Pascher, 1979) as do other molecules of hydration such as acetic acid in the crystals of DLPE (Elder *et al.*, 1977). The polar headgroup dipole appears to be nearly parallel to the bilayer surface in most of these crystals.

In crystals of DMPC (Pearson and Pascher, 1979) the lipids crystallize as stacked bilayers as in the crystals of DLPE, but with the hydrocarbon chains tilted by about 12° relative to the bilayer normal. The tilting of the hydrocarbon chains occurs because the phosphatidylcholine headgroups require a larger surface area than that taken up by two hydrocarbon chains lying perpendicular to the bilayer surface. The cross-sectional area occupied by polar headgroups and hydrocarbon chains can be equal only when the hydrocarbon chains are tilted relative

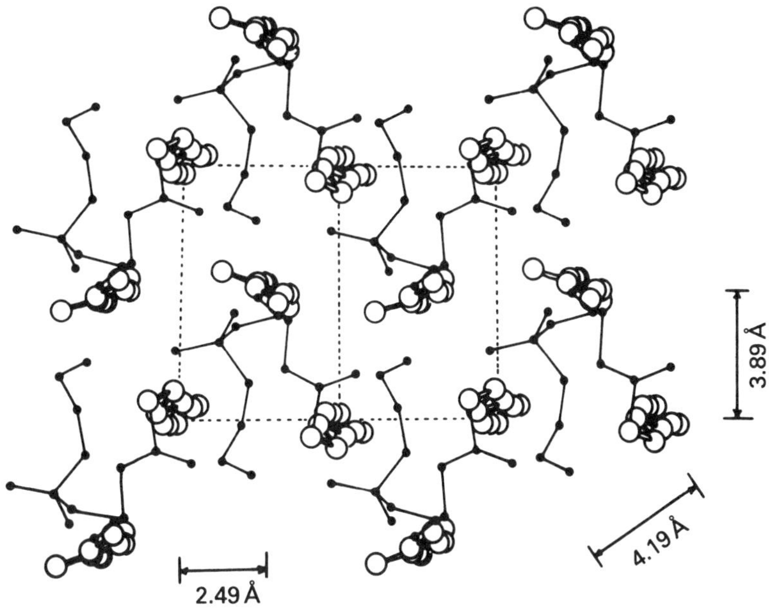

Fig. 9.27 1,2-Dilauroyl-DL-phosphatidylethanolamine–acetic acid: the crystalstructure viewed down the *a* axis, approximately perpendicular to the bilayer surface and parallel to the hydrocarbon chains. Note that the hydrocarbon chains bend slightly. Their packing is not exactly hexagonal because they do not closely resemble circular cylinders at high resolution. From Elder *et al.* (1977).

to the bilayer surface. This phenomenon is also observed in liquid crystals of DMPC in water. The molecular conformations of DLPE and DMPC are very similar in both the crystalline and liquid-crystalline states (Hauser *et al.*, 1981). The hydrocarbon chains of DLPE are not tilted in the crystal because the phosphatidylethanolamine headgroups do not occupy more than 40 Å^2 on the surface of the crystalline bilayer.

(b) Cholesterol

In crystals of both cholesterol monohydrate (Craven, 1976) and anhydrous cholesterol (Shieh *et al.*, 1977) the cholesterol molecules are organized into bilayers with the molecular axes approximately but not exactly perpendicular to the bilayer surface. In both structures, the hydroxyl groups are hydrogen-bonded together in a network stabilizing the crystal. In the monohydrate, none of the molecules has a fully extended hydrocarbon chain. Instead, the chain is partially folded, resulting in a condensed, relatively disordered structure at the centre of the bilayer. This folding may make up the difference in cross-sectional area between the steroid ring system and its short hydrocarbon tail (Rothman and Engelman, 1972), and allow closer packing in the crystal.

(c) Lipid–water liquid crystals; polymorphism

A remarkable variety of structures have been observed in lipid–water systems over a relatively narrow range of conditions. This work has been reviewed, among others, by Luzzati (1968), Shipley (1986), Seddon (1990), Tate *et al.* (1991) and Luzzati *et al.* (1993).

The lamellar phase was discovered over 50 years ago in a hydrated mixture of natural lipids (Bear *et al.*, 1941). Later, Luzzati and co-workers pointed out the large variety of liquid-crystalline phases that could be adopted by lipid–water systems. Among the non-lamellar, the inverted hexagonal (H_{II}) and cubic phases appear to be particularly relevant in biology. A further classification distinguishes between lamellar systems on the grounds of the short-range order of the hydrocarbon chains. If the chains are fluid and give a broad diffraction band centred at about 0.46 nm (4.6 Å), similar to diffraction from liquid paraffins, then the phase is called L_α. If the chains are stiff and parallel and give a sharp reflection at about 0.42 nm (4.2 Å), the phase is called L_β (Luzzati, 1968). In some cases, e.g. PC bilayers, an intermediate phase occurs, with rigid chains and a rippled surface, denoted $P_{\beta'}$. The various lamellar phases constitute an example of thermotropic mesomorphism: the 'gel' (L_β) can be reversibly converted into the 'fluid' (L_α) phase by heating. Occasionally, lamellar phases have been detected consisting of interdigitated bilayers (Slater and Huang, 1988).

Electron density profiles have been obtained from X-ray studies of many lipid–water systems in the lamellar phase. Figure 9.28 shows an electron density profile for a bilayer consisting of lecithin and cholesterol at a 3:2 molar ratio and 100% relative humidity (Franks, 1976). The main features of the profile are two electron density peaks, in this case about 4.4 nm (44 Å) apart, corresponding to the phosphate of the lecithin headgroups, and a pronounced minimum in the electron density at the centre of the bilayer due to the lipid hydrocarbon chain, in particular, the terminal methyl groups. The shoulders, about 1.1 nm (11 Å) from the bilayer centre, are due to cholesterol. Two possible molecular models are shown for the average conformation of lecithin and cholesterol in these bilayers. These models are based on coordinates taken from crystallographic studies of lipid molecules. They are consistent with the X-ray diffraction data and also with neutron diffraction experiments (Worcester and Franks, 1976). The intensities were phased using swelling experiments. In the lamellar array the bilayers are packed 5.89 nm (58.9 Å) apart, and observable diffraction extended to about 0.6 nm (6 Å) spacing. A diffuse band of scattering at about 0.47 nm (4.7 Å) spacing indicated

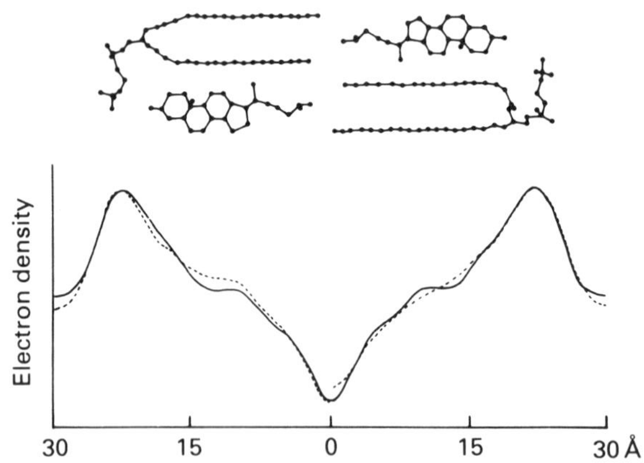

Fig. 9.28 Electron density profile of a lecithin–cholesterol bilayer. The molar ratio of lecithin:cholesterol is 3:2 and the bilayer was at 100% relative humidity. The full curve is that calculated from the X-ray diffraction data. The broken curves are calculated from the two models drawn above. Two distinct molecular models are present, both of which are consistent with the diffraction data. The data, to 0.6 nm (6 Å) resolution, are not of sufficient resolution to allow a distinction to be made between similar lipid molecule conformations. Indeed, in the biolayer, the molecules may take on a large variety of conformations similar to the ones shown here. From Franks (1976).

that the hydrocarbon chains are in the disordered liquid-crystalline state.

In the models presented in Fig. 9.28, the phosphatidylcholine headgroup is oriented approximately parallel to the membrane plane as suggested by the crystallographic studies of L-α-glycerophosphorylcholine (Abrahamsson and Pascher, 1966) and DLPE (Elder *et al.*, 1977). The end of the hydrocarbon chain of cholesterol is very close to the centre of the bilayer in the models in Fig. 9.28, and the cholesterol hydroxyl oxygen is the same distance from the bilayer surface as the fatty acid ester groups (Franks and Lieb, 1979). Very similar electron density profiles are obtained from lamellae in the gel and liquid-crystalline states, as judged from the studies of McIntosh (1978) on dilauroylphosphatidylcholine, except that the bilayer decreases its thickness by about 0.1 nm (1 Å) and the width of the low-density trough occupied by the terminal methyl group of the hydrocarbon chains increases slightly above the transition temperature.

The inverted hexagonal (H_{II}) phase consists of rod-like elements, filled with water (see Fig. 9.22). The H_{II} phase of phospholipids was discovered, and its structure deduced, by Luzzati and Husson (1962), in a lipid extract from human brain containing 52% phosphatidylethanolamine (PE), 35% phosphatidylcholine (PC) and 13% phosphatidylinositol (PI), at 37 °C and water contents below 22 wt%. The structure of this phase, together with information of non-lamellar phase transitions in lipids, has been the object of an excellent review by Seddon (1990).

Liquid-crystalline structures with a cubic symmetry exhibit, like the lamellar and hexagonal phases, long-range order, but have disordered acyl chains. The aggregates formed by the lipid molecules are situated in a cubic three-dimensional lattice. Several different cubic structures exist; an example was shown in Fig. 9.23. Among the liquid-crystalline phases of membrane lipids, the structure of the cubic phases is undoubtedly the most difficult to determine accurately from X-ray diffraction alone (Lindbolm and Rilfors, 1989). Luzzati and Spegt (1967) determined unambiguously, for the first time, the structure of a cubic phase, an anhydrous strontium soap at high temperatures whose structure belonged to the body-centred space group Ia3d. As discussed above, this structure was determined by a trial-and-error approach, which was also used in the resolution of other cubic phases (Luzzati *et al.*, 1987). More recently, a novel pattern recognition approach (Luzzati *et al.*, 1988) has permitted the analysis of the six cubic phases identified so far in lipid-containing systems (Mariani *et al.*, 1988). The theory and biological implications of the cubic phases have been reviewed by Lindblom and Rilfors (1989), Tate *et al.* (1991), Mariani (1991) and Luzzati *et al.* (1993).

Time-resolved X-ray diffraction was first applied to the study of lipid phase transitions by Dupont *et al.* (1972). A rather large number of transitions have been observed since, for which 'characteristic times' have been collected in the review by Caffrey (1989). It is surprising that the range of characteristic times in that collection spans over nine orders of magnitude, from milliseconds to months. Gruner (1987) has also published a detailed review on time-resolved X-ray diffraction as applied to biological samples.

References and further reading

Abrahamsson, S. and Pascher, I. (1966) *Acta Crystallogr.*, **21**, 79.

Bear, R.S., Palmer, K.J. and Schmitt, F.O. (1941) *J. Cell Compar. Physiol.*, **17**, 355.

Caffrey, M. (1985) *Biochemistry*, **24**, 4826.

Caffrey, M. (1989) *Annu. Rev. Biophys. Biophys. Chem.*, **18**, 159.

Chapman, D. (1965) *The Structure of Lipids*, Methuen, London.

Craven, B.M. (1976) *Nature*, **260**, 727.

Dupont, Y., Gabriel, A., Chabre, M., Gulik-Krzywicki, T. and Schechter, E. (1972) *Nature*, **238**, 331.

Elder, M., Hitchcock, P., Mason, R. and Shipley, G.G. (1977) *Proc. R. Soc.*, **A354**, 157.

Franks, N.P. (1976) *J. Mol. Biol.*, **100**, 345.

Franks, N.P. and Lieb, W.R. (1979) *J. Mol. Biol.*, **133**, 469.

Gruner, S.M. (1987) *Science*, **238**, 305.

Harrison, S.C. (1969) *J. Appl. Crystallogr.*, **1**, 84.

Hauser, H., Pascher, I., Pearson, R.H. and Sundell, S. (1981) *Biochim. Biophys. Acta*, **650**, 21.

Hitchcock, P.B., Mason, R., Thomas, K.M. and Shipley, G.G. (1974) *Proc. Natl Acad. Sci. USA*, **71**, 3036.

Huxley, H.E. and Brown, W. (1967) *J. Mol. Biol.*, **30**, 383.

Levine, Y.K. and Wilkins, M.H.F. (1971) *Nature, New Biol.*, **230**, 69.

Lindblom, G. and Rilfors, L. (1989) *Biochim. Biophys. Acta*, **988**, 221.

Luzzati, V. (1968) in *Biological Membranes, Physical Fact and Function* (ed. D. Chapman), Academic Press, New York, p. 71.

Luzzati, V. and Husson, F. (1962) *J. Cell Biol.*, **12**, 207.

Luzzati, V. and Spegt, P.A. (1967) *Nature*, **215**, 701.

Luzzati, V., Mariani, P. and Gulik-Krzywicki, T. (1987) in *Physics of Amphiphilic Layers* (eds J. Meunier, D. Langevin and N. Boccara), Springer, Berlin, p. 131.

Luzzati, V., Mariani, P. and Delacroix, H. (1988) *Makromol. Chem., Macromol. Symp.*, **15**, 1.

Luzzati, V., Vargas, R., Mariani, P., Gulik, A. and Delacroix, H. (1993) *J. Mol. Biol.*, **229**, 540.

Makowski, L. and Li, J. (1983) in *Biomembrane Structure and Function* (ed. D. Chapman), Macmillan, London, p. 43.

Mariani, P. (1991) *Curr. Opin. Struct. Biol.*, **1**, 501.

Mariani, P., Luzzati, V. and Delacroix, H. (1988) *J. Mol. Biol.*, **204**, 165.

McIntosh, T.J. (1978) *Biochim. Biophys. Acta.*, **513**, 43.

McIntosh, T.J., Waldbillig, R.P. and Robertson, J.D. (1976) *Biochim. Biophys. Acta*, **448**, 15.

McIntosh, T.J., Waldbillig, R.P. and Robertson, J.D. (1977) *Biochim. Biophys. Acta*, **466**, 209.

Pascher, I. and Sundell, S. (1977) *Chem. Phys. Lipids*, **20**, 175.

Pascher, I., Sundell, S. and Hauser, H. (1981a) *J. Mol. Biol.*, **153**, 791.

Pascher, I., Sundell, S. and Hauser, H. (1981b) *J. Mol. Biol.*, **153**, 807.

Pearson, R.H. and Pascher, I. (1979) *Nature*, **281**, 499.

Ranck, J.L. (1983) *Chem. Phys. Lipids*, **32**, 251.

Rosenbaum, G., Holmes, K.G. and Witz, J. (1971) *Nature*, **230**, 434.

Rothman, J.E. and Engelman, D.M. (1972) *Nature, New Biol.*, **237**, 42.

Seddon, J.M. (1990) *Biochim. Biophys. Acta*, **1031**, 1.

Seddon, J.M., Cevc, G., Kaye, R.D. and Marsh, D. (1984) *Biochemistry*, **23**, 2634.

Sherwood, D. (1976) *Crystals, X-Rays and Proteins*, Longman, London.

Shieh, H.S., Hoard, L.G. and Nordman, C.E. (1977) *Nature*, **267**, 287.

Shipley, G.G. (1986) in *Handbook of Lipid Research*, vol. 4 (ed. D.M. Small), Plenum, New York.

Slater, J.L. and Huang, C.H. (1988) *Prog. Lipid Res.*, **27**, 325.

Small, D.M. (1986) *Handbook of Lipid Research*, **vol. 4**, *From Alkanes to Phospholipids*, Plenum, New York.

Stamatoff, J.B. and Krimm, S. (1976) *Biophys. J.*, **16**, 503.

Stout, G.H. and Jensen, L.H. (1968) *X-Ray Structure Determination, A Practical Guide*, Macmillan, London.

Tate, M.W., Eikenberry, E.F., Turner, D.C., Shyamsunder, E. and Gruner, S.M. (1991) *Chem. Phys. Lipids*, **57**, 147.

Turner, D.C. and Gruner, S.M. (1990) *Biophys. J.*, **57**, 269a.

Witz, J. (1969) *Acta Crystallogr.*, **A25**, 30.

Woolfson, M.M. (1970) *X-Ray Crystallography*, Cambridge University Press, Cambridge.

Worcester, D.L. and Franks, N.P. (1976) *J. Mol. Biol.*, **100**, 359.

9.6 Mass spectrometry

Of the many analytical techniques now available to the lipid chemist, mass spectrometry (MS) is probably the one that has experienced the fastest growth in the last three decades. This is due both to the development of new techniques (gas and liquid chromatography combined with MS, soft-ionization MS, field desorption MS, atmospheric-pressure MS, etc.) and to the refinement of more traditional methods and their successful application to very complex problems, e.g. the elucidation of glycolipid structure, or the study of structures in lipid mixtures. Much progress has been made since the pioneering work of Ryhage and Stenhagen (1963) on fatty acid methyl esters.

Perhaps the main reason for the success of MS is its ability to detect and characterize organic compounds, giving definite structural information from minute amounts of material. The combination of gas chromatography (GC) and mass spectrometry (GC–MS) is particularly useful for lipid studies because of the ease with which complex mixtures may be separated and identified. In general, any compounds that are volatile enough for GC separation are sufficiently volatile for MS analysis. Compounds that are not sufficiently volatile for GC–MS analysis can usually be studied either after derivatization or with the help of new techniques, such as fast atom bombardment or plasma desorption MS.

9.6.1 Fundamentals of theory and operation

Mass spectra are generally obtained by bombarding the molecules of a compound in the vapour phase at low pressure with electrons of low energy. If the electrons have sufficient energy to cause ionization, the 'molecular ion' is formed:

$$M + e^- \rightarrow M^+ + 2e^- \quad (9.25)$$

As the energy of the bombarding electrons is increased, the probability of ionization increases, and the parent or molecular ion can be formed with excess energy in its electronic and vibrational degrees of freedom. This excess energy may be sufficient to equal the dissociation energy in one of the ion's degrees of freedom, and a fragment ion will be formed. The nature of the ions and their relative abundances depend upon the structure of the molecule being studied. It is common to plot spectra for chemical analysis at bombarding voltages of about 75 eV. At this electron energy the total probability of ionization is high, and there is good sensitivity to detection.

As well as fragments being detected that can be described by assuming simple cleavage of bonds in the parent ion, other fragments arise from atomic rearrangement at the moment of fragmentation. Such rearrangement is known to occur to a certain extent in almost all spectra of molecules containing two or more carbon atoms, and it is very common where migration of hydrogen atoms is involved; it is also common with unsaturated hydrocarbons.

Figure 9.29 is a schematic diagram of a magnetic-field low-resolution mass spectrometer coupled to a gas chromatograph. Modification to this basic design will be discussed in section 9.6.2. The scheme in Fig. 9.29 shows: (1) the gas-chromatograph injection port; (2) gas-chromatograph column and oven; (3) separator; (4) cut-off valve; (5) ion source; (6) entrance slit; (7) magnet; (8) exit slit; (9) multiplier; (10) preamplifier; and (11) amplifier. The ions produced in the ion source are accelerated by a potential of a few thousand volts between two plates and deflected by a magnetic field after passing through a slit. The deflection is usually 60° or 90°. The magnetic field focuses the ion beam onto a collector through a further slit, and the signal is amplified.

The radius of the path described by an ion of mass M and charge e depends upon the accelerating potential V and the magnetic field H. The potential energy eV of the ion will equal the kinetic energy

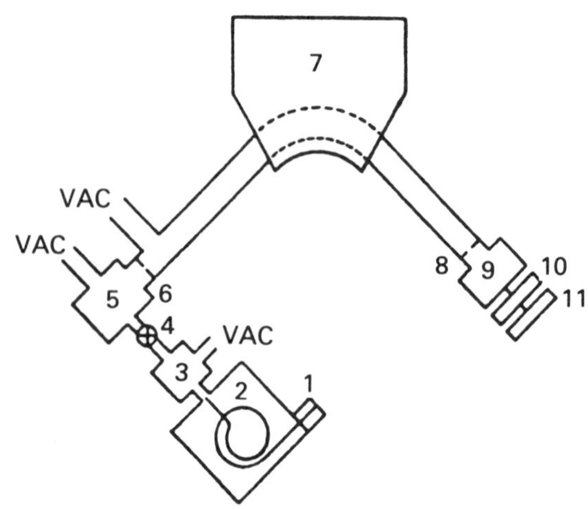

Fig. 9.29 Schematic diagram of a magnetic field low-resolution GC–MS instrument. See text for details.

after full acceleration (where v is the velocity):

$$eV = \tfrac{1}{2} Mv^2 \quad (9.26)$$

In the field H the ion experiences a centripetal force Hev, which is balanced by a centrifugal force Mv^2/r:

$$Hev = Mv^2/r \quad (9.27)$$

Therefore,

$$\frac{M}{e} = \frac{Hr}{v} \quad (9.28)$$

Usually r is kept fixed and, by varying H (or V), ions of differing M/e are allowed to reach the detector. The signal recorded always refers to a mass-to-charge ratio and not to an absolute mass.

A commonly used alternative to magnetic sector MS is the quadrupole mass analyser (Dawson, 1976).

9.6.2 Mass spectrometry techniques

Apart from low-resolution chemical impact mass spectroscopy, which is still the basic standard technique for analytical purposes, more specialized techniques have been developed, including high-resolution (HR), chemical ionization (CI), field ionization (FI) and many others. Some of them will be briefly described here. We shall also discuss the techniques of derivatization and isotope labelling.

(a) High-resolution mass measurements

Accurate mass measurements are usually carried out by the method of peak matching (Saunders and Williams, 1963). Reference compounds that have been used extensively include heptacosafluorotributylamine $C_{12}F_{27}N$ (maximum usable *m/e* 614), perfluorokerosene (maximum *m/e* 700–800) and trisperfluoroheptyl-*sym*-triazine (maximum *m/e* 1185). Other reference compounds have been proposed by Fales (1966) and Olson *et al.* (1977). The accuracy of the mass measurement becomes more critical as the mass increases, in order that the possible atomic combinations may be resolved satisfactorily. The instrumental resolving power needed to separate certain isobaric combinations (e.g. CH_2 and N) rises as the total ionic mass increases. Thus mass measurements of high accuracy (1–2 ppm) are required above *m/e* 500, as, for example, in the identification of the molecular ion in the mass spectrum of 1,2-dioleoyl-3-*sn*-glycerophosphocholine (Klein, 1971a, b, 1978a, b) (Table 9.27). High-resolution work is greatly facilitated by the use of double-focusing instruments, in which the mass analyser contains both an electrical (E) and a magnetic (B) sector. Also, the so-called ion cyclotron resonance Fourier-transform MS (ICR-FTMS) is said to avoid the trade-off between resolution and sensitivity that occurs at high masses in magnetic-sector instruments.

Table 9.27 Common isobaric doublets and the instrumental resolving power needed to separate them (From Klein, 1978a, b)

Doublet	*m/e*	Incremental a.m.u. (atomic mass units) × 10³ (millimass units)	Resolving power *M/M* needed at *m/e* 50	100	500	1000
$^{12}C^{1}H_2$ ^{14}N	14.015 650 14.003 074	12.576	3 976	7 952	39 758	79 517
$^{14}N^{1}H_2$ ^{16}O	16.018 724 15.994 914	23.810	2 100	4 200	21 000	41 999
$^{16}O_2$ ^{32}S	31.989 828 31.972 073	17.755	2 816	5 632	28 161	56 322
$^{14}N_2$ $^{12}C^{16}O$	28.006 148 27.994 914	11.234	4 451	8 902	44 507	89 015
$^{12}C^{1}H_2$ $^{13}C^{1}H$	14.015 650 14.011 179	4.471	11 183	22 366	111 832	223 664
$^{12}C^{1}H_2$ $^{12}C^{2}H(D)$	14.015 650 14.014 102	1.548	32 300	64 600	322 998	645 995
$^{14}C^{1}H_2$ $^{14}N^{1}H_2$	16.018 892 16.018 724	0.168	297 619	595 235	2976 × 10⁶	5 952 × 10⁶

(b) Labelling with stable isotopes or functional groups

Deuterium labelling has been extremely useful in MS studies. For instance, the unexpected formation of the $(M - 43)$ ion in the mass spectra of saturated fatty acid esters, resulting from the intrachain expulsion of C-2, C-3 and C-4, was demonstrated using deuterium labelling (Ryhage and Stenhagen, 1959, 1960). Deuterium has also been used to elucidate the mechanisms by which water is lost under electron impact conditions from glycerides (Morrison *et al.*, 1970; Aasen *et al.*, 1970; Klein and Kemp, 1977).

The introduction of suitable functional groups, usually containing a high-stability charge centre, may be useful in order to direct fragmentation processes, or to produce characteristic mass shifts. The concept of charge localization has been critically reviewed by Williams and Beynon (1976). A recent example of the way in which a functional group may be used to direct fragmentation is the use of fatty acyl pyrrolidides to determine double-bond position and methyl branching (Andersson and Holman, 1975; Andersson *et al.*, 1975; Joseph, 1975). Conversion to the homologous methyl ether or the combined use of capillary GC and MS have been suggested as improved methods for the determination of methyl branching near to the methyl terminus (Karlsson *et al.*, 1973; Apon and Nicolaides, 1975). A large number of other techniques have been examined for the use of derivatives in MS; they are reviewed in Klein and Kemp (1977) and Klein (1978a, b).

The use of combined gas chromatography and mass spectrometry (GS–MS) often requires the preparation of suitable derivatives. From the point of view of GC, derivatives are required in order to improve the velocity and stability of the sample and its chromatographic properties (Drozd, 1975). Derivatives may also give information about functional groups, or provide for selective (non-MS) detection. Reagents used for derivative formation may be general or selective. Among the latter class, reagents that yield cyclic derivatives from bifunctional substrates are of special value; this is the case of the highly versatile boronic acids introduced by Brooks and Watson (1967). A review of derivatives suitable for GC–MS has been published by Brooks *et al.* (1978).

(c) Soft ionization methods

Electron impact (EI), the method outlined in section 9.6.1, is the conventional method of ionizing samples for MS. However, EI is not free from disadvantages, such as inability to provide molecular-weight information from thermally labile samples, difficulty for the study of low-volatility materials, complex spectra arising from interference between molecular and fragment ions, etc. In order to overcome these problems, a large number of alternative ionization methods, collectively known as 'soft ionization methods', have been developed. Reviews of the available techniques have been compiled by Milne and Lacey (1974) and Games (1978). Chemical ionization (CI) and field ionization (FI) are the most useful of these techniques, and both are commercially available. In these methods, less energy is supplied to the molecules in the MS ionization process, and hence the spectra obtained are much simpler than the normal EI spectra since far fewer fragment ions are present (Fig. 9.30).

In chemical ionization (CI) studies the sample is mixed with an excess of reagent gas and this gas is ionized. This initiates ion–molecule reactions, through which the sample itself is eventually ionized, and the ions produced are subsequently analysed (Field, 1972). In FI (Beckey, 1972) the molecule in the gas phase is subjected to a powerful electric field (10^7–10^8 V/cm), and this results in removal of an electron with a much smaller amount of internal energy being supplied, hence fragmentation is reduced. Applications of these methods to lipid analysis may be found in Games (1978); some of them will also be described in section 9.6.3.

(d) Fast atom bombardment and plasma desorption

Fast atom bombardment (FAB) uses a fast atom, usually Ar or Xe, to ionize the sample and transfer the ions to the gas phase. Protonated molecular ions are frequently obtained. The sample is dissolved in a relatively non-volatile liquid matrix such as glycerol. Methodological details on FAB-MS as applied to lipids may be found in Matsubara and Hayashi (1991).

Plasma desorption (PD) is, like FAB, a method for volatilizing large molecules. Samples are bombarded with approximately 100 MeV ions, arising from spontaneous fission of ^{252}Cf, which yields two fragments. One of these hits the detector (starting the so-called time-of-flight timing device) and the other dislodges a number of ions from the sample (Kinsel *et al.*, 1991).

(e) Ionization at atmospheric pressure

Atmospheric-pressure ionization (API) mass spectrometry (Horning *et al.*, 1973; Carroll *et al.*,

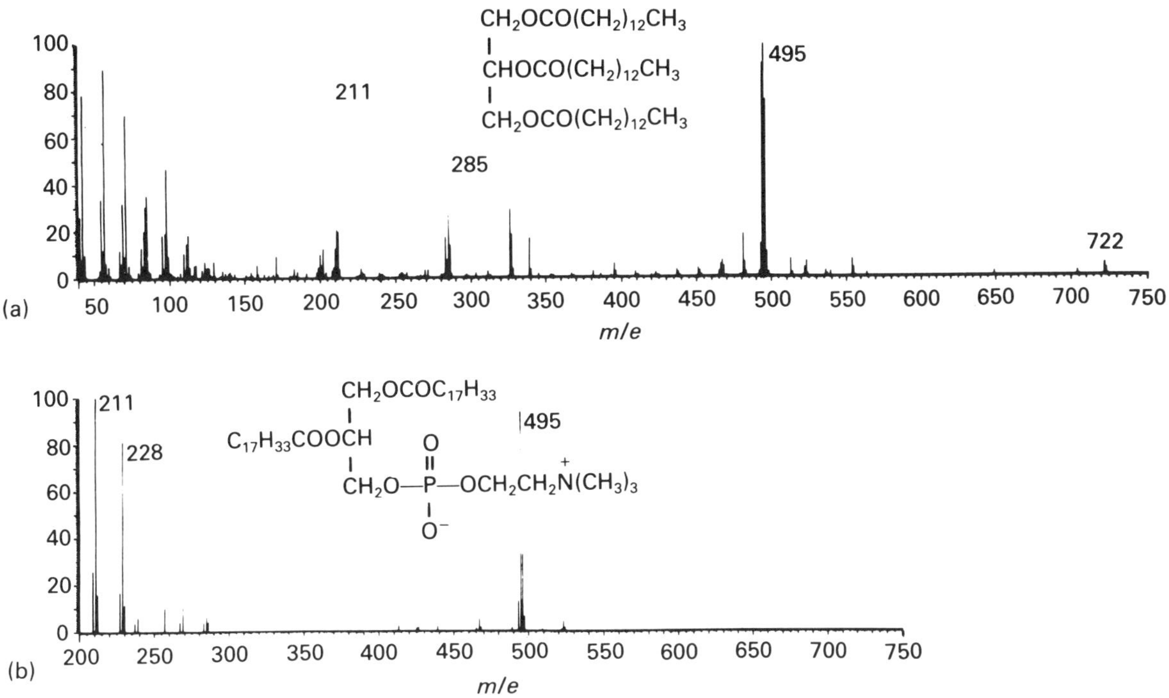

Fig. 9.30 (a) Electron impact mass spectrum of trimyristin. The spectrum was recorded on a Varian CH5D, ionizing voltage 70 eV, source temperature 200 °C, using a gold crucible direct-insertion probe with source operating pressure ≃ 10^6 Torr. The values for *m/e* ratio are given in integral mass units; this also applies to the other spectrum. (b) Methane chemical ionization mass spectrum of distearoylphosphatidylcholine, recorded on a Finnigan 4000 at 0.2 Torr methane, with a source temperature of 220 °C, using in-line combined liquid chromatography–mass spectrometry with a Kapton belt vaporizer at 210 °C. From Games (1978).

1974) is a chemical ionization procedure that differs from the standard technique in a number of ways. The reaction chamber (ion source) is at atmospheric pressure and is external to the high-vacuum region of a quadrupole mass analyser. Ions and molecules enter the mass analyser region through a small aperture. The ions are obtained by ions lenses, while the neutral molecules are removed by pumping. Reagent ions are generated with a ^{63}Ni foil or a corona discharge (Fig. 9.31) rather than by use of a heated filament. Both positive and negative ions are detected. Positive ions are formed from samples by a series of ion–molecule reactions, which usually involve the sequence electrons→ carrier-gas ions→

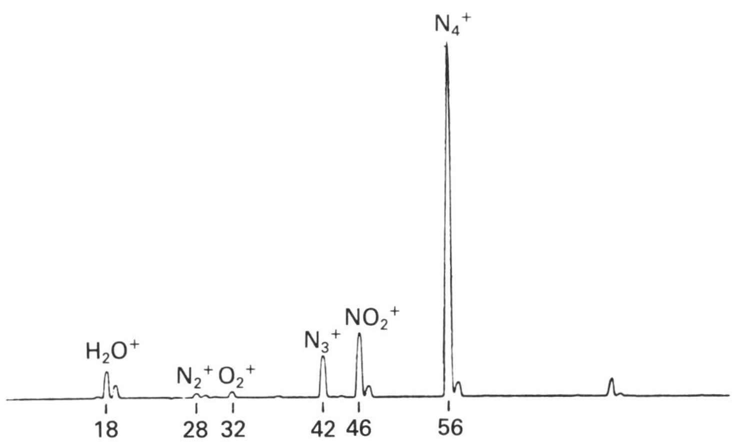

Fig. 9.31 Mass spectrum of the reagent ion present in nitrogen used as a carrier gas with a source assembly of the corona type. From Horning (1978).

reagent ions → sample ions. Negative ions are formed either by direct electron attachment (electron capture) or by ion–molecule reaction (Dzidic *et al.*, 1976). Sub-picogram sensitivity of detection has been achieved for both positive and negative ions. Bioanalytical systems for multicomponent analysis of complex biological samples have been developed, consisting of gas or liquid chromatography systems coupled to MS(API) and a computer (Horning *et al.*, 1973).

(f) Metastable ion techniques

Metastable transitions may be observed for those fragmentations which take place during flight. Such decompositions occur typically between 0.2 and 50 μs after ionization in a double-focusing instrument, and the range of internal energies giving rise to rate constants of this order (2×10^4 to $5 \times 10^6\,s^{-1}$) is known as the 'metastable energy window' (Klein and Kemp, 1977) (Fig. 9.32). Metastable decompositions are useful in structural analysis since their presence

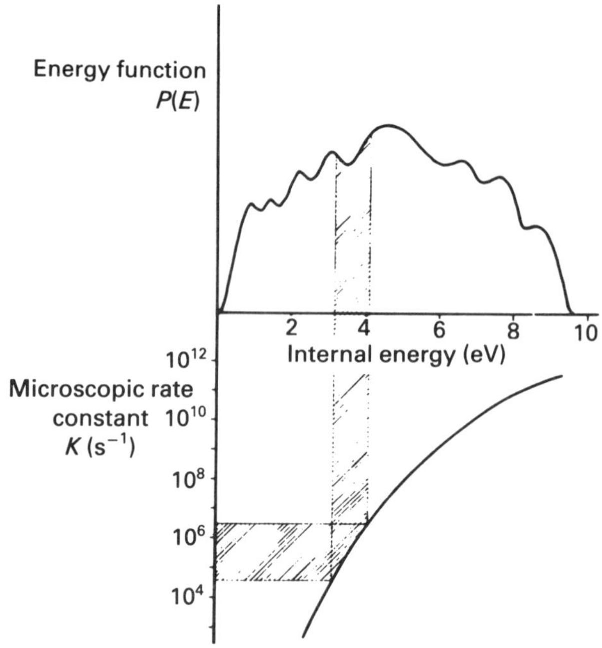

Fig. 9.32 Metastable energy window. From Klein and Kemp (1977).

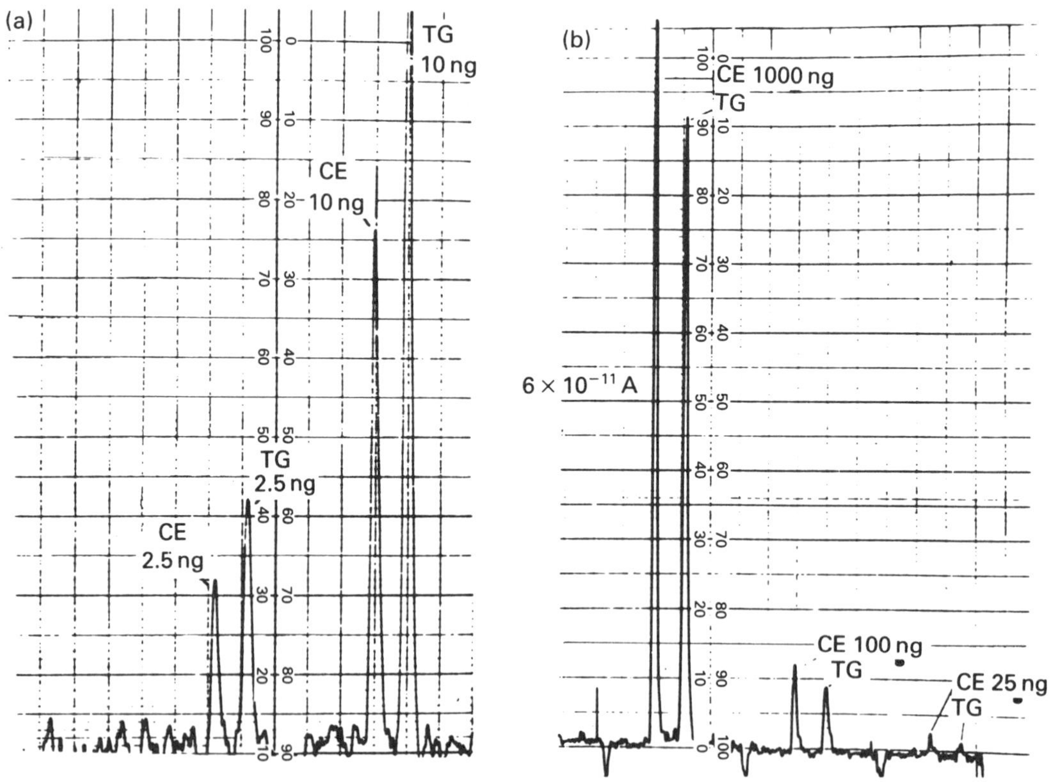

Fig. 9.33 Sensitivities of LC–MS and LC–FID systems. (a) Response of a neutral lipid standard mixture (CE, cholesteryl palmitate; TG, tripalmitin) injected on a 3.2×250 mm column packed with 10 μm silicic acid. Detector: LC–MS interface system, methane chemical ionization, single ion monitors at *m/e* 211. (b) Response of neutral lipid standard mixture (CE, cholesteryl oleate; TG, triolein) injected on a 2.1×250 mm column packed with 6–8 μm Zorbax-Sil. Detector: LC–FID system. From Privett and Erdahl (1978).

indicates that a fragmentation process is at least in part electron-impact-induced, although not necessarily occurring in one step (Holmes and Benoit, 1972). In a double-focusing instrument, transitions may be detected in either the first or second field-free regions (1FFR and 2FFR) (Millington and Smith, 1977). The observation of 1FFR metastable transitions is particularly useful for the detection of very short-lived ions ($\simeq 1\,\mu$s) that leave the source region but fragment before reaching the electrostatic analyser (Ferguson *et al.*, 1977). Daly (1973) have introduced an energy-selective metastable ion detector.

(g) Liquid chromatography–mass spectrometry

Gas chromatography (GC) and MS have long been used in conjunction, as we have repeatedly mentioned. However, GC is not suitable for the separation of complex lipids, which are thermolabile. Liquid chromatography (LC) provides a suitable substitute for MS lipid analysis (Fig. 9.33). Techniques for coupling LC with MS have been reviewed by Privett and Erdahl (1978); these authors have also described an interface for the analysis of lipids by MS. This interface is based on the moving-wire transport principle using an endless stainless-steel belt (Erdahl and Privett, 1977). After evaporation of the solvent, the solute remains as a residue on the belt, which transports it into a reactor where it is volatilized by evaporation or conversion to hydrocarbons. The volatile compounds are then fed into the source of a chemical ionization mass spectrometer. Other commercially available devices (Thermospray, Plasmaspray) work as LC-MS interfaces, even for polar materials, for which the endless belt is unsatisfactory.

9.6.3 Some recent studies

(a) Reviews of MS applications

General reviews on mass spectroscopic studies of fatty acids and lipids include those by Odham and Stenhagen (1972), Klein (1974), Viswanathan (1974), Natale (1977) and Wood (1979). Klein (1978b) has also edited a special issue of *Chemistry and Physics of Lipids* devoted to lipid MS. Studies on the location of double bonds and cyclopropane rings in fatty acids have been reviewed by Minnikin (1978), while data on prostanoids have been compiled by Fisher and Frölich (1982). A review on terpene studies may be found in Enzell *et al.* (1972); steroid MS has been summarized by Budziekievicz (1972). Reviews on glycolipids and glycosphingolipids include ones by Ando *et al.* (1977), Sweeley *et al.* (1978), Karlsson (1976a, b, 1978) and Egge (1978). MS studies of complex lipids of bacterial origin are considered in the review by Batrakov and Bergelson (1978). Carotenoids have been reviewed by Vetter *et al.* (1971). Shackleton (1990) has reviewed the applications of MS to the study of steroids, and Matsubara and Hayashi (1991) have summarized recent studies of FAB-MS of lipids.

(b) A summary of MS studies of lipids

A number of contributions to this field have been indexed in Table 9.28. Reference and, occasionally, remarks on the particular samples under study, or the techniques used, have been included. Two applications are examined in more detail in the following subsections. Examples of methane chemical ionization mass spectra of representative lipids are shown in Figs 9.34–9.37.

(c) MS studies of prostanoids

GC–MS is, together with radioimmunoassay, the most widely used method in the analysis of prostaglandins and related compounds. The main difficulty in this field arises from the fact that a large number of substances (e.g. 12 metabolites in monkey prostacyclin metabolism; Sun and Taylor, 1981) are present in extremely low and widely different concentrations.

Samples for prostanoid analysis by GC–MS must be carefully prepared and purified; high-performance liquid chromatography is particularly useful in this respect (Frölich, 1977). Mass spectrometric quantification is achieved through the use of stable isotopes as internal standards, e.g. $[3,3,4,4\text{-}^2H]_4\text{-}PGF_{2\alpha}$ (Fisher and Frölich, 1982). Gas-chromatographic separation is improved with suitable capillary columns (Rigaud *et al.*, 1981).

Mass spectrometers for prostanoid analysis may belong to either the magnetic focusing or the quadrupole type (Watson, 1976). A low-resolution instrument (RP 500–1000) is enough for qualitative identification of prostanoids; however, high resolution can be very helpful in quantification of a multicomponent sample in the select ion monitoring (SIM) mode, as a high-resolution MS can focus on the exact mass and so excludes impurities with a similar composition. Electron impact or chemical

Table 9.28 A review of mass spectroscopic studies of lipid molecules

Lipid class	References	Remarks[a]
Fatty acids	Graveland (1970)	Hydroxy fatty acids
	Kleiman and Spencer (1973)	Hydroxy fatty acids
	Andersson and Holman (1974)	Pyrrolidine derivatives
	Foglia *et al.* (1976)	α-Substituted esters
	Kuksis and Myher (1976)	
	Kuksis *et al.* (1976)	
	Marai *et al.* (1976)	
	Valicenti *et al.* (1978)	
	Dulaney *et al.* (1978)	Refsum's disease
	Pawlosky *et al.* (1992)	Deuterated fatty acids
Wax esters	Aasen *et al.* (1971)	
	Saito and Gamo (1973)	
Monoacylglycerols	Myher *et al.* (1974)	TMS ethers
	Curstedt (1974)	^{2}H labelling
	Gaskell *et al.* (1976a, b)	Boronate derivatives
Diacylglycerols	Curstedt (1974)	^{2}H labelling
	Myher *et al.* (1978)	t-Butyldimethylsilyl ethers
Triacylglycerols	Lauer *et al.* (1970)	
	Aasen *et al.* (1970)	
	Hites (1970)	
	Rohwedder (1971)	Field ionization
	Beckey (1972)	Field desorption
	Murata and Takahashi (1973)	
	Rohwedder (1975)	Field ionization
	Murata and Takahashi (1977)	Chemical ionization (NH_3)
	Murata (1977)	Chemical ionization (NH_3)
O-Alkylglycerols	Myher *et al.* (1974)	
	Curstedt (1974)	
	Gaskell *et al.* (1976a, b)	
	Satouchi and Saito (1976)	
	Blanck *et al.* (1976)	Isopropylidene derivatives
	Curstedt (1977)	
	Satouchi *et al.* (1978)	TMS ethers
Phosphoglycerides	Klein (1971a, b, 1972)	Metastable ions
	Kemp *et al.* (1972)	New bacterial phospholipid
	Foltz (1972)	Chemical ionization
	Hasegawa and Suzuki (1973, 1975)	Acetolysis
	Karlander *et al.* (1973)	bis-TMS ethers
	Curstedt and Sjövall (1974)	
	Einolf and Fenselau (1974)	Lung lipids
	Wood and Lau (1974)	
	Satouchi and Saito (1976, 1977), Satouchi *et al.* (1978)	Alkyl and alkenyl lipids
	Wood *et al.* (1976, 1977)	Field desorption
	Chebyshev *et al.* (1977)	Field desorption
	Curstedt (1977)	
	Gaskell and Brooks (1977a, b)	
	Ogino *et al.* (1977)	Lung lecithins
	Young *et al.* (1977)	Phosphatidylserine
	Anderson *et al.* (1978)	Sulpholipids
	De Rosa and Gambacorta (1988)	Isoprenoid glycerolipids

Table 9.28 *Continued*

Lipid class	References	Remarks[a]
Glycolipids and sphingolipids	Karlsson (1973)	Glycolipids
	Karlsson *et al.* (1974a, b, c)	Cerebrosides
	Markey and Wenger (1974)	
	Hanfland and Egge (1975, 1976)	Complex glycolipids
	Matsubara (1975)	Sphingophosphonolipids
	Fujino and Ohnishi (1976)	Fatty acid components
	Gaskell and Brooks (1976, 1977a, b)	Boronate derivatives
	Gaskell *et al.* (1976a, b)	Ceramides
	Krisnangkura and Sweeley (1976)	Sphinganine
	Oshima (1976)	
	Breimer *et al.* (1978)	
	Karlsson (1978)	Blood groups
	Karlsson and Larson (1978)	Glycolipids
	Murata *et al.* (1978)	Direct inlet
	Burlingame *et al.* (1992)	Glycolipids
Other complex lipids	Hilker *et al.* (1978)	Ornithinolipids
	Baumann *et al.* (1975)	Diol-derived lipids
	Suzuki and Hasegawa (1976)	Diol-derived lipids
	Sprott (1992)	Archaebacterial lipids
	Kaya (1992)	Taurolipids
Terpenes and steroids	Enzell *et al.* (1972)	
	Coxon *et al.* (1974)	
	Budziekievicz (1972)	
	Burlingame *et al.* (1992)	Steroids, sterols, bile acids, eicosanoids
Prostanoids	Fischer and Frölich (1982)	
Drugs	Horning (1978)	
	Horning *et al.* (1973)	

[a] TMS, trimethylsilyl.

ionization may be used as ionization methods, with no clear-cut advantage of either of them (Claeys *et al.*, 1980).

Derivatization of prostanoids before injection in the GC–MS system is required. Carboxyl groups are converted into methyl or trimethylsilyl (TMS) esters; hydroxy groups into TMS ethers; 1,2- and 1,3-diols into cyclic boronates; and keto groups into oximes (Green *et al.*, 1978; Fischer and Frölich, 1982). GC–MS measurements have also benefited from computer technology, not only for control of instrumental parameters and data processing, but also with the use of specific programs for peak analysis and reconstruction (Rosenkrantz *et al.*, 1982).

When compared to radioimmunoassay, GC–MS requires extensive and time-consuming purification prior to sample injection into the instrument. However, the high specificity and sensitivity of the GC–MS measurement provide reliable results that compensate for the inconvenience of sample preparation.

(d) MS studies of lipid topology in biomembranes

Biomembranes are not static structures, but are rather built on a fluid lipid matrix (Chapman, 1975). Mass spectrometry offers interesting possibilities in the area of affinity labelling, particularly photochemical labelling aimed at the formation of crosslinks between nearest neighbours in membranes. We shall focus our attention on the problem of lipid–lipid interactions, as studied by MS.

Chemical crosslinking as a structural probe was first described by Zahn (1955). The use of MS to characterize lipids labelled with photoactivatable groups was reported by Chakrabarti and Khorana (1975). They showed that the photoaffinity-labelled fatty acids were taken up by an auxotroph of *Escherichia coli* and incorporated into the *sn*-2 position of the glycerol in phospholipids. Nitrene- and carbene-generating precursors have been used to form model membranes; the distribution of cross-

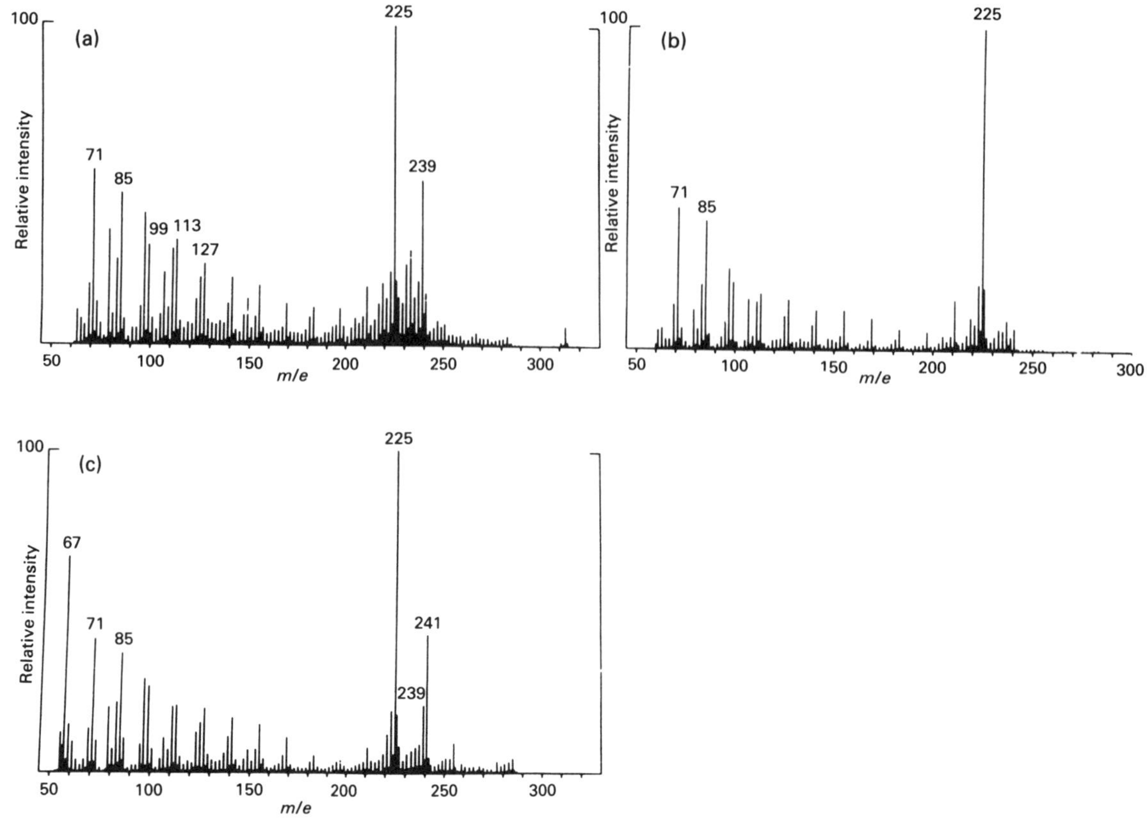

Fig. 9.34 Methane chemical ionization spectra at 2.5 Torr: (a) 1,2-distearoyl-3-hexadecylglycerol, reactor 430 °C; (b) 1-hexadecylglycerol, reactor 430 °C; (c) palmitoyl stearate, reactor 380 °C. Samples were applied to the belt continuously. From Privett and Erdahl (1978).

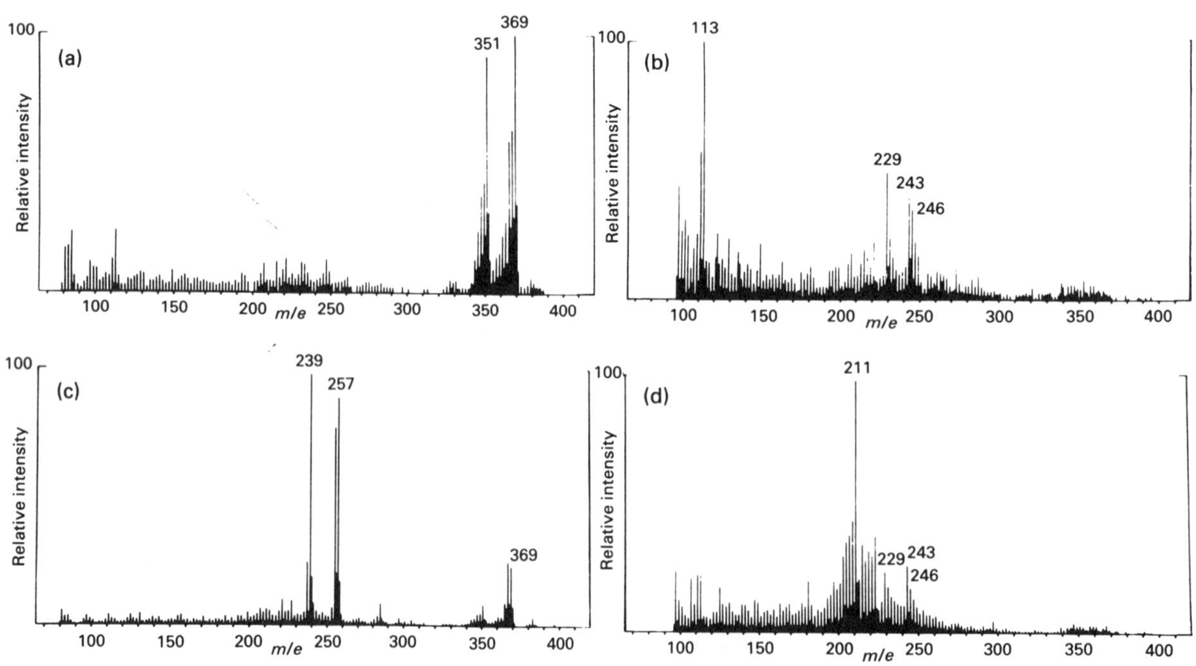

Fig. 9.35 Methane chemical ionization spectra at 2.8 Torr: (a) cholesterol, reactor 380 °C; (b) cholesterol, reactor 430 °C; (c) cholesteryl palmitate, reactor 330 °C; (d) cholesteryl palmitate, reactor 430 °C. Samples were applied to the belt continuously. From Privett and Erdahl (1978).

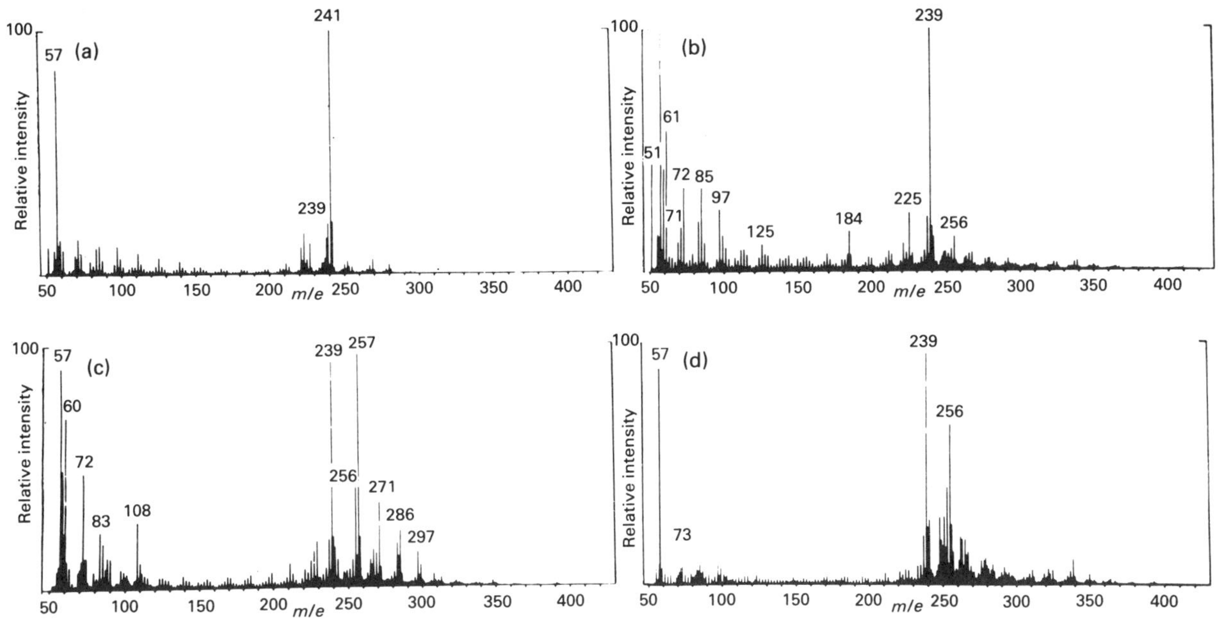

Fig. 9.36 Methane chemical ionization spectra at 2.8 Torr, reactor 339 °C: (a) dihydrosphingosine; (b) ceramides (natural source); (c) sphingomyelin (bovine); (d) cerebrosides (bovine). Samples were applied to the belt continuously. From Privett and Erdahl (1978).

links in the products formed upon irradiation have been determined using MS. Reagents for photochemical crosslinking studies utilizing mass spectral analysis have been discussed by Costello (1982). Some of the results are summarized here.

Radhakrishnan *et al.* (1980) used phospholipids substituted with 2-diazo-3,3,3-trifluoropropionyl or 3*H*-diazirinophenoxy groups in the composition of the fatty acid at *sn*-2; the distribution of insertion points was investigated by examination of mass

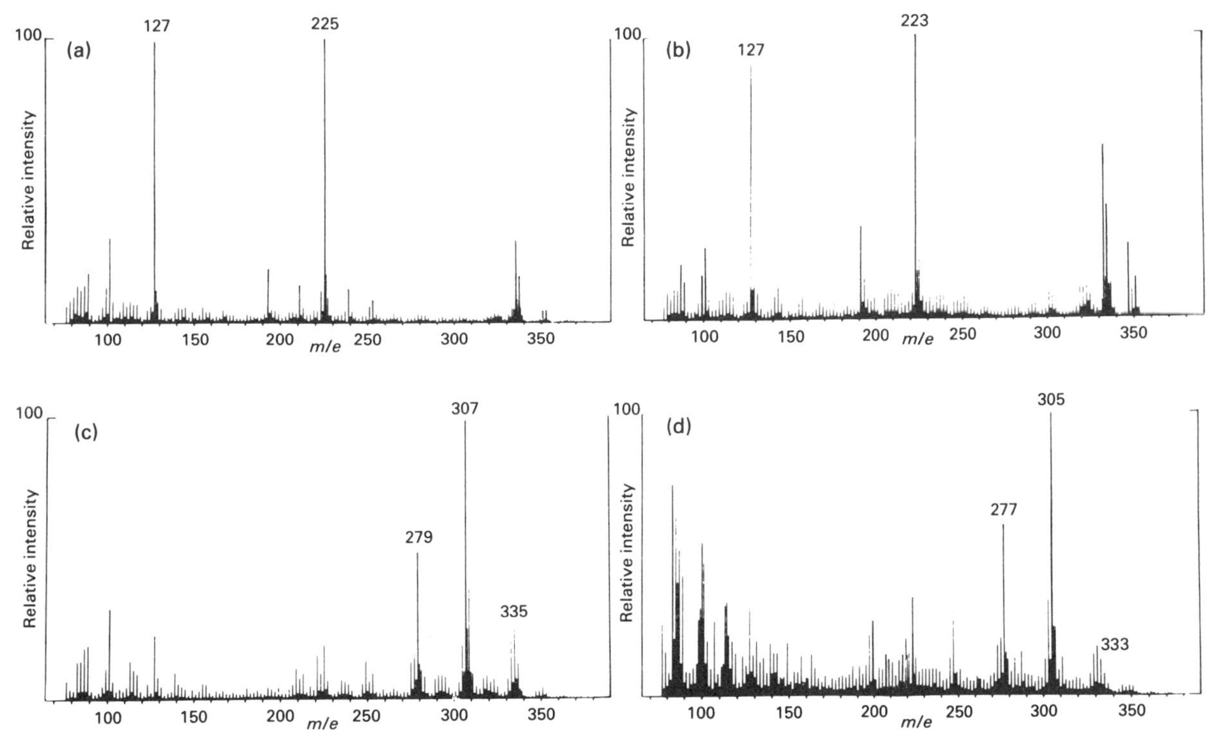

Fig. 9.37 Methane chemical ionization spectra at 2.0 Torr, reactor 225 °C: (a) PGE_1 methyl ester; (b) PGE_2 methyl ester; (c) $PGF_{1\alpha}$ methyl ester. Samples were applied to the belt continuously. From Privett and Erdahl (1978).

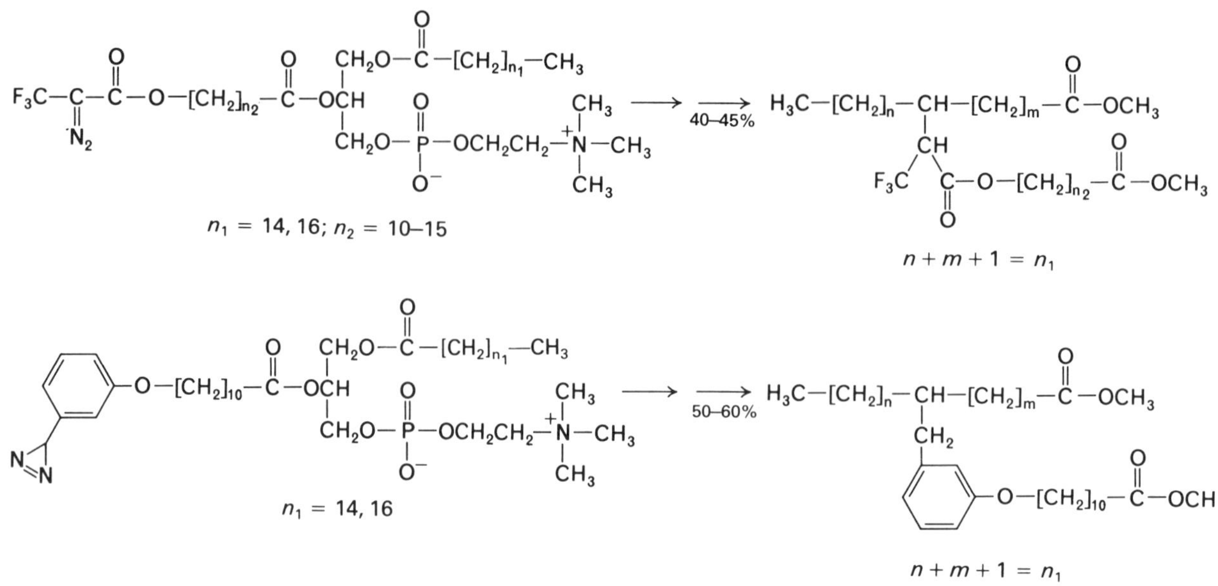

Fig. 9.38 2-Diazo-3,3,3-trifluoropropionyl- and 3*H*-diazirinophenoxy-substituted phospholipids and their respective cross-linked esters.

spectra of the irradiation products. The structures of the synthetic phospholipids and their respective crosslinked esters are shown in Fig. 9.38. The results are consistent with the idea that the molecular conformation of phospholipids in the gel crystalline state is such that the *sn*-2-acyl chain is shorter than the *sn*-1-acyl chain by 2–3 carbon atoms.

Mixtures of a diazirinophenoxy-containing lipid were made with dipalmitoyl- and dioleoylphosphatidylcholines, and irradiated at different temperatures. At low temperatures (i.e. below 30 °C), the diazirino phospholipid partitioned preferentially into the more fluid dioleoyl-rich phase. Thus, the technique may be useful to detect phase separations in biomembranes (Radhakrishnan *et al.*, 1980). Stoffel and co-workers (Stoffel and Metz, 1979; Stoffel and Preissner, 1979) have also carried out a variety of studies with alkylnitrene precursors; they have been able to demonstrate a series of hydrophobic interactions between phospholipids and proteins in membranes. Bayley and Knowles (1980) investigated reactions of phenylnitrene formed within the lipid core of artificial and cell membranes. Interesting results have also been obtained by Breslow *et al.* (1987a, b), who used free radicals generated upon photolysis of benzophenones as probes for molecular conformations in solution and in model membranes. It seems clear that the application of MS to membrane studies will prove fruitful in the elucidation of many topological aspects of the bilayer architecture.

References and further reading

Aasen, A.J., Lauer, W.M. and Holman, R.T. (1970) *Lipids*, **5**, 869.

Aasen, A.M., Hofstetter, H.H., Lyengar, B.T.R. and Holman, R.T. (1971) *Lipids*, **6**, 501.

Anderson, R., Kates, M. and Volcani, B.E. (1978) *Biochim. Biophys. Acta*, **528**, 89.

Andersson, B.A. and Holman, R.T. (1974) *Lipids*, **9**, 185.

Andersson, B.A. and Holman, R.T. (1975) *Lipids*, **10**, 716.

Andersson, B.A., Christie, W.W. and Holman, R.R. (1975) *Lipids*, **10**, 215.

Ando, S., Kon, K., Nagai, Y. and Murata, T. (1977) *J. Biochem. (Tokyo)*, **82**, 1623.

Apon, J.M.B. and Nicolaides, N. (1975) *J. Chromatogr. Sci.*, **13**, 467.

Batrakov, S.G. and Bergelson, L.D. (1978) *Chem. Phys. Lipids*, **21**, 1.

Baumann, W.J., Schupp, E. and Lin, J.T. (1975) *Biochemistry*, **14**, 841.

Bayley, H. and Knowles, J.R. (1980) *Biochemistry*, **19**, 3883.

Beckey, H.D. (1971) *Field Ionization Mass Spectrometry*, Pergamon, New York.

Beckey, H.D. (1972) in *Biochemical Applications of Mass Spectrometry* (ed. G.R. Waller), Interscience, New York, p. 795.

Blanck, M.L., Rainey, W.T., Jr, Christie, W.H., Piantadosi, C. and Snyder, R. (1976) *Chem. Phys. Lipids*, **17**, 201.

Breimer, M.E., Hansson, G.C., Karlsson, K.A., Leffler, H., Pimlott, W. and Samuelsson, B.E. (1978) *FEBS Lett.*, **89**, 42.

Breslow, R., Kitabatake, S. and Rothbard, J. (1978a) *J. Am. Chem. Soc.*, **100**, 8156.

Breslow, R., Rothbard, J., Herman, F. and Rodriguez, M. (1987b) *J. Am. Chem. Soc.*, **100**, 1213.
Brooks, C.J.W. and Watson, J. (1967) *Chem. Commun.*, 952.
Brooks, C.J.W., Edmonds, C.G., Gaskell, S.J. and Smith, A.G. (1978) *Chem. Phys. Lipids*, **21**, 403.
Budziekievicz, H. (1972) in *Biochemical Applications of Mass Spectrometry* (ed. G.R. Waller), Interscience, New York, p. 251.
Burlingame, A.L., Baillie, T.A. and Russell, D.H. (1992) *Anal. Chem.*, **64**, 467R.
Carroll, D.I., Dzidic, I., Stillwell, R.N. and Hörning, E.G. (1974) *Anal. Chem.*, **46**, 706.
Chakrabarti, P. and Khorana, H.G. (1975) *Biochemistry*, **14**, 5021.
Chapman, D. (1965) *The Structure of Lipids*, Methuen, London.
Chapman, D. (1975) *Q. Rev. Biophys.*, **8**, 185.
Chebyshev, A.V., Kabanov, S.P., Petrov, A.A., Serebrennikova, G.A., Kupriyanov, S.E. and Evstigneeva, R.P. (1977) *Bioorg. Khim.*, **3**, 1370.
Claeys, M., Van Hove, C., Duchateau, A. and Herman, A.G. (1980) *Biomed. Mass Spectrom.*, **7**, 544.
Costello, C.E. (1982) *Mass Spectrom. Rev.*, **1**, 169.
Coxon, D.T., Price, K.R., Howard, B., Osman, S.F., Kalan, E.B. and Zacharius, R.M. (1974) *Tetrahedron Lett.*, **34**, 2921.
Curstedt, T. (1974) *Biochim. Biophys. Acta*, **360**, 12.
Curstedt, T. (1977) *Biochim. Biophys. Acta*, **489**, 79.
Curstedt, T. and Sjövall, J. (1974). (1974) *Biochim. Biophys. Acta*, **360**, 24.
Daly, N.R., McCormick, A., Powell, R.E. and Hayes, R. (1973) *Int. J. Mass Spectrom. Ion Phys.*, **11**, 255.
Dawson, P.H. (1976) *Quadrupole Mass Spectrometry and its Applications*, Elsevier, Amsterdam.
De Rosa, M. and Gambacorta, A. (1988) *Prog. Lipid Res.*, **27**, 153.
Drozd, J. (1975) *J. Chromatogr.*, **113**, 303.
Dulaney, J.T., Williams, M., Evans, J.E., Costello, C.E. and Koldny, E.H. (1978) *Biochim. Biophys. Acta*, **529**, 1.
Dzidic, I., Carroll, D.I., Stillwell, R.N. and Horning, E.C. (1976) *Anal. Chem.* **48**, 1768.
Egge, H. (1978) *Chem. Phys. Lipids*, **21**, 349.
Einolf, W.N. and Fenselau, C. (1974) *Biomed. Mass Spectrom.*, **1**, 195.
Enzell, C.R., Appleton, R.A. and Wahlberg, I. (1972) in *Biochemical Applications of Mass Spectrometry* (ed. G.R. Waller), Interscience, New York, p. 351.
Erdahl, W.L. and Privett, O.S. (1977) *Lipids*, **12**, 797.
Fales, H.M. (1966) *Anal. Chem.*, **38**, 1058.
Ferguson, A.M., Gwyn, S.A., Pannell, L.K. and Wright, G.J. (1977) *Anal. Chem.*, **49**, 174.
Field, F.H. (1972) in *International Reviews of Science, Physical Chemistry*, Series 1, vol. 1 (ed. A. Maccoll), Butterworths, London, p. 133.
Fisher, C. and Frölich, J.C. (1982) *Adv. Lipid Res.*, **19**, 185.
Foglia, T., Heller, P. and Dooley, C.J. (1976) *J. Am. Chem. Soc.*, **53**, 4547.
Foltz, R. L. (1972) *Lloydia*, **35**, 344.
Frölich, J.C. (1977) *Prostaglandins*, **3**, 1.
Fujino, Y. and Ohnishi, M. (1976) *J. Chem. Phys. Lipids*, **17**, 275.
Games, D.E. (1978) *Chem. Phys. Lipids*, **21**, 389.
Gaskell, S.J. and Brooks, C.J.W. (1976) *J. Chromatogr.*, **122**, 415.
Gaskell, S.J. and Brooks, C.J.W. (1977a) *J. Chromatogr.*, **142**, 469.
Gaskell, S.J. and Brooks, C.J.W. (1977b) *Org. Mass Spectrom.*, **12**, 651.
Gaskell, S.J., Edmonds, C.G. and Brooks, C.J.W. (1976a) *Anal. Lett.*, **9**, 325.
Gaskell, S.J., Edmonds, C.G. and Brooks, C.J.W. (1976b) *J. Chromatogr.*, **126**, 591.
Graveland, A. (1970) *J. Am. Oil Chem. Soc.*, **47**, 352.
Green, K., Hamberg, M., Samuelsson, B., Smigel, M. and Frölich, J.C. (1978) *Adv. Prostaglandin Thromboxane Res.*, **1**, 47.
Hanfland, P. and Egge, H. (1975) *Chem. Phys. Lipids*, **15**, 243.
Hanfland, P. and Egge, H. (1976) *Chem. Phys. Lipids*, **16**, 201.
Hasegawa, K. and Suzuki, T. (1973) *Lipids*, **8**, 631.
Hasegawa, K. and Suzuki, T. (1975) *Lipids*, **10**, 667.
Hilker, D.R., Gross, M.L., Knocke, H.W. and Shively, J.M. (1978) *Biomed. Mass Spectrom.*, **5**, 64.
Hites, R.A. (1970) *Anal. Chem.*, **42**, 1736.
Holmes, J.L. and Benoit, F.M. (1972) *MITP Int. Rev. Sci.*, **5**, 259.
Horning, E.C., Carroll, D.I., Dzidic, I. and Stillwell, R.N. (1973) *Pure Appl. Chem.*, **50**, 113.
Horning, M.G. (1978) *Methods Enzymol.*, **52**, 331.
Joseph, J.D. (1975) *Lipids*, **10**, 395.
Karlander, S.G., Karlsson, K.R. and Pascjer, I. (1973) *Biochim. Biophys. Acta*, **326**, 174.
Karlsson, K.A. (1973) *FEBS Lett.*, **32**, 317.
Karlsson, K.A. (1976a) in *Ganglioside Function* (eds G. Porcellati *et al.*), Plenum, New York, p. 15.
Karlsson, K.A. (1976b) in *Glycolipid Methodology* (ed. L.A. Witting), American Oil Chemists' Society, Champaign, chap. 5.
Karlsson, K.A. (1978) *Prog. Chem. Fats Other Lipids*, **16**, 207.
Karlsson, K.A. and Larson, G. (1978) *FEBS Lett.*, **87**, 283.
Karlsson, K.A., Samuelsson, B.E. and Stent, G.O. (1973) *Chem. Phys. Lipids*, **11**, 17.
Karlsson, K.A., Leffler, H. and Samuelsson, B.E. (1974a) *J. Biol. Chem.*, **249**, 4819.
Karlsson, K.A., Pascher, I., Pimlott, W. and Samuelsson, B.E. (1974b) *Biomed. Mass Spectrom.*, **1**, 49.
Karlsson, K.A., Pascher, I. and Samuelsson, B.E. (1974c) *Chem. Phys. Lipids*, **12**, 271.
Kaya, K. (1992) *Prog. Lipid Res.*, **31**, 87.
Kemp, P., Dawson, R.M. and Lein, R.A. (1972) *Biochem. J.*, **130**, 221.
Kinsel, G.R., Mowry, C.D., McKown, P.J. and Johnston, M.V. (1991) *Int. J. Mass Spectrom. Ion Processes*, **105**, 35.
Kleiman, R. and Spencer, G.F. (1973) *J. Am. Oil Chem. Soc.*, **50**, 31.
Klein, R.A. (1971a) *J. Lipid Res.*, **12**, 123.

Klein, R.A. (1971b) *J. Lipid Res.*, **12**, 628.
Klein, R.A., (1972) *J. Lipid Res.*, **13**, 672.
Klein, R.A. (1974) *Biochem. Soc., Trans.*, **2**, 1057.
Klein, R.A. (1978a) *Chem. Phys. Lipids,* **21**, 291.
Klein, R.A. (1978b) *Chem. Phys. Lipids,* **21**, 289.
Klein, R.A. and Kemp, P. (1977) in *Methods in Membrane Biology* (ed. E.D. Korn), Plenum, New York, p. 51.
Krisnangkura, K. and Sweeley, C.G. (1976) *Chem. Phys. Lipids,* **13**, 415.
Kuksis, A. and Myher, J.J. (1976) in *Lipids*, vol. 1 (eds R. Paoletti, G. Porcellati and G. Jacini), Raven, New York, p. 23.
Kuksis, A., Myher, J.J., Marai, L. and Geher, D. (1976) *Anal. Biochem.*, **720**, 302.
Lauer, W.M., Aason, A.J., Graff, G. and Holman, R.T. (1970) *Lipids,* **5**, 861.
Marai, L., Myher, J.J. and Kuksis, A. (1976) *Chem. Phys. Lipids,* **17**, 213.
Markey, S.P. and Wenger, D.A. (1974) *Chem. Phys. Lipids,* **12**, 182.
Matsubara, T. (1975) *Chem. Phys. Lipids,* **14**, 247.
Matsubara, T. and Hayashi, A. (1991) *Prog. Lipid Res.*, **30**, 301.
Millington, D.S. and Smith, J.R. (1977) *Org. Mass Spectrom,* **12**, 264.
Milne, G.W.A. and Lacey, M.J. (1974) *Crit. Rev. Anal. Chem.*, **4**, 45.
Minnikin, D.E. (1978) *Chem. Phys. Lipids,* **21**, 313.
Morrison, A., Barratt, M.D. and Aneja, R. (1970) *Chem. Phys. Lipids,* **4**, 47.
Murata, T. (1977) *Anal. Chem.*, **49**, 2209.
Murata, T. and Takahashi, S. (1973) *Anal. Chem.*, **45**, 1816.
Murata, T. and Takahashi, S. (1977) *Anal. Chem.*, **49**, 728.
Murata, T., Ariga, T., Oshima, M. and Miyatake, T. (1978) *J. Lipid Res.*, **19**, 370.
Myher, J.J., Marai, L. and Kuksis, A. (1974) *J. Lipid Res.*, **15**, 586.
Myher, J.J., Kuksis, A., Marai, L. and Yeung, S.K.F. (1978) *Anal. Chem.*, **50**, 557.
Natale, N. (1977) *Lipids,* **12**, 847.
Odham, G. and Stenhagen, E. (1972) in *Biochemical Applications of Mass Spectrometry* (ed. G.R. Waller), Interscience, New York, p. 211.
Ogino, J., Matsumura, T., Satouchi, K. and Saito, K. (1977) *Biomed. Mass Spectrom.*, **4**, 326.
Olson, K.L., Rinehart, K.L. and Cook, J.C. (1977) *Biomed. Mass Spectrom.*, **4**, 248.
Oshima, M. (1976) *J. Biochem. (Tokyo),* **80**, 53.
Pawlosky, R.J., Sprecher, H.W. and Salem, N. (1992) *J. Lipid Res.*, **33**, 1711.
Privett, O.S. and Erdahl, W.L. (1978) *Chem. Phys. Lipids,* **21**, 361.
Radhakrishnan, R., Gupta, C.M., Erni, B., Robson, R.J., Curatolo, W., Majundar, A., Ross, A.J., Takagaki, Y. and Khorana, H.G. (1980) *Ann. NY Acad. Sci.*, **346**, 165.
Rigaud, M., Durand, J., Breton, J.C., Rosello, J. and Gelpi, E. (1981) *Biomed. Mass Spectrom.*, **8**, 149.
Rohwedder, W.K. (1971) *Lipids,* **6**, 906.
Rohwedder, W.K. (1975) in *Analysis of Lipids and Lipoproteins* (ed. E.G. Perkins), American Oil Chemists' Society, Champaign.
Rosenkrantz, B., Hofbauer, K., Fischer, C. and Frölich, J.C. (1982) *Renal Physiol (Basel),* **5**, 186.
Ryhage, R. and Stenhagen, E. (1959) *Ark. Kemi,* **13**, 523.
Ryhage, R. and Stenhagen, E. (1960) *Ark. Kemi,* **15**, 291.
Ryhage, R. and Stenhagen, E. (1963) in *Mass Spectrometry of Organic Ions* (ed. F.W. McLafferty), Academic Press, New York, p. 399.
Saito, K. and Gamo, M. (1973) *Compar. Biochem. Physiol.,* **B45**, 603.
Satouchi, K. and Saito, K. (1976) *Biomed. Mass Spectrom.,* **3**, 122.
Satouchi, K. and Saito, K. (1977) *Biomed. Mass Spectrom.,* **4**, 107.
Satouchi, K., Saito, K. and Kates, M. (1978) *Biomed. Mass Spectrom.,* **5**, 87.
Saunders, R.A. and Williams, A.E. (1963) in *Mass Spectrometry of Organic Ions* (ed. F.W. McLafferty), Academic Press, New York, p. 343.
Shackleton, C.H.L. (1990) *Steroids,* **55**, 139.
Sprott, G.D. (1992) *J. Bioenerg. Biomembr.,* **24**, 555.
Stoffel, W. and Metz, P. (1979) *Hoppe-Seyler's Z. Physiol. Chem.,* **360**, 197.
Stoffel, W. and Pressner, K. (1979) *Hoppe-Seyler's Z. Physiol. Chem.,* **360**, 691.
Sun, F.F. and Taylor, B.M. (1981) *Prostaglandins,* **21**, 307.
Suzuki, T. and Hasegawa, K. (1976) *Agric. Biol. Chem. (Tokyo),* **40**, 221.
Sweeley, C.C., Fung, Y.K., Macher, B.A., Moskal, J.R. and Nunez, H.A. (1978) in *Glycoproteins and Glycolipids in Disease* (ed. E.S. Walborg), American Chemical Society, Washington, DC, p. 47.
Valicenti, A.J., Chapman, C.J., Holman, R.T. and Chipault, J.R. (1978) *Lipids,* **13**, 190.
Vetter, W., Englert, G., Rigassi, N. and Schweiter, U. (1971) in *Carotenoids* (ed. O. Isler), Birkhäuser, Basel, p. 189.
Viswanathan, C.F. (1974) *J. Chromatogr.,* **98**, 105.
Watson, J.T. (1976) *Introduction to Mass Spectrometry: Biomedical, Environmental and Forensic Applications*, Raven, New York.
Williams, D.H. and Beynon, J.H. (1976) *Org. Mass Spectrom.,* **11**, 103.
Wood, G.W. (1979) *Biochemical Applications of Mass Spectrometry,* 1st suppl. vol., Wiley, New York, p. 173.
Wood, G.W. and Lau, P.Y. (1974) *Biomed. Mass Spectrom.,* **1**, 154.
Wood, G.W., Lau, P.Y. and Rao, G.N.S. (1976) *Biomed. Mass Spectrom.,* **3**, 172.
Wood, G.W., Lau, P.Y., Morrow, G., Rao, G.N.S., Schmidt, D.E., Jr and Teubner, J. (1977) *Chem. Phys. Lipids,* **18**, 316.
Young, S.K.F., Kuksis, A., Marai, L. and Myher, J.J. (1977) *Lipids,* **12**, 529.
Zahn, H. (1955) *Angew. Chem.,* **67**, 561.

9.7 Electronic spectra

Electronic spectra involve transitions between different energy levels (states) of the electrons in molecules. Electromagnetic radiation causes the electrons to oscillate in response to the incident electric field. If the induced frequency of oscillation is in resonance with the difference in energy between two different electronic states, the probability of a transition becomes high. The energies involved in electronic transitions correspond to the absorption of a photon in the visible (400–750 nm) and ultraviolet (200–400 nm) regions of the electromagnetic spectrum. Below 200 nm is the far-UV region; in this region, absorption by atmospheric oxygen is considerable, and spectra can be obtained only if the spectrometer is evacuated or purged with He or other suitable gas. Routine instruments therefore extend only to about 185 nm (Campbell and Dwek, 1984).

To describe the intensity of a band, the Beer–Lambert relationship is used:

$$\log (I_0/I) = \varepsilon cl \qquad (9.29)$$

where I_0 is the intensity of the incident radiation,

Table 9.29 Ultraviolet absorption maxima for some chromophoric groups (From Kates, 1986)

Chromophore	Example	λ_{max} (nm)	ε_{max}	Solvent
Unsaturated hydrocarbons				
—C=C—	Octene	177	12600	Heptane
—C≡C—	Octyne	178	10000	Heptane
		196	2100	Heptane
—C=C—C=C—	Butadiene	217	20900	Hexane
—C=C—C=C—C=C—	1,3,5-Hexatriene	247	68000	Chloroform
—(C=C)$_n$—	Conjugated polyenes	217 + 30(n − 2)	20000–100000	Hexane
	Benzene	184	47000	Cyclohexane
		202	7000	
		255	230	
Carbonyl compounds				
—HC=O	Acetaldehyde	290	17	Hexane
C=O	Acetone	275	17	Ethanol
—C=C—CO—C	Methyl vinyl ketone	213	7100	Ethanol
		320	27	
—COOH	Acetic → palmitic acid	208 → 210	32 → 50	Ethanol
—CO_2R	Ethyl acetate	211	50	Ethanol
—CH=CH—COOH	*cis*-Crotonic acid	206	13500	Ethanol
		242	250	
—CH=CH—CH=CH—COOH	Sorbic acid (*trans*)	251	27000	Ethanol
		257	274000	
$H_3C[CH_2]_5[CH{=}CH]_2$—$[CH_2]_7COOH$	Conjugated linoleic acid (all *trans*)	231	29000	Ethanol
$H_3C[CH_2]_3[CH{=}CH]_3$—$[CH_2]_7COOH$	Conjugated linolenic acid (all *trans*)	258	40500	Ethanol
		268	57000	
		279	44000	
—CO—CH_2—COOR	Ethyl acetoacetate	245	6700	Hexane
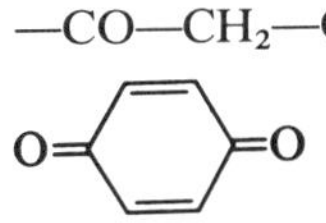	1,4-Benzoquinone	242	24000	Hexane
		281	400	
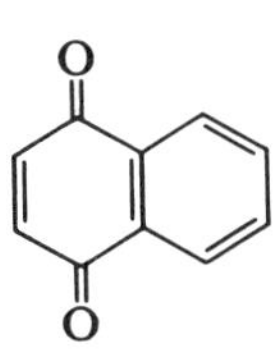	1,4-Naphthoquinone	343	*ca* 24000	Hexane
		249		
		260	*ca* 20000	
		269		

I is the intensity of the transmitted radiation, c is the concentration, ε is the extinction coefficient and l is the length of the optical path (cm). When c is expressed as mol/l, a molar extinction coefficient is required; c may be expressed as g/l, then a specific extinction coefficient is used. The expression $\log(I_0/I)$ is often referred to as absorbance. A related magnitude, percentage transmittance, is defined as $(I/I_0) \times 100$. Recent advances in ultraviolet–visible spectrophotometric instrumentation and applications are reviewed in Lobinski and Marczenko (1992).

Electronic spectra of lipids correspond generally to the UV region of the electromagnetic spectrum. Absorption is generally attributed to the presence of double bonds. Kates (1986) has summarized some UV absorption bands that have diagnostic value for lipids (Table 9.29): λ_{max}, especially in the case of polar compounds, may vary to some extent with the solvent used. The number of double bonds and alkyl or cycloalkyl substituents influence the λ_{max} value for conjugated double-bond systems. The general rule for polyene hydrocarbons or aldehydes is: add 30 nm for each double bond extending the conjugation, and 5 nm for each alkyl or cycloalkyl substituent to a base value of 217 nm (butadiene).

Lipid autoxidation is accompanied by an increase in the absorption at 233 nm, due to increasing diene conjugation (Klein, 1970; Mead, 1976; Logani and Davis, 1980; Goñi *et al.*, 1989). Absorption at 233 nm has been used to evaluate the degree of oxidation of human plasma lipids and commercial oils, and in the assay of antioxidants (Di Luzio, 1973).

When the extent of conjugation in a polyenic molecule reaches a certain stage, absorption occurs in the visible range. Among the lipids, this is the case with porphyrins and carotenoids. Electronic spectral data for porphyrins and haems may be found in Tables 9.30 to 9.32 (Smith and Caughey, 1976). The absorption of the cytochrome c haem group at 695 nm has been related to the ability of the protein to induce lipid oxidation (Valpuesta *et al.*, 1986).

Visible spectra of carotenoids have been extensively studied (Vetter *et al.*, 1971). In general, the shape of the spectrum is determined by the number of double bonds, the substituents, *cis–trans* isomerism and the solvent. Table 9.33 summarizes these influences. As the length of a chain of conjugated double bonds increases, additional double bonds show diminishing effects on the position of the main absorption band. The *apo*-carotenal series in Table 9.34 provides a good example. The bathochromic shift for one additional double bond is 23 nm between

Table 9.30 Electronic spectral data for dipyridine haems in pyridine

Haem derivative	λ_{max} (nm) (ε (mM))			
Haem A[a]	584 (33.5)	535 (8.0)		430 (160)
Haem A monoethyl ether diethyl ester[a]	583 (32.8)	530 (8.4)		429 (164)
Haem B dimethyl ester	557 (36.7)	526 (19.5)	480 (12.7)	420 (196.1)
Haem S/isohaem S[b] dimethyl esters[c]	576.5 (26.0)	536 (10.8)		432 (142.9)
Haem C tetramethyl ester[c]	551 (25.5)	521 (17.5)	472.5 (14.2)	414 (125)

[a] From Caughey *et al.* (1975).
[b] A mixture of the 2,4 positional isomers.
[c] From O'Keeffe (1974).

Table 9.31 Electronic spectral data for μ-oxobishaemins in benzene

μ-Oxo derivative	λ_{max} (nm) (ε (mM))			
Haem A monomethyl ether dimethyl ester	633 (19.4)	586 (12.8)	411 (102.2)	345 (64)
Haemin B dimethyl ester	599 (12.8)	573 (15.2)	397 (134)	357 (88)
Haemin S dimethyl ester[a]	626 (11.3)	583 (11.9)	412 (94.8)	345 (63)
Haemin iso-S dimethyl ester[a]	622 (11.3)	582 (11.9)	412 (94.8)	345 (63)
Haemin C tetramethyl ester[a]	590 (9.8)	579 (11.0)	397 (82.7)	360 (55.8)

[a] From Caughey *et al.* (1975).

Table 9.32 Electronic spectral data for porphyrins in chloroform

Porphyrin derivative	λ_{max} (nm) (ε (mM))				
	I	II	III	IV	Soret
Porphyrin A dimethyl ester[a]	646	584.5	563.5	520	418.5
Porphyrin B dimethyl ester[b]	630 (5.0)	576 (6.5)	541 (11.1)	506 (13.8)	407.5 (161)
Porphyrin S dimethyl ester[c]	642 (2.1)	583.5 (9.1)	559 (14.8)	519 (10.8)	421 (158)
Porphyrin iso-S dimethyl ester[d]	642 (2.0)	584 (9.45)	559 (15.1)	518.5 (10.8)	420 (163)
Porphyrin C tetramethyl ester[c]	625.5 (4.1)	570 (6.2)	538 (8.4)	504 (12.1)	405.5 (116)

[a] From Lemberg and Stewart (1955).
[b] From Caughey *et al.* (1966).
[c] From O'Keeffe (1974).
[d] From Sono and Asakura (1974).

Table 9.33 Effects of substituents and solvents on the main absorption maxima of carotenoids (From Vetter *et al.*, 1971)

Double bond	Chain	+7 to +35 nm
Double bond	Ring	+5 to +9 nm
First carbonyl	Chain	+28 nm
Second carbonyl	Chain	+1 to +7 nm
First carbonyl	Ring	+7 nm
Second carbonyl	Ring	+5 to +9 nm
First carbonyl	Ring *retro*	+9 nm
Second carbonyl	Ring *retro*	+5 nm
First and second	5,6-Epoxide	−8 nm
First and second	5,8-Epoxide	−20 nm
trans → *cis*		−4 nm
Normal → retro		−10 nm
Solvent effect	Ethanol, hexane, petrol, ether	Small effect
	Benzene, chloroform	+15 nm
	Carbon disulphide	+35 nm

Table 9.34 Visible-light absorption maxima of *β-apo*-carotenals in petroleum ether (Modified from Vetter et al., 1971)

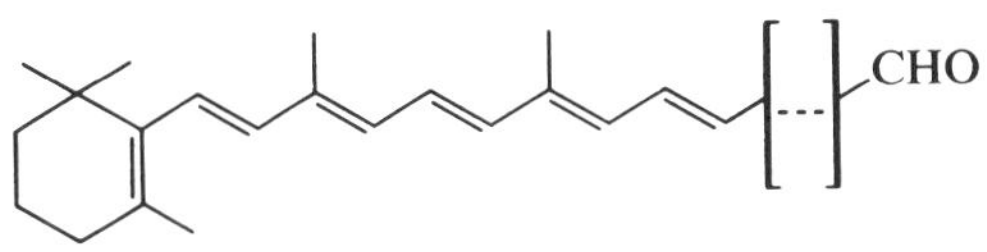

A general formula for *β-apo*-carotenals

No. of C atoms	λ_{max} (nm)
25	414
27	437
30	457
32	473
35	485
37	498
40	508

the first two compounds, in contrast to only 10 nm between the last two members of this series (Vetter et al., 1971).

A recent application of electronic spectra to the lipid field is the study of phospholipid polymers. When arranged in crystalline arrays, compounds that contain diacetylene groups crosslink upon exposure to UV light to form polymers (Wegner, 1972). The resultant polymers are highly coloured owing to the conjugated nature of the crosslinked backbone. Chapman and co-workers have synthesized phosphatidylcholine molecules containing such groupings and these have been polymerized in monolayers and multilayers as well as in liposome form (Hayward and Chapman, 1984). Polymerization is usually monitored by changes in the colour

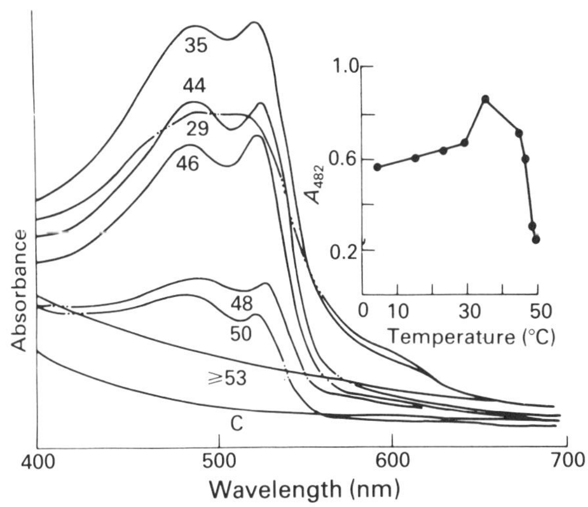

Fig. 9.39 Visible spectra of suspensions of C_{25} diacetylenic phosphatidylcholine liposomes irradiated at various temperatures. Insert shows the effect of temperature of irradiation on the absorbance at 482 nm. Irradiation time was 5 min. Lipid concentration: 1 mg/ml. Light path: 0.2 cm. C: non-irradiated suspension. From Leaver *et al.* (1983.)

of the compounds. The colour is dependent upon the temperature both during and after irradiation (Leaver *et al.*, 1983) (Fig. 9.39).

Cyanine dyes have been frequently used as optical probes in membrane studies. In particular, the anionic dye merocyanine 540 has found widespread application (Smith, 1990). Spectral data processing and curve-fitting techniques have been recently applied to the decomposition of merocyanine 540 absorption spectra in micellar and bilayer environments; the various resolved component bands have been assigned to dye monomers, dimers or larger aggregates, in either polar or non-polar environments (Kaschny and Goñi, 1992). The analysis of spectral parameters of the overall spectra and of each component has revealed that merocyanine 540 is a useful probe in studies of membrane structure and dynamics using visible absorption spectroscopy.

References and further reading

Campbell, I.D. and Dwek, R.A. (1984) *Biological Spectroscopy,* Benjamin/Cummings, Menlo Park, CA.

Caughley, W.S., Alben, J.O., Fujimoto, W.Y. and York, J.L. (1966) *J. Org. Chem.,* **312**, 2631.

Caughey, W.S., Smythe, G.A., O'Keefe, D.H., Maskasky, J.E. and Smith, M.L. (1975) *J. Biol. Chem.,* **250**, 7602.

Di Luzio, N.R. (1973) *Fed. Proc.,* **32**, 1875.

Goñi, F.M., Mañas, J.M.G., Alonso, A. and Arrondo, J.L.R. (1989) in *CRC Handbook of Free Radicals and Antioxidants in Biomedicine* (ed J. Miguel), CRC Press, Boca Raton, FL, p. 123.

Hayward, J.A. and Chapman, D. (1984) *Biomaterials,* **5**, 135.

Kaschny, P. and Goñi, F.M. (1992) *Eur. J. Biochem.* **207**, 1085.

Kates, M. (1986) *Techniques of Lipidology,* 2nd edn, North-Holland, Amsterdam.

Klein, R.A. (1970) *Biochim. Biophys. Acta,* **210**, 486.

Leaver, J., Alonso, A., Durrani, A.R. and Chapman, D. (1983) *Biochim. Biophys. Acta,* **732**, 210.

Lemberg, R. and Stewart, M. (1955) *Austr. J. Exp. Biol. Med. Sci.,* **33**, 451.

Lobinski, R. and Marczenko, Z. (1992) *Crit. Rev. Anal. Chem.,* **23**, 55.

Logani, M.K. and Davies, R.E. (1980) *Lipids,* **15**, 485.

Mead, J.F. (1976) in *Free Radicals in Biology* (ed. W.A. Pryor), Academic Press, New York, p. 51.

O'Keeffe, D.H. (1974) Ph.D. Dissertation, Arizona State University, Tempe.

Smith, J.C. (1990) *Biochim. Biophys. Acta,* **1016**, 1.

Smith, M.L. and Caughey, W.S. (1976) *Methods Enzymol.,* **58**, 421.

Sono, M. and Asakura, T. (1974) *Biochemistry,* **13**, 4386.

Valpuesta, J.M., Ondárroa, M. and Goñi, F.M. (1986) *Biochim. Biophys. Acta,* **878**, 435.

Vetter, W., Englert, G., Rigassi, N. and Schweiter, U. (1971) in *Carotenoids* (ed. O. Isler), Birkhäuser, Basel, p. 189.

Wegner, G. (1972) *Makromol. Chem.,* **154**, 35.

10

CHEMICAL PROPERTIES

F.D. GUNSTONE

10.1 Catalytic hydrogenation, chemical reduction and biohydrogenation

10.1.1 Catalytic hydrogenation*

Catalytic hydrogenation of liquid oils has been extensively studied because this is an important industrial process. Among natural triacylglycerols, liquid oils of vegetable or marine origin predominate, but there is a significant demand for fats of higher melting range for margarine hardstocks and in shortenings. The oils most commonly hydrogenated are soybean and other vegetable oils rich in linoleic (18:2) and, to a lesser extent, linolenic acid (18:3) and fish oils containing significant amounts of more highly unsaturated acids (20:5 and 22:6).

Sometimes it is desirable to effect complete hydrogenation (e.g. fully hydrogenated coconut oil or palmkernel oil), and with present-day procedures iodine values considerably below 1.0 are readily achieved (Hastert, 1990). More usually, however, the desired melting behaviour is achieved by partial hydrogenation combined, when necessary, with blending and/or inter-esterification. The products of partial hydrogenation are more complex than might have been predicted, and understanding of this topic has developed from studies with individual esters such as methyl oleate, linoleate and linolenate. Useful information has also come from deuteration studies. The reduced level of polyunsaturated acids also gives enhanced oxidative stability.

Hydrogenation can be effected with a range of heterogeneous and homogeneous catalysts based on metals such as platinum, palladium, nickel, copper and cobalt. Only nickel is used extensively on an industrial scale, though there is an increasing interest in copper catalysts. About 32 000 tonnes of nickel catalyst containing around 22% nickel is used annually. Spent catalyst contains 3–15% of nickel, and on both economic and environmental grounds there is a case for regeneration and re-use of this metal. This is usually achieved by heating the spent catalyst to 800–1200 °C (Thenee, 1989).

Partial hydrogenation of unsaturated esters is accompanied by double-bond migration and stereomutation, giving rise to a wide range of positional and configurational isomers. The presence of *trans* compounds is important because they have a higher melting point than their *cis* isomers. With the recognition of the dietary importance of linoleic acid (i.e. the 9*c*,12*c* isomer) as an essential fatty acid, solid fats with significant proportions of this acid have been developed. Concern has also been expressed about the ingestion of unnatural isomers – both *cis* and *trans*. Though incorporated into tissue lipids, these appear to be turned over at such a rate that they do not accumulate. Such acids have been part of our dietary intake for many years, with no evidence of harmful effects. Another study suggests that *trans* acids produce no dietary effect so long as the diet contains an adequate quantity of linoleic acid. Nevertheless, there is still disquiet about the nutritional effect of *trans* acids, and a recent report (BNF, 1992) recommends that their intake should not be allowed to rise above the current value, which in the UK is 2% total energy. See also Willett *et al.*, 1993.

The term 'selectivity', frequently used in hydrogenation studies, compares rates of different reactions and may be applied in several ways. It is therefore important to define the term carefully. Linolenic selectivity (S_{Ln}) is the relative rate of conversion of triene to diene compared with the rate of conversion of diene to monoene. It is generally desirable to have this value as high as possible. Linoleic selectivity (S_L) compares the rate of conversion of diene to monoene with that of the conversion of monoene to saturated ester. The specific isomerization index (S_i) compares the number of *trans* double bonds formed with the number of olefinic bonds reduced. It will be shown later that these values can be varied through changes in experimental conditions. The significant parameter is hydrogen availability at the catalyst surface compared with hydrogen demand.

More recently S_{Ln} and S_L have been designated as S_2 and S_1 as indicated below; S_2 is usually between

* See also section 5.5.1.

2 and 3, but S_1 can vary between 5 and 100 (Lips, 1992):

$$18{:}3 \xrightarrow{k_1} \underset{S_2 = k_1/k_2}{18{:}2} \xrightarrow{k_2} \underset{S_1 = k_2/k_3}{18{:}1} \xrightarrow{k_3} 18{:}0$$

Consideration must also be given to triglyceride selectivity, since melting point depends particularly on glycerides with three or two saturated acyl groups. The possibilities are indicated in the case of triolein:

$$O_3 \xrightarrow{H_2} \begin{cases} \text{SOO} & \text{max. selectivity m.p.} \sim 20\,^{\circ}\text{C} \\ \text{SOS, SOO, OOO} & \text{m.p.} \sim 30\,^{\circ}\text{C} \\ \text{SSS, OOO} & \text{min. selectivity m.p.} \sim 50\,^{\circ}\text{C} \end{cases}$$

(a) Methyl oleate

Complete hydrogenation of methyl oleate gives methyl stearate only, but partial hydrogenation gives stearate, oleate and a mixture of 18:1 isomers (iso-oleate). The double bond will no longer be Δ9 only, but may occupy almost any position in the chain. It may also be mainly *trans*. A comparative study of methyl oleate (Δ9*c*) and methyl elaidate (Δ9*t*) showed that these two esters are hydrogenated at about the same rate but that double-bond migration occurred more quickly in the *trans* isomer. Other studies, however, suggest that *trans* double bonds are not reduced as quickly as their *cis* isomers. Catalytic deuteration of methyl oleate shows that replacement of ^{1}H by ^{2}H can be quite extensive, though stearate itself is not deuterated under these conditions. These observations can be explained in terms of the accepted mechanism of catalytic hydrogenation, discussed later.

(b) Methyl linoleate

The isomerization and reduction reactions that may occur with methyl linoleate are outlined in Fig. 10.1. The simplest process would be reduction to stearate via 18:1 (9*c* or 12*c*), but isomerization competes with reduction to produce other diene and monoene esters with *cis* and/or *trans* configuration. It is necessary to distinguish three kinds of dienes depending on the value of *n* in the structural unit —CH=CH[CH_2]$_n$CH=CH—. Some dienes are conjugated ($n = 0$), others are easily converted to conjugated dienes (conjugatable, $n = 1$) and some are not easily converted to conjugated dienes (non-conjugatable, $n > 1$). Partially hydrogenated linoleate is thus a complex mixture, the composition of which depends on the catalyst, the temperature, the pressure and other factors that influence the availability of hydrogen at the catalyst surface where the unsaturated ester is held and where reaction takes place. The availability of hydrogen may decide whether an absorbed species is hydrogenated or isomerized. Also significant are the absorptive/desorptive properties of the various esters on the catalyst and their relative rates of hydrogenation.

The formation of conjugated dienes (9*c*,11*t* and 10*t*,12*c*) is particularly significant, since, with catalysts such as copper chromite, hydrogenation is confined to conjugated dienes produced from methylene-interrupted isomers: non-conjugatable dienes and monoenes are not reduced by this catalyst. This is also a significant reaction with nickel catalysts, but not the only one. Also some diene molecules are reduced directly to stearate, i.e. they undergo hydrogenation twice without desorption between each step.

Methyl linoleate has twice as many double bonds as methyl oleate and might therefore be expected to react twice as quickly, but with a selective catalyst the linoleate might be 5–100 times as reactive. This results from two factors: quick reduction via conjugated diene, which is not possible with oleate; and the fact that polyene esters are more strongly bonded to the nickel surface than are monoene esters. As a consequence of this, diene esters can be largely reduced to the monoene level before there is much reduction of monoene to saturated ester.

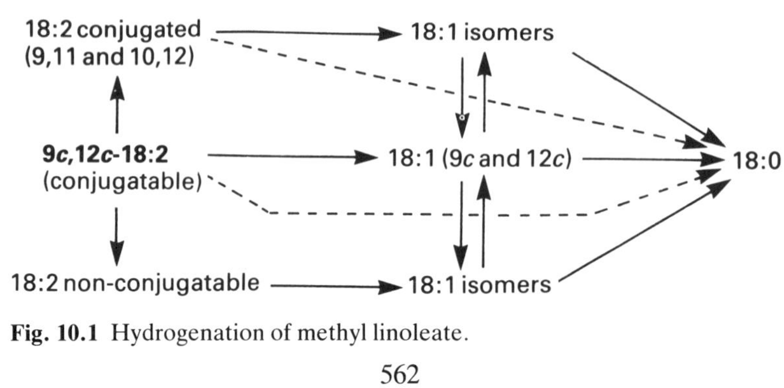

Fig. 10.1 Hydrogenation of methyl linoleate.

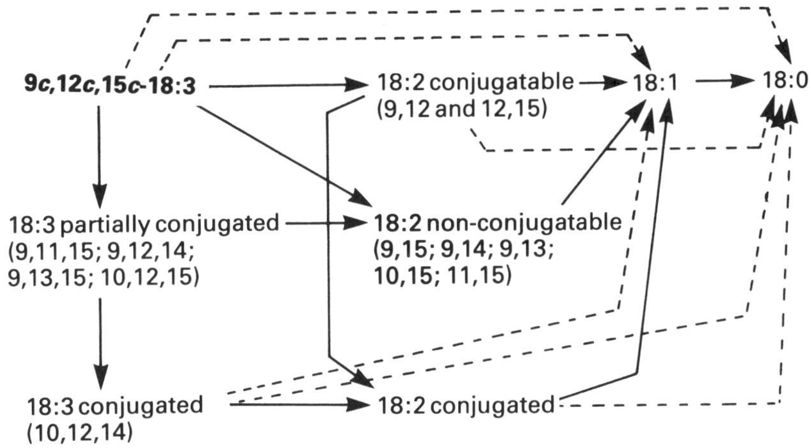

Fig. 10.2 Hydrogenation of methyl linolenate.

(c) *Methyl linolenate*

The reduction with methyl linolenate is even more complex (Fig. 10.2). Direct hydrogenation furnishes, in turn, a number of methylene-interrupted (conjugatable) dienes, then monoenes and finally stearate. Conjugation produces trienes with diene conjugation and, after more extensive double-bond migration, a conjugated triene. The conjugated compounds are usually reduced more quickly than their non-conjugated isomers. Conjugated and conjugatable dienes are readily hydrogenated, but non-conjugatable dienes – like the monoenes – react more slowly and tend to accumulate during partial hydrogenation. All unsaturated intermediates may undergo double-bond migration and/or stereomutation instead of hydrogenation, so that many isomers other than those designated in Fig. 10.2 are actually present. Some molecules may be reduced two or even three times on a single occasion of absorption. The relative importance of these several pathways depends on the choice of catalyst and on other experimental factors.

Because linolenic acid (a minor but significant component of soybean oil) with its $n-3$ ($\Delta 15$) double bond furnishes undesirable flavours after oxidation, and also because linoleic acid has a desirable dietary value, considerable effort has gone into making catalysts with high linolenic/linoleic selectivity. For nickel catalyst this is only 2–3, but with copper catalysts it may be 15–20. Thus soybean oil, reduced to an iodine value of 110 for use as a salad oil, would contain <1% or 4–5% linolenate with copper or nickel catalysts respectively.

Catalytic hydrogenation of triene esters can also produce aromatic compounds, but these can be minimized by appropriate experimental conditions.

The complexity of the hydrogenation process is illustrated by recent studies of the diene esters produced during partial hydrogenation of soybean oil (Mossoba *et al.*, 1991) and of canola oil (Ratnayake and Pelletier, 1992). In the former study the 9*c*,12*c*, 9*c*,12*t* and 9*t*,12*c* dienes were present along with unidentified non-methylene-interrupted dienes. In the latter study 11 diene acids were recognized, though some of them only tentatively: 9*c*,12*t*, 9*t*,12*c*, 9*t*,12*t*, 9*c*,13*t*, 9*c*,13*c*, 9*c*,15*c*, 9*t*,15*c*, 10*t*,15*c*, 8*t*,12*c*, 8*t*,13*c* and 8*c*,13*c*. These structures show clear evidence of double-bond movement and change of configuration.

(d) *Selectivity and isomerization*

The competing processes of double-bond isomerization and hydrogenation (or deuteration) are best discussed in terms of a half-hydrogenated intermediate that may undergo a number of different fates (Fig. 10.3).

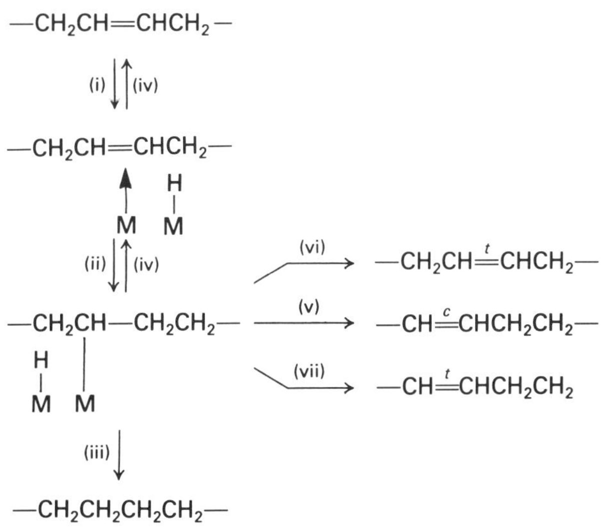

Fig. 10.3 Mechanism of hydrogenation and isomerization (Grau *et al.*, 1990; Simon *et al.*, 1991).

The first step in the reaction sequence (i) is the formation of a π bond between the catalyst and the double bond. Reaction with catalyst-bound hydrogen (ii) then gives the half-hydrogenated intermediate attached to the catalyst by a σ bond. This may react with a second catalyst-bound hydrogen atom (iii) to furnish the hydrogenation product. Alternatively, the half-hydrogenated intermediate may lose hydrogen from either methylene group adjacent to the methine that is attached to the catalyst (iv, v). In one case this will regenerate a *cis* double bond in its original position; in the other it will produce a *cis* double bond shifted from its original position. If rotation occurs before the loss of the hydrogen atom (vi, vii), then a *trans* double bond will be produced. Any unsaturated molecule can enter again into the reaction sequence. Double-bond migration of a polyene may produce a conjugated system.

Hydrogenation will be favoured by a high concentration of hydrogen on the catalyst, which can be enhanced either by raising the pressure or by more vigorous stirring. Alternatively, factors leading to an enhanced hydrogen demand are likely to lead to more isomerization. These include increase of temperature, the amount of catalyst, the activity of the catalyst, or the degree of unsaturation of the oil.

Increasing appreciation of the physical requirements of the catalyst indicate a particle size of ~5 μm for easy filtration and porous particles to provide a large surface area of 50–100 m^2/g of catalyst. The pores should be at least 10 nm in diameter. The overriding factor in determining selectivity in the hydrogenation of oils – as opposed to the simpler methyl esters – is the residence times of triacylglycerol molecules in the pore system of the catalyst. The longer the residence time, the greater the probability of reaction, be it conjugation, isomerization, or hydrogenation.

Though the importance of selectivity of various kinds has been stressed, it should be pointed out that high, intermediate, or low selectivity may be desirable depending on the nature of the hydrogenated product.

Hydrogen is produced by steam reforming of natural gas, by electrolysis, or in other ways. It is desirable to use the purest hydrogen possible, since impurities in the gas can affect the quality of the product and the efficiency of the process (Kubarka *et al.*, 1989; Weise, 1990).

10.1.2 Chemical reduction

The conversion of an unsaturated acid or ester to its perhydro derivative – sometimes a significant step in the identification of new fatty acids – is readily achieved by catalytic hydrogenation using a platinum, palladium, or nickel catalyst. The partial and stereospecific reduction of alkynes to *cis*-alkenes is most commonly effected with Lindlar's catalyst (section 7.1).

Non-catalytic reduction of alkenes is conveniently carried out with hydrazine. Oxygen is a necessary reagent, and the reaction probably involves diimine (N_2H_2). The reduction is stereospecific (*cis*) and occurs without double-bond migration or stereomutation. Partial reduction of a polyene acid or ester thus gives a simpler product than from catalytic processes. A typical experiment with linolenic acid is summarized below, and shows the range of products resulting from partial reduction:

18:3 →	18:2 →	18:1 →	18:0
9,12,15 (26%)	9,12 (15%)	9 (9%)	(5%)
	12,15 (15%)	12 (9%)	
	9,15 (13%)	15 (8%)	

(all unsaturated centres have *cis* configuration)

A recent paper suggests that the various dienes and monoenes are not formed so equally and that the order of reactivity of double bonds is given by the sequence: $CH_2{=}CH$ > double bonds close to the carboxyl group > double bonds close to the methyl group > mid-chain double bonds. Also 1,7-dienes react less quickly than 1,4-dienes. Some of their results are as follows (Ratnayake *et al.*, 1990):

6,9,12-18:3 (60 min)

18:3 →	18:2 →	18:1 →	18:0
6,9,12 (62.7%)	6,9 (7.5%)	6 (1.8%)	(1.4%)
	9,12 (10.1%)	9 (2.5%)	
	6,12 (11.9%)	12 (2.1%)	

11,14,17-20:3 (120 min)

20:3 →	20:2 →	20:1 →	20:0
11,14,17 (5.4%)	11,14 (14.3%)	11 (13.3%)	(20.2%)
	14,17 (9.2%)	14 (13.9%)	
	11,17 (14.8%)	17 (9.0%)	

This *cis* addition also provides the best route to *vic*-dideuterio acids of known stereochemistry:

$$\text{—CH=CH—} \xrightarrow{N_2D_4} \text{-CHDCHD—}$$

cis or *trans* → *erythro* or *threo*

Esters of long-chain acids can be reduced to the corresponding alcohols. In the laboratory, this is most conveniently effected with lithium aluminium hydride. On an industrial scale, reduction is effected by copper chromite at high temperature and pressure, though this reaction may be accompanied by some hydrogenation of double bonds.

New catalysts such as a ruthenium–tin–alumina preparation convert oleic acid to oleyl alcohol with little or no double-bond migration (Cheah *et al.*, 1992).

10.1.3 Biohydrogenation

Ruminant animals such as the cow and the sheep produce relatively saturated depot and milk fats even though their dietary intake is rich in highly unsaturated acids whether in the form of herbage or seed meal concentrate. This situation results from the biohydrogenation that occurs in the rumen through the action of the rumen bacteria. Because of the concern expressed in some quarters about the possible danger of highly saturated dietary fat (section 12.1.3), a special feeding regimen has been developed in which unsaturated acids in the diet of the ruminant animal are protected from bacteria during their passage through the rumen by encapsulation with formaldehyde-treated protein. As a result, both the milk fat and the depot fat are more unsaturated.

Biohydrogenation in the rumen first involves hydrolysis of triacylglycerol (from seed meal concentrate), galactolipid (in grasses) or phospholipid. The liberated acids then undergo double-bond migration and partial or complete hydrogenation. The major products are either stearate or an octadecenoate (mainly $\Delta 11t$, *trans*-vaccenic acid). A few individual bacteria have been isolated and identified, and the biohydrogenation that they effect has been investigated. Some examples are given below but the natural system is more complex:

Butyrivibrio fibrosolvens

18:2 (9*c*,12*c*) → 18:2 (9*c*,11*t*) → 18:1 (9*t* and 11*t*)

Bovine rumen liquor

18:2 (9*c*,12*c*) ⟶ 18:0

18:3 (9*c*,12*c*,15*c*) ⟶ 18:1 (15*c*)

References and further reading

BNF (1992) *Unsaturated Fatty Acids, Nutritional and Physiological Significance* (BNF Report), Chapman & Hall, London, p. 153.

Cheah, K.Y., Tang, T.S., Mizukama, F., Niwa, S., Toba, M. and Choo, Y.M. (1992) *J. Am. Oil Chem. Soc.*, **69**, 410.

Coenen, J.W.E. (1976) *J. Am. Oil Chem. Soc.*, **53**, 382.

Dutton, H.J. (1982) *Chem. Ind.*, 9.

Frankel, E.N. and Dutton, H.J. (1970) *Top. Lipid Chem.*, **1**, 161.

Grau, R.J., Cassano, A.E. and Baltanas, M.A. (1990) *J. Am. Oil Chem. Soc.*, **67**, 226.

Harfoot, C.G. (1980) *Prog. Lipid Res.*, **17**, 21.

Hastert, R.T. (1990) in *World Conference on Oleochemicals* (ed. T.H. Applewhite), American Oil Chemists' Society, Champaign, p. 165.

Hobson, P.N. and Wallace, R.J. (1982) *CRC Crit. Rev. Microbiol.*, **9**, 253.

Kubarka, K.A., Weise, M.K. and Messina, C.A. (1989) *J. Am. Oil Chem. Soc.*, **66**, 1289.

Lips, L. (1992) in *Oils and Fats in the Nineties* (eds V.K.S. Shukla and F.D. Gunstone), IFSC, Lystrup, Denmark, p. 186.

Mossoba, M.M., McDonald, R.E., Armstrong, D.J. and Page, S.W. (1991) *J. Agric. Food Chem.*, **39**, 695.

Ratnayake, W.M.N. and Pelletier, G. (1992) *J. Am. Oil Chem. Soc.*, **69**, 95.

Ratnayake, W.M.N., Grossert, J.S. and Ackman, R.G. (1990) *J. Am. Oil Chem. Soc.*, **67**, 940.

Simon, P., Celkova, A. and Schmidt, S. (1991) *J. Am. Oil Chem. Soc.*, **68**, 74.

Thenee, H. (1989) *Oils Fats Int.*, **5** (4), 30.

Weise, M.K. (1990) *Inform*, **1**, 444.

Willett, W.C., Stampfer, M.J., Manson, J.E., Colditz, G.A., Speizer, F.E., Rosner, B.A., Sampson, L.A. and Hennekens, C.H. (1993) *Lancet*, **341**, 581.

10.2 Autoxidation and photo-oxygenation

Autoxidation and photo-oxygenation are two aspects of the non-enzymic reaction between oxygen and unsaturated fatty acids. The enzymic reactions are discussed in section 10.3. Oxidation of lipids during storage and handling, involving complex substrates and ill-defined reaction conditions, proved difficult to understand. This difficulty is enhanced by the fact that the primary oxidation products are labile and readily converted to secondary oxidation products of several kinds. Understanding of these processes has come from studies of simpler substrates such as methyl oleate or methyl linoleate under clearly defined reaction conditions.

The primary oxidation products are allylic hydroperoxides in which the double bond may have moved and/or changed configuration from its original *cis* form. These hydroperoxides may subsequently be converted to products of lower molecular weight by fission, of similar molecular weight by rearrangement, or of higher molecular weight by dimerization. The oxidations of monoene, diene and triene acids differ slightly, whilst autoxidation and photo-oxygenation usually give similar but not identical products. These points are elaborated in the following discussion.

Recent developments in this field have resulted from improved methods of separation, especially high-performance liquid chromatography, and from additional spectroscopic information, particularly mass spectrometry and ^{1}H and ^{13}C NMR spectroscopy.

This topic has been reviewed by Frankel (1979, 1982, 1983, 1991), Gunstone (1984), Esterbauer and Cheeseman (1987), Chan (1987), Allen and Hamilton (1989) and Hudson (1990).

10.2.1 Autoxidation

This is a radical chain process involving initiation, propagation and termination reactions as set out in Fig. 10.4, where RH represents an unsaturated fatty acid or ester with H attached to an allylic carbon atom.

Initiation $\quad RH \longrightarrow R^{\bullet}$

Propagation $\quad \begin{cases} R^{\bullet} + O_2 \longrightarrow ROO^{\bullet} \\ ROO^{\bullet} + RH \longrightarrow ROOH + R^{\bullet} \end{cases}$

Termination $\quad$ interaction of radicals to produce non-initiating and non-propagating species

Fig. 10.4 Autoxidation of alkene compounds.

In the propagation sequence (given an adequate supply of oxygen), the first step is fast and the second is rate-determining.

The detailed nature of the initiating step is not fully understood, but one important process is the metal-catalysed decomposition of hydroperoxides to produce initiating radicals. (It is very difficult to obtain olefinic compounds entirely free of oxidation products). There is also some evidence that photo-oxygenation may be responsible for the first-formed hydroperoxides, and thermal initiation is possible in heated samples. The structure of the radical $R^{\bullet}$ influences both the composition of the product and the rate of the reaction (see below).

This sequence of reactions is affected by pro-oxidants and by antioxidants already present in the substrate or added to promote some desired effect. Autoxidation is promoted by heat and light, by metals and by radical-producing species. These act mainly on the initiation step.

The most important antioxidants are chain-breaking compounds or termination promoters. These may be synthetic materials – phenols such as gallate esters, butylated hydroxytoluene (BHT), butylated hydroxyanisole (BHA) and t-butylhydroquinone (TBHQ) – or natural compounds – mainly tocopherols and tocotrienols, but also phenols present in rosemary, sesame, tea, etc. The tocopherols, which can also be produced synthetically, are present in animal fats at very low levels and in vegetable oils in higher proportions (see Table 3.169 and 3.170). Other substances sometimes added for their antioxidant effect include sequestrating (chelating) agents (ethylenediaminetetraacetic acid, citric acid, phosphoric acid, amino acids), which help to remove metal ions, oxygen scavengers or reducing agents such as ascorbic acid, which can regenerate spent antioxidant, and singlet oxygen quenchers such as β-carotene.

A more detailed description of the reaction products follows in section 10.2.3.

10.2.2 Photo-oxygenation

This involves interaction between a double bond and singlet oxygen produced from ordinary triplet oxygen by light in the presence of a sensitizer such as chlorophyll, erythrosine, rose Bengal or methylene blue:

$$\text{sens.} + h\nu \xrightarrow{^3O_2} {}^1\text{sens.} \rightarrow {}^3\text{sens.} \rightarrow {}^1O_2$$

The reaction is not a chain reaction; there is no induction period and the reaction is unaffected by antioxidants. It is, however, inhibited by singlet oxygen quenchers such as carotene. Instead, oxidation occurs by the -ene mechanism in which oxygen becomes attached to either of the olefinic carbon atoms with attendant migration and stereomutation of the double bond:

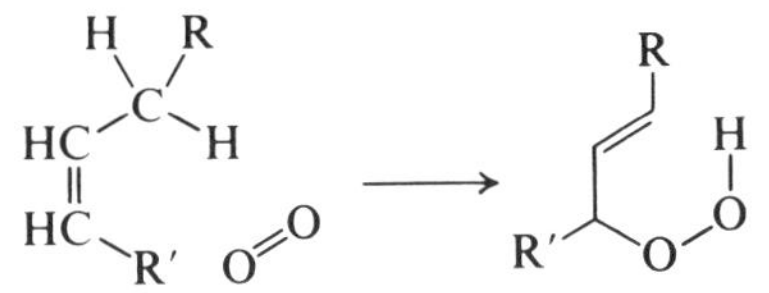

Photo-oxygenation is a quicker reaction than autoxidation, and the relative rates for oleate, linoleate and linolenate are shown in Table 10.1. In autoxidation the methylene-interrupted diene system is significantly more reactive than the isolated double bond of a monoene, but in photo-oxygenation the relative rates of oxidation are more closely related to the number of double bonds. On the basis of these values, photo-oxygenation of methyl oleate can be 30 000 times quicker than autoxidation, and for the polyenes photo-oxygenation can be 1000–1500 times quicker.

Non-methylene-interrupted esters are oxidized more slowly than their methylene-interrupted analogues, so that among C_{18} esters the 5*c*,9*c* diene is less reactive than linoleate and the 5*c*,9*c*,12*c* triene is less reactive than the α-linolenate (Takagi and Miyashita, 1987). The reactivity of methylene-interrupted polyenes rises with the number of double bonds. The induction period for ethyl esters at 5 °C is >60 days, 20 days and 3–4 days for linoleate, for α-linolenate and for EPA and DHA respectively (Cho *et al.*, 1987). The rate of oxidation of EPA and DHA compared to α-linolenate has been given as 5.2 and 8.5 (Cho *et al.*, 1987) and 2 and 2.5 (Cosgrove *et al.*, 1987).

Table 10.1 Relative rates of autoxidation and photo-oxygenation of oleate, linoleate and linolenate

	18:1	18:2	18:3
Autoxidation	1	27	77
Photo-oxygenation	3×10^4	4×10^4	7×10^4

10.2.3 Reaction products

(a) *Methyl oleate*

Oxidation of methyl oleate has been extensively studied and is considered to be typical of all monoene acids/esters. Photo-oxygenation produces only two products – the 9-hydroperoxide (Δ10*t*) and the 10-hydroperoxide (Δ8*t*) – in equal amounts. Reaction is confined to the olefinic carbon atoms and is accompanied by double-bond migration and stereomutation. As shown in Fig. 10.5, autoxidation of methyl oleate forms eight monohydroperoxides, of which two (9-OOH Δ10*c* and 10-OOH Δ8*c*) are only minor products. This range of compounds arises from initial attack at either allylic carbon atom to give a delocalized radical with oxygen finally attached to any one of four carbon atoms.

(b) *Methyl linoleate*

As expected, the diene ester forms a mixture of four hydroperoxides on photo-oxygenation: 9-OOH 10*t*,12*c*-18:2 (35%), 10-OOH 8*t*,12*c*-18:2 (17%),

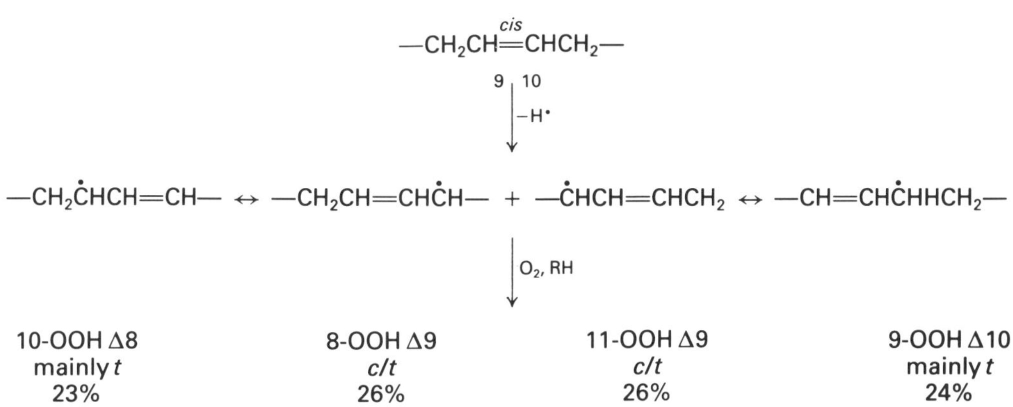

Fig. 10.5 Autoxidation of methyl oleate.

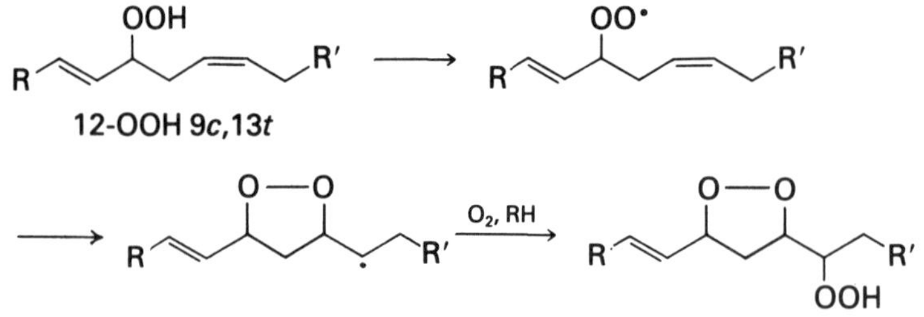

Fig. 10.6 Conversion of an inner hydroperoxide to a hydroperoxyperoxide.

12-OOH 9*c*,13*t*-18:2 (17%) and 13-OOH 9*c*,11*t*-18:2 (32%). Two of these contain a conjugated diene system. In all cases the migrating double bond has become *trans*. The 9- and 13-hydroperoxides (outer hydroperoxides) are present in larger amounts than the 10- and 12-hydroperoxides (inner hydroperoxides). This probably arises from the marked tendency of the inner hydroperoxides, with a β-hydroperoxyalkene system, to undergo autoxidation, yielding hydroperoxyperoxides (Fig. 10.6). This structural feature is absent from the outer hydroperoxides. Any of the products may undergo further photo-oxygenation to give bishydroperoxides so long as the double-bond migration that accompanies this reaction is possible.

In contrast, methyl linoleate gives only the 9- and 13-hydroperoxides as major autoxidation products in equivalent amounts. These can change from *ct* to *tt* dienes with exchange of the hydroperoxide group from C-9 to C-13 or vice versa (see following subsection). Later studies have also shown the presence of the hydroperoxide group on C-8, C-10, C-12 and C-14 at levels between 0.5 and 1.5%. There is still no evidence of a hydroperoxide on C-11.

(c) Methyl linolenate

The monohydroperoxides produced from this triene ester by photo-oxygenation and by autoxidation are set out in Table 10.2. In addition, bishydroperoxides (**1**), hydroperoxyperoxides (**2**) and hydroperoxydiperoxides (**3**) are formed. Typical structures are indicated but in all cases other isomers are also present in the reaction mixture.

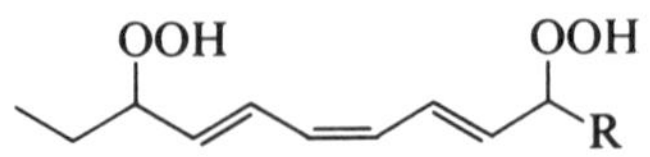

(**1**) Bishydroperoxide

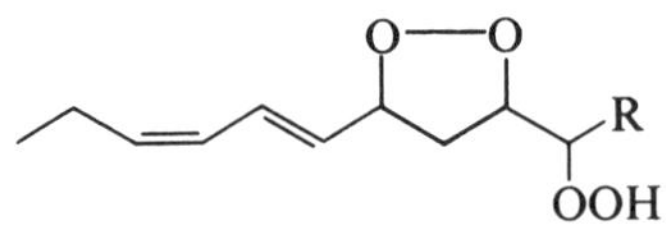

(**2**) Hydroperoxyperoxide

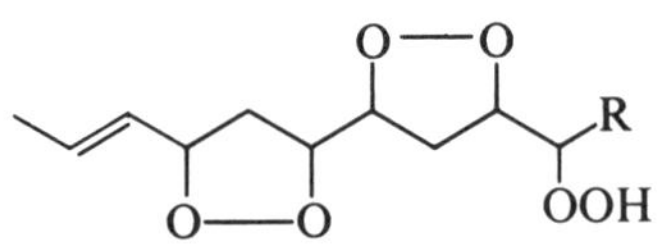

(**3**) Hydroperoxydiperoxide

R = $[CH_2]_7COOMe$

The ability of hydroperoxides from linoleate and linolenate to give cyclic peroxides is summarized in Table 10.3. It will be noted that some of the monohydroperoxides giving rise to these more

Table 10.2 Monohydroperoxides of methyl linolenate formed during autoxidation and photo-oxygenation

Autoxidation	Photo-oxygenation
9-OOH Δ10,12,15 (37%)	9-OOH Δ10,12,15 (23%)
	10-OOH Δ8,12,15 (13%)
12-OOH Δ9,13,15 (8%)	12-OOH Δ9,13,15 (12%)
13-OOH Δ9,11,15 (10%)	13-OOH Δ9,11,15 (14%)
	15-OOH Δ9,12,16 (13%)
16-OOH Δ9,12,14 (45%)	16-OOH Δ9,12,14 (25%)

Table 10.3 The possibility of forming cyclic peroxides from unsaturated hydroperoxides

	Hydroperoxides	→	Cyclic peroxides
18:2	9-OOH	→	None
	10-OOH[a]	→	Monoperoxide
	12-OOH[a]	→	Monoperoxide
	13-OOH	→	None
18:3	9-OOH	→	None
	10-OOH[a]	→	Mono- and diperoxide
	12-OOH	→	Monoperoxide
	13-OOH	→	Monoperoxide
	15-OOH[a]	→	Mono- and diperoxide
	16-OOH	→	None

[a] Not formed by autoxidation.

complex products are not formed during autoxidation but only by photo-oxygenation.

Bishydroperoxides are produced from unsaturated monohydroperoxides by further autoxidation or photo-oxygenation: the former requires an intact pentadiene unit, whilst the latter requires only a double bond that can migrate during the ene oxidation. For example, 9-OOH 10*t*,12*c*,15*c*-18:3 will give the 9,12-, 9,15- and 9,16-bishydroperoxides by photo-oxygenation but only the 9,12- and 9,16-bishydroperoxides by autoxidation:

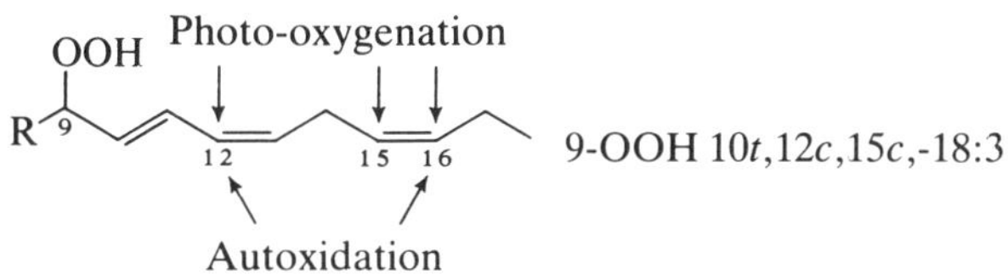

The current explanation of the rearrangement of hydroperoxides with double-bond stereomutation is set out in Fig. 10.7. For example, allylic radical 1A gives rise to two peroxy radicals 2A and 2B. Of these, 2A is the precursor of the 9-hydroperoxide. If, however, bond rotation occurs followed by regeneration of the allyl radical 2C, this, via the 13-peroxy radical, can furnish either the 13-hydroperoxide or a cyclic peroxide.

It has been shown that certain antioxidants, especially α-tocopherol, will inhibit these secondary changes, so that autoxidation of linolenate gives only four products (9-OOH 10*t*,12*c*,15*c*, 12-OOH 9*c*,13*t*,15*c*, 13-OOH 9*c*,11*t*,15*c* and 16-OOH 9*c*,12*c*,14*t*) in approximately equal amounts (cf. Table 10.2). No cyclic peroxides are formed under these conditions.

A study of synthetic triacylglycerols containing only linoleic and/or α-linolenic chains has indicated that the three acyl chains are oxidized progressively, giving hydroperoxides and other products expected from our understanding of the oxidation of the simpler methyl or ethyl esters. The triene chains (α-linolenate) are, of course, more reactive than the diene chains (linoleate). With linoleate glycerol esters there is no difference in reactivity between α and β chains, but with α-linolenate glycerol esters the $n-3$ double bond (Δ15) is slightly more reactive when in the α chain (Neff *et al.*, 1990; Frankel *et al.*, 1990; Miyashita *et al.*, 1990).

(d) Short-chain secondary products

The most important secondary products formed during oxidation are those of lower molecular

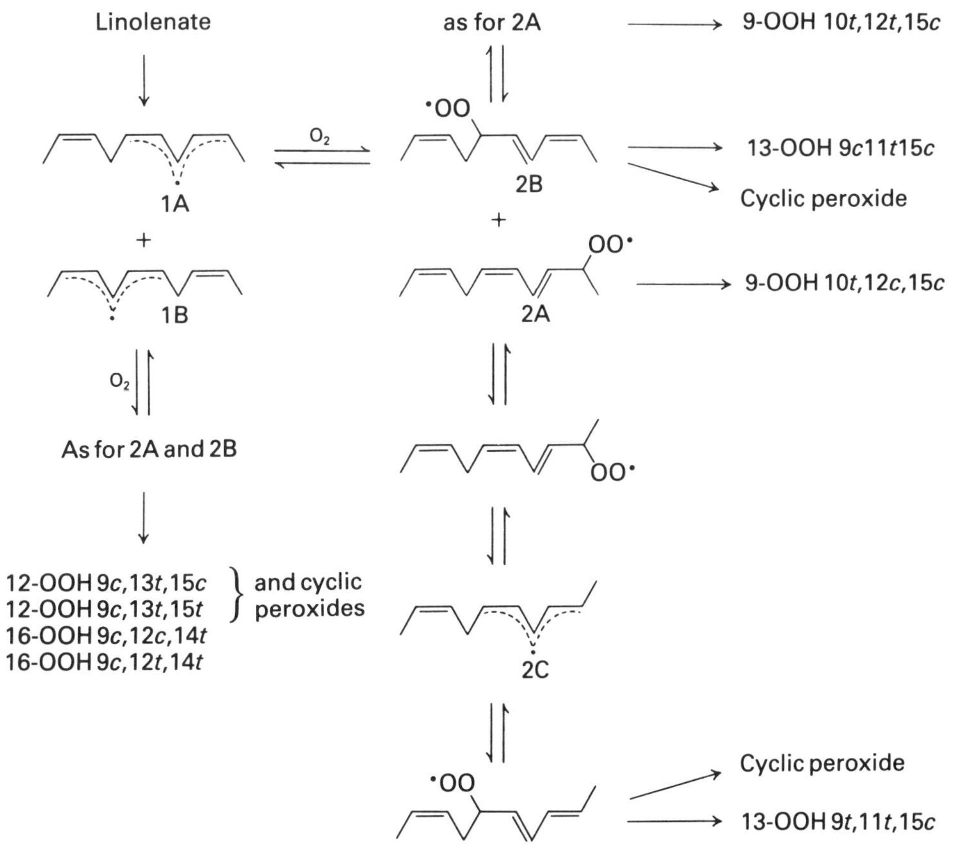

Fig. 10.7 Interconversion of hydroperoxides.

weight resulting from chain fission. Many aldehydes, ketones, alcohols, acids, esters, lactones, ethers and hydrocarbons have been identified. These contribute to odour and flavour – sometimes in a favourable way, imparting characteristic and acceptable flavours to foods both raw (mainly enzymic changes) and cooked (usually non-enzymic), and sometimes in an unfavourable way, imparting the undesirable flavour and odour of rancid fat.

Aldehydes are particularly significant in flavour studies. They result from hydroperoxide decomposition, which is usually homolytic:

$$R^1CH{=}CHCH(OOH)R^2 \longrightarrow R^1CH{=}CHCH(O^\bullet)R^2$$

$$\longrightarrow \begin{cases} R^1CH{=}CHCHO + R^{2\bullet} \\ R^1CH{=}CH^\bullet + R^2CHO \end{cases}$$

but may be heterolytic:

$$R^1CH{=}CHCH(OOH)R^2 \xrightarrow{H^+} R^1CH{=}CHCH(\overset{+}{O}H_2)R^2$$

$$\xrightarrow{-H_2O} R^1CH{=}CHCH(\overset{+}{O})R^2 \longrightarrow R^1CH{=}CHO\overset{+}{C}HR^2$$

$$\xrightarrow{+H_2O} R^1CH{=}CHOCH(OH)R^2 \longrightarrow R^1CH_2CHO + R^2CHO$$

Each hydroperoxide can produce two aldehydes of which the short-chain volatile member is more significant in this context. The other aldehyde is a polyfunctional compound attached to an ester (glyceride) function. Aldehydes produced from natural fats will be complex mixtures because of the large number of hydroperoxides from which they can be produced. This number may be greater still with partially hydrogenated fats because of the large number of double-bond positions possible in such compounds (section 10.1). Most of these aldehydes have a very low flavour threshold level, so they need be present only at minute levels. For example, the deca-2,4-dienal produced from linoleate 9-hydroperoxide produces a deep-fried flavour at a concentration of 5×10^{-10} M.

The four hydroperoxides from autoxidation of oleate give four aldehydes (8:0, 9:0, 2*t*-10:1 and 2*t*-11:1), and the two hydroperoxides from autoxidized linoleate give hexanal and deca-2,4-dienal. The aldehydes produced from other hydroperoxides (Tables 10.2 and 10.3) can be derived from the equation given above. Hept-*cis*-4-enal is reported to give a creamy flavour to butter but a cold-stored flavour to fish. It probably results from one or more of the $n - 3$ acids, but there is no complete explanation of its formation.

The short-chain hydrocarbons include alkanes, alkenes and alkynes. Ethane, ethene, pentane and pentene are sometimes present in exhaled air, and these represent oxidation products of linolenate (C_2 compounds) and linoleate (C_5 compounds).

Other significant short-chain compounds include two C_8 ketones (**4**) and (**5**) and the corresponding secondary alcohols.

(**4**) $CH_2{=}CHCOCH_2CH{=}CHCH_2CH_3$

(**5**) $CH_3COCH{=}CHCH{=}CHCH_2CH_3$

It is still not possible to provide a satisfactory explanation of all the short-chain compounds that have been identified. These could arise from unrecognized hydroperoxides by conventional reactions, from known hydroperoxides by unidentified pathways of breakdown, or from unidentified minor products in the original oil. For example, 10-oxooctadec-8-enoic acid and 9- and 10-hydroxyoctadecanoic acids have recently been identified in soybean oil (Endo *et al.*, 1991).

The formation and decomposition of fatty acid hydroperoxides in aqueous solution under physiological conditions furnishes a range of polyfunctional short-chain compounds such as 4-hydroperoxynon-2-enal and 4-hydroxyoct-2-enal. The latter displays a number of significant physiological effects based on its reaction with thiol groups, including those which are active sites of enzymes:

$$RCH(OH)CH{=}CHCHO \xrightarrow{R'SH}$$

$$[RCH(OH)CH(SR')CH_2CHO] \longrightarrow$$

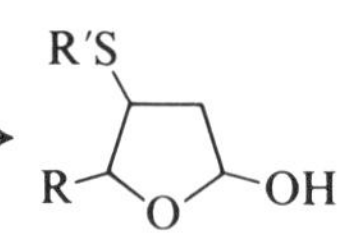

(e) Other secondary products

Other secondary oxidation products may have about the same molecular weight as the starting product or they may be dimers/trimers with a higher molecular weight (Gardner, 1987).

Many of the products that have been identified result from intermediate alkoxy radicals or ions and proceed via epoxides as shown in the following partial structures:

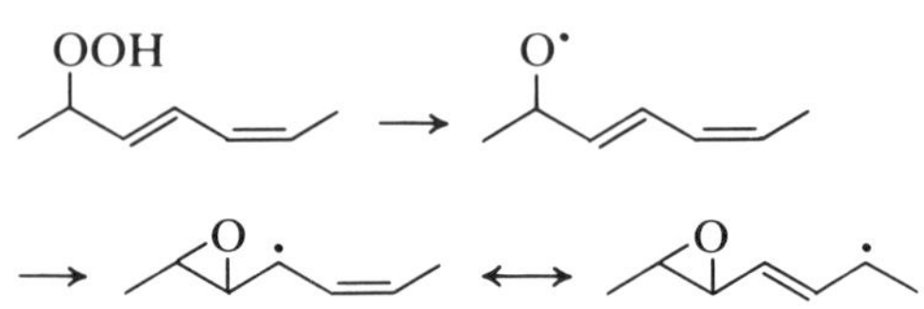

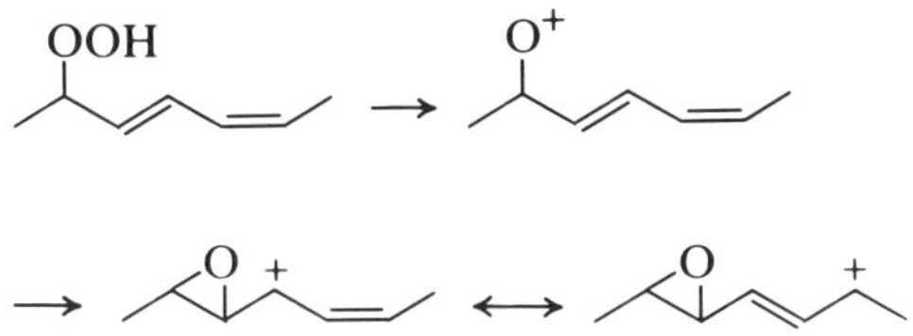

These intermediates can then react, at either reactive site, with oxygen (to give hydroperoxides, ketones, or alcohols) or with solvents or reagents such as alcohols, thiols and carboxylic acids. The epoxide is also a reactive system and may itself undergo further reaction through ring opening. Typical products include epoxyhydroxy, epoxyhydroperoxy and trihydroxy monoenes and hydroxy or keto dienes (Schieberle *et al.*, 1988).

The dimers (and trimers) produced during oxidation are mainly peroxides (C—O—O—C) with some ethers (C—O—C) at around 40 °C. At higher temperatures the C—O—C and C—C dimers are more significant (Frankel, 1991).

References and further reading

Allen, J.C. and Hamilton, R.J. (eds) (1989) *Rancidity in Foods*, 2nd edn, Applied Science, London.

Chan, H.W.S. (ed.) (1987) *Autoxidation of Unsaturated Lipids*, Academic Press, London.

Cho, S.Y., Miyashita, K., Miyazawa, T., Fijimoto, K. and Kaneda, T. (1987) *J. Am. Oil Chem. Soc.*, **64**, 876.

Cosgrove, J.P., Church, D.F. and Pryor, W.A. (1987) *Lipids*, **22**, 299.

Endo, Y., Endo, H., Fujimoto, K. and Kaneda, T. (1991) *J. Am. Oil Chem. Soc.*, **68**, 769.

Esterbauer, M. and Cheeseman, K.H. (1987) *Chem. Phys. Lipids*, **44**, 71–361; **45**, 103–364.

Frankel, E.N. (1979) *Fatty Acids* (ed. E.H. Pryde), American Oil Chemists' Society, Champaign, p. 353.

Frankel, E.N. (1982) *Prog. Lipid Res.*, **19**, 1.

Frankel, E.N. (1983) *Prog. Lipid Res.*, **22**, 1.

Frankel, E.N. (1991) *J. Sci. Food Agric.*, **54**, 495.

Frankel, E.N., Neff, W.E. and Miyashita, K. (1990) *Lipids*, **25**, 40.

Gardner, H. (1987) in *Autoxidation of Unsaturated Lipids* (ed. H.W.S. Chan), Academic Press, London, p. 51.

Gunstone, F.D. (1984) *J. Am. Oil Chem. Soc.*, **61**, 441.

Hudson, B.J.F. (ed.) (1990) *Food Antioxidants*, Elsevier Applied Science, London.

Miyashita, K., Frankel, E.N., Neff, W.E. and Awl, R.A. (1990) *Lipids*, **25**, 48.

Neff, W.E., Frankel, E.N. and Miyashita, K. (1990) *Lipids*, **25**, 33.

Schieberle, P., Trebert, Y., Firl, J. and Grosch, W. (1988) *Chem. Phys. Lipids*, **48**, 281.

Takagi, T. and Miyashita, K. (1987) *J. Am. Oil Chem. Soc.*, **62**, 407.

10.3 Enzymic oxidation

The enzymic oxidation of fatty acids in animal systems involving mainly C_{20} polyene acids and leading to prostaglandins, thromboxanes and leukotrienes, is discussed in section 11.1.7, and attention is now directed to oxidation in plant systems under the influence of lipoxygenase (section 11.1.6) (Veldink *et al.*, 1977; Emken, 1978; Vliegenthart, 1979).

This iron-containing enzyme is widely distributed through the plant kingdom. The natural substrate is linoleic acid (($n-6$)-18:2), which is converted to (*S*)-13-hydroperoxidooctadeca-*cis*-9,*trans*-11-dienoic acid and/or (*R*)-9-hydroperoxidooctadeca-*trans*-10,*cis*-12-dienoic acid. Attack is initiated by removal of hydrogen from C-($n-8$) between the two double bonds, and it has been shown that removal of the pro-*S* hydrogen atoms leads to the (*S*)-13-hydroperoxide, whilst removal of the pro-*R* hydrogen atom leads to the (*R*)-9-hydroperoxide. The mechanism is as follows:

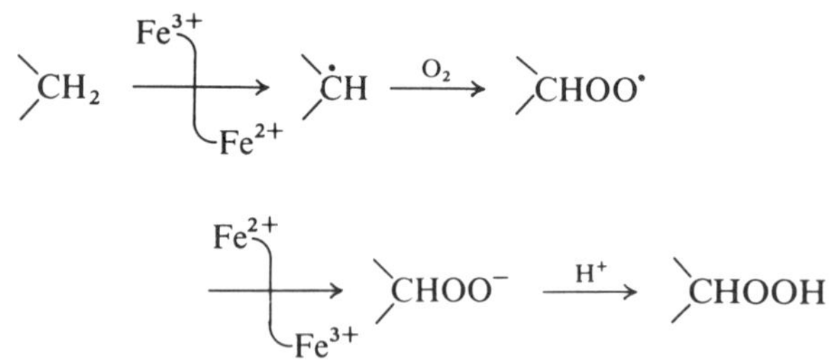

It is not clear whether the two hydroperoxides are produced by a single enzyme or whether two closely related enzymes exist, each leading to a single product. The detailed composition of the hydroperoxide mixture depends on the source of the enzymes and also, to some extent, on the pH and the oxygen pressure. Enzymes from flaxseed, soybean, peanut or eggplant (aubergine) produce the 13-hydroperoxide mainly; the 9-hydroperoxide is the main product with enzyme preparations from rice, corn, potato and tomato. The oxidation process is inhibited by eicosa-5,8,11,14-tetraynoic acid (the acetylenic analogue of arachidonic acid).

Similar hydroperoxides are produced from $n-3$ (18:3, 18:4, 19:3, 20:3, 20:5 and 22:6) and $n-6$ (17:3, 18:3, 19:3, 20:2, 20:3, 20:4, 21:3, 21:4 and 22:3) acids, all of which contain the $n-6$, $n-9$ diene unit, as well as from the ($n-4$)-16:3, ($n-5$)-17:3 and ($n-7$)-18:3 acids. Among the complete series of methylene-interrupted 18:2 acids, the 9,12 isomer is the best substrate, but there is significant oxidation of one other isomer (13*c*,16*c*), which gives the (*S*) 17- (85%) and 13-hydroperoxides (15%). Acids with an appropriately placed 1,4,7-triene unit give a monohydroperoxide fairly quickly and then a bishydroperoxide more slowly as long as the monohydroperoxide still contains a pentadiene unit. Thus α-linolenate gives only the 13-hydroperoxide whereas γ-linolenate gives the 6,13-bishydroperoxide:

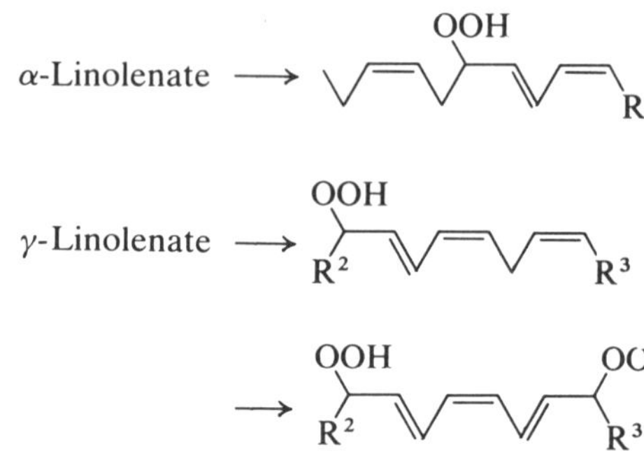

$R^1 = [CH_2]_7COOH$, $R^2 = H_3C[CH_2]_4$,
$R^3 = [CH_2]_4COOH$

Hiruta *et al.* (1988) showed that γ-linolenic acid gives its 9-hydroperoxide (6*c*,8*t*,12*c*) with soybean lipoxygenase in a maximum yield of 35% using a hexane-aqueous two-phase system at pH 6.5 and 10 °C.

Conjugated dienes result from enzymic oxidation of three triene acids (9*c*,12*c*,15*c*, 5*c*,9*c*,12*c* and 5*t*,9*c*,12*c*) almost as efficiently as from linoleic acid. Other polyene acids such as γ-linolenic acid, arachidonic acid, EPA and DHA give lower yields of conjugated diene because the hydroperoxide first formed still contains a pentadiene unit and readily oxidizes to a bishydroperoxide with conjugated triene unsaturation. This further oxidation is apparent at pH 9 but is inhibited at pH 11 (Takagi *et al.*, 1987).

In addition to the hydroperoxides, many other compounds resulting from further reactions have been recognized. Those arising from the 9- and 13-hydroperoxides of linoleic acid are formulated in Fig. 10.8. Each partial structure shows C-8 to C-14 and represents two compounds. The detailed stereochemistry of the 9,10,12- and 9,12,13-trihydroxy acids obtained from the 9- and 13-linoleate hydroperoxides has been reported by Hamberg (1991).

Other products are only produced in compounds having more than two double bonds, such as compound (**6**) produced from linolenic acid by the flaxseed enzyme and the two compounds represented

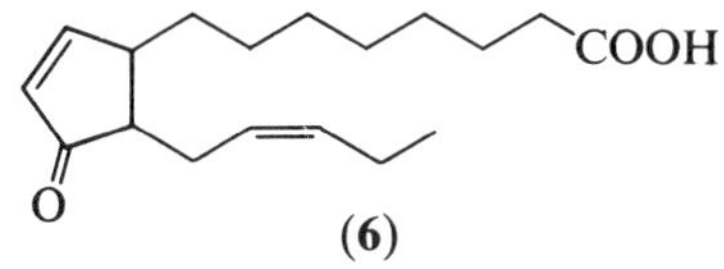

(6)

$CH_3[CH_3]_4CH{=}CHCH{=}CHOCH{=}CH[CH_2]_6COOH$

(8) Colneleic acid

$CH_3CH_2CH{=}CHCH_2CH{=}CHCH{=}CHOCH{=}CH[CH_2]_6COOH$

(9) Colnelenic acid

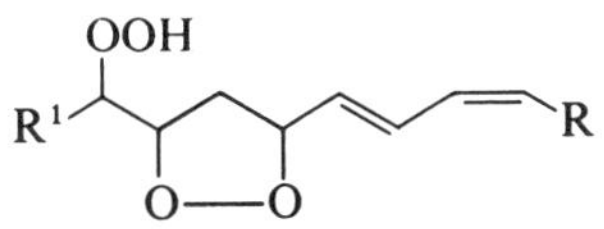

(**7**) (Hydroperoxide on C-9 or C-16)

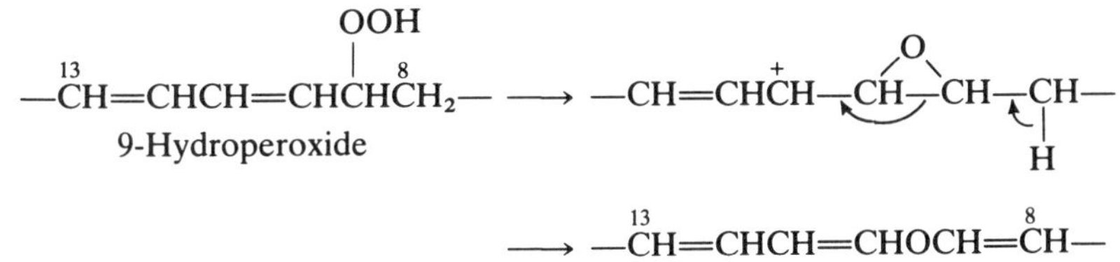

by (**7**) produced from linolenic acid and the soybean flour enzyme(s).

Colneleic (**8**) and colnelenic acids (**9**) are oxidation products of linoleic and α-linolenic acids produced by enzymes in potato tuber. Crombie *et al.* (1991) have shown that these are formed by rearrangement of the appropriate 9-hydroperoxide.

Enzymes present in a range of bacteria convert oleic acid to C_{18} hydroxy acids. *Bacillus pumilus* gives a mixture of 15-, 16- and 17-hydroxyoleic acids (Lanser *et al.*, 1992). *Nocardia cholesterolicum* gives 10-hydroxystearic acid from oleic acid and unsaturated derivatives of this with linoleic and linolenic acids (Knothe *et al.*, 1992). With the bacterial strain PR3 oleic acid is oxidized to 7,10-dihydroxyoctadecenoic acid (8*t*) (Koritala and Bagby, 1992). This is obtained in 60% yield after reaction at pH 7 and 30 °C for two days.

Other plant systems effect degradation of the hydroperoxides to short-chain aldehydes, which have important flavour properties. For example, when cucumber cells are broken and the enzymes

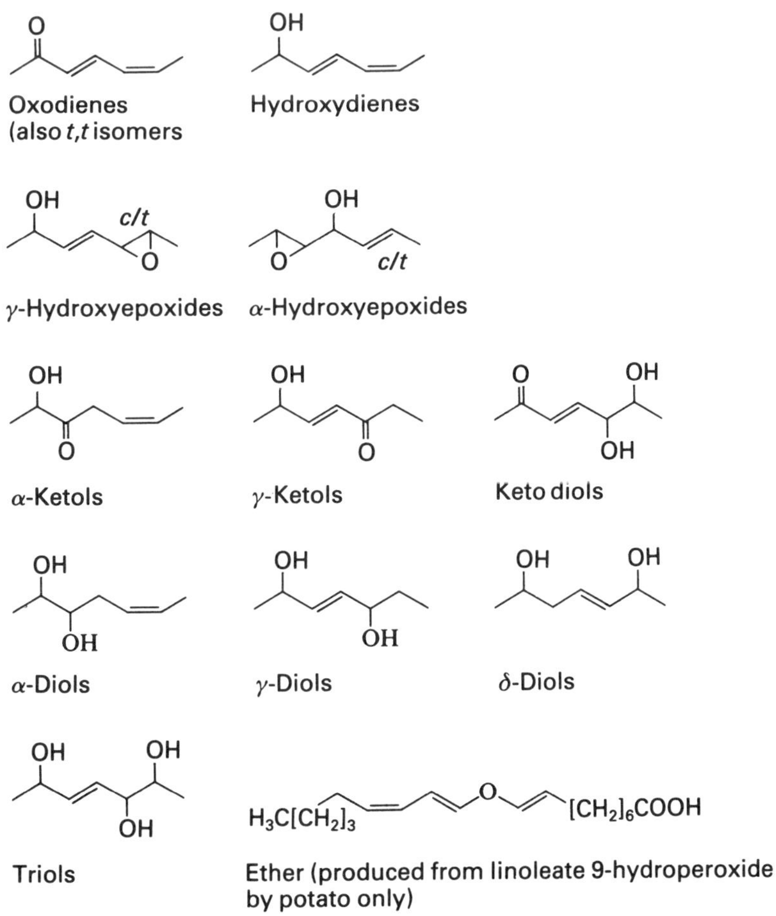

Fig. 10.8 Partial structures of products of enzymic linoleate oxidation (C-8 to C-14; each structure represents two compounds).

liberated, lipolysis, enzymic oxidation and subsequent chain fission give hexanal (from 18:2), non-2-enal (from 18:2), nona-2,6-dienal (from 18:3), $OHC[CH_2]_7COOH$ (from 18:2 and 18:3) and $OHCCH{=}CH[CH_2]_8COOH$ (from 18:2 and 18:3). Similar products have been identified in tomatoes and in tea leaves. (Other enzymic fatty acid oxidations are discussed in section 11.1.5).

References and further reading

Crombie, L., Morgan, D.O. and Smith, E.H. (1991) *J. Chem. Soc., Perkin Trans. I*, 567.

Emken, E.A. (1978) *J. Am. Oil Chem. Soc.*, **55**, 416.

Hamberg, M. (1991) *Lipids*, **26**, 407.

Hiruta, O., Nakahara, T., Yokochi, T., Kamisaka, Y. and Suzuki, O. (1988) *J. Am. Oil Chem. Soc.*, **65**, 1911.

Knothe, G., Bagby, M.O., Peterson, R.E. and Hou, C.T. (1992) *J. Am. Oil Chem. Soc.*, **69**, 367.

Koritala, S. and Bagby, M.O. (1992) *J. Am. Oil Chem. Soc.*, **69**, 1575.

Lanser, A.C., Plattner, R.D. and Bagby, M.O. (1992) *J. Am. Oil Chem. Soc.*, **69**, 363.

Takagi, T., Wakasa, N. and Miyashita, K. (1987) *J. Am. Oil Chem. Soc.*, **64**, 1320.

Veldink, G.A., Vliegenthart, J.F.G. and Boldingh, J. (1977) *Prog. Chem. Fats Other Lipids*, **15**, 131.

Vliegenthart, J.F.G. (1979) *Chem. Ind.*, 241.

10.4 Epoxidation, hydroxylation and oxidative fission

10.4.1 Epoxidation

Epoxidation is the reaction by which a double bond is converted to an epoxide:

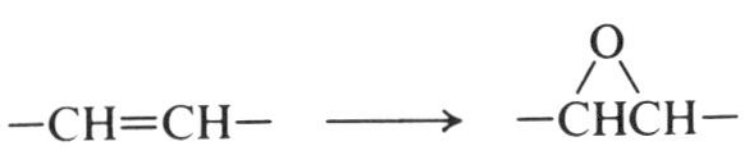

About a dozen natural epoxy acids have been identified (section 1.9).

(a) *Peroxyacid epoxidation*

Epoxidation is extensively employed, both as a laboratory procedure and on a commercial scale. Epoxidized oils are produced at the rate of about 5×10^4 tonne/year and used mainly as a combined stabilizer–plasticizer for poly(vinyl chloride).

Most epoxidations are effected by peroxy acids, which are often preformed and used *in situ* for epoxidation, though some can be stored and used as required. The peroxy acids are made by interaction of carboxylic acids, anhydrides or acid chlorides with hydrogen peroxide, with the first of these requiring an acidic catalyst such as sulphuric acid or a sulphonic acid (Gunstone, 1979):

$$RCOOH + H_2O_2 \xrightleftharpoons{H^+} RCO_3H + H_2O$$

$$(RCO)_2O + H_2O_2 \longrightarrow RCO_3H + RCOOH$$

$$RCOCl + H_2O_2 \longrightarrow RCO_3H + HCl$$

Additionally, peroxyacetic acid is made by oxidation of ethanal.

The most commonly used peroxy acids include aliphatic compounds (peroxyformic, peroxyacetic, peroxytrifluoroacetic, peroxylauric), aromatic compounds (peroxybenzoic, *m*-chloroperoxybenzoic) and monoperoxy acids based on dibasic acids (succinic, maleic, phthalic). These peroxy acids are unlikely to be pure, and should be analysed before use.

When carrying out epoxidation reactions, it should be remembered that the reaction is exothermic, that high concentrations of peroxy acids should be avoided, and that epoxides are acid-labile.

Lie Ken Jie and Cheung (1988) have reported that epoxidation with *m*-chloroperbenzoic acid occurs in 3–5 min in high yield in a microwave oven, and Fringuelli *et al.* (1989) have described smooth epoxidation in aqueous suspension at 0 °C in 30–60 min.

Scheme 10.4(i) Bartlett's proposal for epoxidation mechanism

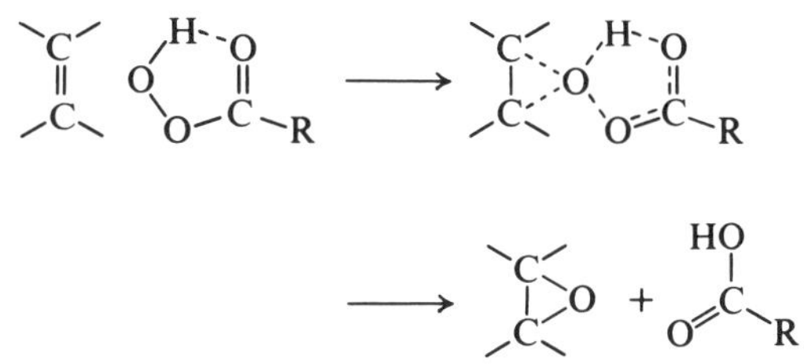

The mechanism of epoxidation remains uncertain. Bartlett's proposal (Scheme 10.4(i)) still finds most favour, but a 1,3 dipole reaction mechanism has also been postulated.

Other epoxidizing reagents that have been proposed but have not been extensively employed with olefinic acids include t-butyl hydroperoxide and transition-metal salts, hydrogen peroxide and isocyanate, sodium chlorate and osmium tetroxide, iodine and silver oxide, and hydrogen peroxide and *ortho* esters.

Epoxidation is a *cis* addition process, so that oleic acid gives *cis*-epoxystearic acid and elaidic acid gives *trans*-epoxystearic acid. Linoleic acid furnishes *cis*-9,10,*cis*-12,3-diepoxystearic acid. This exists in two (racemic) stereochemical forms, which have been separated by crystallization. Partial epoxidation of methyl linoleate gives diepoxide, unreacted linoleate and a mixture of two mono-epoxides (*cis*-9,10-epoxy 12*c*-18:1 and *cis*-12,13-epoxy 9*c*-18:1).

Corey *et al.* (1979) have described routes leading to the selective epoxidation of arachidonic acid. They showed that peroxyarachidonic acid undergoes intramolecular reaction to give *cis*-14,15-epoxy 20:3, possibly via the cyclic intermediate (**10**).

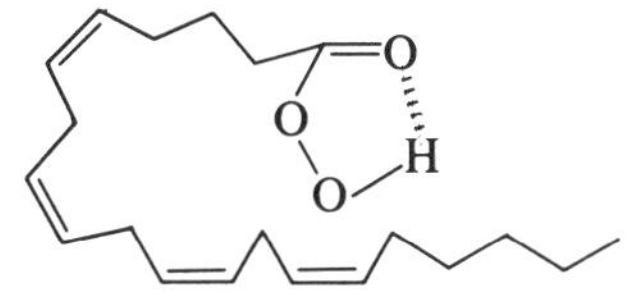

(10)

Selective epoxidation of the 5,6 double bond is based on the iodolactonization reaction, which involves only the 5,6 double bond in arachidonic acid:

RCH=CH[CH$_2$]$_3$COOH
(5*c*,8*c*,11*c*,14*c*-20:4)

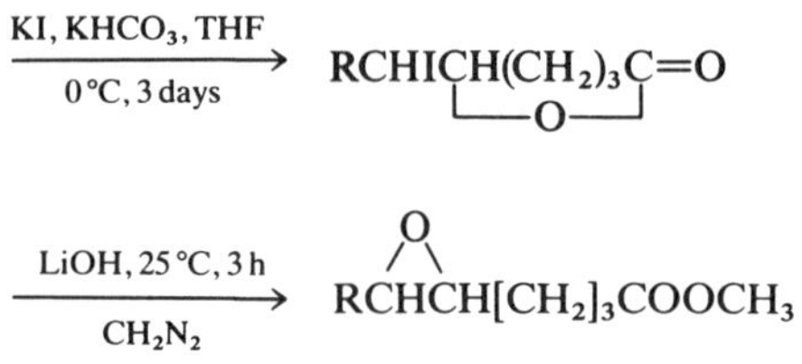

Allene epoxides, produced from allylic hydroperoxides by the enzyme hydroperoxide dehydrase, are among the important metabolites of arachidonic and other polyene acids (Hamberg, 1989):

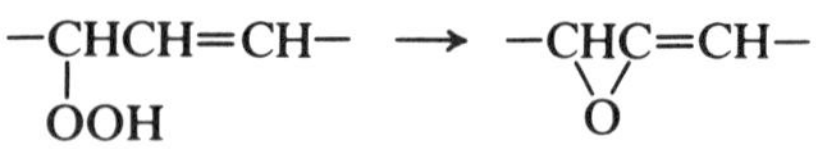

These reactive compounds are the precursors of other metabolites.

Olefinic compounds can be regenerated from epoxides by a number of stereospecific deoxygenation reactions, and these are illustrated in Scheme 10.4(ii) for two sets of conditions giving the *cis*- and *trans*-alkene, respectively, from the *cis*-epoxide.

(b) Spectroscopic properties

The structure of epoxy acids and esters can be derived from the mass spectrum of the epoxide or of some easily obtained derivatives. These are shown in Scheme 10.4(iii), along with the major fragmentations.

Some significant signals (δ values) from the ^{1}H and ^{13}C NMR spectra of *cis*- and *trans*-epoxides are indicated in Scheme 10.4(iv). New ^{13}C NMR information has been provided by Gunstone (1993).

(c) Analysis

Epoxide content is most often determined by the Durbetaki method based on the reaction between epoxide and hydrogen bromide, though gas chromatographic and colorimetric procedures have also been used. Epoxides can be identified by spectroscopic measurements (see subsection (b) above).

(d) Hydrolysis

Epoxides are easily hydrolysed to a diol (or its monoester), especially under acidic conditions. These reactions occur stereospecifically with inversion, so that *cis*- and *trans*-epoxides give the *threo*- and *erythro*-diols respectively. For example, this change is smoothly effected by heating with acetic acid (acetolysis) followed by hydrolysis of the mono-

Scheme 10.4(ii) Regeneration of olefinic compounds

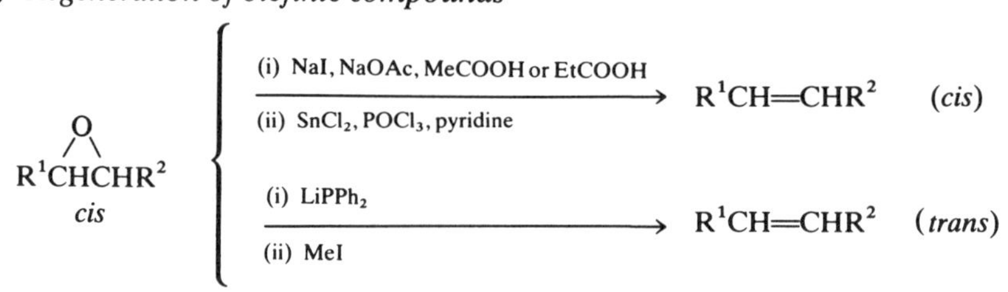

Scheme 10.4(iii) Fragments from mass spectrometry

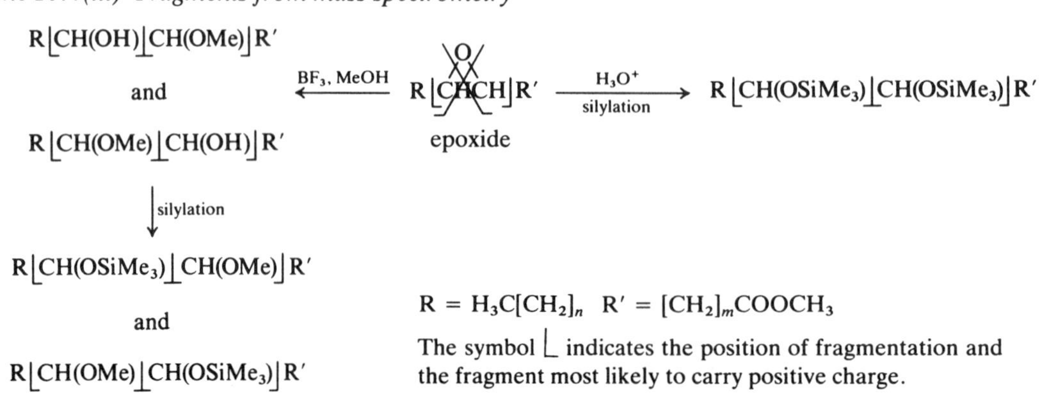

R = H$_3$C[CH$_2$]$_n$ R′ = [CH$_2$]$_m$COOCH$_3$

The symbol ⌊ indicates the position of fragmentation and the fragment most likely to carry positive charge.

Scheme 10.4(iv) Significant chemical shifts

cis isomer	a 1.43	b 2.70
trans isomer	a 1.43	b 2.45

a 1.45 b 2.79 c 2.91 d 1.60

a 1.45 b 2.73 c 2.1–2.3 d 5.41 e 2.03

	a	b	c	d
cis-9,10-epoxide	56.67	27.48	26.25	28.85
trans-9,10-epoxide	58.54	31.89	25.80	28.98

acetate:

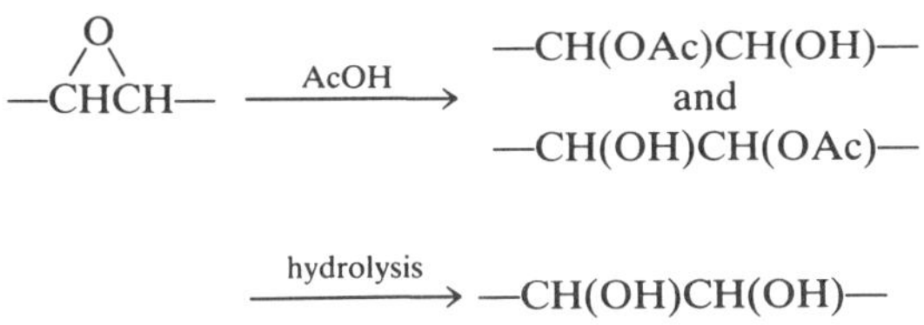

Linoleic acid readily forms a diepoxide, but its conversion to 9,10,12,13-tetrahydroxystearic acid is somewhat less satisfactory. The reaction is reported to take place smoothly with anhydrous dimethylsulphoxide or with 97% formic acid followed by alkaline hydrolysis. With more conventional reagents, a small amount of tetrahydroxy acid is accompanied by cyclic ethers (1,4- and/or 1,5-epoxides) and by the cyclopropanediol (**11**).

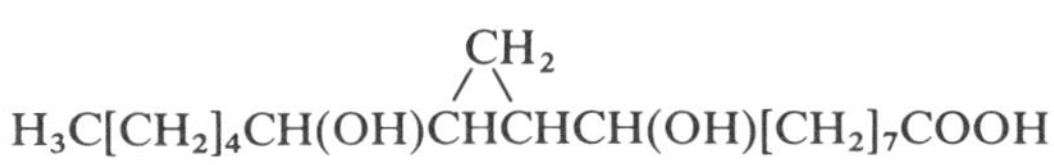

(11)

(e) Other reactions

The epoxide ring is very reactive, especially in acidic solution, and can be converted to a range of other ring systems or to *vic*-disubstituted compounds:

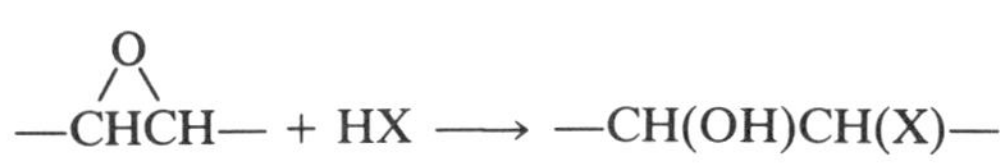

Here HX = H_2 (alcohol), H_2O (diol), ROH (alkoxy alcohol), RCOOH (acyloxy alcohol), $RCONH_2$ (acylamido alcohol), H_2S (mercapto alcohol), R_2NH (amino alcohol) HCN (cyano alcohol), HCl (chlorohydrin) and $NaHSO_3$ (sodium hydroxysulphonate) (Baumann *et al.*, 1988).

(f) Applications

As already indicated, epoxidized oils (especially soybean) are used as a stabilizer–plasticizer for PVC. *Vernonia galamensis* seed oil – a rich source of vernolic acid – has been examined as a source of C_8, C_9 and C_{12} dibasic acids by oxidation with nitric or chromic acids (Ayorinde *et al.*, 1988, 1989). Reaction of epoxide with a polyol gives a polyhydroxy product of high molecular weight, which can be used to make polyurethanes. The triethylamine salt of 10,11-epoxyundecanoic acid shows rust-inhibiting properties in a water-based cutting fluid (Watanabe *et al.*, 1988).

10.4.2 Hydroxylation

The conversion of an alkene to its diol (hydroxylation) can be carried out by several reagents that effect *cis*- or *trans*-hydroxylation to give the *threo*- and *erythro*-diols as shown in Scheme 10.4(v) (Gunstone, 1960).

cis-Hydroxylation is effected by dilute alkaline potassium permanganate, by osmium tetroxide alone, by osmium tetroxide as a catalyst in conjunction with an oxidizing agent such as metal chlorate, hydrogen peroxide/t-butanol, t-butylhydroperoxide or *N*-methylmorpholine *N*-oxide, or by the Woodward reaction, requiring iodine, silver acetate and wet acetic acid.

trans-Hydroxylation results from the Prévost reaction (iodine and silver benzoate under anhydrous conditions) or, more easily, from epoxidation followed by acid-catalysed hydrolysis (section 10.4.1) (Scheme 10.4(vi)).

Oleic acid forms two racemic diols – the *threo* and *erythro* isomers – and the possibilities of

Scheme 10.4(v)

$$\underset{\substack{threo\text{-diol}\\ erythro\text{-diol}}}{\text{—CH(OH)CH(OH)—}} \xleftarrow{trans\text{-hydroxylation}} \underset{\substack{cis\text{-alkene}\\ trans\text{-alkene}}}{\text{CH=CH—}} \xrightarrow{cis\text{-hydroxylation}} \underset{\substack{erythro\text{-diol}\\ threo\text{-diol}}}{\text{—CH(OH)CH(OH)—}}$$

Scheme 10.4(vi)

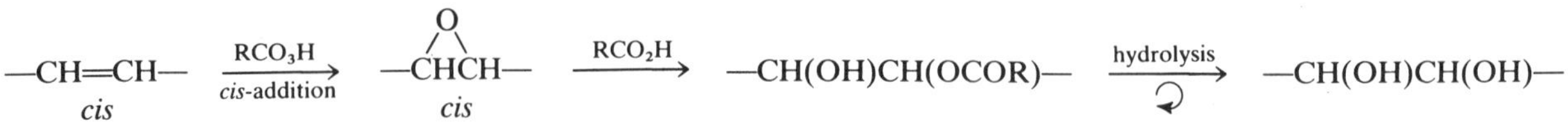

stereoisomerism increase with the number of double bonds. There are eight stereoisomeric 9,10,12,13-tetrahydroxystearic acids and 32 stereoisomeric 9,10,12,13,15,16-hexahydroxystearic acids. Not all these isomers have been identified. Certain difficulties in converting diepoxides to tetrols have already been discussed (section 10.4.1). Ricinoleic acid, which occurs in an enantiomeric form, gives four enantiomeric trihydroxystearic acids on hydroxylation.

The *vic*-diols are cleaved by periodic acid (to aldehydes) and by permanganate (to acids). After suitable derivatization, the hydroxy compounds can be separated by gas chromatography and identified by mass spectrometry (section 10.4.1).

Alkenes can be regenerated stereochemically from diols by reactions such as those set out below:

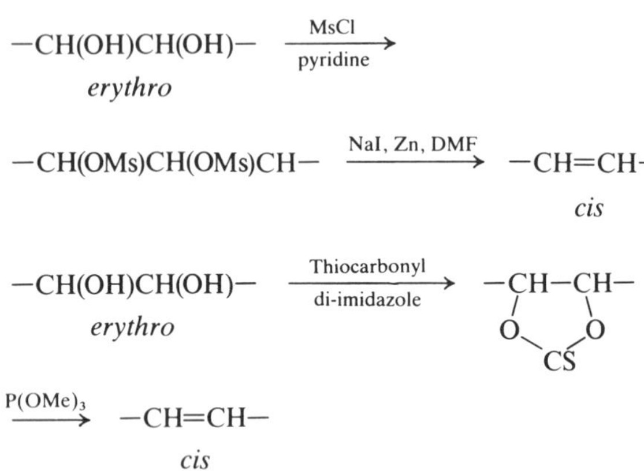

10.4.3 Oxidative fission

Unsaturated acids undergo cleavage at their unsaturated centres, and such processes are valuable for two reasons. On an analytical scale, identification of the cleavage products provides evidence of double-bond position. This may be necessary in the identification of acids of unknown structure and in the analysis of mixtures of acids such as oleate and petroselinate (9*c*- and 6*c*-18:1) or mixtures of monoene isomers resulting from partial hydrogenation. Mixtures of diene esters present even more difficulties. When used for analytical purposes simple procedures giving unequivocal results with small quantities of materials are required. On a preparative and industrial scale, oxidative fission may be used to produce new and interesting compounds. For example, petroselinic acid gives a series of C_6 difunctional and C_{12} monofunctional compounds. Only two procedures have been widely employed: von Rudloff oxidation and ozonolysis.

The classical method of oxidative cleavage – reaction with permanganate – is known to give over-oxidation, i.e. the first-formed oxidation products are subject to chain shortening by further oxidation. This difficulty is overcome in the von Rudloff procedure, which employs a 1:39 mixture of potassium permanganate and sodium metaperiodate. Under these conditions, permanganate is never present in high concentration but is continuously regenerated by oxidation with metaperiodate. Acids are oxidized in aqueous alkaline solution and esters in aqueous t-butanol solution for 6–12 h at room temperature. Under these conditions the reaction takes the following course:

$$R^1CH{=}CHR^2 \xrightarrow{KMnO_4} \left[\begin{array}{c} R^1CH(OH)CH(OH)R^2 \\ R^1COCH(OH)R^2 \\ R^1CH(OH)COR^2 \end{array}\right]$$

$$\xrightarrow{NaIO_4} \left[\begin{array}{c} R^1CHO + R^2CHO \\ R^1COOH + R^2CHO \\ R^1CHO + R^2COOH \end{array}\right]$$

$$\xrightarrow{KMnO_4} R^1COOH + R^2COOH$$

Ozonolysis is now more extensively employed for both analytical and preparative purposes. Reaction occurs via the molozonide, which breaks down to aldehyde and the Criegee zwitterion. The latter then reacts with aldehyde to give ozonide or with solvent to give alkoxyhydroperoxide (alcohol) or acyloxyhydroperoxide (carboxylic acid). All of these subsequently break down to the same ozonolysis products (Privett and Nickell, 1966):

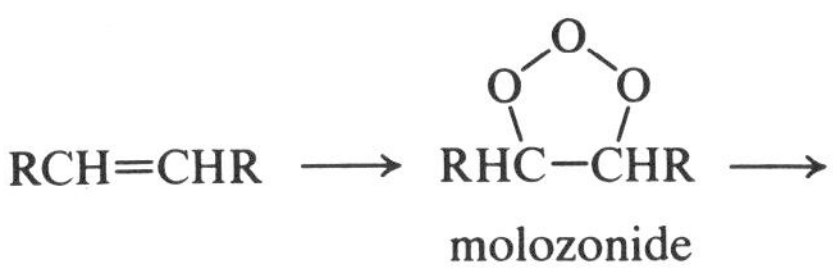

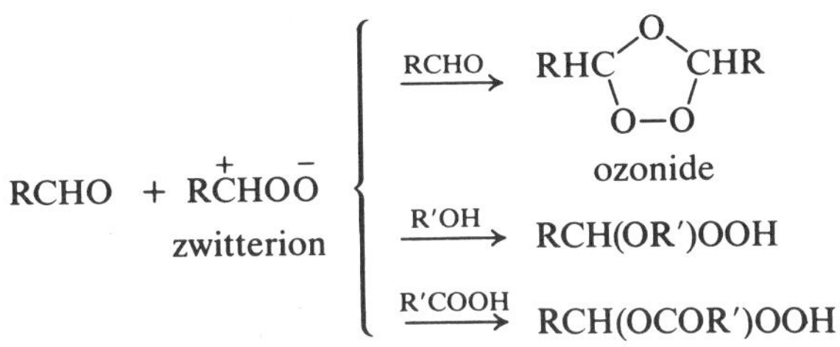

The ozonide exists in *cis* and *trans* forms, and an unsymmetrical alkene will produce symmetrical and unsymmetrical compounds – a total of six ozonides:

$$R^1CH{=}CHR^2 \longrightarrow \begin{matrix} R^1CHO + R^2\overset{+}{C}HO\overset{-}{O} \\ R^1\overset{+}{C}HO\overset{-}{O} + R^2CHO \end{matrix} \longrightarrow$$

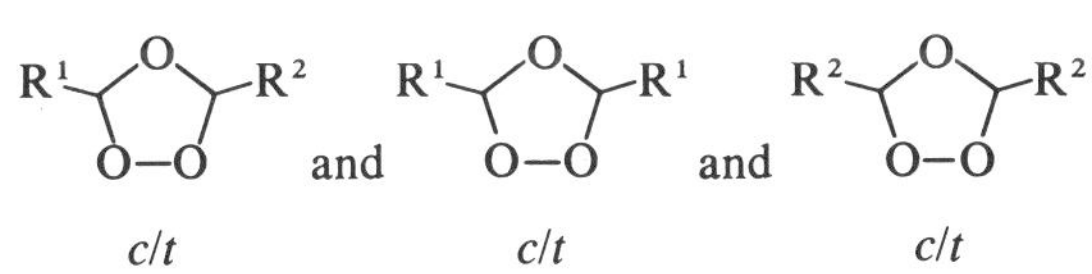

This has been demonstrated for a number of unsymmetrical alkenes including methyl oleate. The six ozonides from methyl oleate were recently separated by HPLC and characterized by their ^{1}H and ^{13}C NMR spectra (Wu *et al.*, 1992).

Of greater importance are the fission products resulting from further reaction of the ozonide or the alkoxy- or acyloxyhydroperoxide. These may be alcohols, aldehydes, acids or amines depending on the method of cleavage. Alcohols are formed by reduction of the ozonide with metal hydrides (lithium aluminium hydride or sodium borohydride) or by catalytic hydrogenation with nickel or platinum catalysts. Aldehydes result under milder reducing conditions, and zinc/acid, triphenylphosphine, dimethylsulphide or Lindlar's catalyst are commonly employed for the purpose. Acids are formed with a wide range of oxidizing agents, including peroxy acids or silver oxide. Amines result from reduction of ozonides over Raney nickel in the presence of ammonia or may be prepared from the aldehydes by reduction of their oximes. Mittelbach and Poklukar (1990) used HCl/MeOH to decompose ozonides, producing methyl esters or methyl acetals. Both EPA and DHA give the acetal ester $(MeO)_2CHCH_2CO_2Me$ as the major C_3 product (about 45% of the theoretical yield). They also studied the partial ozonolysis of C_{18} diene and triene esters.

The ozonolysis of readily available monoene acids is carried out on an industrial scale as a source, in particular, of dibasic acids. The C_9, C_{10}, C_{12}, C_{13} and C_{15} acids of this type are readily available from appropriate monoenes. Modification of the reaction process leads to other difunctional compounds, as indicated in Scheme 10.4(vii) (Klein, 1984).

Two alternatives to ozonolysis have been described. One employs oxygen with aldehydes as the oxidizing agent (Hinze, 1987) and the other uses

Scheme 10.4(vii)

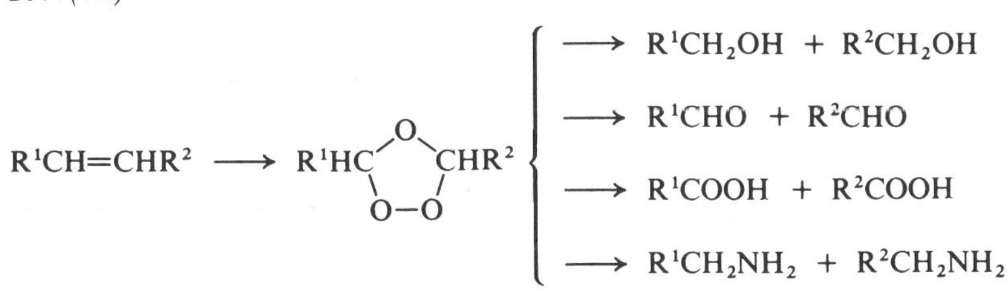

$RCH{=}CH[CH_2]_nCOOH \longrightarrow HOOC[CH_2]_nCOOH$

9*c*-18:1, 10*e*-11:1, 13*c*-22:1 n = 7,8,10,11,13

18:1 (9*c* and 12*c* and 15*c*)

$$RCH{=}CH[CH_2]_nCOOCH_3 \longrightarrow \begin{cases} OHC[CH_2]_nCOOCH_3 \\ HOCH_2[CH_2]_nCOOCH_3 \end{cases}$$

$$H_3C[CH_2]_{10}CH{=}CH[CH_2]_4COOH \longrightarrow H_3C[CH_2]_{10}CH_2NH_2 + H_2NCH_2[CH_2]_4COOH$$

sodium hypochlorite (Zaidman *et al.*, 1988). The latter occurs in three stages: (i) emulsification of oleic acid in water; (ii) non-catalytic oxidation to diol with sodium hypochlorite and sodium hydroxide; and (iii) cleavage of the diol with sodium hypochlorite and ruthenium chloride ($RuCl_3$) as catalyst.

References and further reading

Ayorinde, F.O., Osman, G., Shepard, R.L. and Powers, F.T. (1988) *J. Am. Oil Chem. Soc.*, **65**, 1774.

Ayorinde, F.O., Powers, F.T., Streete, L.D., Shepard, R.L. and Tabi, D.N. (1989) *J. Am. Oil Chem. Soc.*, **66**, 690.

Baumann, H., Bühler, M., Fochem, H., Hirsinger, F., Zoebelein, H. and Falbe, J. (1988) *Angew. Chem., Int. Edn*, **27**, 41.

Corey, E.J., Niwa, H. and Falck, J.R. (1979) *J. Am. Chem. Soc.*, **101**, 1586.

Fringuelli, F., Germani, R., Pizzo, F. and Savelli, G. (1989) *Tetrahedron Lett.*, **30**, 1427.

Gunstone, F.D. (1960) in *Advances in Organic Chemistry: Methods and Results* (eds R.A. Raphael, E.C. Taylor and H. Wynberg), Interscience, New York, p. 103.

Gunstone, F.D. (1979) in *Fatty Acids* (ed. E.H. Pryde), American Oil Chemists' Society, Champaign, p. 379.

Gunstone, F.D. (1993) *J. Am. Oil Chem. Soc.*, **70**, 1139.

Hamberg, M. (1989) *J. Am. Oil Chem. Soc.*, **66**, 1445.

Hinze, A.G. (1987) *Fat Sci. Technol.*, **89**, 339.

Klein, H.P. (1984) *J. Am. Oil Chem. Soc.*, **61**, 306.

Lie Ken Jie, M.S.F. and Cheung, Y.-K. (1988) *Lipids*, **23**, 367.

Mittelbach, M. and Poklukar, N. (1990) *Chem. Phys. Lipids*, **55**, 67.

Privett, O.S. and Nickell, E.C. (1966) *J. Am. Oil Chem. Soc.*, **43**, 393.

Watanabe, S., Fujita, T. and Sakamoto, M. (1988) *J. Am. Oil Chem. Soc.*, **65**, 1311.

Wu, M., Church, D.F., Mahier, T.J., Barker, S.A. and Pryor, W.A. (1992) *Lipids*, **27**, 129.

Zaidman, B., Kisilev, A., Sasson, Y. and Garti, N. (1988) *J. Am. Oil Chem. Soc.*, **65**, 611.

10.5 Halogenation

Halogenation of unsaturated fatty acids has held a significant position in fatty acid chemistry as the basis of standard procedures for measuring total olefinic unsaturation. This is usually recorded as an iodine value (section 6.2.8) but, now that fatty acid composition can be easily determined by gas chromatography, iodine values are less commonly measured, though they are still a useful check on the overall validity of GLC results. Halogenation – especially bromination – generally occurs in a *trans* manner through a polar addition. Bromination of oleic ($\Delta 9c$) and elaidic ($\Delta 9t$) acids gives the *threo* (m.p. 28.5 °C) and *erythro* (m.p. 29.5–30 °C) isomers of 9,10-dibromostearic acid respectively. Linoleic acid furnishes two tetrabromides: a solid product (m.p. 115 °C; 9*R**,10*R**,12*R**,13*R** racemate) and a liquid isomer (9*R**,10*R**,12*S**,13*S** racemate). Linelaidic acid (9*t*,12*t*) gives a solid tetrabromide (m.p. 78 °C) and linolenic acid gives a solid hexabromide (m.p. 185 °C). The chlorides obtained from oleic, linoleic and linolenic acid melt at 37, 123 and 189 °C respectively.

Brominated vegetable oils have been used in the soft-drinks industry as dispersing agents for the flavouring citrus oils and to give the drink a cloudy appearance, but they are no longer permitted in many countries.

The normal halogenation reaction can be diverted in its second stage by reaction with a nucleophile other than Br^-. Reaction in alcohol or acid solution, for example, gives alkoxy or acyloxy bromides:

$$-CH{=}CH- \xrightarrow{Br_2} \underset{-CHCH-}{\overset{Br^+}{\wedge}} \begin{cases} \xrightarrow{Br^-} -CHBrCHBr- \\ \xrightarrow{ROH} -CHBrCH(OR)- \\ \xrightarrow{RCOOH} -CHBrCH(OCOR)- \end{cases}$$

Chlorination of linoleic acid in the presence of water gives tetrachloro (49%), trichloromonohydroxy (38%) and dichlorodihydroxy (13%) compounds. These compounds are formed during the bleaching of wood pulp and related products and may be present in chlorine-bleached flour (McKague and Reeve, 1991).

Alternatively, reaction of an olefinic alcohol or acid can furnish a dimer through interaction of the bromonium ion with a second molecule of reactant acting as nucleophile. Methoxybromo adducts are easily made by reaction of the alkene with t-butoxy bromide in methanol solution.

Olefinic acids react with $KHCO_3 + I_2 + KI$ to form iodolactones if there is a double bond in the 4 (e.g. DHA) or 5 position (e.g. EPA, arachidonic acid). The neutral lactones are easily separated from unreacted acidic material. The $\Delta 4$ acids form γ-lactones and $\Delta 5$ acids form δ-lactones:

$RCH{=}CH[CH_2]_2COOH \longrightarrow$
$\Delta 4$ acids

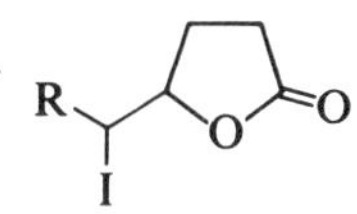

$RCH{=}CH[CH_2]_3COOH \longrightarrow$
$\Delta 5$ acids

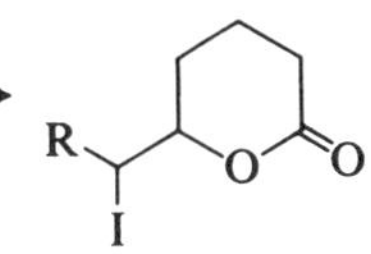

The former are more stable and more readily formed; the latter are less readily formed and react more quickly. The unsaturated acid can be regenerated from the iodolactone by reaction with trimethylsilyl iodide (or $Me_3SiCl + NaI$).

The rate of formation of iodolactone depends on the choice of solvent, on the reaction temperature and on the ratio of KI to I_2; and the regeneration of alkene from iodolactone depends on the amount of Me_3SiI, since the δ-lactone reacts first. The two iodolactones can be separated by HPLC. This reaction can be manipulated in several ways and has been used in an analytical mode (Imbs *et al.*, 1991). It has also been used as a non-chromatographic method to prepare pure DHA, EPA and arachidonic acid (Wright *et al.*, 1987; Gaiday *et al.*, 1991).

vic-Dibromo acids are debrominated by reaction with zinc or sodium iodide (or other reagents such as sodium selenide, thiourea, thiocyanates, chromous sulphate). Both reagents effect *trans* elimination, with the sodium iodide being more stereospecific. The resulting alkene has the same configuration as that from which the dibromide was prepared, and bromination–debromination has been used as a method of protecting double bonds:

$$\underset{\substack{cis \\ trans}}{-CH{=}CH-} \xrightarrow{Br_2} \underset{\substack{threo \\ erythro}}{-CHBrCHBr-} \xrightarrow{NaI} \underset{\substack{cis \\ trans}}{-CH{=}CH-}$$

Dehydrohalogenation of a *vic*-dihalide with base leads to an alkyne or allene depending on which isomer is used. These changes occur by *trans* elimination and the first step occurs easily with both *threo*- and *erythro*-dihalides to form the *Z* and *E* vinyl halides respectively:

$$\underset{cis}{\text{—CH=CH—}} \longrightarrow \underset{threo}{\text{—CHBrCHBr}} \longrightarrow$$

$$\underset{Z}{\text{—CH=CBr—}} \longrightarrow \text{—C}\equiv\text{C—}$$

$$\underset{trans}{\text{—CH}_2\text{CH=CH—}} \longrightarrow \underset{erythro}{\text{—CH}_2\text{CHBrCHBr}} \longrightarrow$$

$$\underset{E}{\text{—CH}_2\text{CBr=CH}} \longrightarrow \text{—CH=C=CH—}$$

Only the *Z* isomer then furnishes the alkyne: the less-reactive *E* isomer gives a mixture of isomeric allenes and alkynes, the latter probably produced by allene rearrangement (Gunstone and Hornby, 1969).

Dehydrohalogenation of the 2-halogeno acid with alcoholic alkali (e.g. KOH–EtOH) gives a mixture of three products: the Δ2, 2-hydroxy and 2-ethoxy acids.

Monohalogenated long-chain acids are most easily prepared from the appropriate hydroxy acid. Reaction with triphenylphosphine and tetrachloromethane or tetrabromomethane occurs smoothly and with complete inversion. D-(−)-12-Hydroxystearic acid is thereby converted to L-(+)-12-chlorostearic acid. Alternatively the alcohol can be converted to its methanesulphonate and then reacted with magnesium bromide in anhydrous ether or magnesium bromide etherate in benzene. Fluorine-containing compounds are also prepared from the corresponding alchols by reaction with 1-chloro-1,2,2-trifluoro-2-diethylaminoethane or DAST (diethylaminosulphur trifluoride) (Delgado *et al.*, 1991) or by interaction of the alcohol toluenesulphonate or methanesulphonate with tetrabutylammonium fluoride.

Carboxylic acids (RCOOH) can be converted to norhalides (RBr or RI) by the Hunsdiecker reaction (silver salt and bromine) or by some modification of this (mercuric oxide and bromine, lead tetraacetate and iodine, or thallium salt and bromine).

Allylic bromination occurs with *N*-bromosuccinimide (NBS) or with t-butyloxy bromide, but in common with other allylic reactions the product is a mixture of isomers. Heavily brominated and highly unsaturated derivatives of jojoba oil have been described (Shani, 1988). These are made by allylic bromination (NBS), removal of hydrogen bromide and bromination.

References and further reading

Delgado, A., Ruiz, M., Camps, F., Hospital, S. and Guerrero, A. (1991) *Chem. Phys. Lipids*, **59**, 127.

Gaiday, N.V., Imbs, A.B., Kuklev, D.V. and Latyshev, N.A. (1991) *J. Am. Oil Chem. Soc.*, **68**, 230.

Gunstone, F.D. and Hornby, G.M. (1969) *Chem. Phys. Lipids*, **3**, 91.

Imbs, A.B., Kuklev, D.V., Vereshchagin, A.D. and Latyshev, N.A. (1991) *Chem. Phys. Lipids*, **60**, 71.

McKague, A.B. and Reeve, D.W. (1991) *Chem. Phys. Lipids*, **58**, 213.

Shani, A. (1988) *J. Am. Oil Chem. Soc.*, **65**, 1318.

Wright, S.W., Kuo, E.Y. and Corey, E.J. (1987) *J. Org. Chem.*, **52**, 4398.

10.6 Oxymercuration

Alkenes react with mercuric acetate (or other mercury salts) and methanol (or other nucleophiles) to give a mercury-containing adduct in a *trans* addition process. This intermediate furnishes the alkene in its original stereoisomeric form by treatment with hydrochloric acid. Alternatively, the mercury group is replaceable by bromine or hydrogen thus:

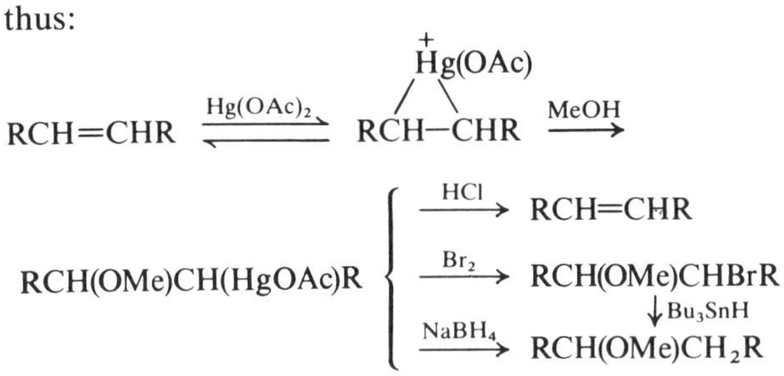

These reactions have been used in a number of ways.

(a) The reversible nature of this reaction and its high stereospecificity underlie its use for the isolation of individual acids from natural mixtures. The adducts from unsaturated acids with one methoxy and one acetoxymercuric group for each double bond vary in polarity and can be separated by partition or by chromatography before being converted back to monoene, diene, etc. Although somewhat overshadowed by silver-ion chromatography, where the adducts do not have to be prepared nor subsequently decomposed, the method still has its attractions and has been recommended as a preparative procedure for isolating pure linoleate from safflower oil and pure linolenate from linseed oil and for the study of the complex mixtures present in partially hydrogenated fish oils. This topic has been reviewed (Sebedio, 1993).

(b) The reaction with mercuric acetate and methanol followed by demercuration with sodium borohydride gives a mixture of isomeric methoxy acids, which can be used to fix double-bond position by mass spectrometry (section 9.6). Oleate, for example, gives an easily identified mixture of 9- and 10-methoxystearate. The method is less satisfactory for polyenes because of the increasing complexity of the spectra of the polymethoxy esters, but good results have been claimed for partial reaction. After incomplete reaction and subsequent hydrogenation, linolenate gives a mixture of six monomethoxystearates (9-, 10-, 12-, 13-, 15- and 16-), which can be identified from its mass spectrum.

(c) With different nucleophiles, the reaction gives a range of interesting products. Oleate, for example, will give a mixture of 9- and 10-substituted stearates, where the substituent can be OMe (MeOH), OEt (EtOH), OAc (AcOH), OH (H_2O), NHAc (CH_3CN) or OOBu^t (Bu^{t}OOH). Interesting intramolecular reactions occur with unsaturated hydroxy acids as starting materials or formed as intermediates in the reaction of a polyene:

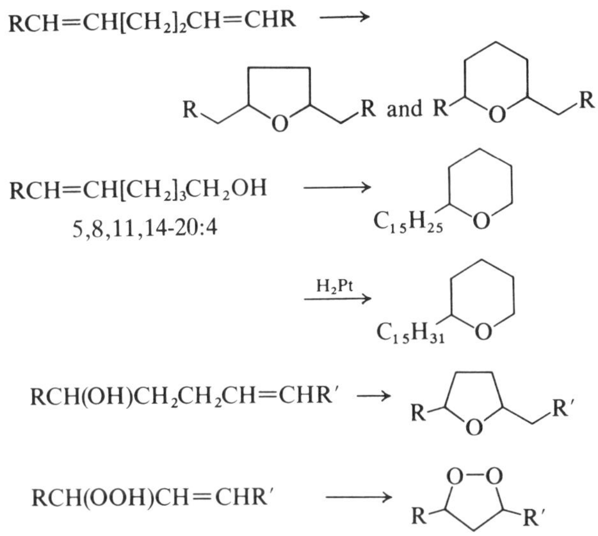

(d) The methoxymercury adducts from oleate and elaidate differ slightly in their ^1H NMR spectrum, and the OCH_3 signals can be distinguished and used as a basis for measuring the *cis/trans* ratio.

Oxymercuration of acetylenic compounds is not reversible and yields oxo or hydroxy compounds depending on the experimental procedure employed. Appropriate diynes yield dioxo compounds, which have been used to prepare furan derivatives.

References and further reading

Christie, W.W. (1989) *Gas Chromatography and Lipids*, Oily Press, Ayr, pp. 80, 141 and 176.
Gunstone, F.D. and Inglis, R.P. (1973) *Chem. Phys. Lipids*, **10**, 73, 89, 105.
Kleiman, R., Bohannon, M.B., Gunstone, F.D. and Barve, J.A. (1976) *Lipids*, **11**, 599.
Lam, C.H. and Lie Ken Jie, M.S.F. (1976) *Chem. Phys. Lipids*, **16**, 181.
Larock, R.C. (1978) *Angew. Chem., Int. Edn*, **17**, 27.
Lie Ken Jie, M.S.F. and Lam, C.H. (1977) *Chem. Phys. Lipids*, **19**, 275.
Plattner, R.E., Spencer, G.F. and Kleiman, R. (1976) *Lipids*, **11**, 222.
Schaumburg, K. (1970) *Lipids*, **5**, 505.
Sebedio, J.L. (1993) in *Advances in Lipid Methodology – Two* (ed. Christie, W.W.), Oily Press, Dundee, p. 139.
Sebedio, J.L. and Ackman, R.G. (1981) *Lipids*, **16**, 461.

10.7 Metathesis

The olefin metathesis reaction is a catalysed equilibrium process, represented most simply by the equation:

$$R^1CH{=}CHR^2 \rightleftharpoons R^1CH{=}CHR^1 + R^2CH{=}CHR^2$$

The number of products increases when the starting material is a polyene or a mixture of olefinic compounds.

The catalyst may be homogeneous or heterogeneous. The former usually consists of a transition-metal compound such as tungsten hexachloride with a Lewis acid or organometallic compound such as ethylaluminium chloride, tetramethyl- or tetrabutyltin, or triethylboron. Heterogeneous catalysts are usually oxides or carbonyls of molybdenum or tungsten on alumina or silica.

At the end of the reaction olefinic compounds, including the starting material(s), are *cis/trans* mixtures. Methyl oleate is converted to a hydrocarbon (octadec-9-ene) and the ester of a dibasic acid (octadec-9-enedioic). Methyl oleate and hex-3-ene give a mixture of hydrocarbons (Δ9-18:1, Δ3-12:1), monoester (Δ9-12:1) and diester (Δ9-18:1).

$$\left.\begin{array}{l} H_3C[CH_2]_7CH{=}CH[CH_2]_7COOCH_3 \\ H_3CCH_2CH{=}CHCH_2CH_3 \end{array}\right] \rightleftharpoons$$

$$\left[\begin{array}{l} H_3C[CH_2]_7CH{=}CH[CH_2]_7CH_3 \\ H_3C[CH_2]_7CH{=}CHCH_2CH_3 \\ H_3CCH_2CH{=}CH[CH_2]_7COOCH_3 \\ H_3COCO[CH_2]_7CH{=}CH[CH_2]_7COOCH_3 \end{array}\right.$$

The polyenes methyl linoleate and methyl linolenate give more complex mixtures. Linoleate is reported to give alkenes (12:1, 15:2, 18:3, 21:4, 24:5), esters of monobasic acids (15:1, 18:2, 21:3, 24:4, 27:5) and esters of dibasic acids (18:1, 21:2, 24:3, 27:4, 30:5), as well as cyclohexa-1,4-diene.

Even more complex products result when the reaction is carried out on glycerides, as a consequence of both inter- and intramolecular reaction.

Metathesis also occurs with olefinic alcohols when these are suitably protected, especially as trimethylsilyl ethers (Warwel *et al.*, 1987).

References and further reading

Boelhouwer, C. and Mol, J.C. (1984) *J. Am. Oil Chem. Soc.*, **61**, 425.

Frankel, E.N. (1979) in *Fatty Acids* (ed. E.H. Pryde), American Oil Chemists' Society, Champaign, p. 447.

Nicolaides, C.P., Opperman, J.H. Scurrell, M.S. and Focke, W.W. (1990) *J. Am. Oil Chem. Soc.*, **67**, 362.

Verkuiljen, E. and Boelhouwer, C. (1979) *Chem. Phys. Lipids*, **24**, 305.

Warwel, S., Döring, N. and Biermanns, F.J. (1987) *Fat Sci. Technol.*, **89**, 335.

Warwel, S. and Jägers, H.J. (1990) in *World Conference on Oleochemicals* (ed. T.H. Applewhite), American Oil Chemists' Society, Champaign, p. 319.

Yarmo, M.A., Alimuniar, A., Rahman, N.A., Ghani, R.A., Yusoff, N.I. and Ghani, M. (1990) in *World Conference on Oleochemicals* (ed. T.H. Applewhite), American Oil Chemists' Society, Champaign, p. 323.

10.8 Stereomutation

Natural unsaturated acids are almost entirely *cis* isomers, and their conversion to *trans* compounds has been well studied. Sometimes this change occurs inadvertently during processing, sometimes, it is promoted with appropriate reagents.

Steam deodorization of linolenic acid above 230 °C produces a number of configurational isomers, among which the 9*t*,12*c*,15*c* and 9*c*,12*c*,15*t* predominate and the 9*t*,12*c*,15*t* and 9*c*,12*t*,15*c* are minor products. Under the same conditions, linoleic acid gives some of the 9*t*,12*c* and 9*c*,12*t* dienes.

Trans isomers of this type have been identified in processed oils and in low-fat spreads (Wijesundera *et al.*, 1989; Wolff and Sebedio, 1991; Wolff, 1992; 1993).

Trans isomers are produced during partial hydrogenation (section 10.1), and reactions involving double-bond migration generally lead to *trans* compounds (sections 10.2 and 10.9).

Cis and *trans* isomers can be interconverted by a sequence of stereospecific reactions that add up to stereomutation. There are a number of procedures for doing this of which the following is typical:

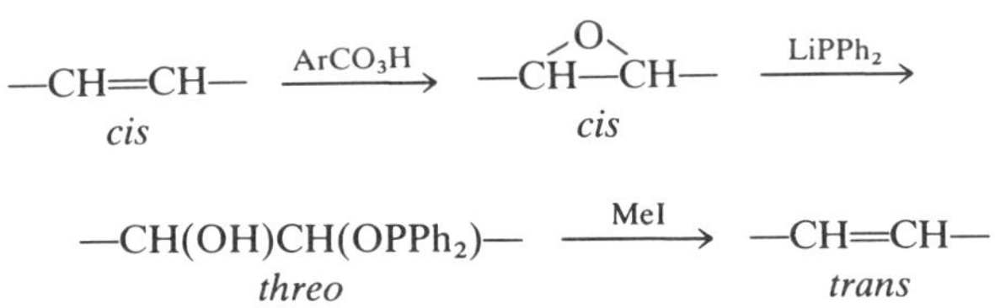

It is more usual, however, to treat the readily available *cis* isomer with a reagent that establishes the *cis*/*trans* equilibrium, and to isolate the *trans* isomer from this mixture by silver-ion chromatography (section 4.4(*b*)) or by crystallization (section 4.1). The higher-melting *trans* acid is less soluble than its *cis* isomer. The reagent should promote stereomutation without double-bond migration or hydrogen transfer. Reaction with selenium at 190–200 °C is not much used now because some double-bond migration also occurs. More often sodium nitrite–nitric acid (a source of nitrogen dioxide) or a sulphur-containing compound (3-mercaptopropionic acid, 2-mercaptoethanoic acid, 2-mercaptoethylamine, thiophenol, or an arylsulphonic acid) is used. *p*-Toluenesulphinic acid also seems to give satisfactory results (Snyder and Scholfield, 1982).

In monoenes and non-conjugated polyenes, each *cis* double bond is converted to the *trans* form in 75–80% yield. Conjugated polyenes isomerize more readily and give a higher proportion of the all-*trans* isomer. This change is simply effected with iodine and light.

The proportion of *trans* isomer in a reaction mixture can be determined by gas chromatography, infrared spectroscopy, ^{13}C NMR, infrared attenuated total reflectance spectrometry, or preparative silver-ion chromatography (section 4.4(*b*)). This topic has recently been discussed by Christie (1991) and by Lanser and Emken (1988).

References and further reading

Christie, W.W. (1991) *Lipid Technol.*, **3**, 65.

Johnson, R.W. and Pryde, E.H. (1979) in *Fatty Acids* (ed. E.H. Pryde), American Oil Chemists' Society, Champaign, p. 319.

Lanser, A.C. and Emken, E.A. (1988) *J. Am. Oil Chem. Soc.*, **65**, 1483.

Litchfield, C., Harlow, R.D., Isbell, A.F. and Reiser, R. (1965) *J. Am. Oil Chem. Soc.*, **42**, 73.

Snyder, J.M. and Scholfield, C.R. (1982) *J. Am. Oil Chem. Soc.*, **59**, 469.

Szczepanska, H. and Chmielarz, B. (1982) *Fette Seifen Anstrichm.*, **84**, 273.

Wijesundera, R.C., Ratnayake, W.M.N. and Ackman, R.G. (1989) *J. Am. Oil Chem. Soc.*, **66**, 1822.

Wolff, R.L. (1992) *J. Am. Oil Chem. Soc.*, **69**, 106.

Wolff, R.L. (1993) *J. Am. Oil Chem. Soc.*, **70**, 219, 425.

Wolff, R.L. and Sebedio, J.L. (1991) *J. Am. Oil Chem. Soc.*, **68**, 719.

10.9 Double-bond migration and cyclization

The movement of double bonds in long-chain unsaturated acids has been known since Varrentrapp converted oleic acid to palmitic acid by fusion with alkali, an observation that led to the invalid conclusion that oleic acid was the Δ2 or Δ3 acid. Double-bond migration unaccompanied by chain fission occurs under milder conditions. The reaction also occurs more easily with methylene-interrupted polyene acids to give products with conjugated unsaturation that are easily recognized by ultraviolet spectroscopy (section 9.7).

Migration occurs via a resonance-stabilized carbanion, and migration of a *cis* double bond is usually accompanied by stereomutation. Linoleic acid (9*c*,12*c*), for example, gives mainly a mixture of the 9*c*,11*t* and 10*t*,12*c* dienes (see Scheme 10.9(i)). Linolenic acid gives a mixture of trienes with diene or triene conjugation.

Isomerization has been effected under a range of alkaline conditions, including potassium hydroxide in ethanediol at 170–180 °C, potassium t-butoxide in t-butanol at 70 °C or in diethyl ether at 25 °C, the potassium salt of the methyl ether of triethylene glycol at 25 or 60 °C, dimsylsodium or dimsylpotassium in dimethylsulphoxide and/or tetrahydrofuran at 25 °C, the potassium salt of 1,3-diaminopropane at 50 °C and tetramethylammonium hydroxide.

Prior to the general use of gas chromatography, alkali isomerization was the basis of a quantitative procedure for determining linoleic and linolenic acids in conjunction with ultraviolet spectroscopy. The reaction was also employed for qualitative analysis. Thus, whilst linolenic acid (Δ9,12,15) gave conjugated diene and triene, the less familiar Δ3,9,12 and Δ5,9,12 isomers gave conjugated diene only. The absence of conjugated triene indicated that one double bond was separated from a methylene-interrupted diene unit by more than one methylene group.

Under acidic conditions, double-bond migration occurs through a sequence of protonation–deprotonation processes. With perchloric acid, oleic acid gives isomeric octadecenoic acids along with the γ-lactone (4-olide) resulting from interaction of carboxyl group and carbonium ion on C-4. With polyphosphoric acid, the product contains lactones and cyclic enones.

Homogeneous organometallic catalysts such as iron and cobalt carbonyls and rhodium complexes effect double-bond migration, as do some metals and their oxides, sulphur dioxide and iodine.

Reactions occurring at allylic sites are generally accompanied by double-bond migration. This is important in oxidation (sections 10.2 and 10.3).

Polyene acids undergo cyclization (without dimerization) at elevated temperatures. This reaction requires double bonds in appropriate positions and of correct configuration, and reagents such as alkali (to promote double-bond migration) and sulphur or iodine (to promote stereomutation) are frequently used. Cyclization of linoleate has been observed, though three or more double bonds are usually required to generate the appropriate intermediates. It is for this reason that linolenic acid (or linseed oil), eleostearic acid (or tung oil) and fish oils have been most extensively studied. The cyclized products contain cyclohexane, cyclohexene, cyclohexadiene, benzene, indane or indene systems, and are easily separated from acyclic compounds by crystallization with urea (section 4.2). The bicyclic acids result from intramolecular diene synthesis requiring a Δ1,3,8-triene unit. Bicyclic acids of this type in tall oil probably result from the Δ5,9,12 acids via a 5,10,12-triene formed under the alkaline conditions of wood pulping. Some typical reaction products are set out in Scheme 10.9(ii), but other products of similar kind are also present since double-bond migration does not give only the isomeric trienes in the sequence shown.

Heat treatment of oils containing linoleic and/or linolenic acid (sunflower, linseed, partially hydrogenated soybean) gives rise to several monomeric cyclic acids. There are *cis* and *trans* isomers of 1,2-disubstituted cyclohexanes and cyclopentanes. Double bonds may still be present in the ring systems or the side chains. Because of their complexity, they have generally been identified only after hydro-

Scheme 10.9(i)

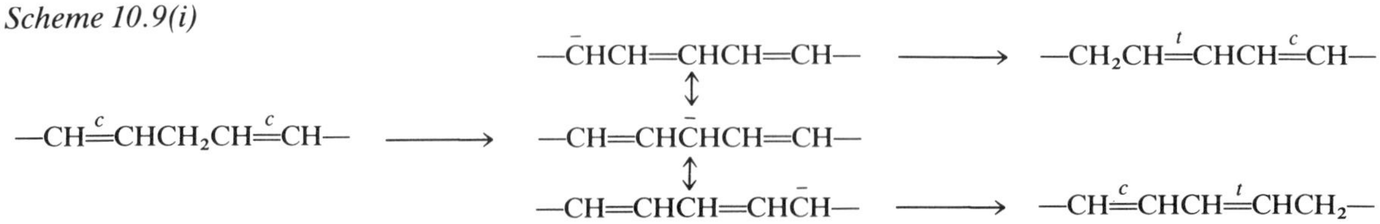

Scheme 10.9(ii)

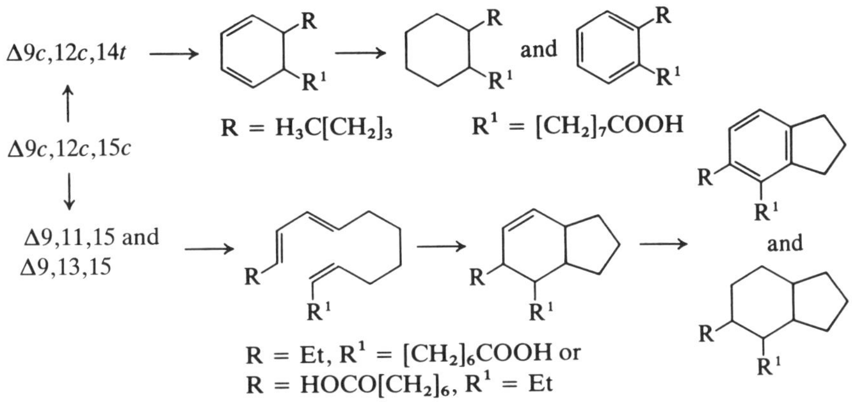

genation of the unsaturated centres, and include those formulated below:

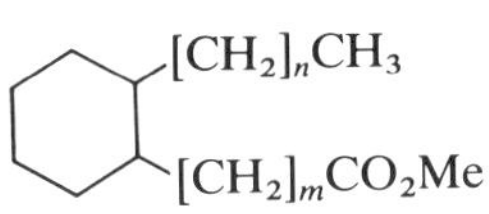

$m = 10-n$
$n = 1,2,3,4$

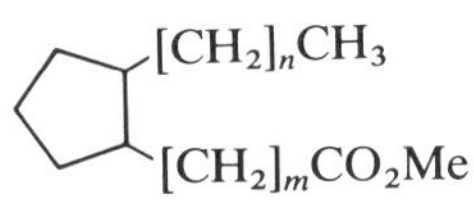

$m = 11-n$
$n = 2,3,5$

Compounds of those types have been identified in used frying oils (Christie *et al.*, 1993). The topic has been reviewed (Sebedio and Grandgirard, 1989).

C_{20} cyclic fatty acids of a slightly different type are produced by diene synthesis from a C_{18} conjugated diene acid and ethene:

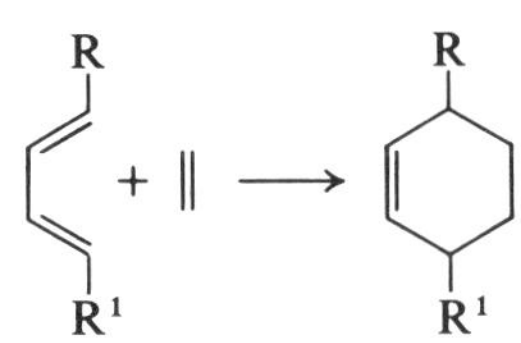

The unsaturated cyclic acids may be used in surface coatings, whilst the hydrogenated products (cyclohexane derivatives) remain liquid at low temperature, and may be useful as low-temperature lubricants.

References and further reading

Christie, W.W., Brechany, E.Y., Sebedio, J.L. and Le Quere, J.L. (1993) *Chem. Phys. Lipids*, **66**, 143.

Frankel, E.N. (1979) in *Fatty Acids* (ed. E.H. Pryde), American Oil Chemists' Society, Champaign, p. 433.

Johnson, R.W. and Pryde, E.H. (1979) in *Fatty Acids* (ed. E.H. Pryde), American Oil Chemists' Society, Champaign, p. 319.

Sebedio, J.L. and Grandgirard, A. (1989) *Prog. Lipid Res.*, **28**, 303.

von Mikusch, J.D. and Sagredos, A.N. (1971) *Fette Seifen Anstrichm.*, **73**, 384.

10.10 Dimerization

Dimerization of fatty acids occurs in the presence of radical sources, on heating and under the influence of clay catalysts, this last being employed on a commercial basis to meet the demand for dimer acids. Typical manufacturing conditions use 4% montmorillonite clay catalyst at 230 °C for 6–8 h. Fractionation of the product into monomer, dimer and trimer involves some form of molecular distillation. Rajadhyaksha *et al.* (1988) have described a new superacid catalyst based on zirconia, which is reported to give a higher dimer/trimer ratio.

In the clay-catalysed reaction, even oleate will furnish a cyclic dimer along with monomer, which is a mixture of saturated and unsaturated (mainly *trans*), straight-chain and branched-chain C_{18} compounds formed by hydrogen transfer and rearrangement. Dimers are formed by diene synthesis (Diels–Alder reaction) between a monoene and a conjugated diene produced from monoene by hydrogen transfer. The cyclohexene derivatives are converted by hydrogen transfer to cyclohexane and benzene derivatives. These monocyclic dimers are accompanied by acyclic and bicyclic dimers such as those formulated:

$R^1CH{=}CHR^2 \longrightarrow R^3CH{=}CHCH{=}CHR^4$

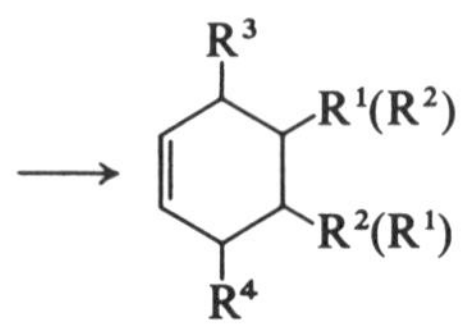

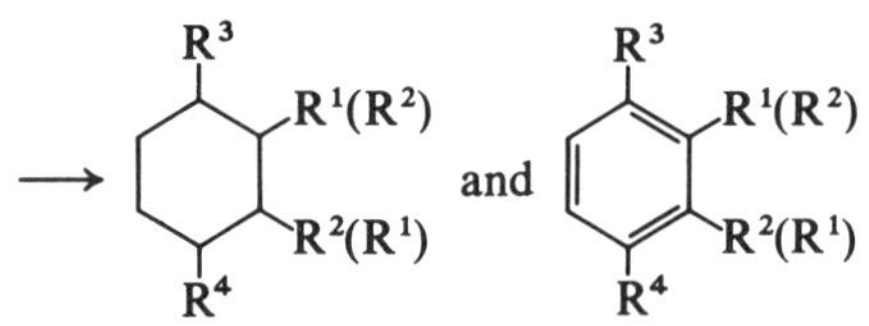

monocyclic dimers

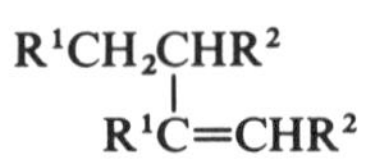

acyclic dimer

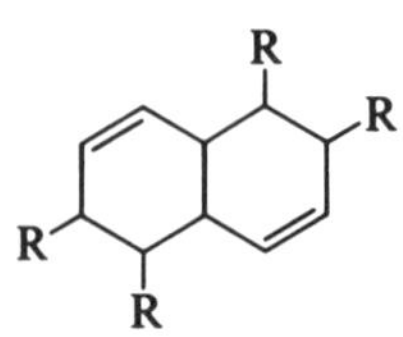

bicyclic dimer

Linoleate reacts in a similar manner. New studies with linoleic acid have provided information about aromatic compounds present in both monomer and dimer fractions (Möhring and Spiteller, 1990; Adelhardt *et al.*, 1991).

Monoene acids produce mainly acyclic (40%) and monocyclic (55%) dimers, but diene acids furnish monocyclic (55%) and bicyclic (40%) dimers. Tall oil fatty acids, containing both oleic and linoleic acid, are frequently used to produce dimers on an industrial scale. A typical product contains 80% dimer acids and 20% trimer acids, with only a trace of monomer. Meadowfoam fatty acids also give a high-quality dimer (Burg and Kleiman, 1991).

Dimer acids are used mainly as polyamides. Non-reactive polyamides, produced from dimer acids and diamines in stoichiometric amounts, are flexible and tough and have excellent adhesive properties. They are used as shoe adhesives, in printing inks, for surface coatings and in textile applications. Reactive polyamides, with free amine groups that can react further, are used as components of thermosetting epoxy systems in surface coatings. Heat-treated vegetable oils are being examined as vehicles for letter-press inks (Erhan and Bagby, 1991; Erhan *et al.*, 1992).

Dimer acids are also used for corrosion inhibition in petroleum-producing equipment (as imidazoles) and as lubricants (as esters).

In the presence of water, and particularly with monoene rather than polyene acids, the yield of dimer declines and is replaced by estolides of the type shown. These also contain the residues of two acyl chains (Anon, 1992).

$$CH_3[CH_2]_{n'}CH[CH_2]_{m'}COOH$$
$$|$$
$$OCO[CH_2]_mCH{=}CH[CH_2)_nCH_3$$

estolide ($n' = n$ or $n+1 = m' = m+1$ or m)

The monomeric acids ('isostearic acid'), usually after hydrogenation, find use as lubricants and in cosmetics.

References and further reading

Adelhardt, R., Link, W. and Spiteller, G. (1991) *Fat Sci. Technol.*, **93**, 277.

Anon (1992) *Oils Fats Int.*, **8** (2), 23.

Burg, D.A. and Kleiman, R. (1991) *J. Am. Oil Chem. Soc.*, **68**, 600.

Christopoulou, C.N. and Perkins, E.G. (1989) *J. Am. Oil Chem. Soc.*, **66**, 1361.

Erhan, S.Z. and Bagby, M.O. (1991) *J. Am. Oil Chem. Soc.*, **68**, 635.

Erhan, S.Z., Bagby, M.O. and Cunningham, H.W. (1992) *J. Am. Oil Chem. Soc.*, **69**, 251.

Figge, K. (1971) *Chem. Phys. Lipids*, **6**, 159, 178.

Johnson, R.W. (1979) in *Fatty Acids* (ed. E.H. Pryde), American Oil Chemists' Society, Champaign, p. 343.

Johnson, R.W. (1988) in *Fatty Acids in Industry* (eds R.W. Johnson and E. Fritz), Dekker, New York, p. 153.
Leonard, E.C. (1979) in *Fatty Acids* (ed. E.H. Pryde), American Oil Chemists' Society, Champaign, p. 520.
Leonard, E.C. (1979) *J. Am. Oil Chem. Soc.*, **56**, 782A.
Möhring, H. and Spiteller, G. (1990) *Fat Sci. Technol.*, **92**, 126.
Rajadhyaksha, R.A., Chaudhari, D.D. and Joshi, G.W. (1988) *J. Am. Oil Chem. Soc.*, **65**, 793.
Sen Gupta, A.K. (1967) *Fette Seifen Anstrichm.*, **69**, 907.
Sen Gupta, A.K. (1968) *Fette Seifen Anstrichm.*, **70**, 86, 153, 265.
Sen Gupta, A.K. (1969) *Fette Seifen Anstrichm.*, **71**, 873.

10.11 Other double-bond reactions

10.11.1 Nitrogen-containing compounds

For a summary, the reader is referred to Maerker (1971), Lie Ken Jie and Lao (1987) and Lie Ken Jie and Syed-Rahmatullah (1991a, b). The best-known nitrogen-containing compounds (other than those associated with the carboxyl group discussed in section 10.15) are the aziridines or epimino compounds and the amino compounds derived from these by ring-opening processes:

$$\text{—CH=CH—} \longrightarrow \text{—CH—CH—}\ (\text{bridged by NH}) \longrightarrow \text{—CH(NH}_2\text{)CH(Z)—}$$

These compounds have a formal resemblance to the epoxides and diols and to the corresponding sulphur compounds.

Aziridines are formed from alkenes in a number of multistep processes involving initial reaction with iodine azide, iodine isocyanate or *N,N*-dichlorourethane (see Scheme 10.11(i)).

Some of the more important reactions of the aziridines are set out in Scheme 10.11(ii). The *vic*-diamine can also be prepared from an epoxide by successive reaction with sodium azide, methanesulphonyl chloride, sodium azide and hydrogen/platinum.

Azides can also be produced from hydroxy compounds via their tosyl or mesyl derivatives, and if the azide is in the γ-position with respect to a double bond then 1-pyrrolino fatty acids are produced:

$$\text{—C=CHCH}_2\text{CH}_2\text{CH(OH)—}$$
isoricinoleate

$$\xrightarrow[\text{NaN}_3]{\text{MsCl}} \text{—CH=CHCH}_2\text{CH}_2\text{CH(N}_3\text{)—}$$

$\longrightarrow$ 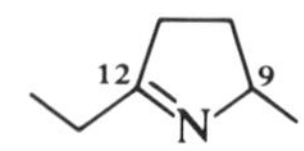

Nitrogen-containing compounds are also prepared by the Ritter reaction (see section 10.11.4).

10.11.2 Sulphur-containing compounds

For details, see Maerker (1971), Silbert and Maxwell (1979), Lower (1984) and Lie Ken Jie and Zheng (1988a, b). Epithio compounds are obtained from epoxides by reaction with sulphur-containing compounds such as thiourea, thiocyanogen, thiocyanates, potassium methyl xanthate and base, or 3-methylbenzothiazol-2-one and trifluoroacetic acid. 9,10-Epithiostearic acid melts at 58 °C (*cis*) and 64 °C (*trans*). Methyl ricinoleate has been converted to a thiophene via the 12-oxo-9,10-epithio ester.

Sulphur-containing compounds can also be made by radical addition (usually photolytic) of hydrogen sulphide or thiols to alkenes (see Scheme 10.11(iii)).

Scheme 10.11(i)

$$\text{—CH=CH—} \left[\begin{array}{lll} \xrightarrow{\text{IN}_3} & \text{—CHICH(N}_3\text{)—} & \xrightarrow{\text{LiAlH}_4} \\ \xrightarrow{\text{INCO}} & \text{—CHICH(NCO)—} & \xrightarrow{\text{MeOH; base}} \\ \xrightarrow{\text{Cl}_2\text{NCO}_2\text{Et}} & \text{—CHClCH(NClCO}_2\text{Et)—} & \xrightarrow{\text{base}} \end{array}\right] \text{—CH—CH—}\ (\text{bridged by NH})$$

Scheme 10.11(ii)

$$\text{—CHCH—}\ (\text{bridged by NH}) \left[\begin{array}{lllll} \xrightarrow{\text{H}^+,\text{MeOH}} & \text{—CH(NH}_2\text{)CH(OMe)—} & & \\ \xrightarrow{\text{HI}} & \text{—CH(NH}_2\text{)CHI—} & & \\ \xrightarrow[\text{or HN}_3]{\text{NaN}_3,\text{NH}_4\text{Cl}} & \text{—CH(NH}_2\text{)CH(N}_3\text{)—} & \xrightarrow{\text{H}_2} & \text{—CH(NH}_2\text{)CH(NH}_2\text{)—} \\ \xrightarrow{\text{RCOOH}} & \text{—CH(NH}_2\text{)CH(OCOR)—} & \underset{\text{acid}}{\overset{\text{base}}{\rightleftharpoons}} & \text{—CH(HNCOR)CH(OH)—} \end{array}\right.$$

Scheme 10.11(iii)

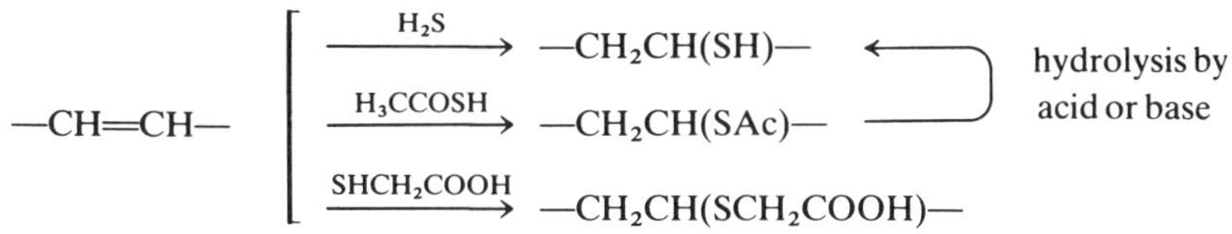

Methyl oleate gives a mixture of the 9- and 10-substituted compounds. The reaction with linoleate is more complex, and the product contains heterocyclic sulphur units.

Certain advantages have been claimed for polar addition reactions using hydrogen sulphide and boron trifluoride at −70 °C.

Sulphur dichloride reacts additively to produce long-chain thioethers containing reactive halogen groups:

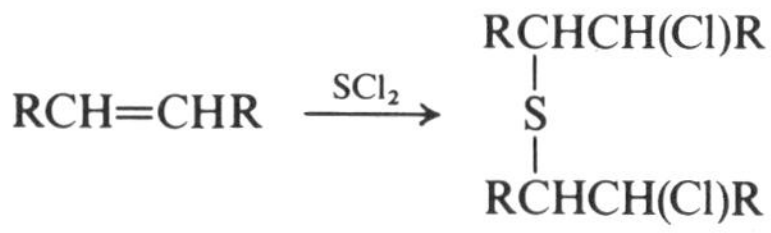

Long-chain thiols and thioethers can be made from hydroxy compounds as their methanesulphonates or from halides as follows:

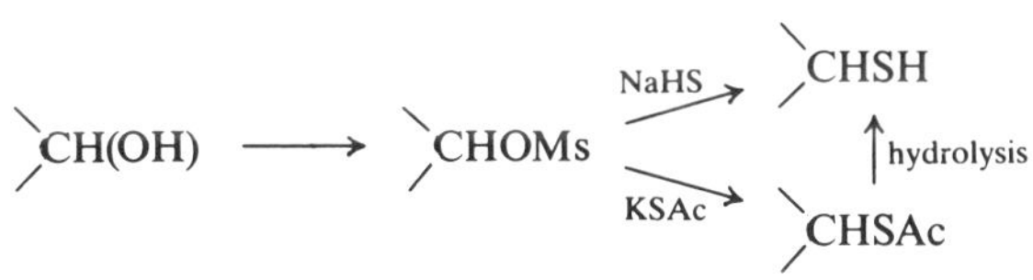

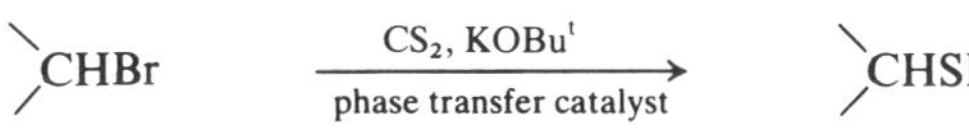

The reaction of thiocyanogen with unsaturated fatty acids, first reported by Kaufmann in 1923, has been the subject of recent re-examination. The addition reaction may occur by a radical or polar mechanism. Under the latter conditions, reaction occurs in a *trans* manner:

$$\text{—CH=CH—} \xrightarrow{\text{XSCN}} \text{—CH(Y)CH(SCN)—}$$

X = NCS,Cl,Br,I Y = NCS,SCN,Cl,Br,I

Vulcanized vegetable oil is known as factice and is used in rubber compounding to improve quality and reduce cost. Reaction with sulphur dichloride gives white factice, whilst sulphur itself gives dark factice. The structures of these compounds have not been fully determined.

10.11.3 Cyclopropanes and cyclopropenes

For more information, consult Christie (1970), Williams and Sgoutas (1972), Simmons *et al.* (1973), Gangadhar *et al.* (1988) and Lie Ken Jie and Lam (1988). The natural occurrence of fatty acids with cyclopropane and cyclopropene systems led to an interest in the synthesis of such acids from appropriate alkenoic and alkynoic acids.

Cyclopropanes are readily prepared from alkenes by the Simmonds–Smith reaction involving diiodomethane and zinc–copper couple. Modifications include the use of diethylzinc, magnesium or zinc films, zinc dust and aqueous copper(I) chloride, zinc–silver couple (benzylmercury)methylene iodine and the more reactive dimethylsulphonium methylide ($Me_2S^+[O]C^-H_2$). For example:

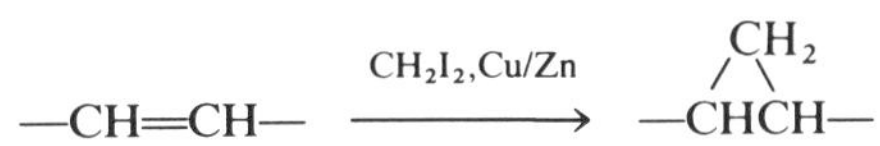

It has been claimed that improved yields result when the reaction is carried out under ultrasonic irradiation. Cyclopropanes can also be produced from olefinic compounds and diazomethane in the presence of palladium acetate.

Cyclopropanes are sometimes produced as by-products in reactions of β-substituted alkenes occurring via a homoallylic carbonium ion.

Cyclopropenes are generated from alkynes by the following reaction sequence:

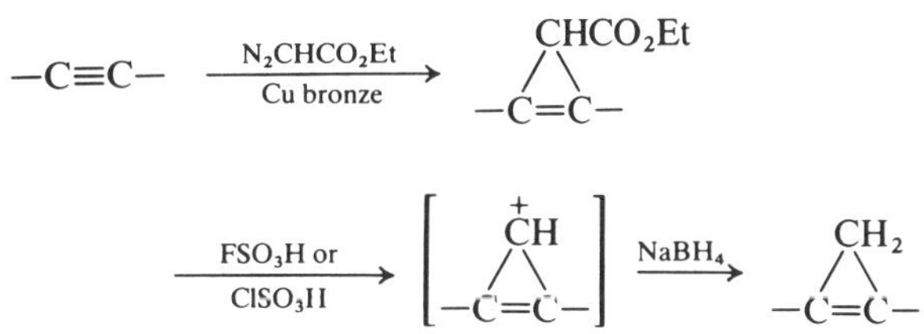

10.11.4 Catalysed addition reactions

For details, the reader is referred to Pryde *et al.* (1972), Frankel and Pryde (1977), Frankel (1979), Madrigal *et al.* (1988), Biermann and Metzger (1990, 1991) and Metzger and Biermann (1992). Many addition reactions of alkenes, which can be represented by the equation

$$\text{—CH=CH—} \xrightarrow{\text{HZ}} \text{—CH}_2\text{CH(Z)—}$$

are of two types: polar addition in the presence of a protonic or Lewis-acid catalyst, and radical reaction usually requiring a radical-initiating species. The products are generally isomeric mixtures, not only because the group Z may be attached to either olefinic carbon atom, but also because the reaction is preceded or accompanied by isomerization of the alkene, so that the ion or radical centre may appear at almost any position in the carbon chain. Isomerization is much reduced with some of the organometallic homogeneous catalysts.

For example, in the Friedel–Crafts reaction between benzene (or other aromatic compound) and oleic acid (or ester or alcohol or nitrile) in the presence of aluminium chloride, the product is a mixture of 5- to 17-phenylstearic acids.

Best known is the hydroformylation (oxo) reaction between unsaturated fatty acids and synthesis gas (hydrogen and carbon monoxide) under pressure (35–135 bar) in the presence of cobalt carbonyl at about 100 °C. The product is mainly a mixture of isomeric formylstearic acids and hydroxymethylstearic acids:

$$\text{—CH=CH—} \xrightarrow{\text{CO, H}_2} \text{—CH}_2\text{CH(CHO)— (major product)} \quad \text{and} \quad \text{—CH}_2\text{CH(CH}_2\text{OH)— (minor product)}$$

The reaction is accompanied by some reduction of double bonds. With homogeneous catalysts based on rhodium, lower pressures are required (~65 bar), higher conversion is achieved, isomer formation is decreased, and with polyenes reaction occurs at each double bond. The formylated product (RCHO) is easily converted to alcohol (RCH_2OH), acid (RCO_2H), acetal ($RCH(OR')_2$), ester (RCOOR′), quaternary diol ($R'C(CH_2OH)_2R''$) and amine (RCH_2NH_2). These polyfunctional molecules have several potential uses.

Reaction between olefinic compounds and formaldehyde in the presence of catalysts such as Me_2AlCl or $EtAlCl_2$ give olefinic hydroxymethyl compounds. Reaction occurs at an olefinic carbon atom and is accompanied by double-bond migration. For example:

$$\text{—CH}_2\text{CH=CH—} \xrightarrow{\text{CH}_2\text{O}} \text{—CH=CHCH(CH}_2\text{OH)—}$$

Other acid-catalysed addition reactions include reaction with nitriles (Ritter reaction), formaldehyde (Prins reaction), and carbon monoxide and water (Koch reaction):

$$\text{—CH=CH—} \begin{cases} \xrightarrow{\text{RCN}} \text{—CH}_2\text{CH(NHCOR)—} \\ \xrightarrow{\text{CH}_2\text{O}} \text{mixed products of uncertain structure} \\ \xrightarrow{\text{CO, H}_2\text{O}} \text{—CH}_2\text{CH(COOH)—} \end{cases}$$

These reactions are normally catalysed by concentrated sulphuric acid. Extensive isomerization occurs and may even lead to quaternary compounds of the type $RC(CH_3)XR'$, where X is the new functional group introduced into the molecule. Homogeneous catalysts have been developed that give simpler products without extensive isomerization.

Primary alcohols (C_1–C_6), but not secondary alcohols, react with olefinic compounds in the presence of sulphuric acid at 140 °C to produce the corresponding ethers:

$$\text{—CH=CH—} \xrightarrow[\text{H}_2\text{SO}_4]{\text{ROH}} \text{—CH}_2\text{CH(OR)—}$$

10.11.5 Diene addition

For a summary, consult Hopkins (1972). Long-chain acids with conjugated diene unsaturation are not readily available, though they can be produced by dehydration of castor oil or by alkali isomerization of linoleic-acid-containing oils. Such products undergo diene synthesis with dienophiles such as ethene or propenoic acid:

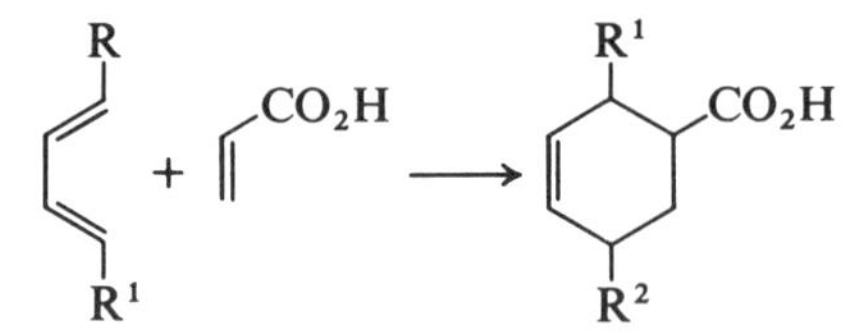

Reaction of a dienophile such as maleic anhydride with the natural conjugated triene acids (section 1.5) was used to assist in their structural identification.

10.11.6 Free-radical additions

The reader can consult Metzger and Linker (1991, 1992) for details. Olefinic compounds such as the methyl esters of undecenoic, oleic and petroselinic acid have been used to study free-radical additions. In the presence of manganese(III) acetate, enolizable compounds such as acetone, acetic acid or malonic acid react at an olefinic carbon atom. The intermediate radical may then give a saturated (H donor) or unsaturated product (Cu(II)). For example:

$$\underset{\text{oleate (etc.)}}{\text{—CH=CHCH}_2\text{—}} \xrightarrow{\text{HZ}} \text{—CHZ}\dot{\text{C}}\text{HCH}_2\text{—} \begin{cases} \xrightarrow{\text{Cu(II)}} \text{—CHZCH=CH—} \\ \xrightarrow{\text{H donor}} \text{—CHZCH}_2\text{CH}_2\text{—} \end{cases}$$

HZ = acetone, acetic acid or malonic acid

Perfluoroalkyl iodides (C_4F_9, C_6F_{13}, C_8F_{17}) react with olefinic esters in the presence of tin(II) chloride and silver acetate to give an iodide, which can be reduced by tributyltin(IV) hydride thus:

$$\underset{\text{oleate}}{\text{—CH=CH—}} \xrightarrow{\text{RfI}} \underset{\text{(+ regioisomer)}}{\text{—CH(Rf)CHI—}}$$

$$\xrightarrow{\text{Bu}_3\text{SnH}} \underset{\text{(+ regioisomer)}}{\text{—CH(Rf)CH}_2\text{—}}$$

References and further reading

Biermann, U. and Metzger, J.O. (1990) *Fat Sci. Technol.*, **92**, 133.

Biermann, U. and Metzger, J.O. (1991) *Fat Sci. Technol.*, **93**, 282.

Christie, W.W. (1970) *Top. Lipid Chem.*, **1**, 1.

Frankel, E.N. (1979) in *Fatty Acids* (ed. E.H. Pryde), American Oil Chemists' Society, Champaign, p. 426.

Frankel, E.N. and Pryde, E.H. (1977) *J. Am. Oil Chem. Soc.*, **54**, 873A.

Gangadhar, A., Subbarao, R. and Lakshminarayana, G. (1988) *J. Am. Oil Chem. Soc.*, **65**, 601.

Hopkins, C.Y. (1972) *Top. Lipid Chem.*, **3**, 37.

Lie Ken Jie, M.S.F. and Lam, W.L.K. (1988) *J. Am. Oil Chem. Soc.*, **65**, 119.

Lie Ken Jie, M.S.F. and Lao, H.B. (1987) *Chem. Phys. Lipids*, **45**, 65.

Lie Ken Jie, M.S.F. and Syed-Rahmatullah, M.S.K. (1991a) *Lipids*, **26**, 842.

Lie Ken Jie, M.S.F. and Syed-Rahmatullah, M.S.K. (1991b) *J. Chem. Soc., Perkin Trans. I*, 421.

Lie Ken Jie, M.S.F. and Zheng, Y.F. (1988a) *Chem. Phys. Lipids*, **49**, 167.

Lie Ken Jie, M.S.F. and Zheng, Y.F. (1988b) *Synthesis*, 467.

Lower, E.S. (1984) *Polym. Paint Colour J.*, **174**, 4125.

Madrigal, R.V., Bagby, M.O. and Pryde, E.H. (1988) *J. Am. Oil Chem. Soc.*, **65**, 1508.

Maerker, G. (1971) *Top. Lipid Chem.*, **2**, 159.

Metzger, J.O. and Biermann, U. (1992) *Synthesis*, 463.

Metzger, J.O. and Linker, U. (1991) *Fat Sci. Technol.*, **93**, 244.

Metzger, J.O. and Linker, U. (1992) *Liebigs Ann. Chem.*, 209.

Pryde, E.H., Frankel, E.N. and Cowan, J.C. (1972) *J. Am. Oil Chem. Soc.*, **49**, 451.

Silbert, L.S. and Maxwell, R.J. (1979) in *Fatty Acids* (ed. E.H. Pryde), American Oil Chemists' Society, Champaign, p. 403.

Simmons, M.E., Cairns, T.L., Vlalduchick, S.A. and Hoiness, C.A. (1973) *Org. React.*, **20**, 1.

Williams, J.L. and Sgoutas, D.S. (1972) *Chem. Phys. Lipids*, **9**, 295.

10.12 Hydrolysis, esterification, inter-esterification, etc.

10.12.1 Hydrolysis

Ester or lipid hydrolysis is most conveniently effected in the laboratory by refluxing with aqueous ethanolic alkali. Acidification of the hydrolysate liberates fatty acids, which can be extracted with hexane or other organic solvent. Non-acidic compounds (unsaponifiables) such as hydrocarbons, long-chain alcohols, sterols and glycerol ethers will also be present in the organic extract, but glycerol remains in the aqueous phase. The hydrolysis may be effected quantitatively for determination of saponification equivalent or saponification value (section 6.2.10).

The production of soap by alkaline hydrolysis (saponification) is usually carried out around 100 °C, with glycerol being recovered as a second commercial product. Sodium and potassium salts are used as soaps (section 5.9). Salts of other metals are used to promote polymerization of drying oils, in the manufacture of grease and lubricants, and as ingredients in plastics formulations.

Fats can also be hydrolysed to free acids in what is probably a homogeneous reaction between fat and water dissolved in the oil phase (fat splitting). This process occurs in the presence of sulphonated long-chain alkylbenzenes (24–48 h, ~100 °C) or of oxides of zinc, magnesium or calcium (2–3 h, 250 °C, 50 bar). These compounds act in part as emulsifying agents to bring fat and water together and in part as catalysts. Continuous high-pressure uncatalysed countercurrent splitting is effected at 20–60 bar and 250 °C. Somewhat lower temperatures are used for highly unsaturated oils (Sonntag, 1979, 1988).

Partial hydrolysis of lipids occurs during digestion under the influence of lipases (sections 11.2 and 11.3). Lipases present in seeds also promote hydrolysis, so that most extracted lipids contain some free acid and some partial glycerides. This is an undesirable change, since removal of free acid during processing is accompanied by some loss of fat. Lipolysis can be minimized by inactivation of the lipase before extraction. Enzymic deacylation is also the basis of some valuable analytical techniques (section 6.2.17).

Serious efforts are being made to carry out lipolysis on a commercial scale to produce fatty acids and glycerol, which are more easily purified (Bilyk *et al.*, 1991).

10.12.2 Esterification

Esters can be prepared by interaction of a carboxylic acid or an acyl derivative with an alcohol or its equivalent (esterification) or by reaction of an ester with an alcohol (alcoholysis), an acid (acidolysis) or another ester (ester–ester interchange, inter-esterification, trans-esterification). These last three processes requiring at least one ester are collectively referred to as trans-esterification reactions. Each process usually requires an acidic, basic or enzymic catalyst (Christie, 1987, 1989; Eisenhard, 1988).

Fatty acids are frequently studied in the form of their esters (e.g. for chromatographic purposes) and most commonly as methyl esters. Methyl esters are made from acids by reaction with a large excess of methanol containing sulphuric acid (1–2%), hydrogen chloride (5%) or boron trifluoride (12–14%) as catalyst, or by reaction with diazomethane. Acid-catalysed procedures are not suitable for acids containing cyclopropane, cyclopropene, epoxide or allylic hydroxy groups, nor for conjugated polyene acids. Such acids, however, can be safely esterified with diazomethane or the ester can be obtained directly from lipid by base-catalysed trans-esterification. Care must also be taken with unsaturated acids. Under more vigorous conditions, both boron trifluoride and sulphuric acid promote the acid-catalysed addition of methanol to a double bond to give a methoxy-substituted acid.

Alternative esterification procedures include the interaction of alkyl halide with the anion of the acid in the form of a salt (silver, sodium, potassium, calcium, tetramethylammonium).

In all these procedures it is convenient to use an excess of alkylating agent. This is not always so, and in glyceride synthesis, for example (sections 7.4 and 7.5), it is more usual to use equivalent or near-equivalent quantities of hydroxy compound and acylating agent, which may be free acid (with appropriate catalyst), acid anhydride or acid chloride.

Enzymes can also be used to catalyse the reaction and have been employed to provide glycerol esters (Padt *et al.*, 1990; Pecnik and Knez, 1992) and wax esters (Trani *et al.*, 1991). The selectivity of enzymes has been exploited to concentrate acids such as γ-linolenic, which is esterified less readily than the more common saturated and unsaturated acids (Hills *et al.*, 1990).

10.12.3 Alcoholysis

Alcoholysis is widely employed to convert lipids directly to methyl esters without first isolating the free acids and to prepare partial glycerides through interaction of a triacylglycerol and free glycerol.

For methanolysis, the fat or oil is dissolved in excess of methanol containing a catalyst along with toluene or dichloromethane as co-solvent:

$$R^1COOR^2 + R^3OH \xrightleftharpoons{\text{catalyst}} R^1COOR^3 + R^2OH$$

The catalyst may be acidic (sulphuric acid, hydrogen chloride, boron trifluoride) or basic (sodium methoxide), with the latter giving the quicker reaction (Christie, 1987, 1989; Eisenhard, 1988):

Acidic methanolysis:

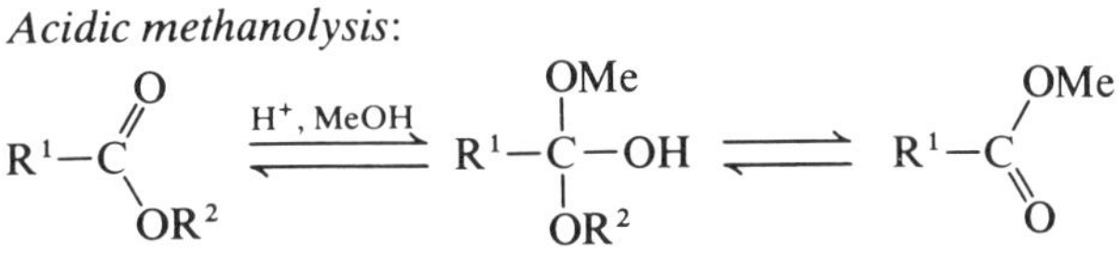

Basic methanolysis:

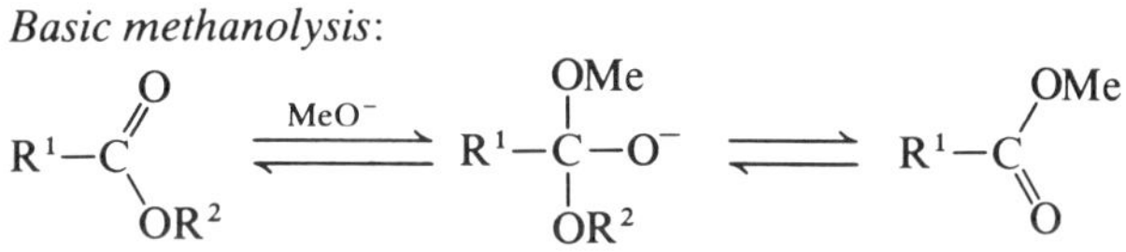

The conversion of glycerol esters to methyl esters can be achieved on a large scale in a continuous process requiring an alkaline catalyst at 240 °C and 100 bar. Direct esterification of acids occurs at 250 °C and 10 bar. Choo *et al.* (1990) have described a procedure for converting crude oils (up to 30% free acid) to methyl esters on a pilot-plant scale (500 kg/h or 3000 tonne/year). Reaction occurs in two stages: esterification of free acid at 80 ± 5 °C using a solid acid catalyst such as a sulphonated ion-exchange resin, followed by methanolysis of glycerol esters using an alkaline catalyst (NaOH) at 70 ± 5 °C.

Tetramethylguanidine has been recommended as a rapid trans-esterification catalyst (2 min at 100 °C). It has the advantage that free acids are also esterified in this reaction (Schuchardt and Lopes, 1988).

When a triacylglycerol and glycerol are heated together in the presence of sodium hydroxide or sodium methoxide as catalyst the following equilibrium is established:

$$\text{triacylglycerol} + \text{glycerol} \rightleftharpoons \text{monoacylglycerol} + \text{diacylglycerol}$$

This is an important method of preparing mono- and diacylglycerols. The composition of the equilibrium mixture of all four components depends on the relative amounts of triacylglycerol and glycerol dissolved in the lipid phase.

There is a growing interest in the preparation of monoacylglycerols on a commercial scale using enzymes as catalysts (section 5.8.1). McNeill *et al.* (1991) and McNeill and Yamane (1991) have emphasized the importance of the reaction temperature, which relates to the melting behaviour of the fat being used. With an appropriate temperature programme, yields of 90% monoacylglycerol were obtained after four days of reaction.

The alcoholysis of sunflower oil with C_1–C_5 alcohols in the presence of *Pseudomonas fluorescens* has been reported (Mittelbach, 1990), and the use of immobilized lipase reactors for modification of oils and fats has been reviewed (Malcata *et al.*, 1990).

10.12.4 Acidolysis

This less common process involves the interaction of ester and carboxylic acid in the presence of sulphuric acid, a metal oxide (zinc, calcium, magnesium, aluminium) or mercuric sulphate at about 150 °C. Applied to natural glycerides and lauric acid, for example, C_{16} and C_{18} acids are replaced by the C_{12} acid:

$$R^1COOR + R^2COOH \rightleftharpoons R^2COOR + R^1COOH$$

See next section for information on enzyme-catalysed acidolysis.

10.12.5 Inter-esterification (trans-esterification)

Natural fats are mixtures of triacylglycerols in which the acyl groups are usually distributed in a non-random manner. Under the influence of an appropriate catalyst, the acyl groups are redistributed, first intramolecularly and then intermolecularly, until a wholly random distribution is finally achieved (Hustedt, 1976; Sreenivasan, 1978; Macrae, 1983). The usual catalysts are sodium hydroxide, sodium methoxide or a sodium–potassium alloy at 0.2–0.4% level. Reaction occurs at about 80 °C during 30–60 min, and the oil or fat should be as free as possible of water, free acid and hydroperoxides, all of which inactivate catalyst. The redistribution of acyl groups leads to a change in the physical properties of the triacylglycerol mixture. Thus the melting point of soybean oil is raised from −7 to +6 °C, that of cottonseed oil from 10 to 34 °C, whilst that of hardened palmkernel oil is reduced from 46 to 35 °C.

This procedure can also be applied to mixtures of oils and fats. All the acids in the two (or more) oils will then be distributed at random, and this provides a method of transferring saturated fatty acids to predominantly unsaturated glycerides and vice versa (section 5.5.2). Methyl glycosides are conveniently acylated by reaction of the fully acetylated glycoside with appropriate methyl esters (Akoh and Swanson, 1989).

These changes can be further modified if the inter-esterification process is carried out at a lower temperature (0–40 °C), so that fully saturated glycerides (S_3) crystallize from the reaction mixture. This is called directed inter-esterification. Crystallization disturbs the equilibrium in the liquid phase, so that more S_3 forms and again separates, producing, finally, more S_3 and U_3 and less of the mixed glycerides:

$$US_2 + U_2S \rightleftharpoons U_3 + S_3$$

This usually causes an increase in the melting range of the fat. Such reactions are slower and may take up to 24 h or more.

Inter-esterification procedures are used industrially to improve the physical properties of lard, to produce cocoa butter substitutes from cheaper oils (usually combined with hydrogenation and fractionation), to produce fats containing acetic acid, and to produce margarine of appropriate melting behaviour with a minimum content of *trans* acids and maximum content of polyene acids. This has been achieved, for example, by inter-esterification of soybean oil (80%) and fully hydrogenated soybean oil (20%).

Enzymic inter-esterification is a recent development of considerable promise (Macrae, 1983). Its advantage over the more conventional procedures lies in the additional control of product composition. The lipase, coated onto a support material (kieselguhr, hydroxyapatite, alumina) in the presence of a little water, supports inter-esterification at about 40 °C and usually requires 16–70 h for complete reaction, but can be shorter depending on catalyst activity. The process may be operated in a batch or continuous manner. The substrate is a natural oil or fat, to which may be added a second oil or fat and/or a particular fatty acid or fatty acid mixture. Three types of enzyme are available. One type (from *Candida cylindricae*, *Corynebacterium acnes* or *Staphylococcus aureus*) is non-specific and leads to complete randomization of all acids at all positions. The product is the same as that obtained from the ordinary catalytic process. A second type of lipase (from *Aspergillus niger*, *Mucor miehei*, *Rhizopus arrhizus*, *R. delemar* and *R. niveus*) is specific for the 1 and 3 positions. Randomization is confined to acids in these positions and any free acid that has been added. Fatty acids in the 2 position remain unchanged. A third type of lipase (*Geotrichum candidum*) will affect only acids with Δ9 unsaturation (oleic, linoleic, linolenic). These acids will be randomized independently of their positions, but saturated acids will not be affected.

An example of the value of this technique is to be found in the upgrading of palm mid-fraction in terms of its potential as a cocoa butter substitute. Before inter-esterification, this contains POP (58%), POS (13%) and SOS (2%). These values are changed to 19, 32 and 13% after inter-esterification with stearic acid in the presence of *A. niger* lipase, which is 1,3 specific. Several other attempts to make cocoa-butter-like fats have been described (Chang *et al.*, 1990; Sridhar *et al.*, 1991; Chong *et al.*, 1992).

A product rich in 2-oleodistearin (79%) has been made from high-oleic sunflower oil and stearic acid by inter-esterification followed by fractionation of the neutral product (Macrae, 1992).

The isomerization of mono- and diacylglycerols depends on a similar acyl migration. The 1- and 2-monoacylglycerols isomerize to a 90:10 mixture of the two isomers under acidic, basic or thermal conditions. Such a change may occur during chromatography, though this is reported to be less serious if the adsorbent is impregnated with boric acid.

Diacylglycerols behave similarly. The equilibrium mixture of 1,2 (40%) and 1,3 (60%) isomers varies somewhat with temperature. Saturated diglycerides isomerize in the solid state to virtually 100% of the 1,3 isomer.

For a detailed review of the use of lipases see Eigtved, 1992.

10.12.6 Preparation and use of enol esters

Enol esters are prepared by interaction of a fatty acid with an alkyne (in the presence of zinc stearate) or with isopropenyl acetate:

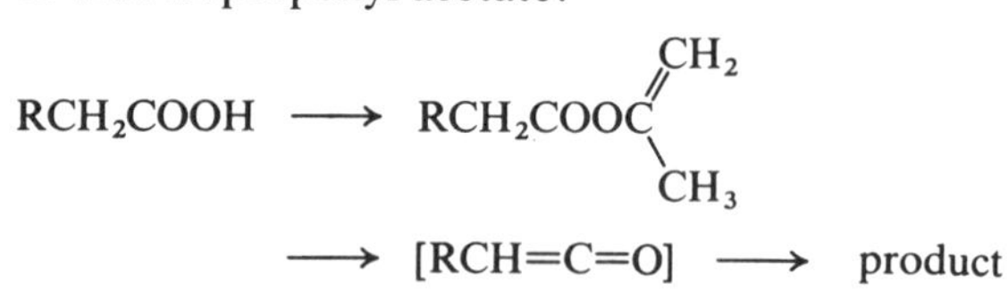

They are powerful acylating agents, and the reaction probably proceeds through the very reactive ketene produced on hydrolysis. Acetone is the only other reaction product. Alcohols (RCH_2COOR'), amines (RCH_2CONHR'), thiols (RCH_2COSR'), sulphonamides ($RCH_2CONHSO_2R'$) and carbamates ($RCH_2CON(R')COOR''$) are readily acylated to furnish the products indicated in parentheses.

References and further reading

Akoh, C.C. and Swanson, B.G. (1989) *J. Am. Oil Chem. Soc.*, **66**, 1295.

Bilyk, A., Bistline, R.G., Haas, M.J. and Feairheller, S.H. (1991) *J. Am. Oil Chem. Soc.*, **68**, 320.

Chang, M.K., Abraham, J. and John, V.T. (1990) *J. Am. Oil Chem. Soc.*, **67**, 832.

Chong, C.N., Hoh, Y.M. and Wang, C.W. (1992) *J. Am. Oil Chem. Soc.*, **69**, 137.

Choo, Y.M., Cheah, K.Y., Ma, A.N., Halim, A. and Ong, A.S.H. (1990) in *World Conference on Oleochemicals* (ed. T.H. Applewhite), American Oil Chemists' Society, Champaign, p. 292.

Christie, W.W. (1987) *High Performance Liquid Chromatography and Lipids*, Pergamon, Oxford.

Christie, W.W. (1989) *Gas Chromatography and Lipids*, Oily Press, Ayr.

Eigtved, P.. (1992) *Adv. Appl. Lipid Res.*, **1**, 1 (ed. F.B. Padley), JAI Press, London, Connecticut.

Eisenhard, W.C. (1988) in *Fatty Acids in Industry* (eds R.W. Johnson and E. Fritz), Dekker, New York, p. 139.

Hills, M.J., Kiewitt, I. and Mukherjee, K.D. (1990) *J. Am. Oil Chem. Soc.*, **67**, 561.

Hustedt, H.H. (1976) *J. Am. Oil Chem. Soc.*, **53**, 390.

Macrae, A.R. (1983) *J. Am. Oil Chem. Soc.*, **60**, 291.

Macrae, A.R. (1992) in *Oils and Fats in the Nineties* (eds V.K.S. Shukla and F.D. Gunstone), IFSC, Lystrup, Denmark, p. 199.

Malcata, F.X., Reyes, H.R., Garcia, H.S., Hill, C.G. and Amundsen, C.H. (1990) *J. Am. Oil Chem. Soc.*, **67**, 890.

McNeill, G.P. and Yamane, T. (1991) *J. Am. Oil Chem. Soc.*, **68**, 6.

McNeill, G.P., Shimizu, S. and Yamane, T. (1991) *J. Am. Oil Chem. Soc.*, **68**, 1.

Mittelbach, M. (1990) *J. Am. Oil Chem. Soc.*, **67**, 168.

Padt, A., Edema, M.J., Sewalt, J.J.W. and Riet, K. (1990) *J. Am. Oil Chem. Soc.*, **67**, 347.

Pecnik, S. and Knez, Z. (1992) *J. Am. Oil Chem. Soc.*, **69**, 261.

Schuchardt, U. and Lopes, D.C. (1988) *J. Am. Oil Chem. Soc.*, **65**, 1940.

Sonntag, N.O.V. (1979) *J. Am. Oil Chem. Soc.*, **56**, 729A.

Sonntag, N.O.V. (1988) in *Fatty Acids in Industry* (eds R.W. Johnson and E. Fritz), Dekker, New York, p. 23.

Sreenivasan, B. (1978) *J. Am. Oil Chem. Soc.*, **55**, 796.

Sridhar, R., Lakshminarayana, G. and Kaimal, T.N.B. (1991) *J. Am. Oil Chem. Soc.*, **68**, 726.

Trani, M., Ergan, F. and André, G. (1991) *J. Am. Oil Chem. Soc.*, **68**, 20.

10.13 Acid chlorides and anhydrides

Acid chlorides are made from fatty acids by reaction with phosphorus trichloride, phosphorus pentachloride, phosphorus oxychloride, phosgene, oxalyl chloride, thionyl chloride or triphenylphosphine and carbon tetrachloride. Phosphorus trichloride or phosgene are probably the most economical reagents for large-scale reaction. Improved yields have been claimed when the reaction is carried out in the absence of oxygen. The most common laboratory procedures require the acid to stand at room temperature for 3–5 days with thionyl chloride (2.4 mol) for saturated acids and with oxalyl chloride (1.8 mol) for unsaturated acids.

Acid anhydrides are prepared by reacting fatty acids or, better, acid chlorides with acetic anhydride at reflux temperature. The least-volatile component of the equilibrium mixture (acetic acid or acetyl chloride) is continuously removed by distillation to shift the equilibrium in the desired direction:

$$(H_3CCO)_2O \xrightleftharpoons{RCOOH} RCOOCOCH_3 \xrightleftharpoons{RCOOH} (RCO)_2O$$

$$2RCOCl + (H_3CCO)_2O \rightleftharpoons (RCO)_2O + 2H_3CCOCl$$

$$2RCOOH + C_6H_{11}N{=}C{=}NC_6H_{11} \longrightarrow (RCO)_2O + C_6H_{11}NHCONHC_6H_{11}$$

Dicyclohexylcarbodiimide will dehydrate acids at room temperature to give the anhydride.

Both the acid chlorides and anhydrides are effective acylating agents used in the synthesis of acylglycerols and phospholipids (sections 7.4 and 7.5).

Mixed anhydrides with methanesulphonic or toluene-4-sulphonic acid are also reported to be good acylating agents. They are made as indicated in the following equations:

$$RCOOC(CH_3){=}CH_2 + CH_3SO_3H \longrightarrow RCOOSO_2CH_3 + (CH_3)_2C{=}O$$

$$RCOCl + AgOSO_2C_6H_4CH_3 \longrightarrow RCOOSO_2C_6H_4CH_3 + AgCl$$

Ketene dimer made from hydrogenated tallow, and therefore mainly a saturated C_{16}/C_{18} mixture, is used for sizing paper. The dimer is made from the acid chloride. This reacts very slowly with water so that its aqueous dispersion is stable for at least one month. The association with paper (cellulose) is partly physical and partly chemical. The latter is thought to result in the formation of a β-oxo ester.

$$RCH_2COOH \longrightarrow RCH_2COCl$$

$\longrightarrow$

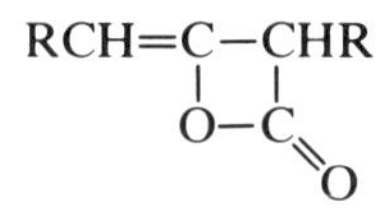

alkyl ketene dimer (AKD)

$$\xrightarrow{\text{cell—OH}} \text{cell—OCHRCOCH}_2\text{R}$$

References and further reading

Grimm, R.A. (1979) in *Fatty Acids* (ed. E.H. Pryde), American Oil Chemists' Society, Champaign, p. 218.

Spence, G.G., Dumas, D.H., Gast, J.C., Ehrhardt, S.M. and Evans, D.B. (1988) in *Fatty Acids In Industry* (eds R.W. Johnson and E. Fritz), Dekker, New York, p. 532.

10.14 Peroxy acids and related compounds

The acyl derivatives of hydrogen peroxide have been widely studied and safely used even on a large scale. Nevertheless, they should always be handled with care. In particular, they should not be ground, heated in a closed container, inhaled as vapour or ingested as dust.

Peroxy acids and diacyl peroxides are mono- and diacyl derivatives of hydrogen peroxide. Peroxy acids form esters, of which the t-butyl compounds are the best known. In contrast to carboxylic acids, which form intermolecular hydrogen-bonded dimers, the peroxy acids exist as intramolecular hydrogen-bonded monomers. This leads to several differences in the properties of these two types of acids.

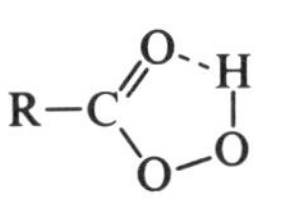
Peroxy acid

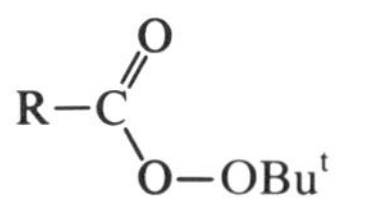
t-Butyl peroxy ester

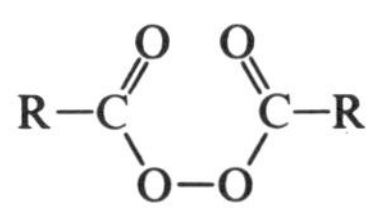
Diacyl peroxide

10.14.1 Peroxy acids

The most important route to the peroxy acids (formerly called peracids) requires establishment of the equilibrium shown in the following equation:

$$\underset{(30\text{–}98\%)}{RCOOH + H_2O_2} \xrightarrow{H^+} RCO_3H + H_2O$$

For acids insoluble in water or hydrogen peroxide, the reaction may be carried out in concentrated sulphuric acid or methanesulphonic acid, both of which then serve as solvent and as catalyst.

Although peroxy acids decompose on heating, they can be stored for quite long times in the dark at sub-zero temperatures. Thermal decomposition occurs mainly by a radical process, giving the lower alcohol, but an alternative concerted pathway giving the carboxylic acid is significant when the radical process is inhibited by antioxidants:

$$RCO_3H \rightarrow ROH + CO_2$$

$$RCO_3H \rightarrow RCO_2H + \tfrac{1}{2}O_2$$

As a consequence of their intramolecular hydrogen bonding, the peroxy acids differ from the carboxylic acids in a number of their physical properties. For example, their acidity is reduced about 1000-fold, the O—H stretching band appears at $3280\,cm^{-1}$ in the infrared spectrum ($3530\,cm^{-1}$ for the unassociated carboxylic acid), and the ^1H NMR signal at $\delta = 11.7\text{–}11.9\,ppm$ is upfield compared with that of the carboxylic acid.

Peroxy acids are useful oxidizing agents and are generally analysed on the basis of their ability to oxidize iodide ion to iodine, which can be determined titrimetrically or colorimetrically. They also oxidize alkenes to epoxides, ketones to esters or lactones, thioethers to sulphoxides and sulphones, and tertiary amines to amine oxides:

$$RCH{=}CHR \longrightarrow R\overset{O}{\widehat{CHCH}}R$$

$$RCOR \longrightarrow RCOOR$$

$$R_2S \longrightarrow R_2SO \longrightarrow R_2SO_2$$

$$R_3N \longrightarrow R_3\overset{+}{N}\text{—}\overset{-}{O}$$

Peroxy acids based on monobasic (C_8 and C_9) and dibasic acids (C_{12}) are used as laundry bleaches.

The melting points of several peroxy acids are given in Table 10.4.

10.14.2 Peroxy esters

t-Butyl esters of peroxy acids result from the reaction of t-butyl hydroperoxide with acyl halides or with carboxylic acids in the presence of appropriate imidazolide derivatives:

$$RCOCl + Bu^tOOH \xrightarrow[\text{petroleum naphtha}]{\text{pyridine}} RCOOOBu^t$$

$$RCOOH + Bu^tOOH \xrightarrow[R'NC(=X)NR']{} RCOOBu^t$$

R′ = 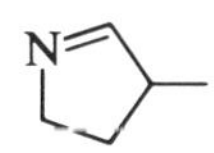 , X = O or S

These peroxy esters undergo smooth thermal decomposition at temperatures around 75 °C and are useful initiators of radical chain reactions:

$$RCOOOBu' \longrightarrow RCOO^{\bullet} + Bu^tO^{\bullet}$$

$$\longrightarrow R^{\bullet} + H_3C^{\bullet} + CO_2 + Me_2CO$$

The melting points of several t-butylperoxy esters are given in Table 10.4.

10.14.3 Diacyl peroxides

Diacyl peroxides are obtained by reaction of acyl halides or anhydrides with sodium peroxide or

Table 10.4 Melting points of peroxy acids, their t-butyl esters and diacyl peroxides

	Melting points (°C)			
$n+2$	Peroxy acids $H_3C[CH_2]_nCO_3H$	Peroxy esters $H_3C[CH_2]_nCO_3Bu^t$	Diacyl peroxides $(H_3C[CH_2]_nCOO)_2$	Diperoxy acids $HO_3C[CH_2]_nCO_3H$
6	15	–	–	117
7	–	–	–	92
8	31	–	22	89
9	35	–	13	90
10	41	−6	41	98
11	48	–	–	104
12	50	8	55	101
13	54	10	49	108
14	56	21	64	106
15	57	–	–	–
16	61	30	72	106
17	62	–	–	–
18	65	39	77	–
19	68.5	–	–	–
20	70.5	–	–	–

hydrogen peroxide or by reaction of carboxylic acid and hydrogen peroxide in the presence of dicyclohexylcarbodiimide (DCC):

$$2RCOCl + Na_2O_2 \xrightarrow{0\text{–}5\,°C} RCOOOCOR$$

$$2RCOOH + H_2O_2 \xrightarrow{DCC} RCOOOCOR$$

These compounds undergo thermal decomposition by various routes, of which the most important involves homolysis of the O—O bond:

$$RCOOOCOR \rightarrow 2RCOO^{\bullet} \rightarrow 2R^{\bullet} + 2CO_2$$

They are used as radical initiators because of their smooth decomposition between 20 and 100 °C. The melting points of several diacyl peroxides are given in Table 10.4.

References and further reading

Swern, D. (1970, 1971) *Organic Peroxides*, vols 1 and 2, Wiley-Interscience, New York.

Swern, D. (1979) in *Fatty Acids* (ed. E.H. Pryde), American Oil Chemists' Society, Champaign, p. 236.

10.15 Nitrogen-containing compounds

Fat-based amides and amines and compounds derived from these are manufactured to the extent of about 450 000 tonnes per annum. Most of their uses are based on their surface-active properties.

Amides are usually made from acids (or esters) and ammonia in a batch operation conducted at 180–200 °C and 3.5–7.0 bar. A catalyst, such as boric acid, alumina or zinc alkoxide, is optional, and urea (at a reaction temperature of 170–240 °C) may replace ammonia:

$$\mathrm{RCOOH + NH_3 \longrightarrow RCONH_2 + H_2O}$$

The amides, especially those based on oleic or erucic acid, are used as anti-slip and anti-block agents in polythene film and as mould-release agents and lubricant additives. They improve the properties of printing inks and are used in the production of water repellents.

Alkylamides have been made from acids, alkyl esters, or glycerol esters and alkylamines with enzymes present in *Candida rugusa*, *Rhizomucor miehei* and porcine pancreas. There is some selectivity for the chain length of the acids and amines (Bistline *et al.*, 1991).

Diamides, made from amides and formaldehyde or from acids and ethanediamine, are used as lubricants:

$$\mathrm{2RCONH_2 + CH_2O \longrightarrow RCONHCH_2NHCOR}$$

$$\mathrm{2RCOOH + H_2N[CH_2]_2NH_2 \longrightarrow RCONH[CH_2]_2NHCOR}$$

Ethoxylated amides result from reaction between amides and ethylene oxide at 150–200 °C in the presence of a basic catalyst or from esters or acids and diethanolamine at 115–160 °C and a basic catalyst:

$$\mathrm{RCONH_2} + n\,\overset{\mathrm{O}}{\widehat{\mathrm{CH_2CH_2}}} \longrightarrow \mathrm{RCONH(CH_2CH_2O)_{\mathit{n}}H}$$

$$\mathrm{RCOOMe + (HOCH_2CH_2)_2NH \longrightarrow RCON(CH_2CH_2OH)_2}$$

Nitriles are made by heating fatty acids with ammonia at 280–360 °C in the presence of a catalyst such as zinc oxide or manganese acetate, probably via the amide:

$$\mathrm{RCOOH \xrightarrow{NH_3} [RCONH_2] \longrightarrow RCN}$$

They are used almost entirely to furnish *amines* by hydrogenation in the presence of ammonia and a nickel or cobalt catalyst. Primary or secondary amines are produced depending on the reaction conditions:

$$\mathrm{RCN \xrightarrow[NH_3]{H_2} RCH_2NH_2 \longrightarrow (RCH_2)_2NH}$$

The amines are used as flotation agents and lubricant additives and serve as emulsifying agents, foaming agents, or corrosion inhibitors after conversion to ethylene oxide adducts of the following types:

$$\mathrm{RNH(CH_2CH_2O)_{\mathit{x}}H} \qquad \mathrm{R_2N(CH_2CH_2O)_{\mathit{x}}H} \qquad \mathrm{RN}\begin{cases}\mathrm{(CH_2CH_2O)_{\mathit{x}}H}\\ \mathrm{(CH_2CH_2O)_{\mathit{y}}H}\end{cases}$$

The primary and secondary amines yield tertiary amines ($\mathrm{RNMe_2}$, $\mathrm{R_2NMe}$) by catalytic reaction with formaldehyde. These are used as corrosion inhibitors, fuel-oil additives, bactericides and fungicides or are converted to *amine oxides* by reaction with hydrogen peroxide or to *quaternary ammonium* compounds (quats) by reaction with methyl chloride (or sulphate) or benzyl chloride. Typical structures include: $\mathrm{RN^+Me_3Cl^-}$, $\mathrm{RN^+Me_2(CH_2Ph)Cl^-}$, $\mathrm{R_2N^+Me_2Cl^-}$ and $\mathrm{R_3N^+MeCl^-}$. These quaternary compounds may have one, two or three long alkyl chains. They are used extensively as fabric softeners, as bactericides and in the production of organic-modified drilling muds.

Diamines, made from amines and acrylonitrile, are valuable as flotation agents, fuel-oil additives and fabric softeners:

$$\mathrm{RNH_2 + CH_2{=}CHCN \longrightarrow RNH[CH_2]_3NH_2}$$

Imidazolines, used as fabric softeners, are made from acids and diamines with diaminoethane or triamines with bisaminoethylamine:

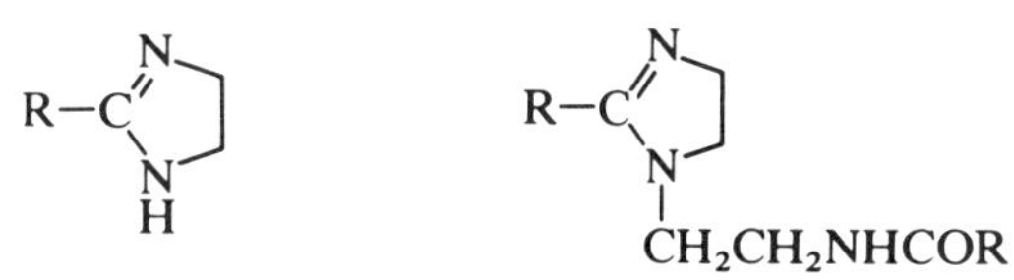

(from diaminoethane) (from bisaminoethylamine)

The hydrolysis of imidazolines in alkaline and acidic solution has been reviewed (Watts, 1990).

Other compounds: several other nitrogen-containing compounds have been made and examined for surface-active properties. Typical examples include pyrimidines, pyrrolidones and derivatives of *N*-acyldiethanolamine. The latter have two ($R' = CH_3$) or three ($R' = C_9H_{19}$ or $C_{11}H_{23}$) long alkyl chains (Tsatsaroni *et al.*, 1987; Takata *et al.*, 1990; Zhu *et al.*, 1991):

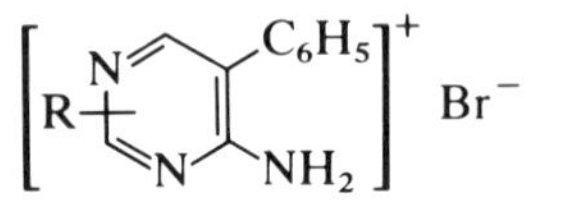

$R = C_6–C_{14}$

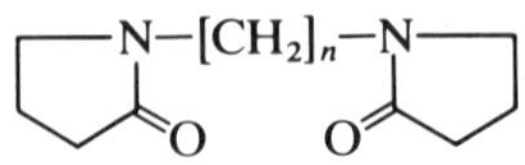

n = 3,5,9,12

$$\begin{array}{c} \qquad\qquad O[CH_2]_3SO_3Na \\ \qquad\qquad | \\ R'CON(CH_2CH_2OCH_2CHCH_2OR)_2 \end{array}$$

$R' = CH_3, C_9H_{19}$ or $C_{11}H_{23}$
$R = C_8H_{17}$ or $C_{10}H_{21}$

On an industrial scale, all these nitrogen-containing compounds are made from mixed fatty acids of natural origin (e.g. coconut, tallow, tall oil) or from individual acids of technical purity.

References and further reading

Beck, R.A. (1979) *J. Am. Oil Chem. Soc.*, **56**, 796A.
Billenstein, S. and Blascke, G. (1984) *J. Am. Oil Chem. Soc.*, **61**, 353.
Bistline, R.G., Bilyk, A. and Feairheller, S.H. (1991) *J. Am. Oil Chem. Soc.*, **68**, 95.
Glankler, C.W. (1979) *J. Am. Oil Chem. Soc.*, **56**, 803A.
House, J.E. (1984) *J. Am. Oil Chem. Soc.*, **61**, 357.
Linfield, W.M. (1984) *J. Am. Oil Chem. Soc.*, **61**, 437.
Reck, R.A. (1988) in *Fatty Acids in Industry* (eds R.W. Johnson and E. Fritz), Dekker, New York, p. 177.
Schaeufele, P.J. (1984) *J. Am. Oil Chem. Soc.*, **61**, 387.
Takata, T., Atobe, I., Kitamura, N. and Endo, T. (1990) *J. Am. Oil Chem. Soc.*, **67**, 739.
Tsatsaroni, E., Pegiadou-Koemtjopoulou, S. and Demertzis, G. (1987) *J. Am. Oil Chem. Soc.*, **64**, 1444.
Watts, M.W. (1990) *J. Am. Oil Chem. Soc.*, **67**, 993.
Zhu, Y.-P., Masuyama, A., Kirito, Y.-I. and Okahara, M. (1991) *J. Am. Oil Chem. Soc.*, **68**, 539.

10.16 Other reactions of the carboxyl group

A carboxyl group raises the activity of the adjacent CH_2 group above that of other CH_2 groups in a long-chain acid. It is thus possible to functionalize the α-methylene group in a number of ways. Some reactions of this type are useful only for saturated acids (or esters) because double bonds sometimes promote similar reactions at their allylic positions, leading to mixed or more complex products. Reference has already been made to α-halogenation (section 10.5), and this account deals with sulphonation and the formation and reactions of α-anions. Some of these reactions are accompanied or followed by decarboxylation, so the whole sequence can be summarized as follows:

$$RCH_2COOH \longrightarrow RCHXCOOH \longrightarrow RCH_2X$$

10.16.1 Sulphonation

Sulphonation of saturated fatty acids (or esters) can be effected with oleum, stabilized sulphur trioxide, dilute vaporized sulphur dioxide or chlorosulphonic acid. Trichloromethane, tetrachloromethane, tetrachloroethene and liquid sulphur dioxide have been used as reaction solvent.

$$RCH_2COOH(CH_3) \longrightarrow RCH(SO_3H)COOH(CH_3)$$

The sulphonic carboxylic acids (α-sulpho fatty acids) have melting points (°C) of 86.5–88 (C_{12}), 85–86.5 (C_{14}), 90–91 (C_{16}), 96–97 (C_{18}) and 96–97 (C_{22}). They form acid salts ($RCH(SO_3Na)COOH$), neutral salts ($RCH(SO_3Na)COONa$) and esters ($RCH(SO_3H)COOR'$), which are more stable than usual to hydrolysis. These compounds are used in detergent bars (as sodium and amine salts), and as viscosity-reducing agents, corrosion inhibitors and ore flotation agents.

The sulphonated products show high levels of surface activity, are less toxic in aqueous environments than other surfactants, and are biodegradable in a few days. Their esters show high hydrolytic stability.

10.16.2 α-Anions

Saturated and monoene acids form dianions when treated with lithium diisopropylamide and hexamethylphosphoramide:

$$RCH_2COOH \xrightarrow[HMPA]{Pr^i_2NLi} RC^-HCOO^-$$

$$\longrightarrow RCHYCOOH \text{ or } RCH_2Y$$

and these undergo further reaction with a wide range of compounds to give α-substituted acids or products arising from these by decarboxylation. The following compounds have been prepared using the reagent indicated in parentheses:

$RCHDCOOH$ (D_2O)
$RCH(R')COOH$ ($R'Br$)
$RCH(CH_2OH)COOH$ and
$RC(=CH_2)COOH$ (CH_2O)
$RCH(OOH)COOH$ and $RCH(OH)COOH$
(O_2 at −78 °C and ambient temperature respectively)
$RCHXCOOH$ (Br_2 or I_2)
$RCH(COOH)_2$ (CO_2 or $ClCO_2Et$ or $(R'O)_2CO$)
$RCH(COR')COOH$ ($R'COCl$)
RCH_2CHO (HCOOH and decarboxylation)
RCH_2NO_2 ($PrONO_2$ and decarboxylation)

$$\begin{array}{l} RCHCH_2CH_2 \\ \;| \quad\quad\; | \qquad \text{(ethylene oxide)} \\ OC\text{——}O \end{array}$$

$RCH(NH_2)COOH$ (NH_2OR')
$RCH(SSH)COOR'$ (R' = H or alkyl, CS_2)

10.16.3 Polyoxyethylene esters

Carboxylic acids react with ethylene oxide and an alkaline catalyst, or with polyoxethylene glycols and an acidic catalyst, to give a hydroxy ester with a range of ethylene oxide units accompanied by the corresponding diester and diol resulting from ester interchange:

$$RCOOH + \overset{O}{\widehat{CH_2CH_2}} \longrightarrow RCO[OCH_2]_nOH$$

$$\longrightarrow \begin{array}{c} RCO[OCH_2]_nOCOR \\ + \\ H[OCH_2CH_2]_nOH \end{array}$$

By this reaction carboxylic acids are converted to non-ionic surfactants, which are used in various ways.

10.16.4 Aldehydes

Esters are readily converted to aldehydes by reaction with lithium aluminium hydride in the presence of excess diethylamine in pentane (Cha and Kwon, 1987). Acids give aldehydes through reaction with thexylbromoborane dimethylsulphide in carbon disulphide (Cha *et al.*, 1987) or in a one-pot reaction via the activated pentacoordinated silyl carboxylate (Corriu *et al.*, 1987).

References and further reading

Bluestein, B.R. and Kapur, B.L. (1979) in *Fatty Acids* (ed. E.H. Pryde), American Oil Chemists' Society, Champaign, p. 260.

Cha, J.S. and Kwon, S.S. (1987) *J. Org. Chem.*, **52**, 5486.

Cha, J.S., Kim, J.E., Oh, S.Y., Lee, J.C. and Lee, K.W. (1987) *Tetrahedron Lett.*, **28**, 2389.

Corriu, R.J.P., Lanneau, G.F. and Perrot, M. (1987) *Tetrahedron Lett.*, **28**, 3941.

Drozd, J.C. (1990) in *World Conference on Oleochemicals* (ed. T.H. Applewhite), American Oil Chemists' Society, Champaign, p. 257.

Inagaki, T. (1990) in *World Conference on Oleochemicals* (ed. T.H. Applewhite), American Oil Chemists' Society, Champaign, p. 269.

Petragnani, N. and Yonashiro, M. (1982) *Synthesis*, 521.

Reck, R.A. (1988) in *Fatty Acids in Industry* (eds R.W. Johnson and E. Fritz), Dekker, New York, p. 201.

Silbert, L.S. and Pfeffer, P.E. (1979) in *Fatty Acids* (ed. E.H. Pryde), American Oil Chemists' Society, Champaign, p. 272.

11

LIPID METABOLISM

J.L. HARWOOD

11.1 Fatty acids

11.1.1 *De novo* synthesis

De novo synthesis of fatty acids requires the concerted action of two multienzyme or multifunctional proteins – acetyl-CoA (coenzyme A) carboxylase and fatty acid synthetase. Acetyl-CoA carboxylase is a type I biotin-containing carboxylase. Its activity depends on the presence of three separate proteins in the case of multienzyme complexes, such as the *Escherichia coli* acetyl-CoA carboxylase, or three domains in the case of multifunctional complexes, such as in plant or animal acetyl-CoA carboxylases. The three proteins/domains are biotin carboxyl carrier protein (BCCP), biotin carboxylase and a carboxyltransferase. In the first part of the reaction, the biotin moiety of BCCP is carboxylated in an (adenosine 5′-triphosphate) ATP-mediated process. Recent evidence has implicated carboxyphosphate as an intermediate. The carboxyl moiety is then transferred to the acceptor acetyl-CoA, probably by the use of carbon dioxide as the electrophile. Study of the half-reactions of acetyl-CoA carboxylase has allowed proposals concerning the mechanism of biotin participation to be made (De Titta *et al.*, 1980). The whole subject of the mechanism of biotin-containing enzymes has been excellently reviewed (Knowles, 1989).

Acetyl-CoA carboxylase has been purified to homogeneity from various tissues including rat liver (Wakil *et al.*, 1983), chicken liver (Wada and Tanabe, 1983), rat and rabbit mammary gland (Wakil *et al.*, 1983) and goose uropygial gland (Rainwater and Kollattukudy, 1982). It is a multifunctional protein of mass 220–260 kDa (kilodaltons)*. By means of combined protein, chemical and molecular cloning techniques, Takai *et al.* (1987) elucidated the primary structure around the biotin-binding site of chicken liver acetyl-CoA carboxylase. Takai *et al.* (1988) described the complete amino acid sequence of the chicken liver enzyme by cloning and sequencing the DNA complementary to the appropriate messenger RNA. The chicken liver acetyl-CoA carboxylase was found to be composed of 2324 amino acid residues with a calculated molecular mass of 262 706 Da. The BCCP domain was in the middle region of the polypeptide. The amino terminal portion showed a homologous primary structure to that of carbamyl phosphate synthase, and was thought to be the site of the biotin carboxylase for which carboxyl phosphate is the postulated reaction intermediate (see Knowles, 1989).

Although animal acetyl-CoA carboxylases have molecular masses of about 250 kDa, they function as polymers of molar masses 4×10^6 to 8×10^6 g/mol (Wakil *et al.*, 1983). The regulation of mammalian acetyl-CoA carboxylase has been well studied owing to the key role of this enzyme in controlling overall fatty acid (and fat) synthesis. Control occurs in two ways: short-term control, which involves allosteric regulation and covalent enzyme modification; or long-term control, where the amounts of the carboxylase are changed. Metabolite control is due to citrate (Moss and Lane, 1971) and long-chain acyl-CoAs. The latter cause depolymerization of mammalian carboxylases. Palmitoyl-CoA, stearoyl-CoA and arachidoyl-CoA are the most effective at inhibiting the carboxylase (Nikawa *et al.*, 1979). Other effectors have been reported to regulate mammalian acetyl-CoA carboxylase (cf. Wakil *et al.*, 1983), but their physiological importance is less certain.

Mammalian acetyl-CoA carboxylase also seems to be subject to regulation by phosphorylation/dephosphorylation. Phosphorylation with one mole of phosphate per mole of rat liver carboxylase subunit causes complete inactivation (Lent and Kim, 1980). The reactions have also been studied using preparations from the rat epididymal fat pad (Krakower and Kim, 1981). Long-term regulation of the carboxylase is caused by diet, thyroxine and insulin. Total enzyme amounts also change during cell differentiation and development (reviewed by

* The dalton (Da) is equal to 1/12 of the mass of an atom of ^{12}C. Many biologists use the dalton as a unit of mass, although it is not a 'recognized' unit. It is especially convenient for structures where the word 'molecule' is incorrect, but is used more generally. One should write that the molecular *mass* (not weight) of a protein, for example, is 220 000 Da or 220 kDa.

Volpe and Vagelos, 1976). Use of antibodies to rat liver acetyl-CoA carboxylase revealed unexpectedly that the mitochondrial outer membrane contained a major pool of rather poorly active enzyme. It was suggested that this protein represented a reservoir of acetyl-CoA carboxylase that could be released and activated under lipogenic conditions (Allred and Roman-Lopez, 1988). The control of mammalian acetyl-CoA carboxylases has been summarized well recently (Goodridge, 1991).

Until fairly recently the molecular nature of plant acetyl-CoA carboxylase was unclear. However, it now seems that the enzyme is also a multifunctional protein of mass 220–240 kDa (see Harwood, 1988). Reference to the purification of higher-molecular-mass forms of the enzyme from parsley and oilseed rape as well as early work will be found in Harwood (1988). A similar form has also been purified to near-homogeneity from the diatom *Cyclotella cryptica* (Roessler, 1990). In addition to acetyl-CoA carboxylase, cell-free extracts from several monocotyledonous (monocot.) and dicotyledonous (dicot.) plant species have been found to contain three additional biotin-containing enzymes – 3-methylcrotonyl-CoA carboxylase, propionyl-CoA carboxylase and pyruvate carboxylase (Wurtele and Nikolau, 1990) – the presence of which can complicate purification protocols making use of avidin affinity columns. Regulation of the plant acetyl-CoA carboxylase is not as well understood as for mammals. Tricarboxylic acids do not seem to function at all, and the enzyme in leaves may be activated, at least in part, through changes in the stroma medium as photosynthesis occurs (see Harwood, 1988). Nevertheless, recent measurement of the pool sizes of intermediates during light-activated fatty acid synthesis in spinach leaves shows clearly that the activity of acetyl-CoA carboxylase is a key regulatory component (Post-Beittenmiller *et al.*, 1991).

In addition to the animal and plant acetyl-CoA carboxylases, that from the yeast *Saccharomyces cerevisiae* also seems to be a multifunctional protein. It has been reported to have a subunit molecular mass of 190 000 Da (Sumper and Riepertinger, 1972).

In contrast to the above, the acetyl-CoA carboxylase from *E. coli* contains three separate proteins – BCCP, biotin carboxylase and carboxyltransferase (Alberts and Vagelos, 1972; Polakis *et al.*, 1974). In this regard, *E. coli* shares the same characteristics as the transcarboxylase from *Propionibacterium shermanii* (Samols *et al.*, 1988). Early work on the structure of the component proteins of *E. coli* acetyl-CoA carboxylase has been summarized (see Volpe and Vagelos, 1976). The DNA sequence of the gene encoding the biotin carboxylase subunit has been reported and encodes a protein of 449 residues. The sequence is strikingly similar to the amino terminal sequence of two biotin-dependent carboxylase proteins, yeast pyruvate carboxylase and the α subunit of rat propionyl-CoA carboxylase (Li and Cronan, 1992). The gene encoding BCCP is located upstream of the biotin carboxylase gene, and the two genes are co-transcribed. The BCCP gene encodes a protein of mass 16.7 kDa in spite of the fact that BCCP appears to have a mass of 22.5 kDa. The *fabE* mutant (acetyl-CoA carboxylase-deficient) was also shown to have a lesion within the BCCP gene (Li and Cronan, 1992). Neither citrate nor phosphorylation play any role in regulating the bacterial enzyme. Instead, guanosine

Table 11.1 Distribution and major properties of type I and type II fatty acid synthetases (From Harwood and Russell, 1984, with permission)

	Distribution	Major characteristics
Type I	Animals Birds Yeast *Euglena gracilis* (cytoplasm) Fungi *Mycobacterium smegmatis* *Corynebacterium diphtheriae*	High-molecular-mass multifunctional proteins Covalently bound ACP[a] Release unesterified fatty acids (animals) or acyl-CoA (yeast)
Type II	Plants *Euglena gracilis* (chloroplasts) Most bacteria Cyanobacteria	Dissociable enzymes in a complex with ACP Dissociable ACP Release acyl thioesters or transfer final products directly to lipid

[a] ACP, acyl carrier protein.

nucleotides, guanosine 3′-diphosphate 5-di- (and tri-) phosphate are used. These inhibit the carboxyltransferase.

Fatty acid synthetases can be categorized as type I, type II and type III. The distribution and some of the properties of the type I and type II enzymes are summarized in Table 11.1. The type I synthetases tend to occur in higher organisms, and all of the purified synthetases of eukaryotes, with the exception of plants, are of this type. Two bacterial genera, *Mycobacterium* and *Corynebacterium*, also contain type I synthetases. These synthetases are large-molecular-mass multifunctional proteins containing covalently bound acyl carrier protein (ACP) (see Wakil *et al.*, 1983). On the other hand, type II synthetases such as that from *E. coli* consist of individual enzymes that can be isolated in an active form. The ACP of type II synthetases also readily dissociates and can be purified and characterized (see Volpe and Vagelos, 1973). Type III synthetases elongate preformed acyl chains, and will be dealt with later.

The partial reactions of fatty acid synthetases are shown in detail in Table 11.2, where assay details for measuring activities in plants are given. A comparison of some features of type I fatty acid synthetases is made in Table 11.3.

The type I fatty acid synthetases from rat liver,

Table 11.2 Partial reactions of fatty acid synthetase (FAS) and assay principles for measuring their activity in plants (From Harwood *et al.*, 1990)

Enzyme	Method
Activities common to all FASs	
Acetyl-CoA:ACP transacylase [^{14}C]Acetyl-CoA + ACP ⇌ [^{14}C]Acetyl-ACP + CoA	Precipitate acetyl-ACP Count
Malonyl-CoA:ACP transacylase [^{14}C]Malonyl-CoA + ACP ⇌ [^{14}C]Malonyl-ACP + CoA	Precipitate malonyl-ACP Count
β-Ketoacyl-ACP synthetase[a] Acetyl-ACP + Malonyl-ACP ⇌ Acetoacetyl-ACP + CO_2 + ACP	Measure absorbance at 303 nm (acetoacetate formation)
or nAcyl-ACP + Malonyl-ACP ⇌ (n + 2)Acyl-ACP + ACP + CO_2	In presence of NADPH[b] and other FAS enzymes, measure absorbance at 340 nm With [^{14}C]malonyl-CoA, measure counts in acyl chains Use of NaH[^{14}C]CO_2 permits a CO_2 exchange assay
β-Ketoacyl-ACP reductase β-Ketoacyl-ACP + NAD(P)H ⇌ β-Hydroxyacyl-ACP + NAD(P)	Change in absorbance at 340 nm
β-Hydroxyacyl-ACP dehydrase β-Hydroxylacyl-ACP ⇌ Enoyl-ACP + H_2O	Back-reaction followed with crotonyl-ACP and decrease in absorbance at 263 nm
Enoyl-ACP Enoyl-ACP + NAD(P)H ⇌ Acyl-ACP + NAD(P)	Change in absorbance at 340 nm
Additional activities present in E. coli and higher plants	
Acetoacetyl-ACP synthetase [^{14}C]Acetyl-CoA + Malonyl-ACP ⇌ [^{14}C]Acetoacetyl-ACP + CoA + CO_2 (+cerulenin)	Precipitate acetoacetyl-ACP Count (reaction conditions not fully defined)
β-Ketoacyl-ACP synthetase II Palmitoyl-ACP + [2-^{14}C]Malonyl-ACP ⇌ [2-^{14}C]β-ketooctadecanoyl-ACP + ACP + CO_2	Reaction products reduced and counts in acyl chain measured

[a] β-Ketoacyl-ACP synthetase I in plants.
[b] NAD(P)(H), nicotinamide adenine dinucleotide (phosphate), (reduced).

Table 11.3 A comparison of type I fatty acid synthetases

Property	Animal	Yeast	Mycobacteria
Molecular mass (Da)	0.4×10^6 to 0.5×10^6	2.4×10^6	2.0×10^6
Subunit arrangement	(A_2)	(A_6B_6)	(A_{6-8})
Malonyl (palmitoyl) transacylase	Separate catalytic sites	Same catalytic site	Same catalytic site
End-product	Unesterified fatty acid	Acyl-CoA	Bimodal distribution of C_{16} and C_{24} fatty acids

adipose and lactating mammary glands, rabbit mammary gland and uropygial gland have been purified and found to have many common features (Table 11.3). The two subunits appear to be identical and of molecular mass 200–250 kDa (cf. Wakil *et al.*, 1983). In contrast, the type I synthetase of yeast is not only larger (2.4×10^6 Da) than those of animals (0.4×10^6 to 5×10^6 Da) but contains two non-identical subunits of molecular masses 213 and 203 kDa. The native yeast synthetase is, thus, an $A_6 B_6$ complex (cf. Schweizer *et al.*, 1973).

Partial reactions for the yeast type I synthetase were studied in a classic series of experiments by Lynen (1967). Peptides capable of these partial reactions have been isolated from yeast and chicken liver enzymes (see Wakil *et al.*, 1983). Particular interest has been focused on the thioesterase. This enzyme, more easily isolated by partial proteolysis than those for most of the other partial reactions, is important in at least partly regulating chain-length termination. Thioesterases isolated from several animal fatty acid synthetases (FASs) have very similar amino acid sequences around the active serine residue (Poulose *et al.*, 1981). The medium-chain fatty acids produced by synthetases from some mammary glands appear to be due to thioesterase II (a second thioesterase) (Libertini and Smith, 1978). The complementary DNA (cDNA) encoding the thioesterase I of the rat mammary gland fatty acid synthetase was isolated and sequenced. It exhibits a low but discernible homology to the medium-chain *S*-acyl fatty acid synthetase thioester hydrolases (thioesterase II) from rat mammary gland and duck uropygial gland, suggesting a distant but common evolutionary ancestry for these proteins (Naggert *et al.*, 1988).

The ACP region of the rat mammary gland FAS was then sequenced. The sequence was flanked at the 3′ end by a region coding for the thioesterase domain and at the 5′ end by a sequence coding for the reductase. The ACP domain only exhibited about a 25% homology with that of the discrete mono-functional ACPs of *E. coli*, spinach and barley (Witkowski *et al.*, 1987). Rat genomic genes encompassing the entire fatty acid synthetase gene have been isolated (Amy *et al.*, 1989) and allowed identification of the promoter region. The 87-nucleotide-long 5′-untranslated messenger RNA (mRNA) sequence was the same for rat liver, lung and mammary gland FAS (Amy *et al.*, 1990). Based on available evidence, the sequence of domains for the mammalian FAS has been proposed (Witkowski *et al.*, 1991) (Fig. 11.1), and this has been confirmed by research on the complete intron-exon organization of rat mammary gland FAS (Amy *et al.*, 1992). Genes for the mammary gland thioesterase I and

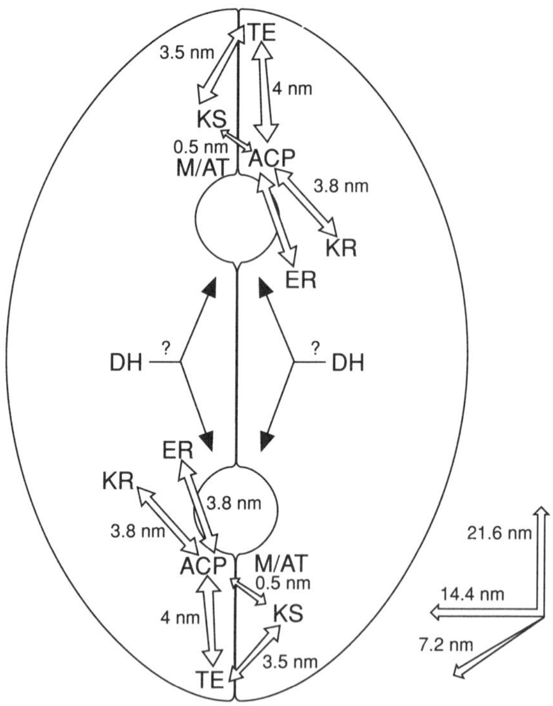

Fig. 11.1 Model of the dimeric animal fatty acid synthetase. The order and location of the functional domains are taken from data presented in the original paper. The overall shape and dimensions of the dimer were established by electron microscopic imaging, and the distances between active sites of the various domains by fluorescence resonance energy transfer measurements. ACP, acyl carrier protein; KS, oxoacyl synthetase; TE, thioesterase; M/AT, malonyl/acetyltransferase; DH, dehydrase; KR, oxoacyl reductase; ER, enoyl reductase. From Witkowski *et al.* (1991).

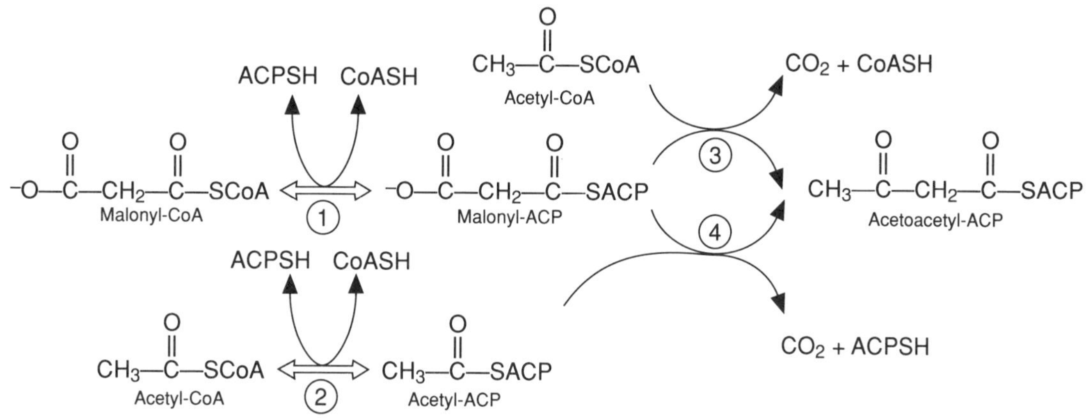

Fig. 11.2 The initiation phase of fatty acid biosynthesis. The initiation of new acyl chains is accomplished by the action of four enzymes: (1) malonyl-CoA:ACP transacylase; (2) acetyl-CoA:ACP transacylase; and (3) acetoacetyl-ACP synthetase or, alternatively, (4) 3-ketoacyl-ACP synthetase. From Jackowski *et al.* (1991).

thioesterase Ii components of FAS have been cloned and expressed under the control of the temperature-sensitive repressor (Naggert *et al.*, 1991).

The type I fatty acid synthetase of *Mycobacterium smegmatis* is unusual in several respects. First, the two reductases have different reduced pyridine nucleotide specificities: the β-ketoacyl-ACP reductase requires NADPH and the enoyl-ACP reductase requires NADH, whereas the reductases of other type I synthetases only use NADPH. The products of the *M. smegmatis* synthetase have a bimodal distribution with peaks at C_{16} and C_{24}, and the overall rate of fatty acid synthesis is increased by two types of polymethylsaccharides found in the mycobacterial cell wall (Bloch, 1977).

The type II synthetase of *E. coli* was the first FAS studied. The individual proteins have all been isolated, purified and characterized (see Volpe and Vagelos, 1973). Many of the genes coding for these proteins have also been identified (Jackowski *et al.*, 1991). The recent demonstration of a new short-chain condensing enzyme (acetoacetyl-ACP synthetase) (Jackowski and Rock, 1987) that by-passes the acetyl-CoA:ACP transacylase reaction (Fig. 11.2) is particularly interesting, since this reaction may regulate the overall rate of fatty acid synthesis. The *E. coli* FAS is also interesting in that it can produce mono-unsaturated (mainly *cis*-vaccenic) acids as well as saturated products. The ratio of the three principal products – palmitate, palmitoleate and *cis*-vaccenate – is controlled by the activity of the enzymes shown in Fig. 11.3. Regulation of the pathway is discussed well by Jackowski *et al.* (1991).

The fatty acid synthetase of *Brevibacterium*

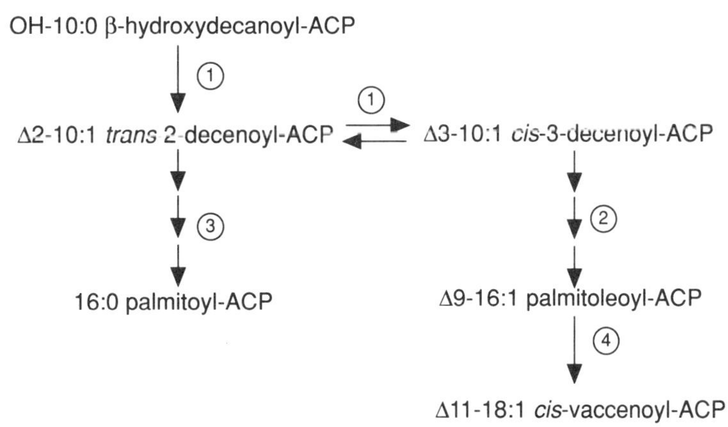

Fig. 11.3 Product diversification in fatty acid biosynthesis. Three main fatty acids are produced by the *E. coli* fatty acid synthetase system. The ratio of these fatty acids is controlled by the activity of three enzymes: (1) 3-hydroxydecanoyl-ACP dehydrase (encoded by the *fabA* gene) is a specific dehydrase that introduces the double bond into the acyl chain; (2) 3-ketoacyl-ACP synthetase I (encoded by the *fabB* gene) catalyses an essential step in the unsaturated fatty acid elongation pathway; (3) both 3-ketoacyl-ACP synthetases I and II can elongate saturated fatty acids; and (4) 3-ketoacyl-ACP synthetase II (encoded by the *fabF* gene) is responsible for the elongation of Δ9-16:1 to Δ11-18:1. From Jackowski *et al.* (1991).

ammoniagenes has a high molecular mass (like type I synthetases) but produces both saturated and unsaturated acids (like *E. coli* type II synthetase). In most Gram-positive and some Gram-negative bacteria, branched-chain fatty acids predominate. These branched-chain acids are made by type II synthetases that use short-chain branched acyl-CoA primers instead of acetyl-CoA and so produce *iso* or *anteiso* fatty acid products (Kaneda, 1977).

Plants also contain a type II FAS. The enzymology of the complex has been reviewed (Stumpf, 1987; Harwood, 1988). Like *E. coli*, there are several condensing enzymes. β-Ketoacyl-ACP synthetase I (KAS I) is responsible for forming keto acids of up to 16 carbons. KAS II then allows the condensation of palmitoyl-ACP with malonyl-ACP and, therefore, controls the production of stearate. These enzymes differ in their sensitivity to cerulenin and arsenite (see Harwood, 1988). A third condensing enzyme KAS III has recently been discovered and, like its counterpart in *E. coli* is responsible for acetoacetyl-ACP formation (Jaworski *et al.*, 1989; Walsh *et al.*, 1990).

Although it is believed generally that *de novo* synthesis of fatty acids in plants is concentrated in plastids (see Harwood, 1988), the recent report of acyl-ACP in plant mitochondria (Chuman and Brody, 1989), together with fatty acid synthesis mediated by ACP in *Neurospora crassa* mitochondria (Mikolajczyk and Brody, 1990), raise the possibility for some activity in this organelle. Much attention has been paid to plant ACP. Earlier work on its purification and properties has been well reviewed (Ohlrogge, 1987). ACP is synthesized in the cytosol and post-translationally imported into plastids, where proteolytic processing to the mature protein takes place. The attachment of the prosthetic group needs holo-ACP synthetase, which is cytosolic (El Hussein *et al.*, 1988). Recent aspects of the biochemistry of plant ACP are covered by Ohlrogge *et al.* (1991) and by Slabas and Fawcett (1992).

All of the proteins catalysing the partial reactions of plant FAS have been purified and, in many cases, genes coding for them have been identified and sequenced. Most of these proteins occur as isoforms (see Harwood, 1988). Recent updates on the genetics of fatty acid synthesis (Ohlrogge *et al.*, 1991) and its molecular biology (Slabas and Fawcett, 1992) have been published. An aspect that has received much attention recently is the mechanism of generation of medium-chain acids such as in coconut, *Cuphea* or palmkernel (see Chapter 3). A medium-chain thioesterase was purified first from the California bay (*Umbellularia california*) (Pollard *et al.*, 1991) and its cDNA cloned. This cDNA has now been expressed in *E. coli* and developing oil seeds, and substantial laurate production induced (Voelker *et al.*, 1992).

The regulation of fatty acid synthesis in plants is not defined at present, although changes in major acyl-ACP intermediates in light- and dark-treated leaf material have shown an increase in acetyl-ACP in the dark, which is consistent with acetyl-CoA carboxylase controlling the overall rate of fatty acid biosynthesis. Other aspects of regulation are discussed by Harwood (1988), Browse and Somerville (1991) and Slabas and Fawcett (1992). By contrast, fatty acid synthetase in animals may be subjected to short-term regulation by stimulation with phosphorylated sugars such as fructose 1,6-diphosphate or by inhibition by palmitoyl-CoA (Wakil *et al.*, 1983). It is also certainly regulated by adaptive changes in enzyme content (cf. Wakil *et al.*, 1983).

Euglena gracilis is an interesting organism because, when it grows heterotrophically (when it is 'animal-like'), it contains a type I synthetase. When it is grown photoautotrophically, it contains, in addition to its cytoplasmic type I enzyme, a type II synthetase in its chloroplasts (Ernst-Fonberg and Bloch, 1971). For more recent work, see Worsham *et al.* (1988).

11.1.2 Elongation systems

As mentioned in section 11.1.1, the elongation of fatty acids is generally carried out by type III fatty acid synthetases, which use malonyl-CoA as the source of C_2 units for addition.

In plants, the elongation of stearate to form very-long-chain saturated fatty acids (which are precursors of the various components of waxes, cutin and suberin: see Kolattukudy, 1980, 1987) takes place via several chain-length-specific systems (Walker and Harwood, 1986). Malonyl-ACP and NADPH are required (see Harwood, 1988) and a cycle of condensation, reduction, dehydration and second reduction reactions using acyl-CoAs established. Proteins catalysing these partial reactions have been demonstrated in leek (Lessire *et al.*, 1989) and *Lunaria annua* (Fehling *et al.*, 1992). In the latter case, mono-unsaturated acyl-CoAs are also substrates, and the elongation of such acids is important in brassicas such as rape, where erucate (Δ13-22:1) is a major component of the seed oil (section 3.3.27). The genetics of the rape elongation system are well described by Ohlrogge *et al.* (1991). Elongation in jojoba (which accumulates lipid as wax esters) uses a system with oleoyl-CoA and malonyl-CoA as substrates, and this plant has been studied in some detail (Pollard and Stumpf, 1980).

Formation of the very-long-chain saturated fatty acids involved in the surface coverings of plants is subject to inhibition by thiocarbamate herbicides (see Harwood, 1990), probably as their sulphoxide metabolites (Abulnaja and Harwood, 1991; see section 12.8).

Very-long-chain fatty acids are important components of the lipids of mammalian brain. Their synthesis has been studied by a number of groups, and key features appear to be the use of acyl-CoA substrates, of NADPH for reduction, of malonyl-CoA as the addition unit and the membranous nature of the elongation systems (Bourre *et al.*, 1975). Several elongases have been detected (Goldberg *et al.*, 1973). Fatty acids can also be elongated in mammals by a mitochondrial system, which uses acetyl-CoA as the unit for C_2 addition and NADH for reduction. In general, the mitochondrial system elongates fatty acids in the range C_{10}–C_{14} whereas the microsomal system uses C_{16} and longer acids. At first it was thought that the mitochondrial elongation system could operate by a reversal of β oxidation, but this is now not considered thermodynamically feasible. Indeed, the (flavin adenine dinucleotide) FAD-dependent acyl-CoA dehydrogenase of β oxidation is substituted by a more thermodynamically favourable enzyme, enoyl-CoA reductase (which is rate-limiting for the overall process in mitochondria; Cook, 1991).

The animal microsomal system contains at least two condensing enzymes – one for saturated and one for unsaturated primers. The dehydrase has been substantially purified and the four reaction intermediates of elongation detected. Although most studies have utilized liver, experiments with mice brains (including those from the Quaking mutant) have revealed chain-length-specific elongation systems. For more details, see Cook (1991) and references therein.

In contrast to animals and plants, bacteria do not usually contain significant amounts of acids longer than C_{18}. A notable exception are the long-chain (up to C_{56}) fatty acids found in mycobacteria. These are formed by an elongation system using acetyl-CoA and NADH, which again may be a reversal of β oxidation (Harwood and Russell, 1984).

11.1.3 Desaturases

Unsaturated fatty acids can be produced by anaerobic or aerobic pathways – the latter being the most usual mechanism. Aerobic desaturases have been studied in a large number of cases, although successful purifications have only been made a few times. Earlier work has been reviewed by Gurr (1974).

The anaerobic method, which is used by many members of the Eubacteriales, including all the anaerobes as well as some aerobes and facultative aerobes, has already been summarized in section 11.1.1. It relies on the activity of the β-hydroxydecanoyl-ACP β,γ-dehydrase, whose isomerase activity allows the synthesis of a *cis*-3-decenoyl-ACP from the usual *trans*-2-decenoyl-ACP. *cis*-3-Decenoyl-ACP cannot be reduced by the enoyl-ACP reductase but can be condensed by the β-keto-acyl-ACP synthetase. This reaction, therefore, allows chain lengthening but preserves the $n-7$ *cis* double bond and results in the synthesis of a series of monoenoic fatty acids (Fig. 11.4).

Details of the enzymes and the position of the genes involved in their synthesis are summarized in Jackowski *et al.* (1991). The ratio of the three main products of *E. coli* FAS is controlled by the relative activity of three enzymes, as illustrated in Fig. 11.3.

Aerobic desaturation involves the stereospecific removal of two hydrogen atoms from an acyl chain. Along with reducing equivalents from NAD(P)H or other reductants, the hydrogens are used to reduce molecular oxygen to water. The desaturases are membrane-bound multienzyme complexes, with the exception of the stearoyl-ACP desaturase found in chloroplasts. The basic details of aerobic desaturation appear to be the same in all organisms in that oxygen and a reduced cofactor are necessary, although the nature of the carriers varies in different systems (Fig. 11.5) as well as their susceptibility to cyanide inhibition. There is also some controversy as to whether the two hydrogens are removed sequentially (with the involvement of a hydroxy intermediate) or in a concerted mechanism. Evidence with *Corynebacterium diphtheria* using tritiated substrates suggested a stepwise mechanism (Schroepfer and Bloch, 1965), whereas experiments with ^{2}H-labelled substrates and *Chlorella vulgaris* indicated concerted hydrogen removal (Morris, 1970).

The first double bond inserted into an acyl chain is usually $\Delta 9$. The stearoyl-CoA $\Delta 9$-desaturase was first purified from rat liver (Strittmatter *et al.*, 1974). The complex consists of three major proteins: NADH-cytochrome b_5 reductase, cytochrome b_5 and a terminal desaturase (or cyanide-sensitive) protein. Usually the activity of the terminal desaturase limits the overall speed of desaturation. The NADH-cytochrome b_5 is a flavoprotein of mass 43 kDa, cytochrome b_5 has a molecular mass of 16.7 kDa and the terminal desaturase is 53 kDa (see Cook, 1991). The approximate location of the three components in the endoplasmic reticulum has been deduced from their amino acid sequence and the use of specific chemical reagents.

Further work on the stearoyl-CoA $\Delta 9$-desaturase

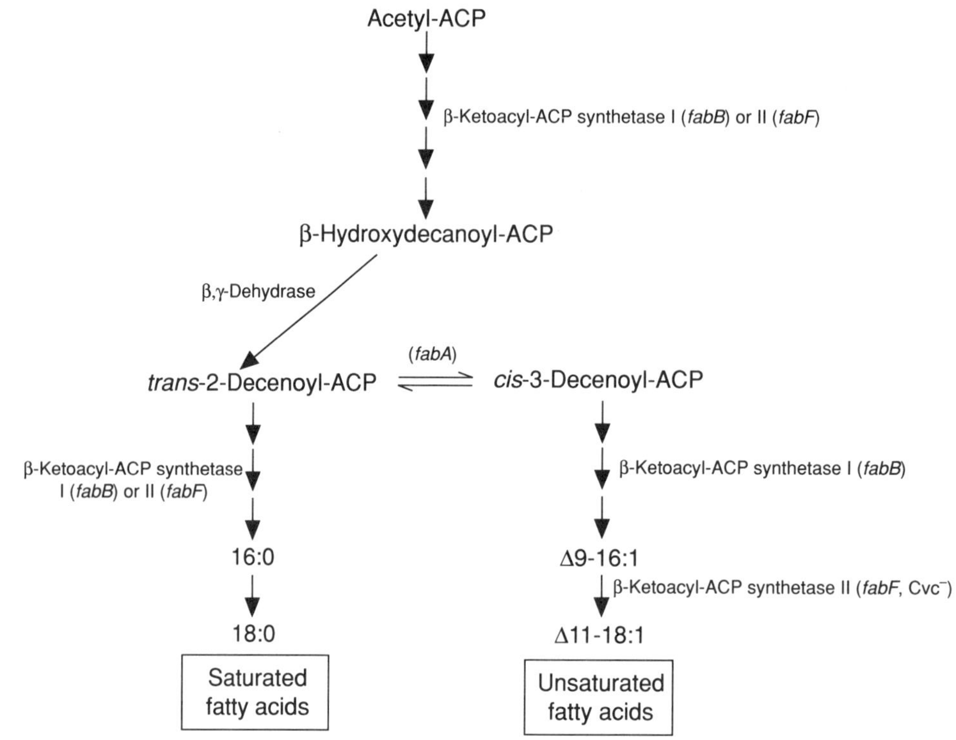

Fig. 11.4 Anaerobic pathway of fatty acid biosynthesis in bacteria showing mutants of *Escherichia coli*. Here *fab* refers to deficient mutants described for *E. coli*. Reproduced from Harwood and Russell (1984) with permission.

has included mechanistic studies on the protein–protein interactions involved in the supply of electrons from NADH (Dailey and Strittmatter, 1979) or NADPH (Dailey and Strittmatter, 1980), induction and characterization of the mRNA coding for the desaturase (Thiede and Strittmatter, 1985), sequencing the cDNA for the enzyme (Thiede *et al.*, 1986) and bacterial synthesis of active liver desaturase (Strittmatter *et al.*, 1988).

The Δ9-desaturases from plants and algae (and

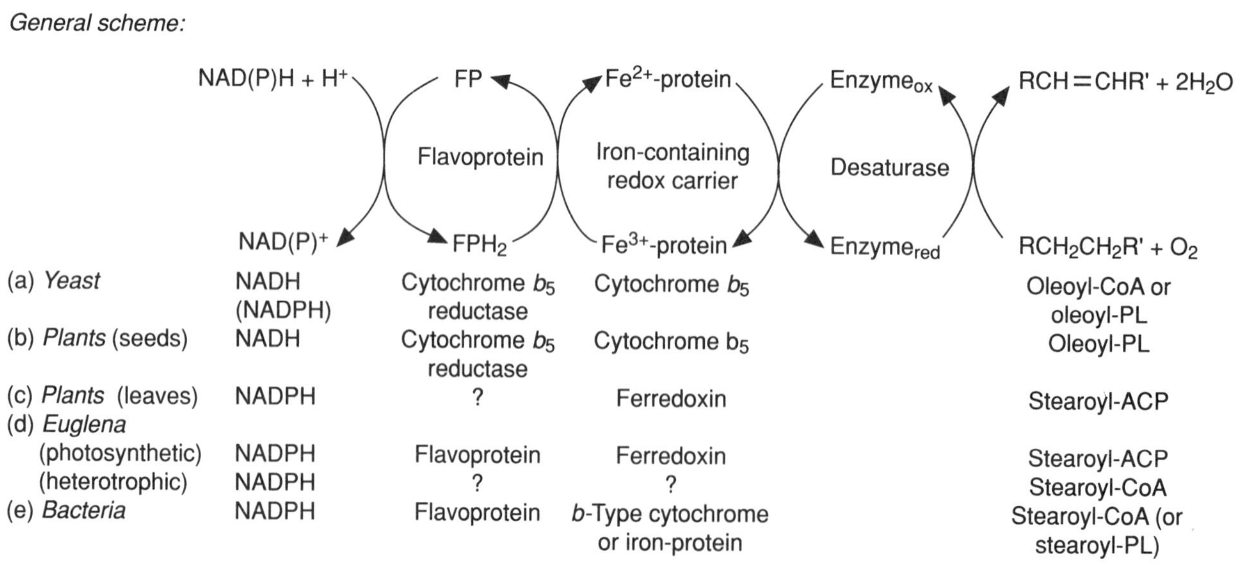

Fig. 11.5 Examples of aerobic desaturation. Abbreviations: FP, flavoprotein; ox, oxidized; red, reduced; R, alkyl chain; R′, acyl chain; PL, phospholipid. Reproduced from Harwood and Russell (1984) with permission.

Euglena gracilis) use stearoyl-ACP (the final product of their FASs) as substrate. The enzyme was first purified substantially from developing safflower seeds by McKeon and Stumpf (1982). The Δ9-(stearoyl)desaturase has been cloned from castor seeds (Knutzon *et al.*, 1991) and from safflower (Thompson *et al.*, 1991). The cDNA from safflower includes a 33-amino-acid transit peptide. Modulation of the stearoyl-ACP desaturase level, using antisense technology in transgenic rapeseed, resulted in a marked decrease in oleate with a commensurate increase in stearate (Kridl *et al.*, 1991). For summaries of the stearoyl-ACP desaturase from plants refer to Jaworski (1987) or Harwood (1988). Mutations affecting the levels of oleic acid in crop plants are discussed in Ohlrogge *et al.* (1991).

In contrast to animals and plants, bacteria are unique in producing Δ10-monoenoic fatty acids. Bacilli commonly possess Δ5- or Δ10-desaturases, and *Bacillus licheniformis* contains both enzymes when it grows at low temperatures (Fulco, 1974). Under these conditions it synthesizes small amounts of Δ5,10-16:2. It will be noted that, although this acid is polyunsaturated, it does not possess the usual methylene-interrupted structure. In fact, a general distinction is often made between bacteria and other organisms in that the former are unable to synthesize methylene-interrupted polyunsaturated fatty acids. This is not always true. For example, the filamentous gliding bacteria (*Flexibacter* spp.) contain considerable amounts of $(n-3)$-20:5, which in some species is the major fatty acid (Johns and Perry, 1977).

Other features of aerobic desaturation in bacteria are discussed by Schweizer (1989). However, it should be noted that the 'anaerobic' pathways utilizing fatty acid synthetase is characteristic of many bacteria, particularly those of the orders Pseudomonadales and Eubacteriales.

Most mammalian tissues can modify acyl chain composition by introducing more than one bond. Like Δ9-desaturation, further desaturation requires molecular oxygen and an associated electron transport system. Animal systems cannot generally insert double bonds beyond the Δ9 position. Consequently, double bonds are inserted at the Δ6, Δ5 and Δ4 positions. However, some species of insect have been shown recently to synthesize the 'essential fatty acid' linoleate from oleoyl-CoA (de Renobales *et al.*, 1987). Polyunsaturated fatty acids in all organisms usually contain methylene-interrupted double bonds and conjugated systems are rare (Cook, 1991). The animal desaturases are also used to further modify polyunsaturated fatty acids from the diet (e.g. linoleate, α-linolenate), and the major pathways are depicted in Fig. 11.6. The metabolism of acids of the $n-6$ family (linoleate, etc.) and the $n-3$ family (α-linolenate) and competition between the pathways are summarized by Cook (1991).

In contrast to the stearate desaturases discussed above, the enzymes forming linoleate and α-linolenate in lower and higher plants use complex lipids as substrates. The use of phosphatidylcholine in this way was proposed originally following experiments with the yeast *Candida albicans* (Pugh and Kates, 1975) and with the green alga *Chlorella vulgaris* (Nichols *et al.*, 1967). However, some yeasts contain both oleoyl-CoA and oleoyl-phospholipid desaturases (Pugh and Kates, 1975).

In plants, desaturation of oleate to linoleate uses phosphatidylcholine (in the endoplasmic reticulum) as the major substrate. The enzyme in safflower has been shown to use cytochrome b_5 as an electron source (Kearns *et al.*, 1991; Smith *et al.*, 1992). However, oleate desaturation in chloroplasts uses ferredoxin (Schmidt and Heinz, 1990). Other complex lipid substrates are also probably used (see Harwood, 1988), especially as demonstrated for chloroplast lipids (see Jones and Harwood, 1980). Further desaturation of linoleate to α-linolenate can take place either using monogalactosyldiacylglycerol (Jones and Harwood, 1980) or with phosphatidylcholine. The relative importance of these two substrates depends on the type of tissue: leaves use monogalactosyldiacylglycerol mainly, but seeds (which have poorly developed plastids) use the non-chloroplastic phosphatidylcholine as the main substrate. These aspects are discussed by Harwood (1988) and by Browse and Somerville (1991). The use of complex lipids for substrates, the cooperation of extraplastidic with plastid pathways and the site of *Arabidopsis* mutations in the pathways are shown in Fig. 11.7.

Some recent exciting work on linoleate desaturase in cyanobacteria has been carried out by Murata and colleagues. They selected four mutants of *Synechocystis* PCC 6803 that were defective in desaturation and were chilling-sensitive (Wada and Murata, 1989). By complementation experiments they cloned a gene that could restore growth at low temperatures and restored Δ12-desaturation. This gene was transferred to *Anacystis nidulans* (which does not normally produce polyunsaturated fatty acids), where it allowed linoleate formation and increased chilling-tolerance (Wada *et al.*, 1990).

Some primitive animals (protozoa) have the ability to desaturate oleate to linoleate amongst their other desaturation reactions. In both *Tetrahymena* (Umeka and Nozawa, 1984) and *Acanthamoeba castellanii* (Jones *et al.*, 1992), phosphatidylcholine appears to be the main substrate.

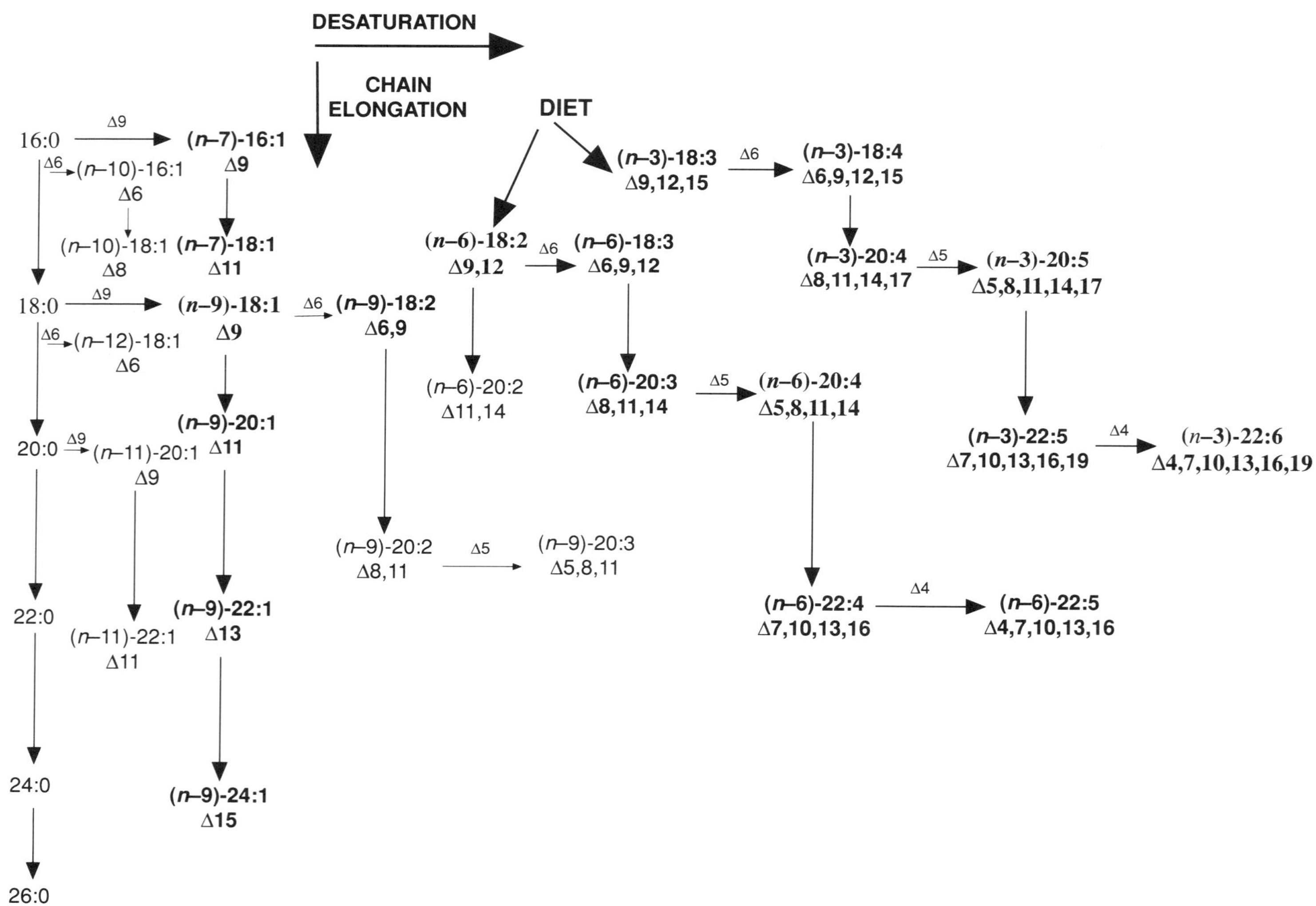

Fig. 11.6 Major pathways of fatty acid biosynthesis by desaturation and chain elongation in animal tissues. Note the alternating sequence of desaturation in the horizontal direction and chain elongation in the vertical direction. Reduced intensity of the arrows distinguishes minor pathways. Boldness of type and size for individual fatty acids reflects, in a general way, relative accumulation in tissues. From Cook (1991).

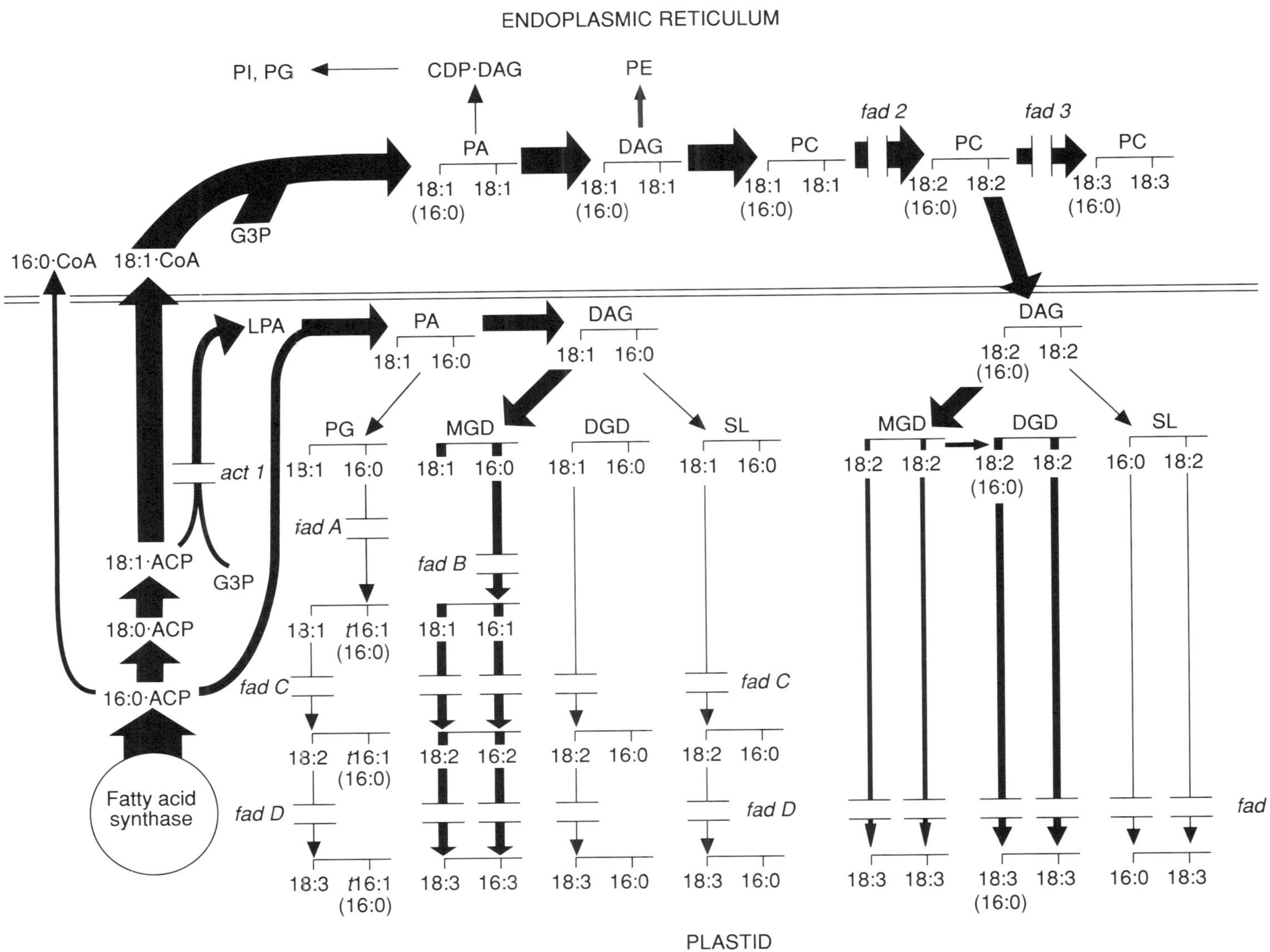

Fig. 11.7 An abbreviated diagram of the two-pathway scheme of glycerolipid biosynthesis in the 16:3 plant *Arabidopsis*. Widths of the lines show the relative fluxes through different reactions. From Browse and Somerville (1991).

11.1.4 Hydroxylation

Hydroxy fatty acids are formed as intermediates in various metabolic sequences (e.g. fatty acid biosynthesis, β-oxidation) as a result of specific hydroxylation reactions and following other activities such as those of lipoxygenase (see section 11.1.6). The hydroxyl group is usually introduced close to one end of the acyl chain (e.g. α, β) and less commonly in the middle of the chain. A good example of the latter in plants is the formation of ricinoleic acid (the major acid of castor oil, section 3.3.4) by hydroxylation of oleoyl-phosphatidylcholine substrate using NADH and oxygen as cofactors (Moreau and Stumpf, 1981; Smith *et al.*, 1992).

Few of the hydroxylases have been characterized, so it is difficult to compare their properties. However, they appear to be mixed-function oxidases and require molecular oxygen and a reduced pyridine nucleotide. Most of the enzymes appear to require cytochrome P_{450}, although ferredoxin (or a related haem protein) may substitute in some bacterial or plant systems (see Galliard, 1980). Cell-free hydroxylating systems have been studied extensively in *Bacillus megaterium* and *Pseudomonas oleovorans* and some enzymes purified (see Schweizer, 1989).

α-Oxidation systems producing 2-hydroxy fatty acids have been demonstrated in yeasts, bacteria and plants (see Harwood and Russell, 1984). ω-Oxidation systems introduce a hydroxyl group to the methyl end of the acyl chain. These oxidations are described more fully in section 11.1.5.

The major hydroxy fatty acids in plants have an ω-OH and an in-chain OH group (e.g. 10,16-dihydroxypalmitic acid). Their synthesis seems to involve ω-hydroxylation with NADPH and O_2 as cofactors, followed by in-chain hydroxylation using the same co-substrates. If the precursor is oleic acid, then the double bond is converted to an epoxide, which is then hydrated to yield 9,10-hydroxy groups. These conversions involve CoA esters. In-chain plant hydroxylase is sensitive to inhibition by *o*-phenanthroline and by CO in a reaction that is reversed by 420–460 nm light (Kolattukudy, 1980, 1987).

It should be noted that a few hydroxylations occur without a mixed-function oxidase enzyme. For example, the ergot fungus *Claviceps purpurea* forms 12-hydroxyoleic (ricinoleic) acid by hydration of linoleic acid (Harwood and Russell, 1984).

11.1.5 Oxidation of fatty acids

Oxidation of fatty acids can occur in a number of ways, depending on the position of oxidation (i.e. α-, β- or ω-oxidation) and the nature of the substrate (e.g. lipoxygenase attack on polyenoic fatty acids).

(a) α-Oxidation

The removal of a single carbon atom from the carbonyl end of a fatty acid is carried out by α-oxidation. This process is particularly active in plants but is also found in mammals (notably brain tissue) and bacteria. The removal of a single carbon may be important when degradation of a fatty acid by β-oxidation (see below) is blocked by the presence of a methyl branch. A medical implication of this is Refsum's disease (section 12.6). Furthermore, α-oxidation is also important because α-hydroxy fatty acids are intermediates in the process and these acids are components of certain sphingolipids (Bowen *et al.*, 1974). Thus, brain cerebroside fatty acids are highly enriched in cerebronic acid (α-OH 24:0). Furthermore, tissues that accumulate large amounts of fatty acids almost invariably also contain significant amounts of odd-chain-length fatty acids (e.g. McIlwain, 1966).

In plants the mechanism of α-oxidation in leaves and seeds is identical. The fatty acid substrate is non-esterified (cf. β-oxidation) and is usually C_{12}–C_{18}. It is attacked by molecular oxygen to generate an unstable 2-hydroperoxy intermediate (Fig. 11.8),

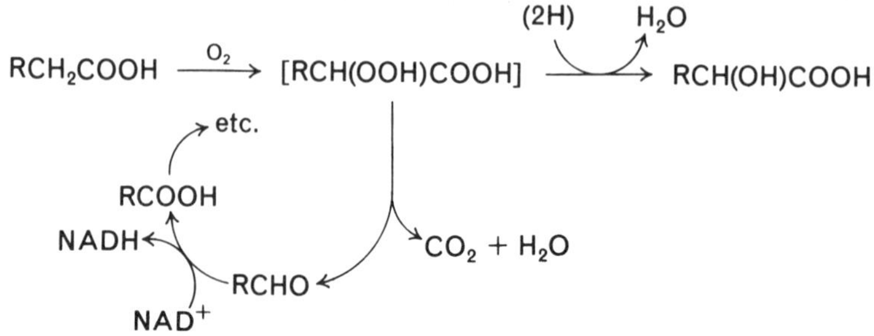

Fig. 11.8 The α-oxidation of fatty acids in plants. Adapted from Galliard (1980).

which decomposes to an aldehyde with release of carbon dioxide. Under certain conditions (for example, in the presence of an enzyme such as glutathione peroxidase, which will reduce peroxides), then a D-2-hydroxy fatty acid may be produced, which cannot be metabolized easily. Under normal conditions, though, the aldehyde is oxidized in the presence of a source of reducing power (pyridine nucleotide or flavoprotein depending on the tissue) to give a fatty acid one carbon atom shorter than the original fatty acid. For a review of plant α-oxidation and the properties of various individual systems, see Galliard (1980).

The occurrence of bacterial fatty acid α-oxidation is not firmly established and, if present, represents a pathway of minor importance (Finnerty, 1989). It has been studied in few bacterial species. However, in *E. coli* the D-2-hydroxy fatty acid is preferentially decarboxylated, unlike in plants and animals where it is the L form that is metabolized preferentially. Thus, the 2-hydroxy acids tend not to accumulate in bacteria (Lekakis, 1977).

(b) β-Oxidation

The β-oxidation of fatty acids was the first metabolic process in which labelled compounds were used for its investigation. Knoop's classic experiments at the turn of the century were later confirmed and extended by others to reveal the details of the process (for references see Greville and Tubbs, 1968; Wakil, 1970; Schulz, 1991a). Key steps in the pathway are the activation of the fatty acid to a coenzyme A thioester, the α,β-dehydrogenation of the acyl-CoA, the hydration of the resultant double bond, oxidation of the β-hydroxyacyl-CoA and thiolytic cleavage of the β-ketoacyl-CoA (Fig. 11.9).

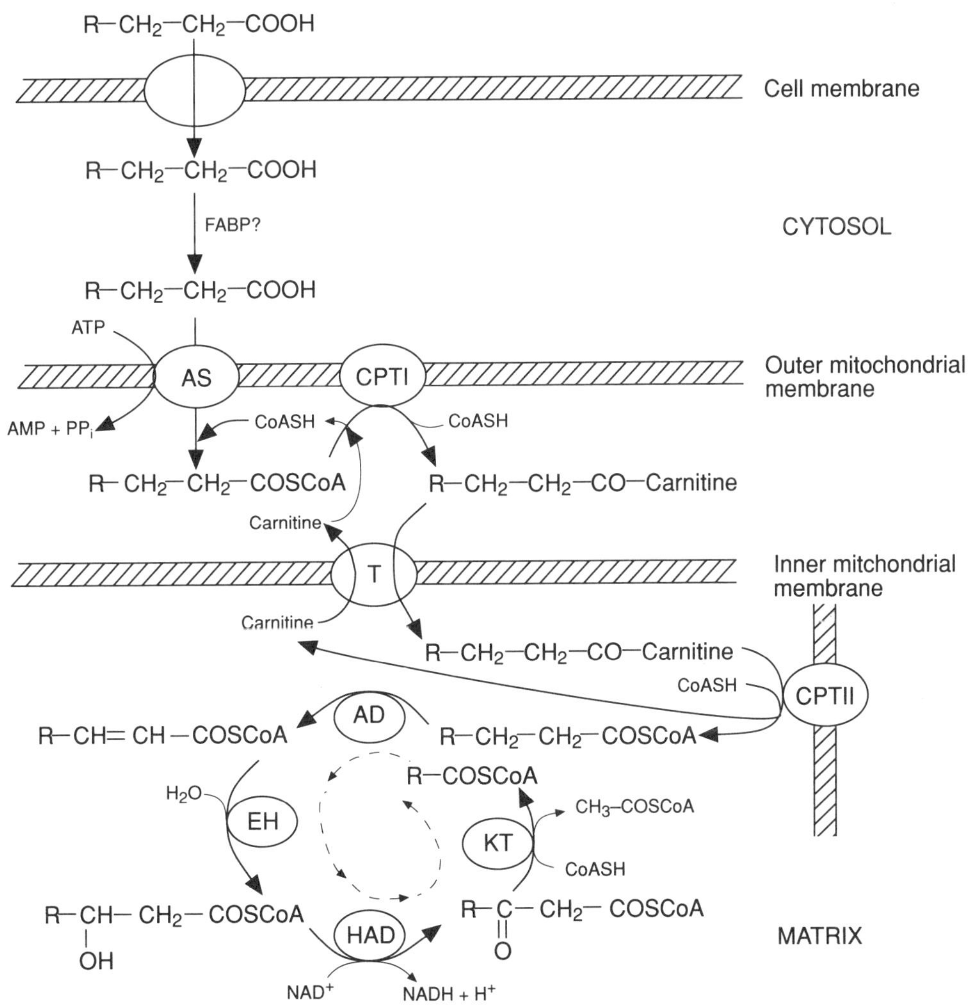

Fig. 11.9 Pathway of mitochondrial fatty acid oxidation. Enzymes of the pathway are: AS, acyl-CoA synthetase; CPT I, carnitine palmitoyltransferase I; T, carnitine:acylcarnitine translocase; CPT II, carnitine palmitoyltransferase II; AD, acyl-CoA dehydrogenase; EH, enoyl-CoA hydratase; HAD, L-3-hydroxyacyl-CoA dehydrogenase; KT, 3-ketoacyl-CoA thiolase. Other abbreviation: FABP, fatty acid binding protein. From Schulz (1991b).

The mechanism of fatty acid uptake by animal cells has not been fully elucidated. Plasma membrane carrier proteins responsible for saturable high-affinity uptake have been identified, but some workers argue that spontaneous and non-specific diffusion of fatty acids across the plasma membrane is also important (see Schulz, 1991b). β-Oxidation is preceded by activation of fatty acids to their coenzyme A (CoA) thioesters. A group of acyl-CoA synthetases have been identified that differ in their chain-length specificities but which catalyse the same type of reaction:

$$\text{R—COOH} + \text{ATP} + \text{CoASH} \rightleftharpoons \text{RCO—SCoA} + \text{AMP} + \text{PP}_i$$

The individual acyl-CoA synthetases are grouped as short-chain, medium-chain or long-chain synthetases depending on their substrate specificities. Short-chain synthetases, which act on acetate or propionate, are present in mitochondria of most animal tissues but, in addition, are found in the cytosol of lipogenic tissues like liver, intestine, adipose tissue and mammary gland. Medium-chain synthetases are also located in the mitochondrial matrix. By contrast, long-chain acyl-CoA synthetases in animals are membrane-bound and associated with the endoplasmic reticulum, mitochondrial outer membrane or peroxisomes (see Schulz, 1991b).

Acyl-CoAs that need to be taken up by mitochondria do so via a carnitine-dependent mechanism. Two carnitine palmitoyltransferases and a carnitine: acylcarnitine translocase are needed, as illustrated in Fig. 11.9.

Three enzymes (differing in chain-length specificity) that catalyse α,β-dehydrogenation of acyl-CoAs have also been purified. The overlapping specificities of these enzymes allow efficient oxidation of fatty acids in the C_4–C_{20} range (Greville and Tubbs, 1968). The acyl-CoA dehydrogenases (EC 1.3.99.3) are flavoproteins containing FAD (flavin adenine dinucleotide) as a prosthetic group. They have molecular masses of 170–190 kDa and are composed of four identical subunits, each of which carries a non-covalently bound FAD. Nucleotide sequences for several rat liver and human acyl-CoA dehydrogenases have been determined, with a greater than 90% homology observed for the same enzyme from rat compared to human (see Schulz, 1991b).

Hydration of the α,β-*trans* double bond is catalysed by enoyl-CoA hydratase (EC 4.2.1.17; crotonase). This enzyme will also attack *cis* double bonds and 3-enoyl-CoAs. An L-OH fatty acid is produced (4- in the latter case and 3- in β oxidation) from the *trans* isomers. For more details, see Stern (1961) and Schulz (1991b).

The L-3-hydroxyacyl-CoA obtained in the step is oxidized by NAD^+ in the presence of 3-hydroxyacyl-CoA dehydrogenase (EC 1.1.1.35). The enzyme is specific for the L form and usually has little chain-length specificity (Wakil, 1970). The enzyme has been purified from pig heart and rat liver. It has a molecular mass of 65 kDa, is composed of two identical subunits and its conformation at 2.8 Å resolution has been determined (Schulz, 1991b). Some enzymes will utilize $NADP^+$ more slowly and these dehydrogenases will also oxidize *S*-3-hydroxy-acyl-*N*-acylthioethanolamine and *S*-3-hydroxyacyl-hydrolipoate.

Thiolytic cleavage of the 3-ketoacyl-CoA is catalysed by acetyl-CoA acyltransferase (EC 2.3.1.16), which liberates acetyl-CoA and an acyl-CoA two carbons shorter than the original substrate (Fig. 11.9). The above enzyme has a broad chain-length specificity, but a second enzyme, acetyl-CoA acetyltransferase (EC 2.3.1.9), catalyses the same reaction with acetoacetyl-CoA as substrate. Kinetic and exchange studies indicate that these thiolase reactions occur in two stages (see Greville and Tubbs, 1968).

Both types of thiolases are homotetramers of mass about 170 kDa. While the 3-ketoacyl-CoA thiolase with its broad chain-length specificity is essential for β oxidation, the acetoacetyl-CoA thiolase probably functions mainly in ketone body metabolism (Schulz, 1991b).

All mitochondrial enzymes of β-oxidation are synthesized in the cytosol on free polysomes (see Ozasa *et al.*, 1984).

Oxidation of saturated straight-chain fatty acids proceeds smoothly using the above series of enzymes, but there are problems in the breakdown of other fatty acids. For example, odd-chain acids finally liberate propionyl-CoA, which is converted to malonyl-CoA for further metabolism. Unsaturated fatty acids must also be modified because their double bonds may be in the wrong position and because of incorrect configuration for normal catabolism. Epimerase and isomerase enzymes are present to cope with these problems (see Greville and Tubbs, 1968; Bressler, 1970; Fig. 11.10).

Apart from β oxidation in mitochondria, peroxisomes are also important. The reactions and enzymes involved have been characterized (see Hashimoto, 1990), with the research being aided by the use of drugs such as clofibrate or di(2-ethyl-hexyl)phthalate, which cause peroxisome proliferation and also induce the enzymes of peroxisomal β-oxidation (Lazarow and Moser, 1989). Although the intermediates of peroxisomal β-oxidation are the same as in mitochondria, there are differences in the enzymes concerned. The first reaction is

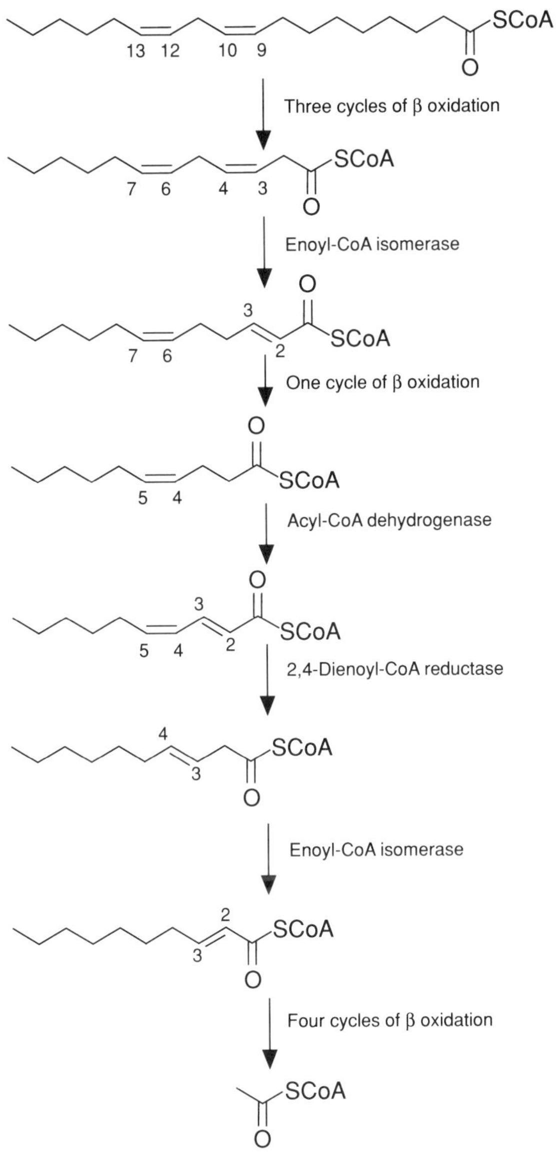

Fig. 11.10 The β oxidation of linoleoyl-CoA. From Schulz (1991b).

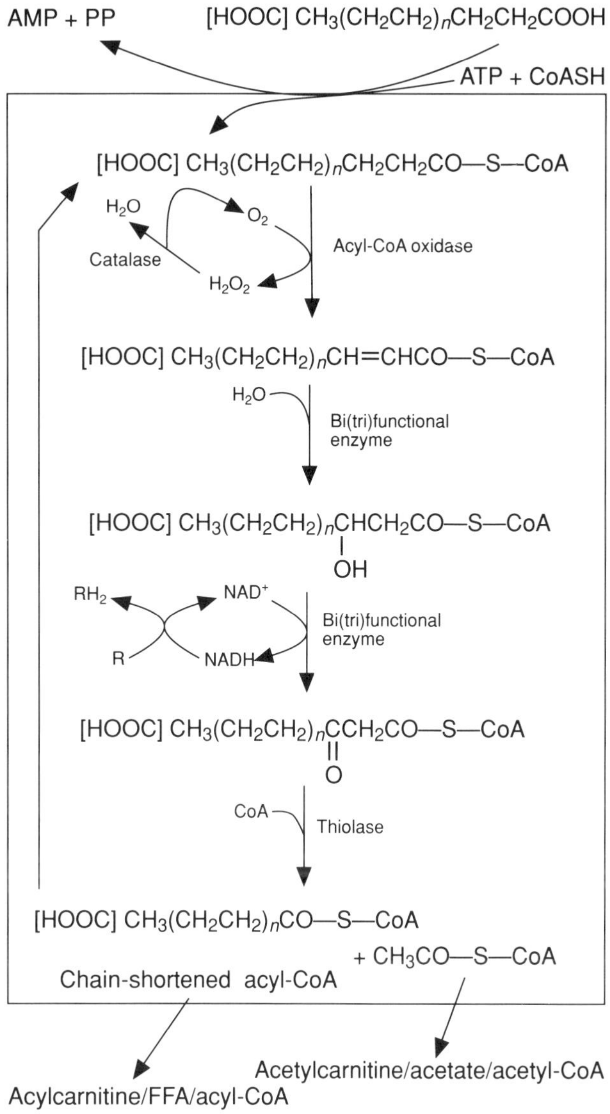

Fig. 11.11 Essential features of peroxisomal β oxidation. A schematic representation of the main metabolic features of peroxisomal β oxidation. The enzymes of peroxisomal β oxidation are contained within the peroxisomal membrane (rectangular box). Possible end-products of β oxidation are indicated at the bottom. The [HOOC] to the left of the substrate acyl-CoA ester indicates that mono-CoA esters of dicarboxylic acids are also substrates for peroxisomal β oxidation. An incorrect stoichiometry of catalase-dependent decomposition of H_2O_2 is used for the sake of brevity. The arrows pointing down to Acylcarnitine/FFA/acyl-CoA, or to Acetylcarnitine/acetate/acetyl-CoA are meant to indicate possible alternative forms of export of chain-shortened fatty acids out of the peroxisome (FFA = free fatty acid). The oxidant (R) of NADH active *in vivo* remains to be established. With isolated peroxisomes, pyruvate (and lactate dehydrogenase) can function in this capacity. From Osmundsen *et al.* (1991).

catalysed by acyl-CoA oxidase, which generates hydrogen peroxide as a product. The latter is rapidly destroyed by peroxisomal catalase activity. The hydration and second dehydrogenation steps are catalysed by a trifunctional enzyme. The essential features are illustrated in Fig. 11.11. Because the acyl-CoA oxidase is almost inactive towards octanoyl-CoA (or shorter chains), peroxisomes are incapable of complete oxidation of fatty acids. For the oxidation of unsaturated fatty acids such as linoleate, a 2,4-dienyl-CoA reductase is present and a Δ^3-*cis*,Δ^2-*trans*-enoyl-CoA isomerase (as the third activity of the trifunctional enzyme) is also needed (see Schulz, 1991b; Osmundsen *et al.*, 1991). The main function of peroxisomal β oxidation in animals seems to be in the chain shortening of very-long-chain fatty acids, prostaglandins, dicarboxylic acids and various xenobiotic compounds (see Schulz, 1991b).

In plants it was observed a number of years ago that peroxisomes, such as the glyoxysomes of germinating oil seeds, contained enzymes necessary for fatty acid β-oxidation (see Galliard, 1980; Kindl, 1987). Early aspects of the work were discussed by Beevers (1980). A thorough review has covered recent aspects of β-oxidation in plants (Gerhardt, 1992). As outlined in this review, there is little evidence for any other location for the process in plants apart from in peroxisomes.

Plants operate a different mechanism from animals when removing the propionate liberated from odd-chain-length fatty acid oxidation. The plant system involves a hydratase and two dehydrogenase enzymes, and the overall scheme is summarized in Fig. 11.12. The pathway is superficially similar to modified β-oxidation in bacteria, but the two systems differ in the fate of the individual carbon atoms (Stumpf, 1970).

Bacteria take up non-esterified fatty acids from the growth medium by a process that probably involves the formation of acyl-CoA. The fatty acid taken up, therefore, is in a form directly available for β-oxidation. In *E. coli* there has been some study of the genes coding for fatty acid degradation (*fad*). These genes, which code for the enzymes of uptake/activation and β-oxidation, are located in three sites on the chromosome and comprise a regulon (Klein *et al.*, 1971). Several types of *fad* mutants are known and another class of mutants, *fadR*, are constitutive for β-oxidation and are probably repressor mutants. In addition to being a repressor for *fad* genes, the *fadR* gene product may also control isocitrate lyase and malate synthetase (Maloy *et al.*, 1980), which are key enzymes of the glyoxylate shunt. There are some differences in the details of β-oxidation enzymes in *E. coli* compared to these from animals. For example, only one acyl-CoA synthetase is present and this can activate both medium-chain and long-chain fatty acids (Schulz, 1991a). Even though *E. coli* does not synthesize polyunsaturated fatty acids, it can easily oxidize them by the reductase-dependent pathway (above). Details of the enzymes involved in *E. coli* fatty acid β-oxidation and mutants available that are defective in individual steps are given by Finnerty (1989). The same author also discusses the features of peroxisomal β-oxidation in microbial eukaryotes such as *Candida tropicalis*.

(c) Inherited diseases of β oxidation

A number of diseases that compromise the functions of liver, muscle and other organs have been ascribed to deficiencies of β-oxidation. These include myopathic carnitine deficiency, deficiencies of short-chain, medium-chain or long-chain acyl-CoA dehydrogenases, and deficiencies in the electron transferring flavoprotein or its dehydrogenase. Several disorders (e.g. Zellweger syndrome) associated with impairment of peroxisomal β-oxidation have also been described. See Schulz (1991a) for more details.

(d) Ketone bodies

In mammalian liver, the excessive amounts of acetyl-CoA liberated by β-oxidation of fatty acid are converted to various ketone bodies. This conversion takes place at high rates when an elevation in the glucagon/insulin ratio occurs, such as during fasting or in uncontrolled diabetes. Several enzymes are involved, as shown in Scheme 11.1(i). These enzymes are also found in extrahepatic tissues such as heart, kidney and intestine.

The ketone bodies produced by various tissues can readily diffuse into the blood and be taken up by other extrahepatic tissues and converted back to acetyl-CoA for complete combustion in the tricarboxylic acid cycle.

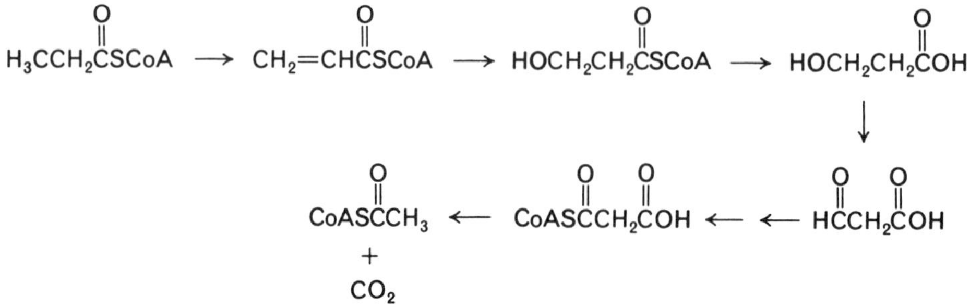

Fig. 11.12 Propionate metabolism in plants.

Scheme 11.1(i)

$$\text{2 Acetyl-CoA} \xrightarrow[\text{thiolase}]{\text{Acetoacetyl-CoA}} \text{Acetoacetyl-CoA + CoA}$$

$$\text{Acetoacetyl-CoA + Acetyl-CoA} \xrightarrow[\text{synthase}]{\text{HMG-CoA}} \text{Hydroxymethylglutaryl-CoA + CoA}$$

$$\text{Hydroxymethylglutaryl-CoA} \xrightarrow[\text{lyase}]{\text{HMG-CoA}} \text{Acetoacetate + Acetyl-CoA}$$

$$\text{Acetoacetate + NADH + } H^+ \xrightarrow[\text{dehydrogenase}]{\text{D-3-Hydroxybutyrate}} \text{3-Hydroxybutyrate + } NAD^+$$

A thorough review of these reactions and of the regulation of ketogenesis (see Fig. 11.13) has been provided by McGarry and Foster (1980).

Ketone bodies provide important alternative fuels to body tissues when carbohydrate is in short supply or cannot be efficiently utilized. A particular example is the central nervous system, which cannot utilize plasma fatty acids for energy. Thus, in prolonged starvation, ketone bodies become more important than glucose as a fuel source. The possibility of the utilization of ketone bodies obviates the harmful degradation of muscle protein for gluconeogenesis. In addition, acetoacetate and 3-hydroxybutyrate are thought to be important precursors for lipid synthesis in neonatal brain (Webber and Edmond, 1979).

Although ketone bodies serve useful functions, excessive accumulation of such compounds in blood can cause clinical problems such as ketoacidosis. This may be severe in diabetes or alcoholism. A comprehensive review of clinical aspects of ketone bodies is that of Soling and Seufert (1978).

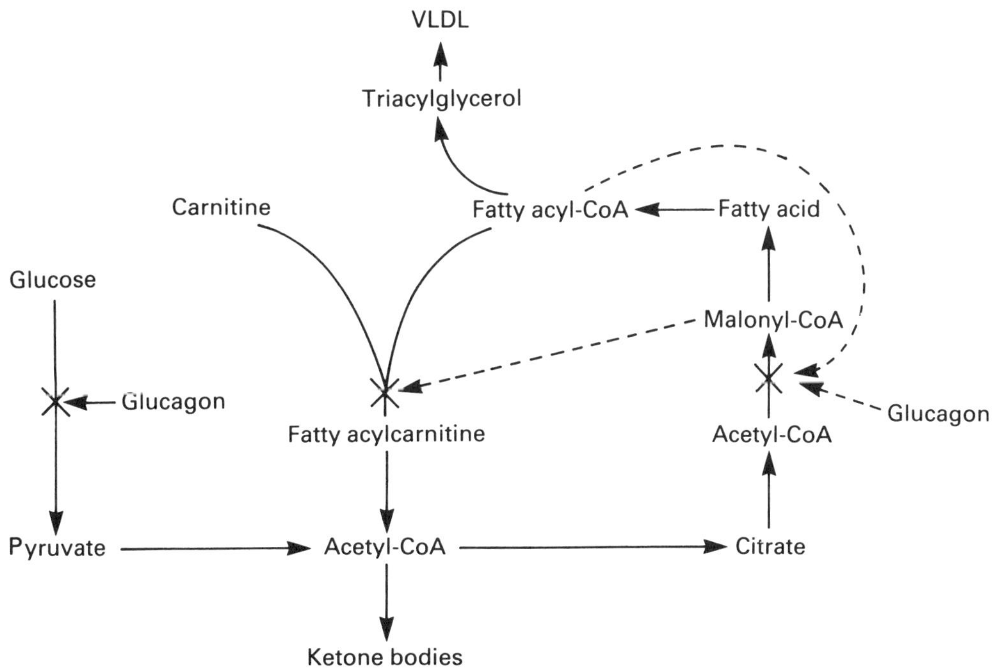

Fig. 11.13 Interactions between fatty acid synthesis and oxidation in liver. In the fed state, malonyl-CoA levels are high. This allows rapid fatty acid synthesis and inhibits β oxidation by lowering carnitine acyltransferase I activity. If triacylglycerol synthesis is impaired, then acyl-CoA will feedback to inhibit acetyl-CoA carboxylase. In the fed state, this does not normally happen, and triacylglycerols are incorporated into very-low-density lipoprotein (VLDL) for export to extrahepatic tissues. Glucagon excess in fasting leads to a suppression of glycolysis, cessation of lipogenesis and activation of β-oxidation and ketogenesis. From McGarry and Foster (1980). Reproduced with permission from *Annual Review of Biochemistry*, **49**, 395. © 1980 by Annual Reviews Inc.

(e) ω-Oxidation

Microsomal preparations from many tissues can oxidize fatty acids in the ω-position. An ω-hydroxy fatty acid is formed first, and this can be oxidized by NAD^+ and cytoplasmic enzymes to yield a dicarboxylic acid, which can be further attacked by β-oxidation (Greville and Tubbs, 1968). The enzyme(s) responsible for the oxidation of the ω-hydroxy fatty acid have been much less studied than the ω-hydroxylation system. There may be an ω-hydroxy acid dehydrogenase, although liver alcohol dehydrogenase can convert such acids to semialdehyde derivatives. Further oxidation of the latter would be catalysed by an NAD-limited aldehyde dehydrogenase (cf. Greville and Tubbs, 1968). ω-Oxidation is important for the further metabolism of fatty acids that are disubstituted in the 2 or 3 positions as well as the catabolism of various xenobiotics that have alkyl chains (see Lenk, 1972).

The role of peroxisomes in the ω-oxidation of fatty acids and xenobiotics is discussed by Osmundsen *et al.* (1991).

The ω-hydroxylase system (alkane 1-monooxygenase, EC 1.14.15.3) has non-haem irons as, apparently, the only prosthetic groups directly involved in the hydroxylation reactions. The system from *Pseudomonas oleovorans* has been isolated by Coon and co-workers (Kusunose *et al.*, 1964). The system was fractionated into three components: a non-haem iron protein (similar to the rubredoxins), a flavoprotein and a final component required for hydroxylase activity called the ω-hydroxylase. The latter has a molecular mass of 42 000 Da with one atom of iron. It also contains a large amount of phospholipid, and activity is lost if this lipid is removed. Other studies have included electron paramagnetic resonance (EPR) experiments on the spectra shown by the non-haem iron protein (for reviews see Coon *et al.*, 1972; Gunsalus *et al.*, 1975).

In plants, the ω-hydroxylase system is responsible for synthesis of ω-hydroxy fatty acyl components of cutin and suberin (sections 1.8 and 2.11). The reaction has been studied in preparations from *Vicia faba* with NADPH and oxygen as the required cofactors (Kolattukudy, 1980). The true substrate for ω-hydroxylation of palmitate is the free acid and the active subcellular preparation is the microsomal fraction. The reaction showed the properties of a classic mixed-function oxidase, being inhibited by *o*-phenanthroline, 8-hydroxyquinoline (metal-ion chelators), sodium azide and thiol-directed reagents. The involvement of cytochrome P_{450} in the *V. faba* system is unproven. Although the hydroxylation is inhibited by carbon monoxide, this inhibition was not reversed by light at 420–460 nm. Thus, if a cytochrome P_{450} is involved in the system, it must have unusual properties when compared to other cytochrome-P_{450}-containing enzymes (Kolattukudy, 1980).

(f) Fatty acid catabolism in ruminants

A specialized situation for lipid breakdown is the rumen of such animals as sheep and ox. A large number of micro-organisms such as bacteria of the genera *Ruminococcus*, *Bacteroides* and *Butyrivibrio* as well as protozoa play a part in the breakdown of leaf lipids (see section 3.10). Thus, α- and β-galactosidases are present to cleave the galactose residues of galactosylglycerides. Active lipases hydrolyse the acyl residues, and the liberated fatty acids are often biohydrogenated to give various mixtures of conjugated fatty acids containing *trans* as well as *cis* double bonds (Table 11.4); see Garton (1977), Harwood and Russell (1984) and Gurr and Harwood (1991). Fermentation results in the production of large quantities of acetic, propionic and butyric acids – the proportions of which vary with the ruminants' diet. These water-soluble short-chain fatty acids are absorbed well, and much of their total quantity is metabolized to ketone bodies in the rumen wall.

Other features of fatty acid metabolism in ruminants are discussed by Garton (1977).

11.1.6 Lipoxygenase

Lipoxygenase (linoleate:oxygen oxidoreductase, EC 1.13.11.12) is a term applied to a group of enzymes that catalyse the oxygenation, by molecular oxygen, of fatty acids containing a *cis*,*cis*-1,4-

Table 11.4 Major products of ruminant biohydrogenation of C_{18} unsaturated fatty acids

Substrate	Products
Oleic acid	18:0
Linoleic acid	9*c*,11*t*-18:2 11*t*-18:1 18:0
α-Linolenic acid	9*c*,11*t*,15*c*-18:3 11*t*,15*c*- and 9*c*,11*t*-18:2 11*t*-18:1 18:0

pentadiene system to produce conjugated hydroperoxydiene derivatives (see also section 10.3):

$$\mathrm{RCH{=}CHCH_2CH{=}CHR'} \xrightarrow{O_2} \mathrm{RCH{=}CHCH{=}CHCH(OOH)R'}$$

Other enzymes of this type have also been termed fat oxidase, carotene oxidase and lipoxidase (see Galliard and Chan, 1980).

Lipoxygenase activity is widespread in the plant kingdom, often in very high amounts. The enzymes are particularly important in the food plants, where they destroy the essential polyunsaturated fatty acids to produce derivatives with characteristic tastes and flavours. They are also used for bleaching natural pigments such as wheat flour carotenoids or alfalfa chlorophyll. General reviews on lipoxygenases have been prepared by Hamberg *et al.* (1974), Galliard (1974), Galliard and Chan (1980) and Veldink *et al.* (1977). A recent comprehensive review of the structure and function of plant lipoxygenase has been published (Siedow, 1991). Eskin *et al.* (1977) have discussed the commercial importance of lipoxygenases, and further references on this aspect will be found in Galliard and Chan (1980).

Plant tissues containing high levels of lipoxygenase activity are shown in Table 11.5. Reviews on the occurrence of the enzyme have been made (see Eskin *et al.*, 1977; Galliard and Chan, 1980). Leguminous seeds generally contain lipoxygenase, but the absence of measured activity in certain plant tissues does not mean that the enzyme is absent, since there are often inhibitors present. The enzyme may be present in plastids (chloroplasts) or in a 'lysosomal' vacuole (see Galliard and Chan, 1980).

Several types of lipoxygenase have been described. These differ in various properties, and the 'classic' soybean lipoxygenase (Theorell *et al.*, 1947) may be atypical in several respects. Two main types are present in soybeans. The Theorell enzyme acts on free acids only and is sometimes called an 'acid' lipoxygenase (Verhue and Francke, 1972). The same enzyme is referred to as an 'alkaline' lipoxygenase by Grosch *et al.* (1977) because its optimal activity is found at pH 9. It is better referred to as a type I enzyme. Similarly, a second lipoxygenase with an optimum pH of 6.5 has been referred to as an 'ester' enzyme, a 'neutral' enzyme or a 'b' enzyme, and it is better referred to as the type II enzyme (Galliard and Chan, 1980).

Isoenzymic forms of lipoxygenase have been purified to homogeneity from many different plants (see Eskin *et al.*, 1977). Most of the enzymes have properties similar to the soybean type II enzyme. Tissues that have been used for study include wheat, alfalfa, potato, barley, pea and various other legumes. Most lipoxygenases appear to be single polypeptides of molecular masses in the region 70 000–100 000 Da. A single atom of non-haem iron per molecule is found (Chan, 1973) and the pH optima are in the range 5.5–7.0. Because lipoxygenase lacks cofactors other than non-haem iron, the number of inhibitors is small. The *trans*-unsaturated fatty acids, acetylenic fatty acids, α-bromo fatty acids and fatty alcohols will all inhibit, and antioxidants or oxygen scavengers are effective in certain cases (see Hamberg *et al.*, 1974; Eskin *et al.*, 1977).

Soybean type I lipoxygenase is particularly stable. Other lipoxygenases are less stable and activity is lost during purification. Heat treatment is often used in the food industry to cause inactivation so as to prevent off-flavours. The purified lipoxygenases are generally unstable at 70 °C, but higher

Table 11.5 Plant tissues containing high levels of lipoxygenase activity (From Galliard and Chan, 1980, with permission)

Plant	Family	Tissue	Lipoxygenase activity (μl O_2 consumed in 10 min per gram fresh weight)
Yellow bean	Leguminosae	Seed	6480
Potato	Solanaceae	Tuber	4560
Eggplant	Solanaceae	Fruit	4320
Soybean	Leguminosae	Seed	4150
Artichoke	Compositae	Heart	3360
Pea	Leguminosae	Seed	1769
Cauliflower	Cruciferae	Floret	1440
Avocado	Lauraceae	Fruit	720
Tomato	Solanaceae	Fruit	360
Lettuce	Compositae	Leaf	120

temperatures may be necessary for inactivating the enzymes in foodstuffs.

A number of assay methods have been used for measuring lipoxygenase. These techniques use radioisotopes or the oxygen electrode and colorimetric methods. These assays and necessary precautions are discussed by Galliard and Chan (1980).

Holman *et al.* (1969) studied the substrate specificity of soybean lipoxygenase. Linoleic acid had the highest activity, but other acids with a double bond at C-13 ($n-6$) were also active. The hydrophobic interaction of the pentadiene system with the enzyme appeared to be more important than the carbonyl group.

Aerobic lipoxygenase reactions give rise to a conjugated *cis,trans*-pentadienyl hydroperoxide, which has its *trans* bond next to the hydroperoxide. The enzymes differ in the position of the hydroperoxide group, which for linoleic acid may be C-9 or C-13, and are enantiospecific. In addition, lipoxygenase also catalyses an anaerobic reaction between the product of the aerobic reaction and linoleic acid. A complex mixture of products results, which include decomposition products of the 13-hydroperoxide (e.g. pentane, oxo acids) and dimeric products (Veldink *et al.*, 1977). Changes in the electron spin resonance (ESR) spectrum during the aerobic reaction have been used to study the involvement of non-haem iron in the overall process (see Galliard and Chan, 1980). Both the anaerobic and the aerobic reactions have also been examined by stopped-flow techniques. It was proposed that the native enzyme catalysed the aerobic reaction before it was converted to the coloured forms (Egmond *et al.*, 1977). Although most of the experiments on the mechanism of lipoxygenase action have utilized the type I soybean enzyme, results with pea lipoxygenase have given essentially similar results (Spaapen *et al.*, 1977). The overall mechanism of lipoxygenase action may, therefore, be similar for the different enzymes.

Lipoxygenases will also catalyse co-oxidation reactions. This is used both in assay methods for the enzyme and in commercial applications. An example of the latter is the addition of soybean or broadbean flours (both rich in lipoxygenase activity) to wheat flour in order to bleach pigments for white bread production. Enzymes from different sources differ in their co-oxidation ability, e.g. soybean type I enzyme has poor activity in this regard while soybean type II enzyme has high co-oxidation activity. The reaction probably proceeds by a free-radical process (Veldink *et al.*, 1977) and requires the presence of a substrate (e.g. linoleic acid) as well as the co-substrate. The extent of the co-oxidation may depend on the lifetime of the radical intermediates and the relative efficiency of the lipoxygenase-mediated radical reduction (Weber and Grosch, 1976).

The hydroperoxide products of lipoxygenase action are not found in healthy plant cells because they are rapidly destroyed by enzymic processes in order to protect cells from membrane or protein damage. The further metabolism of hydroperoxides has been reviewed (Eskin *et al.*, 1977; Veldink *et al.*, 1977; Galliard and Chan, 1980; Gardner, 1975, 1991). Several sorts of reactions can take place. These include reduction to hydroxydienoic acids, isomerization, formation of vinyl ethers and hydroperoxide cleavage. The enzyme that catalyses reduction of hydroperoxides may be lipoxygenase itself (e.g. Streckert and Stan, 1975). A single enzyme system seems to be responsible for the formation of α- and γ-ketols in a number of plants. A second enzyme responsible for isomerization has also been found in alfalfa (Esselman and Clagett, 1974). Vinyl ether formation has been detected in potatoes and is specific for the 9-hydroperoxides (Galliard and Matthew, 1975). These unusual ethers may be present in large amounts in aqueous homogenates of potatoes due to the combined action of acyl hydrolase and lipoxygenase enzymes. It seems possible that vinyl ether formation is catalysed by potato lipoxygenase (Galliard *et al.*, 1974). Cleavage of hydroperoxides is important since the compounds produced contribute to the flavour and smell of many foods. Some of the products of such enzymic attack are shown in Fig. 11.14. Initially, volatile aldehydes (*cis*-3-nonenal and hexanal) and the corresponding C_9 and C_{12} oxo acid fragments are produced. An isomerase enzyme is present in many plants that will convert *cis*-3-enals to *trans*-2 isomers (see Fig. 11.14). The cleavage enzyme is associated with membrane fractions (such as plasmalemma, endoplasmic reticulum and Golgi body) from cucumber fruits (Wardale *et al.*, 1978). In leaves of tea and alfalfa cell cultures the enzyme is associated with chloroplast fragments (e.g. Sekiya and Hatanaka, 1977).

In addition to enzymic reactions, the hydroperoxides formed by lipoxygenase also undergo chemical decomposition under mild conditions. This may give rise to volatile products such as aldehydes. These reactions and the possible physiological functions of lipoxygenase in seed germination and response to wounding are discussed by Galliard and Chan (1980). The product of hydroperoxide dehydrase, 12-oxophytodienoic acid, is metabolized to the plant hormone jasmonic acid. The latter compound (and/or its methyl ester) has often been cited to cause growth inhibition, abscission, senescence and other effects (Gardner, 1991).

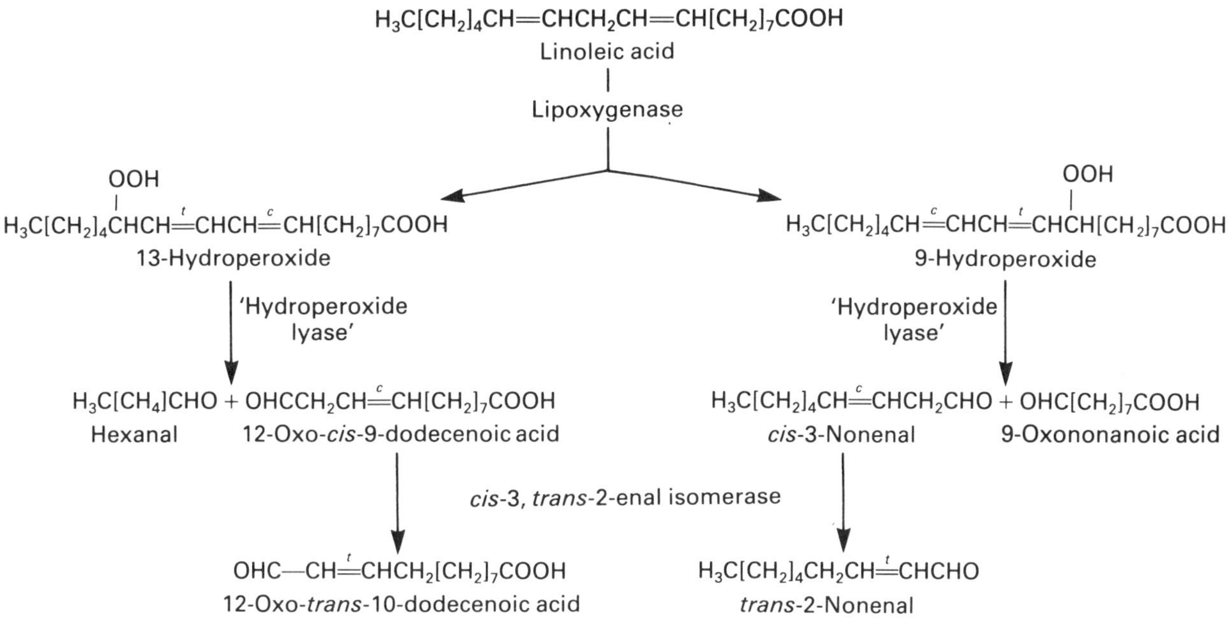

Fig. 11.14 Enzymic formation of aldehydic fragments from linoleic acid. Reproduced from Galliard and Chan (1980) with permission.

Alternatively, wounding appears to activate the hydroperoxide lyase pathway to give 12-oxo-10(E)-dodecenoic acid, which has been reported to have wound-healing properties (Zimmerman and Coudron, 1979). In addition, certain lipoxygenase-pathway products have been found to have anti-fungal properties (see Gardner, 1991).

Other aspects of the physiological role of plant lipoxygenase in relation to growth and development, synthesis of regulatory molecules, senescence, wounding and other stress responses are reviewed by Siedow (1991).

11.1.7 Production and function of the eicosanoids

Eicosanoids are oxygenated 20-carbon fatty acids.

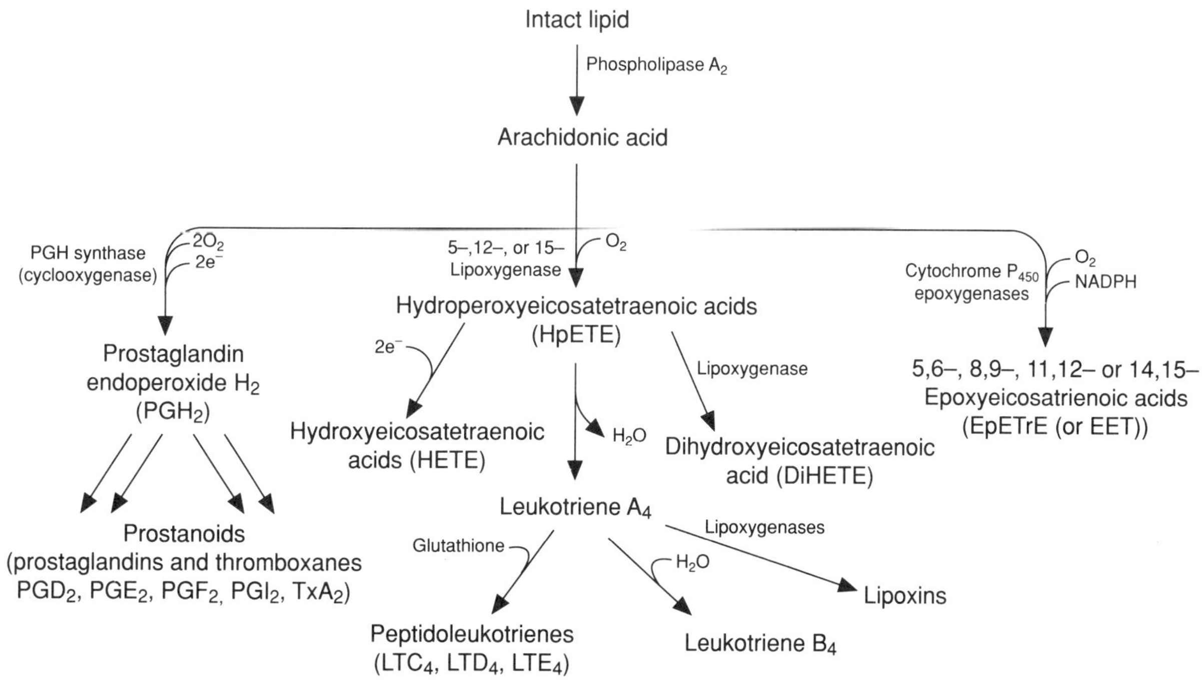

Fig. 11.15 Pathways leading to the formation of eicosanoids from arachidonic acid. From Smith *et al.* (1991a).

Because the major precursor of these compounds in animals is arachidonic acid, the pathways leading to the eicosanoids are also known as the 'arachidonate cascade'. There are three reactions involved in the initial metabolism of arachidonic acid, which itself must be released from membrane lipids through the activity of phospholipases. The enzymes involved are cyclooxygenase, 5-, 12- or 15-lipoxygenases and various cytochrome P_{450} epoxygenases (Fig. 11.15). For prostanoid structures, see section 1.4 and Smith *et al.* (1991a).

Eicosanoid synthesis is initiated following the interaction of a stimulus with the cell's plasma membrane. Interaction of the agonist with its receptor

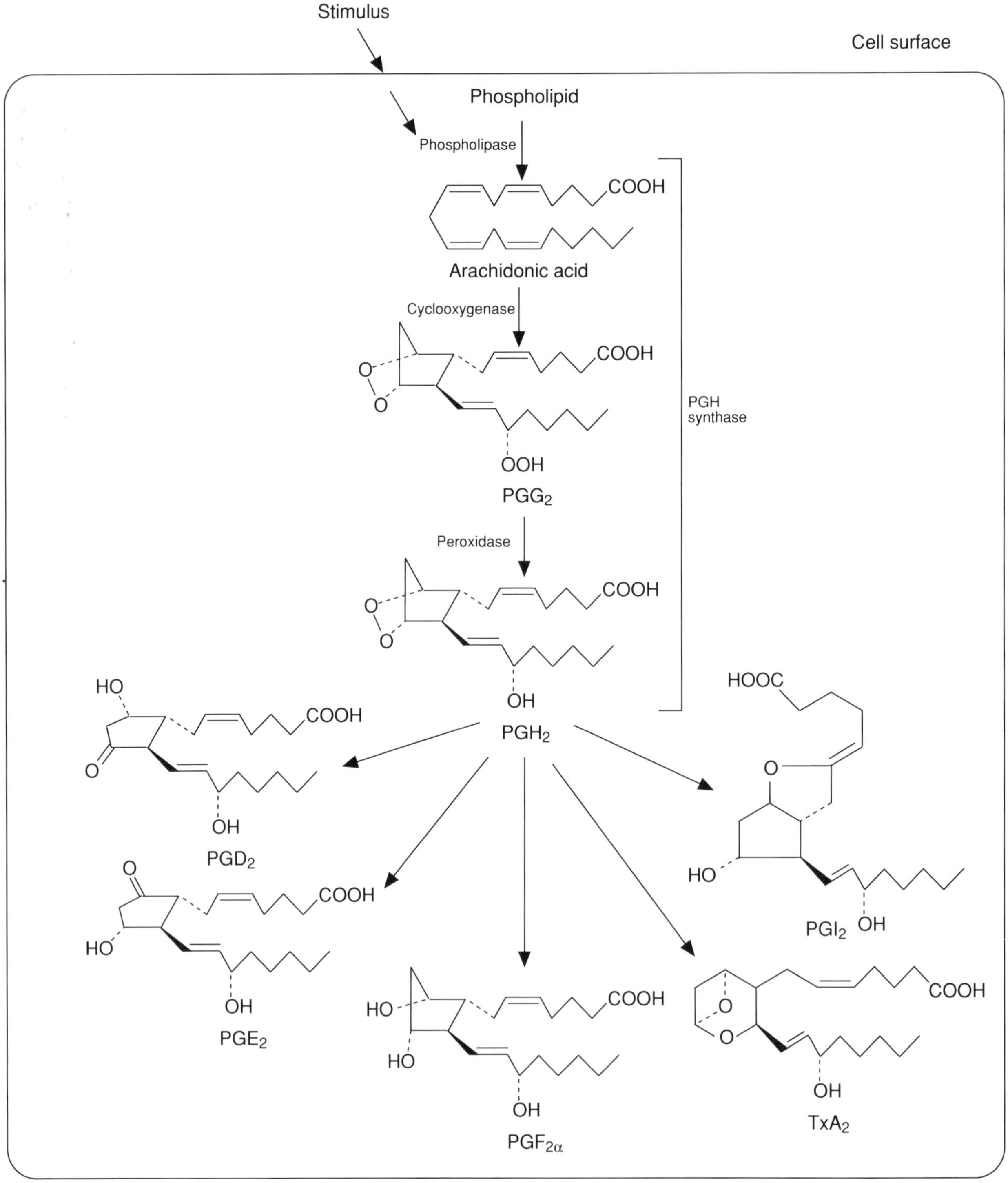

Fig. 11.16 Structures and biosynthetic relationships among prostanoids. From Smith *et al.* (1991a).

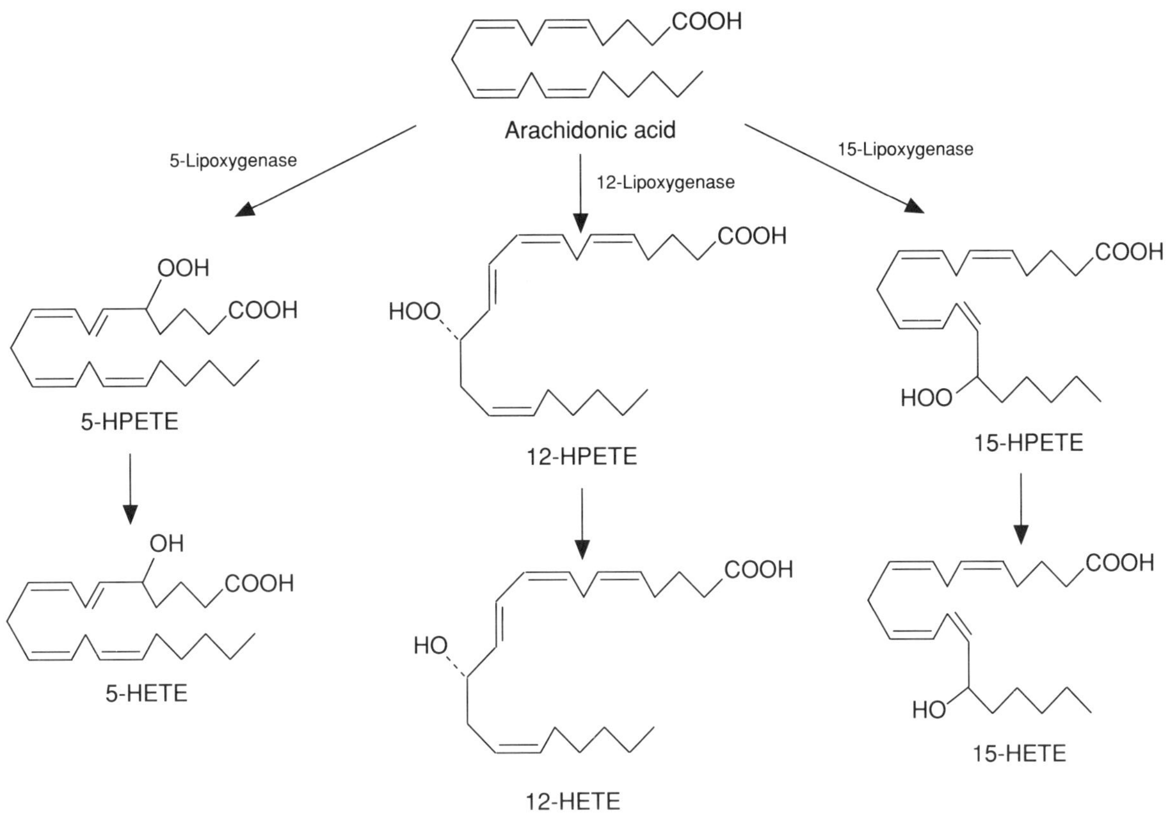

Fig. 11.17 Lipoxygenase pathways for the synthesis of the major HETE isomers. From Spector *et al.* (1988).

leads to the activation of one or more phospholipase. Activation of the phospholipase(s) requires or is accompanied by a significant rise in intracellular Ca^{2+}. The release of arachidonate is selective, and other fatty acids are not hydrolysed (Dennis, 1987).

Two phospholipids are important sources of arachidonate – phosphatidylcholine, which is cleaved by a phospholipase A_2, and phosphatidylinositol, which is catabolized by phospholipase C (see section 11.2.2) and, subsequently, by diacylglycerol (and monoacylglycerol) lipase. Different pools of arachidonate appear to be released for the formation of different eicosanoids (see Smith *et al.*, 1991a).

Once arachidonic acid is released, it can be metabolized by prostaglandin endoperoxide (PGH) synthase. This enzyme has two catalytic activities: a cyclooxygenase, which catalyses the formation of PGG_2; and a peroxidase, which converts this intermediate to PGH_2 (Fig. 11.16). Subsequent metabolism of PGH_2 to one of the major prostanoids then takes place via a cell-specific pathway. Prostanoids with a '2' subscript are derived from arachidonic acid, those with a '1' subscript from 8,11,14-eicosatrienoic acid and those with a '3' subscript from 5,8,11,14,17-eicosapentaenoic acid. The structure and catalytic properties of prostaglandin endoperoxide synthase (Smith and Marnett, 1991) and regulation of its expression (De Witt, 1991) have been reviewed recently.

Once PGH_2 has been formed by PGH synthase, further metabolism is cell-specific (Smith *et al.*, 1991b). For example, platelets form mainly TxA_2, endothelial cells form PGI_2 as their major prostanoid, while PGE_2 is the major product in renal collecting tubule cells. A number of these enzymes have been purified and characterized (see Smith *et al.*, 1991a). Prostanoids are local hormones that act very near their site of synthesis (Smith, 1989). The prostanoids are viewed generally as operating through G-protein-linked receptors, and all have a very short half-life, being rapidly destroyed by catabolism. For PGE_2, the initial reaction is oxidation to the relatively inactive 15-keto derivative. Further catabolism involves reduction of the double bond between C-13 and C-14, ω-oxidation and β-oxidation.

As an alternative to PGH attack, arachidonic acid can be metabolized through the action of one of three lipoxygenases. These enzymes catalyse reactions analogous to the well known plant enzymes (section 11.1.6). The lipoxygenases constitute a family of closely related non-haem iron-containing dioxygenases. The immediate products of their reaction with various eicosaenoic acids are hydroperoxy fatty acids. In the case of arachidonic

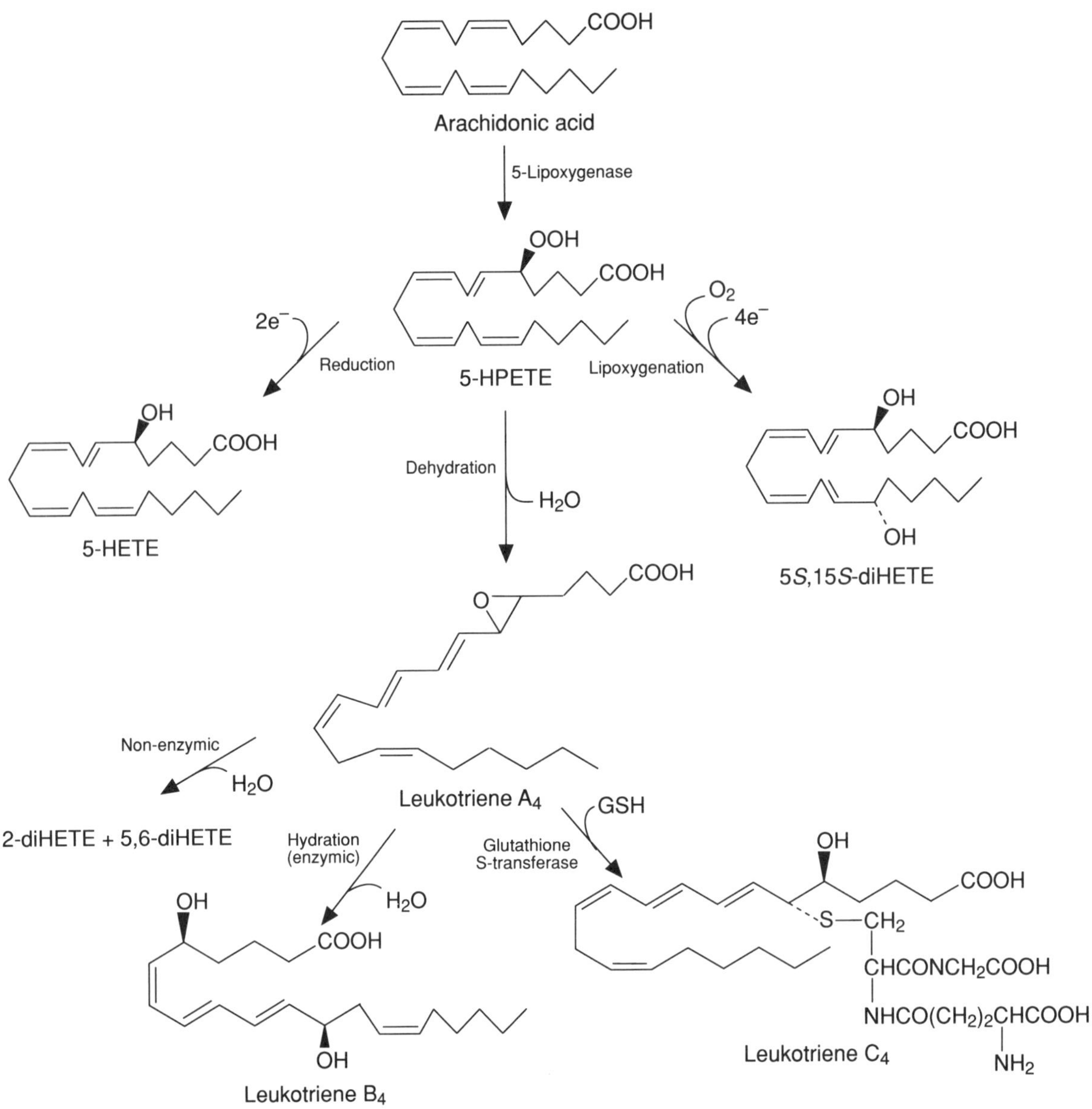

Fig. 11.18 Formation of leukotrienes from arachidonic acid. From Gurr and Harwood (1991).

acid, the products are peroxyeicosatetraenoic acids (HpETEs). Subsequent metabolism generates hydroxyeicosatetraenoic acids (HETEs) (Fig. 11.17). The biochemistry and functional activity of HETEs have been reviewed (Spector *et al.*, 1988).

The activity of the 5-lipoxygenase has been studied more thoroughly than that of the other lipoxygenases, mainly because the leukotrienes (section 1.4) are the end-products of its metabolic pathway. The production of leukotrienes LTA_4, LTB_4 and LTC_4 is shown in Fig. 11.18. For summaries of the actions of leukotrienes, see Smith *et al.* (1991a) and Gurr and Harwood (1991). Depending on the exact chemical structure concerned and the tissue tested, the leukotrienes are potent bronchoconstrictors, arterioconstrictors, vasodilators and chemotactic agents. The function of the 12- and 15-lipoxygenases is much less clear than is that of the 5-lipoxygenase. However, a variety of interesting effects have been noted for 15-lipoxygenase metabolites (Schewe and Kuhn, 1991), and 12-HETE may be important in skin diseases (section 12.4) and in the symptoms caused by sunburn (see Gurr and Harwood, 1991; Spector *et al.*, 1988). Recently, low levels of the 12-lipoxygenase and its mRNA have been found for patients with myeloproliferative disease (Matsuda *et al.*, 1993). For further information on the 5-, 12- and 15-lipoxygenases, see Yamamoto (1989, 1991), and for a review see Yamamoto (1992).

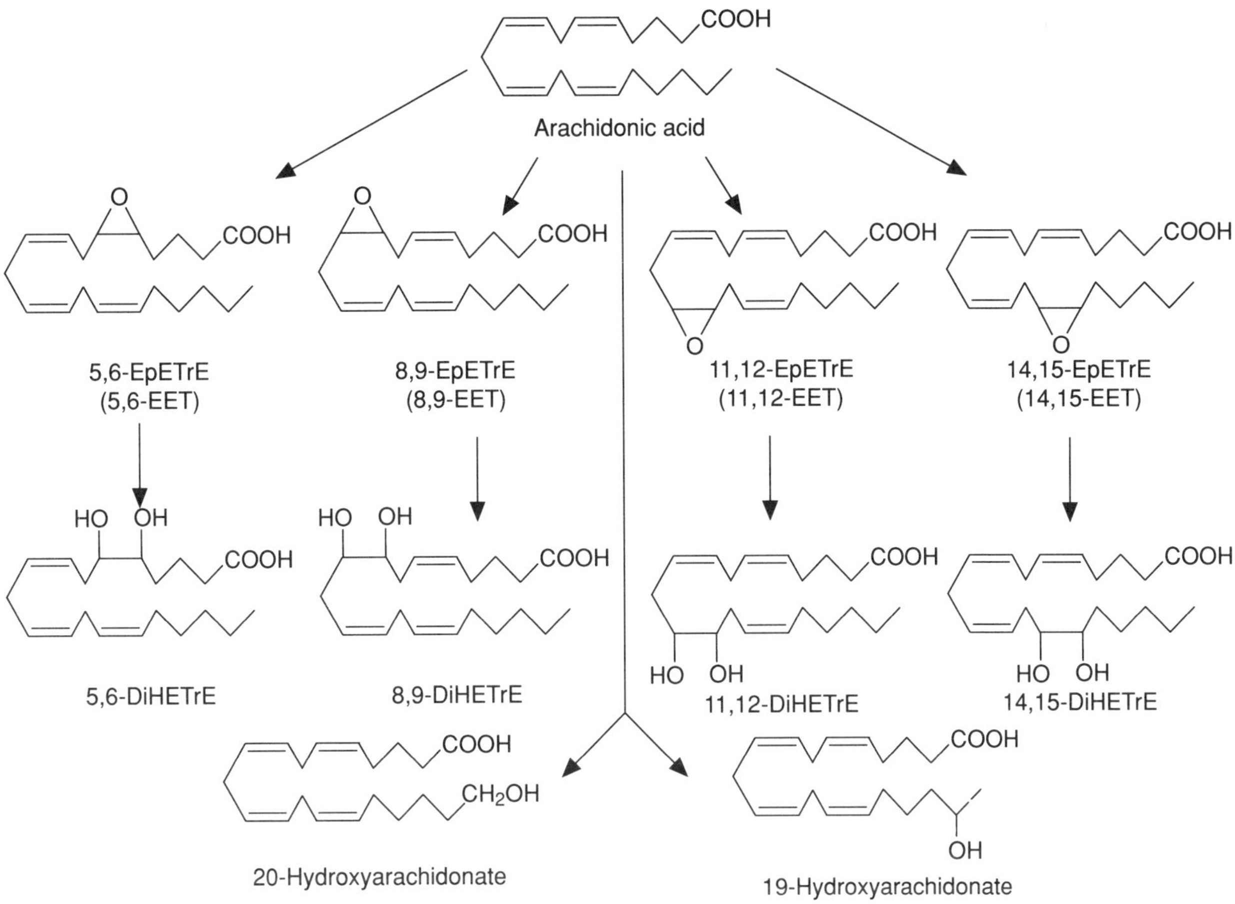

Fig. 11.19 Structures of products of epoxygenase pathways of the arachidonate cascade. From Smith *et al.* (1991a).

The third metabolic pathway for arachidonic acid involves cytochrome P_{450}-mediated epoxygenases. There are two main types of product, *cis*-epoxyeicosatrienoic acids and HETEs (Fig. 11.19). A variety of physiological actions have been noted for metabolites of the epoxygenase pathway, including antagonistic actions with other eicosanoids (see Smith *et al.*, 1991a).

11.1.8 Other conversions

Fatty acids can be modified in a number of other ways. For example, cyclopropane fatty acids are formed by the addition of a methylene group from *S*-adenosylmethionine across the double bond of a mono-unsaturated fatty acid. The latter is esterified to a phospholipid, so that the actual substrate is a membrane lipid. The aldehyde residues in plasmalogens and the alcohol residues in alkyl ether lipids such as those in *Clostridium butyricum* also act as acceptors for the methyl group – in this case forming the corresponding cyclopropane aldehydes and alcohols (Goldfine and Panos, 1971). The control of cyclopropane fatty acid synthetase in bacteria such as *E. coli* has been examined in some detail (see Harwood and Russell, 1984). Cyclopropane fatty acids in plants seem to be made by the same mechanism as for bacteria (Mangold and Spener, 1980). These authors have also reviewed work on the synthesis of cyclopentenyl fatty acids, such as chaulmoogric (13-(2-cyclopentenyl)tridecanoic) acid.

Branched, cyclic and unsaturated hydrocarbons in higher plants are formed from appropriate fatty acids by decarboxylation (see Kolattukudy, 1980, 1987). The process has also been studied in cyanobacteria, insects and other species (see Kolattukudy, 1976). For a review of these and other conversions involved in the synthesis of plant epicuticular waxes, see von Wettstein-Knowles (1979, 1982). These other components include β-diketones, secondary alcohols, primary alcohols and aldehydes. The relationship of these compounds to the fatty acid precursor is depicted in Fig. 11.20.

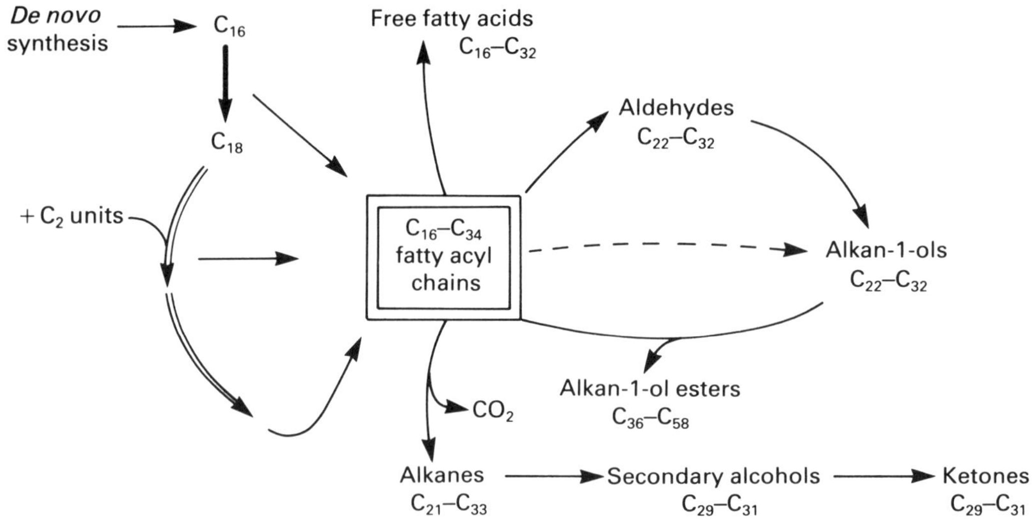

Fig. 11.20 Conversion of fatty acids to other wax components. Reproduced from von Wettstein-Knowles (1979) with permission.

References and further reading

Abulnaja, K.O. and Harwood, J.L. (1991) *Phytochemistry*, **30**, 2883.

Alberts, A. and Vagelos, P.R. (1972) in *The Enzymes*, vol. 6 (ed. P.D. Boyer), Academic Press, New York, p. 37.

Allred, J.B. and Roman-Lopez, C.A. (1988) *Biochem. J.*, **251**, 881.

Amy, C.M., Witkowski, A., Naggert, J., Williams, B., Randhawa, Z. and Smith, S. (1989) *Proc. Natl Acad. Sci. USA*, **86**, 3114.

Amy, C.M., Williams-Ahlf, B., Naggert, J. and Smith, S. (1990) *Biochem. J.*, **271**, 675.

Amy, C.M., Williams-Ahlf, B., Naggert, J. and Smith, S. (1992) *Proc. Natl Acad. Sci. USA*, **89**, 1105.

Beevers, H. (1980) in *The Biochemistry of Plants*, vol. 4 (eds P.K. Stumpf and E.E. Conn), Academic Press, New York, p. 117.

Bloch, K. (1977) *Adv. Enzymol.*, **45**, 1.

Bourre, J.M., Dandu, O.I. and Baumann, N.A. (1975) *Biochem. Biophys. Res. Commun.*, **63**, 1027.

Bowen, D.M., Davison, A.N. and Ramsay, R.B. (1974) *MTP Int. Rev. Sci.*, **4**, 141.

Bressler, R. (1970) in *Lipid Metabolism* (ed. S.J. Wakil), Academic Press, New York, p. 49.

Browse, J. and Somerville, C. (1991) *Annu. Rev. Plant Physiol.*, **42**, 467.

Chan, H.W.S. (1973) *Biochim. Biophys. Acta*, **327**, 32.

Chuman, L. and Brody, S. (1989) *Eur. J. Biochem.*, **184**, 643.

Cook, H.W. (1991) in *Biochemistry of Lipids, Lipoproteins and Membranes*, 2nd edn (eds D.E. Vance and J. Vance), Elsevier, Amsterdam, p. 141.

Coon, M.J., Strobel, H.W., Antor, A.P., Heidema, J. and Duppel, W. (1972) in *Biological Hydroxylation Mechanisms* (eds G.S. Boyd and M.S. Smellie), Academic Press, New York, p. 45.

Cripps, C., Borgeson, C., Blomqvist, G.J. and de Renobales, M. (1990) *Arch. Biochem. Biophys.*, **278**, 46.

Dailey, H.A. and Strittmatter, P. (1979) *J. Biol. Chem.*, **254**, 5388.

Dailey, H.A. and Strittmatter, P. (1980) *J. Biol. Chem.*, **255**, 5184.

Dennis, E.A. (1987) *Biotechnology*, **5**, 1294.

de Renobales, M., Cripps, C., Stanley-Samuelson, D.W., Jurenka, R.A. and Blomqvist, G.J. (1987) *Trends Biochem. Sci.*, **12**, 403.

De Titta, G.T., Parthasarathy, R., Blessing, R.H. and Stalling, W. (1980) *Proc. Natl Acad. Sci. USA*, **77**, 333.

De Witt, D.L. (1991) *Biochim. Biophys. Acta*, **1083**, 121.

Egmond, M.R., Fasella, P.M., Veldink, G.A., Vliegenhart, J.F.G. and Boldingh, J. (1977) *Eur. J. Biochem.*, **76**, 469.

El Hussein, S.A., Miernyk, J.A. and Ohlrogge, J.B. (1988) *Biochem. J.*, **252**, 39.

Ernst-Fonberg, M.L. and Bloch, K. (1971) *Arch. Biochem. Biophys.*, **143**, 392.

Eskin, N., Grossman, S. and Pinsky, A. (1977) *Crit. Rev. Food Sci. Nutr.*, **9**, 1.

Esselman, W.J. and Clagett, C.O. (1974) *J. Lipid Res.*, **15**, 173.

Fehling, E., Lessire, R., Cassagne, C. and Mukherjee, K.D. (1992) *Biochim. Biophys. Acta*, **1126**, 88.

Finnerty, W.R. (1989) in *Microbial Lipids*, vol. 2 (eds C. Ratledge and S.G. Wilkinson), Academic Press, London, p. 525.

Fulco, A.J. (1974) *Annu. Rev. Biochem.*, **43**, 215.

Galliard, T. (1974) *Recent Adv. Phytochem.*, **8**, 209.

Galliard, T. (1980) in *The Biochemistry of Plants*, vol. 4 (eds P.K. Stumpf and E.E. Conn), Academic Press, New York, p. 85.

Galliard, T. and Chan, H.W.S. (1980) in *The Biochemistry of Plants*, vol. 4 (eds P.K. Stumpf and E.E. Conn), Academic Press, New York, p. 131.

Galliard, T. and Matthew, J.W. (1975) *Biochim. Biophys. Acta*, **398**, 1.
Galliard, T., Wardale, D.A. and Matthew, J.W. (1974) *Biochem. J.*, **138**, 23.
Gardner, H.W. (1975) *Agric. Food Chem.*, **23**, 129.
Gardner, H.W. (1991) *Biochim. Biophys. Acta*, **1084**, 221.
Garton, G.A. (1977) *MTP Int. Rev. Sci.*, **14**, 337.
Gerhardt, B. (1992) *Prog. Lipid Res.*, **31**, 417.
Goldberg, I., Schechter, I. and Bloch, K. (1973) *Science*, **182**, 497.
Goldfine, H. and Panos, C. (1971) *J. Lipid Res.*, **12**, 214.
Goodridge, A.G. (1991) in *Biochemistry of Lipids, Lipoproteins and Membranes*, 2nd edn (eds D.E. Vance and J. Vance), Elsevier, Amsterdam, p. 111.
Greville, G.B. and Tubbs, P.K. (1968) *Essays Biochem.*, **4**, 155.
Grosch, W., Laskawy, G. and Kaiser, K.-P. (1977) *Z. Lebensm. Unters. Forsch.*, **163**, 77.
Gunsalus, I.C., Pederson, T.C. and Sligar, S.G. (1975) *Annu. Rev. Biochem.*, **44**, 377.
Gurr, M.I. (1974) in *Biochemistry of Lipids* (ed. T.W. Goodwin), Butterworths, London, p. 181.
Gurr, M.I. and Harwood, J.L. (1991) *Lipid Biochemistry*, 4th edn, Chapman & Hall, London.
Hamberg, M., Samuelsson, B., Bjorkhem, I. and Danielsson, M. (1974) in *Molecular Mechanisms of Oxygen Activation* (ed. O. Hayaishi), Academic Press, New York, p. 29.
Harwood, J.L. (1988) *Annu. Rev. Plant Physiol.*, **39**, 101.
Harwood, J.L. (1990) in *Target Sites for Herbicide Action* (ed. R.C. Kirkwood), Plenum, New York, p. 57.
Harwood, J.L. and Russell, N.J. (1984) *Lipids in Plants and Microbes*, Allen and Unwin, London.
Harwood, J.L., Walsh, M.C. and Walker, K.A. (1990) in *Methods in Plant Biochemistry*, vol. 3 (ed. P.J. Lea), Academic Press, London, p. 193.
Hashimoto, T. (1990) in *Fatty Acid Oxidation: Clinical, Biochemical and Molecular Aspects* (eds K. Tanaka and P.M. Costes), Alan R. Liss, New York, p. 137.
Holman, R.T., Egwin, P.O. and Christie, W.W. (1969) *J. Biol. Chem.*, **244**, 1149.
Jackowski, S. and Rock, C.O. (1987) *J. Biol. Chem.*, **262**, 7927.
Jackowski, S., Cronan, J.E. and Rock, C.O. (1991) in *Biochemistry of Lipids, Lipoproteins and Membranes*, 2nd edn (eds D.E. Vance and J. Vance), Elsevier, Amsterdam, p. 43.
Jaworski, J.G. (1987) in *The Biochemistry of Plants*, vol. 9 (eds P.K. Stumpf and E.E. Conn), Academic Press, New York, p. 159.
Jaworski, J.G., Clough, R.C. and Barnum, S.R. (1989) *Plant Physiol.*, **90**, 41.
Johns, R. and Perry, G.J. (1977) *Arch. Microbiol.*, **114**, 267.
Jones, A.L., Harwood, J.L. and Lloyd, D. (1992) *Biochem. Soc. Trans.*, **20**, 170S.
Jones, A.V.M. and Harwood, J.L. (1980) *Biochem. J.*, **190**, 851.
Kaneda, T. (1977) *Bact. Rev.*, **41**, 391.
Kearns, E.V., Hugly, S. and Somerville, C.R. (1991) *Arch. Biochem. Biophys.*, **284**, 431.
Kindl, H. (1987) in *The Biochemistry of Plants*, vol. 9 (eds P.K. Stumpf and E.E. Conn), Academic Press, New York, p. 31.
Klein, K., Steinberg, R., Fiethen, B. and Overath, P. (1971) *Eur. J. Biochem.*, **19**, 442.
Knowles, J.R. (1989) *Annu. Rev. Biochem.*, **58**, 195.
Knutzon, D.S., Scherer, D.E. and Schreckengost, W.E. (1991) *Plant Physiol.*, **96**, 344.
Kolattukudy, P.E. (ed.) (1976) *Chemistry and Biochemistry of Natural Waxes*, Elsevier, New York.
Kolattukudy, P.E. (1980) in *The Biochemistry of Plants*, vol. 4 (eds P.K. Stumpf and E.E. Conn), Academic Press, New York, p. 571.
Kolattukudy, P.E. (1987) in *The Biochemistry of Plants*, vol. 9 (eds P.K. Stumpf and E.E. Conn), Academic Press, New York, p. 291.
Krakower, G.R. and Kim, K.H. (1981) *J. Biol. Chem.*, **256**, 2408.
Kridl, J.C., Knutzon, D.S., Johnson, W.B., Thompson, G.A., Radke, S.E., Turner, J.C. and Knauf, V.C. (1991) *Int. Soc. for Plant Mol. Biol., 3rd Int. Congr.*, Tucson, AZ, October, Poster 723.
Kusunose, M., Kusunose, E. and Coon, M.J. (1964) *J. Biol. Chem.*, **239**, 2135.
Lazarow, P.B. and Moser, H.W. (1989) in *The Metabolic Basis of Inherited Disease*, 6th edn (eds C.R. Scriver, A.L. Beaudet, W.S. Sly and D. Valle), McGraw-Hill, New York, p. 1479.
Lekakis, N.J. (1977) *FEMS Microbiol. Lett.*, **1**, 289.
Lenk, W. (1972) *Prog. Drug Res.*, **16**, 229.
Lent, B. and Kim, K.H. (1980) *J. Biol. Chem.*, **257**, 1897.
Lessire, R., Bessoule, J.J. and Cassagne, C. (1989) *Biochim. Biophys. Acta*, **1006**, 35.
Li, S.-J. and Cronan, J.E. (1992) *J. Biol. Chem.*, **267**, 855.
Libertini, L.J. and Smith, S. (1978) *J. Biol. Chem.*, **253**, 1393.
Lynen, F.O. (1967) *Biochem. J.*, **102**, 381.
Maloy, S.R., Bohlander, M. and Nunn, W.D. (1980) *J. Bacteriol.*, **143**, 720.
Mangold, H.K. and Spener, F. (1980) in *The Biochemistry of Plants*, vol. 4 (eds P.K. Stumpf and E.E. Conn), Academic Press, New York, p. 647.
Matsuda, S., Murakami, J., Yamamoto, Y., *et al.* (1993) *Biochim. Biophys. Acta*, **1180**, 243.
McGarry, J.D. and Foster, D.W. (1980) *Annu. Rev. Biochem.*, **49**, 395.
McIlwain, H. (1966) *Biochemistry and the Central Nervous System*, Churchill, London.
McKeon, T.A. and Stumpf, P.K. (1982) *J. Biol. Chem.*, **257**, 12141.
Mikolajczyk, S. and Brody, S. (1990) *Eur. J. Biochem.*, **187**, 431.
Moreau, R.A. and Stumpf, P.K. (1981) *Plant Physiol.*, **67**, 672.
Morris, L.J. (1970) *Biochem. J.*, **118**, 681.
Moss, J. and Lane, M.D. (1971) *Adv. Enzymol.*, **35**, 321.
Naggert, J., Witkowski, A., Mikkelsen, J. and Smith, S. (1988) *J. Biol. Chem.*, **263**, 1146.
Naggert, J., Witkowski, A., Wessa, B. and Smith, S. (1991) *Biochem. J.*, **273**, 787.

Nichols, B.W., James, A.T. and Brener, J. (1967) *Biochem. J.*, **104**, 486.
Nikawa, J., Tanabe, T., Ogiwara, H., Shiba, T. and Numa, S. (1979) *FEBS Lett.*, **102**, 223.
Ohlrogge, J.B. (1987) in *The Biochemistry of Plants*, vol. 9 (eds P.K. Stumpf and E.E. Conn), Academic Press, New York, p. 137.
Ohlrogge, J., Browse, J. and Somerville, C.R. (1991) *Biochim. Biophys. Acta*, **1082**, 1.
Osmundsen, H., Bremer, J. and Pedersen, J.I. (1991) *Biochim. Biophys. Acta*, **1085**, 141.
Ozasa, H., Furuta, S., Miyazawa, S., Osumi, T., Hashimoto, T., Mori, M., Miura, S. and Tatibana, M. (1984) *Eur. J. Biochem.*, **144**, 453.
Polakis, S.E., Guchhait, R.B., Zwergel, E.E., Lane, M.D. and Cooper, T.G. (1974) *J. Biol. Chem.*, **244**, 6657.
Pollard, M.R. and Stumpf, P.K. (1980) *Plant Physiol.*, **66**, 649.
Pollard, M.R., Anderson, L., Fan, C., Hawkins, D.J. and Davies, H.M. (1991) *Arch. Biochem. Biophys.*, **284**, 306.
Post-Beittenmiller, D., Jaworski, J.G. and Ohlrogge, J.B. (1991) *J. Biol. Chem.*, **266**, 1858.
Poulose, A.J., Rogers, I. and Kolattukudy, P.E. (1981) *Biochem. Biophys. Res. Commun.*, **103**, 377.
Pugh, E.L. and Kates, M. (1975) *Biochim. Biophys. Acta*, **380**, 442.
Rainwater, D.L. and Kolattukudy, P.E. (1982) *Arch. Biochem. Biophys.*, **213**, 372.
Roessler, P.G. (1990) *Plant Physiol.*, **92**, 73.
Samols, D., Thornton, C.G., Murtif, V.L., Kumar, G.K., Haase, F.C. and Wood, H.G. (1988) *J. Biol. Chem.*, **263**, 6461.
Schewe, T. and Kuhn, H. (1991) *Trends Biochem. Sci.*, **16**, 369.
Schmidt, H. and Heinz, E. (1990) *Plant Physiol.*, **94**, 214.
Schroepfer, G.J. and Bloch, K. (1965) *J. Biol. Chem.*, **240**, 54.
Schulz, H. (1991a) in *Biochemistry of Lipids, Lipoproteins and Membranes*, 2nd edn (eds D.E. Vance and J. Vance), Elsevier, Amsterdam, p. 87.
Schulz, H. (1991b) *Biochim. Biophys. Acta*, **1081**, 109.
Schweizer, E. (1989) in *Microbial Lipids*, vol. 2 (eds C. Ratledge and S.G. Wilkinson), Academic Press, London, p. 3.
Schweizer, E., Kniep, B., Castorph, H. and Holzner, U. (1973) *Eur. J. Biochem.*, **39**, 353.
Sekiya, J. and Hatanaka, A. (1977) *Plant Sci. Lett.*, **10**, 165.
Siedow, J.N. (1991) *Annu. Rev. Plant Physiol.*, **42**, 145.
Sigal, E. (1991) *Am. J. Physiol.*, **260**, L13–28.
Slabas, A.R. and Fawcett, T. (1992) *Plant Mol. Biol.*, **19**, 169.
Smith, M.A., Cross, A.R., Jones, O.T.G., Griffiths, W.T., Stymne, S. and Stobart, A.K. (1990) *Biochem. J.*, **272**, 23.
Smith, M.A., Jonsson, L., Stymne, S. and Stobart, A.K. (1992) *Biochem. J.*, **287**, 141.
Smith, W.L. (1989) *Biochem. J.*, **259**, 315.
Smith, W.L. and Marnett, L.J. (1991) *Biochim. Biophys. Acta*, **1083**, 1.
Smith, W.L., Borgeat, P. and Fitzpatrick, F.A. (1991a) in *Biochemistry of Lipids, Lipoproteins and Membranes*, 2nd edn (eds D.E. Vance and J. Vance), Elsevier, Amsterdam, p. 297.
Smith, W.L., Marnett, L.J. and De Witt, D.L. (1991b) *Pharm. Ther.*, **49**, 153.
Soling, H.D. and Seufert, C.D. (eds) (1978) *Biochemical and Clinical Aspects of Ketone Body Metabolism*, Georg Thiem, Stuttgart.
Spaapen, L.J.M., Vliegenthart, J.F.G. and Boldingh, J. (1977) *Biochim. Biophys. Acta*, **488**, 517.
Spector, A.A., Gordon, J.A. and Moore, S.A. (1988) *Prog. Lipid Res.*, **27**, 271.
Stern, J.R. (1961) in *The Enzymes*, 2nd edn, vol. 5 (eds P.D. Boyer, H. Lardy and K. Myrback), Academic Press, New York, p. 511.
Streckert, G. and Stan, H.J. (1975) *Lipids*, **10**, 847.
Strittmatter, P., Spatz, L., Corcoran, D., Rogers, M.J., Setlow, B. and Redline, R. (1974) *Proc. Natl Acad. Sci. USA*, **71**, 4565.
Strittmatter, P., Thiede, M.A., Hackett, C.S. and Ozols, J. (1988) *J. Biol. Chem.*, **263**, 2532.
Stumpf, P.K. (1970) in *Lipid Metabolism* (ed. S.J. Wakil), Academic Press, New York, p. 79.
Stumpf, P.K. (1987) in *The Biochemistry of Plants*, vol. 9 (eds P.K. Stumpf and E.E. Conn), Academic Press, New York, p. 121.
Sumper, M. and Riepertinger, C. (1972) *Eur. J. Biochem.*, **29**, 237.
Takai, T., Wada, K. and Tanabe, T. (1987) *FEBS Lett.*, **212**, 98.
Takai, T., Yokoyama, C., Wada, K. and Tanabe, T. (1988) *J. Biol. Chem.*, **263**, 2651.
Theorell, H., Bergstrom, S. and Akeson, A. (1947) *Acta Chem. Scand.*, **1**, 571.
Thiede, M.A. and Strittmatter, P. (1985) *J. Biol. Chem.*, **260**, 14459.
Thiede, M.A., Ozols, J. and Strittmatter, P. (1986) *J. Biol. Chem.*, **261**, 13230.
Thompson, G.A., Scherer, D.E., Foxall-Van Aken, S., Kenney, J.W., Young, H.I., Shimani, D.K., Kridl, J.C. and Knauf, V.C. (1991) *Proc. Natl Acad. Sci. USA*, **88**, 2578.
Umeka, S. and Nozawa, Y. (1984) *Biochim. Biophys. Acta*, **792**, 25.
Veldink, G.A., Vliegenthart, J.F.G. and Boldingh, J. (1977) *Prog. Chem. Fats Other Lipids*, **15**, 131.
Verhue, W.H. and Francke, A. (1972) *Biochim. Biophys. Acta*, **285**, 43.
Voelker, T.A., Worrell, A.C., Anderson, L., Bleibaum, J., Fan, C., Hawkins, D.J., Radke, S.E. and Davies, H.M. (1992) *Science*, **257**, 72.
Volpe, J.J. and Vagelos, P.R. (1973) *Annu. Rev. Biochem.*, **42**, 21.
Volpe, J.J. and Vagelos, P.R. (1976) *Physiol. Rev.*, **56**, 339.
von Wettstein-Knowles, P. (1979) *Dev. Plant Biol.*, **3**, 1.
von Wettstein-Knowles, P. (1982) *Dev. Plant Biol.*, **8**, 69.
Wada, H. and Murata, N. (1989) *Plant Cell Physiol.*, **30**, 971.
Wada, H., Gombos, Z. and Murata, N. (1990) *Nature*, **347**, 200.
Wada, K. and Tanabe, T. (1983) *Eur. J. Biochem.*, **135**, 17.

Wakil, S.J. (1970) in *Lipid Metabolism* (ed. S.J. Wakil), Academic Press, New York, p. 1.

Wakil, S.J., Stoops, J.K. and Joshi, V.C. (1983) *Annu. Rev. Biochem.*, **52**, 537.

Walker, K.A. and Harwood, J.L. (1986) *Biochem. J.*, **237**, 41.

Walsh, M.C., Klopfenstein, W.E. and Harwood, J.L. (1990) *Phytochemistry*, **29**, 3797.

Wardale, D.A., Lambert, E.A. and Galliard, T. (1978) *Phytochemistry*, **17**, 205.

Webber, R.J. and Edmond, J. (1979) *J. Biol. Chem.*, **254**, 3912.

Weber, F. and Grosch, W. (1976) *Z. Lebensm. Unters. Forsch.*, **161**, 223.

Witkowski, A., Naggert, J., Mikkelsen, J. and Smith, S. (1987) *Eur. J. Biochem.*, **165**, 601.

Witkowski, A., Rangan, V.S., Randhawa, Z.I., Amy, C.M. and Smith, S. (1991) *Eur. J. Biochem.*, **198**, 571.

Worsham, L.M.S., Tucker, M.M. and Ernst-Fonberg, M.L. (1988) *Biochim. Biophys. Acta*, **963**, 423.

Wurtele, E.S. and Nickolau, B.J. (1990) *Arch. Biochem. Biophys.*, **278**, 179.

Yamamoto, S. (1989) *Prostaglandins, Leukotrienes Essential Fatty Acids*, **35**, 219.

Yamamoto, S. (1991) *Free Radical Biol. Med.*, **10**, 149.

Yamamoto, S. (1992) *Biochim. Biophys. Acta*, **1128**, 117.

Zimmerman, D.L. and Coudron, C.A. (1979) *Plant Physiol.*, **63**, 536.

11.2 Glycerophospholipids

11.2.1 Biosynthesis

The first reaction in glycerophospholipid synthesis can be regarded as the stepwise acylation of glycerol 3-phosphate. The product of these acylations, phosphatidic acid, is a key intermediate in glycerolipid metabolism as illustrated in Fig. 11.21. In addition to providing the backbone of the glycerophospholipids, phosphatidate (via diacylglycerol) also acts as a precursor of triacylglycerols (section 11.3) and the glycosylglycerides (section 11.4).

The biosynthesis of phosphatidate in animals begins with the activation of fatty acids to their acyl-CoAs by one of several chain-length-dependent acyl-CoA ligases (synthetases) (Bloch and Vance, 1977; Groot *et al.*, 1976; Brindley, 1991). The most important locations of the long-chain acyl-CoA ligases are the endoplasmic reticulum and mitochondria (Brindley, 1991). Some ligases show preferences towards saturated or unsaturated fatty acids. Of course, some of the enzymes play a role more in degradation (by providing acyl-CoAs for β-oxidation) than in biosynthesis. The peroxisomal acyl-CoA synthetases come into this category (section 11.1.5). A recent review of the biochemistry of acyl-CoAs is that by Waku (1992).

The main acceptor for acyl-CoA in most tissues is *sn*-glycerol 3-phosphate, derived from glycolysis. Glycerol 3-phosphate acyltransferase is divided equally between mitochondrial and endoplasmic reticulum in mammalian liver, but in other tissues the endoplasmic reticulum is the main site. The latter enzyme is on the cytoplasmic face. The mitochondrial and endoplasmic reticulum enzymes can be distinguished by their relative sensitivities to heat, proteolytic enzymes and –SH reagents (Brindley, 1991). The substrate selectivities of different enzymes are discussed by Waku (1992). In general, a variety of saturated and unsaturated acyl-CoAs are used. The second acyltransferase (lysophosphatidate acyltransferase) has a strong preference for unsaturated fatty acids in animals. In most subcellular fractions (and most animal tissues) the second acyltransferase has much higher activity than the first acyltransferase, so that lysophosphatidate does not accumulate. The substrate specificities of the two enzymes are such that the glycerolipids of animals show a preferential location of saturated fatty acids at the *sn*-1 position and unsaturated fatty acids at the *sn*-2 position. Some properties of the lysophosphatidate acyltransferase are covered in Brindley (1991).

The acylation of dihydroxyacetone phosphate and the subsequent reduction of acyldihydroxyacetone phosphate to monoacylphosphatidic acid provides an alternative route for phosphatidic acid synthesis (Brindley, 1991). Acyltransferases that utilize dihydroxyacetone phosphate have been detected in several tissues (Bell and Coleman, 1980) and purified from guinea-pig liver peroxisomes (Webber and Hajra, 1992). However, the relative contributions of dihydroxyacetone phosphate and glycerophosphate to phosphatidic acid synthesis remain controversial (O'Doherty, 1978).

A possible alternative pathway for phosphatidic acid synthesis (involving diacylglycerol kinase) has been demonstrated in rat liver (see Bell and Coleman, 1980). However, this enzyme is probably not very important quantitatively for phosphatidic acid formation. See also Brindley and Sturton (1982) for a review of phosphatidic acid formation.

Once phosphatidate has been formed, it can be converted to diacylglycerol through the action of phosphatidate phosphohydrolase. This is a key enzyme in glycerolipid metabolism, controlling, on the one hand, the supply of carbon for the major membrane phospholipids, phosphatidylcholine and phosphatidylethanolamine, and the major storage lipid, triacylglycerol, and, on the other, controlling the relative production of anionic phosphoglycerides by a competing cytidylyltransferase (see Fig. 11.21 and below). Phosphatidate phosphohydrolase has been well reviewed (Brindley, 1988; see also Brindley, 1991). Phosphatidate phosphohydrolase in animal tissues seems to be subject to control by translocation between the endoplasmic reticulum and cytosolic compartments (Brindley, 1991). In rat liver it may catalyse the rate-limiting step in glycerolipid synthesis because it has the lowest *in vitro* activity of all the enzymes in the pathway, varies directly with dietary or hormonal regimes that affect triacylglycerol synthesis, and decreases with some drugs that inhibit neutral lipid formation (Bell and Coleman, 1980).

As shown in Fig. 11.21, phosphatidylcholine (the major animal phosphoglyceride) and phosphatidylethanolamine are both synthesized by a (cytidine 5′-diphosphate) CDP-base pathway. This is their major route of formation (Bell and Coleman, 1980; Spanner, 1982). Three enzyme steps are required.

First, the base is phosphorylated by a kinase enzyme. Choline kinase (EC 1.7.1.32) and ethanolamine kinase (EC 2.7.1.82) are soluble enzymes and have been purified from several tissues. The activities may reside in the same protein (Ulane *et al.*, 1977), although separate enzymes have been purified from rat liver (Brophy and Vance, 1976). The purified choline kinase is a dimer of mass 42 kDa (Vance, 1991).

The second enzyme of the pathway is the cytidylyltransferase (EC 2.7.7.15), and this enzyme is partly soluble and partly associated with the endoplasmic reticulum (see Bell and Coleman, 1980). The soluble protein in animals can be aggregated under various conditions and its activity raised by association with membranes (see Vance, 1991). The enzyme seems to be the rate-limiting step in the CDP-choline pathway in animals (Vance and Choy, 1979). The enzyme has been purified (Feldman and Weinhold, 1987), and the cDNA coding for its protein identified and expressed (Vance, 1990). The equivalent enzyme, CDP-ethanolaminephosphate cytidylyltransferase (EC 2.7.7.14), has been partly purified from rat liver (see Vance, 1991).

Choline phosphotransferase (EC 2.7.8.2) and ethanolamine phosphotransferase (EC 2.7.8.1) are located in the endoplasmic reticulum and catalyse the final step in the CDP-base pathway. Their complete purification has yet to be achieved from animals. The two enzymes seem to have somewhat different properties and are distinct proteins (see Vance, 1991).

Phosphatidylcholine can also be synthesized by the stepwise methylation of phosphatidylethanolamine (Fig. 11.21). Methyl groups are transferred

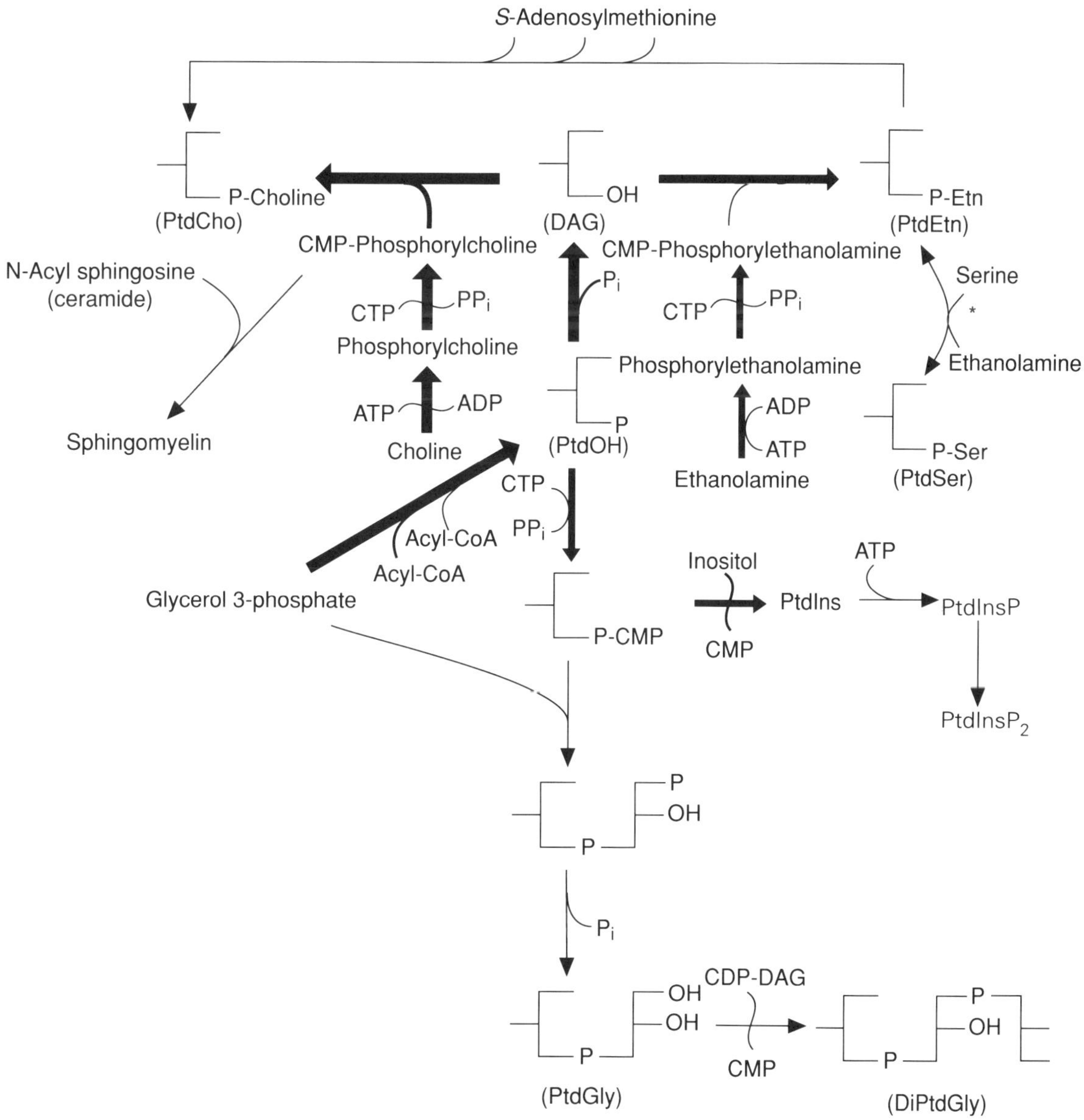

Fig. 11.21 Phospholipid metabolism in animals. The relative thickness of lines shows the approximate carbon flow down each pathway. The asterisk (*) shows that other phospholipids can act as acceptors for serine in base exchange.

from *S*-adenosyl-L-methionine, and this pathway is a minor one in animals (20% of liver phosphatidylcholine synthesis but undetectable in other tissues; Bell and Coleman, 1980). In liver, the three transmethylation reactions are catalysed by a single enzyme with a molecular mass of about 18 kDa. The rate of conversion of phosphatidylethanolamine to phosphatidylcholine appears to be regulated by substrate supply (Vance, 1991). A review of phosphatidylethanolamine methylation has been published (Vance and Ridgway, 1988).

Phosphatidylserine accounts for 5–15% of the total phospholipids in animal cells. It is made by a base exchange reaction (Fig. 11.21) in which the headgroup of a pre-existing phospholipid is exchanged for serine. The enzyme has been purified, and its base exchange activity is not due to phospholipase D activity (see Vance, 1991).

Instead of being dephosphorylated, phosphatidate can be converted to CDP-diacylglycerol. This reaction, catalysed by phosphatidate:CDP-diacylglycerol cytidylyltransferase (EC 2.7.7.41), is important for the synthesis of negatively charged phospholipids such as phosphatidylglycerol, diphosphatidylglycerol and the inositol phospholipids (Fig. 11.21). The cytidylyltransferase has been purified from animals (Bell and Coleman, 1980).

Phosphatidylglycerol can be made by mitochondrial and microsomal fractions from most animal tissues. However, the mitochondrial enzymes are probably more physiologically relevant (Bell and Coleman, 1980), since most of the phosphatidylglycerol (except in lungs; section 12.7) is used for diphosphatidylglycerol formation. The phosphatidylglycerol phosphate synthetase has been localized in the inner mitochondrial membrane, but the phosphatase (Fig. 11.21) is soluble. Together with CDP-diacylglycerol, phosphatidylglycerol is a co-substrate for diphosphatidylglycerol synthesis in animals (Fig. 11.21). This contrasts with the bacterial pathway (see below and Fig. 11.25). A third polyglycerophospholipid, bis(monoacylglycerol) phosphate, comprises less than 1% of the phospholipids of most animal tissues except for alveolar (lung) macrophages, where it represents 14–18% of the phospholipids. Its pathway of formation is under active investigation (Vance, 1991).

Phosphatidylinositol is also formed from CDP-diacylglycerol, though in this case only one enzyme step is involved (Fig. 11.21). Further phosphorylation to phosphatidylinositol 4-phosphate and phosphatidylinositol 4,5-bisphosphate are catalysed by kinase reactions. The first kinase appears to be present in isoforms and these are integral membrane proteins. One enzyme with a low K_m for ATP (20–70 M) and a molecular mass of 55 kDa has been purified to homogeneity (Downes and MacPhee, 1990). Phosphatidylinositol 4-phosphate kinase can be isolated from cytosolic and membrane sources. It appears to be a peripheral membrane protein (Vance, 1991). The above two kinases are important in the 'phosphatidylinositol cycle' (Fig. 11.22), which is connected with the important function of phosphatidylinositol 4,5-bisphosphate as a precursor of second messages that activate many regulatory processes in animal cells (see sections 8.13.2 and 11.2.2; also Vance, 1991).

Phospholipids cannot only be remodelled by base

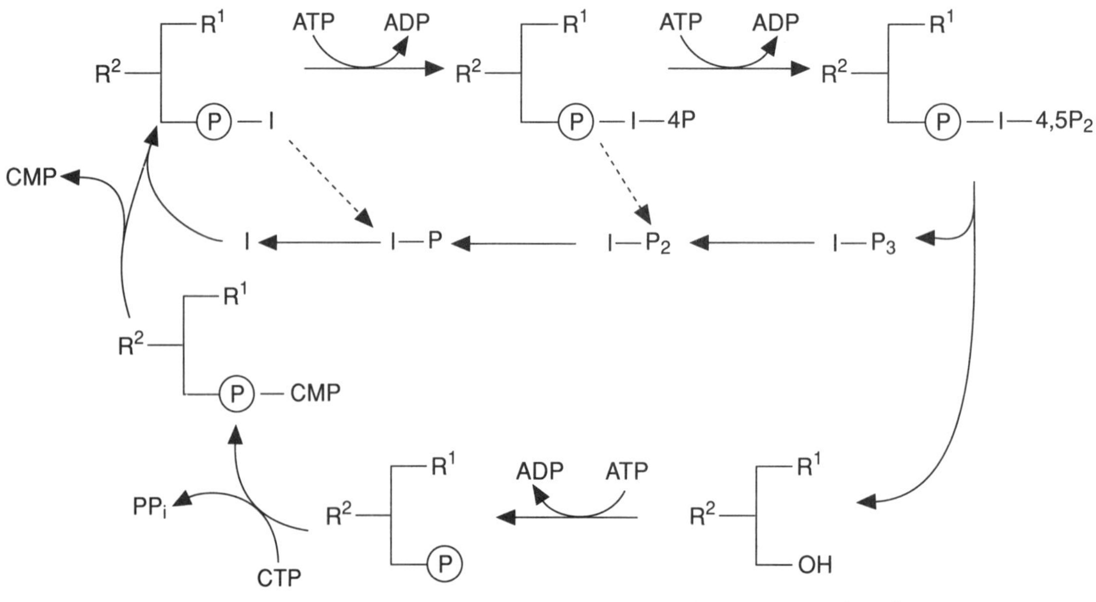

Fig. 11.22 The phosphatidylinositol cycle. Abbreviation: I = inositol. From Vance (1991).

exchange reactions (Taki *et al.*, 1978) but also by reacylation reactions. The latter are extremely important in determining the final molecular species of lipid that are synthesized as well as allowing the participation of specific fatty acids in particular metabolic processes (e.g. release of arachidonate for prostaglandin formation (see van den Bosch, 1982)). The entire subject of reacylation and molecular species turnover has been well reviewed by one of the major research workers in this field (Hill and Lands, 1970). More recent discoveries are covered by Bell and Coleman (1980) and by Hawthorne and Ansell (1982) (see also Vance, 1991; MacDonald and Sprecher, 1991).

The above discussion has centred on the formation of diacylglycerophospholipids, but many phospholipids in Nature have *O*-alkyl bonds. These include such important molecules as the plasmalogens and platelet-activating factor (PAF) (see section 2.3.3). For reviews on the metabolism, regulation and function of ether-linked glycerolipids, refer to Snyder *et al.* (1985) and Snyder (1991).

The overall pathways for phosphoglyceride synthesis in plants have been reviewed (Mudd, 1980; Moore, 1982; Harwood, 1979, 1989). In general, the formation of these lipids is similar to that in animals, but there are some points of difference (Fig. 11.23). First, there are two systems for

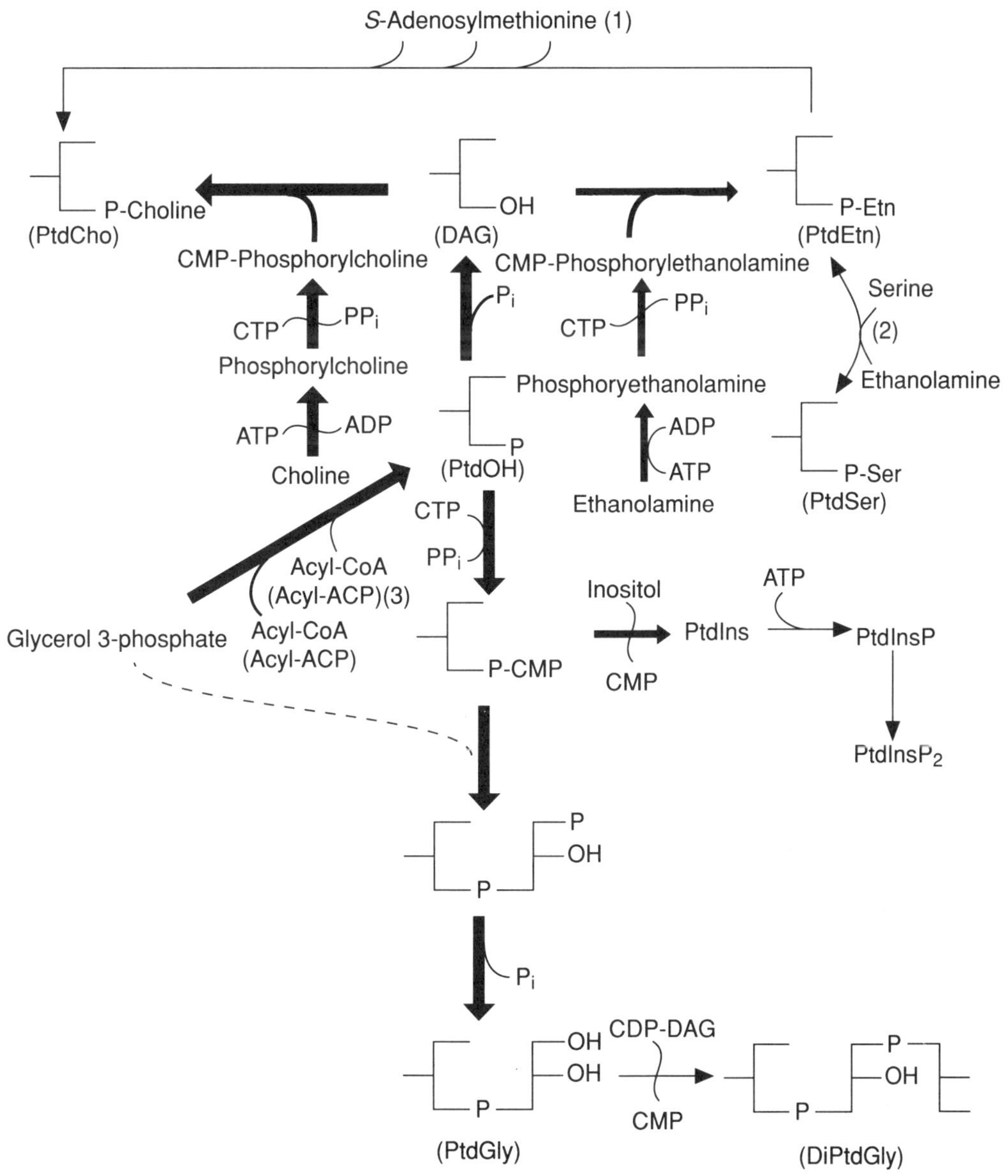

Fig. 11.23 Phospholipid metabolism in plants. The relative thickness of arrows indicates the carbon flux along individual reactions. *Notes*: (1) other pathways for incorporating ethanolamine into PtdChol exist (see text); (2) other phospholipids can serve for base exchange; (3) acyl-ACPs are used by chloroplasts but acyl-CoAs by the endoplasmic reticulum.

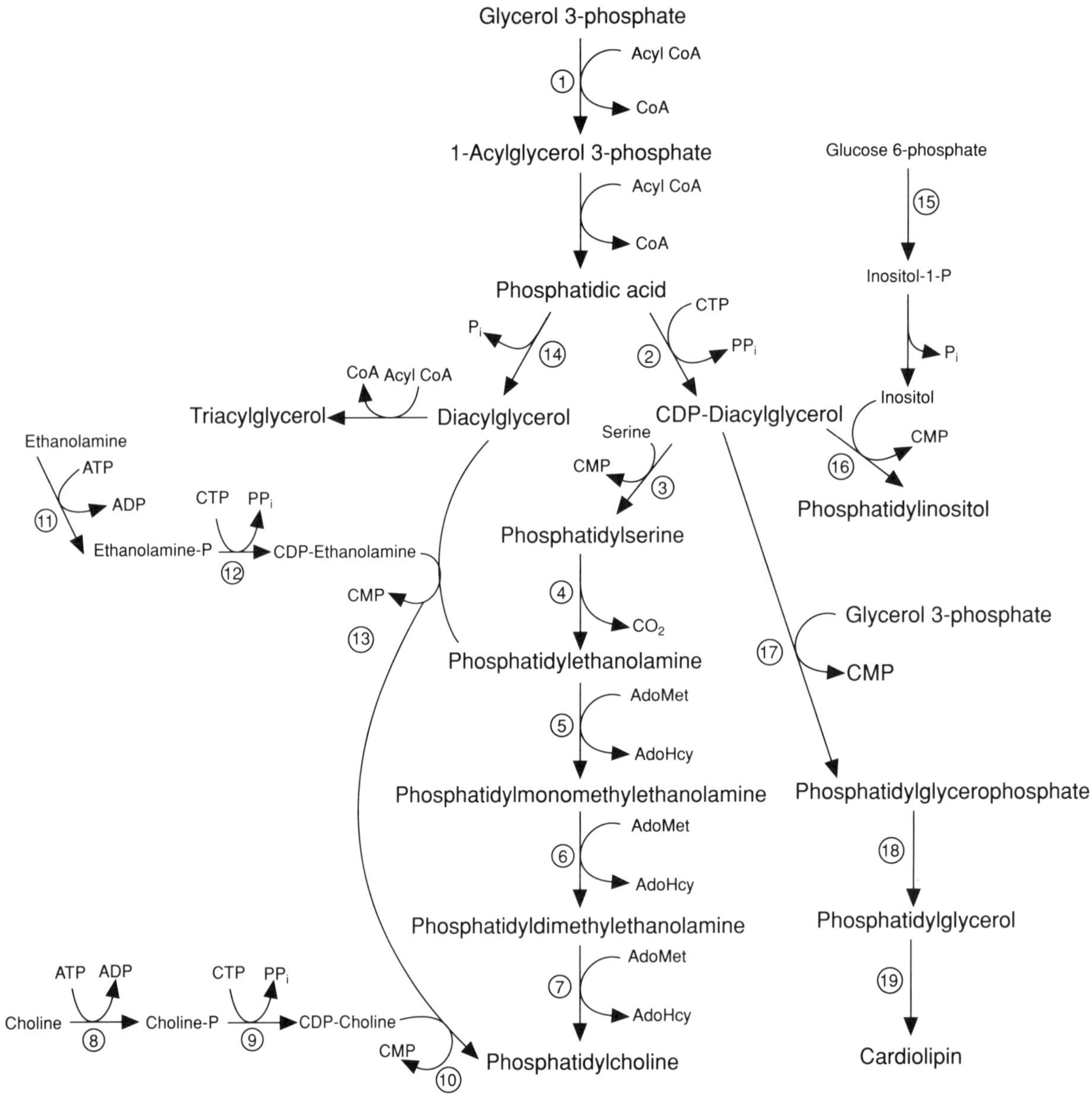

Fig. 11.24 Phospholipid biosynthetic pathways in *Saccharomyces cerevisiae*. The indicated reactions are catalysed by the following enzymes: (1) glycerol 3-phosphate acyltransferase; (2) CDP-DG synthetase; (3) PS synthetase; (4) PS decarboxylase; (5) PE methyltransferase; (6) and (7) phospholipid methyltransferase; (8) choline kinase; (9) choline-phosphate cytidylyltransferase; (10) cholinephosphotransferase; (11) ethanolamine kinase; (12) ethanolaminephosphate cytidylyltransferase; (13) ethanolaminephosphotransferase; (14) PA phosphatase; (15) I-1-P synthetase; (16) PI synthetase; (17) PGP synthetase; (18) PGP phosphatase; (19) CL synthetase. From Carman and Henry (1989).

the initial acylation of glycerol 3-phosphate. In the chloroplast-located pathway, acyl-ACPs are used and the substrate specificity of the enzymes results in the preferential formation of *sn*-1-oleoyl, *sn*-2-palmitoyl molecular species. Subsequent utilization of such phosphatidate for lipid synthesis retains this distribution of fatty acids and gives rise to what have been called 'prokaryotic' molecular species by analogy with cyanobacterial lipid metabolism (see Roughan and Slack, 1984; Harwood, 1989; Joyard and Douce, 1987). A second glycerol 3-phosphate acylation system is located in the endoplasmic reticulum, uses acyl-CoAs and has a substrate specificity such that more saturated fatty acyl groups are usually located at the *sn*-1 position, just as in animals (see Harwood, 1989). The endoplasmic reticulum acylations are responsible for extra-chloroplastic phospholipid synthesis, triacylglycerol formation (Harwood and Page, 1993) as well as some chloroplast lipid synthesis (see Roughan and Slack, 1984; Browse and Somerville, 1991).

The major plant extrachloroplastic lipid, phosphatidylcholine, is mainly made by the CDP-base pathway (Harwood, 1979; Moore, 1982). Separate choline and ethanolamine kinases have been purified from soybean (Harwood, 1979), and the cytidylyltransferase step appears to be rate-limiting (Price-Jones and Harwood, 1986). Separate base-transferases for choline- and for ethanolaminephosphate appear to be present in soybean (see Harwood, 1979).

Although evidence has been produced that plants can convert phosphatidylethanolamine to phosphatidylcholine by stepwise methylation (Mudd, 1980; Moore, 1982), some recent experiments suggest that methylation occurs on various water-soluble intermediates such as ethanolamine-phosphate, which are then incorporated into phosphatidylcholine via phosphatidyldimethyl- (or methyl-) ethanolamine (see Mudd and Datko, 1986).

Phosphatidylserine, phosphatidylglycerol and phosphatidylinositol seem to be synthesized in plants by pathways similar to those discussed above for animals (Mudd, 1980; Moore, 1982; and see Figs 11.21 and 11.23). Phosphatidylglycerol synthesis has been demonstrated in chloroplasts (see Harwood, 1989), in keeping with the major role of this lipid in thylakoid membranes. The metabolism of inositol lipids and their function in cellular regulation in plants has been reviewed recently (Hetherington and Drobak, 1992).

Pathways for the biosynthesis of phospholipids in yeast are generally similar to those demonstrated in other eukaryotes, except that phosphatidylserine is made exclusively from CDP-diacylglycerol (Fig. 11.24) and not by base exchange. Phosphatidylserine is also converted to phosphatidylethanolamine by decarboxylation. Work with yeast has been particularly useful as a model system for other eukaryotes because of a large knowledge base in classical genetics and the ability to elucidate and manipulate individual genes coding for enzymes in the pathways. Purification and properties of CDP-diacylglycerol synthetase, phosphatidylserine synthetase, phosphatidylinositol synthetase and phosphatidylinositol kinase are described by Carman and Henry (1989). These authors also discuss the genetics and molecular biology of phospholipid synthesis in yeasts.

Phospholipid biosynthesis in micro-organisms has been well reviewed by Pieringer (1989). Major pathways for *Escherichia coli* are shown in Fig. 11.25. Several points are of note. First, the acylation of glycerol 3-phosphate uses acyl-ACPs and, thus, phospholipid synthesis is coupled directly to fatty acid formation (Jackowski *et al.*, 1991). Secondly, as in yeasts (Fig. 11.24), phosphatidylserine is made from CDP-diacylglycerol and can then be decarboxylated to phosphatidylethanolamine. Thirdly, the CDP-base pathway for phosphatidylethanolamine is not used (and in bacteria that contain phosphatidylcholine, the latter is made exclusively by the methylation of phosphatidylethanolamine (see Vance and Ridgway, 1988)). Fourthly, diphosphatidylglycerol is made by a reaction involving two phosphatidylglycerol molecules (Fig. 11.21).

Most research in bacteria has concentrated on *E. coli*, a Gram-negative bacterium with a rather restricted habitat. It cannot be emphasized too strongly that pathways and enzymes studied in *E. coli* may not be typical (or even present) in all bacteria (Harwood and Russell, 1984). The overall pathways for phospholipid synthesis in a Gram-positive bacterium, *Streptococcus falcium*, are shown in Fig. 11.26. Individual features of lipid metabolism in bacteria and other prokaryotes or microbes are covered by Pieringer (1989) and by Jackowski *et al.* (1991).

Methods for the isolation and study of numerous enzymes of phospholipid metabolism are given in Dennis and Vance (1992).

11.2.2 Breakdown

The breakdown of glycerophospholipids is catalysed by a series of phospholipases designated A, B, C and D depending on their positions of attack (Fig. 11.27). In addition, phosphatidic acid phosphatase is important, but this has already been dealt with (section 11.2.1). Two phospholipase A specificities are recognized, and these are named A_1 and A_2 depending on the position of the ester hydrolysed in the diacylphosphoglyceride. Phospholipase B is the name usually used for an enzyme that is a monoacylphosphoglyceride acylhydrolase. For reviews of the basic features of phospholipase action, the reader is referred to Hill and Lands (1970), Dawson (1973), van den Bosch (1982), Dennis (1983) and Waite (1987).

In common with many other lipid-metabolizing enzymes, the assay of phospholipases deserves careful consideration. Principally, the potential difficulties are due to the water-insoluble nature of the substrate, the effects of various incubation additions (e.g. solvents, detergents), which can make a big difference to activity, and the problems in kinetic interpretation. Various assay methods have been discussed (Waite, 1987) and an important volume of *Methods in Enzymology* (Dennis, 1991) published. The interpretation of the kinetics of phospholipase inhibition has been the subject of considerable attention (see Harwood and John, 1990), and recently, suicide-inhibitory bifunctionally

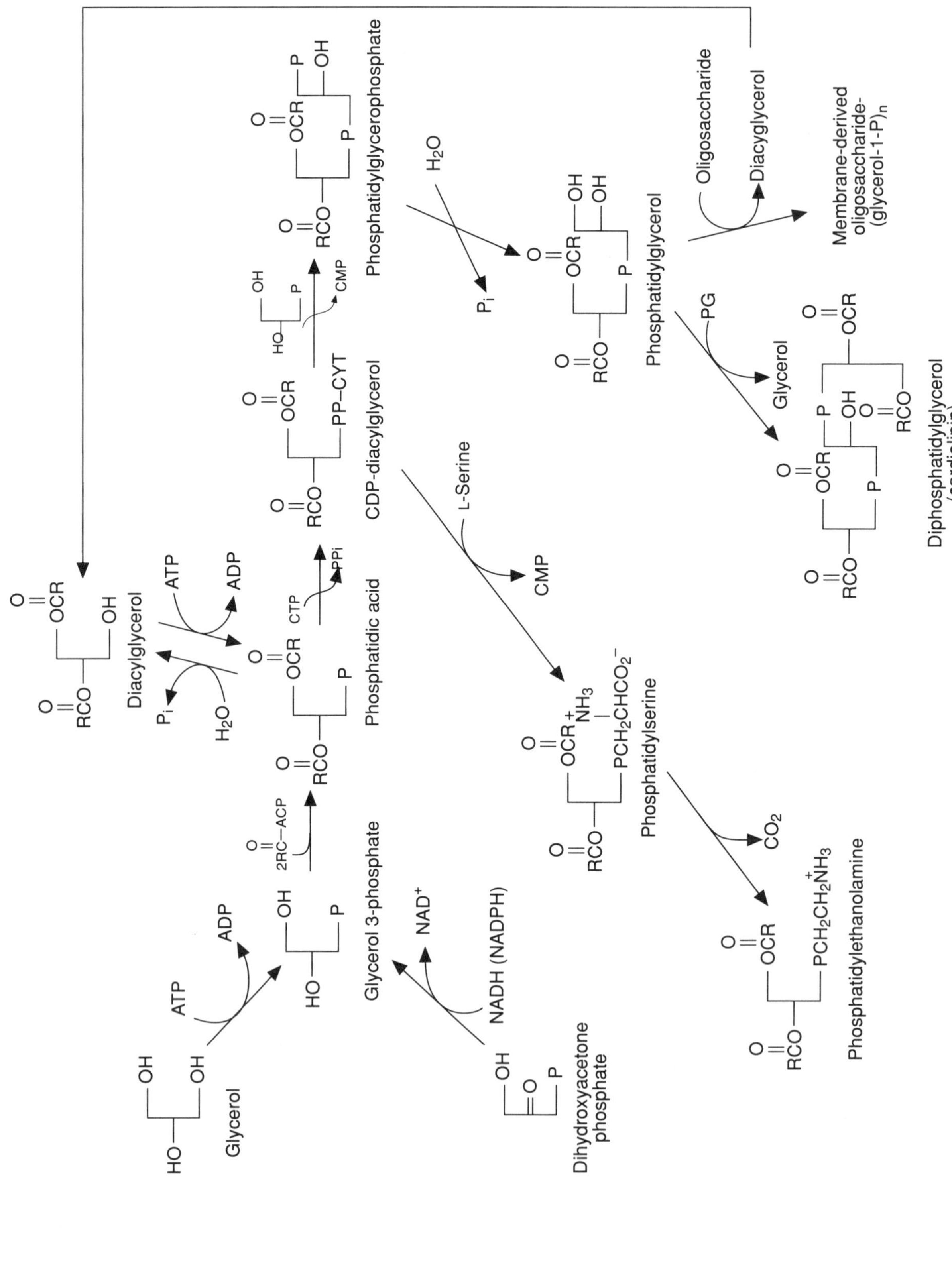

Fig. 11.25 Phospholipid metabolism in *E. coli*. From Pieringer (1989).

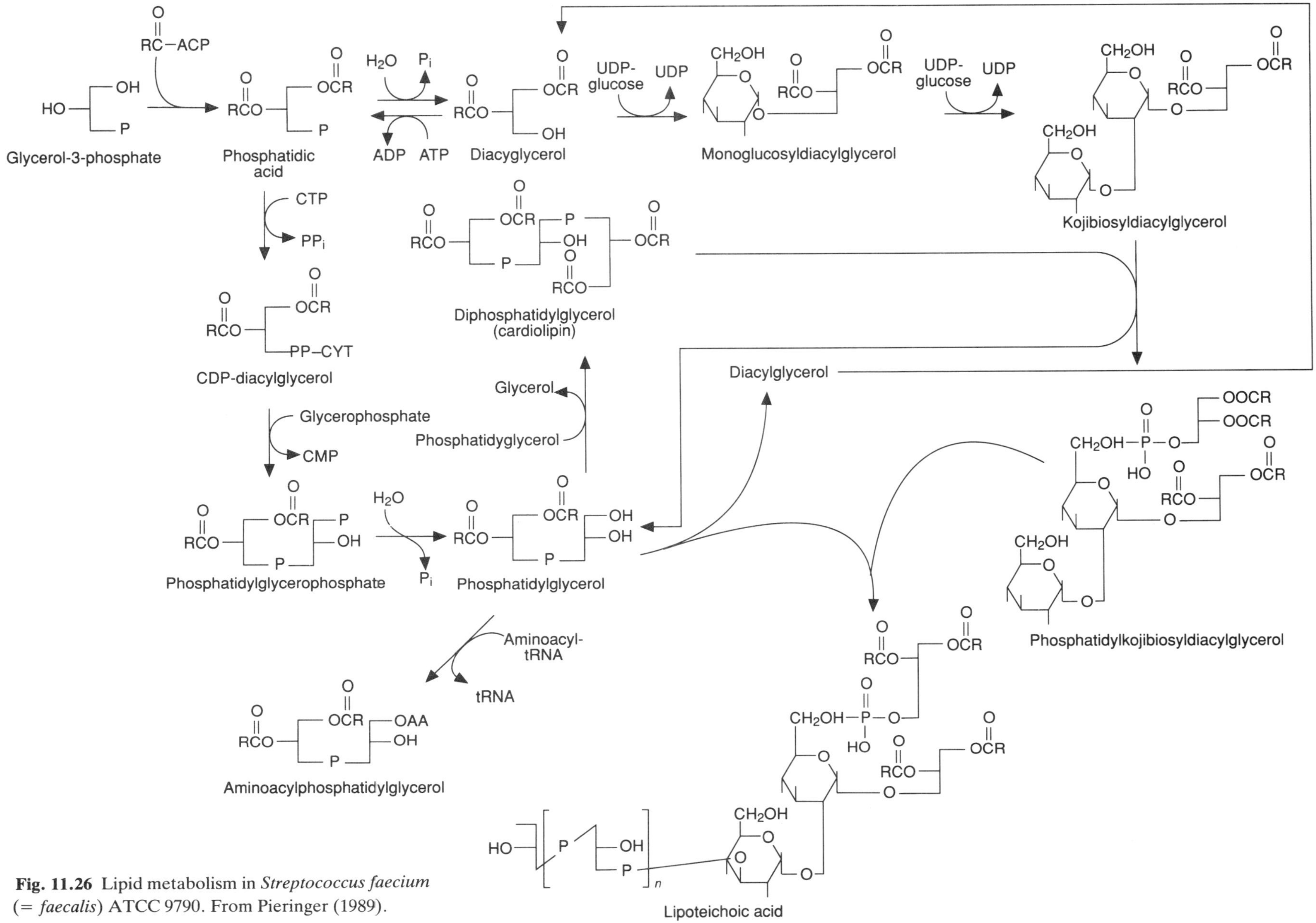

Fig. 11.26 Lipid metabolism in *Streptococcus faecium* (= *faecalis*) ATCC 9790. From Pieringer (1989).

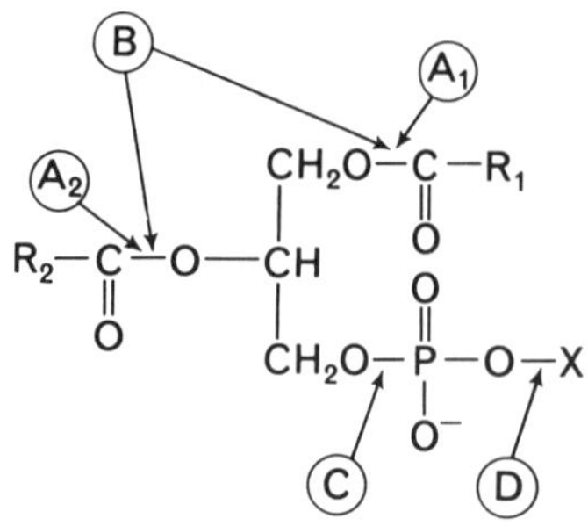

Fig. 11.27 Position of phosphoglyceride hydrolysis for different phospholipases.

linked substrates have been developed as phospholipase A_2 inhibitors (Washburn and Dennis, 1991).

Phospholipase A enzymes – particularly those from snake venoms or digestive secretions – have been widely studied. Phospholipase A_1 is found in microsomal and liposomal fractions (cf. Newkirk and Waite, 1971; Gatt, 1968). It specifically deacylates phosphatidylcholine or phosphatidylethanolamine at the 1-position. Both these substrates are hydrolysed at the same rate by the adrenal medulla lysosomal enzyme, but that from brain prefers phosphatidylcholine as substrate. Detergents will increase phosphatidylethanolamine hydrolysis by the brain enzyme. A phospholipase A_1 (which is relatively specific for phosphatidylglycerol) has been reported from the spores of some bacteria (Raybin *et al.*, 1972), but most bacterial enzymes are unspecific for either the 1- or the 2-positions.

Examples of phospholipase A (unspecific) enzymes from bacteria include the so-called detergent-resistant enzyme from the outer membrane of *E. coli* and a cytoplasmic enzyme from the same bacterium (Nakagawa *et al.*, 1991). These enzymes differ considerably in their properties (Harwood and Russell, 1984). A number of Gram-negative bacteria have phospholipase A activity in their outer membranes, but it is often not determined whether they are phospholipase A_1 or A_2. Several other lipases have been noted to have phospholipase (usually phospholipase A_1) activity. These include rat hepatic lipase (Waite *et al.*, 1991) and milk lipoprotein lipase (Bengtsson-Olivecrona and Olivecrona, 1991).

The specificity of phospholipase A_2 enzymes varies considerably even amongst those from snake venoms (see Dawson, 1973). The enzymes remove a fatty acyl group on the glycerol carbon adjacent to the phosphoryl substituent – i.e. position 2 in a natural phosphoglyceride. If the phosphoryl substituent is at the 2 position, then only a fatty acid with the correct steric configuration is hydrolysed. Substrate specificities also vary markedly with the chain length and degree of unsaturation. For snake venom enzyme hydrolysing phosphatidylcholine, the rate of hydrolysis is in the order: (1-unsat., 2-sat.) > (1-unsat., 2-unsat.) > (1-sat., 2-polyunsat.) > (1-sat., 2-mono-unsat.) > (1-sat., 2-sat.). In contrast, the pancreatic enzyme exhibits no preference for chain length or degree of unsaturation. The hydrolysis of certain phospholipids can be increased in different ways for the various phospholipase A_2 enzymes, e.g. by detergent or diethyl ether addition. This aspect and comments on the underlying physico-chemical mechanisms are discussed by Dawson (1973), Slotboom *et al.* (1982) and Waite (1987), where further references will be found. Specific details of purification and characterization of individual phospholipases are given in Dennis (1991).

Because of the ready availability of the pancreatic phospholipase A_2, this enzyme has been thoroughly examined by enzymologists. The phospholipase is secreted as a zymogen and activated by cleavage of a peptide of three to seven amino acids. The amino acid sequence of the active enzyme has been determined for a number of different preparations (pig, ox, horse and human). The different preparations show a high degree of homology, e.g. the human phospholipase differs from the others in only a single amino acid residue out of 123–125 in each case (Verheij *et al.*, 1983). Further details are given in Waite (1987). Very recently a gene encoding the cobra venom phospholipase A_2 has been expressed in *E. coli* and the recombinant enzyme shown to have the expected activity characteristics (Kelley *et al.*, 1992).

Lysophospholipases are probably distributed as widely as the phospholipases A (Waite, 1987). A true definition of lysophospholipases is difficult, as discussed by van den Bosch (1982). Indeed, a number of enzymes in this group also have activity against diacyl phospholipids and are, therefore, more properly referred to as phospholipase B or A_1 enzymes (van den Bosch *et al.*, 1974). Also, there are several lysophospholipases that show esterase activity and whose classification is difficult. For more discussion see Waite (1987) and specific chapters on individual lysophospholipases or the so-called phospholipase B from *Penicillium notatum* in Dennis (1991).

Phospholipase C enzymes are secreted by several bacteria, particularly pathogens such as *Clostridium* spp. These proteins are zinc-metalloenzymes and some also require Ca^{2+} for activity. The phospholipase C enzymes vary widely in substrate specificity (van den Bosch, 1982). The enzyme from *Clostridium perfringens* attacks choline-containing lipids most readily. Such hydrolysis requires a positive zeta-potential brought about by long-chain cations or by the addition of Ca^{2+} or Mg^{2+}. Phosphatidylcholine breakdown in the presence of Ca^{2+} is greatly increased by the addition of specific concentrations of sodium deoxycholate. The phospholipase C from *Bacillus cereus* attacks a large number of phospholipids (e.g. PG, DPG), which are poorly hydrolysed by the *C. perfringens* enzyme. It will also hydrolyse short-chain phosphoglycerides, which are poor substrates for the latter enzyme (Dawson, 1973). The use of phospholipase C enzymes in the preparation of diacylglycerols for molecular species examination is discussed by Christie (1982).

Mammalian phospholipase C enzymes are now known to be central in regulatory processes (Waite, 1987). Most, if not all, of these enzymes are specific for phosphatidylinositol or its phosphorylated derivatives. Specific enzymes are discussed in Dennis (1991), and their role in transmembrane signalling in various organisms (Berridge and Irvine, 1989) and in higher plants (Hetherington and Drobak, 1992) has been reviewed (see also section 8.13.2). Specific examples of important functional aspects of such phospholipase C activity are given by Turk *et al.* (1987) and by Cockcroft (1992) in relation to insulin secretion by pancreatic islets and in neutrophils, respectively.

There are few examples of phospholipase D (EC 3.1.4.4) in bacteria (Harwood and Russell, 1984), but many plant tissues exhibit high activity. Plant phospholipase D is really a transferase enzyme that catalyses transphosphatidylation to an acceptor molecule. If the enzyme is not fully inactivated, then this reaction will lead to the artifactual formation of phosphatidylmethanol in plant tissues extracted with methanolic solutions (section 6.3.1; Harwood, 1980). A variety of water-soluble primary alcohols can be used as acceptors (e.g. propanol, ethanol, ethylene glycol, glycerol). Most enzymes are stimulated in the presence of linear aliphatic ethers (Dawson, 1973; Galliard, 1980). A possible function in plants for phospholipase D in synthesizing phospholipids has now been largely discounted because the phosphatidylglycerol thus made is a racemic mixture, unlike the naturally occurring phospholipid (Yang *et al.*, 1967). However, Batrakov *et al.* (1975) found evidence for stereospecificity in the transfer. Roughan and Slack (1976) have claimed that phospholipase D is a structural protein that under certain non-physiological conditions possesses enzymic activity.

There has been some recent interest in the possible role of mammalian phospholipase D in signal transduction, in that a wide variety of agents (including hormones, neurotransmitters and growth factors) have been shown to activate a phospholipase D to hydrolyse phosphatidylcholine (see e.g. Hurst *et al.*, 1990; Xie and Dubyak, 1991). A comprehensive review of early data on phospholipase D is that of Heller (1978), and recent information on a glycosylphosphatidylinositol-specific activity (Huang *et al.*, 1991) and an enzyme from rats (Kobayashi and Kanfer, 1991) has been published. Yamane *et al.* (1989) discuss the commercial use of phospholipase D enzymes for transphosphatidylation purposes.

Phosphatidate phosphohydrolase, although a phospholipid degradative enzyme, is mainly involved in the biosynthesis of lipids (Gurr and Harwood, 1991). In mammalian tissues the enzyme may control the overall rate of lipid (triacylglycerol) deposition. A comprehensive review of the enzyme in different organisms and tissues has been published by Brindley (1988) (see also Dennis, 1991).

Many plant tissues also possess high activities of non-specific lipid acyl hydrolases – which can also give rise to problems with the solvent extraction of plants. Acyl hydrolases will hydrolyse fatty acids from a large number of lipids including phosphoglycerides (cf. Galliard, 1980). The enzymes have been purified from several tissues and their substrate specificities and other properties determined. Potato tubers and leaves from *Phaseolus* spp. are particularly rich sources (cf. Galliard, 1980), and the position of bond cleavage has been determined for the latter enzyme to be on the fatty acid side of the oxygen ester bond (Burns *et al.*, 1980).

For an overall review of the functions of phospholipases in such areas as the phosphatidylinositol cycle, the arachidonate cascade, in the digestion of dietary fat in lipoprotein metabolism and in snake venoms, see Waite (1987) and also Vance (1991). Phospholipase A_2 has also been invoked in the protection of membranes from lipid peroxidation damage (van Kuijk *et al.*, 1987).

References and further reading

Batrakov, S.G., Pandsyan, A.G., Kogan, G.A. and Bergelson, L.D. (1975) *Biochem. Biophys. Res. Commun.*, **66**, 755.

Bell, R.M. and Coleman, R.A. (1980) *Annu. Rev. Biochem.*, **49**, 459.

Bengtsson-Olivecrona, G. and Olivecrona, T. (1991) in *Methods in Enzymology*, vol. 197 (ed. E.A. Dennis), Academic Press, New York, p. 345.

Berridge, M.J. and Irvine, R.F. (1989) *Nature*, **341**, 197.

Bloch, K. and Vance, D. (1977) *Annu. Rev. Biochem.*, **46**, 263.

Brindley, D.N. (ed.) (1988) *Phosphatidate Phosphohydrolase*, vols I and II, CRC Press, Boca Raton, FL.

Brindley, D.N. (1991) in *Biochemistry of Lipids*,

Lipoprotein and Membranes (eds D.E. Vance and J. Vance), Elsevier, Amsterdam, p. 171.
Brindley, D.N. and Sturton, R.G. (1982) in *Phospholipids* (eds J.N. Hawthorne and G.B. Ansell), Elsevier, Amsterdam, p. 179.
Brophy, P.J. and Vance, D.E. (1976) *FEBS Lett.*, **62**, 123.
Browse, J. and Somerville, C.R. (1991) *Annu. Rev. Plant Physiol.*, **42**, 467.
Burns, D.D., Galliard, T. and Harwood, J.L. (1980) *Phytochemistry*, **19**, 2281.
Carman, G.M. and Henry, S.A. (1989) *Annu. Rev. Biochem.*, **58**, 635.
Christie, W.W. (1982) *Lipid Analysis*, 2nd edn, Pergamon, Oxford.
Cockcroft, S. (1992) *Biochim. Biophys. Acta*, **1113**, 135.
Dawson, R.M.C. (1973) in *Form and Function of Phospholipids* (eds G.B. Ansell, R.M.C. Dawson and J.N. Hawthorne), Elsevier, Amsterdam, p. 97.
Dennis, E.A. (1983) in *The Enzymes*, vol. XVI (ed. P. Boyer), Academic Press, New York, p. 307.
Dennis, E.A. (ed.) (1991) *Methods in Enzymology*, vol. 197, Academic Press, New York.
Dennis, E.A. and Vance, D.E. (eds) (1992) *Methods in Enzymology*, vol. 209, Academic Press, New York.
Downes, C.P. and MacPhee, C.H. (1990) *Eur. J. Biochem.*, **193**, 1.
Feldman, D.A. and Weinhold, P.A. (1987) *J. Biol. Chem.*, **262**, 9075.
Galliard, T. (1980) in *Biochemistry of Plants*, vol. 4 (eds P.K. Stumpf and E.E. Conn), Academic Press, New York, p. 85.
Gatt, S. (1968) *Biochim. Biophys. Acta*, **159**, 304.
Groot, P.H.E., Scholte, H.R. and Hülsman, W.C. (1976) *Adv. Lipid Res.*, **14**, 75.
Gurr, M.I. and Harwood, J.L. (1991) *Lipid Biochemistry*, 4th edn, Chapman & Hall, London.
Harwood, J.L. (1979) *Prog. Lipid Res.*, **18**, 55.
Harwood, J.L. (1980) in *Biochemistry of Plants*, vol. 4 (eds P.K. Stumpf and E.E. Conn), Academic Press, New York, p. 1.
Harwood, J.L. (1989) *Crit. Rev. Plant Sci.*, **8**, 1.
Harwood, J.L. and John, R.A. (1990) *Trends Biochem. Sci.*, **15**, 409.
Harwood, J.L. and Page, R. (1993) in *Designer Oilseed Crops* (ed. D.J. Murphy), VCH Press, Weinheim.
Harwood, J.L. and Russell, N.J. (1984) *Lipids in Plants and Microbes*, Allen and Unwin, Hemel Hempstead.
Hawthorne, J.N. and Ansell, G.B. (eds) (1982) *Phospholipids*, Elsevier, Amsterdam.
Heller, H. (1978) *Adv. Lipid Res.*, **16**, 267.
Hetherington, A.M. and Drobak, B.K. (1992) *Prog. Lipid Res.*, **31**, 53.
Hill, E.E. and Lands, W.E.M. (1970) in *Lipid Metabolism* (ed. S.J. Wakil), Academic Press, New York, p. 185.
Huang, K.-S., Li, S. and Low, M.G. (1991) *Methods in Enzymology*, vol. 197 (ed. E.A. Dennis), Academic Press, New York, p. 567.
Hurst, K.M., Hughes, B.P. and Barritt, G.J. (1990) *Biochem. J.*, **272**, 749.
Jackowski, S., Cronan, J.E. and Rock, C.O. (1991) in *Biochemistry of Lipids, Lipoproteins and Membranes* (eds D.E. Vance and J. Vance), Elsevier, Amsterdam, p. 43.
Jackson, R.L. and McLean, L.R. (1991) *Methods in Enzymology*, vol. 197 (ed. E.A. Dennis), Academic Press, New York, p. 339.
Joyard, J. and Douce, R. (1987) in *The Biochemistry of Plants*, vol. 9 (eds P.K. Stumpf and E.E. Conn), Academic Press, New York, p. 215.
Kelley, M.J., Crawl, R.M. and Dennis, E.A. (1992) *Biochim. Biophys. Acta*, **1118**, 107.
Kobayashi, M. and Kanfer, J.N. (1991) *Methods in Enzymology*, vol. 197 (ed. E.A. Dennis), Academic Press, New York, p. 575.
MacDonald, J.I.S. and Sprecher, H. (1991) *Biochim. Biophys. Acta*, **1084**, 105.
Moore, T.M. (1982) *Annu. Rev. Plant Physiol.*, **33**, 235.
Mudd, J.B. (1980) in *Biochemistry of Plants*, vol. 4 (eds. P.K. Stumpf and E.E. Conn), Academic Press, New York, p. 249.
Mudd, S.H. and Datko, A.H. (1986) *Plant Physiol.*, **82**, 126.
Nakagawa, Y., Setaka, M. and Nojima, S. (1991) *Methods in Enzymology*, vol. 197 (ed. E.A. Dennis), Academic Press, New York, p. 309.
Newkirk, J.D. and Waite, M. (1971) *Biochim. Biophys. Acta*, **225**, 224.
O'Doherty, P.J.A. (1978) in *Handbook of Lipid Research*, vol. 1 (ed. A. Kuksis), Plenum, New York, p. 289.
Pieringer, R.A. (1989) in *Microbial Lipids* vol. 2 (eds C. Ratledge and S.G. Wilkinson), Academic Press, London, p. 51.
Price-Jones, M.J. and Harwood, J.L. (1986) *Biochem. J.*, **240**, 837.
Raybin, D.M., Bertsch, L.L. and Kornberg, A. (1972) *Biochemistry*, **11**, 1754.
Roughan, P.G. and Slack, C.R. (1976) *Biochim. Biophys. Acta*, **431**, 86.
Roughan, P.G. and Slack, C.R. (1984) *Trends Biochem. Sci.*, **9**, 383–6.
Snyder, F. (1991) in *Biochemistry of Lipids, Lipoproteins and Membranes* (eds D.E. Vance and J. Vance), Elsevier, Amsterdam, p. 241.
Snyder, F., Lee, T.-C. and Wykle, R.L. (1985) in *The Enzymes of Biological Membranes* (ed. A.N. Martonosi), Plenum, New York, p.1.
Slotboom, A.J., Verheij, H.M. and de Haas, G.H. (1982) in *Phospholipids* (eds. J.N. Hawthorne and G.B. Ansell), Elsevier, Amsterdam, p. 359.
Spanner, S. (1982) in *Phospholipids* (eds J.N. Hawthorne and G.B. Ansell), Elsevier, Amsterdam, p. 1.
Taki, T., Miura, T. and Kanfer, J.N. (1978) *Adv. Exp. Med. Biol.*, **101**, 301.
Turk, J., Wolf, B.A. and McDaniel, M.L. (1987) *Prog. Lipid Res.*, **26**, 125.
Ulane, R.E., Stephenson, L.L. and Farell, P.M. (1977) *Anal. Biochem.*, **79**, 526.
Vance, D.E. (1990) *Biochem. Cell Biol.*, **68**, 1151.
Vance, D.E. (1991) in *Biochemistry of Lipids, Lipoproteins and Membranes* (eds D.E. Vance and J. Vance), Elsevier, Amsterdam, p. 205.

Vance, D.E. and Choy, I.C. (1979) *Trends Biochem. Sci.*, **4**, 145.

Vance, D.E. and Ridgway, N.D. (1988) *Prog. Lipid Res.*, **27**, 61.

van den Bosch, H. (1982) in *Phospholipids* (eds J.N. Hawthorne and G.B. Ansell), Elsevier, Amsterdam, p. 313.

van den Bosch, H., Aarsman, A.J. and van Deenen, L.L.M. (1974) *Biochim. Biophys. Acta*, **348**, 197.

van Kuijk, F.J.G.M., Sevanian, A., Handelman, G.J. and Dratz, E.A. (1987) *Trends Biochem. Sci.*, **12**, 31.

Verheij, H.M., Westerman, J., Sternby, B. and de Haas, G.H. (1983) *Biochim. Biophys. Acta*, **747**, 93.

Waite, M. (1987) *Handbook of Lipid Research*, vol. 5, Plenum, New York.

Waite, M., Thuren, T.Y., Wilcox, R.W., Sisson, P.J. and Kucera, G.L. (1991) *Methods in Enzymology*, vol. 197 (ed. E.A. Dennis), Academic Press, New York, p. 331.

Waku, K. (1992) *Biochim. Biophys. Acta*, **1124**, 101.

Washburn, W.N. and Dennis, E.A. (1991) *J. Biol. Chem.*, **266**, 5042.

Webber, K.O. and Hajra, A.K. (1992) *Methods in Enzymology*, vol. 209 (eds E.A. Dennis and D.E. Vance), Academic Press, New York, p. 92.

Xie, M. and Dubyak, G.R. (1991) *Biochem. J.*, **278**, 81.

Yamane, T., Juneja, L.R. and Shimizu, D. (1989) in *Fats for the Future* (ed. R.C. Cambie), Ellis Horwood, Chichester, p. 128.

Yang, S.F., Freer, S. and Benson, A.A. (1967) *J. Biol. Chem.*, **242**, 477.

11.3 Glyceride metabolism

11.3.1 Triacylglycerol synthesis

The lipid stores of animals and almost all plants are triacylglycerols. These compounds, therefore, are of very great importance in physiology as well as to the food industry. One difference between animals and plants lies in the relative amounts of *de novo* synthesized acyl chains in the triacylglycerols. Plants from necessity must make all of the glyceride molecule from simple starting materials (ultimately from photosynthetically fixed CO_2). In contrast, animals make use of dietary fatty acids, which to a large degree determine the composition of the triacylglycerol stores (e.g. copepods consuming phytoplankton; Sargent, 1979).

In oil seeds or oil-rich fruits the triacylglycerol stores are located in oil bodies. These intracellular structures are surrounded by a half-unit phospholipid membrane containing unique proteins called oleosins. The latter have been isolated and sequenced from a number of plants, and probably serve both to stabilize the oil bodies as well as to provide binding sites for lipases to function during germination (for reviews see Murphy, 1990; Huang, 1992).

In animals most triacylglycerol is stored in adipose tissue, though, in times of metabolic stress, significant amounts may build up in liver, heart and skeletal muscle. The formation of triacylglycerols removes the potentially harmful effects of fatty acids or fatty acyl-CoAs (Brindley, 1991). A major function of animal triacylglycerols is to allow the transport of acyl moieties about the body, in the form of the serum lipoproteins. Two major classes of lipoprotein are relevant. Chylomicrons carry absorbed dietary fat from the intestine to other organs, while very-low-density lipoproteins carry triacylglycerol from the liver to other tissues.

There has been considerable recent interest in the use of oleaginous micro-organisms as sources of triacylglycerols, particularly those with unusual fatty acids. In general, yeasts and moulds offer the best possibilities for industrial exploitation, and few bacteria accumulate appreciable amounts of triacylglycerols (Actinomycetes are the most noticeable exceptions). The whole topic of the biotechnology of oils and fats in micro-organisms is reviewed by Ratledge (1989). See also Davies and Holdsworth, 1992.

The major pathways for triacylglycerol synthesis are shown in Fig. 11.28. The formation of phosphatidic acid by the glycerol phosphate pathway or the dihydroxyacetone phosphate pathway has already been discussed (section 11.2.1). In addition, see reviews by Gurr (1980) and O'Doherty (1978). Phosphatidic acid can also be formed from monoacylglycerol or diacylglycerol, but, as we will see, this is not of relevance for triacylglycerol synthesis. Attempts to estimate the relative contribution of different pathways for phosphatidic acid formation (cf. Hill and Lands, 1970; O'Doherty, 1978) indicate that about 50% of lipid glycerol enters by the dihydroxyacetone pathway (e.g. Manning and Brindley, 1972; Pollock *et al.*, 1975). In some tissues (e.g. Ehrlich ascites tumour cells) the acyldihydroxyacetone pathway is the major route of synthesis (Snyder, 1972).

The enzymes responsible for the acylation of glycerol phosphate have been purified and studied many times. For reviews of their purification and properties, see Lennarz (1970), Hill and Lands (1970), O'Doherty (1978), Gurr (1980) and Stymne and Stobart (1987). The glycerol 3-phosphate acyltransferase and monoacylglycerol 3-phosphate acyltransferase enzymes in both animals and plants differ in their substrate specificities. Typically, the glycerol 3-phosphate acyltransferase uses more saturated acyl-CoAs than are subsequently attached to the *sn*-position. The major site of both enzymes is the endoplasmic reticulum, though animal mitochondria (particularly those of liver) also contain a second glycerol 3-phosphate acyltransferase (Brindley, 1991). In plant chloroplasts, glycerol 3-phosphate acylating enzymes (which use acyl-ACP substrates and have different substrate specificities to their endoplasmic reticulum counterparts) are present. Although triacylglycerol synthesis can take place in chloroplasts (particularly under adverse environmental conditions; see Sakaki *et al.*, 1985), the chloroplast enzymes are mainly important for the generation of thylakoid acyl lipids (Ohlrogge *et al.*, 1991). For the dihydroxyacetone pathway (Fig. 11.28), acyltransferases have been found in both the endoplasmic reticulum and in peroxisomes (Brindley, 1991).

As indicated in Fig. 11.28, diacylglycerol can originate from two sources. Either it is formed by phosphatidate phosphohydrolase or it is synthesized from monoacylglycerol. The first demonstration of an enzyme system capable of converting monoacylglycerols to triacylglycerols was the work of Clark and Hübscher (1960) with preparations from rabbit intestine. Indeed, the reactions have been studied most thoroughly in intestine, although activity has been detected in a large number of mammalian

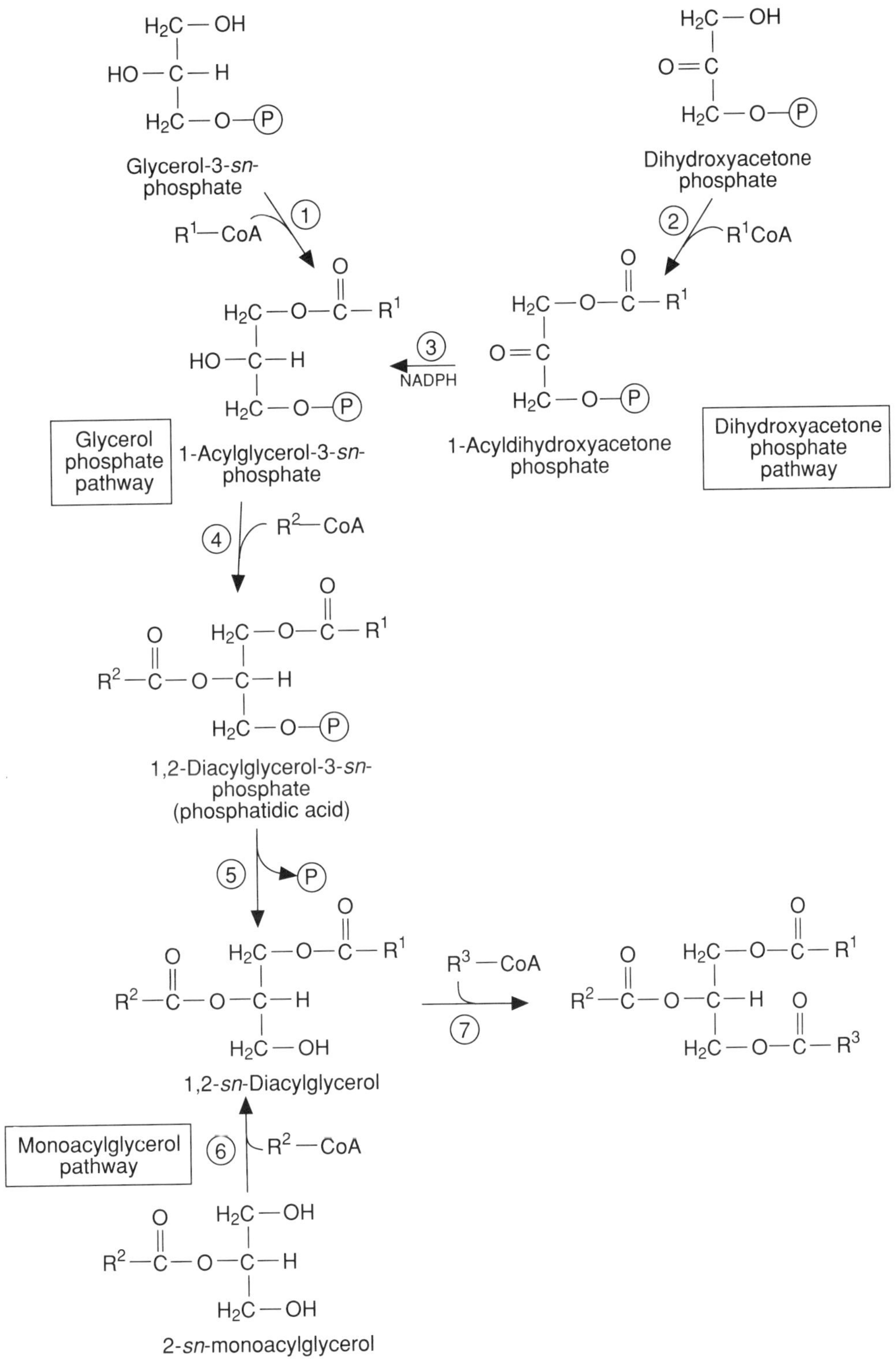

Fig. 11.28 Major pathways of triacylglycerol synthesis. Reactions (1), (4), (5) and (7), glycerol phosphate pathway; reactions (2), (3), (4), (5) and (7), dihydroxyacetone phosphate pathway; reactions (6) and (7), monoacylglycerol pathway. Enzymes: (1) glycerol phosphate 1-acyltransferase; (2) dihydroxyacetone phosphate acyltransferase; (3) acyldihydroxyacetone phosphate reductase; (4) 1-acylglycerol phosphate 2-acyltransferase; (5) phosphatidate phosphohydrolase; (6) monoacylglycerol acyltransferase; (7) diacylglycerol acyltransferase.

tissues including kidney, pancreas, adipose tissue, arterial walls, ascites tumour cells and mammary glands (cf. O'Doherty, 1978; Brindley, 1991). The first enzyme in the pathway is monoacylglycerol acyltransferase (EC 2.3.1.22). The enzyme is particulate and is found in microsomal fractions – probably arising from the endoplasmic reticulum. Evidence for the monoacylglycerol pathway in plants is poor, and it seems probable that diacylglycerol arises from phosphatidate phosphatase in these phyla (Gurr, 1980). For comments on the extensive literature concerning the substrate specificities and other properties of mammalian monoacylglycerol acyltransferases, refer to O'Doherty (1978).

Phosphatidate phosphohydrolase has already been briefly discussed in relation to glycerophospholipid synthesis (section 11.2.1). The enzyme was first discovered in plants (Kates, 1955) and, subsequently, identified in animal tissues. It has been purified from a large number of mammalian tissues including liver, kidney, intestinal mucosa, adipose tissue, erythrocyte membranes and avian salt glands (cf. O'Doherty, 1978). The intracellular distribution of mammalian phosphatidate phosphohydrolase is complicated. It has been found in several particulate fractions – mitochondrial, lysosomal and microsomal (cf. Sedgwick and Hübscher, 1967). However, quantitatively, the most important of these fractions is the endoplasmic reticulum. In addition, there is significant cytosolic activity in animals. The cytosolic phosphatidate phosphohydrolase is able to translocate to the endoplasmic reticulum and mitochondria. The response of such a Mg^{2+}-dependent enzyme to hormones and metabolites, and variations in its intracellular distribution between soluble and particulate compartments, are compatible with an important role in controlling triacylglycerol synthesis; for a full discussion of this function see Brindley (1988, 1991).

In plants, phosphatidate phosphohydrolase has a less obvious role in controlling the rate of triacylglycerol synthesis, since the latter occurs usually at a specific developmental period when it is not in competition with other metabolic pathways such as membrane lipid synthesis (Harwood and Griffiths, 1992). The enzyme has been detected in several subcellular compartments (see Harwood and Price-Jones, 1988), and, moreover, its activity in chloroplast envelopes appears to be important in controlling the flux of lipid carbon between the extrachloroplastic and the chloroplast compartments. This is important in giving rise to different molecular species of lipids in various plants (see Heinz and Roughan, 1983; Roughan, 1987; Ohlrogge *et al.*, 1991). There have been some suggestions that the activity of phosphatidate phosphohydrolase may be important in triacylglycerol accumulation in seeds (Ichihara *et al.*, 1990) even though the enzyme, unlike in animals, is not under acute (hormonal) control.

Synthesis of triacylglycerols requires the enzyme diacylglycerol acyltransferase (EC 2.3.1.20). The enzyme is located in the endoplasmic reticulum (Brindley, 1991) and has been studied in many mammalian tissues (O'Doherty, 1978). In some cases, activity is stimulated by the addition of phosphatidylcholine (O'Doherty *et al.*, 1974), and there is also evidence that the Z protein may be involved in stimulating acyltransferase activity (O'Doherty and Kuksis, 1975). The 1,2 isomers of diacylglycerol are about twice as effective as the 2,3 isomers. The enzymes show little chain-length specificity, however (e.g. Hill *et al.*, 1968).

So far as can be ascertained, diacylglycerol acyltransferase competes for a common pool of diacylglycerol with the choline and ethanolamine phosphotransferases, which form the zwitterionic membrane phospholipids (Brindley, 1991). This enzyme has also been suggested to be rate-limiting (exerts significant flux control) over triacylglycerol synthesis, at least under certain conditions (Mayorek *et al.*, 1989).

The diacylglycerol acyltransferase of plants has been studied in several tissues, though not yet purified. In many plants it seems less specific for its acyl-CoA substrate than are the other acyltransferases (Stymne and Stobart, 1987). It is located in the endoplasmic reticulum and has low activity when measured *in vitro* (e.g. Berneth and Frentzen, 1990). The latter fact, together with the significant build-up of diacylglycerol during triacylglycerol formation *in vivo* (Perry and Harwood, 1990) and *in vitro* (Perry and Harwood, 1991), has led to suggestions that diacylglycerol acyltransferase may exert significant flux control over oil accumulation in crops (see e.g. Griffiths and Harwood, 1990).

Mammalian triacylglycerol synthesis is affected by a large number of factors. These include nutritional, hormonal and pharmacological effects (cf. O'Doherty, 1978, for review). A recent discussion of the control of triacylglycerol synthesis in animals is given by Brindley (1991).

General accounts of triacylglycerol synthesis in plants are given by Gurr (1980), Stymne and Stobart (1987) and Harwood and Page (1993). For specific aspects, including possible control mechanisms and the synthesis of unusual oils, see Quinn and Harwood (1990) and Cherif *et al.* (1992). Another feature worthy of mention is the efforts by breeders to increase oil yields. These have been quite successful (e.g. maize varieties have been changed from an average 4–5% oil content to give IHO lines (20%) or ILO (0.5%) lines). However, the complexity of the phenotype precludes elucidation of the exact genetic basis of these changes (Ohlrogge *et al.*, 1991). On the other hand, a 20% increase in seed fatty acid content (per unit seed weight) has been induced in *Arabidopsis* by a single gene variation (James and Dooner, 1990).

11.3.2 Triacylglycerol breakdown

The breakdown of triacylglycerol is catalysed by lipases. A large number of such enzymes have been purified from animals, plants and microbes (cf. Brockerhoff and Jensen, 1974). It should be noted that the term 'lipase' is frequently misused. A true lipase is one that attacks triacylglycerols and acts only at an oil–water interface. This definition therefore excludes enzymes acting on water-soluble esters (esterases) or those preferentially hydrolysing other lipids (acyl hydrolases).

Since triacylglycerols are important dietary constituents (section 12.1), there is interest in digestive lipases. Although pharyngeal lipases have been found and studied, their importance during overall digestion remains to be established (cf. O'Doherty, 1978). In contrast, pancreatic lipase is the best-known and most investigated of all lipolytic enzymes. It acts on mono-, di- and triacylglycerols, although the reaction rate is slower with partial glycerides (Brockerhoff and Jensen, 1974). It probably lacks stereospecificity. For a thorough review of pancreatic lipase, see Jensen (1971).

Metabolism of triacylglycerols in animals requires the interaction of lipoprotein lipase (involved in uptake of acyl chains from plasma) and hormone-sensitive lipase (involved in release of fatty acids from lipid stores). Some aspects of lipoprotein lipase action are discussed in section 12.3, and the reader is also referred to Brockerhoff and Jensen (1974) and Jensen (1971). The enzyme is also known as clearing factor lipase, requires apo-C_{II} for activity and may be bound via heparan sulphate proteoglycan at the endothelial surface *in vivo* (Williams *et al.*, 1983). Considerable work has been carried out on intracellular processing of the enzyme in active tissues (Cryer, 1981) and on the action of hormones in controlling the adipose and heart tissue enzymes (Ashby and Robinson, 1980; de Gasquet *et al.*, 1975). There is now considerable sequence information of lipoprotein lipase. The sequence is extraordinarily conserved (87–94% with different mammalian enzymes). Comparisons with other lipases have also been made (see Wang *et al.*, 1992). For further information on the structural and functional domains of lipoprotein lipase, the structure and function of apo-C_{II} and the reaction kinetics of the enzyme, see Wang *et al.* (1992).

Hormone-sensitive lipase tissue activity is stimulated by adrenaline (epinephrine), glucagon, ACTH (corticotropin), TSH (thyrotropin) and serotonin (Jensen, 1971). These hormones are presumed to exert their effects on adipose tissue by stimulating adenylate cyclase. Certainly, an increased formation of cAMP (cyclic AMP) is brought about by lipolytic hormones, and cAMP has been shown to stimulate lipase activity in cell-free preparations (cf. O'Doherty, 1978). Phosphorylation/dephosphorylation has been implicated (Belfrage *et al.*, 1983). The cDNA for rat hormone-sensitive lipase has been reported (Holm *et al.*, 1988) and increased expression of the mRNA for this enzyme found in the adipose tissue of cancer patients (Thompson *et al.*, 1993).

Other mammalian lipases that have been studied are milk lipase (cf. Jensen, 1971) and monoacylglycerol lipase. The latter enzyme occurs in several tissues including intestine, liver and adipose tissue. It has high activity with monoacylglycerols when compared to diacylglycerols or triacylglycerols (see O'Doherty, 1978).

Lipase activity has been found in a wide range of plant materials with most work on oil seeds and cereals. One of the best-characterized enzymes is the castor bean acid lipase. Ory and co-workers (Ory, 1969) have characterized the enzyme, which is associated with the spherosome membrane. Huang and Moreau (1978) studied lipolytic activity in a range of germinating oil seeds. The enzyme from peanuts seems to be mainly associated with microsomal fractions (Theimer and Rosnitschek, 1978) and is clearly distinguished from the glyoxysomal monoacylglycerol lipase (Huang and Moreau, 1978). Several lipases have been purified from cereal seeds such as wheat, oat and rice. Galliard (1980) has reviewed this work (cf. also Jensen, 1971) and has discussed the role of plant growth regulators in controlling activity. For a more recent review, see Huang (1987).

Although bacteria do not store energy as triacylglycerol, a number of bacterial lipases have been discovered and studied. An extensive review of microbial lipases (and esterases) has been made by Lawrence (1967). It should be emphasized also that many of the microbial 'lipases' (e.g. the widely studied *Rhizopus arrhizus* enzyme) are, in fact, acyl hydrolases. More recent information is available in Finnerty (1989).

The features of lipases useful for industry in lipid modifications have been discussed by Eigtved (1992), and the monograph by Alberghina *et al.* (1991) contains many specific examples of individual lipases, their characteristics and potential (or actual) industrial use.

References and further reading

Akesson, B., Elovson, J. and Arvidson, G. (1970) *Biochim. Biophys. Acta*, **18**, 44.

Alberghina, L., Schmid, R.D. and Verger, R. (eds) (1991) *Lipases: Structure, Mechanism and Genetic Engineering*, VCH Press, New York.

Appleqvist, L.A. (1975) in *Recent Advances in Chemistry and Biochemistry of Plant Lipids* (eds T. Galliard and E.I. Mercer), Academic Press, London, p. 247.

Ashby, P. and Robinson, D.S. (1980) *Biochem. J.*, **188**, 185.

Belfrage, P., Fredrikson, G., Olsson, H. and Belfrage, P. (1983) in *Cell Function and Differentiation*, Part C (eds G. Akoyunoglou, A.R., Evangelopoulos, J. Georgatsos, G. Pelaiologos, A. Traketellis and C.P. Tsiganos), Alan R. Liss, New York, p. 213.

Berneth, K. and Frentzen, M. (1990) *Plant Sci.*, **67**, 21.

Brindley, D.N. (1978) *Int. J. Obes.*, **2**, 7.

Brindley, D.N. (ed.) (1988) *Phosphatidate Phosphohydrolase*, vols I and II, CRC Press, Boca Raton, FL.

Brindley, D.N. (1991) in *Biochemistry of Lipids, Lipoprotein and Membranes* (eds D.E. Vance and J. Vance), Elsevier, Amsterdam, p. 171.

Brockerhoff, H. and Jensen, R.G. (1974) in *Lipolytic Enzymes*, Academic Press, New York, p. 1.

Cherif, A., Ben Miled-Daoud, D., Marzouk, B., Smaoui, A. and Zarrouk, M. (eds) (1992) *Metabolism, Structure and Utilization of Plant Lipids*, Centre National Pedagogique, Tunis.

Clark, B. and Hübscher, G. (1960) *Nature*, **183**, 35.

Cryer, A. (1981) *Int. J. Biochem.*, **13**, 525.

Davies, R.J. and Holdsworth, J.E. (1992) *Adv. Appl. Lipid Res.*, **1**, 119.

de Gasquet, P., Pequignot-Planche, E., Tonnu, N.T. and Diaby, F.A. (1975) *Horm. Metab. Res.*, **7**, 152.

Douce, R. and Joyard, J. (1980) in *Biochemistry of Plants*, vol. 4 (eds P.K. Stumpf and E.E. Conn), Academic Press, New York, p. 321.

Eigtved, P. (1992) *Appl. Lipid Res.*, **1**, 1.

Finnerty, W.R. (1989) in *Microbial Lipids*, vol. 2 (eds C. Ratledge and S.G. Wilkinson), Academic Press, London, p. 525.

Galliard, T. (1980) in *Biochemistry of Plants*, vol. 4 (eds P.K. Stumpf and E.E. Conn), Academic Press, New York, p. 85.

Griffiths, G. and Harwood, J.L. (1990) in *Plant Lipid Biochemistry, Structure and Utilization* (eds P.J. Quinn and J.L. Harwood), Portland, London, p. 216.

Gurr, M.I. (1980) in *Biochemistry of Plants*, (eds P.K. Stumpf and E.E. Conn), Academic Press, New York, p. 205.

Harwood, J.L. and Griffiths, G. (1992) in *Advances in Plant Cell Biochemistry and Biotechnology* (ed. I.M. Morrison), JAI Press, London, p. 1.

Harwood, J.L. and Page, R. (1993) in *Designer Oilseed Crops* (ed. D.J. Murphy), VCH Press, New York, p. 165.

Harwood, J.L. and Price-Jones, M.J. (1988) in *Phosphatidate Phosphohydrolase*, vol. II (ed. D.N. Brindley), CRC Press, Boca Raton, FL, p. 1.

Heinz, E. (1977) in *Lipids and Lipid Polymers in Higher Plants* (eds M. Tevini and H.K. Lichtenhaler), Springer-Verlag, Berlin, p. 102.

Heinz, E. and Roughan, P.G. (1983) *Plant Physiol.*, **72**, 273.

Hill, E.E. and Lands, W.E.M. (1970) in *Lipid Metabolism* (ed. S.J. Wakil), Academic Press, New York, p. 185.

Hill, E.E., Husbands, D.R. and Lands, W.E.M. (1968) *J. Biol. Chem.*, **243**, 4440.

Hitchcock, C. and Nichols, B.W. (1971) *Plant Lipid Biochemistry*, Academic Press, London.

Holm, C., Kirchgessner, T.G., Lusis, A.J., *et al.* (1988) *Science*, **241**, 1503.

Huang, A.H.C. and Moreau, R.A. (1978) *Planta*, **141**, 111.

Huang, A.H.T. (1987) in *The Biochemistry of Plants*, (eds P.K. Stumpf and E.E. Conn), Academic Press, New York, p. 91.

Huang, A.H.T. (1992) *Annu. Rev. Plant Physiol.*, **43**, 177.

Ichihara, K., Murata, N. and Fujii, S. (1990) *Biochim. Biophys. Acta*, **1043**, 227.

James, D.W. and Dooner, H.K. (1990) *Theor. Appl. Genet.*, **80**, 241.

Jensen, R.G. (1971) *Prog. Chem. Fats Other Lipids*, **11**, 347.

Kanoh, H. and Ohno, K. (1973) *Biochim. Biophys. Acta*, **326**, 17.

Kates, M. (1955) *Can. J. Biochem. Physiol.*, **33**, 575.

Lawrence, R.C. (1967) *Dairy Sci. Abstr.*, **29**, 1, 59.

Lennarz, W.J. (1970) in *Lipid Metabolism* (ed. S.J. Wakil), Academic Press, New York, p. 155.

Manning, R. and Brindley, D.N. (1972) *Biochem. J.*, **130**, 1003.

Mayorek, N., Grinstein, I. and Bar-Tana, J. (1989) *Eur. J. Biochem.*, **182**, 395.

Mitchell, M.P., Brindley, D.N. and Hübscher, G. (1971) *Eur. J. Biochem.*, **18**, 214.

Moore, T.J., Lord, J.M., Kagawa, T. and Beevers, H. (1973) *Plant Physiol.*, **52**, 50.

Murphy, D.J. (1990) *Prog. Lipid Res.*, **29**, 299.

O'Doherty, P.J.A. (1978) in *Fatty Acids and Glycerides* (ed. A. Kuksis), Plenum, New York, p. 289.

O'Doherty, P.J.A. and Kuksis, A. (1975) *FEBS Lett.*, **60**, 256.

O'Doherty, P.J.A., Yousef, I.M. and Kuksis, A. (1974) *Can. J. Biochem.*, **52**, 726.

Olhrogge, J.B., Browse, J. and Somerville, C.R. (1991) *Biochim. Biophys. Acta*, **B1082**, 1.

Ory, R.L. (1969) *Lipids*, **4**, 177.

Perry, H.J. and Harwood, J.L. (1990) in *Plant Lipid Biochemistry, Structure and Utilization* (eds P.J. Quinn and J.L. Harwood), Portland, London, p. 204.

Perry, H.J. and Harwood, J.L. (1991) *Biochem. Soc. Trans.*, **19**, 243S.

Pollock, R.J., Hajra, A.K. and Agranoff, B.W. (1975) *Biochim. Biophys. Acta*, **380**, 421.

Quinn, P.J. and Harwood, J.L. (eds) (1990) *Plant Lipid Biochemistry, Structure and Utilization*, Portland, London.

Ratledge, C. (1989) in *Microbial Lipids*, vol. II (eds C. Ratledge and S.J. Wilkinson), Academic Press, London, p. 567.

Roughan, P.G. (1987) in *The Metabolism, Structure and Function of Plant Lipids* (eds P.K. Stumpf, J.B. Mudd and W.D. Nes), Plenum, New York, p. 247.

Roughan, P.G. and Slack, C.R. (1982) *Annu. Rev. Plant Physiol.*, **33**, 97.

Sakaki, T., Ohnishi, J.-I., Kondo, N. and Yamada, M. (1985) *Plant Cell Physiol.*, **26**, 258.

Sargent, J. (1979) *Trends Biochem. Sci.*, **4**, 209.

Sedgwick, B. and Hübscher, G. (1967) *Biochim. Biophys. Acta*, **144**, 397.

Smith, M.E., Sedgwick, B., Brindley, D.N. and Hübscher, G. (1967) *Eur. J. Biochem.*, **3**, 70.

Snyder, F.L. (1972) *Ether Lipids: Chemistry and Biology*, Academic Press, New York.

Stymne, S. and Stobart, A.K. (1987) in *The Biochemistry of Plants*, vol. 9 (eds P.K. Stumpf and E.E. Conn), Academic Press, New York, p. 175.

Theimer, R.R. and Rosnitschek, I. (1978) *Planta*, **139**, 249.

Thompson, M.P., Coopgerm, S.T., Parry, B.R. and Tuckey, J.A. (1993) *Biochim. Biophys. Acta*, **1180**, 236.

Wang, C.-S., Hartsuck, J. and McConathy, W.J. (1992) *Biochim. Biophys. Acta*, **1123**, 1.

Williams, M.P., Streeter, H.B., Wusteman, F.S. and Cryer, A. (1983) *Biochim. Biophys. Acta*, **756**, 83.

Wilson, R.F. and Rinne, R.W. (1976) *Plant Physiol.*, **57**, 375.

11.4 Glycosylglycerides

11.4.1 Galactosylglycerides

The two galactosylglycerides, monogalactosyldiacylglycerol (MGDG) and digalactosyldiacylglycerol (DGDG), are the major lipid components of the photosynthetic membranes of oxygen-evolving organisms. They are rare or only found in trace amounts in other organisms. Trigalactosyl and tetragalactosyl derivatives are minor components of some chloroplasts, and in marine algae and some bacteria the sugar residue(s) may be glucose.

The pathway for galactosylglyceride synthesis is outlined in Fig. 11.29. UDP-galactose is generated in the cytoplasm of plant leaf cells and is used by two galactosyltransferase enzymes, which are located in the outer half of the envelope membrane. Joyard and Douce (1976) showed that the galactosyltransferase activity was located in the plastid envelope and that any activity in microsomal fractions was due to envelope content. Plastid envelope membranes from non-photosynthetic tissues such as the chromoplasts of *Narcissus pseudonarcissus* (Leidvogel and Kleinig, 1977) and potato tuber amyloplasts (Fishwick and Wright, 1980) are also active.

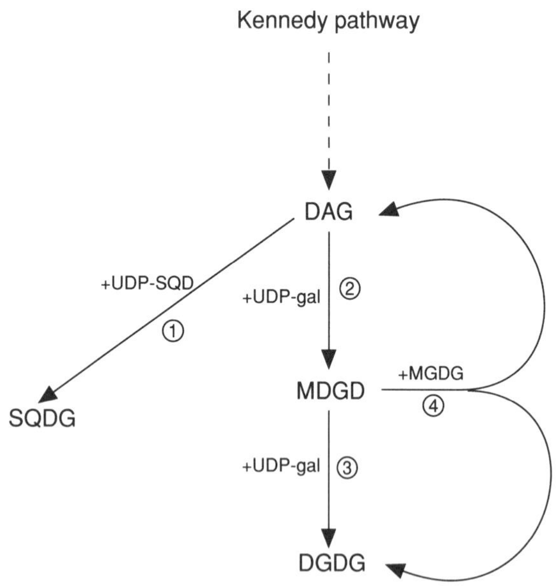

Fig. 11.29 Biosynthesis of plant glycosylglycerides. Abbreviations: DAG, diacylglycerol; MGDG, monogalactosyldiacylglycerol; DGDG, digalactosyldiacylglycerol; SQD, sulphoquinovose; SQDG, sulphoquinovosyldiacylglycerol. Reactions: (1) UDP-SQD: diacylglycerol sulphoquinovosyltransferase; (2) UDP-galactose:diacylglycerol galactosyltransferase; (3) UDP-galactose:MGDG galactosyltransferase; (4) galactolipid:galactolipid galactosyltransferase.

The first galactosyltransferase is specific for the formation of a β-glycosidic bond while the other forms an α-glycosidic bond (Douce and Joyard, 1980). The first galactosyltransferase has been purified from spinach (Teucher and Heinz, 1991).

As an alternative to the use of UDP-galactose, digalactosyldiacylglycerol may be formed by the action of a galactolipid:galactolipid galactosyltransferase (Fig. 11.29), first demonstrated by Von Besouw and Wintermans (1978). Comparative aspects of these alternative pathways have been discussed (Joyard and Douce, 1987), and methods for the analysis of galactolipids and their metabolites reviewed (Douce *et al.*, 1990). Recent results suggest that the galactolipid transferase is the most important enzyme for the formation of digalactosyldiacylglycerol (Heemskerk *et al.*, 1990).

The formation of polyunsaturated fatty acids (mainly α-linolenate) associated with the galactosylglycerides has been reviewed (Harwood, 1988). Depending on the plant type, there seem to be differences between the '16:3 species' and the '18:3 species', as discussed by Heinz and Roughan (1982). The distinctive features of the fatty acid distributions of galactolipids in 16:3 and 18:3 plants is illustrated in section 3.11 and discussed thoroughly by Browse and Somerville (1991). For 18:3 plants the desaturation of oleate in association with phosphatidylcholine is followed by release of the diacylglycerol for galactolipid formation (Joyard and Douce, 1987). Whether diacylglycerol itself or phosphatidylcholine moves from the endoplasmic reticulum to the chloroplast envelope is not known, although Tanaka and Yamada (1982) demonstrated the latter process mediated via a phospholipid transfer protein. After the formation of monogalactosyldiacylglycerol, further desaturation can then take place (Jones and Harwood, 1980; Harwood, 1988; see also section 11.1.3). The fatty acid combinations (Heinz, 1977; Rullkötter *et al.*, 1975) and turnover (Heinz *et al.*, 1979) in galactosylglycerides have been studied thoroughly. In 16:3 plants and cyanobacteria the galactosylglycerides appear to be substrates for successive desaturations of oleate and linoleate to α-linolenate (see Harwood and Jones, 1989).

Initial breakdown of galactosylglycerides is catalysed by non-specific hydrolase enzymes (cf. Galliard, 1980, for review). The enzymes from runner bean leaves (Burns *et al.*, 1979) and potato tubers (Hirayama *et al.*, 1975) have been purified. Two enzymes that differed slightly in substrate specificity were isolated from runner bean leaves (Burns *et al.*, 1979) and the position of hydrolysis identified by mass spectrometry as the bond between the fatty acyl carboxy carbon and the oxygen of glycerol (Burns *et al.*, 1980). Acyl hydrolase enzymes are particularly active, homogenization of potato

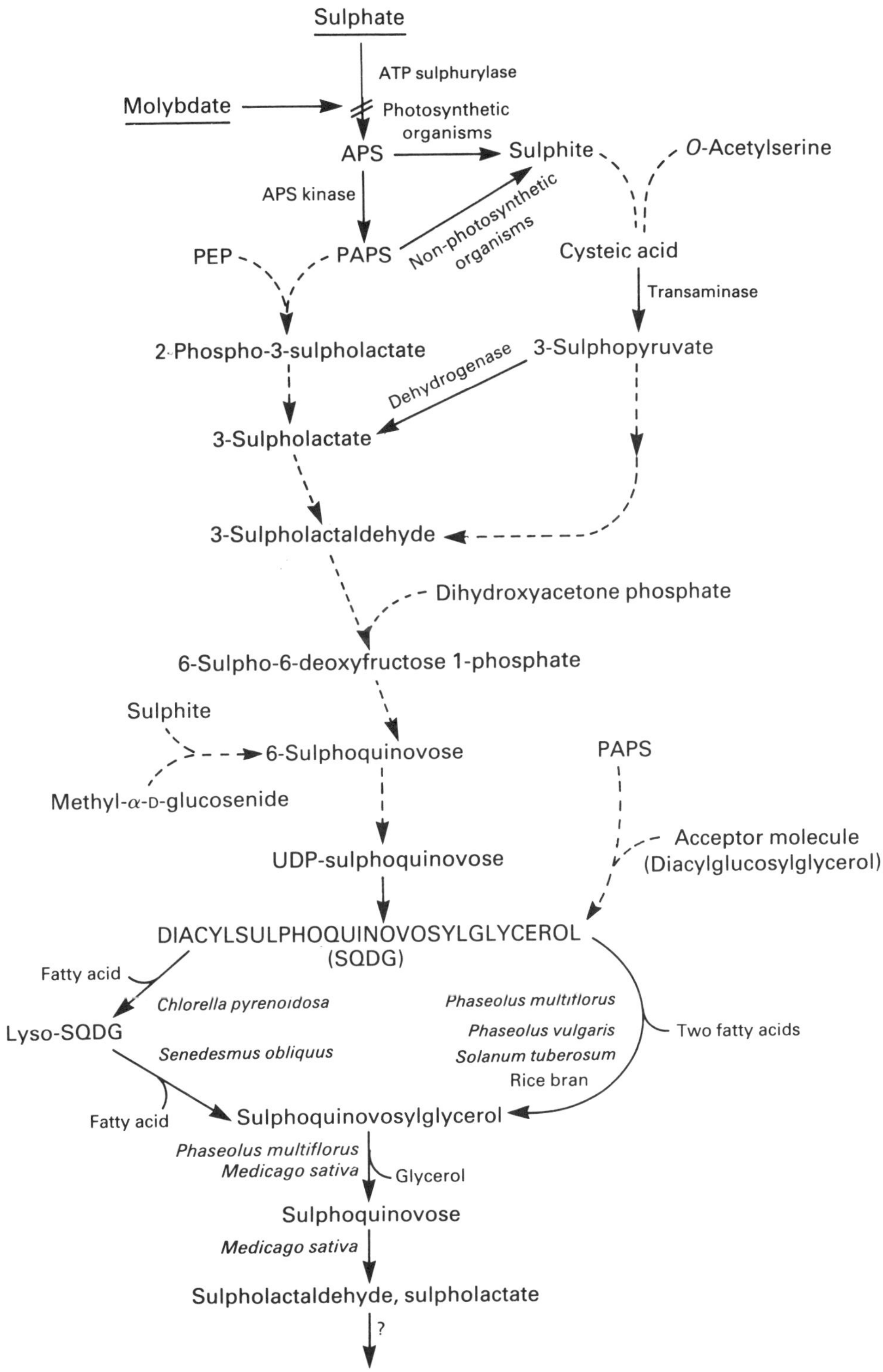

Fig. 11.30 Sulpholipid synthesis and breakdown. Full arrows show reactions that have been demonstrated; broken arrows are possible pathways. Underlined compounds have been used in experiments on sulpholipid synthesis. APS, adenosine phosphosulphate; PAPS, phosphoadenosine phosphosulphate. Adapted from Harwood (1980b).

tubers in water at 0 °C for a few seconds being sufficient for most of the endogenous membrane lipids to be destroyed!

Further hydrolysis of the deacylated moiety probably takes place by the action of α- and β-galactosidases (Sastry and Kates, 1964), but the system is ill-defined at present. However, a β-galactosidase has been purified from spinach leaves that will hydrolyse diacylgalactosylglycerol (Gatt and Baker, 1970).

11.4.2 Sulpholipid (diacylsulphoquinovosylglycerol)

The possible pathways for the formation of diacylsulphoquinovosylglycerol have been reviewed (Harwood and Nichols, 1979; Harwood, 1980b). There are a number of suggestions, each of which can be criticized. The most popular is based on a sulphoglycolytic pathway. Evidence for this has come from the occurrence of intermediates in a number of algae (cf. Benson, 1963) and the incorporation of labelled precursors and intermediates in *Euglena gracilis* (Davies *et al.*, 1966) or higher plants (Harwood, 1975) (Fig. 11.30). An alternative to direct participation of phosphoadenosine phosphosulphate (PAPS) has been proposed by Haines (1973). In his scheme, sulphite is involved in lipid synthesis by nucleophilic displacement on *O*-acetylserine. Harwood and Nichols (1979) have discussed various experiments that are of relevance to these two alternatives. However, Mudd *et al.* (1980), in experiments with spinach seedlings or tobacco cultures, did not show that cysteate was a superior precursor for diacylsulphoquinovosylglycerol when compared with sulphate or cyst(e)ine. These authors also discuss some of the problems inherent in the sulphoglycolytic pathway for the synthesis of plant sulpholipid. Some of the detailed objections are also discussed by Harwood (1980b) and by Mudd and Kleppinger-Sparace (1987). Using the green alga *Chlamydomonas reinhardii*, Hoppe and Schwenn (1981) concluded that the sulphoglycolytic pathway was not utilized but that direct transfer of a sulphite (or sulphate from PAPS) to a suitable acceptor could be. They were, however, unable to identify such a compound. It will be clear that, at present, we do not know how sulphoquinovose is made. However, the nucleotide intermediate has been identified conclusively as UDP-sulphoquinovose (see Seifert and Heinz, 1992).

Plant tissues are capable of, at least, partial catabolism of sulpholipid. Deacylation appears to be the first step and may take place with two enzymes such as in green algae (e.g. Yagi and Benson, 1962) or by the action of a single acyl hydrolase (cf. Harwood, 1980b). Cleavage of sulphoquinovosylglycerol to sulphoquinovose was reported in leaves of *Medicago sativa* (Lee and Benson, 1972) where sulpholactaldehyde and, later, sulpholactic acid accumulated. In contrast, catabolism by cell-free preparations from *Phaseolus multiflorus* stopped at sulphoquinovose (Burns *et al.*, 1980). Further aspects of sulpholipid catabolism are discussed by Harwood (1980b) (cf. Fig. 11.30).

The deacylation step in sulpholipid breakdown is probably catalysed by a non-specific acyl hydrolase (cf. Galliard, 1980; Harwood, 1980b).

Studies on the turnover of molecular species of sulpholipid have been reported in *Vicia faba* and *Hordeum vulgare*. In both these plants, the more saturated species were turned over at high rates whereas the predominant 1-linolenoyl, 2-palmitoyl species has a low rate of turnover. These differences in metabolism may be related to the function of the sulpholipid – the fast-turning-over species being involved in a metabolic function while the trienoic species have a structural role (cf. Harwood, 1980b).

References and further reading

Benson, A.A. (1963) *Adv. Lipid Res.*, **1**, 387.

Browse, J. and Somerville, C. (1991) *Annu. Rev. Plant Physiol.*, **42**, 467.

Burns, D.D., Galliard, T. and Harwood, J.L. (1979) *Phytochemistry*, **18**, 1793.

Burns, D.D., Galliard, T. and Harwood, J.L. (1980) *Phytochemistry*, **19**, 2281.

Davies, W.H., Mercer, E.I. and Goodwin, T.W. (1966) *Biochem. J.*, **98**, 369.

Douce, R. and Joyard, J. (1980) in *Biochemistry of Plants*, vol. 4 (eds P.K. Stumpf and E.E. Conn), Academic Press, New York, p. 321.

Douce, R., Joyard, J., Block, M.A. and Dorne, A.-J. (1990) in *Methods in Plant Biochemistry*, vol. 4 (eds J.L. Harwood and J.R. Boyer), Academic Press, London, p. 71.

Fishwick, M.J. and Wright, A.J. (1980) *Phytochemistry*, **19**, 55.

Galliard, T. (1980) in *Biochemistry of Plants*, vol. 4 (eds P.K. Stumpf and E.E. Conn), Academic Press, New York, p. 85.

Gardner, S.E. and Roughan, P.G. (1983) *Biochem. J.*, **210**, 949.

Gatt, S. and Baker, E.A. (1970) *Biochim. Biophys. Acta*, **206**, 125.

Haines, T.H. (1973) *Annu. Rev. Microbiol.*, **27**, 403.

Harwood, J.L. (1975) *Biochim. Biophys. Acta*, **398**, 224.

Harwood, J.L. (1980a) in *Biogenesis and Function of Plant Lipids* (eds P. Mazliak, P. Benveniste, C. Costes and R. Douce), Elsevier, Amsterdam, p. 143.

Harwood, J.L. (1980b) in *Biochemistry of Plants*, vol. 4 (eds P.K. Stumpf and E.E. Conn), Academic Press, New York, p. 301.
Harwood, J.L. (1988) *Annu. Rev. Plant Physiol.*, **39**, 101.
Harwood, J.L. and Jones, A.L. (1989) *Adv. Bot. Res.*, **16**, 1.
Harwood, J.L. and Nichols, R.G. (1979) *Biochem. Soc. Trans.*, **7**, 440.
Harwood, J.L. and Russell, N.J. (1984) *Lipids in Plants and Microbes*, Allen and Unwin, Hemel Hempstead.
Heemskerk, J.W.M., Storz, T., Schmidt, R.R. and Heinz, E. (1990) *Plant Physiol.*, **93**, 1286.
Heinz, E. (1977) in *Lipids and Lipid Polymers of Higher Plants* (eds M. Tevini and H.K. Lichtenthaler), Springer-Verlag, Berlin, p. 102.
Heinz, E. and Roughan, P.G. (1982) in *Biochemistry and Metabolism of Plant Lipids* (eds J.F.G.M. Wintermans and P.J.C. Kuiper), Elsevier, Amsterdam, p. 169.
Heinz, E., Bertrams, M., Joyard, J. and Douce, R. (1978) *Z. Pflanzenphysiol.*, **87**, 325.
Heinz, E., Siebertz, H.P., Linscheid, M., Joyard, J. and Douce, R. (1979) in *Recent Advances in the Biochemistry and Physiology of Plant Lipids* (eds L.A. Appelqvist and C. Liljenberg), Elsevier, Amsterdam, p. 99.
Hirayama, O., Matsuda, J., Takeda, H., Maenaka, K. and Takatzuka, H. (1975) *Biochim. Biophys. Acta*, **384**, 127.
Hoppe, W. and Schwenn, J.D. (1981) *Z. Naturforsch*, **36c**, 820.
Jones, A.V.M. and Harwood, J.L. (1980) *Biochem. J.*, **190**, 851.
Joyard, J. and Douce, R. (1976) *Physiol. Veg.*, **14**, 31.
Joyard, J. and Douce, R. (1987) in *The Biochemistry of Plants*, vol. 9 (eds P.K. Stumpf and E.E. Conn), Academic Press, New York, p. 275.
Lee, R.F. and Benson, A.A. (1972) *Biochim. Biophys. Acta*, **261**, 35.
Leidvogel, B. and Kleinig, H. (1977) *Planta*, **133**, 249.
Mudd, J.B. and Kleppinger-Sparace, K. (1987) in *The Biochemistry of Plants*, vol. 9 (eds P.K. Stumpf and E.E. Conn), Academic Press, New York, p. 275.
Mudd, J.B., McManus, T.T., Ongun, A. and McCullogh, T.E. (1971) *Plant Physiol.*, **48**, 335.
Mudd, J.B., Dezacks, R. and Smith, J. (1980) in *Biogenesis and Function of Plant Lipids* (eds P. Mazliak, P. Benveniste, C. Costes and R. Douce), Elsevier, Amsterdam, p. 57.
Roughan, P.G. and Slack, C.R. (1982) *Annu. Rev. Plant Physiol.*, **33**, 97.
Rullkötter, J., Heinz, E. and Tulloch, A.P. (1975) *Z. Pflanzenphysiol.*, **76**, 163.
Sastry, P.S. and Kates, M. (1964) *Biochemistry*, **3**, 1280.
Seifert, U. and Heinz, E. (1992) *Bot. Acta.*, **105**, 197.
Tanaka, T. and Yamada, M. (1982) in *Biochemistry and Metabolism of Plant Lipids* (eds J.F.G.M. Wintermans and P.J.C. Kuiper), Elsevier, Amsterdam, p. 99.
Teucher, T. and Heinz, E. (1991) *Planta*, **184**, 319.
Van Besouw, A. and Wintermans, J.F.G.M. (1978) *Biochim. Biophys. Acta*, **529**, 44.
Yagi, T. and Benson, A.A. (1962) *Biochim. Biophys. Acta*, **57**, 601.

11.5 Sphingolipids

11.5.1 Biosynthesis

Sphingolipid chemistry and biosynthesis have been reviewed by Carter *et al.* (1965), Wiegandt (1971), Stoffel (1971), Kanfer and Hakomori (1983) and Sweeley (1991).

Biosynthesis of long-chain bases has been studied in a series of experiments by Snell and co-workers (e.g. Brady *et al.*, 1969) and by Stoffel and co-workers (see Stoffel, 1971). A condensing enzyme, requiring pyridoxal phosphate, is able to condense serine with palmitoyl-CoA to produce 3-oxosphinganine. The reaction, which is probably the rate-limiting step in sphingoid base biosynthesis, proceeds with overall retention of configuration of the C-2 carbon of serine. L-Cycloserine (4-amino-3-isoazolidinone) is an irreversible inhibitor of the palmitoyltransferase and depresses the level of central nervous system sphingolipids if administered to mice (Sweeley, 1991). The next reaction is a reduction using NADPH to produce sphinganine. The reductase responsible is probably closely associated with the palmitoyltransferase in the endoplasmic reticulum. Both the palmitoyltransferase and the reductase exhibit chain-length specificity, with C_{14} to C_{18} CoA esters and C_{14} to C_{20} 3-dehydrosphinganines being utilized. The reaction is shown in Fig. 11.31. The amino acid may form a Schiff-base complex with the pyridoxal phosphate coenzyme and Mg^{2+} during the condensation reaction. Some workers have also suggested that palmitaldehyde is an intermediate in the reaction, although this has not been proven.

Stoffel *et al.* (1968) have studied the direct transformation of sphinganine into 4-hydroxysphinganine in the yeast, *Hansenula ciferri*. The origin of the hydroxyl group is obscure, since Thorpe and Sweeley (1967) concluded that it did not arise from either molecular oxygen or water. For further discussion of the formation of sphingoid bases, see Sweeley (1991).

Ceramides can be rapidly formed from *erythro* or *threo* long-chain bases. The specificity of this acylation by acyl-CoAs has been studied in brain and other tissues (see Stoffel, 1971). The significance of the nature of the fatty acid moiety in ceramides is well seen when the glycosylation of ceramides is considered: hydroxy fatty acid containing ceramides accept predominantly galactose while non-hydroxy fatty acid ceramides accept glucose. Other possible pathways for the synthesis of ceramides are the reversal of the ceramidase reaction and a mechanism involving an active form of the fatty acid, reduced pyridine nucleotide and various other cofactors (see Spence, 1989).

Two pathways are known for the formation of cerebrosides from sphingosine. The quantitative significance of the two synthetic mechanisms is unresolved. In the first pathway, sphingosine is glycosylated with a UDP-sugar (UDP-galactose is usually studied) to yield psychosine. The latter is then acylated with an acyl-CoA (Fig. 11.32). Alternatively, the acylation occurs before the glycosylation (Fig. 11.32).

As expected, the formation of sulphatides involves transfer of sulphate from phosphoadenosine phosphosulphate (PAPS). The latter is formed in a two-step reaction involving ATP and inorganic sulphate via an adenosine phosphosulphate (APS) intermediate. Galactosylceramide and lactosylceramide were both acceptors for the sulphate from PAPS and the reaction is catalysed by the Golgi enzyme, galactosylceramide sulphatotransferase. Sulphate transfer is preceded by receptor-mediated translocation of PAPS from the cytosol across the

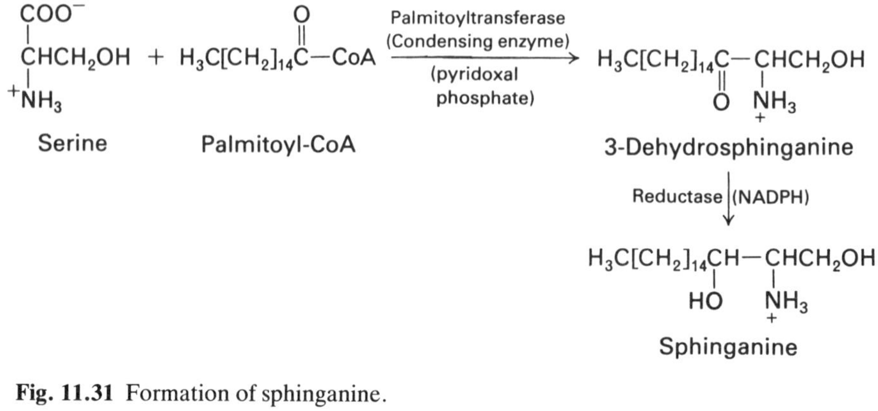

Fig. 11.31 Formation of sphinganine.

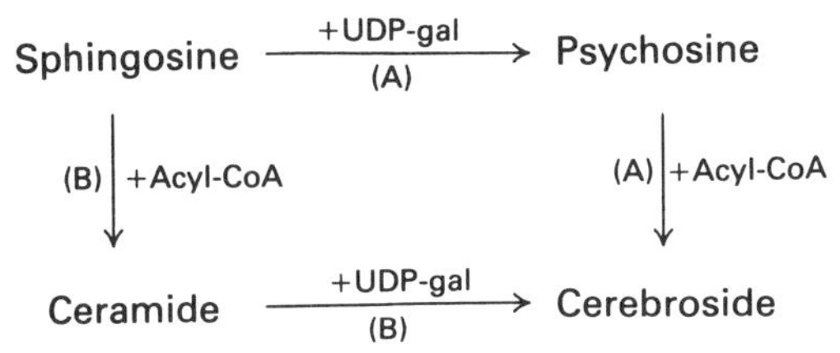

Fig. 11.32 Conversion of sphingosine to cerebroside. The quantitative significance of pathways (A) and (B) in mammalian tissues is unknown.

Golgi membrane, and this process can be inhibited by 3′-P-AMP, palmitoyl-CoA or atractyloside (Sweeley, 1991). Sulphatide synthesis is most rapid in the period 20–25 days after birth in rat brain (Stoffel, 1971). The sulphatides may be involved in the activity of the Na^+/K^+-ATPase (see Stoffel, 1971).

Synthesis of the neutral glycosphingolipids begins from glucosylceramide or galactosylceramide. Specific glycosyltransferases are involved, and the activated forms of the sugar substrates (UDP-Glc, UDP-Gal, UDP-GlcNAc, GDP-Man and GDP-Fuc) are produced in the cytosol from nucleoside triphosphates and hexose 1-phosphates. The formation of different root glycosphingolipids from ceramide is illustrated in Fig. 11.33 and occurs in the lumen of the Golgi apparatus. The active sites of the glycosyltransferases are localized on the lumenal surface of the Golgi membrane and may be organized in several multiglycosyltransferases, which could account for the various types of products that are formed (Sweeley, 1991).

Gangliosides are formed by stepwise elongation of the carbohydrate chain by the action of various glycosyltransferases. Several of the individual enzymes have been studied (see Stoffel, 1971) and assay methods are detailed by Basu *et al.* (1987). The individual sugars are transferred from their UDP derivatives while sialic acid residues are donated by CMP-*N*-acetylneuraminic acid (CMP-NANA). It is important to emphasize here that the glucosylceramide precursor of neutral ceramides and gangliosides is formed from ceramide molecules containing non-hydroxy fatty acids. Thus, the 'higher' sphingolipids do not contain significant amounts of hydroxy fatty acids, which are prevalent in simple ceramides and galactosylceramide (Wiegandt, 1971).

As with neutral glycosphingolipid synthesis, the glycosyltransferases are membrane-bound and predominantly located in the Golgi apparatus. Distribution studies suggest an asymmetric localization of the enzymes in various Golgi in cisternae according to the order in which the sugars are assembled

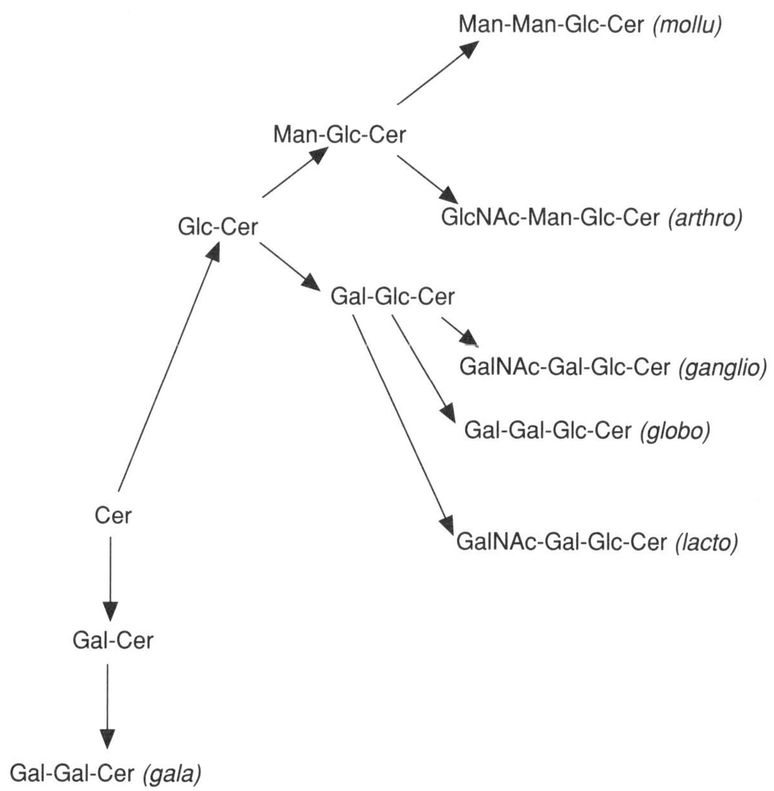

Fig. 11.33 Biosynthesis of different root glycosphingolipids from ceramide. From Sweeley (1991).

Table 11.6 Glycosyltransferases catalysing the biosynthesis of gangliosides (From Sweeley, 1991)

Abbreviation	Name	Linkage created	Substrates
GalT-1	β-Galactosyltransferase	Gal(β1→4)Glc	Glc-Cer
SAT-1	Sialyltransferase	Neu5Ac(α2→3)Gal	Lac-Cer
SAT-2	Sialyltransferase	Neu5Ac(α2→8)Neu5Ac	G_{M3}
SAT-3	Sialyltransferase	Neu5Ac(α2→8)Neu5Ac	G_{D3}
SAT-4	Sialyltransferase	Neu5Ac(α2→3)Gal	G_{A1}, G_{M1a}, G_{D1b}
SAT-5	Sialyltransferase	Neu5Ac(α2→8)Neu5Ac	G_{M1}, G_{D1a}, G_{T1b}
GalNAcT	β-*N*-Acetylgalactos-aminyltransferase	GalNAc(β1→4)Gal	Lac-Cer, G_{M3}, G_{D3}
GalT-2	β-Galactosyltransferase	Gal(β1→3)GalNAc	G_{A2}, G_{M2}, G_{D2}

on the lipid. Enzymes catalysing the various steps involved in ganglioside biosynthesis are shown in Table 11.6. A proposed pathway for the formation of gangliosides involving such enzymes is shown in Fig. 11.34. Regulation of synthesis probably involves transcriptional factors as well as post-translational modification of the glycosyltransferases (see Sweeley, 1991).

Sphingomyelin is a phospholipid as well as a sphingolipid. CDP-choline is the source of the phosphocholine moiety (cf. phosphatidylcholine synthesis, section 11.2.1). As with cerebroside formation, there seem to be two ways of converting a sphingosine base to sphingomyelin. These two pathways differ in the order in which acylation (Fig. 11.35) or phosphocholine transfer take place. The relative importance of these pathways and various features of the reactions are discussed by Thompson (1973) and by Barenholz and Gatt (1982) (see also Sweeley, 1991). A comprehensive review of sphingomyelin metabolism, intracellular transport and various aspects of its biological functions has been published (Koval and Pagano, 1991).

The biological function (Stoffel, 1971) and immunochemistry of the sphingolipids – especially the gangliosides – have been reviewed (Hakomori, 1981).

Kanfer and Hakomori (1983) have published a comprehensive review on sphingolipid biochemistry.

11.5.2 Breakdown

Breakdown of sphingolipids occurs by stepwise hydrolytic cleavage of the various substituents, starting with the terminal hydrophilic portions

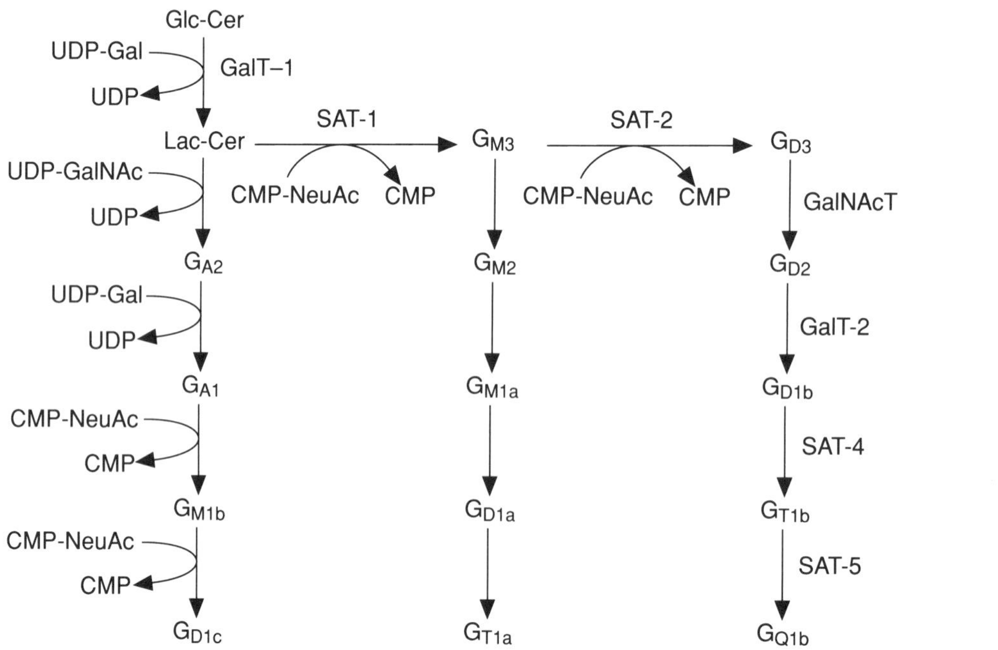

Fig. 11.34 Proposed pathway for biosynthesis of gangliosides (from Sweeley, 1991), using glycosyltransferases detailed in Table 11.6.

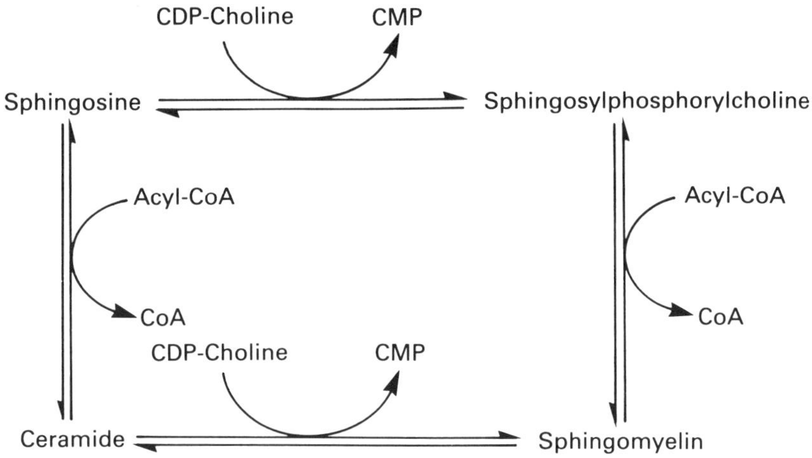

Fig. 11.35 Synthesis of sphingomyelin.

of the molecules. As an example, catabolism of ganglioside G_{M1} [Gal(β1→3)GalNAc(β1→4)Gal-(3←2αAcNeu)(β1→4)Glc(β1→1′)Cer] begins by the action of a β-galactosidase that cleaves the terminal galactose to give ganglioside G_{M2} as the other product. Ganglioside G_{M2} can be cleaved by two pathways. Either a hexosaminidase removes the molecule of GalNAc or a neuroaminidase hydrolyses AcNeu. These reactions have been demonstrated *in vivo*, but because more G_{M2} [GalNAc(β1→4)Gal-(3←2AcNeu)(β1→4)Glc(β1→1′)Cer] accumulates in Tay-Sachs disease than G_{A2} [GalNAc(β1→4)-Gal(β1→4)Glc(β1→1′)Cer], the hexosaminidase reaction appears the more important. The latter enzyme attacks G_{M2} to yield G_{M3} [(AcNeuα2→3)-Gal(β1 → 4)Glc(β1 → 1′)Cer], which is then catalysed further by a neuraminidase to give ceramide lactoside (Fig. 11.36) (see Brady, 1978).

The cleavage of ceramide lactoside needs a β-galactosidase. Two β-galactosidases have been demonstrated in mammalian tissues, both of which have acidic pH optima (pH 4.2 and 4.8). When assayed *in vitro*, either of these enzymes catabolizes ceramide lactoside, depending on the detergent used in the experiment. The pH 4.2 galactosidase also hydrolyses galactose from galactocerebroside. The pH 4.8 enzyme preferentially catalyses the removal of the terminal galactose of G_{M1} as well as catabolizing ceramide lactoside. There are also two liver β-galactosidases with more neutral pH optima that catabolize ceramide lactoside but are inactive with galactocerebroside or G_{M1} (Ben-Yoseph *et al.*, 1977).

Glucocerebroside is hydrolysed by glucocerebrosidase to give glucose and ceramide. Catabolism of ceramide is catalysed by acid (Gatt, 1963) or neutral ceramidase (Sugita *et al.*, 1975) in mammalian tissues. A sphingosine base and a non-esterified fatty acid are liberated. The former is phosphorylated to give a 1-phosphate derivative and then hydrolysed to yield a long-chain aldehyde and phosphoethanolamine (Stoffel, 1971). Sulphatides are catabolized by a sulphatase and the resultant galactocerebroside hydrolysed by a β-galactosidase. In addition, sphingomyelin is hydrolysed by spingo-

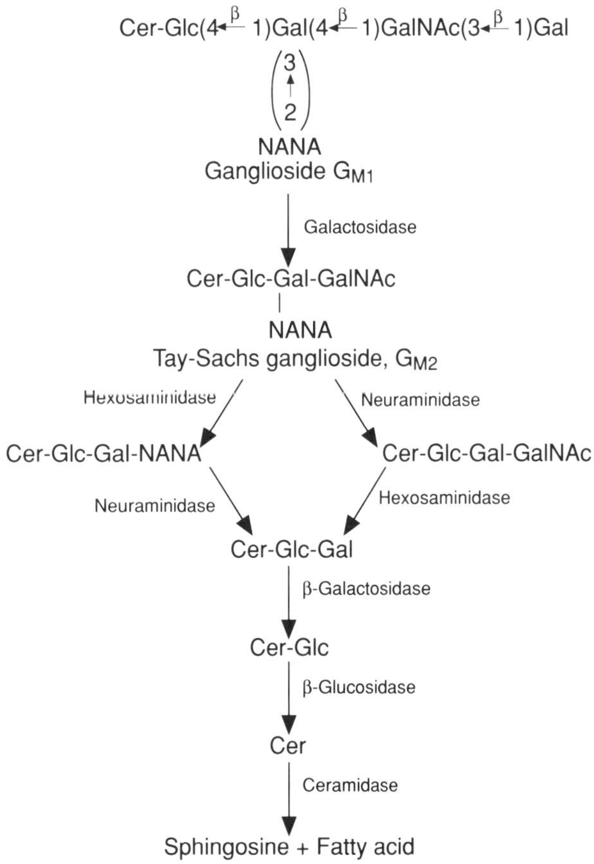

Fig. 11.36 Breakdown of ganglioside G_{M1}-galactosidase.

myelinase to yield ceramide and phosphocholine (see Barenholz and Gatt, 1982).

Many of the above enzymes require detergents for optimal activity *in vitro*. Some workers have suggested that naturally occurring substances may be present in mammalian cells that could serve a similar function. It has also been postulated that the various sphingolipid hydrolases may also be aligned in an ordered fashion on lysosomal membranes. In that case, hydrolysis of sphingolipids should be much more efficient than random access of various substrates to enzymes freely admixed within a lysosomal sac (Brady, 1978).

Sphingolipid breakdown is reviewed by Kanfer and Hakomori (1983) and deficiency diseases of sphingolipid catabolism are covered in section 12.5.

Sweeley (1991) discusses some recent aspects of sphingolipid breakdown and the regulation of their turnover. In the latter connection, glycosphingolipids have important functions during cellular differentiation and oncogenic transformation. Recent interest has also included their function as modulators of transmembrane signalling and as mediators for cellular interactions (Hakomori, 1990).

References and further reading

Barenholz, Y and Gatt, S. (1982) in *Phospholipids* (eds J.N. Hawthorne and G.B. Ansell), Elsevier, Amsterdam, p. 129.

Basu, M., De, T., Das, K., Kyle, J., Chon, H.C., Schaeper, R. and Basu, S. (1987) in *Methods in Enzymology*, vol. 138 (ed. V. Ginsburg), Academic Press, Orlando, FL, p. 575.

Ben-Yoseph, Y., Shapira, E., Edelman, D., Burton, B.K. and Nadler, H.L. (1977) *Arch. Biochem. Biophys.*, **184**, 373.

Brady, R.N., Di Mari, S.J. and Snell, E.E. (1969) *J. Biol. Chem.*, **244**, 491.

Brady, R.O. (1978) *Annu. Rev. Biochem.*, **47**, 687.

Carter, H.E., Johnson, P. and Webber, E.J. (1965) *Annu. Rev. Biochem.*, **34**, 109.

Gatt, S. (1963) *J. Biol. Chem.*, **238**, 3131.

Gurr, M.I. and James, A.T. (1980) *Lipid Biochemistry*, Chapman & Hall, London.

Hakomori, S.-I. (1981) *Annu. Rev. Biochem.*, **50**, 733.

Hakomori, S.-I. (1990) *J. Biol. Chem.*, **265**, 18713.

Kanfer, J.N. and Hakomori, S.-I. (1983) *Sphingolipid Biochemistry*, Plenum, New York.

Koval, M. and Pagano, R.E. (1991) *Biochim. Biophys. Acta*, **1082**, 113.

Spence, M.W. (1989) in *Phosphatidylcholine Metabolism* (ed. D.E. Vance), CRC Press, Boca Raton, FL, p. 186.

Stoffel, W. (1971) *Annu. Rev. Biochem.*, **40**, 57.

Stoffel, W., Sticht, G. and Lekein, D. (1968) *Hoppe-Seyler's Z. Physiol. Chem.*, **349**, 1149.

Sugita, M., Williams, M., Dulaney, J.T. and Moser, H.W. (1975) *Biochim. Biophys. Acta*, **398**, 125.

Sweeley, C.C. (1991) in *Biochemistry of Lipids, Lipoproteins and Membranes* (eds D.E. Vance and J. Vance), Elsevier, Amsterdam, p. 327.

Thompson, G.A. (1973) in *Form and Function of Phospholipids* (eds G.B. Ansell, R.M.C. Dawson and J.N. Hawthorne), Elsevier, Amsterdam, p. 67.

Thorpe, S.R. and Sweeley, C.C. (1967) *Biochemistry*, **6**, 887.

Wiegandt, H. (1971) *Adv. Lipid Res.*, **9**, 249.

11.6 Sterol esters

The esterification of cholesterol in animals has attracted considerable research because of the possible involvement of cholesterol and its ester in various disease states (see Glomset and Norum, 1973; sections 12.1, 12.3 and 12.6). Cholesteryl esters are formed by the action of lecithin cholesterol acyltransferase (LCAT, EC 2.3.1.43), which is particularly active in plasma; see Sabine (1977) for a review of cholesterol metabolism. The reaction involves transfer of a fatty acid from position 2 of lecithin (phosphatidylcholine) to the 3-hydroxyl group of cholesterol with the formation of monoacylphosphatidylcholine. LCAT is a glycoprotein of mass 60 kDa (Fielding, 1990). Although LCAT esterifies plasma cholesterol solely at the interface of high-density lipoprotein (HDL) and very-low-density lipoprotein (VLDL), the cholesteryl esters are transferred to other lipoproteins by a particular transport protein (CETP; cholesteryl ester transfer protein). The CETP catalyses the movement of cholesteryl esters, triacylglycerols and non-polar lipids (such as retinyl esters) between plasma lipoproteins. Physiologically the main effect of CETP may be to promote the transfer of LCAT-derived cholesteryl esters out of HDL (where they were formed) into VLDL and LDL, in exchange for triacylglycerol (see Fielding and Fielding, 1991). CETP is a glycoprotein of mass 53 kDa (Drayna *et al.*, 1987). Its physiological function and mechanism of activity are discussed by Fielding and Fielding (1991). Activity of CETP in plasma is regulated by an inhibitor protein, which acts by displacing it from its lipoprotein binding sites (Morton and Zilverschmidt, 1981). Cholesteryl esters, in contrast to free cholesterol, are taken up by cells mostly via specific receptor pathways (Brown *et al.*, 1981), are hydrolysed by lysosomal enzymes and eventually re-esterified and stored within cells. LCAT may also participate in the movement of cholesterol out of cells by esterifying excess cholesterol in the intravascular circulation (cf. Marcel, 1982).

The purification and properties of LCAT, together with a discussion of its mechanism of reaction, are given by Marcel (1982). A number of disease states involve LCAT activity. Familial LCAT deficiency has been described (Glomset and Norum, 1973), and patients with this rare complaint have been thoroughly investigated. Many of the abnormalities seen in such patients have also been found in those with cholestasis. A discussion of cholesteryl ester metabolism in relation to other liver diseases and dyslipoproteinaemia has been reported (Marcel, 1982). Similarly, the metabolism of cholesteryl esters in relation to arteries and arterial disease has been fully discussed (Kritchevsky and Kothari, 1978). Mammalian steroid sulphates have been reviewed by Farooqui (1981).

Sterol esters and acylated sterol glycosides have been detected in a number of plant tissues and, in some cases, can be quite significant components (Mudd, 1980). For sterol esters any one of palmitate, oleate, linoleate or α-linolenate could be the principal fatty acyl component depending on the tissue. In acylated sterol glycosides, palmitate or linoleate are the most abundant fatty acids.

Synthesis of sterol esters by preparations from spinach leaves have been studied by Mudd's group (cf. Mudd, 1980). Enzyme activity may have been associated with mitochondria, and diacylglycerol was found to be the best acyl donor, although other lipids including phosphatidylcholine could also serve. The acyl donor for sterol ester synthesis in *Phycomyces blakesleanus* was also found to be phosphatidylcholine, as in animal tissues (Bartlett *et al.*, 1974). For a recent summary of plant sterol ester biochemistry, refer to Goad *et al.* (1987).

The acylation of sterol glucosides in plant tissues has been studied by several workers. The research has been reviewed by Eichenberger (1977) and by Mudd (1980). The acylating enzymes are usually particulate and have been partly purified following solubilization. The best purification has been with an enzyme from *Gossypium* spp. (Forsee *et al.*, 1976). Soluble preparations have been studied from carrot roots (cf. Eichenberger, 1977) and bean leaves (Heinz *et al.*, 1975). The acyl donor for acylated sterol glucoside synthesis seems to vary with the preparation being studied. The particulate enzymes tend to use various phosphoglycerides, whereas the soluble enzymes utilize diacyldigalactosylglycerol much better. The bean leaf enzyme would also use diacylgalactosylglycerol efficiently. The overall pathway seems to involve glucosylation of the sterol before acylation (cf. Mudd, 1980).

Work on sterol ester metabolism in insects has been reviewed by Thompson *et al.* (1973).

References and further reading

Bartlett, K., Keat, M.J. and Mercer, E.I. (1974) *Phytochemistry*, **13**, 1107.

Brown, M.S., Kovanen, P.T. and Goldstein, J.L. (1981) *Science*, **212**, 628.

Drayna, D., Jarnagin, A.S., McLean, J., Henzel, W., Kohr, W., Fielding, C. and Lawn, R. (1987) *Nature*, **327**, 632.

Eichenberger, W. (1977) in *Lipids and Lipid Polymers from Higher Plants* (eds M. Tevini and H.K. Lichtenthaler), Springer-Verlag, Berlin, p. 167.

Farooqui, A.A. (1981) *Adv. Lipid Res.*, **18**, 159.

Fielding, C.J. (1990) in *Advances in Cholesterol Research* (eds M. Esfahani and J.B. Swaney), Telford, Caldwell, NJ, p. 271.

Fielding, P.E. and Fielding, C.J. (1991) in *Biochemistry of Lipids* (eds D.E. Vance and J. Vance), Elsevier, Amsterdam, p. 427.

Forsee, W.T., Valkovich, G. and Elbein, A.D. (1976) *Arch. Biochem. Biophys.*, **172**, 410.

Goad, L.J., Zimowski, J., Evershed, R.P. and Male, V.L. (1987) in *The Metabolism, Structure and Function of Plant Lipids* (eds P.K. Stumpf, J.B. Mudd and W.D. Nes), Plenum, New York, p. 95.

Glomset, J.A. and Norum, K.R. (1973) *Adv. Lipid Res.*, **11**, 1.

Heinz, E., Dieler, H.P. and Rullkötter, J. (1975) *Z. Pflanzenphysiol.*, **75**, 78.

Kritchevsky, D. and Kothari, H.V. (1978) *Adv. Lipid Res.*, **16**, 1.

Marcel, Y.L. (1982) *Adv. Lipid Res.*, **19**, 85.

Morton, R.E. and Zilverschmidt, D.B. (1981) *J. Biol. Chem.*, **256**, 11992.

Mudd, J.B. (1980) in *Biochemistry of Plants*, vol. 4 (eds P.K. Stumpf and E.E. Conn), Academic Press, New York, p. 509.

Sabine, J.R. (1977) *Cholesterol*, Marcel Dekker, New York.

Thompson, M.J., Kaplanis, J.N., Robbins, W.E. and Svoboda, J.A. (1973) *Adv. Lipid Res.*, **11**, 219.

11.7 Control mechanisms

The activity of many enzymes alters under the influence of environmental change. There are good reasons for many of these changes in terms of adaptive mechanisms. However, it is not always clear whether the alterations observed in enzyme activities are adaptations to or merely consequences of environmental change.

Individual aspects of control mechanisms are detailed in the earlier sections of this chapter, but a few additional remarks will be made here. For a recent summary of the regulation of different aspects of lipid metabolism, see Vance and Vance (1991); for plant glycerolipid metabolism, refer to Browse and Somerville (1991) and to Quinn and Harwood (1990); and for microbial aspects, see Ratledge and Wilkinson (1988, 1989).

The overall control of lipogenesis has been reviewed by Saggerson (1980). So far as saturated fatty acid synthesis is concerned, the activity of both acetyl-CoA carboxylase and fatty acid synthetase can be altered in various ways (section 11.1.1). Short-term or acute control involves metabolic or allosteric regulation and the covalent modification of enzymes. Long-term control involves alterations in the amounts of enzyme protein (Wakil *et al.*, 1983).

Acetyl-CoA carboxylase activity can be altered by interaction with citrate and other tricarboxylic acids. Much attention has been paid to hepatocyte systems where compounds such as glucagon or dibutyryl cyclic AMP lower cytosolic citrate levels, which cause a lowered acetyl-CoA carboxylase activity (Lane *et al.*, 1979). The latter is also inhibited by raised fatty acyl-CoA concentrations, which cause depolymerization of the mammalian carboxylase. Information is available concerning inhibitor specificity and interactions with other subcellular compounds (see Wakil *et al.*, 1983). The mammalian acetyl-CoA carboxylase has also been reported to undergo covalent modification through phosphorylation/dephosphorylation. Such regulation may involve the simultaneous presence of CoA (Yeh *et al.*, 1981). Various kinase and phosphatase enzymes that may be important in the modification of the mammalian acetyl-CoA carboxylase have been purified (see Wakil *et al.*, 1983).

The mammalian carboxylase also undergoes adaptive changes in protein content. These are caused by changes in the rate of enzyme synthesis and cause long-term regulation under the action of diet, insulin and thyroxine (Volpe and Vagelos, 1976). Yeast carboxylases are also subject to long-term regulation. In these cases the presence of long-chain fatty acids in the media decreases the amount of acetyl-CoA carboxylase by 80%. The decrease results from a depression in mRNA coding for the enzyme and is due to the action of two distinct acyl-CoA synthetases, one of which (synthetase I) provides acyl-CoAs that act as co-repressors (Numa, 1981).

The only bacterial acetyl-CoA carboxylase to have been studied in detail is that from *Escherichia coli*. This enzyme is regulated by guanosine 3′-diphosphate 5′-di- (and tri-) phosphate, which reduce activity by inhibiting the carboxyltransferase (section 11.1.1). In contrast, recent evidence suggests that the plant chloroplastic enzyme is regulated by adenine nucleotides. Thus AMP and ADP were found to be competitive inhibitors, with respect to ATP, of the acetyl-CoA carboxylase. These and other comparative aspects of different acetyl-CoA carboxylases are described by Harwood and Russell (1984).

Recently an AMP-activated protein kinase has been purified that phosphorylates serines 79, 1200 and 1215 on acetyl-CoA carboxylase. The kinase is activated by metabolites, particularly rises in the AMP/ATP ratio. Its characteristics and metabolic roles are discussed by Hardie (1992). Interestingly the kinase also regulates two other important enzymes of lipid metabolism by phosphorylation – (hydroxymethylglutaryl) HMG-CoA reductase and hormone-sensitive lipase.

Fatty acid synthetase is also subject to both adaptive changes in enzyme content as well as short-term metabolic control. The latter effects are ill-defined (Wakil *et al.*, 1983) but diet, triiodothyronine, hydrocortisone and insulin effects have been noted on the amount of synthetase protein. Hydrocortisone and triiodothyronine have no effect alone but potentiate the insulin induction of synthetase (Wakil *et al.*, 1983). The increases in synthetase activity on refeeding or insulin administration are due to an increase in transcription of mRNA, which is elevated 70-fold (Morris *et al.*, 1982).

It is well known that changes in growth temperature lead frequently to a modification in the pattern of fatty acids made and in those accumulated in the membranes of poikilotherms. Typical changes include an increase in unsaturated or of shorter-chain-length fatty acids at lower growth temperatures. The adaptation has been studied in a large number of organisms. In anaerobic bacteria such as *E. coli* it is not possible for desaturases to be induced, so that alterations in saturated/unsaturated fatty acid synthesis has to be controlled via fatty acid synthetase. In the *E. coli* fatty acid synthetase there are β-ketoacyl-ACP synthetases called I and II. Synthetase II is relatively specific for the elongation of palmitoleic to *cis*-vaccenic acid and is more temperature-sensitive that the synthetase I. Thus, at lower growth temperatures the synthetase II is relatively active and this increases the proportion of unsaturated fatty acids produced by the fatty acid synthetase (see section 11.1). Growth temperature has also been reported to alter the fatty acid products of other fatty acid synthetases. For example, the *Saccharomyces cerevisiae* synthetase produces more palmitate rather than stearate at lower growth temperatures (Okuyama *et al.*, 1979).

The control of fatty acid desaturases is not particularly well understood for any organism. The enzymes are under dietary and hormonal control in animals, and the effects of changes in membrane fatty acid unsaturation have been reviewed (Stubbs and Smith, 1984). In higher plants and green algae, it is known that both light and temperature can have profound effects. In the latter case there is some controversy as to whether low temperatures cause an increase in desaturase activity *per se* or whether the increase in activity is a reflection of substrate availability since oxygen is more soluble at lower temperatures (see Harwood, 1994 and Cossins, 1994).

Some of the best-understood desaturases in terms of control are those of bacilli. Two control mechanisms have been demonstrated for the Δ5-desaturase of *Bacillus megaterium*. First, the rate of synthesis of the enzyme is increased at low temperatures owing to a temperature-sensitive modulator, which switches off synthesis at higher temperatures. In addition, the enzyme is thermolabile and at higher growth temperatures is turned over at faster rates (Harwood and Russell, 1984). Membrane fluidity is also known to alter the activity of desaturases in different organisms including plants, protozoa and bacteria. A fuller discussion of these effects is beyond the scope of this section, but more details will be found in Harwood and Russell (1984) and Cossins (1994) together with pertinent references.

In addition to regulation of fatty acid synthesis, the rate of glycerolipid biosynthesis is under careful control. In particular, attention is drawn to the effects of light, temperature and salt stress in photosynthetic organisms (Harwood, 1994). Regulations of triacylglycerol metabolism in mammals (see section 11.3) has been well studied, as well as the net *de novo* synthesis of phosphoglycerides during membrane turnover (Bell and Coleman, 1980). The very fast rates of turnover of certain phosphoglycerides during receptor activation have also attracted considerable research (Hawthorne, 1982). For a comprehensive account of the regulation of phosphoglyceride synthesis and degradation in different organisms, see Hawthorne and Ansell (1982).

References and further reading

Bell, R.M. and Coleman, R.A. (1980) *Annu. Rev. Biochem.*, **49**, 459.

Browse, J. and Somerville, C.R. (1991) *Annu. Rev. Plant Physiol.*, **42**, 467.

Cossins, A. (ed.) (1994) *Adaptation of Biological Membranes*, Portland, London.

Hardie, D.G. (1992) *Biochim. Biophys. Acta*, **1123**, 231

Harwood, J.L. (1994) *Prog. Lipid Res.*, **33**, 193.

Harwood, J.L. and Russell, N.J. (1984) *Lipids in Plants and Microbes*, Allen and Unwin, Hemel Hempstead.

Hawthorne, J.N. (1982) in *Phospholipids* (eds J.N. Hawthorne and G.B. Ansell), Elsevier, Amsterdam, p. 263.

Hawthorne, J.N. and Ansell, G.B. (1982) *Phospholipids, New Comprehensive Biochemistry*, vol. 4, Elsevier, Amsterdam.

Lane, M.D., Watkins, P.A. and Meredith, M.J. (1979) *Crit. Rev. Biochem.*, **7**, 121.

Morris, S.M., Nilson, J.H., Jenik, R.A., Winberry, L.K., McDevitt, M.A. and Goodridge, A.G. (1982) *J. Biol. Chem.*, **257**, 3225.

Numa, S. (1981) *Trends Biochem. Sci.*, **6**, 113.

Okuyama, H., Saito, M., Joshi, V.C., Gunsberg, S. and Wakil, S.J. (1979) *J. Biol. Chem.*, **254**, 12281.

Quinn, P.J. and Harwood, J.L. (eds) (1990) *Plant Lipid Biochemistry, Structure and Utilization*, Portland, London.

Ratledge, C. and Wilkinson, S.G. (eds) (1988) *Microbial Lipids*, vol. 1, Academic Press, London.

Ratledge, C. and Wilkinson, S.G. (eds) (1989) *Microbial Lipids*, vol. 2, Academic Press, London.

Saggerson, E.D. (1980) *Biochemistry of Cellular Regulation*, vol. 2, CRC Press, Boca Raton, FL, p. 207.

Stubbs, C.D. and Smith, A.D. (1984) *Biochim. Biophys. Acta*, **779**, 89.

Vance, D.E. and Vance, J. (eds) (1991) *Biochemistry of Lipids, Lipoproteins and Membranes*, Elsevier, Amsterdam.

Volpe, J.J. and Vagelos, P.R. (1976) *Physiol. Rev.*, **56**, 339.

Wakil, S.J., Stoops, J.K. and Joshi, V.C. (1983) *Annu. Rev. Biochem.*, **52**, 537.

Yeh, L.A., Song, C.S. and Kim, K.H. (1981) *J. Biol. Chem.*, **256**, 2289.

12

MEDICAL AND AGRICULTURAL ASPECTS OF LIPIDS

J.L. HARWOOD, A. CRYER, M.I. GURR and P. DODDS

12.1 Human dietary requirements

12.1.1 Introduction

There are two aspects to lipid requirements in the human diet – these are qualitative and quantitative. First, certain lipids are needed for good health – essential fatty acids and fat-soluble vitamins are good examples. Secondly, it is usually considered that, in the normal diet, some 25–30% of the total calories are conveniently supplied as fat (Jones, 1974). Such lipid (in reasonable amounts) also usually makes food more palatable. Recent comprehensive sources of information on the role of fats in nutrition are Gurr (1992b) and Vergroesen and Crawford (1989).

The human diet has always contained fat, but the amounts and types vary. Typical intakes in Europe and North America are between 80 and 150 g/day, which represents 30–40% of dietary calories. The nutrient fat per capita has been maintained, with a slight overall rise, this century. However, the same foods are not always responsible for the fat consumed (Rizek *et al.*, 1974). Salad and cooking oils have always been major contributors, followed by dairy produce and shortening in the period 1910–30, but by margarine, shortening and meat in the period 1930–60. Since that time, the rise has been due almost entirely to a rise in meat consumption. Moreover, current consumption of fat in different parts of the world varies markedly. In Far Eastern countries there is minimal dietary fat, whereas Eskimos consume 300 g/day – an amount that would nauseate a European.

Because a diet lacking in fat tends to be bulky, then some rough rules can be made: if total intake is <3000 kcal/day, the fat should make up at least 25% of the calories; and for intakes of >3000 kcal/day, then it should represent at least 30% of the calories. This gives minimum intakes of 56–140 g/day. Of course, there are circumstances when dietary fat should be limited (see below and sections 12.2 and 12.3). In a survey in the USA it was found that intake of fat increased with higher incomes. This correlation may reflect a greater consumption of food or, alternatively, the selection of more expensive foods such as meat or dairy products instead of cereals (Rizek *et al.*, 1974). Pearson and Dutson (1990) give information on meat lipids, including several chapters dealing with dietary lipids and health, while Gregory *et al.* (1990) provide data on fat intakes and blood lipid levels for UK subjects.

Major food types that contribute to fat consumption in the United Kingdom are indicated in Table 12.1. For the USA, the intake from fats given by Rizek *et al.* (1974) was 35% for salad and cooking oil, 23% for shortening, 15% for margarine, 8% for butter and the remaining 19% from animal sources (meat, lard, etc.). It will be clearly seen from Table 12.1 that dairy products, meats, spreads and cooking oils are the major contributors. Although not reflected in the table, there has been a significant shift in the sources of fats consumed in recent years. Thus vegetable fats (margarine, cooking oils) are increasingly important when compared to animal fats (dairy products, lard). For example, the contribution of vegetable fats to the total in the USA increased from 17% to 38% in the period 1900–50. This shift has raised the contribution of $n-6$ polyunsaturated fatty acids from 3% to 6% of the total dietary calories. In addition, the reduction in animal fats has lowered cholesterol consumption (cf. section 12.3). Gold *et al.* (1992) have discussed the possible connections between cholesterol and coronary heart disease, and the epidemiological evidence for such interactions is critically reviewed by Gurr (1992a).

Fats are broken down by digestion and absorbed through the alimentary tract. Disorders at this stage lead to impaired fat intake (Sickinger, 1975). Principal fat components of a typical European or North American diet are triacylglycerol >> phospholipid = glycolipid > sterols. In processed food there is often a small amount of monoacylglycerol, and antioxidants such as α-tocopherol may

Table 12.1 Daily per capita fat consumption in the UK (From *Household Food Consumption and Expenditure*, HMSO, London)

Food component	Total fat		Fatty acids			Vitamin A or equivalent	Vitamin D (μg)
	(g)	(% total)	Saturated	Mono-unsaturated	Polyunsaturated	(μg)	
Milk/cream	16.5	15.6	10.4	5.0	0.4	178	0.33
Cheese	4.3	4.1	2.7	1.3	0.1	62	0.05
Meat	29.0	27.4	12.5	12.7	2.5	283	0.02
Fish	1.1	1.0	0.3	0.4	0.4	2	0.48
Eggs	3.2	3.0	1.2	1.4	0.4	41	0.44
Butter	18.7	17.7	11.6	5.6	0.5	228	0.28
Margarine	8.6	8.1	3.1	3.4	1.7	101	0.83
Cooking fats	10.9	10.3	3.9	4.8	1.7	3	0.02
Vegetables	1.5	1.5	0.2	0.6	0.7	280	trace
Fruits	0.5	0.5	0.2	0.2	0.1	13	–
Bread	1.5	1.4	0.5	0.4	0.5	1	–
Cakes, pastries	2.7	2.5	1.0	1.1	0.5	9	0.09
Biscuits	4.5	4.3	2.1	1.9	0.3	–	–
Other cereals	1.3	1.2	0.5	0.5	0.3	6	0.08

be added. Fat-soluble vitamins may be present naturally or have been added. The process of digestion is beyond the scope of this review, but a clear, simple description of it will be found in Gurr and Harwood (1991). The use of medium-chain triacylglycerols in patients with digestive problems is discussed in Babayan (1974), Vergroesen (1975) and Thomas and Holub (1994).

Triacylglycerols are usually 97–100% digested and the products of this process are absorbed into intestinal cells where resynthesis takes place. Unsaturated triacylglycerols are hydrolysed faster than saturated ones. Cholesterol is absorbed as the free sterol and then re-esterified as necessary for transport in the lymph. Bile may act as a cofactor for cholesterol esterase and improves cholesterol absorption, but plant sterols inhibit. Indeed, plant sterols (sitosterol, stigmasterol) and ergosterol are themselves rather poorly absorbed (Boyd, 1975). Although dietary fats increase cholesterol absorption by raising bile flow, saturated lipids are more effective in *net* terms since unsaturated fats also increase cholesterol excretion. The rate of cholesterol absorption is influenced by the age of the subject, their previous dietary history, genetic factors, the amount of cholesterol, the type of fat consumed simultaneously and the frequency of cholesterol intake. Similarly, overall fat absorption is affected by the type and amount of fat consumed, the age of the individual, the presence of emulsifying agents, whether the food has been heated above 250 °C (when fat becomes less digestible) and the presence of calcium (Jones, 1974; Vergroesen, 1975).

Malabsorption of fats can be measured with ^{131}I-labelled fats. Defects in bile salts cause poor digestion and absorption. This may be caused by an inability of the liver to synthesize bile acids or an inability to conjugate such acids with glycine or taurine, by an obstruction in the bile duct or by poor reabsorption from the ileum. Pancreatic insufficiency means that lipase and phospholipase digestive enzymes are decreased or missing. This condition is also caused in cystic fibrosis. Malabsorption is also caused by bacterial infection (tropical sprue, Whipple's disease, small-intestine diverticulosis, blind loop syndrome, scleroderma). In these diseases, symptoms are functional dyspepsia (intolerance of fatty foods) and steatorrhoea. The latter has no pathological significance until the stool fat is greater than 10%. However, it causes distress and inconvenience and, when severe, wasting (loss of salts and liquid, and insufficient uptake of nutrients) and deficiencies (e.g. of fat-soluble vitamins, with vitamin D the first to show). For fuller discussion of these medical aspects, see Petersdorf *et al.* (1983).

12.1.2 Specific fat requirements

Essential fatty acids are necessary since they are utilized as precursors of prostaglandins, thromboxanes and leukotrienes (Samuelsson *et al.*, 1975; 1978; section 11.1.7). There is no doubt that polyunsaturated fatty acids are necessary for normal good health, and, in general, the $n-6,9$ diene unit is required. Of the normal dietary fatty acids, linoleic acid is quantitatively the most important. The role of polyunsaturated fatty acids in human nutrition and metabolism has been reviewed by Soderhjelm *et al.* (1971). For the relationship between structure and efficiency as an essential fatty acid, see Lands *et al.* (1977). Polyunsaturated fatty acids may also be involved in a number of disease states. These include a betalipoproteinaemia, acrodermatitis enteropathica, biliary tract disease, kinky hair disease, hepatoma, kwashiorkor, mycobacterial infections and multiple sclerosis (Soderhjelm *et al.*, 1971). Essential fatty acid deficiency leads to problems in practically every tissue of the body (Holman, 1977). In particular, body weight is decreased, there are changes in skin permeability and in capillary fragility, dermatitis and necrosis are seen, and there are biochemical changes – such as in mitochondrial efficiency – in various tissues (cf. Houtsmuller, 1975; Holman, 1977). Essential fatty acid deficiency is easily recognized because tissue fatty acids of the $n-6$ group are partly replaced by those of the $n-9$ (oleic) group. In particular, arachidonic acid is reduced and eicosatrienoic acid (*cis*-5,7,11-20:3) is increased (Houtsmuller, 1975).

Prostaglandins are cyclopentane derivatives (section 11.1.7) formed from polyunsaturated fatty acids by most mammalian tissues (and by tissues of lower vertebrates and certain invertebrates). These prostaglandins modulate a host of hormonal, neurohormonal and other stimuli, and have a variety of other pharmacological actions (Samuelsson *et al.*, 1975, 1978). Thromboxanes themselves are of pronounced biological activity, but are unstable, with short half-lives (cf. Samuelsson *et al.*, 1978).

The six primary prostaglandins (E_1, E_2, E_3, $F_{1\alpha}$, $F_{2\alpha}$, $F_{3\alpha}$) are formed by the metabolism of three C_{20} acids: Δ8,11,14, Δ5,8,11,14 and Δ5,8,11,14,17 (Lands *et al.*, 1977). The fatty acid precursors are released from phospholipid stores by phospholipases (cf. Hinman, 1972).

There have been two extensive reviews of the role of polyunsaturated fatty acids in the diet. Hoffman and Forster (1981) discussed the influence of dietary linoleic acid on blood-pressure regulation – particularly in relation to salt-loaded individuals. In addition, the mechanism of the hypocholesterolaemic effects of polyunsaturated fatty acids was discussed by Paul *et al.* (1980). These authors classified the hypocholesterolaemic effect as being due to reduced absorption of cholesterol, redistribution of cholesterol from blood to tissues, reduced cholesterol synthesis and increased excretion of cholesterol or its catabolites. These various factors have all been involved in the hypocholesterolaemic effect, but the evidence is equivocal in each case. Other sources of information on dietary polyunsaturated fatty acids are Vergroesen (1975), Kanau and Holman (1977) and Gurr (1992b).

The role of $n-3$ fatty acids in health and disease is covered by Lees and Karel (1990) and Ackman and Cunnane (1992), while a specific review of the possible beneficial effects of γ-linolenic acid is that by Horrobin (1992). For a general review of unsaturated fatty acids and nutrition, see British Nutrition Foundation (1992).

12.1.3 Vitamins

Although somewhat outside the scope of this book, it would be wrong to describe the role of lipids in diets without mentioning fat-soluble vitamins. Vitamins A_1 and A_2 are important in preventing degenerative changes of epithelial surfaces such as keratinization of skin. They are required for normal bone development, and a deficiency in young animals can lead to secondary nervous problems due to compression of the brain and spinal cord. 'Night-blindness' is an early sign of deficiency in humans. Vitamin A activity in foods is frequently due to the presence of precursors such as β-carotene. For a review see H.F. de Luca (1978a).

Vitamin D is needed for the prevention of rickets, and its deficiency is involved in other pathological states (L.M. de Luca, 1978b). A number of different structures have activity, the two most important of which are vitamin D_2 (ergocalciferol) and vitamin D_3 (cholecalciferol). Again, as with vitamin A, the diet is often able to supply provitamins (e.g. ergosterol), which are converted to the active compounds. All of the provitamins require ultraviolet light for their conversion.

Vitamin E is the name given to a number of structurally related compounds, the most important of which is α-tocopherol. The vitamin is needed for mitochondrial electron-transport function and it prevents oxidation of various compounds, including polyunsaturated fatty acids and vitamin A. The dependence of vitamin E requirement on the amount of dietary polyunsaturated fatty acids has been fully discussed by Jager (1975). Traditionally, vitamin E has been known as the 'fertility' vitamin. However, deficiency leads to serious changes in skeletal muscle, the blood system and other tissues before reproduction is impaired. The effects on different animals are described by Jager (1975). Vitamin E is discussed by Scott (1978) and Packer and Fuchs (1993).

Vitamin K, containing the active 2-methyl-1,4-naphthoquinone structure, exists as K_1 in green parts of plants and as K_2 in micro-organisms. Deficiency impairs blood coagulation. As with the other vitamins described above, deficiency of the fat-soluble vitamin

K most frequently results nowadays when general fat absorption is impaired (see above and Sickinger, 1975). Vitamin K is discussed by Suttie (1978). See also Recommendations of the European Atherosclerosis Society (1992) and Willett *et al.* (1993) for data on trans fatty acids, Tholstrup *et al.* (1994) for recent data on saturated acids, Riemersma (1994) for a review of antioxidants in coronary heart disease prevention and Wald (1994) for lipoprotein: CHD relationships.

Certain specific lipids have excited interest because of worries about their actual or potentially harmful effects. Examples of these have been saturated fatty acids (Kanau and Holman, 1977), cholesterol (McGandy and Hegsted, 1975; Sabine, 1977) and erucic acid from some rapeseeds (Vles, 1975). Cyclopropene fatty acids, which occur in some seed oils, are inhibitors of fatty acid desaturation (cf. section 11.1.3).

References and further reading

Ackman, R.G. and Cunnane, S.C. (1992) *Adv. Appl. Lipid Res.*, **1**, 161.

Babayan, V.K. (1974) *J. Am. Oil Chem. Soc.*, **51**, 260.

Boyd, G.S. (1975) in *The Role of Fats in Human Nutrition* (ed. A.J. Vergroesen), Academic Press, London, p. 353.

British Nutrition Foundation (1992) *Report of the Task Force on Unsaturated Fatty Acids*, British Nutrition Foundation, London.

de Luca, H.F. (1978a) in *The Fat-Soluble Vitamins* (ed. H.F. de Luca), Plenum, New York, p. 69.

de Luca, L.M. (1978b) in *The Fat-Soluble Vitamins* (ed. H.F. de Luca), Plenum, New York, p. 1.

Gold, P., Grover, S. and Roncari, D.A.K. (1992) *Cholesterol and Coronary Heart Disease – The Great Debate*, Panthenon, Carnforth and Park Ridge.

Gregory, J., Foster, K., Tyler, H. and Wiseman, M. (1990) *The Dietary and Nutritional Survey of British Adults*, HMSO, London.

Gurr, M.I. (1992a) *Prog. Lipid Res.*, **31**, 195.

Gurr, M.I. (1992b) *The Role of Fat in Food and Nutrition*, Elsevier, London.

Gurr, M.I. and Harwood, J.L. (1991) *Lipid Biochemistry*, 4th edn., Chapman & Hall, London.

Hinman, J.W. (1972) *Annu. Rev. Biochem.*, **41**, 161.

Hoffman, P. and Forster, W. (1981) *Adv. Lipid Res.*, **18**, 203.

Holman, R.T. (1977) in *Polyunsaturated Fatty Acids* (eds W.-H. Kunau and R.T. Holman), American Oil Chemists' Society, Champaign, p. 163.

Horrobin, D.F. (1992) *Prog. Lipid Res.*, **31**, 163.

Houtsmuller, U.M.T. (1975) in *The Role of Fats in Human Nutrition* (ed. A.J. Vergroesen), Academic Press, London, p. 381.

Jager, F.C. (1975) in *The Role of Fats in Human Nutrition* (ed. A.J. Vergroesen), Academic Press, London, p. 2.

Jones, R.J. (1974) *J. Am. Oil Chem. Soc.*, **51**, 251.

Kunau, W.-H. and Holman, R.T. (eds) (1977) *Polyunsaturated Fatty Acids*, American Oil Chemists' Society, Champaign.

Lands, W.E.M., Hemler, M.E. and Crawford, C.G. (1977) in *Polyunsaturated Fatty Acids* (eds W.-H. Kunau and R.T. Holman), American Oil Chemists' Society, Champaign, p. 193.

Lees, R.S. and Karel, M. (eds) (1990) *Omega-3 Fatty Acids in Health and Disease*, Marcel Dekker, New York.

McGandy, R.B. and Hegsted, D.M. (1975) in *The Role of Fats in Human Nutrition* (ed. A.J. Vergroesen), Academic Press, London, p. 211.

Packer, L. and Fuchs, J. (1993) *Vitamin E in Health and Disease*, Marcel Dekker, New York.

Paul, R., Ramesha, C.S. and Ganguly, J. (1980) *Adv. Lipid Res.*, **17**, 155.

Pearson, H.M. and Dutson, T.R. (1990) *Advances in Meat Research*, vol. 6, Elsevier, London.

Petersdorf, R.G., Adams, R.D., Braunwald, E., Isselbacher, K.J., Martin, J.B. and Wilson, J.D. (eds) (1983) *Principles of Internal Medicine*, 10th edn, McGraw-Hill, New York.

Recommendations of the European Atherosclerosis Society (1992) Prevention of coronary heart disease: Scientific background and new clinical guidelines. *Nutri. Metab. Cardiovasc. Dis.*, **2**, 113.

Riesmersma, R.A. (1994) *Proc. Nutri. Soc.*, **53**, 59.

Rizek, R.L., Friend, B. and Page, L. (1974) *J. Am. Oil Chem. Soc.*, **51**, 244.

Sabine, J.R. (1977) *Cholesterol*, Marcel Dekker, New York.

Samuelsson, B., Granström, E., Green, K., Hamberg, M. and Hammerström, S. (1975) *Annu. Rev. Biochem.*, **44**, 669.

Samuelsson, B., Goldyne, M., Granström, E., Hamberg, M., Hammerström, S. and Malmsten, C. (1978) *Annu. Rev. Biochem.*, **47**, 997.

Scott, M.L. (1978) in *The Fat-Soluble Vitamins* (ed. H.F. de Luca), Plenum, New York, p. 133.

Sickinger, K. (1975) in *The Role of Fats in Human Nutrition* (ed. A.J. Vergroesen), Academic Press, London, p. 116.

Soderhjelm, L., Wiese, H.F. and Holman, R.T. (1971) *Prog. Chem. Fats Other Lipids*, **9**, 555.

Suttie, J.W. (1978) in *The Fat-Soluble Vitamins* (ed. H.F. de Luca), Plenum, New York, p. 211.

Tholstrup, T., Marckmann, P., Jespersen, J. and Sandstrom, B. (1994) *Am. J. Clin. Nutri.*, **59**, 371.

Thomas, L.M. and Holub, B.J. (1994) in *Technological Advances in Improved and Alternative Sources of Lipids* (eds B.S. Kamel and Y. Kakude), Blackie, London, p. 16.

Vergroesen, A.J. (ed.) (1975) *The Role of Fats in Human Nutrition*, Academic Press, London.

Vergroesen, A.J. and Crawford, M. (1989) *The Role of Fats in Human Nutrition*, Academic Press, London.

Vles, R.O. (1975) in *The Role of Fats in Human Nutrition* (ed. A.J. Vergroesen), Academic Press, London, p. 434.

Wald, N.J. (1994) *The Lancet*, **343**, 75.

Willett, W.C., Stampfer, M.J., Manson, J.E., Colditz, G.A., Speizer, F.E., Rosner, B.A., Sampson, L. and Hennekens, C.H. (1993) *The Lancet*, **341**, 581.

12.2 Obesity

Obesity can be difficult to define, since individual perceptions of what might be thought 'overfat' vary. Obesity indicates the condition of being overfat, not overweight for height, but carrying excess adipose tissue. Arbitrary definitions of obesity tend to reflect cultural and health attitudes, and in Western populations men who exceed 20% fat ((stored lipid/body weight) × 100) and women who exceed 25% fat might be considered obese. Using these values, up to 40% of adult Western populations could be called obese. Semiquantitative definitions are also used in which body fat is not measured. For example, the most frequently used standard is the body mass index (BMI, weight/height^2), which enjoys consensus support (Burton *et al.*, 1985). When body fat is measured, it is usually deduced from physical measurements, such as skin fold thickness, buoyant weight or from radiotracer dilution data.

Obesity poses a significant health problem in that the risk of certain chronic disease states may be increased up to twofold and life expectancy decreased (US Public Health Service, 1966). Although most prevalent in middle age, individuals of all ages are susceptible. The major conditions associated with obesity include hyperglycaemia, carbohydrate intolerance, an increased incidence of cardiovascular disease (including vascular lesions of the nervous system, hypertensions and coronary heart disease) and certain carcinomas. A small number of grossly obese individuals also develop carbon dioxide insensitivity, hypoventilation, uneven ventilation and somnolence because of the difficulty of maintaining normal respiratory movements. Generally the obese tend to suffer more complications from established disease and have an enhanced risk during surgery and recovery (Mayer, 1968).

Most of the body's stored lipid is found in white adipose tissue in a specialized cell type, the adipocyte. The adult non-obese human may have between 25×10^9 and 50×10^9 of these cells, which are specialized both metabolically and morphologically in order to store large amounts of energy as virtually anhydrous triacylglycerol. Since the principal function of adipocytes is the storage of chemical energy as triacylglycerol during times of nutritional plenty, and the release of this energy in the form of unesterified fatty acid on demand, the expansion of body fat mass must reflect an excessive caloric intake relative to energy expenditure. The mechanism whereby adipose tissue expands therefore depends on the number of adipocytes present and their individual size, and increments in adipose tissue bulk may be the consequence of increases in either adipocyte number or cellular lipid content (size) or a combination of the two. Because recent advances have led to the recognition that the white adipose tissue of both young and mature animals, including humans, can expand by both cellular hypertrophy and hyperplasia, interest in the adipocyte stem cell and in the factors that control its proliferation and differentiation has grown (Grigoriadis *et al.*, 1988). The identification of the adipocyte precursor cell (the stem cell following commitment to the adipocyte lineage) both *in vivo* and *in vitro* has allowed significant progress to be made in the study of both the cell and molecular biology of adipose differentiation and development. Such studies have utilized both cells isolated from the tissue (e.g. Cryer and Cryer, 1990) and useful established cell lines exhibiting appropriate concerted properties (e.g. Ailhaud, 1991). Evidence for the involvement of hormonal and other effectors in the control of the proliferation and differentiation of the adipocyte precursor cell and the molecular mechanisms that exist is providing a fundamental background to the clinical and pathological study of obesity.

Although excess adiposity can be the phenotypic expression of diverse pathogenetic mechanisms, ultimately an expanded body fat mass is the result of excess caloric intake relative to energy expenditure. Thus nutritional intake can, uniquely, affect adipose tissue cellular development during both early and later life. The pattern of adipose tissue expansion is therefore related, at least in part, to the feeding behaviour of the individual. Not only does feeding behaviour include the consumption of food, the seeking out of food and working for food, it also encompasses the response of the individual to those features of a food item that collectively define its 'palatability'. Although the ingestion of excess calories, particularly in the form of a 'high-fat' diet, may increase fat cell numbers, which will contribute to the expansion of adipose tissue mass, the reversal of this cellular hyperplastic response cannot be readily achieved by curtailing dietary intake, although the total adipose mass may decline by a reduction in average cell size (Faust, 1980).

Where excess fat cell production and adiposity are a consequence of diet, reduction in total food intake may be all that is required to effect weight reduction. However, prospectively, it may be possible to gain the same advantage by altering dietary composition (Faust, 1980). Much work has sought to define a metabolic aetiology for obesity, but the comparison of lean and obese individuals does not, in general, allow differences to be identified as either causes or effects of the condition. The differences between genetically obese animals and their lean littermates have also been extensively studied in an attempt to clarify the potential metabolic predispositions that encourage obesity. However, most genetically obese animal models are complicated by the diverse nature of the metabolic lesions in the obese, some of which may not be related to the manifestation of excess adiposity (Bray and York, 1979).

The definition of the precise relationship between the cellular, hormonal, metabolic and neuroendocrinological factors that control energy storage and expenditure in humans is still, however, an unattained goal.

References and further reading

Ailhaud, G. (1991) *Curr. Opin. Cell Biol.*, **2**, 1043.

Bray, G.A. and York, D.A. (1979) *Physiol. Rev.*, **59**, 719.

Burton, B.T., Foster, W.R., Hirsch, J. and van Itallie, T.B. (1985) *Int. J. Obes.*, **9**, 155.

Cryer, A. and Cryer, J. (1990) in *The Control of Body Fat* (eds M.G. Forbes and J. Harvey), Smith Gordon, London, pp. 19–42.

Faust, I.M. (1980) *Int. J. Obes.*, **4**, 314.

Grigoriadis, A.E., Heersche, J.N.M. and Aubin, J.E. (1988) *J. Cell Biol.*, **106**, 2139–51.

Mayer, J. (1968) *Overweight Causes, Cost and Control*, Prentice-Hall, Englewood Cliffs, NJ.

US Public Health Service, Division of Chronic Diseases (1966) *Obesity and Health: A Source Book of Current Information for Professional Health Personnel*, US Government Printing Office, Washington, DC.

12.3 Lipids and cardiovascular disease

12.3.1 Cardiovascular disease: the forms it takes

Of the clinical manifestations associated with cardiovascular disease, the 'heart attack' and the 'stroke' are the most obvious and spectacular. These events, together with other less common symptoms of circulatory malfunction, such as gangrene of the extremities, are the result of progressive deterioration of the circulatory system and its function.

The 'heart attack' occurs when one or more of the coronary arteries, which are responsible for supplying blood to the heart muscle, becomes blocked. The part of the heart that is consequentially robbed of oxygen and nutrients and therefore damaged is called an infarct. From this, the terms myocardial infarct, coronary occlusion and coronary thrombosis (blood clot) all refer, either more or less precisely, to what is called a 'heart attack'.

Coronary arteries may suffer blockage due to the trapping of a blood clot formed elsewhere in the body, the risk being increased by the localized and progressive formation of atherosclerotic plaques. When fully formed, such plaques are composed of accumulated arterial smooth muscle cells, lipid (predominantly cholesterol) and calcium.

Atherosclerotic plaques (atheroma) may promote occlusive clot formation and/or so reduce the size of the blood-vessel lumen that areas of heart muscle become deprived of oxygen. This deprivation may be particularly acute during exercise and lead to the pain characteristic of angina pectoris.

If arteries become blocked in the brain, 'stroke' is the result, with portions of the brain suffering varying degrees of damage. If lesions occur in the limb arteries, blood flow may be impaired and lead to numbness, pain and, eventually, gangrene.

12.3.2 Incidence and distribution of cardiovascular disease

Death rate as a consequence of atherosclerosis and particularly that which contributes to ischaemic heart disease has been increasing, at least until recently, throughout the present century. Even up to the middle of the nineteenth century, descriptions of the condition were rare, and even taking account of any deficiency that might affect earlier clinical judgements, there is no doubt that its related conditions have, in the last 50 years, become the major cause of death in the developed world.

Statistics prepared by WHO indicate a distinct geographical distribution of cardiovascular disease incidence, with Sweden, Scotland, Denmark, USA, England and Wales, Ireland, Finland, Austria, Norway and Australia being the top ten, and with Costa Rica, Japan, Hong Kong, Phillipines, Mexico, Equador, Dominican Republic, Martinique, Honduras and Thailand being the bottom ten, in their incidence rates of atherosclerosis-related deaths.

In susceptible populations, atherosclerosis may start quite early in life. Early (fatty streak) lesions may be seen during the second decade of life and fibrous plaques during the third. These primary pathological changes in arteries therefore antedate, by at least 20 years, the appearance of symptoms above the 'clinical horizon' (Bierman, 1973). In virtually all the populations studied this far, the presence of lesions and the morbidity attributable to underlying atherosclerosis are much more frequent in males than females, and at any given age men have more severe coronary artery disease than women, probably because of its earlier onset.

12.3.3 Epidemiology and the search for risk factors

The analysis of mainly epidemiological data, and most recent longitudinal data, has attempted over a number of years to pinpoint and predict those factors which might be causal in the development of cardiovascular disease. Positive 'risk factors' would therefore be, in one formulation of the concept, 'those habits, traits and abnormalities, with a sizeable (100% or more) increase in susceptibility to atherosclerotic disease'.

One of the major overviews of this (Pooling Project Research Group, 1978) sought to assess the data pertinent to the relation of blood pressure, serum cholesterol, smoking habits, relative weight and ECG abnormalities to the incidence of 'major coronary events'. The identification of a 'risk factor', however, cannot necessarily reveal information on mechanisms, since it may subvert a combination of environmental factors and intrinsic genetic differences.

Hypertension is a strong risk factor predisposing individuals to cardiovascular disease, with the strength of the association being virtually unaffected by age. However, as yet, it is difficult to decide whether hypertension and its haemodynamic consequences per se are contributory, or whether it is other environmental or genetic factors with which it is associated that are more significant.

From epidemiological studies carried out in several countries, smoking has been associated with an increased risk of myocardial infarction in both men and women (Royal College of Physicians, 1971). Although smoking-associated ischaemic heart disease risk is frequently attributed to the effects of carbon monoxide-induced tissue hypoxia, the mutagens present in smoke have also been invoked as significant promoters of vascular lesion initiation (Benditt and Gown, 1980).

Although, as noted above, gender is one of the best-documented risk factors, its associated factors cannot be ignored. For example, women have higher HDL (high-density lipoprotein) cholesterol and lower LDL (low-density lipoprotein) than men of similar ages (see later), and differences exist between the sexes with regard to blood pressure, smoking habits and other risk factors. However, post-menopausal women have twice the risk of myocardial infarction than pre-menopausal women of the same age (Gordon *et al.*, 1978).

12.3.4 Circulating lipids as risk factors

Of the epidemiologically recognized risk factors, plasma lipid concentrations, and particularly plasma cholesterol concentrations, cause the most debate. That an association exists between plasma lipids and cardiovascular disease is based on extensive epidemiological and clinical evidence; however, the exact nature of the interaction is only slowly becoming apparent.

Among the plasma lipid classes cited as atherogenic, or predictive indices of the future clinical manifestations of atherosclerosis, high concentrations of fasting serum cholesterol and low-density lipoproteins (LDL) together with low levels of high-density lipoproteins (HDL) are the most frequently mentioned (Fredrickson *et al.*, 1978; Miller and Miller, 1975), although some investigators rank elevated fasting serum triacylglycerol concentrations as an independent risk factor (Carlson and Bottiger, 1972), and post-prandial lipaemia in those individuals on relatively high-cholesterol diets has also been proposed as important in atherogenesis (Zilversmit, 1979). Although the total dietary intake of lipid, and particularly saturated fats, is related to ischaemic heart disease mortality, there is little firm evidence that the replacement of saturated with unsaturated fats confers protection without a reduction in total fat intake. These considerations do not run contrary to the view that the relative dietary deficiency of essential fatty acid (EFA) may also contribute to the pathogenesis of ischaemic heart disease (e.g. Oliver, 1979). See also Gurr (1992) for a good discussion.

12.3.5 Serum lipids: classification and metabolic roles

All the lipids carried in the plasma exist as specific complexes with protein. Long-chain fatty acids circulate bound to the plasma albumin fraction, the other lipids being integral components of several macromolecular lipid–protein complexes, known as the plasma lipoproteins. Thus the plasma lipoproteins function to transport hydrophobic lipids in an aqueous plasma-soluble form. The structure and function of these particles have been reviewed extensively (e.g. Glickman and Sabesin, 1988). Briefly, the plasma lipoproteins are discrete supramolecular complexes, usually of near-spherical shape with diameters between 8 and 800 nm. The nomenclature applied to lipoproteins is based on their ultracentrifugal flotation properties at specific solvent densities. For each class, the specific density of the lipoproteins is related to the proportions of lipid and protein present in the particles. Using flotation density as the distinguishing criterion, the lipoprotein classes are normally designated very-low-density lipoproteins (VLDL), low-density lipoproteins (LDL) and high-density lipoproteins (HDL). Additionally, chylomicrons also appear in post-prandial plasma, and are large lipoproteins of extremely low density that are derived from the intestinal absorption of dietary lipid. These lipoprotein classes may be further subdivided on the basis of fine differences in density, composition or properties.

The operationally separable classes of lipoproteins have distinct compositions and can be shown to have distinct physiological functions (Table 12.2). Thus chylomicrons are synthesized by mucosal cells of the small intestine and are the plasma transport vehicle for most of the absorbed dietary triacylglycerol and cholesterol. VLDL are also triacylglycerol-rich but are synthesized from endogenous sources and are secreted into the plasma by the liver and to a lesser extent by the small intestine. The endogenous precursors for the synthesis of VLDL triacylglycerol are either non-lipid dietary components or fatty acyl units released from adiposal storage. LDL in the plasma is derived primarily from VLDL, whereas HDL may be secreted as a precursor form from the liver and become modified during VLDL breakdown. A proportion of plasma HDL that is secreted in a nascent or precursor form is converted to native HDL by a process that also involves the plasma

Table 12.2 The average composition of human plasma lipoproteins and their probable general functions

Lipoprotein class	Composition (% total)				Metabolism	Simplified role
	TAG	Cholesterol	Phospholipid	Protein		
Chylomicron	90	5	3	2	Catabolism by extrahepatic lipoprotein lipase to 'remnants' with hepatic fate	Transport of dietary triacylglycerol and cholesterol
VLDL	60	12	18	10	Catabolism by extrahepatic lipoprotein lipase to LDL or remnant	Transport of endogenously synthesized triacylglycerol
LDL	10	50	15	25	Product of VLDL catabolism	Transport of cholesterol to peripheral tissues
HDL	5	20	25	50	Precursor secreted by liver, plasma particle probably derived from action of LPL and LCAT	Donor/acceptor of protein to/from other lipoproteins. Transport of peripheral tissue cholesterol to liver

Abbreviations: LPL = lipoprotein lipase; LCAT = lecithin:cholesterol acyltransferase; TAG = triacylglycerol.

enzyme lecithin:cholesterol acyltransferase (LCAT). Furthermore LDL may regulate cholesterol synthesis *de novo* in non-hepatic tissues, and HDL promotes cholesterol transport from peripheral tissues to the liver.

Apart from their operational distinction on the basis of density and lipid composition, the plasma lipoprotein classes are also distinguished by the nature of the polypeptides present. The apoprotein species have been characterized and designated alphabetic family names. The human plasma lipoprotein apoprotein distribution is shown in Table 12.3. Although the distribution, interchange and metabolism of lipoprotein apoproteins are extremely complicated, some overall aspects of the dynamic situation can be discerned. For example, the apoprotein content of newly synthesized VLDL and chylomicrons change drastically as these lipoproteins enter the circulation and interact with the other lipoproteins present.

Both species pick up apoproteins C and E. Subsequently, as lipolysis proceeds via the action of extra-

Table 12.3 Apoprotein distribution in human plasma lipoprotein classes

	Chylomicron	VLDL	LDL	HDL
Quantitatively major apoproteins	Apo-A_I Apo-B_{48} Apo-C	Apo-B_{100} Apo-C_I Apo-C_{II} Apo-C_{III} Apo-E	Apo-B_{100}	Apo-A_I Apo-A_{II}
Quantitatively minor apoproteins	Apo-A_{II} Apo-E	Apo-A_I Apo-A_{II}	Apo-C	Apo-C_I Apo-C_{II} Apo-C_{III} Apo-D Apo-E

hepatic lipoprotein lipase, the particles lose much of the apoproteins C and E. During VLDL catabolism the apoprotein B_{100} remains with its parent particle and consequently becomes an increasingly larger proportion of the total lipoprotein as the VLDL is metabolized progressively to LDL. The apoprotein B_{48} of chylomicrons is removed from the plasma by the liver as part of a chylomicron remnant particle, whereas the apoprotein A components become associated with the HDL fraction of the plasma apoprotein A pool. The C apoproteins of VLDL and the chylomicrons move rapidly to HDL during lipolysis, and together with apoprotein A contribute to the conversion of nascent HDL into native particles.

12.3.6 Sources and fates of circulating lipids

The overall metabolism of plasma lipoproteins is dependent upon three important enzyme activities: lipoprotein lipase, hepatic triacylglycerol lipase and lecithin:cholesterol acyltransferase. Lipoprotein lipase is widely distributed in extrahepatic tissues and is responsible for the hydrolysis of plasma lipoprotein triacylglycerol, and as such controls the rate-limiting step in the removal of triacylglycerol fatty acids from the circulation. Changes in tissue lipoprotein lipase activities occur in response to a variety of physiological and nutritional stimuli, and alterations in activity correlate closely with modifications in the rate of triacylglycerol fatty acid uptake by tissues such as heart, adipose tissue, mammary gland and skeletal muscle. The generally reciprocal changes in the lipoprotein lipase activities of white adipose tissue and skeletal muscle (the tissues quantitatively most important in the bulk removal of triacylglycerol from the plasma) have led to the view that such changes serve to direct triacylglycerols towards certain tissues preferentially, depending on the physiological circumstances (Cryer, 1987).

The generally held view, for which considerable evidence exists, suggests that lipoprotein lipase exerts its functional activity against chylomicron and VLDL triacylglycerol, following activation by apoprotein C_{II}, while both enzyme and substrate are sequestered at the endothelial cell surface of extrahepatic tissue blood vessels. The enzyme and possibly also the substrate are bound to the lumenal surface of the endothelial cells by specific non-covalent interactions with the glycosaminoglycans and proteoglycans present (Cryer, 1987).

The hepatic lipase has been shown to be localized on the surface of liver endothelial cells, and when the enzyme is specifically inhibited at this site, *in vivo*, although there is no resultant change in VLDL composition or plasma triacylglycerol concentrations, there is specific enrichment in the phospholipid and cholesterol content in the LDL and HDL (and particularly HDL_2) fractions. These findings, and other related information, suggest that the removal of LDL and some HDL phospholipid and cholesterol from the plasma by the liver involves the action of the hepatic lipase.

Hepatic lipase activity, in common with extrahepatic lipoprotein lipase activity, is influenced by a variety of hormones, all of which influence plasma HDL concentrations in an opposite direction. Thus plasma HDL concentrations are determined co-operatively by the activity of two endothelial bound lipolytic activities.

Lecithin:cholesterol acyltransferase is an extracellular enzyme of hepatic and intestinal origin that circulates in plasma and catalyses the transfer of C2 fatty acids from phosphatidylcholine to cholesterol. The enzyme is activated by apoprotein A_I and apoprotein C_I. The transfer of apoprotein E from HDL to VLDL occurs concomitantly with the enzyme's action. The presence of apoprotein A_I in chylomicrons suggests that this enzyme has a role in processing HDL components such as phosphatidylcholine, cholesterol and certain apoproteins that originate during the catabolism of chylomicrons. Thus this enzyme facilitates the formation of HDL from the nascent particles secreted by the liver and allows the uninterrupted metabolism of energy-yielding VLDL triacylglycerol.

The plasma lipoproteins, produced and modified by the actions of the three enzymes mentioned above, and by physical transfer and exchange of components, are ultimately removed from the circulation via uptake by a number of cell types by the process of endocytosis.

Normally the extent to which LDL and HDL are taken up by specific cell types depends on their requirements for cholesterol and cholesteryl ester. Thus uptake is balanced by and contributes to the control of the endogenous rates at which such cells synthesize these components for the maintenance of membrane structure and for other purposes. Derangements of such control exists in certain pathological states where lipoprotein receptor number or function are impaired. In the most thoroughly studied system, LDL has been shown to be interiorized by cells following interaction with a specific cell-surface receptor. This interiorization is then followed by the lysosomal degradation of lipoprotein components, activation of cholesterol-esterifying enzymes and suppression of hydroxymethylglutaryl-CoA reductase activity (Goldstein and Brown, 1977).

12.3.7 Factors affecting the composition and metabolism of circulating lipids

In considering the factors that affect the composition and metabolism of circulating lipids in humans, the nature of the 'normal' subject must be considered, since genetic predispositions, age, sex, diet, exercise and overt or hidden disease may all contribute to discernible differences. For example, sex differences in plasma cholesterol concentrations and distribution are obvious from early ages. Thus, although in girls a gradual decline in total serum cholesterol concentration occurs between five and 17 years, this decline is more marked in boys, although it is only in boys that the ratio of cholesterol in LDL and HDL increases dramatically after puberty. At virtually all ages up to about 50 years, males have higher LDL and VLDL concentrations in their plasma than women. However, in later years the LDL concentration in the plasma of women exceeds that of otherwise comparable men. Plasma VLDL and LDL concentrations also tend to rise in pregnancy.

Where there is no evidence of malnutrition, the dietary components that affect blood lipids and lipoproteins most profoundly are fats and carbohydrates. In the fat component, the nature of the fatty acid composition has attracted most attention. Thus, where conventional diets are modified so as to increase the proportion of fat containing vegetable (unsaturated) oils, significant reductions in total plasma cholesterol concentrations are seen, with the major reduction being in the LDL fraction. When low-fat diets are eaten with the calorie intake kept constant by an additional carbohydrate content, VLDL levels increase without affecting LDL concentrations significantly.

Of the inborn errors of lipid metabolism, those which comprise the familial hyperlipoproteinaemias are probably the most widely studied. These conditions, characterized as types I, IIa, IIb, III, IV and V, are all associated with characteristic derangements of plasma cholesterol and triacylglycerol concentrations.

Type I hyperlipoproteinaemia, known also as familial exogenous hyperlipaemia and familial hyperchylomicronaemia, is a rare recessive condition characterized by lipoprotein lipase deficiency with consequentially elevated plasma triacylglycerol concentrations. Pancreatitis rather than accelerated atherosclerosis is usually the immediate cause of death in these patients.

Type IIa hyperlipoproteinaemia, known also as familial hypercholesterolaemia, is an autosomal dominant condition characterized by an elevated cholesterol concentration in LDL. Patients who are homozygous develop severe cardiovascular disease, normally culminating in death at an early age. Heterozygotes are a more diverse group, who, in general, develop atheromatous lesions similar in appearance to those seen in normocholesterolaemic individuals, but do so over a shorter timescale. Such patients have a 50% probability of myocardial infarction before age 50.

Type IIb hyperlipoproteinaemia represents a larger group of patients than those of type IIa. The condition exhibits elevated levels of both LDL and VLDL, and is thus classified as a mixed hyperlipidaemia.

Type III hyperlipoproteinaemia, or familial dysbetalipoproteinaemia, is characterized by elevated triacylglycerol and cholesterol levels and manifests a broad β band of abnormally migrating VLDL (βVLDL) on electrophoresis. The condition has thus also been referred to as broad β disease. This relatively rare condition too is associated with cardiovascular complications, with premature ischaemic heart disease being clinically detectable.

Type IV hyperlipoproteinaemia, also reported to be associated with an increased risk of atherosclerosis, is characterized by elevated VLDL and plasma triacylglycerol concentrations, and is also known as familial hypertriglyceridaemia. The atherosclerotic lesions seen in those patients appear identical to those found in normolipidaemic individuals.

Type V hyperlipoproteinaemia, which manifests itself as elevated levels of chylomicrons and VLDL and consequentially elevated plasma triacylglycerol and cholesterol concentrations, is generally associated with foam cell lesions of the spleen, skin and bone marrow, but, like type I, patients with this rare condition escape premature coronary atherosclerosis for reasons that remain obscure.

Apart from the hyperlipoproteinaemias that act as determinants of abnormal plasma lipoprotein concentrations, conditions with abnormally low concentrations of certain lipoproteins are also encountered and are generally referred to as the hypolipoproteinaemias. Tangier disease is one such condition that occurs as a rare inherited condition with autosomal recessive characteristics. The condition, a familial HDL deficiency, is characterized by cholesteryl ester accumulation in the reticuloendothelial system, which is similar in pattern to other lipid-storage diseases, such as gangliosidoses. Atheromas are substantially absent from the aorta and coronary vessels of such patients. The plasma cholesterol concentration is reduced and the triacylglycerol concentration is normal or increased in these cases.

In the case of hypoprebeta/betalipoproteinaemia,

both triacylglycerol and cholesterol concentrations may be decreased with either VLDL and/or LDL concentrations being reduced. These again are rare conditions with autosomal dominant inheritance characteristics, in which no structural abnormalities are seen, with the patient being mainly asymptomatic.

A complete absence of LDL in the plasma is seen in patients with the rare autosomal recessive disorder, abetalipoproteinaemia. The patients may also have associated lecithin:cholesterol acyltransferase deficiency, which is also encountered as an independent condition. In this latter case, lipoproteins with abnormal flotation, size and compositional properties exist, with disc-shaped structures (nascent HDL particles?) being present at high concentration.

12.3.8 Circulating lipids and the development of cardiovascular disease

In terms of the nature of processes leading to lipid accumulation during the development of atherosclerosis, the mechanisms of normal uptake of lipoproteins by cells as mentioned above must be considered. Normally, LDL cholesterol is taken up by cells, following its association with high-affinity cell-surface receptors, by endocytosis. It is then delivered to lysosomes, within which the protein part is hydrolysed and the cholesteryl ester is hydrolysed to cholesterol and fatty acids. When sufficient cholesterol has accumulated through this so-called 'LDL pathway' to saturate the cells' requirement for membrane synthesis, the excess cholesterol is re-esterified and stored within the cell. Usually, in a lipoprotein-rich environment, the cell is protected from accumulating excess cholesterol by a mechanism that leads to a decrease in the number of cell-surface LDL receptors. Simultaneously, endogenous cholesterol synthesis is inhibited. It may be considered, therefore, in the normal state, that the LDL pathway serves to protect against atherosclerotic lipid accumulation.

However, in addition to this receptor-mediated LDL uptake, cells can also internalize LDL by a 'bulk-phase pinocytosis', which does not require specific cell-surface binding. In this situation, LDL is taken up at a rate proportional to the environmental concentration. At low extracellular LDL concentrations, the 'bulk' process is insignificant when compared to the receptor-mediated uptake. However, at elevated LDL concentrations, although receptor-mediated uptake becomes rapidly saturated, bulk-phase pinocytosis increases in rate proportionately to the overall LDL concentration. The smooth muscle cells in the 'normal' atherosclerotic lesion accumulate lipid by this receptor-independent process, which is not subject to feedback control. In addition to this, a relative deficiency of lysosomal cholesteryl ester hydrolase activity in the presence of overwhelming concentrations of substrate may also contribute to the aggravated situation.

12.3.9 Treatments used to control circulating lipid concentrations and prospects for the prevention of cardiovascular disease

From the foregoing it is natural that treatments designed to modify plasma lipoprotein concentrations have been sought with vigour. In particular, the aim of reducing serum cholesterol and LDL concentrations has been to induce a decrease in the cholesterol content and size of atheromatous lesions and to restore normal metabolic patterns to the intima and media of affected vessels. However, in advanced lesions the contribution of lowered plasma cholesterol levels to regression is uncertain. Additionally, assessment of such overall trends is well nigh impossible, owing to the variability and inadequacy of disease description *in vivo*. However, in the years since 1965, when coronary heart disease reached its peak in the USA, the proportion of deaths attributable to this cause has declined significantly. The reasons for this decline are still under discussion. In clinical terms, the improved awareness and treatment of high blood pressure and better acute care of heart attack victims have contributed, with improved control of certain risk factors also being of importance. For example, in the USA, among 'at-risk' groups, smoking has declined, as has the per capita consumption of saturated fat and cholesterol-rich foods. The consumption of vegetable-derived fats has increased, together with a more vigorous control of body weight via diet and exercise regimes.

Pharmacological intervention in the prevention and therapy of cardiovascular disease has also been scrutinized. Drug regimes designed to control hypertension, hyperlipoproteinaemia, thrombotic syndromes and stress-induced conditions have all been evaluated for their effect on mortality related to cardiovascular disease. Although the results of such trials have been contradictory and the effectiveness of reducing population cholesterol levels questioned, the major efficacy of such intervention resides in the prevention of further episodes in the survivors of previous non-fatal myocardial infarction.

Drugs affecting lipid metabolism include clofibrate (and its commercial variants), nicotinic acid,

D-thyroxine and oestrogens. These drugs have variable effects both on plasma lipid concentrations and on the survival of patients.

Other drug interventions in survivors of myocardial infarction, such as β blockade, aspirin administration (which affects prostaglandin metabolism) and antithrombotic agents, also reveal uncertain relationships between treatment and improved survival.

Of the non-pharmacological methods used to improve the ratio of plasma cholesterol in HDL compared to LDL, physical exercise and ethanol consumption are the most widely studied. Exercise training raises HDL cholesterol by as much as 50% compared to unexercised controls of similar age, and moderate ethanol consumption is also correlated with a more modest increase in HDL cholesterol, and a possible, increased protection from total coronary mortality. Since a high ratio of HDL/LDL cholesterol produces an enhanced protection against cardiovascular-disease-related mortality, pharmacological agents with the same effect as exercise and ethanol have been sought. Bezafibrate, procetofen and gemfibrozil have been reported to have favourable effects in this regard when used to treat hyperlipidaemic patients.

In the case of primary prevention, a number of trials have sought to test the hypothesis that reduction of raised serum cholesterol concentrations in otherwise healthy adults will reduce risk of cardiovascular disease. Although earlier studies showed a positive correlation, other more recent studies have undermined the certainty (see, for example, Jacobs *et al.*, 1992 and Gurr, 1992).

In the primary prevention studies where drug treatment has been evaluated, clofibrate can produce a modest reduction in serum cholesterol and a reduction in non-fatal myocardial infarctions; however, no reduction in cardiovascular mortality was obvious. Although many of the largest trials are still running, the early indications seem to support, at least in part, lipid involvement in cardiovascular disease hypothesis. However, despite the observed reduction in the incidence of non-fatal myocardial infarctions occurring in conjunction with reductions in raised serum cholesterol levels in some studies, there is no convincing evidence that there has been a consequential reduction in cardiac-related mortality. Indeed, in situations where intervention has led to a reduction in serum cholesterol concentrations, observed increases in non-cardiovascular-related mortality cannot be ignored lightly.

Because several lines of evidence indicate that oxidized LDL promotes atherogenesis through its cytotoxicity, agents that inhibit the oxidation process have also been studied as potential preventive or therapeutic agents. Antioxidants such as tocopherol, butyrated hydroxytoluene, β-carotene, tocotrienols and the drug probucol have been shown to inhibit LDL oxidation both *in vitro* and *in vivo* and to offer the possibility of an additional therapy that would enhance the efficacy of cholesterol-lowering regimes, particularly when such regimes include substituting saturated lipids with polyunsaturated alternatives (Schwartz *et al.*, 1992).

Thus, although reduction of serum cholesterol concentrations by modest amounts, preferably throughout adult life, may help to slow down the progression of cardiovascular disease, once the disease is established and particularly where the 'clinical horizon' has been reached, lipid-lowering regimes appear to have little or no effect.

In terms of dietary counselling, it is advisable, on public health grounds, to reduce the total intake of fat and particularly that which is saturated. A suitable diet may be one that contains no more than 30% of calories from fat and no more than one-third of these from saturated sources. The improvement in diet should be particularly beneficial when aimed at the young, with the possible advantage of preventing normal cholesterol concentrations from rising.

References and further reading

Benditt, E.P. and Gown, A.M. (1980) *Int. Rev. Exp. Pathol.*, **21**, 55.

Bierman, E.L. (1973) *Mech. Age. Dev.*, **2**, 315.

Carlson, L.A. and Bottiger, L.E. (1972) *Lancet*, **i**, 865.

Carlson, L.A., Danielson, M., Ekberg, I., Klintemar, B. and Rosenhamer, G. (1977) *Atherosclerosis*, **28**, 81.

Coronary Drug Project (1972) Research Group Report, *J. Am. Med. Assoc.*, **220**, 996.

Coronary Drug Project (1973) Research Group Report, *J. Am. Med. Assoc.*, **226**, 652.

Coronary Drug Project (1975) Research Group Report, *J. Am. Med. Assoc.*, **231**, 360.

Cryer, A. (1987) in *Lipoprotein Lipase* (ed. J. Borensztajn), Evener, Chicago, pp. 277–327.

Fredrickson, D.S., Goldstein, J.L. and Brown, M.S. (1978) in *The Metabolic Bases of Inherited Disease*, 4th edn (eds J.B. Stanbury, J.B. Wyngaarden and D.S. Fredrickson), McGraw-Hill, New York, p. 604.

Glickman, R.M. and Sabesin, S.M. (1988) in *The Liver: Biology and Pathobiology*, 2nd edn (eds I.M. Arias, W.B. Jakoby, H. Popper, D. Schachter and D.A. Shafnitz), Raven, New York, pp. 331–54.

Goldstein, J.L. and Brown, M.S. (1977) *Annu. Rev. Biochem.*, **46**, 897.

Gordon, T., Kannel, W.B., Hjortland, M.C. and McNamara, P.M. (1978) *Ann. Intern. Med.*, **89**, 157.

Gurr, M.I. (1992) *Prog. Lipid Res.*, **31**, 195.

Jacobs, D., Blackburn, H., Higgins, M., Reed, D., Iso, H.,

McMillan, G., Neaton, J., Nelson, J., Potter, J., Rifkind, B., Rossouw, J., Shekelle, R. and Yusaf, S. (1992) *Circulation*, **86**, 1046–60.

Kannel, W.B., Sorlie, P., Braud, F., Castelli, W.P., McNamara, P.M. and Cherardi, G.J. (1980) in *Atherosclerosis V: Proc. Fifth Int. Symp.* (eds A.M. Gotto, L.C. Smith and B. Allen), Springer, New York, p. 54.

Miller, G.J. and Miller, N.E. (1975) *Lancet*, **i**, 16.

Oliver, M.F. (1979) *Br. Med. J.*, **1**, 839.

Oliver, M.F. (1981) *Lancet*, **ii**, 1090.

Pooling Project Research Group (1978) *J. Chron. Dis.*, **31**, 201.

Royal College of Physicians (1971) *Smoking and Health Now*, Pitman, London.

Schwartz, C.J., Vaente, A.J., Sprague, E.A., Kelley, J.L. and Cayatte, A.J. (1992) *Circulation*, **86**, Suppl. 6, III 116.

WHO Committee of Principal Investigators (1978) *Br. Heart J.*, **10**, 1069.

Zilversmit, D.B. (1979) *Circulation*, **60**, 473.

12.4 Skin lipids and medical implications

The surface cells representing the epidermis play an obvious role in providing a protective permeability barrier between animals and their environment. The mammalian epidermis consists of an undifferentiated germinative layer surmounted by progressively flattened cells that eventually lose recognizable organelles as they transform into corneocytes. Studies on the different layers of the epidermis have shown that, while the basal cells have a lipid composition that is typical of other epithelial cells (see Lampe *et al.*, 1983b), transformation of the stratum granulosum into the stratum corneum is accompanied by a number of dramatic lipid changes. These include depletion of phospholipids with relative retention of neutral lipids, generation and retention of large amounts of sphingolipids, and an enrichment of sphingolipids composed of longer-chain, more saturated fatty acids than are present in lipids in the subjacent viable epidermis (Lampe *et al.*, 1983b; Table 12.4). Possible functions for these changes include both the maintenance of the normal barrier to water loss and the control of stratum corneum desquamation. A full discussion of the role of lamellar bodies with their complement of lipids as well as lipid-metabolizing enzymes in the generation of the unique composition of the stratum corneum is made by Elias and Menon (1991).

A detailed discussion of the biochemistry and functions of stratum corneum lipids is given by Schurer and Elias (1991). Although it is accepted generally that the stratum compactum layer, the lowest region of the stratum corneum, possesses formidable barrier properties (Bowser and White, 1985), most, if not all, of the cells of the stratum corneum also participate. The ability of the intercellular lipids (in the almost complete absence of phosphoglycerides) to do this is dependent on the amphipathic properties of ceramides, non-esterified fatty acids and cholesterol, and on the stabilization of lamellar bilayers (in an aqueous environment) by van der Waals' interactions and hydrogen bonds (Rehfeld *et al.*, 1988, 1990).

When (human) skin lipids are analysed, it should be appreciated that there are two contributing sources. The epidermis itself (discussed above) contributes only a proportion, the remaining originating from the secretion of the sebaceous glands (see below). An approximate composition of lipids from these two sources is shown in Table 12.5. Depending on the site of collection, the contribution of sebum lipids to the total skin lipids will, of course, depend on the number of sebaceous glands. For example, skin lipids from the scalp (where there are 900 pilosebaceous structures/cm^2) have the same overall composition as those from the pilosebaceous glands (Table 12.5), whereas skin lipids from the palms or soles (where sebum is virtually absent) have a composition resembling that of the stratum corneum (Schurer and Elias, 1991). In addition, human stratum corneum lipids display striking regional variations, which could reflect differences in stratum corneum thickness, turnover, desquamation and/or permeability (Elias *et al.*, 1981; Lampe *et al.*, 1983a). In addition to

Table 12.4 Changes in overall lipid composition during epidermal differentiation and cornification (Lampe *et al.*, 1983b)

Lipid	Tissue		
	Basal	Granular	Cornified
Phospholipids	44.5 ± 3.4	25.3 ± 2.6	6.6 ± 2.2
Cholesterol sulphate	2.6 ± 3.4	5.5 ± 1.3	2.0 ± 0.3
Neutrals	51.0 ± 4.5	56.5 ± 2.8	66.9 ± 4.8
Sterols	11.2 ± 1.7	11.5 ± 1.1	18.9 ± 1.5
Fatty acids	7.0 ± 2.1	9.2 ± 1.5	26.0 ± 5.0
Triacylglycerols	12.4 ± 2.9	24.7 ± 4.0	variable
Sterol/wax esters	5.3 ± 1.3	4.7 ± 0.7	7.3 ± 1.2
Squalene	4.9 ± 1.1	4.6 ± 1.0	6.5 ± 2.7
n-Alkanes	3.9 ± 0.3	3.8 ± 0.8	8.2 ± 3.5
Sphingolipids	7.5 ± 1.0	11.7 ± 2.7	24.4 ± 3.8
Glucosylceramides	3.5 ± 0.3	5.8 ± 0.2	trace
Ceramides	3.8 ± 0.2	8.8 ± 0.2	24.4 ± 3.8

Table 12.5 Comparison of typical lipid compositions of human skin lipids derived from pilosebaceous versus epidermal origin (From Schurer and Elias, 1991)

Lipid (%)	Origin	
	Pilosebaceous	Epidermal
Ceramides	–	25
Cholesterol	–	20
Non-esterified fatty acids	16	20
Triacylglycerols	43	10 (variable)
Squalene	12	–
Wax esters	25	–
Cholesterol esters	3	15

endogenous lipids, external sources such as soaps, cosmetics or atmospheric pollutants may also be significant (Burton, 1979).

Sphingolipids are major components of the stratum corneum, and are of special interest because it is this class which provides the major polar lipid from which the extracellular membrane structures of the stratum corneum are constructed. Structures of ceramides and glucosylceramides isolated from human epidermis are shown in Fig. 12.1. One of the most interesting sphingolipids is one that is apparently unique to epidermis, and consists of a glycosylated or non-glycosylated sphingosine base with an amide-linked, long-chain, non-hydroxy or ω-hydroxy acid with an additional ω-esterified non-hydroxy fatty acid (usually linoleate). Because the major ω-esterified acid is linoleic, this may indicate a special role for this component in barrier function (Hansen, 1986). Indeed, the proportion of linoleate esterified in acylsphingolipids varies from 35% in mice, through 50% in humans to 77% in pig (see Schurer and Elias, 1991). Acylated fatty acids containing significant quantities of linoleate are also found (see Hansen, 1986; Schurer and Elias, 1991). The regulation and function of epidermal lipid synthesis have been reviewed by Feingold (1991). Some discussion is also made there about factors such as diet, diabetes and the topical application of detergents and other agents that can affect epidermal lipid synthesis and normal function. In addition, the role of lipid modulators of epidermal proliferation and differentiation has been discussed by Holleran (1991). Particular attention was drawn to platelet-activating factor, phospholipase-generated signalling mechanisms (see sections 8.13.2 and 11.2.2) and the role of sphingolipids, especially gangliosides, in mediating the action of epidermal growth factor. The metabolism and function of skin lipids have also been reviewed by Ziboh and Chapkin (1988).

Because of the importance of the epidermis as a permeability barrier, much research has been made to understand its properties, particularly since such knowledge is needed in order to understand pathological problems (such as found in essential fatty acid deficiency) as well as in finding more efficient ways for the topical delivery of pharmaceutical agents in creams and other vehicles. A useful review is that of Potts *et al.* (1991). In the last two decades, various chemicals (e.g. dimethylsulphoxide, n-decylmethylsulphoxide, 1-dodecylazacycloheptan-2-one (Azone), pyrrolidones, oleic acid) have been investigated for their skin permeability-enhancing action and, hence, their usefulness in the topical administration of therapeutic agents (see Hou *et al.*, 1991).

Normal desquamation and correct maintenance of the stratum corneum are orderly and continuous processes. However, disorders of desquamation are seen: *hyperproliferation* results in excessive numbers of incompletely differentiated cells in the outer layer; and *retention hyperkeratoses* result in increased thickness of the stratum corneum due to a failure of desquamation (Williams, 1991). Apart from inherited disorders, defects in desquamation may be caused as side-effects of drugs. Particular problems are known to occur with hypocholesterolaemic agents (Table 12.6). Metabolic defects associated with disorders of cornification include ichthyosis (vulgaris, recessive X-linked or harlequin), Refsum's disease, neutral lipid storage disease, multiple sulphatase deficiency and CHILD syndrome (see Williams, 1991).

A most obvious connection of skin lipids with medical problems is with acne vulgaris. It is often stated that the severity of acne correlates with the

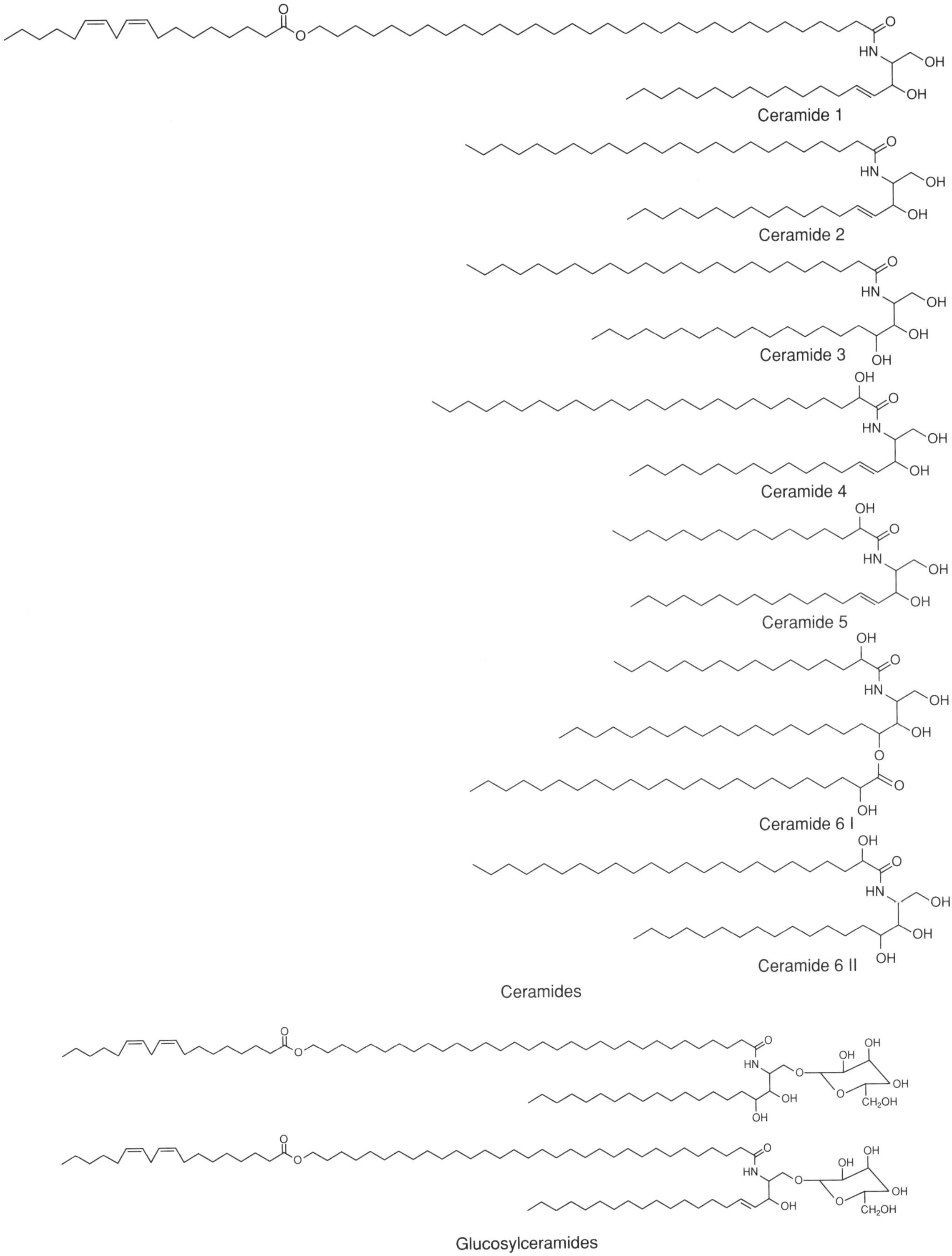

Fig. 12.1 Ceramides and glucosylceramides isolated from human epidermis. See Schurer and Elias (1991) for details.

Table 12.6 Systemic hypocholesterolaemic agents that can cause hyperkeratotic skin conditions (Williams, 1991)

Drug	Therapeutic use	Dermatological side-effects	Effect(s) on cholesterol metabolism
Nicotinic acid	Hypocholesterolaemic agent	Generalized dry skin	Decreases hepatic VLDL synthesis; inhibits HMG-CoA reductase; increases cholesterol oxidation and excretion
Triparanol (Mer-29)	Hypocholesterolaemic agent	Ichthyosis; hair loss and depigmentation	Inhibits Δ^{24} (side chain) reduction
WY-3457 (butyrophenone)	Antipsychotic agent	Ichthyosis; hair loss and depigmentation	Inhibits post-squalene step(s)?
Azacosterol (20,25-diazacholesterol)	Hypocholesterolaemic agent	Hyperkeratosis of palms and soles	Inhibits Δ^{24} (side chain) reduction
Gemfibrozil	Hypocholesterolaemic agent	Exacerbation of psoriasis	Decreases hepatic VLDL secretion; increases peripheral VLDL hydrolysis; inhibits HMG-CoA reductase

amount of seborrhoea (Cunliffe and Shuster, 1969). Moreover, non-esterified fatty acids are thought to play a role in its pathogenesis, since they are both irritant and comedogenic (Stillman *et al.*, 1975). However, the importance of unesterified fatty acids in the pathogenesis of acne has been doubted by others (e.g. Voss, 1974). Alternatively, it has been suggested that the dilutional effect of increased sebaceous gland activity on sebum linoleate may cause acne because local essential fatty acid deficiency is induced (Downing *et al.*, 1986).

Sebum is produced by cells of the sebaceous glands as they move towards the latter's centre and then disintegrate. The lipid cell contents are discharged into the sebaceous duct and then to the pilosebaceous follicle. From there they are excreted to the skin surface. It takes 2–3 weeks for a sebaceous cell to travel from the basal layer to the duct, and another week for the sebum to reach the skin surface. Rates of discharge of sebum are, of course, dependent on changes in sebaceous gland size and mitotic activity, but, normally, the rates remain fairly constant for an individual. Methods for studying sebum secretion rates are discussed by Burton (1979). The most reliable of these techniques is a simple gravimetric method in which sebum is collected onto an absorbent paper in contact with forehead skin (see Burton, 1979). Normal values for the sebum secretion rate for subjects of different ages or sex are available in Pochi and Strauss (1974) and Burton *et al.* (1975), though, because of differences in experimental technique, normal ranges should be obtained by each investigator. Drug treatment and hormonal or other factors that affect sebum secretion rates have been reviewed (e.g. Shuster and Thody, 1974). The selection of 'normal' controls for such testing poses a number of problems (cf. Burton, 1979). The sebum secretion rate is normally expressed in terms of the area of skin studied. It could also be related to the number of functional glands (cf. Burton, 1979).

The chemistry and function of mammalian sebaceous lipids have been reviewed recently (Stewart and Downing, 1991). Apart from details of human sebum, they discuss that from other animals of importance (for example, sheep wool lipids) or from other agriculturally important species. The function of sebum in humans is controversial. It has been suggested to have bactericidal activity and to minimize water loss through the epidermis. However, Kligman (1963) proposed that sebum has none of these effects. Another possible role is to provide a dynamic seal around the hair shaft (Stewart and Downing, 1991).

In acne, a major problem seems to relate to the micro-organisms that are capable of hydrolysing the triacylglycerols of sebum and, hence, generate the irritant and comedogenic fatty acids. Three micro-organisms have been implicated in acne – *Propionibacterium acne*, *Staphylococcus* spp. and *Pityrosporum ovale* – with the former being the most important. However, micrococci, sarcina and yeasts are also present, and may predispose the patient to acne (Holland and Cunliffe, 1979). The hormonal control of sebum secretion, which is critical to the appearance of acne in early puberty, is discussed by Stewart and Downing (1991). Oral medications that reduce sebum secretion, such as oestrogens, antiandrogens and 13-*cis*-retinoic acid, are used to alleviate the symptoms of acne.

In addition to acne, abnormal essential fatty acid deficiency has been cited as a cause of atopic dermatitis and psoriasis. In atopic dermatitis it has been suggested that a deficiency of Δ6-desaturase reduces the amount of γ-linolenic acid available for subsequent eicosanoid synthesis (Horrobin, 1989, 1992). Topical application of this acid is reported to be efficacious in atopic dermatitis treatment. Psoriasis, which is a widespread complaint affecting 2% of the population worldwide, has been suggested to be associated with increased phospholipase A_2 activity. This causes increased levels of unesterified arachidonic acid, and, in addition, there is a relative shift towards lipoxygenase rather than cyclooxygenase activity for eicosanoid production (see section 11.1.7). An imbalance in eicosanoid proportions would then cause changes in cellular metabolism and differentiation (see Kragballe and Voorhees, 1985).

References and further reading

Bowser, P.A. and White, R.J. (1985) *Br. J. Dermatol.*, **112**, 1.

Burton, J.L. (1979) in *Investigative Techniques in Dermatology* (ed. R. Marks), Blackwell, Oxford, p. 1.

Burton, J.L., Pye, R.J., Meyrick, G. and Shuster, S. (1975) *Br. J. Dermatol.*, **92**, 541.

Cunliffe, W.J. and Shuster, S. (1969) *Br. J. Dermatol.*, **81**, 697.

Downing, D.T., Stewart, M.E., Wertz, P.W. and Strauss, J.S. (1986) *J. Am. Acad. Dermatol.*, **14**, 221.

Elias, P.M. and Menon, G.K. (1991) *Adv. Lipid Res.*, **24**, 1.

Elias, P.M., Cooper, E.R., Korc, A. and Brown, B.E. (1981) *J. Invest. Dermatol.*, **76**, 297.

Feingold, K.R. (1991) *Adv. Lipid Res.*, **24**, 57.

Hamanaka, S., Asagami, C., Suzuki, M., Inagaki, F. and Suzuki, A. (1989) *J. Biochem. (Tokyo)*, **105**, 684.

Hansen, H.S. (1986) *Trends Biochem. Sci.*, **11**, 263.

Holland, K.T. and Cunliffe, W.J. (1979) in *Investigative Techniques in Dermatology* (ed. R. Marks), Blackwell, Oxford, p. 21.

Holleran, W.M. (1991) *Adv. Lipid Res.*, **24**, 119.

Horrobin, D.F. (1989) *J. Am. Acad. Dermatol.*, **20**, 1045.

Horrobin, D.F. (1992) *Prog. Lipid Res.*, **31**, 163.

Hou, S.Y.E., Rehfeld, S.J. and Plachy, W.Z. (1991) *Adv. Biol. Skin*, **24**, 141.

Kligman, A.M. (1963) *Adv. Biol. Skin*, **4**, 110.

Kragballe, K. and Voorhees, J.J. (1985) *Curr. Probl. Dermatol.*, **119**, 548.

Lampe, M.A., Burlingame, A.L., Whitney, J., Wilcains, M.L., Brown, B.E., Roitman, E. and Elias, P.M. (1983a) *J. Lipid Res.*, **24**, 120.

Lampe, M.A., Williams, M.L. and Elias, P.M. (1983b) *J. Lipid Res.*, **24**, 131.

Melnik, B.C. (1989) in *The Ichthyoses* (ed. H. Traupe), Springer-Verlag, Berlin, p. 15.

Pochi, P.E. and Strauss, J.S. (1974) *J. Invest. Dermatol.*, **62**, 191.

Potts, R.O., Mak, U.H.W., Guy, R.H. and Francoeur, M.L. (1991) *Adv. Lipid Res.*, **24**, 173.

Rehfeld, S.J., Plachy, W.Z., Williams, M.L. and Elias, P.M. (1988) *J. Invest. Dermatol.*, **91**, 499.

Rehfeld, S.J., Plachy, W.Z., Hou, S.Y. and Elias, P.I. (1990) *J. Invest. Dermatol.*, **95**, 217.

Schurer, N.Y. and Elias, P.M. (1991) *Adv. Lipid Res.*, **24**, 27.

Shuster, S. and Thody, A.J. (1974) *J. Invest. Dermatol.*, **62**, 172.

Stewart, M.E. and Downing, D.T. (1991) *Adv. Lipid Res.*, **24**, 263.

Stillman, M.A., Maibach, H.I. and Shalita, A.R. (1975) *Contact Dermatitis*, **1**, 65.

Voss, J.G. (1974) *Arch. Dermatol.*, **109**, 894.

Williams, M.L. (1991) *Adv. Lipid Res.*, **24**, 211.

Ziboh, V.A. and Chapkin, R.S. (1988) *Prog Lipid Res.*, **27**, 81.

12.5 Sphingolipidoses

There are over a dozen inherited disorders of metabolism that are characterized by the accumulation of different types of sphingolipids in various organs and tissues. Many of these diseases are fatal in the first few years of life, although when the central nervous system is unaffected there are possibilities for treatment (see below). The sphingolipidoses are caused by deficiency of a single lysosomal enzyme, which leads to the accumulation of the enzyme's substrate in affected tissues. The topic of lysosomal storage diseases has been comprehensively reviewed (Scriver *et al.*, 1989).

The structure of sphingolipids has been dealt with in section 2.4 and their metabolism in section 11.5.

The specific metabolic defect in each of the lipid storage diseases has been shown in all cases, with the exceptions of mucolipidosis IV and multiple sulphatase deficiency (Table 12.7) and the possible exception of lactosylceramidosis. The enzymic defects have been reviewed many times (e.g. Neufeld

Table 12.7 Summary of enzyme deficiencies and accumulating lipids in the sphingolipidoses. From Brady (1978). Reproduced with permission from *Annual Review of Biochemistry*, **47**, 687. © 1978 by Annual Reviews Inc. Modified with reference to Neufeld (1991)

Disease	Signs and symptoms	Major lipid accumulation	Enzyme defect
Ceramide lactoside lipidosis	Slowly progressing brain damage	Ceramide lactoside	Neutral β-galactosidase
Fabry's disease	Reddish purple skin rash, kidney failure, pain in lower extremities	Gal-Gal-Glu-ceramide	α-Galactosidase
Farber's disease	Hoarseness, dermatitis, skeletal deformation, mental retardation	Ceramide	Ceramidase
Gaucher's disease	Spleen and liver enlargement, erosion of long bones and pelvis, mental retardation only in infantile form	Glucocerebroside	Glucocerebrosidase
Generalized gangliosidosis (GM_1 gangliosidosis)	Mental retardation, liver enlargement, skeletal deformities, about 50% with red spot in retina	Ganglioside GM_1	β-Galactosidase
Krabbe's disease (globoid leukodystrophy)	Mental retardation, almost total absence of myelin, globoid bodies in white matter of brain	Galactocerebroside	Galactocerebroside β-galactosidase
Metachromatic leukodystrophy (two forms)	Mental retardation, psychological disturbances in adult form, nerves stain yellow brown with cresyl violet dye	Sulphatide	Aryl sulphatase A (sulphatide activator/ saposin inactivator-deficient form)
Niemann–Pick disease	Liver and spleen enlargement, mental retardation, about 30% with red spot in retina	Sphingomyelin	Sphingomyelinase
Tay-Sachs disease	Mental retardation, red spot in retina, blindness, muscular weakness	Ganglioside GM_2	Hexosaminidase A
Tay-Sachs variant	Same as Tay-Sachs disease, but progressing more rapidly	Globoside (and ganglioside GM_2)	Hexosaminidase A and B

et al., 1975; Desnick *et al.*, 1976; Neufeld, 1991), and they are summarized in Table 12.7. Recently, there have been advances in our knowledge concerning the molecular basis of the disorders. It has been proposed that, in these diseases, there is a genetic mutation that leads to a change in the primary structure of the catabolic enzyme. Thus, in patients with adult or juvenile Gaucher's disease, residual glucocerebrosidase activity was found in the tissues. However, the glucocerebrosidase in a pathological spleen was found to have a K_m that was an order of magnitude larger than that in control samples (Brady, 1978). Inactive or less-efficient enzymes have been found in metachromatic leukodystrophy and in G_{M1} gangliosidosis. In the latter disease, the enzyme from the affected liver had a K_m for G_{M1} that was five times that of controls (Norden and O'Brien, 1975).

Much of the progress in our knowledge of the genetics of the sphingolipidoses has come from cell hybridization experiments. Cultured skin fibroblasts have been used to study the isoenzyme abnormalities in Tay-Sachs disease and Sandhoff–Jatzkewitz disease. In the former, only hexosaminidase isoenzyme A is lacking (Okada and O'Brien, 1969), while in the latter, both hexosaminidase A and hexosaminidase B are deficient (Sandhoff *et al.*, 1968). The deficiency of hexosaminidase A in these two disorders is due to different mutations, and activity can be restored by complementation techniques (e.g. Rattazzi *et al.*, 1976). The studies have also provided support for the idea that hexosaminidase A is composed of two non-identical subunits $\{(\alpha\beta)n\}$ and hexosaminidase B is $\{(\beta\beta)n\}$ (cf. Brady, 1978). Hybridization studies have now identified the chromosomes that carry the genetic code for the enzymes. The cistron for the α subunits is on chromosome 5. Also, chromosome 1 seems to carry the cistron for α-L-fucosidase (Brady, 1978).

Pathways for the synthesis, transport and processing of lysosomal enzymes and aspects relevant to the sphingolipidoses have been summarised (Neufeld, 1991). Moreover, we also have information of cDNAs and a number of genes encoding such proteins. Sequence homologies have been noted in several cases and there are families of, for example, α-galactosidases. Molecular analysis of mutations has revealed extensive heterogeneity. Though one or two mutations, often clustered in particular ethnic groups, account for a significant fraction of patients or carriers (Table 12.7), a very large number of rare mutations account for the remainder. Thus patients are very likely to be compound heterozygotes (Neufeld, 1991).

Detailed analysis of genetic mutations has been made in a number of sphingolipidoses. These include GM_2 gangliosidosis, Gaucher's disease, metachromatic leukodystrophy and Fabry's disease (see Neufeld, 1991, for details).

Because of the difficulty of treating a number of the sphingolipidoses, clinical biochemists have been very interested in the development of simple diagnostic tests that can also be used for heterozygous individuals. Fortunately, the enzyme deficiency is the same in all tissues (Table 12.8). Thus, easily available materials can be used for the assay of the enzymes concerned in each disease. Such material includes leukocytes, cultured skin fibroblasts, serum and hair follicles. Urine and tears have also been used. In most cases the enzymes are stable to freezing, so that samples may be shipped to diagnostic centres where automated assays have been developed for many of the determinations (e.g. Peters *et al.*, 1975; cf. Brady, 1978).

Most of the early tests were based on the hydrolysis of the natural sphingolipid that had been labelled in the hydrophilic portion. Thus, the substrate and products could be easily separated by two phase solvent partition or by precipitation of the product. In many cases, artificial chromogenic or fluorogenic glycosides could be used instead of radiolabelled substrates. However, in these cases it is very important to demonstrate that the conditions and specificities of the enzyme reactions are such that the results can be extrapolated to the natural substrate (e.g. Peters *et al.*, 1975). Thus, in the case of Niemann–Pick or Krabbe's diseases, the unsubstituted nitrophenyl derivatives of phosphocholine and galactose, respectively, do not provide the discrimination required. However, analogues with a long-chain aliphatic amide *ortho* to the phenolic hydroxyl of the *p*-nitrophenyl derivative of phosphocholine or β-D-galactopyranoside were found to be entirely suitable for the diagnosis of patients with Niemann–Pick or Krabbe's disease, respectively (Gal *et al.*, 1977). However, even in these cases, special precautions had to be taken so as not to invalidate the assays (Brady, 1978). A β-D-glucopyranosyl derivative of 2-hexadecanoylamino-4-nitrophenol may be useful for diagnosing Gaucher's disease (Gal *et al.*, 1976).

For heterozygous carriers of sphingolipidoses, enzyme assays with leukocytes and especially cultured skin fibroblasts have proved the most useful. However, in a large screening programme for Tay-Sachs disease, a number of mistakes occurred. These problems can be alleviated by the very careful use of the test and maintenance of quality controls (Kaback *et al.*, 1977).

With the exception of Fabry's disease, all of the sphingolipidoses are transmitted as autosomal recessive disorders. In short, both parents must be

Table 12.8 Detection of heterozygous carriers of lipid storage diseases (Taken from Brady, 1972, with permission)

Disorder	Substrate	Enzyme source	Activity (% of normal)	
Ceramide lactoside lipidosis	[^{3}H]Ceramide lactoside	Cultured skin fibroblasts	52	
Fabry's disease	*p*-Nitrophenyl-α-D-galacto-pyranoside	Leukocytes	31	
		Cultured skin fibroblasts	57	
Gaucher's disease	[^{14}C]Glucocerebroside	Circulating leukocytes	59	
		Cultured fibroblasts	47	
Generalized gangliosidosis	*p*-Nitrophenyl-β-D-galacto-pyranoside	Leukocytes	48	
Krabbe's disease	[^{3}H]Galactocerebroside	Serum	26	
Metachromatic leukodystrophy	*p*-nitrocatechol sulphate	Leukocytes	49	
		Cultured skin fibroblasts	19	
Niemann–Pick disease	[^{14}C]Sphingomyelin	Leukocytes	55	
		Cultured skin fibroblasts	64	
Tay-Sachs disease	4-Methylumbelliferyl-β-D-*N*-acetylglucosaminide	Serum	Hex A	Hex B
O variant			56	61
B variant			55	84

carriers in order to produce an affected child. When both parents are heterozygous, then there is a 1 in 4 chance of an affected child, 1 in 4 of a normal child, and 1 in 2 for carriers. In Fabry's disease, only the female need be a carrier to have an affected son; 50% of the sons will be homozygous (affected) and 50% of the daughters will be carriers. When monitoring pregnancies at risk, samples are obtained by transabdominal amniocentesis at around week 14 of gestation. Viable foetus skin cells are obtained from a 20 ml sample of amniotic fluid and cultured for the enzymic tests. Such methods have been shown to be reliable for most of the sphingolipidoses (Milunsky, 1975).

Although the above techniques have proved very useful, occasional problems arise. For example, in the case of Tay-Sachs disease assayed by the use of 4-methylumbelliferyl-β-D-*N*-acetylglucosaminide, certain so-called normal individuals have lowered enzyme activities. Thus, in this case it is important to use radiolabelled G_{M2} substrate in order to prevent the incorrect diagnosis of individuals with unusual hexosaminidase mutations (Tallman *et al.*, 1974).

Some clinicians are critical of the use of amniotic material as described above and prefer to use cells that are more definitely foetal in origin. Therefore, more rapid assay procedures have been developed by increasing the sensitivity of the enzyme assays and reducing the number of tissue culture passages (Brady, 1978).

Alternatives include the isolation of foetal cells from maternal blood, foetal blood sampling, foetal tissue biopsy and somatic cell hybridization. The latter involves the formation of hybrids between deficient parenchymal cells and cultured amniotic fluid cells.

Another possible diagnostic method is based on the fact that alkaline phosphatase activity is increased when the unmetabolizable substrate is added to the medium that bathes cells of an affected individual (Hösli, 1976). This technique has been used for hereditable diseases, including the sphingolipidoses, and may be used to help identify unknown metabolic defects.

Since the discovery that the sphingolipidoses are enzyme deficiency diseases, it has been suggested that a method of treatment would be the administration of exogenous enzymes. Initial trials with cells in culture were encouraging, but early trials in which either purified enzymes or plasma and leukocytes as their source were given to patients with lysosomal storage diseases were disappointing (Neufeld and Muenzer, 1989). Three problems were significant. First, many sphingolipidoses affect brain tissue, and the blood–brain barrier is an important obstacle, although its permeability can be altered under certain conditions. For example α-mannosidase can enter the brain if the barrier is altered by injecting hyperosmotic solutions of mannitol into the external carotid artery. The blood–brain barrier can also be altered by intermittent periods of hyperbaric oxygen (Brady, 1978). Both methods have inherent dangers.

Secondly, it is necessary to have large amounts of enzyme available for therapy. In early trials there was probably not enough administered in the first place (Neufeld, 1991) and, moreover, because of continuous protein turnover in cells, repeated doses are needed. Thirdly, a signal is needed for receptor-mediated endocytosis and, also, a method needs to be found to target the administered liposomes containing the exogenous enzyme to the appropriate tissue(s).

A recent study of enzyme replacement for Gaucher's disease has shown more success (Barton *et al.*, 1990). The enzyme was given in substantial amounts for a two-year period, and the enzyme was modified to expose terminal mannose residues, which aided its binding to mannose receptors present on the surface of reticuloendothelial cells (Stahl *et al.*, 1984).

Because of the failure of early trials with enzyme replacement, other alternatives have been tried, including bone-marrow transplantation. Some limited benefits have been reported in a few cases (see Neufeld, 1991). However, the newer and potentially very beneficial method of gene replacement by transfection of haematopoietic progenitor cells has been tested with considerable success *in vitro*. Thus, for example, retroviral vectors containing human glucocerebrosidase cDNA have been used to express the enzyme in cultured fibroblasts from Gaucher patients (Sorge *et al.*, 1987). Insertion of the human β-glucuronidase gene into the genes line of a completely deficient mouse model resulted in expression of activity in all tissues of the transgenic animal and a completely normal phenotype (Kyle *et al.*, 1990).

These experimental results offer real hope to patients suffering from the sphingolipidoses. Moreover, molecular biology can also help in the use of recombinant DNA technology to produce large quantities of lysosomal enzymes for replacement therapy. In the meantime, genetic counselling remains a way of providing some relief for those families in which these lysosomal storage diseases are known to occur.

References and further reading

Barton, N.W., Furbish, F.S., Murray, G.J., Garfield, M. and Brady, R.O. (1990) *Proc. Natl Acad. Sci. USA*, **87**, 1913.

Brady, R.O. (1972) in *Current Trends in the Biochemistry of Lipids* (eds J. Ganguly and R.M.S. Smellie), Academic Press, New York, p. 113.

Brady, R.O. (1978) *Annu. Rev. Biochem.*, **47**, 687.

Desnick, R.J., Thorpe, S.R. and Fiddler, M.B. (1976) *Physiol. Rev.*, **56**, 57.

Gal, A.E., Pentchev, P.G. and Fash, F.J. (1976) *Proc. Soc. Exp. Biol. Med.*, **153**, 363.

Gal, A.E., Brady, R.O., Pentchev, P.G., Furbish, F.S., Suzuki, K., Tanaka, H. and Schneider, E.L. (1977) *Clin. Chim. Acta*, **77**, 53.

Hösli, P. (1976) *Adv. Exp. Med. Biol.*, **68**, 1.

Kaback, M.M., Rimoin, D.L. and O'Brien, J.S. (1977) *Tay-Sachs Disease: Screening and Prevention*, Alan R. Liss, New York.

Kyle, J.W., Birkenmeier, E.H., Gwynn, B., Vogler, C., *et al.* (1990) *Proc. Natl Acad. Sci. USA*, **87**, 3914.

Milunsky, A. (1975) *The Prevention of Genetic Disease and Mental Retardation*, Saunders, Philadelphia.

Neufeld, E.F. (1991) *Annu. Rev. Biochem.*, **60**, 257.

Neufeld, E.F. and Muenzer, J. (1989) in *The Metabolic Basis of Inherited Disease*, 6th edn (eds C.R. Scriver, A.L. Beaudet, W.S. Sly and D. Valle), McGraw-Hill, New York, p. 1565.

Neufeld, E.F., Lim, L. and Shapiro, L.J. (1975) *Annu. Rev. Biochem.*, **44**, 357.

Norden, A.G.W. and O'Brien, J.S. (1975) *Proc. Natl Acad. Sci. USA*, **72**, 240.

Okada, S. and O'Brien, J.S. (1969) *Science*, **165**, 698.

Peters, S.P., Lee, R.E. and Glew, R.H. (1975) *Clin. Chim. Acta*, **60**, 391.

Rattazzi, M.C., Brown, J.A., Davidson, R.G. and Shows, T.B. (1976) *Am. J. Hum. Genet.*, **28**, 143.

Sandhoff, K., Andrea, U. and Jatzkewitz, H. (1968) *Life Sci.*, **7**, 278.

Scriver, C.R., Beaudet, A.L., Sly, W.S. and Valle, D. (eds) (1989) *The Metabolic Basis of Inherited Disease*, 6th edn, McGraw-Hill, New York.

Sorge, J., Kuhl, W., West, C. and Beutler, E. (1987) *Proc. Natl Acad. Sci. USA*, **84**, 906.

Stahl, P.D., Wileman, T.E., Diment, S. and Shepherd, V.L. (1984) *Biol. Cell.*, **51**, 215.

Tallman, J.F., Brady, R.O., Navon, R. and Padeh, B. (1974) *Nature*, **252**, 254.

12.6 Other disorders of lipid metabolism

In addition to sphingolipidoses, there are a number of other rare disorders where lipids accumulate in tissues – apparently because of enzymic deficiencies.

Phytanic oxidase deficiency is also known as Refsum's disease after the original discoverer. In this disease, there is an accumulation of phytanic acid (3,7,11,15-tetramethylhexadecanoic acid), which may account for 10–30% of the total plasma fatty acids of patients (section 1.6). Phytanic acid originates from dietary phytols (such as those contained in the side chain of chlorophylls) and treatment consists of a low-phytol diet. The familial disease is characterized by ataxic neuropathy, anosmia, retinitis pigmentosa, dry skin, skeletal deformities and ichthyosis, and is commonest in children and young adults of Scandinavian descent. Parents may have high plasma levels of phytanic acid without clinical symptoms. The most important defect is probably a block in the α-oxidation (section 11.1.5) system so that phytanic acid is no longer metabolized. Diagnosis is usually made by chromatographic analysis of plasma (Fredrickson, 1970).

Other rare lipid-storage diseases include Wolman's disease (triacylglycerol and cholesterol accumulation), hepatic cholesterol ester storage disease, ceroid storage disease, histiocytosis X, lipid proteinosis, lipid dermatoarthritis and Farber's disease (see Table 12.7; involves storage of acid mucopolysaccharides also). Many chronic infections are accompanied by exudates and other proliferative processes, which may be associated with collections of foam cells (xanthomas) and sometimes crystals of cholesterol.

A number of adipose tissue disorders have been reported. These include conditions of lipodystrophy (where adipose tissue is abnormal but not necessarily absent) and lipoatrophy. Lipoatrophy may be partial or total. The lipodystrophies are characterized by generalized or partial loss of body fat and various metabolic abnormalities. They are classified as follows (see Foster, 1991):

1. *Generalized*:
   - congenital
   - acquired
2. *Partial*:
   - common (sporadic)
   - dominant
     - limb and trunk
     - with Rieger anomaly
3. *Localized*:
   - inflammatory
   - non-inflammatory

For further details see Frederickson (1970) and Brown and Goldstein (1983), where detailed references are given.

Lipid is an important muscle energy source during rest and prolonged, moderately intense exercise. Oxidation of stored fat involves carnitine as a transport vehicle (see section 11.1.5(b)), and carnitine deficiency (myopathic or systemic) and/or carnitine palmitoyltransferase deficiency have been described (Mendell and Griggs, 1991).

Production of inflammatory mediators by the host in response to bacterial endotoxin is believed generally to be responsible for the pathogenesis of septic shock. Among the many mediators produced by the major effector cells, the macrophages, are arachidonic acid metabolites (Geisel *et al.*, 1991). Changes in polyunsaturated fatty acid metabolism have been suggested to predispose patients to endotoxic shock (Stark *et al.*, 1990).

References and further reading

Brown, M.S. and Goldstein, J.L. (1983) in *Principles of Internal Medicine*, 10th edn (eds R.G. Petersdorf, R.D. Adams, E. Braunwald, K.J. Isselbacher, J.B. Martin and J.D. Wilson), McGraw-Hill, New York, p. 547.

Foster, D.W. (1991) in *Principles of Internal Medicine*, 12th edn (eds J.D. Wilson, *et al.*), McGraw-Hill, New York, p. 1883.

Frederickson, D.S. (1970) in *Principles of Internal Medicine*, 6th edn (eds M.M. Wintrobe, G.W. Thorn, R.D. Adams, I.L. Bennett, K.J. Isselbacher and R.G. Petersdorf), McGraw-Hill, New York, p. 629.

Geisel, J., Cook, J.A., Coffee, K.A., Wise, W.C. and Halushka, P.V. (1991) *Biochim. Biophys. Acta*, **1085**, 15–20.

Mendell, J.R. and Griggs, R.C. (1991) in *Principles of Internal Medicine*, 12th edn (eds J.D. Wilson, *et al.*), McGraw-Hill, New York, p. 2112.

Stark, J.M., Jackson, S.K., Taylor, S., Davies, I. and Harwood, J.L. (1990) *Experientia*, **46**, 472.

12.7 Pulmonary surfactant (lung surfactant)

Pulmonary surfactant is a complex lipoprotein material (King, 1974; Harwood, 1987; van Golde *et al.*, 1988), which is absorbed at the alveolar air/liquid interface and assists in maintaining alveolar stability. It lowers the surface tension to about 10 nN/m and reduces the contractile force of the surface and the work of lung expansion. It also prevents atelectasis because during expiration it forms a solid film on the alveolar surface. Physical requirements for surface tension effects in lung surfactant have been discussed (Hills, 1988).

12.7.1 Composition

The composition and properties of biological materials are, obviously, influenced by the methods used for their isolation. The definition of pulmonary surfactant and a discussion of the properties of various candidate fractions can be found in Clements (1970) and Goerke (1975). Isolation of surfactant requires careful endobronchial infusion and withdrawal of isotonic solutions such as Ringers to yield a cloudy white fluid with the surface-active material in suspension. Pulmonary surfactant has then been purified from the lavage fluid by a number of different procedures using differential and density-gradient centrifugation in either salt (King, 1974; Harwood *et al.*, 1975) or sucrose (Stein *et al.*, 1969; Clements and King, 1976) solutions. The material is isolated at a density of 1.08–1.09 g/ml in salt density gradients. Pulmonary surfactant is a lipid-rich lipoprotein containing little carbohydrate and a large quantity of dipalmitoylphosphatidylcholine (see below and Table 12.9).

Pulmonary surfactant can also be isolated from lung homogenates and from intact inclusion bodies derived from pulmonary type II epithelial cells (Clements and King, 1976).

As shown in Table 12.9, about 80% of isolated pulmonary surfactant is lipid, with phospholipids predominating. Neutral lipids such as cholesterol, triacylglycerol and non-esterified fatty acids make up 10–15% of the total lipids (King, 1974; Clements and King, 1976). The major surface-active component is dipalmitoylphosphatidylcholine (Harwood *et al.*, 1975), but small amounts of unsaturated phospholipids are present, probably to help in the spreading of surfactant at body temperatures (Goerke, 1975). Phosphatidylglycerol is the next most abundant lipid, accounting for up to 10% of the total. It is also highly surface-active (Harwood *et al.*, 1975; Rooney, 1979).

Early preparations of isolated surfactants were shown to contain a number of proteins (e.g. Harwood *et al.*, 1975). Specific surfactant proteins, SP-A, SP-B and SP-C, closely associate with surfactant lipids and contribute to the surfactant properties of the phospholipids. Recently, a fourth protein, SP-D, has been isolated from several preparations. SP-D is a glycoprotein that shows similarities to SP-A in its collagenous structure and carbohydrate-binding properties (Kuroki *et al.*, 1991). SP-A is a major component and has a molecular mass (in humans) of

Table 12.9 Composition of isolated lung surfactant (Data taken from Harwood *et al.*, 1975)

	Rat	Rabbit	Ox	Sheep
	(% w/w)			
Total protein	10	8	18	14
Total lipid	88	90	79	86
	(% w/w of total lipid)			
Triacylglycerol	5	4	15	7
Unesterified fatty acid	3	2	3	10
Cholesterol (total)	3	3	1	3
Phosphatidylethanolamine	4	1	3	2
Phosphatidylcholine	73	83	71	58
Phosphatidylglycerol	5	3	2	4
Phosphatidylinositol	5	2	3	4
Sphingomyelin	2	2	1	3
Lysophosphatidylcholine	tr.	tr.	tr.	11

28–36 kDa depending on oligosaccharides attached (Weaver and Whitsett, 1991). SP-B (8 kDa) and SP-C (about 4 kDa) are extremely hydrophobic. The structures, processing, regulation of expression and functions of these proteins have been well reviewed by Weaver and Whitsett (1991).

12.7.2 Metabolism

There is considerable evidence that the type II alveolar epithelial cell is the site of surfactant production (van Golde, 1976). Methods for the isolation of type II cells from adult rat and rabbit lungs have been developed (cf. Rooney, 1979), but they are difficult to reproduce and there are considerable problems in maintaining the isolated cells in culture for more than short periods. Moreover, cells from adult animals may be of little use in studying hormonally induced changes in lipid metabolism, and therefore procedures for culturing preparations enriched in foetal rat type II cells have been developed (Douglas *et al.*, 1976). As an alternative, some workers have used tumour cells with the morphological characteristics of type II cells as models for study of the synthesis and secretion of surfactant (Rooney, 1979).

Type II cells contain characteristic lamellar inclusion bodies. Disaturated phosphatidylcholine appears to be synthesized in the endoplasmic reticulum and then stored in the lamellar bodies, where it may also be combined with other surfactant components before being released into the alveoli by exocytosis (Goerke, 1975; Rooney, 1979) (Fig. 12.2). Lung synexin may be involved in membrane fusion between the plasma membrane and lamellar bodies during exocytosis (Chander and Wu, 1991). Phosphatidylcholine is synthesized *de novo* in the lung mainly by the CDP-base pathway, and this results in 1-saturated, 2-unsaturated molecular species. These phosphatidylcholines are converted to disaturated species ('remodelling') by deacylation to 1-saturated, 2-lysophosphatidylcholine and then either transacylation with lysophosphatidylcholine or reacylation with palmitoyl-CoA. The latter appears to be more important (Rooney, 1979; Post and van Golde, 1988).

The beginning of pulmonary surfactant synthesis occurs towards the end (last 10%) of gestation, and there is a further increase (10-fold or so) immediately after birth. During these periods the relative amount of sphingomyelin in lung lavage decreases from 38% of the total phospholipids to 7% at full term and 2% after birth. The ratio of phosphatidylcholine/

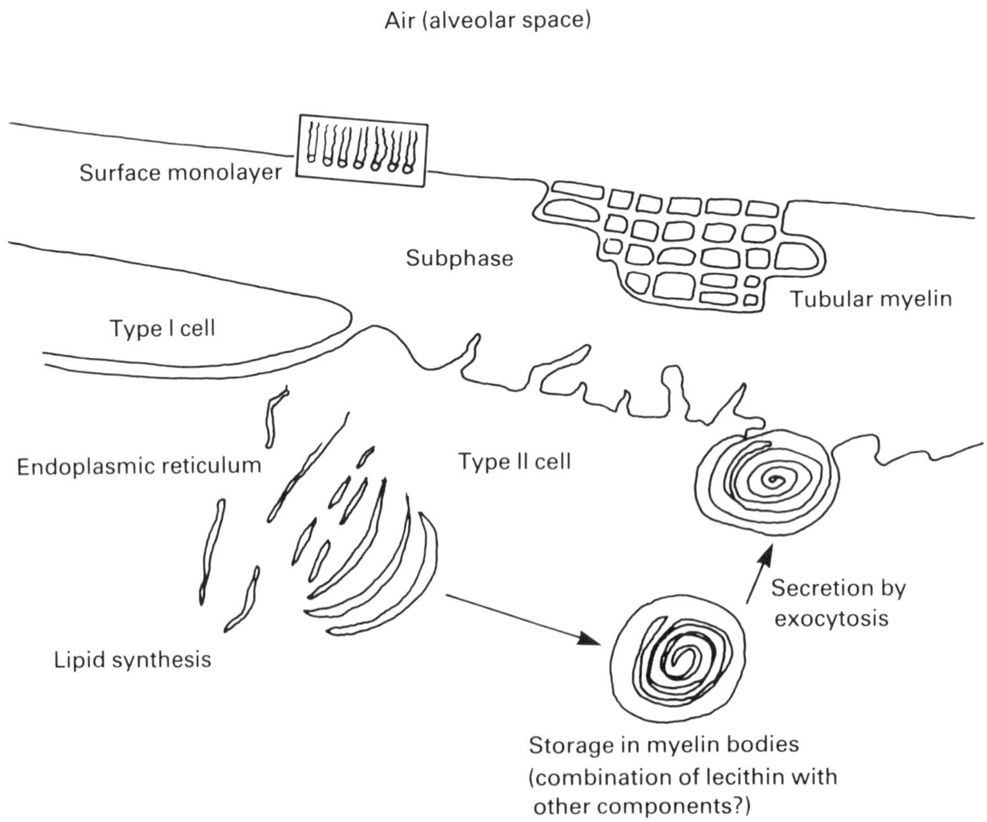

Fig. 12.2 Lung surfactant synthesis and secretion. Adapted from Goerke (1975) with permission.

sphingomyelin in lavage fluid changes, therefore, from less than 1 to 10 at full term to 30–40 after birth. An analysis of the phosphatidylcholine (lecithin)/sphingomyelin (L/S) ratio is used clinically to predict suitable times for elective delivery (Rooney, 1979).

Although changes in a number of enzymes involved in phosphatidylcholine synthesis have been noted during lung development (Rooney, 1979), it seems most likely that the activity of cytidylyltransferase is the rate-limiting step (Vance and Choy, 1979). Other enzymes that show changes around birth and may, in consequence, influence the rate of surfactant synthesis include choline kinase, choline phosphotransferase, phosphatidate phosphatase, phospholipase A_2 and lysolecithin acyltransferase (Rooney, 1979). The production of surfactant has been shown to be stimulated by glucocorticoids, oestrogen, thyroxine and thyrotropin-releasing hormone. It is also influenced by neurohormonal agents and stress factors such as pregnancy and cyclic AMP (Rooney, 1979). Of these effects, those caused by glucocorticoids are the most studied, and the hormones increase the activities of a number of enzymes involved in surfactant lipid synthesis (Table 12.10) and accelerate maturation of foetal lungs. Hormonal regulation of lung development and surfactant synthesis and secretion are reviewed by Post and van Golde (1988) and by van Golde *et al.* (1988).

For general reviews of the biochemistry of lung surfactant lipids, refer to Robertson *et al.* (1984), Harwood (1987), Post and van Golde (1988) and van Golde *et al.* (1988).

12.7.3 Respiratory distress syndrome (hyaline membrane disease)

Respiratory distress syndrome is a developmental disorder that is caused by immaturity of the lungs, which do not produce the required quantity of pulmonary surfactant. The lack of the normal complement of surfactant results in morphological alterations (hyaline membranes, atelectasis) and physiological alterations (decreased compliance, shunting of blood, hypoxaemia) and is a major cause of death in premature infants. The disease afflicts about half of the premature infants in the UK. About 25% of such babies die and some are left with handicaps (Morley *et al.*, 1981). Because surfactant is lacking, the baby has difficulty in expanding his or her lungs and they tend to collapse in expiration, leading to a great increase in the work of breathing, respiratory failure and death unless ventilatory support is given. However, it must be stressed that lung surfactant deficiency may not be the only contributing factor to the disease symptoms (Morley, 1985), thus making treatment more problematic. Indeed, for adult respiratory distress (which is more common and where the prognosis is worse), surfactant deficiency is a complication of lung injury rather than a primary aetiological factor (Lechmann and Danzmann, 1984).

The maturity of foetal lungs can be assessed by obtaining amniotic fluid by transabdominal amniocentesis. It is advisable to perform two types of test on this fluid, including either a determination of the L/S ratio (see above) or phosphatidylcholine concentrations as well as a measurement of surface

Table 12.10 Effects of glucocorticoids and oestrogen on various parameters of surfactant production in the foetal rabbit lung 26–27 days gestation (Taken from Rooney, 1979, with permission)

	Effect (% increase over control)		
	Glucocortoids		17β-Oestrogen to doe
	Cortisol to foetus	Betamethasone to doe	
Phosphatidylcholine content of lung lavage	130	70	320
L/S ratio of lavage	80	78	190
Rate of choline incorporation into PC of lung slices	90	64, 125	–
Enzyme activities:			
Cholinephosphate cytidylyltransferase	22	50	72
Phosphatidate phosphatase	–	50, 94	–
Lysolecithin acyltransferase	33	–	26

characteristics by means of Clement's foam test (Diedrich *et al.*, 1981). A lung maturity test may be positive before week 34 of gestation in cases of gestosis, hypertensive illnesses or previous premature rupture of the membranes. If the mother is diabetic, then lung maturation is delayed. The predictive accuracy of different tests for neonatal respiratory distress syndrome has been discussed (Harwood and Richards, 1985). Normal values for amniotic fluid phosphatidylcholine increase from about 2.5 mg/100 ml at 34 weeks to 8 mg/100 ml at 39 weeks gestation.

There have been considerable efforts to treat neonatal respiratory distress syndrome. Apart from the use of early continuous positive airway pressure and artificial ventilation (Jonson *et al.*, 1984), pharmacological methods have been used to hasten lung maturity and, hence, increase the production of endogenous surfactant (Avery, 1984). Alternatively, surfactant replacement has been used for treatment (Fujiwara, 1984). The efficacy of the latter therapy is now well established both in animal models and in several randomized clinical trials (van Golde *et al.*, 1988).

The biosynthesis and secretion of surfactant are subject to multiple hormonal control (van Golde *et al.*, 1988). First, β-adrenergic stimulation is involved in type II cell development, phosphatidylcholine synthesis and surfactant secretion (Walters, 1985). Secondly, glucocorticoids have extensive activity, particularly with regard to lung maturation and the development of lung surfactant formation (see van Golde *et al.*, 1988; Table 12.10). Glucocorticoids have been used in several clinical trials (see e.g. Avery, 1984). Thyroid hormones may affect lung development and surfactant lipid synthesis and interact with β-adrenergic regulation. Other hormonally related complications are the severer symptoms in male human neonates and for infants of diabetic mothers (van Golde *et al.*, 1988; Post and van Golde, 1988).

Surfactant replacement via the airways can be used prophylactically at birth or as a rescue operation for patients in respiratory failure. A number of different substitutes have been used – nearly all of which have proven clinically efficacious. Most of the preparations are derived from mammalian sources. Their preparation involves organic extraction of lung tissue or more usually lung washings. This process removes SP-A but leaves the hydrophobic proteins SP-B and SP-C still present. The preparations can then be supplemented by addition of lipids and/or SP-A (see van Golde *et al.*, 1988; Cockshutt *et al.*, 1991). Because surfactant-associated proteins are, in principle, antigenic, then efforts have been made to design effective protein-free surfactants. An example is the dipalmitoylphosphatidylcholine–phosphatidylglycerol mixture introduced by Morley *et al.* (1981). This preparation does not have immediate benefit in patients with manifest respiratory distress syndrome (Greenough *et al.*, 1984) but does reduce infant mortality and improve lung function when given prophylactically at birth (Morley, 1987). Another protein-free surfactant substitute is one based on a mixture of dipalmitoylphosphatidylcholine, hexadecanol and tyloxapol (Tooley and Clements, 1984).

In spite of the successful use of artificial surfactants, many questions remain. These particularly concern the dosage, method of administration, speed of removal and possible adverse side-reactions (see van Golde *et al.*, 1988). Other possible problems with replacement therapy are discussed by Morley and Bangham (1981) and by Morley (1985).

12.7.4 Other aspects of pulmonary surfactant

A number of different conditions influence pulmonary surfactant levels in adult lungs. Exposure of animals to silica (Heppleston *et al.*, 1974) or asbestos dusts (Tetley *et al.*, 1977) causes massive increases in pulmonary surfactant and results in pathological conditions similar to human alveolar lipoproteinosis (Sahu *et al.*, 1976). The surfactants from these dust-treated animals also display abnormal surface tension characteristics (McDermott *et al.*, 1977). Asbestos and silica fibres have been shown to be very cytotoxic to cells *in vitro* and these particles are known to be highly damaging *in vivo*. It has been suggested that the increase in pulmonary surfactant is an attempt by animals to neutralize the effect of the inhaled dust. Tests *in vitro* have shown that pulmonary surfactant is strongly bound to the surface of mineral particles and that this surfactant is a very effective material in preventing asbestos-induced lysis of cells (Desai *et al.*, 1975). Alternatively, the increase in surfactant may help in the physical movement of dust particles out of the alveoli (Scarpelli, 1968).

The increase in pulmonary surfactant as a result of environmental factors may be a generalized lung defence mechanism (Witschi, 1976), but exposure to several compounds, such as paraquat, glass powder and anaesthetics, causes a decrease in this material. In addition to stabilizing the alveolar surface, pulmonary surfactant may have a number of vital roles. It has been proposed that it aids in gaseous exchange, helps prevent pulmonary oedema and provides a milieu for movement of macrophages and other cells (Scarpelli, 1968).

Attempts have been made to rationalize the changes brought about in adult pulmonary surfactant levels on the basis of altered rates in synthesis or degradation. The increase following quartz inhalation appears to be due to both an increase in surfactant synthesis as well as a decrease in its degradation (Heppleston *et al.*, 1974). Increases in surfactant following asbestos inhalation may be due to an increase in the activity of the synthetic enzyme, phosphorylcholine glyceride transferase (Tetley *et al.*, 1977).

The massive increases in surfactant levels following dust exposure may have important industrial implications, and changes are also noticed in many other diseases (Scarpelli, 1968) such as cystic fibrosis, sudden unexpected death in infants and for adults suffering from multiple trauma, trauma to the lungs, acute pancreatitis and viral infections. In all of these cases the changes noted in pulmonary surfactant would, at the very least, cause an impairment of lung function.

A general review of surfactant biochemistry in health and disease has been made by Kikkawa and Smith (1983), and the book by Robertson *et al.* (1984) also contains a wealth of information.

References and further reading

Avery, M.E. (1984) in *Pulmonary Surfactant* (eds B. Robertson, L.M.G. van Golde and J.J. Batenburg), Elsevier, Amsterdam, p. 449.

Chander, A. and Wu, R.-D. (1991) *Biochim. Biophys. Acta*, **1086**, 157.

Clements, J.A. (1970) *Am. Rev. Respir. Dis.*, **101**, 984.

Clements, J.A. and King, R.J. (1976) in *The Biochemical Basis of Pulmonary Function* (ed. R.G. Crystal), Marcel Dekker, New York, p. 363.

Cockshutt, A.M., Absolom, D.R. and Possmayer, F. (1991) *Biochim. Biophys. Acta*, **1085**, 248.

Desai, R., Hext, P.M. and Richards, R.J. (1975) *Life Sci.*, **16**, 1931.

Diedrich, K., Lehmann, F. and Krebs, D. (1981) *Prog. Respir. Res.*, **15**, 155.

Douglas, W.H.J., Moorman, G.W. and Teel, R.W. (1976) *In Vitro*, **12**, 373.

Fujiwara, T. (1984) in *Pulmonary Surfactant* (eds B. Robertson, L.M.G. van Golde and J.J. Batenburg), Elsevier, Amsterdam, p. 479.

Goerke, R.J. (1975) *Biochim. Biophys. Acta*, **344**, 241.

Greenough, A., Morley, C.J., Wood, S., Miller, N. and Bangham, A.D. (1984) *Prog. Respir. Res.*, **18**, 253.

Harwood, J.L. (1987) *Prog. Lipid Res.*, **26**, 211.

Harwood, J.L. and Richards, R.J. (1985) *Mol. Aspects Med.*, **8**, 423.

Harwood, J.L., Desai, R., Hext, P., Tetley, T. and Richards, R. (1975) *Biochem. J.*, **151**, 707.

Heppleston, A.G., Fletcher, K. and Wyatt, I. (1974) *Br. J. Exp. Pathol.*, **55**, 384.

Hills, B.A. (1988) *The Biology of Surfactant*, Cambridge University Press, Cambridge.

Jonson, B., Lindroth, M. and Svenningsen, W.W. (1984) in *Pulmonary Surfactant* (eds B. Robertson, L.M.G. van Golde and J.J. Batenburg), Elsevier, Amsterdam, p. 459.

Kikkawa, Y. and Smith, F. (1983) *Lab. Invest.*, **49**, 122.

King, R.J. (1974) *Fed. Proc.*, **33**, 2238.

Kuroki, Y., Shiratori, M., Ogasawara, Y., Tsuzuki, A. and Akino, T. (1991) *Biochim. Biophys. Acta*, **1086**, 185.

Lachmann, B. and Danzmann, E. (1984) in *Pulmonary Surfactant* (eds B. Robertson, L.M.G. van Golde and J.J. Batenburg), Elsevier, Amsterdam, p. 505.

McDermott, M., Wagner, J.C., Tetley, T., Harwood, J.L. and Richards, R.J. (1977) in *Inhaled Particles*, vol. 4 (ed. W.H. Walton), Pergamon, Oxford, p. 415.

Morley, C.J. (1985) *Biochem. Soc. Trans.*, **13**, 1091.

Morley, C.J. (1987) in *Physiology of the Fetal and Neonatal Lung* (eds D.V. Walters, L.B. Strang and F. Geubelle), MTP Press, Lancaster, PA, p. 255.

Morley, C.J. and Bangham, A.D. (1981) *Prog. Respir. Res.*, **15**, 261.

Morley, C.J., Bangham, A.D., Miller, N. and Davis, J.A. (1981) *Lancet*, **i**, 64.

Post, M. and van Golde, L.M.G. (1988) *Biochim. Biophys. Acta*, **947**, 249.

Robertson, B., van Golde, L.M.G. and Batenburg, J.J. (eds) (1984) *Pulmonary Surfactant*, Elsevier, Amsterdam.

Rooney, S.A. (1979) *Trends Biochem. Sci.*, **4**, 189.

Sahu, S., August, R.P. and Lynn, W.S. (1976) *Am. J. Respir. Dis.*, **114**, 177.

Scarpelli, E.H. (1968) *The Surfactant System of the Lung*, Lea and Febiger, Philadelphia.

Stein, J.M., Redding, R.A., Hauck, C.T. and Stein, M. (1969) *Biochem. Biophys. Res. Commun.*, **34**, 434.

Tetley, T.D., Richards, R.J. and Harwood, J.L. (1977) *Biochem. J.*, **166**, 323.

Tooley, W. and Clements, J.A. (1984) in *Clinical Trials of the Use of Surfactant in the Neonatal Respiratory Distress Syndrome* (ed. J.F. Lucey), Ross Labs, Columbus, p. 136.

Vance, D.E. and Choy, P.C. (1979) *Trends Biochem. Sci.*, **4**, 145.

van Golde, L.M.G. (1976) *Am. Rev. Respir. Dis.*, **114**, 977.

van Golde, L.M.G., Batenburg, J.J. and Roberton, B. (1988) *Physiol. Rev.*, **68**, 374.

Walters, D.V. (1985) *Biochem. Soc. Trans.*, **13**, 1089.

Weaver, T.E. and Whitsett, J.A. (1991) *Biochem. J.*, **273**, 249.

Witschi, H. (1976) *Toxicology*, **5**, 267.

12.8 Agricultural aspects

A major purpose of the end-products of agricultural production is to provide human food. Lipids are important components of food, providing dietary energy, fat-soluble vitamins and essential fatty acids (Gurr, 1983; Vergroesen, 1975; Wiseman, 1984). They also enhance the palatability of food by providing flavour and aroma compounds and by influencing 'mouth feel' (Forss, 1972). A wide variety of dietary lipids are contributed by agricultural plant crops and by farm animals.

12.8.1 Plant lipids

The main sources of plant lipids of significance in diets are the leaves, which primarily provide a source of fatty acids for grazing animals, and seed oils, which play a large role in human diets and an increasing one in farm animal diets. These are representative of the 'structural' and 'storage' lipids respectively and differ widely in fat composition (sections 3.3 and 3.11).

(a) Storage lipids: oil seed crops

(i) Commercially important crops
The total world production of seed oils in 1981–82 was 43 million tonnes (Wiseman, 1984). Of the several hundred varieties of plants known to have oil-bearing seeds, only a dozen or so are important commercially (cf. section 3.3).

(ii) Location of the oil
Most oil-bearing plants store their lipid in the form of triacylglycerols, although some, like jojoba, store wax esters. The latter are poorly digested and have little nutritive value (Yaron *et al.*, 1982). Triacylglycerols are stored mainly in the fleshy fruit exocarp (e.g. avocado) or in the seed endosperm (e.g. rape). Some, like palm, store triacylglycerols in both the exocarp (palm oil) and the endosperm (palmkernel oil) (Gurr, 1980). The oil is present in cytoplasmic organelles, the oil bodies, that are surrounded by a half-unit membrane containing protein and phospholipid (Gurr, 1980; section 11.3).

(b) Fatty acid composition

Seed oils vary widely in fatty acid composition (section 3.3). One fatty acid often predominates, is sometimes of unusual structure and is characteristic of a particular plant family (Gurr, 1980). The commercially important seed oils are generally, however, those in which the predominant fatty acids are the common ones, with the exception of coconut and palmkernel oils, which are unusual in containing saturated medium-chain fatty acids. Elsewhere in Nature, only milk contains these fatty acids. The characteristic seed-oil fatty acids are esterified in specific positions on the triacylglycerol molecule and do not occur in the structural lipids (Gurr, 1980). In plants that store triacylglycerols in both exocarp and endosperm, there is usually a marked difference in fatty acid composition of triacylglycerols from the two locations (Gurr, 1980).

(c) Environmental factors influencing lipid composition

The characteristic fatty acid patterns of plant triacylglycerols are to some extent under genetic control. In addition, environmental factors may modify the basic patterns, the extent of modification depending on the species. Thus the seed oils of plants grown in cool climates tend to be more unsaturated than those grown in warm climates. The chief influence seems to be on the characteristic fatty acid of the seed. For example, the proportion of linoleic acid in sunflower seed oil steadily declines between 10 and 30 °C, to be replaced by oleic acid. Yet the linoleic acid content of safflower and the ricinoleic acid of castor are unaffected by the same variation of temperature. Appelqvist (1975) showed that different lines of zero-erucic-acid rape could respond differently to the same climatic variations. Light intensity is another variable that may affect the fatty acid composition of seed-oil triacylglycerols. In Cruciferae, Leguminosae and Linaceae, which have a translucent pericarp and testa and in which the major sites of oil storage are the cotyledons, α-linolenic acid is a major seed-oil component. This may be due in part to the presence of chloroplasts in the cotyledons, which may influence the amount and composition of the stored oil. It may also be mentioned that the presence of functional chloroplasts ensures that a constant supply of oxygen is available to surrounding tissues, so that the effects of light, temperature and oxygen availability may not be entirely independent (Gurr, 1980).

For recent discussions of environmental aspects influencing lipid compositions, refer to Harwood

(1984) and Harwood *et al.* (1992). Other references will be found in Quinn and Harwood (1990), Browse and Somerville (1991) and Cherif *et al.* (1992).

(d) *Improvement of oil composition*

Given the increasing cost of oils for human foods and animal feeds, and the increasing knowledge of the role of dietary fats, it might be expedient to improve the range and availability of edible oils with desirable characteristics. A number of aspects of this subject are discussed by Browse and Somerville (1991).

(i) Agricultural use of little-used species
A search through the available data (Hitchcock and Nichols, 1971; Hilditch and Williams, 1964) reveals seed oils with particularly useful fatty acid composition in species that have hitherto not been used as crops. Recent understanding of the biochemistry of polyunsaturated acids has, for example, highlighted the possible usefulness of γ-linolenic acid, leading to the development of evening primrose (section 3.3.13) as a potential agricultural crop. Examination of the literature reveals that coconut is not the only seed oil containing medium-chain glycerides. *Cuphea* is another provider of these fatty acids that could have agricultural importance (Wiseman, 1984) (section 3.3.12) and there has been considerable work on these plants recently. For example, Dormann *et al.* (1991) have studied the acyl-ACP thioesterase responsible for the release of medium-chain fatty acids, while Heise *et al.* (1991) showed that production of capric acid was cerulenin-independent (thereby implying the importance of a β-ketoacyl-ACP synthetase III enzyme; see section 11.1.1). Lupin seeds contain an oil whose major fatty acids are: palmitic, 11%; oleic, 53%; linoleic, 24%; linolenic, 7% (Hilditch and Williams, 1964). Either lupins or sunflowers (section 3.3.30) could provide an additional oil seed crop in cool climates such as exist in the UK. The latter seed oil is particularly rich in linoleic acid in cool climates. Alternative sources of lipids are discussed by Kamel and Kakuda (1994). The synthesis of uncommon fatty acids and the targeting of such molecules into storage triacylglycerols are discussed by Browse and Somerville (1991).

(ii) Plant breeding
Within the species *Brassica napus*, selection for plants with a low erucic acid content has resulted in a strain that contains no erucic acid in its seed oil. Genetic analysis of the F_2 and F_3 generations and backcrosses between these plants and normal high-erucic-acid types supported the hypothesis that the erucic acid content of the seed oil was controlled by two genes that displayed no dominance and acted in an additive manner. The genes controlled the chain-lengthening pathway from oleic to eicosenoic and erucic acids such that, as the capacity for erucic acid biosynthesis decreased, there was a corresponding increase in the proportion of oleic acid, maintaining a constant oil content of the seed. Three alleles at a single chromosome locus in safflower (*Carthamus tinctorius*) govern the relative amounts of oleic and linoleic acids in the mature seed (Appelqvist, 1975; Gurr, 1980).

The work of the plant breeder, however, is extremely slow, and there is always the possibility that breeding in desired characteristics or breeding out undesired characteristics is accompanied by the elimination of other desirable features, such as cold- or disease-resistance or high yield. Attempts to breed out the high linolenic acid content of soybean oil to improve flavour and oxidative stability have been disappointing but other examples of breeding are given in Murphy (1993).

(iii) Plant tissue culture
Techniques for growing a variety of plant types in tissue culture have been developed. Briefly, sterilized pieces of excised tissue, when incubated on a solid medium containing a full range of nutrients and growth substances, give rise to a disorganized mass of largely undifferentiated cells called a callus. Callus cultures can be maintained on solid media for extended periods or the cells dispersed and grown in liquid culture. Such cultures can be exploited in two main ways: to regenerate clones of plants for use in conventional agriculture; or as a source of low-volume, high-cost products such as drugs, food colourings or flavours. The ability to propagate clones of genetically identical plantlets in such commercially important crops as the oil palm has become a reality. This approach is especially desirable in view of the problems associated with oil palms, for which there are no conventional horticultural methods for making clones. Propagation by tissue culture techniques promises to reduce variability in quality and yield, and should bring plants into production more quickly though at a slightly greater cost than by conventional methods.

(iv) Genetic manipulation
Much of the attention in this field has been devoted to increasing the photosynthetic efficiency of green plants and to developing cereals capable of directly fixing atmospheric nitrogen. Several research groups have been directing their attention to the possibility of inserting into specified plant cells the genes for a desired pathway of lipid metabolism. Such an

approach needs not only improved techniques for the genetic manipulation of higher plant cells, which is lagging behind the microbial technology, but also more detailed knowledge of the genetic control of the pathways of lipid metabolism.

Recent aspects of this area have been reviewed by Browse and Somerville (1991), Ohlrogge *et al.* (1991) and Slabas and Fawcett (1992). A recent example of success has been achieved by using antisense technology to down-regulate the stearoyl-ACP desaturase of *Brassica*, thus producing a storage oil that had a marked increase in its stearate content (Knutzon *et al.*, 1992). Another recent achievement has been the genetic engineering of oil seeds to produce medium-chain fatty acids (Voelker *et al.*, 1992).

(e) Plant leaf lipids

Leaf lipids are discussed fully in section 3.11, and will not be considered further here.

(f) Herbicides that affect acyl lipid metabolism

Some of the most effective modern herbicides are those which interfere with carotenoid formation (Bramley, 1991). These chemicals kill plants because of the chlorophyll photo-oxidation induced by the lack of carotenoids. However, these 'bleaching' herbicides will not be dealt with here, where our attention is focused on chemicals that are known to inhibit acyl lipid synthesis.

Substituted pyridazinones (Fig. 12.3) were originally developed as potential photosynthetic electron-transport inhibitors. However, it was soon discovered that many of them also reduced the formation of thylakoid pigments (both chlorophylls and carotenoids). The mechanism of phytotoxicity has been discussed, as well as the conversion of the herbicide to more toxic metabolites (St John, 1982). In some cases, it was noted that certain herbicides also inhibited fatty acid desaturation (e.g. Hilton *et al.*, 1971). Sandoz 6706 and Sandoz 9785 were both noted to reduce α-linolenate levels, with a commensurate rise in linoleate (see Harwood, 1991b). This action suggested that substituted pyridazinones could inhibit Δ15-desaturation, and that proposal was confirmed by measurement of desaturase activity using radiolabelled precursors (Murphy *et al.*, 1985; Davies and Harwood, 1983). Sandoz 9785 is notable in that its only effect seems to be on Δ15-desaturation and, then, at the level of monogalactosyldiacylglycerol (Norman and St John, 1987) rather than phosphatidylcholine (see section 11.1). Other substituted pyridazinones affect additional metabolic

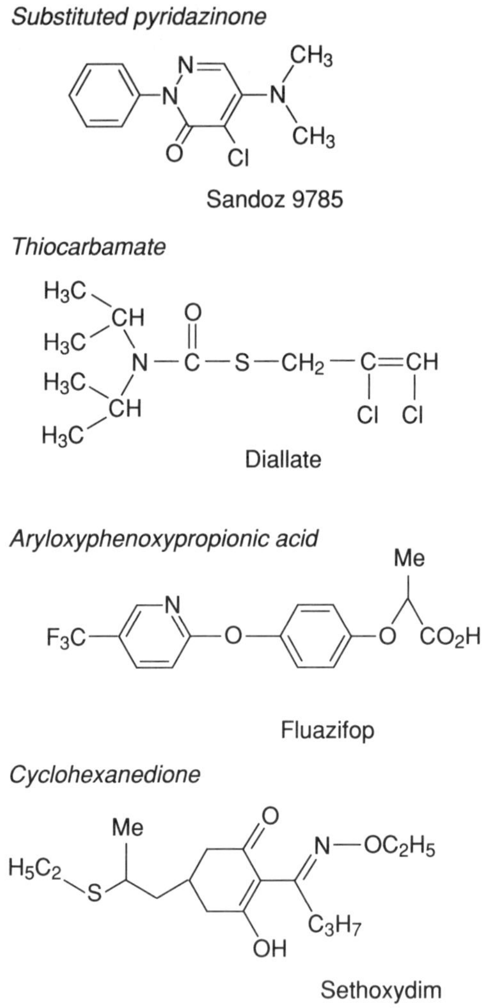

Fig. 12.3 Structures of some herbicides that affect fatty acid synthesis.

activities, and different plants show various susceptibilities (see Harwood, 1991a,b). In particular, algae do not show such clear-cut effects as susceptible higher plants (e.g. Murphy *et al.*, 1985; Henderson *et al.*, 1990).

Another bleaching herbicide is diflufenican, which inhibits phytoene desaturase at very low concentrations and, hence, prevents carotenoid synthesis (Barry and Pallett, 1988). Recently, diflufenican has been shown to inhibit the reductase partial reactions of fatty acid synthetase (Ashton *et al.*, 1992). Although it was not thought that the herbicidal activity of diflufenican was, in any significant way, related to its inhibition of fatty acid synthetase, the observation raises the possibility that other, more effective, inhibitors of fatty acid synthetase may be developed.

The carbamate herbicides, which are normally applied pre-emergence, exert their herbicidal action at early stages of seedling growth (Fuerst, 1987).

Several thiocarbamates (Fig. 12.3), including EPTC, diallate, CDEC and triallate, have been reported to inhibit lipid metabolism. These effects, which are summarized by Harwood (1991b), include alterations in cuticular wax formation, primary alcohol content and the level of very-long-chain fatty acids. The action (and important herbicidal effect) of thiocarbamates on the surface layers of susceptible plants has a rationale in that these chemicals have been shown to be rather specific inhibitors of fatty acid elongation *in vivo*. Since very-long-chain fatty acids are the precursors of many components of wax, cutin and suberin (see sections 2.10, 2.11 and 3.6), it is easy to understand their action. Actually, recent work with subcellular fractions from peas (Abulnaja and Harwood, 1991a) and with safening agents (Abulnaja and Harwood, 1991b) suggests that it is the herbicidally active sulphoxide derivative of thiocarbamates (Fedke, 1982) that is the specific inhibitor of fatty acid elongation. Direct measurement of I_{50} values for specific elongation reactions confirmed this conclusion (Barrett and Harwood, 1993).

Another herbicide that has been reported to interfere with wax formation (Duncan *et al.*, 1982) is ethofumesate. Experiments with this compound were also consistent with its action on waxes being due to inhibition of fatty acid elongation (Abulnaja *et al.*, 1992).

Two important classes of selective (grass-specific) herbicides are the aryloxyphenoxypropionates and the cyclohexanediones (Fig. 12.3). The action of these compounds on lipid metabolism has been reviewed thoroughly (Harwood, 1991a; Hoppe, 1989). The herbicides affect acetyl-CoA carboxylase in the *Poaceae*, but are without action (except at exceptionally high concentrations) on the same enzyme in other monocotyledons or in dicotyledons (see Harwood, 1991b). Where tolerance has been noted in specific *Poaceae*, it has usually been attributed to detoxifying metabolic reactions. Thus, wheat is tolerant to diclofop because of hydroxylation of the herbicide (Walker *et al.*, 1989; Hoppe, 1989). However, there are one or two exceptions in which mutation of the target protein, acetyl-CoA carboxylase, may cause resistance (see Lichtenthaler *et al.*, 1989; Harwood, 1991b).

Examination of the kinetics of acetyl-CoA carboxylase inhibition by aryloxyphenoxypropionates and cyclohexanediones has shown conclusively that it is the carboxyltransferase partial reaction (see section 11.1) that is sensitive (Rendina *et al.*, 1989). Competition experiments have shown that the two classes of herbicide bind to the same site on the carboxyltransferase (Rendina *et al.*, 1989; Burton *et al.*, 1991). A third class of chemical has recently been shown to produce similar effects to the aryloxyphenoxypropionates and cyclohexanediones. This compound, 3-isopropyl-6-[*N*-(2,2-dimethylpropyl)-acetamido]-1,3,5-triazine-2,4(1*H*,3*H*)-dione (PP600), also inhibited *de novo* fatty acid synthesis in monocotyledons but not in dicotyledons. Although acetyl-CoA carboxylase was inhibited by PP600, the enzyme was much less sensitive than whole-tissue preparations, raising the possibility that PP600 is metabolized to a more active metabolite *in vivo* (Walker *et al.*, 1990).

Several other herbicides have been reported to alter lipid metabolism (see Harwood, 1991b), although in most cases this is due to secondary effects following a more acute action elsewhere in metabolism.

12.8.2 Animal lipids

As in plants, animal lipids can be conveniently classified as 'storage' or 'structural'. Storage lipids are, with few exceptions, triacylglycerols, and the structural lipids are mainly phospholipids and cholesterol. The fatty acid compositions of these types of lipids are markedly different.

(a) Storage lipids

(i) Adipose tissue

We shall consider separately adipose tissue from monogastric animals and from ruminant animals.

Monogastric animals. The composition of the adipose tissue of pigs and poultry, the most important monogastric farm animals, is markedly affected by diet (Wiseman, 1984) (Table 12.11). Inclusion of vegetable oils such as soybean oil in the diet results in higher proportions of linoleic acid and lower proportions of oleic and palmitic acids than with the inclusion of tallow. Feeding such unsaturated supplements or including appreciable amounts of copper in pig diets tends to result in a soft back fat (Wiseman, 1984). When these animals are fed low-fat cereal-based diets, adipose tissue fatty acids are synthesized from dietary carbohydrates, and their composition resembles that of animals fed diets containing animal fat (Wiseman, 1984) (Table 12.11). However, the cereals contain structural lipids that influence the adipose tissue fatty acid composition to some extent, depending on the type of cereal (Wiseman, 1984).

Ruminant animals. Because the predominantly polyunsaturated fatty acids in ruminant diets undergo biohydrogenation in the rumen, producing mainly stearic acid and *cis*- and *trans*-mono-unsaturated acids (Wiseman, 1984), the storage fats of grazing ruminants or those fed cereal-based concentrates

Table 12.11 Fatty acid composition of depot (storage) fats of farm animals under different husbandry conditions (g/100 g total fatty acids)

Fatty acid	Pig					Poultry		Cattle			Lamb	
	A	B	C	D	E	F	G	H	I	J	K	L
Myristic (14:0)	1.4	1.4	1.1	1.4	2.0	0.8	1.2	3.3	3.4	–	3.3	4.3
Palmitic (16:0)	28.7	28.5	20.9	21.0	27.7	26.9	21.8	26.0	23.9	19.9	20.9	18.7
Palmitoleic (16:1)	2.5	2.5	2.6	3.8	3.8	8.8	5.0	9.4	10.6	–	4.3	6.0
Stearic (18:0)	15.4	15.3	11.6	16.5	12.9	7.0	6.4	8.2	6.9	9.6	19.6	16.4
Oleic (18:1)	43.3	41.8	45.5	53.7	44.7	44.7	26.9	44.7	48.7	33.3	41.1	36.5
Linoleic (18:2)	9.3	10.7	16.2	3.1	8.9	11.2	35.2	2.1	1.5	23.1	5.1	11.5

A Low-fat cereal-based diet (Wiseman, 1984).
B Selected for lean carcass; low-fat cereal-based diet (Wiseman, 1984).
C High-fat diet containing soybean oil (Brooks, 1971).
D High-fat diet containing tallow (Brooks, 1971).
E Diet containing 250 ppm copper (Christie and Moore, 1969).
F Low-fat cereal-based diet (Bartov *et al.*, 1974).
G High-fat diet containing soybean oil (Bartov *et al.*, 1974).
H Diet based on hay (Wiseman, 1984).
I Diet based on barley (Wiseman, 1984).
J Diet containing 'protected' safflower oil (Cook *et al.*, 1972).
K Cereal-based concentrate diet (Astrup and Nedkvitne, 1975).
L Concentrate diet supplemented with 'protected' soybean oil (Cook *et al.*, 1972).

are on average more saturated and less susceptible to dietary influence than those of monogastric animals (Wiseman, 1984) (Table 12.11). Feeding diets rich in cereals to lambs causes the deposition of large quantities of odd-chain and branched-chain fatty acids. The latter increase the softness of the depot fat (Garton *et al.*, 1972). Feeding fat supplements (15–20% of digestible energy) or whole oil seeds can effect modest changes in depot fat composition (Wiseman, 1984), but more extensive changes are effected by feeding 'protected fat' (McDonald and Scott, 1977). Fat particles are coated with protein that is cross-linked by treatment with formaldehyde. This protects the fat against degradation in the rumen, but, on reaching the acid environment of the abomasum, the protein is digested and the fat can pass unchanged into the duodenum. There it can be digested and absorbed by mechanisms identical to those in monogastric animals (McDonald and Scott, 1977). In this way, large increases in, for example, the proportion of linoleic acid can be brought about in ruminant depot fat (McDonald and Scott, 1977) (Table 12.11) (section 10.1.3).

(ii) Milk

Cow milk and dairy products derived from it are important in human foods, and in the UK 33% of dietary fat intake is from milk and dairy products (Gurr, 1983). Goat milk is increasing in importance in the UK. The fatty acid compositions of cow and goat milk are characterized by a high proportion of short- and medium-chain fatty acids, long-chain saturated and mono-unsaturated acids and small proportions of polyunsaturated acids (Gurr, 1983; section 3.4). Ruminant milks also contain small quantities of a wide variety of branched- and odd-chain fatty acids (Christie, 1983), depending on the feeding system. Milk yields, which have been steadily rising over the years (Oldham and Sutton, 1979), depend on breed, on stage of lactation and on the management, particularly the diet, of the cattle. Strategies to increase yield often result in lower milk fat content and altered fat composition (Oldham and Sutton, 1979). The biggest changes occur when cows are offered highly digestible diets containing large proportions of cereal grains, high-quality forages or fat supplements in order to increase the intake of digestible energy (Oldham and Sutton, 1979). The effect of increasing the concentration of digestible carbohydrate is almost always to decrease milk fat percentage, although not necessarily total milk fat yield because of higher overall milk yield (Oldham and Sutton, 1979). The effects of fat supplements are more variable. Feeding unprotected fat often decreases milk fat content because of inhibitory effects of dietary fatty acids on rumen microorganisms, resulting in decreased fibre digestibility and a decreasing ratio of acetic/propionic acids in the rumen (Oldham and Sutton, 1979; Wiseman, 1984). Feeding protected lipids often overcomes these problems. Higher amounts of lipids can be fed and an increase in milk fat content can be achieved (Oldham and Sutton, 1979; Wiseman, 1984). Feeding fat to ruminants depresses the concentrations of short- and medium-chain fatty acids in milk by inhibiting endogenous synthesis of fatty acids in the mammary gland and allowing the incorporation of dietary long-chain fatty acids into milk. If the protected fat contains linoleic acid, the concentration of this acid in milk can be increased (McDonald and Scott, 1977).

(iii) Fish oils

Fish can be broadly classified into 'lean fish', which store their reserve lipids as triacylglycerols in the liver (e.g. cod), and 'fatty fish' whose triacylglycerols are stored in the flesh (e.g. herring, mackerel) (Wiseman, 1984). They have no adipose tissue. Some fish store their reserve oil as wax esters, e.g. orange roughy (Buisson *et al.*, 1982), which have poor nutritive value. Although fish lipids may differ between species and according to diet, they have some common characteristics that distinguish them from mammalian and avian lipids. They have a high content of fatty acids with 20 or more carbon atoms and are rich in polyunsaturated fatty acids of the $n-3$ family (section 3.5). In partial hydrogenation, a process necessary in the food industry to achieve the desired physical properties and keeping quality, the latter are largely converted into saturated and isomeric (positional and geometric) monoenoic fatty acids (section 10.1.1). The diet of fish is the main factor determining the fatty acid composition of the lipids, and may itself be influenced by factors such as water temperature and whether the fish are wild or farmed (Wiseman, 1984).

(b) Structural lipids

The carcasses of all animals, even when depleted of their energy stores, contain lipids that are components of biological membranes. These are mainly phosphoglycerides, but some phospho- and glycolipids based on sphingosine are also present, especially in nervous tissue (Gurr, 1983). The fatty acid composition of these lipids is less variable and less susceptible to dietary influence than that of the adipose tissue triacylglycerols. There is a high proportion of polyunsaturated fatty acids, even in ruminants, which biohydrogenate a large proportion of their dietary intake. Despite the homeostatic mechanisms tending to keep the composition of membranes within certain bounds consistent with optimal membrane function, these lipids can be manipulated to a degree by feeding to monogastric animals diets containing different fat supplements and to ruminants protected lipids (McDonald and Scott, 1977; Table 12.12).

(c) Lipids in farm animal diets

(i) Growth

The diets of farm animals, especially herbivores, have been traditionally low in fat, not more than 2–5% of digestible energy. It has been known for some time that fat supplements can enhance productive performance in cattle, pigs and poultry, and there is much interest in optimizing the amount and type of fat in the diets of farm animals for performance and quality (Wiseman, 1984). Increasing the fat content, and therefore the nutrient density of diets, results in increased food conversion efficiency and

Table 12.12 Fatty acid composition of some structural lipids occurring in human foods (g/100 g total fatty acids)

Fatty acid	A	B	C	D	E	F
Palmitic (16:0)	18.9	22.5	22.0	15.9	13.7	15.9
Palmitoleic (16:1)	1.6	6.3	2.3	–	–	2.5
Stearic (18:0)	12.0	12.0	13.2	10.8	13.5	2.0
Oleic (18:1)	18.8	33.2	30.3	19.7	5.1	3.4
Linoleic (18:2)	25.5	17.5	18.0	26.2	46.5	13.2
α-Linolenic (18:3)	0.2	0.6	3.9	–	–	61.3
Arachidonic (20:4)	7.7	6.0	6.8	12.7	11.1	–

A Pig muscle (Wood and Lister, 1973).
B Poultry muscle (Bartov *et al.*, 1974).
C Sheep muscle (Crouse *et al.*, 1972).
D Cattle muscle, low-fat diet (Cook *et al.*, 1972).
E Cattle muscle, 'protected' safflower oil diet (Cook *et al.*, 1972).
F Green leaves (pasture grass) (Hitchcock and Nichols, 1971).

faster growth in cattle, pigs and poultry. However, consumers are increasingly demanding leaner carcasses (the fat thickness at the last rib in pigs fell by 0.5 mm each year for the period 1975–85), and, in using fats in feeds, a compromise must be obtained between the advantages of increased efficiency and faster growth and the tendency to lay down excessive carcass fat. In the pig industry, the tendency is towards the increasing use of boars, which are leaner at corresponding weights than castrates. The same relationships hold between bulls and rams on the one hand and steers and wethers on the other (Wiseman, 1984).

(ii) Carcass quality

A consequence of the slower growth that yields a leaner carcass is a relatively more unsaturated fatty acid composition in adipose tissue (in particular, a higher proportion of linoleic acid) and therefore a softer carcass fat (Wiseman, 1984), which is not desirable to the consumer. The same effect is produced by feeding unsaturated fat supplements (Table 12.11). This may be overcome to some extent, while still feeding energy-dense diets, by incorporating more saturated fats such as hydrogenated marine fat and hydrogenated tallow. A disadvantage is low digestibility, but inclusion of some unsaturated fat in a predominantly saturated mix appears to overcome this (Wiseman, 1984). As well as producing fat of undesired appearance and texture, an overabundance of unsaturated fat limits the storage life of meat by oxidative breakdown of unsaturated fatty acids, leading to the production of peroxides and eventually to rancidity (Wiseman, 1984). As stated at the beginning of this section, the fat of meat animals is a major source of dietary fat for man. The oxidative instability and taste problems associated with products from ruminants fed protected oil seeds is a major factor contributing to the failure of such products to become established for use in low-atherogenic diets. Even when the consumer requires leaner carcasses or modified fat products, the associated changes in colour and softness of the fat may be unacceptable.

12.8.3 Xenobiotic lipids

Xenobiotic (or foreign) compounds are a diverse group of chemicals that include drugs, pesticides, food additives and a variety of environmental pollutants. Many xenobiotic compounds are lipophilic and may partition into the fatty tissue of an animal or plant, DDT being a well-known example (Morgan and Roan, 1974). The metabolism of such compounds is usually orchestrated in such a way as to make them less hydrophobic and therefore easier to excrete. The reactions that accomplish the transformation are classified into phase I, where reactive functional groups may be introduced or exposed, and phase II, where the product of phase I can be conjugated with another group such as an amino acid, sugar, glucuronic acid or sulphate (Caldwell, 1986).

Not all xenobiotic compounds behave so conveniently all the time, however. An increasing number are now known to form xenobiotic lipids as a result of being recognized by an enzyme or set of enzymes involved in the biosynthesis of natural lipids. As a simple working definition, *xenobiotic lipids* result when xenobiotic compounds act as substrates for enzymes working in the direction of lipid biosynthesis. In other words, they are all covalently modified xenobiotic compounds that resemble naturally occurring lipids; simple physical association with a lipid phase does not qualify. This definition also excludes compounds that result from degradation by lipid-metabolizing enzymes (for instance, both ω-phenyl fatty acids and some herbicides are known to be degraded by enzymes of β-oxidation (Cole, 1983)), but includes any esters formed by the action of a so-called reverse lipase. Modification of double-bonded lipids by free radicals also falls outside the definition and is covered elsewhere (see sections 10.2 and 10.3).

A number of general reviews cover the whole topic of xenobiotic lipids (Caldwell and Marsh, 1983; Caldwell, 1985; Fears *et al.*, 1984; Hutson, 1982; Niemann, 1990; Quistad and Hutson, 1986). Further information may be obtained by reference to these reviews; additional references are only included where the topic is not covered adequately in the relevant review.

(a) Xenobiotic fatty acids

The simplest class of xenobiotic lipid occurs when a xenobiotic carboxylic acid is elongated, two carbons at a time, to form a xenobiotic *fatty* acid. There are a few examples in the literature of herbicides being so elongated by plants. The most dramatic example of an acid being elongated by more than two carbons is cyclopropanecarboxylic acid (itself a metabolite of the miticide cycloprate), which can be converted by rat tissues into xenobiotic fatty acids 10, 12 or 14 carbons longer than the precursor. The obvious mechanism for such elongation involves the enzyme fatty acid synthetase (see section 11.1.1), an enzyme that is known to accept a variety of primers. This supposition is borne out, to some extent, by the products of chain elongation. For example, the elongated metabolite of benzoic acid has a hydroxy

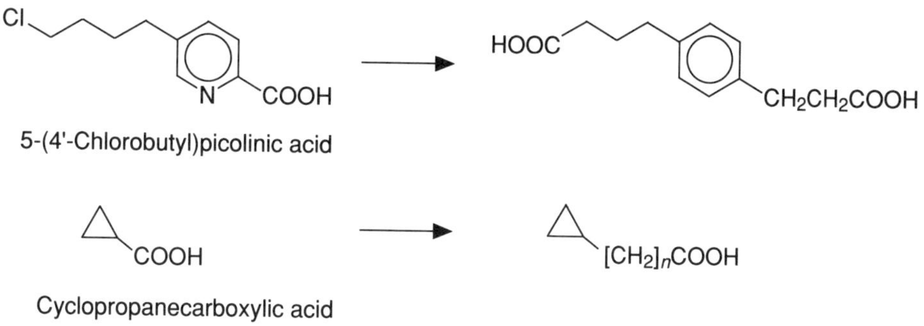

Fig. 12.4 Examples of xenobiotic acids that undergo chain elongation. Chlorobutylpicolinic acid is elongated by two carbons at the 2-carboxylic acid group. The alkyl chlorine also undergoes oxidative metabolism to form the corresponding 5-(3′-carboxypropyl) side chain. Other metabolites are the 2-(2′-unsaturated) acid and the 2-(3′-keto) acid. Cyclopropanecarboxylic acid is elongated by the addition of several two-carbon fragments: $n = 10, 12$, or 14.

group in the 3 (or β) position; 5-(4′-chlorobutyl)-picolinic acid forms enoyl and keto acid derivatives as well as the new saturated acid. All these metabolites represent the predicted intermediates in the fatty acid synthetase pathway. There has been no account of toxicity resulting from xenobiotic fatty acids, and some would appear to undergo normal conjugation, to glycine for instance. Only in the case of the ω-cyclopropyl acids do more complex xenobiotic lipids appear to result. Figure 12.4 shows some of the compounds involved, and further examples can be found in the reviews by Hutson (1982) and Caldwell (1985).

(b) Fatty acid esters of xenobiotic alcohols

It is tempting to refer to this group as 'xenobiotic wax esters' because of their analogy to one class of natural waxes (see section 2.10). In general terms these conjugates occur when the hydroxyl group of a xenobiotic compound, or its phase I metabolite, reacts with a natural fatty acid to form an ester. Probably the simplest examples of the class are the ethyl esters (at least, inasmuch as ethanol may be regarded as xenobiotic), which have been detected in brain tissue from humans (Laposta *et al.*, 1987). There are several other much more elaborate xenobiotic alcohols now known to form esters of this sort: they are recorded in the general reviews on xenobiotic lipids (see above) and in one specialized review (Chang *et al.*, 1986).

Apart from ethanol, there have been reports of fatty acid esters of a number of halogenated pollutants, including 2-chloro-, 2-bromo- and trichloroethanol, pentachlorophenol and 3-chloro-1,2-propanediol (glycerol chloride). Diacyl derivatives of the latter are particularly interesting because of their similarity to natural glycerolipids. Codeine, 7-OH-tetrahydrocannabinol, phencyclidine, dipyridimol, etophenamate, plaunotol, chlorambucil and metabolites of ivermectin and DDT are all widely known and widely used (or abused) compounds that have been shown to form fatty acid esters in animal tissues *in vivo* or *in vitro*. Figure 12.5 shows the structures of a few of the compounds involved. Even the tumour-promoting phorbol diester, 12-tetradecanoylphorbol-13-acetate, can be further acylated at the 20 position (Roeser *et al.*, 1991). This is a phenomenon that is not restricted to animal tissues: the best-documented case occurring in plants is 2-naphthylmethanol (a metabolite of an acaricide), which forms esters with several fatty acids in foliage and fruit from apple trees.

Reports of palmitate esters are more numerous than those of other fatty acids; but, when more exhaustive analysis of metabolites formed *in vivo* or of specificity *in vitro* has been carried out, a wide variety of fatty acid esters have been found. Neither chain length nor degree of unsaturation appears to have much bearing. Many of the fatty acid esters have been detected *in vitro* using a microsomal preparation enriched with coenzyme A, ATP and Mg^{2+} (Chang *et al.*, 1986), strongly suggesting the involvement of an acyl-CoA acyltransferase, possibly acyl-CoA:cholesterol acyltransferase, with the xenobiotic alcohol substituting for the cholesterol. Other reports have shown that cholesterol ester hydrolase (cholesterol esterase) is capable of esterifying the 2-haloethanols (Bhat and Ansari, 1990) and plaunotol. There have been various reports of ethanol acyltransferase activity in the literature (Polokoff and Bell, 1978; Mogelson and Lange, 1984), and these enzymes would also be candidates to esterify at least some of the xenobiotic alcohols.

A number of attempts have been made to determine whether fatty acid esters of xenobiotic alcohols

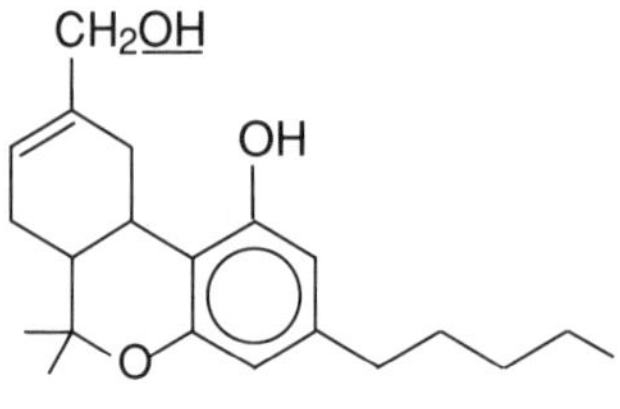

11-Hydroxy-Δ8-tetrahydrocannabinol

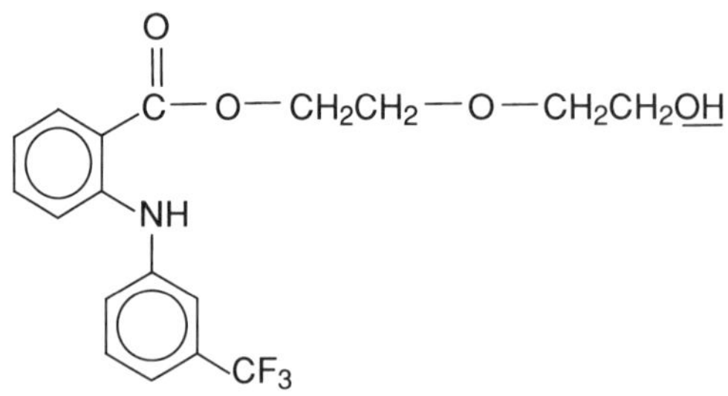

Etofenamate

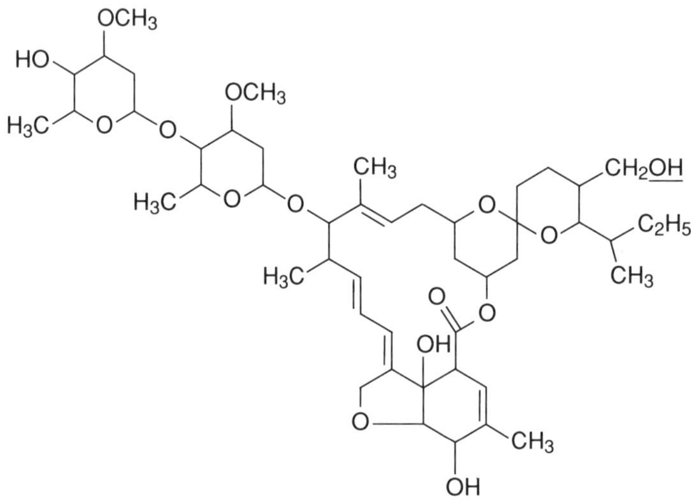

Hydroxy metabolite of ivermectin

Fig. 12.5 Some xenobiotic alcohols that form esters with fatty acids. The alcohol group involved is shown underlined. 11-Hydroxytetrahydrocannabinol forms esters with palmitic, stearic, oleic and linoleic acids. Etofenamate forms esters with all these and palmitoleic, myristic and lauric acids as well. Oleic acid and others of shorter chain length have been found esterified with the ivermectin metabolite.

have any pharmacological or toxic properties over and above those of the parent xenobiotic compound. Fatty acid ethyl esters can be found in membranes, and they have been shown, by fluorescence anisotropy, to disturb the structure of those membranes (Hungund *et al.*, 1988). Palmitoylpentachlorophenol causes selective tissue damage to the exocrine pancreas (Ansari *et al.*, 1987). However, most of the studies conducted suggest that the properties of the conjugate are similar to those of the parent compound (codeine, 7-OH-tetrahydrocannabinol, hydroxy-DDT) or that formation of the conjugate deactivates the parent compound (phorbol diesters). If most of the conjugates produce effects similar to the precursor, they usually do so in a more protracted manner. The results are consistent with the compounds exerting their effects only after they have been released by hydrolysis of the conjugate. The effect of the conjugation is to render the compound more lipophilic, more likely to be stored in adipose tissue and, in consequence, slower to exert its effects and slower to be excreted. It is this tendency to be stored and even concentrated in adipose tissue that causes most general concern amongst toxicologists.

(c) Esters of xenobiotic carboxylic acids

Analogues of natural lipids, mostly cholesterol esters or neutral acylglycerols, are formed when a xenobiotic acid is esterified to a free hydroxy group on a natural acyl acceptor. Most of the published claims that over 38 different xenobiotic carboxylic acids can form such lipids are documented in a specialized review (Dodds, 1991), which also examines the evidence to support the claims. Hypolipidaemic drugs, anti-inflammatory drugs, various pesticides, food additives and some environmental pollutants are all represented in the 38, and a sample of their structures is shown in Fig. 12.6. The lipids they form can be further classified as follows.

(i) Cholesterol esters

Although reports of xenobiotic cholesterol esters are less numerous than those of the following group, there are a variety of structures that are involved. These include substituted benzoic acids, aryl-substituted short-chain acids, analogues of phytanic acid and halogenated acids. The enzymatic mechanism for the synthesis of the cholesterol esters has not been widely studied in spite of a tacit assumption that acyl-CoA:cholesterol acyltransferase may be responsible. In the two cases where studies have been reported, the acyl donor has not been the *S*-acyl coenzyme A derivative but the precursor *O*-acyl ester. For chlorambucil, the precursor was its ester with prednisalone (itself an analogue of cholesterol), and there is evidence that the enzyme involved was lecithin:cholesterol acyltransferase. The acyl donor for the synthesis of the cholesterol ester of 3-(4-chlorophenyl)isovaleric acid is its precursor, fluvalinate. The enzyme responsible for the transfer appears to be highly specific *in vitro* for the acyldonor, if not the acyl acceptor.

The relatively high melting point of a number of xenobiotic cholesterol esters and consequent low fluidity appear to make them poor substrates for enzymes that would otherwise catalyse their hydrolysis. This results in the accumulation of the ester of 1-(4-carboxyphenoxy)-10-(4-chlorophenoxy)decane as its site of synthesis, an occurrence that has been implicated in its mode of action as a hypocholesterolaemic compound. Similarly, the cholesterol ester of 3-(4-chlorophenyl)isovaleric acid forms crystalline structures in a number of tissues. The crystalline structures appear to be the focus for microgranuloma to form and constitute the first 'xenobiotic-lipid storage disease' to be reported. In the light of these findings, the formation of any xenobiotic cholesterol ester must be viewed with concern until such time as its mobilization can be demonstrated.

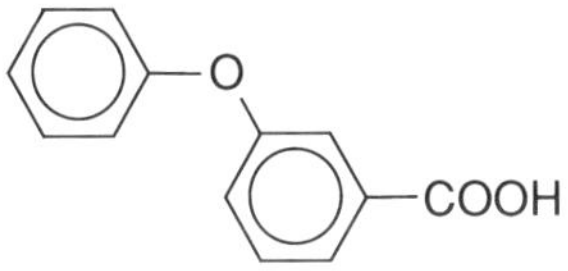

3-Phenoxybenzoic acid

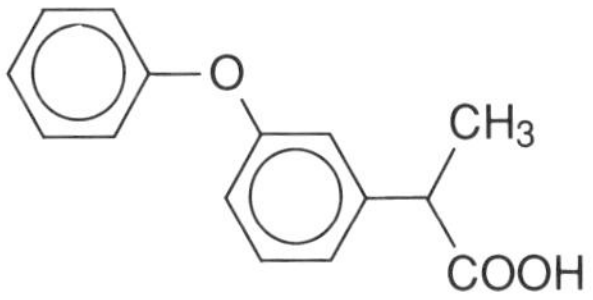

Fenoprofen

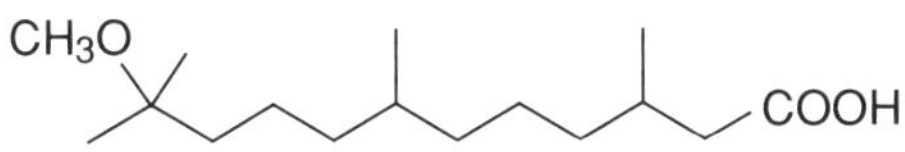

11-Methoxy-3,7,11-trimethyldodecanoic acid

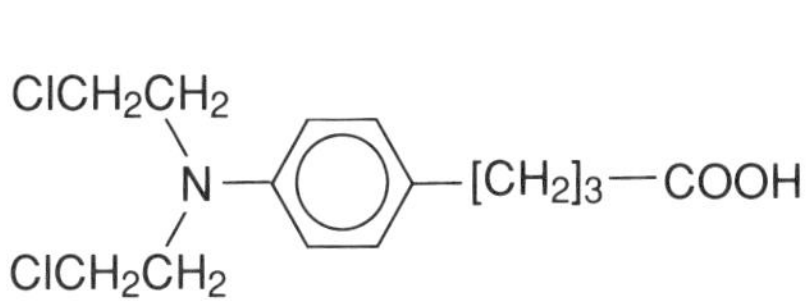

Chlorambucil

Fig. 12.6 Three xenobiotic acids that form xenobiotic lipids: 3-phenoxybenzoic acid (a metabolite of pyrethroid insecticides) is incorporated into di- and triacylglycerol analogues; fenoprofen (a non-steroidal anti-inflammatory drug) also forms triacylglycerol analogues; and 11-methoxy-3,7,11-trimethyldodecanoic acid, a metabolite of the insect growth regulator, methoprene, is incorporated into di- and triacylglycerols as well as cholesterol ester analogues. Chlorambucil is a cytotoxic drug that forms cholesterol esters. Further examples are shown by Dodds (1991).

(ii) Neutral acylglycerols

Reports of xenobiotic acids being incorporated into analogues of triacylglycerol or diacylglycerol easily outnumber those of the other classes. The same general types of xenobiotic acids as encountered amongst the xenobiotic cholesterol esters are found in this group. One of the most interesting groups is the 2-substituted propanoic acids, which includes both the 'profen' non-steroidal anti-inflammatory drugs and the 'fop' herbicides (section 12.8.1(f)). Here, again, the reports derive from a number of different experimental approaches, including analysis of feral tissues, dosing of animals *in vivo*, incubation of whole cells with the acid *in vitro*, and enzyme assays using subcellular preparations.

There is still a shortage of information on the mechanism of synthesis of the xenobiotic acylglycerols. There are a variety of acyltransferase enzymes that are candidates, but, surprisingly, good evidence only for the involvement of monoacylglycerol acyltransferase has been forthcoming. When the acid was 3-phenoxybenzoic acid (3PBA), the activity was shown to require the prior formation of the CoA ester, to be specific for 2- (as opposed to 1-) monoacylglycerol as acyl acceptor, and to show a similar distribution amongst tissues and species as the activity measured with a natural precursor – including the very large increase in hepatic activity found during the suckling period in rats. There is also published evidence that diacylglycerol acyltransferase can be active with some xenobiotic acyl-CoAs or with some xenobiotic diacylglycerols; taken with the established evidence for enzymatic synthesis of xenobiotic acyl-CoA, the entire monoacylglycerol pathway would appear to be operational with at least some xenobiotic acids. Although there have been no reports that other acyltransferases are involved in the synthesis of xenobiotic triacylglycerols, it seems probable that some of the others are involved in at least some cases. For instance, dodecylthioacetic acid is incorporated into both triacylglycerols and phospholipids at rates comparable to its natural counterpart (Skrede *et al.*, 1989), and it would appear likely that glycerol 3-phosphate acyltransferase is the enzyme responsible, especially when, as here, the xenobiotic acid bears a strong structural resemblance to the natural acid. Furthermore, there have been reports of xenobiotic acids occurring in the *sn*-2 position of the triacylglycerol, a position normally accessible to neither monoacylglycerol or diacylglycerol acyltransferase. The involvement here, as elsewhere, of lipolytic or other hydrolytic enzymes working 'in reverse' cannot be ruled out.

Further metabolism of triacylglycerol is normally expected to result in their secretion as lipoprotein or their degradation by lipolytic enzymes. Certainly, there are two reports of xenobiotic triacylglycerols occurring in chylomicrons, and there is circumstantial evidence that 4-benzyloxybenzoic acid is transferred as a triacylglycerol component of VLDL from liver to adipose tissue of rat. The reports of attempts to measure lipolysis of xenobiotic acylglycerols all suggest that the enzymes (both lipoprotein lipase and pancreatic lipase have been used) are considerably less active with xenobiotic substrates than with, say, trioleoylglycerol. The ester bond between the xenobiotic acid and glycerol appears to be particularly resistant: lipolysis of the xenobiotic triacylglycerol may result in a corresponding xenobiotic partial glyceride by the removal of the non-xenobiotic fatty acids.

There does not appear to be any evidence of toxic effects mediated by xenobiotic acylglycerols but speculation that xenobiotic diacylglycerols may be able to mimic the second messenger activities of the natural form were borne out when it was shown that *sn*-1-acyl-2-(3PBA)glycerol was able to stimulate protein kinase C. The further demonstration that the compound was resistant to degradation by lipase suggested a potential to act like the tumour-promoting phorbol diesters – a potential that has yet to be tested in a cell-transformation assay.

On the other hand, it is noticeable that a large number of the compounds involved are either hypolipidaemic agents or have hypolipidaemic properties in addition to their intended properties (e.g. the 'profen' anti-inflammatory drugs and 3PBA). Fears (1985) has shown a direct correlation between the ability of the profens to act as hypolipidaemic agents and their activity as substrates for xenobiotic triacylglycerol synthesis. It was suggested that both effects are mediated by the formation of the corresponding acyl-CoA.

(iii) Glycerophospholipids

There are reports of at least seven xenobiotic acids being incorporated into glycerophospholipids, mostly phosphatidylcholine and phosphatidylethanolamine. All the reports, which are substantiated by reasonable experimental evidence, refer to acids that are close analogues of natural acids. These findings are consistent with the belief that the structures of membrane phospholipids are more tightly controlled than those of the storage lipids.

(iv) Other lipids

ω-Cyclohexyl fatty acids and 2-fluoropalmitate have been shown to form sphingomyelin analogues, and the latter can also be incorporated into ceramide and various glycosphingolipids. These examples differ from the others in the class by being *N*-acyl, rather than *O*-acyl, compounds (see sections 2.3.2 and 2.4).

(d) Xenobiotic modifications to the polar groups of phospholipids

Phospholipids in which the polar headgroup has been replaced with, or modified by, a xenobiotic compound represent a fourth classification of xenobiotic lipids. It remains a classification, the only firmly established member of which results from the *N*-acylation of phosphatidylethanolamine by monochloroacetic acid in tissues of the rat (Bhat *et al.*, 1990). *N*-Acyl esters of phosphatidylethanolamine are found in a number of seeds (see section 2.3) but are rare in animal tissues. Nevertheless, the reaction with chloroacetic acid can be demonstrated in rat liver microsomes and requires coenzyme A and ATP for full activity.

(e) Other related conjugates

Coenzyme A esters, as well as being the substrates for many reactions in lipid biochemistry, are also intermediates in the metabolism of numerous xenobiotic acids, resulting in the formation of amino acid conjugates, for instance. Several enzymes have been implicated in the biosynthesis of xenobiotic acyl-CoA and their role is reviewed elsewhere (Huckle and Millburn, 1982). It is of interest that many xenobiotic acyl-CoAs appear to inhibit fatty acid synthesis, and that this property may be implicated in their action as hypolipidaemic agents or as herbicides (Kemal and Casida, 1992).

Carnitine esters of three small aliphatic acids, including cyclopropanecarboxylic acid, and benzoic acid have been reported in the tissues and urine of a number of species. Although the conjugate is water-soluble, cyclopropanoylcarnitine was stored in considerable quantities in the muscles of dogs; furthermore, a link between its formation and lipid deposition in the heart muscle, mediated by carnitine depletion, has been suggested (Quistad *et al.*, 1986). The same group also detected that the anilino acid of fluvalinate can be conjugated to bile acids and bile salts. These conjugates appear to be excreted with bile and have not caused much concern; however, little is known about their effects on enterohepatic circulation (Quistad *et al.*, 1982).

(f) Concluding comments

Although the compounds known to form xenobiotic lipids of some description probably still number less than 100, one has only to look through a chemical catalogue or the Merck Index to find hundreds more likely candidates. Until more work is done with a wide variety of potential xenobiotic precursors, it will not be possible to infer the rules that will allow us to predict which xenobiotic compounds will be incorporated into which xenobiotic lipids and in what quantities.

Similarly, evidence that xenobiotic lipids mediate toxic effects is rare, and the conversion of the active compound to an inert lipid may be considered a benign or even a protective event. However, it is this very inertia, leading to persistence, storage and concentration of the compound, that causes most general concern. The potential for subsequent release of the concentrated compound under the action of a lipolytic hormone – in a sort of 'flashback' – is one that does not appear to have been investigated. In addition, there are a growing number of cases where the formation of a xenobiotic lipid has physiological or toxic consequences. A short review of the potential toxic implications of xenobiotic lipid formation has appeared (Hutson *et al.*, 1985). Further investigations into the mechanism of synthesis, the rate of degradation and the physiological implications of xenobiotic lipids are required before we can decide whether they are an important 'new' route of metabolism or merely an interesting but inconsequential phenomenon.

References and further reading

Abulnaja, K.O. and Harwood, J.L. (1991a) *Phytochemistry*, **30**, 2883.
Abulnaja, K.O. and Harwood, J.L. (1991b) *Z. Naturforsch.*, **46c**, 931.
Abulnaja, K.O., Tighe, C.R. and Harwood, J.L. (1992) *Phytochemistry*, **31**, 1155.
Ansari, G.A.S., Kaphalia, B.S. and Boor, P.J. (1987) *Toxicology*, **46**, 57.
Appelqvist, L. (1975) in *Recent Advances in the Chemistry and Biochemistry of Plant Lipids* (eds T. Galliard and E.I. Mercer), Academic Press, London, p. 287.
Ashton, I.P., Abulnaja, K.O., Pallett, K.E., Cole, D.E. and Harwood, J.L. (1992) *Pesticide Biochem. Physiol.*, **43**, 14.
Astrup, H.N. and Nedkvitne, J.J. (1975) *Acta Agric. Scand.*, **25**, 49.
Barry, P. and Pallett, K.E. (1988) *Proc. EWRS Symp.*, 51.
Barrett, P.B. and Harwood, J.L. (1993) *Biochem. Soc. Trans.*, **21**, 184S.
Bartov, I., Lipstein, B. and Bornstein, S. (1974) *Poultry Sci.*, **53**, 115.
Bhat, H.K. and Ansari, G.A.S. (1990) *Chem. Res. Toxicol.*, **3**, 311.
Bhat, H.K., Ahmed, A.E. and Ansari, G.A.S. (1990) *Toxicology*, **63**, 35.
Bramley, P.M. (1991) in *Target Sites for Herbicide Action* (ed. R.C. Kirkwood), Plenum, New York, p. 95.
Brooks, C.C. (1971) *J. Animal Sci.*, **33**, 1224.
Browse, J. and Somerville, C.R. (1991) *Annu. Rev. Plant Physiol.*, **42**, 467.
Buisson, D.H., Body, D.R., Dougherty, G.J., Eyres, L. and Vlieg, P. (1982) *J. Am. Oil Chem. Soc.*, **59**, 390.
Burton, J.D., Gronwald, J.W., Keith, R.A., Somers, D.A., Gengenbach, B.G. and Wyse, D.L. (1991) *Pesticide Biochem. Physiol.*, **39**, 100.
Caldwell, J. (1985) *Biochem. Soc. Trans.*, **13**, 852.
Caldwell, J. (1986) *Xenobiotic Conjugation Chemistry, ACS Symp. Ser.* 299 (eds G.D. Paulson, J. Caldwell, D.H. Hutson and J.J. Menn), American Chemical Society, Washington, DC, p. 2.
Caldwell, J. and Marsh, M.V. (1983) *Biochem. Pharmacol.*, **32**, 1667.
Chang, M.J.W., Leighty, E.G., Haggerty, G.C. and

Fentiman, A.F.J. (1986) *Xenobiotic Conjugation Chemistry, ACS Symp. Ser.* **299** (eds G.D. Paulson, J. Caldwell, D.H. Hutson and J.J. Menn), American Chemical Society, Washington, DC, p. 214.

Cherif, A., Ben Miled-Daoud, D., Marzouk, B., Smaoui, A. and Zarrouk, M. (eds) (1992) *Metabolism, Structure and Utilization of Plant Lipids*, Centre National Pedagogique, Tunis.

Christie, W.W. (1983) in *Developments in Dairy Chemistry*, 2: Lipids (ed. P.F. Fox), Applied Science, London, p. 1.

Christie, W.W. and Moore, J.H. (1969) *Lipids*, **4**, 345.

Cole, D. (1983) *Progress in Pesticide Biochemistry*, vol. 3 (eds D.H. Hutson and T.R. Roberts), Wiley, Chichester, p. 199.

Cook, L.J., Scott, T.W., Faichney, G.J. and Davies, H.L. (1972) *Lipids*, **7**, 83.

Crouse, J.D., Kemp, J.D., Fox, J.D., Ely, D.G. and Moody, W.G. (1972) *J. Animal Sci.*, **34**, 388.

Davies, A.O. and Harwood, J.L. (1983) *J. Exp. Bot.*, **34**, 1089.

Dodds, P.F. (1991) *Life Sci.*, **49**, 629.

Dormann, P., Schmid, P.C., Robers, M. and Spener, F. (1991) *Hoppe-Seyler's Z. Biol. Chem.*, **372**, 528.

Duncan, D.N., Meggitt, W.F. and Penner, D. (1982) *Weed Sci.*, **30**, 195.

Fears, R. (1985) *Prog. Lipid Res.*, **24**, 177.

Fears, R., Rush, W.R., Walker, P. and Ferres, H. (1984) *Biochem. Pharmacol.*, **33**, 209.

Fedke, C. (1982) *Biochemistry and Physiology of Herbicide Action*, Springer-Verlag, Heidelberg.

Forss, D.A. (1972) *Prog. Chem. Fats Other Lipids*, **13**, 181.

Fuerst, E.P. (1987) *Weed Technol.*, **1**, 270.

Garton, G.A., Hovell, F.D.D. and Duncan, W.R.H. (1972) *Br. J. Nutr.*, **28**, 409.

Gurr, M.I. (1980) in *Biochemistry of Plants*, vol. 4 (eds P.K. Stumpf and E.E. Conn), Academic Press, New York, p. 205.

Gurr, M.I. (1983) in *Developments in Dairy Chemistry 2: Lipids* (ed. P.F. Fox), Applied Science, London, p. 365.

Harwood, J.L. (1984) in *Structure, Function and Metabolism of Plant Lipids* (eds P.-A. Siegenthaler and W. Eichenberg), Elsevier, Amsterdam, p. 543.

Harwood, J.L. (1991a) in *Topics in Photosynthesis*, vol. 10 (eds N.R. Baker and M.P. Percival), Elsevier, Amsterdam, pp. 209–46.

Harwood, J.L. (1991b) in *Target Sites for Herbicide Action* (ed. R.C. Kirkwood), Plenum, New York, p. 57.

Harwood, J.L., Abulnaja, A.O., Barrett, P.B., Herbert, D., Rutter, A.J. and Williams, M. (1992) in *Metabolism, Structure and Utilization of Plant Lipids* (eds A. Cherif, D.B. Miled-Daoud, B. Marzouk, A. Smaoui and M. Zarrouk), Centre National Pedagogique, Tunis, p. 319.

Heise, K.-P., Fuhrmann, J. and Johnen, T. (1991) *Hoppe-Seyler's Z. Biol. Chem.*, **372**, 532.

Henderson, R.J., MacKinlay, E.E., Hodgson, P. and Harwood, J.L. (1990) *J. Exp. Bot.*, **41**, 729.

Hilditch, T.P. and Williams, P.N. (1964) *The Chemical Constitution of Natural Fats*, Chapman & Hall, London.

Hilton, J.L., St John, J.B., Christiansen, M.N. and Norris, K.H. (1971) *Plant Physiol.*, **48**, 171.

Hitchcock, C. and Nichols, B.W. (1971) *Plant Lipid Biochemistry*, Academic Press, New York.

Hoppe, H.H. (1989) in *Target Sites for Herbicide Action* (eds P. Boger and G. Sandmann), CRC Press, Boca Raton, FL, pp. 65–83.

Huckle, K.R. and Millburn, P. (1982) *Progress in Pesticide Biochemistry*, vol. 2 (ed. D.H. Hutson and T.R. Roberts), Wiley, Chichester, p. 127.

Hungund, B.L., Goldstein, D.B., Villegas, F. and Cooper, T.B. (1988) *Biochem. Pharmacol.*, **37**, 3001.

Hutson, D.H. (1982) *Progress in Pesticide Biochemistry*, vol. 2 (eds D.H. Hutson and T.R. Roberts), Wiley, Chichester, p. 171.

Hutson, D.H., Dodds, P.F. and Logan, C.J. (1985) *Biochem. Soc. Trans.*, **13**, 854.

Kamel, B.S. and Kakuda, Y. (eds) *Technological Advances in Improved and Alternative Sources of Lipids*, Blackie, London.

Kemal, C. and Casida, J.E. (1992) *Life Sci.*, **50**, 533.

Knutzon, D.S., Thompson, G.A., Radke, S.E., Johnson, W.B., Knauf, V.C. and Kridl, J.C. (1992) *Proc. Natl Acad. Sci. USA*, **89**, 2624.

Laposta, E.A., Scherrer, D.E., Mazow, C. and Lange, L.G. (1987) *J. Biol. Chem.*, **262**, 4653.

Lichtenthaler, H.K., Kobek, K. and Focke, M. (1989) *Proc. Brighton Crop Prot. Conf. – Weeds*, p. 173.

McDonald, I.W. and Scott, T.W. (1977) *World Rev. Nutr. Dietet.*, **26**, 144.

Mogelson, S. and Lange, L.G. (1984) *Biochemistry*, **23**, 4075.

Morgan, D.P. and Roan, C.C. (1974) *Essays in Toxicology*, vol. 5 (ed. W.J. Hayes), Academic Press, New York, p. 39.

Murphy, D.J. (ed.) (1993) *Designer Oil Crops*, VCH, Weinheim.

Murphy, D.J., Harwood, J.L., Lee, K.A., Roberto, F., Stumpf, P.K. and St John, J.B. (1985) *Phytochemistry*, **24**, 1923.

Niemann, R. (1990) *Exp. Pathol.*, **39**, 213.

Norman, H.A. and St John, J.B. (1987) *Plant Physiol.*, **85**, 684.

Ohlrogge, J.B., Browse, J. and Somerville, C.R. (1991) *Biochim. Biophys. Acta*, **1082**, 1.

Oldham, J.D. and Sutton, J.D. (1979) in *Feeding Strategy for the High Yielding Dairy Cow* (eds W.H. Broster and H. Swann), Granada, London, p. 114.

Polokoff, M.A. and Bell, R.M. (1978) *J. Biol. Chem.*, **253**, 7173.

Quinn, P.J. and Harwood, J.L. (eds) (1990) *Plant Lipid Biochemistry, Structure and Utilization*, Portland, London.

Quistad, G.B. and Hutson, D.H. (1986) *Xenobiotic Conjugation Chemistry, ACS Symp. Ser.* **299** (eds G.D. Paulson, J. Caldwell, D.H. Hutson and J.J. Menn), American Chemical Society, Washington, DC, p. 204.

Quistad, G.B., Staiger, L.E. and Schooley, D.A. (1982) *Nature*, **296**, 462.

Quistad, G.B., Staiger, L.E. and Schooley, D.A. (1986) *Drug Metab. Disp.*, **14**, 521.

Rendina, A.R., Beaudoin, J.D., Craig-Kennard, A.C. and Breen, M.K. (1989) *Proc. Brighton Crop Prot. Conf. – Weeds*, p. 163.

Roeser, H., Doege, T. and Hecker, E. (1991) *Carcinogen*, **12**, 1563.

Skrede, S., Narce, M., Berseth, S. and Bremer, J. (1989) *Biochim. Biophys. Acta*, **1005**, 296.

Slabas, A.R. and Fawcett, T. (1992) *Plant Mol. Biol.*, **19**, 161.

St John, J.B. (1982) in *Biochemical Responses Induced by Herbicides* (eds D.E. Moreland, J.B. St John and F.D. Hess), American Chemical Society, Washington, DC, p. 97.

Vergroesen, A.J. (ed.) (1975) *The Role of Fats in Human Nutrition*, Academic Press, London.

Voelker, T.A., Worelli, A.C., Radke, S.E., Hawkins, D.J., Dimento, J. and Davies, H.M. (1992) in *Metabolism, Structure and Utilization of Plant Lipids* (eds A. Cherif, D. Ben Miled-Daoud, B. Marzouk, A. Smaoui and M. Zarrouk), Centre National Pedagogique, Tunis.

Walker, K.A., Ridley, S.M., Lewis, T. and Harwood, J.L. (1989) *Rev. Weed Sci.*, **4**, 71.

Walker, K.A., Ridley, S.M., Lewis, T. and Harwood, J.L. (1990) *Phytochemistry*, **29**, 3743.

Wiseman, J. (ed.) (1984) *Fats in Animal Nutrition*, Butterworths, London.

Wood, J.D. and Lister, D. (1973) *J. Sci. Food Agric.*, **24**, 1449.

Yaron, A., Samoiloff, V. and Benzioni, A. (1982) *Lipids*, **17**, 169.

INDEX

Index

DICTIONARY SECTION

Entries taken from the Chapman & Hall Chemical Database

A

Acanthacerebroside A **A-00001**

1-Glucopyranosyl-2-(2-hydroxytetracosanoylamino)-1,3,4-hexadecanetriol

[110744-71-3]

$R = (CH_2)_{21}CH_3$, $R' = (CH_2)_{11}CH_3$

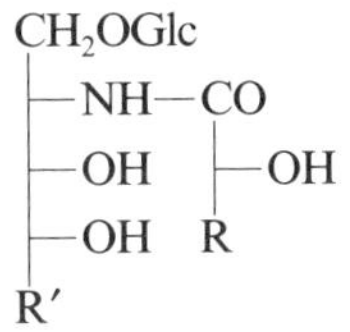

$C_{46}H_{91}NO_{10}$ M 818.226

Cerebroside from the starfish *Acanthaster planci*. Needles + $3H_2O$ (MeOH). Mp 209-210°. $[\alpha]_D$ +2.4° (c, 0.81 in propanol).

Kawano, Y. *et al*, *Justus Liebigs Ann. Chem.*, 1988, 19 (*isol, pmr, cmr, struct*)

Sugiyama, S. *et al*, *Justus Liebigs Ann. Chem.*, 1988, 619; 1990, 1063 (*abs config, synth*)

Acanthacerebroside B **A-00002**

1-Glucopyranosyl-2-(2-hydroxyhexadecanoylamino)-1,3,4-docosanetriol

[110744-72-4]

As Acanthacerebroside A, A-00001 with

$R = -(CH_2)_{13}CH_3$, $R' = (CH_2)_{17}CH_3$

$C_{44}H_{87}NO_{10}$ M 790.172

Cerebroside from the starfish *Acanthaster planci*. Needles + $1H_2O$ (MeOH). Mp 218-219°. $[\alpha]_D$ +7.8° (c, 0.24 in propanol).

Kawano, Y. *et al*, *Justus Liebigs Ann. Chem.*, 1988, 19 (*isol, pmr, cmr, struct*)

Sugiyama, S. *et al*, *Justus Liebigs Ann. Chem.*, 1988, 619 (*abs config*)

Acanthacerebroside C **A-00003**

1-Glucopyranosyl-2-(2-hydroxyhexadecanoylamino)-13-docosene-1,3,4-triol

[110744-73-5]

As Acanthacerebroside A, A-00001 with

$R = (CH_2)_{13}CH_3$, $R' = -(CH_2)_8CH{=}CH(CH_2)_7CH_3$ (*Z*-)

$C_{44}H_{85}NO_{10}$ M 788.156

Cerebroside from the starfish *Acanthaster planci*. Needles (MeOH). Mp 203-204°. $[\alpha]_D$ +18.3° (c, 1.24 in propanol).

Kawano, Y. *et al*, *Justus Liebigs Ann. Chem.*, 1988, 19 (*isol, pmr, cmr, struct*)

Sugiyama, S. *et al*, *Justus Liebigs Ann. Chem.*, 1988, 619 (*abs config*)

Acanthacerebroside D **A-00004**

[110744-74-6]

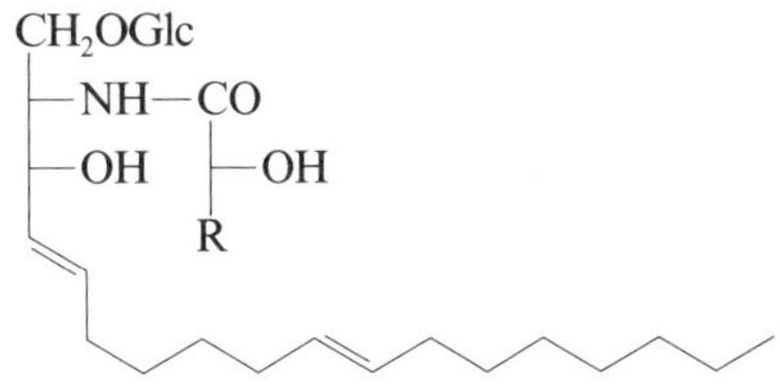

$R = -(CH_2)_{19}CH_3$

$C_{46}H_{87}NO_9$ M 798.195

Cerebroside from the starfish *Acanthaster planci*. Needles + $2H_2O$ (MeOH). Mp 198-199°. $[\alpha]_D$ +1.1° (c, 0.3 in propanol).

Kawano, Y. *et al*, *Justus Liebigs Ann. Chem.*, 1988, 19 (*isol, pmr, cmr, struct*)

Sugiyama, S. *et al*, *Justus Liebigs Ann. Chem.*, 1988, 619 (*abs config*)

Acanthacerebroside E **A-00005**

[110744-75-7]

As Acanthacerebroside D, A-00004 with

$R = -(CH_2)_{20}CH_3$

$C_{47}H_{89}NO_9$ M 812.221

Cerebroside from the starfish *Acanthaster planci*. Needles + $3H_2O$ (MeOH). Mp 194-195°. $[\alpha]_D$ +1.0° (c, 0.3 in propanol).

Kawano, Y. *et al*, *Justus Liebigs Ann. Chem.*, 1988, 19 (*isol, pmr, cmr, struct*)

Sugiyama, S. *et al*, *Justus Liebigs Ann. Chem.*, 1988, 619 (*abs config*)

Acanthacerebroside F **A-00006**

[110744-76-8]

As Acanthacerebroside D, A-00004 with

$R = (CH_2)_{21}CH_3$

$C_{48}H_{91}NO_9$ M 826.248

Cerebroside from the starfish *Acanthaster planci*. Needles + $2H_2O$ (MeOH). Mp 193-194°. $[\alpha]_D$ +1.2° (c, 0.5 in propanol).

Kawano, Y. *et al*, *Justus Liebigs Ann. Chem.*, 1988, 19 (*isol, pmr, cmr, struct*)

Sugiyama, S. *et al*, *Justus Liebigs Ann. Chem.*, 1988, 619 (*abs config*)

Acaranoic acid **A-00007**

*6-Decyltetrahydro-3-methyl-2-oxo-2*H*-pyran-4-carboxylic acid, 9CI*

[20247-99-8]

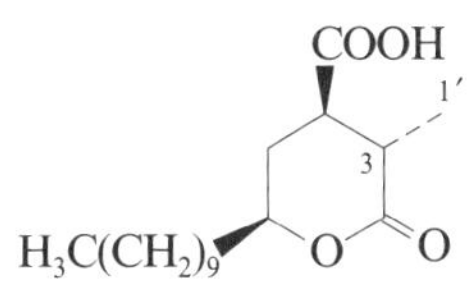

$C_{17}H_{30}O_4$ M 298.422
Constit. of the lichen *Acarospora chlorophana*. Cryst. (Et_2O). Mp 154-155°. $[\alpha]_D^{25}$ −30° (c, 0.29 in $CHCl_3$). Pleopsidic acid was a mixt. of Acaranoic and Acarenoic acids.

Me ester: Needles. Mp 58-59°. $[\alpha]_D^{25}$ −31.1° (c, 1.595 in $CDCl_3$).

1′,3-Didehydro: [20247-98-7]. ***Acarenoic acid.*** *6-Decyltetrahydro-3-methylene-2-oxo-2*H*-pyran-4-carboxylic acid, 9CI*
$C_{17}H_{28}O_4$ M 296.406
Isol. from *A. chlorophana*. Cryst. (C_6H_6). Mp 144-146.5° (132°). $[\alpha]_D^{24}$ −84.3° (c, 0.265 in $CHCl_3$).

Santesson, J., *Acta Chem. Scand.*, 1967, **21**, 1993 (*isol, struct, ir*)
Huneck, S. *et al*, *Phytochemistry*, 1980, **19**, 2713 (*struct*)

6-Acetoxy-7-chloro-3,9,12-pentadecatrien-1-yne A-00008

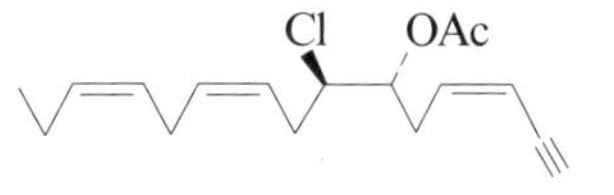

$C_{17}H_{23}ClO_2$ M 294.820
(3Z,6R,7R,9Z,12Z)-form
Metab. of *Laurencia pinnatifida*. Oil. $[\alpha]_D^{25}$ +4.54° (c, 11.0 in $CHCl_3$).

Gonzalez, A.G. *et al*, *Tetrahedron*, 1982, **38**, 1009.

4-Acetoxy-3-hexadecyldihydro-5-methyl-2(3*H*)-furanone A-00009

3-Acetoxy-2-hexadecyl-4-pentanolide

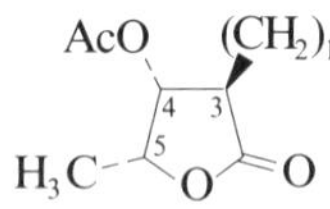

(3*R**,4*R**,5*R**)-*form*

$C_{23}H_{42}O_4$ M 382.582
(3R*,4R*,5R*)-form [81601-89-0]
Constit. of the gorgonian *Plexaura flava*. Oil. $[\alpha]_D$ +31.9° (c, 1.3 in CH_2Cl_2).
(3R*,4S*,5R*)-form [81601-90-3]
Constit. of *P. flava*. $[\alpha]_D^{20}$ −32.5° (c, 1.6 in CH_2Cl_2).

Ravi, B.N. *et al*, *Aust. J. Chem.*, 1982, **35**, 105 (*isol, ir, pmr, ms*)

4-Acetoxy-3-tetradecyl-4,5-dihydro-5-methyl-2(3*H*)-furanone A-00010

3-Acetoxy-2-tetradecyl-4-pentanolide

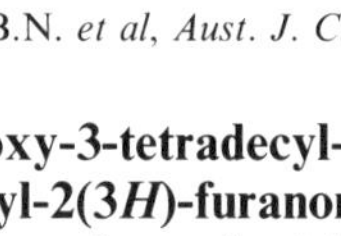

(3*R**,4*R**,5*R**)-*form*

$C_{21}H_{38}O_4$ M 354.529
(3R*,4R*,5R*)-form [81576-10-5]
Constit. of the gorgonian *Plexaura flava*. Oil. $[\alpha]_D^{20}$ +36.1° (c, 1.9 in CH_2Cl_2).
(3R*,4S*,5R*)-form [81576-11-6]
Constit. of *P. flava*. Powder. $[\alpha]_D^{20}$ −36.6° (c, 1.5 in CH_2Cl_2).

Ravi, B.N. *et al*, *Aust. J. Chem.*, 1982, **35**, 105 (*isol, ir, pmr, ms*)

N-Acetylneuraminic acid A-00011

*5-(Acetylamino)-3,5-dideoxy-D-*glycero-*D-*galacto-*2-nonulosonic acid, 9CI.* O-*Sialic acid. NANA. Lactaminic acid. Gynaminic acid. Serolactaminic acid*
[131-48-6]

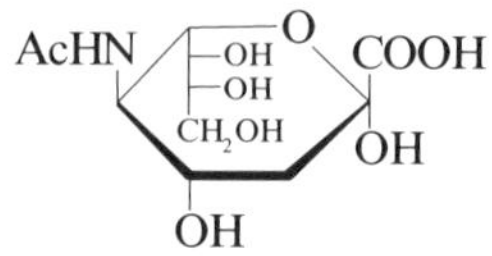

α-D-Pyranose-*form*

$C_{11}H_{19}NO_9$ M 309.272
Isol. from milk, clostrum, submaxillary mucin and meconium by acid or enzymic hydrol. of the constit. sialoproteins and oligosaccharides. Edible bird's nest is also a source. Present in urine of patients with sialuria. Mp 185-187° dec. $[\alpha]_D^{22}$ −32° (H_2O). The only sialic acid formed in human tissues.

4-Ac:
$C_{13}H_{21}NO_{10}$ M 351.310
Mp 199-200°. $[\alpha]_D$ −62° (H_2O).
7-Ac: B-*Sialic acid*
$C_{13}H_{21}NO_{10}$ M 351.310
Mp 138-140°. $[\alpha]_D$ +6°.
4,7,8,9-Tetra-Ac, Me ester:
$C_{20}H_{29}NO_{13}$ M 491.448
Mp 174-175°. $[\alpha]_D^{20}$ −4.8° → −5.2° (c, 0.5 in $CHCl_3$).
*2,4,7,8,9-Penta-Ac, Me ester: Methyl 5-*N*-acetyl-2,4,7,8,9-penta-*O*-acetylneuraminate*
$C_{22}H_{31}NO_{14}$ M 533.485
Mp 156-157°. $[\alpha]_D^{20}$ −3.3° (c, 1.0 in $CHCl_3$).
4,7,8,9-Tetrabenzoyl, Me ester:
$C_{40}H_{37}NO_{13}$ M 739.731
$[\alpha]_D^{20}$ −0.4° (c, 1.0 in $CHCl_3$).

α-D-Pyranose-form
Benzyl glycoside:
$C_{18}H_{25}NO_9$ M 399.397
Mp 194-195°. $[\alpha]_D^{25}$ −7.5° (MeOH).
Benzyl glycoside, Me ester:
$C_{19}H_{27}NO_9$ M 413.424
Mp 85-89°. $[\alpha]_D^{25}$ −3.5° (MeOH).
Me glycoside:
$C_{12}H_{21}NO_9$ M 323.299
Mp 185-188° dec. $[\alpha]_D^{25}$ −16.3° (MeOH).
Me glycoside, Me ester: Mp 166-168° dec. $[\alpha]_D^{25}$ −5.2° (MeOH).
Allyl glycoside:
$C_{14}H_{23}NO_9$ M 349.337
Mp 167-169° dec. $[\alpha]_D$ −14° (MeOH).
Allyl glycoside, Me ester:
$C_{15}H_{25}NO_9$ M 363.364
Mp 154-156°. $[\alpha]_D$ −13° (MeOH).

β-D-Pyranose-form
Me ester: Methyl N*-acetyl-β-D-neuraminate*
$C_{12}H_{21}NO_9$ M 323.299
Prisms + $1H_2O$ (MeOH). Mp 180-182° dec. $[\alpha]_D^{20}$ −28° (c, 1.0 in H_2O).
Me glycoside, Me ester: Mp 115-130°. $[\alpha]_D^{20}$ −46° (c, 0.67 in MeOH).
Me glycoside, Me ester, 4,7,8,9-tetra-Ac: Mp 134-135°. $[\alpha]_D^{20}$ −19.9° (c, 1.0 in $CHCl_3$).

Gottschalk, A., *The Chemistry and Biology of Sialic Acids and Related Substances*, Cambridge Univ. Press, London, 1960 (*rev*)
Blix, G. *et al*, *Methods Carbohydr. Chem.*, 1962, **1**, 246 (*isol*)
Kuhn, R. *et al*, *Chem. Ber.*, 1966, **99**, 611 (*synth*)
O'Connell, A.M., *Acta Crystallogr., Sect. B*, 1973, **29**, 2320 (*cryst struct*)

Codington, J.F. *et al, Methods Carbohydr. Chem.*, 1976, **7**, 226 (*rev*)
Martin, J.E. *et al, Carbohydr. Res.*, 1977, **56**, 432 (*isol*)
Jaques, L.W. *et al, J. Biol. Chem.*, 1977, **252**, 4533 (*cmr*)
Benzing-Nguyen, L. *et al, J. Org. Chem.*, 1978, **43**, 554 (*synth*)
Friebolin, H. *et al, Angew. Chem., Int. Ed. Engl.*, 1980, **19**, 208.
Eschenfelder, V. *et al, Carbohydr. Res.*, 1980, **78**, 190 (*synth*)
Van der Vleugel, D.J.M. *et al, Carbohydr. Res.*, 1982, **102**, 121 (*Me ester, Me glycoside*)
Juliana, R. *et al, Carbohydr. Res.*, 1987, **164**, 415 (*Me ester, synth*)
Csuk, R. *et al, Helv. Chim. Acta*, 1988, **71**, 609 (*synth*)
Kragl, U. *et al, Angew. Chem., Int. Ed. Engl.*, 1991, **30**, 827 (*enzymic synth*)

Aflatoxicol A-00012

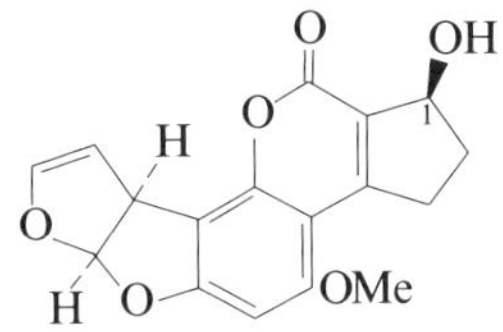

(1*S*)-*form*

$C_{17}H_{14}O_6$ M 314.294

Prod. from biological redn. of aflatoxin B_2.

(*1S*)-*form* [29611-03-8]

Aflatoxicol A. Aflatoxicol R_0

Cryst. (C_6H_6/hexane). Mp 224-226°.

▷ GY1934000.

Et ether: Cryst. (EtOH aq.). Mp 198-200°.

(*1R*)-*form* [61740-00-9]

Aflatoxicol B. Aflatoxicol $R_{0'}$

Cryst. (C_6H_6/hexane). Dec. over broad range starting at 233°.

Et ether: Cryst. (EtOH aq.). Mp 194-196°.

Cole, R.J. *et al, J. Agric. Food Chem.*, 1972, **20**, 1100 (*synth*)
Hisa, M.T. *et al, Experientia*, 1977, **33**, 1132 (*synth*)
Pawlowski, N.E. *et al, J. Agric. Food Chem.*, 1977, **25**, 437 (*synth, uv*)
Newton, M.G. *et al, J. Agric. Food Chem.*, 1979, **27**, 1339 (*stereochem, cryst struct*)
Brumley, W.C. *et al, Anal. Chem.*, 1981, **53**, 2003 (*ms*)
Cole, R.J. *et al, Handbook of Toxic Fungal Metabolites*, Academic Press, N.Y., 1981, 38, 41.
Lewis, R.J., *Sax's Dangerous Properties of Industrial Materials*, 8th Ed., Van Nostrand-Reinhold, 1992, AEW500.

Aflatoxin B_1 A-00013

2,3,6a,9a-Tetrahydro-4-methoxycyclopenta[c]*furo*[*3,2′:4,5*]*furo*[*2,3*-h][*1*]*benzopyran-1,11-dione, 9CI. Aflatoxin FB_1. Aflatoxin B*

[1162-65 8]

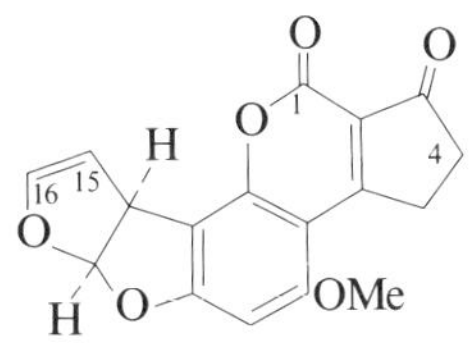

(–)-*form*

$C_{17}H_{12}O_6$ M 312.278

▷ GY1925000.

(–)-*form*

Isol. from *Aspergillus flavus* and *A. parasiticus*. Toxin causing Turkey X disease. Cryst. exhibiting blue fluor. Mp 268-269° dec. $[\alpha]_D^{25}$ –562° (c, 0.115 in $CHCl_3$).

▷ Extremely carcinogenic.

15,16-Dihydro: [7220-81-7]. ***Aflatoxin B_2***

$C_{17}H_{14}O_6$ M 314.294

Metab. of *A. flavus*. Mycotoxin. Yellow cryst. with blue fluor. (MeOH). Mp >310° dec. $[\alpha]_D^{23}$ –492° (c, 0.1 in $CHCl_3$).

▷ Carcinogenic. GY1722000.

15,16-Dihydro, 16-hydroxy: [17878-54-5]. ***Aflatoxin B_{2a}***. *Dihydrohydroxyaflatoxin B_1*

$C_{17}H_{14}O_7$ M 330.293

Metab. of *A. flavus*. Mycotoxin. Cryst. with blue fluor. ($CHCl_3$). Mp 217° (240° dec.).

▷ Carcinogenic.

O-*De-Me:* [32215-02-4]. ***Aflatoxin P_1***

$C_{16}H_{10}O_6$ M 298.251

Metab. of Aflatoxin B_1. Pale-yellow needles (MeOH/C_6H_6/hexane). Mp >320°. $[\alpha]_D^{20}$ –574° (c, 0.08 in MeOH).

▷ Carcinogenic. GY1775000.

15α,16α-Epoxide: [67337-06-8].

$C_{17}H_{12}O_7$ M 328.278

Prob. ultimate carcinogen of Aflatoxin B_1. Cryst. (Me_2CO/CH_2Cl_2). Mp >300° (phase transition at 230°). Stable at –10°, fairly stable at r.t.

▷ Presumed highly toxic and carcinogenic.

15,16-Dihydro, 16-methoxy: [20421-12-9]. ***Aflatoxin Ex_2B_1***

$C_{18}H_{16}O_7$ M 344.320

Metab. of *A. flavus*. Cryst. ($CHCl_3$). Mp 240°.

15,16-Dihydro, 16-ethoxy: [23402-20-2]. ***Aflatoxin ExB_2***

$C_{19}H_{18}O_7$ M 358.347

Isol. from *A. flavus*. Cryst. Mp 245°.

4-Hydroxy: [104700-21-2]. ***Aflatoxin M_4***

$C_{17}H_{12}O_7$ M 328.278

Isol. from cultures of *A. parasiticus*. Cryst.

(±)-*form* [10279-73-9]

Mp 255-256°.

15,16-Dihydro: Cryst. with blue fluor. ($CHCl_3$/MeOH). Mp 303-306° dec.

[42583-46-0]

Asao, T. *et al, J. Am. Chem. Soc.*, 1963, **85**, 1705; 1965, **87**, 882 (*struct, isol, ir, uv, ms, nmr*)
Dutton, M.F. *et al, Biochem. J.*, 1966, **101**, 21P (*deriv*)
Brechbühler, S. *et al, J. Org. Chem.*, 1967, **32**, 2641 (*abs config*)
Roberts, J.C. *et al, J. Chem. Soc. C*, 1968, 22 (*synth, uv, ir, ms*)
Dutton, M.F. *et al, Chem. Ind. (London)*, 1969, 983 (*Aflatoxin Ex_2B_1, Aflatoxin ExB_2*)
Heathcote, J.B. *et al, Tetrahedron*, 1969, **25**, 1497 (*biosynth*)
v. Soest, T.C. *et al, Acta Crystallogr., Sect. B*, 1970, **26**, 1940, 1947 (*cryst struct*)
Büchi, G. *et al, J. Am. Chem. Soc.*, 1971, **93**, 746 (*synth, uv, ms*)
Heathcote, J.B. *et al, Chem. Ind. (London)*, 1976, 270 (*biosynth*)
Pachler, K.G.R. *et al, J. Chem. Soc., Perkin Trans. 1*, 1976, 1182 (*cmr, biosynth*)
Cox, R.H. *et al, J. Org. Chem.*, 1977, **42**, 112 (*cmr*)
Simpson, T.J. *et al, J. Chem. Soc., Chem. Commun.*, 1982, 631; 1983, 338 (*biosynth*)
Pohland, A.E. *et al, Pure Appl. Chem.*, 1982, **54**, 2220 (*uv, ir, pmr, ms, cd*)
Castellino, A.J. *et al, J. Org. Chem.*, 1986, **51**, 1006 (*synth*)
Lafont, P. *et al, Microbiol., Aliments, Nutr.*, 1986, **4**, 65 (*Aflatoxin M_4*)
Townsend, C.A., *Pure Appl. Chem.*, 1986, **58**, 227 (*biosynth*)
Baertschi, S.W. *et al, J. Am. Chem. Soc.*, 1988, **110**, 7929 (*epoxide*)
Weeratunga, G. *et al, J. Chem. Soc., Chem. Commun.*, 1988, 721 (*Aflatoxin B_2*)
Townsend, C.A. *et al, J. Chem. Soc., Perkin Trans. 1*, 1988, 839 (*biosynth*)
Sloan, C.P. *et al, Tetrahedron Lett.*, 1988, **29**, 4685 (*synth*)
Horne, S. *et al, J. Chem. Soc., Chem. Commun.*, 1990, 39 (*Aflatoxin B_2, synth*)
Bhatnager, D. *et al, Biochemistry*, 1991, **30**, 4343 (*biosynth*)
Cole, R.J. *et al, Handbook of Toxic Fungal Metabolites*, Academic Press, N.Y., 1981, 3, 15, 34, 51.
Lewis, R.J., *Sax's Dangerous Properties of Industrial Materials*, 8th Ed., Van Nostrand-Reinhold, 1992, AEU250, AEU750.

Aflatoxin B_3 A-00014

Parasiticol

[23315-33-5]

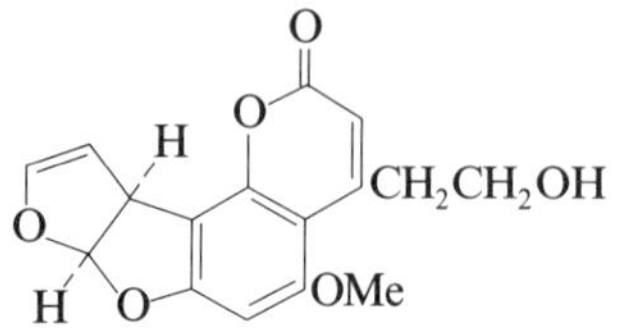

$C_{16}H_{14}O_6$ M 302.283

Isol. from *Aspergillus flavus* and *A. parasiticus*. Cryst. ($CHCl_3$). Mp 217°.

Heathcote, J.G. *et al*, *Tetrahedron*, 1969, **25**, 1497 (*isol, uv, ir, pmr*)
Stubblefield, R.D. *et al*, *J. Agric. Food Chem.*, 1970, **18**, 391 (*isol, uv*)
Cox, R.H. *et al*, *J. Org. Chem.*, 1977, **42**, 112 (*cmr*)
Cole, R.J. *et al*, *Handbook of Toxic Fungal Metabolites*, Academic Press, N.Y., 1981, 10.

Aflatoxin G_1 A-00015

Aflatoxin G

[1165-39-5]

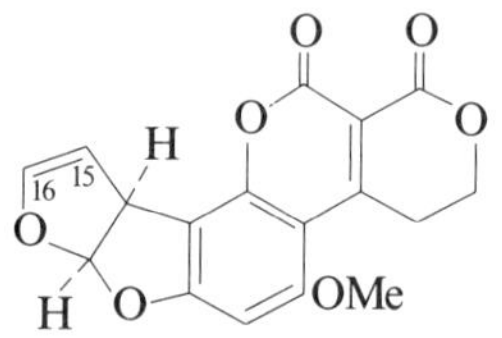

Absolute configuration

$C_{17}H_{12}O_7$ M 328.278

Isol. from *Aspergillus flavus* and *A. parasiticus*. Mycotoxin. Needles (MeOH) exhibiting green fluor. Mp 247-250°. $[\alpha]_D^{23}$ −556° (c, 0.1 in $CHCl_3$).

▷ Carcinogenic. LV1720000.

15,16-Dihydro: [7241-98-7]. ***Aflatoxin G_2***
$C_{17}H_{14}O_7$ M 330.293
Minor constit. of *A. flavus*. Mycotoxin. Cryst. with green fluor. (EtOH). Mp 237-240°. $[\alpha]_D^{23}$ −473° (c, 0.084 in $CHCl_3$).
▷ Carcinogenic, extremely poisonous. LV1700000.

15,16-Dihydro, 16-hydroxy: [20421-10-7]. ***Aflatoxin G_{2a}***
$C_{17}H_{14}O_8$ M 346.293
Metab. of *A. flavus*. Cryst. with green fluor. Mp 190° dec.
▷ Carcinogen.

15,16-Dihydro, 16-ethoxy: [23402-21-3]. *Aflatoxin ExG*
$C_{19}H_{18}O_8$ M 374.346
Mp 203°.

Nesbitt, B.E. *et al*, *Nature (London)*, 1962, **195**, 1062 (*isol*)
Asao, T. *et al*, *J. Am. Chem. Soc.*, 1963, **85**, 1706; 1965, **87**, 882 (*isol, uv, ir, ms, nmr, struct*)
Dutton, M.F. *et al*, *Biochem. J.*, 1966, **101**, 21P (*deriv*)
Brechbühler, S. *et al*, *J. Org. Chem.*, 1967, **32**, 2641 (*abs config*)
Dutton, M.F. *et al*, *Chem. Ind. (London)*, 1969, 983 (*Aflatoxin ExG*)
Büchi, G. *et al*, *J. Am. Chem. Soc.*, 1971, **93**, 746 (*uv, ms, synth*)
Heathcote, J.B. *et al*, *Chem. Ind. (London)*, 1976, 270 (*biosynth*)
Cox, R.H. *et al*, *J. Org. Chem.*, 1977, **42**, 112 (*cmr*)
Cole, R.J. *et al*, *Handbook of Toxic Fungal Metabolites*, Academic Press, N.Y., 1981, 25, 45, 55.
Lewis, R.J., *Sax's Dangerous Properties of Industrial Materials*, 8th Ed., Van Nostrand-Reinhold, 1992, AEV000, AEV500.

Aflatoxin GM_1 A-00016

*3,4,7*a*,10*a*-Tetrahydro-10*a*-hydroxy-5-methoxy-1*H*,12*H*-furo[3′,2′:4,5]furo[2,3-*h*]pyrano[3,4-*c*][1]benzopyran-1,12-dione, 9CI*

[23532-00-5]

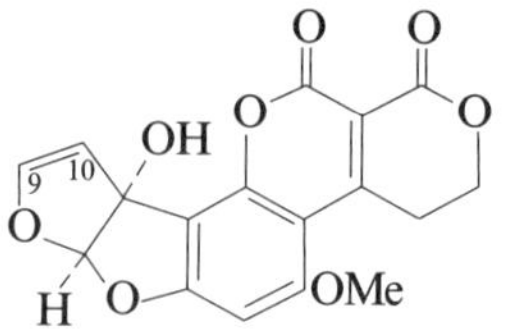

Absolute configuration

$C_{17}H_{12}O_8$ M 344.277

Metab. of *Aspergillus flavus*. Mycotoxin. Cryst. ($CHCl_3$). Mp 276°.

▷ Carcinogenic.

Ac: Mp 280°.

9,10-Dihydro: ***Aflatoxin GM_2***
$C_{17}H_{14}O_8$ M 346.293
Minor metab. of *A. flavus*. Mp 270-272°.

Heathcote, J.B. *et al*, *Tetrahedron*, 1969, **25**, 1497 (*isol, uv, struct*)
Heathcote, J.B. *et al*, *Biochem. Soc. Trans.*, 1974, **2**, 301 (*deriv*)
Heathcote, J.B. *et al*, *Chem. Ind. (London)*, 1976, 270 (*synth*)

Aflatoxin M_1 A-00017

[6795-23-9]

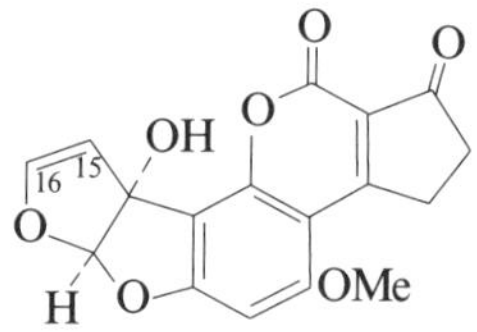

$C_{17}H_{12}O_7$ M 328.278

Minor metab. of *Aspergillus flavus*, also found in the milk of cows and sheep fed toxic meal. Mycotoxin. Cryst. (MeOH) exhibiting blue-violet fluor. Mp 299° dec. $[\alpha]_D^{20}$ −280° (c, 0.1 in DMF).

▷ Carcinogenic. GY1880000.

Ac: Oil.

15,16-Dihydro: [6885-57-0]. ***Aflatoxin M_2***
$C_{17}H_{14}O_7$ M 330.293
Trace metab. of *A. flavus*. Mycotoxin. Cryst. with violet fluor. (MeOH/$CHCl_3$). Mp 293° dec.
▷ Carcinogenic. GY1720000.

15,16-Dihydro, 16-hydroxy: [36601-31-7]. *Aflatoxin M_{2a}*
$C_{17}H_{14}O_8$ M 346.293
Hemiacetal formed from Aflatoxin M_2 by acid-cat. addn. of H_2O.

Holzapfel, C.W. *et al*, *Tetrahedron Lett.*, 1966, 2799 (*isol, struct*)
Büchi, G. *et al*, *J. Am. Chem. Soc.*, 1969, **91**, 5408; 1971, **93**, 746 (*uv, ms, nmr, synth*)
Stack, M.E., *J. Assoc. Off. Anal. Chem.*, 1972, **55**, 313 (*Aflatoxin M_{2a}*)
Heathcote, J.B. *et al*, *Chem. Ind. (London)*, 1976, 270 (*biosynth*)
Büchi, G. *et al*, *J. Am. Chem. Soc.*, 1981, **103**, 3497 (*synth*)
Cole, R.J. *et al*, *Handbook of Toxic Fungal Metabolites*, Academic Press, N.Y., 1981, 29, 49.
Lewis, R.J., *Sax's Dangerous Properties of Industrial Materials*, 8th Ed., Van Nostrand-Reinhold, 1992, AEW000.

11-[(2-Amino-2-carboxyethyl)thio]-12-hydroxy-5,7,9-hexadecatrienedioic acid A-00018

16-Carboxy-17,18,19,20-tetranor-14,15-dihydroleukotriene E_4

[122069-63-0]

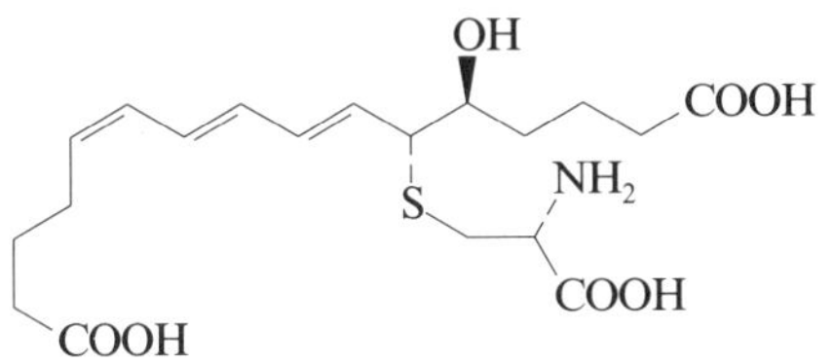

$C_{19}H_{29}NO_7S$ M 415.507

Found in the bile and urine of primates. β-Oxidation prod. of Leukotriene E_4, L-00016. Uv λ_{max} 279 nm (MeOH).

N-*Ac:* [114115-52-5].

$C_{21}H_{31}NO_8S$ M 457.544

Present in rat bile. β-Oxidation prod. of Leukotriene E_4, L-00016. Uv λ_{max} 279 nm (MeOH).

Delorme, D. *et al, Prostaglandins*, 1988, **36**, 291 (*synth, biosynth*)
Keppler, D. *et al, Adv. Enzyme Regul.*, 1989, **28**, 307 (*biosynth*)
Delorme, D. *et al, J. Org. Chem.*, 1989, **54**, 3635 (*synth, uv, pmr*)
Perrin, P. *et al, Prostaglandins*, 1989, **37**, 53 (*synth, biosynth*)
Jedlitschky, G. *et al, J. Biol. Chem.*, 1991, **266**, 24763 (*biosynth*)

13-[(2-Amino-2-carboxyethyl)thio]-14-hydroxy-4,7,9,11-octadecatetraenedioic acid A-00019

*5-Hydroxy-6-cystein-*S*-yl-7,9,11,14-octadecatetraenedioic acid. 18-Carboxydinorleukotriene* E_4

[122069-58-3]

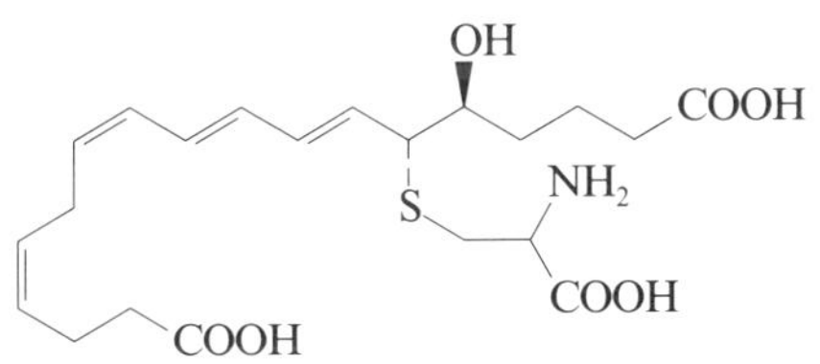

$C_{21}H_{31}NO_7S$ M 441.544

LTE_4 and LTC_4 metab. Uv_{max} 280 nm (MeOH).

N-*Ac:* [114115-51-4].

$C_{23}H_{33}NO_8S$ M 483.582

LTE_4 and LTC_4 metab. Uv λ_{max} 280 nm (MeOH).

Delorme, D. *et al, Prostaglandins*, 1988, **36**, 291; 1989, **37**, 629 (*synth*)
Delorme, D. *et al, J. Org. Chem.*, 1989, **54**, 3635 (*synth, uv, pmr*)
Huber, M. *et al, Eur. J. Biochem.*, 1990, **194**, 309 (*biochem*)

9-[(2-Amino-2-carboxyethyl)thio]-10-hydroxy-3,5,7-tetradecatrienedioic acid A-00020

14-Carboxyhexanorleukotriene E_3

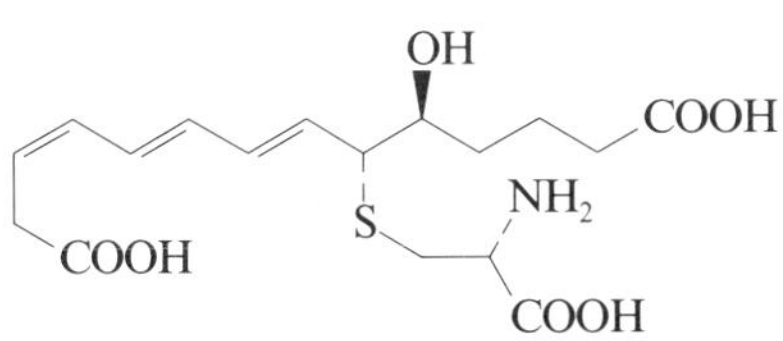

$C_{17}H_{25}NO_7S$ M 387.453

Urinary metab. of Leukotriene E_4, L-00016 in humans.

.*14*N-*Ac:* [114115-54-7].

$C_{19}H_{27}NO_8S$ M 429.490

Metab. of Leukotriene E_4, L-00016.

[137625-92-4]

Stene, D.O. *et al, J. Biol. Chem.*, 1988, **263**, 2773 (*biosynth*)
Sala, A. *et al, J. Biol. Chem.*, 1990, **265**, 21771 (*biosynth*)
Jedlitschky, G. *et al, J. Biol. Chem.*, 1991, **266**, 24763 (*biosynth*)

2-Aminodecanoic acid A-00021

2-Aminocapric acid

[17702-88-4]

$H_3C(CH_2)_7CH(NH_2)COOH$

$C_{10}H_{21}NO_2$ M 187.281

(±)-***form***

Mp 264°.

N-*Benzoyl:*

$C_{17}H_{25}NO_3$ M 291.389

Mp 124°.

Albertson, N.F., *J. Am. Chem. Soc.*, 1946, **68**, 450.

3-Aminodecanoic acid, 9CI A-00022

$H_3C(CH_2)_6CH(NH_2)CH_2COOH$

$C_{10}H_{21}NO_2$ M 187.281

(±)-***form***

Mp 205-206°.

N-*Benzoyl:*

$C_{17}H_{25}NO_3$ M 291.389

Mp 133-134°.

Amide:

$C_{10}H_{22}N_2O$ M 186.297

Cryst. (H_2O). Mp 94-96°.

Lang, K. *et al, Hoppe Seyler's Z. Physiol. Chem.*, 1941, **269**, 236 (*synth*)

9-Aminodecanoic acid A-00023

[134781-68-3]

$H_3CCH(NH_2)(CH_2)_7COOH$

$C_{10}H_{21}NO_2$ M 187.281

(±)-***form***

Solid. Mp 203° dec.

McGhie, J.F. *et al, J. Chem. Soc.*, 1962, 2578.
Bartra, M. *et al, J. Org. Chem.*, 1991, **56**, 5132 (*synth, pmr, ir*)

10-Aminodecanoic acid, 9CI A-00024

[13108-19-5]

$H_2NCH_2(CH_2)_8COOH$

$C_{10}H_{21}NO_2$ M 187.281

Sol. hot H_2O. Mp 187-188°.

B,HCl: Mp 158°.

Me ester:

$C_{11}H_{23}NO_2$ M 201.308

Mp 152° (as hydrochloride).

Blaise, E.-E. *et al, C. R. Hebd. Seances Acad. Sci.*, 1906, **143**, 362.

12-Aminododecanoic acid, 9CI A-00025

[693-57-2]

$H_2NCH_2(CH_2)_{10}COOH$

$C_{12}H_{25}NO_2$ M 215.335

Cryst. (EtOH aq.). Mp 185-187°.

B,HCl: [22543-30-2].

Mp 163-164°.

4-Methylbenzenesulfonyl: Mp 92-93°.

N-tert-*Butoxycarbonyl:* [18934-81-1].
Cryst. (EtOAc/hexane). Mp 80-81°.

Huenig, S. *et al, Angew. Chem., Int. Ed. Engl.*, 1963, **2**, 213.
Genas, M. *et al, Bull. Soc. Chim. Fr.*, 1965, 2833 (*synth*)
Zakharkin, L.I. *et al, CA*, 1965, **62**, 6389d (*synth*)
Huenig, S. *et al, Chem. Ber.*, 1967, **100**, 3039 (*synth*)
Chaturvedi, D.N. *et al, J. Med. Chem.*, 1984, **27**, 1406 (*deriv*)

2-Amino-1,3,4-eicosanetriol A-00026

2-Amino-1,3,4-icosanetriol. 4-Hydroxyeicosasphinganine. C_{20}-*Phytosphingosine*

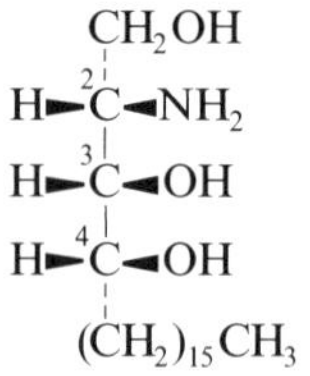

(2*S*,3*S*,4*R*)-*form*

$C_{20}H_{43}NO_3$ M 345.565
Isol. from yeast and fungal cerebrins.

(2S,3S,4R)-form [3530-53-8]
D-ribo-*form*
Cryst. (MeCN). Mp 97-99°. $[\alpha]_D^{17}$ +8.1° (c, 0.34 in $CHCl_3$).
N-*Benzoyl:* Cryst. (EtOAc). Mp 136-137°. $[\alpha]_D$ +3.6° (c, 1 in Py).
N-*Benzoyl, tri*-O-*Ac:* Mp 81-82°. $[\alpha]_D$ +2° (c, 0.7 in $CHCl_3$).

(2RS,3RS,4SR)-form [34506-41-7]
(±)-ribo-*form*
Cryst. (MeCN). Mp 95-105°.

(2RS,3SR,4SR)-form
(±)-xylo-*form*
Cryst. (MeCN). Mp 128-132°.

Gigg, J. *et al, J. Chem. Soc. C*, 1966, 1872, 1876, 1879.
Kisic, A. *et al, Chem. Phys. Lipids*, 1971, **7**, 135.

2-Aminoheptanoic acid A-00027

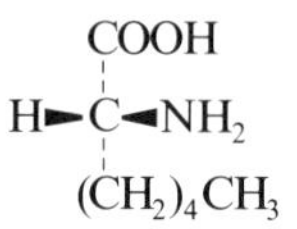

(*R*)-*form*

$C_7H_{15}NO_2$ M 145.201
Prod. by *Claviceps purpurea*.

(R)-form
D-*form*
$[\alpha]_D^{27}$ −24.0° (c, 0.04 in 6*N* HCl).

(S)-form [44902-02-5]
L-*form*
Needles (H_2O). Mp 274° dec. $[\alpha]_D^{20}$ +23.9° (c, 0.04 in 6*N* HCl).

(±)-form [1115-90-8]
Plates (H_2O). Mp 275° dec.
N-*Di-Me, Et ester:*
$C_{11}H_{23}NO_2$ M 201.308
Liq. Bp_{10} 100°.

Helms, A., *Ber.*, 1875, **8**, 1168 (*synth*)
Baker, C.G. *et al, J. Am. Chem. Soc.*, 1951, **73**, 1336 (*resoln, abs config*)
Steiner, M. *et al, Biochem. Z.*, 1964, **340**, 436 (*isol*)

2-Amino-1,3,4-hexadecanetriol A-00028

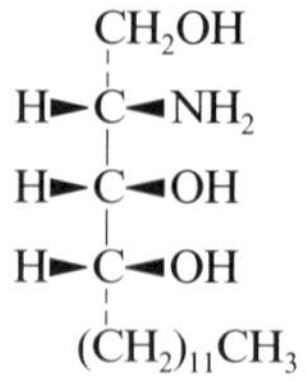

(2*S*,3*S*,4*R*)-*form*

$C_{16}H_{35}NO_3$ M 289.457

(2S,3S,4R)-form [114379-45-2]
D-ribo-*form*
Common phytosphingosine, occurring for example in Acanthocerebrosides (Acanthacerebroside A, A-00001).
N,O,O,O-*Tetra-Ac:*
$C_{24}H_{43}NO_7$ M 457.606
Viscous liq. $[\alpha]_D^{25}$ +27.9° (c, 1.5 in $CHCl_3$).

(2S,3R,4R)-form [128898-88-4]
D-xylo-*form*
N,O,O,O-*Tetra-Ac:* $[\alpha]_D^{27}$ +6.13° (c, 0.56 in $CHCl_3$).

(2S,3R,4S)-form [127061-67-0]
D-arabino-*form*
N,O,O,O-*Tetra-Ac:* $[\alpha]_D^{28}$ −23.3° (c, 1.0 in $CHCl_3$).

(2S,3S,4S)-form [127062-47-9]
D-lyxo-*form*
N,O,O,O-*Tetra-Ac:* $[\alpha]_D^{28}$ −3.3° (c, 0.46 in $CHCl_3$).

Sugiyama, S. *et al, Justus Liebigs Ann. Chem.*, 1988, 619; 1990, 1069 (*abs config, bibl, synth, pmr, cmr*)

2-Aminohexadecanoic acid A-00029

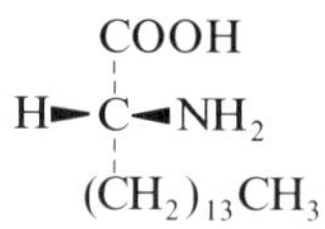

(*R*)-*form*

$C_{16}H_{33}NO_2$ M 271.442

(R)-form [98393-19-2]
Mp 233-236°. $[\alpha]_D^{26}$ −21.0° (c, 0.1 in AcOH).
N-*Chloroacetyl:* [98393-18-1].
Cryst. (hexane). Mp 87-88°. $[\alpha]_D^{21}$ −28.0° (c, 0.5 in $CHCl_3$).

(S)-form [98393-17-0]
Mp 234-236°. $[\alpha]_D^{26}$ +21.8° (c, 0.1 in AcOH).

(±)-form [98320-69-5]
Mp 221-223° dec.
N-*Chloroacetyl:* [98320-70-8].
Cryst. (hexane). Mp 97-99°.

Mori, K. *et al, Tetrahedron*, 1985, **41**, 2369 (*synth, resoln, ir*)

2-Amino-3-hydroxy-4-octadecen-1-sulfonic acid A-00030

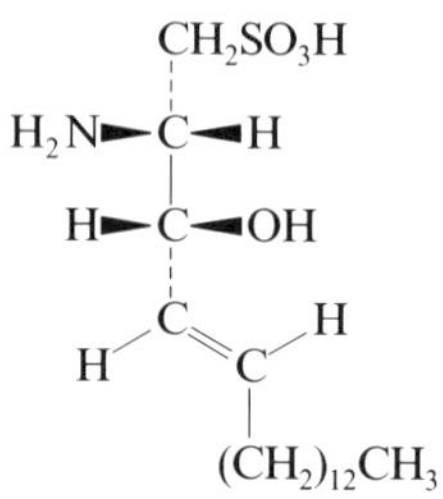

$C_{18}H_{37}NO_4S$ M 363.561

(2S,3R)-form
erythro-*form*

N-*Tetradecanoyl: 1-Deoxydihydroceramide-1-sulfonic acid*
$C_{32}H_{63}NO_5S$ M 573.919
Isol. from lipids of the marine diatom *Nitzschia alba*. The name 1-deoxydihydroceramide-1-sulfonic acid actually refers to the mixt. of alkyl homologues (see also below).
N-*Tetradecenoyl:*
$C_{32}H_{61}NO_5S$ M 571.904
Component of 1-deoxydihydroceramide-1-sulfonic acid from *N. alba*. Posn. of double bond not specified.
N-*Hexadecanoyl:*
$C_{34}H_{67}NO_5S$ M 601.973
Component of 1-deoxydihydroceramide-1-sulfonic acid from *N. alba*.
N-*3-Hexadecenoyl:*
$C_{34}H_{65}NO_5S$ M 599.957
Component of 1-deoxydihydroceramide-1-sulfonic acid from *N. alba*.

Ohashi, K. *et al*, *Tetrahedron*, 1989, **45**, 2557.

3-Aminononanoic acid, 9CI   A-00031

3-Aminopelargonic acid

$H_3C(CH_2)_5CH(NH_2)CH_2COOH$

$C_9H_{19}NO_2$ M 173.255
Mp 205-205.5°.
B,HCl: Mp 135.5-136.0°.
N-*Benzoyl:*
$C_{16}H_{23}NO_3$ M 277.363
Mp 129.5-131°.
Amide, N-*benzoyl:*
$C_{16}H_{24}N_2O_2$ M 276.378
Mp 185-186°.

Rodionov, V. *et al*, *CA*, 1944, **38**, 1473 (*synth*)

9-Aminononanoic acid   A-00032

9-Aminopelargonic acid
[1120-12-3]

$H_2NCH_2(CH_2)_7COOH$

$C_9H_{19}NO_2$ M 173.255
Can be prepd. from soybean oil. Intermed. for nylon-type polymers.
B,HCl: Cryst. (Me_2CO). Mp 115-118°.
N-*Benzenesulfonyl:* Prisms (H_2O). Insol. H_2O, pet. ether. Mp 85°.

Flaschenträger, B. *et al*, *Hoppe Seyler's Z. Physiol. Chem.*, 1930, **192**, 249, 253; *CA*, **25**, 489, 490 (*synth*)
Kohlhase, W. *et al*, *J. Am. Oil Chem. Soc.*, 1974, **51**, 204 (*glc, derivs*)

2-Amino-1,3-octadecanediol, 8CI   A-00033

Octadecacasphinganine. Cerebrin base. Dihydrosphingosine. Sphinganine
[13552-09-5]

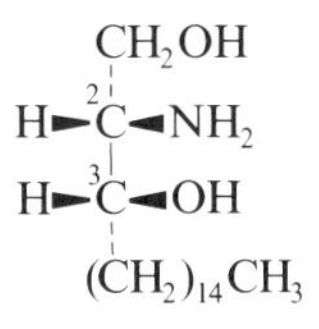

(2*S*,3*R*)-*form*

$C_{18}H_{39}NO_2$ M 301.512
(2*R*,3*R*)-*form* [6036-86-8]
D-threo-*form*
Mp 108°. $[\alpha]^{28}_{546}$ +13° (10:1 $CHCl_3$/MeOH).
N-*Ac:*
$C_{20}H_{41}NO_3$ M 343.549
Mp 96-97°. $[\alpha]^{28}_{546}$ +6.2° (10:1 hexane/EtOH).
N,O,O-*Tri-Ac:*
$C_{24}H_{45}NO_5$ M 427.623
Mp 46°. $[\alpha]^{28}_{546}$ +8.0° (pentane).
(2*S*,3*S*)-*form* [15639-50-6]
L-threo-*form*
Widespread as a component of cerebrosides, in admixture with smaller amts. of homologues. Mp 108°. $[\alpha]^{28}_{D}$ −14.1° ($CHCl_3$).
N-*Ac:* Mp 96-97°. $[\alpha]^{28}_{546}$ −6.5° (10:1 hexane/EtOH).
N,O,O-*Tri-Ac:* Mp 46°. $[\alpha]^{28}_{546}$ −8.0° (pentane).
(2*R*,3*S*)-*form*
L-erythro-*form*
Mp 79°. $[\alpha]^{28}_{546}$ −6.5° (10:1 $CHCl_3$/MeOH).
N-*Ac:* Mp 124-125°. $[\alpha]^{28}_{546}$ −8° (10:1 hexane/EtOH).
N,O,O-*Tri-Ac:* Mp 96-97°. $[\alpha]^{28}_{546}$ −21° (10:1 hexane/EtOH).
(2*S*,3*R*)-*form* [764-22-7]
D-erythro-*form*
Mp 79°, Mp 108-109°. $[\alpha]^{18}_{546}$ +6.5° (10:1 $CHCl_3$/MeOH), $[\alpha]^{28}_{D}$ +13.5° (c, 0.75 in $CHCl_3$).
Sulfate: Cryst. (AcOH). Mp 150°.
N,O,O-*Tri-Ac:* Mp 96-97° (90-93°). $[\alpha]^{19}_{D}$ +16° (c, 0.5 in $CHCl_3$), +21° (10:1 hexane/EtOH).
N-*Ac:* Mp 124-125°. $[\alpha]^{28}_{546}$ +7.7° (10:1 hexane/EtOH).
(2*RS*,3*SR*)-*form* [3102-56-5]
(±)-erythro-*form*
Mp 98-100°.

Grob, C.A. *et al*, *Helv. Chim. Acta*, 1957, **40**, 1145 (*synth*)
Shapiro, D. *et al*, *J. Am. Chem. Soc.*, 1958, **80**, 1194 (*synth*)
Carter, H.E. *et al*, *Biochemistry*, 1963, **2**, 389 (*struct, abs config*)
Gigg, J. *et al*, *J. Chem. Soc. C*, 1966, 1872 (*synth*)
Reist, E.J. *et al*, *J. Org. Chem.*, 1970, **35**, 4127 (*synth*)
Stoffel, W. *et al*, *Hoppe Seyler's Z. Physiol. Chem.*, 1972, **353**, 1962; 1973, **354**, 169 (*synth, cmr*)

2-Amino-1,3,4-octadecanetriol, 8CI   A-00034

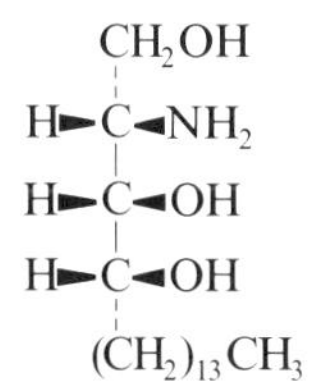

$C_{18}H_{39}NO_3$ M 317.511
Other related bases are also known.
(2*S*,3*S*,4*R*)-*form* [554-62-1]
D-ribo-*form*. *4-Hydroxysphinganine. Phytosphingosine*
Present in cerebrosides and gangliosides.
N-*Benzoyl:* Cryst. (EtOAc). Mp 137.8-138.8°. $[\alpha]_D$ +5° (c, 5.2 in Py).
N-*Benzoyl*, *tri*-O-*Ac:* Cryst. (pet. ether). Mp 78-80°. $[\alpha]_D$ +22° (c, 0.5 in $CHCl_3$).
N,O,O,O-*Tetra-Ac:*
$C_{26}H_{47}NO_7$ M 485.660
Isol. from yeast *Hansenula ciferrii*. Cryst. (pet. ether). Mp 49-50°. $[\alpha]^{25}_{D}$ +5° (c, 4.8 in DMF).

Stodola, F.H. *et al*, *J. Biol. Chem.*, 1960, **235**, 2584 (*isol, deriv*)
Prostenik, M. *et al*, *Tetrahedron*, 1965, **21**, 651.
Gigg, J. *et al*, *J. Chem. Soc. C*, 1966, 1872, 1876.
Gigg, R., *Chem. Phys. Lipids*, 1969, **3**, 106 (*stereochem*)

2-Aminooctadecanoic acid **A-00035**

2-Aminostearic acid, 9CI

[5782-80-9]

$H_3C(CH_2)_{15}CH(NH_2)COOH$

$C_{18}H_{37}NO_2$ M 299.496

(±)-*form*

Cryst. (AcOH). Mp 222°.

Me ester:

$C_{19}H_{39}NO_2$ M 313.523

Needles (EtOAc) (as hydrochloride). Mp 112° (hydrochloride).

Et ester:

$C_{20}H_{41}NO_2$ M 327.549

Needles (EtOAc) (as hydrochloride). Mp 89° (hydrochloride).

Hell, C. *et al*, *Ber.*, 1891, **24**, 2395.
Fischer, E. *et al*, *Justus Liebigs Ann. Chem.*, 1908, **362**, 339.
Huang, Y.-T., *J. Chin. Chem. Soc. (Taipei)*, 1948, **15**, 224; *CA*, **43**, 2166 (*synth*)
Mitsner, B.I. *et al*, *Zh. Org. Khim.*, 1973, **9**, 2010.

12-Aminooctadecanoic acid **A-00036**

12-Aminostearic acid

[18526-74-4]

$H_3C(CH_2)_5CH(NH_2)(CH_2)_{10}COOH$

$C_{18}H_{37}NO_2$ M 299.496

(±)-*form*

Cryst. Mp 141° (136°).

Colonge, J. *et al*, *Bull. Soc. Chim. Fr.*, 1954, 339 (*synth*)
Freedman, B. *et al*, *J. Am. Oil Chem. Soc.*, 1970, **47**, 311 (*synth*)
Caspar, A. *et al*, *Org. Mass Spectrom.*, 1970, **3**, 1351 (*ms*)

2-Amino-4-octadecene-1,3-diol, 9CI **A-00037**

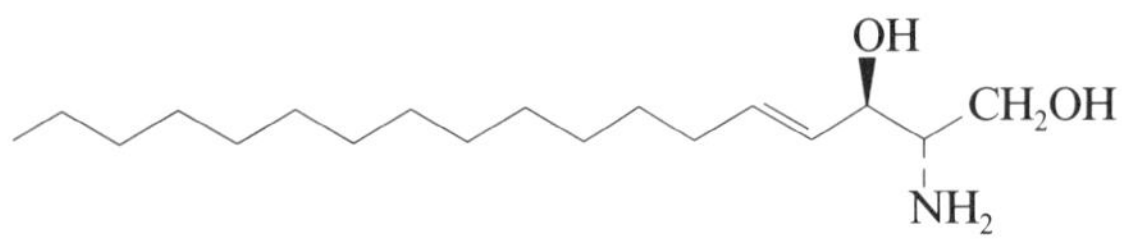

(2*S*,3*R*,4*E*)-*form*

$C_{18}H_{37}NO_2$ M 299.496

(*2S,3R,4E*)-*form* [123-78-4]

D-erythro-trans-*form*, *Sphing-4-enine*. *Sphingosine*

Present in cerebrosides and gangliosides. Cryst. (Et_2O). Mp 80-84° (79-81°). Other Sphingenines having different numbers of carbon atoms are known.

Tri-Ac: [2482-37-3].

$C_{24}H_{43}NO_5$ M 425.607

Mp 103.5-104°. $[\alpha]_D^{19}$ −24.1°, $[\alpha]_D$ −13° (c, 0.5 in $CHCl_3$).

Tribenzoyl:

$C_{39}H_{49}NO_5$ M 611.820

Cryst. (EtOH). Mp 121.5-123.5°. $[\alpha]_D^{27}$ −11.2°.

O[1]-*Glucosyl:* ***Glucopsychosine***

$C_{24}H_{47}NO_7$ M 461.638

Metab. prod. present in Gaucher's disease.

N-*Tetradecanoyl:* N-[*2-Hydroxy-1-(hydroxymethyl)-3-heptadecenyl*]*tetradecenamide, 8CI. 2-Tetradecanoylamino-4-octadecene-1,3-diol.* N-*Tetradecanoyl-4-sphingenine*

$C_{32}H_{63}NO_3$ M 509.855

Present in blood plasma lipids.

N-*Hexadecanoyl:* N-[*2-Hydroxy-1-(hydroxymethyl)-3-heptadecenyl*]*hexadecanamide, 8CI. 2-Hexadecanoylamino-4-octadecene-1,3-diol.* N-*Hexadecanoyl-4-sphingenine*

$C_{34}H_{67}NO_3$ M 537.908

Present in blood and brain lipids.

N-*Octadecanoyl:* [2304-81-6]. N-*Octadecanoylamino-4-octadecene-1,3-diol.* N-*Octadecanoyl-4-sphingenine*

$C_{36}H_{71}NO_3$ M 565.962

Present in blood and brain lipids. Mp 97-98° (91-93°). $[\alpha]_D^{25}$ −3.1° (c, 1.1 in $CHCl_3$).

N-(*9-Octadecenoyl*): [5966-28-9]. N-[*2-Hydroxy-1-(hydroxymethyl)-3-heptadecenyl*]*-9-octadecenamide, 9CI. 2-Octadec-9-enoylamino-4-octadecene-1,3-diol.* N-*Octadec-9-enoylsphingenine.* N-*Oleoyl-4-sphingenine*

$C_{36}H_{69}NO_3$ M 563.946

Present in blood and brain lipids.

N-*Eicosanoyl:* [7344-02-7]. N-[*2-Hydroxy-1-(hydroxymethyl)-3-heptadecenyl*]*eicosanamide, 9CI. 2-Eicosanoylamino-4-octadecene-1,3-diol.* N-*Eicosanoyl-4-sphingenine.* N-*Icosanoyl-4-sphingenine*

$C_{38}H_{75}NO_3$ M 594.015

Present in blood and brain lipids.

N-*Docosanoyl:* [27888-44-4]. N-[*2-Hydroxy-1-(hydroxymethyl)-3-heptadecenyl*]*docosanamide, 9CI. 2-Docosanoylamino-4-octadecene-1,3-diol.* N-*Docosanoyl-4-sphingenine*

$C_{40}H_{79}NO_3$ M 622.069

Present in brain and blood lipids.

N-*13Z-Docosenoyl:* [54135-66-9]. N-[*2-Hydroxy-1-(hydroxymethyl)-3-heptadecenyl*]*-1,3-docosenamide, 9CI.* N-*13-Docosenoyl-4-octadecene-1,3-diol.* N-*13-Docosenoyl-4-sphingenine.* N-*Erucoyl-4-sphingenine*

$C_{40}H_{77}NO_3$ M 620.053

Present in blood lipids.

N-*Tetracosanoyl:* [34435-05-7]. N-[*2-Hydroxy-1-(hydroxymethyl)-3-heptadecenyl*]*tetracosanamide, 9CI. 2-(Tetracosanoylamino)-4-octadecene-1,3-diol.* N-*Tetracosanoyl-4-sphingenine*

$C_{42}H_{83}NO_3$ M 650.123

Present in blood and brain lipids. Cryst. (MeOH). Mp 93-95°.

N-(*15Z-Tetracosenoyl*), *1-*O*-β-*D*-galactopyranoside:* [17283-91-9]. N-*15Z-Tetracosenoylgalactocerebroside. Nervone*

$C_{48}H_{91}NO_8$ M 810.249

Present in brain cerebrosides. Mp 154-156°. $[\alpha]_D^{22}$ −3.70° (c, 10 in Py).

N-(*2-Hydroxy-15Z-tetracosenoyl*), *1-*O*-β-*D*-galactopyranoside:* N-(*2-Hydroxy-15-tetracosenoyl*) *galactocereboside. Oxynervone*

$C_{48}H_{91}NO_9$ M 826.248

Brain cerebroside. Mp 180-181°. $[\alpha]_D^{22}$ +3.80° (c, 10 in Py).

(*2RS,3SR,4E*)-*form* [2733-29-1]

(±)-erythro-trans-*form*

Mp 71-73° (65-68°).

Tri-Ac: Mp 95-95.5° (91-92°).

N-*Hexadecanoyl:* [52019-77-9].

Cryst. (MeOH aq.). Mp 97-98°.

N-*Octadecanoyl:* Mp 101-102°.

N-*Tetracosanoyl:* [26549-48-4].

Cryst. (MeOH aq.). Mp 95-96°.

N-*9-Octadecenoyl:* Mp 68-69°.

(*2S,3R,4Z*)-*form*

D-erythro-cis-*form*

Mp 73-74°. $[\alpha]_D^{21}$ −5° (c, 0.5 in $CHCl_3$).

Tri-Ac: Mp 90-91°.

(*2RS,3SR,4Z*)-*form*

(±)-erythro-cis-*form*
Mp 74-75°.
N-*Ac:*
$C_{20}H_{39}NO_3$ M 341.533
Mp 70-71°.
Tri-Ac: Mp 83-84°.
(2RS,3RS,4E)-form [2304-77-0]
(±)-threo-trans-*form*
N-*Ac:* Mp 69-70°.
N-*Tetradecanoyl:* Mp 88-90°.
(2RS,3RS,4Z)-form [17673-73-3]
(±)-threo-cis-*form*
Mp 80-81°.
N-*Ac:* Mp 97-99°.
Tri-Ac: Mp 40-41°.
N-*Tetradecanoyl:* [17673-78-8].
Mp 96.5-98°.
N-*Tetracosanoyl:* Mp 100-102°.
N-*Hexadecanoyl, 1-*O-*[α-D-galactopyranosyl(1→4)-β-D-galactopyranosyl(1→4)-β-D-glucopyranoside]:* [137896-85-6]. ***Globotriaosylceramide***
$C_{52}H_{97}NO_{18}$ M 1024.334
Isol. from various tissues, e.g. canine intestine. Cell surface receptor. P^k antigen in the P blood-group system. Associated with various carcinomas.
N-*Hexadecanoyl, 1-*O-*[α-D-galactopyranosyl(1→3)-β-D-galactopyranosyl(1→4)-β-D-glucopyranoside]:* [137896-95-8]. ***Isoglobotriaosylceramide***
$C_{52}H_{97}NO_{18}$ M 1024.334
Isol. from various tissues. Associated with rat mammary tumours.

Klenk, E. *et al, Hoppe Seyler's Z. Physiol. Chem.*, 1967, **348**, 1061 (*Nervone, Oxynervone*)
Aver, E. *et al, Monatsh. Chem.*, 1967, **98**, 802 (*synth*)
Samuelsson, B. *et al, J. Lipid Res.*, 1969, **10**, 47 (*occur*)
Karlsson, K.A. *et al, J. Lipid Res.*, 1971, **12**, 466 (*tlc*)
Stoffel, W. *et al, Hoppe Seyler's Z. Physiol. Chem.*, 1972, **353**, 1962 (*cmr*)
Krivit, W. *et al, J. Lipid Res.*, 1972, **13**, 525 (*occur*)
Ong, D.E. *et al, J. Lipid Res.*, 1972, **13**, 819 (*synth*)
Schmidt, R.R. *et al, Angew. Chem., Int. Ed. Engl.*, 1982, **21**, 210; 1986, **25**, 725 (*synth, bibl*)
Garigipati, R.S. *et al, J. Am. Chem. Soc.*, 1983, **105**, 4499 (*synth, ir, pmr, cmr*)
Julina, R. *et al, Helv. Chim. Acta*, 1986, **69**, 368 (*synth*)
Radunz, H.E. *et al, Justus Liebigs Ann. Chem.*, 1988, 1103 (*synth*)
Qiu, D. *et al, Justus Liebigs Ann. Chem.*, 1992, 217 (*Globotriaosylceramide, Isoglobotriaosylceramide*)

2-Amino-8-octadecene-1,3,4-triol, 9CI — A-00038

[81520-97-0]

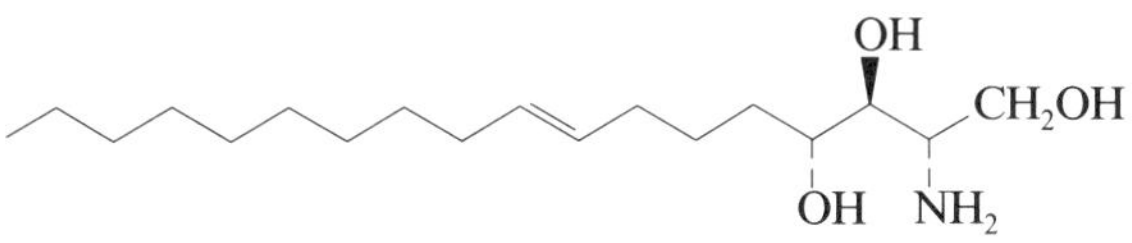

$C_{18}H_{37}NO_3$ M 315.495
(2S,3S,4R,8E)-form [3687-54-5]
D-ribo-trans-*form. Dehydrophytosphingosine*
Isol. from soybean and flax phosphatides. Cryst. (Et_2O). Mp 92-94°. $[\alpha]_D^{25}$ +8.5° (c, 1.2 in EtOH).
N-*Benzoyl:* Cryst. (Me_2CO). Mp 128-129°.
N-*Benzoyl, tri-Ac:* Cryst. (hexane). Mp 77-78°. $[\alpha]_D^{25}$ +10° (c, 0.76 in $CHCl_3$).

[25277-37-6, 51153-53-8]

Prostenik, M. *et al, Naturwissenschaften*, 1961, **48**, 500 (*struct*)
Carter, H.E. *et al, Biochemistry*, 1963, **2**, 389 (*isol, struct*)

3-Aminooctanoic acid, 9CI — A-00039

$H_3C(CH_2)_4CH(NH_2)CH_2COOH$

$C_8H_{17}NO_2$ M 159.228
(±)-form
Cryst. (EtOH). Mp 204-206° dec.
Et ester: [56266-10-5].
$C_{10}H_{21}NO_2$ M 187.281
Bp_{25} 132-133°.

Décombe, J., *C. R. Hebd. Seances Acad. Sci.*, 1930, **190**, 268; **191**, 945 (*synth*)
Lang, K. *et al, Z. Physiol. Chem.*, 1941, **269**, 236 (*synth, props*)

4-Aminopentanoic acid, 9CI — A-00040

4-Aminovaleric acid, 8CI
[13880-74-5]

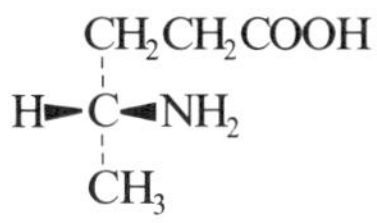

(*R*)-*form*

$C_5H_{11}NO_2$ M 117.147
(R)-form [5937-83-7]
Cryst. (EtOH aq.). Mp 214°. $[\alpha]_D^{22}$ +13.9° (c, 0.8 in H_2O).
N-*Benzoyl:* [22148-85-2].
$C_{12}H_{15}NO_3$ M 221.255
$[\alpha]_D^{20}$ +16.5° (EtOH).
(S)-form [1558-59-4]
$[\alpha]_D^{20}$ −10.7°.
B,HBr: [62504-14-7].
Mp 196-202°. $[\alpha]_D^{20}$ −11.1° (c, 0.27 in H_2O).
N-*Benzoyl:* Needles (H_2O). Mp 133°. $[\alpha]_D^{20}$ −21.9° (EtOH).
(±)-form [627-61-2]
Cryst. Sol. H_2O; prac. insol. EtOH; insol. Et_2O, C_6H_6, petrol. Mp 214°.
B,HCl: [62000-70-8].
Prisms (EtOH). Mp 154°.
Et ester:
$C_7H_{15}NO_2$ M 145.201
Plates (EtOH/Et_2O) (as hydrochloride). Mp 92° (hydrochloride).
N-*Benzoyl:* Cryst. (C_6H_6/pet. ether). Mp 132°.
N-*Di-Et, Et ester:*
$C_{11}H_{23}NO_2$ M 201.308
Bp_8 96°.

Tafel, J. *et al, Ber.*, 1889, **22**, 1861.
Fischer, E. *et al, Justus Liebigs Ann. Chem.*, 1911, **383**, 363 (*synth, resoln*)
Balenović, K. *et al, J. Chem. Soc.*, 1955, 631 (*abs config*)
Lucente, G. *et al, Gazz. Chim. Ital.*, 1965, **95**, 1335.
Kahn, N.H. *et al, J. Org. Chem.*, 1973, **38**, 822.
Kaji, E. *et al, Bull. Chem. Soc. Jpn.*, 1976, **49**, 3181.
Pilipauskas, D.R. *et al, Tetrahedron*, 1976, **32**, 2245 (*synth, resoln*)
Cane, D.E. *et al, J. Am. Chem. Soc.*, 1978, **100**, 3203.
Otsuka, M. *et al, J. Am. Chem. Soc.*, 1990, **112**, 838 (*synth, ir, pmr*)

2-Aminotetradecanoic acid — A-00041

2-Aminomyristic acid

$H_3C(CH_2)_{11}CH(NH_2)COOH$

$C_{14}H_{29}NO_2$ M 243.389
(±)-form
Cryst. (AcOH). Mp 253°.
Me ester; B,HCl: Mp 105°.

Et ester; B,HCl: Cryst. (EtOH). Mp 83°.

Hell, C. *et al*, *Ber.*, 1889, **22**, 1747.
Abderhalden, E. *et al*, *Chem. Zentralbl.*, 1924, **1**, 551.

2-Aminotridecanoic acid A-00042

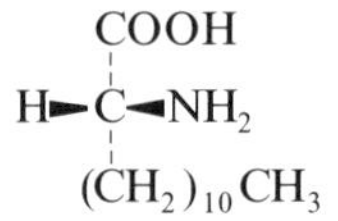

(R)-form

$C_{13}H_{27}NO_2$ M 229.362

(R)-form [96947-52-3]
Solid. Mp 222-223° dec. $[\alpha]_D^{29}$ −21.1° (c, 0.099 in AcOH).
N-*Chloroacetyl:* [96947-51-2].
Cryst. (Me_2CO/hexane). Mp 80-81°. $[\alpha]_D^{21}$ −30.4° (c, 1.39 in $CHCl_3$).

(S)-form [96947-50-1]
Cryst. Mp 223-225° dec. $[\alpha]_D^{19}$ +21.7° (c, 0.105 in AcOH).

(±)*-form* [35237-38-8]
Solid. Mp 226-227° dec.
N-*Chloroacetyl:* [96883-07-7].
Cryst. (Me_2CO/hexane). Mp 91-92°.

Mori, K. *et al*, *Tetrahedron*, 1985, **41**, 547 (*synth, resoln, pmr, ir*)

13-Aminotridecanoic acid, 8CI A-00043

[17437-19-3]

$H_2NCH_2(CH_2)_{11}COOH$

$C_{13}H_{27}NO_2$ M 229.362
Mp 176-178° dec.
B,HCl: Mp *ca.* 151°.
Me ester; B,HCl: Mp 152-154°.
Et ester: Cryst. (EtOH aq.). Mp 73°.
Et ester; B,HCl: Needles (EtOH). Mp 145°.

Baruch, J., *Ber.*, 1893, **26**, 1870.
Müller, A. *et al*, *Ber.*, 1932, **65**, 1357.
Treibs, W. *et al*, *Chem. Ber.*, 1956, **89**, 117 (*synth*)
Green, J.L., *Ind. Eng. Chem. Prod. Res. Dev.*, 1969, **8**, 171 (*synth*)

11-Aminoundecanoic acid, 9CI A-00044

[2432-99-7]

$H_2NCH_2(CH_2)_9COOH$

$C_{11}H_{23}NO_2$ M 201.308
Mp 190-192° (185-187°).
▷ YQ2293000.
Me ester; B,HCl: [29833-32-7].
$C_{12}H_{26}ClNO_2$ M 251.796
Solid (MeOH). Mp 160°.
Et ester: [15351-28-7].
$C_{13}H_{27}NO_2$ M 229.362
$Bp_{0.1}$ 125-130°.

Garmaise, D.L. *et al*, *J. Am. Chem. Soc.*, 1958, **80**, 2332 (*synth*)
Pied, J.P., *Ann. Chim. (Paris)*, 1960, **5**, 469 (*synth*)
Zahn, H. *et al*, *Chem. Ber.*, 1961, **94**, 2470 (*synth*)
East Ger. Pat., 25 925, (1963); *CA*, **60**, 10553f (*synth*)
Tosi, L., *Spectrochim. Acta*, 1963, **19**, 1097 (*ir*)
Lewis, R.J., *Sax's Dangerous Properties of Industrial Materials*, 8th Ed., Van Nostrand-Reinhold, 1992, AMW000.

Annomontacin A-00045

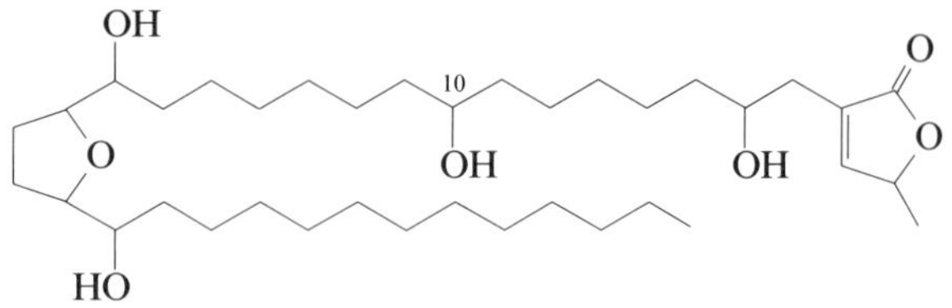

$C_{37}H_{68}O_7$ M 624.940
Constit. of *Annona montana*. Solid. Mp 34-36°. $[\alpha]_D^{20}$ +81° (c, 1 in MeOH).
10-Ketone: [123266-21-7]. ***Annonacinone***
$C_{37}H_{66}O_7$ M 622.924
Constit. of *A. montana.*

Jossang, A. *et al*, *Tetrahedron Lett.*, 1990, **31**, 1861 (*Annonacinone*)
Jossang, A. *et al*, *J. Nat. Prod. (Lloydia)*, 1991, **54**, 967 (*Annomontacin*)

Annonacin† A-00046

[111035-65-5]

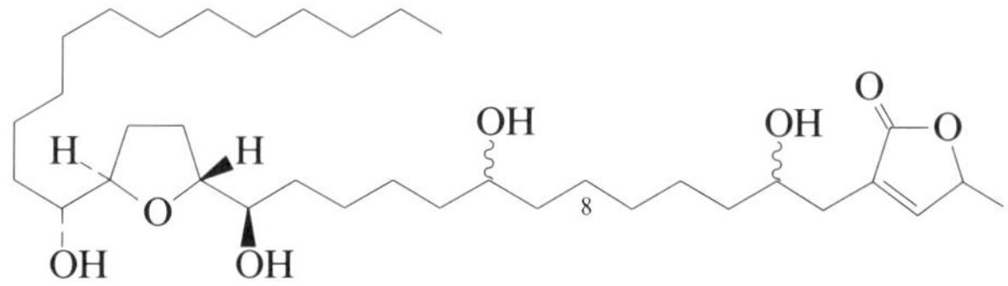

$C_{35}H_{64}O_7$ M 596.886
MF incorr. given as C_{33} in CA. Closely related to Goniothalamicin, G-00272. Constit. of *Goniothalamus giganteus* and *Annona densicoma*. Highly cytotoxic to KB cells and mouse leukaemia cells *in vitro*. Wax.
8-Hydroxy: ***8-Hydroxyannonacin***
$C_{35}H_{64}O_8$ M 612.886
Constit. of *A. densicoma*. Cytotoxic against human tumour cells *in vitro*. Waxy solid. $[\alpha]_D$ +6.1° (c, 0.12 in MeOH).
Stereoisomer: ***Annonacin†***
$C_{35}H_{64}O_7$ M 596.886
Isol. from *A. squamosa*. Amorph. $[\alpha]_D^{25}$ +23.8° (c, 0.4 in $CHCl_3$).

McCloud, T.G. *et al*, *Experientia*, 1987, **43**, 947 (*isol*)
Alkofahi, A. *et al*, *Experientia*, 1988, **44**, 83 (*isol*)
Lieb, F. *et al*, *Planta Med.*, 1990, **56**, 317 (*isol, pmr, cmr, struct*)
Yu, J.G. *et al*, *J. Org. Chem.*, 1992, **57**, 6198 (*8-Hydroxyannonacin*)

Annonin XVI A-00047

[129138-53-0]

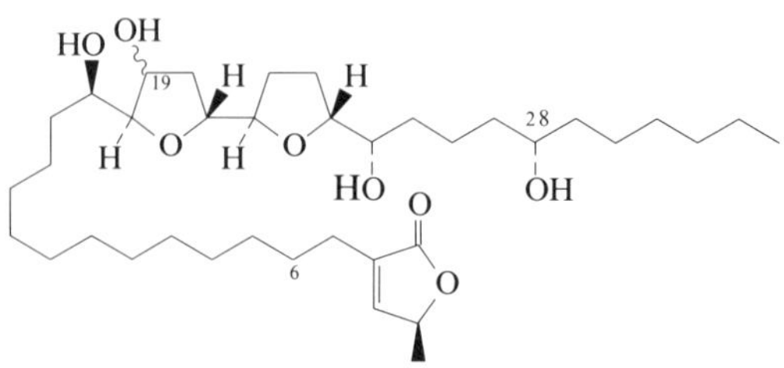

$C_{37}H_{66}O_8$ M 638.924
Constit. of *Annona squamosa*. Amorph. wax. Mp 121-123°. $[\alpha]_D^{25}$ +15.9° (c, 0.21 in CH_2Cl_2).
19-Deoxy: [120298-30-8]. ***Annonin I.*** *Squamocin. Annonin*
$C_{37}H_{66}O_7$ M 622.924

Constit. of *A. squamosa* and *Uvaria narum*. Insecticide. $[\alpha]_D^{22}$ +14.9° (c, 0.94 in MeOH). Identity of Annonin I and Squamocin stereochemically is currently based on CA registry information.

19-Deoxy, 28-ketone: [133594-24-8]. ***Squamocin-28-one***
$C_{37}H_{64}O_7$ M 620.908
Constit. of *U. narum*. Oil.

6-Hydroxy, 19,28-dideoxy: [129212-94-8]. ***Annonin VI***
$C_{37}H_{66}O_7$ M 622.924
Isol. from *A. squamosa*. $[\alpha]_D^{25}$ +15.3° (c, 0.4 in CH_2Cl_2).

[93236-32-9, 93303-16-3]

Fujimoto, Y. *et al, Chem. Pharm. Bull.*, 1988, **36**, 4802 (*isol, pmr, cmr*)
Kawazu, K. *et al, Agric. Biol. Chem.*, 1989, **53**, 2719 (*isol*)
Nonfon, M. *et al, Phytochemistry*, 1990, **29**, 1951 (*isol, pmr, cmr, ms*)
Born, L. *et al, Planta Med.*, 1990, **56**, 312 (*isol, pmr, cmr, cryst struct, derivs*)
Hisham, A. *et al, Phytochemistry*, 1991, **30**, 545 (*isol, pmr, cmr*)

Ascarosides A-00048

Three distinct ascarosides (A-C) have been separated by chromatog. Isol. from the lipid content of *Ascaris* sp.

Ascaroside A [11002-15-6]
Mp 79-80°. $[\alpha]_D$ −55° ($CHCl_3$).

Ascaroside B [11002-16-7]
Mp 83-84°. $[\alpha]_D$ −49° ($CHCl_3$).

Ascaroside C [11002-17-8]
Mp 93-95°. $[\alpha]_D$ −77.5° ($CHCl_3$).

Fouquey, C. *et al, Bull. Soc. Chim. Biol.*, 1957, **39**, 101; 1962, **44**, 69 (*isol, struct*)
Jesyk, P.F. *et al, Comp. Biochem. Physiol.*, 1967, **23**, 691 (*isol, struct*)

Ascidiatrienolide A A-00049

[122752-22-1]

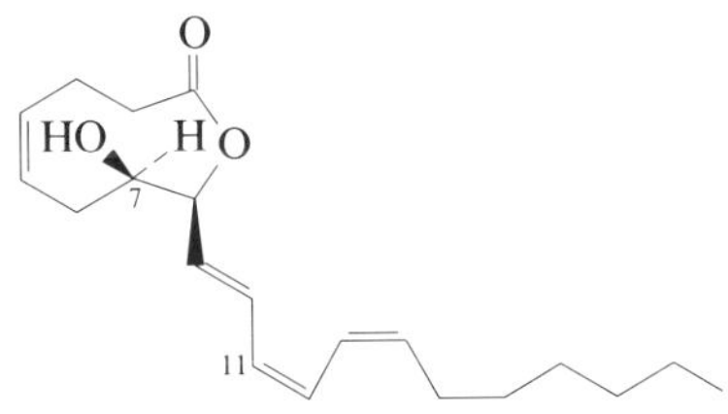

$C_{20}H_{30}O_3$ M 318.455
Constit. of *Didemnum candidum*. $[\alpha]_D$ −14.8° (c, 4.5 in $CHCl_3$).

7-Epimer: [122799-22-8]. ***Ascidiatrienolide B***
$C_{20}H_{30}O_3$ M 318.455
Constit. *D. candidum*. $[\alpha]_D$ −4.1° (c, 3.4 in $CHCl_3$).

11E-Isomer: [122799-23-9]. ***Ascidiatrienolide C***
$C_{20}H_{30}O_3$ M 318.455
Constit. of *D. candidum*. $[\alpha]_D$ −10.6° (c, 11.3 in $CHCl_3$).

Lindquist, N. *et al, Tetrahedron Lett.*, 1989, **30**, 2735.

Asteriacerebroside A A-00050

[134178-88-4]

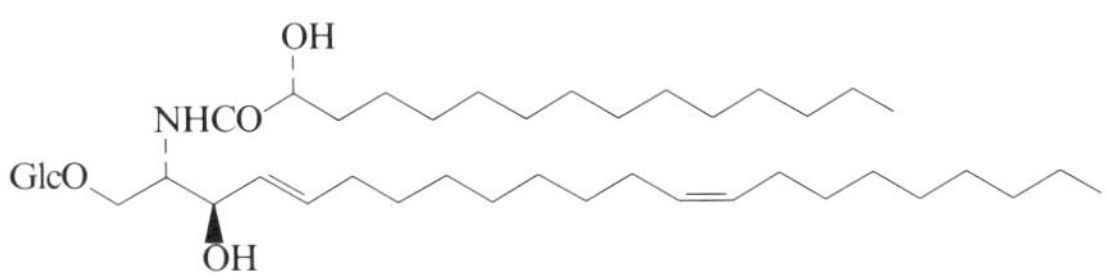

$C_{43}H_{81}NO_9$ M 756.114
Glycosphingolipid. Isol. from the starfish *Asterias amurensis versicolor*. Needles + $4H_2O$ (MeOH). Mp 182-184°. $[\alpha]_D^{25}$ +8.5° (c, 0.27 in 1-propanol).

Higuchi, R. *et al, Justus Liebigs Ann. Chem.*, 1991, 745 (*isol, pmr, cmr, struct*)

Asteriacerebroside B A-00051

[134178-89-5]

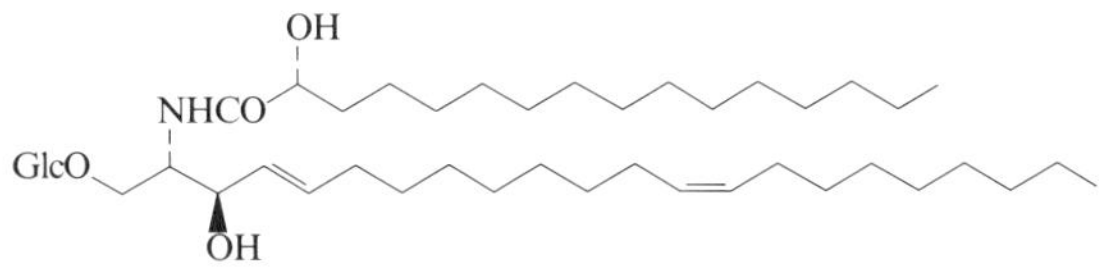

$C_{44}H_{83}NO_9$ M 770.141
Glycosphingolipid. Isol. from the starfish *Asterias amurensis versicolor*. Needles + $3H_2O$ (MeOH). Mp 178-180°. $[\alpha]_D^{25}$ +8.3° (c, 0.28 in 1-propanol).

Higuchi, R. *et al, Justus Liebigs Ann. Chem.*, 1991, 745 (*isol, pmr, cmr, struct*)

Asteriacerebroside C A-00052

[134178-90-8]

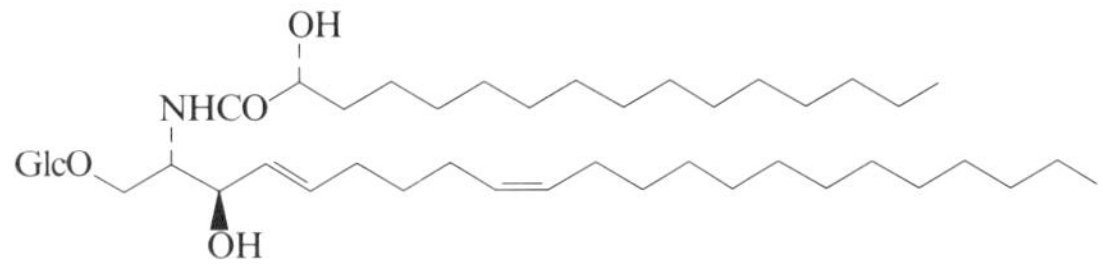

$C_{44}H_{83}NO_9$ M 770.141
Glycosphingolipid. Isol. from the starfish *Asterias amurensis versicolor*. Needles + $3H_2O$ (MeOH). Mp 183-186°. $[\alpha]_D^{25}$ +9.4° (c, 0.28 in 1-propanol).

Higuchi, R. *et al, Justus Liebigs Ann. Chem.*, 1991, 745 (*isol, pmr, cmr, struct*)

Asteriacerebroside D A-00053

[134178-91-9]

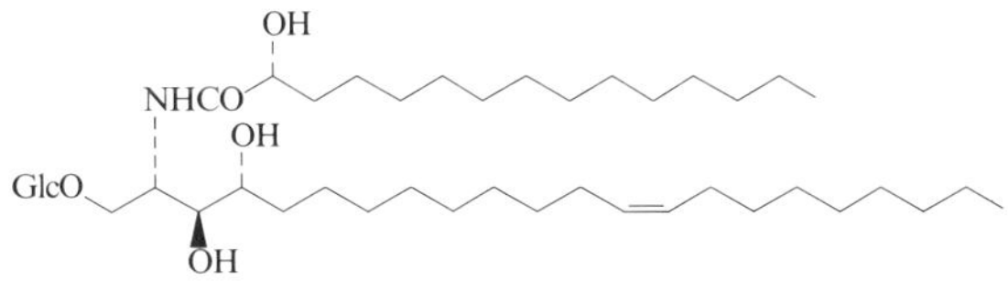

$C_{43}H_{83}NO_{10}$ M 774.129
Glycosphingolipid. Isol. from the starfish *Asterias amurensis versicolor*. Needles + $2H_2O$ (MeOH). Mp 174-176°. $[\alpha]_D^{25}$ +10.5° (c, 0.4 in 1-propanol).

Higuchi, R. *et al, Justus Liebigs Ann. Chem.*, 1991, 745 (*isol, pmr, cmr, struct*)

Asteriacerebroside E A-00054

[134178-92-0]

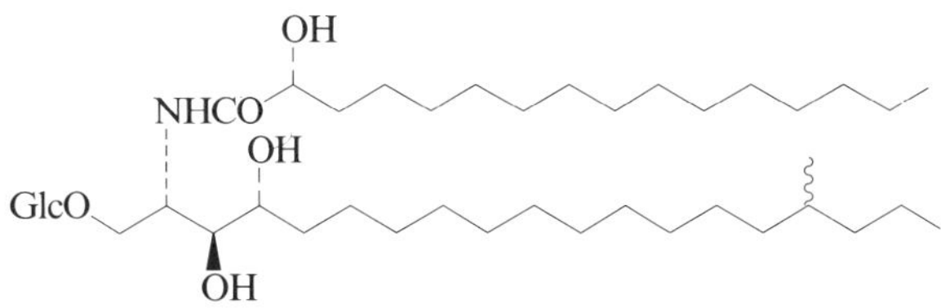

$C_{42}H_{83}NO_{10}$ M 762.118

Glycosphingolipid. Isol. from the starfish *Asterias amurensis versicolor*. Needles + $1H_2O$ (MeOH). Mp 216-218°. $[\alpha]_D^{25}$ +12.2° (c, 0.32 in 1-propanol).

Higuchi, R. *et al*, *Justus Liebigs Ann. Chem.*, 1991, 745 (*isol, pmr, cmr, struct*)

Asteriacerebroside F A-00055

[134178-93-1]

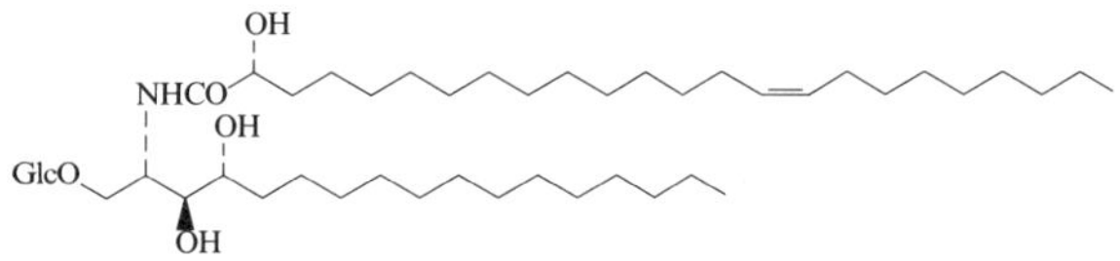

$C_{47}H_{91}NO_{10}$ M 830.237

Glycosphingolipid. Isol. from the starfish *Asterias amurensis versicolor*. Needles + $4H_2O$ (MeOH). Mp 167-169°. $[\alpha]_D^{25}$ +12.6° (c, 0.33 in 1-propanol).

Higuchi, R. *et al*, *Justus Liebigs Ann. Chem.*, 1991, 745 (*isol, pmr, cmr, struct*)

Astrocerebrosides A-00056

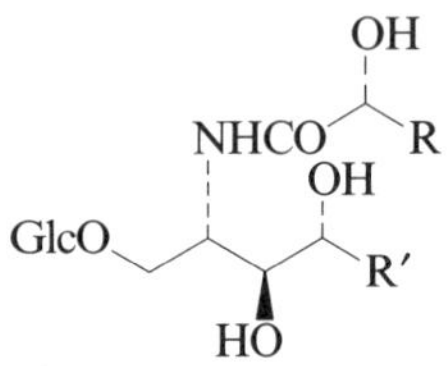

Astrocerebroside

A, R = $(CH_2)_{12}CH_3$, R′ = $(CH_2)_4CH{=}CH(CH_2)_{11}CH_3$ (*Z*-)
B, R = $(CH_2)_{13}CH_3$, R′ = $(CH_2)_4CH{=}CH(CH_2)_{11}CH_3$ (*Z*-)
C, R = $(CH_2)_{21}CH_3$, R′ = $(CH_2)_{10}CH(CH_3)_2$

Isol. from $CHCl_3$/MeOH extract of the starfish *Astropecten latespinosus*.

Astrocerebroside A [126665-07-4]
$C_{43}H_{83}NO_{10}$ M 774.129
Needles + $1H_2O$ (MeOH). Mp 188-192°. $[\alpha]_D^{25}$ +10.3° (c, 1 in 1-propanol).

Astrocerebroside B [122823-41-0]
$C_{44}H_{85}NO_{10}$ M 788.156
Needles + $3H_2O$ (MeOH). Mp 198-202°. $[\alpha]_D^{25}$ +10.8° (c, 1 in 1-propanol).

Astrocerebroside C [122823-47-6]
$C_{47}H_{93}NO_{10}$ M 832.252
Needles + $2H_2O$ (MeOH). Mp 218-219°. $[\alpha]_D^{25}$ +10.7° (c, 1 in 1-propanol).

Higuchi, R. *et al*, *Justus Liebigs Ann. Chem.*, 1990, 659.

14-Azaprostanoic acid A-00057

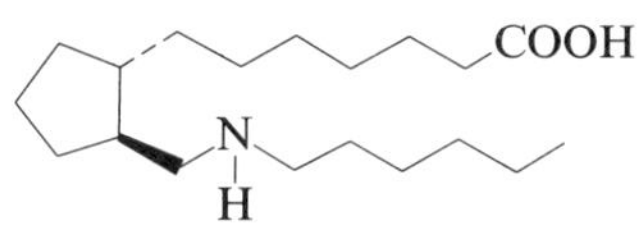

$C_{19}H_{37}NO_2$ M 311.507

Cryst. ($Et_2O/CHCl_3$). Mp 86-88°.

Cho, B.P. *et al*, *J. Org. Chem.*, 1986, **51**, 4279 (*synth, pmr, ms*)

B

Behenic acid tryptamide **B-00001**

N-*[2-(1*H*-Indol-3-yl)ethyl]docosanamide, 9CI. 3-[2-(Docosanoylaminoethyl)]indole*

[7367-79-5]

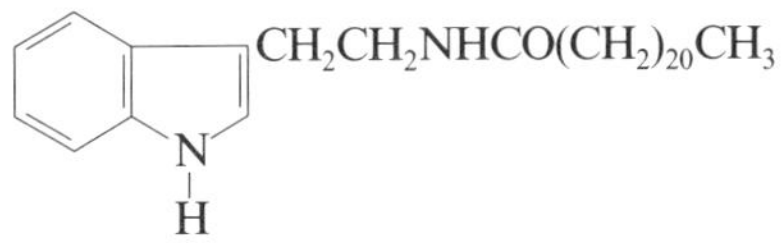

$C_{32}H_{54}N_2O$ M 482.791

Constit. of shell lipids of cocoa beans. Cryst. (Me_2CO/Me_2CO). Mp 116-116.5°.

N-*Ac:* Cryst. (C_6H_6). Mp 95-96°.

Sacher, H., *Z. Lebensm.-Unters. -Forsch.*, 1965, **128**, 264 (*isol, synth, uv, ir*)
Szelag, H. *et al, Nahrung*, 1988, **32**, 285 (*anal*)

Benzyldimethyldodecylammonium(1+) **B-00002**

N-*Dodecyl*-N,N-*dimethylbenzenemethanaminium, 9CI. Benzododecinium. Ajatin. Sterinol*

[10328-35-5]

$H_3C(CH_2)_{11}N^{\oplus}Me_2CH_2Ph$

$C_{21}H_{38}N^{\oplus}$ M 304.538 (ion)

Phase-transfer catalyst. Disinfectant and antiseptic. Sol. H_2O, acids; spar. sol. EtOH; insol. CCl_4.

Bromide: [7281-04-1].
$C_{21}H_{38}BrN$ M 384.442
Used as 10% aq. soln. for photometric detn. of Be (λ_{max} 610 nm, ε 105000); extraction of V(*V*) (C_6H_6/octanol, pH 3.8), $[Fe(CN)_6]^{4\ominus}$; extraction of Prussian blue. Cryst. Sol. H_2O. Mp 47°.
▷ BO7875000.

Chloride: [139-07-1]. *Benzododecinium chloride, INN*
$C_{21}H_{38}ClN$ M 339.991
Mp 31-32°.

Cella, J.A. *et al, J. Am. Chem. Soc.*, 1952, **74**, 2061 (*synth*)
Stepanenko, B.N. *et al, CA*, 1968, **69**, 106057 (*synth*)
Namba, Y. *et al, CA*, 1971, **74**, 125014 (*synth*)
Bartecki, A. *et al, Nukleonika*, 1973, **18**, 71; *CA*, **79**, 35617q (*detn, V*)
Galik, A. *et al, Talanta*, 1974, **21**, 307.
Palacek, J. *et al, Synthesis*, 1976, 550 (*use*)
Fieser and Fieser's Reagents for Organic Synthesis, Wiley, 1979, **7**, 16 (*use*)
Kwapulińska, G. *et al, Mikrochim. Acta*, 1984, **1**, 333 (*detn, Be*)
He, H., *CA*, 1986, **104**, 206795 (*synth*)
Negwer, M., *Organic-Chemical Drugs and their Synonyms*, 6th Ed., Akademie-Verlag, Berlin, 1987, 6402.
Lewis, R.J., *Sax's Dangerous Properties of Industrial Materials*, 8th Ed., Van Nostrand-Reinhold, 1992, BEM000, BEO000.

Benzyldimethyl(tetradecyl)ammonium(1+), 8CI **B-00003**

N,N-*Dimethyl*-N-*tetradecylbenzenemethanaminium, 9CI. Miristalkonium*

$H_3C(CH_2)_{13}N^{\oplus}Me_2CH_2Ph$

$C_{23}H_{42}N^{\oplus}$ M 332.592 (ion)

Antiinfective, surface-active agent.

Chloride: [139-08-2]. *Miristalkonium chloride, INN. Faringets active substance*
$C_{23}H_{42}ClN$ M 368.044
Mp 60-61°.
▷ BO7150000.

U.S. Pat., 3 360 563, (1967); *CA*, **68**, 38525 (*synth, props*)
Namba, Y. *et al, Kogyo Kagaku Zasshi*, 1971, **74**, 413 (*synth*)
Martindale, The Extra Pharmacopoeia, 28th/29th Ed., Pharmaceutical Press, London, 1982/1989, 2259.
Lewis, R.J., *Sax's Dangerous Properties of Industrial Materials*, 8th Ed., Van Nostrand-Reinhold, 1992, TCA500.

Benzylhexadecyldimethylammonium(1+) **B-00004**

N-*Hexadecyl*-N,N-*dimethylbenzenemethanaminium, 9CI. Cetalkonium. Acinol. Armil. Baktonium. Banicol. Bicetonium. Ceetolan. Cetol. Spilan. Tetraseptan. Zettyn. NSC 32942. Win 357*

$PhCH_2N^{\oplus}Me_2(CH_2)_{15}CH_3$

$C_{25}H_{46}N^{\oplus}$ M 360.645 (ion)

Antiinfective and surface-active agent. Cryst. (EtOAc/pet. ether).

Chloride: [122-18-9]. *Cetalkonium chloride, INN, BAN, USAN*
$C_{25}H_{46}ClN$ M 396.098
Mp 68° (59°).

U.S. Pat., 2 075 958, (1937); *CA*, **31**, 3597 (*synth, props*)
Westphal, O. *et al, Ber.*, 1940, **73**, 1011 (*synth*)
Shelton, R.S. *et al, J. Am. Chem. Soc.*, 1946, **68**, 753 (*synth, props*)
Martindale, The Extra Pharmacopoeia, 28th/29th Ed., Pharmaceutical Press, London, 1982/1989, 2214.

Bis(3-oxoundecyl) disulfide **B-00005**

1,1'-Dithiodi-3-undecanone, 8CI

[33796-37-1]

$H_3C(CH_2)_7COCH_2CH_2SSCH_2CH_2CO(CH_2)_7CH_3$

$C_{22}H_{42}O_2S_2$ M 402.705

Constit. of *Dictyopteris plagiogramma*. Cryst. Mp 67-67.5°.

Roller, P. *et al, J. Chem. Soc., Chem. Commun.*, 1971, 503.

Bis(3-oxoundecyl) tetrasulfide **B-00006**

1,1'-Tetrathiobis-3-undecanone, 9CI

[34852-29-4]

$H_3C(CH_2)_7COCH_2CH_2SSSSCH_2CH_2CO(CH_2)_7CH_3$

$C_{22}H_{42}O_2S_4$ M 466.837

Constit. of *Dictyopteris plagiogramma* and *D. australis*. Mp 32-33°.

Moore, R.E., *J. Chem. Soc., Chem. Commun.*, 1971, 1168.

Bis(3-oxoundecyl) trisulfide **B-00007**

1,1'-Trithiobis-3-undecanone, 9CI

[34852-28-3]

$H_3C(CH_2)_7COCH_2CH_2SSSCH_2CH_2CO(CH_2)_7CH_3$

$C_{22}H_{42}O_2S_3$ M 434.771

Constit. of *Dictyopteris plagiogramma* and *D. australis*. Needles (MeOH). Mp 60.5-61°.

Moore, R.E., *J. Chem. Soc., Chem. Commun.*, 1971, 1168.

Bongkrekic acid B-00008

20-(Carboxymethyl)-6-methoxy-2,5,17-trimethyl-2,4,8,10,14,18,20-docosaheptaenedioic acid, 9CI
[11076-19-0]

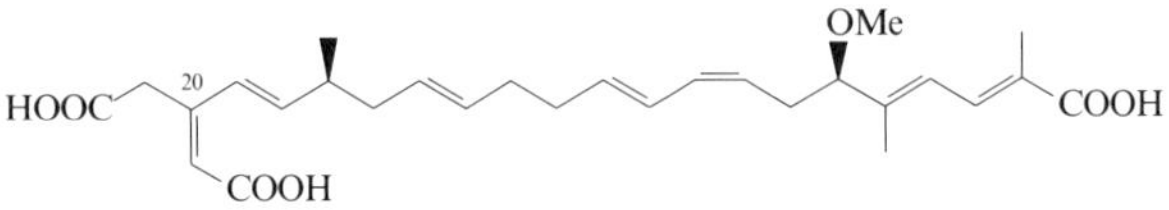

$C_{28}H_{38}O_7$ M 486.604

(20E)-form [60132-21-0]
Isobongkrekic acid
Metab. of *Pseudomonas cocovenenans.*

(20Z)-form
Prod. by *P. cocovenenans* growing on a copra-containing medium. Toxin. Inhibitor of ADP transport by mitochondria. Amorph. solid. Mp 190° (50-60°) dec. $[\alpha]_D^{22}$ +165° (c, 2 in $NaHCO_3$ aq.). λ_{max} 237 (ε 32 000) and 267 nm(36 700) (MeOH).

Nugteren, D.H. *et al, Recl. Trav. Chim. Pays-Bas (J. R. Neth. Chem. Soc.)*, 1957, **76**, 13 (*isol*)
Zylber, J. *et al, Experientia*, 1973, **29**, 387, 648 (*abs config*)
de Bruijn, J. *et al, Tetrahedron*, 1973, **29**, 1541 (*struct*)
Lauquin, G.J.M. *et al, Biochemistry*, 1976, **15**, 2323 (*deriv*)
Corey, E.J. *et al, J. Am. Chem. Soc.*, 1984, **106**, 462 (*synth*)
Chatterjee, S. *et al, J. Org. Chem.*, 1988, **53**, 4883 (*isol, pmr, cmr, Isobongkrekic acid*)

3-Bromobutanoic acid B-00009

[2623-86-1]

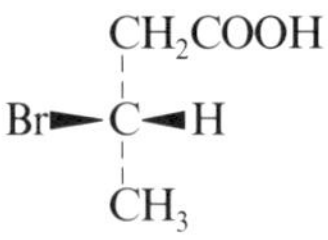

(S)-form

$C_4H_7BrO_2$ M 167.002

(S)-form [77790-08-0]
Mp 17-18°. Bp_{25} 131°. $[\alpha]_D^{25}$ +20.2° (c, 1 in Et_2O) (58.4% e.e.).

(±)-form
Mp 20°. Bp_{16} 122°.
Et ester: [7425-49-2].
$C_6H_{11}BrO_2$ M 195.056
Liq. Bp_{12} 72°.
Nitrile: [20965-20-2] *2-Bromo-1-cyanopropane*
C_4H_6BrN M 148.002
Mp 92-93°.
Amide:
C_4H_8BrNO M 166.017
Cryst. ($CHCl_3$). Mp 92.5-93°.

Lovén, J.M. *et al, Ber.*, 1915, **48**, 1256 (*synth*)
Takebayashi, M., *Nippon Kagaku Kaishi*, 1943, **64**, 41 (*amide*)
Rosenmund, P., *Chem. Ber.*, 1975, **108**, 208 (*ester*)
Tanaka, Y. *et al, J. Org. Chem.*, 1990, **55**, 564 (*synth, pmr, ir*)

2-Bromo-2-butenoic acid, 9CI B-00010

[27243-23-8]

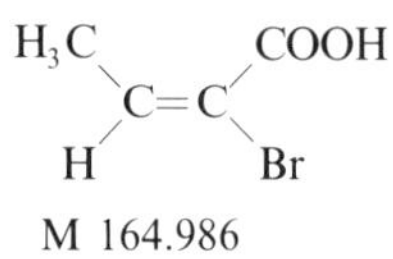

(E)-form

$C_4H_5BrO_2$ M 164.986

(E)-form [36297-22-0]
2-Bromoisocrotonic acid
Needles (H_2O). Mp 92°.
Et ester:
$C_6H_9BrO_2$ M 193.040
Bp_{15} 95-97°.
Amide:
C_4H_6BrNO M 164.002
Mp 60°.

(Z)-form [5405-34-5]
2-Bromocrotonic acid
Needles (H_2O). Mp 107-108°.
Me ester: [22966-48-9].
$C_5H_7BrO_2$ M 179.013
Reagent for bisannellations. Bp_9 61°.
p-*Bromophenacyl ester:*
$C_{12}H_{10}Br_2O_3$ M 362.017
Mp 96-97°.
Amide: Mp 93-94°.

Michael, A. *et al, J. Prakt. Chem.*, 1892, **46**, 257.
James, T.C., *J. Chem. Soc.*, 1910, **97**, 1565.
Bachmann, G.B., *J. Am. Chem. Soc.*, 1933, **55**, 4279.
Backer, H.J. *et al, Recl. Trav. Chim. Pays-Bas (J. R. Neth. Chem. Soc.)*, 1935, **54**, 551.
Mokhnach, V.O. *et al, CA*, 1937, **31**, 5249.
Klein, J. *et al, J. Org. Chem.*, 1970, **35**, 666 (*synth*)
Hagiwara, H. *et al, J. Chem. Soc., Chem. Commun.*, 1976, 413 (*synth, use*)
Heasley, V.L. *et al, J. Org. Chem.*, 1979, **44**, 2608 (*synth*)
Fieser and Fieser's Reagents for Organic Synthesis, Wiley, 1979, **7**, 235 (*use*)

Bromochloroacetic acid B-00011

Bromochloroethanoic acid
[5589-96-8]

ClCHBrCOOH

$C_2H_2BrClO_2$ M 173.393

(±)-form
Mp 31.5°. Bp 215° dec., Bp_{11} 103-104°.
Et ester: [22524-32-9].
$C_4H_6BrClO_2$ M 201.447
Liq. Bp 174° dec., Bp_{12} 65°.
Chloride: [38282-29-0].
C_2HBrCl_2O M 191.839
Bp 138-139°.
Amide: [62872-34-8].
$C_2H_3BrClNO$ M 172.408
Cryst. Sol. hot H_2O. Mp 126° (117°).
Nitrile: [83463-62-1]. *Bromochloroacetonitrile. Bromochlorocyanomethane*
$C_2HBrClN$ M 154.393
Detected in drinking water. Bp 138-140°.

▷ Carcinogenic and mutagenic to mice (skin and lung), rat (liver) and cultured human CCRF-CEM cells.

Compton, H. *et al, J. Chem. Soc.*, 1921, 119, 1874 (*synth*)
Backer, H.J. *et al, J. Chem. Soc.*, 1928, 2125 (*synth*)
McBee, E.T. *et al, J. Am. Chem. Soc.*, 1956, **78**, 4595 (*ester*)
Oliver, B.G. *et al, Environ. Sci. Technol.*, 1983, **17**, 80 (*nitrile*)
Coleman, W.E. *et al, Environ. Sci. Technol.*, 1984, **18**, 674 (*ms, nitrile*)
Lewis, R.J., *Sax's Dangerous Properties of Industrial Materials*, 8th Ed., Van Nostrand-Reinhold, 1992, BMY800.

2-Bromodecanal B-00012

$H_3C(CH_2)_7CHBrCHO$

$C_{10}H_{19}BrO$ M 235.163

(±)-form
Oil.

Satoh, Y. *et al, Synthesis*, 1985, 406 (*synth, ir, pmr, ms*)

1-Bromodecane, 9CI B-00013

Decyl bromide

[112-29-8]

$H_3C(CH_2)_8CH_2Br$

$C_{10}H_{21}Br$ M 221.180

d_4^{20} 1.069. Bp_8 104-104.4°.

▷ HD6850000.

Tuot, M., *Bull. Soc. Chim. Fr.*, 1946, 363 (*synth*)
Tits-Skvortsova, I.N. *et al*, *Zh. Obshch. Khim.*, 1951, **21**, 242 (*synth*)
LeFevre, R.J.W. *et al*, *Aust. J. Chem.*, 1959, **12**, 743 (*ir*)
McLafferty, F.W., *Anal. Chem.*, 1962, **34**, 2 (*ms*)
U.S. Pat., 3 683 037, (1972); *CA*, **77**, 125943 (*synth*)
U.S. Pat., 3 763 241, (1973); *CA*, **79**, 137275 (*synth*)

2-Bromodecanoic acid B-00014

2-Bromocapric acid

[2623-95-2]

$H_3C(CH_2)_7CHBrCOOH$

$C_{10}H_{19}BrO_2$ M 251.163

Used for extraction preconcentration of Pb. Needles. Sol. EtOH, Et_2O, Me_2CO, C_6H_6, $CHCl_3$, alkalis.

(±)-*form*

Has fungicidal props. Sol. Et_2O. Mp 4°.

Me ester:

$C_{11}H_{21}BrO_2$ M 265.190

$Bp_{0.5}$ 89°.

Bagard, P., *Bull. Soc. Chim. Fr.*, 1907, **1**, 307 (*synth*)
Katon, J.E. *et al*, *Appl. Spectrosc.*, 1971, **25**, 229 (*ir*)
Gershon, H. *et al*, *Antimicrob. Agents Chemother.*, 1973, **4**, 435 (*use*)
Onishenko, T.A. *et al*, *Zh. Anal. Khim.*, 1987, **42**, 1616 (*use, Pb*)

9-Bromodecanoic acid B-00015

[134981-66-1]

$H_3CCHBr(CH_2)_7COOH$

$C_{10}H_{19}BrO_2$ M 251.163

(±)-*form*

Oil. $Bp_{0.075}$ 100°.

Moore, B.P. *et al*, *Aust. J. Chem.*, 1976, **29**, 1365.
Bartra, M. *et al*, *J. Org. Chem.*, 1991, **56**, 5132 (*synth, pmr, cmr, ir*)

10-Bromodecanoic acid B-00016

10-Bromocapric acid

[50530-12-6]

$BrCH_2(CH_2)_8COOH$

$C_{10}H_{19}BrO_2$ M 251.163

Plates (pet. ether). Mp 42°. Bp_2 163-165°.

Me ester:

$C_{11}H_{21}BrO_2$ M 265.190

Bp_{12} 161-163°.

Hunsdiecker, H. *et al*, *Chem. Ber.*, 1942, **75**, 291 (*synth*)
Galli, C. *et al*, *J. Am. Chem. Soc.*, 1973, **95**, 8374 (*use*)
Muramoto, Y. *et al*, *Agric. Biol. Chem.*, 1974, **38**, 2197 (*use*)
Schneider, F. *et al*, *Helv. Chim. Acta*, 1974, **57**, 434 (*use*)
Bartra, M. *et al*, *J. Org. Chem.*, 1991, **56**, 5132 (*synth, pmr, cmr, ir*)

10-Bromo-1-decanol B-00017

[53463-68-6]

$BrCH_2(CH_2)_8CH_2OH$

$C_{10}H_{21}BrO$ M 237.179

Liq. $Bp_{0.1}$ 120-122°, $Bp_{0.07}$ 99-101°.

Camps, F. *et al*, *Org. Prep. Proced. Int.*, 1983, **15**, 63.
Schwarz, M. *et al*, *J. Org. Chem.*, 1986, **51**, 260 (*synth*)

1-Bromodocosane B-00018

Docosyl bromide

[6938-66-5]

$H_3C(CH_2)_{20}CH_2Br$

$C_{22}H_{45}Br$ M 389.502

Cryst. (Me_2CO). Mp 44°. $Bp_{0.6}$ 225°.

v. Braun, J. *et al*, *Justus Liebigs Ann. Chem.*, 1929, **472**, 132 (*synth*)
Nakamura, N., *CA*, 1976, **84**, 16340f (*cryst struct*)

16-Bromo-15-docosene-1,3-diyn-5-ol B-00019

18-Dihydrorenierin 1

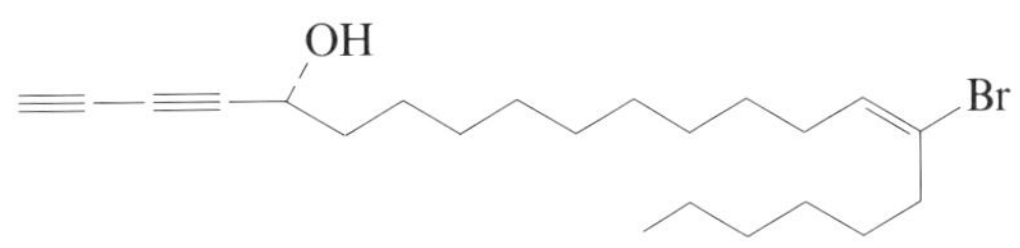

$C_{22}H_{35}BrO$ M 395.422

(*R,E*)-*form* [64010-40-8]

Constit. of *Reniera fulva*. Oil. $[\alpha]_D$ −5.4° (c, 5.4 in $CHCl_3$).

Cimino, G. *et al*, *Tetrahedron Lett.*, 1977, 1325 (*isol*)

16-Bromo-15-docosene-1,3-diyn-5-one B-00020

Renierin 1

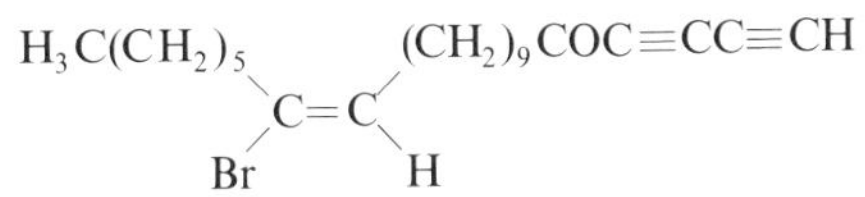

$C_{22}H_{33}BrO$ M 393.406

(*E*)-*form* [63987-87-1]

cis-*form*

Constit. of *Reniera fulva*. Oil.

Cimino, G. *et al*, *Tetrahedron Lett.*, 1977, 1325 (*isol, struct*)

12-Bromododecanal B-00021

[81819-07-0]

$BrCH_2(CH_2)_{10}CHO$

$C_{12}H_{23}BrO$ M 263.217

No phys. props. given, purified as the ethylene acetal. Distn. caused rapid oligomerisation.

Ethylene acetal: [84451-51-4].

$C_{14}H_{27}BrO_2$ M 307.270

Cryst. (pet. ether). Mp 23-24°. $Bp_{0.05}$ 134-136°.

Dinitrophenylhydrazone: Mp 100-102°.

Bidd, I. *et al*, *J. Chem. Soc., Perkin Trans. 1*, 1983, 1369 (*synth, ir*)
Porter, N.A. *et al*, *J. Am. Chem. Soc.*, 1989, **110**, 3554 (*synth, ir, pmr, cmr*)

1-Bromododecane B-00022
Dodecyl bromide. Lauryl bromide
[143-15-7]

$H_3C(CH_2)_{10}CH_2Br$

$C_{12}H_{25}Br$ M 249.233
Liq. d_4^{20} 1.038. Bp_{100} 199.5-201.5°, $Bp_{0.2}$ 92-93°.

Org. Synth., Coll. Vol., 2, 1943, 246 (*synth*)
Hesse, J.H., *J. Chem. Eng. Data*, 1967, **12**, 266.
Kuntz Jr., I.D. *et al*, *J. Am. Chem. Soc.*, 1967, **89**, 6008 (*nmr*)
Stepanenko, B.N. *et al*, *CA*, 1968, **69**, 106057 (*synth*)
Levine, Y.K. *et al*, *J. Chem. Phys.*, 1974, **60**, 2890 (*cmr*)
Salzer, R. *et al*, *Spectrochim. Acta, Part A*, 1974, **30**, 205 (*ir*)
Sax, N.I., *Dangerous Properties of Industrial Materials*, 5th Ed., Van Nostrand-Reinhold, 1979, 764.

2-Bromododecane, 9CI, 8CI B-00023
[13187-99-0]

$H_3C(CH_2)_9CHBrCH_3$

$C_{12}H_{25}Br$ M 249.233
(±)-***form***
Liq. Bp_{10} 125-130°.

Esturo, M. *et al*, *Nagoya Kogyo Daigaku Gakuho*, 1954, **6**, 271; *CA*, **50**, 11943 (*synth*)

2-Bromododecanoic acid B-00024
2-Bromolauric acid

$H_3C(CH_2)_9CHBrCOOH$

$C_{12}H_{23}BrO_2$ M 279.216
(±)-***form*** [111-56-8]
Mp 32°. $Bp_{0.4}$ 161-164°.
Me ester: [617-60-7].
$C_{13}H_{25}BrO_2$ M 293.243
Bp_{12} 169-171°, $Bp_{0.15}$ 113°.
Et ester:
$C_{14}H_{27}BrO_2$ M 307.270
Bp_{12} 175-177°, Bp_3 144-146°.

v. Auwers, K. *et al*, *Ber.*, 1891, **24**, 2224 (*synth*)
Peron, R. *et al*, *C. R. Hebd. Seances Acad. Sci.*, 1962, **254**, 4275 (*ir*)
Reinheckel, H., *J. Prakt. Chem.*, 1962, **15**, 260 (*synth*)
Okamoto, Y. *et al*, *Kogyo Kagaku Zasshi*, 1966, **69**, 1871; *CA*, **66**, 1116934 (*synth*)
Zeigler, H.J., *Synthesis*, 1969, **1**, 39 (*synth*)
Berry, P.J., *J. Org. Chem.*, 1972, **37**, 4396.

11-Bromododecanoic acid B-00025
11-Bromolauric acid

$H_3CCHBr(CH_2)_9COOH$

$C_{12}H_{23}BrO_2$ M 279.216
Mp 38°.

Rountree, J.M. *et al*, *Chem. Ind. (London)*, 1954, 190 (*synth*)

12-Bromododecanoic acid B-00026
12-Bromolauric acid
[73367-80-3]

$BrCH_2(CH_2)_{10}COOH$

$C_{12}H_{23}BrO_2$ M 279.216
Mp 53°.

Rountree, J.M. *et al*, *Chem. Ind. (London)*, 1954, 190 (*synth*)
Mirviss, S.B., *J. Org. Chem.*, 1989, **54**, 1948 (*synth*)

12-Bromo-1-dodecanol B-00027
[3344-77-2]

$BrCH_2(CH_2)_{10}CH_2OH$

$C_{12}H_{25}BrO$ M 265.233
Cryst. (pet. ether). Mp 31-32°. Bp_4 155-156°.

Salmon-Legagneur, F. *et al*, *Bull. Soc. Chim. Fr.*, 1965, 2270 (*synth*)
Bidd, I. *et al*, *J. Chem. Soc., Perkin Trans. 1*, 1983, 1369 (*synth*)
Ainscow, T.A. *et al*, *Tetrahedron*, 1987, **43**, 115 (*synth*)

17-Bromo-6,9-heptadecadiene B-00028

$BrH_2C(CH_2)_6CH{=}CHCH_2CH{=}CH(CH_2)_4CH_3$

$C_{17}H_{31}Br$ M 315.336
Linoleic acid precursor (see 9,12-Octadecadienoic acid, O-00050). Oil. d^{26} 1.02. $Bp_{0.15}$ 131°. n_D^{25} 1.48.

Howton, D.R. *et al*, *J. Am. Chem. Soc.*, 1952, **74**, 1109 (*synth*)

1-Bromoheptadecane B-00029
Heptadecyl bromide
[3508-00-7]

$H_3C(CH_2)_{15}CH_2Br$

$C_{17}H_{35}Br$ M 319.367
Mp 27-29°, Mp 32°. Bp_{10} 193-200°.

Oldham, J.W.H., *J. Chem. Soc.*, 1950, 100 (*synth*)
McKillop, A. *et al*, *J. Org. Chem.*, 1969, **34**, 1172 (*synth*)

1-Bromo-5,8,11-heptadecatriene B-00030

$H_3C(CH_2)_4CH{=}CHCH_2CH{=}CHCH_2CH{=}CHCH_2CH_2CH_2CH_2Br$

$C_{17}H_{29}Br$ M 313.320
(5Z,8Z,11Z)-form
γ-Linolenic acid precursor (see 6,9,12-Octadecatrienoic acid, O-00154). $Bp_{0.0001}$ 120-135°.

Nevenzel, J.C. *et al*, *J. Org. Chem.*, 1958, **23**, 933 (*synth*)

17-Bromo-3,6,9-heptadecatriene B-00031

$BrH_2C(CH_2)_6CH{=}CHCH_2CH{=}CHCH_2CH{=}CHCH_2CH_3$

$C_{17}H_{29}Br$ M 313.320
Linolenic acid precursor (see 9,12,15-Octadecatrienoic acid, O-00158). d^{25} 1.01. n_D^{25} 1.4885.

Nevenzel, J.C. *et al*, *J. Org. Chem.*, 1958, **23**, 933 (*synth*)

1-Bromo-8-heptadecene B-00032
[81861-58-7]

$H_3C(CH_2)_7CH{=}CH(CH_2)_6CH_2Br$

$C_{17}H_{33}Br$ M 317.352
Oleic acid precursor (see 9-Octadecenoic acid, O-00185). Oil.

Dauben, W.G. *et al*, *J. Org. Chem.*, 1989, **54**, 6101 (*synth, pmr, cmr*)

7-Bromo-5-heptynoic acid B-00033
[41300-60-1]

$BrCH_2C{\equiv}C(CH_2)_3COOH$

$C_7H_9BrO_2$ M 205.051
Intermed. in prostagladin synth.

Me ester: [41349-38-6].
$C_8H_{11}BrO_2$ M 219.078
Oil. $Bp_{1.0}$ 90-94°.

Corey, E.J. *et al, J. Am. Chem. Soc.*, 1973, **95**, 8483 (*synth*)
Theil, F. *et al, J. Prakt. Chem.*, 1985, **327**, 917 (*synth*)
Casy, G. *et al, Tetrahedron*, 1986, **42**, 5849 (*synth, ir, pmr*)
Belýk, K. *et al, J. Org. Chem.*, 1992, **57**, 4070 (*Et ester, nitrile, synth, ir, pmr, cmr, ms*)

6-Bromo-5,9-hexacosadienoic acid, 9CI B-00034

[121158-15-4]

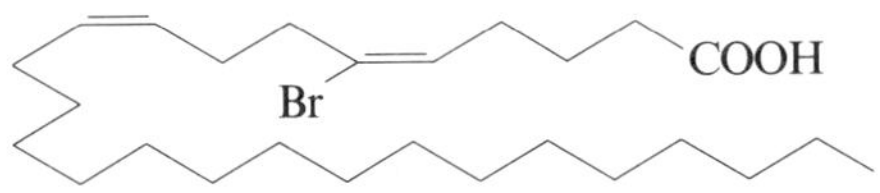

$C_{26}H_{47}BrO_2$ M 471.560
(5E,9Z)-form [121158-12-1]
Isol. from the phospholipids of a marine sponge of the family Hymeniacidonidae.
Me ester: [121158-13-2].
Oil.

Lam, W. *et al, J. Org. Chem.*, 1989, **54**, 3428 (*isol, biosynth, pmr, ms, ir*)

1-Bromohexadecane B-00035

Hexadecyl bromide. Cetyl bromide
[112-82-3]

$H_3C(CH_2)_{14}CH_2Br$

$C_{16}H_{33}Br$ M 305.341
Cryst. (EtOH/MeOH at −50°). Mp 15°, Mp 18°. Bp_{14} 193-196°.
▷ Reacts violently with oxidising materials.

Fürst, H. *et al, Z. Chem.*, 1962, **2**, 338 (*synth*)
McLafferty, F.W., *Anal. Chem.*, 1962, **34**, 2 (*ms*)
Sonnet, P.E., *Synth. Commun.*, 1976, **6**, 21 (*synth*)
Sax, N.I., *Dangerous Properties of Industrial Materials*, 5th Ed., Van Nostrand-Reinhold, 1979, 479.

2-Bromohexadecanoic acid B-00036

2-Bromopalmitic acid

$H_3C(CH_2)_{13}CHBrCOOH$

$C_{16}H_{31}BrO_2$ M 335.324
(±)-form [40305-76-8]
Cryst. (pet. ether). Mp 53°.
Me ester: [115401-57-5].
$C_{17}H_{33}BrO_2$ M 349.350
$Bp_{0.06}$ 134°.
Et ester: [14980-93-9].
$C_{18}H_{35}BrO_2$ M 363.377
Bp_{40} 242°.
Amide:
$C_{16}H_{32}BrNO$ M 334.339
Mp 84.6-84.9°.

Jones, D.B., *J. Am. Chem. Soc.*, 1915, **37**, 586 (*synth*)
Robinet, M., *Bull. Soc. Chim. Belg.*, 1931, **40**, 716 (*synth*)
Weil, J.K. *et al, J. Am. Chem. Soc.*, 1953, **75**, 2526 (*synth*)
Sweet, A.G. *et al, J. Org. Chem.*, 1956, **21**, 1426 (*ir*)
Reinheckel, H., *Chem. Ber.*, 1960, **93**, 2222 (*synth*)
Okamoto, Y. *et al, Kogyo Kagaku Zasshi*, 1966, **69**, 1871; *CA*, **66**, 116934 (*synth*)
Hussain, R. *et al, Justus Liebigs Ann. Chem.*, 1991, 963 (*synth, pmr, ms, ester*)

3-Bromohexadecanoic acid B-00037

3-Bromopalmitic acid

$H_3C(CH_2)_{12}CHBrCH_2COOH$

$C_{16}H_{31}BrO_2$ M 335.324
Cryst. (Et_2O/pet. ether). Mp 44°.

Robinet, M., *Bull. Soc. Chim. Belg.*, 1931, **40**, 710.

16-Bromohexadecanoic acid B-00038

16-Bromopalmitic acid
[2536-35-8]

$BrCH_2(CH_2)_{14}COOH$

$C_{16}H_{31}BrO_2$ M 335.324
Cryst. (pet. ether). Mp 70-70.5°. Bp_1 214-217°.
Me ester:
$C_{17}H_{33}BrO_2$ M 349.351
Cryst. (MeOH). Mp 28-29°, Mp 31°.

Chuit, P. *et al, Helv. Chim. Acta*, 1929, **12**, 463 (*synth*)
Woolford, R.G. *et al, Can. J. Chem.*, 1964, **42**, 1788 (*synth*)

16-Bromo-1-hexadecanol B-00039

[59101-28-9]

$BrCH_2(CH_2)_{14}CH_2OH$

$C_{16}H_{33}BrO$ M 321.340
Cryst. (pet. ether). Mp 53-54°.

Huang, H.C. *et al, J. Org. Chem.*, 1982, **47**, 4018 (*synth, pmr*)

16-Bromo-9-hexadecenoic acid, 9CI B-00040

[56309-90-1]

$BrCH_2(CH_2)_5CH{=}CH(CH_2)_7COOH$

$C_{16}H_{29}BrO_2$ M 333.308
(E)-form
Cryst. (MeOH). Mp 43° (40-42°).
Me ester: [57818-67-4].
Oil. $Bp_{0.3}$ 163-165°.
Et ester: $Bp_{0.01}$ 152-153°.

Blomquist, A.L. *et al, J. Am. Chem. Soc.*, 1948, **70**, 36 (*synth*)
Ger. Pat., 857 636, (1953); *CA*, **47**, 11228b (*synth*)
Mathur, H.H. *et al, J. Chem. Soc.*, 1963, 114.
Jadhav, K.S. *et al, Indian J. Chem., Sect. B*, 1978, **16**, 280 (*synth*)

1-Bromo-12-methyltridecane B-00041

Isomyristyl bromide

$(H_3C)_2CH(CH_2)_{10}CH_2Br$

$C_{14}H_{29}Br$ M 277.287
Bp_3 120-122°.

Fordyce, C.R. *et al, J. Am. Chem. Soc.*, 1933, **55**, 3372.

1-Bromo-4,7,10,13-nonadecatetraene B-00042

$H_3C(CH_2)_4(CH{=}CHCH_2)_4CH_2CH_2Br$

$C_{19}H_{31}Br$ M 339.358
All-(Z)-form [117567-53-0]
Arachidonic acid precursor (see 5,8,11,14-Eicosatetraenoic acid, E-00034). Liq.

Heitz, M.P. *et al, J. Org. Chem.*, 1989, **54**, 500 (*synth, ir, pmr, cmr, ms*)

2-Bromononanoic acid, 9CI **B-00043**

2-Bromopelargonic acid

[2623-83-8]

$H_3C(CH_2)_6CHBrCOOH$

$C_9H_{17}BrO_2$ M 237.136

(±)-form

Liq. Bp_{17} 160°, Bp_2 137-138°.

Et ester: [7425-60-7].

$C_{11}H_{21}BrO_2$ M 265.190

Liq. Bp_{20} 149-154°, $Bp_{0.5}$ 75°.

Chloride:

$C_9H_{16}BrClO$ M 255.582

Liq. Bp_9 108-110°.

Hopwood, A. *et al*, *J. Chem. Soc.*, 1911, **99**, 1577 (*chloride*)
Birnbaum, S.M. *et al*, *J. Biol. Chem.*, 1953, **203**, 333 (*synth*)
Ackerman, B. *et al*, *J. Am. Chem. Soc.*, 1957, **79**, 6524 (*ester*)

9-Bromononanoic acid, 9CI **B-00044**

9-Bromopelargonic acid

[41059-02-3]

$BrCH_2(CH_2)_7COOH$

$C_9H_{17}BrO_2$ M 237.136

Cryst. (pet. ether). Mp 36-36.5°. Bp_2 160-163°.

Et ester: [28598-81-4].

$C_{11}H_{21}BrO_2$ M 265.190

Liq. Bp_1 150-153°.

Chloride: [73674-10-9].

$C_9H_{16}BrClO$ M 255.582

Liq. Bp_{12} 153-155°.

Chuit, P. *et al*, *Helv. Chim. Acta*, 1929, **12**, 469 (*synth*)
Linnell, W.H. *et al*, *J. Pharm. Pharmacol.*, 1952, **4**, 55 (*chloride*)
Salmon-Legagneur, F. *et al*, *Bull. Soc. Chim. Fr.*, 1957, 1463 (*synth, ester*)
Galli, C. *et al*, *J. Am. Chem. Soc.*, 1973, **95**, 8374 (*synth*)

18-Bromo-9,17-octadecadiene-7,15-diynoic acid **B-00045**

$BrCH{=}CHC{\equiv}C(CH_2)_4CH{=}CHC{\equiv}C(CH_2)_5COOH$

$C_{18}H_{23}BrO_2$ M 351.282

Constit. of sponge *Xestospongia testudinaria*. Cryst. (MeOH aq.). Mp 66-67°.

(9E,17E)-form [96175-12-2]

Me ester:

$C_{19}H_{25}BrO_2$ M 365.309

Isol. from the marine sponge *X. testudinaria*.

(9Z,17E)-form

Me ester: [133084-60-3].

Isol. from the marine sponge *X. testudinaria*.

Quinn, R.J. *et al*, *Tetrahedron Lett.*, 1985, **26**, 1671.
Hirsch, S. *et al*, *Tetrahedron*, 1987, **43**, 3257 (*isol*)
Quinn, R.J. *et al*, *J. Nat. Prod. (Lloydia)*, 1991, **54**, 290 (*isol, ms, struct, pmr, cmr*)

1-Bromooctadecane **B-00046**

Octadecyl bromide

[112-89-0]

$H_3C(CH_2)_{16}CH_2Br$

$C_{18}H_{37}Br$ M 333.394

Cryst. (EtOH). Mp 28.5°. $Bp_{1.5}$ 168-169°.

Oskerk, A., *Chem. Zentralbl.*, 1914, **II**, 1264.
Seidel, F. *et al*, *Chem. Ber.*, 1936, **69**, 2567.
Hesse, J.H., *J. Chem. Eng. Data*, 1967, **12**, 266.
Salzer, R. *et al*, *Spectrochim. Acta, Part A*, 1974, **30**, 205 (*ir*)

2-Bromooctadecanoic acid **B-00047**

2-Bromostearic acid

[17108-71-3]

$H_3C(CH_2)_{15}CHBrCOOH$

$C_{18}H_{35}BrO_2$ M 363.377

(±)-form

Cryst. (pet. ether). Mp 60°.

Me ester: [53246-05-2].

$C_{19}H_{37}BrO_2$ M 377.404

Mp 25-25.5°. $Bp_{0.2}$ 160-162°.

Et ester: [5340-56-7].

$C_{20}H_{39}BrO_2$ M 391.431

Mp 35-36°.

Sinclair, R.G. *et al*, *J. Am. Chem. Soc.*, 1952, **74**, 2578 (*ir*)
Reinheckel, H., *Chem. Ber.*, 1960, **93**, 2222 (*synth*)
Okamoto, Y. *et al*, *Kogyo Kagaku Zasshi*, 1966, **69**, 1871; *CA*, **66**, 116934 (*synth*)
Ucciani, E., *Bull. Soc. Chim. Fr.*, 1967, 1913 (*synth*)
Ziegler, H.J. *et al*, *Synthesis*, 1969, **1**, 39 (*synth*)

12-Bromooctadecanoic acid **B-00048**

$H_3C(CH_2)_5CHBr(CH_2)_{10}COOH$

$C_{18}H_{35}BrO_2$ M 363.377

(±)-form

Cryst. (pet. ether). Mp 37.0-37.8°.

Stein, R.A., *J. Chromatogr.*, 1961, **6**, 118 (*synth*)

18-Bromooctadecanoic acid **B-00049**

18-Bromostearic acid

[2536-38-1]

$BrCH_2(CH_2)_{16}COOH$

$C_{18}H_{35}BrO_2$ M 363.377

Cryst. (C_6H_6/pet. ether). Mp 75.2-75.8°.

Chuit, P. *et al*, *Helv. Chim. Acta*, 1929, **12**, 463 (*synth*)
DeVries, V.G. *et al*, *J. Med. Chem.*, 1976, **19**, 946 (*synth*)

18-Bromo-17-octadecene-5,7,15-triynoic acid **B-00050**

$BrCH{=}CHC{\equiv}C(CH_2)_6C{\equiv}CC{\equiv}C(CH_2)_3COOH$

$C_{18}H_{21}BrO_2$ M 349.267

(E)-form [140423-91-2] ***Xestospongic acid***

Isol. from the marine sponge *Xestospongia testudinaria*. Possesses antimicrobial activity. Unstable acid, not obt. pure. Characterised as Me ester.

Me ester: Yellow oil.

Et ester: [140405-24-9].

$C_{20}H_{25}BrO_2$ M 377.320

Isol. from *X. testudinaria*. $Na^{\oplus}/K^{\oplus}$ ATPase inhibitor. Shows weak antimicrobial activity. Pale yellow oil.

Bourguet-Kondracki, M.L. *et al*, *Tetrahedron Lett.*, 1992, **33**, 225 (*isol, pmr, cmr, struct*)

2-Bromooctanal B-00051
[35066-22-9]

$H_3C(CH_2)_5CHBrCHO$

$C_8H_{15}BrO$ M 207.110

(±)-*form* [73368-20-4]
Oil.

Satoh, Y. *et al*, *Synthesis*, 1985, 406 (*synth, ir, pmr, ms*)

2-Bromopentanoic acid, 9CI B-00052
2-Bromovaleric acid, 8CI
[584-93-0]

$H_3CCH_2CH_2CHBrCOOH$

$C_5H_9BrO_2$ M 181.029

▷ YV6191500.

(±)-*form* [2681-92-7]
Liq. d_4^{20} 1.38. Bp_{12} 118°.
Et ester: [79584-43-3].
$C_7H_{13}BrO_2$ M 209.082
Liq. Bp 190-192°, Bp_{16} 84-86°.
Anilide:
$C_{11}H_{14}BrNO$ M 256.142
Cryst. Mp 44-46°.

[615-83-8, 32835-74-8, 42990-12-5]

Juslin, W., *Ber.*, 1884, **17**, 2504 (*ester*)
Favorskii, A.E. *et al*, *CA*, 1936, **30**, 3405 (*synth*)
Smissman, E.E., *J. Am. Chem. Soc.*, 1954, **76**, 5805 (*synth*)
Berry, J.P. *et al*, *J. Org. Chem.*, 1972, **37**, 4396 (*ester*)
Olah, G.A. *et al*, *Helv. Chim. Acta*, 1983, **66**, 1028 (*synth, cmr*)
Lewis, R.J., *Sax's Dangerous Properties of Industrial Materials*, 8th Ed., Van Nostrand-Reinhold, 1992, BOL250.

1-Bromotetradecane, 8CI B-00053
Tetradecyl bromide
[112-71-0]

$H_3C(CH_2)_{12}CH_2Br$

$C_{14}H_{29}Br$ M 277.287
d^{25} 1.018. Mp 4.5°. Bp_{17} 162-165.5°.

U.K. Pat., 629 337, (1949); *CA*, **44**, 4487.
Heston, M. *et al*, *J. Am. Chem. Soc.*, 1950, **72**, 2071.
Kaneko, T. *et al*, *Nippon Kagaku Zasshi*, 1955, **76**, 281; *CA*, **51**, 17722 (*synth*)
McLafferty, F.W., *Anal. Chem.*, 1962, **34**, 2 (*ms*)
Hesse, J.E., *J. Chem. Eng. Data*, 1967, **12**, 266.

1-Bromotridecane B-00054
Tridecyl bromide

$H_3C(CH_2)_{11}CH_2Br$

$C_{13}H_{27}Br$ M 263.260
d_4^{20} 1.026. Mp 6°. $Bp_{9.5}$ 148-149° (Bp_8 156-158°).

Suida, H. *et al*, *Chem. Ber.*, 1942, **75**, 991 (*synth*)
McKillop, A. *et al*, *J. Org. Chem.*, 1969, **34**, 1172 (*synth*)
Cason, J. *et al*, *J. Org. Chem.*, 1972, **37**, 669 (*synth*)
Levine, Y.K. *et al*, *J. Chem. Phys.*, 1974, **60**, 2890 (*synth*)

2-Bromotridecane, 9CI B-00055
[765-09-3]

$H_3C(CH_2)_{10}CHBrCH_3$

$C_{13}H_{27}Br$ M 263.260

▷ YD3100000.

(±)-*form*
Liq. d_4^{20} 1.04. Mp 4-7°. Bp_{10} 148-150°.

Kharasch, M.S. *et al*, *J. Org. Chem.*, 1938, **2**, 196 (*synth*)

2-Bromoundecane, 9CI B-00056
[39563-54-7]

$H_3C(CH_2)_8CHBrCH_3$

$C_{11}H_{23}Br$ M 235.207

(±)-*form*
Liq. Bp_{15} 128°.

Houben, J. *et al*, *Ber.*, 1936, **69**, 1782 (*synth*)
Kharasch, M.S. *et al*, *J. Org. Chem.*, 1937, **2**, 196 (*synth*)

11-Bromoundecanoic acid, 9CI B-00057
[2834-05-1]

$BrCH_2(CH_2)_9COOH$

$C_{11}H_{21}BrO_2$ M 265.190
Cryst. (Me_2CO). Mp 52.5°.
Me ester: [6287-90-7].
$C_{12}H_{23}BrO_2$ M 279.216
$Bp_{0.1}$ 118-119°.
Et ester: [6271-23-4].
$C_{13}H_{25}BrO_2$ M 293.243
$Bp_{0.8}$ 142-143°.

Perrine, T.D., *J. Org. Chem.*, 1953, **18**, 1356 (*synth*)
Chicoisne, A. *et al*, *Bull. Soc. Chim. Fr.*, 1957, 1232 (*synth*)
Woolford, R.G., *Can. J. Chem.*, 1962, **40**, 1846 (*synth*)
Ucciani, E. *et al*, *Bull. Soc. Chim. Fr.*, 1967, 2018 (*synth*)
Schore, N.E. *et al*, *J. Am. Chem. Soc.*, 1975, **97**, 2488 (*synth*)

11-Bromo-1-undecanol B-00058
[1611-56-9]

$BrCH_2(CH_2)_9CH_2OH$

$C_{11}H_{23}BrO$ M 251.206
Mp 45-46°. $Bp_{0.3}$ 134-136°.
Phenylurethane: Mp 64°.

Beets, M.G.J. *et al*, *Perfum. Essent. Oil Rec.*, 1953, **44**, 310 (*synth*)
Pattison, F.L.M. *et al*, *J. Org. Chem.*, 1956, **21**, 739 (*synth*)
Salmon-Ligagneur, F. *et al*, *C. R. Hebd. Seances Acad. Sci.*, 1959, **248**, 2217 (*synth*)

Bromovulone I B-00059
[105343-04-2]

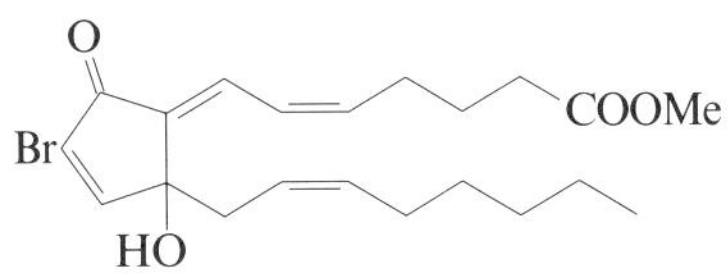

Absolute configuration

$C_{21}H_{29}BrO_4$ M 425.362
Prostanoid isol. from Japanese marine stolonifer *Clavularia viridis*. Shows antitumour activity. Oil.
Iodine analogue: [105343-03-1]. ***Iodovulone I***
$C_{21}H_{29}IO_4$ M 472.362
Isol. from *C. viridis*. Shows antitumour activity. Oil.

Iguchi, K. *et al*, *J. Chem. Soc., Chem. Commun.*, 1986, 981 (*isol, ms, ir, uv, pmr, abs config*)

2-Butenoic acid, 9CI **B-00060**

3-Methylacrylic acid. Ethylideneacetic acid

[3724-65-0]

$H_3CCH{=}CHCOOH$

$C_4H_6O_2$ M 86.090

(*E*)-*form* [107-93-7]

Crotonic acid

Isol. from *Streptomyces collinus*. Used in manuf. of copolymer such as vinyl acetate-crotonic acid adhesives, fungicides and pharmaceuticals. Needles (H_2O or pet. ether). Sol. H_2O. Mp 71.4°. Bp 185° (189°). pK_a 4.61 (25°). n_D^{20} 1.4457.

Me ester: [623-43-8]. *Methyl crotonate*
$C_5H_8O_2$ M 100.117
Liq. Bp 121°.

Et ester: [623-70-1]. *Ethyl crotonate*
$C_6H_{10}O_2$ M 114.144
Liq. Bp 137°, Bp_{15} 38°.

Chloride: [625-35-4].
C_4H_5ClO M 104.536
Liq. Bp 124-125° (120°).

Amide: [625-37-6]. *Crotonamide*
C_4H_7NO M 85.105
Needles. Mp 158°.

Anhydride: [78957-07-0]. *Crotonic anhydride*
$C_8H_{10}O_3$ M 154.165
Reagent for protection of OH groups as crotonate esters. Liq. Bp 246-248°, Bp_{19} 129°.
▷ Mod. toxic.

Anilide: [17645-30-6]. *Crotonanilide*
Needles (H_2O). Mp 115-118°.

Nitrile: [627-26-9]. *Crotononitrile. 1-Cyanopropene*
C_4H_5N M 67.090
Bp 118-119°.

tert-*Butyl ester:*
$C_8H_{14}O_2$ M 142.197
Bp_{15} 49°.
▷ GQ5710000.

(*Z*)-*form*

Isocrotonic acid. β-Crotonic acid

Used in food preservatives. Needles or prisms (petrol). Sol. H_2O, Et_2O, CS_2. Mp 14.6°. Bp 169°, Bp_5 54-55°. Isom. by sunlight or heat to the (*E*)-isomer.

▷ Highly irritant.

Me ester: [4358-59-2].
Liq. Bp 118-119°.

Et ester: [6776-19-8].
$C_6H_{10}O_2$ M 114.144
Pale yellow liq. Bp 128-132°.

Amide: [31110-30-2].
Mp 102°.

Nitrile: [1190-76-7].
Liq. Bp 107.4°.

tert-*Butyl ester:* [73981-49-4].
Bp_{15} 60-65°.

[623-43-8, 18707-60-3]

v. Auwers, K. *et al*, *Ber.*, 1923, **56**, 715 (*config*)
v. Auwers, K. *et al*, *Justus Liebigs Ann. Chem.*, 1923, **432**, 46 (*synth*)
Collin, P.J. *et al*, *Aust. J. Chem.*, 1966, **19**, 317 (*nmr*)
Hocking, M.B., *Synthesis*, 1971, 482 (*synth*)
Fueno, T. *et al*, *Bull. Chem. Soc. Jpn.*, 1972, **45**, 3290 (*uv*)
Brouwer, H. *et al*, *Can. J. Chem.*, 1972, **50**, 601 (*cmr*)
Hocking, M.B., *Can. J. Chem.*, 1972, **50**, 1224 (*isom*)
Org. Synth., 1973, **53**, 123 (*synth*)
Kirk-Othmer Encycl. Chem. Technol., 3rd edn., Wiley, New York, 1978-1984, **7**, 218 (*rev*)
Holmes, J.L. *et al*, *Org. Mass Spectrom.*, 1979, **14**, 204 (*ms*)
Fieser and Fieser's Reagents for Organic Synthesis, Wiley, 1980, **8**, 127 (*use, deriv*)
Abou-Zeid, A. *et al*, *Arab Gulf J. Sci. Res.*, 1984, **2**, 625 (*isol*)
Ger. Pat., 3 245 111, (1984); *CA*, 101, 191148r (*synth*)
Aldrich Library of FT-IR Spectra, 1st edn., 1985, **1**, 498c (*ir*)
Heathcock, C.H. *et al*, *J. Org. Chem.*, 1985, **50**, 1190 (*Et ester, synth*)
Ullmann's Encycl. Ind. Chem., 5th edn., VCH, Weinheim, 1985, **A8**, 85 (*rev*)
Dehmlow, E.V. *et al*, *Chem. Ber.*, 1990, **123**, 583 (*tert-Butyl esters*)
Teixeira-Dias, J.J.C. *et al*, *Vib. Spectrosc.*, 1991, **2**, 107 (*Me ester, ir, raman*)
Lewis, R.J., *Sax's Dangerous Properties of Industrial Materials*, 8th Ed., Van Nostrand-Reinhold, 1992, COB825, COQ750.

2-(2-Butenyl)benzenehexanoic acid, 9CI **B-00061**

Rubrenoic acid A

[95230-66-3]

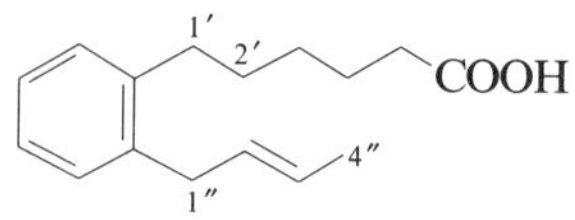

$C_{16}H_{22}O_2$ M 246.349

Metab. of *Ateromonas rubra*.

1′,2′-Didehydro: [95230-67-4]. *6-[2-(2-Butenyl)phenyl]-5-hexenoic acid, 9CI.* ***Rubrenoic acid B***
$C_{16}H_{20}O_2$ M 244.333
From *A. rubra*.

1″,4″-Didehydro: [95230-68-5]. *2-(1,3-Butadienyl) benzenehexanoic acid, 9CI.* ***Rubrenoic acid C***
$C_{16}H_{20}O_2$ M 244.333
From *A. rubra*.

Holland, G.S. *et al*, *Chem. Ind. (London)*, 1984, 850.

5-(1-Butenyl)-dihydro-2(3*H*)-furanone, 9CI **B-00062**

5-Octen-4-olide. 4-Hydroxy-5-octenoic acid lactone

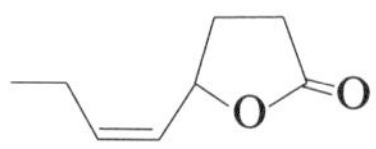

$C_8H_{12}O_2$ M 140.182

(*Z*)-*form* [82934-78-9]

Isol. from tuberose absolute (from *Polianthes tuberosa*). Oil. $Bp_{0.01}$ 85°. Nat. prod. may be opt. active.

Maurer, B. *et al*, *Helv. Chim. Acta*, 1982, **65**, 462 (*isol, synth, spectra*)

2-Butylcyclopropanedodecanoic acid, 9CI **B-00063**

13,14-Methyleneoctadecanoic acid

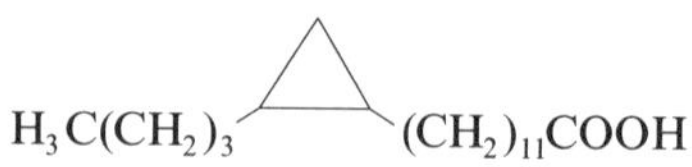

$C_{19}H_{36}O_2$ M 296.492

Both *cis-* and *trans*-forms have been synthesised as Me esters.

Christie, W.W. *et al*, *Chem. Phys. Lipids*, 1968, **2**, 196 (*synth, glc, ir, nmr*)
Christie, W.W., *Top. Lipid Chem.*, 1970, **1**, 4 (*rev*)
Gunstone, F.D. *et al*, *Chem. Phys. Lipids*, 1973, **10**, 303.

5-Butyldihydro-4-methyl-2(3*H*)-furanone B-00064

3-Methyl-4-octanolide. Oaklactone. Quercuslactone. Whisky lactone

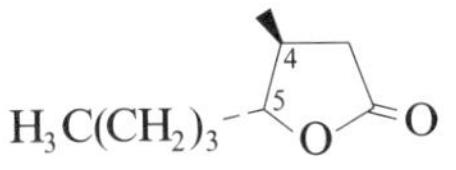

(4*S*,5*R*)-*form*

$C_9H_{16}O_2$ M 156.224

(4S,5R)-form [80041-01-6] ***Quercuslactone a***

Present in oak wood and aged spirits. Aroma constit. $[\alpha]_D^{15}$ +79° (c, 1.04 in MeOH).

(4S,5S)-form [80041-00-5]

Quercuslactone b

Present in oak woods and aged spirits. Aroma constit. $[\alpha]_D^{15}$ −87° (c, 0.92 in MeOH).

Masuda, M. *et al*, *Chem. Lett.*, 1981, 1333 (*struct, uv, cd, ord, bibl*)
Marino, J.P. *et al*, *Tetrahedron Lett.*, 1985, **26**, 5381 (*synth*)
Günther, C. *et al*, *Justus Liebigs Ann. Chem.*, 1986, 2112 (*synth, ir, pmr, cmr*)
Watanabe, M. *et al*, *Chem. Pharm. Bull.*, 1989, **37**, 2914 (*synth*)
Jefford, C.W. *et al*, *Helv. Chim. Acta*, 1989, **72**, 1362 (*synth*)
Miyata, O. *et al*, *Chem. Pharm. Bull.*, 1992, **40**, 2579 (*synth*)
Srikrishna, A. *et al*, *Synth. Commun.*, 1992, **22**, 1127 (*synth*)

2-Butyl-2-octenal, 9CI B-00065

[13019-16-4]

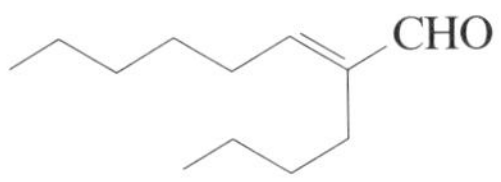

(*E*)-*form*

$C_{12}H_{22}O$ M 182.305

Bp_9 89-91°, Bp_{10} 74-76°.

2,4-Dinitrophenylhydrazone: [49563-02-2].
Cryst. (EtOH). Mp 132-133°.

(E)-form [64935-38-2]

Pheromone isol. from the glands of *Amblypelta nitida* and *Oecophylla longinoda.* Also found in essential oils of *Acorus calamus* and *Lactuca sativa.*

[99915-14-7]

McElvain, S.M. *et al*, *J. Am. Chem. Soc.*, 1954, **76**, 5736 (*synth*)
Lieberman, S.V., *J. Am. Chem. Soc.*, 1955, **77**, 1114 (*synth*)
Damico, R. *et al*, *J. Org. Chem.*, 1967, **32**, 2356 (*synth*)
Baker, J.T. *et al*, *Aust. J. Chem.*, 1972, **25**, 393 (*isol*)
Bradshaw, J.W.S. *et al*, *Nature (London)*, 1975, **258**, 230 (*isol*)
Grandi, R. *et al*, *J. Org. Chem.*, 1977, **42**, 1352 (*synth*)
Kuwajima, I. *et al*, *Chem. Lett.*, 1980, 525 (*synth*)
Pfister, J.R., *Tetrahedron Lett.*, 1980, **21**, 1281 (*synth*)
Casiraghi, G. *et al*, *J. Chem. Soc., Perkin Trans. 1*, 1982, 805 (*synth*)

Butynedioic acid, 9CI B-00066

Acetylenedicarboxylic acid. Ethynedicarboxylic acid

[142-45-0]

HOOCC≡CCOOH

$C_4H_2O_4$ M 114.057

Cryst. + $2H_2O$. Mp 179°. pK_{a1} 1.23; pK_{a2} 2.53 (28°).

Di-Me ester: [762-42-5]. *Dimethyl acetylenedicarboxylate. DMAD*

$C_6H_6O_4$ M 142.111

Useful dienophile in the Diels-Alder reaction. Bp_{20} 98°.
▷ES0175000.

Dichloride:

$C_4Cl_2O_2$ M 150.948

Active dienophile. Obt. in soln. only.

Diamide: [543-21-5]. *Butynediamide. Aquamycin. Cellocidin. Lenamycin*

$C_4H_4N_2O_2$ M 112.088

Isol. from *Streptomyces chibaensis* and *S. reticuli.* Active against gram-positive and -negative bacteria and HeLa cells. Cryst. (MeOH aq.). Mp 216-218° dec. Unstable in alkaline solns., evolving NH_3.
▷AO9900000.

Dinitrile: [1071-98-3]. *Butynedinitrile. Dicyanoacetylene*

C_4N_2 M 76.057

Used in Diels-Alder reactions. Mp 21°. Bp 76°.
▷Toxic, potentially explosive. AO9860000.

Gilman, H. *et al*, *J. Am. Chem. Soc.*, 1945, **67**, 1420 (*synth*)
Saggiomo, A.J., *J. Org. Chem.*, 1957, **22**, 1171.
Suzuki, S. *et al*, *J. Antibiot., Ser. A*, 1958, **11**, 81 (*isol, amide*)
Org. Synth., Coll. Vol., 4, 1963, 329 (*synth*)
Ciganek, E. *et al*, *J. Org. Chem.*, 1968, **33**, 541 (*nitrile*)
Larson, A.C. *et al*, *Acta Crystallogr., Sect. B*, 1973, 1579 (*cryst struct*)
Jones, E.R.H. *et al*, *J. Chem. Soc., Perkin Trans. 1*, 1973, 148 (*biosynth*)
Fieser and Fieser's Reagents for Organic Synthesis, Wiley, 1975, **5**, 205; **7**, 117.
Acheson, R.M. *et al*, *Adv. Heterocycl. Chem.*, 1978, **23**, 263.
Jaeger, P. *et al*, *Chem.-Ing.-Tech.*, 1978, **50**, 787 (*rev, synth*)
Mori, M. *et al*, *Agric. Biol. Chem.*, 1982, **46**, 309 (*pharmacol*)
Maier, G. *et al*, *Chem. Ber.*, 1982, **115**, 804 (*chloride*)
Bretherick, L., *Handbook of Reactive Chemical Hazards*, 2nd Ed., Butterworths, London and Boston, 1979, 518.
Hazards in the Chemical Laboratory, (Bretherick, L., Ed.), 3rd Ed., Royal Society of Chemistry, London, 1981, 221.
Lewis, R.J., *Sax's Dangerous Properties of Industrial Materials*, 8th Ed., Van Nostrand-Reinhold, 1992, ACJ250, DGS000, DOP400.

Butyrospermol B-00067

Eupha-7,24-dien-3β-ol. Basseol

[472-28-6]

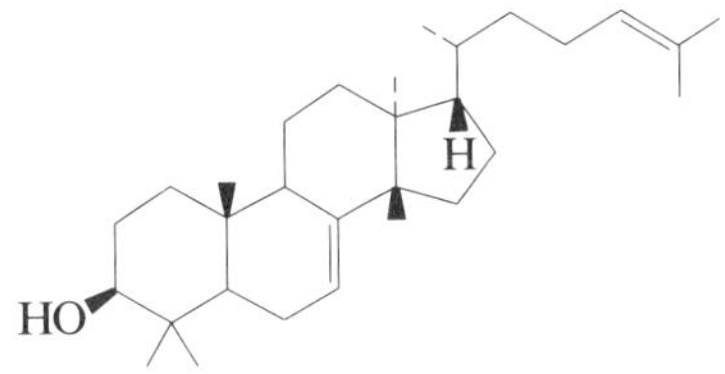

$C_{30}H_{50}O$ M 426.724

Constit. of shea nut fat. Also from *Maclura pomifera, Aesculus hippocastanum, Populus tremuloides* and others. Cryst. (MeOH). Mp 111-113°. $[\alpha]_D^{19}$ −12° ($CHCl_3$).

Ac: [1259-21-8].

$C_{32}H_{52}O_2$ M 468.762

Constit. of *Cudrania javanensis* and *Ficus macrophylla.* Cryst. Mp 136-139°. $[\alpha]_D^{19}$ +14° ($CHCl_3$).

Ketone: [81738-57-0]. ***Butyrospermone***

$C_{30}H_{48}O$ M 424.709

Constit. of *Cassia javanica.* Also found in various vegetable fats.

Lawrie, W. *et al*, *J. Chem. Soc.*, 1956, 2029, 3272 (*struct*)
Galbraith, M.N. *et al*, *Aust. J. Chem.*, 1965, **18**, 226 (*acetate*)
Fedell, E. *et al*, *J. Am. Oil Chem. Soc.*, 1966, **43**, 254 (*occur*)
Pradhan, B.P. *et al*, *J. Indian Chem. Soc.*, 1969, **46**, 331 (*isol, struct*)
Murti, V.V.S. *et al*, *Phytochemistry*, 1972, **11**, 2089 (*acetate*)
Kohle, J.N. *et al*, *Lipids*, 1982, **17**, 166 (*Butyrospermone*)
Ok, H.O., *Diss. Abstr. Int., B*, 1987, **47**, 3775 (*synth*)
Chaudhuri, K. *et al*, *J. Nat. Prod. (Lloydia)*, 1987, **50**, 1183 (*Butyrospermone*)

Byssomeruliol A **B-00068**

12,14-Dihydroxy-1-(2,4,6-trihydroxyphenyl)-1-octadecanone, 9CI

[83212-66-2]

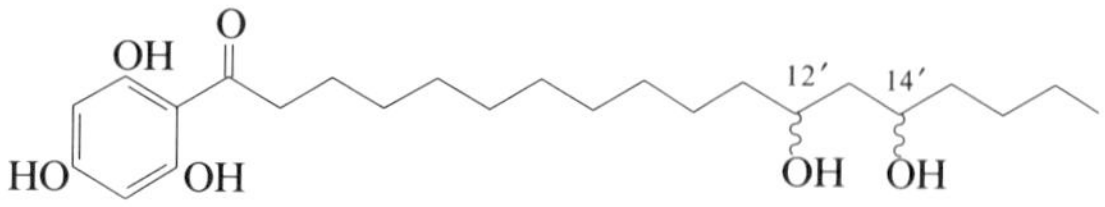

$C_{24}H_{40}O_6$ M 424.576

Isol. from the fruiting bodies and mycelia of *Byssomerulius corium*. Byssomeruliol I was a mixt. of A and B isol. earlier.

12′-O-D-Mannopyranoside: [83162-80-5]. ***Byssomeruliol G***
$C_{30}H_{50}O_{11}$ M 586.718
Isol. from *B. corium*.

14′-O-D-Mannopyranoside: [83162-81-6]. ***Byssomeruliol H***
$C_{30}H_{50}O_{11}$ M 586.718
Isol. from *B. corium*.

12′-Deoxy: [83162-78-1]. *14-Hydroxy-1-(2,4,6-trihydroxyphenyl)-1-octadecanone, 9CI.* ***Byssomeruliol E***
$C_{24}H_{40}O_5$ M 408.577
Isol. from *B. corium*. Stereochem. undetermined.

12′-Deoxy, 14′-O-D-mannopyranoside: [83162-79-2]. ***Byssomeruliol F***
$C_{30}H_{50}O_{10}$ M 570.719
Isol. from *B. corium*.

Stereoisomer: [83213-39-2]. ***Byssomeruliol B***
$C_{24}H_{40}O_6$ M 424.576
Isol. from *B. corium*.

[75656-31-4]

Lunel, M.-C. *et al*, *Tetrahedron*, 1982, **38**, 1235 (*isol, pmr, ms*)

C

Campnospermonol **C-00001**

1-(3-Hydroxyphenyl)-10-nonadecen-2-one

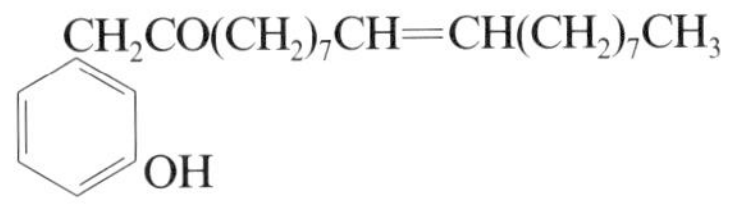

$C_{25}H_{40}O_2$ M 372.590

Isol. from *Campnosperma brevipetiolata*. Oil. Bp_5 260°.

Dihydro: 1-(3-Hydroxyphenyl)-2-nonadecanone. Hydrocampnospermonol

$C_{25}H_{42}O_2$ M 374.606

Needles (pet. ether). Mp 70-71°.

Dihydro, Me ether: Cryst. (EtOH). Mp 53-54°.

Gilbert, T. *et al*, *J. Chem. Soc.*, 1928, 65 (*isol*)

Dalton, L.K. *et al*, *Aust. J. Chem.*, 1958, **11**, 46 (*struct, synth*)

Carbacyclin **C-00002**

Carbaprostacyclin. Carboprostacyclin. 6a-Carbaprostaglandin I_2. Methanoprostacyclin. 6,9-Methylene-11,15-dihydroxy-5,13-prostadienoic acid

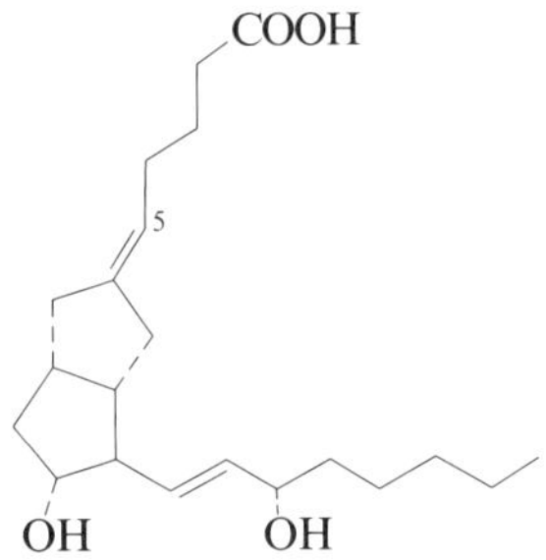

(*5E*)-*form*

$C_{21}H_{34}O_4$ M 350.497

The (*E*)-form is the desired isomer and is properly called Carbacyclin.

(5E)-form

Platelet aggregation inhibitor. Cryst. (Et_2O/hexane). Mp 64.5-66.5°. $[\alpha]_D^{20}$ +92.2° (c, 0.515 in MeOH).

(5Z)-form

Cryst. (Me_2CO/hexane). Mp 113-114°. $[\alpha]_D^{20}$ +40.1° (c, 0.535 in MeOH).

Amemiya, S. *et al*, *Chem. Pharm. Bull.*, 1984, **32**, 1349 (*synth*)

Beierbeck, H. *et al*, *Can. J. Chem.*, 1985, **63**, 1143 (*cmr, pmr*)

Kojima, K. *et al*, *Chem. Pharm. Bull.*, 1985, **33**, 2688 (*synth*)

Miftakhov, M.S. *et al*, *Zh. Org. Khim.*, 1986, **22**, 2235 (*synth*)

Mori, K. *et al*, *Tetrahedron*, 1986, **42**, 435 (*synth, bibl*)

Hutchinson, D.K. *et al*, *J. Am. Chem. Soc.*, 1987, **109**, 4755 (*synth, bibl*)

Carboprost, BAN, INN, USAN **C-00003**

9,11,15-Trihydroxy-15-methylprosta-5,13-dien-1-oic acid, 9CI. 15-Methyl-PGF$_{2\alpha}$. U 32921

[35700-23-3]

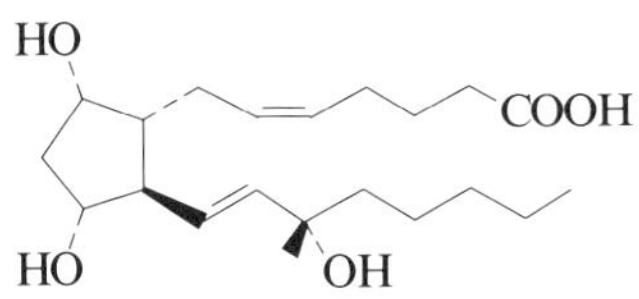

$C_{21}H_{36}O_5$ M 368.512

Smooth muscle stimulant, oxytocic agent, abortifacient. Various stereoisomers also known.

▷ UK8021000.

2-Amino-2-hydroxymethyl-1,3-propanediol salt: [58551-69-2]. *Carboprost trometamol, BAN. Carboprost tromethamine, USAN. Prostin 15M*

▷ UK8026000.

Me ester: [35700-21-1]. *Carboprost methyl. U 36385*

$C_{22}H_{38}O_5$ M 382.539

Cryst. (Et_2O/hexane). Mp 55-56°. $[\alpha]_D$ +24° (c, 81 in EtOH).

[90695-58-2, 90695-59-3, 90695-60-6, 90695-61-7]

Bundy, G.L. *et al*, *Adv. Biosci. (Oxford)*, 1972, **9**, 125 (*synth, pharmacol*)

Yankee, E.W. *et al*, *J. Am. Chem. Soc.*, 1974, **96**, 5865 (*synth, ir, ms, bibl*)

Hansson, G. *et al*, *Biochem. Med.*, 1977, **18**, 420 (*metab*)

Liu, Z. *et al*, *CA*, 1982, **96**, 217527; 1985, **102**, 113106 (*synth*)

Martindale, The Extra Pharmacopoeia, 28th/29th Ed., Pharmaceutical Press, London, 1982/1989, 8064, 8075, 8076.

Weinges, K. *et al*, *Justus Liebigs Ann. Chem.*, 1984, 761 (*stereoisomers*)

Lewis, R.J., *Sax's Dangerous Properties of Industrial Materials*, 8th Ed., Van Nostrand-Reinhold, 1992, CCC100, CCC110, MLO300.

2-(6-Carboxy-2-hexenyl)-3-(3-hydroxy-1-octenyl)bicyclo[3.1.1]heptane **C-00004**

7-[3-(3-Hydroxy-1-octenyl)bicyclo[3.1.1]hept-2-yl]-5-heptenoic acid

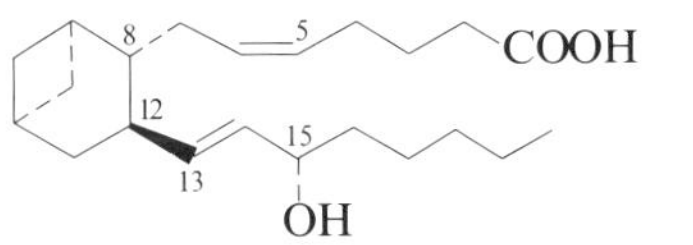

$C_{22}H_{36}O_3$ M 348.525

Prostanoid type numbering shown.

(5Z,8S,12R,13E,15S)-form [74034-57-4]

Carbocyclic thromboxane A_2. CTA_2

Viscous oil. $[\alpha]_D^{30}$ +32.60° (c, 0.95 in MeOH). Selectively inhibits the biosynth. of thromboxanes without compromising PGI_2 production.

Me ester: [74034-54-1].

$[\alpha]_D^{30}$ +25.80° (c, 1.68 in MeOH).

(5Z,8S,12R,13E,15R)-form [74034-56-3]

$[\alpha]_D^{30}$ −23.90° (c, 1.20 in MeOH).

Me ester: [74034-55-2].

$[\alpha]_D^{30}$ −16.70° (c, 0.75 in MeOH).

Ohuchida, S., *Tetrahedron Lett.*, 1979, 3661 (*synth, ir, pmr, ms*)

Nicolaou, K.C. *et al*, *Adv. Prostaglandin, Thromboxane, Leukotriene Res.*, 1980, **6**, 481.

Nicolaou, K.C. *et al*, *J. Am. Chem. Soc.*, 1980, **102**, 1404 (*synth, ir, pmr, ms*)

3-Carboxy-4-methyl-5-pentyl-2-furanpropanoic acid C-00005

Pentylurofuranic acid

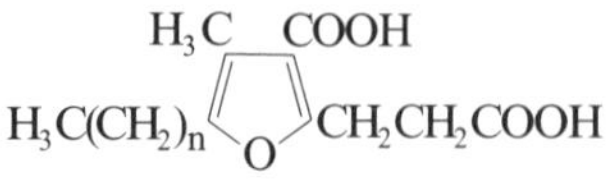

$n = 4$

$C_{14}H_{20}O_5$ M 268.309

Constit. of urine and blood. Characterised as di-Me ester.

[72719-12-1]

Spiteller, M. *et al, Chem. Ber.*, 1980, **113**, 699.
Pfordt, J. *et al, Justus Liebigs Ann. Chem.*, 1980, 175.
Schödel, R. *et al, Justus Liebigs Ann. Chem.*, 1986, 127 (*biosynth*)

3-Carboxy-4-methyl-5-propyl-2-furanpropanoic acid C-00006

Propylurofuranic acid

As 3-Carboxy-4-methyl-5-pentyl-2-furanpropanoic acid, C-00005 with

$n = 2$

$C_{12}H_{16}O_5$ M 240.255

Constit. of urine and blood. Characterised as di-Me ester.

[72719-10-9]

Spiteller, M. *et al, Chem. Ber.*, 1980, **113**, 699 (*isol, ms, synth*)
Pfordt, J. *et al, Justus Liebigs Ann. Chem.*, 1980, 175.

2-(4-Carboxy-2-oxobutyl)-3,5-dihydroxy-2-oxocyclopentaneoctanoic acid C-00007

Dinor-4,13-diketo-7,9-dihydroxy-1,18-prostandioic acid

[68340-29-4]

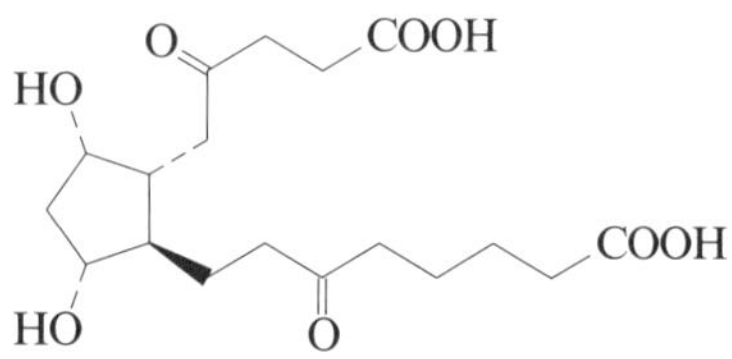

$C_{18}H_{28}O_8$ M 372.414

Prostacyclin, P-00144 metab. in bile. Prostacyclin metab.

Rosenkranz, B. *et al, J. Biol. Chem.*, 1980, **255**, 10194 (*biosynth*)
Taylor, B.M. *et al, J. Pharmacol. Exp. Ther.*, 1980, **214**, 24; 1983, **224**, 692 (*biosynth*)

3-Carboxy-5-pentyl-2-furanpropanoic acid C-00008

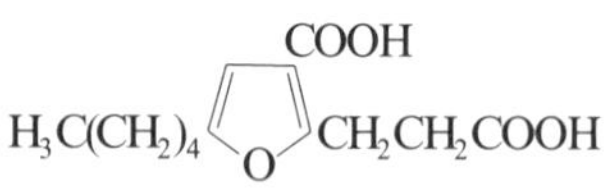

$C_{13}H_{18}O_5$ M 254.282

Found in urine.

Bauer, S. *et al, Justus Liebigs Ann. Chem.*, 1985, 813.

3-Carboxy-5-propyl-2-furanpropanoic acid C-00009

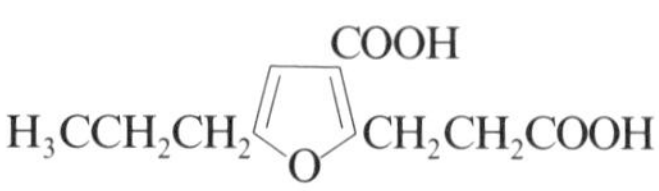

$C_{11}H_{14}O_5$ M 226.229

Found in urine. Characterised as di-Me ester.

Bauer, S. *et al, Justus Liebigs Ann. Chem.*, 1985, 813.

Cardiolipin C-00010

1,3-Diphosphatidylglycerol

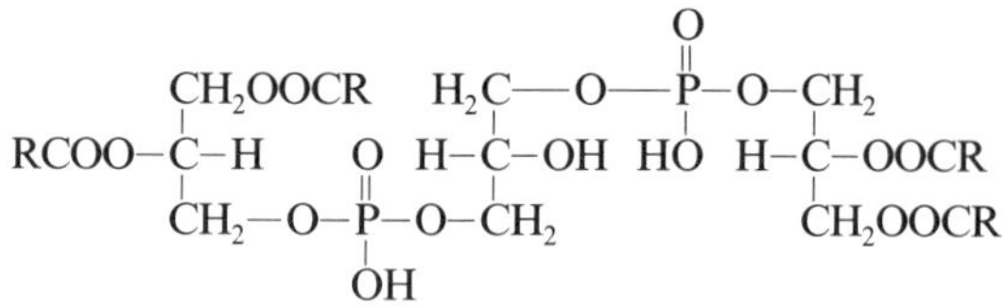

R = alkyl groups

Present in animal (especially heart and liver) and plant tissues as mixtures. Also in the lipids of bacteria. Compds. contg. mixed R groups occur.

Tetramyristoyldiphosphatidylglycerol

$C_{65}H_{126}O_{17}P_2$ M 1241.648

Dihydrate (as NH_4 salt). Mp 181-182° (NH_4 salt dihydrate). $[\alpha]_D^{24}$ +6.9° (c, 2.75 in $CHCl_3$). Has R = $H_3C(CH_2)_{12}$—.

Tetrapalmitoyldiphosphatidylglycerol

$C_{73}H_{142}O_{17}P_2$ M 1353.862

Dihydrate (as NH_4 salt). Mp 177-178°. $[\alpha]_D^{24}$ +6.35° (c, 2.75 in $CHCl_3$). Has R = $H_3C(CH_2)_{14}$—.

Tetrastearoyldiphosphatidylglycerol

$C_{81}H_{158}O_{17}P_2$ M 1466.076

Dihydrate (as NH_4 salt). Mp 182-183° (NH_4 salt). $[\alpha]_D^{24}$ +5.82° (c, 2.75 in $CHCl_3$). Has R = $H_3C(CH_2)_{16}$—.

Macfarlane, M.G., *Biochem. J.*, 1964, **92**, 12C.
Wood, B.J.B. *et al, Biochim. Biophys. Acta*, 1965, **106**, 261 (*isol*)
de Haas, G.H. *et al, Recl. Trav. Chim. Pays-Bas* (*J. R. Neth. Chem. Soc.*), 1965, **84**, 436 (*synth*)
McIlwain, D.L. *et al, Biochemistry*, 1966, **5**, 4054 (*isol*)
Letters, R., *Biochim. Biophys. Acta*, 1966, **116**, 489 (*isol*)
Saunders, R.M. *et al, J. Am. Chem. Soc.*, 1966, **88**, 3844 (*synth*)
Ramirez, F. *et al, Synthesis*, 1976, 769 (*synth*)

α-Carotene C-00011

β,ε-Carotene, 9CI

[432-70-2]

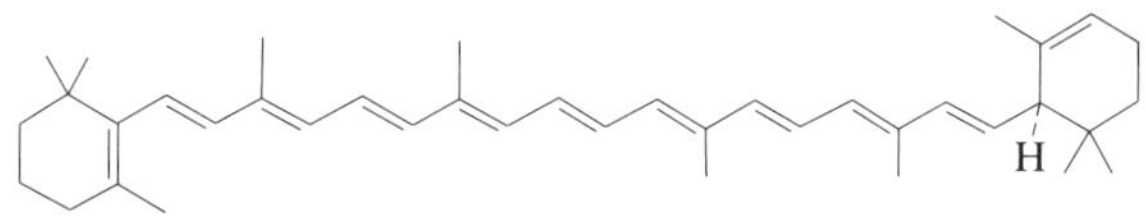

$C_{40}H_{56}$ M 536.882

(R)-form [7488-99-5]

Widespread carotenoid. Has vitamin A activity but less than that of *β*-Carotene. Violet prisms (pet. ether). Mp 187°. $[\alpha]_{643}^{18}$ +385° (C_6H_6). λ_{max} 446, 478, 511 nm (CS_2); 418, 442, 471 (hexane).

(±)-form

Violet platelets (C_6H_6). Mp 160-162°.

Karrer, P. *et al, Helv. Chim. Acta*, 1933, **16**, 975 (*struct*)
Karrer, P. *et al, Helv. Chim. Acta*, 1950, **33**, 1952 (*synth*)
Tscharner, C. *et al, Helv. Chim. Acta*, 1957, **40**, 1676 (*synth*)
Rüegg, R. *et al, Helv. Chim. Acta*, 1961, **44**, 985 (*synth*)
Strain, H.H. *et al, J. Org. Chem.*, 1961, **26**, 5061 (*ir*)
Goodwin, T.W. *et al, Biochem. J.*, 1965, **97**, 28c (*biosynth*)
Eugster, C.H. *et al, Helv. Chim. Acta*, 1969, **52**, 1729 (*abs config*)
Buchecker, R. *et al, Helv. Chim. Acta*, 1971, **54**, 327 (*uv, cd*)
Kjoesen, H. *et al, Acta Chem. Scand.*, 1972, **26**, 3053 (*pmr, ms*)
Karrer, W. *et al, Konstitution und Vorkommen der Organischen Pflanzenstoffe*, 2nd Ed., Birkhäuser Verlag, Basel, 1972-1985, no. 1824 (*occur*)
Straub, O. *et al, Key to Carotenoids*, 2nd Ed., Birkhauser Verlag, Basel and Boston, 1987, 7 (*bibl*)

β-Carotene C-00012

Betacarotene, USAN, INN. Carotaben. Phenoro. Solatene.
β,β-Carotene
[7235-40-7]

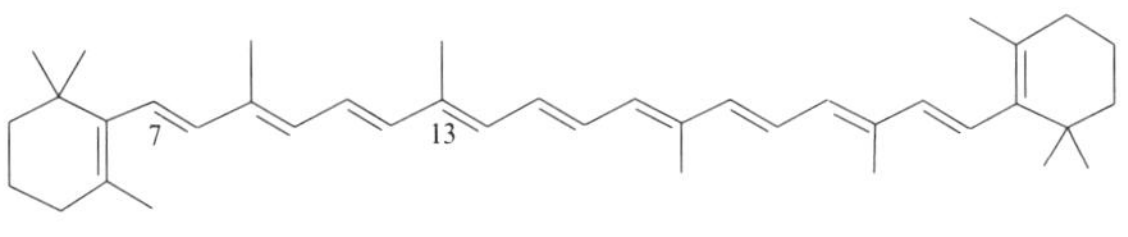

$C_{40}H_{56}$ M 536.882

Widespread, e.g. found in carrots. Exhibits strong vitamin A activity. Used in ultraviolet sunscreen preparations. Yellow food colouring. Deep-purple prisms (C_6H_6/MeOH); red rhombs. (pet. ether). Mp 183° (sealed tube). λ_{max} 452, 484, 520 nm (CS_2). Air-sensitive.

13-cis-form [6811-73-0]

Neo-β-carotene. Pseudo-α-carotene. β,β-Carotene neo B

$C_{40}H_{56}$ M 536.882

Constit. of carrots and other plant spp.

Surmatis, J.D. *et al, J. Org. Chem.*, 1961, **26**, 1171 (*synth*)
Sterling, C., *Acta Crystallogr.*, 1964, **17**, 1224 (*cryst struct*)
Schweiter, U. *et al, Chimia*, 1965, **19**, 294 (*props*)
Surmatis, J.D. *et al, J. Org. Chem.*, 1969, **34**, 559 (*synth*)
Karrer, W. *et al, Konstitution und Vorkommen der Organischen Pflanzenstoffe*, 2nd Ed., Birkhäuser Verlag, Basel, 1972-1985, no. 1821 (*occur*)
Johannes, B. *et al, Org. Mass Spectrom.*, 1974, **9**, 1095 (*ms*)
Fischli, A. *et al, Helv. Chim. Acta*, 1975, **58**, 1584 (*synth*)
Ishida, A. *et al, Chem. Lett.*, 1976, 1127 (*synth*)
Bickel, H. *et al, Phytochemistry*, 1976, **15**, 1253 (*biosynth*)
Bramley, P.M. *et al, Phytochemistry*, 1976, **15**, 1913 (*biosynth*)
Moss, G.P., *Pure Appl. Chem.*, 1976, **47**, 97 (*pmr*)
Ninet, L. *et al, Microb. Technol. (2nd edn.)*, 1979, **1**, 529 (*manuf, rev*)
Wernly, J. *et al, J. Chem. Soc., Chem. Commun.*, 1985, 1221 (*pmr, cmr*)
Koyama, Y. *et al, J. Mol. Struct.*, 1989, **193**, 185 (*pmr*)
Bjoønland, T. *et al, Phytochemistry*, 1989, **28**, 3347 (*isol, pmr*)
Bernhard, K. *et al, Pure Appl. Chem.*, 1991, **63**, 35 (*synth*)
Senge, M.O. *et al, Z. Naturforsch., C*, 1992, **47**, 474 (*cryst struct*)
Lewis, R.J., *Sax's Dangerous Properties of Industrial Materials*, 8th Ed., Van Nostrand-Reinhold, 1992, CCK685.

β,γ-Carotene C-00013

[64129-93-7]

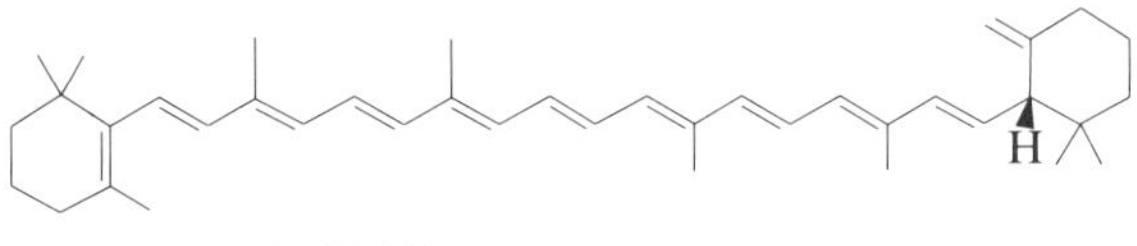

$C_{40}H_{56}$ M 536.882

Constit. of *Caloscypha fulgens*. Cryst. (C_6H_6/MeOH). Mp 176-177°.

Arpin, N. *et al, Phytochemistry*, 1971, **10**, 1595 (*struct*)
Andrewes, A.G. *et al, Acta Chem. Scand.*, 1973, **27**, 1401 (*synth*)
Hallenstvet, M. *et al, Phytochemistry*, 1977, **16**, 583 (*abs config*)

δ-Carotene C-00014

ε,ψ-Carotene
[472-92-4]

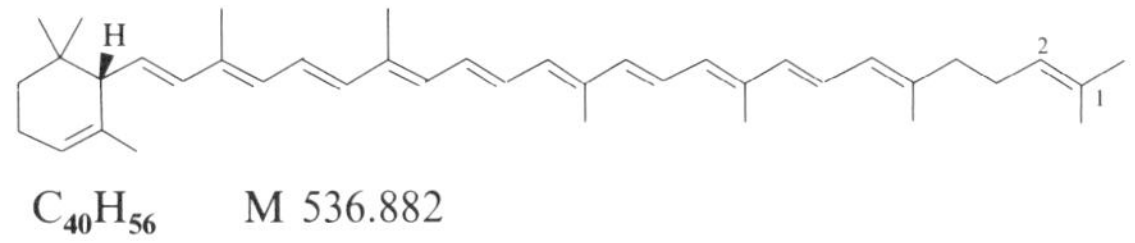

$C_{40}H_{56}$ M 536.882

Constit. of *Gonocaryum pyriforme*, also found in carrots and tomatoes. Red cryst. Mp 172°. $[\alpha]_D^{25}$ +352° (hexane). λ_{max} 280, 431, 456, 488 nm.

1′,2′-Epoxide: [58556-44-8]. ***δ-Carotene-1,2-epoxide.*** *1′,2′-Epoxy-1′,2′-dihydro-ε,ψ-carotene*

$C_{40}H_{56}O$ M 552.882

Isol. from delta tomatoes. Deep red cryst. (CH_2Cl_2/MeOH). Mp 153-155°.

Manchand, P.S. *et al, J. Chem. Soc.*, 1965, 2019 (*synth*)
Buchecker, R. *et al, Helv. Chim. Acta*, 1971, **54**, 327 (*abs config*)
Britton, G. *et al, Phytochemistry*, 1975, **14**, 2530 (*occur, epoxide*)
Pfander, H. *et al, Helv. Chim. Acta*, 1980, **63**, 1792 (*synth, uv, pmr, ms*)
Berset, D. *et al, Helv. Chim. Acta*, 1986, **69**, 340 (*occur, uv, pmr, ms, epoxide*)

ε-Carotene C-00015

ε_1-Carotene. ε,ε-Carotene
[38894-81-4]

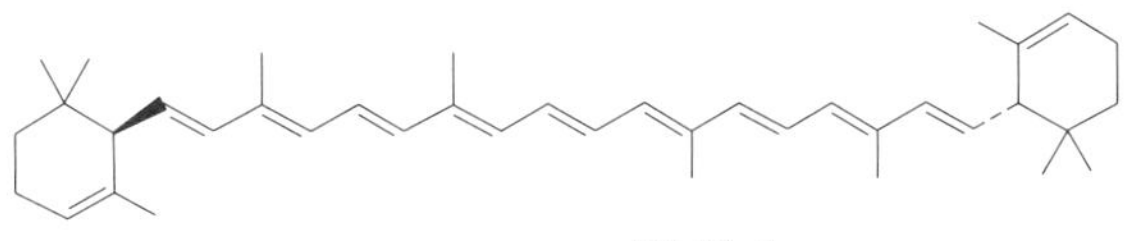

(*R,R*)*-form*

$C_{40}H_{56}$ M 536.882

(*R,R*)*-form* [472-89-9]

Constit. of *Navicula torquatum, Bryopsis corticulans* and *Cryptomonas ovata*. Red cryst. Mp 199-201°, Mp 196° (sealed tube). $[\alpha]_D^{27}$ +806° (C_6H_6). λ_{max} 266, 416, 440, 470 nm.

(*S,S*)*-form*

Mp 197-198° (sealed tube). $[\alpha]_D^{27}$ −786° (C_6H_6).

(*RS,RS*)*-form*

Orange platelets (C_6H_6/MeOH). Mp 190° (sealed tube).

Tscharner, C. *et al, Helv. Chim. Acta*, 1958, **41**, 32 (*synth*)
Manchand, P.S. *et al, J. Chem. Soc.*, 1965, 2019 (*synth*)
Eugster, C.H. *et al, Helv. Chim. Acta*, 1969, **52**, 1729, 1732 (*abs config*)
Gerhard, E., *Helv. Chim. Acta*, 1975, **58**, 2367 (*cmr*)
Bjoørnland, T. *et al, Phytochemistry*, 1989, **28**, 3347 (*isol, pmr, cd*)

γ-Carotene C-00016

β,ψ-Carotene. Sphaerobolin
[472-93-5]

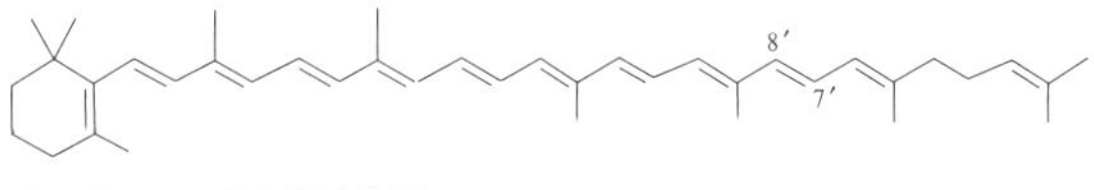

$C_{40}H_{56}$ M 536.882

Widespread carotenoid found esp. in *Gonocaryum pyriforme*. Exhibits vitamin A activity. Violet prisms (C_6H_6/pet. ether). Mp 178° (154°).

7′,8′-Dihydro: [514-90-9]. ***β-Zeacarotene.*** *Carotene X. 7′,8′-Dihydro-β,ψ-carotene*

$C_{40}H_{58}$ M 538.898

Isol. from maize (*Zea mays*), *Citrus* spp., cultures of *Phycomyces blakesleeanus* etc. Orange-yellow platelets (Me_2CO). Mp 96-97°.

Goodwin, T.W., *Annu. Rev. Biochem.*, 1955, **24**, 497 (*rev*)
Rüegg, R. *et al, Helv. Chim. Acta*, 1961, **44**, 994 (*β-Zeacarotene*)
Schweiter, U. *et al, Chimia*, 1965, **19**, 294 (*ms*)
Manchand, P.S. *et al, J. Chem. Soc.*, 1965, 2019 (*synth*)
Williams, R.J.H. *et al, Phytochemistry*, 1965, **4**, 759 (*β-Zeacarotene*)
Weedon, B.C.L., *Prog. Chem. Org. Nat. Prod.*, 1969, **27**, 81 (*pmr*)

Schmidt, K. *et al*, *Acta Chem. Scand.*, 1971, **25**, 2476 (*uv*)
Straub, O. *et al*, *Key to Carotenoids*, 2nd Ed., Birkhauser Verlag, Basel and Boston, 1987, 12 (*bibl*)

γ,γ-Carotene C-00017

[34815-98-0]

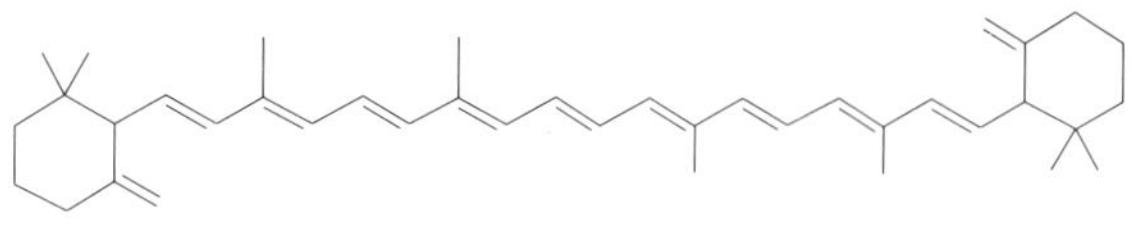

$C_{40}H_{56}$ M 536.882
Constit. of *Macrosiphum liriodendri*. Cryst. (C_6H_6/MeOH). Mp 183°.

Andrewes, A.G. *et al*, *Acta Chem. Scand.*, 1973, **27**, 1401.

Cerebrosides C-00018

RCH(OH)CH(NHCOR′)CH_2O-sugar

Cerebroside is the general name given to monoglycosyl ceramides. On hydrol. they afford a long-chain base such as Sphingenine, a fatty acid (R′COOH) and a sugar (most commonly galactose). These compds. occur as mixtures and are important constits. of lipids of many tissues, esp. brain and nerve. Some members of this class have individual entries.

Rodd's Chem. Carbon Compd. (2nd edn.), 2nd Ed., Vol 1E, 1976, 403.
Schmidt, R.R. *et al*, *Angew. Chem., Int. Ed. Engl.*, 1985, **24**, 65 (*synth*)
Kodato, S. *et al*, *Tetrahedron*, 1989, **45**, 7263 (*synth*)

Cerulenin C-00019

3-(1-Oxo-4,7-nonadienyl)oxiranecarboxamide, 9CI

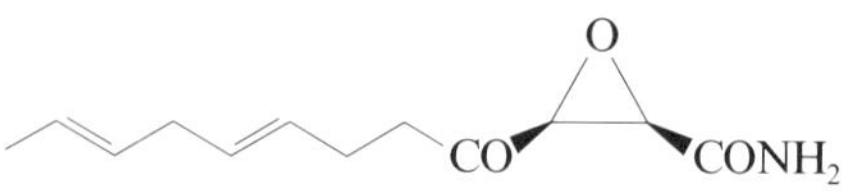

$C_{12}H_{17}NO_3$ M 223.271

(+)-***form*** [17397-89-6]
Isol. from *Cephalosporium caerulens*. Antifungal antibiotic. Inhibitor of the biosynth. of fatty acids and steroids. Needles (CCl_4 or C_6H_6). Mp 93°. $Bp_{0.00001}$ 120°. $[\alpha]_D^{16}$ +63° (c, 2 in MeOH).
▷ JR1670000.
Tetrahydro: [17397-90-9].
Mp 85.5-86.5°. $[\alpha]_D^{25}$ +43° (c, 0.25 in MeOH).

(−)-***form*** [68926-46-5]
Tetrahydro: [68926-45-4].
Mp 84.5°. $[\alpha]_D^{20}$ −52° (c, 1.02 in MeOH).

(±)-***form*** [62964-96-9]
Cryst. (C_6H_6/hexane). Mp 40-43°.

Sano, Y. *et al*, *J. Antibiot., Ser. A*, 1967, **20**, 344.
Awaya, J. *et al*, *J. Antibiot.*, 1975, **28**, 824 (*nmr, biosynth*)
Boeckman, R.K. *et al*, *J. Am. Chem. Soc.*, 1977, **99**, 2805 (*synth*)
Corey, E.J. *et al*, *Tetrahedron Lett.*, 1977, 3847 (*synth*)
Thomas, E.W. *et al*, *CA*, 1978, **89**, 42956j (*synth*)
Pougny, J.R. *et al*, *Tetrahedron Lett.*, 1978, 3301 (*struct, abs config*)
Omura, S. *et al*, *Methods Enzymol.*, 1981, **72**, 520 (*rev*)
Jakubowski, A.A. *et al*, *J. Org. Chem.*, 1982, **47**, 1221 (*synth*)
Furukawa, J. *et al*, *Chem. Pharm. Bull.*, 1988, **36**, 1229 (*synth*)
Yoda, H. *et al*, *Tetrahedron Lett.*, 1991, **32**, 6771 (*synth*)
Morisaki, N. *et al*, *Chem. Pharm. Bull.*, 1992, **40**, 2945, 2954 (*synth*)
Lewis, R.J., *Sax's Dangerous Properties of Industrial Materials*, 8th Ed., Van Nostrand-Reinhold, 1992, ECE500.

1-Chlorodecane, 9CI C-00020

Decyl chloride
[1002-69-3]

$H_3C(CH_2)_8CH_2Cl$

$C_{10}H_{21}Cl$ M 176.729
Bp 223-224.8°.

Rossander, S.S. *et al*, *J. Am. Chem. Soc.*, 1928, **50**, 1495 (*synth*)
Ford, T.A. *et al*, *J. Am. Chem. Soc.*, 1952, **74**, 4323 (*synth*)
Nekrasova, V.A., *Dokl. Akad. Nauk SSSR*, 1953, **88**, 73, 475 (*synth*)
Nekrosova, V.A., *CA*, 1959, **53**, 15949 (*synth*)
Asinger, F. *et al*, *Chem. Ber.*, 1959, **92**, 3085 (*synth*)
Regen, S.L. *et al*, *J. Org. Chem.*, 1975, **40**, 1669 (*synth*)
Sax, N.I., *Dangerous Properties of Industrial Materials*, 5th Ed., Van Nostrand-Reinhold, 1979, 537.

12-Chlorododecanoic acid C-00021

[22075-86-1]

$ClCH_2(CH_2)_{10}COOH$

$C_{12}H_{23}ClO_2$ M 234.765
Cryst. (pet. ether). Mp 43-44°.
Chloride: [78476-69-4].
$C_{12}H_{22}Cl_2O$ M 253.211
$Bp_{0.15}$ 139-145°.

Logemann, E. *et al*, *Chem. Ber.*, 1981, **114**, 2245 (*synth*)
Bidd, I. *et al*, *J. Chem. Soc., Perkin Trans. 1*, 1987, 2455 (*synth*)

10-Chloro-1,16-heptadecadiene-4,6-diyne-3,9-diol C-00022

Ginsenoyne B
[139035-29-3]

$H_2C{=}CHCH(OH)C{\equiv}CC{\equiv}CCH_2CH(OH)CHCl(CH_2)_5CH{=}CH_2$

$C_{17}H_{23}ClO_2$ M 294.820
Isol. from the root of *Panax ginseng*. Oil. $[\alpha]_D$ −33.0° (c, 2.0 in $CHCl_3$). λ_{max} 201, 230, 243, 257 nm ($CHCl_3$).
Di-Ac: [139035-32-8].
Oil. $[\alpha]_D$ +32.8° (c, 2.04 in $CHCl_3$). λ_{max} 200, 220, 232, 244, 257 nm (EtOH).

Hirakura, K. *et al*, *Phytochemistry*, 1991, **30**, 3327 (*isol, ir, uv, ms, pmr, cmr*)

1-Chloroheptadecane C-00023

Heptadecyl chloride

$H_3C(CH_2)_{15}CH_2Cl$

$C_{17}H_{35}Cl$ M 274.916
Mp 24°. Bp_{10} 192-195°.

v. Braun, J. *et al*, *Ber.*, 1911, **44**, 1464.

10-Chloro-1-heptadecene-4,6-diyne-3,9-diol C-00024

Panaxydol chlorohydrin
[111103-92-5]

$H_2C{=}CHCH(OH)C{\equiv}CC{\equiv}CCH_2CH(OH)CHCl(CH_2)_6CH_3$

$C_{17}H_{25}ClO_2$ M 296.836
Isol. from Korean ginseng root. Show cytotoxic activity against L1210 cells. λ_{max} 266nm, 255, 242, 229, 219 (EtOH).

Kitagawa, I. *et al*, *Yakugaku Zasshi*, 1987, **107**, 495 (*synth*)

Ahn, B.Z. *et al*, *Arch. Pharm.* (*Weinheim, Ger.*), 1989, **322**, 223 (*isol, props, cmr, uv, ir*)
Hirakura, K. *et al*, *Phytochemistry*, 1991, **30**, 3327 (*isol, pmr*)

7-Chloro-2-heptenoic acid — C-00025

$ClCH_2(CH_2)_3CH{=}CHCOOH$

$C_7H_{11}ClO_2$ M 162.615

***(E)*-form** [75078-18-1]
Oil. $Bp_{0.5}$ 118-121°.
Et ester: [107408-35-5].
$C_9H_{15}ClO_2$ M 190.669
Bp_{13} 129-131°.

Cooke, M.P. *et al*, *J. Org. Chem.*, 1987, **52**, 1381 (*synth, pmr*)

1-Chlorohexadecane — C-00026

Hexadecyl chloride. Cetyl chloride
[4860-03-1]

$H_3C(CH_2)_{14}CH_2Cl$

$C_{16}H_{33}Cl$ M 260.889
Mp 17.9°. Bp 322°, Bp_{10} 177°.

Sonnet, P.E., *Synth. Commun.*, 1976, **6**, 21 (*synth*)

6-Chloro-2-hexyn-1-ol, 9CI — C-00027

[1002-37-5]

$ClCH_2CH_2CH_2C{\equiv}CCH_2OH$

C_6H_9ClO M 132.589
Liq. $Bp_{0.04}$ 72-74°, $Bp_{0.025}$ 54°.

Rachlin, A.I. *et al*, *J. Org. Chem.*, 1961, **26**, 2688 (*synth*)
Pearson, W.H. *et al*, *Synthesis*, 1990, 156 (*synth, ir, pmr, ms*)

6-Chloro-4-hydroxy-2-naphthoic acid — C-00028

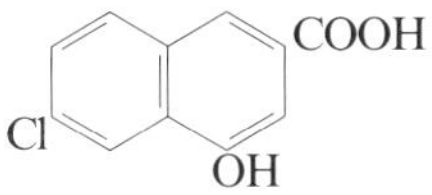

$C_{11}H_7ClO_3$ M 222.627
Cryst. (MeOH or AcOH). Mp 302-304° (297°).

Beech, W.F. *et al*, *J. Chem. Soc.*, 1949, 1887 (*synth*)
El-Abbady, A.M. *et al*, *J. Org. Chem.*, 1961, **26**, 4871 (*synth*)

1-Chlorooctadecane — C-00029

Octadecyl chloride
[3386-33-2]

$H_3C(CH_2)_{16}CH_2Cl$

$C_{18}H_{37}Cl$ M 288.943
Cryst. (Et_2O). Mp 28.6° (18°). Bp 348°, Bp_{10} 199°.

Meyer, K.H. *et al*, *Helv. Chim. Acta*, 1937, **20**, 1179 (*synth*)
McLafferty, F.W., *Anal. Chem.*, 1962, **34**, 2 (*ms*)

7-Chloro-3,9,12-pentadecatrien-1-yn-6-ol — C-00030

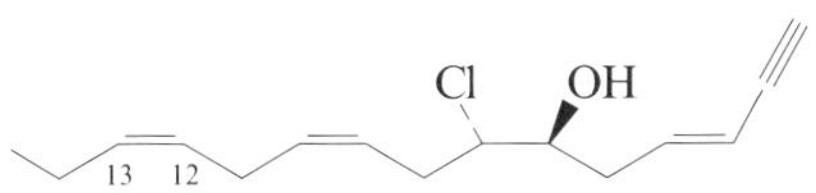

(3*E*,6*S*,7*S*,9*Z*,12*Z*)-*form*

$C_{15}H_{21}ClO$ M 252.783

***(3E,6S,7S,9Z,12Z)*-form** [139026-38-3]
Constit. of *Laurencia pinnatifida.*
Ac: [139026-36-1].
$C_{17}H_{23}ClO_2$ M 294.820
Constit. of *L. pinnatifida*. Oil. $[\alpha]_D^{25}$ −12.9° (c, 0.35 in $CHCl_3$).
12,13-Dihydro: [139026-41-8]. *7-Chloro-3,9-pentadecadien-1-yn-6-ol*
$C_{15}H_{23}ClO$ M 254.799
Constit. of *L. pinnatifida*. Oil. $[\alpha]_D^{25}$ +17.0° (c, 0.18 in $CHCl_3$).
12,13-Dihydro, Ac: [139026-43-0].
$C_{17}H_{25}ClO_2$ M 296.836
Constit. of *L. pinnatifida*. Oil. $[\alpha]_D^{25}$ −12.9° (c, 1.16 in $CHCl_3$).

***(3Z,6S,7S,9Z,12Z)*-form** [139026-37-2]
Ac: [139026-35-0].
Constit. of *L. pinnatifida*. Oil. $[\alpha]_D^{25}$ +18.5° (c, 0.85 in $CHCl_3$).
12,13-Dihydro: [139039-80-8].
Constit. of *L. pinnatifida*. Oil. $[\alpha]_D^{25}$ −9.1° (c, 0.4 in $CHCl_3$).
12,13-Dihydro, Ac: [139026-42-9].
Constit. of *L. pinnatifida*. Oil. $[\alpha]_D^{25}$ +39.7° (c, 0.47 in $CHCl_3$).

Norte, M. *et al*, *Tetrahedron*, 1991, **47**, 9411 (*isol, pmr, cmr*)

2-Chloropropanoic acid, 9CI — C-00031

[598-78-7]

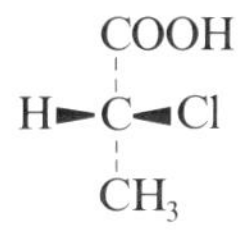

(*R*)-*form*
Absolute configuration

$C_3H_5ClO_2$ M 108.524
▷ UE8575000.

***(R)*-form** [29617-66-1]
Liq. $[\alpha]_D^{25}$ +14.6°.
Me ester: [77287-29-7].
$C_4H_7ClO_2$ M 122.551
Bp 133-134°. $[\alpha]_D^{25}$ +2.15°.

***(S)*-form** [7474-05-7]
Liq. d^{25} 1.249. $[\alpha]_D^{25}$ −14.6°.
Me ester: [73246-45-4].
Bp_{120} 79-80°. $[\alpha]_D$ −26.83°.

(±)-*form* [62138-52-7]
Misc. H_2O, Et_2O. d^0 1.28. Bp 186°, Bp_{12} 84°. pK_{a1} 2.89 (25°).
Me ester: [40705-03-1].
Bp 132-133°.
Chloride: [76248-57-2].
$C_3H_4Cl_2O$ M 126.969
Bp_{744} 110°.
Amide: [27816-36-0].
C_3H_6ClNO M 107.539
Mp 80°.
Nitrile: [1617-17-0]. *α-Chloroethyl cyanide. 1-Chloro-1-cyanoethane*
C_3H_4ClN M 89.524
Liq. Bp 123-124°.

[7623-09-8, 17639-93-9]

Michael, A., *Ber.*, 1901, **34**, 4049 (*synth*)
Frankland, P.F. *et al*, *J. Chem. Soc.*, 1914, **105**, 1110 (*synth*)
Levene, P.A. *et al*, *J. Biol. Chem.*, 1929, **81**, 707 (*synth*)
Fickett, W. *et al*, *J. Am. Chem. Soc.*, 1951, **73**, 5063 (*synth, abs config*)

Fu, J. *et al*, *J. Am. Chem. Soc.*, 1954, **76**, 6054 (*synth*)
Olah, G.A. *et al*, *Helv. Chim. Acta*, 1983, **66**, 1028 (*synth, cmr*)
Lewis, R.J., *Sax's Dangerous Properties of Industrial Materials*, 8th Ed., Van Nostrand-Reinhold, 1992, CKS750, CKT000.

1-Chlorotridecane C-00032

Tridecyl chloride
[822-13-9]

$H_3C(CH_2)_{11}CH_2Cl$

$C_{13}H_{27}Cl$ M 218.809
Bp_9 136°, $Bp_{0.5}$ 103°.

Hartmann, L. *et al*, *J. Am. Oil Chem. Soc.*, 1965, **42**, 65 (*synth*)

1-Chloro-3-tridecene-5,7,9,11-tetrayn-2-ol C-00033

[50818-97-8]

$H_3C(C{\equiv}C)_4CH{=}CHCH(OH)CH_2Cl$

$C_{13}H_9ClO$ M 216.666

(*E*)-*form*
Constit. of *Centaurea ruthenica*. Pale-yellow cryst. Mp 105-106°.
Ac: [50818-98-9].
$C_{15}H_{11}ClO_2$ M 258.703
Constit. of *C. ruthenica*. Mp 53-54°. $[\alpha]_D^{24}$ +105° (c, 0.021 in Et_2O).

Bohlmann, F. *et al*, *Chem. Ber.*, 1973, **106**, 2140.

Chlorovulones C-00034

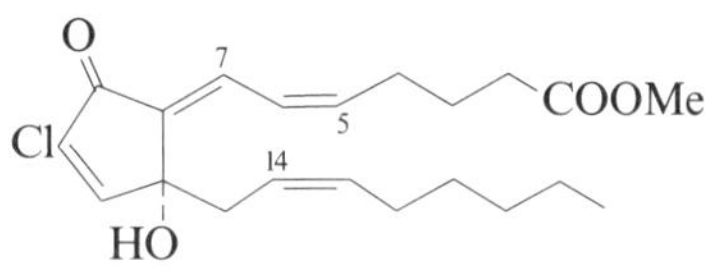

(5*Z*,7*E*,14*Z*)-*form*

$C_{21}H_{29}ClO_4$ M 380.910
Constits. of stolonifer *Clavularia viridis*. Show antitumour activity.

(5Z,7E,14Z)-form [100295-81-6]
Chlorovulone I
Oil. $[\alpha]_D$ −1.2° (c, 0.17 in $CHCl_2$).

(5E,7E,14Z)-form [100295-80-5]
Chlorovulone II
$[\alpha]_D$ +22.7° (c, 0.75 in $CHCl_3$).

(5E,7Z,14Z)-form [100295-79-2]
Chlorovulone III
$[\alpha]_D$ +27.3° (c, 0.033 in $CHCl_3$).

(5Z,7Z,14Z)-form [100201-69-2]
Chlorovulone IV
No phys. props. reported.

Iguchi, K. *et al*, *Tetrahedron Lett.*, 1985, **26**, 5787 (*isol, struct*)
Nagaoka, H. *et al*, *Tetrahedron Lett.*, 1986, **27**, 223 (*abs config, synth*)

Cholesta-5,7-dien-3-ol C-00035

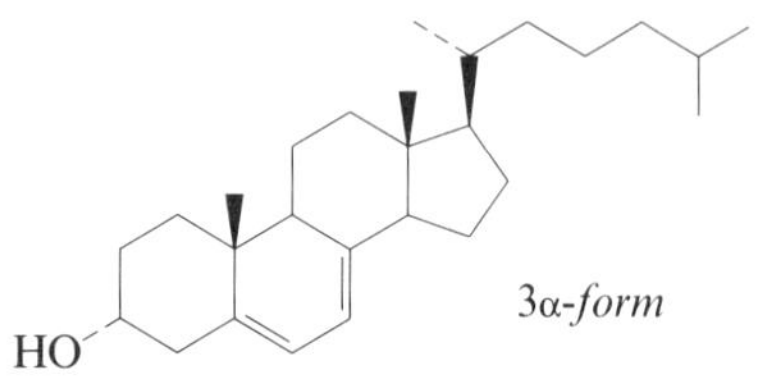

3α-*form*

$C_{27}H_{44}O$ M 384.644

3α-form [57651-84-0]
Needles (MeOH). Mp 125-126°. $[\alpha]_D^{20}$ −70.5° ($CHCl_3$).
Ac:
$C_{29}H_{46}O_2$ M 426.681
Needles (MeOH). Mp 114.5°. $[\alpha]_D^{20}$ −35° ($CHCl_3$).

3β-form [434-16-2] ***7-Dehydrocholesterol.*** *Provitamin D_3*
Occurs in man and higher animals as Vitamin D precursor. Isol. from the horned snail *Buccinum undatum* and pigskin. Irradiation with uv light prod. previtamin D_3. Plates (MeOH aq./Et_2O). Mp 150-151° (anhyd.). $[\alpha]_D^{20}$ −113.6° ($CHCl_3$). Oxid. in air.
▷ FZ5650000.
Ac: [1059-86-5].
Cryst. (MeOH). Mp 129-130°. $[\alpha]_D^{20}$ −85.3° ($CHCl_3$).
▷ FZ5680000.
3-Sulfate: Mp 137.5-138.5° (as Na salt). $[\alpha]_D^{25}$ −48° (DMF) (as Na salt).

(3β,9β)-form [10346-44-8]
Isopyrocholecalciferol. ***Isopyrovitamin D_3***
3,5-Dinitrobenzoyl: Orange needles ($CHCl_3$/MeOH). Mp 170°. $[\alpha]_D^{18}$ +318° ($CHCl_3$).

(3β,9β,10α)-form [5226-01-7] ***Lumisterol$_3$***
Cryst. (Me_2CO).
Ac: Needles (MeOH); plates (Me_2CO). Mp 131-132°. $[\alpha]_D^{18}$ +142° ($CHCl_3$).

(3β,10α)-form [10346-43-7]
Pyrocholecalciferol. ***Pyrovitamin D_3***
3,5-Dinitrobenzoyl: Yellow needles (Me_2O/MeOH). Mp 142°. $[\alpha]_D^{19}$ +221° ($CHCl_3$).
Ac: Needles (Et_2O/pet. ether). Mp 121°. $[\alpha]_D^{19}$ +428° ($CHCl_3$).

(3β,14β)-form [117605-68-2]
Oil.

Windaus, A. *et al*, *Justus Liebigs Ann. Chem.*, 1935, **520**, 98; 1939, **542**, 204 (*synth, 3α-form*)
Windaus, A. *et al*, *Justus Liebigs Ann. Chem.*, 1937, **533**, 118 (*Lumisterol$_3$*)
Windaus, A. *et al*, *Justus Liebigs Ann. Chem.*, 1938, **537**, 1 (*Pyrocholecalciferol, Isopyrocholecalciferol*)
Bernstein, S. *et al*, *J. Org. Chem.*, 1949, **14**, 433 (*synth, 3β-form*)
Handley, G.J., *Aust. J. Pharm.*, 1959, **40**, 1079 (*rev, synth*)
Gaylor, J.L. *et al*, *J. Lipid Res.*, 1964, **5**, 422 (*biosynth*)
Weiss, U. *et al*, *Tetrahedron*, 1965, **21**, 3105 (*ord*)
De Kok, A.J. *et al*, *Acta Crystallogr., Sect. B*, 1974, **30**, 1695 (*cryst struct*)
Tsukida, K. *et al*, *J. Nutr. Sci. Vitaminol.*, 1975, **21**, 411 (*cmr*)
Aberhart, D.J. *et al*, *J. Org. Chem.*, 1976, **41**, 1067 (*synth, 3α-form*)
Iida, T. *et al*, *Steroids*, 1977, **29**, 453 (*pmr, 3β-form*)
Pelc, B. *et al*, *Steroids*, 1978, **31**, 23 (*3β,9β and 3β,10α-forms*)
Confalone, P.N. *et al*, *J. Org. Chem.*, 1981, **46**, 1030 (*synth*)
Anastasia, M. *et al*, *Steroids*, 1987, **49**, 543 (*synth, ir, pmr, 3β,14β-form*)
Kirk, D.N. *et al*, *J. Chem. Soc., Perkin Trans. 2*, 1990, 1567 (*pmr*)
Lewis, R.J., *Sax's Dangerous Properties of Industrial Materials*, 8th Ed., Van Nostrand-Reinhold, 1992, DAK600, DAK800.

Cholesta-5,22-dien-3-ol, 9CI C-00036

$C_{27}H_{44}O$ M 384.644

(3β,22E)-form [34347-28-9]
22E-Dehydrocholesterol
Constit. of a dinoflagellate symbiont of *Zoanthus sociatus* and many other marine organisms. Cryst. (EtOH). Mp 133.5-134°. $[\alpha]_D^{23}$ −60° (c, 1 in $CHCl_3$).
Ac: [26033-11-4].
Cryst. (Et_2O/MeOH). Mp 128-129°. $[\alpha]_D^{23}$ −63.2° (c, 1.15 in $CHCl_3$).

(3β,22Z)-form [26033-10-3]

22Z-Dehydrocholesterol
Constit. of the scallop *Placopecten magellanicus* and the red alga *Hypnea japonica*. Cryst. (MeOH). Mp 134-136°. $[\alpha]_D$ −58.5° ($CHCl_3$).
Ac: [25819-79-8].
Cryst. (Et_2O/MeOH). Mp 117-119°. $[\alpha]_D^{23}$ −68° (c, 1 in $CHCl_3$).
(3β,20S,22E)-form
20-iso form
Mp 154-155°. $[\alpha]_D^{28.5}$ −63.1° (c, 1.01 in $CHCl_3$).
Ac: Mp 107-108°. $[\alpha]_D^{29}$ −66.6° (c, 1.01 in $CHCl_3$).
[566-89-2]

Bergmann, W. *et al*, *J. Org. Chem.*, 1958, **23**, 1245 (*synth, 3β,22E-form*)
Tsuda, K. *et al*, *J. Am. Chem. Soc.*, 1960, **82**, 1442 (*isol*)
Tsuda, K. *et al*, *Chem. Pharm. Bull.*, 1961, **9**, 529 (*synth, 3β,20S,22E-form*)
Tamura, T. *et al*, *Can. J. Biochem.*, 1964, **42**, 1331 (*isol*)
Hutchins, R.F.N. *et al*, *Steroids*, 1970, **15**, 113 (*synth, pmr*)
Rubinstein, I. *et al*, *Phytochemistry*, 1976, **15**, 195 (*pmr*)
Wright, J.L.C. *et al*, *Can. J. Chem.*, 1978, **56**, 1898 (*cmr*)
Kokke, W.C.M.C. *et al*, *Tetrahedron Lett.*, 1979, 3601 (*isol, ir, nmr, ms*)

Cholesta-5,24-dien-3-ol, 9CI C-00037

$C_{27}H_{44}O$ M 384.644
3α-form [67392-80-7]
Cryst. (MeOH). Mp 137-139°. $[\alpha]_D^{28}$ −43.5° ($CHCl_3$).
Ac: [67383-62-4].
$C_{29}H_{46}O_2$ M 426.681
Cryst. (MeOH). Mp 112-115°. $[\alpha]_D^{28}$ −47.5° ($CHCl_3$).
3β-form [313-04-2] ***Desmosterol***
Constit. of the seeds of *Funtumia latifolia*, chick embryos, the skin of rats, barnacles and red algae. Constituent of red algae *Rhodymenia palmata* and *Halosaicion ramentaceum*. Cryst. (Me_2CO/MeOH). Mp 121.5-122.5°. $[\alpha]_D^{20}$ −38.2° (c, 1.143 in $CHCl_3$).
Ac: [2665-04-5].
Cryst. (MeOH/Me_2CO). Mp 99°. $[\alpha]_D^{28}$ −44.8° ($CHCl_3$).
(3β,20S)-form [59532-46-6]
Cryst. (MeOH). Mp 125-127°.

Idler, D.R. *et al*, *Steroids*, 1968, **11**, 465 (*isol, 3β-form*)
Charles, G. *et al*, *C. R. Hebd. Seances Acad. Sci.*, 1969, **268**, 2105 (*isol*)
Ener, M.A. *et al*, *J. Biol. Chem.*, 1973, **248**, 6697 (*ms, 3β-form*)
Dasgupta, S.K. *et al*, *J. Org. Chem.*, 1974, **39**, 1658 (*synth, pmr, 3β-form*)
Chen, S.-M.L. *et al*, *J. Am. Chem. Soc.*, 1975, **97**, 5297 (*biosynth*)
Takeshita, T. *et al*, *Chem. Pharm. Bull.*, 1976, **24**, 1928 (*synth*)
Bu'lock, J.D. *et al*, *Phytochemistry*, 1976, **15**, 1249 (*biosynth*)
Burstein, S. *et al*, *Steroids*, 1976, **27**, 361 (*synth, ir, pmr, 3β,20S-form*)
Ochi, K. *et al*, *Steroids*, 1977, **30**, 795 (*synth, pmr*)
Koreeda, M. *et al*, *J. Org. Chem.*, 1980, **45**, 1172, 1174 (*synth, 3β-form*)
Takano, S. *et al*, *J. Chem. Soc., Chem. Commun.*, 1983, 760 (*synth*)
Kircher, H.W. *et al*, *J. Org. Chem.*, 1987, **52**, 2586 (*synth, 3β-form*)
Ikuina, Y. *et al*, *Chem. Pharm. Bull.*, 1989, **37**, 1755 (*synth*)
Giner, J.L. *et al*, *Phytochemistry*, 1992, **31**, 3865 (*biosynth*)

Cholesta-8(14),24-dien-3-ol, 9CI C-00038

$C_{27}H_{44}O$ M 384.644
(3β,5α)-form [117556-83-9]
Needles (MeOH). Mp 108-109°. $[\alpha]_D$ +16.6° (c, 0.12 in $CHCl_3$).
[68129-47-5]

Apfel, M.A., *J. Org. Chem.*, 1979, **44**, 643 (*synth*)
Dolle, R.E. *et al*, *J. Am. Chem. Soc.*, 1989, **111**, 278 (*synth, ir, pmr, cmr, ms*)

Cholesterol, USAN C-00039

Cholest-5-en-3β-ol. Cholesterin
[57-88-5]

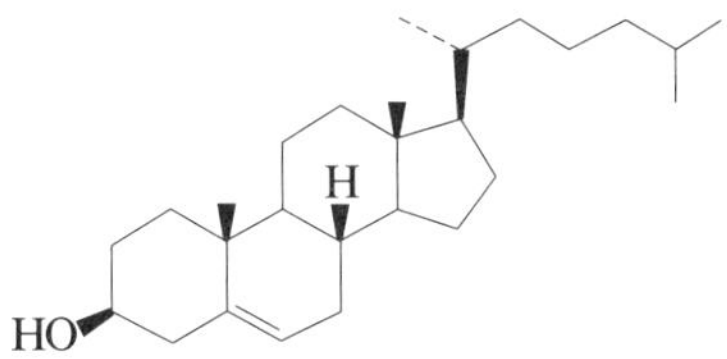

$C_{27}H_{46}O$ M 386.660
Characteristic sterol of higher animals. Constit., either free or as esters, of fish liver oils, egg yolk, bile, bran, gallstones, etc. Constit. of scent material of cotton-top tamarin monkey (*Saguinus oedipus*). Pharmaceutical aid (emulsifying agent). Pearly leaflets (EtOH aq.). Mp 148.5° (anhyd.). $[\alpha]_D$ −31.12° (Et_2O).
▷ Exp. carcinogen. FZ8400000.
Ac: [604-35-3].
$C_{29}H_{48}O_2$ M 428.697
Needles (Me_2CO). Mp 114-115°. $[\alpha]_D^{20}$ −47.4° ($CHCl_3$).
Me ether: 3β-Methoxycholest-5-ene
$C_{28}H_{48}O$ M 400.687
Cryst. (EtOH). Mp 84.5°. $[\alpha]_D^{20}$ −45.8° ($CHCl_3$).
Et ether: 3β-Ethoxycholest-5-ene
$C_{29}H_{50}O$ M 414.713
Mp 88.5°. $[\alpha]_D^{21}$ −39° ($CHCl_3$).
Propanoyl:
$C_{30}H_{50}O_2$ M 442.724
Plates (EtOH/Et_2O). Mp 114°. $[\alpha]_D$ −40.9°.
Butanoyl:
$C_{31}H_{52}O_2$ M 456.751
Mp 97-99°. $[\alpha]_D$ −34.8° ($CHCl_3$). Used as g.c. standard.
2-Methylpropanoyl: Mp 128-130°. $[\alpha]_D$ −37.4° ($CHCl_3$).
Pentanoyl: Cholesteryl valerate
$C_{32}H_{54}O_2$ M 470.777
Mp 87-89°. $[\alpha]_D$ −34.1° ($CHCl_3$).
Heptanoyl: [1182-07-6].
$C_{34}H_{58}O_2$ M 498.831
Mp 106-109°. $[\alpha]_D$ −40° (CCl_4).
Octanoyl: [1182-42-9]. *Cholesteryl caprylate*
$C_{35}H_{60}O_2$ M 512.858
Mp 104-108°. $[\alpha]_D$ −29° ($CHCl_3$).
Nonanoyl: [1182-66-7]. *Cholesteryl pelargonate*
$C_{36}H_{62}O_2$ M 526.885
Mp 58°, Mp 110-124°. $[\alpha]_D$ −33.2° ($CHCl_3$).
Decanoyl:
$C_{37}H_{64}O_2$ M 540.911
Mp 80-81°. $[\alpha]_D$ −27° ($CHCl_3$).
Dodecanoyl: Cholesteryl laurate
$C_{39}H_{68}O_2$ M 568.965
Mp 75-77°. $[\alpha]_D$ −23.4° ($CHCl_3$).
Tetradecanoyl: Cholesteryl myristate
$C_{41}H_{72}O_2$ M 597.019
Mp 69-71°. $[\alpha]_D$ −25° ($CHCl_3$).
Hexadecanoyl: [601-34-3]. *Cholesteryl palmitate*
$C_{43}H_{76}O_2$ M 625.072
Mp 75-77°. $[\alpha]_D$ −25° ($CHCl_3$).
Octadecanoyl: [35602-69-8]. *Cholesteryl stearate*
$C_{45}H_{80}O_2$ M 653.126
Mp 79-80.5°. $[\alpha]_D$ −24.7° ($CHCl_3$).
9Z-Octadecenoyl: Cholesteryl oleate
$C_{45}H_{78}O_2$ M 651.110

Mounting sex pheromone of the American dog or mountain dog tick *Dermacentor variabilis* and the Rocky mountain wood tick *D. andersoni.*

9Z,12Z-Octadecadienoyl: [604-33-1]. *Cholesteryl lineoleate*
$C_{45}H_{76}O_2$ M 649.094
Mp 39-41°. $[\alpha]_D^{20}$ −24° (c, 1 in $CHCl_3$).

Benzoyl: [604-32-0].
$C_{34}H_{50}O_2$ M 490.768
Mp 149-151°. $[\alpha]_D^{20}$ −15.0° (c, 2 in $CHCl_3$).

Cinnamoyl:
$C_{36}H_{52}O_2$ M 516.806
Mp 159°. $[\alpha]_D$ +0.7° ($CHCl_3$).

p-*Toluenesulfonate: Cholesteryl tosylate*
$C_{34}H_{52}O_3S$ M 540.849
Mp 131-132°. $[\alpha]_D$ −42.6° ($CHCl_3$).

3-((+)-14-Methylhexadecanoyl): [19477-24-8]. *Carcinolipin*
$C_{44}H_{78}O_2$ M 639.099
Carcinogenic lipid which stimulates protein synth. Cryst. Mp 75°.
▷ FZ7605000.

Chloroformyl: [7144-08-3]. *Cholesteryl chloroformate*
$C_{28}H_{45}ClO_2$ M 449.115
Reagent for isol. and identification of primary and secondary amines in aq. soln. Mp 119.5-121°. $[\alpha]_D$ −26.7°.

3-(Ethoxyacetyl): ***Chethoxyrol***
$C_{31}H_{52}O_3$ M 472.750
Constit. of *Hilsa ilisha*. Plates (EtOAc/MeOH). Mp 82-83°. $[\alpha]_D^{20}$ −53.8° ($CHCl_3$).

3-O-Glucuronide: Mp 272-274° (242-245°).

3-O-Sulfate: Mp 170-173° (as Na salt). $[\alpha]_D$ −20.8° (MeOH).

(±)-form
Elongated prisms (MeCN). Mp 149-149.5°.

10α-form [23820-70-4]
Needles(MeOH). Mp 118.5-119°. $[\alpha]_D^{25}$ −46° (c, 1.14 in $CHCl_3$).

Ac: [23820-71-5].
Mp 124-124.5°. $[\alpha]_D^{25}$ −47° (c, 1.48 in $CHCl_3$).

14β-form [57759-45-2]
Cryst. (MeOH). Mp 38°. $[\alpha]_D^{21}$ +29° ($CHCl_3$).

Ac: [57759-47-4].
Oil.

(14β,17α)-form [71869-93-7]
Cryst. (hexane). Mp 127-128°. $[\alpha]_D^{23}$ +11° ($CHCl_3$).

(20S)-form [34026-89-6]
Cryst. (MeOH/Me_2CO). Mp 152-153°. $[\alpha]_D^{20}$ −54.2° (c, 0.52 in $CHCl_3$).

Ac: [38774-63-9].
Cryst. (MeOH). Mp 90-92°. $[\alpha]_D^{25}$ −53.9° (c, 1.0 in $CHCl_3$).

ent-form
Mp 146-147°. $[\alpha]_D^{25}$ +40.6° (c, 0.93 in EtOH).

Wagner-Jauregg, T. *et al*, *Z. Physiol. Chem.*, 1932, **213**, 110 (*deriv*)
Bladon, P. *et al*, *J. Chem. Soc.*, 1952, 2737 (*uv, 3β-form*)
Johnson, D.R. *et al*, *J. Am. Chem. Soc.*, 1953, **75**, 52 (*ir, 3β-form*)
Cook, R.P., *Cholesterol* (*Chem., Biochem. Pathol.*), Academic Press, N.Y., 1958 (*rev*)
Org. Synth., Coll. Vol., 4, 1963, 195 (*purifn*)
Keana, J.F.W. *et al*, *Steroids*, 1964, **4**, 457 (*synth*)
Johnson, W.S. *et al*, *Tetrahedron, Suppl.*, No. 8, 1966, 541 (*synth*)
Fieser and Fieser's Reagents for Organic Synthesis, Wiley, 1967, **1**, 140, 141 (*purifn, deriv*)
Edwards, J.T. *et al*, *J. Org. Chem.*, 1970, **35**, 1426 (*synth, ir, pmr, 3β,10α-form*)
Mulheirn, L.J. *et al*, *Chem. Soc. Rev.*, 1972, **1**, 259 (*biosynth, 3β-form*)
Hradec, J., *Prog. Biochem. Pharmacol.*, 1975, **10**, 197 (*Carcinolipin*)
Anastasia, M. *et al*, *J. Org. Chem.*, 1976, **41**, 1064 (*synth, ir, ms, 3β,4β-form*)
Rubinstein, I. *et al*, *Phytochemistry*, 1976, **15**, 195 (*pmr*)
Bu'lock, J.D. *et al*, *Phytochemistry*, 1976, **15**, 1249 (*biosynth*)
Partridge, L.G. *et al*, *J. Org. Chem.*, 1977, **42**, 2799 (*ms*)
Shieh, H.S. *et al*, *Nature (London)*, 1977, **267**, 287 (*cryst struct*)
Joseph-Nathan, P. *et al*, *J. Am. Chem. Soc.*, 1979, **101**, 1289 (*cmr, 3β-form*)
Anastasia, M. *et al*, *J. Org. Chem.*, 1979, **44**, 4983 (*synth, ir, pmr, ms, 3β,14β,17α-form*)
Sawzik, P. *et al*, *Liq. Cryst., Proc. Int. Conf.*, Pittsburgh, USA, 1979, 171 (*esters*)
Nakahara, M. *et al*, *Bull. Chem. Soc. Jpn.*, 1980, **53**, 2499 (*esters*)
Tanabe, M. *et al*, *J. Am. Chem. Soc.*, 1980, **102**, 862 (*synth, pmr, 3β,20S-form*)
Meiboom, S. *et al*, *Phys. Rev. A: Gen. Phys.*, 1981, **24**, 468 (*esters*)
Myant, N.B., *The Biology of Cholesterol and Related Steroids*, Heinemann Medical, London, 1981 (*book*)
Gibbons, G.F. *et al*, *Biochemistry of Cholesterol*, Elsevier, Amsterdam, 1982 (*book*)
Sucrow, W. *et al*, *Justus Liebigs Ann. Chem.*, 1982, 1897 (*synth, ms*)
Ohmori, M. *et al*, *Tetrahedron Lett.*, 1982, **23**, 4709 (*synth*)
Martindale, The Extra Pharmacopoeia, 28th/29th Ed., Pharmaceutical Press, London, 1982/1989, 6412.
Goto, R. *et al*, *Chem. Pharm. Bull.*, 1983, **31**, 3528 (*synth, 3β,20S-form*)
Schmuff, N.R. *et al*, *J. Org. Chem.*, 1983, **48**, 1404 (*synth, pmr, 3β,20S-form*)
Mikami, K. *et al*, *Chem. Lett.*, 1985, 115 (*synth, 20S-form*)
Seo, S. *et al*, *J. Chem. Soc., Perkin Trans. 1*, 1986, 411 (*pmr, 3β-form*)
Haider, S.I. *et al*, *J. Nat. Prod. (Lloydia)*, 1987, **50**, 261 (*Chethoxyrol*)
Tan, F. *et al*, *J. Appl. Cardiol.*, 1988, **3**, 67 (*rev, metab*)
Hamilton, J.G.C. *et al*, *J. Chem. Ecol.*, 1988, **14**, 401 (*oleate*)
Belcher, A. *et al*, *J. Chem. Ecol.*, 1988, **14**, 1367 (*oleate*)
Ramachandran, R., *J. Magn. Reson.*, 1988, **79**, 357 (*cmr*)
Yeagle, P.L.,.Ed., *The Biology of Cholesterol*, CRC Press, 1988 (*book*)
Hamilton, J.G.C. *et al*, *J. Insect Physiol.*, 1989, **35**, 873 (*oleate*)
Rychnovsky, S.D. *et al*, *J. Org. Chem.*, 1992, **57**, 2732 (*ent-form*)
Lewis, R.J., *Sax's Dangerous Properties of Industrial Materials*, 8th Ed., Van Nostrand-Reinhold, 1992, CCJ500, CMD750.

Claviridenones C-00040

4,12-Bis(acetyloxy)-9-oxo-5,7,10,14-prostatetraen-1-oic acid methyl ester, 9CI. Clavulones

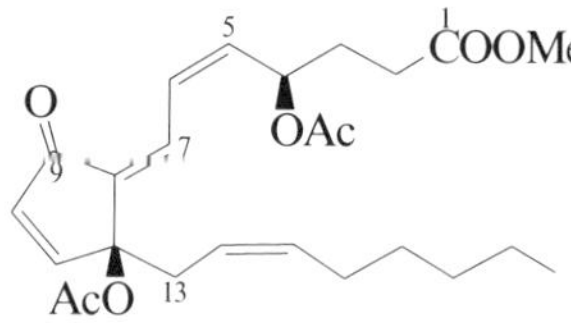

(5Z,7Z)-form

$C_{25}H_{34}O_7$ M 446.539

(5Z,7Z)-form [85611-86-5] ***Claviridenone A.*** *Clavulone IV*
Constit. of *Clavularia viridis*. Yellowish oil. $[\alpha]_D$ −82.2° ($CHCl_3$).

(5E,7Z)-form [85700-44-3] ***Claviridenone B.*** *Clavulone III*
Constit. of *C. viridis*. Yellowish oil. $[\alpha]_D$ +45.5° ($CHCl_3$).

20-Acetoxy: [88216-93-7]. *20-Acetoxyclavulone III*
$C_{27}H_{36}O_9$ M 504.576
Constit. of *C. viridis*. Oil. $[\alpha]_D$ +26.4° (c, 0.86 in $CHCl_3$).

(5E,7E)-form [85700-43-2] ***Claviridenone C.*** *Clavulone II*
Constit. of *C. viridis*. Yellowish oil. $[\alpha]_D$ +10.9° ($CHCl_3$).

20-Acetoxy: [88216-92-6]. *20-Acetoxyclavulone II*
$C_{27}H_{36}O_9$ M 504.576
Constit. of *C. viridis*. Oil. $[\alpha]_D$ +3.7° (c, 0.54 in $CHCl_3$).

(5Z,7E)-form [85700-42-1] ***Claviridenone D.*** *Clavulone I*
Constit. of *C. viridis*. Yellowish oil. $[\alpha]_D$ −28.9° ($CHCl_3$).

20-Acetoxy: [88216-91-5]. *20-Acetoxyclavulone I*
$C_{27}H_{36}O_9$ M 504.576
Constit. of *C. viridis*. Oil. $[\alpha]_D$ −31.1° (c, 0.09 in $CHCl_3$).

Kikuchi, H. *et al*, *Tetrahedron Lett.*, 1982, **23**, 5171 (*isol, struct*)
Kobayashi, M. *et al*, *Tetrahedron Lett.*, 1982, **23**, 5331 (*isol, struct*)
Kikuchi, H. *et al*, *Tetrahedron Lett.*, 1983, **24**, 1549 (*abs config*)
Iguchi, K. *et al*, *Tetrahedron Lett.*, 1983, **24**, 4423 (*derivs*)
Corey, E.J. *et al*, *J. Am. Chem. Soc.*, 1984, **106**, 3384 (*synth*)
Nagaoka, H. *et al*, *Tetrahedron Lett.*, 1984, **25**, 3621 (*synth*)
Kitagawa, I. *et al*, *Tetrahedron*, 1985, **41**, 995 (*isol, struct*)
Hashimoto, S. *et al*, *Tetrahedron Lett.*, 1985, **26**, 2679 (*synth*)
Shibasaki, M. *et al*, *Tetrahedron Lett.*, 1985, **26**, 3841 (*synth*)
Takemoto, M. *et al*, *Chem. Pharm. Bull.*, 1991, **39**, 1106 (*synth*)
Klunder, A.J.H. *et al*, *Tetrahedron Lett.*, 1991, **32**, 3131 (*synth*)

Cloprostenol, BAN, INN C-00041

7-[2-[4-(3-Chlorophenoxy)-3-hydroxy-1-butenyl]-3,5-dihydroxycyclopentyl]-5-heptenoic acid, 9CI. 16-(3-Chlorophenoxy)-9,11,15-trihydroxy-17,18,19,20-tetranorprosta-5,13-dienoic acid
[40665-92-7]

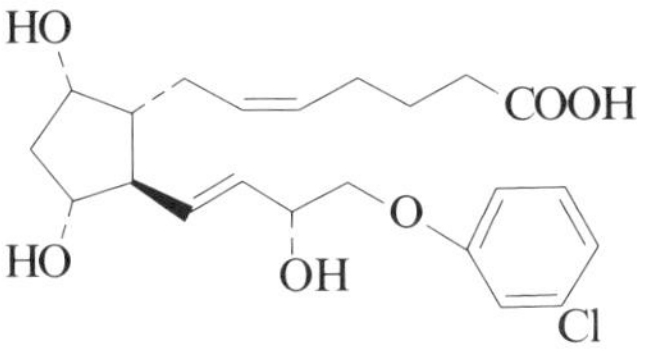

$C_{22}H_{29}ClO_6$ M 424.920
Luteolytic agent. Used as Na salt.
▷ MJ9652000.

Na salt: [55028-72-3]. *Cloprostenol sodium, USAN. Estrumate. Heifex. Oestrophan. Planate. Suimate. ICI 80996*
Amorph. hygroscopic powder. Sol. H_2O, MeOH; insol. Me_2CO.

Ger. Pat., 2 223 365, (1972); *CA*, **78**, 110692v (*synth*)
Binder, D. *et al*, *Prostaglandins*, 1974, **6**, 87; 1978, **15**, 773 (*synth, pharmacol*)
Bowler, J. *et al*, *Prostaglandins*, 1979, **17**, 789 (*synth*)
Bourne, G.R. *et al*, *Xenobiotica*, 1979, **9**, 623 (*metab*)
Baishya, N. *et al*, *Br. Vet. J.*, 1980, **136**, 227 (*props*)
Beeley, N.R.A. *et al*, *Tetrahedron, Suppl.*, No. 9, 1981, **37**, 411 (*synth*)
Martindale, The Extra Pharmacopoeia, 28th/29th Ed., Pharmaceutical Press, London, 1982/1989, 8077.
Liu, Z. *et al*, *CA*, 1985, **103**, 160243 (*synth*)
Lewis, R.J., *Sax's Dangerous Properties of Industrial Materials*, 8th Ed., Van Nostrand-Reinhold, 1992, CMX880.

Colipase C-00042

Lipase cofactor, 9CI
[55126-92-6]

Porcine colipase is a polypeptide of 95 AA residues. The struct. of equine colipase has also been detd. and differs at two positions. Isol. from pancreatic tissue. Assumed to act as a cofactor by anchoring lipase to insoluble triglyceride substrates.

Sakina, K. *et al*, *Chem. Pharm. Bull.*, 1988, **36**, 424 (*struct, bibl*)

Colneleic acid C-00043

9-(1,3-Nonadienyloxy)-8-nonenoic acid, 9CI

$H_3C(CH_2)_4(CH{=}CH)_2OCH{=}CH(CH_2)_6COOH$

$C_{18}H_{30}O_3$ M 294.433
(2′E,4′Z,8E)-form [52761-34-9]
Product of the enzymic oxidation of potato lipids. Isol. as the Me ester by glc.
[52077-21-1]

Galliard, T. *et al*, *Biochem. J.*, 1972, **129**, 743 (*isol*)
Curtis, R.R., *Chem. Phys. Lipids*, 1973, **11**, 11 (*synth*)
Galliard, T. *et al*, *Chem. Phys. Lipids*, 1973, **11**, 173 (*isol*)
Crombie, L. *et al*, *J. Chem. Soc., Chem. Commun.*, 1987, 502, 503 (*biochem*)
Corey, E.J. *et al*, *J. Org. Chem.*, 1990, **55**, 1670 (*synth*)
Crombie, L. *et al*, *J. Chem. Soc., Perkin Trans. 1*, 1991, 567 (*synth, pmr, cmr, ms*)

Colnelenic acid C-00044

9-(1,3,6-Nonatrienyloxy)-8-nonenoic acid, 9CI

$H_3CCH_2CH{=}CHCH_2(CH{=}CH)_2OCH{=}CH(CH_2)_6COOH$

$C_{18}H_{28}O_3$ M 292.417
(2′E,4′Z,7′Z,8E)-form [52591-16-9]
Product of enzymic oxidation of potato lipids. Isol. as the Me ester by glc.
[52077-22-2]

Galliard, T. *et al*, *Chem. Phys. Lipids*, 1973, **11**, 173 (*isol, pmr, uv, ir, ms*)
Crombie, L. *et al*, *J. Chem. Soc., Perkin Trans. 1*, 1991, 567 (*synth, pmr, cmr, ms*)

Conocandin C-00045

3-(6-Methyl-5-dodecyl)-α-methyloxiraneacetic acid, 9CI. ***Antibiotic A32287.*** *A 32287*
[61371-61-7]

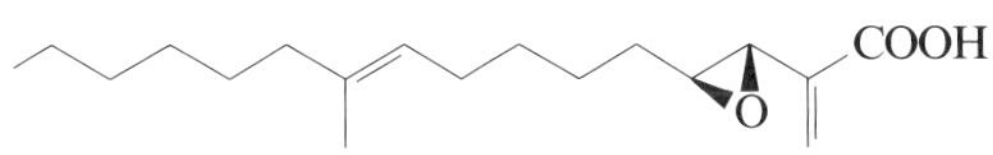

Absolute configuration

$C_{18}H_{30}O_3$ M 294.433
Metab. of *Hormococcus conorum*. Fungicidal antibiotic. Viscous oil. $[\alpha]_D^{20}$ −7° (c, 0.557 in EtOH).
▷ RR0352000.

Müller, J.M. *et al*, *Helv. Chim. Acta*, 1976, **59**, 2506 (*isol, struct*)
Banfi, L. *et al*, *J. Org. Chem.*, 1987, **52**, 5452 (*synth, abs config*)

Constanolactone A C-00046

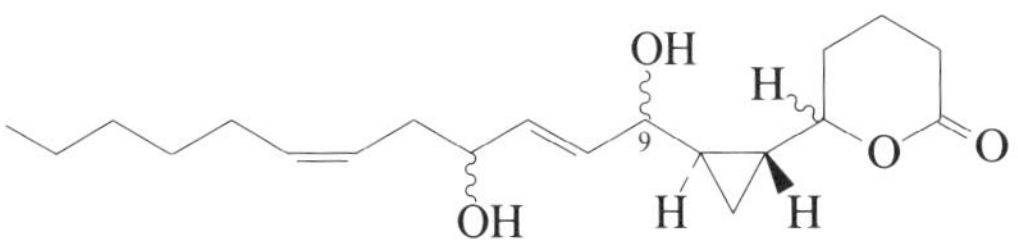

$C_{20}H_{32}O_4$ M 336.470
Metab. of red alga *Constantinea simplex*. Oil (as di-Ac). $[\alpha]_D^{23}$ −5.4° (c, 3.05 in $CHCl_3$) (diacetate).

9-Epimer: ***Constanolactone B***
$C_{20}H_{32}O_4$ M 336.470
Metab. of *C. simplex*. Oil (as di-Ac). $[\alpha]_D^{23}$ −4.8° (c, 2.08 in $CHCl_3$) (diacetate).
[130223-05-1, 130320-78-4]

Nagle, D.G. *et al*, *Tetrahedron Lett.*, 1990, **31**, 2995 (*isol, struct*)

Corynomycolenic acid C-00047

3-Hydroxy-2-tetradecyl-11-octadecenoic acid, 9CI
[578-31-4]

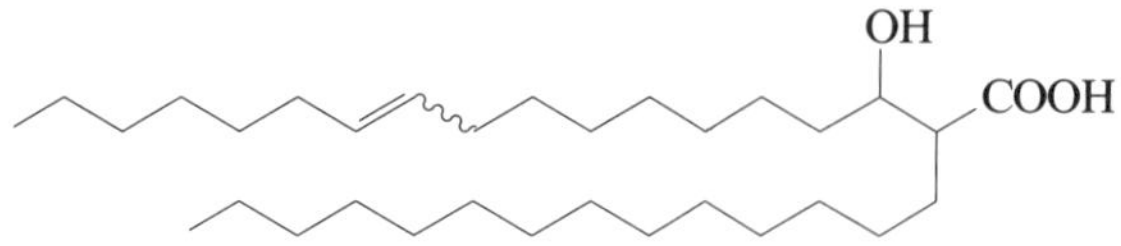

$C_{32}H_{62}O_3$ M 494.840
Isol. from diptheria bacilli, *Corynebacterium diptheriae* and *C. ovis*. Oil. n_D^{19} 1.4758.
Me ester: Oil. d_4^{20} 0.894. $[\alpha]_{546}^{20}$ +9.0°. n_D^{19} 1.4680.

Pudles, J. *et al*, *Biochim. Biophys. Acta*, 1953, **11**, 163 (*isol*)
Senn, M. *et al*, *Eur. J. Biochem.*, 1967, **1**, 353.
Thomas, D.W. *et al*, *Chem. Phys. Lipids*, 1979, **23**, 267 (*isol*)
Inoeda, T. *et al*, *Chem. Phys. Lipids*, 1979, **24**, 1 (*isol*)

Crossoline C-00048

[133352-34-8]

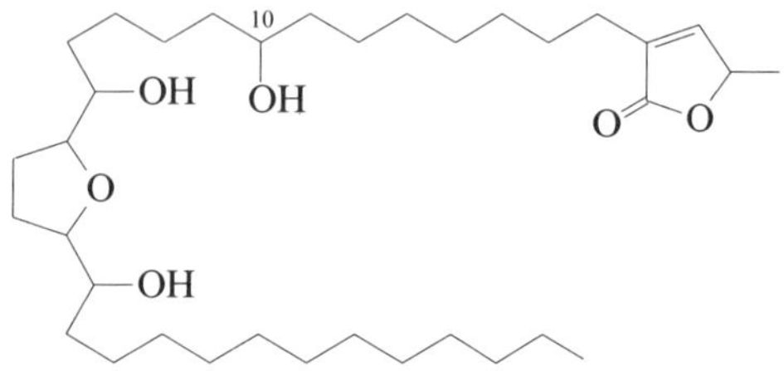

$C_{35}H_{64}O_6$ M 580.887
Constit. of *Annona muricata*. Cytotoxic. Cryst. (EtOAc). Mp 45-40°. $[\alpha]_D$ +19° (c, 0.2 in MeOH).
10-Ketone: [133352-33-7]. ***Crossolone***
$C_{35}H_{62}O_6$ M 578.871
Constit. of *A. muricata*. Cytotoxic. Cryst. (EtOAc). Mp 55-57°. $[\alpha]_D$ +15° (c, 0.13 in MeOH).

Cortes, D. *et al*, *Can. J. Chem.*, 1991, **69**, 8 (*isol, pmr, cmr*)

Cucurbic acid C-00049

3-Hydroxy-2-(2-pentenyl)cyclopentaneacetic acid
[58240-50-9]

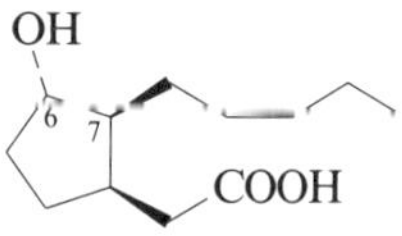

$C_{12}H_{20}O_3$ M 212.288
Constit. of *Cucurbita pepo*. Gum.
6,7-Diepimer: [120330-52-1]. ***6-Epi-7-isocucurbic acid***
$C_{12}H_{20}O_3$ M 212.288
Constit. of *Vicia faba*. Also isol. from *Equisetum* sp. and *Juglans regia*. Oil. $[\alpha]_D^{22}$ +6.1° (c, 0.1 in EtOH).

Fukui, H. *et al*, *Agric. Biol. Chem.*, 1977, **41**, 175 (*isol*)
Kitahara, T. *et al*, *Agric. Biol. Chem.*, 1987, **51**, 1129 (*synth*)
Miersch, O. *et al*, *Phytochemistry*, 1989, **28**, 339 (*isol*)

Cycloartanol C-00050

9β,19-Cyclolanostan-3β-ol. Cycloartan-3β-ol
[4657-58-3]

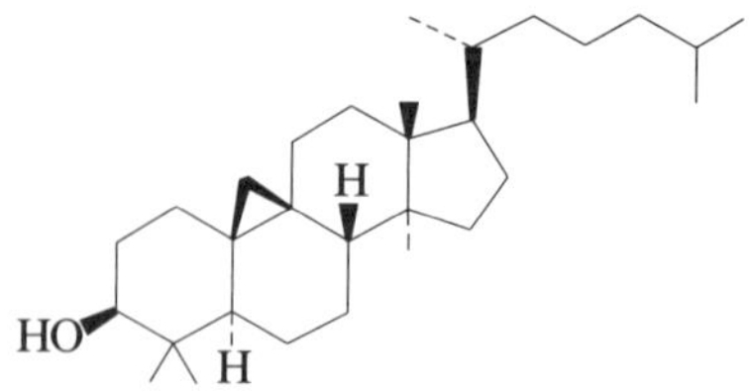

$C_{30}H_{52}O$ M 428.740
Minor constit. of rice bran oil and of the rhizomes of *Polypodium vulgare*. Cryst. (MeOH aq.). Mp 101-102°. $[\alpha]_D$ +45° (c, 3.59 in $CHCl_3$).
Ac: [4575-74-0].
$C_{32}H_{54}O_2$ M 470.777
Isol. from ferns *P.* spp. Cryst. ($CHCl_3$/MeOH). Mp 132-133°. $[\alpha]_D$ +57° (c, 5.31 in $CHCl_3$).

Shimizu, M. *et al*, *Chem. Pharm. Bull.*, 1960, **8**, 108 (*isol*)
Altenburger, E. *et al*, *Helv. Chim. Acta*, 1965, **48**, 704 (*synth*)
Audier, H.E. *et al*, *Tetrahedron Lett.*, 1966, 4341 (*ms*)
Berti, G. *et al*, *Tetrahedron Lett.*, 1967, 125 (*isol*)
Imhof, R. *et al*, *J. Chem. Soc., Chem. Commun.*, 1969, 852 (*synth*)
Khuong-Huu, F. *et al*, *Tetrahedron Lett.*, 1975, 1787 (*cmr*)
Ageta, H. *et al*, *Phytochemistry*, 1984, **23**, 2875 (*isol, acetate*)

Cycloart-24-en-3-ol C-00051

9,19-Cyclo-9β-lanost-24-en-3β-ol

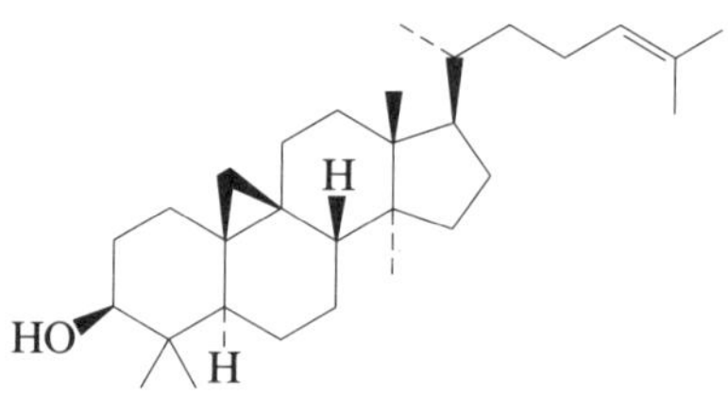

$C_{30}H_{50}O$ M 426.724
3β-form [469-38-5] ***Cycloartenol.*** *Artosenol. Handianol*
Constit. of latex of *Euphorbia balsamifera* and fruits of *Artocarpus integrifolia*. Minor component of linseed oil. Also from *Strychnos nux-vomica* and *Solanum tuberosum* and present in many other plant oils. Cryst. (EtOH). Mp 115° (dry), Mp 99° (solvated). $[\alpha]_D$ +54°.
Ac: Cycloartenyl acetate
$C_{32}H_{52}O_2$ M 468.762
Isol. from latex of *Ficus macrophylla*, also present in other plants. Mp 122°. $[\alpha]_D$ +57° (c, 1.97 in $CHCl_3$).
3-O-(4-Hydroxycinnamoyl) (E-)*:* ***Arundinol***
$C_{39}H_{56}O_3$ M 572.870
Constit. of *Arundina bambusifolia*. Cryst. (EtOAc/pet. ether). Mp 245°. $[\alpha]_D$ +3.52° (MeOH).
3-(4-Hydroxy-3-methoxycinnamoyl): ***Oryzanol A***
$C_{40}H_{58}O_4$ M 602.896
From rice bran oil (*Oryza sativa*). Cryst. Mp 148-150°. $[\alpha]_D$ +40° (c, 1 in $CHCl_3$). Oryzanol B was a mixt. of Oryzanol A and Oryzanol C (see under 24-Methylenecycloartan-3-ol, M-00044).
3-Ketone: [511-63-7]. *9,19-Cyclo-9β-lanost-24-en-3-one.* ***Cycloartenone.*** *Artostenone. Cycloart-24-en-3-one*
$C_{30}H_{48}O$ M 424.709
From *Artocarpus integrifolia*. Mp 109°. $[\alpha]_D$ +24° (c, 3.39 in $CHCl_3$).

Chapon, S. *et al*, *Bull. Soc. Chim. Fr.*, 1952, 456 (*isol*)
Bentley, H.R. *et al*, *J. Chem. Soc.*, 1953, 3673 (*isol, struct*)
Barton, D.H.R., *J. Chem. Soc.*, 1954, 2715 (*struct*)

Irvine, D.S. *et al*, *J. Chem. Soc.*, 1955, 1316 (*struct*)
Shimizü, M. *et al*, *Chem. Pharm. Bull.*, 1960, **8**, 108 (*isol*)
Capella, P., *Nature (London)*, 1961, **190**, 167 (*isol*)
Ardenne, M. *et al*, *Experientia*, 1963, **19**, 178 (*ms*)
Galbraith, M.N. *et al*, *Aust. J. Chem.*, 1965, **18**, 226 (*acetate*)
Fedeli, E. *et al*, *J. Am. Oil Chem. Soc.*, 1966, **43**, 254 (*occur*)
Aplin, R.T. *et al*, *J. Chem. Soc. B*, 1966, 1078 (*ms*)
Barton, D.H.R. *et al*, *J. Chem. Soc., Chem. Commun.*, 1968, 643 (*synth*)
Karrer, W. *et al*, *Konstitution und Vorkommen der Organischen Pflanzenstoffe*, 2nd Ed., Birkhäuser Verlag, Basel, 1972-1985, no. 2035 (*occur*)
Radics, L. *et al*, *Tetrahedron Lett.*, 1975, **48**, 4287 (*cmr*)
Boar, R.B. *et al*, *J. Chem. Soc., Perkin Trans. 1*, 1979, 563 (*synth*)
Seo, S. *et al*, *J. Chem. Soc., Perkin Trans. 1*, 1988, 2407 (*biosynth*)
Milon, A. *et al*, *Helv. Chim. Acta*, 1989, **72**, 1 (*pmr, cmr*)
Majumder, P.L. *et al*, *J. Indian Chem. Soc.*, 1991, **68**, 88 (*Arundinol*)

Cycloeucalenol C-00052

4α,14-Dimethyl-9,19-cyclo-5α,9β-ergost-24(28)-en-3β-ol, 9CI.
4α,14,24-Trimethyl-9,19-cyclo-5α,9β-cholest-24(28)-en-3β-ol.
24-Methylene-29-norcycloartan-3β-ol
[469-39-6]

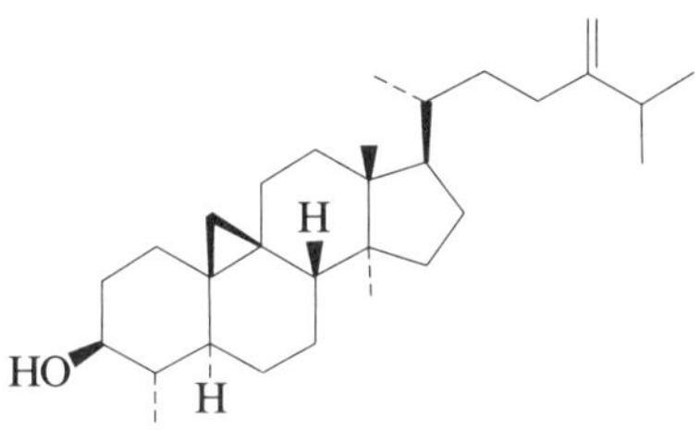

$C_{30}H_{50}O$ M 426.724
Constit. of *Eucalyptus microcorys*, *Erythrophleum guineense*, *Swietenia mahagoni*, *Tristania conferta*, *Cephalosphaera usambarensis* and other plants. Cryst. (pet. ether). Mp 138-139°. $[\alpha]_D$ +45° ($CHCl_3$).

Ac: Cycloeucalenyl acetate
$C_{32}H_{52}O_2$ M 468.762
Isol. from ferns *Polypodium* spp. Mp 108-109°. $[\alpha]_D$ +55.0°.

*24*R*,28-Dihydro:* **Cycloeucalanol.** *24*R*-Methyl-29-norcycloartan-3β-ol*
$C_{30}H_{52}O$ M 428.740
Constit. of *Nervilia purpurea*. Needles (MeOH). Mp 141-142°.

*24*R*,28-Dihydro, Ac:* Cryst. (MeOH). Mp 107.5-108.0°. $[\alpha]_D^{25}$ +70.9° (c, 0.9 in $CHCl_3$).

*24*S*,28-Dihydro:*
$C_{30}H_{52}O$ M 428.740
Constit. of *N. purpurea*. Needles (MeOH). Mp 152-153°.

Cox, J.S.G. *et al*, *J. Chem. Soc.*, 1959, 514 (*struct*)
Amorós-Marín, L. *et al*, *J. Org. Chem.*, 1959, **24**, 411 (*isol*)
Atallah, A.M. *et al*, *Phytochemistry*, 1975, **14**, 1529 (*biosynth*)
Itoh, T. *et al*, *Phytochemistry*, 1977, **16**, 1448 (*isol*)
Conner, A.H. *et al*, *Phytochemistry*, 1981, **20**, 2543 (*isol*)
Ageta, H. *et al*, *Phytochemistry*, 1984, **23**, 2875 (*isol, acetate*)
Kikuchi, T. *et al*, *Chem. Pharm. Bull.*, 1985, **33**, 1914.
Kikuchi, T. *et al*, *Chem. Pharm. Bull.*, 1986, **34**, 2479 (*pmr, cmr*)

11-Cycloheptyl-2-hydroxyundecanoic acid C-00053

α-Hydroxycycloheptaneundecanoic acid, 9CI

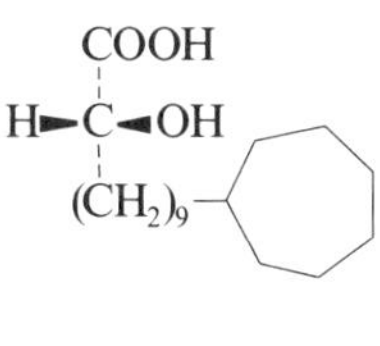

$C_{18}H_{34}O_3$ M 298.465

(R)-form [95896-79-0]
Main component of the fatty acids of a thermoacidic *Bacillus* sp. $[\alpha]_D^{20}$ −4° (c, 0.3 in $CHCl_3$).

Me ester: [95896-80-3].
$[\alpha]_D^{20}$ −7.5° (c, 0.4 in $CHCl_3$).

Allgaier, H. *et al*, *Justus Liebigs Ann. Chem.*, 1985, 378 (*isol, ms, pmr, cmr*)

Cyclohexaneundecanoic acid C-00054

Cyclohexylundecanoic acid

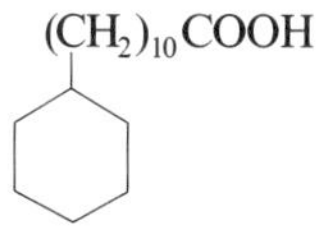

$C_{17}H_{32}O_2$ M 268.439
Minor constit. of butter fat, sheep perinephric fat and bovine rumen bacteria. Mp 55.4-56.6°.

Me ester: [19096-10-7].
$Bp_{0.3}$ 120-124.5°.

Anilide: Mp 84.9-86.6°.

Hiers, G.S. *et al*, *J. Am. Chem. Soc.*, 1926, **48**, 2385 (*synth*)
Schogt, J.C.M. *et al*, *J. Lipid Res.*, 1965, **6**, 466.
Hanson, R.P. *et al*, *Chem. Ind. (London)*, 1967, 1640.
Hanson, R.P. *et al*, *J. Sci. Food Agric.*, 1967, **18**, 225 (*isol*)

7-[2-(3-Cyclohexyl-3-hydroxy-1-propenyl)-3,5-dihydroxycyclopentyl]-5-heptenoic acid C-00055

15-Cyclohexylpentanorprostaglandin $F_{2α}$
[58611-97-5]

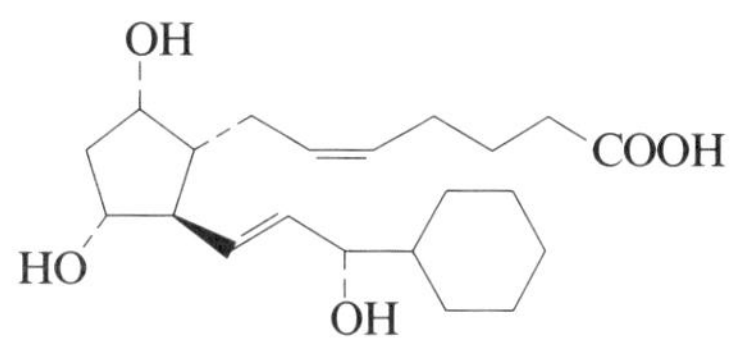

$C_{21}H_{34}O_5$ M 366.497
Shows luteolytic props. $[\alpha]_D^{24}$ +23.6° (c, 0.95 in EtOH).

Ger. Pat., 2 515 770, (1975); *CA*, **84**, 121275m (*synth, ir, pmr, biochem*)
Balapure, A.K. *et al*, *Biochem. Pharmacol.*, 1989, **38**, 2375 (*biochem*)

1-Cyclopentene-1-carboxylic acid, 9CI C-00056

[1560-11-8]

$C_6H_8O_2$ M 112.128
Leaflets or needles (H_2O). Sol. hot H_2O. Mp 123-124°. Bp 210°. Steam-volatile, subl.

Et ester: [10267-94-4].
$C_8H_{12}O_2$ M 140.182
Bp_{25} 92°, Bp_{10} 75°.

Anilide:
$C_{12}H_{13}NO$ M 187.241
Mp 126°.

Nitrile: [3047-38-9]. *1-Cyanocyclopentene*
C_6H_7N M 93.128
Bp_{30} 81°.

Urion, E., *C. R. Hebd. Seances Acad. Sci.*, 1930, **190**, 1512.

Cook, A.H. *et al*, *J. Chem. Soc.*, 1934, 959 (*synth*)
Branner-Jorgensen, S., *Acta Chem. Scand.*, 1966, **20**, 2192 (*synth*)
Odinokov, V.N., *Zh. Org. Khim.*, 1973, **9**, 671; *CA*, **79**, 31536q (*synth*)
Philp, R.P. *et al*, *Aust. J. Chem.*, 1977, **30**, 123 (*synth*)

2-Cyclopentene-1-carboxylic acid, 9CI — C-00057

Aleprolic acid

[2348-89-2]

(*S*)-*form*

$C_6H_8O_2$ M 112.128

(*S*)-*form* [67886-24-2]

Occurs in oil of *Hydnocarpus wightiana*. $[\alpha]_D^{25}$ −262° (c, 3.53 in $CHCl_3$).

(±)-*form*

Liq. Bp_4 65°. n_D^{25} 1.4667.

Me ester: [2258-56-2].
$C_7H_{10}O_2$ M 126.155
Oil. Bp_{12} 50°.

Anilide:
$C_{12}H_{13}NO$ M 187.241
Mp 120°.

Cole, H.I. *et al*, *J. Am. Chem. Soc.*, 1939, **61**, 2350 (*isol*)
Branner-Jorgensen, S., *Acta Chem. Scand.*, 1966, **20**, 2192 (*synth*)
Maercker, A. *et al*, *Chem. Ber.*, 1973, **106**, 773 (*synth*)
Spener, F., *Eur. J. Biochem.*, 1975, **53**, 161 (*biosynth*)
Bunnell, C.A. *et al*, *J. Am. Chem. Soc.*, 1977, **99**, 5184 (*synth*)
Chapman, O.L. *et al*, *J. Am. Chem. Soc.*, 1978, **100**, 4878 (*synth, resoln*)
Coe, D.M. *et al*, *J. Chem. Soc., Perkin Trans. 1*, 1991, 2373 (*synth, pmr*)

3-Cyclopentene-1-carboxylic acid, 9CI — C-00058

[7686-77-3]

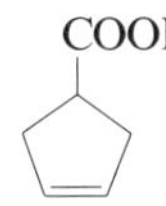

$C_6H_8O_2$ M 112.128

Liq. Bp_{20} 118-121°.

Anilide: [7686-79-5].
$C_{12}H_{13}NO$ M 187.241
Mp 139-140°.

Murdock, K.C. *et al*, *J. Org. Chem.*, 1966, **27**, 2395 (*synth*)
Schmid, G.H. *et al*, *J. Org. Chem.*, 1967, **32**, 254 (*synth*)
Deprés, J.-P. *et al*, *J. Org. Chem.*, 1984, **49**, 928 (*synth, bibl*)

2-Cyclopentene-1-heptanoic acid, 8CI — C-00059

Aleprylic acid

[2348-91-6]

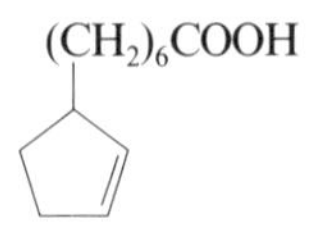

$C_{12}H_{20}O_2$ M 196.289

(+)-*form*

Occurs in *Hydnocarpus wightiana* oil. Mp 32°. $[\alpha]_D^{25}$ +90.8°.

Et ester:
$C_{14}H_{24}O_2$ M 224.342
Bp_{10} 148°. $[\alpha]_D^{25}$ +79.1°.

(±)-*form*

Cryst. (Me_2CO). Mp 32°.

Cole, H.I. *et al*, *J. Am. Chem. Soc.*, 1939, **61**, 2349.
Tanaka, A., *J. Lipid Res.*, 1969, **10**, 681 (*synth*)

2-Cyclopentene-1-nonanoic acid, 9CI — C-00060

Alepric acid

[2519-24-6]

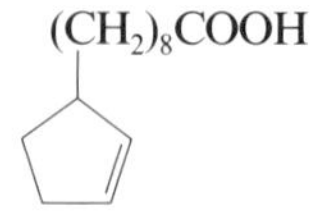

$C_{14}H_{24}O_2$ M 224.342

Occurs in *Hydnocarpus wightiana* oil. Needles (MeOH/pet. ether). Mp 48°. Bp_1 165-168°. $[\alpha]_D^{25}$ +77.1°.

Et ester: Bp_{10} 174°. $[\alpha]_D^{25}$ +66.5°.

Cole, H.I. *et al*, *J. Am. Chem. Soc.*, 1939, **61**, 2349 (*synth*)
Zeman, I. *et al*, *J. Chromatogr.*, 1963, **10**, 15 (*glc*)

2-Cyclopentene-1-nonanol, 9CI — C-00061

3-(9-Hydroxynonyl)-1-cyclopentene

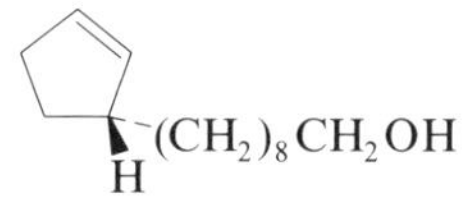

(*R*)-*form*
Absolute configuration

$C_{14}H_{26}O$ M 210.359

(*R*)-*form*

Ac:
$C_{16}H_{28}O_2$ M 252.396
Synthetic insect pheromone for red-banded leaf roller. $[\alpha]_D^{25}$ +70.7° (c, 3.18 in $CHCl_3$).

(*S*)-*form* [67886-29-7]

Ac: Synthetic insect pheromone for European corn borer and red-banded leaf roller. Yellow oil. $[\alpha]_D^{25}$ −70.7° (c, 3.18 in $CHCl_3$).

Chapman, O.L. *et al*, *J. Am. Chem. Soc.*, 1978, **100**, 4878 (*synth, pmr, ir*)

2-Cyclopentene-1-pentadecanoic acid — C-00062

Hormelic acid

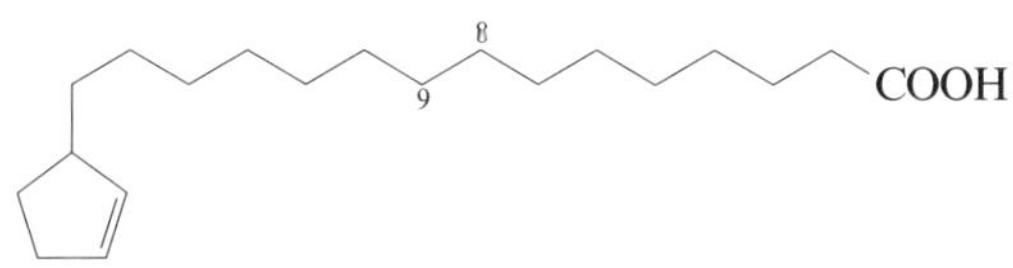

$C_{20}H_{36}O_2$ M 308.503

(+)-*form* [52904-19-5]

Isol. from *Caloncoba echinata* seed oil. Mp 74.5-75°. $[\alpha]_D^{24}$ +52.0° (c, 11 in $CHCl_3$).

Me ester: $[\alpha]_D^{24}$ +49.3° (c, 10.7 in $CHCl_3$).

8,9-Didehydro: [52904-20-8]. *15-(2-Cyclopentenyl)-8-pentadecenoic acid.* ***Oncobic acid***
$C_{20}H_{34}O_2$ M 306.487
Isol. from *C. echinata*, *Hydnocarpus wightiana* and *H. anthelmintica* seed oils. Prob. opt. active and with *Z*-config.

9,10-Didehydro: 15-(2-Cyclopentenyl)-9-pentadecenoic acid
$C_{20}H_{34}O_2$ M 306.487
Isol. from *C. echinata*, *H. anthelmintica* and *H. wightiana*.

Spener, F. *et al*, *Biochemistry*, 1974, **13**, 2241 (*isol, synth*)

2-Cyclopentene-1-tridecanoic acid, 9CI **C-00063**

Chaulmoogric acid. Chaulmoogrina

[502-30-7]

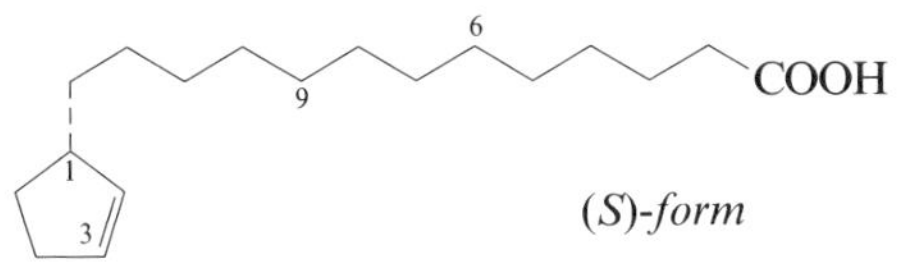

$C_{18}H_{32}O_2$ M 280.450

(S)-form [29106-32-9]

Constit. of oils of *Hydnocarpus* and *Oncola* spp. Also in *Carpotroche brasiliensis*, *Asteriastigma macrocapa*, *Caloncoba* spp., *Lindackeria dentata* and *Buchnerodendron speciosum*. Used in treatment of leprosy. Shiny leaflets (EtOH or pet. ether). Mp 67-68°. $[\alpha]_D^{26}$ +61.7° (c, 4.82 in $CHCl_3$).

Me ester: [24828-59-9].

$C_{19}H_{34}O_2$ M 294.476

Mp 22°. Bp_{20} 227°. $[\alpha]_D$ +50° ($CHCl_3$).

Et ester: [623-32-5]. *Antileprol. Chaulmestrol. Chaulmoogrol. Hyrganol. Moogrol*

$C_{20}H_{36}O_2$ M 308.503

Leprostatic agent. Bp_{20} 230°. $[\alpha]_D$ +50.7°.

Amide:

$C_{18}H_{33}NO$ M 279.465

Mp 106°. $[\alpha]_D$ +57.3° ($CHCl_3$).

Chloride:

$C_{18}H_{31}ClO$ M 298.895

Mp 20-25°.

Nitrile:

$C_{18}H_{31}N$ M 261.450

$Bp_{3.5}$ 158-160°. $[\alpha]_D$ +55.2° (c, 0.3 in $CHCl_3$).

6,7-Didehydro: [502-31-8]. *13-(2-Cyclopentenyl)-6-tridecenoic acid.* ***Gorlic acid***

$C_{18}H_{30}O_2$ M 278.434

Isol. from *Oncoba echinata* and others in Flacourtiaceae. Oil. Mp 6°. Bp 232.5°. $[\alpha]_D^{25}$ +60.7° (c, 25 in $CHCl_3$).

6,7-Didehydro, Me ester: Fp −10°. Bp_{10} 209°. $[\alpha]_D^{25}$ +57.7° (c, 25 in $CHCl_3$).

6,7-Didehydro, amide: Cryst. (Me_2CO). Mp 95°.

9,10-Didehydro: 13-(2-Cyclopentenyl)-9-tridecenoic acid

$C_{18}H_{30}O_2$ M 278.434

Isol. from *Caloncoba echinata, Hydnocarpus antheminthica* and *H. wightiana* seeds. Prob. opt. active and with (*Z*)-config.

Δ^1-Isomer, 3-oxo: 3-Oxo-1-cyclopentenetridecanoic acid. ***Ketochaulmoogric acid***

$C_{18}H_{30}O_3$ M 294.433

Isol. from seed oil of *Carpotroche brasiliensis* (sapucainha oil). Light-yellow granules (EtOH). Mp 116°. Not reported since 1937. Prob. artifact of autoxidn.

(±)-form

Cryst. (EtOAc). Mp 68.5°.

Amide: Cryst. (EtOH). Mp 106°.

Nitrile: $Bp_{2.5}$ 155-158°.

Paget, H., *J. Chem. Soc.*, 1937, 955 (*Ketochaulmoogric acid*)

Cole, H.I. *et al*, *J. Am. Chem. Soc.*, 1938, **60**, 612 (*Gorlic acid*)

Mislow, K. *et al*, *J. Am. Chem. Soc.*, 1955, **77**, 3807 (*synth, abs config, bibl*)

Mani, V.V.S. *et al*, *J. Chromatogr.*, 1969, **39**, 182 (*isol, chromatog*)

Christie, W.W. *et al*, *Lipids*, 1969, **4**, 229 (*ms*)

Karrer, W. *et al*, *Konstitution und Vorkommen der Organischen Pflanzenstoffe*, 2nd Ed., Birkhäuser Verlag, Basel, 1972-1985, no. 978 (*occur*)

Spencer, F. *et al*, *Biochemistry*, 1974, **13**, 2241 (*deriv*)

Levy, L., *Am. Rev. Respir. Dis.*, 1975, **111**, 703 (*props*)

Kaneda, T., *Biochem. Biophys. Res. Commun.*, 1981, **99**, 1226 (*biosynth*)

Martindale, The Extra Pharmacopoeia, 28th/29th Ed., Pharmaceutical Press, London, 1982/1989, 6557.

Cytolipin H **C-00064**

N-*[1-[[(4-O-β-D-Galactopyranosyl-β-D-glucopyranosyl)oxy]methyl]-2-hydroxy-3-heptadecenyl]tetracosanamide. Cerebronylsphingosylglucosidogalactoside. Ganglioside G_{A3}. Lactosylceramide. 1-O-(4-O-β-D-Galactopyranosyl-β-D-glucopyranosyl)ceramide*

[4682-48-8]

$C_{54}H_{103}NO_{13}$ M 974.407

Isol. from human epidermoid cells. Major neutrophil glycolipid serving as a differentiation marker.

Rapport, M.M. *et al*, *Cancer (Philadelphia)*, 1959, **12**, 438 (*isol*)

Rapport, M.M. *et al*, *J. Biol. Chem.*, 1962, **237**, 1056 (*struct*)

Graf, L. *et al*, *Chem. Phys. Lipids*, 1974, **13**, 367.

Symington, F.W. *et al*, *J. Biol. Chem.*, 1987, **262**, 11356.

Nicolaou, K.C. *et al*, *J. Am. Chem. Soc.*, 1988, **110**, 7910 (*synth*)

D

1,9-Decadiene, 9CI — D-00001

[1647-16-1]

$H_2C{=}CH(CH_2)_6CH{=}CH_2$

$C_{10}H_{18}$ M 138.252

Bp 165°. n_D^{20} 1.4318.

Everett, J.L. *et al*, *J. Chem. Soc.*, 1950, 3131 (*synth*)
Blomquist, A.T. *et al*, *J. Am. Chem. Soc.*, 1955, **77**, 6399 (*synth*)
Marvel, C.S. *et al*, *J. Am. Chem. Soc.*, 1959, **81**, 4737 (*synth*)
Hata, G. *et al*, *J. Am. Chem. Soc.*, 1963, **28**, 3457 (*synth*)
Casteignau, G. *et al*, *Bull. Soc. Chim. Fr.*, 1968, 3893 (*synth*)
U.S. Pat., 3 622 646, (1969); *CA*, **76**, 33765a (*synth*)

2,8-Decadiene-4,6-diynedioic acid — D-00002

$HOOCCH{=}CHC{\equiv}CC{\equiv}CCH{=}CHCOOH$

$C_{10}H_6O_4$ M 190.155

(E,E)-form [38329-17-8]

Constit. of *Polyporus anthracophilus*, *P. guttulatus*, *Tricholoma panaeolum* and *Drosophila subatrata*. Amorph. solid (EtOH). Mp 200° dec.

Di-Me ester: [54981-67-8].
$C_{12}H_{10}O_4$ M 218.209
Constit. of *Lepista glaucocana*. Mp 105-107°.

Bu'Lock, J.D. *et al*, *J. Chem. Soc.*, 1957, 1607 (*isol*)
Jones, E.R.H. *et al*, *J. Chem. Soc.*, 1960, 691, 2267 (*isol*)
Hearn, M.T.W. *et al*, *J. Chem. Soc., Perkin Trans. 1*, 1974, 2335 (*isol*)
Hearn, M.T.W. *et al*, *J. Chem. Soc., Perkin Trans. 2*, 1976, 1027 (*cmr*)

2,8-Decadiene-4,6-diyne-1,10-diol — D-00003

[3829-79-6]

$HOCH_2CH{=}CHC{\equiv}CC{\equiv}CCH{=}CHCH_2OH$

$C_{10}H_{10}O_2$ M 162.188

(E,E,)-form

Cryst. (MeOH). Mp 160-161°. Photosensitive.

(Z,Z)-form [24469-82-7]

10-Hydroxymatricarianol
Constit. of *Diplopappus fructicosus* roots. Cryst. (Et_2O/CCl_4 or Et_2O/pet. ether). Mp 64°.

Ac: [39036-41-4].
$C_{12}H_{12}O_3$ M 204.225
Constit. of *D. fructicosus* roots. Oil.

Di-Ac: [39036-42-5].
$C_{14}H_{14}O_4$ M 246.262
Constit. of *D. fructicosus* roots. Oil.

Bohlmann, F. *et al*, *Chem. Ber.*, 1969, **102**, 1679; 1972, **105**, 2421 (*isol*)
Andrewes, A.G., *Acta Chem. Scand.*, 1973, **27**, 2574 (*synth*)

4,5-Decadiene-7,9-diyne-1,3-diol — D-00004

$HC{\equiv}CC{\equiv}CCH{=}C{=}CHCH(OH)CH_2CH_2OH$

$C_{10}H_{10}O_2$ M 162.188

(−)-form

Metab. of *Odontia bicolor* and *Flammula sapinea*. Oil. $[\alpha]_D^{20}$ −210° (c, 0.058 in EtOH).

Isopropylidene deriv.: Oil. $[\alpha]_D^{20}$ −280° (c, 0.032 in EtOH).

Bew, R.E. *et al*, *J. Chem. Soc. C*, 1966, 129 (*isol*)

2,8-Decadiene-4,6-diynoic acid — D-00005

Matricaria acid. Matricaric acid

$H_3CCH{=}CHC{\equiv}CC{\equiv}CCH{=}CHCOOH$

$C_{10}H_8O_2$ M 160.172

(2E,8E)-form

Mp 173° dec.

Me ester: [820-96-2]. trans,trans-*Matricaria ester*
$C_{11}H_{10}O_2$ M 174.199
Constit. of *Matricaria tenuifolia*, *Polyporus anthracophilus* and *Grindelia paludosa*. Mp 63°.

(2E,8Z)-form

Mp 83-90°.

Me ester: [23180-59-8]. trans-*Matricaria ester*
Constit. of *Matricaria inodora*, *Erigeron karwinskyanus* and *Amellus strigosus*. Mp 2°.

2-Methylpropylamide: [113235-92-0]. N-*(2-Methylpropyl)-2,8-decadiene-4,6-diynamide. 2,8-Decadiene-4,6-diynoic acid isobutylamide*
$C_{14}H_{17}NO$ M 215.294
Isol. from underground parts of *Achillea wilhelmsii* (Compositae). Mp 85-89°.

3-Methylbutylamide: [113262-63-8]. N-*(3-Methylbutyl)-2,8-decadiene-4,6-diynamide. 2,8-Decadiene-4,6-diynoic acid isopentylamide*
$C_{15}H_{19}NO$ M 229.321
Isol. from *A. wilhelmsii* (Compositae). Mp 88-91°.

2-Phenylethylamide: [113262-64-9]. N-*(2-Phenylethyl)-2,8-decadiene-4,6-diynamide. 2,8-Decadiene-4,6-diynoic acid 2-phenylethylamide*
$C_{18}H_{17}NO$ M 263.338
Isol. from *A. wilhelmsii* (Compositae). Mp 108-111°.

(2Z,8E)-form

Me ester: [25019-41-4].
Constit. of *Solidago virgaurea* and *E.* spp.

(2Z,8Z)-form [13030-78-9]

Mp 98-99°.

Me ester: [928-36-9]. *Matricaria ester*
Constit. of *M. inodora* and *Erigeron* spp. Widely distributed in the Compositae. Mp 37°.

(2ξ,8ξ)-form

2-Methylpropylamide, N-*Me:* [37064-13-4]. N-*Methyl*-N-*(2-methylpropyl)-2,8-decadiene-4,6-diynamide*
$C_{15}H_{19}NO$ M 229.321
Isol. from the roots of *Anacyclus pyrethrum* (Compositae). Oil. λ_{max} 328, 306.5, 287 nm.

8,9-Epoxide: 8,9-Epoxy-2-decene-4,6-diynoic acid
$C_{10}H_8O_3$ M 176.171
Isol. as Me ester from aerial parts of *E. philadelphicus*. Oil (as Me ester).

Bruun, T. *et al*, *Acta Chem. Scand.*, 1951, **5**, 1244 (*synth*)

Sörensen, J.S. *et al*, *Acta Chem. Scand.*, 1952, **6**, 883; 1953, **7**, 1375; 1954, **8**, 26, 34 (*occur*)
Bu'Lock, J.D. *et al*, *J. Chem. Soc.*, 1957, 1607 (*isol, struct*)
Bell, I. *et al*, *J. Chem. Soc.*, 1958, 1313 (*synth*)
Bohlmann, F. *et al*, *Chem. Ber.*, 1963, **96**, 1485; 1964, **97**, 1193; 1965, **98**, 369; 1966, **99**, 1642, 2096; 1967, **100**, 611; 1982, **21**, 167 (*isol, pmr*)
Jente, R. *et al*, *Chem. Ber.*, 1972, **105**, 1694 (*isol, struct, uv, ms*)
Bohlmann, F. *et al*, *Phytochemistry*, 1982, **21**, 167 (*isol*)
Jakupovic, J. *et al*, *Phytochemistry*, 1986, **25**, 1223 (*epoxides*)
Greger, H. *et al*, *J. Nat. Prod. (Lloydia)*, 1987, **50**, 1100 (*amides, isol, struct, ir, uv, ms, pmr, cmr*)

2,8-Decadiene-4,6-diyn-1-ol D-00006

Matricarianol. Matricariol

[505-98-6]

$H_3CCH{=}CHC{\equiv}CC{\equiv}CCH{=}CHCH_2OH$

$C_{10}H_{10}O$ M 146.188

(E,E)-form [17098-68-9]

Constit. of *Grindelia arenicola* and *G. robusta*. Mp 103-104° (97-99°).

Ac: [1002-64-8].

$C_{12}H_{12}O_2$ M 188.226

Constit. of *G. arenicola*, *G. robusta*, *G. squarrosa* and *Aster sibiricus*. Oil.

3-Methyl-2-butenoyl ester: [29576-67-8].

$C_{15}H_{16}O_2$ M 228.290

Constit. of *Bellium bellidioides*. Oil.

3-Methylbutanoyl ester:

$C_{15}H_{18}O_2$ M 230.306

Constit. of *Osmitopsis asteriscoides*. Oil.

(2E,8Z)-form [13894-67-2]

Constit. of *Diplopappus fruticolosus* and some *G.* spp. Oil.

3-Methylbutanoyl: [29444-87-9].

Constit. of *Carthamus tinctorius*. Oil.

3-Methyl-2-butenoyl: [29576-68-9].

Constit. of *B. bellidioides*. Oil.

(2Z,8E)-form

Constit. of *Polyporus guttulatus*. Cryst. (hexane). Mp 15-16°.

4-Phenylazobenzoyl: Mp 144-145°.

(Z,Z)-form [24486-39-3]

Constit. of *D. fruticolosus* and some *G.* spp. Oil. $Bp_{0.01}$ 100° (bath).

Ac: [24503-75-1].

Constit. of *D. fruticolosus* and *Calotis* sp. $Bp_{0.01}$ 100° (bath).

[104899-48-1]

Bruun, T. *et al*, *Acta Chem. Scand.*, 1951, **5**, 1244 (*synth*)
Gardner, J.N. *et al*, *J. Chem. Soc.*, 1960, 691 (*isol, struct*)
Bohlmann, F. *et al*, *Chem. Ber.*, 1969, **102**, 1679; 1972, **105**, 1919 (*isol, synth, pmr*)
Davies, D.G. *et al*, *J. Chem. Soc., Perkin Trans. 1*, 1978, 1602 (*biosynth, isol*)

3,4-Decadiene-6,8-diyn-1-ol D-00007

9-Methylmarasmin

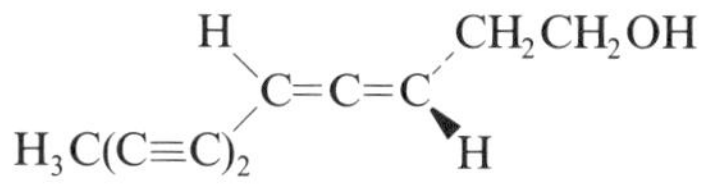

(*S*)-*form*

$C_{10}H_{10}O$ M 146.188

(S)-form

Metab. of *Daedalea juniperina*. Oil. $[\alpha]_D^{21}$ +340° (c, 0.172 in EtOH).

(±)-form

Oil.

Bew, R.E. *et al*, *J. Chem. Soc. C*, 1966, 129 (*isol, synth, uv*)
Landor, S.R. *et al*, *J. Chem. Soc., Perkin Trans. 1*, 1974, 557 (*synth, abs config*)

4,5-Decadiene-7,9-diyn-1-ol D-00008

$HC{\equiv}CC{\equiv}CCH{=}C{=}CHCH_2CH_2CH_2OH$

$C_{10}H_{10}O$ M 146.188

(−)-form

Metab. of *Odontia bicolor* and *Flammula sapinea*. $[\alpha]_D^{21}$ −290° (c, 0.012 in EtOH).

Bew, R.E. *et al*, *J. Chem. Soc. C*, 1966, 129 (*isol, uv*)

6,8-Decadiene-1,3,5-triol D-00009

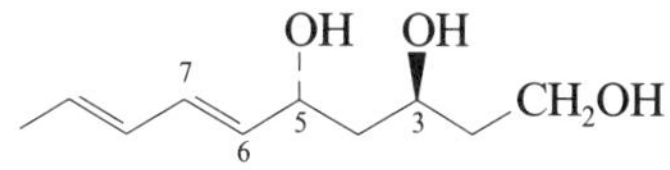

$C_{10}H_{18}O_3$ M 186.250

(3S,5R,6E,8E)-form [133587-33-4] ***Streptenol D***

Isol. from culture filtrates of *Streptomyces fimbriatus*. Mp 35-38.5°. $[\alpha]_D^{23}$ +8.2° (c, 0.98 in $CHCl_3$).

Tri-Ac:

$C_{16}H_{24}O_6$ M 312.362

Oil. $[\alpha]_D^{28}$ +11.4° (c, 1.22 in $CHCl_3$).

5-Ketone: [119188-92-0]. *1,3-Dihydroxy-6,8-decadien-5-one, 9CI.* ***Streptenol C***

$C_{10}H_{16}O_3$ M 184.235

Isol. from *S.* strain C57. Oil. $[\alpha]_D$ +23.4° (c, 0.3 in $CHCl_3$).

6,7-Dihydro, 5-ketone: [86918-62-9]. *1,3-Dihydroxy-8-decen-5-one.* ***Streptenol A***

$C_{10}H_{18}O_3$ M 186.250

Isol. from culture filtrate of *S. fimbriatus*, *S. cirratus* and other *S.* spp. Shows some antitumour and immunostimulating activity. Oil. $[\alpha]_D^{25}$ +23° (c, 1.05 in $CHCl_3$).

Keller-Schierlein, W. *et al*, *Helv. Chim. Acta*, 1983, **66**, 1253 (*isol, uv, pmr, cmr, ms, deriv*)
Arnone, A. *et al*, *Phytochemistry*, 1988, **27**, 3611 (*isol, ms, pmr, deriv*)
Niederer, D. *et al*, *Helv. Chim. Acta*, 1990, **73**, 2129 (*isol, pmr, cmr, ms*)
Grabley, S. *et al*, *J. Antibiot.*, 1991, **44**, 797 (*Streptenols*)
Romeyke, Y. *et al*, *Tetrahedron*, 1991, **47**, 3335 (*synth, bibl*)

2,3-Decadienoic acid, 9CI, 8CI D-00010

[27007-77-8]

$H_3C(CH_2)_5CH{=}C{=}CHCOOH$

$C_{10}H_{16}O_2$ M 168.235

(R)-form [36143-66-5]

$[\alpha]_D^{20}$ −151°.

Me ester: [107865-09-8].

$C_{11}H_{18}O_2$ M 182.262

Liq.

Et ester: [107865-10-1].

$C_{12}H_{20}O_2$ M 196.289

Liq.

(S)-form [36143-65-4]

$[\alpha]_D^{29}$ +141° (Me_2CO).

(±)-form

Cryst. (pentane). Mp 24-25°.

Me ester: [107912-41-4].

$C_{11}H_{18}O_2$ M 182.262
Liq.
Et ester: [107912-42-5].
$C_{12}H_{20}O_2$ M 196.289
Liq.

Jones, V.K. *et al*, *J. Org. Chem.*, 1965, **30**, 3978 (*synth, pmr, ir*)
Endo, K. *et al*, *J. Biol. Chem.*, 1970, **245**, 4293 (*synth, pmr, ir*)
Morisaki, M. *et al*, *Bioorg. Chem.*, 1971, **1**, 188 (*resoln*)
Morisaki, M. *et al*, *Biochemistry*, 1972, **11**, 309 (*synth, ir, uv, pmr*)
Schwab, J.M. *et al*, *J. Am. Chem. Soc.*, 1985, **107**, 6046 (*synth, resoln, cryst struct, abs config*)
Ramaswamy, R. *et al*, *J. Chem. Soc., Chem. Commun.*, 1986, 1545 (*resoln*)

2,4-Decadienoic acid D-00011

[92015-76-4]

$H_3C(CH_2)_4CH{=}CHCH{=}CHCOOH$

$C_{10}H_{16}O_2$ M 168.235
***(E,E)*-form** [30361-33-2]
Mp 49-50°. $Bp_{0.3}$ 121-124°.
Me ester: [7328-33-8].
$C_{11}H_{18}O_2$ M 182.262
$Bp_{0.25}$ 69-70°.
Et ester: [7328-34-9].
$C_{12}H_{20}O_2$ M 196.289
$Bp_{0.05}$ 84-86°.
4-Bromophenacyl ester: Mp 125°.
Chloride:
$C_{10}H_{15}ClO$ M 186.681
Pale yellow liq. Bp_{14} 136°. n_D^{25} 1.5260.
2-Phenylethylamide: [83170-20-1]. *2,4-Decadienoic acid 2-phenylethylamide*
$C_{18}H_{25}NO$ M 271.402
Isol. from underground parts of *Achillea wilhelmsii* (Compositae). Mp 96-98°.
3-Methylbutylamide: [78910-26-6]. *2,4-Decadienoic acid isopentylamide*
$C_{15}H_{27}NO$ M 237.384
Isol. from underground parts of *A. wilhelmsii* (Compositae). Mp 63-65°.
Pyrrolidide: 2,4-Decadienoic acid pyrrolidide. 1-(2,4-Decadienoyl)pyrrolidine
$C_{14}H_{23}NO$ M 221.342
Constit. of pepper (*Piper nigrum*) (Piperaceae). Shows larvicidal activity. Oil.
***(2E,4Z)*-form** [544-48-9]
Stillingic acid
Constit. of Stillingia oil. $Bp_{0.5}$ 122°.
Me ester: [4493-42-9].
Pheromone from spruce engraver beetle *Pityogenes chalcographus*. $Bp_{0.15}$ 71°.
Et ester: [3025-30-7]. *Pear ester*
Constit. of odoriferous principle of Bartlett pears. $Bp_{0.05}$ 84-86°.
4-Bromophenacyl ester: Mp 72°.
2-Methylpropylamide: $Bp_{0.1}$ 142°.
***(2Z,4E)*-form** [68676-77-7]
$Bp_{0.45}$ 112-114°.
Me ester: $Bp_{0.35}$ 73°.
Et ester: Flavouring material.
4-Bromophenacyl ester: Cryst. (EtOH). Mp 82°.
***(Z,Z)*-form**
$Bp_{0.5}$ 112-114°.
Me ester: $Bp_{0.6}$ 78°.
4-Bromophenacyl ester: Cryst. (EtOH). Mp 82°.

Org. Synth., **66**, 22 (*synth, ir, pmr, cmr, chromatog*)
Jacobson, M., *J. Am. Chem. Soc.*, 1953, **75**, 2584 (*synth, derivs*)
Crombie, L., *J. Chem. Soc.*, 1955, 1007 (*synth*)
Burden, R.S. *et al*, *J. Chem. Soc. C*, 1969, 2477 (*isol, synth*)
Näf, F. *et al*, *Helv. Chim. Acta*, 1974, **57**, 1309 (*synth*)
Amos, R.A. *et al*, *J. Org. Chem.*, 1978, **43**, 555 (*synth, pmr*)
Yasuda, I. *et al*, *Chem. Pharm. Bull.*, 1981, **29**, 564 (*isol, cmr*)
Tamura, Y. *et al*, *Tetrahedron Lett.*, 1981, **22**, 1343 (*synth*)
Sato, T. *et al*, *Tetrahedron Lett.*, 1981, **22**, 2375 (*synth*)
Bestmann, H.J. *et al*, *Justus Liebigs Ann. Chem.*, 1982, 363 (*synth*)
Byrne, B. *et al*, *J. Org. Chem.*, 1986, **51**, 2607 (*synth, deriv*)
Greger, H. *et al*, *J. Nat. Prod. (Lloydia)*, 1987, **50**, 1100 (*2-phenylethylamide, isopentylamide*)
Kiuchi, F. *et al*, *Chem. Pharm. Bull.*, 1988, **36**, 2452 (*pyrrolidide*)
Zhengming, L. *et al*, *Synth. Commun.*, 1989, **19**, 91 (*synth, pmr*)
Birgersson, G. *et al*, *J. Insect Physiol.*, 1990, **36**, 391.
Plobeck, N.A. *et al*, *J. Org. Chem.*, 1991, **56**, 4508 (*synth, cmr*)

2,5-Decadienoic acid, 9CI, 8CI D-00012

[35060-78-7]

$H_3C(CH_2)_3CH{=}CHCH_2CH{=}CHCOOH$

$C_{10}H_{16}O_2$ M 168.235
Me ester: [35060-79-8].
$C_{11}H_{18}O_2$ M 182.262
Liq. Bp_{12} 113°.

[41792-25-0]

Adler, K. *et al*, *Justus Liebigs Ann. Chem.*, 1962, **651**, 141.

2,6-Decadienoic acid, 9CI, 8CI D-00013

$H_3CCH_2CH_2CH{=}CHCH_2CH_2CH{=}CHCOOH$

$C_{10}H_{16}O_2$ M 168.235
***(E,E)*-form**
Cryst. Mp 43°. $Bp_{0.4}$ 110°.
p-*Bromophenacyl ester:* Mp 87°.
Chloride:
$C_{10}H_{15}ClO$ M 186.681
Liq. Bp_{17} 118°.
***(2E,6Z)*-form**
$Bp_{0.08}$ 94-95°.
p-*Bromophenacyl ester:* Mp 45°.
Chloride:
$C_{10}H_{15}ClO$ M 186.681
Liq. Bp_{16} 118°.
***(2Z,6E)*-form**
$Bp_{0.15}$ 105°.
p-*Bromophenacyl ester:* Mp 58°.

Jacobson, M., *J. Am. Chem. Soc.*, 1950, **72**, 1489 (*chloride*)
Raphael, R.A. *et al*, *J. Chem. Soc.*, 1950, 120 (*synth*)
Crombie, L., *J. Chem. Soc.*, 1952, 4338 (*synth*)

2,7-Decadienoic acid, 9CI, 8CI D-00014

$H_3CCH_2CH{=}CH(CH_2)_3CH{=}CHCOOH$

$C_{10}H_{16}O_2$ M 168.235
***(E,E)*-form** [66917-14-4]
Me ester: [66800-22-4].
$C_{11}H_{18}O_2$ M 182.262
Liq. Bp_{12} 115-125°.
Benzyl ester: [66917-13-3].
$C_{17}H_{22}O_2$ M 258.360
Liq. $Bp_{0.001}$ 90-110°.
***(2E,7Z)*-form**
Me ester: [66825-29-4].

Liq. Bp_{12} 130-140°, $Bp_{0.6}$ 72-73°.
Et ester: [41547-23-3].
$C_{12}H_{20}O_2$ M 196.289
Liq. Bp_{12} 125-135°.

Waelchi, P.C. *et al, Helv. Chim. Acta*, 1978, **61**, 885 (*synth, ir, pmr, chromatog*)
Waelchi-Schaer, E. *et al, Helv. Chim. Acta*, 1978, **61**, 928 (*synth, pmr*)

3,4-Decadienoic acid, 9CI, 8CI D-00015

[78946-41-5]

$H_3C(CH_2)_4CH{=}C{=}CHCH_2COOH$

$C_{10}H_{16}O_2$ M 168.235
(±)-form
Liq. $Bp_{1.8}$ 131-133°.
Et ester: [36186-28-4].
$C_{12}H_{20}O_2$ M 196.289
Liq. Bp_5 84°.

Amos, R.A. *et al, J. Org. Chem.*, 1978, **43**, 555 (*ester, synth, ir, pmr, ms*)
Sato, T. *et al, Tetrahedron Lett.*, 1981, **22**, 2375 (*synth, ir, pmr*)
Tsuboi, S. *et al, J. Org. Chem.*, 1982, **47**, 4478.
Org. Synth., 1988, **66**, 22 (*ester*)

3,5-Decadienoic acid, 9CI, 8CI D-00016

[41792-19-2]

$H_3C(CH_2)_3CH{=}CHCH{=}CHCH_2COOH$

$C_{10}H_{16}O_2$ M 168.235
(E,E)-form
Me ester: [77761-46-7].
$C_{11}H_{18}O_2$ M 182.262
Constit. of latex of *Euphorbia lathyris*.
Et ester: [93677-59-9].
$C_{12}H_{20}O_2$ M 196.289
Oil.
(3E,5Z)-form
Me ester: [77761-52-5].
Constit. of latex of *E. lathyris*.
(3Z,5E)-form
Me ester: [77761-53-6].
Constit. of latex of *E. lathyris*.

[93677-60-2, 118133-44-1]

Warnaar, F., *Phytochemistry*, 1981, **20**, 89 (*isol*)
Tsuboi, S. *et al, Bull. Chem. Soc. Jpn.*, 1988, **61**, 1410 (*synth, pmr, cmr*)

4,6-Decadienoic acid, 9CI, 8CI D-00017

$H_3CCH_2CH_2CH{=}CHCH{=}CHCH_2CH_2COOH$

$C_{10}H_{16}O_2$ M 168.235
(E,E)-form [60388-65-0]
Mp 39°. Bp_4 138°.
Me ester:
$C_{11}H_{18}O_2$ M 182.262
Liq. Bp_5 103-105°.
Chloride:
$C_{10}H_{15}ClO$ M 186.681
$Bp_{1.8}$ 80°.
N-(*2-Methylpropylamide*): *Isospilanthol*
$C_{14}H_{25}NO$ M 223.358
Mp 60°. $Bp_{1.5}$ 156°.

Yoshioka, T., *Yakugaku Zasshi*, 1955, **75**, 622 (*synth*)
Jacobson, M., *J. Am. Chem. Soc.*, 1956, **78**, 5084 (*synth*)
Vig, B. *et al, J. Indian Chem. Soc.*, 1976, **53**, 303 (*synth*)

4,8-Decadienoic acid, 9CI, 8CI D-00018

[13159-49-4]

$H_3CCH{=}CHCH_2CH_2CH{=}CHCH_2CH_2COOH$

$C_{10}H_{16}O_2$ M 168.235
Me ester: [1191-03-3].
$C_{11}H_{18}O_2$ M 182.262
Constit. of essential oil of hops (*Humulus lupulus*).

[94372-21-1]

Buttery, R.G. *et al, Chem. Ind.* (*London*), 1963, 1981 (*isol, ir, pmr*)
Buttery, R.G. *et al, J. Chromatogr.*, 1965, **18**, 399 (*chromatog*)
Goliaszewski, A. *et al, Tetrahedron*, 1985, **41**, 5779 (*synth, pmr, ms*)

4,9-Decadienoic acid, 9CI, 8CI D-00019

$H_2C{=}CHCH_2CH_2CH_2CH{=}CH(CH_2)_2COOH$

$C_{10}H_{16}O_2$ M 168.235
(E)-form [71859-41-1]
Liq. Bp_1 109-110°.
Me ester: [67140-63-0].
$C_{11}H_{18}O_2$ M 182.262
Liq. Bp_1 64-65°.
Et ester: [83714-45-8].
$C_{12}H_{20}O_2$ M 196.289
Liq. Bp_1 83°.

Tsuji, J. *et al, Bull. Chem. Soc. Jpn.*, 1977, **50**, 2507 (*synth, ir, pmr*)
Zakharkin, L.I. *et al, Zh. Org. Khim.*, 1979, **15**, 1378 (*synth*)

5,8-Decadienoic acid, 9CI, 8CI D-00020

$H_3CCH{=}CHCH_2CH{=}CH(CH_2)_3COOH$

$C_{10}H_{16}O_2$ M 168.235
(E,E)-form
Me ester: [79402-89-4].
$C_{11}H_{18}O_2$ M 182.262
No phys. props. reported.

Hyasi, Y. *et al, Tetrahedron Lett.*, 1981, **22**, 2629 (*synth, ms, pmr*)

5,9-Decadienoic acid, 9CI, 8CI D-00021

$H_2C{=}CHCH_2CH_2CH{=}CH(CH_2)_3COOH$

$C_{10}H_{16}O_2$ M 168.235
(E)-form [84565-16-2]
Liq. $Bp_{0.2}$ 98-100°.
Chloride:
$C_{10}H_{15}ClO$ M 186.681
Liq. $Bp_{0.3}$ 64-65°. n_D 1.4670.
Amide:
$C_{10}H_{17}NO$ M 167.250
Cryst. (pet. ether). Mp 76-77°.
(Z)-form [82315-10-4]
Me ester: [80685-81-0].
$C_{11}H_{18}O_2$ M 182.262
Liq.

Ansell, M.F. *et al, J. Chem. Soc.*, 1960, 5219 (*chloride, amide*)
Tolstikov, G.A. *et al, Zh. Org. Khim.*, 1981, **17**, 2241; 1982, **18**, 721 (*ester, synth, ir*)
Fujisawa, T. *et al, Tetrahedron Lett.*, 1982, **23**, 3583 (*synth*)

6,8-Decadienoic acid, 9CI, 8CI **D-00022**

$H_3CCH=CHCH=CH(CH_2)_4COOH$

$C_{10}H_{16}O_2$ M 168.235

[111876-08-5]

Saxton, H.M. *et al, J. Chem. Soc., Chem. Commun.*, 1987, 1449 (*synth*)

7,9-Decadienoic acid, 9CI, 8CI **D-00023**

[73501-28-7]

$H_2C=CHCH=CH(CH_2)_5COOH$

$C_{10}H_{16}O_2$ M 168.235

[73501-27-6]

Hudlicky, T. *et al, Tetrahedron Lett.*, 1979, 2667.

2,4-Decadienoic isobutylamide **D-00024**

Pellitorine. N-(*2-Methylpropyl*)*-2,4-decadienamide, 9CI.* N-*Isobutyl-2,4-decadienamide, 8CI*

$H_3C(CH_2)_4CH=CHCH=CHCONHCH_2CH(CH_3)_2$

$C_{14}H_{25}NO$ M 223.358

(E,E)-form

Constit. of the seeds of *Piper silvaticum*. Constit. of *Fagara xanthoxyloides, Piper sylvaticum, P. nepalense* and several other *P.* spp., *Anacylus pyrethrum, Achillea millefolium* and *Asiasarum heterotropoides* (Rutaceae, Piperaceae, Compositae, Aristolochiaceae). Produces intense formication and local anaesthesia of the mucous membranes. Insecticide. Needles (pet. ether). Mp 69°, Mp 75°, Mp 90-95°.

▷ Severe irritant.

Crombie, L., *J. Chem. Soc.*, 1955, 1007 (*synth*)
Bowden, K. *et al, J. Chem. Soc.*, 1963, 3503 (*isol, uv, ir*)
Loder, J.W. *et al, Aust. J. Chem.*, 1969, **22**, 1531 (*isol, uv, ir, ms, synth*)
Banerji, A. *et al, Experientia*, 1974, **30**, 223 (*isol, struct*)
Mahanta, P.K. *et al, J. Pharm. Sci.*, 1974, **63**, 1160 (*isol, uv, ir*)
Dasgupta, S. *et al, Indian J. Chem., Sect. B*, 1979, **17**, 538 (*isol, uv, ir, pmr, ms*)
Ono, N. *et al, Chem. Lett.*, 1980, 1365 (*synth*)
Nokami, J. *et al, Tetrahedron Lett.*, 1980, **21**, 4455 (*synth*)
Yasusa, I. *et al, Chem. Pharm. Bull.*, 1981, **29**, 564 (*isol, uv, ir, pmr, cmr, ms*)
Tamura, Y. *et al, Tetrahedron Lett.*, 1981, **22**, 1343 (*synth*)
Sato, T. *et al, Tetrahedron Lett.*, 1981, **22**, 2375 (*synth, ir, pmr*)

2,6-Decadien-4-ynal, 8CI **D-00025**

$H_3CCH_2CH_2CH=CHC\equiv CCH=CHCHO$

$C_{10}H_{12}O$ M 148.204

(2E,6E)-form [29576-71-4]

Constit. of *Grindelia* spp. Oil.

Bohlmann, F. *et al, Chem. Ber.*, 1970, **103**, 2245, 3419 (*synth*)

1,3-Decadien-5-yne **D-00026**

$H_3C(CH_2)_3C\equiv CCH=CHCH=CH_2$

$C_{10}H_{14}$ M 134.221

(E)-form [66426-78-6]

Liq. Bp_{18} 81°.

Mitsudo, T. *et al, J. Org. Chem.*, 1985, **50**, 565 (*synth, pmr, cmr*)

2,6-Decadien-4-ynoic acid, 8CI **D-00027**

$H_3CCH_2CH_2CH=CHC\equiv CCH=CHCOOH$

$C_{10}H_{12}O_2$ M 164.204

(2E,6E)-form

Me ester: [22644-34-4].

$C_{11}H_{14}O_2$ M 178.230

Constit. of the plant *Aster schreberi*. Oil. $Bp_{0.01}$ 60°.

Bohlmann, F. *et al, Chem. Ber.*, 1969, **102**, 1037; 1970, **103**, 3419 (*synth*)

2,6-Decadien-4-yn-1-ol, 8CI **D-00028**

$H_3CCH_2CH_2CH=CHC\equiv CCH=CHCH_2OH$

$C_{10}H_{14}O$ M 150.220

(2E,6E)-form

Constit. of *Grindelia* spp.

Ac: [29576-70-3].

$C_{12}H_{16}O_2$ M 192.257

Constit. of the *G.* spp. Oil. $Bp_{0.1}$ 110°.

Bohlmann, F. *et al, Chem. Ber.*, 1970, **103**, 2245, 3419 (*synth*)

4,9-Decadiynal **D-00029**

[94732-59-9]

$HC\equiv C(CH_2)_3C\equiv CCH_2CH_2CHO$

$C_{10}H_{12}O$ M 148.204

Liq. $Bp_{0.25}$ 48-51°.

Ethylene acetal: [94732-58-8].

$Bp_{0.2}$ 66-68°.

Duñach, E. *et al, J. Am. Chem. Soc.*, 1985, **107**, 1664 (*synth, ir, pmr, ms*)

1,3-Decadiyne **D-00030**

[55682-66-1]

$H_3C(CH_2)_5C\equiv CC\equiv CH$

$C_{10}H_{14}$ M 134.221

Bp_5 85°, $Bp_{2.7}$ 57-60°.

Tikhonova, E.S. *et al, Zh. Org. Khim.*, 1981, **17**, 741; *CA*, **95**, 150544.
Miller, J.A. *et al, Synthesis*, 1983, 128 (*synth, ir, pmr, ms*)

2,8-Decadiyne-1,10-diol, 8CI **D-00031**

[21390-02-3]

$HOCH_2C\equiv C(CH_2)_4C\equiv CCH_2OH$

$C_{10}H_{14}O_2$ M 166.219

Mp 49°.

Mono-Ac: [30427-20-4].

$C_{12}H_{16}O_3$ M 208.257

Oil.

Bohlmann, F. *et al, Chem. Ber.*, 1970, **103**, 3419.

4,6-Decadiyne-1,3,8-triol **D-00032**

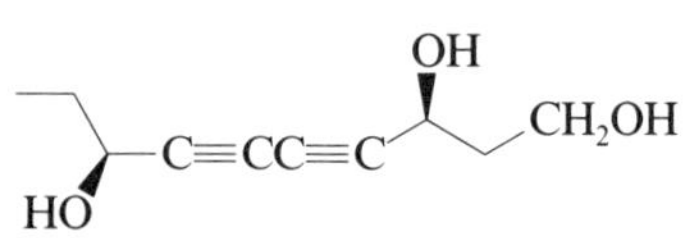

$C_{10}H_{14}O_3$ M 182.219

(3S,8S)-form [105304-52-7]

Isol. from the toxic mushroom *Gymnopilus spectabilis*. Oil. $Bp_{0.4}$ 122°. $[\alpha]_D^{23}$ −5.8° (c, 1.7 in MeOH).

Tribenzoyl: Amorph. solid. Mp 97-102°. $[\alpha]_D^{21}$ +4.70° (c, 1.7 in $CHCl_3$).

Tri-Ac: [105274-17-7].

$C_{16}H_{20}O_6$ M 308.330

Oil. $Bp_{0.4}$ 87°.

Kusano, G. *et al*, *Chem. Pharm. Bull.*, 1986, **34**, 3465 (*isol*)
Hashimoto, M. *et al*, *Bull. Chem. Soc. Jpn.*, 1989, **62**, 2751 (*isol, cmr, abs config*)

2,4-Decadiynoic acid D-00033

$H_3C(CH_2)_4C{\equiv}CC{\equiv}CCOOH$

$C_{10}H_{12}O_2$ M 164.204

Cryst. (pet. ether). Mp 35.5-37°.

Me ester:

$C_{11}H_{14}O_2$ M 178.230

$Bp_{0.15}$ 86°, $Bp_{0.001}$ 40°.

4-Bromophenacyl ester: Cryst. Mp 83°.

Skatteboøl, L., *Acta Chem. Scand.*, 1953, **7**, 1388 (*deriv*)
Crombie, L., *J. Chem. Soc.*, 1955, 1007 (*synth*)

4,6-Decadiyn-1-ol, 9CI D-00034

$H_3CCH_2CH_2C{\equiv}CC{\equiv}CCH_2CH_2CH_2OH$

$C_{10}H_{14}O$ M 150.220

Constit. of *Carthamus tinctorius*. $Bp_{0.001}$ 80-90°.

3-Methylbutanoyl: [29314-16-7]. ***4,6-Decadiyn-1-ol isovalerate***

$C_{15}H_{22}O_2$ M 234.338

Isol. from *C. tinctorius*. Oil.

Bohlmann, F. *et al*, *Chem. Ber.*, 1956, **89**, 1276; 1970, **103**, 2853 (*isol, struct, synth*)

Decanal, 9CI D-00035

Decyl aldehyde. Capraldehyde

[112-31-2]

$H_3C(CH_2)_8CHO$

$C_{10}H_{20}O$ M 156.267

Constit. of Cassia, Neroli and other oils. Used in perfumery and flavour industries. Liq. Mp −5° approx. Bp 208-209°, Bp_7 81°.

▷ HD6000000.

Pickard, K.J., *J. Chem. Soc.*, 1913, **103**, 1947.
Mathïs, C. *et al*, *Phytochemistry*, 1964, **3**, 377 (*isol*)
Attaway, J.A. *et al*, *Phytochemistry*, 1966, **5**, 141 (*isol*)
Morris, W.W., *J. Assoc. Off. Anal. Chem.*, 1973, **56**, 1037 (*ir, pmr*)
Stanley, J.B. *et al*, *J. Food Sci.*, 1975, **40**, 1134 (*ms*)
Org. Synth., 1976, **55**, 84 (*synth*)
Lewis, R.J., *Sax's Dangerous Properties of Industrial Materials*, 8th Ed., Van Nostrand-Reinhold, 1992, DAG000, DAG200.

Decane, 9CI D-00036

[124-18-5]

$H_3C(CH_2)_8CH_3$

$C_{10}H_{22}$ M 142.284

Mp −30°. Bp 174°.

▷ HD6550000.

Simon, I., *CA*, 1929, **23**, 4440 (*synth*)
Maman, A., *CA*, 1936, **30**, 7095 (*synth*)
Mears, T.W. *et al*, *J. Res. Natl. Bur. Stand. (U.S.)*, 1950, **44**, 299.
Coggeshall, N.D., *J. Chem. Phys.*, 1960, **33**, 1247 (*ms*)
McAdie, H.G., *Can. J. Chem.*, 1962, **40**, 2195.
Org. Synth., 1973, **53**, 107 (*synth*)
Sato, F. *et al*, *J. Organomet. Chem.*, 1976, **122**, C25 (*synth*)
Ashby, E.C. *et al*, *Tetrahedron Lett.*, 1977, 4481 (*synth*)
Lewis, R.J., *Sax's Dangerous Properties of Industrial Materials*, 8th Ed., Van Nostrand-Reinhold, 1992, DAG400.

1,10-Decanediamine, 9CI D-00037

Decamethylenediamine. 1,10-Diaminodecane

[646-25-3]

$H_2NCH_2(CH_2)_8CH_2NH_2$

$C_{10}H_{24}N_2$ M 172.313

Used in polymerisations. Mp 62-64°. Bp_{10} 140°.

▷ May cause dermatitis. HD7175000.

B,2HCl: [7408-92-6].

Mp 288-289°.

▷ HD7525000.

Dipicrate: Needles (H_2O). Mp 134°.

N,N-*Di-Me:* [1938-59-6]. *10-Dimethylaminodecylamine*

$C_{12}H_{28}N_2$ M 200.367

Bp_6 128°.

Phookan, R.D. *et al*, *Ber.*, 1892, **25**, 2252 (*synth*)
Brown, R. *et al*, *J. Chem. Soc.*, 1946, 781 (*synth*)
Org. Synth., Coll. Vol., 3, 1955, 229 (*synth*)
Nystrom, R.F., *J. Am. Chem. Soc.*, 1955, **77**, 2544 (*synth*)
McCloskey, J.A., *J. Am. Chem. Soc.*, 1971, **93**, 4955 (*ms*)
Murakami, Y. *et al*, *J. Chem. Soc., Perkin Trans. 1*, 1981, 2800 (*deriv*)

Decanedioic acid, 9CI D-00038

Sebacic acid. Octane-1,8-dicarboxylic acid. Ipomic acid

[111-20-6]

$HOOC(CH_2)_8COOH$

$C_{10}H_{18}O_4$ M 202.250

Isol. from yeasts. Used as a cross-linking agent for epoxy resins and for pptn. separation of Th from lanthanides (pH 2.5). Leaflets. Spar. sol. H_2O. Mp 134.5°. Bp_{100} 294.5°, Bp_{15} 243.5°. pK_{a1} 4.58; pK_{a2} 5.54 (25°, 0.1*M* KNO_3).

▷ VS0875000.

Mono-Me ester: [818-88-2].

$C_{11}H_{20}O_4$ M 216.277

Mp 40-41°. Bp_{20} 208°.

Mono-Me ester, chloride:

$C_{11}H_{19}ClO_3$ M 234.722

Bp_{23} 177°.

Di-Me ester: [106-79-6].

$C_{12}H_{22}O_4$ M 230.303

Mp 26.4° (38°). Bp_{754} 293°, Bp_{20} 175°.

Dichloride:

$C_{10}H_{16}Cl_2O_2$ M 239.141

Bp_{16} 182°. n_D^{18} 1.4684.

Monoamide: Sebacamic acid

$C_{10}H_{19}NO_3$ M 201.265

Mp 170°.

Diamide:

$C_{10}H_{20}N_2O_2$ M 200.280

Mp 210°.

Monoanilide: Sebacanilic acid

$C_{16}H_{23}NO_3$ M 277.363

Leaflets (H_2O). Mp 121-122°.

Dianilide:

$C_{22}H_{28}N_2O_2$ M 352.475

Mp 198°.

Anhydride:

$C_{10}H_{16}O_3$ M 184.235

Cryst. (C_6H_6). Mp 78-79°.
Dihydrazide:
$C_{10}H_{22}N_4O_2$ M 230.309
Mp 184-185°.
Mononitrile: 9-Cyanononanoic acid. ω-Cyanopelargonic acid
$C_{10}H_{17}NO_2$ M 183.250
Mp 50-52°.
Mononitrile, Me ester:
$C_{11}H_{19}NO_2$ M 197.277
Bp_{11} 165-166°.
Dinitrile: Decanedinitrile. Sebaconitrile. 1,8-Dicyanooctane. Octamethylene dicyanide
$C_{10}H_{16}N_2$ M 164.250
Bp_{15} 199-200°.

Boedtker, E., *CA*, 1924, **18**, 3043 (*synth*)
Montonna, R.E., *J. Am. Chem. Soc.*, 1927, **49**, 2114 (*synth*)
Landa, S. *et al, CA*, 1932, **26**, 78 (*synth*)
Stoll, M. *et al, Helv. Chim. Acta*, 1936, **19**, 253 (*synth*)
Morgan, G.T. *et al, J. Chem. Soc.*, 1936, 903 (*synth*)
Carron, M.K. *et al, Anal. Chem.*, 1955, **27**, 1058 (*detn, Th*)
Holmes, J.L. *et al, Org. Mass Spectrom.*, 1970, **3**, 1505 (*ms*)
Freidlin, G.N. *et al, Zh. Org. Khim.*, 1972, **8**, 1162 (*mononitrile*)
Ger. Pat., 2 826 065, (1977); *CA*, **90**, 151587f (*synth*)
Lewis, R.J., *Sax's Dangerous Properties of Industrial Materials*, 8th Ed., Van Nostrand-Reinhold, 1992, SBJ500.

1,2-Decanediol, 9CI — D-00039

[1119-86-4]

$H_3C(CH_2)_7CH(OH)CH_2OH$

$C_{10}H_{22}O_2$ M 174.283

(±)-*form*
Mp 48-49°.

Swern, D. *et al, J. Am. Chem. Soc.*, 1946, **68**, 1504 (*synth*)
Sharpless, K.B. *et al, J. Am. Chem. Soc.*, 1976, **98**, 1986.

1,10-Decanediol, 9CI — D-00040

Decamethylene glycol
[112-47-0]

$HOCH_2(CH_2)_8CH_2OH$

$C_{10}H_{22}O_2$ M 174.283
Mp 72-74°. Bp_8 170°.
Di-Et ether:
$C_{14}H_{30}O_2$ M 230.390
Bp 257-260°.
Cyclic carbonate:
$C_{11}H_{20}O_3$ M 200.277
Mp 55°.
Bis-N-phenylcarbamate: Cryst. (CCl_4). Mp 163-166.5°.

Colonge, J. *et al, Bull. Soc. Chim. Fr.*, 1959, 1248 (*synth*)
Machtlinger, D., *Bull. Soc. Chim. Fr.*, 1961, 1341 (*synth*)
Kraevskii, A.A. *et al, Zh. Org. Khim.*, 1966, **2**, 627 (*synth*)
Diamond, M. *et al, J. Am. Oil Chem. Soc.*, 1967, **44**, 656 (*synth*)
Org. Synth., Coll. Vol., 2, 1943, 154 (*synth*)

2,9-Decanediol — D-00041

[14021-92-2]

$H_3CCH(OH)(CH_2)_6CH(OH)CH_3$

$C_{10}H_{22}O_2$ M 174.283
Mp 33°. Bp_5 114°.
Di-Ac:
$C_{14}H_{26}O_4$ M 258.357
Bp_4 108°.
Bisphenylurethane: Mp 134°.

Heilbron, I.M. *et al, J. Chem. Soc.*, 1943, 268 (*synth*)

3,8-Decanediol — D-00042

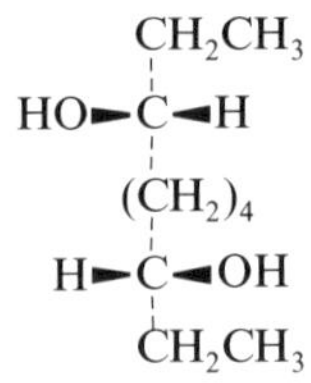

(3*RS*,8*RS*)-*form*

$C_{10}H_{22}O_2$ M 174.283
Needles (Et_2O/pet. ether). Mp 72°. Bp_{11} 143-144°.

(3*RS*,8*RS*)-*form* [13511-17-6]
(±)-*form*
V. viscous liq. $Bp_{0.1}$ 98°.

(3*RS*,8*SR*)-*form* [13307-22-7]
meso-*form*
Cryst. Mp 72°.

Wiemann, J. *et al, C. R. Hebd. Seances Acad. Sci. Ser. C*, 1966, **263**, 313.

4,7-Decanediol, 9CI — D-00043

[4469-89-0]

$H_3CCH_2CH_2CH(OH)CH_2CH_2CH(OH)CH_2CH_2CH_3$

$C_{10}H_{22}O_2$ M 174.283
Mp 79-80°. Bp_{15} 143-147°.

Marvel, C.S. *et al, J. Am. Chem. Soc.*, 1939, **61**, 2709 (*synth*)
Sudweeks, W.B. *et al, J. Org. Chem.*, 1975, **40**, 1131 (*synth*)

Decanoic acid — D-00044

Capric acid. Caprinic acid
[334-48-5]

$H_3C(CH_2)_8COOH$

$C_{10}H_{20}O_2$ M 172.267
Widespread in plant oils, esp. leaf oils and as glycerides in seed oils. Used in corrosion inhibitors and surfactants. Simple esters have perfumery and flavour uses. Used for extraction-preconcentration of Pb. Needles. Sol. EtOH, Et_2O, Me_2CO, C_6H_6, $CHCl_3$, alkalis. d_{40}^{4} 0.8858. Mp 31.5°. Bp 268-270°, Bp_{11} 148-150°. n_D^{40} 1.4288.
▷ HD9100000.
Me ester: [110-42-9].
$C_{11}H_{22}O_2$ M 186.294
Constit. of many plants. Bp 224°, Bp_{15} 114°.
Et ester: [110-38-3].
$C_{12}H_{24}O_2$ M 200.320
Constit. of many plants. Bp 243-245°, Bp_{13} 122-124°.
▷ HD9420000.
Chloride: [112-13-0].
$C_{10}H_{19}ClO$ M 190.712
Bp_{15} 114°.
Amide: [2319-29-1].
$C_{10}H_{21}NO$ M 171.282
Mp 108°, Mp 98°.
Nitrile: [1975-78-6]. *Decanenitrile, 9CI. 1-Cyanononane. Caprinitrile*
$C_{10}H_{19}N$ M 153.267
Bp 235-237°.

Kao, C.H. *et al, J. Chem. Soc.*, 1931, 2046 (*synth*)
Karrer, W. *et al, Konstitution und Vorkommen der Organischen Pflanzenstoffe*, 2nd Ed., Birkhäuser Verlag, Basel, 1972-1985, no. 697 (*occur*)
Tsang, C.W. *et al, J. Chem. Soc., Perkin Trans. 2*, 1975, 1718 (*ms*)
Leovey, E.M.K. *et al, Prostaglandins*, 1975, **10**, 789 (*pmr*)
Gunstone, F.D., *Chem. Phys. Lipids*, 1976, **17**, 1 (*cmr*)
Umemura, J., *J. Chem. Phys.*, 1978, **48**, 42 (*ir, cryst struct*)

Onishenko, T.A. *et al*, *Zh. Anal. Khim.*, 1987, **42**, 1616 (*use*)
Lewis, R.J., *Sax's Dangerous Properties of Industrial Materials*, 8th Ed., Van Nostrand-Reinhold, 1992, DAH400, EHE500.

1-Decanol D-00045

Decyl alcohol

[112-30-1]

$H_3C(CH_2)_8CH_2OH$

$C_{10}H_{22}O$ M 158.283

Isol. from plant sources, eg. citrus oils. Has moderate antifoam activity. Used in manuf. of plasticizers, synthetic lubricants, petroleum additives, herbicides, surface active agents and solvents. Alcohol and esters are perfume and flavour ingredients. Anal. reagent for esterification of fatty acids for glc anal. Oil. Fp 7°. Bp_{700} 232-239°, Bp_7 107-108°.

▷ Mod. toxic. HE4375000.

4-Nitrobenzoyl: Mp 57°.

3,5-Dinitrobenzoyl: Mp 52°.

Schrauth, W. *et al*, *Ber.*, 1931, **64**, 1314 (*synth*)
Komppa, G. *et al*, *J. Prakt. Chem.*, 1932, **135**, 193.
Craig, B.M. *et al*, *J. Am. Oil Chem. Soc.*, 1963, **40**, 61 (*use*)
Hoshino, H., *Bull. Chem. Soc. Jpn.*, 1973, **46**, 3043 (*ms*)
Morris, M.W., *J. Assoc. Off. Anal. Chem.*, 1973, **56**, 1037 (*ir*)
Lewis, R.J., *Sax's Dangerous Properties of Industrial Materials*, 8th Ed., Van Nostrand-Reinhold, 1992, DAI600.

2-Decanol, 9CI D-00046

2-Hydroxydecane

[1120-06-5]

$H_3C(CH_2)_7CH(OH)CH_3$

$C_{10}H_{22}O$ M 158.283

(+)-*form* [33758-16-6]

d_4^{20} 0.825. Bp 210°, Bp_{10} 110-111°. n_D^{20} 1.4344.

Ac:

$C_{12}H_{24}O_2$ M 200.320

Bp_{15} 115°. $[\alpha]_D^{20}$ +5.64°.

(±)-*form* [74742-10-2]

Bp 210-211°, Bp_{10} 110-111°.

3,5-Dinitrobenzoyl: Mp 44°.

[1534-32-3]

Pickard, R.H. *et al*, *J. Chem. Soc.*, 1911, **99**, 55; 1914, **105**, 852 (*synth*)
Heilbron, I.M. *et al*, *J. Chem. Soc.*, 1943, 264 (*synth*)
Petrov, A.D. *et al*, *CA*, 1945, **39**, 4585 (*synth*)
Yamamoto, Y. *et al*, *J. Chem. Soc., Chem. Commun.*, 1976, 672 (*synth*)
Keinan, E. *et al*, *J. Am. Chem. Soc.*, 1986, **108**, 162 (*synth*)

3-Decanol, 9CI D-00047

[1565-81-7]

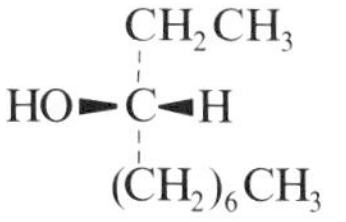

(*R*)-*form*

$C_{10}H_{22}O$ M 158.283

(*R*)-*form*

Bp_{15} 108°. $[\alpha]_D^{20}$ −10.55° (c, 5 in EtOH).

(±)-*form* [74683-67-3]

Mp −7.5°. Bp 213°, Bp_{12} 101°.

1-Naphthylurethane: Mp 60-62°.

Et ether: 3-Ethoxydecane

$C_{12}H_{26}O$ M 186.337

$Bp_{9.5}$ 89.5°.

Pickard, R.H. *et al*, *J. Chem. Soc.*, 1913, **103**, 1923 (*synth, resoln*)
Woods, G.F. *et al*, *J. Am. Chem. Soc.*, 1951, **73**, 2245 (*synth*)
Asinger, F. *et al*, *Chem. Ber.*, 1959, **92**, 3085 (*synth*)
Godsey, C.E., *Anal. Chem.*, 1966, **38**, 842 (*nmr*)
Remberg, G. *et al*, *Chem. Ber.*, 1970, **103**, 3640 (*ms*)

4-Decanol, 9CI D-00048

[2051-31-2]

$H_3C(CH_2)_5CH(OH)CH_2CH_2CH_3$

$C_{10}H_{22}O$ M 158.283

(±)-*form*

Mp −11.6°. Bp_{12} 101°. Opt. active forms are also known.

3,5-Dinitrobenzoyl: Mp 42.3-43.5° (24°).

Levene, P.A. *et al*, *J. Org. Chem.*, 1936, **1**, 76 (*abs config*)
Heilbron, I.M. *et al*, *J. Chem. Soc.*, 1943, 265 (*synth*)
Asinger, F. *et al*, *Chem. Ber.*, 1959, **92**, 3085 (*synth*)
Godsey, C.E., *Anal. Chem.*, 1966, **38**, 842 (*pmr*)
v. Tamelen, E.E. *et al*, *J. Am. Chem. Soc.*, 1974, **96**, 5290 (*synth*)

5-Decanol D-00049

$H_3C(CH_2)_4CH(OH)(CH_2)_3CH_3$

$C_{10}H_{22}O$ M 158.283

(±)-*form*

Mp 8.7°. Bp_{30} 118-120°, Bp_{11} 99°.

3,5-Dinitrobenzoyl: Mp 43.5-44.5°.

1-Naphthylurethane: Mp 55-56°.

Et ether: 4-Ethoxydecane

$C_{12}H_{26}O$ M 186.337

$Bp_{9.5}$ 87.5°.

Woods, G.F. *et al*, *J. Am. Chem. Soc.*, 1951, **73**, 2245 (*synth*)
Asinger, F. *et al*, *Chem. Ber.*, 1959, **92**, 3085 (*synth*)
Godsey, C.E., *Anal. Chem.*, 1966, **38**, 842 (*nmr*)

2-Decanone, 9CI D-00050

[693-54-9]

$H_3C(CH_2)_7COCH_3$

$C_{10}H_{20}O$ M 156.267

Found in oil of rue. Attractant for the olive bane beetle *Phloetribus scarabaeoides*. Needles. d_4^{22} 0.823. Mp 14° (2.5°). Bp_{750} 209°, Bp_{12} 95-97°. n_D^{22} 1.4263.

Semicarbazone: Plates (EtOH). Mp 126° (121°).

Morgan, G.T. *et al*, *J. Soc. Chem. Ind., London*, 1925, **44**, 108T.
Ruzicka, L. *et al*, *Helv. Chim. Acta*, 1926, **9**, 397 (*synth*)
Murakami, K., *CA*, 1930, **24**, 2426 (*synth*)
Rupe, H. *et al*, *Helv. Chim. Acta*, 1932, **15**, 845 (*synth*)
Possel, O. *et al*, *Tetrahedron Lett.*, 1977, 4229 (*synth*)
Sato, F. *et al*, *Chem. Lett.*, 1979, 167 (*synth*)
Pena, A. *et al*, *Entomol. Exp. Appl.*, 1992, **63**, 81.

3-Decanone, 9CI D-00051

3-Oxodecane

[928-80-3]

$H_3C(CH_2)_6COCH_2CH_3$

$C_{10}H_{20}O$ M 156.267

Bp 211°.

Semicarbazone: Mp 100-101°.

2,4-Dinitrophenylhydrazone: [62758-09-2].

Orange needles (MeOH aq.). Mp 53-54°.

Pickard, R.M. *et al*, *J. Chem. Soc.*, 1913, **103**, 1945 (*synth*)
Schwartz, J. *et al*, *J. Am. Chem. Soc.*, 1974, **96**, 4721 (*synth*)
Cahiez, G. *et al*, *Synth. Commun.*, 1979, **9**, 639 (*synth*)

4-Decanone, 9CI — D-00052

[624-16-8]

$H_3C(CH_2)_5COCH_2CH_2CH_3$

$C_{10}H_{20}O$ M 156.267
d_0^{21} 0.824. Mp −9°. Bp 206-207°, Bp_{11} 87-89°.
Semicarbazone: Mp 51-52°.

Karrer, P. *et al*, *Helv. Chim. Acta*, 1930, **13**, 1300 (*synth*)
Georges, S. *et al*, *Bull. Soc. Chim. Fr.*, 1972, 4233 (*synth*)
Waters, W.L. *et al*, *J. Org. Chem.*, 1976, **41**, 889 (*synth*)

2,4,6,8-Decatetraenal, 9CI — D-00053

[40650-87-1]

$H_3C(CH{=}CH)_4CHO$

$C_{10}H_{12}O$ M 148.204
(all-E)-form [39806-05-8]
Mp 104-106°.

Kuhn, R. *et al*, *Ber.*, 1936, **69**, 98.
Blotz, P.E. *et al*, *J. Am. Chem. Soc.*, 1969, **91**, 5930.

2,4,6,8-Decatetraenedioic acid, 9CI, 8CI — D-00054

[6048-86-8]

$HOOCCH{=}CHCH{=}CHCH{=}CHCH{=}CHCOOH$

$C_{10}H_{10}O_4$ M 194.187
Fumagillin degradn. product.
(all-E)-form
Yellow powder. Mp 295-297° dec.
Di-Me ester: [60895-15-0].
$C_{12}H_{14}O_4$ M 222.240
Small yellow prisms. Mp 212-213°.
(E,E,Z,Z)-form [21446-15-1]
Yellow powder. Mp 271-280° dec.
Di-Me ester: [21446-16-2].
Fine yellow needles. Mp 118°.
(E,Z,Z,Z)-form
Di-Me ester: [60860-98-2].
Mp 112°.

Kuhn, R. *et al*, *Ber.*, 1936, **69**, 1757 (*synth*)
Hargie, M.P. *et al*, *J. Am. Chem. Soc.*, 1953, **75**, 2271 (*synth*)
Cantrell, T.S., *J. Am. Chem. Soc.*, 1970, **92**, 5480 (*synth*)
Kaupp, G. *et al*, *Angew. Chem., Int. Ed. Engl.*, 1976, **15**, 779 (*synth, pmr*)

2,4,6,8-Decatetraen-1-ol, 9CI — D-00055

$H_3C(CH{=}CH)_4CH_2OH$

$C_{10}H_{14}O$ M 150.220
(all-E)-form
Metab. of *Daedalea juniperina*. Mp 110-111°.

Bew, R.E. *et al*, *J. Chem. Soc. C*, 1966, 135.

2,4,7-Decatrienal, 9CI — D-00056

[51325-37-2]

$H_3CCH_2CH{=}CHCH_2CH{=}CHCH{=}CHCHO$

$C_{10}H_{14}O$ M 150.220

(2E,4Z,7Z)-form [43108-49-2]
Component of cooked chicken flavour. Also from *n*-3 polyene acyl chains. $Bp_{0.4}$ 65-67°.
[38664-37-8, 66642-86-2]

Meijboom, P.W. *et al*, *J. Am. Oil Chem. Soc.*, 1972, **49**, 555 (*occur, synth*)
Harkes, P.D. *et al*, *J. Am. Oil Chem. Soc.*, 1974, **51**, 356 (*occur*)
Seifert, R.M. *et al*, *J. Agric. Food Chem.*, 1980, **28**, 68 (*synth*)

2,6,8-Decatrienoic acid — D-00057

$H_3CCH{=}CHCH{=}CHCH_2CH_2CH{=}CHCOOH$

$C_{10}H_{14}O_2$ M 166.219
(2E,6Z,8E)-form
N-(*2-Methylpropyl*)*amide:* [25394-57-4]. ***Affinin.*** N-(*2-Methylpropyl*)*-2,6,8-decatrienamide, 9CI.* N-*Isobutyl-2,6,8-decatrienamide. Spilanthol*
$C_{14}H_{23}NO$ M 221.342
Alkaloid from the roots of *Heliopsis longipes*, *Sapilanthes oleraceae* and *S. acmella.* Powerful insecticide. Pale-yellow viscous oil. Mp 23°. $Bp_{0.5}$ 165°, $Bp_{0.001}$ 120-125°. n_D^{25} 1.5135. In the earlier lit. *H. longipes* was incorrectly identified as *Erigeron affinis.*
(all-E)-form
N-(*2-Methylpropyl*)*amide:* [76361-77-8].
Mp 91.5-92.5°. $Bp_{0.25}$ 145°.

Asano, M. *et al*, *Ber.*, 1932, **65**, 1602 (*isol*)
Gokhale, V.G. *et al*, *J. Indian Chem. Soc.*, 1945, **22**, 250 (*isol*)
Acree, F. *et al*, *J. Org. Chem.*, 1945, **10**, 236, 449 (*isol, uv, struct*)
Jacobson, M. *et al*, *J. Org. Chem.*, 1947, **12**, 731.
Jacobson, M., *J. Am. Chem. Soc.*, 1955, **77**, 2461 (*synth*)
Jacobson, M., *Chem. Ind.* (*London*), 1957, 50 (*struct, synth*)
Crombie, L. *et al*, *J. Chem. Soc.*, 1963, 4970 (*ir, uv, synth*)
Correa, J. *et al*, *Org. Magn. Reson.*, 1971, **3**, 1 (*pmr, struct*)
Yasuda, I. *et al*, *Chem. Pharm. Bull.*, 1980, **28**, 2251 (*isol, pmr, cmr, ms, synth, struct*)

2,4,8-Decatrienoic isobutylamide — D-00058

$H_3CCH{=}CHCH_2CH_2CH{=}CHCH{=}CHCONHCH_2CH(CH_3)_2$

$C_{14}H_{23}NO$ M 221.342
(E,E,Z)-form
Constit. of *Otanthus maritimus.* Oil.

Bohlmann, F. *et al*, *Chem. Ber.*, 1974, **107**, 1038 (*isol, struct*)

2,6,8-Decatrien-4-yn-1-ol — D-00059

$H_3CCH{=}CHCH{=}CHC{\equiv}CCH{=}CHCH_2OH$

$C_{10}H_{12}O$ M 148.204
(all-E)-form
Constit. of *Grindelia* spp. and the roots of *Gutierrezia gymnospermoides.* Cryst. (pet. ether). Mp 56°.
Ac: [929-52-2].
$C_{12}H_{14}O_2$ M 190.241
Constit. of *G. gymnospermoides*, *G. robusta* and *G. squarrosa.* Cryst. (pet. ether). Mp 62.5°.
Angeloyl:
$C_{15}H_{18}O_2$ M 230.306
Constit. of *G. gymnospermoides.* Oil. $Bp_{0.001}$ 110°.

Bohlmann, F. *et al*, *Chem. Ber.*, 1965, **98**, 369 (*synth*); 1969, **102**, 1682 (*isol*)
Jente, R. *et al*, *Phytochemistry*, 1976, **15**, 1673 (*biosynth*)

2,4,6-Decatriynedioic acid, 9CI D-00060

[3943-58-6]

HOOCC≡CC≡CC≡CCH$_2$CH$_2$COOH

$C_{10}H_6O_4$ M 190.155

Occurs in *Merulius lachrymans*. Powder. Mp 170° dec.

Di-Me ester:

$C_{12}H_{10}O_4$ M 218.209

Needles. Mp 45°. λ_{max} 201, 217, 226, 257, 272, 288, 307 and 329 nm (CCl_4).

Hodge, P. *et al*, *J. Chem. Soc. C*, 1966, 1216.

4,6,8-Decatriyne-1,3-diol D-00061

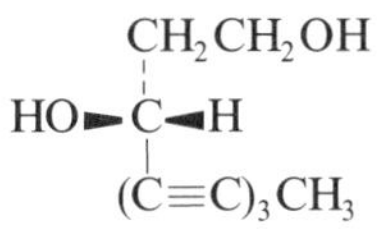

$C_{10}H_{10}O_2$ M 162.188

(R)-form [22733-68-2]

Isol. from lettuce *Lactuea plumieri*. Cryst. (CCl_4). Mp 55-56°. $[\alpha]_D^{20}$ −21° (c, 1 in EtOH).

Bentley, R.K. *et al*, *J. Chem. Soc. C*, 1969, 1096 (*synth*)

4,6,8-Decatriyn-1-ol D-00062

H$_3$CC≡CC≡CC≡CCH$_2$CH$_2$CH$_2$OH

$C_{10}H_{10}O$ M 146.188

Cryst. (hexane). Mp 68-70.5°.

Bew, R.E. *et al*, *J. Chem. Soc. C*, 1966, 129 (*isol, struct*)

2-Decenal D-00063

[3913-71-1]

H$_3$C(CH$_2$)$_6$CH=CHCHO

$C_{10}H_{18}O$ M 154.252

Constit. of essential oil of coriander and oil of *Achasma walang*. Defensive secretion of the bug *Caleotechus sordidus*.

▷ HE2070000.

(E)-form

Bp 229-231°, Bp$_{11.5}$ 107-107.5°.

Oxime:

$C_{10}H_{19}NO$ M 169.266

Mp 55°.

Semicarbazone: [16742-25-9].

Cryst. (EtOH). Mp 168.5°.

(Z)-form

Bp$_{17}$ 112°. 95% pure.

Romburgh, P., *Recl. Trav. Chim. Pays-Bas* (*J. R. Neth. Chem. Soc.*), 1938, **57**, 494 (*isol*)
Swift, C.E. *et al*, *J. Am. Chem. Soc.*, 1949, **71**, 1512 (*synth*)
Tschinkel, W.R., *J. Insect Physiol.*, 1975, **21**, 659 (*isol*)
Dirinck, P. *et al*, *J. Food Sci.*, 1977, **42**, 645 (*isol*)
Selke, E. *et al*, *Lipids*, 1978, **13**, 511 (*glc, ms*)
Smith, R.M., *N.Z. J. Sci.*, 1978, **21**, 121 (*isol*)
Bestmann, H.J. *et al*, *Chem. Ber.*, 1982, **115**, 161 (*synth*)
Lewis, R.J., *Sax's Dangerous Properties of Industrial Materials*, 8th Ed., Van Nostrand-Reinhold, 1992, DAI350.

3-Decenal D-00064

[58474-80-9]

H$_3$C(CH$_2$)$_5$CH=CHCH$_2$CHO

$C_{10}H_{18}O$ M 154.252

(E)-form [68676-85-7]

Liq. Bp$_{0.3}$ 70°.

2,4-Dinitrophenylhydrazone: Mp 126-127°.

(Z)-form [69891-94-7]

Liq.

Di-Me acetal: 1,1-Dimethoxy-3-decene

$C_{12}H_{24}O_2$ M 200.320

Bp$_{11}$ 109-111°.

Stowell, J.C. *et al*, *Synthesis*, 1979, 132 (*synth*)
Achmatowicz, O. *et al*, *Synthesis*, 1987, 413 (*synth, ir, pmr*)

6-Decenal D-00065

H$_3$C(CH$_2$)$_2$CH=CH(CH$_2$)$_4$CHO

$C_{10}H_{18}O$ M 154.252

(E)-form

Bp$_{50}$ 117°.

2,4-Dinitrophenylhydrazone: Cryst. (MeOH). Mp 93°.

Crombie, L. *et al*, *J. Chem. Soc.*, 1952, 2997.

1-Decene, 9CI D-00066

[872-05-9]

H$_3$C(CH$_2$)$_7$CH=CH$_2$

$C_{10}H_{20}$ M 140.268

Liq. d^{20} 0.742. Fp −66.6°. Bp 170.6°, Bp$_3$ 53.5°. n_D^{20} 1.4217.

Vaughn, T.M., *J. Am. Chem. Soc.*, 1934, **56**, 2064 (*synth*)
Maman, A., *CA*, 1936, **30**, 7095 (*synth*)
Holman, R.T. *et al*, *Anal. Chem.*, 1956, **28**, 1533 (*ir*)
McMurry, J.E. *et al*, *J. Org. Chem.*, 1975, **40**, 2555 (*synth*)
Reetz, M.T. *et al*, *Synthesis*, 1977, 135 (*synth*)
Sonnet, P.E. *et al*, *J. Org. Chem.*, 1978, **43**, 1841 (*synth*)
Miller, J.A. *et al*, *J. Org. Chem.*, 1993, **58**, 18 (*synth*)

2-Decene D-00067

[6816-17-7]

H$_3$C(CH$_2$)$_6$CH=CHCH$_3$

$C_{10}H_{20}$ M 140.268

(Z)-form [20348-51-0]

Bp$_{15}$ 62-65°.

Normant, J. *et al*, *Bull. Soc. Chim. Fr.*, 1963, 1888 (*synth*)
Marshall, J.A. *et al*, *Synth. Commun.*, 1975, **5**, 293 (*synth*)
Nixon, J.F. *et al*, *J. Organomet. Chem.*, 1975, **87**, 341 (*synth*)

3-Decene, 9CI D-00068

[19398-37-9]

H$_3$C(CH$_2$)$_5$CH=CHCH$_2$CH$_3$

$C_{10}H_{20}$ M 140.268

Bp 173-173.5°.

Kinney, C.R. *et al*, *J. Org. Chem.*, 1949, **14**, 71 (*synth*)
Katzenellenbogen, J.A. *et al*, *Tetrahedron Lett.*, 1972, 1475 (*synth*)
Dang, H.P. *et al*, *Tetrahedron Lett.*, 1978, 191 (*synth*)

4-Decene, 9CI D-00069

[19689-18-0]

$H_3C(CH_2)_4CH{=}CHCH_2CH_2CH_3$

$C_{10}H_{20}$ M 140.268

Liq. Bp_{761} 170.6°.

Ziegler, K. *et al*, *Justus Liebigs Ann. Chem.*, 1950, **567**, 43 (*synth*)
Baba, S. *et al*, *Tetrahedron Lett.*, 1976, 1927 (*synth*)

5-Decene, 9CI D-00070

[19689-19-1]

$H_3C(CH_2)_3CH{=}CH(CH_2)_3CH_3$

$C_{10}H_{20}$ M 140.268

Bp_{750} 170°.

(E)-form [7433-56-9]

Bp 172-173.5°. n_D^{25} 1.4213.

▷ HE2080000.

(Z)-form [7433-78-5]

Bp 169-170°, Bp_{11} 59°. n_D^{25} 1.4230.

Davis, A. *et al*, *J. Chem. Soc., Perkin Trans. 1*, 1972, 286 (*synth*)
Bestmann, H.J. *et al*, *Synthesis*, 1974, 798 (*synth*)
House, H.O. *et al*, *J. Org. Chem.*, 1974, **39**, 747 (*synth*)
McMurry, J.E. *et al*, *J. Org. Chem.*, 1975, **40**, 2555 (*synth*)
Miyaura, N. *et al*, *Synthesis*, 1975, 669 (*synth*)
Sonnet, P.E. *et al*, *J. Org. Chem.*, 1978, **43**, 1841 (*synth*)
Murahashi, S. *et al*, *J. Org. Chem.*, 1979, **44**, 2408 (*synth, nmr*)

2-Decenedioic acid, 9CI D-00071

1-Octene-1,8-dicarboxylic acid

[6048-93-7]

$HOOC(CH_2)_6CH{=}CHCOOH$

$C_{10}H_{16}O_4$ M 200.234

Isol. from *Candida tropicalis*, fermentations and honeys.

(E)-form [37443-67-7]

Metab. from *Penicillium notatum*. Cryst. (hexane/EtOAc). Mp 172° (168-170°).

[28598-91-6, 37443-69-9]

English, J., *J. Am. Chem. Soc.*, 1941, **63**, 941 (*synth*)
Cram, D.J. *et al*, *J. Am. Chem. Soc.*, 1948, **70**, 4238 (*isol*)
Tsuji, J. *et al*, *J. Organomet. Chem.*, 1977, **131**, 133 (*synth*)
Fujisawa, T. *et al*, *Chem. Lett.*, 1982, 219 (*synth*)
Tan, S.T. *et al*, *J. Agric. Food Chem.*, 1988, **36**, 453 (*isol*)

4-Decenedioic acid D-00072

[67494-01-3]

$HOOC(CH_2)_4CH{=}CHCH_2CH_2COOH$

$C_{10}H_{16}O_4$ M 200.234

Z-form [41221-65-2]

Isol. from urine of patients with Jamaican vomit disease. Solid (hexane/Et_2O). Mp 52-53°.

E-form [128823-57-4]

Solid (hexane/Et_2O). Mp 136-137.5°.

Tanaka, K., *J. Biol. Chem.*, 1972, **247**, 7465 (*isol*)
Kaydos, J.A. *et al*, *J. Org. Chem.*, 1993, **58**, 3463 (*synth, ir, pmr, cmr*)

2-Decene-4,6-diynedioic acid, 9CI D-00073

[28813-56-1]

$HOOCCH_2CH_2C{\equiv}CC{\equiv}CCH{=}CHCOOH$

$C_{10}H_8O_4$ M 192.171

(E)-form [51193-94-3]

Isol. from the fungus *Merulius lachrymans*.

Di-Me ester:

$C_{12}H_{12}O_4$ M 220.224

Occurs in *Polyporus anthracophilus*. Cryst. Mp 56.5-58°. λ_{max} 214.5, 223 (sh), 243 (sh), 255 (sh), 270, 285 and 303 nm (EtOH).

Bu'Lock, J.D. *et al*, *J. Chem. Soc.*, 1957, 1607 (*isol*)
Gardner, J.N. *et al*, *J. Chem. Soc.*, 1960, 691 (*isol*)

2-Decene-4,6-diynoic acid D-00074

Lachnophyllic acid

$H_3CCH_2CH_2C{\equiv}CC{\equiv}CCH{=}CHCOOH$

$C_{10}H_{10}O_2$ M 162.188

(E)-form [80994-68-9]

Mp 123.5-124.5°.

Me ester: [3788-06-5]. (E)-*Lachnophyllum ester*

$C_{11}H_{12}O_2$ M 176.215

Occurs in *Bellis perennis*, *Chrysothamnus nauseosus*, *C. parryi*, *Conyza canadensis*, *Heterotheca grandiflora* and *Baccharis* spp. Cryst. (pet. ether or EtOH). Mp 16-17°.

(Z)-form

Mp 72-74°.

Me ester: [505-01-1]. *Lachnophyllum ester. Dihydromatricaria ester*

Occurs in *Lachnophyllum gossypinium* and many other Compositae spp., also in *Aster spinosus* and *A. sibiricus*. Mp 31-32°.

(ξ)-form

2-Methylpropylamide, N-*Me:* [37064-12-3]. N-*Methyl*-N-*(2-methylpropyl)-2-decene-4,6-diynamide*

$C_{15}H_{21}NO$ M 231.337

Isol. from the roots of *Anacyclus pyrethrum*. Oil. λ_{max} 299, 281, 264, 252 nm. Stereochem. unspecified.

[54497-59-5, 114825-66-0, 114825-67-1]

Sorensen, N.A. *et al*, *Acta Chem. Scand.*, 1950, **4**, 1575 (*uv*)
Tronvold, G.M. *et al*, *Acta Chem. Scand.*, 1953, **7**, 1375 (*isol*)
Bell, I. *et al*, *J. Chem. Soc.*, 1958, 1313 (*synth*)
Bohlmann, F. *et al*, *Chem. Ber.*, 1969, **102**, 1682; 1972, **105**, 3587 (*isol*)
Jente, R. *et al*, *Chem. Ber.*, 1972, **105**, 1694 (*isol, struct, uv, ms*)
Zeisberg, R. *et al*, *Chem. Ber.*, 1974, **107**, 3800 (*cmr*)
Davies, D.G. *et al*, *J. Chem. Soc., Perkin Trans. 1*, 1978, 1602 (*isol, uv, ms, pmr*)
Bohlmann, F. *et al*, *Phytochemistry*, 1979, **18**, 1011, 1367, 1519 (*isol*)
Carpita, A. *et al*, *Gazz. Chim. Ital.*, 1987, **117**, 481 (*synth*)

2-Decene-4,6-diyn-1-ol D-00075

Lachnophyllol

$H_3CCH_2CH_2C{\equiv}CC{\equiv}CCH{=}CHCH_2OH$

$C_{10}H_{12}O$ M 148.204

(E)-form [23180-62-3]

Isol. from *Callistephus chinensis*. Cryst. (pet. ether). Mp 39-40°.

Ac: [24486-26-8].

$C_{12}H_{14}O_2$ M 190.241

Constit. of *Aster sibiricus*. Oil. $Bp_{0.001}$ 90°.

2-Methylpropanoyl:

$C_{14}H_{18}O_2$ M 218.295

Isol. from *Grindelia camporum*. Oil.

3-Methylbutanoyl:

$C_{15}H_{20}O_2$ M 232.322

Constit. of *Osmites hirsuta*. Oil. $Bp_{0.1}$ 120°.

Angeloyl:

$C_{15}H_{18}O_2$ M 230.306
Constit. of *C. chinensis*. Oil.

Bruun, T. *et al*, *Acta Chem. Scand.*, 1950, **4**, 850 (*synth*)
Soerensen, J.S. *et al*, *Aust. J. Chem.*, 1969, **22**, 751 (*isol*)
Bohlmann, F. *et al*, *Chem. Ber.*, 1969, **102**, 1682; 1970, **103**, 2245; 1972, **105**, 2604 (*isol, struct, synth, uv, ir, ms*)
Bohlmann, F. *et al*, *Phytochemistry*, 1982, **21**, 2691 (*isol*)

6-Decene-2,4-diyn-1-ol D-00076

$H_3CCH_2CH_2CH{=}CHC{\equiv}CC{\equiv}CCH_2OH$

$C_{10}H_{12}O$ M 148.204
(E)-form
Cryst. (Et_2O/pet. ether). Mp 38°.

Bohlmann, F. *et al*, *Chem. Ber.*, 1970, **103**, 3419 (*synth*)

8-Decene-4,6-diyn-1-ol D-00077

[30339-12-9]

$H_3CCH{=}CHC{\equiv}CC{\equiv}CCH_2CH_2CH_2OH$

$C_{10}H_{12}O$ M 148.204
(Z)-form
Isol. from above-ground parts of *Chrysanthemum maximum*. Liq. $Bp_{0.05}$ 95°.
Ac:
$C_{12}H_{14}O_2$ M 190.241
Isol. from *C. maximum*. Oil.
3-Methylbutanoyl:
$C_{15}H_{20}O_2$ M 232.322
Constit. of *Carthamus tinctorius*. Oil.

Bohlmann, F. *et al*, *Chem. Ber.*, 1966, **99**, 586; 1970, **103**, 2853 (*isol, uv, ir, pmr, struct, synth*)

8-Decene-1,3,5-triol D-00078

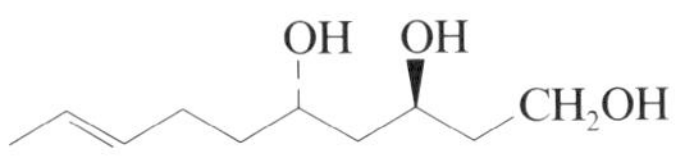

$C_{10}H_{20}O_3$ M 188.266
(3S,5S,8E)-form [135352-00-0] ***Streptenol B***
Prod. by *Streptomyces luteogriseus*. $[\alpha]_D^{20}$ +39° (c, 1 in $CHCl_3$).

Grabley, S. *et al*, *J. Antibiot.*, 1991, **44**, 797 (*isol, pmr, cmr*)

2-Decene-4,6,8-triynoic acid D-00079

Dehydromatricaric acid

$H_3CC{\equiv}CC{\equiv}CC{\equiv}CCH{=}CHCOOH$

$C_{10}H_6O_2$ M 158.156
(E)-form
Me ester: [692-94-4]. *Dehydromatricaria ester*
$C_{11}H_8O_2$ M 172.183
Occurs in *Matricaria inodora* and *M. oreades* and many other Compositae spp. Also isol. from culture medium of *Pleurotus ulmarius*. Light-yellow cryst. (Et_2O/pet. ether). Mp 105-106°.
3-Methylbutylamide: [113235-91-9]. *2-Decene-4,6,8-triynoic acid isopentylamide*
$C_{15}H_{17}NO$ M 227.305
Isol. from the underground parts of *Achillea wilhelmsii* (Compositae). Cryst. Dec. to blue polymers on htg.
2-Phenylethylamide: [113235-93-1]. N-*(2-Phenylethyl)-2-decene-4,6,8-triynamide. 2-Decene-4,6,8-triynoic acid 2-phenylethylamide*
$C_{18}H_{15}NO$ M 261.323
Isol. from the underground parts of *A. wilhelmsii* (Compositae). Cryst. Dec. to blue polymer on htg.
2-Methylpropylamide: [37064-10-1]. ***Dehydromatricaric acid isobutylamide***
$C_{14}H_{15}NO$ M 213.279
Isol. from *Achillea spinulifolia, A. ptarmica, A. impatiens, A. sibirica, Anacyclus pyrethrum* and *Cladanthus arabicus*. Cryst. (CCl_4). Mp 133-139°, Mp 144.5-145.5°.
(Z)-form [23050-77-3]
Mp 150° dec.
Me ester: [2739-57-3].
Occurs in *Artemisia vulgaris* and many other Compositae. Light-yellow needles (pet. ether). Mp 114-115° (112°).

[7329-73-9, 43110-85-6]

Stauholt, K. *et al*, *Acta Chem. Scand.*, 1950, **4**, 1567 (*isol*)
Christensen, P.K. *et al*, *Acta Chem. Scand.*, 1952, **6**, 602 (*synth*)
Sorensen, J.S. *et al*, *Acta Chem. Scand.*, 1954, **8**, 26 (*isol*)
Bell, I. *et al*, *J. Chem. Soc.*, 1958, 1313 (*synth, Me ester*)
Gardner, J.N. *et al*, *J. Chem. Soc.*, 1960, 691 (*isol*)
Bohlmann, F. *et al*, *Chem. Ber.*, 1962, **95**, 1742; 1973, **106**, 1328 (*isol, struct, synth*)
Hodge, P. *et al*, *J. Chem. Soc. C*, 1966, 1216 (*bibl*)
Karrer, W. *et al*, *Konstitution und Vorkommen der Organischen Pflanzenstoffe*, 2nd Ed., Birkhäuser Verlag, Basel, 1972-1985, no. 770.
Kobayashi, A. *et al*, *Agric. Biol. Chem.*, 1975, **39**, 911 (*synth*)
Kawazu, K. *et al*, *Agric. Biol. Chem.*, 1977, **41**, 223.
Greger, H., *Phytochemistry*, 1978, **17**, 86; 1982, **21**, 1071 (*isol, struct*)
Bohlmann, F. *et al*, *Phytochemistry*, 1979, **18**, 1736; 1980, **19**, 841, 2655 (*isol*)
Greger, H. *et al*, *J. Nat. Prod. (Lloydia)*, 1987, **50**, 1100 (*amides, isol, ir, uv, ms, pmr*)

8-Decene-2,4,6-triynoic acid D-00080

$H_3CCH{=}CHC{\equiv}CC{\equiv}CC{\equiv}CCOOH$

$C_{10}H_6O_2$ M 158.156
(E)-form
Me ester:
$C_{11}H_8O_2$ M 172.183
Cryst. (pet. ether). Mp 61°.

Bohlmann, F. *et al*, *Chem. Ber.*, 1955, **88**, 429.

2-Decene-4,6,8-triyn-1-ol D-00081

Dehydromatricarianol
[3930-17-4]

$H_3CC{\equiv}CC{\equiv}CC{\equiv}CCH{=}CHCH_2OH$

$C_{10}H_8O$ M 144.173
(E)-form [6071-16-5]
Constit. of *Tricholoma grammopodium* and other fungi. Cryst. (Et_2O/pentane). Mp 129°.
Ac: [52940-10-0].
$C_{12}H_{10}O_2$ M 186.210
Constit. of *Osmitopsis asteriscoides*. Cryst. (pet. ether). Mp 38.2°.
3-Methylbutanoyl:
$C_{15}H_{16}O_2$ M 228.290
Constit. of *O. asteriscoides*. Oil.
(Z)-form [6071-46-1]
Metab. of *Fistulina hepatica*.
Ac: [115374-19-1].
$C_{12}H_{10}O_2$ M 186.210
Isol. from aerial parts of *Calotis dentex*. Oil. λ_{max} 329, 317, 288, 271 nm (Et_2O).

Jones, J.B., *J. Chem. Soc.*, 1963, 5759 (*synth*)
Jones, E.R.H. *et al*, *J. Chem. Soc. C*, 1966, 139 (*isol*)
Hodge, P. *et al*, *J. Chem. Soc. C*, 1966, 1216 (*biosynth*)
Bohlmann, F. *et al*, *Chem. Ber.*, 1974, **107**, 1409 (*isol*)
Hearn, M.T.W. *et al*, *J. Chem. Soc., Perkin Trans. 2*, 1976, 1027 (*cmr*)
Zdero, C. *et al*, *Phytochemistry*, 1988, **27**, 1105 (*isol, struct, uv, ir, ms, acetate*)

2-Decenoic acid D-00082

[3913-85-7]

$H_3C(CH_2)_6CH{=}CHCOOH$

$C_{10}H_{18}O_2$ M 170.251

Constit. of essential oil of *Achasma walang*. Mosquito (*Aedes aegypti*) repellent. Mp 12°. Bp_{15} 165°, $Bp_{4.5}$ 148-149°.

Chloride:

$C_{10}H_{17}ClO$ M 188.696

Bp_{14} 120-122°, $Bp_{2.5}$ 95-97°.

Amide:

$C_{10}H_{19}NO$ M 169.266

Cryst. (Et_2O). Mp 121-122°.

Zaar, B., *Chem. Zentralbl.*, 1930, **1**, 363.
v. Romburgh, P., *Recl. Trav. Chim. Pays-Bas* (*J. R. Neth. Chem. Soc.*), 1938, **57**, 494 (*isol*)
Skinner, W.A. *et al*, *Experientia*, 1970, **26**, 728.
Quraishi, M.S., *J. Econ. Entomol.*, 1971, **64**, 787.

3-Decenoic acid D-00083

[15469-77-9]

$H_3C(CH_2)_5CH{=}CHCH_2COOH$

$C_{10}H_{18}O_2$ M 170.251

Mp 18°. Bp_{11} 154-163°.

Bromine addn. prod.: 3,4-Dibromodecanoic acid

Mp 44-45°.

(*Z*)-*form* [2430-93-5]

Sex pheromone of the furniture carpet beetle *Athrenus flavipes*.

Tulus, R., *CA*, 1940, **40**, 3722 (*synth*)
Gorge, M., *Ann. Chim.* (*Paris*), 1951, **6**, 648 (*synth*)
Fukui, H. *et al*, *Tetrahedron Lett.*, 1974, 3563 (*isol*)

4-Decenoic acid D-00084

[26303-90-2]

$H_3C(CH_2)_4CH{=}CHCH_2CH_2COOH$

$C_{10}H_{18}O_2$ M 170.251

(*E*)-*form*

Bp_{13} 145-146°.

(*Z*)-*form* [505-90-8]

Obtusilic acid

Constit. of seed oil of *Lindera obtusiloba* and other members of the Lauraceae. Not found in any other family. Bp_{13} 148-150°.

4-Bromophenacyl ester: Mp 43.3°.

[57602-94-5]

Komori, S. *et al*, *Bull. Chem. Soc. Jpn.*, 1937, **12**, 226, 433 (*isol*)
Iwakiri, M., *Yakugaku Zasshi*, 1951, **78**, 1460; *CA*, **53**, 1633 (*synth, config*)
Hopkins, C.Y. *et al*, *Lipids*, 1966, **1**, 118 (*isol*)
Bus, J. *et al*, *Chem. Phys. Lipids*, 1976, **17**, 501 (*cmr*)
Iwakiri, M., *CA*, 1982, **97**, 20619q (*rev*)
Fujisawa, T. *et al*, *Tetrahedron Lett.*, 1982, 3583 (*synth*)
Kawashima, M. *et al*, *Bull. Chem. Soc. Jpn.*, 1988, **61**, 3255 (*synth*)

8-Decenoic acid D-00085

[21968-08-1]

$H_3CCH{=}CH(CH_2)_6COOH$

$C_{10}H_{18}O_2$ M 170.251

Bp_{14} 155-157°.

Me ester:

$C_{11}H_{20}O_2$ M 184.278

Bp_{20} 121-123°.

Chuit, P. *et al*, *Helv. Chim. Acta*, 1927, **10**, 167 (*synth*)
Jacob, J. *et al*, *J. Lipid Res.*, 1968, **9**, 730 (*glc*)
Gunstone, F.D. *et al*, *Chem. Phys. Lipids*, 1977, **18**, 115 (*cmr*)

9-Decenoic acid D-00086

Caproleic acid

[14436-32-9]

$H_2C{=}CH(CH_2)_7COOH$

$C_{10}H_{18}O_2$ M 170.251

Minor constit. of milk fats. Bp_{21} 158-163°.

Me ester:

$C_{11}H_{20}O_2$ M 184.278

Oil. Bp_{20} 120°.

4-Bromophenacyl ester: Cryst. (EtOH aq.). Mp 58°.

Black, H.K. *et al*, *J. Chem. Soc.*, 1953, 1785 (*synth*)
Renner, E. *et al*, *Milchwissenschaft*, 1978, **33**, 489 (*isol*)
Hase, T.A. *et al*, *Synth. Commun.*, 1979, **9**, 63 (*synth*)
Bartra, M. *et al*, *J. Org. Chem.*, 1991, **56**, 5132 (*synth, pmr, cmr, ir*)

2-Decen-1-ol D-00087

$H_3C(CH_2)_6CH{=}CHCH_2OH$

$C_{10}H_{20}O$ M 156.267

(*E*)-*form* [18409-18-2]

Ac: [19487-61-7].

Major component in *Nezara viridula* (green vegetable bug) scent gland secretions.

(*Z*)-*form* [4194-71-2]

$Bp_{0.01}$ 70°.

Bohlmann, F. *et al*, *Chem. Ber.*, 1965, **98**, 3010 (*synth*)
Gilby, A.R. *et al*, *Proc. R. Soc. London, B*, 1965, **162**, 105 (*occur, uv, ms*)
Nokami, J. *et al*, *Tetrahedron Lett.*, 1978, 4903 (*synth*)
Veith, H.J. *et al*, *Experientia*, 1978, **34**, 423 (*isol*)

3-Decen-1-ol D-00088

Slaterol

$H_3C(CH_2)_5CH{=}CHCH_2CH_2OH$

$C_{10}H_{20}O$ M 156.267

(*E*)-*form* [10339-60-3]

Secretion of the terrestrial slater *Porcellio scaber*. Oil. Bp_2 80-81°.

(*Z*)-*form* [10340-22-4]

Secretion of *P. scaber*. Oil. Bp_2 80-81°.

Cavill, G.W.K. *et al*, *Aust. J. Chem.*, 1966, **19**, 1495 (*isol, synth*)
Warwel, S. *et al*, *Synthesis*, 1975, **10**, 632 (*synth*)

4-Decen-1-ol D-00089

$H_3C(CH_2)_4CH{=}CHCH_2CH_2CH_2OH$

$C_{10}H_{20}O$ M 156.267

(*E*)-*form* [10339-62-5]

Constit. of Buchu leaf oil. Liq.

(Z)-form [57074-37-0]
Constit. of Buchu leaf oil.

Kaiser, R. *et al*, *J. Agric. Food Chem.*, 1975, **23**, 943.
Barrett, A.G.M. *et al*, *J. Org. Chem.*, 1991, **56**, 638 (*synth, ir, pmr, cmr, ms*)

5-Decen-1-ol D-00090

$H_3C(CH_2)_3CH{=}CH(CH_2)_3CH_2OH$

$C_{10}H_{20}O$ M 156.267

(E)-form [56578-18-8]
Pheromone of the moth *Anarsia lineatella*. Unpleasant smelling liq. Bp_{14} 114-115°, $Bp_{0.15}$ 83-85°.

(Z)-form [51652-47-2]
Unpleasant-smelling liq. Bp_{14} 114-115°, $Bp_{0.2}$ 100-105°.
Ac: [67446-07-5].
$C_{12}H_{22}O_2$ M 198.305
Pheromone of female moth *Agrotis fucosa*.
3-Methylbutanoyl: [37616-04-9].
$C_{15}H_{28}O_2$ M 240.385
Sex pheromone from the female pine emperor moth *Nudaurelia cytherea cytherea*.

Henderson, H.E. *et al*, *J. Chem. Soc., Chem. Commun.*, 1972, 686.
Roelofs, W. *et al*, *Environ. Entomol.*, 1975, **4**, 580 (*isol*)
Bestmann, H.J. *et al*, *Angew. Chem., Int. Ed. Engl.*, 1978, **17**, 768 (*isol, synth*)
Bestmann, H.J. *et al*, *Justus Liebigs Ann. Chem.*, 1981, 1705 (*synth*)
Buss, A.D. *et al*, *J. Chem. Soc., Perkin Trans. 1*, 1987, 2569 (*synth, ir, pmr, ms*)
Moiseenkov, A.M. *et al*, *J. Chem. Soc., Perkin Trans. 1*, 1991, 2639 (*synth, pmr, cmr*)

6-Decen-1-ol D-00091

$H_3CCH_2CH_2CH{=}CH(CH_2)_4CH_2OH$

$C_{10}H_{20}O$ M 156.267

(E)-form [38421-92-0]
Liq. $Bp_{0.8}$ 78-79°.

(Z)-form [68760-59-8]
Liq. $Bp_{0.3}$ 74-6°.

Svirskaya, P.I. *et al*, *J. Chem. Eng. Data*, 1979, **24**, 152 (*synth, ir, pmr*)
Brown, H.C. *et al*, *J. Org. Chem.*, 1988, **53**, 246 (*synth, ir, pmr, cmr*)

7-Decen-1-ol D-00092

$H_3CCH_2CH{=}CH(CH_2)_5CH_2OH$

$C_{10}H_{20}O$ M 156.267

(E)-form [52957-12-7]
Bp_7 92°, $Bp_{0.02}$ 75°.

(Z)-form [16504-66-8]
Ac:
$C_{12}H_{22}O_2$ M 198.305
Field attractant for many lepidoptera species.

Kita, M., *J. Org. Chem.*, 1957, **22**, 436 (*isol, synth*)
Svirskaya, P.I. *et al*, *J. Chem. Eng. Data*, 1979, **24**, 152 (*synth, ir, pmr*)
Priesner, E., *Z. Naturforsch., C*, 1989, **44**, 1061.

9-Decen-1-ol, 9CI D-00093

[13019-22-2]

$H_2C{=}CH(CH_2)_7CH_2OH$

$C_{10}H_{20}O$ M 156.267

Used in perfumery industry. Bp_8 108°, $Bp_{0.01}$ 60°. n_D^{20} 1.4480.
▷ HE2095000.
Ac: [50816-18-7].
$C_{12}H_{22}O_2$ M 198.305
$Bp_{0.09}$ 95-97°.
▷ AG5230000.
Phenylurethane: Mp 49-50°.

Hill, J.W. *et al*, *J. Am. Chem. Soc.*, 1933, **55**, 5023 (*synth*)
Williams, J.L.R. *et al*, *J. Org. Chem.*, 1958, **23**, 676 (*synth*)
Bestmann, H.J. *et al*, *Chem. Ber.*, 1975, **108**, 3582 (*synth*)
Subramanian, C.S. *et al*, *Tetrahedron Lett.*, 1978, 495 (*synth*)
Lewis, R.J., *Sax's Dangerous Properties of Industrial Materials*, 8th Ed., Van Nostrand-Reinhold, 1992, DAI400, DAI450.

4-Decen-9-olide D-00094

2,3,4,7,8,9,10-Hexahydro-10-methyl-2H-oxecin-2-one, 9CI.
9-Hydroxy-4-decenoic acid lactone

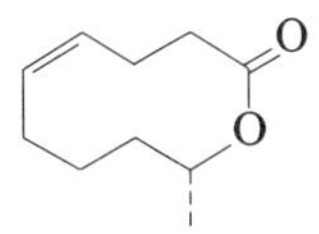

$C_{10}H_{16}O_2$ M 168.235

(R,Z)-form [69980-00-3] ***Phoracantholide J***
Isol. from the gland secretion of *Phorocantha synonyma*.

Moore, B.P. *et al*, *Aust. J. Chem.*, 1976, **29**, 1365 (*isol*)
Petrzilka, M., *Helv. Chim. Acta*, 1978, **61**, 3075 (*synth*)
Wakamatsu, T. *et al*, *J. Org. Chem.*, 1979, **44**, 2008 (*synth*)
Barbier, M., *J. Chem. Soc., Chem. Commun.*, 1982, 668 (*synth*)
Kitahara, T. *et al*, *Agric. Biol. Chem.*, 1983, **47**, 389 (*synth, abs config*)
Malherbe, R. *et al*, *J. Org. Chem.*, 1983, **48**, 860 (*synth*)
Suginome, M. *et al*, *Tetrahedron Lett.*, 1985, **26**, 3715 (*synth*)

3-Decen-2-one, 9CI D-00095

[10519-33-2]

$H_3C(CH_2)_5CH{=}CHCOCH_3$

$C_{10}H_{18}O$ M 154.252
Liq. d_4^{20} 0.847. $Bp_{15.3}$ 101.5-102.5°. n_D^{20} 1.4480.
Semicarbazone: Cryst. + $1H_2O$ (EtOH aq.). Mp 110-111°.
2,4-Dinitrophenylhydrazone: Mp 78°.

Murakami, K., *CA*, 1930, **24**, 2426 (*synth*)
Julia, M. *et al*, *Bull. Soc. Chim. Fr.*, 1975, 1363 (*synth*)

5-(1-Decenyl)dihydro-2(3*H*)-furanone, 9CI D-00096

5-Tetradecen-4-olide

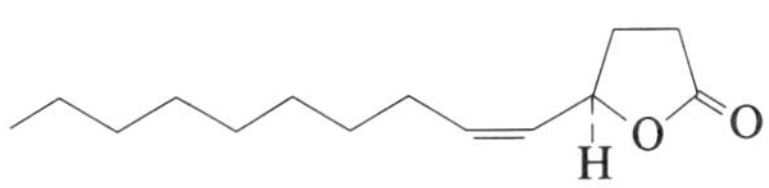

(*R*)-(*Z*)-*form*

$C_{14}H_{24}O_2$ M 224.342

(R)-(Z)-form [64726-91-6] ***Japonilure***
Isol. from the Japanese beetle *Popillia japonica*. Sex pheromone. $[\alpha]_D^{26}$ −69.6° (c, 1.0 in $CHCl_3$).

(±)-*(Z)-form* [72151-71-4]
$Bp_{0.1}$ 110°. Shows no biol. activity owing to the strong inhibition caused by the (*S*)-enantiomer.

Tumlinson, J.H. *et al*, *Science* (*Washington, D.C.*), 1977, **197**, 789 (*isol, synth*)
Sato, K. *et al*, *Agric. Biol. Chem.*, 1979, **43**, 1571 (*synth*)
Nishizawa, M. *et al*, *Tetrahedron Lett.*, 1981, **22**, 247 (*synth*)
Baker, R. *et al*, *J. Chem. Soc., Perkin Trans. 1*, 1982, 69 (*synth, spectra*)

Ramaswamy, S. *et al*, *Tetrahedron*, 1991, **47**, 1145 (*synth, bibl*)
Fukusaki, E. *et al*, *Tetrahedron*, 1991, **47**, 6223 (*synth, bibl*)
Ebata, T. *et al*, *Biosci., Biotechnol., Biochem.*, 1992, **56**, 818 (*synth*)

2-(9-Decenyl)-3-pentyloxirane, 9CI D-00097

11,12-Epoxy-1-heptadecene

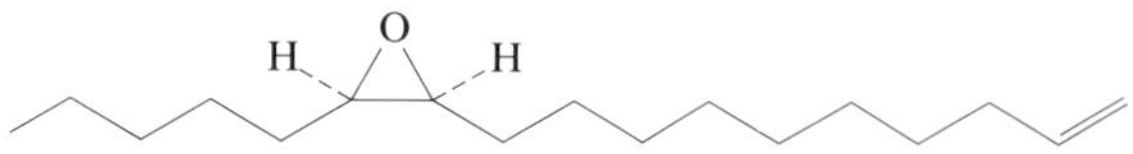

$C_{17}H_{32}O$ M 252.439

(2R,3R)-form [129436-95-9]

Isol. from the brown alga *Notheia anomala*. Unstable oil.

Barrow, R.A. *et al*, *Aust. J. Chem.*, 1990, **43**, 895.

1-Decen-3-yne D-00098

[33622-26-3]

$H_3C(CH_2)_5C{\equiv}CCH{=}CH_2$

$C_{10}H_{16}$ M 136.236

Bp_{15} 66-68°.

Cherkasov, L.N. *et al*, *CA*, 1971, **75**, 140152z.
Negishi, E. *et al*, *J. Org. Chem.*, 1984, **49**, 2629 (*synth, ir, pmr*)

1-Decylamine D-00099

1-Aminodecane. 1-Decanamine, 9CI

[2016-57-1]

$H_3C(CH_2)_8CH_2NH_2$

$C_{10}H_{23}N$ M 157.298

Mp 15.0°. Bp_{747} 218.0-218.5°.

▷ Highly toxic. HD6475000.

B,HCl: Mp 182-183°.

Guyer, A. *et al*, *Helv. Chim. Acta*, 1955, **38**, 1649 (*synth*)
Murr, B.L. *et al*, *J. Am. Chem. Soc.*, 1955, **77**, 1684 (*synth*)
Wolff, H. *et al*, *Ber. Bunsen-Ges. Phys. Chem.*, 1964, **68**, 529 (*ir*)
Malik, A.A. *et al*, *Synthesis*, 1989, 450 (*synth*)
Lewis, R.J., *Sax's Dangerous Properties of Industrial Materials*, 8th Ed., Van Nostrand-Reinhold, 1992, DAG600.

Decylbenzene, 9CI D-00100

1-Phenyldecane, 8CI

[104-72-3]

$Ph(CH_2)_9CH_3$

$C_{16}H_{26}$ M 218.381

Liq. Bp 293-294°, Bp_1 112°.

Bentley, F.F. *et al*, *Spectrochim. Acta*, 1959, 165 (*ir*)
Gray, F.W. *et al*, *J. Org. Chem.*, 1961, **26**, 209 (*synth*)
Lightner, D.A. *et al*, *Appl. Spectrosc.*, 1971, **25**, 253 (*ms*)
Wilkowski, Z. *et al*, *J. Chim. Phys. Phys.-Chim. Biol.*, 1974, **71**, 487 (*cmr*)
Sadykhov, K.I. *et al*, *CA*, 1981, **95**, 80369b (*synth*)
Sax, N.I., *Dangerous Properties of Industrial Materials*, 5th Ed., Van Nostrand-Reinhold, 1979, 531.

2-Decylcyclopropanehexanoic acid, 9CI D-00101

7,8-Methyleneoctadecanoic acid

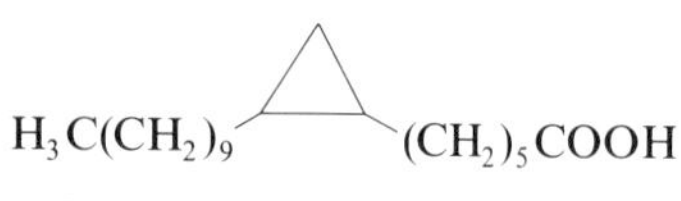

$C_{19}H_{36}O_2$ M 296.492

Both *cis*- and *trans*-forms have been synthesised as Me esters.

Christie, W.W. *et al*, *Chem. Phys. Lipids*, 1968, **2**, 196 (*synth, glc, ir, nmr*)
Christie, W.W., *Top. Lipid Chem.*, 1970, **1**, 4 (*rev*)
Gunstone, F.D. *et al*, *Chem. Phys. Lipids*, 1973, **10**, 303 (*synth, glc, nmr*)

1-Decyne, 9CI D-00102

Octylacetylene

[764-93-2]

$H_3C(CH_2)_7C{\equiv}CH$

$C_{10}H_{18}$ M 138.252

d_0^0 0.791. Mp −36°. Bp_{745} 181-182°, Bp_{10} 69-70°. n_D^{20} 1.4269.

Vaughn, T.M., *J. Am. Chem. Soc.*, 1934, **56**, 2064 (*synth*)
Org. React. (N.Y.), 1949, **5**, 1 (*synth*)
Eisner, D.B. *et al*, *J. Chem. Soc.*, 1951, 893 (*synth*)
Manion, J.J. *et al*, *Spectrochim. Acta*, 1961, **17**, 990 (*ir*)
Klein, J. *et al*, *Tetrahedron*, 1970, **26**, 2127 (*synth*)
Gendreau, V. *et al*, *J. Organomet. Chem.*, 1977, **142**, 1 (*synth*)

2-Decyne D-00103

[2384-70-5]

$H_3C(CH_2)_6C{\equiv}CCH_3$

$C_{10}H_{18}$ M 138.252

Liq. $Bp_{0.8}$ 61°.

Klein, J. *et al*, *Tetrahedron*, 1970, **26**, 2127 (*synth*)

3-Decyne, 9CI D-00104

[2384-85-2]

$H_3C(CH_2)_5C{\equiv}CCH_2CH_3$

$C_{10}H_{18}$ M 138.252

Liq. d^{21} 0.765. Bp 175-176°, Bp_{26} 82°. n_D^{21} 1.433.

Truchet, R., *Ann. Chim. (Paris)*, 1931, **16**, 309.
Wojtkowiak, B., *Ann. Chim. (Paris)*, 1964, **9**, 5.

4-Decyne, 9CI D-00105

[2384-86-3]

$H_3C(CH_2)_4C{\equiv}CCH_2CH_2CH_3$

$C_{10}H_{18}$ M 138.252

Liq. d^{17} 0.772. Bp_{19} 74-75°. n_D^{17} 1.436.

Truchet, R., *Ann. Chim. (Paris)*, 1931, **16**, 309.
Wojtkowiak, B., *Ann. Chim. (Paris)*, 1964, **9**, 5.

5-Decyne, 9CI D-00106

Dibutylacetylene

[1942-46-7]

$H_3C(CH_2)_3C{\equiv}C(CH_2)_3CH_3$

$C_{10}H_{18}$ M 138.252

Liq. d_4^{20} 0.769. Bp 177°, Bp_{48} 94-95°. n_D^{20} 1.4331.

Seifert, M., *Monatsh. Chem.*, 1948, **79**, 198.
Yamada, K. *et al*, *Tetrahedron Lett.*, 1975, 1961 (*synth*)
Tanaka, K. *et al*, *Tetrahedron Lett.*, 1978, 3103 (*synth*)

3-Decynoic acid, 9CI D-00107

[27007-78-9]

$H_3C(CH_2)_5C{\equiv}CCH_2COOH$

$C_{10}H_{16}O_2$ M 168.235

Cryst. (pet. ether). Mp 34-36°.

Gilman, N.W. *et al*, *Chem. Phys. Lipids*, 1974, **13**, 239 (*synth*)

4-Decynoic acid, 9CI D-00108

[16900-59-7]

$H_3C(CH_2)_4C{\equiv}CCH_2CH_2COOH$

$C_{10}H_{16}O_2$ M 168.235

Cryst. (pentane at −78°). Mp 36.5-38.0°.

Et ester:

$C_{12}H_{20}O_2$ M 196.289

$Bp_{0.2}$ 103-104°.

Anilide:

$C_{16}H_{21}NO$ M 243.348

Mp 105.0-105.5°.

Crombie, L., *J. Chem. Soc.*, 1955, 1007.
Kennedy, J. *et al*, *J. Chem. Soc.*, 1961, 4945 (*synth*)
Gilman, N.W. *et al*, *Chem. Phys. Lipids*, 1974, **13**, 239 (*synth*)

5-Decynoic acid, 9CI D-00109

[55182-79-1]

$H_3C(CH_2)_3C{\equiv}C(CH_2)_3COOH$

$C_{10}H_{16}O_2$ M 168.235

$Bp_{0.5}$ 113-116°.

tert-*Butyl ester:* Bp_3 112-114°.

Bohlmann, F. *et al*, *Chem. Ber.*, 1967, **100**, 3706 (*ms*)
Crandall, J.K. *et al*, *J. Organomet. Chem.*, 1973, **51**, 375.
Gilman, N.W. *et al*, *Chem. Phys. Lipids*, 1974, **13**, 239 (*synth*)

6-Decynoic acid, 9CI D-00110

[55182-82-6]

$H_3CCH_2CH_2C{\equiv}C(CH_2)_4COOH$

$C_{10}H_{16}O_2$ M 168.235

$Bp_{0.4}$ 130-135°.

Gilman, N.W. *et al*, *Chem. Phys. Lipids*, 1974, **13**, 239 (*synth*)

7-Decynoic acid D-00111

[54373-85-2]

$H_3CCH_2C{\equiv}C(CH_2)_5COOH$

$C_{10}H_{16}O_2$ M 168.235

Oil. $Bp_{0.1}$ 95-98° (98-100°).

Gilman, N.W. *et al*, *Synth. Commun.*, 1974, **4**, 199 (*synth*)
Gilman, N.W. *et al*, *Chem. Phys. Lipids*, 1974, **13**, 239 (*synth*)
Gunstone, F.D. *et al*, *Chem. Phys. Lipids*, 1976, **17**, 1 (*cmr*)

8-Decynoic acid D-00112

[55182-88-2]

$H_3CC{\equiv}C(CH_2)_6COOH$

$C_{10}H_{16}O_2$ M 168.235

Cryst. (pet. ether). Mp 38.5-39.0°. Mp 41.5-42.5°. $Bp_{0.2}$ 122°.

Et ester:

$C_{12}H_{20}O_2$ M 196.289

$Bp_{3.0}$ 102-104°.

Shishido, K. *et al*, *J. Org. Chem.*, 1962, **27**, 4073 (*synth*)
Gilman, N.W. *et al*, *Chem. Phys. Lipids*, 1974, **13**, 239 (*synth*)
Gunstone, F.D. *et al*, *Chem. Phys. Lipids*, 1976, **17**, 1 (*cmr*)

9-Decynoic acid, 8CI D-00113

[1642-49-5]

$HC{\equiv}C(CH_2)_7COOH$

$C_{10}H_{16}O_2$ M 168.235

Mp 24.5-25.1°. $Bp_{0.08}$ 105-107°.

Me ester:

$C_{11}H_{18}O_2$ M 182.262

$Bp_{0.2}$ 73.5°.

Black, H.K. *et al*, *Chem. Ind. (London)*, 1953, 40 (*synth*)
de Gaudemarais, M. *et al*, *Bull. Soc. Chim. Fr.*, 1962, 315 (*synth*)
Kraevskii, A.A. *et al*, *CA*, 1964, **61**, 5506 (*synth*)
Crombie, L. *et al*, *J. Chem. Soc., Perkin Trans. 1*, 1991, 567 (*synth, ir, pmr*)

1-Decyn-4-ol D-00114

[27907-00-2]

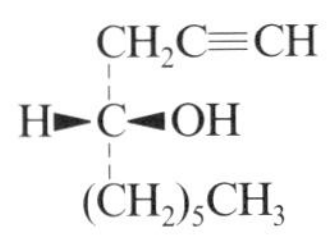

(*R*)-*form*

$C_{10}H_{18}O$ M 154.252

(R)-form

$[\alpha]_D^{21}$ +14.5° (neat), $[\alpha]_D$ 0° ($CHCl_3$).

(±)-form

Golden liq. Bp_{90} 106-110°. Bp_{18} 107-112° also reported.

Lauger, P. *et al*, *Helv. Chim. Acta*, 1959, **42**, 2379 (*synth*)
Buist, P.H. *et al*, *J. Org. Chem.*, 1991, **56**, 3449 (*synth, resoln, pmr*)

4-Decyn-1-ol D-00115

[69222-06-6]

$H_3C(CH_2)_4C{\equiv}CCH_2CH_2CH_2OH$

$C_{10}H_{18}O$ M 154.252

$Bp_{0.25}$ 75-80°.

Svirskaya, P.I. *et al*, *J. Chem. Eng. Data*, 1979, **24**, 152 (*synth*)
Körblová, E. *et al*, *Collect. Czech. Chem. Commun.*, 1985, **50**, 2284 (*synth, ir, ms*)

9-Decyn-1-ol D-00116

[17643-36-6]

$HC{\equiv}C(CH_2)_7CH_2OH$

$C_{10}H_{18}O$ M 154.252

$Bp_{0.5}$ 78-79°. n_D^{25} 1.4851.

Fyles, T.M. *et al*, *Can. J. Chem.*, 1977, **55**, 4135 (*synth, ir, pmr*)
Körblová, E. *et al*, *Collect. Czech. Chem. Commun.*, 1985, **50**, 2284 (*synth, ir, ms*)
Rao, A.V.R. *et al*, *Tetrahedron*, 1986, **42**, 4523 (*synth, ir, pmr*)

1-Decyn-5-one, 9CI D-00117

[55183-17-0]

$H_3C(CH_2)_4COCH_2CH_2C{\equiv}CH$

$C_{10}H_{16}O$ M 152.236

Oil. Bp_{18} 103°.

Ames, D.E. *et al*, *Chem. Phys. Lipids*, 1974, **13**, 223 (*synth*)
Haruta, J. *et al*, *J. Org. Chem.*, 1990, **55**, 4853 (*synth, ir, pmr*)

2-Deoxy-2-(3-hydroxytetradecanoylamino) glucose D-00118

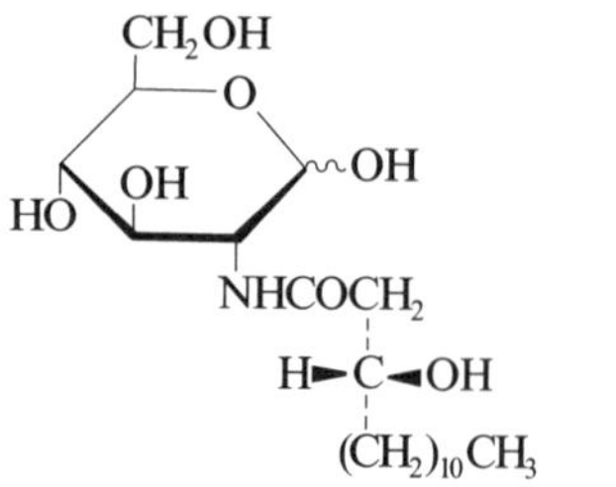

(R)-form

$C_{20}H_{39}NO_7$ M 405.531
Lipid A component analog.

(D,R)-form [68297-06-3]
$[\alpha]_D$ +35.0° (c, 0.24 in oxolane aq.).

(D,S)-form [77448-49-8]
$[\alpha]_D$ +23.7° (c, 0.25 in oxolane aq.).

Kiso, M. *et al*, *Carbohydr. Res.*, 1981, **88**, C5.

Diacetyl peroxide, 9CI D-00119

Acetyl peroxide
[110-22-5]

$H_3CCO—O–O–COCH_3$

$C_4H_6O_4$ M 118.089
Cryst. with very pungent odour. Mp 30°. Bp_{21} 63°. Aq. soln. slowly evolves O_2.

▷ Exceedingly explosive. Highly irritant. AP8500000.

Walker, O.J., *J. Chem. Soc.*, 1928, 2041.
Hawkins, E.G.E., *Org. Peroxides*, Spon, London, 1961 (*synth, props*)
Bretherick, L., *Handbook of Reactive Chemical Hazards*, 2nd Ed., Butterworths, London and Boston, 1979, 484.
Hazards in the Chemical Laboratory, (Bretherick, L., Ed.), 3rd Ed., Royal Society of Chemistry, London, 1981, 266.
Lewis, R.J., *Sax's Dangerous Properties of Industrial Materials*, 8th Ed., Van Nostrand-Reinhold, 1992, ACV500, ACV750.

4-*O*-(1,2-Diacylglyceryl)-*N,N,N*-trimethylhomoserine D-00120

DGTS

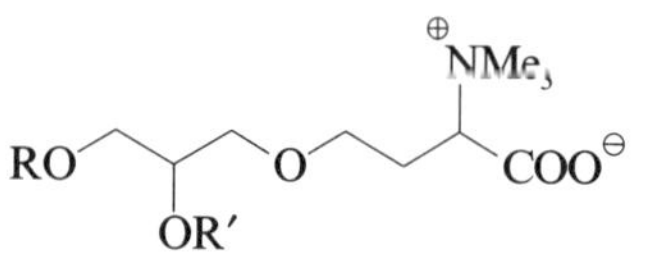

Betaine lipid present in most ferns and in some green algae, not found in higher plants.

Sato, N. *et al*, *Phytochemistry*, 1984, **23**, 1625 (*bibl*)

2,5-Dialkylpyrrolidines D-00121

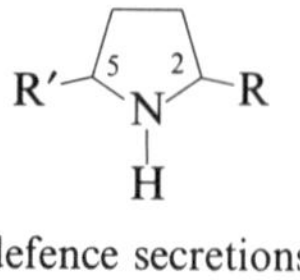

Components of the defence secretions of the ants *Solenopsis* and *Monomorium* spp. R, R′ = Alkyl groups, various dehydro analogues also found.

2-Ethyl-5-pentylpyrrolidine
$C_{11}H_{23}N$ M 169.309
1,2-Didehydro: [61772-94-9]. *5-Ethyl-3,4-dihydro-2-pentyl-2H-pyrrole, 9CI.* ***2-Ethyl-5-pentyl-Δ¹-pyrroline***
$C_{11}H_{21}N$ M 167.294
From *S. punctaticeps.*
1,5-Didehydro: [61772-95-0]. *2-Ethyl-3,4-dihydro-5-pentyl-2H-pyrrole, 9CI.* ***5-Ethyl-2-pentyl-Δ¹-pyrroline***
$C_{11}H_{21}N$ M 167.294
From *S. punctaticeps.*

2-Ethyl-5-heptylpyrrolidine
$C_{13}H_{27}N$ M 197.363
From *S. punctaticeps.*
1,2-Didehydro: ***2-Ethyl-5-heptyl-Δ¹-pyrroline***
$C_{13}H_{25}N$ M 195.347
From *S. punctaticeps.*
1,5-Didehydro: [61772-93-8]. *2-Ethyl-5-heptyl-3,4-dihydro-2H-pyrrole, 9CI.* ***5-Ethyl-2-heptyl-Δ¹-pyrroline***
$C_{13}H_{25}N$ M 195.347
From *S. punctaticeps.*

2-Ethyl-5-undecylpyrrolidine, 9CI
Dendrobates *Alkaloid 197B*
$C_{17}H_{35}N$ M 253.470
Constit. of the venom of *Monomorium* sp. Also isol. from skin extracts of the Colombian poison frog *Dendrobates histrionicus*: traces constit. of one population of *D. pumilio*. Oil. $[\alpha]_D^{27}$ +5.8° (c, 0.79 in $CHCl_3$). The *D.* sp. alkaloid to which the phys. props. refer, is the (2*S*,5*S*)-isomer.

2-Ethyl-5-(12-tridecenyl)pyrrolidine, 9CI
$C_{19}H_{37}N$ M 279.508
Constit. of the venom of *Monomorium* sp.

2-Butyl-5-pentylpyrrolidine [19501-71-4]
$C_{13}H_{27}N$ M 197.363
From *S. punctaticeps* and *M.* sp.

2-Butyl-5-heptylpyrrolidine [61772-92-7]
$C_{15}H_{31}N$ M 225.417
From *S. fugax*, *S. punctaticeps* and *M. latinode.*
N-*Me:* [83688-93-1]. ***2-Butyl-5-heptyl-1-methylpyrrolidine, 9CI***
$C_{16}H_{33}N$ M 239.443
Venom from *M. latinode.*
1,5-Didehydro: [83688-87-3]. *2-Butyl-5-heptyl-3,4-dihydro-2H-pyrrole, 9CI.* ***5-Butyl-2-heptyl-Δ¹-pyrroline***
$C_{15}H_{29}N$ M 223.401
Venom from *M. latinode.*

2-Butyl-5-nonylpyrrolidine
8″,9″-Didehydro: [120091-01-2]. ***2-Butyl-5-(8-nonenyl) pyrrolidine***
$C_{17}H_{33}N$ M 251.454
From *M.* sp.

2-Hexyl-5-pentylpyrrolidine [73427-38-0]
$C_{15}H_{31}N$ M 225.417
From *S. molesta* and *S. texanas.*
5′,6′-Didehydro: [58623-41-9]. ***2-(5-Hexenyl)-5-pentylpyrrolidine, 9CI***
$C_{15}H_{29}N$ M 223.401
Venom from *M. pharaonis.*

2-Hexyl-5-nonylpyrrolidine [81943-71-7]
$C_{19}H_{39}N$ M 281.524
From *M.* spp.
5′,6′,8″,9″-Tetradehydro: [81943-70-6]. ***2-(5-Hexenyl)-5-(8-nonenyl)pyrrolidine***
$C_{19}H_{35}N$ M 277.492
Component of venom from *M.* sp. and from *Chelaner antarcticus*. The *C. antarcticus* isolate was shown to have the *trans*-config.
5′,6′,8″,9″-Tetradehydro, N-*Me:* ***2-(5-Hexenyl)-1-methyl-5-(8-nonenyl)pyrrolidine***
$C_{20}H_{37}N$ M 291.519
Venom from *M.* sp.

2-Heptyl-5-(5-hexenyl)pyrrolidine
$C_{17}H_{33}N$ M 251.454
From *M.* sp.

2-(5-Hexenyl)-5-nonylpyrrolidine [100594-88-5]
$C_{19}H_{37}N$ M 279.508
From *M.* sp.

N-*Me:* ***2-(5-Hexenyl)-1-methyl-5-nonylpyrrolidine***
$C_{20}H_{39}N$ M 293.535
From *M.* sp.

1,2,8″,9″-Tetradehydro: ***2-(5-Hexenyl)-5-(8-nonenyl)-Δ^1-pyrroline***
$C_{19}H_{33}N$ M 275.476
From *M.* sp.

1,5,8″,9″-Tetradehydro: ***5-(5-Hexenyl)-2-(8-nonenyl)-Δ^1-pyrroline***
$C_{19}H_{33}N$ M 275.476
From *M.* sp.

[71732-77-9, 71732-78-0, 75372-27-9, 75372-32-6, 81943-78-4, 95018-45-4, 95018-46-5, 103833-14-3, 103833-64-3, 114640-28-7, 114640-29-8, 114640-32-3, 114640-33-4, 116558-86-2, 116558-88-4, 131682-66-1]

Pedder, D.J. *et al*, *Tetrahedron*, 1976, **32**, 2275.
Jones, T.H. *et al*, *Tetrahedron Lett.*, 1979, 1031.
Blum, M.S. *et al*, *Naturwissenschaften*, 1980, **67**, 144.
Jones, T.H. *et al*, *J. Chem. Ecol.*, 1982, **8**, 285.
Jones, T.H. *et al*, *Tetrahedron*, 1982, **38**, 1949.
Attygalle, A.B. *et al*, *Chem. Soc. Rev.*, 1984, **13**, 245.
Daly, J.W. *et al*, *J. Nat. Prod. (Lloydia)*, 1986, **49**, 265 (*isol, ms, struct, Dendrobates alkaloid 197B*)
Jones, T.H. *et al*, *J. Org. Chem.*, 1986, **51**, 2712.
Jones, T.H. *et al*, *J. Chem. Ecol.*, 1988, **14**, 35 (*isol*)
Jones, T.H. *et al*, *J. Nat. Prod. (Lloydia)*, 1990, **53**, 375 (*isol, pmr, cmr*)
Bäckvall, J.-E. *et al*, *J. Org. Chem.*, 1990, **55**, 826 (*synth*)
Machinaga, N. *et al*, *J. Org. Chem.*, 1991, **56**, 1386 (*synth*)

1,10-Dibromodecane, 9CI D-00122

Decamethylene dibromide
[4101-68-2]

$BrCH_2(CH_2)_8CH_2Br$

$C_{10}H_{20}Br_2$ M 300.076
Mp 27° (25.5-26°). $Bp_{0.3}$ 110°.

Org. Synth., 1940, **20**, 24 (*synth*)
Woolford, R.G., *Can. J. Chem.*, 1962, **40**, 1846 (*synth*)
Conin, J.M. *et al*, *Bull. Soc. Chim. Fr.*, 1963, 1925 (*synth*)
Woolford, R.G. *et al*, *Can. J. Chem.*, 1964, **42**, 1788 (*synth*)

1,12-Dibromododecane, 9CI D-00123

[3344-70-5]

$BrCH_2(CH_2)_{10}CH_2Br$

$C_{12}H_{24}Br_2$ M 328.130
Mp 40.5-41°. Bp_{15} 215°.
▷ JR2295000.

Nineham, A.W., *J. Chem. Soc.*, 1953, 2601 (*synth*)
Colonge, J. *et al*, *Bull. Soc. Chim. Fr.*, 1960, 763 (*synth*)
Woolford, R.G., *Can. J. Chem.*, 1962, **40**, 1846 (*synth*)
Ainscow, T.A. *et al*, *Tetrahedron*, 1987, **43**, 115 (*synth*)

1,20-Dibromoeicosane D-00124

[14296-16-3]

$BrCH_2(CH_2)_{18}CH_2Br$

$C_{20}H_{40}Br_2$ M 440.344
Mp 62-64°. $Bp_{0.05}$ 170-180°.

Woolford, R.G., *Can. J. Chem.*, 1962, **40**, 1846 (*synth*)
Natrajan, A. *et al*, *J. Am. Chem. Soc.*, 1987, **109**, 7477 (*synth, pmr, cmr*)

N,N-Dibromoformamide, 9CI D-00125

[42144-85-4]

$HCONBr_2$

$CHBr_2NO$ M 202.833
Yellow cryst. (CCl_4). Mp 96-98°.

Gottardi, W. *et al*, *Monatsh. Chem.*, 1973, **104**, 421 (*synth*)

1,17-Dibromoheptadecane D-00126

$BrCH_2(CH_2)_{15}CH_2Br$

$C_{17}H_{34}Br_2$ M 398.264
Needles (EtOH). Mp 38-38.4°. Bp_3 208-210°.

Chuit, P. *et al*, *Helv. Chim. Acta*, 1929, **12**, 854 (*synth*)

1,16-Dibromohexadecane, 9CI D-00127

[45223-18-5]

$BrCH_2(CH_2)_{14}CH_2Br$

$C_{16}H_{32}Br_2$ M 384.237
Needles (EtOH). Mp 56.2-56.7°. Bp_4 204-205°.

Chuit, P. *et al*, *Helv. Chim. Acta*, 1929, **12**, 852 (*synth*)
Woolford, R.G., *Can. J. Chem.*, 1962, **40**, 1846 (*synth*)

14,16-Dibromo-7,13,15-hexadecatrien-5-ynoic acid D-00128

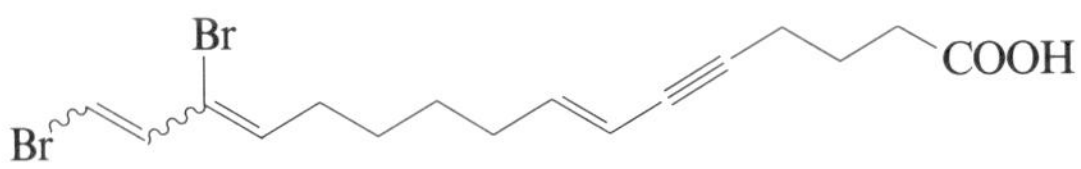

$C_{16}H_{20}Br_2O_2$ M 404.141
(7E,15Z)-form [70147-24-9]
Extract of the sponge *Xestospongia muta*. Shows tumour-inhibitory and CNS activity. Oil.

Schmitz, F.J. *et al*, *Tetrahedron Lett.*, 1978, 3637 (*isol*)

1,18-Dibromooctadecane, 9CI D-00129

Octadecamethylene bromide
[31772-06-2]

$BrCH_2(CH_2)_{16}CH_2Br$

$C_{18}H_{36}Br_2$ M 412.290
Needles (EtOH). Mp 63.5-64°. $Bp_{1.5}$ 205-207°.

Chuit, P. *et al*, *Helv. Chim. Acta*, 1929, **12**, 854 (*synth*)

9,10-Dibromooctadecanoic acid, 9CI D-00130

9,10-Dibromostearic acid
[19117-94-3]

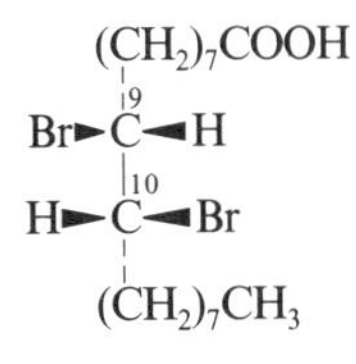

(9*R**,10*R**)-*form*
Relative configuration

$C_{18}H_{34}Br_2O_2$ M 442.273
(9R*,10R*)-form [18416-28-9]
(−)-threo-*form*
$[\alpha]_D$ −18.5°.

Strychnine salt: $[\alpha]_D$ −122°.

(9R*,10S*)-form [18485-62-6]

(−)-erythro-*form*

$[\alpha]_D$ −27.75°.

Strychnine salt: $[\alpha]_D$ −128°.

(9RS,10RS)-form [53323-97-0]

(±)-threo-*form. Elaidodibromostearic acid. Elaidic acid dibromide*

Mp 31° (27°).

(9RS,10SR)-form

(±)-erythro-*form. Oleodibromostearic acid. Oleic acid dibromide*

Mp 28.5-29.5°.

Manuyama, T. *et al, Chem. Zentralbl.*, 1931, **2**, 2594; 1932, **1**, 2307 (*config*)
Ucciani, U. *et al, Bull. Soc. Chim. Fr.*, 1967, 2018 (*synth*)

1,15-Dibromopentadecane, 9CI — D-00131

Pentadecamethylene bromide

$BrCH_2(CH_2)_{13}CH_2Br$

$C_{15}H_{30}Br_2$ M 370.210

Mp 27.2-27.5°. Bp_2 192°.

Chuit, P. *et al, Helv. Chim. Acta*, 1929, **12**, 852 (*synth*)

1,14-Dibromotetradecane, 9CI — D-00132

Tetradecamethylene bromide

[37688-96-3]

$BrCH_2(CH_2)_{12}CH_2Br$

$C_{14}H_{28}Br_2$ M 356.183

Cryst. (EtOH). Mp 50.4°. $Bp_{0.05}$ 135°.

Chuit, P., *Helv. Chim. Acta*, 1926, **9**, 268, 272 (*synth*)
Müller, A. *et al, Ber.*, 1938, **71**, 691 (*synth*)

1,13-Dibromotridecane, 9CI — D-00133

Tridecamethylene bromide

[31772-05-1]

$BrCH_2(CH_2)_{11}CH_2Br$

$C_{13}H_{26}Br_2$ M 342.156

d^{15} 1.276. Mp 8-10°. Bp_9 185-187°.

Chuit, P., *Helv. Chim. Acta*, 1926, **9**, 271 (*synth*)
Müller, A., *Ber.*, 1934, **67**, 297 (*synth*)

1,11-Dibromoundecane, 9CI — D-00134

1,11-Dibromohendecane

[16696-65-4]

$BrCH_2(CH_2)_9CH_2Br$

$C_{11}H_{22}Br_2$ M 314.103

d^{15} 1.332. Mp −10.6°. Bp_{12} 179°.

Chuit, P., *Helv. Chim. Acta*, 1926, **9**, 267 (*synth*)

2,3-Dichlorobutanedioic acid, 9CI — D-00135

2,3-Dichlorosuccinic acid, 8CI

[19922-87-3]

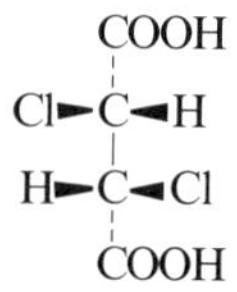

(2*R*,3*R*)-*form*
Absolute configuration

$C_4H_4Cl_2O_4$ M 186.979

(2R,3R)-form

Di-Me ester:

$C_6H_8Cl_2O_4$ M 215.032

Cryst. (MeOH). Mp 62-63°. $Bp_{0.13}$ 117-118°. $[\alpha]_D^{21}$ −114.1° (c, 6.63 in CS_2).

(2RS,3RS)-form [1114-09-6]

(±)-*form*

Prisms. Mp 175° dec.

Anhydride: 3,4-Dichloro-3,4-dihydro-2,5-furandione

$C_4H_2Cl_2O_3$ M 168.963

Cryst. Mp 95°.

(2RS,3SR)-form [3856-37-9]

meso-*form*

Prisms. Mp 217-218° dec.

Di-Me ester: [1114-22-3].

Cryst. Mp 31.5-32°.

[20688-20-4, 62173-55-1]

Kirchoff, H., *Justus Liebigs Ann. Chem.*, 1894, **280**, 211 (*synth*)
van der Riet, B., *Justus Liebigs Ann. Chem.*, 1894, **280**, 219 (*anhydride*)
Kuhn, R. *et al, Ber.*, 1928, **61**, 504 (*synth*)
Walther, H. *et al, Chem. Ber.*, 1958, **91**, 630 (*synth*)
Fuer, H. *et al, J. Org. Chem.*, 1959, **24**, 81 (*anhydride*)
Auram, M. *et al, Chem. Ber.*, 1964, **97**, 372 (*synth*)
Larsson, E., *Org. Magn. Reson.*, 1974, **6**, 103 (*pmr*)

1,10-Dichlorodecane, 9CI — D-00136

Decamethylene chloride

[2162-98-3]

$ClCH_2(CH_2)_8CH_2Cl$

$C_{10}H_{20}Cl_2$ M 211.173

Liq. Mp 15.6°. Bp_{28} 167-168°.

Huber, N.F., *J. Am. Chem. Soc.*, 1951, **73**, 2730 (*synth*)
Grogan, C.H. *et al, J. Org. Chem.*, 1953, **18**, 728 (*synth*)

1,12-Dichlorododecane, 9CI — D-00137

Dodecamethylene dichloride

[3922-28-9]

$ClCH_2(CH_2)_{10}CH_2Cl$

$C_{12}H_{24}Cl_2$ M 239.227

Mp 29°. Bp_1 170 172°.

v. Braun, J. *et al, Ber.*, 1911, **44**, 1474 (*synth*)
U.S. Pat., 3 111 545, (1963); *CA*, **60**, 4008 (*synth*)
Netherlands Pat., 6 411 864, (1965); *CA*, **63**, 11358 (*synth*)
Kazno, S. *et al, Bull. Chem. Soc. Jpn.*, 1966, **39**, 480 (*synth*)

Dicranenone A — D-00138

[87403-25-6]

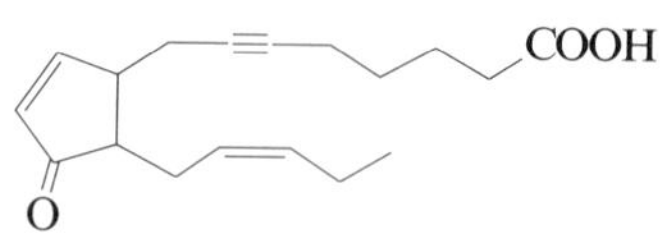

$C_{18}H_{24}O_3$ M 288.386

Isol. from the Japanese moss *Dicranum scoparium*. Shows some antibiotic activity. Oil. $[\alpha]_D^{25.5}$ +191° (c, 0.3 in MeOH). Stereochem. prob. as in the prostaglandin series.

Ichikawa, T. *et al, Tetrahedron Lett.*, 1983, **24**, 3337 (*isol, ir, uv, pmr, cmr, cd*)

Dicranenone B$_1$ **D-00139**

[87403-24-5]

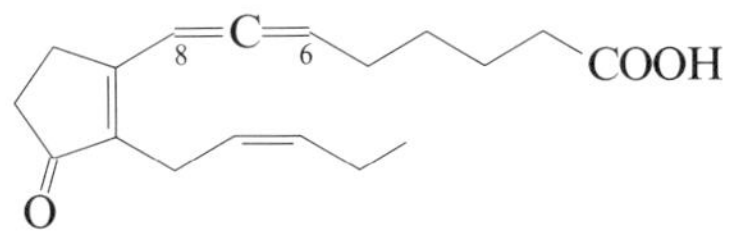

$C_{18}H_{24}O_3$ M 288.386

Tentative struct. Isol. from the Japanese moss *Dicranum japonicum*. Shows some antibiot. activity.

6,7,7,8-Tetrahydro: [93633-61-5]. *Tetrahydrodicranenone B$_1$*
$C_{18}H_{28}O_3$ M 292.417
Isol. from Japanese mosses. Shows antimicrobial activity. Pale yellow oil.

[72458-41-4]

Vick, B.A. *et al*, *Lipids*, 1979, **14**, 734.
Ichikawa, T. *et al*, *Tetrahedron Lett.*, 1983, **24**, 3337 (*isol, pmr, ir*)
Sakai, K. *et al*, *Tetrahedron Lett.*, 1985, **26**, 2089 (*isol*)
Casy, G. *et al*, *J. Chem. Soc., Chem. Commun.*, 1988, 454 (*synth*)
Moody, C.J. *et al*, *J. Chem. Soc., Perkin Trans. 1*, 1988, 1401 (*synth, ir, pmr, ms*)

Didecyl sulfate **D-00140**

Sulfuric acid didecyl ester

[66186-16-1]

$H_3C(CH_2)_9OSO_2O(CH_2)_9CH_3$

$C_{20}H_{42}O_4S$ M 378.615

Mp 37.6-37.8°.

Barkenbus, C. *et al*, *J. Am. Chem. Soc.*, 1934, **56**, 1204 (*synth*)

1,3-Didodecanoyl-5-tetradecanoylbenzene **D-00141**

$C_{44}H_{76}O_3$ M 653.083

Constit. of root bark of *Cochlospermum planchonii*. Cryst. Mp 31-33°.

Addae-Mensak, I. *et al*, *Justus Liebigs Ann. Chem.*, 1985, 1284.

Didodecylamine, 8CI **D-00142**

N-*Dodecyl-1-dodecanamine, 9CI*

[3007-31-6]

$H_3C(CH_2)_{11}NH(CH_2)_{11}CH_3$

$C_{24}H_{51}N$ M 353.674

Mp 55-56°. Bp$_{27}$ 263-265°, Bp$_{0.7}$ 195°.

B,HCl: [2486-85-3].
Cryst. (2-propanol). Mp 207.8-210.8° dec.
B,HI: Mp 232.8-234° dec.

Wojcik, B. *et al*, *J. Am. Chem. Soc.*, 1934, **56**, 2419 (*synth*)
Petrillo, V.C., *J. Am. Chem. Soc.*, 1951, **73**, 2381 (*derivs*)
Sax, N.I., *Dangerous Properties of Industrial Materials*, 5th Ed., Van Nostrand-Reinhold, 1979, 569.

Didodecyl phosphate **D-00143**

Didodecyl hydrogen phosphate. Dilauryl phosphate. Didodecyl phosphoric acid

[7057-92-3]

$[H_3C(CH_2)_{11}O]_2P(O)OH$

$C_{24}H_{51}O_4P$ M 434.638

Cryst. (MeOH or Et_2O). Mp 59°.

Brown, D.A. *et al*, *J. Chem. Soc.*, 1955, 1584 (*synth*)
Ramirez, F. *et al*, *Phosphorus Sulfur Relat. Elem.*, 1978, **4**, 43 (*synth*)

Didodecyl sulfate **D-00144**

Sulfuric acid didodecyl ester, 9CI. Dilauryl sulfate

[2649-11-8]

$H_3C(CH_2)_{11}OSO_2O(CH_2)_{11}CH_3$

$C_{24}H_{50}O_4S$ M 434.723

Cryst. (MeOH). Mp 48.4-48.5°.

Barkenbus, C. *et al*, *J. Am. Chem. Soc.*, 1934, **56**, 1204 (*synth*)
East Ger. Pat., 32 780, (1965); *CA*, **63**, 6863a (*synth*)

Didodecyl sulfide **D-00145**

1,1'-Thiobisdodecane, 9CI. Dodecyl sulfide, 8CI

[2469-45-6]

$H_3C(CH_2)_{11}S(CH_2)_{11}CH_3$

$C_{24}H_{50}S$ M 370.725

Antioxidant, stabiliser. Cryst. (Me_2CO). Mp 38-40°. Bp$_4$ 260-263°.

Bezzi, S. *et al*, *Gazz. Chim. Ital.*, 1950, **80**, 180 (*synth, bibl, use*)
Mori, K. *et al*, *J. Org. Chem.*, 1969, **34**, 4170 (*synth*)

2,3-Dihydro-5,7-dihydroxy-2-(4,7,10,13,16-nonadecapentaenylidene)-4*H*-1-benzopyran-4-one **D-00146**

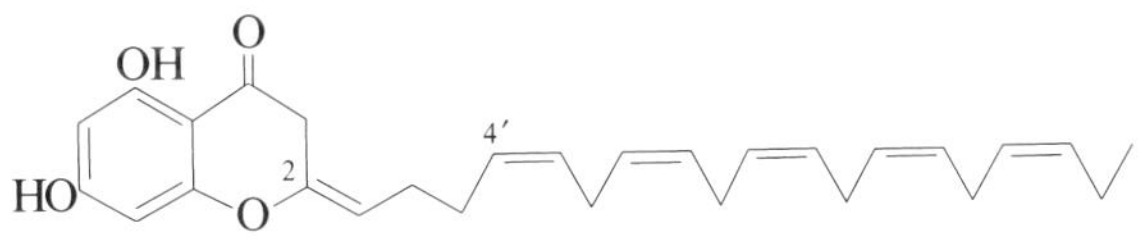

$C_{28}H_{34}O_4$ M 434.574

Mp 45-46°.

(2E,4'Z,7'Z,10'Z,13'Z,16'Z)-form
Isol. from the brown alga *Zonaria tournefortii*.

Tringali, C. *et al*, *Gazz. Chim. Ital.*, 1982, **112**, 465 (*isol, struct*)

2,3-Dihydro-5,7-dihydroxy-2-pentadecylidene-4*H*-1-benzopyran-4-one **D-00147**

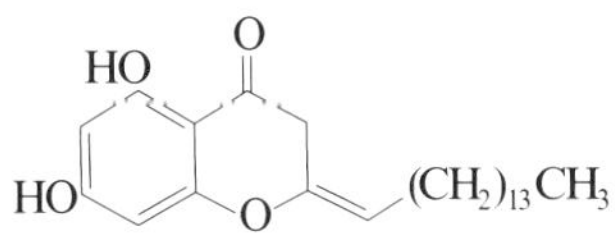

$C_{24}H_{36}O_4$ M 388.546

(E)-form
Isol. from the brown alga *Zonaria tournefortii*. Mp 67-68°.

Tringali, C. *et al*, *Gazz. Chim. Ital.*, 1982, **112**, 465.

3,4-Dihydro-6,8-dihydroxy-3-undecyl-1*H*-2-benzopyran-1-one, 9CI **D-00148**

3,4-Dihydro-6,8-dihydroxy-3-undecylisocoumarin

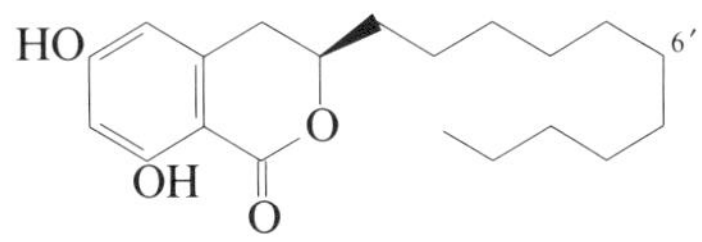

(*R*)-*form*

$C_{20}H_{30}O_4$ M 334.455

(R)-form [128232-78-0]
Constit. of *Ononis natrix* ssp. *hispanica*. Cryst. (hexane/CH_2Cl_2). Mp 104-105°. $[\alpha]_D$ −17.6° (c, 1.12 in $CHCl_3$).
6-Me ether: [88510-01-4]. ***3,4-Dihydro-8-hydroxy-6-methoxy-3-undecylisocoumarin***
$C_{21}H_{32}O_4$ M 348.481
Constit. of *O. natrix* ssp. *hispanica*. Cryst. (MeOH). Mp 98°. $[\alpha]_D^{24}$ −23.2° (c, 0.31 in $CHCl_3$).
6′S-Hydroxy: [129277-47-0]. ***3,4-Dihydro-6,8-dihydroxy-3-(6-hydroxyundecyl)isocoumarin***
$C_{20}H_{30}O_5$ M 350.454
Constit. of *O. natrix* spp. *hispanica*. Oil. $[\alpha]_D$ −2.4° (c, 1 in $CHCl_3$). Diagram in ref. inconsistent with name.
6′-Hydroxy, 6-Me ether: ***3,4-Dihydro-8-hydroxy-3-(6-hydroxyundecyl)-6-methoxyisocoumarin***
$C_{21}H_{32}O_5$ M 364.481
Constit. of *O. natrix*. Cryst. (hexane/CH_2Cl_2). Mp 65-66°. $[\alpha]_D$ −15.3° (c, 0.51 in $CHCl_3$).
6′-Oxo: [128232-79-1]. ***3,4-Dihydro-6,8-dihydroxy-3-(6-oxoundecyl)isocoumarin***
$C_{20}H_{28}O_5$ M 348.438
Constit. of *O. natrix*. Cryst. (hexane/CH_2Cl_2). Mp 86-87°.
6′-Oxo, 6-Me ether: [128232-81-5]. ***3,4-Dihydro-8-hydroxy-6-methoxy-3-(6-oxoundecyl)isocoumarin***
$C_{21}H_{30}O_5$ M 362.465
Constit. of *O. natrix*. Cryst. (hexane). Mp 73-74°. $[\alpha]_D$ −20.7° (c, 0.58 in $CHCl_3$).

[128232-80-4, 128232-82-6]

Feliciano, A.S. *et al*, *Phytochemistry*, 1983, **22**, 2031; 1990, **29**, 945 (*isol, pmr, cmr*)
Barrero, A.F. *et al*, *Phytochemistry*, 1990, **29**, 1967 (*isol*)

4,5-Dihydro-4-hydroxy-5-(9-hydroxy-1,3,5,7-decatetraynyl)-2(3*H*)-furanone, 9CI D-00149

3,4,13-Trihydroxy-5,7,9,11-tetradecatetraynoic acid γ-lactone. 3,13-Dihydroxy-5,7,9,11-tetradecatetrayne-1,4-olide
[83475-38-1]

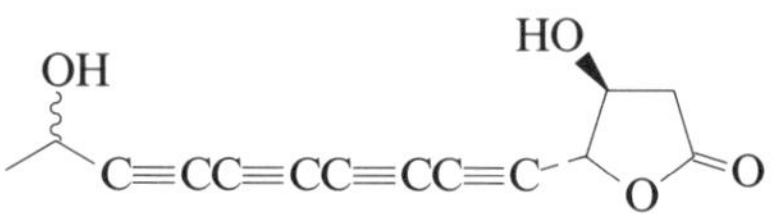

$C_{14}H_{10}O_4$ M 242.231
Minor metabolite isol. from *Mycena viridimarginata*.

Jente, R. *et al*, *Phytochemistry*, 1985, **24**, 553 (*isol, struct, spectra*)

7,8-Dihydrolycopene D-00150

7,8-Dihydro-ψ,ψ-carotene, 9CI. ***Neurosporene.*** *Neurosporin. Flavorhodin. Proneurosporene*
[502-64-7]

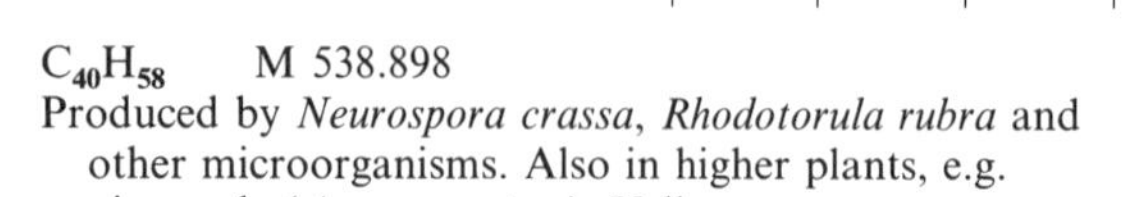

$C_{40}H_{58}$ M 538.898
Produced by *Neurospora crassa*, *Rhodotorula rubra* and other microorganisms. Also in higher plants, e.g. pineapple (*Ananas sativus*). Yellow-orange cryst. (EtOH). Mp 124° (115-116°).
(9Z,7′Z,9′Z)-form [10467-46-6]
Isol. from *Lycopersicon esculentum* var. 'Tangella'. Dark red viscous oil. λ_{max} 410, 433 nm (hexane).

Davis, J.B. *et al*, *J. Chem. Soc. C*, 1966, 2154 (*synth, pmr*)
Suzue, G. *et al*, *Biochim. Biophys. Acta*, 1969, **176**, 653 (*biosynth*)
Karrer, W. *et al*, *Konstitution und Vorkommen der Organischen Pflanzenstoffe*, 2nd Ed., Birkhäuser Verlag, Basel, 1972-1985, no. 1817b (*occur*)
Clough, J.M. *et al*, *J. Chem. Soc., Perkin Trans. 1*, 1983, 3011 (*cmr, ms*)
Frecknell, E.A. *et al*, *Phytochemistry*, 1984, **23**, 1707 (*isol, occur*)
Straub, O. *et al*, *Key to Carotenoids*, 2nd Ed., Birkhauser Verlag, Basel and Boston, 1987, 34 (*bibl*)

Dihydro-4-methyl-5-pentyl-2(3*H*)-furanone D-00151

Tetrahydro-4-methyl-5-pentyl-2-furanone. 3-Methyl-4-nonanolide. Cognac lactone

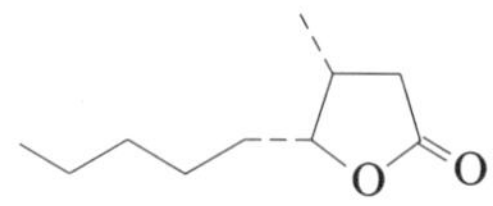

$C_{10}H_{18}O_2$ M 170.251
Higher homologue of 5-Butyldihydro-4-methyl-2(3*H*)-furanone, B-00064 found in cognac, presumably in an enantiomeric form. Oil.

Ortuño, R.M. *et al*, *Tetrahedron*, 1987, **43**, 4497 (*synth*)
Jefford, C.W. *et al*, *Helv. Chim. Acta*, 1989, **72**, 1362 (*synth, ir, pmr, cmr, bibl*)
Miyata, O. *et al*, *Chem. Pharm. Bull.*, 1992, **40**, 2579 (*synth*)

2,5-Dihydro-2-(6,8-nonadien-2,4-diynylidene)furan D-00152

10,13-Epoxy-1,3,9,11-tridecatetraene-5,7-diyne
[78876-47-8]

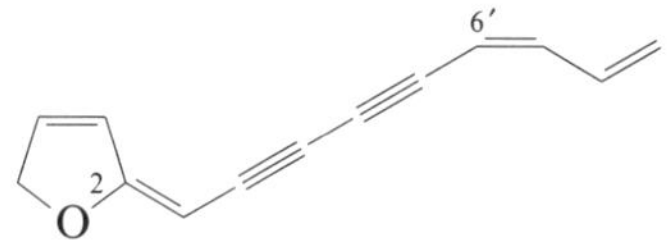

$C_{13}H_{10}O$ M 182.221
(2′E,6Z)-form [15400-64-3]
Constit. of the root of *Carlina vulgaris* and *C. diae*. Yellow oil. $Bp_{0.01}$ 120° sl. dec.

[16548-48-4]

Bohlmann, F. *et al*, *Chem. Ber.*, 1967, **100**, 1507 (*isol, struct, ir, uv, pmr, ms*)
Bohlmann, F. *et al*, *Phytochemistry*, 1981, **20**, 823 (*isol*)

Dihydro-5-(2,5-octadienyl)-2(3*H*)-furanone, 9CI D-00153

6,9-Dodecadien-4-olide. 4-Hydroxy-6,9-dodecadienoic acid lactone

$C_{12}H_{18}O_2$ M 194.273
(Z,Z)-form [82934-81-4]
Isol. from tuberose absolute (from *Polianthes tuberosa*). Oil.
5′,6′-Dihydro: see *Dihydro-5-(2-octenyl)-2(3H)-furanone, D-00155*

Maurer, B. *et al*, *Helv. Chim. Acta*, 1982, **65**, 462 (*isol, synth*)

Dihydro-5-(1-octenyl)-2(3*H*)-furanone D-00154

5-Dodecen-4-olide

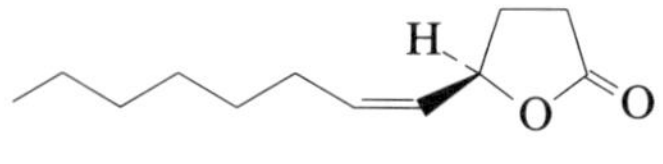

$C_{12}H_{20}O_2$ M 196.289

(R,Z)-form [139477-52-4]

Isol. from the beetle *Anomala cuprea*. Sex pheromone. $Bp_{0.2}$ 115°. $[\alpha]_D^{24}$ −80.9° (c, 0.99 in $CHCl_3$), $[\alpha]_D^{24}$ −61.8° (c, 1 in hexane).

[139477-53-5]

Leal, W.S., *Naturwissenschaften*, 1991, **78**, 521 (*isol, struct*)
Fukusaki, E. *et al*, *Biosci., Biotechnol., Biochem.*, 1992, **56**, 1160 (*synth*)

Dihydro-5-(2-octenyl)-2(3*H*)-furanone, 9CI **D-00155**

5-(2-Octenyl)tetrahydrofuran-2-one. 6-Dodecen-4-olide

[15456-69-6]

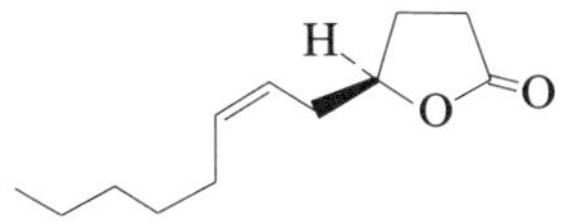

(*R,Z*)-*form*

$C_{12}H_{20}O_2$ M 196.289

Isol. from butterfat and from the tarsal organs of deer. Intermed. in synth. of 5,12-Dihydroxy-6,8,10,14-eicosatetraenoic acid, D-00171.

(R,Z)-form

$[\alpha]_D^{20}$ −16.1° (c, 0.3 in MeOH).

(S,Z)-form [63357-98-2]

$[\alpha]_D^{20}$ +15.0° (c, 0.1 in MeOH).

[18679-18-0]

Brownlee, R.G. *et al*, *Nature (London)*, 1969, **221**, 284 (*isol*)
Sucrow, W. *et al*, *Chem. Ber.*, 1975, **108**, 3518 (*synth*)
Meuller-Schwarze, D. *et al*, *J. Chem. Ecol.*, 1976, **2**, 389; 1978, **4**, 247.
Burger, B.V. *et al*, *Z. Naturforsch., C*, 1977, **32**, 49 (*synth*)
Ravid, U. *et al*, *Tetrahedron*, 1978, **34**, 1449 (*synth*)
Dormán, G. *et al*, *J. Chem. Soc., Perkin Trans. 1*, 1989, 1543 (*use*)

Dihydro-5-octyl-2(3*H*)-furanone, 9CI **D-00156**

4-Hydroxydodecanoic acid γ-lactone. 4-Dodecanolide. γ-Dodecalactone. γ-Dodecanolactone

[2305-05-7]

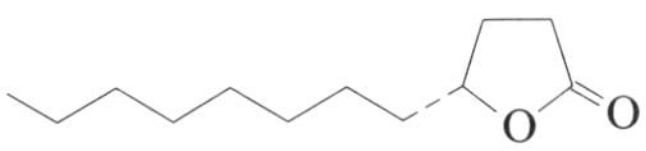

(*R*)-*form*

$C_{12}H_{22}O_2$ M 198.305

Found in the pygidial glands of the beetles *Bledius mandibularis* and *B. spectabilis*. Also present in several food aromas.

▷ LU3600000.

(R)-form [69830-91-7]

$[\alpha]_D$ +33° (c, 0.73 in MeOH).

(S)-form [69830-92-8]

$[\alpha]_D$ −32.8° (c, 1.6 in MeOH).

(±)-*form* [57084-18-1]

Used in perfumery and flavour industries. Oil. Bp_{11} 170-171°.

Muys, G.T. *et al*, *Nature (London)*, 1962, **194**, 995.
Wheeler, J.M. *et al*, *Tetrahedron Lett.*, 1972, 4635 (*isol*)
Sucrow, W. *et al*, *Chem. Ber.*, 1975, **108**, 3518 (*synth*)
Opdyke, D.L.J., *Food Cosmet. Toxicol.*, 1976, **14**, 751 (*rev, tox*)
Nakamura, E. *et al*, *J. Am. Chem. Soc.*, 1977, **99**, 7360 (*synth*)
Vigneron, J.P. *et al*, *Tetrahedron Lett.*, 1980, **21**, 1735.
Solladie, G. *et al*, *J. Org. Chem.*, 1982, **47**, 91.
Chattopadhyay, S. *et al*, *Tetrahedron*, 1990, **46**, 3667 (*synth*)
Fronza, G. *et al*, *J. Chem. Soc., Perkin Trans. 1*, 1991, 2977 (*synth*)
Cardillo, R. *et al*, *J. Org. Chem.*, 1991, **56**, 5237 (*synth*)

Dihydro-5-(2-pentenyl)-2(3*H*)-furanone **D-00157**

6-Nonen-4-olide

[82934-80-3]

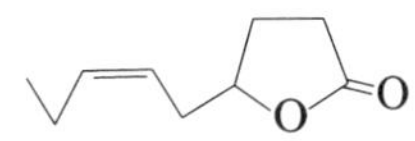

$C_9H_{14}O_2$ M 154.208

(Z)-form

Isol. from tuberose absolute (from *Polianthes tuberosa*). Oil.

Maurer, B. *et al*, *Helv. Chim. Acta*, 1982, **65**, 462 (*isol, synth*)

5,6-Dihydro-6-(2-pentenyl)-2*H*-pyran-2-one, 9CI **D-00158**

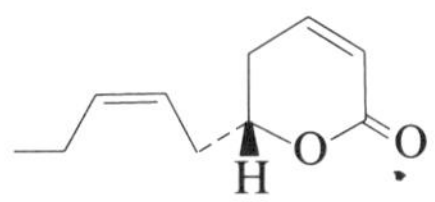

$C_{10}H_{14}O_2$ M 166.219

(R,Z)-form [60503-15-3] ***Tuberolactone***

Constit. of leaves of *Polianthes tuberosa*. $[\alpha]_D$ −77.3° (c, 0.67 in $CHCl_3$).

Kaiser, R. *et al*, *Tetrahedron Lett.*, 1976, 1659 (*struct*)
Fehr, C. *et al*, *Helv. Chim. Acta*, 1981, **64**, 1247 (*synth*)
Cassani, G. *et al*, *Tetrahedron Lett.*, 1983, **24**, 2513 (*synth*)
Hoffmann, R.W. *et al*, *Tetrahedron Lett.*, 1983, **24**, 3209 (*synth*)

Dihydro-5-pentyl-2(3*H*)-furanone, 9CI, 8CI **D-00159**

*γ-Nonalactone. 4-Hydroxynonanoic acid γ-lactone. γ-Amylbutyrolactone. 1,4-Nonanolide. 5-Pentyldihydro-2(3*H*)-*furanone. Apricolin. Prunolide. γ-Pelargonolactone. Cocos aldehyde* (*misleading*). *Aldehyde* C_{18} (*misleading*)

[104-61-0]

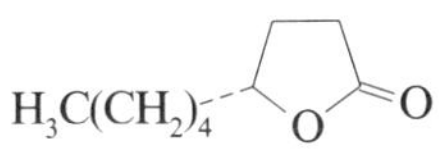

(*R*)-*form*

$C_9H_{16}O_2$ M 156.224

Isol. from ethereal oil of *Rosa rugosa*.

▷ LU3675000.

(R)-form [63357-96-0]

Repellant for cats and dogs: attractant for rice weevil *Sitophilus zeamais*. Oil. Bp_{12} 108°. $[\alpha]_D^{24}$ +47.5° (c, 1.82 in MeOH). ν_{max} 1770 cm^{-1}.

(S)-form [63357-97-1]

$[\alpha]_D^{24}$ −47.9° (c, 1.70 in MeOH).

(±)-*form* [57084-16-9]

Perfumery and flavouring ingredient. Bp_{12} 134-134.5°, $Bp_{1.5}$ 93-94°.

[82373-92-0]

Terai, Y. *et al*, *Bull. Chem. Soc. Jpn.*, 1956, **29**, 822 (*synth*)
Sakai, T. *et al*, *CA*, 1962, **58**, 8845 (*isol*)
Goldwhite, H. *et al*, *Tetrahedron*, 1964, **20**, 1657 (*synth, ir*)
Honkanen, E. *et al*, *Acta Chem. Scand.*, 1965, **19**, 370 (*ms*)
Oser, B.L. *et al*, *Food Cosmet. Toxicol.*, 1965, **3**, 563 (*rev*)
Opdyke, D.L.J., *Food Cosmet. Toxicol.*, 1975, **13**, 889 (*rev, tox*)
Sataro, M. *et al*, *Bull. Chem. Soc. Jpn.*, 1980, **53**, 770 (*synth*)
Miyashita, Y. *et al*, *Agric. Biol. Chem.*, 1981, **45**, 2521 (*synth*)
Cardellach, J. *et al*, *J. Heterocycl. Chem.*, 1984, **21**, 327 (*synth*)
Ballini, R. *et al*, *Synthesis*, 1987, 711 (*synth, ir, pmr*)
Cardillo, R. *et al*, *J. Org. Chem.*, 1991, **56**, 5237 (*synth*)
Lewis, R.J., *Sax's Dangerous Properties of Industrial Materials*, 8th Ed., Van Nostrand-Reinhold, 1992, CNF250.

5,15-Dihydroperoxy-6,8,11,13-eicosatetraenoic acid D-00160

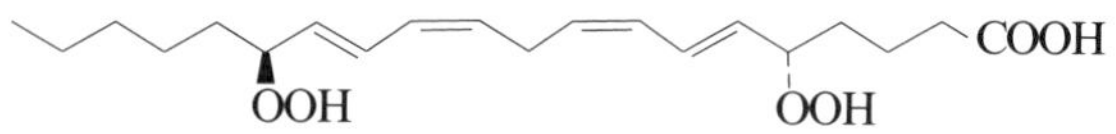

$C_{20}H_{32}O_6$ M 368.469

(5S,6E,8Z,11Z,13E,15S)-form [76964-60-8]
Metab. of arachidonic acid.

van Os, C.P.A. *et al*, *Biochim. Biophys. Acta*, 1981, **663**, 177.

5,6-Dihydro-6-(2,4,6-trihydroxyheneicosyl)-2*H*-pyran-2-one, 9CI D-00161

7,9,11-Trihydroxy-2-hexacosen-1,5-olide
[116561-97-8]

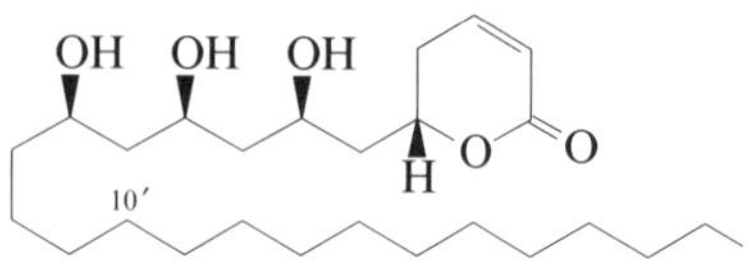

$C_{26}H_{48}O_5$ M 440.662
Constit. of *Eupatorium pilosum*. Cryst. (MeOH). Mp 101°.

10′-Hydroxy: [69616-80-4]. *5,6-Dihydro-6-(2,4,6,10-tetrahydroxyheneicosyl)-2*H*-pyran-2-one. 7,9,11,15-Tetrahydroxy-2-hexacosen-1,5-olide*
$C_{26}H_{48}O_6$ M 456.662
Constit. of *E. pilosum*. Needles (MeOH). Mp 109°. $[\alpha]_D^{22}$ −33.6° (c, 0.0250 in MeOH).

[69616-81-5]

Herz, W. *et al*, *Phytochemistry*, 1978, **17**, 1327 (*isol*)
Nakata, T. *et al*, *Tetrahedron Lett.*, 1987, **28**, 5661 (*synth, abs config*)

3,6-Dihydroxycholan-24-oic acid D-00162

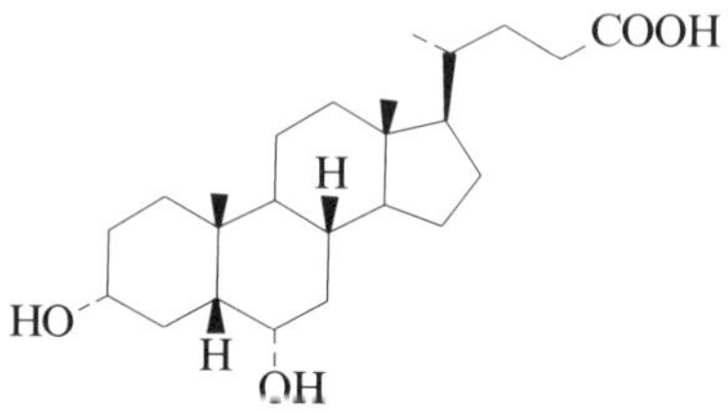

$C_{24}H_{40}O_4$ M 392.578

(3α,5β,6α)-form [83-49-8] ***Hyodeoxycholic acid***
Isol. from pig bile. Cryst. Mp 197-198°. $[\alpha]_D$ +8° (MeOH).

Glycine amide: [13042-33-6]. ***Glycohyodeoxycholic acid.*** *Hyodeoxycholylglycine*
$C_{26}H_{43}NO_5$ M 449.629
Occurs in bile of pig and wild boar.

(3α,5β,6β)-form
Cryst. (MeOH). Mp 209-210°. $[\alpha]_D$ +37° (MeOH).

Wieland, H. *et al*, *Z. Phys. Chem. (Leipzig)*, 1933, **215**, 18 (*isol*)
Moffett, R.B. *et al*, *J. Am. Chem. Soc.*, 1947, **69**, 1995 (*struct*)
Leibfritz, D. *et al*, *J. Am. Chem. Soc.*, 1973, **95**, 4996 (*cmr*)
Hall, S.R. *et al*, *Acta Crystallogr., Sect. B*, 1974, **30**, 1441 (*cryst struct*)
Zhou, W.-S. *et al*, *J. Chem. Soc., Perkin Trans. 1*, 1990, 1 (*synth*)
Hoehm, W.M. *et al*, *J. Am. Chem. Soc.*, 1946, **68**, 1855.
Tsukamoto, L.T., *J. Biochem. (Tokyo)*, 1940, **32**, 451.

3,12-Dihydroxycholan-24-oic acid D-00163

$C_{24}H_{40}O_4$ M 392.578

(3α,5α,12α)-form [1912-55-6] ***Allodeoxycholic acid.*** *β-Lagodeoxycholic acid*
Constit. of rabbit bile and faeces. Cryst. (MeOH). Mp 214-215°. $[\alpha]_D^{27}$ +42° (c, 0.75 in 90% EtOH aq.).

Me ester:
$C_{25}H_{42}O_4$ M 406.604
Mp 174-176°. $[\alpha]_D^{23}$ +35.6° (c, 1 in MeOH).

(3α,5β,12α)-form [83-44-3] ***Deoxycholic acid.*** *Choleic acid. Cholorebic acid. Droxolan. Pyrochol. Septochol*
Occurs in bile of man, ox, goat and sheep. Choleretic, antiinflammatory agent. Cryst. (EtOH). Mp 187-189°. $[\alpha]_D^{19}$ +47.7° (c, 1.7 in dioxan), $[\alpha]_D$ +53° ($CHCl_3$).
Adduct with menthol is marketed as Degalol.
▷ Exp. carcinogen. FZ2100000.

12-Ac:
$C_{26}H_{42}O_5$ M 434.615
Cryst. (MeOH). Mp 174-175°.

Me ester:
$C_{25}H_{42}O_4$ M 406.604
Cryst. (butanone). Mp 105-106°. $[\alpha]_D^{20.5}$ +48° (butanone).

Et ester:
$C_{26}H_{44}O_4$ M 420.631
Needles. Mp 81°.

Amide:
$C_{24}H_{41}NO_3$ M 391.593
Needles. Mp 186°. Sinters at 162°.

Glycine amide: [360-65-6]. ***Glycodeoxycholic acid.*** *Deoxycholylglycine*
$C_{26}H_{43}NO_5$ M 449.629
Bile acid. Cryst. Mp 187-188°. $[\alpha]_D^{23}$ +45.9° (c, 1 in EtOH), $[\alpha]_D$ +53° ($CHCl_3$).
▷ MB9670000.

(3α,5β,12β)-form
12-epi-*Deoxycholic acid*
Cryst. Mp 188°. $[\alpha]_D$ +38° (dioxan).

(3β,5α,12α)-form
Needles (MeOH). Mp 230°. $[\alpha]_D$ +38° (MeOH).
12-Ac: Cryst. (MeOH). Mp 174-175°.

(3β,5β,12α)-form [570-63-8]
Constit. of rabbit faeces. Cryst. (MeOH). Mp 176°. $[\alpha]_D$ +4.7° ($CHCl_3$).

(3β,5β,12β) form [84848-09-9]
Cryst. (MeOH). Mp 204-206°. $[\alpha]_D$ +27° ($CHCl_3$).

Reichstein, T. *et al*, *Helv. Chim. Acta*, 1942, **25**, 797 (*synth*)
Bergström, S. *et al*, *Acta Chem. Scand.*, 1953, **7**, 481 (*synth*)
Fieser, L.F. *et al*, *Steroids*, Reinhold, N.Y., 1959 (*rev*)
Danielsson, H. *et al*, *J. Biol. Chem.*, 1963, **238**, 3846 (*isol*)
Chang, F.C. *et al*, *J. Org. Chem.*, 1965, **30**, 1718 (*synth*)
Leibfritz, D. *et al*, *J. Am. Chem. Soc.*, 1973, **95**, 4996 (*cmr*)
Inoue, M. *et al*, *Chem. Pharm. Bull.*, 1974, **22**, 1949 (*ms*)
Iida, T. *et al*, *J. Lipid Res.*, 1985, **26**, 874 (*synth*)
Coiro, V.M. *et al*, *Acta Crystallogr., Sect. B*, 1979, **35**, 2941 (*cryst struct*)
Matsumoto, K., *J. Biochem. (Tokyo)*, 1944, **36**, 173.
Lewis, R.J., *Sax's Dangerous Properties of Industrial Materials*, 8th Ed., Van Nostrand-Reinhold, 1992, DAQ400.

2,10-Dihydroxy-4,6,8-decatriynoic acid D-00164

$C_{10}H_8O_4$ M 192.171

(+)-***form***
Metab. of *Poria sinuosa.*
Me ester:
$C_{11}H_{10}O_4$ M 206.198
Cryst. (CS_2 or CH_2Cl_2/hexane). Mp 102-103° dec. $[\alpha]_D^{17}$ +44° (c, 0.45 in EtOH).
(±)-***form***
Cryst. (Et_2O). Mp 152-153° dec. Turns purple on exp. to light.
Me ester: Cryst. (CH_2Cl_2/hexane). Mp 108-109° dec.

Cambie, R.C. *et al, J. Chem. Soc.*, 1963, 2056, 3466 (*isol, synth*)

2,10-Dihydroxy-4-decene-6,8-diynoic acid D-00165

$HOCH_2C{\equiv}CC{\equiv}CCH{=}CHCH_2CH(OH)COOH$

$C_{10}H_{10}O_4$ M 194.187
(***E***)-***form***
Metab. of *Poria sinuosa.*
Me ester: Oil.
(±)-(***E***)-***form***
Cryst. (Et_2O/hexane). Mp 163-166° dec.
Me ester: Cryst. (Et_2O/hexane). Mp 85-87° dec.

Cambie, R.C. *et al, J. Chem. Soc.*, 1963, 2056, 3466 (*isol, synth*)

11,15-Dihydroxy-16,16-dimethyl-9-methylene-5,13-prostadienoic acid D-00166

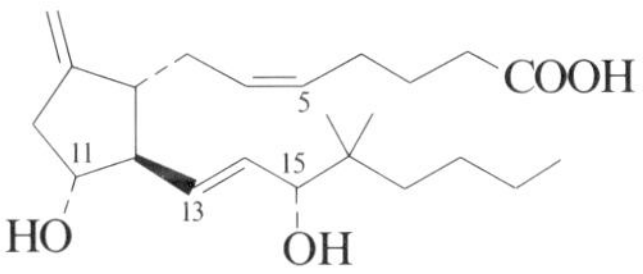

$C_{23}H_{38}O_4$ M 378.551
(***5Z,11R,13E,15R***)-***form*** [61263-35-2] ***Meteneprost, INN.*** *U 46785*
Oxytocic agent, shows high abortifacient activity.
Adamantamine salt: [65470-21-5].
Mp 115-117°. $[\alpha]_D$ −31° (c, 1 in MeOH).

Bygdeman, M. *et al, Lancet*, 1979, **8126**, 1136 (*pharmacol*)
Kimball, F.A. *et al, Prostaglandins*, 1979, **17**, 657 (*ir, pmr, ms, synth, props*)
Steffenrud, S., *Drug Metab. Dispos.*, 1983, **11**, 255 (*metab*)
Martindale, The Extra Pharmacopoeia, 28th/29th Ed., Pharmaceutical Press, London, 1982/1989, 8090.

11,15-Dihydroxy-16,16-dimethyl-9-oxo-5,13-prostadienoic acid, 9CI D-00167

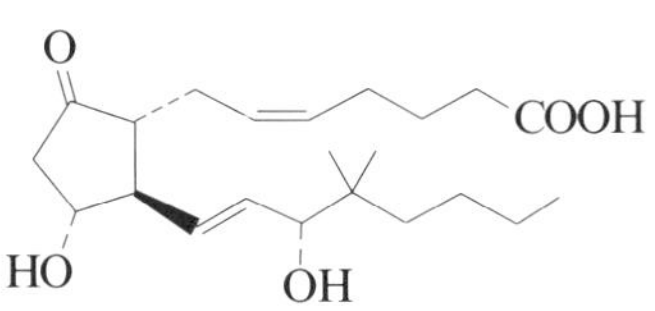

$C_{22}H_{36}O_5$ M 380.523
(***5Z,11R,13E,15R***)-***form*** [39746-25-3]
16,16-Dimethyl-PGE₂. 16,16-Dimethyldinoprostone
Has abortive and contraceptive activity. Also affects stomach secretion. Viscous oil.
▷ UK7894000.
11-Deoxy: [53658-98-3]. *15-Hydroxy-16,16-dimethyl-9-oxo-5,13-prostadienoic acid. 11-Deoxy-16,16-dimethyl PGE₂*
$C_{22}H_{36}O_4$ M 364.524
Inhibitor of gastric acid secretion. Potential therapeutic agent for the treatment of peptic ulcers.
(***5Z,11R,13E,15S***)-***form***
11-Deoxy: [53658-99-4]. *AY 24696*
Less active than (15*R*)-form.

[55309-96-1, 55309-97-2]

Magerlein, B.J. *et al, Prostaglandins*, 1973, **4**, 143 (*synth*)
Lippmann, W., *Prostaglandins*, 1974, **7**, 231 (*biochem*)
U.S. Pat., 3 903 131, (1975); *CA*, **83**, 192667 (*synth*)
U.S. Pat., 3 987 081, (1976); *CA*, **86**, 139483d (*synth*)
Martindale, The Extra Pharmacopoeia, 28th/29th Ed., Pharmaceutical Press, London, 1982/1989, 8080.

13,14-Dihydroxydocosanoic acid, 8CI D-00168

13,14-Dihydroxybehenic acid
[616-01-3]

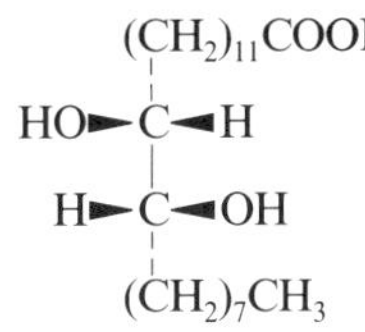

(13*R*,14*R*)-*form*

$C_{22}H_{44}O_4$ M 372.587
(***13R,14R***)-***form*** [125876-57-5]
(+)-threo-*form*
$[\alpha]_D$ +19.8° (MeOH).
Me ester:
$C_{23}H_{46}O_4$ M 386.614
$[\alpha]_D$ +18.8° (MeOH).
(***13S,14S***)-***form*** [33447-89-1]
(−)-threo-*form*
$[\alpha]_D$ −19.8° (MeOH).
(***13RS,14RS***)-***form*** [10489-64-2]
(±)-threo-*form*
Cryst. (EtOH). Mp 101°.
Me ester: [6088-35-3].
Cryst. (MeOH). Mp 78°.
(***13RS,14SR***)-***form*** [13028-35-8]
(±)-erythro-*form*
Isol. from seed oil of *Cardamine impatiens*. Plates (EtOH). Mp 132-133°. Nat. prod. had no detectable opt. activity.
Me ester: [6088-34-2].
Cryst. (MeOH). Mp 111°.

[33447-88-0, 125876-58-6]

Doreé, C. *et al, J. Chem. Soc.*, 1942, 477 (*synth*)
Bader, A.R., *J. Am. Chem. Soc.*, 1948, **70**, 3939 (*synth, config*)
Mikolajczak, C.R. *et al, J. Am. Oil Chem. Soc.*, 1965, **42**, 939 (*isol, synth*)
Roomi, M.W. *et al, J. Chromatogr.*, 1966, **24**, 93 (*chromatog*)
McGhie, J.F. *et al, Chem. Ind. (London)*, 1971, 1074 (*resoln*)

5,12-Dihydroxy-6,8,10,14,17-eicosapentaenoic acid D-00169

Leukotriene B₅. LTB₅
[80445-66-5]

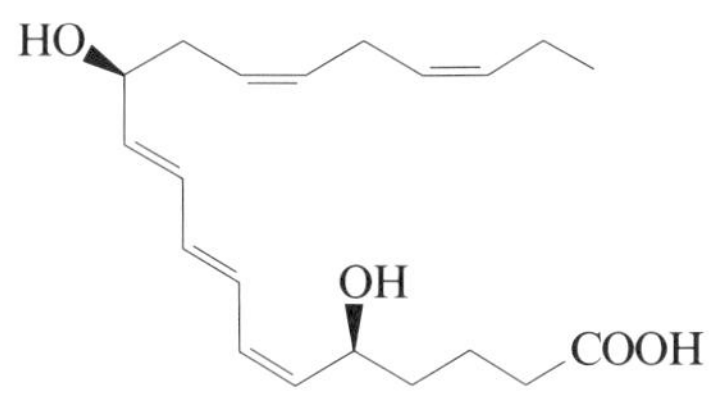

$C_{20}H_{30}O_4$ M 334.455
Prod. *in vivo* from eicosapentaenoic acid. No phys. props. reported.

Murphy, R.A. *et al, Prostaglandins*, 1981, **22**, 613 (*isol*)
Corey, E.J. *et al, Proc. Natl. Acad. Sci. U.S.A.*, 1983, **80**, 3581 (*biosynth*)
Corey, E.J. *et al, Tetrahedron Lett.*, 1983, **24**, 4883 (*synth, pmr, uv, ms*)

5,6-Dihydroxy-7,9,11,14-eicosatetraenoic acid D-00170

5,6-DiHETE

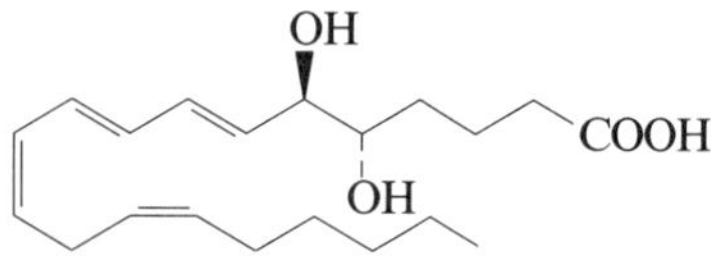

$C_{20}H_{32}O_4$ M 336.470

(5S,6R,7E,9E,11Z,14Z)-form
Metab. of arachidonic acid derived from Leukotriene A_4.
Me ester:
$C_{21}H_{34}O_4$ M 350.497
Oil. $[\alpha]_D^{20}$ +0.91° (c, 0.44 in $CHCl_3$). Other diastereoisomers also prepd.

(5S,6S,7E,9E,11Z,14Z)-form [82948-87-6]
$[\alpha]_D^{22}$ −65° (c, 0.05 in EtOH).

Cristol, J.P. *et al, Res. Commun. Chem. Pathol. Pharmacol.*, 1988, **59**, 423.
Nicolaou, K.C. *et al, Angew. Chem., Int. Ed. Engl.*, 1989, **28**, 587 (*synth, ir, uv, pmr*)
Muller, A. *et al, Prostaglandins*, 1989, **38**, 635 (*synth, biochem*)
Kugel, C. *et al, Tetrahedron Lett.*, 1989, **30**, 4947 (*synth, pmr, ms*)

5,12-Dihydroxy-6,8,10,14-eicosatetraenoic acid D-00171

[73151-67-4]

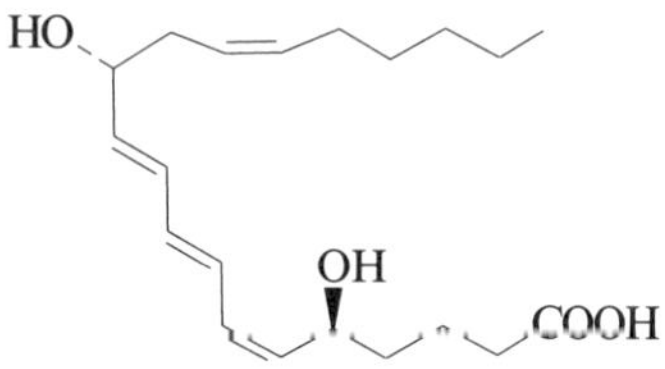

Absolute configuration

$C_{20}H_{32}O_4$ M 336.470

(5S,12R,6Z,8E,10E,14Z)-form
***Leukotriene B_4**. LTB$_4$*
Powerful chemotactic substance for macrophages and neutrophils, involved in tissue response to inflammation. Oil. $[\alpha]_D^{25}$ +12.6° (c, 0.46 in $CDCl_3$).
Me ester: [83058-42-8].
$[\alpha]_D^{20}$ +6.5° (c, 1.8 in CCl_4). Uv λ_{max} 260, 270, 281.5 nm (MeOH).
5,12-Di-Ac:
$C_{24}H_{36}O_6$ M 420.545
Selective, competitive and reversible inhibitor of LTB_4.
19-Hydroxy: [81994-54-9]. *5,12,19-Trihydroxy-6,8,10,14-eicosatetraenoic acid*
$C_{20}H_{32}O_5$ M 352.470
LTB_4 metab. Uv λ_{max} 281, 270.5, 260 nm (MeOH).
20-Hydroxy: [79516-82-8]. *5,12,20-Trihydroxy-6,8,10,14-eicosatetraenoic acid*
$C_{20}H_{32}O_5$ M 352.470
Primary metab. of LTB_4.
12-Oxo: [136696-10-1]. *5-Hydroxy-12-oxo-6,8,10,14-eicosatetraenoic acid*
$C_{20}H_{30}O_4$ M 334.455
Biosynth. from LTB_4. Formed from LTB_4 by polymorphonuclear leukocytes.
20-Aldehyde:
$C_{20}H_{30}O_5$ M 350.454
LTB_4 metab.
20-Carboxylic acid: [80434-82-8]. *5,12-Dihydroxy-6,8,10,14-eicosatetraenedioic acid*
$C_{20}H_{30}O_6$ M 366.453
Primary LTB_4 metab.
δ-Lactone: [85082-09-3]. *Tetrahydro-6-(7-hydroxy-1,3,5,9-pentadecatetraenyl)-2*H*-pyran-2-one. Leukotriene B_4 δ-lactone*
$C_{20}H_{30}O_3$ M 318.455
Useful intermed. giving LTB_4 covalently bound to a variety of ligands. Ir (neat) 3380, 1720, 1238, 1034 cm^{-1}.
14,15-Didehydro: [114616-11-4]. *5,12-Dihydroxy-6,8,10-eicosatrien-14-ynoic acid*
$C_{20}H_{30}O_4$ M 334.455
Precursor of tritiated LTB_4. Inhibits LTB_4. Ir (neat) 3380, 1712, 1385 cm^{-1}.
6,7,14,15-Tetradehydro: [90108-38-6]. *5,12-Dihydroxy-8,10-eicosadiene-6,14-diynoic acid*
$C_{20}H_{28}O_4$ M 332.439
LTB_4 antagonist. Deuteration or tritiation intermed. No phys. props. reported.
1-Methylamide: [83024-92-4]. *5,12-Dihydroxy-*N,N*-dimethyl-6,8,10,14-eicosatetraenamide*
$C_{21}H_{35}NO_3$ M 349.512
Inhibits neutrophil degranulation induced by LTB_4.
1-Dimethylamide: LTB$_4$-DMA
$C_{22}H_{37}NO_3$ M 363.539
Inhibits binding of tritiated LTB_4. Antagonises the degranulation response to LTB_4. Uv λ_{max} 281.5, 270, 260 nm (MeOH).
3-Aminopropylamide: [89596-43-0]. N*-(3-Aminopropyl)-5,12-dihydroxy-6,8,10,14-eicosatetraenamide. LTB$_4$-APA*
$C_{23}H_{40}N_2O_3$ M 392.581
Affinity label for the LTB_4 receptor. Selectively binds to high affinity leuckocyte LTB_4 receptors and induces receptor densensitisation.
3-Aminopropylamidefluorescein deriv.: Fluorescent LTB_4 receptor ligand. Used for studying LTB_4 receptors.
20, 20, 20-Trifluoro: [115178-97-7]. *20,20,20-Trifluoro-5,12-dihydroxy-6,8,10,14-eicosatetraenoic acid*
$C_{20}H_{29}F_3O_4$ M 390.442
As potent as LTB_4 at inducing chemotaxis, adhesion and aggregation. Antagonist for degranulation. Used in pharmacological studies of LTB_4.
20, 20, 20-Trifluoro, Me ester: [121145-72-0].
$C_{21}H_{31}F_3O_4$ M 404.469
$[\alpha]_D^{22}$ +17.89° (c, 0.833 in $CHCl_3$).

(5S,12S,6E,8Z,10E,14Z)-form
5S,12S-DiHETE. Leukotriene B_x. LTB$_x$
Metab. of arachidonic acid found in human leucocytes. Has weak chemotactic activity. Cryst. Mp 101-103°. Biosynth. markedly diff. from LTB_4.

(5S,12R,6E,8E,10E,14Z)-form [71652-82-9]
*(5S,12R)-DiHETE. 6-*trans*-Leukotriene B_4*
LTB_4 metab. Biosynth. from LTA_4 and cysteinyl leukotrienes.

(5S,12S,6Z,8E,10E,14Z)-form [83709-73-3]
*12-*epi*-Leukotriene B_4*
Leukotriene metab. Less active than LTB_4.

(5S,12S,6E,8E,10E,14Z)-form [71548-19-1]
*6-*trans*-12-*epi*-Leukotriene B_4*
Leukotriene metab.

[71160-24-2, 79056-01-2, 82978-32-3, 82978-33-4, 98056-93-0, 98056-94-1, 98103-85-6, 98167-17-0]

Borgeat, P. *et al, J. Biol. Chem.*, 1979, **254**, 2643 (*struct*)
Goetzl, J.E. *et al, J. Exp. Med.*, 1981, **153**, 482 (*Di-Ac*)
Ford-Hutchinson, A.W. *et al, Prostaglandins*, 1981, **21**, 143; 1983, **25**, 29 (*biochem*)
Corey, E.J. *et al, Tetrahedron Lett.*, 1981, **22**, 1077, 1587, 3339 (*synth, uv, bibl*)
Showell, H.J. *et al, Biochem. Biophys. Res. Commun.*, 1982, **106**, 741 (*synth, biochem*)
Zamboni, R. *et al, Tetrahedron Lett.*, 1982, **23**, 2631, 4751 (*synth, bibl*)
Gúindon, Y. *et al, Tetrahedron Lett.*, 1982, **23**, 739 (*synth*)
Lee, C.W. *et al, J. Biol. Chem.*, 1983, **258**, 15004 (*biosynth*)
Young, R.N. *et al, Prostaglandins*, 1983, **26**, 605 (*biochem*)
Green, R.H. *et al, Tetrahedron*, 1983, **39**, 1687 (*rev*)
Mills, L.S. *et al, Tetrahedron Lett.*, 1983, **24**, 409 (*synth*)
Fuganti, C. *et al, Tetrahedron Lett.*, 1983, **24**, 5285 (*synth*)
Nicolaou, K.C. *et al, J. Am. Chem. Soc.*, 1984, **106**, 3548 (*synth*)
Sugiura, M. *et al, J. Am. Chem. Soc.*, 1984, **106**, 4021 (*conformn*)
Adams, J. *et al, Tetrahedron Lett.*, 1984, **25**, 1227 (*synth*)
Han, C.-O. *et al, J. Org. Chem.*, 1986, **51**, 1253 (*synth, uv, pmr*)
Nicolaou, K.C. *et al, Tetrahedron Lett.*, 1986, **27**, 1881 (*synth, bibl*)
Le Merrer, Y. *et al, Tetrahedron Lett.*, 1986, **27**, 4161 (*synth*)
Leblanc, Y. *et al, Prostaglandins*, 1987, **33**, 617 (*synth*)
Tanaka, Y. *et al, Arch. Biochem. Biophys.*, 1988, **263**, 178 (*synth, biosynth, biochem*)
Kitamura, S. *et al, Eur. J. Biochem.*, 1988, **176**, 725 (*synth*)
Guindon, Y. *et al, J. Org. Chem.*, 1988, **53**, 267 (*synth, pmr, uv*)
Tsai, B.S. *et al, Prostaglandins*, 1989, **37**, 287 (*synth, biochem, ir, uv, pmr*)
Avignon-Tropis, M. *et al, Tetrahedron Lett.*, 1989, **30**, 6335 (*synth*)
Kiel, D. *et al, Adv. Prostaglandin, Thromboxane, Leukotriene Res.*, 1990, **21A**, 403 (*rev*)
Le Merver, Y. *et al, Bull. Soc. Chim. Fr.*, 1990, 798 (*synth, uv, pmr*)
Kobayashi, Y. *et al, J. Org. Chem.*, 1990, **55**, 5324 (*synth, ir, pmr*)
Shimazaki, T. *et al, Prostaglandins*, 1990, **39**, 459 (*biochem*)
Wainwright, S.L. *et al, J. Biol. Chem.*, 1991, **266**, 20899.
Babudri, F. *et al, J. Org. Chem.*, 1991, **56**, 6245 (*synth, pmr*)
Kerdesky, F.A.J. *et al, J. Org. Chem.*, 1993, **58**, 3516 (*synth, pmr, ir, uv*)

5,15-Dihydroxy-6,8,11,13-eicosatetraenoic acid D-00172

5,15-DiHETE

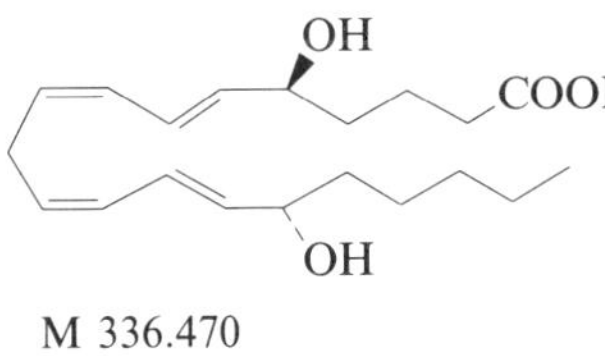

$C_{20}H_{32}O_4$ M 336.470

(5S,6E,8Z,11Z,13E,15S)-form

Metab. of arachidonic acid. Biosynth. in human eosinophils. Enhances neutrophil degranulation.

[81994-56-1]

Van Os, P.A. *et al, Biochim. Biophys. Acta*, 1981, **663**, 177 (*isol*)
Maas, R.L. *et al, J. Biol. Chem.*, 1982, **257**, 7056 (*isol, ms, uv, props*)
Borgeat, B. *et al, Lipids*, 1982, **17**, 676 (*metab*)
Nicolaou, K.C. *et al, J. Am. Chem. Soc.*, 1984, **106**, 5734 (*synth*)

5,20-Dihydroxy-6,8,11,14-eicosatetraenoic acid D-00173

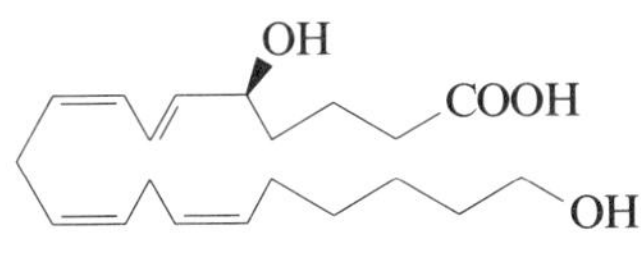

$C_{20}H_{32}O_4$ M 336.470

(5S,6E,8Z,11Z,14Z)-form [107373-23-9]

Hydroxylated eicosanoid metab. $[\alpha]_D^{23}$ +10.2° (c, 1.53 in Me_2CO).

Lumin, S. *et al, J. Chem. Soc., Chem. Commun.*, 1987, 389 (*synth, pmr*)

8,15-Dihydroxy-5,9,11,13-eicosatetraenoic acid D-00174

8,15-DiHETE

[77667-08-4]

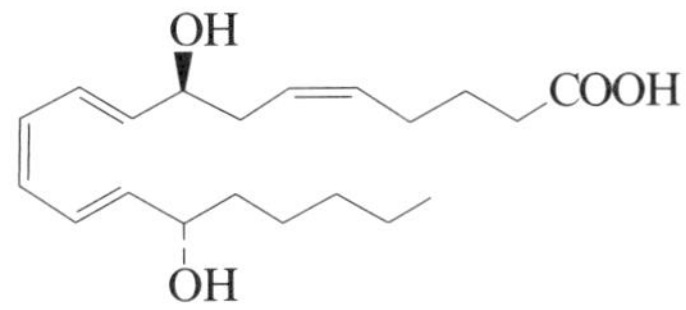

$C_{20}H_{32}O_4$ M 336.470

(5Z,8R,9E,11E,13E,15S)-form

Metab. of 15-Hydroperoxy-5,8,11,13-eicosatetraenoic acid, H-00340. No phys. props. reported.

(5Z,8R,9E,11Z,13E,15S)-form

HPETE metab. No phys. props. reported.

(5Z,8S,9E,11E,13E,15S)-form [80234-66-8]

Metab. of arachidonic acid. No phys. props. reported.

(5Z,8S,9E,11Z,13E,15S)-form

Metab. of arachidonic acid. Biosynth. in human leukocytes, eosinophils and platelets.

Van Os, C.P.A. *et al, Biochim. Biophys. Acta*, 1981, **663**, 177 (*isol*)
Sok, D.E. *et al, Biochem. Biophys. Res. Commun.*, 1983, **110**, 273 (*biosynth*)
Shak, S. *et al, J. Biol. Chem.*, 1983, **258**, 14948 (*synth, isol, uv, ms, props, activity*)
Nicolaou, K.C. *et al, J. Am. Chem. Soc.*, 1984, **106**, 5734 (*synth*)
Hopkins, N.K. *et al, J. Biol. Chem.*, 1984, **259**, 14048 (*biosynth*)
Fitzsimmons, B.J. *et al, Tetrahedron Lett.*, 1984, **25**, 3043 (*synth*)

12,19-Dihydroxy-5,8,10,14-eicosatetraenoic acid D-00175

12,19-DiHETE

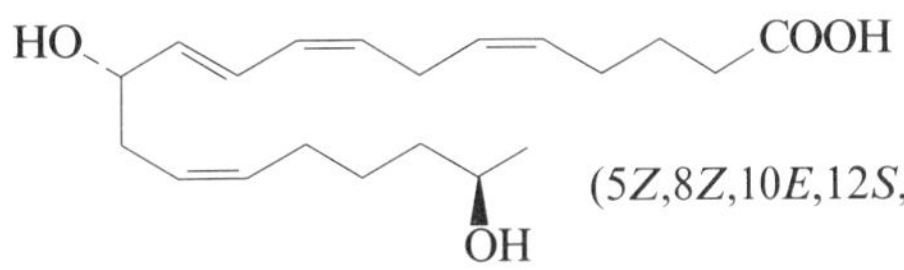

(5Z,8Z,10E,12S,14Z,19R)-form

$C_{20}H_{32}O_4$ M 336.470

(5Z,8Z,10E,12S,14Z,19R)-form [110066-02-9]

Probable metabolite of 12-Hydroxy-5,8,10,14-eicosatetraenoic acid, H-00404.

(5Z,8Z,10E,12S,14Z,19S)-form [110066-03-0]

Probable metabolite of 12-Hydroxy-5,8,10,14-eicosatetraenoic acid, H-00404.

Manna, S. *et al, Tetrahedron Lett.*, 1986, **27**, 2679 (*synth*)

12,20-Dihydroxy-5,8,10,14-eicosatetraenoic acid D-00176

12,20-DiHETE

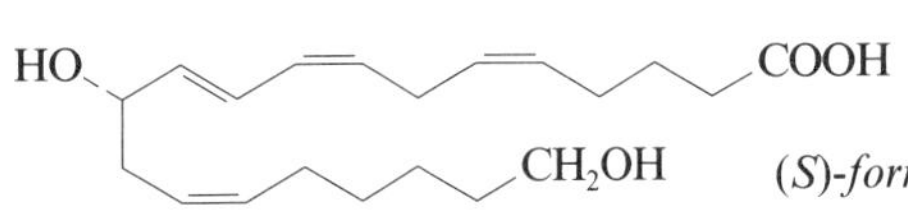

(S)-form

$C_{20}H_{32}O_4$ M 336.470

(5Z,8Z,10E,12S,14Z)-form [89614-44-8]
Metab. of arachidonic acid. $[\alpha]_D^{22}$ +10.3° (c, 1.36 in Me_2CO).

Marcus, A.J. *et al*, *Proc. Natl. Acad. Sci. U.S.A.*, 1984, **81**, 903 (*isol, biochem, ms*)
Leblanc, Y. *et al*, *J. Org. Chem.*, 1986, **51**, 789 (*synth, pmr*)
Manna, S. *et al*, *Tetrahedron Lett.*, 1986, **27**, 2679 (*synth*)

14,15-Dihydroxy-5,8,10,12-eicosatetraenoic acid **D-00177**

14,15-DiHETE
[77667-09-5]

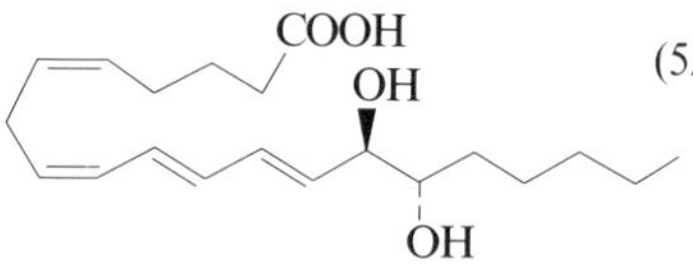

(5*Z*,8*Z*,10*E*,12*E*,14*R*,15*S*)-*form*

$C_{20}H_{32}O_4$ M 336.470

(5Z,8Z,10E,12E,14R,15S)-form [82263-61-4]
Formed by human blood platelets and swine leukocytes. No phys. props. given.

(5Z,8Z,10E,12E,14S,15S)-form [82263-60-3]
Formed by arachidonate lipoxygenase of leukocytes. No phys. props. given.

[89410-77-5, 89410-78-6]

Maas, R.L. *et al*, *Proc. Natl. Acad. Sci. U.S.A.*, 1983, **80**, 2884 (*biosynth*)
Falck, J.R. *et al*, *Tetrahedron Lett.*, 1983, **24**, 5719 (*synth, pmr, bibl*)
Radmark, O. *et al*, *J. Biol. Chem.*, 1984, **259**, 13011 (*synth, biochem*)
Corey, E.J. *et al*, *Tetrahedron Lett.*, 1985, **26**, 1919 (*synth*)
Brash, A.R. *et al*, *Arch. Biochem. Biophys.*, 1989, **273**, 414.

15,20-Dihydroxy-5,8,11,13-eicosatetraenoic acid, 9CI **D-00178**

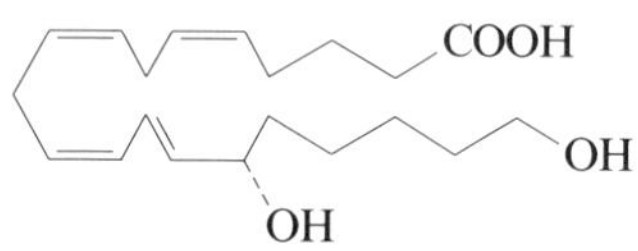

$C_{20}H_{32}O_4$ M 336.470

(5Z,8Z,11Z,13E,15S)-form [112168-45-3]
Hydroxylated eicosanoid metabolite. $[\alpha]_D^{23}$ +12.0° (c, 1.76 in Me_2CO).

Lumin, S. *et al*, *J. Chem. Soc., Chem. Commun.*, 1987, 389 (*synth, pmr*)

5,6-Dihydroxy-8,11,14-eicosatrienoic acid **D-00179**

[81920-20-9]

$H_3CCH_2(CH_2)_3(CH{=}CHCH_2)_3CH(OH)CH(OH)CH_2CH_2CH_2COOH$

$C_{20}H_{34}O_4$ M 338.486

Metab. of arachidonic acid. 5*R*,6*R*-form synth. No phys. props. reported.

δ-Lactone: [98050-32-9]. *Tetrahydro-6-(1-hydroxy-3,6,9-pentadecatrienyl)-2*H*-pyran-2-one*
No phys. props. reported.

[98050-31-8]

Corey, E.J. *et al*, *Tetrahedron Lett.*, 1983, **24**, 265 (*synth*)
Moustakis, C.A. *et al*, *J. Am. Chem. Soc.*, 1985, **107**, 5283 (*synth*)
Nishimura, M. *et al*, *Prostaglandins*, 1989, **38**, 413 (*biosynth*)
Balazy, M., *J. Biol. Chem.*, 1991, **266**, 23561 (*biosynth*)

5,12-Dihydroxy-6,8,10-eicosatrienoic acid, 9CI **D-00180**

Leukotriene B₃
[88099-35-8]

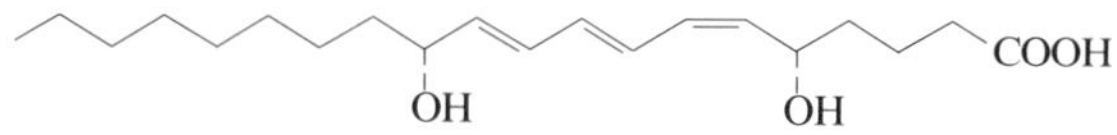

$C_{20}H_{34}O_4$ M 338.486

Metab. of arachidonic acid. Cryst. (CH_2Cl_2/Et_2O). Mp 101-103°. $[\alpha]_D^{21}$ +7.8° (c, 0.23 in $CDCl_3$).

Cotterill, I. *et al*, *J. Chem. Soc., Chem. Commun.*, 1990, 1661 (*synth*)
Kobayashi, Y. *et al*, *J. Org. Chem.*, 1990, **55**, 5324 (*synth, uv, pmr, cmr*)
Babudri, F. *et al*, *J. Org. Chem.*, 1991, **56**, 6245 (*synth, pmr*)

8,9-Dihydroxy-5,11,14-eicosatrienoic acid **D-00181**

[81943-03-5]

$H_3C(CH_2)_4CH{=}CHCH_2CH{=}CHCH_2CH(OH)CH(OH)CH_2CH{=}CHCH_2CH_2CH_2COOH$

$C_{20}H_{34}O_4$ M 338.486

Arachidonic acid metab.

Oliw, E.H. *et al*, *Biochim. Biophys. Acta*, 1982, **721**, 135 (*biosynth*)
Oliw, E.H. *et al*, *J. Biol. Chem.*, 1982, **257**, 3771 (*biosynth*)
Nishimura, M. *et al*, *Prostaglandins*, 1989, **38**, 413 (*biosynth*)

14,15-Dihydroxy-5,8,11-eicosatrienoic acid **D-00182**

[79551-81-8]

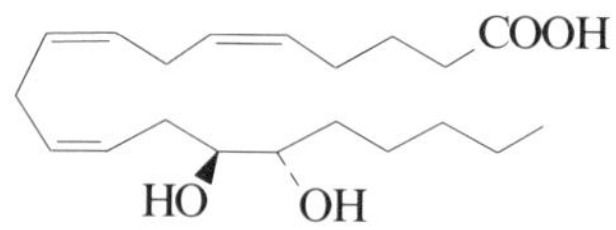

(14*S*,15*S*)-*form*

$C_{20}H_{34}O_4$ M 338.486

(5Z,8Z,11Z,14R,15S)-form [89410-73-1]
Prod. of arachidonic acid metab.

(5Z,8Z,11Z,14S,15S)-form [89410-74-2]
Prod. of arachidonic acid metab.

[98050-20-5]

Oliw, E.H. *et al*, *J. Biol. Chem.*, 1981, **256**, 9924 (*biosynth*)
Manna, S. *et al*, *Tetrahedron Lett.*, 1983, **24**, 33 (*synth*)
Falck, J.R. *et al*, *Tetrahedron Lett.*, 1983, **24**, 5715 (*synth, bibl*)
Moustakis, C.A. *et al*, *J. Am. Chem. Soc.*, 1985, **107**, 5283 (*synth*)
Nishimura, M. *et al*, *Prostaglandins*, 1989, **38**, 413 (*biosynth*)

9,10-Dihydroxy-4,6-heptadecadiyn-3-one **D-00183**

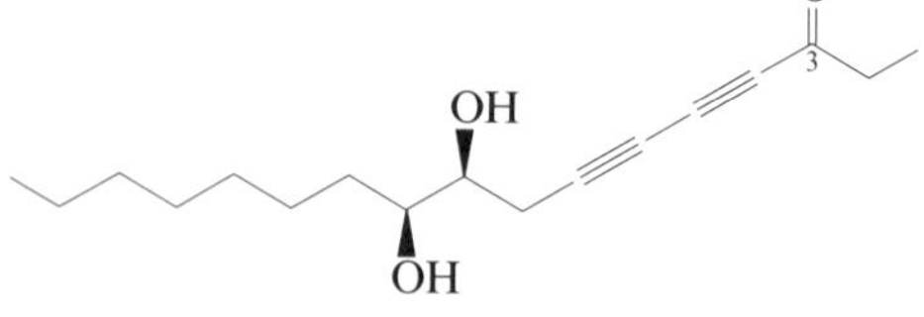

$C_{17}H_{26}O_3$ M 278.391

(9S,10S)-form [106828-96-0] ***Panaxacol***
Isol. from callus of *Panax ginseng*. $[\alpha]_D$ +18.3° (c, 1.0 in MeOH).

3S(?)-Alcohol: [113122-25-1]. *4,6-Heptadecadiyne-3,9,10-triol.* ***Dihydropanaxacol***
$C_{17}H_{28}O_3$ M 280.406
Isol. from *P. ginseng*.

[114687-50-2]

Fujimoto, Y. *et al*, *Phytochemistry*, 1987, **26**, 2850.
Fujimoto, Y. *et al*, *Chem. Pharm. Bull.*, 1990, **38**, 1447 (*synth, abs config*)

1,2-Dihydroxy-16-heptadecen-4-one D-00184

Avocadenone

$H_2C{=}CH(CH_2)_{11}COCH_2CH(OH)CH_2OH$

$C_{17}H_{32}O_3$ M 284.438

1-Ac: [25346-18-3].
$C_{19}H_{34}O_4$ M 326.475
Mp 47°. $[\alpha]_D^{25}$ +20° ($CHCl_3$).

Kashman, Y. *et al*, *Isr. J. Chem.*, 1969, **7**, 173; *Tetrahedron*, 1969, **25**, 4617 (*isol*)
Alves, H.M. *et al*, *Ann. Acad. Bras. Ciénc.*, 1970, **42S**, 45 (*isol*)
Neeman, L. *et al*, *Appl. Microbiol.*, 1970, **19**, 470 (*isol*)

1,2-Dihydroxy-16-heptadecyn-4-one D-00185

Avocadynone

$HC{\equiv}C(CH_2)_{11}COCH_2CH(OH)CH_2OH$

$C_{17}H_{30}O_3$ M 282.422

1-Ac: [24607-10-1].
$C_{19}H_{32}O_4$ M 324.459
Constit. of the Avocado pear. Mp 47°. $[\alpha]_D^{25}$ +26° ($CHCl_3$).

Kashman, Y. *et al*, *Isr. J. Chem.*, 1969, **7**, 173; *Tetrahedron*, 1969, **25**, 4617 (*isol, nmr*)
Alves, H.M. *et al*, *Ann. Acad. Bras. Ciénc.*, 1970, **42S**, 45 (*isol*)

2,3-Dihydroxyhexadecanoic acid D-00186

2,3-Dihydroxypalmitic acid

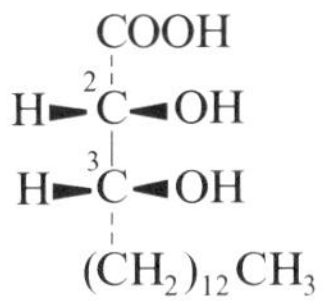

(2*RS*,3*RS*)-*form*
Relative configuration

$C_{16}H_{32}O_4$ M 288.426

(2RS,3RS)-form [59612-64-5]
(±)-erythro-*form*
Cryst. (EtOH). Mp 103-104°.

(2RS,3SR)-form [59612-63-4]
(±)-threo-*form*
Cryst. (EtOH). Mp 124°.

Ansari, A.A. *et al*, *J. Am. Oil Chem. Soc.*, 1976, **53**, 118 (*synth*)

3,12-Dihydroxyhexadecanoic acid D-00187

3,12-Dihydroxypalmitic acid. Operculinolic acid. Turpetholic acid D

[66675-73-8]

$H_3C(CH_2)_3CH(OH)(CH_2)_8CH(OH)CH_2COOH$

$C_{16}H_{32}O_4$ M 288.426
Mp 83.4°.

Me ester: [53781-47-8].
$C_{17}H_{34}O_4$ M 302.453
Mp 81-82°. $[\alpha]_D$ +0.9°.

Et ester:
$C_{18}H_{36}O_4$ M 316.480
Mp 72-73°.

12-O-[β-D-Glucopyranosyl-(1→3)-α-L-rhamnopyranosyl(1→3)-β-D-glucopyranosyl(1→3)-β-D-glucopyranoside]: [66895-64-5]. ***Turpethinic acid D***
$C_{40}H_{72}O_{23}$ M 920.995
Isol. from *Ipomoea turpethum*.

12-O-[α-L-Rhamnopyranosyl-(1→6)-[α-D-glucopyranosyl-(1→4)]-α-D-glucopyranosyl-(1→4)-L-rhamnopyranosyl-(1→2)[β-D-glucopyranosyl-(1→3)]-β-D-glucopyranoside]: [34437-89-3]. ***Operculinic acid.*** *Operculic acid. Rhamnoconvolvulinolic acid*
$C_{52}H_{92}O_{32}$ M 1229.280
Constit. of *Ipomoea operculata*. Mp 182-185°. $[\alpha]_D^{24}$ −48.6° (c, 1.2 in Py).

[66965-42-2]

Shellard, E.J., *Planta Med.*, 1961, **9**, 141 (*isol*)
Ciraf, E. *et al*, *Arch. Pharm. (Weinheim, Ger.)*, 1974, **307**, 628 (*synth*)
Wagner, H. *et al*, *Phytochemistry*, 1977, **16**, 711 (*struct, Operculinic acid*)
Wagner, H. *et al*, *Planta Med.*, 1978, **33**, 144 (*Turpethinic acid D*)
Piatak, D.M. *et al*, *Lipids*, 1985, **20**, 903 (*synth*)

4,12-Dihydroxyhexadecanoic acid D-00188

4,12-Dihydroxypalmitic acid. Turpetholic acid C

[66895-60-1]

$H_3C(CH_2)_3CH(OH)(CH_2)_7CH(OH)CH_2CH_2COOH$

$C_{16}H_{32}O_4$ M 288.426

Me ester: [66940-06-5].
$C_{17}H_{34}O_4$ M 302.453
Mp 70°.

12-O-[β-D-Glucopyranosyl-(1→3)-α-L-rhamnopyranosyl-(1→3)-β-D-glucopyranosyl-(1→3)-β-D-glucopyranoside]: [66895-63-4]. ***Turpethinic acid C***
$C_{40}H_{72}O_{23}$ M 920.995
Isol. from *Ipomoea turpethum*.

Wagner, H. *et al*, *Planta Med.*, 1978, **33**, 144 (*isol*)

5,6-Dihydroxyhexadecanoic acid D-00189

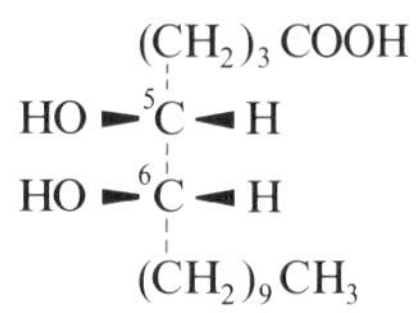

$C_{16}H_{32}O_4$ M 288.426

(5R*,6S*)-form
erythro-*form*

Di-Ac:
$C_{20}H_{36}O_6$ M 372.501
Present in mosquito eggs (possibly in an enantiomeric form). Oil.

(5RS,6SR)-form
(±)-erythro-*form*
Cryst. (Et_2O/pet. ether). Mp 123-124°.

Starratt, A.N., *Chem. Phys. Lipids*, 1976, **16**, 215 (*synth*)

9,10-Dihydroxyhexadecanoic acid D-00190

[29242-09-9]

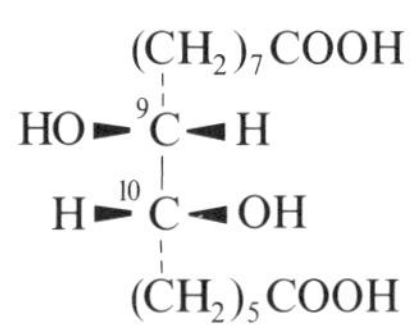

$C_{16}H_{30}O_6$ M 318.409

(9R*,10R*)-form

threo-*form*

Minor constit. of shellac.

Me ester: Mp 62-65°.

Christie, W.W. *et al*, *J. Chem. Soc.*, 1964, 5833.

15,16-Dihydroxyhexadecanoic acid D-00191

$HOCH_2CH(OH)(CH_2)_{13}COOH$

$C_{16}H_{32}O_4$ M 288.426

(R)-form

Ustilic acid A

Component of the antibiotic Ustilagic acid. Mp 112-113°. $[\alpha]_D$ −8° (MeOH).

Me ester: Mp 85.5-86°. $[\alpha]_D$ −7° (MeOH).

Lemieux, R.U., *Can. J. Chem.*, 1953, **31**, 396.

7-[3,5-Dihydroxy-2-(3-hydroxy-5-phenyl-1-pentenyl)cyclopentyl]-5-heptenoic acid D-00192

17-Phenyltrinorprostaglandin $F_{2\alpha}$

[38344-08-0]

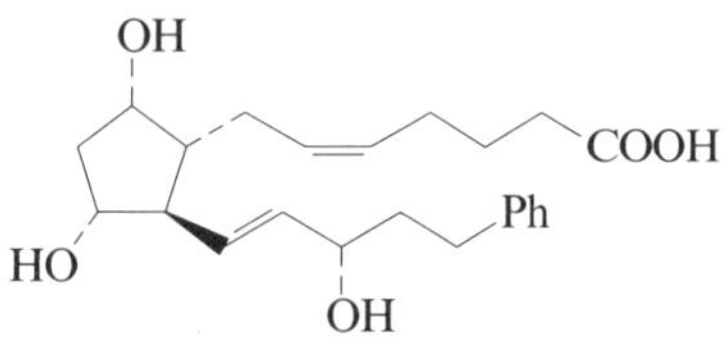

$C_{23}H_{32}O_5$ M 388.503

Shows anti-fertility and luteolytic activity. Mp 79-80°.

[41639-71-8, 88293-69-0, 88293-70-3, 130273-90-4]

Powell, W.S. *et al*, *Eur. J. Biochem.*, 1975, **59**, 271 (*biochem*)
Magerlein, B.J. *et al*, *Prostaglandins*, 1975, **9**, 5 (*synth*)
Miller, W.L. *et al*, *Prostaglandins*, 1975, **9**, 9 (*biochem*)
Schaaf, T.K. *et al*, *J. Med. Chem.*, 1983, **26**, 328 (*biochem*)
Balapure, A.K. *et al*, *Biochem. Pharmacol.*, 1989, **38**, 2375 (*biochem*)

6,8-Dihydroxy-11-isopropyl-4,8-dimethyl-14-oxo-4,9-pentadecadienoic acid D-00193

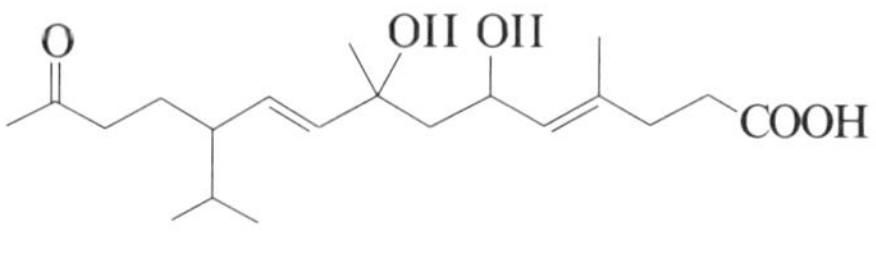

$C_{20}H_{34}O_5$ M 354.486

Two diastereoisomers characterised, configs. undetd.

Isomer A

Isol. from tobacco. Solanone precursor. Oil (as Me ester). $[\alpha]_D^{25}$ +14° (Me ester).

Isomer B

Isol. from tobacco. Solanone precursor. Oil (as Me ester). $[\alpha]_D^{25}$ −36.7° (Me ester).

Kinzer, G.W. *et al*, *J. Org. Chem.*, 1966, **31**, 1797.

11,15-Dihydroxy-11-methyl-9-oxo-5,13-prostadienoic acid D-00194

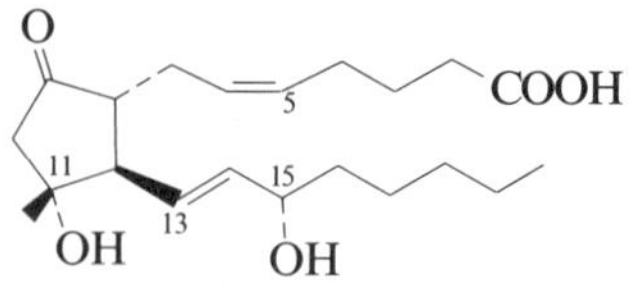

$C_{21}H_{34}O_5$ M 366.497

(5Z,11R,13E,15S)-form [55517-06-1]

Oil.

Guzman, A. *et al*, *Prostaglandins*, 1974, **8**, 85 (*synth*)
Lin, C.H., *Chem. Ind. (London)*, 1976, 994 (*synth, ir, pmr*)

11,15-Dihydroxy-15-methyl-9-oxo-5,13-prostadienoic acid, 9CI D-00195

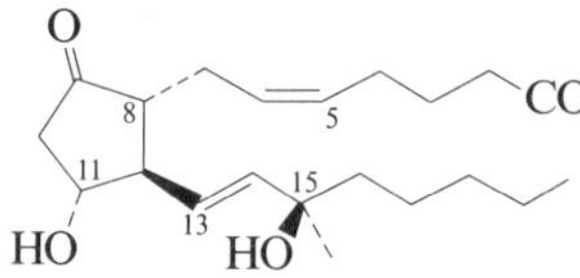

(5*Z*,11*R*,13*E*,15*R*)-*form*

$C_{21}H_{34}O_5$ M 366.497

(5Z,11R,13E,15R)-form [55028-70-1]

(15R)-Methyl-PGE$_2$. ***Arbaprostil, INN.*** *Arbacet. U 42842*

Gastric antisecretory and antiulcer agent.

▷ UK7895000.

Me ester: [39219-61-9].

Oil. $[\alpha]_D$ −74° (c, 1.0 in $CHCl_3$).

(5Z,11R,13E,15S)-form [35700-27-7]

(15S)-Methyl-PGE$_2$

Has vasodepressor activity.

▷ UK7896000.

Me ester: [36210-00-1].

Has vasodepressor activity. Oil. $[\alpha]_D$ −79° (c, 1.3 in $CHCl_3$).

(5Z,8R,11R,12S,13E,15R)-form

(15R)-Methyl-ent-*PGE*$_2$

Me ester: [53317-79-6].

Oil. $[\alpha]_D$ +78° (c, 0.84 in $CHCl_3$).

(5Z,8R,11R,12S,13E,15S)-form

(15S)-Methyl-ent-*PGE*$_2$

Me ester: [53317-78-5].

Oil. $[\alpha]_D$ +77° (c, 0.75 in $CHCl_3$).

Bundy, G.L. *et al*, *Ann. N.Y. Acad. Sci.*, 1971, **180**, 76 (*synth*)
Cooper, E.L. *et al*, *J. Am. Chem. Soc.*, 1974, **96**, 5876 (*synth, pmr, ir, uv, ms*)
Yankee, E.W. *et al*, *J. Am. Chem. Soc.*, 1974, **96**, 5865 (*synth, ir, pmr, uv*)
Thornburgh, B.A. *et al*, *Eur. J. Drug Metab. Pharmacokinet.*, 1986, **11**, 61 (*metab*)
Martindale, The Extra Pharmacopoeia, 28th/29th Ed., Pharmaceutical Press, London, 1982/1989, 8072, 8073.
Lewis, R.J., *Sax's Dangerous Properties of Industrial Materials*, 8th Ed., Van Nostrand-Reinhold, 1992, AQT575, MOV800.

11,15-Dihydroxy-15-methyl-9-oxo-13-prostenoic acid, 9CI D-00196

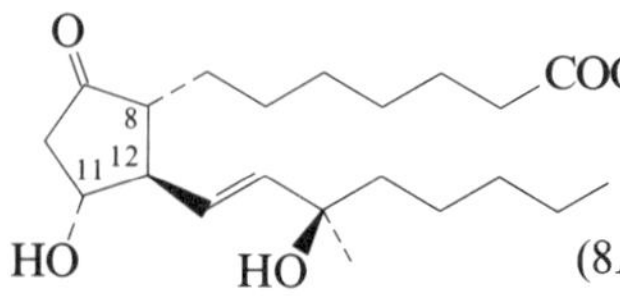

(8*R*,11*R*,12*R*,13*E*,15*S*)-*form*

$C_{21}H_{36}O_5$ M 352.513

(8R,11R,12R,13E,15S)-form [35700-26-6]

15-Methylprostaglandin E₁. (8α,11α,15α)-form

Shows potent antiinflammatory activity (see 11,15-Dihydroxy-9-oxo-13-prostenoic acid, D-00232).

Me ester: [51411-13-3].

$C_{22}H_{38}O_5$ M 366.540

Off-white small needles (hexane/Et_2O). Mp 45-47°. Uv λ_{max} 278 (ε 27250) nm (basic EtOH).

[93566-59-7]

Yankee, E.W. *et al*, *J. Am. Chem. Soc.*, 1974, **96**, 5865 (*synth*)
Sih, C.J. *et al*, *J. Am. Chem. Soc.*, 1975, **97**, 857 (*synth, ir, pmr*)
Kleunder, H.C. *et al*, *Tetrahedron Lett.*, 1977, **24**, 2063 (*synth, biochem*)
Kunkel, S.L. *et al*, *J. Clin. Invest.*, 1979, **64**, 1525 (*biochem*)
Gardiner, P.J. *et al*, *Prostaglandins*, 1980, **19**, 349 (*biochem*)
Marino, J.P. *et al*, *J. Org. Chem.*, 1987, **52**, 4898 (*synth, ir, cmr, pmr, ms*)

11,16-Dihydroxy-16-methyl-9-oxo-13-prostenoic acid, 9CI — D-00197

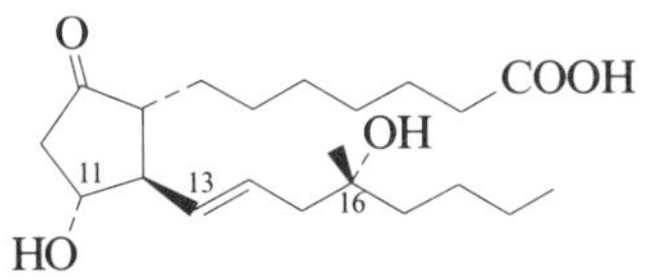

(11*R*,13*E*,16*R*)-*form*

$C_{21}H_{36}O_5$ M 368.512

(11R,13E,16R)-form

Me ester: [59122-48-4].

Inhibitor of gastric secretion. $[\alpha]_D^{25}$ −55.5° (MeOH).

(11S,13E,16S)-form

Me ester: [59122-50-8].

$[\alpha]_D^{25}$ +56.5° (MeOH).

(11S,13E,16R)-form

Me ester: [59122-51-9].

$[\alpha]_D^{25}$ +47.5° (MeOH).

(11R,13E,16S)-form

Me ester: [59122-49-5].

Inhibits gastric secretion. $[\alpha]_D^{25}$ −54.0° (MeOH).

Collins, P.W. *et al*, *Tetrahedron Lett.*, 1975, 4217 (*synth*)
Chen, S.-M.L. *et al*, *J. Org. Chem.*, 1978, **43**, 3450 (*synth, pmr, cmr, ms*)

2,4-Dihydroxy-3-methyl-6-undecylbenzoic acid — D-00198

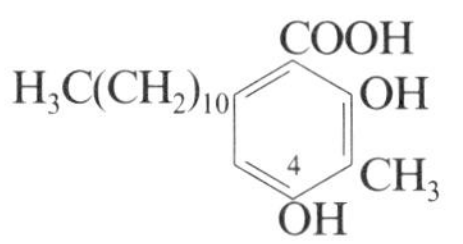

$C_{19}H_{30}O_4$ M 322.444

4-Me ether: [135607-81-7]. *2-Hydroxy-4-methoxy-3-methyl-6-undecylbenzoic acid.* ***Grammitic acid***

$C_{20}H_{32}O_4$ M 336.470

Constit. of the exudate of *Grammitis argyrata*.

Wollenweber, E. *et al*, *Phytochemistry*, 1991, **30**, 2307 (*isol, pmr, cmr*)

5,7-Dihydroxy-2-(4,7,10,13,16-nonadecapentaenyl)-4*H*-1-benzopyran-4-one — D-00199

5,7-Dihydroxy-2-(4,7,10,13,16-nonadecapentaenyl)chromone

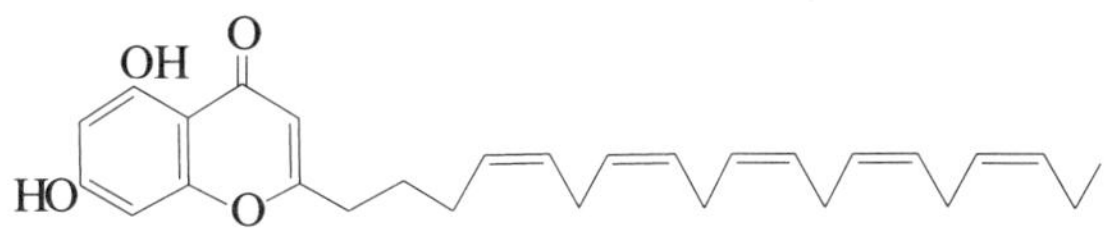

$C_{28}H_{34}O_4$ M 434.574

(all-Z)-form

Constit. of the brown alga *Zanaria tournefortii*. Mp 53-54°.

Tringali, C. *et al*, *Tetrahedron Lett.*, 1982, **23**, 1509 (*isol, ir, uv, ms, pmr, cmr*)

5,8-Dihydroxy-9,12-octadecadienoic acid — D-00200

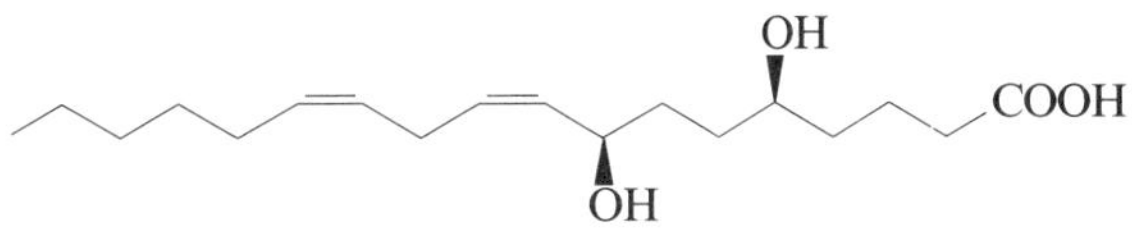

$C_{18}H_{32}O_4$ M 312.448

(5S,8R,9Z,12Z)-form [138078-48-5]

Sporogenic Psi Factor Cα. Psi Cα

Prod. by *Aspergillus nidulans*. Possesses sporogenic activity.

Mazur, P. *et al*, *J. Chem. Soc., Chem. Commun.*, 1991, 1486 (*isol, synth*)

9,10-Dihydroxyoctadecanedioic acid, 9CI — D-00201

Phloionic acid. Floionic acid

[137-21-3]

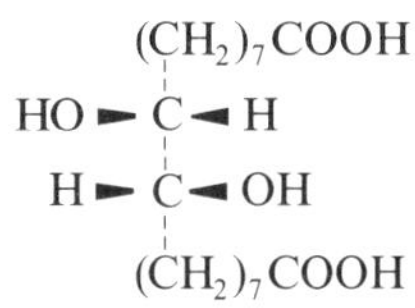

(9*R*,10*R*)-*form*

$C_{18}H_{34}O_6$ M 346.463

(9R,10R)-form [23843-53-0]

(+)-threo-*form*

Isol. from cork and oak. Mp 134-135°. $[\alpha]_D$ +21.45°.

Di-Me ester: [37428-14-1].

$C_{20}H_{38}O_6$ M 374.517

Mp 63-64°. $[\alpha]_D$ +20.45°.

(9S,10S)-form [23843-38-1]

(−)-threo-*form*

Isol. from cork and oak. Mp 132-134°. $[\alpha]_D$ −21.5°.

Di-Me ester: [37179-66-1].

Mp 63.5-64°. $[\alpha]_D$ −20.45°.

(9RS,10RS)-form [23843-52-9]

(±)-*form*

Isol. from cork and oak. Mp 124°. pK_{a2} 5.74.

Di-Me ester: [23200-14-8].

Mp 77-78°.

(9RS,10SR)-form [18287-34-8]

meso-*form*. *Isophloionic acid*

Mp 159.5-160.5° (153°).

Di-Me ester: [23200-17-1].

Mp 102-103°.

Zetsche, F. *et al*, *Helv. Chim. Acta*, 1931, **14**, 846 (*ester*)
Zetsche, F. *et al*, *J. Prakt. Chem.*, 1938, **150**, 140.

McGhie, J.F. *et al*, *J. Chem. Soc. C*, 1968, 1 (*synth, bibl*)
McGhie, J.F. *et al*, *Chem. Ind. (London)*, 1972, 536 (*abs config*)
Seoane, E. *et al*, *Chem. Ind. (London)*, 1978, 165 (*synth*)

2,3-Dihydroxyoctadecanoic acid D-00202

2,3-Dihydroxystearic acid

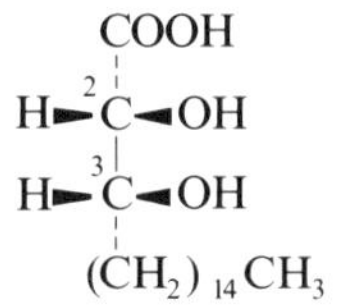

(2*RS*,3*RS*)-*form*
Relative configuration

$C_{18}H_{36}O_4$ M 316.480

(2RS,3RS)-form

(±)-erythro-*form*
Mp 108°.

Et ester:

$C_{20}H_{40}O_4$ M 344.534
Cryst. (pet. ether). Mp 69°.

(2RS,3SR)-form

(±)-threo-*form*
Mp 127°.

Et ester: Cryst. (MeCN). Mp 71-72°.

Palmeta, B. *et al*, *Tetrahedron*, 1963, **19**, 1463 (*synth*)
Palmeta, B. *et al*, *J. Org. Chem.*, 1964, **29**, 1026 (*config*)

4,14-Dihydroxyoctadecanoic acid D-00203

$H_3C(CH_2)_3CH(OH)(CH_2)_9CH(OH)CH_2CH_2COOH$

$C_{18}H_{36}O_4$ M 316.480

14-O-Diglucoside, Et ester: ***Muricatin A***

$C_{32}H_{21}O_{10}$ M 565.512
Isol. from *Ipomoea muricata*. Mp 117-118°. $[\alpha]_D^{28}$ −29.8° (EtOH). Complete struct. not known. Muricatin A is not homogeneous.

Khanna, S.N. *et al*, *Phytochemistry*, 1967, **6**, 735.

6,7-Dihydroxyoctadecanoic acid D-00204

6,7-Dihydroxystearic acid

[583-84-6]

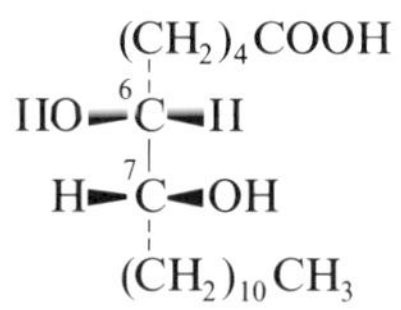

(6*RS*,7*RS*)-*form*
Relative configuration

$C_{18}H_{36}O_4$ M 316.480

(2RS,3RS)-form [13985-16-5]

(±)-threo-*form*
Mp 116-117°.

(2RS,3SR)-form [14073-16-6]

(±)-erythro-*form*
Mp 122-123°.

Kannan, R. *et al*, *Fette, Seifen, Anstrichm.*, 1974, **76**, 344 (*pmr*)
Osman, S.M. *et al*, *Fette, Seifen, Anstrichm.*, 1975, **77**, 473 (*synth*)

7,8-Dihydroxyoctadecanoic acid D-00205

7,8-Dihydroxystearic acid

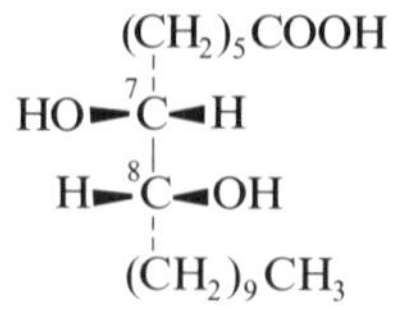

(7*RS*,8*RS*)-*form*
Relative configuration

$C_{18}H_{36}O_4$ M 316.480

(7RS,8RS)-form

(±)-threo-*form*
Cryst. (EtOH). Mp 94-94.5°.

(7RS,8SR)-form

(±)-erythro-*form*
Cryst. (EtOH). Mp 130.5-131°.

Fusari, S.A. *et al*, *J. Am. Oil Chem. Soc.*, 1951, **28**, 416 (*synth*)
Lutton, E.S. *et al*, *J. Am. Chem. Soc.*, 1951, **73**, 5206 (*struct*)

8,9-Dihydroxyoctadecanoic acid D-00206

8,9-Dihydroxystearic acid

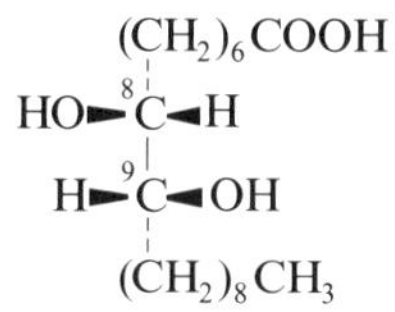

(8*RS*,9*RS*)-*form*
Relative configuration

$C_{18}H_{36}O_4$ M 316.480

(8RS,9RS)-form

(±)-threo-*form*
Cryst. (EtOH). Mp 94-94.3°.

(8RS,9SR)-form

(±)-erythro-*form*
Cryst. (EtOH). Mp 118-119°.

Fusari, S.A. *et al*, *J. Am. Oil Chem. Soc.*, 1951, **28**, 416 (*synth*)
Lutton, E.S. *et al*, *J. Am. Chem. Soc.*, 1951, **73**, 5206 (*struct*)

9,10-Dihydroxyoctadecanoic acid D-00207

9,10-Dihydroxystearic acid

[120-87-6]

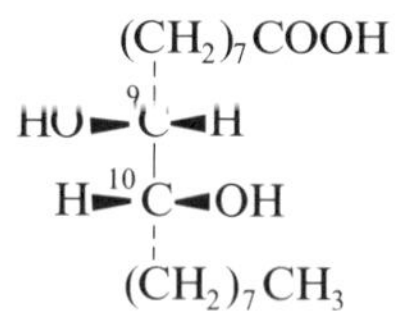

(9*R*,10*R*)-*form*

$C_{18}H_{36}O_4$ M 316.480

(9R,10R)-form [10067-09-1]

(+)-threo-*form*
Constit. of the sclerotial lipids of the fungus *Claviceps sulcata*. Mp 99.5°. $[\alpha]_D^{20}$ +23.5° (c, 5 in MeOH).

Me ester:

$C_{19}H_{38}O_4$ M 330.507
Mp 58-60°. $[\alpha]_D^{26}$ +22.5° (c, 0.1 in MeOH).

(9S,10R)-form [38003-70-2]

(−)-erythro-*form*
Prepared from *Shorea robusta* seed oil. Mp 136°. $[\alpha]_D$ −0.5° (c, 1 in $CHCl_3$/MeOH).

Me ester: [38172-86-0].

Mp 105°. $[\alpha]_D$ −2° (c, 1 in $CHCl_3$/MeOH).

Et ester:

$C_{20}H_{40}O_4$ M 344.534
Mp 98-99°. $[\alpha]_D$ −2.1°.

(9RS,10RS)-form [2391-05-1]
(±)-threo-*form*
Obt. by performic acid oxidn. of Oleic acid. Cryst. (EtOH). Mp 94-95°.
Me ester: [3639-31-4].
Mp 71°.
(9RS,10SR)-form [3639-32-5]
(±)-erythro-*form*
Obt. by performic acid oxidn. of Elaidic acid. Cryst. (EtOH). Mp 130-131°.
Me ester: [3639-30-3].
Mp 105-106°.
Et ester: Mp 99-100°.

Org. Synth., Coll. Vol., 4, 1963, 317 (*synth*)
Morris, L.J. *et al*, *Lipids*, 1972, **7**, 372 (*config*)
Korver, O. *et al*, *Recl. Trav. Chim. Pays-Bas* (*J. R. Neth. Chem. Soc.*), 1973, **92**, 127 (*isol, config*)
Kannan, R. *et al*, *Fette, Seifen, Anstrichm.*, 1974, **76**, 344 (*pmr*)

9,14-Dihydroxyoctadecanoic acid D-00208

[65370-66-3]

$H_3C(CH_2)_3CH(OH)(CH_2)_4CH(OH)(CH_2)_7COOH$

$C_{18}H_{36}O_4$ M 316.480
Constit. of *Peganum harmala* oil. Mp 89-90°.
Me ester:
$C_{19}H_{38}O_4$ M 330.507
Mp 65-66°.

Ahmad, I. *et al*, *Phytochemistry*, 1977, **16**, 1761 (*isol, pmr, ms*)

10,11-Dihydroxyoctadecanoic acid D-00209

10,11-Dihydroxystearic acid

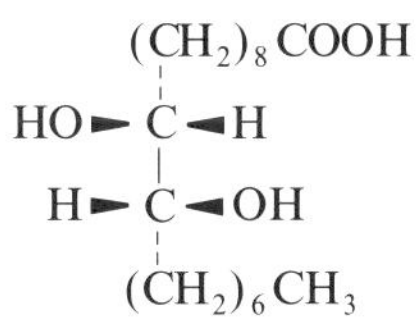

(10*RS*,11*RS*)-*form*

$C_{18}H_{36}O_4$ M 316.480
(10RS,11RS)-form
(±)-threo-*form*
Cryst. (EtOH). Mp 98-99.5°.
(10RS,11SR)-form
(±)-erythro-*form*
Cryst. (EtOH). Mp 120-121°.

Huber, W.F., *J. Am. Chem. Soc.*, 1951, **73**, 2730.

11,12-Dihydroxyoctadecanoic acid D-00210

11,12-Dihydroxystearic acid

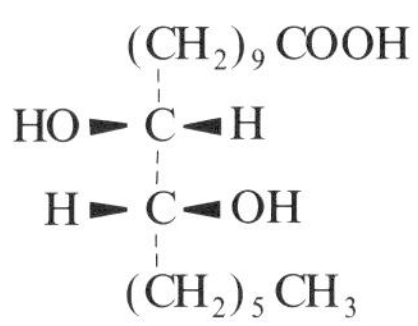

(11*RS*,12*RS*)-*form*

$C_{18}H_{36}O_4$ M 316.480
(11RS,12RS)-form
(±)-threo-*form*
Cryst. (EtOH aq.). Mp 95-96°.
(11RS,12SR)-form
(±)-erythro-*form*
Cryst. (EtOH). Mp 129-130°.

Ahmad, K. *et al*, *J. Am. Chem. Soc.*, 1948, **70**, 3391.
Bounds, D.G. *et al*, *J. Chem. Soc.*, 1954, 4219.

12,13-Dihydroxyoctadecanoic acid D-00211

12,13-Dihydroxystearic acid

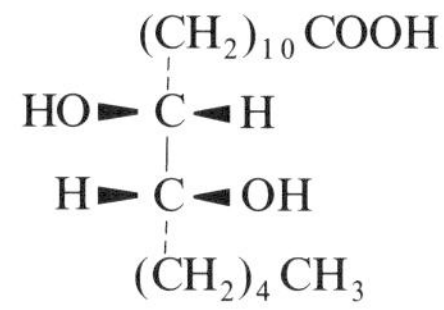

(12*R**,13*R**)-*form*
Relative configuration

$C_{18}H_{36}O_4$ M 316.480
(12R*,13R*)-form
(+)-threo-*form*
Mp 97.8-98°. $[\alpha]_D^{27}$ +23.8°.
(12S*,13S*)-form
(−)-threo-*form*
Mp 98.5°. $[\alpha]_D^{27}$ −23.4°.
(12RS,13RS)-form
(±)-threo-*form*
Cryst. (EtOH). Mp 96.5-97°.
Me ester: Mp 67.5-69°.
Et ester: Mp 61-63°.
4-Bromophenacyl ester: Mp 102-106°.
(12RS,13SR)-form
(±)-erythro-*form*
Cryst. (EtOH or MeOH). Mp 119-120°.
Me ester: Mp 102-103°.
Et ester: Mp 97-98°.
4-Bromophenacyl ester: Mp 136.5-137.5°.

Huber, W.F., *J. Am. Chem. Soc.*, 1951, **73**, 2730.
Bharucha, K.E. *et al*, *J. Chem. Soc.*, 1956, 1611.
Chisholm, M.C. *et al*, *Chem. Ind.* (*London*), 1960, 1134.
Ames, D.E. *et al*, *J. Chem. Soc.*, 1961, 4409.
Osbond, J.M., *J. Chem. Soc.*, 1961, 5270.
Scott, W.E. *et al*, *Chem. Ind.* (*London*), 1962, 2038.

15,16-Dihydroxyoctadecanoic acid D-00212

15,16-Dihydroxystearic acid

$H_3CCH_2CH(OH)CH(OH)(CH_2)_{13}COOH$

$C_{18}H_{36}O_4$ M 316.480
(+)-*form*
Prepared from *Camelina sativa* seed oil. Cryst. (pet. ether). Mp 96-97°. $[\alpha]_D^{20}$ +3.3° (c, 1 in EtOH).
Me ester:
$C_{19}H_{38}O_4$ M 330.507
Mp 89.5-90.5°.
Et ester:
$C_{20}H_{40}O_4$ M 344.534
Mp 74-74.5°.

Gunstone, F.D. *et al*, *J. Chem. Soc.*, 1959, 2127 (*synth*)

7,14-Dihydroxy-4,8,10,12-octadecatetraenedioic acid D-00213

18-Carboxy-19,20-dinor LTB$_4$. 18-Carboxyleukotriene B$_4$

[102674-12-4]

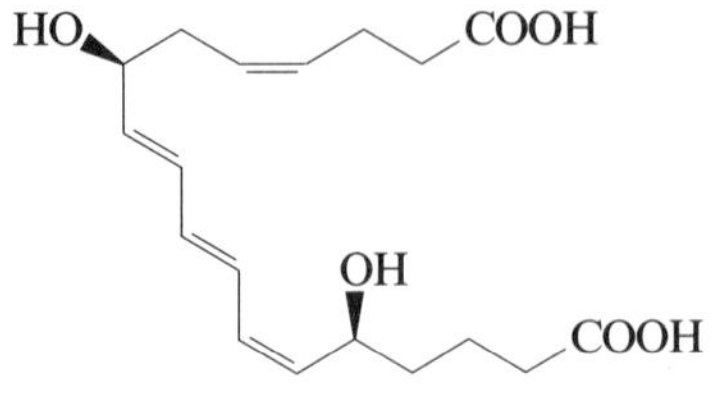

$C_{18}H_{26}O_6$ M 338.400

Formed in hepatocyte metab. of LTB$_4$ (see 5,12-Dihydroxy-6,8,10,14-eicosatetraenoic acid, D-00171).

Harper, T.W. *et al*, *J. Biol. Chem.*, 1986, **261**, 5414 (*biosynth*)
Jedlitschky, G. *et al*, *J. Biol. Chem.*, 1991, **266**, 24763 (*biosynth*)

5,8-Dihydroxy-9-octadecenoic acid D-00214

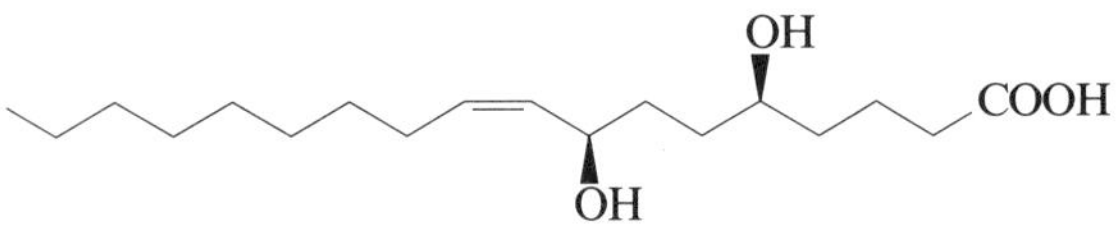

$C_{18}H_{34}O_4$ M 314.464

(5S,8R,9Z)-form [138078-50-9]

Sporogenic Psi Factor Cβ. Psi Cβ

Prod. by *Aspergillus nidulans*. Possesses sporogenic activity.

Mazur, P. *et al*, *J. Chem. Soc., Chem. Commun.*, 1991, 1486 (*isol, struct*)

9,10-Dihydroxy-12-octadecenoic acid D-00215

[53734-70-6]

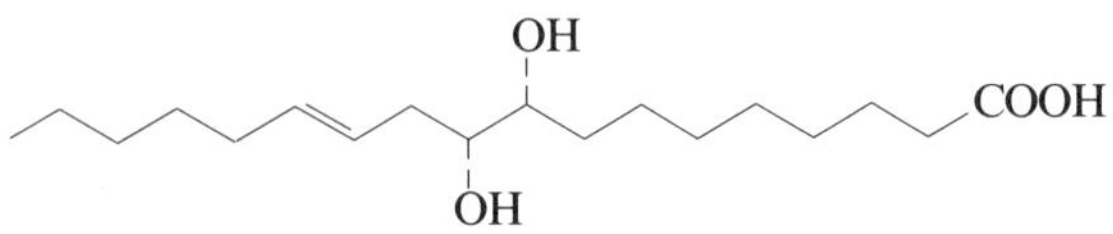

(9*RS*,10*RS*,12*E*)-*form*

$C_{18}H_{34}O_4$ M 314.464

(9RS,10RS,12E)-form [59959-43-2]

(±)-trans-threo-*form*

Cryst. (Et_2O/pet. ether). Mp 71-74°.

(9RS,10RS,12Z)-form [59959-42-1]

(±)-cis-threo-*form*

Cryst. (Et_2O/pet. ether). Mp 62-63°.

(9RS,10SR,12E)-form [59959-50-1]

(±)-trans-erythro-*form*

Cryst. (Et_2O/pet. ether). Mp 119-120.5°.

(9RS,10SR,12Z)-form [59959-49-8]

(±)-cis-erythro-*form*

Cryst. (Et_2O/pet. ether). Mp 96-99°.

Gunstone, F.D. *et al*, *Chem. Phys. Lipids*, 1975, **15**, 174 (*synth*;) 198 (*ms*)
Fukui, H. *et al*, *Agric. Biol. Chem.*, 1977, **41**, 175 (*isol*)

12,13-Dihydroxy-9-octadecenoic acid D-00216

[53734-71-7]

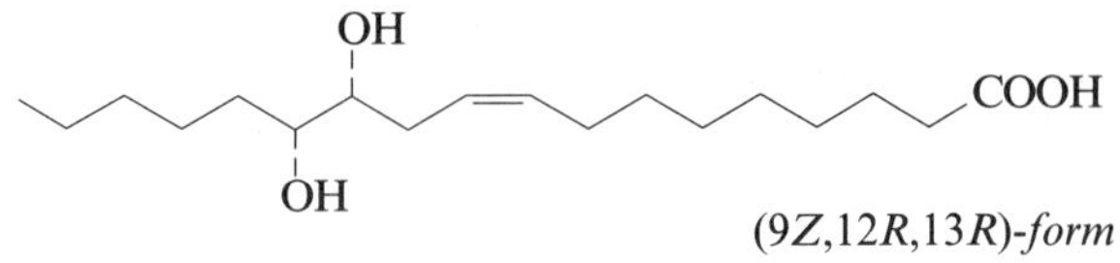

(9*Z*,12*R*,13*R*)-*form*

$C_{18}H_{34}O_4$ M 314.464

(9Z,12R,13R)-form [23201-88-9]

(+)-threo-*Dihydroxyoleic acid*

Prepared by enzymic action on *Vernonia* oils or chemical acetolysis of *Malope trifida* oil. Cryst. (EtOAc). Mp 63°. $[\alpha]_D^{20}$ +19° (EtOH).

(9Z,12S,13S)-form

(−)-threo-*Dihydroxyoleic acid*

Prepared by chemical acetolysis of *Vernonia* oils. Mp 63°. $[\alpha]_D^{20}$ −19° (EtOH).

1,13-Lactone: [63872-94-6]. *12-Hydroxy-9-octadecen-13-olide*

$C_{18}H_{32}O_3$ M 296.449

Isol. from *Crepis conyzaefolia* seed oils. Mp 49-50°. $[\alpha]_D^{26}$ −38.0° (c, 0.584 in EtOH).

(12RS,13RS,9Z)-form [59981-82-7]

(±)-threo-*Dihydroxyoleic acid*

Cryst. (Et_2O/pet. ether). Mp 50-52°.

(9E,12RS,13RS)-form [59981-83-8]

(±)-threo-*Dihydroxyelaidic acid*

Cryst. (Et_2O/pet. ether). Mp 65-68°.

(9Z,12RS,13SR)-form [59981-84-9]

(±)-erythro-*Dihydroxyoleic acid*

Cryst. (Et_2O/pet. ether). Mp 73-76°.

(9E,12RS,13SR)-form [59981-85-0]

(±)-erythro-*Dihydroxyelaidic acid*

Cryst. (Et_2O/pet. ether). Mp 103-105°.

Gunstone, F.D. *et al*, *J. Chem. Soc.*, 1954, 1611 (*synth*)
Scott, W.E. *et al*, *Chem. Ind. (London)*, 1962, 2038 (*isol*)
Miwa, J.K. *et al*, *J. Am. Oil Chem. Soc.*, 1963, **40**, 225 (*biosynth*)
Morris, L.J. *et al*, *Lipids*, 1972, **7**, 372 (*config*)
Gunstone, F.D. *et al*, *Chem. Phys. Lipids*, 1975, **15**, 174 (*synth*)
Spencer, G.F. *et al*, *Phytochemistry*, 1976, **16**, 764 (*lactone*)

2,3-Dihydroxy-1-octadec-9′-enyloxypropane D-00217

Glyceryl 1-octadec-9′-enyl ether. α-Oleyl glyceryl ether. Selachyl alcohol. Glyceryl α-oleyl ether

[15863-83-9]

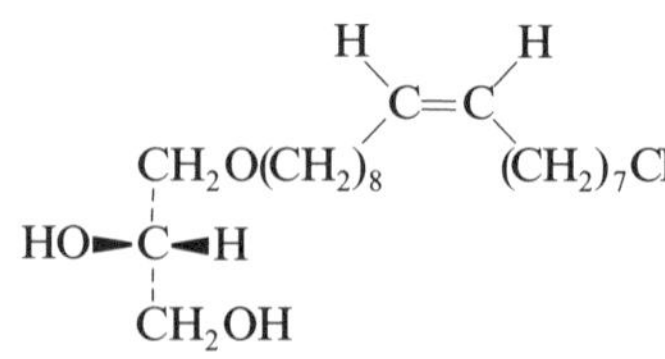

$C_{21}H_{42}O_3$ M 342.561

(S,Z)-form [34043-91-9]

D-cis-*form. 1-O-Octadec-(Z)-9′-enyl-*sn-*glycerol*

Constit. of shark liver and other elasmobranch fish oils. Oil. Opt. rotn. concn.-dependent.

Isopropylidene deriv.:

$C_{24}H_{46}O_3$ M 382.626

$[\alpha]_D^{22}$ −15.2° (c, 0.34 in hexane).

(RS,Z)-form [34783-94-3]

D,L-cis-*form*

Cryst. (hexane). Mp 17.6-18.5°.

Di-Ac: [21994-81-0].
$C_{25}H_{46}O_5$ M 426.635
Mp 32.9-33.7°.

Hallgren, B. *et al*, *J. Lipid Res.*, 1962, **3**, 31 (*occur*)
Wood, R. *et al*, *Lipids*, 1967, **2**, 161 (*synth, chromatog, pmr*)
Chacko, G.K. *et al*, *Biochim. Biophys. Acta*, 1968, **164**, 252 (*synth*)

2,3-Dihydroxy-1-octadecyloxypropane — D-00218

Batyl alcohol. Glyceryl 1-octadecyl ether. Glyceryl α-stearyl ether. α-Stearyl glyceryl ether. ***Batilol,*** **INN**
[544-62-7]

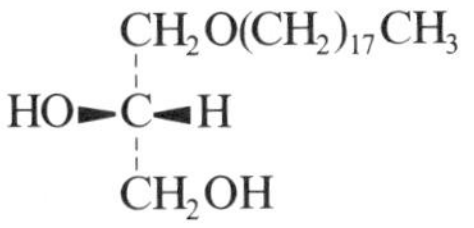

(S)-form

$C_{21}H_{44}O_3$ M 344.577
▷ TZ0960000.

(S)-form [10567-22-3]
*1-O-Octadecyl-*sn*-glycerol.* D*-form*
Constit. of shark liver and other elasmobranch fish oils. Radioprotective agent. Mp 70.5-71.5°. Opt. rotn. concn. dependent.
Di-Ac: [18330-40-0].
$C_{25}H_{48}O_5$ M 428.651
Mp 39-40°. $[\alpha]_D^{22}$ −11.8° (c, 0.27 in hexane).
Isopropylidene deriv.:
$C_{24}H_{48}O_3$ M 384.641
$[\alpha]_D^{22}$ −16.2° (c, 0.46 in hexane).

Heilbron, I.M. *et al*, *J. Chem. Soc.*, 1928, 942 (*isol*)
Baer, E. *et al*, *J. Biol. Chem.*, 1941, **140**, 397 (*synth, config*)
Kind, C.A. *et al*, *J. Org. Chem.*, 1942, **7**, 424 (*isol*)
Edlund, T., *Nature (London)*, 1954, **174**, 1102 (*props*)
Bodman, J. *et al*, *Clin. Chim. Acta*, 1958, **3**, 253 (*rev, occur*)
Hallgren, B. *et al*, *J. Lipid Res.*, 1962, **3**, 31 (*occur*)
Wood, R. *et al*, *Lipids*, 1967, **2**, 161 (*synth, chromatog, pmr*)
Chacko, G.K. *et al*, *Biochim. Biophys. Acta*, 1968, **164**, 252 (*synth*)
Suemune, H. *et al*, *Chem. Pharm. Bull.*, 1987, **35**, 3112 (*synth, ir, pmr*)

6,8-Dihydroxyoctanoic acid, 9CI — D-00219

[74903-53-0]

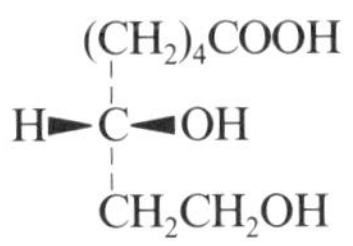

(S)-form

$C_8H_{16}O_4$ M 176.212

(S)-form
Me ester: [116349-04-3].
$C_9H_{18}O_4$ M 190.239
Lipoic acid precursor. $Bp_{0.005}$ 120-125°. $[\alpha]_D^{25}$ −4.1° ($CHCl_3$).

(±)-form
Me ester: [66859-07-2].
Oil.

Tsuji, J. *et al*, *J. Org. Chem.*, 1978, **43**, 3606 (*synth, pmr*)
White, R.H., *J. Am. Chem. Soc.*, 1980, **102**, 6605 (*synth*)
Brookes, M.H. *et al*, *J. Chem. Soc., Perkin Trans. 1*, 1988, 9 (*synth, pmr, ir, ms*)
Dasaradhi, L. *et al*, *J. Chem. Soc., Chem. Commun.*, 1990, 729 (*synth, pmr, use*)

9,11-Dihydroxy-7-oxaprostenoic acid — D-00220

6-[[3,5-Dihydroxy-2-(3-hydroxy-1-octenyl)cyclopentyl]oxy]hexanoic acid, 8CI

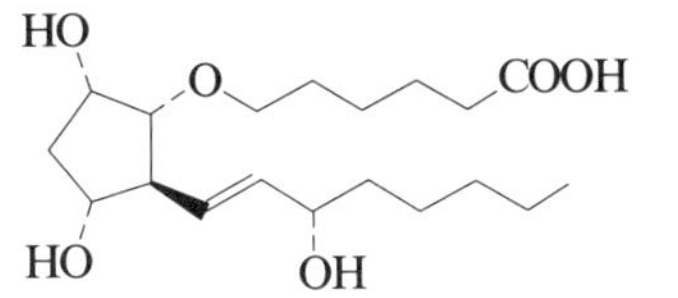

(+)-form

$C_{19}H_{34}O_6$ M 358.474

(+)-form [34153-08-7]
Mp 65-67°. $[\alpha]_D$ +6.8°.

(−)-form [34047-91-1]
Mp 65-67°. $[\alpha]_D$ −6.2°. Unnatural stereochem. No prostaglandin activity.

(±)-form [20476-32-8]
Mp 90.5-90.7°.

Fried, J. *et al*, *J. Chem. Soc., Chem. Commun.*, 1968, 634 (*synth, ir, pmr*)
Fried, J. *et al*, *J. Am. Chem. Soc.*, 1971, **93**, 5594 (*synth*)

6,7-Dihydroxy-4-oxo-2-heptenoic acid — D-00221

$HOCH_2CH(OH)CH_2COCH{=}CHCOOH$

$C_7H_{10}O_5$ M 174.153

(E)-form
7-O-Benzoyl, Me ester: [135626-21-0]. *Methyl 7-(benzoyloxy)-6-hydroxy-4-oxo-2-heptenoate.* ***7-Hydroxy-6-hydromelodienone***
$C_{15}H_{16}O_6$ M 292.288
Isol. from *Melodorum fruticosum*. Cytotoxic. Yellow cryst. Mp 96-97°.

Jung, J.H. *et al*, *J. Nat. Prod. (Lloydia)*, 1991, **54**, 500 (*isol, pmr, cmr*)

14,18-Dihydroxy-12-oxo-9,13,15-octadecatrienoic acid — D-00222

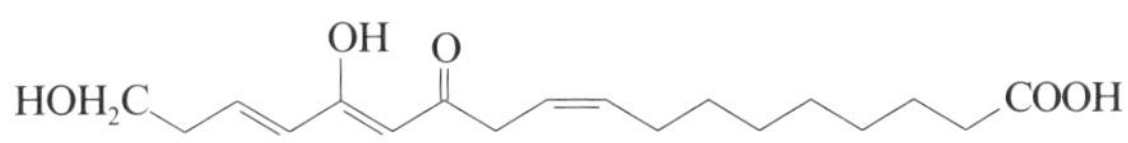

$C_{18}H_{28}O_5$ M 324.416

(9Z,13Z,15E)-form [130523-93-2] ***Cibaric acid***
Isol. from damaged fruit bodies of the mushroom *Cantharellus cibarius*. Cryst. (EtOAc). Mp 69.5-70.5°.

Pang, Z. *et al*, *J. Org. Chem.*, 1991, **56**, 1233 (*struct, ir, uv, ms, pmr, cmr*)

9,11-Dihydroxy-15-oxo-5,13-prostadienoic acid, 9CI — D-00223

7-[3,5-Dihydroxy-2-(3-oxo-1-octenyl)cyclopentyl]-5-heptenoic acid, 8CI
[35850-13-6]

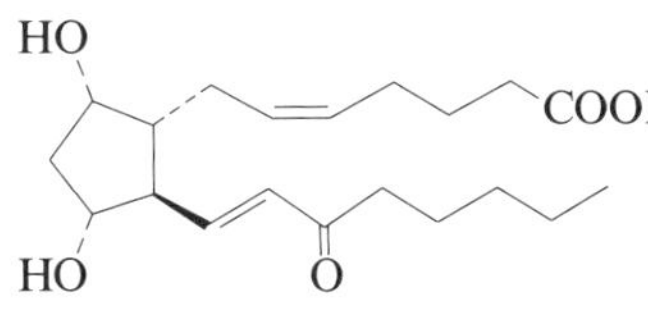

(5Z,9*S*,11*R*,13*E*)*-form*

$C_{20}H_{32}O_5$ M 352.470

(5Z,9S,11R,13E)-form
15-Oxo-PGF$_{2\alpha}$
Metab. of $PGF_{2\alpha}$ (see 9,11,15-Trihydroxyprosta-5,13-dienoic acid, T-00297).

13,14-Dihydro: see *9,11-Dihydroxy-15-oxo-5-prostenoic acid, D-00229*

Bundy, G.L. *et al*, *Adv. Biosci. (Oxford)*, 1972, **9**, 125 (*synth*)
Pace-Asciak, C. *et al*, *Prostaglandins*, 1972, **1**, 469 (*biosynth*)
Hansen, H.S., *Prostaglandins*, 1976, **12**, 647 (*rev*)
Schneider, W.P. *et al*, *J. Am. Chem. Soc.*, 1977, **99**, 1222 (*synth*)
Speziale, E. *et al*, *Biochem. Biophys. Res. Commun.*, 1984, **124**, 69 (*biochem*)

9,15-Dihydroxy-11-oxo-5,13-prostadienoic acid D-00224

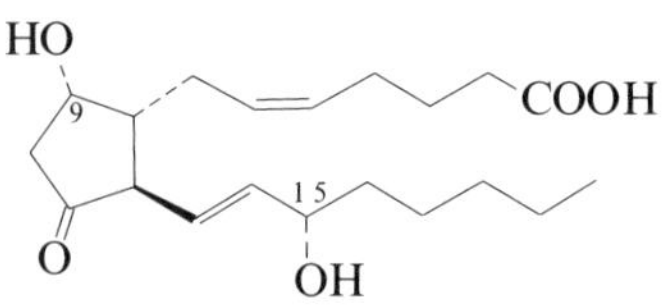

(5Z,9*S*,13*E*,15*S*)-*form*

$C_{20}H_{32}O_5$ M 352.470

(5Z,9S,13E,15S)-form [41598-07-6] ***Prostaglandin D$_2$***. *PGD$_2$*
Metab. of PGH_2, bronchoconstrictor, inhibits platelet aggregation. Cryst. Mp 68°.
▷ UK7930000.

13,14-Dihydro, 15-Oxo: [59894-07-4]. *9-Hydroxy-11,15-dioxo-5-prostenoic acid*
$C_{20}H_{32}O_5$ M 352.470
PGD_2 metab.

[118594-01-7, 118594-29-9, 118594-30-2, 118594-31-3]

Hayashi, M. *et al*, *J. Org. Chem.*, 1973, **38**, 2115 (*synth, ir, pmr*)
Jones, R.L. *et al*, *Br. J. Pharmacol.*, 1976, **56**, 339p (*biochem*)
Anderson, N.H. *et al*, *Prostaglandins*, 1977, **14**, 61 (*synth*)
Watanabe, T. *et al*, *Arch. Biochem. Biophys.*, 1982, **216**, 372 (*biosynth*)
Stehle, R.G., *Methods Enzymol.*, 1982, **86**, 436 (*rev*)
Martindale, The Extra Pharmacopoeia, 28th/29th Ed., Pharmaceutical Press, London, 1982/1989, 8094.
Bundy, G.L. *et al*, *J. Med. Chem.*, 1983, **26**, 790 (*synth, pharmacol*)
Collington, E.W. *et al*, *Tetrahedron Lett.*, 1983, **24**, 3125 (*synth*)
Cepa, S.R. *et al*, *Prostaglandins*, 1984, **27**, 645 (*ms*)
Suzuki, M. *et al*, *Tetrahedron Lett.*, 1984, **25**, 1383 (*synth*)
Whittle, B.J.R. *et al*, *Adv. Exp. Med. Biol.*, 1985, **192**, 109 (*rev, props*)
Cainelli, G. *et al*, *Tetrahedron*, 1985, **41**, 1385 (*synth*)
Ogawa, Y. *et al*, *J. Org. Chem.*, 1986, **51**, 1625 (*synth*)
Eur. Pat., 281 239, (1988); *CA*, **110**, 75151y (*synth, pmr*)
Giles, H. *et al*, *Prostaglandins*, 1988, **35**, 277 (*rev*)
Lewis, R.J., *Sax's Dangerous Properties of Industrial Materials*, 8th Ed., Van Nostrand Reinhold, 1992, POC275.

11,15-Dihydroxy-9-oxo-5,13-prostadienoic acid, 9CI D-00225

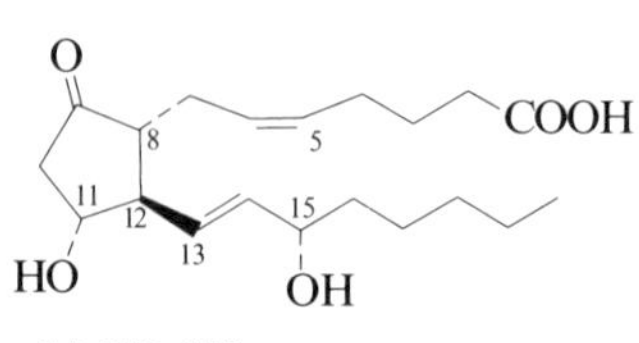

$C_{20}H_{32}O_5$ M 352.470

(5Z,8R,11R,12R,13E,15R)-form [38873-82-4]
15(R)-*Prostaglandin E$_2$*
Isol. from *Plexaura homomalla*.

Me ester: [31687-35-1].
Oil.

(5Z,8R,11R,12R,13E,15S)-form [363-24-6]
Prostaglandin E$_2$. PGE$_2$. ***Dinoprostone, BAN, INN, JAN, USAN.*** *Dinopron EM. Enzaprost E. Medullin. Minprostin E$_2$. Prostarmon E. Prostenon. Prostin E$_2$. U 12062*
Most common and biologically active of mammalian prostaglandins. Oxytocic, abortifacient and vasodilator. Cryst. Mp 66-68°. $[\alpha]_D^{26}$ −61° (c, 1 in THF).
▷ UK8000000.

(5E,8R,11R,12R,13E,15S)-form [36150-00-2]
(*5*E)-*PGE$_2$*
Plates (Et_2O/hexane). Mp 76-77°. $[\alpha]_D$ −66° (c, 0.983 in EtOH), −95° (c, 0.903 in $CHCl_3$).
▷ Exp. teratogen.

(5Z,8R,11S,12R,13E,15S)-form [38310-90-6]
11-epi-*PGE$_2$*
Oil. $[\alpha]_D^{25}$ −26° (c, 0.0076 in EtOH).

(5Z,8SR,11RS,12RS,13E,15SR)-form [31660-17-0]
(±)-*8*-iso-*PGE$_2$*
Cryst. (EtOAc/heptane). Mp 90-92°.

(5Z,8RS,11RS,12RS,13E,15RS)-form [31660-13-6]
(±)-*15*-epi-*PGE$_2$*

[38873-84-6]

Bergström, S. *et al*, *Biochim. Biophys. Acta*, 1964, **90**, 207 (*biosynth*)
Corey, E.J. *et al*, *J. Am. Chem. Soc.*, 1969, **91**, 5675; *Tetrahedron Lett.*, 1970, 307 (*synth*)
Light, R.J. *et al*, *Eur. J. Biochem.*, 1972, **28**, 232 (*isol*)
Bundy, G.L. *et al*, *J. Am. Chem. Soc.*, 1972, **94**, 2124 (*synth*)
Horton, E.W., *Proc. R. Soc. London, B*, 1972, **182**, 411 (*rev*)
Floyd, D.M. *et al*, *Tetrahedron Lett.*, 1972, 3269 (*synth*)
Von Euler, U.S., *Arch. Int. Pharmacodyn. Ther.*, 1973, **202**, 295 (*rev, pharmacol*)
Heather, J.B. *et al*, *Tetrahedron Lett.*, 1973, 2313 (*synth*)
Sih, C.J. *et al*, *J. Am. Chem. Soc.*, 1975, **97**, 865 (*synth*)
Uekama, K. *et al*, *Chem. Lett.*, 1977, 1389 (*cd*)
Schneider, W.P. *et al*, *J. Am. Chem. Soc.*, 1977, **99**, 1222 (*synth, ms*)
Andersen, N.H. *et al*, *Prostaglandins*, 1977, **14**, 61 (*synth*)
Van Dyk, J.M., *Drugs of Today (Barcelona)*, 1978, **14**, 74 (*rev*)
Chen, S.-M.L. *et al*, *J. Org. Chem.*, 1978, **43**, 3450 (*synth, ir, pmr, cmr, ms*)
Nakamura, N. *et al*, *Tetrahedron Lett.*, 1978, 1549 (*synth, pmr*)
Newton, R.F. *et al*, *J. Chem. Soc., Perkin Trans. 1*, 1979, 2789 (*synth*)
de Titta, G.T. *et al*, *Acta Crystallogr., Sect. B*, 1980, **36**, 638 (*cryst struct, conformn*)
Howard, C. *et al*, *J. Chem. Soc., Perkin Trans. 1*, 1981, 2049 (*Me ester, synth, ir, pmr*)
Stehle, R.G., *Methods Enzymol.*, 1982, **86**, 436 (*rev*)
Martindale, The Extra Pharmacopoeia, 28th/29th Ed., Pharmaceutical Press, London, 1982/1989, 8082.
Donaldson, R.E. *et al*, *J. Org. Chem.*, 1983, **48**, 2167 (*synth*)
Tanaka, T. *et al*, *Tetrahedron Lett.*, 1985, **26**, 5575 (*synth*)
Suzuki, M. *et al*, *J. Am. Chem. Soc.*, 1988, **110**, 4718 (*synth*)
Johnson, C.R. *et al*, *J. Am. Chem. Soc.*, 1988, **110**, 4726 (*synth, pmr, cmr*)
Murray, C.K. *et al*, *J. Am. Chem. Soc.*, 1990, **112**, 5660 (*synth*)
Lewis, R.J., *Sax's Dangerous Properties of Industrial Materials*, 8th Ed., Van Nostrand-Reinhold, 1992, DVJ200.

9,11-Dihydroxy-15-oxoprostanoic acid, 9CI D-00226

2,4-Dihydroxy-5-(3-oxooctyl)cyclopentaneheptanoic acid, 8CI
[56649-46-8]

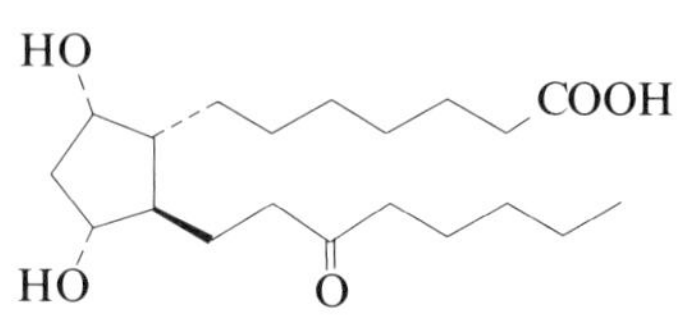

$C_{20}H_{36}O_5$ M 356.501

(9S,15R)-form [29044-75-5]
13,14-Dihydro-15-oxo-PGF$_{1\alpha}$
Metab. of $PGF_{1\alpha}$. Pale-yellow oil.

Pace-Asciak, C. *et al*, *Biochim. Biophys. Acta*, 1970, **218**, 288.
Lincoln, F.H. *et al*, *J. Org. Chem.*, 1973, **38**, 951 (*synth, ir, pmr, ms*)

11,15-Dihydroxy-9-oxoprostanoic acid D-00227

2-(6-Carboxyhexyl)-4-hydroxy-3-(3-hydroxyoctyl)-1-cyclopentanone

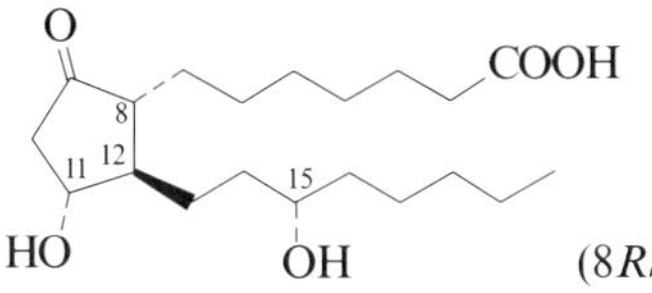

(8*RS*,11*RS*,12*RS*,15*SR*)-*form*

$C_{20}H_{36}O_5$ M 356.501

(8RS,11RS,12RS,15SR)-form [28896-13-1]
(±)-*(11α,15α)-form*. (±)-*Dihydro-PGE*$_1$
Oil.

(8R,11S,12R,15S)-form [25140-29-8]
(+)-*(11β,15α)-form*. (+)-*11*-epi-*Dihydro-PGE*$_1$
$[\alpha]_D$ +12° ($CDCl_3$).

(8RS,11SR,12SR,15SR)-form [34389-02-1]
(±)-*(11β,15α)-form*. (±)-*11*-epi-*Dihydro-PGE*$_1$
Oil.

(8RS,11RS,12RS,15RS)-form [34389-01-0]
(±)-*(11α,15β)-form*. (±)-*15*-epi-*Dihydro-PGE*$_1$
Oil.

(8RS,11SR,12RS,15RS)-form [34389-00-9]
(±)-*(11β,15β)-form*. (±)-*11,15*-diepi-*PGE*$_1$
Cryst. (EtOAc/hexane). Mp 76-77°.

Änggard, E. *et al*, *J. Biol. Chem.*, 1964, **239**, 4097 (*ms*)
Beal, P.F. *et al*, *J. Am. Chem. Soc.*, 1966, **88**, 3131 (*synth*)
Klok, R. *et al*, *Recl. Trav. Chim. Pays-Bas* (*J. R. Neth. Chem. Soc.*), 1970, **89**, 1043 (*synth, pmr, ir, uv*)
Miyano, M. *et al*, *J. Org. Chem.*, 1972, **37**, 1818 (*synth, pmr*)
Lincoln, F.H. *et al*, *J. Org. Chem.*, 1973, **38**, 951 (*synth, ir, pmr, ms*)

11,15-Dihydroxy-9-oxo-5,13,17-prostatrienoic acid, 9CI D-00228

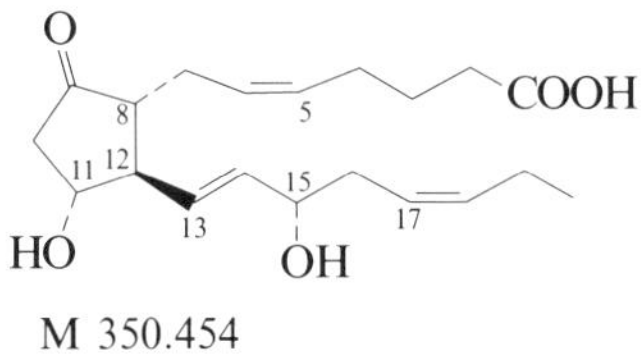

$C_{20}H_{30}O_5$ M 350.454

(5Z,8R,11R,12R,13E,15S,17Z)-form
Prostaglandin E$_3$. *PGE*$_3$
Constit. of sheep prostate gland and human menstrual fluid. Viscous oil. $[\alpha]_D^{24}$ −48.9° (c, 1.2 in THF).

Bergström, S. *et al*, *Ark. Kemi*, 1963, **19**, 563 (*isol, ms*)
Samuelsson, B., *J. Am. Chem. Soc.*, 1963, **85**, 1878 (*struct, pmr*)
Axen, U. *et al*, *J. Chem. Soc., Chem. Commun.*, 1970, 602 (*synth*)
Corey, E.J. *et al*, *J. Am. Chem. Soc.*, 1971, **93**, 1490 (*synth*)

9,11-Dihydroxy-15-oxo-5-prostenoic acid, 9CI D-00229

7-[3,5-Dihydroxy-2-(3-oxooctyl)cyclopentyl]-5-heptenoic acid, 8CI

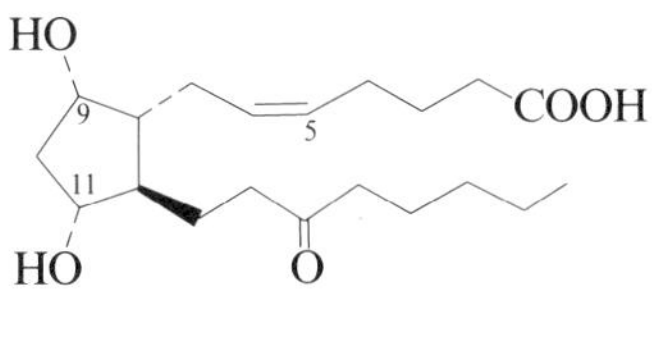

$C_{20}H_{34}O_5$ M 354.486

(5Z,9S,11R,13E)-form [27376-76-7]
Metab. of $PGF_{2\alpha}$ and PGE_2 occurring in blood, urine and semen. Viscous oil.

Hamberg, M. *et al*, *Ann. N.Y. Acad. Sci.*, 1971, **180**, 164.

9,11-Dihydroxy-15-oxo-13-prostenoic acid, 9CI D-00230

2,4-Dihydroxy-5-(3-oxo-1-octenyl)cyclopentaneheptanoic acid, 8CI. 3,5-Dihydroxy-2-(3-oxo-1-octenyl) cyclopentaneheptanoic acid, 8CI. 15-Oxoprostaglandin F$_{1\alpha}$. *15-Oxo-PGF*$_{1\alpha}$
[52087-42-0]

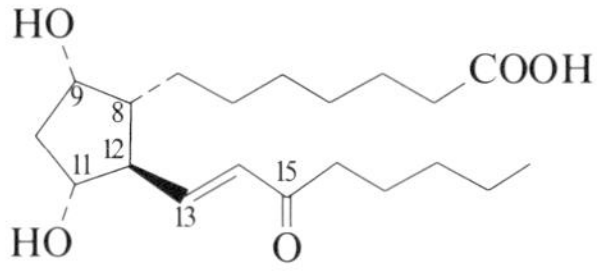

(8*RS*,9*SR*,11*RS*,13*E*)-*form*

$C_{20}H_{34}O_5$ M 354.486
Metab. of $PGF_{1\alpha}$ (in enantiomeric form). Mp 81-82° (racemate).

[52087-41-9]

Pike, J.E. *et al*, *J. Org. Chem.*, 1969, **34**, 3512 (*synth, ir, uv*)
Pace-Asciak, C. *et al*, *Biochim. Biophys. Acta*, 1970, **218**, 288.
Miyano, M. *et al*, *J. Org. Chem.*, 1975, **40**, 1748 (*synth, pmr, uv*)

9,15-Dihydroxy-11-oxo-13-prostenoic acid, 9CI D-00231

5-Hydroxy-2-(3-hydroxy-1-octenyl)-3-oxocyclopentaneheptanoic acid, 8CI

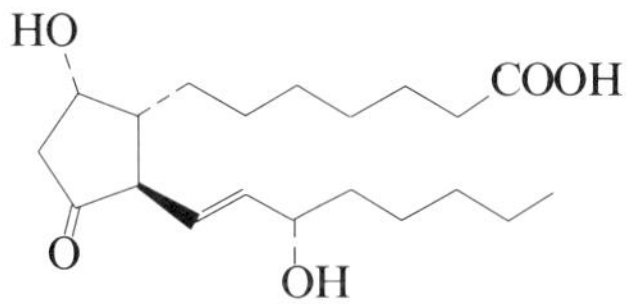

$C_{20}H_{34}O_5$ M 354.486

(9S,13E,15S)-form [71121-05-6]
Prostaglandin D$_1$. *PGD*$_1$
Formed by sheep seminal vesicles.

Granstrom, E. *et al*, *J. Biol. Chem.*, 1968, **243**, 4104.
Foss, P.S., *Biochemistry*, 1972, **11**, 2271 (*ms, pmr, biosynth*)
Cameron, A.G. *et al*, *J. Chem. Soc., Perkin Trans. 1*, 1984, 2337 (*synth, ir, pmr*)
Suzuki, M. *et al*, *Tetrahedron Lett.*, 1984, **25**, 1383 (*synth*)

11,15-Dihydroxy-9-oxo-13-prostenoic acid, 9CI D-00232

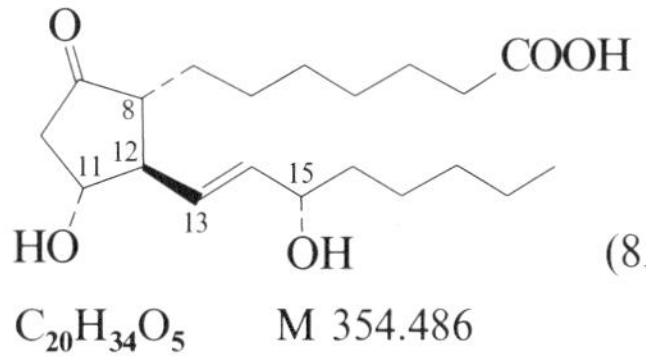

(8*R*,11*R*,12*R*,13*E*,15*S*)-*form*

$C_{20}H_{34}O_5$ M 354.486

(8R,11R,12R,13E,15S)-form [745-65-3]
(8α,11α,15α)-form. Prostaglandin E$_1$. (−)-*PGE*$_1$.
Alprostadil, BAN, USAN, INN. *Minprog. Prostin VR. U 10136*
Constit. of sheep's prostate and calf thymus glands. Occurs in semen. Vasodilator, oxytocic and abortifacient. Cryst. (EtOAc). Mp 114-116.5°. $[\alpha]_D^{24}$ −53.2° (c, 0.977 in THF).

▷ GY4569800.

Et ester: [35900-16-4]. *Prostaglandin E_1 ethyl ester*
Inhibits human platelet aggregation.

(8S,11S,12S,13E,15R)-form
(+)-*PGE_1*. ent-*PGE_1*
Mp 114-117°. $[\alpha]_D$ +57° (c, 0.50 in THF).

(8RS,11RS,12RS,13E,15SR)-form [20348-58-7]
(±)-*PGE_1*
Mp 113.5-115°.

Me ester: (±)-*PGE_1 methyl ester*
Oil.

(8S,11R,12R,13E,15S)-form [21003-46-3]
(8β,11α,15α)-form. 8-epi-*Prostaglandin E_1*. (−)-*8*-epi-*PGE_1*. *Ovinonic acid*
Cryst. (EtOAc/pet. ether). Mp 87-88°. Polymorphic.

(8RS,11SR,12SR,13E,15RS)-form [23203-65-8]
(±)-*8*-epi-*Prostaglandin E_1*. (±)-*8*-epi-*PGE_1*
Mp 101-102°.

Me ester: Mp 52-53°.

(8RS,11SR,12RS,13E,15SR)-form [20348-68-9]
(8α,11β,15α)-form. (±)-*11*-epi-*Prostaglandin E_1*. (±)-*11*-epi-*PGE_1*
Cryst. (Et_2O). Mp 92.5-93°.

(8RS,11RS,12RS,13E,15RS)-form [20897-96-5]
(8α,11α,15β)-form. (±)-*15*-epi-*Prostaglandin E_1*. (±)-*15*-epi-*PGE_1*
Oil.

Me ester: (±)-*15*-epi-*PGE_1 methyl ester*
Oil.

(8RS,11SR,12SR,13E,15SR)-form [55869-97-1]
(8β,11α,15β)-form. (±)-*8,15*-diepi-*Prostaglandin E_1*. (±)-*8,15*-diepi-*PGE_1*
Oil.

(8RS,11SR,12RS,13E,15RS)-form [20348-69-0]
(8α,11β,15β)-form. (±)-*11,15*-diepi-*Prostaglandin E_1*. (±)-diepi-*PGE_1*
Has twice the biological activity of (±)-Prostaglandin E_1. Cryst. (EtOAc). Mp 91-92°.

Berti, F. *et al*, *Prog. Biochem. Pharmacol.*, 1967, **3**, 110 (*rev, props*)
Corey, E.J. *et al*, *J. Am. Chem. Soc.*, 1968, **90**, 3245, 3247; 1969, **91**, 535 (*synth*)
Daniels, E.G. *et al*, *J. Am. Chem. Soc.*, 1968, **90**, 5894 (*isol*)
Schneider, W.P. *et al*, *J. Am. Chem. Soc.*, 1968, **90**, 5895 (*synth*)
Hamberg, M. *et al*, *J. Biol. Chem.*, 1971, **246**, 6713 (*metab*)
Weeks, J.R. *et al*, *Annu. Rev. Pharmacol.*, 1972, **12**, 317 (*rev*)
Sih, C.J. *et al*, *J. Am. Chem. Soc.*, 1972, **94**, 3643 (*synth*)
Lincoln, F.H. *et al*, *J. Org. Chem.*, 1973, **38**, 951 (*synth*)
Lukacs, G. *et al*, *Tetrahedron Lett.*, 1973, 515 (*cmr*)
Sih, C.J. *et al*, *J. Am. Chem. Soc.*, 1975, **97**, 857 (*synth, ir, pmr*)
Miyano, M. *et al*, *J. Org. Chem.*, 1975, **40**, 1740, 1748 (*synth, bibl, ord, cd*)
Spek, A.L., *Acta Crystallogr., Sect. B*, 1977, **33**, 816 (*cryst struct*)
Murakami, A. *et al*, *Chem. Pharm. Bull.*, 1977, **25**, 2870; 1979, **27**, 548 (*conformn*)
Nakamura, N. *et al*, *Tetrahedron Lett.*, 1978, 1549 (*synth*)
Bernady, K.F. *et al*, *J. Org. Chem.*, 1979, **44**, 1438 (*synth*)
Mai, J. *et al*, *Prostaglandins*, 1980, **20**, 187 (*isol*)
Nokami, J. *et al*, *Bull. Chem. Soc. Jpn.*, 1982, **55**, 3043 (*synth*)
Martindale, The Extra Pharmacopoeia, 28th/29th Ed., Pharmaceutical Press, London, 1982/1989, 8071.
Tanaka, T. *et al*, *Tetrahedron Lett.*, 1983, **24**, 4103 (*synth*)
Cepa, S.P. *et al*, *Prostaglandins*, 1984, **27**, 645 (*ms*)
Whittle, B.J.R. *et al*, *Adv. Exp. Med. Biol.*, 1985, **192**, 109 (*rev, props*)
Suzuki, M. *et al*, *J. Am. Chem. Soc.*, 1988, **110**, 4718 (*Me ester, synth, ir, pmr*)
Johnson, C.R. *et al*, *J. Org. Chem.*, 1991, **56**, 3352 (*synth, ir, pmr, cmr*)
Lewis, R.J., *Sax's Dangerous Properties of Industrial Materials*, 8th Ed., Van Nostrand-Reinhold, 1992, POC350.

11,16-Dihydroxy-9-oxo-13-prostenoic acid, 9CI D-00233

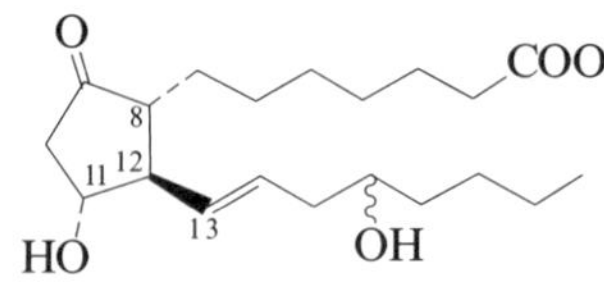

$C_{20}H_{34}O_5$ M 354.486

(8RS,11RS,12RS,13E)-form
Bronchodilator.

Me ester: [59122-45-1].
Inhibits gastric secretion.

Collins, P.W. *et al*, *Tetrahedron Lett.*, 1975, 4217 (*synth, pmr*)
Collins, P.W. *et al*, *J. Med. Chem.*, 1977, **20**, 1152 (*synth*)

11,15-Dihydroxy-9-oxo-5-prosten-13-ynoic acid D-00234

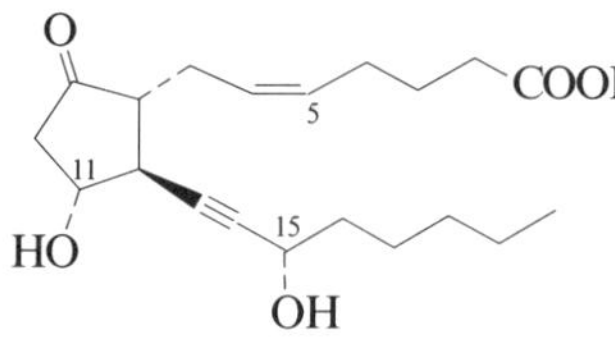

$C_{20}H_{30}O_5$ M 350.454

(5Z,11R,15S)-form [39918-02-0]
13,14-Dehydro-PGE_2

Cooper, G.F. *et al*, *Proc. Int. Conf. Stable Isot. Chem., Biol., Med., 1st*, 1973, 72 (*cmr*)
Fried, J. *et al*, *Tetrahedron Lett.*, 1973, 3899 (*synth*)

3,6-Dihydroxy-2-(1-oxo-10-tetradecenyl)-2-cyclohexen-1-one, 9CI D-00235

[112727-26-1]

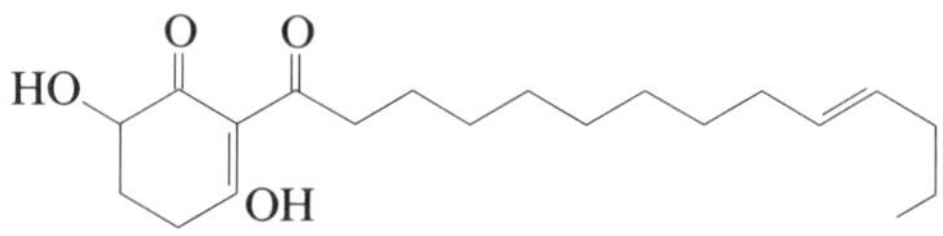

$C_{20}H_{32}O_4$ M 336.470
Constit. of *Corythucha ciliata*. Oil.

Lusby, W.R. *et al*, *J. Nat. Prod. (Lloydia)*, 1987, **50**, 1126.

9,11-Dihydroxy-15-oxo-2,3,18,19-tetranorprost-5-ene-1,20-dioic acid D-00236

2-(4-Carboxy-2-butenyl)-3,5-dihydroxy-γ-oxocyclopentanehexanoic acid, 9CI
[99295-34-8]

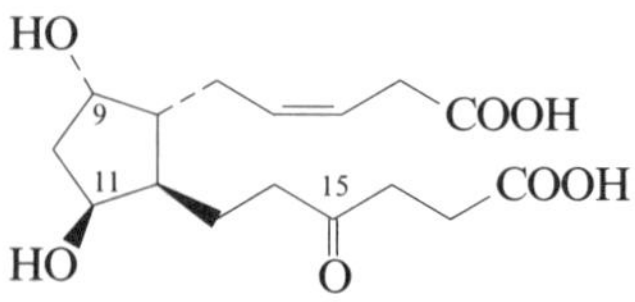

$C_{16}H_{24}O_7$ M 328.361

(5Z,9S,11S)-form
Major urinary metabolite of prostaglandin D_2.
[118908-30-8, 118908-31-9]

Liston, T.E. *et al*, *J. Biol. Chem.*, 1985, **260**, 13172 (*isol, struct*)
Prakash, C. *et al*, *J. Chem. Soc., Perkin Trans. 1*, 1988, 2821 (*synth, pmr, ms*)

3,12-Dihydroxypentadecanoic acid D-00237

Turpetholic acid A

[66895-58-7]

$H_3CCH_2CH_2CH(OH)(CH_2)_8CH(OH)CH_2COOH$

$C_{15}H_{30}O_4$ M 274.400

Me ester: [66884-49-9].
$C_{16}H_{32}O_4$ M 288.426
Cryst. (Et_2O). Mp 49-50°.

12-O-[β-D-Glucopyranosyl-(1→3)-α-L-rhamnopyranosyl-(1→3)-β-D-glucopyranosyl-(1→3)-β-D-glucopyranoside]: [66895-61-2]. ***Turpethinic acid A***
$C_{39}H_{70}O_{23}$ M 906.968
Isol. from *Ipomoea turpethum. I. turpethum* resin used as a purgative. Mp 75° (as Me ester).

[66965-43-3, 130548-12-8]

Wagner, H. *et al*, *Planta Med.*, 1978, **33**, 144 (*isol, synth*)

4,12-Dihydroxypentadecanoic acid D-00238

Turpetholic acid B

[66895-59-8]

$H_3CCH_2CH_2CH(OH)(CH_2)_7CH(OH)CH_2CH_2COOH$

$C_{15}H_{30}O_4$ M 274.400

Me ester: [66884-50-2].
$C_{16}H_{32}O_4$ M 288.426
Mp 65°.

12-O-[β-D-Glucopyranosyl-(1→3)-α-L-rhamnopyranosyl-(1→3)-β-D-glucopyranosyl-(1→3)-β-D-glucopyranoside]: [66895-62-3]. ***Turpethinic acid B***
$C_{39}H_{70}O_{23}$ M 906.968
Isol. from *Ipomoea turpethum.*

Wagner, H. *et al*, *Planta Med.*, 1978, **33**, 144 (*isol*)

2,4-Dihydroxy-6-(8-pentadecenyl)benzoic acid D-00239

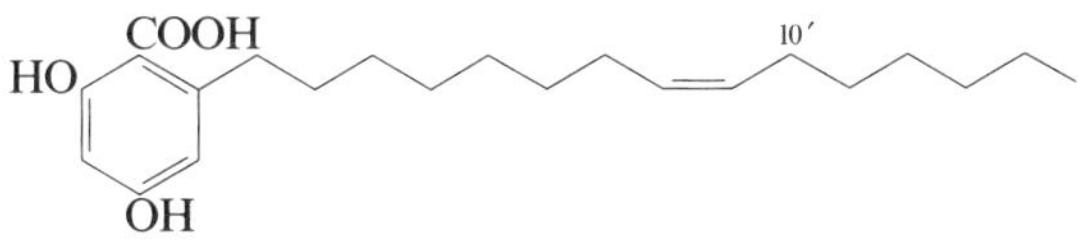

$C_{22}H_{34}O_4$ M 362.508

(Z)-form [62071-06-1]
Constit. of *Ginkgo biloba* and *Ononis speciosa.*

10′-Acetoxy: [120727-07-3]. *6-(10-Acetoxy-8-pentadecenyl)-2,4-dihydroxybenzoic acid*
$C_{24}H_{36}O_6$ M 420.545
Constit. of *O. speciosa.*

[120727-09-5]

Gellerman, J.L. *et al*, *Phytochemistry*, 1976, **15**, 1959 (*isol, uv, pmr*)
Barrero, A.F. *et al*, *Phytochemistry*, 1989, **28**, 161 (*isol, cmr*)

1-(2,6-Dihydroxyphenyl)-4-hexadecen-1-one D-00240

1-(4-Hexadecenoyl)-2,6-dihydroxybenzene. 2-(4-Hexadecenyl)-1,3-benzenediol

[96820-26-7]

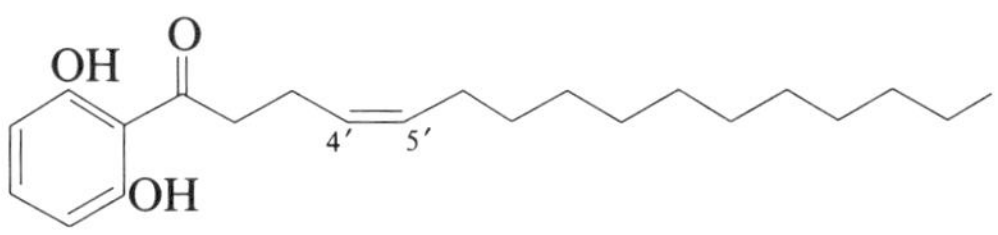

$C_{22}H_{34}O_3$ M 346.509

Constit. of *Virola* spp. Viscous oil.

4′,5′-Dihydro: [96820-25-6]. *2-Hexadecanoyl-1,3-benzenediol. 1-Hexadecanoyl-2,6-dihydroxybenzene*
$C_{22}H_{36}O_3$ M 348.525
Constit. of *V. elongata.* Amorph.

Kato, M.J. *et al*, *Phytochemistry*, 1985, **24**, 533.

25-(3,5-Dihydroxyphenyl)-9,10-pentacosanediol D-00241

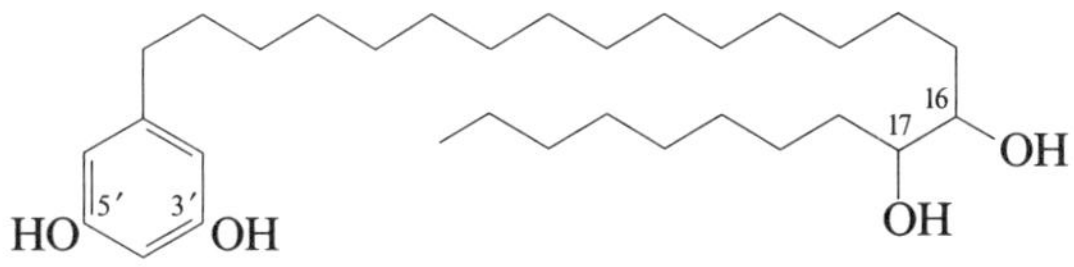

$C_{31}H_{56}O_4$ M 492.781

16(17)-Ac, 3′,5′-di-O-sulfate: [85528-23-0]. ***Antibiotic A 32724A.*** *A 32724A*
$C_{33}H_{58}O_{11}S_2$ M 694.947
Isol. from *Chaetomella raphigera.* Active against gram-positive and -negative bacteria, fungi and yeasts. Possesses hypotensive activity and inhibits glucosyl transferase. Amorph. powder (as di-Na salt). Mp 175-176° dec. (di-Na salt). Posn. of Ac group not certain.

17(16)-Ac, 3′,5′-di-O-sulfate: [85537-32-2]. ***Antibiotic A32724B.*** *A 32724B*
$C_{33}H_{58}O_{11}S_2$ M 694.947
Isol. from *C. raphigera.* Amorph. powder (as di-Na salt). Mp 160-162° dec. (di-Na salt).

Japan. Pat., 83 188 851, (1983); *CA*, **100**, 119333 (*isol*)
U.S. Pat., 4 375 462, (1983); *CA*, **98**, 177504 (*isol*)

1-(2,6-Dihydroxyphenyl)-1-tetradecanone, 9CI D-00242

2-Tetradecanoyl-1,3-benzenediol

[113201-68-6]

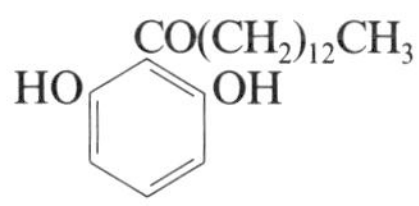

$C_{20}H_{32}O_3$ M 320.471

Constit. of the stem bark of *Myristica dactyloides.* Pale-yellow needles (CH_2Cl_2/pet. ether). Mp 91-92°.

Mono-Me ether: [114226-24-3]. *1-(2-Hydroxy-6-methoxyphenyl)-1-tetradecanone*
$C_{21}H_{34}O_3$ M 334.498
Constit. of the stem bark of *M. dactyloides.* Needles (MeOH). Mp 51-52°.

Kumar, N.S. *et al*, *Phytochemistry*, 1988, **27**, 465.

11,15-Dihydroxy-5,13-prostadienoic acid, 9CI D-00243

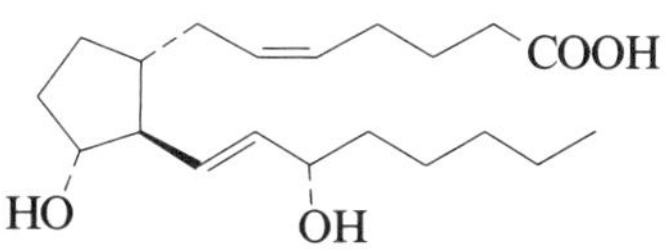

$C_{20}H_{34}O_4$ M 338.486

(5Z,11R,13E,15S)-form [51424-50-1]
9-Deoxy PGE₂

Guzman, A. *et al*, *Chem. Lett.*, 1973, 1073 (*synth*)

9,15-Dihydroxy-13-prostenoic acid D-00244

[16471-84-4]

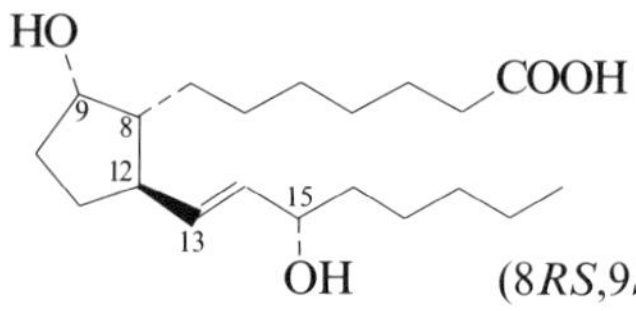

(8*RS*,9*SR*,12*RS*,13*E*,15*SR*)-*form*

$C_{20}H_{36}O_4$ M 340.502

(8RS,9SR,12RS,13E,15SR)-form [40098-35-9]

(±)-*11-Deoxy-PGF*$_{1\alpha}$

Has hypertensive activity. Mp 97-98.5°.

(8RS,9SR,12RS,13E,15RS)-form [40098-37-1]

(±)-*11-Deoxy-15*-epi-*PGF*$_{1\alpha}$

Mp 102-104°.

(8RS,9RS,12RS,13E,15SR)-form

(±)-*11-Deoxy-PGF*$_{1\beta}$

Mp 69-70.5°.

(8RS,9RS,12RS,13E,15RS)-form [40098-38-2]

(±)-*11-Deoxy-15*-epi-*PGF*$_{1\beta}$

Mp 59-60.5°.

[37785-98-1, 37785-99-2]

Bagli, J.F. *et al*, *Tetrahedron Lett.*, 1966, 465 (*synth*)
Alvarez, F.S. *et al*, *J. Am. Chem. Soc.*, 1972, **94**, 7823 (*synth, pmr, ir, ms*)
Lincoln, F.H. *et al*, *J. Org. Chem.*, 1973, **38**, 951 (*synth*)
Alvarez, F.S. *et al*, *Tetrahedron Lett.*, 1973, 569 (*synth, ir, pmr*)
Burton, T.S. *et al*, *J. Chem. Soc., Perkin Trans. 1*, 1976, 2550 (*synth*)

11,15-Dihydroxy-13-prostenoic acid, 9CI D-00245

9-Deoxyprostaglandin E$_1$. *9-Deoxy-PGE*$_1$

[53317-71-8]

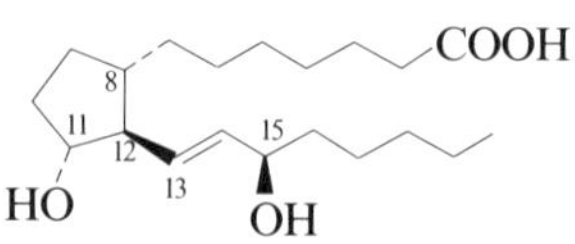

(8*SR*,11*RS*,12*SR*,13*E*,15*RS*)-*form*

$C_{20}H_{36}O_4$ M 340.502

Several racemates prepd.

[53317-72-9, 53317-74-1, 53317-75-2]

Finch, N. *et al*, *J. Org. Chem.*, 1975, **40**, 206 (*synth, pmr, ir, ms*)

3,9-Dihydroxytetradecanoic acid D-00246

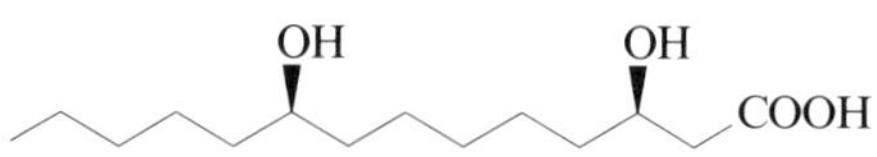

(3*R*,9*R*)-*form*

$C_{14}H_{28}O_4$ M 260.373

(3R,9R)-form

Me ester: [115482-43-4].

$C_{15}H_{30}O_4$ M 274.400

Solid. Mp 57-59°. $[\alpha]_D^{27}$ −10.55° (c, 1.45 in $CHCl_3$).

Schreiber, S.L. *et al*, *J. Am. Chem. Soc.*, 1988, **110**, 6210 (*synth, pmr, cmr, ms, ir*)

3,11-Dihydroxytetradecanoic acid D-00247

3,11-Dihydroxymyristic acid. Ipurolic acid

$H_3C(CH_2)_2CH(OH)(CH_2)_7CH(OH)CH_2COOH$

$C_{14}H_{28}O_4$ M 260.373

Occurs in *Ipomoea purpurea*, *I. purga*, *I. fistulosa* and *Pharbitis nil*. Mp 100-101°.

Ag salt: Mp 160°.

Me ester:

$C_{15}H_{30}O_4$ M 274.400

Mp 68-69°. $[\alpha]_D$ +1.59° (EtOH).

11-[O-6-Deoxy-β-D-glucopyranosyl-(1→4)-O-6-deoxy-α-L-mannopyranosyl-(1→6)-O-[6-deoxy-α-L-mannopyranosyl-(1→2)-O-β-D-glucopyranosyl-(1→2)]-β-D-glucopyranoside: Pharbitic acid C, 9CI. 11-[O-6-Deoxy-β-D-glucopyranosyl-(1→4)-O-6-deoxy-α-L-mannopyranosyl-(1→6)-O-[6-deoxy-α-L-mannopyranosyl-(1→2)-O-β-D-glucopyranosyl-(1→2)]-β-D-glucopyranosyl)oxy]-3-hydroxytetradecanoic acid, 9CI

$C_{44}H_{78}O_{26}$ M 1023.085

Constit. of seeds of *P. nil*. Cryst. +1H_2O. Mp 120-129°. $[\alpha]_D$ −541° (MeOH).

Power, F.B. *et al*, *Chem. Zentralbl.*, 1908, **2**, 887 (*isol*)
Asahina, Y. *et al*, *CA*, 1925, **19**, 3479 (*isol*)
Legler, G., *Phytochemistry*, 1965, **4**, 29 (*isol, ir, pmr*)
Okabe, H. *et al*, *Tetrahedron Lett.*, 1970, 3123 (*Pharbitic acid C*)
Okabe, H. *et al*, *Chem. Pharm. Bull.*, 1971, **19**, 2394; 1972, **20**, 514 (*Pharbitic acid C*)
Miyahari, K. *et al*, *Chem. Pharm. Bull.*, 1974, **47**, 1737 (*pmr*)

9,10-Dihydroxytetradecanoic acid D-00248

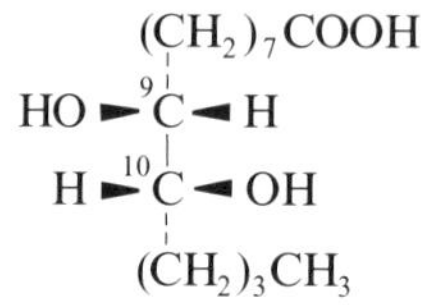

(9*R**,10*R**)-*form*

$C_{14}H_{28}O_4$ M 260.373

(9R*,10R*)-form

threo-*form*

Minor constit. of shellac. Cryst. (EtOAc). Mp 78.5-79.5°. Opt. activity of isol. sample too small to measure.

Me ester:

$C_{15}H_{30}O_4$ M 274.400

Mp 58-60°.

(9RS,10SR)-form

(±)-threo-*form*

Cryst. (EtOH). Mp 123-124°.

Boughton, B.W. *et al*, *J. Chem. Soc.*, 1952, 671 (*synth*)
Ames, D.E. *et al*, *J. Chem. Soc.*, 1963, 775 (*synth*)
Christie, W.W. *et al*, *J. Chem. Soc.*, 1964, 5833 (*isol*)

6-(2,12-Dihydroxytridecyl)-2,4-dihydroxybenzoic acid D-00249

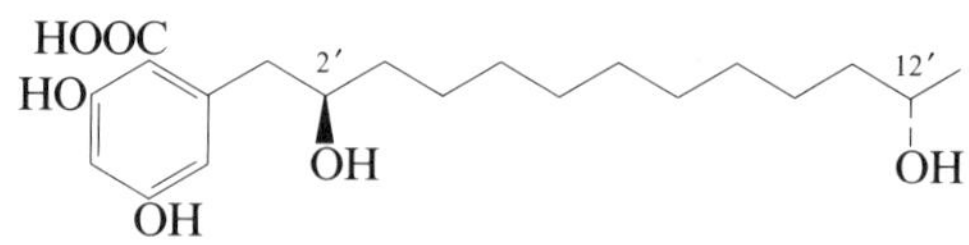

$C_{20}H_{32}O_6$ M 368.469

(2′R,12′S)-form

2′-Ac: [133361-20-3]. *2-(2-Acetoxy-12-hydroxytridecyl)-4,6-dihydroxybenzoic acid*

$C_{22}H_{34}O_7$ M 410.506

Constit. of *Ononis viscosa*. Oil (as Me ester). $[\alpha]_D^{20}$ −5.5° (c, 1.01 in $CHCl_3$)(Me ester).

2′,12′-Di-Ac: [133361-19-0]. *2-(2,12-Diacetoxytridecyl)-4,6-dihydroxybenzoic acid*

$C_{24}H_{36}O_8$ M 452.544

Constit. of *O. viscosa*. Oil (as Me ester). $[\alpha]_D^{20}$ +11.9° (c, 1.2 in $CHCl_3$)(Me ester).

12′-Deoxy, 2′-Ac, 4-Me ether: [129277-42-5]. *2-(2-Acetoxytridecyl)-6-hydroxy-4-methoxybenzoic acid*

$C_{23}H_{36}O_6$ M 408.534

Constit. of *O. natrix* ssp. *hispanica*. Oil (as Me ester). $[\alpha]_D^{22}$ −15.2° (c, 0.99 in MeOH)(Me ester).

[129277-48-1, 133361-22-5, 133361-24-7]

Barrero, A.F. *et al*, *Phytochemistry*, 1990, **29**, 1967; 1991, **30**, 641 (*isol*)

3,6-Dihydroxy-2-(1,3,5-undecatrienyl) benzaldehyde, 9CI — D-00250

Aurocitrin

[73151-59-4]

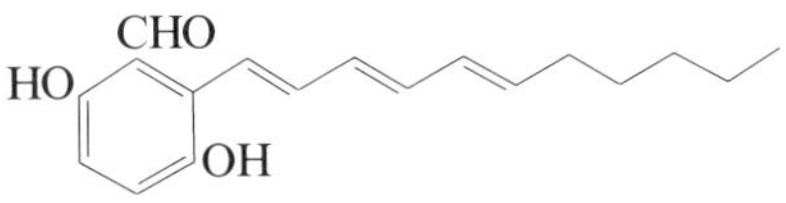

$C_{18}H_{22}O_3$ M 286.370

Polyketide antibiotic. Prod. by *Hypocrea citrina*. Shows broad spectrum antimicrobial activity. Orange-red cryst. Mp 106-107°. λ_{max} 413, 314, 266 nm.

Nair, M.S. *et al*, *Tetrahedron Lett.*, 1979, 3233 (*isol*)
Ronald, R.C. *et al*, *J. Org. Chem.*, 1982, **47**, 2541 (*synth*)
Harwood, L.M., *J. Chem. Soc., Perkin Trans. 1*, 1984, 2577 (*synth*)

4,8-Dimethyldecanal — D-00251

[75983-36-7]

(4*R*,8*R*)-*form*

$C_{12}H_{24}O$ M 184.321

(4R,8R)-form [85880-35-9]

Aggregation pheromone of the red flour beetle *Tribolium castaneum* and the confused flour beetle *T. confusum*. $Bp_{2.5}$ 65°. $[\alpha]_D^{22.5}$ −7.37° (c, 2.04 in $CHCl_3$).

(4S,8S)-form [85880-36-0]

Bp_1 72°. $[\alpha]_D^{22}$ +7.33° (c, 3.42 in $CHCl_3$).

(4R,8S)-form [85880-37-1]

Poss. synergistic activity with (4*R*,8*R*)-isomer. Bp_3 60°. $[\alpha]_D^{22}$ +9.94° (c, 2.40 in $CHCl_3$). n_D^{21} 1.4320.

(4S,8R)-form [85880-38-2]

Bp_1 75°. $[\alpha]_D^{22}$ −9.92° (c, 2.51 in $CHCl_3$). n_D^{21} 1.4320.

Suzuki, T., *Agric. Biol. Chem.*, 1981, **45**, 1357 (*isol*)
Suzuki, T. *et al*, *Agric. Biol. Chem.*, 1983, **47**, 869 (*synth*)
Levinson, H.Z. *et al*, *Naturwissenschaften*, 1983, **70**, 190 (*activity*)
Mori, K. *et al*, *Tetrahedron*, 1983, **39**, 2439 (*synth*)
Mori, K. *et al*, *Justus Liebigs Ann. Chem.*, 1985, 861; 1991, 497 (*synth, pmr*)

4,6-Dimethyl-4-decen-3-one — D-00252

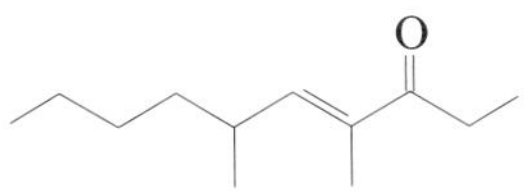

$C_{12}H_{22}O$ M 182.305

Constit. of the essential oil of *Evernia prunastri*.

(*E*)-*form* [111187-59-8] ***Bishomomanicone***

Constit. of the mandibular gland secretion of the ant *Manica rubida*.

Corbier, B. *et al*, *Recherches*, 1974, **19**, 289 (*isol*)
Bestmann, H.J. *et al*, *Justus Liebigs Ann. Chem.*, 1988, 55 (*isol, pmr*)

2,2-Dimethyl-1,3-dioxolane-4-methanol, 9CI, 8CI — D-00253

1,2-O-Isopropylideneglycerol. 4-Hydroxymethyl-2,2-dimethyl-1,3-dioxolan. Acetoneglycerol. Glycerol acetonide. Solketal

[100-79-8]

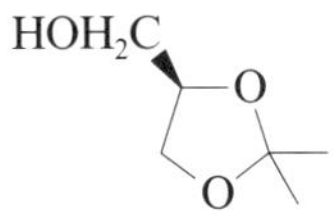

(*R*)-*form*

$C_6H_{12}O_3$ M 132.159

Important synthetic intermed. for chiral compds. Use limited by instability and hydrophilicity.

▷ JI0400000.

(R)-form [14347-78-5]

D-form

Intermed. for synth. of lipids etc. Liq. $Bp_{11\text{-}12}$ 80-80.5°, $Bp_{0.5}$ 24°. Opt. rotation is solv. dependent. The abs. config. is wrongly descr. as (*S*) by Hirth *et al.*

4-Methylbenzenesulfonyl: $[\alpha]_D$ +4.5° (EtOH). Has (*S*)-chirality.

Triphenylmethyl ether: [16975-61-4]. *D-1,2-O-Isopropylidene-3-O-tritylglycerol*

$C_{25}H_{26}O_3$ M 374.479

Mp 89-90°. $[\alpha]_D^{20}$ +15.5° (c, 8.4 in dioxan).

(S)-form [22323-82-6]

L-form

Intermed. for synth. of lipids etc. Bp_8 72-73°. $[\alpha]_D$ +10.7° (MeOH), $[\alpha]_D$ −14.4° (pure liq.) (o.p. >99.4%).

4-Methylbenzenesulfonyl: $[\alpha]_D$ −4.6° (EtOH). Has (*R*)-chirality.

(±)-***form***

Solvent, plasticiser. Liq. Sol. H_2O, org. solvs. Bp_{31} 104-106°.

O-Ac:

$C_8H_{14}O_4$ M 174.196

Bp_9 84°.

3,5-Dinitrobenzoyl: Mp 85°.

4-Methylbenzenesulfonyl: [23737-51-1].

Mp 47°.

Baer, E. *et al*, *J. Biol. Chem.*, 1939, **128**, 463.
Fischer, H.O.L. *et al*, *Chem. Rev.*, 1941, **29**, 287 (*use*)
Baer, E. *et al*, *J. Am. Chem. Soc.*, 1945, **67**, 944; 1948, **70**, 609 (*synth, abs config*)
Gresham, W.F. *et al*, *J. Org. Chem.*, 1949, **14**, 1103 (*synth*)
Nelson, W.L. *et al*, *J. Org. Chem.*, 1977, **42**, 1006.
Chirila, T. *et al*, *Rev. Chim. (Bucharest)*, 1977, **28**, 827 (*synth*)
Oelenschlaeger, J. *et al*, *Lipids*, 1978, **13**, 557 (*synth, ms*)
Jung, M.E. *et al*, *J. Am. Chem. Soc.*, 1980, **102**, 6304 (*synth, bibl*)
Hirth, G. *et al*, *Helv. Chim. Acta*, 1985, **68**, 1863 (*synth, props*)
Suemune, H. *et al*, *Chem. Pharm. Bull.*, 1986, **34**, 3440 (*synth*)
Lewis, R.J., *Sax's Dangerous Properties of Industrial Materials*, 8th Ed., Van Nostrand-Reinhold, 1992, DVR600.

2,4-Dimethyldocosanoic acid — D-00254

$H_3C(CH_2)_{17}CH(CH_3)CH_2CH(CH_3)COOH$

$C_{24}H_{48}O_2$ M 368.642

(+)-***form***

Constit. of the lipids of tubercle bacilli. Mp 43-45°. $[\alpha]_D^{25}$ +2.2° (c, 0.72 in $CHCl_3$).

(±)-***form***

Mp 50-53°.

Cason, J. *et al*, *Tetrahedron*, 1964, **20**, 1955.

3,15-Dimethyldocosanoic acid D-00255

$H_3C(CH_2)_6CH(CH_3)(CH_2)_{11}CH(CH_3)CH_2COOH$

$C_{24}H_{48}O_2$ M 368.642

Me ester: $Bp_{0.7}$ 195°.

Minnikin, D.E., *J. Chem. Soc. C*, 1967, 575.

2,5-Dimethyldodecanoic acid D-00256

2,5-Dimethyllauric acid

[100675-91-0]

$H_3C(CH_2)_6CH(CH_3)CH_2CH_2CH(CH_3)COOH$

$C_{14}H_{28}O_2$ M 228.374

Isol. from the cyanobacterium *Lyngbya aestuarii*. Shows herbicidal activity. $[\alpha]_D^{22}$ −9.4° (c, 4.4 in MeOH). Config. not determined.

Me ester: [100675-92-1].

$C_{15}H_{30}O_2$ M 242.401

Oil. $[\alpha]_D^{22}$ −15° (c, 0.77 in CCl_4).

(Z)-*3-Acetoxy-3-bromo-1-propenyl ester:* [139116-17-9].

$C_{19}H_{33}BrO_4$ M 405.371

From a *L.* sp. Pale yellow oil. $[\alpha]_D$ −4.7° (c, 0.51 in $CHCl_3$).

Entzeroth, M. *et al*, *Phytochemistry*, 1985, **24**, 2875 (*isol*)
Hodder, A.R. *et al*, *J. Nat. Prod. (Lloydia)*, 1991, **54**, 1668 (*deriv*)

7,7-Dimethyl-5,8-eicosadienoic acid D-00257

DEDA

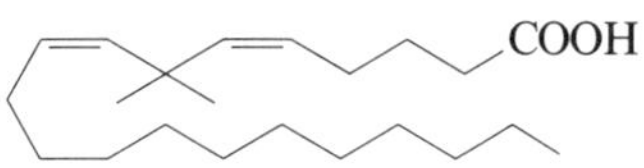

(Z,Z)-form

$C_{22}H_{40}O_2$ M 336.557

(5Z,8Z)-form [89560-01-0]

Inhibits phospholipase A_2 and 5-lipoxygenase. Viscous oil.

Ackroyd, J. *et al*, *Tetrahedron Lett.*, 1983, **24**, 5139 (*synth, pmr, biochem*)
Cohen, N. *et al*, *Prostaglandins*, 1984, **27**, 553 (*synth, biochem*)
Menon, S.D. *et al*, *Lipids*, 1990, **25**, 321 (*biochem*)

7,7-Dimethyl-5,8,11-eicosatrienoic acid D-00258

$H_3C(CH_2)_7CH{=}CHCH_2CH{=}CHC(CH_3)_2CH{=}CH(CH_2)_3COOH$

$C_{22}H_{38}O_2$ M 334.541

(all-Z)-form [86982-61-8]

5-Lipoxygenase inhibitor. Oil.

Perchonock, C.D. *et al*, *Tetrahedron Lett.*, 1983, **24**, 2457 (*synth*)

10,10-Dimethyl-5,8,11-eicosatrienoic acid D-00259

$H_3C(CH_2)_7CH{=}CHC(CH_3)_2CH{=}CHCH_2CH{=}CH(CH_2)_3COOH$

$C_{22}H_{38}O_2$ M 334.541

(all-Z)-form [86982-62-9]

5-Lipoxygenase inhibitor.

Perchonock, C.D. *et al*, *Tetrahedron Lett.*, 1983, **24**, 2457 (*synth*)

8,14-Dimethyl-2,4,6-heptadecatriyne, 9CI D-00260

Trichodermene A

[34285-15-9]

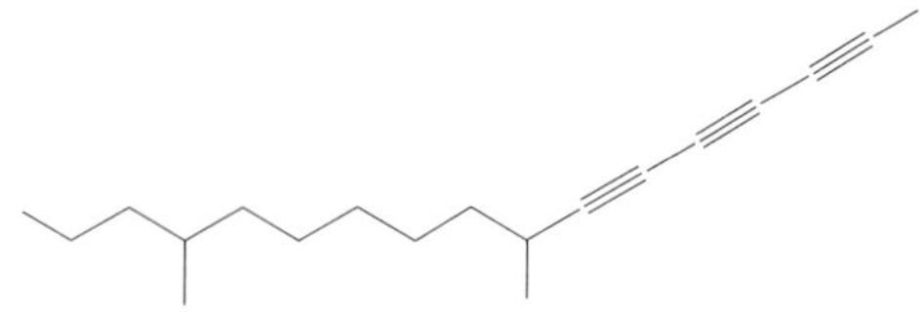

$C_{19}H_{28}$ M 256.430

Isol. from *Trichoderma pseudokoningii*.

Kamal, A. *et al*, *Pak. J. Sci. Ind. Res.*, 1971, **14**, 71; *CA*, **75**, 126877 (*isol*)

2,4-Dimethyl-2,4-heptadien-1-ol D-00261

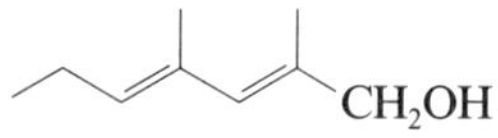

$C_9H_{16}O$ M 140.225

(E,E)-form [62332-67-6]

Constit. of the defense secretions of *Leiobunum leiopenis*. Liq. Bp_8 96-99°.

Jones, T.H. *et al*, *Proc. Natl. Acad. Sci. U.S.A.*, 1977, **74**, 419 (*isol*)

15,19-Dimethylheptatriacontane, 9CI D-00262

[56987-91-8]

$H_3C(CH_2)_{13}CH(CH_3)CH_2CH_2CH_2CH(CH_3)(CH_2)_{17}CH_3$

$C_{39}H_{80}$ M 549.061

A component of pheromones from female stable fly *Stomoxys calcitrans* and from female tsetse fly *Glossina morsitans morsitans*.

Uebel, E.C. *et al*, *J. Chem. Ecol.*, 1975, **1**, 377; *CA*, **83**, 189262 (*isol*)
Sonnet, P.E. *et al*, *CA*, 1976, **84**, 122071 (*synth*)
Carlson, D.A. *et al*, *Science (Washington, D.C.)*, 1978, **201**, 750 (*isol*)

17,21-Dimethylheptatriacontane, 9CI D-00263

$H_3C(CH_2)_{15}CH(CH_3)CH_2CH_2CH_2CH(CH_3)(CH_2)_{15}CH_3$

$C_{39}H_{80}$ M 549.061

A component of pheromones isol. from female tsetse fly *Glossina morsitans morsitans*.

Carlson, D.A. *et al*, *Science (Washington, D.C.)*, 1978, **201**, 750 (*isol, struct, synth*)
Ade, E. *et al*, *Tetrahedron Lett.*, 1980, **21**, 1137.

3,7-Dimethyl-2-hexadecanol D-00264

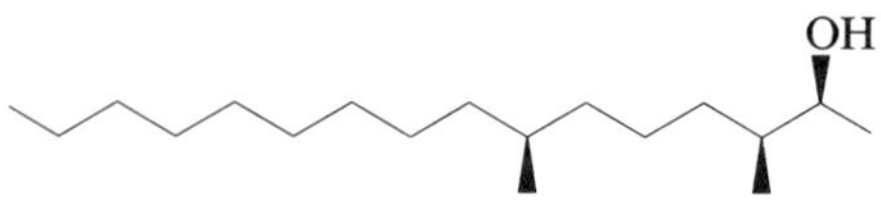

$C_{18}H_{38}O$ M 270.498

(2S,3S,7S)-form

Minor component of the sex pheromone of European pine sawfly *Neodiprion sertifer*. $Bp_{0.7}$ 130°. $[\alpha]_D^{22}$ −12.1° (c, 1 in hexane).

Hedenstrom, E. *et al*, *Tetrahedron*, 1992, **48**, 3139 (*isol, synth*)

2,4-Dimethyl-2,4-hexadien-1-ol D-00265

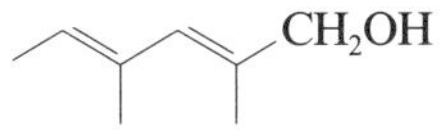

$C_8H_{14}O$ M 126.198

(E,E)-form [62332-66-5]

Constit. of the defence secretions of some *Leiobunum* spp. Liq. Bp_8 83-85°, $Bp_{0.05}$ 48-50°.

Aldehyde: [62332-65-4]. *2,4-Dimethyl-2,4-hexadienal*

$C_8H_{12}O$ M 124.182

Constit. of the defense secretion of some *L.* spp. Oil. Bp_{39} 110-120°.

Aldehyde, 2,4-dinitrophenylhydrazone: [136464-82-9]. Cryst. + $\frac{1}{2}H_2O$ (EtOH). Mp 174-176°.

Jones, T.H. *et al*, *Proc. Natl. Acad. Sci. U.S.A.*, 1977, **74**, 419 (*isol*)
Bartelt, R.J. *et al*, *J. Agric. Food Chem.*, 1990, **38**, 2192 (*synth*)
Patel, P. *et al*, *J. Chem. Soc., Perkin Trans. 1*, 1991, 1941 (*synth*)

7,11-Dimethyl-3-methylene-1,6,10-dodecatriene D-00266

β-Farnesene

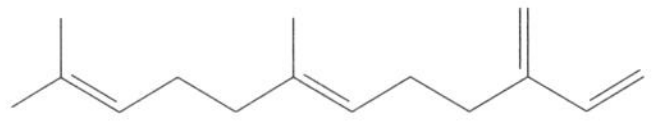

(*E*)-*form*

$C_{15}H_{24}$ M 204.355

(E)-form [18794-84-8]

Constit. of hop, camomile and other essential oils. Alarm pheromone of aphids; found in male mouse urine. Oil. Bp_9 121-122°.

(Z)-form [28973-97-9]

Oil.

Naves, Y.R., *Helv. Chim. Acta*, 1966, **49**, 1029 (*uv, ir, pmr*)
Brieger, G. *et al*, *J. Org. Chem.*, 1969, **34**, 3789 (*synth*)
Anet, E.F.L.J., *Aust. J. Chem.*, 1970, **23**, 2101 (*synth*)
Tanaka, S. *et al*, *J. Am. Chem. Soc.*, 1975, **97**, 3252 (*synth*)
Vig, O.P. *et al*, *J. Indian Chem. Soc.*, 1975, **13**, 1244 (*synth*)
Mandai, T. *et al*, *J. Org. Chem.*, 1983, **48**, 2601 (*isol*)
Harvey, S. *et al*, *J. Chem. Ecol.*, 1989, **15**, 2061.

3,11-Dimethyl-2-nonacosanone D-00267

[53623-10-2]

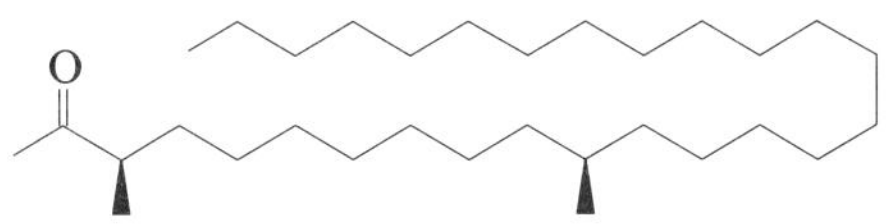

(3*R*,11*R*)-*form*

$C_{31}H_{62}O$ M 450.830

Sex pheromone of the German cockroach *Blattella germanica*. The (3*S*,11*S*)-isomer appears to be the predominant natural isomer, but all 4 stereoisomers show activity.

(3R,11R)-form [69274-88-0]

Mp 44.5-45°. $[\alpha]_D$ −5.63° (c, 4.1 in hexane).

(3S,11S)-form [69274-90-4]

Natural isomer. Mp 45-46°. $[\alpha]_D$ +5.1° (c, 3.54 in hexane).

(3R,11S)-form [69274-89-1]

Mp 39-39.5°. $[\alpha]_D$ −5.68° (c, 4.0 in hexane).

(3S,11R)-form [69274-91-5]

Mp 38-38.5°. $[\alpha]_D$ +5.73° (c, 2.04 in hexane).

Nishida, R. *et al*, *J. Chem. Ecol.*, 1976, **2**, 449; 1979, **5**, 289 (*occur*)
Burgstahler, A.W. *et al*, *Synthesis*, 1977, **6**, 405 (*synth*)
Place, P. *et al*, *Tetrahedron*, 1978, **34**, 1931 (*synth, ir, pmr, ms*)
Mori, K. *et al*, *Tetrahedron*, 1981, **37**, 1329; 1990, **46**, 4473 (*synth, ir, pmr*)

3,7-Dimethylnonadecane D-00268

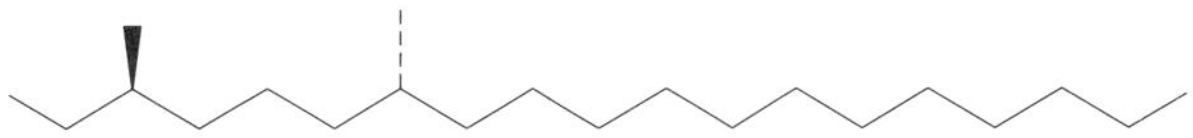

(3*R*,7*R*)-*form*

$C_{21}H_{44}$ M 296.579

Female sex pheromone of the alfalfa blotch leafminer *Agromyza frontella*.

(3R,7R)-form

$[\alpha]_D^{26}$ −21.4° (c, 1.85 in $CHCl_3$).

(3R,7S)-form [130797-69-2]

$[\alpha]_D^{26}$ −7.0° (c, 1.50 in $CHCl_3$).

(3S,7R)-form [130797-70-5]

$[\alpha]_D^{26}$ +6.9° (c, 1.50 in $CHCl_3$).

(3S,7S)-form

$[\alpha]_D^{26}$ +22.3° (c, 0.71 in $CHCl_3$). n_D^{19} 1.4400.

Carriere, Y. *et al*, *J. Chem. Ecol.*, 1988, **14**, 947.
Mori, K. *et al*, *Justus Liebigs Ann. Chem.*, 1991, 213 (*synth, bibl*)

9,10-Dimethyloctacosanoic acid D-00269

Nodoic acid

[37577-48-3]

$$H_3C(CH_2)_{17}CH(CH_3)CH(CH_3)(CH_2)_7COOH$$

$C_{30}H_{60}O_2$ M 452.803

Prepared from nodolidate. Cryst. (MeOH). Mp 54-55°. $[\alpha]_D^{30}$ +12.2° ($CHCl_3$).

Me ester: [37577-50-7].

$C_{31}H_{62}O_2$ M 466.830

Cryst. (Me_2CO at 0°). Mp 23-24°.

Amide:

$C_{30}H_{61}NO$ M 451.818

Cryst. (EtOH aq.). Mp 95-96°.

Rigvi, S.A.I. *et al*, *Phytochemistry*, 1972, **11**, 1823 (*synth, pmr, ms*)

7,11-Dimethyloctadecane D-00270

[65431-89-2]

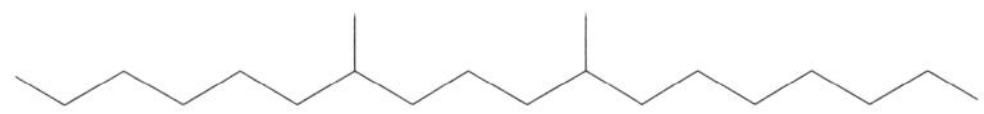

$C_{20}H_{42}$ M 282.552

Prod. by *Pseudomonas aeruginosa*. Ovipositional attractant for the mosquito *Aedes aegypti*.

Ikeshoji, T. *et al*, *CA*, 1979, **91**, 135529 (*isol*)
Odiokov, V.N. *et al*, *Dokl. Akad. Nauk SSSR, Ser. Khim.*, 1983, **271**, 261 (*synth*)
Odinokov, V.N. *et al*, *Khim. Prir. Soedin.*, 1990, **26**, 818; *Chem. Nat. Compd. (Engl. Transl.)*, 697 (*synth*)

2,4-Dimethyl-2-pentacosenoic acid D-00271

$$H_3C(CH_2)_{20}CH(CH_3)CH{=}C(CH_3)COOH$$

$C_{27}H_{52}O_2$ M 408.707

(E)-(?)-form

Mp 69.5-70.3°.

Cason, J. *et al*, *J. Org. Chem.*, 1954, **19**, 1836.

5,9-Dimethylpentadecane D-00272

[119163-96-1]

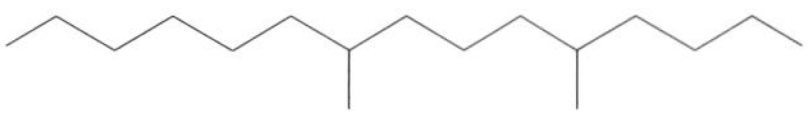

$C_{17}H_{36}$ M 240.471

Constit. of the sex pheromone of female *Perileucoptera coffeella.*

[135461-23-3, 135904-84-6]

Francke, W. *et al, Z. Naturforsch., C*, 1988, **43**, 787 (*isol*)
Poppe, L. *et al, Tetrahedron Lett.*, 1991, **32**, 2643 (*synth*)

3,7-Dimethyl-2-pentadecanol, 9CI D-00273

Diprionol

[59056-73-4]

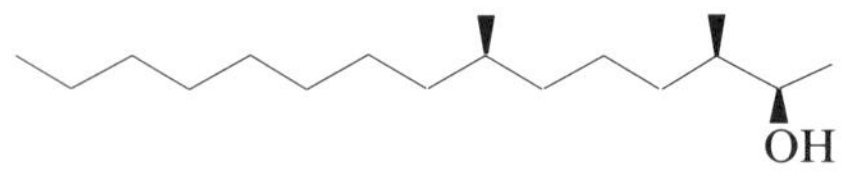

(2*R*,3*R*,7*R*)-*form*

$C_{17}H_{36}O$ M 256.471

Various stereoisomers and their acetates and propionates are sex pheromones of diprionid sawflies *Neodiprion, Diprion* and *Gilpinia* spp. The exact composition of the pheromone mix is species-variable.

(2R,3R,7R)-form [67253-82-1]
$Bp_{0.4}$ 115-117°. $[\alpha]_D^{22}$ +11.50° (neat). n_D^{20} 1.4513.
Ac:
$C_{19}H_{38}O_2$ M 298.508
$[\alpha]_D$ +6.4° (c, 1.5 in hexane). n_D^{20} 1.4409.

(2R,3R,7S)-form [67253-84-3]
$[\alpha]_D^{22}$ +13.04° (neat). n_D^{20} 1.4513.
Ac: $[\alpha]_D^{22}$ +7.1° (c, 0.8 in hexane). n_D^{20} 1.4410.

(2S,3S,7S)-form [67253-83-2]
Preferred stereoisomer (as esters) for most *N.* spp. $[\alpha]_D^{22}$ −11.8° (neat). n_D^{20} 1.4512.
Ac: $[\alpha]_D$ −5.6° (c, 1.0 in hexane). n_D^{20} 1.4407.

(2S,3S,7R)-form [67315-12-2]
$[\alpha]_D^{22}$ −13.05° (neat). n_D^{20} 1.4513.
Ac: $[\alpha]_D$ −6.7° (c, 1.4 in hexane). n_D^{20} 1.4513.

(2R,3S,7S)-form
$[\alpha]_D$ −15.5° (neat). n_D^{20} 1.4513.
Ac: $[\alpha]_D$ −8.4° (c, 1.1 in hexane). n_D^{20} 1.4411.

(2S,3R,7R)-form
Preferred stereoisomer for some *N.* spp. for *D. similis* and for *G. frutetorum.* $[\alpha]_D$ +15.11° (neat). n_D^{20} 1.4512.
Ac: $[\alpha]_D$ +7.8° (c, 1.2 in hexane). n_D^{20} 1.4411.

(2R,3S,7R)-form
$[\alpha]_D^{22}$ −16.33° (neat). n_D^{22} 1.4512.
Ac: $[\alpha]_D$ −8.3° (c, 1.4 in hexane). n_D^{20} 1.4410.

(2S,3R,7S)-form
$[\alpha]_D^{22}$ +16.07° (neat). n_D^{20} 1.4512.
Ac: $[\alpha]_D$ +8.7° (c, 1.4 in hexane). n_D^{23} 1.4408.

[59056-74-5, 59056-75-6]

Kramer, M.E. *et al, J. Chem. Ecol.*, 1984, **10**, 983.
Olaifa, J.I. *et al, J. Chem. Ecol.*, 1988, **14**, 1131.
Högberg, H.E. *et al, Tetrahedron*, 1990, **46**, 3007 (*synth, pmr, bibl*)

6,12-Dimethyl-2-pentadecanone D-00274

[110606-12-7]

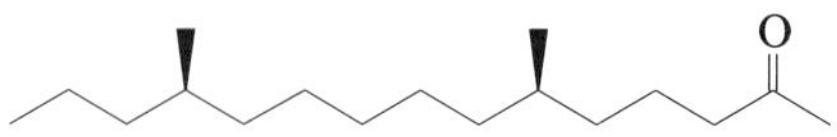

(6*R*,12*R*)-*form*

$C_{17}H_{34}O$ M 254.455

(6R,12R)-form [114907-67-4]
Sex pheromone of the female banded cucumber beetle *Diabrotica balteata.* $Bp_{0.29}$ 108-111°. $[\alpha]_D^{22}$ −0.5° (c, 1.13 in $CHCl_3$). n_D^{22} 1.4388.

(6R,12S)-form [114907-68-5]
$Bp_{0.29}$ 104-107°. $[\alpha]_D^{20}$ +1.9° (c, 1.16 in $CHCl_3$). n_D^{20} 1.4392.

(6S,12R)-form [114907-69-6]
$Bp_{0.28}$ 114-117°. $[\alpha]_D^{20}$ −1.9° (c, 1.14 in $CHCl_3$). $n_D^{19.5}$ 1.4392.

(6S,12S)-form [114907-70-9]
$Bp_{0.2}$ 108-113°. $[\alpha]_D^{21}$ +0.5° (c, 1.06 in $CHCl_3$). n_D^{21} 1.4393.

Chuman, T. *et al, J. Chem. Ecol.*, 1987, **13**, 1601 (*isol*)
Mori, K. *et al, Justus Liebigs Ann. Chem.*, 1988, 717 (*synth, pmr, cmr, abs config, bibl*)
McLaughlin, J.R. *et al, J. Econ. Entomol.*, 1991, **84**, 99 (*activity*)

3,4-Dimethyl-5-pentyl-2-furannonanoic acid D-00275

F_3 Acid

[57818-40-3]

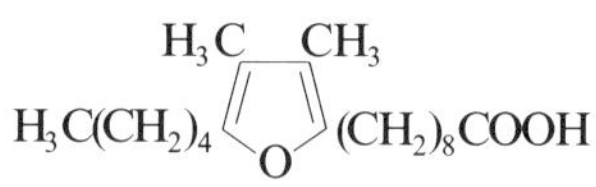

$C_{20}H_{34}O_3$ M 322.487

Present in fish oils. Also found in human and bovine blood plasma as phospholipid.

Glass, R.L. *et al, Lipids*, 1974, **9**, 1004; 1975, **10**, 695; 1977, **12**, 828.
Gunstone, F.D. *et al, J. Sci. Food Agric.*, 1978, **29**, 539.
Puchta, V. *et al, Justus Liebigs Ann. Chem.*, 1987, 25 (*occur*)
Scheinkönig, J. *et al, Justus Liebigs Ann. Chem.*, 1991, 451 (*biosynth*)
Batna, A. *et al, Justus Liebigs Ann. Chem.*, 1991, 861 (*biosynth*)

3,4-Dimethyl-5-pentyl-2-furanundecanoic acid, 9CI D-00276

12,15-Epoxy-13,14-dimethyleicosa-12,14-dienoic acid. F_6 Acid

[57818-36-7]

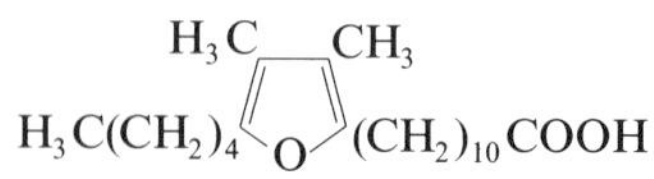

$C_{22}H_{38}O_3$ M 350.540

The most common member of a series of furanoid acids varying in chain length and in the number of methyl groups; see also 3-Methyl-5-pentyl-2-furannonanoic acid, M-00134. Present in fish lipids and reaching high proportions in starved fish. Component of F acid fraction present in human and beef blood serum.

Glass, R.L. *et al, Lipids*, 1977, **12**, 828.
Gunstone, F.D. *et al, J. Sci. Food Agric.*, 1978, **29**, 539.
Puchta, V. *et al, Justus Liebigs Ann. Chem.*, 1988, 25 (*occur*)
Batna, A. *et al, Justus Liebigs Ann. Chem.*, 1991, 861 (*biosynth*)

3,4-Dimethyl-5-propyl-2-furannonanoic acid, 9CI D-00277

F_1 Acid

[57818-38-9]

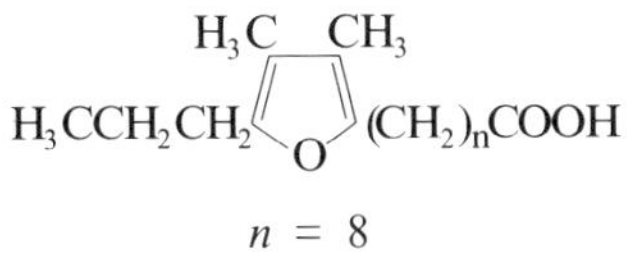

$n = 8$

$C_{18}H_{30}O_3$ M 294.433

Present in fish oils. Also found in human and bovine blood plasma as phospholipid.

Glass, R.L. *et al*, *Lipids*, 1974, **9**, 1004; 1975, **10**, 695; 1977, **12**, 828.
Gunstone, F.D. *et al*, *J. Sci. Food Agric.*, 1978, **29**, 539.
Puchta, V. *et al*, *Justus Liebigs Ann. Chem.*, 1988, 25 (*occur*)
Scheinkönig, J. *et al*, *Justus Liebigs Ann. Chem.*, 1991, 451 (*biosynth*)
Batna, A. *et al*, *Justus Liebigs Ann. Chem.*, 1991, 861 (*biosynth*)

3,4-Dimethyl-5-propyl-2-furanundecanoic acid, 9CI D-00278

F_4 Acid

[57818-41-4]

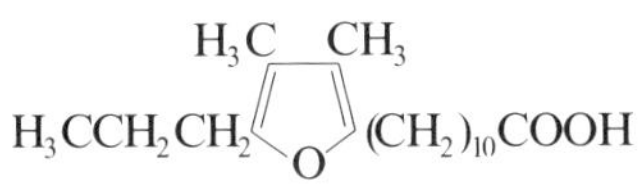

$C_{20}H_{34}O_3$ M 322.487

Present in fish oils. Also found in human and bovine blood plasma as phospholipids.

Glass, R.L. *et al*, *Lipids*, 1974, **9**, 1004; 1975, **10**, 695; 1977, **12**, 828.
Gunstone, F.D. *et al*, *J. Sci. Food Agric.*, 1978, **29**, 539.
Puchta, V. *et al*, *Justus Liebigs Ann. Chem.*, 1988, 25 (*occur*)
Scheinkönig, J. *et al*, *Justus Liebigs Ann. Chem.*, 1991, 451 (*biosynth*)
Batna, A. *et al*, *Justus Liebigs Ann. Chem.*, 1991, 858 (*biosynth*)

3,7-Dimethyl-2-tetradecanol D-00279

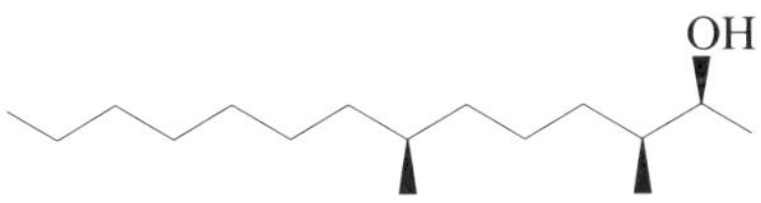

$C_{16}H_{34}O$ M 242.444

(2S,3S,7S)-form

Minor component of the sex pheromone of European pine sawfly *Neodiprion sertifer*. $Bp_{0.7}$ 100°. $[\alpha]_D^{22}$ −13.5° (c, 1 in hexane).

Hedenstrom, E. *et al*, *Tetrahedron*, 1992, **48**, 3139 (*isol, synth*)

3,7-Dimethyl-2-tridecanol D-00280

$C_{15}H_{32}O$ M 228.417

(2S,3S,7S)-form

Minor component of the sex pheromone of European pine sawfly *Neodiprion sertifer*. $Bp_{0.7}$ 90°. $[\alpha]_D^{22}$ −14.2° (c, 1 in hexane).

Hedenstrom, E. *et al*, *Tetrahedron*, 1992, **48**, 3139 (*isol, synth*)

2,3-Dinor-6-oxoprostaglandin $F_{1\alpha}$ D-00281

3,5-Dihydroxy-2-(3-hydroxy-1-octenyl)-γ-oxocyclopentanepentanoic acid, 9CI. Dinor-4-keto-7,9,13-trihydroxyprost-11-enoic acid

[64700-71-6]

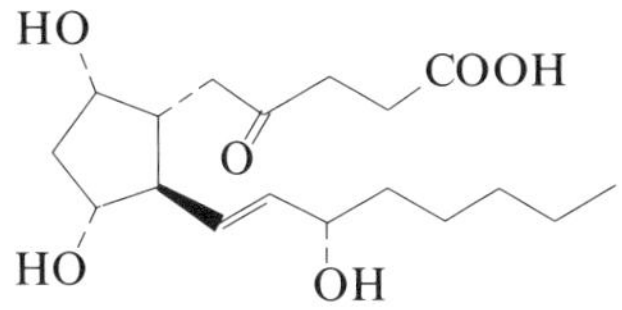

$C_{18}H_{30}O_6$ M 342.431

Major enzymatic metab. of Prostacyclin, P-00144 and 6-ketoprostaglandin $F_{1\alpha}$ in man. Viscous oil.

Rosencranz, B. *et al*, *Biochim. Biophys. Acta*, 1980, **619**, 207.
Rosencranz, B. *et al*, *J. Biol. Chem.*, 1980, **255**, 10194 (*isol*)
Sun, F.F. *et al*, *Prostaglandins*, 1980, **20**, 729 (*synth*)
Bundy, G.L. *et al*, *Tetrahedron*, 1981, **37**, 4419.
Brash, A.R. *et al*, *J. Pharmacol. Exp. Ther.*, 1983, **226**, 78.
Meese, C.O., *Tetrahedron Lett.*, 1984, **25**, 2199 (*synth*)

Dioctadecyl phosphate D-00282

Distearyl phosphate. Dioctadecyl hydrogen phosphate. Dioctadecyl phosphoric acid

$[H_3C(CH_2)_{17}O]_2P(O)OH$

$C_{36}H_{75}O_4P$ M 602.960

Cryst. (MeOH). Mp 82°.

Brown, D.A. *et al*, *J. Chem. Soc.*, 1955, 1584 (*synth*)
Ramirez, F. *et al*, *Phosphorus Sulfur Relat. Elem.*, 1978, **4**, 43 (*synth*)

Dioctyl sulfide D-00283

Octyl sulfide

[2690-08-6]

$H_3C(CH_2)_7S(CH_2)_7CH_3$

$C_{16}H_{34}S$ M 258.511

Used in extraction-separation of Pd. Liq. with strong odour. Bp_{10} 180°.

Vogel, A.I. *et al*, *J. Chem. Soc.*, 1943, 16 (*purifn, props*)
U.S. Pat., 2 551 813, (1951); *CA*, **45**, 9559 (*synth*)
Mojski, M., *Talanta*, 1978, **25**, 163 (*use*)

6,9-Dioxodecanoic acid D-00284

$H_3CCOCH_2CH_2CO(CH_2)_4COOH$

$C_{10}H_{16}O_4$ M 200.234

Mp 73° (70-71°).

Stetter, H. *et al*, *Chem. Ber.*, 1978, **111**, 431; 1985, **118**, 1115 (*synth, ir, pmr*)

3,6-Dioxododecanoic acid D-00285

$H_3C(CH_2)_5COCH_2CH_2COCH_2COOH$

$C_{12}H_{20}O_4$ M 228.288

Et ester: [86254-29-7].

$C_{14}H_{24}O_4$ M 256.341

Oil.

Brown, P.J. *et al*, *J. Chem. Soc., Perkin Trans. 1*, 1984, 2049 (*synth, ir, pmr*)

9,12-Dioxododecanoic acid, 9CI D-00286

[51551-01-0]

$OHCCH_2CH_2CO(CH_2)_7COOH$

$C_{12}H_{20}O_4$ M 228.288

Cryst. (Et_2O/pentane). Mp 55-62°.

Me ester: [50266-44-9].

$C_{13}H_{22}O_4$ M 242.314

Oil. n_D^{25} 1.4788.

Me ester, ethylene acetal: [96202-77-6].

$C_{15}H_{26}O_5$ M 286.367

Solid (pentane). Mp 27-28°.

Brown, E. *et al, Tetrahedron*, 1975, **31**, 1047 (*synth, ir, pmr*)
Boga, C. *et al, Synthesis*, 1986, 212 (*deriv, synth, pmr, ir*)

9,10-Dioxooctadecanoic acid D-00287

9,10-Dioxostearic acid

[656-73-5]

$H_3C(CH_2)_7COCO(CH_2)_7COOH$

$C_{18}H_{32}O_4$ M 312.448

Yellow cryst. Mp 86°.

Monoxime:

$C_{18}H_{33}NO_4$ M 327.463

Mp 76-81°.

Dioxime:

$C_{18}H_{34}N_2O_4$ M 342.478

Mp 153-154°.

2,4-Dinitrophenylosazone: Orange cryst. Mp 146.5°.

Spieckermann, A., *Ber.*, 1895, **28**, 276.
Grindley, D.N. *et al, J. Am. Oil Chem. Soc.*, 1972, **49**, 338.
Jenson, H.P. *et al, J. Org. Chem.*, 1974, **39**, 2314 (*synth*)

4,7-Dioxooctanoic acid D-00288

$H_3CCOCH_2CH_2COCH_2CH_2COOH$

$C_8H_{12}O_4$ M 172.180

Mp 77-78° (75-76°).

Me ester: [61426-50-4].

$C_9H_{14}O_4$ M 186.207

Liq. $Bp_{0.2}$ 120°.

Stetter, H. *et al, Chem. Ber.*, 1985, **118**, 1115 (*synth, ir, pmr*)
Abeysekera, A.M. *et al, J. Chem. Soc., Perkin Trans. 1*, 1991, 2021 (*synth, pmr*)
Zimmer, R. *et al, Justus Liebigs Ann. Chem.*, 1992, 709 (*synth, pmr, ester*)

Dipalmitoylphosphatidylcarnitine D-00289

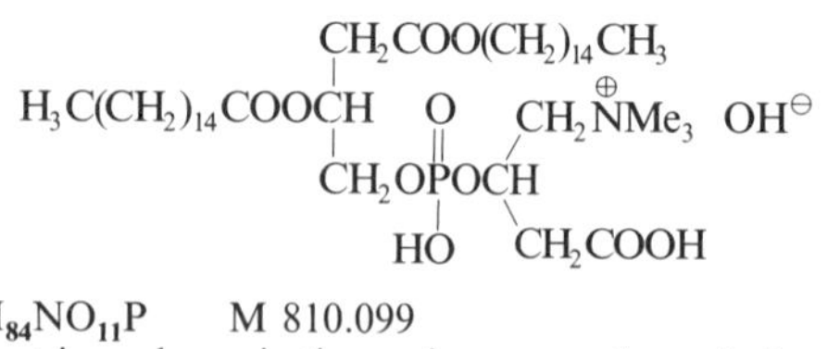

$C_{42}H_{84}NO_{11}P$ M 810.099

Present in embryonic tissues in an enantiomeric form.

rac-form

Cryst. ($CHCl_3$/Me_2CO). Mp 212°.

Benzyl ester: Mp 110°.

Mehlman, M.A. *et al, Arch. Biochem. Biophys.*, 1963, **102**, 346.
Hintze, U. *et al, Lipids*, 1975, **10**, 20 (*synth, cmr*)

Dipropanoyl peroxide D-00290

Bis(1-oxopropyl) peroxide, 9CI. Propionyl peroxide, 8CI

[3248-28-0]

$H_3CCH_2CO—O-O-COCH_2CH_3$

$C_6H_{10}O_4$ M 146.143

Cryst.

▷ Explodes on standing at r.t.. JL8750000.

Goldschmidt, S. *et al, Justus Liebigs Ann. Chem.*, 1952, **577**, 153 (*synth*)
Renbaum, A. *et al, J. Chem. Phys.*, 1955, **23**, 909 (*synth*)
Pacansky, J. *et al, J. Chem. Phys.*, 1979, **71**, 2811 (*synth, ir*)
Pacansky, J. *et al, J. Am. Chem. Soc.*, 1982, **104**, 415 (*ir*)
Bretherick, L., *Handbook of Reactive Chemical Hazards*, 2nd Ed., Butterworths, London and Boston, 1979, 596.
Hazards in the Chemical Laboratory, (Bretherick, L., Ed.), 3rd Ed., Royal Society of Chemistry, London, 1981, 315.
Lewis, R.J., *Sax's Dangerous Properties of Industrial Materials*, 8th Ed., Van Nostrand-Reinhold, 1992, DWQ800.

Disparlure D-00291

2-Decyl-3-(5-methylhexyl)oxirane, 9CI. 7,8-Epoxy-2-methyloctadecane

[29804-22-6]

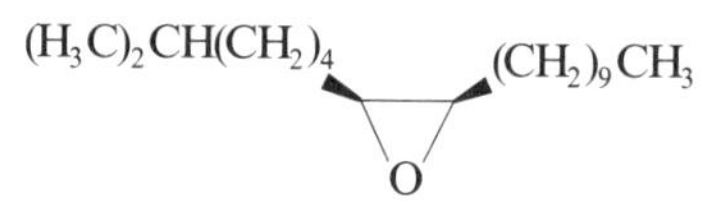

$C_{19}H_{38}O$ M 282.509

Sex attractant of *Porthetria dispar* and *Lymantria monachia*.

Bierl, B.A. *et al, Science (Washington, D.C.)*, 1970, **170**, 87 (*isol*)
Kasang, G. *et al, Naturwissenschaften*, 1974, **61**, 130 (*biosynth*)
Bestmann, H.J. *et al, Chem. Ber.*, 1976, **109**, 3375 (*synth*)
Mori, K. *et al, Tetrahedron*, 1976, 3953; 1986, **42**, 3471 (*synth*)
Farnum, D.G. *et al, Tetrahedron Lett.*, 1977, 4009 (*synth*)
Tolstikov, G.A. *et al, Tetrahedron Lett.*, 1978, 1857 (*synth*)
Brown, H.C. *et al, Synthesis*, 1983, 283 (*synth*)
Vig, O.P. *et al, Indian J. Chem., Sect. B*, 1985, **24**, 860 (*synth*)
Mori, K. *et al, Tetrahedron*, 1986, **42**, 3471 (*synth, ir, pmr, cmr, ms*)
Tsuboi, S. *et al, Bull. Chem. Soc. Jpn.*, 1987, **60**, 2475 (*synth*)
Satoh, T. *et al, Tetrahedron Lett.*, 1988, **29**, 313 (*synth*)
Marczak, S. *et al, Justus Liebigs Ann. Chem.*, 1990, 345 (*synth*)
Fukusaki, E. *et al, J. Ferment. Bioeng.*, 1992, **73**, 284 (*manuf*)

Diundecylamine, 8CI D-00292

N-*Undecyl-1-undecanamine, 9CI. Dihendecylamine*

[16165-33-6]

$H_3C(CH_2)_{10}NH(CH_2)_{10}CH_3$

$C_{22}H_{47}N$ M 325.620

Needles (C_6H_6/EtOH under N_2). Mp 51.5-52.5°. $Bp_{0.8}$ 190-194°.

Wright, J.B. *et al, J. Org. Chem.*, 1946, **11**, 111 (*synth*)

5,13-Docosadienoic acid D-00293

$H_3C(CH_2)_7CH{=}CH(CH_2)_6CH{=}CH(CH_2)_3COOH$

$C_{22}H_{40}O_2$ M 336.557

(Z,Z)-form [676-39-1]

Constit. of the seed fat of *Limnanthes douglasii*. Fp −4° (sealed tube). $Bp_{0.15}$ 183-185°.

Bagby, M.O. *et al, J. Org. Chem.*, 1961, **26**, 1261 (*isol*)
Ames, D.E. *et al, J. Chem. Soc.*, 1965, 894 (*synth*)
Blum, S. *et al, J. Org. Chem.*, 1972, **37**, 3121 (*ms*)
Pierce, R.O. *et al, Crop Sci.*, 1977, **17**, 521.

4,7,10,13,16,19-Docosahexaenoic acid D-00294

[2091-24-9]

$C_{22}H_{32}O_2$ M 328.494

(all-Z)-form [6217-54-5]

Cervonic acid

Metabolic prod. of α-Linolenic acid present in fish oils and in many phospholipids. Tuna eyeballs are a major source. Reduces blood cholesterol and blood pressure, claimed to increase intelligence in rats. Extensively marketed as a dietary supplement in Japan.

Me ester: [2566-90-7].

$C_{23}H_{34}O_2$ M 342.520

Oil.

Hands, A.R. *et al*, *Biochem. J.*, 1963, **87**, 263.
Beach, D.H. *et al*, *Biochim. Biophys. Acta*, 1974, **369**, 16 (*biosynth*)
Morales, R.W. *et al*, *Biochim. Biophys. Acta*, 1976, **431**, 206 (*isol*)
Sanders, T.A.B. *et al*, *Am. J. Clin. Nutr.*, 1978, **31**, 805.
Moreno, V.J. *et al*, *Lipids*, 1979, **14**, 15 (*biosynth*)
Selivonchick, D.P. *et al*, *Lipids*, 1979, **14**, 66 (*isol*)

4,7,10,13,16,19-Docosahexaynoic acid D-00295

$H_3CCH_2(C{\equiv}CCH_2)_6CH_2COOH$

$C_{22}H_{20}O_2$ M 316.399

Cryst. (EtOH). Mp 138-139°.

Kunau, W.H., *Hoppe Seyler's Z. Physiol. Chem.*, 1971, **352**, 542.

Docosane D-00296

[629-97-0]

$H_3C(CH_2)_{20}CH_3$

$C_{22}H_{46}$ M 310.605

Constit. of green alga *Scenedesmus acutus* and moth scales of the corn earworm *Heliothis zea.* Also isol. from higher plant sources. Has kairomone activity. Cryst. (Et_2O). Mp 47°. Bp_{15} 224°.

Levene, P.A., *J. Biol. Chem.*, 1915, **20**, 528.
Minami, Y. *et al*, *CA*, 1953, 1422e (*synth*)
Hunter, G.L.K. *et al*, *Phytochemistry*, 1966, **5**, 807 (*isol*)
Jones, R.L. *et al*, *Environ. Entomol.*, 1973, **2**, 593 (*isol*)
Snyder, R.G., *J. Chem. Phys.*, 1978, **68**, 4156 (*ir*)
Kakemoto, M. *et al*, *Nippon Kagaku Kaishi*, 1979, 80 (*glc, ms*)

11,12-Docosanedicarboxylic acid D-00297

$H_3C(CH_2)_9CH(COOH)CH(COOH)(CH_2)_9CH_3$

$C_{24}H_{46}O_4$ M 398.625

Di-Me ester: Cryst. (EtOH). Mp 54-54.5°.

Macfarlane, J.J. *et al*, *Nature (London)*, 1965, **207**, 520 (*ms, nmr*)

Docosanedioic acid D-00298

Phellogenic acid

[505-56-6]

$HOOC(CH_2)_{20}COOH$

$C_{22}H_{42}O_4$ M 370.571

Constit. of cork, bark and Japan wax. Cryst. (Me_2CO). Mp 122-124°, Mp 126-128°.

Mono-Me ester:

$C_{23}H_{44}O_4$ M 384.598

Cryst. (pet. ether). Mp 82-84°, Mp 86-87°.

Di-Me ester: [22399-98-0].

$C_{24}H_{46}O_4$ M 398.625

Cryst. (pet. ether). Mp 69-70°.

Di-Et ester: [26818-50-8].

$C_{26}H_{50}O_4$ M 426.679

Cryst. (pet. ether). Mp 61°.

Drake, N.L. *et al*, *J. Am. Chem. Soc.*, 1941, **63**, 617 (*isol, synth*)
Anderson, J.D. *et al*, *J. Org. Chem.*, 1966, **31**, 3890 (*synth*)

1,22-Docosanediol D-00299

[22513-81-1]

$HOCH_2(CH_2)_{20}CH_2OH$

$C_{22}H_{46}O_2$ M 342.604

Constit. of bark and carnauba wax. Cryst. (Me_2CO/C_6H_6). Mp 105-106°.

Di-Ac:

$C_{26}H_{50}O_4$ M 426.679

Mp 68.4°. Bp_1 229°.

Drake, N.L. *et al*, *J. Am. Chem. Soc.*, 1941, **63**, 617 (*synth*)
Murray, K.E. *et al*, *Aust. J. Chem.*, 1955, **8**, 432 (*isol*)

11,12-Docosanediol D-00300

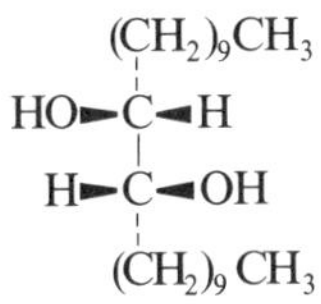

(11*RS*,12*RS*)-*form*

$C_{22}H_{46}O_2$ M 342.604

(10RS,11RS)-form [88454-88-0]

(±)-threo-*form*

Cryst. (C_6H_6/hexane). Mp 85-87°.

(10RS,11SR)-form [88454-87-9]

erythro-*form*

Cryst. (pentane/Me_2CO). Mp 135-137°.

Fyles, T.M. *et al*, *J. Org. Chem.*, 1984, **49**, 753 (*synth, ir, pmr, cmr, ms*)

Docosanoic acid D-00301

Behenic acid

[112-85-6]

$H_3C(CH_2)_{20}COOH$

$C_{22}H_{44}O_2$ M 340.588

Occurs in *Lophira alata* kernel fat, carnauba wax and *Mimusops elengi* seed oil. Mp 81-82°.

Me ester: [929-77-1].

$C_{23}H_{46}O_2$ M 354.615

Constit. of *Cassia javanica* leaves. Mp 54°.

Et ester: [5908-87-2].

$C_{24}H_{48}O_2$ M 368.642

Mp 50°. Bp_{10} 240-242°, $Bp_{0.2}$ 185°.

Chloride:

$C_{22}H_{43}ClO$ M 359.034

Cryst. Mp 73-75°.

Amide:

$C_{22}H_{45}NO$ M 339.604

Mp 111°.

Anilide:

$C_{28}H_{49}NO$ M 415.701

Mp 101-102°.

Levene, P.A. *et al*, *J. Biol. Chem.*, 1924, **59**, 905.
Bleyberg, W. *et al*, *Ber.*, 1931, **64**, 2504 (*synth*)
Dinh-Nguyen, N. *et al*, *Acta Chem. Scand.*, 1966, **20**, 1423 (*synth*)
Holmbom, B., *J. Am. Oil Chem. Soc.*, 1977, **54**, 289 (*glc*)

Pesic, Z. *et al*, *Phytopathol. Z.*, 1978, **91**, 121.
Conti, G. *et al*, *J. Mol. Struct.*, 1984, **118**, 237 (*cryst struct*)
Joshi, U.R. *et al*, *Synth. Commun.*, 1984, **14**, 681 (*synth*)
Chaudhuri, K. *et al*, *Indian J. Pharm. Sci.*, 1985, **47**, 172; *CA*, **105**, 21726t (*occur, Me ester*)

1-Docosanol D-00302

1-Hydroxydocosane. Behenyl alcohol

[661-19-8]

$H_3C(CH_2)_{20}CH_2OH$

$C_{22}H_{46}O$ M 326.605

Constit. of several waxes. Mp 87° (70.8°, 73.4°).

Phenylurethane: Mp 86-86.5°.

Ac:

$C_{24}H_{48}O_2$ M 368.642

Minor constit. of female sex pheromones from *Sparganothis pilleriana* and *Agapeta zoegana*.

Francis, F. *et al*, *Proc. R. Soc. London, A*, 1937, **158**, 695 (*synth*)
Estrada, H., *CA*, 1967, **66**, 108163g (*nmr*)
Cavero, I. *et al*, *Ann. Chim. (Rome)*, 1970, **60**, 469 (*isol*)
Precht, D., *Fette, Seifen, Anstrichm.*, 1976, **78**, 189 (*cryst struct*)
Willuhn, G. *et al*, *Planta Med.*, 1977, **31**, 328 (*isol*)
Tóth, M. *et al*, *Can. Entomol.*, 1985, **117**, 1163 (*isol*)

Docosanoylbenzene D-00303

Docosanophenone. Behenophenone. Heneicosyl phenyl ketone

[5667-66-3]

$PhCO(CH_2)_{20}CH_3$

$C_{28}H_{48}O$ M 400.687

Cryst. (Me_2CO). Mp 74-76°.

Mikeska, L.A. *et al*, *J. Org. Chem.*, 1941, **6**, 788 (*synth*)
Achaya, K.T. *et al*, *Indian J. Chem.*, 1967, **5**, 109 (*synth*)

2-Docosanoylnaphthalene D-00304

Docosanonaphthone. Behenonaphthone. Heneicosyl 2-naphthyl ketone

$C_{32}H_{50}O$ M 450.746

Cryst. (Me_2CO/pet. ether). Mp 67-69°.

Mikeska, L.A. *et al*, *J. Org. Chem.*, 1941, **6**, 787 (*synth*)

4,7,10,13,16-Docosapentaenoic acid D-00305

[2313-14-6]

$H_3C(CH_2)_3(CH_2CH{=}CH)_5CH_2CH_2COOH$

$C_{22}H_{34}O_2$ M 330.509

(all-Z)-form

Metab. of Linoleic acid.

Me ester: $Bp_{0.001}$ 138-146°.

Et ester: $Bp_{0.01}$ 147-152°.

Osbond, J.M. *et al*, *Chem. Ind. (London)*, 1959, 1288 (*synth*)
Osbond, J.M. *et al*, *J. Chem. Soc.*, 1961, 2779 (*synth*)

4,8,12,15,19-Docosapentaenoic acid D-00306

Clupanodonic acid

[2548-85-8]

$H_3CCH_2CH{=}CHCH_2CH_2CH{=}CHCH_2CH{=}CHCH_2$
$CH_2CH{=}CHCH_2CH_2CH{=}CHCH_2CH_2COOH$

$C_{22}H_{34}O_2$ M 330.509

Constit. of many fish oils e.g. sardine, mackerel and herring.

Me ester: d_{15} 0.936. $Bp_{0.07}$ 169-170°. n_D^{15} 1.4933.

Tsujimoto, M., *Chem. Zentralbl.*, 1923, **1**, 38.
Toyama, Y. *et al*, *Bull. Chem. Soc. Jpn.*, 1935, **10**, 433, 441 (*isol, struct*)
Baudent, P., *Compt. Rend. Trav. Faculté. Sci. Marseille*, 1941, **1**, 79; 1942, **1**, 86; 1943, **1**, 127 (*isol, struct*)
Varanasi, U. *et al*, *Biochim. Biophys. Acta*, 1975, **409**, 304 (*isol*)

7,10,13,16,19-Docosapentaenoic acid, 9CI D-00307

[2234-74-4]

$H_3C(CH_2CH{=}CH)_5(CH_2)_5COOH$

$C_{22}H_{34}O_2$ M 330.509

Metab. of Linolenic acid.

Myagkova, G.I. *et al*, *CA*, 1967, **66**, 94666 (*synth*)

4,7,10,13,16-Docosapentaynoic acid D-00308

[16326-33-3]

$H_3C(CH_2)_4(C{\equiv}CCH_2)_5CH_2COOH$

$C_{22}H_{24}O_2$ M 320.430

Cryst. (EtOH). Mp 100-101.5°.

Osbond, J.M. *et al*, *J. Chem. Soc.*, 1961, 2779 (*synth*)
Kunau, W.H., *Hoppe Seyler's Z. Physiol. Chem.*, 1971, **352**, 542 (*synth*)

7,10,13,16,19-Docosapentaynoic acid D-00309

[3507-86-6]

$H_3CCH_2C{\equiv}CCH_2C{\equiv}CCH_2C{\equiv}CCH_2C{\equiv}CCH_2C{\equiv}C(CH_2)_5COOH$

$C_{22}H_{24}O_2$ M 320.430

Cryst. (EtOH). Mp 109-110°.

Kunau, W.H., *Hoppe Seyler's Z. Physiol. Chem.*, 1971, **352**, 542 (*synth*)

3-(3,6,9,12-Docosatetraenyl)-5-hydroxy-5-methyl-2(5*H*)-furanone D-00310

2-(3,6,9,12-Docosatetraenyl)-4-hydroxy-4-methyl-2-buten-4-olide

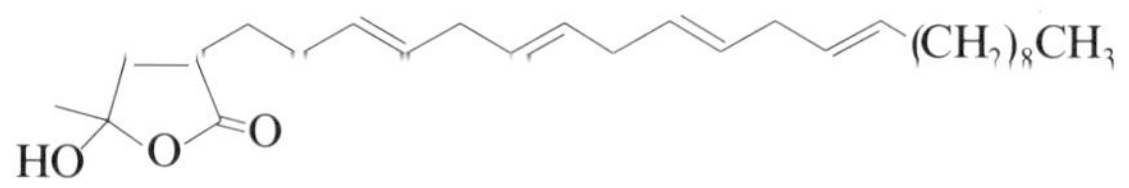

$C_{27}H_{42}O_3$ M 414.627

Obt. from the Japanese gorgonian *Euplexaura flava*.

3′,4′-Dihydro: 3-(6,9,12-Docosatrienyl)-5-hydroxy-5-methyl-2(5H)-furanone

$C_{27}H_{44}O_3$ M 416.643

From *E. flava*.

Kituchi, H. *et al*, *Chem. Lett.*, 1982, 233 (*isol*)

4,7,10,13-Docosatetraynoic acid D-00311

[31820-13-0]

$H_3C(CH_2)_7C{\equiv}CCH_2C{\equiv}CCH_2C{\equiv}CCH_2C{\equiv}CCH_2CH_2COOH$

$C_{22}H_{28}O_2$ M 324.462

Cryst. (EtOH). Mp 86.5-87.5°.

Kunau, W.H., *Hoppe Seyler's Z. Physiol. Chem.*, 1971, **352**, 542 (*synth*)

5,8,11,14-Docosatetraynoic acid D-00312

[32895-88-8]

$H_3C(CH_2)_6C{\equiv}CCH_2C{\equiv}CCH_2C{\equiv}CCH_2C{\equiv}C(CH_2)_3COOH$

$C_{22}H_{28}O_2$ M 324.462

Cryst. (EtOAc). Mp 85-86°.

Beerthuis, R.K. *et al*, *Recl. Trav. Chim. Pays-Bas* (*J. R. Neth. Chem. Soc.*), 1971, **90**, 943 (*synth*)

7,10,13,16-Docosatetraynoic acid D-00313

[10434-01-2]

$H_3C(CH_2)_4C{\equiv}CCH_2C{\equiv}CCH_2C{\equiv}CCH_2C{\equiv}C(CH_2)_5COOH$

$C_{22}H_{28}O_2$ M 324.462

Cryst. (hexane). Mp 82-82.5°.

Struijk, C.B. *et al*, *Recl. Trav. Chim. Pays-Bas* (*J. R. Neth. Chem. Soc.*), 1966, **85**, 1233 (*synth*)
Myagkova, G.I. *et al*, *CA*, 1967, **66**, 75631 (*synth*)
Kunau, W.H., *Hoppe Seyler's Z. Physiol. Chem.*, 1971, **352**, 542 (*synth*)
Bus, J. *et al*, *Recl. Trav. Chim. Pays-Bas* (*J. R. Neth. Chem. Soc.*), 1972, **91**, 394 (*ir*)

3,6,9-Docosatriene, 9CI D-00314

$H_3C(CH_2)_{11}CH{=}CHCH_2CH{=}CHCH_2CH{=}CHCH_2CH_3$

$C_{22}H_{40}$ M 304.558

(3Z,6Z,9Z)-form [97689-04-8]

Minor component of the sex pheromones of several Lepidoptera species including *Mocis megas* and *Ectropis obliqua*.

9S,10R-Epoxide: [97719-68-1]. *2-Dodecyl-3-(2,5-octadienyl) oxirane, 9CI. 9,10-Epoxy-3,6-docosadiene*

$C_{22}H_{40}O$ M 320.557

Sex attractant for several geometrid/noctuid moths.

[139405-38-2]

Wong, J.W. *et al*, *J. Chem. Ecol.*, 1985, **11**, 727 (*activity*)
Frerot, B. *et al*, *C. R. Hebd. Seances Acad. Sci. Ser. 3*, 1988, **306**, 157 (*activity*)
Renou, M. *et al*, *J. Insect Physiol.*, 1988, **34**, 481 (*synth, activity*)
Descoins, C. *et al*, *Experientia*, 1990, **46**, 536 (*isol, activity*)
Yao, E. *et al*, *Prog. Nat. Sci.*, 1991, **1**, 566 (*isol, activity*)

3,9,15-Docosatrienoic acid D-00315

$H_3C(CH_2)_5CH{=}CH(CH_2)_4CH{=}CH(CH_2)_4CH{=}CHCH_2COOH$

$C_{22}H_{38}O_2$ M 334.541

(all-Z)-form [35087-15-1]

Oil. $Bp_{0.01}$ 180°.

Ames, D.E. *et al*, *J. Chem. Soc., Perkin Trans. 1*, 1972, 255 (*synth*)

5,11,17-Docosatrienoic acid D-00316

$H_3C(CH_2)_3CH{=}CH(CH_2)_4CH{=}CH(CH_2)_4CH{=}CH(CH_2)_3COOH$

$C_{22}H_{38}O_2$ M 334.541

(all-Z)-form [35087-10-6]

Oil.

Ames, D.E. *et al*, *J. Chem. Soc., Perkin Trans. 1*, 1972, 255 (*synth*)

7,10,13-Docosatrienoic acid D-00317

$H_3C(CH_2)_7CH{=}CHCH_2CH{=}CHCH_2CH{=}CH(CH_2)_5COOH$

$C_{22}H_{38}O_2$ M 334.541

(all-Z)-form [28874-30-8]

Metab. of Oleic acid prod. particularly under conditions of essential fatty acid deficiency.

Kunau, W.H. *et al*, *Hoppe Seyler's Z. Physiol. Chem.*, 1971, **352**, 1297.

13,16,19-Docosatrienoic acid D-00318

[59708-86-0]

$H_3CCH_2(CH{=}CHCH_2)_3(CH_2)_9CH_2COOH$

$C_{22}H_{38}O_2$ M 334.541

All-(Z)-form [28845-86-5]

Isol. from rape seed oil. Widespread among marine organisms. Found in lipids of egg yolk, embryo of dogfish and in lipids and phospholipids of pancreatic islets.

[108698-01-7]

Haeffner, E., *Lipids*, 1970, **5**, 430 (*isol, struct*)
Moreno, V.J. *et al*, *Lipids*, 1979, **14**, 15, 313 (*biosynth*)
Diaz, G.B. *et al*, *Lipids*, 1988, **23**, 1125; 1990, **25**, 724 (*occur*)
Berger, A. *et al*, *Lipids*, 1990, **25**, 473 (*biosynth*)

2,4,10-Docosatrienoic isobutylamide D-00319

$H_3C(CH_2)_{10}CH{=}CH(CH_2)_4CH{=}CHCH{=}CHCONHCH_2CH(CH_3$

$C_{26}H_{47}NO$ M 389.663

(E,E,Z)-form [62726-18-5] ***Filfiline***

Constit. of the fruits of *Piper officinarum*. Has antifertility props. Cryst. (pet. ether). Mp 66-66.5°.

Gupta, O.P. *et al*, *Indian J. Chem., Sect. B*, 1976, **14**, 912 (*isol*)
Gupta, O.P. *et al*, *Indian J. Pharm. Sci.*, 1978, **40**, 113 (*use*)

3,9,15-Docosatriynoic acid D-00320

$H_3C(CH_2)_5C{\equiv}C(CH_2)_4C{\equiv}C(CH_2)_4C{\equiv}CCH_2COOH$

$C_{22}H_{32}O_2$ M 328.494

Cryst. (pet. ether). Mp 61.5-62°.

Ames, D.E. *et al*, *J. Chem. Soc., Perkin Trans. 1*, 1972, 255 (*synth*)

8,11,14-Docosatriynoic acid D-00321

[19960-56-6]

$H_3C(CH_2)_6C{\equiv}CCH_2C{\equiv}CCH_2C{\equiv}C(CH_2)_6COOH$

$C_{22}H_{32}O_2$ M 328.494

Mp 56-57.5°.

Beerthuis, R.K. *et al*, *Recl. Trav. Chim. Pays-Bas* (*J. R. Neth. Chem. Soc.*), 1968, **87**, 461 (*synth*)
Bus, J. *et al*, *Recl. Trav. Chim. Pays-Bas* (*J. R. Neth. Chem. Soc.*), 1972, **91**, 394 (*ir*)

10,13,16-Docosatriynoic acid D-00322

[60323-48-0]

$H_3C(CH_2)_4C{\equiv}CCH_2C{\equiv}CCH_2C{\equiv}C(CH_2)_8COOH$

$C_{22}H_{32}O_2$ M 328.494

Weak inhibitor of prostaglandin biosynth.

Goetz, J.M. *et al*, *Prostaglandins*, 1976, **12**, 187.

1-Docosene, 9CI D-00323

[1599-67-3]

$H_3C(CH_2)_{19}CH{=}CH_2$

$C_{22}H_{44}$ M 308.590

Isol. in small amt. from rose oil wax (*Rosa* sp.). Mp 58-60°.

Touhiana, R. *et al*, *Ann. Chim. (Paris)*, 1962, **7**, 593 (*synth*)
Wollrab, V. *et al*, *Collect. Czech. Chem. Commun.*, 1965, **30**, 1654 (*isol*)
Field, F.H., *J. Am. Chem. Soc.*, 1968, **90**, 5649 (*ms*)
Sharpless, K.B. *et al*, *J. Chem. Soc., Chem. Commun.*, 1972, 370 (*synth*)

11-Docosene D-00324

[62978-77-2]

$H_3C(CH_2)_9CH{=}CH(CH_2)_9CH_3$

$C_{22}H_{44}$ M 308.590

(E)-form [61759-29-3]
Obt. cont. 15% (*Z*)-isomer.

(Z)-form [61759-30-6]
Obt. cont. <10° (*E*)-isomer.

Léonard-Goppens, A.M. *et al*, *Tetrahedron Lett.*, 1976, 3227 (*synth*)
Lucchetti, J. *et al*, *C. R. Hebd. Seances Acad. Sci. Ser. C*, 1979, **288**, 553 (*synth*)
Odinokov, V.N. *et al*, *Zh. Org. Khim.*, 1979, **15**, 1403; *CA*, **92**, 22315h (*synth*)
Fyles, T.M. *et al*, *J. Org. Chem.*, 1984, **49**, 753 (*synth, pmr, cmr, ir, ms*)

15-Docosene-1,3-diyn-5-one, 9CI D-00325

$H_3C(CH_2)_5CH{=}CH(CH_2)_9COC{\equiv}CC{\equiv}CH$

$C_{22}H_{34}O$ M 314.510

(Z)-form [63987-88-2]
Constit. of *Reniera fulva*. Oil. λ_{max} 236 (ε 5 800), 246 (9 320), 262 (12 900) and 276 nm (8 700) (MeOH).

Cimino, G. *et al*, *Tetrahedron Lett.*, 1977, 1325 (*ir, uv, pmr*)

5-Docosenoic acid D-00326

$H_3C(CH_2)_{15}CH{=}CH(CH_2)_3COOH$

$C_{22}H_{42}O_2$ M 338.573

(Z)-form [64777-02-2]
Unusual fatty acid isol. from *Limnanthes douglasii* and *L. alba* seed oil.

Smith, C.R. *et al*, *J. Org. Chem.*, 1960, **25**, 1770 (*isol*)
Bergelson, L.D. *et al*, *Tetrahedron*, 1963, **19**, 149 (*synth*)
Pollard, M.R. *et al*, *Plant Physiol.*, 1980, **66**, 649; *CA*, **94**, 2061k (*biosynth*)
Sebedio, J.L. *et al*, *Lipids*, 1982, **17**, 469 (*isol*)

11-Docosenoic acid D-00327

[506-36-5]

$H_3C(CH_2)_9CH{=}CH(CH_2)_9COOH$

$C_{22}H_{42}O_2$ M 338.573

(E)-form
Catelaidic acid
Constit. of partially hydrogenated fish oil.

(Z)-form [1002-96-6]
Cetoleic acid
Esters occur commonly in oil from marine mammals and fish and rapeseed. Mp 33-33.7°.

Toyama, Y., *CA*, 1928, **22**, 575 (*isol*)
Ackman, R.G. *et al*, *Fette, Seifen, Anstrichm.*, 1977, **79**, 15.
Gudbjarnason, S. *et al*, *Biochim. Biophys. Acta*, 1977, **487**, 10.
Kirk, R.S. *et al*, *J. Sci. Food Agric.*, 1978, **29**, 880 (*glc*)
Loew, F.M. *et al*, *Nutr. Metab.*, 1978, **22**, 207.
Thomas, D.C. *et al*, *Arch. Microbiol.*, 1978, **117**, 239.

13-Docosenoic acid D-00328

$H_3C(CH_2)_7CH{=}CH(CH_2)_{11}COOH$

$C_{22}H_{42}O_2$ M 338.573

(E)-form [506-33-2]
Brassidic acid
Cryst. (EtOH). Mp 61.5°. Bp_{30} 282°, Bp_{10} 256°.
Me ester: [7439-44-3].
Mp 34-35°.
Et ester: Mp 30.5°.
Nitrile: Mp 22°. Bp_{17} 257°.
Anhydride: Mp 64°.

(Z)-form [112-86-7]
Erucic acid
An important vegetable fatty acid, mainly confined to seed fats of the families Cruciferae and Tropaelaceae (e.g. rape, mustard and wallflower seeds). Cryst. (CH_3CN). Mp 33.5°. Undergoes solid-state phase transitions.
Chloride: Mp 14°.
Amide: Mp 94° (65-66°).
Me ester: [1120-34-9].
Oil. Bp_5 221-222°. Ir ν 1744 cm^{-1}.
Et ester: [37910-77-3].
Bp_5 229-230°.
Anhydride: Mp 47.5-48°.
Nitrile: Bp 238-240°.

Holde, D. *et al*, *Ber.*, 1923, **56**, 2052.
Org. Synth., Coll. Vol., 2, 1943, 258 (*synth*)
Bowman, R.E., *J. Chem. Soc.*, 1950, 177 (*synth*)
Lakizo, V.I., *CA*, 1968, **69**, 2475g (*synth*)
Niewiadomski, H., *Przem. Chem.*, 1969, **48**, 659; *CA*, **72**, 78, 293 (*rev*)
Chang, S.P. *et al*, *J. Am. Oil Chem. Soc.*, 1972, **49**, 422 (*synth*)
Mitcham, D. *et al*, *J. Am. Oil Chem. Soc.*, 1973, **50**, 446 (*ir*)
Barrette, J.P. *et al*, *J. Assoc. Off. Anal. Chem.*, 1976, **59**, 855 (*glc*)
Yamada, K. *et al*, *J. Org. Chem.*, 1978, **43**, 2076 (*synth*)
Kaneko, F. *et al*, *Acta Crystallogr., Sect. C*, 1992, **48**, 1060 (*cryst struct*)
Painuly, P. *et al*, *J. Chromatogr.*, 1992, **590**, 139 (*purifn*)
Kling, M.R. *et al*, *J. Chem. Soc., Perkin Trans. 1*, 1993, 1183 (*synth, ir, ms*)

15-Docosenoic acid D-00329

[14134-54-4]

$H_3C(CH_2)_5CH{=}CH(CH_2)_{13}COOH$

$C_{22}H_{42}O_2$ M 338.573

(Z)-form [17735-97-6]
Found in pig brain sphingolipids.

[62732-79-0, 82683-31-6]

Kishimoto, Y. *et al*, *J. Lipid Res.*, 1963, **4**, 437; 1964, **5**, 98 (*isol, struct*)
Sebedio, S.L. *et al*, *Lipids*, 1982, **17**, 469 (*isol*)

19-Docosenoic acid D-00330

$H_3CCH_2CH{=}CH(CH_2)_{17}COOH$

$C_{22}H_{42}O_2$ M 338.573

(Z)-form [55833-27-7]
Cryst. (Me_2CO). Mp 60-62°.

Klok, R. *et al, Recl. Trav. Chim. Pays-Bas (J. R. Neth. Chem. Soc.)*, 1974, **93**, 222 (*synth*)

13-Docosen-1-ol D-00331

Docosenyl alcohol

$H_3C(CH_2)_7CH{=}CH(CH_2)_{11}CH_2OH$

$C_{22}H_{44}O$ M 324.589

(Z)-form [629-98-1]
Erucyl alcohol
Isol. from seed wax of *Simmondsia californica* (prob. the Z-form). Cryst. (Me_2CO or MeOH). Mp 34.5-35.5°. Bp_{10} 240.5-241.5°, $Bp_{0.2}$ 199°.
Dibromide: 13,14-Dibromo-1-docosanol
$C_{22}H_{44}Br_2O$ M 484.397
Mp 45°.
Phenylurethane: Mp 86-86.5°.

Willstätter, R., *Justus Liebigs Ann. Chem.*, 1911, **378**, 101 (*synth*)
Bleyberg, W. *et al, Ber.*, 1931, **64**, 2504 (*synth*)
Green, T.G. *et al, J. Chem. Soc.*, 1936, 1750 (*isol*)
Kameoka, H. *et al, Yukagaku*, 1970, **19**, 70.

2-Docosyl-3-hydroxy-21-tetracontenoic acid D-00332

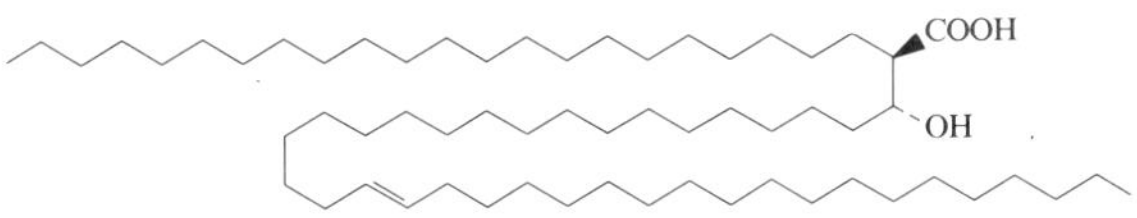

$C_{62}H_{122}O_3$ M 915.644
Component of mycolic acids prod. by mycobacteria.

(2RS,3RS,21E)-form
(±)-trans-threo-*form*
Et ester: [82741-55-7].
Cryst. Mp 74-77°.

(2RS,3RS,21Z)-form
(±)-cis-threo-*form*
Et ester: [82741-56-8].
Cryst. (Me_2CO). Mp 73-76°.

[82741-71-7, 82741-72-8]

Huang, H.C. *et al, J. Org. Chem.*, 1982, **47**, 4018 (*synth, pmr*)

11-Docosyne D-00333

[88454-86-8]

$H_3C(CH_2)_9C{\equiv}C(CH_2)_9CH_3$

$C_{22}H_{42}$ M 306.574
Bp_1 138-140°.

Fyles, T.M. *et al, J. Org. Chem.*, 1984, **49**, 753 (*synth, pmr, cmr, ir, ms*)

13-Docosynoic acid D-00334

Behenolic acid
[506-35-4]

$H_3C(CH_2)_7C{\equiv}C(CH_2)_{11}COOH$

$C_{22}H_{40}O_2$ M 336.557
Cryst. (MeOH). Mp 57-57.5°.

Me ester:
$C_{23}H_{42}O_2$ M 350.584
Mp 22°.
Et ester:
$C_{24}H_{44}O_2$ M 364.610
Mp 15-16°.
Chloride:
$C_{22}H_{39}ClO$ M 355.002
Mp 29-30°.
Amide:
$C_{22}H_{41}NO$ M 335.572
Mp 90°.
Anilide:
$C_{28}H_{45}NO$ M 411.670
Mp 73°.

Bhattacharya, R. *et al, J. Chem. Soc.*, 1928, 2678 (*synth*)
Osman, F. *et al, J. Chromatogr.*, 1967, **26**, 286 (*tlc*)
Churacek, J. *et al, CA*, 1969, **71**, 4743h (*glc*)

2,6-Dodecadienal, 9CI D-00335

$H_3C(CH_2)_4CH{=}CHCH_2CH_2CH{=}CHCHO$

$C_{12}H_{20}O$ M 180.289
Component of cooked chicken flavour. Also from *n*-6 polyene acyl chains.

[21662-13-5, 21662-14-6]

Meijboom, P.W. *et al, Fette, Seifen, Anstrichm.*, 1968, **70**, 477.
Harkes, P.D. *et al, J. Am. Oil Chem. Soc.*, 1974, **51**, 356 (*occur*)

5,7-Dodecadienal D-00336

$H_3C(CH_2)_3CH{=}CHCH{=}CH(CH_2)_3CHO$

$C_{12}H_{20}O$ M 180.289

(5E,7Z)-form [75539-65-0]
Pheromone of western tent caterpillar *Malacosoma californicum*. $Bp_{0.1}$ 70-90°.

(5Z,7E)-form [75983-34-5]
Pheromone of *Dendrolimus pini* and forest tent caterpillar *M. disstria*.

(5Z,7Z)-form [75983-33-4]
$Bp_{0.01}$ 55-60°.

Chisholm, M.D. *et al, Environ. Entomol.*, 1980, **9**, 278.
Bestmann, H.J. *et al, Justus Liebigs Ann. Chem.*, 1982, 1359 (*synth*)

6,9-Dodecadienal D-00337

$H_3CCH_2CH{=}CHCH_2CH{=}CH(CH_2)_4CHO$

$C_{12}H_{20}O$ M 180.289

(6Z,9Z)-form [21662-12-4]
$Bp_{1.5}$ 101-102°.
2,4-Dinitrophenylhydrazone: [22644-18-4].
Mp 58-59°.

Ward, J.P. *et al, Recl. Trav. Chim. Pays-Bas (J. R. Neth. Chem. Soc.)*, 1969, **88**, 177 (*synth*)

8,10-Dodecadienal D-00338

$H_3CCH{=}CHCH{=}CH(CH_2)_6CHO$

$C_{12}H_{20}O$ M 180.289

(8E,10E)-form [69775-58-2]
$Bp_{0.01}$ 70°.

Bestmann, H.J. *et al, Justus Liebigs Ann. Chem.*, 1981, 2117 (*synth, pmr*)

9,11-Dodecadienal, 9CI D-00339

$H_2C{=}CHCH{=}CH(CH_2)_7CHO$

$C_{12}H_{20}O$ M 180.289

Both E and Z forms identified as components (E:Z ratio 10:1) of the sex pheromone of *Tortricidia testacea*.

[97475-08-6, 97475-11-1]

Ando, T. *et al*, *Agric. Biol. Chem.*, 1985, **49**, 413; 1988, **52**, 1415 (*isol, synth, ms, glc*)
Chisholm, M.D. *et al*, *J. Chem. Ecol.*, 1985, **11**, 1341 (*synth, isol, activity*)
Trost, B.M. *et al*, *J. Org. Chem.*, 1988, **53**, 915 (*synth*)

1,11-Dodecadiene D-00340

[5876-87-9]

$H_2C{=}CH(CH_2)_8CH{=}CH_2$

$C_{12}H_{22}$ M 166.306

Casteignau, G. *et al*, *Bull. Soc. Chim. Fr.*, 1968, 3893 (*glc*)

4,8-Dodecadiene D-00341

$H_3CCH_2CH_2CH{=}CHCH_2CH_2CH{=}CHCH_2CH_2CH_3$

$C_{12}H_{22}$ M 166.306

(E,E)-form [26907-85-7]
Bp_{16} 87-88°.

(Z,E)-form [38926-02-2]
Bp_8 75-76°.

Mochel, V.D. *et al*, *J. Am. Chem. Soc.*, 1972, **94**, 6202 (*synth, cmr*)
Pinazzi, C. *et al*, *Bull. Soc. Chim. Fr.*, 1973, 1027 (*synth*)
Pinazzi, C., *Nuova Chim.*, 1974, **50**, 77 (*synth*)
Deneke, M. *et al*, *Makromol. Chem.*, 1975, **176**, 1471 (*cmr*)
Pinazzi, C. *et al*, *Tetrahedron Lett.*, 1976, **13**, 989 (*synth*)

2,4-Dodecadienoic acid D-00342

$H_3C(CH_2)_6CH{=}CHCH{=}CHCOOH$

$C_{12}H_{20}O_2$ M 196.289

Probable constit. of *Sebastiana lingustrina* seed oil.

(2E,4E)-form [24738-48-5]
Cryst. (pet. ether). Mp 49-51°.
Me ester: [24738-46-3].
$C_{13}H_{22}O_2$ M 210.316
$Bp_{0.3}$ 106-108°.
2-Methylpropylamide: [24738-51-0]. ***2,4-Dodecadienoic acid isobutylamide***
$C_{16}H_{29}NO$ M 251.411
Alkaloid from *Anacyclus pyrethrum* and *Leucocyclus formosus* (Compositae) (in admixture with homologues). Mp 89-90°.
Pyrrolidide: [117137-69-6]. *2,4-Dodecadienoic acid pyrrolidide. 1-(2,4-Dodecadienoyl)pyrrolidine*
$C_{16}H_{27}NO$ M 249.395
Constit. of pepper (*Piper nigrum*) (Piperaceae). Shows larvicidal activity. Not obt. completely pure.

Holman, R.T. *et al*, *J. Am. Oil Chem. Soc.*, 1955, **32**, 356 (*isol*)
Burden, R.S. *et al*, *J. Chem. Soc. C*, 1969, 2477 (*isol, synth*)
Greger, H. *et al*, *Phytochemistry*, 1981, **20**, 2579 (*isol, ms, struct*)
Kiuichi, F. *et al*, *Chem. Pharm. Bull.*, 1988, **36**, 2452 (*pyrrolidide*)

2,6-Dodecadienoic acid D-00343

$H_3C(CH_2)_4CH{=}CHCH_2CH_2CH{=}CHCOOH$

$C_{12}H_{20}O_2$ M 196.289

(2E,6Z)-form
Me ester: [28369-22-4].
$C_{13}H_{22}O_2$ M 210.316
Constit. of Barlett pears.
Et ester: [28380-08-7].
$C_{14}H_{24}O_2$ M 224.342
Constit. of Barlett pears.

Creveling, R.K. *et al*, *J. Agric. Food Chem.*, 1970, **18**, 19.
Naf, F. *et al*, *Helv. Chim. Acta*, 1972, **55**, 82 (*synth*)

2,8-Dodecadienoic acid D-00344

$H_3C(CH_2)_2CH{=}CH(CH_2)_4CH{=}CHCOOH$

$C_{12}H_{20}O_2$ M 196.289

(E,E)-form
Mp 35°. $Bp_{0.7}$ 137°.
S-*Benzylthiuronium salt:* Mp 159°.
N-*Isobutylamide:*
$C_{16}H_{29}NO$ M 251.411
Mp 54°. Bp 004° 145.

(2E,8Z)-form
$Bp_{0.00001}$ 98-100°.
S-*Benzylthiuronium salt:* Cryst. Mp 143-144°.

(2Z,8E)-form
$Bp_{0.5}$ 115-117°.
S-*Benzylthiuronium salt:* Mp 148°.
N-*Isobutylamide:*
$C_{16}H_{29}NO$ M 251.411
$Bp_{0.01}$ 142-143°.

(2Z,8Z)-form
$Bp_{0.0001}$ 108-109°.
S-*Benzylthiuronium salt:* Cryst. (EtOAc). Mp 129-130°.

Raphael, R.A. *et al*, *J. Chem. Soc.*, 1950, 115 (*synth*)
Raphael, R.A. *et al*, *J. Chem. Soc.*, 1951, 2693 (*synth*)
Crombie, L., *J. Chem. Soc.*, 1952, 2997.
Dobson, N.A. *et al*, *J. Chem. Soc.*, 1955, 3558 (*synth*)

5,7-Dodecadienoic acid D-00345

$H_3C(CH_2)_3CH{=}CHCH{=}CH(CH_2)_3COOH$

$C_{12}H_{20}O_2$ M 196.289

(E,E)-form [97827-04-8]
Oil.
Me ester: [97827-02-6]
$C_{13}H_{22}O_2$ M 210.316
Oil.

[97827-03-7]

Watanabe, Y. *et al*, *J. Org. Chem.*, 1989, **54**, 4088 (*synth, ir, pmr, cmr, ms*)

7,9-Dodecadienoic acid D-00346

$H_3CCH_2CH{=}CHCH{=}CH(CH_2)_5COOH$

$C_{12}H_{20}O_2$ M 196.289

(7E,9Z)-form
Me ester: [72553-56-1].
$C_{13}H_{22}O_2$ M 210.316
$Bp_{0.01}$ 80-85°.

Bestmann, H.J. *et al*, *Justus Liebigs Ann. Chem.*, 1981, 2117 (*synth, pmr*)

8,10-Dodecadienoic acid D-00347

$H_3CCH{=}CHCH{=}CH(CH_2)_6COOH$

$C_{12}H_{20}O_2$ M 196.289

(8E,10E)-form

Me ester: [60099-82-3].
Pale-yellow liq. $Bp_{0.4}$ 107-110°.

(8E,10Z)-form

Me ester: [80625-47-4].
$C_{13}H_{22}O_2$ M 210.316
$Bp_{0.01}$ 75-80°.

Naoshima, Y. *et al, Agric. Biol. Chem.*, 1980, **44**, 1429 (*pmr, synth, ir, ms*)
Bestmann, H.J. *et al, Justus Liebigs Ann. Chem.*, 1981, 227 (*synth, pmr*)

2,3-Dodecadien-1-ol D-00348

[34656-66-1]

$H_3C(CH_2)_7CH{=}C{=}CHCH_2OH$

$C_{12}H_{22}O$ M 182.305

Landor, P.D. *et al*, 1971, 1638 (*synth, ir*)
Djahanbini, D. *et al, Tetrahedron*, 1987, **43**, 3441 (*synth, ir, pmr*)

3,5-Dodecadien-1-ol, 9CI D-00349

$H_3C(CH_2)_5CH{=}CHCH{=}CHCH_2CH_2OH$

$C_{12}H_{22}O$ M 182.305

(3E,5Z)-form [32504-51-1]
Disruptive to the trail-following pheromone of southern subterranean termite, *Reticulitermes virginicus.*

(3Z,5Z)-form [32451-96-0]
Minor component of the essential oil of *Cyperus iria* (Kogomegayatsuri).

Ac: [69285-94-5].
$C_{14}H_{24}O_2$ M 224.342
Major component of the essential oil of *C. iria*.

[32451-97-1]

Tai, A. *et al, J. Insect Physiol.*, 1971, **17**, 181 (*synth, activity*)
Matsumura, F. *et al, J. Econ. Entomol.*, 1972, **65**, 600 (*activity*)
Iwamura, J. *et al, Nippon Nogei Kagaku Kaishi*, 1978, **52**, 379 (*isol*)
Reed, D.W. *et al, J. Chem. Ecol.*, 1985, **11**, 1645 (*activity*)
Golovanov, I.B. *et al, Mol. Biol.*, 1991, **25**, 1594 (*activity*)

3,6-Dodecadien-1-ol D-00350

$H_3C(CH_2)_4CH{=}CHCH_2CH{=}CHCH_2CH_2OH$

$C_{12}H_{22}O$ M 182.305
Synthetic trail-following substance for termites.

(Z,Z)-form [29125-78-8]
Yellowish liq.

Carvalho, J.F. *et al, J. Org. Chem.*, 1984, **49**, 1251 (*synth, ir, pmr, cmr, bibl*)

5,7-Dodecadien-1-ol D-00351

$H_3C(CH_2)_3CH{=}CHCH{=}CH(CH_2)_3CH_2OH$

$C_{12}H_{22}O$ M 182.305

(5Z,7E)-form [73416-71-4]
Sex pheromone component of the forest tent caterpillar *Malacosoma disstria*. $Bp_{0.05}$ 90-91°.

Chisholm, M.D. *et al, Environ. Entomol.*, 1980, **9**, 278.
Rossi, R. *et al, Tetrahedron*, 1983, **39**, 287 (*synth*)
Stille, J.K. *et al, J. Am. Chem. Soc.*, 1987, **109**, 2138 (*synth, pmr, cmr, ir*)

7,9-Dodecadien-1-ol, 9CI D-00352

$H_3CCH_2CH{=}CHCH{=}CH(CH_2)_5CH_2OH$

$C_{12}H_{22}O$ M 182.305

(E,E)-form [54364-61-3]
Pheromone for insect spp. $Bp_{0.01}$ 98°.

Ac: [54364-63-5]. *12-Acetoxy-3,5-dodecadiene*
$C_{14}H_{24}O_2$ M 224.342
Pheromone for insect spp. $Bp_{0.01}$ 88°.

(7E,9Z)-form [66471-35-0]
$Bp_{0.33}$ 88°.

Ac: [55774-32-8]. ***Eudemone***
Isol. from the abdomen of the grapevine moth *Lobesia botrana* females. Pheromone for various moths, particularly *L. botrana*. $Bp_{0.2}$ 86-86.5°.

Buser, H.R. *et al, Z. Naturforsch., C*, 1974, **29**, 781.
Buser, H.R. *et al, J. Chromatogr.*, 1975, **106**, 83 (*glc, ms*)
Labovitz, J.N. *et al, Tetrahedron Lett.*, 1975, 4209.
Descoius, C. *et al, Bull. Soc. Chim. Fr.*, 1977, 941 (*ir, nmr, glc, ms*)
Warthen, J.D. *et al, Chromatographia*, 1977, **10**, 720.
Nigishi, E. *et al, Tetrahedron Lett.*, 1977, 411 (*ir, nmr, glc*)
Bestmann, H.J. *et al, Justus Liebigs Ann. Chem.*, 1981, 2117 (*synth, pmr*)
Rossi, R. *et al, Tetrahedron*, 1981, **37**, 2617 (*synth, pmr, cmr, ms*)
Tsuboi, S. *et al, J. Org. Chem.*, 1982, **47**, 4478 (*synth*)
Rossi, R. *et al, Tetrahedron*, 1982, **38**, 639 (*cmr*)
Cassani, G *et al, Tetrahedron Lett.*, 1983, **24**, 2513 (*synth*)

7,10-Dodecadien-1-ol D-00353

[70654-43-2]

$H_3CCH{=}CHCH_2CH{=}CH(CH_2)_5CH_2OH$

$C_{12}H_{22}O$ M 182.305

(E,E)-form
$Bp_{0.01}$ 85-90°.

Ac: [70654-46-5].
$C_{14}H_{24}O_2$ M 224.342
$Bp_{0.01}$ 85-90°.

Bestmann, H.J. *et al, Tetrahedron Lett.*, 1979, 245.
Bestmann, H.J. *et al, Justus Liebigs Ann. Chem.*, 1982, 536 (*synth, pmr*)

8,10-Dodecadien-1-ol D-00354

[57002-06-9]

$H_3CCH{=}CHCH{=}CH(CH_2)_6CH_2OH$

$C_{12}H_{22}O$ M 182.305

(8E,10E)-form [33956-49-9] ***Codlemone***
Sex attractant of codling moth *Laspeyresia pomonella*. Cryst. (pentane). Mp 29-30°. $Bp_{0.01}$ 80-85°.

Ac: [53880-51-6].
$C_{14}H_{24}O_2$ M 224.342
$Bp_{0.01}$ 90°.

(8E,10Z)-form [33956-50-2]
$Bp_{0.01}$ 80-85°.

Ac: $Bp_{0.01}$ 85-90°.

(8Z,10Z)-form [39616-21-2]
$Bp_{0.01}$ 85-90°.

Ac: [67992-61-4].
$Bp_{0.01}$ 85-90°.

McDonough, L.M. *et al, J. Econ. Entomol.*, 1969, **62**, 62.
Roelofs, W. *et al, Science (Washington, D.C.)*, 1971, **174**, 297.

Descoins, C. *et al*, *Tetrahedron Lett.*, 1972, 2999 (*synth*)
Samain, D. *et al*, *Synthesis*, 1978, 388 (*synth*)
Barabas, A., *Tetrahedron*, 1978, **34**, 2191 (*cmr*)
Bestmann, H.J. *et al*, *Justus Liebigs Ann. Chem.*, 1981, 2117 (*synth, pmr*)
Babler, J.H. *et al*, *J. Org. Chem.*, 1982, **47**, 4801 (*synth*)
Rossi, R. *et al*, *Tetrahedron*, 1982, **38**, 639 (*cmr*)
Yamada, S. *et al*, *Chem. Lett.*, 1983, 1003 (*synth*)
Bloch, R. *et al*, *Tetrahedron Lett.*, 1983, **24**, 1247 (*synth*)
Takayama, H. *et al*, *J. Org. Chem.*, 1986, **51**, 4934 (*synth, pmr, cmr, ir, uv*)

8,11-Dodecadien-1-ol — D-00355

[65954-18-9]

$H_2C{=}CHCH_2CH{=}CH(CH_2)_6CH_2OH$

$C_{12}H_{22}O$ M 182.305

(E)-form

$Bp_{0.01}$ 70-75°.

Ac:

$C_{14}H_{24}O_2$ M 224.342

$Bp_{0.01}$ 80-85°.

Bestmann, H.J. *et al*, *Justus Liebigs Ann. Chem.*, 1982, 536 (*synth, pmr*)

9,11-Dodecadien-1-ol — D-00356

$H_2C{=}CHCH{=}CH(CH_2)_7CH_2OH$

$C_{12}H_{22}O$ M 182.305

(E)-form [55110-79-7]

$Bp_{0.01}$ 85-90°.

Ac: [50767-78-7].

$C_{14}H_{24}O_2$ M 224.342

Major sex pheromone of the red bollworm moth *Diparopsis castanea*. $Bp_{0.5}$ 108-110°, $Bp_{0.01}$ 80-85°.

(Z)-form

Ac: Liq. $Bp_{0.5}$ 80-100°.

Nesbitt, B.F. *et al*, *J. Insect Physiol.*, 1975, **21**, 1091 (*isol*)
Bestmann, H.J. *et al*, *Justus Liebigs Ann. Chem.*, 1981, 2117 (*synth, pmr*)
Rossi, R. *et al*, *Tetrahedron*, 1981, **37**, 2617 (*synth, pmr, cmr, ms*)
Rossi, R. *et al*, *Tetrahedron*, 1982, **38**, 639 (*cmr*)
Yamada, S. *et al*, *Chem. Lett.*, 1983, 1003 (*synth*)
Ochiai, M. *et al*, *Chem. Pharm. Bull.*, 1983, **31**, 1641 (*synth*)
Michelot, D., *Synthesis*, 1983, 130 (*synth*)
Hayashi, T. *et al*, *Tetrahedron Lett.*, 1983, **24**, 2665 (*synth*)
Takayama, H. *et al*, *J. Org. Chem.*, 1986, **51**, 4934 (*synth, deriv, pmr, cmr, ir*)

1,3-Dodecadiyne, 9CI — D-00357

[79965-25-6]

$HC{\equiv}CC{\equiv}C(CH_2)_7CH_3$

$C_{12}H_{18}$ M 162.274

Bp_2 90-92°. n_D^{20} 1.4743.

Remizova, L.A. *et al*, *Zh. Org. Khim.*, 1986, **22**, 2459; *J. Org. Chem. USSR* (*Engl. Transl.*), 2209 (*synth*)

1,4-Dodecadiyne, 9CI — D-00358

[20334-65-0]

$H_3C(CH_2)_6C{\equiv}CCH_2C{\equiv}CH$

$C_{12}H_{18}$ M 162.274

Liq. Bp_8 100-102°. n_D^{23} 1.4550.

Klenberg, D., *Ark. Kemi*, 1968, **29**, 23; *CA*, **69**, 18543p (*synth*)

1,5-Dodecadiyne, 9CI — D-00359

$H_3C(CH_2)_5C{\equiv}CCH_2CH_2C{\equiv}CH$

$C_{12}H_{18}$ M 162.274

Pleasant-smelling liq. Bp_{20} 100°. n_D^{20} 1.4545.

Gunstone, F.D. *et al*, *J. Chem. Soc.*, 1962, 3055 (*synth*)

1,6-Dodecadiyne, 9CI — D-00360

[3515-92-2]

$H_3C(CH_2)_4C{\equiv}C(CH_2)_3C{\equiv}CH$

$C_{12}H_{18}$ M 162.274

Liq. Bp_{14} 106-107°. n_D^{21} 1.4597.

Ames, D.E. *et al*, *J. Chem. Soc.*, 1965, 4373 (*synth*)
Gunstone, F.D. *et al*, *Chem. Phys. Lipids*, 1970, **4**, 1 (*synth*)

1,7-Dodecadiyne — D-00361

[35087-24-2]

$H_3C(CH_2)_3C{\equiv}C(CH_2)_4C{\equiv}CH$

$C_{12}H_{18}$ M 162.274

$Bp_{0.7}$ 94-96°.

Ames, D.E. *et al*, *J. Chem. Soc.*, *Perkin Trans. 1*, 1972, 255.

1,10-Dodecadiyne, 9CI — D-00362

$H_3CC{\equiv}C(CH_2)_7C{\equiv}CH$

$C_{12}H_{18}$ M 162.274

Mp 27°.

Nineham, A.W., *J. Chem. Soc.*, 1953, 2601 (*synth*)

1,11-Dodecadiyne, 9CI — D-00363

[20521-44-2]

$HC{\equiv}C(CH_2)_8C{\equiv}CH$

$C_{12}H_{18}$ M 162.274

Liq. Bp_{15} 104°.

Dale, J. *et al*, *J. Chem. Soc.*, 1963, 73 (*synth*)
Butere, R.J. *et al*, *Polym. Mater. Sci. Eng.*, 1986, **54**, 218 (*cryst struct*)

5,7-Dodecadiyne, 9CI — D-00364

[1120-29-2]

$H_3C(CH_2)_3C{\equiv}CC{\equiv}C(CH_2)_3CH_3$

$C_{12}H_{18}$ M 162.274

Liq. d 1.489. Bp_{10} 112-113°. n_D^{20} 1.4890.

Black, H.K. *et al*, *J. Chem. Soc.*, 1954, 1704 (*synth, uv*)
Evans, D.F., *J. Chem. Soc.*, 1960, 1735 (*uv*)
Traficante, D.D. *et al*, *J. Phys. Chem.*, 1965, **69**, 1348 (*pmr, cmr*)
Heilbronner, E. *et al*, *Helv. Chim. Acta*, 1977, **60**, 1697 (*pe*)
Barebas, A. *et al*, *Tetrahedron*, 1978, **34**, 2191 (*cmr*)

5,11-Dodecadiynoic acid — D-00365

[35087-17-3]

$HC{\equiv}C(CH_2)_4C{\equiv}C(CH_2)_3COOH$

$C_{12}H_{16}O_2$ M 192.257

$Bp_{0.1}$ 118°.

Ames, D.E. *et al*, *J. Chem. Soc.*, *Perkin Trans. 1*, 1972, 255 (*synth*)

3,6-Dodecadiyn-1-ol D-00366

[65090-68-8]

$H_3C(CH_2)_4C{\equiv}CCH_2C{\equiv}CCH_2CH_2OH$

$C_{12}H_{18}O$ M 178.274

Yellow liq.

Carvalho, J.F. *et al*, *J. Org. Chem.*, 1984, **49**, 1251 (*synth, ir, pmr*)

Dodecanal D-00367

Lauric aldehyde. Laurylaldehyde

[112-54-9]

$H_3C(CH_2)_{10}CHO$

$C_{12}H_{24}O$ M 184.321

Occurs in oil from *Chamaecyparis lawsonia*, *Citrus* spp. and a common component of lepidopteran pheromones. Perfumery ingredient. Cryst. Mp 44.5°. Bp_{100} 184-185°, Bp_{22} 142-143°.

▷ JR1910000.

Oxime:

$C_{12}H_{25}NO$ M 199.336

Mp 73°.

2,4-Dinitrophenylhydrazone: Mp 106°.

Kraft, F., *Ber.*, 1880, **13**, 1413.
Davies, R.R. *et al*, *J. Chem. Soc.*, 1943, 84 (*synth*)
Ikeda, R.M. *et al*, *J. Agric. Food Chem.*, 1962, **10**, 98 (*isol*)
Boss, B.D. *et al*, *Anal. Chem.*, 1976, **48**, 417 (*glc, ms*)
Tengoe, J. *et al*, *Comp. Biochem. Physiol., B: Comp. Biochem.*, 1976, **55**, 179 (*isol*)
Lewis, R.J., *Sax's Dangerous Properties of Industrial Materials*, 8th Ed., Van Nostrand-Reinhold, 1992, DXT000.

Dodecane, 9CI D-00368

[112-40-3]

$H_3C(CH_2)_{10}CH_3$

$C_{12}H_{26}$ M 170.337

Constit. of petroleum. Detergent feedstock, in admixture with other alkanes in C_{11}-C_{16} range. Fp −12°. Bp 214.5°, Bp_{100} 145°.

▷ JR2125000.

Krafft, F. *et al*, *Ber.*, 1882, **15**, 1698 (*synth*)
Schrauth, W. *et al*, *Ber.*, 1931, **64**, 1316 (*synth*)
Duric, R.A. *et al*, *Spectrochim. Acta*, 1962, **18**, 1505 (*ir*)
Eastham, J.F. *et al*, *J. Am. Chem. Soc.*, 1963, **85**, 2171 (*synth*)
Org. Synth., 1973, **53**, 107 (*synth*)
Lewis, R.J., *Sax's Dangerous Properties of Industrial Materials*, 8th Ed., Van Nostrand-Reinhold, 1992, DXT200.

1,12-Dodecanediamine, 9CI D-00369

1,12-Diaminododecane

[2783-17-7]

$H_2NCH_2(CH_2)_{10}CH_2NH_2$

$C_{12}H_{28}N_2$ M 200.367

Fp 66-67°.

▷ JR2200000.

Ziegler, K. *et al*, *Justus Liebigs Ann. Chem.*, 1959, **623**, 9 (*synth*)
Saotome, K. *et al*, *Bull. Chem. Soc. Jpn.*, 1966, **39**, 480 (*synth*)
Mayerl, F. *et al*, *Helv. Chim. Acta*, 1976, **59**, 127 (*ms*)
Lewis, R.J., *Sax's Dangerous Properties of Industrial Materials*, 8th Ed., Van Nostrand-Reinhold, 1992, DXW800.

Dodecanedioic acid D-00370

[693-23-2]

$HOOC(CH_2)_{10}COOH$

$C_{12}H_{22}O_4$ M 230.303

Isol. from leaf wax of *Chamaecyparis obtusa*. Cryst. (EtOAc). Mp 129°. Bp_{10} 245°.

Mono-Me ester: [3903-40-0].

$C_{13}H_{24}O_4$ M 244.330

Cryst. (EtOH). Mp 51-52°.

Di-Me ester: [1731-79-9].

$C_{14}H_{26}O_4$ M 258.357

Mp 31-31.5°. Bp_9 167-169°.

Di-Et ester: [10471-28-0].

$C_{16}H_{30}O_4$ M 286.411

Mp 16.7-17.2°. Bp_{15} 204-205°.

Diamide: [6224-99-3].

$C_{12}H_{24}N_2O_2$ M 228.334

Mp 189°.

Dinitrile: [4543-66-2].

$C_{12}H_{20}N_2$ M 192.303

Bp_{17} 225°.

Chuit, P., *Helv. Chim. Acta*, 1926, **9**, 264 (*synth*)
Bhattacharya, R. *et al*, *J. Chem. Soc.*, 1928, 2678 (*synth*)
Kariyone, T. *et al*, *Yakugaku Zasshi*, 1959, **79**, 54 (*isol*)
Marosi, L. *et al*, *Justus Liebigs Ann. Chem.*, 1973, 584 (*props, cryst struct*)
Vanier, M. *et al*, *Acta Crystallogr., Sect. B*, 1982, **38**, 643 (*cryst struct*)

1,2-Dodecanediol D-00371

1,2-Dihydroxydodecane

[1119-87-5]

$H_3C(CH_2)_9CH(OH)CH_2OH$

$C_{12}H_{26}O_2$ M 202.336

Cryst. Mp 56-58°, Mp 60-61°.

Swern, D. *et al*, *J. Am. Chem. Soc.*, 1946, **68**, 1504 (*synth*)
Emmons, W.D. *et al*, *J. Am. Chem. Soc.*, 1954, **76**, 3472 (*synth*)
Mysak, A.E. *et al*, *Zh. Anal. Khim.*, 1971, **26**, 1225 (*glc*)
Pliev, T.N. *et al*, *Zh. Prikl. Khim. (Leningrad)*, 1977, **50**, 157 (*ir, nmr*)

1,9-Dodecanediol D-00372

$H_3CCH_2CH_2CHOH(CH_2)_7CH_2OH$

$C_{12}H_{26}O_2$ M 202.336

Cryst. (Me_2CO/hexane). Mp 38-39°.

Cason, J. *et al*, *J. Org. Chem.*, 1949, **14**, 137 (*synth*)

1,10-Dodecanediol D-00373

[39516-27-3]

$H_3CCH_2CHOH(CH_2)_8CH_2OH$

$C_{12}H_{26}O_2$ M 202.336

Cryst. (hexane/Me_2CO). Mp 39-40°. Bp_{10} 172.5-173°.

Cason, J. *et al*, *J. Org. Chem.*, 1949, **14**, 137 (*synth*)
Baser, I.A., *Fette, Seifen, Anstrichm.*, 1972, **74**, 524.

1,12-Dodecanediol D-00374

Dodecamethylenediol

[5675-51-4]

$HOCH_2(CH_2)_{10}CH_2OH$

$C_{12}H_{26}O_2$ M 202.336

Constit. of the leaf wax of *Chamaecyparis obtusa* and of *Pinus thunbergii*. Cryst. (EtOH aq. or C_6H_6). Mp 80-81°. Bp_{12} 189°.

Di-Me ether:
$C_{14}H_{30}O_2$ M 230.390
Mp 11.5°. Bp_{15} 155-156°.

Di-Ac:
$C_{16}H_{30}O_4$ M 286.411
Mp 36.5-37.5°.

Di-Ph ether: [61575-03-9]. *1,1'-[1,12-Dodecanediylbis(oxy)] bisbenzene, 9CI. 1,2-Diphenoxydodecane*
$C_{24}H_{34}O_2$ M 354.531
Isol. from the leaves of *Arachis hypogaea* infected with *Puccinia arachidis*. Phytoalexin.

Chuit, P., *Helv. Chim. Acta*, 1926, **9**, 264 (*synth*)
Guyer, A. *et al*, *Helv. Chim. Acta*, 1955, **38**, 976.
Colonge, J. *et al*, *Bull. Soc. Chim. Fr.*, 1959, 1248 (*synth*)
White, R.W. *et al*, *Tetrahedron Lett.*, 1971, 3587 (*synth*)
Drewes, S.E. *et al*, *J. Chem. Soc., Perkin Trans. 1*, 1974, 2578.
Rosen, M.J. *et al*, *J. Am. Oil Chem. Soc.*, 1976, **53**, 742 (*di-Ph ether*)
Rao, P.V.S. *et al*, *Oleagineux*, 1991, **46**, 501; *CA*, **116**, 252152 (*di-Ph ether*)

2,3-Dodecanediol D-00375

$H_3C(CH_2)_8CH(OH)CH(OH)CH_3$

$C_{12}H_{26}O_2$ M 202.336
Mp 68-69°.

Luttringhaus, H. *et al*, *Justus Liebigs Ann. Chem.*, 1945, **557**, 46.
Krief, A. *et al*, *Angew. Chem.*, 1977, **89**, 103 (*nmr*)

2,11-Dodecanediol D-00376

[33666-71-6]

$H_3CCHOH(CH_2)_8CHOHCH_3$

$C_{12}H_{26}O_2$ M 202.336
Mp 54°. Bp_{15} 165-170°.

Hawkins, E.G.E. *et al*, *J. Chem. Soc.*, 1950, 2804 (*synth*)

4,9-Dodecanediol D-00377

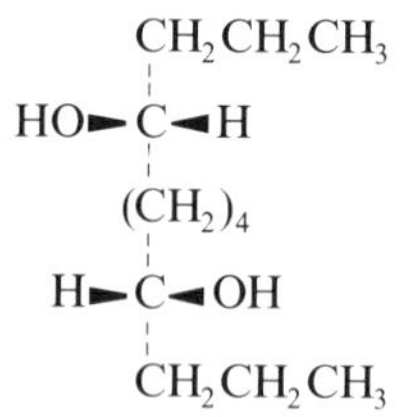

(*4RS,9RS*)*-form*

$C_{12}H_{26}O_2$ M 202.336

(4RS,9RS)-form [13511-19-8]
(±)-*form*
Mp 89°.

(4RS,9SR)-form
meso-*form*
Cryst. (pet. ether). Mp 54°.

Bowden, K. *et al*, *J. Chem. Soc.*, 1947, 1579 (*synth*)
Soussan, G. *et al*, *Bull. Soc. Chim. Fr.*, 1972, 4228.

1-Dodecanesulfonic acid, 9CI D-00378

[1510-16-3]

$H_3C(CH_2)_{11}SO_3H$

$C_{12}H_{26}O_3S$ M 250.402
Detergent. Mp 49-52°. $Bp_{0.1}$ 220°.

Me ester: [7560-61-4].
$C_{13}H_{28}O_3S$ M 264.428
Cryst. (Me_2CO). Mp 34-34.5°. $Bp_{0.3}$ 130-133°.

Et ester:
$C_{14}H_{30}O_3S$ M 278.455
$Bp_{0.3}$ 132-135°.

Chloride: Mp 42-43°.

Klamann, D. *et al*, *Monatsh. Chem.*, 1952, **83**, 463 (*synth*)
Showell, J.S. *et al*, *J. Org. Chem.*, 1962, **27**, 2853 (*synth*)
Nerasin, A. *et al*, *J. Appl. Polym. Sci.*, 1965, **9**, 1653 (*ir*)

Dodecanoic acid D-00379

Lauric acid. Laurostearic acid
[143-07-7]

$H_3C(CH_2)_{10}COOH$

$C_{12}H_{24}O_2$ M 200.320
Occurs as glyceride in coconut oil and palm kernel oil. Widespread in plants, esp. in the Palmae and Lauraceae. Simple esters are flavour ingredients. Cryst. (EtOH). Mp 44°. Bp_{100} 225°, $Bp_{0.6-0.7}$ 141-142°.

▷ OE9800000.

Me ester: [111-82-0].
$C_{13}H_{26}O_2$ M 214.347
Mp 5°. Bp_{15} 141°.

Et ester: [106-33-2].
$C_{14}H_{28}O_2$ M 228.374
Bp_{25} 163°.

Chloride:
$C_{12}H_{23}ClO$ M 218.766
Fp −17°. Bp_{18} 145°, Bp_{10} 135-140°.

Amide:
$C_{12}H_{25}NO$ M 199.336
Cryst. Mp 110°. $Bp_{12.5}$ 199-200°.

Nitrile:
$C_{12}H_{23}N$ M 181.320
Mp 3.9°. Bp_{100} 198°, Bp_{30} 160°.

Anilide:
$C_{18}H_{29}NO$ M 275.433
Cryst. (EtOH aq.). Mp 78°.

Anhydride:
$C_{24}H_{46}O_3$ M 382.626
Cryst. Mp 41.8°.

Org. Synth., 1950, **30**, 106.
Warren, C.H. *et al*, *Can. J. Chem.*, 1973, **51**, 3901 (*raman*)
Kameoka, H. *et al*, *Yukagaku*, 1976, **25**, 490 (*isol*)
Goto, M. *et al*, *Bull. Chem. Soc. Jpn.*, 1978, **51**, 70 (*cryst struct*)
Barton, F.E., *J. Am. Oil Chem. Soc.*, 1978, **55**, 574 (*cmr*)
Shostenko, A.G. *et al*, *CA*, 1979, **90**, 103343b (*synth*)
Lewis, R.J., *Sax's Dangerous Properties of Industrial Materials*, 8th Ed., Van Nostrand-Reinhold, 1992, ELY700, LBL000.

1-Dodecanol D-00380

Dodecyl alcohol. Lauryl alcohol
[112-53-8]

$H_3C(CH_2)_{10}CH_2OH$

$C_{12}H_{26}O$ M 186.337
Isol. from a few plant sources, eg. oil of *Furcraea gigantea*. Occurs also as ester. Used in perfumery. Sulphuric esters are used as wetting agents and detergents. Cryst. (EtOH aq.). Mp 24°, Mp 26°. Bp_{18} 145-148°.

▷ JR5775000.

Formyl:
$C_{13}H_{26}O_2$ M 214.347
Bp_{15} 145-146°.

Ac: [112-66-3].

$C_{14}H_{28}O_2$ M 228.374
Sex pheromone of processionary moth *Thanmetopoea pityocampa*.
4-Methylbenzenesulfonyl: Cryst. (pet. ether). Mp 30°.

Adkins, H. *et al*, *J. Am. Chem. Soc.*, 1931, **53**, 1095.
Jerchal, D. *et al*, *Z. Naturforsch., B*, 1954, **96**, 169.
Salgues, R., *C. R. Hebd. Seances Acad. Sci.*, 1956, **242**, 2392 (*isol*)
Morris, W.W., *J. Assoc. Off. Anal. Chem.*, 1973, **56**, 1037 (*ir*)
Nagai, T. *et al*, *Yukagaku*, 1974, **23**, 43 (*nmr*)
Precht, D., *Fette, Seifen, Anstrichm.*, 1976, **78**, 145, 189 (*cryst struct*)
Rossi, R. *et al*, *Tetrahedron*, 1982, **38**, 639 (*cmr*)
Lewis, R.J., *Sax's Dangerous Properties of Industrial Materials*, 8th Ed., Van Nostrand-Reinhold, 1992, DXV600.

3-Dodecanol D-00381

[10203-30-2]

$H_3C(CH_2)_8CH(OH)CH_2CH_3$

$C_{12}H_{26}O$ M 186.337

(−)-form

Mp 25°. Bp_{15} 130°. $[\alpha]_D^{20}$ −6.10° (EtOH).

Acid phthalate: Mp 25°. $[\alpha]_D^{20}$ −15.6° (EtOH).

(±)-form

Mp 12°. Bp_{14} 133°.

Acid phthalate: Mp 31-32°.

Pickard, R.H. *et al*, *J. Chem. Soc.*, 1913, **103**, 19.
Singliar, M., *Petrochemia*, 1971, **11**, 99 (*glc*)
Krupcik, J. *et al*, *Petrochemia*, 1975, **15**, 137 (*glc*)
Boss, B.D. *et al*, *Anal. Chem.*, 1976, **48**, 417 (*glc, ms*)
Holotik, S. *et al*, *Chromatographia*, 1976, **9**, 443 (*glc, ms*)

6-Dodecanol D-00382

[6836-38-0]

$H_3C(CH_2)_5CH(OH)(CH_2)_4CH_3$

$C_{12}H_{26}O$ M 186.337
Cryst. (pet. ether). Mp 30°. Bp_9 119°.

Krupcik, J. *et al*, *Petrochemia*, 1975, **15**, 137 (*glc*)
Boss, B.D. *et al*, *Anal. Chem.*, 1976, **48**, 417 (*glc, ms*)
Holotik, S. *et al*, *Chromatographia*, 1976, **9**, 443 (*glc, ms*)

2-Dodecanone D-00383

$H_3C(CH_2)_9COCH_3$

$C_{12}H_{24}O$ M 184.321
Detected in rue oils, fish volatiles, tomato leaf oil and *Cannabis sativa* oil. Also in hop oil (*Humulus lupulus*) and oil of *Schisandra nigra*. Mp 21°. Bp_{100} 177-178°, $Bp_{3.5}$ 101°.

Semicarbazone: Needles (EtOH). Mp 122-123°.
2,4-Dinitrophenylhydrazone: Mp 81°.

Pickard, R.H. *et al*, *J. Chem. Soc.*, 1911, **99**, 57 (*synth*)
Asinger, F., *Ber.*, 1944, **77**, 73 (*synth*)
U.S. Pat., 3 365 499, (1968); *CA*, **69**, 35440 (*manuf*)
Cooke, M.P., *J. Org. Chem.*, 1973, **38**, 4082 (*synth*)
Bente, P.F. *et al*, *J. Phys. Chem.*, 1975, **79**, 713 (*ms*)
Kabalka, G.W. *et al*, *J. Chem. Educ.*, 1976, **53**, 549 (*synth*)

3-Dodecanone D-00384

[1534-27-6]

$H_3C(CH_2)_8COCH_2CH_3$

$C_{12}H_{24}O$ M 184.321
Mp 19°. Bp_{18} 134°.

Semicarbazone: Cryst. (EtOH). Mp 89°.

Pickard, R.H. *et al*, *J. Chem. Soc.*, 1913, **103**, 1936 (*synth*)
Lawrence, B. *et al*, *Phytochemistry*, 1972, **11**, 2639 (*occur*)
Prestwich, G.D. *et al*, *Tetrahedron Lett.*, 1975, 4701 (*isol, synth*)
Boss, B.D. *et al*, *Anal. Chem.*, 1976, **48**, 417 (*glc*)
Savio, D. *et al*, *J. Org. Chem.*, 1978, **43**, 2907 (*synth*)

6-Dodecanone D-00385

[6064-27-3]

$H_3C(CH_2)_5CO(CH_2)_4CH_3$

$C_{12}H_{24}O$ M 184.321
Mp 9°. Bp_3 88-90°.

Oxime: Bp_{10} 147°.

Bouveault, L. *et al*, *C. R. Hebd. Seances Acad. Sci.*, 1905, **140**, 1699 (*synth*)
Okamoto, Y. *et al*, *Kogyo Kagaku Zasshi*, 1967, **70**, 1178; *CA*, **68**, 39725u (*synth*)
Pelter, A. *et al*, *J. Chem. Soc., Perkin Trans. 1*, 1976, 2435 (*synth*)
Sato, F. *et al*, *Chem. Lett.*, 1978, 1337 (*synth*)

2,4,6,8,10-Dodecapentaenal D-00386

$H_3C(CH{=}CH)_5CHO$

$C_{12}H_{14}O$ M 174.242
Cryst. Mp 164-166°.

Blatz, P.E. *et al*, *J. Am. Chem. Soc.*, 1969, **91**, 5930 (*synth, uv*)

3,5,7,9,11-Dodecapentaenoic acid, 8CI D-00387

C_{12}-*Ajenoic acid*

[30409-26-8]

$H_2C{=}CH(CH{=}CH)_4CH_2COOH$

$C_{12}H_{14}O_2$ M 190.241
Constit. of aje, the body fat of *Llaveia axia*.

Cason, J. *et al*, *J. Org. Chem.*, 1971, **36**, 2621.

2-(1,3,5,7,9-Dodecapentaenyloxy)-1,3-propanediol, 9CI D-00388

2-O-(1,3,5,7,9-Dodecapentaenyl)glycerol

$H_3CCH_2(CH{=}CH)_5OCH(CH_2OH)_2$

$C_{15}H_{22}O_3$ M 250.337

(all-E)-form [89229-66-3]

Isomer of 3-(1,3,5,7,9-Dodecapentaenyloxy)-1,2-propanediol, D-00389. Of interest in carcinogenicity studies.

▷ Mutagenic.

Gunatilaka, A.A.L. *et al*, *Tetrahedron Lett.*, 1983, **24**, 5457 (*synth*)

3-(1,3,5,7,9-Dodecapentaenyloxy)-1,2-propanediol, 9CI D-00389

Fecapentaene 12

[84000-59-9]

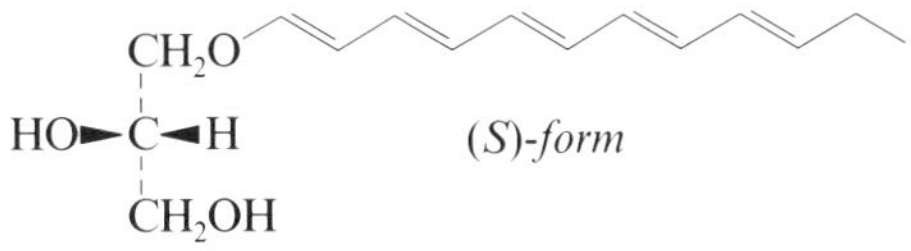

(*S*)-*form*

$C_{15}H_{22}O_3$ M 250.337
Natural fecapentaene 12 is a mixt. of geom. isomers in which the all *E*-form predominates. Isol. from faeces and urine of humans.

(R,all-E)-form
Mp 190.5-191° (sealed tube under Ar). $[\alpha]_D^{20}$ +9° (c, 0.5 in DMSO).

(S,all-E)-form [83248-46-8]
1-O-(1,3,5,7,9-Dodecapentaenyl)-sn-glycerol
Mutagen with suspected implication in human colonic cancer. Pale-yellow plates. Mp 191-192° (sealed tube under Ar). $[\alpha]_D^{20}$ −7° (c, 0.5 in DMSO). This is the main biological isomer. It is labile and unstable and many biol. studies may have been carried out on isomeric mixts. Mutagenicity is however independent of chirality.
▷ Highly mutagenic.

((±)-all-trans)-form
Mp 194.5-197.5° (sealed tube under Ar).

[89300-15-2, 89300-16-3, 91423-46-0]

Hirai, N. *et al*, *J. Am. Chem. Soc.*, 1982, **104**, 6149 (*struct, ms, uv, pmr*)
Bruce, W.R. *et al*, *Naturwissenschaften*, 1982, **69**, 557.
Gunatilaka, A.A.L. *et al*, *Biochemistry*, 1983, **22**, 241.
Gunatilaka, A.A.L. *et al*, *Tetrahedron Lett.*, 1983, **24**, 5457 (*synth*)
de Wit, P.P. *et al*, *Recl. Trav. Chim. Pays-Bas* (*J. R. Neth. Chem. Soc.*), 1985, **104**, 307 (*synth*)
Pfaendler, H.R. *et al*, *J. Am. Chem. Soc.*, 1986, **108**, 1338 (*synth, pmr*)
Baptista, J. *et al*, *Angew. Chem., Int. Ed. Engl.*, 1987, **26**, 1186 (*config*)
Pfaendler, R. *et al*, *Justus Liebigs Ann. Chem.*, 1988, 499 (*synth, ir, pmr, cmr, ms*)
Kingston, D.G.I. *et al*, *Chem. Res. Toxicol.*, 1990, **3**, 391 (*rev*)
Povey, A.C. *et al*, *Mutat. Res.*, 1991, **259**, 387 (*rev, props*)

2,4,8,10-Dodecatetraene-1,12-dioic acid D-00390

HOOCCH=CHCH=CHCH$_2$CH$_2$CH=CHCH=CHCOOH

$C_{12}H_{14}O_4$ M 222.240

(all-E)-form [120828-06-0]
Solid. Darkens >245° without melting.

Aurell, M.J. *et al*, *Synthesis*, 1990, 317 (*synth, pmr, cmr*)

2,4,6,8-Dodecatetraene-1,12-diol D-00391

1,12-Dihydroxy-2,4,6,8-dodecatetraene

HOCH$_2$CH$_2$CH$_2$(CH=CH)$_4$CH$_2$OH

$C_{12}H_{18}O_2$ M 194.273

(all-E)-form
Metab. of *Poria pearsonii* and *Coprinus flocculosus*. Plates (pet. ether/Et$_2$O). Mp 125-128°.

Bew, R.E. *et al*, *J. Chem. Soc. C*, 1966, 135.

2,4,8,10-Dodecatetraenoic acid D-00392

H$_3$CCH=CHCH=CHCH$_2$CH$_2$CH=CHCH=CHCOOH

$C_{12}H_{16}O_2$ M 192.257

N-*Isobutylamide:*
$C_{16}H_{25}NO$ M 247.380
Isol. from roots of *Asiasarum heterotropoides*. The (2*E*,4*E*,8*Z*,10*E*)- and (2*E*,4*E*,8*Z*,10*Z*)-isomers were isol.

N-*2-Hydroxy-2-methylpropylamide:* ***Sanshoamide***
$C_{16}H_{25}NO_2$ M 263.379
Alkaloid from fresh unripe fruits of *Zanthoxylum piperitum* (Rutaceae). Needles (C_6H_6). Mp 110-111°.

Aihara, T. *et al*, *Yakugaku Zasshi*, 1951, **71**, 1112.
Yasuda, I. *et al*, *Chem. Pharm. Bull.*, 1981, **29**, 564 (*Isobutylamide*)

2,6,8,10-Dodecatetraenoic acid D-00393

H$_3$CCH=CHCH=CHCH=CHCH$_2$CH$_2$CH=CHCOOH

$C_{12}H_{16}O_2$ M 192.257

(all-E)-form
N-(*2-Methylpropyl*)*amide:* *β-Sanshool*
Produced by isomerisation of α-Sanshool. Insecticide. Cryst. (hexane). Mp 100-101°.

(2E,6Z,8E,10E)-form [18744-21-3]
Cryst. (hexane). Mp 93.5-95°. Material contaminated with some *all-E*-isomer.

2-Methylpropylamide: [504-97-2]. ***Neoherculin.*** N-*Isobutyl-2,6,8,10-dodecatetraenamide. Echinacein. α-Sanshool*
$C_{16}H_{25}NO$ M 247.380
Constit. of *Echinacea angustifolia* roots and *Zanthoxylum* spp. Insecticide. Cryst. (pet. ether). Mp 69° (61-63°). Unstable in air.

Crombie, L., *J. Chem. Soc.*, 1952, 2997; 1955, 995; 1957, 2760 (*isol, uv, ir, struct*)
Crombie, L., *Nature* (*London*), 1954, **174**, 833.
Sonnet, P.E., *J. Org. Chem.*, 1969, **34**, 1147 (*synth, uv*)

1,8,10-Dodecatriene D-00394

H$_3$CCH=CHCH=CH(CH$_2$)$_5$CH=CH$_2$

$C_{12}H_{20}$ M 164.290

(8E,10Z)-form [69775-60-6]
Bp$_2$ 67-69°.

Bestmann, H.J. *et al*, *Justus Liebigs Ann. Chem.*, 1981, 2117 (*synth, pmr*)

3,5,7-Dodecatriene-9,11-diyn-1-ol, 9CI D-00395

[27285-85-4]

HC≡CC≡CCH=CHCH=CHCH=CHCH$_2$CH$_2$OH

$C_{12}H_{12}O$ M 172.226

(all-E)-form
Constit. of *Chrysanthemum leptophyllum*. Unstable oil.

Bohlmann, F. *et al*, *Chem. Ber.*, 1970, **103**, 1879.

3,6,8-Dodecatrien-1-ol D-00396

H$_3$CCH$_2$CH$_2$CH=CHCH=CHCH$_2$CH=CHCH$_2$CH$_2$OH

$C_{12}H_{20}O$ M 180.289

(3Z,6Z,8E)-form [19926-64-8]
Termite trail-following pheromone. Produced by the fungi *Lenzites trabea* and *Glaeophyllum trabeum*. All 8 geom. isomers have been synthesised.

Matsumura, F. *et al*, *Nature* (*London*), 1968, **219**, 963 (*isol, synth*)
Tai, A. *et al*, *J. Org. Chem.*, 1969, **34**, 2180 (*synth*)
Eya, B.K. *et al*, *Tetrahedron*, 1990, **46**, 2695 (*synth, pmr, cmr, bibl*)

1-*O*-(4,6,8-Dodecatrienyl)glycerol D-00397

3-(4,6,8-Dodecatrienyloxy)-1,2-propanediol

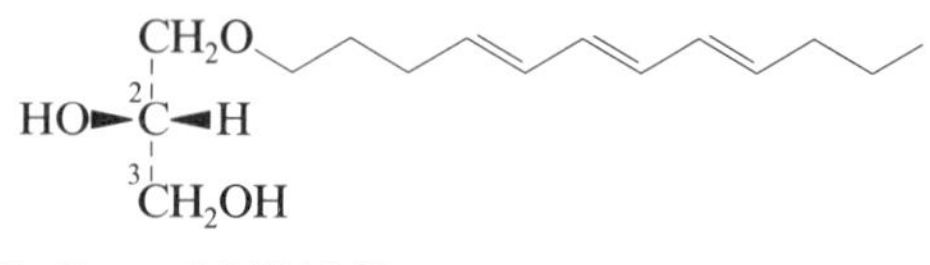

$C_{15}H_{26}O_3$ M 254.369
Closely related to the fecapentaenes.

(S, all-E)-form
2-O-(4-Hydroxybenzoyl): [123231-45-8]. ***Bretonin A***
$C_{22}H_{30}O_5$ M 374.476

Isol. from an unidentified marine demosponge. Oil. Opt. rotn. too small to measure.

2-O-(4-Hydroxybenzoyl), di-Ac: $[\alpha]_D^{25}$ −7.0° (c, 0.13 in $CHCl_3$).

3-O-(4-Hydroxybenzoyl): ***Isobretonin A***

$C_{22}H_{30}O_5$ M 374.476

Isol. from an unidentified marine demosponge. Oil. $[\alpha]_D^{25}$ +7.3° ($CHCl_3$). Has (*R*-) config.

Guella, G. *et al*, *Helv. Chim. Acta*, 1989, **72**, 1121 (*isol, uv, pmr, cmr, struct*)
Mancini, I. *et al*, *Helv. Chim. Acta*, 1991, **74**, 941 (*synth*)

5,8,11-Dodecatriynoic acid D-00398

[53292-96-9]

HC≡CCH$_2$C≡CCH$_2$C≡C(CH$_2$)$_3$COOH

$C_{12}H_{12}O_2$ M 188.226

Beige plates (hexane). Mp 57-58°.

Fryer, R.I. *et al*, *J. Org. Chem.*, 1975, **40**, 348 (*synth, ir, pmr*)

2,5,8-Dodecatriyn-1-ol D-00399

[32895-85-5]

H$_3$CCH$_2$CH$_2$C≡CCH$_2$C≡CCH$_2$C≡CCH$_2$OH

$C_{12}H_{14}O$ M 174.242

$Bp_{0.01}$ 115-125°.

Beerthuis, R.K. *et al*, *Recl. Trav. Chim. Pays-Bas* (*J. R. Neth. Chem. Soc.*), 1971, **90**, 943 (*synth*)

2-Dodecenal D-00400

[4826-62-4]

H$_3$C(CH$_2$)$_8$CH=CHCHO

$C_{12}H_{22}O$ M 182.305

▷JR5150000.

(E)-form [20407-84-5]

Present in many foods including citrus, ginger, carrots, milk, peanuts. Also found in the millipede *Rhinocricus insulatus*. Bp_{10} 125-128°, $Bp_{0.5}$ 73.4°.

Semicarbazone: [16742-26-0].

Mp 165.5-166°.

Bedoukian, P.Z., *J. Am. Chem. Soc.*, 1957, **79**, 889.
Wheeler, J.W. *et al*, *Science* (*Washington, D.C.*), 1964, **144**, 540.
Naves, Y.R. *et al*, *Helv. Chim. Acta*, 1967, **50**, 1461 (*ir*)
Chang, Y.H. *et al*, *Synthesis*, 1990, 1033 (*synth*)
Lewis, R.J., *Sax's Dangerous Properties of Industrial Materials*, 8th Ed., Van Nostrand-Reinhold, 1992, DXU280.

9-Dodecenal, 9CI D-00401

[121052-29-7]

H$_3$CCH$_2$CH=CH(CH$_2$)$_7$CHO

$C_{12}H_{22}O$ M 182.305

(Z)-form [56219-03-5]

Component of the sex pheromone of the plume moth, *Oidaematophorus monodactylus* ((*E*)-form exerts synergistic effect). Liq. $Bp_{0.02}$ 62° (⩾95% geom. purity).

Bestmann, H.J. *et al*, *Chem. Ber.*, 1975, **108**, 3582 (*synth, pmr, ir, ms*)
Priesner, E. *et al*, *Z. Naturforsch., C*, 1975, **30**, 283 (*activity*)
Kamm, J.A. *et al*, *Pan-Pac Entomol.*, 1979, **55**, 284; *CA*, **94**, 115938v (*isol, activity*)
Meighen, E.A. *et al*, *J. Chem. Ecol.*, 1982, **8**, 911 (*isol, activity*)
Horiike, M. *et al*, *Biomed. Environ. Mass Spectrom.*, 1988, **17**, 301 (*ms*)

1-Dodecene, 9CI D-00402

[112-41-4]

H$_3$C(CH$_2$)$_9$CH=CH$_2$

$C_{12}H_{24}$ M 168.322

Bp_{15} 92-95°.

McMurry, J.E. *et al*, *J. Org. Chem.*, 1975, **40**, 2555 (*synth*)
Remion, J.R. *et al*, *Tetrahedron Lett.*, 1976, 1385 (*synth*)
Utimoto, K. *et al*, *Tetrahedron*, 1977, **33**, 1945 (*synth*)
Welch, S.C. *et al*, *J. Org. Chem.*, 1978, **43**, 2715 (*synth*)
Yamada, K. *et al*, *J. Org. Chem.*, 1978, **43**, 2076 (*synth*)

6-Dodecene D-00403

[29493-00-3]

H$_3$C(CH$_2$)$_4$CH=CH(CH$_2$)$_4$CH$_3$

$C_{12}H_{24}$ M 168.322

(E)-form [7206-17-9]

Liq. Mp −34° to −33°. Bp_3 55-56°.

(Z)-form [7206-29-3]

Liq. Bp_3 53-54°.

Ziegler, K. *et al*, *Justus Liebigs Ann. Chem.*, 1950, **567**, 43 (*synth*)
Hamatani, T. *et al*, *Tetrahedron*, 1988, **44**, 2875 (*synth, ir, pmr, ms*)

2-Dodecenedioic acid, 9CI D-00404

[124-00-5]

HOOC(CH$_2$)$_8$CH=CHCOOH

$C_{12}H_{20}O_4$ M 228.288

(E)-form [6402-36-4]

Traumatic acid

Isol. from string bean pods. Plant wound hormone. Spar. sol. H_2O. Mp 166-167°. $Bp_{0.001}$ 150-160°.

(Z)-form

Spar. sol. H_2O. Mp 67-68°.

English, J. *et al*, *J. Am. Chem. Soc.*, 1939, **61**, 3434.
Lauer, W.M. *et al*, *J. Am. Chem. Soc.*, 1945, **67**, 1171 (*synth*)
Rao, A.S.C.P. *et al*, *Synthesis*, 1975, 608 (*synth*)
Villemin, D., *Chem. Ind.* (*London*), 1986, 36 (*synth*)

1-Dodecene-5,11-diyne D-00405

[94732-61-3]

HC≡C(CH$_2$)$_4$C≡CCH$_2$CH$_2$CH=CH$_2$

$C_{12}H_{16}$ M 160.258

Liq. Bp_{120} 92-94°.

Duñach, E. *et al*, *J. Am. Chem. Soc.*, 1985, **107**, 1664 (*synth, ir, pmr, ms*)

4-Dodecene-6,8,10-triyn-3-ol D-00406

H$_3$CC≡CC≡CC≡CCH=CHCH(OH)CH$_2$CH$_3$

$C_{12}H_{12}O$ M 172.226

(E)-form

Constit. of the fungus *Reniophera resinosa*. Mp 37-38°. $[\alpha]_D^{20}$ +12.5° (c, 0.7 in EtOH).

Ahmed, M. *et al*, *J. Chem. Soc., Perkin Trans. 1*, 1978, 1487 (*ir, pmr, ms*)

2-Dodecenoic acid D-00407

[4412-16-2]

H$_3$C(CH$_2$)$_8$CH=CHCOOH

$C_{12}H_{22}O_2$ M 198.305

(E)-form [32466-54-9]
Mp 13-18°. Bp_3 155-158°, $Bp_{0.15}$ 127-130°.
Me ester: [6208-91-9].
$C_{13}H_{24}O_2$ M 212.331
Occurs in volatile component of pears.
Et ester: [28290-90-6].
$C_{14}H_{26}O_2$ M 226.358
Occurs in pears.
(Z)-form [55928-65-9]

Org. Synth., Coll. Vol., 4, 1963, 398 (*synth*)
Creveling, R.K. *et al*, *J. Agric. Food Chem.*, 1970, **18**, 19 (*ir, pmr, ms, uv*)
Bus, J. *et al*, *Chem. Phys. Lipids*, 1976, **17**, 501; 1977, **18**, 130 (*cmr*)

4-Dodecenoic acid, 8CI — D-00408

[2430-94-6]

$H_3C(CH_2)_6CH{=}CHCH_2CH_2COOH$

$C_{12}H_{22}O_2$ M 198.305
(E)-form
Mp 27-27.5°. Bp_{13} 166.5-168.5°.
Me ester:
$C_{13}H_{24}O_2$ M 212.331
Bp_{12} 125-126°.
p-*Bromophenacyl ester:* Mp 75.5-76°.
(Z)-form
Linderic acid
Constit. of oil of *Lindera obtusiloba* and *L. umbellata*. $Bp_{0.13}$ 170-172°. n_D^{15} 1.4545. Incorr. abstracted in CA as 5-dodecenoic acid.

Toyama, Y. *et al*, *Chem. Zentralbl.*, 1938, **1**, 2647.
Komori, S. *et al*, *Chem. Zentralbl.*, 1938, **1**, 2855.
Iwakiri, M., *Nippon Kagaku Zasshi*, 1958, **79**, 910; *CA*, **54**, 4363d (*synth*)
Hopkins, C.Y. *et al*, *Lipids*, 1966, **1**, 118 (*isol, struct*)

6-Dodecenoic acid — D-00409

$H_3C(CH_2)_4CH{=}CH(CH_2)_4COOH$

$C_{12}H_{22}O_2$ M 198.305
Two forms known, said to be probable geom. isomers. $Bp_{0.04}$ 100-106°.
α-form
Mp 41-42°.
4-Bromophenacyl ester: Mp 61-62°.
β-form
Mp 49-50°.
4-Bromophenacyl ester: Mp 49-50°.

Baudart, P., *C. R. Hebd. Seances Acad. Sci.*, 1945, **220**, 404 (*synth*)
Kovaleva, A.S. *et al*, *Zh. Org. Khim.*, 1974, **10**, 696 (*synth*)
Bus, J. *et al*, *Chem. Phys. Lipids*, 1977, **18**, 130 (*cmr*)
Scholz, D., *Justus Liebigs Ann. Chem.*, 1984, 264 (*synth*)

7-Dodecenoic acid, 9CI — D-00410

[21968-10-5]

$H_3C(CH_2)_3CH{=}CH(CH_2)_5COOH$

$C_{12}H_{22}O_2$ M 198.305
(E)-form [16695-39-9]
Occurs in human depot fat.
(Z)-form [52956-99-7]
Occurs in human depot fat.
Amide:
$C_{12}H_{23}NO$ M 197.320
Mp 69-70°.

Taylor, W.R. *et al*, *J. Am. Chem. Soc.*, 1950, **72**, 4263 (*synth*)
Gunstone, F.D. *et al*, *Chem. Phys. Lipids*, 1977, **18**, 115 (*cmr*)
Seidel, W. *et al*, *Chem. Ber.*, 1977, **110**, 3544 (*synth*)

9-Dodecenoic acid — D-00411

[2382-40-3]

$H_3CCH_2CH{=}CH(CH_2)_7COOH$

$C_{12}H_{22}O_2$ M 198.305
(E)-form [21091-01-0]
Occurs in human depot fat.
(Z)-form [22032-47-9]
Occurs in human depot fat.

Kono, M. *et al*, *Nippon Nogei Kagaku Kaishi*, 1939, **15**, 177; *CA*, **33**, 6239.
Kovalevea, A.S. *et al*, *Zh. Org. Khim.*, 1974, **10**, 696 (*synth*)
Gunstone, F.D. *et al*, *Chem. Phys. Lipids*, 1977, **18**, 115 (*cmr*)

10-Dodecenoic acid — D-00412

$H_3CCH{=}CH(CH_2)_8COOH$

$C_{12}H_{22}O_2$ M 198.305
(Z)-form [18654-82-5]
Mp 18°. Bp_9 166-168°.

Chuit, P. *et al*, *Helv. Chim. Acta*, 1927, **10**, 113 (*synth*)
Gunstone, F.D. *et al*, *Chem. Phys. Lipids*, 1967, **1**, 209 (*synth*); 1977, **18**, 115 (*cmr*)

11-Dodecenoic acid, 9CI — D-00413

[65423-25-8]

$H_2C{=}CH(CH_2)_9COOH$

$C_{12}H_{22}O_2$ M 198.305
Mp 20°. Bp_{13} 171-172°, Bp_3 143-144°.
Me ester: [29972-79-0].
$C_{13}H_{24}O_2$ M 212.331
Bp_{13} 138-139°, Bp_3 121°.
Et ester:
$C_{14}H_{26}O_2$ M 226.358
Bp_{18} 154.5-155.5°.

Chuit, P. *et al*, *Helv. Chim. Acta*, 1927, **10**, 113.
Rountree, J.M. *et al*, *Chem. Ind. (London)*, 1954, 190 (*synth, ir*)
Scarborough, R.M., *Tetrahedron Lett.*, 1977, 4361 (*synth*)

2-Dodecen-1-ol — D-00414

[22104-81-0]

$H_3C(CH_2)_8CH{=}CHCH_2OH$

$C_{12}H_{24}O$ M 184.321
(E)-form [69064-36-4]
$Bp_{0.8}$ 95-97°.

Patterson, J.W. *et al*, *J. Org. Chem.*, 1983, **48**, 2572 (*synth, ir, pmr*)

3-Dodecen-1-ol — D-00415

$H_3C(CH_2)_7CH{=}CHCH_2CH_2OH$

$C_{12}H_{24}O$ M 184.321
(Z)-form [32451-95-9]
Liq.

Carvalho, J.F. *et al*, *J. Org. Chem.*, 1984, **49**, 1251 (*synth, ir, pmr*)

5-Dodecen-1-ol D-00416

$H_3C(CH_2)_5CH{=}CH(CH_2)_3CH_2OH$

$C_{12}H_{24}O$ M 184.321

(Z)-form

Ac: [16676-96-3].

$C_{14}H_{26}O_2$ M 226.358

Constit. of the oil of *Hibiscus abelmoschus*. Oil.

Warthen, J.D. *et al, J. Chromatogr. Sci.*, 1976, **14**, 513.
Maurer, B. *et al, Helv. Chim. Acta*, 1977, **60**, 1155.
Ando, T. *et al, CA*, 1979, **80**, 120969 (*synth*)

6-Dodecen-1-ol D-00417

$H_3C(CH_2)_4CH{=}CH(CH_2)_4CH_2OH$

$C_{12}H_{24}O$ M 184.321

$Bp_{0.1}$ 75-76°.

(E)-form

Liq. $Bp_{0.7}$ 98-99°.

(Z)-form [40642-39-5]

$Bp_{0.4}$ 90-92°.

Ames, D.E. *et al, J. Chem. Soc.*, 1963, 4363 (*synth, derivs*)
Kovaleva, A.S. *et al, J. Org. Chem. USSR* (*Engl. Transl.*), 1974, **10**, 700 (*synth*)
Horiike, M. *et al, Agric. Biol. Chem.*, 1978, **42**, 1963 (*synth*)
Brown, H.C. *et al, J. Org. Chem.*, 1988, **53**, 246 (*synth, ir, pmr, cmr*)

7-Dodecen-1-ol D-00418

$H_3C(CH_2)_3CH{=}CH(CH_2)_5CH_2OH$

$C_{12}H_{24}O$ M 184.321

(E)-form

Ac: [14959-86-5].

Sex pheromone of the false codling moth *Argyroploce leuctotreta*. $Bp_{0.05}$ 78-82°.

(Z)-form [20056-92-2]

Liq. n_D^{20} 1.4540.

▷ JR5255000.

Ac: [16695-41-3].

Sex pheromone of the female cabbage looper moth *Trichoplusia ni*. $Bp_{0.08}$ 85-90°, $Bp_{0.05}$ 98-100°.

Green, N. *et al, J. Med. Chem.*, 1967, **10**, 533 (*isol, synth*)
Read, J.S. *et al, J. Chem. Soc., Chem. Commun.*, 1968, 792 (*isol*)
Leznoff, C.C. *et al, Can. J. Chem.*, 1977, **55**, 1143 (*synth*)
Seidel, W. *et al, Chem. Ber.*, 1977, **110**, 3544 (*synth*)
Brown, H.C. *et al, J. Org. Chem.*, 1986, **51**, 4512 (*synth, ir, pmr, cmr*)
Lewis, R.J., *Sax's Dangerous Properties of Industrial Materials*, 8th Ed., Van Nostrand-Reinhold, 1992, DXU800.

8-Dodecen-1-ol D-00419

$H_3CCH_2CH_2CH{=}CH(CH_2)_6CH_2OH$

$C_{12}H_{24}O$ M 184.321

(Z)-form [40642-40-8]

Pheromone component of the Oriental fruit moth *Grapholitha molesta* and several other insect spp. $Bp_{1.1}$ 99-100°.

Ac: [28079-04-1]. *Orfamone. Orfralure*

Pheromone component for *G. molesta*. $Bp_{1.2}$ 97-98°.

Roelofs, W.L. *et al, Nature (London)*, 1969, **224**, 723 (*occur*)
Holan, G. *et al, Tetrahedron Lett.*, 1973, 673 (*synth*)
Mori, K. *et al, Tetrahedron*, 1977, **33**, 385 (*synth*)
Peking Institute of Zoology, *CA*, 1978, **89**, 196930 (*synth*)
Chisholm, M.D. *et al, J. Chem. Ecol.*, 1978, **4**, 657 (*synth*)
Aukrust, A. *et al, Acta Chem. Scand., Ser. B*, 1985, **39**, 267 (*synth*)

9-Dodecen-1-ol, 9CI D-00420

$H_3CCH_2CH{=}CH(CH_2)_7CH_2OH$

$C_{12}H_{24}O$ M 184.321

(E)-form [35237-62-8]

Pheromone for various insect spp. $Bp_{0.03}$ 81-83°. n_D^{21} 1.4534.

Ac: [35148-19-7].

$C_{14}H_{26}O_2$ M 226.358

Pheromone for various insect spp.

(Z)-form [35148-18-6]

Pheromone for varous insect spp. $Bp_{0.01}$ 83°.

Ac: [16974-11-1].

Pheromone for various insect spp. $Bp_{0.1}$ 101-102°.

Formyl: [56218-77-0].

$C_{13}H_{24}O_2$ M 212.331

Pheromone for various insect spp. $Bp_{0.05}$ 78-80°.

Propanoyl: [56218-84-9].

$C_{15}H_{28}O_2$ M 240.385

Pheromone for various insect spp. $Bp_{0.05}$ 96-97°.

Kovaleva, A.S. *et al, Zh. Org. Khim.*, 1974, **10**, 696.
Bestmann, H.J. *et al, Chem. Ber.*, 1975, **108**, 3582; 1976, **109**, 1694 (*synth, ir*)
Barabas, A. *et al, Tetrahedron*, 1978, **34**, 2191 (*nmr*)
Michelot, D., *Synthesis*, 1983, 130 (*synth*)
Körblová, E. *et al, Collect. Czech. Chem. Commun.*, 1985, **50**, 2284 (*synth, ir*)
Szurdoki, F. *et al, Org. Prep. Proced. Int.*, 1988, **20**, 475 (*synth, pmr, cmr, bibl*)
Moiseenkov, A.M. *et al, J. Chem. Soc., Perkin Trans. 1*, 1991, 2639 (*synth, Ac, pmr*)

10-Dodecen-1-ol, 9CI D-00421

$H_3CCH{=}CH(CH_2)_8CH_2OH$

$C_{12}H_{24}O$ M 184.321

(E)-form [35237-63-9]

Sex pheromone for the red banded leafroller *Argyrotaenia velutinana*. Bp_{13} 143-144°.

Ac: [35153-09-4].

Sex pheromone for various spp. including the spotted tentiform leaf miner moth *Lithocolletis blancardella*.

(Z)-form [35289-30-6]

Pheromone for various spp.

Ac: [35148-20-0].

Pheromone for various spp.

Phenoxyacetyl: [71255-21-5].

Cryst. Mp 40-41°.

Ando, T. *et al, Agric. Biol. Chem.*, 1977, **41**, 1485 (*synth*)
Polina, I. *et al, J. Chem. Eng. Data*, 1979, **24**, 152 (*synth*)
Mandai, T. *et al, Tetrahedron*, 1979, **35**, 309 (*synth, ir, ms, nmr*)

11-Dodecen-1-ol, 9CI D-00422

[35289-31-7]

$H_2C{=}CH(CH_2)_9CH_2OH$

$C_{12}H_{24}O$ M 184.321

Pheromone for the red banded leafroller *Argyrotaenia velutinana*.

Ac: [35153-10-7].

$C_{14}H_{26}O_2$ M 226.358

Pheromone for various moths, esp. *A. orana* and the red bollworm *Diparopsis castanea*.

Nesbitt, B.F. *et al*, *Nature (London)*, 1973, **244**, 208.
Nesbitt, B.F. *et al*, *J. Insect Physiol.*, 1975, **21**, 1091.

8-Dodecen-11-olide D-00423

12-Methyloxacyclodec-9-en-2-one, 9CI. *3-Methyl-2-oxacyclododec-5-en-1-one*. *11-Hydroxy-8-dodecenoic acid lactone*

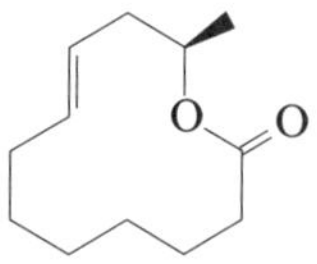

(*R*)-(*E*)-*form*

$C_{12}H_{20}O_2$ M 196.289

(*R*)-(*E*)-*form* [59780-28-8] ***Recifeiolide***

Metab. of *Cephalosporium recifei*. Shows antibiotic props. Liq. $Bp_{0.01}$ 110°. $[\alpha]_D^{25}$ +73.2° (c, 7.5 in $CHCl_3$).

(±)-(*E*)-*form* [58654-20-9]

Oil.

Vesonder, R.F. *et al*, *Can. J. Chem.*, 1971, **49**, 2029 (*isol, struct, ir, uv, ms, pmr*)
Wassermann, H.H. *et al*, *Tetrahedron*, 1981, **37**, 4059 (*synth*)
Bestmann, H.J. *et al*, *Angew. Chem., Int. Ed. Engl.*, 1983, **22**, 780 (*synth*)
Ahmed, A. *et al*, *Bull. Chem. Soc. Jpn.*, 1984, **57**, 781 (*synth*)
Rao, K.V. *et al*, *Synth. Commun.*, 1984, **14**, 321 (*synth, bibl*)
Inoue, S. *et al*, *CA*, 1985, **103**, 215039 (*synth*)
Rao, A.V.R. *et al*, *Synth. Commun.*, 1986, **16**, 1149 (*synth*)

5-Dodecen-11-yne D-00424

7-Dodecen-1-yne

$H_3C(CH_2)_3CH{=}CH(CH_2)_4C{\equiv}CH$

$C_{12}H_{20}$ M 164.290

(*E*)-*form* [35087-11-7]

Bp_1 55°.

Ames, D.E. *et al*, *J. Chem. Soc., Perkin Trans. 1*, 1972, 255 (*synth*)

8-Dodecen-10-ynoic acid D-00425

$H_3CC{\equiv}CCH{=}CH(CH_2)_6COOH$

$C_{12}H_{18}O_2$ M 194.273

(*E*)-*form*

Me ester: [80625-37-2].

$C_{13}H_{20}O_2$ M 208.300

$Bp_{0.01}$ 63-68°.

Bestmann, H.J. *et al*, *Justus Liebigs Ann. Chem.*, 1981, 2117 (*synth, pmr*)

3-Dodecen-1-yn-5-ol D-00426

$H_3C(CH_2)_6CH(OH)CH{=}CHC{\equiv}CH$

$C_{12}H_{20}O$ M 180.289

(±)-(*E*)-*form*

$Bp_{0.01}$ 80°.

Bohlmann, F. *et al*, *Chem. Ber.*, 1971, **104**, 1362 (*synth*)

8-Dodecen-10-yn-1-ol D-00427

$H_3CC{\equiv}CCH{=}CH(CH_2)_6CH_2OH$

$C_{12}H_{20}O$ M 180.289

(*Z*)-*form*

Ac: [80625-38-3].

$C_{14}H_{22}O_2$ M 222.327

$Bp_{0.05}$ 90-95°.

Bestmann, H.J. *et al*, *Justus Liebigs Ann. Chem.*, 1981, 2117 (*synth, pmr*)

9-Dodecen-7-yn-1-ol D-00428

$H_3CCH_2CH{=}CHC{\equiv}C(CH_2)_5CH_2OH$

$C_{12}H_{20}O$ M 180.289

(*Z*)-*form*

Tetrahydropyranyl ether: [80625-40-7].

$C_{17}H_{28}O_2$ M 264.407

$Bp_{0.01}$ 100-110°.

Bestmann, H.J. *et al*, *Justus Liebigs Ann. Chem.*, 1981, 2117 (*synth, pmr*)

10-Dodecen-8-yn-1-ol D-00429

$H_3CCH{=}CHC{\equiv}C(CH_2)_6CH_2OH$

$C_{12}H_{20}O$ M 180.289

(*Z*)-*form*

Tetrahydropyranyl ether: [80625-39-4].

$C_{17}H_{28}O_2$ M 264.407

$Bp_{0.01}$ 100-110°.

Bestmann, H.J. *et al*, *Justus Liebigs Ann. Chem.*, 1981, 2117 (*synth, pmr*)

11-Dodecen-9-yn-1-ol D-00430

[82334-20-1]

$H_2C{=}CHC{\equiv}C(CH_2)_7CH_2OH$

$C_{12}H_{20}O$ M 180.289

$Bp_{0.05}$ 85-86°. n_D^{25} 1.4825.

Rossi, R. *et al*, *Tetrahedron*, 1982, **38**, 631 (*synth, ir, pmr*)

Dodecylamine D-00431

1-Dodecanamine, 9CI. *Laurylamine*. *1-Aminododecane*

[124-22-1]

$H_3C(CH_2)_{10}CH_2NH_2$

$C_{12}H_{27}N$ M 185.352

Used as an anal. reagent for aldehydes, and as a 2% soln. in BuOH for extraction-photometric detn. of Nb; extraction of Mn(*III*) and Fe(*III*) pyrophosphate complexes. Cryst. (C_6H_6). Sol. EtOH, Et_2O, C_6H_6, $CHCl_3$. Mp 27-28°. Bp 247-249°, Bp_{10} 126°. n_D^{20} 1.4421.

▷ JR6475000.

N-*Me:* [7311-30-0].

$C_{13}H_{29}N$ M 199.379

Oil. $Bp_{1.2\text{-}1.5}$ 110-115°.

N-*Di-Me*, N-*oxide:* [1643-20-5].

$C_{14}H_{31}NO$ M 229.405

Used in detergents. V. hygroscopic needles (dry toluene). Mp 130-131°.

▷ JR6650000.

Krafft, F., *Ber.*, 1890, **23**, 2364.
Adkins, H. *et al*, *J. Am. Chem. Soc.*, 1934, **56**, 2419.
v. Braun, J. *et al*, *Ber.*, 1937, **70**, 979.
Heierman, F. *et al*, *Recl. Trav. Chim. Pays-Bas (J. R. Neth. Chem. Soc.)*, 1938, **57**, 456.
Siggia, S. *et al*, *Anal. Chem.*, 1953, **25**, 830 (*use*)
Org. Synth., Coll. Vol., 4, 1963, 564 (*deriv*)
Shevchuk, I.A. *et al*, *Zavod. Lab.*, 1967, **33**, 288 (*use, Fe*)
Shevchuk, I.A. *et al*, *Zh. Anal. Khim.*, 1968, **23**, 1386 (*use, Mn*)

Bedenbaugh, A.O. *et al*, *J. Am. Chem. Soc.*, 1970, **92**, 5774 (*deriv*)
Org. Synth., 1970, **50**, 56 (*deriv*)
Gilson, D.F.R., *Can. J. Chem.*, 1974, **52**, 2421 (*nmr*)
Ivanov, N. *et al*, *Fresenius' Z. Anal. Chem.*, 1976, **280**, 223 (*use, Nb*)
Lewis, R.J., *Sax's Dangerous Properties of Industrial Materials*, 8th Ed., Van Nostrand-Reinhold, 1992, DRS200, DXW000.

2-Dodecylbenzenesulfonic acid, 9CI D-00432

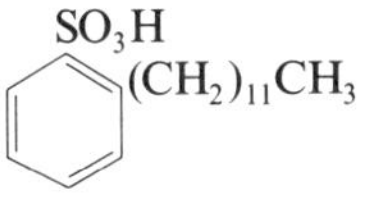

$C_{18}H_{30}O_3S$ M 326.499
Detergent.
S-*Benzylisothiouronium salt:* Mp 100-101°.

Truce, W.E. *et al*, *J. Am. Chem. Soc.*, 1951, **73**, 126 (*synth*)

3-Dodecylbenzenesulfonic acid, 9CI D-00433

[16577-13-2]

$C_{18}H_{30}O_3S$ M 326.499
Detergent, flotation agent.
Na salt: [19589-59-4].
Detergent.
S-*Benzylisothiouronium salt:* Mp 97-98°.
p-*Toluidine salt:* Mp 103-104°.

Truce, W.E. *et al*, *J. Am. Chem. Soc.*, 1951, **73**, 126 (*synth*)

4-Dodecylbenzenesulfonic acid, 9CI D-00434

[121-65-3]

$C_{18}H_{30}O_3S$ M 326.499
Detergent.
Na salt: [2211-98-5].
Detergent. Mp 144-145°.
S-*Benzylisothiouronium salt:* Mp 117-118°.
p-*Toluidine salt:* Mp 144-145° (138-139.5°).

Ogawa, M., *Nippon Kagaku Kaishi*, 1951, **54**, 779; *CA*, **48**, 3918 (*synth*)
Truce, W.E. *et al*, *J. Am. Chem. Soc.*, 1951, **73**, 126 (*synth*)
Pueschel, F. *et al*, *Tenside Deterg.*, 1970, **7**, 249; *CA*, **74**, 99561 (*synth, ir*)

2-Dodecylcyclopropanebutanoic acid D-00435

5,6-Methyleneoctadecanoic acid

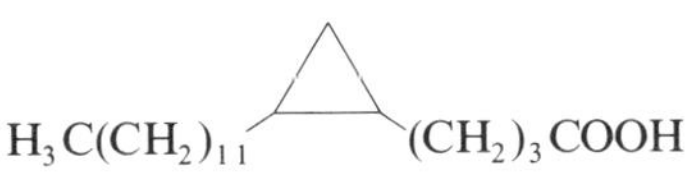

$C_{19}H_{36}O_2$ M 296.492
Both *cis*- and *trans*-forms have been synthesised as Me esters.

Christie, W.W. *et al*, *Chem. Phys. Lipids*, 1968, **2**, 196 (*synth, glc, ir, nmr*)
Christie, W.W., *Top. Lipid Chem.*, 1970, **1**, 4 (*rev*)
Gunstone, F.D. *et al*, *Chem. Phys. Lipids*, 1973, **10**, 303 (*synth, glc, nmr*)

5-Dodecyldihydro-2(3*H*)-furanone, 9CI D-00436

4-Hexadecanolide. γ-Palmitolactone. 4-Dodecyl-γ-butyrolactone

[730-46-1]

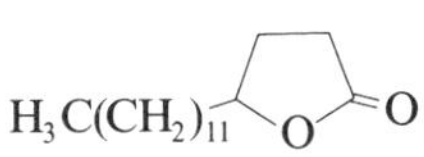

$C_{16}H_{30}O_2$ M 254.412
Found in pineapple essence and essential oil of *Pastinaca sativa*. Mp 54°.

Näf-Müller, R. *et al*, *Helv. Chim. Acta*, 1971, **54**, 1880 (*isol*)
Kubeczka, K.H. *et al*, *Planta Med.*, 1977, **31**, 173; *CA*, **86**, 177157 (*isol*)

Dodecyldimethylsulfonium, 9CI D-00437

Lauryldimethylsulfonium

[45164-82-7]

$$H_3C(CH_2)_{11}S^{\oplus}Me_2$$

$C_{14}H_{31}S^{\oplus}$ M 231.465 (ion)
Synthetic reagent.
Chloride: [41580-87-4].
$C_{14}H_{31}ClS$ M 266.918
Cryst. (Me_2CO). Mp 64-66°.
Bromide: [41580-88-5].
$C_{14}H_{31}BrS$ M 311.369
Cryst. (Me_2CO). Mp 68-70°.
Iodide: [18412-81-2].
$C_{14}H_{31}IS$ M 358.369
Cryst. (Me_2CO). Mp 87°.

Yano, Y. *et al*, *J. Org. Chem.*, 1973, **38**, 3912 (*synth, use*)
Fieser and Fieser's Reagents for Organic Synthesis, Wiley, 1975, **5**, 363 (*synth, use*)

2-Dodecyl-3-methylbutanedioic acid, 9CI D-00438

2-Dodecyl-3-methylsuccinic acid

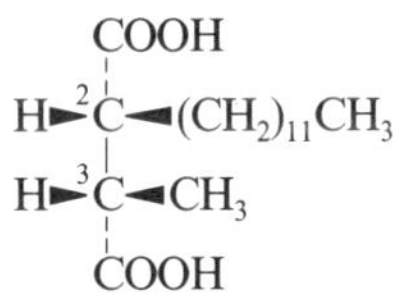

$C_{17}H_{32}O_4$ M 300.437
(*2R,3S*)-*form* [22139-54-4] ***Roccellic acid***
Occurs in *Lecanora* spp., *Roccella tinctoria*, *Lepraria latebrarum*, *Lecidea aglaeotara*, *Pannaria languinosa* and other lichens. Cryst. (pet. ether). Mp 131-132°. $[\alpha]_D$ +17.4° (EtOH).
Di-Me ester: Cryst. (Et_2O). Mp 28-29°.
Phenylimide: Cryst. (MeOH). Mp 57-58°.

Åkermark, B., *Acta Chem. Scand.*, 1962, **16**, 599 (*abs config*)
Huneck, S. *et al*, *Phytochemistry*, 1969, **8**, 1301 (*isol*)
Huneck, S. *et al*, *Z. Naturforsch., B*, 1969, **24**, 750 (*isol*)
Åkermark, B., *Acta Chem. Scand.*, 1970, **24**, 1456 (*isol*)
Ollis, W.D. *et al*, *J. Chem. Soc. C*, 1971, 1318 (*isol*)

1-Dodecyl sulfate D-00439

Sulfuric acid monododecyl ester, 9CI. Lauryl sulfate

[151-41-7]

$$H_3C(CH_2)_{11}OSO_2OH$$

$C_{12}H_{26}O_4S$ M 266.401
Detergent, surface-active agent. Used as 1% aq. soln. in photometric detns. of metals (with some azo reagents). Cryst. Sol. H_2O.
▷ WT0700000.
Glycine salt: Mp 110°.
Urea salt: Mp 110°.

Jaeger, D.A. *et al*, *J. Org. Chem.*, 1977, **42**, 3298 (*synth*)
Hung, S.C. *et al*, *Talanta*, 1982, **29**, 85 (*detn, Ag*)
Tan Zhe, *et al*, *Talanta*, 1984, **31**, 624 (*detn, Zn*)
Lewis, R.J., *Sax's Dangerous Properties of Industrial Materials*, 8th Ed., Van Nostrand-Reinhold, 1992, MRH250.

2-Dodecyl sulfate D-00440

$H_3C(CH_2)_9CH(OSO_2OH)CH_3$

$C_{12}H_{26}O_4S$ M 266.401

Na salt: Cryst. (H_2O or EtOH). Sinters at 142-5°.

Baumgarten, P., *Ber.*, 1942, **75**, 977; 1943, **76**, 213 (*synth*)

Dodecyltriphenylphosphonium(1+), 9CI D-00441

$Ph_3P^{\oplus}(CH_2)_{11}CH_3$

$C_{30}H_{40}P^{\oplus}$ M 431.620 (ion)

Topical antifungal agent.

Bromide: [15510-55-1]. *Nycal. Myxal-Losung. DTPB. D 305*

$C_{30}H_{40}BrP$ M 511.524

Hygroscopic cryst. (EtOAc/Et_2O). Mp 87-88°, Mp 98-99°. NaH or butyllithium → ylide.

Pentachlorophenol salt (1:1): [24360-58-5]. ***Pentaphonate.*** *Claraphos*

Antibacterial agent. Viscous brown oil.

Ylide: [54208-04-7]. *Dodecylidenetriphenylphosphorane*

$C_{30}H_{39}P$ M 430.612

Used in Wittig reaction. Red.

Jerchel, M. *et al*, *Chem. Ber.*, 1950, **83**, 277 (*synth*)
Kimmig, J. *et al*, *Klin. Wochenschr.*, 1950, **28**, 429 (*pharmacol*)
Ger. Pat., 920 726, (1954); *CA*, **53**, 291 (*Pentaphonate*)
Jerchel, D. *et al*, *Z. Naturforsch., B*, 1961, **16**, 85 (*props*)
Chasin, D.G. *et al*, *Chem. Phys. Lipids*, 1971, **6**, 8 (*synth, use*)
Bestmann, H.J. *et al*, *Angew. Chem., Int. Ed. Engl.*, 1976, **15**, 298 (*use*)
Lie Ken Jie, M.S.F. *et al*, *Chem. Phys. Lipids*, 1978, **21**, 275 (*use*)
Ku, T.W. *et al*, *J. Med. Chem.*, 1985, **28**, 1847 (*synth*)

2-Dodecyne D-00442

Methylnonylacetylene

[629-49-2]

$H_3C(CH_2)_8C{\equiv}CCH_3$

$C_{12}H_{22}$ M 166.306

Fp −9°. Bp_{15} 105°.

Krafft, F. *et al*, *Ber.*, 1892, **25**, 2243 (*synth*)

6-Dodecyne D-00443

Diamylacetylene

[6975-99-1]

$H_3C(CH_2)_4C{\equiv}C(CH_2)_4CH_3$

$C_{12}H_{22}$ M 166.306

Mp −41° to −40°. Bp_{30} 115-116°.

Bried, E.A. *et al*, *J. Am. Chem. Soc.*, 1937, **59**, 1310 (*synth*)
Matsuda, H. *et al*, *Tetrahedron*, 1988, **44**, 2865 (*ir, pmr, ms*)

3-Dodecynoic acid, 9CI D-00444

[55182-75-7]

$H_3C(CH_2)_7C{\equiv}CCH_2COOH$

$C_{12}H_{20}O_2$ M 196.289

Cryst. (pet. ether). Mp 47-49°.

Me ester:

$C_{13}H_{22}O_2$ M 210.316

$Bp_{0.05}$ 70°.

Williams, J.L. *et al*, *Chem. Phys. Lipids*, 1972, **9**, 295 (*synth, ir, nmr*)
Gilman, N.W. *et al*, *Chem. Phys. Lipids*, 1974, **13**, 239 (*synth*)
Gunstone, F.D. *et al*, *Chem. Phys. Lipids*, 1976, **17**, 1 (*cmr*)

4-Dodecynoic acid, 9CI D-00445

[55182-77-9]

$H_3C(CH_2)_6C{\equiv}CCH_2CH_2COOH$

$C_{12}H_{20}O_2$ M 196.289

Mp 48-50°.

Bohlmann, F. *et al*, *Chem. Ber.*, 1967, **100**, 3706 (*ms*)
Gilman, N.W. *et al*, *Chem. Phys. Lipids*, 1974, **13**, 239 (*synth*)

5-Dodecynoic acid, 9CI D-00446

[19307-20-1]

$H_3C(CH_2)_5C{\equiv}C(CH_2)_3COOH$

$C_{12}H_{20}O_2$ M 196.289

Cryst. (pet. ether). Mp 20.2-22.4°. $Bp_{0.3}$ 122-123°.

Hofmann, K. *et al*, *J. Biol. Chem.*, 1959, **234**, 1672 (*synth*)
Gunstone, F.D. *et al*, *Chem. Phys. Lipids*, 1967, **1**, 337 (*nmr*)
Gilman, N.W. *et al*, *Chem. Phys. Lipids*, 1974, **13**, 239 (*synth*)
Gunstone, F.D. *et al*, *Chem. Phys. Lipids*, 1976, **17**, 1 (*cmr*)

6-Dodecynoic acid, 9CI D-00447

[676-17-5]

$H_3C(CH_2)_4C{\equiv}C(CH_2)_4COOH$

$C_{12}H_{20}O_2$ M 196.289

Oil. Mp <20°. $Bp_{0.15}$ 130-135° (Bp_1 112-114°).

Taylor, W.R. *et al*, *J. Am. Chem. Soc.*, 1950, **72**, 4263 (*synth*)
Ames, D.E. *et al*, *J. Chem. Soc.*, 1965, 894 (*synth*)
Gunstone, F.D. *et al*, *Chem. Phys. Lipids*, 1967, **1**, 337 (*nmr*)
Gilman, N.W. *et al*, *Chem. Phys. Lipids*, 1974, **13**, 239 (*synth*)

7-Dodecynoic acid, 9CI D-00448

[19307-22-3]

$H_3C(CH_2)_3C{\equiv}C(CH_2)_5COOH$

$C_{12}H_{20}O_2$ M 196.289

Mp 18.5-19°. Bp_1 142-143°, $Bp_{0.4}$ 118-120°.

tert-*Butyl ester:* $Bp_{0.5}$ 108-111°.

Anilide:

$C_{18}H_{25}NO$ M 271.402

Mp 40-41°.

Taylor, W.R. *et al*, *J. Am. Chem. Soc.*, 1950, **72**, 4263 (*synth*)
Gunstone, F.D. *et al*, *Chem. Phys. Lipids*, 1967, **1**, 209, 337 (*synth, nmr*)
Crandall, J.K. *et al*, *J. Organomet. Chem.*, 1973, **51**, 375.
Gilman, N.W. *et al*, *Chem. Phys. Lipids*, 1974, **13**, 239 (*synth*)
Gunstone, F.D. *et al*, *Chem. Phys. Lipids*, 1976, **17**, 1 (*cmr*)

8-Dodecynoic acid, 9CI D-00449

[18545-07-8]

$H_3C(CH_2)_2C{\equiv}C(CH_2)_6COOH$

$C_{12}H_{20}O_2$ M 196.289

Mp 39.5-41.0°. $Bp_{0.4}$ 118.5-119.5°.

Gunstone, F.D. *et al*, *Chem. Phys. Lipids*, 1967, **1**, 209, 337 (*synth, pmr*)
Kovaleva, A.S. *et al*, *CA*, 1973, **79**, 18006x.
Gilman, N.W. *et al*, *Chem. Phys. Lipids*, 1974, **13**, 239 (*synth*)

9-Dodecynoic acid, 9CI D-00450

[19307-24-5]

$H_3CCH_2C{\equiv}C(CH_2)_7COOH$

$C_{12}H_{20}O_2$ M 196.289

Mp 39.5-41.0°.

Gunstone, F.D. *et al*, *Chem. Phys. Lipids*, 1967, **1**, 209, 337 (*synth, nmr*)
Gilman, N.W. *et al*, *Chem. Phys. Lipids*, 1974, **13**, 239 (*synth*)

10-Dodecynoic acid, 9CI — D-00451

[19307-25-6]

$H_3CC{\equiv}C(CH_2)_8COOH$

$C_{12}H_{20}O_2$ M 196.289
Mp 54.5-56.5°.

Gunstone, F.D. *et al*, *Chem. Phys. Lipids*, 1967, **1**, 209, 337 (*synth, pmr*)
Gilman, N.W. *et al*, *Chem. Phys. Lipids*, 1974, **13**, 239 (*synth*)
Gunstone, F.D. *et al*, *Chem. Phys. Lipids*, 1976, **17**, 1 (*cmr*)

11-Dodecynoic acid — D-00452

$HC{\equiv}C(CH_2)_9COOH$

$C_{12}H_{20}O_2$ M 196.289
Cryst. (pet. ether). Mp 44-46°. Bp_1 155-161° ($Bp_{0.6}$ 132-136°).

Me ester:
$C_{13}H_{22}O_2$ M 210.316
$Bp_{0.3}$ 49-50°.

Gunstone, F.D. *et al*, *Chem. Phys. Lipids*, 1967, **1**, 209, 337 (*synth, pmr*)

2-Dodecyn-1-ol, 9CI — D-00453

[69064-46-6]

$H_3C(CH_2)_8C{\equiv}CCH_2OH$

$C_{12}H_{22}O$ M 182.305
Mp 30-31°. $Bp_{0.3}$ 102-107°.

Christie, W.W. *et al*, *Chem. Phys. Lipids*, 1967, **1**, 407 (*synth*)
Jain, S.C. *et al*, *J. Org. Chem.*, 1983, **48**, 2274 (*synth, ir, pmr, ms*)
Mori, K. *et al*, *Synthesis*, 1991, 1125 (*synth, ir, pmr*)

3-Dodecyn-1-ol — D-00454

[55182-73-5]

$H_3C(CH_2)_7C{\equiv}CCH_2CH_2OH$

$C_{12}H_{22}O$ M 182.305
Liq. $Bp_{0.25}$ 106-107°.

Gilman, N.W. *et al*, *Chem. Phys. Lipids*, 1974, **13**, 239 (*synth*)
Carvalho, J.F. *et al*, *J. Org. Chem.*, 1984, **49**, 1251 (*synth, ir, pmr*)

Dohexacontane, 9CI — D-00455

[7719-83-7]

$H_3C(CH_2)_{60}CH_3$

$C_{62}H_{126}$ M 871.677
Isol. from wax of *Leptochloa digitata*. Cryst. (pet. ether). Mp 99.5-100°.

Kranz, Z.H. *et al*, *Aust. J. Chem.*, 1961, **14**, 264 (*isol*)
Kudchadker, A.P. *et al*, *J. Chem. Eng. Data*, 1966, **11**, 253 (*isol*)

Dolichols — D-00456

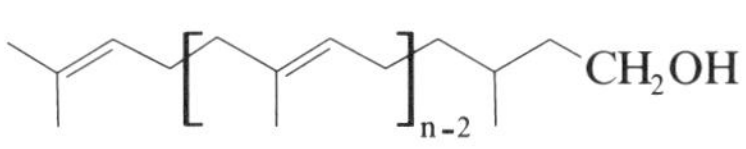

A homologous family of isoprene alcohols. The best-characterised member appears to be Dolichol 20. Occur, sometimes in large amounts, in unsaponifiable fractions of plant and animal tissues. Of largely unknown biol. function; act as cofactors in biosynth. of secondary glycoproteins in the endoplasmic reticulum. The phosphates are involved in transport of *N*-glycoproteins.

Dolichol 11 [12777-41-2]
Bactoprenol
$C_{55}H_{92}O$ M 769.331
Isol. from lactobacilli.

Dolichol 20 [2067-66-5]
$C_{100}H_{164}O$ M 1382.395
Viscous oil. Mp −10°. $[\alpha]^{25}_{360}$ −1.8° (dioxan).

[26113-10-0, 37839-88-6, 42436-66-8, 57229-24-0, 68688-74-4, 73742-37-7, 73742-38-8, 124898-59-5, 124989-36-2]

Burgos, J. *et al*, *Biochem. J.*, 1963, **88**, 470 (*isol, uv, ir, pmr, ms*)
Morton, R.A., *Biochem. J.*, 1972, **128**, 11P (*rev*)
Ekström, T. *et al*, *Acta Chem. Scand., Ser. B*, 1982, **36**, 411 (*biosynth*)
Suzuki, S. *et al*, *Tetrahedron Lett.*, 1983, **24**, 5103 (*synth*)
Eggers, I., *Acta Chem. Scand., Ser. B*, 1987, **41**, 67 (*biosynth*)
Appelkvist, E.-L., *Acta Chem. Scand., Ser. B*, 1987, **41**, 73 (*biosynth*)
Jaenicke, L. *et al*, *Chem. Phys. Lipids*, 1989, **51**, 159 (*synth, bibl*)

Dotriacontane, 9CI — D-00457

Dicetyl. Lacceran. Bicetyl

[544-85-4]

$H_3C(CH_2)_{30}CH_3$

$C_{32}H_{66}$ M 450.873
Constit. of higher plants, eg. *Bacopa monnieri*, *Mentha aquatica*, *Agave sisalana*. Plates (Et_2O). Mp 74-75°. $Bp_{1.5}$ 245°.

▷ JT2360000.

Levene, P.A. *et al*, *J. Biol. Chem.*, 1915, **20**, 530 (*synth*)
Brown, I. *et al*, *J. Chem. Soc.*, 1960, 2783 (*isol*)
Razafindrazaka, J. *et al*, *Bull. Soc. Chim. Fr.*, 1963, 1633 (*isol*)

Dotriacontanoic acid — D-00458

Lacceric acid

[3625-52-3]

$H_3C(CH_2)_{30}COOH$

$C_{32}H_{64}O_2$ M 480.856
Constit. of wool fat and of several leaf waxes. Cryst. (C_6H_6). Mp 96°.

Me ester: [41755-79-7].
$C_{33}H_{66}O_2$ M 494.883
$Bp_{0.01}$ 192°.

Et ester: [29030-82-8].
$C_{34}H_{68}O_2$ M 508.910
Mp 76°.

Gascard, A., *C. R. Hebd. Seances Acad. Sci.*, 1914, **159**, 260 (*isol*)
Francis, F. *et al*, *J. Am. Chem. Soc.*, 1939, **61**, 577 (*isol*)
Schuette, H.A. *et al*, *Oil Soap* (*Chicago*), 1945, **22**, 107 (*props*)
Liu, S.-L., *CA*, 1955, **49**, 164 (*isol*)
Hashimoto, A., *CA*, 1968, **68**, 76049.

1-Dotriacontanol — D-00459

Laccerol

$H_3C(CH_2)_{30}CH_2OH$

$C_{32}H_{66}O$ M 466.873
Minor constit. of wax of *Arbutus unedo* and obt. from several other sources. Plates (C_6H_6). Mp 89°.

Ac: [23811-94-1].
$C_{34}H_{68}O_2$ M 508.910
Found in *Aniba gardneri*. Mp 76-77°.

p-*Hydroxycinnamoyl:* [92533-57-8]. ***Randilongin***
$C_{41}H_{72}O_3$ M 613.018
Isol. from leaves of *Randia longiflora* and from *Cinnamomum cassia*. Cryst. ($CHCl_3$/petrol). Mp 82°.

Gascard, A., *C. R. Hebd. Seances Acad. Sci.*, 1914, **159**, 258 (*isol*)
Sosa, A., *C. R. Hebd. Seances Acad. Sci.*, 1950, **230**, 995 (*isol*)
Cocker, W. *et al*, *J. Chem. Soc.*, 1962, 5194 (*isol*)
Kuo, Y. *et al*, *J. Chin. Chem. Soc.* (*Taipei*), 1984, **31**, 159 (*Randilongin*)
Talapatra, S.K. *et al*, *J. Indian Chem. Soc.*, 1989, **66**, 694 (*Randilongin*)

2-Dotriacontanone, 9CI D-00460

[77327-15-2]

$H_3C(CH_2)_{29}COCH_3$

$C_{32}H_{64}O$ M 464.857
Isol. from the snakes *Boiga irregularis* and *Thamnophis sirtalis parietalis*. Sex attractant pheromone.

Mason, R.T. *et al*, *J. Chem. Ecol.*, 1990, **16**, 2353 (*isol, ms*)
Murata, Y. *et al*, *J. Nat. Prod.* (*Lloydia*), 1991, **54**, 233 (*isol*)

Duryne D-00461

4,15,26-Triacontatriene-1,29-diyne-3,28-diol
[108641-87-8]

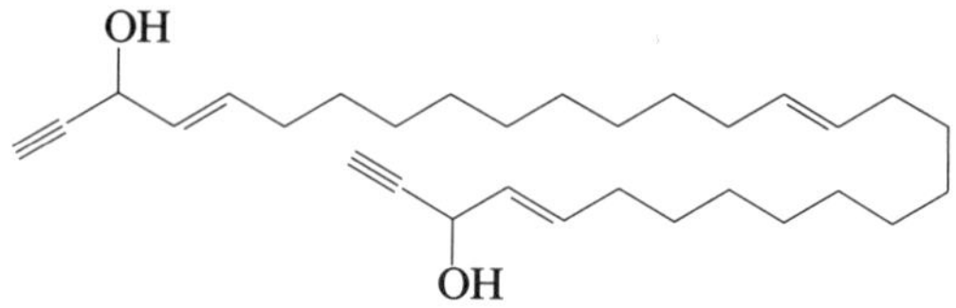

$C_{30}H_{48}O_2$ M 440.708
Isol. from marine sponge *Cribrochalina dura*. Potent cytotoxic metabolite. Solid. Mp 44-45°. Nomenclature error in CA.

Wright, A.E. *et al*, *Tetrahedron Lett.*, 1987, **28**, 1377 (*isol, ir, ms, pmr, cmr*)

E

1,19-Eicosadiene — E-00001

1,19-Icosadiene

[14811-95-1]

$H_2C{=}CH(CH_2)_{16}CH{=}CH_2$

$C_{20}H_{38}$ M 278.520

Cryst. (EtOH). Mp 24.5° (19.5-20°). $Bp_{0.1}$ 135-45° ($Bp_{0.2}$125-30°).

Canonica, L. *et al*, *CA*, 1953, **47**, 11131 (*synth*)
Bounds, D.G. *et al*, *J. Chem. Soc.*, 1953, 2393 (*synth*)
Corey, E.J. *et al*, *Tetrahedron Lett.*, 1975, 4091 (*synth*)

5,11-Eicosadienoic acid — E-00002

5,11-Icosadienoic acid

$H_3C(CH_2)_7CH{=}CH(CH_2)_4CH{=}CH(CH_2)_3COOH$

$C_{20}H_{36}O_2$ M 308.503

Isol. from Gymnospermae seed lipids.

Takagi, T. *et al*, *Lipids*, 1982, **17**, 716 (*occur, glc, cmr*)

7,13-Eicosadienoic acid — E-00003

7,13-Icosadienoic acid

[13344-56-4]

$H_3C(CH_2)_5CH{=}CH(CH_2)_4CH{=}CH(CH_2)_5COOH$

$C_{20}H_{36}O_2$ M 308.503

Low-melting solid.

Tetrabromide: 7,8,13,14-Tetrabromoeicosanoic acid
$C_{20}H_{36}Br_4O_2$ M 628.119
Mp 132-134°.

Gunstone, F.D. *et al*, *J. Chem. Soc.*, 1962, 3055 (*synth*)

8,11-Eicosadienoic acid, 9CI — E-00004

8,11-Icosadienoic acid

[3039-63-2]

$H_3C(CH_2)_7CH{=}CHCH_2CH{=}CH(CH_2)_6COOH$

$C_{20}H_{36}O_2$ M 308.503

(all-Z)-form [29428-99-7]
Metab. of Oleic acid.

Castuma, J.C. *et al*, *CA*, 1972, **76**, 112613.

11,14-Eicosadienoic acid — E-00005

11,14-Icosadienoic acid

[60362-00-7]

$H_3C(CH_2)_4CH{=}CHCH_2CH{=}CH(CH_2)_9COOH$

$C_{20}H_{36}O_2$ M 308.503

$Bp_{0.1}$ 198°.

(Z,Z)-form
Isol. from lipids of *Ginkgo biloba*, *Podocarpus nageia* and other plants.

Karrer, P. *et al*, *Helv. Chim. Acta*, 1943, **26**, 619 (*synth*)
Ito, S. *et al*, *Bull. Chem. Soc. Jpn.*, 1963, **36**, 1439 (*isol, uv, ir, struct*)
Williams, M.A. *et al*, *Biochem. Biophys. Res. Commun.*, 1973, **54**, 1560 (*isol*)
Spener, F. *et al*, *Chem. Phys. Lipids*, 1973, **11**, 215 (*synth*)
Sung-Kee, C. *et al*, *Tetrahedron Lett.*, 1974, 3023 (*synth*)
Bardyshev, I.I. *et al*, *CA*, 1975, **83**, 166073q (*isol*)

11,15-Eicosadienoic acid — E-00006

11,15-Icosadienoic acid

[133530-13-9]

$H_3C(CH_2)_3CH{=}CHCH_2CH_2CH{=}CH(CH_2)_9COOH$

$C_{20}H_{36}O_2$ M 308.503

Constit. of the sponge *Amphimedon complanata*. Isol. as Me ester.

Carballeira, N.M. *et al*, *J. Nat. Prod. (Lloydia)*, 1991, **54**, 315 (*isol*)

1,19-Eicosadiyne — E-00007

1,19-Icosadiyne

$HC{\equiv}C(CH_2)_{16}C{\equiv}CH$

$C_{20}H_{34}$ M 274.489

Mp 58-58.5°.

Canonica, L. *et al*, *CA*, 1953, **47**, 11131 (*synth*)
Beckmann, W. *et al*, *Synthesis*, 1975, 423 (*synth*)

7,13-Eicosadiynoic acid, 9CI — E-00008

7,13-Icosadiynoic acid

[26192-08-5]

$H_3C(CH_2)_5C{\equiv}C(CH_2)_4C{\equiv}C(CH_2)_5COOH$

$C_{20}H_{32}O_2$ M 304.472

Cryst. (EtOH). Mp 45.5-46°.

Gunstone, F.D. *et al*, *J. Chem. Soc.*, 1962, 3055 (*synth*)
Ames, D.E. *et al*, *J. Chem. Soc., Perkin Trans. 1*, 1972, 255 (*synth*)

8,11-Eicosadiynoic acid — E-00009

8,11-Icosadiynoic acid

$H_3C(CH_2)_7C{\equiv}CCH_2C{\equiv}C(CH_2)_6COOH$

$C_{20}H_{32}O_2$ M 304.472

Cryst. Mp 40-41°.

Sprecher, H., *Prog. Chem. Fats Other Lipids*, 1978, **15**, 219.

10,13-Eicosadiynoic acid — E-00011

10,13-Icosadiynoic acid

$H_3C(CH_2)_5C{\equiv}CCH_2C{\equiv}C(CH_2)_8COOH$

$C_{20}H_{32}O_2$ M 304.472

Mp 40-41°.

Sprecher, H., *Biochim. Biophys. Acta*, 1971, **231**, 122 (*synth*)
Sprecher, H., *Prog. Chem. Fats Other Lipids*, 1977, **15**, 219.

1-Eicosanamine, 9CI E-00012

1-Icosylamine. 1-Aminoeicosane

[10525-37-8]

$H_3C(CH_2)_{18}CH_2NH_2$

$C_{20}H_{43}N$ M 297.566

Mp 58°. Bp 196°.

Zvejnieks, A., *Sven. Kem. Tidskr.*, 1954, **66**, 316.

Eicosane, 9CI E-00013

Icosane

[112-95-8]

$H_3C(CH_2)_{18}CH_3$

$C_{20}H_{42}$ M 282.552

Isol. from waxes and ethereal oils and from wood of *Manilkara bidentata*. Cryst. (EtOH). Mp 36.7°. Bp_{15} 205°.

Carothers, W.H. *et al*, *J. Am. Chem. Soc.*, 1930, **52**, 5280 (*synth*)
La Face, D., *Helv. Chim. Acta*, 1950, 33, 249 (*isol*)
Cocker, W. *et al*, *J. Chem. Soc.*, 1963, 677 (*isol*)
Wagner, J.R. *et al*, *J. Chem. Eng. Data*, 1968, **13**, 22.
Yasuo, S. *et al*, *Bull. Chem. Soc. Jpn.*, 1970, **43**, 3924 (*synth*)
Karl, B. *et al*, *Helv. Chim. Acta*, 1970, **53**, 1833.
Crob, K. *et al*, *J. Chromatogr.*, 1971, **62**, 1 (*glc, ms*)
Sax, N.I., *Dangerous Properties of Industrial Materials*, 5th Ed., Van Nostrand-Reinhold, 1979, 639.

4,5-Eicosanediamine E-00014

4,5-Icosanediamine. 4,5-Diaminoeicosane. ***Necrosamine***

$H_3C(CH_2)_{14}CH(NH_2)CH(NH_2)CH_2CH_2CH_3$

$C_{20}H_{44}N_2$ M 312.581

Prod. in the phospholipid fraction of *E. coli*.

B,2HCl: Cryst. Mp *ca.* 275° dec.

Dibenzoyl: Cryst. (EtOH aq.). Mp 72.5-73.5°.

Dipicrate: Cryst. (EtOH). Mp 159-161°.

Ikawa, M. *et al*, *J. Am. Chem. Soc.*, 1953, **75**, 3439, 6314 (*struct, synth*)

Eicosanedioic acid E-00015

Octadecane-1,18-dicarboxylic acid. Icosanedioic acid

[2424-92-2]

$HOOC(CH_2)_{18}COOH$

$C_{20}H_{38}O_4$ M 342.518

Constit. of conifer bark waxes and angiosperms, eg. *Rhus* spp. Mp 125-126.5°. Bp_2 233-234°.

Di-Me ester: [42235-38-1].

$C_{22}H_{42}O_4$ M 370.571

Cryst. (MeOH). Mp 65-65.8°.

Di-Et ester: [42235-39-2].

$C_{24}H_{46}O_4$ M 398.625

Mp 54.5-55.2°.

Dianilide:

$C_{32}H_{48}N_2O_2$ M 492.743

Cryst. (EtOH). Mp 143-146°.

Dinitrile:

$C_{20}H_{36}N_2$ M 304.518

Mp 70-71°.

Günthard, H.H. *et al*, *Helv. Chim. Acta*, 1953, **36**, 1147 (*synth*)
Aleby, S. *et al*, *Ark. Kemi*, 1969, **31**, 267 (*ir, polymorphism*)
Marosi, L. *et al*, *Justus Liebigs Ann. Chem.*, 1973, 584 (*cryst struct*)

2,19-Eicosanedione E-00016

2,19-Icosanedione

$H_3CCO(CH_2)_{16}COCH_3$

$C_{20}H_{38}O_2$ M 310.519

Mp 92-93°.

Dioxime:

$C_{20}H_{40}N_2O_2$ M 340.548

Mp 110-114°.

Bis(2,4-dinitrophenylhydrazone): Mp 106-108°.

Stoll, M., *Helv. Chim. Acta*, 1951, **34**, 1817 (*synth*)
Ames, D. *et al*, *J. Chem. Soc.*, 1952, 1057 (*synth*)
Canonica, L. *et al*, *CA*, 1953, **47**, 11131 (*synth*)
U.S. Pat., 2 671 810, (1955); *CA*, **49**, 5522 (*synth*)

4,5-Eicosanedione E-00017

4,5-Icosanedione

$H_3C(CH_2)_2COCO(CH_2)_{14}CH_3$

$C_{20}H_{38}O_2$ M 310.519

Cryst. (EtOH). Mp 56.5-57.5°.

Phenylhydrazone: Mp 71-73°.

Dioxime:

$C_{20}H_{40}N_2O_2$ M 340.548

Mp 136.5-137.5°.

Ikawa, M. *et al*, *J. Am. Chem. Soc.*, 1953, **75**, 6314 (*synth*)

7,14-Eicosanedione, 9CI E-00018

7,14-Icosanedione

[31335-10-1]

$H_3C(CH_2)_5CO(CH_2)_6CO(CH_2)_5CH_3$

$C_{20}H_{38}O_2$ M 310.519

Cryst. (EtOH). Mp 83.5-84°.

Bissemicarbazone: Mp 146-148°.

Samour, C.M. *et al*, *J. Am. Chem. Soc.*, 1954, **76**, 441 (*synth*)

1,2,3,4-Eicosanetetrol, 9CI E-00019

1,2,3,4-Icosanetetrol

[42415-56-5]

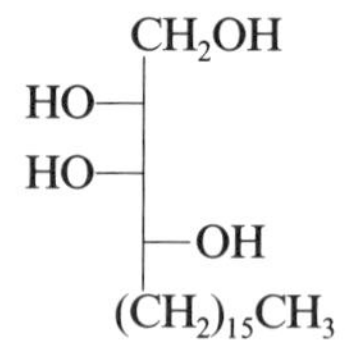

(*2R,3S,4R*)-*form*

$C_{20}H_{42}O_4$ M 346.549

(2R,3S,4R)-form [105368-57-8]

D-lyxo-*form*

$[\alpha]_D$ +2.5° (c, 0.28 in Py).

Tetra-Ac: [105368-58-9].

Needles (hexane). Mp 84-86°. $[\alpha]_D^{20}$ +20.3° (c, 1 in $CHCl_3$).

(2S,3R,4R)-form [105368-50-1]

D-ribo-*form*

Cryst. (MeOH). $[\alpha]_D^{20}$ +6.8° (c, 0.6 in Py).

Tetra-Ac: [105265-44-9].

Oil. $[\alpha]_D^{20}$ +3.3° (c, 1.2 in $CHCl_3$).

(2S,3R,4S)-form [105368-53-4]

L-lyxo-*form*

$[\alpha]_D$ −2.1° (c, 0.3 in Py).

(2S,3S,4R)-form [105368-60-3]

D-xylo-*form.* ***Guggultetrol 20***

Isol. from the saponified resin of *Commiphora mukul*. Cryst. (EtOH). $[\alpha]_D^{20}$ +9.2° (c, 0.28 in EtOH).

Tetra-Ac: [105368-61-4].
Cryst. (hexane). Mp 54-56°. $[\alpha]_D^{20}$ +2.3° (c, 1 in $CHCl_3$).

Kjaer, A. *et al, Tetrahedron*, 1986, **42**, 1439 (*synth*)
Kumar, V. *et al, Tetrahedron*, 1987, **43**, 5933 (*synth, abs config*)

Eicosanoic acid — E-00020

Arachidic acid. Icosanoic acid. Arachic acid
[506-30-9]

$$H_3C(CH_2)_{18}COOH$$

$C_{20}H_{40}O_2$ M 312.535
Constit. of arachis (*Arachis hypogaea*) (ground-nut) oil glycerides. Widely distributed in other seed oils. Cryst. (EtOH). Mp 77°. Bp_1 203-205°.
▷ JX3780000.

Me ester: [1120-28-1].
$C_{21}H_{42}O_2$ M 326.562
Mp 45.8-46.3°. Bp_{10} 215-216°, Bp_2 188°.

Et ester: [18281-05-5].
$C_{22}H_{44}O_2$ M 340.588
Constit. of *Cassia javanica* leaves. Mp 41.4-42°. Bp_{100} 295-297°, Bp_2 186-187°.

Amide:
$C_{20}H_{41}NO$ M 311.550
Mp 109°.

Nitrile:
$C_{20}H_{39}N$ M 293.535
Mp 49.5°.

Hexacosyl ester: [17318-45-5]. *Hexacosyl eicosanoate*
$C_{46}H_{92}O_2$ M 677.232
Constit. of the waxes of *Desmodium* sp., *Nardostachys* sp., *Scutellaria* sp. and green tobacco leaves.

Adam, N.K. *et al, J. Chem. Soc.*, 1925, **127**, 72 (*synth*)
Sy, M. *et al, C. R. Hebd. Seances Acad. Sci.*, 1954, **239**, 1813 (*synth*)
Radler, F. *et al, Aust. J. Chem.*, 1965, **18**, 1059 (*isol*)
Marosi, L. *et al, Justus Liebigs Ann. Chem.*, 1973, 584 (*props, cryst struct*)
Lewis, R.J., *Sax's Dangerous Properties of Industrial Materials*, 8th Ed., Van Nostrand-Reinhold, 1992, EAF000.

1-Eicosanol, 9CI — E-00021

Arachidic alcohol. 1-Icosanol. Arachic alcohol
[629-96-9]

$$H_3C(CH_2)_{18}CH_2OH$$

$C_{20}H_{42}O$ M 298.551
Constit. of conifer waxes. Cryst. (Me_2CO). Mp 72.5-73°. Bp_3 220°.

Formyl:
$C_{21}H_{42}O_2$ M 326.562
Mp 74°.

Ac:
$C_{22}H_{44}O_2$ M 340.588
Constit. of *Bombus* spp. head secretions. Cryst. (Me_2CO aq.). Mp 40°.

Levene, P.A. *et al, J. Biol. Chem.*, 1924, **59**, 905 (*synth*)
Heilbron, I.M. *et al, J. Chem. Soc.*, 1941, 344 (*isol*)
Karrer, W. *et al, Konstitution und Vorkommen der Organischen Pflanzenstoffe*, 2nd Ed., Birkhäuser Verlag, Basel, 1972-1985, no. 90.
Bergström, G. *et al, Chem. Scr.*, 1973, **4**, 231 (*isol*)
Precht, V.D., *Fette, Seifen, Anstrichm.*, 1976, **78**, 189 (*cryst struct*)

2-Eicosanol, 9CI — E-00022

2-Icosanol
[4340-76-5]

$$H_3C(CH_2)_{17}CH(OH)CH_3$$

$C_{20}H_{42}O$ M 298.551

(+)-*form*
Isol. from bacterial lipids *Mycobacterium* spp. Needles (MeOH). $[\alpha]_D$ +4.2° ($CHCl_3$), $[\alpha]_D$ +6.79° (Et_2O).
[34019-45-9]

Anderson, R.J. *et al, J. Am. Chem. Soc.*, 1936, **58**, 10; 1937, **59**, 858 (*isol*)
Serck-Hanssen, K. *et al, CA*, 1954, **48**, 5071 (*synth*)
Elémadi, A.H. *et al, Bull. Soc. Chim. Biol.*, 1965, **47**, 2095 (*biosynth*)
Pearce, P.J. *et al, J. Chem. Soc., Perkin Trans. 1*, 1972, 1655 (*synth*)

3-Eicosanol, 9CI — E-00023

3-Icosanol

$$H_3C(CH_2)_{16}CH(OH)CH_2CH_3$$

$C_{20}H_{42}O$ M 298.551
Mp 57.4-58.5°.

Breusch, F.L. *et al, Chem. Ber.*, 1953, **86**, 678 (*synth*)

2-Eicosanone — E-00024

Methyl octadecyl ketone. 2-Oxoicosane. 2-Icosanone
[29703-52-4]

$$H_3C(CH_2)_{17}COCH_3$$

$C_{20}H_{40}O$ M 296.535
Constit. of *Euonymus latifolius* leaves. Cryst. (EtOH). Mp 58-59°.

Oxime:
$C_{20}H_{41}NO$ M 311.550
Cryst. (EtOH). Mp 73-74°.

Semicarbazone: Cryst. (EtOH). Mp 128°.

Pangborn, M.C. *et al, J. Am. Chem. Soc.*, 1936, **58**, 10 (*synth*)
Breusch, F.L. *et al, Chem. Ber.*, 1953, **86**, 684 (*synth*)
Ulubelen, A. *et al, Phytochemistry*, 1973, **12**, 1824 (*isol*)

3-Eicosanone — E-00025

3-Icosanone

$$H_3C(CH_2)_{16}COCH_2CH_3$$

$C_{20}H_{40}O$ M 296.535
Cryst. (EtOH). Mp 60-61°.

Oxime:
$C_{20}H_{41}NO$ M 311.550
Cryst. (EtOH). Mp 55.5-56.5°.

Semicarbazone: Mp 76°, Mp 89-91°.

Gilman, H. *et al, Recl. Trav. Chim. Pays-Bas (J. R. Neth. Chem. Soc.)*, 1936, **55**, 518 (*synth*)
Sherk, K.W. *et al, J. Am. Chem. Soc.*, 1945, **67**, 2239 (*synth*)
Breusch, F.L. *et al, Chem. Ber.*, 1953, **86**, 684 (*synth*)

7-Eicosanone — E-00026

7-Icosanone

$$H_3C(CH_2)_{12}CO(CH_2)_5CH_3$$

$C_{20}H_{40}O$ M 296.535
Mp 52-53°. Bp_{11} 210-211°.

Breusch, F.L. *et al, Chem. Ber.*, 1953, **86**, 684 (*synth*)

2,5,8,11,14-Eicosapentaenoic acid **E-00027**

2,5,8,11,14-Icosapentaenoic acid

$H_3C(CH_2)_4(CH{=}CHCH_2)_4CH{=}CHCOOH$

$C_{20}H_{30}O_2$ M 302.456

(2E,5Z,8Z,11Z,14Z)-form [41245-85-6]

Heslinga, L. *et al*, *Recl. Trav. Chim. Pays-Bas* (*J. R. Neth. Chem. Soc.*), 1973, **92**, 287 (*synth*)
Egmond, G.J.N. *et al*, *Recl. Trav. Chim. Pays-Bas* (*J. R. Neth. Chem. Soc.*), 1977, **96**, 172 (*synth*)

5,8,11,14,16-Eicosapentaenoic acid, 9CI **E-00028**

5,8,11,14,16-Icosapentaenoic acid. 16,17-Dehydroarachidonic acid

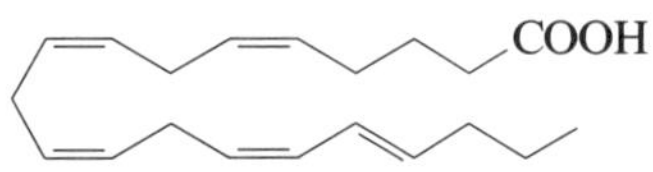

$C_{20}H_{30}O_2$ M 302.456

(5Z,8Z,11Z,14Z,16E)-form

Substrate for soybean lipoxygenase.

Corey, E.J. *et al*, *Tetrahedron Lett.*, 1987, **28**, 5391 (*synth, uv, ms*)

5,8,11,14,17-Eicosapentaenoic acid **E-00029**

5,8,11,14,17-Icosapentaenoic acid. EPA

[1553-41-9]

$H_3CCH_2(CH{=}CHCH_2)_5CH_2CH_2COOH$

$C_{20}H_{30}O_2$ M 302.456

(all-Z)-form [10417-94-4] ***Timnodonic acid***

Present in fish oils as an acylglycerol and in animal phospholipids. Metab. of α-Linolenic acid. Precursor of PG_3 series of prostaglandins.

Et ester: [86227-47-6]. *Ethyl EPA*

$C_{22}H_{34}O_2$ M 330.509

Isol. from fish oils. Antithrombotic agent.

[84494-70-2]

Teshima, S. *et al*, *CA*, 1978, **89**, 193306 (*isol, nmr*)
Von Schacky, C. *et al*, *J. Clin. Invest.*, 1985, **76**, 2446 (*metab, pharmacol*)
Eur. Pat., 292 846, (1988); *CA*, **111**, 63962 (*isol, Et ester*)
Corey, E.J. *et al*, *J. Org. Chem.*, 1988, **53**, 5980 (*synth, ir, pmr, ms*)

5,8,11,14,17-Eicosapentaynoic acid **E-00030**

5,8,11,14,17-Icosapentaynoic acid, 9CI

[5871-07-8]

$H_3CCH_2(C{\equiv}CCH_2)_5CH_2CH_2COOH$

$C_{20}H_{20}O_2$ M 292.377

Cryst. (Et_2O). Mp 116-116.5°.

Pabon, H.J.J. *et al*, *Recl. Trav. Chim. Pays-Bas* (*J. R. Neth. Chem. Soc.*), 1965, **84**, 1319 (*synth*)
Kunau, W.H. *et al*, *Hoppe Seyler's Z. Physiol. Chem.*, 1971, **352**, 542 (*synth*)
Bus, J. *et al*, *Recl. Trav. Chim. Pays-Bas* (*J. R. Neth. Chem. Soc.*), 1972, **91**, 394 (*ir*)

5,8,11,14-Eicosatetraenedioic acid **E-00031**

5,8,11,14-Icosatetraenedioic acid

$HOOC(CH_2)_3(CH{=}CHCH_2)_4(CH_2)_3COOH$

$C_{20}H_{30}O_4$ M 334.455

Metab. produced during NADPH dependent enzymatic oxidation of arachidonic acid.

Manna, S. *et al*, *Tetrahedron Lett.*, 1983, **24**, 33 (*synth, bibl*)

4,7,10,13-Eicosatetraenoic acid **E-00032**

4,7,10,13-Icosatetraenoic acid

[32907-97-4]

$H_3C(CH_2)_5(CH{=}CHCH_2)_4CH_2COOH$

$C_{20}H_{32}O_2$ M 304.472

(all-Z)-form

Present in some animal phospholipids. Metab. of 9-Hexadecenoic acid.

Kunau, W.H. *et al*, *Eur. J. Biochem.*, 1974, **48**, 311.

4,8,11,14-Eicosatetraenoic acid **E-00033**

4,8,11,14-Icosatetraenoic acid

$H_3C(CH_2)_4(CH{=}CHCH_2)_3CH_2CH{=}CHCH_2CH_2COOH$

$C_{20}H_{32}O_2$ M 304.472

(all-Z)-form

Beerthius, R.K. *et al*, *Recl. Trav. Chim. Pays-Bas* (*J. R. Neth. Chem. Soc.*), 1971, **90**, 943 (*synth*)

5,8,11,14-Eicosatetraenoic acid **E-00034**

5,8,11,14-Icosatetraenoic acid

[7771-44-0]

$H_3C(CH_2)_4CH{=}CHCH_2CH{=}CHCH_2$
$CH{=}CHCH_2CH{=}CH(CH_2)_3COOH$

$C_{20}H_{32}O_2$ M 304.472

▷ JX3850000.

(all-Z)-form [506-32-1]

Arachidonic acid

Constit. of many animal phospholipids, also of some ferns and mosses. Essential fatty acid. Precursor of Prostaglandin G_2, P-00146 series prostaglandins. Mp −49.5°. Bp_1 163°.

▷ Potentially explosive. CE6675000.

Me ester: [2566-89-4].

$C_{21}H_{34}O_2$ M 318.498

$Bp_{0.7}$ 194-196°.

(2-Hydroxyethyl)amide: [94421-68-8]. *Arachnidonylethanolamide.* ***Anandamide***

$C_{22}H_{37}NO_2$ M 347.540

Isol. from porcine brain. Cannabinoid receptor ligand.

20,20,20-Trifluoro: [115178-89-7]. *20-Trifluoroarachidonic acid*

$C_{20}H_{29}F_3O_2$ M 358.443

Used in studying physiological conditions of leukotriene B_4 synth (see 5,12-Dihydroxy-6,8,10,14-eicosatetraenoic acid, D-00171).

[77297-91-7]

Osbond, J.M. *et al*, *J. Chem. Soc.*, 1961, 2779 (*synth*)
Schlenk, H. *et al*, *J. Am. Oil Chem. Soc.*, 1965, **42**, 481 (*occur*)
Wagner, H. *et al*, *Naturwissenschaften*, 1965, **52**, 305 (*occur*)
Frost, D.J. *et al*, *Chem. Phys. Lipids*, 1975, **15**, 53 (*pmr*)
Fryer, R.I. *et al*, *J. Org. Chem.*, 1975, **40**, 348 (*synth*)
Gunstone, F.D. *et al*, *Chem. Phys. Lipids*, 1976, **17**, 1 (*cmr*)
Henderson, J.D. *et al*, *Chem. Eng. News*, 1983, 2 (*haz*)
Corey, E.J. *et al*, *Tetrahedron Lett.*, 1984, **25**, 2729 (*purifn*)
Tanaka, Y. *et al*, *Arch. Biochem. Biophys.*, 1988, **263**, 178 (*synth*)
Devane, W.A. *et al*, *Science* (*Washington, D.C.*), 1992, **258**, 1946 (*Anandamide*)
Lewis, R.J., *Sax's Dangerous Properties of Industrial Materials*, 8th Ed., Van Nostrand-Reinhold, 1992, AQS750.

5,11,14,17-Eicosatetraenoic acid E-00035

5,11,14,17-Icosatetraenoic acid

[2271-31-0]

$H_3C(CH_2CH{=}CH)_3(CH_2)_4CH{=}CH(CH_2)_3COOH$

$C_{20}H_{32}O_2$ M 304.472

Occurs in Gymnospermae seed lipids.

(5Z,11Z,14Z,17Z)-form [18016-45-0]

Constit. of *Ephedra* spp. and *Caltha palustris* seed oils.

Kleiman, R. *et al*, *Chem. Ind.* (*London*), 1967, 1326 (*isol*)
Smith, C.R. *et al*, *Lipids*, 1968, **3**, 37 (*isol*)
Plattner, R.D. *et al*, *Lipids*, 1976, **11**, 222 (*struct*)
Takagi, T. *et al*, *Lipids*, 1982, **17**, 716 (*occur, glc, cmr*)

8,11,14,17-Eicosatetraenoic acid E-00036

8,11,14,17-Icosatetraenoic acid

[2091-26-1]

$H_3CCH_2(CH{=}CHCH_2)_4(CH_2)_5COOH$

$C_{20}H_{32}O_2$ M 304.472

(all-Z)-form [24880-40-8]

Present in fish oils and in animal phospholipids. Metab. of α-Linolenic acid.

Nicolasev, V. *et al*, *CA*, 1974, **81**, 117622.
Dhopeshwarkor, G. *et al*, *Lipids*, 1976, **11**, 689.

8,11,14,18-Eicosatetraenoic acid E-00037

8,11,14,18-Icosatetraenoic acid

$H_3CCH{=}CHCH_2(CH_2CH{=}CH)_3(CH_2)_6COOH$

$C_{20}H_{32}O_2$ M 304.472

Beerthius, R.K. *et al*, *Recl. Trav. Chim. Pays-Bas* (*J. R. Neth. Chem. Soc.*), 1971, **90**, 943 (*synth*)

4,8,11,14-Eicosatetraynoic acid E-00038

4,8,11,14-Icosatetraynoic acid, 9CI

[32895-77-5]

$H_3C(CH_2)_4(C{\equiv}CCH_2)_3CH_2C{\equiv}CCH_2CH_2COOH$

$C_{20}H_{24}O_2$ M 296.408

Cryst. (EtOH or EtOAc/pet. ether). Mp 83.5-85.5°.

Beerthuis, R.K. *et al*, *Recl. Trav. Chim. Pays-Bas* (*J. R. Neth. Chem. Soc.*), 1971, **90**, 943 (*synth*)
Bus, J. *et al*, *Recl. Trav. Chim. Pays-Bas* (*J. R. Neth. Chem. Soc.*), 1972, **91**, 394 (*ir*)
Sprecher, H., *Prog. Chem. Fats Other Lipids*, 1978, **15**, 219.

5,8,11,14-Eicosatetraynoic acid E-00039

5,8,11,14-Icosatetraynoic acid, 9CI

[1191-85-1]

$H_3C(CH_2)_4C{\equiv}CCH_2C{\equiv}CCH_2C{\equiv}CCH_2C{\equiv}C(CH_2)_3COOH$

$C_{20}H_{24}O_2$ M 296.408

Cryst. (pet. ether or MeOH). Mp 81-82°.

Me ester: Mp 10-12°. Bp_3 180-185°.

Et ester: $Bp_{0.5}$ 138-140°.

Amide: Cryst. (MeOH). Mp 104-106.5°.

Pleshakov, M.G. *et al*, *Zh. Obshch. Khim.*, 1960, **30**, 2983.
Osbond, J.M. *et al*, *J. Chem. Soc.*, 1961, 2779 (*synth*)
Bus, J. *et al*, *Recl. Trav. Chim. Pays-Bas* (*J. R. Neth. Chem. Soc.*), 1972, **91**, 394 (*ir*)
Sprecher, H., *Prog. Chem. Fats Other Lipids*, 1978, **15**, 219.

8,11,14,18-Eicosatetraynoic acid E-00040

8,11,14,18-Icosatetraynoic acid, 9CI

[32895-82-2]

$H_3CC{\equiv}CCH_2CH_2C{\equiv}CCH_2C{\equiv}CCH_2C{\equiv}C(CH_2)_6COOH$

$C_{20}H_{24}O_2$ M 296.408

Cryst. (EtOAc or EtOH). Mp 90-91.5°.

Beerthuis, R.K. *et al*, *Recl. Trav. Chim. Pays-Bas* (*J. R. Neth. Chem. Soc.*), 1971, **90**, 943 (*synth*)
Bus, J. *et al*, *Recl. Trav. Chim. Pays-Bas* (*J. R. Neth. Chem. Soc.*), 1972, **91**, 394 (*ir*)
Sprecher, H., *Prog. Chem. Fats Other Lipids*, 1978, **15**, 219.

3,6,9-Eicosatriene, 9CI E-00041

3,6,9-Icosatriene

[134370-60-8]

$H_3CCH_2CH{=}CHCH_2CH{=}CHCH_2CH{=}CH(CH_2)_9CH_3$

$C_{20}H_{36}$ M 276.504

(3Z,6Z,9Z)-form [87255-14-9]

Component of the sex pheromones of several Lepidoptera species; principal component of the sex pheromone of the Spring cankerworm moth, *Paleacrita vernata*. Major component (with 3,6,9-Heneicosatriene, H-00016) of the sex pheromones of the velvetbean caterpillar moth, *Anticarsia gemmatalis*, and of forage looper *Caenurgina erechtea*.

*9*S,*10*R*-Epoxide:* [97719-76-1]. *2-Decyl-3-(2,5-octadienyl) oxirane, 9CI. 9,10-Epoxy-3,6-eicosadiene*

$C_{20}H_{36}O$ M 292.504

Component of the sex pheromone of moths *Euclidea cuspidea*, *C. distincta*, *Eupithecia annulata* and *Tyria jacobaeae*. Synergistic activity shown with 3,6,9-Eicosatriene, and 2-(2,5-Octadienyl)-3-undecyloxirane, O-00235.

[97668-91-2, 97719-66-9, 125535-23-1, 139405-37-1]

Heath, R.R. *et al*, *J. Chem. Ecol.*, 1983, **9**, 645 (*isol, synth*)
Underhill, E.W. *et al*, *J. Chem. Ecol.*, 1983, **9**, 1413 (*isol, synth, activity*)
Wong, J.W. *et al*, *J. Chem. Ecol.*, 1985, **11**, 727 (*isol, synth, epoxide*)
Mitchell, E.C. *et al*, *J. Econ. Entomol.*, 1986, **79**, 289 (*activity*)
Bestmann, H.J. *et al*, *Justus Liebigs Ann. Chem.*, 1987, 417 (*synth, pmr, ir, ms*)
Yu, H.M. *et al*, *Chem. Ind.* (*London*), 1988, 787, 836 (*synth*)
Frerot, B. *et al*, *Entomol. Exp. Appl.*, 1988, **46**, 281 (*isol, synth, activity, epoxide*)
Millar, J.G. *et al*, *J. Chem. Ecol.*, 1990, **16**, 3393; 1991, **17**, 2095 (*isol, activity*)

2,4,8-Eicosatrienoic acid E-00042

2,4,8-Icosatrienoic acid

$H_3C(CH_2)_{10}CH{=}CHCH_2CH_2(CH{=}CH)_2COOH$

$C_{20}H_{34}O_2$ M 306.487

(2E,4E,8Z)-form

2-Methylpropylamide: [64543-30-2]. ***2,4,8-Eicosatrienoic acid isobutylamide.*** N*-Isobutyl-2,4,8-eicosatrienamide*

$C_{24}H_{43}NO$ M 361.610

Alkaloid from *Piper officinarum* (Piperaceae). Mp 67-67.5°.

Gupta, O.P. *et al*, *Phytochemistry*, 1977, **16**, 1436 (*uv, ir, pmr, ms, isol, struct*)

5,8,11-Eicosatrienoic acid **E-00043**

5,8,11-Icosatrienoic acid

$H_3C(CH_2)_7CH{=}CHCH_2CH{=}CHCH_2CH{=}CH(CH_2)_3COOH$

$C_{20}H_{34}O_2$ M 306.487

(all-Z)-form [20590-32-3]

Mead acid

Metab. of Oleic acid.

Struijk, C.B. *et al, Recl. Trav. Chim. Pays-Bas (J. R. Neth. Chem. Soc.)*, 1966, **85**, 1233, 1241 (*synth*)
Ghosh, A. *et al, Lipids*, 1982, **17**, 314 (*synth, pmr*)

5,11,14-Eicosatrienoic acid **E-00044**

5,11,14-Icosatrienoic acid

$H_3C(CH_2)_4CH{=}CHCH_2CH{=}CH(CH_2)_4CH{=}CH(CH_2)_3COOH$

$C_{20}H_{34}O_2$ M 306.487

(all-Z)-form

Podocarpic acid (*obsol.*)

Isol. from tall oil (*Pinus sylvestris*), seed oil of *Podocarpus nageia* and from *Ginkgo biloba*.

Me ester:

$C_{21}H_{36}O_2$ M 320.514

n_D^{32} 1.4660.

Ito, S. *et al, Bull. Chem. Soc. Jpn.*, 1963, **36**, 1439 (*isol, uv, ir*)
Gellerman, J.L. *et al, Experientia*, 1963, **19**, 522 (*isol*)
Takagi, T. *et al, Lipids*, 1982, **17**, 716 (*occur, glc, cmr*)

7,10,13-Eicosatrienoic acid **E-00045**

7,10,13-Icosatrienoic acid

[34025-13-3]

$H_3C(CH_2)_5CH{=}CHCH_2CH{=}CHCH_2CH{=}CH(CH_2)_5COOH$

$C_{20}H_{34}O_2$ M 306.487

(all-Z)-form [32907-94-1]

Metab. of 9-Hexadecenoic acid.

Struijk, C.B. *et al, Recl. Trav. Chim. Pays-Bas (J. R. Neth. Chem. Soc.)*, 1966, **85**, 1233 (*synth*)
Budny, J. *et al, Biochim. Biophys. Acta*, 1971, **239**, 190.

7,11,14-Eicosatrienoic acid **E-00046**

7,11,14-Icosatrienoic acid

[73675-34-0]

$H_3C(CH_2)_4CH{=}CHCH_2CH{=}CHCH_2$
$CH_2CH{=}CH(CH_2)_5COOH$

$C_{20}H_{34}O_2$ M 306.487

Isol. from *Picea abies*. Stereochem. undetermined.

(7Z,11Z,14E)-form [81861-75-8]

Bishomocolumbinic acid

(7Z,11Z,14Z)-form [107167-33-9]

Isol. from *P. orientalis*.

Ekman, R., *Phytochemistry*, 1980, **19**, 147 (*isol*)
Houtsmuller, U.M.T., *Prog. Lipids Res.*, 1981, **20**, 889.
Demirbas, A., *Chim. Acta Turc.*, 1985, **13**, 143 (*isol*)

8,11,14-Eicosatrienoic acid **E-00047**

8,11,14-Icosatrienoic acid

[7324-41-6]

$H_3C(CH_2)_4CH{=}CHCH_2CH{=}CHCH_2CH{=}CH(CH_2)_6COOH$

$C_{20}H_{34}O_2$ M 306.487

(all-Z)-form [1783-84-2]

Dihomogammalinolenic acid

Minor component of animal phospholipids. Metab. of Linoleic acid. Precursor of PG_1 series of prostaglandins.

Struijk, C.B. *et al, Recl. Trav. Chim. Pays-Bas (J. R. Neth. Chem. Soc.)*, 1966, **85**, 1241 (*synth*)

8,12,14-Eicosatrienoic acid **E-00048**

8,12,14-Icosatrienoic acid

$H_3C(CH_2)_4CH{=}CHCH{=}CHCH_2CH_2CH{=}CH(CH_2)_6COOH$

$C_{20}H_{34}O_2$ M 306.487

(8Z,12E,14Z)-form [26322-26-9]

Cryst. (MeOH at −30°). Mp 8°.

Et ester: [33001-41-1].

Oil. $Bp_{0.005}$ 135-140°.

Beerthius, R.K. *et al, Recl. Trav. Chim. Pays-Bas (J. R. Neth. Chem. Soc.)*, 1971, **90**, 943 (*synth*)

5,8,14-Eicosatrien-11-ynoic acid **E-00049**

5,8,14-Icosatrien-11-ynoic acid

$H_3C(CH_2)_4CH{=}CHCH_2C{\equiv}CCH_2CH{=}CH$
$CH_2CH{=}CH(CH_2)_3COOH$

$C_{20}H_{30}O_2$ M 302.456

Irreversibly inhibits leukotriene and prostaglandin biosynth.

Corey, E.J. *et al, Tetrahedron Lett.*, 1982, **23**, 1651 (*synth*)

5,11,14-Eicosatrien-8-ynoic acid **E-00050**

5,11,14-Icosatrien-8-ynoic acid

$H_3C(CH_2)_4CH{=}CHCH_2CH{=}CHCH_2$
$C{\equiv}CCH_2CH{=}CH(CH_2)_3COOH$

$C_{20}H_{30}O_2$ M 302.456

Corey, E.J. *et al, Tetrahedron Lett.*, 1982, **23**, 1651 (*synth*)

8,11,14-Eicosatrien-5-ynoic acid **E-00051**

8,11,14-Icosatrien-5-ynoic acid. 5,6-Dehydroarachidonic acid

$H_3C(CH_2)_4CH{=}CHCH_2CH{=}CHCH_2$
$CH{=}CHCH_2C{\equiv}C(CH_2)_3COOH$

$C_{20}H_{30}O_2$ M 302.456

(all-Z)-form [58688-54-3]

Irreversibly inhibits leukotriene and prostaglandin biosynth. Oil.

Hedlinga, L. *et al, Recl. Trav. Chim. Pays-Bas (J. R. Neth. Chem. Soc.)*, 1975, **94**, 262 (*synth, ir, pmr*)
Corey, E.J. *et al, Tetrahedron Lett.*, 1980, **21**, 4243; 1982, **23**, 1651; 1984, **25**, 2419 (*synth, bibl*)
Corey, E.J. *et al, J. Am. Chem. Soc.*, 1982, **104**, 1752 (*synth, biochem*)

5,8,11-Eicosatriynoic acid **E-00052**

5,8,11-Icosatriynoic acid, 9CI

[13488-22-7]

$H_3C(CH_2)_7C{\equiv}CCH_2C{\equiv}CCH_2C{\equiv}C(CH_2)_3COOH$

$C_{20}H_{28}O_2$ M 300.440

Mp 70.5-71.5°.

Struijk, C.B. *et al, Recl. Trav. Chim. Pays-Bas (J. R. Neth. Chem. Soc.)*, 1966, **85**, 1233 (*synth*)

Bus, J. *et al*, *Recl. Trav. Chim. Pays-Bas* (*J. R. Neth. Chem. Soc.*), 1972, **91**, 394 (*ir*)
Sprecher, H., *Prog. Chem. Fats Other Lipids*, 1978, **15**, 219.

6,9,12-Eicosatriynoic acid E-00053

6,9,12-Icosatriynoic acid, 9CI
[13488-21-6]

$H_3C(CH_2)_6C{\equiv}CCH_2C{\equiv}CCH_2C{\equiv}C(CH_2)_4COOH$

$C_{20}H_{28}O_2$ M 300.440
Mp 50-51°.

Struijk, C.B. *et al*, *Recl. Trav. Chim. Pays-Bas* (*J. R. Neth. Chem. Soc.*), 1966, **85**, 1233 (*synth*)
Bus, J. *et al*, *Recl. Trav. Chim. Pays-Bas* (*J. R. Neth. Chem. Soc.*), 1972, **91**, 394 (*ir*)
Sprecher, H., *Prog. Chem. Fats Other Lipids*, 1978, **15**, 219.

7,10,13-Eicosatriynoic acid E-00054

7,10,13-Icosatriynoic acid, 9CI
[13488-20-5]

$H_3C(CH_2)_5C{\equiv}CCH_2C{\equiv}CCH_2C{\equiv}C(CH_2)_5COOH$

$C_{20}H_{28}O_2$ M 300.440
Mp 63.5-65.5°.

Struijk, C.B. *et al*, *Recl. Trav. Chim. Pays-Bas* (*J. R. Neth. Chem. Soc.*), 1966, **85**, 1233 (*synth*)
Bus, J. *et al*, *Recl. Trav. Chim. Pays-Bas* (*J. R. Neth. Chem. Soc.*), 1972, **91**, 394 (*ir*)
Sprecher, H., *Prog. Chem. Fats Other Lipids*, 1978, **15**, 219.

9,12,15-Eicosatriynoic acid E-00055

9,12,15-Icosatriynoic acid, 9CI
[13488-19-2]

$H_3C(CH_2)_3C{\equiv}CCH_2C{\equiv}CCH_2C{\equiv}C(CH_2)_7COOH$

$C_{20}H_{28}O_2$ M 300.440
Mp 65-66°.

Struijk, C.B. *et al*, *Recl. Trav. Chim. Pays-Bas* (*J. R. Neth. Chem. Soc.*), 1966, **85**, 1233 (*synth*)
Bus, J. *et al*, *Recl. Trav. Chim. Pays-Bas* (*J. R. Neth. Chem. Soc.*), 1972, **91**, 394 (*ir*)
Sprecher, H., *Prog. Chem. Fats Other Lipids*, 1978, **15**, 219.

1-Eicosene, 9CI E-00056

1-Icosene
[3452-07-1]

$H_3C(CH_2)_{17}CH{=}CH_2$

$C_{20}H_{40}$ M 280.536
Present in small amount in rose oil. Fp 28.5°. $Bp_{1.5}$ 151°. n_D^{30} 1.4440.

Niemann, C. *et al*, *J. Org. Chem.*, 1942, **7**, 227 (*synth*)
Wollrab, V. *et al*, *Collect. Czech. Chem. Commun.*, 1965, **30**, 1654 (*isol*)

5-Eicosenoic acid, 9CI E-00057

5-Icosenoic acid
[7329-42-2]

$H_3C(CH_2)_{13}CH{=}CH(CH_2)_3COOH$

$C_{20}H_{38}O_2$ M 310.519
(E)-form [35237-01-5]
Waxy solid. Mp 52.5-54°.
(Z)-form [7050-07-9]
cis-*form*
Isol. from seed oils of *Limnanthes douglasii*, *Thalictrum venulosum* and *Delphinium consolida*. Waxy solid or cryst. (MeOH aq.). Mp 26-27°.
[20839-34-3, 69119-84-2, 116430-55-8]

Smith, C.R. *et al*, *J. Org. Chem.*, 1960, **25**, 1770 (*isol*)
Bhatty, M.K. *et al*, *Can. J. Biochem.*, 1966, **44**, 311 (*isol*)
Mitcham, D. *et al*, *J. Am. Oil Chem. Soc.*, 1973, **50**, 446 (*ir*)
Bailey, A.V. *et al*, *J. Am. Oil Chem. Soc.*, 1975, **52**, 196 (*cryst struct*)
Young, R.N. *et al*, *Tetrahedron Lett.*, 1981, 4933 (*synth*)
Okuyama, S. *et al*, *Chem. Pharm. Bull.*, 1982, **30**, 2453 (*pmr*)
Levin, D. *et al*, *J. Chem. Soc., Perkin Trans. 1*, 1988, 1799 (*synth, ir, pmr*)

9-Eicosenoic acid, 9CI E-00058

9-Icosenoic acid
[506-31-0]

$H_3C(CH_2)_9CH{=}CH(CH_2)_7COOH$

$C_{20}H_{38}O_2$ M 310.519
(E)-form
Gadelaidic acid
Cryst. (MeOH). Mp 54°.
Amide:
$C_{20}H_{39}NO$ M 309.534
Mp 90-91°.
(Z)-form [29204-02-2] ***Gadoleic acid***
Constit. of fish oil glycerides. Mp 23-23.5°. $Bp_{0.1}$ 170°.
Et ester:
$C_{22}H_{42}O_2$ M 338.573
$Bp_{0.1}$ 154-155°.
Amide: Mp 78-79°.

Boughton, B.W. *et al*, *J. Chem. Soc.*, 1952, 671 (*synth*)

11-Eicosenoic acid, 9CI E-00059

11-Icosenoic acid
[2462-94-4]

$H_3C(CH_2)_7CH{=}CH(CH_2)_9COOH$

$C_{20}H_{38}O_2$ M 310.519
(E)-form
Mp 49-51°.
(Z)-form [5561-99-9] ***Gondoic acid***
Constit. of rape oil and fish oils as glyceride. Also in other plant oils, eg. *Camelina sativa*, *Brassica napobrassica*. Cryst. (Me_2CO/hexane). Mp 24-25°.
Me ester: [2390-09-2].
$C_{21}H_{40}O_2$ M 324.546
Mp −45°.

Fieser, L.F. *et al*, *J. Am. Chem. Soc.*, 1948, **70**, 71 (*synth*)
Argoudelis, C.J. *et al*, *Lipids*, 1968, **3**, 379 (*struct, ms*)
Karrer, W. *et al*, *Konstitution und Vorkommen der Organischen Pflanzenstoffe*, 2nd Ed., Birkhäuser Verlag, Basel, 1972-1985, no. 742 (*occur*)
Spencer, F. *et al*, *Chem. Phys. Lipids*, 1973, **11**, 215 (*synth*)

13-Eicosenoic acid, 9CI E-00060

13-Icosenoic acid
[14134-51-1]

$H_3C(CH_2)_5CH{=}CH(CH_2)_{11}COOH$

$C_{20}H_{38}O_2$ M 310.519
(Z)-form [17735-94-3]
Isol. from herring oil and rapeseed oil.
[69119-92-2, 69120-02-1, 82683-17-8]

Kishimoto, Y. *et al*, *J. Lipid Res.*, 1964, **5**, 98 (*isol, struct*)
Morales, R.W. *et al*, *Biochim. Biophys. Acta*, 1976, **431**, 206 (*isol*)
Richter, I. *et al*, *Z. Naturforsch., C*, 1978, **33**, 629 (*synth*)
Yu, Q.T. *et al*, *Lipids*, 1989, **24**, 79 (*ms*)

14-Eicosenoic acid E-00061

14-Icosenoic acid

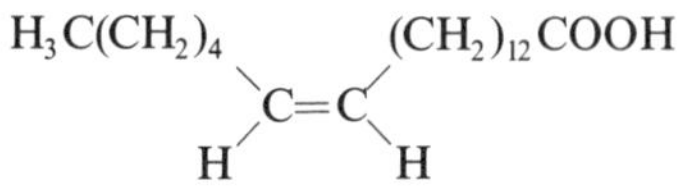

(*Z*)-*form*

$C_{20}H_{38}O_2$ M 310.519

(*Z*)-*form* [17735-95-4]

Cryst. (EtOH). Mp 42.5°.

Baker, C.D. *et al*, *J. Chem. Soc.*, 1963, 489 (*synth*)
Ackman, R.G. *et al*, *J. Gas Chromatogr.*, 1967, **5**, 489 (*chromatog*)

15-Eicosenoic acid E-00062

15-Icosenoic acid

[2654-73-1]

$H_3C(CH_2)_3CH{=}CH(CH_2)_{13}COOH$

$C_{20}H_{38}O_2$ M 310.519

(*Z*)-*form* [34739-80-5]

A minor component of lipids from marine sources.

Ackman, R.G. *et al*, *Lipids*, 1966, **1**, 341.
Vickery, J.R., *Phytochemistry*, 1971, **10**, 123.

17-Eicosenoic acid E-00063

17-Icosenoic acid

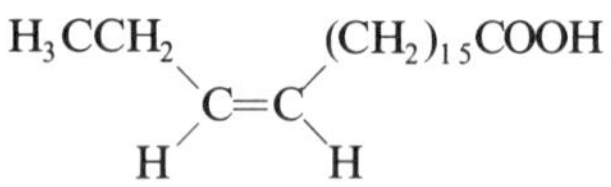

(*Z*)-*form*

$C_{20}H_{38}O_2$ M 310.519

Cryst. (Me_2CO). Mp 50-51°.

(*Z*)-*form* [55833-23-3]

Klok, R. *et al*, *Recl. Trav. Chim. Pays-Bas (J. R. Neth. Chem. Soc.)*, 1974, **93**, 222 (*synth, pmr*)

11-Eicosen-1-ol, 9CI E-00064

11-Icosen-1-ol

[87268-65-3]

$H_3C(CH_2)_7CH{=}CH(CH_2)_9CH_2OH$

$C_{20}H_{40}O$ M 296.535

(*E*)-*form* [68760-58-7]

Cryst. (hexane). Mp 43-44°.

(*Z*)-*form* [62442-62-0]

Present in seeds of *Simmondsia chinensis* (jojoba), *S. californica*, *Halobates hayanus*, *Rheumatobates aestuarius*, and other natural sources. Pheromone isolated from thoracic glands of male carpenter bees *Xylocopa micheneri*. Plates (pentane at −10°). Mp 25-26°.

Ac:

$C_{22}H_{42}O_2$ M 338.573

Cryst. (pentane at −10°). Mp 19.5°.

Kobayashi, A. *et al*, *Agric. Biol. Chem.*, 1978, **42**, 1973 (*synth, ir, glc, pmr, acetate*)
Shani, A., *J. Chem. Ecol.*, 1979, **5**, 557 (*pmr*)
Pickett, J.A. *et al*, *J. Chem. Ecol.*, 1982, **8**, 163 (*glc, ms*)
Tagabi, T. *et al*, *Lipids*, 1985, **20**, 675 (*glc*)
Andersen, J.F. *et al*, *J. Chem. Ecol.*, 1988, **14**, 1153.

13-Eicosen-10-one E-00065

13-Icosen-10-one

$H_3C(CH_2)_8COCH_2CH_2CH{=}CH(CH_2)_5CH_3$

$C_{20}H_{38}O$ M 294.520

(*Z*)-*form* [63408-44-6]

Pheromone of Japanese peach fruit moth *Carposia niponensis*. Pale-yellow liq.

Tamaki, Y. *et al*, *Appl. Entomol. Zool.*, 1977, **12**, 60.
Yoshida, T. *et al*, *Bull. Chem. Soc. Jpn.*, 1982, **55**, 3047 (*synth*)
Sviridov, A.F. *et al*, *Tetrahedron Lett.*, 1983, **24**, 4359 (*synth*)
Yamamoto, T. *et al*, *J. Org. Chem.*, 1989, **54**, 747 (*synth, ir, pmr, cmr*)
Kang, S.K. *et al*, *Org. Prep. Proced. Int.*, 1990, **22**, 122 (*synth*)

Eicosylpropanedioic acid E-00066

Heneicosane-1,1-dicarboxylic acid. Icosylmalonic acid

$H_3C(CH_2)_{19}CH(COOH)_2$

$C_{23}H_{44}O_4$ M 384.598

Spar. sol. $CHCl_3$. Mp 119°. Evolves CO_2 at 150°.

Mononitrile: 2-Cyanodocosanoic acid

$C_{23}H_{43}NO_2$ M 365.598

Mp 87-89°.

Fileti, M., *Gazz. Chim. Ital.*, 1910, **27**, 302.

1-Eicosyne E-00067

1-Icosyne

[765-27-5]

$H_3C(CH_2)_{17}C{\equiv}CH$

$C_{20}H_{38}$ M 278.520

Mp 34-35°.

Gensler, W.J. *et al*, *J. Am. Chem. Soc.*, 1956, **78**, 163 (*synth*)
Streibl, M. *et al*, *Fette, Seifen, Anstrichm.*, 1968, **70**, 343 (*isol*)
Luftmann, H. *et al*, *Org. Mass Spectrom.*, 1971, **5**, 1073 (*ms*)
Huang, M.C. *et al*, *J. Org. Chem.*, 1982, **47**, 4018 (*synth, pmr*)

9,11-Epidioxy-15-hydroperoxy-13-prosten-1-oic acid, 9CI E-00068

6-(3-Hydroperoxy-1-octenyl)-2,3-dioxabicyclo[2.2.1]heptane-5-heptanoic acid, 8CI

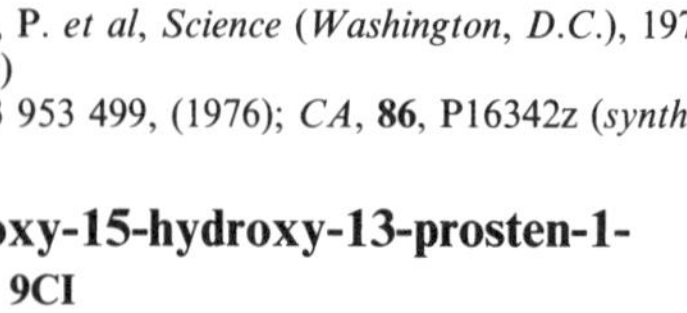

$C_{20}H_{34}O_6$ M 370.485

(*9S,11R,13E,15S*)-*form* [52162-11-5]

PGG$_1$

Formed by enzymes of seminal vesicles.

Needleman, P. *et al*, *Science (Washington, D.C.)*, 1976, **193**, 163 (*biosynth*)
U.S. Pat., 3 953 499, (1976); *CA*, **86**, P16342z (*synth*)

9,11-Epidioxy-15-hydroxy-13-prosten-1-oic acid, 9CI E-00069

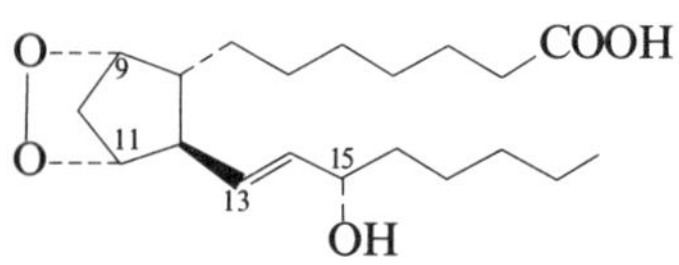

$C_{20}H_{34}O_5$ M 354.486

(9S,11R,13E,15S)-form

PGH_1

Formed by enzymes of seminal vesicles. Induces contraction of bovine coronary strips.

Needleman, P. *et al*, *Science (Washington, D.C.)*, 1976, **193**, 163 (*biosynth*)
Gorman, R.R. *et al*, *Prostaglandins*, 1977, **13**, 1043 (*synth, ms, ir*)

3,6-Epidioxy-6-methoxy-4,16,18-eicosatrienoic acid **E-00070**

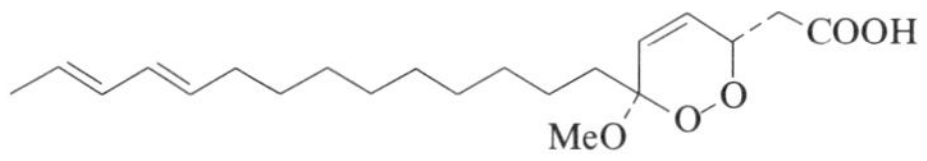

$C_{21}H_{34}O_5$ M 366.497

Me ester: Methyl 3,6-epidioxy-6-methoxy-4,16,18-eicosatrienoate

$C_{22}H_{36}O_5$ M 380.523

Constit. of sponge *Plakortis lita*. Antitumour agent. Solid. Mp 47.5°. $[\alpha]_D^{20}$ +36.4° (c, 1.1 in MeOH).

Sakemi, S. *et al*, *Tetrahedron*, 1987, **43**, 263.

3,6-Epidioxy-6-methoxy-4-octadecenoic acid **E-00071**

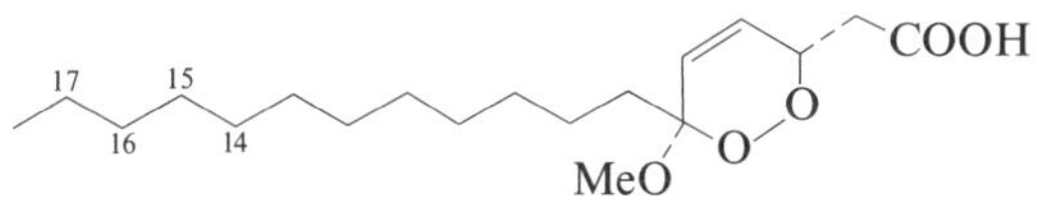

$C_{19}H_{34}O_5$ M 342.475

Me ester: Methyl 3,6-epidioxy-6-methoxy-4-octadecenoate

$C_{20}H_{36}O_5$ M 356.501

Constit. of sponge *Plakortis lita*. Antitumour agent. Solid. Mp 49°. $[\alpha]_D^{20}$ +38.1° (c, 5.3 in MeOH).

16,17-Didehydro, Me ester: Methyl 3,6-epidioxy-6-methoxy-4,16-octadecadienoate

$C_{20}H_{34}O_5$ M 354.486

Constit. of *P. lita*. Antitumour agent. Solid. Mp 37.5°. $[\alpha]_D^{20}$ +41.4° (c, 2.9 in MeOH).

14,15,16,17-Tetradehydro, Me ester: Methyl 3,6-epidioxy-6-methoxy-4,14,16-octadecatrienoate

$C_{20}H_{32}O_5$ M 352.470

Constit. of *P. lita*. Antitumour agent. Solid. Mp 38-39°. $[\alpha]_D^{20}$ +40.8° (c, 4.9 in MeOH).

Sakemi, S. *et al*, *Tetrahedron*, 1987, **43**, 263.

6,9-Epithio-11,15-dihydroxy-5,13-prostadienoic acid **E-00072**

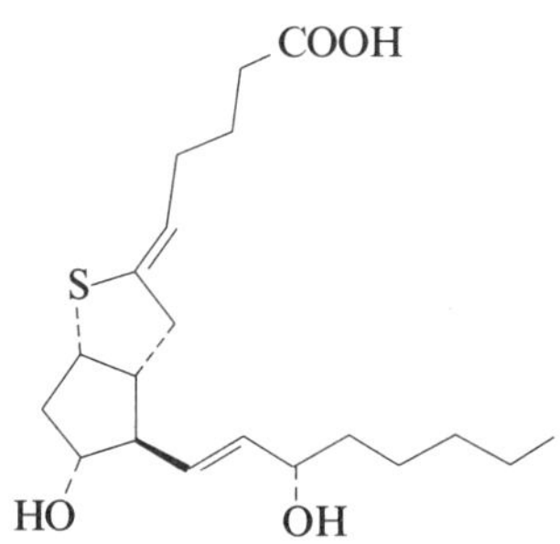

$C_{20}H_{30}O_4S$ M 366.521

(5Z,9S,11R,13E,15S)-form [64868-63-9]

6,9-Thiaprostacyclin. PGI_2-S. $TPGI_2$

Sulphur analogue of Prostacyclin. Inhibits blood platelet aggregation and is a potent vasoconstrictor. Antihypertensive agent. Oil.

Me ester: Oil. $[\alpha]_D^{25}$ +58.4° (c, 0.011 in MeOH).

Shibasaki, M. *et al*, *Tetrahedron Lett.*, 1978, 559 (*synth, ir, pmr, ms*)
Shimoji, K. *et al*, *Chem. Lett.*, 1978, 1375 (*synth*)
Lefer, A.M. *et al*, *Life Sci.*, 1979, **25**, 259 (*pharmacol*)
Nicolaou, K.C. *et al*, *J. Am. Chem. Soc.*, 1981, **103**, 3472 (*synth, pmr*)
Yokomori, H. *et al*, *Heterocycles*, (Spec. Issue), 1982, **18**, 251 (*synth*)

6,9-Epoxy-11,15-dihydroxy-13-prostenoic acid, 9CI **E-00073**

5,6-Dihydroprostacyclin. Prostaglandin I_1

[67008-19-9]

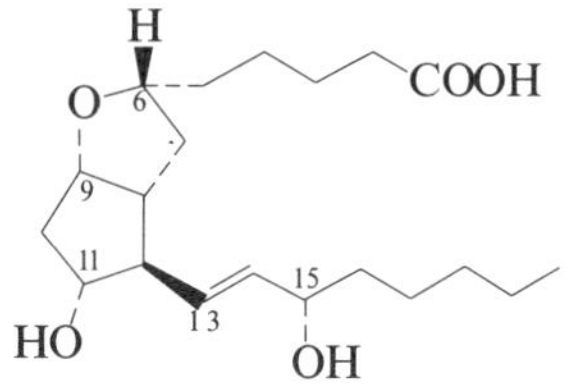

(6*R*,9*S*,11*R*,13*E*,15*S*)-*form*

$C_{20}H_{34}O_5$ M 354.486

Shows antilipolytic activity.

(6R,9S,11R,13E,15S)-form [62777-90-6]

Synthetic stable analogue of PGI_2 which inhibits blood platelet aggregation. Mp 97-99°.

Me ester: Silky needles. Mp 71-73°. $[\alpha]_D$ +13° (c, 0.8245 in $CHCl_3$).

(6S,9S,11R,13E,15S)-form [62770-50-7]

Stable synthetic analogue of PGI_2 which inhibits blood platelet aggregation. Mp 78-81°.

Me ester: [63557-52-8].

Mp 43°. $[\alpha]_D$ +23° (c, 0.8715 in $CHCl_3$).

Corey, E.J. *et al*, *J. Am. Chem. Soc.*, 1977, **99**, 2006 (*synth*)
Johnson, R.A. *et al*, *J. Am. Chem. Soc.*, 1977, **99**, 4182; 1978, **100**, 7690 (*synth, pmr*)
Olson, C.V. *et al*, *Biochim. Biophys. Acta*, 1985, **845**, 411 (*biochem*)
Tan, S. *et al*, *Prostaglandins, Leukotrienes Med.*, 1985, **17**, 117 (*biochem*)

3,4-Epoxy-2-dodecanone **E-00074**

1-(3-Octyloxiranyl)ethanone, 9CI. 2-Acetyl-3-octyloxirane

[126912-72-9]

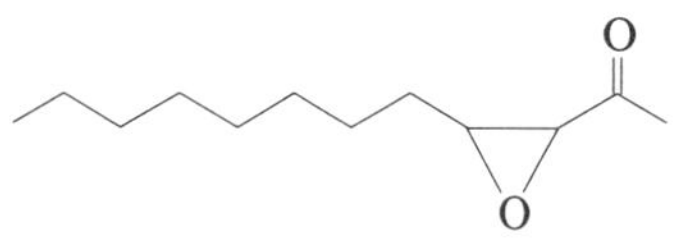

$C_{12}H_{22}O_2$ M 198.305

Sex pheromone constit. of the preorbital gland secretion of the Grey Duiker *Sylvicapra grimmia* (antelope).

Burger, B.V. *et al*, *J. Chem. Ecol.*, 1990, **16**, 397.

## 5,6-Epoxy-7,9,11,14,17-eicosapentaenoic acid	E-00075

3-(1,3,5,8,11-Tetradecapentaenyl)-oxiranebutanoic acid. ***Leukotriene*** $\boldsymbol{A_5}$
[83952-39-0]

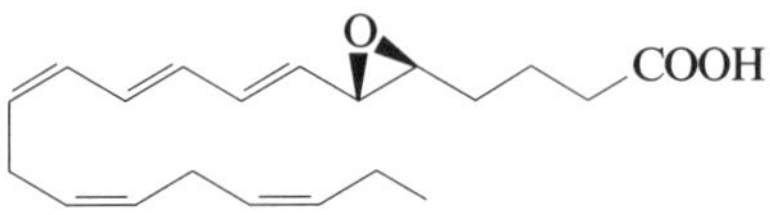

$C_{20}H_{28}O_3$ M 316.439
Substrate and inhibitor of rat and human neutrophil Leukotriene A_4 (see 5,6-Epoxy-7,9,11,14-eicosatetraenoic acid, E-00076). Unstable.
Me ester: [90121-06-5].
$C_{21}H_{30}O_3$ M 330.466

Hammarstroem, S., *J. Biol. Chem.*, 1983, **258**, 1427 (*biosynth*)
Spur, B. *et al*, *Arch. Pharm. (Weinheim, Ger.)*, 1984, **317**, 280 (*synth*)
Nathaniel, D.J. *et al*, *Biochem. Biophys. Res. Commun.*, 1985, **131**, 827 (*synth, biochem*)

## 5,6-Epoxy-7,9,11,14-eicosatetraenoic acid	E-00076

3-(1,3,5,8-Tetradecatetraenyl)-oxiranebutanoic acid, 9CI
[71548-17-9]

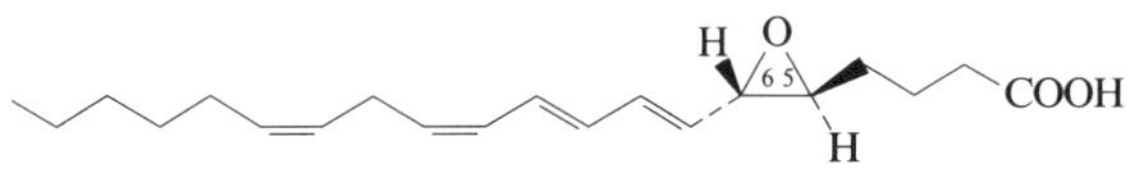

(5*S*,6*S*,7*E*,9*E*,11*Z*,14*Z*)-*form*

$C_{20}H_{30}O_3$ M 318.455
(5S,6S,7E,9E,11Z,14Z)-form [72059-45-1] ***Leukotriene*** $\boldsymbol{A_4}$**.** *LTA*$_4$
$[\alpha]^{25}_{D}$ $-2.19°$ (c, 0.32 in cyclohexane).
Me ester: [73466-12-3].
$C_{21}H_{32}O_3$ M 332.482
Mp 28-32°. $[\alpha]^{25}_{D}$ $-27°$ (hexane). Uv λ_{max} 290 (ε 43100), 278 (ε 56700), 270 (ε 43900) nm.
(5S,6S,7E,9Z,11Z,14Z)-form
Metab. of arachidonic acid.
(5R,6S,7E,9E,11Z,14Z)-form
5-epi-*Leukotriene* A_4
Me ester: [79356-68-6].
$[\alpha]_D$ $+18.1°$ (c, 0.5 in hexane).
(5S,6R,7E,9E,11Z,14Z)-form
6-epi-*Leukotriene* A_4
Me ester: [76738-49-3].
$[\alpha]_D$ $-21.5°$ (c, 0.2 in hexane).
(5R,6R,7E,9E,11Z,14Z)-form
5-epi, 6-epi-*Leukotriene* A_4
Me ester: [79356-65-3].
$[\alpha]_D$ $+28.0°$ (c, 1.0 in hexane).
(5RS,6RS,7Z,9E,11Z,14Z)-form
Oil.

Corey, E.J. *et al*, *J. Am. Chem. Soc.*, 1980, **102**, 1436 (*synth*)
Rokach, J. *et al*, *Tetrahedron Lett.*, 1981, **22**, 2759 (*synth*)
Atrache, V. *et al*, *Tetrahedron Lett.*, 1981, **22**, 3443 (*synth*)
Ernest, I. *et al*, *Tetrahedron Lett.*, 1982, **23**, 167 (*synth*)
Cohen, N. *et al*, *J. Am. Chem. Soc.*, 1983, **105**, 3661 (*synth, uv, pmr, bibl*)
Wang, Y. *et al*, *Tetrahedron Lett.*, 1986, **27**, 4583 (*synth*)
Baker, S.R. *et al*, *Tetrahedron Lett.*, 1988, **29**, 991 (*synth*)
Corey, E.J. *et al*, *J. Am. Chem. Soc.*, 1989, **111**, 1452 (*biosynth*)

## 11,12-Epoxy-5,7,9,14-eicosatetraenoic acid	E-00077

[3-(2-Octenyl)oxiranyl]-5,7,9-decatrienoic acid
[75290-57-2]

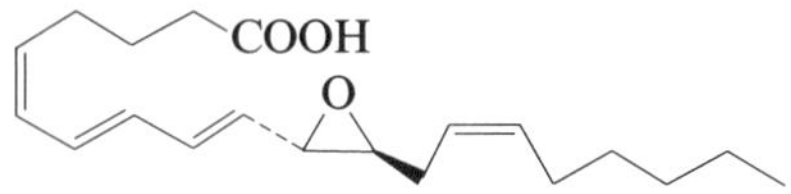

$C_{20}H_{30}O_3$ M 318.455
(5Z,7E,9E,14Z,11S,12S)-form [95722-34-2]
11,12-Leukotriene A_4. *11,12-LTA*$_4$
Metab. of arachidonic acid. More unstable than 5,6-Epoxy-7,9,11,14-eicosatetraenoic acid, E-00076 and 14,15-Epoxy-5,8,10,12-eicosatetraenoic acid, E-00078.
Me ester: [95645-85-5].
$[\alpha]_D$ $-14.6°$ (c, 4 in hexane).

Corey, E.J. *et al*, *J. Am. Chem. Soc.*, 1980, **102**, 6607 (*synth*)
Zamboni, R. *et al*, *Tetrahedron Lett.*, 1984, **25**, 5835 (*synth*)

## 14,15-Epoxy-5,8,10,12-eicosatetraenoic acid	E-00078

13-(3-Pentyloxiranyl)-5,8,10,12-tridecatetraenoic acid. 14,15-EPETE

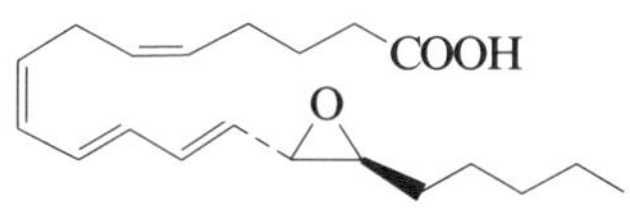

$C_{20}H_{30}O_3$ M 318.455
(5Z,8Z,10E,12E,14S,15S)-form [81918-96-9]
14,15-Leukotriene A_4. *14,15-LTA*$_4$
Intermed. in 15-lipoxygenase-initiated leukotriene biosynth.
Me ester: [75290-58-3].
$C_{21}H_{32}O_3$ M 332.482
$[\alpha]^{22}_{D}$ $-5.0°$ (c, 0.3 in cyclohexane).
[82263-59-0, 88852-50-0]

Corey, E.J. *et al*, *J. Am. Chem. Soc.*, 1980, **102**, 6607 (*synth*)
Attrache, V. *et al*, *Tetrahedron Lett.*, 1981, **22**, 3443 (*synth*)
Zamboni, R. *et al*, *Tetrahedron Lett.*, 1983, **24**, 4899 (*synth*)
Corey, E.J. *et al*, *Tetrahedron Lett.*, 1983, **24**, 4917 (*synth*)

## 5,6-Epoxy-7,9,11-eicosatrienoic acid	E-00079

3-(1,3,5-Tetradecatrienyl)oxiranebutanoic acid. ***Leukotriene*** $\boldsymbol{A_3}$
[87791-73-9]

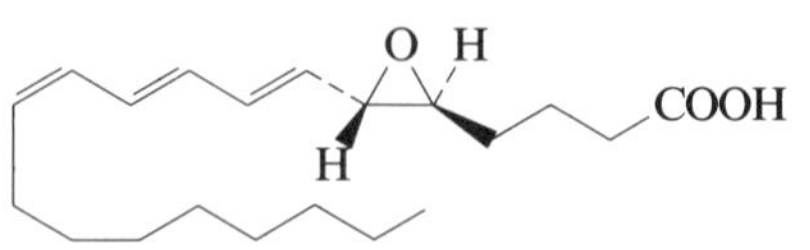

$C_{20}H_{32}O_3$ M 320.471
Derived from 5,8,11-Eicosatrienoic acid, E-00043 by the 5-lipoxygenase leukotriene pathway. Potent inhibitor of rat and human neutrophil LTA_4 hydrolase.
Me ester: [83851-38-1].
$C_{21}H_{34}O_3$ M 334.498
Ir (film) 1740, 1425 cm^{-1}; uv 271, 280 (ε 50000), 290 nm (EtOH).

Okuyama, S. *et al*, *Chem. Pharm. Bull.*, 1982, **30**, 2453 (*Me ester, synth, ir, uv, pmr*)
Spur, B. *et al*, *Arch. Pharm. (Weinheim, Ger.)*, 1983, **316**, 789 (*synth, cmr*)
Jakschik, B.A. *et al*, *J. Biol. Chem.*, 1983, **258**, 12797 (*biosynth*)
Evans, J.F. *et al*, *J. Biol. Chem.*, 1985, **260**, 10966 (*biochem*)

5,6-Epoxy-8,11,14-eicosatrienoic acid E-00080

3-(2,5,8-Tetradecatrienyl)oxiranebutanoic acid. 5,6-EET

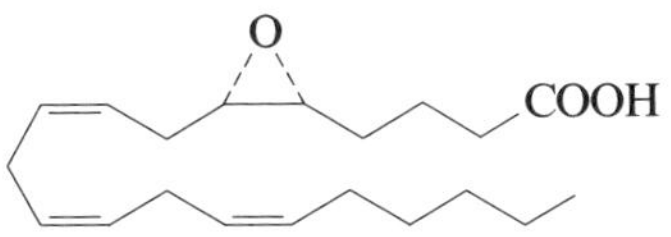

(5*R*,6*S*,8*Z*,11*Z*,14*Z*)-*form*

$C_{20}H_{32}O_3$ M 320.471

Metab. of cytochrome P-450 epoxygenase (as mixt. of 5*R*,6*S* and 5*S*,6*R* isomers).

(5R,6S,8Z,11Z,14Z)-form [98103-46-9]

Me ester: [90080-06-1].

$C_{21}H_{34}O_3$ M 334.498

Metab. of arachidonic acid (by cytochrome P-450 monooxygenase). No phys. props. reported.

(5S,6R,8Z,11Z,14Z)-form [98103-47-0]

Me ester: [98103-49-2].

Metab. of arachidonic acid (by cytochrome P-450 monooxygenase). No phys. props. reported.

Corey, E.J. *et al*, *J. Am. Chem. Soc.*, 1979, **101**, 1586 (*synth*)
Falck, J.R. *et al*, *J. Am. Chem. Soc.*, 1984, **106**, 3334 (*biosynth*)
Moustakis, C.A. *et al*, *J. Am. Chem. Soc.*, 1985, **107**, 5283 (*synth*)

8,9-Epoxy-5,11,14-eicosatrienoic acid E-00081

7-[3-(2,5-Undecadienyl)oxiranyl]-5-heptenoic acid, 9CI. 8,9-EET

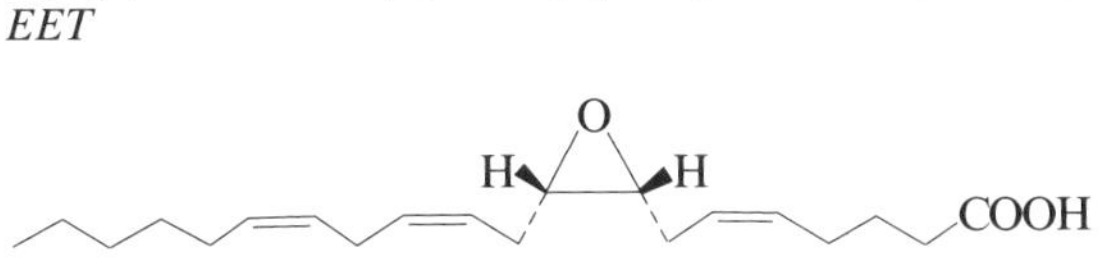

$C_{20}H_{32}O_3$ M 320.471

(5Z,8R*,9R*,11Z,14Z)-form [82864-43-5]

Arachidonic acid metab.

(5Z,8R,9S,11Z,14Z)-form [129172-62-9]

Arachidonic acid metab.

Me ester: [90080-07-2].

$C_{21}H_{34}O_3$ M 334.498

No phys. props. reported.

(5Z,8S,9R,11Z,14Z)-form [123931-39-5]

Arachidonic acid metab. Major isomer prod.

Me ester: [110901-55-8].

No phys. props. reported.

Falck, J.R. *et al*, *Tetrahedron Lett.*, 1982, **23**, 1755 (*synth, pmr*)
Mosset, P. *et al*, *Tetrahedron Lett.*, 1986, **27**, 6035 (*synth*)
Capdevila, J.H. *et al*, *J. Biol. Chem.*, 1990, **265**, 10865 (*biosynth*)
Karara, A. *et al*, *J. Biol. Chem.*, 1991, **266**, 7561 (*biosynth*)

11,12-Epoxy-5,8,14-eicosatrienoic acid E-00082

10-[3-(2-Octenyl)oxiranyl]-5,8-decadienoic acid, 9CI. 11,12-EET

[81276-02-0]

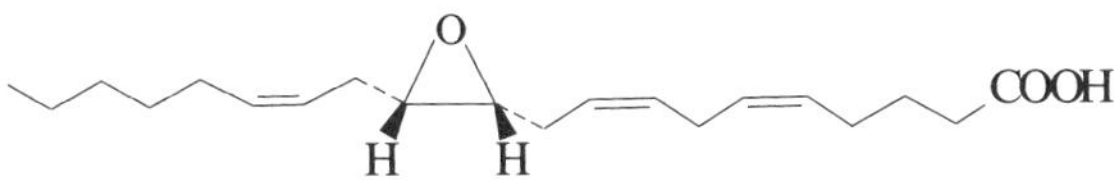

$C_{20}H_{32}O_3$ M 320.471

(5Z,8Z,11R,12S,14Z)-form [123931-38-4]

Arachidonic acid metab.

Me ester: [110901-54-7].

No phys. props. reported.

(5Z,8Z,11S,12R,14Z)-form [123931-40-8]

Arachidonic acid metab.

Me ester: [90080-08-3].

$[\alpha]_D^{24}$ −2.34° (c, 0.66 in $CHCl_3$). No phys. props. reported.

(5Z,8Z,11RS,12SR,14Z)-form

Oil.

Corey, E.J. *et al*, *J. Am. Chem. Soc.*, 1980, **102**, 1433 (*synth*)
Falck, J.R. *et al*, *J. Am. Chem. Soc.*, 1984, **106**, 3334 (*biosynth*)
Falck, J.R. *et al*, *Tetrahedron Lett.*, 1984, **25**, 2443 (*synth*)
Mosset, P. *et al*, *Tetrahedron Lett.*, 1986, **27**, 6035 (*synth*)
Karara, A. *et al*, *J. Biol. Chem.*, 1989, **264**, 19822; 1991, **266**, 7561 (*biosynth*)
Capdevila, J.H. *et al*, *J. Biol. Chem.*, 1990, **265**, 10865 (*biosynth*)

14,15-Epoxy-5,8,11-eicosatrienoic acid E-00083

13-(3-Pentyloxiranyl)-5,8,11-tridecatrienoic acid, 9CI. 14,15-EET

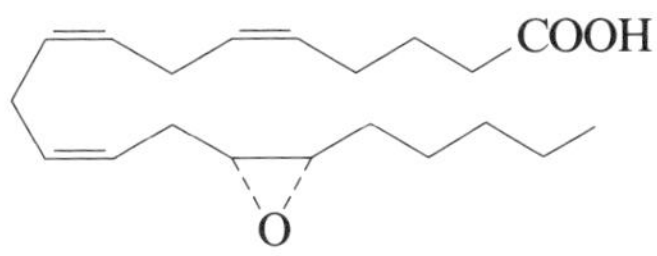

(5*Z*,8*Z*,11*Z*,14*R*,15*S*)-*form*

$C_{20}H_{32}O_3$ M 320.471

(5Z,8Z,11Z,14R,15S)-form [74868-37-4]

Cytochrome P-450 epoxygenase metab.

Me ester: [92143-14-1].

Inhibits platelet aggregation. $[\alpha]_D^{24}$ −2.73° (c, 0.90 in $CHCl_3$). Other isomers were prepd.

(5Z,8Z,11Z,14S,15R)-form [105370-31-8]

Inhibits platelet aggregation.

Me ester: [110901-53-6].

$[\alpha]_D^{24}$ +2.95° (c, 0.95 in $CHCl_3$).

[73799-13-0, 105370-30-7]

Corey, E.J. *et al*, *J. Am. Chem. Soc.*, 1979, **101**, 1586 (*synth*)
Falck, J.R. *et al*, *Tetrahedron Lett.*, 1984, **25**, 2443 (*synth, bibl*)
Moustakis, C.A. *et al*, *J. Am. Chem. Soc.*, 1985, **107**, 5283 (*synth*)
Fitzpatrick, F.A. *et al*, *J. Biol. Chem.*, 1986, **261**, 15334 (*biochem*)
Ennis, M.D. *et al*, *Tetrahedron Lett.*, 1986, **27**, 6031 (*synth*)

14,15-Epoxy-11-eicosenoic acid E-00084

13-(3-Pentyloxiranyl)-11-tridecenoic acid

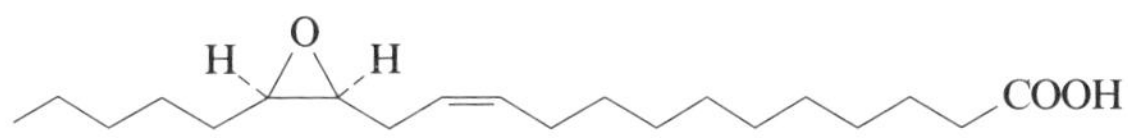

$C_{20}H_{36}O_3$ M 324.503

(14S,15R)-(Z)-form [64019-95-0]

Alchornoic acid

A major acid in the seed fat of *Alchornea cordifolia*.

Kleiman, R. *et al*, *Lipids*, 1977, **12**, 610 (*isol, ord, ir, nmr, ms*)

14,15-Epoxy-5-hydroxy-6,8,10,12-eicosatetraenoic acid E-00085

5-Hydroxy-13(3-pentyloxiranyl)-6,8,10,12-tridecatetraenoic acid, 9CI. 5-Hydroxy-14,15-LTA$_4$

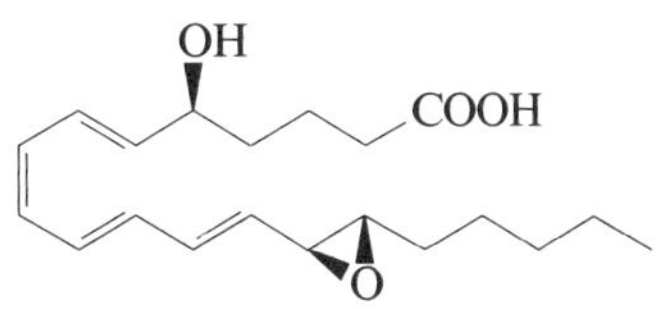

$C_{20}H_{30}O_4$ M 334.455

(5S,6E,8Z,10E,12E)-form [114144-28-4]

Intermed. in lipoxin biosynth. $t_{1/2}$ 2 min. in pH 7.4 buffer.

Leblanc, Y. *et al*, *Tetrahedron Lett.*, 1987, **28**, 3449 (*synth, biochem*)

5,6-Epoxy-20-hydroxy-8,11,14-eicosatrienoic acid E-00086

3-(14-Hydroxy-2,5,8-tetradecatrienyl)oxiranebutanoic acid, 9CI

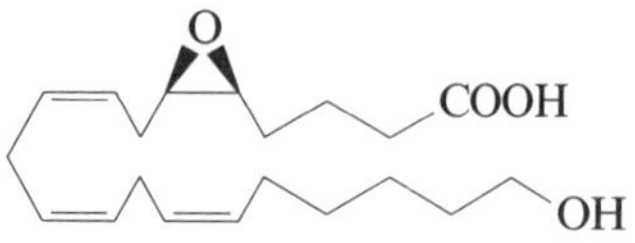

$C_{20}H_{32}O_4$ M 336.470

(5S,6R,8Z,11Z,14Z)-form [112168-43-1]

Hydroxylated eicosanoid metab. $[\alpha]_D^{23}$ −3.77° (c, 1.28 in Me_2CO).

Lumin, S. *et al, J. Chem. Soc., Chem. Commun.*, 1987, 389 (*synth, pmr*)

14,15-Epoxy-20-hydroxy-5,8,11-eicosatrienoic acid E-00087

13-[3-(5-Hydroxypentyl)oxiranyl]-5,8,11-tridecatrienoic acid, 9CI

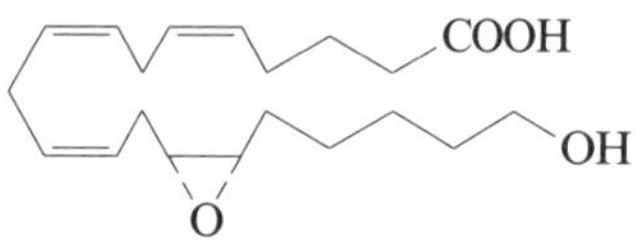

$C_{20}H_{32}O_4$ M 336.470

(5Z,8Z,11Z,14R,15S)-form [112168-44-2]

Hydroxylated eicosanoid metabolite. $[\alpha]_D^{23}$ +4.66° (c, 1.01 in Me_2CO).

Lumin, S. *et al, J. Chem. Soc., Chem. Commun.*, 1987, 389 (*synth, pmr*)

9,10-Epoxy-8-hydroxy-9-methyl-2,4,6-decatrienoic acid E-00088

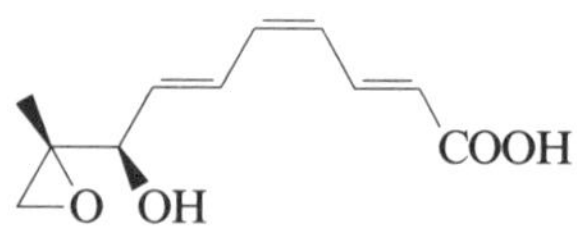

$C_{11}H_{14}O_4$ M 210.229

(2E,4Z,6E,8R,9S)-form [127702-86-7]

Prod. by *Alternaria alternata*. Biological precursor of AK-toxins.

[125555-66-0, 125582-64-1]

Feng, B. *et al, Agric. Biol. Chem.*, 1990, **54**, 845 (*isol*)
Nakatsuka, S. *et al, Phytochemistry*, 1990, **29**, 1529 (*biosynth*)

9,10-Epoxy-18-hydroxyoctadecanoic acid E-00089

3-(8-Hydroxyoctyl)oxiraneoctanoic acid, 9CI

[3233-92-9]

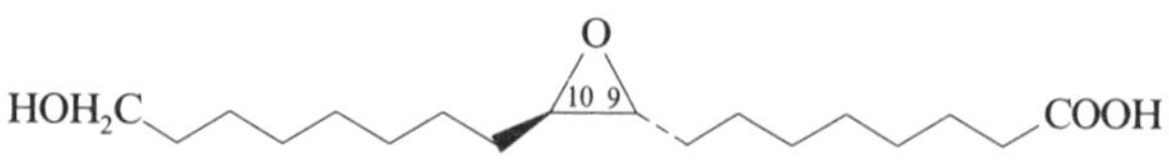

(9*R**,10*R**)-*form*

$C_{18}H_{34}O_4$ M 314.464

(9R*,10R*)-form [36693-06-8]

(+)-trans-*form*

Mp 70-70.5°. $[\alpha]_D$ +20.84°.

(9S*,10S*)-form [36693-07-9]

(−)-trans-*form*

Mp 70-71°. $[\alpha]_D$ −20.81°.

(9RS,10RS)-form

(±)-trans-*form*

Me ester: [73334-96-0].

$C_{19}H_{36}O_4$ M 328.491

Cryst. (hexane/Et_2O). Mp 50-51°.

(9R*,10S*)-form [35688-45-0]

cis-*form*

Present in several cutins. Mp 82°.

Me ester: [65492-89-9].

Cryst. (Et_2O). Mp 36°.

Breiskorn, C.H. *et al, Fette, Seifen, Anstrichm.*, 1964, **66**, 925; *CA*, **62**, 8318 (*isol, ir*)
McGhie, J.F. *et al, Chem. Ind. (London)*, 1972, 463 (*resoln*)
Croteau, R. *et al, Arch. Biochem. Biophys.*, 1975, **170**, 61, 73 (*biosynth*)
Seoane, E. *et al, Chem. Ind. (London)*, 1977, 662 (*occur*)
Gonzalez, B. *et al, Chem. Ind. (London)*, 1979, 477 (*synth, ir, pmr, ms*)
Velcheva, M.P. *et al, Phytochemistry*, 1981, **20**, 2225 (*chromatog, ms*)

9,10-Epoxy-18-hydroxy-12-octadecenoic acid E-00090

3-(8-Hydroxy-2-octenyl)oxiraneoctanoic acid, 9CI

[50632-63-8]

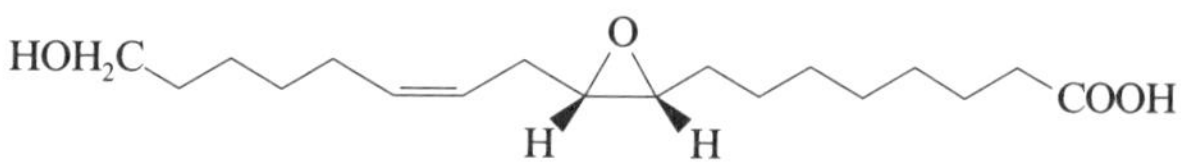

$C_{18}H_{32}O_4$ M 312.448

(9R*,10S*,Z)-form

Present in some cutins.

Holloway, P.J. *et al, Phytochemistry*, 1973, **12**, 1721, 2913 (*occur*)
Croteau, R. *et al, Arch. Biochem. Biophys.*, 1974, **162**, 458 (*biosynth*)

6,9-Epoxy-18-nonadecene-7,10-diol E-00091

Tetrahydro-4-hydroxy-α-(8-nonenyl)-5-pentyl-2-furanmethanol, 9CI

[74350-84-8]

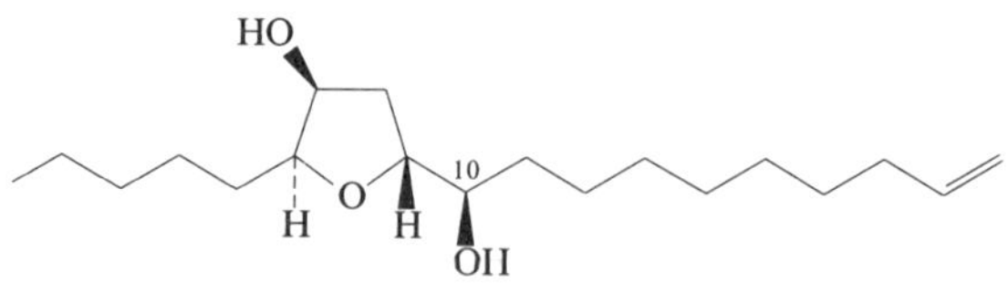

$C_{19}H_{36}O_3$ M 312.492

Isol. from the brown alga *Notheia anomala*. Plates or needles. Mp 54.5-55°. $[\alpha]_D^{21}$ +15.0° (c, 1 in $CHCl_3$). Also present as alkanoyl esters.

Warren, R.G. *et al, Aust. J. Chem.*, 1980, **33**, 891 (*isol, pmr, cmr, ms*)
Barrow, R.A. *et al, Aust. J. Chem.*, 1990, **43**, 895 (*isol, pmr, cmr*)

2,3-Epoxy-6,8-nonadiynoic acid E-00092

3-(3,5-Hexadiynyl)oxiranecarboxylic acid

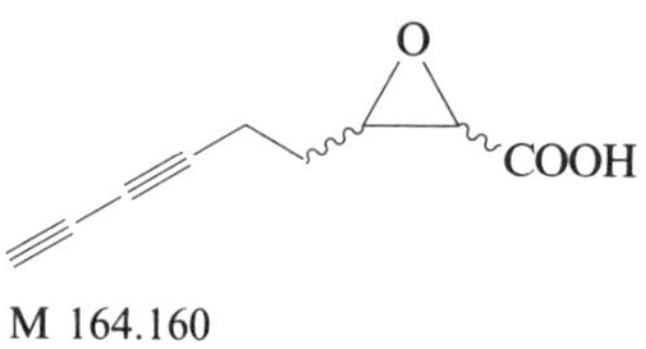

$C_9H_8O_3$ M 164.160

2-Phenylethylamide: [96917-25-8]. *3-(3,5-Hexadiynyl)-*N*-(2-phenylethyl)oxiranecarboxamide. 2,3-Epoxy-6,8-nonadiynoic acid 2-phenylethylamide*
$C_{17}H_{17}NO_2$ M 267.327
Isol. from *Acmella ciliata* and *Salmea scandens*. Oil.
(Z)-*2-Phenylethenylamide:* [75872-78-5]. *3-(3,5-Hexadiynyl)-*N*-(2-phenylethenyl)oxiranecarboxamide, 9CI. 1,2-Epoxy-6,8-nonadiynoic acid styrylamide*
$C_{17}H_{15}NO_2$ M 265.311
Isol. from *Spilanthes alba*. Oil.

Bohlmann, F. *et al*, *Phytochemistry*, 1980, **19**, 1535; 1985, **24**, 595 (*isol, struct, pmr, ir, ms*)
Martin, R. *et al*, *Phytochemistry*, 1985, **24**, 2295 (*isol, struct, pmr, ms*)

9,10-Epoxyoctadecanedioic acid **E-00093**

2,3-Oxiranedioctanoic acid, 9CI
[50632-64-9]

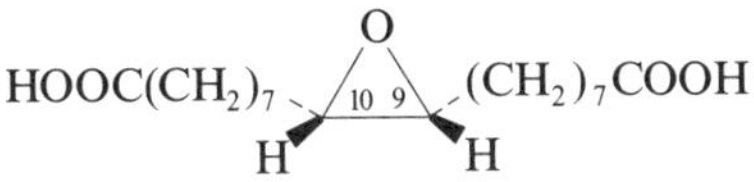

$C_{18}H_{32}O_5$ M 328.448
(9RS,10SR)-form
cis-*form*
Present in some cutins.

Holloway, J.P., *Phytochemistry*, 1974, **13**, 229, 1721 (*occur*)
Seoane, E. *et al*, *Chem. Ind. (London)*, 1977, 662 (*occur*)

2,3-Epoxyoctadecanoic acid **E-00094**

3-Pentadecyloxiranecarboxylic acid, 9CI. 2,3-Epoxystearic acid

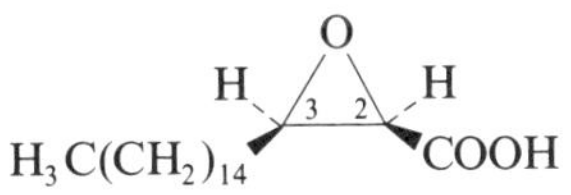

(2*RS*,3*RS*)-*form*

$C_{18}H_{34}O_3$ M 298.465
(2RS,3RS)-form
(±)-cis-*form*
Cryst. (pet. ether). Mp 84.0-86.5°, Mp 90.5-91.0°.
Me ester:
$C_{19}H_{36}O_3$ M 312.492
Liq.
(2RS,3SR)-form
(±)-trans-*form*
Cryst. (pet. ether). Mp 85.0-88.0°, Mp 87.5-88.0°.
Me ester: [38511-87-4].
Liq.

Myers, G.S., *J. Am. Chem. Soc.*, 1952, **74**, 1390.
Gunstone, F.D. *et al*, *Chem. Phys. Lipids*, 1972, **9**, 26 (*glc, pmr, ms*)

3,4-Epoxyoctadecanoic acid **E-00095**

3-Tetradecyloxiraneacetic acid, 9CI. 3,4-Epoxystearic acid

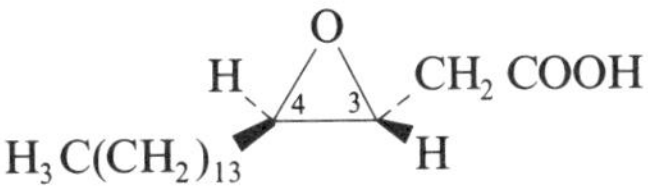

(3*RS*,4*RS*)-*form*

$C_{18}H_{34}O_3$ M 298.465
(3RS,4RS)-form
(±)-trans-*form*
Liq.
Me ester: [38520-13-7].
$C_{19}H_{36}O_3$ M 312.492
Liq.
(3RS,4SR)-form
(±)-cis-*form*
Me ester: [38520-12-6].
Liq.

Gunstone, F.D. *et al*, *Chem. Phys. Lipids*, 1972, **9**, 26 (*glc, pmr, ms*)

4,5-Epoxyoctadecanoic acid **E-00096**

3-Tridecyloxiranepropanoic acid, 9CI. 4,5-Epoxystearic acid

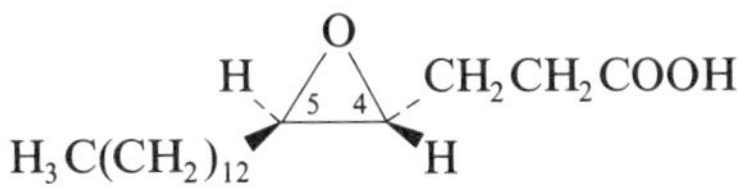

(4*RS*,5*RS*)-*form*

$C_{18}H_{34}O_3$ M 298.465
(4RS,5RS)-form
(±)-trans-*form*
Me ester: [38520-15-9].
$C_{19}H_{36}O_3$ M 312.492
Cryst. Mp 37.0-39.5°.
(4RS,5SR)-form
(±)-cis-*form*
Me ester: [38520-14-8].
Cryst. Mp 33.0-34.0°.

Gunstone, F.D. *et al*, *Chem. Phys. Lipids*, 1972, **9**, 26 (*glc, pmr, ms*)

5,6-Epoxyoctadecanoic acid **E-00097**

3-Dodecyloxiranebutanoic acid, 9CI. 5,6-Epoxystearic acid

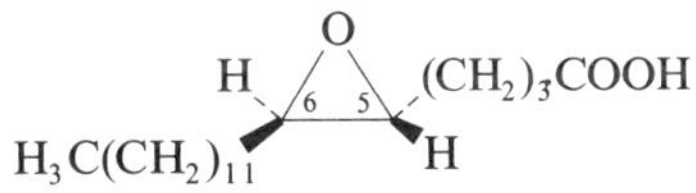

(5*RS*,6*RS*)-*form*

$C_{18}H_{34}O_3$ M 298.465
(5RS,6RS)-form
(±)-trans-*form*
Me ester: [38520-17-1].
$C_{19}H_{36}O_3$ M 312.492
Liq.
(5RS,6SR)-form
(±)-cis-*form*
Me ester: [38520-16-0].
Liq.

Gunstone, F.D. *et al*, *Chem. Phys. Lipids*, 1972, **9**, 26 (*glc, pmr, ms*)

6,7-Epoxyoctadecanoic acid **E-00098**

3-Undecyloxiranepentanoic acid, 9CI. 6,7-Epoxystearic acid

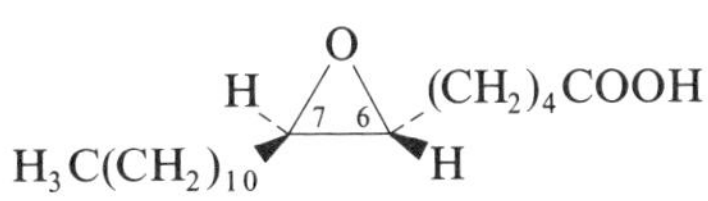

(6*RS*,7*RS*)-*form*

$C_{18}H_{34}O_3$ M 298.465
(6RS,7RS)-form
(±)-trans-*form*
Cryst. (pet. ether). Mp 64.5-65.5°.
Me ester: [6175-11-7].
$C_{19}H_{36}O_3$ M 312.492
Liq.
(6RS,7SR)-form
(±)-cis-*form*
Cryst. (pet. ether). Mp 56.5-58.0°, Mp 59.5-60.0°.

Me ester: [1041-25-4].
Liq.

Farooq, M.O. *et al*, *Fette, Seifen, Anstrichm.*, 1959, **61**, 636.
Gunstone, F.D. *et al*, *Chem. Phys. Lipids*, 1972, **9**, 26 (*glc, pmr, ms*)

7,8-Epoxyoctadecanoic acid E-00099

3-Decyloxiranehexanoic acid, 9CI. 7,8-Epoxystearic acid

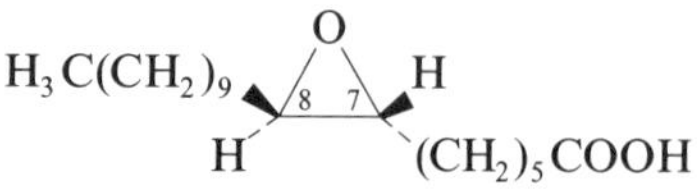

$C_{18}H_{34}O_3$ M 298.465

(7RS,8RS)-form
(±)-trans-*form*
Cryst. (pet. ether). Mp 54.0-55.0°.
Me ester: [38520-21-7].
$C_{19}H_{36}O_3$ M 312.492
Liq.

(7RS,8SR)-form
(±)-cis-*form*
Cryst. (pet. ether). Mp 46.5-48.0°.
Me ester: [38520-20-6].
Liq.

Gunstone, F.D. *et al*, *Chem. Phys. Lipids*, 1972, **9**, 26 (*glc, pmr, ms*)

8,9-Epoxyoctadecanoic acid E-00100

3-Nonyloxiraneheptanoic acid, 9CI. 8,9-Epoxystearic acid

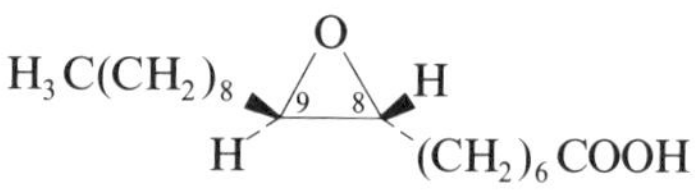

(8*RS*,9*RS*)-*form*

$C_{18}H_{34}O_3$ M 298.465

(8RS,9RS)-form
(±)-trans-*form*
Cryst. (pet. ether). Mp 52.5-54.0°.
Me ester: [38520-23-9].
$C_{19}H_{36}O_3$ M 312.492
Liq.

(8RS,9SR)-form
(±)-cis-*form*
Cryst. (pet. ether). Mp 55.0-56.0°.
Me ester: [38520-22-8].
Liq.

Gunstone, F.D. *et al*, *Chem. Phys. Lipids*, 1972, **9**, 26 (*glc, pmr, ms*)

9,10-Epoxyoctadecanoic acid E-00101

9,10-Epoxystearic acid. 3-Octyloxiraneoctanoic acid

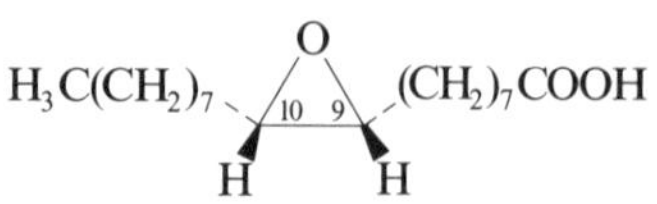

(9*R*,10*S*)-*form*

$C_{18}H_{34}O_3$ M 298.465

(9R,10S)-form [16833-55-9]
(+)-cis-*form*
Constit. of the uredospores of *Puccinia graminis* var. *tritici* and from the spore oil of other fungi and of *Lycopodium* spp. Also in seed oil of *Ongokea gore*. Mp 58.5-59.5°. $[\alpha]_D^{25}$ +1.61° (MeOH).
▷ Exp. carcinogen.

(9RS,10SR)-form [24560-98-3]
(±)-cis-*form*
Cryst. (Me_2CO). Mp 59.5-59.8°.
Me ester: [2566-91-8].
$C_{19}H_{36}O_3$ M 312.492
Mp 15-16.5°.
Et ester:
$C_{20}H_{38}O_3$ M 326.518
Mp 18°.

(9RS,10RS)-form [17682-85-8]
(±)-trans-*form*
Cryst. (MeOH or Me_2CO). Mp 56-57°.
Me ester: [6084-76-0].
Cryst. (pet. ether or EtOH). Mp 32-33.5°.
Et ester: Mp 21°.

Swern, D. *et al*, *J. Am. Chem. Soc.*, 1944, **66**, 1925 (*synth*)
Tulloch, A.P. *et al*, *Can. J. Microbiol.*, 1959, **5**, 485 (*isol*)
Vioque, E. *et al*, *J. Am. Oil Chem. Soc.*, 1961, **38**, 489 (*isol*)
Morris, L.J., *J. Chem. Soc.*, 1963, 5779 (*isol*)
Tulloch, A.P., *Can. J. Chem.*, 1965, **43**, 415 (*isol*)
Powell, R.G. *et al*, *Lipids*, 1967, **2**, 172.
Gunstone, F.D. *et al*, *Chem. Phys. Lipids*, 1972, **9**, 26 (*synth*)
Korver, O. *et al*, *Recl. Trav. Chim. Pays-Bas* (*J. R. Neth. Chem. Soc.*), 1973, **92**, 127 (*abs config*)
Sax, N.I., *Dangerous Properties of Industrial Materials*, 5th Ed., Van Nostrand-Reinhold, 1979, 643.

10,11-Epoxyoctadecanoic acid E-00102

3-Heptyloxiranenonanoic acid, 9CI. 10,11-Epoxystearic acid

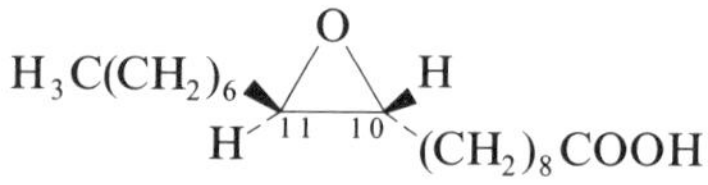

(10*RS*,11*RS*)-*form*

$C_{18}H_{34}O_3$ M 298.465

(10RS,11RS)-form
(±)-trans-*form*
Cryst. (pet. ether). Mp 51.0-52.5°.
Me ester: [38520-27-3].
$C_{19}H_{36}O_3$ M 312.492
Liq.

(10RS,11SR)-form
(±)-cis-*form*
Cryst. (pet. ether). Mp 55.0-56.0°.
Me ester: [38520-26-2].
Liq.

Gunstone, F.D., *Chem. Phys. Lipids*, 1972, **9**, 26 (*glc, pmr, ms*)

11,12-Epoxyoctadecanoic acid E-00103

3-Hexyloxiranedecanoic acid, 9CI. 11,12-Epoxystearic acid

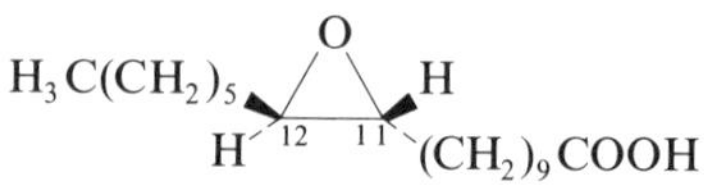

(11*RS*,12*RS*)-*form*

$C_{18}H_{34}O_3$ M 298.465

(11RS,12RS)-form
(±)-trans-*form*
Cryst. (pet. ether). Mp 51.0-52.0°.
Me ester: [38520-29-5].
$C_{19}H_{36}O_3$ M 312.492
Cryst. Mp 29.5-30.5°.

(11RS,12SR)-form
(±)-cis-*form*
Cryst. (pet. ether). Mp 41.5-44.0°, Mp 46.0-47.6°.
Me ester: [38520-28-4].
Liq.

Hofmann, K. *et al*, *J. Biol. Chem.*, 1953, **205**, 55.
Gunstone, F.D. *et al*, *Chem. Phys. Lipids*, 1972, **9**, 26 (*glc, pmr, ms*)

12,13-Epoxyoctadecanoic acid E-00104

3-Pentyloxiraneundecanoic acid, 9CI. 12,13-Epoxystearic acid

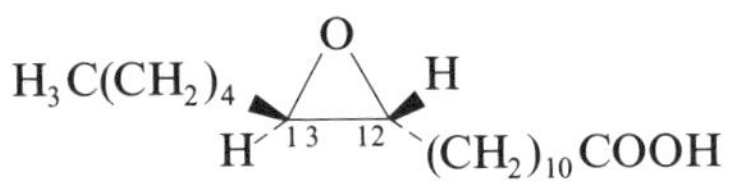

(12*RS*,13*RS*)-*form*

$C_{18}H_{34}O_3$ M 298.465

(12RS,13RS)-form
(±)-trans-*form*
Cryst. (pet. ether). Mp 51.0-53.5°.
Me ester: [38520-31-9].
$C_{19}H_{36}O_3$ M 312.492
Liq.

(12RS,13SR)-form
(±)-cis-*form*
Cryst. (pet. ether). Mp 52.0-53.5°.
Me ester: [38520-30-8].
Liq.

Gunstone, F.D. *et al*, *Chem. Phys. Lipids*, 1972, **9**, 26 (*glc, pmr, ms*)

13,14-Epoxyoctadecanoic acid E-00105

3-Butyloxiranedodecanoic acid, 9CI. 13,14-Epoxystearic acid

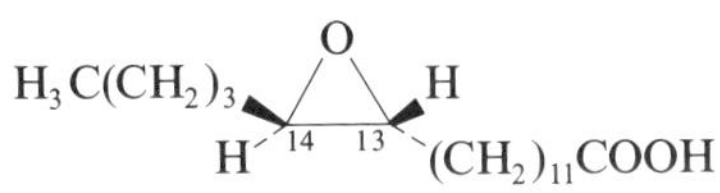

(13*RS*,14*RS*)-*form*

$C_{18}H_{34}O_3$ M 298.465

(13RS,14RS)-form
(±)-trans-*form*
Cryst. (pet. ether). Mp 53.0-55.0°.
Me ester: [38520-33-1].
$C_{19}H_{36}O_3$ M 312.492
Cryst. Mp 32.5-34.0°.

(13RS,14SR)-form
(±)-cis-*form*
Cryst. (pet. ether). Mp 46.5-48.0°.
Me ester: [38520-32-0].
Liq.

Gunstone, F.D. *et al*, *Chem. Phys. Lipids*, 1972, **9**, 26 (*glc, pmr, ms*)

14,15-Epoxyoctadecanoic acid E-00106

3-Propyloxiranetridecanoic acid, 9CI. 14,15-Epoxystearic acid

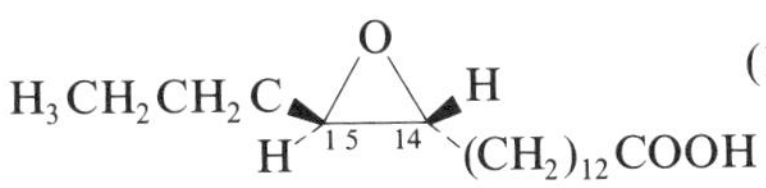

(14*RS*,15*RS*)-*form*

$C_{18}H_{34}O_3$ M 298.465

(14RS,15RS)-form
(±)-trans-*form*
Cryst. (pet. ether). Mp 56.0-58.5°.
Me ester: [38520-35-3].
$C_{19}H_{36}O_3$ M 312.492
Mp 33.0-34.0°.

(14RS,15SR)-form
(±)-cis-*form*
Cryst. (pet. ether). Mp 54.5-56.5°.
Me ester: [38520-34-2].
Liq.

Gunstone, F.D., *Chem. Phys. Lipids*, 1972, **9**, 26 (*glc, pmr, ms*)

15,16-Epoxyoctadecanoic acid E-00107

3-Ethyloxiranetetradecanoic acid, 9CI. 15,16-Epoxystearic acid

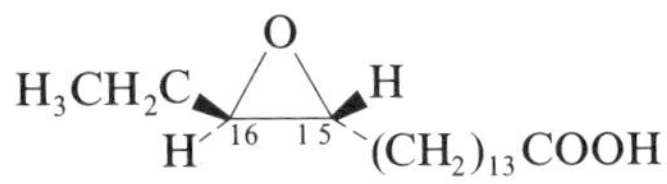

(15*RS*,16*RS*)-*form*

$C_{18}H_{34}O_3$ M 298.465

(15RS,16RS)-form
(±)-trans-*form*
Cryst. (pet. ether). Mp 59.5-62.0°.
Me ester: [38520-37-5].
$C_{19}H_{36}O_3$ M 312.492
Mp 42.0-44.0°.

(15RS,16SR)-form
(±)-cis-*form*
Cryst. (pet. ether). Mp 47.0-49.0°.
Me ester: [38520-36-4].
Liq.

Gunstone, F.D. *et al*, *Chem. Phys. Lipids*, 1972, **9**, 26 (*glc, pmr, ms*)

16,17-Epoxyoctadecanoic acid E-00108

3-Methyloxiranepentadecanoic acid, 9CI. 16,17-Epoxystearic acid

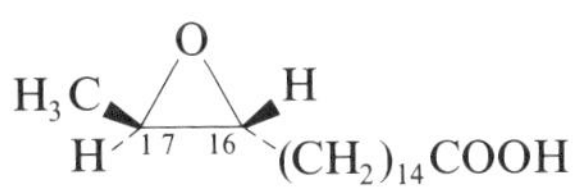

(16*RS*,17*RS*)-*form*

$C_{18}H_{34}O_3$ M 298.465

(16RS,17RS)-form
(±)-trans-*form*
Cryst. (pet. ether). Mp 80.0-82.0°.
Me ester: [38520-39-7].
$C_{19}H_{36}O_3$ M 312.492
Mp 40.0-43.0°.

(16RS,17SR)-form
(±)-cis-*form*
Cryst. (pet. ether). Mp 57.0-58.5°.
Me ester: [38520-38-6].
Mp 37.0-38.0°.

Gunstone, F.D. *et al*, *Chem. Phys. Lipids*, 1972, **9**, 26 (*glc, pmr, ms*)

17,18-Epoxyoctadecanoic acid E-00109

Oxiranehexadecanoic acid, 9CI. 17,18-Epoxystearic acid

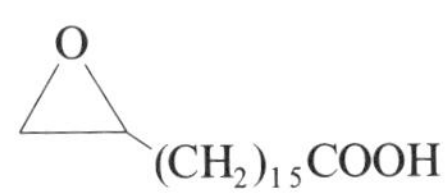

$C_{18}H_{34}O_3$ M 298.465

(±)-*form*
Mp 62.5-63.5°.
Me ester: [38520-40-0].
$C_{19}H_{36}O_3$ M 312.492
Mp 38.0-40.0°.

Jennen, A. *et al*, *C. R. Hebd. Seances Acad. Sci.*, 1960, **251**, 91.
Gunstone, F.D. *et al*, *Chem. Phys. Lipids*, 1972, **9**, 26 (*glc, pmr, ms*)

5,6-Epoxy-12-octadecenoic acid **E-00110**

3-(6-Dodecenyl)oxiranebutanoic acid, 9CI

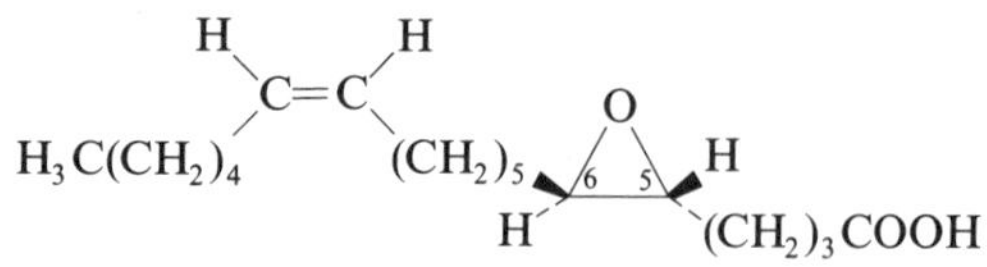

$C_{18}H_{32}O_3$ M 296.449

(5RS,6SR,12Z)-form

cis,cis-*form*

threo-*Dihydroxy acid:* (Z)-threo-*5,6-Dihydroxy-12-octadecenoic acid*

$C_{18}H_{34}O_4$ M 314.464

Cryst. (Et_2O/pet. ether). Mp 51-54°.

Gunstone, F.D. *et al*, *Chem. Phys. Lipids*, 1975, **15**, 174, 189, 198 (*synth, glc, tlc, pmr, ms*)

6,7-Epoxy-9-octadecenoic acid **E-00111**

3-(2-Undecenyl)oxiranepentanoic acid, 9CI

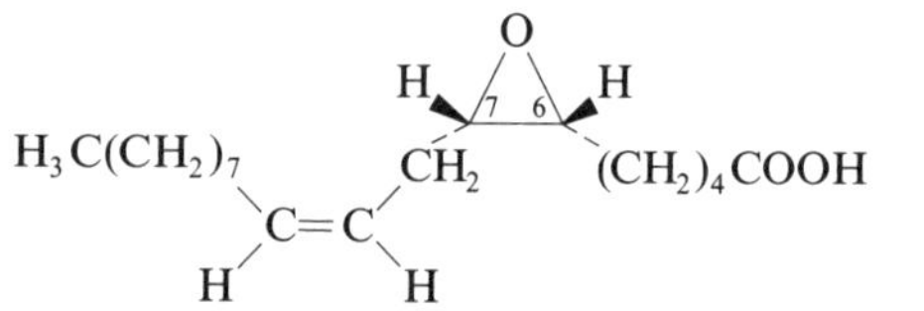

$C_{18}H_{32}O_3$ M 296.449

(6RS,7SR)-(Z)-form

(±)-cis,cis-*form*

threo-*Dihydroxy acid:* (Z)-threo-*6,7-Dihydroxy-9-octadecenoic acid*

$C_{18}H_{34}O_4$ M 314.464

Cryst. (Et_2O/pet. ether). Mp 71.5-72.5°.

Gunstone, F.D. *et al*, *Chem. Phys. Lipids*, 1975, **15**, 174, 189, 198 (*synth, glc, tlc, pmr, ms*)

6,7-Epoxy-10-octadecenoic acid **E-00112**

3-(3-Undecenyl)oxiranepentanoic acid, 9CI

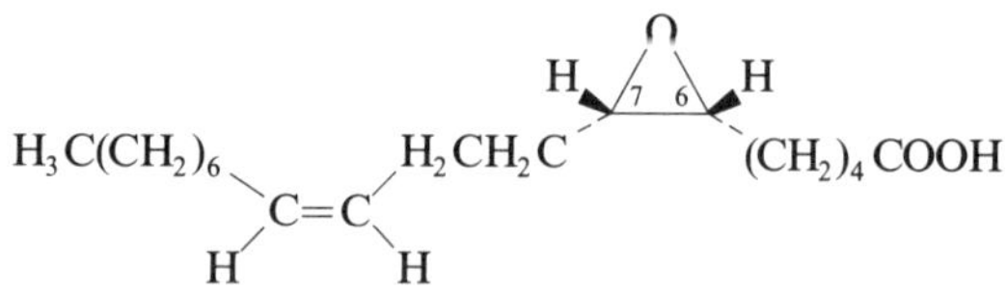

$C_{18}H_{32}O_3$ M 296.449

(6RS,7SR)-(Z)-form

(±)-cis,cis-*form*

threo-*Dihydroxy acid:* (Z)-threo-*6,7-Dihydroxy-10-octadecenoic acid*

$C_{18}H_{34}O_4$ M 314.464

Cryst. (Et_2O/pet. ether). Mp 58.5-60.0°.

Gunstone, F.D. *et al*, *Chem. Phys. Lipids*, 1975, **15**, 174, 189, 198 (*synth, glc, tlc, pmr, ms*)

6,7-Epoxy-11-octadecenoic acid **E-00113**

3-(4-Undecenyl)oxiranepentanoic acid, 9CI

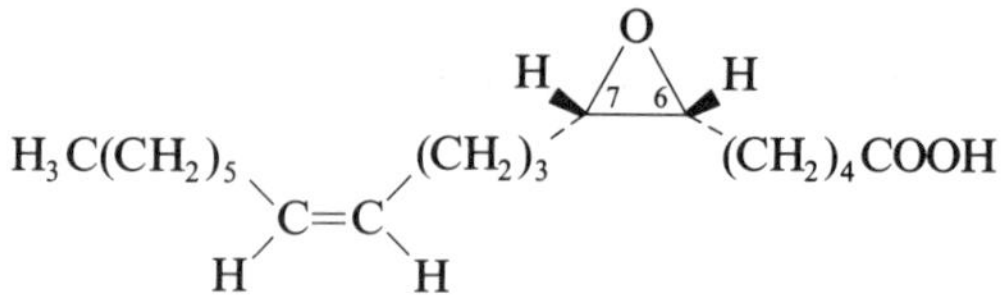

$C_{18}H_{32}O_3$ M 296.449

(6RS,7SR)-(Z)-form

(±)-cis,cis-*form*

threo-*Dihydroxy acid:* (Z)-threo-*6,7-Dihydroxy-11-octadecenoic acid*

$C_{18}H_{34}O_4$ M 314.464

Cryst. (Et_2O/pet. ether). Mp 65-66°.

Gunstone, F.D. *et al*, *Chem. Phys. Lipids*, 1975, **15**, 174, 189, 198 (*synth, glc, tlc, pmr, ms*)

6,7-Epoxy-12-octadecenoic acid **E-00114**

3-(5-Undecenyl)oxiranepentanoic acid, 9CI

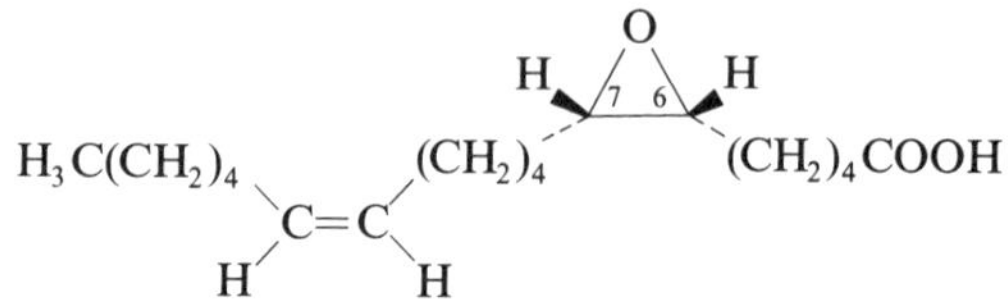

$C_{18}H_{32}O_3$ M 296.449

(6RS,7SR)-(Z)-form

(±)-cis,cis-*form*

threo-*Dihydroxy acid:* (Z)-threo-*6,7-Dihydroxy-12-octadecenoic acid*

$C_{18}H_{34}O_4$ M 314.464

Cryst. (Et_2O/pet. ether). Mp 72-73°.

Gunstone, F.D., *Chem. Phys. Lipids*, 1975, **15**, 174, 189, 198 (*synth, glc, tlc, pmr, ms*)

7,8-Epoxy-12-octadecenoic acid **E-00115**

3-(4-Decenyl)oxiranehexanoic acid, 9CI

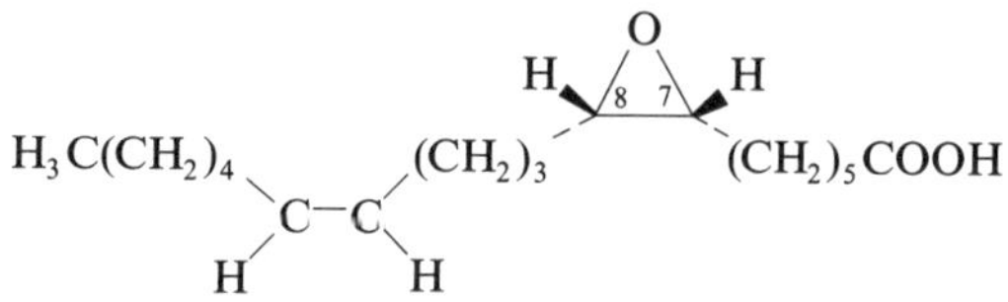

$C_{18}H_{32}O_3$ M 296.449

(7RS,8SR)-(Z)-form

(±)-cis,cis-*form*

threo-*Dihydroxy acid:* (Z)-threo-*7,8-Dihydroxy-12-octadecenoic acid*

$C_{18}H_{34}O_4$ M 314.464

Cryst. (Et_2O/pet. ether). Mp 52-54°.

Gunstone, F.D. *et al*, *Chem. Phys. Lipids*, 1975, **15**, 174, 189, 198.

8,9-Epoxy-12-octadecenoic acid **E-00116**

3-(3-Nonenyl)oxiraneheptanoic acid, 9CI

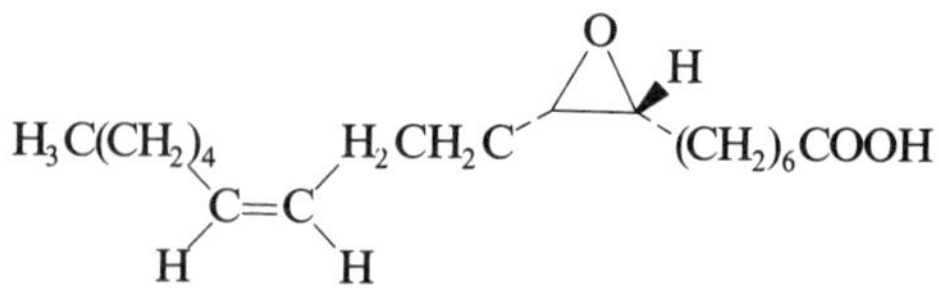

$C_{18}H_{32}O_3$ M 296.449

(8RS,9SR)-(Z)-form
(±)-cis,cis-*form*
threo-*Dihydroxy acid:* (Z)-threo-*8,9-Dihydroxy-12-octadecenoic acid*
$C_{18}H_{34}O_4$ M 314.464
Cryst. (Et_2O/pet. ether). Mp 62-63°.

Gunstone, F.D. *et al*, *Chem. Phys. Lipids*, 1975, **15**, 174, 189, 198 (*glc, tlc, pmr, ms*)

9,10-Epoxy-6-octadecenoic acid E-00117

8-(3-Octyloxiranyl)-6-octenoic acid, 9CI

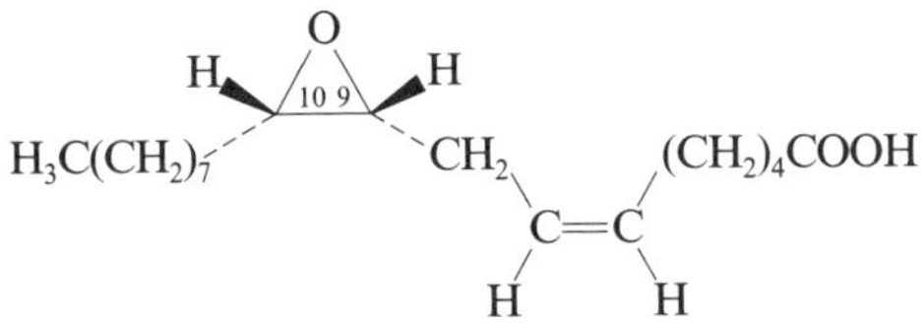

$C_{18}H_{32}O_3$ M 296.449
(9RS,10SR)-(Z)-form
(±)-cis,cis-*form*
threo-*Dihydroxy acid:* (Z)-threo-*9,10-Dihydroxy-6-octadecenoic acid*
$C_{18}H_{34}O_4$ M 314.464
Cryst. (Et_2O/pet. ether). Mp 64.5-65°.

Gunstone, F.D. *et al*, *Chem. Phys. Lipids*, 1975, **15**, 174, 189, 198 (*synth, glc, tlc, pmr, ms*)

9,10-Epoxy-12-octadecenoic acid E-00118

3-(2-Octenyl)oxiraneoctanoic acid, 9CI. Leukotoxin A
[65167-83-1]

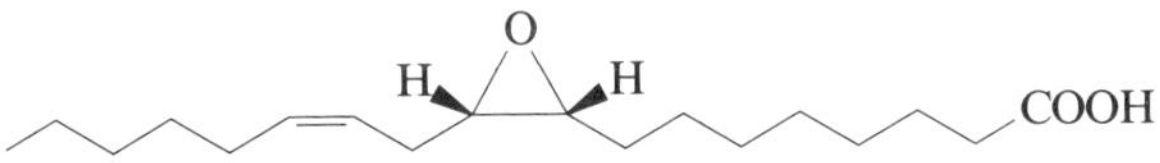

$C_{18}H_{32}O_3$ M 296.449
(9R,10S,12Z)-form [16833-56-0] ***Coronaric acid***
Found in *Chrysanthemum coronarium* and some other seed oil. Isol. from rice plant, Fukuyuki. Inhibits rice blast fungus. $[\alpha]_D$ +3.44°, $[\alpha]_{350}$ +8.94° (c, 9.54 in hexane). Racemic diastereoisomers known synthetically and characterised as the corresp. diols.

[61949-82-4]

Smith, C.R. *et al*, *J. Org. Chem.*, 1960, **25**, 218 (*isol*)
Powell, R.G. *et al*, *Lipids*, 1967, **2**, 172 (*abs config*)
Gunstone, F.D. *et al*, *Chem. Phys. Lipids*, 1975, **15**, 174, 189, 198 (*synth, glc, tlc, pmr, ms*)
Sen Gupta, A.K., *Chem. Ind.* (*London*), 1976, 257.
Kato, T. *et al*, *Tetrahedron Lett.*, 1983, **24**, 4715 (*isol*)

10,11-Epoxy-6-octadecenoic acid E-00119

9-(3-Heptyloxiranyl)-6-nonenoic acid, 9CI

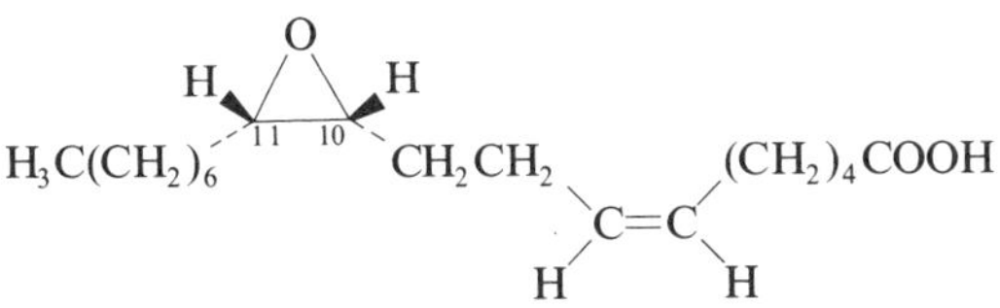

$C_{18}H_{32}O_3$ M 296.449
(10RS,11SR)-(Z)-form
(±)-cis,cis-*form*
threo-*Dihydroxy acid:* (Z)-threo-*10,11-Dihydroxy-6-octadecenoic acid*
$C_{18}H_{34}O_4$ M 314.464
Cryst. (Et_2O/pet. ether). Mp 53-55.5°.

Gunstone, F.D. *et al*, *Chem. Phys. Lipids*, 1975, **15**, 174, 189, 198 (*synth, glc, tlc, pmr, ms*)

11,12-Epoxy-6-octadecenoic acid E-00120

10-(3-Hexyloxiranyl)-6-decenoic acid, 9CI

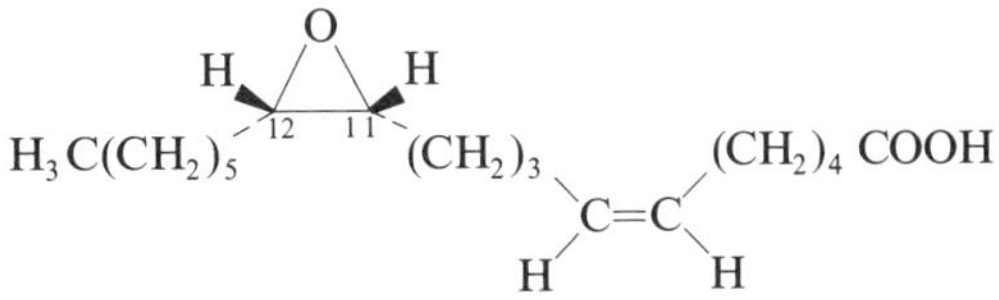

$C_{18}H_{32}O_3$ M 296.449
(11RS,12SR)-(Z)-form
(±)-cis,cis-*form*
threo-*Dihydroxy acid:* (Z)-threo-*11,12-Dihydroxy-6-octadecenoic acid*
$C_{18}H_{34}O_4$ M 314.464
Cryst. (Et_2O/pet. ether). Mp 59.5-61°.

Gunstone, F.D. *et al*, *Chem. Phys. Lipids*, 1975, **15**, 174, 189, 198 (*synth, glc, tlc, pmr, ms*)

12,13-Epoxy-5-octadecenoic acid E-00121

11-(3-Pentyloxiranyl)-5-undecenoic acid, 9CI

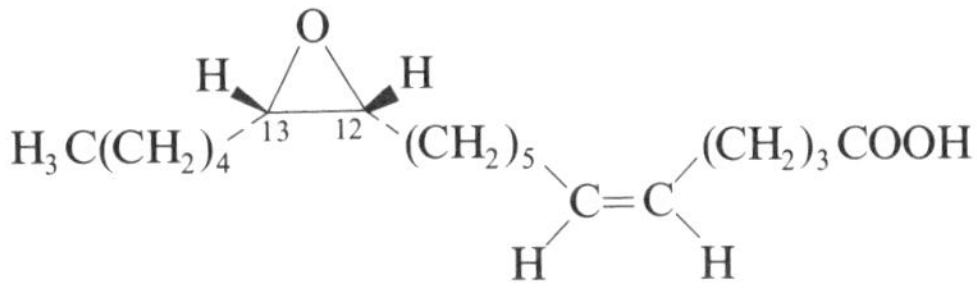

$C_{18}H_{32}O_3$ M 296.449
(12RS,13SR)-form
(±)-cis,cis-*form*
threo-*Dihydroxy acid:* (Z)-threo-*12,13-Dihydroxy-5-octadecenoic acid*
$C_{18}H_{34}O_4$ M 314.464
Cryst. (Et_2O/pet. ether). Mp 45-50°.

Gunstone, F.D. *et al*, *Chem. Phys. Lipids*, 1975, **15**, 174, 189, 198 (*synth, glc, tlc, pmr, ms*)

12,13-Epoxy-6-octadecenoic acid E-00122

11-(3-Pentyloxiranyl)-6-undecenoic acid, 9CI

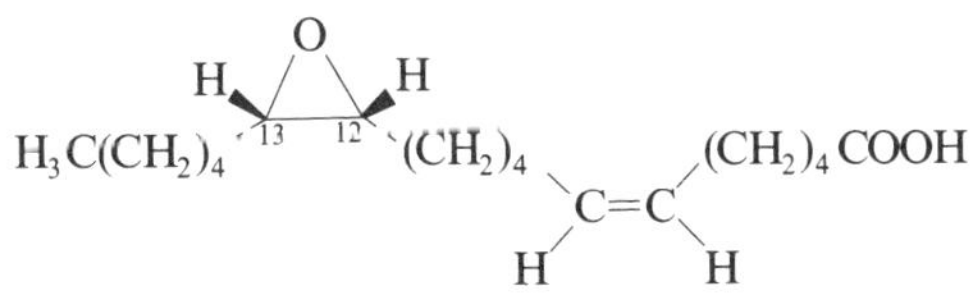

$C_{18}H_{32}O_3$ M 296.449
(12RS,13RS)-(Z)-form
(±)-cis,cis-*form*
threo-*Dihydroxy acid:* (Z)-threo-*12,13-Dihydroxy-6-octadecenoic acid*
$C_{18}H_{34}O_4$ M 314.464
Cryst. (Et_2O/pet. ether). Mp 53-54°.

Gunstone, F.D. *et al*, *Chem. Phys. Lipids*, 1975, **15**, 174, 189, 198 (*synth, glc, tlc, pmr, ms*)

12,13-Epoxy-7-octadecenoic acid E-00123

11-(3-Pentyloxiranyl)-7-undecenoic acid, 9CI

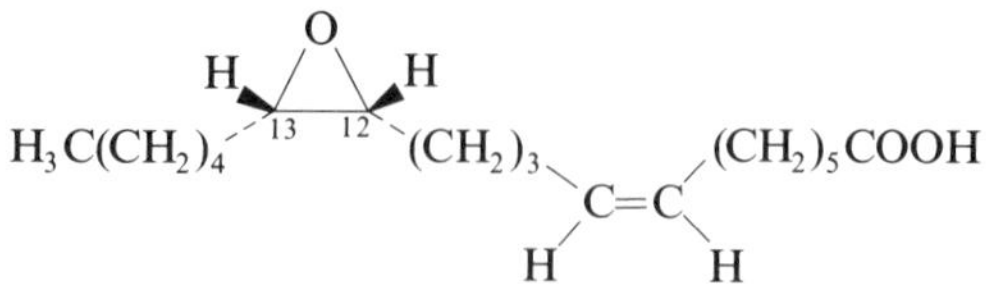

$C_{18}H_{32}O_3$ M 296.449

(12RS,13SR)-(Z)-form

(±)-cis,cis-*form*

threo-*Dihydroxy acid:* (Z)-threo-*12,13-Dihydroxy-7-octadecenoic acid*

$C_{18}H_{34}O_4$ M 314.464

Cryst. (Et_2O/pet. ether). Mp 40.5-42°.

Gunstone, F.D. *et al, Chem. Phys. Lipids*, 1975, **15**, 174, 189, 198 (*synth, glc, tlc, pmr, ms*)

12,13-Epoxy-8-octadecenoic acid E-00124

11-(3-Pentyloxiranyl)-8-undecenoic acid, 9CI

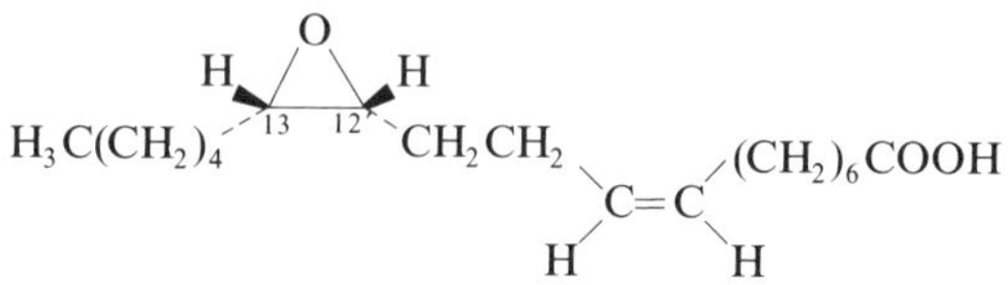

$C_{18}H_{32}O_3$ M 296.449

(12RS,13SR)-(Z)-form

(±)-cis,cis-*form*

threo-*Dihydroxy acid:* (Z)-threo-*12,13-Dihydroxy-8-octadecenoic acid*

$C_{18}H_{34}O_4$ M 314.464

Cryst. (Et_2O/pet. ether). Mp 51-52°.

Gunstone, F.D. *et al, Chem. Phys. Lipids*, 1975, **15**, 174, 189, 198 (*synth, glc, tlc, pmr, ms*)

12,13-Epoxy-9-octadecenoic acid E-00125

Leukotoxin B

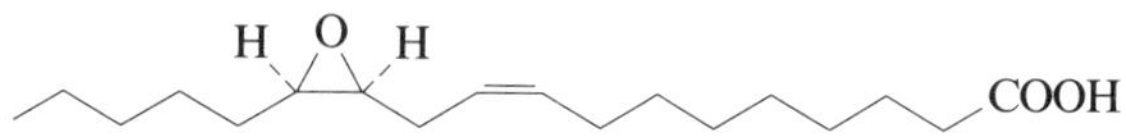

(9Z,12S,13R)-*form*

$C_{18}H_{32}O_3$ M 296.449

(9Z,12S,13R)-form [38003-78-0]

(+)-*Vernolic acid*

Constit. of *Vernonia anthelmintica* and some other oils. Mp 32.5°. $[\alpha]_D^{20}$ +2.03° (hexane).

(9Z,12R,13S)-form [503-07-1]

(−)-*Vernolic acid*

Constit. of several Malvaceae seed oils.

(9Z,12RS,13SR)-form [58800-38-7]

(±)-*Vernolic acid.* cis-*12,13-Epoxyoleic acid*

Cryst. (pet. ether at −10°). Mp 35-36°.

(9E,12R*,13S*)-form

cis-*12,13-Epoxyelaidic acid*

Isol. from *Mucuna prurita*. No phys. props. reported.

[21019-42-1]

Hopkins, C.Y. *et al, J. Am. Oil Chem. Soc.*, 1960, **37**, 682 (*isol*)
Osbond, J.M., *J. Chem. Soc.*, 1961, 5270 (*synth*)
Tallent, W.H. *et al, Lipids*, 1966, **1**, 335 (*occur*)
Gunstone, F.D. *et al, Chem. Phys. Lipids*, 1975, **15**, 174, 189, 198 (*synth, chromatog, pmr, ms*)
Hasan, S.Q. *et al, J. Indian Chem. Soc.*, 1980, **57**, 920 (*isol*)
Kato, T. *et al, Tetrahedron Lett.*, 1983, **24**, 4715 (*isol*)
Moustakis, C.A. *et al, Tetrahedron Lett.*, 1986, **27**, 303 (*synth*)

9,10-Epoxy-12-octadecynoic acid E-00126

3-(2-Octynyl)oxiraneoctanoic acid, 9CI

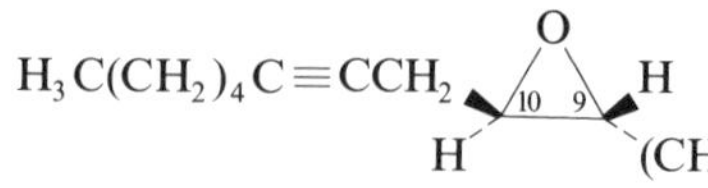

(9*RS*,10*RS*)-*form*

$C_{18}H_{30}O_3$ M 294.433

(9RS,10RS)-form

(±)-trans-*form*

erythro-*Dihydroxy acid:* erythro-*9,10-Dihydroxy-12-octadecynoic acid*

$C_{18}H_{32}O_4$ M 312.448

Cryst. (Et_2O/pet. ether). Mp 105.5-107°.

(9R*,10S*)-form

cis-*form*

Present in *Helichrysum bracteatum* seed oil probably as the 9*S*,10*R*-form.

threo-*Dihydroxy acid:* threo-*9,10-Dihydroxy-12-octadecynoic acid*

Cryst. (Et_2O/pet. ether). Mp 68-70°, Mp 72-72.5°.

Mikolajczak, K.L. *et al, J. Org. Chem.*, 1964, **29**, 318.
Conacher, M.B.S. *et al, Lipids*, 1970, **5**, 137 (*occur*)
Gunstone, F.D. *et al, Chem. Phys. Lipids*, 1975, **15**, 174, 189, 198 (*synth, glc, tlc, pmr, ms*)

11,12-Epoxy-9-octadecynoic acid E-00127

10-(3-Hexyloxiranyl)-9-decynoic acid, 9CI

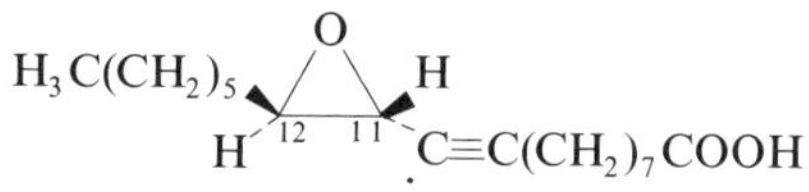

$C_{18}H_{30}O_3$ M 294.433

(11RS,12RS)-form

Abbot, G.G. *et al, Chem. Phys. Lipids*, 1971, **7**, 290.
Gunstone, F.D. *et al, Chem. Phys. Lipids*, 1979, **24**, 193.

12,13-Epoxy-9-octadecynoic acid E-00128

11-(3-Pentyloxiranyl)-9-undecynoic acid, 9CI

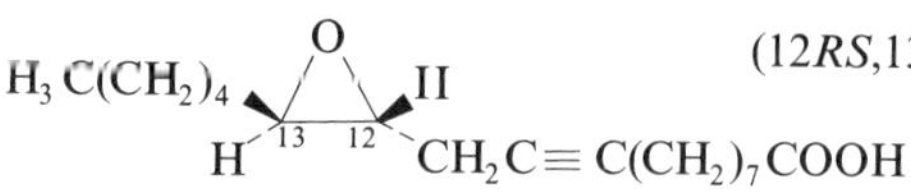

(12*RS*,13*RS*)-*form*

$C_{18}H_{30}O_3$ M 294.433

(12RS,13RS)-form

(±)-trans-*form*

erythro-*Dihydroxy acid:* erythro-*12,13-Dihydroxy-9-octadecynoic acid*

$C_{18}H_{32}O_4$ M 312.448

Cryst. (Et_2O/pet. ether). Mp 93.5-95°.

(12RS,13SR)-form

(±)-cis-*form*

threo-*Dihydroxy acid:* threo-*12,13-Dihydroxy-9-octadecynoic acid*

Cryst. (Et_2O/pet. ether). Mp 59-62°.

Gunstone, F.D. *et al, Chem. Phys. Lipids*, 1975, **15**, 174, 189, 198 (*synth, glc, tlc, pmr, ms*)

9,10-Epoxy-18-oxooctadecanoic acid E-00129

3-(8-Oxooctyl)oxiraneoctanoic acid, 9CI

[41607-37-8]

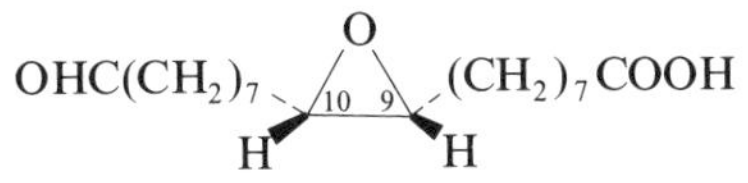

$C_{18}H_{32}O_4$ M 312.448

(9R*,10S*)-form

cis-*form*

Present in some cutins.

Kolattukudy, P.E., *Lipids*, 1973, **8**, 90 (*occur*)

5,9-Epoxy-4-tetradecen-1-yn-3-one E-00130

Tetrahydro-2-(2-oxobut-3-ynylidene)-6-pentylpyran

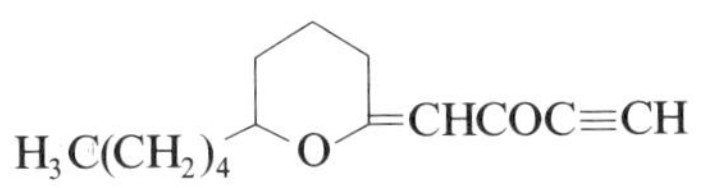

$C_{14}H_{20}O_2$ M 220.311

Constit. of *Seseli Lippomarathrum* roots. Oil. $[\alpha]_D^{22}$ −88°, $[\alpha]_{436}^{22}$ −207° (c, 5.71 in Et_2O).

Bohlmann, F. *et al*, *Chem. Ber.*, 1971, **104**, 2354 (*isol, ir, uv, ms*)

12,13-Epoxy-9,10,18-trihydroxyoctadecanoic acid E-00131

8,9-Dihydroxy-3-(5-hydroxypentyl)oxiraneundecanoic acid, 9CI

[52744-08-8]

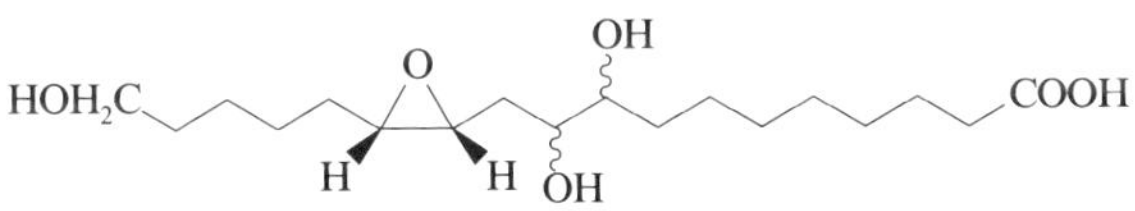

$C_{18}H_{34}O_6$ M 346.463

(12R*,13S*)-form

cis-*form*

Present in some cutins. The OH groups are probably *threo*.

Croteau, R., *Arch. Biochem. Biophys.*, 1974, **162**, 458 (*biosynth*)

Ergosta-5,7-dien-3-ol E-00132

24-Methylcholesta-5,7-dien-3-ol

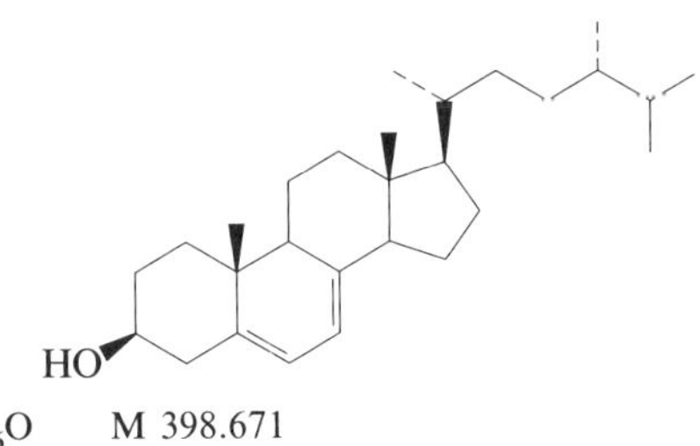

$C_{28}H_{46}O$ M 398.671

(3β,24S)-form [516-79-0]

22,23-Dihydroergosterol. Provitamin D_4

Constit. of *Saccharomyces cerevisiae*. Also from lipids of *Claviceps purpurea*, seed oil of *Vitis vinifera* and from *Polyporus* spp. Cryst. + H_2O (EtOAc/MeOH). Mp 152-153°. $[\alpha]_D$ −128.7° (c, 0.4 in $CHCl_3$).

Barton, D.H.R. *et al*, *J. Chem. Soc., Perkin Trans. 1*, 1974, 1326 (*isol*)
Kircher, H.W. *et al*, *Lipids*, 1975, **10**, 517 (*synth*)
Barton, D.H.R. *et al*, *J. Chem. Soc., Perkin Trans. 1*, 1976, 821 (*synth*)
Brynjolffssen, J. *et al*, *J. Chem. Soc., Perkin Trans. 1*, 1976, 826 (*synth*)
Mercer, E.I. *et al*, *Phytochemistry*, 1976, **15**, 283 (*biosynth*)
Curry, D.J. *et al*, *J. Chem. Soc., Perkin Trans. 1*, 1977, 822 (*synth*)

Ergosta-7,24(28)-dien-3-ol E-00133

24-Methylenecholest-7-en-3-ol

$C_{28}H_{46}O$ M 398.671

(3β,5α)-form [474-68-0] ***Episterol***

Constit. of yeasts, *Neurospora crassa* and *Asterias amurensis*. Also present in lichens. Cryst. (MeOH). Mp 127.5-128.5° (150-151°). $[\alpha]_D$ +4.4° (c, 2.3 in $CHCl_3$) (−5°).

Barton, D.H.R. *et al*, *J. Chem. Soc.*, 1945, 813; 1946, 512 (*struct*)
Kobayashi, M. *et al*, *Tetrahedron*, 1973, **29**, 1193 (*isol*)
Renaud, R. *et al*, *Phytochemistry*, 1976, **15**, 977 (*isol*)

Ergosta-7,22-dien-3-one E-00134

24-Methylcholesta-7,22-dien-3-one

$C_{28}H_{44}O$ M 396.655

(5α,24S)-form [32507-77-0]

Campesta-7,22-dien-3-one

Constit. of *Fomes fomentarius*, *F. applanatus* and *Coriolus sanguineus*. Cryst. (Me_2CO). Mp 184-187°. $[\alpha]_D$ +6° ($CHCl_3$).

Munro, H.D. *et al*, *J. Chem. Soc. C*, 1971, 685.

Ergosta-5,8,17(20)-trien-3-ol E-00135

24-Methylcholesta-5,8,17(20)-trien-3-ol

$C_{28}H_{44}O$ M 396.655

(3β,17(20)ξ,24ξ)-form [20362-29-2]

Constit. of *Edgeworthia papyrifera*. Cryst. Mp 136-137°.

Sugiyama, N. *et al*, *Nippon Kagaku Zasshi*, 1968, **89**, 710; *CA*, **70**, 862.

Ergosta-5,8,22-trien-3-ol E-00136

24-Methylcholesta-5,8,22-trien-3-ol

$C_{28}H_{44}O$ M 396.655

(3β,22E,24R)-form [50657-31-3] ***Lichesterol***

Constit. of *Xanthoria parietina*. Cryst. (Et_2O/MeOH). Mp 114-115°. $[\alpha]_D^{25}$ −26.6° (c, 0.3 in $CHCl_3$).

Ac: Mp 125-126°.

Lenton, J.R. *et al*, *Phytochemistry*, 1973, **12**, 1135.

Ergosta-6,8(14),22-trien-3-ol E-00137

24-Methylcholesta-6,8(14),22-trien-3-ol

$C_{28}H_{44}O$ M 396.655

(3β,5α,22E)-form [516-89-2]

Ergosterol B_2. Isoergosterol

Cryst. (MeOH). Mp 113-114°. $[\alpha]_D^{25}$ −115° (c, 2.10 in $CHCl_3$). λ_{max} 252 nm (EtOH).

Ac: [5226-38-0].

$C_{30}H_{46}O_2$ M 438.692

Cryst. (MeOH). Mp 115-117°. $[\alpha]_D$ −102° (c, 0.44 in $CHCl_3$).

(3β,5β,22E)-form [63039-38-3]

Mp 88°. $[\alpha]_D$ +75° (c, 0.81 in $CHCl_3$).

Ac: [63180-63-2].

Mp 115-116°. $[\alpha]_D$ +73° (c, 0.89 in $CHCl_3$).

[69610-90-8]

Fieser, M. *et al*, *J. Am. Chem. Soc.*, 1952, **74**, 5397 (*synth*)
Andrieux, J. *et al*, *J. Chem. Soc., Perkin Trans. 1*, 1977, 359 (*synth*)
Kaneko, C. *et al*, *Chem. Pharm. Bull.*, 1978, **26**, 3582 (*synth*)

Barton, D.H.R. *et al*, *Synthesis*, 1979, 265 (*synth*)
Nes, W.D. *et al*, *J. Chem. Soc., Chem. Commun.*, 1985, 1431 (*biosynth*)
Uomori, A. *et al*, *J. Chem. Soc., Chem. Commun.*, 1984, 1176 (*biosynth*)
Seo, S. *et al*, *J. Chem. Soc., Perkin Trans. 1*, 1992, 569 (*biosynth*)

Ergosta-7,9(11),22-trien-3-ol, 9CI — E-00138

24-Methylcholesta-7,9(11),22-trien-3-ol

$C_{28}H_{44}O$ M 396.655

(3β,5α,22E)-form [516-84-7]

Ergosterol D

Mp 167°. $[\alpha]_D$ +25° ($CHCl_3$).

Ac:

$C_{30}H_{46}O_2$ M 438.692

Cryst. Mp 178-181°. $[\alpha]_D$ +32° (c, 1 in $CHCl_3$).

[102916-84-7]

Anderson, R.C. *et al*, *J. Chem. Soc.*, 1952, 2901.
Crawshaw, A. *et al*, *J. Chem. Soc.*, 1955, 3420 (*synth*)
Tsuda, M. *et al*, *Chem. Phys. Lipids*, 1979, **25**, 49 (*cmr*)

Ergosta-7,14,22-trien-3-ol — E-00139

24-Methylcholesta-7,14,22-trien-3-ol

$C_{28}H_{44}O$ M 396.655

(3β,5α,22E,24R)-form [516-87-0] ***Ergosterol B₃***

Isol. from *Zoophagus insidians*. Cryst. (MeOH). Mp 139°. $[\alpha]_D$ −238° ($CHCl_3$).

Ac: Cryst. ($CHCl_3$/MeOH). Mp 139°. $[\alpha]_D$ −217° ($CHCl_3$).

(3β,5β,22E,24R)-form

Ac: Cryst. Mp 140-141°. $[\alpha]_D^{20}$ −19.3° (c, 0.86 in $CHCl_3$).

Windaus, A. *et al*, *Justus Liebigs Ann. Chem.*, 1931, **488**, 91 (*synth*)
Barton, D.H.R. *et al*, *J. Chem. Soc.*, 1951, 257 (*synth*)
Paterson, J. *et al*, *J. Chem. Soc.*, 1954, 325.
Warner, S.A. *et al*, *Phytochemistry*, 1982, **21**, 2135 (*isol*)
Dolle, R.E. *et al*, *J. Org. Chem.*, 1986, **51**, 4047 (*deriv*)
Dolle, R.E. *et al*, *Tetrahedron Lett.*, 1988, **29**, 1581 (*synth*)

Ergosta-7,22,24(28)-trien-3-ol — E-00140

24-Methylenecholesta-7,22-dien-3-ol

$C_{28}H_{44}O$ M 396.655

(3β,5α)-form [36680-39-4]

Constit. of yeast. Cryst. (MeOH/Me_2CO). Mp 119-121°. $[\alpha]_D$ +36.5° (c, 0.85 in $CHCl_3$).

Barton, D.H.R. *et al*, *J. Chem. Soc., Perkin Trans. 1*, 1972, 513, 1231 (*isol, struct, biosynth*)

Ergost-5-en-3-ol, 9CI — E-00141

24-Methylcholest-5-en-3-ol

$C_{28}H_{48}O$ M 400.687

(3β,24R)-form [474-62-4] ***Campesterol***

Constit. of *Brassica* spp., *Glycine max* and *Triticum* spp. Found in virtually all plant oils. Cryst. (Me_2CO). Mp 157-158°. $[\alpha]_D^{23}$ −33° ($CHCl_3$).

(3β,24S)-form

Dihydrobrassicasterol

Occurs in lipids of *Chlorella ellipsoidea* and in *Physalis peruviana*. Needles (Me_2CO). Mp 157-158°. $[\alpha]_D^{24}$ −46.3° (c, 1.2 in $CHCl_3$).

Thompson, M.J. *et al*, *Steroids*, 1965, **5**, 745 (*synth*)
Fedeli, E. *et al*, *J. Am. Oil Chem. Soc.*, 1966, **43**, 254 (*occur*)
Tarzia, G. *et al*, *Gazz. Chim. Ital.*, 1967, **97**, 102 (*synth*)
Thompson, M.J. *et al*, *Phytochemistry*, 1972, **11**, 1781 (*pmr*)
Karrer, W. *et al*, *Konstitution und Vorkommen der Organischen Pflanzenstoffe*, 2nd Ed., Birkhäuser Verlag, Basel, 1972-1985, no. 2053 (*occur*)
Lenton, J.R. *et al*, *Phytochemistry*, 1975, **14**, 1523 (*biosynth*)
Rubinstein, I. *et al*, *Phytochemistry*, 1976, **15**, 195 (*cmr*)

Ergosterol — E-00142

Ergosta-5,7,22E-trien-3β-ol, 9CI. 24R-Methylcholesta-5,7,22E-trien-3β-ol. Ergosterin. Provitamin D₂. Ertron

[57-87-4]

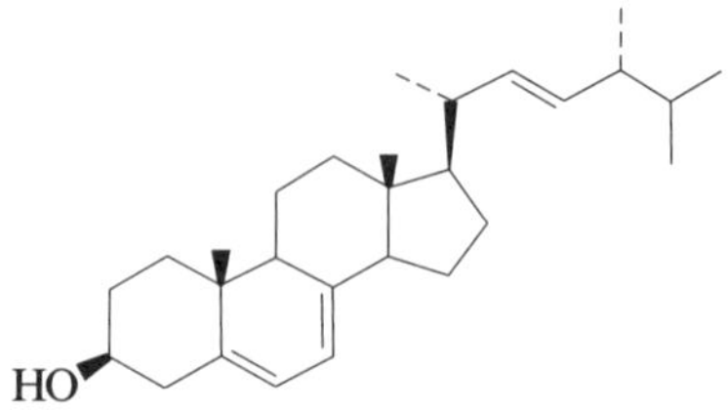

$C_{28}H_{44}O$ M 396.655

Occurs in yeast and fungi. The main fungal steroid. Also found in small amts. in higher plant prods., e.g. palm oil. Small hydrated plates (EtOH). Mp 168° (165°). $[\alpha]_D$ −133° ($CHCl_3$). λ_{max} 280 nm (EtOH).

Ac: [2418-45-3].

$C_{30}H_{46}O_2$ M 438.692

Plates (Et_2O/EtOH). Mp 175-176°.

Benzoyl:

$C_{35}H_{48}O_2$ M 500.763

Mp 168°. $[\alpha]_D$ −68°.

Me ether:

$C_{29}H_{46}O$ M 410.682

Mp 151-152°.

O-*Hexadecanoyl: Ergosteryl palmitate*

$C_{44}H_{74}O_2$ M 635.067

Isol. from microorganisms, *Penicillium* spp. and *Aspergillus itaconicus*. Cryst. Mp 107-108°. $[\alpha]_D^{20}$ −57° ($CHCl_3$).

9,10-Diepimer: see *Lumisterol, L-00029*

9-Epimer: [474-70-4]. *Isopyrocalciferol*

$C_{28}H_{44}O$ M 396.655

Cryst. (EtOH). Mp 112-120°. $[\alpha]_D$ +332° ($CHCl_3$).

3-O-β-D-Galactopyranoside:

$C_{34}H_{54}O_6$ M 558.797

Mp 204-205° (as 2′,3′,4′-Tri-Ac, Me ester). $[\alpha]_D^{25}$ −27.9° ($CHCl_3$).

3-Sulfate: Mp 186° (as Na salt).

Sell, H.M. *et al*, *J. Biol. Chem.*, 1938, **125**, 235 (*galactopyranoside*)
Kennedy, I. *et al*, *J. Chem. Soc.*, 1939, 250 (*struct*)
Castells, J. *et al*, *J. Chem. Soc.*, 1959, 1159 (*synth*)
Jacobs, H.J.C. *et al*, *Recl. Trav. Chim. Pays-Bas* (*J. R. Neth. Chem. Soc.*), 1965, **84**, 932 (*ord*)
Karrer, W. *et al*, *Konstitution und Vorkommen der Organischen Pflanzenstoffe*, 2nd Ed., Birkhäuser Verlag, Basel, 1972-1985, no. 2060 (*occur*)
Fryberg, M. *et al*, *J. Am. Chem. Soc.*, 1973, **95**, 5747 (*biosynth*)
Hull, S.E. *et al*, *Acta Crystallogr., Sect. B*, 1976, 2370 (*struct*)
Mercer, E.I. *et al*, *Phytochemistry*, 1976, **15**, 283 (*biosynth*)
Adler, J.H. *et al*, *Lipids*, 1977, **12**, 364 (*pmr, struct*)
Smith, W.B. *et al*, *Org. Magn. Reson.*, 1977, **9**, 644 (*cmr*)
Kirk, D.N. *et al*, *J. Chem. Soc., Perkin Trans. 2*, 1990, 1567 (*pmr*)

(2-Ethoxy-2-oxoethyl) triphenylphosphonium, 9CI — E-00143

(Carboxymethyl)triphenylphosphonium ethyl ester, 8CI. Carbethoxymethyltriphenylphosphonium

[42809-80-3]

$Ph_3P^{\oplus}CH_2COOEt$

$C_{22}H_{22}O_2P^{\oplus}$ M 349.388 (ion)

pK_{a1} 9.2.
Chloride: [17577-28-5].
$C_{22}H_{22}ClO_2P$ M 384.841
Cryst. (CH_2Cl_2/CCl_4). Mp 87° dec., 144° dec. (after intensive drying). Forms a dihydrate.
Bromide: [1530-45-6].
$C_{22}H_{22}BrO_2P$ M 429.292
Cryst. (C_6H_6). Mp 158° dec. pK_{a1} 8.85 (80%, EtOH aq.).

Isler, O. *et al, Helv. Chim. Acta*, 1957, **40**, 1242 (*bromide*)
Considine, W., *J. Org. Chem.*, 1962, **27**, 647 (*chloride, ir*)
Quemeneur, F. *et al, J. Chem. Res., Synop.*, 1979, 187 (*bromide*)

2-Ethylcyclopropanetetradecanoic acid, 9CI E-00144

15,16-Methyleneoctadecanoic acid

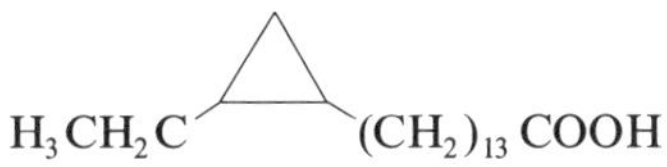

$C_{19}H_{36}O_2$ M 296.492
Both *cis*- and *trans*-forms have been synthesised as Me esters.

Christie, W.W. *et al, Chem. Phys. Lipids*, 1968, **2**, 196 (*synth, glc, ir, nmr*)
Christie, W.W., *Top. Lipid Chem.*, 1970, **1**, 4 (*rev*)
Gunstone, F.D. *et al, Chem. Phys. Lipids*, 1973, **10**, 303 (*synth, glc, nmr*)

3-Ethyl-2-hydroxynonanoic acid E-00145

3-Ethyl-2-hydroxypelargonic acid

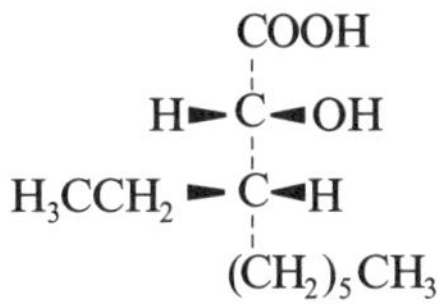

Relative configuration

$C_{11}H_{22}O_3$ M 202.293
(2RS,3SR)-form
Cryst. (pet. ether). Mp 47°.
Et ester:
$C_{13}H_{26}O_3$ M 230.347
Liq. Bp_{15} 148-150°.

Bagard, P., *Bull. Soc. Chim. Fr.*, 1907, 307, 346 (*synth*)

F

Ficulinic acid A **F-00001**
2-Heptyl-10-oxo-11-nonadecenoic acid, 9CI

$H_3C(CH_2)_6CH{=}CHCO(CH_2)_7CH(COOH)(CH_2)_6CH_3$

$C_{26}H_{48}O_3$ M 408.663
***(E)*-form** [102791-30-0]
Isol. from the sponge *Ficulina ficus*. Shows weak cytotoxic props. Mp 33-35°.

Guyot, M. *et al*, *J. Nat. Prod.* (*Lloydia*), 1986, **49**, 307 (*isol*)

Ficulinic acid B **F-00002**
2-Heptyl-12-oxo-13-heneicosenoic acid, 9CI

$H_3C(CH_2)_6CH{=}CHCO(CH_2)_9CH(COOH)(CH_2)_6CH_3$

$C_{28}H_{52}O_3$ M 436.717
***(E)*-form** [102791-31-1]
Isol. from the sponge *Ficulina ficus*. Shows weak cytotoxic props. Mp 31-32°.

Guyot, M. *et al*, *J. Nat. Prod.* (*Lloydia*), 1986, **49**, 307 (*isol*)

16-Fluorohexadecanoic acid **F-00003**
16-Fluoropalmitic acid
[3109-58-8]

$FCH_2(CH_2)_{14}COOH$

$C_{16}H_{31}FO_2$ M 274.418
Constit. of the seeds of *Dichapetalum toxicarium*. Cryst. (pet. ether). Mp 71-73°. Toxic.

Ward, P.F.V. *et al*, *Nature* (*London*), 1964, **201**, 611 (*isol*)
Cross, B.E. *et al*, *J. Chem. Soc., Perkin Trans. 1*, 1975, **23**, 2523 (*synth*)

3-Fluorononanoic acid **F-00004**

$H_3C(CH_2)_5CHFCH_2COOH$

$C_9H_{17}FO_2$ M 176.230
(±)-*form* [134056-75-0]
Cryst. Mp 43-45°.

Buist, P.H. *et al*, *J. Org. Chem.*, 1991, **56**, 3449 (*synth*, *pmr*)

18-Fluorooctadecanoic acid **F-00005**
18-Fluorostearic acid

$FCH_2(CH_2)_{16}COOH$

$C_{18}H_{35}FO_2$ M 302.472
Cryst. (MeOH). Mp 76-77° (68.5-69°). Toxic.
Me ester:
$C_{19}H_{37}FO_2$ M 316.499
Cryst. (pet. ether). Mp 33-34°.

Pattison, F.L.M. *et al*, *J. Am. Chem. Soc.*, 1956, **78**, 2255 (*synth*)
Pattison, F.L.M. *et al*, *Nature* (*London*), 1961, **192**, 1284 (*synth*)

18-Fluoro-9-octadecenoic acid **F-00006**
ω-Fluorooleic acid

$FCH_2(CH_2)_7CH{=}CH(CH_2)_7COOH$

$C_{18}H_{33}FO_2$ M 300.456
***(E)*-form**
Cryst. (Me_2CO). Mp 53-54°.
***(Z)*-form**
Isol. from seeds of *Dichapetalum toxicarium*. Mp 12.4-13.5°, Mp 15.8-16.8°.

Peters, R.A. *et al*, *Biochem. J.*, 1960, **77**, 17 (*isol*)
Pattison, A.M. *et al*, *Nature* (*London*), 1961, **192**, 1284 (*synth*)
Dear, R.E.A. *et al*, *J. Am. Chem. Soc.*, 1963, **85**, 622 (*synth*)

12-Fluoro-9,11,15-trihydroxy-5,13-prostadienoic acid **F-00007**

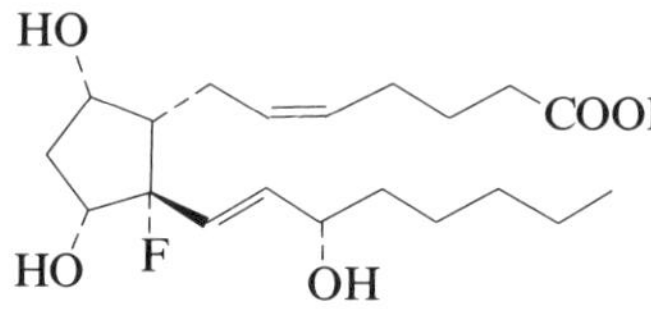

$C_{20}H_{33}FO_5$ M 372.476
***(5Z,9S,11R,13E,15S)*-form** [61111-02-2]
12-Fluoro-PGF$_{2\alpha}$

Wang, C.-L.J. *et al*, *J. Chem. Soc., Chem. Commun.*, 1976, 468 (*synth*)

Fuligorubin A **F-00008**
[108343-55-1]

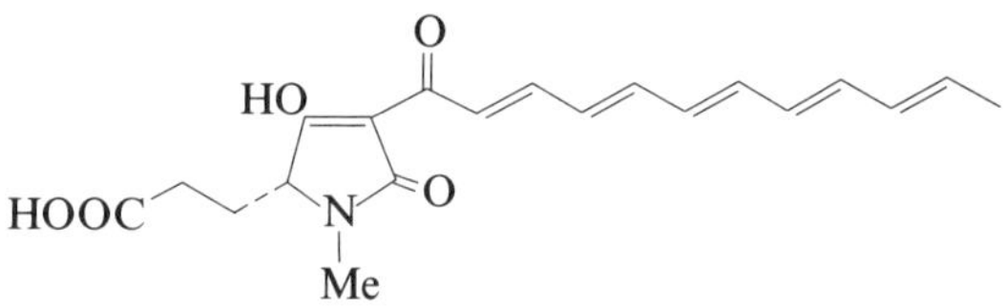

$C_{20}H_{23}NO_5$ M 357.405
Tetramic acid deriv. Isol. from the slime mold *Fuligo septica* (Myxomycetes). Isol. from the sponge *Topsentia roquensis*. Red cryst. + $2H_2O$. Mp 150° dec.

Casser, I. *et al*, *Angew. Chem., Int. Ed. Engl.*, 1987, **26**, 586 (*isol*, *uv*, *ir*, *pmr*, *cmr*, *ms*, *cd*, *struct*)
Ley, S.V. *et al*, *Tetrahedron*, 1992, **48**, 1145 (*synth*)

2,5-Furandiacetic acid **F-00009**
[50738-75-5]

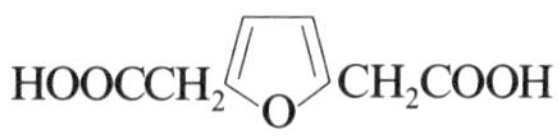

$C_8H_8O_5$ M 184.148
Found in urine. Cryst. (PhOMe). Mp 154-155°.
Di-Me ester:
$C_{10}H_{12}O_5$ M 212.202
Liq. d_4^{20} 1.205. Bp_1 117-118°. n_D^{20} 1.4788.
Diamide:
$C_8H_{10}N_2O_3$ M 182.179
Needles (H_2O). Mp 194-195°.
Dinitrile: 2,5-Bis(cyanomethyl)furan
$C_8H_6N_2O$ M 146.148
Platelets (EtOH/pet. ether). Mp 56-57°. $Bp_{0.5}$ 140-143°.

Novitskii, K.Yu. *et al*, *J. Gen. Chem. USSR* (*Engl. Transl.*), 1962, **32**, 3245 (*synth*)
Bauer, S. *et al*, *Justus Liebigs Ann. Chem.*, 1985, 813 (*isol*)

G

Gigantecin — G-00001

[130818-48-3]

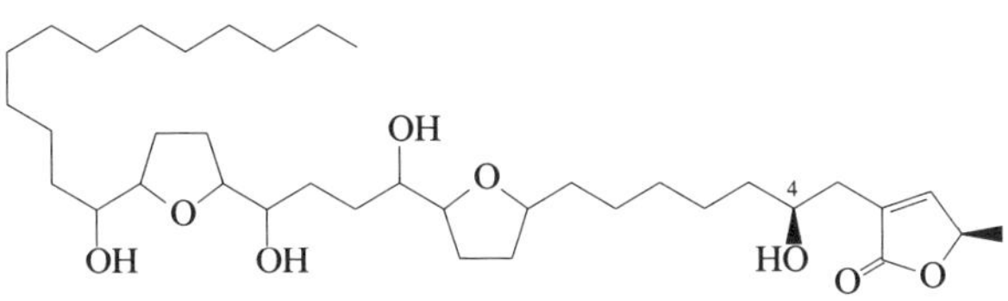

$C_{37}H_{66}O_8$ M 638.924

Constit. of *Goniothalamus giganteus*. Wax. Mp 96-98°. $[\alpha]_D$ +3.15° (c, 0.25 in MeOH).

4-Deoxy: ***4-Deoxygigantecin***

$C_{37}H_{66}O_7$ M 622.924

Isol. from the bark of *G. giganteus*. Wax. Mp 97-99°. $[\alpha]_D^{25}$ +15.5° (c, 0.2 in MeOH).

Alkofahi, A. *et al*, *Experientia*, 1990, **46**, 539 (*isol, pmr, cmr, ms*)
Fang, X.-P. *et al*, *Heterocycles*, 1992, **34**, 105 (*4-Deoxygigantecin*)
Fang, X.-P. *et al*, *Heterocycles*, 1992, **34**, 1075 (*4-Deoxygigantecin*)

Gigantetrocin — G-00002

[134955-48-9]

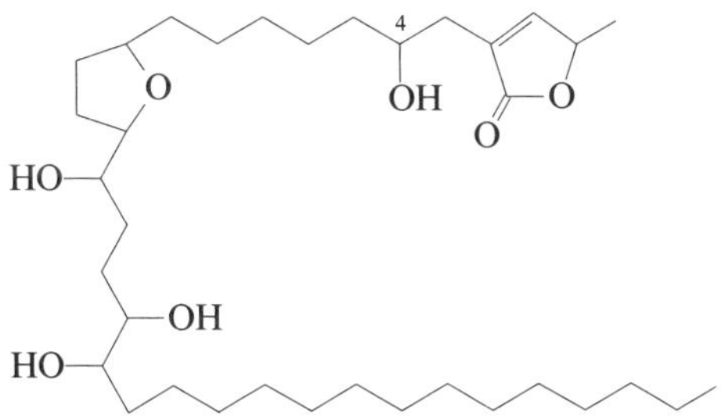

$C_{35}H_{64}O_7$ M 596.886

Acetogenin. Constit. of *Goniothalamus giganteus*. Cytotoxic. Amorph. solid. Mp 80-81°. $[\alpha]_D^{25}$ +10.3° ($CHCl_3$).

4-Deoxy: [134955-49-0]. ***Gigantriocin***

$C_{35}H_{64}O_6$ M 580.887

Constit. of *G. giganteus*. Cytotoxic. Cryst. Mp 69-71°. $[\alpha]_D$ +18° ($CHCl_3$).

Stereoisomer(?): [144239-83-8]. *Densicomacin*

$C_{35}H_{64}O_7$ M 596.886

Isol. from stembark of *Annona densicoma*. Cytotoxic against human tumour cells *in vitro*. Cryst. Mp 83-84°. $[\alpha]_D$ +26° (c, 0.05 in MeOH). Stereochemical relationship with Gigantetrocin not yet defined. Props. are similar.

[144239-84-9]

Fang, X.-P. *et al*, *Heterocycles*, 1991, **32**, 11 (*isol, pmr, cmr*)
Yu, J.G. *et al*, *J. Org. Chem.*, 1992, **57**, 6198 (*Densicomacin*)

Gingerdiones — G-00003

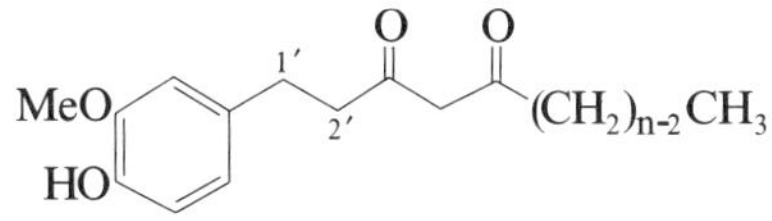

A series of homologous compds. from ginger (*Zingiber officinale*). Enolised *β*-diketones.

[***6***]***Gingerdione*** [61871-71-4]

1-(4-Hydroxy-3-methoxyphenyl)-3,5-decanedione

$C_{17}H_{24}O_4$ M 292.374

Constit. of *Z. officinale*. Prostaglandin inhibitor. Oil.

1′,2′-Didehydro: [76060-35-0]. [***6***]***-Dehydrogingerdione***

$C_{17}H_{22}O_4$ M 290.358

Isol. from *Z. officinale*. Yellow needles. Mp 83.5-84.5°. Enolised.

[***8***]***Gingerdione*** [77334-06-6]

1-(4-Hydroxy-3-methoxyphenyl)-3,5-dodecanedione, 9CI

$C_{19}H_{28}O_4$ M 320.428

Constit. of *Z. officinale*.

1,2-Didehydro: [77334-08-8]. [***8***]***-Dehydrogingerdione***

$C_{19}H_{26}O_4$ M 318.412

From *Z. officinale*. Mp 48-49°.

[***10***]***Gingerdione***

1-(4-Hydroxy-3-methoxyphenyl)-3,5-tetradecanedione

$C_{21}H_{32}O_4$ M 348.481

Isol. from *Z. officinale*. Strong prostaglandin inhibitor. Solid.

1′,2′-Didehydro: [82206-04-0]. [***10***]***-Dehydrogingerdione***

$C_{21}H_{30}O_4$ M 346.466

From *Z. officinale*. Prostaglandin inhibitor. Yellow needles. Mp 65-69.5°.

[***12***]***Gingerdione*** [91815-31-5]

1-(4-Hydroxy-3-methoxyphenyl)-3,5-hexadecanedione, 9CI

$C_{23}H_{36}O_4$ M 376.535

From *Z. officinale*. Mp 44-45°.

Denniff, P. *et al*, *J. Chem. Soc., Perkin Trans. 1*, 1981, 82 (*synth*)
Harvey, D.J., *J. Chromatogr.*, 1981, **212**, 75.
Kiuchi, F. *et al*, *Chem. Pharm. Bull.*, 1982, **30**, 754.

Glisoprenin A — G-00004

[144376-62-5]

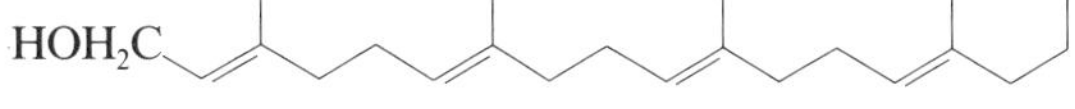

$C_{45}H_{82}O_5$ M 703.140

Prod. by *Gliocladium* sp. FO-1513. Inhibitor of acyl CoA: cholesterol acyltransferase. Oil. $[\alpha]_D^{18}$ +0.4° (c, 1 in $CHCl_3$).

Tomoda, H. *et al*, *J. Antibiot.*, 1992, **45**, 1202, 1669 (*isol, pmr, cmr, struct*)

Glisoprenin B G-00005

[144376-63-6]

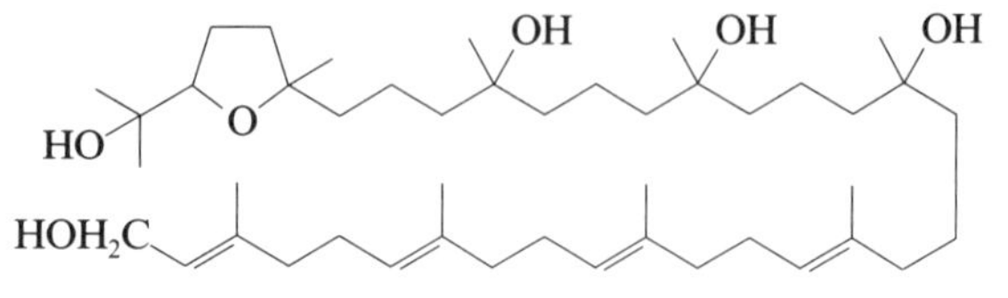

$C_{45}H_{82}O_6$ M 719.139

Prod. by *Gliocladium* sp. FO-1513. Inhibitor of acyl CoA: cholesterol acyltransferase. Oil. $[\alpha]_D^{18}$ −0.8° (c, 1 in $CHCl_3$).

Tomoda, H. *et al*, *J. Antibiot.*, 1992, **45**, 1202, 1669 (*isol, pmr, cmr, struct*)

Glycerol G-00006

1,2,3-Propanetriol, 9CI. 1,2,3-Trihydroxypropane. Glycerin. Amylac. Babylax. Bulboid. Cristal. Dagralax. Glyrol. Luxoral. Meprolax. Osmoglyn. Glyceritol

[56-81-5]

$HOCH_2CH(OH)CH_2OH$

$C_3H_8O_3$ M 92.094

Found extensively in esterified form in animal and plant glycerides. Obt. on large scale by alkaline hydrol. of fats during soap manuf. and by other routes. Used in cosmetics, foods, tobacco processing and extensively in numerous industrial and domestic prods. Component of alkyd resins and polyurethanes. Mild laxative, diuretic. Used for detn. of B; photometric detn. of Co. Syrup with sweet taste. Misc. H_2O, EtOH; insol. C_6H_6, $CHCl_3$, CCl_4. d 1.262, d_{15}^{15} 1.265. Mp 17.8°. Bp 290° part. dec., Bp_{20} 182°. pK_{a1} 14.4. V. hygroscopic. Steam-volatile. Numerous glycerides are. listed separately. Skin and eye irritant. Ingestion can cause headache and vomiting. Exp. reprod. effects.

▷ Reacts violently or explosively with many oxid. agents. MA8050000.

Bismuth salt: [17140-54-4]. *Bismutoral*
Oral antiseptic.

Complex with aluminium hydroxide: [12182-48-8]. ***Glucalox*, INN.** *Glycalox*, BAN. *Manalox AG*
Antacid.

Ac: see *Glycerol 2-acetate, G 00008*

Tribenzoyl: [614-33-5].
$C_{24}H_{20}O_6$ M 404.418
Mp 72°.

Tris-4-nitrobenzoyl: Mp 197.5°.

1-Me ether: [623-39-2]. *3-Methoxy-1,2-propanediol*
$C_4H_{10}O_3$ M 106.121
Bp 220°, Bp_{40} 136°.

2-Me ether: [761-06-8]. *2-Methoxy-1,3-propanediol*
$C_4H_{10}O_3$ M 106.121
Bp 232°, Bp_{40} 148°.

1,2-Di-Me ether: [40453-77-8]. *2,3-Dimethoxy-1-propanol*
$C_5H_{12}O_3$ M 120.148
Bp 180°, Bp_{40} 100°.

1,3-Di-Me ether: [623-69-8]. *1,3-Dimethoxy-2-propanol*
$C_5H_{12}O_3$ M 120.148
Bp 169°, Bp_{40} 88°.

Tri-Me ether: [20637-49-4]. *1,2,3-Trimethoxypropane*
$C_6H_{14}O_3$ M 134.175
Bp 148°.

2-O-Benzyl:
$C_{10}H_{14}O_3$ M 182.219
Oil.

[18673-08-0, 28959-04-8, 36887-04-4]

Welcher, F.J., *Organic Analytical Reagents*, Van Nostrand, New York, 1947, **1**, 91 (*use*)
Gidez, L.I. *et al*, *J. Am. Chem. Soc.*, 1952, **74**, 2413 (*synth*)
Miner, C.S. *et al*, *Glycerol, Am. Chem. Soc. Monograph No. 117*, N.Y., 1953 (*rev*)
Nessonova, G.D. *et al*, *Zavod. Lab.*, 1959, **25**, 786 (*detn, Co*)
U.S. Pat., 3 352 895, (1967); *CA*, **68**, 33186x (*synth, Glucalox*)
Newman, A.A., *Glycerol, Am. Chem. Soc. Monograph No. 117*, Morgan-Grampian, London, 1968 (*rev*)
Karrer, W. *et al*, *Konstitution und Vorkommen der Organischen Pflanzenstoffe*, 2nd Ed., Birkhäuser Verlag, Basel, 1972-1985, no. 139.
Budden, R. *et al*, *Arzneim.-Forsch.*, 1978, **28**, 1579 (*pharmacol*)
Kern, J.C., *Kirk-Othmer Encycl. Chem. Technol.*, 3rd Ed., Wiley, N.Y., 1978-1984, **11**, 921 (*rev*)
Carmaniu, S. *et al*, *Arch. Int. Physiol. Biochim.*, 1980, **88**, 255 (*metab*)
Frank, M.S. *et al*, *Pharmacotherapy (Carlisle, Mass.)*, 1981, **1**, 147 (*rev, pharmacol*)
Negwer, M., *Organic-Chemical Drugs and their Synonyms*, 6th Ed., Akademie-Verlag, Berlin, 1987, 142.
Lewis, R.J., *Food Additives Handbook*, Van Nostrand Reinhold International, New York, 1989, GGA000, GGU000.
Marinier, A. *et al*, *Can. J. Chem.*, 1992, **70**, 2350 (*benzyl ether*)
Martindale, The Extra Pharmacopoeia, 30th edn., Pharmaceutical Press, London, 1993, 1374.
Lewis, R.J., *Sax's Dangerous Properties of Industrial Materials*, 8th Ed., Van Nostrand-Reinhold, 1992, GGA000, GGU000.
Bretherick, L., *Handbook of Reactive Chemical Hazards*, 4th edn., Butterworth, London and Boston, 1990, 1209.
Luxon, S.G., *Hazards in the Chemical Laboratory*, 5th edn., Royal Society of Chemistry, Cambridge, 1992, 648.

Glycerol 1-acetate G-00007

1,2,3-Propanetriol 1-acetate, 9CI. 3-Acetoxy-1,2-propanediol. α-Monoacetin

[106-61-6]

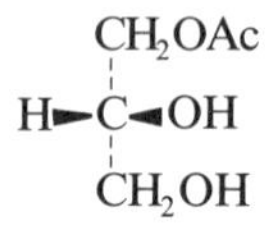

(R)-form

$C_5H_{10}O_4$ M 134.132

▷ AK3600000.

(R)-form [57416-01-3]
L-form
$[\alpha]_D$ −10.5° (dry Py).

(±)-form [93713-40-7]
Used in manuf. of smokeless powder and dynamite, solvent for basic dyes and tanning leather. d_4^{20} 1.206. Bp_4 134-135°, $Bp_{0.4}$ 103°. n_D^{20} 1.4157.

Di-Me ether:
$C_7H_{14}O_4$ M 162.185
Bp_{18} 100-110°.

Bis(4-nitrobenzoyl): Mp 129-130°.

[105498-20-2]

Baer, E. *et al*, *J. Am. Chem. Soc.*, 1945, **67**, 2031 (*synth, abs config*)
Williams, P.H. *et al*, *J. Am. Chem. Soc.*, 1960, **82**, 4883 (*synth*)
Ger. Pat., 1 810 568, (1969); *CA*, **71**, 90859 (*synth*)
Dahlhoff, W.V. *et al*, *Justus Liebigs Ann. Chem.*, 1975, 1914 (*synth*)
Gronowitz, S. *et al*, *Chem. Phys. Lipids*, 1976, **17**, 244 (*synth, cd*)
Suemune, H. *et al*, *Chem. Pharm. Bull.*, 1986, **34**, 3440 (*synth*)
Lewis, R.J., *Sax's Dangerous Properties of Industrial Materials*, 8th Ed., Van Nostrand-Reinhold, 1992, GGO000.

Glycerol 2-acetate G-00008

Propanetriol 2-acetate, 9CI. 2-Acetoxy-1,3-propanediol. β-Monoacetin

[100-78-7]

$HOCH_2CH(OAc)CH_2OH$

$C_5H_{10}O_4$ M 134.132

Viscous oil. $Bp_{0.3}$ 177-178°.

1,3-Benzylidene:

$C_{12}H_{14}O_4$ M 222.240

Mp 101°.

Averill, H.P. *et al*, *J. Am. Chem. Soc.*, 1929, **51**, 866 (*synth*)
Bergmann, M. *et al*, *Hoppe Seyler's Z. Physiol. Chem.*, 1930, **191**, 211 (*synth*)
McElroy, O.E. *et al*, *J. Am. Chem. Soc.*, 1934, **56**, 1191 (*synth*)
Artamonova, N. *et al*, *CA*, 1979, **91**, 56320 (*synth*)

Glycerol 1,2-diacetate 3-octadecanoate G-00009

Octadecanoic acid 2,3-bis(acetyloxy)propyl ester. α,β-Diacetostearin

[33599-07-4]

$C_{25}H_{46}O_6$ M 442.635

(R)-form [57416-08-7]

*1,2-Diacetyl-3-stearoyl-*sn*-glycerol*

Cryst. (EtOH). Mp 39-40°. $[M]_D^{25}$ +10.2° (C_6H_6).

(±)-form [60562-71-2]

Mp 48°. Exhibits polymorphism.

Gronowitz, S. *et al*, *Chem. Phys. Lipids*, 1975, **14**, 174 (*synth, ord*)
v. Menz, H.-U., *Fette, Seifen, Anstrichm.*, 1975, **77**, 170 (*synth*)
Bus, J. *et al*, *Chem. Phys. Lipids*, 1976, **16**, 123 (*pmr*)

Glycerol 1,3-didecanoate G-00010

Decanoic acid 2-hydroxy-1,3-propanediyl ester, 9CI. α,α'-Dicaprin

[17598-93-5]

$C_{23}H_{44}O_5$ M 400.598

Cryst. (hexane). Mp 44°.

Mank, A.P.J. *et al*, *Chem. Phys. Lipids*, 1976, **16**, 107 (*synth*)

Glycerol 1,2-didecanoate 3-dodecanoate G-00011

Dodecanoic acid 2,3-bis[(1-oxodecyl)oxy]propyl ester, 9CI. α,β-Dicaprolaurin

[60138-10-5]

$C_{35}H_{66}O_6$ M 582.903

(±)-form

Cryst. (C_6H_6). Mp 30°. Exhibits polymorphism.

Carter, M.G.R. *et al*, *J. Chem. Soc.*, 1939, 577 (*synth*)
v. Soest, T.C. *et al*, *Recl. Trav. Chim. Pays-Bas (J. R. Neth. Chem. Soc.)*, 1976, **95**, 162.

Glycerol 1,3-didecanoate 2-dodecanoate G-00012

Dodecanoic acid 1-[[(1-oxodecyl)oxy]methyl]-2-[(1-oxodecyl)oxy]ethyl ester, 9CI. α,α'-Dicaprolaurin

[60138-21-8]

$C_{35}H_{66}O_6$ M 582.903

Cryst. (EtOH). Mp 37.5°. Exhibits polymorphism.

Malkin, T. *et al*, *J. Chem. Soc.*, 1939, 103 (*synth*)
v. Soest, T.C. *et al*, *Recl. Trav. Chim. Pays-Bas (J. R. Neth. Chem. Soc.)*, 1976, **95**, 162.

Glycerol 1,2-didecanoate 3-hexadecanoate G-00013

α,β-Dicapropalmitin

$C_{39}H_{74}O_6$ M 639.010

(±)-form

Cryst. (C_6H_6). Mp 35°. Exhibits polymorphism.

Carter, M.G.R. *et al*, *J. Chem. Soc.*, 1939, 1518.

Glycerol 1,3-didecanoate 2-hexadecanoate G-00014

α,α'-Dicapropalmitin

$C_{39}H_{74}O_6$ M 639.010

Cryst. (EtOH). Mp 40°. Exhibits polymorphism.

Malkin, T. *et al*, *J. Chem. Soc.*, 1939, 1141.

Glycerol 1,3-didecanoate 2-octadecanoate G-00015

α,α'-Dicaprostearin

$C_{41}H_{78}O_6$ M 667.064

Cryst. (EtOH). Mp 44.5°. Exhibits polymorphism.

Malkin, T. *et al*, *J. Chem. Soc.*, 1939, 1141.

Glycerol 1,3-didecanoate 2-(9-octadecenoate) G-00016

$C_{41}H_{76}O_6$ M 665.048

(Z)-form

α,α'-Dicaproolein

Mattson, F.H. *et al*, *J. Lipid Res.*, 1964, **5**, 374 (*synth*)

Glycerol 1,2-didecanoate 3-phosphocholine G-00017

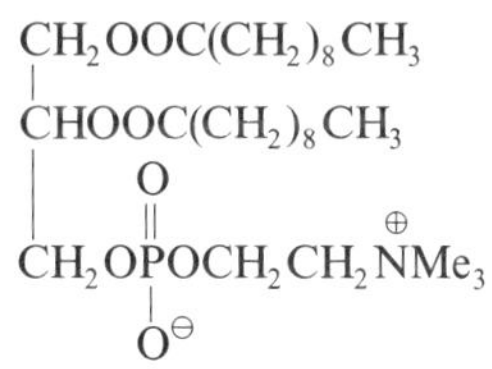

$C_{28}H_{56}NO_8P$ M 565.726

(±)-form [13699-47-3]

Dicaproylphosphatidylcholine

Chapman, D. *et al*, *J. Biol. Chem.*, 1966, **241**, 5044 (*nmr*)

Glycerol 1,2-didecanoate 3-tetradecanoate G-00018

Tetradecanoic acid 2,3-bis[(1-oxodecyl)oxy]propyl ester, 9CI. α,β-Dicapromyristin

$C_{37}H_{70}O_6$ M 610.956

(R)-form [57416-11-2]

*1,2-Didecanoyl-3-tetradecanoyl-*sn*-glycerol*

Cryst. (EtOH). Mp 26-26.5°. $[M]_{300}^{27}$ +2.9° (C_6H_6).

(±)-form

Cryst. (C_6H_6). Mp 34.5°. Exhibits polymorphism.

Carter, M.G.R. *et al*, *J. Chem. Soc.*, 1939, 1518 (*synth*)
Gronowitz, S. *et al*, *Chem. Phys. Lipids*, 1975, **14**, 174 (*synth, ord*)

Glycerol 1,3-didecanoate 2-tetradecanoate G-00019

α,α'-Dicapromyristin

$C_{37}H_{70}O_6$ M 610.956

Cryst. (EtOH). Mp 34°. Exhibits polymorphism.

Malkin, T. *et al*, *J. Chem. Soc.*, 1939, 1141.

Glycerol 1,2-didocosanoate 3-octadecanoate G-00020

α,β-Dibehenostearin

$C_{65}H_{126}O_6$ M 1003.707

(±)-***form***
Cryst. (pet. ether/Et_2O). Mp 73.5°. Exhibits polymorphism.

Jackson, F.L. *et al*, *J. Am. Chem. Soc.*, 1950, **72**, 4519.

Glycerol 1,2-didodecanoate **G-00021**

Dodecanoic acid 1-(hydroxymethyl)-1,2-ethanediyl ester, 9CI. αβ-Dilaurin

[17598-94-6]

$C_{27}H_{52}O_5$ M 456.705

(*S*)-***form*** [60562-15-4]
*1,2-Dilauroyl-*sn*-glycerol*
Cryst. ($CHCl_3$/pet. ether). Mp 45-47°. $[\alpha]_D^{23}$ −3.24° (c, 7.3 in $CHCl_3$).

Benzyl ether: Oil. $[\alpha]_D^{23}$ +6.90° (c, 7.33 in $CHCl_3$).

(±)-***form*** [59540-17-9]
Cryst. (hexane). Mp 20°, Mp 39°. Exhibits polymorphism.

Howe, R.J. *et al*, *J. Chem. Soc.*, 1951, 2663 (*synth, props*)
Ioannou, P.V. *et al*, *Synthesis*, 1979, 939 (*synth*)
Pascher, I. *et al*, *J. Mol. Biol.*, 1981, **153**, 791 (*cryst struct*)
Michelsen, P. *et al*, *Chem. Phys. Lipids*, 1983, **32**, 27 (*synth*)

Glycerol 1,3-didodecanoate **G-00022**

Dodecanoic acid 2-hydroxy-1,3-propanediyl ester, 9CI. α,α′-Dilaurin

[539-93-5]

$C_{27}H_{52}O_5$ M 456.705

Cryst. (EtOH/hexane). Mp 57.8°.

Baur, F.J. *et al*, *J. Am. Chem. Soc.*, 1951, **73**, 3926 (*synth*)

Glycerol 1,2-didodecanoate 3-decanoate **G-00023**

Dodecanoic acid 1-[[(1-oxodecyl)oxy]methyl]-1,2-ethanediyl ester, 9CI. α,β′-Dilaurocaprin

$C_{37}H_{70}O_6$ M 610.956

(±)-***form***
Cryst. (C_6H_6). Mp 35.5°. Exhibits polymorphism.

Carter, M.G.R. *et al*, *J. Chem. Soc.*, 1939, 577 (*synth*)
v. Soest, T.C. *et al*, *Recl. Trav. Chim. Pays-Bas (J. R. Neth. Chem. Soc.)*, 1976, **95**, 162.

Glycerol 1,3-didodecanoate 2-decanoate **G-00024**

Dodecanoic acid 2-[(1-oxodecyl)oxy]-1,3-propanediyl ester, 9CI. α,α′-Dilaurocaprin. 2-Caprinyldilaurin

[60138-22-9]

$C_{37}H_{70}O_6$ M 610.956

Isol. from seeds of *Cinnamomum laureiri* and cocoa butter. Cryst. (C_6H_6). Mp 38.5°. Exhibits polymorphism (Lower Mp's reported 8°, 23°, 33°).

Malkin, T. *et al*, *J. Chem. Soc.*, 1939, 103 (*synth*)
v. Soest, T.C. *et al*, *Recl. Trav. Chim. Pays-Bas (J. R. Neth. Chem. Soc.)*, 1976, **95**, 162.

Glycerol 1,3-didodecanoate 2-hexadecanoate **G-00025**

α,α′-Dilauropalmitin

$C_{43}H_{82}O_6$ M 695.117

Cryst. (EtOH). Mp 45.5°. Exhibits polymorphism.

Malkin, T. *et al*, *J. Chem. Soc.*, 1939, 1141 (*synth, props*)

Glycerol 1,3-didodecanoate 2-octadecanoate **G-00026**

α,α′-Dilaurostearin

$C_{45}H_{86}O_6$ M 723.171

Cryst. (EtOH). Mp 47°. Exhibits polymorphism.

Malkin, T. *et al*, *J. Chem. Soc.*, 1939, 1141 (*synth*)

Glycerol 1,2-didodecanoate 3-(9-octadecenoate) **G-00027**

9-Octadecenoic acid 2,3-bis[(1-oxododecyl)oxy]propyl ester, 9CI

$C_{45}H_{84}O_6$ M 721.155

(*R*)-(*Z*)-***form*** [67254-18-6]
*1,2-Dilauroyl-3-oleoyl-*sn*-glycerol*
Oil. $[M]_{300}$ +4.5° (C_6H_6).

(±)-(*E*)-***form***
α,β-Dilauroelaidin
Cryst. (EtOH). Mp 27°.

(±)-(*Z*)-***form*** [20299-66-5]
α,β-Dilauroolein
Cryst. (Et_2O). Mp 16° (20°). Exhibits polymorphism.

Daubert, B.F. *et al*, *J. Am. Chem. Soc.*, 1943, **65**, 2142 (*synth*)
Daubert, B.F., *J. Am. Chem. Soc.*, 1944, **66**, 290 (*synth*)
Åkesson, B. *et al*, *Lipids*, 1978, **13**, 338 (*synth*)
Gronowitz, S. *et al*, *Chem. Phys. Lipids*, 1979, **23**, 101 (*ord*)

Glycerol 1,3-didodecanoate 2-(9-octadecenoate) **G-00028**

$C_{45}H_{84}O_6$ M 721.155

(*Z*)-***form***
α,α′-Dilauroolein
Cryst. (pet. ether at −30°). Mp 15°. Exhibits polymorphism.

Jackson, F.L. *et al*, *J. Am. Chem. Soc.*, 1944, **66**, 289 (*synth*)
Daubert, B.F. *et al*, *J. Am. Chem. Soc.*, 1944, **66**, 690.

Glycerol 1,2-didodecanoate 3-octanoate **G-00029**

α,β-Dilaurocaprylin

$C_{35}H_{66}O_6$ M 582.903

(±)-***form***
Cryst. (EtOH/Me_2CO). Mp 28.5°.

McElroy, O.E. *et al*, *J. Am. Chem. Soc.*, 1934, **56**, 1191 (*synth*)

Glycerol 1,2-didodecanoate 3-phosphocholine **G-00030**

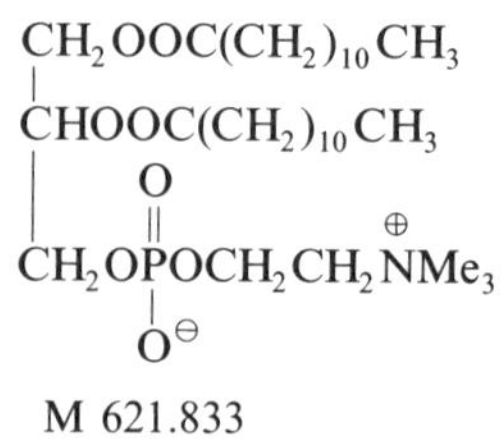

$C_{32}H_{64}NO_8P$ M 621.833

(*R*)-L-***form*** [18194-25-7]
Dilauroylphosphatidylcholine
$[\alpha]_D^{26}$ +6.1° (c, 1.5 in 1:1 $CHCl_3$/MeOH).

Baer, E. *et al*, *Biochemistry*, 1962, **1**, 518 (*synth*)
Brandt, A.E. *et al*, *Biochim. Biophys. Acta*, 1967, **144**, 605 (*synth*)

Glycerol 1,2-didodecanoate 3-phosphoethanolamine G-00031

Dodecanoic acid 1-[[[(2-aminoethoxy)hydroxyphosphinyl]oxy]methyl]-1,2-ethanediyl ester, 9CI.
Dilauroylphosphatidylethanolamine

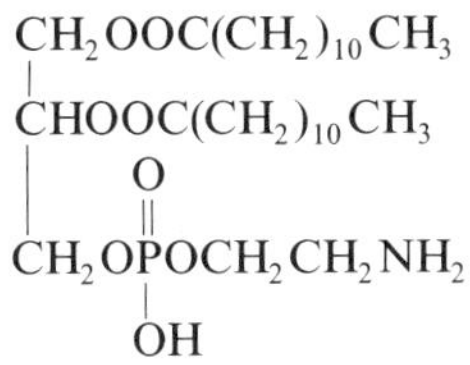

$C_{29}H_{58}NO_8P$ M 579.753

(±)-form [42436-56-6]
Mp 208°.

Bevan, T.H. *et al, J. Chem. Soc.*, 1951, 2667.
Chapman, D. *et al, Trans. Faraday Soc.*, 1966, **62**, 2607 (*pmr, props*)
v. Dijck, P.W.M., *Biochim. Biophys. Acta*, 1976, **455**, 576.
Wood, G.W. *et al, Chem. Phys. Lipids*, 1977, **18**, 316 (*ms*)
Cullis, P.R. *et al, Biochim. Biophys. Acta*, 1978, **513**, 31 (*nmr*)
Blume, A., *Biochim. Biophys. Acta*, 1979, **557**, 32.

Glycerol 1,2-didodecanoate 3-tetradecanoate G-00032

Tetradecanoic acid 2,3-bis[(1-oxododecyl)oxy]propyl ester, 9CI. α,β-Dilauromyristin. 1-Myristodilaurin

$C_{41}H_{78}O_6$ M 667.064

Isol. from lipids of *Attalea speciosa*.

(R)-form [57416-12-3]
*1,2-Dilauroyl-3-myristoyl-*sn-*glycerol*
Cryst. (EtOH). Mp 39.5-40°. $[M]^{27}_{300}$ +0.8° (C_6H_6).

(±)-form [60175-30-6]
Cryst. (C_6H_6). Mp 43.5°. Exhibits polymorphism (Lower Mp's reported 19°, 33.5°, 39°).

Carter, M.G.R. *et al, J. Chem. Soc.*, 1939, 577 (*synth*)
Gronowitz, S. *et al, Chem. Phys. Lipids*, 1975, **14**, 174 (*synth, ord*)
v. Soest, T. *et al, Recl. Trav. Chim. Pays-Bas (J. R. Neth. Chem. Soc.)*, 1976, **95**, 162.

Glycerol 1,3-didodecanoate 2-tetradecanoate G-00033

Tetradecanoic acid 2-[(1-oxododecyl)oxy]-1-[[(1-oxododecyl)oxy]methyl]ethyl ester, 9CI. α,α'-Dilauromyristin

[51464-76-7]

$C_{41}H_{78}O_6$ M 667.064

Cryst. (EtOH). Mp 48°. Exhibits polymorphism.

Malkin, T. *et al, J. Chem. Soc.*, 1939, 103 (*synth*)
v. Soest, T. *et al, Recl. Trav. Chim. Pays-Bas (J. R. Neth. Chem. Soc.)*, 1976, **95**, 162.

Glycerol 1,2-di-5,8,11,14-eicosatetraenoate 3-phosphoethanolamine G-00034

1,2-Dieicosa-5,8,11,14-tetraenoylglycero-3-phosphoethanolamine

[14994-07-1]

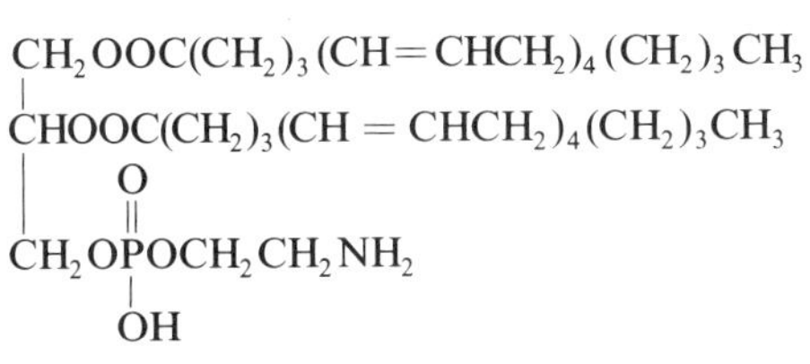

$C_{45}H_{74}NO_8P$ M 788.055

(R)-(all-Z)-form
Diarachidonoylphosphatidylethanolamine
Renin inhibitor. $[\alpha]^{20}_D$ +6.0°.

Bova, L.M. *et al, CA*, 1966, **64**, 3343 (*synth, ir*)
Miyazaki, M. *et al, Proc. Soc. Exp. Biol. Med.*, 1977, **155**, 468.

Glycerol 1,2-diheptadecanoate 3-phosphocholine G-00035

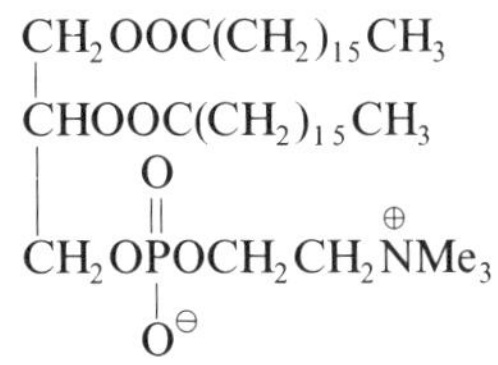

$C_{42}H_{84}NO_8P$ M 762.101

(R)-L-form [70897-27-7]
Diheptadecanoylphosphatidylcholine

Sylvius, J.R. *et al, Biochim. Biophys. Acta*, 1979, **555**, 175.

Glycerol 1,2-di-10,12-hexacosadiynoate 3-phosphoethanolamine G-00036

Di-10,12-hexacosadiynoylphosphatidylethanolamine

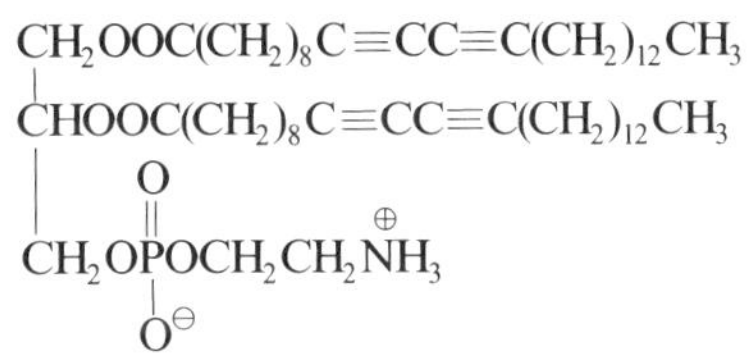

$C_{57}H_{98}NO_8P$ M 956.377

Polymerisable phospholipid.

(±)-form [77074-02-3]
Mp 180°.

N-*Tri-Me:* [77074-00-1]. *Glycerol 1,2-di-10,12-hexacosadiynoate-3-phosphocholine. Di-10,12-hexacosadiynoylphosphatidylcholine*
$C_{60}H_{104}NO_8P$ M 998.457
Polymerisable phospholipid. Monohydrate. Mp 215°.

Hupfer, B. *et al, Chem. Phys. Lipids*, 1983, **33**, 355 (*synth, ir, ms*)

Glycerol 1,2-dihexadecanoate G-00037

Hexadecanoic acid 1-(hydroxymethyl)-1,2-ethanediyl ester, 9CI. α,β-Dipalmitin. Glycerol 1,2-palmitate. 1,2-Dipalmitin

[761-35-3]

$C_{35}H_{68}O_5$ M 568.919

(S)-form [30334-71-5]
*1,2-Dipalmitoyl-*sn-*glycerol*
Cryst. ($CHCl_3$/pet. ether). Mp 68-69°. $[\alpha]^{23}_D$ −2.75° (c, 7.62 in $CHCl_3$).

Benzyl ether: Mp 42-42.5°. $[\alpha]_D$ +6.3° (c, 8.5 in $CHCl_3$).
4-Nitrobenzoyl: Mp 60-61°. $[\alpha]_D$ −1.4° (c, 10.7 in $CHCl_3$).

(±)-form [40290-32-2]
Cryst. (hexane). Mp 50° (65-66°). Exhibits polymorphism.

Ac: Mp 51-52°.
Benzyl ether: Mp 45-46°.

Howe, R.J. *et al, J. Chem. Soc.*, 1951, 2663 (*synth, props*)
Dorset, D.L. *et al, Z. Naturforsch., C*, 1978, **33**, 39 (*cryst struct*)
Ioannou, P.V. *et al, Synthesis*, 1979, 939 (*synth*)

Glycerol 1,3-dihexadecanoate G-00038

Hexadecanoic acid 2-hydroxy-1,3-propanediyl ester, 9CI. α,α'-Dipalmitin. Glycerol 1,3-dipalmitate. 1,3-Dipalmitin

[502-52-3]

$C_{35}H_{68}O_5$ M 568.919

Cryst. (Me_2CO/hexane at 0°). Mp 73-74°.

Ac: Mp 54°.

O-*Benzyl:* Mp 40°.

Mank, A.P.J. *et al, Chem. Phys. Lipids*, 1976, **16**, 107 (*synth*)

Glycerol 1,2-dihexadecanoate 3-decanoate G-00039

α,β-Dipalmitocaproin

$C_{45}H_{86}O_6$ M 723.171

(±)-***form***

Cryst. ($EtOH/C_6H_6$). Mp 45.5°. Exhibits polymorphism.

Carter, M.G.R. *et al, J. Chem. Soc.*, 1939, 1518.

Glycerol 1,3-dihexadecanoate 2-decanoate G-00040

Hexadecanoic acid 2-[(1-oxodecyl)oxy]-1,3-propanediyl ester, 9CI. α,α'-Dipalmitocaprin

[67826-14-6]

$C_{45}H_{86}O_6$ M 723.171

Cryst. (EtOH). Mp 52°. Exhibits polymorphism.

Malkin, T. *et al, J. Chem. Soc.*, 1939, 1141.
Gray, M.S. *et al, J. Am. Oil Chem. Soc.*, 1978, **55**, 601.

Glycerol 1,3-dihexadecanoate 2-docosanoate G-00041

α,α'-Dipalmitobehenin

$C_{57}H_{110}O_6$ M 891.492

Cryst. (pet. ether/Et_2O). Mp 66.6°. Exhibits polymorphism.

Jackson, F.L. *et al, J. Am. Chem. Soc.*, 1950, **72**, 4519.

Glycerol 1,2-dihexadecanoate 3-(9,12-octadecadienoate) G-00042

9,12-Octadecadienoic acid 2,3-bis[(1-oxohexadecyl)oxy] propyl ester, 9CI

$C_{53}H_{98}O_6$ M 831.354

(±)-(***all-Z***)-***form*** [2535-35-5]

α,β-Dipalmitolinolein

Cryst. Mp 36.5-38°.

Serebrennikova, G.A. *et al, CA*, 1963, **58**, 8898b (*synth*)

Glycerol 1,3-dihexadecanoate 2-(9,12-octadecadienoate) G-00043

9,12-Octadecadienoic acid 2-[(1-oxohexadecyl)oxy]-1-[[(1-oxohexadecyl)oxy]methyl]ethyl ester, 9CI

$C_{53}H_{98}O_6$ M 831.354

(***all-Z***)-***form*** [2442-56-0]

α,α'-Dipalmitolinolein

Cryst. Mp 25.5-26°.

Serebrennikova, G.A. *et al, CA*, 1963, **59**, 435b (*synth*)
Tamaki, Y. *et al, J. Insect Physiol.*, 1971, **17**, 1239 (*synth*)

Glycerol 1,2-dihexadecanoate 3-octadecanoate G-00044

Octadecanoic acid 2,3-bis[(1-oxohexadecyl)oxy]propyl ester, 9CI

$C_{53}H_{102}O_6$ M 835.385

(***S***)-***form*** [5281-84-5]

2,3-Dipalmitoyl-1-stearoyl-sn-glycerol

Cryst. (C_6H_6/EtOH). Mp 58°.

(±)-***form*** [5568-22-9]

α,β-Dipalmitostearin

Cryst. (C_6H_6/EtOH). Mp 62.5-63.5°. Exibits polymorphism.

Carter, M.G.R. *et al, J. Chem. Soc.*, 1939, 577.
Mitchell, L.C., *J. Am. Oil Chem. Soc.*, 1972, **49**, 281.
Lok, C.M. *et al, Chem. Phys. Lipids*, 1976, **16**, 115.

Glycerol 1,3-dihexadecanoate 2-octadecanoate G-00045

Octadecanoic acid 2[(1-oxohexadecyl)oxy]-1-[[(1-oxohexadecyl)oxy]methyl]ethyl ester, 9CI. α,α'-Dipalmitostearin

[2177-97-1]

$C_{53}H_{102}O_6$ M 835.385

Cryst. (C_6H_6/EtOH). Mp 68°. Exibits polymorphism.

Malkin, T. *et al, J. Chem. Soc.*, 1939, 103.
Mitchell, L.C., *J. Am. Oil Chem. Soc.*, 1972, **49**, 281.
Gray, M.S. *et al, J. Am. Oil Chem. Soc.*, 1978, **55**, 601.

Glycerol 1,2-dihexadecanoate 3-(9-octadecenoate) G-00046

9-Octadecenoic acid 2,3-bis[(1-oxohexadecyl)oxy]propyl ester, 9CI

[1867-91-0]

$C_{53}H_{100}O_6$ M 833.369

Found in palm oil and other oils.

(±)-(***E***)-***form***

α,β-Dipalmitoelaidin

Cryst. Mp 51.7°.

(***R***)-(***Z***)-***form*** [53023-11-3]

Mp 35.5-36°. $[\alpha]_D$ 0° (c, 10.2 in $CHCl_3$).

(***S***)-(***Z***)-***form*** [53023-10-2]

1-Oleoyl-2,3-dipalmitoyl-sn-glycerol

Cryst. (Me_2CO or C_6H_6/EtOH). Mp 37°. $[\alpha]_D$ 0° (c, 10.2 in $CHCl_3$). $[M]_{300}$ −3.6° (C_6H_6).

(±)-(***Z***)-***form*** [37179-82-1]

α,β-Dipalmitoolein. 2,3-Dipalmito-1-olein, 8CI. 3-Oleo-1,2-dipalmitin

Cryst. (Et_2O). Mp 34.5°. n_D^{40} 1.4556.

Bömer, A. *et al, CA*, 1938, **32**, 495 (*synth*)
Daubert, B.F. *et al, J. Am. Chem. Soc.*, 1943, **65**, 2143 (*synth*)
Tattrie, N.H. *et al, Arch. Biochem. Biophys.*, 1958, **78**, 319 (*synth*)
Chapman, D., *J. Am. Oil Chem. Soc.*, 1960, **37**, 73 (*ir*)
Chapman, D., *J. Chem. Soc.*, 1960, 436 (*pmr, polymorphism*)
Lok, C.M. *et al, Chem. Phys. Lipids*, 1976, **16**, 115 (*synth*)
Bus, J. *et al, Chem. Phys. Lipids*, 1978, **21**, 253 (*pmr*)
Gronowitz, S. *et al, Chem. Phys. Lipids*, 1979, **23**, 101 (*synth, ord, cd*)

Glycerol 1,3-dihexadecanoate 2-(9-octadecenoate) G-00047

9-Octadecenoic acid 2-[(1-oxohexadecyl)oxy]-1-[[(1-oxohexadecyl)oxy]methyl]ethyl ester, 9CI

$C_{53}H_{100}O_6$ M 833.369

(***E***)-***form*** [32221-51-5]

α,α'-Dipalmitoelaidin

Cryst. (Me_2CO). Mp 56.5°. Exibits polymorphism.

(***Z***)-***form*** [2190-25-2]

α,α'-Dipalmitoolein. 2-Oleodipalmitin

Constit. of cocoa butter and of *Stillingia* spp. Cryst. (Me_2CO). Mp 37.2°. Exibits polymorphism.

Lutton, E.S. *et al*, *J. Am. Chem. Soc.*, 1950, **72**, 3254 (*isol, synth, polymorphism*)
Lovegren, N.V. *et al*, *J. Am. Oil Chem. Soc.*, 1971, **48**, 116 (*synth, props*)

Glycerol 1,2-dihexadecanoate 3-phosphate G-00048

Hexadecanoic acid 1-[(phosphonoxy)methyl]-1,2-ethanediyl ester, 9CI. Dipalmitoylphosphatidic acid

[19698-29-4]

$C_{35}H_{69}O_8P$ M 648.899

(R)-form [7091-44-3]

*1,2-Dihexadecanoyl-*sn*-glycerol 3-phosphate*

Mp 70-71°. $[\alpha]_D^{26}$ +4.0° (c, 9.6 in $CHCl_3$).

Di-Me ester: Mp 42.5-43.5°. $[\alpha]_D$ +2.0° (c, 7.6 in $CHCl_3$).

Diphenyl ester: Mp 47-48°. $[\alpha]_D$ +2.4° (c, 2.4 in $CHCl_3$).

(±)-form [5129-68-0]

Di-Me ester: Mp 46°.

Dibenzyl ester: Mp 50-50.5°.

Monophenyl ester: Mp 54.5-55.5°.

Baer, E., *J. Biol. Chem.*, 1951, **189**, 235.
Hessel, L.W. *et al*, *Recl. Trav. Chim. Pays-Bas* (*J. R. Neth. Chem. Soc.*), 1954, **73**, 150.
Baer, E. *et al*, *J. Biol. Chem.*, 1955, **212**, 39.
Baer, E., *Prog. Chem. Fats Other Lipids*, 1963, **6**, 33 (*rev*)
Aneja, R. *et al*, *Chem. Phys. Lipids*, 1974, **12**, 39 (*pmr*)
Avenado, R. *et al*, *CA*, 1975, **83**, 113590.

Glycerol 1,2-dihexadecanoate 3-phosphocholine G-00049

Dipalmitoylphosphatidylcholine. 1,2-Dihexadecanoylglycero-3-phosphocholine

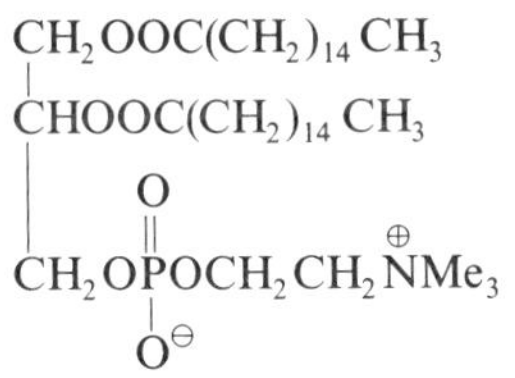

$C_{40}H_{80}NO_8P$ M 734.048

(R)-L-form [63-89-8]

Mp 235-236°. $[\alpha]_D^{25}$ +7.0° (c, 5.6 in $CHCl_3$).

Baer, E. *et al*, *J. Am. Chem. Soc.*, 1952, **74**, 158 (*synth*)
Brandt, A.E. *et al*, *Biochim. Biophys. Acta*, 1967, **144**, 605 (*synth*)
Klein, R.A., *J. Lipid Res.*, 1971, **12**, 123 (*ms*)
Ghosh, D. *et al*, *Biochim. Biophys. Acta*, 1972, **266**, 41.
Gordon, D.T. *et al*, *Lipids*, 1972, **7**, 261 (*synth*)
Bunow, M.R. *et al*, *Biochim. Biophys. Acta*, 1977, **489**, 191 (*raman*)

Glycerol 1,2-dihexadecanoate 3-phosphoethanolamine G-00050

Hexadecanoic acid 1-[[[(2-aminoethoxy)hydroxyphosphinyl]oxy]methyl]-1,2-ethanediyl ester, 9CI. Dipalmitoylphosphatidylethanolamine. 1,2-Dihexadecanoylglycero-3-phosphoethanolamine. β,γ-Dipalmitoyl-α-cephalin

[3026-45-7]

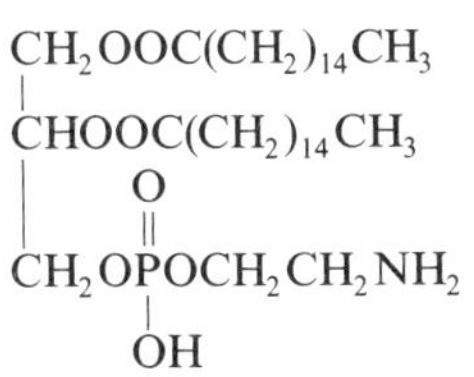

$C_{37}H_{74}NO_8P$ M 691.967

(R)-form [923-61-5]

Mp 185-186°, Mp 210-211°. $[\alpha]_D$ +6.55°.

(±)-form [5681-36-7]

Mp 196°.

v. Veen, A. *et al*, *Recl. Trav. Chim. Pays-Bas* (*J. R. Neth. Chem. Soc.*), 1960, **79**, 1085 (*synth*)
Chapman, D. *et al*, *Trans. Faraday Soc.*, 1966, **62**, 2607 (*pmr, props*)
Billimoria, J.D. *et al*, *J. Chem. Soc. C*, 1968, 1404 (*synth*)
Billimoria, J.D. *et al*, *Chem. Phys. Lipids*, 1974, **12**, 327 (*synth*)
Wood, G.W. *et al*, *Chem. Phys. Lipids*, 1977, **18**, 316 (*ms*)
Mendelsohn, R., *Biochemistry*, 1978, 3944 (*raman*)
Skarjune, R. *et al*, *Biochemistry*, 1979, **18**, 5903 (*conformn*)

Glycerol 1,2-dihexadecanoate 3-phospho-1′-glycerol G-00051

Hexadecanoic acid 1-[[[(2,3-dihydroxypropoxy)hydroxyphosphinyl]oxy]methyl]-1,2-ethanediyl ester, 9CI. Dipalmitoylphosphatidylglycerol

[4537-77-3]

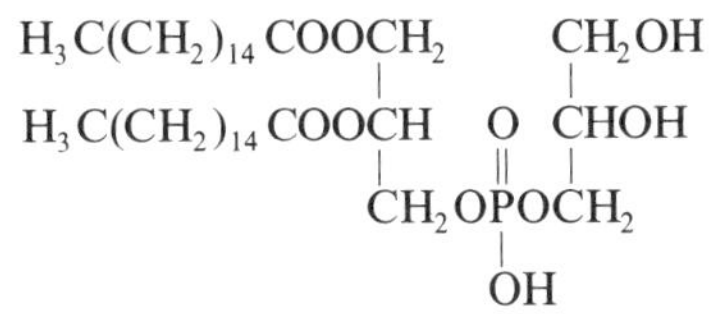

$C_{38}H_{75}O_{10}P$ M 722.978

Lammers, J.G. *et al*, *Recl. Trav. Chim. Pays-Bas* (*J. R. Neth. Chem. Soc.*), 1979, **98**, 243 (*synth, cmr*)

Glycerol 1,2-dihexadecanoate 3-phosphoinositol G-00052

Dipalmitoylphosphatidylinositol

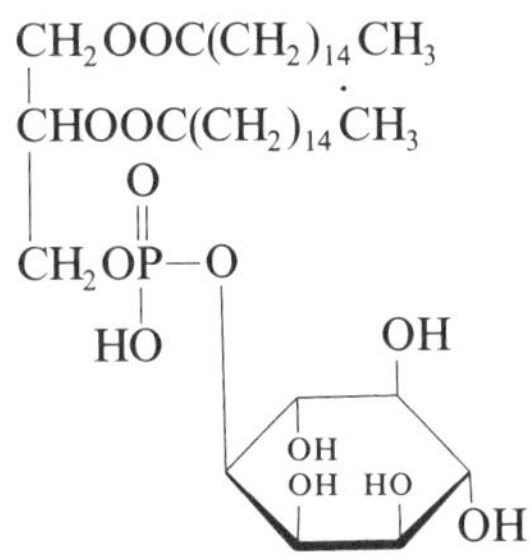

$C_{41}H_{79}O_{13}P$ M 811.041

Myoinositol-form

myo *Inositol mono[[2,3-bis(1-oxohexadecyl)oxy]propyl] 1-(dihydrogen phosphate) ester, 9CI*

NH_4 salt: Mp 169-172°. $[\alpha]_D^{20}$ +7.48° (c, 0.3 in $CHCl_3$).

Penta-O-benzyl: Mp 53-54°.

Klyashehitskii, B.A. *et al*, *Tetrahedron Lett.*, 1970, 587 (*synth*)
Zhelvakova, E.G., *CA*, 1970, **72**, 133119 (*synth*)
Lyutik, A.Y. *et al*, *CA*, 1975, **82**, 86519 (*synth*)
Sadovnikova, M.S. *et al*, *CA*, 1975, **83**, 131855 (*synth*)
Sukhanov, V.A. *et al*, *CA*, 1978, **88**, 62556 (*synth*)

Glycerol 1,2-dihexadecanoate 3-phosphoserine G-00053

Serine 2,3-bis[(1-oxohexadecyl)oxy]propyl hydrogen phosphate ester, 9CI. Dipalmitoylphosphatidylserine

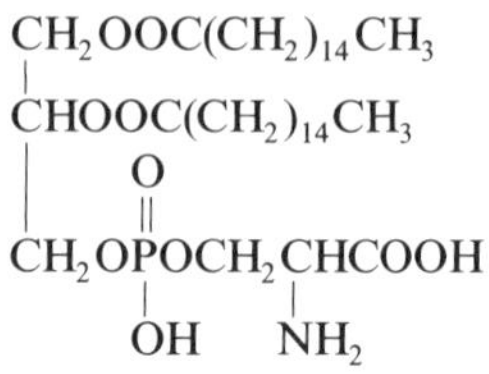

$C_{38}H_{74}NO_{10}P$ M 735.977

(R)-L-form [40290-42-4]

*1,2-Dipalmitoyl-*sn*-glycero-L-serine*

Mp 155-158°. $[\alpha]_D$ +5.5° ($CHCl_3$, $MeOH/H_2O$).

Browning, J. *et al, Chem. Phys. Lipids*, 1979, **24**, 103 (*pmr*)
Browning, J. *et al, Biochemistry*, 1980, **19**, 1262 (*nmr*)

Glycerol 1,3-dihexadecanoate 2-tetradecanoate G-00054

Hexadecanoic acid 2-[(1-oxotetradecyl)oxy]-1,3-propanediyl ester, 9CI. α,α′-Dipalmitomyristin

[56599-89-4]

$C_{49}H_{94}O_6$ M 779.278

Cryst. ($EtOH/C_6H_6$). Mp 60°. Exhibits polymorphism.

Malkin, T. *et al, J. Chem. Soc.*, 1939, 103 (*synth*)
Gray, M.S. *et al, J. Am. Oil Chem. Soc.*, 1978, **55**, 601.

Glycerol 1,2-di-11-methyldodecanoate 3-phosphocholine G-00055

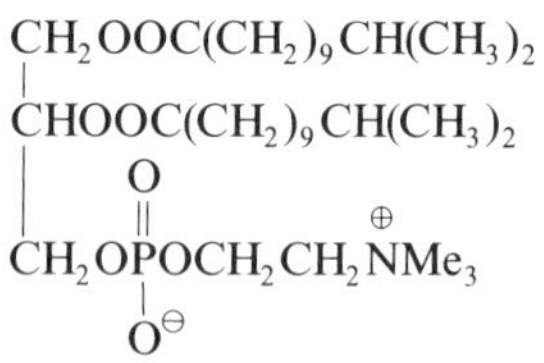

$C_{34}H_{68}NO_8P$ M 649.887

(R)-L-form [71368-25-7]

1,2-Diisotridecanoylphosphatidylcholine

Sylvius, J.R. *et al, Chem. Phys. Lipids*, 1979, **24**, 287 (*synth*)

Glycerol 1,2-di-16-methylheptadecanoate 3-phosphocholine G-00056

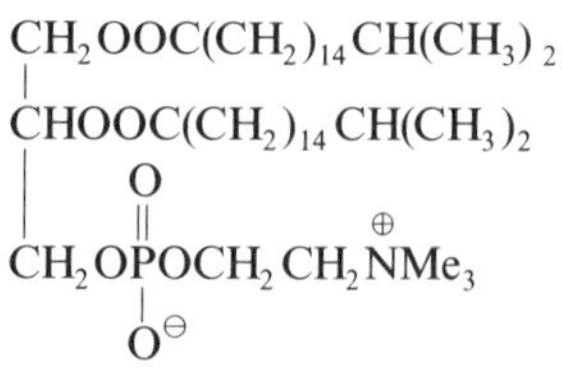

$C_{44}H_{88}NO_8P$ M 790.155

(R)-L-form [60683-79-6]

1,2-Diisooctadecanoylphosphatidylcholine

Sylvius, J.R. *et al, Chem. Phys. Lipids*, 1979, **24**, 287.

Glycerol 1,2-di-15-methylhexadecanoate 3-phosphocholine G-00057

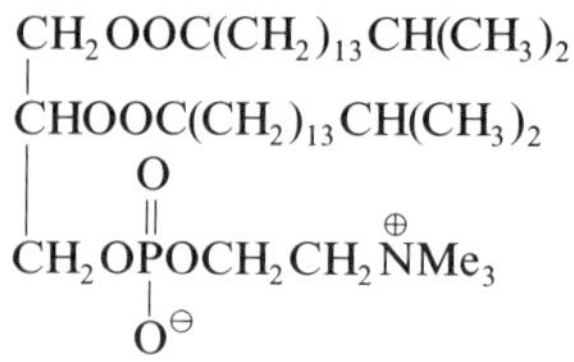

$C_{42}H_{84}NO_8P$ M 762.101

(R)-L-form [71368-21-3]

Diisoheptadecanoylphosphatidylcholine

Sylvius, J.R. *et al, Chem. Phys. Lipids*, 1979, **24**, 287.

Glycerol 1,2-di-14-methylpentadecanoate 3-phosphocholine G-00058

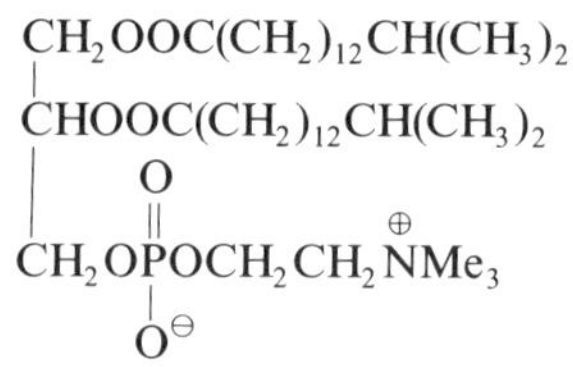

$C_{40}H_{80}NO_8P$ M 734.048

(R)-L-form [71368-22-4]

Diisohexadecanoylphosphatidylcholine

Sylvius, J.R. *et al, Chem. Phys. Lipids*, 1979, **24**, 287.

Glycerol 1,2-di-13-methyltetradecanoate 3-phosphocholine G-00059

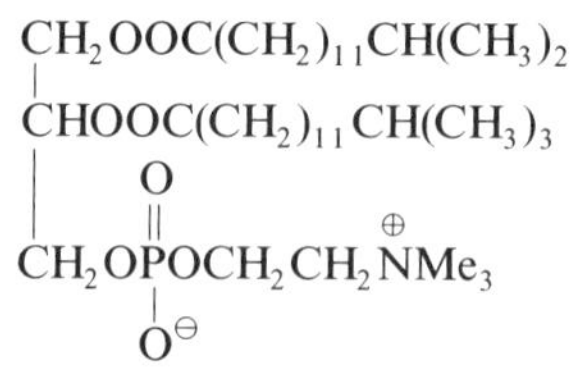

$C_{38}H_{76}NO_8P$ M 705.994

(R)-L-form [71368-23-5]

1,2-Diisopentadecanoylphosphatidylcholine

Sylvius, J.R. *et al, Chem. Phys. Lipids*, 1979, **24**, 287.

Glycerol 1,2-di-12-methyltridecanoate 3-phosphocholine G-00060

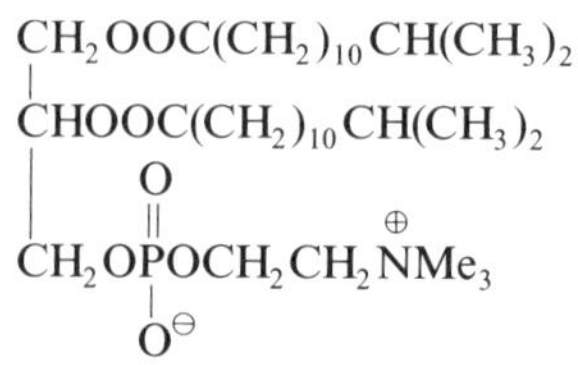

$C_{36}H_{72}NO_8P$ M 677.940

(R)-L-form [71368-24-6]

1,2-Diisotetradecanoylphosphatidylcholine

Sylvius, J.R. *et al, Chem. Phys. Lipids*, 1979, **24**, 287 (*synth*)

Glycerol 1,2-di-10-methylundecanoate 3-phosphocholine G-00061

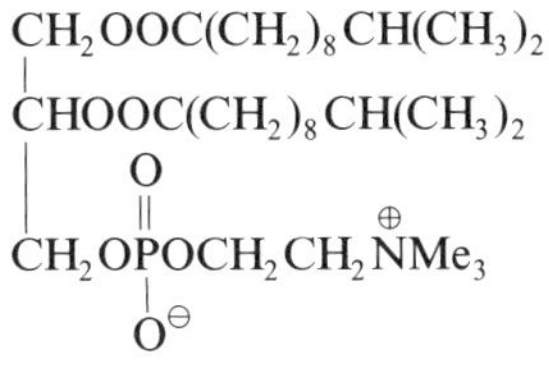

$C_{32}H_{64}NO_8P$ M 621.833

***(R)*-L-form** [71368-26-8]

Diisododecanoylphosphatidylcholine

Sylvius, J.R. *et al, Chem. Phys. Lipids*, 1979, **24**, 287 (*synth*)

Glycerol 1,2-dinonadecanoate 3-phosphocholine G-00062

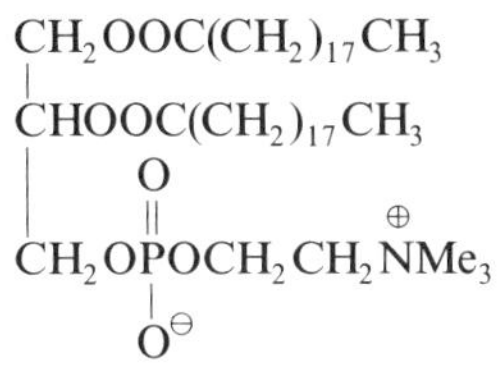

$C_{46}H_{92}NO_8P$ M 818.208

***(R)*-L-form**

Dinonadecanoylphosphatidylcholine

Sylvius, J.R. *et al, Biochim. Biophys. Acta*, 1979, **555**, 175.

Glycerol 1,2-dioctacosanoate G-00063

Octacosanoic acid 1-(hydroxymethyl)-1,2-ethanediyl ester, 9CI. α,β-Dimontanylglycerol

[52363-45-8]

$C_{59}H_{116}O_5$ M 905.562

Constit. of *Morus alba* root bark. Cryst. (EtOAc). Mp 82-82.5°. Stereochemistry not known.

Ac: Cryst. (MeOH/$CHCl_3$). Mp 68-69.5°.

Kondo, K. *et al, Chem. Pharm. Bull.*, 1973, **21**, 2265 (*isol, ir, pmr*)

Glycerol 1,2-di-9,12-octadecadienoate G-00064

9,12-Octadecadienoic acid 1-(hydroxymethyl)-1,2-ethanediyl ester, 9CI

$C_{39}H_{68}O_5$ M 616.963

***(S)*-(*all-Z*)-form** [24529-89-3]

1,2-Dilinoleoyl-sn-*glycerol*

Oil. $[\alpha]_D^{25}$ −2.0° (c, 1.1 in $CHCl_3$).

(±)-(*all-Z*)-form [68069-29-4]

α,β-Dilinolein

Oil.

Octabromide: Mp 103-104°.

Jackson, J.E. *et al, J. Am. Oil Chem. Soc.*, 1963, **40**, 502 (*synth*)
Rakhit, S. *et al, Can. J. Chem.*, 1969, **47**, 2906 (*synth*)
Pfeiffer, F.R. *et al, J. Org. Chem.*, 1970, **35**, 221 (*synth*)

Glycerol 1,3-di-9,12-octadecadienoate G-00065

9,12-Octadecadienoic acid 2-hydroxy-1,3-propanediyl ester, 9CI

[59925-29-0]

$C_{39}H_{68}O_5$ M 616.963

(*all-Z*)-form [15818-46-9]

α,α′-Dilinolein

Oil. Mp −2.6°.

Octabromide: Mp 90-93°.

Daubert, B.F. *et al, J. Am. Chem. Soc.*, 1947, **69**, 1449 (*synth*)
Jackson, J.E. *et al, J. Am. Oil Chem. Soc.*, 1963, **40**, 502 (*synth*)

Glycerol 1,2-di-9,12-octadecadienoate 3-dodecanoate G-00066

9,12-Octadecadienoic acid 1-[[(1-oxododecyl)oxy]methyl]-1,2-ethanediyl ester, 9CI

$C_{51}H_{90}O_6$ M 799.268

(±)-(*all-Z*)-form [61824-39-3]

α,β-Dilinoleolaurin

Oil. Mp −12° to −11.

Daubert, B.F. *et al, J. Am. Chem. Soc.*, 1944, **66**, 1507 (*synth*)

Glycerol 1,2-di-9,12-octadecadienoate 3-hexadecanoate G-00067

9,12-Octadecadienoic acid 2-[[(1-oxohexadecyl)oxy]methyl]-1,2-ethanediyl ester, 9CI

$C_{55}H_{98}O_6$ M 855.376

(±)-(*all-Z*)-form [2190-15-0]

α,β-Dilinoleopalmitin

Oil. Mp −4° to −3°.

Serebrennikova, G.A. *et al, CA*, 1962, **56**, 11732d (*synth*)

Glycerol 1,3-di-9,12-octadecadienoate 2-hexadecanoate G-00068

9,12-Octadecadienoic acid 2-[(1-oxohexadecyl)oxy]-1,3-propanediyl ester, 9CI

$C_{55}H_{98}O_6$ M 855.376

(*all-Z*)-form [2190-16-1]

α,α′-Dilinoleopalmitin

Oil. Mp −3° to −1.5°.

Serebrennikova, G.A. *et al, CA*, 1963, **59**, 435b.

Glycerol 1,2-di-9,12-octadecadienoate 3-octadecanoate G-00069

$C_{57}H_{102}O_6$ M 883.429

(±)-(*all-Z*)-form

α,β-Dilinoleostearin

Oil.

Octabromide: Mp 43.5-44°.

Jackson, J.E. *et al, J. Am. Oil Chem. Soc.*, 1963, **40**, 502 (*synth*)

Glycerol 1,3-di-9,12-octadecadienoate 2-octadecanoate G-00070

9,12-Octadecadienoic acid 2-[(1-oxooctadecyl)oxy]-1,3-propanediyl ester, 9CI

$C_{57}H_{102}O_6$ M 883.429

(*all-Z*)-form [2190-18-3]

α,α′-Dilinoleostearin

Oil.

Octabromide: Mp 40-41°.

Jackson, J.E. *et al, J. Am. Oil Chem. Soc.*, 1963, **40**, 502 (*synth*)

Glycerol 1,2-di-9,12-octadecadienoate 3-phosphate G-00071

$C_{39}H_{69}O_8P$ M 696.943

Shvets, V.I. *et al, CA*, 1966, **65**, 20000 (*synth*)

Glycerol 1,2-di-9,12-octadecadienoate 3-phosphocholine G-00072

1,2-Dioctadeca-9,12-dienoylglycero-3-phosphocholine

$CH_2OOC(CH_2)_6CH_2(CH{=}CHCH_2)_2CH_2CH_2CH_2CH_3$
$CHOOC(CH_2)_6CH_2(CH{=}CHCH_2)_2CH_2CH_2CH_2CH_3$
$CH_2OP(O)(O^{\ominus})OCH_2CH_2\overset{\oplus}{N}H_2Me_3$

$C_{44}H_{80}NO_8P$ M 782.092

***(R)*-L-(*all-Z*)-form** [998-06-1]

Dilinoleoylphosphatidylcholine

Gelatinous. Mp 232°. $[\alpha]_D^{20}$ +6.02°.

Demel, R.A. *et al*, *Biochim. Biophys. Acta*, 1972, **266**, 26.
Gordon, D.T. *et al*, *Lipids*, 1972, **7**, 261 (*synth*)
Wood, G.W. *et al*, *Chem. Phys. Lipids*, 1977, **18**, 316 (*ms*)
Warner, T.G. *et al*, *J. Lipid Res.*, 1977, **18**, 548 (*synth*)
Lammers, J.G. *et al*, *Chem. Phys. Lipids*, 1978, **22**, 293 (*nmr, ir*)

Glycerol 1,2-di-9,12-octadecadienoate 3-phosphoethanolamine G-00073

9,12-Octadecadienoic acid 1-[[[(2-aminoethoxy)hydroxyphosphinyl]oxy]methyl]-1,2-ethanediyl ester, 9CI. 1,2-Di-9,12-octadecadienoylphosphatidylcholine

$CH_2OOC(CH_2)_6CH_2(CH{=}CHCH_2)_2CH_2CH_2CH_2CH_3$
$CHOOC(CH_2)_6CH_2(CH{=}CHCH_2)_2CH_2CH_2CH_2CH_3$
$CH_2OP(O)(OH)OCH_2CH_2NH_2$

$C_{41}H_{74}NO_8P$ M 740.011

***(R)*-(*all-Z*)-form** [20707-71-5]

Dilinoleoylphosphatidylethanolamine

Mp 25°. $[\alpha]_D^{25}$ +6.2° (c, 1 in $CHCl_3$).

(±)-(*all-Z*)-form

Waxy solid.

Shvets, V.I. *et al*, *CA*, 1962, **56**, 4611 (*synth*)
Pfeiffer, F.R. *et al*, *J. Org. Chem.*, 1969, **34**, 2795 (*synth*)
Billimoria, J.D. *et al*, *Chem. Phys. Lipids*, 1974, **12**, 327 (*synth*)

Glycerol 1,2-di-9,12-octadecadienoate 3-phosphoinositol G-00074

Dilinoleoylphosphatidylinositol

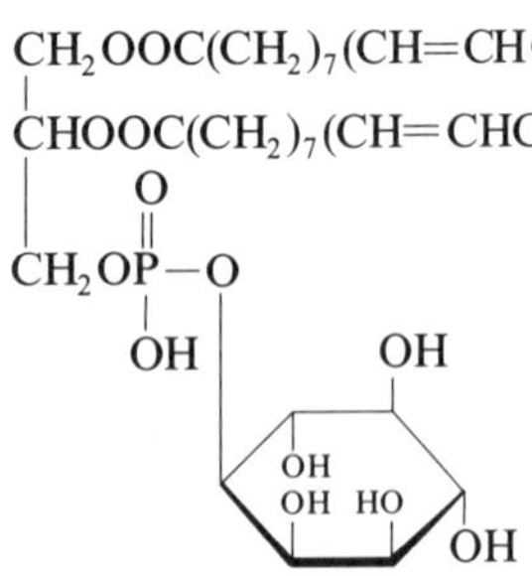

$C_{45}H_{79}O_{13}P$ M 859.085

Shvets, V.I. *et al*, *CA*, 1966, **64**, 14261 (*synth*)
Petrova, M.K. *et al*, *CA*, 1966, **65**, 20207 (*synth*)

Glycerol 1,2-di-9,12-octadecadienoate 3-phosphoserine G-00075

$CH_2OOC(CH_2)_6CH_2(CH{=}CHCH_2)_2CH_2CH_2CH_2CH_3$
$CHOOC(CH_2)_6CH_2(CH{=}CHCH_2)_2CH_2CH_2CH_2CH_3$
$CH_2OP(O)(OH)OCH_2CH(NH_2)COOH$

$C_{42}H_{74}NO_{10}P$ M 784.021

(±)-(*all-Z*)-form

Dilinoleoylphosphatidylserine

Turner, D.L. *et al*, *Lipids*, 1966, **1**, 439 (*synth*)

Glycerol 1,2-dioctadecanoate G-00076

Octadecanoic acid 1-(hydroxymethyl)-1,2-ethanediyl ester, 9CI. α,β-Distearin. Glycerol 1,2-distearate

[1188-58-5]

$C_{39}H_{76}O_5$ M 625.026

***(S)*-form** [10567-21-2]

*1,2-Distearoyl-*sn*-glycerol*

Cryst. ($CHCl_3$/pet. ether). Mp 74-76°. $[\alpha]_D^{23}$ −2.91° (c, 6.18 in $CHCl_3$).

3-Benzyl ether: Mp 50-51°. $[\alpha]_D^{23}$ +6.1° (c, 7.48 in $CHCl_3$).

(±)-form [51063-97-9]

Cryst. (hexane). Mp 71-72° (59.5°). Exhibits polymorphism.

3-Benzyl ether: Mp 53.5-54°.

Gigg, J. *et al*, *J. Chem. Soc. C*, 1967, 431 (*synth*)
Ioannou, P.N. *et al*, *Synthesis*, 1979, 939 (*synth*)

Glycerol 1,3-dioctadecanoate G-00077

Octadecanoic acid 2-hydroxy-1,3-propanediyl ester, 9CI. α,α′-Distearin

[504-40-5]

$C_{39}H_{76}O_5$ M 625.026

Cryst. (Me_2CO). Mp 80-81°.

2-Ac: Mp 64°.

2-Benzyl ether: Mp 49.5°.

Mank, A.P.J. *et al*, *Chem. Phys. Lipids*, 1976, **16**, 107 (*synth*)

Glycerol 1,2-dioctadecanoate 3-butanoate G-00078

Octadecanoic acid 1-[(1-oxobutoxy)methyl]-1,2-ethanediyl ester, 9CI. α,β-Distearobutyrin

[60562-74-5]

$C_{43}H_{82}O_6$ M 695.117

***(S)*-form** [60456-31-7]

*3-Butyryl-1,2-stearoyl-*sn*-glycerol*

Cryst. (Me_2CO). Mp 49°.

Lok, C.M. *et al*, *Chem. Phys. Lipids*, 1976, **16**, 115 (*synth*)
Bus, J. *et al*, *Chem. Phys. Lipids*, 1976, **16**, 123 (*pmr*)

Glycerol 1,3-dioctadecanoate 2-decanoate G-00079

Octadecanoic acid 2-[(1-oxodecyl)oxy]-1,3-propanediyl ester, 9CI. α,α′-Distearocaprin

[56968-93-5]

$C_{49}H_{94}O_6$ M 779.278

Cryst. (EtOH). Mp 57°. Exhibits polymorphism.

Malkin, T. *et al*, *J. Chem. Soc.*, 1939, 1141 (*synth*)
Lovegren, N.V. *et al*, *J. Am. Oil Chem. Soc.*, 1978, **55**, 310.

Glycerol 1,2-dioctadecanoate 3-docosanoate G-00080

Docosanoic acid 2,3-bis[(1-oxooctadecyl)oxy]propyl ester, 9CI. α,β-Distearobehenin

[38280-88-5]

$C_{61}H_{118}O_6$ M 947.600

(±)-form

Cryst. (hexane/Me_2CO). Mp 49°, Mp 75.5°. Exhibits polymorphism.

Lutton, E.S. *et al*, *J. Am. Oil Chem. Soc.*, 1972, **49**, 336.

Glycerol 1,2-dioctadecanoate 3-dodecanoate G-00081

Octadecanoic acid 3-[(1-oxododecyl)oxy]-1,2-propanediyl ester, 9CI. α,β-Distearolaurin

$C_{51}H_{98}O_6$ M 807.332

Cryst. (EtOH/C_6H_6). Mp 52°. Exibits polymorphism.

Carter, M.G.R. *et al*, *J. Chem. Soc.*, 1939, 1518.

Glycerol 1,3-dioctadecanoate 2-dodecanoate G-00082

Octadecanoic acid 2-[(1-oxododecyl)oxy]-1,3-propanediyl ester, 9CI. α,α'-Distearolaurin

[56183-45-0]

$C_{51}H_{98}O_6$ M 807.332

Cryst. (EtOH/C_6H_6). Mp 60.5°. Exibits polymorphism.

Malkin, T. *et al*, *J. Chem. Soc.*, 1939, 1141 (*synth*)
Lovegren, N.V. *et al*, *J. Am. Oil Chem. Soc.*, 1978, **55**, 310.

Glycerol 1,2-dioctadecanoate 3-ethanoate G-00083

Octadecanoic acid 1-[(acetyloxy)methyl]-1,2-ethanediyl ester, 9CI. α,β-Distearoacetin

[60562-72-3]

$C_{41}H_{78}O_6$ M 667.064

(S)-form [20236-49-1]

*3-Acetyl-1,2-distearoyl-*sn*-glycerol*

Constit. of *Euonymus* spp. seed oils. Cryst. (Me_2CO). Mp 49°, Mp 53°. $[M]^{27}_{300}$ −60° (C_6H_6).

Kleiman, R. *et al*, *Lipids*, 1966, **1**, 286 (*isol*)
Gronowitz, S. *et al*, *Chem. Phys. Lipids*, 1975, **14**, 174 (*synth*, *ord*)
Lok, C.M. *et al*, *Chem. Phys. Lipids*, 1976, **16**, 115 (*synth*)
Bus, J. *et al*, *Chem. Phys. Lipids*, 1976, **16**, 123 (*pmr*)

Glycerol 1,3-dioctadecanoate 2-hexadecanoate G-00084

Octadecanoic acid 2-[(1-oxohexadecyl)oxy]-1,3-propanediyl ester, 9CI. α,α'-Distearopalmitin. 2-Palmitodistearin

[2190-24-1]

$C_{55}H_{106}O_6$ M 863.439

Isol. from cocoa butter. Cryst. (EtOH/pet. ether). Mp 68°. Exhibits polymorphism.

Lutton, E.S. *et al*, *J. Am. Chem. Soc.*, 1948, **70**, 2441.
Lovegren, N.V. *et al*, *J. Am. Oil Chem. Soc.*, 1978, **55**, 310.

Glycerol 1,3-dioctadecanoate 2-hexanoate G-00085

Octadecanoic acid 2-[(1-oxohexyl)oxy]-1,3-propanediyl ester, 9CI. α,α'-Distearocaproin

[66411-62-9]

$C_{45}H_{86}O_6$ M 723.171

Cryst. (EtOH/Et_2O). Mp 53°.

Jackson, F.L. *et al*, *J. Am. Chem. Soc.*, 1951, **73**, 4280.
Lovegren, N.V. *et al*, *J. Am. Oil Chem. Soc.*, 1978, **55**, 310.

Glycerol 1,2-dioctadecanoate 3-(9,12-octadecadienoate) G-00086

9,12-Octadecadienoic acid 2,3-bis[(1-oxooctadecyl)oxy] propyl ester, 9CI

$C_{57}H_{106}O_6$ M 887.461

(±)-(Z,Z)-form [2442-53-7]

α,β-Distearolinolein

Cryst. Mp 39.5-40.8°.

Tetrabromide: Cryst. Mp 50.5-51.5°.

Jackson, J.E. *et al*, *J. Am. Oil Chem. Soc.*, 1963, **40**, 502 (*synth*)

Glycerol 1,3-dioctadecanoate 2-(9,12-octadecadienoate) G-00087

9,12-Octadecadienoic acid 2-[(1-oxooctadecyl)oxy]-1-[[(1-oxooctadecyl)oxy]methyl]ethyl ester, 9CI

$C_{57}H_{106}O_6$ M 887.461

(Z,Z)-form [2190-13-8]

α,α'-Distearolinolein

Cryst. Mp 36.3-37.0°.

Tetrabromide: Cryst. Mp 53.5-54.5°.

Jackson, J.E. *et al*, *J. Am. Oil Chem. Soc.*, 1963, **40**, 502 (*synth*)

Glycerol 1,2-dioctadecanoate 3-(9-octadecenoate) G-00088

9-Octadecenoic acid 2,3-bis[(1-oxooctadecyl)oxy]propyl ester, 9CI. 1-Oleodistearin

$C_{57}H_{108}O_6$ M 889.477

Isol. from cocoa butter.

(R,Z)-form

*1,2-Distearoyl-3-oleoyl-*sn*-glycerol*

Mp 45-46°. $[M]_{300}$ +3.4° (C_6H_6).

(±)-(Z)-form [51195-71-2]

α,β-Distearoolein

Cryst. (Me_2CO). Mp 42.1°. Exhibits polymorphism.

Landmann, W. *et al*, *J. Am. Oil Chem. Soc.*, 1961, **38**, 681 (*synth*, *props*)
Barber, M. *et al*, *Tetrahedron Lett.*, 1964, 1063 (*ms*)
Gronowitz, S. *et al*, *Chem. Phys. Lipids*, 1979, **23**, 101 (*synth*, *ord*)

Glycerol 1,3-dioctadecanoate 2-(9-octadecenoate) G-00089

9-Octadecenoic acid 2-[(1-oxooctadecyl)oxy]-1-[[(1-oxooctadecyl)oxy]methyl]ethyl ester, 9CI

$C_{57}H_{108}O_6$ M 889.477

(E)-form

α,α'-Distearoelaidin

Cryst. (Me_2CO, pet. ether, Et_2O). Mp 59.5°. Exhibits polymorphism.

(Z)-form [2846-04-0]

α,α'-Distearoolein. 2-Oleodistearin

Present in oils of *Dipterocarpus* spp., *Allanblackia* spp. and others and cocoa butter. Cryst. (hexane). Mp 40.5-42.5°. Exhibits polymorphism.

Meara, M.L., *J. Chem. Soc.*, 1945, 22 (*isol*, *bibl*)
Minor, J.E. *et al*, *J. Am. Chem. Soc.*, 1953, **75**, 2685 (*synth*, *polymorphism*)
Barber, M. *et al*, *Tetrahedron Lett.*, 1964, 1063 (*ms*)
Mitchell, L.C. *et al*, *J. Am. Oil Chem. Soc.*, 1972, **49**, 281 (*synth*)

Glycerol 1,2-dioctadecanoate 3-phosphate G-00090

Octadecanoic acid 1-[(phosphonoxy)methyl]-1,2-ethanediyl ester, 9CI. Distearoylphosphatidic acid

$C_{39}H_{77}O_8P$ M 705.006

(R)-form [17966-16-4]

*1,2-Dioctadecanoyl-*sn*-glycerol 3-phosphate*
Mp 75.5-76.5°. $[\alpha]_D^{26}$ +3.8° (c, 9.3 in $CHCl_3$).
Di-Me ester: Mp 52-53°. $[\alpha]_D$ +1.9° (c, 9.9 in $CHCl_3$).
Diphenyl ester: Mp 54.5-55°. $[\alpha]_D$ +2.0° (c, 10.2 in $CHCl_3$).
(±)-form [13563-93-4]
Mp 70-71°.
Phenyl ester: Mp 58-59°.
Dibenzyl ester: Mp 56-57°.

Baer, E., *J. Biol. Chem.*, 1951, **189**, 235.
Hessel, L.W. *et al*, *Recl. Trav. Chim. Pays-Bas* (*J. R. Neth. Chem. Soc.*), 1954, **73**, 150.
Baer, E. *et al*, *J. Biol. Chem.*, 1955, **212**, 39.
Baer, E. *et al*, *Arch. Biochem. Biophys.*, 1958, **78**, 294.
Baer, E., *Prog. Chem. Fats Other Lipids*, 1963, **6**, 33 (*rev*)
Avenado, R. *et al*, *CA*, 1975, **83**, 113590.

Glycerol 1,2-dioctadecanoate 3-phospho-2-amino-1-propanol G-00091

Distearoyl-α-glycerylphosphoryl-2-amino-1-propanol.
Dioctadecanoyl-α-glycerylphosphoryl-2-amino-1-propanol

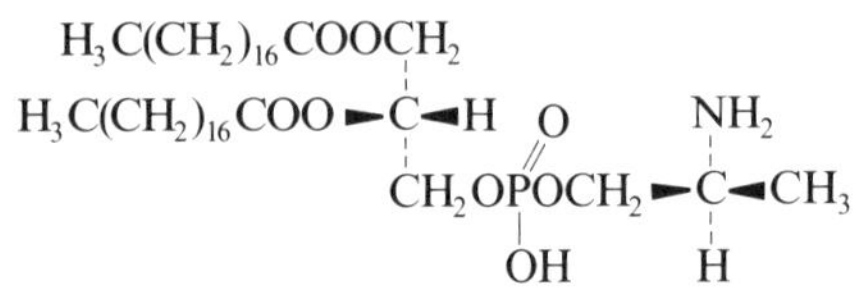

$C_{42}H_{84}NO_8P$ M 762.101
(R)-L-form
Mp 182-183°. $[\alpha]_D^{23}$ +7° (c, 2.7 in $CHCl_3$).

Baer, E. *et al*, *J. Biol. Chem.*, 1963, **238**, 3591.

Glycerol 1,2-dioctadecanoate 3-phosphocholine G-00092

1,2-Dioctadecanoylglycero-3-phosphocholine.
Distearoylphosphatidylcholine
[4539-70-2]

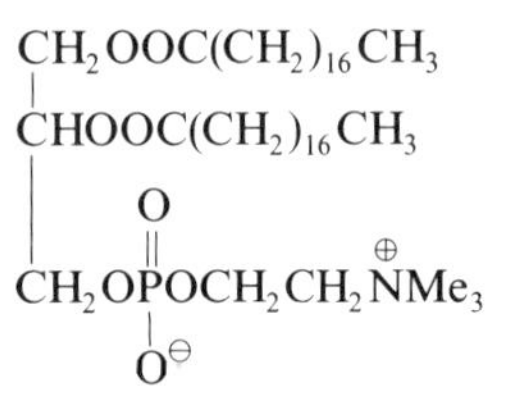

$C_{44}H_{88}NO_8P$ M 790.155
(R)-form [816-94-4]
Mp 231-232°. $[\alpha]_D^{29}$ +6.2° (c, 10 in $CHCl_3$/MeOH).

Baer, E. *et al*, *Can. J. Biochem. Physiol.*, 1959, **37**, 453 (*synth*)
Chapman, D. *et al*, *J. Biol. Chem.*, 1966, 241, 5044 (*nmr*)
Aneja, R. *et al*, *Biochim. Biophys. Acta*, 1971, **248**, 455 (*synth*)
Ghosh, D. *et al*, *Biochim. Biophys. Acta*, 1972, **266**, 41.
Klein, R.A., *J. Lipid Res.*, 1972, **13**, 672 (*ms*)
Stoffel, W. *et al*, *Hoppe Seyler's Z. Physiol. Chem.*, 1972, **253**, 1962 (*cmr*)
Wood, G.W., *Chem. Phys. Lipids*, 1977, **18**, 316 (*ms*)

Glycerol 1,2-dioctadecanoate 3-phosphoethanolamine G-00093

Octadecanoic acid 1-[[[(2-aminoethoxy)hydroxyphosphinyl] oxy]methyl]-1,2-ethanediyl ester, 9CI.
Distearoylphosphatidylethanolamine
[4537-76-2]

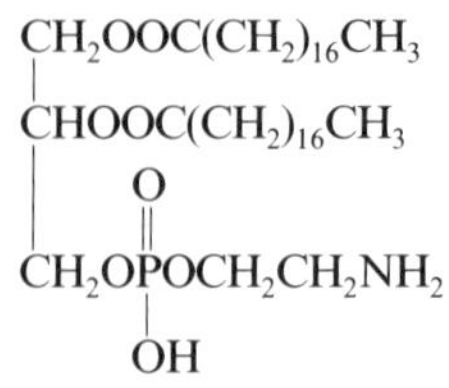

$C_{41}H_{82}NO_8P$ M 748.074
(R)-form
Cryst. Mp 185° (196°). $[\alpha]_D$ +6.2° (c, 3.4 in 9:1 $CHCl_3$/AcOH).
(±)-form
Cryst. Mp 180-181° (193°).

Hoefnagel, M.A. *et al*, *Recl. Trav. Chim. Pays-Bas* (*J. R. Neth. Chem. Soc.*), 1960, **79**, 330 (*synth*)
Chapman, D. *et al*, *Proc. R. Soc. London, A*, 1966, **290**, 1151 (*ir, pmr, props*)
Chapman, D. *et al*, *Trans. Faraday Soc.*, 1966, **62**, 2607 (*pmr, props*)
Aneja, R. *et al*, *Biochim. Biophys. Acta*, 1969, **187**, 439; 1970, **218**, 102 (*synth*)
Wood, G.W. *et al*, *Chem. Phys. Lipids*, 1977, **18**, 316 (*ms*)
Lammers, J.G. *et al*, *Chem. Phys. Lipids*, 1978, **22**, 293 (*ir, pmr*)

Glycerol 1,2-dioctadecanoate 3-phospho-1′-glycerol G-00094

Octadecanoic acid 1-[[[(2,3-dihydroxypropoxy) hydroxyphosphinyl]oxy]methyl]-1,2-ethanediyl ester, 9CI.
Distearoylphosphatidylglycerol
[4537-78-4]

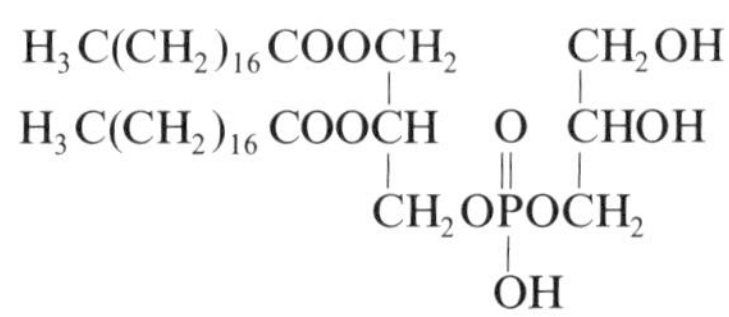

$C_{42}H_{83}O_{10}P$ M 779.085
Mp 66.5-67° $[\alpha]_D^{22}$ +2.0° (c, 10 in $CHCl_3$).
Ba salt: Mp 166°. $[\alpha]_D^{25}$ +9.2° (c, 1. in Py).

Baer, E. *et al*, *J. Biol. Chem.*, 1958, **232**, 895 (*synth*)
Baer, E., *Prog. Chem. Fats Other Lipids*, 1963, **6**, 33 (*rev*)
Saunders, R.M. *et al*, *J. Am. Chem. Soc.*, 1966, **88**, 3844 (*synth*)
Lammers, J.G. *et al*, *Recl. Trav. Chim. Pays-Bas* (*J. R. Neth. Chem. Soc.*), 1977, **96**, 216 (*synth*)
Lammers, J.G. *et al*, *Recl. Trav. Chim. Pays-Bas* (*J. R. Neth. Chem. Soc.*), 1979, **98**, 243 (*synth, cmr*)

Glycerol 1,2-dioctadecanoate 3-phosphoinositol G-00095

Distearoylphosphatidylinositol

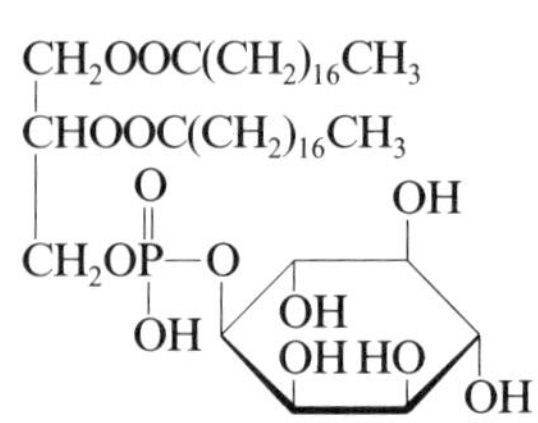

$C_{45}H_{87}O_{13}P$ M 867.148

Myoinositol-form

Mp 135-137°. Other stereoisomers have been synthesised.

Penta-Ac: Mp 119-122°.

Davies, J.M. *et al*, *Chem. Ind. (London)*, 1959, 1155 (*synth*)
Shevchenko, V.P. *et al*, *CA*, 1977, **86**, 190385 (*synth*)
Krylova, V.N. *et al*, *CA*, 1979, **91**, 19354 (*synth*)

Glycerol 1,2-dioctadecanoate 3-phosphoserine G-00096

Serine 2,3-bis[(1-oxooctadecyl)oxy]propyl hydrogen phosphate ester, 9CI. Distearoylphosphatidylserine

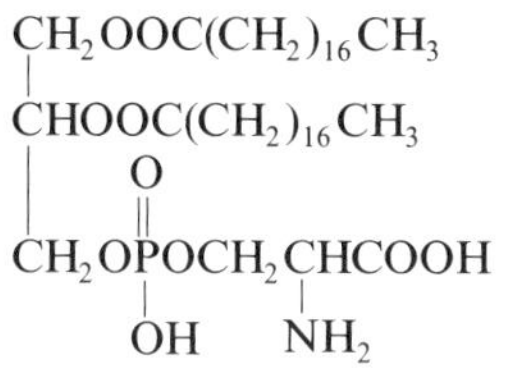

$C_{42}H_{82}NO_{10}P$ M 792.084

(R)-L-form [51446-62-9]

*1,2-Distearoyl-*sn-*glycero-3-phospho-L-serine*

Mp 159-161°.

(R)-DL-form

Mp 162°. $[\alpha]_D^{55}$ +7.4° (c, 0.6 in 9:1 $CHCl_3/Me_2CO$).

Baer, E. *et al*, *J. Biol. Chem.*, 1955, **212**, 25.
Baer, E., *Prog. Chem. Fats Other Lipids*, 1963, **6**, 33 (*rev*)
Aneja, R. *et al*, *Biochim. Biophys. Acta*, 1970, **218**, 102.

Glycerol 1,2-dioctadecanoate 3-propanoate G-00097

Octadecanoic acid 1-[(1-oxopropoxy)methyl]-1,2-ethanediyl ester, 9CI. α,β-Distearopropionin

[60562-73-4]

$C_{42}H_{80}O_6$ M 681.090

(S)-form [60456-30-6]

*3-Propionyl-1,2-distearoyl-*sn-*glycerol*

Cryst. (Me_2CO). Mp 48°.

Lok, C.M. *et al*, *Chem. Phys. Lipids*, 1976, **16**, 115 (*synth*)
Bus, J. *et al*, *Chem. Phys. Lipids*, 1976, **16**, 123 (*pmr*)

Glycerol 1,2-dioctadecanoate 3-tetradecanoate G-00098

Octadecanoic acid 1-[[(1-oxotetradecyl)oxy]methyl]-1,2-ethanediyl ester, 9CI

$C_{53}H_{102}O_6$ M 835.385

(S)-form [57416-14-5]

*1,2-Distearoyl-3-myristoyl-*sn-*glycerol*

Cryst. (EtOH). Mp 60-61°. $[M]_{300}^{27}$ −1.4° (C_6H_6).

(±)-form [60138-19-4]

α,β-Distearomyristin

Cryst. (C_6H_6/EtOH). Mp 62°. Exibits polymorphism.

Carter, M.G.R. *et al*, *J. Chem. Soc.*, 1939, 1518 (*synth*)
Gronowitz, S. *et al*, *Chem. Phys. Lipids*, 1975, **14**, 174 (*synth, ord*)

Glycerol 1,3-di-9,12,15-octadecatrienoate G-00099

$C_{39}H_{64}O_5$ M 612.932

(all-Z)-form

α,α′-Dilinolenin

Oil. Mp −12.3°.

Daubert, B.F. *et al*, *J. Am. Chem. Soc.*, 1947, **69**, 1449 (*synth*)

Glycerol 1,2-di-9,12,15-octadecatrienoate 3-phosphate G-00100

9,12,15-Octadecatrienoic acid 1-[(phosphonoxy)methyl]-1,2-ethanediyl ester, 9CI

$C_{39}H_{65}O_8P$ M 692.911

(R)-(all-Z)-form

*1,2-Dilinolenoyl-*sn-*glycerol 3-phosphate*

$[\alpha]_D^{20}$ +31° (c, 3.5 in $CHCl_3$).

Rakhit, S. *et al*, *Can. J. Chem.*, 1969, **47**, 2906 (*synth*)

Glycerol 1,2-di-9,12,15-octadecatrienoate 3-phosphocholine G-00101

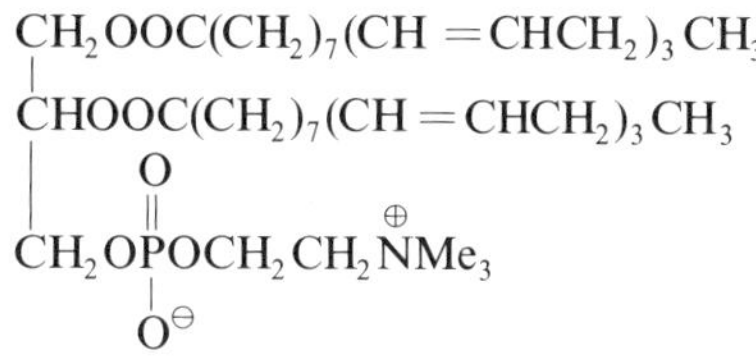

$C_{44}H_{76}NO_8P$ M 778.060

(R)-L-(all-Z)-form

Dilinolenoylphosphatidylcholine

Demel, R.A. *et al*, *Biochim. Biophys. Acta*, 1972, **266**, 26.
Warner, T.G. *et al*, *J. Lipid Res.*, 1977, **18**, 548 (*nmr*)
Wood, G.W. *et al*, *Chem. Phys. Lipids*, 1977, **18**, 316 (*ms*)

Glycerol 1,2-di-9,12,15-octadecatrienoate 3-phosphoethanolamine G-00102

9,12,15-Octadecatrienoic acid 1-[[[(2-aminoethoxy)hydroxyphosphinyl]oxy]methyl]-1,2-ethanediyl ester, 9CI

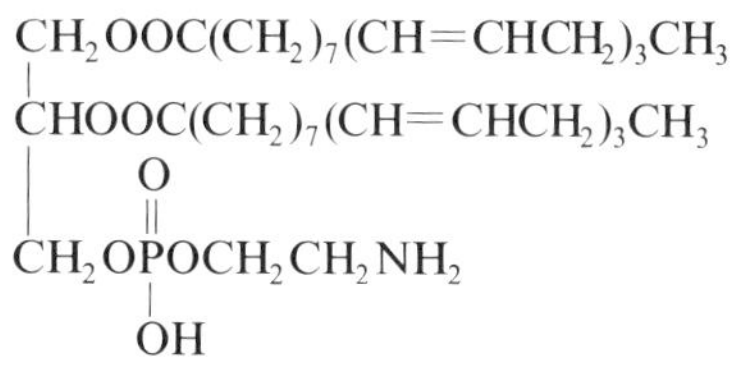

$C_{41}H_{70}NO_8P$ M 735.980

(R)-(all-Z)-form [34813-40-6]

Dilinolenoylphosphatidylethanolamine

$[\alpha]_D$ +6.1° (c, 2 in $CHCl_3$).

Shvets, V.I., *CA*, 1964, **60**, 1579 (*synth*)
Rakhit, S. *et al*, *Can. J. Chem.*, 1969, **47**, 2906 (*synth*)

Glycerol 1,2-di-9-octadecenoate G-00103

9-Octadecenoic acid 1-(hydroxymethyl)-1,2-ethanediyl ester, 9CI

$C_{39}H_{72}O_5$ M 620.995

(S)-(E,E)-form

*1,2-Dielaidoyl-*sn-*glycerol*

Mp *ca.* 25°. $[\alpha]_D^{25}$ −1.85° (c, 1.0 in $CHCl_3$).

(S)-(Z,Z)-form [24529-88-2]

*1,2-Dioleoyl-*sn-*glycerol*

Oil. $[\alpha]_D^{25}$ −1.91° (c, 6.2 in $CHCl_3$).

(±)-(Z,Z)-form [3738-74-7]

α,β-Diolein

Oil.

Krabisch, L. *et al*, *J. Lipid Res.*, 1965, **6**, 156 (*synth*)
Rakhit, S. *et al*, *Can. J. Chem.*, 1969, **47**, 2906 (*synth*)
Pfeiffer, F.R. *et al*, *J. Org. Chem.*, 1970, **35**, 221 (*synth*)

Glycerol 1,3-di-9-octadecenoate G-00104

9-Octadecenoic acid 2-hydroxy-1,3-propanediyl ester, 9CI

$C_{39}H_{72}O_5$ M 620.995

(E,E)-form

α,α′-Dielaidin

Cryst. (Et_2O/EtOH at 10°). Mp 50.1°.

(±)-***(E,Z)-form***

α-Elaido-α′-olein

Cryst. (Et_2O/MeOH). Mp 28.8°. Exhibits polymorphism.

(Z,Z)-form [2465-32-9]

α,α′-Diolein

Cryst. (Me_2CO). Mp 26-28°.

Benedict, J.H. *et al*, *J. Am. Oil Chem. Soc.*, 1950, **27**, 91 (*synth*)
Baur, J.F. *et al*, *J. Am. Chem. Soc.*, 1951, **73**, 3926 (*synth*)
Mank, A.P.J. *et al*, *Chem. Phys. Lipids*, 1976, **16**, 107 (*synth*)

Glycerol 1,2-di-9-octadecenoate 3-dodecanoate G-00105

*9-Octadecenoic acid 1-[[(1-oxododecyl)oxy]methyl]-1,2-ethanediyl ester, 9CI. 5,7-Dimethoxy-δ-(3-methyl-but-2-enoyl)coumarin. Angelicone. 5,7-Dimethoxy-δ-(3-methyl-1-oxo-2-butenyl)-2*H*-1-benzopyran-2-one, 9CI*

$C_{51}H_{94}O_6$ M 803.300

(±)-***(E,E)-form*** [51067-89-1]

α,β-Dielaidolaurin

Cryst. (EtOH). Mp 35.5°.

(±)-***(Z,Z)-form*** [4016-52-8]

α,β-Dioleolaurin

Oil, Cryst. (Et_2O at −20°). Mp 5-6°.

Daubert, B.F. *et al*, *J. Am. Chem. Soc.*, 1944, **66**, 290 (*synth*)
Cook, P.F.E. *et al*, *J. Chem. Soc.*, 1965, 4594 (*synth*)

Glycerol 1,2-di-9-octadecenoate 3-hexadecanoate G-00106

9-Octadecenoic acid 1-[[(1-oxohexadecyl)oxy]methyl]-1,2-ethanediyl ester, 9CI. 1-Palmitodiolein

$C_{55}H_{102}O_6$ M 859.407

Isol. from cocoa butter, olive oil and palm oil.

(S)-(Z,Z)-form [14960-35-1]

sn-*Glyceryl 1,2-dioleate 3-palmitate*

Oil. Mp 18.5-19°. $[M]_{300}$ −3.3 (C_6H_6).

(±)-***(E,E)-form*** [68069-77-2]

α,β-Dielaidopalmitin

Cryst. Mp 40.3°, Mp 46°. Exhibits polymorphism.

(±)-***(Z,Z)-form*** [65390-75-2]

α,β-Dioleopalmitin

Oil.

Meara, M.L., *J. Chem. Soc.*, 1948, 724 (*isol*)
Minor, J.E. *et al*, *J. Am. Chem. Soc.*, 1953, 2685 (*synth*)
Tamaki, Y. *et al*, *J. Insect Physiol.*, 1971, **17**, 1239 (*synth*)
Quinn, J.G. *et al*, *Lipids*, 1971, **6**, 692 (*synth*)
Gronowitz, S. *et al*, *Chem. Phys. Lipids*, 1979, **23**, 101 (*synth, ord*)

Glycerol 1,3-di-9-octadecenoate 2-hexadecanoate G-00107

9-Octadecenoic acid 2-[(1-oxohexadecyl)oxy]-1,3-propanediyl ester, 9CI

$C_{55}H_{102}O_6$ M 859.407

(E,E)-form

α,α′-Dielaidopalmitin

Cryst. Mp 44.5°. Exhibits polymorphism.

(Z,Z)-form [1716-07-0]

α,α′-Dioleopalmitin

Oil. Mp 5-6°, Mp 19°. Exhibits polymorphism.

Minor, J.E. *et al*, *J. Am. Chem. Soc.*, 1953, **75**, 2685 (*synth*)
Cook, P.F.E. *et al*, *J. Chem. Soc.*, 1965, 4594 (*synth*)

Glycerol 1,2-di-9-octadecenoate 3-(9,12-octadecadienoate) G-00108

9,12-Octadecadienoic acid 2,3-bis[(1-oxo-9-octadecenyl)oxy]propyl ester, 9CI

$C_{57}H_{102}O_6$ M 883.429

(±)-***(all-Z)-form*** [2190-20-7]

α,β-Dioleolinolein

Tamaki, Y. *et al*, *J. Insect Physiol.*, 1971, **17**, 1239 (*synth*)

Glycerol 1,3-di-9-octadecenoate 2-(9,12-octadecadienoate) G-00109

9,12-Octadecadienoic acid 2-[(1-oxo-9-octadecenyl)oxy]-1-[[(1-oxo-9-octadecenyl)oxy]methyl]ethyl ester, 9CI

$C_{57}H_{102}O_6$ M 883.429

(all-Z)-form [2190-19-4]

α,α′-Dioleolinolein

Tamaki, Y. *et al*, *J. Insect Physiol.*, 1971, **17**, 1239 (*synth*)

Glycerol 1,2-di-9-octadecenoate 3-octadecanoate G-00110

9-Octadecenoic acid 2-[(1-oxo-9-octadecenyl)oxy]-3-[(1-oxooctadecyl)oxy]propyl ester, 9CI

$C_{57}H_{106}O_6$ M 887.461

(±)-***(E,E)-form***

α,β-Dielaidostearin

Cryst. (Me_2CO). Mp 49.9°. Exhibits polymorphism.

(±)-***(E,Z)-form*** [30465-26-0]

α-Elaido-β-oleostearin

Cryst. Mp 29-31°.

(±)-***(Z,E)-form*** [35219-91-1]

α-Oleo-β-elaidostearin

Cryst. Mp 27-28.5°.

(±)-***(Z,Z)-form*** [2410-28-8]

α,β-Dioleostearin

Cryst. Mp 22.5-23.5°. Exhibits polymorphism.

Minor, J.E. *et al*, *J. Am. Chem. Soc.*, 1953, **75**, 2685 (*synth, props*)
Cook, P.F.E. *et al*, *J. Chem. Soc.*, 1965, 4594 (*synth*)
Carreau, J.-P., *Bull. Soc. Chim. Fr.*, 1970, 4111 (*synth*)

Glycerol 1,3-di-9-octadecenoate 2-octadecanoate G-00111

9-Octadecenoic acid 2-[(1-oxooctadecyl)oxy]-1,3-propanediyl ester, 9CI

$C_{57}H_{106}O_6$ M 887.461

(E,E)-form

α,α′-Dielaidostearin

Cryst. (Me_2CO). Mp 50.1°. Exhibits polymorphism.

(E,Z)-form [30465-28-2]

α-Elaido-α′-oleostearin

Cryst. Mp 26-28°.

(Z,Z)-form

α,α′-Dioleostearin

Cryst. (Et_2O). Mp 10.5-12.5° (14-15°). Exhibits polymorphism.

Minor, J.E. *et al*, *J. Am. Chem. Soc.*, 1953, **75**, 2685 (*synth, props*)
Cook, P.F.E. *et al*, *J. Chem. Soc.*, 1965, 4595 (*synth*)
Carreau, J.P., *Bull. Soc. Chim. Fr.*, 1970, 4111 (*synth*)

Glycerol 1,2-di-9-octadecenoate 3-phosphate G-00112

9-Octadecenoic acid 1-[(phosphonoxy)methyl]-1,2-ethanediyl ester, 9CI

$C_{39}H_{73}O_8P$ M 700.975

(R)-(Z)-form

*1,2-Dioleoyl-*sn*-glycerol 3-phosphate. Dioleoylphosphatidic acid*

$[\alpha]_D^{22}$ +3.8° (c, 5 in $CHCl_3$).

(±)-(Z)-form [5487-64-9]

Baer, E. *et al*, *Arch. Biochem. Biophys.*, 1958, **78**, 294.
Baer, E., *Prog. Chem. Fats Other Lipids*, 1963, **6**, 33 (*rev*)
Shvets, V.I. *et al*, *CA*, 1966, **65**, 20000 (*synth*)
Browning, J. *et al*, *Chem. Phys. Lipids*, 1979, **24**, 103 (*pmr*)

Glycerol 1,2-di-9-octadecenoate 3-phosphocholine G-00113

1,2-Di-9-octadecenoylglycerol-3-phosphocholine

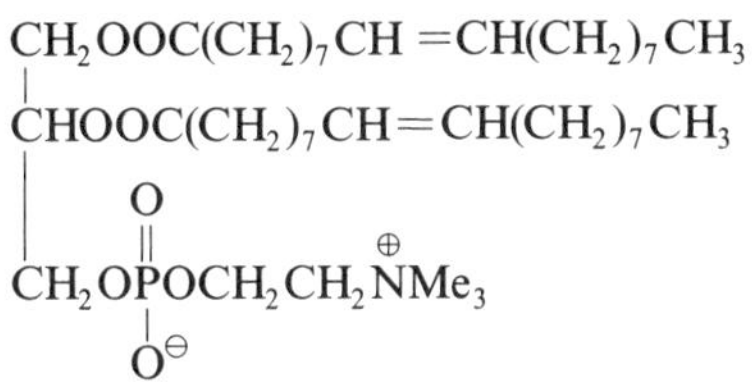

$C_{44}H_{84}NO_8P$ M 786.123

(R)-(Z,Z)-form [4235-95-4]

Dioleoylphosphatidylcholine

$[\alpha]_D$ +6.1° (c, 10 in $CHCl_3$/MeOH).

Baer, E. *et al*, *Can. J. Biochem. Physiol.*, 1959, **37**, 953 (*synth*)
Stoffel, W. *et al*, *Hoppe Seyler's Z. Physiol. Chem.*, 1969, **350**, 1385 (*props*)
Klein, R.A., *J. Lipid Res.*, 1971, **12**, 123, 628; 1972, **13**, 672 (*ms*)
Lippert, J., *Biochim. Biophys. Acta*, 1972, **282**, 8 (*raman*)
Barton, P.G. *et al*, *J. Biol. Chem.*, 1975, **250**, 4470 (*isomers*)
Wood, G.W. *et al*, *Chem. Phys. Lipids*, 1977, **18**, 316 (*ms*)
Seelig, J., *Biochemistry*, 1978, **17**, 3310 (*nmr*)

Glycerol 1,2-di-9-octadecenoate 3-phosphoethanolamine G-00114

9-Octadecenoic acid 1-[[[(2-aminoethoxy)hydroxyphosphinyl]oxy]methyl]-1,2-ethanediyl ester, 9CI. 1,2-Di-9-octadecenoylphosphatidylethanolamine

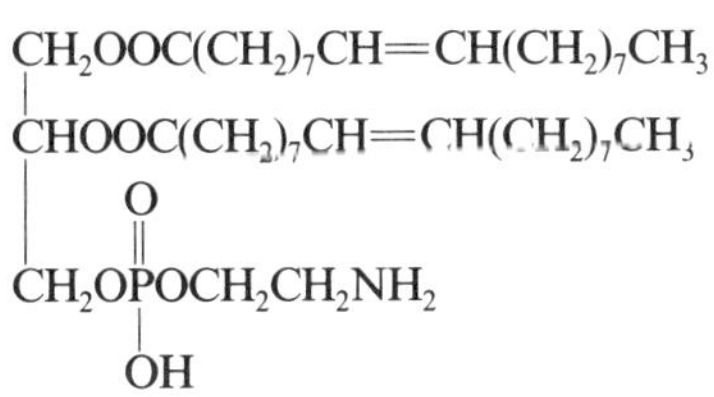

$C_{41}H_{78}NO_8P$ M 744.043

(R)-(all-E)-form

Dielaidylphosphatidylethanolamine

Cryst. ($CHCl_3$/MeOH). Mp 193°. $[\alpha]_D^{22}$ +6.1° ($CHCl_3$).

(±)-(all-E)-form [16777-83-6]

Cryst. Mp 188°. Exhibits polymorphism.

(R)-(all-Z)-form [4004-05-1]

Dioleoylphosphatidylethanolamine

Cryst. (MeOH). Mp 195-200°. $[\alpha]_D^{20}$ +6.0° ($CHCl_3$).

(±)-(all-Z)-form [2462-63-7]

Exhibits polymorphism.

Chapman, D. *et al*, *Biochim. Biophys. Acta*, 1966, **120**, 148.
Chapman, D. *et al*, *Proc. R. Soc. London, A*, 1966, **290**, 1151 (*ir, nmr, props*)
Chapman, D. *et al*, *Trans. Faraday Soc.*, 1966, **62**, 607.
Billimoria, J.D. *et al*, *J. Chem. Soc. C*, 1968, 1404 (*synth*)
Stoffel, W. *et al*, *Hoppe Seyler's Z. Physiol. Chem.*, 1969, **350**, 1385.
v. Dijck, P.W.M., *Biochim. Biophys. Acta*, 1976, **455**, 576.
Wood, G.W. *et al*, *Chem. Phys. Lipids*, 1977, **18**, 316 (*ms*)
Cullis, P.R. *et al*, *Biochim. Biophys. Acta*, 1978, **513**, 31 (*nmr*)
Lammers, J.G. *et al*, *Recl. Trav. Chim. Pays-Bas (J. R. Neth. Chem. Soc.)*, 1979, **98**, 243 (*synth*)

Glycerol 1,2-di-9-octadecenoate 3-phospho-1′-glycerol G-00115

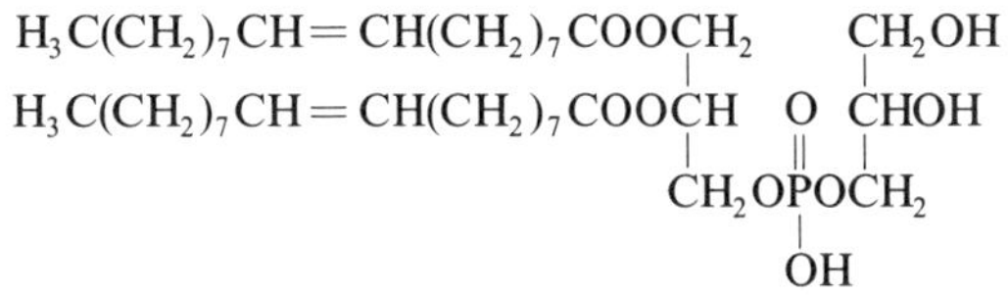

$C_{42}H_{79}O_{10}P$ M 775.054

(all-Z)-form

*1,2-Dioleoyl-*sn*-glycerol-3-phospho-1′-*sn*-glycerol*

$[\alpha]_D^{22}$ +2.35° (c, 11 in EtOH), $[\alpha]_D^{21}$ +2.0° (c, 10 in $CHCl_3$).

Baer, E. *et al*, *J. Biol. Chem.*, 1958, **232**, 895 (*synth*)
Baer, E., *Prog. Chem. Fats Other Lipids*, 1963, **6**, 33 (*rev*)

Glycerol 1,2-di-9-octadecenoate 3-phosphoserine G-00116

$C_{42}H_{78}NO_{10}P$ M 788.053

(R)-D-form

*1,2-Dioleoyl-*sn*-glycero-3-phospho-D-serine*

$[\alpha]_D^{22}$ +21.3° (C_6H_6), $[\alpha]_D^{22}$ +13.4° ($CHCl_3$).

(R)-L-form

*1,2-Dioleoyl-*sn*-glycero-3-phospho-L-serine*

$[\alpha]_D^{22}$ −7.8° ($CHCl_3$), $[\alpha]_D^{22}$ +5.1° ($CHCl_3$/MeOH/H_2O), $[\alpha]_D^{22}$ −17.8° (C_6H_6).

Ba salt: Mp 165-169°.

(R)-DL-form

*1,2-Dioleoyl-*sn*-glycero-3-phospho-DL-serine*

$[\alpha]_D^{22}$ +4.7° ($CHCl_3$), $[\alpha]_D^{22}$ +7.0° ($CHCl_3$/MeOH/H_2O), $[\alpha]_D^{22}$ +4.9° (C_6H_6).

Browning, J. *et al*, *Chem. Phys. Lipids*, 1979, **24**, 103 (*synth, pmr*)
Browning, J. *et al*, *Biochemistry*, 1980, **19**, 1262 (*cmr*)

Glycerol 1,3-dipentadecanoate G-00117

Pentadecanoic acid 2-hydroxy-1,3-propanediyl ester, 9CI. α,α′-Dipentadecanoin

[59891-25-7]

$C_{33}H_{64}O_5$ M 540.866

Cryst. (hexane). Mp 70°.

Mank, A.P.J. *et al*, *Chem. Phys. Lipids*, 1976, **16**, 107 (*synth*)

Glycerol 1,2-dipentadecanoate 3-phosphocholine G-00118

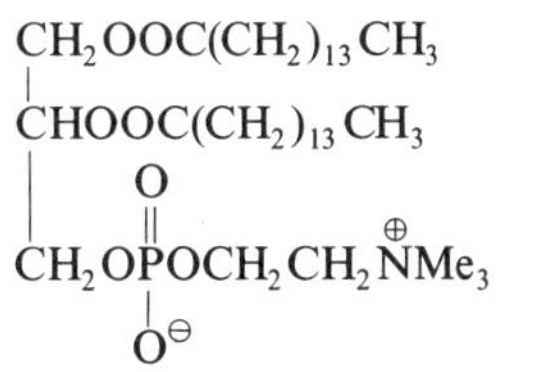

$C_{38}H_{76}NO_8P$ M 705.994

(R)-L-form [3355-27-9]

Dipentadecanoylphosphatidylcholine

Sylvius, J.R. *et al*, *Biochim. Biophys. Acta*, 1979, **555**, 175.

Glycerol 1,2-ditetradecanoate G-00119

Tetradecanoic acid 1-(hydroxymethyl)-1,2-ethanediyl ester, 9CI. αβ-Dimyristin

[20255-94-1]

$C_{31}H_{60}O_5$ M 512.812

(S)-form [60562-16-5]

*1,2-Dimyristoyl-*sn*-glycerol*

Cryst. ($CHCl_3$/pet. ether). Mp 58.5-60°. $[\alpha]_D^{23}$ −3.05° (c, 6.95 in $CHCl_3$).

Me ether:

$C_{32}H_{62}O_5$ M 526.839

Cryst. Mp 34-35°. [M] +925° (EtOH).

Hexadecyl ether:

$C_{47}H_{92}O_5$ M 737.241

Mp 49-49.5°. [M] +1747° (EtOH).

Benzyl ether:

$C_{38}H_{66}O_5$ M 602.936

Cryst. Mp 34-35°. $[\alpha]_D^{23}$ +5.90° (c, 7.1 in $CHCl_3$).

(±)-form [56270-93-0]

Cryst. (hexane). Mp 37.5°, Mp 54°, 59°. Exhibits polymorphism.

Howe, R.J. *et al*, *J. Chem. Soc.*, 1951, 2663 (*synth, props*)
Ioannou, P.V. *et al*, *Synthesis*, 1979, 939 (*synth*)
Michelson, P. *et al*, *Chem. Phys. Lipids*, 1983, **33**, 27 (*synth, cd*)

Glycerol 1,3-ditetradecanoate G-00120

Tetradecanoic acid 2-hydroxy-1,3-propanediyl ester, 9CI. α,α′-Dimyristin

[7770-09-4]

$C_{31}H_{60}O_5$ M 512.812

Cryst. (hexane). Mp 66-67°.

Mank, A.P.J. *et al*, *Chem. Phys. Lipids*, 1976, **16**, 107 (*synth*)

Glycerol 1,2-ditetradecanoate 3-decanoate G-00121

Tetradecanoic acid 1-[[(1-oxodecyl)oxy]methyl]-1,2-ethanediyl ester, 9CI. α,β-Dimyristocaprin

[60138-16-1]

$C_{41}H_{78}O_6$ M 667.064

(±)-form

Cryst. (C_6H_6). Mp 43.5°. Exhibits polymorphism.

Carter, M.G.R. *et al*, *J. Chem. Soc.*, 1939, 1518 (*synth*)
v. Soest, T. *et al*, *Recl. Trav. Chim. Pays-Bas* (*J. R. Neth. Chem. Soc.*), 1976, **95**, 162.

Glycerol 1,3-ditetradecanoate 2-decanoate G-00122

α,α′-Dimyristocaprin

$C_{41}H_{78}O_6$ M 667.064

Cryst. (EtOH). Mp 43.5°. Exhibits polymorphism.

Malkin, T. *et al*, *J. Chem. Soc.*, 1939, 1141.

Glycerol 1,2-ditetradecanoate 3-dodecanoate G-00123

Tetradecanoic acid 1-[[(1-oxododecyl)oxy]methyl]-1,2-ethanediyl ester, 9CI. α,β-Dimyristolaurin. Laurodimyristin

$C_{43}H_{82}O_6$ M 695.117

Major constit. of the seed oil of *Virola surinamensis*.

(±)-form

Cryst. (C_6H_6). Mp 46.5°, Mp 48.5°. Exhibits polymorphism.

Carter, M.G.R. *et al*, *J. Chem. Soc.*, 1939, 577 (*synth*)
Culp, T.W. *et al*, *J. Am. Oil Chem. Soc.*, 1965, **42**, 974 (*isol*)
v. Soest, T.C. *et al*, *Recl. Trav. Chim. Pays-Bas* (*J. R. Neth. Chem. Soc.*), 1976, **95**, 162.

Glycerol 1,3-ditetradecanoate 2-dodecanoate G-00124

Tetradecanoic acid 2-[(1-oxododecyl)oxy]-1,3-propanediyl ester, 9CI. α,α′-Dimyristolaurin

[60138-24-1]

$C_{43}H_{82}O_6$ M 695.117

Cryst. (EtOH). Mp 50°. Exhibits polymorphism.

Malkin, T. *et al*, *J. Chem. Soc.*, 1939, 103 (*synth*)
v. Soest, T.C. *et al*, *Recl. Trav. Chim. Pays-Bas* (*J. R. Neth. Chem. Soc.*), 1976, **95**, 162.

Glycerol 1,3-ditetradecanoate 2-hexadienoate G-00125

$C_{37}H_{66}O_6$ M 606.925

(E,E)-form

2-Sorbo-1,3-dimyristin

Triglyceride present in the aphid *Dactynotus jaceae*. Cryst. (MeOH). Mp 54-54.5°.

Bowie, J.H. *et al*, *J. Chem. Soc.*, 1965, 5651 (*struct*)

Glycerol 1,2-ditetradecanoate 3-octadecanoate G-00126

α,β-Dimyristostearin

$C_{49}H_{94}O_6$ M 779.278

(±)-form

Cryst. (EtOH/C_6H_6). Mp 56°. Exhibits polymorphism.

Carter, M.G.R. *et al*, *J. Chem. Soc.*, 1939, 1518 (*synth*)
Chapman, D., *J. Chem. Soc.*, 1958, 3186 (*synth, ir*)

Glycerol 1,3-ditetradecanoate 2-octadecanoate G-00127

α,α′-Dimyristostearin

$C_{49}H_{94}O_6$ M 779.278

Cryst. (EtOH/C_6H_6). Mp 55.5°. Exibits polymorphism.

Malkin, T. *et al*, *J. Chem. Soc.*, 1939, 1141.

Glycerol 1,3-ditetradecanoate 2-(9-octadecenoate) G-00128

$C_{49}H_{92}O_6$ M 777.262

(Z)-form

α,α′-Dimyristoolein

Cryst. (EtOH/C_6H_6). Mp 28.5°. Exhibits polymorphism.

Malkin, T. *et al*, *J. Chem. Soc.*, 1949, 369 (*synth*)

Glycerol 1,2-ditetradecanoate 3-phosphate G-00129

Tetradecanoic acid 1-[(phosphonoxy)methyl]-1,2-ethanediyl ester, 9CI. Dimyristoylphosphatidic acid

[30170-00-4]

$C_{31}H_{61}O_8P$ M 592.792

(R)-form [28874-52-4]

*1,2-Ditetradecanoyl-*sn*-glycerol 3-phosphate*

Mp 61.5-62.5°. $[\alpha]_D^{24}$ +4.4° (c, 11 in $CHCl_3$).

Ba salt: Mp 213-215°.

Di-Me ester: Mp 32-33°. $[\alpha]_D$ +2.3°.

Diphenyl ester: Mp 38-39°. $[\alpha]_D$ +2.6° (c, 10 in $CHCl_3$).

(±)-form

Mp 54-55°.

Phenyl ester: Mp 48.5-49.5°.

Dibenzyl ester: Mp 40-41°.

Baer, E., *J. Biol. Chem.*, 1951, **189**, 235.
Hessel, L.W. *et al*, *Recl. Trav. Chim. Pays-Bas* (*J. R. Neth. Chem. Soc.*), 1954, **73**, 150.
Baer, E. *et al*, *J. Biol. Chem.*, 1955, **212**, 39.
Baer, E., *Prog. Chem. Fats Other Lipids*, 1963, **6**, 33 (*rev*)
Browning, J. *et al*, *Chem. Phys. Lipids*, 1979, **24**, 103.

Glycerol 1,2-ditetradecanoate 3-phosphocholine — G-00130

Dimyristoylphosphatidylcholine. 1,2-Ditetradecanoylglycero-3-phosphocholine

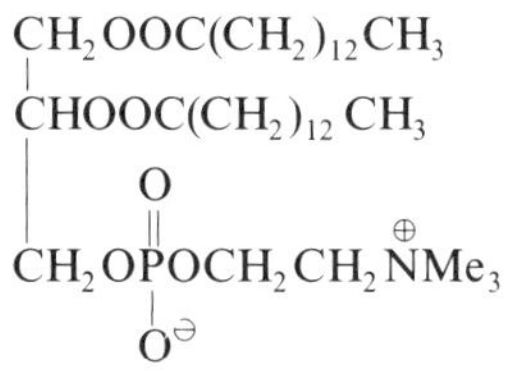

$C_{36}H_{72}NO_8P$ M 677.940

***(R)*-form** [18194-24-6]

Mp 236-237°. $[\alpha]_D^{22}$ +7.0° (c, 10 in 1:1 $CHCl_3$/MeOH).

Baer, E. *et al*, *Can. J. Biochem.*, 1959, **37**, 953.
Klein, R.A., *J. Lipid Res.*, 1972, **13**, 672 (*ms*)
Dorset, D.L., *Biochim. Biophys. Acta*, 1975, **380**, 257.
Koyama, Y. *et al*, *Chem. Phys. Lipids*, 1977, **19**, 74 (*ir, raman*)
Wood, G.W. *et al*, *Chem. Phys. Lipids*, 1977, **18**, 316 (*ms*)

Glycerol 1,2-ditetradecanoate 3-phosphoethanolamine — G-00131

Tetradecanoic acid 1-[[[(2-aminoethoxy)hydroxyphosphinyl]oxy]methyl]-1,2-ethanediyl ester, 9CI. Dimyristoylphosphatidylethanolamine. 1,2-Ditetradecanoylglycero-3-phosphoethanolamine. 1,2-Dimyristoylcephalin

[20255-95-2]

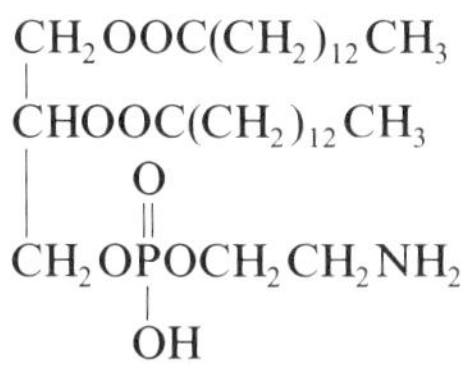

$C_{33}H_{66}NO_8P$ M 635.860

***(R)*-form** [998-07-2]

Mp 175-177°. $[\alpha]_D^{26}$ +6.7° (c, 8.4 in $CHCl_3$).

(±)-form [5683-46-5]

Cryst. (EtOH). Mp 200°.

Baer, E. *et al*, *J. Am. Chem. Soc.*, 1952, **74**, 152 (*synth*)
Hoefnagel, M.A. *et al*, *Recl. Trav. Chim. Pays-Bas* (*J. R. Neth. Chem. Soc.*), 1960, **79**, 330 (*synth*)
Chapman, D. *et al*, *Proc. R. Soc. London, A*, 1966, **290**, 1151 (*pmr, ir, props*)
Klein, R.A., *J. Lipid Res.*, 1972, **13**, 672 (*ms*)
v. Dijck, P.W.M., *Biochim. Biophys. Acta*, 1976, **455**, 576.
Eibl, H. *et al*, *Chem. Phys. Lipids*, 1978, **22**, 1 (*synth*)
Blume, A., *Biochim. Biophys. Acta*, 1979, **557**, 32.

Glycerol 1,2-ditetradecanoate 3-phosphono-2-amino-1-propanol — G-00132

Ditetradecanoyl-α-glycerylphosphoryl-2-amino-1-propanol. Dimyristoyl-α-glycerylphosphoryl-2-amino-1-propanol

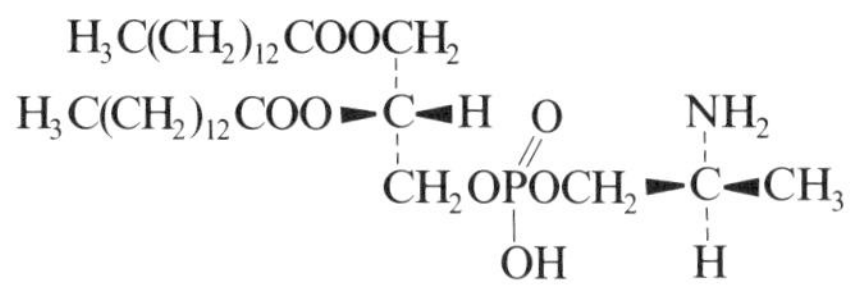

$C_{34}H_{68}NO_8P$ M 649.887

***(R)*-L-form**

Mp 181-182°. $[\alpha]_D^{24}$ +7.4° (c, 5 in $CHCl_3$).

Baer, E. *et al*, *J. Biol. Chem.*, 1963, **238**, 3591.

Glycerol 1,2-ditetradecanoate 3-phosphoserine — G-00133

Dimyristoylphosphatidylserine

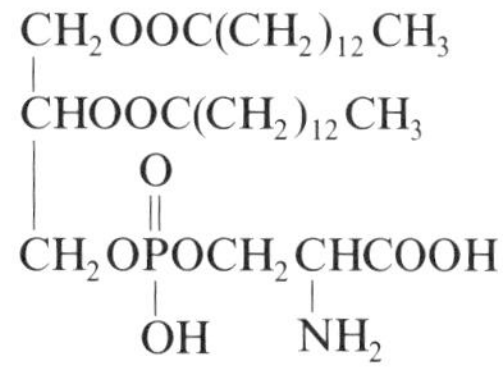

$C_{34}H_{66}NO_{10}P$ M 679.870

***(R)*-D-form**

*1,2-Dimyristoyl-*sn*-glycero-3-phospho-D-serine*

$[\alpha]_D^{22}$ +11.5° ($CHCl_3$), $[\alpha]_D^{22}$ +6.3° ($CHCl_3$/MeOH/H_2O), $[\alpha]_D^{22}$ +21.5° (C_6H_6).

***(R)*-L-form**

*1,2-Dimyristoyl-*sn*-glycero-3-phospho-L-serine*

$[\alpha]_D^{22}$ −6.0° ($CHCl_3$), $[\alpha]_D^{22}$ +6.2° ($CHCl_3$/MeOH/H_2O), $[\alpha]_D^{22}$ −13.5° (C_6H_6).

***(R)*-DL-form**

*1,2-Dimyristoyl-*sn*-glycero-3-phospho-DL-serine*

$[\alpha]_D^{22}$ +3.6° ($CHCl_3$), $[\alpha]_D^{22}$ +7.3° ($CHCl_3$/MeOH/H_2O), $[\alpha]_D^{22}$ +4.2° (C_6H_6).

Browning, J. *et al*, *Chem. Phys. Lipids*, 1979, **24**, 103 (*synth*)
Browning, J. *et al*, *Biochemistry*, 1980, **19**, 1262 (*nmr*)

Glycerol 1,3-ditridecanoate — G-00134

Tridecanoic acid 2-hydroxy-1,3-propanediyl ester, 9CI. α,α′-Ditridecanoin

[59891-27-9]

$C_{29}H_{56}O_5$ M 484.758

Cryst. (hexane). Mp 62.0°.

Mank, A.P.J. *et al*, *Chem. Phys. Lipids*, 1976, **16**, 107 (*synth*)

Glycerol 1,2-ditridecanoate 3-phosphocholine — G-00135

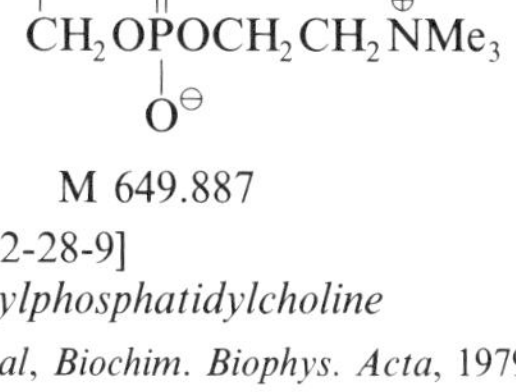

$C_{34}H_{68}NO_8P$ M 649.887

***(R)*-L-form** [71242-28-9]

Ditridecanoylphosphatidylcholine

Sylvius, J.R. *et al*, *Biochim. Biophys. Acta*, 1979, **555**, 175.

Glycerol 1-docosanoate 3-eicosanoate G-00136

Docosanoic acid 2-hydroxy-3-[(1-oxoeicosyl)oxy]propyl ester, 9CI. α,α′-Arachidobehenin

[57799-98-1]

$C_{45}H_{88}O_5$ M 709.187

Constit. of *Flourensia cernua*. Cryst. (cyclohexane). Mp 70.5-71.5°. Stereochem. not known.

Kingston, D.G.I. *et al*, *Phytochemistry*, 1975, **14**, 2033 (*isol, ms*)

Glycerol 1-docosanoate 3-hexadecanoate G-00137

Docosanoic acid 2-hydroxy-3-[(1-oxohexadecyl)oxy]propyl ester, 9CI. α,α′-Behenopalmitin

$C_{41}H_{80}O_5$ M 653.080

(±)-form [59891-35-9]

Cryst. (Me_2CO). Mp 76-77°.

Mank, A.P.J. *et al*, *Chem. Phys. Lipids*, 1976, **16**, 107 (*synth*)

Glycerol 1-dodecanoate G-00138

2,3-Dihydroxypropyl dodecanoate, 9CI. 1-Monolaurin, 8CI. α-Monolaurin. Glycerol 1-laurate

[142-18-7]

$HOH_2CCH(OH)CH_2OOC(CH_2)_{10}CH_3$

$C_{15}H_{30}O_4$ M 274.400

▷ OF1600000.

***(S)*-form**

sn-*Glycerol 1-dodecanoate*

Mp 54-55°. $[\alpha]_D$ −4.90° (dry Py).

(±)-form

Has antimicrobial activity. Mp 63°. Bp_1 186°.

Averill, H.P. *et al*, *J. Am. Chem. Soc.*, 1929, **51**, 866 (*synth*)
Baer, E. *et al*, *J. Am. Chem. Soc.*, 1945, **67**, 2031 (*synth*)
Gronowitz, S. *et al*, *Chem. Phys. Lipids*, 1975, **14**, 174 (*stereochem*)
Artamov, A.F. *et al*, *Izv. Akad. Nauk Kaz. SSR, Ser. Khim.*, 1977, **27**, 54 (*synth*)
Kabara, J.J. *et al*, *Lipids*, 1977, **12**, 753.
Lewis, R.J., *Sax's Dangerous Properties of Industrial Materials*, 8th Ed., Van Nostrand-Reinhold, 1992, MRJ000.

Glycerol 2-dodecanoate G-00139

2-Hydroxy-1-(hydroxymethyl)ethyl dodecanoate, 9CI. β-Monolaurin. Glycerol 2-laurate

[1678-45-1]

$C_{15}H_{30}O_4$ M 274.400

Has antimicrobial activity. Mp 51°.

1,3-Benzylidene:

$C_{22}H_{34}O_4$ M 362.508

Mp 46.6°.

Averill, H.P. *et al*, *J. Am. Chem. Soc.*, 1929, **51**, 866 (*synth*)
Stimmel, B.F. *et al*, *J. Am. Chem. Soc.*, 1934, **56**, 1724.

Glycerol 1-dodecanoate 2-octadecanoate 3-phosphocholine G-00140

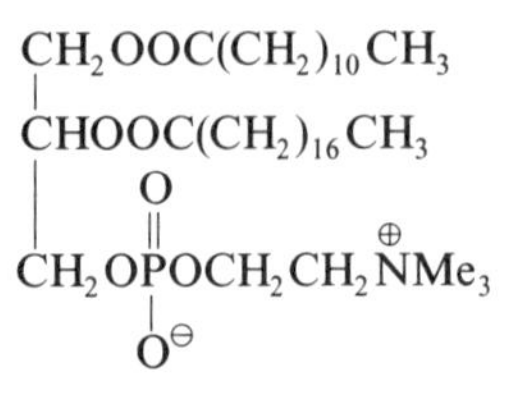

$C_{38}H_{76}NO_8P$ M 705.994

(±)-form

Lauroylstearoylphosphatidylcholine

Mp 225°. The (*R*)-form has also been synthesised.

de Haas, G.H. *et al*, *Tetrahedron Lett.*, 1960, 1 (*synth*)
de Haas, G.H. *et al*, *Recl. Trav. Chim. Pays-Bas* (*J. R. Neth. Chem. Soc.*), 1961, **80**, 951 (*synth, ir*)

Glycerol 1-dodecanoate 2-(9-octadecenoate) 3-phosphoinositol G-00141

myo-*Inositol 1-[2-[(1-oxo-9-octadecenyl)oxy]]-3-[(1-oxododecyl)oxy]propyl hydrogen phosphate, 9CI*

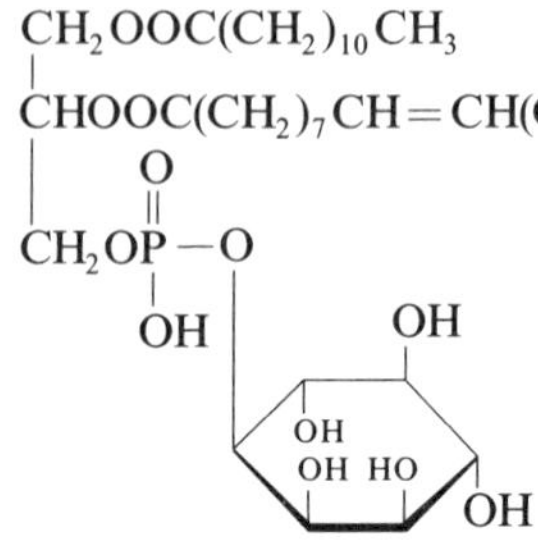

$C_{39}H_{73}O_{13}P$ M 780.972

Molotkovskii, Y.G. *et al*, *CA*, 1971, **75**, 88864 (*synth*)

Glycerol 1-dodecanoate 2-tetradecanoate 3-octanoate G-00142

2-Myristocaprylolaurin. α-Caprylo-β-myristo-α′-laurin

$C_{37}H_{70}O_6$ M 610.956

Isol. from coconut oil. Cryst. Mp 14.1°.

Bomer, A., *Chem.-Ztg.*, 1914, **38**, 844 (*isol*)
Heiduschka, A. *et al*, *J. Prakt. Chem.*, 1928, **120**, 145 (*synth*)

Glycerol 1-eicosanoate 3-octadecanoate G-00143

Eicosanoic acid 2-hydroxy-3-[(1-oxooctadecyl)oxy]propyl ester, 9CI. αα′-Arachidostearin

$C_{41}H_{80}O_5$ M 653.080

(±)-form [59891-28-0]

Cryst. (Me_2CO). Mp 76°.

Mank, A.P.J. *et al*, *Chem. Phys. Lipids*, 1976, **16**, 107 (*synth*)

Glycerol 1-(5,8,11,14-eicosatetraenoate) 2-octadecanoate 3-phosphoethanolamine G-00144

5,8,11,14-Eicosatetraenoic acid 3-[[(2-aminoethoxy)hydroxyphosphinyl]oxy]-2-[(1-oxooctadecyl)oxy]propyl ester, 9CI. 1-Eicosa-5,8,11,14-tetraenoyl 2 octadecanoylglycero-3-phosphoethanolamine

[22259-33-2]

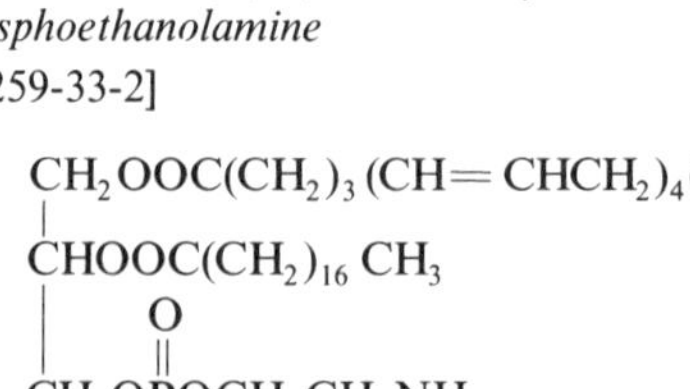

$C_{43}H_{78}NO_8P$ M 768.065

***(R)-(all-Z)*-form**

Arachidonoylstearoylphosphatidylethanolamine

Renin inhibitor.

Miyazaki, M. *et al*, *Proc. Soc. Exp. Biol. Med.*, 1977, **155**, 468 (*synth*)

Glycerol 1-hexadecanoate G-00145

Hexadexanoic acid 2,3-dihydroxypropyl ester, 9CI. α-Monopalmitin. Glycerol 1-palmitate

[542-44-9]

$$H_3C(CH_2)_{14}COOCH_2CH(OH)CH_2OH$$

$C_{19}H_{38}O_4$ M 330.507

(R)-form [5309-46-6]

3-Palmitoyl-sn-glycerol

Cryst. (pet. ether). Mp 71-72°. $[\alpha]_D$ −4.3° (c, 10 in Py).

(±)-form [19670-51-0]

Cryst. (pet. ether). Mp 77°.

Baer, E. *et al*, *J. Am. Chem. Soc.*, 1945, **67**, 2031 (*synth*)
Schlenk, W.J., *J. Am. Oil Chem. Soc.*, 1965, **42**, 945 (*synth*)
Quinn, J.G. *et al*, *J. Am. Oil Chem. Soc.*, 1967, **44**, 439 (*synth*)

Glycerol 2-hexadecanoate G-00146

Hexadecanoic acid 2-hydroxy-1-(hydroxymethyl)ethyl ester, 9CI. β-Monopalmitin. Glycerol 2-palmitate

[23470-00-0]

$$H_3C(CH_2)_{14}COOCH(CH_2OH)_2$$

$C_{19}H_{38}O_4$ M 330.507

Cryst. (pet. ether). Mp 69°.

1,3-Benzylidene: Mp 63.5°.

Cook, P.F.E. *et al*, *J. Chem. Soc.*, 1965, 4594 (*synth*)

Glycerol 1-hexadecanoate 3-decanoate G-00147

Hexadecanoic acid 2-hydroxy-3-[(1-oxodecyl)oxy]propyl ester, 9CI. α,α′-Palmitocaprin

$C_{29}H_{56}O_5$ M 484.758

(±)-form [59891-26-8]

Cryst. (hexane). Mp 54.5°.

Mank, A.P.J. *et al*, *Chem. Phys. Lipids*, 1976, **16**, 107 (*synth*)

Glycerol 1-hexadecanoate 2,3-didodecanoate G-00148

Hexadecanoic acid 2,3-bis[(1-oxododecyl)oxy]propyl ester, 9CI. α,β-Dilauropalmitin

$C_{43}H_{82}O_6$ M 695.117

(R)-form [57416-16-7]

1,2-Dilauroyl-3-palmitoyl-sn-glycerol

Cryst. Mp 42.5-43°. $[M]^{27}_{300}$ +2.0° (C_6H_6).

(±)-form

Cryst. (EtOH/C_6H_6). Mp 46.5°. Exhibits polymorphism.

Carter, M.G.R. *et al*, *J. Chem. Soc.*, 1939, 1518 (*synth, props*)
Gronowitz, S. *et al*, *Chem. Phys. Lipids*, 1975, **14**, 174 (*synth, ord*)

Glycerol 1-hexadecanoate 2,3-dioctadecanoate G-00149

Octadecanoic acid 1-[[(1-oxohexadecyl)oxy]methyl]-1,2-ethanediyl ester, 9CI. α,β-Distearopalmitin

[2177-99-3]

$C_{55}H_{106}O_6$ M 863.439

(S)-form [5281-85-6]

1,2-Distearoyl-3-palmitoyl-sn-glycerol

Mp 61°.

(±)-form [5281-50-5]

Cryst. (EtOH/pet. ether). Mp 65.2°. Exhibits polymorphism.

Lutton, E.S. *et al*, *J. Am. Chem. Soc.*, 1948, **70**, 2441 (*synth, props*)
Lok, C.M. *et al*, *Chem. Phys. Lipids*, 1976, **16**, 115 (*synth*)

Glycerol 1-hexadecanoate 3-dodecanoate G-00150

Hexadecanoic acid 2-hydroxy-3-[(1-oxododecyl)oxy]propyl ester, 9CI. α-Palmito-α′-laurin

[51604-53-6]

$C_{31}H_{60}O_5$ M 512.812

(±)-form

Cryst. (hexane). Mp 54.5°.

Mank, A.P.J. *et al*, *Chem. Phys. Lipids*, 1976, **16**, 107 (*synth*)

Glycerol 1-hexadecanoate 2-(5,8,11,14-eicosatetraenoate) 3-phosphocholine G-00151

1-Hexadecanoyl-2-eicosa-5,8,11,14-tetraenoylglycero-3-phosphocholine

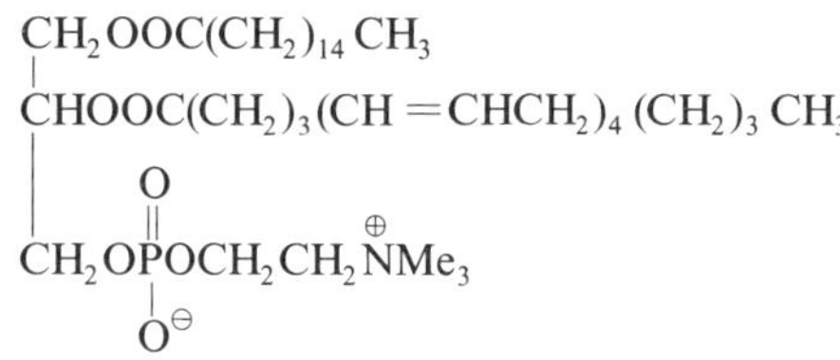

$C_{44}H_{80}NO_8P$ M 782.092

(R)-L-(all-Z)-form [35418-58-7]

Palmitoylarachidonoylphosphatidylcholine

Evans, R.W. *et al*, *Chem. Phys. Lipids*, 1978, **22**, 207.

Glycerol 1-hexadecanoate 2-(9,12-octadecadienoate) G-00152

9,12-Octadecadienoic acid 1-(hydroxymethyl)-2-[(1-oxohexadecyl)oxy]ethyl ester, 9CI

$C_{37}H_{68}O_5$ M 592.941

(S)-(all-Z)-form [51621-26-2]

1-Palmitoyl-2-linoleoyl-sn-glycerol

Oil. $[\alpha]^{25}_D$ +2.5° (neat).

Buchnea, D., *Lipids*, 1974, **9**, 55 (*synth*)

Glycerol 1-hexadecanoate 2-(9,12-octadecadienoate) 3-phosphocholine G-00153

1-Hexadecanoyl-2-octadeca-9,12-dienoylglycero-3-phosphocholine

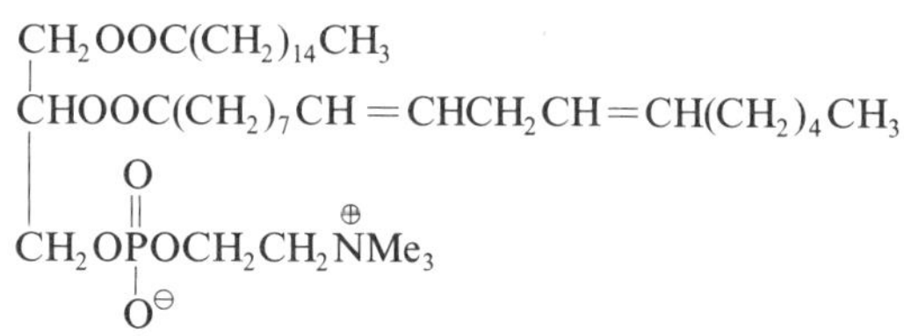

$C_{42}H_{80}NO_8P$ M 758.070

(R)-L-(Z,Z)-form [17708-90-6]

Palmitoyllinoleoylphosphatidylcholine

Demel, R.A. *et al*, *Biochim. Biophys. Acta*, 1972, **266**, 26.
Ghosh, D. *et al*, *Biochim. Biophys. Acta*, 1972, **266**, 41.
Wood, G.W. *et al*, *Chem. Phys. Lipids*, 1977, **18**, 316 (*ms*)
Porter, N.A. *et al*, *Lipids*, 1979, **14**, 20 (*hplc*)

Glycerol 1-hexadecanoate 2-octadecanoate G-00154

Octadecanoic acid 1-(hydroxymethyl)-2-[(1-oxohexadecyl)oxy]ethyl ester, 9CI. α-Palmito-β-stearin

$C_{37}H_{72}O_5$ M 596.973

(±)-form [20296-26-8]

Cryst. (hexane). Mp 59-61°.

Gigg, J. *et al*, *J. Chem. Soc. C*, 1967, 431 (*synth*)

Glycerol 1-hexadecanoate 2-octadecanoate 3-phosphocholine G-00155

Palmitoylstearoylphosphatidylcholine. 1-Hexadecanoyl-2-octadecanoylglycero-3-phosphocholine

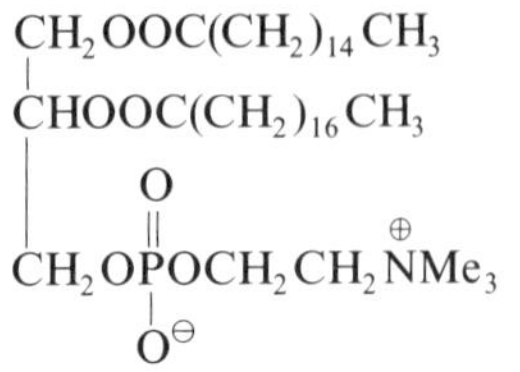

$C_{42}H_{84}NO_8P$ M 762.101

Keough, K.M.W. *et al*, *Biochemistry*, 1979, **18**, 1453.
Porter, N.A. *et al*, *Lipids*, 1979, **14**, 20 (*hplc*)

Glycerol 1-hexadecanoate 2-octadecanoate 3-tetradecanoate G-00156

Octadecanoic acid 1-[[(1-oxohexadecyl)oxy]methyl]-2-[(1-oxotetradecyl)oxy]ethyl ester, 9CI. α-Palmito-β-stearomyristin

[55256-01-4]

$C_{51}H_{98}O_6$ M 807.332

(±)-form
Cryst. (EtOH/C_6H_6). Mp 58.5-59°.

Verkade, P.E. *et al*, *Recl. Trav. Chim. Pays-Bas* (*J. R. Neth. Chem. Soc.*), 1937, **56**, 365.

Glycerol 1-hexadecanoate 2-(9,12,15-octadecatrienoate) 3-phosphocholine G-00157

[17118-61-5]

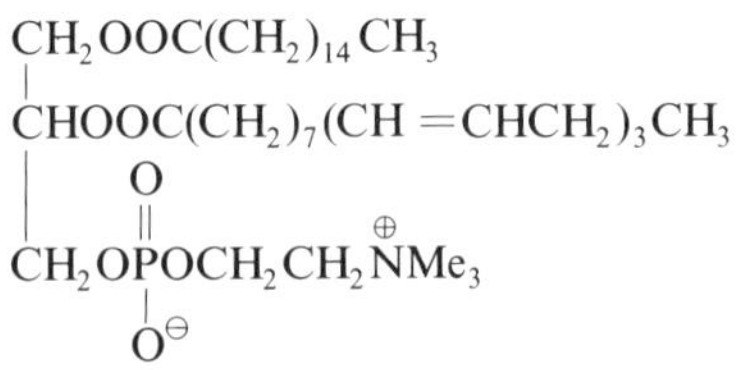

$C_{42}H_{78}NO_8P$ M 756.054

(R)-L-(all-Z)-form
Palmitoyllinolenoylphosphatidylcholine

Ghosh, D. *et al*, *Chem. Phys. Lipids*, 1971, **7**, 173.
Demel, R.A. *et al*, *Biochim. Biophys. Acta*, 1972, **266**, 26.
Ghosh, D. *et al*, *Biochim. Biophys. Acta*, 1972, **266**, 41.

Glycerol 1-hexadecanoate 2-(9,12,15-octadecatrienoate) 3-phosphoethanolamine G-00158

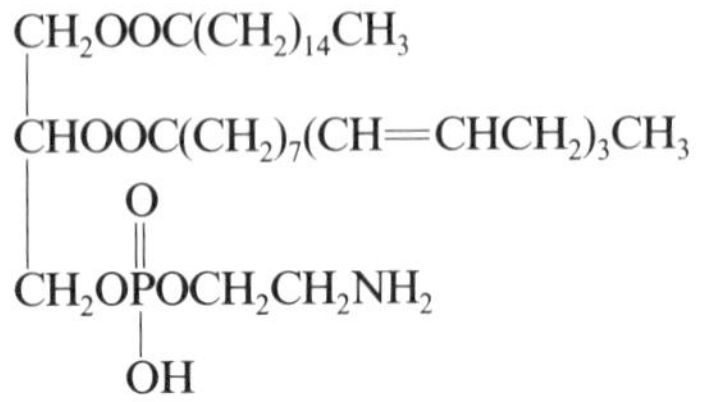

$C_{39}H_{72}NO_8P$ M 713.973

(R)-(all-Z)-form
Palmitoyllinoleoylphosphatidylethanolamine
$[\alpha]_D^{20}$ +6.0° (c, 18 in $CHCl_3$).

de Haas, G.H. *et al*, *Biochim. Biophys. Acta*, 1962, **65**, 260.

Glycerol 1-hexadecanoate 2-(9-octadecenoate) G-00159

9-Octadecenoic acid 1-(hydroxymethyl)-2-[(1-oxohexadecyl)oxy]ethyl ester, 9CI

$C_{37}H_{70}O_5$ M 594.957

(S)-(Z)-form [29541-66-0]
*1-Palmitoyl-2-oleoyl-*sn*-glycerol*
Oil. $[\alpha]_D^{25}$ +2.5° (neat).

(±)-(Z)-form [3123-73-7]
α-Palmito-β-olein
Mp 31-32°.

Zapesochnaya, G.G. *et al*, *CA*, 1963, **58**, 11207d (*synth*)
Buchnea, D., *Lipids*, 1974, **9**, 55 (*synth*)

Glycerol 1-hexadecanoate 2-(9-octadecenoate) 3-phosphate G-00160

$C_{37}H_{71}O_8P$ M 674.937

(R)-(Z)-form
*1-Palmitoyl-2-oleoyl-*sn*-glycerol 3-phosphate*
Di-NH$_4$ salt: Mp 192-194°. $[\alpha]_D$ +4.5° (c, 3 in $CHCl_3$).

Molotkovsky, J.G. *et al*, *Chem. Phys. Lipids*, 1976, **17**, 108 (*synth*)

Glycerol 1-hexadecanoate 2-(6-octadecenoate) 3-phosphocholine G-00161

Palmitoylpetroselinoylphosphatidylcholine

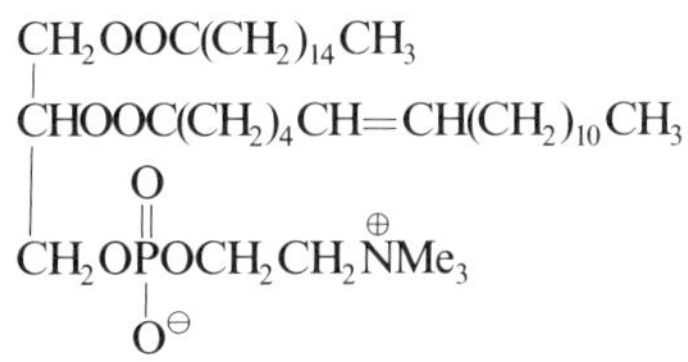

$C_{42}H_{82}NO_8P$ M 760.085
Cryst. (Me_2CO/$CHCl_3$).

Guyer, W. *et al*, *Chem. Phys. Lipids*, 1983, **33**, 313 (*synth*)

Glycerol 1-hexadecanoate 2-(9-octadecenoate) 3-phosphocholine G-00162

1-Hexadecanoyl-2-octadec-9-enoylglycero-3-phosphocholine

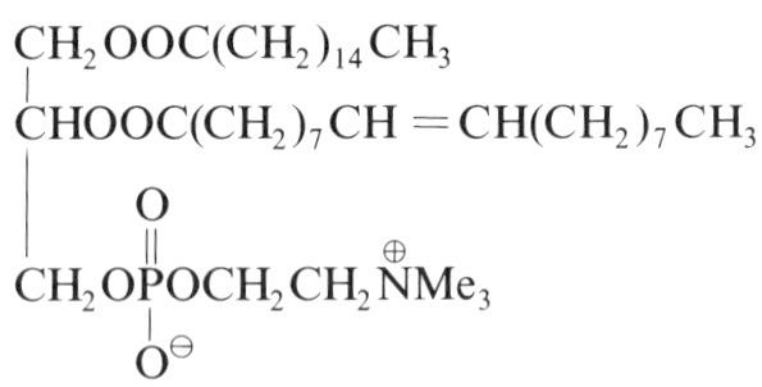

$C_{42}H_{82}NO_8P$ M 760.085

(R)-L-(Z)-form [26853-31-6]
Palmitoyloleoylphosphatidylcholine

Slotboom, A.J. *et al*, *Chem. Phys. Lipids*, 1970, **4**, 15.
Ghosh, D. *et al*, *Biochim. Biophys. Acta*, 1972, **266**, 41.
Wood, G.W. *et al*, *Chem. Phys. Lipids*, 1977, **18**, 316 (*ms*)
Seelig, J. *et al*, *Biochemistry*, 1978, **17**, 3310 (*nmr*)
Porter, N.A., *Lipids*, 1979, **14**, 20 (*hplc*)
Guyer, W. *et al*, *Chem. Phys. Lipids*, 1983, **33**, 313 (*synth*)

Glycerol 1-hexadecanoate 2-(11-octadecenoate) 3-phosphocholine G-00163

Palmitoylvaccenoylphosphatidylcholine

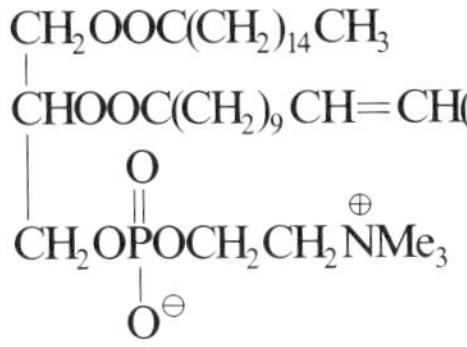

$C_{42}H_{82}NO_8P$ M 760.085

Cryst. ($Me_2CO/CHCl_3$).

Guyer, W. *et al, Chem. Phys. Lipids*, 1983, **33**, 313 (*synth*)

Glycerol 1-hexadecanoate 2-(9-octadecenoate) 3-phosphoethanolamine G-00164

9-Octadecenoic acid 1-[[[(2-aminoethoxy)hydroxyphosphinyl] oxy]methyl]-2-[(1-oxohexadecyloxy)ethyl] ester, 9CI. 1-Hexadecanoyl-2-octadec-9-enoylglycero-3-phosphoethanolamine

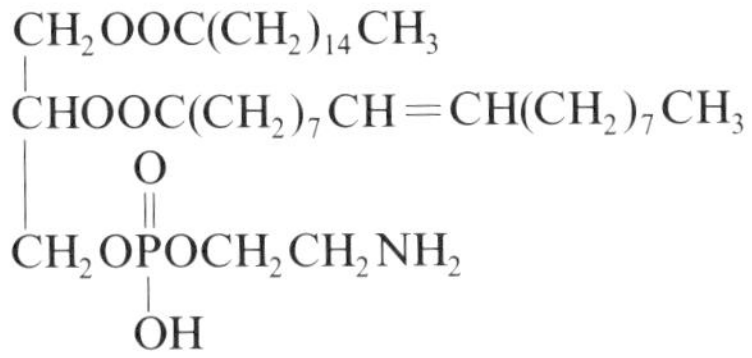

$C_{39}H_{76}NO_8P$ M 718.005

(*R*)-(*Z*)-*form* [39749-44-5]

Palmitoyloleoylphosphatidylethanolamine

Mp 184-185° (194-196°). $[\alpha]_D^{22}$ +6.35° (c, 3 in $CHCl_3$).

Daemen, F.J.M. *et al, Chem. Phys. Lipids*, 1967, **1**, 476 (*synth*)
Billimoria, J.D. *et al, Chem. Phys. Lipids*, 1974, **12**, 327 (*synth*)
Molotkovksy, I.G. *et al, Chem. Phys. Lipids*, 1976, **17**, 108 (*synth*)
Eibl, H. *et al, Chem. Phys. Lipids*, 1978, **22**, 1 (*synth*)

Glycerol-1-hexadecanoate 2-(9-octadecenoate) 3-phosphoinositol G-00165

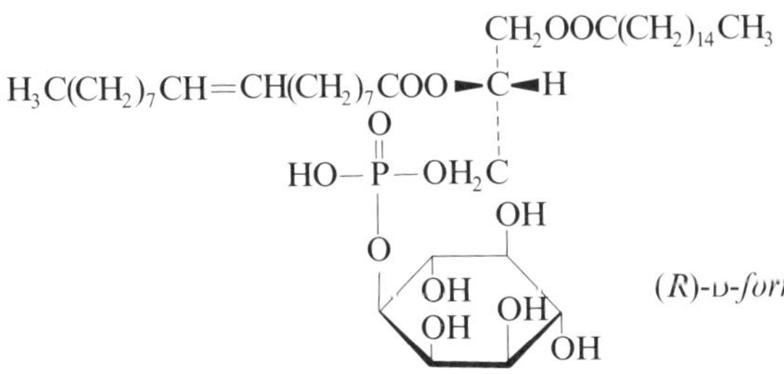

(*R*)-D-*form*

$C_{43}H_{81}O_{13}P$ M 837.079

The (*R*)-symbol designates the chirality of the glyceride moiety (≡ 1-phosphate acc. to the *sn*-system) and the D or L symbol refers to the abs. config. of the inositol ring system.

(*R*)-D-*form*

*D-1-O-(1′-Palmitoyl-2′-oleoyl-*sn*-glycero-3′-phosphate) myoinositol*

Cryst. ($CHCl_3$/MeOH). Mp 172-174°. $[\alpha]_D$ +6.1° (c, 12 in $CHCl_3$). Natural config.

(*R*)-L-*form*

*L-1-O-(1′-Palmitoyl-2′-oleoyl-*sn*-glycero-3′-phosphate) myoinositol*

Cryst. ($CHCl_3$/MeOH). Mp 169-172°. $[\alpha]_D$ +9.2° (c, 0.4 in $CHCl_3$).

(*R*)-DL-*form*

*DL-1-O-(1′-Palmitoyl-2′-oleoyl-*sn*-glycero-3′-phosphate) myoinositol*

4-Phosphate triammonium salt: Cryst. ($CHCl_3$/MeOH). Mp 187°. $[\alpha]_D$ +1.8° (c, 0.3 in $CHCl_3$).

Molotkovsky, J.G. *et al, Chem. Phys. Lipids*, 1973, **11**, 135.
Shevchenko, V.P. *et al, Chem. Phys. Lipids*, 1975, **15**, 95.
Sadovnika, M.S. *et al, CA*, 1976, **85**, 17835.

Glycerol 1-hexadecanoate 3-tetradecanoate G-00166

α-Palmito-α′-myristin

[17637-37-5]

$C_{33}H_{64}O_5$ M 540.866

(±)-*form*

Cryst. (pet. ether). Mp 63.5-64°.

Verkade, P.E. *et al, Recl. Trav. Chim. Pays-Bas* (*J. R. Neth. Chem. Soc.*), 1937, **56**, 365 (*synth*)

Glycerol 1-hexadecanoate 2-tetradecanoate 3-dodecanoate G-00167

Hexadecanoic acid 3-[(1-oxododecyl)oxy]-2-[(1-oxotetradecyl)oxy]propyl ester, 9CI. α-Palmito-β-myristolaurin

[60138-25-2]

$C_{45}H_{86}O_6$ M 723.171

(±)-*form*

Cryst. (Me_2CO). Mp 49°. Exhibits polymorphism.

Sidhu, S.S. *et al, J. Am. Chem. Soc.*, 1947, **69**, 1451.

Glycerol 1-hexadecanoate 2-tetradecanoate 3-phosphocholine G-00168

Palmitoylmyristoylphosphatidylcholine. 1-Hexadecanoyl-2-tetradecanoylglycero-3-phosphocholine

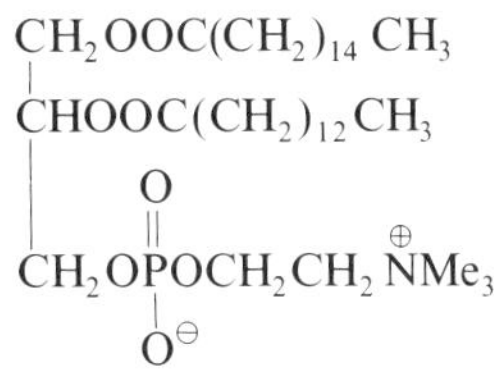

$C_{38}H_{76}NO_8P$ M 705.994

Wood, G.W. *et al, Chem. Phys. Lipids*, 1977, **18**, 316 (*ms*)
Keough, K.M.W. *et al, Biochemistry*, 1979, **18**, 1453.

Glycerol 1 (9,12-octadecadienoate) G-00169

2,3-Dihydroxypropyl-9,12-octadecadienoic acid, 9CI. 1-Monolinolein, 8CI. α-Monolinolein

$C_{21}H_{38}O_4$ M 354.529

(*R*)-(*Z,Z*)-*form* [2258-92-6]

Sporogenic compd. of *Sclerotinia fructicola*. $[\alpha]_D$ −5.2° (c, 1.56 in MeOH).

(*S*)-(*Z,Z*)-*form* [67968-46-1]

$[\alpha]_D$ +5.4° (c, 1.16 in MeOH).

(±)-(*Z,Z*)-*form* [2277-28-3]

Stimulates germination of pollen in *Chrysanthemum leucanthemum* (*in vitro*). Cryst. (C_6H_6 at −26°). Mp 14-15°.

Daubert, B.F. *et al, J. Am. Chem. Soc.*, 1944, **66**, 998 (*synth*)
Koichi, K. *et al, Bull. Inst. Chem. Res., Kyoto Univ.*, 1972, **50**, 142.
Katayama, M. *et al, Agric. Biol. Chem.*, 1978, **42**, 1431 (*synth, resoln, ms*)

Glycerol 2-(9,12-octadecadienoate) G-00170

$C_{21}H_{38}O_4$ M 354.529

(Z,Z)-form [3443-82-1]

β-Monolinolein

Oil, cryst. (pet. ether at −35°). Mp 8.9°.

Martin, J.B., *J. Am. Chem. Soc.*, 1953, **75**, 5482 (*synth*)

Glycerol 1-(9,12-octadecadienoate) 3-hexadecanoate G-00171

9,12-Octadecadienoic acid 2-hydroxy-3-[(1-oxohexadecyl)oxy]propyl ester, 9CI

$C_{37}H_{68}O_5$ M 592.941

(±)-***(all-Z)-form***

α-Linoleo-α′-palmitin

Cryst. Mp 27-28°.

Serebrennikova, G.A. *et al*, *CA*, 1961, **55**, 22145g (*synth*)

Glycerol 1-(9,12-octadecadienoate) 2-hexadecanoate 3-octadecanoate G-00172

$C_{55}H_{102}O_6$ M 859.407

(±)-***(all-Z)-form*** [2680-57-1]

α-Linoleo-β-palmitostearin

Cryst. Mp 34-34.5°.

Serebrennikova, G.A. *et al*, *CA*, 1963, **59**, 435c.

Glycerol 1-(9,12-octadecadienoate) 2-hexadecanoate 3-(9-octadecenoate) G-00173

9,12-Octadecadienoic acid 2-[(1-oxohexadecyl)oxy]-3-[(1-oxo-9-octadecenyl)oxy]propyl ester, 9CI

$C_{55}H_{100}O_6$ M 857.391

(±)-***(all-Z)-form*** [2534-97-6]

α-Linoleo-β-palmitoolein

Oil. Mp −10° to −9.

Serebrennikova, G.A. *et al*, *CA*, 1963, **59**, 435b (*synth*)

Glycerol 1-(9,12-octadecadienoate) 2-hexadecanoate 3-phosphocholine G-00174

1-Octadeca-9,12-dienoyl-2-hexadecanoylglycero-3-phosphocholine

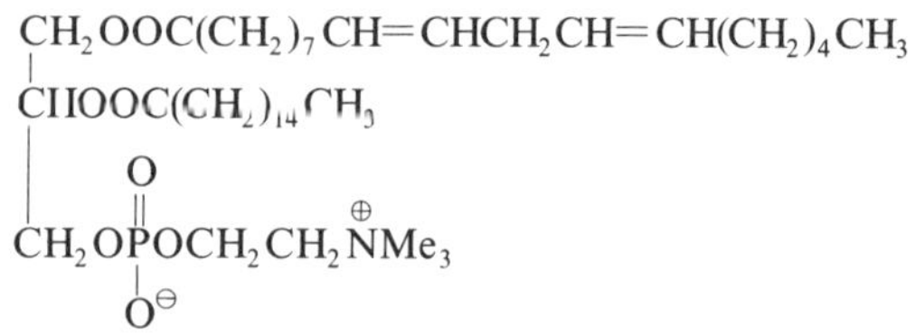

$C_{42}H_{80}NO_8P$ M 758.070

(R)-L-(Z,Z)-form

Linoleoylpalmitoylphosphatidylcholine

Demel, R.A. *et al*, *Biochim. Biophys. Acta*, 1972, **266**, 26.

Glycerol 1-(9,12-octadecadienoate) 2-hexadecanoate 3-phosphoethanolamine G-00175

1-Octadeca-9,12-dienoyl-2-hexadecanoylglycero-3-phosphoethanolamine

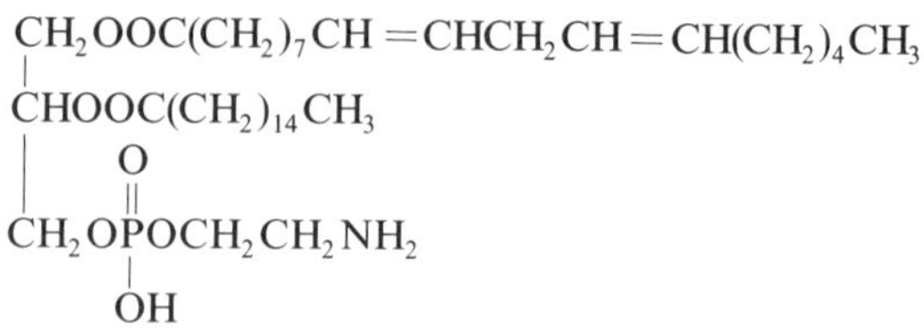

$C_{39}H_{74}NO_8P$ M 715.989

(R)-(all-Z)-form [26662-95-3]

Linoleoylpalmitoylphosphatidylethanolamine

Finean, J.B., *Biochim. Biophys. Acta*, 1953, **10**, 371.

Glycerol 1-(9,12-octadecadienoate) 2-octadecanoate G-00176

$C_{39}H_{72}O_5$ M 620.995

(S)-(Z,Z)-form

*1-Linoleoyl-2-stearoyl-*sn*-glycerol*

Oil. Mp 18.5°.

(±)-***(Z,Z)-form***

α-Linoleo-β-stearin

Oil. Mp 9-10°.

Bogoslovskii, N.A. *et al*, *CA*, 1962, **57**, 14154i (*synth*)
Dorofeeva, L.T. *et al*, *CA*, 1965, **63**, 6853a (*synth*)

Glycerol 1-(9,12-octadecadienoate) 2-octadecanoate 3-hexadecanoate G-00177

$C_{55}H_{102}O_6$ M 859.407

(±)-***(all-Z)-form*** [2534-96-5]

α-Linoleo-β-stearopalmitin

Cryst. Mp 36-37°.

Serebrennikova, G.A. *et al*, *CA*, 1963, **59**, 435c (*synth*)

Glycerol 1-(9,12-octadecadienoate) 2-octadecanoate 3-(9-octadecenoate) G-00178

9,12-Octadecadienoic acid 3-[(1-oxo-9-octadecenyl)oxy]-2-[(1-oxooctadecyl)oxy]propyl ester, 9CI

$C_{57}H_{104}O_6$ M 885.445

(all-Z)-form [2410-30-2]

α-Linoleo-β-stearoolein

Mp −20° to −19°, −12°.

Serebrennikova, G.A. *et al*, *CA*, 1963, **58**, 8898b (*synth*)

Glycerol 1-(9,12-octadecadienoate) 2-octadecanoate 3-phosphocholine G-00179

1-Octadeca-9,12-dienoyl-2-octadecanoylglycero-3-phosphocholine

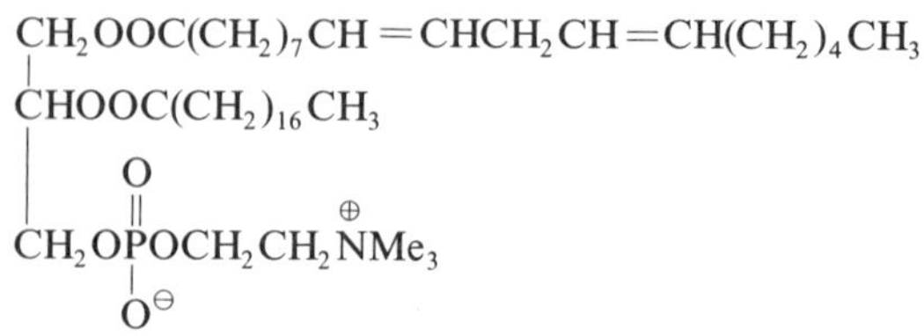

$C_{44}H_{84}NO_8P$ M 786.123

(±)-***(Z,Z)-form***

Hygroscopic powder (EtOH/Me_2CO).

Bogolovskii, N.A. *et al*, *CA*, 1963, **58**, 8895 (*synth, ir*)

Glycerol 1-(9,12-octadecadienoate) 2-octadecanoate 3-phosphoethanolamine G-00180

1-Octadec-9,12-dienoyl-2-octadecanoylglycero-3-phosphocholine

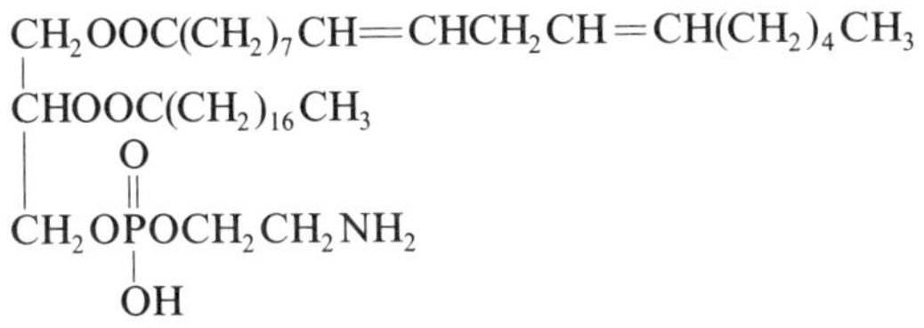

$C_{41}H_{78}NO_8P$ M 744.043

(±)-***(all-Z)-form*** [18468-68-3]

Linoleoylstearoylphosphatidylethanolamine

Mp 181-183° (194-195°).

Shvets, V.I. *et al*, *CA*, 1962, **56**, 4611 (*synth*)
Bogoslovskii, N.A. *et al*, *CA*, 1962, **57**, 14151 (*synth*)
Dorofeeva, L.T. *et al*, *CA*, 1965, **63**, 6852 (*synth*)
Chapman, D. *et al*, *Proc. R. Soc. London, A*, 1966, **290**, 1151 (*ir, nmr, props*)

Glycerol 1-(9,12-octadecadienoate) 3-(9-octadecenoate) G-00181

$C_{39}H_{70}O_5$ M 618.979

(±)-***form***

α-Linoleo-α′-olein

Oil.

Serebrennikova, G.A. *et al*, *CA*, 1961, **55**, 22145h (*synth*)

Glycerol 1-(9,12-octadecadienoate) 2-(9-octadecenoate) G-00182

9,12-Octadecadienoic acid 3-hydroxy-2-[(1-oxo-9-octadecenyl)oxy]propyl ester, 9CI

$C_{39}H_{70}O_5$ M 618.979

(***R***)-***(all-Z)-form***

*2-Oleoyl-3-linoleoyl-*sn*-glycerol*

Oil. $[\alpha]_D^{20}$ −4.47°.

Shvets, V.I. *et al*, *CA*, 1965, **62**, 10329g (*synth*)

Glycerol 1-(9,12-octadecadienoate) 2-(9-octadecenoate) 3-hexadecanoate G-00183

9,12-Octadecadienoic acid 3-[(1-oxohexadecyl)oxy]-2-[(1-oxo-9-octadecenyl)oxy]propyl ester, 9CI

$C_{55}H_{100}O_6$ M 857.391

(±)-***(all-Z)-form*** [2680-59-3]

α-Linoleo-β-oleopalmitin

Mp 13-13.5°. Exhibits polymorphism.

Zapesochnaya, G.G. *et al*, *CA*, 1963, **58**, 11207b (*synth*)

Glycerol 1-(9,12-octadecadienoate) 2-(9-octadecenoate) 3-octadecanoate G-00184

9,12-Octadecadienoic acid 2-[(1-oxo-9-octadecenyl)oxy]-3-[(1-oxooctadecyl)oxy]propyl ester, 9CI

$C_{57}H_{104}O_6$ M 885.445

(all-Z)-form [2190-14-9]

α-Linoleo-β-oleostearin

Mp −17° to −15°, −4° to −2°.

Serebrennikova, G.A. *et al*, *CA*, 1963, **58**, 8898b (*synth*)

Glycerol 1-(9,12-octadecadienoate) 2-(9-octadecenoate) 3-phosphoinositol G-00185

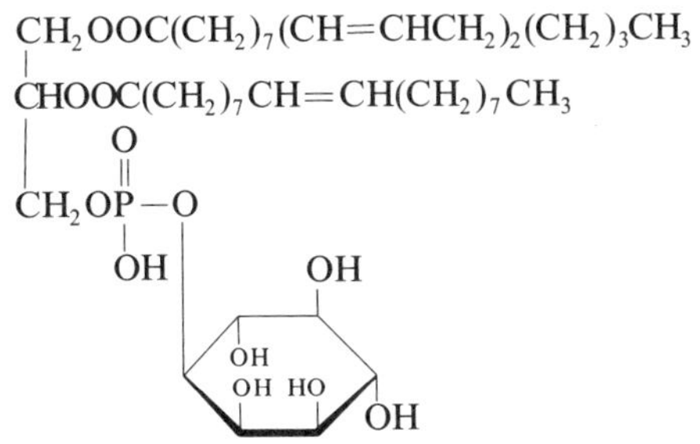

$C_{45}H_{81}O_{13}P$ M 861.101

Shvets, V.I. *et al*, *CA*, 1966, **64**, 14261 (*synth*)

Glycerol 1-octadecanoate G-00186

Octadecanoic acid 2,3-dihydroxypropyl ester, 9CI. α-Monostearin. Glycerol 1-stearate. 1-O-Stearoylglycerol

[123-94-4]

$$H_3C(CH_2)_{16}COOCH_2CH(OH)CH_2OH$$

$C_{21}H_{42}O_4$ M 358.560

Isol. from cultures of *Penicillium* sp.

(***R***)-***form*** [14811-92-8]

*3-Stearoyl-*sn*-glycerol*

Cryst. ($CHCl_3$ at −20°). Mp 74-75°. $[\alpha]_{300}^{27}$ −34.1° (EtOH).

Isopropylidene: Cryst. (EtOH). Mp 42-43°.

(±)-***form*** [22610-63-5]

Cryst. (hexane/EtOH). Mp 77-79°. Exhibits polymorphism.

Isopropylidene: Cryst. (pet. ether). Mp 39-40°.

Härri, E. *et al*, *Helv. Chim. Acta*, 1963, **46**, 1235 (*isol*)
Segredos, A.N., *Justus Liebigs Ann. Chem.*, 1973, 87 (*synth*)
Gronowitz, S. *et al*, *Chem. Phys. Lipids*, 1975, **14**, 174 (*synth, ord*)
Bus, J. *et al*, *Chem. Phys. Lipids*, 1976, **16**, 123 (*pmr*)
Sax, N.I., *Dangerous Properties of Industrial Materials*, 5th Ed., Van Nostrand-Reinhold, 1979, 706.

Glycerol 2-octadecanoate G-00187

Octadecanoic acid 2-hydroxy-1-(hydroxymethyl)ethyl ester, 9CI. β-Monostearin. Glycerol 2-stearate

[621-61-4]

$$H_3C(CH_2)_{16}COOCH(CH_2OH)_2$$

$C_{21}H_{42}O_4$ M 358.560

Cryst. (pet. ether). Mp 74°.

1,3-Benzylidene: Mp 69°.

Cook, P.F.E. *et al*, *J. Chem. Soc.*, 1965, 4594 (*synth*)

Glycerol 1-octadecanoate 2-butanoate 3-hexadecanoate G-00188

Octadecanoic acid 2-(1-oxobutoxy)-3-[(1-oxohexadecyl)oxy]propyl ester, 9CI. α-Stearo-β-butyropalmitin

$C_{41}H_{78}O_6$ M 667.064

(***S***)-***form*** [60456-32-8]

*2-Butyryl-3-palmitoyl-1-stearoyl-*sn*-glycerol*

Cryst. (Me_2CO). Mp 49°.

(±)-***form*** [62833-22-1]

Cryst. (C_6H_6/EtOH). Mp 45.5-46°.

Verkade, P.E., *Recl. Trav. Chim. Pays-Bas (J. R. Neth. Chem. Soc.)*, 1943, **62**, 393 (*synth*)
Lok, C.M. *et al*, *Chem. Phys. Lipids*, 1976, **16**, 115 (*synth*)

Glycerol 1-octadecanoate 2-decanoate 3-dodecanoate G-00189

α-Stearo-β-caprolaurin

$C_{43}H_{82}O_6$ M 695.117

Cryst. (Me_2CO). Mp 41.8°. Exhibits polymorphism.

Chen, C. *et al*, *J. Am. Chem. Soc.*, 1945, **67**, 1256.
Filer, L.J. *et al*, *J. Am. Chem. Soc.*, 1945, **67**, 2085.

Glycerol 1-octadecanoate 2-decanoate 3-tetradecanoate G-00190

α-Stearo-β-capromyristin

$C_{45}H_{86}O_6$ M 723.171

(±)-form

Cryst. (Me_2CO). Mp 52.5°. Exhibits polymorphism.

Chen, C. *et al*, *J. Am. Chem. Soc.*, 1945, **67**, 1256.
Filer, L.J. *et al*, *J. Am. Chem. Soc.*, 1945, **67**, 2085.

Glycerol 1-octadecanoate 3-dodecanoate G-00191

α-Stearo-α′-laurin

$C_{33}H_{64}O_5$ M 540.866

(±)-form

Cryst. Mp 62.0°.

Daubert, B.F. *et al*, *J. Am. Chem. Soc.*, 1944, **66**, 53 (*synth*)

Glycerol 1-octadecanoate 2-dodecanoate 3-hexadecanoate G-00192

α-Stearo-β-lauropalmitin

$C_{49}H_{94}O_6$ M 779.278

(±)-form

Cryst. (Me_2CO). Mp 57.5°. Exhibits polymorphism.

Chen, C. *et al*, *J. Am. Chem. Soc.*, 1945, **67**, 1256 (*synth*)

Glycerol 1-octadecanoate 2-dodecanoate 3-phosphocholine G-00193

[15071-88-2]

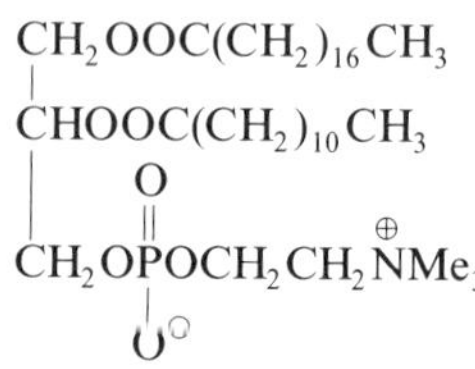

$C_{38}H_{76}NO_8P$ M 705.994

***(R)*-L-form**

Stearoyllauroylphosphatidylcholine

Powder (Et_2O). Mp 211-212°.

de Haas, G.H. *et al*, *Tetrahedron Lett.*, 1960, 1 (*synth*)

Glycerol 1-octadecanoate 2-(5,8,11,14-eicosatetraenoate) 3-phosphoethanolamine G-00194

5,8,11,14-Eicosatetraenoic acid 1-[[[(2-aminoethoxy)hydroxyphosphinyl]oxy]methyl]-2-[(1-oxooctadecyl)oxy]ethyl ester. 1-Octadecanoyl-2-eicosa-5,8,11,14-tetraenoylglycero-3-phosphoethanolamine

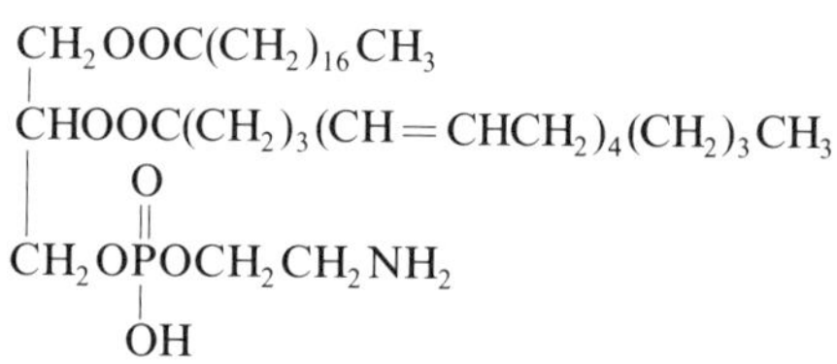

$C_{43}H_{78}NO_8P$ M 768.065

***(R)*-(*all-Z*)-form**

Stearoylarachidonylphosphatidylethanolamine

Renin inhibitor.

Stoffel, W. *et al*, *Hoppe Seyler's Z. Physiol. Chem.*, 1969, **350**, 1385.
Miyazaki, M. *et al*, *Proc. Soc. Exp. Biol. Med.*, 1977, **155**, 468 (*synth*)

Glycerol 1-octadecanoate 2-hexadecanoate G-00195

α-Stearo-β-palmitin

$C_{37}H_{72}O_5$ M 596.973

(±)-form

Cryst. (hexane). Mp 69-71°.

Gigg, J. *et al*, *J. Chem. Soc. C*, 1967, 431 (*synth*)

Glycerol 1-octadecanoate 3-hexadecanoate G-00196

Octadecanoic acid 2-hydroxy-3-[(1-oxohexadecyl)oxy]propyl ester, 9CI. α-Stearo-α′-palmitin

[17708-08-6]

$C_{37}H_{72}O_5$ M 596.973

***(S)*-form** [5281-87-8]

*1-Stearoyl-3-palmitoyl-*sn*-glycerol*

Cryst. (Me_2CO). Mp 74-75°.

(±)-form [5281-52-7]

Cryst. (Me_2CO). Mp 71°.

Mank, A.P.J. *et al*, *Chem. Phys. Lipids*, 1976, **16**, 107 (*synth*)
Lok, C.M. *et al*, *Chem. Phys. Lipids*, 1976, **16**, 115 (*synth*)
Bus, J. *et al*, *Chem. Phys. Lipids*, 1976, **16**, 123 (*pmr*)

Glycerol 1-octadecanoate 2-hexadecanoate 3-butanoate G-00197

Octadecanoic acid 3-(1-oxobutoxy)-2-[(1-oxohexadecyl)oxy]propyl ester, 9CI. α-Stearo-β-palmitobutyrin

$C_{41}H_{78}O_6$ M 667.064

***(S)*-form** [60456-33-9]

*3-Butyryl-2-palmitoyl-1-stearoyl-*sn*-glycerol*

Cryst. (Me_2CO). Mp 49°.

(±)-form [62833-21-0]

Cryst. (C_6H_6/EtOH). Mp 47.5-48°.

Verkade, P.F., *Recl. Trav. Chim. Pays-Bas* (*J. R. Neth. Chem. Soc.*), 1943, **62**, 393 (*synth*)
Lok, C.M. *et al*, *Chem. Phys. Lipids*, 1976, **16**, 115 (*synth*)

Glycerol 1-octadecanoate 2-hexadecanoate 3-dodecanoate G-00198

α-Stearo-β-palmitolaurin

$C_{49}H_{94}O_6$ M 779.278

(±)-form

Cryst. (Me_2CO). Mp 52°. Exhibits polymorphism.

Chen, C. *et al*, *J. Am. Chem. Soc.*, 1945, **67**, 1256 (*synth*)

Glycerol 1-octadecanoate 2-hexadecanoate 3-phosphocholine G-00199

Stearoylpalmitoylphosphatidylcholine. 1-Octadecyl-2-hexadecanoylglycerol-3-phosphocholine

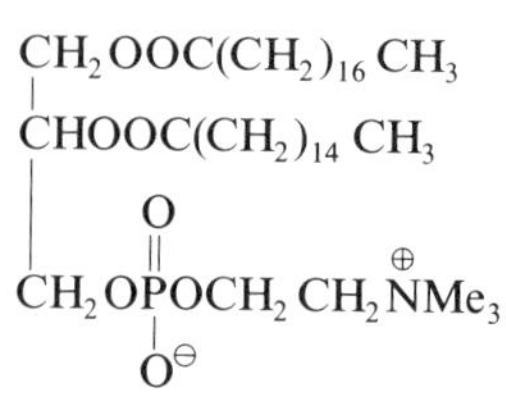

$C_{42}H_{84}NO_8P$ M 762.101

Nakajima, T. *et al*, *CA*, 1976, **85**, 176859 (*synth*)
Keough, K.M.W. *et al*, *Biochemistry*, 1979, **18**, 1453.
Porter, N.A. *et al*, *Lipids*, 1979, **14**, 20 (*hplc*)

Glycerol 1-octadecanoate 2-hexadecanoate 3-tetradecanoate G-00200

Octadecanoic acid 2-[(1-oxohexadecyl)oxy]-3-[(1-oxotetradecyl)oxy]propyl ester, 9CI. α-Stearo-β-palmitomyristin

[60138-20-7]

$C_{51}H_{98}O_6$ M 807.332
Cryst. (Me_2CO). Mp 56°. Exibits polymorphism.

Chen, C. *et al*, *J. Am. Chem. Soc.*, 1945, **67**, 1256 (*synth*)
Filer, L.J. *et al*, *J. Am. Chem. Soc.*, 1945, **67**, 2085.
Barber, M. *et al*, *Tetrahedron Lett.*, 1964, 1063 (*ms*)

Glycerol 1-octadecanoate 2-(9,12-octadecadienoate) G-00201

9,12-Octadecadienoic acid 1-(hydroxymethyl)-2-[(1-oxooctadecyl)oxy]ethyl ester, 9CI

$C_{39}H_{72}O_5$ M 620.995

(S)-(Z,Z)-form [34487-26-8]
*1-Stearoyl-2-linoleoyl-*sn*-glycerol*
Oil. Mp 17.5°.

(±)-(Z,Z)-form
α-Stearo-β-linolein
Oil. Mp 13.5°.

Dorofeeva, L.T. *et al*, *CA*, 1964, **60**, 1577b; 1965, **63**, 6853b (*synth*)

Glycerol 1-octadecanoate 2-(9,12-octadecadienoate) 3-hexadecanoate G-00202

9,12-Octadecandienoic acid 1-[[(1-oxohexadecyl)oxy]methyl]-2-[[(1-oxooctadecyl)oxy]ethyl] ester, 9CI

$C_{55}H_{102}O_6$ M 859.407

(±)-(Z,Z)-form [2190-12-7]
α-Stearo-β-linoleopalmitin

Tamaki, Y. *et al*, *J. Insect Physiol.*, 1971, **17**, 1239 (*synth*)

Glycerol 1-octadecanoate 2-(9,12-octadecadienoate) 3-phosphocholine G-00203

1-Octadecanoyl-2-octadeca-9,12-dienoylglycero-3-phosphocholine

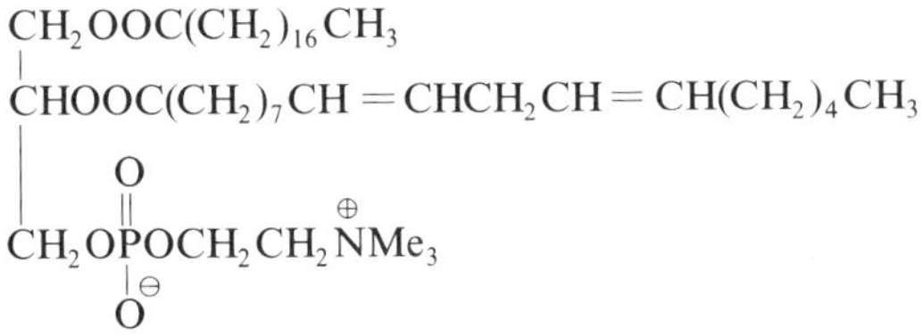

$C_{44}H_{84}NO_8P$ M 786.123

(R)-L-(all-Z)-form [27098-24-4]
Stearoyllinoleoylphosphatidylcholine
Gelatinous. $[\alpha]_D^{20}$ +5.02°.

Shvets, V.I., *CA*, 1965, **62**, 9001 (*synth*)
Stoffel, W. *et al*, *Hoppe Seyler's Z. Physiol. Chem.*, 1969, **350**, 1385 (*props*)
Ghosh, D. *et al*, *Biochim. Biophys. Acta*, 1972, **266**, 41.
Stoffel, W. *et al*, *Hoppe Seyler's Z. Physiol. Chem.*, 1972, **353**, 1962 (*cmr*)

Glycerol 1-octadecanoate 2-(9,12-octadecadienoate) 3-phosphoethanolamine G-00204

9,12-Octadecadienoic acid 1-[[[(2-aminoethoxy)hydroxyphosphinyl]oxy]methyl]-2-[(1-oxooctadecyl)oxy]ethyl ester, 9CI. 1-Octadecanoyl-2-octadeca-9,12-dienoylglycero-3-phosphoethanolamine

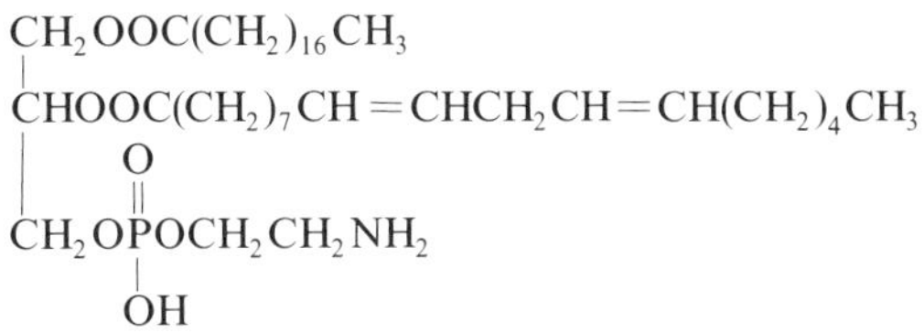

$C_{41}H_{78}NO_8P$ M 744.043

(R)-(all-Z)-form [7266-53-7]
Stearoyllinoleoylphosphatidylethanolamine

(±)-(all-Z)-form
Mp 193-194°.

Dorofeeva, L.T. *et al*, *CA*, 1965, **63**, 6851 (*synth*)
Chapman, D. *et al*, *Proc. R. Soc. London, A*, 1966, **290**, 1151 (*ir, nmr, props*)
Stoffel, W. *et al*, *Hoppe Seyler's Z. Physiol. Chem.*, 1969, **350**, 1385.

Glycerol 1-octadecanoate 2-(9,12,15-octadecatrienoate) 3-phosphocholine G-00205

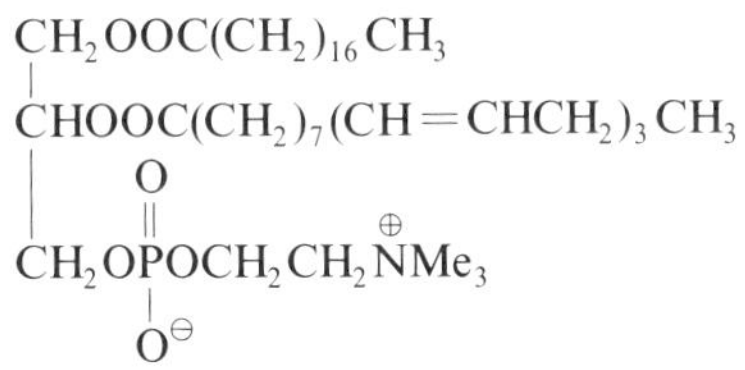

$C_{44}H_{82}NO_8P$ M 784.107

(R)-L-(all-Z)-form [35418-57-6]
Stearoyllinolenoylphosphatidylcholine

Stoffel, W. *et al*, *Hoppe Seyler's Z. Physiol. Chem.*, 1969, **350**, 1385.
Ghosh, D. *et al*, *Chem. Phys. Lipids*, 1971, **7**, 173.
Ghosh, D. *et al*, *Biochim. Biophys. Acta*, 1972, **266**, 41.

Glycerol 1-octadecanoate 2-(9,12,15-octadecatrienoate) 3-phosphoethanolamine G-00206

9,12,15-Octadecatrienoic acid 1-[[[(2-aminoethoxy)hydroxyphosphinyl]oxy]methyl]-2-[(1-oxooctadecyl)oxy]ethyl ester, 9CI

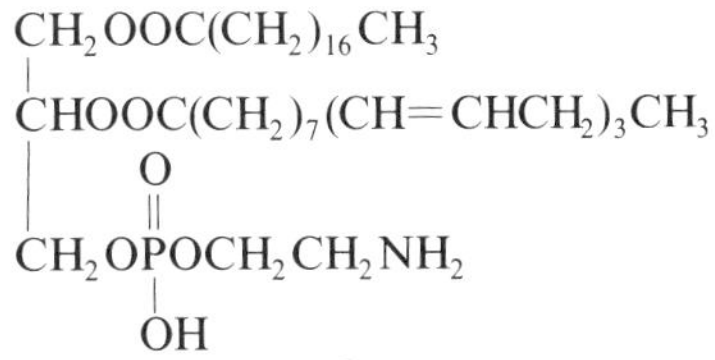

$C_{41}H_{76}NO_8P$ M 742.027

(±)-(all-Z)-form [55822-08-7]
Stearoyllinolenoylphosphatidylethanolamine
Renin inhibitor.

Shvets, V.I., *CA*, 1964, **60**, 179 (*synth*)
Stoffel, W. *et al*, *Hoppe Seyler's Z. Physiol. Chem.*, 1969, **350**, 1385.
Miyazaki, M., *Proc. Soc. Exp. Biol. Med.*, 1977, **155**, 468.

Glycerol 1-octadecanoate 2-(9-octadecenoate) G-00207

9-Octadecenoic acid 1-(hydroxymethyl)-2-[(1-oxooctadecyl)oxy]ethyl ester, 9CI

$C_{39}H_{74}O_5$ M 623.011

(S)-(Z)-form [53702-48-0]

1-Stearoyl-2-oleoyl-sn-glycerol

$[\alpha]_D$ −2.8° (c, 9.2 in $CHCl_3$).

(±)-*(Z)-form* [21059-30-3]

α-Stearo-β-olein

Mp 39.5-40°.

Volkova, L.V. *et al*, *CA*, 1963, **58**, 4415a (*synth*)
Baer, E., *Can. J. Biochem.*, 1974, **52**, 570 (*synth*)

Glycerol 1-octadecanoate 2-(9-octadecenoate) 3-eicosanoate G-00208

Eicosanoic acid 2-[(1-oxo-9-octadecenyl)oxy]-3-[(1-oxooctadecyl)oxy]propyl ester, 9CI

$C_{59}H_{112}O_6$ M 917.530

(±)-*(Z)-form* [59891-33-7]

α-Stearo-β-oleoarachidin

Cryst. (Me_2CO). Mp 41-42°.

Mank, A.P.J. *et al*, *Chem. Phys. Lipids*, 1976, **16**, 107 (*synth*)

Glycerol 1-octadecanoate 2-(9-octadecenoate) 3-hexadecanoate G-00209

9-Octadecenoic acid 1-[[(oxohexadecyl)oxy]methyl]-2-[(1-oxooctadecyl)oxy]ethyl ester, 9CI

$C_{55}H_{104}O_6$ M 861.423

(±)-*(Z)-form* [2190-27-4]

α-Stearo-β-oleopalmitin

Constit. of cocoa butter and confectionery fats. Cryst. (Me_2CO/hexane). Mp 32-33°, Mp 37-38°. Exhibits polymorphism.

Carreau, J.P., *Bull. Soc. Chim. Fr.*, 1970, 4111 (*synth*)
Mank, A.P.J. *et al*, *Chem. Phys. Lipids*, 1976, **16**, 107 (*synth*)

Glycerol 1-octadecanoate 2-(9-octadecenoate) 3-phosphocholine G-00210

1-Octadecanoyl-2-octadec-9-enoylglycero-3-phosphocholine

[6753-56-6]

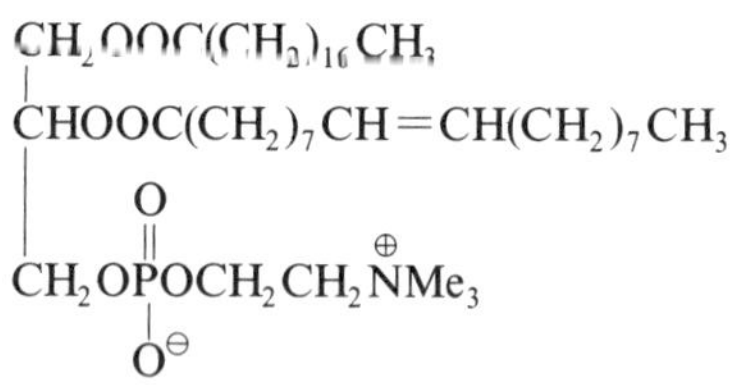

$C_{44}H_{86}NO_8P$ M 788.139

(R)-form [56421-10-4]

Stearoyloleoylphosphatidylcholine

Mp 230-231°. $[\alpha]_D^{20}$ +6.0° (c, 9 in $CHCl_3$).

de Haas, G.H. *et al*, *Recl. Trav. Chim. Pays-Bas* (*J. R. Neth. Chem. Soc.*), 1961, **80**, 951 (*synth, ir*)
Chapman, D. *et al*, *J. Biol. Chem.*, 1966, **241**, 5044 (*nmr*)
Stoffel, W. *et al*, *Hoppe Seyler's Z. Physiol. Chem.*, 1969, **350**, 1385.
Klein, R.A., *J. Lipid Res.*, 1971, **12**, 123, 628; 1972, **13**, 672 (*ms*)
Demel, R.A., *Biochim. Biophys. Acta*, 1972, **266**, 26.

Glycerol 1-octadecanoate 2-(9-octadecenoate) 3-phosphoethanolamine G-00211

9-Octadecenoic acid 1-[[[(2-aminoethoxy)hydroxyphosphinyl]oxy]methyl]-2-[(1-oxooctadecyl)oxy]ethyl ester, 9CI. 1-Octadecanoyl-2-octadec-9-enoylglycero-3-phosphoethanolamine

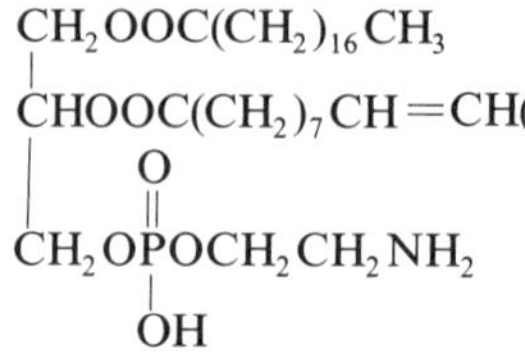

$C_{41}H_{80}NO_8P$ M 746.059

(R)-(Z)-form [6418-95-7]

Stearoyloleoylphosphatidylethanolamine

Mp 182-183°. $[\alpha]_D$ +6.0° (c, 10 in $CHCl_3$).

(±)-*(E)-form*

Stearoylelaidoylphosphatidylethanolamine

Mp 190-192°.

(±)-*(Z)-form*

Mp 189-190°.

Baer, E. *et al*, *Can. J. Biochem.*, 1961, **39**, 1471 (*synth*)
Chapman, D. *et al*, *Biochim. Biophys. Acta*, 1966, **120**, 148.
Chapman, D. *et al*, *Proc. R. Soc. London, A*, 1966, **290**, 1151 (*ir, nmr, props*)
Chapman, D. *et al*, *Trans. Faraday Soc.*, 1966, **62**, 2607.
Stoffel, W. *et al*, *Hoppe Seyler's Z. Physiol. Chem.*, 1969, **350**, 1385.

Glycerol 1-octadecanoate 2-(9-octadecenoate) 3-phosphoinositol G-00212

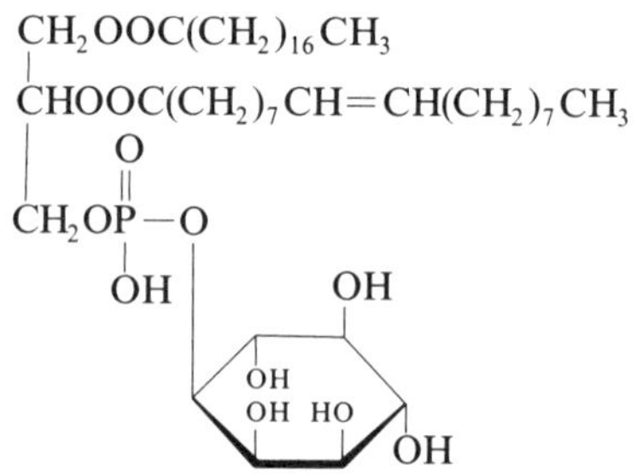

$C_{45}H_{85}O_{13}P$ M 865.132

(R)-(Z)-form [58116-17-9]

Lyutik, A.Y. *et al*, *CA*, 1975, **82**, 86520; 1976, **84**, 59900 (*synth*)

Glycerol 1-octadecanoate 3-tetradecanoate G-00213

α-Stearo-α′-myristin

$C_{35}H_{68}O_5$ M 568.919

(±)-*form*

Cryst. (Et_2O/EtOH). Mp 66.5-67°.

Sidhu, S.S. *et al*, *J. Am. Chem. Soc.*, 1946, **68**, 2603 (*synth*)

Glycerol 1-octadecanoate 2-tetradecanoate 3-hexadecanoate G-00214

Octadecanoic acid 3-[(1-oxohexadecyl)oxy]-2-[(1-oxotetradecyl)oxy]propyl ester, 9CI. α-Stearo-β-myristopalmitin

[56554-26-8]

$C_{51}H_{98}O_6$ M 807.332

Cryst. (Me_2CO). Mp 59.5°. Exibits polymorphism.

Chen, C. *et al*, *J. Am. Chem. Soc.*, 1945, **67**, 1256 (*synth*)
Filer, J.L. *et al*, *J. Am. Chem. Soc.*, 1945, **67**, 2085.

Glycerol 1-(9,12,15-octadecatrienoate) G-00215

(2,3-Dihydroxypropyl)-9,12,15-octadecatrienoic acid, 9CI

[18465-99-1]

$C_{21}H_{36}O_4$ M 352.513

(all-Z)-form

α-Monolinolenin. 1-Monolinolenin

Present in *Tulipa gesneriana* flowers, stimulates (*in vitro*) germination of pollen of *Chrysanthemum leucanthemum*. Cryst. in two forms (EtOH/C_6H_6 at −20°). Mp −15.7° (to −13.5°).

Daubert, B.F. *et al*, *J. Am. Chem. Soc.*, 1944, **66**, 997 (*synth*)
Koichi, K. *et al*, *Bull. Inst. Chem. Res., Kyoto Univ.*, 1972, **50**, 142; *CA*, **78**, 80736t (*isol*)

Glycerol 1-(9,12,15-octadecatrienoate) 2-hexadecanoate 3-phosphoethanolamine G-00216

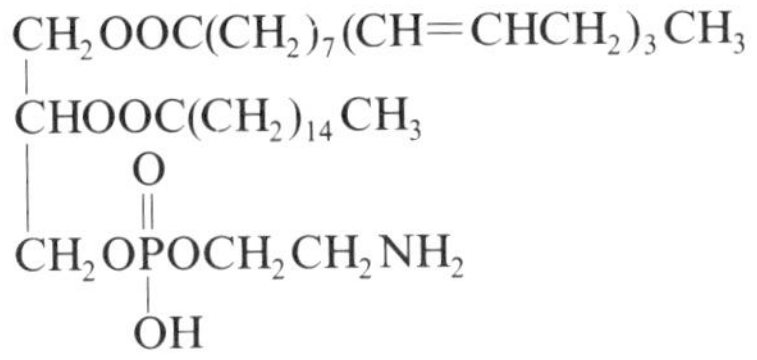

$C_{39}H_{72}NO_8P$ M 713.973

(R)-(all-Z)-form

Linolenoylpalmitoylphosphatidylethanolamine

$[\alpha]_D^{20}$ +6.2° (c, 8 in $CHCl_3$).

de Haas, G.H. *et al*, *Biochim. Biophys. Acta*, 1962, **65**, 260.

Glycerol 1-(9-octadecenoate) G-00217

9-Octadecenoic acid 2,3-dihydroxypropyl ester, 9CI

$H_3C(CH_2)_7CH{=}CH(CH_2)_7COOCH_2CH(OH)CH_2OH$

$C_{21}H_{40}O_4$ M 356.545

(±)-(*E*)-*form* [2716-53-2]

α-Monoelaidin

Cryst. Mp 58.5°. Exhibits polymorphism.

(R,Z)-form [34487-30-4]

3-Oleoyl-sn-glycerol

Paste. $[\alpha]_D$ −3.6° (c, 10 in Py).

(±)-(*Z*)-*form* [30836-40-9]

α-Monoolein

Cryst. Mp 35°. Exhibits polymorphism.

Carter, M.G.R. *et al*, *J. Chem. Soc.*, 1947, 554 (*synth, props*)
Anfinsen, J.R. *et al*, *J. Am. Oil Chem. Soc.*, 1964, **41**, 779 (*synth*)
Buchnea, D., *Lipids*, 1971, **6**, 734 (*synth*)

Glycerol 2-(9-octadecenoate) G-00218

2-Hydroxy-1-(hydroxymethyl)ethyl 9-octadecenoate, 9CI

$C_{21}H_{40}O_4$ M 356.545

(E)-form

2-Monoelaidin

Mp 53.7°.

(Z)-form [3443-84-3]

2-Monoolein, 8CI. β-Monoolein

Insect attractant present in mushrooms. Mp 26°.

Bournat, K., *Biochem. Z.*, 1914, **65**, 156 (*synth*)
Martin, J.B., *J. Am. Chem. Soc.*, 1953, **75**, 5482 (*synth*)
Muto, T. *et al*, *Agric. Biol. Chem.*, 1968, **32**, 624.
Buchnea, D., *Lipids*, 1974, **9**, 55 (*synth*)

Glycerol 1-(9-octadecenoate) 3-dodecanoate G-00219

$C_{33}H_{62}O_5$ M 538.850

(±)-(*Z*)-*form*

α-Oleo-α′-laurin

Cryst. (Et_2O/EtOH). Mp 32°.

Daubert, B.F. *et al*, *J. Am. Chem. Soc.*, 1944, **66**, 53 (*synth*)
Hartman, L., *J. Chem. Soc.*, 1957, 3572 (*synth*)

Glycerol 1-(9-octadecenoate) 2-hexadecanoate G-00220

9-Octadecenoic acid 3-hydroxy-2-[(1-oxohexadecyl)oxy] propyl ester, 9CI

$C_{37}H_{70}O_5$ M 594.957

(R)-(Z)-form [38991-91-2]

*3-Oleoyl-2-palmitoyl-*sn*-glycerol*

Oil.

O'Doherty, P.J.A. *et al*, *Can. J. Biochem.*, 1972, **50**, 881 (*synth*)

Glycerol 1-(9-octadecenoate) 3-hexadecanoate G-00221

9-Octadecenoic acid 2-hydroxy-3-[(1-oxohexadecyl)oxy] propyl ester

$C_{37}H_{70}O_5$ M 594.957

(S)-(Z)-form [36805-77-3]

*1-Oleoyl-3-palmitoyl-*sn*-glycerol*

Cryst. (Me_2CO). Mp 45-46° (48.5-49°).

(±)-(*Z*)-*form* [3343-30-4]

α-Oleo-α′-palmitin

Cryst. (EtOH/pet. ether). Mp 44.5-45°.

Hartman, L., *J. Chem. Soc.*, 1957, 3572 (*synth*)
Lok, C.M. *et al*, *Chem. Phys. Lipids*, 1976, **16**, 115 (*synth*)

Glycerol 1-(9-octadecenoate) 2-hexadecanoate 3-dodecanoate G-00222

$C_{49}H_{92}O_6$ M 777.262

(±)-(*Z*)-*form*

α-Oleo-β-palmitolaurin

Cryst. (Me_2CO/pet. ether). Mp 29.5°.

Hartman, L., *J. Chem. Soc.*, 1957, 3572 (*synth*)

Glycerol 1-(9-octadecenoate) 2-hexadecanoate 3-octadecanoate G-00223

9-Octadecenoic acid 2-[(1-oxohexadecyl)oxy]-3-[(1-oxooctadecyl)oxy]propyl ester, 9CI. α-Oleo-β-palmitostearin

$C_{55}H_{104}O_6$ M 861.423

Isol. from cocoa butter.

(±)-(*Z*)-*form* [59891-29-1]

Cryst. Mp 40-41°.

Mank, A.P.J. *et al*, *Chem. Phys. Lipids*, 1976, **16**, 107 (*synth*)

Glycerol 1-(9-octadecenoate) 2-hexadecanoate 3-phosphate G-00224

9-Octadecenoic acid 1-[[(1-oxohexadecyl)oxy]methyl]-2-(phosphonoxy)ethyl ester, 9CI

[10015-87-9]

$C_{37}H_{71}O_8P$ M 674.937

(±)-*form*

Bonsen, P.P.M. *et al*, *Chem. Phys. Lipids*, 1966, **1**, 100 (*synth*)

Glycerol 1-(9-octadecenoate) 2-hexadecanoate 3-phosphocholine G-00225

1-Octadec-9-enyl-2-hexadecanoylglycero-3-phosphocholine

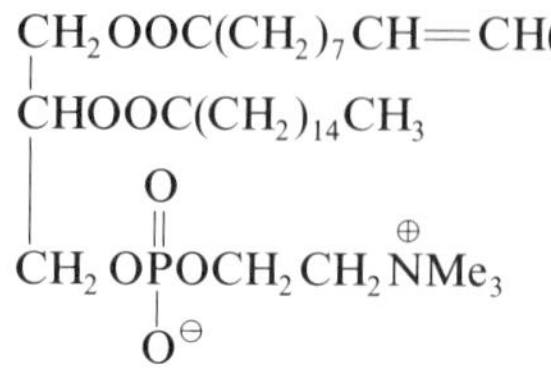

$C_{42}H_{82}NO_8P$ M 760.085

(R)-L-(Z)-form [59491-62-2]

Oleoylpalmitoylphosphatidylcholine

Wood, G.W. *et al*, *Chem. Phys. Lipids*, 1977, **18**, 316 (*ms*)

Glycerol 1-(9-octadecenoate) 2-hexadecanoate 3-phospho-1′-glycerol G-00226

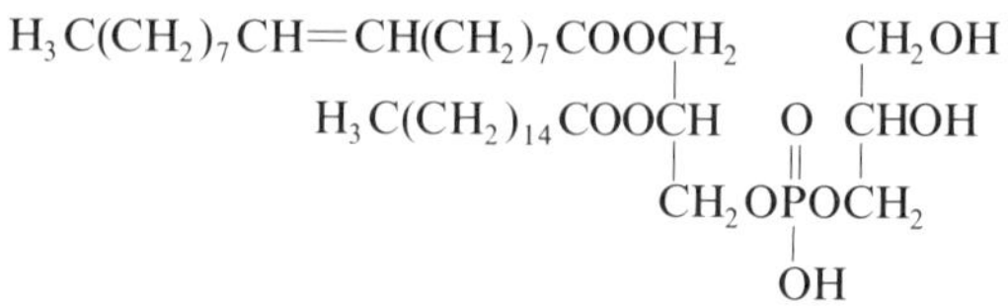

$C_{40}H_{77}O_{10}P$ M 749.016

(Z)-form

Olein 2-palmito-1-(2,3-dihydroxypropyl)hydrogen phosphate, 8CI

Na salt: [13879-80-6].

Mp 176-179°. $[\alpha]_D^{20}$ +1.02° (c, 10 in $CHCl_3$).

Bonsen, P.P.M. *et al*, *Chem. Phys. Lipids*, 1966, **1**, 33 (*synth*)

Glycerol 1-(9-octadecenoate) 2-hexadecanoate 3-phosphoserine G-00227

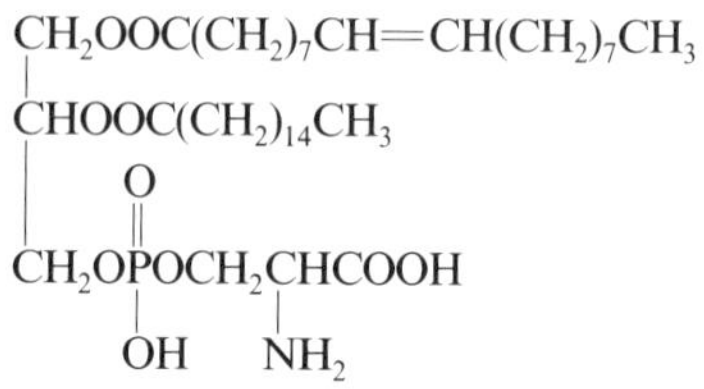

$C_{40}H_{76}NO_{10}P$ M 762.015

(R)-DL-form

*1-Oleoyl-2-palmitoyl-*sn*-glycero-3-phospho-DL-serine*

Cryst. (pet. ether/Me_2CO). $[\alpha]_D^{20}$ +5.8° (c, 12 in $CHCl_3$).

de Haas, G.M. *et al*, *Recl. Trav. Chim. Pays-Bas (J. R. Neth. Chem. Soc.)*, 1964, **83**, 99 (*synth, ir*)

Glycerol 1-(9-octadecenoate) 2-(9,12-octadecadienoate) G-00228

$C_{39}H_{70}O_5$ M 618.979

(R)-(all-Z)-form

*2-Linoleoyl-3-oleoyl-*sn*-glycerol*

Oil. $[\alpha]_D^{20}$ −2.6° (c, 2.0 in $CHCl_3$).

(±)-(all-Z)-form

α-Oleo-β-linolein

Oil.

Shvets, V.I. *et al*, *CA*, 1963, **59**, 9781g (*synth*)
Volkova, L.V. *et al*, *CA*, 1963, **59**, 9782a (*synth*)

Glycerol 1-(9-octadecenoate) 2-(9,12-octadecadienoate) 3-hexadecanoate G-00229

9,12-Octadecadienoic acid 1-[[(1-oxohexadecyl)oxy]methyl]-2-[(1-oxo-9-octadecenyl)oxy]ethyl ester, 9CI

$C_{55}H_{100}O_6$ M 857.391

(±)-(all-Z)-form [2579-00-2]

α-Oleo-β-linoleopalmitin

Mp 13-14°.

Zapesochnaya, G.G. *et al*, *CA*, 1963, **58**, 11207d (*synth*)
Tamaki, Y. *et al*, *J. Insect Physiol.*, 1971, **17**, 1239 (*synth*)

Glycerol 1-(9-octadecenoate) 2-(9,12-octadecadienoate) 3-octadecanoate G-00230

9,12-Octadecadienoic acid 1-[[(1-oxo-9-octadecenyl)oxy]methyl]-2-[(1-oxooctadecyl)oxy]ethyl ester, 9CI

$C_{57}H_{104}O_6$ M 885.445

(all-Z)-form [7162-26-7]

α-Oleo-β-linoleostearin

Mp 8.5-9° (14-14.5°).

Maurit, M.E. *et al*, *CA*, 1960, **54**, 20889e (*synth*)
Serebrennikova, G.A. *et al*, *CA*, 1963, **58**, 8898b (*synth*)

Glycerol 1-(9-octadecenoate) 2-(9,12-octadecadienoate) 3-phosphoethanolamine G-00231

1-Octadec-9-enoyl-2-octadeca-9,12-dienoylglycero-3-phosphoethanolamine

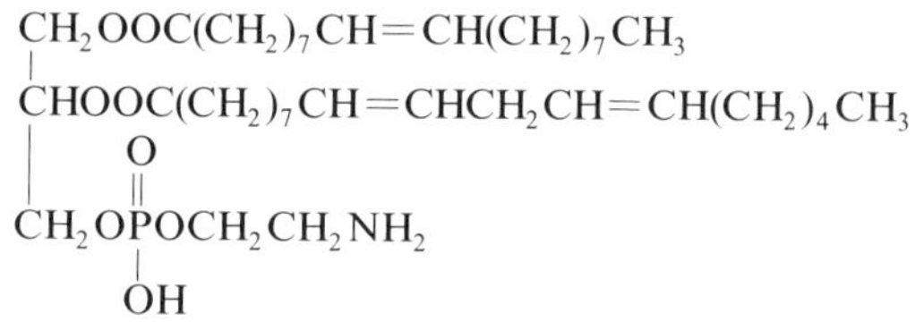

$C_{41}H_{76}NO_8P$ M 742.027

(R)-(all-Z)-form

Oleoyllinoleoylphosphatidylethanolamine

Mp 155°. $[\alpha]_D^{20}$ +6.0° (c, 5 in $CHCl_3$).

Daemen, F.J.M. *et al*, *Recl. Trav. Chim. Pays-Bas (J. R. Neth. Chem. Soc.)*, 1963, **82**, 487 (*synth*)
Volkova, L.V. *et al*, *CA*, 1965, **62**, 13039 (*synth*)
Shvets, V.I. *et al*, *CA*, 1966, **65**, 13534 (*synth*)

Glycerol 1-(9-octadecenoate) 2-(9,12-octadecadienoate) 3-phosphoserine G-00232

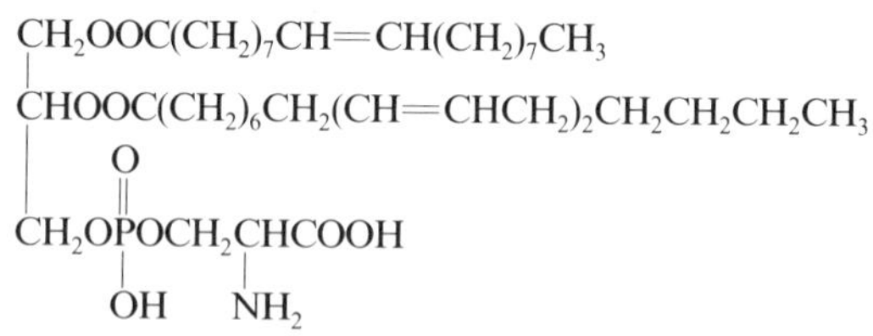

$C_{42}H_{76}NO_{10}P$ M 786.037

(±)-form

Oleoyllinoleoylphosphatidylserine

Shvets, V.I. *et al*, *CA*, 1968, **68**, 3169 (*synth*)

Glycerol 1-(9-octadecenoate) 2-octadecanoate G-00233

9-Octadecenoic acid 3-hydroxy-2-[(1-oxooctadecyl)oxy]propyl ester, 9CI

[38635-46-0]

$C_{39}H_{74}O_5$ M 623.011

(±)-(*Z*)-*form* [73788-95-1]

α-Oleo-β-stearin

Mp 19-20°.

Bogoslovskii, N.A. *et al*, *CA*, 1961, **55**, 24578c (*synth*)

Glycerol 1-(9-octadecenoate) 3-octadecanoate G-00234

9-Octadecenoic acid 2-hydroxy-3-[(1-oxooctadecyl)oxy] propyl ester, 9CI

$C_{39}H_{74}O_5$ M 623.011

(*E*)-*form*

α-Elaido-α'-stearin

Cryst. (Et_2O/EtOH). Mp 65.5-66°. Exhibits polymorphism.

(±)-(*Z*)-*form* [5568-23-0]

α-Oleo-α'-stearin

Cryst. (Me_2CO). Mp 48-49°. Exhibits polymorphism.

Benedict, J.H. *et al*, *J. Am. Oil Chem. Soc.*, 1950, **27**, 91 (*synth*)
Mank, A.P.J. *et al*, *Chem. Phys. Lipids*, 1976, **16**, 107 (*synth*)

Glycerol 1-(9-octadecenoate) 2-octadecanoate 3-hexadecanoate G-00235

9-Octadecenoic acid 3-[(1-oxohexadecyl)oxy]-2-[(1-oxooctadecyl)oxy]propyl ester, 9CI

$C_{55}H_{104}O_6$ M 861.423

(±)-(*Z*)-*form* [2190-28-5]

α-Oleo-β-stearopalmitin

Cryst. Mp 41-41.5°. Exhibits polymorphism.

Chapman, D. *et al*, *J. Chem. Soc.*, 1957, 1502 (*synth*)

Glycerol 1-(9-octadecenoate) 2-octadecanoate 3-phosphocholine G-00236

2-Octadecanoyl-1-octadec-9-enoylglycero-3-phosphocholine

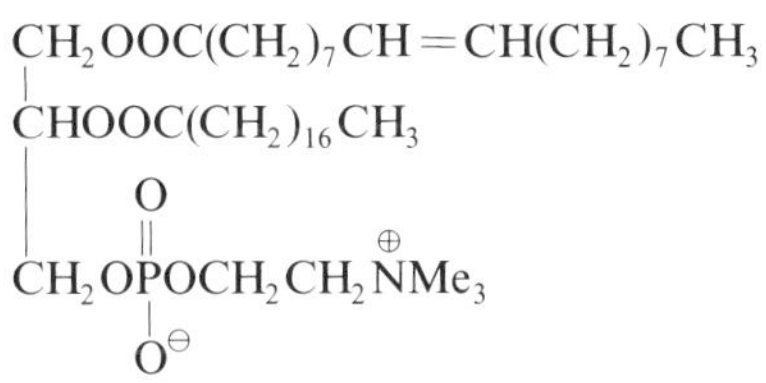

$C_{44}H_{86}NO_8P$ M 788.139

(*R*)-(*Z*)-*form* [7319-55-3]

Oleoylstearoylphosphatidylcholine

Mp 235°. $[\alpha]_D^{20}$ +6.2° (c, 10 in $CHCl_3$).

de Haas, G.H. *et al*, *Recl. Trav. Chim. Pays-Bas* (*J. R. Neth. Chem. Soc.*), 1961, **80**, 951 (*synth, ir*)
Infante, R. *et al*, *Biochim. Biophys. Acta*, 1968, **164**, 436 (*synth*)
Demel, R.A., *Biochim. Biophys. Acta*, 1972, **266**, 26 (*props*)
Klein, R.A., *J. Lipid Res.*, 1972, **13**, 672 (*ms*)

Glycerol 1-(9-octadecenoate) 2-octadecanoate 3-phosphoethanolamine G-00237

9-Octadecenoic acid 3-[[[(2-Aminoethoxy)hydroxyphosphinyl]oxy]methyl]-2-[(1-oxooctadecyl)oxy]propyl ester, 9CI. 1-Octadec-9-enoyl-2-octadecanoylglycero-3-phosphoethanolamine

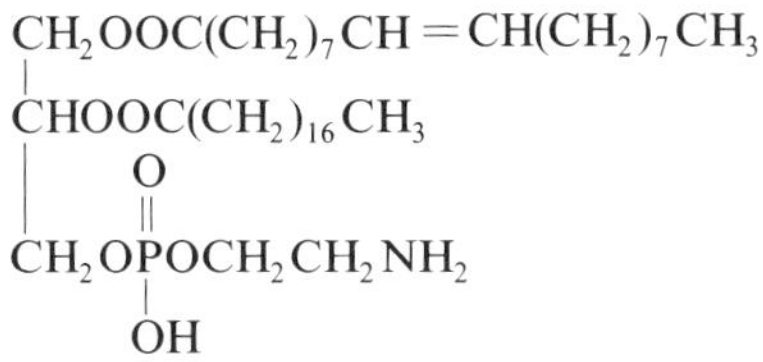

$C_{41}H_{80}NO_8P$ M 746.059

(*R*)-(*Z*)-*form*

Oleoylstearoylphosphatidylethanolamine

Cryst. Mp 181-182°. $[\alpha]_D$ +6.0° (c, 10 in $CHCl_3$).

(±)-(*E*)-*form*

Elaidoylstearoylphosphatidylethanolamine

Mp 190-192°.

(±)-(*Z*)-*form* [55176-55-1]

Mp 188-190°.

Baer, E. *et al*, *Can. J. Biochem.*, 1961, **39**, 1477 (*synth*)
Daemen, F.J.M. *et al*, *Recl. Trav. Chim. Pays-Bas* (*J. R. Neth. Chem. Soc.*), 1962, **81**, 348 (*synth*)
Chapman, D. *et al*, *Biochim. Biophys. Acta*, 1966, **120**, 148.
Chapman, D. *et al*, *Proc. R. Soc. London, A*, 1966, **290**, 1151 (*ir, nmr, props*)
Chapman, D. *et al*, *Trans. Faraday Soc.*, 1966, **62**, 2607 (*pmr, props*)

Glycerol 1-(9-octadecenoate) 2-octadecanoate 3-phosphoserine G-00238

Serine 2-[(1-oxooctadecenyl)oxy]-3-[(1-oxooctadecyl)oxy] propyl ester, dihydrogen phosphate ester, 9CI

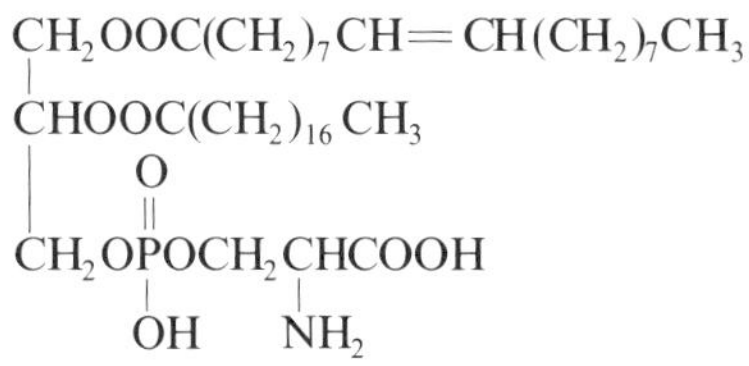

$C_{42}H_{80}NO_{10}P$ M 790.068

(±)-*form*

Oleoylpalmitoylphosphatidylserine

Turner, D.L. *et al*, *J. Lipid Res.*, 1964, **5**, 616 (*synth*)
Turner, D.L. *et al*, *J. Med. Chem.*, 1966, **9**, 771 (*synth*)

Glycerol 1-(9-octadecenoate) 2-octanoate 3-tetradecanoate G-00239

2-Capryloleomyristin. α-Oleo-β-caprylo-α'-myristin

$C_{43}H_{80}O_6$ M 693.101

Isol. from palm kernel oil. Cryst. Mp 14.8°.

Heiduschka, A. *et al*, *J. Prakt. Chem.*, 1928, **120**, 145 (*synth*)

Glycerol 1-(9-octadecenoate) 3-tetradecanoate G-00240

$C_{35}H_{66}O_5$ M 566.903

(±)-(*Z*)-*form*

α-Oleo-α'-myristin

Cryst. (Et_2O/EtOH). Mp 41°.

Daubert, B.F. *et al*, *J. Am. Chem. Soc.*, 1944, **66**, 53 (*synth*)

Glycerol 1-(9-octadecenoate) 2-tetradecanoate 3-phosphate G-00241

$C_{35}H_{67}O_8P$ M 646.883

(R)-form

1-Oleoyl-2-palmitoyl-sn-glycerol 3-phosphate

Cryst. (Me_2CO). Mp 61-63°. $[\alpha]_D^{18}$ +2.05° (c, 6 in $CHCl_3$).

Ba salt: Mp >360°. $[\alpha]_D^{20}$ +12.4° (c, 8 in $CHCl_3$/MeOH).

Di-Na salt: Mp 208-210°. $[\alpha]_D^{20}$ +5.70° (c, 11 in $CHCl_3$).

Bonsen, P.P.M. *et al, Chem. Phys. Lipids*, 1966, **1**, 100.

Glycerol 1-(9-octadecenoate) 2-tetradecanoate 3-phosphoglycerophosphate G-00242

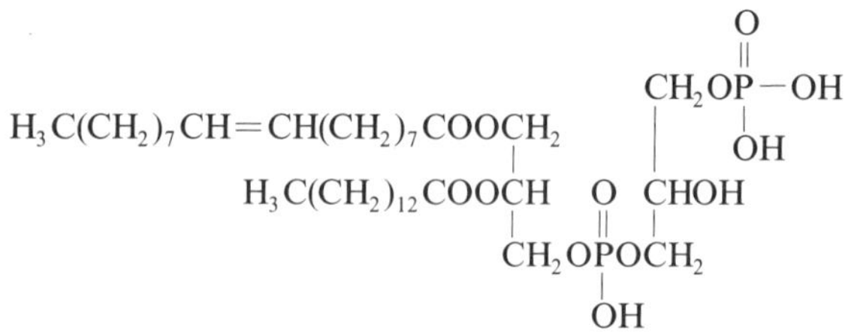

$C_{38}H_{74}O_{13}P_2$ M 800.942

(Z)-form

*Olein 2-myristo-1-(2,3-dihydroxypropyl) hydrogen phosphate 3-dihydrogen phosphate, 8CI. 1-Oleoyl-2-myristoylglycerol-3-phospho-*rac*-1′-glycerol-3′-phosphate*

Ba salt: [17708-99-5].

Mp >300°. $[\alpha]_D^{20}$ +4.70° (c, 7 in $CHCl_3$/MeOH).

Bonson, P.P.M. *et al, Chem. Phys. Lipids*, 1966, **1**, 100.

Glycerol 1-tetradecanoate G-00243

2,3-Dihydroxypropyl tetradecanoate, 9CI. 1-Monomyristin, 8CI. α-Monomyristin. Glycerol 1-myristate

[589-68-4]

$H_3C(CH_2)_{12}COOCH_2CH(OH)CH_2OH$

$C_{17}H_{34}O_4$ M 302.453

(±)-form

Mp 70-71°.

Averill, H.P. *et al, J. Am. Chem. Soc.*, 1929, **51**, 866 (*synth*)
Johnson, C.B. *et al, Lipids*, 1966, **1**, 371 (*ms*)
Gronowitz, S. *et al, Chem. Phys. Lipids*, 1975, **14**, 174.
Artamov, A.F. *et al, Izv. Akad. Nauk Kaz. SSR, Ser. Khim.*, 1977, **27**, 54 (*synth*)

Glycerol 2-tetradecanoate G-00244

β-Monomyristin. Glycerol 2-myristate

$H_3C(CH_2)_{12}COOCH(CH_2OH)_2$

$C_{17}H_{34}O_4$ M 302.453

Cryst. (pet. ether). Mp 60-61°.

1,3-Benzylidene: Mp 62°.

Cook, P.F.E. *et al, J. Chem. Soc.*, 1965, 4594 (*synth*)

Glycerol 1-tetradecanoate 2,3-di-9,12-octadecadienoate G-00245

$C_{53}H_{94}O_6$ M 827.322

(±)-(all-Z)-form

α,β-Dilinoleomyristin

Oil. Mp −9° to −8°.

Daubert, B.F. *et al, J. Am. Chem. Soc.*, 1944, **66**, 1507 (*synth*)

Glycerol 1-tetradecanoate 2,3-di-9-octadecenoate G-00246

9-Octadecenoic acid 1-[[(1-oxotetradecyl)oxy]methyl]-1,2-ethanediyl ester, 9CI

$C_{53}H_{98}O_6$ M 831.354

(±)-(all-E)-form

α,β-Dielaidomyristin

Cryst. (pet. ether or EtOH). Mp 40°.

(±)-(all-Z)-form [35804-98-9]

α,β-Dioleomyristin

Cryst. (Et_2O at −20°). Mp 12-13°.

Daubert, B.F. *et al, J. Am. Chem. Soc.*, 1944, **66**, 290 (*synth*)
Cook, P.F.E. *et al, J. Chem. Soc.*, 1965, 4594 (*synth*)

Glycerol 1-tetradecanoate 2-hexadecanoate 3-phosphocholine G-00247

Myristoylpalmitoylphosphatidylcholine. 1-Tetradecanoyl-2-hexadecanoylglycero-3-phosphocholine

[10589-50-1]

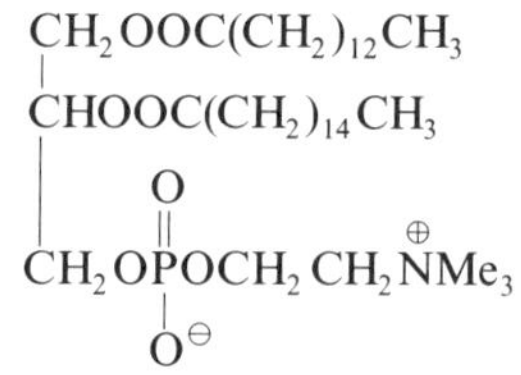

$C_{38}H_{76}NO_8P$ M 705.994

Keough, K.M.W. *et al, Biochemistry*, 1978, **18**, 1453.

Glycerol triacrylate, 8CI G-00248

2-Propenoic acid 1,2,3-propanetriyl ester, 9CI. Glyceryl triacrylate. Triacrylin

[5459-38-1]

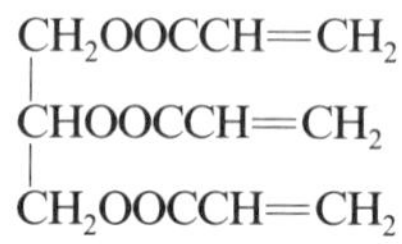

$C_{12}H_{14}O_6$ M 254.239

Monomer for cross-linked polymers. Liq. Bp_3 118°.

Sadtler Standard Infrared Spectra, No. 40091 (*ir*)
Reppe, W., *Justus Liebigs Ann. Chem.*, 1953, **582**, 1 (*synth*)
Kirk-Othmer Encycl. Chem. Technol., 3rd Ed., Wiley, N.Y., 1978-1984, **1**, 330, 386 (*rev*)

Glycerol tridecanoate G-00249

Decanoic acid 1,2,3-propanetriyl ester, 9CI. Tricaprin. Caprin

[11139-88-1]

$C_{33}H_{62}O_6$ M 554.849

Cryst. (pet. ether/EtOH). Mp 32°. Exhibits polymorphism.

Lutton, E.S. *et al, Lipids*, 1970, **5**, 90 (*synth*)

Glycerol tridocosanoate G-00250

Docosanoic acid 1,2,3-propanetriyl ester, 9CI. Tribehenin

[18641-57-1]

$C_{69}H_{134}O_6$ M 1059.814

Cryst. (hexane). Mp 82.5°. Exhibits polymorphism.

Lutton, E.S. *et al, Lipids*, 1970, **5**, 90 (*synth*)
Hagemann, J.W. *et al, J. Am. Oil Chem. Soc.*, 1972, **49**, 118.

Glycerol tri-13-docosenoate G-00251

13-Docosenoic acid 1,2,3-propanetriyl ester, 9CI

$C_{69}H_{128}O_6$ M 1053.767

(all-E)-form [37635-44-2]

Tribrassidin

Mp 58°. Exhibits polymorphism.

(all-Z)-form [2752-99-0]

Trierucin. Erucin†

Widespread in seed oils. Mp 32°. Exhibits polymorphism.

Hagemann, J.W. *et al*, *J. Am. Oil Chem. Soc.*, 1972, **49**, 118.

Glycerol trieicosanoate G-00252

Eicosanoic acid 1,2,3-propanetriyl ester, 9CI. Triarachidin

[620-64-4]

$C_{63}H_{122}O_6$ M 975.653

Cryst. (hexane). Mp 78°. Exhibits polymorphism.

Lutton, E.S. *et al*, *Lipids*, 1970, **5**, 90 (*synth*)
Hagemann, J.W. *et al*, *J. Am. Oil Chem. Soc.*, 1972, **49**, 118.

Glycerol triheneicosanoate G-00253

Triheneicosanoin

[26536-14-1]

$C_{66}H_{128}O_6$ M 1017.734

Cryst. (hexane). Mp 75.9°. Exhibits polymorphism.

Lutton, E.S. *et al*, *Lipids*, 1970, **5**, 90.

Glycerol triheptadecanoate G-00254

Triheptadecanoin, 8CI. Trimargarin

[2438-40-6]

$C_{54}H_{104}O_6$ M 849.412

Cryst. (hexane). Mp 64°. Exhibits polymorphism.

Lutton, E.S. *et al*, *Lipids*, 1970, **5**, 90.
Simpson, T.D., *J. Am. Oil Chem. Soc.*, 1983, **60**, 19 (*polymorphism*)

Glycerol trihexadecanoate G-00255

Glycerol tripalmitate. Tripalmitin. Palmitin

[555-44-2]

$C_{51}H_{98}O_6$ M 807.332

Present in palm oil and other plant waxes, e.g. *Bombax malabaricum*, *Dipterocarpus* spp. and wax of *Mycobacterium* spp. Cryst. (Et_2O). Mp 66.4°. Polymorphic.

Hilditch, T.P. *et al*, *The Chemical Constitution of Natural Fats*, Chapman and Hall, London, 4th Ed., 1964, 664.
Chapman, D., *The Structure of Lipids*, Methuen and Co., London, 1965, 268.
Dorset, D.L. *et al*, *Z. Naturforsch., C*, 1978, **33**, 50 (*cryst struct*)

Glycerol trihexanoate G-00256

Hexanoic acid 1,2,3-propanetriyl ester, 9CI. Tricaproin

[621-70-5]

$C_{21}H_{38}O_6$ M 386.528

Mp −25°.

▷ MB2700000.

Scheij, L.T.C., *Recl. Trav. Chim. Pays-Bas (J. R. Neth. Chem. Soc.)*, 1899, **18**, 169.
Lewis, R.J., *Sax's Dangerous Properties of Industrial Materials*, 8th Ed., Van Nostrand-Reinhold, 1992, GGK000.

Glycerol tri-12-hydroxy-9-octadecenoate G-00257

$CH_2OOC(CH_2)_7CH{=}CHCH_2CH(OH)(CH_2)_5CH_3$
|
$CHOOC(CH_2)_7CH{=}CHCH_2CH(OH)(CH_2)_5CH_3$
|
$CH_2OOC(CH_2)_7CH{=}CHCH_2CH(OH)(CH_2)_5CH_3$

$C_{57}H_{104}O_9$ M 933.443

all-[R-(Z)-]-form [2540-54-7]

Triricinolein. Ricinolein

Isol. from castor oil. Oil. d^{29} 0.962. n_D^{29} 1.4763.

Meyer, H., *Arch. Pharm. (Weinheim, Ger.)*, 1897, **235**, 189 (*synth*)
Subrahmanyan, V.V.R. *et al*, *J. Chem. Eng. Data*, 1961, **6**, 548 (*props*)
Subrahmanyan, V.V.R. *et al*, *J. Sci. Ind. Res.*, 1961, **20D**, 45 (*isol*)
Rashkes, Y.V. *et al*, *Khim. Prir. Soedin.*, 1986, 551 (*ms*)

Glycerol trimethacrylate, 8CI G-00258

2-Methyl-2-propenoic acid 1,2,3-propanetriyl ester, 9CI. Glyceryl trimethacrylate

[7401-88-9]

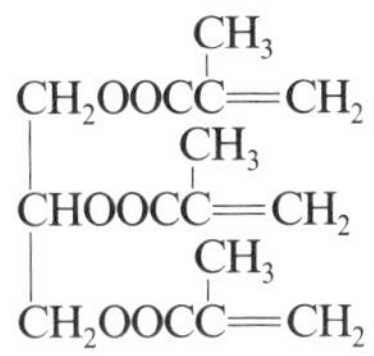

$C_{15}H_{20}O_6$ M 296.319

Monomer for cross-linked polymers. Liq. d_4^{20} 1.096. Mp −14°. Bp_1 156-160°. n_D 1.467.

Sadtler Standard Infrared Spectra, No. 40092 (*ir*)
U.K. Pat., 461 979, (1937); *CA*, **31**, 5378 (*synth*)
Dvořák, J. *et al*, *Collect. Czech. Chem. Commun.*, 1957, **22**, 379 (*synth*)
Kirk-Othmer Encycl. Chem. Technol., 3rd Ed., Wiley, N.Y., 1978-1984, **1**, 330, 386 (*rev*)
Encyclopedia of Polymer Science and Engineering, Wiley, 1985, **1**, 234 (*rev*)

Glycerol trinonadecanoate G-00259

Trinonadecanoin, 8CI

$C_{60}H_{116}O_6$ M 933.573

Cryst. (hexane). Mp 71°. Exhibits polymorphism.

Lutton, E.S. *et al*, *Lipids*, 1970, **5**, 90.

Glycerol tri-9,12-octadecadienoate G-00260

$C_{57}H_{98}O_6$ M 879.398

(9Z,12Z)-form [537-40-6]

Glycerol trilinoleate. Trilinolein

Constit. of seed oils rich in linoleic acid, e.g., sunflower oil. Isol. from lichens. Oil. Mp −11°. Exists in polymorphic forms.

Hilditch, T.P. *et al*, *The Chemical Constitution of Natural Fats*, Chapman and Hall, London, 4th Ed., 1964.
Hagemann, J.W. *et al*, *J. Am. Oil Chem. Soc.*, 1972, **49**, 118 (*polymorphism*)
Frost, D.J. *et al*, *Chem. Phys. Lipids*, 1975, **14**, 189 (*pmr*)

Glycerol trioctadecanoate G-00261

Glycerol tristearate. Tristearin. Stearin

[555-43-1]

$C_{57}H_{110}O_6$ M 891.492

Present in many vegetable fats after complete hydrogenation. Cryst. (Et_2O or hexane). Mp 73.5°. Exhibits polymorphism (lower Mp's reported 54.9°, 64°, 54.5°).

Lutton, E.S., *J. Am. Chem. Soc.*, 1955, **77**, 2646 (*synth*)
Markley, K.S., *Fatty Acids*, 2nd Ed,. Interscience, London, 1960, 326.
Barber, M. *et al*, *Tetrahedron Lett.*, 1964, 1063 (*ms*)
Hilditch, T.P. *et al*, *The Chemical Constitution of Natural Fats*, 4th Ed., Chapman and Hall, London, 1964, 664.
Lewis, R.J., *Sax's Dangerous Properties of Industrial Materials*, 8th Ed., Van Nostrand-Reinhold, 1992, GGU400.

Glycerol tri-9,11,13,15-octadecatetraenoate G-00262

Triparinarin

$C_{57}H_{86}O_6$ M 867.303

Isol. from the seed fat of *Parinarium laurinium*. Cryst. Mp 49-50°.

Riley, J.P., *J. Chem. Soc.*, 1951, 291 (*isol*)

Glycerol tri-9,12,15-octadecatrienoate G-00263

$C_{57}H_{92}O_6$ M 873.350

(9Z,12Z,15Z)-form

Glycerol trilinolenate. Trilinolenin. Linolenin

Constit. of linseed oil. Isol. from *Pinus taeda* pollen. Oil. Mp −24.2°. Exists in polymorphic forms.

Scott, R.W. *et al*, *Phytochemistry*, 1962, **1**, 189 (*isol*)
Hilditch, T.P. *et al*, *The Chemical Constitution of Natural Fats*, Chapman and Hall, London, 4th ed., 1964.
Frost, D.J. *et al*, *Chem. Phys. Lipids*, 1975, **14**, 189 (*pmr*)

Glycerol tri-6-octadecenoate G-00264

6-Octadecenoic acid 1,2,3-propanetriyl ester, 9CI. Tripetroselin

$C_{57}H_{104}O_6$ M 885.445

(all-E)-form [35017-28-8]

Tripetroselaidin

Mp 52°. Exibits polymorphism.

(all-Z)-form [3296-43-3]

Tripetroselinin

Constit. of some Umbelliferae seed oils, e.g. *Petroselinum* spp. Mp 28°. Exibits polymorphism.

Hagemann, J.W. *et al*, *J. Am. Oil Chem. Soc.*, 1972, **49**, 118 (*synth*)
Batchelor, J.G. *et al*, *J. Org. Chem.*, 1974, **39**, 1698 (*cmr*)

Glycerol tri-9-octadecenoate G-00265

$C_{57}H_{104}O_6$ M 885.445

(E,E,E)-form [537-39-3]

Glycerol trielaidate. Trielaidin

Mp 41°. Exists in polymorphic forms.

(Z,Z,Z)-form [122-32-7]

Glycerol trioleate. Triolein. Olein

Constit. of olive oil and other vegetable oils. Oil. Mp 5°. Bp_{18} 235-240° part. dec. Exists in polymorphic forms.

Hilditch, T.P. *et al*, *The Chemical Constitution of Natural Fats*, Chapman and Hall, London, 4th Ed., 1964.
Hagemann, J.W. *et al*, *J. Am. Oil Chem. Soc.*, 1972, **49**, 118 (*polymorphism*)

Glycerol tritetracosanoate G-00266

Tetracosanoic acid 1,2,3-propenetriyl ester, 9CI. Trilignocerin

[56933-26-7]

$C_{75}H_{146}O_6$ M 1143.975

Cryst. (hexane). Mp 86°. Exhibits polymorphism.

Simpson, T.D. *et al*, *J. Am. Oil Chem. Soc.*, 1975, **52**, 303 (*struct*)

Glycerol tritetradecanoate G-00267

Glycerol trimyristate. Trimyristin. Myristin

$C_{45}H_{86}O_6$ M 723.171

Constit. of nutmegs and other seed fats rich in myristic acid. Mp 58.5°. Exists in polymorphic forms.

Hilditch, T.P. *et al*, *The Chemical Constitution of Natural Fats*, Chapman and Hall, London, 4th Ed., 1964.
Lutton, E.S. *et al*, *Lipids*, 1970, **5**, 90 (*polymorphism*)

Glycerol tritridecanoate G-00268

Tritridecanoin, 8CI

[26536-12-9]

$C_{42}H_{80}O_6$ M 681.090

Cryst. (hexane). Mp 44.5°. Exibits polymorphism.

Lutton, E.S. *et al*, *Lipids*, 1970, **5**, 90.

Glycerol triundecanoate G-00269

Triundecanoin

[13552-80-2]

$C_{36}H_{68}O_6$ M 596.930

Cryst. (hexane). Mp 31.2°. Exhibits polymorphism.

Lutton, E.S. *et al*, *Lipids*, 1970, **5**, 90.

Glycocholic acid G-00270

Cholylglycine

[475-31-0]

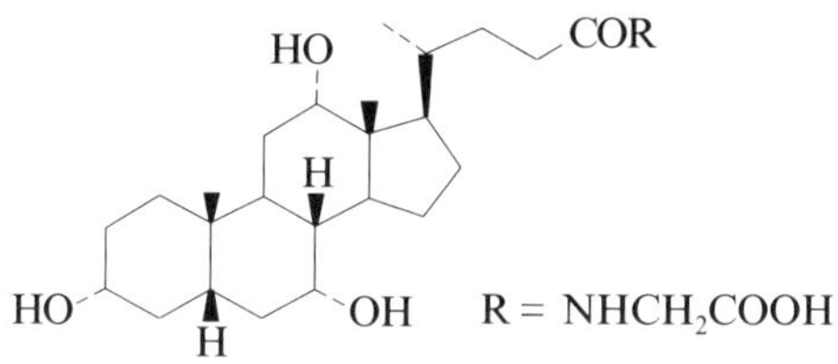

$C_{26}H_{43}NO_6$ M 465.629

Constit. of bile. Small needles (EtOH aq.). Mp 140-142°, Mp 154-155° dec. $[\alpha]_D^{13}$ +27.8° (EtOH). pK_a 4.40.

Na salt: [863-57-0].

Cryst. (EtOH/Et_2O). Mp 230-240°. $[\alpha]_D^{23}$ +29.2° (H_2O).

Cortese, F. *et al*, *J. Am. Chem. Soc.*, 1935, **57**, 1393 (*synth*)
Cortese, F., *J. Am. Chem. Soc.*, 1937, **59**, 2532 (*synth*)
Begemann, F., *Z. Klin. Chem. Klin. Biochem.*, 1972, **10**, 29 (*isol*)
Games, D.E. *et al*, *Tetrahedron Lett.*, 1974, **27**, 2377 (*ms*)
Tserng, K.-Y. *et al*, *J. Lipid Res.*, 1977, **18**, 404 (*synth*)
Shaw, R. *et al*, *Biomed. Mass Spectrom.*, 1978, **5**, 433 (*ms*)

N-Glycolylneuraminic acid G-00271

*3,5-Dideoxy-5-[(hydroxyacetyl)amino]-*D-glycero-D-galacto-*2-nonulosonic acid, 9CI. Porcine sialic acid. NGNA*

[1113-83-3]

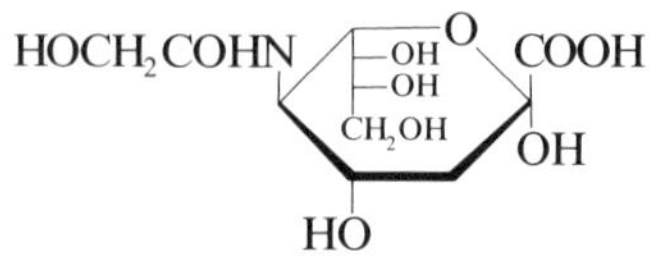

α-D-Pyranose-*form*

$C_{11}H_{19}NO_{10}$ M 325.272

Isol. from hog submaxillary mucin. Also found together with *N*-Acetylneuraminic acid in certain gangliosides and sialoproteins of other animal spp., e.g. horse erythrocytes, horse kidney and beef serum. Mp 185-187°. $[\alpha]_D$ −32° (H_2O).

Blix, G. *et al*, *Nature (London)*, 1955, **175**, 340 (*isol*)
Zilliken, F. *et al*, *Adv. Carbohydr. Chem.*, 1958, **13**, 237 (*rev*)

Gottschalk, A., *The Chemistry and Biology of Sialic Acids and Related Substances*, Cambridge Univ. Press, 1960 (*rev*)
Schoop, H.J. *et al*, *Hoppe Seyler's Z. Physiol. Chem.*, 1969, **350**, 155.
Schauer, R., *Angew. Chem., Int. Ed. Engl.*, 1973, **12**, 127.
Codington, J.F. *et al*, *Methods Carbohydr. Chem.*, 1976, **7**, 226 (*occur, purifn*)
Corfield, A.P. *et al*, *Eur. J. Biochem.*, 1976, **68**, 597 (*biosynth*)

Goniothalamicin G-00272

[113817-64-4]

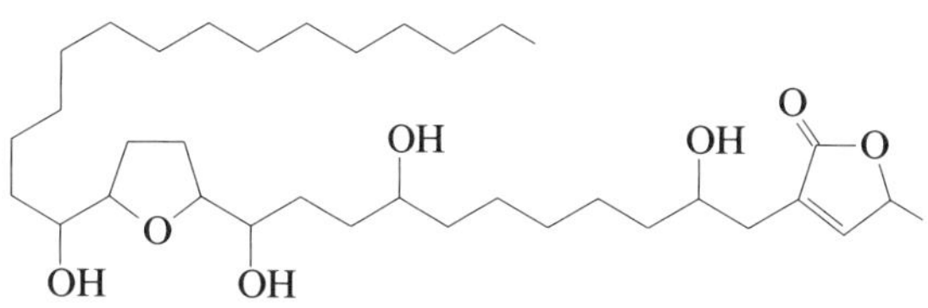

$C_{35}H_{64}O_7$ M 596.886

Closely related to Annonacin†, A-00046. Constit. of *Goniothalamus giganteus* and the stem bark of *Annona densicoma*. Cytotoxic against human tumour cells *in vitro*. Wax. Mp 86-88°, Mp 91-92°. $[\alpha]_D$ +1.6°, $[\alpha]_D$ +10.4° (c, 0.08 in MeOH).

Alkofahi, A. *et al*, *Experientia*, 1988, **44**, 83.
Yu, J.G. *et al*, *J. Org. Chem.*, 1992, **57**, 6198 (*isol, pmr, cmr, ir, uv, ms*)

Gossypol G-00273

1,1′,6,6′,7,7′-Hexahydroxy-3,3′-dimethyl-5,5′-bis(1-methylethyl)-[2,2′-binaphthalene]-8,8′-dicarboxaldehyde, 9CI. Thespesin

[303-45-7]

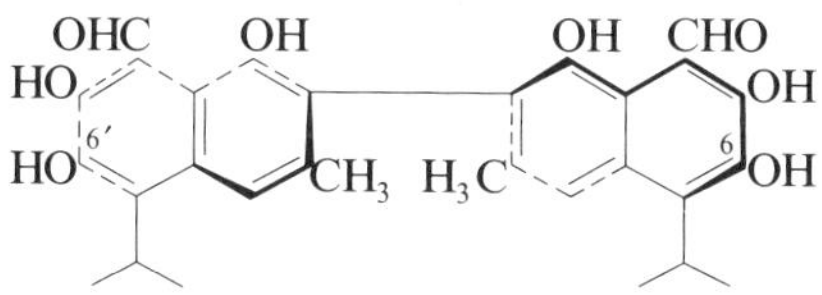

Absolute configuration

$C_{30}H_{30}O_8$ M 518.562

Atropisomeric compd., occurs naturally in both (+)- and (±)-forms. Tautomeric in polar solvents with intramolecular bis-lactol and bis-ketol tautomers.

▷ Toxic. DU3100000.

(+)-***form*** [20300-26-9]

Constit. of *Thespesia populnea*. Pale-yellow needles (pet. ether), deep-yellow prisms + Me_2CO (Me_2CO), large elongated plates (Me_2CO aq.). Mp 181-183°. $[\alpha]_D^{19}$ +445° (c, 0.15 in $CHCl_3$).

▷ DU3101000.

Hexa-Me ether: [17273-30-2].
Mp 242-244°. $[\alpha]_D$ +177° ($CHCl_3$).

(±)-***form*** [40112-23-0]

Toxic component of cotton boll cavities. Used as 0.2% EtOH soln. for detn. of Sn(*IV*); for photometric detn. of Ge. Male antifertility agent, undergoing widespread trials in the Peoples' Republic of China. Cryst. in three forms (Et_2O, $CHCl_3$, pet. ether). Mp 184°, Mp 199°, Mp 214°.

Hexa-Ac: [30719-67-6].
Mp 276-279°.
▷ DU3103000.

Bisphenylhydrazone: Yellow plates (C_6H_6). Mp 303°.

6-Me ether: [54302-42-0].
$C_{31}H_{32}O_8$ M 532.589
Constit. of the roots of *Gossypium* spp. Yellow cryst. (C_6H_6/hexane). Mp 146-149°.

6,6′-Di-Me ether: [1110-58-3].
$C_{32}H_{34}O_8$ M 546.616
From *G.* spp. Golden-yellow cryst. (C_6H_6/hexane). Mp 181-184°.

Hexa-Me ether: Cryst. in three forms, two colourless and one red. Mp 158-160° (red), Mp 221°, Mp 231-232°.

2-Aminoethylsulfonate: [87606-98-2]. *GSN (as disodium salt). Metaphin (as disodium salt)*
Immunosuppressive, antiviral agent.

[62770-38-1, 90141-22-3]

Vioque-Pizarro, A., *Anal. Chim. Acta*, 1951, **5**, 529 (*detn, Sn*)
Adams, R. *et al*, *Chem. Rev.*, 1960, **60**, 555 (*rev*)
Bell, A., *Phytopathology*, 1967, **57**, 759.
Bhakuni, D.S. *et al*, *Experientia*, 1968, **24**, 109 (*isol*)
King, T.J. *et al*, *Tetrahedron Lett.*, 1968, 261 (*isol*)
Wood, A.B. *et al*, *Chem. Ind. (London)*, 1969, 1738 (*conformn*)
Edwards, J.D., *J. Am. Oil Chem. Soc.*, 1970, **47**, 441 (*synth*)
Abou-Donia, M.B. *et al*, *Lipids*, 1970, **5**, 938 (*metab*)
Talipov, S.T. *et al*, *Zh. Anal. Khim.*, 1970, **25**, 1420 (*detn, Ge*)
Stipanovic, R.D. *et al*, *Phytochemistry*, 1975, **14**, 1077.
O'Brien, D.H. *et al*, *J. Org. Chem.*, 1978, **43**, 1105 (*nmr*)
Martindale, The Extra Pharmacopoeia, 28th/29th Ed., Pharmaceutical Press, London, 1982/1989, 12798.
Aripov, U.A. *et al*, *Khim.-Farm. Zh.*, 1983, **17**, 908 (*pharmacol, deriv*)
Sega, S.J. *et al*, *Gossypol*, Plenum Press, N.Y., 1985 (*book*)
Masciadri, R. *et al*, *J. Chem. Soc., Chem. Commun.*, 1985, 1573 (*biosynth*)
Talipov, S.A. *et al*, *Khim. Prir. Soedin.*, 1985, **21**, 835; 1986, **22**, 112; *Chem. Nat. Compd. (Engl. Transl.)*, 797, 108 (*cryst struct*)
Ibragimov, B.T. *et al*, *Khim. Prir. Soedin.*, 1985, **21**, 837; 1986, **22**, 113; *Chem. Nat. Compd. (Engl. Transl.)*, 799, 110 (*cryst struct*)
Wu, D.F. *et al*, *Clin. Pharmacol. Ther. (St. Louis)*, 1986, **39**, 613 (*pharmacol*)
Stipanovic, R.D. *et al*, *J. Chem. Soc., Chem. Commun.*, 1986, 100 (*biosynth*)
Sampath, D.S. *et al*, *J. Chem. Soc., Chem. Commun.*, 1986, 649 (*cd, resoln*)
Lacombe, L. *et al*, *J. Nat. Prod. (Lloydia)*, 1987, **50**, 277 (*cmr*)
Tyson, R., *Chem. Ind. (London)*, 1988, 118 (*hplc, resoln*)
Huang, L. *et al*, *Collect. Czech. Chem. Commun.*, 1988, **53**, 2664 (*cd, abs config*)
Brzezinski, B. *et al*, *J. Mol. Struct.*, 1990, **239**, 23 (*tautom, ir*)
Marciniak, B. *et al*, *J. Chem. Soc., Perkin Trans. 2*, 1991, 1359 (*pmr, tautom*)
Jaroszewski, J.W. *et al*, *Chirality*, 1992, **4**, 216 (*props*)
Jaroszewski, J.W. *et al*, *Planta Med.*, 1992, **58**, 454 (*occur*)
Akhila, A. *et al*, *Phytochemistry*, 1993, **33**, 335 (*biosynth*)
Lewis, R.J., *Sax's Dangerous Properties of Industrial Materials*, 8th Ed., Van Nostrand-Reinhold, 1992, GJM000, GJM025, GJM030, GJM035.

H

6,9-Heneicosadiene **H-00001**

$H_3C(CH_2)_4CH{=}CHCH_2CH{=}CH(CH_2)_{10}CH_3$

$C_{21}H_{40}$ M 292.547

(6Z,9Z)-form [85613-23-6]

Pheromone of *Utetheisa ornatrix* and *Mocis latipes*.

Jain, S.C. *et al*, *J. Org. Chem.*, 1983, **48**, 2266 (*isol, synth, pmr*)
O'Connor, B. *et al*, *J. Org. Chem.*, 1987, **52**, 1801 (*synth*)

1,6-Heneicosadien-11-one **H-00002**

$H_3C(CH_2)_9CO(CH_2)_3CH{=}CH(CH_2)_3CH{=}CH_2$

$C_{21}H_{38}O$ M 306.531

(E)-form [65956-72-1]

Mp 30°.

(Z)-form [65956-73-2]

Additional sex pheromone of Douglas fir tussock moth. Liq.

Smith, L.M. *et al*, *J. Org. Chem.*, 1978, **43**, 2361 (*synth, pmr, ir, ms,cmr, uv*)

1,6,9,12,15,18-Heneicosahexaene, 9CI, 8CI **H-00003**

$C_{21}H_{32}$ M 284.484

(all-Z)-form [33439-58-6]

Isol. from *Fucus vesiculosus* and the brown alga *Notheia anomala*. Highly unstable oil.

18,19-Dihydro: [33439-59-7]. *1,6,9,12,15-Heneicosapentaene, 9CI, 8CI*
$C_{21}H_{34}$ M 286.500
Isol. from *N. anomala* and *F. vesiculosus*. Highly unstable oil.

6,7:15,16-Diepoxide: 6,7:15,16-Diepoxy-1,9,12,18-heneicosatetraene. 6,7:15,16-Bisepoxy-3,9,12,20-heneicosatetraene (incorr.)
$C_{21}H_{32}O_2$ M 316.483
Isol. from *N. anomala*. Yellow oil. $[\alpha]_D$ +9.1° (c, 0.3 in $CHCl_3$).

9,10:15,16-Diepoxide: 9,10:15,16-Diepoxy-1,6,12,18-heneicosatetraene. 6,7:12,13-Bisepoxy-3,9,15,20-heneicosatetraene (incorr.)
$C_{21}H_{32}O_2$ M 316.483
Isol. from *N. anomala*. Oil. $[\alpha]_D$ +15.6° (c, 0.3 in $CHCl_3$).

9,10:15,16-Diepoxide, 18,19-dihydro: 9,10:15,16-Diepoxy-1,6,12-heneicosatriene. 6,7:12,13-Bisepoxy-9,15,20-heneicosatriene (incorr.)
$C_{21}H_{34}O_2$ M 318.498
From *N. anomala*. Unstable oil. $[\alpha]_D$ +8.6° (c, 0.5 in $CHCl_3$).

6,7:12,13:15,16-Triepoxide: 6,7:12,13:15,16-Triepoxy-1,9,18-heneicosatriene. 6,7:9,10:15,16-Trisepoxy-3,12,20-heneicosatriene (incorr.)
$C_{21}H_{32}O_3$ M 332.482
From *N. anomala*. Yellow oil. $[\alpha]_D$ +12.3° (c, 1.7 in $CHCl_3$).

6,7:12,13:15,16-Triepoxide,18,19-dihydro: 6,7:12,13:15,16-Triepoxy-1,9-heneicosadiene. 6,7:9,10:15,16-Trisepoxy-12,20-heneicosadiene (incorr.)
$C_{21}H_{34}O_3$ M 334.498
From *N. anomala*. Yellow oil. $[\alpha]_D$ +5.9° (c, 2.3 in $CHCl_3$).

12S,13R:15S,16R-Diepoxide, 18,19-dihydro: [129436-93-7]. *12,13:15,16-Diepoxy-1,6,9-heneicosatriene*
$C_{21}H_{34}O_2$ M 318.498
Isol. from *N. anomala*. Unstable oil. $[\alpha]_D$ +25.6° (c, 0.7 in $CHCl_3$).

Halsall, T.G. *et al*, *J. Chem. Soc., Chem. Commun.*, 1971, 448 (*isol, pmr*)
Caccamuse, S. *et al*, *Experientia*, 1978, **34**, 1129 (*ms*)
Barrow, R.A. *et al*, *Aust. J. Chem.*, 1990, **43**, 895 (*isol, pmr, ms*)
Murray, L.M. *et al*, *Aust. J. Chem.*, 1991, **44**, 843.

3,6,9,12,15,18-Heneicosahexaene **H-00004**

[127702-72-1]

$H_3C(CH_2CH{=}CH)_6CH_2CH_3$

$C_{21}H_{32}$ M 284.484

All-(Z)-form [29118-86-3]

Isol. from marine algae and plankton. Ir ν 3020, 715 (C=C) cm^{-1}.

[32779-95-6]

Lee, R.F. *et al*, *Biochim. Biophys. Acta*, 1970, **202**, 386 (*isol, struct, ms*)
Blumer, M. *et al*, *Mar. Biol. (Berlin)*, 1970, **6**, 226 (*isol, struct, uv, ir, ms*)
Wright, J.L.C., *Phytochemistry*, 1980, **19**, 143 (*cmr*)

Heneicosane, 9CI **H-00005**

Henicosane

[629-94-7]

$H_3C(CH_2)_{19}CH_3$

$C_{21}H_{44}$ M 296.579

Common constit. of plant waxes. Constit. of female sex pheromone of *Hypsilpyla robusta*. Wax. Mp 40.5°. Bp 356.5°, Bp_{15} 215°, $Bp_{0.05}$ 129°.

Schmidt, A.W. *et al*, *Ber.*, 1941, **74**, 1313.
Karrer, W. *et al*, *Konstitution und Vorkommen der Organischen Pflanzenstoffe*, 2nd Ed., Birkhäuser Verlag, Basel, 1972-1985, no. 107 (*occur*)
Sevrin, M. *et al*, *Tetrahedron Lett.*, 1976, 2643 (*synth*)
Bosson, G.A. *et al*, *C. R. Hebd. Seances Acad. Sci.*, 1982, **294**, 819 (*isol*)

Heneicosanedioic acid **H-00006**

[505-55-5]

$HOOC(CH_2)_{19}COOH$

$C_{21}H_{40}O_4$ M 356.545

Little known. Was descr. (1900) as nat. prod. under the name Japanic acid but this was erroneous.

Lamberton, J.A., *Aust. J. Chem.*, 1961, **14**, 323.

Heneicosanoic acid, 9CI **H-00007**

[2363-71-5]

$H_3C(CH_2)_{19}COOH$

$C_{21}H_{42}O_2$ M 326.562

Isol. from wood of *Ulex europaeus*, bark of *Pinus contorta*, from *Shorea maranti*, olive oil (*Olea europaea*) etc. Needles (Me_2CO). Mp 73-74°.

[28898-67-1, 59252-40-3]

Le Sueur, H.R. *et al*, *J. Chem. Soc.*, 1915, **107**, 736 (*synth*)
Cocker, W. *et al*, *Perfum. Essent. Oil Rec.*, 1963, **54**, 235 442 (*isol*)
Org. Synth., 1981, **60**, 11 (*synth*)

2-Heneicosanone H-00008

[22589-04-4]

$H_3C(CH_2)_{18}COCH_3$

$C_{21}H_{42}O$ M 310.562
Mp 100-100.5°.

Semicarbazone: Mp 128-129°.

Stenhagen, E., *Ark. Kemi*, 1951, **3**, 381 (*synth*)
Bailey, A.V. *et al*, *J. Chem. Eng. Data*, 1970, **15**, 542 (*cryst struct*)

4-Heneicosanone H-00009

[32822-45-0]

$H_3C(CH_2)_{16}COCH_2CH_2CH_3$

$C_{21}H_{42}O$ M 310.562
Mp 55-56°.

Ohlson, J.L. *et al*, *CA*, 1956, **50**, 1893 (*synth*)

7-Heneicosanone H-00010

$H_3C(CH_2)_{13}CO(CH_2)_5CH_3$

$C_{21}H_{42}O$ M 310.562
Mp 55.6-56.4°.

Breusch, F.L. *et al*, *Chem. Ber.*, 1953, **86**, 684 (*synth*)

8-Heneicosanone H-00011

$H_3C(CH_2)_{12}CO(CH_2)_6CH_3$

$C_{21}H_{42}O$ M 310.562
Mp 55.1-56.4°.

Breusch, F.L. *et al*, *Chem. Ber.*, 1953, **86**, 684 (*synth*)

9-Heneicosanone H-00012

$H_3C(CH_2)_{11}CO(CH_2)_7CH_3$

$C_{21}H_{42}O$ M 310.562
Mp 55.4-56.2°.

Breusch, F.L., *Chem. Ber.*, 1953, **86**, 684 (*synth*)

10-Heneicosanone H-00013

[22026-16-0]

$H_3C(CH_2)_{10}CO(CH_2)_8CH_3$

$C_{21}H_{42}O$ M 310.562
Mp 55.5-56.9°.

Breusch, F.L., *Chem. Ber.*, 1953, **86**, 684.
Arai, K. *et al*, *CA*, 1969, **70**, 16862p (*glc*)

11-Heneicosanone H-00014

[19781-72-7]

$H_3C(CH_2)_9CO(CH_2)_9CH_3$

$C_{21}H_{42}O$ M 310.562
Cryst. (EtOH aq.). Mp 64°.

Oxime:
$C_{21}H_{43}NO$ M 325.577
Mp 27.5°.

Breusch, F.L. *et al*, *Fette, Seifen, Anstrichm.*, 1968, **70**, 864.
Matsota, N.P., *Zh. Prikl. Khim. (Leningrad)*, 1971, **44**, 1823.
Schill, G. *et al*, *Synthesis*, 1974, 117 (*synth*)

1,3,6,9-Heneicosatetraene H-00015

$H_2C{=}CHCH{=}CHCH_2CH{=}CHCH_2CH{=}CH(CH_2)_{10}CH_3$

$C_{21}H_{36}$ M 288.515

(all-Z)-form [85612-05-1]
Pheromone of the bella moth *Utetheisa ornatrix*.

Jain, S.C. *et al*, *J. Org. Chem.*, 1983, **48**, 2266 (*isol, synth, spectra*)
Huang, W. *et al*, *J. Org. Chem.*, 1983, **48**, 2270 (*synth*)

3,6,9-Heneicosatriene H-00016

$H_3CCH_2CH{=}CHCH_2CH{=}CHCH_2CH{=}CH(CH_2)_{10}CH_3$

$C_{21}H_{38}$ M 290.531

(all-Z)-form [87255-15-0]
Pheromone of *Utetheisa ornatrix* and *Anticarsia gemmatalis*; isol. from various insects incl. *Arctia* spp., *Anticarsia* spp. and *Caenurgina* spp.

Jain, S.C. *et al*, *J. Org. Chem.*, 1983, **48**, 2266 (*isol*)
Bell, T.W. *et al*, *J. Chem. Ecol.*, 1986, **12**, 385 (*occur*)
Yu, H.M. *et al*, *Chem. Ind. (London)*, 1988, 787, 836 (*synth*)
Heath, R.R. *et al*, *J. Chem. Ecol.*, 1988, **14**, 1121.

9-Heneicosene H-00017

[629-95-8]

$H_3C(CH_2)_{10}CH{=}CH(CH_2)_7CH_3$

$C_{21}H_{42}$ M 294.563
Mp 3°. Bp_{12} 201-202°. The (*Z*)-form is an insect sex attractant.

Mai, J., *Ber.*, 1889, **22**, 2133.
Schaal, R., *Ber.*, 1907, **40**, 4787.
Mansingh, A. *et al*, *Can. Entomol.*, 1972, **104**, 1963.

8-Heneicosen-1-ol H-00018

$H_3C(CH_2)_{11}CH{=}CH(CH_2)_6CH_2OH$

$C_{21}H_{42}O$ M 310.562

Ac:
$C_{23}H_{44}O_2$ M 352.599
Constit. of cuticular lipid of mature male little house fly *Fannia canicularis*.

Uebel, E.C. *et al*, *J. Chem. Ecol.*, 1977, **3**, 269 (*isol, glc, ir, nmr, ms*)

6-Heneicosen-11-one H-00019

$H_3C(CH_2)_9CO(CH_2)_3CH{=}CH(CH_2)_4CH_3$

$C_{21}H_{40}O$ M 308.546

(E)-form [54844-66-5]
Sex attractant for Tussock moth *Orgyia pseudotsugata*. Mp 36-38°.

(Z)-form [54844-65-4]
Sex attractant for *O. pseudotsugata*, other *O.* spp. and honeybee *Apis mellifera*. Oil. $Bp_{0.4}$ 118° ($Bp_{0.2}$ 138-140°).

Smith, R.G. *et al*, *J. Org. Chem.*, 1975, **40**, 1593 (*synth, ir, pmr, ms*)
Smith, R.G. *et al*, *Science (Washington, D.C.)*, 1975, **188**, 63 (*occur*)

Mori, K. *et al*, *Tetrahedron*, 1977, **33**, 385 (*synth, ir, pmr, ms*)
Akermark, B. *et al*, *J. Org. Chem.*, 1978, **43**, 4387 (*synth, ir, pmr*)
Murata, Y. *et al*, *Bull. Chem. Soc. Jpn.*, 1983, **56**, 2539 (*synth*)
Naoshima, Y. *et al*, *Agric. Biol. Chem.*, 1984, **48**, 2151 (*isol, ms, synth*)
Nishiyama, H. *et al*, *Tetrahedron Lett.*, 1984, **25**, 223 (*synth*)
Sodeoka, M. *et al*, *J. Org. Chem.*, 1985, **50**, 1147 (*synth, activity*)
Dasaradhi, L. *et al*, *Synth. Commun.*, 1987, **17**, 1845 (*synth*)
Ballini, R., *J. Chem. Soc., Perkin Trans. 1*, 1991, 1491 (*synth, pmr, cmr, ms, ir, glc*)

2-Heneicosyl-3,6-dihydroxy-5-methyl-1,4-benzoquinone H-00020

2-Heneicosyl-3,6-dihydroxy-5-methyl-2,5-cyclohexadiene-1,4-dione, 9CI. ***Polygonaquinone***

[750-71-0]

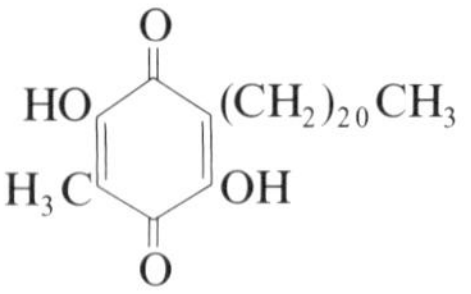

$C_{28}H_{48}O_4$ M 448.685

Constit. of the roots of *Polygonatum falcatum*. Orange prisms (EtOH). Mp 133-134°.

Di-Ac: Yellow needles (EtOH). Mp 92-93°.

Di-Me ether: Yellow needles (EtOH). Mp 79-80°.

Nakata, H. *et al*, *Tetrahedron*, 1964, **20**, 2319 (*isol*)
Yoshihira, K. *et al*, *Chem. Pharm. Bull.*, 1966, **14**, 1052 (*synth*)

6-Heneicosyl-3,4,5,6-tetrahydro-2*H*-pyran-2-one H-00021

5-Hexacosanolide. 5-Hydroxyhexacosanoic acid lactone

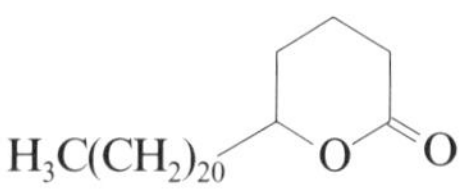

$C_{26}H_{50}O_2$ M 394.680

(+)-*form* [63389-34-4]

Constit. of *Heliotropium curassavicum*. Mp 76°. $[\alpha]_D$ +25.1° ($CHCl_3$).

Subramanian, P.S. *et al*, *Experientia*, 1977, **33**, 707 (*isol*)

26-Henpentacontanone H-00022

Cerotone

[542-48-3]

$H_3C(CH_2)_{24}CO(CH_2)_{24}CH_3$

$C_{51}H_{102}O$ M 731.366

Mp 93°.

Oura, H. *et al*, *Yakugaku Zasshi*, 1958, **78**, 141; *CA*, **52**, 10885f (*synth*)

Hentriacontane, 9CI H-00023

[630-04-6]

$H_3C(CH_2)_{29}CH_3$

$C_{31}H_{64}$ M 436.847

Constit. of many waxes. Mp 68°. Bp 458°, Bp_{15} 302°.

Heilbron, I.H. *et al*, *J. Chem. Soc.*, 1934, 1573.
Keller, A. *et al*, *J. Polym. Sci., Polym. Phys. Ed.*, 1970, **8**, 19 (*synth, ir*)
Karrer, W. *et al*, *Konstitution und Vorkommen der Organischen Pflanzenstoffe*, 2nd Ed., Birkhäuser Verlag, Basel, 1972-1985, no. 18 (*occur*)

Hentriacontanoic acid, 9CI H-00024

[38232-01-8]

$H_3C(CH_2)_{29}COOH$

$C_{31}H_{62}O_2$ M 466.830

Isol. from leaf wax of *Agave sisalana* and other plants. Cryst. (Me_2CO). Mp 93-93.2°.

[77630-51-4]

Piper, S.H. *et al*, *Biochem. J.*, 1934, **28**, 2175 (*synth*)
Razafindrazaka, J. *et al*, *Bull. Soc. Chim. Fr.*, 1963, 1633 (*isol*)
Rao, S.J. *et al*, *Indian J. Chem., Sect. B*, 1987, **26**, 208 (*synth*)

1-Hentriacontanol, 9CI H-00025

[544-86-5]

$H_3C(CH_2)_{29}CH_2OH$

$C_{31}H_{64}O$ M 452.846

Found in several plant species incl. *Phyllarthron comorense* and *Agave sisalana* (leafwax). Lustrous plates ($CHCl_3$/pet. ether). Mp 87.1-87.2°.

Jones, R.G. *et al*, *J. Am. Chem. Soc.*, 1947, **69**, 2350 (*synth*)
Ställberg, G. *et al*, *Acta Chem. Scand.*, 1952, **6**, 313.
Razafindrazaka, J. *et al*, *Bull. Soc. Chim. Fr.*, 1963, 1633 (*isol*)
Rao, S.J. *et al*, *Indian J. Chem., Sect. B*, 1987, **26**, 208 (*synth*)

16-Hentriacontanol H-00026

[1070-54-8]

$H_3C(CH_2)_{14}CH(OH)(CH_2)_{14}CH_3$

$C_{31}H_{64}O$ M 452.846

Constit. of natural waxes from several sources. Cryst. (EtOH). Mp 84-85°.

Ac:

$C_{33}H_{66}O_2$ M 494.883

Cryst. Mp 47-49°.

Kipping, F.S., *J. Chem. Soc.*, 1890, **57**, 986 (*synth*)
Pudles, J. *et al*, *Bull. Soc. Chim. Biol.*, 1954, **36**, 754.
Bordet, C. *et al*, *Bull. Soc. Chim. Biol.*, 1964, **46**, 1101.
Hirose, M. *et al*, *Nippon Kagaku Zasshi*, 1968, **89**, 887 (*isol*)
Kolattukudy, P.E., *Lipids*, 1970, **5**, 398 (*isol, ms, ir*)
Macey, M.J.K. *et al*, *Phytochemistry*, 1970, **9**, 5 (*isol*)
Mahato, S.B. *et al*, *Phytochemistry*, 1971, **10**, 2847 (*isol*)

16-Hentriacontanone H-00027

Palmitone

[502-73-8]

$H_3C(CH_2)_{14}CO(CH_2)_{14}CH_3$

$C_{31}H_{62}O$ M 450.830

Constit. of *Adiantum caudatum*, *Platanus orientalis*, *Piper nigrum* and *Gymnelaea cunninghamii*. Also in other lipids from bacteria, fungi and higher plants. Cryst. (EtOH). Mp 82-83°.

Oxime:

$C_{31}H_{63}NO$ M 465.845

Cryst. (EtOH or AcOH). Mp 59°.

Chibnall, A.C. *et al*, *Biochem. J.*, 1937, **31**, 1981 (*isol*)
Asselineau, J., *Biochim. Biophys. Acta*, 1961, **54**, 359 (*isol*)
Hayashi, N. *et al*, *J. Indian Chem. Soc.*, 1971, **48**, 288 (*isol*)
Singh, J.K. *et al*, *J. Indian Chem. Soc.*, 1976, **53**, 1162 (*isol*)
Dhav, D.N. *et al*, *Planta Med.*, 1976, **29**, 91 (*isol*)

11-Hentriacontene H-00028

[69086-62-0]

$H_3C(CH_2)_{18}CH{=}CH(CH_2)_9CH_3$

$C_{31}H_{62}$ M 434.831

(Z)-form [66648-68-8]

Mating stimulant pheromone and cuticular lipid constit. for the fly *Fannia femoralis*.

Uebel, E.C. *et al*, *J. Chem. Ecol.*, 1978, **4**, 73, 83 (*synth*)

7-Hentriaconten-16-one H-00029

Palmitenone

$H_3C(CH_2)_5CH{=}CH(CH_2)_7CO(CH_2)_{14}CH_3$

$C_{31}H_{60}O$ M 448.814

(Z)-form

Produced by *Corynebacterium diphtheriae*. Cryst. Mp 40°.

Pudles, J. *et al*, *Biochim. Biophys. Acta*, 1953, **11**, 602 (*isol*)
Pudles, J. *et al*, *Bull. Soc. Chim. Biol.*, 1954, **36**, 759.

5,9-Heptacosadiene H-00030

[96314-01-1]

$H_3C(CH_2)_{16}CH{=}CHCH_2CH_2CH{=}CH(CH_2)_3CH_3$

$C_{27}H_{52}$ M 376.708

Constit. of the cuticle of *Drosophila melanogaster* and *D. pseudoobscura*. (*Z,Z*)-isomer predominates.

Blomquist, G.J. *et al*, *Insect Biochem.*, 1985, **15**, 25 (*isol*)
Jallon, J.M. *et al*, *C. R. Hebd. Seances Acad. Sci. Ser. 2*, 1989, **309**, 1551 (*isol*)

7,11-Heptacosadiene H-00031

$H_3C(CH_2)_5CH{=}CHCH_2CH_2CH{=}CH(CH_2)_{14}CH_3$

$C_{27}H_{52}$ M 376.708

(Z,Z)-form

Major sex pheromone of the fruit fly *Drosophila melanogaster*.

Antony, C. *et al*, *Insects Physiol.*, 1982, **28**, 873.
Davis, T.L. *et al*, *Synthesis*, 1989, 936 (*synth*)

12,14-Heptacosadiynoic acid H-00032

[106510-41-2]

$H_3C(CH_2)_{11}C{\equiv}CC{\equiv}C(CH_2)_{10}COOH$

$C_{27}H_{46}O_2$ M 402.659

Solid. Mp 62-63°.

Singh, Z. *et al*, *Synth. Commun.*, 1986, **16**, 847 (*synth*)
Xu, Z. *et al*, *J. Org. Chem.*, 1991, **56**, 7183 (*synth, pmr*)

Heptacosane, 9CI H-00033

[593-49-7]

$H_3C(CH_2)_{25}CH_3$

$C_{27}H_{56}$ M 380.739

Constit. of tobacco oil and many waxes. Pearly leaflets (EtOAc). Spar. sol. Et_2O, insol. EtOH. Mp 59.5°. Bp 422°, Bp_{15} 270°.

Glund, W., *Ber.*, 1919, **52**, 1051.
Lipp, A. *et al*, *J. Prakt. Chem.*, 1919, **99**, 256.
Karrer, W. *et al*, *Konstitution und Vorkommen der Organischen Pflanzenstoffe*, 2nd Ed., Birkhäuser Verlag, Basel, 1972-1985, no. 14 (*occur*)

Heptacosanoic acid, 9CI H-00034

[7138-40-1]

$H_3C(CH_2)_{25}COOH$

$C_{27}H_{54}O_2$ M 410.722

Present as ester in Chinese and Montan waxes. Mp 82°.

Me ester:

$C_{28}H_{56}O_2$ M 424.749

Mp 64°.

Gascard, A., *C. R. Hebd. Seances Acad. Sci.*, 1920, **170**, 1328 (*isol*)
Tropsch, H. *et al*, *CA*, 1922, **16**, 2111 (*isol*)

1-Heptacosanol, 9CI H-00035

[2004-39-9]

$H_3C(CH_2)_{25}CH_2OH$

$C_{27}H_{56}O$ M 396.739

Constit. of peel and pulp of *Citullus colocynthis*. Mp 81.5° (76°).

Ac: [1873-38-7].

$C_{29}H_{58}O_2$ M 438.776

Mp 62-64°.

Piper, S.H. *et al*, *Biochem. J.*, 1934, **28**, 2175 (*synth*)
El Khadem, H. *et al*, *J. Chem. Soc.*, 1965, 3488 (*isol, ir, pmr*)

9-Heptacosene H-00036

[69188-49-4]

$H_3C(CH_2)_7CH{=}CH(CH_2)_{16}CH_3$

$C_{27}H_{54}$ M 378.724

(Z)-form [36258-12-5]

Constit. of cuticular lipids of various insect spp.

Uebel, E.C. *et al*, *J. Chem. Ecol.*, 1977, **3**, 269 (*isol, glc, ir, nmr, ms*)
Prestwich, G.D. *et al*, *J. Chem. Ecol.*, 1977, **3**, 579 (*isol, glc, ir, nmr, ms*)

13-Heptacosene H-00037

[61906-95-4]

$H_3C(CH_2)_{12}CH{=}CH(CH_2)_{11}CH_3$

$C_{27}H_{54}$ M 378.724

(Z)-form [54863-75-1]

Pheromone for *Musca autumnalis*.

Küpper, F. *et al*, *Z. Naturforsch.*, *B*, 1976, **31**, 1256 (*synth*)
Subramanian, G.B.V. *et al*, *Tetrahedron*, 1986, **42**, 3967 (*synth, ir, pmr*)

20-Heptacosenoic acid H-00038

$H_3C(CH_2)_5CH{=}CH(CH_2)_{18}COOH$

$C_{27}H_{52}O_2$ M 408.707

(Z)-form [140197-98-4]

Constit. of the sponge *Amphimedon compressa*.

Carballeira, N.M. *et al*, *J. Nat. Prod. (Lloydia)*, 1992, **55**, 333.

1,8-Heptadecadiene, 9CI H-00039

Tetrahydroaplotaxene

[58045-14-0]

$H_3C(CH_2)_7CH{=}CH(CH_2)_5CH{=}CH_2$

$C_{17}H_{32}$ M 236.440

(Z)-form [56134-02-2]
Constit. of roots of *Centaurea scabiosa* and *Cirsium japonicum*, of oakmoss essential oil and flour-beetle volatiles.

Suzuki, T. *et al*, *Agric. Biol. Chem.*, 1975, **39**, 2207 (*pmr, ms, ir*)
Yano, K., *Phytochemistry*, 1977, **16**, 263.

1,12-Heptadecadiene H-00040

[81500-71-2]

$H_3C(CH_2)_3CH{=}CH(CH_2)_9CH{=}CH_2$

$C_{17}H_{32}$ M 236.440

(Z)-form
Shows strong sex pheromone activity for *Heliothis zea*.

Carlson, D.A. *et al*, *Experientia*, 1982, **38**, 309.

6,9-Heptadecadiene, 9CI H-00041

[81265-03-4]

$H_3C(CH_2)_6CH{=}CHCH_2CH{=}CH(CH_2)_4CH_3$

$C_{17}H_{32}$ M 236.440
Found in Australian formicine ants.

(6Z,9Z)-form [133410-67-0]
Constit. of *Dermatophagoides* sp.

[132833-56-8, 132833-58-0]

Brophy, J.J. *et al*, *Insect Biochem.*, 1982, **12**, 215 (*isol*)
Hutzinger, M.W. *et al*, *J. Org. Chem.*, 1991, **56**, 2918 (*synth, isomers, pmr, cmr, ms*)

2,9-Heptadecadiene-4,6-diyne, 8CI H-00042

$H_3C(CH_2)_6CH{=}CHCH_2C{\equiv}CC{\equiv}CCH{=}CHCH_3$

$C_{17}H_{24}$ M 228.377

(2E,6Z)-form [20311-60-8]
Constit. of *Oenanthe crocata*. $Bp_{0.001}$ 110°.

Bohlmann, F. *et al*, *Chem. Ber.*, 1968, **101**, 1163 (*uv*); 1971, **104**, 1362 (*synth*)

1,8-Heptadecadiene-4,6-diyne-3,10-diol, 9CI H-00043

[63910-76-9]

$H_2C{=}CHCH(OH)C{\equiv}CC{\equiv}CCH{=}CHCH(OH)(CH_2)_6CH_3$

$C_{17}H_{24}O_2$ M 260.375
Isol. from roots and rhizomes of *Aegopodium podagraria* and *Pituranthos tortuosus*. Also from *Panax quinquefolium*.

(+)-(Z)-form [132075-01-5] ***Seselidiol***
Isol. from the roots of *Seseli mairei*. Shows cytotoxic activity. Yellowish oil. $[\alpha]_D^{23}$ +192° (c, 0.53 in Et_2O).

Di-Ac: [131989-87-2].
$C_{21}H_{28}O_4$ M 344.450
Yellowish oil.

Schulte, K.E. *et al*, *Arch. Pharm.* (*Weinheim, Ger.*), 1977, **310**, 285, 945 (*isol, struct, uv, ir, ms, pmr*)
Shim, S.C. *et al*, *Phytochemistry*, 1987, **26**, 2849 (*isol, pmr, cmr*)
Hu, C.-Q. *et al*, *J. Nat. Prod.* (*Lloydia*), 1990, **53**, 932 (*isol, struct*)
Fujimoto, Y. *et al*, *Chem. Pharm. Bull.*, 1991, **39**, 521 (*isol*)

1,9-Heptadecadiene-4,6-diyne-3,8-diol, 9CI H-00044

[30779-95-4]

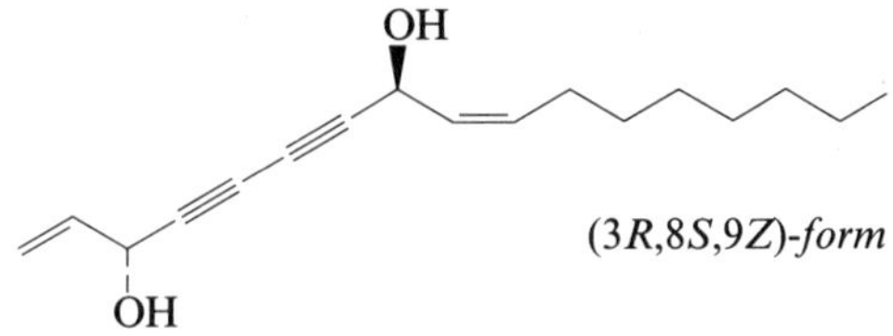

(3*R*,8*S*,9*Z*)-*form*

$C_{17}H_{24}O_2$ M 260.375

(3R,8S,9Z)-form [55297-87-5] ***Falcarindiol***
Constit. of roots of several plants including the common carrot *Daucus carota*. Has antifungal and analgesic activity.

3-Ac: [22526-53-0].
$C_{19}H_{26}O_3$ M 302.413
Constit. of *D. carota*.

8-Ac: [55249-27-9].
$C_{19}H_{26}O_3$ M 302.413
Constit. of *Centella* spp.

3-Ketone: [18089-23-1]. *8-Hydroxy-1,9-heptadecadien-4,6-diyn-3-one*. ***Falcarinolone***
$C_{17}H_{22}O_2$ M 258.360
Isol. from *Carum carvi*, *Ligusticum mucronatum*, *Aegopodium*, *Angelica* spp . and *D. carota*. Unstable pale yellow oil. $[\alpha]_D^{20}$ +255° (c, 5.1 in Et_2O).

3-Ketone, 8-Ac: [55249-28-0]. *8-Acetoxy-1,9-heptadecadien-4,6-diyn-3-one*
$C_{19}H_{24}O_3$ M 300.397
Isol. from *Centella* spp. and *Pituranthos tortuosus*.

Diketone: [65892-84-4]. *1,9-Heptadecadiene-4,6-diyne-3,8-dione*. ***Falcarindione***
$C_{17}H_{20}O_2$ M 256.344
Isol. from roots of *Carum carvi*. Unstable yellowish oil.

(3R*,8R*)-form

9,10-Epoxide: ***9,10-Epoxy-1-heptadecene-4,6-diyne-3,8-diol.*** *PQ6*
$C_{17}H_{24}O_3$ M 276.375
Constit. of *Panax quinquefolium*. Oil. $[\alpha]_D$ −27° (c, 1.0 in $CHCl_3$).

Bohlmann, F. *et al*, *Chem. Ber.*, 1961, **94**, 958, 3189; 1967, **100**, 3450, 3454; 1971, **104**, 2033; 1975, **108**, 511 (*isol, derivs*)
Bentley, R.K. *et al*, *J. Chem. Soc. C*, 1969, 685 (*isol, ir, uv, pmr*)
Schulte, K.E. *et al*, *Arch. Pharm.* (*Weinheim, Ger.*), 1977, **310**, 945 (*synth, ir, uv, pmr*)
Garrod, B. *et al*, *Physiol. Plant Pathol.*, 1978, **13**, 241 (*ir, uv, pmr, ms*)
Kemp, M.S., *Phytochemistry*, 1978, **17**, 1002 (*isol, props*)
Lemmich, E., *Phytochemistry*, 1981, **20**, 1419 (*abs config*)
Fujimoto, Y. *et al*, *Phytochemistry*, 1992, **31**, 3499 (*PQ6*)

2,8-Heptadecadiene-4,6-diyne-1,10-diol, 8CI H-00045

$H_3C(CH_2)_6CH(OH)CH{=}CHC{\equiv}CC{\equiv}CCH{=}CHCH_2OH$

$C_{17}H_{24}O_2$ M 260.375

(all-E)-form [18229-50-0]
Constit. of *Oenanthe crocata*. Mp 60°.

Bohlmann, F. *et al*, *Chem. Ber.*, 1968, **101**, 525; 1971, **104**, 1362 (*synth*)

2,9-Heptadecadiene-4,6-diyne-1,8-diol, 8CI **H-00046**

$H_3C(CH_2)_6CH{=}CHCH(OH)C{\equiv}CC{\equiv}CCH{=}CHCH_2OH$

$C_{17}H_{24}O_2$ M 260.375

(all-Z)-form [32768-89-1]

Constit. of *Azorella trifurcata*. Oil.

1-Ac: [32768-88-0].

$C_{19}H_{26}O_3$ M 302.413

Constit. of *A. trifurcata*. Oil.

Bohlmann, F. *et al*, *Chem. Ber.*, 1971, **104**, 1322.

8,10-Heptadecadiene-4,6-diyne-1,12-diol, 8CI **H-00047**

$H_3C(CH_2)_4CH(OH)$
$CH{=}CHCH{=}CHC{\equiv}CC{\equiv}CCH_2CH_2CH_2OH$

$C_{17}H_{24}O_2$ M 260.375

(all-E)-form [25525-19-3]

Constit. of roots and rhizomes of *Cicuta virosa*. Oil.

Bohlmann, F. *et al*, *Chem. Ber.*, 1969, **102**, 3293.

1,16-Heptadecadiene-4,6-diyne-3,9,10-triol, 9CI **H-00048**

Ginsenoyne C

[139163-35-2]

$H_2C{=}CHCH(OH)C{\equiv}CC{\equiv}CCH_2CH(OH)CH(OH)$
$(CH_2)_5CH{=}CH_2$

$C_{17}H_{24}O_3$ M 276.375

Isol. from the root of *Panax ginseng*. Amorph. powder. $[\alpha]_D$ −16.5° (c, 1.5 in $CHCl_3$). λ_{max} 201, 231, 243, 257 nm (EtOH).

Tri-Ac: [139035-33-9].

Oil. $[\alpha]_D$ +13.3° (c, 1.49 in $CHCl_3$). λ_{max} 200, 220, 232, 244, 258 nm (EtOH).

Acetonide: Oil. $[\alpha]_D$ −21.2° (c, 1.26 in $CHCl_3$). λ_{max} 200, 229, 242, 257 nm (EtOH).

Hirakura, K. *et al*, *Phytochemistry*, 1991, **30**, 3327 (*isol, ir, uv, ms, pmr, cmr*)

1,9-Heptadecadiene-4,6-diyn-3-ol **H-00049**

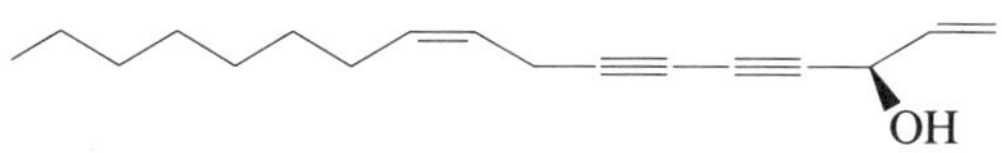

$C_{17}H_{24}O$ M 244.376

(R)-(Z)-form [21852-80-2] ***Falcarinol.*** *Carotatoxin. Carrotatoxin. Falcalinol*

Constit. of *Falcaria vulgaris* and some other plants. $Bp_{0.02}$ 115°. $[\alpha]_D^{20}$ −22.5° (c, 1.0 in Et_2O). n_D^{18} 1.5018.

▷ Toxic.

(S)-(Z)-form [81203-57-8] ***Panaxynol***

Isol. from *Panax ginseng* and *Saposhnikovia divaricata*.

Bohlmann, F. *et al*, *Chem. Ber.*, 1965, **98**, 3010 (*synth*)
Bohlmann, F. *et al*, *Chem. Ber.*, 1966, **99**, 3552 (*isol*)
Crosby, D.G. *et al*, *Tetrahedron*, 1967, **23**, 465 (*ir, uv*)
Larsen, P.K. *et al*, *Acta Chem. Scand.*, 1969, **23**, 2552 (*abs config*)
Bentley, R.K. *et al*, *J. Chem. Soc. C*, 1969, 685 (*isol, ir, uv, nmr*)
Shim, S.C. *et al*, *Tetrahedron Lett.*, 1985, **26**, 5775 (*abs config*)
Terada, A. *et al*, *Bull. Chem. Soc. Jpn.*, 1989, **62**, 2977 (*isol, bibl*)
Fujimoto, Y. *et al*, *Chem. Pharm. Bull.*, 1991, **39**, 521 (*isol*)

2,9-Heptadecadiene-4,6-diyn-1-ol, 8CI **H-00050**

[22531-01-7]

$H_3C(CH_2)_6CH{=}CHCH_2C{\equiv}CC{\equiv}CCH{=}CHCH_2OH$

$C_{17}H_{24}O$ M 244.376

(2E,9Z)-form [18836-70-9]

Constit. of *Oenanthe crocata*. Oil.

Ac: [18836-71-0].

$C_{19}H_{26}O_2$ M 286.413

Constit. of *O. crocata*.

(all-Z)-form [63767-82-8]

Constit. of *Bupleurum ranunculoides* and *Pituranthos tortuosus*.

Ac: [65892-82-2].

Constit. of *P. tortuosus*.

Bohlmann, F. *et al*, *Chem. Ber.*, 1968, **101**, 525 (*isol, uv*); 1971, **104**, 1362, 2030 (*synth*)
Schulte, K., *Arch. Pharm. (Weinheim, Ger.)*, 1977, **310**, 285 (*pmr, ms, ir, uv*)

2,9-Heptadecadiene-4,6-diyn-8-ol, 9CI **H-00051**

$H_3C(CH_2)_6CH{=}CHCH(OH)C{\equiv}CC{\equiv}CCH{=}CHCH_3$

$C_{17}H_{24}O$ M 244.376

(all-E)-form [32768-72-2]

Constit. of *Oenanthe crocata*.

(2Z,9E)-form [50816-77-8]

Constit. of roots of *Heracleum mantegazzianum*.

(all-Z)-form [32768-91-5]

Constit. of *Heracleum sphondylium*. Oil.

Bohlmann, F. *et al*, *Chem. Ber.*, 1971, **104**, 1322, 1362.
Vanhaelen-Fastré, R. *et al*, *Phytochemistry*, 1973, **12**, 2687.

8,15-Heptadecadiene-11,13-diyn-1-ol, 9CI **H-00052**

$H_3CCH{=}CHC{\equiv}CC{\equiv}CCH_2CH{=}CH(CH_2)_6CH_2OH$

$C_{17}H_{24}O$ M 244.376

Ac: [83905-71-9].

$C_{19}H_{26}O_2$ M 286.413

Isol. from *Pleiotaxis rugosa*.

Bohlmann, F. *et al*, *Phytochemistry*, 1982, **21**, 1434 (*isol*)

9,15-Heptadecadiene-11,13-diyn-8-ol, 8CI **H-00053**

$H_3C(CH_2)_6CH(OH)CH{=}CHC{\equiv}CC{\equiv}CCH{=}CHCH_3$

$C_{17}H_{24}O$ M 244.376

(all-E)-form [32768-70-0]

Constit. of *Oenanthe crocata*. Oil.

Bohlmann, F. *et al*, *Chem. Ber.*, 1968, **101**, 1163 (*isol, uv*); 1971, **104**, 1362 (*synth*)

9,16-Heptadecadiene-4,6-diyn-3-ol, 8CI **H-00054**

$H_2C{=}CH(CH_2)_5CH{=}CHCH_2C{\equiv}CC{\equiv}CCH(OH)CH_2CH_3$

$C_{17}H_{24}O$ M 244.376

(Z)-form [24587-16-4]

Constit. of *Coreopsis gigantea* and aerial parts of *Diotis maritima*. Oil.

9,10-Epoxide, Ac: [142465-54-1]. ***Ginsenoyne H***

$C_{19}H_{26}O_3$ M 302.413

Isol. from the roots of *Panax ginseng*. Oil. $[\alpha]_D$ −208.9° (c, 2.27 in $CHCl_3$).

Bohlmann, F. *et al*, *Chem. Ber.*, 1969, **102**, 1691.
de Pascual, T.J., *An. Quim.*, 1977, **73**, 1525; *CA*, **89**, 176303c.
Hirakura, K. *et al*, *Phytochemistry*, 1991, **30**, 4053 (*Ginsenoyne H*)

1,9-Heptadecadiene-4,6-diyn-3-one H-00055

$H_3C(CH_2)_6CH{=}CHCH_2C{\equiv}CC{\equiv}CCOCH{=}CH_2$

$C_{17}H_{22}O$ M 242.360

(Z)-form [4117-11-7] ***Falcarinone***
Isol. from *Falcaria vulgaris*, *Oenanthe* spp. *Sium sisarum*, *Chaerophyllum temulum*, *Eryngium planum*, *Galinsoga paviflora*, *Hedera helix* etc. Pale-yellow oil.
9,10-Epoxide: see *Panaxydol, P-00003*

Bohlmann, F. *et al*, *Chem. Ber.*, 1961, **94**, 958; 1962, **95**, 1320; 1965, **98**, 3010.

9,16-Heptadecadiene-4,6-diyn-3-one H-00056

$H_2C{=}CH(CH_2)_5CH{=}CHCH_2C{\equiv}CC{\equiv}CCOCH_2CH_3$

$C_{17}H_{22}O$ M 242.360

(Z)-form [24587-18-6]
Constit. of *Coreopsis gigantea*. Oil.

Bohlmann, F. *et al*, *Chem. Ber.*, 1969, **102**, 1691.

8,10-Heptadecadienoic acid H-00057

$H_3C(CH_2)_5CH{=}CHCH{=}CH(CH_2)_6COOH$

$C_{17}H_{30}O_2$ M 266.423

(all-E)-form [5375-38-2]

Crundwell, E. *et al*, *J. Med. Chem.*, 1972, **15**, 754 (*synth*)

10,12-Heptadecadienoic acid H-00058

$H_3C(CH_2)_3CH{=}CHCH{=}CH(CH_2)_8COOH$

$C_{17}H_{30}O_2$ M 266.423

(10E,12Z)-form
Me ester: [63024-92-0].
$C_{18}H_{32}O_2$ M 280.450
$Bp_{0.03}$ 121-122°.

Bestmann, H.J. *et al*, *Justus Liebigs Ann. Chem.*, 1981, 2117 (*synth, pmr*)

9,11-Heptadecadien-1-ol H-00059

$H_3C(CH_2)_4CH{=}CHCH{=}CH(CH_2)_7CH_2OH$

$C_{17}H_{32}O$ M 252.439

(9E,11Z)-form [80625-58-7]
$Bp_{0.01}$ 120-125°.
Ac: [80625-71-4].
$C_{19}H_{34}O_2$ M 294.476
$Bp_{0.01}$ 120-125°.

Bestmann, H.J. *et al*, *Justus Liebigs Ann. Chem.*, 1981, 2117 (*synth, pmr*)

10,12-Heptadecadien-1-ol H-00060

$H_3C(CH_2)_3CH{=}CHCH{=}CH(CH_2)_8CH_2OH$

$C_{17}H_{32}O$ M 252.439

(10E,12Z)-form [63024-97-5]
$Bp_{0.01}$ 122-130°.
Ac: [63025-07-0].
$C_{19}H_{34}O_2$ M 294.476
$Bp_{0.01}$ 132°.

Bestmann, H.J. *et al*, *Justus Liebigs Ann. Chem.*, 1981, 2117 (*synth, pmr*)

11,13-Heptadecadien-1-ol H-00061

$H_3CCH_2CH_2CH{=}CHCH{=}CH(CH_2)_9CH_2OH$

$C_{17}H_{32}O$ M 252.439

(11E,13Z)-form [80625-63-4]
$Bp_{0.01}$ 120-125°.
Ac: [80625-75-8].
$C_{19}H_{34}O_2$ M 294.476
$Bp_{0.01}$ 120-125°.

Bestmann, H.J. *et al*, *Justus Liebigs Ann. Chem.*, 1981, 2117 (*synth, pmr*)

10,16-Heptadecadien-8-ynoic acid H-00062

$H_2C{=}CH(CH_2)_4CH{=}CHC{\equiv}C(CH_2)_6COOH$

$C_{17}H_{26}O_2$ M 262.391

(E)-form
Constit. of the seed oil of *Acanthosyris spinescens*.
Me ester:
$C_{18}H_{28}O_2$ M 276.418
Oil.

Powell, R.G. *et al*, *Biochemistry*, 1966, **5**, 625.

5,8-Heptadecadiynoic acid, 8CI H-00063

[22117-03-9]

$H_3C(CH_2)_7C{\equiv}CCH_2C{\equiv}C(CH_2)_3COOH$

$C_{17}H_{26}O_2$ M 262.391
Cryst. Mp 45.5-46.5°.

Marcel, Y.L. *et al*, *Chem. Phys. Lipids*, 1968, **2**, 173 (*synth*)

6,9-Heptadecadiynoic acid, 8CI H-00064

[22117-04-0]

$H_3C(CH_2)_6C{\equiv}CCH_2C{\equiv}C(CH_2)_4COOH$

$C_{17}H_{26}O_2$ M 262.391
Mp 15-17°.

Marcel, Y.L. *et al*, *Chem. Phys. Lipids*, 1968, **2**, 173 (*synth*)

7,10-Heptadecadiynoic acid, 8CI H-00065

[22117-05-1]

$H_3C(CH_2)_5C{\equiv}CCH_2C{\equiv}C(CH_2)_5COOH$

$C_{17}H_{26}O_2$ M 262.391
Cryst. Mp 35-37°.

Marcel, Y.L. *et al*, *Chem. Phys. Lipids*, 1968, **2**, 173 (*synth*)

Heptadecanal H-00066

[629-90-3]

$H_3C(CH_2)_{15}CHO$

$C_{17}H_{34}O$ M 254.455
Isol. from lemon oil (*Citrus limon*). Cryst. (pet. ether). Mp 36°. Bp_{26} 204°.
Oxime: [24435-10-7].
$C_{17}H_{35}NO$ M 269.470
Mp 89.5°.
Semicarbazone: Mp 108°.

Ikeda, R.M. *et al*, *J. Agric. Food Chem.*, 1962, **10**, 98 (*isol*)
Rao, P.V. *et al*, *J. Lipid Res.*, 1967, **8**, 380 (*synth, glc*)
Org. Synth., 1971, **51**, 39 (*synth*)

Heptadecane H-00067

[629-78-7]

$H_3C(CH_2)_{15}CH_3$

$C_{17}H_{36}$ M 240.471

Occurs in plant waxes, e.g. rose wax and in wood of *Manilkara bidentata*. Widespread in biol. materials. Constit. of Dufour's gland secretion of the ant *Harpagoxenus sublaevis*. Mp 22°. Bp 303°, Bp_{10} 162°.

Schmidt, A.W. *et al*, *Ber.*, 1941, **74**, 1313 (*synth*)
Cocker, W. *et al*, *J. Chem. Soc.*, 1963, 677 (*isol*)
Ollett, D.G. *et al*, *Z. Naturforsch., C*, 1987, **42**, 140.

1,14-Heptadecanediol H-00068

$H_3CCH_2CH_2CH(OH)(CH_2)_{12}CH_2OH$

$C_{17}H_{36}O_2$ M 272.470

Opt. active-form

Prepared by hydrogenation of Oenanthotoxin or Cicutoxin. Cryst. ($CHCl_3$/pet. ether). Mp 71°. $[\alpha]_D^{12}$ 0° ± 1° (c, 2.0 in MeOH).

(±)-***form*** [16339-74-5]

Cryst. ($CHCl_3$/pet. ether). Mp 64-65°.

Anet, E.F.L.J. *et al*, *J. Chem. Soc.*, 1953, 309 (*synth*)
Lüning, B. *et al*, *Acta Chem. Scand.*, 1967, **21**, 829 (*synth, ms*)

1,2,3-Heptadecanetricarboxylic acid, 9CI H-00069

Norrangiformic acid

[26534-85-0]

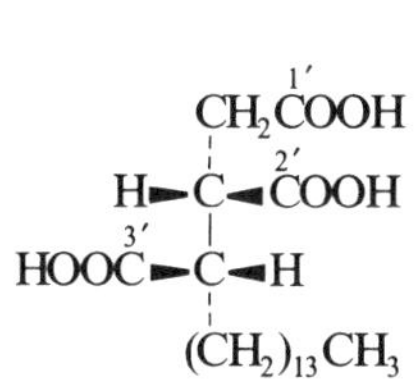

$C_{20}H_{36}O_6$ M 372.501

Cryst. (MeOH). Mp 105-108°. $[\alpha]_D^{24}$ +75° (c, 1.87 in MeOH).

2′-Me ester: ***Isorangiformic acid***

$C_{21}H_{38}O_6$ M 386.528

Constit. of *Lecanora stenotropa*. Cryst. Mp 83-85°. $[\alpha]_D^{24}$ +5.2° (c, 2.5 in MeOH).

2′-Me ester, 1′,3′-cyclic anhydride:

$C_{21}H_{36}O_5$ M 368.512

Cryst. (hexane). Mp 92-94°. $[\alpha]_D^{24}$ +13.4° (c, 1.74 in $CHCl_3$).

1′- or 3′-Me ester: ***Rangiformic acid***

$C_{21}H_{38}O_6$ M 386.528

Metab. of *Cladonia rangiformis*, *C. mitis* and *Lecanora polytropa*. Needles (C_6H_6). Mp 106°. $[\alpha]_D^{24}$ +16.2° (EtOH).

Tri-Me ester:

$C_{23}H_{42}O_6$ M 414.581

Cryst. (MeOH). Mp 48-49°. $[\alpha]_D^{24}$ +9.6° (c, 1.7 in $CHCl_3$).

Asahina, Y. *et al*, *Bull. Chem. Soc. Jpn.*, 1942, **17**, 495 (*isol*)
Huneck, S., *Z. Naturforsch., B*, 1966, **21**, 888 (*isol*)
Huneck, S., *Phytochemistry*, 1982, **21**, 2407 (*isol, struct*)

Heptadecanoic acid H-00070

Margaric acid. Daturinic acid

[506-12-7]

$H_3C(CH_2)_{15}COOH$

$C_{17}H_{34}O_2$ M 270.454

Constit. of *Erythrina crista-galli* trunkwood and bark. Common constit. of lipids. Used (with octadecanoic acid) for amino acid sequencing in peptides. Cryst. (pet. ether). Mp 62-63°.

▷ LD_{50} 36 mg/kg (mouse, i.v.). MI3850000.

Me ester: [1731-92-6].

$C_{18}H_{36}O_2$ M 284.481

Cryst. (EtOH). Mp 30°. Bp_9 184-187°.

Et ester: [14010-23-2].

$C_{19}H_{38}O_2$ M 298.508

Cryst. (EtOH aq.). Mp 28°. Bp_5 185°.

Amide: [25844-13-7].

$C_{17}H_{35}NO$ M 269.470

Cryst. (EtOH). Mp 108°.

Nitrile:

$C_{17}H_{33}N$ M 251.454

Cryst. (EtOH). Mp 34°. Bp_{10} 185°.

Bricas, E. *et al*, *Biochemistry*, 1965, **4**, 2254 (*use*)
Buchta, E. *et al*, *Justus Liebigs Ann. Chem.*, 1966, **698**, 93 (*synth*)
Imamura, H. *et al*, *Gigu Daigaku Nogakubu Kenkyu Hokoku*, 1981, 77; *CA*, **96**. 196524y (*occur*)
Bengsch, E. *et al*, *J. Magn. Reson.*, 1986, **68**, 1 (*cmr*)
Lewis, R.J., *Sax's Dangerous Properties of Industrial Materials*, 8th Ed., Van Nostrand-Reinhold, 1992, HAS500.

1-Heptadecanol H-00071

Heptadecyl alcohol. 1-Hydroxyheptadecane

[1454-85-9]

$H_3C(CH_2)_{15}CH_2OH$

$C_{17}H_{36}O$ M 256.471

Cryst. (Me_2CO). Mp 54°. Bp 308°.

4-Nitrobenzyl: Mp 54°.

Watanabe, A., *Bull. Chem. Soc. Jpn.*, 1954, **32**, 1295 (*synth*)

2-Heptadecanol H-00072

[16813-18-6]

$H_3C(CH_2)_{14}CH(OH)CH_3$

$C_{17}H_{36}O$ M 256.471

(±)-***form***

Cryst. Mp 34-35°, Mp 54°. $Bp_{0.5}$ 140°.

Gascard, A., *Justus Liebigs Ann. Chem.*, 1921, **15**, 332.
Kuhn, R. *et al*, *Hoppe Seyler's Z. Physiol. Chem.*, 1936, **242**, 171.
Chang, N.C. *et al*, *Biochim. Biophys. Acta*, 1973, **306**, 437.

3-Heptadecanol H-00073

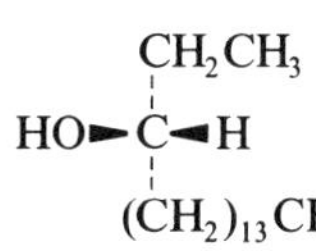

(*R*)-*form*

$C_{17}H_{36}O$ M 256.471

(***R***)-***form*** [26432-14-4]

Constit. of umbelliferous plants. Mp 54-55°. $[\alpha]_D^{18.2}$ −5.7° (c, 1.4 in $CHCl_3$).

(***S***)-***form*** [26432-15-5]

Constit. of umbelliferous plants. Mp 54-55°. $[\alpha]_D^{18.2}$ +5.8° (c, 1.4 in $CHCl_3$).

(±)-form
Mp 45.6-47.5°.

Breusch, F.L. *et al*, *Chem. Ber.*, 1953, **86**, 678 (*synth*)
Larsen, P.K. *et al*, *Acta Chem. Scand.*, 1969, **23**, 2552 (*isol, abs config*)

4-Heptadecanol, 9CI H-00074

$H_3C(CH_2)_{12}CH(OH)CH_2CH_2CH_3$

$C_{17}H_{36}O$ M 256.471
(±)-form
Mp 44.7-46.1°, Mp 53°.

Anet, E. *et al*, *Chem. Ind. (London)*, 1952, 757.
Breusch, F.L. *et al*, *Chem. Ber.*, 1953, **86**, 678 (*synth*)

5-Heptadecanol, 9CI H-00075

$H_3C(CH_2)_{11}CH(OH)(CH_2)_3CH_3$

$C_{17}H_{36}O$ M 256.471
(±)-form
Mp 45.8-46.7°.

Breusch, F.L. *et al*, *Chem. Ber.*, 1953, **86**, 678 (*synth*)

6-Heptadecanol, 9CI H-00076

$H_3C(CH_2)_{10}CH(OH)(CH_2)_4CH_3$

$C_{17}H_{36}O$ M 256.471
(±)-form
Mp 46.4-47.4°.

Breusch, F.L. *et al*, *Chem. Ber.*, 1953, **86**, 678 (*synth*)

7-Heptadecanol, 9CI H-00077

$H_3C(CH_2)_9CH(OH)(CH_2)_5CH_3$

$C_{17}H_{36}O$ M 256.471
(±)-form
Mp 47.2-48.9°.

Breusch, F.L., *Chem. Ber.*, 1953, **86**, 678.

8-Heptadecanol, 9CI H-00078

[2541-75-5]

$H_3C(CH_2)_8CH(OH)(CH_2)_6CH_3$

$C_{17}H_{36}O$ M 256.471
Mp 52.5-53°.

Breusch, F.L., *Chem. Ber.*, 1953, **86**, 678.

9-Heptadecanol H-00079

[624-08-8]

$H_3C(CH_2)_7CH(OH)(CH_2)_7CH_3$

$C_{17}H_{36}O$ M 256.471
Cryst. (EtOH aq.). Mp 61°.

Kipping, F.S., *J. Chem. Soc.*, 1893, **63**, 452.
Breusch, F.L. *et al*, *Chem. Ber.*, 1953, **86**, 678 (*synth*)
Spiteller, G., *Monatsh. Chem.*, 1968, **99**, 1839.
Francis, G.W. *et al*, *Acta Chem. Scand., Ser. B*, 1977, **31**, 721.

2-Heptadecanone H-00080

[2922-51-2]

$H_3C(CH_2)_{14}COCH_3$

$C_{17}H_{34}O$ M 254.455
Constit. of aroma compounds of white bread. Cryst. Mp 48°. Bp 319-320°, Bp_{110} 246°.

Keskin, H., *CA*, 1946, **40**, 1822[7] (*synth*)
Bente, P.F. *et al*, *J. Phys. Chem.*, 1975, **79**, 713 (*synth, ms*)
Obretenov, T. *et al*, *Z. Lebensm.-Unters. -Forsch.*, 1977, **165**, 195 (*isol, glc, ms*)
Sax, N.I., *Dangerous Properties of Industrial Materials*, 5th Ed., Van Nostrand-Reinhold, 1979, 828.

9-Heptadecanone H-00081

Dioctyl ketone
[540-08-9]

$H_3C(CH_2)_7CO(CH_2)_7CH_3$

$C_{17}H_{34}O$ M 254.455
Cryst. (MeOH). Mp 53°. Bp 250-253°.
Oxime:
$C_{17}H_{35}NO$ M 269.470
Mp 11-12°.
4-Nitrophenylhydrazone: Mp 54°.

Briese, R.R. *et al*, *J. Am. Chem. Soc.*, 1933, **55**, 1697 (*synth*)
Pelter, A. *et al*, *J. Chem. Soc., Perkin Trans. 1*, 1975, 129 (*synth*)
Reznikov, S.A. *et al*, *Zh. Fiz. Khim.*, 1975, **49**, 356.
Savoia, D. *et al*, *J. Org. Chem.*, 1978, **43**, 2907 (*synth*)

2,4,8,10,16-Heptadecapentaen-6-yn-1-al H-00082

$H_2C{=}CH(CH_2)_4CH{=}CHCH{=}CH$
$C{\equiv}CCH{=}CHCH{=}CHCHO$

$C_{17}H_{20}O$ M 240.344
(all-E)-form [29739-54-6]
Constit. of *Tridax trilobata*. Oil.

Bohlmann, F. *et al*, *Justus Liebigs Ann. Chem.*, 1970, **739**, 135 (*isol*)

1,3,7,9,13-Heptadecapentaen-11-yne H-00083

$H_3CCH_2CH_2CH{=}CHC{\equiv}CCH{=}CH$
$CH{=}CHCH_2CH_2CH{=}CHCH{=}CH_2$

$C_{17}H_{22}$ M 226.361
(all-E)-form [22644-33-3]
Constit. of *Aster schreberi*. Oil.

Bohlmann, F. *et al*, *Chem. Ber.*, 1969, **102**, 1037 (*isol*)

1,7,9,13,15-Heptadecapentaen-11-yne H-00084

$H_3CCH{=}CHCH{=}CHC{\equiv}CCH{=}CH$
$CH{=}CH(CH_2)_4CH{=}CH_2$

$C_{17}H_{22}$ M 226.361
Constit. of *Dahlia merckii*. Oil.

Neuschild, K. *et al*, *Phytochemistry*, 1992, **31**, 4353 (*isol, pmr*)

2,4,8,10,16-Heptadecapentaen-6-yn-1-ol H-00085

$H_2C{=}CH(CH_2)_4CH{=}CHCH{=}CH$
$C{\equiv}CCH{=}CHCH{=}CHCH_2OH$

$C_{17}H_{22}O$ M 242.360

(all-E)-form [29739-59-1]
Constit. of *Tridax trilobata*. Oil.

Bohlmann, F. *et al, Justus Liebigs Ann. Chem.*, 1970, **739**, 135.

4,8,10,14,16-Heptadecapentaen-6-yn-3-ol H-00086

$H_2C{=}CHCH{=}CHCH_2CH_2CH{=}CHCH{=}CHC{\equiv}CCH{=}CHCH(OH)CH_2CH_3$

$C_{17}H_{22}O$ M 242.360

(all-E)-form [22644-41-3]
Constit. of *Aster schreberi*. Oil.

Bohlmann, F. *et al, Chem. Ber.*, 1969, **102**, 1037.

1,4,8,10,16-Heptadecapentaen-6-yn-3-one H-00087

$H_2C{=}CH(CH_2)_4CH{=}CHCH{=}CHC{\equiv}CCH{=}CHCOCH{=}CH_2$

$C_{17}H_{20}O$ M 240.344

(all-E)-form [29739-56-8]
Constit. of *Tridax trilobata* and *Calea scabra*. Oil.

Bohlmann, F. *et al, Justus Liebigs Ann. Chem.*, 1970, **739**, 135 (*isol*)
Bohlmann, F. *et al, Phytochemistry*, 1977, **16**, 1065 (*isol*)

1,8,11,14-Heptadecatetraene H-00088

Aplotaxene
[10482-53-8]

$H_3CCH_2CH{=}CHCH_2CH{=}CHCH_2CH{=}CH(CH_2)_5CH{=}CH_2$

$C_{17}H_{28}$ M 232.408

(all-Z)-form
Constit. of Compositae spp. Bp_{11} 153-155° (Bp_8 110-115°).

Mathur, S.B., *Phytochemistry*, 1972, **11**, 1513 (*isol*)
Binder, R.G. *et al, Phytochemistry*, 1975, **14**, 2085 (*ms, nmr, ir, uv*)
Yano, K., *Phytochemistry*, 1977, **16**, 263 (*ms, nmr, ir*)
Andersen, A.B. *et al, Phytochemistry*, 1977, **16**, 1829 (*isol*)
Bohlmann, F. *et al, Phytochemistry*, 1979, **18**, 79 (*isol*)

1,7,9,15-Heptadecatetraene-11,13-diyn-6-ol H-00089

$H_3CCH{=}CHC{\equiv}CC{\equiv}C(CH{=}CH)_2CH(OH)(CH_2)_3CH{=}CH_2$

$C_{17}H_{20}O$ M 240.344

(all-E)-form [24587-23-3]
Constit. of *Coreopsis gigantea*. Cryst. (Et_2O/pet. ether). Mp 41-42°. Unstable.
Ac: [26131-04-4].
$C_{19}H_{22}O_2$ M 282.382
Constit. of the roots of *Dahlia coccinea*. Oil.

Bohlmann, F. *et al, Chem. Ber.*, 1969, **102**, 1691 (*isol*)
Chin, C. *et al, J. Chem. Soc. C*, 1970, 314 (*synth*)

2,8,10,16-Heptadecatetraene-4,6-diyn-1-ol H-00090

[13081-27-1]

$H_2C{=}CH(CH_2)_4CH{=}CHCH{=}CHC{\equiv}CC{\equiv}CCH{=}CHCH_2OH$

$C_{17}H_{20}O$ M 240.344

(all-E)-form [4117-04-8]
Constit. of *Isostigma peucedanifolium, Onopordum* spp., *Tetragonotheca repanda* and *Coreopsis gigantea*. Cryst. (pet. ether). Mp 77-78°.
Ac: [5059-56-3].
$C_{19}H_{22}O_2$ M 282.382
Isol. from *Calea scabra, C. zacatechichi, Senecio* sp. and *Coreopsis longula*.

(2Z,8E,10E)-form [1540-88-1]
Isol. from *Tridax trilobata, Coreopsis gigantea, Anthemis tinctoria* and *Tetragonotheca repanda*. Cryst. (pet. ether). Mp 32°.
Ac: [73488-98-9].
Isol. from roots of *Tetragonotheca* sp.

(2E,8Z,10E)-form [88126-59-4]
Isol. from the roots of *S. microglossus*.

[64185-69-9, 69905-03-9]

Bohlmann, F. *et al, Chem. Ber.*, 1964, **97**, 2135; 1965, **98**, 1225, 3010; 1966, **99**, 2096, 3201 (*isol, uv, ir, pmr, struct*)
Soørensen, J.S. *et al, Acta Chem. Scand.*, 1966, **20**, 992 (*isol*)
Bohlmann, F. *et al, Phytochemistry*, 1979, **18**, 79; 1983, **22**, 1675 (*isol*)
Quijano, L. *et al, Phytochemistry*, 1979, **18**, 1529 (*isol*)
Metwally, M.A. *et al, Phytochemistry*, 1985, **24**, 182 (*isol*)
Jakupovic, J. *et al, Phytochemistry*, 1988, **27**, 3881 (*isol*)

1,7,9,15-Heptadecatetraene-11,13-diyn-6-one H-00091

$H_3CCH{=}CHC{\equiv}CC{\equiv}CCH{=}CHCH{=}CHCO(CH_2)_3CH{=}CH_2$

$C_{17}H_{18}O$ M 238.329

(all-E)-form [24587-24-4]
Constit. of *Dahlia coccinea*. Needles (Et_2O/pet. ether). Mp 108°.

Bohlmann, F. *et al, Chem. Ber.*, 1969, **102**, 1691 (*isol*)
Chin, C. *et al, J. Chem. Soc. C*, 1970, 314 (*synth*)

5,7,9,15-Heptadecatetraene-11,13-diyn-4-one H-00092

$H_3CCH{=}CHC{\equiv}CC{\equiv}CCH{=}CHCH{=}CHCH{=}CHCO(CH_2)_2CH_3$

$C_{17}H_{18}O$ M 238.329

(all-E)-form [25525-20-6]
Constit. of the roots and rhizomes of *Cicuta virosa*. Oil.

Bohlmann, F. *et al, Chem. Ber.*, 1969, **102**, 3293 (*synth*)

1,7,9,13-Heptadecatetraen-11-yne H-00093

$H_3CCH_2CH_2CH{=}CHC{\equiv}CCH{=}CHCH{=}CH_2(CH_2)_4CH{=}CH_2$

$C_{17}H_{24}$ M 228.377

(all-E)-form [22645-00-7]
Isol. from *Aster schreberi*. Oil.

Bohlmann, F. *et al, Chem. Ber.*, 1969, **102**, 1037 (*isol;*)1970, **103**, 3419 (*synth*)

1,7,9,13-Heptadecatetraen-11-yn-3-ol H-00094

$H_3CCH_2CH_2CH{=}CHC{\equiv}CCH{=}CHCH{=}CH(CH_2)_3CH(OH)CH{=}CH_2$

$C_{17}H_{24}O$ M 244.376

(all-E)-form [22644-37-7]
Constit. of *Aster schreberi*. Oil.

Bohlmann, F. *et al, Chem. Ber.*, 1969, **102**, 1037 (*isol;*)1970, **103**, 3424 (*synth*)

2,7,9,13-Heptadecatetraen-11-yn-1-ol H-00095

$H_3CCH_2CH_2CH{=}CHC{\equiv}CCH{=}CH$
$CH{=}CH(CH_2)_3CH{=}CHCH_2OH$

$C_{17}H_{24}O$ M 244.376

(2Z,7E,9E,13E)-form [22644-35-5]

Occurs in *Aster schreberi*. Oil.

Bohlmann, F. *et al*, *Chem. Ber.*, 1969, **102**, 1037 (*uv*, *isol*); 1970, **103**, 3419 (*synth*)

4,8,10,16-Heptadecatetraen-6-yn-3-ol H-00096

$H_2C{=}CH(CH_2)_4CH{=}CHCH{=}CHC{\equiv}CCH{=}CHCH(OH)$
CH_2CH_3

$C_{17}H_{24}O$ M 244.376

(all-E)-form [22644-39-9]

Occurs in *Aster schreberi*. Oil.

Bohlmann, F. *et al*, *Chem. Ber.*, 1969, **102**, 1037 (*isol*, *uv*); 1970, **103**, 3419 (*synth*)

8,10,14,16-Heptadecatetraen-6-yn-3-one H-00097

$H_2C{=}CHCH{=}CHCH_2CH_2CH{=}CH$
$CH{=}CHC{\equiv}CCH_2CH_2COCH_2CH_3$

$C_{17}H_{22}O$ M 242.360

(all-E)-form [22644-43-5]

Occurs in *Aster schreberi*. Oil.

Bohlmann, F. *et al*, *Chem. Ber.*, 1969, **102**, 1037 (*isol*, *uv*)

1,11,13-Heptadecatriene H-00098

$H_3CCH_2CH_2CH{=}CHCH{=}CH(CH_2)_8CH{=}CH_2$

$C_{17}H_{30}$ M 234.424

(11E,13Z)-form [80625-35-0]

$Bp_{0.01}$ 90-95°.

Bestmann, H.J. *et al*, *Justus Liebigs Ann. Chem.*, 1981, 2117 (*synth*, *pmr*)

3,6,9-Heptadecatriene, 9CI H-00099

[81265-04-5]

$H_3C(CH_2)_6CH{=}CHCH_2CH{=}CHCH_2CH{=}CHCH_2CH_3$

$C_{17}H_{30}$ M 234.424

(All-Z)-form [109877-47-6]

Found in *Semiothisa* sp. and female *Tephrina arenacearria*. Sex pheromone.

3R,4S*-Epoxide:* [109877-48-7]. *3,4-Epoxy-6,9-heptadecadiene. 2-Ethyl-3-(2,5-tridecadienyl)oxirane, 9CI*

Constit. of the sex pheromones of *S.* spp. and *T. arenacearia.*

6R,7S*-Epoxide:* [109877-49-8]. *6,7-Epoxy-3,9-heptadecadiene. 2-(2-Decenyl)-3-(2-pentenyl)oxirane, 9CI*

$C_{17}H_{30}O$ M 250.423

Constit. of the sex pheromones of *S.* spp. and *T. arenacearia.*

9R,10S*-Epoxide:* [109877-51-2]. *9,10-Epoxy-6,9-heptadecadiene. 2-Heptyl-3-(2,5-octadienyl)oxirane, 9CI*

Constit. of sex pheromone of *T. arenacearia.*

[129938-28-9, 130193-58-7, 130194-36-4]

Millar, J.G. *et al*, *J. Chem. Ecol.*, 1987, **13**, 1371; 1990, **16**, 2317 (*isol*, *synth*, *ms*, *pmr*, *ir*)
Sabharwal, a. *et al*, *J. Indian Chem. Soc.*, 1990, **67**, 173 (*synth*)
Toth, M. *et al*, *Z. Naturforsch., C*, 1991, **46**, 257 (*isol*, *synth*)

1,7,9-Heptadecatriene-11,13-diyne H-00100

$H_3CCH_2CH_2C{\equiv}CC{\equiv}CCH{=}CHCH{=}CH(CH_2)_4CH{=}CH_2$

$C_{17}H_{22}$ M 226.361

(all-E)-form [24587-22-2]

Constit. of *Coreopsis gigantea*. Oil. $Bp_{0.001}$ 110°.

Bohlmann, F. *et al*, *Chem. Ber.*, 1969, **102**, 1691 (*isol*, *uv*)

1,8,15-Heptadecatriene-11,13-diyne, 9CI H-00101

$H_3CCH{=}CHC{\equiv}CC{\equiv}CCH_2CH{=}CH(CH_2)_5CH{=}CH_2$

$C_{17}H_{22}$ M 226.361

8,9-Epoxide: [95925-05-6]. *8,9-Epoxy-1,15-heptadecadiene-11,13-diyne. 2-(6-Heptenyl)-3-(6-octene-2,4-diynyl) oxirane, 9CI*

$C_{17}H_{22}O$ M 242.360

Isol. from *Solidago virgaurea* and *S. multiradiata.*

(8E,15E)-form [28834-02-8]

Constit. of the plant *Silybum marianum.*

(8Z,15E)-form

Constit. of *Coreopsis gigantea*. Oil.

(8Z,15Z)-form [6386-16-9]

Occurs in *Aster bellidiastrum* and *Chrysanthemum frutescens*. Also in *C. maximum* and *S. marianum.*

Bohlmann, F. *et al*, *Chem. Ber.*, 1966, **99**, 586, 3201; 1968, **101**, 532; 1969, **102**, 1691 (*isol*, *uv*, *ir*, *pmr*, *ms*, *struct*, *synth*)
Schulte, K.E. *et al*, *Arch. Pharm. (Weinheim, Ger.)*, 1970, **303**, 7 (*isol*, *pmr*, *ir*, *uv*)
Lam, J., *Phytochemistry*, 1971, **10**, 647 (*epoxide*)
Bohlmann, F. *et al*, *Planta Med.*, 1984, **50**, 201 (*epoxide*)

2,8,10-Heptadecatriene-4,6-diyne H-00102

$H_3C(CH_2)_5CH{=}CHCH{=}CHC{\equiv}CC{\equiv}CCH{=}CHCH_3$

$C_{17}H_{22}$ M 226.361

(all-E)-form [20311-59-5]

Constit. of *Oenanthe crocata*. Oil. $Bp_{0.001}$ 100°.

Bohlmann, F. *et al*, *Chem. Ber.*, 1968, **101**, 1163 (*isol*, *uv*); 1971, **104**, 362 (*synth*)

2,8,10-Heptadecatriene-4,6-diyne-1,14-diol H-00103

[20311-78-8]

$H_3CCH_2CH_2CH(OH)$
$CH_2CH_2CH{=}CHCH{=}CHC{\equiv}CC{\equiv}CCH{=}CHCH_2OH$

$C_{17}H_{22}O_2$ M 258.360

▷ JY8750000.

(E,E,E)-form

Enanthotoxin. Oenanthotoxin

Isol. from roots of *Oenanthe crocata*. Cryst. (C_6H_6 or MeOH). Mp 87°. $[\alpha]_D^{15}$ +30.5° (c, 2 in MeOH).

▷ Poisonous.

1-Ac: [20311-73-3].

$C_{19}H_{24}O_3$ M 300.397

Isol. from *O. crocata*. Oil.

Di-Ac: ***Centaur Z***

$C_{21}H_{26}O_4$ M 342.434

Isol. from roots of *Centaurea montana*. Needles (pentane). Mp 33°. Probable structure.

(E,E,E)-(±)-form

Cryst. (pet. ether). Mp 68°.

Anet, E. *et al*, *Chem. Ind. (London)*, 1952, 757 (*isol*)
Bohlmann, F. *et al*, *Chem. Ber.*, 1955, **88**, 1245.

Lofrgen, N. *et al*, *Acta Chem. Scand.*, 1963, **17**, 1065 (*isol, struct, Centaur Z*)
Bohlmann, F. *et al*, *Chem. Ber.*, 1968, **101**, 1162 (*isol*)
Lewis, R.J., *Sax's Dangerous Properties of Industrial Materials*, 8th Ed., Van Nostrand-Reinhold, 1992, EAO500.

8,10,12-Heptadecatriene-4,6-diyne-1,14-diol H-00104

[82905-09-7]

$H_3CCH_2CH_2CH(OH)(CH{=}CH)_3(C{\equiv}C)_3CH_2CH_2CH_2OH$

$C_{17}H_{22}O_2$ M 258.360

(all-E)-(−)-form [505-75-9] ***Cicutoxin***
Isol. from *Cicuta virosa* roots and from *C. maculata*. Shows antileukaemic activity. Cryst. (Et_2O/pet. ether). Mp 54°. $[\alpha]_D^{15}$ −14.5° (c, 1.7 in EtOH).

(all-E)-(±)-form
Cryst. (Et_2O/pet. ether). Mp 67°.

[107882-70-2, 107882-71-3]

Anet, E.F.L.J. *et al*, *J. Chem. Soc.*, 1953, 309 (*isol, struct, uv*)
Hill, B.E. *et al*, *J. Chem. Soc.*, 1955, 1770 (*synth*)
Konoshima, T. *et al*, *J. Nat. Prod. (Lloydia)*, 1986, **49**, 1117 (*isol, struct, bibl, pmr, ms, ir*)

7,9,15-Heptadecatriene-11,13-diynoic acid H-00105

$H_3CCH{=}CHC{\equiv}CC{\equiv}CCH{=}CHCH{=}CH(CH_2)_5COOH$

$C_{17}H_{20}O_2$ M 256.344

(all-E)-form [137217-71-1]
Isol. from aerial parts of *Baccharis pedunculata*.

Me ester: [29708-53-0].
$C_{18}H_{22}O_2$ M 270.371
Occurs in *B. trinervis* and aerial parts of *B. pedunculata*. Oil.

Bohlmann, F. *et al*, *Chem. Ber.*, 1970, **103**, 2327 (*isol, pmr, ir, uv, ms, struct*)
Bohlmann, F. *et al*, *Phytochemistry*, 1979, **18**, 1011 (*isol*)
Jakupovic, J. *et al*, *Phytochemistry*, 1991, **30**, 2785 (*isol*)

1,8,10-Heptadecatriene-4,6-diyn-3-ol H-00106

$H_3C(CH_2)_5CH{=}CHCH{=}CHC{\equiv}CC{\equiv}CCH(OH)$
$CH{=}CH_2$

$C_{17}H_{22}O$ M 242.360

(all-E)-form [18836-79-8]
Constit. of *Pittosporum buchanani*. Oil.

Bohlmann, F. *et al*, *Chem. Ber.*, 1968, **101**, 525, 1889 (*uv, isol*; 1971, **104**, 1362 (*synth*)

1,9,16-Heptadecatriene-4,6-diyn-3-ol H-00107

[88153-63-3]

$H_2C{=}CH(CH_2)_5CH{=}CHCH_2C{\equiv}CC{\equiv}CCH(OH)$
$CH{=}CH_2$

$C_{17}H_{22}O$ M 242.360

(−)-(Z)-form [36150-08-0] ***Dehydrofalcarinol***
Isol. from roots of *Artemisia atrata*, *A. borealis* and *Nidorella* sp. Pale-yellow oil. $[\alpha]_D^{23}$ −5.0° (c, 13.9 in Et_2O).

Ac: [78516-98-0].
$C_{19}H_{24}O_2$ M 284.397
Isol. from *Viguiera* sp. Oil.

9,10-Epoxide: [139163-34-1]. *9,10-Epoxy-1,16-heptadecadiene-4,6-diyn-3-ol.* ***Ginsenoyne A***
$C_{17}H_{22}O_2$ M 258.360
Isol. from the root of *Panax ginseng*. Oil. $[\alpha]_D$ −121.9° (c, 1.0 in $CHCl_3$). λ_{max} 201, 230, 243, 256 nm (EtOH).

9,10-Epoxide, Ac: [142449-71-6]. ***Ginsenoyne F***
$C_{19}H_{24}O_3$ M 300.397
Isol. from the roots of *P. ginseng*. Oil. $[\alpha]_D$ −54.2° (c, 2.35 in $CHCl_3$).

[13894-97-8]

Bohlmann, F. *et al*, *Chem. Ber.*, 1966, **99**, 3552 (*isol, ir, pmr, struct*)
Bohlmann, F. *et al*, *Phytochemistry*, 1981, **20**, 113 (*isol*)
Harada, R. *et al*, *Phytochemistry*, 1982, **21**, 2009 (*isol*)
Wang, Y. *et al*, *Phytochemistry*, 1990, **29**, 3101 (*isol, struct, bibl, pmr, cmr*)
Hirakura, K. *et al*, *Phytochemistry*, 1991, **30**, 3327, 4053 (*isol, ir, uv, ms, pmr, cmr*)

2,8,10-Heptadecatriene-4,6-diyn-1-ol H-00108

[18836-75-4]

$H_3C(CH_2)_5CH{=}CHCH{=}CHC{\equiv}CC{\equiv}CCH{=}CHCH_2OH$

$C_{17}H_{22}O$ M 242.360

(E,E,E)-form
Enanthetol. *Oenanthetol*
Constit. of roots of *Oenanthe crocata*. Cryst. (pet. ether). Mp 71°.

Anet, E. *et al*, *Chem. Ind. (London)*, 1952, 757 (*isol*)
Hill, B.E. *et al*, *J. Chem. Soc.*, 1955, 1770 (*synth, config*)
Bohlmann, F. *et al*, *Chem. Ber.*, 1968, **101**, 1163; 1969, **102**, 1702 (*isol, biosynth*)

2,9,16-Heptadecatriene-4,6-diyn-1-ol H-00109

$H_2C{=}CH(CH_2)_5CH{=}CHCH_2C{\equiv}CC{\equiv}CCH{=}CHCH_2OH$

$C_{17}H_{22}O$ M 242.360

(2E,9Z)-form [61102-22-5]
Isol. from *Calea integrifolia*, *Senecio* sp. and roots of *Bidens graveolens*.

2-Methylpropanoyl: [87710-43-8].
$C_{21}H_{28}O_2$ M 312.451
Isol. from *B. graveolens*. Gum. λ_{max} 280, 266, 252 nm (Et_2O).

(2Z,9Z)-form
Isol. from roots of *Anthemis tinctoria*. Oil. λ_{max} 251.5, 267, 282.5 nm.

[71392-32-0]

Bohlmann, F. *et al*, *Chem. Ber.*, 1966, **99**, 2096.
Bohlmann, F. *et al*, *Phytochemistry*, 1976, **15**, 1177; 1983, **22**, 1281 (*isol, ir, uv, ms*)

8,10,12-Heptadecatriene-4,6-diyn-1-ol H-00110

$H_3C(CH_2)_3(CH{=}CH)_3C{\equiv}CC{\equiv}CCH_2CH_2CH_2OH$

$C_{17}H_{22}O$ M 242.360

(all-E)-form
Cicutol
Constit. of *Cicuta virosa* roots. Cryst. (CCl_4/pet. ether). Mp 65-66°.

Anet, E.F.L.J. *et al*, *J. Chem. Soc.*, 1953, 309 (*isol, struct, uv*)
Bohlmann, F. *et al*, *Chem. Ber.*, 1955, **88**, 1347 (*synth*)

8,10,16-Heptadecatriene-4,6-diyn-3-ol H-00111

$H_2C{=}CH(CH_2)_4CH{=}CHCH{=}CHC{\equiv}CC{\equiv}CCH(OH)$
CH_2CH_3

$C_{17}H_{22}O$ M 242.360

(all-E)-form [24587-15-3]
Constit. of *Coreopsis gigantea*. Oil.

Bohlmann, F. *et al*, *Chem. Ber.*, 1969, **102**, 1691 (*isol, uv*); 1970, **103**, 3419 (*synth*)

1,8,16-Heptadecatriene-4,6-diyn-3-one H-00112

$H_2C{=}CH(CH_2)_6CH{=}CHC{\equiv}CC{\equiv}CCOCH{=}CH_2$

$C_{17}H_{20}O$ M 240.344

(E)-form [29739-55-7]
Constit. of *Tridax trilobata*. Oil.

Bohlmann, F. *et al*, *Justus Liebigs Ann. Chem.*, 1970, **739**, 135 (*isol, synth, uv*)

7,9,15-Heptadecatriene-11,13-diyn-4-one H-00113

[18836-73-2]

$CH_3(CH_2)_2CO(CH_2)_2CH{=}CHCH{=}CHC{\equiv}CC{\equiv}CCH{=}CH_2CH_3$

$C_{17}H_{20}O$ M 240.344

(E,E,E)-form
Enanthetone. Oenanthetone
Constit. of roots of *Oenanthe crocata*. Cryst. (pet. ether). Mp 46°.

Anet, E.F.L.J. *et al*, *J. Chem. Soc.*, 1953, 309 (*isol, struct*)
Bohlmann, F. *et al*, *Chem. Ber.*, 1955, **88**, 1245; 1968, **101**, 1162 (*synth, isol*)

8,10,16-Heptadecatriene-4,6-diyn-3-one H-00114

$H_2C{=}CH(CH_2)_4CH{=}CHCH{=}CHC{\equiv}CC{\equiv}CCOCH_2CH_3$

$C_{17}H_{20}O$ M 240.344

(E,E)-form
Constit. of *Coreopsis gigantea*. Oil.

Bohlmann, F. *et al*, *Chem. Ber.*, 1969, **102**, 1691 (*isol, uv*)

1,7,9-Heptadecatriene-11,13,15-triyne, 9CI H-00115

Centaur X_3
[20252-54-4]

$H_3CC{\equiv}CC{\equiv}CC{\equiv}CCH{=}CHCH{=}CH(CH_2)_4CH{=}CH_2$

$C_{17}H_{18}$ M 222.329

(all-E)-form [41688-30-6]
Constit. of *Chrysanthemum leucanthemum* and *Dahlia* spp. Also from *Centaurea cyanus* and *Artemisia vulgaris*. Mp 17.5°.

(7Z,9E)-form
Cryst. (pet. ether). Mp −12°.

Bohlmann, F. *et al*, *Chem. Ber.*, 1957, **90**, 1661 (*synth*)
Bentley, R.K. *et al*, *J. Chem. Soc. C*, 1969, 1096 (*isol*)
Bedford, C.T. *et al*, *J. Chem. Soc., Perkin Trans. 1*, 1976, 735 (*synth*)

1,7,9-Heptadecatriene-11,13,15-triyn-6-ol H-00116

$H_3CC{\equiv}CC{\equiv}CC{\equiv}CCH{=}CHCH{=}CHCH(OH)(CH_2)_3CH{=}CH_2$

$C_{17}H_{18}O$ M 238.329

(all-E)-form [16697-23-7]
Constit. of *Artemisia selengensis* and *Dahlia* spp. Unstable cryst. (Et_2O/pet. ether). Mp 46-47°. $[\alpha]_D^{20}$ −9° (c, 0.65 in $CHCl_3$).

Ac: [59950-64-0].
$C_{19}H_{20}O_2$ M 280.366
Constit. of *Dahlia* spp.

Bohlmann, F. *et al*, *Chem. Ber.*, 1967, **100**, 1940 (*isol, struct*)
Bedford, C.T. *et al*, *J. Chem. Soc., Perkin Trans. 1*, 1976, 735 (*isol, synth*)

3-(8,11,14-Heptadecatrienyl)-1,2-benzenediol H-00117

3-(8,11,14-Heptadecatrienyl)catechol
[4807-34-5]

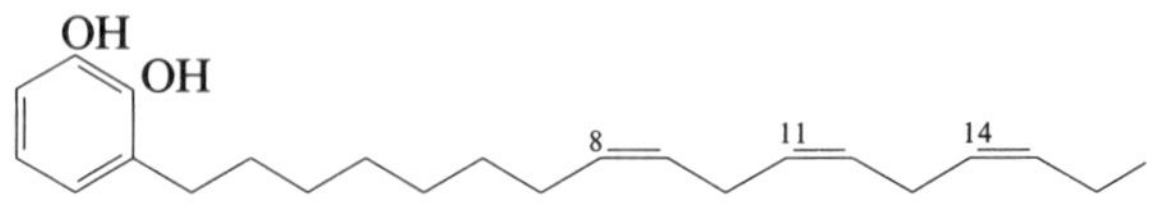

$C_{23}H_{34}O_2$ M 342.520
Constit. of poison oak (*Toxicodendron diversilobum*). $Bp_{0.022}$ 209-217°.

14,15-Dihydro: [5130-56-3]. *3-(8,11-Heptadecadienyl)-1,2-benzenediol. 3-(8,11-Heptadecadienyl)catechol*
$C_{23}H_{36}O_2$ M 344.536
Major vesicant principle of *Smodingium argutum*, also isol. from *T. radicans* and *Melanorrhoea diversilobum* (Burmese lac). $Bp_{0.03}$ 212-221°.
▷ Irritant.

[83258-40-6, 108006-51-5]

Murakami, M. *et al*, *Bull. Chem. Soc. Jpn.*, 1963, **36**, 897 (*synth*)
Jefferson, A. *et al*, *J. Chromatogr.*, 1986, **367**, 145 (*isol*)
El Sohly, M.A. *et al*, *J. Nat. Prod. (Lloydia)*, 1987, **45**, 532 (*isol, pmr, cmr*)
Gorst-Allman, C.P. *et al*, *S. Afr. J. Chem.*, 1987, **40**, 82 (*isol*)

6-(5,8,11-Heptadecatrienyl)-5,6-dihydro-2*H*-pyran-2-one, 9CI H-00118

2,10,13,16-Docosatetraen-5-olide

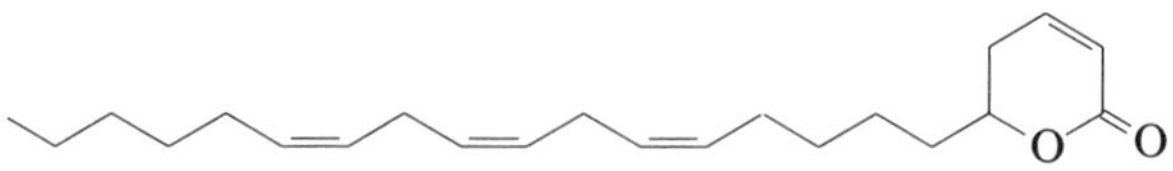

$C_{22}H_{34}O_2$ M 330.509

(all-Z)-form [81575-28-2]
Obt. from the red alga *Phacelocarpus labillardieri*. Oil. $[\alpha]_D^{25}$ −51° (c, 0.7 in $CHCl_3$).

Kazlauskas, R. *et al*, *Aust. J. Chem.*, 1982, **35**, 113 (*ms, pmr, cmr*)

8,10,16-Heptadecatrien-6-yn-3-one H-00119

$H_2C{=}CH(CH_2)_4CH{=}CHCH{=}CHC{\equiv}CCH_2CH_2COCH_2CH_3$

$C_{17}H_{24}O$ M 244.376

(all-E)-form [22644-44-6]
Constit. of *Aster schreberi*. Oil.

Bohlmann, F. *et al*, *Chem. Ber.*, 1969, **102**, 1037 (*isol, uv*)

5,8,11-Heptadecatriynoic acid, 8CI H-00120

[22117-02-8]

$H_3C(CH_2)_4C{\equiv}CCH_2C{\equiv}CCH_2C{\equiv}C(CH_2)_3COOH$

$C_{17}H_{22}O_2$ M 258.360
Cryst. Mp 52-54°.

Marcel, Y.L. *et al*, *Chem. Phys. Lipids*, 1968, **2**, 173 (*synth*)

1-Heptadecene, 9CI H-00121

[6765-39-5]

$H_3C(CH_2)_{14}CH{=}CH_2$

$C_{17}H_{34}$ M 238.456

Occurs in roots of *Cirsium japonicum* and *Centaurea scabiosa*. Constit. of the sex pheromone of the red flour beetle *Tribolium castaneum*. Mp 11.2°. Bp 300°, Bp_{10} 157°.

Kozacik, A.P. *et al*, *J. Am. Chem. Soc.*, 1938, **60**, 2436.
Schmidt, A.W. *et al*, *Ber.*, 1941, **74**, 1313.
Suzuki, T. *et al*, *Agric. Biol. Chem.*, 1975, **39**, 2207 (*ir, pmr, uv*)
Anderson, A.B. *et al*, *Phytochemistry*, 1977, **16**, 1829.

8-Heptadecene, 9CI H-00122

$H_3C(CH_2)_7CH{=}CH(CH_2)_6CH_3$

$C_{17}H_{34}$ M 238.456

(E)-form [16416-42-5]

Bp_{20} 164-168°.

(Z)-form [16369-12-3]

Occurs in the ant *Myrmecia gulosa* and the flowers of *Citrus bigaradia*.

Cavill, E.W.K. *et al*, *J. Insect Physiol.*, 1967, **13**, 1097 (*ir, pmr*)
Villieras, J., *Bull. Soc. Chim. Fr.*, 1967, 1511 (*synth, ir*)
Cavill, E.W.K. *et al*, *J. Insect Physiol.*, 1974, **20**, 2049 (*ms*)
Corbier, B. *et al*, *Recherches*, 1974, **19**, 289 (*isol, pmr, ir*)
Francis, E.W. *et al*, *Acta Chem. Scand., Ser. B*, 1977, **31**, 721 (*synth*)

8-Heptadecenedioic acid H-00123

$HOOC(CH_2)_7CH{=}CH(CH_2)_6COOH$

$C_{17}H_{30}O_4$ M 298.422

(E)-form

Cryst. (C_6H_6/AcOH). Mp 94-95°.

Di-Et ester:

$C_{21}H_{38}O_4$ M 354.529

$Bp_{0.03}$ 143-145°.

Mathur, H.H. *et al*, *J. Chem. Soc.*, 1963, 114 (*synth*)

8-Heptadecene-4,6-diyne-1,10-diol, 8CI H-00124

$H_3C(CH_2)_6CH(OH)CH{=}CHC{\equiv}CC{\equiv}CCH_2CH_2CH_2OH$

$C_{17}H_{26}O_2$ M 262.391

(E)-form [18829-51-1]

Occurs in *Oplopanax chironium*. Mp 44°.

1-Ac: [20311-72-2].

$C_{19}H_{28}O_3$ M 304.428

Occurs in *Oenanthe crocata*. Oil.

Bohlmann, F. *et al*, *Chem. Ber.*, 1968, **101**, 525 (*isol*); 1968, **101**, 1163 (*struct*); 1971, **104**, 1362 (*synth*)

1-Heptadecene-4,6-diyne-3,9,10-triol, 9CI H-00125

Panaxytriol. *Falcarintriol*

[87005-03-6]

$H_2C{=}CHCH(OH)C{\equiv}CC{\equiv}CCH_2CH(OH)CH(OH)(CH_2)_6CH_3$

$C_{17}H_{26}O_3$ M 278.391

Isol. from *Panax ginseng*. Also from *P. quinquefolium*. Shows antitumour activity. Needles (H_2O).

10-Me ether: [133921-57-0]. *10-Methoxy-1-heptadecene-4,6-diyne-3,9-diol, 9CI. PQ 1*

$C_{18}H_{28}O_3$ M 292.417

Isol. from *P. quinquefolium*. Shows some antileukaemic props. Oil. $[\alpha]_D$ −21.7° (c, 0.58 in MeOH).

10-Ac: [122855-48-5]. ***10-Acetylpanaxytriol***

$C_{19}H_{28}O_4$ M 320.428

Constit. of the roots of *Panax ginseng*. Cytotoxic.

Matsunaga, H. *et al*, *Chem. Pharm. Bull.*, 1989, **37**, 1279 (*isol, uv, pmr, cmr, synth*)
Kim, S.I. *et al*, *Yakhak Hoechi*, 1989, **33**, 118; *CA*, **111**, 146363 (*10-Acetyl panaxytriol*)
Fujimoto, Y. *et al*, *Chem. Pharm. Bull.*, 1991, **39**, 521 (*isol, ms, cmr, struct*)

9-Heptadecene-4,6-diyn-1-ol, 8CI H-00126

$H_3C(CH_2)_6CH{=}CHCH_2C{\equiv}CC{\equiv}CCH_2CH_2CH_2OH$

$C_{17}H_{26}O$ M 246.392

(Z)-form [20311-69-7]

Constit. of *Oenanthe crocata* and from *Falcaria vulgaris*. Oil. $Bp_{0.001}$ 110°.

Bohlmann, F. *et al*, *Chem. Ber.*, 1966, **99**, 3552; 1968, **101**, 1163; 1971, **104**, 1362 (*isol, struct, synth, ir, pmr*)

9-Heptadecene-4,6-diyn-8-ol, 9CI H-00127

$H_3C(CH_2)_6CH{=}CHCH(OH)C{\equiv}CC{\equiv}CCH_2CH_2CH_3$

$C_{17}H_{26}O$ M 246.392

(Z)-form [32768-90-4]

Constit. of *Heracleum mantegazzianum* and *H. sphondylium* roots. Oil.

Bohlmann, F. *et al*, *Chem. Ber.*, 1971, **104**, 1322.
Vanhaelen-Fastré, R. *et al*, *Phytochemistry*, 1973, **12**, 2687.

9-Heptadecene-4,6-diyn-3-one H-00128

1,2-Dihydrofalcarinone

$H_3C(CH_2)_6CH{=}CHCH_2C{\equiv}CC{\equiv}CCOCH_2CH_3$

$C_{17}H_{24}O$ M 244.376

(Z)-form [32768-86-8]

Constit. of the root of *Conium maculatum*. Oil.

9R,10S-Epoxide, 3-alcohol: 9,10-Epoxy-4,6-heptadecadiyn-3-ol. ***Ginsenoyne D***

$C_{17}H_{26}O_2$ M 262.391

Isol. from roots of *Panax ginseng*. Oil. $[\alpha]_D$ −90.7° (c, 3.0 in $CHCl_3$). λ_{max} 200, 230, 241, 255 nm (EtOH).

9R,10S-Epoxide, 3-alcohol, Ac: [142449-72-7]. ***Ginsenoyne G***

$C_{19}H_{28}O_3$ M 304.428

Isol. from the roots of *P. ginseng*. Oil. $[\alpha]_D$ −182.6° (c, 3.77 in $CHCl_3$).

Bohlmann, F. *et al*, *Chem. Ber.*, 1971, **104**, 1322 (*isol*)
Schulte, K.E. *et al*, *Arch. Pharm. (Weinheim, Ger.)*, 1977, **310**, 945 (*synth*)
Hirakawa, K. *et al*, *Phytochemistry*, 1991, **30**, 3327, 4053 (*Ginsenoynes*)

16-Heptadecene-1,2,4-triol, 9CI H-00129

Avocadene

[24607-08-7]

$H_2C{=}CH(CH_2)_{11}CH(OH)CH_2CH(OH)CH_2OH$

$C_{17}H_{34}O_3$ M 286.454

Constit. of avocado pear. Mp 68°. $[\alpha]_D^{25}$ −6.7° ($CHCl_3$).

1-Ac: [24607-09-8].

$C_{19}H_{36}O_4$ M 328.491

Constit. of avocado pear. Mp 58-59°. $[\alpha]_D^{25}$ −6.7° ($CHCl_3$).

2-Ac: [51352-69-3].
$C_{19}H_{36}O_4$ M 328.491
Detected in avocado pear.
4-Ac: [59499-18-2].
$C_{19}H_{36}O_4$ M 328.491
Detected in avocado pear.

Kashman, Y. *et al, Tetrahedron*, 1969, **25**, 4617.
Alves, H.M. *et al, Ann. Acad. Bras. Ciénc.*, 1970, **425**, 45 (*ir, ms, pmr, uv*)
Bittner, S. *et al, Phytochemistry*, 1971, **10**, 1417.
Brown, B.I. *et al, J. Chromatogr.*, 1973, **86**, 239.

2-Heptadecenoic acid, 9CI **H-00130**

[20753-89-3]

$H_3C(CH_2)_{13}CH{=}CHCOOH$

$C_{17}H_{32}O_2$ M 268.439
(E)-form [2825-78-7]
Cryst.(Me_2CO). Mp 56.6-58.5°. $Bp_{0.4}$ 145-150°.
Me ester: [42607-02-3].
$C_{18}H_{34}O_2$ M 282.465
No. phys. props. reported.
Amide:
$C_{17}H_{33}NO$ M 267.454
Mp 110-110.5°.
(Z)-form
Cryst. Mp 51.8-53.7°.

Lauer, W.M. *et al, J. Am. Chem. Soc.*, 1941, **63**, 1153 (*synth*)
Rusoff, I.I. *et al, J. Am. Chem. Soc.*, 1945, **67**, 673 (*uv*)
Gernot, G. *et al, Justus Liebigs Ann. Chem.*, 1965, **685**, 154 (*synth, ir*)
Dolezal, S. *et al, Collect. Czech. Chem. Commun.*, 1968, **33**, 3176 (*synth*)
Siouffi, A.M. *et al, Chem. Phys. Lipids*, 1973, **11**, 66 (*ester*)

3-Heptadecenoic acid, 9CI **H-00131**

$H_3C(CH_2)_{12}CH{=}CHCH_2COOH$

$C_{17}H_{32}O_2$ M 268.439
Constit. of *Grindelia oxylepis.*

[75656-23-4]

Kleiman, R. *et al, Lipids*, 1966, **1**, 301.

7-Heptadecenoic acid, 9CI **H-00132**

[71997-05-2]

$H_3C(CH_2)_8CH{=}CH(CH_2)_5COOH$

$C_{17}H_{32}O_2$ M 268.439
(Z)-form [87268-66-4]
Cryst. Mp 3.5-4.5°.

Gernot, G. *et al, Chem. Ber.*, 1963, **96**, 3370 (*synth*)

8-Heptadecenoic acid, 9CI **H-00133**

Civetic acid
[1975-86-6]

$H_3C(CH_2)_7CH{=}CH(CH_2)_6COOH$

$C_{17}H_{32}O_2$ M 268.439
Constit. of cod liver oil and animal lipids.
(E)-form [69695-25-6]
No Mp reported.
(Z)-form [7432-41-9]
Mp 5.5-7.0°.
Me ester: [65092-95-7].
$C_{18}H_{34}O_2$ M 282.465
Liq. Bp_6 159-165°.

Weitkamp, A.W. *et al, J. Am. Chem. Soc.*, 1947, **69**, 1938 (*isol*)
Gernot, G. *et al, Chem. Ber.*, 1963, **96**, 3370 (*synth*)
Boon, J.J. *et al, Lipids*, 1977, **12**, 717 (*synth, ms*)
Hase, T.A. *et al, Synth. Commun.*, 1979, **9**, 63 (*synth*)
Christie, W.W. *et al, Lipids*, 1986, **21**, 657 (*isol, chromatog*)

9-Heptadecenoic acid **H-00134**

[10136-52-4]

$H_3C(CH_2)_6CH{=}CH(CH_2)_7COOH$

$C_{17}H_{32}O_2$ M 268.439
(E)-form
Mp 38°.
Et ester:
$C_{19}H_{36}O_2$ M 296.492
$Bp_{0.5}$ 150°.
(Z)-form [1981-50-6]
Minor constit. of several animal and vegetable fats. Mp 11.4-12.2°, Mp 14.5°. $Bp_{0.5}$ 175°.
Me ester: [14101-91-8].
$C_{18}H_{34}O_2$ M 282.465
Mp −26.3° to −25.2°. Bp_2 149-152°.
Et ester: $Bp_{0.1}$ 139°.

Ames, D.H. *et al, J. Chem. Soc.*, 1951, 1079.
Vauer, V.A. *et al, Izv. Akad. Nauk SSSR, Ser. Khim.*, 1966, 2241 (*synth, ir*)
Boon, J.J. *et al, Lipids*, 1977, **12**, 717 (*synth, ms*)

16-Heptadecenoic acid, 9CI **H-00135**

[65119-97-3]

$H_2C{=}CH(CH_2)_{14}COOH$

$C_{17}H_{32}O_2$ M 268.439
Cryst. (butanone). Mp 52-54°.

Sadtler Standard Infrared Spectra, No. 23432.
Hunig, S. *et al, Chem. Ber.*, 1962, **95**, 2493 (*synth*)
Mirviss, S.B., *J. Org. Chem.*, 1989, **54**, 1948 (*synth*)
Mach, R.H. *et al, Tetrahedron Lett.*, 1989, **30**, 4069 (*synth*)

11-Heptadecen-1-ol, 9CI **H-00136**

$H_3C(CH_2)_4CH{=}CH(CH_2)_9CH_2OH$

$C_{17}H_{34}O$ M 254.455
(Z)-form [73461-68-4]
Ac: [73461-65-1]. *11-Heptadecenyl acetate*
$C_{19}H_{36}O_2$ M 296.492
Key component of the sex pheromone of the cabbage armyworm *Mamestra brassicae.*

[84801-19-4, 87436-91-7]

Novak, L. *et al, Acta Chim. Acad. Sci. Hung.*, 1979, **102**, 135 (*synth, isol*)
Maryanoff, B.E. *et al, J. Am. Chem. Soc.*, 1985, **107**, 217 (*synth*)
Attygalle, A.B. *et al, J. Chem. Ecol.*, 1987, **13**, 1299 (*isol*)
Deng, Y. *et al, Chin. Sci. Bull.*, 1989, **34**, 203 (*synth, isol*)
Den Otter, C.J. *et al, Insect Sci. Its Appl.*, 1989, **10**, 235 (*isol, activity*)

3-(8-Heptadecenyl)-1,2-benzenediol, 9CI **H-00137**
3-(8-Heptadecenyl)catechol
[5159-28-4]

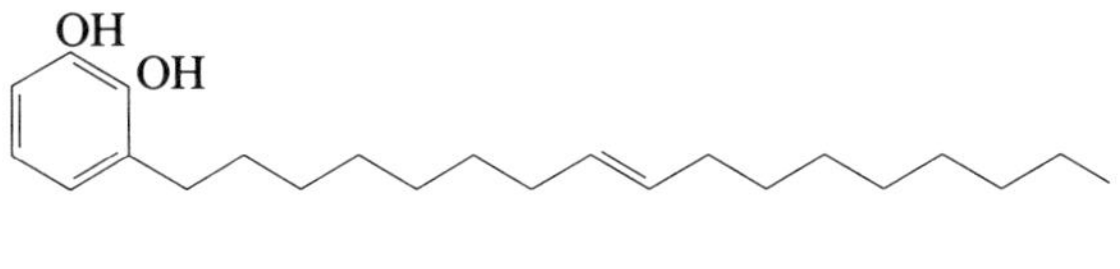

(E)-form

$C_{23}H_{38}O_2$ M 346.552
Mp 33-36°. $Bp_{0.8}$ 215-220°.
(E)-form [122850-19-5]
Constit. of *Melanorrhoea usitata* (Burmese lac).
(Z)-form [54954-20-0]
Constit. of *M. usitata* and of *Toxicodendron radicans* (poison oak).

Murakami, M. *et al*, *Bull. Chem. Soc. Jpn.*, 1963, **36**, 263 (*synth*)
El Sohly, M.A. *et al*, *J. Nat. Prod.* (*Lloydia*), 1982, **45**, 532 (*isol, cmr, pmr*)
Sargent, M.V. *et al*, *J. Chem. Soc., Perkin Trans. 1*, 1989, 431 (*isol, synth*)
Bestmann, H.J. *et al*, *Synlett*, 1990, 555 (*synth*)

3-(10-Heptadecenyl)-1,2-benzenediol, 9CI **H-00138**
3-(10-Heptadecenyl)catechol
$C_{23}H_{38}O_2$ M 346.552
(Z)-form [89838-23-3]
Constit. of *Rhus vernicifera* (Japanese lac) and *Melanorrhoea usitata* (Burmese lac).

Du, Y. *et al*, *J. Chromatogr.*, 1984, **284**, 463; **295**, 179 (*occur, pmr*)
Du, Y. *et al*, *Phytochemistry*, 1986, **25**, 2211 (*occur*)
Miyagoshi, T. *et al*, *Yukagaku*, 1990, **39**, 660; *CA*, **114**, 81132s (*synth*)

5-(2-Heptadecenyl)-1,3-benzenediol, 9CI **H-00139**
5-(2-Heptadecenyl)resorcinol
$C_{23}H_{38}O_2$ M 346.552
(Z)-form [94751-34-5]
Isol. from mango (*Mangifera indica*) latex.

Mamdapur, V.R., *J. Agric. Food Chem.*, 1985, **33**, 377 (*isol*)

5-(8-Heptadecenyl)-1,3-benzenediol, 9CI **H-00140**
5-(8-Heptadecenyl)resorcinol
$C_{23}H_{38}O_2$ M 346.552
(Z)-form
Isol. from *Cardwellia sublimis*, *Grevillea hilliana*, *Hakea persiehana* and *H. amplexicaulis*.
[54954-20-0, 117940-49-5]

Cirigottis, K.A. *et al*, *Aust. J. Chem.*, 1974, **27**, 345 (*isol*)

5-(12-Heptadecenyl)-1,3-benzenediol **H-00141**
5-(12-Heptadecenyl)resorcinol
$C_{23}H_{38}O_2$ M 346.552
(Z)-form [103462-06-2]
Constit. of peel of mango fruit (*Mangifera indica*). Antifungal agent. Oil.
[103462-07-3]

Cojocaru, M. *et al*, *Phytochemistry*, 1986, **25**, 1093.

5-(8-Heptadecenyl)dihydro-3-hydroxy-2(3*H*)-furanone, 9CI **H-00142**
[83725-45-5]

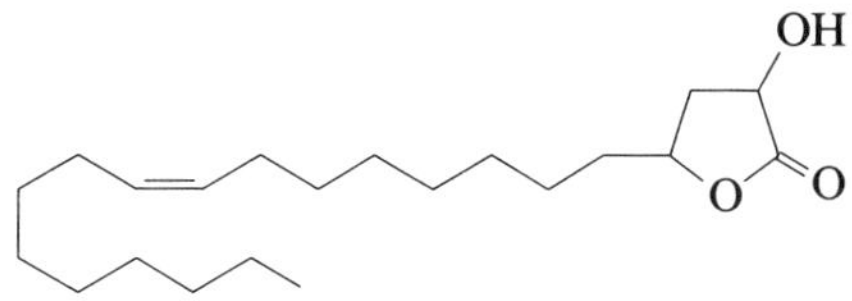

$C_{21}H_{38}O_3$ M 338.529
Constit. of *Garcinia mannii*. Oil.

Hussain, R.A. *et al*, *Phytochemistry*, 1982, **21**, 1393.

10-Heptadecen-8-ynoic acid, 8CI **H-00143**

$$H_3C(CH_2)_5CH{=}CHC{\equiv}C(CH_2)_6COOH$$

$C_{17}H_{28}O_2$ M 264.407
(E)-form [16714-84-4]
Pyrulic acid
Constit. of the seed oil of *Pyrularia pubera*. Mp 33-34°.
Me ester: [16714-85-5].
$C_{18}H_{30}O_2$ M 278.434
Oil.

Hopkins, C.Y. *et al*, *Chem. Ind.* (*London*), 1967, 998.

Heptadecylamine **H-00144**
1-Aminoheptadecane. 1-Heptadecanamine, 9CI
[4200-95-7]

$$H_3C(CH_2)_{15}CH_2NH_2$$

$C_{17}H_{37}N$ M 255.486
Cryst. Mp 49°. Bp 335-360°, $Bp_{0.5}$ 128°.
B,HCl: [1838-07-9].
Cryst. (EtOH). Mp 162-164° (158°).
▷MI4230000.
Benzoyl: Leaflets (C_6H_6). Mp 91°.
N-*Ac:* Cryst. Mp 62°.

Naegeli, C. *et al*, *Helv. Chim. Acta*, 1929, **12**, 227 (*synth*)
Briggs, L.H. *et al*, *J. Chem. Soc.*, 1942, 61 (*deriv*)
Borrows, E.T. *et al*, *J. Chem. Soc.*, 1947, 197 (*synth*)
Murr, B.L. *et al*, *J. Am. Chem. Soc.*, 1955, **77**, 1684 (*synth*)
Gohlke, R.S. *et al*, *Anal. Chem.*, 1962, **34**, 1281 (*ms*)
Goto, R. *et al*, *Nippon Kagaku Zasshi*, 1966, **87**, 1220; *CA*, **66**, 75599 (*synth*)

16-Heptadecyne-1,2,4-triol, 9CI **H-00145**
Avocadyne
[34524-38-4]

$$HC{\equiv}C(CH_2)_{11}CH(OH)CH_2CH(OH)CH_2OH$$

$C_{17}H_{32}O_3$ M 284.438
Constit. of avocado pear. Mp 76°. $[\alpha]_D^{25}$ −4.9° ($CHCl_3$).
1-Ac: [24607-06-5].
$C_{19}H_{34}O_4$ M 326.475
Constit. of avocado pear. Mp 73°. $[\alpha]_D^{25}$ −4.6° ($CHCl_3$).
2-Ac: [37139-04-1].
$C_{19}H_{34}O_4$ M 326.475
Constit. of avocado pear.
4-Ac: [28884-46-0].
$C_{19}H_{34}O_4$ M 326.475
Constit. of the avocado pear.

Kashman, Y. *et al*, *Tetrahedron*, 1969, **25**, 4617.

Alves, H.M. *et al, Ann. Acad. Bras. Ciénc.*, 1970, **425**, 45.
Kashman, Y. *et al, Tetrahedron*, 1970, **26**, 1943 (*struct*)
Brown, B.I., *J. Agric. Food Chem.*, 1972, **20**, 753.

7-Heptadecynoic acid H-00146

$H_3C(CH_2)_8C{\equiv}C(CH_2)_5COOH$

$C_{17}H_{30}O_2$ M 266.423
Mp 43.5-44.5°.

Grimmer, G. *et al, Chem. Ber.*, 1963, **96**, 3370 (*synth*)

8-Heptadecynoic acid, 9CI H-00147

[5556-50-3]

$H_3C(CH_2)_7C{\equiv}C(CH_2)_6COOH$

$C_{17}H_{30}O_2$ M 266.423
Mp 34.0-34.8°.
Me ester:
$C_{18}H_{32}O_2$ M 280.450
Mp −11° to −13°. $Bp_{0.03}$ 125-128°.

Grimmer, G. *et al, Chem. Ber.*, 1963, **96**, 3370 (*synth*)
Gensler, W.J. *et al, J. Org. Chem.*, 1970, **35**, 2301 (*synth, ir, nmr, raman*)
Williams, J.L. *et al, Chem. Phys. Lipids*, 1972, **9**, 320 (*pmr*)

9-Heptadecynoic acid, 9CI H-00148

[10269-71-3]

$H_3C(CH_2)_6C{\equiv}C(CH_2)_7COOH$

$C_{17}H_{30}O_2$ M 266.423
Mp 42.8-43.5°.
Me ester:
$C_{18}H_{32}O_2$ M 280.450
$Bp_{0.025}$ 114°.

Grimmer, G. *et al, Chem. Ber.*, 1963, **96**, 3370 (*synth*)
Pawlowski, N.E. *et al, J. Org. Chem.*, 1972, **37**, 3245 (*ir, ms*)

12-Heptadecynoic acid H-00149

$H_3C(CH_2)_3C{\equiv}C(CH_2)_{10}COOH$

$C_{17}H_{30}O_2$ M 266.423
Mp 35-36°. $Bp_{0.15}$ 158-163°.

Ames, D.E. *et al, J. Chem. Soc.*, 1963, 3775 (*synth*)

16-Heptadecynoic acid H-00150

$HC{\equiv}C(CH_2)_{14}COOH$

$C_{17}H_{30}O_2$ M 266.423
Cryst. Mp 66-68°.
4-Bromophenacyl ester: Mp 76°.

Schulte, K.E. *et al, Arch. Pharm. (Weinheim, Ger.)*, 1957, **290**, 483.
Shishido, K. *et al, J. Org. Chem.*, 1964, **29**, 2968.

2,4-Heptadienal, 9CI H-00151

[4313-03-5]

$H_3CCH_2CH{=}CHCH{=}CHCHO$

$C_7H_{10}O$ M 110.155
(E,E)-form
Bp_{12} 78°, $Bp_{0.06}$ 27-29°.
Di-Et acetal:
$C_{11}H_{20}O_2$ M 184.278
Bp_{12} 96.5-97.5°.
2,4-Dinitrophenylhydrazone: Purple needles. Mp 157-159°.

Nishida, S. *et al, Chem. Lett.*, 1976, 1297 (*synth*)
Zimmermann, B. *et al, Chem. Ber.*, 1986, **119**, 2848 (*synth, ir, pmr*)

3,5-Heptadien-1-ol, 9CI H-00152

$H_3CCH{=}CHCH{=}CHCH_2CH_2OH$

$C_7H_{12}O$ M 112.171
(E,E)-form [37944-01-7]
Liq. Bp_{21} 98°.

Normant, H., *C. R. Hebd. Seances Acad. Sci.*, 1955, **240**, 1111 (*synth*)
Mauldin, C.H. *et al, Tetrahedron Lett.*, 1972, 2955 (*synth, pmr*)
Tolstikov, G.A. *et al, Zh. Obshch. Khim.*, 1975, **45**, 1322; *CA*, **83**, 96363 (*synth*)

3,6-Heptadien-1-ol H-00153

$H_2C{=}CHCH_2CH{=}CHCH_2CH_2OH$

$C_7H_{12}O$ M 112.171
(E)-form [80502-28-9]
No phys. props. reported.
(Z)-form [80502-29-0]
Oil. $Bp_{0.5}$ 60°.

Snider, B.B. *et al, J. Org. Chem.*, 1983, **48**, 3003 (*synth, pmr, ir, ms*)
Sargent, M.V. *et al, J. Chem. Soc., Perkin Trans. 1*, 1990, 1429 (*synth, pmr, cmr, ir*)

1,4-Heptanediol, 9CI H-00154

[40646-07-9]

$H_3C(CH_2)_2CH(OH)(CH_2)_2CH_2OH$

$C_7H_{16}O_2$ M 132.202
(±)-form
Misc. with H_2O. Bp_{750} 240-245°, Bp_3 112-115°.
Di-Ac: [39846-58-7].
$C_{11}H_{20}O_4$ M 216.277
Bp_{748} 249-252°, Bp_1 113.0-113.5°.
α-Naphthylurethane: Cryst. (petrol). Mp 81-82°.

Utimoto, K. *et al, Tetrahedron*, 1977, **33**, 1945 (*synth, spectra*)

1,7-Heptanediol, 9CI H-00155

1,7-Dihydroxyheptane
[629-30-1]

$HOH_2C(CH_2)_5CH_2OH$

$C_7H_{16}O_2$ M 132.202
Hygroscopic solid. Sol. H_2O, spar. sol. Et_2O. Mp 22.5°. 262° (Bp_3 112-115°).
Di-Ac: [52819-34-8].
$C_{11}H_{20}O_4$ M 216.277
Bp 274°.
Bisphenylurethane: Mp 137°.
Di-Me ether: 1,7-Dimethoxyheptane
$C_9H_{20}O_2$ M 160.256
201° (Bp_{38} 108°).

Huber, W.F., *J. Am. Chem. Soc.*, 1951, **73**, 2730 (*synth*)
Blanc, P.A. *et al, Org. Mass Spectrom.*, 1978, **13**, 135 (*ms*)

2,5-Heptanediol H-00156

[70444-25-6]

$H_3CCH_2CH(OH)CH_2CH_2CH(OH)CH_3$

$C_7H_{16}O_2$ M 132.202

$Bp_{18.5}$ 132°.

Di-Ac:

$C_{11}H_{20}O_4$ M 216.277

Bp_{11} 121-124°.

Bisphenylurethane: Mp 147°.

Helferich, B. *et al*, *Ber.*, 1925, **58**, 1246.

6,28-Heptatriacontadien-2-one, 9CI H-00157

$H_3C(CH_2)_7CH{=}CH(CH_2)_{20}CH{=}CH(CH_2)_3COCH_3$

$C_{37}H_{70}O$ M 530.959

(Z,Z)-form [133530-21-9]

Isol. from the snake *Boiga irregularis.*

Murata, Y. *et al*, *J. Nat. Prod.* (*Lloydia*), 1991, **54**, 233.

8,28-Heptatriacontadien-2-one, 9CI H-00158

$H_3C(CH_2)_7CH{=}CH(CH_2)_{18}CH{=}CH(CH_2)_5COCH_3$

$C_{37}H_{70}O$ M 530.959

(Z,Z)-form [133530-22-0]

Isol. from the snake *Boiga irregularis.*

Murata, Y. *et al*, *J. Nat. Prod.* (*Lloydia*), 1991, **54**, 233 (*isol, pmr, cmr*)

28-Heptatriaconten-2-one, 9CI H-00159

$H_3C(CH_2)_7CH{=}CH(CH_2)_{25}COCH_3$

$C_{37}H_{72}O$ M 532.975

(Z)-form [130385-26-1]

Isol. from the snakes *Boiga irregularis* and *Thamnophis sirtalis parietalis.* Sex attractant pheromone.

Mason, R.T. *et al*, *J. Chem. Ecol.*, 1990, **16**, 2353 (*isol*)
Murata, Y. *et al*, *J. Nat. Prod.* (*Lloydia*), 1991, **54**, 233 (*isol, pmr, cmr*)

2-Heptenal, 9CI H-00160

[2463-63-0]

$H_3C(CH_2)_3CH{=}CHCHO$

$C_7H_{12}O$ M 112.171

Flavouring constit. of many foods. Isol. from soya bean oil (*Glycine max*).

(E)-form [18829-55-5]

Bp 165-167°, Bp_{15} 61-62°.

2,4-Dinitrophenylhydrazone: [2122-10-3]. Mp 132-133°.

(Z)-form [57266-86-1]

Bp_{16} 64-65°.

Di-Et acetal: 1,1-Diethoxy-1-heptene

$C_{11}H_{22}O_2$ M 186.294

Bp_{15} 90-92°.

[20810-22-4]

Fujimaki, M. *et al*, *Agric. Biol. Chem.*, 1965, **29**, 855 (*isol*)
Tadema, G. *et al*, *CA*, 1976, **85**, 5141m (*synth*)
Wada, M. *et al*, *Chem. Lett.*, 1977, 345 (*synth*)
Yamamoto, K. *et al*, *Synthesis*, 1977, 721 (*synth, ir, pmr*)
Bestmann, H.J. *et al*, *Chem. Ber.*, 1982, **115**, 161 (*synth*)
Huang, Y.-Z. *et al*, *J. Org. Chem.*, 1987, **52**, 3558 (*synth, ir, ms, pmr*)

4-Heptenal, 9CI H-00161

[62238-34-0]

$H_3CCH_2CH{=}CHCH_2CH_2CHO$

$C_7H_{12}O$ M 112.171

(E)-form [929-22-6]

Constit. of black Hemiptera bugs. Oil with nauseous odour. Bp 112-113°.

2,4-Dinitrophenylhydrazone: Dull-orange needles (EtOH). Mp 120°.

(Z)-form [6728-31-0]

Component of butter. Flavouring agent. Liq. with creamy odour. Bp_{12} 42-44°.

Mukerji, S.K. *et al*, *Tetrahedron Lett.*, 1966, 2479 (*synth*)
Stetter, H. *et al*, *Synthesis*, 1975, 379 (*synth*)

5-Heptenal H-00162

$H_3CCH{=}CHCH_2CH_2CH_2CHO$

$C_7H_{12}O$ M 112.171

(E)-form [21662-19-1]

Bp_{85} 90-92°.

2,4-Dinitrophenylhydrazone: Cryst. (EtOH aq.). Mp 109.5-110°.

(Z)-form [21661-98-3]

Bp_{20} 58°. Cont. 20% *E*-isomer.

2,4-Dinitrophenylhydrazone: Cryst. (EtOH). Mp 94-95°.

LeBel, N.A. *et al*, *J. Am. Chem. Soc.*, 1964, **86**, 3759 (*synth, ir*)
Oppolzer, W. *et al*, *Tetrahedron*, 1985, **41**, 3497 (*synth, ir, pmr, ms*)

6-Heptenal H-00163

[17206-61-0]

$H_2C{=}CH(CH_2)_4CHO$

$C_7H_{12}O$ M 112.171

Liq. Bp_{80} 88-90°.

2,4-Dinitrophenylhydrazone: Cryst. (EtOH aq.). Mp 95-95.8°.

LeBel, N.A. *et al*, *J. Am. Chem. Soc.*, 1964, **86**, 3759 (*synth, ir*)

4-Heptenoic acid, 9CI H-00164

Hex-3-ene-1-carboxylic acid

[35194-37-7]

$H_3CCH_2CH{=}CH(CH_2)_2COOH$

$C_7H_{12}O_2$ M 128.171

Isol. as the Me ester from oil of *Humulus lupulus.* d^{15} 0.949. Bp_5 104-109°. n_D^{15} 1.4418. This data prob. refers to the (*E*)-form.

Chloride: [74132-57-3].

$C_7H_{11}ClO$ M 146.616

Bp_{14} 66.5°.

(Z)-form [41653-95-6]

Me ester: [39924-30-6].

$C_8H_{14}O_2$ M 142.197

Bp_4 106.5°.

Hunsdiecker, H., *Chem. Ber.*, 1942, **75**, 460.
Buttery, R.G. *et al*, *CA*, 1965, **62**, 1050 (*isol*)
Furukawa, Y. *et al*, *CA*, 1976, **84**, 150153a.
Sato, T. *et al*, *Bull. Chem. Soc. Jpn.*, 1981, **54**, 505.
Kawashima, M. *et al*, *Bull. Chem. Soc. Jpn.*, 1988, **61**, 3255 (*synth, pmr*)

5-Heptenoic acid, 9CI **H-00165**

[3593-00-8]

$H_3CCH{=}CH(CH_2)_3COOH$

$C_7H_{12}O_2$ M 128.171

(E)-form [18776-90-4]

Bp_{16} 120-121°.

Nitrile: [55999-04-7]. *5-Heptenenitrile, 9CI. 6-Cyano-2-pentene*

$C_7H_{11}N$ M 109.171

Bp_{25} 90°.

4-Bromophenacyl ester: Plates (EtOH aq.). Mp 62-63°.

(Z)-form

Bp_{13} 116-117°.

4-Bromophenacyl ester: Cryst. (MeOH). Mp 49-50°.

Ansell, M.F. *et al, J. Chem. Soc.*, 1958, 1788 (*synth*)
Ansell, M.F. *et al, J. Chem. Soc. C*, 1968, 217 (*synth*)
Chiusoli, G.P. *et al, Gazz. Chim. Ital.*, 1973, **103**, 569.
Denmark, S.E. *et al, J. Org. Chem.*, 1987, **52**, 877 (*deriv, synth, pmr*)

6-(1-Heptenyl)-5,6-dihydro-2*H*-pyran-2-one, 9CI **H-00166**

2,6-Dodecadien-5-olide

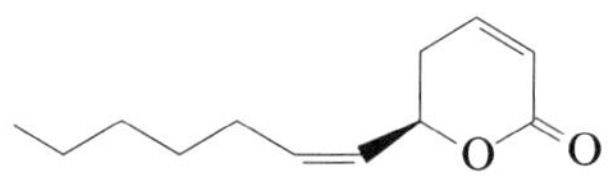

$C_{12}H_{18}O_2$ M 194.273

(R,Z)-form [64543-31-3] ***Argentilactone***

Constit. of *Aristolochia argentina*. $Bp_{0.6}$ 139-143°. $[\alpha]_D$ −21.1° (c, 2.25 in EtOH).

Priestap, H.A. *et al, Phytochemistry*, 1977, **16**, 1579 (*isol, uv, ir, pmr, ms*)
Fehr, C. *et al, Helv. Chim. Acta*, 1981, **64**, 1247 (*synth*)
O'Connor, B. *et al, Tetrahedron Lett.*, 1986, **27**, 5201 (*synth*)

2-Heptylcyclopropanenonanoic acid, 9CI **H-00167**

10,11-Methyleneoctadecanoic acid

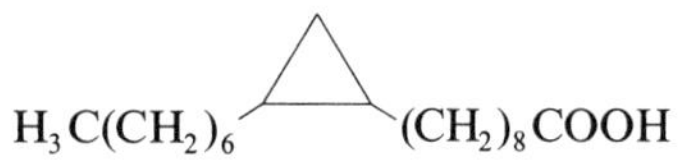

$C_{19}H_{36}O_2$ M 296.492

Both *cis-* and *trans-*forms have been synthesised as Me esters.

Christie, W.N. *et al, Chem. Phys. Lipids*, 1968, **2**, 168 (*synth, glc, ir, nmr*)
Christie, W.W., *Top. Lipid Chem.*, 1970, **1**, 4 (*rev*)
Gunstone, F.D. *et al, Chem. Phys. Lipids*, 1973, **10**, 303 (*synth, glc, nmr*)

6-Heptyltetrahydro-2*H*-pyran-2-one, 9CI **H-00168**

δ-Dodecalactone. Dodecan-5-olide

[16400-73-0]

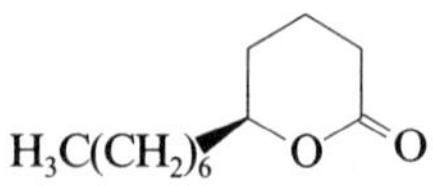

(S)-form

$C_{12}H_{22}O_2$ M 198.305

Isol. from coconut oil.

(S)-form [108943-47-1]

$[\alpha]_D$ −48.8°.

(±)-form [3051-22-7]

Liq. Mp −12°. $Bp_{0.03}$ 100-103°. n_D^{25} 1.4684 (1.4571).

Taylor, R.J. *et al, CA*, 1961, **55**, 390 (*synth*)
Allen, R.R. *et al, Chem. Ind. (London)*, 1965, 1560 (*isol, ir*)
Parliament, T.H. *et al, Chem. Ind. (London)*, 1966, 1845 (*synth*)
Lamberti, V. *et al, Recl. Trav. Chim. Pays-Bas (J. R. Neth. Chem. Soc.)*, 1967, **86**, 504 (*synth*)
Kohver, O., *Tetrahedron*, 1970, **26**, 2391 (*abs config, ord, cd*)
Bianco, L. *et al, Tetrahedron Lett.*, 1988, **29**, 1915.

2-Heptynal **H-00169**

[1846-67-9]

$H_3CCH_2CH_2CH_2C{\equiv}CCHO$

$C_7H_{10}O$ M 110.155

Liq. Bp_5 45°.

Semicarbazone: Cryst. Mp 72°.

2,4-Dinitrophenylhydrazone: Cryst. (EtOH). Mp 73°.

Di-Et acetal: 1,1-Diethoxy-2-heptyne

$C_{11}H_{20}O_2$ M 184.278

Bp_{12} 102°.

Durand, M.-H., *Bull. Soc. Chim. Fr.*, 1961, 2387 (*acetal*)
Kanakam, C.C. *et al, J. Chem. Soc., Perkin Trans. 1*, 1989, 1907 (*synth, ir, pmr*)

3-Heptynoic acid **H-00170**

[59862-93-0]

$H_3CCH_2CH_2C{\equiv}CCH_2COOH$

$C_7H_{10}O_2$ M 126.155

Liq. Bp_{22} 140-141°.

Bigley, D.B. *et al, J. Chem. Soc., Perkin Trans. 2*, 1972, 592 (*synth*)

4-Heptynoic acid **H-00171**

[42441-83-8]

$H_3CCH_2C{\equiv}CCH_2CH_2COOH$

$C_7H_{10}O_2$ M 126.155

Needles (pet. ether). Mp 57-59°.

4-Bromophenacyl ester: Plates (EtOH). Mp 99-100°.

Nitrile: [18719-33-0]. *4-Heptynenitrile, 9CI. 1-Cyano-3-hexyne*

C_7H_9N M 107.155

Bp_{16} 82-83°.

Ansell, M.F. *et al, J. Chem. Soc. C*, 1968, 217 (*synth*)

6-Heptynoic acid **H-00172**

[30964-00-2]

$HC{\equiv}C(CH_2)_4COOH$

$C_7H_{10}O_2$ M 126.155

Liq. Bp_1 93-94°.

Me ester:

$C_8H_{12}O_2$ M 140.182

Bp_{11} 80°.

Nitrile: [15295-69-9]. *6-Heptynenitrile, 9CI. 6-Cyano-1-hexyne*

C_7H_9N M 107.155

Bp_{14} 87-93°.

Taylor, W.R. *et al, J. Am. Chem. Soc.*, 1950, **72**, 4263 (*synth*)
Ferrier, R.J. *et al, J. Chem. Soc.*, 1957, 1435 (*synth, nitrile*)
Gautier, J.A. *et al, Bull. Soc. Chim. Fr.*, 1967, 1551 (*nitrile*)
Moody, C.J. *et al, J. Chem. Soc., Perkin Trans. 1*, 1988, 3249 (*synth, ir, pmr, ms*)

2-Heptyn-1-ol, 9CI, 8CI **H-00173**

[1002-36-4]

$H_3C(CH_2)_3C{\equiv}CCH_2OH$

$C_7H_{12}O$ M 112.171

Bp_{56} 113-116°.

▷ Distillation residue may explode.

3,5-Dinitrobenzoyl: Mp 62°.

Newman, M.S. *et al*, *J. Am. Chem. Soc.*, 1949, **71**, 1292 (*synth*)
Bus, J. *et al*, *Recl. Trav. Chim. Pays-Bas* (*J. R. Neth. Chem. Soc.*), 1972, **91**, 229 (*ir*)
Kanakam, C.C. *et al*, *J. Chem. Soc., Perkin Trans. 1*, 1989, 1907 (*synth, ir, pmr*)
Bretherick, L., *Handbook of Reactive Chemical Hazards*, 2nd Ed., Butterworths, London and Boston, 1979, 640.
Hazards in the Chemical Laboratory, (Bretherick, L., Ed.), 3rd Ed., Royal Society of Chemistry, London, 1981, 344.
Lewis, R.J., *Sax's Dangerous Properties of Industrial Materials*, 8th Ed., Van Nostrand-Reinhold, 1992, HBS500.

6-Heptyn-1-ol **H-00174**

[63478-76-2]

$HC{\equiv}C(CH_2)_4CH_2OH$

$C_7H_{12}O$ M 112.171

Liq. Bp_{20} 105°.

Knapp, F.F. *et al*, *J. Med. Chem.*, 1984, **27**, 57 (*synth, pmr*)
Theobald, P.G. *et al*, *J. Org. Chem.*, 1990, **55**, 741 (*synth*)

9,10,12,13,15,16-Hexabromooctadecanoic acid **H-00175**

Hexabromostearic acid. Linolenic acid hexabromide

[4167-08-2]

$H_3C(CH_2CHBrCHBr)_3(CH_2)_7COOH$

$C_{18}H_{30}Br_6O_2$ M 757.858

3 forms known.

α-form

Cryst. (dioxan). Insol. Et_2O. Mp 181.9°.

Me ester:

$C_{19}H_{32}Br_6O_2$ M 771.885

Mp 157-158°.

Et ester:

$C_{20}H_{34}Br_6O_2$ M 785.911

Mp 151.5-152.6°.

β-form

Mp 169-170°.

Et ester: Mp 114-115°.

γ-form

Needles (C_6H_6). Mp 195-196° dec.

Vincente, M.L.A. *et al*, *CA*, 1928, **22**, 4105 (*synth*)
Stanfield, K.E. *et al*, *J. Am. Chem. Soc.*, 1932, **54**, 4358.
Kass, J.P. *et al*, *J. Am. Chem. Soc.*, 1941, **63**, 1060.
Org. Synth., 1942, **22**, 82 (*synth*)

Hexacontane, 9CI **H-00176**

[7667-80-3]

$H_3C(CH_2)_{58}CH_3$

$C_{60}H_{122}$ M 843.624

Mp 98.5-99.3°.

Carothers, W.H. *et al*, *J. Am. Chem. Soc.*, 1930, **52**, 5279 (*synth*)
Heitz, W. *et al*, *Makromol. Chem.*, 1972, **162**, 63 (*synth*)

5,9-Hexacosadienoic acid **H-00177**

[59708-84-8]

$H_3C(CH_2)_{15}CH{=}CHCH_2CH_2CH{=}CH(CH_2)_3COOH$

$C_{26}H_{48}O_2$ M 392.664

(Z,Z)-form

Constit. of marine sponges *Microciona prolifera* and *Xestospongia halichondroides.*

Jefferts, E. *et al*, *Lipids*, 1974, **9**, 244 (*ir, nmr, glc*)
Morales, R.W. *et al*, *Lipids*, 1977, **12**, 570.
Litchfield, C. *et al*, *Lipids*, 1978, **13**, 199 (*glc*)
Bäckvall, J.E. *et al*, *Acta Chem. Scand., Ser. B*, 1988, **42**, 397 (*synth*)
Hahn, S. *et al*, *J. Am. Chem. Soc.*, 1988, **110**, 8117 (*synth, pmr, ms*)

Hexacosanal **H-00178**

[26627-85-0]

$H_3C(CH_2)_{24}CHO$

$C_{26}H_{52}O$ M 380.696

Isol. from *Vitis vinifera* var. *sultana.* Cryst. (MeOH). Mp 73-73.5°.

Crabbé, P. *et al*, *Bull. Soc. Chim. Belg.*, 1961, **70**, 168 (*synth, ir*)
Radler, F. *et al*, *Aust. J. Chem.*, 1965, **18**, 1059 (*isol*)

Hexacosane, 9CI **H-00179**

Cerane

[630-01-3]

$H_3C(CH_2)_{24}CH_3$

$C_{26}H_{54}$ M 366.713

Occurs in many biological systems including *Chrysanthemum indicum.* Mp 56.4°, Mp 59-60°. $Bp_{0.05}$ 169°.

Whitmore, F.C. *et al*, *J. Am. Chem. Soc.*, 1945, **67**, 2059 (*synth*)
Reipe, W. *et al*, *Adv. Mass Spectrom. Biochem. Med., Proc. Int. Symp. Mass Spectrom. Biochem. Med. 2nd, 3rd, 1974, 1975*, 1976, **1**, 401 (*ms*)

Hexacosanedioic acid **H-00180**

[3365-67-1]

$HOOC(CH_2)_{24}COOH$

$C_{26}H_{50}O_4$ M 426.679

Mp 123.5°.

Di-Et ester: [28353-88-0].

$C_{30}H_{58}O_4$ M 482.786

Mp 66°.

Fairweather, D.A., *Proc. R. Soc. London*, 1926, **46**, 71 (*synth*)
Bongini, A. *et al*, *Ann. Chim. (Rome)*, 1970, **60**, 155.

1,26-Hexacosanediol **H-00181**

[15541-01-2]

$HOCH_2(CH_2)_{24}CH_2OH$

$C_{26}H_{54}O_2$ M 398.711

Constit. of *Spartium junceum* flowers and carnauba wax. Also from apple and in esterified form from *Avena sativa.* Cryst. (C_6H_6/EtOAc). Mp 110.5° (102-105°).

Di-Ac: [15541-04-5].

$C_{30}H_{58}O_4$ M 482.786

Mp 77.5°.

Dibenzoyl:

$C_{40}H_{62}O_4$ M 606.927

Cryst. (MeOH). Mp 67-70°.

Musgrave, O.C. *et al*, *J. Chem. Soc.*, 1952, 4393 (*isol, synth*)
Murray, K.E. *et al*, *Aust. J. Chem.*, 1955, **8**, 432 (*isol*)
Mazliak, P., *Phytochemistry*, 1962, **1**, 79 (*isol*)
Daniels, D.G.H. *et al*, *Chem. Ind.* (*London*), 1965, 1763 (*isol*)

Hexacosanoic acid, 9CI — H-00182

Cerinic acid. Cerotic acid. Ceric acid. Ceratinic acid

[506-46-7]

$H_3C(CH_2)_{24}COOH$

$C_{26}H_{52}O_2$ M 396.696

Constit. of various oils, waxes and bacterial lipids. Mp 88-89°.

Me ester: [5802-82-4].
$C_{27}H_{54}O_2$ M 410.722
Mp 63-68°. Bp_{15} 286°, $Bp_{3.5}$ 248°.

Et ester: [29030-81-7].
$C_{28}H_{56}O_2$ M 424.749
Mp 59.5-59.8°.

Amide: [22715-86-2].
$C_{26}H_{53}NO$ M 395.711
Mp 105-107°.

2-Methylpropylamide: [119736-87-7]. N-(*2-Methylpropyl*) *hexacosanamide, 9CI. Hexacosanoic acid isobutylamide.* ***Longamide***
$C_{30}H_{61}NO$ M 451.818
Minor constit. of fruits of *Piper longum* (Piperaceae). Amorph. powder. Mp 72°.

Levene, P.A. *et al*, *J. Biol. Chem.*, 1924, **59**, 905 (*synth*)
Helz, G.E. *et al*, *J. Biol. Chem.*, 1936, **116**, 203.
Bernhard, K. *et al*, *Helv. Chim. Acta*, 1948, **31**, 977.
Asselineau, J., *C. R. Hebd. Seances Acad. Sci.*, 1953, **237**, 1804.
Tember, G.A. *et al*, *CA*, 1974, **80**, 47410a (*synth*)
Watanabe, S. *et al*, *Z. Naturforsch., C*, 1975, **30**, 825 (*ms*)
Koul, S.K. *et al*, *Phytochemistry*, 1988, **27**, 3523 (*Longamide*)

1-Hexacosanol, 9CI — H-00183

Ceryl alcohol

[506-52-5]

$H_3C(CH_2)_{24}CH_2OH$

$C_{26}H_{54}O$ M 382.712

Occurs as fatty acid esters in Chinese insect wax and many other waxes. Mp 80°. Bp_{70} 305° dec.

Ac: [822-32-2].
$C_{28}H_{56}O_2$ M 424.749
Mp 65° (62-63°).

Rabate, J., *Bull. Soc. Chim. Biol.*, 1930, **12**, 758.
Chibnall, A.C. *et al*, *Biochem. J.*, 1934, **28**, 2189 (*isol*)
Chandra, G. *et al*, *J. Chem. Soc.*, 1964, 3648 (*isol*)
Karrer, W. *et al*, *Konstitution und Vorkommen der Organischen Pflanzenstoffe*, 2nd Ed., Birkhäuser Verlag, Basel, 1972-1985, no. 95 (*occur*)
Precht, D., *Fette, Seifen, Anstrichm.*, 1976, **78**, 145 (*cryst struct*)

9-Hexacosanone — H-00184

9-Oxohexacosane

$H_3C(CH_2)_{16}CO(CH_2)_7CH_3$

$C_{26}H_{52}O$ M 380.696

Cryst. (MeOH). Mp 67-68°.

Oxime:
$C_{26}H_{53}NO$ M 395.711
Mp 40-41°.

Whitmore, F.C. *et al*, *J. Am. Chem. Soc.*, 1945, **67**, 2059 (*synth*)

5,9,17-Hexacosatrienoic acid, 9CI — H-00185

[59708-88-2]

$H_3C(CH_2)_7CH{=}CH(CH_2)_6CH{=}CH$
$CH_2CH_2CH{=}CH(CH_2)_3COOH$

$C_{26}H_{46}O_2$ M 390.648

(all-Z)-form

Component of lipids of the sponge *Microciona prolifera*.

Morales, R.W. *et al*, *Biochim. Biophys. Acta*, 1976, **431**, 206 (*occur*)

5,9,19-Hexacosatrienoic acid, 8CI — H-00186

[59708-89-3]

$H_3C(CH_2)_5CH{=}CH(CH_2)_8CH{=}CH$
$CH_2CH_2CH{=}CH(CH_2)_3COOH$

$C_{26}H_{46}O_2$ M 390.648

(Z,Z,Z)-form [52715-56-7]

Constit. of marine sponge *Microciona prolifera*.

Jefferts, E. *et al*, *Lipids*, 1974, **9**, 244 (*ir, nmr, glc*)
Morales, R.W. *et al*, *Lipids*, 1977, **12**, 570.

19-Hexacosenal — H-00187

$H_3C(CH_2)_5CH{=}CH(CH_2)_{17}CHO$

$C_{26}H_{50}O$ M 378.680

(Z)-form [140163-49-1]

Constit. of the sponge *Amphimedon compressa*.

Carballeira, N.M. *et al*, *J. Nat. Prod.* (*Lloydia*), 1992, **55**, 333.

9-Hexacosenoic acid — H-00188

[59708-77-9]

$H_3C(CH_2)_{15}CH{=}CH(CH_2)_7COOH$

$C_{26}H_{50}O_2$ M 394.680

(Z)-form [86901-41-9]

Isol. from the marine sponge *Microciona prolifera*.

Morales, R.W. *et al*, *Biochim. Biophys. Acta*, 1976, **431**, 206 (*isol*)
Cervilla, M. *et al*, *Anal. Chem.*, 1983, **55**, 2100 (*ms*)
Lam, W. *et al*, *J. Org. Chem.*, 1989, **54**, 3428 (*isol*)

17-Hexacosenoic acid — H-00189

[544-84-3]

$H_3C(CH_2)_7CH{=}CH(CH_2)_{15}COOH$

$C_{26}H_{50}O_2$ M 394.680

(Z)-form [66274-43-9]

Ximenic acid

Isol. from seed fat of *Tropaeolum speciosum* and *Ximenia* spp. Also in fish oil lipids and sponges. Mp 50.5-50.9°.

Me ester: Oil. Ir ν 1744 cm^{-1}.

Lightelm, S.P. *et al*, *J. Sci. Food Agric.*, 1954, **5**, 281 (*isol, config, bibl*)
Litchfield, C. *et al*, *Lipids*, 1970, **5**, 144; 1978, **13**, 199 (*isol*)
Kling, M.R. *et al*, *J. Chem. Soc., Perkin Trans. 1*, 1993, 1183 (*synth, ir, ms*)

6,11-Hexadecadienal, 9CI — H-00190

[114915-41-2]

$H_3C(CH_2)_3CH{=}CH(CH_2)_3CH{=}CH(CH_2)_4CHO$

$C_{16}H_{28}O$ M 236.397

(6E,11Z)-form [58701-06-7]
Pheromone for the moths *Antheraea polyphemus* and *A. pernyi*. Oil. $Bp_{0.1}$ 120°.
(6Z,11E)-form [83483-58-3]
$Bp_{0.1}$ 120-126°.

[92232-34-3, 104837-06-1]

Kochansky, J.P. *et al*, *J. Chem. Ecol.*, 1977, **3**, 419 (*synth, ir, glc*)
Tomida, I. *et al*, *Agric. Biol. Chem.*, 1986, **50**, 2143 (*synth*)
Bestmann, H.-J., *Z. Naturforsch., C*, 1987, **42**, 631.
Rastegaeva, V.H. *et al*, *Zh. Org. Khim.*, 1989, **25**, 2066.
Bestmann, H.-J. *et al*, *Synthesis*, 1990, 1039.

7,11-Hexadecadienal, 9CI H-00191

$H_3C(CH_2)_3CH{=}CHCH_2CH_2CH{=}CH(CH_2)_5CHO$

$C_{16}H_{28}O$ M 236.397
(7Z,11Z)-form [96883-53-3]
Component of the sex pheromone of *Phyllocnistis wampella*, and citrus leafminer moth, *P. citrella*. Liq.

[96883-54-4, 131392-75-1, 131400-46-9]

Andelic, I. *et al*, *Acta Chem. Scand., Ser. B*, 1985, **39**, 231 (*synth, pmr, ir, ms, glc*)
Ando, T. *et al*, *Agric. Biol. Chem.*, 1985, **49**, 3633 (*isol, activity, synth, pmr, ir, ms, glc*)
Odinokov, V.N. *et al*, *Zh. Org. Khim.*, 1990, **26**, 1619 (*synth*)
Yen, Y.P. *et al*, *J. Chin. Chem. Soc. (Peking)*, 1991, **38**, 293; *CA*, **115**, 207457t (*isol, synth, activity*)

9,11-Hexadecadienal, 9CI H-00192

$H_3C(CH_2)_3CH{=}CHCH{=}CH(CH_2)_7CHO$

$C_{16}H_{28}O$ M 236.397
(9Z,11E)-form [77822-24-3]
Component of the sex pheromone of the leek moth, *Acrolepiopsis assectella* and *Diatraea saccharalis*.

[74728-46-4, 84643-63-0, 92080-87-0]

Renou, M. *et al*, *Entomol. Exp. Appl.*, 1981, **29**, 198 (*activity*)
U.S. Pat., 4 357 474, (1983); *CA*, **98**, 125738b.
Svirskaya, P.L. *et al*, *J. Chem. Ecol.*, 1984, **10**, 795 (*synth, glc, cmr*)
Ando, T. *et al*, *Agric. Biol. Chem.*, 1988, **52**, 1415 (*ms*)

10,12-Hexadecadienal H-00193
Bombykal

$H_3CCH_2CH_2CH{=}CHCH{=}CH(CH_2)_8CHO$

$C_{16}H_{28}O$ M 236.397
Trimerises on standing.
(10E,12Z)-form [63024-98-6]
Minor pheromone of the silkworm *Bombyx mori* and tobacco hornworm *Manduca sexta*. $Bp_{0.01}$ 135-145°.
(10E,12E)-form
Pheromone of spiny bollworm *Earias insulana*.

Bestmann, H.J. *et al*, *Tetrahedron Lett.*, 1977, 121 (*isol*)
Kasang, G. *et al*, *Angew. Chem., Int. Ed. Engl.*, 1978, **17**, 60 (*isol*)
Hall, D.R. *et al*, *Experientia*, 1980, **36**, 152.
Bestmann, H.J. *et al*, *Justus Liebigs Ann. Chem.*, 1981, 2117 (*synth, pmr*)
Klug, J.T. *et al*, *Chem. Ind. (London)*, 1982, 372 (*synth*)
Dunkelblum, E. *et al*, *J. Chem. Ecol.*, 1984, **10**, 421 (*trimer*)
Yadav, J.S. *et al*, *Synth. Commun.*, 1989, **19**, 125 (*synth*)
Lee, J.K. *et al*, *J. Neurocytol.*, 1990, **69**, 519.
Kanzaki, R. *et al*, *J. Comp. Physiol., A*, 1991, **168**, 281.

11,13-Hexadecadienal H-00194

$H_3CCH_2CH{=}CHCH{=}CH(CH_2)_9CHO$

$C_{16}H_{28}O$ M 236.397
Oil.
(E,E)-form [73264-91-2]
Sex pheromone of the cabbage webworm *Hellulla undalis fabricius*.
(Z,Z)-form [71317-73-2]
Pheromone of the navel orangeworm *Pamyelois transitella*. Liq. $Bp_{0.1}$ 140° (bath).

Bishop, C.E. *et al*, *J. Org. Chem.*, 1983, **48**, 657 (*synth*)
Michelot, D., *Synthesis*, 1983, 130 (*synth*)
Takayama, H. *et al*, *J. Org. Chem.*, 1986, **51**, 4934 (*synth, ms, pmr, cmr*)
Lo, V.M. *et al*, *Synth. Commun.*, 1986, **16**, 1647 (*synth, pmr, bibl*)

1,15-Hexadecadiene, 8CI H-00195
[21964-51-2]

$H_2C{=}CH(CH_2)_{12}CH{=}CH_2$

$C_{16}H_{30}$ M 222.413

Casteignau, G. *et al*, *Bull. Soc. Chim. Fr.*, 1968, 3893, 3904 (*synth, glc*)

3,13-Hexadecadiene-1,15-diyne H-00196

$HC{\equiv}CCH{=}CH(CH_2)_8CH{=}CHC{\equiv}CH$

$C_{16}H_{22}$ M 214.350
(3Z,13Z)-form [137319-24-5] ***Callydiyne***
Isol. from the sponge *Callyspongia flammea*. Oil.

Miao, S. *et al*, *J. Nat. Prod. (Lloydia)*, 1991, **54**, 1433 (*isol, pmr, cmr, ms, struct*)

7,14-Hexadecadiene-10,12-diyn-1-ol, 8CI H-00197

$H_3CCH{=}CHC{\equiv}CC{\equiv}CCH_2CH{=}CH(CH_2)_5CH_2OH$

$C_{16}H_{22}O$ M 230.349
(7E,14Z)-form [16697-15-7]
Constit. of *Dahlia merckii*.
(7Z,14E)-form
Constit. of *Centaurea macrocephala*. Oil.

Bohlmann, F. *et al*, *Chem. Ber.*, 1967, **100**, 1936; 1970, **103**, 2100.

6,8-Hexadecadiene-10,12,14-triynoic acid, 9CI H-00198

$H_3CC{\equiv}CC{\equiv}CC{\equiv}CCH{=}CHCH{=}CH(CH_2)_4COOH$

$C_{16}H_{16}O_2$ M 240.301
(all-E)-form
Me ester: [40241-96-1].
$C_{17}H_{18}O_2$ M 254.328
Constit. of *Chrysocoma* spp. Mp 77°.

Bohlmann, F. *et al*, *Chem. Ber.*, 1972, **105**, 3587.

1,8-Hexadecadiene-10,12,14-triyn-7-ol, 9CI H-00199

$H_3CC{\equiv}CC{\equiv}CC{\equiv}CCH{=}CHCH(OH)(CH_2)_4CH{=}CH_2$

$C_{16}H_{18}O$ M 226.318
(E)-form [113762-93-9]
Isol. from aerial parts of *Leucanthemum pallens*. Oil.
λ_{max} 330, 308, 290, 273, 258, 242, 232 nm.

Bellido, I.S. *et al*, *Phytochemistry*, 1988, **27**, 282 (*isol, struct, ir, uv, pmr, cmr*)

2,4-Hexadecadienoic acid, 9CI H-00200

$H_3C(CH_2)_{10}CH{=}CHCH{=}CHCOOH$

$C_{16}H_{28}O_2$ M 252.396

(all-E)-form [59404-48-7]

2-Methylpropylamide: ***N-Isobutyl-2,4-hexadecadienamide***
$C_{20}H_{37}NO$ M 307.518
Constit. of the fruits of *Piper guineense* and *Evodia huipephensis* (Piperaceae, Rutaceae). Noncryst.

(2E,4Z)-form

Me ester: [54977-89-8].
$C_{17}H_{30}O_2$ M 266.423
Used as aroma compd. in the food industry.

Et ester:
$C_{18}H_{32}O_2$ M 280.450
Aroma compd.

Okogun, J.I. *et al*, *J. Chem. Soc., Perkin Trans. 1*, 1974, 2195 (*pmr, ms*)
Swiss Pat., 557 645, (1974); *CA*, **82**, 138023 (*synth*)
Vig, O.P. *et al*, *Indian J. Chem.*, 1975, **13**, 1358 (*synth*)
Reisch, J. *et al*, *J. Nat. Prod. (Lloydia)*, 1985, **48**, 862 (*isol, uv, ir, pmr, ms, struct, isobutylamide*)

3,9-Hexadecadienoic acid, 9CI H-00201

$H_3C(CH_2)_5CH{=}CH(CH_2)_4CH{=}CHCH_2COOH$

$C_{16}H_{28}O_2$ M 252.396

(all-Z)-form [35087-13-9]
Cryst. + $\frac{1}{2}$ H_2O (pet. ether). Mp 20-22°. $Bp_{0.1}$ 140°.

Ames, D.E. *et al*, *J. Chem. Soc., Perkin Trans. 1*, 1972, 255.

6,8-Hexadecadienoic acid H-00202

$H_3C(CH_2)_6CH{=}CHCH{=}CH(CH_2)_4COOH$

$C_{16}H_{28}O_2$ M 252.396

(6E,8Z)-form

Me ester: [80625-49-6].
$C_{17}H_{30}O_2$ M 266.423
$Bp_{0.01}$ 110-115°.

Bestmann, H.J. *et al*, *Justus Liebigs Ann. Chem.*, 1981, 2117 (*synth, pmr*)

6,9-Hexadecadienoic acid H-00203
[4780-50-1]

$H_3C(CH_2)_5CH{=}CHCH_2CH{=}CH(CH_2)_4COOH$

$C_{16}H_{28}O_2$ M 252.396

(Z,Z)-form
A minor component of rat liver lipids and of some fish lipids. Metab. of 9-Hexadecenoic acid.

Schmitte, B. *et al*, *Lipids*, 1977, **12**, 307.
Moreno, V.J. *et al*, *Lipids*, 1979, **14**, 15.

8,10-Hexadecadienoic acid H-00204

$H_3C(CH_2)_4CH{=}CHCH{=}CH(CH_2)_6COOH$

$C_{16}H_{28}O_2$ M 252.396

(all-Z)-form
Low-melting solid.

4-Bromophenacyl ester: Mp 74-75°.

Gunstone, F.D. *et al*, *J. Chem. Soc.*, 1962, 3455 (*synth*)

9,12-Hexadecadienoic acid H-00205
[20261-45-4]

$H_3CCH_2CH_2CH{=}CHCH_2CH{=}CH(CH_2)_7COOH$

$C_{16}H_{28}O_2$ M 252.396
Isol. from the seed fat and seed pod fat of *Acacia giraffae* and from seed oil of *Asclepias syriaca*.

(Z,Z)-form [5070-03-1]
Isol. from the butterfly *Prodenia litura*.

Harrison, G.S. *et al*, *J. S. Afr. Chem. Inst.*, 1952, **5**, 23; *CA*, **48**, 7320h (*isol*)
Stoffel, W. *et al*, *J. Am. Chem. Soc.*, 1958, **80**, 6604 (*isol, struct*)
Chisholm, M.J. *et al*, *Can. J. Chem.*, 1960, **38**, 805 (*isol, struct*)
Zayed, S.M.A. *et al*, *Hoppe Seyler's Z. Physiol. Chem.*, 1965, **341**, 91 (*isol, struct*)
Schlenk, H. *et al*, *Biochim. Biophys. Acta*, 1967, **144**, 305.

10,12-Hexadecadienoic acid H-00206

$H_3CCH_2CH_2CH{=}CHCH{=}CH(CH_2)_8COOH$

$C_{16}H_{28}O_2$ M 252.396

(10E,12Z)-form

Me ester: [63024-91-9].
$C_{17}H_{30}O_2$ M 266.423
$Bp_{0.03}$ 116-117°.

Bestmann, H.J. *et al*, *Justus Liebigs Ann. Chem.*, 1981, 2117 (*synth, pmr*)

6,8-Hexadecadien-1-ol H-00207

$H_3C(CH_2)_6CH{=}CHCH{=}CH(CH_2)_4CH_2OH$

$C_{16}H_{30}O$ M 238.412

(6E,8Z)-form
$Bp_{0.01}$ 110-115°.

Ac: [80625-65-6].
$C_{18}H_{32}O_2$ M 280.450
$Bp_{0.01}$ 110-115°.

Bestmann, H.J. *et al*, *Justus Liebigs Ann. Chem.*, 1981, 2117 (*synth, pmr*)

6,11-Hexadecadien-1-ol, 9CI H-00208

$H_3C(CH_2)_3CH{=}CH(CH_2)_3CH{=}CH(CH_2)_4CH_2OH$

$C_{16}H_{30}O$ M 238.412

(6E,11Z)-form [65370-61-8]
Yellow oil.

Ac: [58701-05-6].
$C_{18}H_{32}O_2$ M 280.450
Pheromone for the moths *Antheraea polyphemus* and *A. pernyi*. Pale-yellow oil.

(6Z,11Z)-form

Ac: Pheromone for *Eudia pavonia* and *Drosophila melanogaster*.

Kochansky, J.P. *et al*, *J. Chem. Ecol.*, 1977, **3**, 419 (*synth, ir, glc*)
Bestmann, H.-J., *Z. Naturforsch., C*, 1986, **42**, 631.
Ducoux, J.P. *et al*, *Tetrahedron*, 1992, **48**, 6403 (*synth*)

7,11-Hexadecadien-1-ol, 9CI H-00209
[59741-58-1]

$H_3C(CH_2)_3CH{=}CHCH_2CH_2CH{=}CH(CH_2)_5CH_2OH$

$C_{16}H_{30}O$ M 238.412
The acetate (mixt. of *E*- and *Z*-isomers) is known as Gossyplure. Sex pheromone of *Sitotroga cerealella*.

(all-E)-form [53963-09-0]
Ac: [53042-81-2].
$C_{18}H_{32}O_2$ M 280.450
$Bp_{0.12}$ 97-99°, $Bp_{0.05}$ 118°.
(7Z,11E)-form [56879-24-4]
$Bp_{0.35}$ 121-122°.
Ac: [51607-94-4].
Pheromone of the pink bollworm *Pectinophora gossypiella*. $Bp_{0.5}$ 135-136°, $Bp_{0.08}$ 92-94°.
(7Z,11Z)-form [53963-06-7]
$Bp_{0.2}$ 118°.
Ac: [52207-99-5].
Pheromone for the pink bollworm *P. gossypiella*. $Bp_{0.05}$ 86-89°.
[50933-33-0]

Mori, K., *Agric. Biol. Chem.*, 1974, **38**, 1551 (*synth*)
Vick, K.W. *et al*, *Experientia*, 1974, **30**, 17.
Mori, K. *et al*, *Tetrahedron*, 1975, **31**, 1846 (*synth, ir, pmr*)
Disselnkötter, H. *et al*, *Tetrahedron*, 1976, **32**, 1591 (*synth*)
Rossi, R. *et al*, *Tetrahedron*, 1982, **38**, 639 (*cmr*)
Michelot, D., *Synthesis*, 1983, 130 (*synth*)
Andelic, I. *et al*, *Acta Chem. Scand., Ser. B*, 1985, **39**, 231 (*synth*)
Lewis, R.J., *Sax's Dangerous Properties of Industrial Materials*, 8th Ed., Van Nostrand-Reinhold, 1992, GJK000.

8,10-Hexadecadien-1-ol **H-00210**

$H_3C(CH_2)_4CH=CHCH=CH(CH_2)_6CH_2OH$

$C_{16}H_{30}O$ M 238.412
(8E,10Z)-form [80625-56-5]
$Bp_{0.01}$ 110-115°.
Ac: [80625-69-0].
$C_{18}H_{32}O_2$ M 280.450
$Bp_{0.01}$ 110-115°.

Bestmann, H.J. *et al*, *Justus Liebigs Ann. Chem.*, 1981, 2117.

9,11-Hexadecadien-1-ol **H-00211**

$H_3C(CH_2)_3CH=CHCH=CH(CH_2)_7CH_2OH$

$C_{16}H_{30}O$ M 238.412
(9E,11Z)-form [63025-04-7]
$Bp_{0.01}$ 114-124°.
Ac: [63025-09-2].
$C_{18}H_{32}O_2$ M 280.450
$Bp_{0.01}$ 110-115°.

Bestmann, H.J. *et al*, *Justus Liebigs Ann. Chem.*, 1981, 2117 (*synth, pmr*)

9,12-Hexadecadien-1-ol **H-00212**

$H_3CCH_2CH_2CH=CHCH_2CH=CH(CH_2)_7CH_2OH$

$C_{16}H_{30}O$ M 238.412
(E,E)-form
$Bp_{0.03}$ 115-120°.
Ac:
$C_{18}H_{32}O_2$ M 280.450
$Bp_{0.01}$ 115-120°.

Bestmann, H.J. *et al*, *Justus Liebigs Ann. Chem.*, 1982, 536 (*synth, pmr*)

10,12-Hexadecadien-1-ol, 9CI **H-00213**

$H_3CCH_2CH_2CH=CHCH=CH(CH_2)_8CH_2OH$

$C_{16}H_{30}O$ M 238.412
(10E,12E)-form [965-19-5]
Minor component of female sex pheromone of the spiny bollworm *Earias insulana*. Mp 37-38°.
4-Nitrophenylazobenzoyl: Mp 112.5-113.5°.
(10E,12Z)-form [765-17-3] ***Bombykol***
Sex pheromone of the silk worm moth. The first lepidopteran pheromone to be identified. Oil. $Bp_{0.005}$ 130-133°.
Ac:
$C_{18}H_{32}O_2$ M 280.450
$Bp_{0.01}$ 124-125°.
4-Nitrophenylazobenzoyl: Mp 95-96°.
(10Z,12E)-form [1002-94-4] ***Isobombykol***
Oil. $Bp_{0.005}$ 128-130°.
4-Nitrophenylazobenzoyl: Mp 85.5-86.5°.
(10Z,12Z)-form [765-18-4]
Mp 24.5-25°.
4-Nitrophenylazobenzoyl: Mp 96.5-97°.

Butenandt, A. *et al*, *Angew. Chem.*, 1961, **73**, 349.
Truscheit, E. *et al*, *Justus Liebigs Ann. Chem.*, 1962, **658**, 65.
Normant, J.F. *et al*, *Tetrahedron Lett.*, 1975, 1465 (*ir, pmr*)
Hall, D.R. *et al*, *Experientia*, 1980, **36**, 152 ((*E,E*)-*form*)
Bestmann, H.J. *et al*, *Justus Liebigs Ann. Chem.*, 1981, 2117 (*synth, pmr*)
Miyaura, N. *et al*, *Tetrahedron*, 1983, **39**, 3271 (*synth*)
Ando, T. *et al*, *Agric. Biol. Chem.*, 1988, **52**, 141, 473, 881 (*biosynth, Bombykol*)

11,13-Hexadecadien-1-ol **H-00214**

$H_3CCH_2CH=CHCH=CH(CH_2)_9CH_2OH$

$C_{16}H_{30}O$ M 238.412
(11E,13Z)-form [80625-62-3]
$Bp_{0.01}$ 115-120°.
Ac: [80625-74-7].
$C_{18}H_{32}O_2$ M 280.450
$Bp_{0.01}$ 115-120°.
(E,E)-form [98010-23-2]
No phys. props. reported.

Bestmann, H.J. *et al*, *Justus Liebigs Ann. Chem.*, 1981, 2117 (*synth, pmr*)
Takayama, H. *et al*, *J. Org. Chem.*, 1986, **51**, 4934 (*synth, ms, cmr, uv*)

11,14-Hexadecadien-1-ol **H-00215**

$H_3CCH=CHCH_2CH=CH(CH_2)_9CH_2OH$

$C_{16}H_{30}O$ M 238.412
(E,E)-form [70654-45-4]
$Bp_{0.03}$ 115-120°.
Ac: [69775-64-0].
$C_{18}H_{32}O_2$ M 280.450
$Bp_{0.01}$ 120-125°.

Bestmann, H.J. *et al*, *Justus Liebigs Ann. Chem.*, 1982, 536 (*synth, pmr*)

7,9-Hexadecadiynedioic acid H-00216

[33128-29-9]

$HOOC(CH_2)_5(C{\equiv}C)_2(CH_2)_5COOH$

$C_{16}H_{22}O_4$ M 278.347

Cryst. (Me_2CO/pet. ether). Mp 129.5-130°.

Morris, S.G., *J. Am. Oil Chem. Soc.*, 1971, **48**, 376 (*synth*)

3,9-Hexadecadiynoic acid, 9CI H-00217

[35087-12-8]

$H_3C(CH_2)_5C{\equiv}C(CH_2)_4C{\equiv}CCH_2COOH$

$C_{16}H_{24}O_2$ M 248.364

Cryst. + $\frac{1}{2}$ H_2O (pet. ether). Mp 51.5-52.5°.

Ames, D.E. *et al*, *J. Chem. Soc., Perkin Trans. 1*, 1972, 255.

7,10-Hexadecadiynoic acid, 9CI H-00218

[19181-12-5]

$H_3C(CH_2)_4C{\equiv}CCH_2C{\equiv}C(CH_2)_5COOH$

$C_{16}H_{24}O_2$ M 248.364

Cryst. (pet. ether). Mp 35-36°.

Sprecher, H., *Lipids*, 1968, **3**, 14 (*synth*)

8,10-Hexadecadiynoic acid H-00219

$H_3C(CH_2)_4C{\equiv}CC{\equiv}C(CH_2)_6COOH$

$C_{16}H_{24}O_2$ M 248.364

Mp 32-34°. Becomes red on exp. to daylight.

Gunstone, F.D. *et al*, *J. Chem. Soc.*, 1962, 3056 (*synth*)

3,9-Hexadecadiyn-1-ol, 9CI H-00220

[35182-39-9]

$H_3C(CH_2)_5C{\equiv}C(CH_2)_4C{\equiv}CCH_2CH_2OH$

$C_{16}H_{26}O$ M 234.381

$Bp_{0.1}$ 140-144°.

Ames, D.E. *et al*, *J. Chem. Soc., Perkin Trans. 1*, 1972, 255.

2,4,6,8,10,12,14-Hexadecaheptaenedioic acid H-00221

$HOOC(CH{=}CH)_7COOH$

$C_{16}H_{16}O_4$ M 272.300

Red tablets. Mp *ca.* 300°.

Di-Me ester:

$C_{18}H_{20}O_4$ M 300.354

Orange cryst. (AcOH). Mp 236°.

Di-Et ester:

$C_{20}H_{24}O_4$ M 328.407

Orange cryst. (AcOH). Mp 217°.

Kuhn, R. *et al*, *Ber.*, 1937, **70**, 1318 (*synth*)

Hexadecanal, 9CI H-00222

Palmitic aldehyde. Palmitaldehyde

[629-80-1]

$H_3C(CH_2)_{14}CHO$

$C_{16}H_{32}O$ M 240.428

Occurs as an enol ether of diacylglycerol (plasmalogens). Isol. from lemon oil (*Citrus limon*) and bacterial lipids. Common constit. of lepidopteran sex pheromones. Mp 34°. Bp_{29} 200-202°. Polymerises on standing to the trimer (Mp 73°).

Oxime:

$C_{16}H_{33}NO$ M 255.443

Mp 88°.

Di-Me acetal: [2791-29-9]. *1,1-Dimethoxyhexadecane*

$C_{18}H_{38}O_2$ M 286.497

Mp 10°. Bp_{15} 139-146°.

Le Sueur, H.R., *J. Chem. Soc.*, 1905, **87**, 1888 (*synth*)
Ikeda, R.M. *et al*, *J. Agric. Food Chem.*, 1962, **10**, 98 (*isol*)
Goldfine, H., *J. Biol. Chem.*, 1964, **239**, 2130 (*isol*)
Christiansen, K. *et al*, *Lipids*, 1969, **4**, 421 (*ms, acetal*)
Degani, I. *et al*, *J. Chem. Soc., Perkin Trans. 1*, 1976, 323 (*synth*)

Hexadecane H-00223

Cetane

[544-76-3]

$H_3C(CH_2)_{14}CH_3$

$C_{16}H_{34}$ M 226.445

Isol. from rose petal oil, oil of *Piper longum* and wood of *Manilkara bidentata*. Standard for rating diesel-fuel ignition quality. Mp 18.17°. Bp_{160} 287°, $Bp_{0.1}$ 105-110°.

Corey, P.C. *et al*, *J. Chem. Soc.*, 1933, 346 (*synth*)
Org. Synth., 1935, **15**, 27 (*synth*)
Cocher, W. *et al*, *J. Chem. Soc.*, 1963, 677 (*isol*)
Niwa, Y. *et al*, *Bull. Chem. Soc. Jpn.*, 1968, **41**, 1746 (*ms*)
Tosi, C. *et al*, *Spectrochim. Acta, Part A*, 1972, **28**, 585 (*ir*)
Fritz, H. *et al*, *Chem. Ber.*, 1976, **109**, 1258 (*cmr, pmr*)

Hexadecanedioic acid H-00224

Thapsic acid

[505-54-4]

$HOOC(CH_2)_{14}COOH$

$C_{16}H_{30}O_4$ M 286.411

Occurs as esters in *Thapsia garganica* and in *Juniperus sabina*. Cryst. (EtOH or EtOAc). Mp 126°.

Mono-Me ester: [18451-85-9].

$C_{17}H_{32}O_4$ M 300.437

Cryst. (MeOH). Mp 45-46°.

Di-Me ester: [19102-90-0].

$C_{18}H_{34}O_4$ M 314.464

Mp 51-52°. $Bp_{0.3}$ 150-160°.

Mono-Et ester: [18017-66-8].

$C_{18}H_{34}O_4$ M 314.464

Bp_3 220°.

Di-Et ester: [15786-31-9].

$C_{20}H_{38}O_4$ M 342.518

Cryst. Mp 39°. $Bp_{0.3}$ 160-165°.

Anhydride:

$C_{16}H_{28}O_3$ M 268.395

Cryst. (C_6H_6). Mp 71°.

Bougault, J., *C. R. Hebd. Seances Acad. Sci.*, 1910, **150**, 875.
Loveland, P.M. *et al*, *Phytochemistry*, 1972, **11**, 430 (*isol*)
Pettersen, J.E. *et al*, *Biochim. Biophys. Acta*, 1973, **326**, 305.
Tember, G.A. *et al*, *Zh. Prikl. Khim. (Leningrad)*, 1974, **47**, 477 (*synth*)
Gershen, H. *et al*, *Can. J. Microbiol.*, 1976, **22**, 1198.
Espelie, K.E. *et al*, *Plant Physiol.*, 1979, **63**, 433.

1,16-Hexadecanediol, 9CI H-00225

[7735-42-4]

$HOCH_2(CH_2)_{14}CH_2OH$

$C_{16}H_{34}O_2$ M 258.443

Isol. from waxes of conifers, eg. *Pinus thunbergii*, *Thuja orientalis* and from *Cycas revoluta*. Needles (C_6H_6), platelets (EtOH). Mp 91.5°.

Chiut, P., *Helv. Chim. Acta*, 1926, **9**, 264, 274 (*synth*)
Kariyone, T. *et al*, *CA*, 1952, **46**, 5869; 1956, **50**, 13748; 1959, **53**, 10031 (*isol*)
Tanaka, A., *Yakugaku Zasshi*, 1959, **79**, 1327 (*synth*)
Taits, S.Z. *et al*, *Izv. Akad. Nauk SSSR, Ser. Khim.*, 1963, 1289; *CA*, **59**, 13990 (*synth*)
Saotome, K. *et al*, *Bull. Chem. Soc. Jpn.*, 1966, **39**, 480 (*synth*)

1-Hexadecanesulfonic acid, 9CI H-00226

[6140-88-1]

$H_3C(CH_2)_{15}SO_3H$

$C_{16}H_{34}O_3S$ M 306.509

Cryst. (pet. ether). Mp 72.5-73°.

Me ester:

$C_{17}H_{36}O_3S$ M 320.536

Cryst. (EtOH). Mp 47.7-47.9°.

S-Benzylthiuronium salt: Mp 122-123°.

Stowell, J.S. *et al*, *J. Org. Chem.*, 1962, **27**, 2853 (*synth*)
U.S. Pat., 3 518 299, (1970); *CA*, **73**, 98388 (*synth*)

Hexadecanoic acid H-00227

Palmitic acid. Aethalic acid

[57-10-3]

$H_3C(CH_2)_{14}COOH$

$C_{16}H_{32}O_2$ M 256.428

Occurs in the form of esters (glycerides) in oils and fats of vegetable and animal origin. Usually obt. from palm oil. Widely distributed in plants. Used in detn. of water hardness. Cryst. Mp 63-64°. Bp 390°, Bp_{100} 268.5°, Bp_{15} 215°.

▷ LD_{50} 57 mg/kg (mouse, i.v.). RT4550000.

Me ester: [112-39-0].

$C_{17}H_{34}O_2$ M 270.454

Constit. of trunkwood of *Machaerium kuhlmanii*. Mp 30.5°. Bp_{147} 415-418°, Bp_{15} 196°, Bp_2 148°.

Et ester: [628-97-7].

$C_{18}H_{36}O_2$ M 284.481

Mp 19.3°, Mp 24° (dimorph.). Bp_{10} 191°

Tetratriacontanyl ester: [84461-48-3]. ***Tetratriacontanyl palmitate***

$C_{50}H_{100}O_2$ M 733.339

Isol. from seeds of *Cassia spectabilis* (Leguminosae) and flowers of *Melastoma candidum albiflorum*. Cryst. (C_6H_6). Mp 69°.

Chloride: [112-67-4].

$C_{16}H_{31}ClO$ M 274.873

Mp 12°. Bp_{20} 199°.

Anhydride: [623-65-4].

$C_{32}H_{62}O_3$ M 494.840

Mp 64°.

Amide: [629-54-9]. *Hexadecanamide, 9CI. Palmitamide, 8CI*

$C_{16}H_{33}NO$ M 255.443

Isol. from seeds of *Casimiroa edulis*. Mp 106-107°. Bp_{12} 235-236°.

Nitrile:

$C_{16}H_{31}N$ M 237.428

Mp 33°. Bp 333°, Bp_{100} 251°, Bp_1 142°.

Whitby, G.S., *J. Chem. Soc.*, 1926, 1458 (*synth*)
Smith, J.C., *J. Chem. Soc.*, 1931, 802.
Welcher, F.J., *Organic Analytical Reagents*, Van Nostrand, NY, 1947, **2** (*use*)
Kincl, F.A. *et al*, *J. Chem. Soc.*, 1956, 4163 (*isol, amide*)
Bailey, A.V. *et al*, *J. Am. Oil Chem. Soc.*, 1971, **48**, 775 (*pmr*)
Karrer, W. *et al*, *Konstitution und Vorkommen der Organischen Pflanzenstoffe*, 2nd Ed., Birkhäuser Verlag, Basel, 1972-1985, no. 703 (*occur*)
Hayashi, S., *J. Chem. Phys.*, 1975, **63**, 775 (*ir*)
Ollis, W.D. *et al*, *Phytochemistry*, 1978, **17**, 1383 (*Me ester*)
Singh, M. *et al*, *Z. Naturforsch.*, *B*, 1984, **39**, 1425 (*Tetratriacontanyl palmitate*)
Lewis, R.J., *Sax's Dangerous Properties of Industrial Materials*, 8th Ed., Van Nostrand-Reinhold, 1992, PAE250.

1-Hexadecanol H-00228

Cetyl alcohol, **USAN.** *Cetanol. Palmityl alcohol*

[36653-82-4]

$H_3C(CH_2)_{14}CH_2OH$

$C_{16}H_{34}O$ M 242.444

Occurs chiefly in the form of esters in certain waxes of which spermaceti (from the head oil of the sperm whale) is the most important. Used in cosmetics industry and (as sulphate) as detergent and wetting agent. Pharmaceutic aid (emulsifying and stiffening agent). Mp 50°. Bp 334°, Bp_{12} 178-182°.

▷ Mod. toxic. MM0225000.

Ac:

$C_{18}H_{36}O_2$ M 284.481

Component of the hair pencil secretion of male butterflies *Lycorea ceres ceres*. Dimorphous. Mp 24.2°. Bp_{205} 220-225°, Bp_{15} 200°.

4-Nitrobenzoyl: Mp 68.4°.

[124-29-8]

Org. Synth., 1930, **10**, 62 (*synth*)
Meinwald, J. *et al*, *J. Am. Chem. Soc.*, 1966, **88**, 1305.
Dolejs, L. *et al*, *Org. Mass Spectrom.*, 1968, **1**, 563 (*ms*)
Geiseler, G. *et al*, *Spectrochim. Acta, Part A*, 1968, **24**, 1007 (*ir*)
Rossi, R. *et al*, *Tetrahedron*, 1982, **38**, 639 (*cmr*)
Lewis, R.J., *Sax's Dangerous Properties of Industrial Materials*, 8th Ed., Van Nostrand-Reinhold, 1992, HCP000.

2-Hexadecanol, 9CI H-00229

2-Hydroxyhexadecane

[14852-31-4]

$H_3C(CH_2)_{13}CH(OH)CH_3$

$C_{16}H_{34}O$ M 242.444

(+)-form

Mp 42-44°. $[\alpha]_D^{22}$ +4.2° (c, 2.8 in $CHCl_3$).

(±)-form [34019-44-8]

Mp 43-44°. Bp_1 135°.

Phenylurethane: Mp 58-59°.

Asinger, F. *et al*, *Ber.*, 1943, **76**, 579 (*synth*)
Jones, D.F. *et al*, *J. Chem. Soc. C*, 1968, 2801.
Hamberg, M., *Chem. Phys. Lipids*, 1971, **6**, 152 (*glc*)
Il'ina, A.I. *et al*, *Khim. Promst. (Moscow)*, 1974, 329.
Brown, N.M.D. *et al*, *J. Chem. Soc., Perkin Trans. 2*, 1976, 1769.
Bouvier, J.M. *et al*, *Bull. Soc. Chim. Fr.*, 1977, 1093.

3-Hexadecanol, 9CI H-00230

3-Hydroxyhexadecane

[593-03-3]

$H_3C(CH_2)_{12}CH(OH)CH_2CH_3$

$C_{16}H_{34}O$ M 242.444

(−)-form

Cryst. (EtOH). Mp 50°. Bp_4 152°. $[\alpha]_D^{20}$ −16.47° (EtOH).

2-Carboxybenzoyl: Cryst. (pet. ether). Mp 51°. $[\alpha]_D^{20}$ −16.47° (EtOH).

(±)-form

Mp 42-42.4°. Bp_{16} 176°, Bp_4 156°.

2-Carboxybenzoyl: Cryst. (pet. ether). Mp 51-52°.

Pickard, R.H. *et al*, *J. Chem. Soc.*, 1913, **103**, 1923.
Asinger, F. *et al*, *Ber.*, 1943, **76**, 579 (*synth*)
Klein, D.A. *et al*, *Appl. Microbiol.*, 1969, **17**, 676.
Zinbo, M. *et al*, *Anal. Lett.*, 1977, **10**, 119 (*glc*)

4-Hexadecanol, 9CI — H-00231

[19781-43-2]

$H_3C(CH_2)_{11}CH(OH)CH_2CH_2CH_3$

$C_{16}H_{34}O$ M 242.444

(±)-form

Mp 38-39°. Bp_1 128°.

Phenylurethane: Mp 45°.

Asinger, F. *et al*, *Ber.*, 1943, **76**, 579 (*synth*)
Klein, D.A. *et al*, *Appl. Microbiol.*, 1969, **17**, 676.
Il'ina, A.I. *et al*, *Khim. Promst.* (*Moscow*), 1974, 329.

5-Hexadecanol, 9CI — H-00232

[21078-87-5]

$H_3C(CH_2)_{10}CH(OH)(CH_2)_3CH_3$

$C_{16}H_{34}O$ M 242.444

(±)-form

Mp 36°. $Bp_{1.5}$ 134°.

Asinger, F. *et al*, *Ber.*, 1943, **76**, 579 (*synth*)
Il'ina, A.I. *et al*, *Khim. Promst.* (*Moscow*), 1974, 329.
Zinbo, M. *et al*, *Anal. Lett.*, 1977, **10**, 119 (*glc*)

6-Hexadecanol, 9CI — H-00233

[591-73-1]

$H_3C(CH_2)_9CH(OH)(CH_2)_4CH_3$

$C_{16}H_{34}O$ M 242.444

(±)-form

Mp 34-35°. $Bp_{3.5}$ 148°.

Asinger, F. *et al*, *Ber.*, 1943, **76**, 579 (*synth*)
Il'ina, A.I. *et al*, *Khim. Promst.* (*Moscow*), 1974, 329.
Zinbo, M. *et al*, *Anal. Lett.*, 1977, **10**, 119 (*glc*)

7-Hexadecanol, 9CI — H-00234

[591-72-0]

$H_3C(CH_2)_8CH(OH)(CH_2)_5CH_3$

$C_{16}H_{34}O$ M 242.444

(±)-form

Mp 32°. Bp_3 155°.

Asinger, F. *et al*, *Ber.*, 1943, **76**, 579 (*synth*)
Il'ina, A.I. *et al*, *Khim. Promst.* (*Moscow*), 1974, 329.
Zinbo, M. *et al*, *Anal. Lett.*, 1977, **10**, 119 (*glc*)

8-Hexadecanol, 9CI — H-00235

[19781-83-0]

$H_3C(CH_2)_7CH(OH)(CH_2)_6CH_3$

$C_{16}H_{34}O$ M 242.444

(±)-form

Mp 48°. $Bp_{1.5}$ 128°.

Asinger, F. *et al*, *Ber.*, 1943, **76**, 579 (*synth*)
Il'ina, A.I. *et al*, *Khim. Promst.* (*Moscow*), 1974, 329.

2-Hexadecanone — H-00236

Methyl tetradecyl ketone

[18787-63-8]

$H_3C(CH_2)_{13}COCH_3$

$C_{16}H_{32}O$ M 240.428

Isol. from hop oil (*Humulus lupulus*). Cryst. Mp 43-43.5°. Bp_{100} 130-131°.

▷ MM0300000.

Semicarbazone: Cryst. (EtOH). Mp 117-118°.

Jahnsen, V.J., *Nature* (*London*), 1962, **196**, 474 (*isol*)
Klein, D.A. *et al*, *Appl. Microbiol.*, 1969, **17**, 676.
Bailey, A.V. *et al*, *J. Chem. Eng. Data*, 1970, **15**, 542.
Wyrick, S.D. *et al*, *J. Med. Chem.*, 1976, **19**, 219 (*synth*)

3-Hexadecanone — H-00237

Ethyl tridecyl ketone

[18787-64-9]

$H_3C(CH_2)_{12}COCH_2CH_3$

$C_{16}H_{32}O$ M 240.428

Cryst. (pet. ether). Mp 42°. Bp_{17} 184°, Bp_2 140°.

Semicarbazone: Cryst. (EtOH aq.). Mp 86°.

Pickard, R.H. *et al*, *J. Chem. Soc.*, 1913, **103**, 1923.
Asinger, F. *et al*, *Chem. Ber.*, 1943, **76**, 579.
Klein, D.A. *et al*, *Appl. Microbiol.*, 1969, **17**, 676 (*biosynth*)
Gilbertson, J.R. *et al*, *Lipids*, 1976, **11**, 172.

2-Hexadecanoylamino-1,3-octadecanediol — H-00238

N-*[2-Hydroxy-1-(hydroxymethyl)heptadecyl] hexadecanamide, 8CI*

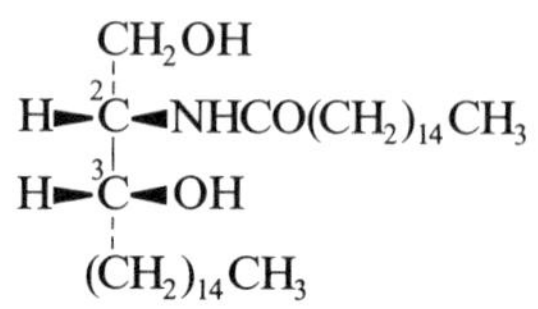

$C_{34}H_{69}NO_3$ M 539.924

***(2RS,3SR)* form** [21275-74-1]

(±)-erythro-*form.* N-*Hexadecanoylsphinganine*

Mp 104-105°.

Mitsner, B.I. *et al*, *CA*, 1968, **68**, 95290k (*synth*)
Ong, D.E. *et al*, *J. Lipid Res.*, 1972, **13**, 819 (*synth*)
Alpes, H., *Chem. Phys. Lipids*, 1974, **13**, 109 (*synth*)
Gaskell, S.J. *et al*, *J. Chromatogr.*, 1976, **126**, 591.

3-Hexadecanoyloxy-2-oxopropyl phosphate — H-00239

Hexadecanoic acid 2-oxo-3-(phosphonooxy)propyl ester, 9CI. 1-O-Hexadecanoyldihydroxyacetone 3-phosphate

[17378-38-0]

$H_3C(CH_2)_{14}COOCH_2COCH_2OPO_3H_2$

$C_{19}H_{37}O_7P$ M 408.471

Mp 80-81°.

Hajra, A.K. *et al*, *Chem. Phys. Lipids*, 1983, **33**, 179 (*synth, ir, pmr*)

2,6,8,12-Hexadecatetraenoic acid H-00240

$H_3CCH_2CH_2CH{=}CHCH_2CH_2(CH{=}CH)_2$
$CH_2CH_2CH{=}CHCOOH$

$C_{16}H_{24}O_2$ M 248.364

(2E,6Z,8Z,12E)-form
Oil.
Isobutylamide:
$C_{20}H_{33}NO$ M 303.487
Oil.

(2E,6Z,8Z,12Z)-form
Oil.
Isobutylamide: Oil.

(2Z,6Z,8Z,12E)-form
Oil.
Isobutylamide: Oil.

(all-Z)-form
Oil.
Isobutylamide: Oil.

Meisters, A. *et al*, *Aust. J. Chem.*, 1966, **19**, 1207.

4,7,10,13-Hexadecatetraenoic acid H-00241
[3209-28-7]

$H_3CCH_2CH{=}CHCH_2CH{=}CHCH_2CH{=}CH$
$CH_2CH{=}CHCH_2CH_2COOH$

$C_{16}H_{24}O_2$ M 248.364

(all-Z)-form [29259-52-7]
Isol. from algae *Chlorella* spp., *Euglena gracilis*, *Scenedesmus* spp. Liq. $Bp_{0.0001}$ 76°.
[25377-57-5]

Paschke, R.F. *et al*, *J. Am. Oil Chem. Soc.*, 1954, **31**, 81 (*isol, struct*)
Klenk, E. *et al*, *Hoppe Seyler's Z. Physiol. Chem.*, 1959, **317**, 243 (*isol, struct*)

6,8,12,14-Hexadecatetraen-10-ynoic acid, 9CI H-00242

$H_3CCH{=}CHCH{=}CHC{\equiv}CCH{=}CHCH{=}CH(CH_2)_4COOH$

$C_{16}H_{20}O_2$ M 244.333

(all-E)-form
Me ester: [40241-97-2].
$C_{17}H_{22}O_2$ M 258.360
Constit. of *Chrysocoma* spp. Oil.

Bohlmann, F. *et al*, *Chem. Ber.*, 1972, **105**, 3587.

6,8,14-Hexadecatrien-10,12-diynoic acid, 9CI H-00243

$H_3CCH{=}CHC{\equiv}CC{\equiv}CCH{=}CHCH{=}CH(CH_2)_4COOH$

$C_{16}H_{18}O_2$ M 242.317

(all-E)-form
Me ester: [40241-94-9].
$C_{17}H_{20}O_2$ M 256.344
Constit. of *Chrysocoma* spp. Oil.

(6E,8E,14Z)-form
Me ester: Constit. of *Chrysocoma* spp. Oil.

Bohlmann, F. *et al*, *Chem. Ber.*, 1972, **105**, 3587.

1,8,10-Hexadecatriene H-00244

$H_3C(CH_2)_4CH{=}CHCH{=}CH(CH_2)_5CH{=}CH_2$

$C_{16}H_{28}$ M 220.397

(8E,10Z)-form [80625-33-8]
$Bp_{0.01}$ 86-87°.

Bestmann, H.J. *et al*, *Justus Liebigs Ann. Chem.*, 1981, 2117 (*synth, pmr*)

1,11,13-Hexadecatriene H-00245

$H_3CCH_2CH{=}CHCH{=}CH(CH_2)_8CH{=}CH_2$

$C_{16}H_{28}$ M 220.397

(11E,13Z)-form [80634-97-5]
$Bp_{0.3}$ 114°.

Bestmann, H.J. *et al*, *Justus Liebigs Ann. Chem.*, 1981, 2117 (*synth, pmr*)

7,10,13-Hexadecatrienoic acid H-00246
[2271-35-4]

$H_3CCH_2CH{=}CHCH_2CH{=}CHCH_2CH{=}CH(CH_2)_5COOH$

$C_{16}H_{26}O_2$ M 250.380

(Z,Z,Z)-form [7561-64-0]
Present in a wide variety of angiosperm leaves, also in lipids of ferns and green algae.

Heyes, J.K. *et al*, *Biochem. J.*, 1951, **49**, 503 (*struct*)
Radunz, A., *Phytochemistry*, 1967, **6**, 399 (*isol*)
Jamieson, G.R. *et al*, *Phytochemistry*, 1971, **10**, 1837 (*occur, bibl*)

4,6,10-Hexadecatrien-1-ol, 9CI H-00247

$H_3C(CH_2)_4CH{=}CHCH_2CH_2CH{=}CH$
$CH{=}CHCH_2CH_2CH_2OH$

$C_{16}H_{28}O$ M 236.397

(4*E*,6*E*,10*Z*)- and (4*E*,6*Z*,10*Z*)-forms are components of the sex pheromone of the cocoa pod borer moth, *Conopomorpha cramerella*.
Ac:
$C_{18}H_{30}O_2$ M 278.434
Component of the sex pheromone of *C. cramerella*.
[101372-99-0, 101373-00-6, 101373-01-7, 101405-31-6]

Beevor, P.S. *et al*, *J. Chem. Ecol.*, 1986, **12**, 1 (*isol, props, synth, glc*)
Malosse, C. *et al*, *Anal. Chem.*, 1990, **62**, 287 (*ms*)
Yen, Y.P. *et al*, *Synth. Commun.*, 1992, **22**, 1567 (*synth*)

6,8,12-Hexadecatrien-10-ynoic acid H-00248

$H_3CCH_2CH_2CH{=}CHC{\equiv}CCH{=}CHCH{=}CH(CH_2)_4COOH$

$C_{16}H_{22}O_2$ M 246.349

(E,E,E)-form
Me ester: [85571-37-5].
$C_{17}H_{24}O_2$ M 260.375
Constit. of *Chrysocoma tenuifolia*. Gum.

Bohlmann, F. *et al*, *Phytochemistry*, 1982, **21**, 2742.

2,5,8-Hexadecatriyn-1-ol, 8CI H-00249
[32895-86-6]

$H_3C(CH_2)_6C{\equiv}CCH_2C{\equiv}CCH_2C{\equiv}CCH_2OH$

$C_{16}H_{22}O$ M 230.349

Cryst. (EtOH at −20°). Mp 43-45°. $Bp_{0.005}$ 138°.

Beerthuis, R.K. *et al*, *Recl. Trav. Chim. Pays-Bas* (*J. R. Neth. Chem. Soc.*), 1971, **90**, 943 (*synth*)

2-Hexadecenal H-00250

$H_3C(CH_2)_{12}CH{=}CHCHO$

$C_{16}H_{30}O$ M 238.412

(E)-form [22644-96-8]

Mp 34°. Bp_1 135°.

Hino, T. *et al*, *J. Chem. Soc., Perkin Trans. 1*, 1986, 1687 (*synth, ir, pmr, ms*)
Chang, Y.-H. *et al*, *Synthesis*, 1990, 1033 (*synth, ir, pmr, ms*)

7-Hexadecenal H-00251

$H_3C(CH_2)_7CH{=}CH(CH_2)_5CHO$

$C_{16}H_{30}O$ M 238.412

(Z)-form [56797-40-1]

Active component of trail-following pheromone of the Argentine ant *Iridomyrmex humilis*. Liq. $Bp_{0.05}$ 110-112°. n_D^{20} 1.4525.

Van Vorhiskey, S.E. *et al*, *J. Chem. Ecol.*, 1982, **8**, 3.
Brown, H.C. *et al*, *J. Org. Chem.*, 1986, **51**, 4518 (*synth, pmr, cmr*)

9-Hexadecenal H-00252

$H_3C(CH_2)_5CH{=}CH(CH_2)_7CHO$

$C_{16}H_{30}O$ M 238.412

(Z)-form [56219-04-6]

Found in *Heliothis* spp. and other moths. Pheromone. Oil.

Klun, J.A. *et al*, *J. Chem. Ecol.*, 1980, **6**, 165, 177 (*isol, glc, ms*)

10-Hexadecenal, 9CI H-00253

[86416-31-1]

$H_3C(CH_2)_4CH{=}CH(CH_2)_8CHO$

$C_{16}H_{30}O$ M 238.412

(E)-form [72698-30-7]

Major component of the sex pheromone of the yellow peach moth, *Dichocrocis punctiferalis*, and minor component of the sex pheromones of *D. chlorophanata* and the cabbage webworm, *Hellula undalis*.

(Z)-form

Synergistic minor component of the sex pheromones of *D. punctiferalis*, and the diamondback moth, *Plutella xylostella*.

Ando, T. *et al*, *Appl. Entomol. Zool.*, 1979, **14**, 362 (*isol, activity*)
Arai, K. *et al*, *Agric. Biol. Chem.*, 1982, **46**, 2395 (*isol, activity, synth, tlc*)
Konno, Y. *et al*, *Appl. Entomol. Zool.*, 1982, **17**, 207; 1986, **21**, 622 (*isol, activity, ms, glc*)

11-Hexadecenal H-00254

$H_3C(CH_2)_3CH{=}CH(CH_2)_9CHO$

$C_{16}H_{30}O$ M 238.412

(Z)-form [53939-28-9]

Sex pheromone of the rice stem borer, old world bollworm *Heliothis armigera* and spotted bollworm *Earias vittella*. $Bp_{0.05}$ 115-120° (bath).

Bestmann, H.J. *et al*, *Chem. Ber.*, 1978, **111**, 248 (*synth*)
Leadbetter, G. *et al*, *J. Chem. Ecol.*, 1979, **5**, 101 (*synth*)
Cork, A. *et al*, *J. Chem. Ecol.*, 1988, **14**, 929.

12-Hexadecenal, 9CI H-00255

[106608-75-7]

$H_3CCH_2CH_2CH{=}CH(CH_2)_{10}CHO$

$C_{16}H_{30}O$ M 238.412

(E)-form [72698-32-9]

Possible minor component of the sex pheromone of the cabbage webworm, *Hellula undalis*.

(Z)-form [72698-31-8]

Shows synergistic effect on the male attractive properties of the female sex pheromone of the diamondback moth *Plutella xylostella*. Possible minor component of the sex pheromone of the cabbage webworm, *H. undalis*.

Ando, T. *et al*, *Appl. Entomol. Zool.*, 1979, **14**, 362 (*activity*)
Hall, D.R. *et al*, *Experientia*, 1980, **36**, 152 (*isol*)
Arai, K. *et al*, *Agric. Biol. Chem.*, 1982, **46**, 2395 (*activity*)
Verba, G.G. *et al*, *Khim. Prir. Soedin.*, 1990, 252 (*synth*)

1-Hexadecene H-00256

Cetene

[629-73-2]

$H_3C(CH_2)_{13}CH{=}CH_2$

$C_{16}H_{32}$ M 224.429

Polymer substrate. Mp 4.1°. Bp 284°, Bp_{15} 157.5°.

Roberti, G. *et al*, *CA*, 1942, **36**, 5767 (*synth*)
Field, F.H., *J. Am. Chem. Soc.*, 1968, **90**, 5649 (*ms*)
Vinuits'kii, O.M. *et al*, *Neftekhimiya*, 1969, **9**, 767 (*synth*)

2-Hexadecenoic acid, 9CI H-00257

[629-56-1]

$H_3C(CH_2)_{12}CH{=}CHCOOH$

$C_{16}H_{30}O_2$ M 254.412

(E)-form [929-79-3]

Mp 46.5-47.6°.

Me ester: [2825-81-2].

$C_{17}H_{32}O_2$ M 268.439

Bp_{16} 199°.

(Z)-form [2825-68-5]

Mp 40.5-41.7°.

Et ester:

$C_{18}H_{34}O_2$ M 282.465

Mp 9-10°.

Grimmer, G. *et al*, *Justus Liebigs Ann. Chem.*, 1965, **685**, 154 (*synth*)
Falk, F. *et al*, *Fette, Seifen, Anstrichm.*, 1968, **70**, 709 (*synth*)
Hino, T. *et al*, *J. Chem. Soc., Perkin Trans. 1*, 1986, 1687 (*synth, pmr, ms*)

3-Hexadecenoic acid, 9CI H-00258

$H_3C(CH_2)_{11}CH{=}CHCH_2COOH$

$C_{16}H_{30}O_2$ M 254.412

(E)-form [1686-10-8]

Constit. of many plant systems including spinach leaves, red clover (as phosphatidylglycerol ester) and *Aster* spp.. Other major sources are seed oils of *Grindelia oxylepis* and *Helenium bigelovii*. Also isol. from algae. Mp 53-54°.

Klenk, E. *et al*, *Hoppe Seyler's Z. Physiol. Chem.*, 1963, **334**, 44 (*isol*)
Weenik, R.O. *et al*, *Biochim. Biophys. Acta*, 1964, **84**, 613.
Hopkins, C.Y. *et al*, *Can. J. Chem.*, 1964, **42**, 2224 (*isol*)
Haverkate, F. *et al*, *Experientia*, 1964, **20**, 511.
Kleiman, R. *et al*, *Lipids*, 1966, **1**, 301 (*isol*)
Knipprath, W. *et al*, *Lipids*, 1966, **1**, 81 (*synth*)
Kannan, R. *et al*, *Fette, Seifen, Anstrichm.*, 1974, **76**, 344 (*pmr*)

6-Hexadecenoic acid, 9CI H-00259

[2197-46-8]

$H_3C(CH_2)_8CH{=}CH(CH_2)_4COOH$

$C_{16}H_{30}O_2$ M 254.412

(E)-form [28290-76-8]

Constit. of seed oil of *Picramnia sellowii*. Also present in adipose tissue, skin and parsley oil.

(Z)-form [17004-51-2]

Present in seed oils of *Thunbergia alata* and *Picramnia sellowii*. Also in sea anemones and other marine animals.

Spencer, G.F. *et al*, *Lipids*, 1971, **6**, 712 (*isol*)

9-Hexadecenoic acid, 9CI H-00260

Hypogaic acid

[2091-29-4]

$H_3C(CH_2)_5CH{=}CH(CH_2)_7COOH$

$C_{16}H_{30}O_2$ M 254.412

(E)-form [10030-73-6]

Palmitelaidic acid

Constit. of many plant and animal systems. Pheromone of hide beetle. Mp 32-33°.

(Z)-form [373-49-9]

Palmitoleic acid. Zoomaric acid. Physetoleic acid

A major component of lipids of marine plants and animals, also found in plant oils, eg. *Macadamia ternifolia* seed oil (20%). Mp 0.5°. Bp_1 180-183°, $Bp_{0.6}$ 162°. n_D^{20} 1.4582.

Me ester:

$C_{17}H_{32}O_2$ M 268.439

Liq. Bp_5 140-141°.

Armstrong, E.F. *et al*, *J. Soc. Chem. Ind., London*, 1925, **44**, 182T.
Bridge, R.E. *et al*, *J. Chem. Soc.*, 1950, 2396 (*isol*)
Boughton, B.W. *et al*, *J. Chem. Soc.*, 1952, 671 (*synth*)
Hofmann, K. *et al*, *J. Biol. Chem.*, 1955, **213**, 415.
Karrer, W. *et al*, *Konstitution und Vorkommen der Organischen Pflanzenstoffe*, 2nd Ed., Birkhäuser Verlag, Basel, 1972-1985, no. 734 (*occur*)
Suzuki, T. *et al*, *Agric. Biol. Chem.*, 1974, **38**, 2269 (*ir, ms*)
Batchelor, J.G. *et al*, *J. Org. Chem.*, 1974, **39**, 1698 (*cmr*)
Mannan, A. *et al*, *Chem. Ind. (London)*, 1984, 851 (*isol, occur*)

10-Hexadecenoic acid, 9CI H-00261

[25976-13-0]

$H_3C(CH_2)_4CH{=}CH(CH_2)_8COOH$

$C_{16}H_{30}O_2$ M 254.412

(Z)-form [2511-97-9]

Present in adipose tissue, brain and skin (in psoriasis). Mp 15-16°.

Baker, C.D. *et al*, *J. Chem. Soc.*, 1963, 489 (*synth*)
Bus, J. *et al*, *Lipids*, [*Invited Lect. Symp. Int. Congr. Fat Res.*], *1974* (*1976*), 1976, **2**, 243 (*cmr, pmr*)

11-Hexadecenoic acid H-00262

Lycopodiumoleic acid. Tanacetumoleic acid. Palmitvaccenic acid

[2271-34-3]

$H_3C(CH_2)_3CH{=}CH(CH_2)_9COOH$

$C_{16}H_{30}O_2$ M 254.412

(E)-form [73292-39-4]

Isol. from *Chrysanthemum vulgare*, *Lycopodium vulgare*, etc. Bp_{18} 235-240°.

(Z)-form [2416-20-8]

Present along with the more common 9*E*-isomer in many bacterial and animal lipids.

Ackman, R.G. *et al*, *Lipids*, 1966, **1**, 341 (*occur*)
Walker, R.W., *Lipids*, 1969, **4**, 15 (*occur*)
Nichols, P.D. *et al*, *J. Microbiol. Methods*, 1986, **5**, 49.

13-Hexadecenoic acid H-00263

[14134-46-4]

$H_3CCH_2CH{=}CH(CH_2)_{11}COOH$

$C_{16}H_{30}O_2$ M 254.412

(Z)-form

Isol. from herring oil.

[92661-08-0]

Ackman, R.G. *et al*, *Lipids*, 1966, **1**, 341.

7-Hexadecen-1-ol H-00264

$H_3C(CH_2)_7CH{=}CH(CH_2)_5CH_2OH$

$C_{16}H_{32}O$ M 240.428

(E)-form [51824-10-3]

Pheromone of Lepidoptera spp.

Ac: [23192-83-8].

$C_{18}H_{34}O_2$ M 282.465

Pheromone of Lepidoptera spp.

(Z)-form [24880-48-6]

Isol. from the heads of male bumblebees, *Bombus agrorum*. Pheromone of many insect spp.

Formyl: [65954-23-6].

$C_{17}H_{32}O_2$ M 268.439

Sex attractant for cotton pests.

Ac: [23192-42-9]. ***Hexalure***

Pheromone for many insect spp. $Bp_{0.08}$ 121.5-124.5°.

Butanoyl:

$C_{20}H_{38}O_2$ M 310.519

Constit. of cephalic secretions of the stingless bee *Scaptotrigona postica*.

Green, N. *et al*, *Experientia*, 1969, **25**, 682.
Calam, D.H., *Nature (London)*, 1969, **221**, 856 (*isol*)
Barabas, A. *et al*, *Tetrahedron*, 1978, **34**, 2191 (*nmr*)
Engels, E. *et al*, *J. Chem. Ecol.*, 1987, **13**, 371.
Subramanian, G.B.V. *et al*, *Chem. Ind. (London)*, 1988, 664 (*synth*)

9-Hexadecen-1-ol H-00265

$H_3C(CH_2)_5CH{=}CH(CH_2)_7CH_2OH$

$C_{16}H_{32}O$ M 240.428

(Z)-form [10378-01-5]

Palmitoleyl alcohol

Isol. from the heads of male bumblebees *Bombus lapidarius*. Pheromone for many insect spp.

Formyl: [56218-71-4].

$C_{17}H_{32}O_2$ M 268.439

Pheromone for many insect spp.

Ac: [34010-20-3].
$C_{18}H_{34}O_2$ M 282.465
Pheromone for many insect spp.

Calam, D.H., *Nature (London)*, 1969, **221**, 856 (*isol*)
Bestman, H.J. *et al, Chem. Ber.*, 1975, **108**, 3582.

11-Hexadecen-1-ol H-00266

$H_3C(CH_2)_3CH{=}CH(CH_2)_9CH_2OH$

$C_{16}H_{32}O$ M 240.428

(E)-form [61301-56-2]

Ac: [56218-72-5].
Sex pheromone produced by female sweet potato leaf folder moth (*Brachmia macroscopa*).

(Z)-form

Sex pheromone component of the diamond back moth *Plutella xylostella* and the clover cutworm moth *Scotogramma trifolii*. Female sex pheromone of cotton bollworm *Heliothis armigera*. $Bp_{0.15}$ 140°.

Ac: [34010-21-4].
Major component of the sex pheromone of the bertha armyworm *Mamestra configurata*; female sex pheromone of the purple stem borer moth *Sesamia inferens* and rice green caterpillar *Natanga aenescens*; sex pheromone component of diamond back moth and cabbage armyworm moth *Mamestra brassicae* and corn stalk borer *Sesamia nonagroides*. $Bp_{0.1}$ 108°.

Chisholm, M.D. *et al, Can. Entomol.*, 1975, **107**, 361 (*isol*)
Underhill, E.W. *et al, Environ. Entomol.*, 1976, **5**, 307 (*isol*)
Hirano, C. *et al, Naturwissenschaften*, 1976, **63**, 439 (*isol*)
Tamaki, Y. *et al, Appl. Entomol. Zool.*, 1977, **12**, 208 (*isol*)
Hirai, Y. *et al, Appl. Entomol. Zool.*, 1978, **13**, 136 (*isol*)
Bestmann, H.J. *et al, Chem. Ber.*, 1978, **111**, 248 (*synth, pmr, ir*)
Mazomenos, B.E., *J. Chem. Ecol.*, 1989, **15**, 1241.

7-Hexadecen-16-olide H-00267

Oxacycloheptadec-8-en-2-one, 9CI. Ambrettolic acid lactone

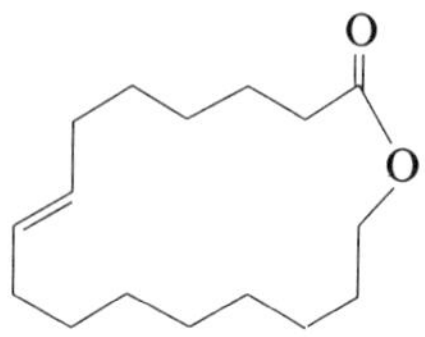

$C_{16}H_{28}O_2$ M 252.396

(E)-form [51155-12-5]

trans-*Ambrettolide*
Fragrance compd. Oil.

(Z)-form [123-69-3]

Ambrettolide
Occurs in the plant *Abelmoschus moschatus* and the Dufour's gland of bees. Used in the perfume industry. Has a strong musk-like odour. Oil. Bp_1 154-156°.

Baudart, P., *C. R. Hebd. Seances Acad. Sci.*, 1945, **221**, 205 (*synth*)
Sabnis, S.D. *et al, J. Chem. Soc.*, 1963, 2477 (*synth, ir*)
Morris, W.W., *J. Assoc. Off. Anal. Chem.*, 1973, **56**, 1037 (*ir, pmr*)
Bergström, G., *Chem. Scr.*, 1974, **5**, 39.
Sanz, V. *et al, J. Chem. Soc., Perkin Trans. 1*, 1982, 1837 (*synth*)

3-(7-Hexadecenyl)-4-hydroxy-5-methylene-2(5*H*)-furanone H-00268

2-(7-Hexadecenyl)-3-hydroxy-2,4-pentadien-4-olide

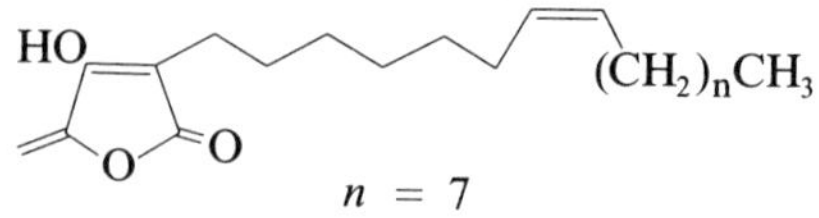

$C_{21}H_{34}O_3$ M 334.498

(Z)-form

Constit. of *Lomatium dissectum*. Shows antimicrobial props. Unstable oil. Occurs as mixt. with its lower homologue (n = 5).

Vanwagenen, B.C. *et al, Tetrahedron*, 1986, **42**, 1117.

3-(7-Hexadecenyl)-4-methyl-2,5-furandione, 9CI H-00269

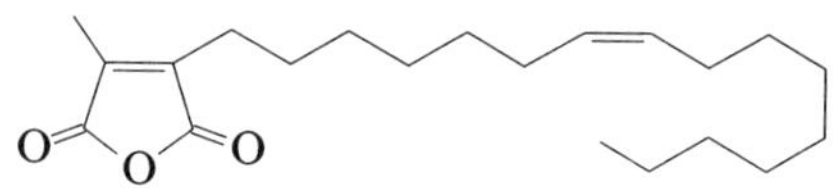

$C_{21}H_{34}O_3$ M 334.498

(Z)-form [84306-79-6]

Constit. of *Piptoporus australiensis*. Oil. $Bp_{0.1}$ 105°.

Gill, M., *Phytochemistry*, 1982, **21**, 1786.

3-(1-Hexadecenyloxy)-1,2-propanediol, 9CI H-00270

1-O-(1-Hexadecenyl)glycerol
[1559-76-8]

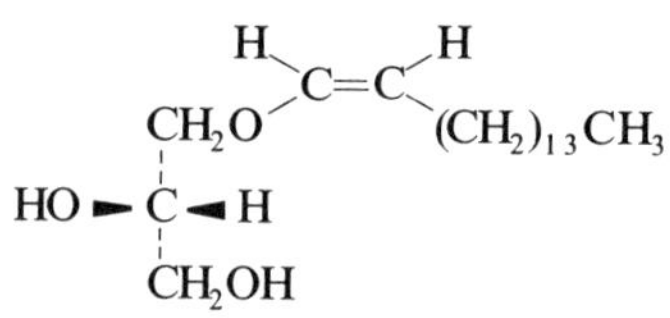

$C_{19}H_{38}O_3$ M 314.507

(2S,1'Z)-form [34677-26-4]

1-Hexadec-cis-*1'-enyl*-sn-*glycerol*
Isol. from pig's heart plasmalogen. Mp 37-39°. $[\alpha]_D^{20}$ −2.1° (c, 1 in $CHCl_3$). Acid hydrol. gives Glycerol and Hexadecanal.

Cymerman-Craig, J. *et al, Chem. Ind. (London)*, 1965, 1559 (*synth*)
Vtorov, I.B. *et al, Tetrahedron Lett.*, 1971, 4603 (*synth*)

13-Hexadecen-11-yn-1-ol H-00271

$H_3CCH_2CH{=}CHC{\equiv}C(CH_2)_9CH_2OH$

$C_{16}H_{28}O$ M 236.397

(Z)-form [75089-05-3]

Ac: [78617-58-0]. ***Pityolure***
$C_{18}H_{30}O_2$ M 278.434
Sex pheromone of the processionary moth *Thaumetopoea pityocampa*. The first acetylenic compd. found as an insect pheromone.

Camps, F. *et al, Chem. Lett.*, 1981, 703.
Guerrero, A. *et al, Tetrahedron Lett.*, 1981, **22**, 2013 (*isol, nmr, ms*)
Cardillo, G. *et al, Gazz. Chim. Ital.*, 1982, **112**, 231 (*synth*)
Michelot, D. *et al, J. Chem. Res., Synop.*, 1982, 93 (*synth*)
Rossi, R. *et al, Tetrahedron*, 1983, **39**, 287 (*synth*)

1-*O*-Hexadecyl-2-*O*-acetyl-*sn*-glycero-3-phospho-(*N,N,N*-trimethyl) hexanolamine — H-00272

*11-(Acetyloxy)-8-hydroxy-*N,N,N*-trimethyl-7,9,13-trioxa-8-phosphanonacosan-1-aminium hydroxide, inner salt, 8-oxide. Hexanolamino-PAF*

[137566-83-7]

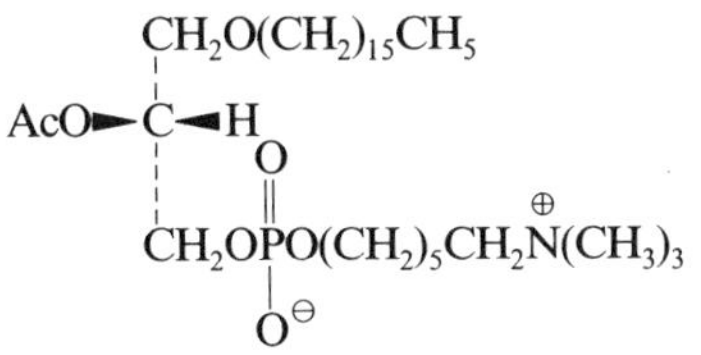

$C_{30}H_{62}NO_7P$ M 579.796

Inhibits PAF induced platelet aggregation. White powder + $1H_2O$. Ir 1730 cm^{-1} (KBr).

[99885-01-5]

Wissner, A. *et al*, *J. Med. Chem.*, 1986, **29**, 328 (*synth, ir*)
Rouis, M. *et al*, *Biochem. Biophys. Res. Commun.*, 1988, **156**, 1293 (*biochem*)
Grigoriadis, G. *et al*, *Br. J. Pharmacol.*, 1991, **104**, 171 (*biochem*)

Hexadecylamine — H-00273

1-Hexadecanamine, 9CI. 1-Aminohexadecane. Cetylamine

[143-27-1]

$H_3C(CH_2)_{14}CH_2NH_2$

$C_{16}H_{35}N$ M 241.459

Mp 46.2°. $Bp_{5.2}$ 162-165°. pK_a 10.61 (25°).

▷ML9100000.

B,HF: [3151-59-5]. ***Hetaflur*, BAN, USAN, INN.** *GA 242. SKF 2208*

Dental caries prophylactic.

▷MM1120200.

B,HCl: Mp 178°.

N-*Di-Me:* Bp_2 135-136°.

Markley, K.S., *Fatty Acids*, 2nd Ed., Interscience, N.Y., 1654 (*synth*)
Koyama, T. *et al*, *Kumamoto Pharm. Bull.*, 1954, 52; *CA*, **50**, 3207 (*synth*)
Zvejnieks, A., *Sven. Kem. Tidskr.*, 1954, **66**, 316 (*synth*)
Oktawiec, M., *CA*, 1959, **53**, 14920 (*synth*)
Gohlke, R.S. *et al*, *Anal. Chem.*, 1962, **34**, 1281 (*ms*)
Goto, R. *et al*, *Nippon Kagaku Zasshi*, 1966, **87**, 1220 (*synth*)
Martindale, The Extra Pharmacopoeia, 28th/29th Ed., Pharmaceutical Press, London, 1982/1989, 12551.
Lewis, R.J., *Sax's Dangerous Properties of Industrial Materials*, 8th Ed., Van Nostrand-Reinhold, 1992, CDF400, HCO500.

1-*O*-Hexadecyl-2-*O*-arachidonoyl-*sn*-glycero-3-phosphocholine — H-00274

*7-[(Hexadecyloxy)methyl]-4-hydroxy-*N,N,N*-trimethyl-9-oxo-3,5,8-trioxa-4-phosphaoctacosa-13,16,19,22-tetraen-1-aminium hydroxide, inner salt, 4-oxide. Arachidonyl-PAF*

[86288-11-1]

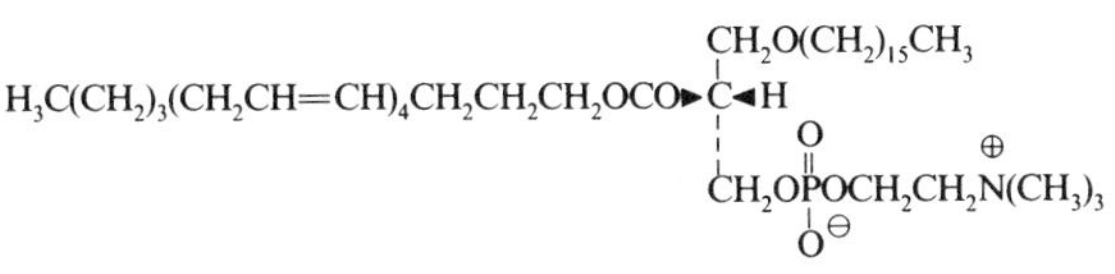

$C_{44}H_{82}NO_7P$ M 768.108

Main source of arachidonic acid from human neutrophils and macrophages.

Robinson, M. *et al*, *J. Biol. Chem.*, 1985, **260**, 7889 (*biosynth*)
Kramer, R.M. *et al*, *Biochim. Biophys. Acta*, 1988, **959**, 269 (*synth, biochem*)
Leslie, C.C., *Methods Enzymol.*, 1990, **187**, 216 (*synth, props*)

1-*O*-Hexadecyl-2,3-di-*O*-hexadecanoylglycerol — H-00275

1-Hexadecyloxy-2,3-dihexadecanoyloxypropane

[1116-45-6]

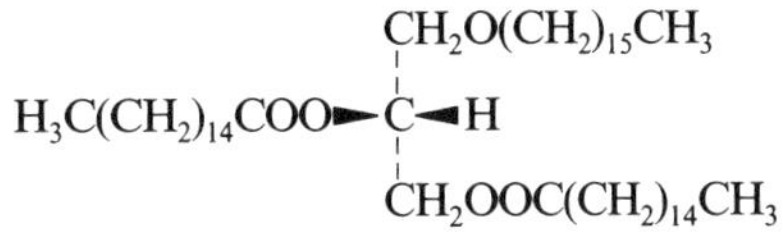

(S)-form

$C_{51}H_{100}O_5$ M 793.348

Isol. from the soft coral *Lobophytum* spp. Commercially available. Solid. Mp 56-57°. $[\alpha]_D^{20}$ +7.40° (c, 1.0 in $CHCl_3$).

Subrahmanyam, C. *et al*, *Tetrahedron*, 1992, **48**, 3111 (*isol, pmr, cmr*)
Chenevert, R. *et al*, *J. Org. Chem.*, 1993, **58**, 1054 (*synth, ir, pmr, cmr*)

1-*O*-Hexadecyl-2-*O*-dihomogammalinolenoyl-*sn*-glycero-3-phosphocholine — H-00276

DHLA-PAF

[131907-73-8]

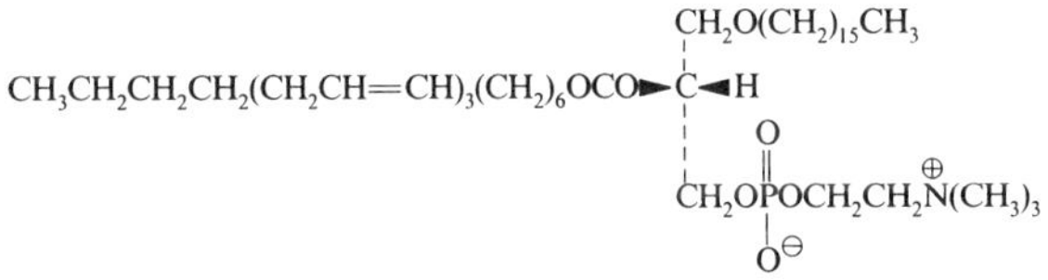

$C_{44}H_{84}NO_7P$ M 770.124

Isol. from the Japanese oyster *Crassostrea gigas*. PAF precursor.

Jeong, B.Y., *Lipids*, 1990, **25**, 624 (*isol*)

3-Hexadecyldihydro-4-hydroxy-5-methylene-2(3*H*)-furanone, 9CI — H-00277

2-Hexadecyl-3-hydroxy-4-penten-4-olide

[81576-08-1]

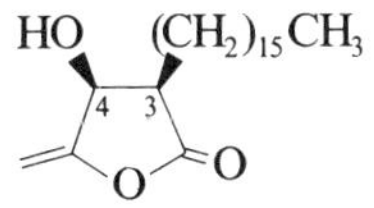

$C_{21}H_{38}O_3$ M 338.529

(3R*,4S*)-form

Isol. from the gorgonian *Plexaura flava*. Wax. $[\alpha]_D^{20}$ +2.8° (c, 1.4 in CH_2Cl_2).

Ravi, B.N. *et al*, *Aust. J. Chem.*, 1982, **35**, 105 (*isol, ir, pmr, ms*)

1-*O*-Hexadecyl-2-*O*-docosahexaenoyl-*sn*-glycero-3-phosphocholine **H-00278**

DCHA-PAF

[97717-93-6]

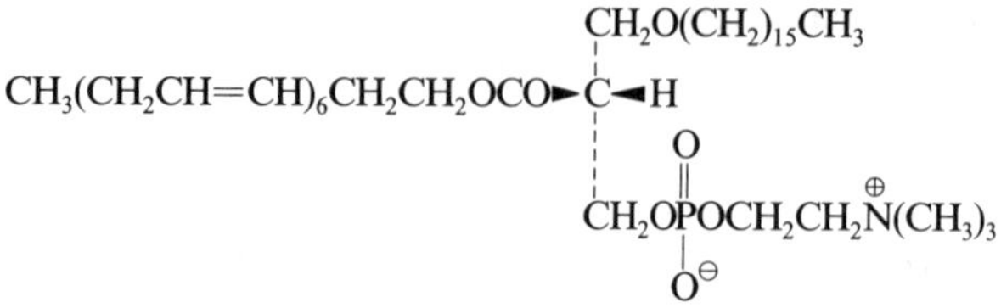

$C_{46}H_{82}NO_7P$ M 792.130

Isol. from the Japanese oyster *Crassostrea gigas*. PAF precursor.

Robinson, M. *et al*, *J. Biol. Chem.*, 1985, **260**, 7889 (*synth*)
Jeong, B.Y. *et al*, *Lipids*, 1990, **25**, 624 (*isol*)

1-*O*-Hexadecyl-2-*O*-eicosapentaenoyl-*sn*-glycero-3-phosphocholine **H-00279**

EPA-PAF

[132196-28-2]

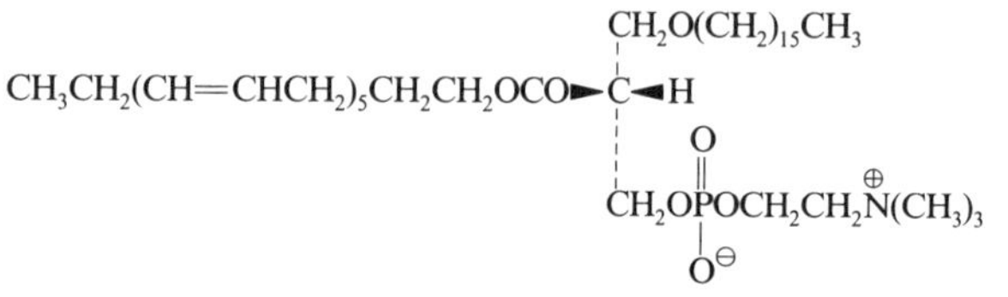

$C_{44}H_{80}NO_7P$ M 766.092

Isol. from the Japanese oyster *Crassostrea gigas*. PAF precursor.

Jeong, B.Y., *Lipids*, 1990, **25**, 624 (*isol*)

1-*O*-Hexadecyl-*sn*-glycero-3-phosphocholine **H-00280**

*4,7-Dihydroxy-*N,N,N*-trimethyl-3,5,9-trioxa-4-phosphapentacosan-1-aminium hydroxide, inner salt, 4-oxide. Lyso-PAF (C-16)*

[52691-62-0]

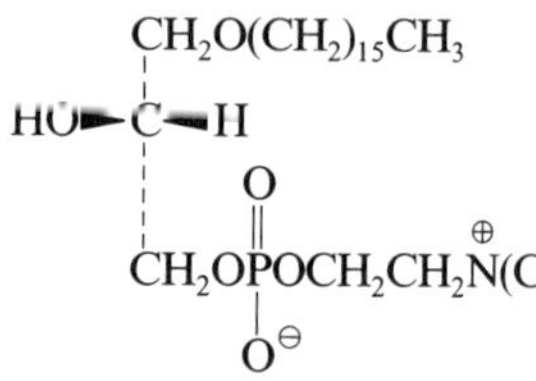

(R)-form

$C_{24}H_{52}NO_6P$ M 481.652

White powder. Mp 250° dec. $[\alpha]_D^{25}$ −6.09° (c, 1.04 in $CHCl_3$/MeOH).

2-Ac: [74389-68-7]. *PAF(C-16). 1-*O*-Hexadecyl-2-*O*-acetyl-*sn*-glycero-3-phosphonocholine. 1-*O*-Hexadecyl-PAF*

$C_{26}H_{54}NO_7P$ M 523.689

Major molecular form of PAF (see Platelet activating factor, P-00136). Powerful mediator in physiological processes such as anaphylaxis and inflammation. Amorph. solid. Mp 247° dec. $[\alpha]_D^{20}$ −3.30° (c, 0.53 in $CHCl_3$/MeOH).

2-Me ether: [78858-44-3]. *1-*O*-Hexadecyl-2-*O*-methyl-*sn*-glycero-3-phosphocholine. 2-*O*-Methyl PAF (C-16)*

$C_{25}H_{54}NO_6P$ M 495.678

Potential antimetastatic and antitumour agent. Cryst. $[\alpha]_D^{20}$ −5.41° (c, 0.95 in $CHCl_3$/MeOH).

2-Et ether: [78858-42-1]. *1-*O*-Hexadecyl-2-*O*-ethyl-*sn*-glycero-3-phosphocholine. 7-Ethoxy-4-hydroxy-*N,N,N*-trimethyl-3,5,9-trioxa-4-phosphapentacosan-1-aminium hydroxide, inner salt, 4-oxide. AEGPC. E-PAF*

$C_{26}H_{56}NO_6P$ M 509.705

PAF-analogue. Degranulates human neutrophils and desensitises neutrophils to PAF. $[\alpha]_D^{23}$ −0.19° (c, 1.3 in MeOH/$CHCl_3$).

2-Benzyl ether: [91326-84-0]. *1-*O*-Hexadecyl-2-*O*-benzyl-*sn*-glycero-3-phosphocholine. 2-*O*-Benzyl PAF (C-16)*

$C_{31}H_{58}NO_6P$ M 571.776

Stable PAF analogue. Amorph. solid. Mp 200°. $[\alpha]_D^{20}$ +2.41° (c, 2.32 in MeOH).

*2-*N*-Methylcarbamoyl:* [110406-02-5]. *1-*O*-Hexadecyl-2-*N*-methylcarbamoyl-*sn*-glycero-3-phosphocholine. 7-[(Aminocarbonyl)oxy]-4-hydroxy-*N,N,N*-trimethyl-3,5,9-trioxa-4-phosphahexacosan-1-aminium hydroxide, inner salt, 4-oxide. C-PAF*

$C_{26}H_{55}N_2O_7P$ M 538.703

Biologically potent non-metabolizable PAF analogue. Stimulates human neutrophils to synthesise PAF. No phys. props. reported.

N-*Methylpyrrolidine, 2-Ac:* [91021-63-5]. *1-*O*-Hexadecyl-2-*O*-acetyl-*sn*-glycero-3-phospho-(*N*-methylpyrrolidino) ethanolamine. 1-[7-(Acetyloxy)-4-hydroxy-3,5,9-trioxa-4-phosphapentacos-1-yl]-1-methylpyrrolidinium hydroxide, inner salt, *P*-oxide. Pyrrolidino-PAF*

$C_{28}H_{56}NO_7P$ M 549.727

Induces platelet aggregation. Hypotensive agent. Ten times more active than parent PAF. $[\alpha]_D^{20}$ −4.12° (c, 0.38 in $CHCl_3$/MeOH).

(S)-form [77249-80-0]

Enantio-PAF (C-16)

$C_{26}H_{54}NO_7P$ M 523.689

Biologically inactive enantiomer of PAF. $[\alpha]_D^{20}$ +3.12° (c, 2.09 in $CHCl_3$/MeOH).

Wykle, R.L. *et al*, *Biochem. Biophys. Res. Commun.*, 1981, **100**, 1651 (*synth, biochem*)
Hirth, G. *et al*, *Helv. Chim. Acta*, 1982, **65**, 1059 (*synth*)
Kasuya, Y. *et al*, *Can. J. Physiol. Pharmacol.*, 1984, **62**, 457 (*deriv*)
Ohno, M. *et al*, *Chem. Pharm. Bull.*, 1985, **33**, 572 (*synth, ir, pmr, deriv, biochem*)
Kertscher, H.P. *et al*, *Pharmazie*, 1986, **41**, 596 (*synth, lyso PAF*)
O'Flaherty, J.T. *et al*, *Biochem. Biophys. Res. Commun.*, 1987, **147**, 18 (*synth, biochem*)
Bhatia, S.K. *et al*, *J. Org. Chem.*, 1988, **53**, 5034 (*synth, ir, pmr*)
Wolfgang, J., *Synform*, 1988, **6**, 115; 1990, **8**, 313 (*rev, synth*)
Hofmann, B. *et al*, *Thromb. Res.*, 1988, **49**, 415 (*biochem*)
Benveniste, J., *Adv. Prostaglandin, Thromboxane, Leukotriene Res.*, 1989, **19**, 355 (*rev, props*)
Ostermann, G. *et al*, *Biomed. Biochim. Acta.*, 1989, **47**, 5233 (*biochem*)
Tessner, T.G. *et al*, *J. Biol. Chem.*, 1989, **264**, 4794 (*biochem*)
Guivisdalsky, P.N. *et al*, *J. Org. Chem.*, 1989, **54**, 4637, 4643 (*synth, ir, pmr*)
Murari, M.P. *et al*, *Lipids*, 1990, **25**, 606 (*synth, lyso PAF*)
Nakamura, N. *et al*, *Tetrahedron Lett.*, 1990, **31**, 699 (*synth*)

Hexadecyl hexadecanoate **H-00281**

Hexadecyl palmitate. Cetyl palmitate. Cetin

[540-10-3]

$$H_3C(CH_2)_{14}COO(CH_2)_{15}CH_3$$

$C_{32}H_{64}O_2$ M 480.856

Constit. of plant and animal waxes. Cryst. (Et_2O or AcOH). Mp 54°.

▷ RT4750000.

Iyengar, B.T.R. *et al*, *Lipids*, 1969, **4**, 28 (*synth*)

Aasen, A.J. *et al, Lipids*, 1971, **6**, 502 (*ms*)
Marosi, L. *et al, Justus Liebigs Ann. Chem.*, 1973, 584 (*props, cryst struct*)

3-Hexadecyl-5-hydroxy-5-methyl-2(5*H*)-furanone H-00282

2-Hexadecyl-4-hydroxy-4-methyl-2-buten-4-olide
[81346-95-4]

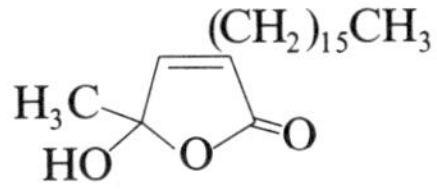

$C_{21}H_{38}O_3$ M 338.529
Isol. from the Japanese gorgonian *Euplexaura flava*.

Kituchi, H. *et al, Chem. Lett.*, 1982, 233 (*isol*)

Hexadecyl octadecanoate H-00283

Hexadecyl stearate. Cetyl stearate
[1190-63-2]

$H_3C(CH_2)_{16}COO(CH_2)_{15}CH_3$

$C_{34}H_{68}O_2$ M 508.910
Constit. of plant and animal waxes. Cryst. (Et_2O or MeOH). Mp 58°.

Iyengar, B.T.R. *et al, Lipids*, 1969, **4**, 28 (*synth*)
Aasen, A.J. *et al, Lipids*, 1971, **6**, 502 (*ms*)
Marosi, L. *et al, Justus Liebigs Ann. Chem.*, 1973, 584 (*props, cryst struct*)

1-(Hexadecyloxy)-3-(hexyloxy)-2-propanol, 9CI H-00284

$CH_2O(CH_2)_5CH_3$
HO►C◄H (*R*)-*form*
$CH_2O(CH_2)_{15}CH_3$

$C_{25}H_{52}O_3$ M 400.684
(*R*)-*form* [86158-15-8]
*3-O-Hexadecyl-1-O-hexyl-*sn*-glycerol*
Mp 35.5°.
(*S*)-*form* [86158-16-9]
*1-O-Hexadecyl-3-O-hexyl-*sn*-glycerol*
Mp 35-36°. [M] +2.54° (C_6H_6).

Michelsen, P. *et al, Chem. Phys. Lipids*, 1983, **32**, 27 (*synth, ord*)

3-Hexadecyloxy-2-oxopropyl phosphate H-00285

1-(Hexadecyloxy)-3-(phosphonooxy)-2-propanone, 9CI. 1-O-Hexadecyldihydroxyacetone-3-phosphate
[32113-53-4]

$H_3C(CH_2)_{15}OCH_2COCH_2OPO_3H_2$

$C_{19}H_{39}O_6P$ M 394.487
Mp 85-86°.

Hajra, A.K. *et al, Chem. Phys. Lipids*, 1983, **33**, 179 (*synth, ir, pmr*)

3-Hexadecyloxy-1,2-propanediol H-00286

α-Palmityl glyceryl ether. Glyceryl 1-hexadecyl ether. Glyceryl α-palmityl ether. ***Chimyl alcohol***

$CH_2O(CH_2)_{15}CH_3$
HO►C◄H (*S*)-*form*
CH_2OH

$C_{19}H_{40}O_3$ M 316.523
(*R*)-*form*
*3-O-Hexadecyl-*sn*-glycerol*
Mp 64-65°. [M] +14° (EtOH).
(*S*)-*form* [25378-11-4]
*D-form. 1-O-Hexadecyl-*sn*-glycerol*
Constit. of shark liver and other elasmobranch fish oils. Mp 64.5-65.5°. Opt. rotn. concn. dependent.
2-Acetate: [77133-35-8]. *1-O-Hexadecyl-2-O-acetyl-*sn*-glycerol. 3-Hexadecyloxy-1,2-propanediol 2-acetate*
$C_{21}H_{42}O_4$ M 358.560
Biosynth. precursor to Platelet activating factor, P-00136. Stimulates cell differentiation and possesses hypotensive activity. Ir 3420, 1720 (CO), 1180 (COC), 1060 (OH) cm^{-1}.
Di-Ac: [10152-42-8].
$C_{23}H_{44}O_5$ M 400.598
$[\alpha]_D^{20}$ −12.8° (c, 0.2 in hexane).
2-Me ether: [96960-92-8]. *1-O-Hexadecyl-2-O-methyl-*sn*-glycerol. 3-(Hexadecyloxy)-2-methoxy-1-propanol*
Inhibitor of protein kinase. Mp 29-30° (35-36°). $[\alpha]_D^{20}$ −9.92° (c, 1.64 in $CHCl_3$).
Isopropylidene deriv.:
$C_{22}H_{44}O_3$ M 356.588
$[\alpha]_D^{22}$ −14.0° (c, 0.29 in hexane).
(±)-*form* [6145-69-3]
Cryst. (hexane). Mp 59.9-60.6°.

Hallgren, B. *et al, J. Lipid Res.*, 1962, **3**, 31 (*occur*)
Wood, R. *et al, Lipids*, 1967, **2**, 161 (*synth, chromatog, pmr*)
Chacko, G.K. *et al, Biochim. Biophys. Acta*, 1968, **164**, 252 (*synth*)
Michelson, P. *et al, Chem. Phys. Lipids*, 1983, **32**, 27 (*synth, cd*)
Moschidis, M.C. *et al, Chem. Phys. Lipids*, 1983, **33**, 87 (*synth, biochem*)
Ohno, M. *et al, Chem. Pharm. Bull.*, 1985, **33**, 572 (*synth, ir, pmr*)
Bass, D.A. *et al, J. Biol. Chem.*, 1988, **263**, 19610 (*synth, biochem*)
Bhatia, S.K. *et al, J. Org. Chem.*, 1988, **53**, 5034 (*synth, ir, pmr*)
Krammer, I.M. *et al, J. Biol. Chem.*, 1989, **264**, 5876 (*biochem*)
Guivisdalsky, P.N. *et al, J. Org. Chem.*, 1989, **54**, 4637 (*synth, ir, pmr*)
Bauer, F. *et al, Justus Liebigs Ann. Chem.*, 1991, 765 (*synth*)
Kumar, A. *et al, Synth. Commun.*, 1991, **21**, 1763 (*synth, ir*)

1-Hexadecylpyridinium(1+) H-00287

Cetylpyridinium. Other proprietary names
[7773-52-6]

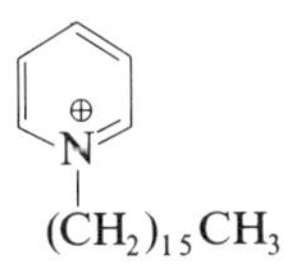

$C_{21}H_{38}N^{\oplus}$ M 304.538 (ion)
Surface active agent with bactericidal properties; used in throat pastilles etc. Used as a 0.2-0.3% aq. soln. to increase sensitivity in photometric detn. of metals; forms ternary complexes with metal ions and some chromogenic reagents.
Chloride: [123-03-5]. *Cetylpyridinium chloride, BAN, INN. Merocet. Ceepryn. Cepacol*
$C_{21}H_{38}ClN$ M 339.991
Sol. H_2O. Mp 87-88°. Monohydrate, Mp 77-83°.
▷ UU4900000.
Bromide: [140-72-7].
$C_{21}H_{38}BrN$ M 384.442
Microcryst. Sol. H_2O, EtOH. Mp 61-62°, Mp 66-68°.
Iodide: [2349-55-5].
$C_{21}H_{38}IN$ M 431.442
Mp 101°.

Warren, M.R. *et al*, *J. Pharm. Pharmacol.*, 1942, **74**, 401 (*pharmacol, tox*)
Huyck, C.L. *et al*, *Am. J. Pharm.*, 1944, **116**, 50 (*rev*)
Adam, N.K. *et al*, *Trans. Faraday Soc.*, 1946, **42**, 523 (*synth*)
Ziegler, M., *Angew. Chem.*, 1956, **68**, 411, 620 (*synth*)
Chumakov, Y.I. *et al*, *CA*, 1966, **64**, 19549d (*synth*)
Svoboda, V., *Talanta*, 1966, **13**, 237 (*use*)
Tikhonov, B.N., *Zh. Anal. Khim.*, 1977, **32**, 1435 (*rev*)
Martindale, The Extra Pharmacopoeia, 28th/29th Ed., Pharmaceutical Press, London, 1982/1989, 2217.
Marczenko, Z., *Separation and Spectrophotometric Determination of Elements*, Horwood, Chichester, 1986, 78 (*use*)
Negwer, M., *Organic-Chemical Drugs and their Synonyms*, 6th Ed., Akademie-Verlag, Berlin, 1987, 6403 (*synonyms*)
Lewis, R.J., *Sax's Dangerous Properties of Industrial Materials*, 8th Ed., Van Nostrand-Reinhold, 1992, CCX000.

Hexadecyltrimethylammonium(1+), 8CI — H-00288

N,N,N-*Trimethyl-1-hexadecanaminium, 9CI. Cetyltrimethylammonium. Cetrimonium*
[6899-10-1]

$H_3C(CH_2)_{15}N^{\oplus}Me_3$

$C_{19}H_{42}N^{\oplus}$ M 284.547 (ion)
Topical antiseptic, surfactant. Reagent for isol. and purifn. of acidic polysaccharides and sulphonated carbohydrates. Used as 0.2-0.3% aq. soln. Increases sensitivity in photometric detn. of metals (e.g. Al, Be, In, Th, Zr) with some chelating reagents (e.g. chromazurol S, bromopyrogallol red, phenylfluorone); forms ternary complexes.

Chloride: [112-02-7]. *Cetrimonium chloride, BAN*
$C_{19}H_{42}ClN$ M 320.000

Bromide: [57-09-0]. *Cetrimonium bromide, INN. Cetrimide. Cetab. Other synonyms*
$C_{19}H_{42}BrN$ M 364.451
Microcryst. Sol. H_2O, EtOH, insol. C_6H_6, Et_2O. Mp 237-243°.

▷Toxic, exp. teratogen. Emits highly toxic fumes when heated. BQ7875000.

Pentachlorophenolate: [87-76-3]. *TCAP. T.S.P.*
Topical antiseptic, fungicide.

4-Methylbenzenesulfonate: [138-32-9]. *Cetrimonium tosilate. Intol*

Fieser and Fieser's Reagents for Organic Synthesis, Wiley, 1967, **1**, 121.
Borisova, N.N. *et al*, *Zh. Prikl. Khim. (Leningrad)* 1967, **40**, 2838 (*synth, pharmacol*)
Uno, T. *et al*, *Spectrochim. Acta, Part A*, 1968, **24**, 1749 (*ir*)
Chojnacki, J. *et al*, *Pol. J. Chem.* (*Rocz. Chem.*), 1971, **45**, 1997 (*cryst struct*)
LoSurdo, A. *et al*, *J. Phys. Chem.*, 1972, **76**, 130 (*pmr*)
Kalyanasundaram, K. *et al*, *J. Phys. Chem.*, 1976, **80**, 1462 (*conformn*)
Tikhonov, B.N., *Zh. Anal. Khim.*, 1977, **32**, 1435 (*rev*)
Isomaa, B. *et al*, *Acta Pharmacol. Toxicol.*, 1980, **47**, 17 (*pharmacol, toxicity*)
Martindale, The Extra Pharmacopoeia, 28th/29th Ed., Pharmaceutical Press, London, 1982/1989, 2216.
Callahan, J.H. *et al*, *Anal. Chem.*, 1984, **56**, 1632 (*theory*)
Chernova, R.K. *et al*, *Zh. Anal. Khim.*, 1984, **39**, 1019 (*use*)
Antonovich, V.P. *et al*, *Zh. Anal. Khim.*, 1984, **39**, 1157 (*use*)
Ohashi, M. *et al*, *Chem. Pharm. Bull.*, 1985, **33**, 2499 (*ms*)
Marczenko, Z., *Separation and Spectrophotometric Determination of Elements*, Horwood, Chichester, 1986 (*use*)
Negwer, M., *Organic-Chemical Drugs and their Synonyms*, 6th Ed., Akademie-Verlag, Berlin, 1987, 5535.
Lewis, R.J., *Sax's Dangerous Properties of Industrial Materials*, 8th Ed., Van Nostrand-Reinhold, 1992, HCQ500, TLN150.

1-Hexadecyne — H-00289

[629-74-3]

$H_3C(CH_2)_{13}C{\equiv}CH$

$C_{16}H_{30}$ M 222.413
Mp 15°. Bp_{15} 155°, $Bp_{0.3}$ 114-116°.

Kraft, F., *Ber.*, 1896, **29**, 2232.
Klein, J. *et al*, *Tetrahedron*, 1970, **26**, 2127 (*synth*)

5-Hexadecyne — H-00290

[71899-37-1]

$H_3C(CH_2)_9C{\equiv}C(CH_2)_3CH_3$

$C_{16}H_{30}$ M 222.413
Liq. $Bp_{0.04}$ 71-73°.

Sikorski, J.A. *et al*, *J. Org. Chem.*, 1986, **51**, 4521 (*synth, pmr, cmr*)

2-Hexadecynoic acid — H-00291

$H_3C(CH_2)_{12}C{\equiv}CCOOH$

$C_{16}H_{28}O_2$ M 252.396
Cryst. Mp 54-55° (49.1-50.2°).

Sunko, D.E. *et al*, *Ark. Kemi*, 1955, **27**, 217; *CA*, **50**, 11943 (*synth*)
Wailes, P.C., *Aust. J. Chem.*, 1959, **12**, 173 (*deriv*)
Grimmer, G. *et al*, *Justus Liebigs Ann. Chem.*, 1965, **685**, 154 (*synth*)

4-Hexadecynoic acid — H-00292

$H_3C(CH_2)_{10}C{\equiv}CCH_2CH_2COOH$

$C_{16}H_{28}O_2$ M 252.396
Cryst. (pet. ether). Mp 68-70°.

Ames, D.E. *et al*, *J. Chem. Soc.*, 1963, 5889 (*synth*)

7-Hexadecynoic acid — H-00293

Palmitolic acid
[629-57-2]

$H_3C(CH_2)_7C{\equiv}C(CH_2)_5COOH$

$C_{16}H_{28}O_2$ M 252.396
Cryst. Mp 38-39°, Mp 47°. $Bp_{0.6}$ 160-163°.

Me ester: [39077-31-1].
$C_{17}H_{30}O_2$ M 266.423
$Bp_{0.05}$ 102°.

Ames, D.E. *et al*, *J. Chem. Soc.*, 1963, 775 (*synth*)
Williams, J.L. *et al*, *Chem. Phys. Lipids*, 1972, **9**, 295, 320 (*synth, pmr*)

10-Hexadecynoic acid, 8CI — H-00294

[676-25-5]

$H_3C(CH_2)_4C{\equiv}C(CH_2)_8COOH$

$C_{16}H_{28}O_2$ M 252.396
Cryst. (MeOH or MeOH aq.). Mp 35-36°.

Baker, C.D. *et al*, *J. Chem. Soc.*, 1963, 489 (*synth*)

3-(15-Hexadecynylidene)dihydro-4-hydroxy-5-methyl-2(3*H*)-furanone H-00295

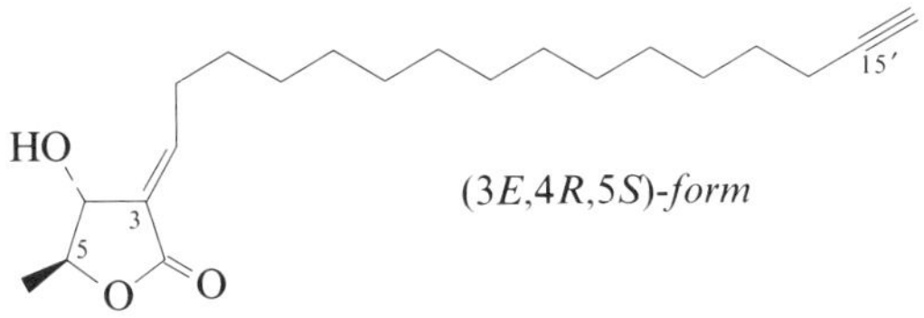

$C_{21}H_{34}O_3$ M 334.498

(3E,4R,5S)-form [71339-54-3] ***Isodihydromahubynolide A***

Isol. from the trunkwood of *Clinostemon mahuba*. Mp 64-65°. $[\alpha]_D^{24}$ +42.4°.

15′,16′-Dihydro: [71339-52-1]. ***Isodihydromahubenolide A***

$C_{21}H_{36}O_3$ M 336.514

From *C. mahuba*. Mp 45-46°. $[\alpha]_D^{24}$ +37.1°.

15′,15′,16′,16′-Tetrahydro: [71359-05-2]. ***Isodihydromahubanolide A***

$C_{21}H_{38}O_3$ M 338.529

From *C. mahuba*. Mp 49-50°. $[\alpha]_D^{24}$ +36.6°.

(3E,4S,5S)-form [71325-96-7] ***Isodihydromahubynolide B***

From *C. mahuba*. Mp 79-80°. $[\alpha]_D^{24}$ −90.2°.

15′,16′-Dihydro: [71325-94-5]. ***Isodihydromahubenolide B***

$C_{21}H_{36}O_3$ M 336.514

From *C. mahuba*. Mp 66-67°. $[\alpha]_D^{24}$ −96.5°.

15′,15′,16′,16′-Tetrahydro: [71358-20-8]. ***Isodihydromahubanolide B***

$C_{21}H_{38}O_3$ M 338.529

From *C. mahuba*. Mp 70-71° (78-79°). $[\alpha]_D^{24}$ −93.3°, $[\alpha]_D^{21}$ −102° (dioxan).

(3Z,4R,5S)-form [71339-53-2] ***Dihydromahubynolide A***

From *C. mahuba*. Mp 61-63°. $[\alpha]_D^{24}$ +9.0°.

15′,16′-Dihydro: [71339-51-0]. ***Dihydromahubenolide A***

$C_{21}H_{36}O_3$ M 336.514

From *C. mahuba*. Mp 48-50°. $[\alpha]_D^{24}$ +8.5°.

15′,15′,16′,16′-Tetrahydro: [71359-04-1]. ***Dihydromahubanolide A***

$C_{21}H_{38}O_3$ M 338.529

From *C. mahuba*.

(3Z,4S,5S)-form [71325-95-6] ***Dihydromahubynolide B***

From *C. mahuba*. Mp 55-56°. $[\alpha]_D^{24}$ −32.6°.

15′,16′-Dihydro: [71325-93-4]. ***Dihydromahubenolide B***

$C_{21}H_{36}O_3$ M 336.514

From *C. mahuba*. Mp 46-47°. $[\alpha]_D^{24}$ −35.1°.

15′,16′-Tetrahydro: [71325-97-8]. ***Dihydromahubanolide B***

$C_{21}H_{38}O_3$ M 338.529

From *C. mahuba*. Mp 66-67°. $[\alpha]_D^{21}$ −37° (dioxan)

[71358-21-9, 78339-87-4, 78339-88-5, 78339-89-6, 121703-31-9]

Rollinson, S.W. *et al*, *J. Am. Chem. Soc.*, 1981, **103**, 4114 (*synth*)
Amos, R.A. *et al*, *J. Am. Chem. Soc.*, 1981, **103**, 4114 (*Dihydromahubanolide B*)
Martinez, J.C. *et al*, *Phytochemistry*, 1981, **20**, 459 (*isol*)
Tanaka, K. *et al*, *Bull. Chem. Soc. Jpn.*, 1982, **55**, 3935 (*Dihydromahubanolide B, Isodihydromahubanolide B*)
Ortuno, R.M. *et al*, *Tetrahedron*, 1988, **44**, 5139 (*abs config*)
Wood, W.W. *et al*, *J. Chem. Soc., Perkin Trans. 1*, 1990, 3201 (*synth*)

2,4-Hexadienal, 9CI H-00296

Sorbic aldehyde. Sorbaldehyde

[142-83-6]

$H_3CCH{=}CHCH{=}CHCHO$

C_6H_8O M 96.129

▷ WG1925000.

(E,E)-form

Bp 173-174°, Bp_{11} 64-66°.

▷ Toxic by skin absorption and inhalation.

Oxime: [1515-97-5].

C_6H_9NO M 111.143

Needles (EtOH). Mp 159.5-160.5° dec.

Semicarbazone: Plates (EtOH). Mp 206°.

Phenylhydrazone: Yellow plates (EtOH). Mp 101-102°.

(2Z,4E)-form [54716-12-0]

Pale yellow malodorous liq. Bp *ca.* 180°.

2,4-Dinitrophenylhydrazone: Mp 161-163°.

[80466-34-8]

Albriktsen, P. *et al*, *Acta Chem. Scand.*, 1973, **27**, 3993 (*pmr, config*)
Shiess, P. *et al*, *Helv. Chim. Acta*, 1974, **57**, 2583 (*synth*)
Corey, E.J. *et al*, *Chem. Ber.*, 1978, **111**, 1362 (*synth*)
Furber, M. *et al*, *J. Chem. Soc., Perkin Trans. 1*, 1989, 683 (*synth, ir, pmr, ms*)
Lewis, R.J., *Sax's Dangerous Properties of Industrial Materials*, 8th Ed., Van Nostrand-Reinhold, 1992, SKT500.

3,5-Hexadienoic acid, 9CI H-00297

[29949-29-9]

$H_2C{=}CHCH{=}CHCH_2COOH$

$C_6H_8O_2$ M 112.128

(E)-form [73670-87-8]

Bp_{20} 79-80°.

Me ester: [32775-94-3].

$C_7H_{10}O_2$ M 126.155

Bp_{17} 66-68°.

Et ester: [74054-58-3].

$C_8H_{12}O_2$ M 140.182

Bp_{30} 87-88°.

Chloride: [73670-84-5].

C_6H_7ClO M 130.573

Liq. Bp_{50} 79-81°.

p-*Nitrobenzyl ester:* [87785-65-7].

Yellow cryst. (Et_2O/pentane). Mp 40-41°.

(Z)-form [2196-20-5]

Oil. Bp_{70} 100-103°.

[40338-61-2]

Torssell, K., *Tetrahedron Lett.*, 1974, 623 (*synth, deriv*)
Stevens, R.V. *et al*, *J. Am. Chem. Soc.*, 1976, **98**, 6317 (*synth, deriv, ir, pmr*)
Martin, S.F. *et al*, *J. Am. Chem. Soc.*, 1980, **102**, 5274 (*deriv, synth, ir, pmr, ms*)
Ueda, Y. *et al*, *Can. J. Chem.*, 1983, **61**, 1996 (*synth, deriv, ir, pmr*)

5,9-Hexaeicosadienoic acid H-00298

$H_3C(CH_2)_{15}CH{=}CHCH_2CH_2CH{=}CH(CH_2)_3COOH$

$C_{26}H_{48}O_2$ M 392.664

(5Z,9Z)-form [52715-55-6]

Isol. from *Axinella verrucosa* and *Perrosia ficiformis*. Other stereoisomers have been prepd.

[90913-50-1, 90913-54-5, 90913-61-4, 90913-68-1]

Ayanoglu, E. *et al*, *Lipids*, 1982, **17**, 617 (*isol*)
Lankelma, J. *et al*, *Lipids*, 1983, **18**, 853 (*isol*)
Mena, P.L. *et al*, *J. Org. Chem.*, 1984, **49**, 3260 (*synth, ir, pmr, ms*)

7,7′,8,8′,11,12-Hexahydrolycopene H-00299

7,7′,8,8′,11,12-Hexahydro-ψ,ψ-carotene. ***Phytofluene***

[540-05-6]

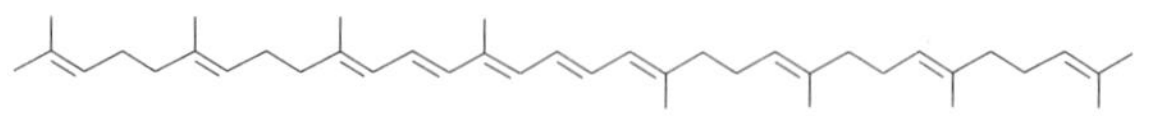

$C_{40}H_{62}$ M 542.930

Constit. of *Neurospora* spp. and other microorganisms. Widespread in plants. Pale-yellow oil with brilliant green fluorescence. The (all-*E*) and several mono- and di-Z isomers appear to occur naturally.

9Z-form

9-cis-*Phytofluene*

$C_{40}H_{62}$ M 542.930

Constit. of *Dunaliella bardawil*. Pale yellow liq.

(15Z,9′Z)-form [72746-34-0]

Isol. from *Lycopersicon esculentum* var. 'Tangella'. Yellow oil with intense blue-green fluorescence. λ_{max} 249, 257, 331, 348, 367 nm (hexane).

Zechmeister, L. *et al*, *Arch. Biochem. Biophys.*, 1953, **47**, 160 (*occur*)
Davis, J.B. *et al*, *J. Chem. Soc. C*, 1966, 2154 (*isol, struct*)
Karrer, W. *et al*, *Konstitution und Vorkommen der Organischen Pflanzenstoffe*, 2nd Ed., Birkhäuser Verlag, Basel, 1972-1985, no. 1817 (*occur*)
Brown, D.J. *et al*, *Biochem. Soc. Trans.*, 1975, **3**, 741 (*biosynth*)
Clough, J.M. *et al*, *J. Chem. Soc., Perkin Trans. 1*, 1983, 3011 (*15Z, 9′Z-form, ir, pmr, cmr*)
Frecknall, E.A. *et al*, *Phytochemistry*, 1984, **23**, 1707 (*occur*)
Ebenezer, W.J. *et al*, *J. Chem. Soc., Perkin Trans. 1*, 1993, 1869 (*9Z-isomer*)

8,16,18,26,34,36-Hexahydroxy-2,6,10,14,24,28,32-hentetracontaneheptone, 9CI H-00300

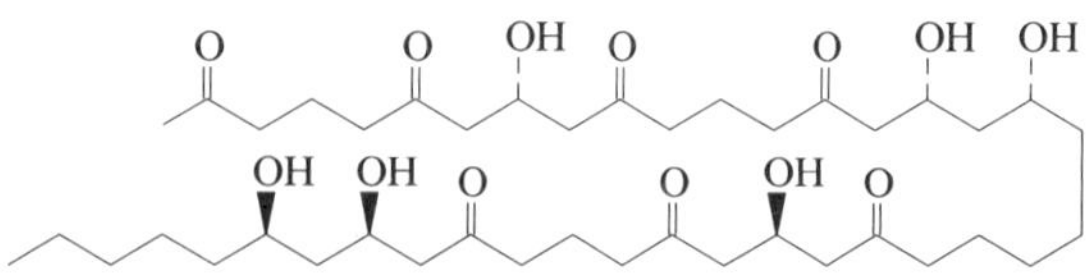

$C_{41}H_{70}O_{13}$ M 770.996

(8S,16R,18R,26S,34R,36R)-form [73356-03-3] ***HMT Band 2 toxin***

Isol. from *Helminthosporium maydis* race T. Host-specific pathotoxin. Powder. Mp 127-128°. $[\alpha]_D^{26}$ +17.5° (c, 0.6 in MeOH).

Kono, Y. *et al*, *Bioorg. Chem.*, 1981, **10**, 206 (*isol, struct, props*)
Kono, Y. *et al*, *Agric. Biol. Chem.*, 1985, **49**, 559 (*abs config*)

9,10,12,13,15,16-Hexahydroxyoctadecanoic acid H-00301

$H_3C(CH_2CHOHCHOH)_3(CH_2)_7COOH$

$C_{18}H_{36}O_8$ M 380.478

Prepared by $KMnO_4$ oxidn. of α-Linolenic acid. Cryst. (H_2O). Mp 198-200°.

Bayzer, H. *et al*, *Monatsh. Chem.*, 1958, **89**, 15 (*synth, ir*)
Niehaus, W.G. *et al*, *Tetrahedron Lett.*, 1967, 5021 (*synth, ms*)

6,14,22,30,32,34-Hexahydroxy-8,16,24-pentatriacontanetrione, 9CI H-00302

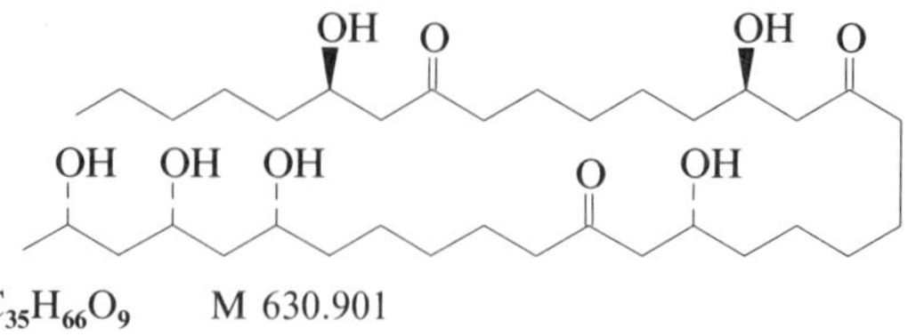

$C_{35}H_{66}O_9$ M 630.901

(6R,14R,22R,30R,32S,34S)-form

PM Toxin C

Isol. from *Phyllosticta maydis*. Host-specific pathotoxin. Powder. Mp 108-109°. $[\alpha]_D^{27}$ −6° (c, 0.13 in MeOH).

Kono, Y. *et al*, *Biochemistry*, 1984, **23**, 759 (*isol, struct, props*)
Kono, Y. *et al*, *Agric. Biol. Chem.*, 1985, **49**, 149, 559 (*synth, abs config*)

4,12,20,24,26,28-Hexahydroxy-2,10,18-tritriacontanetrione, 9CI H-00303

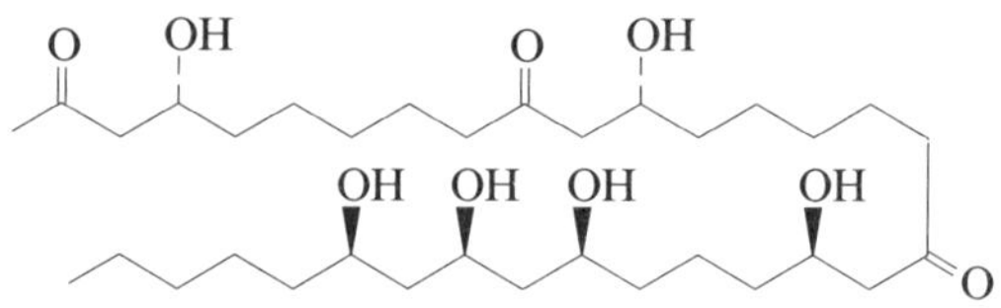

$C_{33}H_{62}O_9$ M 602.847

(4R,12R,20R,24S,26R,28R)-form [88212-13-9] ***PM Toxin D***

Isol. from *Phyllosticta maydis*. Host-specific pathotoxin. Mp 110-111°. $[\alpha]_D^{27}$ −8° (MeOH).

Kono, Y. *et al*, *Biochemistry*, 1984, **23**, 759 (*isol, struct, props*)
Kono, Y. *et al*, *Agric. Biol. Chem.*, 1985, **49**, 149, 559 (*synth, abs config*)

Hexatetracontanoic acid H-00304

$H_3C(CH_2)_{44}COOH$

$C_{46}H_{92}O_2$ M 677.232

Cryst. Mp 107°.

Et ester:

$C_{48}H_{96}O_2$ M 705.285

Mp 90.5°.

Francis, F. *et al*, *J. Chem. Soc.*, 1937, 999 (*synth*)

4,12,23,27,42-Hexatetracontapentaene-1,18,21,45-tetrayne-3,20,44,?-tetrol H-00305

Petroformyne 5

[128646-05-9]

$C_{46}H_{68}O_4$ M 685.041

Posn. of one hydroxyl grp. not known. Constit. of the sponge *Petrosia ficiformis*.

Cimino, G. *et al*, *J. Nat. Prod. (Lloydia)*, 1990, **53**, 345.

6,27-Hexatriacontadien-2-one, 9CI H-00306

$H_3C(CH_2)_7CH{=}CH(CH_2)_{19}CH{=}CH(CH_2)_3COCH_3$

$C_{36}H_{68}O$ M 516.933

(Z,Z)-form [133530-19-5]

Isol. from the snake *Boiga irregularis*.

Murata, Y. *et al*, *J. Nat. Prod. (Lloydia)*, 1991, **54**, 233.

8,27-Hexatriacontadien-2-one, 9CI H-00307

$H_3C(CH_2)_7CH{=}CH(CH_2)_{17}CH{=}CH(CH_2)_5COCH_3$

$C_{36}H_{68}O$ M 516.933

(Z,Z)-form [133530-20-8]

Isol. from the snake *Boiga irregularis.*

Murata, Y. *et al*, *J. Nat. Prod. (Lloydia)*, 1991, **54**, 233 (*isol, pmr, cmr*)

Hexatriacontane H-00308

[630-06-8]

$H_3C(CH_2)_{34}CH_3$

$C_{36}H_{74}$ M 506.981

Cryst. (pet. ether). Mp 76°.

Morgat, J.-L. *et al*, *C. R. Hebd. Seances Acad. Sci.*, 1965, **260**, 5579 (*synth*)
Boistelle, R. *et al*, *Acta Crystallogr., Sect. B*, 1976, **32**, 1240 (*cryst struct*)

1-Hexatriacontanol H-00309

[82741-64-8]

$H_3C(CH_2)_{34}CH_2OH$

$C_{36}H_{74}O$ M 522.980

Constit. of *Erythrina stricta* bark. Mp 93.5-94.5°.

Singh, H. *et al*, *J. Nat. Prod. (Lloydia)*, 1981, **44**, 526 (*occur*)
Huang, H.C. *et al*, *J. Org. Chem.*, 1982, **47**, 4018 (*synth, pmr*)

5-Hexatriacontanone H-00310

$H_3C(CH_2)_3CO(CH_2)_{30}CH_3$

$C_{36}H_{72}O$ M 520.964

Constit. of *Solanum torvum*. Cryst. (Me_2CO). Mp 78°.

Mahmood, U. *et al*, *Phytochemistry*, 1983, **22**, 167.

17-Hexatriaconten-1-ol H-00311

$H_3C(CH_2)_{17}CH{=}CH(CH_2)_{15}CH_2OH$

$C_{36}H_{72}O$ M 520.964

(E)-form [82741-68-2]

Cryst. (Me_2CO). Mp 82-83°.

Methanesulfonate: [82741-69-3].

Cryst. ($CHCl_3$/EtOH). Mp 78-79°.

(Z)-form [82741-65-9]

Cryst. (EtOH). Mp 61.5 63°.

Methanesulfonate: [82741-66-0].

Cryst. ($CHCl_3$/EtOH). Mp 75-76°.

Huang, H.C. *et al*, *J. Org. Chem.*, 1982, **47**, 4018 (*synth, pmr*)

17-Hexatriacontyn-1-ol H-00312

[82741-61-5]

$H_3C(CH_2)_{17}C{\equiv}C(CH_2)_{15}CH_2OH$

$C_{36}H_{70}O$ M 518.948

Cryst. (EtOH). Mp 77-78.5°.

Methanesulfonyl: Cryst. ($CHCl_3$/EtOH). Mp 65-66°.

Huang, H.C. *et al*, *J. Org. Chem.*, 1982, **47**, 4018 (*synth, pmr*)

1,3,5-Hexatriene, 9CI H-00313

sym-*Divinylethylene*

[2235-12-3]

$H_2C{=}CHCH{=}CHCH{=}CH_2$

C_6H_8 M 80.129

▷ MP5425000.

(E)-form [821-07-8]

Bp_{764} 77-78.5°. Polymerises.

(Z)-form [2612-46-6]

Bp 78.5°. Polymerises.

v. Romburgh, P. *et al*, *J. Chem. Soc.*, 1906, 130 (*synth*)
Woods, G.F. *et al*, *J. Am. Chem. Soc.*, 1955, **77**, 1800 (*synth*)
Figeys, H.P. *et al*, *Bull. Soc. Chim. Belg.*, 1974, **83**, 381 (*synth*)
Spangler, C.W. *et al*, *Synthesis*, 1976, 108 (*synth*)
Yamamoto, K. *et al*, *Bull. Chem. Soc. Jpn.*, 1981, **54**, 2541 (*synth*)
Bloch, R. *et al*, *Tetrahedron Lett.*, 1985, **26**, 1301 (*synth, pmr*)
Lewis, R.J., *Sax's Dangerous Properties of Industrial Materials*, 8th Ed., Van Nostrand-Reinhold, 1992, HEZ000.

2-Hexenal, 9CI H-00314

3-Propylacrolein

[505-57-7]

$H_3CCH_2CH_2CH{=}CHCHO$

$C_6H_{10}O$ M 98.144

Constit. of many foods, hornbeam leaves, and of scent gland of many bugs, e.g., *Pternistria bispina*; alarm pheromone of e.g. *Cimex lectularius.*

▷ MP5880000.

(E)-form [6728-26-3]

Leaf aldehyde

Used in perfumery and flavourings. Bp 146-147°, Bp_{12} 43°.

▷ MP5900000.

2,4-Dinitrophenylhydrazone: [2122-02-3].

Red needles. Mp 147°.

Semicarbazone: Cryst. (MeOH). Mp 179°.

Curtius, T. *et al*, *Justus Liebigs Ann. Chem.*, 1913, **404**, 93 (*isol*)
Elkik, E., *Bull. Soc. Chim. Fr.*, 1968, 283 (*synth*)
Hatanaka, A. *et al*, *Phytochemistry*, 1976, **15**, 1125 (*biosynth*)
Ito, Y. *et al*, *J. Org. Chem.*, 1978, **43**, 1011 (*synth*)
Zimmermann, B. *et al*, *Chem. Ber.*, 1986, **119**, 2848 (*synth, ir, pmr*)
Lewis, R.J., *Sax's Dangerous Properties of Industrial Materials*, 8th Ed., Van Nostrand-Reinhold, 1992, HFA500, PNC500.

4-Hexenal H-00315

$H_3CCH{=}CHCH_2CH_2CHO$

$C_6H_{10}O$ M 98.144

(E)-form [25166-87-4]

Oil. Bp_{50} 34°.

Oxime:

$C_6H_{11}NO$ M 113.159

Bp_{13} 80°.

Oppolzer, W. *et al*, *Tetrahedron*, 1985, **41**, 3497 (*synth, ir, pmr, ms*)

5-Hexenal H-00316

[764-59-0]

$H_2C{=}CHCH_2CH_2CH_2CHO$

$C_6H_{10}O$ M 98.144

Liq. Bp 118-121°.

2,4-Dinitrophenylhydrazone: Cryst. (EtOH aq.). Mp 99-100°.

Kharasch, M.S. *et al, J. Org. Chem.*, 1953, **18**, 1225.
LeBel, N.A. *et al, J. Am. Chem. Soc.*, 1964, **86**, 3759 (*synth*)

5-Hexenoic acid, 9CI H-00317

4-Pentene-1-carboxylic acid

[1577-22-6]

$H_2C{=}CH(CH_2)_3COOH$

$C_6H_{10}O_2$ M 114.144

Bp 203°. pK_a 4.72.

Me ester: [2396-80-7].
$C_7H_{12}O_2$ M 128.171
Bp_{12} 48°.

Et ester: [54653-25-7].
$C_8H_{14}O_2$ M 142.197
Bp_4 40°.

Hydrazide:
$C_6H_{12}N_2O$ M 128.174
Mp 46.5-47.5°.

Nitrile: 5-Cyano-1-pentene
C_6H_9N M 95.144
Liq. Bp 160-162°.

Anhydride:
$C_{12}H_{18}O_3$ M 210.272
No phys. props. given.

Wallach, O., *Justus Liebigs Ann. Chem.*, 1905, **343**, 28 (*synth*)
Bailey, W.J. *et al, J. Org. Chem.*, 1977, **42**, 3895 (*synth*)
Abd El Samii, Z.K.M. *et al, J. Chem. Soc., Perkin Trans. 1*, 1988, 2523 (*nitrile*)
Moody, C.J. *et al, J. Chem. Soc., Perkin Trans. 1*, 1988, 3249 (*anhydride*)
Juaristi, E. *et al, J. Org. Chem.*, 1991, **56**, 1623 (*synth, pmr, cmr, ir*)

5-(1-Hexenyl)-dihydro-2(3*H*)-furanone H-00318

5-Decen-4-olide. 4-Hydroxy-5-decenoic acid lactone

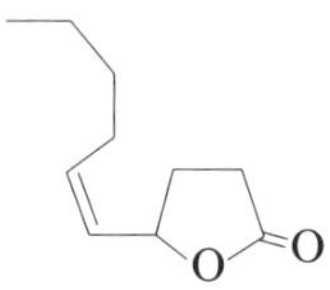

$C_{10}H_{16}O_2$ M 168.235

(Z)-form [82934-79-0]

Constit. of *Polianthes tuberosa*. Oil.

Maurer, B. *et al, Helv. Chim. Acta*, 1982, **65**, 45, 462 (*isol, synth, spectra*)

2-Hexen-4-ynedioic acid H-00319

1-Butene-3-yne-1,4-dicarboxylic acid

[20011-47-6]

$HOOCC{\equiv}CCH{=}CHCOOH$

$C_6H_4O_4$ M 140.095

(E)-form

Di-Me ester: [54299-80-8].
$C_8H_8O_4$ M 168.149
Dimer of methyl propiolate. Solid. Mp 53-54°.

Dinitrile: [129422-27-1]. *1,4-Dicyano-1-buten-3-yne*
$C_6H_2N_2$ M 102.095
Cryst. by subl. Mp 64°.

[129422-28-2]

Wenkert, E. *et al, Can. J. Chem.*, 1963, **41**, 1844 (*synth*)
Acheson, R.M. *et al, J. Chem. Soc., Perkin Trans. 1*, 1982, 1905 (*synth, use*)
Ferris, J.P. *et al, J. Org. Chem.*, 1990, **55**, 5601 (*dinitrile*)

2-Hexyl-1-cyclopropaneacetic acid H-00320

3,4-Methylenedecanoic acid. Cascarillic acid

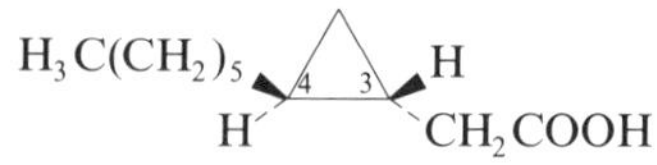

$C_{11}H_{20}O_2$ M 184.278

(3R*,4S*)-form

trans-*form*

Component of cascarilla essential oil.

Me ester:
$C_{12}H_{22}O_2$ M 198.305
Oil. $[\alpha]_D^{20}$ −10.5° (c, 0.55 in $CHCl_3$). The (±)-*cis*- and *trans*-forms have both been synthesised as their Me esters.

Motil, O. *et al, Phytochemistry*, 1972, **11**, 407 (*isol*)
Vig, B. *et al, J. Indian Chem. Soc.*, 1973, **50**, 557 (*synth*)
Wilson, S.R. *et al, Tetrahedron Lett.*, 1976, 4231 (*synth, ir, nmr, ms*)

2-Hexylcyclopropanedecanoic acid, 9CI H-00321

11,12-Methyleneoctadecanoic acid

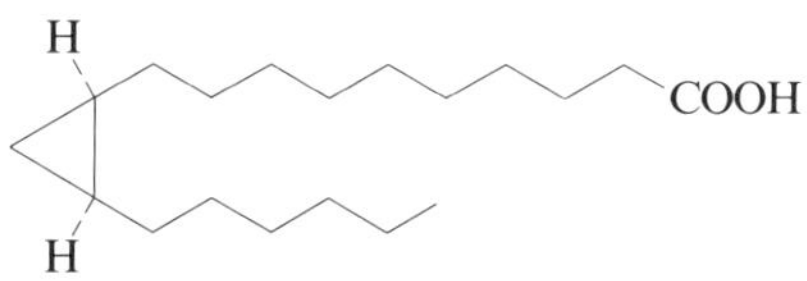

$C_{19}H_{36}O_2$ M 296.492

(1R,2S)-form [19625-10-6]

cis-*form*. ***Lactobacillic acid.*** *Phytomonic acid*

Produced by *Lactobacillus arabinosus*, *L. casei*, *Agrobacterium* (*Phytomonos*) *tumefaciens* and *Streptococcus* spp. Mp 33.6-35°.

Hofmann, K. *et al, J. Am. Chem. Soc.*, 1954, **76**, 1799; 1958, **80**, 5717 (*isol*)
Christie, W.W. *et al, Chem. Phys. Lipids*, 1968, **2**, 196 (*synth, ms, chromatog*)
Gunstone, F.D. *et al, Chem. Phys. Lipids*, 1971, **6**, 147; 1973, **10**, 303 (*synth, nmr, chromatog*)
Tocanne, J.F., *Tetrahedron*, 1972, **28**, 363 (*abs config*)
Buist, P.H. *et al, Can. J. Chem.*, 1981, **59**, 828 (*biosynth*)

2-Hexylcyclopropaneoctanoic acid, 9CI H-00322

9,10-Methylenehexadecanoic acid

[5618-00-8]

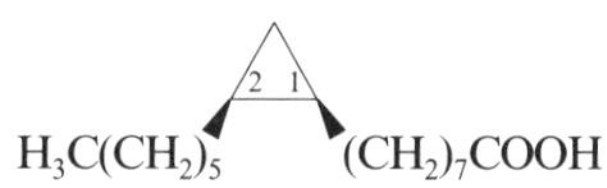

$C_{17}H_{32}O_2$ M 268.439

(1RS,2SR)-form [4675-60-9]

(±)-cis-*form*

Constit. of some bacterial lipids (poss. in an enantiomeric form). Mp 8.5-9.5°.

Me ester:
$C_{18}H_{34}O_2$ M 282.465
d_4^{22} 0.894. n_D^{22} 1.4518.

Amide:
$C_{17}H_{33}NO$ M 267.454
Cryst. Mp 79-81° (71.5-73°).

[36254-90-7]

Kaneshiro, T. *et al*, *J. Biol. Chem.*, 1961, **236**, 2615 (*isol*)
Toubiana, R. *et al*, *C. R. Hebd. Seances Acad. Sci.*, 1962, **254**, 369 (*isol, synth*)
Panos, C. *et al*, *J. Gen. Microbiol.*, 1974, **80**, Pt.1, 93 (*metab*)
Sullivan, K.H. *et al*, *J. Bacteriol.*, 1979, **138**, 133.

5-Hexyldihydro-2(3*H*)-furanone, 9CI H-00323

4-Hydroxydecanoic acid lactone, 8CI. γ-Hexylbutyrolactone. γ-Decalactone. 4-Decanolide
[706-14-9]

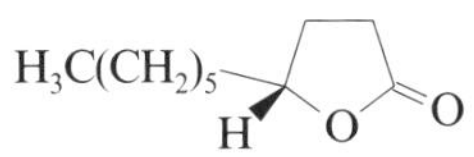

$C_{10}H_{18}O_2$ M 170.251

(***R***)***-form***
Formed by *Sporobolomyces odorus*, also found in oil from *Pityosporum* spp. and musk ox *Ovibos moschatus*. Flavorant and perfume ingredient. Component of apricot, plum and strawberry aromas. Oil with peach-like odour. Bp 281°, $Bp_{7.7}$ 140-145°. $[\alpha]_D^{20}$ +31° (c, 1.2 in $CHCl_3$).
▷ LU4600000.

Honkanen, E. *et al*, *Acta Chem. Scand.*, 1965, **19**, 370 (*ms*)
Heiba, E.I. *et al*, *J. Am. Chem. Soc.*, 1974, **96**, 7977 (*synth*)
Pyysalo, H. *et al*, *Finn. Chem. Lett.*, 1975, 136; *CA*, **84**, 69088m (*synth, pmr, cmr, ms*)
Opdyke, D.L.J., *Food Cosmet. Toxicol.*, (Suppl.), 1976, **14**, 741; *CA*, **91**, 198665r (*rev*)
Grieco, P.A. *et al*, *Tetrahedron Lett.*, 1978, 419 (*synth*)
Labows, J.N. *et al*, *Appl. Environ. Microbiol.*, 1979, **38**, 412; *CA*, **91**, 189438s (*isol*)
Bravo, P. *et al*, *Tetrahedron*, 1987, **43**, 4635 (*synth*)
Flood, P.F. *et al*, *J. Chem. Ecol.*, 1989, **15**, 2207.
Lewis, R.J., *Sax's Dangerous Properties of Industrial Materials*, 8th Ed., Van Nostrand-Reinhold, 1992, HKA500.

5-Hexyl-2-furanoctanoic acid H-00324

9,12-Epoxy-9,11-octadecadienoic acid

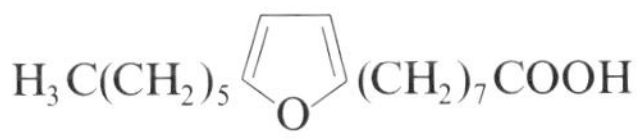

$C_{18}H_{30}O_3$ M 294.433
Isol. from the seed oil of *Exocarpus cupressiformis*. Cryst. Mp 27-29° (20-21°). $Bp_{0.01}$ 159°. Poss. artifact.
Me ester: Oil. $Bp_{0.5}$ 174-176°.

Buchta, E. *et al*, *Justus Liebigs Ann. Chem.*, 1965, **686**, 77 (*synth*)
Morris, L.J. *et al*, *Tetrahedron Lett.*, 1966, 4249 (*isol, uv, ir, pmr, ms*)
Elix, J.A *et al*, *J. Chem. Soc. C*, 1968, 595 (*synth*)
Lie Ken Jie, M.S.F. *et al*, *Chem. Phys. Lipids*, 1978, **21**, 275 (*synth, ms*)

2-Hexyl-3-methylenebutanedioic acid H-00325

Hexylmethylenebutanedioic acid, 9CI. 1-Nonene-2,3-dicarboxylic acid. Hexylitaconic acid

$H_2C{=}C(COOH)CH(COOH)(CH_2)_5CH_3$

$C_{11}H_{18}O_4$ M 214.261

(+)-***form*** [94513-51-6]
Metab. of *Aspergillus niger*. Plant-growth regulator, precursor of Canadensolide. Cryst. Mp 46-47°. $[\alpha]_D$ +15.3° (c, 2 in MeOH).

Isogai, A. *et al*, *Agric. Biol. Chem.*, 1984, **48**, 2607.

2-Hexyl-5-methyl-3(2*H*)-furanone, 9CI H-00326

2-Hexyl-5-methyl-3-oxo-2,3-dihydrofuran. Norcepanone
[33922-66-6]

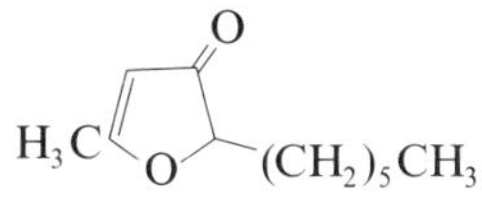

$C_{11}H_{18}O_2$ M 182.262
Constit. of onions, leeks and shallots.

(±)-***form***
Bp_3 79-81° ($Bp_{0.15}$ 78°).
[59462-25-8]

Boelens, M. *et al*, *J. Agric. Food Chem.*, 1971, **19**, 984 (*isol, ms, pmr, ir*)
de Rijke, D. *et al*, *Recl. Trav. Chim. Pays-Bas* (*J. R. Neth. Chem. Soc.*), 1973, **92**, 731 (*synth, ir, pmr*)
Galetto, W.G. *et al*, *J. Agric. Food Chem.*, 1976, **24**, 854 (*synth*)
Thomas, A.F. *et al*, *Tetrahedron Lett.*, 1986, **27**, 505 (*synth*)

5-Hexyltetrahydro-2-furanoctanoic acid, 9CI H-00327

9,12-Epoxyoctadecanoic acid
[61781-98-4]

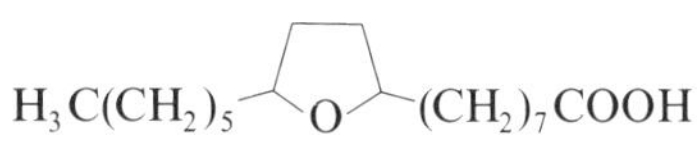

$C_{18}H_{34}O_3$ M 298.465
Prepared from linoleic acid, from 9,12-dihydroxystearic acid and from 9- and 12-hydroxystearic acids. Present in heat-treated soybean oil. Diastereoisomeric mixt.

Abbot, G.G. *et al*, *Chem. Phys. Lipids*, 1970, **4**, 351; 1971, **7**, 303 (*synth, ms*)
Doi, H. *et al*, *Yukagaku*, 1976, **25**, 831 (*isol*)

3-Hexynoic acid H-00328

[17814-72-1]

$H_3CCH_2C{\equiv}CCH_2COOH$

$C_6H_8O_2$ M 112.128
Needles (pet. ether). Mp 61-63°.

Jones, E.R.H. *et al*, *J. Chem. Soc.*, 1954, 3201 (*synth*)
Cymerman Craig, J. *et al*, *J. Chem. Soc.*, 1963, 4402 (*synth*)

4-Hexynoic acid H-00329

[41143-12-8]

$H_3CC{\equiv}CCH_2CH_2COOH$

$C_6H_8O_2$ M 112.128
Needles (pet. ether). Mp 100-101°.
Me ester: [41143-13-9].
$C_7H_{10}O_2$ M 126.155
Bp_{21} 77-78°, Bp_{10} 61.5°.

Jones, E.R.H. *et al*, *J. Chem. Soc.*, 1954, 3201 (*synth*)
Gravestock, M.B. *et al*, *J. Am. Chem. Soc.*, 1978, **100**, 4274 (*synth, pmr*)

5-Hexynoic acid H-00330

[53293-00-8]

$HC{\equiv}C(CH_2)_3COOH$

$C_6H_8O_2$ M 112.128
Liq. Bp_{20} 120-125°, Bp_6 95-96°. pK_a 4.59 (25°, 0.1*M* NaCl). Claimed in one paper to be cryst., Mp 41-43.5°

Me ester: [77758-51-1].
$C_7H_{10}O_2$ M 126.155
Bp_{10} 56°.
Chloride: [55183-45-4].
C_6H_7ClO M 130.573
Liq. $Bp_{0.5}$ 50-55°.
Nitrile: [14918-21-9]. *5-Cyano-1-pentyne*
C_6H_7N M 93.128
Bp_7 65.5°.
Anhydride:
$C_{12}H_{14}O_3$ M 206.241
Liq. $Bp_{0.6}$ 122-124°.

Eglington, G. *et al*, *J. Chem. Soc.*, 1953, 3052 (*synth*)
Just, G. *et al*, *Synth. Commun.*, 1979, **9**, 613 (*synth*)
Gerlach, H. *et al*, *Helv. Chim. Acta*, 1980, **63**, 2312 (*synth, pmr, ir, ms*)
Krafft, G.A. *et al*, *J. Am. Chem. Soc.*, 1981, **103**, 5459 (*synth, pmr*)
Millar, J.G. *et al*, *J. Org. Chem.*, 1984, **49**, 2332 (*synth, pmr, ms*)
Earl, R.A. *et al*, *J. Org. Chem.*, 1984, **49**, 4786 (*synth*)
Atkinson, R.S. *et al*, *J. Chem. Soc., Perkin Trans. 1*, 1986, 1215 (*synth*)
Moody, C.J. *et al*, *J. Chem. Soc., Perkin Trans. 1*, 1988, 3249 (*anhydride*)

Hurgadenyne H-00331

[132412-60-3]

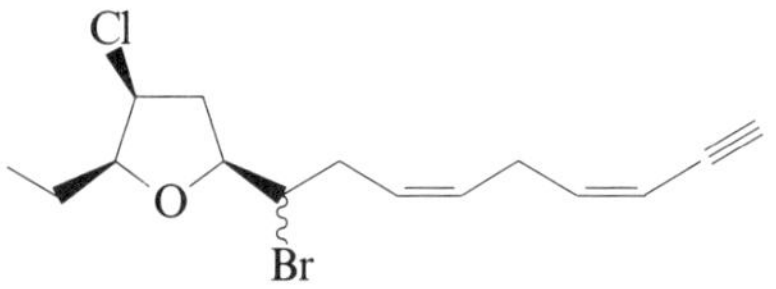

$C_{15}H_{20}BrClO$ M 331.679
Metab. of *Laurencia obtusa*. Oil. $[\alpha]_D^{24}$ −110° (c, 5 in $CHCl_3$).

Ayyad, S.-E.N. *et al*, *Phytochemistry*, 1990, **29**, 3193 (*isol, pmr, cmr*)

Hydnocarpic acid H-00332

2-Cyclopentene-1-undecanoic acid, 9CI. 11-(2-Cyclopentenyl) undecanoic acid
[459-67-6]

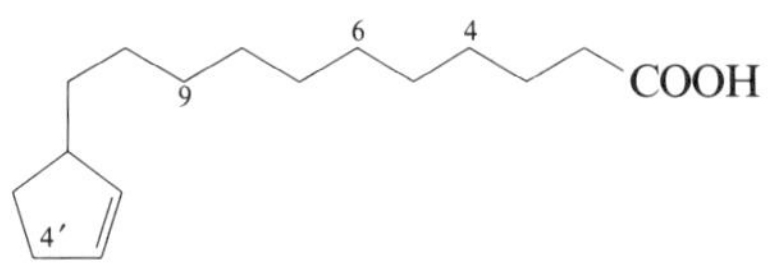

$C_{16}H_{28}O_2$ M 252.396
(+)-*form*
Constit. of chaulmoogra oil and seed oils of other Flacourtiaceae spp., *Hydnocarpus*, *Asteriastigma*, *Carpotroche*, *Caloncoba* and *Lindackeria* spp. Cryst. (EtOH). Mp 58-59°. $[\alpha]_D^{25}$ +68° ($CHCl_3$).
Me ester: [28046-99-3].
$C_{17}H_{30}O_2$ M 266.423
Mp 8°. Bp_{19} 200-203°. $[\alpha]_D^{25}$ +62.4° ($CHCl_3$).
Et ester:
$C_{18}H_{32}O_2$ M 280.450
Bp_{19} 211°. $[\alpha]_D$ +51.6° ($CHCl_3$).
Amide:
$C_{16}H_{29}NO$ M 251.411
Mp 111-112°. $[\alpha]_D^{25}$ +69.4°.
4,5-Didehydro: 11-(2-Cyclopentenyl)-4-undecenoic acid
$C_{16}H_{26}O_2$ M 250.380
Isol. from *Calocoba echinata* seeds. Prob. opt. active and with (Z)-config.
6,7-Didehydro: [52904-22-0]. *11-(2-Cyclopentenyl)-6-undecenoic acid.* ***Manoaic acid***
$C_{16}H_{26}O_2$ M 250.380
Isol. from seeds of *C. echinata*. Prob. opt. active and with *Z*-config.
9,10-Didehydro: 11-(2-Cyclopentenyl)-9-undecenoic acid
$C_{16}H_{26}O_2$ M 250.380
Isol. from *C. echinata* seeds. Prob. opt. active and with *Z*-config.
4′-Oxo: Ketohydnocarpic acid
$C_{16}H_{26}O_3$ M 266.380
Minor constit. of oil of *Carpotroche brasiliensis*. Presumably artifact of autoxidn.
(±)-*form*
Cryst. (pet. ether). Mp 58-59°.
Nitrile:
$C_{16}H_{27}N$ M 233.396
Bp_3 155-156°.

Hinegardner, W.S., *J. Am. Chem. Soc.*, 1933, **55**, 2831 (*synth*)
Paget, H., *J. Chem. Soc.*, 1937, 955 (*Ketohydnocarpic acid*)
Adriaens, L., *CA*, 1946, **40**, 6754 (*isol*)
Christie, W.W. *et al*, *Lipids*, 1969, **4**, 229 (*ms*)
Sengupta, A., *J. Sci. Food Agric.*, 1973, **24**, 669 (*isol*)
Spener, F. *et al*, *Biochemistry*, 1974, **13**, 2241 (*derivs*)
Spener, F. *et al*, *Chem. Phys. Lipids*, 1974, **12**, 344 (*isol*)

5-Hydroperoxy-6,8,11,14-eicosatetraenoic acid H-00333

5-HPETE
[74581-83-2]

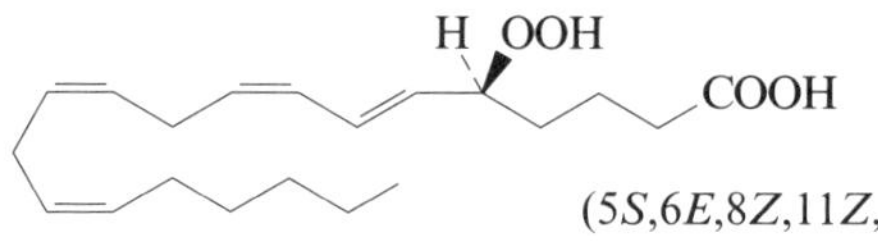

(5*S*,6*E*,8*Z*,11*Z*,14*Z*)-*form*

$C_{20}H_{32}O_4$ M 336.470
(5S,6E,8Z,11Z,14Z)-form [71774-08-8]
Metab. of arachidonic acid.
(5RS,6E,8Z,11Z,14Z)-form [70968-82-0]
Obt. from arachidonic acid by autoxidn. or photooxygenation.

Porter, N.A. *et al*, *J. Org. Chem.*, 1979, **44**, 3177 (*synth, ms*)
Corey, E.J. *et al*, *J. Am. Chem. Soc.*, 1980, **102**, 1435; 1989, **111**, 1452 (*synth, ir, pmr, uv*)
Boeynaems, J.M. *et al*, *Prostaglandins*, 1980, **19**, 87 (*synth*)
Terao, J. *et al*, *Agric. Biol. Chem.*, 1981, **45**, 587 (*synth*)
Porter, N.A. *et al*, *J. Am. Chem. Soc.*, 1981, **103**, 6447 (*synth*)

6-Hydroperoxy-4,8,11,14-eicosatetraenoic acid H-00334

6-HPETE

$$H_3C(CH_2)_3(CH_2CH{=}CH)_3CH_2CH(OOH)CH{=}CHCH_2CH_2COOH$$

$C_{20}H_{32}O_4$ M 336.470
(4E,6RS,8Z,11Z,14Z)-form
Obt. from arachidonic acid by photooxidn.

Porter, N.A. *et al*, *J. Org. Chem.*, 1979, **44**, 3177 (*ms*)
Terao, J. *et al*, *Agric. Biol. Chem.*, 1981, **45**, 587.

8-Hydroperoxy-5,9,11,14-eicosatetraenoic acid H-00335

8-HPETE

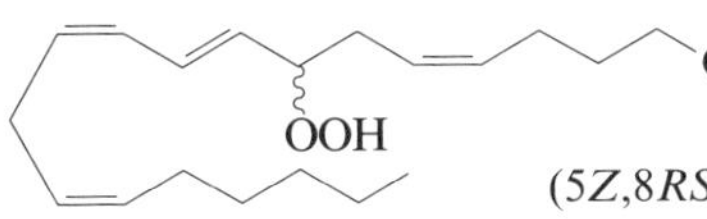

(5Z,8RS,9E,11Z,14Z)-form

$C_{20}H_{32}O_4$ M 336.470

(5Z,8RS,9E,11Z,14Z)-form [70968-80-8]

Autoxidn. or photooxygenation prod. of arachidonic acid.

Porter, N.A. *et al*, *J. Org. Chem.*, 1979, **44**, 3177 (*synth, ms*)
Porter, N.A. *et al*, *J. Am. Chem. Soc.*, 1980, **102**, 5912 (*synth*)
Boeynaems, J.M. *et al*, *Prostaglandins*, 1980, **19**, 87 (*synth*)
Terao, J. *et al*, *Agric. Biol. Chem.*, 1981, **45**, 587 (*synth*)
Porter, N.A. *et al*, *J. Am. Chem. Soc.*, 1981, **103**, 6447 (*synth*)

9-Hydroperoxy-5,7,11,14-eicosatetraenoic acid H-00336

9-HPETE

[77913-06-5]

$H_3C(CH_2)_4(CH{=}CHCH_2)_2CH(OOH)$
$(CH{=}CH)_2(CH_2)_3COOH$

$C_{20}H_{32}O_4$ M 336.470

(5Z,7E,9RS,11Z,14Z)-form

Autoxidn. or photooxygenation prod. of arachidonic acid. Metab. (in chiral form) of arachidonic acid.

Porter, N.A. *et al*, *J. Org. Chem.*, 1979, **44**, 3177 (*synth, ms*)
Boeynaems, J.M. *et al*, *Prostaglandins*, 1980, **19**, 87 (*synth*)
Terao, J. *et al*, *Agric. Biol. Chem.*, 1981, **45**, 587.
Porter, N.A. *et al*, *J. Am. Chem. Soc.*, 1981, **103**, 6447.

11-Hydroperoxy-5,8,12,14-eicosatetraenoic acid H-00337

11-HPETE

$H_3C(CH_2)_4(CH{=}CH)_2CH(OOH)$
$(CH_2CH{=}CH)_2(CH_2)_3COOH$

$C_{20}H_{32}O_4$ M 336.470

(5Z,8Z,11RS,12E,14Z)-form [73959-85-0]

Obt. from arachidonic acid by autoxidn. or photooxygenation.

Porter, N.A. *et al*, *J. Org. Chem.*, 1979, **44**, 3177 (*synth, ms*)
Corey, E.J. *et al*, *J. Am. Chem. Soc.*, 1980, **102**, 1433 (*synth*)
Boeynaems, J.M. *et al*, *Prostaglandins*, 1980, **19**, 87 (*synth*)
Terao, J. *et al*, *Agric. Biol. Chem.*, 1981, **45**, 587 (*synth*)
Porter, N.A. *et al*, *J. Am. Chem. Soc.*, 1981, **103**, 6447 (*synth*)

12-Hydroperoxy-5,8,10,14-eicosatetraenoic acid H-00338

12-HPETE

$H_3C(CH_2)_4CH{=}CHCH_2CH(OOH)$
$(CH{=}CH)_2CH_2CH{=}CH(CH_2)_3COOH$

$C_{20}H_{32}O_4$ M 336.470

(5Z,8Z,10E,12RS,14Z)-form [71030-35-8]

Obt. from arachidonic acid by photooxygenation. A metab. of arachidonic acid (presumably in chiral form).

Porter, N.A. *et al*, *J. Org. Chem.*, 1979, **44**, 3177 (*synth, ms*)
Corey, E.J. *et al*, *J. Am. Chem. Soc.*, 1980, **102**, 1433 (*synth*)
Porter, N.A. *et al*, *J. Am. Chem. Soc.*, 1980, **102**, 5912; 1981, **103**, 6447 (*synth*)
Boeynaems, J.M. *et al*, *Prostaglandins*, 1980, **19**, 87 (*synth*)
Terao, J. *et al*, *Agric. Biol. Chem.*, 1981, **45**, 587 (*synth*)

14-Hydroperoxy-5,8,11,15-eicosatetraenoic acid H-00339

14-HPETE

$H_3C(CH_2)_3CH{=}CHCH(OOH)$
$(CH_2CH{=}CH)_3(CH_2)_3COOH$

$C_{20}H_{32}O_4$ M 336.470

(5Z,8Z,11Z,14RS,15E)-form [77913-07-6]

Obt. from arachidonic acid by photooxygenation.

Porter, N.A. *et al*, *J. Org. Chem.*, 1979, **44**, 3177 (*ms*)
Terao, J. *et al*, *Agric. Biol. Chem.*, 1981, **45**, 587.

15-Hydroperoxy-5,8,11,13-eicosatetraenoic acid H-00340

15-HPETE

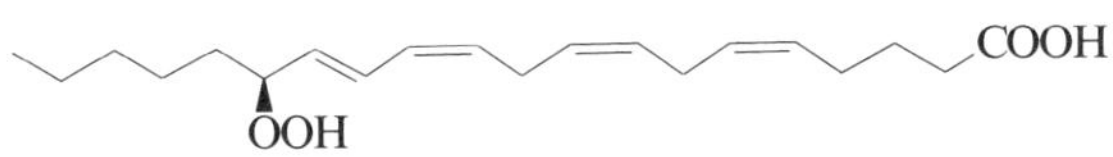

$C_{20}H_{32}O_4$ M 336.470

(15S,5Z,8Z,11Z,13E)-form [70981-96-3]

Metab. of arachidonic acid. $[\alpha]_D$ −4.6° (MeOH).

Me ester: [77026-90-5].

$C_{21}H_{34}O_4$ M 350.497

$[\alpha]_D$ −3.5° (MeOH).

(15RS,5Z,8Z,11Z,13E)-form [73804-66-7]

Obt. from arachidonic acid by autoxidn. or photooxygenation.

Baldwin, J.E. *et al*, *J. Chem. Soc., Perkin Trans. 1*, 1979, 115 (*synth*)
Porter, N.A. *et al*, *J. Org. Chem.*, 1979, **44**, 3177 (*synth, ms*)
Corey, E.J. *et al*, *J. Am. Chem. Soc.*, 1980, **102**, 1433 (*synth*)
Boeynaems, J.M. *et al*, *Prostaglandins*, 1980, **19**, 87 (*synth*)
Terao, J. *et al*, *Agric. Biol. Chem.*, 1981, **45**, 587 (*synth*)
Porter, N.A. *et al*, *J. Am. Chem. Soc.*, 1981, **103**, 6447 (*synth*)
Dussault, P. *et al*, *J. Org. Chem.*, 1992, **57**, 1952 (*synth*)

3-Hydroxybutanoic acid, 9CI H-00341

3-Hydroxybutyric acid, 8CI

[300-85-6]

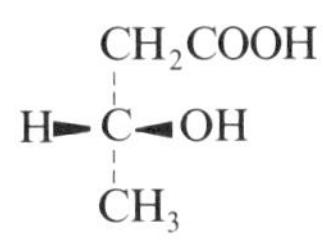

(R)-form

$C_4H_8O_3$ M 104.105

Isol. from bacteria, in polymeric form.

(R)-form [625-72-9]

V. hygroscopic cryst. V. sol. H_2O, EtOH, insol. C_6H_6. Mp 49-50°. $[\alpha]_D^{16}$ −24.9°.

Me ester: [3976-69-0].

$C_5H_{10}O_3$ M 118.132

d_{20}^{20} 1.06. Bp_{13} 67-68.5°. $[\alpha]_D^{20}$ −21.09°.

Amide:

$C_4H_9NO_2$ M 103.121

Cryst. (EtOAc). Mp 99-100°. $[\alpha]_D^{20}$ −22.49° (MeOH).

Nitrile: 1-Cyano-2-propanol

C_4H_7NO M 85.105

Bp_{15} 99-100°. $[\alpha]_D^{18}$ −10.03° (H_2O).

(S)-form [6168-83-8]

Na salt: Cryst. (EtOH).

Nitrile: Bp_{15} 99-100°. $[\alpha]_D^{18}$ +8.78°.

(±)-form [625-71-8]

Hygroscopic syrup. $Bp_{12\text{-}14}$ 130°. pK_a 4.29. Steam-volatile.

Ac: [24621-58-7].
$C_6H_{10}O_4$ M 146.143
d^{18} 1.13. $Bp_{0.5}$ 93-94°.
Me ester: [1487-49-6].
Bp_{12-13} 67-68°.
▷ Irritant. ET4700000.

Me ether, 4-bromophenacyl ester: Mp 59-60°.
Amide: [40482-53-9].
Prisms (H_2O). Mp 84-87°.
Nitrile: [4368-06-3].
Bp 220-221°, Bp_{22} 123-125°.
Nitrile, Ac:
$C_6H_9NO_2$ M 127.143
Bp 210°.

[5864-66-4, 25654-45-9]

Wislicenus, J., *Justus Liebigs Ann. Chem.*, 1869, **149**, 205 (*synth*)
Levene, P.A. *et al, J. Biol. Chem.*, 1926, **68**, 418 (*abs config*)
Forsyth, W.G.C. *et al, Nature (London)*, 1958, **182**, 800 (*isol*)
Ottaway, J.H., *Biochem. J.*, 1962, **84**, 11 (*synth*)
Gottschulk, G. *et al, Nature (London)*, 1965, **205**, 308.
Meyers, A.I. *et al, Tetrahedron Lett.*, 1974, 1333 (*synth*)
Wipf, B. *et al, Helv. Chim. Acta*, 1983, **66**, 485.
Bernardi, R. *et al, J. Chem. Soc., Chem. Commun.*, 1984, 460 (*resoln*)
Lewis, R.J., *Sax's Dangerous Properties of Industrial Materials*, 8th Ed., Van Nostrand-Reinhold, 1992, MKP250.

3-Hydroxycholan-24-oic acid H-00342

3-Hydroxycholanic acid
[35498-67-0]

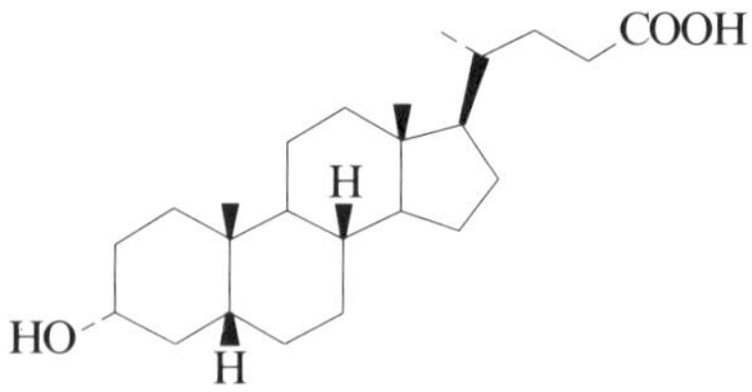

$C_{24}H_{40}O_3$ M 376.578

(3α,5α)-form [2276-94-0] ***Allolithocholic acid***
Isol. from human faeces. Mp 204-208°. $[\alpha]_D^{26}$ +26.0°.
Me ester: [15074-01-8].
$C_{25}H_{42}O_3$ M 390.605
Mp 160-161°.

(3α,5β)-form [434-13-9] ***Lithocholic acid***
Occurs in human, cattle and rabbit bile, and in cattle and pig gallstones. Secondary, faecal, bile acid. Prisms (EtOH aq.). Mp 186°. $[\alpha]_D^{20}$ +33.7° (c, 1.5 in EtOH).
▷ Exp. reprod. and teratogenic effects. LD_{50} (mus, orl) 3.9 mg/kg. FZ2275000.
Me ester: [1249-75-8].
Cryst. (MeOH). Mp 90-93°, 130° (dimorph.).
Et ester: [6112-82-9].
$C_{26}H_{44}O_3$ M 404.632
Cryst. Mp 145-148°.
Ac: [4057-84-5].
$C_{26}H_{42}O_4$ M 418.615
Mp 170-171°.
Ac, Me ester: [3253-69-8].
$C_{27}H_{44}O_4$ M 432.642
Cryst. (MeOH). Mp 133-135°.
Glycine conjugate: ***Glycolithocholic acid.*** N-(*3-Hydroxy-24-oxocholan-24-yl)glycine, 9CI*
$C_{26}H_{43}NO_4$ M 433.630
Plates (EtOH aq.). Mp 212-214° (184-186°). $[\alpha]_D$ +30.6° (c, 1 in EtOH).
Taurine conjugate: [516-90-5]. ***Taurolithocholic acid.*** *2-[(3-Hydroxy-24-oxocholan-24-yl)amino]ethanesulfonic acid*
$C_{26}H_{45}NO_5S$ M 483.711
Needles (EtOH/EtOAc). Mp 222-223° (212-213°). $[\alpha]_D$ +27.2° (c, 1 in EtOH).

(3β,5α)-form [2276-93-9]
Needles. Mp 219-220° (208-211°). $[\alpha]_D^{20}$ +33.3° (c, 0.12 in EtOH).
3-Ac: [33628-54-5].
Needles (EtOAc/hexane). Mp 161-163°. $[\alpha]_D^{20}$ +21.7° (c, 0.286 in $CHCl_3$).
Me ester, Ac: [1255-52-3].
Mp 164-165°. $[\alpha]_D^{22}$ +21° ($CHCl_3$).

(3β,5β)-form [1534-35-6] ***Isolithocholic acid***
Constit. of the bile of man, ox, rabbit and hog. Mp 176-176.5°. $[\alpha]_D$ +32° ($CHCl_3$).
Me ester: [5405-42-5].
Laths (MeOH). Mp 117.1-118°. $[\alpha]_D$ +21.3° ($CHCl_3$).
Ac, Me ester: [1255-51-2].
Hexagons (EtOAc). Mp 168.5-171.5°. $[\alpha]_D$ +21.1° ($CHCl_3$).

Pearlman, W.H. *et al, J. Am. Chem. Soc.*, 1949, **71**, 4126 (*synth, Lithocholic acid*)
Blickenstaff, R.T. *et al, J. Am. Chem. Soc.*, 1958, **80**, 2726 (*synth, Isolithocholic acid, derivs*)
Sarel, S. *et al, J. Org. Chem.*, 1959, **24**, 2018 (*synth*)
Ushizawa, I., *CA*, 1961, **55**, 6534g (*derivs*)
Anderson, I.G. *et al, Biochem. J.*, 1962, **85**, 236 (*synth, Allolithocholic acid*)
Norita, D.A. *et al, Acta Crystallogr.*, 1965, **19**, 477 (*cryst struct, ester*)
Egger, H., *Monatsh. Chem.*, 1968, **99**, 1163 (*ms*)
Leibfritz, D. *et al, J. Am. Chem. Soc.*, 1973, **95**, 4996 (*cmr*)
Cowen, A.E. *et al, Gastroenterology*, 1975, **69**, 59, 67, 77 (*metab*)
Arora, S.K. *et al, Acta Crystallogr., Sect. B*, 1976, **32**, 415 (*cryst struct*)
Tserng, K.-Y. *et al, J. Lipid Res.*, 1977, **18**, 404 (*synth, Glyco- and Tauro-conjugates*)
Famaey, J.P. *et al, Arch. Int. Physiol. Biochim.*, 1978, **86**, 577 (*props conjugates*)
Shaw, R. *et al, Biomed. Mass Spectrom.*, 1978, **5**, 433 (*ms conjugates*)
Shaw, R. *et al, Lipids*, 1980, **15**, 805 (*synth, Allolithocholic acid, conjugates*)
Fukumoto, K. *et al, Tetrahedron*, 1982, **38**, 3701 (*synth, 3β,5α-form*)
Setchell, K.D.R. *et al, J. Lipid Res.*, 1983, **24**, 1085 (*analysis of bile acids in faeces*)
Iida, T. *et al, Synthesis*, 1984, 957 (*ester*)
Iida, T. *et al, J. Chromatogr.*, 1986, **366**, 396 (*hplc, esters*)
Radominska-Pyrek, A. *et al, J. Lipid Res.*, 1986, **27**, 102 (*synth*)
Seldes, A.M. *et al, Magn. Reson. Chem.*, 1986, **24**, 185 (*cmr, esters*)
Kirk, D.N. *et al, J. Chem. Soc., Perkin Trans. 2*, 1990, 1567 (*pmr*)
Lewis, R.J., *Sax's Dangerous Properties of Industrial Materials*, 8th Ed., Van Nostrand-Reinhold, 1992, LHW000.

10-Hydroxy-2,8-decadiene-4,6-diynoic acid H-00343

10-Hydroxymatricaric acid

$$HOCH_2CH{=}CHC{\equiv}CC{\equiv}CCH{=}CHCOOH$$

$C_{10}H_8O_3$ M 176.171
Isol. as an (*E,Z*)-mixt. from the fungus *Camarophyllus virgineus*.

(E,E)-form [51193-98-7]
Constit. of *Fistulina hepatica*.
Me ester: [51193-95-4].
$C_{11}H_{10}O_3$ M 190.198
Isol. from *Polyporus anthracophilus*. Pale-yellow prisms (Et_2O/pet. ether). Mp 80-81°.
10-Angeloyl, Me ester: [77355-58-9].

$C_{16}H_{16}O_4$ M 272.300
Isol. from the roots of *Solidago odora* and *S. nemoralis*.

(2E,8Z)-form

Me ester: [65367-66-0].
Isol. from *P. anthracophilus* and *Chrysothamnus parryi*. Cryst. (Et_2O/pet. ether). Mp 38°.

(Z,Z)-form

Me ester: [65367-59-1].
Isol. from *Camarophyllus virgineus*, *Chrysothamnus nauseosus* and *C. parryi*. Mp 38.5-40.5°. λ_{max} 330, 310, 294, 258, 246, 233 nm (MeOH).

Ac, Me ester: [73149-56-1].
$C_{13}H_{12}O_4$ M 232.235
Isol. from *C. nauseosus* and *C. parryi*. Oil. λ_{max} 334, 312, 294, 259, 247 nm (Et_2O).

10-Angeloyl, Me ester: [58969-60-1].
Isol. from the roots of *Solidago altissima*, *C. parryi* and *Baccharis eleagnoides*. λ_{max} 246.5, 259, 291, 310, 331.5 nm.

O-*(2-Methylpropanoyl), Me ester:* [73149-55-0].
$C_{15}H_{16}O_4$ M 260.289
Isol. from *C. parryi*. Oil.

8,9-Epoxide: 8,9-Epoxy-10-hydroxy-2-decene-4,6-diynoic acid. 7-[3-(Hydroxymethyl)oxiranyl]-2-heptene-4,6-diynoic acid
$C_{10}H_8O_4$ M 192.171

8,9-Epoxide, Me ester: [74728-17-9].
$C_{11}H_{10}O_4$ M 206.198
Isol. from *C. nauseosus*. Mp 53-54°. λ_{max} (MeOH) 307 (11200), 288 (12300), 273 (8100), 260sh (5300), 227 (2400) nm.

8,9-Epoxide, Ac, Me ester: [73149-58-3].
$C_{13}H_{12}O_5$ M 248.235
Isol. from *C. nauseosus* and *C. parryi*. Oil. λ_{max} (MeOH) 304 (9600), 287 (1100), 272 (7200), 285sh (4000), 225 (21600) nm.

[62000-93-5, 65367-67-1, 106111-45-9]

Farrell, I.W. *et al*, *J. Chem. Soc., Perkin Trans. 1*, 1973, 2642; 1977, 1886 (*isol, struct, synth, ir, uv, pmr, ms*)
Ichihara, K. *et al*, *Agric. Biol. Chem.*, 1976, **40**, 353 (*isol, struct, uv, ir, ms, pmr*)
Kobayashi, A. *et al*, *Agric. Biol. Chem.*, 1976, **40**, 2257 (*synth, ir, uv, ms, pmr*)
Davies, D.G. *et al*, *J. Chem. Soc., Perkin Trans. 1*, 1978, 1602 (*isol, struct, ir, uv, ms, pmr*)
Bohlmann, F. *et al*, *Phytochemistry*, 1979, **18**, 1519; 1980, **19**, 2655 (*epoxide, isol, struct, ms, pmr, ir, uv*)
Rose, A.F. *et al*, *Phytochemistry*, 1980, **19**, 563 (*epoxide, isol, struct, activity, ir, uv, pmr, ms*)
Hnneck, S. *et al*, *Pharmazie*, 1986, **41**, 673 (*isol*)
Rehder, V.L.G. *et al*, *J. Nat. Prod. (Lloydia)*, 1990, **53**, 692 (*isol, cmr*)

3-Hydroxy-4,6-decadiynoic acid H-00344

$H_3CCH_2CH_2C{\equiv}CC{\equiv}CCH(OH)CH_2COOH$

$C_{10}H_{12}O_3$ M 180.203

(Z)-form

O-*Angeloyl, Me ether:*
$C_{16}H_{20}O_4$ M 276.332
Constit. of *Bellis perennis*. Oil.

Bohlmann, F. *et al*, *Chem. Ber.*, 1969, **102**, 1682; 1970, **103**, 561 (*pmr, ms, isol, struct, synth*)

2-Hydroxydecanoic acid H-00345

2-Hydroxycapric acid
[5393-81-7]

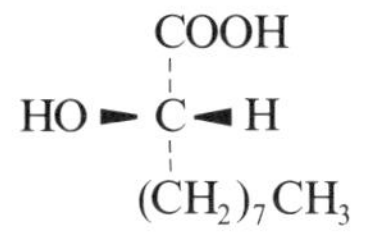

$C_{10}H_{20}O_3$ M 188.266

(S)-form

Cryst. (isopropyl ether). Mp 77-78°. $[\alpha]_D^{24}$ +5.3° (c, 8.6 in $CHCl_3$).

Me ester:
$C_{11}H_{22}O_3$ M 202.293
Oil. Bp_{22} 152-155°. $[\alpha]_D^{22}$ +3.0° (neat).

Horn, D.H.S. *et al*, *J. Chem. Soc.*, 1954, 1460 (*synth, config*)
Haga, T., *CA*, 1962, **56**, 311e (*synth*)

3-Hydroxydecanoic acid H-00346

[33044-91-6]

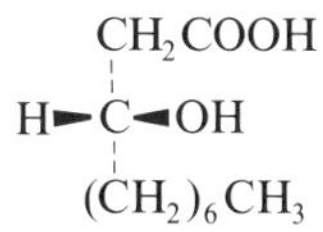

(R)-form

$C_{10}H_{20}O_3$ M 188.266

(R)-form

Myrmicacin
Constit. of the secretion of the metathoracic glands of Myrmicine ants. Occurs in bacterial lipids, residue present in the antibiotic viscosin. Mp 48-49°. $[\alpha]_D^{20}$ −20.8° (c, 1 in $CHCl_3$).

O-*Rhamnoside:* ***Pyolipic acid***
$C_{16}H_{30}O_7$ M 334.409
Isol. from cultures of *Pseudomonas pyocyanea*.

(S)-form

Me ester:
$C_{11}H_{22}O_3$ M 202.293
$Bp_{0.05}$ 74-76°. $[\alpha]_D^{20}$ +25.8° (c, 1 in cyclohexane).

(±)-form [5561-87-5]
Mp 54.6-56.6°.

Bergström, S. *et al*, *Arch. Biochem.*, 1946, **10**, 165 (*Pyolipic acid*)
Skogh, M., *Acta Chem. Scand.*, 1952, **6**, 809 (*synth*)
Serck-Hanssen, K. *et al*, *Acta Chem. Scand.*, 1955, **9**, 866 (*synth*)
Bishop, D.G. *et al*, *Biochem. Biophys. Res. Commun.*, 1962, **7**, 337 (*isol*)
Shemyakin, M.M. *et al*, *Tetrahedron Lett.*, 1964, 47 (*synth*)
Schildnecht, H. *et al*, *Angew. Chem.*, 1971, **10**, 124 (*isol*)
Furukawa, K. *et al*, *Yukagaku*, 1976, **25**, 358.
Tahara, S. *et al*, *Agric. Biol. Chem.*, 1978, **42**, 879 (*synth*)
Kawahara, K. *et al*, *Biochim. Biophys. Acta*, 1979, **572**, 1 (*biosynth*)
Ketelaar, P.E.F. *et al*, *Tetrahedron Lett.*, 1985, **26**, 4665 (*synth*)

4-Hydroxydecanoic acid H-00347

[17369-51-6]

$H_3C(CH_2)_5CH(OH)CH_2CH_2COOH$

$C_{10}H_{20}O_3$ M 188.266
Cryst. Readily dehydrates to lactone.

Lactone: 5-Hexyl-3,4-dihydro-2(3H)-furanone
$C_{10}H_{18}O_2$ M 170.251

Spar. sol. H_2O. Bp 281°. Steam-volatile.

Fittig, R. *et al*, *Justus Liebigs Ann. Chem.*, 1885, **227**, 93.
Wyatt, C.J. *et al*, *Lipids*, 1967, **2**, 208 (*ms, glc*)

5-Hydroxydecanoic acid H-00348

5-Hydroxycapric acid
[624-00-0]

$H_3C(CH_2)_4CH(OH)(CH_2)_3COOH$

$C_{10}H_{20}O_3$ M 188.266
Constit. of massory bark (*Cryptocarya massoia*).
Lactone: see *Tetrahydro-6-pentyl-2*H*-pyran-2-one, T-00130*

Meyer, T.M., *Recl. Trav. Chim. Pays-Bas* (*J. R. Neth. Chem. Soc.*), 1940, **59**, 191 (*isol*)
Wyatt, C.J. *et al*, *Lipids*, 1967, **2**, 208 (*ms, glc*)

6-Hydroxydecanoic acid, 9CI H-00349

[16899-10-8]

$H_3C(CH_2)_3CH(OH)(CH_2)_4COOH$

$C_{10}H_{20}O_3$ M 188.266
Present in lipid fraction of royal jelly.
Lactone:
$C_{10}H_{18}O_2$ M 170.251
$Bp_{1.3}$ 96°. n_D^{20} 1.4601.

Netherlands Pat., 100 368, (1962); *CA*, **60**, 2783c (*synth, lactone*)
Japan. Pat., 74 85 276, (1974); *CA*, **82**, 71817a (*use*)
Lercker, G. *et al*, *Lipids*, 1981, **16**, 912 (*anal*)

7-Hydroxydecanoic acid, 9CI H-00350

[2034-55-1]

$H_3CCH_2CH_2CH(OH)(CH_2)_5COOH$

$C_{10}H_{20}O_3$ M 188.266
Present in hydrolysate of leaf resin of *Ipomoea fistulosa* and saponified cells of *Mucor* spp. Liq.
[73276-73-0]

Legler, G., *Phytochemistry*, 1965, **4**, 29 (*isol*)
Satoshi, T. *et al*, *Agric. Biol. Chem.*, 1980, **44**, 193 (*synth, ms, isol, ir, glc*)

9-Hydroxydecanoic acid, 9CI H-00351

[1422-27-1]

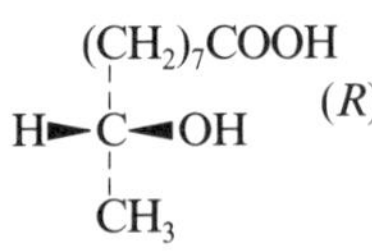
(*R*)-*form*

$C_{10}H_{20}O_3$ M 188.266
Present in *Agaricus campestris*, *Mucor hiemalis* and royal jelly.
(*R*)-*form* [35433-73-9]
Oil. Mp 32°. $[\alpha]_D$ −7.5° (MeOH), $[\alpha]_D^{23}$ −8.1° (c, 0.11 in $CHCl_3$).
Me ester:
$C_{11}H_{22}O_3$ M 202.293
$Bp_{0.005}$ 78-80°. $[\alpha]_D$ −9.5° (c, 1.45 in cyclohexane).
tert-*Butyl ester:*
$C_{14}H_{28}O_3$ M 244.373
$[\alpha]_D$ −7.0° (c, 1.73 in cyclohexane).
Lactone: [74183-94-1]. *10-Methyl-2-oxecanone, 9CI.* ***Phoracantholide I***
$C_{10}H_{18}O_2$ M 170.251
Present in metasternal gland of *Phoracantha synonyma*. $[\alpha]_D^{22}$ −37.4° (c, 0.02 in $CHCl_3$).
(*S*)-*form* [35433-72-8]
$[\alpha]_D^{22}$ +8.14° (c, 0.14 in $CHCl_3$).
Me ester: Oil. $Bp_{0.002}$ 80°. $[\alpha]_D$ +9.3° (c, 2.0 in cyclohexane).
Lactone: $[\alpha]_D^{22}$ +38.38° (c, 0.16 in $CHCl_3$).
(±)-*form* [40151-97-1]
Oil. Mp *ca.* 0°.
Me ester: $Bp_{0.01}$ 85°.
tert-*Butyl ester:* $Bp_{0.01}$ 72-74°.
Lactone: [65371-24-6].
Oil. $Bp_{0.75}$ 60°.

Moore, B.P. *et al*, *Aust. J. Chem.*, 1976, **29**, 1365 (*isol*)
Gerlach, H. *et al*, *Helv. Chim. Acta*, 1978, **61**, 1226 (*synth, ir, pmr, ms*)
Fink, M. *et al*, *Helv. Chim. Acta*, 1982, **65**, 2563 (*derivs*)
Tressl, R. *et al*, *J. Agric. Food Chem.*, 1982, **30**, 89 (*glc, ms*)
Voss, G. *et al*, *Helv. Chim. Acta*, 1983, **66**, 2294 (*synth*)
Clarke, M.J. *et al*, *J. Liq. Chromatogr.*, 1986, **9**, 1711 (*hplc*)
Naoshima, Y. *et al*, *Chem. Lett.*, 1987, 2379 (*synth, lactone*)
Bantra, M. *et al*, *J. Org. Chem.*, 1991, **56**, 5132 (*synth*)
Nagumo, S. *et al*, *Tetrahedron*, 1992, **48**, 8667 (*synth*)

10-Hydroxydecanoic acid H-00352

10-Hydroxycapric acid
[1679-53-4]

$HOCH_2(CH_2)_8COOH$

$C_{10}H_{20}O_3$ M 188.266
Cryst. (Et_2O/pet. ether). Mp 75.5-76.5°.
Me ester: [2640-94-0].
$C_{11}H_{22}O_3$ M 202.293
Mp 34-35°. Bp_7 154°, Bp_3 145-147°.
Ac:
$C_{12}H_{22}O_4$ M 230.303
Mp 36-37°. Bp_{15} 213°, Bp_2 172-174°.
Lactone:
$C_{10}H_{18}O_2$ M 170.251
Mp 6°. Bp_{15} 113-115°.

Grün, A. *et al*, *Ber.*, 1922, **55**, 2206 (*synth*)
Chuit, P. *et al*, *Helv. Chim. Acta*, 1929, **12**, 463 (*synth*)
Lycan, W.H. *et al*, *J. Am. Chem. Soc.*, 1929, **51**, 625 (*synth*)
Elliger, C.A. *et al*, *J. Am. Oil Chem. Soc.*, 1972, **49**, 278 (*synth*)
Goto, G. *et al*, *Chem. Pharm. Bull.*, 1985, **33**, 4422 (*synth*)

1-Hydroxy-4,6,8-decatriyn-3-one H-00353

[22692-42-8]

$H_3CC{\equiv}CC{\equiv}CC{\equiv}CCOCH_2CH_2OH$

$C_{10}H_8O_2$ M 160.172
Constit. of the lettuce *Lactuca plumieri*.

Bentley, R.K. *et al*, *J. Chem. Soc. C*, 1969, 1096 (*synth*)

2-Hydroxy-8-decene-4,6-diynoic acid H-00354

$H_3CCH{=}CHC{\equiv}CC{\equiv}CCH_2CH(OH)COOH$

$C_{10}H_{10}O_3$ M 178.187
(*Z*)-*form*
O-*Angeloyl, Me ester:* [24486-35-9].
$C_{16}H_{18}O_4$ M 274.316
Constit. of *Bellis perennis*. Oil.

Bohlmann, F. *et al*, *Chem. Ber.*, 1969, **102**, 1682 (*isol, pmr, ms*)

8-Hydroxy-2-decene-4,6-diynoic acid, 9CI H-00355

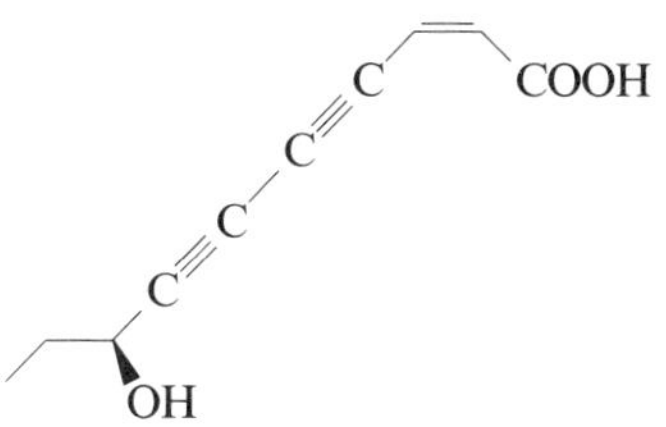

$C_{10}H_{10}O_3$ M 178.187

***(S,E)*-form** [56085-68-8]

Isol. from *Aster novi-belgii* and *Clitocybe rhizophora*. Needles (Et_2O/pet. ether). Mp 97-100°. $[\alpha]_D^{20}$ −27.4° (c, 0.36 in H_2O).

Me ester: $[\alpha]_D^{20}$ −11.5° (c, 0.76 in EtOH).

***(S,Z)*-form**

Me ester: [3347-86-2].

$C_{11}H_{12}O_3$ M 192.214

Isol. from *Aster radula*. Oil.

O-*Angeloyl, Me ester:* Isol. from the roots of *A. novi-belgii*, *A. radula* and *Baccharis eleagnoides*. $Bp_{0.0001}$ 120-130° dec. $[\alpha]_D^{20}$ −164°. λ_{max} 303, 284.5, 269, 255, 225, 218 nm.

[131899-90-6]

Bohlmann, F. *et al*, *Chem. Ber.*, 1964, **97**, 3469; 1965, **98**, 2608 (*isol, uv, ir, pmr, ms, abs config*)
Ord, M.R. *et al*, *J. Chem. Soc., Perkin Trans. 1*, 1975, 687 (*synth, uv, ir, pmr*)
Render, V.L.G. *et al*, *J. Nat. Prod.* (*Lloydia*), 1990, **53**, 692 (*isol, cmr*)

10-Hydroxy-8-decene-4,6-diynoic acid H-00356

$HOCH_2CH{=}CHC{\equiv}CC{\equiv}CCH_2CH_2COOH$

$C_{10}H_{10}O_3$ M 178.187

***(E)*-form** [54541-69-4]

Constit. of *Fistulina pallida*.

***(Z)*-form** [51193-97-6]

Constit. of *F. hepatica*.

Farrell, I.W. *et al*, *J. Chem. Soc., Perkin Trans. 1*, 1973, 2642 (*isol*)
Ahmed, M. *et al*, *J. Chem. Soc., Perkin Trans. 1*, 1974, 1981 (*isol*)
Kobayashi, A. *et al*, *Agric. Biol. Chem.*, 1975, **39**, 2247 (*synth*)

10-Hydroxy-8-decene-2,4,6-triynoic acid, 9CI H-00357

$HOCH_2CH{=}CHC{\equiv}CC{\equiv}CC{\equiv}CCOOH$

$C_{10}H_6O_3$ M 174.156

***(E)*-form**

Me ester: [80050-30-2].

$C_{11}H_8O_3$ M 188.182

Isol. from cultures of the fungus *Fayodia bisphaerigera*. Mp 85° dec. λ_{max} 352, 329.5, 309, 291, 271 infl., 257, 245, 236 nm (Et_2O).

Glycyl methyl ester: [80050-33-5]. N-(*10-Hydroxy-1-oxo-8-decene-2,4,6-triynyl*)*glycine methyl ester*. N-(*10-Hydroxy-8-decene-2,4,6-triynoyl*)*glycine methyl ester*

$C_{13}H_{11}NO_4$ M 245.234

Isol. from *F. bisphaerigera*. Needles. Mp 80° dec. λ_{max} 349, 326, 306, 288, 246 nm (EtOH).

L-*Alanyl methyl ester:* [80050-31-3]. N-(*10-Hydroxy-1-oxo-8-decene-2,4,6-triynyl*)*alanine methyl ester*. N-(*10-Hydroxy-8-decene-2,4,6-triynoyl*)*alanine methyl ester*

$C_{14}H_{13}NO_4$ M 259.261

Isol. from *F. bisphaerigera*. Needles. Mp 95° dec. $[\alpha]_D^{20}$ +64° (c, 0.2 in EtOH). λ_{max} 349, 326.5, 306.5, 288, 246 nm (EtOH).

Ahmed, M. *et al*, *J. Chem. Res., Synop.*, 1981, 262 (*isol, synth, uv, ir, pmr, ms*)

9-Hydroxy-2-decenoic acid, 9CI H-00358

[1422-28-2]

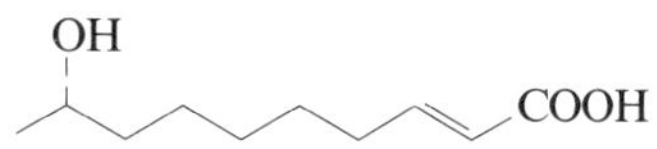

(R,E)-form

$C_{10}H_{18}O_3$ M 186.250

***(R)*-*(E)*-form**

$[\alpha]_D$ −5.4° (c, 2.1 in EtOH).

***(S)*-*(E)*-form**

$[\alpha]_D$ +5.2° (c, 7.61 in EtOH).

***(E)*-form** [4448-33-3]

Minor constit. of royal jelly. A pheromone stabilizing honey-bee swarms. Oil.

Butler, C.G. *et al*, *Nature* (*London*), 1964, **201**, 733.
Butler, C.G. *et al*, *Proc. R. Entomol. Soc. B*, 1968, **43**, 62 (*isol*)
Kovalev, B.G. *et al*, *Zh. Org. Khim.*, 1971, **7**, 667 (*synth*)
Pain, J. *et al*, *Apidologie*, 1974, **5**, 319.
Boch, R. *et al*, *J. Chem. Ecol.*, 1975, **1**, 133.
Kandil, A.A. *et al*, *Can. J. Chem.*, 1982, **61**, 1166 (*synth*)

10-Hydroxy-2-decenoic acid H-00359

[765-01-5]

$HOCH_2(CH_2)_6CH{=}CHCOOH$

$C_{10}H_{18}O_3$ M 186.250

***(E)*-form** [14113-05-4]

Constit. of royal jelly. Mp 52-53°, Mp 64.5-65°.

***(Z)*-form**

Mp 73.5-74.5°.

Fray, G.I. *et al*, *Tetrahedron*, 1961, **15**, 18 (*synth*)
Barker, S.A. *et al*, *Tetrahedron*, 1962, **18**, 177 (*synth*)
Vig, O.P. *et al*, *J. Indian Chem. Soc.*, 1975, **52**, 538 (*synth*)
Tsuji, J. *et al*, *Bull. Chem. Soc. Jpn.*, 1977, **50**, 2507 (*synth*)
Pain, J. *et al*, *Apidologie*, 1978, **9**, 263 (*glc*)
Dixon, S.E. *et al*, *Can. J. Zool.*, 1978, **56**, 592.

10-Hydroxy-8-decenoic acid H-00360

[79728 54-4]

$HOCH_2CH{=}CH(CH_2)_6COOH$

$C_{10}H_{18}O_3$ M 186.250

***(E)*-form** [106541-97-3]

Isol. from the injured fruit bodies of *Cantharellus tubaeformis*.

Me ester: [67803-47-8].

$C_{11}H_{20}O_3$ M 200.277

$Bp_{0.04}$ 115°.

Ac, Me ester: [67803-46-7].

$Bp_{0.13}$ 115°.

[69152-90-5]

Ranganathan, D. *et al*, *Tetrahedron*, 1980, **36**, 1869 (*synth*)
Baker, S.R. *et al*, *Tetrahedron Lett.*, 1983, **24**, 4469 (*synth*)
Pang, Z. *et al*, *Acta Chem. Scand.*, 1992, **46**, 301 (*isol, synth, pmr*)

11-Hydroxy-14-(3,5-dihydroxy-2-methylcyclopentyl)-9-tetradecen-12-ynoic acid H-00361

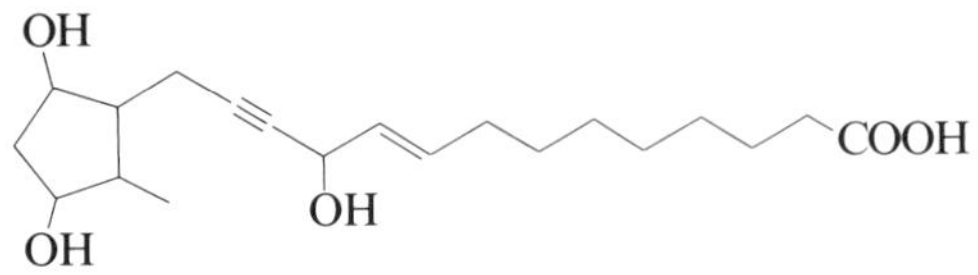

$C_{20}H_{32}O_5$ M 352.470

Isol. from the aquatic sedge *Eleocharis microcarpa*.

van Aller, R.T. *et al*, *Lipids*, 1983, **18**, 617 (*isol, ms, pmr, cmr*)

8-Hydroxy-2,7-dimethyl-2,4-decadienedioic acid, 8CI H-00362

[16515-82-5]

$HOOCCH_2CH(OH)CH(CH_3)CH_2CH{=}CHCH{=}C(CH_3)$
COOH

$C_{12}H_{18}O_5$ M 242.271

***(E,E)*-form**

Constit. of the seed of *Phaseolus multiflorus*. Needles (EtOAc/pet. ether). Mp 125-127°.

Di-Me ester: [16515-83-6].

$C_{14}H_{22}O_5$ M 270.325

Oil.

MacMillan, J. *et al*, *Tetrahedron*, 1967, **23**, 2417 (*isol, struct*)

29-Hydroxy-3,11-dimethyl-2-nonacosanone H-00363

[60789-53-9]

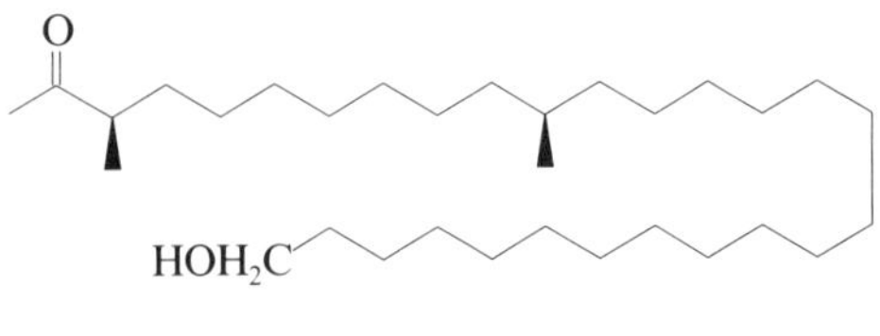

(3*R*,11*R*)-*form*

$C_{31}H_{62}O_2$ M 466.830

***(3R,11R)*-form** [79016-06-1]

Mp 38-40°. $[\alpha]_D$ −6.5° (c, 0.38 in hexane).

***(3S,11S)*-form** [79016-03-8]

Natural isom. in the cuticular wax of sexually mature females of the German cockroach (*Blattella germanica*). Mp 42.5-43°.

***(3S,11R)*-form** [79016-05-0]

Mp 40-40.5°. $[\alpha]_D$ +6.4° (c, 0.55 in hexane).

***(3R,11S)*-form** [79016-04-9]

Mp 39-40°. $[\alpha]_D$ −6.8° (c, 1.0 in hexane).

Nishida, R. *et al*, *J. Chem. Ecol.*, 1976, **2**, 449 (*occur*)
Mori, K. *et al*, *Tetrahedron*, 1981, **37**, 1329 (*synth*)

9-Hydroxy-2,8-dimethyl-3-oxo-4,6-undecadienoic acid H-00364

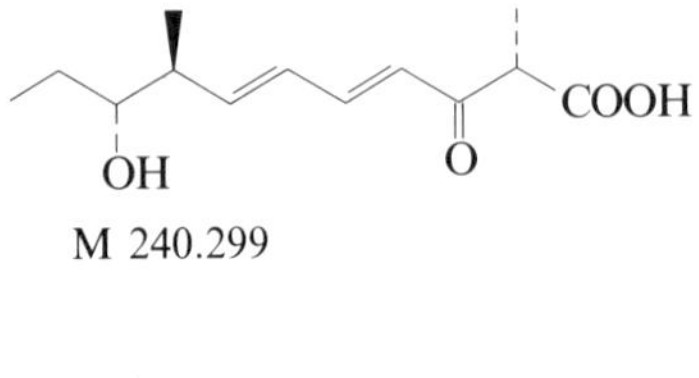

$C_{13}H_{20}O_4$ M 240.299

(2S,4E,6E,8S,9R)*-form** [138258-53-4] ***Mycinonic acid III

Prod. by *Micromonospora griseorubida*. Proposed biosynth. intermed. of mycinamycins. Oil (as Me ester).

[138223-71-9]

Kinoshita, K. *et al*, *J. Chem. Soc., Perkin Trans. 1*, 1991, 2547 (*isol, cryst struct*)

11-Hydroxy-9,15-dioxo-5,13-prostadienoic acid, 9CI H-00365

7-[3-Hydroxy-5-oxo-2-(3-oxo-1-octenyl)cyclopentyl]-5-heptenoic acid, 8CI

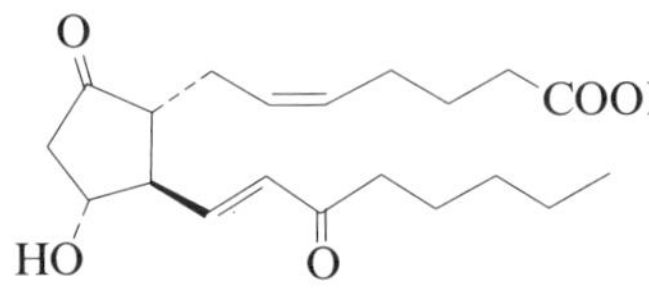

$C_{20}H_{30}O_5$ M 350.454

***(5Z,11R,13E)*-form** [26441-05-4]

15-Oxo-PGE$_2$

Induces blood platelet aggregation. Metab. of $PGE_{2\alpha}$. Cryst.

Angaard, E. *et al*, *J. Biol. Chem.*, 1964, **239**, 4097 (*synth*)
Miyano, M. *et al*, *J. Org. Chem.*, 1972, **37**, 1810 (*synth*)

11-Hydroxy-9,15-dioxoprostanoic acid, 9CI H-00366

3-Hydroxy-5-oxo-2-(3-oxooctyl)cyclopentaneheptanoic acid, 8CI

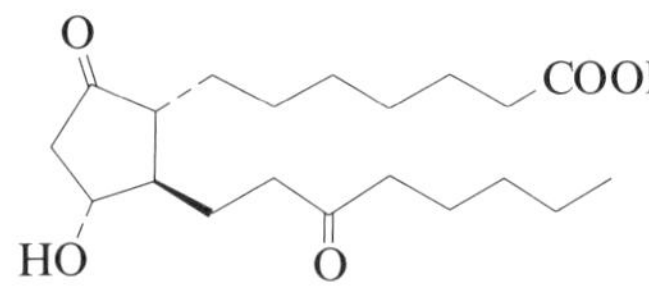

$C_{20}H_{34}O_5$ M 354.486

***(11R)*-form** [5094-14-4]

13,14-Dihydro-15-oxo-PGE$_1$

Metab. of PGE_1. Pale-yellow oil. $[\alpha]_D$ −9° ($CHCl_3$).

Angaard, E. *et al*, *Mem. Soc. Endocrinol.*, 1966, 107 (*struct, ir, uv*)
Pike, J.E. *et al*, *Prostaglandins, Proc. Nobel Symp., 2nd., 1966*, 1967, 161 (*synth, ms, pmr, ir*)
Lincoln, F.H. *et al*, *J. Org. Chem.*, 1973, **38**, 951 (*synth, ms, pmr, ir*)

11-Hydroxy-9,15-dioxo-5-prostenoic acid H-00367

7-[3-Hydroxy-5-oxo-2-(3-oxooctyl)cyclopentyl]-5-heptenoic acid, 8CI

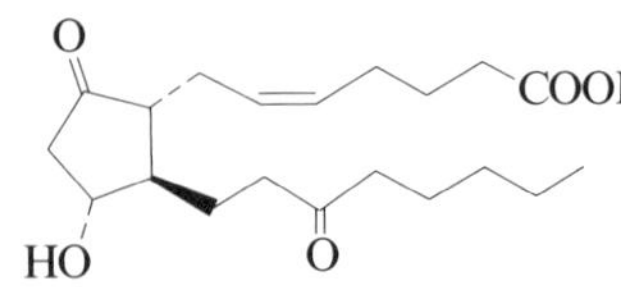

$C_{20}H_{32}O_5$ M 352.470

***(5Z,11R)*-form**

13,14-Dihydro-15-oxo-prostaglandin E$_2$

Metab. of PGE_2 occurring in blood. Induces platelet aggregation. Viscous oil.

Hamberg, M. *et al*, *Ann. N.Y. Acad. Sci.*, 1971, **180**, 164.

11-Hydroxy-9,15-dioxo-13-prostenoic acid, 9CI H-00368

3-Hydroxy-5-oxo-2-(3-oxo-1-octenyl)cyclopentaneheptanoic acid, 8CI

[22973-19-9]

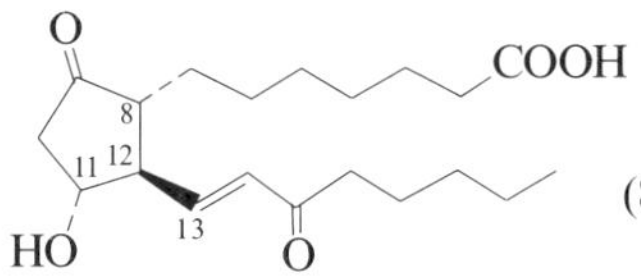

(8*RS*,11*RS*,12*RS*,13*E*)-*form*

$C_{20}H_{32}O_5$ M 352.470

(8R,11R,12R,13E)-form [34402-60-3]

15-Oxo-PGE$_1$

Metab. of PGE$_1$. Oil. $[\alpha]_D^{25}$ −41.6° (c, 1.009 in MeOH).

(8RS,11RS,12RS,13E)-form [54984-02-0]

(±)-15-Oxo-PGE$_1$

Mp 67°.

(8RS,11SR,12RS,13E)-form [34388-82-4]

(±)-*11*-epi-*15-Oxo-PGE$_1$*

Mp 60°.

Korver, O., *Recl. Trav. Chim. Pays-Bas* (*J. R. Neth. Chem. Soc.*), 1969, **88**, 1070 (*ord, cd*)
Miyano, M. *et al*, *J. Org. Chem.*, 1972, **37**, 1810; 1975, **40**, 1748 (*synth, uv, pmr*)

9-Hydroxy-11,15-dioxo-2,3,18,19-tetranor-5-prosten-1,20-dioic acid H-00369

2-(4-Carboxy-2-butenyl)-3-hydroxy-γ,5-dioxocyclopentanehexanoic acid, 9CI

[70803-92-8]

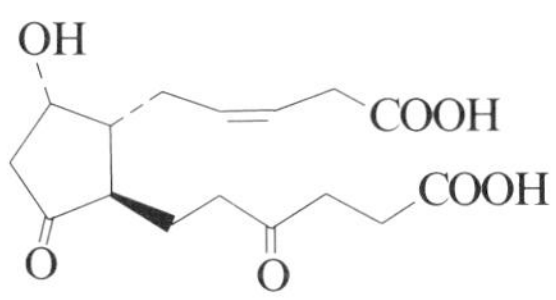

$C_{16}H_{22}O_7$ M 326.346

Human urinary metab. of prostaglandin D$_2$.

Ellis, C.K. *et al*, *J. Biol. Chem.*, 1979, **254**, 4152 (*isol, ms*)
Corey, E.J. *et al*, *J. Am. Chem. Soc.*, 1983, **105**, 1662 (*synth, uv*)

2-Hydroxydocosanoic acid H-00370

2-Hydroxybehenic acid

[13980-14-8]

$H_3C(CH_2)_{19}CH(OH)COOH$

$C_{22}H_{44}O_3$ M 356.588

Constit. of cork. Cryst. ($CHCl_3$). Mp 97°.

Me ester: [13980-17-1].

$C_{23}H_{46}O_3$ M 370.615

Cryst. (Et_2O/pet. ether). Mp 75°.

Et ester:

$C_{24}H_{48}O_3$ M 384.641

Cryst. Mp 70-71°.

Ac:

$C_{24}H_{46}O_4$ M 398.625

Cryst. (EtOH). Mp 79°.

Zetzsche, F. *et al*, *Helv. Chim. Acta*, 1931, **14**, 632.
Ansari, A.A. *et al*, *J. Am. Oil Chem. Soc.*, 1976, **53**, 118 (*synth, pmr*)

3-Hydroxydocosanoic acid H-00371

3-Hydroxybehenic acid

$H_3C(CH_2)_{18}CH(OH)CH_2COOH$

$C_{22}H_{44}O_3$ M 356.588

(±)-form

Cryst. (pet. ether). Mp 95.8-96°.

Skogh, M., *Acta Chem. Scand.*, 1952, **6**, 809 (*synth, cryst struct*)
de Pascual, T.J. *et al*, *CA*, 1967, **66**, 460w (*isol*)

13-Hydroxydocosanoic acid H-00372

[13980-16-0]

$H_3C(CH_2)_8CH(OH)(CH_2)_{11}COOH$

$C_{22}H_{44}O_3$ M 356.588

(S)-form

Prod. by the yeast *Candida bogoriensis*. Cryst. (Me_2CO). Mp 85-86°.

Me ester: [23560-01-2].

$C_{23}H_{46}O_3$ M 370.615

Mp 69-70°.

(±)-form

Me ester: Mp 63-63.5°.

Roomi, M.W. *et al*, *J. Chromatogr.*, 1966, **24**, 93 (*tlc*)
Tulloch, A.P., *Can. J. Chem.*, 1968, **46**, 3727 (*abs config*)
Esders, T.W. *et al*, *J. Lipid Res.*, 1972, **13**, 663 (*glc, ms*)

22-Hydroxydocosanoic acid H-00373

22-Hydroxybehenic acid. Phellonic acid

[506-45-6]

$HOCH_2(CH_2)_{20}COOH$

$C_{22}H_{44}O_3$ M 356.588

Present in cork, birch bark and carnauba wax. Cryst. ($CHCl_3$/Me_2CO). Mp 99.2-100.5°.

Me ester: [38646-51-4].

$C_{23}H_{46}O_3$ M 370.615

Cryst. (C_6H_6). Mp 73.5-74°.

Ac:

$C_{24}H_{46}O_4$ M 398.625

Mp 79°.

Lactone: [58296-51-8]. *Oxacyclotricosan-2-one, 9CI. 22-Docosanolide*

$C_{22}H_{42}O_2$ M 338.573

Constit. of the secretion of Dufour's gland of various bee spp. and from *Armitermes* spp. Pheromone or defence substance.

Guillemonat, M.A. *et al*, *Bull. Soc. Chim. Fr.*, 1950, 860 (*struct*)
Murry, K.E. *et al*, *Aust. J. Chem.*, 1955, **8**, 437 (*isol*)
Ames, D.E. *et al*, *J. Chem. Soc. C*, 1967, 1556 (*synth*)
Holloway, P.J., *Chem. Phys. Lipids*, 1972, **9**, 158 (*isol, glc, ms*)
Duffield, R.M. *et al*, *J. Chem. Ecol.*, 1982, **8**, 535 (*isol, lactone*)
Prestwich, G.D., *Tetrahedron*, 1982, **38**, 1911 (*isol, struct, lactone*)
Hefetz, A. *et al*, *J. Chem. Ecol.*, 1986, **12**, 197 (*occur, lactone*)

10-Hydroxy-7,11,13,16,19-docosapentaenoic acid H-00374

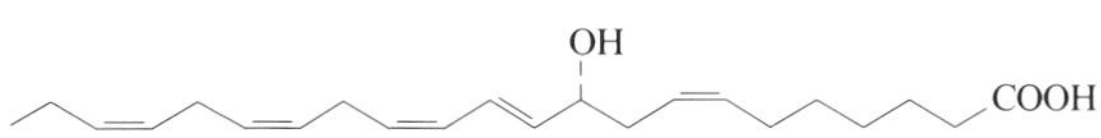

$C_{22}H_{34}O_3$ M 346.509

(7Z,10R,11E,13Z,16Z,19Z)-form [117332-99-7] ***Leiopathic acid***

Isol. from the antipatharian hexacoral, *Leiopathes* sp. $[\alpha]_D^{20}$ +3.1° (c, 1.85 in $CHCl_3$).

Et ester: [117333-00-3].
$C_{24}H_{38}O_3$ M 374.562
Isol. from *L.* sp. $[\alpha]_D^{20}$ +5.3° (c, 0.53 in $CHCl_3$).

Guerriero, A. *et al, Helv. Chim. Acta*, 1988, **71**, 1094 (*isol, struct, uv, pmr, cmr, ms, abs config*)

12-Hydroxy-3,6-dodecadienoic acid H-00375

$HOCH_2(CH_2)_4CH{=}CHCH_2CH{=}CHCH_2COOH$

$C_{12}H_{20}O_3$ M 212.288

(Z,Z)-form
Lactone: [90134-39-7]. *3,6-Dodecadienolide*
$C_{12}H_{18}O_2$ M 194.273
No phys. props. reported.

Pierce, A.M. *et al, J. Agric. Food Chem.*, 1983, **12**, 1367.
Millar, J.G. *et al, J. Org. Chem.*, 1984, **49**, 2332 (*synth, ir, pmr, ms*)

2-Hydroxydodecanoic acid H-00376

2-Hydroxylauric acid
[2984-55-6]

$H_3C(CH_2)_9CH(OH)COOH$

$C_{12}H_{24}O_3$ M 216.320
Present in lipids of *Azotobacter agilis.*

(±)-form
Cryst. ($CHCl_3$). Mp 73-74°.
Et ester:
$C_{14}H_{28}O_3$ M 244.373
Cryst. (pet. ether). Mp 43°. Bp_{15} 165°.
Et ester, Ac:
$C_{16}H_{30}O_4$ M 286.411
Bp_{13} 172-173°.

Davies, R.R. *et al, J. Soc. Chem. Ind., London*, 1943, **62**, 128 (*synth*)
Kaneshiro, T. *et al, Biochim. Biophys. Acta*, 1963, **70**, 271 (*isol*)
Breusch, F.L. *et al, Fette, Seifen, Anstrichm.*, 1971, **73**, 635.

3-Hydroxydodecanoic acid H-00377

3-Hydroxylauric acid
[1883-13-2]

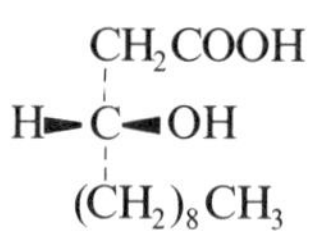
(R)-form

$C_{12}H_{24}O_3$ M 216.320

(R)-form [28254-78-6]
Hydrol. prod. of Isariin. Present in lipids of *Serratia marcescens* and *Azotobacter agilis.* Mp 62.5-63.3°. $[\alpha]_D^{25}$ −15.2° (c, 1.6 in $CHCl_3$).

(S)-form [45162-48-9]
Mp 61-63.3°. $[\alpha]_D^{25}$ +14.4° (c, 2.8 in $CHCl_3$).

(±)-form [53941-38-1]
Mp 70-70.5°.

Vining, L.C., *Can. J. Chem.*, 1962, **40**, 1579.
Barbosa, H.R. *et al, Rev. Microbiol.*, 1975, **6**, 1 (*isol*)
Rietschel, E.T. *et al, J. Bacteriol.*, 1975, **122**, 1180.
Hiramatsu, T. *et al, Can. J. Microbiol.*, 1976, **22**, 731 (*isol*)
Tahara, S. *et al, Agric. Biol. Chem.*, 1978, **42**, 879 (*biosynth*)
Kawahara, K. *et al, Biochim. Biophys. Acta*, 1979, **572**, 1 (*biosynth*)

4-Hydroxydodecanoic acid H-00378

4-Hydroxylauric acid
[16899-07-3]

$H_3C(CH_2)_7CH(OH)CH_2CH_2COOH$

$C_{12}H_{24}O_3$ M 216.320
Cryst. (C_6H_6/pet. ether). Mp 62.5-63.5°.
Lactone: see *Dihydro-5-octyl-2(3H)-furanone, D-00156*

Chuit, P. *et al, Helv. Chim. Acta*, 1927, **10**, 113 (*synth*)

7-Hydroxydodecanoic acid, 9CI H-00379

[70393-62-3]

$H_3C(CH_2)_4CH(OH)(CH_2)_5COOH$

$C_{12}H_{24}O_3$ M 216.320

(±)-form [73276-74-1]
Mp 55.5-56.5°. (*R*)-form also known.

[58262-37-6, 78737-61-8]

Tahera, S. *et al, Agric. Biol. Chem.*, 1980, **44**, 193 (*synth, ir, ms, glc*)
Midland, M.M. *et al, J. Org. Chem.*, 1981, **46**, 3933 (*synth, pmr*)

9-Hydroxydodecanoic acid H-00380

$H_3CCH_2CH_2CH(OH)(CH_2)_7COOH$

$C_{12}H_{24}O_3$ M 216.320
Constit. of *Blepharis sindica.* Cryst. (Me_2CO/pet. ether). Mp 46-47°.

Ahmad, I. *et al, Phytochemistry*, 1983, **22**, 493.

11-Hydroxydodecanoic acid, 9CI H-00381

[32459-66-8]

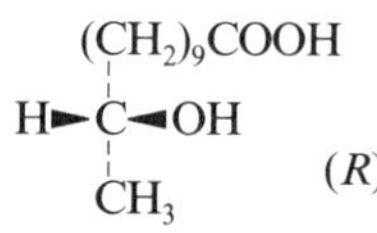
(R)-form

$C_{12}H_{24}O_3$ M 216.320
Constit. of royal jelly. Metab. of *Fusarium oxysporum.*

(R)-form
Me ester: [88785-25-5].
Oil. $Bp_{0.002}$ ~100°. $[\alpha]_D$ −7.7° (c, 2.15 in cyclohexane).

(±)-form [83515-89-3]
Cryst. (Et_2O/pet. ether). Mp 54°.
Me ester: [109718-22-1].
$C_{13}H_{26}O_3$ M 230.347
Liq. or cryst. Bp_7 160°.
Anilide: Small prisms (pet. ether). Mp 87°.
Lactone: 11-Decanolide. Dihydrorecifeiolide
$C_{12}H_{22}O_2$ M 198.305
Oil. $Bp_{0.001}$ 65-70°.

[66605-95-6]

Ellin, A. *et al, Arch. Biochem. Biophys.*, 1973, **158**, 597 (*synth, derivs*)
Voss, G. *et al, Helv. Chim. Acta*, 1983, **66**, 2294 (*synth, ir, pmr, cmr*)
Hesse, B. *et al, Helv. Chim. Acta*, 1984, **67**, 1713; 1986, **69**, 1323 (*lactone*)
Makita, A. *et al, J. Antibiot.*, 1986, **39**, 1257 (*synth, derivs*)

12-Hydroxydodecanoic acid **H-00382**

Sabinic acid. 12-Hydroxylauric acid

[505-95-3]

$HOCH_2(CH_2)_{10}COOH$

$C_{12}H_{24}O_3$ M 216.320

Constit. of *Juniperus oxycedrus* leaves. Also from *J. sabina*, *Thuja occidentalis*, *Pinus* spp., *Chamaecyparis* spp. and *Cycas revoluta*. Widespread in conifer waxes. Cryst. (MeOH aq., C_6H_6 or EtOAc). Mp 78-79°, Mp 84-85°.

Me ester:

$C_{13}H_{26}O_3$ M 230.347

Mp 34-35°, Mp 44.5°. Bp_7 160°.

Ac:

$C_{14}H_{26}O_4$ M 258.357

Mp 49-49.2° (43°). Bp 202-203°.

Anilide: Cryst. (pet. ether). Mp 87°.

Lactone: [947-05-7]. *Oxacyclotridecan-2-one*

$C_{12}H_{22}O_2$ M 198.305

No phys. props. reported.

Chuit, P. *et al*, *Helv. Chim. Acta*, 1929, **12**, 463 (*synth*)
Lycan, W.H. *et al*, *J. Am. Chem. Soc.*, 1929, **51**, 625 (*synth*)
De Pascual, T.J. *et al*, *An. Quim.*, 1973, **69**, 941 (*isol*)
Fan, L.L. *et al*, *Anal. Biochem.*, 1976, **71**, 265 (*hplc*)
Kurihara, T. *et al*, *Tetrahedron Lett.*, 1976, 2455 (*synth*)
Bidd, I. *et al*, *J. Chem. Soc., Perkin Trans. 1*, 1983, 1369 (*lactone*)

3-Hydroxy-5-dodecenoic acid **H-00383**

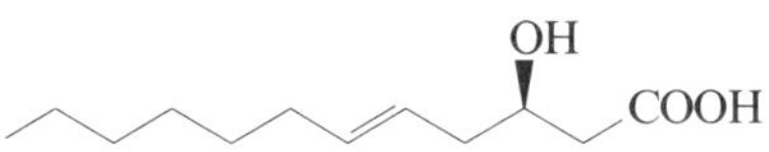

$C_{12}H_{22}O_3$ M 214.304

(R,Z)-form [88456-81-9]

Isol. from *Pythium ultimum* and *Serratia marcescens*. Phytotoxin. Yellow oil. $[\alpha]_D^{22}$ −10.7° (c, 1.55 in $CHCl_3$).

Bishop, D.G. *et al*, *Biochem. Biophys. Res. Commun.*, 1962, **7**, 337 (*isol*)
Ichihara, A. *et al*, *Agric. Biol. Chem.*, 1985, **49**, 2207 (*isol, abs config*)

10-Hydroxy-11-dodecenoic acid, 9CI **H-00384**

[85288-97-7]

$H_2C{=}CHCH(OH)(CH_2)_8COOH$

$C_{12}H_{22}O_3$ M 214.304

(±)-form

Cryst. (Et_2O/pet. ether). Mp 43-44°.

Me ester: [106753-87-1].

$C_{13}H_{24}O_3$ M 228.331

Oil.

Cameron, A.G. *et al*, *J. Chem. Soc., Perkin Trans. 1*, 1986, 161 (*synth, pmr, ir*)

12-Hydroxy-3-dodecenoic acid **H-00385**

$HOCH_2(CH_2)_7CH{=}CHCH_2COOH$

$C_{12}H_{22}O_3$ M 214.304

(Z)-form

Lactone: [87371-99-1]. *Oxacyclotridec-4-en-2-one, 9CI. 3-Dodecenolide*

$C_{12}H_{20}O_2$ M 196.289

Aggregation pheromone of the flat grain beetle *Cryptolestes pusillus*.

Millar, J.G. *et al*, *J. Org. Chem.*, 1983, **48**, 4404 (*synth, ir, pmr, ms*)

12-Hydroxy-10-dodecenoic acid, 9CI **H-00386**

[85288-99-9]

$HOCH_2CH{=}CH(CH_2)_8COOH$

$C_{12}H_{22}O_3$ M 214.304

(E)-form

Cryst. (Et_2O/pet. ether). Mp 39-40°.

Me ester: [106753-90-6].

$C_{13}H_{24}O_3$ M 228.331

Oil. $Bp_{0.1}$ 175° (oven).

Lactone: 10-Dodecen-12-olide

$C_{12}H_{20}O_2$ M 196.289

$Bp_{0.1}$ 85° (oven).

Cameron, A.G. *et al*, *J. Chem. Soc., Perkin Trans. 1*, 1986, 161 (*synth, pmr, ir*)

5-Hydroxy-9-dotriacontanone, 9CI **H-00387**

[132923-40-1]

$H_3C(CH_2)_{22}CO(CH_2)_3CH(OH)(CH_2)_3CH_3$

$C_{32}H_{64}O_2$ M 480.856

Isol. from bulbs of *Crinum augustum*. Mp 80-82°.

Abd El-Hafiz, M.A., *Phytochemistry*, 1990, **29**, 3936 (*isol, ir, ms, pmr*)

7-Hydroxy-2-dotriacontanone **H-00388**

[146356-94-7]

$H_3C(CH_2)_{24}CH(OH)(CH_2)_4COCH_3$

$C_{32}H_{64}O_2$ M 480.856

Constit. of the shoots of *Leucas aspera*. Cryst. (C_6H_6/EtOAc). Mp 84-85°.

Ac: Mp 67-68°.

Misra, T.N. *et al*, *Phytochemistry*, 1993, **32**, 199 (*isol, pmr*)

25-Hydroxy-3-dotriacontanone, 9CI **H-00389**

8-Hydroxy-30-dotriacontanone

[94413-31-7]

$H_3C(CH_2)_6CH(OH)(CH_2)_{21}COCH_2CH_3$

$C_{32}H_{64}O_2$ M 480.856

Isol. from leaves and stems of *Duboisia myoporoides*. Mp 76-77°.

Shukla, Y.N. *et al*, *Phytochemistry*, 1984, **23**, 1516 (*isol, ir, ms*)

25-Hydroxy-6-dotriacontanone, 9CI **H-00390**

[87314-41-8]

$H_3C(CH_2)_6CH(OH)(CH_2)_{18}CO(CH_2)_4CH_3$

$C_{32}H_{64}O_2$ M 480.856

Isol. from aerial parts of *Artemisia roxburghiana*.

Prasad, A.V.K. *et al*, *Indian J. Chem., Sect. B*, 1983, **22**, 610 (*isol, pmr, ir, ms*)

30-Hydroxy-5-dotriacontanone, 9CI **H-00391**

[92122-80-0]

$H_3CCH_2CH(OH)(CH_2)_{24}CO(CH_2)_3CH_3$

$C_{32}H_{64}O_2$ M 480.856

Isol. from leaves and stems of *Duboisia myoporoides*. Mp 75-76°.

Shukla, Y.N. *et al*, *Phytochemistry*, 1984, **23**, 799 (*isol, ir, pmr, ms*)

11-Hydroxy-12,14-eicosadienoic acid H-00392

$H_3C(CH_2)_4CH{=}CHCH{=}CHCH(OH)(CH_2)_9COOH$

$C_{20}H_{36}O_3$ M 324.503

(11S,12E,14Z)-form

Prepared from 11,14-Eicosadienoic acid in the presence of sheep vesicular glands. Cryst. (pet. ether). Mp 48°. $[\alpha]_D^{20}$ −10° (c, 1.5 in MeOH), $[\alpha]_D$ −5° (c, 1.7 in $CHCl_3$).

Nugteren, D.H. *et al*, *Nature (London)*, 1966, **212**, 38 (*synth*)

2-Hydroxyeicosanoic acid H-00393

2-Hydroxyarachidic acid

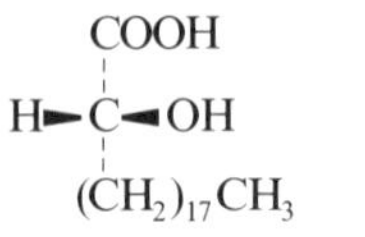

(*R*)-*form*

$C_{20}H_{40}O_3$ M 328.534

(*R*)-*form* [26632-14-4]

Constit. of brain cerebrosides.

(±)-*form* [16742-48-6]

Cryst. (C_6H_6/pet. ether). Mp 91-92°.

Me ester: [16742-49-7].
$C_{21}H_{42}O_3$ M 342.561
Mp 62-64°.

Et ester:
$C_{22}H_{44}O_3$ M 356.588
Mp 62-63°.

Hammarström, S., *FEBS Lett.*, 1969, **5**, 192 (*config*)

3-Hydroxyeicosanoic acid H-00394

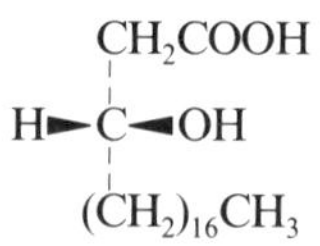

(*R*)-*form*

$C_{20}H_{40}O_3$ M 328.534

(*R*)-*form*

Isol. from hydrol. of peptidolipin of *Nocardia asteroides*.

(±)-*form*

Constit. of gum resin of *Citrus ladaniferus*. Mp 93°.

Skogh, M., *Acta Chem. Scand.*, 1952, **6**, 809 (*synth, cryst struct*)
Guinand, M. *et al*, *Biochim. Biophys. Acta*, 1966, **125**, 75 (*isol, ir, struct*)
de Pascual, J., *CA*, 1967, **66**, 460w (*isol*)

19-Hydroxyeicosanoic acid, 9CI H-00395

19-Hydroxyicosanoic acid. 19-Hydroxyarachidic acid

[34383-23-8]

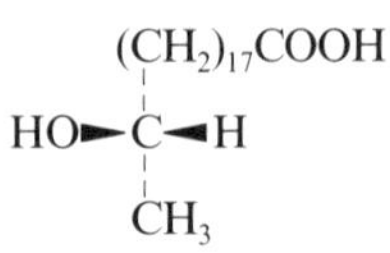

$C_{20}H_{40}O_3$ M 328.534

(*S*)-*form* [34370-60-0]

Isol. from oil of *Torulopsis magnoliae*. Cryst. (Me_2CO). Mp 86.5-87.5°. $[\alpha]_D^{25}$ +5.2° (c, 1.4 in AcOH).

Me ester:
$C_{21}H_{42}O_3$ M 342.561
Cryst. (pet. ether). Mp 58-60°. $[\alpha]_D^{25}$ +3.6° (c, 2.4 in MeOH).

Tulloch, A.P. *et al*, *Can. J. Chem.*, 1962, **40**, 1362.

20-Hydroxyeicosanoic acid, 9CI H-00396

[62643-46-3]

$HOCH_2(CH_2)_{18}COOH$

$C_{20}H_{40}O_3$ M 328.534

Isol. from *Carnauba* wax; stem and root of *Malus pumila*; *Dicranum elongatum*, *Eriophorum vaginatum*, *Pseudotsuga menziesii* bark. Mp 97.4-97.8°.

Me ester: [37477-29-5].
$C_{21}H_{42}O_3$ M 342.561
Mp 68-68.5°.

Ac: [104966-87-2].
$C_{22}H_{42}O_4$ M 370.571
Needles (hexane). Mp 78°.

Ac, Et ester: [104967-06-8].
$C_{24}H_{46}O_4$ M 398.625
Needles (hexane). Mp 53°.

Lactone: [31844-01-6]. *Oxacycloheneicosan-2-one, 9CI. 20-Eicosanolide, 9CI. 20-Icosanolide*
$C_{20}H_{38}O_2$ M 310.519
Pheromone from social bee *Lasioglossum zephyrum* and from other bees.

Murry, K.E. *et al*, *Aust. J. Chem.*, 1955, **8**, 437 (*isol*)
Loveland, P.M. *et al*, *Phytochemistry*, 1972, **11**, 3080 (*isol*)
Corey, E.J. *et al*, *Tetrahedron Lett.*, 1976, 3405 (*synth, lactone*)
Holloway, P.J., *Phytochemistry*, 1982, **21**, 2517 (*isol*)
Goto, G. *et al*, *Chem. Pharm. Bull.*, 1985, **33**, 4422 (*synth, derivs, ir, pmr*)
Hefetz, A. *et al*, *J. Chem. Ecol.*, 1986, **12**, 197 (*occur, lactone*)
Karunen, P. *et al*, *Phytochemistry*, 1987, **26**, 1728; 1988, **27**, 2045 (*isol*)
Smith, B.H. *et al*, *J. Chem. Ecol.*, 1988, **14**, 87.

8-Hydroxy-5,9,11,14,17-eicosapentaenoic acid, 9CI H-00397

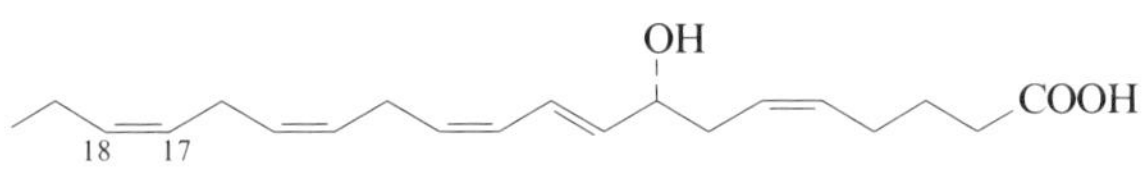

$C_{20}H_{30}O_3$ M 318.455

(5Z,8R,9E,11Z,14Z,17Z)-form [117407-06-4]

Constit. of the antipatharian hexacoral *Leiopathes* sp. $[\alpha]_D^{20}$ +3.0° (c, 0.43 in $CHCl_3$).

Et ester: [117333-01-4].
$C_{22}H_{34}O_3$ M 346.509
Constit. of *L.* sp. $[\alpha]_D^{20}$ +3.7° (c, 0.33 in $CHCl_3$).

17,18-Dihydro: see *8-Hydroxy-5,9,11,14-eicosatetraenoic acid, H-00401*

Guerriero, A. *et al*, *Helv. Chim. Acta*, 1988, **71**, 1094 (*isol, struct, uv, pmr, cmr, abs config*)

9-Hydroxy-2,5,7,11,14-eicosapentaenoic acid H-00398

$H_3C(CH_2)_4(CH{=}CHCH_2)_2CH(OH)(CH{=}CH)_2CH_2CH{=}CHCOOH$

$C_{20}H_{30}O_3$ M 318.455

(2Z,5Z,7E,11Z,14Z)-form

Antimicrobial component of the red alga *Laurencia hybrida*.

Me ester: [81892-93-5].
$C_{21}H_{32}O_3$ M 332.482
$[\alpha]_D^{20}$ −4.5° (c, 0.65 in MeOH).

Higgs, M.D., *Tetrahedron*, 1981, **37**, 4255 (*isol, pmr, ms*)

5-Hydroxy-6,8,11,14-eicosatetraenoic acid, 9CI H-00399

5-HETE

[71030-39-2]

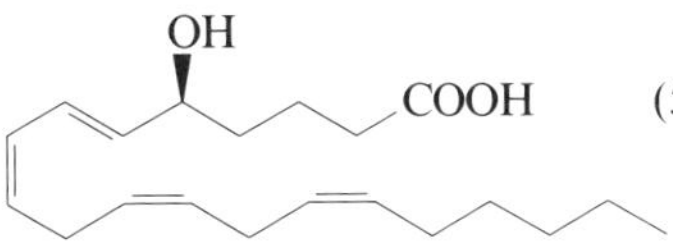

(5*R*,6*E*,8*Z*,11*Z*,14*Z*)-*form*

$C_{20}H_{32}O_3$ M 320.471

(5R,6E,8Z,11Z,14Z)-form

δ-Lactone: [81517-87-5]. *Tetrahydro-6-(1,3,6,9-pentadecatetraenyl)-2*H*-pyran-2-one. 5-HL*

$C_{20}H_{30}O_2$ M 302.456

5-Lipoxygenase inhibitor.

(5S,6E,8Z,11Z,14Z)-form [70608-72-9]

Metab. of arachidonic acid.

Me ester: [78037-99-7].

$C_{21}H_{34}O_3$ M 334.498

$[\alpha]^{23}_{436}$ +12.42° (c, 0.99 in EtOH), $[\alpha]^{23}_{D}$ +4.73° (c, 0.99 in EtOH).

(6E,8Z,11Z,14Z)-form

5-Oxo: [106154-18-1]. *5-Keto-6,11,14-eicosatetraenoic acid. 5-Oxo-ETE. 5-KETE*

5-Lipoxygenase inhibitor.

5-Oxo, Me ester: [74785-00-5].

5-Lipoxygenase inhibitor. Oil.

δ-Lactone: [73279-37-5].

Ir 985 cm^{-1} (*trans* HC=CH).

[126432-17-5, 127708-42-3]

Porter, N.A. *et al, J. Org. Chem.*, 1979, **44**, 3177 (*ms*)
Boeynaems, J.M. *et al, Anal. Biochem.*, 1980, **104**, 259 (*synth*)
Corey, E.J. *et al, J. Am. Chem. Soc.*, 1980, **102**, 1435 (*synth, pmr, ir*)
Rabinovitch, H. *et al, Lipids*, 1981, **16**, 518 (*metab*)
Atrache, V. *et al, Tetrahedron Lett.*, 1981, **22**, 3443 (*synth*)
Arai, Y. *et al, J. Med. Chem.*, 1983, **26**, 72 (*synth, ms, ir, pmr, uv*)
Rokach, J. *et al, Tetrahedron Lett.*, 1983, **24**, 5185 (*synth*)
Gunn, B.P., *Tetrahedron Lett.*, 1985, **26**, 2869 (*synth*)
Nicolaou, K.C. *et al, Synthesis*, 1986, 344 (*synth*)
Haviv, F. *et al, J. Med. Chem.*, 1987, **30**, 254 (*synth, biochem*)
Kerdesky, F.A.J. *et al, J. Med. Chem.*, 1987, **30**, 1177 (*synth, biochem*)
Schulam, P.G. *et al, J. Immunol.*, 1990, **144**, 2696 (*biosynth*)

6-Hydroxy-4,8,11,14-eicosatetraenoic acid H-00400

6-HETE

[70968-94-4]

$H_3C(CH_2)_4(CH{=}CHCH_2)_3CH(OH)$
$CH{=}CHCH_2CH_2COOH$

$C_{20}H_{32}O_3$ M 320.471

Porter, N.A. *et al, J. Org. Chem.*, 1979, **44**, 3177 (*synth, ms*)

8-Hydroxy-5,9,11,14-eicosatetraenoic acid, 9CI H-00401

8-HETE

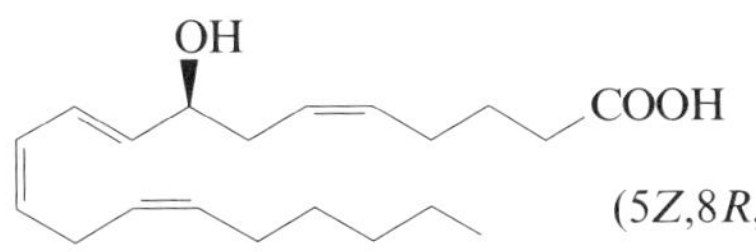

(5*Z*,8*R*,9*E*,11*Z*,14*Z*)-*form*

$C_{20}H_{32}O_3$ M 320.471

(5Z,8R,9E,11Z,14Z)-form [105500-09-2]

Isol. from the black coral *Leiopathes* sp. $[\alpha]^{20}_{D}$ +4.0° (c, 0.48 in $CHCl_3$). Uv 236.5 (21150) nm (95% EtOH).

Et ester: [117333-02-5].

$C_{22}H_{36}O_3$ M 348.525

Isol. from *L.* sp. $[\alpha]^{20}_{D}$ +4.5° (c, 0.32 in $CHCl_3$).

8-Oxo: [116539-62-9]. *8-Keto-5,9,11,14-eicosatetraenoic acid. 8-Oxo-ETE. 8-KETE*

$C_{20}H_{30}O_3$ M 318.455

Arachidonic acid metab.

(8S,5Z,9E,11Z,14Z)-form [98462-03-4]

Metab. of arachidonic acid.

Me ester:

$C_{21}H_{34}O_3$ M 334.498

Oil. $[\alpha]^{22}_{D}$ −4.75° (c, 0.4 in $CHCl_3$).

[70968-93-3, 79495-84-4, 100896-77-3]

Porter, N.A. *et al, J. Org. Chem.*, 1979, **44**, 3177 (*synth, ms*)
Boeynaems, J.M. *et al, Anal. Biochem.*, 1980, **104**, 259 (*synth*)
Rabinovitch, H. *et al, Lipids*, 1981, **16**, 518 (*metab*)
Adams, J. *et al, Tetrahedron Lett.*, 1984, **25**, 35 (*synth*)
Just, G. *et al, J. Org. Chem.*, 1986, **51**, 4796 (*synth, ms, pmr*)
Yadagiri, P. *et al, Tetrahedron Lett.*, 1986, **27**, 6039 (*synth*)
Wiseman, J.S. *et al, Biochem. Biophys. Res. Commun.*, 1988, **154**, 544 (*synth*)
Guerriero, A. *et al, Helv. Chim. Acta*, 1988, **71**, 1094 (*synth, uv, ir, pmr, cmr*)
Kuehn, H. *et al, Eicosanoids*, 1991, **4**, 9 (*biosynth*)

9-Hydroxy-5,7,11,14-eicosatetraenoic acid H-00402

9-HETE

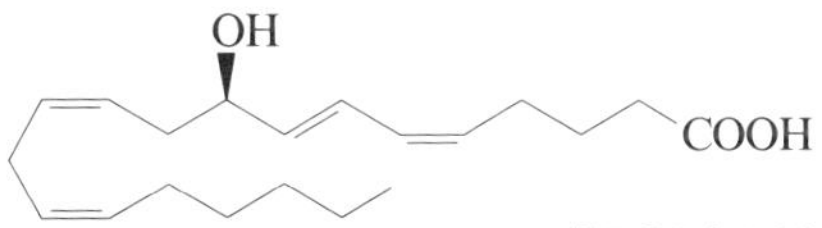

(5*Z*,7*E*,9*R*,11*Z*,14*Z*)-*form*

$C_{20}H_{32}O_3$ M 320.471

(5Z,7E,9R,11Z,14Z)-form [107656-14-4]

Metab. of arachidonic acid.

(5Z,7E,9S,11Z,14Z)-form [70968-92-2]

Metab. of arachidonic acid (poss. as the 9*R*-enantiomer). Isol. from rat brain.

Me ester: [70968-96-6].

$C_{21}H_{34}O_3$ M 334.498

$[\alpha]^{22}_{D}$ −7.1° (c, 2.05 in $CHCl_3$).

(5Z,7E,11Z,14Z)-form

9-Oxo: [135200-66-7]. *9-Keto-5,7,11,14-eicosatetraenoic acid. 9-KETE. 9-Oxo-ETE*

Metab. of arachidonic acid.

[117957-82-1]

Porter, N.A. *et al, J. Org. Chem.*, 1977, **44**, 3177 (*synth, ms*)
Boeynaems, J.M. *et al, Anal. Biochem.*, 1980, **104**, 259 (*synth*)
Rabinovitch, M. *et al, Lipids*, 1981, **16**, 518 (*metab*)
Capdevila, J. *et al, Proc. Natl. Acad. Sci. U.S.A.*, 1982, **79**, 767 (*synth, ms*)
Adams, J. *et al, Tetrahedron Lett.*, 1984, **25**, 35 (*synth*)
Capdevila, J. *et al, Biochem. Biophys. Res. Commun.*, 1986, **141**, 1007 (*biosynth*)
Just, G. *et al, J. Org. Chem.*, 1986, **51**, 4796 (*synth, pmr, uv, ms*)
Saniere, M. *et al, Tetrahedron*, 1989, **45**, 7317 (*synth, pmr*)
Kuehn, H. *et al, Eicosanoids*, 1991, **4**, 9 (*biosynth*)

11-Hydroxy-5,8,12,14-eicosatetraenoic acid H-00403

11-HETE

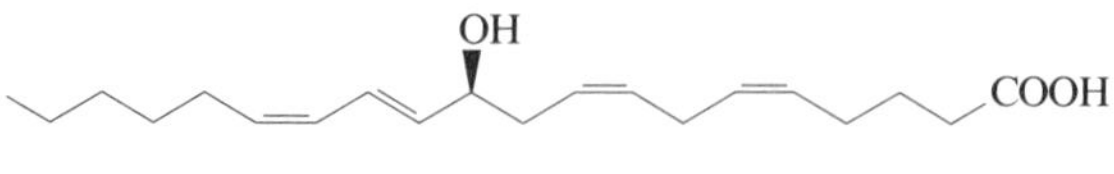

(5*Z*,8*Z*,11*R*,12*E*,14*Z*)-*form*

$C_{20}H_{32}O_3$ M 320.471

Metab. of arachidonic acid.

(5Z,8Z,11R,12E,14Z)-form [73347-43-0]

$[\alpha]_D^{21}$ +11.2° (CH_2Cl_2).

Me ester: [79083-18-4].

$C_{21}H_{34}O_3$ M 334.498

$[\alpha]_D^{23}$ +10.17° (c, 1.0 in $CHCl_3$).

(5Z,8Z,11S,12E,14Z)-form

$[\alpha]_D^{21}$ −11.3° (CH_2Cl_2).

Me ester: $[\alpha]_D^{22}$ −10.3° (c, 1.2 in $CHCl_3$).

(5E,8Z,12Z,14Z)-form

11-Oxo: [135200-65-6]. *11-Keto-5,8,12,14-eicosatetraenoic acid. 11-Oxo-ETE. 11-KETE*

Metab. of arachidonic acid.

Porter, N.A. *et al*, *J. Org. Chem.*, 1979, **44**, 3177 (*synth, ms*)
Boeynaems, J.M., *Anal. Biochem.*, 1980, **104**, 259 (*synth*)
Corey, E.J. *et al*, *J. Am. Chem. Soc.*, 1980, **102**, 1433; 1981, **103**, 4618 (*synth, pmr*)
Rabinovitch, H., *Lipids*, 1981, **16**, 518 (*metab*)
Capdevilla, J. *et al*, *Proc. Natl. Acad. Sci. U.S.A.*, 1982, **79**, 767 (*synth*)
Just, G. *et al*, *Tetrahedron Lett.*, 1982, **23**, 1331, 2285 (*synth*)
Just, G. *et al*, *J. Org. Chem.*, 1986, **51**, 4796 (*deriv, synth, uv, pmr, ms*)
Kuehn, H. *et al*, *Eicosanoids*, 1991, **4**, 9 (*synth*)

12-Hydroxy-5,8,10,14-eicosatetraenoic acid, 9CI H-00404

12-HETE

[59985-28-3]

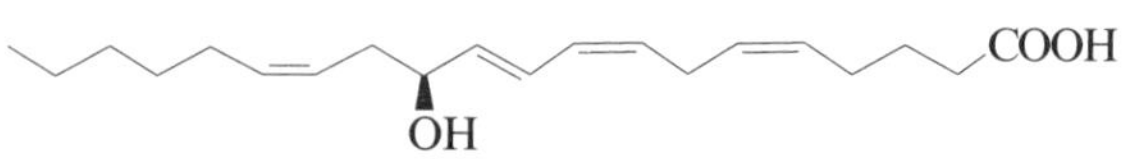

(5*Z*,8*Z*,10*E*,12*S*,14*Z*)-*form*

$C_{20}H_{32}O_3$ M 320.471

(5Z,8Z,10E,12S,14Z)-form [54397-83-0]

Metab. of arachidonic acid. Isol. from rat brain.

Me ester: [57872-14-7].

$C_{21}H_{34}O_3$ M 334.498

$[\alpha]_D^{22}$ +1.3° (c, 0.3 in $CHCl_3$).

12-Oxo: [108437-64-5]. *12-Keto-5,8,10,14-eicosatetraenoic acid. 12-Oxo-ETE. 12-KETE*

Formed in the nervous system of *Aplysia*. Metab. of arachidonic acid.

(5Z,8Z,10Z,12S,14Z)-form

$[\alpha]_D^{21}$ −1.87° (c, 6.1 in $CHCl_3$).

Me ester: [81623-72-5].

$[\alpha]_D^{23}$ −2.06° (c, 5.4 in $CHCl_3$).

(5Z,8Z,10E,12R,14Z)-form

Metab. of arachidonic acid. Biochem. diff. from 12*S*-form.

Me ester: Oil. $[\alpha]_D^{20}$ −1.25° (c, 0.25 in $CHCl_3$).

[108437-55-4]

Corey, E.J. *et al*, *J. Am. Chem. Soc.*, 1978, **100**, 1942 (*synth, pmr*)
McGuire, J.C. *et al*, *Prep. Biochem.*, 1978, **8**, 147 (*metab*)
Porter, N.A. *et al*, *J. Org. Chem.*, 1979, **44**, 3177 (*synth, ms*)
Boeynaems, J.M. *et al*, *Anal. Biochem.*, 1980, **104**, 259 (*synth*)
Corey, E.J. *et al*, *J. Am. Chem. Soc.*, 1980, **102**, 1433 (*synth*)
Rabinovitch, H. *et al*, *Lipids*, 1981, **16**, 518 (*metab*)
Russell, S.W. *et al*, *J. Chem. Soc., Perkin Trans. 1*, 1982, 545 (*synth*)
Corey, E.J. *et al*, *Tetrahedron Lett.*, 1984, **25**, 5115 (*synth*)
Woollard, P.M., *Biochem. Biophys. Res. Commun.*, 1986, **136**, 169.
Just, G. *et al*, *J. Org. Chem.*, 1986, **51**, 4796 (*deriv, synth, pmr, ir, ms*)
Nicolaou, K.C. *et al*, *Synthesis*, 1986, 344 (*synth*)
Yadagiri, P. *et al*, *Tetrahedron Lett.*, 1986, **27**, 6039 (*synth*)
Fruteau de Laclos, B. *et al*, *Prostaglandins*, 1987, **33**, 315 (*synth, ms*)
Taffer, I.M. *et al*, *Tetrahedron Lett.*, 1987, **28**, 6543 (*synth*)
Saniere, M. *et al*, *Angew. Chem., Int. Ed. Engl.*, 1989, **28**, 614 (*synth*)
Piomelli, D. *et al*, *Ann. N.Y. Acad. Sci.*, 1989, **559**, 202 (*rev*)
Nicolaou, K.C. *et al*, *Synthesis*, 1989, 898 (*synth, ir, uv, pmr, bibl*)

14-Hydroxy-5,8,11,15-eicosatetraenoic acid H-00405

14-HETE

[70968-91-1]

$H_3C(CH_2)_3CH{=}CHCH(OH)(CH_2CH{=}CH)_3(CH_2)_3COOH$

$C_{20}H_{32}O_3$ M 320.471

Porter, N.A. *et al*, *J. Org. Chem.*, 1979, **44**, 3177 (*synth, ms*)

15-Hydroxy-5,8,11,13-eicosatetraenoic acid H-00406

15-HETE

[71030-36-9]

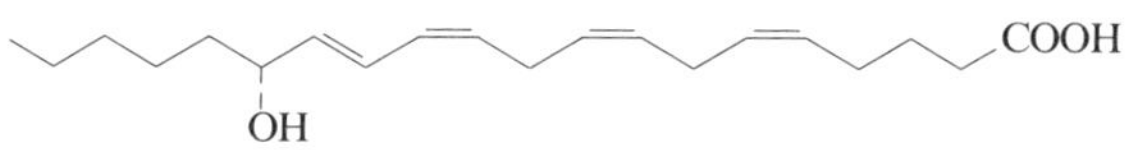

(5*Z*,8*Z*,11*Z*,13*E*,15*R*)-*form*

$C_{20}H_{32}O_3$ M 320.471

(5Z,8Z,11Z,13E,15R)-form [83603-31-0]

Metab. of arachidonic acid.

(5Z,8Z,11Z,13E,15S)-form [54845-95-3]

Metab. of arachidonic acid.

(5Z,8Z,11Z,13E)-form

15-Oxo: [81416-72-0]. *15-Oxo-5,8,11,13-eicosatetraenoic acid. 15-Oxo-ETE. 15-KETE*

$C_{20}H_{30}O_3$ M 318.455

Inhibits 5- and 12-lipoxygenase.

[80574-96-5, 118013-72-2]

Baldwin, J.E. *et al*, *J. Chem. Soc., Perkin Trans. 1*, 1979, 115 (*synth*)
Porter, N.A. *et al*, *J. Org. Chem.*, 1979, **44**, 3177 (*synth, ms*)
Boeynaems, J.M. *et al*, *Anal. Biochem.*, 1980, **104**, 259 (*synth*)
Corey, E.J. *et al*, *J. Am. Chem. Soc.*, 1980, **102**, 1433 (*synth*)
Funk, M.O. *et al*, *Lipids*, 1980, **15**, 1051 (*synth, pmr, cmr*)
Rabinovitch, H. *et al*, *Lipids*, 1981, **16**, 158 (*metab*)
Atrache, V. *et al*, *Tetrahedron Lett.*, 1981, **22**, 13443 (*synth*)
Vanderhoek, J.Y. *et al*, *Biochem. Pharmacol.*, 1982, **31**, 3463 (*biochem*)
Claeys, M. *et al*, *Biochim. Biophys. Acta*, 1982, **713**, 160 (*biosynth*)
Nicolaou, K.C. *et al*, *J. Chem. Soc., Chem. Commun.*, 1985, 1580 (*synth*)
Haviv, F. *et al*, *J. Med. Chem.*, 1987, **30**, 254 (*synth*)

19-Hydroxy-5,8,11,14-eicosatetraenoic acid H-00407

19-HETE

$H_3CCH(OH)CH_2CH_2(CH_2CH{=}CH)_4(CH_2)_3COOH$

$C_{20}H_{32}O_3$ M 320.471
Metab. produced during enzymatic NADPH dependent oxidation of arachidonic acid.

Manna, S. *et al*, *Tetrahedron Lett.*, 1983, **24**, 33 (*synth, bibl*)

20-Hydroxy-5,8,11,14-eicosatetraenoic acid H-00408

20-Hydroxyarachidonic acid. 20-HETE

[41376-14-1]

$HOCH_2(CH_2)_4(CH{=}CHCH_2)_4CH_2CH_2COOH$

$C_{20}H_{32}O_3$ M 320.471
Metab. produced during NADPH dependent enzymatic oxidation of arachidonic acid.

Manna, S. *et al*, *Tetrahedron Lett.*, 1983, **24**, 33 (*synth, bibl*)
Corey, E.J. *et al*, *Tetrahedron Lett.*, 1983, **24**, 37 (*ir, pmr*)

14-Hydroxy-11-eicosenoic acid H-00409

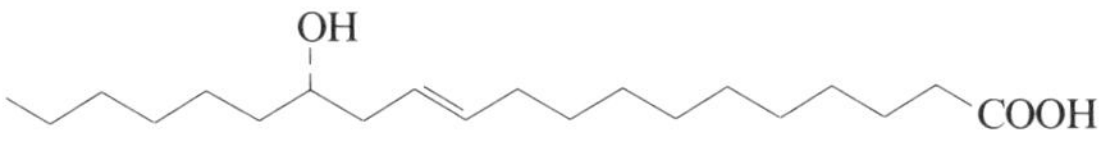

(*R,E*)*-form*

$C_{20}H_{38}O_3$ M 326.518

(R,E)-form

Obt. by isomerisation of the (*Z*)-form.

Me ester: [7706-02-7].
$C_{21}H_{40}O_3$ M 340.545
Cryst. (Me_2CO aq.). Mp 47.5-48.5°. $[\alpha]_D$ +14° (c, 0.4 in cyclohexane).

(R,Z)-form [4103-20-2] ***Lesquerolic acid***

Constit. of *Lesquerella* spp. seed oils. $[\alpha]_D$ +6° (c, 1.6 in $CHCl_3$).

Me ester: [4102-96-9].
$[\alpha]_D$ +10° (c, 1.7 in cyclohexane).

Applewhite, T.H., *Tetrahedron Lett.*, 1965, 3391 (*synth*)
Applewhite, T.H. *et al*, *J. Org. Chem.*, 1967, **32**, 1173 (*config*)

8-Hydroxy-11,12-epoxy-5,9,14-eicosatrienoic acid H-00410

Hepoxilin A. Hepoxilin A_3

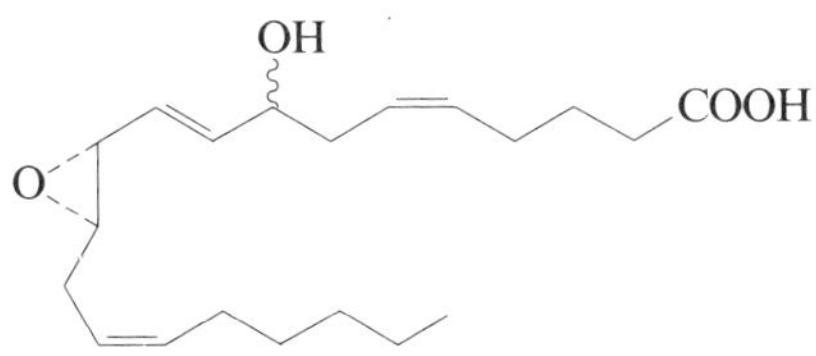

$C_{20}H_{32}O_4$ M 336.470

(5Z,9E,11S,12S,14Z)-form

Metab. of arachidonic acid also prod. in pancreatic islet cells. 8-Epimers sepd. but no phys. props. given.

Me ester: Methyl 8-hydroxy-10-[3-(2-octenyl)oxiranyl]-5,9-decadienoate. Hepoxilin A_3 methyl ester
$C_{21}H_{34}O_4$ M 350.497
Two diastereoisomers sepd. by flash chromatog.

Pace-Asciak, C.R. *et al*, *J. Biol. Chem.*, 1983, **258**, 6835 (*isol*)
Corey, E.J. *et al*, *Tetrahedron Lett.*, 1984, **25**, 5119 (*synth, pmr*)
Chabert, P. *et al*, *Tetrahedron Lett.*, 1989, **30**, 2545 (*synth*)

10-Hydroxy-11,12-epoxy-5,8,14-eicosatrienoic acid H-00411

10-Hydroxy-10-[3-(2-octenyl)oxiranyl]-5,8-decadienoic acid. Hepoxilin B. Hepoxilin B_3

[94161-10-1]

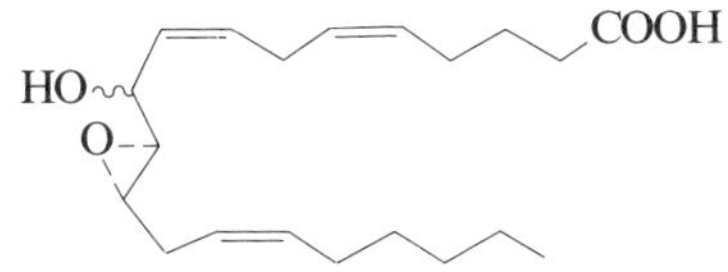

$C_{20}H_{32}O_4$ M 336.470
Metab. of arachidonic acid in mammalian blood platelets.

(5Z,8Z,10R,11S,12S,14Z)-form

Me ester: $[\alpha]_D$ −68.2° (c, 0.5 in Me_2CO).

(5Z,8Z,10S,11S,12S,14Z)-form

Me ester: [89461-47-2]. *Hepoxilin B_3 methyl ester*
$[\alpha]_D$ +34.1° (c, 0.5 in Me_2CO).

[89461-48-3, 89461-49-4]

Pace-Asciak, C.R. *et al*, *J. Biol. Chem.*, 1983, **258**, 6835 (*synth*)
Corey, E.J. *et al*, *Tetrahedron Lett.*, 1983, **24**, 4913 (*synth*)
Wu, W.-L. *et al*, *J. Chem. Soc., Perkin Trans. 1*, 1992, 2705 (*synth, ir, pmr, ms*)
Wu, W.-L. *et al*, *J. Org. Chem.*, 1993, **58**, 2760 (*synth, ir, pmr, ms*)

11-Hydroxy-12,13-epoxy-9,15-octadecadienoic acid H-00412

11-Hydroxy-1-[3-(2-pentenyl)oxiranyl]-9-undecenoic acid, 9CI

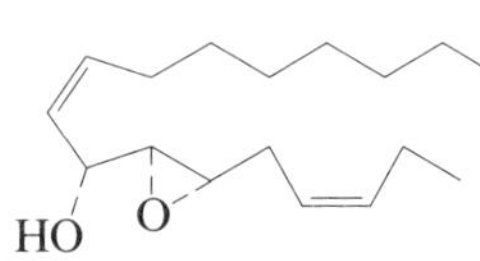

(9*Z*,11*R*,12*S*,13*S*,15*Z*)*-form*

$C_{18}H_{30}O_4$ M 310.433

(9Z,11R,12S,13S,15Z)-form [106034-49-5]

Isol. from rice plants with rice blast disease (*Pyricularia oryzae*).

O-(p-*Bromophenacyl*)*:* $[\alpha]_D$ +26.0° (c, 0.493 in $CHCl_3$).

(9Z,11S,12S,13S,12Z)-form [105977-35-3]

Isol. from rice plants with rice blast disease (*P. oryzae*).

O-(p-*Bromophenacyl*)*:* $[\alpha]_D$ −41.2° (c, 0.131 in $CHCl_3$).

[106034-56-4, 106034-57-5]

Kato, T. *et al*, *J. Chem. Soc., Chem. Commun.*, 1986, 743 (*isol, ms, pmr, abs config*)

5-Hydroxy-6-(glycinylcysteinyl)-7,9-eicosadienoic acid H-00413

N-[S-[*1-(4-Carboxy-1-hydroxybutyl)-2,4-pentadecadienyl*] *cysteinyl*]*glycine, 9CI*

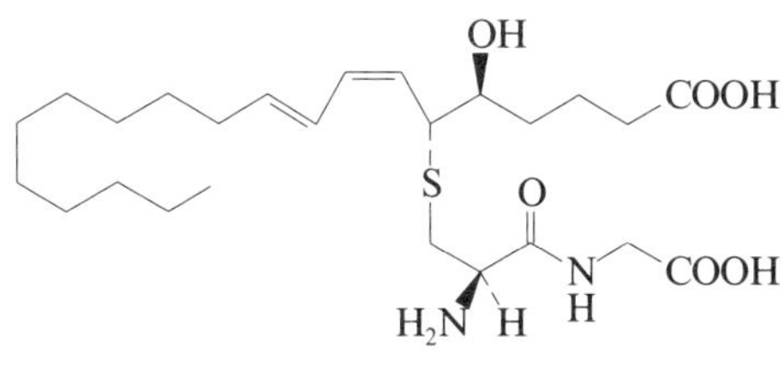

(7*Z*,9*E*)*-form*

$C_{25}H_{44}N_2O_6S$ M 500.698

(7Z,9E)-form

Fine powder. $[\alpha]_D^{20}$ +107° (c, 0.003 in MeOH).

(7E,9E)-form [146610-26-6]
Fine powder. $[\alpha]_D^{20}$ −23° (c, 0.0012 in MeOH).

Klotz, P. *et al*, *J. Org. Chem.*, 1993, **58**, 1076 (*synth, pmr, ms, uv*)

2-Hydroxyheneicosanoic acid H-00414

[93361-62-7]

$H_3C(CH_2)_{18}CH(OH)COOH$

$C_{21}H_{42}O_3$ M 342.561
Constit. of various microalgae. Also found as constit. of phospholipids and sphingolipids.

Matsumoto, G.I. *et al*, *Phytochemistry*, 1984, **23**, 1421 (*occur*)
Carballeria, N.M. *et al*, *Lipids*, 1989, **24**, 229 (*occur*)
Hayashi, A. *et al*, *Biochim. Biophys. Acta*, 1991, **1083**, 179 (*occur*)
Yasugi, E. *et al*, *J. Biochem.* (*Tokyo*), 1991, **110**, 202.

14-Hydroxyheneicosanoic acid, 9CI H-00415

[97534-12-8]

$H_3C(CH_2)_6CH(OH)(CH_2)_{12}COOH$

$C_{21}H_{42}O_3$ M 342.561
Constit. of the rhizomes of *Panax pseudo-ginseng*. Mp 84°.

Shukla, Y.N. *et al*, *Phytochemistry*, 1985, **24**, 1091 (*isol*)

15-Hydroxyheneicosanoic acid, 9CI H-00416

[79097-24-8]

$H_3C(CH_2)_5CH(OH)(CH_2)_{13}COOH$

$C_{21}H_{42}O_3$ M 342.561
Constit. of the leaves and stems of *Hyoscyamus muticus*. Cryst. (C_6H_6/hexane). Mp 80-82°.

Goswami, A. *et al*, *Phytochemistry*, 1981, **20**, 1315 (*isol*)

8-Hydroxy-14,16-hentriacontanedione H-00417

[10368-07-7]

$H_3C(CH_2)_{14}COCH_2CO(CH_2)_5CH(OH)(CH_2)_6CH_3$

$C_{31}H_{60}O_3$ M 480.813
Isol. from leaf-surface wax of wheat *Triticum compactum* var. Little Club. Obt. as a mixt. with the 9-hydroxy compd. Mp 69.5-70.5°.

Tulloch, A.P. *et al*, *Can. J. Chem.*, 1969, **47**, 3119 (*isol, uv, pmr*)

9-Hydroxy-3-hentriacontanone, 9CI H-00418

[132923-41-2]

$H_3C(CH_2)_{21}CH(OH)(CH_2)_5COCH_2CH_3$

$C_{31}H_{62}O_2$ M 466.830
Isol. from bulbs of *Crinum augustum*. Mp 76-78°.

Abd El-Hafiz, M.A., *Phytochemistry*, 1990, **29**, 3936 (*isol, ir, ms, pmr*)

13-Hydroxy-16-hentriacontanone H-00419

[64118-77-0]

$H_3C(CH_2)_{11}CH(OH)CH_2CH_2CO(CH_2)_{14}CH_3$

$C_{31}H_{62}O_2$ M 466.830
Constit. of *Neolitsea sericea*. Mp 83°.

Takaoka, D. *et al*, *Phytochemistry*, 1977, **16**, 1042 (*isol, ir, pmr*)

10-Hydroxy-2,8-heptadecadiene-4,6-diyn-1-al H-00420

$H_3C(CH_2)_6CH(OH)CH{=}CHC{\equiv}CC{\equiv}CCH{=}CHCHO$

$C_{17}H_{22}O_2$ M 258.360
***(E,E)*-form** [20311-70-0]
Constit. of *Oenanthe crocata*. Oil.

Bohlmann, F. *et al*, *Chem. Ber.*, 1968, **101**, 1163 (*isol*); 1971, **104**, 1362 (*synth*)

17-Hydroxy-7,9-heptadecadiene-11,13-diyn-4-one H-00421

$HOCH_2CH_2CH_2C{\equiv}CC{\equiv}CCH{=}CHCH{=}CH$
$CH_2CH_2COCH_2CH_2CH_3$

$C_{17}H_{22}O_2$ M 258.360
***(E,E)*-form** [20311-76-6]
Constit. of *Oenanthe crocata* and *Cicuta virosa*. Cryst. (Et_2O/pet. ether). Mp 35.5°.

Bohlmann, F. *et al*, *Chem. Ber.*, 1968, **101**, 1163; 1969, **102**, 3293.

17-Hydroxy-7,9-heptadecadiene-11,13-diyn-6-one H-00422

$HOCH_2CH_2CH_2C{\equiv}CC{\equiv}CCH{=}CHCH{=}CHCO(CH_2)_4CH_3$

$C_{17}H_{22}O_2$ M 258.360
***(E,E)*-form** [25525-17-1]
Constit. of the roots and rhizomes of *Cicuta virosa*. Oil.

Bohlmann, F. *et al*, *Chem. Ber.*, 1969, **102**, 3293.

2-Hydroxy-6-(8,10-heptadecadienyl)benzoic acid H-00423

Pelandjauic acid. Pentaspadonic acid

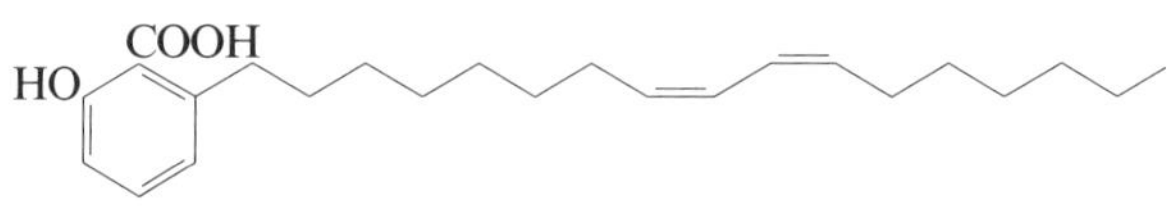

$C_{24}H_{36}O_3$ M 372.547
Isol. from oils of *Pentaspadon officinalis* and *P. motleyi*. Cryst. (Et_2O). Mp 23-26°.
Dihydro: 2-Hydroxy-6-(8-heptadecenyl)benzoic acid. 6-(8-Heptadecenyl)salicylic acid
$C_{24}H_{38}O_3$ M 374.562
Minor constit. of Pelandjauic acid from *P.* spp.

Lamberton, J.A., *Aust. J. Chem.*, 1959, **12**, 234.

7-Hydroxy-10,16-heptadecadien-8-ynoic acid H-00424

$H_2C{=}CH(CH_2)_4CH{=}CHC{\equiv}CCH(OH)(CH_2)_5COOH$

$C_{17}H_{26}O_3$ M 278.391
***(E)*-form**
Constit. of the oil of *Acanthosyris spinescens*. Probably opt. active (see ester).
Me ester: Oil. $[\alpha]_D^{27}$ −3° (c, 3.7 in cyclohexane).

Powell, R.G. *et al*, *Chem. Ind.* (*London*), 1965, 470.
Powell, R.G. *et al*, *J. Org. Chem.*, 1966, **31**, 528 (*uv*)
Kleiman, R. *et al*, *J. Am. Oil Chem. Soc.*, 1973, **50**, 31 (*ms*)

8-Hydroxy-1,9,16-heptadecatriene-4,6-diyne-3-one, 9CI H-00425

$H_2C{=}CH(CH_2)_5CH{=}CHCH(OH)$
$C{\equiv}CC{\equiv}CCOCH{=}CH_2$

$C_{17}H_{20}O_2$ M 256.344

(Z)-form [13894-98-9] ***Dehydrofalcarinolone.***
Dehydrofalcarinonol
Constit. of *Artemisia* spp. Unstable.
Ac: [70387-29-0].
Isol. from *Nidorella agria.* Oil. $[\alpha]_D^{24}$ +86.8° (c, 0.8). λ_{max} 289, 271, 258 nm (Et_2O).

Bohlmann, F. *et al, Chem. Ber.*, 1966, **99**, 3552 (*isol, struct, uv, ir*)
Bohlmann, F. *et al, Phytochemistry*, 1978, **17**, 1769; 1979, **18**, 334 (*isol*)
Greger, H., *Phytochemistry*, 1979, **18**, 1319 (*isol*)

12-Hydroxy-5,8,10-heptadecatrienoic acid H-00426
12-HHT
[50683-78-8]

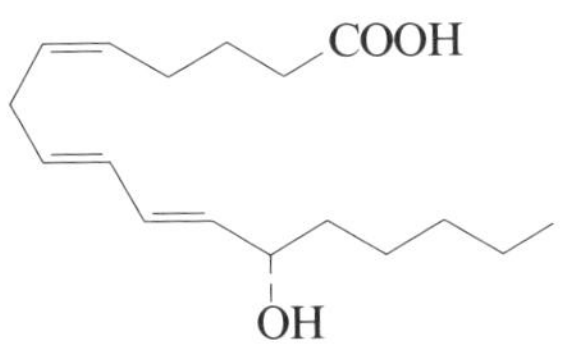

(*S*)-*form*

$C_{17}H_{28}O_3$ M 280.406
Stimulates prostacyclin production.
(5Z,8E,10E,12S)-form [54397-84-1]
Major metab. of arachidonic acid.
Me ester: [79171-56-5].
$C_{18}H_{30}O_3$ M 294.433
Oil. $[\alpha]_D^{25}$ +7.5° (c, 0.2 in $CHCl_3$). Uv λ_{max} 240 nm (MeOH).
12-Ketone: [103374-38-5]. *12-Oxo-5,8,10-heptadecatrienoic acid. 12-KHT*
Arachidonic acid metab. produced by human leukaemia cells. Formed from keto-PGH_2 by prostacyclin synthase and thromboxane synthase.

[81585-98-0, 81610-84-6, 100679-07-0]

Hamberg, M. *et al, Proc. Natl. Acad. Sci. U.S.A.*, 1974, **71**, 3400, 3824 (*biosynth, isol, uv, ms*)
Russell, S.W. *et al, J. Chem. Soc., Perkin Trans. 1*, 1982, 545 (*synth, pmr, ir, ms*)
Sirois, P. *et al, Prostaglandins*, 1982, **24**, 405 (*biosynth*)
Agins, A.P. *et al, Biochem. Pharmacol.*, 1987, **36**, 1799 (*biosynth*)
Hecker, M. *et al, Eicosanoids*, 1988, **1**, 19 (*biosynth*)
De Montarby, L. *et al, Bull. Soc. Chim. Fr.*, 1989, 419 (*synth, pmr, cmr, ir*)
Nicolaou, K.C. *et al, J. Chem. Soc., Perkin Trans. 1*, 1989, 2131 (*synth, pmr*)
Suzuki, M. *et al, J. Org. Chem.*, 1989, **54**, 5292 (*synth, ir, pmr*)

17-Hydroxy-9-heptadecene-11,13-diyn-4-one, 8CI H-00427

$HOCH_2CH_2CH_2C{\equiv}CC{\equiv}CCH{=}CH(CH_2)_4COCH_2CH_2CH_3$

$C_{17}H_{24}O_2$ M 260.375
(E)-form [20311-77-7]
Constit. of *Oenanthe crocata* and *Cicuta virosa.* Mp 87°.

Bohlmann, F. *et al, Chem. Ber.*, 1968, **101**, 1163; 1969, **102**, 3293.

7-Hydroxy-10-heptadecen-8-ynoic acid, 9CI H-00428

$H_3C(CH_2)_5CH{=}CHC{\equiv}CCH(OH)(CH_2)_5COOH$

$C_{17}H_{28}O_3$ M 280.406
(E)-form
Constit. of the oil of *Acanthosyris spinescens.*

Powell, R.G. *et al, J. Org. Chem.*, 1966, **31**, 528.
Kleiman, R. *et al, J. Am. Oil Chem. Soc.*, 1973, **50**, 31 (*ms*)

6-Hydroxyheptanoic acid H-00429

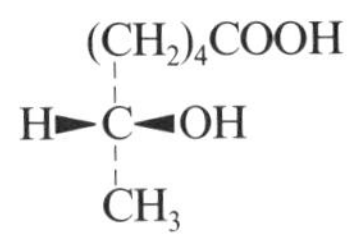

(*R*)-*form*

$C_7H_{14}O_3$ M 146.186
(R)-form
Lactone: [69765-34-0]. *7-Methyl-2-oxepanone, 9CI. 6-Heptanolide. ε-Caprolactone*
$C_7H_{12}O_2$ M 128.171
$[\alpha]_D^{20}$ +25.0° (c, 1.8 in $CHCl_3$).
(±)-form
Lactone: [69854-30-4].
Liq. $Bp_{0.08}$ 46-47°.

[2549-61-3]

Starcher, R.S. *et al, J. Am. Chem. Soc.*, 1958, **80**, 4079 (*synth*)
Pirkle, W.H. *et al, J. Org. Chem.*, 1979, **44**, 2169 (*synth, ir, pmr*)
Fouque, E. *et al, Synthesis*, 1989, 661 (*synth, pmr, cmr, ms*)

7-Hydroxy-2-heptynoic acid H-00430

$HOCH_2CH_2CH_2CH_2C{\equiv}CCOOH$

$C_7H_{10}O_3$ M 142.154
Me ester: [52500-26-2].
$C_8H_{12}O_3$ M 156.181
Liq. $Bp_{0.5}$ 120-122°, $Bp_{0.15}$ 90-95°.
O-Methanesulfonyl, Me ester: Oil. $Bp_{0.25}$ 155°.

Tufariello, J.J. *et al, J. Org. Chem.*, 1974, **39**, 3378 (*synth, ir, pmr*)
Kita, Y. *et al, J. Org. Chem.*, 1991, **56**, 119 (*synth, ir, pmr*)

7-Hydroxy-5-heptynoic acid H-00431
[41300-59-8]

$HOCH_2C{\equiv}C(CH_2)_3COOH$

$C_7H_{10}O_3$ M 142.154
Intermed. in prostagladin synth.
Me ester: [50781-91-4].
$C_8H_{12}O_3$ M 156.181
Oil. $Bp_{0.05}$ 150°.
Nitrile: [69285-47-8]. *6-Cyano-2-hexyn-1-ol*
C_7H_9NO M 123.154
Oil.

Corey, E.J. *et al, J. Am. Chem. Soc.*, 1973, **95**, 8483 (*synth*)
Ger. Pat., 2 313 868, (1974); *CA*, **80**, 59566u.
Casy, G. *et al, Tetrahedron*, 1986, **42**, 5849 (*synth, ir, pmr*)
Belyk, K. *et al, J. Org. Chem.*, 1992, **57**, 4070 (*nitrile*)

2-Hydroxyhexacosanoic acid H-00432

[14176-13-7]

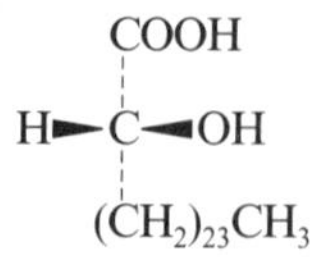

$C_{26}H_{52}O_3$ M 412.695

(R)-form [26631-99-2]

Major component of yeast cerebrin. Also found in brain cerebrosides. Mp 104-105°. $[\alpha]_D^{20}$ +2.1° (c, 2.8 in Py).

Ac:

$C_{28}H_{54}O_4$ M 454.732

Cryst. (Me_2CO/pet. ether). Mp 84.84.5°.

[72741-91-4]

Chibnall, A.C. *et al, Biochem. J.*, 1953, **55**, 711 (*isol*)
Hammarström, S. *et al, FEBS Lett.*, 1969, **5**, 192 (*resoln*)
Nurminen, T. *et al, Biochem. J.*, 1971, **125**, 963 (*isol*)
Muralidharan, F.N. *et al, Chem. Phys. Lipids*, 1984, **34**, 257 (*synth*)

26-Hydroxyhexacosanoic acid, 9CI H-00433

26-Hydroxycerotic acid

[506-47-8]

$HOCH_2(CH_2)_{24}COOH$

$C_{26}H_{52}O_3$ M 412.695

Isol. from Carnauba wax.

Me ester: [79162-71-3].

$C_{27}H_{54}O_3$ M 426.722

Mp 82.5-83.0°.

Ac, Me ester:

$C_{29}H_{56}O_4$ M 468.759

Mp 68°.

Murry, K.E. *et al, Aust. J. Chem.*, 1955, **8**, 437 (*isol*)
Abrams, S.R., *Chem. Phys. Lipids*, 1981, **28**, 379 (*synth*)
Ekman, R. *et al, Phytochemistry*, 1982, **21**, 121 (*isol*)

2-Hydroxyhexadecanoic acid, 9CI H-00434

2-Hydroxypalmitic acid

[764-67-0]

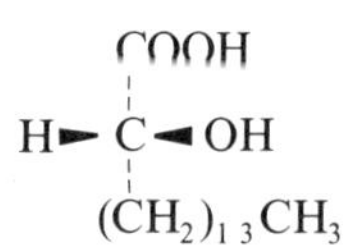

(*R*)-*form*

$C_{16}H_{32}O_3$ M 272.427

(R)-form [16452-51-0]

Occurs in wool fat. Metab. of *Hansenula sydowiorum*. Mp 93.3-93.6°. $[\alpha]_D$ −1° (EtOH), $[\alpha]_D$ −2.7° ($CHCl_3$).

Me ester: [36451-47-5].

$C_{17}H_{34}O_3$ M 286.454

Mp 45-46°. Bp_5 250-251°, Bp_1 153.5-154°.

(S)-form [16452-52-1]

Mp 93.3-93.6°. $[\alpha]_D^{26}$ +2.7° ($CHCl_3$).

(±)-form [10067-06-8]

Mp 86-87° (82-83°).

Me ester: [16742-51-1].

Mp 59-60°.

Et ester:

$C_{18}H_{36}O_3$ M 300.481

Mp 55.5-56.5°.

Amide:

$C_{16}H_{33}NO_2$ M 271.442

Mp 150°.

Levene, P.A. *et al, J. Biol. Chem.*, 1914, **18**, 466.
Kuwata, T., *J. Am. Chem. Soc.*, 1938, **60**, 559.
Horn, D.H.S. *et al, J. Chem. Soc.*, 1954, 1460.
Eglinton, G. *et al, Org. Mass Spectrom.*, 1968, **1**, 593 (*ms*)
Annett, R.G. *et al, Anal. Biochem.*, 1972, **47**, 638 (*resoln*)
Sokolov, V.P. *et al, Zh. Obshch. Khim.*, 1972, **42**, 2119 (*ir*)

3-Hydroxyhexadecanoic acid H-00435

3-Hydroxypalmitic acid

[2398-34-7]

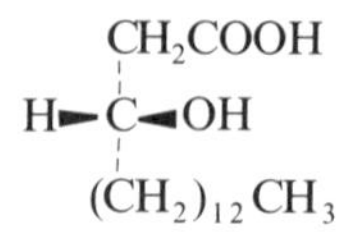

(*R*)-*form*

$C_{16}H_{32}O_3$ M 272.427

(R)-form [20595-04-4]

Present in extracellular glycolipids of *Rhodotorula* spp. Cryst. (Me_2CO). Mp 78-79°. $[\alpha]_D^{22}$ −19.2° (c, 1.18 in $CHCl_3$), $[\alpha]_D^{20}$ −13.8° (c, 1 in $CHCl_3$).

Me ester: [51883-36-4].

$C_{17}H_{34}O_3$ M 286.454

Cryst. (Me_2CO). Mp 48-49°. $[\alpha]_D$ −13.8° (c, 5.6 in $CHCl_3$), $[\alpha]_D$ +2.1° (c, 6.8 in MeOH).

(±)-form [928-17-6]

Cryst. ($CHCl_3$). Mp 84-85°.

Ac:

$C_{18}H_{34}O_4$ M 314.464

Mp 58°.

Robinet, M., *Bull. Soc. Chim. Fr.*, 1931, **40**, 710 (*synth*)
Tulloch, A.P., *Can. J. Chem.*, 1964, **42**, 830 (*isol*)
Davidoff, F. *et al, J. Biol. Chem.*, 1964, **239**, 2496 (*biosynth*)
Ammers, M. *et al, Recl. Trav. Chim. Pays-Bas (J. R. Neth. Chem. Soc.)*, 1964, **83**, 708 (*isol*)
Mayberry, W.R. *et al, J. Bacteriol.*, 1973, **116**, 1091 (*glc, ms*)
Kurtzman, C.P. *et al, Mycologia*, 1974, **66**, 580 (*biosynth*)
Yano, I. *et al, Lipids*, 1976, **11**, 685 (*isol*)
Kellner, B.M.J. *et al, Chem. Phys. Lipids*, 1979, **23**, 41.
Nakahata, M. *et al, Bull. Chem. Soc. Jpn.*, 1982, **55**, 2186 (*synth*)
Utaka, M. *et al, J. Org. Chem.*, 1990, **55**, 3917 (*synth, ir, pmr*)

4-Hydroxyhexadecanoic acid H-00436

4-Hydroxypalmitic acid

$H_3C(CH_2)_{11}CH(OH)CH_2CH_2COOH$

$C_{16}H_{32}O_3$ M 272.427

Cryst. (Me_2CO). Mp 79-79.6°.

γ-Lactone:

$C_{16}H_{30}O_2$ M 254.412

Mp 40.7-41.3°.

Houston, D.F. *et al, J. Am. Chem. Soc.*, 1947, **69**, 517 (*synth*)

5-Hydroxyhexadecanoic acid H-00437

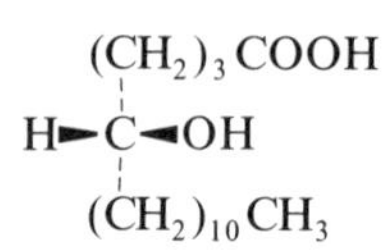

(*R*)-*form*

$C_{16}H_{32}O_3$ M 272.427

(R)-form [96895-71-5]

Needles (Me_2CO/hexane). Mp 67-68°. $[\alpha]_D^{20}$ −1.07° (c, 1.03 in MeOH). Opt. pure.

Ac: [96856-91-6].

$C_{18}H_{34}O_4$ M 314.464

Solid. Mp 31-32°. $[\alpha]_D^{21}$ −2.54° (c, 1.01 in $CHCl_3$).

(S)-form [96895-70-4]

Needles (Et_2O/hexane). Mp 67-68°. $[\alpha]_D^{20}$ +1.07° (c, 1.01 in MeOH).

[96856-90-5]

Mori, K. *et al*, *Tetrahedron*, 1985, **41**, 547 (*synth, ir, pmr*)

8-Hydroxyhexadecanoic acid H-00438

8-Hydroxypalmitic acid

$H_3C(CH_2)_7CH(OH)(CH_2)_6COOH$

$C_{16}H_{32}O_3$ M 272.427

(+)-form

Constit. of the spores of *Lycopodium complanatum* and from fern *Lygodium japonicum*. Spore germination inhibitor. Cryst. (Me_2CO). Mp 78.5-79.5°. $[\alpha]_D$ +0.3° (c, 19.5 in $CHCl_3$).

Me ester:

$C_{17}H_{34}O_3$ M 286.454

Cryst. (pet. ether). Mp 44-45.5°. $[\alpha]_D$ +0.8° (c, 14 in MeOH).

(±)-form

Mp 73.5-74.5°.

Me ester: Mp 45-46°.

Tulloch, A.P., *Can. J. Chem.*, 1965, **43**, 415 (*isol*)
Yamane, H. *et al*, *Agric. Biol. Chem.*, 1980, **44**, 1697 (*isol*)

11-Hydroxyhexadecanoic acid H-00439

11-Hydroxypalmitic acid. Jalapinolic acid. Scammonolic acid. Turpetholic acid E

[502-75-0]

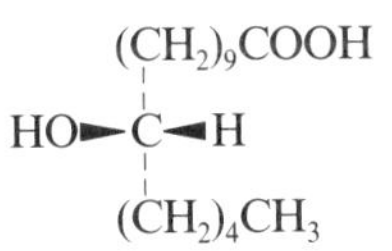

(S)-form

$C_{16}H_{32}O_3$ M 272.427

(S)-form [13717-17-4]

Occurs in jalap and Scammony resins (*Ipomoea orizabensis* and *Convolvulus scammonia*) and other *I.* spp., in glycosidic form. Cryst. (EtOAc). Mp 68-69°. $[\alpha]_D$ +0.79° ($CHCl_3$).

Me ester: [60368-18-5].

$C_{17}H_{34}O_3$ M 286.454

Cryst. (pet. ether or EtOAc). Mp 40.5-41.5°, Mp 47-49°. Bp_3 183-186°.

Et ester:

$C_{18}H_{36}O_3$ M 300.481

Cryst. Mp 47-48°.

Ac:

$C_{18}H_{34}O_4$ M 314.464

Oil. Bp_{50} 224-225°.

O-[*α-L-Rhamnopyranosyl(1→4)-α-L-rhamnopyranoside*]: [68124-11-8]. ***Muricatin B***

$C_{28}H_{52}O_{11}$ M 564.712

Isol. from *I. muricata*. Syrup. $[\alpha]_D^{26}$ −56.6° (c, 0.495 in EtOH)(synthetic).

O-*Glycoside: Jalapinic acid. Scammoninic acid. Scammonic acid*

Isol. from *I. orizabensis* and *C. scammonia*. Hygroscopic solid. Mp 66°. Gives Jalapinolic acid, glucose, rhamnose and rhodeose on hydrol. May not be homogeneous.

11-O-[β-D-Glucopyranosyl-(1→3)-α-L-rhamnopyranosyl-(1→3)-β-D-glucopyranosyl-(1→3)-β-D-glucopyranoside]: [66895-65-6]. ***Turpethinic acid E***

$C_{40}H_{72}O_{22}$ M 904.996

Isol. from *Ipomoea turpethum*.

O-[*α-L-Rhamnopyranosyl(1→?)-α-L-rhamnopyranosyl(1→?)-α-L-rhamnopyranosyl(1→?)-α-L-rhamnopyranosyl(1→?)-α-L-rhamnopyranoside*]: [89382-36-5]. ***Quamoclitic acid***

$C_{46}H_{82}O_{23}$ M 1003.140

Isol. from *Ipomoea quamoclit, I. lacunosa, I. pandurata* and *Convolvulus al-sirensis*. Mp 129-131° (as Me ester). $[\alpha]_D^{22}$ −67.6° (c, 0.93 in H_2O) (Me ester). Struct. not fully determined. Glycosidic linkages are (1→3) or (1→4).

O-[*β-D-Glucopyranosyl-(1→3)[α-L-rhamnnopyranosyl-(1→4)]-α-L-rhamnopyranosyl-(1→4)-α-L-rhamnopyranosyl-(1→2)-β-D-fucopyranoside*]: [120583-62-2]. ***Operculinic acid A***. *Mammoside J*

$C_{46}H_{82}O_{24}$ M 1019.139

Constit. of the roots of *I. operculata*. Powder (MeOH aq.). Mp 172-174° dec. $[\alpha]_D^{18}$ −63.6° (c, 1 in MeOH).

O-[*β-D-Glucopyranosyl-(1→3)[α-L-rhamnopyranosyl-(1→4)]-α-L-rhamnopyranosyl-(1→4)-α-L-rhamnopyranosyl-(1→2)-β-D-glucopyranoside*]: [126572-20-1]. ***Operculinic acid B***

$C_{46}H_{82}O_{25}$ M 1035.139

Constit. of the roots of *I. operculata*. Powder (MeOH aq.). Mp 169-171° dec. $[\alpha]_D^{24}$ −80.9° (c, 2.6 in MeOH).

O-[*α-L-Rhamnopyranosyl-(1→4)-α-L-rhamnopyranosyl-(1→4)-α-L-rhamnopyranosyl-(1→2)-β-D-fucopyranoside*]: [126642-32-8]. ***Operculinic acid C***

$C_{40}H_{72}O_{19}$ M 856.997

Constit. of the roots of *I. operculata*. Powder (MeOH aq.). Mp 116-120°. $[\alpha]_D^{18}$ −84.0° (c, 1.9 in MeOH).

O-[*β-D-Glucopyranosyl-(1→3)[α-L-rhamnopyranosyl-(1→4)]-α-L-rhamnopyranosyl-(1→4)-α-L-rhamnopyranosyl-(1→2)-β-D-xylopyranoside*]: [126465-32-5]. ***Operculinic acid D***

$C_{45}H_{80}O_{24}$ M 1005.113

Constit. of the roots of *I. operculata*. Powder (MeOH aq.). Mp 175-177° dec. $[\alpha]_D^{16}$ −88.6° (c, 0.7 in MeOH).

O-[*α-L-Rhamnopyranosyl-(1→4)-α-L-rhamnopyranosyl-(1→4)-α-L-rhamnopyranosyl-(1→2)-β-D-glucopyranoside*]: [126465-33-6]. ***Operculinic acid E***

$C_{40}H_{72}O_{20}$ M 872.997

Constit. of the roots of *I. operculata*. Powder. Mp 114-116° dec. $[\alpha]_D^{18}$ −76.8° (c, 1.6 in MeOH).

O-[*α-L-Rhamnopyranosyl-(1→4)-α-L-rhamnopyranosyl-(1→4)-α-L-rhamnopyranosyl-(1→2)-β-D-xylopyranoside*]: [132333-81-4]. ***Operculinic acid F***

$C_{39}H_{70}O_{19}$ M 842.971

Constit. of the roots of *I. operculata*. Powder. Mp 104-109° dec. $[\alpha]_D^{32}$ −97.9° (c, 2.2 in MeOH).

O-[2-O-*Methyl-β-D-glucopyranosyl-(1→3)[α-L-rhamnopyranosyl-(1→4)]-α-L-rhamnopyranosyl-(1→4)-α-L-rhamnopyranosyl-(1→2)-β-D-fucopyranoside*]: [132333-82-5]. ***Operculinic acid G***

$C_{47}H_{84}O_{24}$ M 1033.166

Constit. of the roots of *I. operculata*. Powder. Mp 110-114° dec. $[\alpha]_D^{24}$ −62.0° (c, 1.3 in MeOH).

(±)-form

Cryst. (EtOAc). Mp 68-69°.

Me ester: Cryst. (pet. ether). Mp 40.5-41.5°. Bp_3 183-186°.

[114926-99-7, 121043-05-8, 121153-17-1]

Power, F.B. *et al*, *J. Chem. Soc.*, 1912, **101**, 398 (*isol*)
Davies, C.A. *et al*, *J. Am. Chem. Soc.*, 1928, **50**, 1754 (*synth*)
Bauer, K.H. *et al*, *Arch. Pharm. (Weinheim, Ger.)*, 1934, **272**, 841 (*Jalapinic acid*)
Smith, C.R. *et al*, *Phytochemistry*, 1964, **3**, 289 (*isol*)
Khanna, S.N. *et al*, *Phytochemistry*, 1967, **6**, 735.
Tulloch, A.P., *Phytochemistry*, 1976, **15**, 1153 (*isol*)
Wagner, H. *et al*, *Phytochemistry*, 1977, **16**, 715 (*glc, ms*)
Liptak, A. *et al*, *Phytochemistry*, 1978, **17**, 997 (*Muricatin B*)
Wagner, H. *et al*, *Planta Med.*, 1978, **33**, 144 (*Turpethinic acid E*)
Wagner, H. *et al*, *Planta Med.*, 1983, **49**, 154 (*isol, ms*)

Shibuya, H. *et al*, *Chem. Pharm. Bull.*, 1989, **37**, 260 (*synth, abs config*)
Ono, M. *et al*, *Chem. Pharm. Bull.*, 1989, **37**, 3209; 1990, **38**, 2650 (*Operculinic acids*)

14-Hydroxyhexadecanoic acid — H-00440

14-Hydroxypalmitic acid

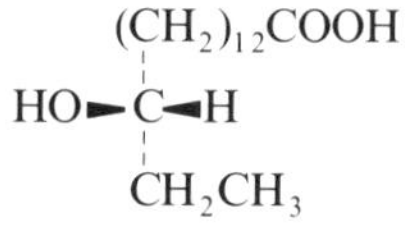

(*S*)-*form*

$C_{16}H_{32}O_3$ M 272.427

(*S*)-*form* [34370-57-5]

Me ester:
$C_{17}H_{34}O_3$ M 286.454
Constit. of beeswax. Cryst. (hexane). Mp 47-48°.

Lactone: 14-Ethyltetradecanolide
$C_{16}H_{30}O_2$ M 254.412
Liq. with pleasant, musky odour. Bp_3 140-150°.

Toyama, Y. *et al*, *Fette, Seifen, Anstrichm.*, 1951, **53**, 556.
Tulloch, A.P., *Chem. Phys. Lipids*, 1971, **6**, 235 (*isol, struct*)
Ho, P.P., *Biochim. Biophys. Acta*, 1976, **431**, 249.

15-Hydroxyhexadecanoic acid — H-00441

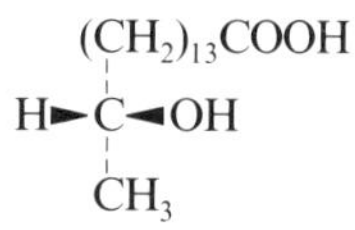

(*R*)-*form*

$C_{16}H_{32}O_3$ M 272.427

(*R*)-*form*

Lactone: [69297-56-9]. *15-Hexadecanolide. 16-Methyloxacyclohexadecan-2-one, 9CI*
$C_{16}H_{30}O_2$ M 254.412
Constit. of Galbanum absolute (from *Ferula galbaniflua* and *F. rubicaulis*). $[\alpha]_D^{20}$ −16.5° (c, 1.03 in $CHCl_3$).

(*S*)-*form*

Lactone: Oil. $Bp_{0.05}$ 110-115°. $[\alpha]_D^{20}$ +2.41° (c, 1.66 in $CHCl_3$).

(±)-*form*

Lactone: $Bp_{0.06}$ 130°.

Becker, J. *et al*, *Helv. Chim. Acta*, 1971, **54**, 2889 (*synth*)
Kaiser, R. *et al*, *Helv. Chim. Acta*, 1978, **61**, 2671 (*isol, synth*)
Kostova, K. *et al*, *Helv. Chim. Acta*, 1984, **67**, 1713 (*synth, pmr, cmr*)
Stanchev, S. *et al*, *Helv. Chim. Acta*, 1989, **72**, 1052 (*synth, abs config, bibl*)

16-Hydroxyhexadecanoic acid, 9CI — H-00442

16-Hydroxypalmitic acid. Juniperic acid. Juniperinic acid
[506-13-8]

$HOCH_2(CH_2)_{14}COOH$

$C_{16}H_{32}O_3$ M 272.427
Isol. from many conifer waxes, eg. *Juniperus*, *Picea* spp. and other plant waxes. Cryst. (C_6H_6/Et_2O). Mp 95°.

Ac: [66146-69-8].
$C_{18}H_{34}O_4$ M 314.464
Leaflets (EtOH aq.). Mp 63°. Bp_2 215-218°.

Me ester: [36575-67-4].
$C_{17}H_{34}O_3$ M 286.454
Cryst. (pet. ether/EtOH). Mp 55-55.5°. Bp_2 194-196°.

Lactone: [109-29-5]. *Oxacycloheptadecan-2-one, 9CI. Dihydroambrettolide*
$C_{16}H_{30}O_2$ M 254.412
Cryst. (EtOH) with musky odour. Mp 33-34°. Bp_{15} 188°.
▷ RN9730000.

Kerschbaum, *Ber.*, 1927, **60**, 902.
Jeger, O. *et al*, *Helv. Chim. Acta*, 1946, **29**, 684 (*synth*)
Sabnis, S.D. *et al*, *J. Chem. Soc.*, 1965, 4580 (*synth*)
Rohwedder, W.K., *Lipids*, 1971, **6**, 906 (*ms*)
Caldicott, A.B. *et al*, *Phytochemistry*, 1975, **14**, 2223 (*glc, ms*)
Agrawal, V.P. *et al*, *Arch. Biochem. Biophys.*, 1978, **191**, 452.
Espelie, K.E. *et al*, *Plant Physiol.*, 1979, **63**, 433 (*isol*)
Majee, R.N. *et al*, *Chem. Ind. (London)*, 1983, 43 (*synth*)
Steliou, K. *et al*, *J. Am. Chem. Soc.*, 1983, **105**, 7130 (*lactone*)
Lewis, R.J., *Sax's Dangerous Properties of Industrial Materials*, 8th Ed., Van Nostrand-Reinhold, 1992, OKU000.

2-(2-Hydroxyhexadecanoylamino)-1,3-octadecanediol — H-00443

*2-Hydroxy-*N*-[2-hydroxy-1-(hydroxymethyl)heptadecanyl] hexadecanamide*

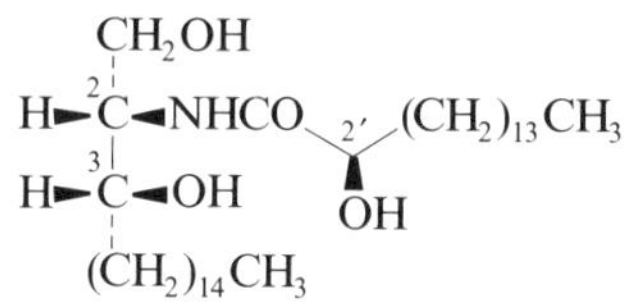

$C_{34}H_{69}NO_4$ M 555.923

(*2S,3R,2'R*)-*form* [34839-06-0]
N-*2-Hydroxyhexadecanoylsphinganine*
Present in lipids of *Acetobacter* spp. Mp 108-114°.

Hammerström, S., *J. Lipid Res.*, 1971, **12**, 760 (*synth*)
Fujino, Y., *CA*, 1978, **89**, 176022k (*occur*)

2-(2-Hydroxyhexadecanoylamino)-4-octadecene-1,3-diol — H-00444

*2-Hydroxy-*N*-[2-hydroxy-1-(hydroxymethyl)-3-heptadecenyl] hexadecanamide, 9CI*

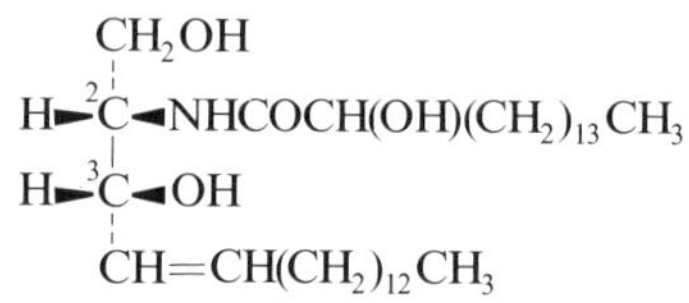

$C_{34}H_{67}NO_4$ M 553.908

(*2S,3R*)-*form*
D-erythro-*form*. N-*2-Hydroxyhexadecanoyl-4-sphingenine*
Mp 100-101°.

(*2RS,3SR*)-*form*
(±)-erythro-*form*
Mp 99-100°.

Hammarström, S., *J. Lipid Res.*, 1971, **12**, 760 (*synth, glc*)
Alpes, H., *Chem. Phys. Lipids*, 1974, **13**, 109 (*synth*)

12-Hydroxy-9-hexadecenoic acid — H-00445

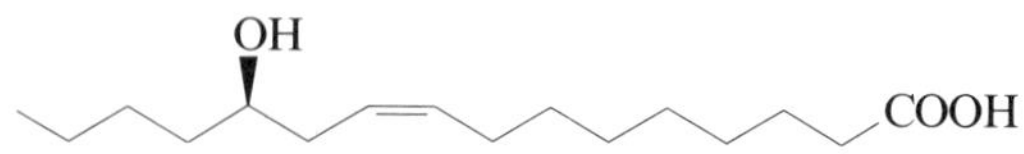

$C_{16}H_{30}O_3$ M 270.411

(*R*)-(*Z*)-*form*
Minor constit. of the seed oil of *Lesquerella densipila*.

Me ester:
$C_{17}H_{32}O_3$ M 284.438
$[\alpha]_D^{26}$ +6.2° (c, 3.6 in MeOH).

Binder, R.G. *et al*, *J. Org. Chem.*, 1966, **31**, 1477.

16-Hydroxy-5-hexadecenoic acid **H-00446**

Δ^5-Isoambrettolic acid

[4444-15-9]

$$HOCH_2(CH_2)_9CH{=}CH(CH_2)_3COOH$$

$C_{16}H_{30}O_3$ M 270.411

Cryst. (C_6H_6). Mp 61-62°.

Formyl:

$C_{17}H_{30}O_4$ M 298.422

Cryst. (pentane). Mp 43-44°.

Lactone: Δ^5-Isoambrettolide

$C_{16}H_{28}O_2$ M 252.396

Bp_2 143°.

Collaud, C., *Helv. Chim. Acta*, 1943, **26**, 849.

16-Hydroxy-6-hexadecenoic acid **H-00447**

Δ^6-Isoambrettolic acid

[4444-16-0]

$$HOCH_2(CH_2)_8CH{=}CH(CH_2)_4COOH$$

$C_{16}H_{30}O_3$ M 270.411

Cryst. (C_6H_6). Mp 77-77.5°.

Formyl:

$C_{17}H_{30}O_4$ M 298.422

Cryst. (pentane). Mp 53-54°.

Lactone: Δ^6-Isoambrettolide

$C_{16}H_{28}O_2$ M 252.396

Mp 28.5°. $Bp_{0.15}$ 117°.

Collaud, C., *Helv. Chim. Acta*, 1943, **26**, 849 (*synth*)

16-Hydroxy-7-hexadecenoic acid **H-00448**

Ambrettolic acid

[506-14-9]

$$HOCH_2(CH_2)_7CH{=}CH(CH_2)_5COOH$$

$C_{16}H_{30}O_3$ M 270.411

Obt. from musk seed oils and *Satureia calamintha* plant oils. Mp 53-55° (α-form), Mp 26-27° (β-form).

Lactone: see *7-Hexadecen-16-olide, H-00267*

Stoll, M. *et al*, *Helv. Chim. Acta*, 1934, **17**, 1609.
Baudart, P., *C. R. Hebd. Seances Acad. Sci.*, 1945, **221**, 205 (*synth*)
Gupta, Y.N. *et al*, *Indian Perfum.*, 1969, **13**, 26 (*isol*)

16-Hydroxy-9-hexadecenoic acid **H-00449**

Δ^9-Isoambrettolic acid

[17278-80-7]

$$HOCH_2(CH_2)_5CH{=}CH(CH_2)_7COOH$$

$C_{16}H_{30}O_3$ M 270.411

(E)-form [18951-79-6]

Cryst. in 2 forms. Mp 55-55.5°, Mp 70° (dimorph.).

Et ester:

$C_{18}H_{34}O_3$ M 298.465

Cryst. (Et_2O). Mp 50-55°. Bp_4 198-200°.

Ac:

$C_{18}H_{32}O_4$ M 312.448

Bp_2 190°.

Lactone: Δ^9-Isoambrettolide

$C_{16}H_{28}O_2$ M 252.396

$Bp_{0.7}$ 131°.

(Z)-form [1619-68-7]

Minor constit. of shellac; fermentation prod. of *Torulopsis apicola*. Mp 17-19°.

Hunsdiecker, H., *Naturwissenschaften*, 1942, **30**, 587 (*synth*)
Christie, W.W. *et al*, *J. Chem. Soc.*, 1964, 5833 (*isol*)
Ames, D.E. *et al*, *J. Chem. Soc. C*, 1968, 268.
Tulloch, A.P. *et al*, *Can. J. Chem.*, 1968, **46**, 1523 (*isol*)
Hevesi, L. *et al*, *Bull. Soc. Chim. Belg.*, 1975, **84**, 709 (*synth*)
Singh, A.N. *et al*, *Tetrahedron*, 1978, **34**, 595 (*synth*)
Majee, R.N. *et al*, *Chem. Ind. (London)*, 1983, 43 (*synth*)

1-Hydroxy-10-hexadecen-7-one, 9CI **H-00450**

Disparolone

$$H_3C(CH_2)_4CH{=}CHCH_2CH_2CO(CH_2)_5CH_2OH$$

$C_{16}H_{30}O_2$ M 254.412

(Z)-form [49831-68-7]

Tumour inhibitor from *Portheria dispar*. Waxy solid. Mp 25°.

Jacobson, M. *et al*, *Experientia*, 1973, **29**, 783.

(15-Hydroxyhexadecyl) methylenebutanedioic acid, 9CI **H-00451**

18-Hydroxy-1-nonadecene-2,3-dicarboxylic acid. α-(15-Hydroxyhexyl)itaconic acid

[62722-96-7]

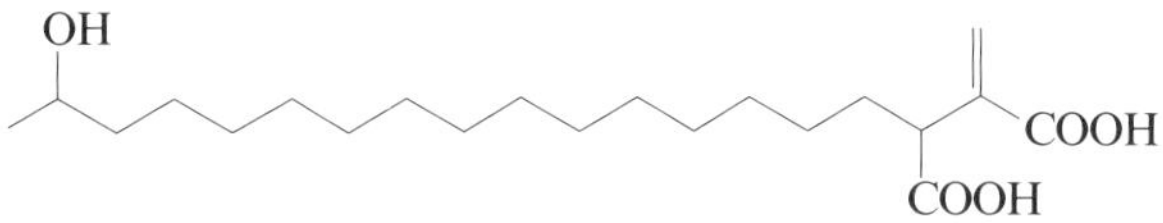

$C_{21}H_{38}O_5$ M 370.528

Constit. of *Usnea aliphatica*. Cryst. (AcOH aq.). Mp 77-79°.

Keogh, M.F. *et al*, *Phytochemistry*, 1977, **16**, 134 (*isol, struct*)

2-Hydroxyhexanedioic acid, 9CI **H-00452**

2-Hydroxyadipic acid

[18294-85-4]

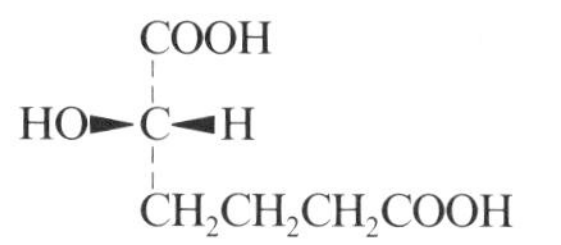

(S)-form

$C_6H_{10}O_5$ M 162.142

(S)-form

Di-Et ester: [136173-89-2].

$C_{10}H_{18}O_5$ M 218.249

Synthon for chiral lactones and leucotrienes. Pale yellow oil. $[\alpha]_D^{21}$ 3.46° (c, 4.38 in EtOH) (≥95% e.e.).

(±)-form

Cryst. (EtOAc/C_6H_6 or H_2O). Sol. H_2O. Mp 101-104°, Mp 151°.

Di-Me ester: [22191-05-5].

$C_8H_{14}O_5$ M 190.196

$Bp_{0.5}$ 99-100°.

Me ether, di-Me ester: [18333-67-0]. *Dimethyl 2-methoxyadipate*

$C_9H_{16}O_5$ M 204.222

Bp_{11} 157-160°.

[136113-68-3]

Ince, W.H., *J. Chem. Soc.*, 1895, **67**, 155 (*synth*)
Borsche, W. *et al*, *Justus Liebigs Ann. Chem.*, 1933, **505**, 177 (*synth*)
Morin, R.B. *et al*, *J. Am. Chem. Soc.*, 1969, **91**, 1396 (*synth*)
Blaser, F. *et al*, *Helv. Chim. Acta*, 1991, **74**, 141 (*synth, use*)

3-Hydroxyhexanoic acid, 9CI, 8CI H-00453

3-Hydroxycaproic acid

[10191-24-9]

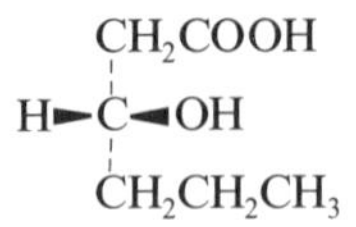

(R)-form

$C_6H_{12}O_3$ M 132.159

(R)-form [77877-35-1]

Cryst. (petrol). Mp 41-41.6°. $[\alpha]_D^{24}$ −27.9° ($CHCl_3$).

Me ester:

$C_7H_{14}O_3$ M 146.186

$[\alpha]_D$ −27.1° ($CHCl_3$).

(±)-form

V. sol. H_2O. Mp 13°.

Et ester: [2305-25-1].

$C_8H_{16}O_3$ M 160.213

Food flavourant. Bp_{12} 95-98°.

[66997-60-2]

Baker, C.F. *et al*, *Justus Liebigs Ann. Chem.*, 1894, **283**, 124 (*synth*)
Lemieux, R.U. *et al*, *Can. J. Chem.*, 1951, **29**, 678 (*abs config*)
Serck-Hanssen, K., *Ark. Kemi*, 1956, **10**, 136 (*abs config*)
Vuitel, L. *et al*, *Helv. Chim. Acta*, 1974, **57**, 1713 (*nmr*)
Ketelaar, P.E.F. *et al*, *Tetrahedron Lett.*, 1985, **26**, 4665 (*synth*)

6-Hydroxyhexanoic acid, 9CI H-00454

[1191-25-9]

$HOCH_2(CH_2)_4COOH$

$C_6H_{12}O_3$ M 132.159

Liq. Polymerises on heating to cryst. polymer, Mp 53-5°.

Et ester: [5299-60-5].

$C_8H_{16}O_3$ M 160.213

d_4^{20} 0.978. Bp_{15} 134°. n_D^{20} 1.4390.

Robinson, R. *et al*, *J. Chem. Soc.*, 1937, 371 (*synth*)
Heilbron, I. *et al*, *J. Chem. Soc.*, 1947, 1588 (*synth*)
U.K. Pat., 1 129 292, (1968); *CA*, **70**, 11132r (*synth*)
Ger. Pat., 2 013 524, (1970); *CA*, **74**, 12620t (*synth*)

6-Hydroxy-2-hexynoic acid H-00455

$HOCH_2CH_2CH_2C{\equiv}CCOOH$

$C_6H_8O_3$ M 128.127

Cryst. (Et_2O/pet. ether). Mp 65-66°.

Me ester: [123368-72-9].

$C_7H_{10}O_3$ M 142.154

Oil. $Bp_{0.3}$ 118-124°, $Bp_{0.08}$ 115-136°.

Jones, E.R.H. *et al*, *J. Chem. Soc.*, 1954, 3208 (*synth, ir, uv*)
Pflieger, D. *et al*, *Tetrahedron*, 1989, **45**, 2031 (*synth, pmr, ir*)
Kita, Y. *et al*, *J. Org. Chem.*, 1991, **56**, 119 (*synth, pmr, ir*)

12-Hydroxy-13-(hydroxymethyl)-3,5,7-trimethyl-2,4-tetradecadiene-1,14-dioic acid, 9CI H-00456

Antibiotic 1233B

[34668-58-1]

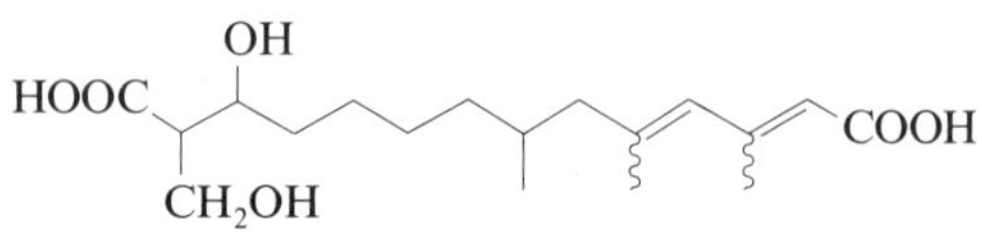

$C_{18}H_{30}O_6$ M 342.431

Metab. of *Cephalosporium* spp. Shows antibiotic props. Cryst. (Et_2O). Mp 88-94°.

Aldridge, D.C. *et al*, *J. Chem. Soc. C*, 1971, 3888 (*isol, uv, ir, ms, nmr, struct*)

7-[3-Hydroxy-2-(3-hydroxy-5-phenyl-1-pentenyl)-5-oxocyclopentyl]-5-heptenoic acid H-00457

17-Phenyltrinorprostaglandin E_2

[38315-43-4]

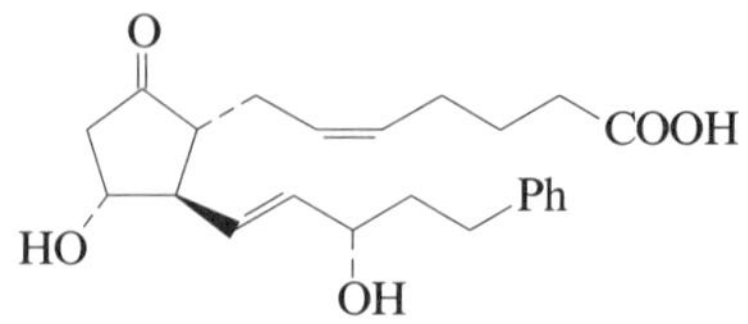

$C_{23}H_{30}O_5$ M 386.487

Shows similar activity and potency to 11,15-Dihydroxy-9-oxo-5,13-prostadienoic acid, D-00225. Mp 95-96°.

Magerlein, B.J. *et al*, *Prostaglandins*, 1975, **9**, 5 (*synth*)
Miller, W.L. *et al*, *Prostaglandins*, 1975, **9**, 9 (*biochem*)
Schaaf, T.K. *et al*, *J. Med. Chem.*, 1983, **26**, 328 (*biochem*)

3-(4-Hydroxy-3-methoxyphenyl)-2-propenoic acid, 9CI H-00458

4-Hydroxy-3-methoxycinnamic acid. ***Ferulic acid.*** *Caffeic acid 3-methyl ether*

[1135-24-6]

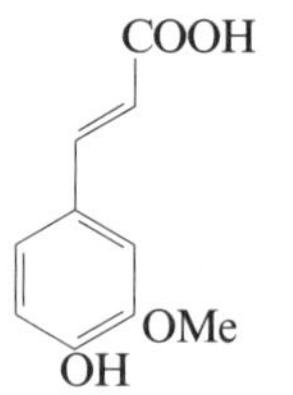

(E)-form

$C_{10}H_{10}O_4$ M 194.187

(E)-form [537-98-4]

Widely distributed in plants. Metab. of a *Penicillium* sp. Orthorhombic needles (H_2O). Mp 174°.

▷ GD9275000.

Me ester:

$C_{11}H_{12}O_4$ M 208.213

Mp 63-64°.

Hexadecyl ester: [64190-80-3]. *Hexadecyl ferulate*

$C_{26}H_{42}O_4$ M 418.615

Isol. from *Euchresta horsfieldii* and *Ipomoea batatas*.

Eicosanyl ester: Eicosanoyl ferulate. Icosanoyl ferulate

$C_{30}H_{50}O_4$ M 474.723

Isol. from heartwood of *Larix laricina*. Cryst. (cyclohexane). Mp 61-62°.

Docosyl ester: [101927-24-6]. *Docosyl* (E)-*ferulate*

$C_{32}H_{54}O_4$ M 502.776

Isol. from the stems of *Bauhinia manca*.

Tricosyl ester: [101959-30-2]. *Tricosyl* (E)-*ferulate*

$C_{33}H_{56}O_4$ M 516.803

Isol. from the stems of *B. manca*.

Tetracosyl ester: [101927-25-7]. *Tetracosyl* (E)-*ferulate. Lignoceryl ferulate*

$C_{34}H_{58}O_4$ M 530.830

Isol. from the stems of *B. manca*, *Hedysarum polybotrys* and from *Pygeum africanum*.

Pentacosyl ester: [101959-33-5]. *Pentacosyl* (E)-*ferulate*

$C_{35}H_{60}O_4$ M 544.857

Isol. from the stems of *B. manca*.

Hexacosyl ester: [63034-29-7]. *Hexacosyl* (E)-*ferulate*
$C_{36}H_{62}O_4$ M 558.883
Isol. from the stems of *B. manca* and from *Pinus roxburghii*.
Heptacosyl ester: [101959-35-7]. *Heptacosyl* (E)-*ferulate*
$C_{37}H_{64}O_4$ M 572.910
Isol. from the stems of *B. manca*.
Octacosyl ester: [101959-37-9]. *Octacosyl* (E)-*ferulate*
$C_{38}H_{66}O_4$ M 586.937
Isol. from the stems of *B. manca* and *Tecomella undulata*.
Ac:
$C_{12}H_{12}O_5$ M 236.224
Mp 124°.
Ac, octacosyl ester: [102686-01-1]. *Octacosyl 4-O-acetyl-*(E)-*ferulate*
$C_{40}H_{68}O_5$ M 628.974
Isol. from *T. undulata*. Granules. Mp 96-97°.
4-O-β-D-Glucopyranoside: [14364-12-6]. ***Glucoferulic acid***
$C_{16}H_{20}O_9$ M 356.329
Present in various plant sources, e.g. soft fruits, needles of *Larix leptolepis*. Pale yellow granular cryst. (EtOH). Mp 245-247°.
O-(*3,7-Dimethyl-2,6-octadienyl*): p-*Geranyloxyferulic acid*
$C_{20}H_{26}O_4$ M 330.423
Cryst. (pet. ether). Mp 109-111°. The name *p*-Geranyloxyferulic acid is misleading.
O-(*3,7-Dimethyl-2,6-octadienyl*), *Me ester:*
$C_{21}H_{28}O_4$ M 344.450
Constit. of the bark of *Acronychia baueri*. Prisms (MeOH aq.). Mp 60-62°.

(*Z*)-*form* [1014-83-1]
Oil. λ_{max} 316 nm (EtOH).

Tanaka, S., *CA*, 1930, **24**, 2453.
Pearl, I.A. *et al*, *J. Org. Chem.*, 1951, **16**, 216.
Comte, P. *et al*, *C. R. Hebd. Seances Acad. Sci.*, 1957, **245**, 1144.
Herrmann, K., *Naturwissenschaften*, 1958, **45**, 111 (*occur*)
Nair, G.V. *et al*, *Can. J. Chem.*, 1959, **37**, 1608 (*eicosanoyl ester*)
Prager, R.H. *et al*, *Aust. J. Chem.*, 1966, **19**, 451 (*p*-Geranyloxyferulic acid)
Challice, J.S. *et al*, *J. Chromatogr.*, 1966, **21**, 357 (*isom*, *uv*)
Karrer, W. *et al*, *Konstitution und Vorkommen der Organischen Pflanzenstoffe*, 2nd Ed., Birkhäuser Verlag, Basel, 1972-1985, no. 959 (*occur*)
Von Massow, F. *et al*, *J. Chem. Soc., Perkin Trans. 2*, 1976, 977 (*cmr*)
Kelley, C.J. *et al*, *J. Org. Chem.*, 1976, **41**, 449 (*cmr*)
Ohmoto, T. *et al*, *Chem. Pharm. Bull.*, 1983, **31**, 919 (*Glucoferulic acid*)
Agarwal, V.K., *Phytochemistry*, 1984, **23**, 1342 (*Glucoferulic acid*)
Chenault, J. *et al*, *Synth. Commun.*, 1984, **14**, 1059 (*synth*)
Joshi, K.C. *et al*, *Planta Med.*, 1986, **52**, 71 (*Octacosyl acetylferulate*)
Achenbach, H. *et al*, *Z. Naturforsch., C*, 1986, **41**, 164 (*esters*)
Nethaji, M. *et al*, *Acta Crystallogr., Sect. C*, 1988, **44**, 275 (*cryst struct*)
Tillett, R. *et al*, *Arch. Mikrobiol.*, 1990, **154**, 206 (*isol*)
Mizuno, M. *et al*, *Phytochemistry*, 1990, **29**, 2663 (*Hexadecyl ferulate*)
Lewis, R.J., *Sax's Dangerous Properties of Industrial Materials*, 8th Ed., Van Nostrand-Reinhold, 1992, FBP200.

3-Hydroxy-16-methylheptadecanoic acid H-00459

$(H_3C)_2CH(CH_2)_{12}CH(OH)CH_2COOH$

$C_{18}H_{36}O_3$ M 300.481
Plates (hexane). Mp 70-71°.
Ac: 3-Acetoxy-16-methylheptadecanoic acid
$C_{20}H_{38}O_4$ M 342.518
Constit. of *Lavandula gibsonii*. Viscous liq.
Ac, Me ester: Methyl-3-acetoxy-16-methylheptadecanoate
$C_{21}H_{40}O_4$ M 356.545
Constit. of *L. gibsonii*. Viscous liq. $Bp_{0.6}$ 195-200°.

Patwardhan, S.A. *et al*, *Phytochemistry*, 1983, **22**, 165.

34-Hydroxy-8-methyl-5-heptatriacontanone, 9CI H-00460

[133956-31-7]

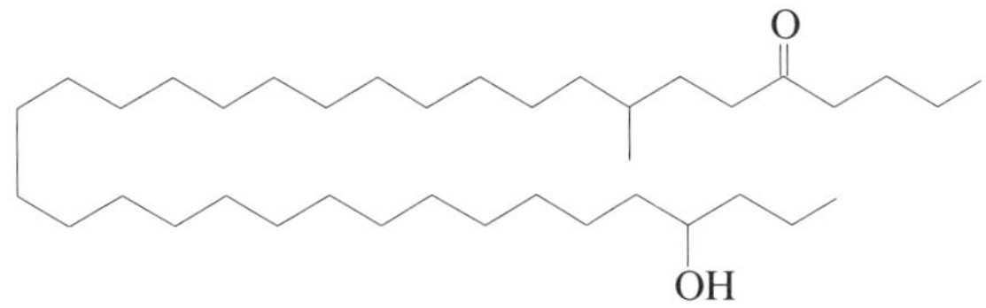

$C_{38}H_{76}O_2$ M 565.017
Constit. of the leaves of *Adenocalymma alliaceum*. Cryst. (MeOH). Mp 91-92°.
Ac: [133956-33-9].
Mp 83-84°.

Misra, T.N. *et al*, *Phytochemistry*, 1991, **30**, 541 (*isol*)

15-Hydroxy-11-methyl-9-oxo-5,10,13-prostatrienoic acid, 9CI H-00461

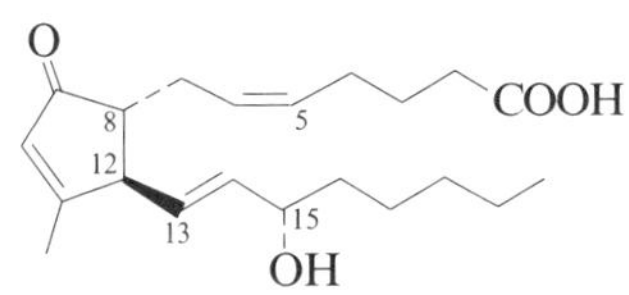

$C_{21}H_{32}O_4$ M 348.481
(*5Z,13E,15S)-*form*** [41416-42-6]
11-Methyl-PGA$_2$
Synthetic prostaglandin analogue.

Vogel, P. *et al*, *Helv. Chim. Acta*, 1973, **56**, 557 (*synth, uv, ir, pmr, ms*)
Lin, C.H., *Chem. Ind. (London)*, 1976, 994 (*synth, ir, pmr, ms*)

25-Hydroxy-33-methyl-6-pentatriacontanone H-00462

[129355-87-9]

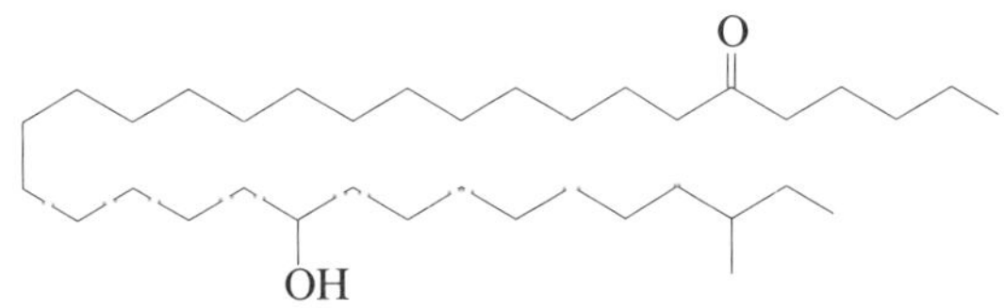

$C_{36}H_{72}O_2$ M 536.964
Constit. of the rhizomes of *Curculigo orchioides*. Cryst. (C_6H_6). Mp 92°.
2,4-Dinitrophenylhydrazone: [129355-88-0].
Cryst. (MeOH). Mp 99-101°.

Mehta, B.K. *et al*, *Indian J. Chem., Sect. B*, 1990, **29**, 493 (*isol, pmr*)

2-Hydroxy-1,2,3-nonadecanetricarboxylic acid, 8CI **H-00463**

Agaric acid. Agaricic acid. Agaricin. Laricic acid

[666-99-9]

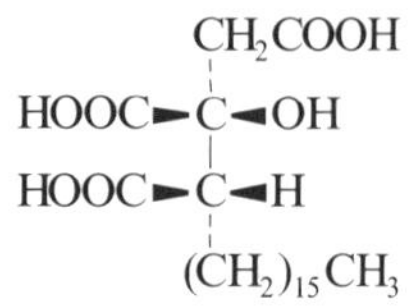

$C_{22}H_{40}O_7$ M 416.554

▷ RA5365000.

(2S,3S)-form

Constit. of *Polyporus officinalis*. Leaflets + $1\frac{1}{2}$ H_2O (EtOH aq.). Anhyd. at 100°. Spar. sol. EtOH, Et_2O. Mp 141.5-142°. $[\alpha]_D^{19}$ −8.84° (NaOH).

Tri-Me ester: Mp 63-64°.

Tri-Et ester: [42065-60-1].

Mp 36-37°.

Thoms, H. *et al*, *Justus Liebigs Ann. Chem.*, 1907, **357**, 145 (*struct*)
Evans, D.W.S., *J. Chem. Soc.*, 1959, 1313.
Brandange, S. *et al*, *Acta Chem. Scand., Ser. B*, 1977, **31**, 307 (*abs config*)

2-Hydroxynonanoic acid, 9CI **H-00464**

2-Hydroxypelargonic acid

[15896-36-3]

$H_3C(CH_2)_6CH(OH)COOH$

$C_9H_{18}O_3$ M 174.239

(±)-*form*

Mp 70°.

Ac:

$C_{11}H_{20}O_4$ M 216.277

Oil. Bp_{10} 171-174°.

Et ester: [18295-07-3].

$C_{11}H_{22}O_3$ M 202.293

Mp 23-24°.

Chloride, Ac:

$C_{11}H_{19}ClO_3$ M 234.722

Oil. $Bp_{0.5}$ 94°.

Ames, D.E. *et al*, *J. Chem. Soc.*, 1951, 1079 (*synth*)
Moersch, G.W. *et al*, *Synthesis*, 1970, 647 (*synth*)

3-Hydroxynonanoic acid, 9CI **H-00465**

3-Hydroxypelargonic acid

[40165-87-5]

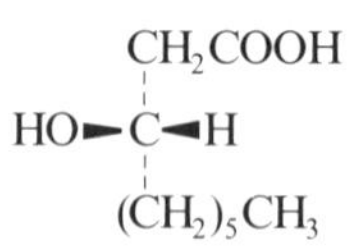

(*S*)-*form*

$C_9H_{18}O_3$ M 174.239

(S)-form [45023-83-4]

Mp 47-48°, Mp 58-59°. $[\alpha]_D^{25}$ +3.3° (c, 12.9 in $CHCl_3$).

Me ester: [66997-65-7].

$C_{10}H_{20}O_3$ M 188.266

$[\alpha]_D^{23}$ −1.00° (c, 1.97 in $CHCl_3$).

Me ester, Me ether:

$C_{11}H_{22}O_3$ M 202.293

$[\alpha]_D^{23}$ −0.07° (c, 5.00 in $CHCl_3$).

(±)-*form*

Isol. from milk. Mp 61°.

Et ester: [26257-80-7].

$C_{11}H_{22}O_3$ M 202.293

Oil. Bp_{13} 145°.

Hydrazide: [33824-92-9].

$C_9H_{20}N_2O_2$ M 188.269

Mp 111-112°.

Harding, V.J. *et al*, *J. Chem. Soc.*, 1910, **97**, 299 (*synth*)
Kuhn, M., *Ber.*, 1938, **71**, 1119 (*synth*)
Meyers, A.I. *et al*, *Tetrahedron Lett.*, 1974, 1333 (*synth*)

4-Hydroxynonanoic acid, 9CI **H-00466**

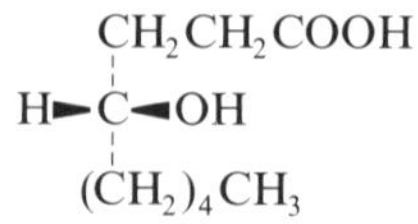

(*R*)-*form*

$C_9H_{18}O_3$ M 174.239

(R)-form

Lactone: see *Dihydro-5-pentyl-2(3*H*)-furanone, D-00159*

Hydrazide:

$C_9H_{20}N_2O_2$ M 188.269

Mp 79-80°.

(±)-*form*

Oil. Bp_{10} 117-120°.

Hydrazide: [22094-44-6].

Mp 79-80°.

Lactone: see *Dihydro-5-pentyl-2(3*H*)-furanone, D-00159*

[104-61-0]

Yumoto, T. *et al*, *CA*, 1956, **50**, 16674 (*synth*)
Allen, C.F.H. *et al*, *J. Heterocycl. Chem.*, 1969, **6**, 349 (*synth*)

6-Hydroxynonanoic acid, 9CI **H-00467**

[75544-91-1]

$H_3CCH_2CH_2CH(OH)(CH_2)_4COOH$

$C_9H_{18}O_3$ M 174.239

(±)-*form*

Et ester:

$C_{11}H_{22}O_3$ M 202.293

Liq. d^{25} 0.940. Bp_{15} 151-152°. n_D^{20} 1.4412, n_D^{25} 1.4540.

Lactone:

$C_9H_{16}O_2$ M 156.224

Liq. $Bp_{2.4}$ 100°. n_D^{20} 1.4577, n_D^{25} 1.4582.

Hydrazide: [16425-24-4].

Solid. Mp 79-81°.

[84770-42-3]

Kemeoka, H. *et al*, *Nippon Kagaku Zasshi*, 1964, **85**, 60 (*ester*)
Mornet, R. *et al*, *Bull. Soc. Chim. Fr.*, 1965, 3043 (*synth*)
Lardelli, G. *et al*, *Recl. Trav. Chim. Pays-Bas (J. R. Neth. Chem. Soc.)*, 1967, **86**, 481 (*hydrazide, lactone*)

7-Hydroxynonanoic acid, 9CI **H-00468**

[70478-77-2]

$H_3CCH_2CH(OH)(CH_2)_5COOH$

$C_9H_{18}O_3$ M 174.239

Isol. from *Poecilocerus pictus*.

(±)-*form*

Liq. Bp_{25} 204°.

Et ester:

$C_{11}H_{22}O_3$ M 202.293

Liq. Bp_{98} 151-152°.

Blaise, E. *et al*, *Bull. Soc. Chim. Fr.*, 1910, **7**, 415 (*synth*)
Panda, R.K. *et al*, *Indian J. Biochem. Biophys.*, 1978, **15**, 483; *CA*, **91**, 16964g (*isol*)

8-Hydroxynonanoic acid, 9CI **H-00469**
8-Hydroxypelargonic acid

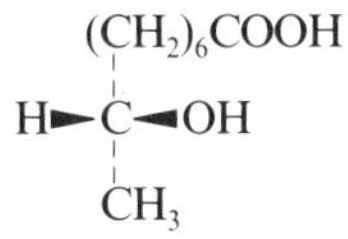

(R)-form

$C_9H_{18}O_3$ M 174.239
Isol. from Calluna heathland soil. Has strong phytotoxic and fungitoxic props.

(R)-form [126230-50-0]
$[\alpha]_D^{20}$ −8.6° (c, 1 in THF).
Me ester: [126372-28-9].
$C_{10}H_{20}O_3$ M 188.266
$[\alpha]_D^{20}$ −5.4° (c, 1.15 in THF).

(S)-form
Lactone:
$C_9H_{16}O_2$ M 156.224
$[\alpha]_D^{20}$ +45.5° (c, 1 In THF).

(±)-form [75544-92-2]
Me ester: Oil. $Bp_{0.01}$ 68°.
Me ester, Ac:
$C_{12}H_{22}O_4$ M 230.303
Oil. Bp_5 142°.
Et ester: [25482-96-6].
$C_{11}H_{22}O_3$ M 202.293
Liq. Bp_{11} 137-140°.
Lactone: Liq. $Bp_{0.65}$ 50-51°.

Kameoka, H. *et al, Kogyo Kagaku Zasshi*, 1969, **72**, 1204; *CA*, **71**, 80615h (*synth*)
Voss, G. *et al, Helv. Chim. Acta*, 1983, **66**, 2294 (*synth*)
Jalal, M.A.F. *et al, Plant Soil*, 1983, **70**, 257, 273 (*isol, glc, ms*)
Fouque, E. *et al, Synthesis*, 1989, 661 (*synth, pmr, cmr, ir, ms*)
Birch, A.J. *et al, J. Chem. Soc., Perkin Trans. 1*, 1990, 1423 (*synth, ir, pmr*)

9-Hydroxynonanoic acid, 9CI **H-00470**
[3788-56-5]

$HOCH_2(CH_2)_7COOH$

$C_9H_{18}O_3$ M 174.239
Mp 53-54°.
Ac:
$C_{11}H_{20}O_4$ M 216.277
Mp 1°. $Bp_{1.5}$ 156-160°.
Me ester: [34957-73-8].
$C_{10}H_{20}O_3$ M 188.266
Oil. Bp_3 137-139°.
Me ester, phenylurethane: Mp 53-54°.
Lactone: [6008-27-1]. *2-Oxecanone, 9CI. Nonanolactone*
$C_9H_{16}O_2$ M 156.224
Prisms (MeOH). Mp 34°. Bp_3 73°.

Bestmann, H.J. *et al, Chem. Ber.*, 1971, **104**, 65 (*synth*)
White, R.W. *et al, Tetrahedron Lett.*, 1971, 3587 (*synth*)
Pryde, E.H. *et al, J. Am. Oil Chem. Soc.*, 1976, **53**, 90 (*synth*)
Wiberg, K.B. *et al, J. Am. Chem. Soc.*, 1991, **113**, 971 (*lactone, synth, cryst struct*)

9-Hydroxy-4-nonanone, 9CI **H-00471**
6-Oxo-1-nonanol (incorr.)
[4066-93-7]

$H_3CCH_2CH_2CO(CH_2)_4CH_2OH$

$C_9H_{18}O_2$ M 158.240
Sex pheromone constit. of the rectal gland secretions of various fruit flies, *Dacus* (*Bactrocera*) spp. Bp_{12} 142-143°.

Mornet, R. *et al, Bull. Soc. Chim. Fr.*, 1965, 3043 (*synth*)
Gol'dfarb, Y.L. *et al, Izv. Akad. Nauk SSSR, Ser. Khim.*, 1973, 2160 (*synth*)
Kitching, W. *et al, J. Org. Chem.*, 1989, **54**, 3893 (*occur*)
Perkins, M.V. *et al, J. Chem. Soc., Perkin Trans. 1*, 1990, 1111 (*occur*)

9-Hydroxy-7-nonene-3,5-diynoic acid, 9CI **H-00472**

$HOCH_2CH{=}CHC{\equiv}CC{\equiv}CCH_2COOH$

$C_9H_8O_3$ M 164.160

(E)-form
Found in oil of *Poria selecta*. Isol. from extracts of culture fluids of *Fistulina pallida*. Oil.
Me ester:
$C_{10}H_{10}O_3$ M 178.187
Mp 56° dec.

(Z)-form [58065-04-6]
Isol. from culture fluids of *F. pallida*. Mp 113-114°.
[58065-05-7]

Bew, R.E. *et al, J. Chem. Soc. C*, 1966, 135 (*isol, synth*)
Ahmed, M. *et al, J. Chem. Soc., Perkin Trans. 1*, 1974, 1981 (*isol, synth, pmr, ms*)
Kobayashi, A. *et al, Agric. Biol. Chem.*, 1975, **39**, 2247 (*synth*)

9-Hydroxy-5-nonenoic acid **H-00473**

$HOCH_2CH_2CH_2CH{=}CH(CH_2)_3COOH$

$C_9H_{16}O_3$ M 172.224

(Z)-form [109997-30-0]
Yellow oil.
Me ester: [75964-05-5].
$C_{10}H_{18}O_3$ M 186.250
Oil.

Bergman, N.-A. *et al, J. Org. Chem.*, 1987, **52**, 4449 (*synth, pmr*)

28-Hydroxyoctacosanoic acid, 9CI **H-00474**
[52900-17-1]

$HOCH_2(CH_2)_{26}COOH$

$C_{28}H_{56}O_3$ M 440.749
Isol. from Carnauba wax.
Me ester: [79162-66-6].
$C_{29}H_{58}O_3$ M 454.775
Mp 86-87°.
Ac, Me ester:
$C_{31}H_{60}O_4$ M 496.813
Mp 72°.

Murry, K.E. *et al, Aust. J. Chem.*, 1955, **8**, 437 (*isol*)
Abrams, S.R., *Chem. Phys. Lipids*, 1981, **28**, 379 (*synth*)

8-Hydroxy-13,17-octadecadiene-9,11-diynoic acid **H-00475**

$H_2C{=}CHCH_2CH_2CH{=}CHC{\equiv}CC{\equiv}CCH(OH)(CH_2)_6COOH$

$C_{18}H_{24}O_3$ M 288.386

(Z)-form
Constit. of Isano oil (*Ongokea gore*).

Badami, R.C. *et al, J. Sci. Food Agric.*, 1963, **14**, 863 (*uv, struct*)
Gunstone, F.D. *et al, J. Chem. Soc.*, 1963, 5772 (*isol*)

6-Hydroxy-7,9-octadecadiene-11,13,15,17-tetraynoic acid, 9CI — H-00476

H(C≡C)$_4$CH=CHCH=CHCH(OH)(CH$_2$)$_4$COOH

$C_{18}H_{16}O_3$ M 280.323
Exhibits potent antimicrobial activity.

(7E,9E)-form [111051-87-7] ***Caryoynencin A***
Isol. from *Pseudomonas caryophilli.*

(7E,9Z)-form [111138-75-1] ***Caryoynencin C***
From *P. caryophilli.*

(7Z,9E)-form [111138-74-0] ***Caryoynencin B***
From *P. caryophilli.*

Kusumi, T. *et al, Tetrahedron Lett.*, 1987, **28**, 3981 (*isol, struct*)

6-Hydroxy-9,12-octadecadienoic acid — H-00477

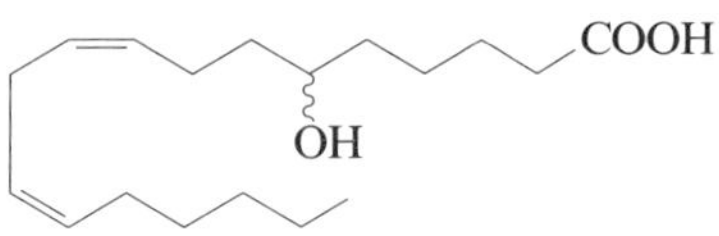

$C_{18}H_{32}O_3$ M 296.449

(9Z,12Z)-form
6-Hydroxylinoleic acid

6-Ac: [137052-51-8]. ***6-Acetoxylinoleic acid***
$C_{20}H_{34}O_4$ M 338.486
Isol. from the marine alga *Spatoglossum pacificum.* Pollen growth inhibitor. $[\alpha]_D^{23}$ −1.04° (c, 0.5 in $CHCl_3$). C-6 config. not determined.

Tazaki, H. *et al, Agric. Biol. Chem.*, 1991, **55**, 2149 (*isol*)

9-Hydroxy-10,12-octadecadienoic acid, 9CI — H-00478

[15514-85-9]

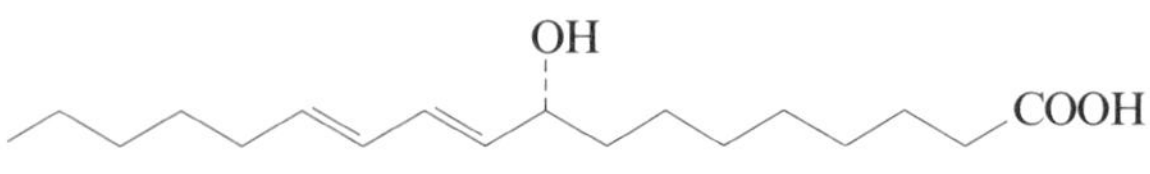

(9*S*,10*E*,12*E*)-*form*

$C_{18}H_{32}O_3$ M 296.449

(*S*)-(*10E,12E*)-*form* [6114-34-7]
β-Dimorphecolic acid
Constit. of the seed fat of *Dimorphotheca aurantiaca, D. sinuata, D. pluvialis* and *Osteospermum ecklonis.* Cryst. (Me_2CO). Mp 36-37°. $[\alpha]_D^{25}$ +15.2° (c, 5.2 in MeOH).

Me ester:
$C_{19}H_{34}O_3$ M 310.476
$[\alpha]_D^{25}$ +5° (c, 5.0 in $CHCl_3$).

(*S*)-(*10E,12Z*)-*form* [10075-11-3]
α-Dimorphecolic acid
Occurs in *Xeranthemum annuum.* Present in plant and animal lipids as autoxidn. or lipoxyenase oxidn. prod. of linoleic acid. Prostaglandin agonist/antagonist, adenylcyclase regulator in thrombocytes. $[\alpha]_{400}$ +1.5° (c, 1.6 in hexane).

Me ester: $[\alpha]_{540}^{20}$ +4.2° (c, 4.17 in EtOH).

Smith, C.R. *et al, J. Am. Chem. Soc.*, 1960, **82**, 1417.
Applewhite, T.H. *et al, J. Chem. Soc., Chem. Commun.*, 1965, 255 (*abs config*)
Powell, R.C. *et al, J. Org. Chem.*, 1967, **32**, 1442.
Chan, H.W.S. *et al, Lipids*, 1977, **12**, 99 (*synth*)
Sessa, D.J. *et al, Lipids*, 1977, **12**, 613 (*isol, uv, ir, pmr, ms*)
Tsuboi, S. *et al, Bull. Chem. Soc. Jpn.*, 1986, **59**, 2050 (*synth*)
Rao, A.V.R. *et al, Tetrahedron*, 1986, **42**, 4523 (*synth, ir, pmr*)
Kraus, R. *et al, Justus Liebigs Ann. Chem.*, 1991, 335 (*isol, synth, bibl*)

12-Hydroxy-9,15-octadecadienoic acid, 9CI — H-00479

Densipolic acid

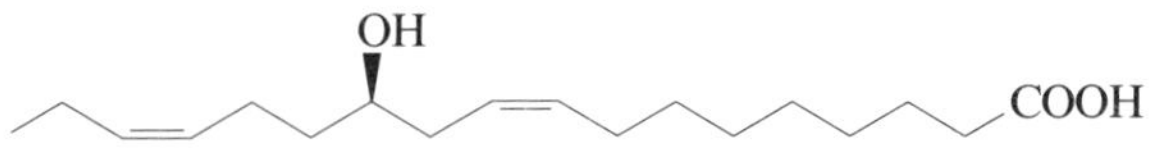

$C_{18}H_{32}O_3$ M 296.449

(9Z,12R,15Z)-form [7121-47-3]
Constit. of the seed oils of *Lesquerella densipila* and *L. auriculata.*

Me ester: [6114-36-9].
$[\alpha]_D^{26}$ −0.58° (c, 13.1 in MeOH).

Smith, C.R. *et al, J. Org. Chem.*, 1962, **27**, 3112 (*occur*)
Applewhite, T.H. *et al, J. Chem. Soc., Chem. Commun.*, 1965, 255 (*abs config*)

13-Hydroxy-9,11-octadecadienoic acid, 9CI — H-00480

[5204-88-6]

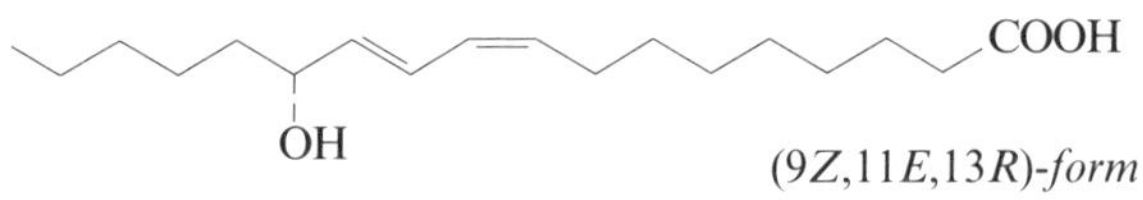

(9*Z*,11*E*,13*R*)-*form*

$C_{18}H_{32}O_3$ M 296.449

(9Z,11E,13R)-form [10219-69-9]
(−)-Coriolic acid. α-Artemesiolic acid
Occurs in *Coriaria nepalensis* and *Xeranthemum annuum* seed oils. $[\alpha]_{400}$ −0.5° (c, 1.6 in hexane). Abs. config. is not assigned to α-Artemesiolic acid.

Me ester:
$C_{19}H_{34}O_3$ M 310.476
Oil. $[\alpha]_D^{23}$ −7.5° (c, 2.3 in hexane).

(9Z,11E,13S)-form [29623-28-7]
(+)-Coriolic acid
Occurs in *Monnina emarginata.*

Me ester: $[\alpha]_{546}^{20}$ +15.2° (0.86 in EtOH).

Lactone: [29473-40-3]. ***Coriolide.*** *Coriolic acid lactone*
$C_{18}H_{30}O_2$ M 278.434
From *M. emarginata* seed oil. Mp 39.5-42°. $[\alpha]_D^{24}$ +32° (c, 2.56 in hexane)

(±)-***(9E,11E)-form*** [32819-31-1]
Occurs in *Absinthium* oils.

[73804-64-5]

Tallent, W.H. *et al, Tetrahedron Lett.*, 1966, 4329 (*struct, uv*)
Powell, R.C. *et al, J. Org. Chem.*, 1967, **32**, 1442.
Conacher, H.B.S. *et al, Chem. Phys. Lipids*, 1969, **3**, 191 (*synth*)
Phillips, B.E. *et al, J. Org. Chem.*, 1970, **35**, 1916 (*Coriolide*)
Kleiman, R. *et al, J. Am. Oil Chem. Soc.*, 1973, **50**, 31 (*ms*)
Chan, H.W.S. *et al, Lipids*, 1977, **12**, 99 (*synth*)
Rao, A.V.R. *et al, J. Org. Chem.*, 1986, **51**, 4158 (*synth, ir, pmr, ms*)
Moustakis, C.A. *et al, Tetrahedron Lett.*, 1986, **27**, 303 (*synth*)
Tsuboi, S. *et al, Bull. Chem. Soc. Jpn.*, 1987, **60**, 1103 (*synth*)
Kobayashi, Y. *et al, Tetrahedron Lett.*, 1987, **28**, 3959 (*synth, biochem*)
Chan, C. *et al, J. Chem. Soc., Chem. Commun.*, 1988, 971 (*synth*)
Tranchepain, I. *et al, Tetrahedron*, 1989, **45**, 2057 (*synth*)

8-Hydroxy-11,17-octadecadien-9-ynoic acid **H-00481**

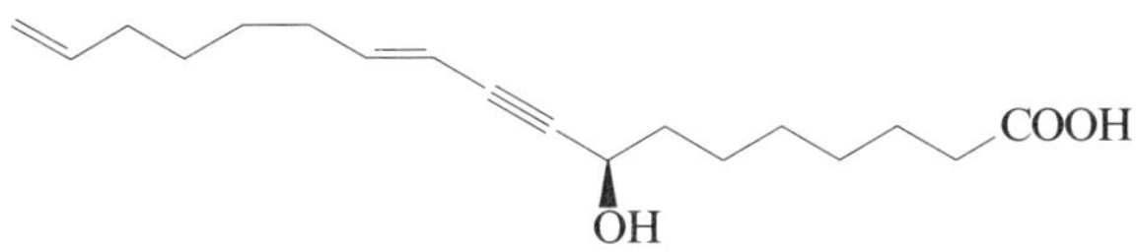

$C_{18}H_{28}O_3$ M 292.417

(R)-(E)-form [20714-97-0]

Occurs in the seed oil of *Pyrularia pubera* and *Acanthosyris spinescens*.

Me ester: [20715-00-8].

Oil. $[\alpha]_D$ −3.0° (c, 3.0 in cyclohexane).

Powell, R.G. *et al*, *J. Org. Chem.*, 1966, **31**, 528.
Hopkins, C.Y. *et al*, *J. Chem. Soc. C*, 1968, 2462 (*uv, ir*)
Kleiman, R. *et al*, *J. Am. Oil Chem. Soc.*, 1973, **50**, 31 (*ms*)

9-Hydroxy-10,14-octadecadien-12-ynoic acid **H-00482**

[130523-94-3]

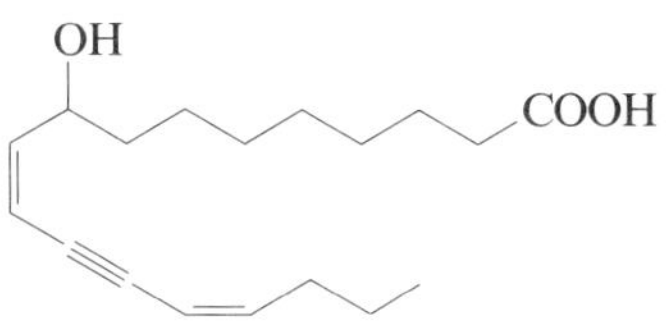

$C_{18}H_{28}O_3$ M 292.417

Prod. of autoxidation of 9,14-Octadecadien-12-ynoic acid, O-00075. Oil.

Pang, Z. *et al*, *J. Org. Chem.*, 1991, **56**, 1233 (*struct, ir, uv, ms, pmr*)

8-Hydroxy-9,11-octadecadiynoic acid **H-00483**

[64144-77-0]

$H_3C(CH_2)_5C{\equiv}CC{\equiv}CCH(OH)(CH_2)_6COOH$

$C_{18}H_{28}O_3$ M 292.417

Constit. of Isano oil (*Ongokea gore*).

Miller, R.W. *et al*, *Phytochemistry*, 1977, **16**, 947 (*glc, nmr, ms, uv*)

8-Hydroxy-10,12-octadecadiynoic acid **H-00484**

[64144-73-6]

$H_3C(CH_2)_4C{\equiv}CC{\equiv}CCH_2CH(OH)(CH_2)_6COOH$

$C_{18}H_{28}O_3$ M 292.417

Isol. from Isano oil (*Ongokea gore*).

Miller, R.W. *et al*, *Phytochemistry*, 1977, **16**, 947.

2-Hydroxyoctadecanoic acid, 9CI **H-00485**

2-Hydroxystearic acid

[629-22-1]

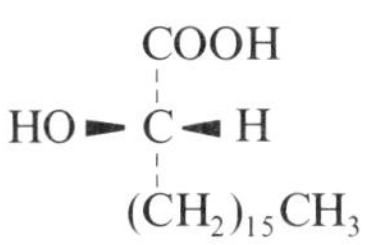

(S)-form

$C_{18}H_{36}O_3$ M 300.481

Occurs in Actinomycetales bacteria and the sphingolipids of the influenza virus. Has mycobactericidal activity.

(S)-form [26531-80-6]

Occurs in wool wax. Mp 76.5-77.5°.

Me ester:

$C_{19}H_{38}O_3$ M 314.507

Mp 38.5-39.5°. $[\alpha]_D$ +3.8° ($CHCl_3$).

(±)-*form*

Mp 93°.

Me ester: [2420-35-1].

Mp 65.7-65.9°.

Et ester:

$C_{20}H_{40}O_3$ M 328.534

Mp 62-63°.

Amide:

$C_{18}H_{37}NO_2$ M 299.496

Mp 148-149°.

Darzens, G., *C. R. Hebd. Seances Acad. Sci.*, 1933, **196**, 348.
Fischmeister, I., *Ark. Kemi*, 1967, **27**, 167 (*ir*)
Yano, I. *et al*, *FEBS Lett.*, 1969, **4**, 96.
Hammerstrom, S. *et al*, *Anal. Biochem.*, 1973, **52**, 169 (*resoln*)
Karlsson, K.A. *et al*, *Chem. Phys. Lipids*, 1974, **12**, 65 (*synth*)
Tulloch, A.P., *Org. Magn. Reson.*, 1978, **11**, 109 (*cmr*)

3-Hydroxyoctadecanoic acid, 9CI **H-00486**

3-Hydroxystearic acid

[17773-30-7]

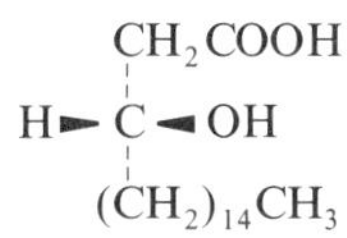

(R)-form

$C_{18}H_{36}O_3$ M 300.481

(R)-form [14531-50-1]

Occurs in extracellular lipid of Rhodotorula yeasts. Cryst. (Me_2CO). Mp 83-85°. $[\alpha]_D$ −12.2° (c, 1.4 in $CHCl_3$).

Me ester:

$C_{19}H_{38}O_3$ M 314.507

Mp 56.5-57.5°. $[\alpha]_D$ −12.9° (c, 4.1 in $CHCl_3$).

(±)-*form* [14530-89-3]

Occurs in gum resin of *Cistus lodaniferus*. Cryst. ($CHCl_3$). Mp 89°.

Me ester: Mp 51.1-54.4°.

Et ester: [65144-03-8].

$C_{20}H_{40}O_3$ M 328.534

Mp 45-46°.

Bergstrom, S. *et al*, *Acta Chem. Scand.*, 1952, **6**, 1157 (*synth*)
Tulloch, A.P. *et al*, *Can. J. Chem.*, 1964, **42**, 830.
Fischmeister, I., *Ark. Kemi*, 1967, **27**, 167 (*ir*)
Eglinton, G. *et al*, *Org. Mass Spectrom.*, 1968, **1**, 593 (*ms*)
Santaniello, E. *et al*, *Synthesis*, 1977, 698 (*synth*)
Tulloch, A.P., *Org. Magn. Reson.*, 1978, **11**, 109 (*cmr*)

4-Hydroxyoctadecanoic acid **H-00487**

4-Hydroxystearic acid

[2858-39-1]

$H_3C(CH_2)_{13}CH(OH)CH_2CH_2COOH$

$C_{18}H_{36}O_3$ M 300.481

(±)-*form* [40152-01-0]

Mp 89°.

Me ester: [2420-38-4].

$C_{19}H_{38}O_3$ M 314.507

Mp 58.6-59.3°.

Lactone: 4-Octadecanolide. Stearolactone

$C_{18}H_{34}O_2$ M 282.465

Pheromone for *Lasioglossum zephyrum*. Mp 47-48°.

Bergstrom, S. *et al*, *Acta Chem. Scand.*, 1952, **6**, 1157 (*synth*)

Fischmeister, I., *Ark. Kemi*, 1967, **27**, 167 (*ir*)
Tulloch, A.P., *Org. Magn. Reson.*, 1978, **11**, 109 (*cmr*)
Smith, B.H. *et al*, *J. Chem. Ecol.*, 1988, **14**, 87.

5-Hydroxyoctadecanoic acid H-00488

5-Hydroxystearic acid
[2858-38-0]

$H_3C(CH_2)_{12}CH(OH)CH_2CH_2CH_2COOH$

$C_{18}H_{36}O_3$ M 300.481
(±)-*form* [36377-30-7]
Mp 81.9-82.1°.
Me ester: [2420-40-8].
$C_{19}H_{38}O_3$ M 314.507
Mp 51.7-52.1°.

Yegorov, M.A., *Chem. Zentralbl.*, 1915, **1**, 934.
Bergstrom, S. *et al*, *Acta Chem. Scand.*, 1952, **6**, 1157 (*synth*)
Fischmeister, I., *Ark. Kemi*, 1967, **27**, 167 (*ir*)
Tulloch, A.P., *Org. Magn. Reson.*, 1978, **11**, 109 (*cmr*)

6-Hydroxyoctadecanoic acid, 9CI H-00489

6-Hydroxystearic acid
[3158-47-2]

$H_3C(CH_2)_{11}CH(OH)(CH_2)_4COOH$

$C_{18}H_{36}O_3$ M 300.481
(±)-*form* [36377-31-8]
Mp 83°.
Me ester:
$C_{19}H_{38}O_3$ M 314.507
Mp 58.2-58.5°.
Ac:
$C_{20}H_{38}O_4$ M 342.518
Mp 52-53°.

Bougault, J. *et al*, *C. R. Hebd. Seances Acad. Sci.*, 1911, **153**, 573.
Bergström, S. *et al*, *Acta Chem. Scand.*, 1952, **6**, 1157 (*synth*)
Fischmeister, I., *Ark. Kemi*, 1967, **27**, 167 (*ir*)
Tulloch, A.P., *Org. Magn. Reson.*, 1978, **11**, 109 (*cmr*)

7-Hydroxyoctadecanoic acid, 9CI H-00490

7-Hydroxystearic acid
[3158-48-3]

$H_3C(CH_2)_{10}CH(OH)(CH_2)_5COOH$

$C_{18}H_{36}O_3$ M 300.481
(±)-*form* [36377-32-9]
Mp 77.6-77.8°.
Me ester: [2379-96-6].
$C_{19}H_{38}O_3$ M 314.507
Mp 50.7-51.0°.
Et ester:
$C_{20}H_{40}O_3$ M 328.534
Mp 37.5°.

Bergström, S. *et al*, *Acta Chem. Scand.*, 1952, **6**, 1157 (*synth*)
Fischmeister, I., *Ark. Kemi*, 1967, **27**, 167 (*ir*)
Tulloch, A.P., *Org. Magn. Reson.*, 1978, **11**, 109 (*cmr*)

8-Hydroxyoctadecanoic acid, 9CI H-00491

8-Hydroxystearic acid
[6807-91-6]

$H_3C(CH_2)_9CH(OH)(CH_2)_6COOH$

$C_{18}H_{36}O_3$ M 300.481
(±)-*form*
Mp 81.5-81.7°.
Me ester: [2379-98-8].
$C_{19}H_{38}O_3$ M 314.507
Mp 55.3-55.6°.

Bergström, S. *et al*, *Acta Chem. Scand.*, 1952, **6**, 1157 (*synth*)
Fischmeister, I., *Ark. Kemi*, 1967, **27**, 167 (*ir*)
Tulloch, A.P., *Org. Magn. Reson.*, 1978, **11**, 109 (*cmr*)

9-Hydroxyoctadecanoic acid, 9CI H-00492

9-Hydroxystearic acid
[3384-24-5]

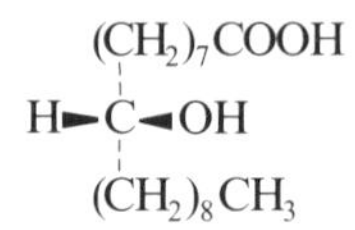

(*R*)-*form*

$C_{18}H_{36}O_3$ M 300.481
(*R*)-*form* [6114-37-0]
Obt. by hydrogenation of natural unsaturated analogues or by synth. Cryst. (MeOH). Mp 83.5-84.5°.
Me ester: Cryst. (pentane). Mp 51.5-52.8°. $[\alpha]_{250}$ −1.24°, $[\alpha]_D^{27}$ −0.17° (MeOH).
(*S*)-*form* [64144-76-9]
Constit. of isano oil.
(±)-*form*
Mp 75.4-75.9°.
Me ester: [2447-53-2].
Mp 45-45.3°, Mp 50.3-50.6°.

Bergström, S. *et al*, *Acta Chem. Scand.*, 1952, **6**, 1157 (*synth*)
Baker, C.D. *et al*, *J. Chem. Soc.*, 1963, 759 (*synth*)
Fischmeister, I., *Ark. Kemi*, 1967, **27**, 167 (*ir*)
Applewhite, T.H. *et al*, *J. Org. Chem.*, 1967, **32**, 1173 (*abs config*)
Miller, R.W. *et al*, *Phytochemistry*, 1977, **16**, 947 (*isol, ms, pmr, ord*)
Tulloch, A.P., *Org. Magn. Reson.*, 1978, **11**, 109 (*cmr*)

10-Hydroxyoctadecanoic acid, 9CI H-00493

10-Hydroxystearic acid. Rosilic acid

$H_3C(CH_2)_7CH(OH)(CH_2)_8COOH$

$C_{18}H_{36}O_3$ M 300.481
Constit. of leaf cutins of rosemary. Mp 84°.
Me ester:
$C_{19}H_{38}O_3$ M 314.507
Mp 54.1-54.2°. Bp_4 212-216°.

Bergström, S. *et al*, *Acta Chem. Scand.*, 1952, **6**, 1157 (*synth*)
Fischmeister, I., *Ark. Kemi*, 1967, **27**, 127 (*ir*)
Rohwedder, K.W., *Lipids*, 1971, **6**, 906 (*ms*)
Brieskorn, C.H. *et al*, *Phytochemistry*, 1971, **10**, 3195.
Tulloch, A.P., *Org. Magn. Reson.*, 1978, **11**, 109 (*cmr*)

11-Hydroxyoctadecanoic acid, 9CI H-00494

11-Hydroxystearic acid
[13419-91-5]

$H_3C(CH_2)_6CH(OH)(CH_2)_9COOH$

$C_{18}H_{36}O_3$ M 300.481
(±)-*form*
Mp 81-82°.
Me ester: [2380-03-2].
$C_{19}H_{38}O_3$ M 314.507
Mp 49.6-49.8°. Bp_4 213-217°.
Et ester:
$C_{20}H_{40}O_3$ M 328.534
Mp 44°.

Tomecko, C.E. *et al*, *J. Am. Chem. Soc.*, 1927, **49**, 525 (*synth*)

Bergström, S. *et al*, *Acta Chem. Scand.*, 1952, **6**, 1157 (*synth*)
Fischmeister, I., *Ark. Kemi*, 1967, **27**, 167 (*ir*)
Tulloch, A.P., *Org. Magn. Reson.*, 1978, **11**, 109 (*cmr*)

12-Hydroxyoctadecanoic acid, 9CI — H-00495

12-Hydroxystearic acid
[106-14-9]

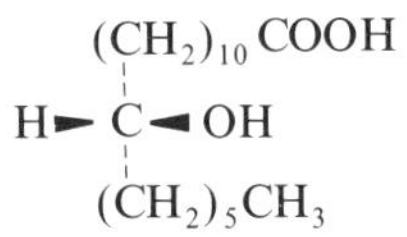

(*R*)-*form*

$C_{18}H_{36}O_3$ M 300.481
▷ Exp. neoplastic agent. WI3850000.

(*R*)-*form* [5762-36-7]
Produced by hydrogenation of (natural) 12-Hydroxy-9-octadecenoic acid (Ricinoleic acid). Used in surface-active and lubricating formulations. (+)-form has been prepared by anodic synth.
Me ester: [6114-39-2].
$C_{19}H_{38}O_3$ M 314.507
$[\alpha]_{300}$ −1.5° (MeOH), $[\alpha]_{400}$ −0.8° (MeOH).

(±)-*form*
Mp 78.5-78.9°.
Me ester: [141-23-1].
Mp 53.4-53.6°. Bp_4 204-206°.
▷ WI4200000.

Bergström, S. *et al*, *Acta Chem. Scand.*, 1956, **6**, 1157 (*synth*)
Applewhite, T.H. *et al*, *J. Org. Chem.*, 1967, **32**, 1173 (*abs config, ord*)
Fischmeister, I., *Ark. Kemi*, 1967, **27**, 167 (*ir*)
Rohwedder, K.H., *Lipids*, 1971, **6**, 906 (*ms*)
Kannan, R. *et al*, *Fette, Seifen, Anstrichm.*, 1974, **76**, 344 (*pmr*)
Tulloch, A.P., *Org. Magn. Reson.*, 1978, **11**, 109 (*cmr*)
Lewis, R.J., *Sax's Dangerous Properties of Industrial Materials*, 8th Ed., Van Nostrand-Reinhold, 1992, HOG000, HOG500.

13-Hydroxyoctadecanoic acid, 9CI — H-00496

13-Hydroxystearic acid
[17773-34-1]

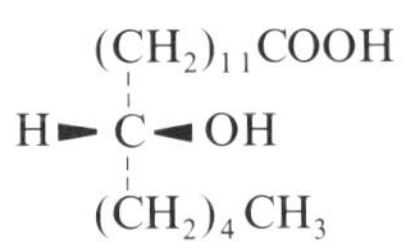

(*R*)-*form*

$C_{18}H_{36}O_3$ M 300.481

(*R*)-*form* [10075-10-2]
Mp 80.5-81°. $[\alpha]_D^{18}$ −0.41° (Py).
Me ester:
$C_{19}H_{38}O_3$ M 314.507
Mp 56.6-57.0°. Bp_4 202-204°. $[\alpha]_D^{20}$ −0.32° (Py).

(±)-*form* [36377-34-1]
Mp 77.2-77.5°.
Me ester: [2540-76-3].
Mp 53.3-53.5°.
Et ester:
$C_{20}H_{40}O_3$ M 328.534
Mp 50.3-51.6°.

Tomecko, C.G. *et al*, *J. Am. Chem. Soc.*, 1927, **49**, 527.
Straus, F. *et al*, *Ber.*, 1933, **66**, 632.
Bergström, S. *et al*, *Acta Chem. Scand.*, 1952, **6**, 1157 (*synth*)
Fischmeister, I., *Ark. Kemi*, 1967, **27**, 167 (*ir*)
Chan, H.W.S. *et al*, *Biochim. Biophys. Acta*, 1975, **380**, 141 (*ms*)
Tulloch, A.P., *Org. Magn. Reson.*, 1975, **11**, 109 (*cmr*)

14-Hydroxyoctadecanoic acid, 9CI — H-00497

14-Hydroxystearic acid
[17773-35-2]

$H_3C(CH_2)_3CH(OH)(CH_2)_{12}COOH$

$C_{18}H_{36}O_3$ M 300.481

(±)-*form* [36377-35-2]
Mp 76.6-76.8°, Mp 79.6-79.9° (dimorph.).
Me ester: [2380-08-7].
$C_{19}H_{38}O_3$ M 314.507
Mp 54°, Mp 54.5-54.8° (dimorph.). Bp_2 185-189°.

Tomecko, C.G. *et al*, *J. Am. Chem. Soc.*, 1927, **49**, 527.
Bergström, S. *et al*, *Acta Chem. Scand.*, 1952, **6**, 1157 (*synth*)
Fischmeister, I., *Ark. Kemi*, 1967, **27**, 167 (*ir*)
Tulloch, A.P., *Org. Magn. Reson.*, 1978, **11**, 109 (*cmr*)

15-Hydroxyoctadecanoic acid — H-00498

15-Hydroxystearic acid
[17773-36-3]

$H_3CCH_2CH_2CH(OH)(CH_2)_{13}COOH$

$C_{18}H_{36}O_3$ M 300.481

(±)-*form* [36377-36-3]
Mp 75.3-75.5°.
Me ester: [2380-10-1].
$C_{19}H_{38}O_3$ M 314.507
Mp 51°, Mp 58.7-58.9° (dimorph.).

Bergström, S. *et al*, *Acta Chem. Scand.*, 1952, **6**, 1157 (*synth*)
Fischmeister, I., *Ark. Kemi*, 1967, **27**, 167 (*ir*)
Hammerstrom, S. *et al*, *Anal. Biochem.*, 1973, **52**, 169 (*resoln*)
Tulloch, A.P., *Org. Magn. Reson.*, 1978, **11**, 109 (*cmr*)

16-Hydroxyoctadecanoic acid — H-00499

16-Hydroxystearic acid
[17773-37-4]

$H_3CCH_2CH(OH)(CH_2)_{14}COOH$

$C_{18}H_{36}O_3$ M 300.481
Occurs in beeswax (probably in an enantiomeric form).

(±)-*form* [36377-37-4]
Mp 78.4-78.6°.
Me ester:
$C_{19}H_{38}O_3$ M 314.507
Mp 56°, Mp 61.2-61.5° (dimorph.).

Bergström, S. *et al*, *Acta Chem. Scand.*, 1952, **6**, 1157 (*synth*)
Fischmeister, I., *Ark. Kemi*, 1967, **27**, 167 (*ir*)
Tulloch, A.P., *Chem. Phys. Lipids*, 1971, **6**, 235.
Hammerstrom, S. *et al*, *Anal. Biochem.*, 1973, **52**, 169 (*resoln*)
Frankel, E.W. *et al*, *Lipids*, 1977, **12**, 1055 (*ms*)
Tulloch, A.P., *Org. Magn. Reson.*, 1978, **11**, 109 (*cmr*)

17-Hydroxyoctadecanoic acid — H-00500

17-Hydroxystearic acid
[4552-19-6]

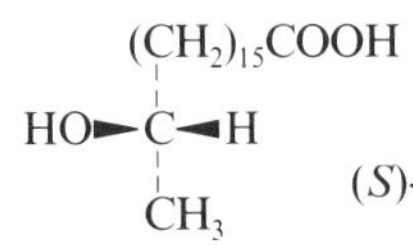

(*S*)-*form*

$C_{18}H_{36}O_3$ M 300.481
Occurs in beeswax (probably in an enantiomeric form).

(*S*)-*form*
Found as sophorose glycoside in *Torulopsis magnoliae*. Cryst. (Me_2CO). Mp 78-80°. $[\alpha]_D$ +4.4° (c, 7.9 in AcOH).

(±)-form

Mp 76.4-76.6°, Mp 81-81.4° (dimorph.).

Me ester: [2380-14-5].

$C_{19}H_{38}O_3$ M 314.507

Mp 52.0-52.5°, Mp 64.1-64.3° (dimorph.).

Bergström, S. *et al*, *Acta Chem. Scand.*, 1952, **6**, 1157 (*synth*)
Gorin, P.A.J. *et al*, *Can. J. Chem.*, 1961, **39**, 846 (*isol*)
Tulloch, A.P., *Chem. Phys. Lipids*, 1971, **6**, 235 (*isol*)
Hammerstrom, S. *et al*, *Anal. Biochem.*, 1973, **52**, 169 (*resoln*)
Tulloch, A.P. *et al*, *Org. Magn. Reson.*, 1978, **11**, 109 (*cmr*)

18-Hydroxyoctadecanoic acid, 9CI H-00501

18-Hydroxystearic acid

[3155-42-8]

$HOCH_2(CH_2)_{16}COOH$

$C_{18}H_{36}O_3$ M 300.481

Occurs in a number of waxes, eg. carnauba wax, oil of *Torulopsis magnoliae*. Mp 87-88°, Mp 99.3-99.5° (dimorph.).

Me ester: [2380-16-7].

$C_{19}H_{38}O_3$ M 314.507

Mp 62.4-62.6°.

Ac:

$C_{20}H_{38}O_4$ M 342.518

Mp 48.5-49.1°, Mp 64°.

Lactone: [4444-85-3]. *Oxacyclononadecan-2-one, 9CI, 8CI. 18-Octadecanolide*

$C_{18}H_{34}O_2$ M 282.465

Major pheromone constit. of Dufour's gland secretion of various bee spp.

Bergström, S. *et al*, *Acta Chem. Scand.*, 1952, **6**, 1157 (*synth*)
Crombie, L. *et al*, *J. Chem. Soc.*, 1954, 2816.
Tulloch, A.P. *et al*, *Can. J. Chem.*, 1962, **14**, 1326 (*isol*)
Fischmeister, I., *Ark. Kemi*, 1967, **27**, 167 (*ir*)
Scott, L.T. *et al*, *Synthesis*, 1976, 738 (*synth, lactone*)
Corey, E.J. *et al*, *Tetrahedron Lett.*, 1976, 3405 (*synth, lactone*)
Tulloch, A.P., *Org. Magn. Reson.*, 1978, **11**, 109 (*cmr*)
Hefetz, A. *et al*, *Science (Washington, D.C.)*, 1979, **204**, 415 (*ms*)
Goto, G. *et al*, *Chem. Pharm. Bull.*, 1985, **33**, 4422 (*synth*)
Hefetz, A. *et al*, *J. Chem. Ecol.*, 1986, **12**, 197 (*occur, lactone*)
Gangadhar, A. *et al*, *J. Am. Oil Chem. Soc.*, 1987, **64**, 1135 (*synth, lactone*)

1-Hydroxy-8,10,14,16-octadecatetraen-12-yn-3-one H-00502

$H_3CCH{=}CHCH{=}CHC{\equiv}CCH{=}CH$
$CH{=}CH(CH_2)_4COCH_2CH_2OH$

$C_{18}H_{24}O_2$ M 272.386

(8E,10E,14E,16E)-form

Ac: 1-Acetoxy-8,10,14,16-octadecatetraen-12-yn-3-one

$C_{20}H_{26}O_3$ M 314.424

Isol. from *Cosmos sulphureus*. Cryst. (Et_2O/pet. ether). Mp 73°.

Bohlmann, F. *et al*, *Chem. Ber.*, 1966, **99**, 142 (*isol, uv, ir, synth*)

17-Hydroxy-9,11,13,15-octadecatetraynoic acid H-00503

Minquartynoic acid

$H_3CCH(OH)C{\equiv}CC{\equiv}CC{\equiv}CC{\equiv}C(CH_2)_7COOH$

$C_{18}H_{20}O_3$ M 284.354

(−)-form [123154-43-8]

Grey-yellow needles (CH_2Cl_2). Isol. from the stem bark of *Minquartia guianensis*. Shows cytotoxic activity. Mp 95°. $[\alpha]_D^{25}$ −29° (c, 0.1 in MeOH). λ_{max} 239, 227, 217, 208 nm (MeOH).

Marles, R.J. *et al*, *J. Nat. Prod. (Lloydia)*, 1989, **52**, 261 (*isol, struct, pmr, cmr, ir, uv, ms*)

2-Hydroxy-9,12,15-octadecatrienoic acid H-00504

$H_3CCH_2CH{=}CHCH_2CH{=}CHCH_2CH{=}CH(CH_2)_6CH(OH)$
$COOH$

$C_{18}H_{30}O_3$ M 294.433

(R)-(all-Z)-form

2-Hydroxylinolenic acid

Constit. of *Salvia nilotica* and *Thymus vulgaris* seed oils. $[\alpha]_{400}$ −24.3° (c, 0.16 in $CHCl_3$).

Smith, C.R. *et al*, *Lipids*, 1969, **4**, 9 (*isol, struct, ord*)
Bohannon, M.B. *et al*, *Lipids*, 1975, **10**, 703 (*isol, ms, ir, nmr*)

18-Hydroxy-9,11,13-octadecatrienoic acid H-00505

18-Hydroxyelaeostearic acid. Kamlolenic acid. Camlolenic acid

[4444-93-3]

$HOCH_2(CH_2)_3CH{=}CHCH{=}CHCH{=}CH(CH_2)_7COOH$

$C_{18}H_{30}O_3$ M 294.433

(all-E)-form

β-Kamlolenic acid

Formed from the α-isomer under the influence of I_2/uv light. Mp 89°.

Me ester:

$C_{19}H_{32}O_3$ M 308.460

Mp 45°.

Ac:

$C_{20}H_{32}O_4$ M 336.470

Mp 58-59°.

(9Z,11E,13E)-form

α-Kamlolenic acid

Present in kamala oil (*Mallotus philippensis* seed fat). Also in *M. discolor*, *M. elaoxyloides* and *Trewia nudiflora*. Mp 78°.

Me ester: Mp 24°.

Ac: Mp 43-44°.

Ac, maleic anhydride adduct. Cryst. (Et_2O/pet. ether). Mp 70-71°.

Crombie, L. *et al*, *J. Chem. Soc.*, 1954, 2816.
Calderwood, R.C. *et al*, *J. Sci. Food Agric.*, 1954, **5**, 382.
Gupta, S. *et al*, *J. Am. Oil Chem. Soc.*, 1955, **32**, 501.
Hatt, H.H. *et al*, *Aust. J. Chem.*, 1961, **14**, 321 (*isol*)
Bergel'son, L.D. *et al*, *Izv. Akad. Nauk SSSR, Ser. Khim.*, 1964, 2003 (*synth*)
Chisholm, M.J. *et al*, *J. Am. Oil Chem. Soc.*, 1966, **43**, 390 (*isol*)
Hopkins, C.Y. *et al*, *Lipids*, 1969, **4**, 89.

13-Hydroxy-9,11,15-octadecatrien-6-ynoic acid H-00506

$H_3CCH_2CH{=}CHCH_2CH(OH)$
$CH{=}CHCH{=}CHCH_2C{\equiv}C(CH_2)_4COOH$

$C_{18}H_{26}O_3$ M 290.402

(9Z,11E,15Z)-form [93633-58-0]

Isol. from Japanese moss *Dicranum scoporium*.

Ichikawa, T. *et al*, *Tetrahedron Lett.*, 1983, **24**, 3337 (*isol, pmr, cmr*)

8-Hydroxy-13-octadecene-9,11-diynoic acid H-00507

$H_3C(CH_2)_3CH{=}CHC{\equiv}CC{\equiv}CCH(OH)(CH_2)_6COOH$

$C_{18}H_{26}O_3$ M 290.402

(Z)-form

Constit. of Isano oil (*Ongokea gore*).

Badami, R.C. *et al*, *J. Sci. Food Agric.*, 1963, **14**, 863 (*uv, struct*)
Gunstone, F.D. *et al*, *J. Chem. Soc.*, 1963, 5772 (*isol*)

8-Hydroxy-17-octadecene-9,11-diynoic acid H-00508

Isanolic acid

[64144-78-1]

$H_2C{=}CH(CH_2)_4C{\equiv}CC{\equiv}CCH(OH)(CH_2)_6COOH$

$C_{18}H_{26}O_3$ M 290.402

Constit. of Isano oil (*Ongokea gore*).

Miller, R.W. *et al*, *Phytochemistry*, 1977, **16**, 947 (*glc, nmr, ms, uv*)

8-Hydroxy-17-octadecene-10,12-diynoic acid, 9CI H-00509

[64144-75-8]

$H_2C{=}CH(CH_2)_3C{\equiv}CC{\equiv}CCH_2CH(OH)(CH_2)_6COOH$

$C_{18}H_{26}O_3$ M 290.402

Isol. from Isano oil (*Ongokea gore*).

Miller, R.W. *et al*, *Phytochemistry*, 1977, **16**, 947.

5-Hydroxy-2-octadecenoic acid H-00510

$H_3C(CH_2)_{12}CH(OH)CH_2CH{=}CHCOOH$

$C_{18}H_{34}O_3$ M 298.465

(E)-(?)-form

Mp 76.3-77.1°.

Breusch, F.L. *et al*, *Istanbul Univ. Fen Fak. Mecm., Seri A*, 1953, **18**, 218; *CA*, **48**, 10578.

8-Hydroxy-9-octadecenoic acid H-00511

[91999-05-2]

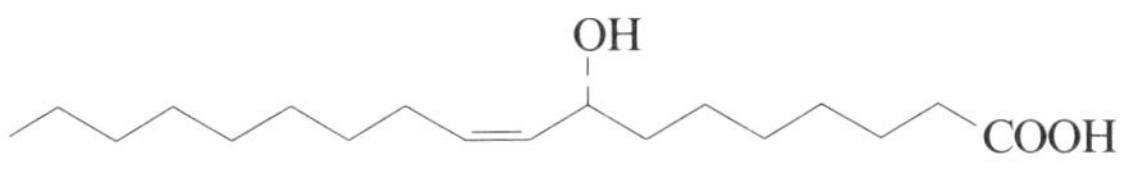

(R,Z)-form

$C_{18}H_{34}O_3$ M 298.465

(8R,9Z)-form

Sporogenic Psi Factor Bβ. psi Bβ

Prod. by *Aspergillus nidulans*. Possesses sporogenic activity.

Me ester: $[\alpha]_D^{26}$ +13.7° (c, 0.012 in $CHCl_3$).

Ross, J. *et al*, *J. Am. Chem. Soc.*, 1949, **18**, 282.
Chan, H.W.S. *et al*, *Chem. Ind. (London)*, 1977, 692.
Garwood, R.F. *et al*, *J. Chem. Soc., Chem. Commun.*, 1977, 364 (*synth*)
Frankel, E.N. *et al*, *Lipids*, 1977, **12**, 901.
Mazur, P. *et al*, *J. Chem. Soc., Chem. Commun.*, 1991, 1486 (*isol, synth*)

9-Hydroxy-10-octadecenoic acid H-00512

$H_3C(CH_2)_6CH{=}CHCH(OH)(CH_2)_7COOH$

$C_{18}H_{34}O_3$ M 298.465

Present in autoxidised Oleic acid after redn. of the hydroperoxide.

(R)-(E)-form

Isol. from stromata of *Epichloe typhina*. Shows antifungal props.

Me ester: $[\alpha]_D^{23}$ −2.14° (c, 0.28 in EtOH).

(±)-(E)-form

Me ester:

$C_{19}H_{36}O_3$ M 312.492

Oil.

(±)-(Z)-form

Me ester: Oil.

Ross, J. *et al*, *J. Am. Chem. Soc.*, 1949, **71**, 282.
Chan, H.W.S. *et al*, *Chem. Ind. (London)*, 1977, 692.
Garwood, R.F. *et al*, *J. Chem. Soc., Chem. Commun.*, 1977, 364 (*synth*)
Frankel, E.N. *et al*, *Lipids*, 1977, **12**, 901.
Frankel, E.N. *et al*, *J. Chem. Soc., Perkin Trans. 1*, 1984, 2233 (*deriv, synth, ir, pmr, ms*)
Koshino, T. *et al*, *Tetrahedron Lett.*, 1987, **28**, 73 (*isol, props*)

9-Hydroxy-12-octadecenoic acid H-00513

[38231-95-7]

$H_3C(CH_2)_4CH{=}CHCH_2CH_2CH(OH)(CH_2)_7COOH$

$C_{18}H_{34}O_3$ M 298.465

(R)-(E)-form

Mp 57.5-59.5°. $[\alpha]_D$ +1.3° (AcOH).

(R)-(Z)-form

Constit. of the seed fat of *Strophanthus* spp. and some other seed fats. Mp 30-32°. $[\alpha]_D$ +3.5° (AcOH).

(S)-(Z)-form [26806-10-0]

Isol. from Apocynaca seed oils. $[\alpha]_{400}$ −2.54°.

(±)-(Z)-form [35017-84-6]

Waxy plates. Mp 28-32°.

Gunstone, F.D., *J. Chem. Soc.*, 1952, 1274.
Bharucha, K.E. *et al*, *J. Sci. Food Agric.*, 1955, **6**, 373.
Gunstone, F.D. *et al*, *J. Sci. Food Agric.*, 1959, **10**, 522.
Kennedy, J. *et al*, *J. Chem. Soc.*, 1961, 4945 (*synth*)
Applewhite, T. *et al*, *J. Org. Chem.*, 1967, **32**, 1173 (*abs config*)
Powell, R.G. *et al*, *Lipids*, 1969, **4**, 450 (*occur*)

10-Hydroxy-8-octadecenoic acid, 8CI H-00514

[94376-50-8]

$H_3C(CH_2)_7CH(OH)CH{=}CH(CH_2)_6COOH$

$C_{18}H_{34}O_3$ M 298.465

Present in autoxidised Oleic acid after reduction of the hydroperoxide.

(8E,10R)-form

Isol. from stromata of *Epichloe typhina*. Shows antifungal props.

Me ester:

$C_{19}H_{36}O_3$ M 312.492

$[\alpha]_D^{23}$ −1.48° (c, 0.27 in EtOH).

Ross, J. *et al*, *J. Am. Chem. Soc.*, 1949, **71**, 282.
Chan, H.W.S. *et al*, *Chem. Ind. (London)*, 1977, 692.
Garwood, R.F. *et al*, *J. Chem. Soc., Chem. Commun.*, 1977, 364 (*synth*)
Frankel, E.N. *et al*, *Lipids*, 1977, **12**, 901.
Mosset, P. *et al*, *Tetrahedron Lett.*, 1986, **27**, 299 (*synth*)
Koshino, T. *et al*, *Tetrahedron Lett.*, 1987, **28**, 73 (*isol, props*)

11-Hydroxy-9-octadecenoic acid H-00515

$H_3C(CH_2)_6CH(OH)CH{=}CH(CH_2)_7COOH$

$C_{18}H_{34}O_3$ M 298.465

Present in autoxidised Oleic acid after redn. of the hydroperoxide.

Ross, J. *et al*, *J. Am. Chem. Soc.*, 1949, **71**, 282.
Chan, M.W.S. *et al*, *Chem. Ind. (London)*, 1977, 692.
Frankel, E.N. *et al*, *Lipids*, 1977, **12**, 901.
Garwood, R.F. *et al*, *J. Chem. Soc., Chem. Commun.*, 1977, 364 (*synth*)

12-Hydroxy-9-octadecenoic acid H-00516

[7431-95-0]

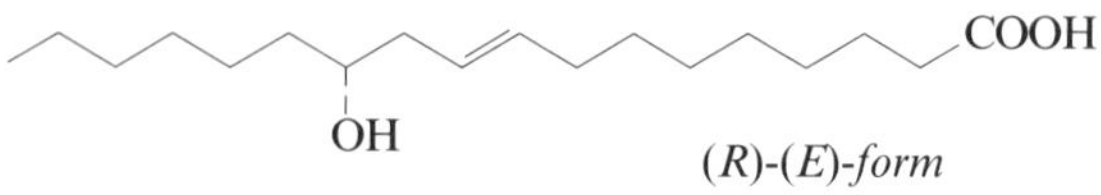

(*R*)-(*E*)-*form*

$C_{18}H_{34}O_3$ M 298.465

***(R,E)*-form** [540-12-5]

Ricinelaidic acid

Made from Ricinoleic acid by double-bond stereomutation. Mp 51-52°. Bp_{10} 240-242°. $[\alpha]_D^{20}$ +6.67° (EtOH).

Me ester:

$C_{19}H_{36}O_3$ M 312.492

$[\alpha]_{400}$ +12.7°.

Et ester:

$C_{20}H_{38}O_3$ M 326.518

Mp 16°.

Amide:

$C_{18}H_{35}NO_2$ M 297.480

Mp 91-93°.

Lactone:

$C_{18}H_{32}O_2$ M 280.450

Oil. $[\alpha]_D^{23}$ +40.4° (c, 0.98 in $CHCl_3$).

Lactone dimer: Mp 38-40°. $[\alpha]_D^{23}$ +1.2° (c, 0.85 in $CHCl_3$).

(R,Z)*-form** [141-22-0] ***Ricinoleic acid. *12-Hydroxyoleic acid. Ricinusoleic acid. Ricinic acid. Ricinolic acid*

Occurs in castor oil and other oils eg. grape, *Lesquerella densipila* and ergot (*Claviceps purpurea*). Used in polymer manuf., textile finishing and contraceptive jellies. Liq. Mp 5.5°. Bp_{10} 245°. $[\alpha]_D^{22}$ +6.67° (c, 5 in Me_2CO), $[\alpha]_D^{26}$ +7.15° (c, 5 in Me_2CO).

▷ VJ3150000.

Ac: [14936-79-9].

$C_{20}H_{36}O_4$ M 340.502

Bp_{13} 210°. $[\alpha]_D^{22}$ +15.25°.

Me ester: [141-24-2].

Bp_{15} 225-227°. $[\alpha]_{400}$ +14.9°.

Et ester: Bp_{13} 258°. $[\alpha]_D^{22}$ +5.28°.

***(S,E)*-form**

$[\alpha]_D$ −5.02° (c, 1.120 in MeOH).

Me ester: $[\alpha]_D$ −5.20° (c, 1.120 in MeOH).

***(S,Z)*-form**

$[\alpha]_D$ −4.98° (c, 1.0 in MeOH).

Me ester: $[\alpha]_D$ −5.11° (c, 1.027 in MeOH).

James, A.T. *et al*, *Biochem. J.*, 1965, **95**, 448 (*biosynth*)
Applewhite, T.H. *et al*, *J. Org. Chem.*, 1967, **32**, 1173 (*abs config*)
Suzuki, O. *et al*, *Yakugaku Zasshi*, 1970, **19**, 140 (*pmr*)
Chakrabarti, P. *et al*, *Biochemistry*, 1975, **14**, 5021 (*synth*)
Ariga, T. *et al*, *Chem. Phys. Lipids*, 1977, **19**, 14 (*ms*)
Rakoff, H. *et al*, *Lipids*, 1979, **14**, 81 (*cmr*)
McGhie, J.F. *et al*, *Chem. Ind. (London)*, 1982, 719.
Steliou, K. *et al*, *J. Am. Chem. Soc.*, 1983, **105**, 7130 (*synth, pmr, cmr, ms*)
Lewis, R.J., *Sax's Dangerous Properties of Industrial Materials*, 8th Ed., Van Nostrand-Reinhold, 1992, RJP000.

12-Hydroxy-10-octadecenoic acid H-00517

$H_3C(CH_2)_5CH(OH)CH{=}CH(CH_2)_8COOH$

$C_{18}H_{34}O_3$ M 298.465

(±)-*(E)*-form [95189-43-8]

Cryst. Mp 50-50.5° (45-46°).

Me ester: [95080-12-9].

$C_{19}H_{36}O_3$ M 312.492

Oil.

(±)-*(Z)*-form

Me ester: [95080-13-0].

Oil.

Frankel, E.N. *et al*, *J. Chem. Soc., Perkin Trans. 1*, 1984, 2233 (*synth, ir, pmr, ms*)

17-Hydroxy-9-octadecenoic acid, 8CI H-00518

$H_3CCH(OH)(CH_2)_6CH{=}CH(CH_2)_7COOH$

$C_{18}H_{34}O_3$ M 298.465

(+)-*(Z)*-form

Metab. of *Torulopsis magnoliae*. Mp 5°. $[\alpha]_D$ +4.3° (c, 6.1 in MeOH).

Me ester: [16104-93-1].

$C_{19}H_{36}O_3$ M 312.492

$Bp_{0.04}$ 145°. $[\alpha]_D$ +3.5° (c, 6.5 in MeOH).

Gorin, P.A.J. *et al*, *Can. J. Chem.*, 1961, **39**, 846.

18-Hydroxy-9-octadecenoic acid H-00519

$HOCH_2(CH_2)_7CH{=}CH(CH_2)_7COOH$

$C_{18}H_{34}O_3$ M 298.465

***(Z)*-form**

Constit. of cork (*Quercus suber*) and *Betula verrucosa*.

Me ester: [3329-38-2].

$C_{19}H_{36}O_3$ M 312.492

$Bp_{0.7}$ 185-190°, $Bp_{0.1}$ 130°.

Riba, I. *et al*, *An. R. Soc. Esp. Fis. Quim., Ser. B*, 1954, **50**, 971; *CA*, **50**, 806 (*isol*)
Jensen, W. *et al*, *CA*, 1958, **52**, 1067.
Duhamel, L., *Bull. Soc. Chim. Fr.*, 1965, 399 (*config*)
Eglinton, L. *et al*, *Phytochemistry*, 1968, **7**, 313 (*ms*)
Seoane, E. *et al*, *An. Quim.*, 1977, **73**, 1336 (*synth*)

8-Hydroxy-11-octadecen-9-ynoic acid H-00520

$H_3C(CH_2)_5CH{=}CHC{\equiv}CCH(OH)(CH_2)_6COOH$

$C_{18}H_{30}O_3$ M 294.433

(±)-*(E)*-form

8-Hydroxyximenynic acid

Isol. from oil of *Ximenia caffra*, prob. in an enantiomeric form. Brown semisolid. Mp 29-34°.

Me ester: $Bp_{0.01}$ 150-152°.

(±)-*(Z)*-form

Pale-yellow oil.

Me ester: $Bp_{0.01}$ 152-156°.

Crombie, L. *et al*, *J. Chem. Soc.*, 1958, 4435 (*synth*)
Gunstone, F.D. *et al*, *J. Chem. Soc.*, 1963, 5772 (*occur*)

9-Hydroxy-10-octadecen-12-ynoic acid, 8CI **H-00521**

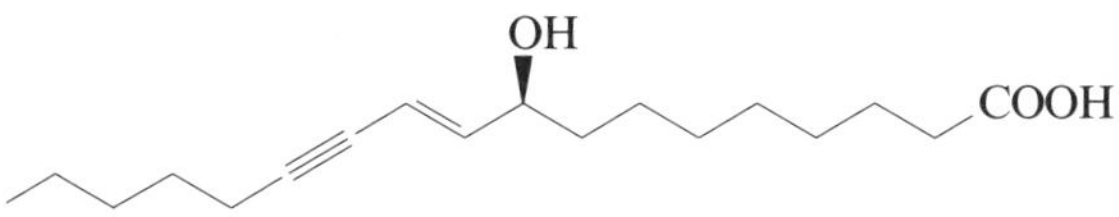

$C_{18}H_{30}O_3$ M 294.433

(S)-(E)-form

Helenynolic acid

Constit. of the seeds of *Helichrysum bracteatum*. $[\alpha]^{26}_{600}$ −7° (c, 3.6 in EtOH).

Me ester: Oil. $[\alpha]^{26}_{600}$ −7° (c, 3.6 in EtOH).

Powell, R.G. *et al*, *J. Am. Oil Chem. Soc.*, 1965, **42**, 165 (*uv*)
Powell, R.G. *et al*, *J. Org. Chem.*, 1965, **30**, 610 (*ir, pmr*)
Craig, J.C. *et al*, *J. Org. Chem.*, 1965, **30**, 4342 (*abs config*)
Patrick, T.B. *et al*, *J. Org. Chem.*, 1979, **44**, 645 (*synth*)

9-Hydroxy-10-octadecynoic acid **H-00522**

$H_3C(CH_2)_6C{\equiv}CCH(OH)(CH_2)_7COOH$

$C_{18}H_{32}O_3$ M 296.449

(±)-form

Me ester: [95080-10-7].

$C_{19}H_{34}O_3$ M 310.476

Oil.

Frankel, E.N. *et al*, *J. Chem. Soc., Perkin Trans. 1*, 1984, 2233 (*synth, ir, pmr, ms*)

12-Hydroxy-9-octadecynoic acid, 9CI **H-00523**

Ricinstearolic acid

[540-13-6]

$H_3C(CH_2)_5CH(OH)CH_2C{\equiv}C(CH_2)_7COOH$

$C_{18}H_{32}O_3$ M 296.449

(−)-form

Mp 52°. Bp_{10} 260°. $[\alpha]_D$ −5.54° (c, 1.123 in MeOH).

Me ester: $[\alpha]_D$ −5.08° (c, 1.82 in MeOH).

(±)-form

Me ester: [71692-12-1].

$C_{19}H_{34}O_3$ M 310.476

Oil. Bp_{12} 225°.

Crombie, L. *et al*, *J. Chem. Soc.*, 1955, 1740 (*synth*)
Jadhav, K.S. *et al*, *Indian J. Chem., Sect. B*, 1978, **16**, 275 (*synth*)
Jadhav, K.S. *et al*, *Indian J. Chem., Sect. B*, 1978, **16**, 280 (*lactone*)
McGhie, J.F. *et al*, *Chem. Ind. (London)*, 1982, 719.

12-Hydroxy-10-octadecynoic acid **H-00524**

$H_3C(CH_2)_5CH(OH)C{\equiv}C(CH_2)_8COOH$

$C_{18}H_{32}O_3$ M 296.449

(±)-form

Me ester: [95080-08-3].

$C_{19}H_{34}O_3$ M 310.476

Oil.

Frankel, E.N. *et al*, *J. Chem. Soc., Perkin Trans. 1*, 1984, 2233 (*synth, ir, pmr, ms*)

2-Hydroxyoctanoic acid, 9CI, 8CI **H-00525**

α-Hydroxycaprylic acid

[617-73-2]

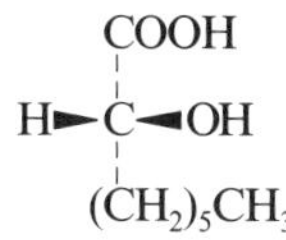

(R)-form

$C_8H_{16}O_3$ M 160.213

(R)-form [30117-44-3]

Cryst.

Ba salt: $[\alpha]^{26}_D$ +9.1° (1M NaOH).

Me ester: [92572-96-8].

$C_9H_{18}O_3$ M 174.239

Liq. Bp_{10} 110°. $[\alpha]^{20}_D$ −9.91° (c, 1.95 in $CHCl_3$).

(S)-form [70267-27-5]

Cryst. (diisopropyl ether). Mp 70.2-70.6°. $[\alpha]^{19}_D$ +6.2° (c, 9 in $CHCl_3$).

Ba salt: Cryst. $[\alpha]^{26}_D$ −9.3° (c, 1 in aq. NaCl).

Me ester: Liq. d^{16}_4 0.9686. Bp_{22} 118°. $[\alpha]^{16}_D$ +11° (c, 10 in $CHCl_3$). n^{18}_D 1.4342.

(±)-form [6482-96-8]

Cryst. (C_6H_6). Mp 70-71°.

Et ester: [4219-47-0].

$C_{10}H_{20}O_3$ M 188.266

Liq. Bp_{715} 229-230°, Bp_{20} 130°.

Benzyl ester: [95513-13-6].

$C_{15}H_{22}O_3$ M 250.337

Bp_{400} 110°.

Amide:

$C_8H_{17}NO_2$ M 159.228

Plates. Mp 150°.

Nitrile: 1-Cyanoheptanol

$C_8H_{15}NO$ M 141.213

Liq. d^{17} 0.905. Mp −10°. Bp_{19} 143.5-144°.

[6482-96-8]

Boeseken, M.J. *et al*, *Recl. Trav. Chim. Pays-Bas (J. R. Neth. Chem. Soc.)*, 1918, **37**, 165 (*synth, props*)
Verhulst, J., *Bull. Soc. Chim. Belg.*, 1931, **40**, 486 (*amide*)
Bruylants, P. *et al*, *Bull. Soc. Chim. Belg.*, 1934, **43**, 211 (*nitrile*)
Baker, C.G. *et al*, *J. Am. Chem. Soc.*, 1951, **73**, 1336 (*synth*)
Horn, D.H.S. *et al*, *J. Chem. Soc.*, 1954, 1460 (*synth*)
Solokov, V.P. *et al*, *Zh. Obshch. Khim.*, 1972, **42**, 219 (*resoln*)
Ger. Pat., 2 832 230, (1979); *CA*, **90**, 202244b (*resoln*)
Ko, K.-Y. *et al*, *Tetrahedron*, 1984, **40**, 1333 (*esters, synth, ir, pmr, cmr*)
Shiosaki, K. *et al*, *J. Org. Chem.*, 1985, **50**, 1229 (*synth, ir, pmr*)
Larcheveque, M. *et al*, *Bull. Soc. Chim. Fr.*, 1989, 130 (*ester*)

3-Hydroxyoctanoic acid, 9CI **H-00526**

3-Hydroxycaprylic acid

[14292-27-4]

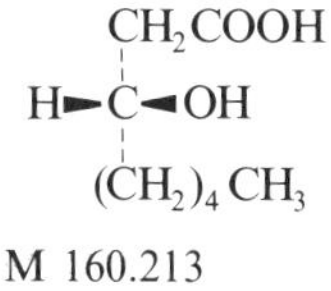

(R)-form

$C_8H_{16}O_3$ M 160.213

(R)-form [44987-72-6]

Isol. from secretion of the metathoracic glands of myrmicine ants. Also found in milk, strawberries and pineapple concentrate. $[\alpha]^{24}_D$ −21° (c, 1.9 in $CHCl_3$).

Me ester: [78672-90-9].

$C_9H_{18}O_3$ M 174.239

$[\alpha]^{24}_D$ −18.4° (c, 1 in $CHCl_3$).

Hydrazide:

$C_8H_{18}N_2O_2$ M 174.242
Mp 127.5-128°. $[\alpha]_D^{24}$ −11.4° (c, 1 in H_2O).

***(S)*-form** [33796-86-0]
Mp 20-22°. $Bp_{0.15}$ 123°. $[\alpha]_D^{18}$ +23.0° (c, 2 in $CHCl_3$).
Me ester: [66997-64-6].
Oil. $[\alpha]_D^{18}$ −27.1° (c, 1 in $CHCl_3$).

(±)-form
Mp 38-38.5°.
Et ester: [7367-90-0].
$C_{10}H_{20}O_3$ M 188.266
Oil. Bp_5 101-104°.
Hydrazide: [33824-91-8].
Mp 105-107°.

Lemieux, R.U. *et al*, *Can. J. Chem.*, 1951, **29**, 678 (*synth*)
Schildknecht, H. *et al*, *Angew. Chem., Int. Ed. Engl.*, 1971, **10**, 124 (*isol*)
Kuwajima, I., *Tetrahedron Lett.*, 1976, 2253 (*synth*)
Tahara, S. *et al*, *Agric. Biol. Chem.*, 1978, **42**, 879 (*synth*)
Ketelaar, P.E.F. *et al*, *Tetrahedron Lett.*, 1985, **26**, 4665 (*synth*)
Schreiber, S.L. *et al*, *J. Am. Chem. Soc.*, 1988, **110**, 6210 (*Me ester, synth, pmr, cmr, ms, ir*)

4-Hydroxyoctanoic acid, 9CI — H-00527

4-Hydroxycaprylic acid
[7779-55-7]

$H_3C(CH_2)_3CH(OH)CH_2CH_2COOH$

$C_8H_{16}O_3$ M 160.213
Free acid immediately lactonises.

***(R)*-form**
Et ester: [107736-67-4].
$C_{10}H_{20}O_3$ M 188.266
Liq.

(±)-form
Hydrazide: [24535-12-4].
$C_8H_{18}N_2O_2$ M 174.242
Cryst. Mp 79-80°.
Lactone: [104-50-7]. *5-Butyldihydro-2(3*H*)furanone, 9CI. γ-Octalactone. γ-Butyl-γ-butyrolactone*
Liq. Bp_{20} 132-133°, $Bp_{5.5}$ 102-105°.
▷ LU3562000.

Sadtler Standard Infrared Spectra, No. 20213 (*lactone*)
Sadtler Standard NMR Spectra, Nos. 2500, 2362 (*lactone, pmr, cmr*)
Terai, Y. *et al*, *Bull. Chem. Soc. Jpn.*, 1956, **29**, 822 (*lactone, hydrazide*)
Schmidt, U. *et al*, *Justus Liebigs Ann. Chem.*, 1963, **666**, 201 (*synth*)
McFadden, W.H. *et al*, *Anal. Chem.*, 1965, **37**, 89 (*lactone, ms*)
Sedavkina, V., *CA*. 1976, **84**, 30362 (*synth*)
Thijs, L. *et al*, *Recl. Trav. Chim. Pays-Bas* (*J. R. Neth. Chem. Soc.*), 1986, **105**, 332 (*ester*)
Lewis, R.J., *Sax's Dangerous Properties of Industrial Materials*, 8th Ed., Van Nostrand-Reinhold, 1992, OCE000.

5-Hydroxyoctanoic acid, 9CI, 8CI — H-00528

[17369-50-5]

$H_3CCH_2CH_2CH(OH)CH_2CH_2CH_2COOH$

$C_8H_{16}O_3$ M 160.213
Present in milk fat lipids.

(±)-form
Lactonises on dist.
Lactone:
$C_8H_{14}O_2$ M 142.197
Liq. $Bp_{1.9}$ 95-97°. n_D^{25} 1.4535.
Hydrazide: [16425-11-9].
$C_8H_{18}N_2O_2$ M 174.242
Cryst. Mp 97-98°.

Honkanenea, E., *Acta Chem. Scand.*, 1965, **19**, 370 (*synth, ms*)
McFadden, W.H. *et al*, *Anal. Chem.*, 1965, **37**, 89 (*ms*)
Wyatt, C.J., *Lipids*, 1967, **2**, 208 (*isol*)
Lardelli, G. *et al*, *Recl. Trav. Chim. Pays-Bas* (*J. R. Neth. Chem. Soc.*), 1967, **86**, 481 (*synth*)

6-Hydroxyoctanoic acid, 9CI, 8CI — H-00529

6-Hydroxycaprylic acid

$H_3CCH_2CH(OH)CH_2CH_2CH_2CH_2COOH$

$C_8H_{16}O_3$ M 160.213

(±)-form
Viscous liq. Lactonises on dist.
Et ester:
$C_{10}H_{20}O_3$ M 188.266
Liq. d_4^{20} 0.96. Bp_{14} 140-141°. n_D^{20} 1.4405.
Hydrazide:
$C_8H_{18}N_2O_2$ M 174.242
Cryst. Mp 83-84°.
Lactone:
$C_8H_{14}O_2$ M 142.197
Liq. Bp_{10} 114-115°.

Mornet, R. *et al*, *Bull. Soc. Chim. Fr.*, 1965, 3043 (*ester*)
Lardelli, G., *Recl. Trav. Chim. Pays-Bas* (*J. R. Neth. Chem. Soc.*), 1967, **86**, 481 (*synth, lactone, hydrazide*)

7-Hydroxyoctanoic acid, 9CI, 8CI — H-00530

[17173-14-7]

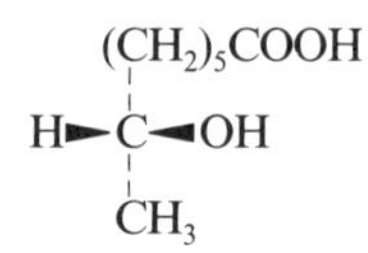

(*R*)-form

$C_8H_{16}O_3$ M 160.213

***(R)*-form** [119434-02-5]
Me ester: [119434-03-6].
$[\alpha]_D^{22}$ −8.3° (c, 1.8 in $CHCl_3$).

***(S)*-form**
Me ester: [88851-70-1].
$C_9H_{18}O_3$ M 174.239
Liq. Bp_{12} 128-132°. $[\alpha]_D$ +10.7° (c, 2.7 in cyclohexane).
Lactone: [126372-23-4]. *8-Methyl-7-oxocanone, 9CI. 7-Octanolide*
$C_8H_{14}O_2$ M 142.197
$[\alpha]_D^{20}$ +67.8° (c, 2.2 in THF) (99% e.e.).

(±)-form [77842-06-9]
Bp_4 150-160°. May dec. on dist.
Me ester: [77758-48-6].
Oil.
Et ester:
$C_{10}H_{20}O_3$ M 188.266
Liq. d_{25}^{25} 0.962. Bp_2 111-113°. n_D^{25} 1.4387.
Benzyl ester: [16209-10-2].
$C_{15}H_{22}O_3$ M 250.337
Liq. Bp_3 145-149°.
Ac: [17173-15-8].
$C_{10}H_{18}O_4$ M 202.250
Bp_1 128°.
Lactone: [119366-72-2].
Liq. $Bp_{0.1}$ 34-36°.
Me ether: [821-68-1]. *7-Methoxyoctanoic acid*
$C_9H_{18}O_3$ M 174.239
d 0.98. Bp_1 121-122°.
Me ether, Me ester:

$C_{10}H_{20}O_3$ M 188.266
d 0.94. Bp_{20} 118-119°.
Chloride, Ac:
$C_{10}H_{17}ClO_3$ M 220.695
Liq. Bp_5 158°.

[126372-27-8]

Lease, E.J. *et al, J. Am. Chem. Soc.*, 1933, **55**, 807 (*Et ester*)
Michel, C. *et al, Bull. Soc. Chim. Fr.*, 1964, 2230 (*deriv.*)
Sugiyama, N. *et al, Bull. Chem. Soc. Jpn.*, 1967, **40**, 2713 (*synth, ir*)
Baker, P.M. *et al, J. Chem. Soc. C*, 1967, 1913 (*benzyl ester*)
Musgrave, O.C. *et al, J. Chem. Soc. C*, 1968, 250 (*synth, ir*)
Voss, G., *Helv. Chim. Acta*, 1983, **66**, 2294 (*synth*)
Lefevre, M.F., *J. Chromatogr. Sci.*, 1989, **27**, 23 (*chromatog, ms*)
Ngooi, T.K. *et al, J. Org. Chem.*, 1989, **54**, 911 (*synth, ir, pmr, resoln*)
Fouque, E. *et al, Synthesis*, 1989, 661 (*lactone*)
Birch, A.J. *et al, J. Chem. Soc., Perkin Trans. 1*, 1990, 1423 (*synth, ir, pmr*)

8-Hydroxyoctanoic acid H-00531

8-Hydroxycaprylic acid
[764-89-6]

$HOCH_2(CH_2)_6COOH$

$C_8H_{16}O_3$ M 160.213
Isol. from *Apis mellifera* royal jelly. Sol. H_2O, spar. sol. pet. ether. Mp 62-63° (58°).
Me ester: [20257-95-8].
$C_9H_{18}O_3$ M 174.239
Oil. Bp_8 137-138°.
Ac: [871-66-9].
$C_{10}H_{18}O_4$ M 202.250
Mp 9-10°. $Bp_{1.5}$ 155-158°.
Amide:
$C_8H_{17}NO_2$ M 159.228
Cryst. Mp 76-78°.
Hydrazide:
$C_8H_{18}N_2O_2$ M 174.242
Cryst. (Et_2O). Mp 130.5-132°.
Lactone:
$C_8H_{14}O_2$ M 142.197
Liq.
Me ether: [90677-38-6]. *8-Methoxyoctanoic acid*
$C_9H_{18}O_3$ M 174.239
$Bp_{0.5}$ 98-100°.
Me ether, chloride: [95978-39-5].
$C_9H_{17}ClO_2$ M 192.685
$Bp_{0.5}$ 80-85°.

Chuit, P. *et al, Helv. Chim. Acta*, 1929, **12**, 463 (*synth*)
Palomaa, M.H., *Chem. Ber.*, 1941, **74**, 294 (*Me ether*)
Friess, S.L. *et al, J. Am. Chem. Soc.*, 1952, **74**, 2679 (*hydrazide*)
Bulock, J.D. *et al, Chem. Ind. (London)*, 1954, 990 (*amide*)
Bestmann, H.J. *et al, Justus Liebigs Ann. Chem.*, 1966, **699**, 33 (*synth*)
Weaver, N. *et al, Lipids*, 1968, **3**, 535 (*isol*)
Mukaiyama, T. *et al, Chem. Lett.*, 1976, 49 (*lactone*)
Anol, R.A. *et al, Lipids*, 1982, **17**, 414 (*synth*)
Goodman, M.M. *et al, J. Med. Chem.*, 1985, **28**, 807 (*deriv, synth, pmr, ms*)

2-Hydroxy-3-octanone H-00532

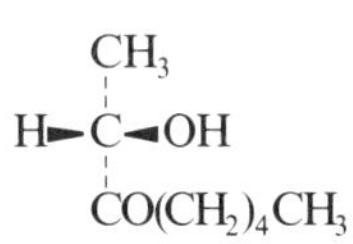

$C_8H_{16}O_2$ M 144.213

(*S*)-form [84435-13-2]
Constit. of male sex pheromone of the grape borer *Xylotrechus pyrrhoderus* and *X. chinensis.*

Sakai, T. *et al, Chem. Lett.*, 1984, 263 (*isol, synth*)
Bel-Rhlid, R. *et al, J. Org. Chem.*, 1989, **54**, 322 (*synth*)

7-[6-(3-Hydroxy-1-octenyl)-2,3-diazabicyclo[2.2.1]hept-2-en-5-yl]-5-heptenoic acid, 9CI H-00533

9,11-Diazo-15-hydroxyprosta-5,13-dienoic acid
[57712-08-0]

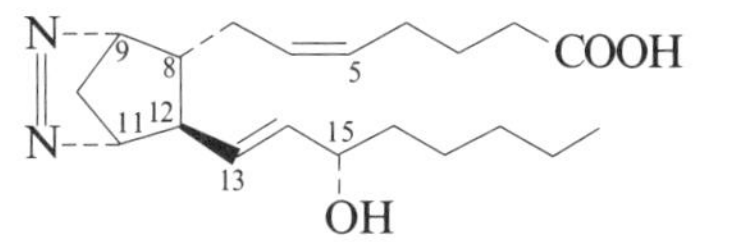

Prostanoid numbering shown

$C_{20}H_{32}N_2O_3$ M 348.484
(5Z,9S,11R,13E,15S)-form
(9α,11α,15α)-(5Z,13E)-form
Powerful mimic of the endoperoxides PGH_2 and PGG_2 with reference to platelet aggregation and Serotonin release, about seven times more active than PGG_2 in stimulating muscle contraction. Cryst. (pentane/Et_2O). Mp 74-75°. $[\alpha]_D^{22}$ −17.7° ($CHCl_3$). Stable indefinitely at 0° in aq. soln. or as the pure solid.

Corey, E.J. *et al, Proc. Natl. Acad. Sci. U.S.A.*, 1975, **72**, 3355 (*synth, uv, ir*)
Corey, E.J. *et al, J. Am. Chem. Soc.*, 1976, **98**, 6417 (*synth*)
Fitzpatrick, F.A. *et al, Nature (London)*, 1978, **275**, 764 (*pharmacol*)

5-(3-Hydroxy-1-octenyl)-3-methyl-2-oxo-1-imidazolidineheptanoic acid, 9CI H-00534

[77580-27-9]

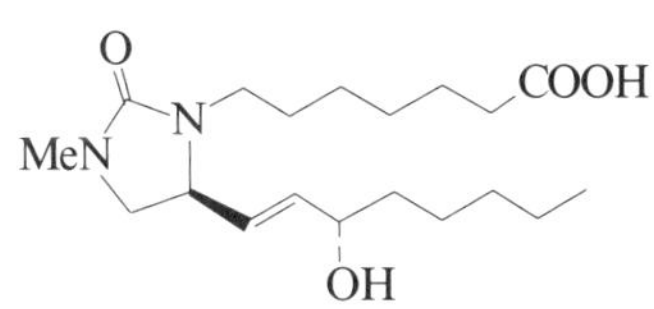

$C_{19}H_{34}N_2O_4$ M 354.489
(12R,13E,15S)-form
11-Deoxy-10-methyl-8,10-diaza-PGE$_1$
Cryst. (EtOAc/hexane). Mp 85-86°.
Me ester:
$C_{20}H_{36}N_2O_4$ M 368.515
Cryst. (EtOAc/pet. ether). Mp 58-59°.
(12R,13E,15R)-form
*11-Deoxy-8,10-diaza-10-methyl-15-*epi-*PGE$_1$*
Oil.
Me ester: Oil.

Saijo, S. *et al, Chem. Pharm. Bull.*, 1980, **28**, 1459 (*synth, ir, pmr, deriv*)

4-(3-Hydroxy-1-octenyl)-1-methyl-2-oxo-3-pyrrolidineheptanoic acid, 9CI H-00535

10-Aza-15-hydroxy-10-methyl-9-oxo-13-prostenoic acid
[63599-19-9]

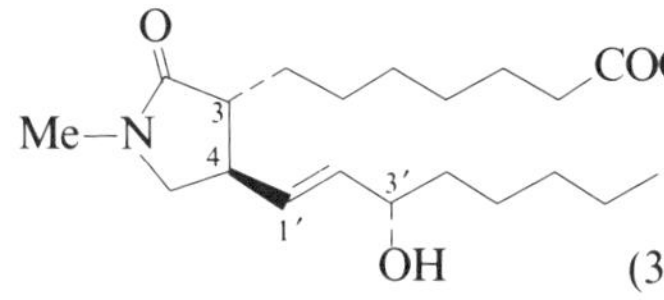

(3RS,4RS,1′E,3′SR)-form

$C_{20}H_{35}NO_4$ M 353.501

(3RS,4RS,1'E,3'SR)-form

(±)-*10-Aza-11-deoxy-PGE*$_1$

Cryst. (MeOH/Et$_2$O). Mp 123-124°.

Me ester: [63599-22-4].

Cryst. (MeOH/Et$_2$O). Mp 54-54.5°.

(3RS,4RS,1'E,3'RS)-form [63599-20-2]

(±)-*10-Aza-11-deoxy-15-*epi-*PGE*$_1$

Cryst. (MeOH/Et$_2$O). Mp 83-84°.

Me ester: [63599-21-3].

Cryst. (MeOH/Et$_2$O). Mp 50-51°.

Zoretic, P.A. *et al*, *J. Org. Chem.*, 1980, **45**, 810 (*synth, ir, pmr, ms*)

7-[6-(3-Hydroxy-1-octenyl)-2-oxabicyclo[2.2.1]hept-5-yl]-5-heptenoic acid, 9CI — H-00536

11,9-Epoxymethano-15-hydroxy-prosta-5,13-dienoic acid

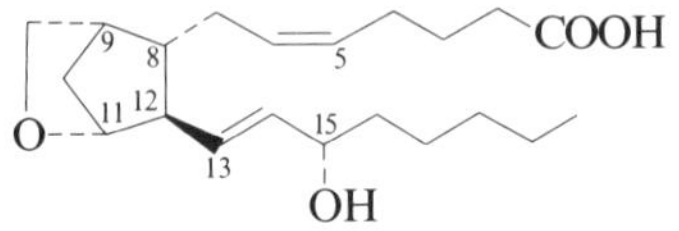

$C_{21}H_{34}O_4$ M 350.497

Prostanoid numbering shown.

(5Z,9R*,11R*,13E,15S*)-form [56985-40-1]

(9α,11α,15α)-(5Z,13E)-form

Potent bronchoconstrictor.

▷ MJ9681090.

Bundy, G.L., *Tetrahedron Lett.*, 1975, 1957 (*synth, pmr, ms*)

2-(3-Hydroxy-1-octenyl)-4-oxo-1-azetidineheptanoic acid, 9CI — H-00537

8-Aza-15-hydroxy-9-oxo-11-nor-13-prostenoic acid. 8-Aza-11-nor-PGE$_1$

[68254-96-6]

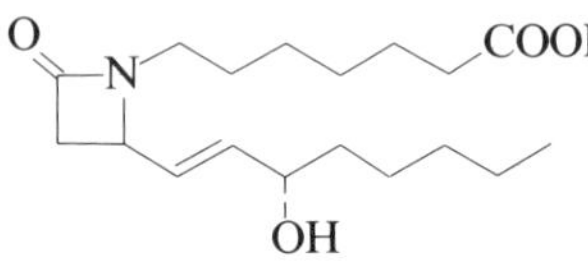

$C_{18}H_{31}NO_4$ M 325.447

Synthetic prostaglandin analogue.

Duprès, J.-P. *et al*, *Tetrahedron Lett.*, 1978, 2191 (*synth, ir, pmr*)

2-(3-Hydroxy-1-octenyl)-4-oxo-3-azetidineheptanoic acid, 9CI — H-00538

10-Aza-15-hydroxy-9-oxo-11-nor-5,13-prostadienoic acid. 10-Aza-11-nor-PGE$_1$

[68254-98-8]

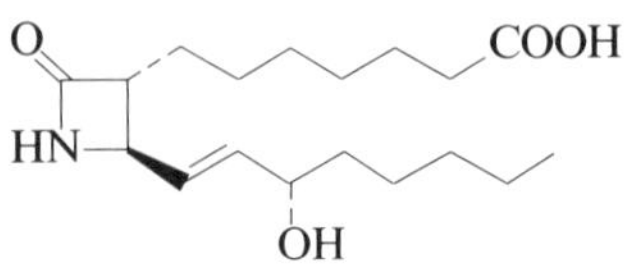

$C_{18}H_{31}NO_4$ M 325.447

Synethic prostaglandin analogue.

Duprès, J.-P. *et al*, *Tetrahedron Lett.*, 1978, 2191 (*synth, ir, pmr*)

7-[5-(3-Hydroxy-1-octenyl)-2-oxobicyclo[2.2.1]hept-6-yl]-5-heptenoic acid, 9CI — H-00539

9,11-Epoxymethano-15-hydroxyprosta-5,13-dienoic acid

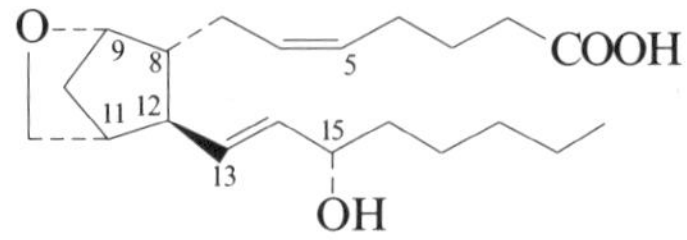

Prostanoid numbering shown

$C_{21}H_{34}O_4$ M 350.497

(5Z,9S*,11S*,13E,15S*)-form [56985-32-1]

(9α,11α,15α)-(5Z,13E)-form

Potent bronchoconstrictor.

Bundy, G.L., *Tetrahedron Lett.*, 1975, 1957 (*synth, pmr*)

5-(3-Hydroxy-1-octenyl)-2-oxo-1-imidazolidineheptanoic acid, 9CI — H-00540

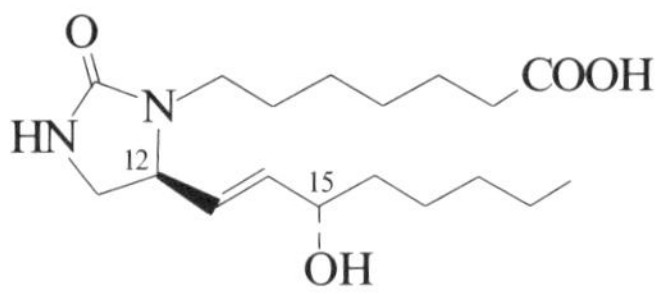

$C_{18}H_{32}N_2O_4$ M 340.462

(12RS,13E,15SR)-form [76453-44-6]

(±)-*11-Deoxy-8,10-diaza-PGE*$_1$

Cryst. (EtOAc). Mp 122-123°.

Me ester:

$C_{19}H_{34}N_2O_4$ M 354.489

Cryst. (EtOAc/hexane). Mp 114-115°.

(12RS,13E,15RS)-form [76453-43-5]

(±)-*11-Deoxy-8,10-diaza-15-*epi-*PGE*$_1$

Cryst. (EtOAc/hexane). Mp 69-70°.

Me ester: Cryst. (EtOAc/hexane). Mp 96-97°.

Saijo, S. *et al*, *Chem. Pharm. Bull.*, 1980, **28**, 1459 (*synth, ir, pmr*)

2-(3-Hydroxy-1-octenyl)-ζ-oxo-1-piperidineheptanoic acid — H-00541

7-Oxo-8-aza-10a-homo-13-prostenoic acid

[65950-27-8]

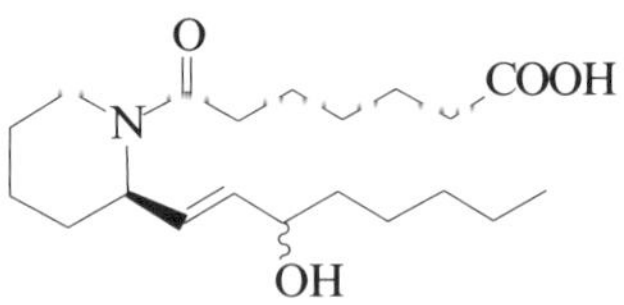

$C_{20}H_{35}NO_4$ M 353.501

Inhibits gastric acid secretion.

Zoretic, P.A. *et al*, *J. Heterocycl. Chem.*, 1977, **14**, 1293 (*synth, ir, pmr, ms*)

7-[2-(3-Hydroxy-1-octenyl)-5-oxo-1-pyrrolidinyl]-5-heptenoic acid — H-00542

11-Deoxy-8-aza-PGE$_2$

[60380-19-0]

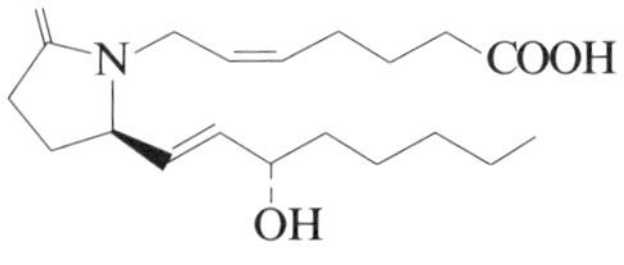

$C_{19}H_{31}NO_4$ M 337.458
Synthetic prostaglandin analogue.

Bruin, J.W. *et al*, *Tetrahedron Lett.*, 1975, 4599 (*synth, ir, pmr, ms*)

7-Hydroxy-4-oxo-2,5-heptadienoic acid H-00543

$HOCH_2CH=CHCOCH=CHCOOH$

$C_7H_8O_4$ M 156.138

(E,E)-form

O-Benzoyl, Me ester: [130518-24-0]. ***Melodienone***
$C_{15}H_{14}O_5$ M 274.273
Constit. of stem bark of *Melodorum fruticosum*. Cytotoxic. Light yellow cryst. (hexane/Me_2CO). Mp 69-70°.

O-Benzoyl, Et ester: [135626-20-9]. ***Homomelodienone***
$C_{16}H_{16}O_5$ M 288.299
Constit. of *M. fruticosum*. Cytotoxic. Oil.

(2Z,5E)-form

O-Benzoyl, Me ester: [130518-23-9]. ***Isomelodienone***
$C_{15}H_{14}O_5$ M 274.273
Constit. of *M. fruticosum*. Cytotoxic. Light yellow liq.

O-Benzoyl, Et ester: [135626-22-1]. ***Homoisomelodienone***
$C_{16}H_{16}O_5$ M 288.299
Constit. of *M. fruticosum*. Cytotoxic. Oil.

Jung, J.H. *et al*, *Tetrahedron*, 1990, **46**, 5043 (*isol, struct*)
Jung, J.H. *et al*, *J. Nat. Prod. (Lloydia)*, 1991, **54**, 500 (*isol, pmr, cmr*)

4-Hydroxy-2-oxoheptanedioic acid, 9CI H-00544

4-Hydroxy-2-ketopimelic acid
[53795-98-5]

$HOOCCH_2CH_2CH(OH)CH_2COCOOH$

$C_7H_{10}O_6$ M 190.152
Present in the fern *Asplenium septentrionale*.

(±)-form

1,4-Lactone:
$C_7H_8O_5$ M 172.137
Cryst. (EtOAc). Mp 116-117°.

Virtanen, A.I. *et al*, *Acta Chem. Scand.*, 1954, **8**, 1720 (*isol*)
Leung, P. *et al*, *J. Bacteriol.*, 1974, **120**, 168 (*synth*)

3-Hydroxy-2-(1-oxo-12,14-hexadienyl)-2-cyclohexen-1-one H-00545

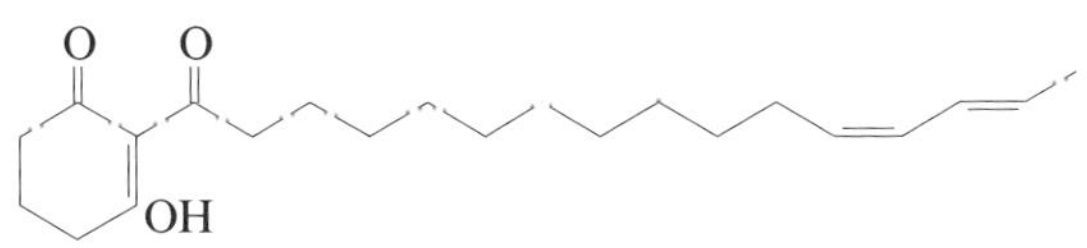

$C_{22}H_{34}O_3$ M 346.509

(12′Z,14′E)-form [79053-03-5]
Isol. from larval mandibular glands of *Ephestia kühnellia*.

Mudd, A., *J. Chem. Soc., Perkin Trans. 1*, 1981, 2357 (*isol, ms, pmr, cmr*)

2-Hydroxy-18-oxo-1,2,3-nonadecanetricarboxylic acid, 9CI H-00546

3,4-Dicarboxy-3-hydroxy-19-oxoeicosanoic acid

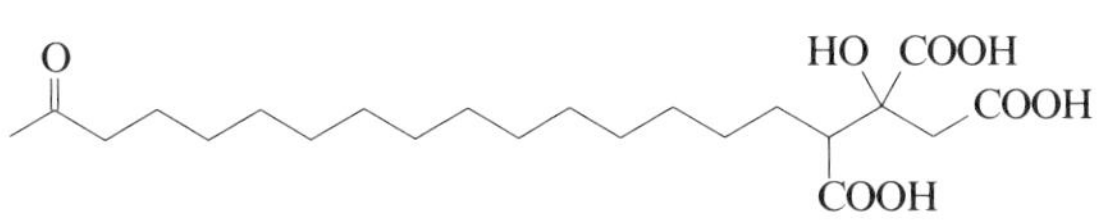

$C_{22}H_{38}O_8$ M 430.537

Tri-Me ester: [64767-77-7].
$C_{25}H_{44}O_8$ M 472.618
Constit. of *Usnea meridensis*. Cryst. (MeOH aq.). Mp 125°. $[\alpha]_D^{25}$ −15.4° (c, 0.06 in dioxan).

Keogh, M.F. *et al*, *Phytochemistry*, 1977, **16**, 1605.

3-Hydroxy-2-(1-oxo-9-octadecenyl)-2-cyclohexen-1-one H-00547

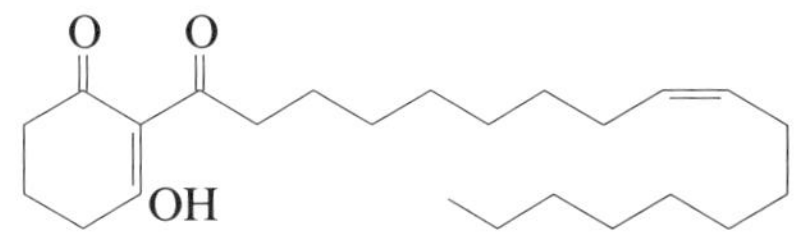

$C_{24}H_{40}O_3$ M 376.578

(Z)-form [79048-56-9]
Isol. from larval mandibular glands of *Ephestia kühnellia*.

Mudd, A., *J. Chem. Soc., Perkin Trans. 1*, 1981, 2357 (*isol, ms, pmr, cmr*)

15-Hydroxy-9-oxo-4,13-prostadienoic acid, 9CI H-00548

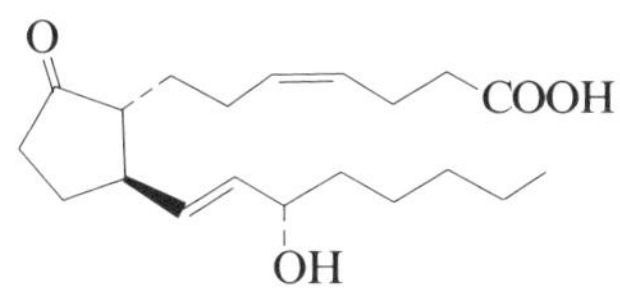

$C_{20}H_{32}O_4$ M 336.470

(4Z,13E)-form

U.S. Pat., 4 098 815, (1978); *CA*, **90**, P38588u (*synth*)

15-Hydroxy-9-oxo-5,13-prostadienoic acid, 9CI H-00549

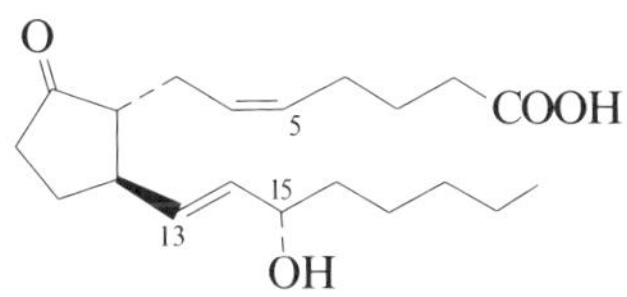

$C_{20}H_{32}O_4$ M 336.470

(5Z,13E,15S)-form [35183-82-5]
11-Deoxy-PGE$_2$
Inhibits blood platelet aggregation.

Abraham, M.A., *Tetrahedron Lett.*, 1973, 451 (*synth*)
Cooper, G.F. *et al*, *Proc. Int. Conf. Stable Isot. Chem., Biol., Med., 1st*, 1973, 72; *CA*, **81**, 48947p (*cmr*)
Corey, E.J. *et al*, *J. Org. Chem.*, 1975, **40**, 2265 (*synth*)

15-Hydroxy-9-oxo-8(12),13-prostadienoic acid, 9CI H-00550

2-(3-Hydroxy-1-octenyl)-5-oxo-1-cyclopentene-1-heptanoic acid, 8CI

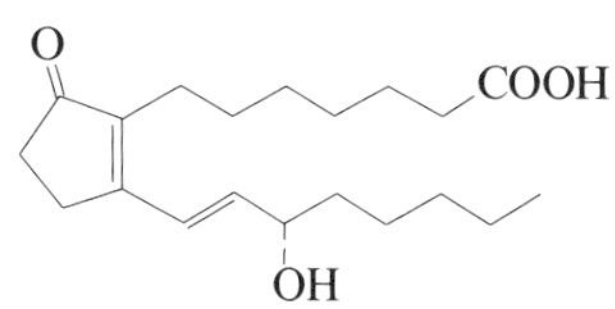

$C_{20}H_{32}O_4$ M 336.470

(13E,15S)-form [13345-51-2]

PGB_1. Prostaglandin B_1. Prostaglandin E_1-278
Occurs in human seminal fluid. Inhibits uterine motility (*in vitro*). Cryst.

Hamberg, M. *et al, Prostaglandins, Proc. Nobel Symp., 2nd., 1966*, 1967, 63.
Collins, P. *et al, Isr. J. Chem.*, 1968, **6**, 839 (*synth, pmr, uv*)
Pappo, R. *et al, Ann. N.Y. Acad. Sci.*, 1971, **180**, 64 (*synth*)

15-Hydroxy-9-oxo-10,13-prostadienoic acid, 9CI — H-00551

2-(3-Hydroxy-1-octenyl)-5-oxo-2-cyclopentene-1-heptanoic acid, 8CI

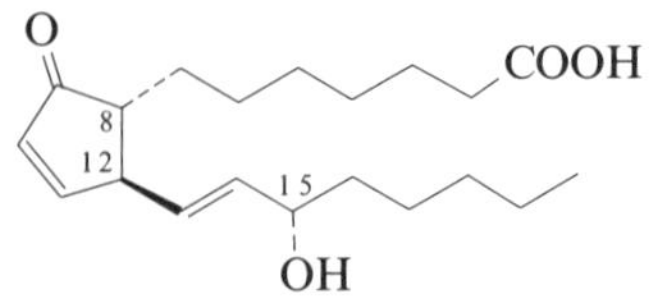

$C_{20}H_{32}O_4$ M 336.470

(8R,12S,13E,15S)-form [14152-28-4]
Prostaglandin A_1. PGA_1
Occurs in semen. Biosynth. from 8,11,14-Eicosatrienoic acid, E-00047. Inhibits blood platelet aggregation and smooth muscle contraction. Stimulates renin release. Cryst. Mp 42-44°.
▷ GY5977900.

(8RS,12SR,13E,15SR)-form [20348-61-2]
(±)-*PGA_1*
Oil.

Pike, J.E. *et al, Prostaglandins, Proc. Nobel Symp., 2nd., 1966*, 1967 (*synth, ms, pmr, ir*)
Corey, E.J. *et al, J. Am. Chem. Soc.*, 1968, **90**, 3245 (*synth*)
Middleditch, B.S. *et al, Prostaglandins*, 1973, **4**, 31 (*ms*)
Hahn, J. *et al, Prostaglandins*, 1980, **20**, 15 (*biosynth*)
Lewis, R.J., *Sax's Dangerous Properties of Industrial Materials*, 8th Ed., Van Nostrand-Reinhold, 1992, POC250.

15-Hydroxy-9-oxo-5,10,13,17-prostatetraenoic acid — H-00552

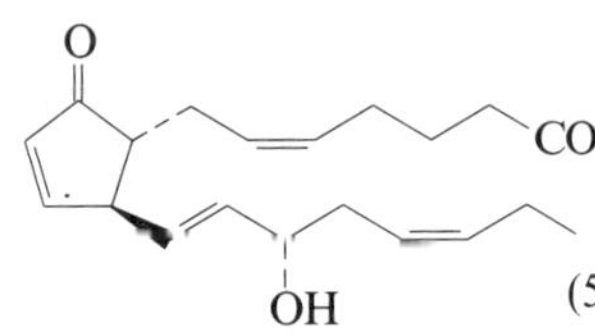

(5*Z*,13*E*,15*S*,17*Z*)-*form*

$C_{20}H_{28}O_4$ M 332.439

(5Z,13E,15S,17Z)-form [36614-31-0] ***Prostaglandin A_3***
Isol. from seminal fluid of sheep.

Ger. Pat., 2 314 519, (1973); *CA*, **80**, 595645 (*isom*)
Schevchenko, V.P. *et al, J. Labelled Compd. Radiopharm.*, 1989, **27**, 1177 (*synth*)

15-Hydroxy-9-oxo-5,8(12),13-prostatrienoic acid, 9CI — H-00553

7-[2-(3-Hydroxy-1-octenyl)-5-oxo-1-cyclopenten-1-yl]-5-heptenoic acid, 8CI

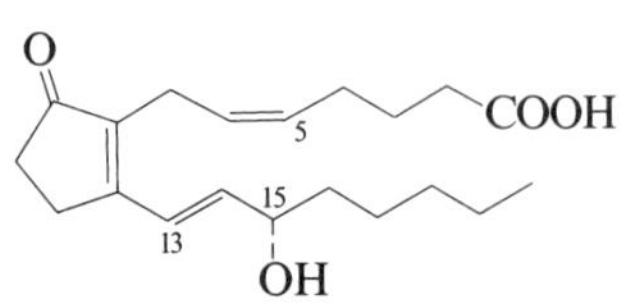

$C_{20}H_{30}O_4$ M 334.455

(5Z,13E,15S)-form [13367-85-6]
PGB_2. Prostaglandin B_2
Occurs in human seminal plasma. Most abundant prostaglandin released from osteoblasts. Inhibits uterine motility (*in vitro*). Cryst.

Hamberg, M. *et al, Prostaglandins, Proc. Nobel Symp., 2nd., 1966*, 1967, 63 (*occur*)
Ger. Pat., 2 011 969, (1970) (*synth*)
Middleditch, B.S. *et al, Lipids*, 1973, **8**, 267 (*ms*)
Feyen, J.H.M. *et al, Prostaglandins*, 1984, **28**, 769 (*isol*)

15-Hydroxy-9-oxo-5,10,13-prostatrien-1-oic acid, 9CI — H-00554

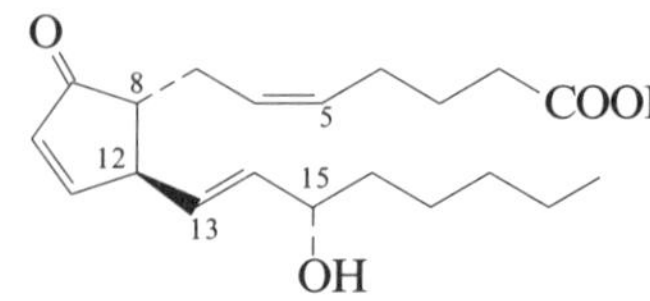

$C_{20}H_{30}O_4$ M 334.455

(5Z,8R,12S,13E,15S)-form [13345-50-1]
15α-(5Z,13E)-form. Prostaglandin A_2. PGA_2
Constit. of *Plexaura homomalla* and human seminal plasma. Oil. $[\alpha]_D^{20}$ +140° (c, 1.15 in $CHCl_3$).
▷ UK8050000.

Me ester: [31753-19-2].
Pale-yellow oil. $[\alpha]_D$ +148°.
15-Ac: [36323-03-2].
$[\alpha]_D$ +102°.
13,14-Dihydro, 15-oxo: [74872-89-2]. *9,15-Dioxo-5,10-prostadienoic acid*
$C_{20}H_{30}O_4$ M 334.455
Metab. of prostaglandin biosynth. Viscous oil.

(5Z,8R,12S,13E,15R)-form
15β-(5Z,13E)-form. 15-epi-PGA_2
Constit. of *P. homomalla.*

(5E,8R,12S,13E,15S)-form
15α-(5E,13E)-form. (5E)-PGA_2
Constit. of *P. homomalla.* Pale-yellow oil. $[\alpha]_D$ +128° ($CHCl_3$), $[\alpha]_D$ +137° ($CHCl_3$).

Hamberg, M. *et al, J. Biol. Chem.*, 1966, **241**, 257.
Weinheimer, A.J. *et al, Tetrahedron Lett.*, 1969, 5158 (*isol, struct*)
Bundy, G.L. *et al, J. Am. Chem. Soc.*, 1972, **94**, 2124 (*isol, uv*)
Corey, E.J. *et al, J. Am. Chem. Soc.*, 1973, **95**, 6831 (*synth*)
Lukacs, G. *et al, Tetrahedron Lett.*, 1973, 515 (*cmr*)
Stork, G. *et al, J. Am. Chem. Soc.*, 1976, **98**, 1583 (*synth, uv*)
Hansen, H.S., *Prostaglandins Ser.*, 1976, **12**, 647 (*rev*)
Uekama, K. *et al, Chem. Lett.*, 1977, 1389 (*cd*)
Schneider, W.P. *et al, J. Am. Chem. Soc.*, 1977, **99**, 1222 (*isol, ir, uv, ms*)
Buendia, J. *et al, Bull. Soc. Chim. Fr.*, 1978, 140 (*synth, uv, ir, pmr, ms*)
Greene, A.E. *et al, J. Org. Chem.*, 1978, **43**, 4377 (*isol*)
Finch, M.A.W. *et al, J. Chem. Soc., Chem. Commun.*, 1979, 677 (*synth*)
Peskar, B.M. *et al, FEBS Lett.*, 1980, **115**, 123 (*biochem*)
Greene, A.E. *et al, J. Org. Chem.*, 1982, **47**, 2553 (*synth*)
Martindale, The Extra Pharmacopoeia, 28th/29th Ed., Pharmaceutical Press, London, 1982/1989, 8093.
Newton, R.F. *et al, J. Chem. Soc., Perkin Trans. 1*, 1983, 683 (*synth*)
Suemune, H. *et al, Chem. Pharm. Bull.*, 1988, **36**, 15 (*synth*)
Lewis, R.J., *Sax's Dangerous Properties of Industrial Materials*, 8th Ed., Van Nostrand-Reinhold, 1992, MCA025.

15-Hydroxy-9-oxo-13-prostenoic acid, 9CI H-00555

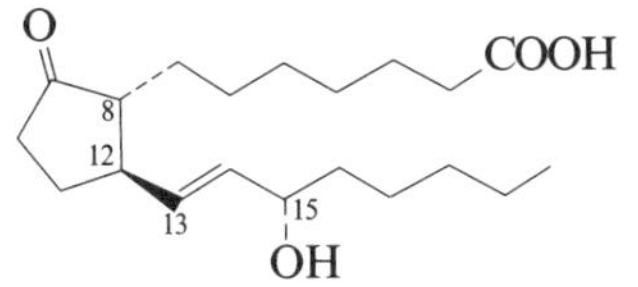

$C_{20}H_{34}O_4$ M 338.486

(8R,12R,13E,15S)-form
11-Deoxy-PGE$_1$
Mp 91-92°. $[\alpha]_D^{24}$ −45.73° (c, 0.73 in $CHCl_3$).

(8RS,12RS,13E,15SR)-form [34603-80-0]
(±)-11-Deoxy-PGE$_1$
Needles (Et_2O/pet. ether). Mp 85-86°.

(8RS,12RS,13E,15RS)-form [40098-34-8]
*(±)-11-Deoxy-15-*epi-*PGE$_1$*
Cryst. (EtOAc/hexane). Mp 53-56°.

(8R,12R,13Z,15R)-form
Mp 54-55°. $[\alpha]_D$ −75° (c, 1.0 in MeOH).

(8S,12S,13E,15S)-form [38380-61-9]
*11-Deoxy-15-*epi-ent-*PGE$_1$*
Cryst. (EtOAc/pentane). Mp 52-54°. $[\alpha]_D^{24}$ +43.02° (c, 1.47 in $CHCl_3$).

(8S,12S,13Z,15S)-form [39765-13-4]
Mp 54.5-55.5°. $[\alpha]_D$ +82° (c, 0.81 in MeOH).

Alvarez, F.S. *et al*, *J. Am. Chem. Soc.*, 1972, **94**, 7823 (*synth, ir, pmr, bibl*)
Alvarez, F.S. *et al*, *Tetrahedron Lett.*, 1973, 569 (*synth, ir, pmr, bibl*)
Sih, C.J. *et al*, *J. Am. Chem. Soc.*, 1975, **97**, 857 (*synth*)
Bernady, K.F. *et al*, *Tetrahedron Lett.*, 1975, 765 (*synth*)
Greene, A. *et al*, *Tetrahedron Lett.*, 1975, 2215 (*synth, bibl*)
Burton, T.S. *et al*, *J. Chem. Soc., Perkin Trans. 1*, 1976, 2550 (*synth*)
Mel'nikova, V.I. *et al*, *J. Gen. Chem. USSR* (*Engl. Transl.*), 1976, **46**, 1403 (*synth*)
Kondo, K. *et al*, *Tetrahedron Lett.*, 1977, 113 (*synth*)
Bernady, K.F. *et al*, *J. Org. Chem.*, 1979, **44**, 1438 (*synth, ir, pmr*)
Greene, A.E. *et al*, *J. Org. Chem.*, 1982, **47**, 2553 (*synth*)

15-Hydroxypentacosanoic acid H-00556

[136051-38-2]

$H_3C(CH_2)_9CH(OH)(CH_2)_{13}COOH$

$C_{25}H_{50}O_3$ M 398.668
Constit. of *Butea monosperma*. Needles (C_6H_6). Mp 76-78°.

Sharma, S. *et al*, *Indian J. Chem., Sect. B*, 1991, **30**, 715 (*isol*)

2-Hydroxy-6-(8,11-pentadecadienyl) benzoic acid H-00557

6-Pentadecadienylsalicylic acid
[18654-17-6]

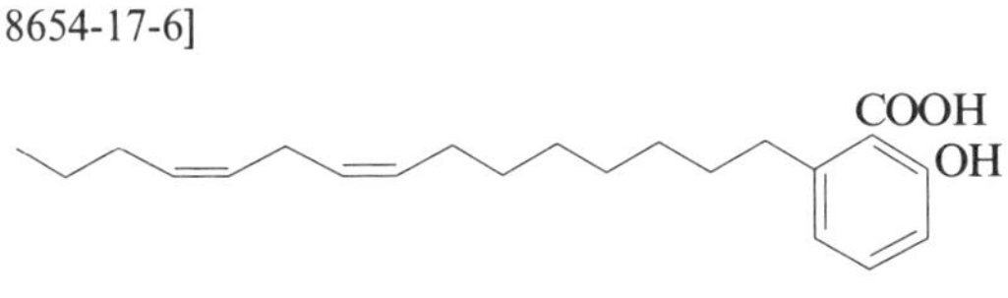

$C_{22}H_{32}O_3$ M 344.493

(all-Z)-form
Anacardic acid
Found in cashew nut shell. Mp 25-26°.
11′,12′-Dihydro: [22910-60-7]. *2-Hydroxy-6-(8-pentadecenyl) benzoic acid. 6-(8-Pentadecenyl)salicylic acid.* ***Ginkgoic acid.*** *Ginkgolic acid*
$C_{22}H_{34}O_3$ M 346.509
Constit. of *Ginkgo biloba* and minor constit. of cashew nut shell. Needles by subl. Mp 45.3-48° (42-43°).
11′,12′-Dihydro, Me ester: Bp_2 230-233°.
8′,9′-Dihydro, Δ$^{10'}$-isomer: 2-Hydroxy-6-(10-pentadecenyl) benzoic acid. 6-(10-Pentadecenyl)salicylic acid
$C_{22}H_{34}O_3$ M 346.509
Constit. of *Ozoroa mucronata*. Inhibits prostaglandin synthetase. Needles. Mp 45.8-46.2°.
14′,15′-Didehydro: 2-Hydroxy-6-(8,11,14-pentadecatrienyl) benzoic acid
$C_{22}H_{30}O_3$ M 342.477
Minor constit. of cashew nut shell. Geom. isom. not specified.

[25377-74-6]

Paul, V.J. *et al*, *Nature (London)*, 1954, **174**, 604 (*deriv*)
Tyman, J.H.P., *J. Org. Chem.*, 1976, **41**, 894 (*synth*)
Tyman, J.H.P., *Chem. Soc. Rev.*, 1979, **8**, 499 (*rev*)
Kubo, I. *et al*, *Chem. Lett.*, 1987, 1101 (*struct*)
Yamigawa, Y. *et al*, *Tetrahedron*, 1987, **43**, 3387 (*synth*)

2-Hydroxypentadecanoic acid, 9CI H-00558

[2507-54-2]

$H_3C(CH_2)_{12}CH(OH)COOH$

$C_{15}H_{30}O_3$ M 258.400

(R)-form [50849-29-1]
Formed by *Endomycopsis* spp. Mp 89.5-90°. $[\alpha]_D$ −0.8° (EtOH).

(±)-form
Mp 84.5°.
Et ester:
$C_{17}H_{34}O_3$ M 286.454
$Bp_{0.3}$ 165-167°.
Ac:
$C_{17}H_{32}O_4$ M 300.437
Mp 51-52°.
Amide:
$C_{15}H_{31}NO_2$ M 257.415
Cryst. (EtOH). Mp 149-150°.
Chloride:
$C_{15}H_{29}ClO_2$ M 276.846
$Bp_{0.6}$ 160°.

Broughton, B.W. *et al*, *J. Chem. Soc.*, 1952, 671 (*synth*)
Lemieux, R.U., *Can. J. Chem.*, 1953, **31**, 396.
Sokolov, V.P. *et al*, *Zh. Obshch. Khim.*, 1972, **42**, 2119 (*ir*)

3-Hydroxypentadecanoic acid H-00559

[32602-70-3]

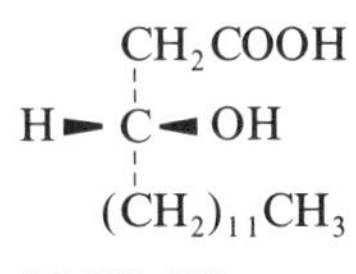

(*R*)-*form*

$C_{15}H_{30}O_3$ M 258.400

(R)-form [61365-61-5]
$[\alpha]_D^{20}$ −15.0° (c, 1 in $CHCl_3$).

(S)-form [61365-60-4]
Mp 75-76°. $[\alpha]_D^{20}$ +14.5° (c, 1 in $CHCl_3$).

(±)-form [61365-59-1]
Mp 76-77°, Mp 86.7-88.0°.
Et ester: [52315-82-9].
$C_{17}H_{34}O_3$ M 286.454
$Bp_{1.3}$ 155-159°.

Skogh, M., *Acta Chem. Scand.*, 1952, **6**, 809 (*synth*)

Borodina, G.M., *Zh. Obshch. Khim.*, 1954, **24**, 235; *CA*, **49**, 4518 (*synth*)
Lammek, B. *et al*, *Pol. J. Chem.* (*Rocz. Chem.*), 1976, **50**, 997 (*resoln, ord, cd*)

4-Hydroxypentadecanoic acid — H-00560

$H_3C(CH_2)_{10}CH(OH)CH_2CH_2COOH$

$C_{15}H_{30}O_3$ M 258.400

(±)-*form*

Lactone: 4-Pentadecanolide
$C_{15}H_{28}O_2$ M 240.385
Mp 25°. $Bp_{0.8}$ 155-157°.

Kawasaki, T., *Yakugaku Zasshi*, 1950, **70**, 485.
Bowman, R.E. *et al*, *J. Chem. Soc.*, 1951, 2753 (*synth*)

11-Hydroxypentadecanoic acid — H-00561

$H_3C(CH_2)_3CH(OH)(CH_2)_9COOH$

$C_{15}H_{30}O_3$ M 258.400

(±)-*form*
Mp 63.5-64.0°.

Me ester:
$C_{16}H_{32}O_3$ M 272.427
Mp 29-32°. Bp_2 166°.

Davies, L.A. *et al*, *J. Am. Chem. Soc.*, 1928, **50**, 1754.

15-Hydroxypentadecanoic acid, 9CI — H-00562

[4617-33-8]

$HOCH_2(CH_2)_{13}COOH$

$C_{15}H_{30}O_3$ M 258.400
Occurs as ester in musk-seed oil. Use in polyester formulations. Mp 82-84°.

Ac:
$C_{17}H_{32}O_4$ M 300.437
Mp 59°.

Lactone: [106-02-5]. *Oxacyclohexadecan-2-one, 9CI, 8CI. Pentadecanolide.* ***Exaltolide***
$C_{15}H_{28}O_2$ M 240.385
Constit. of angelica root oil (*Angelica archangelica*). Important musk odorant. Mp 34°, Mp 52-53°. Bp_{15} 176°

Dhekne, V.V. *et al*, *J. Chem. Soc.*, 1962, 2348 (*lactone*)
Klouwen, M.H. *et al*, *Parfuem. Kosmet.*, 1962, **43**, 35 (*rev*)
Mathur, H.H. *et al*, *J. Chem. Soc.*, 1963, 3505.
McCrae, W., *Tetrahedron*, 1964, **20**, 1773 (*lactone*)
McFadden, W.H. *et al*, *Anal. Chem.*, 1965, **37**, 89 (*ms, lactone*)
Nozaki, H. *et al*, *Can. J. Chem.*, 1968, **46**, 3767 (*synth*)
Scott, L.T. *et al*, *Synthesis*, 1976, 738 (*lactone*)
Narasaka, K. *et al*, *Chem. Lett.*, 1978, 885 (*lactone*)
Sato, T. *et al*, *Bull. Chem. Soc. Jpn.*, 1981, **54**, 945 (*synth*)
Subramanian, G.B.V. *et al*, *Chem. Ind.* (*London*), 1985, 379 (*synth*)
Milenkov, B. *et al*, *Helv. Chim. Acta*, 1987, **70**, 760 (*synth*)

5-(10-Hydroxy-8-pentadecenyl)-1,3-benzenediol — H-00563

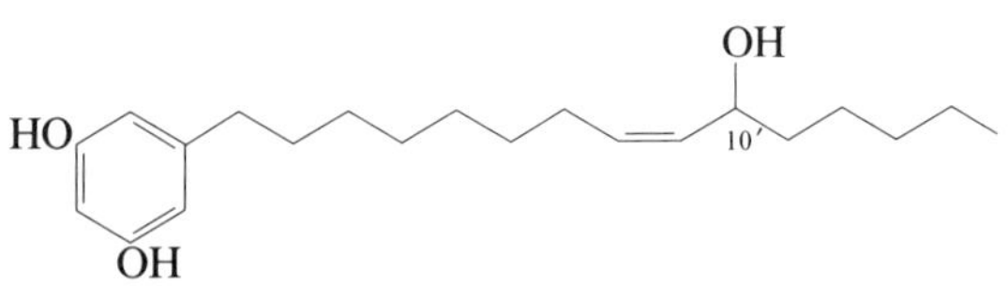

$C_{21}H_{34}O_3$ M 334.498

(*Z*)-*form* [120727-08-4]
Cryst. ($CHCl_3$). Mp 87-88°.

10′-Ac: [120727-06-2]. *5-(10-Acetoxy-8-pentadecenyl)-1,3-benzenediol*
$C_{23}H_{36}O_4$ M 376.535
Constit. of *Ononis speciosa*. Oil.

[120727-11-9]

Barrero, A.F. *et al*, *Phytochemistry*, 1989, **28**, 161 (*isol*)

4-Hydroxy-6-pentadecyl-2*H*-pyran-2-one, 9CI — H-00564

Conrauanalactone

[75666-76-1]

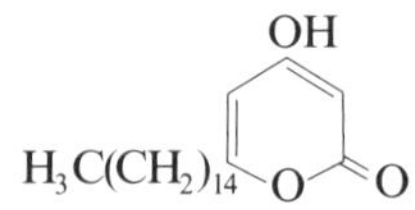

$C_{20}H_{34}O_3$ M 322.487
Isol. from *Garcinia conrauana* bark. Off-white needles (EtOAc/pet. ether). Mp 90°.

Waterman, P.G. *et al*, *Phytochemistry*, 1980, **19**, 1187 (*isol, ir, uv, pmr, cmr, ms, struct*)

4-Hydroxy-2-pentanone — H-00565

[4161-60-8]

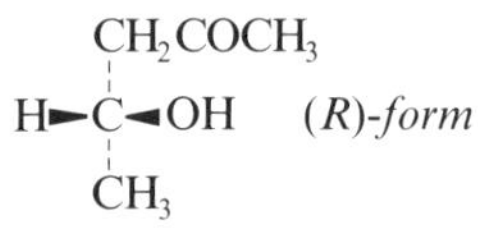

$C_5H_{10}O_2$ M 102.133

(*R*)-*form* [63315-69-5]
Liq. $[\alpha]_D^{20}$ −10.2° (neat, part. resolved), $[\alpha]_D$ −18° (calcd.).

Ac: [51953-37-8].
$C_7H_{12}O_3$ M 144.170
Liq. Bp_{20} 92°. $[\alpha]_D^{20}$ +1.33° (neat).

Benzyl ether: [87841-68-7].
$C_{12}H_{16}O_2$ M 192.257
$[\alpha]_D^{20}$ −20° (c, 1 in EtOH).

(±)-*form* [74339-90-5]
Liq. d_4^{20} 1.01. Bp 177°, Bp_{28} 77°.

Ac: [55577-75-8].
Bp_{12} 84°.

Ac, phenylhydrazone: Plates (EtOH). Mp 115°.

Oxime:
$C_5H_{11}NO_2$ M 117.147
Bp_{13} 123-124°.

Phenylhydrazone: Mp 102-103°.

[73836-68-7]

Tanabe, T., *Bull. Chem. Soc. Jpn.*, 1973, **46**, 2233 (*abs config*)
Eichenauer, H. *et al*, *Synthesis*, 1977, **12**, 893 (*synth*)
Fuganti, C. *et al*, *J. Org. Chem.*, 1984, **49**, 543 (*deriv, synth, pmr*)
Cardillo, G. *et al*, *J. Org. Chem.*, 1984, **49**, 701 (*synth, ir, pmr, ms*)

43-Hydroxy-2,44-pentatetracontapentaenediynoic acid — H-00566

$C_{45}H_{72}O_3$ M 661.062

Me ester: [128646-24-2]. ***Petroformyne B***
$C_{46}H_{74}O_8$ M 755.086
Constit. of the sponge *Petrosia ficiformis*. Posn. of ethylenic groups not determined.

Cimino, G. *et al*, *J. Nat. Prod.* (*Lloydia*), 1990, **53**, 345.

28-Hydroxy-7-pentatriacontanone H-00567

[146356-93-6]

$H_3C(CH_2)_6CH(OH)(CH_2)_{20}CO(CH_2)_5CH_3$

$C_{35}H_{70}O_2$ M 522.937

Constit. of the shoots of *Leucas aspera*. Cryst. (C_6H_6/hexane). Mp 63-64°.

Ac: Mp 55°.

Misra, T.N. *et al*, *Phytochemistry*, 1993, **32**, 199 (*isol, pmr*)

1-(4-Hydroxyphenyl)-1-nonanone, 9CI H-00568

p-*Hydroxynonanophenone*

[14392-69-9]

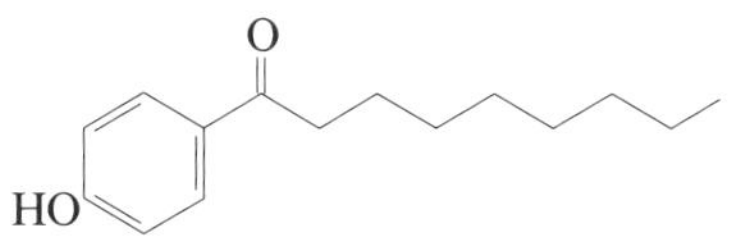

$C_{15}H_{22}O_2$ M 234.338

Constit. of *Humulus lupulus*. Oil.

Chandra, A. *et al*, *Phytochemistry*, 1991, **30**, 495 (*isol, pmr, cmr*)

7-[2-(3-Hydroxy-5-phenylpentyl)-5-oxo-3-cyclopenten-1-yl]-5-heptenoic acid H-00569

17-Phenyltrinor-13,14-dihydroprostaglandin A_2

[130209-80-2]

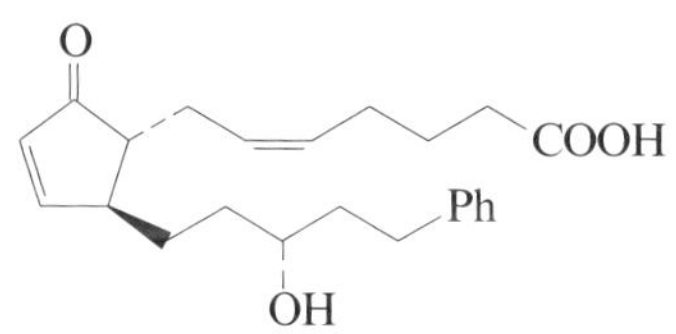

$C_{23}H_{30}O_4$ M 370.488

Interm. in prostaglandin synth.

Isopropyl ester: [130209-79-9].

Used in the treatment of glaucoma.

Pat. Coop. Treaty (*WIPO*), 90 02 553, (1990); *CA*, **113**, 205515v (*synth, pmr*)

4-Hydroxy-2-(11-phenylundecanoyl)-1,3-cyclohexanedione H-00570

[96820-28-9]

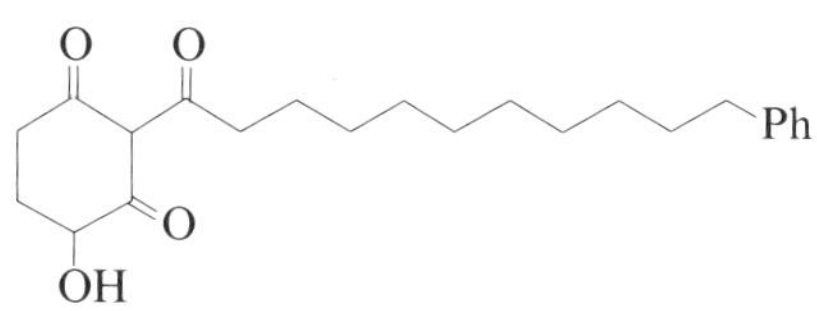

$C_{23}H_{32}O_4$ M 372.503

Enolised β-diketone. Constit. of *Virola elongata*. Oil.

Kato, M.J. *et al*, *Phytochemistry*, 1985, **24**, 533.

2-Hydroxysterculic acid H-00571

α-Hydroxy-2-octyl-1-cyclopropene-1-octanoic acid, 8CI. 2-Hydroxy-8-(2-octyl-1-cycylpropenyl)octanoic acid

[18273-00-2]

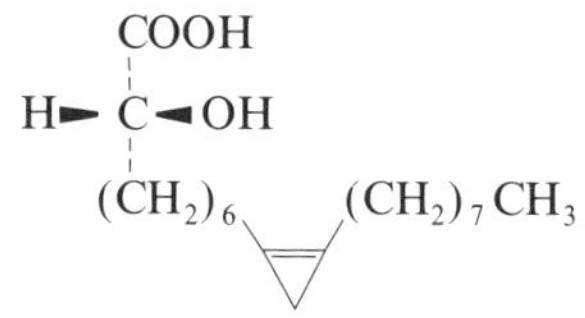

$C_{19}H_{34}O_3$ M 310.476

***(R)*-form** [14602-38-1]

Constit. of seed oils of *Bombacopsis glabra*, *Pachira insignis* and *P. aquatica*.

Me ester: [13487-77-9].

$C_{20}H_{36}O_3$ M 324.503

Oil. $[\alpha]^{24}_{546}$ −5.1° (c, 3 in $CHCl_3$), $[\alpha]^{24}_D$ −10° (c, 0.083 in $CHCl_3$). Ir ν 1740 cm^{-1}.

Morris, L.J. *et al*, *Chem. Ind.* (*London*), 1967, 32.
Recourt, J.H. *et al*, *J. Chromatogr.*, 1967, **30**, 35 (*ms, ir, pmr*)
Nakatani, M. *et al*, *Bull. Chem. Soc. Jpn.*, 1991, **64**, 3084 (*isol, ir, pmr, cmr, ms, abs config*)
Baird, M.S. *et al*, *J. Chem. Soc., Perkin Trans. 1*, 1993, 1547 (*synth*)

2-Hydroxytetracosanoic acid H-00572

Cerebronic acid. *Phrenosic acid. Phrenosinic acid*

[544-57-0]

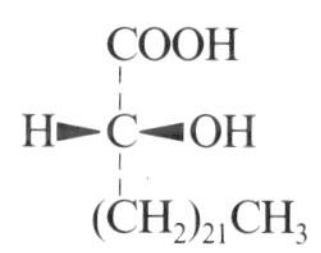

(*R*)-*form*

$C_{24}H_{48}O_3$ M 384.641

Constit. of various glycosphingolipids. Isol. from wheat, corn, other plant spp., lichens and sponges.

***(R)*-form** [26632-11-1]

Isol. from *Penicillium* cultures; *Polyporus umbllatus* and *Aspergillus flavus*. Cryst. ($CHCl_3$); plates (Me_2CO). Mp 104.5-105.5° (99.5-100°). $[\alpha]^{25}_D$ +3.4° (c, 1.6 in Py).

Me ester:

$C_{25}H_{50}O_3$ M 398.668

Mp 64-65°. $[\alpha]_D$ −3.25° (c, 0.92 in $CHCl_3$).

Ac: [127061-75-0].

$C_{26}H_{50}O_4$ M 426.679

Cryst. (pet. ether). Mp 81.5-82.5°. $[\alpha]^{20}_D$ +8.46° (c, 1.01 in $CHCl_3$).

(±)-*form* [139237-67-5]

Cryst. (C_6H_6). Mp 99.5-100°.

[2433-95-6, 54563-85-8, 54563-86-9, 54563-87-0, 54563-88-1, 73580-21-9, 132842-33-2]

Ashton, R. *et al*, *J. Chem. Soc.*, 1936, 283 (*synth*)
Muller, A., *Ber.*, 1939, **72B**, 615 (*synth*)
Yoshioka, I. *et al*, *Yakugaku Zasshi*, 1964, **84**, 742 (*isol, deriv*)
Vacheron, H.J. *et al*, *Phytochemistry*, 1968, **7**, 1645 (*isol*)
Hoshi, M. *et al*, *J. Biol. Chem.*, 1973, **248**, 4123 (*pharmacol*)
Tatsumi, K. *et al*, *Arch. Biochem. Biophys.*, 1974, **165**, 656 (*abs config*)
Florentina, N. *et al*, *Chem. Phys. Lipids*, 1984, **34**, 257 (*synth*)
Koike, K. *et al*, *Carbohydr. Res.*, 1987, **162**, 237 (*synth*)
Yamagata, K. *et al*, *J. Chem. Soc., Perkin Trans. 1*, 1990, 3355 (*synth*)
Sugiyama, S. *et al*, *Justus Liebigs Ann. Chem.*, 1990, 1063 (*synth*)
Sugai, T. *et al*, *Tetrahedron Lett.*, 1991, **32**, 7063 (*synth, bibl*)

3-Hydroxytetracosanoic acid H-00573

[91297-89-1]

$H_3C(CH_2)_{20}CH(OH)CH_2COOH$

$C_{24}H_{48}O_3$ M 384.641

Constit. of the bark and latex of *Parahancornia arnapa*. Also found in marine sediments.

Sobrinho, D.C. *et al*, *J. Braz. Chem. Soc.*, 1991, **2**, 15 (*isol*)

24-Hydroxytetracosanoic acid, 9CI H-00574

[75912-18-4]

$HOCH_2(CH_2)_{22}COOH$

$C_{24}H_{48}O_3$ M 384.641

Isol. from Carnauba wax and from the wood of *Angophora subvelutina*. Mp 97-98°.

Me ester: [37477-30-8].
$C_{25}H_{50}O_3$ M 398.668
Cryst. (pet. ether). Mp 79-80°.

Me ester, Ac:
$C_{27}H_{52}O_4$ M 440.705
Mp 63-65°.

Lactone: [58296-52-9]. *Oxacyclopentacosan-2-one, 9CI. 24-Tetracosanolide*
$C_{24}H_{46}O_2$ M 366.626
Pheromone from the Dufour's gland secretion of various bees and from termites.

Ritchie, E. *et al*, *Aust. J. Chem.*, 1961, **14**, 473 (*isol*)
Duhamel, L., *Ann. Chim. (Paris)*, 1963, **8**, 315 (*synth*)
Abrams, S.R., *Chem. Phys. Lipids*, 1981, **28**, 379 (*synth*)
Duffield, R.M. *et al*, *J. Chem. Ecol.*, 1982, **8**, 535 (*isol, lactone*)
Prestwich, G.D., *Tetrahedron*, 1982, **38**, 1911 (*isol, struct, lactone*)
Hefetz, A. *et al*, *J. Chem. Ecol.*, 1986, **12**, 197 (*occur, lactone*)

24-Hydroxy-10-tetracosenoic acid H-00575

$HOCH_2(CH_2)_{12}CH{=}CH(CH_2)_8COOH$

$C_{24}H_{46}O_3$ M 382.626

***(Z)*-form** [136068-52-5]

(Z)*-7-Heneicosenyl ester:* [136068-49-0]. *7Z-Heineicosenyl 10Z-tetracosenoate*
$C_{45}H_{86}O_3$ M 675.173
Constit. of *Albizzia lebbek*.

[136068-50-3]

Agrawal, P.K. *et al*, *Indian J. Pharm. Sci.*, 1991, **53**, 24 (*isol*)

2-Hydroxytetradecanoic acid H-00576

2-Hydroxymyristic acid

[2507-55-3]

$H_3C(CH_2)_{11}CH(OH)COOH$

$C_{14}H_{28}O_3$ M 244.373

***(R)*-form** [26632-17-7]
Present in wool wax and in brain cerebrosides. Mp 88.2-88.5°. $[\alpha]_D$ −3.1° ($CHCl_3$).

Me ester:
$C_{15}H_{30}O_3$ M 258.400
Mp 34.8-35.0°. Bp_1 133.5-136.5°.

***(S)*-form** [67253-09-2]
Occurs in *Salmonella* lipopolysaccharides.

(±)-*form* [37639-48-8]
Occurs in the *Actinomycetales* bacteria. Mp 81-82°.

4-Nitrobenzyl ester: Mp 67-68°.

Amide:
$C_{14}H_{29}NO_2$ M 243.389
Mp 150°.

Nitrile:
$C_{14}H_{27}NO$ M 225.373
Cryst. (pet. ether). Mp 44.5°.

Horn, D.H.S. *et al*, *J. Chem. Soc.*, 1954, 177.
Annett, R.L. *et al*, *Anal. Biochem.*, 1972, **47**, 638 (*resoln*)
Sokolov, V.P. *et al*, *Zh. Obshch. Khim.*, 1972, **42**, 2119 (*ir*)
Konen, D.A. *et al*, *J. Org. Chem.*, 1975, **40**, 3253 (*synth*)
Dahlen, B. *et al*, *Acta Crystallogr., Sect. B*, 1976, **32**, 2059 (*cryst struct*)
Bryn, K. *et al*, *Eur. J. Biochem.*, 1978, **86**, 311.

3-Hydroxytetradecanoic acid, 9CI H-00577

3-Hydroxymyristic acid

[1961-72-4]

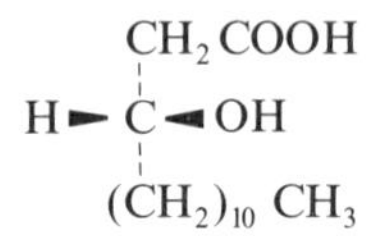

(R)-form

$C_{14}H_{28}O_3$ M 244.373

▷XB8585000.

***(R)*-form** [28715-21-1]
Occurs in phospholipids of *E. coli*. Mp 73-74°. $[\alpha]_D^{25}$ −9.6° (Py), $[\alpha]_D^{22}$ −16.2° (c, 1 in $CHCl_3$).

4-Bromophenacyl ester: Mp 112.5-113.0°.

***(S)*-form** [35683-15-9]
Formed by heart mitochondria. Mp 71-73°. $[\alpha]_D^{25}$ +6.5° (Py).

(±)-*form*
Mp 78-79°.

Et ester:
$C_{16}H_{32}O_3$ M 272.427
$Bp_{2\text{-}3}$ 164-167°.

Ikawa, M. *et al*, *J. Am. Chem. Soc.*, 1953, **75**, 1035.
Lammek, B. *et al*, *Pol. J. Chem. (Rocz. Chem.)*, 1976, **50**, 997 (*ord, resoln, abs config*)
Leibrand, R.J. *et al*, *Adv. Mass Spectrom. Biochem. Med., Proc. Int. Symp. Mass Spectrom. Biochem. Med. 2nd, 3rd, 1974, 1975*, 1977, **2**, 505 (*ms*)
Nakahata, M. *et al*, *Bull. Chem. Soc. Jpn.*, 1982, **55**, 2186 (*synth*)
Utaka, M. *et al*, *J. Org. Chem.*, 1990, **55**, 3917 (*synth, ir, pmr*)

5-Hydroxytetradecanoic acid H-00578

$H_3C(CH_2)_8CH(OH)CH_2CH_2CH_2COOH$

$C_{14}H_{28}O_3$ M 244.373

(±)-*form*

Lactone: [2721-22-4]. *Tetrahydro-6-nonyl-2*H*-pyran-2-one, 9CI. 5-Tetradecanolide. δ-Tetradecalactone*
$C_{14}H_{26}O_2$ M 226.358
Found in coconut oil, butter, cheese, pork fat, tuberose oil and others (poss. in an enantiomeric form). Liq. Mp 14°. Bp_1 148°. n_D^{25} 1.4592.

[92446-06-5]

Allen, R.R., *Chem. Ind. (London)*, 1965, 1560 (*isol, ir*)
Landelli, G. *et al*, *Recl. Trav. Chim. Pays-Bas (J. R. Neth. Chem. Soc.)*, 1967, **86**, 481 (*synth*)
Ijima, J. *et al*, *Chem. Pharm. Bull.*, 1973, **21**, 215 (*synth*)
Oser, B.L. *et al*, *Food Technol. (Chicago)*, *(Chicago)*, 1978, **32**, 60, 64, 68 (*rev, tox*)

6-Hydroxytetradecanoic acid, 9CI H-00579

6-Hydroxymyristic acid

[1747-18-8]

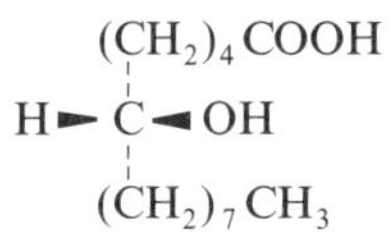

(R)-form

$C_{14}H_{28}O_3$ M 244.373

(R)-form [24205-54-7]

Butolic acid

Constit. of lac. Mp 58-59°. $[\alpha]_D$ −1.3° ($CHCl_3$).

Me ester:

$C_{15}H_{30}O_3$ M 258.400

Mp 26-27°. Bp_1 120-122°. $[\alpha]_D$ −2.2° ($CHCl_3$).

(±)-*form*

Mp 65-66°.

Christie, W.W. *et al*, *J. Chem. Soc.*, 1963, 5768 (*synth*)
Khurana, R.G. *et al*, *Tetrahedron Lett.*, 1964, 1537.
Wadia, M.S. *et al*, *Tetrahedron*, 1969, 3841 (*pmr, ir*)

10-Hydroxytetradecanoic acid, 9CI H-00580

10-Hydroxymyristic acid

[16899-08-4]

$H_3C(CH_2)_3CH(OH)(CH_2)_8COOH$

$C_{14}H_{28}O_3$ M 244.373

Metab. of *E. coli*. Mp 56-57°. Probably an opt. active form.

Me ester:

$C_{15}H_{30}O_3$ M 258.400

Occurs in oat (*Avena sativa*) leaf wax.

Papa, D. *et al*, *J. Org. Chem.*, 1949, **14**, 723.
Michiano, M. *et al*, *J. Biochem.* (*Tokyo*), 1968, **63**, 390 (*synth*)
Tulloch, A.P. *et al*, *Lipids*, 1973, **8**, 617.

11-Hydroxytetradecanoic acid H-00581

11-Hydroxymyristic acid. Convolvulinolic acid. β-Desoxyipuronic acid

$H_3CCH_2CH_2CH(OH)(CH_2)_9COOH$

$C_{14}H_{28}O_3$ M 244.373

(+)-*form*

Present in *Ipomoea jalapa* and *I. fistulosa*. Mp 51-52°.

(±)-*form*

Mp 54.5-55.5° (51-52°).

Kawasaki, T., *Yakugaku Zasshi*, 1950, **70**, 485; *CA*, **45**, 5624a (*struct, synth*)
Shellard, E.J., *Planta Med.*, 1961, **9**, 141.
Legler, G., *Phytochemistry*, 1965, **4**, 29 (*isol*)
Graf, E. *et al*, *Arch. Pharm.* (*Weinheim, Ger.*), 1974, **307**, 628.

14-Hydroxytetradecanoic acid, 9CI H-00582

[17278-74-9]

$HOCH_2(CH_2)_{12}COOH$

$C_{14}H_{28}O_3$ M 244.373

Isol. from needle lipids of *Picea pungens*, *Sphagnum fuscum* and stem cutin of *Pinus radiata*. Cryst. (C_6H_6/Et_2O or hexane). Mp 93-95° (88-89°).

Me ester: [50515-98-5].

$C_{15}H_{30}O_3$ M 258.400

Cryst. (hexane). Mp 41-42°. Bp_{10} 196-198°.

Hydrazide: [24535-05-5].

Mp 141-142°.

Ruzicka, L. *et al*, *Helv. Chim. Acta*, 1928, **11**, 1159 (*synth*)
Chuit, P. *et al*, *Helv. Chim. Acta*, 1929, **12**, 463 (*synth*)
von Rudloff, E., *Can. J. Chem.*, 1959, **37**, 1038 (*isol*)
Ames, D.E. *et al*, *J. Chem. Soc.*, 1963, 5889 (*synth*)
Allen, C.F.H. *et al*, *J. Heterocycl. Chem.*, 1969, **6**, 349 (*synth, ir*)
Ellin, A. *et al*, *Arch. Biochem. Biophys.*, 1973, **158**, 597 (*synth*)
Nimitz, J.S. *et al*, *Tetrahedron Lett.*, 1978, 3523 (*synth*)
Zakharkin, L.I. *et al*, *Zh. Org. Khim.*, 1981, **17**, 755 (*synth*)
Ekman, R. *et al*, *Phytochemistry*, 1982, **21**, 121 (*isol*)
Villemin, D. *et al*, *Synthesis*, 1984, 230 (*synth, ir, pmr*)

6-Hydroxy-2,4,8-tetradecatrienoic acid, 9CI H-00583

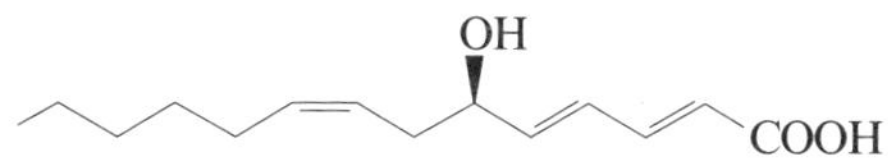

$C_{14}H_{22}O_3$ M 238.326

(2E,4E,6R,8Z)-form

Et ester: [82493-54-7].

$C_{16}H_{26}O_3$ M 266.380

Important intermed. in synth. of leukotrienes. $[\alpha]_D$ +11.4° (c, 1.31 in $CDCl_3$).

Et ester, tert-*butyldiphenylsilyl ether:* [82498-75-7].

Oil. $[\alpha]_D^{25}$ +40.6° (c, 2.0 in $CHCl_3$).

Guindon, Y. *et al*, *Tetrahedron Lett.*, 1982, **23**, 739 (*synth*)
Sih, C.J. *et al*, *J. Org. Chem.*, 1986, **51**, 1253 (*synth, ir, pmr*)

12-Hydroxy-13-tetradecenoic acid, 9CI H-00584

[85288-98-8]

$H_2C{=}CHCH(OH)(CH_2)_{10}COOH$

$C_{14}H_{26}O_3$ M 242.358

(±)-*form*

Mp 51-52°.

Me ester: [106753-88-2].

$C_{15}H_{28}O_3$ M 256.384

Cryst. Mp 36-38°.

Cameron, A.G. *et al*, *J. Chem. Soc., Perkin Trans. 1*, 1986, 161 (*synth, ir, pmr*)

14-Hydroxy-5-tetradecenoic acid H-00585

$HOCH_2(CH_2)_7CH{=}CH(CH_2)_3COOH$

$C_{14}H_{26}O_3$ M 242.358

Lactone: [63958-52-1]. *5-Tetradecen-14-olide. Oxacyclopentadec-6-en-2-one*

$C_{14}H_{24}O_2$ M 224.342

Constit. of the oil of *Hibiscus abelmoschus*. Perfumery component. Oil. $Bp_{0.005}$ 100°.

Maurer, B. *et al*, *Helv. Chim. Acta*, 1977, **60**, 1155 (*synth*)
Midland, M.M. *et al*, *J. Org. Chem.*, 1981, **46**, 4107 (*synth*)
Ramaswamy, S. *et al*, *Tetrahedron*, 1991, **47**, 1145 (*synth*)

14-Hydroxy-12-tetradecenoic acid, 9CI H-00586

[85289-00-5]

$HOCH_2CH{=}CH(CH_2)_{10}COOH$

$C_{14}H_{26}O_3$ M 242.358

(E)-form [106753-93-9]

Cryst. (Et_2O/pet. ether). Mp 47-48°.

Lactone: 12-Tetradecen-14-olide

$C_{14}H_{24}O_2$ M 224.342

$Bp_{0.01}$ 110° (oven).

(Z)-form [106541-96-2]
Cryst. (Et_2O/pet. ether). Mp 42.5-43.0°.
Lactone: $Bp_{0.03}$ 110° (oven).

Ames, D.E. *et al*, *J. Chem. Soc.*, 1963, 5889 (*synth*)
Cameron, A.G. *et al*, *J. Chem. Soc., Perkin Trans. 1*, 1986, 161 (*synth, ir, pmr*)

22-Hydroxy-13-tetratriacontanone H-00587

Grewinol
[61774-47-8]

$H_3C(CH_2)_{11}CH(OH)(CH_2)_8CO(CH_2)_{11}CH_3$

$C_{34}H_{68}O_2$ M 508.910
Constit. of the flowers of *Grewia asiatica*. Cryst. ($MeOH/CHCl_3$). Mp 81-82°.
Ac:
$C_{36}H_{70}O_3$ M 550.947
Cryst. (MeOH). Mp 70-71°.
2,4-Dinitrophenylhydrazone: Mp 140-141°.

Lakshmi, V. *et al*, *J. Nat. Prod. (Lloydia)*, 1976, **39**, 372 (*isol*)

32-Hydroxy-5-tetratriacontanone H-00588

[79105-74-1]

$H_3CCH_2CH(OH)(CH_2)_{26}CO(CH_2)_3CH_3$

$C_{34}H_{68}O_2$ M 508.910
Constit. of the leaves and stems of *Hyoscyamus muticus*. Cryst. (C_6H_6/hexane). Mp 76-77°.

Goswami, A. *et al*, *Phytochemistry*, 1981, **20**, 1315 (*isol*)

12-Hydroxy-4,7-triacontanedione H-00589

[85708-50-5]

$H_3C(CH_2)_{17}CH(OH)(CH_2)_4COCH_2CH_2COCH_2CH_2CH_3$

$C_{30}H_{58}O_3$ M 466.786
Constit. of the bark of *Wisteria sinensis*.
[85708-53-8, 85708-54-9]

Tiwari, K.P. *et al*, *Proc. Natl. Acad. Sci., India*, 1981, **51**, 263 (*isol*)

30-Hydroxytriacontanoic acid H-00590

[52900-18-2]

$HOCH_2(CH_2)_{28}COOH$

$C_{30}H_{60}O_3$ M 468.802
Isol. from Carnauba wax.
Me ester: [79162-70-2].
$C_{31}H_{62}O_3$ M 482.829
Mp 89.0-89.5°.
Ac, Me ester:
$C_{33}H_{64}O_4$ M 524.866
Mp 73.8-74.3°.

Murry, K.E. *et al*, *Aust. J. Chem.*, 1955, **8**, 437 (*isol*)
Abrams, S.R., *Chem. Phys. Lipids*, 1981, **28**, 379 (*synth*)

2-Hydroxytricosanoic acid, 9CI H-00591

[2718-37-8]

$H_3C(CH_2)_{20}CH(OH)COOH$

$C_{23}H_{46}O_3$ M 370.615
Present in cerebrosides.
[26632-12-2]

Downing, D.T., *Aust. J. Chem.*, 1961, **14**, 150 (*isol*)

3-Hydroxytridecanoic acid H-00592

[32602-69-0]

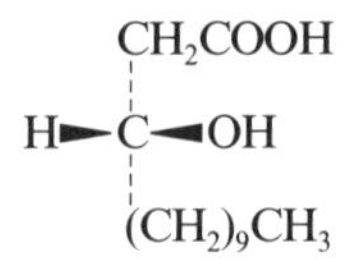

(*R*)-*form*

$C_{13}H_{26}O_3$ M 230.347
(R)-form [14664-26-7]
Cryst. (EtOAc/hexane). Mp 79°. $[\alpha]_D$ −16.6° (c, 0.6 in $CHCl_3$).
(±)-*form* [19516-31-5]
Cryst. (EtOAc). Mp 79.5-80°.
[14664-25-6]

Skogh, M., *Acta Chem. Scand.*, 1952, **6**, 809 (*synth*)
Ovchinnikov, Y.A. *et al*, *Tetrahedron Lett.*, 1966, 5285 (*synth, resoln*)
Fahlstadius, P. *et al*, *J. Chem. Soc., Perkin Trans. 1*, 1990, 2027 (*synth, resoln, ms*)

13-Hydroxy-9-tridecenoic acid H-00593

$HOCH_2CH_2CH_2CH{=}CH(CH_2)_7COOH$

$C_{12}H_{22}O_3$ M 214.304
(Z)-form
Lactone: [79894-05-6]. *Oxacyclotridec-10-en-2-one, 9CI. 9-Tridecenolide*
Constit. of the leaves of *Medicago rugosa*.
[79894-06-7]

Inoue, S. *et al*, *Nippon Kagaku Kaishi*, 1985, 425; *CA*, **103**, 215039 (*synth*)
Doss, R.P. *et al*, *Phytochemistry*, 1989, **28**, 3311 (*isol, pmr, cmr*)
Doss, R.P. *et al*, *J. Chem. Ecol.*, 1991, **17**, 535 (*props*)

5-Hydroxy-4,6,8-trimethyl-6-undecen-3-one H-00594

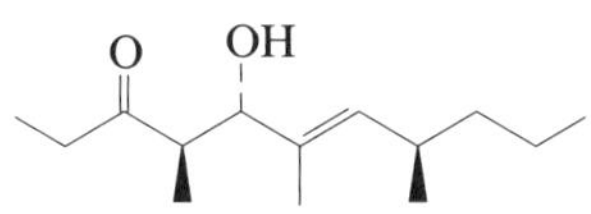

$C_{14}H_{26}O_2$ M 226.358
(4R,5S,8R)-form
Propanoyl: [91466-59-0]. *4,6,8-Trimethyl-5-(propanoyloxy)-6-undecen-3-one*
$C_{17}H_{30}O_3$ M 282.422
Isol. from the mollusc *Siphonaria australis*. $[\alpha]_D$ −19.4° ($CHCl_3$).

Hochlowski, J.E. *et al*, *J. Org. Chem.*, 1984, **49**, 3838.
Sundram, U.N. *et al*, *Tetrahedron Lett.*, 1992, **33**, 437.

41-Hydroxy-2,42-tritetracontapentaenediynoic acid H-00595

$C_{43}H_{68}O_3$ M 633.008
Me ester: [128646-22-0]. ***Petroformyne A***
$C_{44}H_{70}O_3$ M 647.035
Constit. of the sponge *Petrosia ficiformis*. Posns. of ethylenic groups not determined.

Cimino, G. *et al*, *J. Nat. Prod. (Lloydia)*, 1990, **53**, 345.

4-Hydroxy-16,18-tritriacontanedione **H-00596**

[97191-41-8]

$H_3C(CH_2)_{14}COCH_2CO(CH_2)_{11}CH(OH)CH_2CH_2CH_3$

$C_{33}H_{64}O_3$ M 508.867

Constit. of the leaf waxes of *Eucalyptus globulus*.

Osawa, T. *et al*, *J. Agric. Food Chem.*, 1985, **33**, 777 (*occur*)

18-Hydroxy-16-tritriacontanone **H-00597**

[97191-42-9]

$H_3C(CH_2)_{14}CH(OH)CH_2CO(CH_2)_{14}CH_3$

$C_{33}H_{66}O_2$ M 494.883

Constit. of the leaf waxes of *Eucalyptus globulus*.

Osawa, T. *et al*, *J. Agric. Food Chem.*, 1985, **33**, 777 (*occur*)

2-Hydroxyundecanoic acid, 9CI **H-00598**

[19790-86-4]

$H_3C(CH_2)_8CH(OH)COOH$

$C_{11}H_{22}O_3$ M 202.293

(±)-*form*

Cryst. (EtOAc/hexane). Mp 104-105°.

Me ester: [14919-24-5].

$C_{12}H_{24}O_3$ M 216.320

$Bp_{0.3}$ 80-82°.

Et ester:

$C_{13}H_{26}O_3$ M 230.347

Mp 38°.

Amide:

$C_{11}H_{23}NO_2$ M 201.308

Cryst. (EtOAc/hexane). Mp 143-144°.

Anilide:

$C_{17}H_{27}NO_2$ M 277.406

Mp 80°.

Bagard, P., *Bull. Soc. Chim. Fr.*, 1907, **1**, 310, 354.
Thompson, J.E., *J. Org. Chem.*, 1967, **32**, 3947 (*synth, pmr*)
Sokolov, V.P. *et al*, *Zh. Obshch. Khim.*, 1972, **42**, 2119 (*ir*)
Satoh, T. *et al*, *J. Org. Chem.*, 1991, **56**, 4129 (*synth, derivs, pmr, ir, ms*)

3-Hydroxyundecanoic acid, 9CI **H-00599**

[40165-88-6]

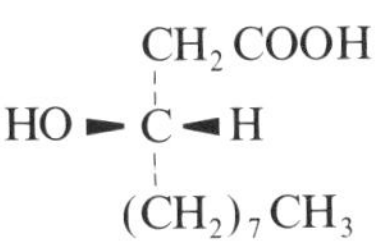

(*S*)-*form*

$C_{11}H_{22}O_3$ M 202.293

(*S*)-*form* [66997-62-4]

Mp 64-65°. $[\alpha]_D^{18}$ +17.1° (c, 2.1 in $CHCl_3$).

(±)-*form*

Occurs in *Pseudomonas maltophilia*. Cryst. (pet. ether). Mp 73-73.5°.

Skogh, M., *Acta Chem. Scand.*, 1952, **6**, 809 (*synth*)
Moss, C.W. *et al*, *J. Bacteriol.*, 1973, **114**, 1018 (*ms, ir*)
Tahara, S. *et al*, *Agric. Biol. Chem.*, 1978, **42**, 879 (*synth*)

4-Hydroxyundecanoic acid **H-00600**

$H_3C(CH_2)_6CH(OH)CH_2CH_2COOH$

$C_{11}H_{22}O_3$ M 202.293

(+)-*form*

Lactone: 4-Undecanolide

$C_{11}H_{20}O_2$ M 184.278

Bp_9 145-146°. $[\alpha]_D^{27}$ +29°.

(±)-*form*

Mp 34°.

Lactone: Bp 286°, Bp_{13} 162°.

Zhukov, A.A. *et al*, *Chem.-Ztg.*, 1908, **2**, 1415.

10-Hydroxyundecanoic acid, 9CI **H-00601**

[6336-28-3]

$H_3CCH(OH)(CH_2)_8COOH$

$C_{11}H_{22}O_3$ M 202.293

(±)-*form*

Cryst. (pet. ether). Mp 49.0-49.5°.

Et ester:

$C_{13}H_{26}O_3$ M 230.347

$Bp_{0.08}$ 129-131°.

Kinoshita, M. *et al*, *Bull. Chem. Soc. Jpn.*, 1971, **44**, 3355 (*synth*)
Osman, S.M. *et al*, *Fette, Seifen, Anstrichm.*, 1975, **77**, 473 (*synth*)

11-Hydroxyundecanoic acid, 9CI **H-00602**

[3669-80-5]

$HOCH_2(CH_2)_9COOH$

$C_{11}H_{22}O_3$ M 202.293

Needles (H_2O). Mp 65-66°, Mp 76°. Bp_4 148-150°.

Me ester: [24724-07-0].

$C_{12}H_{24}O_3$ M 216.320

Mp 27-27.5°. Bp_8 168-169°.

Ac:

$C_{13}H_{24}O_4$ M 244.330

Mp 34°. Bp_2 184-185°.

Chuit, P. *et al*, *Helv. Chim. Acta*, 1929, **12**, 476.
Lycan, W.H. *et al*, *J. Am. Chem. Soc.*, 1929, **51**, 628 (*synth*)
Tamao, K. *et al*, *J. Am. Chem. Soc.*, 1978, **100**, 2268 (*synth*)

2-Hydroxy-10-undecenoic acid, 9CI **H-00603**

[69695-27-8]

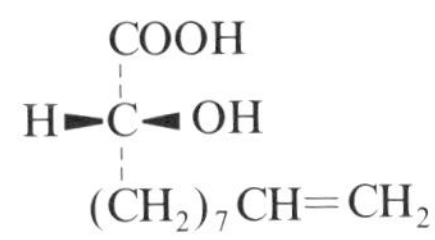

(*R*)-*form*
Absolute configuration

$C_{11}H_{20}O_3$ M 200.277

(*R*)-*form* [34456-31-0]

Mp 65.0-65.5°. $[\alpha]_D^{25}$ +6.3°.

(*S*)-*form* [34456-30-9]

Mp 64.9-65.5°. $[\alpha]_D^{23}$ −6.3°.

(±)-*form* [34456-29-6]

Mp 60.4-60.9°.

Eliasson, B. *et al*, *Acta Chem. Scand.*, 1971, **25**, 2217 (*synth, resoln, ms*)
Hase, T.A. *et al*, *Synth. Commun.*, 1979, **9**, 63 (*synth*)

2-Hydroxy-6-undecylbenzoic acid, 9CI **H-00604**

Anagigantic acid

[14155-32-9]

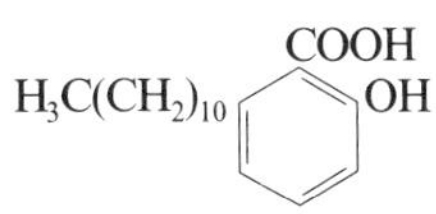

$C_{18}H_{28}O_3$ M 292.417

Isol. from the pericarp of *Anacardium giganteum* and from *Schistochila appendiculata*. Needles (hexane). Mp 81-82°.

K salt: [108049-75-8].
Mp 201-204° dec.

Sharma, P.V.K. *et al*, *Indian J. Chem.*, 1966, **4**, 99 (*isol, struct*)
Durrani, A.A. *et al*, *J. Chem. Soc., Perkin Trans. 1*, 1979, 2069 (*synth, pmr, ir*)
Asakawa, Y. *et al*, *Phytochemistry*, 1987, **26**, 735 (*isol*)

I

muco-Inositol, 9CI, 8CI **I-00001**

(1α,2α,3β,4α,5α,6β)-Cyclohexanehexol. 1,2,4,5/3,6-Inositol. Mucoinositol

[488-55-1]

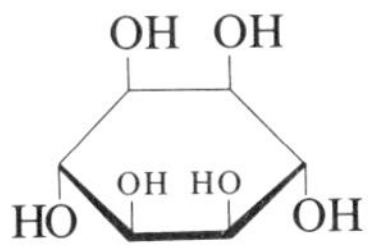

$C_6H_{12}O_6$ M 180.157

Occurs as the methyl ether in conifers (exc. Pinaceae), *Taxopsida* and Cistaceae. Mp 285-290°. Opt. inactive (*meso-*).

*1,2:4,5-Di-O-isopropylidene: 1,2:4,5-Di-O-isopropylidene-*muco*-inositol*
$C_{12}H_{20}O_6$ M 260.286
Mp 184°.

1,2:4,5-Di-O-isopropylidene, 3-brosyl: [58706-28-8].
Mp 114-115°.

1,2:4,5-Di-O-isopropylidene, 3-benzyl: [58706-24-4].
$C_{19}H_{26}O_6$ M 350.411
Mp 121-122°.

3,6-Di-Ac:
$C_{10}H_{16}O_8$ M 264.232
Mp 201-202°.

*Hexa-Ac: 1,2,3,4,5,6-Hexa-O-acetyl-*muco*-inositol*
$C_{18}H_{24}O_{12}$ M 432.380
Mp 177-178°.

3,6-Dimesyl: Mp 191-193°.

1-Me:
$C_7H_{14}O_6$ M 194.184
Occurs in Conifers (exc. Pinaceae), *Taxopsida*, Cistaceae.
$[\alpha]_D^{20}$ −53° (EtOH).

3,6-Dibenzyl:
$C_{20}H_{24}O_6$ M 360.406
Mp 147-148°.

Dangschat, G. *et al*, *Naturwissenschaften*, 1939, **27**, 756 (*synth*)
Angyal, S.J. *et al*, *J. Chem. Soc.*, 1962, 2985.
Foxall, C.D. *et al*, *J. Chem. Soc.*, 1963, 5573 (*isol*)
Suami, T. *et al*, *Bull. Chem. Soc. Jpn.*, 1967, **40**, 1488.
Espelie, K.E. *et al*, *Carbohydr. Res.*, 1976, **46**, 53.
Dangschat, G. *et al*, *Carbohydr. Res.*, 1987, **164**, 343 (*synth*)

myo-Inositol, 8CI **I-00002**

(1α,2α,3α,4β,5α,6β)-Cyclohexanehexol. meso-*Inositol. Dambose. Nucitol. 1,2,3,5/4,6-Inositol.* i-*Inositol. Inositol. Mesoinositol. Phaseomannitol. Myoinositol. Other proprietary names*

[87-89-8]

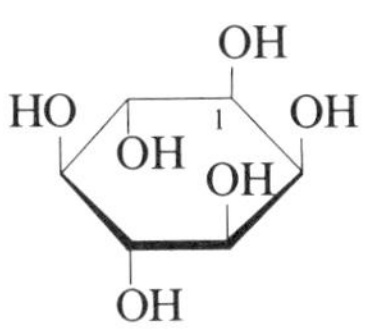

$C_6H_{12}O_6$ M 180.157

The most widely distributed member of the group of stereoisomers. Of common occurrence in plants and animals. Obt. comly. from phytic acid in corn steep liquor. Growth factor for animals and microorganisms. Lipotropic agent. The phosphates are important cellular second messengers. Mp 225°. Opt. inactive (*meso-*).

▷ Toxic.

1,2-O-Cyclohexylidene: [6763-47-9].
$C_{12}H_{20}O_6$ M 260.286
Mp 179°.

1,2:3,4-Di-O-isopropylidene: 1,2:3,4-Di-O-isopropylidene-myo-inositol
$C_{12}H_{20}O_6$ M 260.286
(Pet. ether - EtOAc, 10:1). Mp 150-152°.

1,2:3,4-Di-O-isopropylidene, di-Ac:
$C_{16}H_{24}O_8$ M 344.361
Mp 140-142°.

1,2:4,5-Di-O-isopropylidene: [34379-30-1].
$C_{12}H_{20}O_6$ M 260.286
Mp 174°.

1,2:5,6-Di-O-isopropylidene: 1,2:5,6-Di-O-isopropylidene-myo-inositol
$C_{12}H_{20}O_6$ M 260.286
Mp 172-175°.

1,2:5,6-Di-O-isopropylidene, di-Ac:
$C_{16}H_{24}O_8$ M 344.361
Mp 165-168°.

1,2:5,6-Di-O-isopropylidene, dibenzoyl:
$C_{26}H_{28}O_8$ M 468.502
(MeOH). Mp 197-200°.

1,2:5,6-Di-O-cyclohexylidene: [40773-57-7].
$C_{18}H_{28}O_6$ M 340.416
Mp 133°, Mp 153-156°. $[\alpha]_D^{20}$ −7.4° (c, 0.5 in C_6H_6).
▷ LD_{50} (rat, scu) 1180 mg/kg.

1,4,5,6-Tetra-Ac:
$C_{14}H_{20}O_{10}$ M 348.306
Mp 100°.

1,2,3,4,6-Penta-Ac:
$C_{16}H_{22}O_{11}$ M 390.343
Mp 178°.

Hexa-Ac: [1254-38-2].
$C_{18}H_{24}O_{12}$ M 432.380
Mp 216-217°.

Hexabenzoyl:
$C_{48}H_{36}O_{12}$ M 804.805
Mp 258°.

1,2:3,4-Tetrabenzyl, di-Ac: Mp 132-134°.

1,2:5,6-Tetrabenzyl, di-Ac:
$C_{38}H_{40}O_8$ M 624.729
(Pet. ether 60-80°). Mp 106-108°.

Hexa(3-pyridinecarboxyl): [6556-11-2]. ***Inositol nicotinate*, BAN, INN.** *Inositol niacinate, USAN. Hexanicotinoylinositol. Hexopal. Mesonex. Lipoflavonoid. Numerous proprietary names*
$C_{42}H_{30}N_6O_{12}$ M 810.732
Peripheral vasodilator, antilipidaemic agent. Mp 254-255°.
▷ NM7535400.

1-Phosphate: [573-35-3].
$C_6H_{13}O_9P$ M 260.137
Wheat bran phytase hydrol. prod. of sodium phytate. Mp 195-197° dec.

1-Phosphate, biscyclohexylammonium salt: Mp 202-204°. $[\alpha]_D^{20}$ +3.4° (c, 1.0 in H_2O).

1,4-Diphosphate:

$C_6H_{14}O_{12}P_2$ M 340.117

Mp 206-207° (as bis(cyclohexylammonium) salt).

1,4,5-Triphosphate:

$C_6H_{15}O_{15}P_3$ M 420.097

Prod. *in vivo* (as the D-enantiomer) by hydrol. of phosphatidylinositol 4,5-diphosphate. Present in mammalian brain tissue. Important cellular second messenger. Controls $Ca^{2\oplus}$ release from intracellular stores. $[\alpha]_D$ −27.4° (as cyclohexylamine salt) (D-form).

Hexaphosphate: [83-86-3]. ***Phytic acid.*** *Fytic acid, INN. Alkalovert*

$C_6H_{18}O_{24}P_6$ M 660.036

Widely distributed in the Gramineae and seeds of many other higher plants. Complexing agent for the removal of traces of heavy metals; therapeutic hypocalcaemic agent, freq. employed as nona-Na salt (Sodium phytate, USAN). Straw-coloured syrup.

▷ NM7525000.

1-O-β-D-Glucopyranoside: [4539-73-5]. *1-O-β-D-Glucopyranosyl-*myo-*inositol.* ***Glucinol***

$C_{12}H_{22}O_{11}$ M 342.299

Occurs in vascular tissues of higher plants. Mp 252-260°. $[\alpha]_D^{25}$ −17.5° (H_2O). Has (1*R*-config).

1-O-β-D-Glucopyranoside, nona-Me:

$C_{21}H_{40}O_{11}$ M 468.540

Mp 95-98°. $[\alpha]_D^{25}$ −10° (c, 1.0 in EtOH).

*1,2:5,6-Tetrabenzyl: 1,2:5,6-Tetra-O-benzyl-*myo-*inositol*

$C_{34}H_{36}O_6$ M 540.655

(EtOH). Mp 169-170°.

[7205-52-9, 17211-15-3]

Badgett, C.O. *et al, J. Am. Chem. Soc.*, 1947, **69**, 2907 (*nicotinate*)
Biochem. Prep., 1952, **2**, 65.
Angyal, S.J. *et al, Adv. Carbohydr. Chem.*, 1959, **14**, 135 (*rev*)
Ballou, L.E. *et al, J. Biol. Chem.*, 1961, **236**, 54, 1902 (*isol, triphosphate*)
Angyal, S.J. *et al, J. Chem. Soc.*, 1961, 4116, 4122, 4718.
Gorin, P.A.J. *et al, Can. J. Chem.*, 1965, **43**, 2259 (*Glucinol*)
Karrer, W. *et al, Konstitution und Vorkommen der Organischen Pflanzenstoffe*, 2nd Ed., Birkhäuser Verlag, Basel, 1972-1985, no. 280 (*bibl*)
Shvets, V.I. *et al, Tetrahedron*, 1973, **29**, 331.
Isbrandt, L.R. *et al, J. Am. Chem. Soc.*, 1980, **102**, 3144.
Hopf, H. *et al, CA*, 1981, **94**, 13612n (*occur, Glucinol*)
Wells, W.W., *Kirk-Othmer Encycl. Chem. Technol.*, 3rd Ed., Wiley, N.Y., 1983, **24**, 50 (*rev*)
Berridge, M.J. *et al, Nature (London)*, 1984, **312**, 315 (*rev, triphosphate*)
Bleasdale, J.E. *et al*, Eds., *Inositol and Phosphoinositides:Metabolism and Regulation*, Humane Press,Clifton, N.J., 1985 (*book*)
Holub, B.J., *Annu. Rev. Nutr.*, 1986, **6**, 563 (*rev*)
Gigg, J. *et al, J. Chem. Soc., Perkin Trans. 1*, 1987, 2411 (*isopropylidene, benzyl derivs*)
Negwer, M., *Organic-Chemical Drugs and their Synonyms*, 6th Ed., Akademie-Verlag, Berlin, 1987, 540, 588, 8197 (*synonyms*)
Ley, S.V. *et al, Tetrahedron*, 1990, **46**, 4995 (*synth, derivs*)
Martindale, The Extra Pharmacopoeia, 30th edn., Pharmaceutical Press, London, 1993, 1056-z, 7877-p, 9243-y.
Lewis, R.J., *Sax's Dangerous Properties of Industrial Materials*, 8th Ed., Van Nostrand-Reinhold, 1992, HFG550, IDE300, PIB250.

1-Iododecane, 9CI, 8CI — I-00003

Decyl iodide

[2050-77-3]

$H_3C(CH_2)_8CH_2I$

$C_{10}H_{21}I$ M 268.180

Liq. d_4^{20} 1.2567. Bp_{15} 132°.

Krafft, F., *Ber.*, 1892, **19**, 2218 (*synth*)
Komppa, G. *et al, J. Prakt. Chem.*, 1932, **135**, 193 (*synth*)
Sevrin, M. *et al, Tetrahedron Lett.*, 1976, 2647 (*synth*)
Jung, M.E. *et al, Tetrahedron Lett.*, 1977, 2659 (*synth*)

1-Iododocosane — I-00004

Docosyl iodide

$H_3C(CH_2)_{20}CH_2I$

$C_{22}H_{45}I$ M 436.502

Cryst. (pet. ether). Mp 41.5°, Mp 48.2° (dimorph.). $Bp_{0.23}$ 198°.

Francis, F. *et al, Proc. R. Soc. London, A*, 1937, **158**, 691 (*synth*)

1-Iodododecane — I-00005

Dodecyl iodide. Lauryl iodide

[4292-19-7]

$H_3C(CH_2)_{10}CH_2I$

$C_{12}H_{25}I$ M 296.234

$Bp_{0.7}$ 145-150°.

Thiouronium picrate: Mp 139°.

Levene, P.A. *et al, J. Biol. Chem.*, 1914, **18**, 478.

1-Iodoeicosane — I-00006

Eicosyl iodide. Icosyl iodide

[41380-77-2]

$H_3C(CH_2)_{18}CH_2I$

$C_{20}H_{41}I$ M 408.448

Cryst. (Me_2CO). Mp 42-43°.

Levene, P.A. *et al, J. Biol. Chem.*, 1924, **59**, 916.
Li, J.C.M. *et al, J. Chem. Eng. Data*, 1961, **6**, 268.
Brady, G.W. *et al, J. Chem. Phys.*, 1973, **58**, 3535.

1-Iodoheptadecane — I-00007

Heptadecyl iodide

[26825-83-2]

$H_3C(CH_2)_{15}CH_2I$

$C_{17}H_{35}I$ M 366.368

Cryst. (Me_2CO/EtOH). Mp 34.8-35.3°. $Bp_{0.5}$ 158-159°.

Levene, P.A. *et al, J. Biol. Chem.*, 1915, **20**, 531.
Webster, V.S. *et al, CA*, 1940, **34**, 2784.
Silbert, L.S. *et al, J. Am. Oil Chem. Soc.*, 1969, **46**, 615 (*synth*)
Silbert, L.S., *CA*, 1970, **73**, 120055a (*synth*)

7-Iodo-2-heptynoic acid — I-00008

$ICH_2(CH_2)_3C{\equiv}CCOOH$

$C_7H_9IO_2$ M 252.051

Me ester: [121823-95-8].

$C_8H_{11}IO_2$ M 266.078

Oil. $Bp_{0.2}$ 120-122°.

Piers, E. *et al, Tetrahedron*, 1989, **45**, 363 (*synth, ir, pmr*)
Kita, Y. *et al, J. Org. Chem.*, 1991, **56**, 119 (*synth, pmr, ir*)

1-Iodohexacosane I-00009

Hexacosyl iodide. Ceryl iodide

[52644-81-2]

$H_3C(CH_2)_{24}CH_2I$

$C_{26}H_{53}I$ M 492.609

Mp 54.4°, Mp 59.1° (dimorph.).

Bleyberg, W. *et al*, *Chem. Ber.*, 1931, **64**, 2504 (*synth*)
Francis, F. *et al*, *Proc. R. Soc. London, A*, 1937, **158**, 691 (*synth*)

1-Iodohexadecane I-00010

Cetyl iodide. Hexadecyl iodide

[40474-98-4]

$H_3C(CH_2)_{14}CH_2I$

$C_{16}H_{33}I$ M 352.341

Cryst. (EtOH). Mp 22°. Bp_{15} 211°.

Org. Synth., 1935, **15**, 29 (*synth*)
Hauser, E.A. *et al*, *J. Phys. Chem.*, 1941, **45**, 954.
Bentley, F.F. *et al*, *Spectrochim. Acta*, 1964, **20**, 105 (*ir*)

2-Iodohexadecanoic acid I-00011

2-Iodopalmitic acid

$H_3C(CH_2)_{13}CHICOOH$

$C_{16}H_{31}IO_2$ M 382.324

Cryst. (pet. ether). Mp 60-61°.

Amide:

$C_{16}H_{32}INO$ M 381.339

Cryst. (EtOH). Mp 109°.

Jones, D.B., *J. Am. Chem. Soc.*, 1915, **37**, 589.
Sweet, R.S. *et al*, *J. Org. Chem.*, 1956, **21**, 1426 (*ir*)

3-Iodohexadecanoic acid I-00012

$H_3C(CH_2)_{12}CHICH_2COOH$

$C_{16}H_{31}IO_2$ M 382.324

Mp 50.5°.

Robinet, M., *Bull. Soc. Chim. Belg.*, 1931, **40**, 710.

16-Iodohexadecanoic acid I-00013

***Iodocetilic acid,* INN.** *RA-C-384*

[2536-36-9]

$ICH_2(CH_2)_{14}COOH$

$C_{16}H_{31}IO_2$ M 382.324

Mp 76°. ^{123}I labelled cpd. is used as a diagnostic aid.

[54510-20-2]

Bougault, J., *C. R. Hebd. Seances Acad. Sci.*, 1910, **150**, 876.
Hunsdiecker, H. *et al*, *Chem. Ber.*, 1942, **75**, 291.
Fox, R.B. *et al*, *J. Chem. Eng. Data*, 1963, **8**, 612.
Laufer, P. *et al*, *Eur. J. Nucl. Med.*, 1981, **18**, 1205 (*synth, props*)
Okada, R.D. *et al*, *J. Labelled Compd. Radiopharm.*, 1983, **8**, 211 (*synth*)
Knapp, F.F. *et al*, *J. Med. Chem.*, 1984, **27**, 94 (*synth*)

1-Iodohexatriacontane I-00014

[62154-87-4]

$H_3C(CH_2)_{34}CH_2I$

$C_{36}H_{73}I$ M 632.877

Cryst. (Me_2CO). Mp 73-74°.

Huang, H.C. *et al*, *J. Org. Chem.*, 1982, **47**, 4018 (*synth, pmr*)

1-Iodo-17-hexatriacontene I-00015

$H_3C(CH_2)_{17}CH{=}CH(CH_2)_{15}CH_2I$

$C_{36}H_{71}I$ M 630.861

(E)-form [82741-70-6]

Cryst. (EtOH). Mp 61-62°.

(Z)-form [82741-67-1]

Cryst. ($CHCl_3$/EtOH). Mp 55-56.5°.

Huang, H.C. *et al*, *J. Org. Chem.*, 1982, **47**, 4018 (*synth, pmr*)

1-Iodo-17-hexatriacontyne I-00016

[82741-63-7]

$H_3C(CH_2)_{17}C{\equiv}C(CH_2)_{15}CH_2I$

$C_{36}H_{69}I$ M 628.846

Cryst. ($CHCl_3$/EtOH). Mp 44-46°.

Huang, H.C. *et al*, *J. Org. Chem.*, 1982, **47**, 4018 (*synth, pmr*)

6-Iodo-2-hexynoic acid I-00017

$ICH_2CH_2CH_2C{\equiv}CCOOH$

$C_6H_7IO_2$ M 238.025

Me ester: [51755-86-3].

$C_7H_9IO_2$ M 252.051

Oil. $Bp_{0.5}$ 125-132°.

Kita, Y. *et al*, *J. Org. Chem.*, 1991, **56**, 119 (*synth*)

1-Iodononadecane I-00018

Nonadecyl iodide

$H_3C(CH_2)_{17}CH_2I$

$C_{19}H_{39}I$ M 394.422

Mp 42.5-43.5°. $Bp_{0.22}$ 174-174.5°.

Levene, P.A. *et al*, *J. Biol. Chem.*, 1924, **59**, 905.
Shiina, S., *CA*, 1935, **29**, 13854 (*synth*)

1-Iodooctacosane I-00019

Octacosyl iodide

$H_3C(CH_2)_{26}CH_2I$

$C_{28}H_{57}I$ M 520.663

Mp 62.8-63.2°.

Bleyberg, W. *et al*, *Chem. Ber.*, 1931, **64**, 2504 (*synth*)

1-Iodooctadecane I-00020

Octadecyl iodide. Stearyl iodide

[629-93-6]

$H_3C(CH_2)_{16}CH_2I$

$C_{18}H_{37}I$ M 380.395

Mp 34.5-35°. $Bp_{0.5}$ 160-170°.

Bleyberg, W. *et al*, *Ber.*, 1931, **64**, 2504 (*synth*)
Frewing, J.J., *Proc. R. Soc. London, A*, 1944, **182**, 270 (*synth*)
Shirley, D.A. *et al*, *J. Org. Chem.*, 1953, **18**, 1591.
Juergen, H. *et al*, *CA*, 1973, **78**, 128785n.

2-Iodooctadecanoic acid I-00021

2-Iodostearic acid

$H_3C(CH_2)_{15}CHICOOH$

$C_{18}H_{35}IO_2$ M 410.378

Mp 66°.

Amide:

$C_{18}H_{36}INO$ M 409.393

Cryst. (EtOH). Mp 112°.

Ponzio, G., *Gazz. Chim. Ital.*, 1911, **41**, 786.
Frewing, J.J., *Proc. R. Soc. London, A*, 1944, **182**, 270 (*synth*)
Myers, S.G., *J. Am. Chem. Soc.*, 1951, **73**, 2100 (*synth*)
Robertson, R.F. *et al*, *Can. J. Chem.*, 1956, **34**, 716.

3-Iodooctadecanoic acid I-00022

3-Iodostearic acid

$H_3C(CH_2)_{14}CHICH_2COOH$

$C_{18}H_{35}IO_2$ M 410.378
Cryst. (AcOH). Mp 60-61°.

Me ester:
$C_{19}H_{37}IO_2$ M 424.405
Mp 41°.

Eckert, A. *et al*, *Monatsh. Chem.*, 1913, **34**, 1817.
Pigulevskii, G.V. *et al*, *CA*, 1943, **37**, 2716 (*synth*)

1-Iodotetracosane I-00023

Tetracosyl iodide

$H_3C(CH_2)_{22}CH_2I$

$C_{24}H_{49}I$ M 464.556
Mp 54.5-55.5°. $Bp_{0.35}$ 207-209°.

Levene, P.A. *et al*, *J. Biol. Chem.*, 1924, **59**, 917.
Bleyberg, W. *et al*, *Chem. Ber.*, 1931, **64**, 2504 (*synth*)
Schildknecht, H. *et al*, *Fette, Seifen, Anstrichm.*, 1964, **66**, 176 (*synth*)

1-Iodotetradecane I-00024

Tetradecyl iodide. Myristyl iodide

[19218-94-1]

$H_3C(CH_2)_{12}CH_2I$

$C_{14}H_{29}I$ M 324.288
$Bp_{17.5}$ 192-195°, $Bp_{0.5}$ 128°.

Majima, R. *et al*, *Ber.*, 1913, **46**, 4089 (*synth*)
Levene, P.A. *et al*, *J. Biol. Chem.*, 1915, **20**, 529.
Li, J.C.M. *et al*, *J. Chem. Eng. Data*, 1961, **6**, 268.
McLafferty, F.W., *Anal. Chem.*, 1962, **34**, 2 (*ms*)

Isocarbacyclin I-00025

1,3a,4,5,6,6a-Hexahydro-5-hydroxy-6-(3-hydroxy-1-octenyl)-2-pentalenepentanoic acid

[88931-59-3]

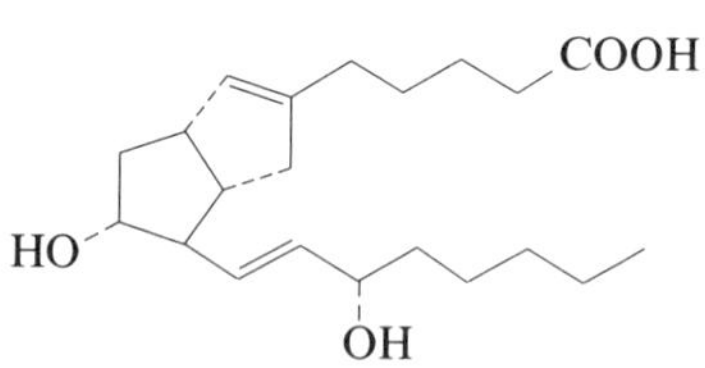

$C_{21}H_{34}O_4$ M 350.497
Inhibitor of platelet aggregation. Mp 73-79°. $[\alpha]_D^{25}$ +16.0° (c, 0.25 in MeOH).

Shibasaki, M. *et al*, *Tetrahedron Lett.*, 1983, **24**, 3493, 3497 (*synth, ir, pmr*)
Mase, T. *et al*, *Tetrahedron Lett.*, 1984, **25**, 5087 (*synth*)
Bannai, K. *et al*, *Tetrahedron Lett.*, 1986, **27**, 6353 (*synth*)
Suzuki, M. *et al*, *J. Org. Chem.*, 1987, **52**, 5583 (*synth*)
Hashimoto, S. *et al*, *Tetrahedron Lett.*, 1987, **28**, 637 (*synth*)
Hashimoto, S. *et al*, *J. Chem. Soc., Chem. Commun.*, 1988, 1137 (*synth*)
Bannai, K. *et al*, *Tetrahedron*, 1990, **46**, 6689 (*synth*)

Isodactylyne I-00026

*3-Bromo-2-(3-bromo-2-pentenyl)-5-chlorotetrahydro-6-(2-penten-4-ynyl)-2*H*-pyran, 9CI. 9,13-Dibromo-7-chloro-6,10-epoxy-3,12-pentadecadien-1-yne*

[58001-90-4]

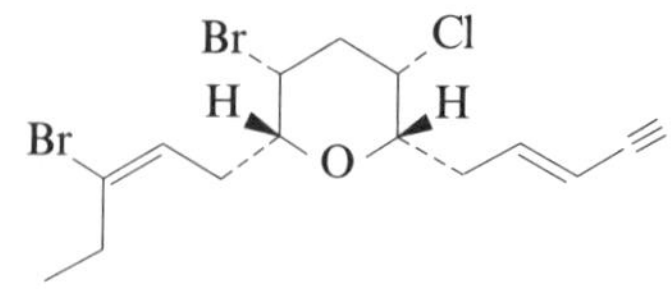

$C_{15}H_{19}Br_2ClO$ M 410.575
Constit. of the sea hare *Aplysia dactylomela*. Shows antibiotic props. Oil. $[\alpha]_D^{24}$ −8.06° (c, 7.97 in $CHCl_3$).

Vanderah, D.J. *et al*, *J. Org. Chem.*, 1976, **41**, 3480 (*isol, ir, uv, ms, nmr, struct*)
Gao, L.-X. *et al*, *Tetrahedron Lett.*, 1992, **33**, 4349 (*synth*)

Isolaureatin I-00027

7-Bromo-4-(1-bromopropyl)-2-(2-penten-4-ynyl)-3,9-dioxabicyclo[4.2.1]nonane, 9CI. 9,13-Dibromo-6,12;7,10-diepoxypentadec-3-en-1-yne

[19897-64-4]

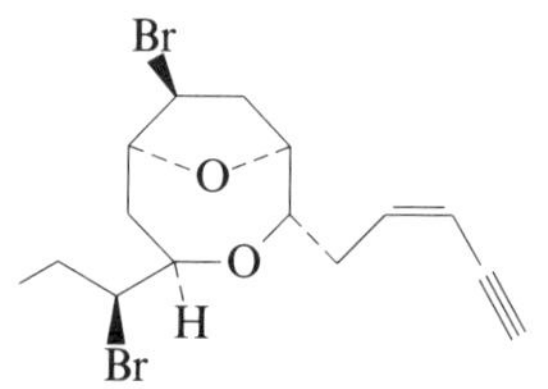

Absolute configuration

$C_{15}H_{20}Br_2O_2$ M 392.130
Constit. of *Laurencia nipponica*. Mp 83-84°. $[\alpha]_D$ +40° (CCl_4).

E-*Isomer:* [112710-66-4]. ***(3E)-Isolaureatin***
$C_{15}H_{20}Br_2O_2$ M 392.130
Constit. of *L. nipponica*. Oil. $[\alpha]_D^{15}$ −8.73° (c, 1.11 in $CHCl_3$).

Irie, T. *et al*, *Tetrahedron Lett.*, 1968, 2735 (*isol, pmr, ms*)
Kurosawa, E. *et al*, *Tetrahedron Lett.*, 1973, 3857 (*cryst struct*)
Suzuki, M. *et al*, *Bull. Chem. Soc. Jpn.*, 1987, **60**, 3791 (*isol*)

Isomaneonene A I-00028

[62583 57 7]

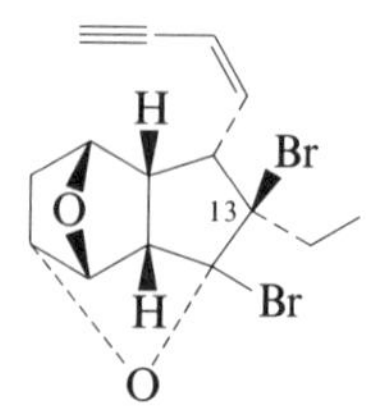

$C_{15}H_{16}Br_2O_2$ M 388.098
Constit. of *Laurencia nidifica*. Yellow cryst. Mp 114.5-115.5°. $[\alpha]_D^{21}$ +106° (c, 1.72 in $CHCl_3$).

13-Epimer: [62623-94-3]. ***Isomanoenene B***
Isol. from *L. nidifica*. Cryst. Mp 136.5-137°. $[\alpha]_D$ +87° (c, 0.83 in $CHCl_3$).

Waraszkiewicz, S.M. *et al*, *J. Org. Chem.*, 1978, **43**, 3194.

Isoprelaurefucin I-00029

5-Bromo-3-(1-bromopropyl)-8-(2-penten-4-ynyl)-2,7-dioxabicyclo[4.2.1]nonane, 9CI. 10,13-Dibromo-6,9;7,12-diepoxypentadec-3-en-1-yne

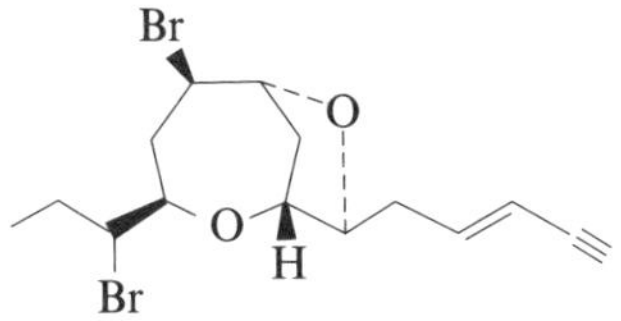

$C_{15}H_{20}Br_2O_2$ M 392.130

Constit. of the seaweed *Laurencia nipponica* and *L. subopposita*. Oil. $[\alpha]_D$ −54.4°.

Kurosawa, E. *et al*, *Tetrahedron Lett.*, 1973, 4135 (*isol, uv, ir, pmr, ms*)
Suzuki, M. *et al*, *Bull. Chem. Soc. Jpn.*, 1986, **59**, 2953 (*abs config, bibl*)

1-Isothiocyanato-4,5,6,7-decanetetrol, 9CI I-00030

4,5,6,7-Tetrahydroxydecyl isothiocyanate
[57103-43-2]

$H_3CCH_2CH_2CH(OH)CH(OH)CH(OH)CH(OH)$
$(CH_2)_3NCS$

$C_{11}H_{21}NO_4S$ M 263.357

Constit. of *Capparis grandis* roots. Cryst. (MeOH). Mp 178-180° dec. $[\alpha]_D^{24}$ +6.4° (c, 1 in MeOH).

Tetra-Ac: [57103-44-3].
$C_{19}H_{29}NO_8S$ M 431.506
Cryst. ($CHCl_3$/hexane). Mp 140-141°.

Gaind, K.N. *et al*, *Phytochemistry*, 1975, **14**, 1415 (*isol, pmr, ms*)

J

Jasmonic acid **J-00001**

3-Oxo-2-(2-pentenyl)-cyclopentaneacetic acid, 9CI, 8CI

[6894-38-8]

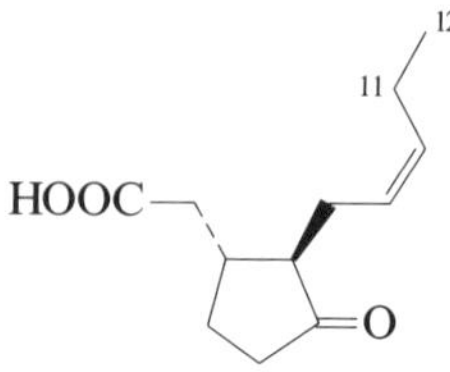

$C_{12}H_{18}O_3$ M 210.272

Esters are present in *Jasminum grandiflorum* and are responsible for its odour. Viscous oil. $Bp_{0.001}$ 125°. $[\alpha]_D$ −83.5° (c, 0.97 in $CHCl_3$). n_D^{19} 1.4885.

2,4-Dinitrophenylhydrazone: Pale citron-yellow cryst. (Et_2O/pet. ether). Mp 152-154°, Mp 157-159° (double Mp).

Me ester: [20073-13-6]. *Methyl jasmonate*

$C_{13}H_{20}O_3$ M 224.299

From *J. grandiflorum*. Has characteristic odour of jasmine; used in perfumery. Oil. $Bp_{0.001}$ 81-84°. $[\alpha]_D$ −76.5° (c, 3.4 in MeOH). n_D^{22} 1.4730.

Me ester, 2,4-dinitrophenylhydrazone: Mp 50-54°. Mixture of geom. isomers.

2-Epimer: ***7-Isojasmonic acid***

$C_{12}H_{18}O_3$ M 210.272

Metab. of *Botryodiplodia theobromae*. Plant growth regulator. Oil. $[\alpha]_D$ +64° (c, 0.1 in MeOH).

2′,3′-Dihydro: [98674-52-3]. *3-Oxo-2-pentylcyclopentaneacetic acid. 9,10-Dihydrojasmonic acid*

$C_{12}H_{20}O_3$ M 212.288

Isol. from *Vicia faba*. Oil. $[\alpha]_D^{22}$ −28.3° (c, 0.1 in MeOH).

2′,3′-Dihydro, Me ester: [2630-39-9]. *Methyl dihydrojasmonate. Hedione*

$C_{13}H_{22}O_3$ M 226.315

Perfumery ingredient.

8ξ-Hydroxy: ***8-Hydroxyjasmonic acid***

$C_{12}H_{18}O_4$ M 226.272

Metab. of *B. theobromae*.

11R-Hydroxy: ***11R-Hydroxyjasmonic acid***

$C_{12}H_{18}O_4$ M 226.272

Metab. of *B. theobromae*. $[\alpha]_D$ −88°.

11S-Hydroxy: ***11S-Hydroxyjasmonic acid***

$C_{12}H_{18}O_4$ M 226.272

Metab. of *B. theobromae*. $[\alpha]_D$ −44.1°.

12-Hydroxy: ***12-Hydroxyjasmonic acid***

$C_{12}H_{18}O_4$ M 226.272

Metab. of *B. theobromae*.

[76968-33-7]

Hill, R.K. *et al*, *Tetrahedron*, 1965, **21**, 1501 (*abs config*)
Fukui, H. *et al*, *Agric. Biol. Chem.*, 1977, **41**, 189 (*synth*)
Dubs, P. *et al*, *Helv. Chim. Acta*, 1978, **61**, 990.
Gerlach, H. *et al*, *Helv. Chim. Acta*, 1978, **61**, 2503.
Näf, F. *et al*, *Helv. Chim. Acta*, 1978, **61**, 2524.
Sato, T. *et al*, *Bull. Chem. Soc. Jpn.*, 1981, **54**, 505 (*synth*)
Kitahara, T. *et al*, *Agric. Biol. Chem.*, 1982, **46**, 1369; 1984, **48**, 1731; 1987, **51**, 1129; 1991, **55**, 1013 (*synth*)
Johnson, F. *et al*, *J. Org. Chem.*, 1982, **47**, 4254 (*synth*)
Oppolzer, W. *et al*, *Helv. Chim. Acta*, 1983, **66**, 2140 (*synth*)
Nishiyama, H. *et al*, *Tetrahedron Lett.*, 1984, **25**, 2487 (*synth*)
Posner, G.H. *et al*, *J. Org. Chem.*, 1985, **50**, 2589 (*synth*)
Smith, A.B. *et al*, *J. Org. Chem.*, 1985, **50**, 3239 (*synth*)
Miersch, O. *et al*, *Phytochemistry*, 1986, **26**, 1037; 1989, **28**, 339 (*isol, derivs*)
Kataoka, H. *et al*, *Tetrahedron*, 1987, **43**, 4107 (*synth*)
Billington, D.C. *et al*, *J. Chem. Res., Miniprint*, 1988, 2601 (*synth*)
Knöfel, H.-D. *et al*, *Z. Naturforsch., C*, 1988, **43**, 29 (*synth*)
Montforts, F.-P. *et al*, *Helv. Chim. Acta*, 1989, **72**, 1852 (*synth*)
Miersch, O. *et al*, *Phytochemistry*, 1991, **30**, 4049 (*Hydroxyjasmonic acids*)
Lee, W.Y. *et al*, *Synth. Commun.*, 1992, **22**, 1283 (*synth*)

Juruenolide B **J-00002**

[88206-70-6]

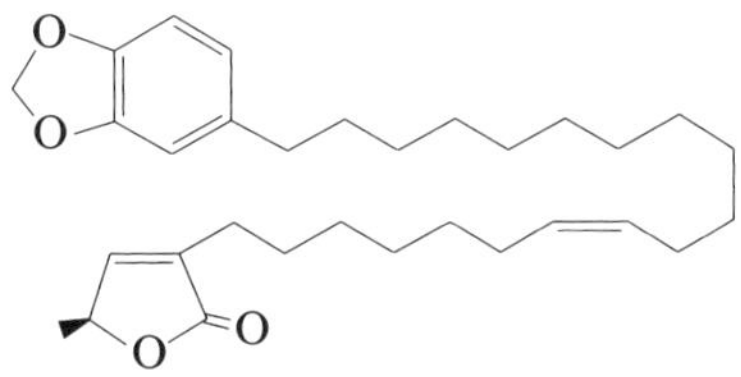

$C_{31}H_{46}O_4$ M 482.702

Constit. of *Iryanthera juruensis*. Oil. $[\alpha]_D^{25}$ +15° (MeOH).

Vieira, P.C. *et al*, *Phytochemistry*, 1983, **22**, 711.

K

Kumausyne **K-00001**

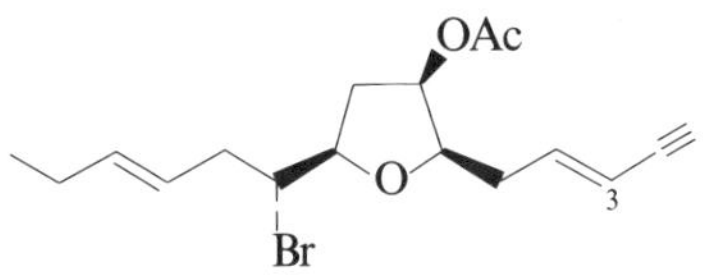

$C_{17}H_{23}BrO_3$ M 355.271

(3E)-form [88034-94-0]

Constit. of the red alga *Laurencia nipponica*. Oil. $[\alpha]_D^{26}$ −2.3° (c, 0.62 in $CHCl_3$).

O-*De-Ac:* [88034-95-1]. ***Deacetylkumausyne***

$C_{15}H_{21}BrO_2$ M 313.234

Constit. of *L. nipponica*. Oil. $[\alpha]_D^{20}$ +6.5° (c, 1.08 in $CHCl_3$).

(3Z)-form [87980-57-2]

Constit. of *L. nipponica*. Oil. $[\alpha]_D^{25}$ −3.2° (c, 1.03 in $CHCl_3$).

O-*De-Ac:* [87980-61-8].

Constit. of *L. nipponica*. Oil. $[\alpha]_D^{20}$ +10.8° (c, 1.88 in $CHCl_3$).

Suzuki, T. *et al*, *Chem. Lett.*, 1983, 1643.
Brown, M.J. *et al*, *J. Am. Chem. Soc.*, 1991, **113**, 5378 (*synth*)

L

Lancifolide **L-00001**

[121748-17-2]

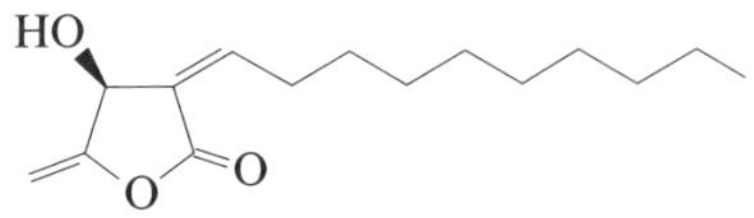

$C_{15}H_{24}O_3$ M 252.353

Constit. of *Actinodaphne lancifolia*. Oil. $[\alpha]_D$ −49.0° (c, 0.58 in $CHCl_3$).

(E)-*Isomer:* ***Isolancifolide***

$C_{15}H_{24}O_3$ M 252.353

Constit. of *A. lancifolia*. Oil. $[\alpha]_D$ −59.0° (c, 0.5 in $CHCl_3$).

Tanaka, H. *et al*, *Phytochemistry*, 1989, **28**, 626.

Lanosta-9(11),24-dien-3-ol **L-00002**

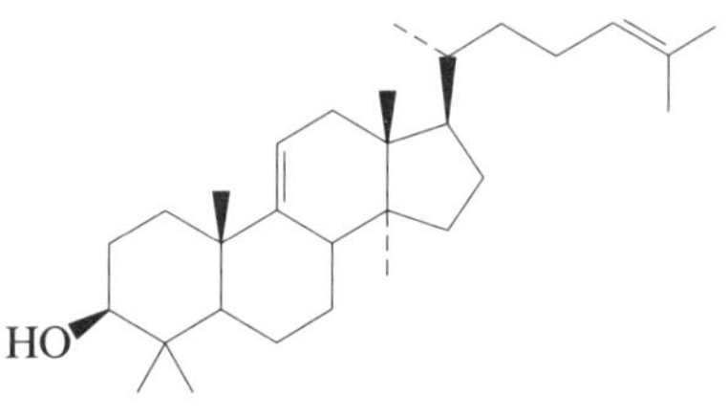

$C_{30}H_{50}O$ M 426.724

3β-form [514-45-4] ***Parkeol***

Constit. of *Butyrospermum parkii*. Cryst. (Et_2O/MeOH). Mp 162-165°. $[\alpha]_D$ +76.8° (c, 1 in $CHCl_3$).

Schreiber, K. *et al*, *Tetrahedron*, 1964, **20**, 1803 (*struct*)
v. Tamelen, E.E. *et al*, *J. Am. Chem. Soc.*, 1972, **94**, 8225 (*synth*)
Itoh, T. *et al*, *Steroids*, 1976, **27**, 275 (*pmr*)

Lanosterol **L-00003**

Lanosta-8,24-dien-3β-ol. Cryptosterol. Isocholesterol. Kryptosterol

[79-63-0]

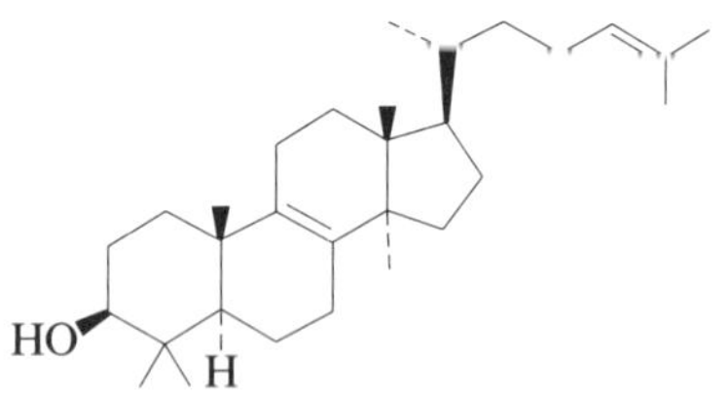

$C_{30}H_{50}O$ M 426.724

Constit. of wool fat. Also from yeast and some other fungi and *Euphorbia regis-jubae* and some other plants. Cryst. (MeOH/Me_2CO). Mp 140-141°. $[\alpha]_D^{17}$ +58° ($CHCl_3$).

Ac: Mp 131-133°. $[\alpha]_D^{20}$ +62.5° (c, 1.12 in $CHCl_3$).

3-Sulfate: Mp 199-200° (as K salt).

Sobel, A.E. *et al*, *J. Am. Chem. Soc.*, 1941, **63**, 1259 (*sulfate*)
Voser, W. *et al*, *Helv. Chim. Acta*, 1952, **35**, 66, 2414 (*struct*)
Hemmert, F. *et al*, *Bull. Soc. Chim. Fr.*, 1966, 976 (*pmr*)
Legrand, M. *et al*, *C. R. Hebd. Seances Acad. Sci. Ser. C*, 1966, **262**, 1290 (*cd*)
Barton, D.H.R. *et al*, *J. Chem. Soc. C*, 1969, **332** (*synth*)
Shishibori, T. *et al*, *Chem. Lett.*, 1973, 1289 (*biosynth*)
Boar, R.B. *et al*, *J. Chem. Soc., Perkin Trans. 1*, 1973, 1583 (*synth*)

Laurallene **L-00004**

[72719-97-2]

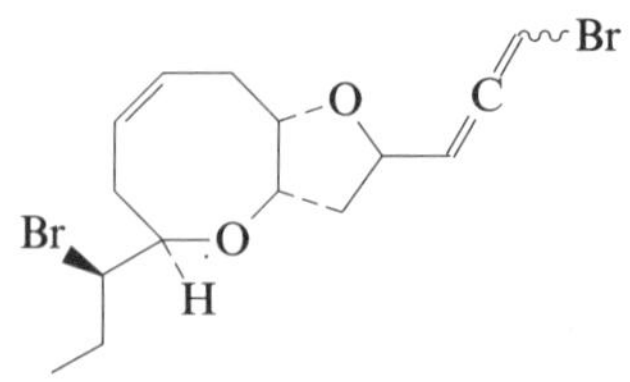

$C_{15}H_{20}Br_2O_2$ M 392.130

Constit. of *Laurencia nipponica*. Cryst. (hexane). Mp 53-54°. $[\alpha]_D$ +173.6° (c, 1.13 in $CHCl_3$).

Fukuzawa, A. *et al*, *Tetrahedron Lett.*, 1979, 2797.

Laureatin **L-00005**

6-Bromo-4-(1-bromopropyl)-2-(2-penten-4-ynyl)-3,8-dioxabicyclo[5.1.1]nonane, 9CI. 10,13-Dibromo-6,12;7,9-diepoxypentadec-3-en-1-yne

[18762-30-6]

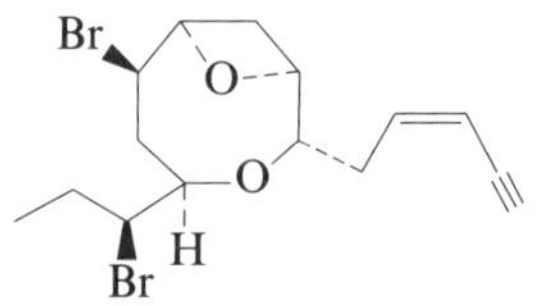

Absolute configuration

$C_{15}H_{20}Br_2O_2$ M 392.130

Constit. of *Laurencia nipponica*. Mp 82-83°. $[\alpha]_D$ +96° (CCl_4).

(E)-*Isomer:* **(3E)-*Laureatin***

$C_{15}H_{20}Br_2O_2$ M 392.130

Constit. of *L. nipponica*. Cryst. (MeOH). Mp 77-78°. $[\alpha]_D^{18}$ +36.6° (c, 0.95 in $CHCl_3$).

Irie, T. *et al*, *Tetrahedron Lett.*, 1968, 2091 (*isol, pmr, ms*)
Kurosawa, E. *et al*, *Tetrahedron Lett.*, 1973, 3875 (*cryst struct*)
Suzuki, M. *et al*, *Bull. Chem. Soc. Jpn.*, 1987, **60**, 3791 (*isol*)
Fukuzawa, A. *et al*, *Tetrahedron Lett.*, 1992, **33**, 2017 (*synth*)

Laureepoxide **L-00006**

[73328-63-9]

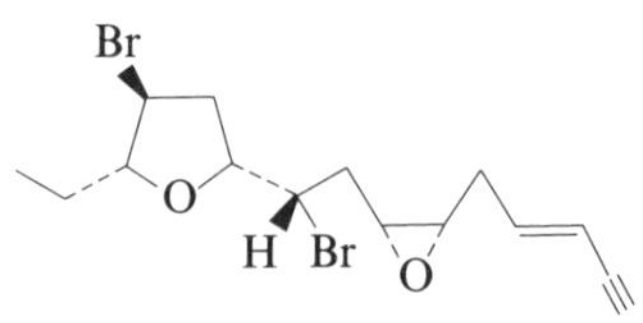

$C_{15}H_{20}Br_2O_2$ M 392.130

Constit. of *Laurencia nipponica*. Oil. $[\alpha]_D$ +25° (c, 1.0 in $CHCl_3$).

Fukuzawa, A. *et al*, *Tetrahedron Lett.*, 1980, 1471.

Laurefucin L-00007

12-Bromo-10-hydroxy-6,9;7,13-diepoxypentadec-3-en-1-yne

[36431-73-9]

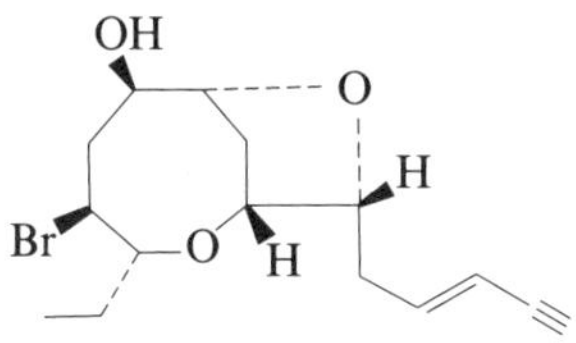

$C_{15}H_{21}BrO_3$ M 329.233

Constit. of *Laurencia nipponica.* Mp 107-108°. $[\alpha]_D$ −80°.

Ac:

$C_{17}H_{23}BrO_4$ M 371.270

Constit. of *Laurencia nipponica.* Oil. $[\alpha]_D$ −126.5°.

Hexahydro:

$C_{15}H_{27}BrO_3$ M 335.280

Mp 63.5°.

Furusaki, A. *et al, Tetrahedron Lett.*, 1973, 4579 (*struct*)

Laurencianol L-00008

[84323-26-2]

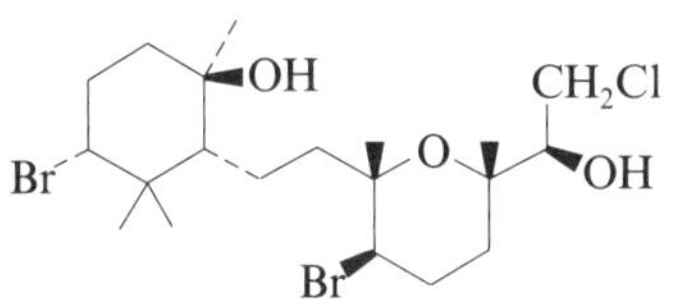

Absolute configuration

$C_{20}H_{35}Br_2ClO_3$ M 518.755

Constit. of marine alga *Laurencia obtusa.* Antibacterial agent. Cryst. (C_6H_6/hexane). Mp 114-116°.

Caccamese, S. *et al, Tetrahedron Lett.*, 1982, **23**, 3415 (*isol, cryst struct*)

Laurencienyne L-00009

[75179-72-5]

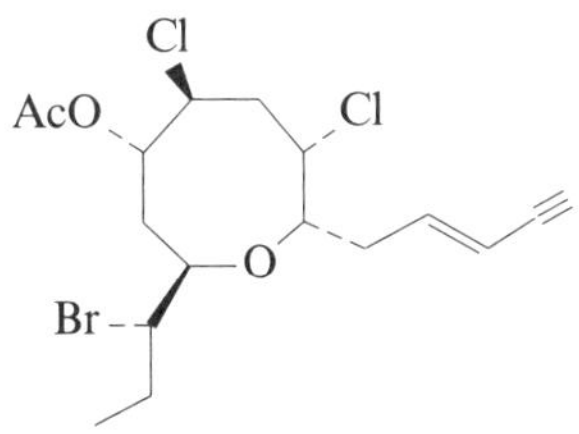

$C_{17}H_{23}BrCl_2O_3$ M 426.176

Constit. of *Laurencia obtusa.* Cryst. (C_6H_6/hexane). Mp 119-120°. $[\alpha]_D^{24}$ +50.3° (c, 1.1 in $CHCl_3$).

Caccamese, S. *et al, Tetrahedron Lett.*, 1980, 2299 (*cryst struct, abs config*)

Laurencin L-00010

7-Bromo-8-ethyl-3,6,7,8-tetrahydro-α-(2-penten-4-ynyl)-2H-oxocin-2-methanol acetate, 8CI. 6-Acetoxy-12-bromo-7,13-epoxy-3,9-pentadecadien-1-yne

[3442-58-8]

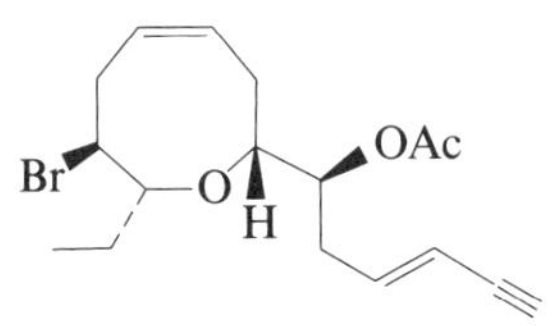

(+)-*form*

$C_{17}H_{23}BrO_3$ M 355.271

(+)-*form*

Constit. of *Laurencia glandulifera.* Cryst. (MeOH). Mp 73-74°. $[\alpha]_D^{23}$ +70.2° (c, 1.0 in $CHCl_3$).

(±)-*form*

Mp 45-47°.

Irie, T. *et al, Tetrahedron*, 1968, **24**, 4193 (*isol, struct, pmr, ms*)
Cameron, A.F. *et al, J. Chem. Soc. B*, 1969, 559 (*cryst struct*)
Murai, A. *et al, Tetrahedron Lett.*, 1977, 2507 (*synth*)
Tsushima, K. *et al, Tetrahedron Lett.*, 1992, **33**, 4345 (*synth*)

Laurenyne L-00011

[77182-65-1]

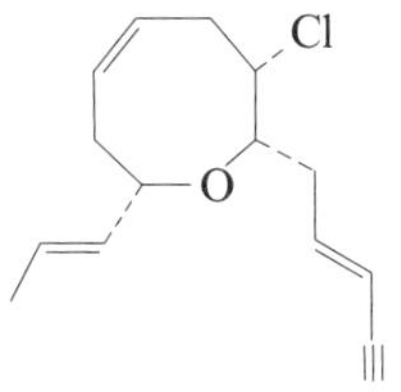

$C_{15}H_{19}ClO$ M 250.767

Constit. of *Laurencia obtusa.* Cryst. (pet. ether). Mp 78-80°. $[\alpha]_D^{17}$ +22.64° (c, 2.35 in $CHCl_3$).

Falshaw, C.P. *et al, Tetrahedron Lett.*, 1980, 4951 (*cryst struct, abs config*)

Leukotriene C_4 L-00012

N-[S-[*1-(4-Carboxy-1-hydroxybutyl)-2,4,6,9-pentadecatetraenyl*]-N-*L-γ-glutamyl-L-cysteinyl*]*glycine, 9CI. Leukotriene C. Leukotriene C_1*

[72025-60-6]

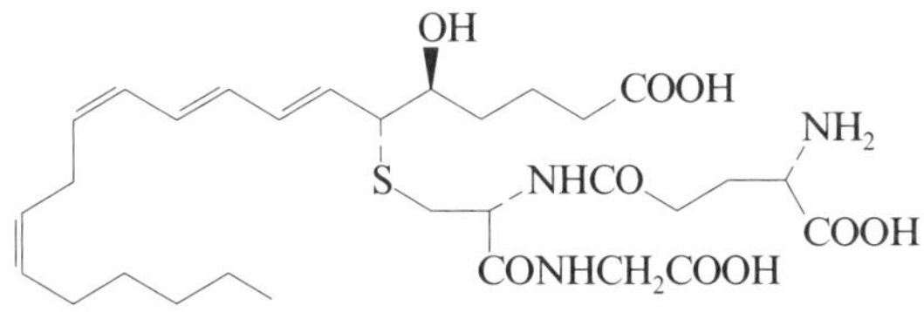

$C_{30}H_{47}N_3O_9S$ M 625.782

Constit. of the slow reacting substance of anaphylaxis. Isol. from mouse mastocytoma cells and human polymorphonuclear leukocytes. Metab. of arachidonic acid. Important agonist in asthma and various forms of hypersensitivity.

(*11E*)-*form* [74841-69-3]

*11-*trans*-Leukotriene C_4. Leukotriene C_2*

Constit. of the slow reacting substance of anaphylaxis. Isol. from human polymorphonuclear leukocytes. Slow reacting substance. Uv λ_{max} 278 (ε 40000) nm.

N-*Methyl:* [131391-65-6]. N-*Methylleukotriene C_4*

$C_{31}H_{49}N_3O_9S$ M 639.809

Metabolically stable LTC_4 mimetic. Solid.

Me-ester: [89286-84-0].

$C_{31}H_{49}N_3O_9S$ M 639.809

No phys. props. reported.

[112924-15-9]

Murphy, R.C. *et al, Proc. Natl. Acad. Sci. U.S.A.*, 1979, **76**, 4275 (*isol*)
Hammarstroem, S. *et al, Biochem. Biophys. Res. Commun.*, 1980, **92**, 946 (*struct*)
Hansson, G. *et al, FEBS Lett.*, 1980, **122**, 87 (*isol*)
Corey, E.J. *et al, J. Am. Chem. Soc.*, 1980, **102**, 1436, 4278 (*synth*)
Corey, E.J. *et al, Tetrahedron Lett.*, 1980, **21**, 3143; 1982, **23**, 2351 (*synth*)
Koenig, W. *et al, Monogr. Allergy*, 1983, **18**, 201 (*Me ester*)

Green, R.H. *et al*, *Tetrahedron*, 1983, **39**, 1687 (*rev*)
Baker, S.R. *et al*, *Prostaglandins*, 1990, **40**, 445 (*N-Me*)
Gareau, Y. *et al*, *J. Org. Chem.*, 1993, **58**, 1582 (*synth, pmr*)

Leukotriene C_5 L-00013

N-[S-[*1-(4-Carboxy-1-hydroxybutyl)-2,4,6,9,12-pentadecapentaenyl*]-N-*L-γ-glutamyl-L-cysteinyl*]*glycine*, *9CI*
[75207-09-9]

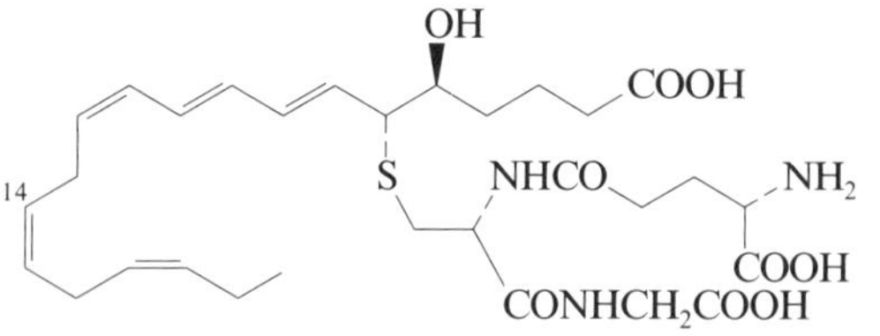

$C_{30}H_{45}N_3O_9S$ M 623.766

Prod. by incubation of 5,8,11,14,17-eicosapentaenoic acid with rat basophilic leukaemia cells and with mouse mastocytoma cells. Slow reacting substance.

14,15,17,18-Tetrahydro: [77009-77-9]. ***Leukotriene C$_3$***
$C_{30}H_{49}N_3O_9S$ M 627.798
Prod. from 5,8,11-eicosatrienoic acid by incubation with rat leukaemia or mouse mastocytoma cells. Slow reacting substance.

Hammarstroem, S., *J. Biol. Chem.*, 1980, **255**, 7093; 1981, **256**, 2275.
Oerning, L. *et al*, *Eur. J. Biochem.*, 1981, **120**, 41.
Okuyama, S. *et al*, *Chem. Pharm. Bull.*, 1982, **30**, 2453 (*synth*)
Spur, B. *et al*, *Arch. Pharm. (Weinheim, Ger.)*, 1983, **316**, 789; 1984, **317**, 280 (*synth*)

8,9-Leukotriene C_3 L-00014

N-[S-[*1-(7-Carboxy-1-hydroxyheptyl)-2,4,6-dodecatrienyl*]-N-*L-γ-glutamyl-L-cysteinyl*]*glycine*, *9CI*
[79162-55-3]

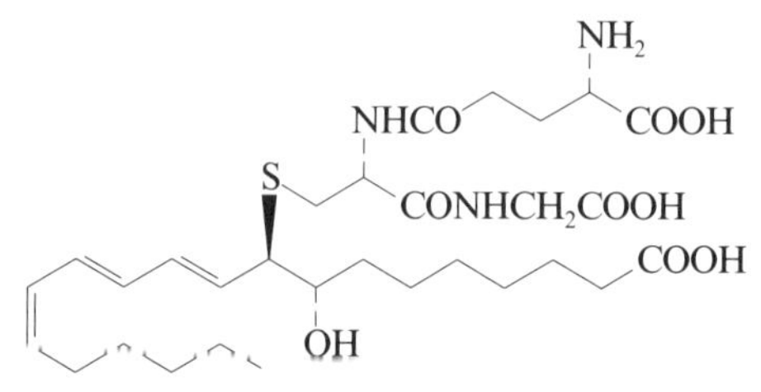

$C_{30}H_{49}N_3O_9S$ M 627.798

Prod. from dihomo-γ-linolenic acid by incubation with mouse mastocytoma cells or with purified lipoxygenase from potato tubers.

Hammarstroem, S., *J. Biol. Chem.*, 1981, **256**, 7712.
Baker, R.S. *et al*, *Tetrahedron Lett.*, 1983, **24**, 4469 (*synth*)
Reddy, C.C. *et al*, *Arch. Biochem. Biophys.*, 1990, **279**, 211.

Leukotriene D_4 L-00015

Leukotriene D. N-[S-[*1-(4-Carboxy-1-hydroxybutyl)-2,4,6,9-pentadecatetraenyl*]-*L-cysteinyl*]*glycine*, *9CI*
[73836-78-9]

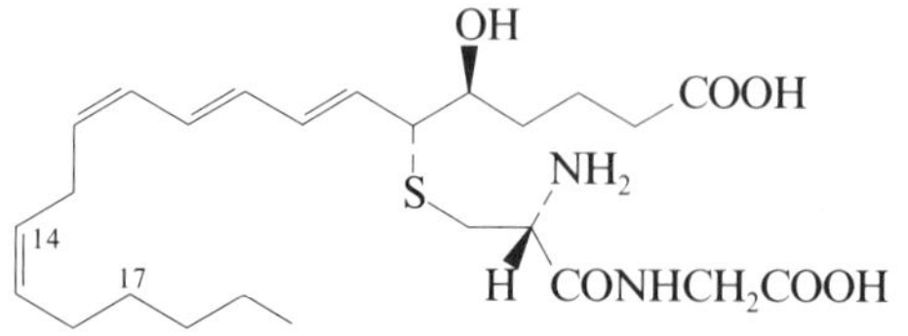

$C_{25}H_{40}N_2O_6S$ M 496.667

Constit. of the slow reacting substance of anaphylaxis. Isol. from cat basophilic leukaemic cells and from human lung. Metab. of arachidonic acid. Agonist in various forms of hypersensitivity of the lung including asthma and allergic rhinitis.

17,18Z-Didehydro: [79695-13-9]. *Leukotriene D$_5$*
$C_{25}H_{38}N_2O_6S$ M 494.651
Prod. by incubation of 5,8,11,14,17-Eicosapentaenoic acid, E-00029 with rat leukaemia cells. Slow reacting substance.

14,15-Dihydro: [77165-74-3]. *Leukotriene D$_3$*
$C_{25}H_{42}N_2O_6S$ M 498.683
Prod. from 5,8,11,14,17-Eicosapentaenoic acid, E-00029 by incubation with rat leukaemia or mouse mastocytoma cells. Slow reacting substance.

Me ester: [89286-83-9].
$C_{26}H_{42}N_2O_6S$ M 510.694
Induces blood vessel contraction.

11-trans-form [79768-40-4]
11-trans-*Leukotriene D$_4$*
Biologically less active than 11-*cis*-LTD$_4$. No phys. props. reported.

[122331-00-4]

Oerning, L. *et al*, *Proc. Natl. Acad. Sci. U.S.A.*, 1980, **77**, 2014 (*isol, struct*)
Lewis, R.A. *et al*, *Proc. Natl. Acad. Sci. U.S.A.*, 1980, **77**, 3710 (*isol*)
Corey, E.J. *et al*, *Tetrahedron Lett.*, 1980, **21**, 3143; 1982, **23**, 2351 (*synth*)
Baker, S.R. *et al*, *Biochem. Biophys. Res. Commun.*, 1981, **103**, 1258 (*synth, biochem*)
Oerning, L. *et al*, *Eur. J. Biochem.*, 1981, **120**, 41 (*derivs*)
Hammerstroem, S., *J. Biol. Chem.*, 1981, **256**, 2275.
Krell, R.D. *et al*, *Prostaglandins*, 1981, **22**, 387 (*biochem*)
Baker, S.R. *et al*, *Prostaglandins*, 1982, **23**, 569 (*biochem*)
Spur, B. *et al*, *Arch. Pharm. (Weinheim, Ger.)*, 1983, **316**, 289; 1984, **317**, 280 (*synth, deriv*)
Green, R.H. *et al*, *Tetrahedron*, 1983, **39**, 1687 (*rev*)
Kimura, I. *et al*, *Jpn. J. Pharmacol.*, 1989, **50**, 253 (*Me ester, biochem*)

Leukotriene E_4 L-00016

6-[2-Amino-2-(carboxyethyl)thio]-5-hydroxy-7,9,11,14-eicosatetraenoic acid, *9CI*. *Leukotriene E*
[75715-89-8]

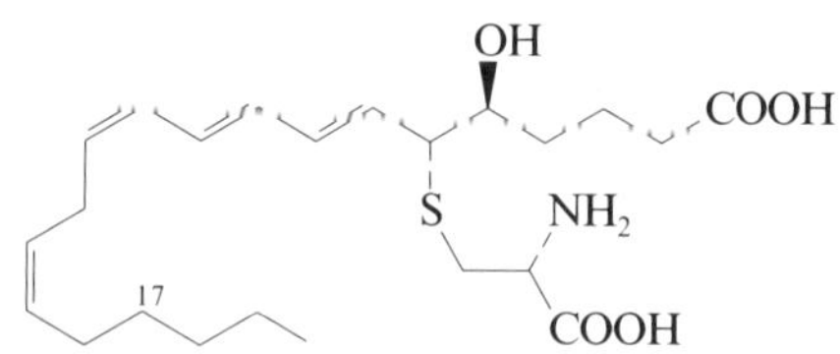

$C_{23}H_{37}NO_5S$ M 439.615

Constit. of the slow reacting substance of anaphylaxis. Derived from the oxidative metabolism of arachidonic acid. Highly spasmogenic material, which possibly plays an important role in asthma and other diseases of the respiratory system.

14,15-Dihydro: [79494-05-6]. ***Leukotriene E$_3$***
$C_{23}H_{39}NO_5S$ M 441.631
Prod. from 5,8,11-Eicosatrienoic acid, E-00043 by incubation with rat leukaemia cells. Metab. of Leukotriene C$_3$ in pig and rat kidney. Slow-reacting substance.

17,18-Didehydro(Z−): [79695-14-0]. ***Leukotriene E$_5$***
$C_{23}H_{35}NO_5S$ M 437.599

Prod. by incubation of 5,8,11,14,17-Eicosapentaenoic acid, E-00029 with rat leukaemia cells. Metab. of Leukotriene D_5 in pig kidney. Slow-reacting substance.
20-Hydroxy: [111844-33-8].
$C_{23}H_{37}NO_6S$ M 455.614
LTC_4 and LTE_4 metab.
N-*Ac:* [80115-95-3].
$C_{25}H_{39}NO_6S$ M 481.652
LTC_4 and LTE_4 metab.
20-Hydroxy, N-*Ac:* [107701-63-3].
$C_{25}H_{39}NO_7S$ M 497.652
LTE_4 metab.
20-Carboxylic acid: [111844-34-9]. *15-[(2-Amino-2-carboxyethyl)thio]-16-hydroxy-6,9,11,13-eicosatetraenedioic acid*
$C_{23}H_{35}NO_7S$ M 469.598
LTC_4 and LTE_4 metab. in humans and monkey.
20-Carboxylic acid, N-*Ac:* [112066-58-7].
$C_{25}H_{37}NO_8S$ M 511.635
LTC_4 and LTE_4 metab. found in bile.
S,S-*Dioxide:* [82850-11-1].
$C_{23}H_{37}NO_7S$ M 471.614
LTE_4 metab. Shows similar biol. activity to LTE_4. Uv λ_{max} 292, 280, 270 nm (MeOH).
S-*Oxide:* [88196-63-8].
$C_{23}H_{37}NO_6S$ M 455.614
LTE_4 metab.
20,20,20-Trifluoro: [131074-64-1]. *6-[(2-Amino-2-carboxyethyl)thio]-20,20,20-trifluoro-5-hydroxy-7,9,11,14-eicosatetraenoic acid*
$C_{23}H_{34}F_3NO_5S$ M 493.586
Inhibits oxidation of LTE_4.

11-(trans)-form [75715-88-7]
11-(trans)*-Leukotriene E_4*
LTE_4 metab.

Lewis, R.A. *et al, Biochem. Biophys. Res. Commun.*, 1980, **96**, 271 (*struct*)
Oerning, L. *et al, Eur. J. Biochem.*, 1981, **120**, 41 (*derivs*)
Bernstroem, K. *et al, J. Biol. Chem.*, 1981, **256**, 9579.
Atrache, V. *et al, Proc. Natl. Acad. Sci. U.S.A.*, 1981, **78**, 1523 (*biosynth 11-trans*)
Lewis, R.A. *et al, Proc. Natl. Acad. Sci. U.S.A.*, 1981, **78**, 4579 (*synth, activity N-Ac*)
Ormstad, K. *et al, Biochem. Biophys. Res. Commun.*, 1982, **104**, 1434.
Jones, T. *et al, Prostaglandins*, 1982, **24**, 279 (*sulfone synth, pmr, uv*)
Rosenberger, M. *et al, J. Am. Chem. Soc.*, 1983, **105**, 3656 (*synth*)
Lee, C.W. *et al, J. Biol. Chem.*, 1983, **258**, 15004 (*sulfoxide*)
Green, R.H. *et al, Tetrahedron*, 1983, **39**, 1687 (*rev*)
Spuri, B. *et al, Arch. Pharm. (Weinheim, Ger.)*, 1984, **317**, 280.
Foster, A. *et al, Prostaglandins*, 1986, **31**, 1077 (*synth, N-Ac, biochem*)
Oerning, L., *Eur. J. Biochem.*, 1987, **170**, 77 (*isol, biosynth, biochem*)
Hammarstroem, S. *et al, Ann. N.Y. Acad. Sci.*, 1988, **524**, 43 (*biosynth 20-OH, N-Ac*)
Delorme, D. *et al, Prostaglandins*, 1988, **36**, 291 (*biosynth*)
Keppler, D. *et al, Adv. Enzyme Regul.*, 1989, **28**, 307 (*biosynth*)
Tagari, P. *et al, Prostaglandins*, 1989, **37**, 629 (*synth, biosynth 20-OH*)
Jedlitschky, G. *et al, Arch. Biochem. Biophys.*, 1990, **282**, 333 (*biochem trifluoro*)
Huker, M. *et al, Eur. J. Biochem.*, 1990, **194**, 309 (*biosynth, metab*)

Leukotriene F_4 L-00017

S-*[1-(4-Carboxy-1-hydroxybutyl)-2,4,6,9-pentadecatetraenyl]*-N-*L-γ-glutamyl-L-cysteine, 9CI*
[83851-42-7]

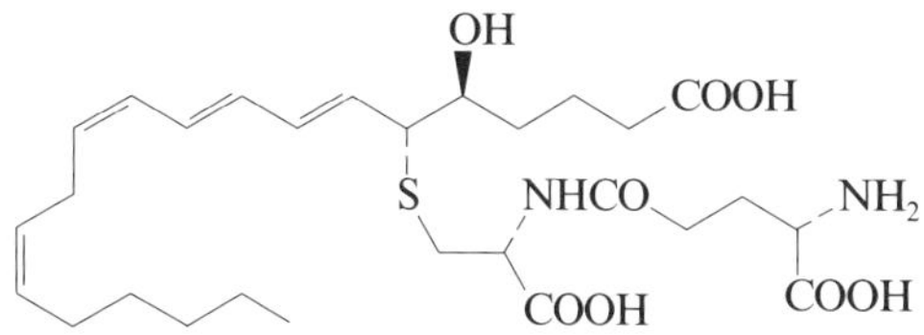

$C_{28}H_{44}N_2O_8S$ M 568.730
Formed from leukotriene E_4 by incubation with γ-glutamyltranspeptidase in the presence of glutathione. Induces contractions in isolated guinea pig ileum (less potent than leukotriene E_4).

Bernstroem, K. *et al, Biochem. Biophys. Res. Commun.*, 1982, **109**, 800.
Okuyama, S. *et al, Chem. Pharm. Bull.*, 1982, **30**, 2453 (*synth, pmr, ir, uv, ms*)
Ellis, F. *et al, Tetrahedron Lett.*, 1982, **23**, 3735 (*synth*)

Leukotriene G_4 L-00018

6-[(2-Carboxy-2-oxoethyl)thio]-5-hydroxy-7,9,11,14-eicosatetraenoic acid
[117675-20-4]

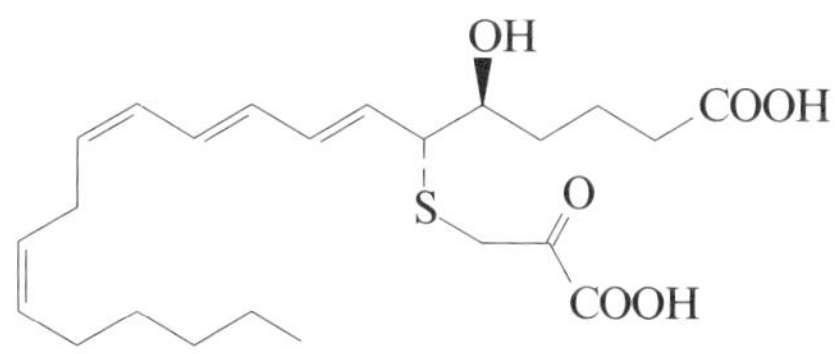

$C_{23}H_{34}O_6S$ M 438.584
Formed by transamination of Leukotriene E_4, L-00016 by cysteine conjugate aminotransferase. Leukotriene E_4, L-00016 metab. in kidney and liver.

Tomisawa, H. *et al, Biochem. Biophys. Res. Commun.*, 1988, **155**, 1119.

Levuglandin D_2 L-00019

[91712-44-6]

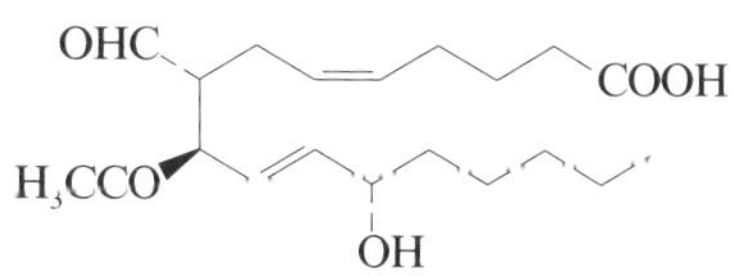

$C_{20}H_{32}O_5$ M 352.470
H_2O-induced rearrangement prod. of Prostaglandin H_2. Less stable than Levuglandin E_2, L-00020.

Salomon, R.G. *et al, J. Am. Chem. Soc.*, 1984, **106**, 6049 (*pmr*)

Levuglandin E_2 L-00020

8-Acetyl-9-formyl-12-hydroxy-5,10-heptadecadienoic acid, 9CI. LGE$_2$
[91712-41-3]

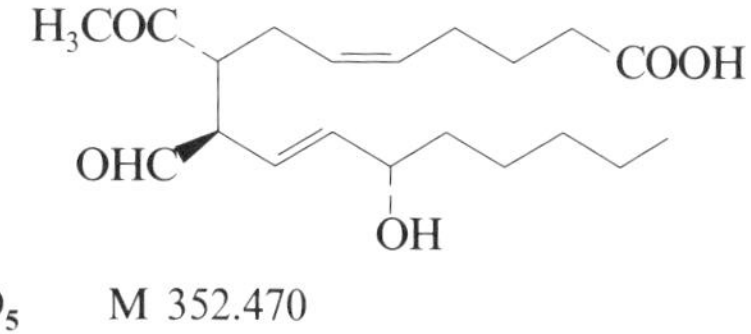

$C_{20}H_{32}O_5$ M 352.470

H_2O-induced rearrangement prod. of Prostaglandin H_2.

Salomon, R.G. *et al*, *J. Am. Chem. Soc.*, 1984, **106**, 6049, 8296 (*synth, pmr*)
Miller, D.B. *et al*, *J. Org. Chem.*, 1990, **55**, 3164 (*synth*)
Iyer, R.S. *et al*, *J. Org. Chem.*, 1990, **55**, 3175 (*props, pmr*)

Lichesteric acid L-00021

2,5-Dihydro-4-methyl-5-oxo-2-tridecyl-3-furancarboxylic acid, 9CI. Lichesterinic acid

[22800-25-5]

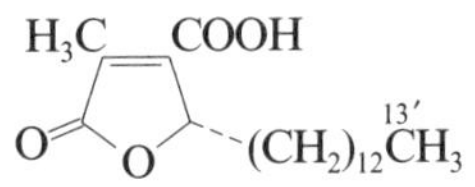

$C_{19}H_{32}O_4$ M 324.459

▷ LV1820000.

(S)-form

Isol. from Iceland moss (*Cetraria islandica*), *Nephromopsis stracheyi*, *Parmelia kamtschadalis* and other lichens. Mp 124.5-125° (121-122°). $[\alpha]_D^{25}$ −32.66°.

Me ester: Prisms. Mp 53-54°. $[\alpha]_D^{14}$ −28.07° ($CHCl_3$).

(R)-(?)-form

13′-Acetoxy: 13-Acetoxylichesterinic acid

$C_{21}H_{34}O_6$ M 382.496

Metab. of *Neuropogon trachycarpus*. Cryst. (hexane). Mp 98-99°.

Asano, M. *et al*, *Ber.*, 1932, **65**, 1175 (*isol, struct*)
Asahina, Y. *et al*, *Ber.*, 1936, **69**, 120 (*struct*)
Boll, P.M., *Acta Chem. Scand.*, 1968, **22**, 3245 (*abs config*)
Huneck, S. *et al*, *J. Hattori Bot. Lab.*, 1979, **45**, 1 (*cd, abs config*)
Ghogomu, R.T. *et al*, *Phytochemistry*, 1982, **21**, 2355 (*deriv*)

Lipid A L-00022

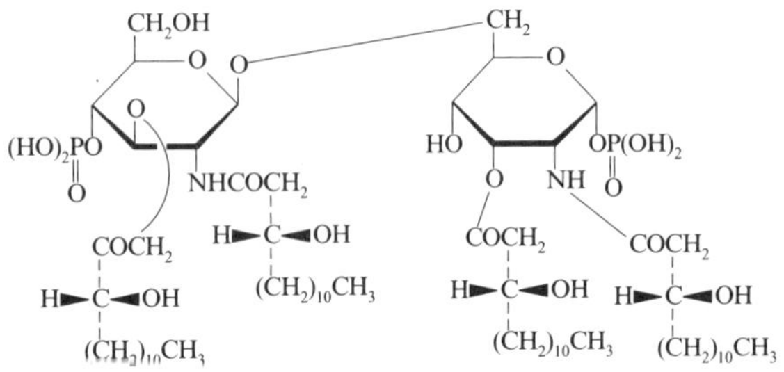

$C_{68}H_{130}N_2O_{23}P_2$ M 1405.722

Struct. of lipid A from a *Salmonella* mutant is shown. Structs. of lipid A's from *E. coli*, *S. minnesota* and *Proteus mirabilis* have been detd. and are derived from the struct. shown by acylation at two or more of the 3-hydroxytetradecanoyl residues. Component of gram-negative bacteria lipopolysaccharide. Plays an important role in toxicity, pyrogenicity and adjuvant activity.

[95991-05-2]

Imoto, M. *et al*, *Tetrahedron Lett.*, 1985, **26**, 907, 1545 (*struct, synth*)
Takayama, K. *et al*, *Microbiology* (*Washington, D.C.*), 1986, 5 (*rev*)
Imoto, M. *et al*, *Bull. Chem. Soc. Jpn.*, 1987, **60**, 2205 (*synth*)
Ikeda, K. *et al*, *Chem. Pharm. Bull.*, 1987, **35**, 4436, 4517 (*bibl*)
Kusumoto, S. *et al*, *Pure Appl. Chem.*, 1989, **61**, 461.

α-Lipoic acid L-00023

1,2-Dithiolane-3-pentanoic acid, 9CI. Protogen A. Thioctic acid. Other proprietary names

[62-46-4]

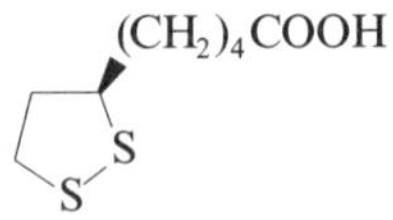

(R)-form

$C_8H_{14}O_2S_2$ M 206.329

▷ LD_{50} (rat, orl) 1130 mg/kg. Exp. reprod. effects. JP1190000.

(R)-form [1200-22-2]

Widely distributed in nature. Bacterial and protozoal growth factor. Cryst. Mp 46-48°. $[\alpha]_D^{23}$ +104° (c, 0.88 in C_6H_6). pK_a 5.4.

S^1-*Oxide:* [6992-30-9]. ***β-Lipoic acid.*** *Protogen B. Lipoic thiosulfinate*

$C_8H_{14}O_3S_2$ M 222.329

Prod. by bacteria. $[\alpha]_D^{25}$ +105°. Stereochem. apparently not established.

(S)-form [1077-27-6]

Cryst. (cyclohexane). Mp 45-47.5°. $[\alpha]_D^{23}$ −113° (c, 0.88 in C_6H_6).

(±)-form [1077-28-7]

Used to treat liver disease and claimed for the treatment of *Amanita* toadstool poisoning (liver necrosis). Pale yellow plates (cyclohexane or by subl.). Prac. insol. H_2O. Mp 59-61°. Bp_{25} 85-90°. pK_a 4.7.

Hornberger, C.S. *et al*, *J. Am. Chem. Soc.*, 1952, **74**, 2382 (*synth*)
Reed, L.J. *et al*, *J. Am. Chem. Soc.*, 1952, **74**, 2383 (*struct*)
Bullock, M.W. *et al*, *J. Am. Chem. Soc.*, 1954, **76**, 1828 (*oxide*)
Yurugi, S. *et al*, *Yakugaku Zasshi*, 1960, **80**, 1686, 1691 (*synth*)
Schmidt, U. *et al*, *Angew. Chem., Int. Ed. Engl.*, 1965, **4**, 846 (*rev, bibl*)
Neubert, L.A. *et al*, *Tetrahedron Lett.*, 1974, 3543 (*cd*)
Koike, M. *et al*, *Metab. Pathways, 3rd edn.*, 1975, **7**, 87 (*rev, metab*)
Paukstelis, J.V. *et al*, *J. Org. Chem.*, 1977, **42**, 3941 (*cmr*)
Koike, M., *Method. Chim.*, 1977, **11**, 112 (*rev, props*)
Furr, H.C. *et al*, *Arch. Biochem. Biophys.*, 1978, **185**, 576 (*oxide*)
Parry, R.J. *et al*, *J. Am. Chem. Soc.*, 1978, **100**, 5243 (*biosynth*)
Tsuji, J. *et al*, *J. Org. Chem.*, 1978, **43**, 3606 (*synth*)
Brookes, M.H. *et al*, *J. Chem. Soc., Chem. Commun.*, 1983, 1051 (*synth, abs config*)
Elliot, J.D. *et al*, *Tetrahedron Lett.*, 1985, **26**, 2535 (*synth*)
Rao, A.V.R. *et al*, *Carbohydr. Res.*, 1986, **148**, 51 (*synth*)
Negwer, M., *Organic-Chemical Drugs and their Synonyms*, 6th Ed., Akademie-Verlag, Berlin, 1987, 996 (*synonyms*)
Rao, A.V.R. *et al*, *Synth. Commun.*, 1987, **17**, 1095, 1339 (*synth*)
Menon, R.B. *et al*, *Tetrahedron Lett.*, 1987, **28**, 5313 (*synth*)
Bulman Page, P.C. *et al*, *J. Chem. Soc., Perkin Trans. 1*, 1990, 1615 (*synth, ir, pmr, cmr*)
Gopalan, A.S. *et al*, *J. Chem. Soc., Perkin Trans. 1*, 1990, 1897 (*synth*)
Martindale, The Extra Pharmacopoeia, 30th edn., Pharmaceutical Press, London, 1993, 1420.
Lewis, R.J., *Sax's Dangerous Properties of Industrial Materials*, 8th Ed., Van Nostrand-Reinhold, 1992, DXN800.

Litsenolide A L-00024

3-(11-Dodecenylidene)dihydro-4-hydroxy-5-methyl-2(3H)-furanone, 9CI

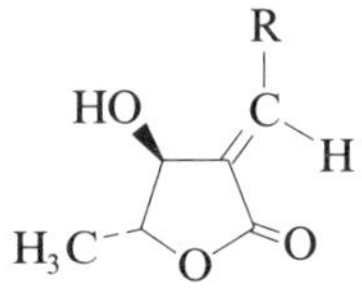

(E)-form
R = $(CH_2)_9CH{=}CH_2$

$C_{17}H_{28}O_3$ M 280.406

(E)-form [38965-58-1]

Litsenolide A_2

Obt. from root of *Litsea japonica*. $[\alpha]_D^{23}$ −40.4° (c, 0.67 in dioxan).

(Z)-form [38965-57-0]

Litsenolide A_1

Obt. from *L. japonica*. Oil. $[\alpha]_D^{23}$ −2.4° (c, 0.62 in dioxan).

Takeda, K. *et al*, *Tetrahedron*, 1972, **28**, 3757.
Nii, M. *et al*, *Agric. Biol. Chem.*, 1979, **43**, 1349.
Kende, A.S. *et al*, *J. Org. Chem.*, 1982, **47**, 163 (*synth*)

Litsenolide B L-00025

3-(11-Dodecynyldiene)dihydro-4-hydroxy-5-methyl-2(3H)-furanone

As Litsenolide A, L-00024 with

R = -$(CH_2)_9C{\equiv}CH$

$C_{17}H_{26}O_3$ M 278.391

(E)-form [38965-60-5]

Litsenolide B_2

Obt. from *Litsea japonica* root. Mp 42-43°. $[\alpha]_D^{24}$ −44.9° (c, 0.53 in dioxan).

(Z)-form [38965-59-2]

Litsenolide B_1

Obt. from *L. japonica*. Mp 40-41°. $[\alpha]_D^{24}$ −5.9° (c, 0.71 in dioxan).

Takeda, K. *et al*, *Tetrahedron*, 1972, **28**, 3757.
Nii, H. *et al*, *Agric. Biol. Chem.*, 1979, **43**, 1349.
Kende, A.S. *e' al*, *J. Org. Chem.*, 1982, **47**, 163 (*synth*)

Litsenolide C L-00026

Dihydro-4-hydroxy-5-methyl-3-tetradecylidene-2(3H)-furanone

As Litsenolide A, L-00024 with

R = -$(CH_2)_{12}CH_3$

$C_{19}H_{34}O_3$ M 310.476

(E)-form [38965-62-7]

Litsenolide C_2

Obt. from *Litsea japonica* root. Mp 44-45°. $[\alpha]_D^{26}$ −45.2° (c, 0.99 in dioxan).

(Z)-form [38965-61-6]

Litsenolide C_1

Obt. from *L. japonica*. Mp 60-62°. $[\alpha]_D^{24}$ −9.4° (c, 0.62 in dioxan).

Takeda, K. *et al*, *Tetrahedron*, 1972, **28**, 3757.
Kende, A.S. *et al*, *J. Org. Chem.*, 1982, **47**, 163 (*synth*)
Barbier, P. *et al*, *Tetrahedron Lett.*, 1982, **23**, 3513 (*synth*)
Chen, S.-Y. *et al*, *Tetrahedron Lett.*, 1983, **24**, 5027 (*synth*)

Litsenolide D_1 L-00027

[56820-22-5]

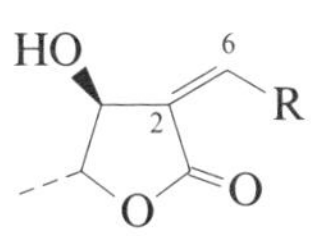

R = $(CH_2)_{10}CH_3$

$C_{17}H_{30}O_3$ M 282.422

Constit. of *Litsea japonica*. Prisms (Et_2O). Mp 49-51°. $[\alpha]_D$ −8.4° (c, 0.37 in $CHCl_3$).

(E)-*Isomer:* [128396-32-7]. ***Litsenolide D_2***

$C_{17}H_{30}O_3$ M 282.422

Constit. of *L. japonica*. Prisms (Et_2O). Mp 33-34°. $[\alpha]_D$ −21.6° (c, 0.49 in $CHCl_3$).

Niwa, M. *et al*, *Tetrahedron Lett.*, 1975, 1539 (*synth, pmr, ms*)
Tanaka, H. *et al*, *Phytochemistry*, 1990, **29**, 857 (*isol, pmr, cmr*)

Litsenolide E_1 L-00028

[128443-47-0]

As Litsenolide D_1, L-00027 with

R = $-(CH_2)_9CH{=}CHCH_2CH_3$ (*E*-)

$C_{19}H_{32}O_3$ M 308.460

Constit. of *Litsea japonica*. Oil. $[\alpha]_D$ −6.0° (c, 0.4 in $CHCl_3$).

(E)-*Isomer:* [128396-33-8]. ***Litsenolide E_2***

$C_{19}H_{32}O_3$ M 308.460

Constit. of *L. japonica*. Oil. $[\alpha]_D$ −42.0° (c, 0.05 in $CHCl_3$).

Tanaka, H. *et al*, *Phytochemistry*, 1990, **29**, 857 (*isol, pmr, cmr*)

Lumisterol L-00029

(9β,10α)-Ergosta-5,7,22E-trien-3β-ol, 9CI

[474-69-1]

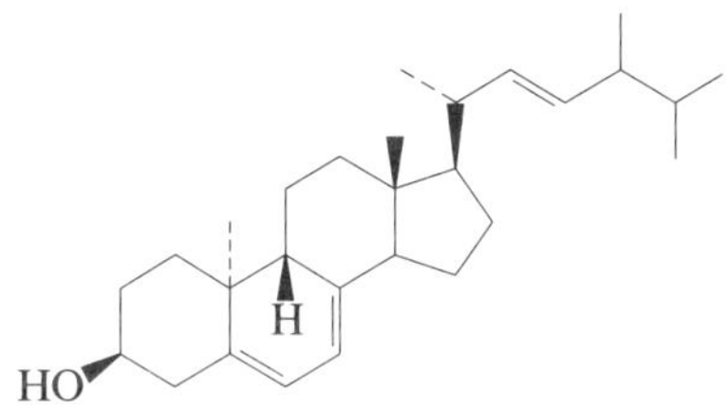

$C_{28}H_{44}O$ M 396.655

Initial prod. of uv irrad. of, and stereoisomeric with, Ergosterol, E-00142. Further irrad. → Tachysterol. Cryst. (Me_2O/MeOH). Mp 118°. $[\alpha]_D^{18}$ +197° ($CHCl_3$).

Ac:

$C_{30}H_{46}O_2$ M 438.692

Needles (Me_2CO/MeOH). Mp 100°. $[\alpha]_D^{18}$ +142° ($CHCl_3$).

Heilbron, I.M. *et al*, *J. Chem. Soc.*, 1935, 1221; 1937, 411; 1938, 869 (*synth, struct*)
Castells, J. *et al*, *J. Chem. Soc.*, 1959, 1159 (*stereochem*)
Jones, E.R.H. *et al*, *J. Chem. Soc.*, 1965, 2054.
de Kok, A.J. *et al*, *Acta Crystallogr., Sect. B*, 1974, **30**, 1695 (*struct*)

α-Lysolecithin L-00030

4,7-Dihydroxy-N,N,N-trimethyl-10-oxo-3,5,9-trioxa-4-phosphapentacosan-1-aminium hydroxide, inner salt, 4-oxide, 9CI. Choline hydroxide dihydrogen phosphate inner salt 3-ester with 1-monopalmitin, 8CI. 1-Palmitoylglycero-3-phosphocholine. Choline phosphate 3-ester with 1-monopalmitin

[14863-27-5]

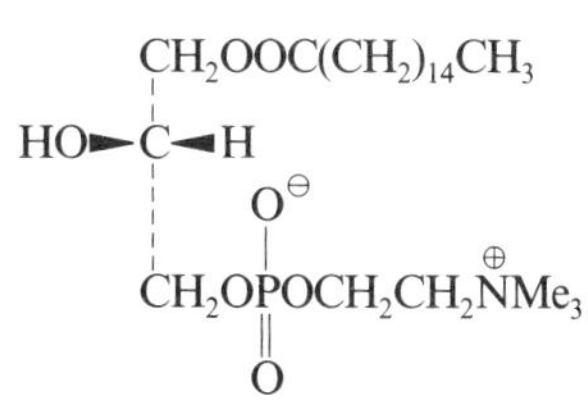

$C_{24}H_{50}NO_7P$ M 495.635

(R)-form [17364-16-8]

L-form

Constit. of biomembranes. $[\alpha]^{22}_{546}$ −4.3° (c, 3.5 in $CHCl_3$/MeOH).

[17364-17-9, 17364-18-0]

Arnold, D. *et al*, *Justus Liebigs Ann. Chem.*, 1967, **709**, 234 (*synth*)
Eibl, H. *et al*, *Justus Liebigs Ann. Chem.*, 1970, **738**, 161 (*synth*)
Dorset, D.L., *Biochim. Biophys. Acta*, 1975, **380**, 257 (*cryst struct*)
Spiker, R.C. *et al*, *Biochim. Biophys. Acta*, 1975, **388**, 361 (*raman*)
Yabusaki, K.K. *et al*, *Chem. Phys. Lipids*, 1976, **17**, 120 (*synth*)
Smith, N.B. *et al*, *Can. J. Biochem.*, 1978, **56**, 1149 (*synth, pmr*)
Tokumura, A. *et al*, *Chem. Pharm. Bull.*, 1983, **31**, 4425 (*ms*)

M

Macrolactinic acid **M-00001**

7,13,15,23-Tetrahydroxy-2,4,8,10,16,18-tetracosahexaenoic acid

[122540-33-4]

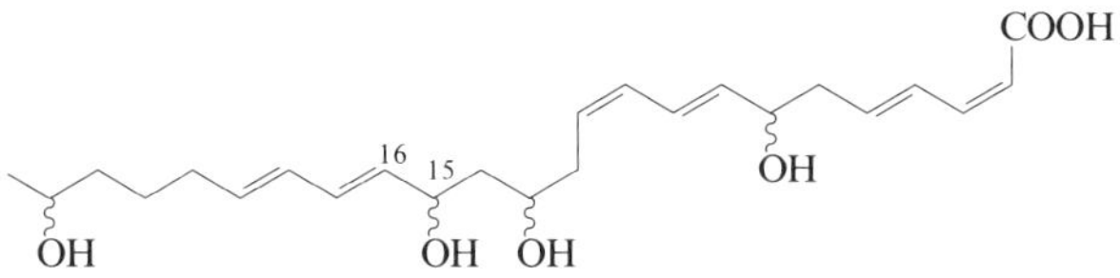

$C_{24}H_{36}O_6$ M 420.545

Isol. from an unclassifiable marine bacterium. $[\alpha]_D$ −13.9° (c, 0.58 in MeOH).

16,17-Dihydro, 15-ketone: [122540-34-5]. ***Isomacrolactinic acid***

$C_{24}H_{36}O_6$ M 420.545

From a marine bacterium. $[\alpha]_D$ −9.3° (c, 0.22 in MeOH).

Gustafson, K. *et al, J. Am. Chem. Soc.*, 1989, **111**, 7519 (*isol, pmr, cmr, struct*)

Maesanin **M-00002**

2-Hydroxy-5-methoxy-3-(10-pentadecenyl)-1,4-benzoquinone

[82380-21-0]

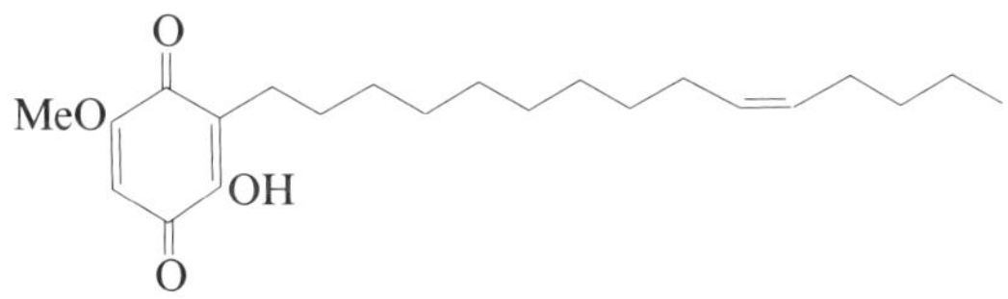

$C_{22}H_{34}O_4$ M 362.508

Constit. of fruits of *Maesa lanceolata*. Host defence stimulant. Inhibits KDO transferase, antibacterial, potential antiasthmatic. Yellow needles. Mp 77°.

Kubo, I. *et al, Tetrahedron*, 1987, **43**, 2653 (*struct, synth*)
Danheiser, R.L. *et al, Tetrahedron Lett.*, 1990, **31**, 1527 (*synth*)
Yadav, J.S. *et al, J. Org. Chem.*, 1992, **57**, 3242 (*synth*)

Majorynolide **M-00003**

[126622-45-5]

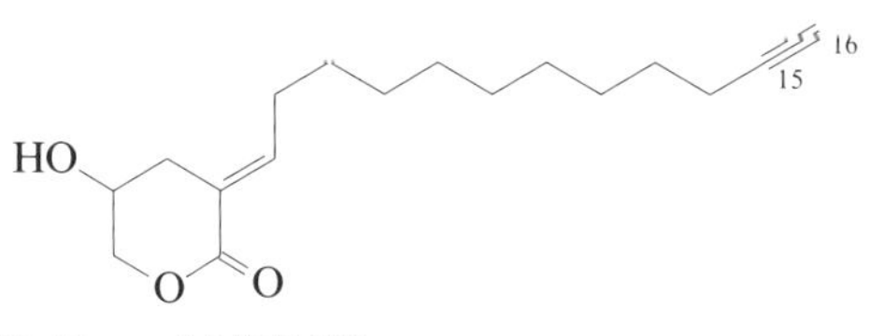

$C_{17}H_{26}O_3$ M 278.391

Constit. of *Persea major*. Cryst. (CH_2Cl_2/hexane). Mp 39°. $[\alpha]_D$ −17° (c, 0.1 in CH_2Cl_2/hexane).

15,16-Dihydro: [126622-46-6]. ***Majorenolide***

$C_{17}H_{28}O_3$ M 280.406

From *P. major*. Oil.

Ma, W.-W. *et al, J. Nat. Prod. (Lloydia)*, 1989, **52**, 1263 (*isol, pmr, cmr*)

Malvalic acid **M-00004**

2-Octyl-1-cyclopropene-1-heptanoic acid, 9CI. 8,9-Methylene-8-heptadecenoic acid. Malvalinic acid. Halphenic acid

[503-05-9]

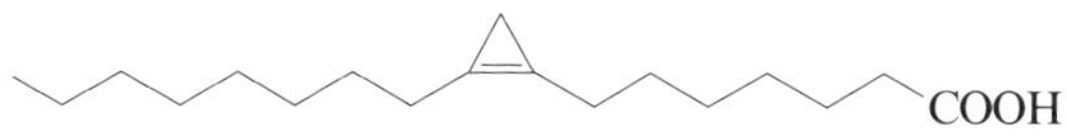

$C_{18}H_{32}O_2$ M 280.450

Occurs in seed oils of the Malvaceae, Bombacaceae and Sterculiaceae families. Inhibitor of insect metabolism. Oil. Mp 10.3-10.5°.

Me ester: [5026-66-4].

Oil. Mp −18° to −17°. $Bp_{0.04}$ 136°.

Shenstone, F.S. *et al, Nature (London)*, 1956, **177**, 94 (*isol*)
Craven, B.M. *et al, Nature (London)*, 1959, **183**, 676 (*struct*)
Fogarty, A.C. *et al, J. Am. Oil Chem. Soc.*, 1965, **42**, 885 (*isol*)
Gensler, W.J. *et al, J. Org. Chem.*, 1970, **35**, 2301 (*synth, ir, pmr*)
Boudreaux, G.J. *et al, J. Am. Oil Chem. Soc.*, 1972, **49**, 278 (*pmr*)
Pawlowski, N.E. *et al, Chem. Phys. Lipids*, 1975, **13**, 164 (*ms*)
Ahmad, M.S. *et al, Chem. Phys. Lipids*, 1979, **25**, 29 (*occur*)
Nakatani, M. *et al, Bull. Chem. Soc. Jpn.*, 1991, **64**, 3084 (*isol, bibl*)

Malyngamide E **M-00005**

N-*[2-Hydroxy-2-(4-hydroxy-3,5,5-trimethyl-6-oxo-1-cyclohexen-1-yl)-1-(methoxymethyl)ethyl]-7-methoxy-9-methyl-4-hexadecenamide, 9CI*

[67488-05-5]

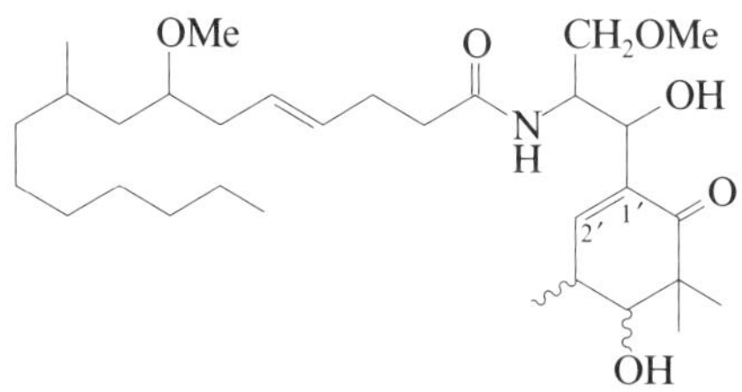

$C_{31}H_{55}NO_6$ M 537.779

Constit. of *Lyngbya majuscula*. Shows mild antibiotic activity against *Mycobacterium smegmatis* and *Bacillus subtilis*. $[\alpha]_D^{24}$ +24.2° (c, 0.6 in $CHCl_3$).

Di Ac: $[\alpha]_D^{25}$ +37.5° (c, 0.12 in $CHCl_3$).

1′,2′-Dihydro, 2′-hydroxy: [67488-04-4]. ***Malyngamide D***†

$C_{31}H_{57}NO_7$ M 555.794

Metab. from *L. majuscula*. Shows similar prop. as parent. $[\alpha]_D^{25}$ −33.0° (c, 0.53 in $CHCl_3$).

Mynderse, J.S. *et al, J. Org. Chem.*, 1978, **43**, 4359 (*pmr, ms, cmr, ir, uv, struct*)

Maneonenes M-00006

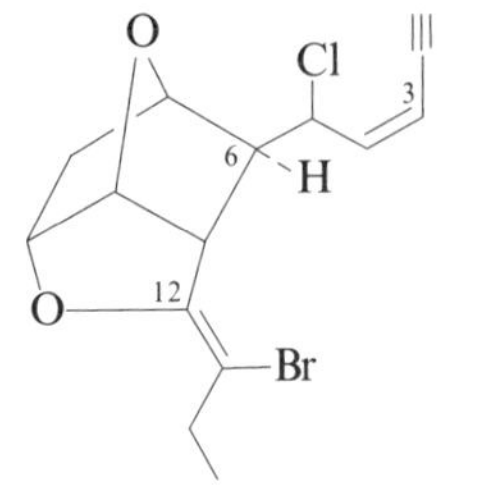

(3Z,6S,12E)-form

$C_{15}H_{16}BrClO_2$ M 343.647

(3Z,6R,12E)-form [62624-86-6]

Maneonene C

Constit. of *Laurencia nidifica*. Oil. $[\alpha]_D^{21}$ +336° (c, 1.46 in $CHCl_3$).

(3Z,6S,12E)-form [61661-24-3]

Maneonene A

From *L. nidifica*. Oil. $[\alpha]_D^{21}$ +39° (c, 1 in $CHCl_3$).

(3E,6S,12Z)-form [61688-65-1]

(E)-*Maneonene B*

From *L. nidifica*. Oil. $[\alpha]_D^{21}$ −25° (c, 0.4 in $CHCl_3$).

(3Z,6S,12Z)-form [61661-25-4]

(Z)-*Maneonene B*

From *L. nidifica*. Oil. $[\alpha]_D^{21}$ −49° (c, 3.6 in $CHCl_3$).

Waraskiewicz, S.M. *et al*, *J. Org. Chem.*, 1978, **43**, 3194 (*isol, struct*)
Holmes, A.B. *et al*, *J. Chem. Soc., Chem. Commun.*, 1983, 415; 1984, 1594 (*synth*)

2-Mercaptooctadecanoic acid, 9CI M-00007

2-Mercaptostearic acid

[13595-54-5]

$H_3C(CH_2)_{15}CH(SH)COOH$

$C_{18}H_{36}O_2S$ M 316.547

(±)-form

Used in photosensitive films, in soaps as a germicide and in the dyeing of polypropylene fibres. Mp 80° (74°).

Eckert, A. *et al*, *Monatsh. Chem.*, 1913, **34**, 1811 (*synth*)
Nicolet, B.H. *et al*, *J. Am. Chem. Soc.*, 1927, **49**, 2065 (*synth*)
Schinzel, T. *et al*, *Bull. Soc. Chim. Fr.*, 1939, **6**, 501.
Belg. Pat., 710 798, (1968); *CA*, **71**, 22929c (*synth*)

3-Mercaptopropanoic acid, 9CI M-00008

Thiohydracrylic acid

[107-96-0]

$HSCH_2CH_2COOH$

$C_3H_6O_2S$ M 106.145

Used for photometric detn. of Re(*VII*) (λ_{max} 470 nm, ε 13500). Cryst. Sol. H_2O, EtOH, Et_2O. Mp 16.8°. Bp_{13} 114-115.5°, Bp_3 85-86°. pK_{a1} 4.38; pK_{a2} 10.38 (25°).

▷ UF5270000.

Me ester: [2935-90-2].
$C_4H_8O_2S$ M 120.172
Bp_{20} 67-68°.

S-*Me:* [646-01-5]. *3-Methylmercaptopropanoic acid. 3-Methylthiopropionic acid. Methylthiohydracrylic acid*
$C_4H_8O_2S$ M 120.172
$Bp_{0.5}$ 102-104°.

S-*Me, Me ester:* [13532-18-8].
$C_5H_{10}O_2S$ M 134.199
Bp_{11} 69°.

Anilide: [10220-66-3]. *3-Mercapto-*N*-phenylpropanamide, 9CI*
$C_9H_{11}NOS$ M 181.258
Used as a 5% soln. in MeOH for extraction-photometric detn. of Pd (λ_{max} 400 nm, ε 4500, $CHCl_3$). Cryst.

Billmann, E., *Justus Liebigs Ann. Chem.*, 1906, **348**, 120 (*synth*)
Holmberg, B., *Ark. Kemi, Sect. B*, No. 7, 1945, **21** (*deriv*)
Cheney, L.C. *et al*, *J. Am. Chem. Soc.*, 1945, **67**, 731 (*synth*)
Danehy, J.P. *et al*, *J. Org. Chem.*, 1967, **32**, 1491 (*synth*)
Talipova, L.L. *et al*, *Zh. Anal. Khim.*, 1976, **31**, 106 (*detn, Re*)
Nacu, A. *et al*, *Rev. Roum. Chim.*, 1977, **28**, 1091 (*detn, Pd*)
Lewis, R.J., *Sax's Dangerous Properties of Industrial Materials*, 8th Ed., Van Nostrand-Reinhold, 1992, MCQ000.

(2-Methoxy-2-oxoethyl) triphenylphosphonium(1+), 9CI M-00009

(Carboxymethyl)triphenylphosphonium methyl ester, 8CI. Carbomethoxymethyltriphenylphosphonium(1+)

$Ph_3P^{\oplus}CH_2COOMe$

$C_{21}H_{20}O_2P^{\oplus}$ M 335.361 (ion)

Treatment of salts with 2M NaOH yields the ylide.

Chloride: [2181-97-7].
$C_{21}H_{20}ClO_2P$ M 370.814
Solid. Mp 155° dec.

Bromide: [1779-58-4].
$C_{21}H_{20}BrO_2P$ M 415.266
Cryst. (H_2O). Mp 164-165° dec.

Iodide: [39720-65-5].
$C_{21}H_{20}IO_2P$ M 462.266
Cryst. (CH_2Cl_2/Et_2O). Mp 162-163°.

Tetrafluoroborate:
$C_{21}H_{20}BF_4O_2P$ M 422.166
Cryst. (CH_2Cl_2/Et_2O). Mp 130-131°.

Isler, O. *et al*, *Helv. Chim. Acta*, 1957, **40**, 1242 (*synth*)
Grim, S.O. *et al*, *J. Phys. Chem.*, 1966, **70**, 581 (*nmr*)
Gallagher, M.J. *et al*, *Aust. J. Chem.*, 1968, **21**, 1197 (*pmr*)
Albright, T.A. *et al*, *J. Am. Chem. Soc.*, 1976, **98**, 6249 (*cmr*)
Nesterov, L.V. *et al*, *Zh. Obshch. Khim.*, (*Engl. transl.* p. 1161), 1977, **47**, 1259 (*bromide, iodide, nmr*)
Ayrey, G. *et al*, *J. Labelled Compd. Radiopharm.*, 1978, **14**, 935 (*bromide, ylide*)
Lang, R.W. *et al*, *Helv. Chim. Acta*, 1980, **63**, 438 (*bromide, use*)

5-Methoxy-3-(8,11,14-pentadecatrienyl)-1,2,4-benzenetriol M-00010

1,2,4-Trihydroxy-5-methoxy-3-(8,11,14-pentadecatrienyl) benzene

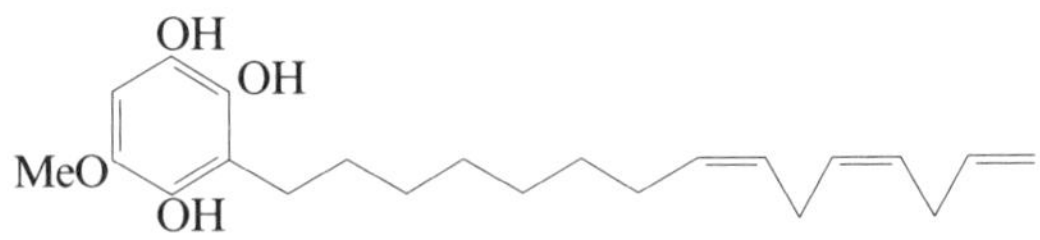

$C_{22}H_{32}O_4$ M 360.492

(8Z,11Z)-form [105018-77-7]

Germination stimulant for *Striga asiatica*, an obligate parasitic plant. V. unstable, rapidly converts to the 1,4-benzoquinone.

Lynn, D.G. *et al*, *J. Am. Chem. Soc.*, 1986, **108**, 7858 (*isol, nmr*)

Methyl acetate, 9CI, 8CI M-00011

[79-20-9]

$H_3CCOOMe$

$C_3H_6O_2$ M 74.079

Industrial solv. for gums, lacquers etc. Pleasant-smelling, volatile liq. Sol. H_2O. d_4^{20} 0.934. Mp −98.7°. Bp 56.3°. n_D^{20} 1.3614. Dissolves many metallic salts.

▷ Irritant, narcotic in high concs. TLV610. Highly flammable, fl.p. −9°. AI9100000.

Mumford, S.A. *et al*, *J. Chem. Soc.*, 1950, 75 (*synth, props*)
Ger. Pat., 1 939 286, (1968); *CA*, **74**, 126332 (*manuf*)
Techniques of Chemistry, (Weisberger, A., Ed.), 3rd Ed., 1970, 276, 748 (*props, purifn*)
Hazards in the Chemical Laboratory, (Bretherick, L., Ed.), 3rd Ed., Royal Society of Chemistry, London, 1981, 388.
Lewis, R.J., *Sax's Dangerous Properties of Industrial Materials*, 8th Ed., Van Nostrand-Reinhold, 1992, MFW100.

Methyl arachidonate hydroperoxides M-00012

$H_3C(CH_2)_3(CH_2CH{=}CH)_3CH{=}CHCH(OOH)(CH_2)_3COOH$ (5-isomer)

$C_{20}H_{28}O_4$ M 332.439

A mixt. of isomeric prods. All eight hydroperoxides are formed by reaction with singlet oxygen. Free radical autoxidn. probably gives the 5-, 8-, 9-, 11-, 12- and 15-isomers only. Enzymic oxidn. can give rise to the 15- or 12-hydroperoxides depending on the enzyme source.

Cobern, D. *et al*, *J. Chem. Soc. C*, 1966, 1897.
Hamberg, M. *et al*, *J. Biol. Chem.*, 1967, **242**, 5329.
Nugteren, D.H. *et al*, *Biochim. Biophys. Acta*, 1975, **380**, 299.
Funk, M.O. *et al*, *Lipids*, 1976, **11**, 113.
Salomon, J.A. *et al*, *Biochim. Biophys. Acta*, 1978, **523**, 250.
Porter, N.A. *et al*, *J. Org. Chem.*, 1979, **44**, 3179.
Boeynaems, J.M. *et al*, *Prostaglandins*, 1980, **19**, 87.
Corey, E.J. *et al*, *J. Am. Chem. Soc.*, 1980, **102**, 1433, 1435.

24-Methylcholesta-5,22-dien-3-ol M-00013

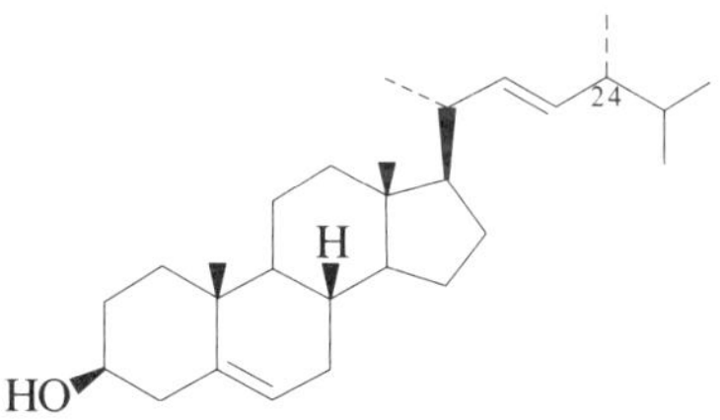

(3β,22*E*,24*R*)-*form*

$C_{28}H_{46}O$ M 398.671

(3β,22E,24R)-form [474-67-9]
Ergosta-5,22-dien-3-ol. ***Brassicasterol***
Constit. of *Brassica rapa* oil and of various sponges. Cryst. Mp 157-158° (148°). $[\alpha]_D^{19}$ −39.4°, $[\alpha]_D^{19}$ −54° ($CHCl_3$). Prob. identical with Shakosterol isol. from a mollusc.

(3β,22E,24S)-form [17472-78-5] ***Crinosterol.*** *Campesta-5,22-dien-3-ol*
Constit. of a crinoid (*Comutala* sp.), the sponges *Jaspis stellifera, Phaeodactylum tricornutum* and *Chlorella ellipsoidea*. Cryst. (MeOH or Me_2CO). Mp 147-148°. $[\alpha]_D^{26}$ −47.2°.

Bolker, H.I. *et al*, *Nature (London)*, 1967, **213**, 905 (*isol*)
Rubinstein, I. *et al*, *Phytochemistry*, 1974, **13**, 485 (*isol*)
Sheikh, Y.M. *et al*, *Tetrahedron*, 1974, **30**, 4095 (*isol*)
Sheikh, Y.M. *et al*, *Steroids*, 1975, **26**, 129 (*synth*)
Kobayashi, M. *et al*, *Steroids*, 1975, **26**, 605 (*isol*)
Adler, J.H. *et al*, *Lipids*, 1976, **11**, 634 (*biosynth*)
Rubinstein, I. *et al*, *Phytochemistry*, 1976, **15**, 195 (*pmr*)
Tsai, L.B. *et al*, *Phytochemistry*, 1976, **15**, 1131 (*biosynth*)
Wright, J.L.C. *et al*, *Can. J. Chem.*, 1978, **56**, 1898 (*cmr*)
Theobald, N. *et al*, *J. Am. Chem. Soc.*, 1978, **100**, 7677 (*isol*)
Lang, R.W. *et al*, *Helv. Chim. Acta*, 1982, **65**, 407 (*synth*)
Anastasia, M. *et al*, *J. Chem. Soc., Perkin Trans. 1*, 1983, 379, 2365 (*synth*)
Seo, S. *et al*, *J. Chem. Soc., Chem. Commun.*, 1984, 1174 (*biosynth*)

4-Methylcholest-7-en-3-ol M-00014

$C_{28}H_{48}O$ M 400.687

(3β,4α,5α)-form [481-25-4] ***Lophenol.*** *Methosterol*
Constit. of *Lophocereus schottii, Solanum tuberosum* and *S. demissum*. Cryst. (MeOH). Mp 149-151°. $[\alpha]_D$ +5° ($CHCl_3$).

Djerassi, C. *et al*, *J. Am. Chem. Soc.*, 1958, **80**, 6284 (*struct*)
Scallen, T.J. *et al*, *J. Lipid Res.*, 1968, **9**, 120 (*ir*)
Sucrow, W. *et al*, *Chem. Ber.*, 1977, **110**, 1523 (*synth*)
Iida, T. *et al*, *Steroids*, 1977, **29**, 453 (*pmr*)
Schreiber, K. *et al*, *Tetrahedron*, 1964, **20**, 2575 (*isol*)

24-Methylcycloart-25-en-3-ol M-00015

24-Methyl-9,19-cyclolanost-25-en-3-ol

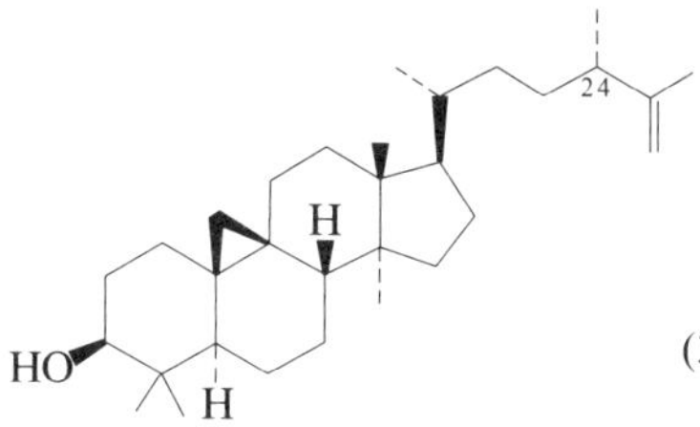

(3β,9β,24*S*)-*form*

$C_{31}H_{52}O$ M 440.751

(3α,9β,24S)-form
3-Epicyclolaudenol
Constit. of *Euphorbia caducifolia*. Cryst. Mp 140°. $[\alpha]_D$ −10° (c, 0.9 in $CHCl_3$).

(3β,9β,24R)-form
Constit. of *Polypodium formosanum*. Cryst. Mp 123-124°. $[\alpha]_D^{24}$ +36.5° ($CHCl_3$).

Ac:
$C_{33}H_{54}O_2$ M 482.788
Constit. of *P. formosanum*. Cryst. Mp 127-128°. $[\alpha]_D^{24}$ +53.5° ($CHCl_3$).

3-Ketone: [2315-13-1]. *24*R*-Methylcycloart-25-en-3-one. 24*R*-Methyl-9,19-cyclo-9β-lanost-25-en-3-one.* ***Cyclolaudenone***
$C_{31}H_{50}O$ M 438.735
Constit. of *P. formosanum*. Cryst. Mp 105°. $[\alpha]_D^{24}$ +14.2° ($CHCl_3$).

(3β,9β,24S)-form [511-61-5] ***Cyclolaudenol***
Found in opium and roots of *P. vulgare*. Also from *Manilkara bidentata, Ceterach officinarum, Phyllitis scolopendrium* and others. Cryst. (MeOH). Mp 125°. $[\alpha]_D$ +46° (c, 1.5 in $CHCl_3$).

Ac: [2315-12-0].
Cryst. ($CHCl_3$/MeOH). Mp 120-121°. $[\alpha]_D$ +55° (c, 1.5 in $CHCl_3$).

Berti, G. *et al*, *Bull. Soc. Chim. Fr.*, 1964, 2359 (*isol*)
Ghisalberti, E.L. *et al*, *J. Chem. Soc., Chem. Commun.*, 1969, 1401 (*biosynth*)
Ageta, H. *et al*, *Chem. Lett.*, 1982, 881 (*isol, struct*)
Govardhan, C. *et al*, *Phytochemistry*, 1984, **23**, 411 (*isol*)

2-Methylcyclopropanepentadecanoic acid, 9CI M-00016

16,17-Methyleneoctadecanoic acid

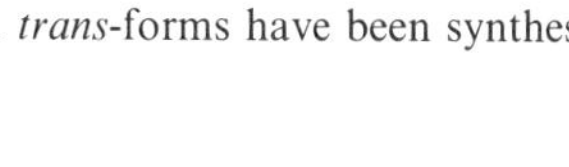

$C_{19}H_{36}O_2$ M 296.492

Both *cis*- and *trans*-forms have been synthesised as Me esters.

Christie, W.W. *et al*, *Chem. Phys. Lipids*, 1968, **2**, 196 (*synth, glc, ir, nmr*)
Christie, W.W., *Top. Lipid Chem.*, 1970, **1**, 4 (*rev*)
Gunstone, F.D. *et al*, *Chem. Phys. Lipids*, 1973, **10**, 303 (*synth, glc, nmr*)

2-Methyldecane — M-00017

[6975-98-0]

$H_3C(CH_2)_7CH(CH_3)_2$

$C_{11}H_{24}$ M 156.311
Constit. of *Angelica* spp., *Cicer arietinum* and *Hypericum* spp. d_4^{20} 0.7365. Bp 190°, Bp_{10} 74-75°. n_D^{20} 1.4152.

Francis, A.W., *Ind. Eng. Chem.*, 1943, **35**, 442.
Streif, A.J. *et al*, *Anal. Chem.*, 1955, **27**, 411.
Mathis, C. *et al*, *Phytochemistry*, 1964, **3**, 115 (*isol*)
Mann, G. *et al*, *Tetrahedron*, 1967, **23**, 3393 (*conformn, props*)
Lehmkuhl, H. *et al*, *Justus Liebigs Ann. Chem.*, 1975, 145, 1176 (*synth*)

4-Methyldecane — M-00018

[2847-72-5]

$H_3C(CH_2)_5CH(CH_3)CH_2CH_2CH_3$

$C_{11}H_{24}$ M 156.311
Constit. of the secretion of the Dufour's glands of ants. Also isol. from *Cicer arietinum* and coconut.

(±)-*form*
d_4^{20} 0.742. Bp 187°, Bp_{12} 71-72°. n_D^{20} 1.4175.

Letsinger, R.L. *et al*, *J. Am. Chem. Soc.*, 1948, **70**, 3342; 1950, **72**, 849 (*synth*)
Mann, G. *et al*, *Tetrahedron*, 1967, **23**, 3393 (*conformn, props*)
Lehmkuhl, H. *et al*, *Justus Liebigs Ann. Chem.*, 1975, 145 (*synth*)
Netzel, D.A. *et al*, *Fuel*, 1981, **60**, 307 (*cmr*)
Lipshutz, B.H. *et al*, *J. Org. Chem.*, 1984, **49**, 3928 (*synth*)
Bagneres, A.G. *et al*, *Biochem. Syst. Ecol.*, 1991, **19**, 25 (*isol*)

5-Methyldecane — M-00019

[13151-35-4]

$H_3C(CH_2)_4CH(CH_3)CH_2CH_2CH_2CH_3$

$C_{11}H_{24}$ M 156.311
Constit. of the pheromones of *Platynereis dumerilii* and certain ants. Also isol. from *Juncus roemerianus*.

Miles, D.H. *et al*, *Phytochemistry*, 1973, **12**, 1399 (*isol*)
Zeeck, E. *et al*, *Tetrahedron Lett.*, 1990, **31**, 5613 (*isol, synth*)
Bagneres, A.G. *et al*, *Biochem. Syst. Ecol.*, 1991, **19**, 25 (*isol*)

2-Methyldecanoic acid, 9CI — M-00020

2-Methylcapric acid
[24323-23-7]

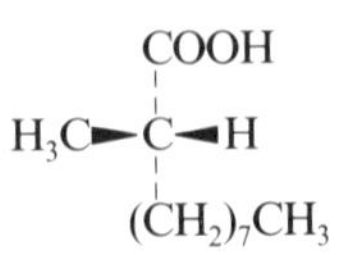

(*S*)-*form*

$C_{11}H_{22}O_2$ M 186.294

(*S*)-*form*
Et ester: [16177-05-2].
$C_{13}H_{26}O_2$ M 214.347
Occurs in preen gland of water birds.

(±)-*form*
Bp_{14} 137-139°, Bp_2 103°.
Me ester: [29619-64-5].
$C_{12}H_{24}O_2$ M 200.320
Bp 105-107°.
Amide:
$C_{11}H_{23}NO$ M 185.309
Cryst. (pet. ether). Mp 76.5-77.3°.

Ställberg-Stenhagen, S. *et al*, *Ark. Kemi*, No. 9, 1947, **24B** (*abs config*)
Schulte, R.E. *et al*, *Hoppe Seyler's Z. Physiol. Chem.*, 1951, **288**, 69.
Cason, J. *et al*, *J. Org. Chem.*, 1953, **18**, 1594 (*synth*)
Odham, E., *Fette, Seifen, Anstrichm.*, 1967, **69**, 164.
Karlsson, H. *et al*, *Ark. Kemi*, 1969, **31**, 143 (*ms*)
Japan. Pat., 73 81 810, (1973); *CA*, **80**, 107992d (*synth*)

3-Methyldecanoic acid, 9CI — M-00021

3-Methylcapric acid
[60308-82-9]

$H_3C(CH_2)_6CH(CH_3)CH_2COOH$

$C_{11}H_{22}O_2$ M 186.294

(±)-*form*
Perfume ingredient. Bp_{14} 153-154°.
Me ester:
$C_{12}H_{24}O_2$ M 200.320
Bp_{18} 110-113°.

Keil, W., *Hoppe Seyler's Z. Physiol. Chem.*, 1942, **274**, 175 (*synth*)
Polgar, N. *et al*, *J. Chem. Soc.*, 1945, 389.
Ger. Pat., 2 558 657, (1967); *CA*, **85**, 108310v 8synth).

4-Methyldecanoic acid, 9CI — M-00022

4-Methylcapric acid

$H_3C(CH_2)_5CH(CH_3)CH_2CH_2COOH$

$C_{11}H_{22}O_2$ M 186.294

(+)-*form*
Et ester: [16177-06-3].
$C_{13}H_{26}O_2$ M 214.347
Occurs in preen gland of water birds.

(±)-*form* [24323-24-8]
Oil. Bp_{12} 150-152°.
Me ester:
$C_{12}H_{24}O_2$ M 200.320
Bp_{13} 111°.

Staudinger, H. *et al*, *Helv. Chim. Acta*, 1924, **7**, 255.
Polgar, N. *et al*, *J. Chem. Soc.*, 1945, 389.
Odham, G., *Fette, Seifen, Anstrichm.*, 1967, **69**, 164.
Karlssen, H., *Ark. Kemi*, 1969, **31**, 143 (*ms*)

5-Methyldecanoic acid — M-00023

5-Methylcapric acid

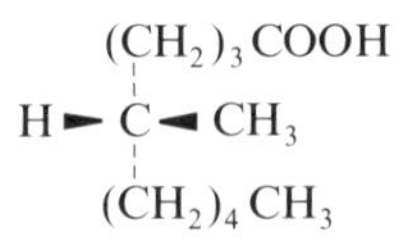

(*R*)-*form*

$C_{11}H_{22}O_2$ M 186.294

(*R*)-*form*
Bp_3 135°. $[\alpha]_D^{25}$ −0.18°.
Et ester:
$C_{13}H_{26}O_2$ M 214.347
Bp_{25} 140°.

(±)-*form*
Bp_{20} 167-168°, Bp_2 117-119°.

Levene, P.A. *et al*, *J. Biol. Chem.*, 1932, **95**, 153; 1933, **103**, 299.
Polgar, N. *et al*, *J. Chem. Soc.*, 1945, 389.

6-Methyldecanoic acid, 9CI M-00024

6-Methylcapric acid

$H_3C(CH_2)_3CH(CH_3)(CH_2)_4COOH$

$C_{11}H_{22}O_2$ M 186.294

(+)-***form***

Et ester: [16177-07-4].

$C_{13}H_{26}O_2$ M 214.347

Occurs in preen gland of water birds.

(±)-***form*** [53696-14-3]

Bp_{18} 164-165°, $Bp_{1.5}$ 127°.

Me ester: [51771-68-7].

$C_{12}H_{24}O_2$ M 200.320

Occurs in preen gland of birds.

Polgar, N. *et al*, *J. Chem. Soc.*, 1945, 389.
Linstead, R.P. *et al*, *J. Chem. Soc.*, 1953, 1538.
Odham, G., *Fette, Seifen, Anstrichm.*, 1967, **69**, 164.

7-Methyldecanoic acid M-00025

7-Methylcapric acid

$H_3CCH_2CH_2CH(CH_3)(CH_2)_5COOH$

$C_{11}H_{22}O_2$ M 186.294

(±)-***form***

Bp_{15} 161-162°.

Polgar, N. *et al*, *J. Chem. Soc.*, 1945, 389.

8-Methyldecanoic acid M-00026

8-Methylcapric acid

[5601-60-5]

$H_3CCH_2CH(CH_3)(CH_2)_6COOH$

$C_{11}H_{22}O_2$ M 186.294

Occurs in *Propionibacterium freudenreichii* and tobacco and mutton flavours. Mp −11.6°. Bp_{25} 175°, $Bp_{0.5}$ 111-112°.

Me ester:

$C_{12}H_{24}O_2$ M 200.320

Bp_{22} 129-132°, $Bp_{1.95}$ 88°.

Weitkamp, A.W. *et al*, *J. Am. Chem. Soc.*, 1945, **67**, 447.
Nunn, J.R., *J. Chem. Soc.*, 1951, 1740.
Laneele, M.A., *C. R. Hebd. Seances Acad. Sci.*, 1968, **266**, 1901 (*isol*)

9-Methyldecanoic acid, 9CI M-00027

9-Methylcapric acid

[1119-63-7]

$(H_3C)_2CH(CH_2)_7COOH$

$C_{11}H_{22}O_2$ M 186.294

Occurs in *Propionibacterium freudenreichii* and mutton flavour. Bp_{23} 174-174.5°, Bp_3 122-124°.

Levene, P.A. *et al*, *J. Biol. Chem.*, 1916, **27**, 449.
Laneele, M.A., *C. R. Hebd. Seances Acad. Sci. Ser. D*, 1968, **266**, 1901 (*isol*)
Takahashi, M. *et al*, *Yakugaku Zasshi*, 1976, **96**, 1000 (*synth*)
Kaga, H. *et al*, *Synthesis*, 1989, 864 (*synth, ir, pmr*)

8-Methyl-2-decanol, 9CI M-00028

[110072-54-3]

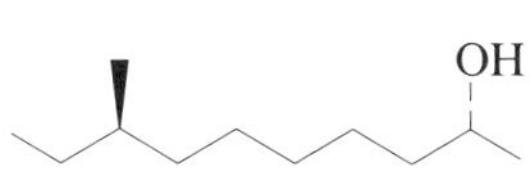

(2*R*,8*R*)-*form*

$C_{11}H_{24}O$ M 172.310

(***2R,8R***)-***form*** [90359-38-9]

Liq. Bp_5 87-95° (bath). $[\alpha]_D^{23.5}$ −14.9° (c, 1.09 in $CHCl_3$) (o.p. 100%). $n_D^{23.5}$ 1.4337.

Propanoyl: [90410-62-1].

$C_{14}H_{28}O_2$ M 228.374

Sex pheromone of the female western corn rootworm *Diabrotica virgifera*. Liq. Bp_5 95-98° (bath). $[\alpha]_D^{23}$ −7.57° (c, 1.05 in $CHCl_3$). $n_D^{23.1}$ 1.4248. Abs. config. of natural pheromone apparently not yet known but males prefer the (8*R*) isomers.

(***2S,8R***)-***form*** [90359-40-3]

Liq. Bp_{10} 104° (bath). $[\alpha]_D^{21.5}$ −1.32° (c, 1.19 in $CHCl_3$) (o.p. 100%). $n_D^{21.5}$ 1.4339.

Propanoyl: [90410-63-2].

Sex pheromone of *D. virgifera* and *D. porracea*. Bp_5 104° (bath). $[\alpha]_D^{21.5}$ −4.25° (c, 1.1 in $CHCl_3$). $n_D^{21.5}$ 1.4249.

(***2S,8S***)-***form*** [104435-11-2]

Propanoyl: [95840-36-1].

$[\alpha]_D$ +8.597° (c, 12.7 in $CHCl_3$).

[81931-28-4, 86736-98-3, 91853-82-6, 91853-83-7, 91853-84-8, 95840-37-2, 104435-71-4, 104486-91-1, 104487-04-9, 104487-05-0]

Guss, P.L. *et al*, *J. Chem. Ecol.*, 1984, **10**, 1123; 1985, **11**, 21 (*activity*)
Mori, K. *et al*, *Tetrahedron*, 1984, **40**, 299 (*synth, pmr, ir*)
Sonnet, P.E. *et al*, *J. Chem. Ecol.*, 1987, **13**, 1279; 1985, **11**, 1371 (*isol*)
Ferreira, J.T.B. *et al*, *Tetrahedron*, 1990, **46**, 6311 (*synth, pmr, cmr, ir, bibl*)
Naoshima, Y. *et al*, *J. Chem. Soc., Perkin Trans. 1*, 1991, 549 (*synth, 2R,8R-form, ir, pmr, ms*)

21-Methyldocosanoic acid, 9CI M-00029

Isotricosanoic acid

[59708-74-6]

$(H_3C)_2CH(CH_2)_{19}COOH$

$C_{23}H_{46}O_2$ M 354.615

Constit. of the marine sponge *Microciona prolifera*. Mp 73.5°.

Et ester:

$C_{25}H_{50}O_2$ M 382.669

Mp 55.5°.

Levene, P.A. *et al*, *J. Biol. Chem.*, 1922, **52**, 227 (*synth*)
Morales, R.W. *et al*, *Biochim. Biophys. Acta*, 1976, **431**, 206.

21-Methyldocosanol M-00030

Isotricosanol

$(H_3C)_2CH(CH_2)_{19}CH_2OH$

$C_{23}H_{48}O$ M 340.632

Mp 69°.

Levene, P.A. *et al*, *J. Biol. Chem.*, 1922, **52**, 227 (*synth*)

2-Methyldodecanedioic acid M-00031

$HOOC(CH_2)_9CH(CH_3)COOH$

$C_{13}H_{24}O_4$ M 244.330

(±)-***form***

Needles (C_6H_6/pet. ether). Mp 76-76.5°. Bp 210-215°.

Di-Me ester: Bp_{13} 187-188°.

Di-Et ester: Bp_{12} 197°.

Chuit, P. *et al*, *Helv. Chim. Acta*, 1927, **10**, 167 (*synth*)

3-Methyldodecanedioic acid M-00032

$HOOC(CH_2)_8CH(CH_3)CH_2COOH$

$C_{13}H_{24}O_4$ M 244.330

(±)-*form*

Cryst. (C_6H_6). Mp 82°. Bp_1 210-211°.

Di-Me ester: Bp_8 175-176°.

Di-Et ester: Bp_8 187-189°.

Chuit, P. *et al*, *Helv. Chim. Acta*, 1927, **10**, 113, 167 (*synth*)

4-Methyldodecanedioic acid M-00033

$HOOC(CH_2)_7CH(CH_3)CH_2CH_2COOH$

$C_{13}H_{24}O_4$ M 244.330

(±)-*form*

Cryst. (C_6H_6/pet. ether). Mp 71°.

Di-Et ester: $Bp_{0.5}$ 140° *ca.*.

Ruzicka, P. *et al*, *Helv. Chim. Acta*, 1927, **10**, 680 (*synth*)

6-Methyldodecanedioic acid M-00034

$HOOC(CH_2)_5CH(CH_3)(CH_2)_4COOH$

$C_{13}H_{24}O_4$ M 244.330

(±)-*form*

Di-Me ester: Bp_{10} 180-187°.

Ruzicka, L. *et al*, *Helv. Chim. Acta*, 1934, **17**, 1308 (*synth*)

2-Methyldodecanoic acid, 9CI M-00035

2-Methyllauric acid

[2874-74-0]

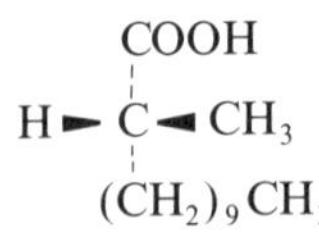

(*R*)-*form*

$C_{13}H_{26}O_2$ M 214.347

(*R*)-*form*

$[\alpha]_D^{25}$ −6.38° (Et_2O).

Amide:

$C_{13}H_{27}NO$ M 213.362

Mp 77°. $[\alpha]_D^{25}$ −3.01° (Et_2O).

(*S*)-*form*

Lustrous plates. Mp 22°. Bp_1 153°, $Bp_{0.4}$ 124-12725°, +8.47. Shows no opt. rotation after neutralisation.

Amide: Mp 94°. $[\alpha]_D^{20}$ +12.6° ($CHCl_3$).

(±)-*form*

Occurs in uropygial gland of the oyster catcher and ducks. Mp 24°. $Bp_{4.4}$ 159-161°, $Bp_{0.2}$ 119-122°.

Amide: Mp 89°.

Levene, P. *et al*, *J. Biol. Chem.*, 1929, **24**, 590.
Ställberg-Stenhagen, S., *Ark. Kemi*, No. 15, 1947, **23A** (*abs config*)
Org. Synth., Coll. Vol., 4, 1963, 616.
Karlsson, A. *et al*, *Ark. Kemi*, 1969, **31**, 143 (*ms*)
DeKlein, W.J., *Recl. Trav. Chim. Pays-Bas* (*J. R. Neth. Chem. Soc.*), 1975, **94**, 48 (*synth*)

3-Methyldodecanoic acid, 9CI M-00036

[13490-36-3]

$H_3C(CH_2)_8CH(CH_3)CH_2COOH$

$C_{13}H_{26}O_2$ M 214.347

(±)-*form*

Mp 22°. Bp_7 162°.

Ulusoy, E., *CA*, 1952, **46**, 3946.

10-Methyldodecanoic acid M-00037

Anteisotridecanoic acid

[7416-57-1]

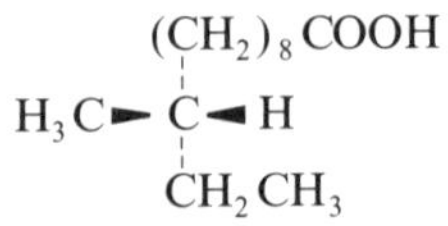

(*S*)-*form*

$C_{13}H_{26}O_2$ M 214.347

(*S*)-*form* [32726-07-1]

Occurs in wool wax.

(±)-*form*

Occurs in sprucewood, mandibular gland of ants and some bacteria. Mp 5.6°. $Bp_{0.3}$ 124-126°.

Me ester: [5129-65-7].

$C_{14}H_{28}O_2$ M 228.374

Bp_{22} 156-158°.

Amide:

$C_{13}H_{27}NO$ M 213.362

Mp 91.6-92.6°.

Nunn, J.R., *J. Chem. Soc.*, 1951, 1740.
Laneelle, M.A. *et al*, *C. R. Hebd. Seances Acad. Sci.*, 1968, **266**, 1961.
Goodrich, B. *et al*, *Aust. J. Chem.*, 1971, **24**, 153.

10-Methyl-1-dodecanol M-00038

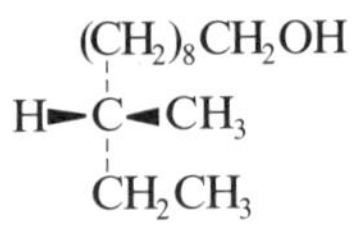

(*R*)-*form*

$C_{13}H_{28}O$ M 200.364

(*R*)-*form* [71777-32-7]

Oil. $Bp_{0.04}$ 84-86°.

Ac: [71777-34-9].

$C_{15}H_{30}O_2$ M 242.401

Constit. of the sex pheromone of the lesser tea tortrix *Adoxophyes* sp. Oil. $Bp_{0.05}$ 120-125°. $[\alpha]_D^{24}$ −5.57° (c, 21.8 in $CHCl_3$).

(*S*)-*form* [71777-33-8]

Oil.

Ac: [71777-35-0].

Minor component of the pheromone of *A.* spp. Somewhat less active than the (*R*)-enantiomer. Oil. $[\alpha]_D^{24}$ +5.6° (c, 21.8 in $CHCl_3$).

Sonnet, P.E., *J. Org. Chem.*, 1982, **47**, 3793 (*synth*)
Hjalmarsson, M. *et al*, *Acta Chem. Scand., Ser. B*, 1985, **39**, 793 (*synth, bibl*)
Bestmann, H.J. *et al*, *Justus Liebigs Ann. Chem.*, 1990, 829 (*synth*)

2-Methyl-2-dodecenoic acid M-00039

$H_3C(CH_2)_8CH{=}C(CH_3)COOH$

$C_{13}H_{24}O_2$ M 212.331

Mp 28.5-32°. $Bp_{1.4}$ 146-147°. Described as the *trans*-isomer but this term is ambiguous.

Me ester:

$C_{14}H_{26}O_2$ M 226.358

$Bp_{14.5}$ 153-154°, $Bp_{5.2}$ 135°.

Cason, J. *et al*, *J. Org. Chem.*, 1953, **18**, 842 (*synth, uv*)
Freeman, N.K., *J. Am. Chem. Soc.*, 1953, **75**, 1859 (*ir*)
Org. Synth., Coll. Vol., 4, 1963, 608 (*synth*)

2-Methyldotriacontane, 9CI M-00040

[1720-11-2]

$H_3C(CH_2)_{29}CH(CH_3)_2$

$C_{33}H_{68}$ M 464.900

Isol. from leaf wax of *Aeonium lindleyi* and *Nicotiana* sp.. Present in surface lipids of pupal tobacco budworms (*Heliothis virescens*), in sex pheromones of insect spp. and in camomile wax.

[69067-11-4]

Eglington, D. *et al*, *Phytochemistry*, 1966, **5**, 1349 (*isol, ir, ms*)
Guelz, P.G. *et al*, *Phytochemistry*, 1968, **7**, 1009 (*isol, ms*)
Gamou, K. *et al*, *Agric. Biol. Chem.*, 1980, **44**, 2119 (*isol, glc*)
Heeman, V. *et al*, *Phytochemistry*, 1983, **22**, 133 (*isol, ms*)

18-Methyl-5,9-eicosadienoic acid M-00041

[133530-15-1]

$H_3CCH_2CH(CH_3)$
$(CH_2)_7CH{=}CHCH_2CH_2CH{=}CH(CH_2)_3COOH$

$C_{21}H_{38}O_2$ M 322.530

Constit. of the sponge *Erylus formosus*. Isol. as Me ester.

Carballeira, N.M. *et al*, *J. Nat. Prod.* (*Lloydia*), 1991, **54**, 305 (*isol*)

19-Methyl-5,9-eicosadienoic acid M-00042

$(H_3C)_2CH(CH_2)_8CH{=}CHCH_2CH_2CH{=}CH(CH_2)_3COOH$

$C_{21}H_{38}O_2$ M 322.530

(E,E)-form [132171-20-1]
Constit. of the sponge *Plakortis halichondrioides*.

(Z,Z)-form [133530-16-2]
Constit. of the sponge *Erylus formosus*. Isol. as Me ester.

Carballeira, N.M. *et al*, *Lipids*, 1990, **25**, 835 (*isol*)
Carballeira, N.M. *et al*, *J. Nat. Prod.* (*Lloydia*), 1991, **54**, 305 (*isol*)

2-Methyl-2-eicosenoic acid M-00043

$H_3C(CH_2)_{16}CH{=}C(CH_3)COOH$

$C_{21}H_{40}O_2$ M 324.546

Cryst. Mp 66.3-67.6°.

Cason, J. *et al*, *J. Org. Chem.*, 1954, **19**, 1836 (*synth, uv*)

24-Methylenecycloartan-3-ol M-00044

24-Methylene-9β,19-cyclolanostan-3-ol

[1449-09-8]

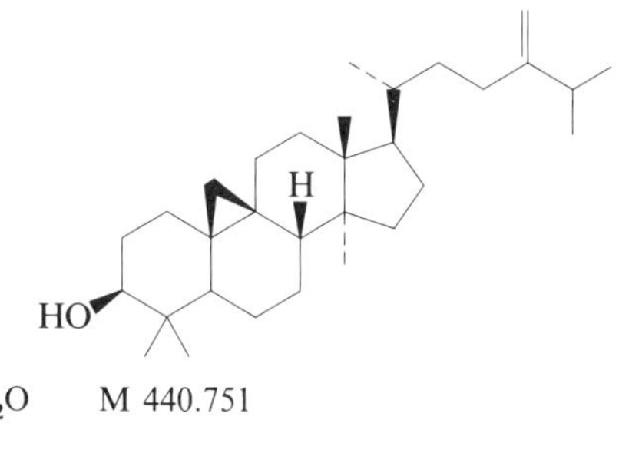

$C_{31}H_{52}O$ M 440.751

3β-form
Constit. of rice bran oil and the woods from *Tristania conferta*, *Angophora subvelutina* and *Cephalosphaera usambarensis*. Also from *Populus*, *Solanum*, *Betula*, *Euphorbia* and others. Cryst. (MeOH). $[\alpha]_D^{14.5}$ +48.8° (c, 0.26 in $CHCl_3$).

Ac:
$C_{33}H_{54}O_2$ M 482.788
Isol. from *Polypodium niponicum* and other ferns. Mp 118-120°. $[\alpha]_D$ +31° ($CHCl_3$).

*3-(4-Hydroxy-*E*-cinnamoyl):* [99132-92-0]. ***Pholidotin***
$C_{40}H_{58}O_3$ M 586.896
Constit. of *Pholidota rubra* and *Cirrhopetalum elatum*. Cryst. Mp 196°. $[\alpha]_D$ +5.54° ($CHCl_3$).

3-(4-Hydroxy-3-methoxycinnamoyl): ***Oryzanol C***
$C_{41}H_{60}O_4$ M 616.923
From rice bran oil. Cryst. ($CHCl_3$/MeOH). Mp 193-194°. $[\alpha]_D$ +36° (c, 1.87 in $CHCl_3$).

3-Ketone: 24-Methylenecycloartan-3-one
$C_{31}H_{50}O$ M 438.735
Isol. from *Microsorium fortunei*. Mp 110-112°. $[\alpha]_D$ +11° ($CHCl_3$).

Ohta, G., *Chem. Pharm. Bull.*, 1960, **8**, 5, 9 (*Oryzanol C*)
Cocker, W. *et al*, *J. Chem. Soc.*, 1965, 1692 (*synth*)
Alpin, R.T. *et al*, *J. Chem. Soc. B*, 1966, 1078 (*ms*)
Malhotra, H.C. *et al*, *J. Biol. Chem.*, 1972, **247**, 6243 (*biosynth*)
Lenton, J.R. *et al*, *Phytochemistry*, 1975, **14**, 1523 (*biosynth*)
Itoh, T. *et al*, *Steroids*, 1976, **27**, 275 (*pmr*)
Ageta, H. *et al*, *Phytochemistry*, 1984, **23**, 2875 (*isol, acetate*)
Murakami, T. *et al*, *Yakugaku Zasshi*, 1985, **105**, 655 (*isol, ketone*)
Majumder, P.L. *et al*, *Indian J. Chem., Sect. B*, 1987, **26**, 297 (*Pholidotin*)
Seo, S. *et al*, *J. Chem. Soc., Perkin Trans. 1*, 1988, 2407 (*biosynth*)

2-Methylenedodecanoic acid, 9CI M-00045

2-Decyl-2-propenoic acid

[52756-21-5]

$H_3C(CH_2)_9C(=CH_2)COOH$

$C_{13}H_{24}O_2$ M 212.331

Cryst. Mp 33.3-34.2°. $Bp_{1.7}$ 149-151°.

tert-*Butyl ester:* Bp_3 129-130°.

Cason, J. *et al*, *J. Org. Chem.*, 1953, **18**, 842 (*synth*)
Reich, H.J. *et al*, *J. Org. Chem.*, 1986, **51**, 2981 (*synth, pmr*)

24-Methylenelanost-8-en-3-ol M-00046

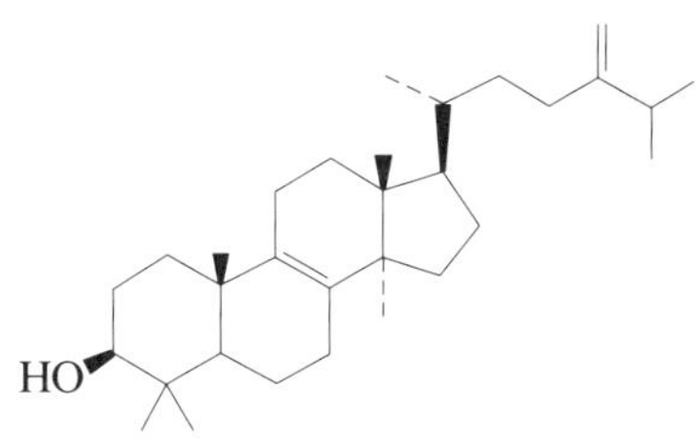

$C_{31}H_{52}O$ M 440.751

3β-form [6890-88-6] ***Eburicol.*** *Obtusifoldienol*
Isol. from *Fomes officinalis* and *Mucor pusillus*. Biogenetic precursors of steroids. Cryst. Mp 158-159°. $[\alpha]_D$ +66° (c, 0.3 in $CHCl_3$).

Lahey, F.N. *et al*, *J. Chem. Soc.*, 1951, 873 (*synth*)
González González, A. *et al*, *CA*, 1959, **53**, 20123 (*isol*)
Barton, D.H.R. *et al*, *J. Chem. Soc., Chem. Commun.*, 1966, 595 (*biosynth*)

Epstein, W.W. *et al, J. Org. Chem.*, 1966, **31**, 3434 (*isol*)
Anding, C. *et al, C. R. Hebd. Seances Acad. Sci. Ser. D*, 1973, **276**, 205 (*biosynth*)

24-Methylenelanost-9(11)-en-3-ol — M-00047

$C_{31}H_{52}O$ M 440.751

3β-form [52222-75-0]

Constit. of shea butter. Cryst. Mp 178.5-179.5°. $[\alpha]_D^{20}$ +60° (c, 0.92 in $CHCl_3$).

Itoh, T. *et al, Lipids*, 1975, **10**, 454 (*isol*)
Itoh, T. *et al, Steroids*, 1976, **27**, 275 (*pmr*)

10-Methyleneoctadecanoic acid, 9CI — M-00048

10-Methylenestearic acid. 10-Octyl-10-undecenoic acid

[7091-52-3]

$$H_3C(CH_2)_7C(=CH_2)(CH_2)_8COOH$$

$C_{19}H_{36}O_2$ M 296.492

Intermediate in the biosynth. of 10-Methyloctadecanoic acid. Mp 28-30°.

Et ester: [18432-34-3].

$C_{21}H_{40}O_2$ M 324.546

Oil. $Bp_{0.1}$ 130°.

Jauréguiberry, G. *et al, J. Chem. Soc., Chem. Commun.*, 1966, 855 (*biosynth*)
Toubiana, R. *et al, Bull. Soc. Chim. Fr.*, 1967, 94 (*synth*)

5-Methylene-3-(6,9,12,15,18,21-tetracosahexaenyl)-2(5*H*)-furanone, 9CI — M-00049

2-(6,9,12,15,18,21-Tetracosahexaenyl)-2,4-pentadien-4-olide

$C_{29}H_{40}O_2$ M 420.634

(all-Z)-form [81576-09-2]

Isol. from the gorgonian *Plexaura flava*. Incorrectly abstracted as (*all-E*)-form in *CA*.

Ravi, B.N. *et al, Aust. J. Chem.*, 1982, **35**, 105 (*isol, ir, pmr, ms*)

4-Methylergosta-7,24(28)-dien-3-ol — M-00050

4-Methyl-24-methylenecholest-7-en-3-ol

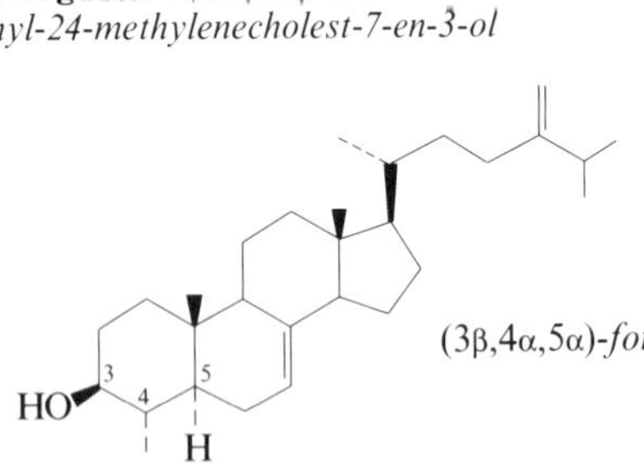

(3β,4α,5α)-*form*

$C_{29}H_{48}O$ M 412.698

(3β,4α,5α)-form [1176-52-9]

24-Methylenelophenol. Gramisterol. Gramisterin

Constit. of *Saccharum officinarum* and *Solanum tuberosum*. Also *Citrus paradisi, Solanum demissum, Pisum sativum* and *Nicotiana tabacum*. Cryst. Mp 172-173°. $[\alpha]_D^{19}$ +6° (c, 1.4 in $CHCl_3$).

(3β,4β,5α)-form [21490-25-5]

Isol. from Marigold (*Calendula officinalis*) flowers. Mp 151°.

(3β,4α,5α,20S)-form

20-Epi-24-methylenelophenol

Constit. of *Typha latifolia*. Possesses antialgal props. Cryst. (EtOH). Mp 176-178°. $[\alpha]_D$ +3° (c, 0.8 in $CHCl_3$).

Schreiber, K. *et al, Tetrahedron*, 1964, **20**, 2575 (*isol*)
Osske, G. *et al, Tetrahedron*, 1965, **21**, 1559 (*isol*)
Pyrek, J.St., *J. Chem. Soc., Chem. Commun.*, 1969, 107 (*isol*)
Lenton, J.R. *et al, Phytochemistry*, 1975, **14**, 1523 (*biosynth*)
Della Greca, M. *et al, Phytochemistry*, 1990, **29**, 1797 (*isol, pmr*)

3-Methyl-2,5-furandipropanoic acid — M-00051

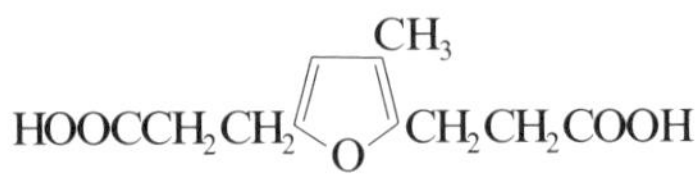

$C_{11}H_{14}O_5$ M 226.229

Found in urine. Characterised as di-Me ester.

Bauer, S. *et al, Justus Liebigs Ann. Chem.*, 1985, 813.

3-Methylheneicosanoic acid — M-00052

[66344-19-2]

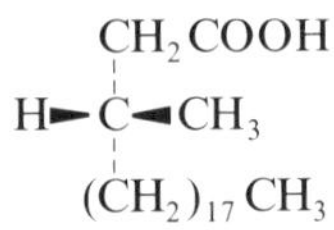

(*R*)-*form*

$C_{22}H_{44}O_2$ M 340.588

(R)-form [63473-62-1]

D-form

Cryst. (AcOH). Mp 56.5-57.5°. $[\alpha]_D^{10}$ +3.66° (c, 3.00 in $CHCl_3$).

(S)-form

$[\alpha]_D^{10}$ −3.8° (c, 3.0 in $CHCl_3$).

Amide:

$C_{22}H_{45}NO$ M 339.604

Needles (EtOH). Mp 92°.

Bailey, A.S. *et al, J. Chem. Soc.*, 1959, 661 (*synth*)
Rossi, R. *et al, Tetrahedron*, 1985, **41**, 627 (*synth*)

20-Methyl-4,7,10,13,16-heneicosapentaenoic acid — M-00053

[81607-73-0]

$$(H_3C)_2CHCH_2CH_2(CH{=}CHCH_2)_5CH_2COOH$$

$C_{22}H_{34}O_2$ M 330.509

Obt. from *Phytomonas* and *Herpatomonas* flagellates. Probably (*all-Z*)-form.

Fish, W.R. *et al, Mol. Biochem. Parasitol.*, 1982, **5**, 1; *CA*, **96**, 177577 (*ms, pmr*)

16-Methyl-6,9-heptadecadienoic acid — M-00054

[81607-68-3]

$$(H_3C)_2CH(CH_2)_5CH{=}CHCH_2CH{=}CH(CH_2)_4COOH$$

$C_{18}H_{32}O_2$ M 280.450

Obt. from *Phytomonas* and *Herpatomonas* flagellates. Probably (all-*Z*)-form.

Fish, W.R. *et al, Mol. Biochem. Parasitol.*, 1982, **5**, 1; *CA*, **96**, 177577 (*ms, pmr*)

16-Methyl-9,12-heptadecadienoic acid M-00055

[81607-69-4]

$(H_3C)_2CHCH_2CH_2CH{=}CHCH_2CH{=}CH(CH_2)_7COOH$

$C_{18}H_{32}O_2$ M 280.450

Obt. from *Phytomonas* and *Herpatomonas* flagellates. Probably (all-*Z*)-form.

Fish, W.R. *et al*, *Mol. Biochem. Parasitol.*, 1982, **5**, 1; *CA*, **96**, 177577 (*ms, pmr*)

2-Methylheptadecane, 9CI M-00056

[1560-89-0]

$H_3C(CH_2)_{14}CH(CH_3)_2$

$C_{18}H_{38}$ M 254.498

Pheromone from Tiger moth (Arctiidae family). Mp 5.7°. Bp 311° dec., Bp_{15} 178.5°.

Topchiev, A.V. *et al*, *Zh. Neorg. Khim.*, 1956, **1**, 1185; *CA*, **51**, 2570 (*synth*)
Roelofs, W.L. *et al*, *Science (Washington, D.C.)*, 1971, **171**, 684 (*isol*)
Rossi, R., *Chim. Ind. (Milan)*, 1975, **57**, 242 (*isol, synth*)
Dasaradhi, L. *et al*, *Synth. Commun.*, 1987, **17**, 1845 (*synth*)
Charlton, R.E. *et al*, *Arch. Insect Biochem. Physiol.*, 1991, **18**, 81 (*biosynth*)

9-Methylheptadecane, 8CI M-00057

[18869-72-2]

$H_3C(CH_2)_7CH(CH_3)(CH_2)_7CH_3$

$C_{18}H_{38}$ M 254.498

Bp_{10} 172-174°.

Petrov, A.D. *et al*, *CA*, 1945, **39**, 1619 (*synth*)

2-Methylheptadecanoic acid, 9CI M-00058

[5638-12-0]

$H_3C(CH_2)_{14}CH(CH_3)COOH$

$C_{18}H_{36}O_2$ M 284.481

Constit. of uropygial wax of the night heron (presumably in an enantiomeric form).

Me ester: [2490-28-0].
$C_{19}H_{38}O_2$ M 298.508
Occurs in the photosynthetic bacterium *Rhodospirillum rubrum*.

(±)-form
Mp 34-35°, Mp 51°, Bp_5 179-183°.

Morgan, G.T. *et al*, *J. Soc. Chem. Ind., London*, 1927, **46**, 152T (*synth*)
Stanley, W.M. *et al*, *J. Am. Chem. Soc.*, 1929, **51**, 1265 (*synth*)
Jacob, J., *Hoppe Seyler's Z. Physiol. Chem.*, 1975, **356**, 1823 (*ms*)
Maudinas, B. *et al*, *Phytochemistry*, 1977, **16**, 1299 (*ms*)

16-Methylheptadecanoic acid M-00059

Isostearic acid

[2724-58-5]

$(H_3C)_2CH(CH_2)_{14}COOH$

$C_{18}H_{36}O_2$ M 284.481

Occurs in adipose tissue, milk fat, liver and tobacco smoke. Has surfactant props. Mp 69.5°.

Amide:
$C_{18}H_{37}NO$ M 283.496
Mp 107.3°.

Fordyce, C.R., *J. Am. Chem. Soc.*, 1933, **55**, 3371.
Abrahamsson, S. *et al*, *Acta Crystallogr., Sect. B*, 1972, **28**, 2562 (*cryst struct*)

16-Methyl-6,9,12-heptadecatrienoic acid M-00060

[81607-70-7]

$(H_3C)_2CHCH_2CH_2CH{=}CHCH_2CH{=}CH$
$CH_2CH{=}CH(CH_2)_4COOH$

$C_{18}H_{30}O_2$ M 278.434

Obt. from *Phytomonas* and *Herpatomonas* flagellates. Probably (all-*Z*)-form.

Fish, W.R. *et al*, *Mol. Biochem. Parasitol.*, 1982, **5**, 1; *CA*, **96**, 177597 (*ms, pmr*)

2-Methyl-2-heptadecene M-00061

$H_3C(CH_2)_{13}CH{=}C(CH_3)_2$

$C_{18}H_{36}$ M 252.482

Mp −2.5°. Bp 314°, Bp_{10} 277°.

Landa, S. *et al*, *Collect. Czech. Chem. Commun.*, 1930, **2**, 540; *CA*, **25**, 67 (*synth*)

2-Methyl-16-heptadecenoic acid M-00062

$H_2C{=}CH(CH_2)_{13}CH(CH_3)COOH$

$C_{18}H_{34}O_2$ M 282.465

Mp 43-43.5°. Bp_3 186-187°.

Me ester: Bp_3 158-159°.

Chuit, P. *et al*, *Helv. Chim. Acta*, 1927, **10**, 131.

6-Methylheptanoic acid, 9CI M-00063

Isooctanoic acid. Isocaprylic acid

[929-10-2]

$(H_3C)_2CH(CH_2)_4COOH$

$C_8H_{16}O_2$ M 144.213

Obt. by hydrol. of polymyxin antibiotics. Also present in hop oil (*Humulus lupulus*) as the Me ester. Liq. Mp 0°. Bp_{762} 232°, Bp_{14} 126-127°.

Et ester: [62337-58-0].
$C_{10}H_{20}O_2$ M 172.267
Liq. Bp 200.3°.

Amide: [2979-83-1].
$C_8H_{17}NO$ M 143.228
Plates. Mp 114°.

Nitrile: 1-Cyano-5-methylhexane
$C_8H_{15}N$ M 125.213
Liq. Bp 194°.

Levene, P.A. *et al*, *J. Biol. Chem.*, 1916, **27**, 452.
Wilkinson, S. *et al*, *Nature (London)*, 1963, **200**, 1008 (*isol*)

2-Methyl-2-hexacosenoic acid M-00064

$H_3C(CH_2)_{22}CH{=}C(CH_3)COOH$

$C_{27}H_{52}O_2$ M 408.707

Mp 85.4-86.2°.

Cason, J. *et al*, *J. Org. Chem.*, 1954, **19**, 1836.

2-Methylhexadecanedioic acid, 9CI M-00065

2-Methylthapsic acid

$HOOC(CH_2)_{13}CH(CH_3)COOH$

$C_{17}H_{32}O_4$ M 300.437

(±)-*form*

Cryst. (EtOAc). Mp 89-90°. $Bp_{0.2}$ 223-225°.

Di-Me ester: [19102-97-7].

Mp 19-21°. Bp_8 210-213°.

Chuit, P. *et al*, *Helv. Chim. Acta*, 1927, **10**, 167 (*synth*)
U.K. Pat., 1 096 326, (1967); *CA*, **68**, 48302 (*synth*)

3-Methylhexadecanedioic acid, 9CI M-00066

3-Methylthapsic acid

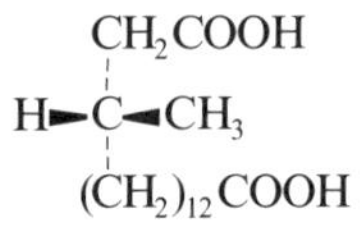

(*R*)-*form*

$C_{17}H_{32}O_4$ M 300.437

(*R*)-*form*

Mp 78.6-79°. $[\alpha]_D^{24}$ +0.9° ($CHCl_3$).

Di-Me ester: $[\alpha]_D^{24}$ +2.27°.

(*S*)-*form*

Mp 79-79.4°. $[\alpha]_D^{24}$ −0.91° (c, 10.1 in $CHCl_3$).

Di-Me ester: $Bp_{0.15}$ 143-152°. $[\alpha]_D^{25}$ −2.31°.

(±)-*form*

Cryst. (EtOAc). Mp 78-79.5°. $Bp_{0.45}$ 112-114°.

Di-Me ester: $Bp_{3.5}$ 198-200°.

Di-Et ester: $Bp_{3.5}$ 209-211°.

Chuit, P. *et al*, *Helv. Chim. Acta*, 1927, **10**, 167 (*synth*)
Ställberg-Stenhagen, S., *Ark. Kemi*, 1951, **3**, 517 (*synth, abs config*)
Sarycheva, I.K. *et al*, *Zh. Obshch. Khim.*, 1960, **30**, 2539 (*synth*)

4-Methylhexadecanedioic acid, 9CI M-00067

4-Methylthapsic acid

$HOOC(CH_2)_{11}CH(CH_3)CH_2CH_2COOH$

$C_{17}H_{32}O_4$ M 300.437

(±)-*form*

Cryst. (EtOAc). Mp 78-78.4°. Bp_1 238-240°.

Di-Me ester:

$C_{19}H_{36}O_4$ M 328.491

Mp 15°. Bp_4 203-204°.

Di-Et ester:

$C_{21}H_{40}O_4$ M 356.545

Bp_8 223-225°.

Chuit, P. *et al*, *Helv. Chim. Acta*, 1927, **10**, 167 (*synth*)

5-Methylhexadecanedioic acid M-00068

5-Methylthapsic acid

$HOOC(CH_2)_{10}CH(CH_3)(CH_2)_3COOH$

$C_{17}H_{32}O_4$ M 300.437

(±)-*form*

Cryst. (EtOH aq.). Mp 78.8-79°. Bp_3 252-253°.

Di-Me ester: Bp_3 200-202°.

Di-Et ester: Bp_9 234-236°.

Chuit, P. *et al*, *Helv. Chim. Acta*, 1927, **10**, 167 (*synth*)

8-Methylhexadecanedioic acid, 9CI M-00069

8-Methylthapsic acid

$HOOC(CH_2)_7CH(CH_3)(CH_2)_6COOH$

$C_{17}H_{32}O_4$ M 300.437

(±)-*form*

Mp 77-78°.

Di-Me ester: [2490-52-0].

Bp_1 187-193°.

Ruzicka, L. *et al*, *Helv. Chim. Acta*, 1934, **17**, 1308 (*synth*)

2-Methylhexadecanoic acid, 9CI M-00070

2-Methylpalmitic acid

[27147-71-3]

$H_3C(CH_2)_{13}CH(CH_3)COOH$

$C_{17}H_{34}O_2$ M 270.454

Occurs in uropygial gland of night heron and in the bacterium *Rhodospirillum rubrum*. Mp 47.5-48°, Mp 54°. Bp_{1-2} 174-177°.

Me ester: [2490-53-1].

$C_{18}H_{36}O_2$ M 284.481

Isol. from *Convolvulus* and other plant species. Shows insecticidal props.

Et ester: [14146-53-3].

$C_{19}H_{38}O_2$ M 298.508

Liq. Bp_2 17°.

Amide:

$C_{17}H_{35}NO$ M 269.470

Cryst (MeOH or Me_2CO). Mp 101.5-102°.

Greer, C.M. *et al*, *J. Am. Chem. Soc.*, 1930, **52**, 2542.
Weitzel, G., *Hoppe Seyler's Z. Physiol. Chem.*, 1951, **287**, 296 (*ester, amide*)
Gerson, T. *et al*, *Chem. Phys. Lipids*, 1968, **2**, 213 (*synth*)
Ubik, K. *et al*, *Collect. Czech. Chem. Commun.*, 1975, **40**, 2826 (*ms*)
Jacob, J. *et al*, *Hoppe Seyler's Z. Physiol. Chem.*, 1975, **356**, 1823.
Dobner, B. *et al*, *Pharmazie*, 1984, **39**, 69 (*anhydride*)
Adams, J. *et al*, *Org. Mass Spectrom.*, 1988, **23**, 307 (*ms*)

3-Methylhexadecanoic acid, 9CI M-00071

3-Methylpalmitic acid

[42172-35-0]

$H_3C(CH_2)_{12}CH(CH_3)CH_2COOH$

$C_{17}H_{34}O_2$ M 270.454

(±)-*form*

Cryst. (Me_2CO). Mp 42.5-43° (41-42°). $Bp_{0.5}$ 174-176°.

Et ester: [66344-25-0].

$C_{19}H_{38}O_2$ M 298.508

Liq. $Bp_{0.7}$ 160-162°.

Amide:

$C_{17}H_{35}NO$ M 269.470

Cryst. (EtOAc). Mp 90.5-91°.

[65128-49-6, 91357-11-8]

Weitzel, G. *et al*, *Hoppe Seyler's Z. Physiol. Chem.*, 1951, **287**, 65 (*synth, ester, amide*)
Hwang, Y.S. *et al*, *J. Agric. Food Chem.*, 1978, **26**, 557 (*synth*)
Dobner, B. *et al*, *Pharmazie*, 1984, **39**, 69 (*synth*)

4-Methylhexadecanoic acid, 9CI M-00072

4-Methylpalmitic acid

[53696-23-4]

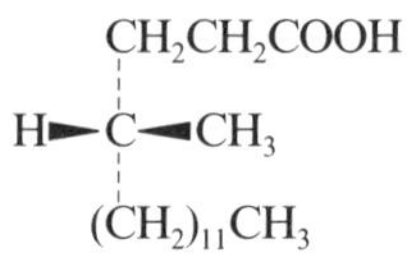

(*R*)-*form*

$C_{17}H_{34}O_2$ M 270.454

Constit. of animal lipids.

(*R*)-*form* [89051-73-0]

No phys. props. reported.

(±)-form
Cryst. (pet. ether). Mp 44.5-45° (41°). $Bp_{0.2}$ 159-162°.
Et ester:
$C_{19}H_{38}O_2$ M 298.508
Liq. Bp_{15} 200-202°.
Amide:
$C_{17}H_{35}NO$ M 269.470
Cryst. (MeOH). Mp 74-74.5°.
Anilide:
$C_{23}H_{39}NO$ M 345.567
Cryst. or needles. Mp 60°.

[2490-14-4]

Asano, M. *et al*, *Yakugaku Zasshi*, 1950, **70**, 202 (*synth*)
Weitzel, G. *et al*, *Hoppe Seyler's Z. Physiol. Chem.*, 1951, **287**, 65 (*ester, amide*)
Ackman, R.G. *et al*, *Lipids*, 1972, **7**, 683 (*isol*)
Kuwahara, S. *et al*, *Agric. Biol. Chem.*, 1983, **47**, 2599 (*synth*)

5-Methylhexadecanoic acid, 9CI M-00073

5-Methylpalmitic acid

$H_3C(CH_2)_{10}CH(CH_3)(CH_2)_3COOH$

$C_{17}H_{34}O_2$ M 270.454
(±)-form
Cryst. (Me_2CO or pet. ether). Mp 34.5-35°. $Bp_{0.2}$ 158-160°.
Et ester:
$C_{19}H_{38}O_2$ M 298.508
Liq. $Bp_{0.7}$ 160°.
Amide:
$C_{17}H_{35}NO$ M 269.470
Cryst. (EtOAc). Mp 72-72.5°.

Weitzel, G. *et al*, *Hoppe Seyler's Z. Physiol. Chem.*, 1951, **287**, 65; 1951, **288**, 189.

7-Methylhexadecanoic acid, 9CI M-00074

7-Methylpalmitic acid
[56796-91-9]

$H_3C(CH_2)_8CH(CH_3)(CH_2)_5COOH$

$C_{17}H_{34}O_2$ M 270.454
(±)-form
Cryst. (pet. ether). Mp 31-31.5°.
Et ester:
$C_{19}H_{38}O_2$ M 298.508
Liq. $Bp_{0.4}$ 152°.
Amide:
$C_{17}H_{35}NO$ M 269.470
Cryst. (EtOAc). Mp 69-69.5°.

[18720-85-9]

Weitzel, G. *et al*, *Hoppe Seyler's Z. Physiol. Chem.*, 1951, **287**, 65, 296; 1951, **288**, 189 (*synth*)
Dobner, B. *et al*, *Z. Chem.*, 1987, **27**, 63 (*synth*)

8-Methylhexadecanoic acid, 9CI M-00075

8-Methylpalmitic acid
[17301-65-4]

$H_3C(CH_2)_7CH(CH_3)(CH_2)_6COOH$

$C_{17}H_{34}O_2$ M 270.454
Constit. of animal and bacterial lipids.
(±)-form
Cryst. (Me_2CO). Mp 21-21.5°. $Bp_{0.5}$ 168-170°.
Et ester:
$C_{19}H_{38}O_2$ M 298.508
Liq. $Bp_{0.5}$ 156°.
Amide:
$C_{17}H_{35}NO$ M 269.470
Cryst. (EtOAc). Mp 71-71.5°.

[2490-52-0]

Weitzel, G. *et al*, *Hoppe Seyler's Z. Physiol. Chem.*, 1951, **287**, 65; 1951, **288**, 189 (*synth*)
Carson, J. *et al*, *J. Biol. Chem.*, 1963, **238**, 883 (*isol*)
Ryhage, R., *J. Dairy Res.*, 1967, **34**, 115 (*ester, chromatog, ms*)
Ackman, R.G. *et al*, *Lipids*, 1972, **7**, 683 (*synth*)

9-Methylhexadecanoic acid, 9CI M-00076

9-Methylpalmitic acid
[17001-27-3]

$H_3C(CH_2)_6CH(CH_3)(CH_2)_7COOH$

$C_{17}H_{34}O_2$ M 270.454
Constit. of animal lipids. Cryst. (Me_2CO). Mp 21.5-22°. $Bp_{0.4}$ 188°.
Et ester:
$C_{19}H_{38}O_2$ M 298.508
Liq. Bp_6 183°.
Amide:
$C_{17}H_{35}NO$ M 269.470
Cryst. (EtOAc). Mp 62.5-63°.

Weitzel, G. *et al*, *Hoppe Seyler's Z. Physiol. Chem.*, 1951, **287**, 65, 296; 1951, **288**, 189 (*synth*)
Ryhage, R., *J. Dairy Res.*, 1967, **34**, 115 (*chromatog, ms*)
Dobner, B. *et al*, *Z. Chem.*, 1987, **27**, 63 (*synth*)

10-Methylhexadecanoic acid, 9CI M-00077

10-Methylpalmitic acid
[17001-26-2]

$H_3C(CH_2)_5CH(CH_3)(CH_2)_8COOH$

$C_{17}H_{34}O_2$ M 270.454
Constit. of animal, sponge and bacterial lipids.
(±)-form
Cryst. (Me_2CO). Mp 24-24.5°. $Bp_{1.5}$ 189°.
Et ester:
$C_{19}H_{38}O_2$ M 298.508
Liq. Bp_1 163°.
Amide:
$C_{17}H_{35}NO$ M 269.470
Cryst. (EtOAc). Mp 80.5-81°.

[2490-51-9]

Weitzel, G. *et al*, *Hoppe Seyler's Z. Physiol. Chem.*, 1951, **287**, 65, 296; 1951, **288**, 189 (*synth*)
Cason, J. *et al*, *J. Biol. Chem.*, 1963, **238**, 883 (*isol*)
Ryhage, R., *J. Dairy Res.*, 1967, **34**, 115 (*chromatog, ms*)
Ackman, R.G. *et al*, *Lipids*, 1972, **7**, 683 (*chromatog*)
Raederstorff, D. *et al*, *J. Org. Chem.*, 1987, **52**, 2337.

11-Methylhexadecanoic acid, 9CI M-00078

11-Methylpalmitic acid
[17001-25-1]

$H_3C(CH_2)_4CH(CH_3)(CH_2)_9COOH$

$C_{17}H_{34}O_2$ M 270.454
Constit. of butter fat. Cryst. (Me_2CO). Mp 18.5-19°. $Bp_{0.4}$ 187-188°.
Et ester:
$C_{19}H_{38}O_2$ M 298.508
Liq. $Bp_{0.7}$ 160°.
Amide:
$C_{17}H_{35}NO$ M 269.470

Cryst. (EtOAc). Mp 71-71.5°.

Weitzel, G. *et al*, *Hoppe Seyler's Z. Physiol. Chem.*, 1951, **287**, 65; 1951, **288**, 189 (*synth*)
Ryhage, R., *J. Dairy Res.*, 1967, **34**, 115 (*chromatog, ms*)
Dobner, B. *et al*, *Z. Chem.*, 1987, **27**, 63 (*synth*)

12-Methylhexadecanoic acid, 9CI — M-00079

12-Methylpalmitic acid

[17001-24-0]

$H_3C(CH_2)_3CH(CH_3)(CH_2)_{10}COOH$

$C_{17}H_{34}O_2$ M 270.454

Constit. of animal lipids. Cryst. (Me_2CO). Mp 28-28.5°. Bp_2 189°.

Et ester:
$C_{19}H_{38}O_2$ M 298.508
Liq. $Bp_{0.5}$ 197°.

Amide:
$C_{17}H_{35}NO$ M 269.470
Cryst. (EtOAc). Mp 85.5-86°.

[2490-50-8]

Weitzel, G. *et al*, *Hoppe Seyler's Z. Physiol. Chem.*, 1951, **287**, 65; 1951, **288**, 189 (*synth*)
Ryhage, R., *J. Dairy Res.*, 1967, **34**, 115 (*chromatog*)
Ackman, R.G. *et al*, *Lipids*, 1972, **7**, 633 (*chromatog*)

13-Methylhexadecanoic acid, 9CI — M-00080

13-Methylpalmitic acid

[103386-25-0]

$H_3CCH_2CH_2CH(CH_3)(CH_2)_{11}COOH$

$C_{17}H_{34}O_2$ M 270.454

(±)-*form*
Cryst. (Me_2CO). Mp 30-30.5°. $Bp_{0.4}$ 186°.

Et ester:
$C_{19}H_{38}O_2$ M 298.508
Liq. Bp_2 172°.

Amide:
$C_{17}H_{35}NO$ M 269.470
Cryst. (EtOAc). Mp 78.5-79°.

Weitzel, G. *et al*, *Hoppe Seyler's Z. Physiol. Chem.*, 1951, **287**, 65; **288**, 189 (*synth, ester, amide*)
Dobner, B. *et al*, *Z. Chem.*, 1987, **27**, 63 (*synth*)

14-Methylhexadecanoic acid, 9CI — M-00081

14-Methylpalmitic acid

[5918-29-6]

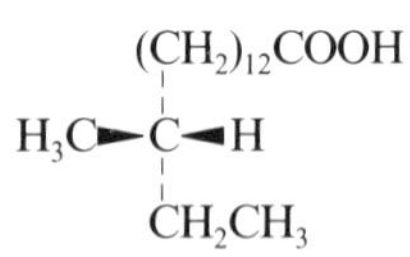

(*S*)-*form*

$C_{17}H_{34}O_2$ M 270.454

(*S*)-*form* [5746-59-8]
Occurs in several animal fats, bacteria and softwoods. Mp 39.5-40.0°. $[\alpha]_D^{29}$ +5.2° (c, 14 in Me_2CO).

Me ester:
$C_{18}H_{36}O_2$ M 284.481
Liq. Bp_2 156.5°. $[\alpha]_D^{27}$ +4.49°.

Anilide:
$C_{23}H_{39}NO$ M 345.567
Cryst. Mp 61.3-61.6°.

(±)-*form* [73641-04-0]
Cryst. (Me_2CO). Mp 39-40°.

Me ester: [2490-49-5].
Liq. $Bp_{1.8}$ 156°.

Et ester:
$C_{19}H_{38}O_2$ M 298.508
Liq. Bp_2 169-174°.

Amide:
$C_{17}H_{35}NO$ M 269.470
Mp 89-90°.

[73641-07-3]

Velick, S.F. *et al*, *J. Biol. Chem.*, 1945, **160**, 473 (*synth, resoln, derivs*)
Velick, S.F., *J. Am. Chem. Soc.*, 1947, **69**, 2321 (*cryst struct*)
Sobotka, H. *et al*, *J. Am. Chem. Soc.*, 1950, **72**, 5139 (*ir*)
Cason, J. *et al*, *J. Org. Chem.*, 1951, **16**, 1187 (*ester*)
Milburn, A.H. *et al*, *J. Chem. Soc.*, 1954, 3344 (*amide, anilide*)
Campbell, I.M. *et al*, *J. Lipid Res.*, 1969, **10**, 593 (*ms*)
Silvins, J.R., *Chem. Phys. Lipids*, 1980, **26**, 67 (*synth*)

15-Methylhexadecanoic acid, 9CI — M-00082

15-Methylpalmitic acid. Subtiloheptadecanoic acid

[1603-03-8]

$(H_3C)_2CH(CH_2)_{13}COOH$

$C_{17}H_{34}O_2$ M 270.454

Occurs widely in nature. Cryst. (C_6H_6 or Me_2CO). Mp 60.3-60.5°.

Me ester: [6929-04-0].
$C_{18}H_{36}O_2$ M 284.481
Liq. Mp 18.3-18.9°. $Bp_{1.4}$ 151-153°.

Et ester:
$C_{19}H_{38}O_2$ M 298.508
Liq. $Bp_{0.8}$ 162°.

Amide:
$C_{17}H_{35}NO$ M 269.470
Cryst. (EtOH/C_6H_6). Mp 100.5-101°.

Weitzel, G. *et al*, *Hoppe Seyler's Z. Physiol. Chem.*, 1951, **287**, 65 (*synth*)
Horn, D.H.S. *et al*, *J. Chem. Soc.*, 1954, 177 (*synth*)
Hansen, A.E. *et al*, *Biochem. J.*, 1955, **61**, 141 (*synth*)
Crossley, A.T. *et al*, *Can. J. Chem.*, 1955, **33**, 1426 (*synth*)
Saito, K., *J. Biochem. (Tokyo)*, 1960, **47**, 710 (*isol*)
Ryhage, R., *J. Dairy Res.*, 1967, **34**, 115 (*synth, chromatog, ms*)
Ubik, K. *et al*, *Collect. Czech. Chem. Commun.*, 1975, **40**, 2826 (*ms*)
Carballeira, N. *et al*, *J. Org. Chem.*, 1986, **51**, 2751 (*isol, biosynth*)

14-Methyl-8-hexadecenal — M-00083

Trogodermal

$H_3CCH_2CH(CH_3)(CH_2)_4CH{=}CH(CH_2)_6CHO$

$C_{17}H_{32}O$ M 252.439

Pheromone from *Trogoderma* spp.

(*R*)-(*Z*)-*form* [70224-30-5]
$Bp_{0.65}$ 127-128°. $[\alpha]_D^{21}$ −5.94° (c, 1.06 in $CHCl_3$).

(*R*)-(*E*)-*form* [70144-70-6]
$Bp_{0.8}$ 125-128°. $[\alpha]_D^{21}$ −5.99°.

(*S*)-(*Z*)-*form* [66007-17-8]
$Bp_{1.5}$ 140-141°. $[\alpha]_D^{25}$ +6.54° (c, 4.96 in Et_2O).

(*S*)-(*E*)-*form* [66007-19-0]
Bp_1 131-133°. $[\alpha]_D^{25}$ +6.06° (c, 3.55 in Et_2O).

Cross, J.H. *et al*, *J. Chem. Ecol.*, 1976, **2**, 457 (*isol*)
Rossi, R. *et al*, *Tetrahedron*, 1977, **33**, 2447 (*synth*)
Mori, Y. *et al*, *Tetrahedron*, 1978, **34**, 3119 (*synth, ir, ms*)
Suguro, T. *et al*, *Agric. Biol. Chem.*, 1979, **43**, 409 (*synth, ir, nmr*)
Mori, K. *et al*, *Tetrahedron*, 1982, **38**, 2291 (*synth*)
Sato, T. *et al*, *Tetrahedron Lett.*, 1982, **23**, 3587 (*synth*)
Bestmann, H.-J. *et al*, *Justus Liebigs Ann. Chem.*, 1988, 877 (*synth*)
Tolstikov, A.G. *et al*, *Khim. Prir. Soedin.*, 1991, 707 (*synth*)

7-Methyl-6-hexadecenoic acid, 9CI **M-00084**

[56796-89-5]

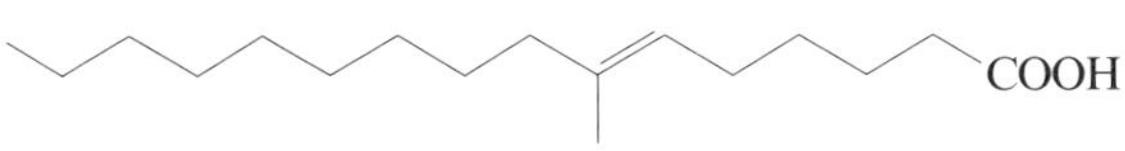

$C_{17}H_{32}O_2$ M 268.439

(E)-form [18689-89-9]

Isol. from whale oil.

Sano, Y., *Yukagaku*, 1967, **16**, 605; *CA*, **68**, 31327 (*isol, ir, pmr, ms*)
Pascal, J.L. *et al, Lipids*, 1975, **10**, 478 (*chromatog, pmr, ms*)

10-Methyl-9-hexadecenoic acid, 9CI **M-00085**

$C_{17}H_{32}O_2$ M 268.439

Constit. of lipids of *Mycobacterium tuberculosis.*

Carson, J. *et al, J. Biol. Chem.*, 1963, **238**, 883 (*isol*)

14-Methyl-8-hexadecenoic acid **M-00086**

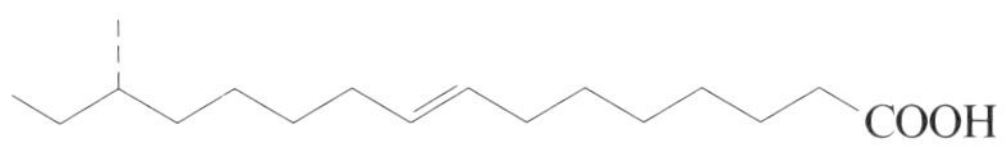

(R,Z)-form

$C_{17}H_{32}O_2$ M 268.439

(R)-(Z)-form

Me ester: [51018-59-8].

$C_{18}H_{34}O_2$ M 282.465

$Bp_{0.1}$ 125-127°. $[\alpha]_D^{25}$ +3.75° (c, 1.975 in $CHCl_3$).

(S)-(Z)-form

Me ester: [25227-40-1].

Sex pheromone for the Khapra beetle *Trogoderma inclusum*. Also weak attractant for *T. granarium*.

(E)-form

Me ester: [56941-91-4].

Component of the sex pheromone of *T. glabrum*.

Rodin, J.O. *et al, Science (Washington, D.C.)*, 1969, **165**, 904.
de Graw, J.I. *et al, J. Org. Chem.*, 1971, **36**, 2902 (*synth*)
Levinson, H.Z. *et al, Naturwissenschaften*, 1974, **61**, 685.
Mori, K., *Tetrahedron*, 1974, **30**, 3817 (*synth, ir, pmr, ms*)
Yarger, R.G. *et al, J. Chem. Ecol.*, 1975, **1**, 323.

14-Methyl-8-hexadecen-1-ol, 9CI **M-00087**

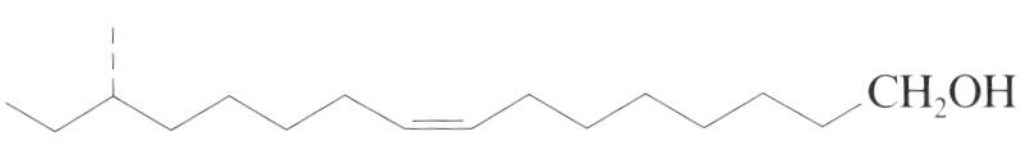

(R,Z)-form

$C_{17}H_{34}O$ M 254.455

(R)-(Z)-form [25227-39-8]

Sex pheromone of the Khapra beetle (*Trogoderma inclusum*) and *T. granarium*.

(S)-(Z)-form [51018-58-7]

$Bp_{0.1}$ 150-151°, $Bp_{0.2}$ 132-133°.

(S)-(E)-form [66007-18-9]

$Bp_{0.03}$ 108-109°. $[\alpha]_D^{25}$ +5.52° (c, 2.823 in $CHCl_3$). An (*E*)-form is reported to be a sex pheromone of *T. glabrum*.

Rodin, J.O. *et al, Science (Washington, D.C.)*, 1969, **165**, 904.
Levinson, H.Z. *et al, Naturwissenschaften*, 1974, **61**, 685.
Mori, K., *Tetrahedron*, 1974, **30**, 3817 (*synth, abs config, pmr, ir, ms*)
Yarger, R.G. *et al, J. Chem. Ecol.*, 1975, **1**, 323.
Rossi, R. *et al, Tetrahedron*, 1977, **33**, 2447 (*synth, abs config, ir, ms, pmr*)

3-[(13-Methylhexadecyl)oxy]-1,2-propanediol, 9CI **M-00088**

1-O-(13-Methylhexadecyl)glycerol

[88948-43-0]

$H_3CCH_2CH_2CH(CH_3)(CH_2)_{12}OCH_2CH(OH)CH_2OH$

$C_{20}H_{42}O_3$ M 330.550

Isol. from a marine sponge *Aaptos* sp.

Do Muu, N. *et al, Tetrahedron Lett.*, 1983, **24**, 5699 (*isol, synth, ms, pmr, cmr*)

Methyl α-linolenate hydroperoxides **M-00089**

Methyl hydroperoxidooctadecatrienoates

$H_3CCH_2CH{=}CHCH_2CH{=}CHCH{=}CHCH(OOH)$
$(CH_2)_7COOCH_3$ (9-isomer)

$C_{19}H_{32}O_4$ M 324.459

Reaction with singlet oxygen produces the 9-, 10-, 12-, 13-, 15- and 16-hydroperoxides. Free radical autoxidn. produces the 9-, 12-, 13- and 16-hydroperoxides. The 13-isomer is formed stereospecifically (13*S*-) by lipoxygenase which occurs in flax seed and cabbage leaves.

Frankel, E.N. *et al, J. Org. Chem.*, 1961, **26**, 4663.
Cobern, D. *et al, J. Chem. Soc. C*, 1966, 1897.
Chan, H.W.-S., *J. Am. Oil Chem. Soc.*, 1977, **54**, 100; *Lipids*, 1977, **12**, 837.
Frankel, E.N. *et al, Lipids*, 1977, **12**, 1055.
Terao, J. *et al, J. Am. Oil Chem. Soc.*, 1977, **54**, 234.

Methyl δ-linolenate hydroperoxides **M-00090**

$H_3C(CH_2)_4CH{=}CHCH_2CH{=}CHCH{=}CHCH(OOH)$
$(CH_2)_4COOMe$ (6-isomer)

$C_{19}H_{28}O_4$ M 320.428

Singlet oxygen gives the 6-, 7-, 9-, 10-, 12-, and 13-hydroperoxides. Free radical autoxidn. gives 6-, 9-, 10- and 12-hydroperoxides. In each case the prod. is a mixt. of configurational and positional isomers. The (6*Z*,10*E*,12*Z*)-9-isomer and the (6*Z*,9*Z*,11*E*)-13-isomer can be obt. in an enantiomeric form using soybean lipoxygenase.

Porter, N.A. *et al, J. Org. Chem.*, 1975, **40**, 3614.
Funk, M.O. *et al, Lipids*, 1976, **11**, 113.

2-Methyl-6-methylene-1,7-octadiene, 9CI **M-00091**

α-Myrcene

[1686-30-2]

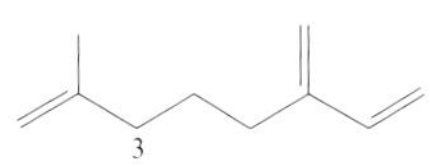

$C_{10}H_{16}$ M 136.236

Found in many essential oils, usually as a mixt. with *β*-myrcene. Intermed. in manuf. of terpene alcohols for perfumery etc. Oil. Bp_{10} 44°.

3-Hydroperoxy: 3-Hydroperoxy-7-methyl-6-methylene-1,7-octadiene. ***α-Myrcene hydroperoxide***

$C_{10}H_{16}O_2$ M 168.235

Constit. of *Artemisia annua*. Unstable oil.

Wrolstad, R.E. *et al*, *J. Agric. Food Chem.*, 1964, **12**, 507 (*ir*)
Ohloff, G. *et al*, *Justus Liebigs Ann. Chem.*, 1964, **675**, 83 (*synth, spectra*)
Ryhage, R. *et al*, *Acta Chem. Scand.*, 1965, **17**, 2025 (*ms*)
Mitzner, B.M. *et al*, *J. Org. Chem.*, 1965, **30**, 646 (*synth, spectra*)
Karrer, W. *et al*, *Konstitution und Vorkommen der Organischen Pflanzenstoffe*, 2nd Ed., Birkhäuser Verlag, Basel, 1972-1985, no. 32 (*occur*)
Rücker, G. *et al*, *J. Nat. Prod.* (*Lloydia*), 1987, **50**, 287 (*hydroperoxide*)

3-Methyl-6-(1-methylethenyl)-3,9-decadien-1-ol, 9CI — M-00092

6-Isopropenyl-3-methyl-3,9-decadien-1-ol
[66348-55-8]

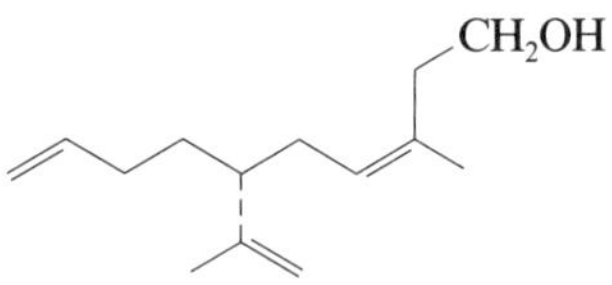

(3*Z*,6*R*)-*form*

$C_{14}H_{24}O$ M 208.343

(3Z,6R)-form [72257-34-2]
$[\alpha]_D$ −6° (c, 1.7 in $CHCl_3$).
Ac: [67225-47-2].
$C_{16}H_{26}O_2$ M 250.380
Constit. of the sex pheromone of California red scale (*Aonidiella aurantii*). $[\alpha]_D$ −7.14° (c, 2.1 in $CHCl_3$).

[64309-04-2, 67169-08-8, 67225-48-3, 67225-49-4, 67225-50-7, 67225-51-8, 72244-15-6]

Roelofs, W.L. *et al*, *Nature* (*London*), 1977, **267**, 698 (*isol, activity, synth, pmr, glc*)
Cooke, M.P. *et al*, *J. Org. Chem.*, 1982, **47**, 4955 (*synth, pmr, cmr, ir, glc*)
Oppolzer, W. *et al*, *Tetrahedron Lett.*, 1986, **27**, 1139 (*synth*)
Becker, D. *et al*, *Tetrahedron*, 1988, **44**, 4541 (*synth*)
Baudouy, R. *et al*, *Tetrahedron*, 1989, **45**, 2067 (*synth, pmr, cmr, ir, ms*)

2-Methylnonacosane — M-00093

Isotriacontane. Melissane

$(H_3C)_2CH(CH_2)_{26}CH_3$

$C_{30}H_{62}$ M 422.820
Leaflets. Mp 73-74°. $Bp_{0.3}$ 222°.

Levene, P.A. *et al*, *J. Biol. Chem.*, 1915, **20**, 533.

18-Methyl-5,8,11,14-nonadecatetraenoic acid — M-00094

[81607-72-9]

$(H_3C)_2CHCH_2CH_2CH{=}CHCH_2CH{=}CH$
$CH_2CH{=}CHCH_2CH{=}CH(CH_2)_3COOH$

$C_{20}H_{32}O_2$ M 304.472
Obt. from *Phytomonas* and *Herpatomonas* flagellates. Probably (all-*Z*)-form.

Fish, W.R. *et al*, *Mol. Biochem. Parasitol.*, 1982, **5**, 1; *CA*, **96**, 177577 (*ms, pmr*)

18-Methyl-8,11,14-nonadecatrienoic acid — M-00095

[81607-71-8]

$(H_3C)_2CHCH_2CH_2CH{=}CHCH_2CH{=}CHCH_2CH{=}CH(CH_2)_6COOH$

$C_{20}H_{34}O_2$ M 306.487
Obt. from *Phytomonas* and *Herpatomonas* flagellates. Probably (all-*Z*)-form.

Fish, W.R. *et al*, *Mol. Biochem. Parasitol.*, 1982, **5**, 1; *CA*, **96**, 177577 (*ms, pmr*)

2-Methylnonanoic acid, 9CI — M-00096

2-Methylpelargonic acid. Nonane-2-carboxylic acid
[24323-21-5]

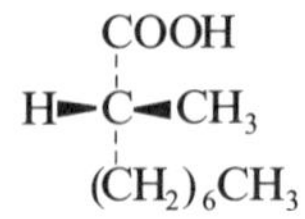

(*R*)-*form*

$C_{10}H_{20}O_2$ M 172.267

(R)-form
Bp_1 115°. $[\alpha]^{25}_D$ −8.72° (Et_2O).
Et ester:
$C_{12}H_{24}O_2$ M 200.320
$[\alpha]^{25}_D$ −8.60°.

(S)-form
Bp_4 145-147°. $[\alpha]^{25}_D$ +8.34°.
Chloride:
$C_{10}H_{19}ClO$ M 190.712
Bp_1 73-74°. $[\alpha]^{25}_D$ +4.89°.
Amide:
$C_{10}H_{21}NO$ M 171.282
Cryst. (EtOH aq.). Mp 78°. $[\alpha]^{25}_D$ +7.07° (95% EtOH).
Nitrile: 2-Cyanononane
$C_{10}H_{19}N$ M 153.267
Bp_7 85-94°. $[\alpha]^{25}_D$ +13.61°.

(±)-form
Bp 261-265°.
Amide: Needles (EtOH). Mp 76°.
Nitrile: Bp_{10} 100°.

Guerbot, M., *C. R. Hebd. Seances Acad. Sci.*, 1902, **135**, 174 (*synth*)
Levene, P.A. *et al*, *J. Biol. Chem.*, 1929, **84**, 571; 1932, **98**, 1 (*resoln*)

3-Methylnonanoic acid, 9CI — M-00097

3-Methylpelargonic acid
[35205-79-9]

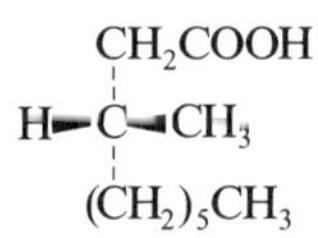

(*R*)-*form*

$C_{10}H_{20}O_2$ M 172.267

(R)-form [52075-16-8]
Bp_1 105-106°. $[\alpha]^{24}_D$ +5.25° (neat).
Et ester:
$C_{12}H_{24}O_2$ M 200.320
Bp_{36} 135°.

(±)-form
Bp_{12} 147-148°.
Et ester: Bp_{13} 115°.
p-*Toluidide:* Mp 76-77°.

[59614-86-7]

Levene, P.A. *et al*, *J. Biol. Chem.*, 1922, **54**, 351; 1931, **91**, 77 (*synth*)
Cymerman, J. *et al*, *J. Chem. Soc.*, 1944, 144 (*synth*)
Meyers, A.I. *et al*, *J. Am. Chem. Soc.*, 1976, **98**, 2290 (*synth*)
Meyers, A.I. *et al*, *J. Org. Chem.*, 1979, **44**, 2247, 2250.
Jackman, L.M. *et al*, *J. Org. Chem.*, 1983, **48**, 4789 (*synth*)

7-Methyl-6-nonen-3-ol M-00098

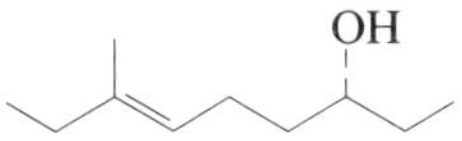

(R,E)-form

$C_{10}H_{20}O$ M 156.267

(R,E)-form [123807-63-6]

Bp_{18} 99-102°. $[\alpha]_D^{21}$ −10.8° (c, 0.30 in $CHCl_3$). n_D^{21} 1.4489. *ca.* 100% opt. pure.

Ac: [100429-35-4]. ***Quadrilure***

$C_{12}H_{22}O_2$ M 198.305

Aggregation pheromone of the square-necked grain beetle *Cathartus quadricollis*. Bp_{13} 63-64°. $[\alpha]_D^{20}$ +9.59° (c, 1.09 in $CHCl_3$). n_D^{20} 1.4331.

(S,E)-form [100429-33-2]

Oil. Bp_{14} 94-95°. $[\alpha]_D^{21}$ +10.4° (c, 0.96 in $CHCl_3$). n_D^{21} 1.4491. *ca.* 100% opt. pure.

Ac: [100429-32-1].

Bp_{12} 54-55°. $[\alpha]_D^{20}$ −9.05° (c, 1.05 in $CHCl_3$). n_D^{20} 1.4332.

[122923-49-3]

Johnston, B.D. *et al*, *J. Org. Chem.*, 1986, **51**, 760 (*synth, pmr, ir, ms*)
Kocienski, P. *et al*, *Tetrahedron Lett.*, 1988, **29**, 2353 (*synth*)
Mori, K. *et al*, *Justus Liebigs Ann. Chem.*, 1990, 159 (*synth, ir, pmr, cmr, ord, bibl*)
Tomioka, H. *et al*, *Biosci., Biotechnol., Biochem.*, 1992, **56**, 999 (*synth*)

22-Methyl-5,9-octacosadienoic acid M-00099

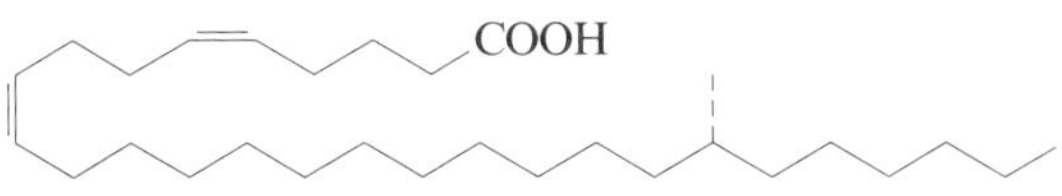

$C_{29}H_{54}O_2$ M 434.744

(5Z,9Z,22R)-form [108644-87-7]

Isol. from the marine sponge *Aplysina fistularis*. $[\alpha]_D^{20}$ −0.12° (c, 6.7 in $CHCl_3$).

[79605-33-7]

Walkup, R.D. *et al*, *Lipids*, 1981, **16**, 631 (*isol*)
Raederstorff, D. *et al*, *J. Org. Chem.*, 1987, **52**, 2337 (*synth, abs config*)

2-Methyloctadecanoic acid, 9CI M-00100

2-Methylstearic acid

[7217-83-6]

$H_3C(CH_2)_{15}CH(CH_3)COOH$

$C_{19}H_{38}O_2$ M 298.508

Occurs in the uropygial gland of the night heron and in *Corvus frugilegus*, probably in an enantiomeric form.

(R)-form

Mp 54.7-55°. $[\alpha]_D^{25}$ −9.4° (c, 9.08 in $CHCl_3$).

Me ester: $[\alpha]_D^{25}$ −12.1° (c, 7.54 in $CHCl_3$).

(S)-form

Mp 54.7-55°. $[\alpha]_D^{25}$ +9.3° (c, 8.63 in $CHCl_3$).

Me ester: $[\alpha]_D^{25}$ +12.1° (c, 7.17 in $CHCl_3$).

(±)-form

Has weak larvicidal activity in the mosquito. Mp 55°, Mp 68°. $Bp_{0.1}$ 135-140°.

Amide: Cryst. (MeOH). Mp 105°.

Morgan, G.T. *et al*, *J. Soc. Chem. Ind., London*, 1927, **46**, 152T (*synth*)
Weitzel, G., *Hoppe Seyler's Z. Physiol. Chem.*, 1950, **285**, 220 (*synth*)
Ställberg, G., *Ark. Kemi*, 1958, **12**, 153 (*synth, ir*)
Hwang, Y.-S. *et al*, *J. Agric. Food Chem.*, 1974, **22**, 400.
Jacob, J., *Hoppe Seyler's Z. Physiol. Chem.*, 1975, **356**, 1823.

3-Methyloctadecanoic acid, 9CI M-00101

3-Methylstearic acid

[52304-07-1]

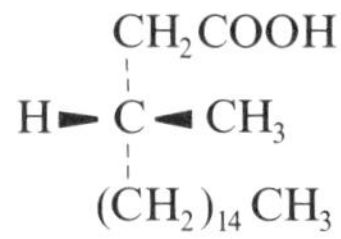

(R)-form

$C_{19}H_{38}O_2$ M 298.508

(R)-form

Mp 44°. $[\alpha]_D^{20}$ +4.84°.

Amide:

$C_{19}H_{39}NO$ M 297.523

Mp 91.5°.

(S)-form

Mp 44°. $[\alpha]_D^{20}$ −4.75°.

Amide: Mp 91.6°.

(±)-form

Has high larvicidal activity against the mosquito. Mp 51.0-51.5°.

Me ester: [45267-60-5].

$C_{20}H_{40}O_2$ M 312.535

Has high larvicidal activity against the mosquito. $Bp_{0.4}$ 148-149°.

Amide: Mp 93.5-94.0°.

Ställberg-Stenhagen, S., *Ark. Kemi*, 1949, **1**, 187.
Weitzel, G., *Hoppe Seyler's Z. Physiol. Chem.*, 1951, **287**, 210 (*synth*)
Hwang, Y.-S. *et al*, *J. Agric. Food Chem.*, 1978, **26**, 557.

4-Methyloctadecanoic acid, 9CI M-00102

4-Methylstearic acid

[63060-52-6]

$H_3C(CH_2)_{13}CH(CH_3)CH_2CH_2COOH$

$C_{19}H_{38}O_2$ M 298.508

(±)-form

Amide:

$C_{19}H_{39}NO$ M 297.523

Mp 79.5-80.0°.

Weitzel, G., *Hoppe Seyler's Z. Physiol. Chem.*, 1951, **287**, 310 (*synth*)
Ackman, R.G. *et al*, *Lipids*, 1972, **7**, 683.

5-Methyloctadecanoic acid M-00103

5-Methylstearic acid

$H_3C(CH_2)_{12}CH(CH_3)(CH_2)_3COOH$

$C_{19}H_{38}O_2$ M 298.508

(±)-form

Mp 44.0-44.5°. $Bp_{0.1}$ 174°.

Amide:

$C_{19}H_{39}NO$ M 297.523

Mp 79.0-79.5°.

Cason, J. *et al*, *J. Org. Chem.*, 1950, **15**, 139 (*synth*)
Weitzel, G., *Hoppe Seyler's Z. Physiol. Chem.*, 1951, **287**, 310.

6-Methyloctadecanoic acid M-00104

6-Methylstearic acid

$H_3C(CH_2)_{11}CH(CH_3)(CH_2)_4COOH$

$C_{19}H_{38}O_2$ M 298.508

(±)-*form*

Mp 40.0-40.5°. $Bp_{0.1}$ 172°.

Amide:

$C_{19}H_{39}NO$ M 297.523

Mp 79.5-80.0°.

Cason, J. *et al*, *J. Org. Chem.*, 1950, **15**, 139 (*synth*)
Weitzel, G., *Hoppe Seyler's Z. Physiol. Chem.*, 1951, **287**, 310 (*synth*)
Ackman, R.G. *et al*, *Lipids*, 1972, **7**, 683.

7-Methyloctadecanoic acid M-00105

7-Methylstearic acid

$H_3C(CH_2)_{10}CH(CH_3)(CH_2)_5COOH$

$C_{19}H_{38}O_2$ M 298.508

(±)-*form*

Mp 37°. $Bp_{0.1}$ 173°.

Amide:

$C_{19}H_{39}NO$ M 297.523

Mp 70°.

Cason, J. *et al*, *J. Org. Chem.*, 1950, **15**, 139 (*synth*)
Weitzel, G., *Hoppe Seyler's Z. Physiol. Chem.*, 1951, **287**, 310 (*synth*)

8-Methyloctadecanoic acid, 8CI M-00106

8-Methylstearic acid

[17670-89-2]

$H_3C(CH_2)_9CH(CH_3)(CH_2)_6COOH$

$C_{19}H_{38}O_2$ M 298.508

(±)-*form*

Mp 33.5°. $Bp_{0.1}$ 174°.

Amide:

$C_{19}H_{39}NO$ M 297.523

Mp 71.0-72.6°.

Cason, J. *et al*, *J. Org. Chem.*, 1950, **15**, 135 (*synth*)
Weitzel, G., *Hoppe Seyler's Z. Physiol. Chem.*, 1951, **287**, 310 (*synth*)
Ackman, R.G. *et al*, *Lipids*, 1972, **7**, 683.

9-Methyloctadecanoic acid, 9CI M-00107

9-Methylstearic acid

$H_3C(CH_2)_8CH(CH_3)(CH_2)_7COOH$

$C_{19}H_{38}O_2$ M 298.508

(±)-*form* [37963-53-4]

Mp 38.5-39.0°. $Bp_{0.1}$ 171°.

Amide:

$C_{19}H_{39}NO$ M 297.523

Mp 69.5-70.0°.

Cason, J. *et al*, *J. Org. Chem.*, 1950, **15**, 139 (*synth*)
Weitzel, G., *Hoppe Seyler's Z. Physiol. Chem.*, 1951, **287**, 310 (*synth*)

10-Methyloctadecanoic acid, 9CI M-00108

10-Methylstearic acid. Tuberculostearic acid

[542-47-2]

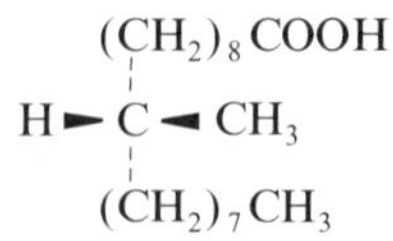

(*R*)-*form*

$C_{19}H_{38}O_2$ M 298.508

(*R*)-*form*

Isol. from the wax of the tubercle bacillus. Mp 12.8-13.4°. $Bp_{0.7}$ 175-178°. $[\alpha]_D^{22}$ −0.11° (Me_2CO), $[\alpha]_D^{19}$ −0.41° (c, 17 in Py).

Amide:

$C_{19}H_{39}NO$ M 297.523

Mp 75.1-76.3°. $[\alpha]_D^{24}$ −0.09° ($CHCl_3$).

(±)-*form*

Mp 25.5-26.0°.

Amide: Mp 79.0-79.5°.

Prout, F.S. *et al*, *J. Am. Chem. Soc.*, 1948, **70**, 298 (*synth*)
Weitzel, G., *Hoppe Seyler's Z. Physiol. Chem.*, 1951, **287**, 310 (*synth*)
Sy, M. *et al*, *C. R. Hebd. Seances Acad. Sci.*, 1954, **239**, 1813 (*synth*)
Hünig, S. *et al*, *Angew. Chem.*, 1959, **71**, 339 (*synth*)
Jaureguiberry, G. *et al*, *J. Chem. Soc., Chem. Commun.*, 1966, 855 (*biosynth*)
Ackman, R.G. *et al*, *Lipids*, 1972, **7**, 683.
O'Neill, H.J. *et al*, *Trans. Ill. State Acad. Sci.*, 1976, **69**, 344; *CA*, **87**, 196944z (*ms*)
Saito, K., *GC-MS News*, 1978, **6**, 2 (*rev*)

11-Methyloctadecanoic acid M-00109

11-Methylstearic acid

$H_3C(CH_2)_6CH(CH_3)(CH_2)_9COOH$

$C_{19}H_{38}O_2$ M 298.508

(±)-*form* [40811-98-1]

Mp 21°. $Bp_{0.1}$ 174°.

Amide:

$C_{19}H_{39}NO$ M 297.523

Mp 73.5-74.0°.

Cason, J. *et al*, *J. Org. Chem.*, 1950, **15**, 139 (*synth*)
Weitzel, G., *Hoppe Seyler's Z. Physiol. Chem.*, 1951, **287**, 310 (*synth*)

12-Methyloctadecanoic acid, 8CI M-00110

12-Methylstearic acid

$H_3C(CH_2)_5CH(CH_3)(CH_2)_{10}COOH$

$C_{19}H_{38}O_2$ M 298.508

(±)-*form*

Mp 36.5-37.0°. $Bp_{0.6}$ 194°.

Amide:

$C_{19}H_{39}NO$ M 297.523

Mp 87.5-88.0°.

Cason, J. *et al*, *J. Org. Chem.*, 1950, **15**, 135 (*synth*)
Weitzel, G., *Hoppe Seyler's Z. Physiol. Chem.*, 1951, **287**, 310 (*synth*)
Perkins, E.G. *et al*, *Chem. Phys. Lipids*, 1971, **6**, 8 (*synth*, *ms*, *pmr*)

13-Methyloctadecanoic acid M-00111

13-Methylstearic acid

$H_3C(CH_2)_4CH(CH_3)(CH_2)_{11}COOH$

$C_{19}H_{38}O_2$ M 298.508

(±)-form

Mp 30.5-31.0°. $Bp_{0.5}$ 190°.

Amide:

$C_{19}H_{39}NO$ M 297.523

Mp 78°.

Cason, J. *et al*, *J. Org. Chem.*, 1950, **15**, 135 (*synth*)
Weitzel, G., *Hoppe Seyler's Z. Physiol. Chem.*, 1951, **287**, 310 (*synth*)

14-Methyloctadecanoic acid M-00112

14-Methylstearic acid

$H_3C(CH_2)_3CH(CH_3)(CH_2)_{12}COOH$

$C_{19}H_{38}O_2$ M 298.508

(±)-form

Lustrous prisms (Me_2CO). Mp 37.5°. $Bp_{0.4}$ 182°.

Me ester:

$C_{20}H_{40}O_2$ M 312.535

$Bp_{0.9}$ 159-162°.

Amide:

$C_{19}H_{39}NO$ M 297.523

Mp 79.0-79.3°.

Ställberg-Stenhagen, S., *CA*, 1947, **41**, 5444.
Cason, J. *et al*, *J. Org. Chem.*, 1950, **15**, 135 (*synth*)
Weitzel, G., *Hoppe Seyler's Z. Physiol. Chem.*, 1951, **287**, 310 (*synth*)

15-Methyloctadecanoic acid M-00113

15-Methylstearic acid

$H_3CCH_2CH_2CH(CH_3)(CH_2)_{13}COOH$

$C_{19}H_{38}O_2$ M 298.508

(+)-form

Tribromoanilide:

$C_{25}H_{40}Br_3NO$ M 610.309

Mp 104.4-105.5°. $[\alpha]_D^{25}$ +0.49° ($CHCl_3$).

(−)-form

Mp 35.2-36.2°.

Tribromoanilide: Mp 104.4-105.3°. $[\alpha]_D^{25}$ −0.67° ($CHCl_3$).

(±)-form

Mp 40.9-42.4°. $Bp_{0.4}$ 183°.

Amide:

$C_{19}H_{39}NO$ M 297.523

Mp 80.0-80.5°.

Tribromoanilide: Mp 103.6-104.4°.

Cason, J., *J. Am. Chem. Soc.*, 1950, **72**, 4695.
Weitzel, G., *Hoppe Seyler's Z. Physiol. Chem.*, 1951, **287**, 310 (*synth*)

16-Methyloctadecanoic acid, 9CI M-00114

16-Methylstearic acid. Anteisononadecanoic acid

[17001-28-4]

$H_3CCH_2CH(CH_3)(CH_2)_{14}COOH$

$C_{19}H_{38}O_2$ M 298.508

Occurs in butterfat, several woods, tobacco and Baltic salmon. Mp 49.9-50.6°. The natural compd. is probably enantiomeric.

Me ester: [2490-16-6].

$C_{20}H_{40}O_2$ M 312.535

Occurs in *Staphylococcus aureus* membrane. $Bp_{1.2}$ 176°.

Amide:

$C_{19}H_{39}NO$ M 297.523

Mp 93.2°.

Cason, J. *et al*, *J. Am. Chem. Soc.*, 1944, **66**, 46.
Ryhage, R., *J. Dairy Res.*, 1967, **34**, 115 (*ms*)

17-Methyloctadecanoic acid, 9CI M-00115

17-Methylstearic acid. Isononadecanoic acid

[2724-59-6]

$H_3CCH(CH_3)(CH_2)_{15}COOH$

$C_{19}H_{38}O_2$ M 298.508

Occurs in *Microciona prolifera*, seeds of *Asphodelus tennifolus* and Baltic salmon. Mp 67.3-67.8°. $Bp_{0.3}$ 180°.

Me ester:

$C_{20}H_{40}O_2$ M 312.535

Mp 26.0-27.5°. Bp_{1-2} 171-172°.

Amide:

$C_{19}H_{39}NO$ M 297.523

Mp 100.2-101.3°.

Cason, J., *J. Am. Chem. Soc.*, 1942, **64**, 1106.
Weitzel, G., *Hoppe Seyler's Z. Physiol. Chem.*, 1951, **287**, 310 (*synth*)
Golovnya, R.V. *et al*, *Prikl. Biokhim. Mikrobiol.*, 1978, **14**, 609 (*ms*)

3-[(15-Methyloctadecyl)oxy]-1,2-propanediol, 9CI M-00116

[88948-44-1]

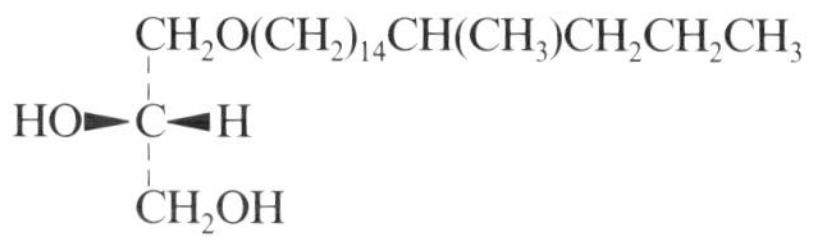

$C_{22}H_{46}O_3$ M 358.604

(2S)-form

1-O-(15-Methylhexadecyl)-sn-glycerol

Isol. from a marine sponge *Aaptos* sp.

Do, M.N. *et al*, *Tetrahedron Lett.*, 1983, **24**, 5699 (*isol, ir, pmr, ms*)

6-Methyl-2,4-octadienoic acid, 9CI M-00117

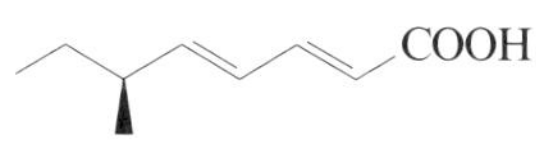

$C_9H_{14}O_2$ M 154.208

(2E,4E,6S)-form [121661-45-8] ***Dendryphiellic acid A***

Metab. of *Dendryphiella salina*. Oil. $[\alpha]_D^{20}$ +52.2° (c, 0.17 in EtOH aq.).

Guerriero, A. *et al*, *Helv. Chim. Acta*, 1989, **72**, 438 (*isol, pmr, cmr, synth*)

2-Methyloctanoic acid, 9CI M-00118

2-Methylcaprylic acid

[3004-93-1]

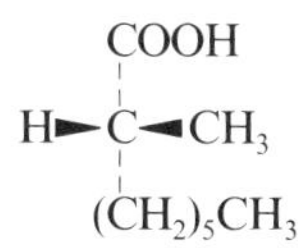

(R)-form

$C_9H_{18}O_2$ M 158.240

(R)-form

$Bp_{0.5}$ 93-95°. $[\alpha]_D$ −16.53° (neat) (97.8% e.e.).

(S)-form [61866-40-8]

$[\alpha]_D$ +16.12° (c, 1.179 in $CHCl_3$) (>97% e.e.).

(±)-form

Bp_{14} 137-139°. n_D^{25} 1.4281.

Me ester: [2177-86-8].

$C_{10}H_{20}O_2$ M 172.267

Mp 70-71°.

Amide:

$C_9H_{19}NO$ M 157.255

Mp 80-81°.

Nitrile: [2570-96-9]. *2-Cyanooctane*

$C_9H_{17}N$ M 139.240

Bp 206°, $Bp_{17.5}$ 95-97°.

Kullhem, H.A., *Justus Liebigs Ann. Chem.*, 1874, **173**, 322 (*synth*)
Cason, J. *et al*, *J. Org. Chem.*, 1953, **18**, 1594 (*synth*)
Sonnet, P.E. *et al*, *Org. Prep. Proced. Int.*, 1990, **22**, 203 (*synth*)

3-Methyloctanoic acid, 9CI M-00119

3-Methylcaprylic acid

[6061-10-5]

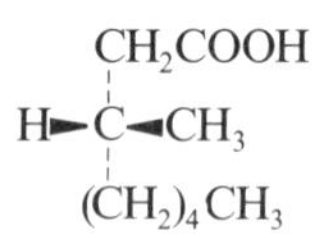

(*R*)-form

$C_9H_{18}O_2$ M 158.240

(*R*)-form

$Bp_{0.5}$ 90-95°. $[\alpha]_D$ +5.50° (c, 1.237 in $CHCl_3$).

Chloride:

$C_9H_{17}ClO$ M 176.685

Bp_{20} 95°. $[\alpha]_D^{24}$ +1.36° (neat).

(*S*)-form

Bp_{16} 135°. $[\alpha]_D$ −6.5° (c, 1.135 in $CHCl_3$).

Et ester:

$C_{11}H_{22}O_2$ M 186.294

Bp_{35} 117°. $[\alpha]_D^{23}$ −2.03° (neat).

Amide:

$C_9H_{19}NO$ M 157.255

Mp 85.6-86.0°. $[\alpha]_D^{25}$ −5.03° (EtOH).

(±)-form

Bp 234-236°, Bp_{14} 130-131°.

S-*Benzylisothiouronium salt:* Cryst. (EtOH aq.). Mp 138°.

Et ester: Bp_{18} 100-102°, $Bp_{0.1}$ 93-95°.

Amide: Mp 82-84.5°.

Helg, R. *et al*, *Helv. Chim. Acta*, 1956, **39**, 1269 (*synth*)
Prout, F.S. *et al*, *J. Org. Chem.*, 1962, **27**, 1488 (*synth*)
Descotes, G. *et al*, *Bull. Soc. Chim. Fr.*, 1968, 3395 (*synth*)
Meyers, A.I. *et al*, *J. Am. Chem. Soc.*, 1975, **97**, 6266 (*abs config*)
Sonnet, P.E. *et al*, *Org. Prep. Proced. Int.*, 1990, **22**, 203 (*synth*)

4-Methyloctanoic acid, 9CI M-00120

4-Methylcaprylic acid

[54947-74-9]

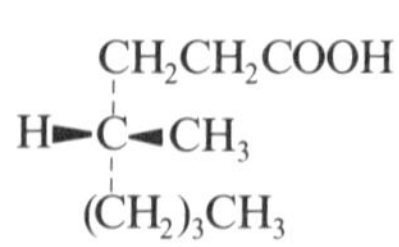

(*R*)-form

$C_9H_{18}O_2$ M 158.240

(*R*)-form

Bp_{25} 166-172°. $[\alpha]_D$ −1.53° (c, 1.114 in $CHCl_3$) (*ca.* 95% ee).

(*S*)-form

$[\alpha]_D$ +1.45° (c, 1.106 in $CHCl_3$) (*ca.* 95% ee).

(±)-form

Bp_{10} 140-142°, Bp_2 110.5-111.5°.

S-*Benzylthiouronium salt:* Needles. Mp 135°.

Et ester: [56196-53-3].

$C_{11}H_{22}O_2$ M 186.294

Bp_7 140°, Bp_2 108°.

Levene, P.A. *et al*, *J. Org. Chem.*, 1936, **1**, 76 (*abs config*)
Weitzel, G. *et al*, *Hoppe Seyler's Z. Physiol. Chem.*, 1951, **287**, 65 (*synth*)
Linstead, R.P. *et al*, *J. Chem. Soc.*, 1953, 1538 (*synth*)
Kenyon, J. *et al*, *J. Chem. Soc.*, 1953, 2129 (*synth*)
Wynberg, H. *et al*, *J. Am. Chem. Soc.*, 1960, **82**, 1447 (*synth*)
Sonnet, P.E. *et al*, *Org. Prep. Proced. Int.*, 1990, **22**, 203 (*synth*)

5-Methyloctanoic acid, 9CI M-00121

5-Methylcaprylic acid

[60218-42-0]

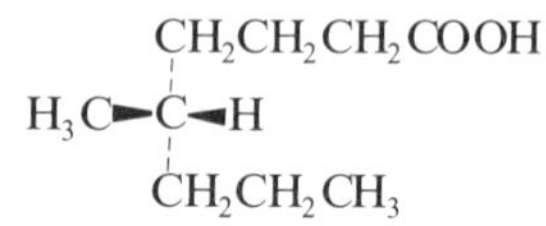

(*S*)-form

$C_9H_{18}O_2$ M 158.240

(*S*)-form

Bp_5 127°. $[\alpha]_D^{25}$ −0.37° (neat).

Et ester: Bp_{30} 112°.

(±)-form

Oil. Bp_{30} 150°.

4-Bromophenacyl ester: Plates (EtOH aq.). Mp 46.0-46.5°.

Levene, P.A. *et al*, *J. Biol. Chem.*, 1932, **95**, 162 (*synth*)
Levene, P.A. *et al*, *J. Org. Chem.*, 1936, **1**, 76 (*abs config*)
Ames, D.E. *et al*, *J. Chem. Soc.*, 1950, 329 (*synth*)

6-Methyloctanoic acid, 9CI M-00122

6-Methylcaprylic acid. Isopelargonic acid

[504-99-4]

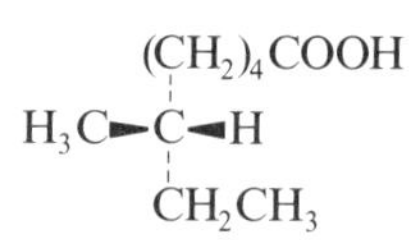

(*S*)-form

$C_9H_{18}O_2$ M 158.240

(*S*)-form [928-30-3]

Degradn. prod. from Polymyxins. Bp_{21} 148-149°. $[\alpha]_D^{19}$ +7.9°.

Et ester: [16177-02-9].

$C_{11}H_{22}O_2$ M 186.294

Bp_{25} 110°.

4-Bromophenacyl ester: Plates (EtOH aq.). Mp 58-59°.

Amide:

$C_9H_{19}NO$ M 157.255

Mp 90-91°.

(±)-form

Bp_{23} 148-150°. n_D^{20} 1.4337.

Me ester:

$C_{10}H_{20}O_2$ M 172.267

$Bp_{1.95}$ 60°.

4-Bromophenacyl ester: Plates (EtOH aq.). Mp 57-58°.

Amide: Plates (EtOH/pet. ether). Mp 91°.

Levene, P.A. *et al*, *J. Biol. Chem.*, 1932, **95**, 163; 1933, **103**, 299 (*synth*)
Cason, J. *et al*, *J. Am. Chem. Soc.*, 1944, **66**, 46 (*synth*)
Wilkinson, S. *et al*, *Nature (London)*, 1949, **164**, 622; 1963, **200**, 1008 (*isol*)
Crombie, L. *et al*, *J. Chem. Soc.*, 1950, 2685 (*synth, abs config*)
Hayashi, K. *et al*, *Experientia*, 1966, **22**, 354 (*isol*)

7-Methyloctanoic acid, 9CI M-00123

7-Methylcaprylic acid. Isononanoic acid

[693-19-6]

$(H_3C)_2CH(CH_2)_5COOH$

$C_9H_{18}O_2$ M 158.240

V. spar. sol. H_2O. Bp 248°, Bp_2 132-134°.

Et ester: [29490-61-7].

$C_{11}H_{22}O_2$ M 186.294

Bp 220.5°, Bp_{12} 102°.

Amide:

$C_9H_{19}NO$ M 157.255

Cryst. Mp 106.5°.

Levene, P.A. *et al*, *J. Biol. Chem.*, 1916, **27**, 433 (*synth*)
Cason, J., *J. Am. Chem. Soc.*, 1942, **64**, 1106 (*synth*)

2-Methyl-4-oxoheptadecanoic acid M-00124

Lichesterylic acid

[29595-87-7]

$H_3C(CH_2)_{12}COCH_2CH(CH_3)COOH$

$C_{18}H_{34}O_3$ M 298.465

(±)-form

Cryst. (AcOH). Mp 81-83°.

Semicarbazone: Mp 126°.

Kaufmann, H.P. *et al*, *J. Prakt. Chem.*, 1962, **15**, 121 (*synth*)
Bloomer, J.L. *et al*, *J. Chem. Soc. C*, 1970, 1848.

16-Methyl-10-oxoheptadecanoic acid M-00125

10-Oxoisostearic acid. 10-Ketoisostearic acid

$(H_3C)_2CH(CH_2)_5CO(CH_2)_8COOH$

$C_{18}H_{34}O_3$ M 298.465

Cryst. (EtOAc). Mp 71-72°.

Fordyce, C.R. *et al*, *J. Am. Chem. Soc.*, 1933, **55**, 3368.
McGhie, J.F. *et al*, *J. Chem. Soc.*, 1962, 350.

7-Methyl-4-oxooctanoic acid M-00126

$(H_3C)_2CHCH_2CH_2COCH_2CH_2COOH$

$C_9H_{16}O_3$ M 172.224

Mp 47-49.5°.

Me ester:

$C_{10}H_{18}O_3$ M 186.250

Bp_{22} 145°, Bp_8 117°.

Cronnyn, M.W. *et al*, *J. Am. Chem. Soc.*, 1952, **94**, 1225 (*synth*)

7-Methyl-6-oxooctanoic acid M-00127

[59210-01-4]

$(H_3C)_2CHCO(CH_2)_4COOH$

$C_9H_{16}O_3$ M 172.224

Oil. $Bp_{0.05}$ 99-101°.

Kaga, H. *et al*, *Synthesis*, 1989, 864 (*synth, ir, pmr*)

2-Methyl-4-oxopentadecanoic acid, 9CI M-00128

Nephrosterylic acid

[75232-41-6]

$H_3C(CH_2)_{10}COCH_2CH(CH_3)COOH$

$C_{16}H_{30}O_3$ M 270.411

Present in several spp. of lichen. Shows antitumour activity. Mp 78°.

Nitrile:

$C_{16}H_{29}NO$ M 251.411

Mp 24°.

Kaufmann, H.P. *et al*, *Fette, Seifen, Anstrichm.*, 1962, **64**, 434; *CA*, **57**, 4783a (*chromatog*)
Kaufmann, H.P. *et al*, *J. Prakt. Chem.*, 1962, **15**, 121 (*synth, nitrile*)
Hinayama, T. *et al*, *Yakugaku Zasshi*, 1980, **100**, 755 (*isol*)

14-Methyl-10-oxopentadecanoic acid M-00129

10-Oxoisopalmitic acid. 10-Ketoisopalmitic acid

$(H_3C)_2CH(CH_2)_3CO(CH_2)_8COOH$

$C_{16}H_{30}O_3$ M 270.411

Cryst. (EtOAc). Mp 68-69°.

Semicarbazone: Mp 108-111°.

Fordyce, C.R. *et al*, *J. Am. Chem. Soc.*, 1933, **55**, 3370.
Akiya, S. *et al*, *Yakugaku Zasshi*, 1957, **76**, 1401.

12-Methyl-10-oxotridecanoic acid M-00130

10-Oxoisomyristic acid. 10-Ketoisomyristic acid

$(H_3C)_2CHCH_2CO(CH_2)_8COOH$

$C_{14}H_{26}O_3$ M 242.358

Cryst. (pet. ether). Mp 54-55°.

Fordyce, C.R. *et al*, *J. Am. Chem. Soc.*, 1933, **55**, 3368.

14-Methylpentadecanoic acid, 9CI M-00131

Isopalmitic acid

[4669-02-7]

$(H_3C)_2CH(CH_2)_{12}COOH$

$C_{16}H_{32}O_2$ M 256.428

Occurs in many natural systems including tall oil, butterfat, adipose tissue, pine resins and bacteria such as *Mycobacterium phlei* and *Arthrobacter simplex*. Mp 61.8-62.4°.

Amide:

$C_{16}H_{33}NO$ M 255.443

Mp 102°.

Fordyce, C.R. *et al*, *J. Am. Chem. Soc.*, 1933, **55**, 3370 (*synth*)
Fujii, K. *et al*, *Eur. J. Biochem.*, 1970, **17**, 552 (*ms*)

2-Methyl-1-pentadecanol M-00132

[25354-98-7]

$H_3C(CH_2)_{12}CH(CH_3)CH_2OH$

$C_{16}H_{34}O$ M 242.444

Minor component of the wax alcohols of the uropygial gland wax of tufted duck and night heron. Mp 12-13°. Bp_{15} 185-187°.

Saus, A. *et al*, *Tenside Deterg.*, 1969, **6**, 129 (*synth*)
Jacob, J. *et al*, *Z. Naturforsch., B*, 1970, **25**, 1438 (*glc, ms*)
Jacob, J., *Hoppe Seyler's Z. Physiol. Chem.*, 1975, **356**, 1823 (*ms*)

3-Methyl-1-pentadecanol M-00133

$H_3C(CH_2)_{11}CH(CH_3)CH_2CH_2OH$

$C_{16}H_{34}O$ M 242.444

(±)-form

Bp_9 171-174°.

Cason, J. *et al*, *J. Org. Chem.*, 1953, **18**, 842.

3-Methyl-5-pentyl-2-furannonanoic acid, 9CI M-00134

10,13-Epoxy-11-methyl-10,12-octadecadienoic acid. F_2 Acid

[57818-39-0]

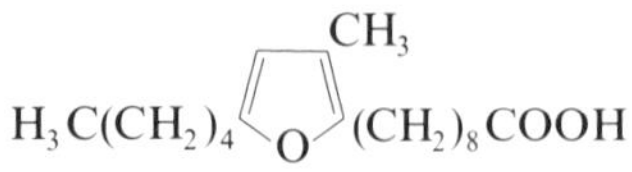

$C_{19}H_{32}O_3$ M 308.460

Comprises 90% of triacylglycerols of *Hevea brasiliensis* (rubber) latex, a minor component among fish furanoid acids. Component of F acid fraction present in human and beef blood serum.

Glass, R.L. *et al, Lipids*, 1977, **12**, 828 (*occur*)
Hasma, H. *et al, Lipids*, 1978, **13**, 905 (*occur, ms, nmr*)
Lie Ken Jie, M.S.F. *et al, Chem. Phys. Lipids*, 1981, **28**, 99 (*cmr*)
Puchta, V. *et al, Justus Liebigs Ann. Chem.*, 1988, 25 (*occur*)
Batna, A. *et al, Justus Liebigs Ann. Chem.*, 1991, 861 (*biosynth*)

3-Methyl-5-pentyl-2-furantridecanoic acid, 9CI M-00135

F_7 Acid

[57818-42-5]

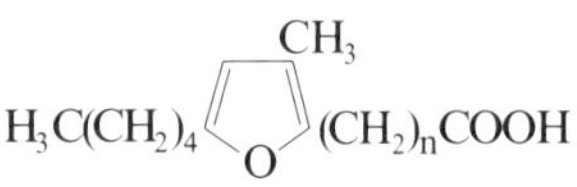

$n = 12$

$C_{23}H_{40}O_3$ M 364.567

Present in fish oils.

Glass, R.L. *et al, Lipids*, 1974, **9**, 1004; 1975, **10**, 695; 1977, **12**, 828.
Gunstone, F.D. *et al, J. Sci. Food Agric.*, 1978, **29**, 539.
Scheinkönig, J. *et al, Justus Liebigs Ann. Chem.*, 1991, 451 (*biosynth*)

3-Methyl-5-pentyl-2-furanundecanoic acid, 9CI M-00136

F_5 Acid

[57818-37-8]

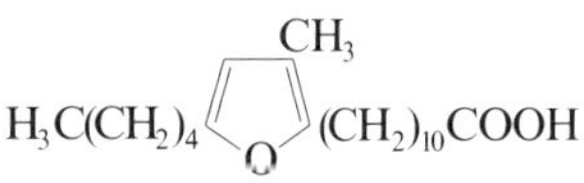

$C_{21}H_{36}O_3$ M 336.514

Constit. of fats of the liver and gonads of fishes, e.g. pike (*Esox lucius*).

Glass, R.L. *et al, Lipids*, 1974, **9**, 1004; 1975, **10**, 695; 1977, **12**, 828.
Scheinkönig, J. *et al, Justus Liebigs Ann. Chem.*, 1991, 451 (*biosynth*)

3-Methyl-5-propyl-2-furannonanoic acid M-00137

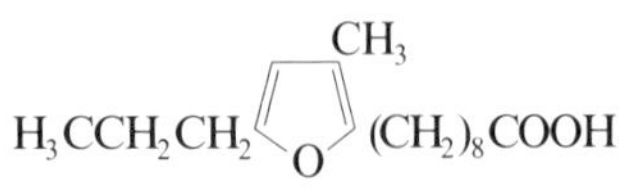

$C_{17}H_{28}O_3$ M 280.406

Et ester:

$C_{19}H_{32}O_3$ M 308.460

Isol. from the brown alga *Acrocarpia paniculata*. Viscous liq.

Kazlauskas, R. *et al, Aust. J. Chem.*, 1982, **35**, 165.

7-Methyl-3-propyl-2,6-nonadien-1-ol M-00138

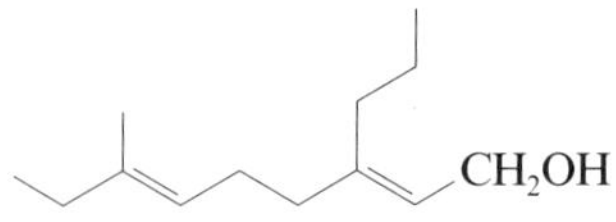

$C_{13}H_{24}O$ M 196.332

(2E,6E)-form

Ac: [142947-40-8].

$C_{15}H_{26}O_2$ M 238.369

Constit. of *Cathartus quadricollis*. Oil.

[119342-00-6]

Pierce, H.D. *et al, J. Chem. Ecol.*, 1988, **14**, 2169 (*isol*)
Dodd, D.S. *et al, J. Org. Chem.*, 1992, **57**, 5250 (*isol, pmr, cmr, ir, ms, synth*)

4-Methylstigmasta-7,24(28)-dien-3-ol M-00139

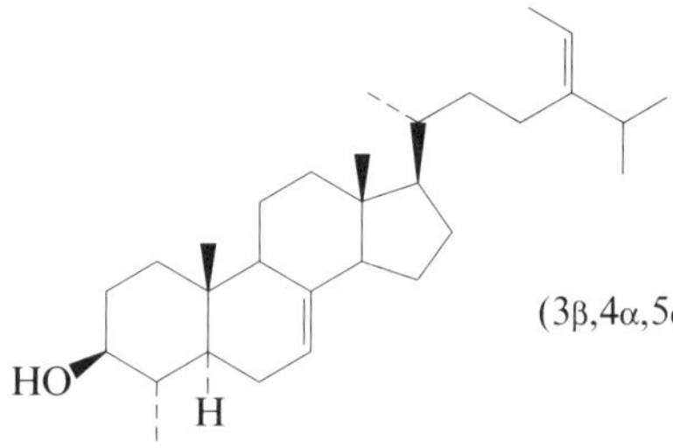

(3β,4α,5α,24*E*)-*form*

$C_{30}H_{50}O$ M 426.724

(3β,4α,5α,24E)-form [39006-52-5] ***Epicitrostadienol***

Constit. of marigold flowers. Cryst. (MeOH).

(3β,4α,5α,24Z)-form [474-40-8]

α_1-Sitosterol. Citrostadienol. 24-Ethylidenelophenol

Constit. of *Solanum tuberosum, S. demissum, Betula verrucosa, Populus* spp., *Pisum sativum*, citrus oils and other plants. Cryst. (Et_2O/EtOH). Mp 162-164°. $[\alpha]_D$ +24° (c, 1 in $CHCl_3$).

(3β,4β,5α,24Z)-form [21490-23-3]

Constit. of *Calendula officinalis*. Cryst. Mp 165°. $[\alpha]_D$ +6° ($CHCl_3$).

[11040-28-1]

Mazur, Y. *et al, J. Am. Chem. Soc.*, 1958, **80**, 1008, 6293, 6296.
Bates, R.B. *et al, Tetrahedron Lett.*, 1968, 6163 (*struct*)
St. Pyrek, J., *J. Chem. Soc., Chem. Commun.*, 1969, 107 (*isol*)
Brooks, C.J.W. *et al, Steroids*, 1972, **20**, 487 (*ms*)
Lenton, J.R. *et al, Phytochemistry*, 1975, **14**, 1523 (*biosynth*)
Sucrow, W. *et al, Chem. Ber.*, 1977, **110**, 1523 (*synth*)

4-Methylstigmast-7-en-3-ol M-00140

$C_{30}H_{52}O$ M 428.740

(3β,4α,5α,24R)-form [36735-29-2]

24-Ethyllophenol

Isol. from seeds of *Bryonia alba, Lycopersicon esculentum* and numerous vegetable oils. Cryst. (Me_2CO/MeOH). Mp 175.5-176.5°.

Ac: Cryst. (Me_2CO/MeOH). Mp 153-155°.

[33903-18-3, 65437-62-9, 77610-47-0]

Panosyan, A.G. *et al, Khim. Prir. Soedin.*, 1977, 353; 1980, 781 (*isol*)
Itoh, T. *et al, Fette, Seifen, Anstrichm.*, 1978, **80**, 382.
Itoh, T. *et al, Phytochemistry*, 1980, **19**, 2491.
Itoh, T. *et al, Yukagaku*, 1981, **30**, 586 (*hplc*)

23-Methyl-5,9-tetracosadienoic acid M-00141

$(H_3C)_2CH(CH_2)_{12}CH{=}CHCH_2CH_2CH{=}CH(CH_2)_3COOH$

$C_{25}H_{46}O_2$ M 378.637

(Z,Z)-form [140245-76-7]

Constit. of the sponge *Ircinia* spp.

Carballeira, N.M. *et al*, *Comp. Biochem. Physiol., B: Comp. Biochem.*, 1991, **100**, 489 (*isol*)

2-Methyltetracosane M-00142

Isopentacosane

[1560-78-7]

$(H_3C)_2CH(CH_2)_{21}CH_3$

$C_{25}H_{52}$ M 352.686

Constit. of the cuticular hydrocarbons of the termite *Reticulitermes flavipes*. Also in hop (*Humulus lupulus*) and orange (*Citrus sinensis*) oils. Mp 56°.

Levene, P.A. *et al*, *J. Biol. Chem.*, 1922, **52**, 227.
Streibl, M. *et al*, *Fette, Seifen, Anstrichm.*, 1968, **70**, 543 (*glc*)
Howard, R.W. *et al*, *J. Chem. Ecol.*, 1978, **4**, 233 (*glc, ms*)

23-Methyltetracosanoic acid M-00143

$(H_3C)_2CH(CH_2)_{21}COOH$

$C_{25}H_{50}O_2$ M 382.669

Mp 78.5°.

Et ester:

$C_{27}H_{54}O_2$ M 410.722

Mp 57°.

Levene, P.A. *et al*, *J. Biol. Chem.*, 1922, **52**, 227.

23-Methyltetracosanol M-00144

$(H_3C)_2CH(CH_2)_{21}CH_2OH$

$C_{25}H_{52}O$ M 368.685

Mp 75°.

Levene, P.A. *et al*, *J. Biol. Chem.*, 1922, **52**, 227.

3-Methyltetradecanedioic acid M-00145

$HOOC(CH_2)_{10}CH(CH_3)CH_2COOH$

$C_{15}H_{28}O_4$ M 272.384

(±)-form

Cryst. (EtOH). Mp 75.2°. Bp_4 239-241°.

Dinitrile: Bp_{16} 229-232°.

Chuit, P. *et al*, *Helv. Chim. Acta*, 1927, **10**, 178.

5-Methyltetradecanedioic acid M-00146

$HOOC(CH_2)_8CH(CH_3)(CH_2)_3COOH$

$C_{15}H_{28}O_4$ M 272.384

(±)-form

Cryst. (C_6H_6/pet. ether). Mp 74-75°.

Ruzicka, L. *et al*, *Helv. Chim. Acta*, 1927, **10**, 689.

12-Methyltetradecanoic acid M-00147

Sarcinic acid. *Aseonostatin P5. Anteisopentadecanoic acid*

[5502-94-3]

$H_3CCH_2CH(CH_3)(CH_2)_{10}COOH$

$C_{15}H_{30}O_2$ M 242.401

Isol. from *Bacillus* sp., *Sarcina* phospholipids and other microbial lipids. β-Lactamase inhibitor. Liq. Mp 23.7°. $[\alpha]_D^{18}$ +4.8°.

2,3-Dihydroxypropyl ester: [19207-26-2]. ***Aggreceride A***

$C_{18}H_{36}O_4$ M 316.480

Prod. by *Streptomyces* sp. Om 3209. Platelet aggregation inhibitor. $[\alpha]_D^{21}$ +9.7° (c, 0.33 in $CHCl_3$). Strain also prod. Aggreceride B and Aggreceride C which contain 14-methylpentadecanoic acid and 15-methylhexadecanoic acid, resp.

Akashi, S. *et al*, *J. Biochem.* (*Tokyo*), 1960, **47**, 222 (*isol*)
Song, Y. *et al*, *J. Antibiot.*, 1981, **34**, 980 (*isol, props, bibl*)
Omura, S. *et al*, *J. Antibiot.*, 1986, **39**, 1180 (*Aggreceride A*)
Carballeira, N. *et al*, *J. Org. Chem.*, 1986, **51**, 2751 (*synth*)
Ishida-Okawara, A. *et al*, *J. Antibiot.*, 1991, **44**, 524 (*Aseonostatin*)

12-Methyl-1-tetradecanol M-00148

[50313-71-8]

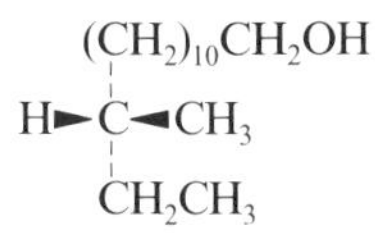

(*R*)-*form*

$C_{15}H_{32}O$ M 228.417

The acetate (chirality currently unspecified) is a component of the female sex pheromone of the tortrix moth *Adoxophyes* spp.

(R)-form [127995-63-5]

$[\alpha]_D$ −5.73° (c, 15 in $CHCl_3$).

Ac: [127914-36-7].

$C_{17}H_{34}O_2$ M 270.454

$[\alpha]_D$ −5.13° (c, 11.7 in $CHCl_3$).

(S)-form [10159-99-6]

$[\alpha]_D$ +5.73° (c, 15.7 in $CHCl_3$).

Ac: [127914-37-8].

$[\alpha]_D$ +5.84° (c, 15.4 in $CHCl_3$).

(±)-form [127914-35-6]

$Bp_{0.05}$ 92-94°.

Ac: [127995-64-6].

Bp_5 123-125°.

Horiike, M. *et al*, *Agric. Biol. Chem.*, 1982, **46**, 1925 (*isol, synth*)
Bestmann, H.J. *et al*, *Justus Liebigs Ann. Chem.*, 1990, 829 (*synth, pmr, bibl*)

13-Methyl-1-triacontene, 9CI M-00149

[35857-60-4]

$H_2C{=}CH(CH_2)_{10}CH(CH_3)(CH_2)_{16}CH_3$

$C_{31}H_{62}$ M 434.831

Synthetic stimulant for parasite of corn earworm *Heliothis zea*.

Jones, R.L. *et al*, *Science* (*Washington, D.C.*), 1971, **173**, 842.

2-Methyltricosane M-00150

[1928-30-9]

$H_3C(CH_2)_{20}CH(CH_3)_2$

$C_{24}H_{50}$ M 338.659

Minor constit. of rose-petal leaf wax and of hops (*Humulus lupulus*). Mp 42°.

Wollrab, V. *et al*, *Collect. Czech. Chem. Commun.*, 1965, **30**, 1654, 1670 (*glc, ir*)

12-Methyltricosane M-00151

Isotetracosane

[22331-52-8]

$H_3C(CH_2)_{10}CH(CH_3)(CH_2)_{10}CH_3$

$C_{24}H_{50}$ M 338.659

Cryst. (Et_2O). Mp 138-138.5°. Bp_9 222-225°.

Levene, P.A. *et al*, *J. Biol. Chem.*, 1914, **18**, 480.
Streibl, M. *et al*, *Fette, Seifen, Anstrichm.*, 1968, **70**, 543 (*glc*)

2-Methyl-2-tricosanol M-00152

$(H_3C)_2C(OH)(CH_2)_{20}CH_3$

$C_{24}H_{50}O$ M 354.658

Cryst. (Et_2O). Mp 58.8-60°.

Breusch, F.L. *et al*, *CA*, 1964, **61**, 6910h.

12-Methyl-1-tricosanol M-00153

Isotetracosanol

$H_3C(CH_2)_{10}CH(CH_3)(CH_2)_{10}CH_2OH$

$C_{24}H_{50}O$ M 354.658

Cryst. ($CHCl_3$). Mp 72°. $Bp_{0.8}$ 220°.

Levene, P.A. *et al*, *J. Biol. Chem.*, 1914, **18**, 480.

2-Methyl-2-tricosene M-00154

$H_3C(CH_2)_{19}CH{=}C(CH_3)_2$

$C_{24}H_{48}$ M 336.643

Mp 41.5°. Bp_{10} 234-237°.

Landa, S. *et al*, *Collect. Czech. Chem. Commun.*, 1930, **2**, 520.
Landa, S. *et al*, *Chem. Zentralbl.*, 1931, **1**, 2454.

2-Methyltridecanedioic acid M-00155

2-Methylbrassylic acid

[37443-08-6]

$HOOC(CH_2)_{10}CH(CH_3)COOH$

$C_{14}H_{26}O_4$ M 258.357

Cryst. (C_6H_6/pet. ether). Mp 87.5-88.5°.

Di-Me ester: [18451-90-6].
Bp_9 185°.

Chuit, P. *et al*, *Helv. Chim. Acta*, 1927, **10**, 175.
Lefort, D. *et al*, *Izv. Akad. Nauk SSSR, Ser. Khim.*, 1967, 648.
Lefort, D. *et al*, *Bull. Soc. Chim. Fr.*, 1968, 4613.

3-Methyltridecanedioic acid M-00156

3-Methylbrassylic acid

$HOOC(CH_2)_9CH(CH_3)CH_2COOH$

$C_{14}H_{26}O_4$ M 258.357

Cryst. (EtOH aq. or C_6H_6/pet. ether). Mp 68.5-69.5°.

Di-Me ester: Bp_8 182-185°.

Chuit, P. *et al*, *Helv. Chim. Acta*, 1927, **10**, 175.

4-Methyltridecanedioic acid M-00157

4-Methylbrassylic acid

$HOOC(CH_2)_8CH(CH_3)CH_2CH_2COOH$

$C_{14}H_{26}O_4$ M 258.357

Cryst. (C_6H_6/pet. ether). Mp 47°. $Bp_{0.5}$ 190-200°.

Ruzicka, L. *et al*, *Helv. Chim. Acta*, 1927, **10**, 689.

3-Methyltridecanoic acid M-00158

[58054-74-3]

$H_3C(CH_2)_9CH(CH_3)CH_2COOH$

$C_{14}H_{28}O_2$ M 228.374

Present in uropygial gland of birds.

Me ester: [2412-84-2].
$C_{15}H_{30}O_2$ M 242.401
Bp_{10} 147-149°.

Grimwood, P.D. *et al*, *J. Chem. Soc. C*, 1971, 870 (*synth*)
Zeman, A. *et al*, *Fette, Seifen, Anstrichm.*, 1973, **75**, 667.
Ubik, K. *et al*, *Collect. Czech. Chem. Commun.*, 1975, **40**, 2826 (*ms*)

12-Methyltridecanoic acid, 9CI M-00159

Isomyristic acid. Aseonostatin P1. Aseanostatin P1

[2724-57-4]

$(H_3C)_2CH(CH_2)_{10}COOH$

$C_{14}H_{28}O_2$ M 228.374

Occurs in several natural sources including wool wax, pine resin, Baltic salmon and sperm whale. Also prod. by various bacteria. Has bactericidial activity. Mp 53°. Bp_2 158°.

Et ester:
$C_{16}H_{32}O_2$ M 256.428
Bp_5 140-142°.

Amide:
$C_{14}H_{29}NO$ M 227.389
Mp 107°.

Fordyce, C.R. *et al*, *J. Am. Chem. Soc.*, 1933, **55**, 3371 (*synth*)
Tyrrell, D., *Lipids*, 1968, **3**, 368 (*ir, ms*)
Goodrich, B. *et al*, *Aust. J. Chem.*, 1971, **24**, 153.
Carballeira, N. *et al*, *J. Org. Chem.*, 1986, **51**, 2751 (*synth*)
Ishida-Okawara, A. *et al*, *J. Antibiot.*, 1991, **44**, 524 (*Aseonostatin*)

2-Methyl-3-tridecanol M-00160

$H_3C(CH_2)_9CH(OH)CH(CH_3)_2$

$C_{14}H_{30}O$ M 214.390

(+)-*form*

Needles. Mp 20°. Bp_{29} 164°. $[\alpha]_D$ +16.15° (EtOH), $[\alpha]_D$ +18.34° (EtOH).

Hydrogen phthalate: Cryst. (pet. ether). Mp 65-66°. $[\alpha]_D$ +16.8° (EtOH), $[\alpha]_D$ +14.2° ($CHCl_3$).

(±)-*form*

Bp 274°.

Hydrogen phthalate: Cryst. (pet. ether). Mp 58-59°.

Pickard, R.H. *et al*, *J. Chem. Soc.*, 1912, **101**, 636.

2-Methyl-4-tridecanol M-00161

$H_3C(CH_2)_8CH(OH)CH_2CH(CH_3)_2$

$C_{14}H_{30}O$ M 214.390

Bp_{14} 147-148°.

Allophanate: Mp 90-91.5°.

Pratt, E.F. *et al*, *J. Am. Chem. Soc.*, 1954, **76**, 52.

4-Methyl-4-tridecanol M-00162

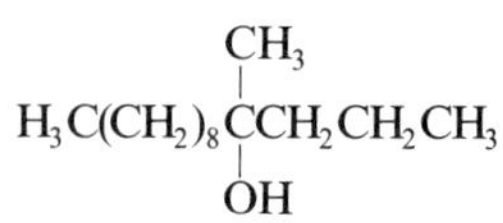

$C_{14}H_{30}O$ M 214.390

(±)-form
Bp_{13} 140-142°.

Thoms, H. *et al*, *Arch. Pharm.* (*Weinheim, Ger.*), 1925, **263**, 263.

12-Methyl-1-tridecanol M-00163

Isomyristyl alcohol
[21987-21-3]

$(H_3C)_2CH(CH_2)_{10}CH_2OH$

$C_{14}H_{30}O$ M 214.390
Mp 10-11°. Bp_6 145-150°.

Fordyce, C.R. *et al*, *J. Am. Chem. Soc.*, 1933, **55**, 3372 (*synth*)
Dolejs, L. *et al*, *Org. Mass Spectrom.*, 1968, **1**, 563 (*ms*)

12-Methyl-13-tridecanolide M-00164

13-Methyloxacyclotetradecan-2-one
[57092-32-7]

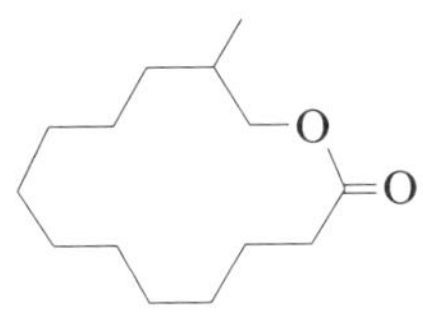

$C_{14}H_{26}O_2$ M 226.358
Constit. of root of *Angelica archangelica*.
(±)-form
Yellow oil with musk-like odour. $Bp_{0.005}$ 100°.

Taskinen, J., *Acta Chem. Scand., Ser. B*, 1975, **29**, 637 (*isol, ir, ms, pmr*)
Stanchev, S. *et al*, *Helv. Chim. Acta*, 1987, **70**, 1389 (*synth, ir, pmr, cmr, ms*)

10-Methyl-2-tridecanone M-00165

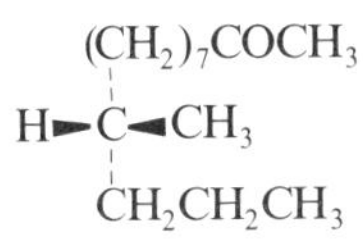

$C_{14}H_{28}O$ M 212.375
*(**R**)-form* [82621-53-2]
Pheromone of the Southern corn rootworm or spotted cucumber beetle *Diabrotica undecimpunctata howardi*. Oil. Bp_3 105-115°. $[\alpha]_D^{23}$ −1.4° (c, 3 in $CHCl_3$).

Sonnett, P.E., *J. Org. Chem.*, 1982, **47**, 3793 (*synth, pmr*)
Senda, S. *et al*, *Agric. Biol. Chem.*, 1983, **47**, 795 (*synth*)
Gus, P.L. *et al*, *J. Chem. Ecol.*, 1983, **8**, 545 (*isol*)
Rossi, R. *et al*, *Tetrahedron*, 1985, **41**, 627 (*synth, ir, pmr*)

3-Methyl-8-tridecenoic acid, 8CI M-00166

$H_3C(CH_2)_3CH{=}CH(CH_2)_4CH(CH_3)CH_2COOH$

$C_{14}H_{26}O_2$ M 226.358
(±)-(?)-form
Me ester: [31535-17-8].
$C_{15}H_{28}O_2$ M 240.385
Bp_{13} 151-153°. Double-bond config. not indicated.

Grimwood, P.D. *et al*, *J. Chem. Soc. C*, 1971, 870.

2-Methyl-5-(8-tridecenyl)-1,3-benzenediol M-00167

2-Methyl-5-(8-tridecenyl)resorcinol. 2-Methylroupelliol

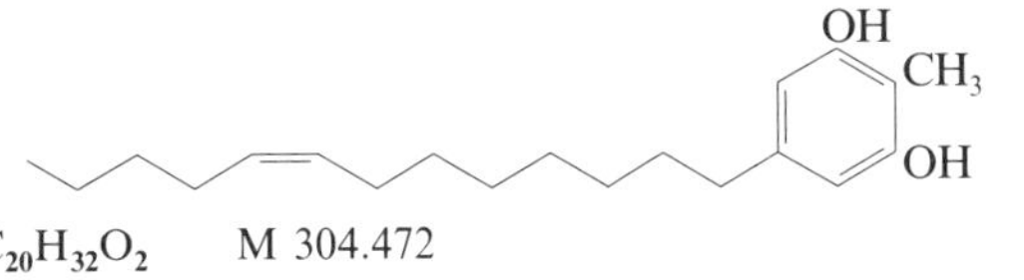

$C_{20}H_{32}O_2$ M 304.472
*(**Z**)-form* [72629-62-0] ***Ardisin.*** *Ardisinol I*
Isol. from *Ardisia japonica* and *Protea roupelliae*. Shows antitubercular activity. Oil.
▷ Skin irritant.
Di-Ac: [125187-40-8].
Yellow oil.
Di-Me ether: [125187-41-9]. *1,3-Dimethoxy-2-methyl-5-(8-tridecenyl)benzene*
$C_{22}H_{36}O_2$ M 332.525
Isol. from *P. roupelliae*. High boiling oil.
8′,9′-Dihydro: [66284-53-5]. *2-Methyl-5-tridecyl-1,3-benzenediol, 9CI. 2-Methyl-5-tridecylresorcinol. 2-C-Methylgrevillol*
$C_{20}H_{34}O_2$ M 306.487
Isol. from *Hakea saligna* wood. Prisms (pet. ether). Mp 109-110°.

[66284-54-6]

Manju, M. *et al*, *Indian J. Chem., Sect. B*, 1977, **15**, 1090 (*isol*)
Hu, Y. *et al*, *CA*, 1980, **92**, 124878m (*isol*)
Varma, M. *et al*, *Gazz. Chim. Ital.*, 1980, **110**, 641 (*cmr*)
Bullock, L.M. *et al*, *S. Afr. J. Chem.*, 1989, **42**, 76 (*isol*)

2-Methyl-3-tridecylbutanedioic acid M-00168

Pedicellic acid

$H_3C(CH_2)_{12}CH(COOH)CH(COOH)CH_3$

$C_{18}H_{34}O_4$ M 314.464
Constit. of the leaves of *Didymocarpus pedicellata*. Cryst. Mp 92-93°. This natural compd. is probably opt. active (see anhydride).
Di-Me ester: Oil.
Anhydride: Plates ($CHCl_3$). Mp 63°. $[\alpha]_D^{29}$ +9.9° (c, 0.6 in $CHCl_3$).

Rao, K.V. *et al*, *Tetrahedron*, 1966, **22**, 1495.

2-Methyl-6-tridecylpiperidine, 9CI M-00169

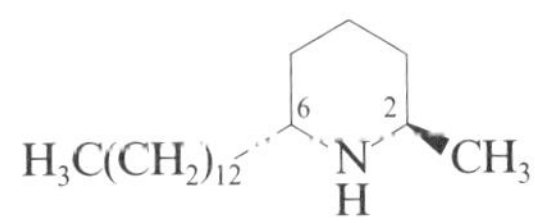

(2R,6R*)-form*

$C_{19}H_{39}N$ M 281.524
*(**2R*,6R***)-form* [32778-77-1]
trans-*form.* ***Solenopsin B***
Minor component of the venom of the red form of the fire ant (*Solenopsis saevissima*). Also present in the venom of *S. invicta*. Oil.
4′,5′-Didehydro (Z-)*:* [32778-76-0]. ***2-Methyl-6-(4-tridecenyl)piperidine.*** *Dehydrosolenopsin B*
$C_{19}H_{37}N$ M 279.508
Major component of venom of *S. saevissima*, also found in *S. xyloni* and *S. geminata*. Dehydrosolenopsin B is the 2,6-*trans*-form but the *cis*-form also occurs.
*(**2R*,6S***)-form* [35285-26-8]
cis-*form*
Minor component of the venoms of both red and black forms of *S. saevissima*, *S. xyloni* and *S. geminata*. Trace component of the venom of *S. aurea*. Oil.

MacConnell, J.G. *et al*, *Tetrahedron*, 1971, **27**, 1129 (*isol*, *synth*, *ms*)
Maruoka, K. *et al*, *J. Am. Chem. Soc.*, 1983, **105**, 2831 (*synth*)
Taber, D.F. *et al*, *J. Org. Chem.*, 1988, **53**, 2968 (*synth*, *ir*, *pmr*, *cmr*, *ms*, *Solenopsin B*)

11-Methyl-3,6-undecadienolide M-00170

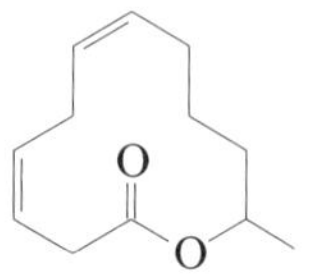

$C_{12}H_{18}O_2$ M 194.273

(3Z,6Z)-form [86583-51-9]
Isol. from the rusty grain beetle *Cryptolestes ferrugineus*, aggregation pheromone for the merchant grain beetle *Oryzaephilus mercator*.

Wong, J.W. *et al*, *J. Chem. Ecol.*, 1983, **9**, 451 (*isol*)
Millar, J.G. *et al*, *J. Org. Chem.*, 1984, **49**, 2332 (*synth*, *ir*, *pmr*, *ms*)

2-Methylundecanal, 9CI M-00171

2-Methylnonylacetaldehyde. Aldehyde MNA
[110-41-8]

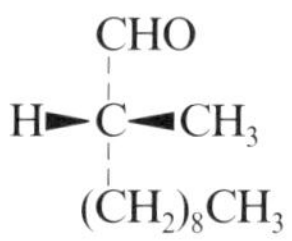

(R)-form

$C_{12}H_{24}O$ M 184.321
Important perfumery ingredient. Enantiomers show little difference in odour props.

(R)-form
$Bp_{9.75}$ 114°. $[\alpha]_D^{25}$ −13.7° (neat) (e.e. 96%).

(S)-form [128871-20-5]
Bp_{10} 115°. $[\alpha]_D^{25}$ +13.7° (neat).

(±)-*form* [135026-96-9]
Bp 171°, Bp_{10} 114°.
Semicarbazone: [16742-16-8].
Mp 86°.

Wright, R.H., *Am. Perfum. Cosmet.*, 1968, **83**, 43; *CA*, **69**, 91487c.
Khulesza, J. *et al*, *Riechst., Aromen, Koerperpflegem.*, 1969, **19**, 8; *CA*, **71**, 49167j (*synth*)
Shutikova, L.A. *et al*, *CA*, 1974, **74**, 12552x (*synth*)
Groenewegen, F. *et al*, *Tetrahedron Lett.*, 1979, 2817.
Enders, D. *et al*, *Justus Liebigs Ann. Chem.*, 1990, 1107 (*synth*, *pmr*, *cmr*)
Lewis, R.J., *Sax's Dangerous Properties of Industrial Materials*, 8th Ed., Van Nostrand-Reinhold, 1992, MQI550.

3-Methylundecane, 9CI M-00172

[1002-43-3]

$H_3C(CH_2)_7CH(CH_3)CH_2CH_3$

$C_{12}H_{26}$ M 170.337

(±)-*form*
Bp_{32} 109-110°.

Franklin, S. *et al*, *J. Am. Chem. Soc.*, 1948, **70**, 298 (*synth*)
Posner, G.H. *et al*, *Tetrahedron Lett.*, 1973, 1815 (*synth*)
Lehnskuhl, H. *et al*, *Justus Liebigs Ann. Chem.*, 1975, 1176 (*synth*)

10-Methylundecanoic acid M-00173

Isododecanoic acid
[2724-56-3]

$(H_3C)_2CH(CH_2)_8COOH$

$C_{12}H_{24}O_2$ M 200.320
Isol. from wool fat, from lipids of *Sarcina lutea* and resin of *Shorea maranti*. Tablets (Me_2CO or C_6H_6). Mp 41.2°. Bp_3 140-145°.
[5129-56-6]

Weitkamp, A.W., *J. Am. Chem. Soc.*, 1945, **67**, 447 (*isol*)
Fieser, L.F. *et al*, *J. Am. Chem. Soc.*, 1948, **70**, 3174 (*synth*)
Cocker, W. *et al*, *Perfum. Essent. Oil Rec.*, 1964, **55**, 442 (*isol*)

2-Methyl-2-undecanol M-00174

[32836-42-3]

$(H_3C)_2C(OH)(CH_2)_8CH_3$

$C_{12}H_{26}O$ M 186.337
Bp_{15} 134-135° (Bp_{23} 118°).

Prévost, Ch. *et al*, *Bull. Soc. Chim. Fr.*, 1950, 1068.

2-Methyl-3-undecanol M-00175

$H_3C(CH_2)_7CH(OH)CH(CH_3)_2$

$C_{12}H_{26}O$ M 186.337

(+)-*form* [60671-36-5]
Bp_{26} 135°. $[\alpha]_D^{20}$ +18.55° (EtOH), $[\alpha]_D^{20}$ +21.50° (EtOH).
Hydrogen phthalate: Needles. Mp 47-48°. $[\alpha]_D$ +17.2° (EtOH).

(±)-*form*
Bp 236°.
Hydrogen phthalate: Cryst. (pet. ether). Mp 55°.

Pickard, R.H. *et al*, *J. Chem. Soc.*, 1912, **101**, 633.
Baratchart, M. *et al*, *Bull. Soc. Chim. Fr.*, 1976, 953.

2-Methyl-5-undecanol M-00176

[33978-71-1]

$H_3C(CH_2)_5CH(OH)CH_2CH_2CH(CH_3)_2$

$C_{12}H_{26}O$ M 186.337

(±)-*form*
Bp_{20} 128-130°.

Petrov, A.D. *et al*, *Zh. Obshch. Khim.*, 1948, **18**, 859.
Seibl, J., *Arch. Mass Spectral Data*, 1971, **2**, 196, 256, 300, 306 (*ms*)

6-Methyl-6-undecanol M-00177

[5340-31-8]

$H_3C(CH_2)_4C(CH_3)(OH)(CH_2)_4CH_3$

$C_{12}H_{26}O$ M 186.337
Bp_2 80-83°.

Whitmore, F.C. *et al*, *J. Am. Chem. Soc.*, 1933, **55**, 406.

11-Methyl-3-undecenolide M-00178

12-Methyloxacyclododec-4-en-2-one, 9CI

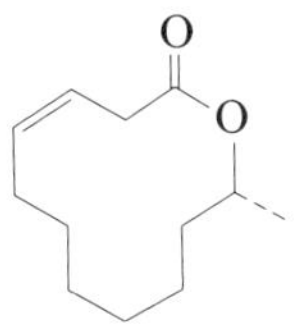

(R,Z)-form

$C_{12}H_{20}O_2$ M 196.289

(R,Z)-form [87583-38-8] ***Cucujolide II***
Aggregation pheromone of the grain beetle. Attractant for *Oryzaephilus mercator*.

(S,Z)-form [86578-99-6] ***Ferrulactone II***
Pheromone of the rusty grain beetle *Cryptolestes ferrugineus*. Aggregating pheromone for *O. mercator*. $[\alpha]_D^{22.5}$ +70.5° (c, 0.96 in $CHCl_3$).

(±)-*(Z)-form* [87583-37-7]
Liq. $Bp_{0.1}$ 60-70°. Contains *ca.* 5% of the *E*-isomer.

[86631-33-6, 87583-37-7, 97372-89-9, 126415-33-6, 126415-34-7, 126841-00-7, 126841-01-8, 127888-78-2]

Oehlschlager, A.C. *et al*, *J. Org. Chem.*, 1983, **48**, 5009 (*synth*)
U.S. Pat., 4 560 551, (1985); *CA*, **104**, 125076b (*synth*)
Pierce, A.M. *et al*, *Environ. Entomol.*, 1989, **18**, 747; *CA*, **112**, 195479z (*struct*)
Keinan, E. *et al*, *Tetrahedron*, 1991, **47**, 4631 (*synth*)

2-Methyl-6-undecylpiperidine, 9CI M-00179

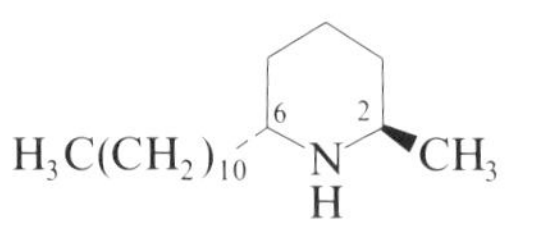

(2R,6R*)-form*

$C_{17}H_{35}N$ M 253.470

(2R*,6R*)-form [28720-60-7]
trans-*form*. ***Solenopsin A***
One of the major components of the venoms of fire ants spp. *Solenopsis xyloni*, *S. geminata*, *S. aurea*. Minor component of venoms of both red and black forms of *S. saevissima*. Also present in *S. invicta*. Oil.

B,HCl: Mp 114°.

1,2-Didehydro: 2,3,4,5-Tetrahydro-6-methyl-2-undecylpyridine. ***1-Methyl-6-undecyl-Δ²-piperidine***
$C_{17}H_{33}N$ M 251.454
From *S.* spp.

(2R*,6S*)-form [35285-24-6]
cis-*form*. ***Isosolenopsin A***
One of the major components of the venoms of *S. xyloni*, *S. geminata* and *S. aurea*. Minor component in venoms of both red and black forms of *S. saevissima*. Oil. $Bp_{0.02}$ 91-96°.

B,HCl: Mp 155°.

MacConnell, J.G. *et al*, *Tetrahedron*, 1971, **27**, 1129 (*isol, synth, ms, ir, pmr*)
Moriyama, Y. *et al*, *Tetrahedron Lett.*, 1977, 825 (*synth, ms, cmr*)
Maruoka, K. *et al*, *J. Am. Chem. Soc.*, 1983, **105**, 2831 (*synth*)
Wasserman, H.H. *et al*, *Tetrahedron Lett.*, 1988, **29**, 4977 (*synth*)

Moenocinol M-00180

[19953-93-6]

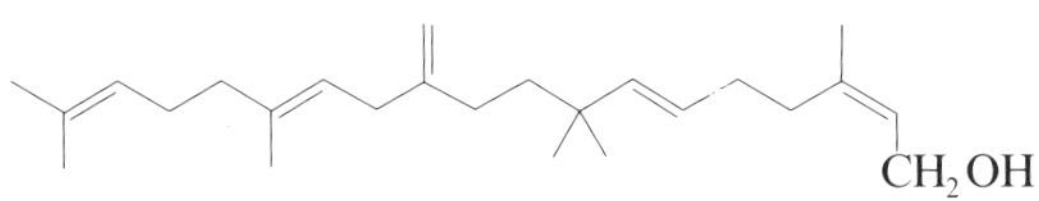

$C_{25}H_{42}O$ M 358.606

Produced by various *Streptomyces* strains. Lipid component of Moenomycin and Prasinomycin. Oil.

Slusarchyk, W.A. *et al*, *Tetrahedron Lett.*, 1969, 659 (*isol*)
Tschesche, R., *Justus Liebigs Ann. Chem.*, 1974, 853 (*synth*)
Grieco, P.A. *et al*, *J. Am. Chem. Soc.*, 1975, **97**, 1597 (*synth*)
Coates, R.M. *et al*, *J. Org. Chem.*, 1980, **45**, 2685 (*synth, cmr*)
Kocienski, P. *et al*, *J. Chem. Soc., Perkin Trans. 1*, 1983, 1777 (*synth*)
Böttger, D. *et al*, *Tetrahedron Lett.*, 1983, **24**, 5201 (*synth*)
Stumpp, M.C. *et al*, *Tetrahedron*, 1986, **42**, 5941 (*synth*)

Murolic acid M-00181

Tetrahydro-2-(14-hydroxypentadecyl)-4-methylene-5-oxo-3-furancarboxylic acid, 9CI

[70579-58-7]

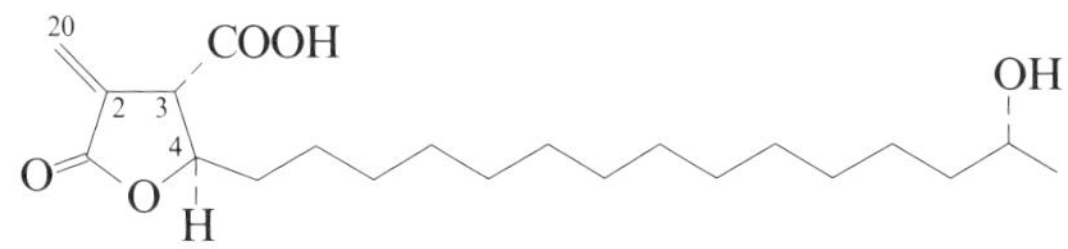

Absolute configuration

$C_{21}H_{36}O_5$ M 368.512

Isol. from the lichen *Lecanora muralis*, also in *L. melanophthalma* and *L. rubina*. Platelets (MeOH). Mp 110-111°. $[\alpha]_D^{24}$ +5.13° (c, 1.11 in MeOH).

2S,20-Dihydro: [70579-57-6]. ***Neodihydromurolic acid***
$C_{21}H_{38}O_5$ M 370.528
Isol. from *L. muralis*, also in *L. melanophthalma* and *L. rubina*. Platelets (MeOH). Mp 122-123°. $[\alpha]_D^{24}$ +17.14° (c, 1.59 in MeOH).

14′-Ketone: [70579-63-4]. ***Muronic acid***
$C_{21}H_{34}O_5$ M 366.497
Isol. from *Neuropogon trachycarpus*. Cryst. (Me_2CO aq.). Mp 93-94°. $[\alpha]_D^{20}$ +8.7° (c, 0.085 in $CHCl_3$).

14′-Ketone, 4-epimer: [101899-62-1]. ***Allopertusaric acid***
$C_{21}H_{34}O_5$ M 366.497
Isol. from *Pertusaria albescens*. Seedling growth inhibitor. Plates. Mp 76-78°. $[\alpha]_D^{24}$ −95.5° (c, 1.525 in MeOH). Pertusaric acid appears to be unknown.

▷ Pertussive agent, nasal irritant.

2S,20-Dihydro, 14′-ketone, 3,4-diepimer: [101899-68-7]. ***Dihydropertusaric acid***
$C_{21}H_{36}O_5$ M 368.512
Isol. from *P. albescens*. Small needles. Mp 105-107°. $[\alpha]_D^{24}$ −74.9° (c, 2.15 in MeOH).

Huneck, S. *et al*, *J. Hattori Bot. Lab.*, 1979, **45**, 1 (*isol, pmr, cd, struct, abs config*)
Ghogomu, R.T. *et al*, *Phytochemistry*, 1982, **21**, 2355 (*Muronic acid*)
Huneck, S. *et al*, *Phytochemistry*, 1986, **25**, 453 (*Allopertusaric acid, Dihydropertusaric acid*)

Mycolic acid M-00182

Corynomycolic acid. Nocardic acid. Smegmamycolic acid

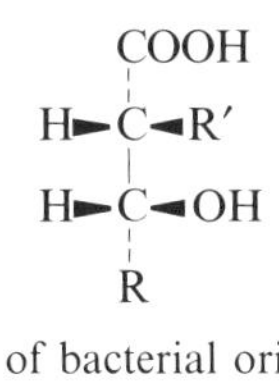

Absolute configuration

Glycolipids of bacterial origin which are homologous mixts. varying with the origin of the sample. On pyrol. they yield RCHO + $R'CH_2COOH$. R′ is a saturated alkyl residue. R may be saturated or unsaturated. The following classes have been recognised. (i) From

Corynobacterium diphtheriae and *Norcardia opaca*: R = C_{15}; R′ = C_{14} R = *n*-C_{31}, *n*-C_{33}, *n*-C_{35}; R′ = C_8, C_{10}, C_{12} (ii) Acids containing unsaturated alkyl chains (and in some cases branching methyl groups) from *C. diphtheriae*, *N.* spp. and *Mycobacterium smegmatis*: R = 8-pentadecenyl; R′ = $C_{14}H_{29}$ R = C_{37}, C_{39}, C_{41}; R′ = $C_{22}H_{45}$ *also* dienes, R = $C_{49}H_{95}$—$C_{58}H_{113}$ (10 members), some of which have a branching methyl group. (iii) Acids containing one cyclopropane unit, one double bond and one branching methyl, from *M. smegmatis*: e.g. R = $H_3C(CH_2)_{17}$-cyclopropyl-$(CH_2)_{13}CH{=}CHCH(CH_3)$; R′ = $C_{22}H_{45}$ (iv) Acids containing two cyclopropane units from *M. tuberculosis* and *M. paratuberculosis*.(v) Acids with additional oxygenated functions (-OMe, C=O and -COOH) in R′: e.g. R′ = $H_3C(CH_2)_{17}CH(CH_3)CO(CH_2)_{16}$-cyclopropyl-$(CH_2)_{19}$-; R = $C_{24}H_{49}$ R′ = HO.

[23599-54-4, 23642-94-6, 23725-22-6, 23950-11-0]

Polgar, N., *Top. Lipid Chem.*, 1971, **2**, 207.
Gensler, W.J., *Chem. Phys. Lipids*, 1977, **19**, 128.
Asselineau, C. *et al*, *Prog. Chem. Fats Other Lipids*, 1978, **16**, 59.
Wong, M.Y.H., *J. Biol. Chem.*, 1979, **254**, 5734, 5741.
Kitano, Y. *et al*, *J. Chem. Soc., Chem. Commun.*, 1985, 498 (*synth*)
Utaka, M. *et al*, *J. Chem. Soc., Chem. Commun.*, 1987, 1368 (*synth*)
Fujisawa, T. *et al*, *Bull. Chem. Soc. Jpn.*, 1988, **61**, 1273 (*synth*)

Mycophenolic acid, USAN, INN **M-00183**

6-(1,3-Dihydro-4-hydroxy-6-methoxy-7-methyl-3-oxo-5-isobenzofuranyl)-4-methyl-4-hexenoic acid, 9CI. Melbex. Lilly 68618. NSC 129185

[24280-93-1]

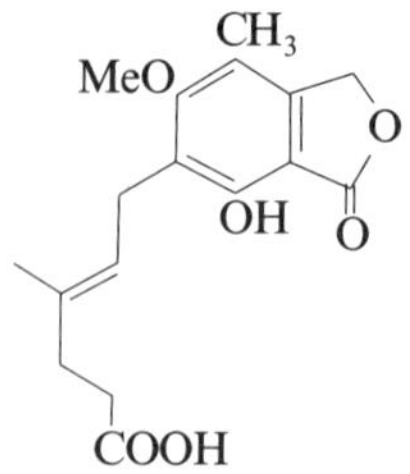

$C_{17}H_{20}O_6$ M 320.341

Metab. of *Penicillium brevi-compactum*, *P. stoloniferum* and other *P.* spp. Shows antitumour and antiviral props. Useful in treating psoriasis and leishmaniasis. Needles (H_2O). Mp 141°.

▷ MP8050000.

Clutterbuck, P.W. *et al*, *Biochem. J.*, 1932, **26**, 1451 (*isol*)
Canonica, L. *et al*, *J. Chem. Soc., Perkin Trans. 1*, 1972, 2639 (*ir, nmr, biosynth*)
Canonica, L. *et al*, *Tetrahedron*, 1972, **28**, 4395 (*synth, pmr, ir*)
Harrison, W. *et al*, *J. Chem. Soc., Perkin Trans. 2*, 1972, 1542 (*cryst struct*)
Bowen, L. *et al*, *J. Chem. Soc., Chem. Commun.*, 1977, 949, 950 (*ms, pmr, biosynth*)
Colombo, L. *et al*, *J. Chem. Soc., Perkin Trans. 1*, 1982, 365 (*biosynth*)
Danheiser, R.L. *et al*, *J. Am. Chem. Soc.*, 1986, **108**, 806 (*synth, bibl*)
Martindale, The Extra Pharmacopoeia, 28th/29th Ed., Pharmaceutical Press, London, 1982/1989, 12984.
Watanabe, M. *et al*, *Chem. Pharm. Bull.*, 1989, **37**, 2948 (*synth*)
Cole, R.J. *et al*, *Handbook of Toxic Fungal Metabolites*, Academic Press, N.Y., 1981, 866.
Lewis, R.J., *Sax's Dangerous Properties of Industrial Materials*, 8th Ed., Van Nostrand-Reinhold, 1992, MRX000.

N

Narumicin I — N-00001

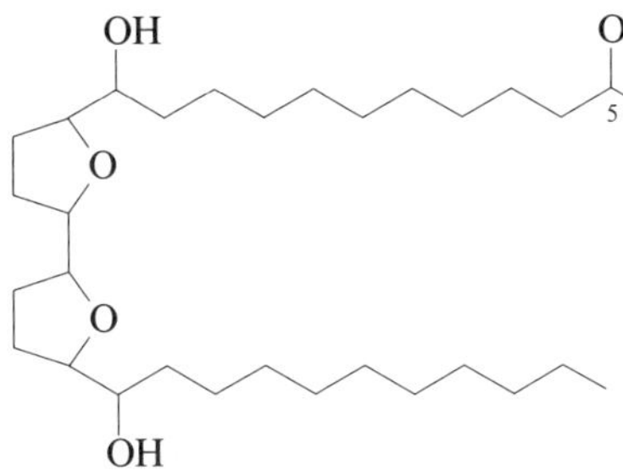

$C_{37}H_{66}O_7$ M 622.924
Constit. of *Uvaria narum*.
Stereoisomer: ***Narumicin II***
$C_{37}H_{66}O_7$ M 622.924
Constit. of *U. narum*.
5-Deoxy: [136033-39-1]. ***Isodesacetyluvaricin***
$C_{37}H_{66}O_6$ M 606.925
Isol. from *U. narum*. No stereochem. detd., may not have the same stereochem. as Narumicin I.

[136005-76-0, 136005-77-1]

Hisham, A. *et al*, *Phytochemistry*, 1991, **30**, 2373 (*isol, pmr, cmr*)

Neoannonin — N-00002

[125205-43-8]

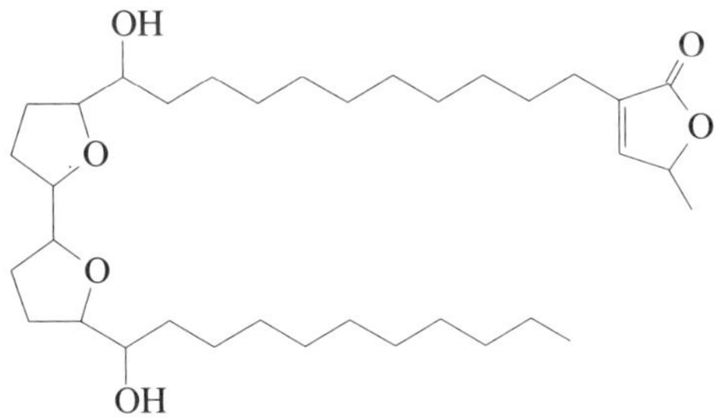

$C_{35}H_{62}O_6$ M 578.871
Constit. of *Annona squamosa*. Insecticide. Wax. $[\alpha]_D^{22}$ +18.8° (c, 1.35 in MeOH).

Kawaza, K. *et al*, *Agric. Biol. Chem.*, 1989, **53**, 2719 (*isol, pmr, cmr*)

Nephrosterinic acid — N-00003

Tetrahydro-4-methylene-5-oxo-2-undecyl-3-furancarboxylic acid, 9CI

[70630-67-0]

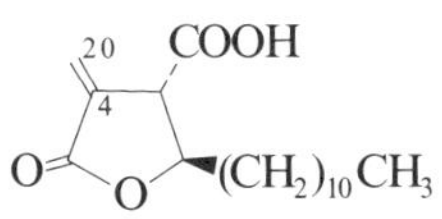

Absolute configuration

$C_{17}H_{28}O_4$ M 296.406
Prod. by *Nephromopsis endocrocea*. Leaflets (AcOH aq.). Mp 96°. $[\alpha]_D^{10}$ +10.81° (c, 6.785 in $CHCl_3$).
Semicarbazone: Leaflets (EtOH). Mp 183-184° dec.
4S,20-Dihydro: [70579-56-5]. ***Nephrosteranic acid***
$C_{17}H_{30}O_4$ M 298.422
Prod. by *N. endocrocea*. Plates (pet. ether). Mp 95°. $[\alpha]_D^{21}$ +38.4° ($CHCl_3$).

Asahina, Y. *et al*, *Ber.*, 1937, **70**, 227 (*isol, struct*)
Bendz, G. *et al*, *Acta Chem. Scand.*, 1966, **20**, 1181 (*tlc*)
Carlson, R.M. *et al*, *J. Org. Chem.*, 1976, **41**, 4065 (*synth*)
Huneck, S. *et al*, *J. Hattori Bot. Lab.*, 1979, **45**, 1 (*cd, abs config*)

1-Nitro-1-pentadecene — N-00004

$H_3C(CH_2)_{12}CH{=}CHNO_2$

$C_{15}H_{29}NO_2$ M 255.400
(E)-form [53520-53-9]
Defence compd. of the termite *Prorhintermes simplex*. Oil.
▷ Contact poison.

Vrkoc, J. *et al*, *Tetrahedron Lett.*, 1974, 1463.
Spanton, S.G. *et al*, *Science (Washington, D.C.)*, 1981, **214**, 1363.
Naya, Y. *et al*, *Tetrahedron Lett.*, 1982, **23**, 3047.

7,11-Nonacosadiene — N-00005

$H_3C(CH_2)_5CH{=}CHCH_2CH_2CH{=}CH(CH_2)_{16}CH_3$

$C_{29}H_{56}$ M 404.761
(7Z,11Z)-form
Pheromone of *Drosophila melanogaster*.

Bac, N.V. *et al*, *Tetrahedron Lett.*, 1986, **27**, 841 (*synth*)
Kelkar, S.V. *et al*, *Synth. Commun.*, 1989, **19**, 1369 (*synth*)

Nonacosane — N-00006

[630-03-5]

$H_3C(CH_2)_{27}CH_3$

$C_{29}H_{60}$ M 408.793
Constit. of the leaves of *Brassica oleracea* (Brussels sprouts) and other plants. Cryst. (pet. ether). Mp 64°. Bp_{40} 346-348°, $Bp_{0.05}$ 196-198°.

Gluud, W., *Ber.*, 1919, **52**, 1039 (*synth*)
Karrer, W. *et al*, *Konstitution und Vorkommen der Organischen Pflanzenstoffe*, 2nd Ed., Birkhäuser Verlag, Basel, 1972-1985, no. 16 (*occur*)
Kami, T. *et al*, *J. Agric. Food Chem.*, 1975, **23**, 795 (*glc, ms*)
Behari, M. *et al*, *Curr. Sci.*, 1976, **45**, 481 (*isol*)
Ulubelen, L. *et al*, *Planta Med.*, 1976, **29**, 258 (*isol*)
Koehlen, L. *et al*, *Z. Pflanzenphysiol.*, 1976, **77**, 99 (*isol*)

8,10-Nonacosanedione, 9CI — N-00007

[58170-84-6]

$H_3C(CH_2)_{18}COCH_2CO(CH_2)_6CH_3$

$C_{29}H_{56}O_2$ M 436.760
Constit. of leaf wax of many *Rhododendron* spp.

Evans, D. *et al*, *Phytochemistry*, 1975, **14**, 2447 (*isol, chromatog, ms*)

12,14-Nonacosanedione, 9CI — N-00008

[58141-94-9]

$H_3C(CH_2)_{14}COCH_2CO(CH_2)_{10}CH_3$

$C_{29}H_{56}O_2$ M 436.760

Constit. of leaf wax of *Eucalyptus* and *Rhododendron* spp.

Horn, D.H.S. *et al*, *Aust. J. Chem.*, 1964, **17**, 464 (*isol*)
Evans, D. *et al*, *Phytochemistry*, 1975, **14**, 2447 (*isol, chromatog, ms*)

Nonacosanoic acid N-00009

[4250-38-8]

$H_3C(CH_2)_{27}COOH$

$C_{29}H_{58}O_2$ M 438.776
Cryst. (toluene). Mp 90.3° (89-90°).
Me ester: [4082-55-7].
$C_{30}H_{60}O_2$ M 452.803
Cryst. (petrol). Mp 68.8° (69-70°). Ir ν 1730 cm^{-1}.
Et ester:
$C_{31}H_{62}O_2$ M 466.830
Mp 66.6°.

Francis, F. *et al*, *J. Am. Chem. Soc.*, 1939, **61**, 577 (*synth*)
Flores, S.E. *et al*, *Acta Cient. Venez.*, 1971, **22**, 147 (*isol, ir*)
Rao, S.J., *Indian J. Chem., Sect. B*, 1987, **26**, 208 (*synth*)
Brown, H.C. *et al*, *Tetrahedron*, 1992, **48**, 9187 (*synth*)
Kling, M.R. *et al*, *J. Chem. Soc., Perkin Trans. 1*, 1993, 1183 (*synth, ir, ms*)

1-Nonacosanol N-00010

[6624-76-6]

$H_3C(CH_2)_{27}CH_2OH$

$C_{29}H_{60}O$ M 424.792
Isol. from leaf wax of *Agave sisalana*. Cryst. (C_6H_6). Mp 85°.

Piper, S.H. *et al*, *Biochem. J.*, 1934, **28**, 2175 (*synth*)
Razafindrazaka, J. *et al*, *Bull. Soc. Chim. Fr.*, 1963, 1633 (*isol*)
Rao, S.J. *et al*, *Indian J. Chem., Sect. B*, 1987, **26**, 208 (*synth*)

10-Nonacosanol N-00011

Ginnol. Celidoniol
[2606-50-0]

$H_3C(CH_2)_{18}CH(OH)(CH_2)_8CH_3$

$C_{29}H_{60}O$ M 424.792
(+)-*form*
Wax from *Ginkgo biloba*, *Chelidonium majus*, *Bazzania pompeana*, *Pinus caribaea*, *Chamaecyparis* spp., *Cycas revoluta* and others. Cryst. (EtOH, Me_2CO or hexane). Mp 83-84°.
Ac:
$C_{31}H_{62}O_2$ M 466.830
Cryst. (Me_2CO). Mp 44.5-45.5°.

Bentley, H.R. *et al*, *J. Chem. Soc.*, 1955, 596 (*isol*)
Kariyone, T. *et al*, *Yakugaku Zasshi*, 1959, **79**, 51, 54, 58 (*isol, struct*)
Seone, E., *Chem. Ind. (London)*, 1961, 1080 (*isol, struct*)
Naruto, S. *et al*, *Phytochemistry*, 1972, **11**, 2644 (*isol*)
De Pascual, T. *et al*, *An. Quim.*, 1975, **71**, 110 (*isol*)
Caldicott, A.B. *et al*, *Phytochemistry*, 1975, **14**, 1799 (*isol*)
Isono, H. *et al*, *Yakugaku Zasshi*, 1976, **96**, 957 (*glc, ms*)

6-Nonacosanone N-00012

[79913-69-2]

$H_3C(CH_2)_4CO(CH_2)_{22}CH_3$

$C_{29}H_{58}O$ M 422.777
Isol. from pod husks of *Cassia auriculata* (Leguminosae). Prisms. Mp 81°.

Lohar, D.R. *et al*, *J. Indian Chem. Soc.*, 1981, **58**, 989.

10-Nonacosanone N-00013

Ginnone. Celidonione
[504-56-3]

$H_3C(CH_2)_{18}CO(CH_2)_8CH_3$

$C_{29}H_{58}O$ M 422.777
Isol. from plant waxes, e.g. *Cycas revoluta*, *Diplopterygium glaucum*, *Ginkgo biloba*. Cryst. (EtOH aq.). Mp 74-75°.
Oxime:
$C_{29}H_{59}NO$ M 437.791
Cryst. (EtOH). Mp 49-50°.
Semicarbazone: Cryst. (EtOH aq.). Mp 45-46°.

Piper, H. *et al*, *Biochem. J.*, 1932, **25**, 2072 (*synth*)
Seoane, E., *Chem. Ind. (London)*, 1961, 1080 (*isol*)
Dodova, A.M. *et al*, *Phytochemistry*, 1973, **12**, 2239 (*isol, ms, struct*)
Ulubelen, A. *et al*, *Planta Med.*, 1973, **24**, 290 (*isol*)
Soeder, R.W. *et al*, *Phytochemistry*, 1975, **14**, 2508 (*isol, glc, ms*)

9-Nonacosene N-00014

[69203-98-1]

$H_3C(CH_2)_7CH{=}CH(CH_2)_{18}CH_3$

$C_{29}H_{58}$ M 406.777
(*Z*)-*form* [36258-10-3]
Obt. from cuticle of the cockroach *Periplaneta japonica* and from secretions of the soldiers of the East African fungus-growing termite *Macrotermes subhyalinus*. Mating stimulant pheromone of the little house fly *Fannia canicularis*.

Jackson, L.L., *Comp. Biochem. Physiol., B: Comp. Biochem.*, 1972, **41**, 331 (*ms, bibl*)
Prestwick, G.D. *et al*, *J. Chem. Ecol.*, 1977, **3**, 579 (*bibl*)

13-Nonacosene N-00015

[61906-97-6]

$H_3C(CH_2)_{14}CH{=}CH(CH_2)_{11}CH_3$

$C_{29}H_{58}$ M 406.777
(*Z*)-*form* [54863-79-5]
Sex pheromone of the face fly *Musca autumnalis*.

Kuepper, F. *et al*, *Chem.-Ztg.*, 1975, **99**, 464 (*synth*)
Sonnett, P. *et al*, *Environ. Entomol.*, 1975, **4**, 761.
Uebel, E.C., *J. Chem. Ecol.*, 1975, **1**, 195.
Lockey, K.M., *Insect Biochem.*, 1978, **8**, 237.

14-Nonacosene N-00016

[61906-96-5]

$H_3C(CH_2)_{12}CH{=}CH(CH_2)_{13}CH_3$

$C_{29}H_{58}$ M 406.777
(*Z*)-*form* [54863-80-8]
Sex pheromone of the face fly *Musca autumnalis*. n_D^{25} 1.4514.

Sonnet, P.E. *et al*, *Org. Prep. Proced. Int.*, 1974, **6**, 269 (*synth*)
Küpper, F.W. *et al*, *Chem.-Ztg.*, 1975, **99**, 464; *Z. Naturforsch., B*, 1976, **31**, 1256 (*synth*)
Uebel, E.C. *et al*, *J. Chem. Ecol.*, 1975, **1**, 195 (*synth*)

2-Nonacosyl-3-tridecene-1,10-diol, 9CI N-00017

*12-(Hydroxymethyl)-10-hentetraconten-4-ol. **Drechslerol A***

[108793-61-9]

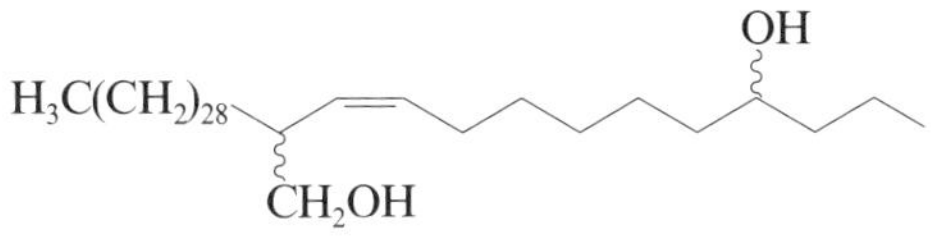

$C_{42}H_{84}O_2$ M 621.124

Prod. by *Drechslera maydis*. Phytotoxin.

Shukla, R.S. *et al, Plant Sci.* (*Limerick, Irel.*), 1987, **48**, 159 (*isol*)

6,9-Nonadecadiene, 9CI N-00018

[52223-55-9]

$$H_3C(CH_2)_4CH{=}CHCH_2CH{=}CH(CH_2)_8CH_3$$

$C_{19}H_{36}$ M 264.493

(6Z,9Z)-form [95663-63-1]

Component of sex pheromone of Lepidoptera spp., including *Alsophila quadripunctata*, *Bupalus piniarius*, *Utetheisa ornatrix* and *Paleacrita vernata*. Viscous oil. $Bp_{0.03}$ 95-99°.

[98291-75-9, 98291-76-0]

Szoes, G. *et al, Entomol. Exp. Appl.*, 1984, **36**, 287.
Bogenschuetz, H. *et al, Z. Angew. Entomol.*, 1985, **100**, 349.
McDonough, L.M., *J. Chem. Ecol.*, 1986, **12**, 2107 (*isol, props*)
Subchev, M. *et al, Z. Naturforsch., C*, 1986, **41**, 1082 (*synth, props*)
Bestmann, H.J. *et al, Justus Liebigs Ann. Chem.*, 1987, **5**, 417 (*synth, pmr, glc, ir, ms*)
Mangold, H.K. *et al, Z. Naturforsch., C*, 1987, **42**, 1035 (*synth, props*)
Becker, H. *et al, Chem. Ind.* (*London*), 1989, 39 (*synth, props*)
Millar, J.G. *et al, J. Chem. Ecol.*, 1990, **16**, 3393 (*synth, ms, glc, props*)
Viala, J. *et al, Tetrahedron*, 1991, **47**, 3347 (*synth, pmr, cmr, ir, ms*)

9,18-Nonadecadiene-6,7-diol, 9CI N-00019

[129436-89-1]

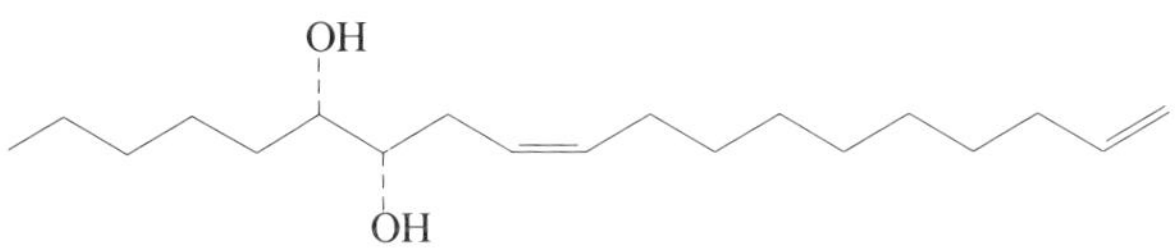

$C_{19}H_{36}O_2$ M 296.492

(6S,7S,9Z)-form

Isol. from the brown alga *Notheia anomala*. Unstable oil. $[\alpha]_D$ −7.5° (c, 0.3 in $CHCl_3$).

Barrow, R.A. *et al, Aust. J. Chem.*, 1990, **43**, 895.

10,13-Nonadecadiynoic acid, 8CI N-00020

[4184-92-3]

$$H_3C(CH_2)_4C{\equiv}CCH_2C{\equiv}C(CH_2)_8COOH$$

$C_{19}H_{30}O_2$ M 290.445

Mp 39.5-41.5°.

Et ester: [16326-29-7].

$C_{21}H_{34}O_2$ M 318.498

Cryst. (pet. ether). Mp 38-39°. $Bp_{0.0028}$ 130-150°.

Osbond, J.M. *et al, J. Chem. Soc.*, 1961, 2779 (*synth*)
Swiss Pat., 414 595, (1966); *CA*, **67**, 2772 (*synth*)
Sprecher, H., *Prog. Chem. Fats Other Lipids*, 1978, **15**, 219.

Nonadecane N-00021

[629-92-5]

$$H_3C(CH_2)_{17}CH_3$$

$C_{19}H_{40}$ M 268.525

Isol. from apple wax, *Manilkara bidentata* and many other plant sources. Constit. of sex pheromone of female fall cankerworm *Alsophila pometaria*, male southern green stinkbug *Nezara viridula* and female tigermoth *Holomelina lamae*. Mp 32°. Bp 330°, Bp_{30} 212°.

Hildebrandt, J.H. *et al, J. Am. Chem. Soc.*, 1929, **51**, 2487.
Cocker, W. *et al, J. Chem. Soc.*, 1963, 677 (*isol*)
Carnduff, J. *et al, Phytochemistry*, 1966, **5**, 1029 (*isol*)
Roelofs, W.L. *et al, Science* (*Washington, D.C.*), 1971, **171**, 684.
Olf, H.G. *et al, J. Chem. Phys.*, 1973, **59**, 534 (*raman*)
Balinova-Tsvetkova, A. *et al, Dokl. Skh. Akad.* (*Sofia*), 1974, **7**, 45; *CA*, **83**, 32911c (*glc*)
Wong, J.W. *et al, J. Chem. Ecol.*, 1984, **10**, 463.

Nonadecanoic acid N-00022

[646-30-0]

$$H_3C(CH_2)_{17}COOH$$

$C_{19}H_{38}O_2$ M 298.508

Isol. from wood of *Ulex europaeus* and *Shorea maranti* and other plant sources. Cryst. (EtOH). Mp 68.7°. Bp_{100} 297-298°, Bp_{10} 227-230°.

Me ester: [1731-94-8].

$C_{20}H_{40}O_2$ M 312.535

Mp 38.9°.

Tetratriacontanyl ester: [94410-18-1]. *Tetratriacontanyl nonadecanoate*

$C_{53}H_{106}O_2$ M 775.419

Isol. from seeds of *Cassia spectabilis* (Leguminosae). Cryst. (C_6H_6). Mp 65°.

Jones, R.N., *Can. J. Chem.*, 1962, **40**, 321 (*ir*)
Ames, D.E. *et al, J. Chem. Soc.*, 1963, 775 (*synth*)
Cocker, W. *et al, Perfum. Essent. Oil Rec.*, 1963, **54**, 235; 1964, **55**, 442 (*isol*)
Matucha, M. *et al, J. Chromatogr.*, 1972, **65**, 371 (*glc*)
Marosi, L.S. *et al, Justus Liebigs Ann. Chem.*, 1973, 584 (*cryst struct*)
Morales, R.W. *et al, Biochim. Biophys. Acta*, 1976, **431**, 206 (*isol*)
Singh, M. *et al, Z. Naturforsch., B*, 1984, **39**, 1425 (*Tetratriacontanyl nonadecanoate*)

1-Nonadecanol N-00023

Nonadecyl alcohol

[1454-84-8]

$$H_3C(CH_2)_{17}CH_2OH$$

$C_{19}H_{40}O$ M 284.524

Mp 62-63°. $Bp_{0.32}$ 166-167°, $Bp_{0.2}$ 179-184°.

Formyl:

$C_{20}H_{40}O_2$ M 312.535

Mp 66.9°.

4-Nitrobenzoyl: Mp 58.9°.

Levene, P.A. *et al, J. Biol. Chem.*, 1924, **59**, 905 (*synth*)
Shiina, S., *Chem. Zentralbl.*, 1936, **1**, 755 (*synth*)
Baarschers, W.H. *et al, Chem. Scr.*, 1974, **5**, 233 (*isol*)
Mace, A.W., *J. Pharm. Pharmacol.*, 1975, **27**, 209 (*glc*)

2-Nonadecanol, 9CI N-00024

[26533-36-8]

$$H_3C(CH_2)_{16}CH(OH)CH_3$$

$C_{19}H_{40}O$ M 284.524

(±)-form
Mp 51-51.8°.

Breusch, F.L. *et al*, *Chem. Ber.*, 1953, **86**, 669 (*synth*)

2-Nonadecanone N-00025

Heptadecyl methyl ketone
[629-66-3]

$H_3C(CH_2)_{16}COCH_3$

$C_{19}H_{38}O$ M 282.509
Cryst. (EtOH). Mp 55-56°. Bp_{110} 266.5°.
Oxime:
$C_{19}H_{39}NO$ M 297.523
Cryst. (EtOH). Mp 76°.
Semicarbazone: Mp 125.5-126° (119°).

Thoms, H. *et al*, *Justus Liebigs Ann. Chem.*, 1907, **357**, 161.
Bailey, A.V. *et al*, *J. Chem. Eng. Data*, 1970, **15**, 542 (*cryst struct*)
Golemba, F.J. *et al*, *Macromolecules*, 1972, **5**, 63.
Middleditch, B.S. *et al*, *Org. Mass Spectrom.*, 1972, **6**, 179 (*ms*)
Simoneit, B.R. *et al*, *Adv. Org. Geochem., Proc. Int. Meet., 6th*, 1974, 191 (*isol*)

4-Nonadecanone N-00026

$H_3C(CH_2)_{14}CO(CH_2)_2CH_3$

$C_{19}H_{38}O$ M 282.509
Spar. sol. EtOH. Mp 50.5°. Bp_{11} 211°, part. dec.
Oxime:
$C_{19}H_{39}NO$ M 297.523
Cryst. in 2 forms. Mp 25.5-26.5°, Mp 43.5-44.5° (dimorph.).

Furukawa, S., *CA*, 1933, **27**, 2131 (*synth*)
Breuch, F.L. *et al*, *Chem. Ber.*, 1953, **86**, 684 (*synth*)

10-Nonadecanone N-00027

Dinonyl ketone. Caprinone
[504-57-4]

$H_3C(CH_2)_8CO(CH_2)_8CH_3$

$C_{19}H_{38}O$ M 282.509
Mp 58°, Mp 65.5°.

Briese, R.R. *et al*, *J. Am. Chem. Soc.*, 1933, **55**, 1697 (*synth*)
Komarewsky, V.I. *et al*, *J. Am. Chem. Soc.*, 1941, **63**, 3269 (*synth*)
Yakerson, V.I. *et al*, *Zh. Anal. Khim.*, 1968, **23**, 293 (*glc*)
Meyer, E.F. *et al*, *J. Phys. Chem.*, 1973, **77**, 2140 (*glc*)

3,6,9,12,15-Nonadecapentaene, 9CI N-00028

$H_3CCH_2(CH{=}CHCH_2)_5CH_2CH_3$

$C_{19}H_{30}$ M 258.446
All-(Z)-form [33426-21-0]
Found in marine benthic algae.

Youngblood, W.W. *et al*, *Mar. Biol. (Berlin)*, 1971, **8**, 190 (*isol, struct, ir, ms*)

1,3,6,9-Nonadecatetraene N-00029

$H_2C{=}CHCH{=}CHCH_2CH{=}CHCH_2CH{=}CH(CH_2)_8CH_3$

$C_{19}H_{32}$ M 260.462
(Z,Z,Z)-form [82970-94-3]
Pheromone from the geometrid moth *Operophtera brumata*. $Bp_{0.25}$ 110-115°.

Roelofs, W.L. *et al*, *Science (Washington, D.C.)*, 1982, **217**, 657 (*isol*)
Huang, W. *et al*, *J. Org. Chem.*, 1983, **48**, 2270 (*synth*)
Jain, S.C. *et al*, *J. Org. Chem.*, 1983, **48**, 2274 (*synth*)
Bestmann, H.J. *et al*, *Tetrahedron Lett.*, 1982, **23**, 4007 (*isol, struct, synth*)
Baker, R. *et al*, *J. Chem. Res., Synop.*, 1984, 190 (*synth*)

1,10,13,16-Nonadecatetraene, 9CI N-00030

$H_2C{=}CH(CH_2)_7CH{=}CHCH_2CH{=}CHCH_2CH{=}CHCH_2CH_3$

$C_{19}H_{32}$ M 260.462
(all-Z)-form [129436-92-6]
Isol. from the brown alga *Nothaia anomala*. Unstable oil. Not obt. pure.
16,17-Dihydro: [106799-43-3]. *1,10,13-Nonadecatriene, 9CI*
$C_{19}H_{34}$ M 262.478
Isol. from *N. anomala*. Unstable oil.
16,17-Dihydro,10S,11R:13S,14R-diepoxide: [129436-85-7]. *10,11:13,14-Diepoxy-1-nonadecene*
$C_{19}H_{34}O_2$ M 294.476
Isol. from *N. anomala*. Unstable oil.

Bestmann, H.J. *et al*, *Justus Liebigs Ann. Chem.*, 1987, 417 (*synth*)
Barrow, R.A. *et al*, *Aust. J. Chem.*, 1990, **43**, 895 (*isol, pmr, cmr, ms*)

3,6,9,11-Nonadecatetraene, 9CI N-00031

$H_3C(CH_2)_6CH{=}CHCH{=}CHCH_2$
$CH{=}CHCH_2CH{=}CHCH_2CH_3$

$C_{19}H_{32}$ M 260.462
(3*Z*,6*Z*,9*Z*,11*E*)- and (3*Z*,6*Z*,9*Z*,11*Z*)-isomers, together with 3,6,9-Nonadecatriene, N-00032, are the principal components of the sex pheromone of the fall cankerworm moth, *Alsophila pometaria*.
[90075-17-5, 90130-70-4, 94537-06-1]

Mitter, C. *et al*, *J. Chem. Ecol.*, 1984, **10**, 463, 1579; 1985, **11**, 727; 1987, **13**, 1823 (*isol, synth*)
Palaniswamy, P. *et al*, *Environ. Entomol.*, 1986, **15**, 943.

3,6,9-Nonadecatriene N-00032

$H_3CCH_2CH{=}CHCH_2CH{=}CHCH_2CH{=}CH(CH_2)_8CH_3$

$C_{19}H_{34}$ M 262.478
(3Z,6Z,9Z)-form [89353-62-8]
Pheromone component of the giant looper *Boarmia selenaria* and *B. rhomboidaria*. $Bp_{0.05}$ 110-130° (bath).
6R,7S-Epoxide: [105228-40-8]. *2-(2-Dodecenyl)-3-(2-pentenyl)oxirane. 6,7-Epoxy-3,9-nonadecadiene*
Sex pheromone constit. of secretions found in glands of *Colotois penuaria* and *Eramis defoliaria* (as stereoisomeric mixt.). $[\alpha]_D^{21}$ −1.3° (c, 11.3 in $CHCl_3$).
6S,7R-Epoxide: [105228-35-1].
$[\alpha]_D^{22}$ +1.2° (c, 7.1 in $CHCl_3$) (~ 100% e.e).
9S,10R-Epoxide: [97668-96-7]. *2-Nonyl-3-(2,5-octadienyl) oxirane, 9CI. 9,10-Epoxy-3,6-nonadecadiene*
Synergistic component of the sex pheromones of several geometrid and noctuid moths. Antagonistic to *Sicya macularia*.
3R,4S-Epoxide: [129895-16-5]. *3,4-Epoxy-6,9-nonadecadiene*
$C_{19}H_{34}O$ M 278.477
Behavioural antagonist to the male-attractive property of the 3*S*,4*R*-form. Liq. $[\alpha]_D^{22}$ −21° (c, 2.83 in CH_2Cl_2).
3S,4R-Epoxide: [129895-14-3].
Component of female sex pheromones of several Lepidoptera species. Liq. $[\alpha]_D^{22}$ +2.5° (c, 2.7 in CH_2Cl_2).
[97697-78-4, 97697-79-5, 97718-63-3, 129895-12-1, 129939-19-1]

Becker, D. *et al*, *Tetrahedron Lett.*, 1983, **24**, 5505; 1990, **31**, 4923 (*isol, synth*)

Baker, R. *et al*, *J. Chem. Res., Synop.*, 1984, 190 (*synth*)
Wong, J.W. *et al*, *J. Chem. Ecol.*, 1985, **11**, 727 (*9,10-epoxide*)
Bestmann, H. *et al*, *Justus Liebigs Ann. Chem.*, 1987, 417 (*synth, pmr*)
Mangold, E.W. *et al*, *Z. Naturforsch., C*, 1987, **42**, 1035 (*synth*)
Dogra, V. *et al*, *J. Indian Chem. Soc.*, 1989, **66**, 169 (*synth*)
Millar, J.G. *et al*, *J. Chem. Ecol.*, 1990, **16**, 2153, 2307, 2317, 3393 (*epoxides, deriv, isol, synth, ms*)
Mori, K. *et al*, *Synthesis*, 1991, 1125 (*synth, epoxides*)
Cosse, A.A. *et al*, *J. Chem. Ecol.*, 1992, **18**, 165 (*epoxides, isol, synth, ms*)

8,11,14-Nonadecatriynoic acid **N-00033**

[19960-54-4]

$H_3C(CH_2)_3C{\equiv}CCH_2C{\equiv}CCH_2C{\equiv}C(CH_2)_6COOH$

$C_{19}H_{26}O_2$ M 286.413
Cryst. Mp 48.5-49.5°.

Struijk, C.B. *et al*, *Recl. Trav. Chim. Pays-Bas (J. R. Neth. Chem. Soc.)*, 1966, **85**, 1233 (*ir*)
Beerthuis, R.K. *et al*, *Recl. Trav. Chim. Pays-Bas (J. R. Neth. Chem. Soc.)*, 1968, **87**, 461 (*synth, ir, pmr*)
Netherlands Pat., 6 802 212, (1968); *CA*, **70**, 106059 (*synth*)

1-Nonadecene **N-00034**

[18435-45-5]

$H_3C(CH_2)_{16}CH{=}CH_2$

$C_{19}H_{38}$ M 266.509
Defensive secretion of tenebrionid beetles. Mp 21.7°. Bp_{10} 177°.

Kozacik, A.P. *et al*, *J. Am. Chem. Soc.*, 1938, **60**, 2436 (*synth*)
Zaitsev, B.A. *et al*, *CA*, 1970, **72**, 61041n.
Van Putte, K., *J. Chem. Soc., Faraday Trans. 2*, 1972, **68**, 357.
Rogers, D.W. *et al*, *J. Phys. Chem.*, 1974, **78**, 2569.
Tschinkel, W.R. *et al*, *J. Insect Physiol.*, 1975, **21**, 753 (*glc*)

9-Nonadecene, 9CI **N-00035**

[31035-07-1]

$H_3C(CH_2)_7CH{=}CH(CH_2)_8CH_3$

$C_{19}H_{38}$ M 266.509
Pheromone in male bees of several spp. of *Bombus* and *Psithyrus*. Liq.

Kullenberg, B. *et al*, *Zoon, Suppl.*, 1973, 23; *CA*, **80**, 93309x (*isol*)

12-Nonadecen-9-one **N-00036**

$H_3C(CH_2)_5CH{=}CHCH_2CH_2CO(CH_2)_7CH_3$

$C_{19}H_{36}O$ M 280.493
(E)-form [63408-51-5]
Active component of female sex pheromone of the peach fruit moth *Carposina nipponensis*.
(Z)-form [63408-45-7]
Active component of female sex pheromone of *C. nipponensis*.

Tamaki, Y. *et al*, *Appl. Entomol. Zool.*, 1977, **12**, 60.
Yoshida, T. *et al*, *Bull. Chem. Soc. Jpn.*, 1982, **55**, 3047 (*synth*)
Hernandez, J.E. *et al*, *Synth. Commun.*, 1983, **13**, 191 (*synth*)
Sviridov, A.F. *et al*, *Tetrahedron Lett.*, 1983, **24**, 4359 (*synth*)
Wenkert, E. *et al*, *J. Org. Chem.*, 1985, **50**, 719 (*synth, pmr, ir*)
Yamamoto, I. *et al*, *J. Org. Chem.*, 1989, **54**, 747 (*synth, pmr, ir, cmr*)
Savoia, D. *et al*, *J. Org. Chem.*, 1991, **56**, 1822 (*synth, ir, pmr*)

18-Nonadecynoic acid **N-00037**

[95978-51-1]

$HC{\equiv}C(CH_2)_{16}COOH$

$C_{19}H_{34}O_2$ M 294.476
Incorrectly referred to as 19-nonadecynoic acid. Cryst. (pet. ether). Mp 74-75°.
Me ester: [95978-52-2].
$C_{20}H_{36}O_2$ M 308.503
Cryst. (MeOH). Mp 42-43°.

Goodman, M.M. *et al*, *J. Med. Chem.*, 1985, **28**, 807 (*synth, pmr, ms*)

3,6-Nonadienal, 9CI **N-00038**

$H_3CCH_2CH{=}CHCH_2CH{=}CHCH_2CHO$

$C_9H_{14}O$ M 138.209
(3E,6Z)-form [57386-95-5]
Bp_{22} 94-95°.
(3Z,6Z)-form [21944-83-2]
Constit. of cucumber flavour also derived from α-linolenic acid. Bp_{14} 86-91°.

Kajiwara, T. *et al*, *Agric. Biol. Chem.*, 1975, **39**, 1617; 1977, **41**, 1481 (*synth, nmr*)
Phillips, D.R. *et al*, *Phytochemistry*, 1978, **17**, 355.

5,7-Nonadienal **N-00039**

$H_3CCH{=}CHCH{=}CH(CH_2)_3CHO$

$C_9H_{14}O$ M 138.209
(E,E)-form [96556-50-2]
Light-yellow oil.

Kurth, M.J. *et al*, *J. Org. Chem.*, 1985, **50**, 2626 (*synth, pmr, ir*)

1,4-Nonadiene, 9CI **N-00040**

[58688-15-6]

$H_2C{=}CHCH_2CH{=}CH(CH_2)_3CH_3$

C_9H_{16} M 124.225
(E)-form [60835-96-3]
Bp 128°.
(Z)-form [54068-77-8]
Bp_{14} 39-40°.

Normant, J.F. *et al*, *J. Organomet. Chem.*, 1974, 269 (*synth*)
Larcock, R.C. *et al*, *J. Organomet. Chem.*, 1978, **156**, 45 (*synth, pmr, ir*)
Yatagai, H. *et al*, *J. Org. Chem.*, 1980, **45**, 1640 (*synth*)
Bestmann, H.J. *et al*, *Justus Liebigs Ann. Chem.*, 1981, 1705 (*synth, pmr*)

1,6-Nonadiene, 9CI **N-00041**

[60024-69-3]

$H_2C{=}CH(CH_2)_3CH{=}CHCH_2CH_3$

C_9H_{16} M 124.225

Byron, N.T. *et al*, *J. Chem. Soc., Chem. Commun.*, 1976, 216 (*synth*)
Rossi, R., *Chim. Ind. (Milan)*, 1977, **59**, 591 (*synth*)

2,7-Nonadiene-1,9-diol N-00042

$HOH_2CCH{=}CH(CH_2)_3CH{=}CHCH_2OH$

$C_9H_{16}O_2$ M 156.224

***(E,E)*-form** [84117-77-1]

Liq.

***(Z,Z)*-form** [81077-36-3]

Liq.

Lennon, P. *et al*, *J. Am. Chem. Soc.*, 1983, **105**, 1233 (*synth*)

3,4-Nonadiene-6,8-diyn-1-ol, 9CI N-00043

$HC{\equiv}CC{\equiv}CHC{=}C{=}CHCH_2CH_2OH$

C_9H_8O M 132.162

***(R)*-form** [505-97-5]

Marasin

Antibiotic from *Murasmius ramealis*. Also prod. by *Cortinellus berkeleyanus*. Oil. Sol. EtOH, Et_2O, pentane. $[\alpha]_D^{25}$ −325° (c, 0.2 in EtOH).

***(S)*-form** [52611-52-6]

Antibiotic from *Aleurodiscus roseus*. Oil. Sol. Et_2O, pentane. $[\alpha]_D^{22}$ +360° (c, 1.6 in EtOH).

(±)-*form*

Oil.

Cambie, R.C. *et al*, *J. Chem. Soc.*, 1963, 4120 (*isol, ir, uv*)
Bew, R.E. *et al*, *J. Chem. Soc. C*, 1966, 129 (*isol, ir, uv*)
Landor, S.R. *et al*, *J. Chem. Soc., Chem. Commun.*, 1966, 585; *J. Chem. Soc., Perkin Trans. 1*, 1974, 557 (*synth, abs config*)
Landor, S.R. *et al*, *J. Chem. Soc. C*, 1967, 189 (*synth, ir, uv*)
Evans, R.J.D. *et al*, *J. Chem. Soc., Perkin Trans. 1*, 1974, 552 (*synth, uv*)
Runge, W. *et al*, *J. Am. Chem. Soc.*, 1977, **99**, 5597.

2,4-Nonadienoic acid N-00044

$H_3C(CH_2)_3CH{=}CHCH{=}CHCOOH$

$C_9H_{14}O_2$ M 154.208

***(2Z,4E)*-form** [121742-69-6]

Pale yellow oil.

Me ester:

$C_{10}H_{16}O_2$ M 168.235

Pale yellow oil.

Furber, M. *et al*, *J. Chem. Soc., Perkin Trans. 1*, 1989, 683 (*synth, pmr, ir, cmr, ms*)

2,6-Nonadienoic acid, 9CI N-00045

[60026-10-0]

$H_3CCH_2CH{=}CHCH_2CH_2CH{=}CHCOOH$

$C_9H_{14}O_2$ M 154.208

***(2E,6Z)*-form** [23605-13-2]

Constit. of tobacco.

Koualev, B.G., *Zh. Org. Khim.*, 1969, **5**, 1771; *CA*, **72**, 11996 (*synth*)
Kimland, B. *et al*, *Phytochemistry*, 1973, **12**, 835 (*pmr, ms*)
Chunan, T. *et al*, *Agric. Biol. Chem.*, 1977, **41**, 1021 (*isol*)
Snider, B.B. *et al*, *J. Org. Chem.*, 1988, **53**, 5320 (*synth*)

5,7-Nonadienoic acid N-00046

$H_3CCH{=}CHCH{=}CH(CH_2)_3COOH$

$C_9H_{14}O_2$ M 154.208

***(E,E)*-form** [77494-72-5]

Liq. $Bp_{0.2}$ 85°.

Hudlicky, T. *et al*, *J. Org. Chem.*, 1981, **46**, 2911 (*synth, ir, nmr*)

2,6-Nonadien-1-ol, 9CI N-00047

[7786-44-9]

$H_3CCH_2CH{=}CH(CH_2)_2CH{=}CHCH_2OH$

$C_9H_{16}O$ M 140.225

***(2E,6E)*-form** [5820-89-3]

Bp_{2-3} 58°.

Benzoyl:

$C_{16}H_{20}O_2$ M 244.333

Liq. $Bp_{0.3}$ 137°.

***(2E,6Z)*-form** [28069-72-9]

Violet alcohol. Cucumber alcohol. Violet-leaf alcohol

Found in *Cucumis sativus* (cucumbers), watermelons and prickly pears. Used in perfumes and as a flavouring agent in tobacco substitutes. Liq. Bp_{11} 98-100°.

▷ RA5394000.

***(2Z,6Z)*-form**

Isol. from violet leaves and peppermint oil. Oily liq. Bp_{24} 108-110°.

Ruzicka, L. *et al*, *Helv. Chim. Acta*, 1944, **27**, 1561 (*struct*)
Zobrist, F. *et al*, *Helv. Chim. Acta*, 1952, **35**, 2380 (*synth*)
Sondheimer, F., *J. Am. Chem. Soc.*, 1952, **74**, 4040 (*synth*)
Vig, O.P. *et al*, *J. Indian Chem. Soc.*, 1965, **42**, 637 (*synth*)
Ali, M.E. *et al*, *Sci. Res. (Dacca)*, 1968, **5**, 91 (*isol*)
Kemp, T.R., *J. Agric. Food Chem.*, 1974, **22**, 717; *Phytochemistry*, 1975, **14**, 2637 (*isol*)
Lewis, R.J., *Sax's Dangerous Properties of Industrial Materials*, 8th Ed., Van Nostrand-Reinhold, 1992, NMV780.

3,6-Nonadien-1-ol N-00048

$H_3CCH_2CH{=}CHCH_2CH{=}CHCH_2CH_2OH$

$C_9H_{16}O$ M 140.225

***(3Z,6E)*-form** [53250-56-9]

Isol. from *Citrullus vulgaris* and *Cucumis melo*.

***(all-Z)*-form** [53046-97-2]

Isol. from *Citrullus vulgaris* and *Cucumis melo*. Constit. of sex pheromone of Caribbean and Mexican fruit flies (*Anastrepha suspensa* and *A. ludens*).

Kemp, T.R. *et al*, *J. Agric. Food Chem.*, 1974, **22**, 717 (*isol*)
Kemp, T.R. *et al*, *Phytochemistry*, 1974, **13**, 1167 (*isol, synth*)
Kajiwara, T. *et al*, *Agric. Biol. Chem.*, 1975, **39**, 1617; 1977, **41**, 1481 (*synth*)
Nation, J.L., *Environ. Entomol.*, 1975, **4**, 27.
Hatanaka, A. *et al*, *Phytochemistry*, 1975, **14**, 2589 (*isol*)
Flath, R.A. *et al*, *J. Agric. Food Chem.*, 1978, **26**, 835 (*isol*)

2,7-Nonadiyne-1,9-diol N-00049

[81077-35-2]

$HOCH_2C{\equiv}C(CH_2)_3C{\equiv}CCH_2OH$

$C_9H_{12}O_2$ M 152.193

Waxy solid. Mp ~20°.

Lennon, P. *et al*, *J. Am. Chem. Soc.*, 1983, **105**, 1233 (*synth*)

3,5-Nonadiyne-1,2,7-triol N-00050

[3858-81-9]

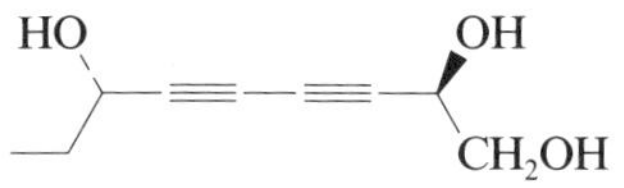

Absolute configuration

$C_9H_{12}O_3$ M 168.192

(2R,7S)-form [104516-02-1]

Metab. of *Clitocybe rhizophora*. Needles (Et_2O/hexane). Mp 71.5-72.5°. $[\alpha]_D^{26}$ −21° (c, 0.53 in EtOH).

1,2-O-Isopropylidene:

$C_{12}H_{16}O_3$ M 208.257

Oil. $[\alpha]_D^{24}$ −61° (c, 0.23 in EtOH).

(2RS,7SR)-form

Mp 68°.

Jones, E.R.H. *et al*, *J. Chem. Soc.*, 1964, 1478 (*isol, synth*)
Jones, E.R.H. *et al*, *J. Chem. Soc. C*, 1966, 144; 1971, 3308 (*biosynth*)
Ord, M.R. *et al*, *J. Chem. Soc., Perkin Trans. 1*, 1975, 687 (*abs config*)

Nonanal, 9CI N-00051

Nonaldehyde. Pelargonaldehyde

[124-19-6]

$H_3C(CH_2)_7CHO$

$C_9H_{18}O$ M 142.241

Found in various plant sources incl. cassava, rice, *Cannabis sativa*. Also found in wood ant secretions. Acts as an insect pheromone and may also be a pheromone in humans. Perfumery ingredient. Liq. with sharp, pleasant odour. Bp 190-192°, Bp_{22} 91-92°. Polymerises readily with H_2SO_4. Oxidised fairly rapidly in air.

▷ RA5700000.

Trimer: [50876-89-6]. *2,4,6-Trioctyl-1,3,5-trioxane*

$C_{27}H_{54}O_3$ M 426.722

Solid. Mp 33.5°.

Di-Me acetal: [18824-63-0]. *1,1-Dimethoxynonane*

$C_{11}H_{24}O_2$ M 188.309

Liq. Bp_{15} 96-98°.

Oxime: [2243-24-5].

$C_9H_{19}NO$ M 157.255

Leaflets (EtOH aq.). Mp 64°. Volatile in steam.

Semicarbazone: [16742-11-3].

Mp 100°.

2,4-Dinitrophenylhydrazone: Golden-yellow needles (EtOH). Mp 106-107° (100°).

Scanlan, J.T. *et al*, *J. Am. Chem. Soc.*, 1940, **62**, 2305 (*synth*)
Harrison, A.G., *Org. Mass Spectrom.*, 1970, **3**, 549 (*ms*)
Fleet, G.W.J. *et al*, *Tetrahedron Lett.*, 1972, 3749 (*synth*)
Botteghi, C. *et al*, *Chim. Ind. (Milan)*, 1977, **59**, 839 (*synth*)
Johnstone, R.A.W. *et al*, *J. Chem. Soc., Chem. Commun.*, 1978, 354 (*synth*)
Sato, F. *et al*, *Chem. Lett.*, 1979, 167 (*synth*)
Katritzky, A.R. *et al*, *J. Org. Chem.*, 1991, **56**, 2143 (*synth*)
Lewis, R.J., *Sax's Dangerous Properties of Industrial Materials*, 8th Ed., Van Nostrand-Reinhold, 1992, NMW500.

1-Nonanamine, 9CI N-00052

Nonylamine, 8CI. 1-Aminononane

[112-20-9]

$H_3C(CH_2)_7CH_2NH_2$

$C_9H_{21}N$ M 143.272

Bp 201°, $Bp_{38.5}$ 107-108.5°. pK_a 10.64 (25°).

N-*Ac:*

$C_{11}H_{23}NO$ M 185.309

Leaflets. Mp 34-35°.

Von Braun, J. *et al*, *Ber.*, 1911, **44**, 1469 (*synth*)
Maihle, A., *Bull. Soc. Chim. Fr.*, 1918, **23**, 235 (*synth*)
Eggert, H. *et al*, *J. Am. Chem. Soc.*, 1973, **95**, 3710 (*cmr*)
Kliger, G.A. *et al*, *Neftekhimiya*, 1977, **17**, 602 (*synth*)
Sax, N.I., *Dangerous Properties of Industrial Materials*, 5th Ed., Van Nostrand-Reinhold, 1979, 871.

Nonane, 9CI N-00053

[111-84-2]

$H_3C(CH_2)_7CH_3$

C_9H_{20} M 128.257

Has been isol. from numerous plant sources incl. pine oils, petroleum and olive oils. Has slight anaesthetic properties. An alarm pheromone in ants. Liq. Fp −51°. Bp 149.5°.

▷ Narcotic in high concs., TLV 1050. RA6115000.

Fischer, W. *et al*, *Z. Phys. Chem., Abt. A*, 1930, **147**, 275 (*rev*)
Averin, V.A. *et al*, *CA*, 1970, **72**, 2733 (*purifn*)
Kajdas, C., *Pol. J. Chem. (Rocz. Chem.)*, 1971, **45**, 1771 (*ms*)
Norman, N. *et al*, *Acta Chem. Scand.*, 1972, **26**, 3913 (*cryst struct*)
Deshayes, H. *et al*, *J. Chem. Soc., Chem. Commun.*, 1975, 439 (*synth*)
Baker, A. *et al*, *Helv. Chim. Acta*, 1977, **60**, 2958 (*synth*)
Hazards in the Chemical Laboratory, (Bretherick, L., Ed.), 3rd Ed., Royal Society of Chemistry, London, 1981, 421.
Lewis, R.J., *Sax's Dangerous Properties of Industrial Materials*, 8th Ed., Van Nostrand-Reinhold, 1992, NMX000.
Low, L.K. *et al*, *Ethel Browning's Toxicity and Metabolism of Industrial Solvents*, (Snyder, R., Ed.), 2nd Ed., Elsevier, 1987, **1**, 318 (*rev, tox*)

Nonanedioic acid, 9CI N-00054

1,7-Heptanedicarboxylic acid. ***Azelaic acid,*** **INN.** *Anchoic acid. Lepargylic acid. Emerox 1144. ZK 62498*

[123-99-9]

$HOOC(CH_2)_7COOH$

$C_9H_{16}O_4$ M 188.223

Occurs in rancid fats. Isol. from *Lycopodium* spp. and a few higher plants. Used to treat acne. Antineoplastic agent. Needles or leaflets. Spar. sol. H_2O. Mp 106.5°. Bp >360° part. dec., Bp_{10} 226°. pK_{a1} 4.56; pK_{a2} 5.53 (25°, 0.1*M* KNO_3). Not steam-volatile.

▷ CM1980000.

Di-Me ester: [1732-10-1].

$C_{11}H_{20}O_4$ M 216.277

Bp_8 140°.

Dichloride:

$C_9H_{14}Cl_2O_2$ M 225.114

Bp_{18} 166°.

Anhydride: [4196-95-6]. *2,10-Oxecanedione, 9CI. Azelaic anhydride*

$C_{18}H_{30}O_7$ M 358.431

Mp 53-55°.

Diamide:

$C_9H_{18}N_2O_2$ M 186.253

Prisms. Mp 175°.

Dinitrile: [1675-69-0]. *1,7-Dicyanoheptane*

$C_9H_{14}N_2$ M 150.223

Liq. Bp_{20} 195°. n_D^{25} 1.4426.

Ger. Pat., 565 158, (1933); *CA*, **27**, 1008 (*synth*)
Org. Synth., 1933, **13**, 4 (*synth*)
Mowry, D.T. *et al*, *J. Am. Chem. Soc.*, 1950, **72**, 4439 (*synth*)
Org. Synth., Coll. Vol., 4, 1963, 62 (*nitrile*)
Housty, J. *et al*, *Acta Crystallogr.*, 1967, **22**, 288 (*cryst struct*)

Martindale, The Extra Pharmacopoeia, 28th/29th Ed., Pharmaceutical Press, London, 1982/1989, 14036.
Mingrone, G. *et al*, *Drugs Exp. Clin. Res.*, 1983, **9**, 447 (*tox*)
Breathnach, A.S. *et al*, *Br. J. Dermatol.*, 1984, **111**, 115 (*rev, pharmacol*)
Thesen, R., *Pharm. Ztg.*, 1992, **137**, 30, 34 (*rev*)
Lewis, R.J., *Sax's Dangerous Properties of Industrial Materials*, 8th Ed., Van Nostrand-Reinhold, 1992, ASB750.

1,3-Nonanediol, 9CI, 8CI — N-00055

[23433-07-0]

$H_3C(CH_2)_5CH(OH)CH_2CH_2OH$

$C_9H_{20}O_2$ M 160.256

Constit. of rectal gland secretions of Melon fly *Dacus cucurbitae*.

(−)-*form*

Liq. $Bp_{0.1}$ 97-99°. $[\alpha]_D^{22}$ −6.00° (neat), $[\alpha]_D$ −6.30° (c, 10.0 in EtOH).

(±)-*form*

Cosmetic preservative and bactericide. The diacetate is a perfumery ingredient. Oil. Bp_{12} 144-146°, $Bp_{0.8}$ 90-92°.

N-*Phenylcarbamate:* Cryst. (2,2,4-trimethylpentane). Mp 83-84.5°.

Bis-N-*phenylcarbamate:* Cryst. (2,2,4-trimethylpentane). Mp 85-85.5°.

Bendz, G. *et al*, *Acta Chem. Scand.*, 1957, **11**, 903 (*synth*)
Diaper, D.G.M. *et al*, *Can. J. Chem.*, 1960, **38**, 1976 (*synth*)
Kuleszi, J. *et al*, *Riechst., Aromen, Koerperpflegem.*, 1969, **19**, 157; *CA*, **71**, 49865r (*synth*)
Akedo, M. *et al*, *J. Food Sci.*, 1977, **42**, 699; *CA*, **87**, 48469h (*use*)
Kitching, W. *et al*, *J. Org. Chem.*, 1989, **54**, 3893.
Perkins, M.V. *et al*, *J. Chem. Soc., Perkin Trans. 1*, 1990, 1111.

1,9-Nonanediol, 9CI — N-00056

Nonamethylene glycol

[3937-56-2]

$HOCH_2(CH_2)_7CH_2OH$

$C_9H_{20}O_2$ M 160.256

Acts as a synergist to the pheromone of the moth *Adoxophyes orana*. Cryst. (C_6H_6). V. sol. EtOH, Et_2O; sol. hot C_6H_6; insol. petrol. Mp 45.8°. Bp_{20} 173-175°.

Di-Ac: [4944-60-9].
$C_{13}H_{24}O_4$ M 244.330
Pheromone synergist. Liq. $Bp_{0.4}$ 124-128°.

Di-Me ether: 1,9-Dimethoxynonane
$C_{11}H_{24}O_2$ M 188.309
Liq. Bp_{10} 114-115°.

Chuit, P., *Helv. Chim. Acta*, 1926, **9**, 264 (*synth*)
Polyakova, S.G. *et al*, *Zh. Prikl. Khim. (Leningrad)*, 1970, **43**, 1144 (*synth*)
Hendry, L.B. *et al*, *J. Chem. Ecol.*, 1975, **1**, 317 (*synth*)
Blanc, P.A. *et al*, *Org. Mass Spectrom.*, 1978, **13**, 135 (*ms*)

1-Nonanesulfonic acid, 9CI — N-00057

$H_3C(CH_2)_8SO_3H$

$C_9H_{20}O_3S$ M 208.321

Anhyd. cryst. (H_2O). Mp 46°.

Benzylaniline salt: Cryst. ($EtOH/Et_2O$). Mp 90.5-91°.

tert-*Butyl ester:* [52075-30-6].
$C_{13}H_{28}O_3S$ M 264.428
Liq. $Bp_{0.15}$ 123°.

Diethylamide: [52075-35-1].
Liq. $Bp_{0.35}$ 120°.

Noller, C.R. *et al*, *J. Am. Chem. Soc.*, 1933, **55**, 1090 (*synth*)
McBain, E.L. *et al*, *J. Am. Chem. Soc.*, 1939, **61**, 3210 (*props*)
Vol'fson, L.G. *et al*, *Zh. Obshch. Khim.*, 1950, **20**, 2085.
Truce, W.E. *et al*, *J. Org. Chem.*, 1974, **39**, 1449 (*derivs*)

Nonanoic acid, 9CI — N-00058

Nonoic acid. Pelargonic acid. Octane-1-carboxylic acid. Nonylic acid

[112-05-0]

$H_3C(CH_2)_7COOH$

$C_9H_{18}O_2$ M 158.240

Occurs as esters in oil of pelargonium. Acid isol. from *Ajania, Rubus, Rhamnus, Paulownia, Artemisia* spp., numerous other plant sources and human skin. Acts as bactericide and fungicide. Is used in snail repellents. Simple esters used as flavouring ingedients. Oily, irritant liq. with characteristic odour. Mp 15°, Fp 12.24°. Bp 255°, Bp_1 98.9°. pK_a 4.95. Spar. steam-volatile.

▷ RA6650000.

NH_4 salt: [63718-65-0].
Solid (H_2O). Mp 115°.

Me ester: [1731-84-6].
$C_{10}H_{20}O_2$ M 172.267
Isolated from *Cistus, Hyssopus* and *Pinus* spp., and from civet, cloves, hops, tobacco, vanilla and certain wines. Liq. Bp_8 88-91°.

4-Bromophenacyl ester: [35300-70-0].
Cryst. (EtOH). Mp 68.5°.

Chloride: [764-85-2].
$C_9H_{17}ClO$ M 176.685
Liq. Fp −60.5°. Bp 215°, Bp_{15} 98°.

Anhydride: [1680-36-0].
$C_{18}H_{34}O_3$ M 298.465
Liq. Fp 5° (16°). Bp_{15} 207°.

Amide: [1120-07-6].
$C_9H_{19}NO$ M 157.255
Cryst. Insol. cold H_2O. Mp 92-93° (98.9°).

Nitrile: [2243-27-8]. *Nonanenitrile. 1-Cyanooctane. Pelargononitrile. Octyl cyanide*
$C_9H_{17}N$ M 139.240
Liq. Fp −34.2°. Bp 224°. n_D^{20} 1.4260.
▷ RA6606000.

Deffet, L., *Bull. Soc. Chim. Belg.*, 1931, **40**, 385 (*props*)
Foglia, T.A. *et al*, *J. Am. Oil Chem. Soc.*, 1977, **54**, 858A (*synth, ester*)
Krapcho, A.P. *et al*, *J. Org. Chem.*, 1977, **42**, 3749 (*synth*)
Miyazaki, S. *et al*, *J. Am. Oil Chem. Soc.*, 1978, **55**, 536 (*synth*)
Evans, D.A. *et al*, *J. Am. Chem. Soc.*, 1979, **101**, 371 (*synth, ir, pmr, ester*)
Jiricny, J. *et al*, *J. Chem. Soc., Perkin Trans. 1*, 1981, 1487 (*nitrile*)
Lewis, R.J., *Sax's Dangerous Properties of Industrial Materials*, 8th Ed., Van Nostrand-Reinhold, 1992, NMY000.

2-Nonanol, 9CI, 8CI — N-00059

[628-99-9]

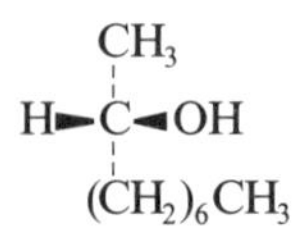

(*S*)-*form*

$C_9H_{20}O$ M 144.256

Isol. from various essential oils, e.g. *Larix laricina, L. lyallii, Citrus sinensis, Zingiber officinale, Anthocephalus cadamba*.

(*S*)-*form* [70419-06-6]
Bp_{19} 105°. $[\alpha]_D^{20}$ +11.90° (C_6H_6).

(±)-*form* [74683-66-2]
Fp −35°. Bp 193-194°, Bp_{10} 84-86°.

Ac: [14936-66-4].
$C_{11}H_{22}O_2$ M 186.294
Bp_{30} 108-109°.
3,5-Dinitrobenzoyl: Mp 43-45°.
Me ether: [54894-33-6]. *2-Methoxynonane*
$C_{10}H_{22}O$ M 158.283
Bp 186-188°.

Pickard, R.H. *et al, J. Chem. Soc.*, 1911, **99**, 56, 70 (*synth*)
Keil, W., *Hoppe Seyler's Z. Physiol. Chem.*, 1942, **274**, 175 (*synth*)
Nair, G.V. *et al, Can. J. Chem.*, 1959, **37**, 1608; 1960, **38**, 177 (*isol*)
McKillop, A. *et al, Tetrahedron*, 1974, **30**, 2467 (*synth*)
Millon, J. *et al, Tetrahedron Lett.*, 1976, 1095 (*synth*)
Normant, J.F. *et al, Bull. Soc. Chim. Fr.*, 1980, 289 (*synth*)
Keinan, E. *et al, J. Am. Chem. Soc.*, 1986, **108**, 162 (*synth*)

3-Nonanol, 9CI N-00060

Ethylhexylcarbinol
[624-51-1]

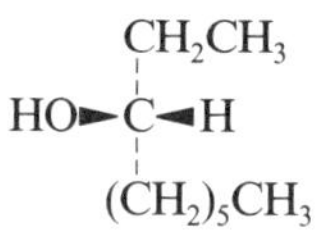

$C_9H_{20}O$ M 144.256
Constit. of Scotch spearmint oil and yarrow oil from *Achillea fragrantissima*. Also a Myrmincine worker ant pheromone.
(R)-form [61925-50-6]
$[\alpha]_D^{24}$ −7.40° (c, 8.1 in $CHCl_3$).
(S)-form
$[\alpha]_D^{24}$ +9.60° (c, 8.3 in $CHCl_3$).
(±)-form
Bp_{20} 94°. n_D^{20} 1.4279.
3,5-Dinitrobenzoyl: Mp 44-45°.

Shalaby, A.F. *et al, J. Pharm. Sci.*, 1964, **53**, 1502; *CA*, **62**, 3879h (*isol*)
Thewalt, K. *et al, J. Prakt. Chem.*, 1964, **26**, 237 (*synth*)
Mukaiyama, T. *et al, Chem. Lett.*, 1976, 893 (*synth*)
Cammert, M.C. *et al, J. Insect Physiol.*, 1978, **24**, 207; *CA*, **90**, 19298u.
Japan. Pat., 78 25 502, (1978); *CA*, **89**, 42412d (*synth*)

4-Nonanol, 9CI, 8CI N-00061

Amylpropylcarbinol. Pentylpropylcarbinol
[5932-79-6]

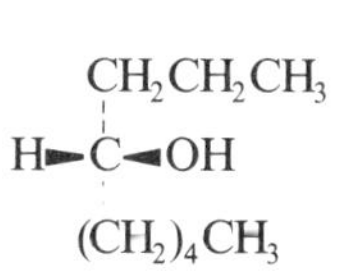

(S)-form

$C_9H_{20}O$ M 144.256
(S)-form [52708-03-9]
Bp_{18} 94-95°. $[\alpha]_D^{25}$ +0.57° (neat).
(±)-form
Bp 191-192.5°.
Ac: [5990-73-8].
$C_{11}H_{22}O_2$ M 186.294
Bp 199-200°.

Pearce, P.J. *et al, J. Chem. Soc., Perkin Trans. 1*, 1972, 1655 (*synth*)
Schlosser, M. *et al, Chem. Ber.*, 1974, **107**, 1162 (*synth*)
Ohta, S., *Bull. Chem. Soc. Jpn.*, 1986, **59**, 1181 (*synth*)

2-Nonanone, 9CI N-00062

Heptyl methyl ketone
[821-55-6]

$H_3C(CH_2)_6COCH_3$

$C_9H_{18}O$ M 142.241
Present in attar of rose, clove oil, passionflowers, sorghum, asparagus, tomato, corn, cheese, beer etc. Acts as an alarm pheromone in ants, hornets and honeybees. Constit. of temporal gland secretions of *Elephas maximas* (Asian elephant). Has weak bactericidal activity. Liq. Fp −15°. Bp 194-196°, Bp_{12} 75-77°.
▷ Mod. toxic. RA8225000.
Oxime: [52435-37-7].
$C_9H_{19}NO$ M 157.255
Liq. or cryst. Mp 16°. Bp_{15} 131°.
Semicarbazone: [818-06-4]. *2-(1-Methyloctylidene) hydrazinecarboxamide*
Solid. Mp 119-120°.

Morgan, G.T. *et al, J. Soc. Chem. Ind., London*, 1925, **44**, 108T (*synth*)
Carpenter, W. *et al, J. Am. Chem. Soc.*, 1967, **89**, 6167 (*ms*)
Olah, G.A. *et al, Synthesis*, 1977, 419 (*synth*)
Demuth, M., *Helv. Chim. Acta*, 1978, **61**, 3136 (*synth*)
Sonnek, G. *et al, J. Prakt. Chem.*, 1978, **320**, 341 (*synth*)
Sato, F. *et al, Chem. Lett.*, 1979, 167 (*synth*)
Rasmussen, L.E.L. *et al, J. Chem. Ecol.*, 1990, **16**, 2167.
Lewis, R.J., *Sax's Dangerous Properties of Industrial Materials*, 8th Ed., Van Nostrand-Reinhold, 1992, NMY500.

3-Nonanone N-00063

[925-78-0]

$H_3C(CH_2)_5COCH_2CH_3$

$C_9H_{18}O$ M 142.241
Insect alarm pheromone; component of some beetle defence secretions. Liq. Fp −8°. Bp 190°, Bp_{20} 86°.
Oxime: [22418-72-0].
$C_9H_{19}NO$ M 157.255
Liq. Bp_1 70°.
Phenylhydrazone: Amber oil. Bp_{15} 190-192°.
Semicarbazone: [1669-37-0]. *2-(1-Ethylheptylidene) hydrazinecarboxamide*
Cryst. (EtOH aq.). Mp 111-112°.

Pickard, R.H. *et al, J. Chem. Soc.*, 1913, **103**, 1923 (*synth*)
Kraft, M. *et al, Org. Mass Spectrom.*, 1968, **1**, 617 (*ms*)
Yamamoto, Y. *et al, Bull. Chem. Soc. Jpn.*, 1975, **48**, 3682 (*synth, ir, pmr*)
Masuyama, Y. *et al, Tetrahedron Lett.*, 1976, 2967 (*synth*)
Cammaerts, M.C. *et al, J. Insect Physiol.*, 1978, **24**, 207 (*isol*)

4-Nonanone, 9CI N-00064

Amyl propyl ketone. Pentyl propyl ketone
[4485-09-0]

$H_3C(CH_2)_4CO(CH_2)_2CH_3$

$C_9H_{18}O$ M 142.241
Isol. from lemongrass. Alarm pheromone in ants and hornets. Liq. Bp 187-188°, Bp_{20} 75-76°.
Semicarbazone: Cryst. (EtOH or ligroin). Mp 73-74° (rapid heat) (67°, 145°).
4-Nitrophenylhydrazone: Unstable yellow cryst. (EtOH) or light-yellow plates (ligroin). Mp 84-85°.

Bryant, S.A. *et al, J. Chem. Soc.*, 1931, 2080 (*synth*)
Brodman, B.W. *et al, Sep. Sci. Technol.*, 1967, **2**, 139 (*cryst struct*)
Soussan, G. *et al, Bull. Soc. Chim. Fr.*, 1972, 4228 (*synth*)

Sheikh, M.Y. *et al*, *Tetrahedron Lett.*, 1972, 257 (*synth*)
McAdoo, D.J. *et al*, *Org. Mass Spectrom.*, 1974, **9**, 525 (*ms*)
Kende, A.S. *et al*, *Synth. Commun.*, 1978, **8**, 59 (*synth*)

Nonatriacontane N-00065

[7194-86-7]

$H_3C(CH_2)_{37}CH_3$

$C_{39}H_{80}$ M 549.061
Cryst. (C_6H_6). Mp 80-80.2°.

Stenhagen, E. *et al*, *J. Am. Chem. Soc.*, 1944, **66**, 845 (*synth*)
Nelson, D.R. *et al*, *J. Lipid Res.*, 1972, **13**, 413 (*glc*)

1,3,9,15,21,27,33,39-Nonatriacontaneoctacarboxylic acid N-00066

Octacidomycin

[120636-98-8]

$HOOC(CH_2)_6[CH(COOH)(CH_2)_5]_5CH(COOH)CH_2CH_2COOH$

$C_{47}H_{80}O_{16}$ M 901.139
Oligocarboxylic antibiotic. Isol. from a *Streptomyces* sp. Red-brown oil.

Octa-Me ester: [120665-58-9].
Amorph. Mp 92°. $[\alpha]_D^{20}$ +0.7° (c, 1 in $CHCl_3$).

Kerber, K.-H. *et al*, *Justus Liebigs Ann. Chem.*, 1989, 712, 719 (*isol, struct, synth*)

4,6,8-Nonatriyne-1,2-diol N-00067

HC≡CC≡CC≡$CCH_2CH(OH)CH_2OH$

$C_9H_8O_2$ M 148.161
Isol. from *Polyporus biformis*. Shows antibacterial and antifungal activity. Needles. Highly unstable. Darkens rapidly in light.

Mono-3,5-dinitrobenzoyl: Needles. Mp 50-55°.
Bis-3,5-dinitrobenzoyl, 1-naphthylamine complex: Orange needles (C_6H_6/EtOH). Mp 111-112°.

Anchel, M. *et al*, *J. Biol. Chem.*, 1954, **208**, 319 (*isol, uv, props*)
Bohlmann, F. *et al*, *Chem. Ber.*, 1961, **94**, 948 (*synth, uv, ir*)
Jones, E.R.H. *et al*, *J. Chem. Soc.*, 1963, 2048 (*synth, uv, ir*)

3-Nonenal N-00068

$H_3C(CH_2)_4CH{=}CHCH_2CHO$

$C_9H_{16}O$ M 140.225
(*Z*)-*form* [31823-43-5]
Unstable oil.

Viala, J. *et al*, *Synthesis*, 1988, 395 (*synth, pmr, cmr*)
Treilhou, M. *et al*, *J. Org. Chem.*, 1992, **57**, 3203 (*synth, ir*)

4-Nonenal, 9CI N-00069

$H_3C(CH_2)_3CH{=}CHCH_2CH_2CHO$

$C_9H_{16}O$ M 140.225
(*E*)-*form* [2277-16-9]
Bp_2 80°.
(*Z*)-*form* [2277-15-8]
Bp_2 76°.

Bestmann, H.J. *et al*, *Justus Liebigs Ann. Chem.*, 1981, 1705 (*synth*)
Takai, K. *et al*, *Bull. Chem. Soc. Jpn.*, 1984, **57**, 446 (*synth, ir, pmr, ms*)

1-Nonene, 9CI N-00070

[124-11-8]

$H_3C(CH_2)_6CH{=}CH_2$

C_9H_{18} M 126.241
Isol. from tars, *Oenanthe*, *Petasites* and *Polygonum* spp., peaches, beef and from beetle defensive secretions. Liq. d^{21} 0.730. Fp −88° (−82.2°). Bp 146°, Bp_{15} 50°. n_D^{21} 1.4140.

Tschinkel, W.R., *J. Insect Physiol.*, 1975, **21**, 659, 753 (*isol*)
Westmijze, H. *et al*, *Recl. Trav. Chim. Pays-Bas* (*J. R. Neth. Chem. Soc.*), 1976, **95**, 299 (*synth*)
Miyaura, N. *et al*, *Bull. Chem. Soc. Jpn.*, 1977, **50**, 2199 (*synth*)
Roberts, J.L. *et al*, *Tetrahedron Lett.*, 1977, 1299 (*synth*)
Sato, F. *et al*, *J. Organomet. Chem.*, 1978, **157**, C30 (*synth*)
Muurisepp, A. *et al*, *Zh. Org. Khim.*, 1978, **14**, 2017 (*ms*)

8-Nonene-3,4-diol N-00071

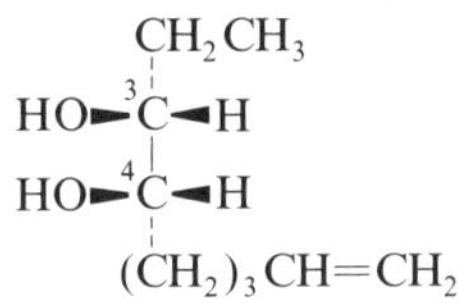

(3*R*,4*S*)-*form*

$C_9H_{18}O_2$ M 158.240
(3*R*,4*S*)-*form* [101833-81-2]
Cryst. (hexane). Mp 81-82°. $[\alpha]_D^{21}$ −11.5° (c, 0.92).
(3*S*,4*R*)-*form* [100762-81-0]
Cryst. (hexane). Mp 81-82°. $[\alpha]_D^{21}$ +11.6° (c, 1.0 in $CHCl_3$).

Mori, K. *et al*, *Tetrahedron*, 1985, **41**, 3429 (*synth, ir, pmr*)
Mori, K. *et al*, *Synthesis*, 1991, 487 (*synth, ir, pmr*)
Mori, K. *et al*, *Justus Liebigs Ann. Chem.*, 1992, 989 (*synth*)

2-Nonene-6,8-diynoic acid, 9CI N-00072

HC≡CC≡$CCH_2CH_2CH{=}CHCOOH$

$C_9H_8O_2$ M 148.161
(*E*)-*form*
2-Phenylethylamide: [99615-84-6]. ***N-(2-Phenylethyl)-2-nonene-6,8-diynamide***
$C_{17}H_{17}NO$ M 251.327
Isol. from *Acmella ciliata*.
(*Z*)-*form*
2-Methylpropylamide: [96602-65-2]. ***N-(2-Methylpropyl)-2-nonene-6,8-diynamide.*** *2-Nonene-6,8-diynoic acid isobutylamide*
$C_{13}H_{17}NO$ M 203.283
Isol. from roots of *Spilanthes oleracea*. Oil.
2-Phenylethylamide: [75872-74-1].
$C_{17}H_{17}NO$ M 251.327
Isol. from *Spilanthes alba* and *Salmea scandens*. Oil. (*E*-) and (*Z*-)forms characterised from *S. alba*.

[75872-75-2, 75872-77-4]

Bohlmann, F. *et al*, *Phytochemistry*, 1980, **19**, 1535; 1985, **24**, 595 (*isol, struct*)
Greger, H. *et al*, *Monatsh. Chem.*, 1985, **116**, 273 (*isol, struct, uv, ir, cmr, pmr*)
Martin, R. *et al*, *Phytochemistry*, 1985, **24**, 2295 (*isol, pmr*)

2-Nonenoic acid, 9CI, 8CI N-00073

2-Nonylenic acid

[3760-11-0]

$H_3C(CH_2)_5CH{=}CHCOOH$

$C_9H_{16}O_2$ M 156.224
Occurs in mushrooms, mutton and tobacco. Used in mosquito repellants, and as a fungicide. Me ester used in perfumery and flavourings. Liq. Spar. sol. H_2O. Fp 0.3°. Bp_{20} 173°. Steam-volatile.
Et ester: [17463-01-3].
$C_{11}H_{20}O_2$ M 184.278
Used as a food preservative. Liq. Bp_{12} 118-120°.
4-Bromophenacyl ester:
$C_{17}H_{21}BrO_3$ M 353.255
Cryst. Mp 77.5-78°.
Chloride: [764-86-3].
$C_9H_{15}ClO$ M 174.670
Liq. Bp_{12} 90-91°.
Amide:
$C_9H_{17}NO$ M 155.239
Cryst. (C_6H_6) or plates (MeOH aq.). Insol. petrol. Mp 118-119° (126-127°).
(*E*)-*form* [14812-03-4]
Isol. from tobacco. Acts as a pheromone in the honeybee. Liq. Mp 1-2° (4°). Bp_{14} 146°.
Et ester: [38112-59-3].
$C_{11}H_{20}O_2$ M 184.278
Liq. Sol. Et_2O, THF, hexane. Bp_{20} 114°.
Chloride: [60784-27-2].
$C_9H_{15}ClO$ M 174.670
Liq. $Bp_{0.4}$ 70-72°.
Amide: [14952-05-7].
$C_9H_{17}NO$ M 155.239
Cryst. (C_6H_6). Mp 131.5°. $Bp_{0.73}$ 120° subl.
(*Z*)-*form* [1577-98-6]
Liq. at −15°. Fp 2-3°. Bp_{15} 140°.
Et ester: [72284-17-4].
Liq. Bp_{15} 108°.
Amide: Cryst. (C_6H_6). Mp 74-74.5°.

Blum, M.S. *et al*, *J. Insect Physiol.*, 1971, **17**, 349 (*use*)
Org. Synth., 1973, **53**, 123 (*synth*)
Katzenellenbogen, J.A. *et al*, *J. Am. Chem. Soc.*, 1974, **96**, 6153 (*synth, ir, ms, pmr, ester*)
Yamada, K. *et al*, *Bull. Chem. Soc. Jpn.*, 1977, **50**, 3431 (*synth, ms, pmr, ester*)
Lombardo, L. *et al*, *Synthesis*, 1978, 131; *Synth. Commun.*, 1978, **8**, 463 (*synth*)
Lucchetti, J. *et al*, *Tetrahedron Lett.*, 1978, 2693 (*synth*)

3-Nonenoic acid, 9CI, 8CI N-00074

[4124-88-3]

$H_3C(CH_2)_4CH{=}CHCH_2COOH$

$C_9H_{16}O_2$ M 156.224
Isol. from gardenia, grapes, tobacco and strawberries. Liq. Fp −4.4°. Bp_2 114-115°.
Me ester: [13481-87-3].
$C_{10}H_{18}O_2$ M 170.251
Isol. from guava.
Chloride: [764-87-4].
$C_9H_{15}ClO$ M 174.670
Liq. $Bp_{4.5}$ 80-81°.
Amide:
$C_9H_{17}NO$ M 155.239
Cryst. (EtOH). Mp 74.5-75°.
(*E*)-*form* [28163-88-4]
Liq. Sol. Et_2O, insol. H_2O. Bp_1 106°.
4-Bromophenacyl ester: Cryst. Mp 68.2-69.3°.
(*Z*)-*form* [41653-98-9]
Isol. from tobacco.
Me ester: [41654-16-4].
$C_{10}H_{18}O_2$ M 170.251
Isol. from tobacco.

Morton, A.A. *et al*, *J. Am. Chem. Soc.*, 1950, **72**, 3785 (*synth*)
Groff, T.M. *et al*, *Ark. Kemi*, 1968, **29**, 179 (*ms, ester*)
Koyama, H. *et al*, *Agric. Biol. Chem.*, 1972, **36**, 793 (*synth, pmr, ester*)
Kimland, B. *et al*, *Phytochemistry*, 1973, **12**, 835 (*isol*)
Hattori, R. *et al*, *Agric. Biol. Chem.*, 1978, **42**, 1351 (*isol*)
Shiota, H., *Koryo*, 1978, **121**, 23 (*isol, ester*)
Kawashima, M. *et al*, *Bull. Chem. Soc. Jpn.*, 1988, **61**, 3255 (*synth*)

4-Nonenoic acid N-00075

[28482-04-4]

$H_3C(CH_2)_3CH{=}CHCH_2CH_2COOH$

$C_9H_{16}O_2$ M 156.224
(*E*)-*form* [35329-50-1]
Et ester: [69361-38-2].
$C_{11}H_{20}O_2$ M 184.278
Liq. Bp_2 83-86°, $Bp_{0.3}$ 60°.

Hammoud, A. *et al*, *Bull. Soc. Chim. Fr.*, 1978, 299.
Evans, D.A. *et al*, *J. Am. Chem. Soc.*, 1982, **104**, 3695.

6-Nonenoic acid N-00076

$H_3CCH_2CH{=}CH(CH_2)_4COOH$

$C_9H_{16}O_2$ M 156.224
(*E*)-*form* [31502-23-5]
$Bp_{0.1}$ 93-95°.

Jacobson, M. *et al*, *J. Med. Chem.*, 1971, **14**, 236 (*synth*)
Scholz, D., *Justus Liebigs Ann. Chem.*, 1983, 98 (*synth*)

8-Nonenoic acid, 9CI N-00077

8-Nonylenic acid
[31642-67-8]

$H_2C{=}CH(CH_2)_6COOH$

$C_9H_{16}O_2$ M 156.224
Liq. with waxy odour. Mp 5° (Fp 10°). Bp 248-249°, Bp_{15} 147°.
p-*Toluidide:* Cryst. (pet. ether). Mp 68°.
Me ester: [20731-23-1].
$C_{10}H_{18}O_2$ M 170.251
Bp_{15} 110° (bath).
Et ester: [35194-39-9].
$C_{11}H_{20}O_2$ M 184.278
Liq. Sol. EtOH, Et_2O, spar. sol. H_2O. Bp 226-228°, Bp_{15} 114-116°.
Chloride:
$C_9H_{15}ClO$ M 174.670
Liq. Bp_{30} 110°.
Nitrile: [5048-34-0]. *8-Nonenenitrile, 9CI. 8-Cyanooctene*
$C_9H_{15}N$ M 137.224
Liq.

Gaubert, P. *et al*, *J. Chem. Soc.*, 1937, 1971 (*synth*)
Dias, J.R. *et al*, *J. Am. Chem. Soc.*, 1972, **94**, 473 (*ms*)
Ogibin, Yu.N. *et al*, *Izv. Akad. Nauk SSSR, Ser. Khim.*, 1976, 2518 (*synth*)
Liu, S.-H., *J. Org. Chem.*, 1977, **42**, 3209 (*synth*)
Giese, B. *et al*, *Chem. Ber.*, 1984, **117**, 3160 (*ester, nitrile, synth, ir, pmr*)

1-Nonen-3-ol, 9CI N-00078

Hexylvinylcarbinol
[21964-44-3]

$H_3C(CH_2)_5CH(OH)CH{=}CH_2$

$C_9H_{18}O$ M 142.241
Isol. from *Petasites japonicus*. Also reported as the (−)-enantiomer from water fennel oil (*Phellandrium aquaticum*).
(±)-form [79605-61-1]
Bp 193-194°, Bp_{20} 92.5°.
4-Nitrobenzoyl: Plates (EtOH). Mp 36-37°.
[79646-41-6]

Dumont, W. *et al, Angew. Chem., Int. Ed. Engl.*, 1974, **13**, 804 (*synth*)
Ueno, Y. *et al, Synthesis*, 1980, 1011 (*synth, pmr*)
Martin, V.S. *et al, J. Am. Chem. Soc.*, 1981, **103**, 6237 (*resoln*)
Halazy, S. *et al, Tetrahedron Lett.*, 1981, 2135 (*synth*)
Boughdady, N.M. *et al, Aust. J. Chem.*, 1987, **40**, 767 (*synth, pmr, ir*)

4-Nonen-1-ol N-00079

$H_3C(CH_2)_3CH{=}CHCH_2CH_2CH_2OH$

$C_9H_{18}O$ M 142.241
(E)-form [16695-34-4]
Liq. Bp_4 55° (Bp_3 83°).

Hammoud, A. *et al, Bull. Soc. Chim. Fr.*, 1978, 299.
Evans, D.A. *et al, J. Am. Chem. Soc.*, 1982, **104**, 3695.

6-Nonen-1-ol N-00080

$H_3CCH_2CH{=}CH(CH_2)_4CH_2OH$

$C_9H_{18}O$ M 142.241
Sex pheromone of the Mediterranean fruit fly *Ceratitus captitata*.
(E)-form [31502-19-9]
$Bp_{0.1}$ 75-80°.
Ac: [30412-53-4].
$C_{11}H_{20}O_2$ M 184.278
Bp_3 89-90°.

Jacobson, M. *et al, J. Med. Chem.*, 1971, **14**, 236 (*synth*)
Scholz, D., *Justus Liebigs Ann. Chem.*, 1983, 98 (*synth*)
Buss, A.D. *et al, J. Chem. Soc., Perkin Trans. 1*, 1987, 2569 (*synth, ir, pmr, ms*)
Brown, H.C. *et al, J. Org. Chem.*, 1988, **53**, 246 (*synth, ir, pmr, cmr*)

8-Nonen-1-ol, 9CI, 8CI N-00081

[13038-21-6]

$H_2C{=}CH(CH_2)_6CH_2OH$

$C_9H_{18}O$ M 142.241
Liq. Bp_{13} 103-104°.
Ac: [13038-22-7].
$C_{11}H_{20}O_2$ M 184.278
Acts as a sex attractant to some *Lepidoptera*. Liq. Bp_{24} 151-155°, $Bp_{0.15}$ 64-67°.

Shishido, K. *et al, J. Org. Chem.*, 1962, **27**, 4073 (*synth, ir*)
Buendia, J., *Bull. Soc. Chim. Fr.*, 1966, 2778 (*synth*)
Dolezal, S., *Collect. Czech. Chem. Commun.*, 1966, **31**, 3765 (*synth*)
Jacobsen, M. *et al, J. Med. Chem.*, 1971, **14**, 236 (*synth, pharmacol*)
Adler, V.E. *et al, J. Econ. Entomol.*, 1972, **65**, 1582 (*occur*)
Cahiez, G. *et al, Tetrahedron Lett.*, 1978, 3013 (*synth*)

1-Nonen-3-one, 9CI N-00082

[24415-26-7]

$H_3C(CH_2)_5COCH{=}CH_2$

$C_9H_{16}O$ M 140.225
Minor constit. in defensive secretion of tenebrionid beetle, *Eleodes beamerii*. Liq. Bp_{35} 85-87°, Bp_{12} 80°.
[95831-13-3]

Cookson, R. *et al, J. Chem. Soc., Chem. Commun.*, 1978, 821 (*synth*)
Takahashi, T. *et al, Tetrahedron Lett.*, 1978, 799 (*synth, ir, pmr*)
Stetter, H. *et al, Chem. Ber.*, 1979, **112**, 1410 (*synth, ir, pmr*)
Shono, T. *et al, J. Am. Chem. Soc.*, 1979, **101**, 984 (*synth, ir, pmr*)
Adams, D.R. *et al, J. Chem. Soc., Perkin Trans. 1*, 1984, 2061 (*synth*)
Ahlbrecht, H. *et al, Synthesis*, 1988, 210 (*synth, pmr*)

4-Nonen-2-ynoic acid N-00083

$H_3C(CH_2)_3CH{=}CHC{\equiv}CCOOH$

$C_9H_{12}O_2$ M 152.193
(E)-form
Cryst. (pentane). Mp 45-47°.

Vlahov, R. *et al, Synth. Commun.*, 1986, **16**, 509 (*synth, ir, pmr*)

2-Nonylcyclopropaneheptanoic acid, 9CI N-00084

8,9-Methyleneoctadecanoic acid

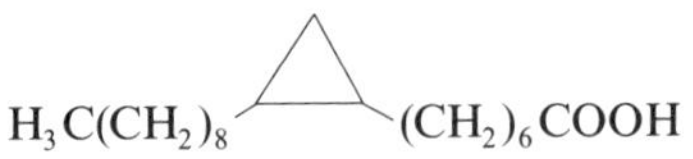

$C_{19}H_{36}O_2$ M 296.492
Both *cis*- and *trans*-forms have been synthesised as Me esters.

Christie, W.W., *Chem. Phys. Lipids*, 1968, **2**, 196 (*synth, glc, ir, nmr*)
Christie, W.W., *Top. Lipid Chem.*, 1970, **1**, 4 (*rev*)
Gunstone, F.D. *et al, Chem. Phys. Lipids*, 1973, **10**, 303 (*synth, glc, nmr*)

4-Nonylphenol N-00085

[104-40-5]

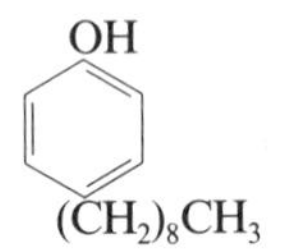

$C_{15}H_{24}O$ M 220.354
Constit. of the temporal gland secretions of elephants (*Elephas* spp.). Also found in the pheromone of *Anthonomus grandis* (boll weevil). Mp 41-42.5°. Bp_{10} 180-181°, $Bp_{1.3}$ 140-142°.
Benzoyl: [6316-62-7].
$C_{22}H_{28}O_2$ M 324.462
Mp 36.2°.

Sandulesco, G. *et al, Bull. Soc. Chim. Fr.*, 1930, **47**, 1300 (*synth*)
Lindemann, L.P. *et al, Anal. Chem.*, 1964, **36**, 2414 (*pmr*)
Hedin, P.A. *et al, J. Insect Physiol.*, 1974, **20**, 2135 (*occur*)
Itokawa, H. *et al, Chem. Pharm. Bull.*, 1989, **37**, 1619 (*synth*)
Krzyzanowska, E. *et al, J. Prakt. Chem.*, 1989, **331**, 617 (*synth*)
Rasmussen, L.E.L. *et al, J. Chem. Ecol.*, 1990, **16**, 2167 (*occur*)
Kiedik, M. *et al, Przem. Chem.*, 1991, **70**, 257 (*synth*)

3-Nonynal N-00086

[87745-64-0]

$H_3C(CH_2)_4C{\equiv}CCH_2CHO$

$C_9H_{14}O$ M 138.209
$Bp_{0.05}$ 48°.

Barbot, F. *et al, Synthesis*, 1983, 651 (*synth, pmr*)
Corey, E.J. *et al, J. Org. Chem.*, 1990, **55**, 1670 (*synth, pmr, ir*)

2-Nonynoic acid, 9CI, 8CI N-00087

[1846-70-4]

$H_3C(CH_2)_5C{\equiv}CCOOH$

$C_9H_{14}O_2$ M 154.208

Possesses some antifungal activity. Liq. Mp −8°. Bp_{18} 154-156°, $Bp_{2.2}$ 122°.

Me ester: [111-80-8]. *Methyl octynecarboxylate*

$C_{10}H_{16}O_2$ M 168.235

Possesses some antifungal activity. Is used in the manuf. of perfumes and cosmetics. Liq. $Bp_{2.3}$ 85°.

▷ RB3450000.

Amide:

$C_9H_{15}NO$ M 153.224

Cryst. (Et_2O/petrol). Mp 94°.

Wotiz, J.H. *et al*, *J. Org. Chem.*, 1954, **19**, 1580 (*synth*)
Groff, T.M. *et al*, *Ark. Kemi*, 1968, **29**, 179 (*ms, ester*)
Silveira, A. Jr. *et al*, *J. Am. Oil Chem. Soc.*, 1971, **48**, 661 (*synth*)
Opdyke, D.L.J., *Food Cosmet. Toxicol., Suppl.*, 1975, **13**, 871 (*rev, bibl, ester*)
Org. Synth., 1976, **55**, 73 (*ester*)
Gershon, H. *et al*, *Can. J. Microbiol.*, 1978, **24**, 593 (*use*)
Lewis, R.J., *Sax's Dangerous Properties of Industrial Materials*, 8th Ed., Van Nostrand-Reinhold, 1992, MNC000.

4-Nonynoic acid N-00088

$H_3C(CH_2)_3C{\equiv}CCH_2CH_2COOH$

$C_9H_{14}O_2$ M 154.208

Mp 48°. $Bp_{5.0}$ 135°. pK_a 6.47.

Me ester:

$C_{10}H_{16}O_2$ M 168.235

$Bp_{3.0}$ 85°.

Amide:

$C_9H_{15}NO$ M 153.224

Mp 110°.

Wotiz, J.H. *et al*, *J. Org. Chem.*, 1954, **19**, 1580 (*synth*)

5-Nonynoic acid N-00089

[56630-34-3]

$H_3CCH_2CH_2C{\equiv}C(CH_2)_3COOH$

$C_9H_{14}O_2$ M 154.208

Mp −3°. $Bp_{1.0}$ 115°. pK_a 6.80.

Me ester:

$C_{10}H_{16}O_2$ M 168.235

$Bp_{1.5}$ 79°.

Amide:

$C_9H_{15}NO$ M 153.224

Mp 102°.

Wotiz, J.H. *et al*, *J. Org. Chem.*, 1954, **19**, 1580 (*synth*)

6-Nonynoic acid N-00090

$H_3CCH_2C{\equiv}C(CH_2)_4COOH$

$C_9H_{14}O_2$ M 154.208

Mp 14°. $Bp_{3.0}$ 130°. pK_a 6.79.

Me ester:

$C_{10}H_{16}O_2$ M 168.235

$Bp_{4.0}$ 85°.

Amide:

$C_9H_{15}NO$ M 153.224

Mp 96°. $Bp_{4.0}$ 85°.

Wotiz, J.H. *et al*, *J. Org. Chem.*, 1954, **19**, 1580.

7-Nonynoic acid N-00091

$H_3CC{\equiv}C(CH_2)_5COOH$

$C_9H_{14}O_2$ M 154.208

Mp 52°. $Bp_{1.0}$ 118°. pK_a 6.83.

Me ester:

$C_{10}H_{16}O_2$ M 168.235

$Bp_{1.5}$ 85°.

Amide:

$C_9H_{15}NO$ M 153.224

Mp 85°.

Wotiz, J.H. *et al*, *J. Org. Chem.*, 1954, **19**, 1580 (*synth*)

8-Nonynoic acid, 9CI, 8CI N-00092

[30964-01-3]

$HC{\equiv}C(CH_2)_6COOH$

$C_9H_{14}O_2$ M 154.208

Cryst. or liq. Mp 19°. Bp_6 100° (Bp_4 128°). pK_a 6.84.

Me ester: [7003-48-7].

$C_{10}H_{16}O_2$ M 168.235

Liq. Bp_4 87°.

Amide:

$C_9H_{15}NO$ M 153.224

Cryst. (Et_2O/petrol). Mp 85°.

Wotiz, J.H. *et al*, *J. Org. Chem.*, 1954, **19**, 1580 (*synth*)
Groff, T.M. *et al*, *Ark. Kemi*, 1968, **29**, 179 (*ms, ester*)
Ueno, A. *et al*, *Yakugaku Zasshi*, 1970, **90**, 1578 (*synth*)
Kovaleva, A.S. *et al*, *Zh. Org. Khim.*, 1972, **8**, 2474 (*synth*)
Gunstone, F.D. *et al*, *Chem. Phys. Lipids*, 1976, **17**, 1 (*cmr*)

8-Nonyn-1-ol N-00093

[10160-28-8]

$HC{\equiv}C(CH_2)_6CH_2OH$

$C_9H_{16}O$ M 140.225

Liq. $Bp_{0.5}$ 68-69°. n_D^{21} 1.4532.

Riemschneider, R. *et al*, *Monatsh. Chem.*, 1965, **96**, 1766 (*synth, ir*)
Knapp, F.F. *et al*, *J. Med. Chem.*, 1984, **27**, 57 (*synth, pmr*)
Mori, K. *et al*, *Justus Liebigs Ann. Chem.*, 1990, 1257 (*synth, pmr*)
Ma, D. *et al*, *Tetrahedron*, 1990, **46**, 6319 (*synth*)

29-Norcycloartan-3-ol N-00094

[17320-15-9]

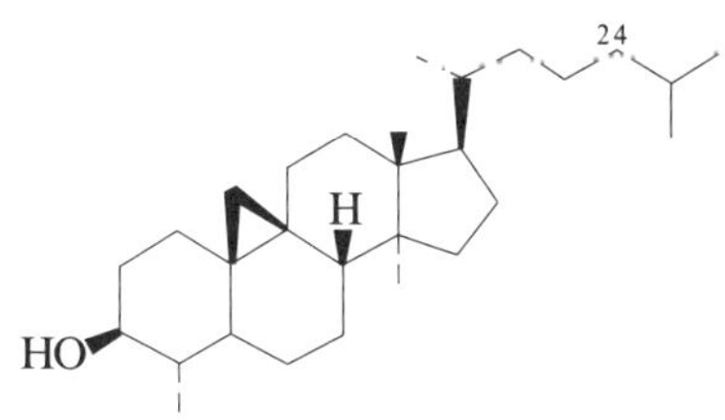

$C_{29}H_{50}O$ M 414.713

3β-form

29-Norcycloartanol. *31-Norcycloartanol* (*incorr.*)

Constit. of *Polypodium vulgare* and *Smilax aspera*. Cryst. (MeOH or Me_2CO). Mp 132°. $[\alpha]_D$ +42° ($CHCl_3$).

Ac:

$C_{31}H_{52}O_2$ M 456.751

Isol. from *P.* spp. Mp 91-92°. $[\alpha]_D$ +45.6°.

24,25-Didehydro: [28288-92-8]. *4α,14α-Dimethyl-9,19-cyclo-5α,9β-cholest-24-en-3β-ol. 29-Norcycloart-24-en-3β-ol. 29-Norcycloartenol*

$C_{29}H_{48}O$ M 412.698

Isol. from *Carnegiea gigantea*. Cryst. (MeOH)(as 3-Ac). Mp 93-95° (acetyl). $[\alpha]_D^{20}$ +63° ($CHCl_3$)(acetyl).

Ayengar, K.N.N., *Tetrahedron Lett.*, 1967, 3567 (*isol*)
Ghisalberti, E.L. *et al*, *J. Chem. Soc., Chem. Commun.*, 1969, 1403 (*biosynth*)
Devys, M. *et al*, *Tetrahedron Lett.*, 1970, 4621 (*deriv*)
Ageta, H. *et al*, *Phytochemistry*, 1984, **23**, 2875 (*isol, acetate*)

29-Norlanosterol N-00095

29-Norlanosta-8,24-dien-3β-ol

[32162-65-5]

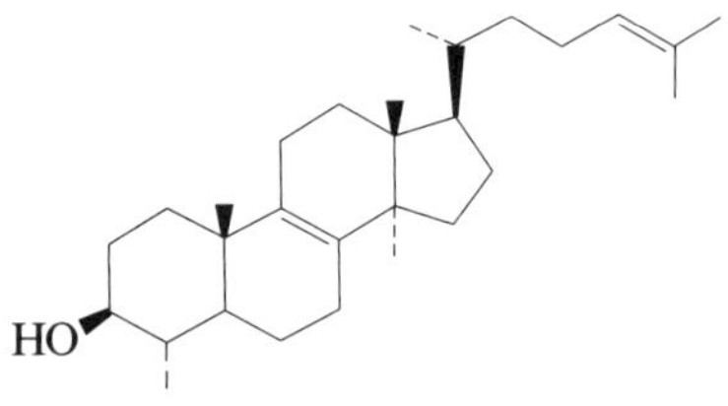

$C_{29}H_{48}O$ M 412.698

Constit. of *Funtumia latifolia*. Cryst. Mp 110°. $[\alpha]_D$ +60°.

Charles, G. *et al*, *C. R. Hebd. Seances Acad. Sci.*, 1969, **268**, 2105 (*struct*)
v. Tamelen, E.E. *et al*, *J. Am. Chem. Soc.*, 1971, **93**, 5279 (*biosynth*)

O

Obtusenol O-00001

[79406-07-8]

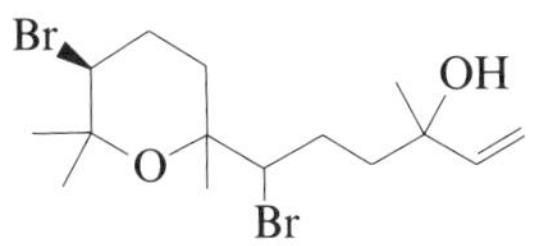

$C_{15}H_{26}Br_2O_2$ M 398.177

Constit. of *Laurencia obtusa*. Oil. $[\alpha]_D^{16}$ −50.2° (c, 1.7 in $CHCl_3$).

Imre, S. *et al*, *Phytochemistry*, 1981, **20**, 833.
Gonzalez, A.G. *et al*, *Tetrahedron Lett.*, 1981, **22**, 5071 (*synth*)

Obtusenyne O-00002

[71939-43-0]

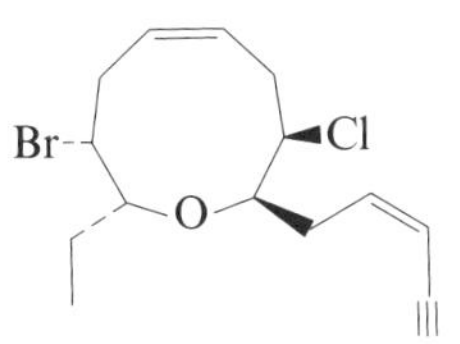

$C_{15}H_{20}BrClO$ M 331.679

Constit. of *Laurencia* spp. Oil. $[\alpha]_D^{20}$ +111.4° (c, 2.8 in $CHCl_3$).

King, T.J. *et al*, *Tetrahedron Lett.*, 1979, 1453 (*cryst struct*)
Howard, B.M. *et al*, *Tetrahedron*, 1980, **36**, 1747.

Obtusilactone O-00003

*3-(11-Dodecenylidene)dihydro-4-hydroxy-5-methylene-2(3*H*)-furanone, 9CI*

[56799-51-0]

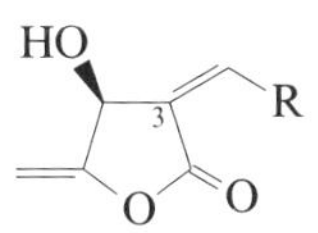

$R = (CH_2)_9CH{=}CH_2$

$C_{17}H_{26}O_3$ M 278.391

Constit. of *Lindera obtusiloba*. Viscous liq. $[\alpha]_D^{23}$ −53° (c, 0.35 in MeOH).

(E)-*Isomer:* [56522-14-6]. ***Isoobtusilactone***

$C_{17}H_{26}O_3$ M 278.391

Isol from *L. obtusiloba*. Viscous liq. $[\alpha]_D^{23}$ −56° (c, 0.67 in $CHCl_3$).

Niwa, M. *et al*, *Chem. Lett.*, 1975, 655 (*Isoobtusilactone*)
Niwa, M. *et al*, *Tetrahedron Lett.*, 1975, 1539 (*isol, struct*)
Rollinson, S.W. *et al*, *J. Am. Chem. Soc.*, 1981, **103**, 4114 (*synth*)

Obtusilactone A O-00004

*Dihydro-4-hydroxy-5-methylene-3-tetradecylidene-2(3*H*)-furanone, 9CI. Borbonol*

[56522-15-7]

As Obtusilactone, O-00003 with

$R = (CH_2)_{12}CH_3$

$C_{19}H_{32}O_3$ M 308.460

Constit. of *Lindera obtusiloba*, *Persea borbonia* and other *P.* spp. Antifungal agent. Viscous liq. $[\alpha]_D^{23}$ −46° (c, 0.45 in $CHCl_3$). Stereochem. identity of Borbonol with Obtusilactone A is not certain.

(E)-*Isomer:* [56522-16-8]. ***Isoobtusilactone A. Borbonol 2***

$C_{19}H_{32}O_3$ M 308.460

Isol. from *L. obtusiloba*, also from *Persea borbonia* and other *P.* spp. Viscous liq. $[\alpha]_D^{23}$ −54° (c, 0.50 in $CHCl_3$).

[71242-34-7]

Niwa, M. *et al*, *Chem. Lett.*, 1975, 655 (*isol, struct*)
Zaki, A.I. *et al*, *Physiol. Plant Pathol.*, 1980, **16**, 205 (*isol*)
Rollinson, S.W. *et al*, *J. Am. Chem. Soc.*, 1981, **103**, 4114 (*synth*)

Obtusilactone B O-00005

*3-(7-Hexadecenylidene)dihydro-4-hydroxy-5-methylene-2(3*H*)-furanone, 9CI*

[58940-65-1]

As Obtusilactone, O-00003 with

$R = (CH_2)_5CH{=}CH(CH_2)_7CH_3$ (*Z*−)

$C_{21}H_{34}O_3$ M 334.498

Constit. of *Lindera obtusiloba*. Oil.

(3E)-*Isomer:* [58940-66-2]. ***Isoobtusilactone B***

$C_{21}H_{34}O_3$ M 334.498

Isol. from *L. obtusiloba*.

Niwa, M. *et al*, *Tetrahedron Lett.*, 1975, 4395.

Obtusin† O-00006

[73618-74-3]

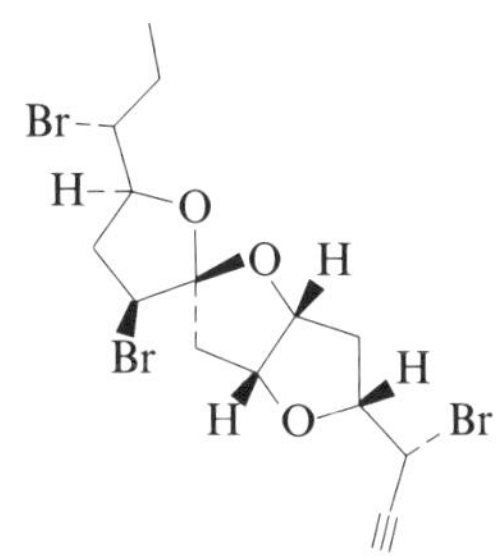

$C_{15}H_{19}Br_3O_3$ M 487.025

Constit. of *Laurencia obtusa*. Cryst. (hexane). Mp 93-95°. $[\alpha]_D$ −107° (c, 0.84 in $CHCl_3$).

Howard, B.M. *et al*, *Tetrahedron Lett.*, 1979, 2841 (*cryst struct, abs config*)
González, A.G. *et al*, *Tetrahedron Lett.*, 1983, **24**, 4143 (*cryst struct*)

Octacosanal O-00007

[22725-64-0]

$H_3C(CH_2)_{26}CHO$

$C_{28}H_{56}O$ M 408.750

Isol. from wax of grapes (*Vitis vinifera*) and from cabbage leaves (*Brassica oleracea*).

Radler, F. *et al*, *Aust. J. Chem.*, 1965, **18**, 1059 (*isol*)
Schmid, H.H.O. *et al*, *Hoppe Seyler's Z. Physiol. Chem.*, 1969, **350**, 462 (*isol*)

Octacosane O-00008

[630-02-4]

$H_3C(CH_2)_{26}CH_3$

$C_{28}H_{58}$ M 394.766

Isol. from plant waxes, e.g. *Coriolus sanguineus*, *Bacopa monniera*. Mp 61.2-61.4°, Fp 64-65°. $Bp_{1.1}$ 224°.

Hidebrand, J.H. *et al*, *J. Am. Chem. Soc.*, 1929, **51**, 2487 (*synth*)
Brown, I. *et al*, *J. Chem. Soc.*, 1960, 2783 (*isol*)
Cambie, R.C. *et al*, *J. Chem. Soc. C*, 1966, 72 (*isol*)
George, R.A. *et al*, *J. Phys. Chem.*, 1972, **5**, 871 (*uv*)
Boistelle, R. *et al*, *Acta Crystallogr., Sect. B*, 1976, **32**, 1240 (*cryst struct*)
Prakash, A. *et al*, *Curr. Sci.*, 1976, **45**, 705 (*isol, ir*)

1,28-Octacosanediol O-00009

$HOCH_2(CH_2)_{26}CH_2OH$

$C_{28}H_{58}O_2$ M 426.765

Constit. of the unsaponifiable fraction of carnauba wax. Also from apple (*Pyrus malus*) and oat (*Avena sativa*) in esterified form. Cryst. (C_6H_6/Me_2CO). Mp 112.2-112.3°.

Di-Ac:
$C_{32}H_{62}O_4$ M 510.839
Mp 80.9-81°. $Bp_{0.5}$ 258°.

Murray, K.E. *et al*, *Aust. J. Chem.*, 1955, **8**, 832.
Mazliak, P. *et al*, *Phytochemistry*, 1962, **1**, 79 (*isol*)
Daniels, D.G.H. *et al*, *Chem. Ind. (London)*, 1965, 1763 (*isol*)

Octacosanoic acid O-00010

Montanic acid
[506-48-9]

$H_3C(CH_2)_{26}COOH$

$C_{28}H_{56}O_2$ M 424.749

Found in carnauba wax, agave leaf wax, in wood of *Vitex divaricata*, etc. Rosettes (AcOH). Mp 90.3-90.5°.

Me ester:
$C_{29}H_{58}O_2$ M 438.776
Cryst. (MeOH/pet. ether). Mp 66-68°.

Et ester:
$C_{30}H_{60}O_2$ M 452.803
Mp 64.8-65°.

Bleyberg, W. *et al*, *Ber.*, 1931, **64**, 2504 (*synth*)
Cocker, W. *et al*, *J. Chem. Soc.*, 1962, 5194 (*isol*)
Razafindrazaka, J. *et al*, *Bull. Soc. Chim. Fr.*, 1963, 1633 (*isol*)
Satish, S. *et al*, *Phytochemistry*, 1972, **11**, 2888 (*isol*)
Misra, G. *et al*, *Planta Med.*, 1975, **28**, 165 (*isol, ms*)

1-Octacosanol O-00011

Montanyl alcohol. Cluytyl alcohol
[557-61-9]

$H_3C(CH_2)_{26}CH_2OH$

$C_{28}H_{58}O$ M 410.766

Present in leaf waxes. Mp 83.2-83.4°.

Ac: [18206-97-8]. *1-Octacosanyl acetate*
$C_{30}H_{60}O_2$ M 452.803
Isol. from various plant spp. incl. *Prosopis spicigera*. Also obt. from the cuticle and labial glands of the bumblebee. Bp_1 245°.

Bleyberg, W. *et al*, *Ber.*, 1931, **64**, 2504 (*synth*)
Pollard, A. *et al*, *Biochem. J.*, 1933, **27**, 1889 (*isol*)
Murray, K.E. *et al*, *Aust. J. Chem.*, 1955, **8**, 424.
Mathis, C. *et al*, *Phytochemistry*, 1964, **3**, 379 (*isol*)
Radler, F. *et al*, *Aust. J. Chem.*, 1965, **18**, 1059 (*isol*)
Joshi, K.C. *et al*, *Planta Med.*, 1975, **28**, 190 (*isol*)

9-Octacosenoic acid O-00012

[68593-79-3]

$H_3C(CH_2)_{17}CH{=}CH(CH_2)_7COOH$

$C_{28}H_{54}O_2$ M 422.733

(Z)-form [86901-44-2]
Isol. from *Mycobacterium tuberculosis* and *M. phlei*.

Takayama, K. *et al*, *Lipids*, 1978, **13**, 575 (*isol, ms*)
Cervilla, M. *et al*, *Anal. Chem.*, 1983, **55**, 2100 (*ms*)

19-Octacosenoic acid O-00013

$H_3C(CH_2)_7CH{=}CH(CH_2)_{17}COOH$

$C_{28}H_{54}O_2$ M 422.733

(Z)-form [86901-42-0]
Isol. from kernel oils of *Ximenia* spp. Cryst. Mp 57.8-58.2°.

[108045-06-3]

Ligthelm, S.P. *et al*, *J. Sci. Food Agric.*, 1954, **5**, 281 (*isol, struct*)
Cervilla, M. *et al*, *Anal. Chem.*, 1983, **55**, 2100 (*ms*)

21-Octacosenoic acid O-00014

$H_3C(CH_2)_5CH{=}CH(CH_2)_{19}COOH$

$C_{28}H_{54}O_2$ M 422.733

(Z)-form [140163-41-3]
Constit. of the sponge *Amphimedon compressa*.

Carballeira, N.M. *et al*, *J. Nat. Prod. (Lloydia)*, 1992, **55**, 333.

23-Octacosenoic acid O-00015

$H_3C(CH_2)_3CH{=}CH(CH_2)_{21}COOH$

$C_{28}H_{54}O_2$ M 422.733

(Z)-form [97761-97-2]
Constit. of the seed oil of *Grevillea decora*.

Kleiman, R. *et al*, *Lipids*, 1985, **20**, 373 (*isol*)

2,13-Octadecadienal, 9CI O-00016

$H_3C(CH_2)_3CH{=}CH(CH_2)_9CH{=}CHCHO$

$C_{18}H_{32}O$ M 264.450

(2E,13Z)-form [99577-57-8] ***Koiganal II***
Principal component of the sex pheromones of the webbing clothes moth, *Tiniella bisselliella* (together with Koiganal I, 2-Octadecenal, O-00164) and *Tinea pellionella*.

Yamaoka, R. *et al*, *Shitsuryo Bunseki*, 1985, **33**, 189; *CA*, **104**, 17841p (*isol, activity*)
Szocs, G. *et al*, *Biochem. Syst. Ecol.*, 1989, **17**, 417 (*synth*)
Japan. Pat., 91 41 003, (1991); *CA*, **115**, 108581x (*synth, activity*)

9,12-Octadecadienal O-00017

[26537-70-2]

$H_3C(CH_2)_4CH{=}CHCH_2CH{=}CH(CH_2)_7CHO$

$C_{18}H_{32}O$ M 264.450

(Z,Z)-form [2541-61-9]
Present in glycerolipids as an enol ether, sex pheromone in the female fall webworm moth *Hyphantria cunea*. Mp −32.3°. Readily forms a dimethyl acetal.

2,4-Dinitrophenylhydrazone: Mp 50-51°.

Mahadevan, V. *et al*, *Lipids*, 1966, **1**, 183 (*synth, ir*)
Christiansen, K. *et al*, *Lipids*, 1969, **4**, 421 (*ms*)

Mahadevan, V., *Prog. Chem. Fats Other Lipids*, 1972, **11**, 81 (*rev*)
Valicenti, A.J. *et al*, *Chem. Phys. Lipids*, 1976, **17**, 389 (*synth*)
Maneuso, A.J. *et al*, *J. Org. Chem.*, 1978, **43**, 2480 (*synth*)
Hill, A.S. *et al*, *J. Chem. Ecol.*, 1982, **8**, 383 (*pharmacol*)

1,9-Octadecadiene-4,6-diyne-3,8,18-triol O-00018

$H_2C{=}CHCH(OH)C{\equiv}CC{\equiv}CCH(OH)$
$CH{=}CH(CH_2)_7CH_2OH$

$C_{18}H_{26}O_3$ M 290.402

(Z)-form

18-Ac:
$C_{20}H_{28}O_4$ M 332.439
Obt. from roots of *Smyrnium olusatrum*. Oil.

Bohlmann, F. *et al*, *Chem. Ber.*, 1973, **106**, 3614.

9,16-Octadecadiene-12,14-diynoic acid O-00019

$H_3CCH{=}CHC{\equiv}CC{\equiv}CCH_2CH{=}CH(CH_2)_7COOH$

$C_{18}H_{24}O_2$ M 272.386

(9Z,16E)-form

Me ester: [41345-50-0].
$C_{19}H_{26}O_2$ M 286.413
Cryst. (pet. ether). Mp <0°.

(9Z,16Z)-form

Me ester: [41345-48-6].
Cryst. (pet. ether). Mp <0°.

Fallis, A.G. *et al*, *J. Chem. Soc., Perkin Trans. 1*, 1973, 743 (*synth, nmr, ms*)

13,17-Octadecadiene-9,11-diynoic acid O-00020

$H_2C{=}CHCH_2CH_2CH{=}CHC{\equiv}CC{\equiv}C(CH_2)_7COOH$

$C_{18}H_{24}O_2$ M 272.386

(Z)-form

Found in *Ongokea gore* seed oil.

Gunstone, F.D. *et al*, *J. Chem. Soc.*, 1963, 5772.
Hopkins, C.Y., *Top. Lipid Chem.*, 1972, **3**, 37.

2,4-Octadecadienoic acid O-00021

[76282-17-2]

$H_3C(CH_2)_{12}CH{=}CHCH{=}CHCOOH$

$C_{18}H_{32}O_2$ M 280.450

(2E,4E)-form [59404-49-8]

Me ester: [86120-41-4].
$C_{19}H_{34}O_2$ M 294.476
Oil.

2-Methylpropylamide: N-*Isobutyl-2,4-octadecadienamide*
$C_{22}H_{41}NO$ M 335.572
Constit. of *Piper guineense* fruits.

Okogun, J.I. *et al*, *J. Chem. Soc., Perkin Trans. 1*, 1974, 2195 (*isol, ms*)
Vig, O.P. *et al*, *Indian J. Chem.*, 1975, **13**, 1358 (*synth, deriv*)
Akimoto, A. *et al*, *Angew. Chem., Int. Ed. Engl.*, 1981, **20**, 90 (*synth*)
Cardillo, G. *et al*, *Tetrahedron*, 1986, **42**, 917 (*deriv, synth, ir, pmr, cmr, ms*)

2,5-Octadecadienoic acid O-00022

$H_3C(CH_2)_{11}CH{=}CHCH_2CH{=}CHCOOH$

$C_{18}H_{32}O_2$ M 280.450

(all-Z)-form

Mp 34-36°.

Christie, W.W. *et al*, *Chem. Phys. Lipids*, 1967, **1**, 407 (*synth*)
Sprecher, H., *Prog. Chem. Fats Other Lipids*, 1977, **15**, 219.

2,6-Octadecadienoic acid O-00023

$H_3C(CH_2)_{10}CH{=}CHCH_2CH_2CH{=}CHCOOH$

$C_{18}H_{32}O_2$ M 280.450

Lam, C.H. *et al*, *J. Chromatogr.*, 1975, **115**, 559; 1976, **117**, 365; 1976, **121**, 303 (*synth, glc*)

2,7-Octadecadienoic acid O-00024

$H_3C(CH_2)_9CH{=}CH(CH_2)_3CH{=}CHCOOH$

$C_{18}H_{32}O_2$ M 280.450

Lie Ken Jie, M.J.F. *et al*, *J. Chromatogr.*, 1975, **109**, 81 (*synth, glc*)

3,6-Octadecadienoic acid O-00025

$H_3C(CH_2)_{10}CH{=}CHCH_2CH{=}CHCH_2COOH$

$C_{18}H_{32}O_2$ M 280.450

(all-Z)-form

Mp 22-23°.

Christie, W.W. *et al*, *Chem. Phys. Lipids*, 1967, **1**, 407 (*synth*)
Kajiwara, T. *et al*, *Agric. Biol. Chem.*, 1977, **41**, 1481.
Sprecher, H., *Prog. Chem. Fats Other Lipids*, 1977, **15**, 219.

3,7-Octadecadienoic acid O-00026

$H_3C(CH_2)_9CH{=}CHCH_2CH_2CH{=}CHCH_2COOH$

$C_{18}H_{32}O_2$ M 280.450

Lam, C.H. *et al*, *J. Chromatogr.*, 1975, **115**, 559; 1976, **117**, 365; 1976, **121**, 303 (*synth, glc*)

3,8-Octadecadienoic acid O-00027

$H_3C(CH_2)_8CH{=}CH(CH_2)_3CH{=}CHCH_2COOH$

$C_{18}H_{32}O_2$ M 280.450

Lie Ken Jie, M.S.F. *et al*, *J. Chromatogr.*, 1975, **109**, 81 (*synth, glc*)

3,12-Octadecadienoic acid, 9CI O-00028

$H_3C(CH_2)_4CH{=}CH(CH_2)_7CH{=}CHCH_2COOH$

$C_{18}H_{32}O_2$ M 280.450

(3E,12Z)(?)-form

Constit. of seed oil of *Grindelia oxylepis*.

Kleiman, R. *et al*, *Lipids*, 1966, **1**, 301.

4,7-Octadecadienoic acid O-00029

$H_3C(CH_2)_9CH{=}CHCH_2CH{=}CHCH_2CH_2COOH$

$C_{18}H_{32}O_2$ M 280.450

(all-Z)-form

Mp 11-12°.

Christie, W.W. *et al*, *Chem. Phys. Lipids*, 1967, **1**, 407 (*synth*)
Sprecher, H., *Prog. Chem. Fats Other Lipids*, 1978, **15**, 219.

4,8-Octadecadienoic acid **O-00030**

$H_3C(CH_2)_8CH{=}CHCH_2CH_2CH{=}CHCH_2CH_2COOH$

$C_{18}H_{32}O_2$ M 280.450

Lam, C.H. *et al*, *J. Chromatogr.*, 1975, **115**, 559; 1976, **117**, 365; 1976, **121**, 303 (*synth, glc*)

4,9-Octadecadienoic acid, 8CI **O-00031**

[20194-44-9]

$H_3C(CH_2)_7CH{=}CH(CH_2)_3CH{=}CHCH_2CH_2COOH$

$C_{18}H_{32}O_2$ M 280.450

Present in the hardwood of *Pinus sibirica* as wax esters.

Lisina, A.I. *et al*, *CA*, 1968, **69**, 41702 (*isol*)
Lie Ken Jie, M.J.F. *et al*, *J. Chromatogr.*, 1975, **109**, 81 (*synth, glc*)

5,6-Octadecadienoic acid **O-00032**

Laballenic acid

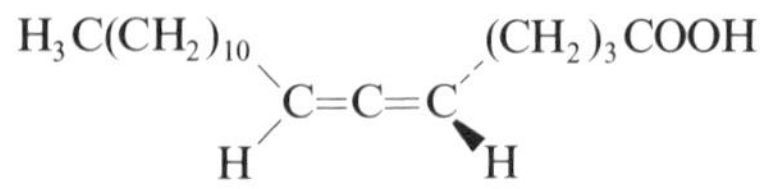

(*R*)-*form*

$C_{18}H_{32}O_2$ M 280.450

(R)-form [5204-84-2]

Constit. of *Leonotis nepetaefolia* seed oil. $[\alpha]_D^{20}$ −3.0°.

Me ester:

$C_{19}H_{34}O_2$ M 294.476

$[\alpha]_D^{26}$ −47.3° (c, 1.8 in EtOH).

4-Bromophenacyl ester: Mp 47-47.5°.

(±)-form [37973-79-8]

Me ester: [37973-80-1].

$Bp_{0.05}$ 144-146.5°.

4-Bromophenacyl ester: Mp 46-46.5°.

Bagby, M.O. *et al*, *J. Org. Chem.*, 1965, **30**, 4227 (*isol, struct*)
Cowie, J.S. *et al*, *J. Chem. Soc., Perkin Trans. 1*, 1972, 2197 (*synth*)

5,8-Octadecadienoic acid **O-00033**

$H_3C(CH_2)_8CH{=}CHCH_2CH{=}CH(CH_2)_3COOH$

$C_{18}H_{32}O_2$ M 280.450

(all-Z)-form

Mp −8° to −9°.

Christie, W.W. *et al*, *Chem. Phys. Lipids*, 1967, **1**, 407 (*synth*)
Sprecher, H., *Prog. Chem. Fats Other Lipids*, 1977, **15**, 219.

5,9-Octadecadienoic acid **O-00034**

$H_3C(CH_2)_7CH{=}CHCH_2CH_2CH{=}CH(CH_2)_3COOH$

$C_{18}H_{32}O_2$ M 280.450

Isol. from Gymnospermae seed lipids.

(5Z,9E)-form [7049-91-4]

Metabolic prod. of (*E*)-9-Octadecenoic acid.

(5Z,9Z)-form [26549-54-2]

Occurs in *Larix leptolepis* and *Taxus baccata* and other seed oils.

(5E,9Z)-form

Isol. from seed oil of *Thalictrum venulosum*.

Bhatty, M.K. *et al*, *Can. J. Biochem.*, 1966, **44**, 311 (*isol*)
Guo, L.S.S. *et al*, *Nutr. Metab.*, 1974, **16**, 51.
Lam, C.H. *et al*, *J. Chromatogr.*, 1975, **115**, 559; 1976, **117**, 365; 1976, **121**, 303 (*synth, glc*)
Plattner, R.D. *et al*, *Lipids*, 1975, **10**, 413 (*occur*)
Madrigal, R.V. *et al*, *Lipids*, 1975, **10**, 502 (*occur*)
Takagi, T. *et al*, *Lipids*, 1982, **17**, 716 (*occur, glc, cmr*)

5,10-Octadecadienoic acid, 9CI **O-00035**

[57568-22-6]

$H_3C(CH_2)_6CH{=}CH(CH_2)_3CH{=}CH(CH_2)_3COOH$

$C_{18}H_{32}O_2$ M 280.450

Lie Ken Jie, M.J.F. *et al*, *J. Chromatogr.*, 1975, **109**, 81 (*synth, glc*)

5,11-Octadecadienoic acid **O-00036**

[62528-83-0]

$H_3C(CH_2)_5CH{=}CH(CH_2)_4CH{=}CH(CH_2)_3COOH$

$C_{18}H_{32}O_2$ M 280.450

(all-Z)-form [71541-85-0]

Obt. from the seaweed *Cladophora rupestris*.

Ratnayake, W.N. *et al*, *Lipids*, 1979, **14**, 580.

5,12-Octadecadienoic acid **O-00037**

$H_3C(CH_2)_4CH{=}CH(CH_2)_5CH{=}CH(CH_2)_3COOH$

$C_{18}H_{32}O_2$ M 280.450

(E,E)-form

Mp 16-19°, Mp 26.5-27°.

Et ester:

$C_{20}H_{36}O_2$ M 308.503

$Bp_{0.4}$ 153-154°.

(Z,Z)-form [50499-21-3]

Mp −30°. $Bp_{0.4}$ 172-174°.

Et ester: $Bp_{0.2}$ 200°.

(5E,12Z)-form

Mp −12°. $Bp_{0.3}$ 168-170°.

Et ester: $Bp_{0.2}$ 150-152°.

(5Z,12E)-form

Mp −4°. $Bp_{0.3}$ 170°.

Et ester: $Bp_{0.2}$ 148-151°.

Ames, D.E. *et al*, *J. Chem. Soc.*, 1963, 4363 (*synth*)
Gunstone, F.D. *et al*, *Chem. Phys. Lipids*, 1969, **3**, 297 (*pmr*)
Gunstone, F.D. *et al*, *Chem. Phys. Lipids*, 1970, **4**, 1 (*synth*)
Gunstone, F.D. *et al*, *Chem. Phys. Lipids*, 1977, **18**, 115 (*cmr*)

6,8-Octadecadienoic acid **O-00038**

$H_3C(CH_2)_8CH{=}CHCH{=}CH(CH_2)_4COOH$

$C_{18}H_{32}O_2$ M 280.450

(E,E)-form [26181-82-8]

Cryst. (pet. ether). Mp 52-52.5°.

Gunstone, F.D. *et al*, *Chem. Phys. Lipids*, 1969, **3**, 297 (*pmr*)
Gunstone, F.D. *et al*, *Chem. Phys. Lipids*, 1970, **4**, 1 (*synth*)

6,9-Octadecadienoic acid **O-00039**

$H_3C(CH_2)_7CH{=}CHCH_2CH{=}CH(CH_2)_4COOH$

$C_{18}H_{32}O_2$ M 280.450

(all-E)-form

Mp 14-16°.

(all-Z)-form

Mp −11.5° to −10.5°.

Christie, W.W. *et al*, *Chem. Phys. Lipids*, 1967, **1**, 407 (*synth*)

Gunstone, F.D. *et al*, *Chem. Phys. Lipids*, 1969, **3**, 297; 1970, **4**, 1, 131 (*synth, pmr, glc, tlc*)
Sprecher, H., *Prog. Chem. Fats Other Lipids*, 1978, **15**, 219.

6,10-Octadecadienoic acid O-00040

$H_3C(CH_2)_6CH{=}CHCH_2CH_2CH{=}CH(CH_2)_4COOH$

$C_{18}H_{32}O_2$ M 280.450

(all-E)-form
Mp 40.5-41°.

Gunstone, F.D. *et al*, *Chem. Phys. Lipids*, 1969, **3**, 297; 1970, **4**, 1 (*synth, nmr*)
Lam, H. *et al*, *J. Chromatogr.*, 1976, **117**, 365; 1976, **121**, 303 (*synth*)
Sprecher, H., *Prog. Chem. Fats Other Lipids*, 1978, **15**, 219.

6,11-Octadecadienoic acid O-00041

[61628-30-6]

$H_3C(CH_2)_5CH{=}CH(CH_2)_3CH{=}CH(CH_2)_4COOH$

$C_{18}H_{32}O_2$ M 280.450

(E,E)-form
Mp 27.0-27.5°.

(6E,11Z)-form [88216-90-4]
Formed by *Eubacterium lentum*.

(Z,Z)-form [50381-82-3] ***Cilienic acid***
Found in the ciliated protozoan *Tetrahymena pyriformis* and related spp. Occurs in the sponge *Euryspongia rosea*.

Me ester: [26181-68-0].
$C_{19}H_{34}O_2$ M 294.476
Bp_{20} 80-82°.

[26181-77-1]

Gunstone, F.D. *et al*, *Chem. Phys. Lipids*, 1969, **3**, 297; 1970, **4**, 1; 1977, **18**, 115 (*cmr, pmr, synth*)
Lie Ken Jie, M.S.F., *J. Chromatogr.*, 1975, **109**, 81 (*synth, glc*)
Koroly, M.J. *et al*, *J. Biol. Chem.*, 1976, **251**, 7588.
Schick, B.P. *et al*, *Biochim. Biophys. Acta*, 1979, **575**, 475 (*metab*)
Watanabe, T. *et al*, *Biochim. Biophys. Acta*, 1980, **620**, 133; 1981, **665**, 66; **666**, 36 (*isol*)
Janssen, G. *et al*, *Biomed. Mass Spectrom.*, 1985, **12**, 134 (*synth*)
Carballeira, N.M. *et al*, *Lipids*, 1989, **24**, 665 (*isol, ms*)

6,12-Octadecadienoic acid O-00042

$H_3C(CH_2)_4CH{=}CH(CH_2)_4CH{=}CH(CH_2)_4COOH$

$C_{18}H_{32}O_2$ M 280.450

(E,E)-form
Mp 37-37.5°.

Gunstone, F.D. *et al*, *Chem. Phys. Lipids*, 1970, **4**, 1 (*synth*)
Frost, D.J. *et al*, *Chem. Phys. Lipids*, 1975, **15**, 48 (*pmr*)
Gunstone, F.D. *et al*, *Chem. Phys. Lipids*, 1977, **18**, 115 (*cmr*)

7,10-Octadecadienoic acid O-00043

$H_3C(CH_2)_6CH{=}CHCH_2CH{=}CH(CH_2)_5COOH$

$C_{18}H_{32}O_2$ M 280.450

(all-Z)-form [34143-92-5]
Mp −15° to −16°.

Christie, W.W. *et al*, *Chem. Phys. Lipids*, 1967, **1**, 407 (*synth*)
Sprecher, H., *Prog. Chem. Fats Other Lipids*, 1977, **15**, 219.

7,11-Octadecadienoic acid O-00044

$H_3C(CH_2)_5CH{=}CHCH_2CH_2CH{=}CH(CH_2)_5COOH$

$C_{18}H_{32}O_2$ M 280.450

(7Z,11Z)-form
Low-melting solid.

Tetrabromide: 7,8,11,12-Tetrabromooctadecanoic acid
$C_{18}H_{32}Br_4O_2$ M 600.066
Mp 120-122°.

Gunstone, F.D. *et al*, *J. Chem. Soc.*, 1962, 3055 (*synth*)
Lam, C.H. *et al*, *J. Chromatogr.*, 1975, **115**, 559; 1976, **117**, 365; 1976, **121**, 303 (*synth, glc*)

7,12-Octadecadienoic acid O-00045

$H_3C(CH_2)_4CH{=}CH(CH_2)_3CH{=}CH(CH_2)_5COOH$

$C_{18}H_{32}O_2$ M 280.450

(all-E)-form
Mp 26.5-27°.

Gunstone, F.D. *et al*, *Chem. Phys. Lipids*, 1969, **3**, 297; 1970, **4**, 1 (*synth, nmr*)
Davies, J.E. *et al*, *Chem. Phys. Lipids*, 1975, **15**, 48 (*raman*)
Sprecher, H., *Prog. Chem. Fats Other Lipids*, 1978, **15**, 219.

8,10-Octadecadienoic acid O-00046

[57568-21-5]

$H_3C(CH_2)_6CH{=}CHCH{=}CH(CH_2)_6COOH$

$C_{18}H_{32}O_2$ M 280.450

(E,E)-form
Cryst. (Et_2O). Mp 56-56.5°.

Maleic anhydride adduct:
$C_{22}H_{34}O_5$ M 378.508
Mp 110°.

Me ester, maleic anhydride adduct:
$C_{23}H_{36}O_5$ M 392.534
Mp 60.5-61.5°.

Gupta, S.C. *et al*, *J. Am. Oil Chem. Soc.*, 1960, **37**, 32 (*synth*)

8,11-Octadecadienoic acid O-00047

[2197-52-6]

$H_3C(CH_2)_5CH{=}CHCH_2CH{=}CH(CH_2)_6COOH$

$C_{18}H_{32}O_2$ M 280.450

(Z,Z)-form [4906-90-5]
Mp 16° (−12.5° to −19.5°).

Gensler, W.J. *et al*, *J. Org. Chem.*, 1963, **28**, 1254 (*synth*)
Christie, W.W. *et al*, *Chem. Phys. Lipids*, 1967, **1**, 407 (*synth*)
Christie, W.W. *et al*, *J. Chromatogr.*, 1968, **37**, 27 (*chromatogr*)

8,12-Octadecadienoic acid, 8CI O-00048

[34588-42-6]

$H_3C(CH_2)_4CH{=}CHCH_2CH_2CH{=}CH(CH_2)_6COOH$

$C_{18}H_{32}O_2$ M 280.450

(all-E)-form
Mp 38.5-39°.

Gunstone, F.D. *et al*, *Chem. Phys. Lipids*, 1969, **3**, 297; 1970, **4**, 1 (*synth, nmr*)
Davies, J.E. *et al*, *Chem. Phys. Lipids*, 1975, **15**, 48 (*raman*)
Lam, C.H. *et al*, *J. Chromatogr.*, 1976, **117**, 365; 1976, **121**, 305 (*synth*)
Sprecher, H., *Prog. Chem. Fats Other Lipids*, 1978, **15**, 219.

9,11-Octadecadienoic acid O-00049

[1839-11-8]

$H_3C(CH_2)_5CH{=}CHCH{=}CH(CH_2)_7COOH$

$C_{18}H_{32}O_2$ M 280.450

(E,E)-form [544-71-8]

Mangold's acid

Cryst. (EtOH). Mp 54°.

Maleic anhydride adduct:

$C_{22}H_{34}O_5$ M 378.508

Mp 94.5°.

Amide:

$C_{18}H_{33}NO$ M 279.465

Mp 98-99°.

(9Z,11E)-form [872-23-1]

Mp 20°.

(Z,Z)-form [544-70-7]

Ricinenic acid

Mp 42-43.2°.

Me ester: [822-10-6].

$C_{19}H_{34}O_2$ M 294.476

Mp −8° to −7°.

Nichols, P.L. *et al*, *J. Am. Chem. Soc.*, 1951, **73**, 247 (*synth*)
Alder, K. *et al*, *Justus Liebigs Ann. Chem.*, 1957, **609**, 19 (*synth*)
Gunstone, F.D. *et al*, *Chem. Phys. Lipids*, 1971, **7**, 121 (*synth*)
Batchelor, J.G. *et al*, *J. Org. Chem.*, 1974, **39**, 1698 (*cmr*)

9,12-Octadecadienoic acid O-00050

[2197-37-7]

$H_3C(CH_2)_4CH{=}CHCH_2CH{=}CH(CH_2)_7COOH$

$C_{18}H_{32}O_2$ M 280.450

(E,E)-form [506-21-8]

Linolelaidic acid

Constit. of *Chilopsis linearis* seed fat. Readily obt. by stereomutation of Linoleic acid. Cryst. (MeOH). Mp 28-29°. $Bp_{0.8}$ 179-183°.

Me ester: [2566-97-4].

$C_{19}H_{34}O_2$ M 294.476

Mp −7.8°. Bp_2 151-152°.

(9E,12Z)-form [2420-55-5]

Obt. by stereomutation of Linoleic acid. Mp 1°.

(9Z,12E)-form [2420-42-0]

Constit. of *Dimorphotheca sinuata* and *Crepis rubra* seed oils, also obt. from Linoleic acid. Mp 1°.

(Z,Z)-form [60-33-3]

Linoleic acid. Leinolic acid. Telfairic acid. Linolic acid

Constit. of most vegetable oils and animal fats. Occurs in lichens. Constit. of mandibular gland secretions of *Philanthus basilaris* and *P. bicinctus*. Essential fatty acid. Mp −5°. Bp_{16} 229-230°, $Bp_{0.02}$ 129°.

▷ RF9990000.

Ca salt: [19704-83-7]. *Linocalcium*

Me ester: [112-63-0].

Mp −35°. Bp_1 168-170°.

Et ester:

$C_{20}H_{36}O_2$ M 308.503

$Bp_{2.5}$ 175°.

Cyclohexylamide: [25167-65-1]. ***Clinolamide*, INN.**

Linolexamide

$C_{24}H_{43}NO$ M 361.610

Used in manuf. of paints, emulsifiers, vitamins. Antipruritic agent. Mp 34-35°. $Bp_{0.02}$ 180-181°.

[3207-50-9]

Harlow, R.D. *et al*, *J. Am. Oil Chem. Soc.*, 1963, **40**, 505.
Japan. Pat., 63 625 966, (1963); *CA*, **65**, 7066a.
Nakatani, H. *et al*, *Science (Washington, D.C.)*, 1966, **153**, 1267 (*Clinolamide, metab, activity*)
Gunstone, F.D. *et al*, *Chem. Phys. Lipids*, 1972, **9**, 112 (*synth, pmr*)
Karrer, W. *et al*, *Konstitution und Vorkommen der Organischen Pflanzenstoffe*, 2nd Ed., Birkhäuser Verlag, Basel, 1972-1985, no. 754 (*occur*)
Ackman, R.G. *et al*, *J. Chromatogr. Sci.*, 1974, **12**, 131 (*chromatog*)
Gunstone, F.D. *et al*, *J. Sci. Food Agric.*, 1976, **27**, 675 (*isol*)
McDaniel, C.A. *et al*, *J. Chem. Ecol.*, 1987, **13**, 227.
Lewis, R.J., *Sax's Dangerous Properties of Industrial Materials*, 8th Ed., Van Nostrand-Reinhold, 1992, LGG000.

9,13-Octadecadienoic acid O-00051

$H_3C(CH_2)_3CH{=}CHCH_2CH_2CH{=}CH(CH_2)_7COOH$

$C_{18}H_{32}O_2$ M 280.450

Lam, C.H. *et al*, *J. Chromatogr.*, 1975, **115**, 559; 1976, **117**, 365; 1976, **121**, 303 (*synth, glc*)

10,12-Octadecadienoic acid O-00052

[22880-03-1]

$H_3C(CH_2)_4CH{=}CHCH{=}CH(CH_2)_8COOH$

$C_{18}H_{32}O_2$ M 280.450

(E,E)-form [1072-36-2]

Constit. of *Chilopsis linearis* seed fat. Cryst. (EtOH). Mp 55.5-56°.

Me ester: [13038-48-7].

$C_{19}H_{34}O_2$ M 294.476

Cryst. (MeOH). Mp 25°. Bp_9 207-208°.

Maleic anhydride adduct:

$C_{22}H_{34}O_5$ M 378.508

Mp 102.5-103°.

(10E,12Z)-form

Mp 22-23°.

Me ester: [13058-53-2].

Mp −12° to −9°.

(Z,Z)-form

Mp 38-39°.

Chipault, J.R. *et al*, *J. Am. Oil Chem. Soc.*, 1959, **36**, 535 (*synth*)
Hopkins, C.Y. *et al*, *J. Am. Oil Chem. Soc.*, 1964, **41**, 42 (*isol, struct*)
Gunstone, F.D. *et al*, *Chem. Phys. Lipids*, 1969, **3**, 297; 1970, **4**, 1 (*pmr, synth*)
Bestmann, H.J. *et al*, *Justus Liebigs Ann. Chem.*, 1981, 2117 (*synth, pmr*)

10,13-Octadecadienoic acid O-00053

$H_3C(CH_2)_3CH{=}CHCH_2CH{=}CH(CH_2)_8COOH$

$C_{18}H_{32}O_2$ M 280.450

(Z,Z)-form [5027-60-1]

Mp −10.5° to −11°. $Bp_{0.0001}$ 121-128°.

Christie, W.W. *et al*, *Chem. Phys. Lipids*, 1967, **1**, 407 (*synth*)
Christie, W.W., *J. Chromatogr.*, 1968, **37**, 27 (*chromatog*)

10,14-Octadecadienoic acid O-00054

$H_3CCH_2CH_2CH{=}CHCH_2CH_2CH{=}CH(CH_2)_8COOH$

$C_{18}H_{32}O_2$ M 280.450

Lam, C.H. *et al*, *J. Chromatogr.*, 1975, **115**, 559; 1976, **117**, 365; 1976, **121**, 303 (*synth, glc*)

11,14-Octadecadienoic acid O-00055

$H_3C(CH_2)_2CH{=}CHCH_2CH{=}CH(CH_2)_9COOH$

$C_{18}H_{32}O_2$ M 280.450

(Z,Z)-form [17027-31-5]

Mp 4.5-5.5°. Bp 0001° 112-115°.

Christie, W.W. *et al*, *Chem. Phys. Lipids*, 1967, **1**, 407 (*synth*)
Christie, W.W., *J. Chromatogr.*, 1968, **37**, 27 (*chromatog*)

11,15-Octadecadienoic acid O-00056

$H_3CCH_2CH{=}CHCH_2CH_2CH{=}CH(CH_2)_9COOH$

$C_{18}H_{32}O_2$ M 280.450

(11Z,15Z)-form [18541-76-9]

Mp *ca.* 11°.

Pabon, H.J.J. *et al*, *J. Am. Oil Chem. Soc.*, 1969, **46**, 269 (*synth, ir, nmr*)
Lam, C.H. *et al*, *J. Chromatogr.*, 1975, **115**, 559; 1976, **117**, 365; 1976, **121**, 303 (*synth, glc*)

12,15-Octadecadienoic acid O-00057

$H_3CCH_2CH{=}CHCH_2CH{=}CH(CH_2)_{10}COOH$

$C_{18}H_{32}O_2$ M 280.450

(all-Z)-form

Mp 18-18.5°.

Christie, W.W. *et al*, *Chem. Phys. Lipids*, 1967, **1**, 407 (*synth*)
Sprecher, H., *Prog. Chem. Fats Other Lipids*, 1977, **15**, 219.

12,16-Octadecadienoic acid O-00058

$H_3CCH{=}CHCH_2CH_2CH{=}CH(CH_2)_{10}COOH$

$C_{18}H_{32}O_2$ M 280.450

Lam, C.H. *et al*, *J. Chromatogr.*, 1975, **115**, 559; 1976, **117**, 365; 1976, **121**, 303 (*synth, glc*)

13,16-Octadecadienoic acid O-00059

$H_3CCH{=}CHCH_2CH{=}CH(CH_2)_{11}COOH$

$C_{18}H_{32}O_2$ M 280.450

(all-Z)-form

Mp 20.5-21.5°.

Christie, W.W. *et al*, *Chem. Phys. Lipids*, 1967, **1**, 407 (*synth*)
Sprecher, H., *Prog. Chem. Fats Other Lipids*, 1977, **15**, 219.

13,17-Octadecadienoic acid O-00060

$H_2C{=}CHCH_2CH_2CH{=}CH(CH_2)_{11}COOH$

$C_{18}H_{32}O_2$ M 280.450

Lam, C.H. *et al*, *J. Chromatogr.*, 1975, **115**, 559; 1976, **117**, 365; 1976, **121**, 303 (*synth, glc*)

14,17-Octadecadienoic acid O-00061

$H_2C{=}CHCH_2CH{=}CH(CH_2)_{12}COOH$

$C_{18}H_{32}O_2$ M 280.450

(all-Z)-form

Mp 37-37.5°.

Christie, W.W. *et al*, *Chem. Phys. Lipids*, 1967, **1**, 407 (*synth*)
Sprecher, H., *Prog. Chem. Fats Other Lipids*, 1977, **15**, 219.

2,13-Octadecadien-1-ol O-00062

$H_3C(CH_2)_3CH{=}CH(CH_2)_9CH{=}CHCH_2OH$

$C_{18}H_{34}O$ M 266.466

(2E,13Z)-form

Ac:

$C_{20}H_{36}O_2$ M 308.503

Pheromone of *Vitacea polistiformis*, *Synanthedon acerrubri* and *Melittia satyriniformis*, moths of the Sesiidae family.

Schwarz, M. *et al*, *Tetrahedron Lett.*, 1983, **24**, 1007.

3,13-Octadecadien-1-ol O-00063

$H_3C(CH_2)_3CH{=}CH(CH_2)_8CH{=}CHCH_2CH_2OH$

$C_{18}H_{34}O$ M 266.466

(3E,13Z)-form [66410-28-4]

Ac: [53120-26-6].

Pheromone of the female lesser peach tree borer *Synanthedon exitiosa*.

(Z,Z)-form [66410-24-0]

Ac: [53120-27-7].

$C_{20}H_{36}O_2$ M 308.503

Pheromone of the female peach tree borer *S. pictipes* and *S. exitiosa*.

Tumlinson, J.H. *et al*, *Science (Washington, D.C.)*, 1974, **185**, 614 (*isol*)
Heath, R.R. *et al*, *J. Chromatogr. Sci.*, 1977, **15**, 10 (*hplc*)
Uchida, M. *et al*, *Agric. Biol. Chem.*, 1978, **42**, 1067 (*synth*)
Ebata, T. *et al*, *Agric. Biol. Chem.*, 1979, **43**, 1567 (*synth*)
Uchida, M. *et al*, *Agric. Biol. Chem.*, 1979, **43**, 1919 (*synth*)
Voerman, S., *J. Chem. Ecol.*, 1979, **5**, 759 (*synth*)
Heath, R.R. *et al*, *J. Chromatogr.*, 1980, **189**, 199 (*chromatog*)
Vinczer, P. *et al*, *Tetrahedron Lett.*, 1984, **25**, 2701 (*synth*)
Moiseenkov, A.M. *et al*, *J. Chem. Soc., Perkin Trans. 1*, 1991, 2639 (*synth, pmr*)

3,14-Octadecadien-1-ol O-00064

$H_3CCH_2CH_2CH{=}CH(CH_2)_9CH{=}CHCH_2CH_2OH$

$C_{18}H_{34}O$ M 266.466

Schwarz, M. *et al*, *Tetrahedron Lett.*, 1983, **24**, 1007 (*synth*)

4,15-Octadecadien-1-ol O-00065

$H_3CCH_2CH{=}CH(CH_2)_9CH{=}CHCH_2CH_2CH_2OH$

$C_{18}H_{34}O$ M 266.466

Schwarz, M. *et al*, *Tetrahedron Lett.*, 1983, **24**, 1007 (*synth*)

5,16-Octadecadien-1-ol O-00066

$H_3CCH{=}CH(CH_2)_9CH{=}CH(CH_2)_3CH_2OH$

$C_{18}H_{34}O$ M 266.466

Schwarz, M. *et al*, *Tetrahedron Lett.*, 1983, **24**, 1007 (*synth*)

6,17-Octadecadien-1-ol O-00067

$H_2C{=}CH(CH_2)_9CH{=}CH(CH_2)_4CH_2OH$

$C_{18}H_{34}O$ M 266.466

Schwarz, M. *et al*, *Tetrahedron Lett.*, 1983, **24**, 1007 (*synth*)

9,12-Octadecadien-1-ol O-00068

$H_3C(CH_2)_4CH{=}CHCH_2CH{=}CH(CH_2)_7CH_2OH$

$C_{18}H_{34}O$ M 266.466

(Z,Z)-form [506-43-4]

Linoleyl alcohol

Prepared by reduction of linoleic acid Me ester. Mp −5° to −2°. Bp_1 148-150°.

Jamieson, G.R. *et al*, *J. Chromatogr.*, 1967, **26**, 8 (*chromatog*)
Wood, R., *J. Gas Chromatogr.*, 1968, **6**, 94 (*synth, chromatog*)

10,12-Octadecadien-1-ol O-00069

$H_3C(CH_2)_4CH{=}CHCH{=}CH(CH_2)_8CH_2OH$

$C_{18}H_{34}O$ M 266.466

(10E,12Z)-form [80625-60-1]

$Bp_{0.01}$ 125-130°.

Ac: [80625-72-5].

$C_{20}H_{36}O_2$ M 308.503

$Bp_{0.01}$ 120-125°.

Bestmann, H.J. *et al*, *Justus Liebigs Ann. Chem.*, 1981, 2117 (*synth, pmr*)

11,13-Octadecadien-1-ol O-00070

$H_3C(CH_2)_3CH{=}CHCH{=}CH(CH_2)_9CH_2OH$

$C_{18}H_{34}O$ M 266.466

(11E,13Z)-form [80625-64-5]

$Bp_{0.01}$ 125-130°.

Ac: [80625-76-9].

$C_{20}H_{36}O_2$ M 308.503

$Bp_{0.01}$ 125-130°.

Bestmann, H.J. *et al*, *Justus Liebigs Ann. Chem.*, 1981, 2117 (*synth, pmr*)

2-(9,12-Octadecadienoylamino)-4,8-octadecadiene-1,3-diol O-00071

N-[*2-Hydroxy-1-(hydroxymethyl)-3,7-heptadecadienyl*]*-9,12-octadecadienamide*

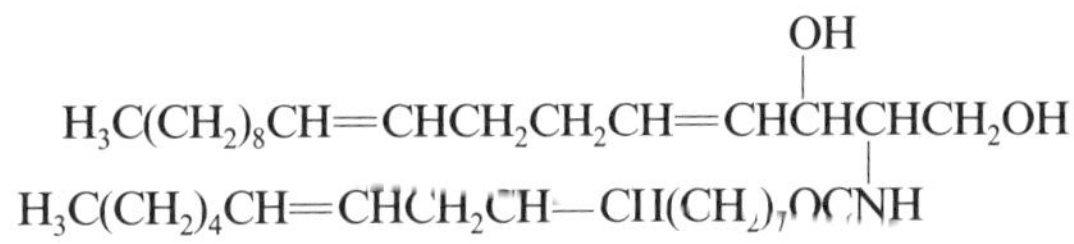

$C_{36}H_{65}NO_3$ M 559.914

Karlsson, K.A., *J. Lipid Res.*, 1971, **12**, 466 (*tlc*)

2-(9,12-Octadecadienoylamino)-1,3-octadecanediol O-00072

N-[*2-Hydroxy-1-(hydroxymethyl)heptadecyl*]*-9,12-octadecadienamide, 9CI*

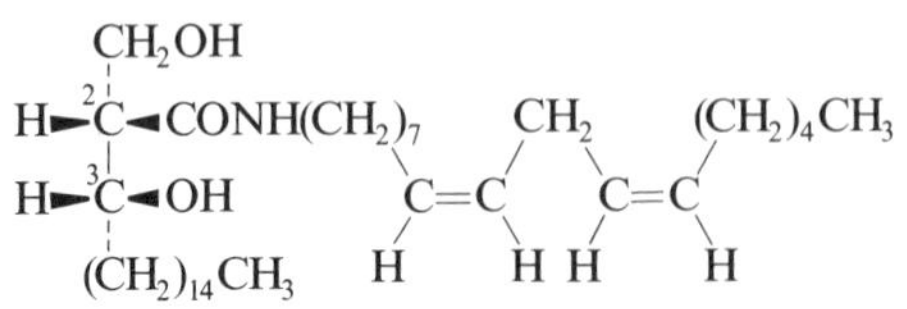

$C_{36}H_{69}NO_3$ M 563.946

(2S,3R,Z,Z)-form [34227-66-2]

D-erythro-trans,trans-*form. Linoleoylsphinganine.* N-*Octadeca-9,12-dienoylsphinganine*

Karlsson, K.A., *J. Lipid Res.*, 1971, **12**, 466 (*tlc*)
Gaskell, S.J. *et al*, *J. Chromatogr.*, 1976, **126**, 591 (*glc, ms*)

2-Octadeca-9,12-dienoylamino-4-octadecene-1,3-diol O-00073

N-[*2-Hydroxy-1-(hydroxymethyl)-3-heptadecenyl*]*-9,12-octadecadienamide*

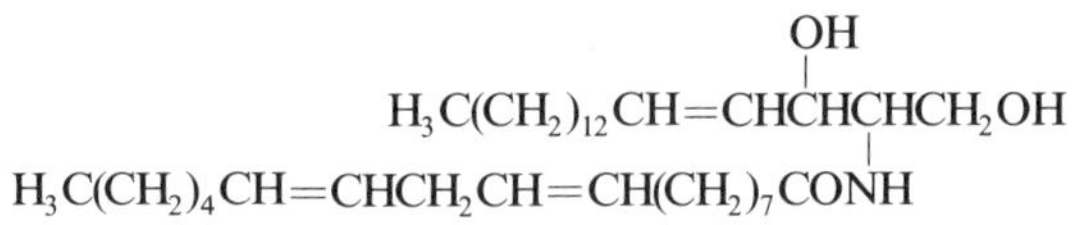

$C_{36}H_{67}NO_3$ M 561.930

(E,Z,Z)-D-erythro-form [34249-39-3]

N-*Linoleoyl-4-sphingenine.* N-*Octadeca-9,12-dienoyl-4-sphingenine*

Karlsson, K.A. *et al*, *J. Lipid Res.*, 1971, **12**, 466 (*tlc*)
Löfgren, H. *et al*, *Chem. Phys. Lipids*, 1977, **20**, 273 (*props*)

9,12-Octadecadien-6-ynoic acid O-00074

$H_3C(CH_2)_4CH{=}CHCH_2CH{=}CHCH_2C{\equiv}C(CH_2)_4COOH$

$C_{18}H_{28}O_2$ M 276.418

(9Z,12Z)-form [56795-52-9]

Isol. from various mosses e.g. *Dicranum scoporium.*

Ichikawa, T. *et al*, *Tetrahedron Lett.*, 1983, **24**, 3337 (*isol, bibl*)

9,14-Octadecadien-12-ynoic acid O-00075

$H_3CCH_2CH_2CH{=}CHC{\equiv}CCH_2CH{=}CH(CH_2)_7COOH$

$C_{18}H_{28}O_2$ M 276.418

(Z,Z)-form [4154-44-3]

14,15-Dehydrocrepenynic acid

Constit. of the seed oils of *Afzelia* spp. and the fungus *Tricholoma grammopodium.*

Bu'Lock, J.D. *et al*, *J. Chem. Soc. C*, 1967, 332.
Gunstone, F.D. *et al*, *J. Chem. Soc., Chem. Commun.*, 1967, 295 (*isol*)
Gunstone, F.D. *et al*, *Chem. Phys. Lipids*, 1977, **18**, 115 (*cmr*)
Farrell, I.W. *et al*, *J. Chem. Res., Synop.*, 1987, 234 (*synth, ir, uv, ms, pmr*)

11,13-Octadecadien-9-ynoic acid O-00076

$H_3C(CH_2)_3CH{=}CHCH{=}CHC{\equiv}C(CH)_7COOH$

$C_{18}H_{28}O_2$ M 276.418

(E,E)-form

Constit. of the root lipids of *Ximenia americana.* Also from *Santalum acuminatum, S acuminata, Leptomeria* spp. Cryst. (pentane). Mp 45-46°.

4-Bromophenacyl ester: Cryst. (EtOH). Mp 62-63°.

Crombie, L.C. *et al*, *J. Chem. Soc.*, 1957, 1632 (*synth*)
Hatt, H.H. *et al*, *Aust. J. Chem.*, 1960, **13**, 488 (*isol*)
Bu'lock, J.D. *et al*, *Phytochemistry*, 1963, **2**, 289 (*isol*)
Ketola, M., *Tetrahedron*, 1974, **30**, 2717.

8,10-Octadecadiynedioic acid O-00077

[33128-30-2]

$HOOC(CH_2)_6C{\equiv}CC{\equiv}C(CH_2)_6COOH$

$C_{18}H_{26}O_4$ M 306.401

Mp 135-135.5°.

Morris, S.G., *J. Am. Oil Chem. Soc.*, 1971, **48**, 376.

2,5-Octadecadiynoic acid, 8CI O-00078

[18202-14-7]

$H_3C(CH_2)_{11}C{\equiv}CCH_2C{\equiv}CCOOH$

$C_{18}H_{28}O_2$ M 276.418

Mp 53-54°.

Christie, W.W. *et al*, *Chem. Phys. Lipids*, 1967, **1**, 407 (*synth*)
Sprecher, H., *Prog. Chem. Fats Other Lipids*, 1978, **15**, 219.

2,6-Octadecadiynoic acid O-00079

[58443-88-2]

$H_3C(CH_2)_{10}C{\equiv}CCH_2CH_2C{\equiv}CCOOH$

$C_{18}H_{28}O_2$ M 276.418

Cryst. (hexane). Mp 64-64.5°.

Lam, C.H. *et al*, *J. Chromatogr.*, 1975, **115**, 559 (*synth, glc*)

2,7-Octadecadiynoic acid O-00080

[56682-43-0]

$H_3C(CH_2)_9C{\equiv}C(CH_2)_3C{\equiv}CCOOH$

$C_{18}H_{28}O_2$ M 276.418

Lie Ken Jie, M.S.F., *J. Chromatogr.*, 1975, **109**, 81 (*synth, glc*)

3,6-Octadecadiynoic acid, 8CI O-00081

[18202-15-8]

$H_3C(CH_2)_{10}C{\equiv}CCH_2C{\equiv}CCH_2COOH$

$C_{18}H_{28}O_2$ M 276.418

Mp 84-85°.

Christie, W.W. *et al*, *Chem. Phys. Lipids*, 1967, **1**, 407 (*synth*)
Sprecher, H., *Prog. Chem. Fats Other Lipids*, 1978, **15**, 219.

3,7-Octadecadiynoic acid O-00082

[58443-94-0]

$H_3C(CH_2)_9C{\equiv}CCH_2CH_2C{\equiv}CCH_2COOH$

$C_{18}H_{28}O_2$ M 276.418

Cryst. Mp 93-94°.

Lam, C.H. *et al*, *J. Chromatogr.*, 1975, **115**, 559 (*synth, glc*)

3,8-Octadecadiynoic acid O-00083

[56682-46-3]

$H_3C(CH_2)_8C{\equiv}C(CH_2)_3C{\equiv}CCH_2COOH$

$C_{18}H_{28}O_2$ M 276.418

Lie Ken Jie, M.S.F., *J. Chromatogr.*, 1975, **109**, 81 (*synth, glc*)

4,6-Octadecadiynoic acid O-00084

[136667-35-1]

$H_3C(CH_2)_{10}C{\equiv}CC{\equiv}CCH_2CH_2COOH$

$C_{18}H_{28}O_2$ M 276.418

Solid. Mp 74-75°.

Xu, Z. *et al*, *J. Org. Chem.*, 1991, **56**, 7183 (*synth, pmr*)

4,7-Octadecadiynoic acid O-00085

$H_3C(CH_2)_9C{\equiv}CCH_2C{\equiv}CCH_2CH_2COOH$

$C_{18}H_{28}O_2$ M 276.418

Christie, W.W. *et al*, *Chem. Phys. Lipids*, 1967, **1**, 407 (*synth*)
Sprecher, H., *Prog. Chem. Fats Other Lipids*, 1978, **15**, 219.

4,8-Octadecadiynoic acid O-00086

[58443-95-1]

$H_3C(CH_2)_8C{\equiv}CCH_2CH_2C{\equiv}CCH_2CH_2COOH$

$C_{18}H_{28}O_2$ M 276.418

Cryst. Mp 95-95.5°.

Lam, C.H. *et al*, *J. Chromatogr.*, 1975, **115**, 559 (*synth, glc*)

4,9-Octadecadiynoic acid O-00087

[56682-50-9]

$H_3C(CH_2)_7C{\equiv}C(CH_2)_3C{\equiv}CCH_2CH_2COOH$

$C_{18}H_{28}O_2$ M 276.418

Lie Ken Jie, M.S.F., *J. Chromatogr.*, 1975, **109**, 81 (*synth, glc*)

5,7-Octadecadiynoic acid O-00088

[136667-54-4]

$H_3C(CH_2)_9C{\equiv}CC{\equiv}C(CH_2)_3COOH$

$C_{18}H_{28}O_2$ M 276.418

Solid. Mp 47-49°.

Xu, Z. *et al*, *J. Org. Chem.*, 1991, **56**, 7183 (*synth, pmr*)

5,8-Octadecadiynoic acid O-00089

$H_3C(CH_2)_8C{\equiv}CCH_2C{\equiv}C(CH_2)_3COOH$

$C_{18}H_{28}O_2$ M 276.418

Christie, W.W. *et al*, *Chem. Phys. Lipids*, 1967, **1**, 407 (*synth*)
Sprecher, H., *Prog. Chem. Fats Other Lipids*, 1978, **15**, 219.

5,9-Octadecadiynoic acid O-00090

[58443-99-5]

$H_3C(CH_2)_7C{\equiv}CCH_2CH_2C{\equiv}C(CH_2)_3COOH$

$C_{18}H_{28}O_2$ M 276.418

Cryst. (hexane). Mp 51-51.5°.

Lam, C.H. *et al*, *J. Chromatogr.*, 1975, **115**, 559 (*synth, glc*)

5,10-Octadecadiynoic acid O-00091

$H_3C(CH_2)_6C{\equiv}C(CH_2)_3C{\equiv}C(CH_2)_3COOH$

$C_{18}H_{28}O_2$ M 276.418

Lie Ken Jie, M.S.F., *J. Chromatogr.*, 1975, **109**, 81 (*synth, glc*)

5,12-Octadecadiynoic acid O-00092

$H_3C(CH_2)_4C{\equiv}C(CH_2)_5C{\equiv}C(CH_2)_3COOH$

$C_{18}H_{28}O_2$ M 276.418

Mp 42.5-43°.

Gunstone, F.D. *et al*, *Chem. Phys. Lipids*, 1970, **4**, 1 (*synth*)

6,8-Octadecadiynoic acid O-00093

$H_3C(CH_2)_8C{\equiv}CC{\equiv}C(CH_2)_4COOH$

$C_{18}H_{28}O_2$ M 276.418
Cryst. (Me_2CO/pet. ether). Mp 56.5-57°.

Gunstone, F.D. *et al*, *Chem. Phys. Lipids*, 1970, **4**, 1 (*synth*)
Morris, S.G., *J. Am. Oil Chem. Soc.*, 1971, **48**, 376 (*synth*)
Morris, S.G. *et al*, *J. Am. Oil Chem. Soc.*, 1972, **49**, 505.
Gunstone, F.D. *et al*, *Chem. Phys. Lipids*, 1975, **15**, 48 (*raman*)

6,9-Octadecadiynoic acid O-00094

$H_3C(CH_2)_7C{\equiv}CCH_2C{\equiv}C(CH_2)_4COOH$

$C_{18}H_{28}O_2$ M 276.418
Mp 33-34°, Mp 32.5-33°.

Gunstone, F.D. *et al*, *Chem. Phys. Lipids*, 1970, **4**, 1 (*synth*)

6,10-Octadecadiynoic acid O-00095
[26192-15-4]

$H_3C(CH_2)_6C{\equiv}C(CH_2)_2C{\equiv}C(CH_2)_4COOH$

$C_{18}H_{28}O_2$ M 276.418
Mp 56-56.5°.

Gunstone, F.D. *et al*, *Chem. Phys. Lipids*, 1970, **4**, 1 (*synth*)
Gunstone, F.D. *et al*, *Chem. Phys. Lipids*, 1975, **15**, 48 (*raman*)
Lam, C.H. *et al*, *J. Chromatogr.*, 1975, **115**, 559 (*synth, glc*)
Lam, C.H. *et al*, *J. Chromatogr.*, 1976, **121**, 303 (*glc*)
Lam, C.H., *J. Chromatogr.*, 1976, **124**, 147 (*tlc*)

6,11-Octadecadiynoic acid O-00096
[26192-12-1]

$H_3C(CH_2)_5C{\equiv}C(CH_2)_3C{\equiv}C(CH_2)_4COOH$

$C_{18}H_{28}O_2$ M 276.418
Mp 31-31.5°.

Gunstone, F.D. *et al*, *Chem. Phys. Lipids*, 1970, **4**, 1 (*synth*)
Gunstone, F.D. *et al*, *Chem. Phys. Lipids*, 1975, **15**, 48 (*raman*)
Lie Ken Jie, M.S.F., *J. Chromatogr.*, 1975, **109**, 81 (*synth, glc*)

6,12-Octadecadiynoic acid O-00097

$H_3C(CH_2)_4C{\equiv}C(CH_2)_4C{\equiv}C(CH_2)_4COOH$

$C_{18}H_{28}O_2$ M 276.418
Mp 32.5-33°.

Gunstone, F.D. *et al*, *Chem. Phys. Lipids*, 1970, **4**, 1 (*synth*)

7,9-Octadecadiynoic acid O-00098
[33128-25-5]

$H_3C(CH_2)_7C{\equiv}CC{\equiv}C(CH_2)_5COOH$

$C_{18}H_{28}O_2$ M 276.418
Isol. from root bark of *Paramacrolobium caeruleum*. Cryst. (Me_2CO/pet. ether). Mp 51.5-52° (41-43°).

Me ester: [35042-74-1].
Oil.

Morris, S.G., *J. Am. Oil Chem. Soc.*, 1971, **48**, 376 (*synth*)
Morris, S.G., *J. Am. Oil Chem. Soc.*, 1972, **49**, 92, 505.
Patil, A.D. *et al*, *J. Nat. Prod.* (*Lloydia*), 1989, **52**, 153 (*isol, struct, ir, pmr, cmr, ms*)

7,10-Octadecadiynoic acid O-00099

$H_3C(CH_2)_6C{\equiv}CCH_2C{\equiv}C(CH_2)_5COOH$

$C_{18}H_{28}O_2$ M 276.418

Christie, W.W., *Chem. Phys. Lipids*, 1967, **1**, 407 (*synth*)
Sprecher, H., *Prog. Chem. Fats Other Lipids*, 1978, **15**, 219.

7,11-Octadecadiynoic acid O-00100
[26192-14-3]

$H_3C(CH_2)_5C{\equiv}CCH_2CH_2C{\equiv}C(CH_2)_5COOH$

$C_{18}H_{28}O_2$ M 276.418
Cryst. (hexane). Mp 48-48.5°. $Bp_{0.001}$ 158-160°.

Gunstone, F.D. *et al*, *J. Chem. Soc.*, 1962, 3055 (*synth*)
Lam, C.H. *et al*, *J. Chromatogr.*, 1975, **115**, 559 (*synth, glc*)

7,12-Octadecadiynoic acid O-00101
[26192-11-0]

$H_3C(CH_2)_4C{\equiv}C(CH_2)_3C{\equiv}C(CH_2)_5COOH$

$C_{18}H_{28}O_2$ M 276.418
Mp 42.5-43°.

Gunstone, F.D. *et al*, *Chem. Phys. Lipids*, 1970, **4**, 1 (*synth*)
Gunstone, F.D. *et al*, *Chem. Phys. Lipids*, 1975, **15**, 48 (*raman*)

8,10-Octadecadiynoic acid O-00102

$H_3C(CH_2)_6C{\equiv}CC{\equiv}C(CH_2)_6COOH$

$C_{18}H_{28}O_2$ M 276.418
Cryst. (Me_2CO/pet. ether). Mp 41.5-42° (42-43°).

Morris, S.G., *J. Am. Oil Chem. Soc.*, 1971, **48**, 376 (*synth*)
Morris, S.G., *J. Am. Oil Chem. Soc.*, 1972, **49**, 505.
Xu, Z. *et al*, *J. Org. Chem.*, 1991, **56**, 7183 (*synth, pmr*)

8,11-Octadecadiynoic acid, 8CI O-00103
[24397-37-3]

$H_3C(CH_2)_5C{\equiv}CCH_2C{\equiv}C(CH_2)_6COOH$

$C_{18}H_{28}O_2$ M 276.418

Christie, W.W. *et al*, *Chem. Phys. Lipids*, 1967, **1**, 407 (*synth*)
Sprecher, H., *Prog. Chem. Fats Other Lipids*, 1978, **15**, 219.

8,12-Octadecadiynoic acid O-00104
[26192-13-2]

$H_3C(CH_2)_4C{\equiv}C(CH_2)_2C{\equiv}C(CH_2)_6COOH$

$C_{18}H_{28}O_2$ M 276.418
Mp 52.5-53°.

Gunstone, F.D. *et al*, *Chem. Phys. Lipids*, 1970, **4**, 1; 1975, **15**, 48 (*synth, raman*)
Lam, C.H. *et al*, *J. Chromatogr.*, 1975, **115**, 559; 1976, **124**, 147 (*synth, glc, tlc*)

9,11-Octadecadiynoic acid O-00105

$H_3C(CH_2)_5C{\equiv}CC{\equiv}C(CH_2)_7COOH$

$C_{18}H_{28}O_2$ M 276.418
Major component of isano oil (*Ongokea gore*). Needles (EtOH aq.). Mp 46.5-47°. $Bp_{0.001}$ 145°.

Gunstone, F.D. *et al*, *J. Chem. Soc.*, 1963, 5772 (*isol*)
Morris, S.G., *J. Am. Oil Chem. Soc.*, 1971, **48**, 376; 1972, **49**, 505 (*synth*)

9,12-Octadecadiynoic acid O-00106

[2012-14-8]

$H_3C(CH_2)_4C{\equiv}CCH_2C{\equiv}C(CH_2)_7COOH$

$C_{18}H_{28}O_2$ M 276.418

Cryst. Mp 43.5-44°. $Bp_{0.1}$ 165-168°.

[62439-44-5]

de Gaudemaris, M. *et al*, *Bull. Soc. Chim. Fr.*, 1962, 315 (*synth*)
Gunstone, F.D. *et al*, *Chem. Phys. Lipids*, 1970, **4**, 1 (*synth*)
Bus, J. *et al*, *Recl. Trav. Chim. Pays-Bas* (*J. R. Neth. Chem. Soc.*), 1972, **91**, 394 (*ir*)
Gunstone, F.D. *et al*, *Chem. Phys. Lipids*, 1975, **15**, 48 (*raman*)
Gunstone, F.D. *et al*, *Chem. Phys. Lipids*, 1976, **17**, 1 (*cmr*)
Crombie, L. *et al*, *J. Chem. Soc., Perkin Trans. 1*, 1985, 2425; 1991, 567 (*synth, ir, pmr, cmr*)

9,13-Octadecadiynoic acid O-00107

[58444-08-9]

$H_3C(CH_2)_3C{\equiv}CCH_2CH_2C{\equiv}C(CH_2)_7COOH$

$C_{18}H_{28}O_2$ M 276.418

Cryst. (hexane). Mp 53.5-54.5°.

Lam, C.H. *et al*, *J. Chromatogr.*, 1975, **115**, 559 (*synth, glc*)

10,12-Octadecadiynoic acid O-00108

$H_3C(CH_2)_4C{\equiv}CC{\equiv}C(CH_2)_8COOH$

$C_{18}H_{28}O_2$ M 276.418

Mp 43-44°.

Allen, R.R., *J. Org. Chem.*, 1956, **21**, 143 (*synth*)
Gunstone, F.D. *et al*, *Chem. Phys. Lipids*, 1970, **4**, 1 (*synth*)
Gunstone, F.D. *et al*, *Chem. Phys. Lipids*, 1975, **15**, 48 (*raman*)

10,13-Octadecadiynoic acid O-00109

$H_3C(CH_2)_3C{\equiv}CCH_2C{\equiv}C(CH_2)_8COOH$

$C_{18}H_{28}O_2$ M 276.418

Christie, W.W. *et al*, *Chem. Phys. Lipids*, 1967, **1**, 407 (*synth*)
Sprecher, H., *Prog. Chem. Fats Other Lipids*, 1978, **15**, 219.

10,14-Octadecadiynoic acid O-00110

[61056-46-0]

$H_3CCH_2CH_2C{\equiv}CCH_2CH_2C{\equiv}C(CH_2)_8COOH$

$C_{18}H_{28}O_2$ M 276.418

Cryst. (hexane). Mp 61-62°.

Lam, C.H. *et al*, *J. Chromatogr.*, 1975, **115**, 559 (*synth, glc*)

11,14-Octadecadiynoic acid, 8CI O-00111

[18202-16-9]

$H_3CCH_2CH_2C{\equiv}CCH_2C{\equiv}C(CH_2)_9COOH$

$C_{18}H_{28}O_2$ M 276.418

Needles (pet. ether). Mp 63-64°.

Christie, W.W. *et al*, *Chem. Phys. Lipids*, 1967, **1**, 407 (*synth*)
Sprecher, H., *Prog. Chem. Fats Other Lipids*, 1978, **15**, 219.

11,15-Octadecadiynoic acid O-00112

[24397-41-9]

$H_3CCH_2C{\equiv}CCH_2CH_2C{\equiv}C(CH_2)_9COOH$

$C_{18}H_{28}O_2$ M 276.418

Cryst. (hexane). Mp 75-76°.

Pabon, H.J.J. *et al*, *J. Am. Oil Chem. Soc.*, 1969, **46**, 269 (*synth*)
Lam, C.H. *et al*, *J. Chromatogr.*, 1975, **115**, 559 (*synth, glc*)

12,14-Octadecadiynoic acid O-00113

[136667-55-5]

$H_3C(CH_2)_2C{\equiv}CC{\equiv}C(CH_2)_{10}COOH$

$C_{18}H_{28}O_2$ M 276.418

Solid. Mp 58-59°.

Xu, Z. *et al*, *J. Org. Chem.*, 1991, **56**, 7183 (*synth, pmr*)

12,15-Octadecadiynoic acid, 8CI O-00114

[18202-17-0]

$H_3CCH_2C{\equiv}CCH_2C{\equiv}C(CH_2)_{10}COOH$

$C_{18}H_{28}O_2$ M 276.418

Mp 55-57°.

Christie, W.W. *et al*, *Chem. Phys. Lipids*, 1967, **1**, 407 (*synth*)
Sprecher, H., *Prog. Chem. Fats Other Lipids*, 1978, **15**, 219.

12,16-Octadecadiynoic acid O-00115

[61056-47-1]

$H_3CC{\equiv}CCH_2CH_2C{\equiv}C(CH_2)_{10}COOH$

$C_{18}H_{28}O_2$ M 276.418

Cryst. (hexane). Mp 78-79°.

Lam, C.H. *et al*, *J. Chromatogr.*, 1975, **115**, 559 (*synth, glc*)

13,16-Octadecadiynoic acid, 8CI O-00116

[18379-07-2]

$H_3CC{\equiv}CCH_2C{\equiv}C(CH_2)_{11}COOH$

$C_{18}H_{28}O_2$ M 276.418

Mp 83-85°.

Christie, W.W. *et al*, *Chem. Phys. Lipids*, 1967, **1**, 407 (*synth*)
Sprecher, H., *Prog. Chem. Fats Other Lipids*, 1978, **15**, 219.

13,17-Octadecadiynoic acid O-00117

[58444-10-3]

$HC{\equiv}CCH_2CH_2C{\equiv}C(CH_2)_{11}COOH$

$C_{18}H_{28}O_2$ M 276.418

Cryst. Mp 62-62.5°.

Lam, C.H. *et al*, *J. Chromatogr.*, 1975, **115**, 559 (*synth, glc*)

14,17-Octadecadiynoic acid O-00118

$HC{\equiv}CCH_2C{\equiv}C(CH_2)_{12}COOH$

$C_{18}H_{28}O_2$ M 276.418

Christie, W.W. *et al*, *Chem. Phys. Lipids*, 1967, **1**, 407 (*synth*)
Sprecher, H., *Prog. Chem. Fats Other Lipids*, 1978, **15**, 219.

2,4,8,10,12,16-Octadecahexaenoic acid O-00119

$H_3CCH{=}CHCH_2CH_2CH{=}CHCH{=}CH$
$CH{=}CHCH_2CH_2CH{=}CHCH{=}CHCOOH$

$C_{18}H_{24}O_2$ M 272.386

2-Methylpropylamide: ***Heliopsin.*** *2,4,8,10,12-Octadecapentaenoic isobutylamide*
$C_{22}H_{33}NO$ M 327.509

Isol. from seeds of *Heliopsis helianthoides* var. *scabra*. Light-yellow viscous oil. $Bp_{0.08}$ 198-200° part dec. Struct. not certain.

Jacobson, M., *J. Am. Chem. Soc.*, 1957, **79**, 356.

Octadecanal **O-00120**

Stearaldehyde

[638-66-4]

$H_3C(CH_2)_{16}CHO$

$C_{18}H_{36}O$ M 268.482

Constit. of female sex pheromones of the large pine weevil *Hylobius abietis*, the rice stem borer *Chilo suppressalis* and *Cacoecimorpha pronubana*. Mp 38°, Mp 55°. Polymerises rapidly.

Oxime:

$C_{18}H_{37}NO$ M 283.496

Mp 89°.

Semicarbazone: Mp 108-109°.

Stephen, H., *J. Chem. Soc.*, 1925, 1876 (*synth*)
Flora, T. *et al*, *Proc. Symp. Coord. Chem.*, *3rd*, 1970, **1**, 517.
Auslander, D.E. *et al*, *J. Pharm. Sci.*, 1973, **62**, 167.
Kalo, P., *Finn. Chem. Lett.*, 1979, 189 (*isol*)
Tatsuki, S. *et al*, *Appl. Entomol. Zool.*, 1983, **18**, 443.

Octadecane **O-00121**

[593-45-3]

$H_3C(CH_2)_{16}CH_3$

$C_{18}H_{38}$ M 254.498

Isol. from plant oils and waxes, e.g. *Piper longum*, *Ulex europaeus*, hop and rose oil. Cryst. (Me_2CO). Mp 28°. Bp 305-307°, Bp_{15} 180-183°.

Schrauth, W. *et al*, *Chem. Ber.*, 1931, **64**, 1314 (*synth*)
Carey, P.C. *et al*, *J. Chem. Soc.*, 1933, 346 (*synth*)
Nyburg, S.C. *et al*, *Acta Crystallogr., Sect. B*, 1972, **28**, 2992 (*cryst struct*)
Trenner, N.R. *et al*, *J. Chromatogr.*, 1972, **71**, 415 (*glc*)
Pacakova, V. *et al*, *Chromatographia*, 1974, **7**, 171 (*glc*)
Erkelens, J. *et al*, *J. Catal.*, 1975, **39**, 173 (*ir*)

1,18-Octadecanediamine, 9CI **O-00122**

1,18-Diaminooctadecane

[10341-25-0]

$H_2NCH_2(CH_2)_{16}CH_2NH_2$

$C_{18}H_{40}N_2$ M 284.527

Mp 93°.

B,2HCl: Leaflets. Mp >225°.

N,N′-*Dibenzoyl:*

$C_{32}H_{48}N_2O_2$ M 492.743

Leaflets (EtOH). Mp 150°.

Pfeiffer, P. *et al*, *J. Prakt. Chem.*, 1933, **136**, 321.

Octadecanedioic acid, 9CI **O-00123**

[871-70-5]

$HOOC(CH_2)_{16}COOH$

$C_{18}H_{34}O_4$ M 314.464

Cryst. (Me_2CO). Mp 123-124°.

Sengupta, A. *et al*, *J. Indian Chem. Soc.*, 1956, **33**, 804 (*synth*)
Adamov, A.A. *et al*, *Elektrokhimiya*, 1975, **11**, 1773 (*synth*)

1,2-Octadecanediol **O-00124**

[20294-76-2]

$H_3C(CH_2)_{15}CH(OH)CH_2OH$

$C_{18}H_{38}O_2$ M 286.497

(±)-*form*

Cryst. (MeOH/petrol). Mp 81-82°.

Di-Ac: [20294-92-2].

$C_{22}H_{42}O_4$ M 370.571

Mp 40°. Bp_1 191-195°.

Nieman, C. *et al*, *J. Org. Chem.*, 1942, **7**, 227 (*synth*)
Swern, D. *et al*, *J. Am. Chem. Soc.*, 1946, **68**, 1504 (*synth*)

1,18-Octadecanediol **O-00125**

[3155-43-9]

$HOCH_2(CH_2)_{16}CH_2OH$

$C_{18}H_{38}O_2$ M 286.497

Isol. from *Spartium japonicum*. Cryst. (dioxan or C_6H_6). Mp 97-98°.

Bennett, G.M. *et al*, *J. Chem. Soc.*, 1938, 1679 (*synth*)
Drake, N.L. *et al*, *J. Am. Chem. Soc.*, 1941, **63**, 617 (*synth*)
Musgrave, O.C. *et al*, *J. Chem. Soc.*, 1952, 4373 (*isol*)
Ueno, M. *et al*, *J. Am. Oil Chem. Soc.*, 1974, **51**, 373 (*synth*)

1,2,3,4-Octadecanetetrol **O-00126**

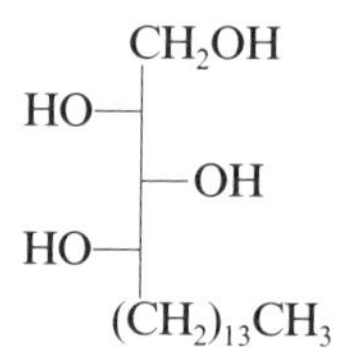

(2*R*,3*R*,4*S*)-*form*

$C_{18}H_{38}O_4$ M 318.496

(2*R*,3*R*,4*S*)-*form* [116002-57-4]

L-xylo-*form*

Cryst. (EtOH). Mp 81-83°. $[\alpha]_D$ −5.7° (c, 0.42 in EtOH).

(2*R*,3*S*,4*R*)-*form* [116002-56-3]

D-lyxo-*form*

Cryst. (EtOH). Mp 138-140°. $[\alpha]_D$ +5.1° (c, 0.1 in EtOH).

(2*R*,3*S*,4*S*)-*form* [116002-55-2]

L-ribo-*form*

Needles (EtOH). Mp 101-103°. $[\alpha]_D$ −8.9° (c, 0.44 in EtOH).

(2*S*,3*S*,4*R*)-*form* [105368-62-5]

D-xylo-*form*. ***Guggultetrol 18***

Constit. of the saponified resin of *Commiphora mukul*. Mp 80-82°. $[\alpha]_D^{25}$ +11.4° (c, 0.34 in EtOH).

Kumar, V. *et al*, *Tetrahedron*, 1987, **43**, 5933 (*isol, synth, struct*)

1-Octadecanethiol **O-00127**

1-Mercaptooctadecane. Octadecyl mercaptan

[2885-00-9]

$H_3C(CH_2)_{16}CH_2SH$

$C_{18}H_{38}S$ M 286.564

Waxy solid with strong odour. Mp 32-32.5°. Bp_4 170-175°.

Org. Synth., 1941, **21**, 36 (*synth*)
U.S. Pat., 2 551 813, (1951); *CA*, **45**, 9559 (*synth*)
Shirley, D.A. *et al*, *J. Org. Chem.*, 1953, **18**, 1591 (*synth*)
U.S. Pat., 2 998 455, (1960); *CA*, **56**, 2333 (*synth*)

Octadecanoic acid O-00128

Stearic acid. Talgic acid. Stearophanic acid. Bassinic acid. Lactaric acid

[57-11-4]

$H_3C(CH_2)_{16}COOH$

$C_{18}H_{36}O_2$ M 284.481

Constit. of most vegetable and animal fats. Acid and derivatives widely used as additives to industrial preparations. Used as EtOH soln. for nephelometric detn. of Ca. Used (with heptadecanoic acid) for amino acid sequencing in peptides. Leaflets. Sol. EtOH, C_6H_6, $CHCl_3$; sl. sol. H_2O. Mp 69.7°. Bp_5 213°.

▷ WI2800000.

Zn salt: [557-05-1]. ***Zinc stearate,*** **USAN**

Dusting powder, pharmaceutical acid (tablet lubricant).

Me ester: [112-61-8].

$C_{19}H_{38}O_2$ M 298.508

Mp 39.1°. Bp_{15} 215°.

▷ WI4460000.

Et ester: [111-61-5].

$C_{20}H_{40}O_2$ M 312.535

Mp 33.4°, Mp 30.9°. Bp_{10} 199°.

▷ WI3600000.

Chloride: [112-76-5].

$C_{18}H_{35}ClO$ M 302.927

Mp 23°. Bp_6 202-203°.

Amide: [124-26-5].

$C_{18}H_{37}NO$ M 283.496

Mp 109°. Bp_{12} 250-251°.

▷ RG0182000.

Anhydride:

$C_{36}H_{70}O_3$ M 550.947

Reagent for anal. detn. of alcohols and phenols.

Nitrile: [638-65-3].

$C_{18}H_{35}N$ M 265.481

Mp 41°. Bp_{10} 193°.

▷ WI5500000.

[822-16-2, 1592-23-0]

Welcher, F.J., *Organic Analytical Reagents*, Van Nostrand, NY, 1947, **2**, 82 (*use*)
Bricas, E. *et al*, *Biochemistry*, 1965, **4**, 2254 (*use*)
Fieser and Fieser's Reagents for Organic Synthesis, Wiley, 1967, **1**, 1117.
Malta, V. *et al*, *J. Chem. Soc. B*, 1971, 548 (*cryst struct*)
Karrer, W. *et al*, *Konstitution und Vorkommen der Organischen Pflanzenstoffe*, 2nd Ed., Birkhäuser Verlag, Basel, 1972-1985, no. 705 (*occur*)
Gunstone, F.D. *et al*, *Chem. Phys. Lipids*, 1976, **17**, 1 (*cmr*)
Lewis, R.J., *Sax's Dangerous Properties of Industrial Materials*, 8th Ed., Van Nostrand-Reinhold, 1992, MJW000, OAR000, SLK000, SLL500.

1-Octadecanol O-00129

Stearyl alcohol

[112-92-5]

$H_3C(CH_2)_{16}CH_2OH$

$C_{18}H_{38}O$ M 270.498

Obt. by redn. of stearic acid or methyl stearate. Isol. from fungal and higher plant lipids. Constit. of sex pheromone of female silkworm *Bombyx mori*. Mp 59.4-59.8°, Fp 57.95°.

▷ RG2010000.

Ac:

$C_{20}H_{40}O_2$ M 312.535

Common constit. of lepidopteran pheromones. Mp 33°.

Me ether: 1-Methoxyoctadecane

$C_{19}H_{40}O$ M 284.524

Cryst. (Et_2O/EtOH). Mp 30-31°.

Precht, V.D., *Fette, Seifen, Anstrichm.*, 1976, **78**, 145, 189 (*cryst struct*)
Yamaoka, R. *et al*, *Nippon Oyo Dobutsu Konchu Gakkaishi*, 1983, **27**, 77.
Lewis, R.J., *Sax's Dangerous Properties of Industrial Materials*, 8th Ed., Van Nostrand-Reinhold, 1992, OAX000.

3-Octadecanol O-00130

$H_3C(CH_2)_{14}CH(OH)CH_2CH_3$

$C_{18}H_{38}O$ M 270.498

(+)-*form*

Isol. from diphtheria bacterial lipids.

(−)-*form*

Cryst. (EtOH). Mp 56°. Bp_2 172°. $[\alpha]_D^{20}$ −4.78° (EtOH).

Hydrogen phthalate: Mp 32-33°. $[\alpha]_D^{20}$ −15.6° (EtOH).

(±)-*form*

Mp 50-51.8°. Bp_{13} 202°.

Hydrogen phthalate: Mp 39-41°.

Pickard, R.H. *et al*, *J. Chem. Soc.*, 1913, **103**, 1953 (*synth*)
Kanchukh, A.A., *CA*, 1955, **49**, 1142 (*isol*)

2-Octadecanone O-00131

[7373-13-9]

$H_3C(CH_2)_{15}COCH_3$

$C_{18}H_{36}O$ M 268.482

Constit. of the leaves of *Euonymus latifolius*. Mp 52°. Bp_{100} 251-252°.

Semicarbazone: Mp 125.5-126°.

Morgan, G.T. *et al*, *Chem. Ind. (London)*, 1925, **44**, 108.
Ulubelen, A. *et al*, *Phytochemistry*, 1973, **12**, 1824 (*isol*)
Brown, N.M.D. *et al*, *J. Chem. Soc., Perkin Trans. 2*, 1976, 1769.
Bergbreiter, D.E. *et al*, *J. Org. Chem.*, 1976, **41**, 2750 (*synth*)
Gawronski, J., *Tetrahedron Lett.*, 1976, 3845.

3-Octadecanone O-00132

3-Oxooctadecane

[18261-92-2]

$H_3C(CH_2)_{14}COCH_2CH_3$

$C_{18}H_{36}O$ M 268.482

Cryst. (EtOH/Et_2O). Mp 53° (51-52°). Bp_{14} 198°, Bp_{11} 197.5°.

Oxime:

$C_{18}H_{37}NO$ M 283.496

Cryst. in 2 forms. Mp 44°, Mp 58-59°.

Semicarbazone: Mp 76°.

Pickard, R.H. *et al*, *J. Chem. Soc.*, 1913, **103**, 1923.
Okamoto, Y. *et al*, *CA*, 1968, **68**, 39725u (*synth*)

2-(Octadecanoylamino)-4,8-octadecadiene-1,3-diol O-00133

N-[*2-Hydroxy-1-(hydroxymethyl)-3,7-heptadecadienyl*] *octadecanamide, 8CI*

$H_3C(CH_2)_8CH{=}CHCH_2CH_2CH{=}CHCH(OH)$
$CH(CH_2OH)NHCO(CH_2)_{16}CH_3$

$C_{36}H_{69}NO_3$ M 563.946

Karlsson, K.A., *J. Lipid Res.*, 1971, **12**, 466 (*tlc*)

2-(Octadecanoylamino)-1,3-octadecanediol O-00134

N-[*2-Hydroxy-1-(hydroxymethyl)heptadecyl*]*octadecanamide, 8CI*

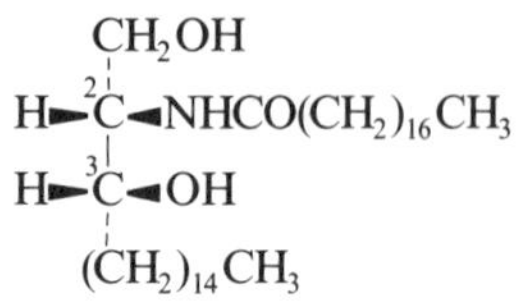

$C_{36}H_{73}NO_3$ M 567.978

(2RS,3SR)-form [21481-56-1]

($\pm$)-erythro-*form.* N-*Octadecanoylsphinganine*

Mp 107-108°.

Karlsson, K.A., *J. Lipid Res.*, 1971, **12**, 466 (*tlc*)
Ong, D.E. *et al*, *J. Lipid Res.*, 1972, **13**, 819 (*synth*)
Evenkova, E.N. *et al*, *CA*, 1975, **82**, 97613f (*synth*)
Gaskell, S.J. *et al*, *J. Chromatogr.*, 1976, **126**, 591 (*glc, ms*)

2-(Octadecanoylamino)-1,3,4-octadecanetriol O-00135

N-[*2,3-Dihydroxy-1-(hydroxymethyl)heptadecyl*] *octadecanamide, 8CI*

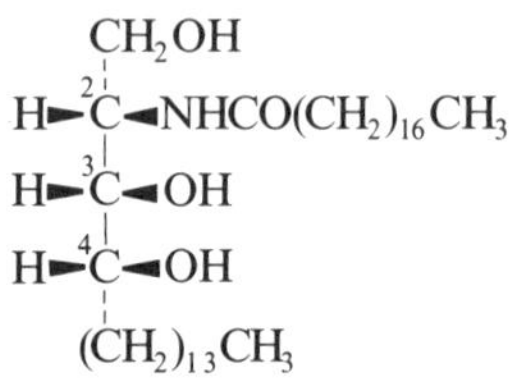

$C_{36}H_{73}NO_4$ M 583.977

(2S,3R,4R)-form

D-ribo-*form.* N-*Octadecanoylphytosphingosine*

Mp 99-100°.

Karlsson, K.A., *J. Lipid Res.*, 1971, **12**, 466 (*tlc*)
Hammarström, S., *J. Lipid Res.*, 1971, **12**, 760 (*synth*)

1-*O*-Octadecanoyl-2-*O*-arachidonoyl-*sn*-glycero-3-phosphocholine O-00136

*1-Acyl-2-arachidonyl-GPC. 1-Stearoyl-2-arachidonoyl-*sn-*glycero-3-phosphorylcholine. 4-Hydroxy-*N,N,N-*trimethyl-9-oxo-7-[[(1-oxooctadecyl)oxy]methyl]-3,5,8-trioxa-4-phosphaoctacosa-13,16,19,22-tetraen-1-aminiumhydroxide, inner salt, 4-oxide*

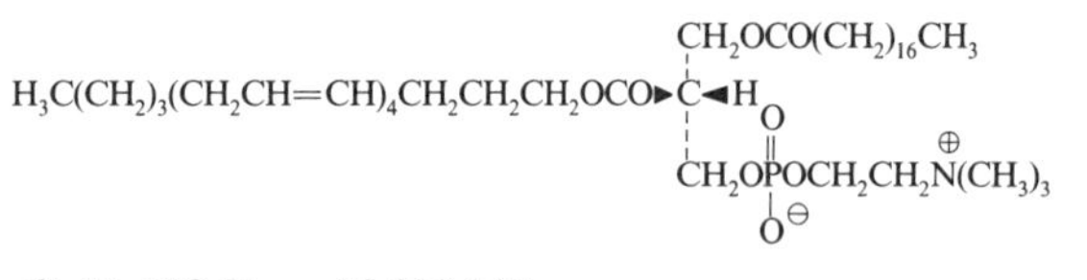

$C_{46}H_{84}NO_8P$ M 810.145

(R)-(all-Z)-form [35418-59-8]

Stearoylarachidonylphosphatidylcholine

Main source of arachidonic acid from human platelets and endothelial cells.

[18892-74-5, 128114-96-5]

Ghosh, D. *et al*, *Biochim. Biophys. Acta*, 1972, **266**, 41.
Cronholm, T. *et al*, *Eur. J. Biochem.*, 1977, **77**, 337 (*biosynth*)
Porter, N.A. *et al*, *Lipids*, 1979, **14**, 20 (*hplc*)
Santaren, J.F. *et al*, *Org. Magn. Reson.*, 1982, **18**, 98 (*synth, cmr*)
Chilton, F.H. *et al*, *J. Biol. Chem.*, 1986, **261**, 7771 (*metab*)
Chilton, F.H. *et al*, *Methods Enzymol.*, 1990, **187**, 157 (*biochem*)
Cubre, F. *et al*, *Biochim. Biophys. Acta*, 1992, **1124**, 297 (*biochem*)

Octadecanoylbenzene O-00137

1-Phenyl-1-octadecanone, 9CI. Octadecanophenone. Stearophenone. Heptadecyl phenyl ketone

[6786-36-3]

$$PhCO(CH_2)_{16}CH_3$$

$C_{24}H_{40}O$ M 344.579

Cryst. (EtOAc). Mp 63-64°.

Oxime:

$C_{24}H_{41}NO$ M 359.594

Cryst. (EtOH). Mp 53°.

Phenylhydrazone: Cryst. (EtOH). Mp 54°.

2,4-Dinitrophenylhydrazone: Mp 100-101°.

Breusch, F.L. *et al*, *Chem. Ber.*, 1954, **87**, 1225 (*synth*)
Maynard, J.A., *Aust. J. Chem.*, 1962, **15**, 867.

N-Octadecanoylgalactocerebroside O-00138

N-[*1-(β-D-Galactopyranosyloxy)methyl*]*-2-hydroxy-3-heptadecenyloctadecanamide, 9CI*

$$H_3C(CH_2)_{12}CH{=}CHCH(OH)CH(CH_2OGal)$$
$$NHCO(CH_2)_{16}CH_3$$

$C_{42}H_{81}NO_8$ M 728.104

Abrahamsson, S. *et al*, *Chem. Phys. Lipids*, 1972, **8**, 152 (*props*)
Andersson, B.A. *et al*, *Chem. Phys. Lipids*, 1972, **9**, 89 (*ms*)
Karlsson, K.A. *et al*, *Chem. Phys. Lipids*, 1972, **9**, 230 (*ms*)
Wood, G.W. *et al*, *Chem. Phys. Lipids*, 1977, **18**, 316 (*ms*)

N-Octadecanoylphytogalactocerebroside O-00139

N-[*1-[(β-D-Galactopyranosyloxy)methyl*]*-2,3-dihydroxyheptadecyl*]*octadecanamide, 9CI*

$$H_3C(CH_2)_{13}CH(OH)CH(OH)CH(CH_2OGal)$$
$$NHCO(CH_2)_{16}CH_3$$

$C_{42}H_{83}NO_9$ M 746.119

Cryst. ($CHCl_3$/MeOH). Mp 197-198°. $[\alpha]_D^{20}$ −1.4° (c, 3 in Py), $[\alpha]_D^{20}$ +10.4° (c, 3 in EtOH/$CHCl_3$).

Andersson, B.A. *et al*, *Chem. Phys. Lipids*, 1972, **9**, 89 (*ms*)
Karlsson, K.A. *et al*, *Chem. Phys. Lipids*, 1972, **9**, 230 (*ms*)
Pascher, I., *Chem. Phys. Lipids*, 1974, **12**, 303 (*synth, ir*)

3,6,9,12,15-Octadecapentaenoic acid O-00140

[56683-48-8]

$$H_3C(CH_2CH{=}CH)_5CH_2COOH$$

$C_{18}H_{26}O_2$ M 274.402

(all-Z)-form [51592-59-7]

Found in photosynthetic dinoflagellates and in zooxanthellae (*Gymnodinium microadriaticum*) isol. from clam tissue.

Joseph, J.D., *Lipids*, 1975, **10**, 395 (*ir, config, bibl*)
Bishop, D.G. *et al*, *Aust. J. Plant Physiol.*, 1976, **3**, 33 (*occur*)
Kuklev, D.V. *et al*, *Phytochemistry*, 1992, **31**, 2401 (*isol*)

3,9,12,15-Octadecatetraenoic acid O-00141

$$H_3CCH_2CH{=}CHCH_2CH{=}CHCH_2$$
$$CH{=}CH(CH_2)_4CH{=}CHCH_2COOH$$

$C_{18}H_{28}O_2$ M 276.418

(3E,9Z,12Z,15Z)-form

Constit. of the seed oil of *Tecoma stans*.

Me ester: [59149-03-0].

$C_{19}H_{30}O_2$ M 290.445

Fp −30°.

Hopkins, C.Y. *et al*, *J. Chem. Soc.*, 1965, 907 (*isol*)
Plattner, R.D. *et al*, *Lipids*, 1976, **11**, 222 (*isol, glc, ms*)

5,8,11,14-Octadecatetraenoic acid, 9CI O-00142

[75236-11-2]

$H_3CCH_2(CH_2CH{=}CH)_4(CH_2)_3COOH$

$C_{18}H_{28}O_2$ M 276.418

All-(Z)-form [57193-03-0]

Arachidonoyl-CoA inhibitor.

[69938-89-2]

Jakschik, B.A. *et al*, *Prostaglandins*, 1980, **20**, 401 (*activity*)
Neufield, E.J. *et al*, *J. Lipid Res.*, 1984, **25**, 288 (*synth*)
Garcia, M.C. *et al*, *Lipids*, 1990, **25**, 211 (*metab*)

5,9,12,15-Octadecatetraenoic acid O-00143

$H_3CCH_2(CH{=}CHCH_2)_3CH_2CH{=}CH(CH_2)_3COOH$

$C_{18}H_{28}O_2$ M 276.418

Isol. from Gymnospermae seed lipids.

Takagi, T. *et al*, *Lipids*, 1982, **17**, 716 (*occur, glc, cmr*)

6,9,12,15-Octadecatetraenoic acid O-00144

[2091-28-3]

$H_3CCH_2CH{=}CHCH_2CH{=}CH$
$CH_2CH{=}CHCH_2CH{=}CH(CH_2)_4COOH$

$C_{18}H_{28}O_2$ M 276.418

(6Z,9Z,12Z,15Z)-form

Isol. from *Lithospermum officinale*, *Onosmodium occidentale*, and from herring and other fish oils. Pale-yellow oil. Mp *ca.* −57°. n_D^{16} 1.4888.

Klenk, E. *et al*, *Hoppe Seyler's Z. Physiol. Chem.*, 1957, **307**, 272 (*isol*)
Craig, B.M. *et al*, *J. Am. Oil Chem. Soc.*, 1964, **41**, 209 (*isol*)
De Gomez Dumm, I.N.T. *et al*, *J. Lipid Res.*, 1975, **16**, 264.
De Gomez Dumm, I.N.T. *et al*, *Lipids*, 1975, **10**, 315.

9,11,13,15-Octadecatetraenoic acid O-00145

Parinaric acid

[18427-44-6]

$H_3CCH_2CH{=}CHCH{=}CHCH{=}CHCH{=}CH(CH_2)_7COOH$

$C_{18}H_{28}O_2$ M 276.418

(9E,11E,13E,15E)-form [18841-21-9]

β-Parinaric acid

Mp 95-96°.

(9Z,11E,13E,15Z)-form [593-38-4]

α-Parinaric acid

Acid of the kernel fat of *Parinarium laurinum*, *P. glaberrimum* and *Impatiens* spp. Cryst. (pet. ether). Mp 85-86°. Isom. by light or I_2 → *β-form*. A further labile stereoisomer (?) Mp. 72-74° known as Impatiensic acid was isol. from *I.* seed oil.

Farmer, E.H. *et al*, *J. Chem. Soc.*, 1935, 759 (*isol*)
Kaufmann, H. *et al*, *Chem. Ber.*, 1948, **81**, 152 (*isol*)
Kaufmann, H.P. *et al*, *Chem. Ber.*, 1959, **92**, 2797 (*Impatiensic acid*)
Bagby, M.O. *et al*, *Lipids*, 1966, **1**, 263 (*isol, config*)
Sklar, L.A. *et al*, *Biochemistry*, 1977, **16**, 5100.
Yabusaki, K.K. *et al*, *Proc. Natl. Acad. Sci. U.S.A.*, 1978, **75**, 691.

5,8,11,14-Octadecatetraynoic acid, 9CI O-00146

[32895-87-7]

$H_3CCH_2CH_2C{\equiv}CCH_2C{\equiv}CCH_2C{\equiv}CCH_2C{\equiv}C(CH_2)_3COOH$

$C_{18}H_{20}O_2$ M 268.355

Cryst. (pet. ether/EtOAc). Mp 84-85°.

Beerthuis, R.K. *et al*, *Recl. Trav. Chim. Pays-Bas (J. R. Neth. Chem. Soc.)*, 1971, **15**, 219 (*synth*)
Sprecher, H., *Prog. Chem. Fats Other Lipids*, 1978, **15**, 219.

9,12,15-Octadecatrienal O-00147

$H_3CCH_2CH{=}CHCH_2CH{=}CHCH_2CH{=}CH(CH_2)_7CHO$

$C_{18}H_{30}O$ M 262.434

(Z,Z,Z)-form [2423-13-4]

Linolenic aldehyde

Present in glycerolipids as an enol ether, sex pheromone in the female fall webworm moth *Hyphantria cunea*. Readily forms a dimethyl acetal.

2,4-Dinitrophenylhydrazone: Mp 42-43°.

Mahadevan, V. *et al*, *Lipids*, 1966, **1**, 183 (*synth, ir*)
Christiansen, K. *et al*, *Lipids*, 1969, **4**, 421 (*ms*)
Mahadeven, V., *Prog. Chem. Fats Other Lipids*, 1972, **11**, 81 (*rev*)
Valicenti, A.J., *Chem. Phys. Lipids*, 1976, **17**, 389 (*synth*)
Hill, A.S. *et al*, *J. Chem. Ecol.*, 1982, **8**, 383 (*pharmacol*)

1,11,13-Octadecatriene O-00148

$H_3C(CH_2)_3CH{=}CHCH{=}CH(CH_2)_8CH{=}CH_2$

$C_{18}H_{32}$ M 248.451

(11E,13Z)-form [80625-36-1]

$Bp_{0.01}$ 100-104°.

Bestmann, H.J. *et al*, *Justus Liebigs Ann. Chem.*, 1981, 2117 (*synth, pmr*)

2,9,12-Octadecatrienoic acid O-00149

$H_3C(CH_2)_4CH{=}CHCH_2CH{=}CH(CH_2)_5CH{=}CHCOOH$

$C_{18}H_{30}O_2$ M 278.434

(2E,9Z,12Z)-form [22684-06-6]

Pollen attractant of honey bees.

Me ester: [30959-44-5].

$Bp_{0.04}$ 98°.

Hopkins, C.Y. *et al*, *Can. J. Biochem.*, 1969, **47**, 433 (*isol, ir, pmr*)
Starratt, A.N. *et al*, *Can. J. Biochem.*, 1971, **49**, 251 (*synth*)
Sharpless, K.B. *et al*, *J. Am. Chem. Soc.*, 1973, **95**, 6137 (*synth*)
Trost, B.M. *et al*, *J. Org. Chem.*, 1975, **40**, 148 (*synth*)
Bus, J. *et al*, *Chem. Phys. Lipids*, 1976, **17**, 501; 1977, **18**, 130 (*cmr*)
Trost, B.M., *J. Am. Chem. Soc.*, 1976, **98**, 4887 (*synth*)

3,9,12-Octadecatrienoic acid, 9CI O-00150

$H_3C(CH_2)_4CH{=}CHCH_2CH{=}CH(CH_2)_4CH{=}CHCH_2COOH$

$C_{18}H_{30}O_2$ M 278.434

(3E,9Z,12Z)-form [2277-04-5]

Isol. from various seed oil.

(9R,10S*)-Epoxide: 9,10-Epoxy-3,12-octadecadienoic acid*

$C_{18}H_{30}O_3$ M 294.433

Present in *Stenachaenium macrocephalum* seed oil.

Bagby, M.O. *et al*, *J. Am. Oil Chem. Soc.*, 1965, **42**, 50 (*isol, uv, ir, pmr*)
Kleiman, R. *et al*, *Lipids*, 1966, **1**, 301; 1971, **6**, 617 (*isol, pmr*)
Morris, L.J. *et al*, *Lipids*, 1968, **3**, 91 (*isol*)

5,6,16-Octadecatrienoic acid O-00151

Lamenallenic acid

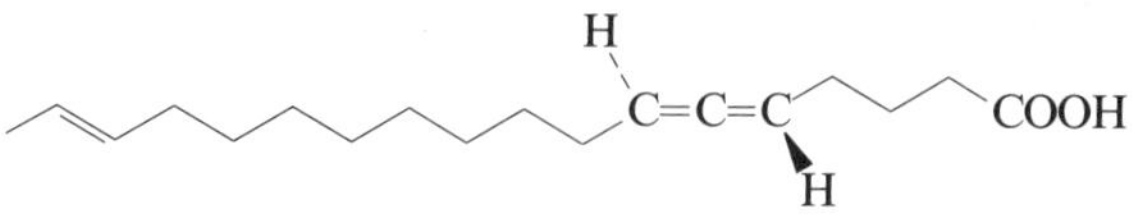

(*R*)-(*E*)-*form*

$C_{18}H_{30}O_2$ M 278.434

***(R,E)*-form** [18689-88-8]

Isol. from *Laminium purpureum* oil. $[\alpha]_D^{20}$ −6.21° (c, 3.6 in EtOH).

Me ester: [37973-92-5].

$C_{19}H_{32}O_2$ M 292.461

$[\alpha]_D^{25}$ −50° (c, 0.82 in EtOH).

4-Bromophenacyl ester: Mp 40.5-41.5°.

(±)-(*E*)-*form*

Me ester: $Bp_{0.0015}$ 136-140°.

Mikolajczak, K.L. *et al*, *Biochem. J.*, 1967, **105**, 1245 (*isol*)
Cowie, J.S. *et al*, *J. Chem. Soc., Perkin Trans. 1*, 1972, 2197 (*synth, abs config*)

5,8,11-Octadecatrienoic acid, 8CI O-00152

[17206-72-3]

$H_3C(CH_2)_5CH{=}CHCH_2CH{=}CHCH_2CH{=}CH(CH_2)_3COOH$

$C_{18}H_{30}O_2$ M 278.434

Sprecher, H.W., *Biochim. Biophys. Acta*, 1971, **231**, 122 (*synth*)

5,9,12-Octadecatrienoic acid O-00153

[13237-97-3]

$H_3C(CH_2)_4CH{=}CHCH_2CH{=}CH$
$CH_2CH_2CH{=}CH(CH_2)_3COOH$

$C_{18}H_{30}O_2$ M 278.434

Occurs in Gymnospermae seed lipids.

***(E,Z,Z)*-form** [2441-53-4]

Ranunculeic acid. Columbinic acid. Δ^5-trans-*Linoleic acid. Aquilegic acid*

Constit. of *Thalictrum polycarpum, T. venulosum, Aquilegia vulgaris* and other seed oils. λ_{max} 233 nm.

***(Z,Z,Z)* form**

Constit. of *Teucrium depressum,* other seed oils, and tall oil. n_D^{20} 1.4804.

Hexabromide: 5,6,9,10,12,13-Hexabromooctadecanoic acid

$C_{18}H_{30}Br_6O_2$ M 757.858

Mp 145.5-147.5°.

Bagby, M.O. *et al*, *Biochemistry*, 1962, **1**, 632 (*isol*)
Rowe, J.W. *et al*, *J. Org. Chem.*, 1964, **29**, 1554 (*isol*)
Max, F.H. *et al*, *Anal. Chem.*, 1967, **39**, 1118 (*chromatog*)
Smith, C.R. *et al*, *Lipids*, 1969, **4**, 462 (*isol*)
Freiman, R.E. *et al*, *Khim. Prir. Soedin.*, 1970, **6**, 167; *CA*, **73**, 63191 (*isol*)
Bus, J. *et al*, *Lipids*, 1976, **12**, 343 (*cmr, pmr*)
Takagi, T. *et al*, *Lipids*, 1982, **17**, 716 (*occur, glc, cmr*)
Elliot, W.J. *et al*, *J. Org. Chem.*, 1983, **48**, 5378 (*synth*)

6,9,12-Octadecatrienoic acid, 9CI O-00154

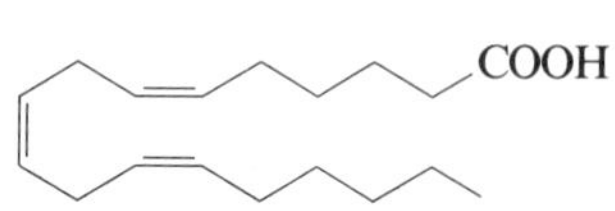

$C_{18}H_{30}O_2$ M 278.434

***(Z,Z,Z)*-form** [506-26-3]

γ-*Linolenic acid.* ***Gamolenic acid, BAN, INN***

Isol. from evening primrose seed oil. A minor component of many animal lipids. Antihypercholesterolaemic drug.

Hexabromide:

$C_{18}H_{30}Br_6O_2$ M 757.858

Mp 196° dec.

Bernhard, K. *et al*, *Helv. Chim. Acta*, 1948, **31**, 977, 981 (*props*)
Riley, J.P., *J. Chem. Soc.*, 1949, 2728 (*struct*)
Osbond, J.M. *et al*, *J. Chem. Soc.*, 1961, 2779, 5270 (*synth*)
Bus, J. *et al*, *Chem. Phys. Lipids*, 1976, **17**, 501 (*cmr*)
Lie Ken Jie, M., *Chem. Phys. Lipids*, 1976, **18**, 130 (*cmr*)
Fr. Pat., 1 603 386, (1976); *CA*, **76**, 84681 (*isol*)
Gunstone, F.D., *Prog. Lipids Res.*, 1992, **31**, 145 (*rev*)

7,9,12-Octadecatrienoic acid O-00155

$H_3C(CH_2)_4CH{=}CHCH_2CH{=}CHCH{=}CH(CH_2)_5COOH$

$C_{18}H_{30}O_2$ M 278.434

Stoffel, W. *et al*, *Hoppe Seyler's Z. Physiol. Chem.*, 1966, **345**, 52 (*synth*)

8,10,12-Octadecatrienoic acid O-00156

$H_3C(CH_2)_4CH{=}CHCH{=}CHCH{=}CH(CH_2)_6COOH$

$C_{18}H_{30}O_2$ M 278.434

***(E,E,E)*-form** [822-19-5]

Formed by stereomutation of the natural isomers. Mp 77-78°.

Maleic anhydride adduct: Cryst. (pet. ether). Mp 71°.

(8E,10E,12Z)*-form** [28872-28-8] ***Calendic acid. *Calendulic acid*

Isol. from *Calendula officinalis* seed oil. Also from *C. stellata, Osteospermum spinescens* and *O. hyoseroides.* Mp 40-40.5°.

Tetracyanoethylene adduct: Mp 112.5-113.5°.

Maleic anhydride adduct: Mp 73-74°.

(8Z,10E,12Z)*-form** [18961-59-6] ***Jacaric acid. *Jacaranda acid*

Isol. from the seed oil of *Jacaranda mimosifolia* and *J. chelonia.* Mp 43.5-44°.

McLean, J. *et al*, *J. Chem. Soc.*, 1956, 777 (*isol*)
Chisholm, M.J. *et al*, *Can. J. Chem.*, 1960, **38**, 2500 (*isol*)
Chisholm, M.J. *et al*, *J. Org. Chem.*, 1962, **27**, 3137 (*isol*)
Bergel'son, L.D. *et al*, *CA*, 1967, **67**, 99692 (*synth*)
Conacher, H.B.S. *et al*, *Lipids*, 1970, **5**, 434.
Crombie, L. *et al*, *J. Chem. Soc., Chem. Commun.*, 1984, 953 (*biosynth*)

9,11,13-Octadecatrienoic acid O-00157

[13296-76-9]

$H_3C(CH_2)_3CH{=}CHCH{=}CHCH{=}CH(CH_2)_7COOH$

$C_{18}H_{30}O_2$ M 278.434

***(E,E,E)*-form** [544-73-0]

β-Eleostearic acid

Can be formed by stereomutation of any of the 9,11,13-Octadecatrienoic acids. Cryst. (EtOH). Mp 72°. Bp_1 188°.

Me ester: [4238-03-3].

$C_{19}H_{32}O_2$ M 292.461

Mp 12.5-13.2°. Bp_1 162°.

Et ester: [6114-25-6].

$C_{20}H_{34}O_2$ M 306.487

Bp_{14} 232°.

Maleic anhydride adduct: Mp 75-78°. A mixt. of two isomeric adducts.

(9Z,11E,13E)-form [506-23-0]
α-Eleostearic acid. Elaeostearinic acid. Margarolic acid. Elaeomargarinic acid
Occurs in tung oil and in some other fats. The most common acid of this group. Isol. from seed oil of *Momordica charantia* and *Centranthus macrosiphon*. Cryst. (MeOH). Mp 49-49.2°. Rapidly polymerises. λ_{max} 262, 272, 283 nm (cyclohexane).
Me ester: [4175-47-7].
Bp_1 148°.
Et ester: [42021-86-3].
Mp −26.5°. Bp_1 158°.
Maleic anhydride adduct: Mp 64°.

(9E,11E,13Z)-form [4337-71-7]
Catalpic acid
Constit. of the seed oil of *Catalpa ovata*, *C. speciosa* and *C. bignonioides*. Mp 31.5-32°. λ_{max} 262, 272, 283 nm (cyclohexane).
Maleic anhydride adduct: Cryst. (Et_2O/pet. ether). Mp 73-73.5°.

(9Z,11Z,13E)-form [544-72-9]
Cryst. (pentane). Mp 61.5-62.5°. λ_{max} 261, 270, 280 nm.

(9Z,11E,13Z)-form [24022-76-2]
Punicic acid. Trichosanic acid. Punicinic acid
Found in pomegranate oil, (*Punica granatum*) seed oils of *Momordica* spp., *Cucurbita* spp., *Trichosanthes cucumeroides* and others. Mp 44-45°. λ_{max} 265, 275, 287 nm (cyclohexane).

Crombie, L. *et al*, *J. Chem. Soc.*, 1957, 1632; 1962, 2449 (*isol*)
Alder, H. *et al*, *Justus Liebigs Ann. Chem.*, 1957, **609**, 24, 36 (*isol*)
Hopkins, C.Y. *et al*, *Can. J. Chem.*, 1962, **40**, 2078; 1965, **43**, 3160 (*isol, struct*)
Chisholm, M.J. *et al*, *Can. J. Chem.*, 1964, **42**, 560 (*isol*)
Hopkins, C.Y., *Top. Lipid Chem.*, 1973, **3**, 37 (*rev*)
Tulloch, A.P., *Lipids*, 1979, **14**, 996 (*cmr*)
Tsuboi, S. *et al*, *Bull. Chem. Soc. Jpn.*, 1987, **60**, 1103 (*synth*)

9,12,15-Octadecatrienoic acid O-00158

[1955-33-5]

$H_3CCH_2CH{=}CHCH_2CH{=}CHCH_2CH{=}CH(CH_2)_7COOH$

$C_{18}H_{30}O_2$ M 278.434

(E,E,E)-form [28290-79-1] ***Linolenelaidic acid.*** *Elaidolinolenic acid*
Isol. from seed oils of *Carthamus tinctorius*, *Chilopsis linearis*, *Catalpa bignonioides*, etc. Cryst. (C_6H_6). Mp 29.5-30°. $Bp_{0.8}$ 179-183°. n_D^{27} 1.4655.
Me ester:
$C_{19}H_{32}O_2$ M 292.461
Mp −3.5°.
Et ester:
$C_{20}H_{34}O_2$ M 306.487
Bp_2 140°.
Hexabromide:
$C_{18}H_{30}Br_6O_2$ M 757.858
Mp 169-170°.

(Z,Z,Z)-form [463-40-1] ***α-Linolenic acid***
Constit. of linseed and other drying oils. An essential fatty acid. Mp −11.3° to −11°. $Bp_{0.07}$ 137°, $Bp_{0.05}$ 125°.
Me ester: [301-00-8].
Mp −57°, Mp −52.4° to − 51.8°. $Bp_{0.018}$ 109°.
Et ester: [1191-41-9].
$Bp_{2.5}$ 174°.
Chloride:
$C_{18}H_{29}ClO$ M 296.879
$Bp_{2.5}$ 170-171°.
Hexabromide: Mp 180-181°.

(9E,12Z,15E)-form
Me ester: Mp −45.5°.

(9Z,12E,15E)-form
Me ester: Mp −43.5°.

(9E,12E,15Z)-form
Me ester: Mp −50.5°.

(9Z,12Z,15E)-form
Me ester: Mp −60°.

(9E,12Z,15Z)-form
Me ester: Mp −57°.

(9Z,12E,15Z)-form
Me ester: Mp −52°.

Nigam, S.S. *et al*, *J. Chem. Soc.*, 1956, 4049 (*synth*)
Chisholm. M.J., *et al*, *Can. J. Chem.*, 1963, **41**, 1888; 1965, **43**, 2566 (*isol, uv, ir*)
Strocchi, A. *et al*, *Rev. Fr. Corps. Gras*, 1969, **13**, 3 (*isom*)
Karrer, W. *et al*, *Konstitution und Vorkommen der Organischen Pflanzenstoffe*, 2nd Ed., Birkhäuser Verlag, Basel, 1972-1985, no. 759 (*occur*)
Ackman, R.G. *et al*, *J. Chromatogr. Sci.*, 1974, **12**, 131 (*cmr*)
Gunstone, F.D. *et al*, *Chem. Phys. Lipids*, 1977, **18**, 115 (*cmr*)
Rakoff, M. *et al*, *Chem. Phys. Lipids*, 1982, **31**, 213 (*synth, cmr, glc*)
Lewis, R.J., *Sax's Dangerous Properties of Industrial Materials*, 8th Ed., Van Nostrand-Reinhold, 1992, OAX100.

10,12,14-Octadecatrienoic acid O-00159

$H_3CCH_2CH_2(CH{=}CH)_3(CH_2)_8COOH$

$C_{18}H_{30}O_2$ M 278.434

(E,E,E)-form
Pseudoeleostearic acid
Product of alkaline isomerisation of α-linolenic acid. Mp 77-77.5°.
Me ester:
$C_{19}H_{32}O_2$ M 292.461
Cryst. (MeOH or C_6H_6). Mp 40.6-41°.
Tetrabromide:
$C_{18}H_{30}Br_4O_2$ M 598.050
Cryst. (EtOH). Mp 104-104.5°.

Kass, J.P. *et al*, *J. Am. Chem. Soc.*, 1939, **61**, 3293 (*synth*)
Ault, W.C. *et al*, *Ind. Eng. Chem.*, 1942, **34**, 1120.
Holman, R.T. *et al*, *J. Am. Chem. Soc.*, 1945, **67**, 1392 (*uv*)
v. Mikusch, J.D., *Farbe + Lack*, 1954, **60**, 178 (*synth*)
Strocchi, A. *et al*, *Rev. Fr. Corps. Gras*, 1969, **16**, 3; *CA*, **71**, 4751.

9,12,15-Octadecatrien-6-ynoic acid O-00160

[61481-30-9]

$H_3CCH_2CH{=}CHCH_2CH{=}CHCH_2$
$CH{=}CHCH_2C{\equiv}C(CH_2)_4COOH$

$C_{18}H_{26}O_2$ M 274.402

(all-Z)-form [53766-42-0]
Component of the glycerides of several mosses, e.g. *Fortinalis*, *Dicranum*, *Ceratodon*, *Sulacomnium*, *Bryum* spp.

Andersson, B. *et al*, *Lipids*, 1974, **9**, 506 (*ir, bibl*)
Reid, E.H. *et al*, *Phytochemistry*, 1976, **15**, 1731.
Ichikawa, T. *et al*, *Tetrahedron Lett.*, 1983, **24**, 3337 (*isol, bibl*)

5,8,11-Octadecatriynoic acid O-00161

$H_3C(CH_2)_5C{\equiv}CCH_2C{\equiv}CCH_2C{\equiv}C(CH_2)_3COOH$

$C_{18}H_{24}O_2$ M 272.386
Cryst. (pet. ether). Mp 62-63°.

Sprecher, H.W., *Biochim. Biophys. Acta*, 1971, **231**, 122 (*synth*)

6,9,12-Octadecatriynoic acid O-00162

$H_3C(CH_2)_4C{\equiv}CCH_2C{\equiv}CCH_2C{\equiv}C(CH_2)_4COOH$

$C_{18}H_{24}O_2$ M 272.386

Cryst. (pet. ether). Mp 53-55°.

Me ester: [16326-31-1].

$C_{19}H_{26}O_2$ M 286.413

$Bp_{35.5}$ 140-156°.

Osbond, J.M. *et al*, *J. Chem. Soc.*, 1961, 2779 (*synth*)
Swiss Pat., 414 595, (1966); *CA*, **67**, 2772 (*synth*)
Sprecher, H., *Prog. Chem. Fats Other Lipids*, 1978, **15**, 219.

8,11,14-Octadecatriynoic acid O-00163

[19960-53-3]

$H_3CCH_2CH_2C{\equiv}CCH_2C{\equiv}CCH_2C{\equiv}C(CH_2)_6COOH$

$C_{18}H_{24}O_2$ M 272.386

Cryst. Mp 47.5-49°.

Beerthuis, R.K. *et al*, *Recl. Trav. Chim. Pays-Bas* (*J. R. Neth. Chem. Soc.*), 1968, **87**, 461 (*synth, ir, nmr*)

2-Octadecenal, 9CI O-00164

[56554-96-2]

$H_3C(CH_2)_{14}CH{=}CHCHO$

$C_{18}H_{34}O$ M 266.466

(E)-form [51534-37-3] ***Koiganal I***

Major component (together with 2,13-Octadecadienal, O-00016) of the sex pheromone of the webbing clothes moth, *Tiniella bisselliella*. Light yellow solid. Mp 44.5-46.5°. $Bp_{0.001}$ 123-125°.

Stoffel, W. *et al*, *Hoppe Seyler's Z. Physiol. Chem.*, 1973, **354**, 1626 (*synth*)
Yamaoka, R. *et al*, *Shitsuryo Bunseki*, 1985, **33**, 189; *CA*, **104**, 17841p (*isol, activity, synth*)
Pohmakotr, M. *et al*, *Tetrahedron Lett.*, 1985, **26**, 3613 (*synth*)
Shi, L. *et al*, *Justus Liebigs Ann. Chem.*, 1988, 377 (*synth, pmr, glc*)

9-Octadecenal O-00165

[5090-41-5]

$H_3C(CH_2)_7CH{=}CH(CH_2)_7CHO$

$C_{18}H_{34}O$ M 266.466

(Z)-form [2423-10-1]

Oleic aldehyde

Constit. of the sex pheromone of the Southwestern corn borer *Diatraea grandiosella* and the yellow stem borer *Scirpophaga incertulas*. Isol. from the oil of hops (*Humulus lupulus* and *Elaeagnus moorcraftii*). Also found in the aromas of various cooked meats. Liq. $Bp_{0.001}$ 108-110°.

2,4-Dinitrophenylhydrazone: Yellow cryst. (EtOH). Mp 67-68°.

[10009-79-7]

Valicenti, A.J. *et al*, *Chem. Phys. Lipids*, 1976, **17**, 389.
Wedmid, Y. *et al*, *J. Org. Chem.*, 1977, **42**, 3624.
Mancuso, A.J. *et al*, *J. Org. Chem.*, 1978, **43**, 2480.
Tatsuki, S. *et al*, *Appl. Entomol. Zool.*, 1985, **20**, 357.
Hedin, P.A. *et al*, *J. Chem. Ecol.*, 1986, **12**, 2051 (*isol*)
Tsay, S.-C. *et al*, *J. Chem. Soc., Perkin Trans. 1*, 1990, 757 (*synth*)

11-Octadecenal O-00166

$H_3C(CH_2)_5CH{=}CH(CH_2)_9CHO$

$C_{18}H_{34}O$ M 266.466

(Z)-form [4273-95-4]

Sex attractant of the lesser waxmoth *Achroia grisella* and spotted bollworm *Earias vittella*.

Dehm, K.H. *et al*, *Naturwissenschaften*, 1971, **58**, 265.
Cork, A. *et al*, *J. Chem. Ecol.*, 1988, **14**, 929 (*isol*)

13-Octadecenal, 9CI O-00167

[56554-90-6]

$H_3C(CH_2)_5CH{=}CH(CH_2)_9CHO$

$C_{18}H_{34}O$ M 266.466

Component of the labial gland secretion (pathway marker) of the male cuckoo bumblebee, *Psithyrus bohemicus*.

(Z)-form [58594-45-9]

Component of sex pheromones of several Lepidoptera spp., including rice stem borer *Chilo suppressalis*, *Eoreuma loftini*, tobacco hornworm *Manduca sexta*, corn earworm *Heliothis zea*, southwestern corn borer *Diatraea grandiosella*. Oil. $Bp_{0.01}$ 110-112°.

Ohta, K. *et al*, *Agric. Biol. Chem.*, 1976, **40**, 1897 (*synth, pmr, ir, ms, glc, activity*)
Kanno, H. *et al*, *Appl. Entomol. Zool.*, 1978, **13**, 321 (*activity*)
Bestmann, H.J. *et al*, *Chem. Ber.*, 1979, **112**, 3740 (*synth, pmr, ir, ms, activity*)
McDonough, L.M. *et al*, *Environ. Entomol.*, 1982, **11**, 711 (*activity*)
Hedin, P.A. *et al*, *J. Chem. Ecol.*, 1986, **12**, 2051 (*activity*)
Mangold, H.K. *et al*, *Z. Naturforsch., C*, 1986, **41**, 555 (*synth*)
Lanne, B.S. *et al*, *Comp. Biochem. Physiol., B: Comp. Biochem.*, 1987, **88**, 631 (*isol*)
Robacker, D.C. *et al*, *J. Chem. Ecol.*, 1989, **15**, 105 (*activity*)
Zagatti, P. *et al*, *J. Chem. Ecol.*, 1991, **17**, 1399 (*isol, activity*)

1-Octadecene O-00168

[112-88-9]

$H_3C(CH_2)_{15}CH{=}CH_2$

$C_{18}H_{36}$ M 252.482

Mp 18° (17.5°). Bp_3 144-146°.

Dibromide: 1,2-Dibromoctadecane

Cryst. (EtOH). Mp 24°.

Krafft, F., *Ber.*, 1884, **17**, 1371 (*deriv*)
Asinger, F., *Ber.*, 1942, **75**, 1247.
Nieman, C. *et al*, *J. Org. Chem.*, 1942, **7**, 227 (*synth*)
Yano, I. *et al*, *Eur. J. Biochem.*, 1971, **23**, 220.
Krapcho, A.P. *et al*, *J. Org. Chem.*, 1977, **42**, 3749.
Voloshin, N.L. *et al*, *CA*, 1978, **89**, 179638b.
Sax, N.I., *Dangerous Properties of Industrial Materials*, 5th Ed., Van Nostrand-Reinhold, 1979, 874.

9-Octadecenedioic acid, 9CI O-00169

[4494-16-0]

$HOOC(CH_2)_7CH{=}CH(CH_2)_7COOH$

$C_{18}H_{32}O_4$ M 312.448

Present in *Sphagnum*, pine, spruce and birch bark; root and stem corks of *Malus pumilia* and suberin from sapwood of oak.

(E)-form [20701-67-1]

Solid (EtOAc). Mp 99-101° (95.5-98°).

Mono-Me ester: [24753-51-3].

$C_{19}H_{34}O_4$ M 326.475

Solid (pet. ether). Mp 46-47°.

Di-Me ester: [24753-49-9].
$C_{20}H_{36}O_4$ M 340.502
Solid. Mp 33-34° (30°).

(Z)-form [20701-68-2]
Manuf. from hexadecenoate by *Candidum tropicalis*. Solid. Mp 67-68°.

[13481-97-5, 40393-46-2, 70445-23-7, 75722-65-5, 80060-75-9, 80060-80-6]

Blomquist, A.T. *et al*, *J. Am. Chem. Soc.*, 1948, **70**, 34 (*synth*)
Dupont, G. *et al*, *Bull. Soc. Chim. Fr.*, 1956, 819 (*synth*)
Ribas, J. *et al*, *CA*, 1956, **50**, 806 (*synth*)
Van Dam, P.V. *et al*, *J. Chem. Soc., Chem. Commun.*, 1972, 1221 (*synth*)
Aberca, B.G. *et al*, *Chem. Ind. (London)*, 1979, 477 (*synth*)
Tsuji, J. *et al*, *J. Organomet. Chem.*, 1981, **218**, 69 (*synth, ir, pmr*)

3-Octadecene-1,2-diol O-00170

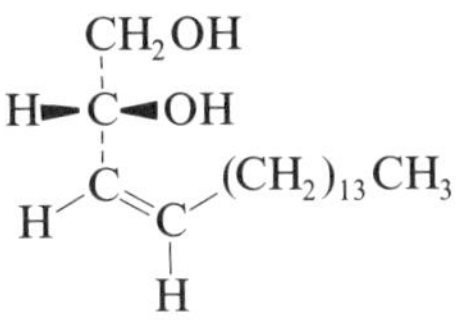

(R,Z)-form

$C_{18}H_{36}O_2$ M 284.481

(R,Z)-form
Cryst. (hexane). Mp 56-57°. $[\alpha]_D^{20}$ −7.5° (c, 2.0 in $CHCl_3$).
Dibenzoyl:
$C_{32}H_{44}O_4$ M 492.697
$[\alpha]_D^{20}$ +38.0° (c, 3.25 in $CHCl_3$).

(S,Z)-form
Platelets (hexane). Mp 56-57°. $[\alpha]_D^{20}$ +8.1° (c, 3.23 in $CHCl_3$).
1-Benzoyl:
$C_{25}H_{40}O_3$ M 388.589
$[\alpha]_D^{20}$ +11.6° (c, 2.2 in Py).
Dibenzoyl: $[\alpha]_D^{20}$ −38.2° (c, 3.25 in $CHCl_3$).

Mulzer, J. *et al*, *Tetrahedron*, 1986, **42**, 5961 (*synth, ir, pmr, cmr*)

9-Octadecene-1,12-diol O-00171

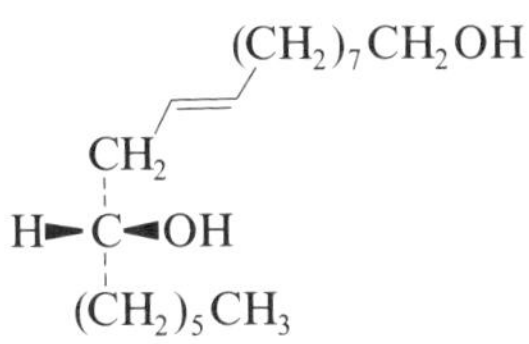

(R,E)-form

$C_{18}H_{36}O_2$ M 284.481

(R,E)-form [35732-93-5]
(+)-trans-*form. Ricinelaidyl alcohol*
Mp 51.2-52.4°. $[\alpha]_D^{27}$ +6.0° (c, 1.51 in MeOH).
12-Ac: Yellow oil. $Bp_{0.5}$ 178°.

(R,Z)-form [540-11-4]
(+)-cis-*form. Ricinoleyl alcohol*
$Bp_{0.5}$ 175°. $[\alpha]_D^{27}$ +6.7° (c, 1.65 in MeOH).
12-Ac: [2581-27-3]. *Gyplure*
$C_{20}H_{38}O_3$ M 326.518
Insect sex attractant. Oil. $Bp_{0.5}$ 182°. $[\alpha]_D^{30}$ +7.4° (c, 1 in $CHCl_3$).
Di-Ac:
$C_{22}H_{40}O_4$ M 368.556
Liq. $Bp_{1.3}$ 180°. $[\alpha]_D^{30}$ +8.7° (c, 1 in $CHCl_3$).

Jacobson, M. *et al*, *J. Org. Chem.*, 1962, **27**, 2523 (*synth*)
Applewhite, T.H. *et al*, *J. Org. Chem.*, 1967, **32**, 1173 (*synth, ord*)
Eiter, K. *et al*, *Justus Liebigs Ann. Chem.*, 1967, **709**, 29 (*synth*)

5-Octadecene-7,9-diynoic acid, 9CI O-00172

$$H_3C(CH_2)_7C{\equiv}CC{\equiv}CCH{=}CH(CH_2)_3COOH$$

$C_{18}H_{26}O_2$ M 274.402
Inhibitor of HMG-CoA reductase.

(E)-form [120193-28-4]
Isol. from the root bark of *Paramacrolobium caeruleum*. Gum. λ_{max} 282, 266, 252, 240, 217 nm (hexane).
Me ester: [120193-35-3].
Oil. λ_{max} 282, 267, 252, 240, 218 nm (hexane).
5ξ,6ξ-Epoxide: [120193-30-8]. *3-(1,3-Dodecadiynyl)oxiranebutanoic acid, 9CI. 5,6-Epoxy-7,9-octadecadiynoic acid*
$C_{18}H_{26}O_3$ M 290.402
Isol. from root bark of *P. caeruleum*. Yellow gum. λ_{max} 283, 269, 258, 245, 207 nm (MeOH).

(Z)-form
Isol. from the root bark of *P. caeruleum*. Gum. λ_{max} 284, 268, 253, 241, 217 nm (hexane).

[120193-36-4]

Patil, A.D. *et al*, *J. Nat. Prod. (Lloydia)*, 1989, **52**, 153 (*isol, struct, ir, uv, pmr, cmr*)
Mhaskar, S.Y. *et al*, *Synth. Commun.*, 1990, **20**, 2001 (*synth, pmr, ester*)

9-Octadecene-12,14-diynoic acid O-00173

$$H_3C(CH_2)_2C{\equiv}CC{\equiv}CCH_2CH{=}CH(CH_2)_7COOH$$

$C_{18}H_{26}O_2$ M 274.402

(Z)-form

Fallis, A.G. *et al*, *J. Chem. Soc., Perkin Trans. 1*, 1973, 743 (*synth, pmr*)

13-Octadecene-9,11-diynoic acid O-00174

$$H_3C(CH_2)_3CH{=}CHC{\equiv}CC{\equiv}C(CH_2)_7COOH$$

$C_{18}H_{26}O_2$ M 274.402

(E)-form [13089-73-1] ***Exocarpic acid***
Present in the root fat of *Exocarpus cupressiformis, E. stricta, Buckleya distichophylla*. Cryst. (pentane) which turn blue on exposure to light. Mp 41-42°, Mp 42.5-43.5°.

(Z)-form
Constit. of *Ongokea gore* seed oil. Cryst. (pentane) which rapidly turn pink in air. Mp 19-19.5°.

Hatt, H.H. *et al*, *Aust. J. Chem.*, 1959, **12**, 190 (*isol*)
Crombie, L. *et al*, *J. Chem. Soc.*, 1962, 2449 (*synth*)
Gunstone, F.D. *et al*, *J. Chem. Soc.*, 1963, 5772 (*isol*)
Hopkins, C.Y. *et al*, *Chem. Ind. (London)*, 1966, 1533 (*isol*)
Hatt, H.H. *et al*, *Aust. J. Chem.*, 1967, **20**, 2285 (*isol*)

17-Octadecene-9,11-diynoic acid O-00175

Erythrogenic acid. Isanic acid

$$H_2C{=}CH(CH_2)_4C{\equiv}CC{\equiv}C(CH_2)_7COOH$$

$C_{18}H_{26}O_2$ M 274.402
Constit. of *Ongokea gore*. Cryst. (pet. ether). Mp 38.4-39°.

Armitage, J.P. *et al*, *J. Chem. Soc.*, 1952, 1998 (*synth*)
Black, H.K. *et al*, *J. Chem. Soc.*, 1953, 1753 (*synth*)
Gunstone, F.D. *et al*, *J. Chem. Soc.*, 1963, 5772 (*isol, struct*)
Vieux, A.S. *et al*, *Oleagineux*, 1968, **23**, 325 (*isol*)

9-Octadecene-12,14,16-triynoic acid, 9CI O-00176

[56319-30-3]

$H_3C(C{\equiv}C)_3CH_2CH{=}CH(CH_2)_7COOH$

$C_{18}H_{22}O_2$ M 270.371

(Z)-form

Me ester:

$C_{19}H_{24}O_2$ M 284.397

Cryst. (pet. ether at −40°), liq. at r.t. Mp *ca.* 10°.

Fallis, A.G. *et al, J. Chem. Soc., Perkin Trans. 1*, 1973, 743 (*synth, nmr, ms*)

15-Octadecene-9,11,13-triynoic acid O-00177

$H_3CCH_2CH{=}CHC{\equiv}CC{\equiv}CC{\equiv}C(CH_2)_7COOH$

$C_{18}H_{22}O_2$ M 270.371

Major component of the lipids of *Santalum acuminatum*. λ_{max} 212, 223, 232, 244, 258, 274, 291, 310, 332 nm.

Bu'Lock, J.D. *et al, Phytochemistry*, 1963, **2**, 289.

2-Octadecenoic acid O-00178

[5340-63-6]

$H_3C(CH_2)_{14}CH{=}CHCOOH$

$C_{18}H_{34}O_2$ M 282.465

(E)-form [2825-79-8]

Cryst. (EtOH or pet. ether). Mp 58.5°.

Me ester: [14663-11-7].

$C_{19}H_{36}O_2$ M 296.492

Mp 36°.

Et ester:

$C_{20}H_{38}O_2$ M 310.519

Cryst. (MeOH). Mp 25-26°. Bp >360°.

Amide:

$C_{18}H_{35}NO$ M 281.481

Cryst. (EtOH). Mp 107-108°.

Dibromide: Cryst. (pet. ether). Mp 72°.

(Z)-form [2825-66-3]

Cryst. (pet. ether). Mp 50.5°.

Barve, J.A. *et al, Chem. Phys. Lipids*, 1971, **7**, 311 (*synth*)
Davies, J.E.D. *et al, J. Chem. Soc., Perkin Trans. 2*, 1972, 1557 (*ir, raman*)
Gershon, H. *et al, Antimicrob. Agents Chemother.*, 1973, **4**, 435.
Howton, D.R., *Org. Prep. Proced. Int.*, 1974, **6**, 175 (*synth*)
Schmitz, B. *et al, Chem. Phys. Lipids*, 1975, **15**, 248.
Gunstone, F.D. *et al, Chem. Phys. Lipids*, 1977, **18**, 115 (*cmr*)

3-Octadecenoic acid O-00179

[21735-77-3]

$H_3C(CH_2)_{13}CH{=}CHCH_2COOH$

$C_{18}H_{34}O_2$ M 282.465

(E)-form [13161-77-8]

Isol. from seed oil of *Grindelia oxylepis*. Cryst. (pet. ether). Mp 64.5-65.5°.

(Z)-form [17257-45-3]

Cryst. (pet. ether). Mp 49.5-50.5°.

Gunstone, F.D. *et al, Chem. Phys. Lipids*, 1967, **1**, 264 (*synth*)
Barve, J.A. *et al, Chem. Phys. Lipids*, 1971, **7**, 311 (*synth*)
Davies, J.E.D. *et al, J. Chem. Soc., Perkin Trans. 2*, 1972, 1557 (*ir, raman*)
Gunstone, F.D. *et al, Chem. Phys. Lipids*, 1973, **11**, 43.
Schwartz, D.P., *Anal. Biochem.*, 1976, **74**, 320.
Gunstone, F.D. *et al, Chem. Phys. Lipids*, 1977, **18**, 115 (*cmr*)
Saito, Y. *et al, Biochim. Biophys. Acta*, 1978, **529**, 224.

4-Octadecenoic acid O-00180

$H_3C(CH_2)_{12}CH{=}CHCH_2CH_2COOH$

$C_{18}H_{34}O_2$ M 282.465

(E)-form [34450-19-6]

Mp 58.5-59.5°.

(Z)-form [19308-06-6]

Mp 45.5-46.5°.

James, A.T. *et al, Biochem. J.*, 1956, **63**, 144 (*glc*)
Barve, J.A. *et al, Chem. Phys. Lipids*, 1971, **7**, 311 (*synth*)
Davies, J.E.D. *et al, J. Chem. Soc., Perkin Trans. 2*, 1972, **11**, 1557 (*ir, raman*)
Welles, H.L. *et al, Adv. Chem. Ser.*, 1975, **144**, 135.
Schwartz, D.P., *Anal. Biochem.*, 1976, **74**, 320.
Saito, Y. *et al, Biochim. Biophys. Acta*, 1978, **529**, 224.

5-Octadecenoic acid O-00181

$H_3C(CH_2)_{11}CH{=}CH(CH_2)_3COOH$

$C_{18}H_{34}O_2$ M 282.465

(E)-form [7056-85-1]

Constit. of *Thalictrum polycarpum* seed oil, also *T. simplex* and *T. venulosum*. Mp 47.5° (43-44°).

(Z)-form [676-29-9]

Isol. from seed oil of *T. venulosum*. Mp 12.5-13.5°.

Bagby, M.O. *et al, Biochemistry*, 1962, **1**, 632 (*isol*)
Barve, J.A. *et al, Chem. Phys. Lipids*, 1971, **7**, 311 (*synth*)
Davies, J.E.D. *et al, J. Chem. Soc., Perkin Trans. 2*, 1972, 1557 (*ir, raman*)
Schwartz, D.P., *Anal. Biochem.*, 1976, **74**, 320.
Bus, J. *et al, Chem. Phys. Lipids*, 1977, **18**, 130 (*cmr*)
Saito, Y. *et al, Biochim. Biophys. Acta*, 1978, **529**, 224.

6-Octadecenoic acid O-00182

$H_3C(CH_2)_{10}CH{=}CH(CH_2)_4COOH$

$C_{18}H_{34}O_2$ M 282.465

(E)-form [593-40-8]

Petroselaidic acid. Tarelaidinic acid

Minor constit. of plant oils accompanying the major *Z*-isomer. Cryst. ($CHCl_3$ or MeOH). Mp 52.7-53.4°, Mp 54-59°.

Me ester: [14620-36-1].

$C_{19}H_{36}O_2$ M 296.492

Bp_{10} 208-210°, Bp_4 185-186°.

Chloride:

$C_{18}H_{33}ClO$ M 300.911

Bp_2 144-146°.

(Z)-form [593-39-5]

Petroselinic acid

Found in umbelliferous seed oils eg. major constit. of oils of parsley, ivy, fennel, celery, *Seseli indicum* and others. Cryst. (MeOH or EtOH aq.). Mp 33°. $Bp_{2\text{-}3}$ 215-217°, $Bp_{0.15}$ 148-152°.

Baker, B.W. *et al, J. Chem. Soc.*, 1952, 1804.
Lamb, P.B. *et al, J. Chem. Soc.*, 1952, 5032 (*synth, config*)
Swern, D. *et al, J. Am. Oil Chem. Soc.*, 1955, **32**, 539 (*cryst struct*)
Barve, J.A. *et al, Chem. Phys. Lipids*, 1971, **7**, 311 (*synth*)
Batchelor, J.G. *et al, J. Org. Chem.*, 1974, **39**, 1698 (*cmr*)
Parodi, P.W., *J. Dairy Sci.*, 1976, **59**, 1870 (*synth, glc*)
Saito, Y. *et al, Biochim. Biophys. Acta*, 1978, **529**, 224.
Conacher, H.B.S. *et al, J. Assoc. Off. Anal. Chem.*, 1978, **61**, 702 (*glc*)

7-Octadecenoic acid O-00183

[14122-42-0]

$H_3C(CH_2)_9CH{=}CH(CH_2)_5COOH$

$C_{18}H_{34}O_2$ M 282.465

(E)-form [13126-32-4]

Found in partially hydrogenated fats. Mp 44.2-44.5°.

(Z)-form [13126-31-3]

Found in partially hydrogenated fats. Mp 11.8-12.5°, Mp 12.5-13.1°.

Huber, W.F., *J. Am. Chem. Soc.*, 1951, **73**, 2732 (*cryst struct*)
Swern, D. *et al*, *J. Am. Oil Chem. Soc.*, 1955, **32**, 539 (*cryst struct*)
Barve, J.A. *et al*, *Chem. Phys. Lipids*, 1971, **7**, 311 (*synth*)
Schwartz, D.P., *Anal. Biochem.*, 1976, **74**, 320.
Parodi, P.W. *et al*, *J. Dairy Sci.*, 1976, **59**, 1870 (*glc*)
Saito, Y. *et al*, *Biochim. Biophys. Acta*, 1978, **529**, 224.

8-Octadecenoic acid O-00184

$H_3C(CH_2)_8CH{=}CH(CH_2)_6COOH$

$C_{18}H_{34}O_2$ M 282.465

(E)-form [5684-81-1]

Found in partially hydrogenated fat as a glyceride. Mp 51.5-52.3°, Mp 52-52.6°.

(Z)-form [5684-71-9]

Found in partially hydrogenated fat as a glyceride. Mp 22.7-23.8°.

Huber, W.F., *J. Am. Chem. Soc.*, 1951, **73**, 2730 (*synth*)
Swern, D. *et al*, *J. Am. Oil Chem. Soc.*, 1955, **32**, 539 (*cryst struct*)
Barve, J.A. *et al*, *Chem. Phys. Lipids*, 1971, **7**, 311 (*synth*)
Davies, J.E.D. *et al*, *J. Chem. Soc., Perkin Trans. 2*, 1972, 1557 (*ir, raman*)
Gunstone, F.D. *et al*, *Chem. Phys. Lipids*, 1977, **18**, 115 (*cmr*)
Saito, Y. *et al*, *Biochim. Biophys. Acta*, 1978, **529**, 224.

9-Octadecenoic acid O-00185

[2027-47-6]

$H_3C(CH_2)_7CH{=}CH(CH_2)_7COOH$

$C_{18}H_{34}O_2$ M 282.465

(E)-form [112-79-8]

Elaidic acid

Minor constit. of plant oils. Cryst. (EtOH). Mp 45-45.5°. Bp_{15} 234°, $Bp_{0.8}$ 180-185°.

▷ JX6125000.

Me ester: [1937-62-8].
$C_{19}H_{36}O_2$ M 296.492
Bp_{15} 213-215°.

Et ester:
$C_{20}H_{38}O_2$ M 310.519
Bp_{15} 217-219°.

Chloride:
$C_{18}H_{33}ClO$ M 300.911
Bp_{11} 200°.

Amide:
$C_{18}H_{35}NO$ M 281.481
Mp 93-94°.

Anhydride:
$C_{36}H_{66}O_3$ M 546.916
Mp 51°.

Nitrile:
$C_{18}H_{33}N$ M 263.465
Fp −1°. Bp_{16} 213°.

(Z)-form [112-80-1]

Oleic acid. Rapinic acid

Major constit. of plant oils e.g. olive oil (*ca.* 80%), almond oil (*ca.* 80%) and many others, mainly as glyceride. Mp 12° (labile form), Mp 16° (stable form). Bp_5 203-205°. ^{125}I and ^{131}I labelled cpds. are used as radioactive agents (Oleotope).

▷ RG2275000.

Me ester: [112-62-9].
Bp_{15} 212-213°.
▷ An exp. carcinogen. RK0895000.

Et ester: [111-62-6].
Bp_{151} 216-217°.

Vinyl ester: Bp_2 173°.

Chloride: Bp_{11} 200°.

Anhydride: [24909-72-6].
Mp 22-24°.

Amide: Mp 75-76°.

Nitrile: Bp 330-335° dec.

Gensler, W.J. *et al*, *J. Am. Chem. Soc.*, 1951, **73**, 1071 (*synth*)
Org. Synth., 1957, **37**, 66 (*synth*)
Karrer, W. *et al*, *Konstitution und Vorkommen der Organischen Pflanzenstoffe*, 2nd Ed., Birkhäuser Verlag, Basel, 1972-1985, nos. 738, 739 (*occur*)
Bonfand, A. *et al*, *C. R. Hebd. Seances Acad. Sci. Ser. C*, 1975, **281**, 723 (*synth*)
Bailey, A.W., *J. Am. Oil Chem. Soc.*, 1976, **53**, 89 (*cryst struct*)
Ackman, R.G. *et al*, *J. Chromatogr. Sci.*, 1978, **16**, 204 (*glc*)
Miller, R.A. *et al*, *J. Liq. Chromatogr.*, 1978, **1**, 291 (*hplc*)
Bretherick, L., *Handbook of Reactive Chemical Hazards*, 2nd Ed., Butterworths, London and Boston, 1979, 738.
Lewis, R.J., *Sax's Dangerous Properties of Industrial Materials*, 8th Ed., Van Nostrand-Reinhold, 1992, EAF500, OHU000, OHW000.

10-Octadecenoic acid O-00186

Isooleic acid

[3386-63-8]

$H_3C(CH_2)_6CH{=}CH(CH_2)_8COOH$

$C_{18}H_{34}O_2$ M 282.465

(E)-form [5684-82-2]

Found in partially hydrogenated fats as a glyceride. Mp 52-52.6°. Bp_3 208-216°.

Amide:
$C_{18}H_{35}NO$ M 281.481
Mp 83°.

(Z)-form [2442-70-8]

Found in partially hydrogenated fats as a glyceride. Mp 22.2-22.8°.

Huber, F., *J. Am. Chem. Soc.*, 1951, **73**, 2730 (*synth*)
Swern, D. *et al*, *J. Am. Oil Chem. Soc.*, 1955, **32**, 539 (*cryst struct*)
Barve, J.A. *et al*, *Chem. Phys. Lipids*, 1971, **7**, 311 (*synth*)
Gunstone, F.D. *et al*, *Chem. Phys. Lipids*, 1977, **18**, 115 (*cmr*)
Saito, Y. *et al*, *Biochim. Biophys. Acta*, 1978, **529**, 224.

11-Octadecenoic acid O-00187

[143-25-9]

$H_3C(CH_2)_5CH{=}CH(CH_2)_9COOH$

$C_{18}H_{34}O_2$ M 282.465

(E)-form [693-72-1]

Vaccenic acid

Occurs in small proportions in ruminant fats (e.g., butter) *via* biohydrogenation of dietary polyene acids. Main component of *Streptomyces pyogenes* protoplast lipids. Mp 43.5-44.1°.

Me ester: [6198-58-9].
$C_{19}H_{36}O_2$ M 296.492
Bp_3 172-173°.

(Z)-form [506-17-2]

Asclepic acid

A minor component of various natural fats. Cryst. (Me_2CO). Mp 14.5-15.5°.

Bounds, D.G. *et al*, *J. Chem. Soc.*, 1954, 4219 (*synth*)
Swern, D. *et al*, *J. Am. Oil Chem. Soc.*, 1955, **32**, 539 (*cryst struct*)
Cohen, M. *et al*, *Biochemistry*, 1966, **5**, 2385 (*isol*)
Barve, J.A. *et al*, *Chem. Phys. Lipids*, 1971, **7**, 311 (*synth*)
Noble, R.C. *et al*, *Br. J. Nutr.*, 1974, **31**, 99.
Bus, J. *et al*, *Recl. Trav. Chim. Pays-Bas* (*J. R. Neth. Chem. Soc.*), 1974, **93**, 213 (*cmr*)
Morales, R.W. *et al*, *Biochim. Biophys. Acta*, 1976, **431**, 206 (*isol*)
Saito, Y. *et al*, *Biochim. Biophys. Acta*, 1978, **529**, 224 (*isol*)

12-Octadecenoic acid O-00188

[7378-88-3]

$H_3C(CH_2)_4CH{=}CH(CH_2)_{10}COOH$

$C_{18}H_{34}O_2$ M 282.465

(E)-form [13126-38-0]

Present in partially hydrogenated fat as a glyceride. Cryst. (pet. ether). Mp 39.7-40.1°, Mp 52-53°.

(Z)-form [13126-37-9]

Mp 26.5-27.5°, Mp 26.8-27.6°. $Bp_{1.5}$ 196°.

Et ester:

$C_{20}H_{38}O_2$ M 310.519

$Bp_{0.4}$ 184°.

Huber, F., *J. Am. Chem. Soc.*, 1951, **73**, 2730 (*synth*)
Baker, C.D. *et al*, *J. Chem. Soc.*, 1963, 489 (*synth*)
Barve, J.A. *et al*, *Chem. Phys. Lipids*, 1971, **7**, 311 (*synth*)
Schwartz, D.P. *et al*, *Anal. Biochem.*, 1976, **74**, 320.
Gunstone, F.D., *Chem. Phys. Lipids*, 1977, **18**, 115 (*cmr*)
Conacher, H.B.S. *et al*, *J. Assoc. Off. Anal. Chem.*, 1978, **61**, 702 (*glc*)

15-Octadecenoic acid O-00189

[15737-21-0]

$H_3CCH_2CH{=}CH(CH_2)_{13}COOH$

$C_{18}H_{34}O_2$ M 282.465

(E)-form [13126-44-8]

A minor component of milk fat. Cryst. (EtOH). Mp 59°, Mp 61-61.5°.

(Z)-form [13126-43-7]

Cryst. (Me_2CO). Mp 41-42°.

Barve, J.A. *et al*, *Chem. Phys. Lipids*, 1971, **7**, 311 (*synth*)
Klok, R. *et al*, *Recl. Trav. Chim. Pays-Bas* (*J. R. Neth. Chem. Soc.*), 1974, **93**, 222 (*synth*)
Parodi, P.W. *et al*, *J. Dairy Sci.*, 1976, **59**, 1870 (*isol, synth, glc*)
Gunstone, F.D. *et al*, *Chem. Phys. Lipids*, 1977, **18**, 115 (*cmr*)
Conacher, H.B.S. *et al*, *J. Am. Oil Chem. Soc.*, 1978, **61**, 702 (*glc*)
Saito, Y. *et al*, *Biochim. Biophys. Acta*, 1978, **529**, 224.

16-Octadecenoic acid O-00190

[21739-76-4]

$H_3CCH{=}CH(CH_2)_{14}COOH$

$C_{18}H_{34}O_2$ M 282.465

(E)-form [10440-93-4]

Occurs in hydrolysates of ruminant fats. Cryst. (MeOH). Mp 65.6-66.2°.

Me ester: [42199-41-7].

$C_{19}H_{36}O_2$ M 296.492

Mp 28.3-28.9°.

(Z)-form [13126-45-9]

Mp 54°.

Kapp, R. *et al*, *J. Am. Chem. Soc.*, 1943, **65**, 2062 (*synth*)
Barve, J.A. *et al*, *Chem. Phys. Lipids*, 1971, **7**, 311 (*synth*)
Frost, D.J. *et al*, *Chem. Phys. Lipids*, 1975, **15**, 53 (*pmr*)
Bus, J. *et al*, *Chem. Phys. Lipids*, 1976, **17**, 501 (*cmr*)
Schwartz, D.P. *et al*, *Anal. Biochem.*, 1976, **74**, 320.
Saito, Y. *et al*, *Biochim. Biophys. Acta*, 1978, **529**, 224.

17-Octadecenoic acid O-00191

[19307-16-5]

$H_2C{=}CH(CH_2)_{15}COOH$

$C_{18}H_{34}O_2$ M 282.465

Cryst. (MeOH aq.). Mp 55-55.5°.

Kapp, R. *et al*, *J. Am. Chem. Soc.*, 1943, **65**, 2062 (*synth*)
Barve, J.A. *et al*, *Chem. Phys. Lipids*, 1971, **7**, 311 (*synth*)
Davies, J.E.D. *et al*, *J. Chem. Soc., Perkin Trans. 2*, 1972, 1557 (*raman*)
Gunstone, F.D. *et al*, *Chem. Phys. Lipids*, 1977, **18**, 115 (*cmr*)
Saito, Y. *et al*, *Biochim. Biophys. Acta*, 1978, **529**, 224.

2-Octadecen-1-ol O-00192

[22104-84-3]

$H_3C(CH_2)_{14}CH{=}CHCH_2OH$

$C_{18}H_{36}O$ M 268.482

(E)-form [41207-34-5]

Oil.

(Z)-form [2831-86-9]

Low-melting solid.

Borgini, A. *et al*, *J. Chem. Soc., Perkin Trans. 1*, 1986, 1339, 1345 (*synth, ir, pmr, cmr*)

9-Octadecen-1-ol O-00193

[593-47-5]

$H_3C(CH_2)_7CH{=}CH(CH_2)_7CH_2OH$

$C_{18}H_{36}O$ M 268.482

(E)-form [506-42-3]

Elaidyl alcohol. Elaidic alcohol

Mp 34°, Mp 36-37°. Bp_{18} 216°.

(Z)-form [143-28-2]

Oleyl alcohol. Oleic alcohol

Occurs in fish oils. Also obt. from wood of *Eperua falcata*. d_4^{20} 0.849. Bp_{15} 205-210°. n_D^{20} 1.4607.

▷ RG4120000.

Noller, C.R. *et al*, *J. Am. Chem. Soc.*, 1934, **56**, 1563 (*synth*)
Org. Synth., 1935, **15**, 72 (*synth*)
Nyström, R.F. *et al*, *J. Am. Chem. Soc.*, 1947, **69**, 1197.
Blake, S. *et al*, *J. Chem. Soc.*, 1963, 430 (*isol*)
Lewis, R.J., *Sax's Dangerous Properties of Industrial Materials*, 8th Ed., Van Nostrand-Reinhold, 1992, OBA000.

11-Octadecen-1-ol O-00194

$H_3C(CH_2)_5CH{=}CH(CH_2)_9CH_2OH$

$C_{18}H_{36}O$ M 268.482

(E)-form [57716-88-8]

Vaccenyl alcohol

(Z)-form

Ac: [6186-98-7].

$C_{20}H_{38}O_2$ M 310.519

Component of the hairpencil secretion of male butterflies *Lycorea ceres ceres* and male *Drosophila melanogaster*. $Bp_{0.2}$ 128-140°.

Meinwald, J. *et al*, *J. Am. Chem. Soc.*, 1966, **88**, 1308 (*synth, struct*)
Brieger, G. *et al*, *Science* (*Washington, D.C.*), 1970, **167**, 1262 (*isol*)

13-Octadecen-1-ol O-00195

[62803-20-7]

$H_3C(CH_2)_3CH{=}CH(CH_2)_{11}CH_2OH$

$C_{18}H_{36}O$ M 268.482

(Z)-form [69820-27-5]

Pheromone component of the rice stem borer *Chilo suppressalis*. Mp 14-15°. $Bp_{0.01}$ 127-128°.

Ac: [60037-58-3].

Pheromone component of *Eoreuma loftini*. $Bp_{0.01}$ 125-127°.

Bestmann, H.J. *et al*, *Chem. Ber.*, 1979, **112**, 3740 (*synth, pmr, ir, ms*)
Shaver, T.N. *et al*, *J. Chem. Ecol.*, 1988, **14**, 391.

2-(9-Octadecenoylamino)-4,8-octadecadiene-1,3-diol O-00196

N-[*2-Hydroxy-1-(hydroxymethyl)-3,7-heptadecadienyl*]-*9-octadecenamide*

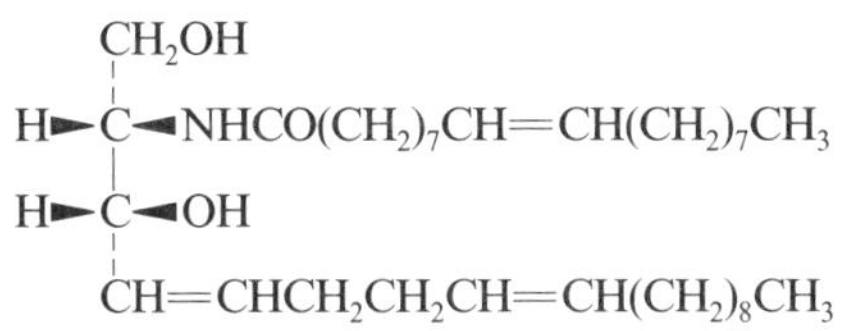

$C_{36}H_{67}NO_3$ M 561.930

Karlsson, K.A., *J. Lipid Res.*, 1971, **12**, 466 (*tlc*)

2-(9-Octadecenoylamino)-1,3-octadecanediol O-00197

N-[*2-Hydroxy-1-(hydroxymethyl)heptadecyl*]-*9-octadecenamide, 9CI*

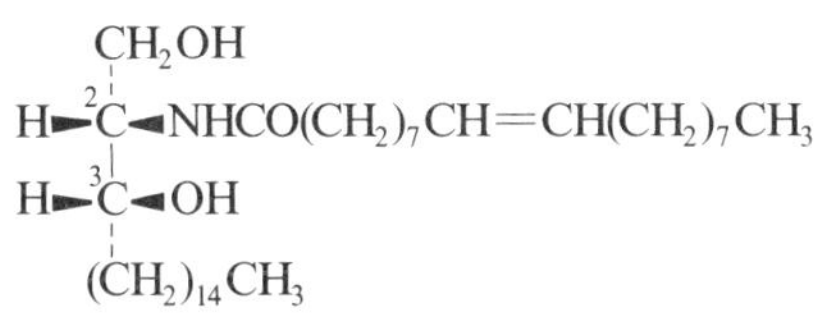

$C_{36}H_{71}NO_3$ M 565.962

(2RS,3SR)-form

(±)-erythro-*form*. N-*Oleoylsphinganine*. N-*Octadec-9-enoylsphinganine*

Mp 86-87°.

Karlsson, K.A., *J. Lipid Res.*, 1971, **12**, 466 (*tlc*)
Alpes, H., *Chem. Phys. Lipids*, 1974, **13**, 109 (*synth*)
Gaskell, S.J. *et al*, *J. Chromatogr.*, 1976, **126**, 591 (*glc, ms*)

9-Octadecen-2,4,6-triynedioic acid O-00198

$HOOCC{\equiv}CC{\equiv}CC{\equiv}CCH_2CH{=}CH(CH_2)_7COOH$

$C_{18}H_{20}O_4$ M 300.354

(Z)-form

Di-Me ester:

$C_{20}H_{24}O_4$ M 328.407

Needles (pet. ether). Mp *ca.* 0°.

Fallis, A.G. *et al*, *J. Chem. Soc., Perkin Trans. 1*, 1973, 743 (*bibl*)

9-Octadecenylamine, 8CI O-00199

9-Octadecen-1-amine, 9CI. 1-Amino-9-octadecene

[1838-19-3]

$H_3C(CH_2)_7CH{=}CH(CH_2)_7CH_2NH_2$

$C_{18}H_{37}N$ M 267.497

B,HF: [36505-83-6]. ***Dectaflur***, **INN.** *SKF 38094*

Dental caries prophylactic.

(E)-form

Elaidylamine

Cryst. Mp 39°. Bp_2 149°.

Picrate: Mp 89.5°.

(Z)-form [112-90-3]

Oleylamine

Oil. Mp 25° (21°). Bp_2 147°.

Picrate: Mp 76.4°.

Ucciani, E. *et al*, *Chem. Phys. Lipids*, 1968, **2**, 240, 256 (*synth, pmr*)

3-(11-Octadecenyloxy)-1,2-propanediol, 9CI O-00200

1-O-(11-Octadecenyl)glycerol. Paramecyl alcohol

[76971-20-5]

$H_3C(CH_2)_5CH{=}CH(CH_2)_{10}OCH_2CH(OH)CH_2OH$

$C_{21}H_{42}O_3$ M 342.561

Constit. of phospholipids of *Paramecium tetraurelia* and *P. caudatum*.

Rhoads, D.E. *et al*, *Biochem. Biophys. Res. Commun.*, 1981, **98**, 858 (*isol, struct*)

9-Octadecen-12-ynoic acid O-00201

$H_3C(CH_2)_4C{\equiv}CCH_2CH{=}CH(CH_2)_7COOH$

$C_{18}H_{30}O_2$ M 278.434

(E)-form [38681-47-9]

Mp 128°. $Bp_{0.001}$ 120-140°.

(Z)-form [2277-31-8]

Crepenynic acid

Constit. of the seed oil of *Crepis foetida* and *Afzelia cuanzensis*. Mp 10.5-11°.

Mikolajczak, K.L. *et al*, *J. Org. Chem.*, 1964, **29**, 318 (*isol*)
Gunstone, F.D. *et al*, *J. Chem. Soc., Chem. Commun.*, 1967, 295 (*isol*)
Bradshaw, R.W. *et al*, *J. Chem. Soc. C*, 1971, 1156 (*synth*)
Gunstone, F.D. *et al*, *Chem. Phys. Lipids*, 1972, **9**, 112 (*nmr*)
Smith, C.R., *Lipids*, 1974, **9**, 640 (*isol*)
Macdonald, T.L., *Tetrahedron Lett.*, 1978, 4201.

11-Octadecen-9-ynoic acid O-00202

$H_3C(CH_2)_5CH{=}CHC{\equiv}C(CH_2)_7COOH$

$C_{18}H_{30}O_2$ M 278.434

(E)-form [557-58-4]

Ximenynic acid. Santalbic acid

Occurs widely in the seed lipids of many spp. of the Santalaceae and Olacaceae. Cryst. (Me_2CO or pet. ether). Mp 39-40°. $Bp_{0.075}$ 198°.

Me ester: $Bp_{0.2}$ 148°.

Et ester: $Bp_{0.2}$ 168-169°.

Amide: Mp 105-105.5°.

4-Bromophenacyl ester: Mp 53.8-54.2°.

(Z)-form

Mp 25.5-26.5°.

Lighthelm, S.P. *et al*, *J. Chem. Soc.*, 1952, 1088.
Gunstone, F.D. *et al*, *J. Chem. Soc.*, 1955, 3782.
Crombie, L. *et al*, *J. Chem. Soc.*, 1957, 1622; 1958, 4435 (*synth*)
Hopkins, C.Y. *et al*, *Phytochemistry*, 1969, **8**, 161 (*occur*)
Hopkins, C.Y., *Top. Lipid Chem.*, 1972, **3**, 37 (*rev*)

17-Octadecen-9-ynoic acid — O-00203

[5916-03-0]

$$H_2C{=}CH(CH_2)_6C{\equiv}C(CH_2)_7COOH$$

$C_{18}H_{30}O_2$ M 278.434

Fatty acid from the seed of *Acanthosyris spinescens*. Also from the seed oil of *Scleropyrum wallichianum*.

Me ester:
$C_{19}H_{32}O_2$ M 292.461
Oil.

Powell, R.G. *et al*, *Biochemistry*, 1966, **5**, 625.
Wang, J. *et al*, *CA*, 1992, **117**, 66648 (*isol*)

Octadecylamine — O-00204

1-Octadecanamine, 9CI. 1-Aminooctadecane. Stearylamine

[124-30-1]

$$H_3C(CH_2)_{16}CH_2NH_2$$

$C_{18}H_{39}N$ M 269.513

Cryst. Mp 49-52° (46°). Bp_5 183.0-183.1°.

▷ RG4150000.

B,HCl: [1838-08-0].
Cryst. Mp 160-161°.
▷ RG4375000.

N-*Ac:*
$C_{20}H_{41}NO$ M 311.550
Cryst. ($EtOH/Et_2O$). Mp 84-85°.

N-*Di-Me:* [124-28-7]. N,N-*Dimethyl-1-octadecanamine, 9CI.* ***Dimantine*, INN.** *Dymanthine*
$C_{20}H_{43}N$ M 297.566
Anthelmintic agent. Fp 22.9°.
▷ RG4200000.

N-*Di-Me; B,HCl:* [1613-17-8]. *Dymanthine hydrochloride, USAN. Doda. GS 1339. NSC 5547. Thelmesan*
▷ RG4210000.

[2190-04-7]

Ralston, A.W., *Oil Soap (Chicago)*, 1940, **17**, 89; *CA*, **34**, 3674 (*synth*)
Reck, R.A. *et al*, *J. Org. Chem.*, 1947, **12**, 517 (*deriv*)
Clemo, G.R. *et al*, *J. Chem. Soc.*, 1952, 4684 (*synth*)
Murr, B.L. *et al*, *J. Am. Chem. Soc.*, 1955, **77**, 1684 (*synth*)
Gohlke, R.S. *et al*, *Anal. Chem.*, 1962, **34**, 1281 (*ms*)
Cavier, R. *et al*, *Therapie*, 1964, **19**, 265, 947 (*activity, deriv*)
Goto, R. *et al*, *Nippon Kagaku Zasshi*, 1966, **87**, 1220; *CA*, **66**, 75599 (*synth*)
Kreuger, P.J. *et al*, *Can. J. Chem.*, 1967, **45**, 1605 (*ir*)
Narayanan, C.R. *et al*, *Indian J. Chem., Sect. B*, 1980, **19**, 209 (*nmr*)
Douvres, F.W. *et al*, *Vet. Parasitol.*, 1980, **7**, 195 (*activity*)
Martindale, The Extra Pharmacopoeia, 28th/29th Ed., Pharmaceutical Press, London, 1982/1989, 776.
Lewis, R.J., *Sax's Dangerous Properties of Industrial Materials*, 8th Ed., Van Nostrand-Reinhold, 1992, DTC400, OBC000, OBE000.

Octadecylbenzene — O-00205

1-Phenyloctadecane

[4445-07-2]

$$Ph(CH_2)_{17}CH_3$$

$C_{24}H_{42}$ M 330.596

Mp 32-33°, Mp 36°. $Bp_{0.25}$ 155-160°.

Gilman, H. *et al*, *J. Org. Chem.*, 1943, **8**, 126 (*synth*)
Shirley, D.A. *et al*, *J. Am. Chem. Soc.*, 1953, **75**, 6333 (*synth*)

2-Octadecyleicosanoic acid — O-00206

Dioctadecylacetic acid

[108293-08-9]

$$[H_3C(CH_2)_{17}]_2CHCOOH$$

$C_{38}H_{76}O_2$ M 565.017

Cryst. (EtOAc). Mp 81-83°.

Adam, N.K. *et al*, *J. Chem. Soc.*, 1925, 70 (*synth*)
Belletire, J.L. *et al*, *J. Org. Chem.*, 1987, **52**, 2549 (*synth, pmr, cmr, ir, ms*)

1-*O*-Octadecyl-*sn*-glycero-3-phosphocholine — O-00207

*4,7-Dihydroxy-*N,N,N-*trimethyl-3,5,9-trioxa-4-phosphaheptacosan-1-aminium hydroxide, inner salt, 4-oxide. Lyso-PAF (C-18)*

[74430-89-0]

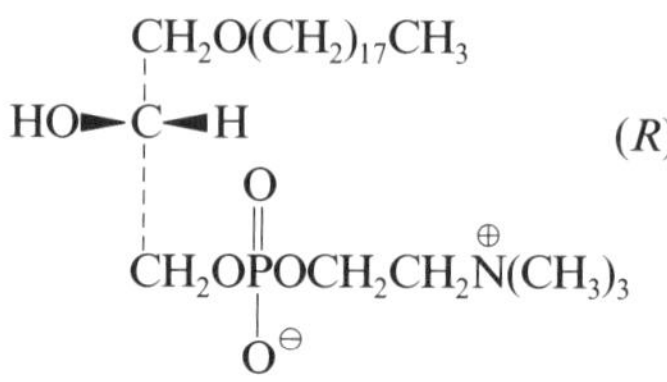

(*R*)-*form*

$C_{26}H_{56}NO_6P$ M 509.705

Mp 240° dec. $[\alpha]_D^{20}$ −4.65° (c, 1.00 in $CHCl_3$/MeOH).

2-Ac: [74389-69-8]. *1-*O*-Octadecyl-2-*O*-acetyl-*sn*-glycero-3-phosphocholine. PAF (C-18). 1-*O*-Octadecyl-PAF*
$C_{28}H_{58}NO_7P$ M 551.742
Major molecular form of PAF (see Platelet activating factor, P-00136). Powerful mediator in physiological processes such as anaphylaxis and inflammation. Mp 260° (214-215° dec.). $[\alpha]_D^{20}$ −3.46° (c, 0.90 in $CHCl_3$/MeOH).

9,10-Didehydro, 2-Ac: [85966-90-1]. *1-*O*-Oleyl-2-*O*-acetyl-*sn*-glycero-3-phosphocholine. 2-*O*-Acetyl-1-*O*-(9-octadecenyl)-*sn*-glycero-3-phosphocholine. Oleyl-PAF*
$C_{28}H_{56}NO_7P$ M 549.727
$[\alpha]_D^{20}$ −0.25° (c, 5.0 in MeOH/$CHCl_3$).

2-Me ether: [77286-66-9]. *1-*O*-Octadecyl-2-*O*-methyl-*sn*-glycero-3-phosphocholine. 4-Hydroxy-7-methoxy-*N,N,N*-trimethyl-3,5,9-trioxa-4-phosphaheptacosan-1-aminium hydroxide, inner salt, 4-oxide. 2-*O*-Methyl PAF (C-18)*
$C_{27}H_{58}NO_6P$ M 523.732
Antitumour and antimetastatic agent. Glass-like substance (Me_2CO). Mp 80°. $[\alpha]_D^{20}$ +2.8° (10% in $CHCl_3$). Sinters at 230°.

2-Benzyl ether: [80483-86-9]. *1-*O*-Octadecyl-2-*O*-benzyl-*sn*-glycero-3-phosphocholine. 2-*O*-Benzyl PAF (C-18)*
Mp 220° dec. $[\alpha]_D^{20}$ +2.24° (c, 2.18 in MeOH).

(*S*)-*form* [80707-96-6]
*3-*O*-Octadecyl-2-acetyl-*sn*-glycero-1-phosphocholine. Enantio-PAF (C-18)*
$[\alpha]_D^{20}$ +3.38° (c, 1.60 in $CHCl_3$/MeOH).

[34240-68-1, 72490-82-5]

Lartigue-Mattel, C. *et al*, *Agents Actions*, 1982, **12**, 703 (*pharmacokin*)
Tsushima, S. *et al*, *Chem. Pharm. Bull.*, 1982, **30**, 3260 (*synth, Me ether*)
Berchtold, R., *Chem. Phys. Lipids*, 1982, **30**, 389 (*synth, Me ether*)
Hirth, G. *et al*, *Helv. Chim. Acta*, 1982, **65**, 1059 (*synth, ir, pmr*)
Keraly, C. *et al*, *Br. J. Haematol*, 1983, **53**, 513 (*biochem, struct*)

Hirth, G. *et al*, *Helv. Chim. Acta*, 1983, **66**, 1210 (*synth, oleyl PAF, ir, pmr*)
Hadvary, P. *et al*, *Thromb. Res.*, 1983, **30**, 143 (*biochem*)
Ohno, M. *et al*, *Chem. Pharm. Bull.*, 1985, **33**, 572 (*synth, pmr, ir, biochem*)
Sturk, A. *et al*, *Thromb. Res.*, 1985, **40**, 359 (*biochem*)
Cimetiere, B. *et al*, *Tetrahedron Lett.*, 1986, **27**, 6329 (*synth, PAF*)
Wolfgang, J., *Synform*, 1988, **6**, 115; 1990, **8**, 313 (*rev, synth*)
Benveniste, J., *Adv. Prostaglandin, Thromboxane, Leukotriene Res.*, 1989, **19**, 355 (*rev, props*)
Nakamura, N. *et al*, *Tetrahedron Lett.*, 1990, **31**, 699 (*synth*)
Kumar, A. *et al*, *Synth. Commun.*, 1991, **21**, 1763 (*synth, enantio-PAF, pmr*)

Octadecyl hydrogen sulfate O-00208

[143-03-3]

$H_3C(CH_2)_{17}OSO_3H$

$C_{18}H_{38}O_4S$ M 350.562
Hygroscopic cryst. Mp 54.1-55.5° (sealed tube).

Maurer, E.W. *et al*, *J. Am. Oil Chem. Soc.*, 1960, **37**, 34 (*synth*)
Maurer, E.W. *et al*, *Biochem. Prep.*, 1966, **11**, 96 (*synth*)

4-(4-Octadecylphenyl)-4-oxo-2-butenoic acid O-00209

3-(4-Octadecyl)benzoylacrylic acid. OBAA
[134531-42-3]

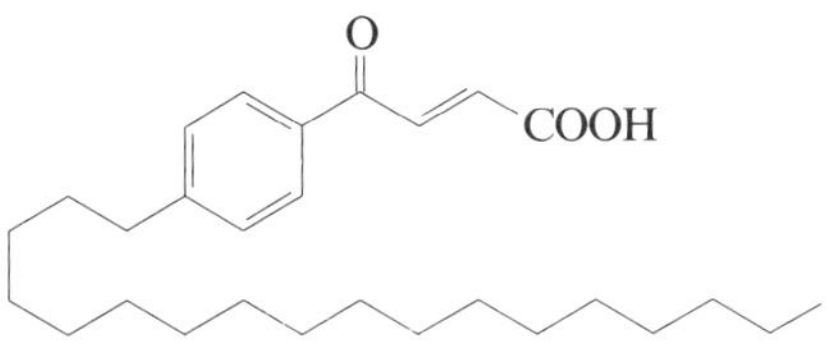

$C_{28}H_{44}O_3$ M 428.654
Inhibits phospholipase A_2 of snake venom.

Koehler, T. *et al*, *Agents Actions*, 1991, **32**, 70 (*struct, biochem*)
Ger. Pat., DD 287 253, (1991); *CA*, **115**, 28900k (*synth*)

1-Octadecyne O-00210

[629-89-0]

$H_3C(CH_2)_{15}C{\equiv}CH$

$C_{18}H_{34}$ M 250.467
Cryst. (EtOH). Mp 22.5°, Mp 28°. Bp_{15} 180°.

Krafft, F. *et al*, *Ber.*, 1892, **25**, 2248 (*synth*)
Van Putte, K., *J. Chem. Soc., Faraday Trans. 2*, 1972, **68**, 357.

2-Octadecynoic acid O-00211

[2834-01-7]

$H_3C(CH_2)_{14}C{\equiv}CCOOH$

$C_{18}H_{32}O_2$ M 280.450
Cryst. (pet. ether). Mp 57°.

Barve, J.A. *et al*, *Chem. Phys. Lipids*, 1971, **7**, 311 (*synth*)
Frost, D.J. *et al*, *Chem. Phys. Lipids*, 1975, **15**, 53 (*pmr*)
Gunstone, F.D. *et al*, *Chem. Phys. Lipids*, 1976, **17**, 1 (*cmr*)

3-Octadecynoic acid O-00212

$H_3C(CH_2)_{13}C{\equiv}CCH_2COOH$

$C_{18}H_{32}O_2$ M 280.450
Cryst. Mp 74°.

Barve, J.A. *et al*, *Chem. Phys. Lipids*, 1971, **7**, 311 (*synth*)
Frost, D.J. *et al*, *Chem. Phys. Lipids*, 1975, **15**, 53 (*pmr*)

4-Octadecynoic acid O-00213

[19307-18-7]

$H_3C(CH_2)_{12}C{\equiv}CCH_2CH_2COOH$

$C_{18}H_{32}O_2$ M 280.450
Cryst. Mp 75°.

Barve, J.A. *et al*, *Chem. Phys. Lipids*, 1971, **7**, 311 (*synth*)
Frost, D.J. *et al*, *Chem. Phys. Lipids*, 1975, **15**, 53 (*pmr*)
Gunstone, F.D. *et al*, *Chem. Phys. Lipids*, 1976, **17**, 1 (*cmr*)

5-Octadecynoic acid O-00214

[676-30-2]

$H_3C(CH_2)_{11}C{\equiv}C(CH_2)_3COOH$

$C_{18}H_{32}O_2$ M 280.450
Cryst. Mp 52°.

Barve, J.A. *et al*, *Chem. Phys. Lipids*, 1971, **7**, 311 (*synth*)
Frost, D.J. *et al*, *Chem. Phys. Lipids*, 1975, **15**, 53 (*pmr*)
Gunstone, F.D. *et al*, *Chem. Phys. Lipids*, 1976, **17**, 1 (*cmr*)

6-Octadecynoic acid O-00215

Tariric acid
[544-74-1]

$H_3C(CH_2)_{10}C{\equiv}C(CH_2)_4COOH$

$C_{18}H_{32}O_2$ M 280.450
Occurs as glyceride in fruit of *Picramia* spp. Cryst. (EtOH). Mp 50.4-51°.

Dibromide:
$C_{18}H_{32}Br_2O_2$ M 440.258
Mp 32°.

Diiodide:
$C_{18}H_{32}I_2O_2$ M 534.259
Mp 48.5°.

Tetrabromide: 6,6,7,7-Tetrabromooctadecanoic acid
$C_{18}H_{32}Br_4O_2$ M 600.066
Mp 125°.

Robertson, W.B. *et al*, *J. Am. Chem. Soc.*, 1944, **66**, 1894.
Lumb, P.B. *et al*, *J. Chem. Soc.*, 1952, 5032 (*synth*)
Barve, J.A. *et al*, *Chem. Phys. Lipids*, 1971, **7**, 311 (*synth*)
Frost, D.J. *et al*, *Chem. Phys. Lipids*, 1975, **15**, 53 (*pmr*)
Gunstone, F.D. *et al*, *Chem. Phys. Lipids*, 1976, **17**, 1 (*cmr*)

7-Octadecynoic acid O-00216

[19220-35-0]

$H_3C(CH_2)_9C{\equiv}C(CH_2)_5COOH$

$C_{18}H_{32}O_2$ M 280.450
Cryst. Mp 49°.

Barve, J.A. *et al*, *Chem. Phys. Lipids*, 1971, **7**, 311 (*synth*)
Frost, D.J. *et al*, *Chem. Phys. Lipids*, 1975, **15**, 53 (*pmr*)
Gunstone, F.D. *et al*, *Chem. Phys. Lipids*, 1976, **17**, 1 (*cmr*)

8-Octadecynoic acid O-00217

[19220-36-1]

$H_3C(CH_2)_8C{\equiv}C(CH_2)_6COOH$

$C_{18}H_{32}O_2$ M 280.450
Cryst. Mp 47°.

Barve, J.A. *et al*, *Chem. Phys. Lipids*, 1971, **7**, 311 (*synth*)
Frost, D.J. *et al*, *Chem. Phys. Lipids*, 1975, **15**, 53 (*pmr*)
Gunstone, F.D. *et al*, *Chem. Phys. Lipids*, 1976, **17**, 1 (*cmr*)

9-Octadecynoic acid O-00218

Stearolic acid

[506-24-1]

$H_3C(CH_2)_7C{\equiv}C(CH_2)_7COOH$

$C_{18}H_{32}O_2$ M 280.450

Constit. of the seed oil of Santalaceae spp., e.g. *Pyrularia pubera*, *Sterculia foetida*, *Exocarpus cupressiformis*, *Santalum* spp. and *Buckleya distichophylla*. Cryst. (EtOH aq. or EtOH). Mp 47-48°. $Bp_{1.8}$ 189-190°.

Et ester:

$C_{20}H_{36}O_2$ M 308.503

Bp_3 195-197°.

Chloride:

$C_{18}H_{31}ClO$ M 298.895

Bp_{15} 210°.

Tetrabromide: 9,9,10,10-Tetrabromooctadecanoic acid

$C_{18}H_{32}Br_4O_2$ M 600.066

Cryst. (EtOH). Mp 70°.

Org. Synth., 1947, **27**, 76 (*synth*)
Hopkins, C.Y. *et al*, *Tetrahedron Lett.*, 1964, 3011 (*isol, glc*)
Barve, J.A. *et al*, *Chem. Phys. Lipids*, 1971, **7**, 311 (*synth*)
Hirabayashi, Y. *et al*, *Bull. Chem. Soc. Jpn.*, 1971, **44**, 2733 (*ir*)
Frost, D.J. *et al*, *Chem. Phys. Lipids*, 1975, **15**, 53 (*pmr*)
Gunstone, F.D. *et al*, *Chem. Phys. Lipids*, 1976, **17**, 1 (*cmr*)
Knapp, F.F. *et al*, *J. Med. Chem.*, 1984, **27**, 94 (*synth, pmr*)

10-Octadecynoic acid O-00219

[19220-39-4]

$H_3C(CH_2)_6C{\equiv}C(CH_2)_8COOH$

$C_{18}H_{32}O_2$ M 280.450

Cryst. Mp 46°.

Barve, J.A. *et al*, *Chem. Phys. Lipids*, 1971, **7**, 311 (*synth*)
Frost, D.J. *et al*, *Chem. Phys. Lipids*, 1975, **15**, 53 (*pmr*)
Gunstone, F.D. *et al*, *Chem. Phys. Lipids*, 1976, **17**, 1 (*cmr*)

11-Octadecynoic acid O-00220

[19220-40-7]

$H_3C(CH_2)_5C{\equiv}C(CH_2)_9COOH$

$C_{18}H_{32}O_2$ M 280.450

Cryst. (Me_2CO). Mp 45.5-46.5°.

Ahmad, K. *et al*, *J. Am. Chem. Soc.*, 1948, **70**, 3391 (*synth*)
Barve, J.A. *et al*, *Chem. Phys. Lipids*, 1971, **7**, 311 (*synth*)
Frost, D.J. *et al*, *Chem. Phys. Lipids*, 1975, **15**, 53 (*pmr*)
Gunstone, F.D. *et al*, *Chem. Phys. Lipids*, 1976, **17**, 1 (*cmr*)
Kleiman, R. *et al*, *Lipids*, 1976, **11**, 599 (*ms*)

12-Octadecynoic acid O-00221

[19220-41-8]

$H_3C(CH_2)_4C{\equiv}C(CH_2)_{10}COOH$

$C_{18}H_{32}O_2$ M 280.450

Cryst. Mp 47°.

Barve, J.A. *et al*, *Chem. Phys. Lipids*, 1971, **7**, 311 (*synth*)
Frost, D.J. *et al*, *Chem. Phys. Lipids*, 1975, **15**, 53 (*pmr*)
Gunstone, F.D. *et al*, *Chem. Phys. Lipids*, 1976, **17**, 1 (*cmr*)

13-Octadecynoic acid O-00222

[19220-42-9]

$H_3C(CH_2)_3C{\equiv}C(CH_2)_{11}COOH$

$C_{18}H_{32}O_2$ M 280.450

Cryst. Mp 49°.

Barve, J.A. *et al*, *Chem. Phys. Lipids*, 1971, **7**, 311 (*synth*)
Frost, D.J. *et al*, *Chem. Phys. Lipids*, 1975, **15**, 53 (*pmr*)
Gunstone, F.D. *et al*, *Chem. Phys. Lipids*, 1976, **17**, 1 (*cmr*)

14-Octadecynoic acid O-00223

[34494-26-3]

$H_3CCH_2CH_2C{\equiv}C(CH_2)_{12}COOH$

$C_{18}H_{32}O_2$ M 280.450

Cryst. Mp 64°.

Barve, J.A. *et al*, *Chem. Phys. Lipids*, 1971, **7**, 311 (*synth*)
Frost, D.J. *et al*, *Chem. Phys. Lipids*, 1975, **15**, 53 (*pmr*)
Gunstone, F.D. *et al*, *Chem. Phys. Lipids*, 1976, **17**, 1 (*cmr*)

15-Octadecynoic acid O-00224

[19220-43-0]

$H_3CCH_2C{\equiv}C(CH_2)_{13}COOH$

$C_{18}H_{32}O_2$ M 280.450

Cryst. (pet. ether/C_6H_6). Mp 66-67°.

Barve, J.A. *et al*, *Chem. Phys. Lipids*, 1971, **7**, 311 (*synth*)
Klok, R. *et al*, *Recl. Trav. Chim. Pays-Bas* (*J. R. Neth. Chem. Soc.*), 1974, **93**, 222.
Frost, D.J. *et al*, *Chem. Phys. Lipids*, 1975, **15**, 53 (*pmr*)
Gunstone, F.D. *et al*, *Chem. Phys. Lipids*, 1976, **17**, 1 (*cmr*)

16-Octadecynoic acid O-00225

[34450-17-4]

$H_3CC{\equiv}C(CH_2)_{14}COOH$

$C_{18}H_{32}O_2$ M 280.450

Cryst. Mp 72-73°.

Barve, J.A. *et al*, *Chem. Phys. Lipids*, 1971, **7**, 311 (*synth*)
Frost, D.J. *et al*, *Chem. Phys. Lipids*, 1975, **15**, 53 (*pmr*)

17-Octadecynoic acid O-00226

[34450-18-5]

$HC{\equiv}C(CH_2)_{15}COOH$

$C_{18}H_{32}O_2$ M 280.450

Mp 67°.

Me ester: [68950-90-3].

$C_{19}H_{34}O_2$ M 294.476

Oil or cryst. Mp 29°.

Barve, J.A. *et al*, *Chem. Phys. Lipids*, 1971, **7**, 311 (*synth*)
Frost, D.J. *et al*, *Chem. Phys. Lipids*, 1975, **15**, 53 (*pmr*)
Gunstone, F.D. *et al*, *Chem. Phys. Lipids*, 1976, **17**, 1 (*cmr*)
Knapp, F.F. *et al*, *J. Med. Chem.*, 1984, **27**, 94 (*synth, pmr, ir, ms*)
Augustin, K.E. *et al*, *Justus Liebigs Ann. Chem.*, 1991, 1037 (*synth*)

2,6-Octadienoic acid O-00227

$H_3CCH{=}CHCH_2CH_2CH{=}CHCOOH$

$C_8H_{12}O_2$ M 140.182

(E,E)-form [116565-36-7]

Waxy solid. Mp 44-46°.

Me ester: [25172-05-8].

$C_9H_{14}O_2$ M 154.208

Liq.

Crandall, J.K. *et al*, *J. Org. Chem.*, 1970, **35**, 3049 (*synth, ir, pmr*)
Snider, B.B., *J. Org. Chem.*, 1988, **53**, 5320 (*synth, pmr, cmr, ir*)

4,7-Octadienoic acid O-00228

$H_2C{=}CHCH_2CH{=}CHCH_2CH_2COOH$

$C_8H_{12}O_2$ M 140.182

(Z)-form

Et ester: [69925-33-3].

$C_{10}H_{16}O_2$ M 168.235

Aroma component of purple passion fruit *Passiflora edulis*. Bp_{7-10} 90-95°.

Furrer, A. *et al*, *Helv. Chim. Acta*, 1979, **62**, 135 (*isol*)
Bhanot, R.K. *et al*, *Indian J. Chem., Sect. B*, 1981, **20**, 970 (*synth*)
Kawashima, M. *et al*, *Bull. Chem. Soc. Jpn.*, 1988, **61**, 3255 (*synth*)

5,7-Octadienoic acid O-00229

$H_2C{=}CHCH{=}CHCH_2CH_2CH_2COOH$

$C_8H_{12}O_2$ M 140.182

(E)-form [73501-16-3]

Liq. $Bp_{0.5}$ 80°.

(Z)-form [74199-16-9]

Mp 82.5-85°.

Hudlicky, T. *et al*, *J. Org. Chem.*, 1981, **46**, 2911 (*synth, ir, nmr*)

1,5-Octadien-3-ol, 9CI O-00230

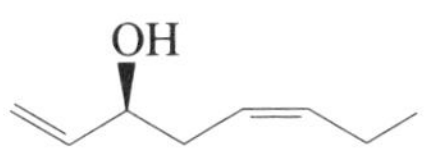

(*S,Z*)-*form*

$C_8H_{14}O$ M 126.198

(S,Z)-form [56994-74-2]

Constit. of essential oil of *Chondrococcus hornemanni*. $[\alpha]_D$ −8° (c, 0.5 in CH_2Cl_2).

(E)-form [50306-14-4]

Found in the fungus *Trichothecium roseum*. Insect attractant for *Tyrophagus putrescentiae*.

(Z)-form [50306-18-8]

From *T. roseum*. Off-flavour component of cooked prawns and sand-lobsters. Insect attractant.

Woolard, F.X. *et al*, *J. Chem. Soc., Chem. Commun.*, 1975, 486 (*pmr*)
Vanhaelen, M., *J. Chromatogr.*, 1977, **144**, 108.
Whitfield, F.B. *et al*, *Aust. J. Chem.*, 1982, **35**, 373 (*isol*)

2,4-Octadien-1-ol O-00231

$H_3CCH_2CH_2CH{=}CHCH{=}CHCH_2OH$

$C_8H_{14}O$ M 126.198

(2E,4E)-form [18409-20-6]

Mobile liq. with pleasant fruity odour. Bp_2 70-73°. n_D^{25} 1.4865.

(2E,4Z)-form [56904-85-9]

$Bp_{0.1}$ 48°. n_D^{20} 1.4912.

Jacobson, M., *J. Am. Chem. Soc.*, 1956, **78**, 5084 (*synth*)
Commerçon, A. *et al*, *Tetrahedron*, 1980, **36**, 1215 (*synth, ir, pmr*)
Kuroda, S. *et al*, *Tetrahedron Lett.*, 1987, **28**, 803 (*synth, pmr*)

2,6-Octadien-1-ol O-00232

$H_3CCH{=}CHCH_2CH_2CH{=}CHCH_2OH$

$C_8H_{14}O$ M 126.198

(2Z,6E)-form [124562-08-9]

Oil.

Aebi, J.D. *et al*, *J. Med. Chem.*, 1990, **33**, 999 (*synth, ir, pmr*)

2,7-Octadien-1-ol O-00233

[23578-51-0]

$H_2C{=}CHCH_2CH_2CH_2CH{=}CHCH_2OH$

$C_8H_{14}O$ M 126.198

▷ RG5425000.

(E)-form [62179-18-4]

Bp_{14} 95-97°.

(Z)-form

Bp_{12} 96°.

Saloman, R.G. *et al*, *J. Am. Chem. Soc.*, 1982, **104**, 998 (*synth, pmr*)
Bulman Page, P.C. *et al*, *J. Chem. Soc., Perkin Trans. 1*, 1990, 1375 (*synth, ir, pmr, cmr, ms*)

4,7-Octadien-1-ol O-00234

$H_2C{=}CHCH_2CH{=}CHCH_2CH_2CH_2OH$

$C_8H_{14}O$ M 126.198

(E)-form [81651-45-8]

Bp_{14} 92-98°.

Ac: [81651-48-1].

$C_{10}H_{16}O_2$ M 168.235

Bp_{15} 80-100°.

(Z)-form [72820-69-0]

Bp_{15} 90-91°.

Alexakis, A. *et al*, *Synthesis*, 1979, 826 (*synth*)
Bestmann, H.J. *et al*, *Justus Liebigs Ann. Chem.*, 1982, 536 (*synth, pmr*)

2-(2,5-Octadienyl)-3-undecyloxirane, 9CI O-00235

9,10-Epoxy-3,6-heneicosadiene

[107596-95-2]

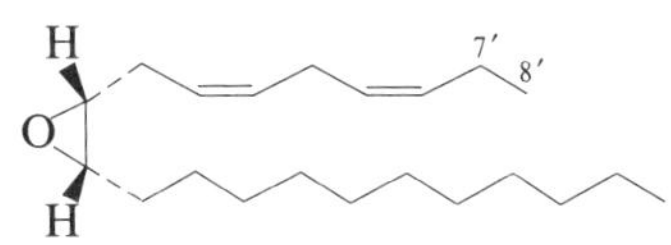

$C_{21}H_{38}O$ M 306.531

(2S,3R)-form [81309-90-2]

Component of sex pheromone of the saltmarsh caterpillar moth *Estigmene acrea* and of the fall webworm moth *Hyphantria cunea*. $[\alpha]_{230}$ −43.3°.

7′,8′-Didehydro: [119072-10-5]. *2-(2,5,7-Octatrienyl)-3-undecyloxirane. 9,10-Epoxy-1,3,6-heneicosatriene*

$C_{21}H_{36}O$ M 304.515

Component of the sex pheromone of *H. cunea*. Amorph. solid. Mp 14-15°. $[\alpha]_D^{16}$ −0.41° (c, 1.97 in $CHCl_3$).

[81369-84-8]

Mori, K. *et al*, *Tetrahedron*, 1986, **42**, 3471 (*synth, ir, pmr, cmr, ms*)
Mori, K. *et al*, *Justus Liebigs Ann. Chem.*, 1989, 453 (*deriv*)

2,6-Octadien-4-ynedioic acid, 9CI O-00236

$HOOCCH{=}CHC{\equiv}CCH{=}CHCOOH$

$C_8H_6O_4$ M 166.133

(E,E)-form

Mp 120-121°.

Di-Me ester: [74149-28-3].

$C_{10}H_{10}O_4$ M 194.187

Occurs in *Polyporus anthracophilus*. Plates. Mp 117-119.5°.

Bu'Lock, J.D. *et al*, *J. Chem. Soc.*, 1957, 1607 (*isol, synth*)
Julia, M. *et al*, *C. R. Hebd. Seances Acad. Sci.*, 1958, **246**, 3648 (*synth*)

2,6-Octadiyne-1,8-diol O-00237

[58471-75-3]

$HOCH_2C{\equiv}CCH_2CH_2C{\equiv}CCH_2OH$

$C_8H_{10}O_2$ M 138.166
Solid (C_6H_6). Mp 88.5-89.5°.

Lespieau, R., *C. R. Hebd. Seances Acad. Sci.*, 1914, 707.
Lennon, P. *et al*, *J. Am. Chem. Soc.*, 1983, **105**, 1233.

3,5-Octadiyne-1,8-diol O-00238

[15808-23-8]

$HOCH_2CH_2C{\equiv}CC{\equiv}CCH_2CH_2OH$

$C_8H_{10}O_2$ M 138.166
Metab. of *Collybia peronata*. Plates (Et_2O). Mp 41.5-43°. $Bp_{0.0001}$ 135°.
Di-Ac: [54448-17-8].
$C_{12}H_{14}O_4$ M 222.240
Mp 63°.

Bowden, K. *et al*, *J. Chem. Soc.*, 1947, 1579 (*uv*)
Kuznetsova, T.S., *Zh. Org. Khim.*, 1973, **9**, 1806 (*synth*)
Zeisberg, R., *Chem. Ber.*, 1974, **107**, 3800 (*cmr*)
Higham, C.A. *et al*, *J. Chem. Soc., Perkin Trans. 1*, 1974, 1991 (*pmr*)

7,7′,8,8′,11,11′,12,12′-Octahydrolycopene O-00239

7,7′,8,8′,11,11′,12,12′-Octahydro-ψ,ψ-carotene. ***Phytoene***
[540-04-5]

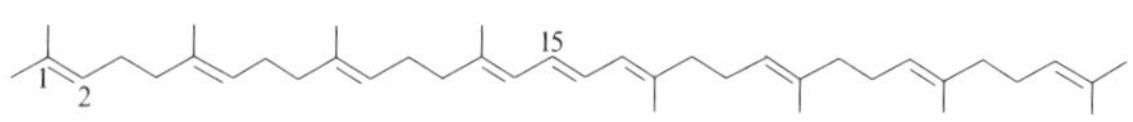

$C_{40}H_{64}$ M 544.946
Constit. of *Mycobacterium phlei* and other bacteria. Widespread in plants. Viscous oil with strong uv fluorescence.
1,2-Epoxide: [26107-96-0]. ***Phytoene 1,2-epoxide.*** *1,2-Epoxy-1,2,7,7′,8,8′,11,11′,12,12′-decahydro-ψ,ψ-carotene*
$C_{40}H_{64}O$ M 560.945
Isol. from ripe tomatoes. λ_{max} 275, 285, 296 nm.

(15Z)-form [13920-14-4] ***15-cis-Pytoene***
Isol. from *Lycopersicon esculentum* var. 'Tangella', and *Dunaliella bardawil.* Clear branch point for poly-Z-carotene formation. Oil. λ_{max} 286 nm (hexane).

Britton, G. *et al*, *Phytochemistry*, 1969, **8**, 2257 (*epoxide*)
Karrer, W. *et al*, *Konstitution und Vorkommen der Organischen Pflanzenstoffe*, 2nd Ed., Birkhäuser Verlag, Basel, 1972-1985, no. 1817a (*occur*)
Gregonis, D.E. *et al*, *Biochemistry*, 1974, **13**, 1538 (*biosynth*)
Khan, N. *et al*, *J. Chem. Soc., Perkin Trans. 1*, 1975, 1457 (*struct, synth*)
Moss, G.P., *Pure Appl. Chem.*, 1976, **47**, 97 (*cmr*)
Clough, J.M. *et al*, *J. Chem. Soc., Perkin Trans. 1*, 1983, 3011 (*15Z-Phytoene, ir, pmr, cmr*)
Frecknell, E.A. *et al*, *Phytochemistry*, 1984, **23**, 1707 (*isol, occur*)
Straub, O. *et al*, *Key to Carotenoids*, 2nd Ed., Birkhauser Verlag, Basel and Boston, 1987, 44 (*bibl*)
Ebenezer, W.J. *et al*, *J. Chem. Soc., Perkin Trans. 1*, 1993, 1869 (*isol, Z-form*)

Octanedioic acid, 9CI O-00240

Suberic acid, 8CI
[505-48-6]

$HOOC(CH_2)_6COOH$

$C_8H_{14}O_4$ M 174.196
Isol. from stems and leaves of *Anthyllis sericea* and roots of *Phaseolus vulgaris*. Used in plastics industry. Cryst. (H_2O or C_6H_6/EtOH). Mp 144°. pK_{a1} 4.53; pK_{a2} 5.52 (25°, 0.1*M* KNO_3). Subl. >300°.
Mono-Me ester: [3946-32-5].
$C_9H_{16}O_4$ M 188.223
Mp 14-15°. Bp_{18} 185-186°.
Di-Me ester: [1732-09-8].
$C_{10}H_{18}O_4$ M 202.250
Fp −3.1°. Bp 268°, Bp_9 130-131°.
Dichloride: [10027-07-3].
$C_8H_{12}Cl_2O_2$ M 211.087
Bp_{12} 149-150°.
Anhydride: Suberic anhydride
$C_8H_{12}O_3$ M 156.181
Needles (C_6H_6). Mp 65-66°.
Monoamide: [73427-53-9]. *8-Amino-8-oxooctanoic acid, 9CI. Suberamic acid*
$C_8H_{15}NO_3$ M 173.211
Cryst. (H_2O). Mp 125-127°.
Diamide: [3891-73-4].
$C_8H_{16}N_2O_2$ M 172.227
Cryst. (H_2O). Mp 216-217°.
Dinitrile: [629-40-3]. *Suberonitrile. 1,6-Dicyanohexane*
$C_8H_{12}N_2$ M 136.196
Mp −3.5°. Bp_{11} 176-178°.

Green, T.G. *et al*, *J. Chem. Soc.*, 1937, 766 (*synth*)
Cason, J. *et al*, *J. Org. Chem.*, 1949, **14**, 37 (*synth*)
Hopff, H. *et al*, *Justus Liebigs Ann. Chem.*, 1966, **691**, 61 (*synth*)
Angelo, B., *C. R. Hebd. Seances Acad. Sci.*, 1971, **273**, 1767 (*synth*)
Takasugi, M. *et al*, *Chem. Lett.*, 1973, 445 (*isol*)
Chern, C.-I. *et al*, *J. Org. Chem.*, 1976, **41**, 1077 (*synth*)
Marco, J.A. *et al*, *Phytochemistry*, 1978, **17**, 1438 (*isol*)
Ansell, M.F. *et al*, *J. Chem. Soc., Perkin Trans. 1*, 1984, 1061 (*synth, pmr*)

2,3-Octanediol O-00241

[20653-90-1]

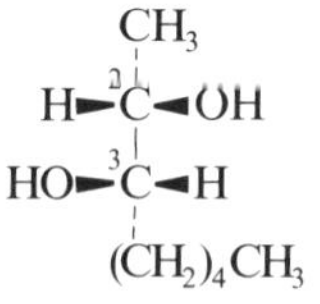

(2*S*,3*S*)-*form*

$C_8H_{18}O_2$ M 146.229

(2S,3S)-form
(−)-erythro-*form*
Constit. of male sex pheromone of the grape borer *Xylotrechus pyrrhoderus* and *X. chinensis*. $[\alpha]_D^{21}$ −19.2° (c, 1.45 in $CHCl_3$).

(2RS,3RS)-form [37163-97-6]
(±)-erythro-*form*
Mp 42.5-44°.

(2RS,3SR)-form [37163-98-7]
(±)-threo-*form*
Liq.

Katzenellenbogen, J.A. *et al*, *J. Org. Chem.*, 1973, **38**, 627 (*synth, pmr*)

Sakai, T. *et al*, *Chem. Lett.*, 1984, 263 (*isol, synth*)
Mori, K. *et al*, *Tetrahedron*, 1985, 553 (*synth*)
Iwabuchi, K., *Appl. Entomol. Zool.*, 1987, **22**, 110.

Octanoic acid, 9CI **O-00242**

n-*Caprylic acid*. n-*Octoic acid*. n-*Octylic acid*
[124-07-2]

$H_3C(CH_2)_6COOH$

$C_8H_{16}O_2$ M 144.213
Widespread in plant oils, free and as glycerides. Used in perfume and dye manuf. Simple esters have flavour use. Antifungal agent. Used as a soln. in $CHCl_3$/butanol for extraction-separation of Co, Fe, Rh, Ti, Al, Ga, In. Liq. or cryst. Sol. alkalis, EtOH, $CHCl_3$; spar. sol. hot H_2O. Mp 16°. Bp 239.7°, Bp_{10} 124°. pK_a 4.89 (25°).
▷ Skin irritant. LD_{50} (rat, orl) approx. 10g/kg. RH0175000.

Cu salt: Mp 264-266°.
Me ester: [111-11-5].
$C_9H_{18}O_2$ M 158.240
Bp 193-194°, Bp_{15} 83°.
4-Bromophenacyl ester: Mp 67.4°.
Chloride: [111-64-8].
$C_8H_{15}ClO$ M 162.659
Bp 195-196°, Bp_{15} 83°.
Anhydride: [623-66-5].
$C_{16}H_{30}O_3$ M 270.411
Mp −1°. Bp 280-285°, Bp_{15} 186°.
Amide: [629-01-6].
$C_8H_{17}NO$ M 143.228
Mp 110°.
Nitrile: [124-12-9]. *Octanenitrile, 9CI. Caprylonitrile. 1-Cyanoheptane. Heptyl cyanide*
$C_8H_{15}N$ M 125.213
Bp 198-200°. n_D^{20} 1.4200.
▷ LD_{50} (mus, orl) 1764 mg/kg. RG9700300.

Langenbeck, W. *et al*, *Chem. Ber.*, 1956, **89**, 202 (*synth*)
Markley, K.S., *Fatty Acids*, Part 1, Interscience, N.Y., 2nd Ed., 1960 (*rev*)
Tripett, S. *et al*, *J. Chem. Soc.*, 1960, 2976 (*deriv*)
Nakagawa, K. *et al*, *Chem. Pharm. Bull.*, 1963, **11**, 296 (*deriv*)
Karrer, W. *et al*, *Konstitution und Vorkommen der Organischen Pflanzenstoffe*, 2nd Ed., Birkhäuser Verlag, Basel, 1972-1985, no. 694 (*occur*)
U.S. Pat., 3 770 782, (1973); *CA*, **80**, 14572 (*synth*)
Cahiez, G. *et al*, *J. Organomet. Chem.*, 1975, **94**, 463 (*synth*)
Baer, T.A. *et al*, *Tetrahedron Lett.*, 1976, 4697 (*synth*)
Dinizo, S.E. *et al*, *J. Am. Chem. Soc.*, 1977, **99**, 182 (*synth*)
Belova, V.V. *et al*, *CA*, 1979, **90**, 128252j (*detn, Rh*)
Muehl, P. *et al*, *Talanta*, 1979, **26**, 227 (*detn, Fe*)
Kopacz, S. *et al*, *Zh. Neorg. Khim.*, 1981, **26**, 1625; *CA*, **95**, 50325s (*detn, Co*)
Sokhan, V.F. *et al*, *Zh. Prikl. Khim. (Leningrad)*, 1981, **54**, 1023; *CA*, **95**, 100983c (*detn, Ti*)
Kolomiets, L.L. *et al*, *Ukr. Khim. Zh. (Russ. edn.)*, 1982, **48**, 51 (*detn, Al, In, Ga*)
Pyatnitskii, I.V. *et al*, *Zh. Anal. Khim.*, 1985, **40**, 115 (*detn, Ga*)
Lewis, R.J., *Food Additives Handbook*, Van Nostrand Reinhold International, New York, 1989, OCY000.
Martindale, The Extra Pharmacopoeia, 30th edn., Pharmaceutical Press, London, 1993, 319.
Lewis, R.J., *Sax's Dangerous Properties of Industrial Materials*, 8th Ed., Van Nostrand-Reinhold, 1992, OCY000, ODE200.

2-Octanol, 9CI **O-00243**

Capryl alcohol
[123-96-6]

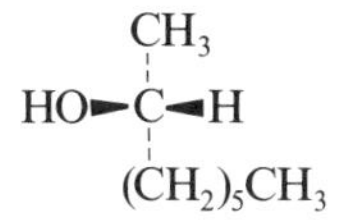

(*R*)-*form*

$C_8H_{18}O$ M 130.230
Isol. from geranium oil. Used as fragrance ingredient. Reagent for the chromatographic sepn. of amino acids.

(*R*)-*form*
Bp_{20} 86°. $[\alpha]_D^{17}$ −9.9°.
Hydrogen phthalate: Needles (pet. ether). Mp 75°. $[\alpha]_D^{24}$ −48.4°.
Benzoyl: [78687-05-5].
$C_{15}H_{22}O_2$ M 234.338
$Bp_{0.6}$ 90°. $[\alpha]_D$ −34.3° (c, 1.5 in CH_2Cl_2).

(*S*)-*form* [6169-06-8]
Bp_{20} 86°. $[\alpha]_D^{17}$ +9.9°.
Benzoyl: [34881-29-3].
$Bp_{0.6}$ 90°. $[\alpha]_D^{18}$ +33.5° (MeCl).

(±)-*form* [4128-31-8]
Used in manuf. of perfumes and disinfectant soaps. Antifoaming agent, solv. for waxes and fats. Oily refractive liq. with aromatic yet unpleasant odour. d_4^{25} 0.819. n_D^{20} 1.4203.
Ac: [2051-50-5].
$C_{10}H_{20}O_2$ M 172.267
Bp_{744} 194.5°, Bp_{15} 84°.
Me ether: [1541-09-9].
$C_9H_{20}O$ M 144.256
Bp_{44} 76-77°.

Kenyon, J. *et al*, *J. Chem. Soc.*, 1923, **123**, 18.
Org. Synth., 1926, **6**, 68 (*synth*)
Bohnsack, H., *Ber.*, 1942, **75**, 502 (*isol*)
Doering von E., W. *et al*, *J. Am. Chem. Soc.*, 1952, **74**, 2997 (*abs config*)
Cristol, S.J., *J. Am. Chem. Soc.*, 1955, **77**, 2512.
Goering, H.L., *J. Am. Chem. Soc.*, 1956, **78**, 2270.
Gil-Av, E., *J. Chromatogr.*, 1965, **17**, 408 (*use*)
Fruwert, J., *Z. Phys. Chem. (Leipzig)*, 1968, **238**, 262 (*ir*)
Whitney, T.A. *et al*, *Polym. Prepr., (Am. Chem. Soc., Div. Polym. Chem.)*, 1971, **13**, 688 (*synth*)
Morris, W.W., *J. Assoc. Off. Anal. Chem.*, 1973, **56**, 1037 (*ir*)
Keinan, E. *et al*, *J. Am. Chem. Soc.*, 1986, **108**, 162 (*synth*)
Sax, N.I., *Dangerous Properties of Industrial Materials*, 5th Ed., Van Nostrand-Reinhold, 1979, 467.

3-Octanol, 9CI **O-00244**

[589-98-0]

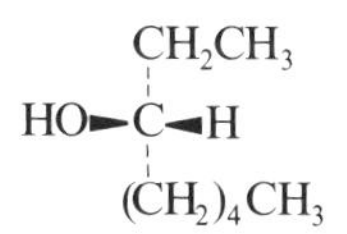

(*R*)-*form*

$C_8H_{18}O$ M 130.230
Released by *Curvularia falcata* and *Mucor* spp.
▷ RH0855000.

(*R*)-*form* [70492-66-9]
Bp_{24} 82°. $[\alpha]_D^{27}$ −7.40° (EtOH).
1-Naphthylurethane: Cryst. (EtOH aq.). Mp 79-80°. $[\alpha]_D^{20}$ −2.48° (EtOH).
Ac: [4864-61-3].
Bp_{10} 90°. $[\alpha]_D^{20}$ −5°.
Hydrogen phthalate: Plates (pet. ether). Mp 62-63°.

(S)-form [22658-92-0]
Present in Japanese peppermint oil and many other essential oils. Bp 178.5-179.5°, Bp_{16} 76°.
Hydrogen phthalate: Plates (pet. ether). Mp 66-68°. $[\alpha]_D$ +21.67° (EtOH).
(±)-form [20296-29-1]
d_{15}^{15} 0.829. Bp 176-177.5°. n_D^{20} 1.4279.

Pickard, R.H. *et al, J. Chem. Soc.*, 1913, **103**, 1944 (*synth*)
Woods, G.F. *et al, J. Am. Chem. Soc.*, 1956, **78**, 4380 (*synth*)
Cason, J. *et al, J. Org. Chem.*, 1961, **26**, 3645 (*synth*)
Spiteller, G. *et al, Angew. Chem.*, 1965, **77**, 393 (*ms*)
Fruwert, J., *Z. Phys. Chem. (Leipzig)*, 1968, **238**, 262 (*ir*)
Fried, J. *et al, J. Am. Chem. Soc.*, 1972, **94**, 4342 (*abs config*)
Karrer, W. *et al, Konstitution und Vorkommen der Organischen Pflanzenstoffe*, 2nd Ed., Birkhäuser Verlag, Basel, 1972-1985, no. 78 (*occur*)
Tang, K.-C. *et al, J. Med. Chem.*, 1981, **24**, 1277 (*synth, pmr*)
Brand, J.M. *et al, J. Chem. Ecol.*, 1987, **13**, 357 (*isol*)
Lewis, R.J., *Sax's Dangerous Properties of Industrial Materials*, 8th Ed., Van Nostrand-Reinhold, 1992, OCY100.

2-Octanone, 9CI — O-00245

[111-13-7]

$H_3CCH_2(CH_2)_4COCH_3$

$C_8H_{16}O$ M 128.214
Trace constit. of plant oils. Perfumery and flavouring ingredient. Liq. d_4^{20} 0.820. Fp −16°. Bp 172.92°, Bp_{11} 59-60°. n_D^{20} 1.4151.
▷ RH1484000.
Di-Et acetal:
$C_{12}H_{26}O_2$ M 202.336
Bp_{16} 101-103°.
Oxime: [7207-49-0].
$C_8H_{17}NO$ M 143.228
Oil. Bp_{725} 213-214°, Bp_{15} 116.5°.
Semicarbazone: Cryst. (EtOH aq.). Mp 123°.
2,4-Dinitrophenylhydrazone: Orange cryst. (EtOH). Mp 58°.

Moureu, C. *et al, C. R. Hebd. Seances Acad. Sci.*, 1920, **171**, 652 (*synth*)
Verhulst, G. *et al, Bull. Soc. Chim. Belg.*, 1932, **41**, 501 (*synth*)
Fruwert, J., *Z. Phys. Chem. (Leipzig)*, 1968, **238**, 262, 401 (*ir*)
Lewis, R.J., *Sax's Dangerous Properties of Industrial Materials*, 8th Ed., Van Nostrand-Reinhold, 1992, ODG000.

3-Octanone, 9CI — O-00246

[106-68-3]

$H_3C(CH_2)_4COCH_2CH_3$

$C_8H_{16}O$ M 128.214
Essential constit. of *Mentha arvensis* and present in many fragrant oils. Food flavouring. Alarm pheromone for myrmicine ants. Liq. d^{15} 0.826. Bp 165-166° (170°). n_D^{23} 1.4156.
▷ RH1485000.
Di-Me ketal: Bp_{26} 90-92°.
Semicarbazone: Cryst. (EtOH aq.). Mp 112° (slow heat) (117-117.5°).
2,4-Dinitrophenylhydrazone: Mp 67°.

Pickard, R.H. *et al, J. Chem. Soc.*, 1913, **103**, 1936 (*derivs*)
Asinger, F. *et al, Chem. Ber.*, 1957, **90**, 485 (*synth, bibl*)
Shuikin, N.I. *et al, Zh. Obshch. Khim.*, 1957, **27**, 402; *CA*, **51**, 15489i (*synth*)
Shimizu, S. *et al, Agric. Biol. Chem.*, 1961, **25**, 263 (*isol*)
Fruwert, J., *Z. Phys. Chem. (Leipzig)*, 1968, **238**, 262 (*ir*)
Lewis, R.J., *Sax's Dangerous Properties of Industrial Materials*, 8th Ed., Van Nostrand-Reinhold, 1992, ODI000.

1,3,5-Octatriene — O-00247

$H_3CCH_2CH{=}CHCH{=}CHCH{=}CH_2$

C_8H_{12} M 108.183
(3E,5E)-form [33580-04-0] ***Sarohornene B***
Constit. of male-attractant pheromone from *Sargassum horneri*. Bp_{60} 67°.
(3E,5Z)-form [40087-61-4] ***Fucoserratene***
Female sex attractant from the ova of *Fucus serratus* and *F. vesiculosus*. Bp_{40} 56°.
(3Z,5E)-form [33580-05-1] ***Sarohornene C***
Constit. of male-attractant pheromone from *S. horneri*.
(3Z,5Z)-form [40087-62-5]
No phys. props. reported.

Alder, K. *et al, Justus Liebigs Ann. Chem.*, 1957, **608**, 195 (*synth*)
Jaenicke, L. *et al, Chem. Ber.*, 1975, **108**, 225 (*synth*)
Mueller, D.G. *et al, Naturwissenschaften*, 1978, **65**, 389 (*isol*)
Kajiwara, T. *et al, Naturwissenschaften*, 1980, **67**, 612 (*Sarohornenes*)
Janicke, L. *et al, Angew. Chem., Int. Ed. Engl.*, 1982, **21**, 643 (*rev*)
Baldwin, J.E. *et al, J. Am. Chem. Soc.*, 1987, **109**, 8051 (*synth, pmr, props*)
Kishi, N. *et al, Tetrahedron*, 1992, **48**, 4087 (*synth*)

2,4,6-Octatrienedioic acid, 9CI — O-00248

Hexatriene-1,6-dicarboxylic acid
[5698-41-9]

$HOOCCH{=}CHCH{=}CHCH{=}CHCOOH$

$C_8H_8O_4$ M 168.149
Cryst. Mp 316-318°.
Di-Me ester: [77185-98-9].
$C_{10}H_{12}O_4$ M 196.202
Cryst. (MeOH). Mp 172°.
Dichloride:
$C_8H_6Cl_2O_2$ M 205.040
Mp 98-100°.

Kuhn, R., *Angew. Chem.*, 1937, **50**, 703 (*synth*)
Sadeh, T. *et al, J. Am. Chem. Soc.*, 1962, **84**, 3970 (*cryst struct*)
Lapitskii, G.A. *et al, CA*, 1964, **61**, 14520 (*synth*)
Kaupp, G. *et al, Chem. Ber.*, 1977, **110**, 3084 (*synth*)

2-(2,5,7-Octatrienyl)-3-decyloxirane — O-00249

9,10-Epoxy-1,3,6-eicosatriene

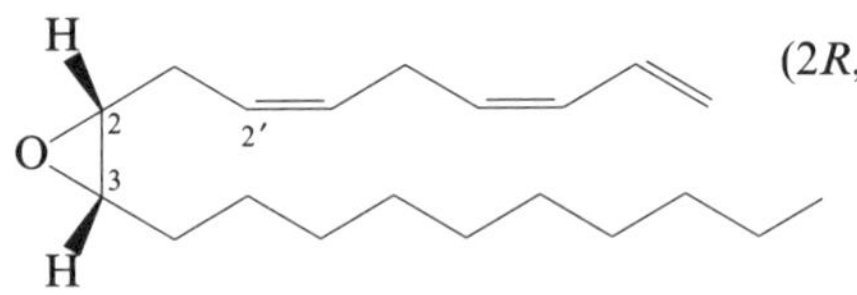

(2R,2′Z,3S,5′Z)-form

$C_{20}H_{34}O$ M 290.488
(2R,2′Z,3S,5′Z)-form [119181-53-2]
Mp 2-3°. $[\alpha]_D^{16}$ +0.62° (c, 2.34 in $CHCl_3$). n_D^{16} 1.4779. Unstable.
(2S,2′Z,3R,5′Z)-form [119072-11-6]
Isol. from the fall webworm moth *Hyphantria cunea*. Active component of the sex pheromone. Mp 2-3°. $[\alpha]_D^{16}$ −0.57° (c, 1.23 in $CHCl_3$). n_D^{16} 1.4787. Unstable.

Carriere, Y. *et al, J. Chem. Ecol.*, 1988, **14**, 947.
Mori, K. *et al, Justus Liebigs Ann. Chem.*, 1989, 453 (*synth, ir, pmr, cmr, bibl*)

3,5,7-Octatriyn-1-ol, 9CI O-00250

[6071-20-1]

HC≡CC≡CC≡CCH$_2$CH$_2$OH

C_8H_6O M 118.135

Isol. from *Lentinus edodes*. λ_{max} 209 nm.

Bew, R.E. *et al*, *J. Chem. Soc. C*, 1966, 129 (*synth, ir, uv*)
Tokimoto, K. *et al*, *Proc. Jpn. Acad., Ser. B*, 1987, **63**, 277 (*isol*)

2-Octenal, 9CI O-00251

[2363-89-5]

H$_3$C(CH$_2$)$_4$CH=CHCHO

$C_8H_{14}O$ M 126.198

(E)-form

Occurs in many insect and plant systems including the scent gland of the nymph of *Pternistria bispina*. Also isol. from essential oil of *Achasma walang*. Liq.

2,4-Dinitrophenylhydrazone: Mp 125-126°.

(Z)-form

Bp_{15} 81-82°. 92% pure.

Di-Et acetal: 1,1-Diethoxy-2-octene

$C_{12}H_{24}O_2$ M 200.320

Bp_{14} 104-106°.

v. Romburgh, P., *Recl. Trav. Chim. Pays-Bas* (*J. R. Neth. Chem. Soc.*), 1938, **57**, 494 (*isol*)
Baker, J.T. *et al*, *Aust. J. Chem.*, 1969, **22**, 1793.
Riehl, J.J. *et al*, *Tetrahedron Lett.*, 1969, 3139 (*synth*)
Corey, E.J. *et al*, *J. Am. Chem. Soc.*, 1971, **93**, 1724 (*ir, pmr*)
Makin, S.M. *et al*, *Zh. Org. Khim.*, 1974, **10**, 2044 (*synth*)
Bestmann, H.J. *et al*, *Chem. Ber.*, 1982, **115**, 161 (*synth*)

3-Octenal O-00252

[60671-71-8]

H$_3$C(CH$_2$)$_3$CH=CHCH$_2$CHO

$C_8H_{14}O$ M 126.198

Used in pheromone synthesis.

(E)-form [76595-71-6]

Liq. $Bp_{0.02}$ 60°.

2,4-Dinitrophenylhydrazone: Cryst. (EtOH). Mp 80-81°.

(Z)-form [78693-34-2]

Liq. Bp_{14} 65-75° (bath).

Bestmann, H.J. *et al*, *Justus Liebigs Ann. Chem.*, 1981, 1705 (*synth, ir, pmr*)
Achmatowicz, O. *et al*, *Synthesis*, 1987, 413 (*synth, ir, pmr*)

1-Octene, 9CI O-00253

Caprylene

[111-66-0]

H$_3$C(CH$_2$)$_5$CH=CH$_2$

C_8H_{16} M 112.214

d_4^{20} 0.716. Fp −102°. Bp 122°, Bp_{100} 61.5-61.7°. n_D^{25} 1.4062.

▷ Highly flammable, flash pt. 8°.

Geiseler, G. *et al*, *Chem. Ber.*, 1962, **95**, 96 (*synth, props*)
Blouri, B. *et al*, *Bull. Soc. Chim. Fr.*, 1963, 1670 (*synth*)
Lundeen, A.J. *et al*, *J. Am. Chem. Soc.*, 1963, **85**, 2180 (*synth*)
Hauser, C.F. *et al*, *J. Org. Chem.*, 1963, **28**, 372 (*synth*)
Hashimoto, H. *et al*, *J. Organomet. Chem.*, 1967, **10**, 518 (*synth*)
Olah, G.A. *et al*, *Synthesis*, 1976, 607 (*synth*)
Miller, J.A. *et al*, *J. Org. Chem.*, 1993, **58**, 18 (*synth*)
Sax, N.I., *Dangerous Properties of Industrial Materials*, 5th Ed., Van Nostrand-Reinhold, 1979, 875.

4-Octene-1,8-diol O-00254

[56506-08-2]

HOCH$_2$CH$_2$CH$_2$CH=CHCH$_2$CH$_2$CH$_2$OH

$C_8H_{16}O_2$ M 144.213

(Z)-form [62422-45-1]

Intermed. in nat. prod. synth. $Bp_{0.001}$ 78°.

Di-Ac: [67161-27-7].

$C_{12}H_{20}O_4$ M 228.288

Bp_1 102-103°.

Nagarkatti, J.P. *et al*, *J. Indian Chem. Soc.*, 1976, **53**, 666 (*synth*)
Tolstikov, G.A. *et al*, *Tetrahedron Lett.*, 1978, 1857 (*synth*)
Odinokov, V.N. *et al*, *J. Org. Chem. USSR* (*Engl. Transl.*), 1979, **15**, 1252 (*synth, ir, pmr*)
Raederstorff, D. *et al*, *J. Org. Chem.*, 1987, **52**, 2337 (*synth, pmr, ms*)

6-Octene-2,4-diynoic acid, 9CI O-00255

H$_3$CCH=CHC≡CC≡CCOOH

$C_8H_6O_2$ M 134.134

(E)-form [132368-12-8]

Isol. from *Siphocampylus sulfureus*.

Me ester: Needles (pentane). Mp 34-35°.

(Z)-form

Needles (pet. ether). Mp 50-60°. λ_{max} 303, 285.5, 269.5, 257, 248, 233 nm.

Jones, J.B., *J. Chem. Soc.*, 1963, 5759 (*synth, uv, ir*)
Bohlmann, F. *et al*, *Chem. Ber.*, 1964, **97**, 2586 (*synth, ir, uv*)
Magalhaes, A.F. *et al*, *Quim. Nova*, 1990, **13**, 65 (*isol*)

3-Octenoic acid, 9CI O-00256

2-Heptene-1-carboxylic acid

[1577-19-1]

H$_3$C(CH$_2$)$_3$CH=CHCH$_2$COOH

$C_8H_{14}O_2$ M 142.197

$Bp_{0.7-1.0}$ 84-87°.

(E)-form [5163-67-7]

Me ester: [35234-16-3].

$C_9H_{16}O_2$ M 156.224

Bp_{11} 70-73°.

Et ester: [26553-47-9].

$C_{10}H_{18}O_2$ M 170.251

Bp_{11} 80-89°.

Nitrile:

$C_8H_{13}N$ M 123.197

Bp_{20} 93-95°.

DiPietro, J. *et al*, *Can. J. Chem.*, 1966, **44**, 2241 (*synth*)
Kharasch, N. *et al*, *J. Chem. Soc., Chem. Commun.*, 1967, 435 (*synth*)
Brown, H.C. *et al*, *J. Am. Chem. Soc.*, 1970, **92**, 1761 (*synth*)
Näf-Müller, R. *et al*, *Helv. Chim. Acta*, 1971, **54**, 1880 (*synth, spectra*)

5-Octenoic acid O-00257

[63892-00-2]

H$_3$CCH$_2$CH=CHCH$_2$CH$_2$CH$_2$COOH

$C_8H_{14}O_2$ M 142.197

(E)-form [16424-53-6]

d^{25} 0.93. Mp −9°. $Bp_{1.5}$ 120°.

(Z)-form [41653-97-8]

Oil. d_4^{24} 0.96. $Bp_{0.4}$ 88.5°.

Me ester: [41654-15-3].

$C_9H_{16}O_2$ M 156.224
Fragrant oil. $Bp_{5.5}$ 63°.
4-Bromophenacyl ester: Fine needles (pet. ether). Mp 47.2-48°.

[41653-97-8]

Howton, D.R. *et al*, *J. Org. Chem.*, 1951, **16**, 1405 (*synth, ir*)
Scholz, D., *Justus Liebigs Ann. Chem.*, 1984, 264.
Kawashima, M. *et al*, *Bull. Chem. Soc. Jpn.*, 1988, **61**, 3255.
Hdrlik, P.F. *et al*, *Tetrahedron*, 1988, **44**, 3791 (*synth, pmr, ms*)

7-Octenoic acid, 9CI O-00258

1-Heptene-7-carboxylic acid
[18719-24-9]

$H_2C{=}CH(CH_2)_5COOH$

$C_8H_{14}O_2$ M 142.197
Bp_{12} 126°.
Me ester: [15766-90-2].
$C_9H_{16}O_2$ M 156.224
Bp 190°, Bp_{15} 84°.
Et ester: [35194-38-8].
$C_{10}H_{18}O_2$ M 170.251
Bp 210°, Bp_{22} 106°.
4-Bromophenacyl ester: Mp 58°.

Colonge, J. *et al*, *Bull. Soc. Chim. Fr.*, 1962, 177 (*synth*)
Rubottom, G.M. *et al*, *Tetrahedron Lett.*, 1977, 4013 (*synth*)
Matsuda, H. *et al*, *J. Org. Chem.*, 1978, **43**, 4567 (*derivs*)
Scholz, D., *Justus Liebigs Ann. Chem.*, 1984, 264 (*synth*)

1-Octen-2-ol, 9CI O-00259

$H_3C(CH_2)_5C(OH){=}CH_2$

$C_8H_{16}O$ M 128.214
Ac: [26735-84-2].
$C_{10}H_{18}O_2$ M 170.251
Bp_{42} 108°.
Me ether: 2-Methoxy-1-octene
$C_9H_{18}O$ M 142.241
Bp 166-168°.

Moureu, C., *C. R. Hebd. Seances Acad. Sci.*, 1904, **138**, 287.
Hudrlik, P.F. *et al*, *J. Org. Chem.*, 1973, **38**, 4254 (*synth, spectra*)

1-Octen-3-ol, 9CI O-00260

Amyl vinyl carbinol. Matsutake alcohol. Matsutakeol
[3391-86-4]

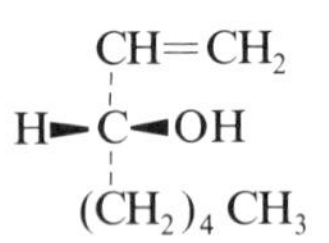

(*R*)*-form*

$C_8H_{16}O$ M 128.214
Constit. of the defense secretions of *Leiobunum nigripalpi*. Oil. Mp Bp43° 73.
▷ RH3300000.
(R)-form [22658-80-6]
Isol. from a number of essential oils, e.g. lavender, leek, mint and mushrooms. $[\alpha]_D^{17}$ −13.1° (EtOH).
Ac:
$C_{10}H_{18}O_2$ M 170.251
Found in lavender oil. Used in perfumery. Bp 190°. $[\alpha]_D^{20}$ +3.5°.
Hydrogen phthalate: Mp 57°.
(S)-form [62247-46-5]
$[\alpha]_D^{17}$ +10.7° (EtOH).
(±)*-form* [74072-82-5]
Bp 173.5°.
Hydrogen phthalate: Mp 79°.

[3687-48-7, 24587-53-9, 50999-79-6]

Honkanen, E. *et al*, *Acta Chem. Scand.*, 1963, **17**, 858 (*isol*)
Tsuji, J. *et al*, *Bull. Chem. Soc. Jpn.*, 1976, **49**, 1701 (*synth, spectra*)
Roumestant, M.L. *et al*, *Synthesis*, 1976, 755 (*synth*)
McGahren, W.J. *et al*, *J. Org. Chem.*, 1977, **42**, 1659 (*abs config*)
Nakai, T. *et al*, *Tetrahedron Lett.*, 1977, 2425 (*synth*)
Takabe, K. *et al*, *Synth. Commun.*, 1980, **10**, 89 (*synth*)
Whitfield, F.B. *et al*, *Aust. J. Chem.*, 1982, **35**, 373 (*isol*)
Shimazaki, M. *et al*, *Chem. Pharm. Bull.*, 1992, **40**, 3073 (*synth*)
Lewis, R.J., *Sax's Dangerous Properties of Industrial Materials*, 8th Ed., Van Nostrand-Reinhold, 1992, ODW000.

2-Octen-1-ol, 9CI O-00261

[22104-78-5]

$H_3C(CH_2)_4CH{=}CHCH_2OH$

$C_8H_{16}O$ M 128.214
(E)-form [18409-17-1]
Found in grapes, ripe bananas and mushroom volatiles. Bp_{11} 87-89°.
Ac: [3913-80-2].
$C_{10}H_{18}O_2$ M 170.251
Bp_{13} 82-84°.
(Z)-form [26001-58-1]
Bp_{10} 85-87°.
3,5-Dinitrobenzoyl: Mp 60°.

[26806-12-2]

Colonge, J., *Bull. Soc. Chim. Fr.*, 1955, 953 (*synth*)
Osbond, J., *J. Chem. Soc.*, 1961, 5270 (*synth*)
Schlosser, M. *et al*, *Synthesis*, 1972, 575 (*synth*)
Sood, R. *et al*, *Tetrahedron Lett.*, 1974, 423 (*synth*)
Pyysalo, H., *Acta Chem. Scand., Ser. B*, 1976, **30**, 235 (*isol*)
Tsuji, J. *et al*, *Bull. Chem. Soc. Jpn.*, 1976, **49**, 1701.
Tsuji, J. *et al*, *Tetrahedron Lett.*, 1979, 39 (*synth, spectra*)
Porter, N.A. *et al*, *J. Org. Chem.*, 1985, **50**, 2252 (*synth, pmr, cmr*)
Nicolaou, K.C. *et al*, *Synthesis*, 1986, 453 (*synth, ir, pmr*)
Binns, M.R. *et al*, *Aust. J. Chem.*, 1987, **40**, 281 (*synth*)

2-Octen-4-ol, 9CI O-00262

[4798-61-2]

$H_3C(CH_2)_3CH(OH)CH{=}CHCH_3$

$C_8H_{16}O$ M 128.214
(±)-*(E)-form* [20125-81-9]
Bp 175°, Bp_{20} 83°.
3,5-Dinitrobenzoyl: Plates (EtOH aq.). Mp 54°.
(±)-*(Z)-form* [22287-00-9]
Ac:
$C_{10}H_{18}O_2$ M 170.251
Bp_{45} 105-107°.

Hargreaves, G.H. *et al*, *J. Chem. Soc.*, 1947, 756 (*synth*)
Dauben, W.G. *et al*, *J. Org. Chem.*, 1970, **35**, 374 (*synth*)
Gocmen, M. *et al*, *Bull. Soc. Chim. Fr.*, 1973, 1310 (*synth*)
Umbreit, M.A. *et al*, *J. Am. Chem. Soc.*, 1977, **99**, 5526 (*synth*)
Ishida, A. *et al*, *Bull. Chem. Soc. Jpn.*, 1978, **51**, 2077 (*synth*)

3-Octen-1-ol, 9CI O-00263

[18185-81-4]

$H_3C(CH_2)_3CH{=}CHCH_2CH_2OH$

$C_8H_{16}O$ M 128.214
(E)-form [20125-85-3]
Found in ripe bananas. Bp_{12} 86°.

(Z)-form [20125-84-2]
Liq. Bp_{15} 95-96°.

Normant, H., *C. R. Hebd. Seances Acad. Sci.*, 1948, **226**, 733.
Rossi, R. *et al*, *Gazz. Chim. Ital.*, 1978, **108**, 709.
Alexakis, A. *et al*, *Tetrahedron*, 1980, **36**, 1961 (*synth, ir, pmr*)
Mitra, R.B. *et al*, *Synthesis*, 1989, 694 (*synth, ir, pmr*)

4-Octen-1-ol, 9CI O-00264

[6770-26-9]

$H_3CCH_2CH_2CH{=}CHCH_2CH_2CH_2OH$

$C_8H_{16}O$ M 128.214

(E)-form [31502-21-3]
Bp_{17} 95°.
Ac:
$C_{10}H_{18}O_2$ M 170.251
Bp_{19} 101°.
3,5-Dinitrobenzoyl: Mp 44°.

(Z)-form [54393-36-1]
Bp_{20} 100°.

Riobé, O., *C. R. Hebd. Seances Acad. Sci.*, 1947, **225**, 334.
Jacabson, M. *et al*, *J. Med. Chem.*, 1971, **14**, 236 (*synth*)
Schlosser, M. *et al*, *Helv. Chim. Acta*, 1974, **57**, 1567 (*synth, spectra*)
Staehle, M. *et al*, *Helv. Chim. Acta*, 1977, **60**, 1730 (*synth*)

6-Octen-1-ol, 9CI O-00265

[63768-12-7]

$H_3CCH{=}CH(CH_2)_4CH_2OH$

$C_8H_{16}O$ M 128.214
Bp 187-193°.

[67133-87-3]

Sumitani, M. *et al*, *Bull. Chem. Soc. Jpn.*, 1978, **51**, 1511 (*synth, ir, pmr*)
De Medeiros, E.F. *et al*, *J. Chem. Soc., Perkin Trans. 1*, 1991, 2725 (*synth*)

1-Octen-3-one, 9CI O-00266

[4312-99-6]

$H_3C(CH_2)_4COCH{=}CH_2$

$C_8H_{14}O$ M 126.198
Mushroom flavour component. Bp_{14} 63-65°.

Sasson, Y. *et al*, *Can. J. Chem.*, 1974, **52**, 3825 (*synth*)
Grieco, P.A. *et al*, *J. Chem. Soc., Chem. Commun.*, 1974, 497 (*synth*)
Pyysalo, H., *Acta Chem. Scand., Ser. B*, 1976, **30**, 235 (*isol*)
Venturello, C. *et al*, *J. Org. Chem.*, 1991, **56**, 5924 (*synth, ir, pmr*)

3-Octen-2-one O-00267

[1669-44-9]

$H_3C(CH_2)_3CH{=}CHCOCH_3$

$C_8H_{14}O$ M 126.198
Vanilla volatile, mushroom flavour component.

(E)-form [18402-82-9]
Oil. Bp_{50} 100°.
2,4-Dinitrophenylhydrazone: Mp 93°.

(Z)-form [51193-77-2]
Oil. Bp_{50} 68-69°.
2,4-Dinitrophenylhydrazone: Mp 93°.

Heilman, R. *et al*, *Bull. Soc. Chim. Fr.*, 1957, 119 (*synth, uv, ir*)
Clinet, J.C. *et al*, *Tetrahedron Lett.*, 1978, 1137 (*synth*)
Beak, P. *et al*, *J. Am. Chem. Soc.*, 1980, **102**, 3848 (*synth*)
Grayson, D.H. *et al*, *J. Chem. Soc., Perkin Trans. 1*, 1986, 2137 (*synth, ir, pmr*)

7-[5-(1-Octenyl)-4-imidazolidinyl]-5-heptenoic acid O-00268

U 51605
[77331-77-2]

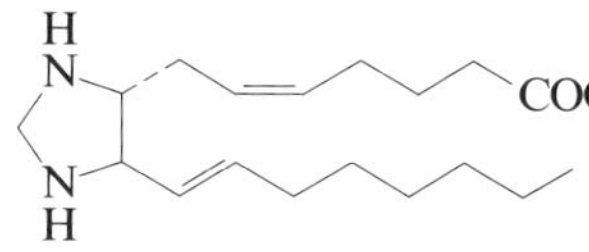

$C_{18}H_{32}N_2O_2$ M 308.463
Thromboxane synthetase inhibitor.

Rao, G.H. *et al*, *Prostaglandins Med.*, 1981, **6**, 75 (*biochem*)
Rittenhouse, S.E. *et al*, *J. Clin. Invest.*, 1982, 70, 1216 (*biochem*)

7-[5-(1-Octenyl)-2-oxa-3-azabicyclo[2.2.1]hept-6-yl]-5-heptenoic acid O-00269

9,11-Epoxyimino-5,13-prostadienoic acid

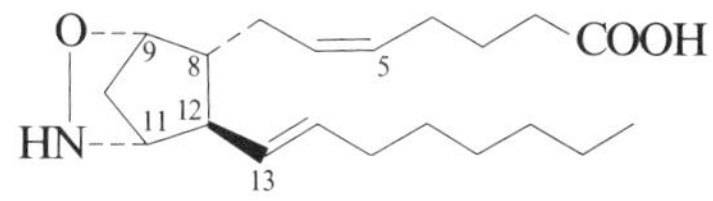

$C_{20}H_{33}NO_3$ M 335.486
Prostanoid numbering shown.

(9α,11α)-(5Z,13E)-form [66464-51-5]
Potent thromboxane synthetase inhibitor.

Bundy, G.L. *et al*, *Tetrahedron Lett.*, 1978, 41 (*synth, pmr*)

7-[6-(1-Octenyl)-2-oxa-3-azabicyclo[2.2.1]hept-5-yl]-5-heptenoic acid O-00270

11,9-Epoxyimino-5,13-prostadienoic acid

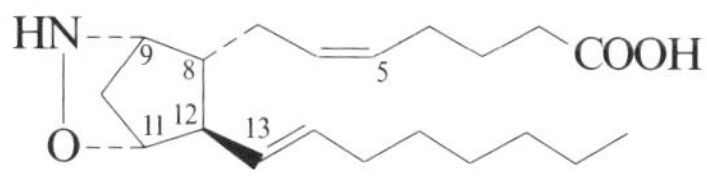

$C_{20}H_{33}NO_3$ M 335.486
Prostanoid numbering shown.

(9α,11α)-(5Z,13E)-form [66464-50-4]
Potent thromboxane synthetase inhibitor. Cryst. (Et_2O/hexane). Mp 53-54°.

Bundy, G.L. *et al*, *Tetrahedron Lett.*, 1978, 41 (*synth, pmr*)

7-Octen-2-yn-1-ol O-00271

[79972-66-0]

$H_2C{=}CHCH_2CH_2CH_2C{\equiv}CCH_2OH$

$C_8H_{12}O$ M 124.182
Bp_{15} 100-101°.

Saloman, R.G. *et al*, *J. Am. Chem. Soc.*, 1982, **104**, 998 (*synth, pmr*)
Bulman Page, P.C. *et al*, *J. Chem. Soc., Perkin Trans. 1*, 1990, 1375 (*synth, ir, pmr, cmr, ms*)

2-Octylcyclopropaneoctanoic acid O-00272

9,10-Methyleneoctadecanoic acid

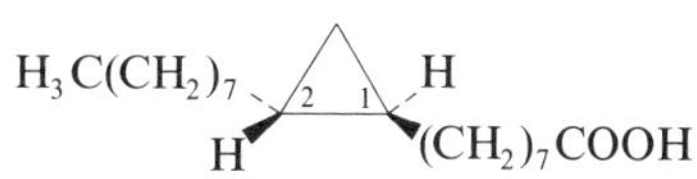

$C_{19}H_{36}O_2$ M 296.492

(1R*,2R*)-form [29203-99-4]

trans-*form*

Mp 28.4-29.4°, Mp 33.6-35.0°.

(1R*,2S*)-form [4675-61-0]

Dihydrosterculic acid

Occurs in a number of seed oils, frequently with sterculic acid, and in some microorganisms. Mp 35.6-36.5°, Mp 38.6-39.0°.

Wood, R. *et al*, *J. Am. Oil Chem. Soc.*, 1965, **42**, 315 (*synth*)
Christie, W.W. *et al*, *Chem. Phys. Lipids*, 1968, **2**, 196 (*synth*, *glc*, *ir*, *nmr*)
Christie, W.W., *Top. Lipid Chem.*, 1970, **1**, 4 (*rev*)
Gunstone, F.D. *et al*, *Chem. Phys. Lipids*, 1973, **10**, 303 (*synth*, *glc*, *nmr*)
Gensler, W.J. *et al*, *J. Org. Chem.*, 1977, **42**, 126 (*ms*)
White, J.L. *et al*, *J. Am. Oil Chem. Soc.*, 1977, **54**, 1345.

2-Octyl-1-cyclopropene-1-acetic acid, 9CI O-00273

3,4-Methylene-3-dodecenoic acid

$C_{13}H_{22}O_2$ M 210.316

Williams, J.L. *et al*, *Chem. Phys. Lipids*, 1972, **9**, 295 (*synth*, *nmr*)

2-Octyl-1-cyclopropene-1-butanoic acid, 9CI O-00274

5,6-Methylene-5-tetradecenoic acid

[39647-67-1]

$C_{15}H_{26}O_2$ M 238.369

Williams, J.L. *et al*, *Chem. Phys. Lipids*, 1972, **9**, 295 (*synth*, *nmr*)

2-Octyl-1-cyclopropene-1-carboxylic acid, 9CI O-00275

2,3-Methylene-2-undecenoic acid

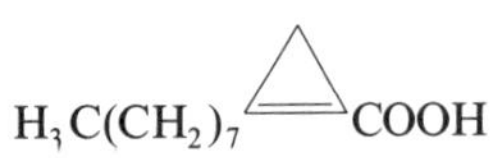

$C_{12}H_{20}O_2$ M 196.289

Williams, J.L. *et al*, *Chem. Phys. Lipids*, 1972, **9**, 295 (*synth*, *nmr*)

2-Octyl-1-cyclopropene-1-hexanoic acid, 9CI O-00276

7,8-Methylene-7-hexadecenoic acid

[39647-65-9]

$C_{17}H_{30}O_2$ M 266.423

Williams, J.L. *et al*, *Chem. Phys. Lipids*, 1972, **9**, 295 (*synth*, *nmr*)

2-Octyl-1-cyclopropene-1-propanoic acid, 9CI O-00277

4,5-Methylene-4-tridecenoic acid

[39647-68-2]

$C_{14}H_{24}O_2$ M 224.342

Williams, J.L. *et al*, *Chem. Phys. Lipids*, 1972, **9**, 295 (*synth*, *nmr*)

9-Octyl-9-heptadecanol, 9CI O-00278

Trioctylcarbinol

[17687-72-8]

$(H_3CCH_2CH_2CH_2CH_2CH_2CH_2CH_2)_3COH$

$C_{25}H_{52}O$ M 368.685

Mp 8-9°. $Bp_{0.02}$ 110-120°.

Challenger, F. *et al*, *CA*, 1955, **49**, 6813e (*synth*)
Friedrich, L. *et al*, *Chem. Ber.*, 1955, **88**, 913 (*synth*)

2-(Octylthio)ethanol, 9CI O-00279

2-Hydroxyethyl octyl sulfide

[3547-33-9]

$H_3C(CH_2)_7SCH_2CH_2OH$

$C_{10}H_{22}OS$ M 190.349

Insect repellent. Pale-amber liq. Sl. sol. H_2O, misc. most org. solvs. d_4^{20} 0.93. Mp 0°. $Bp_{0.1}$ 98°. n_D^{20} 1.470-1.478.

▷ KL9700000.

U.S. Pat., 2 863 799, (1955); *CA*, **53**, 6518g.
Pesticide Manual, 6th Ed., 1979, 301.
Lewis, R.J., *Sax's Dangerous Properties of Industrial Materials*, 8th Ed., Van Nostrand-Reinhold, 1992, OGA000.

2-Octynoic acid, 9CI O-00280

1-Heptyne-1-carboxylic acid

[5663-96-7]

$H_3C(CH_2)_4C{\equiv}CCOOH$

$C_8H_{12}O_2$ M 140.182

Fungicide. Liq. d_4^{13} 0.962. Fp 2-5°. Bp 180-220° dec., Bp_{19} 148-149°. n_D^{20} 1.4595.

▷ RI2725000.

Me ester: [111-12-6]. *Methyl heptyne carbonate*

$C_9H_{14}O_2$ M 154.208

Perfumery ingredient. Bp_{20} 107°, Bp_{10} 94°.

▷ RI2735000.

Et ester: [10519-20-7].

$C_{10}H_{16}O_2$ M 168.235

$Bp_{0.2}$ 51-52°.

Chloride:

$C_8H_{11}ClO$ M 158.627

Bp_{17} 88-90°.

Amide:

$C_8H_{13}NO$ M 139.197

Plates (EtOH). Mp 91°.

Nitrile:

$C_8H_{11}N$ M 121.182

Bp 194-196°, Bp_{13} 80-81°.

Zoss, A.O. *et al*, *J. Am. Chem. Soc.*, 1941, **63**, 1152 (*synth*)
Nametkin, I.S.S. *et al*, *CA*, 1942, **36**, 3783 (*synth*)
Newman, M.S. *et al*, *J. Am. Chem. Soc.*, 1949, **71**, 1292 (*synth*)
Wotiz, J.H. *et al*, *J. Am. Chem. Soc.*, 1949, **71**, 3441 (*ir*)
Bohlmann, F. *et al*, *Chem. Ber.*, 1967, **100**, 3706 (*ms*)
Gershon, H. *et al*, *Can. J. Microbiol.*, 1978, **24**, 593 (*use*)

Sprecher, H., *Prog. Chem. Fats Other Lipids*, 1978, **15**, 219 (*rev*)
Wuts, P.G.M. *et al*, *J. Org. Chem.*, 1989, **53**, 5989 (*Me ester synth, pmr, cmr, ir*)
Crombie, L. *et al*, *J. Chem. Soc., Perkin Trans. 1*, 1992, 1929 (*synth, ir, pmr*)
Lewis, R.J., *Sax's Dangerous Properties of Industrial Materials*, 8th Ed., Van Nostrand-Reinhold, 1992, MND275.

7-Octynoic acid, 9CI — O-00281

HC≡C$(CH_2)_5$COOH

$C_8H_{12}O_2$ M 140.182
Mp 20°. Bp_2 123°, $Bp_{0.25}$ 89-93.5°.
Me ester: [18458-50-9].
$C_9H_{14}O_2$ M 154.208
Bp_2 64°, $Bp_{0.1}$ 44-46°.
Amide:
$C_8H_{13}NO$ M 139.197
Cryst. (pet. ether). Mp 90.5-91°.
Anilide:
$C_{14}H_{17}NO$ M 215.294
Mp 59-60°.
Anhydride: [119837-88-6].
$C_{16}H_{22}O_3$ M 262.348
Bp_1 160-166°.

Newman, M.S. *et al*, *J. Am. Chem. Soc.*, 1949, **71**, 1292 (*synth*)
Wotiz, J.H. *et al*, *J. Am. Chem. Soc.*, 1949, **71**, 3441 (*ir*)
Taylor, W.R. *et al*, *J. Am. Chem. Soc.*, 1950, **50**, 4263 (*synth*)
Eglinton, G. *et al*, *J. Chem. Soc.*, 1953, 3052 (*synth*)
Kraevskii, A.A. *et al*, *Zh. Obshch. Khim.*, 1964, **34**, 1180 (*synth*)
Myagkova, G.I. *et al*, *Zh. Org. Khim.*, 1967, **3**, 1415 (*synth*)
Gunstone, F.D. *et al*, *Chem. Phys. Lipids*, 1976, **17**, 1 (*cmr*)
Sprecher, H., *Prog. Chem. Fats Other Lipids*, 1978, **15**, 219.
Moody, C.J. *et al*, *J. Chem. Soc., Perkin Trans. 1*, 1988, 3249 (*anhydride*)

2-Octyn-1-ol, 9CI — O-00282

Pentylpropargyl alcohol
[20739-58-6]

$H_3C(CH_2)_4$C≡CCH_2OH

$C_8H_{14}O$ M 126.198
Liq. d_4^{17} 0.884. Bp_8 90°. n_D^{17} 1.4585.
Ac: [16491-64-8].
$C_{10}H_{16}O_2$ M 168.235
Bp_6 79-81°.

Gelin, R. *et al*, *Bull. Soc. Chim. Fr.*, 1971, 4146 (*synth, spectra*)
Sood, R. *et al*, *Tetrahedron Lett.*, 1974, 423 (*synth*)
Egmond, G.J.N. *et al*, *Recl. Trav. Chim. Pays-Bas* (*J. R. Neth. Chem. Soc.*), 1977, **96**, 172 (*synth*)
Rickards, G. *et al*, *J. Org. Chem.*, 1978, **43**, 3607 (*synth, spectra*)
Porter, N.A. *et al*, *J. Org. Chem.*, 1985, **50**, 2252 (*synth, pmr*)
Kanakan, C.C. *et al*, *J. Chem. Soc., Perkin Trans. 1*, 1989, 1907 (*synth, ir, pmr*)
Zhengming, L. *et al*, *Synth. Commun.*, 1989, **19**, 91 (*synth, pmr*)
Crombie, L. *et al*, *J. Chem. Soc., Perkin Trans. 1*, 1991, 567 (*synth, ir, pmr*)

3-Octyn-1-ol, 9CI — O-00283

[14916-80-4]

$H_3CCH_2CH_2CH_2$C≡CCH_2CH_2OH

$C_8H_{14}O$ M 126.198
Oil. Bp_6 86°.

Schlosser, M. *et al*, *Helv. Chim. Acta*, 1973, **56**, 2166 (*synth, pmr*)
Flahaut, J. *et al*, *Helv. Chim. Acta*, 1978, **61**, 2275 (*synth*)
Mitra, R.B. *et al*, *Synthesis*, 1989, 694 (*synth, ir, pmr*)

Ohchinin — O-00284

[67023-80-7]

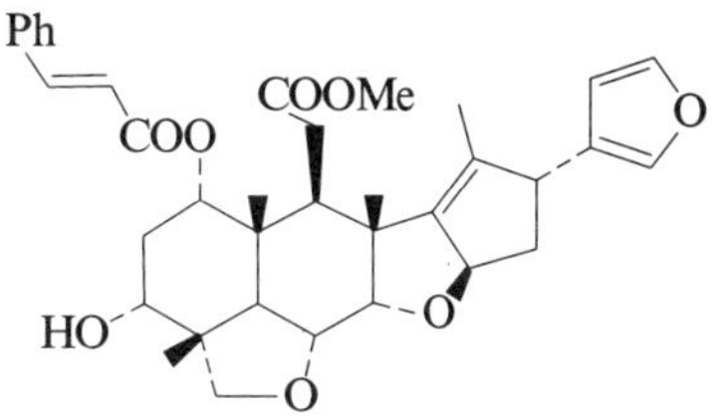

$C_{36}H_{42}O_8$ M 602.723
Constit. of *Melia azedarach*. Cryst. (MeOH). Mp 184-185°. $[\alpha]_D^{23}$ +64° (c, 0.17 in EtOH).
Ac: ***Ohchinin acetate***
$C_{38}H_{44}O_9$ M 644.760
Constit. of *M. azedarach*. Cryst. (Et_2O/CH_2Cl_2). Mp 223-226°. $[\alpha]_D^{23}$ +227° (c, 0.23 in EtOH).
3-O-Tigloyl:
$C_{41}H_{48}O_9$ M 684.825
Isol. from *M. azedarach* fruits.

Fukuyama, Y. *et al*, *Bull. Chem. Soc. Jpn.*, 1983, **56**, 1139.
Kraus, W., *Stud. Org. Chem. (Amsterdam)*, 1986, **26**, 237 (*tigloyl*)

12-Oleanen-3-ol — O-00285

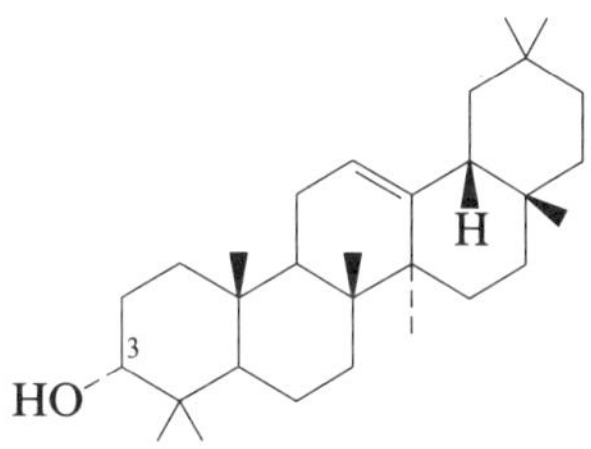

3α-*form*

$C_{30}H_{50}O$ M 426.724
3α-form [6811-63-8] ***Pulcherrol.*** *3-Epi-β-amyrin. 3-epi-β-Amyrin*
Constit. of *Euphorbia pulcherrima*, *Excoecaria agallocha* and *Neopringlea integrifolia*. Cryst. Mp 228°.
Ac: [33055-28-6]. ***Pulcherryl acetate***
$C_{32}H_{52}O_2$ M 468.762
Isol. from *Eupatorium havanense*. Mp 125-126°. $[\alpha]_D$ +70°.
3β-form [559-70-6] ***β-Amyrin.*** *β-Amyrenol. α-Viscol*
Occurs in many plants. Mp 197-197.5°. $[\alpha]_D^{19}$ +88.4° ($CHCl_3$).
Ac: [1616-93-9].
$C_{32}H_{52}O_2$ M 468.762
Constit. of *Alstonia boonei*, *Vernonia cinerea* and many other plants. Cryst. (pet. ether). Mp 244-245° (238-239°). $[\alpha]_D^{17}$ +81.4° ($CHCl_3$).
O-*Hexadecanoyl:* [5973-06-8]. ***Palmitoyl-β-amyrin***
$C_{46}H_{80}O_2$ M 665.137
Constit. of many plants. Cryst. ($CHCl_3$/MeOH). Mp 75-76°. $[\alpha]_D^{20}$ +67.9° ($CHCl_3$).
Me ether: [14021-26-2]. *3β-Methoxy-12-oleanene.* ***Isosawamilletin***
$C_{31}H_{52}O$ M 440.751
Constit. of *Cortaderia toetoe*, *Chionchloa*, *Arundo conspicua* and others. Cryst. Mp 247-248°. $[\alpha]_D$ −92° (c, 4.8 in $CHCl_3$).
3-Ketone: [638-97-1]. *12-Oleanen-3-one.* ***β-Amyrone.*** *β-Amyrenone. Pulcherrone*
$C_{30}H_{48}O$ M 424.709

Constit. of *Mimusops globosa*, *Ilex*, *Quercus* and some other plants. Cryst. (MeOH). Mp 177-179°. $[\alpha]_D$ +107.2° ($CHCl_3$).

O-*Cinnamoyl:* [21593-81-7].
$C_{39}H_{56}O_2$ M 556.870
Isol. from *Cnidoscolus urens*. Cryst. (pet. ether). Mp 240-242°. $[\alpha]_D$ +66.2°.

Benzoyl: [3607-93-0].
$C_{37}H_{54}O_2$ M 530.832
Isol. from *Lippia citriodora* and *Calotropis gigantea*. Cryst. (EtOAc). Mp 234-235°. $[\alpha]_D$ +98°.

O-*Decanoyl:* [64142-76-3].
$C_{40}H_{68}O_2$ M 580.976
Isol. from members of the Sapotaceae. Cryst. (MeOH). Mp 103-104°. $[\alpha]_D^{27}$ +37.9°.

O-(*9Z-Hexadecenoyl*)*:* [65118-59-4].
$C_{46}H_{78}O_2$ M 663.121
Isol. from *Wedelia rugosa*.

O-*Hexanoyl:* [64166-12-7]. ***β-Amyrin caproate***
$C_{36}H_{60}O_2$ M 524.869
Isol. from *Mimusops littoralis*. Fine needles (EtOH). Mp 158-160°.

O-*Octadecanoyl:* [62736-38-3]. ***β-Amyrin stearate***
$C_{48}H_{84}O_2$ M 693.190
Isol. from *Balanophora indica*. Mp 52-54°. $[\alpha]_D$ +62° ($CHCl_3$).

O-*Eicosanoyl:* [35286-33-0].
$C_{50}H_{88}O_2$ M 721.244
Isol. from *Leuconotis eugenifolia*. Mp 79-83°.

O-(*9Z,12Z-Octadecadienoyl*)*:* [57403-22-2].
$C_{48}H_{80}O_2$ M 689.159
Isol. from *Spilanthes acmella*.

O-*Docosanoyl:* [35241-37-3].
$C_{52}H_{92}O_2$ M 749.298
Isol. from *L. eugenifolia*.

O-*Dodecanoyl:* [55803-92-4].
$C_{42}H_{72}O_2$ M 609.030
Isol. from *S. acmella*.

O-(*9Z,12Z,15Z-Octadecatrienoyl*)*:* [57403-24-4].
$C_{48}H_{78}O_2$ M 687.143
From *S. acmella*.

O-(*5-Phenyl-2,4-pentadienoyl*)*:* [54303-80-9].
$C_{41}H_{58}O_2$ M 582.908
Isol. from *Marsdenia pringlei*.

O-*Tetradecanoyl:* [57403-20-0].
$C_{44}H_{76}O_2$ M 637.083
Isol. from *S. acmella*.

(3β,18α)-form [109838-28-0]
Reported from *Pernettya poeppigii*.

Ac: [64200-19-7].
Reported from *Vernonia* spp. and *Podachaenium eminens*. No details given other than struct. diagram. No comment is made on the abnormal stereochem.

[32164-06-0, 52690-48-9]

Savoir, R. *et al*, *Bull. Soc. Chim. Belg.*, 1967, **76**, 335 (*pmr*)
Corey, E.J. *et al*, *J. Am. Chem. Soc.*, 1967, **89**, 3362 (*biosynth*)
Barton, D.H.R. *et al*, *J. Chem. Soc. C*, 1968, 1031 (*synth*)
Lawrie, W. *et al*, *Phytochemistry*, 1970, **9**, 1669 (*isol*)
Dominguez, X.A. *et al*, *Rev. Latinoam. Quim.*, 1970, **1**, 68 (*isol*)
Buckley, D.G. *et al*, *Chem. Ind.* (*London*), 1971, 298 (*pmr*)
Kim, K.Y. *et al*, *Phytochemistry*, 1971, **10**, 2803 (*derivs*)
Kawashima, T. *et al*, *Phytochemistry*, 1971, **10**, 3308 (*isol*)
van Tamelen, E.E. *et al*, *J. Am. Chem. Soc.*, 1972, **94**, 8229 (*synth*)
Karrer, W. *et al*, *Konstitution und Vorkommen der Organischen Pflanzenstoffe*, 2nd Ed., Birkhäuser Verlag, Basel, 1972-1985, no. 1967 (*occur*)
Shinga, T. *et al*, *Chem. Pharm. Bull.*, 1973, **21**, 2252 (*pmr*)
Dominguez, X.A. *et al*, *Phytochemistry*, 1973, **12**, 2060 (*Pulcherryl acetate*)
Boar, R.B. *et al*, *Phytochemistry*, 1973, **12**, 2571 (*rev*)
Barton, D.H.R. *et al*, *J. Chem. Soc., Chem. Commun.*, 1974, 861 (*biosynth*)
Knight, S.A., *Org. Magn. Reson.*, 1974, **6**, 603 (*cmr*)
Dominguez, X.A. *et al*, *Phytochemistry*, 1974, **13**, 2617 (*derivs*)
Doddrell, D.M. *et al*, *Tetrahedron Lett.*, 1974, 2381 (*cmr*)
Atallah, A.M. *et al*, *Phytochemistry*, 1975, **14**, 1529 (*biosynth*)
Krishnaswamy, N.R. *et al*, *Phytochemistry*, 1975, **14**, 1666 (*derivs*)
Narayanan, C.R. *et al*, *Indian J. Chem., Sect. B*, 1976, **14**, 906 (*stearate*)
Gunasekera, S.P. *et al*, *Phytochemistry*, 1977, **16**, 923 (*decanoate*)
Bohlmann, F. *et al*, *Phytochemistry*, 1977, **16**, 1304 (*18α-acetate*)
Aleman, E. *et al*, *CA*, 1978, **88**, 3076g (*isol, esters*)
Rao, C. *et al*, *Curr. Sci.*, 1979, **48**, 534 (*benzoate*)
Banerji, R. *et al*, *Indian J. Chem., Sect. B*, 1979, **18**, 293 (*caproate*)
Reyes, J. *et al*, *Rev. Latinoam. Quim.*, 1983, **14**, 67 (*cryst struct*)
Pradhan, B.P. *et al*, *Tetrahedron*, 1985, **41**, 2513 (*synth*)
Battacharyya, J. *et al*, *Phytochemistry*, 1986, **25**, 274 (*Cinnamoyl, cmr*)
Martinez, J.R. *et al*, *Rev. Latinoam. Quim.*, 1987, **18**, 69 (*3β,18α-form*)
Yan, X.-Z. *et al*, *Phytochemistry*, 1989, **28**, 1541 (*cryst struct*)
Corey, E.J. *et al*, *J. Am. Chem. Soc.*, 1993, **115**, 8871 (*synth*)

13(18)-Oleanen-3-ol O-00286

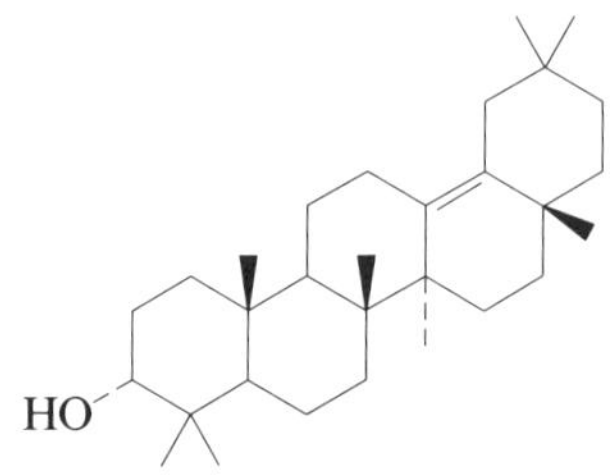

3α-*form*

$C_{30}H_{50}O$ M 426.724

3α-form [52647-56-0] ***Gymnorhizol***
Isol. from leaves of *Bruguiera gymnorhiza*. Mp 208-210°. $[\alpha]_D$ −48°.

3-Ketone: [20248-08-2]. *13(18)-Oleanen-3-one*. ***δ-Amyrone***. *Lupenone II*. *δ-Amyrenone*
$C_{30}H_{48}O$ M 424.709
Constit. of *Alnus* spp., *Euphorbia* spp., *Cistus quadrangularis* and others. Cryst. (EtOH). Mp 203° (198-201°). $[\alpha]_D^{20}$ −9.7° ($CHCl_3$).

3β-form [508-04-3] ***δ-Amyrin***
Constit. of *Spartium junceum* and *C. quadrangularis*. Cryst. (MeOH). Mp 212-213.5°. $[\alpha]_D^{20}$ −50.5° ($CHCl_3$).

3-O-Formyl: ***δ-Amyrin formate***
$C_{31}H_{50}O_2$ M 454.735
Constit. of *E. supina*. Needles. Mp 254-256°. $[\alpha]_D^{23}$ +15.2° (c, 0.85 in $CHCl_3$).

Ac: [51361-60-5].
$C_{32}H_{52}O_2$ M 468.762
Isol. from roots of *Vernonia cinerea* and *Scaevola* spp. Cryst. (MeOH/$CHCl_3$). Mp 208°. $[\alpha]_D$ −21° (c, 1 in $CHCl_3$).

[39776-31-3]

Musgrave, O.C. *et al*, *J. Chem. Soc.*, 1952, 4393 (*isol*)
Brownlie, G. *et al*, *J. Chem. Soc.*, 1956, 1377 (*synth*)
Fischer, F.G. *et al*, *Justus Liebigs Ann. Chem.*, 1961, **644**, 162 (*isol*)
van Tamelen, E.E. *et al*, *J. Am. Chem. Soc.*, 1972, **94**, 8229 (*synth*)
Wollenweber, E., *Z. Naturforsch., C*, 1974, **29**, 362 (*isol*)
Bohlmann, F. *et al*, *Chem. Ber.*, 1975, **108**, 433 (*δ-Amyrin acetate*)
Sarkar, A. *et al*, *Indian J. Chem., Sect. B*, 1978, **16**, 742 (*Gymnorhizol*)
Singh, R.S. *et al*, *J. Nat. Prod.* (*Lloydia*), 1984, **47**, 368 (*δ-Amyrin acetate*)
Bhutani, K.K. *et al*, *Phytochemistry*, 1984, **23**, 407 (*isol*)

3-Oxobutanoic acid, 9CI O-00287

Acetoacetic acid. Acetonecarboxylic acid. Acetylacetic acid
[541-50-4]

$$H_3CCOCH_2COOH \rightleftharpoons H_3CC(OH){=}CHCOOH$$

$C_4H_6O_3$ M 102.090
Syrup or cryst. Misc. H_2O. Mp 36-37°. Bp <100° dec. pK_{a1} 3.6 (25°). Strong but unstable acid. On warming loses $CO_2 \rightarrow Me_2CO$.
▷ Irritant.
Li salt: [3483-11-2].
Cryst. (MeOH/Et_2O).
Me ester: [105-45-3]. *Methyl acetoacetate*
$C_5H_8O_3$ M 116.116
Liq. Misc. H_2O. Bp_{20} 169-170°. pK_{a1} 10.72 (21°). *Cis-* and *trans-*enol forms detected spectroscopically.
▷ Mod. toxic. AK5775000.
tert-*Butyl ester:* [1694-31-1]. tert-*Butyl acetoacetate*
$C_8H_{14}O_3$ M 158.197
Reagent for acyloin and $\alpha\beta$-unsatd. ketone synth. Bp_{20} 85°.
Fluoride: [2343-88-6].
$C_4H_5FO_2$ M 104.081
Acetoacetylating agent for alcohols and amines. Bp 132-134°. Dec. slowly at r.t., can be stored at 0°.
Amide: [5977-14-0]. *3-Oxobutanamide, 9CI. Acetoacetamide*
$C_4H_7NO_2$ M 101.105
Reagent for synth. of heterocycles. Cryst. (Me_2CO/pet. ether). Sol. H_2O, EtOH, AcOH. Mp 54°. Dec. on attempted dist.
Methylamide:
$C_5H_9NO_2$ M 115.132
Liq.
Nitrile: [2469-99-0]. *3-Oxobutanenitrile, 9CI. Cyanoacetone. Acetoacetonitrile*
C_4H_5NO M 83.090
Liq. Bp >230° dec. Polymerises.
Phenylhydrazone: Mp 128°.

Enol-form
3-Hydroxy-2-butenoic acid. 3-Hydroxycrotonic acid
Ac: [38550-63-9].
$C_6H_8O_4$ M 144.127
Bp 212-214°.
Me ether: [67107-76-0]. *3-Methoxy-2-butenoic acid*
$C_5H_8O_3$ M 116.116
Mp 129-130°.
Me ether, Me ester: [35217-21-1].
$C_6H_{10}O_3$ M 130.143
Bp 175.8°.
Et ether: [38624-58-7].
$C_6H_{10}O_3$ M 130.143
Mp 141°.

Krampitz, L.O., *Arch. Biochem.*, 1948, **17**, 82 (*synth*)
Olah, G.A. *et al*, *J. Org. Chem.*, 1961, **26**, 225 (*fluoride*)
Biochem. Prep., 1963, **10**, 1 (*synth*)
Fieser and Fieser's Reagents for Organic Synthesis, Wiley, 1967, **1**, 4, 83.
Matusch, R., *Angew. Chem., Int. Ed. Engl.*, 1975, **14**, 260 (*tautom*)
Kato, T. *et al*, *Heterocycles*, 1975, **3**, 413 (*amide, rev*)
DeShong, P. *et al*, *J. Org. Chem.*, 1988, **53**, 1356 (*deriv, synth, ir, pmr, ms*)
Lewis, R.J., *Sax's Dangerous Properties of Industrial Materials*, 8th Ed., Van Nostrand-Reinhold, 1992, MFX250.

5-Oxo-1-cyclopentene-1-heptanoic acid, 9CI O-00288

2-(6-Carboxyhexyl)-2-cyclopenten-1-one
[5239-43-0]

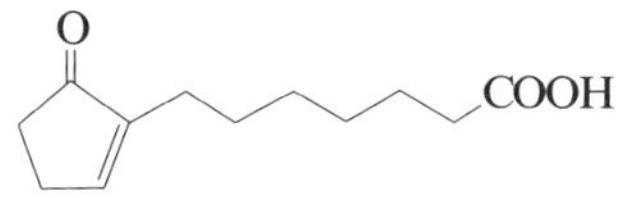

$C_{12}H_{18}O_3$ M 210.272
Me ester: [34546-57-1].
$C_{13}H_{20}O_3$ M 224.299
Prostanoid synthon. Liq. Bp_1 152°.

Valcavi, U. *et al*, *Synthesis*, 1983, 124 (*synth, bibl*)

10-Oxo-5,8-decadienoic acid O-00289

9-Formyl-5,8-nonadienoic acid

$$OHCCH{=}CHCH_2CH{=}CH(CH_2)_3COOH$$

$C_{10}H_{14}O_3$ M 182.219
Et ester, di-Et acetal: [81585-93-5].
$C_{16}H_{28}O_4$ M 284.395
Oil.

Russell, S.W. *et al*, *J. Chem. Soc., Perkin Trans. 1*, 1982, 545 (*synth, pmr*)

2-Oxodecanoic acid, 9CI O-00290

2-Oxocapric acid
[333-60-8]

$$H_3C(CH_2)_7COCOOH$$

$C_{10}H_{18}O_3$ M 186.250
Mp 48-49°. Bp_6 118-120°.
2,4-Dinitrophenylhydrazone: [18281-52-2].
Yellow cryst. Mp 134°.
Oxime:
$C_{10}H_{19}NO_3$ M 201.265
Mp 85-86°.

Adickes, F. *et al*, *Justus Liebigs Ann. Chem.*, 1943, **555**, 41 (*synth*)
Schreiber, J., *Bull. Soc. Chim. Fr.*, 1956, 1361 (*synth*)
Lapkin, H. *et al*, *CA*, 1962, **57**, 11010 (*synth*)
Jain, S. *et al*, *Acta Crystallogr.*, 1966, **21**, 445 (*cryst struct*)
Reinheckel, H., *Monatsh. Chem.*, 1967, **98**, 1217 (*synth*)

3-Oxodecanoic acid, 9CI O-00291

3-Oxocapric acid. Octanoylacetic acid
[13283-92-6]

$$H_3C(CH_2)_6COCH_2COOH$$

$C_{10}H_{18}O_3$ M 186.250
Leaflets. Mp 83-84°. pK_a 5.22.
Me ester: [22348-96-5].
$C_{11}H_{20}O_3$ M 200.277
$Bp_{2.8}$ 107-109°.

Breusch, F.L. *et al*, *CA*, 1946, **40**, 5400 (*synth*)
Mitz, M.A. *et al*, *J. Am. Chem. Soc.*, 1950, **72**, 1231 (*synth*)
Stenhagen, E., *Ark. Kemi*, 1951, **3**, 381; *CA*, **46**, 7050 (*synth*)
Epstein, J. *et al*, *J. Org. Chem.*, 1977, **42**, 759 (*synth*)

4-Oxodecanoic acid, 9CI O-00292

4-Oxocapric acid
[4144-54-1]

$$H_3C(CH_2)_5COCH_2CH_2COOH$$

$C_{10}H_{18}O_3$ M 186.250

Cryst. (EtOH aq.). Mp 69-70°.

Patrick, T.M., *J. Org. Chem.*, 1952, **17**, 1009 (*synth*)
Org. Synth., Coll. Vol., 4, 1963, 430 (*synth*)
Bohlmann, F. *et al*, *Chem. Ber.*, 1961, **94**, 3193 (*synth*)
Reinheckel, H., *Angew. Chem., Int. Ed. Engl.*, 1965, **4**, 785 (*synth*)
Takeda, A. *et al*, *J. Org. Chem.*, 1966, **31**, 616; *Bull. Chem. Soc. Jpn.*, 1971, **44**, 1342 (*synth*)
Wineburg, J.P. *et al*, *J. Heterocycl. Chem.*, 1975, **12**, 749 (*synth*)

5-Oxodecanoic acid, 9CI — O-00293

5-Oxocapric acid
[624-01-1]

$H_3C(CH_2)_4CO(CH_2)_3COOH$

$C_{10}H_{18}O_3$ M 186.250
Mp 58°. $Bp_{0.03}$ 125°.
Me ester: [6093-95-4].
$C_{11}H_{20}O_3$ M 200.277
Mp *ca.* 5°. Bp_5 116-121°.
Semicarbazone: Mp 126°.

Rosenmund, K.W. *et al*, *Chem. Ber.*, 1961, **94**, 2401 (*synth*)
Lardelli, G. *et al*, *Recl. Trav. Chim. Pays-Bas* (*J. R. Neth. Chem. Soc.*), 1967, **86**, 481 (*synth*)
Nobuhara, A. *et al*, *Agric. Biol. Chem.*, 1969, **33**, 1264 (*synth*)
Ijima, A. *et al*, *Chem. Pharm. Bull.*, 1973, **21**, 215 (*synth*)

6-Oxodecanoic acid, 9CI — O-00294

6-Oxocapric acid
[4144-60-9]

$H_3C(CH_2)_3CO(CH_2)_4COOH$

$C_{10}H_{18}O_3$ M 186.250
Cryst. (pet. ether). Mp 45-46°.
Et ester: [4144-61-0].
$C_{12}H_{22}O_3$ M 214.304
Bp_{15} 151°.
Semicarbazone: Cryst. (MeOH). Mp 159°.

Cason, J. *et al*, *J. Org. Chem.*, 1963, **28**, 552 (*synth*)
Campbell, I.D. *et al*, *J. Chem. Soc.*, 1964, 1092 (*synth*)
Graf, E. *et al*, *Chem. Ber.*, 1964, **97**, 2785 (*synth*)
Lamant, M. *et al*, *Bull. Soc. Chim. Fr.*, 1965, 3041 (*synth*)

7-Oxodecanoic acid, 9CI — O-00295

7-Ketocapric acid. 6-Butanoylhexanoic acid
[35383-65-4]

$H_3CCH_2CH_2CO(CH_2)_5COOH$

$C_{10}H_{18}O_3$ M 186.250
Mp 42°. $Bp_{2.5}$ 157-158°.
Semicarbazone: Mp 158-159°.
Me ester:
$C_{11}H_{20}O_3$ M 200.277
Bp_{18} 106°.
Me ester, 2,4-dinitrophenylhydrazone: Mp 121-122°.

Manyik, R.M. *et al*, *J. Am. Chem. Soc.*, 1953, **75**, 5030 (*synth*)
Keskin, H., *CA*, 1963, **59**, 15173 (*synth*)
U.S. Pat., 3 197 488, (1965); *CA*, **63**, 13083 (*synth*)
Gol'dfarb, Y.L. *et al*, *CA*, 1972, **76**, 139896 (*synth*)
Tahara, S. *et al*, *Agric. Biol. Chem.*, 1980, **44**, 193 (*synth, ester*)

8-Oxodecanoic acid, 9CI — O-00296

8-Ketocapric acid. 7-Propionylheptanoic acid
[3006-51-7]

$H_3CCH_2CO(CH_2)_6COOH$

$C_{10}H_{18}O_3$ M 186.250
Mp 64° (59.4-60.9°).
Et ester:
$C_{12}H_{22}O_3$ M 214.304
Bp_{15} 157°.
Semicarbazone: Cryst. (AcOH). Mp 184°.

Blaise, E.E. *et al*, *Bull. Soc. Chim. Fr.*, 1910, **7**, 225 (*synth*)
Brockman, J.A. *et al*, *J. Am. Chem. Soc.*, 1957, **79**, 5027 (*synth, ester*)
Breusch, F.L. *et al*, *CA*, 1964, **60**, 9139 (*synth*)
Odinokov, V.N. *et al*, *Zh. Org. Khim.*, 1980, **16**, 524; *CA*, **93**, 94515 (*synth, pmr, ms*)

9-Oxodecanoic acid, 9CI — O-00297

9-Oxocapric acid
[1422-26-0]

$H_3CCO(CH_2)_7COOH$

$C_{10}H_{18}O_3$ M 186.250
Intermed. for prostaglandin synth. Mp 48°.
Me ester: [2575-07-7].
$C_{11}H_{20}O_3$ M 200.277
Bp_3 98-100°.
Semicarbazone: Cryst. (MeOH aq.). Mp 115-116°.

Cason, J. *et al*, *J. Org. Chem.*, 1958, **23**, 1497 (*synth*)
Barbier, M. *et al*, *Bull. Soc. Chim. Fr.*, 1961, 951 (*synth*)
Omura, S. *et al*, *Chem. Pharm. Bull.*, 1976, **24**, 3139 (*synth, cmr*)
Subramaniam, C.S. *et al*, *Synthesis*, 1978, 468 (*synth*)
Hase, T.A. *et al*, *Synth. Commun.*, 1979, **9**, 63.
Karim, M.R. *et al*, *J. Org. Chem.*, 1990, **55**, 598 (*synth, pmr, cmr*)

10-Oxodecanoic acid, 9CI — O-00298

9-Formylnonanoic acid. 9-Aldehydononanoic acid. Sebacic semialdehyde. 9-Formylpelargonic acid
[5578-80-3]

$OHC(CH_2)_8COOH$

$C_{10}H_{18}O_3$ M 186.250
Glistening plates (Me_2CO/pet. ether). Mp 56-57°. $Bp_{0.1}$ 150°.
Semicarbazone: Nodules (EtOH). Mp 170°.
Oxime:
$C_{10}H_{19}NO_3$ M 201.265
Prisms (MeOH). Mp 111°.
Me ester: [14811-73-5].
$C_{11}H_{20}O_3$ M 200.277
Pale-yellow oil. Bp_2 119-121°, $Bp_{0.6}$ 108-110°.
Me ester, semicarbazone: Mp 98-100°.
Me ester, 2,4-Dinitrophenylhydrazone: Orange prisms (MeOH). Mp 86-87° (83.8-84.6°).
Me ester, Di-Me acetal: [65157-90-6].
$C_{13}H_{26}O_4$ M 246.346
$Bp_{0.25}$ 137-138°.

Noller, C.R. *et al*, *J. Am. Chem. Soc.*, 1926, **48**, 1074 (*deriv, synth*)
Hargreaves, G.H. *et al*, *J. Chem. Soc.*, 1947, 753 (*synth*)
Kraevskii, A.A. *et al*, *Zh. Obshch. Khim.*, 1964, **34**, 1180.
Gokhale, P.D. *et al*, *Synthesis*, 1974, 718 (*synth*)
Adlof, R.O. *et al*, *J. Am. Oil Chem. Soc.*, 1977, **54**, 414 (*synth*)
Cameron, A.G. *et al*, *J. Chem. Soc., Perkin Trans. 1*, 1986, 161 (*deriv, synth, pmr*)

9-Oxo-2,4,5,7-decatetraenoic acid, 9CI — O-00299

Antibiotic LA 1. *LA 1*
[61947-93-1]

$H_3CCOCH{=}CHCH{=}C{=}CHCH{=}CHCOOH$

$C_{10}H_{10}O_3$ M 178.187

Isol. from culture broth of *Streptomyces kitasatoensis*. Antibiotic active against gram-positive organisms. Unstable solid. $[\alpha]_D^{20}$ +50° ($CHCl_3$). λ_{max} 270 nm (MeOH).

Omura, S. *et al*, *Chem. Pharm. Bull.*, 1976, **24**, 3139.

5-Oxo-7-decenoic acid O-00300

$H_3CCH_2CH{=}CHCH_2CO(CH_2)_3COOH$

$C_{10}H_{16}O_3$ M 184.235

(*E*)-*form* [87086-54-2]
$Bp_{1.5}$ 130°.
[87086-53-1]

Utaka, M. *et al*, *J. Org. Chem.*, 1986, **51**, 935 (*synth, pmr*)

9-Oxo-2-decenoic acid, 9CI O-00301

[2575-01-1]

$H_3CCO(CH_2)_5CH{=}CHCOOH$

$C_{10}H_{16}O_3$ M 184.235

(*E*)-*form* [334-20-3]
Queen substance. Queen bee pheromone
Queen substance of honey bee *Apis mellifera*. Reputed to have medicinal benefits. Cryst. (MeOH aq. or Me_2CO). Mp 54-55.5°. $Bp_{0.001}$ 37-39°.
Me ester: [1189-64-6].
$C_{11}H_{18}O_3$ M 198.261
$Bp_{0.6}$ 107.5-108°.
2,4-Dinitrophenylhydrazone: Mp 126-130°.
Semicarbazone: Cryst. (MeOH aq.). Mp 165-168°.

(*Z*)-*form* [29282-32-4]
Oil. $Bp_{0.2}$ 110-150°.

Van der Plas, H.C. *et al*, *Recl. Trav. Chim. Pays-Bas* (*J. R. Neth. Chem. Soc.*), 1964, **83**, 701 (*synth, bibl*)
Cromer, D.T. *et al*, *Acta Crystallogr., Sect. B*, 1972, **28**, 2128 (*cryst struct*)
Boch, R. *et al*, *J. Chem. Ecol.*, 1975, **1**, 133 (*field trials*)
Trost, B.M. *et al*, *J. Org. Chem.*, 1975, **40**, 148 (*synth*)
Tsuji, J. *et al*, *Tetrahedron Lett.*, 1977, 2267 (*synth*)
Subramaniam, C.S. *et al*, *Indian J. Chem., Sect. B*, 1978, **16**, 318 (*synth, ir, pmr*)
Lombardi, L. *et al*, *Synthesis*, 1978, 131.
Hase, T.A. *et al*, *Synth. Commun.*, 1979, **9**, 63 (*synth*)
Tamaru, Y. *et al*, *Tetrahedron*, 1979, **35**, 329 (*synth, ir, pmr*)
Villemin, D., *Chem. Ind. (London)*, 1986, 69 (*synth*)
Dhokte, V.P. *et al*, *Synth. Commun.*, 1987, **17**, 355 (*synth*)
Bestmann, H.J. *et al*, *Synthesis*, 1988, 49 (*synth, ir, pmr, ms*)

5-Oxo-7-decynoic acid O-00302

[72666-29-6]

$H_3CCH_2C{\equiv}CCH_2CO(CH_2)_3COOH$

$C_{10}H_{14}O_3$ M 182.219
Cryst. Mp 50.0-50.5°.

Utaka, M. *et al*, *J. Org. Chem.*, 1986, **51**, 935 (*synth, ir, pmr, cmr*)

4-Oxodocosanoic acid O-00303

4-Oxobehenic acid

$H_3C(CH_2)_{17}COCH_2CH_2COOH$

$C_{22}H_{42}O_3$ M 354.572
Cryst. (AcOH). Mp 103°.
Me ester:
$C_{23}H_{44}O_3$ M 368.599
Mp 62.6°.

Shukowski, G. *et al*, *Chem. Zentralbl.*, 1908, **11**, 1415.
Stenhagen, S.S. *et al*, *CA*, 1947, **41**, 4105 (*synth*)

10-Oxodocosanoic acid O-00304

10-Oxobehenic acid

$H_3C(CH_2)_{11}CO(CH_2)_8COOH$

$C_{22}H_{42}O_3$ M 354.572
Cryst. (MeOH). Mp 94°.
Amide:
$C_{22}H_{43}NO_2$ M 353.587
Cryst. (MeOH). Mp 99°.

Robinson, G.M. *et al*, *J. Chem. Soc.*, 1926, 2204 (*synth*)

9-Oxododecanoic acid O-00305

9-Oxolauric acid. 9-Ketolauric acid

$H_3CCH_2CH_2CO(CH_2)_7COOH$

$C_{12}H_{22}O_3$ M 214.304
Cryst. Mp 50°.
Semicarbazone: Cryst. Mp 131°.

Asano, M., *CA*, 1924, **18**, 1645 (*synth*)

12-Oxododecanoic acid, 9CI O-00306

11-Formylundecanoic acid
[3956-80-7]

$OHC(CH_2)_{10}COOH$

$C_{12}H_{22}O_3$ M 214.304
Mp 65-66°. $Bp_{0.05}$ 151-153°.
Me ester: [2009-59-8].
$C_{13}H_{24}O_3$ M 228.331
Pale-yellow oil, solidifies on refrigeration. Bp_5 154°, $Bp_{0.1}$ 109-110°.
Me ester, Di-Me acetal: [1931-67-5].
$C_{15}H_{30}O_4$ M 274.400
Liq. $Bp_{0.05}$ 129-130°.

Tomecko, C.G. *et al*, *J. Am. Chem. Soc.*, 1927, **49**, 522 (*deriv, synth*)
Adlof, R.O. *et al*, *J. Am. Oil Chem. Soc.*, 1977, **54**, 414 (*synth*)
Cameron, A.G. *et al*, *J. Chem. Soc., Perkin Trans. 1*, 1986, 161 (*deriv, synth, pmr*)

12-Oxo-5,8,10-dodecatrienoic acid, 9CI O-00307

12-Formyl-5,8,10-undecatrienoic acid

$OHCCH{=}CHCH{=}CHCH_2CH{=}CH(CH_2)_3COOH$

$C_{12}H_{16}O_3$ M 208.257

(*5Z,7E,8E*)-*form* [81892-90-2]
Antimicrobial component of the red alga *Laurencia hybrida*.

Higgs, M.D., *Tetrahedron*, 1981, **24**, 4255 (*isol, pmr, ms*)

12-Oxo-10-dodecenoic acid O-00308

11-Formyl-10-undecenoic acid

$OHCCH{=}CH(CH_2)_8COOH$

$C_{12}H_{20}O_3$ M 212.288

(*E*)-*form* [65410-38-0]
Traumatin. Traumatic half aldehyde
Arises in plants by cleavage of unsatd. C_{18} fatty acids. Found for example in *Thea sinensis* chloroplasts. Wound hormone. Solid (cyclohexane/Et_2O). Mp 65-66°.

Me ester: [63024-87-3].
Amber liq. Mp 0.7°. $Bp_{0.05}$ 109-111°.

(Z)-form

Me ester:
$C_{13}H_{22}O_3$ M 226.315
$Bp_{0.23}$ 126-131°.

Kajiwara, T. *et al*, *Agric. Biol. Chem.*, 1977, **41**, 8 (*synth, bibl*)
Hatanaka, A. *et al*, *Phytochemistry*, 1977, **16**, 1828 (*isol*)
Bestmann, H.J. *et al*, *Tetrahedron Lett.*, 1977, 121 (*synth, pmr*)
Rakoff, H. *et al*, *Synth. Commun.*, 1979, **9**, 185 (*synth, pmr, ir*)
Bestmann, H.J. *et al*, *Justus Liebigs Ann. Chem.*, 1981, 1705 (*synth, ir, pmr*)
Gaede, B.J., *Org. Prep. Proced. Int.*, 1987, **19**, 461 (*synth, pmr, cmr, bibl*)

19-Oxo-5,8,11,14-eicosatetraenoic acid O-00309

$H_3CCOCH_2CH_2(CH_2CH{=}CH)_4(CH_2)_3COOH$

$C_{20}H_{30}O_3$ M 318.455
Metab. produced during NADPH dependent enzymatic oxidation of arachidonic acid.

Manna, S. *et al*, *Tetrahedron Lett.*, 1983, **24**, 33 (*synth, bibl*)

5-Oxoheneicosanoic acid O-00310

[95007-77-5]

$H_3C(CH_2)_{15}COCH_2CH_2CH_2COOH$

$C_{21}H_{40}O_3$ M 340.545
Constit. of the lichen *Usnea hirta*.

Solberg, Y., *J. Hattori Bot. Lab.*, 1987, **63**, 357; *CA*, **108**, 218962.

9-Oxoheptadecanoic acid O-00311

9-Oxomargaric acid

$H_3C(CH_2)_7CO(CH_2)_7COOH$

$C_{17}H_{32}O_3$ M 284.438
Cryst. ($CHCl_3$/pet. ether). Mp 77.8-78.8°.

Me ester:
$C_{18}H_{34}O_3$ M 298.465
Cryst. (MeOH). Mp 45.5°.

Et ester:
$C_{19}H_{36}O_3$ M 312.492
Cryst. (EtOH aq.). Mp 38°.

Amide:
$C_{17}H_{33}NO_2$ M 283.453
Cryst. ($CHCl_3$/pet. ether). Mp 119°.

Semicarbazone: Cryst. (Me_2CO). Mp 111°.

LeSueur, H.R. *et al*, *J. Chem. Soc.*, 1914, **105**, 2800 (*synth*)
Oezeris, S., *Fette, Seifen, Anstrichm.*, 1961, **63**, 805 (*synth*)

17-Oxo-20-hexacosenoic acid O-00312

$H_3C(CH_2)_4CH{=}CHCH_2CH_2CO(CH_2)_{15}COOH$

$C_{26}H_{48}O_3$ M 408.663

(Z)-form

Constit. of the seed oil of *Cuspidaria pterocarpa*.

Me ester:
$C_{27}H_{50}O_3$ M 422.690
Mp 44-45°.

Smith, C.R. *et al*, *Lipids*, 1966, **1**, 268.

2-Oxohexadecanoic acid O-00313

2-Oxopalmitic acid. 2-Ketopalmitic acid

$H_3C(CH_2)_{13}COCOOH$

$C_{16}H_{30}O_3$ M 270.411
Cryst. (pet. ether). Mp 65-66°.

Me ester: [55836-30-1].
$C_{17}H_{32}O_3$ M 284.438
Mp 47°.

Oxime:
$C_{16}H_{31}NO_3$ M 285.426
Mp 81-82°.

Kuwata, T., *J. Am. Chem. Soc.*, 1938, **60**, 559.
Reinheckel, H., *Monatsh. Chem.*, 1967, **98**, 1217.

3-Oxohexadecanoic acid O-00314

3-Oxopalmitic acid. 3-Ketopalmitic acid

[928-19-8]

$H_3C(CH_2)_{12}COCH_2COOH$

$C_{16}H_{30}O_3$ M 270.411
Mp 98°.

Me ester: [14427-53-3].
$C_{17}H_{32}O_3$ M 284.438
Mp 40.1°.

Et ester: [62148-72-5].
$C_{18}H_{34}O_3$ M 298.465
Bp_3 165-170°.

Asano, M. *et al*, *Chem. Ber.*, 1939, **72**, 35.
Franzke, C. *et al*, *Fette, Seifen, Anstrichm.*, 1968, **70**, 638.

4-Oxohexadecanoic acid O-00315

4-Oxopalmitic acid. 4-Ketopalmitic acid

[4144-56-3]

$H_3C(CH_2)_{11}CO(CH_2)_2COOH$

$C_{16}H_{30}O_3$ M 270.411
Cryst. (pet. ether). Mp 91-92°.

Oxime:
$C_{16}H_{31}NO_3$ M 285.426
Cryst. (pet. ether). Mp 54°.

Robinson, G.M. *et al*, *J. Chem. Soc.*, 1925, **127**, 175 (*synth*)
Houston, D.F., *J. Am. Chem. Soc.*, 1947, **69**, 517.
Ahmad, M.U. *et al*, *J. Am. Oil Chem. Soc.*, 1978, **55**, 491 (*synth*)

5-Oxohexadecanoic acid O-00316

5-Oxopalmitic acid. 5-Ketopalmitic acid

$H_3C(CH_2)_{10}CO(CH_2)_3COOH$

$C_{16}H_{30}O_3$ M 270.411
Trace constit. of milk fat. Cryst. (MeOH, C_6H_6 or pet. ether). Mp 88°.

Robinson, G.M., *J. Chem. Soc.*, 1930, 745 (*synth*)
Weihrauch, J.L. *et al*, *Lipids*, 1974, **9**, 883 (*isol*)

7-Oxohexadecanoic acid O-00317

7-Oxopalmitic acid. 7-Ketopalmitic acid

[54527-29-6]

$H_3C(CH_2)_8CO(CH_2)_5COOH$

$C_{16}H_{30}O_3$ M 270.411
Trace constit. of milk fat. Cryst. (MeOH or pet. ether). Mp 78°.

Robinson, G.M., *J. Chem. Soc.*, 1930, 745 (*synth*)
Weihrauch, J.L. *et al*, *Lipids*, 1974, **9**, 883 (*isol*)

8-Oxohexadecanoic acid O-00318

8-Oxopalmitic acid. 8-Ketopalmitic acid

[2777-52-8]

$H_3C(CH_2)_7CO(CH_2)_6COOH$

$C_{16}H_{30}O_3$ M 270.411

Trace constit. of milk fat. Constit. of spore lipids of *Lycopodium* spp. Cryst. (MeOH). Mp 77-78°.

Amide:

$C_{16}H_{31}NO_2$ M 269.426

Mp 58°.

Bodenstein, M., *Ber.*, 1894, **27**, 3397.
Robinson, G.M., *J. Chem. Soc.*, 1930, 745 (*synth*)
Tulloch, A.P., *Can. J. Chem.*, 1965, **43**, 415 (*isol*)
Weihrauch, J.L. *et al*, *Lipids*, 1974, **9**, 883 (*isol*)

9-Oxohexadecanoic acid O-00319

9-Oxopalmitic acid. 9-Ketopalmitic acid

[54527-30-9]

$H_3C(CH_2)_6CO(CH_2)_7COOH$

$C_{16}H_{30}O_3$ M 270.411

Trace constit. of milk fat. Cryst. (EtOAc). Mp 73.5-74.5°.

Davies, L.A. *et al*, *J. Am. Chem. Soc.*, 1928, **50**, 1749 (*synth*)
Weihrauch, J.L. *et al*, *Lipids*, 1974, **9**, 883 (*isol, glc, ms*)

10-Oxohexadecanoic acid O-00320

10-Oxopalmitic acid. 10-Ketopalmitic acid

[818-26-8]

$H_3C(CH_2)_5CO(CH_2)_8COOH$

$C_{16}H_{30}O_3$ M 270.411

Cryst. (EtOAc). Mp 75-75.8°.

Semicarbazone: Cryst. (EtOH). Mp 154-155°.

Fordyce, C.R. *et al*, *J. Am. Chem. Soc.*, 1933, **55**, 3368 (*synth*)

3-Oxo-9-hexadecenal O-00321

$H_3C(CH_2)_5CH{=}CH(CH_2)_5COCH_2CHO$

$C_{16}H_{28}O_2$ M 252.396

Unstable β-ketoaldehyde.

(Z)-form [76963-23-0]

Major component of the defence secretion of the termite *Acorhinotermes subfusciceps.*

Prestwich, G.D. *et al*, *J. Org. Chem.*, 1981, **46**, 2383 (*isol, struct, synth*)

4-Oxohexanoic acid, 9CI O-00322

4-Oxocaproic acid. Homolevulinic acid

[1117-74-4]

$H_3CCH_2COCH_2CH_2COOH$

$C_6H_{10}O_3$ M 130.143

Leaflets (Et_2O/pet. ether). Mp 40°. Bp_{16} 143-144°, $Bp_{0.4}$ 89°. Hygroscopic.

Phenylhydrazone: Mp 152°.

Semicarbazone: Prisms (EtOH). Mp 186°.

Et ester: [3249-33-0].

$C_8H_{14}O_3$ M 158.197

Bp_{14} 91°.

Et ester, semicarbazone: Needles (EtOAc/pet. ether). Mp 142° (106°).

Nitrile: [14274-30-7]. *1-Cyano-3-pentanone*

C_6H_9NO M 111.143

Oil. $Bp_{0.5}$ 75-80°.

Taylor, H.T., *J. Chem. Soc.*, 1958, 3922 (*synth*)
Noltes, A.W. *et al*, *Recl. Trav. Chim. Pays-Bas* (*J. R. Neth. Chem. Soc.*), 1961, **80**, 1334 (*synth*)
Ried, W. *et al*, *Justus Liebigs Ann. Chem.*, 1964, **678**, 113 (*synth*)
Reinheckel, H. *et al*, *Angew. Chem., Int. Ed. Engl.*, 1965, **4**, 785; 1966, **5**, 511 (*synth*)
Vita, G. *et al*, *Chem. Ber.*, 1966, **99**, 3387 (*synth*)
Kawashima, M. *et al*, *Bull. Chem. Soc. Jpn.*, 1988, **61**, 3255 (*synth*)
Coveney, D.J. *et al*, *J. Chem. Soc., Perkin Trans. 1*, 1990, 2721 (*nitrile*)

5-Oxo-2-hexenoic acid, 8CI O-00323

$H_3CCOCH_2CH{=}CHCOOH$

$C_6H_8O_3$ M 128.127

(E)-form [32794-09-5]

Oil.

Harris, C.M. *et al*, *J. Org. Chem.*, 1971, **36**, 2181 (*synth*)

2-Oxo-4-methylenepentanedioic acid O-00324

α-Keto-γ-methyleneglutamic acid

[55601-65-5]

$H_2C{=}C(COOH)CH_2COCOOH$

$C_6H_6O_5$ M 158.110

Found in tulip *Tulipa gesneriana*, *Lilium* spp. and ground nut (*Arachis hypogaea*).

2,4-Dinitrophenylhydrazone: Mp 228° dec.

5-Amide:

$C_6H_7NO_4$ M 157.126

Found in *T. gesneriana.*

Towers, G.H.N. *et al*, *J. Am. Chem. Soc.*, 1954, **76**, 1959.
Fowden, L. *et al*, *Biochem. J.*, 1955, **59**, 228.

10-Oxononadecanoic acid O-00325

$H_3C(CH_2)_8CO(CH_2)_8COOH$

$C_{19}H_{36}O_3$ M 312.492

Cryst. (MeOH). Mp 86-87°.

Amide:

$C_{19}H_{37}NO_2$ M 311.507

Cryst. (MeOH). Mp 83°.

Robinson, G.M. *et al*, *J. Chem. Soc.*, 1926, 2204 (*synth*)

3-Oxononanoic acid, 9CI, 8CI O-00326

3-Ketopelargonic acid. Heptanoylacetic acid

[40435-57-2]

$H_3C(CH_2)_5COCH_2COOH$

$C_9H_{16}O_3$ M 172.224

Mp 68-68.5° dec.

Et ester: [6622-36-2].

$C_{11}H_{20}O_3$ M 200.277

$Bp_{0.8}$ 87°.

Amide:

$C_9H_{17}NO_2$ M 171.239

Leaflets (H_2O). Mp 106-107°.

Nitrile: 1-Cyano-2-octanone

$C_9H_{15}NO$ M 153.224

Bp_{15} 137-141°.

Stenhagen, E., *Ark. Kemi*, 1951, **3**, 381 (*synth, cryst struct*)
Angelo, B., *C. R. Hebd. Seances Acad. Sci. Ser. C*, 1973, **276**, 293 (*synth*)

Katzenellenbogen, J.A. *et al*, *J. Am. Chem. Soc.*, 1974, **96**, 6153 (*ester*)
Oikawa, Y. *et al*, *J. Org. Chem.*, 1978, **43**, 2087 (*ester*)
Barnick, J.W.F.K. *et al*, *Synthesis*, 1979, 787 (*synth*)

4-Oxononanoic acid, 9CI — O-00327

3-Caproylpropionic acid
[6064-52-4]

$H_3C(CH_2)_4COCH_2CH_2COOH$

$C_9H_{16}O_3$ M 172.224
Cryst. (pet. ether). Mp 69-70°.
Et ester: [37174-92-8].
$C_{11}H_{20}O_3$ M 200.277
Bp_2 90-92°.
Semicarbazone: Mp 157-159° dec.
Phenylhydrazone: Mp 85-87°.

Ernest, I. *et al*, *Collect. Czech. Chem. Commun.*, 1959, **24**, 3341 (*synth*)
Takeda, A. *et al*, *J. Org. Chem.*, 1966, **31**, 616 (*synth*)
Steglich, W. *et al*, *Angew. Chem., Int. Ed. Engl.*, 1971, **10**, 655 (*synth*)
Wineburg, J.P. *et al*, *J. Heterocycl. Chem.*, 1975, **12**, 749 (*synth*)
Stetter, H. *et al*, *Chem. Ber.*, 1976, **109**, 534 (*synth*)
Kawashima, M. *et al*, *Bull. Chem. Soc. Jpn.*, 1988, **61**, 3255 (*synth*)

5-Oxononanoic acid, 9CI, 8CI — O-00328

5-Ketopelargonic acid. 4-Valerylbutyric acid
[3637-15-8]

$H_3C(CH_2)_3CO(CH_2)_3COOH$

$C_9H_{16}O_3$ M 172.224
Mp 46°. Bp_2 140-141°.
Semicarbazone: Mp 140-141°.
Et ester: [24071-99-6].
$C_{11}H_{20}O_3$ M 200.277
Bp_{20} 150-155°.
Amide:
$C_9H_{17}NO_2$ M 171.239
Mp 91-92°.

Keskin, H., *CA*, 1963, **59**, 15173 (*synth*)
Lardelli, G. *et al*, *Recl. Trav. Chim. Pays-Bas* (*J. R. Neth. Chem. Soc.*), 1967, **86**, 481 (*synth*)
LeGuillanton, G., *Bull. Soc. Chim. Fr.*, 1969, 2871 (*synth*)
Ijima, A. *et al*, *Chem. Pharm. Bull.*, 1973, **21**, 215 (*synth*)

6-Oxononanoic acid, 9CI, 8CI — O-00329

6-Ketopelargonic acid. 5-Butyrylvaleric acid
[4144-58-5]

$H_3CCH_2CH_2CO(CH_2)_4COOH$

$C_9H_{16}O_3$ M 172.224
Mp 40-41°. Bp_{15} 170-175°.
Semicarbazone: Mp 153°.
Et ester: [4144-59-6].
$C_{11}H_{20}O_3$ M 200.277
Bp_{10} 139°.

Lamant, M. *et al*, *Bull. Soc. Chim. Fr.*, 1965, 3041 (*synth*)
Shostakovskii, V.M. *et al*, *Dokl. Akad. Nauk SSSR, Ser. Khim.*, 1967, **176**, 610; *CA*, **68**, 114003 (*synth, ester*)
Gol'dfarb, Y.L. *et al*, *CA*, 1972, **76**, 139896 (*synth*)

7-Oxononanoic acid, 9CI, 8CI — O-00330

7-Ketopelargonic acid. 6-Propionylcaproic acid
[20356-92-7]

$H_3CCH_2CO(CH_2)_5COOH$

$C_9H_{16}O_3$ M 172.224
Leaflets (Et_2O/pet. ether). Mp 41.5-42.5°. Bp_2 150-151°.
Oxime: [20356-93-8].
$C_9H_{17}NO_3$ M 187.238
Mp 77°.
Semicarbazone: Cryst. (EtOH). Mp 168-169.5°.
Me ester: [4219-20-9].
$C_{10}H_{18}O_3$ M 186.250
Bp_{21} 143°.
Me ester, 2,4-dinitrophenylhydrazone: Mp 128°.

Hauser, C.R. *et al*, *J. Am. Chem. Soc.*, 1948, **70**, 4023 (*synth*)
Huenig, S. *et al*, *Chem. Ber.*, 1958, **91**, 129 (*synth*)
Suyama, T. *et al*, *CA*, 1968, **69**, 51501 (*synth*)

8-Oxononanoic acid, 9CI — O-00331

7-Acetylheptanoic acid
[25542-64-7]

$H_3CCO(CH_2)_6COOH$

$C_9H_{16}O_3$ M 172.224
Needles (hexane). Mp 42.5-44°. $Bp_{10.5}$ 165-169°.
Semicarbazone: Mp 137-138°.
Me ester: [34455-70-4].
$C_{10}H_{18}O_3$ M 186.250
Oil.
Et ester: [25482-95-5].
$C_{11}H_{20}O_3$ M 200.277
$Bp_{15.5}$ 139-142°.
Et ester, semicarbazone: [25542-65-8].
Mp 108°.

Feugas, C., *Bull. Soc. Chim. Fr.*, 1963, 2568 (*synth*)
Burgstahler, A.W. *et al*, *J. Org. Chem.*, 1975, **40**, 3456 (*synth*)
Tsuji, J. *et al*, *Bull. Chem. Soc. Jpn.*, 1978, **51**, 1915 (*synth*)
Bosone, E. *et al*, *Synthesis*, 1983, 942 (*synth*)

9-Oxononanoic acid, 9CI — O-00332

Azelaialdehydic acid. 8-Formylcaprylic acid. Azelaic hemialdehyde
[2553-17-5]

$OHC(CH_2)_7COOH$

$C_9H_{16}O_3$ M 172.224
Insol. H_2O. Mp 57-63° (70°). Bp_{15} 181-182°.
Me ester: [1931-63-1].
$C_{10}H_{18}O_3$ M 186.250
Bp_3 111-112°.
Me ester, di-Me acetal: [1599-18-0].
$C_{12}H_{24}O_4$ M 232.319
Bp_{14} 148-150°.
Me ester, semicarbazone: Mp 104-105° (107°).
Me ester, 2,4-dinitrophenylhydrazone: Mp 67-68°.

Noller, C.R. *et al*, *J. Am. Chem. Soc.*, 1926, **48**, 1074 (*synth*)
Fischer, F.G. *et al*, *Ber.*, 1932, **65**, 1471 (*synth*)
Crombie, L. *et al*, *J. Chem. Soc., Perkin Trans. 1*, 1991, 567 (*synth, ir*)

19-Oxo-22-octacosenoic acid — O-00333

$H_3C(CH_2)_4CH{=}CHCH_2CH_2CO(CH_2)_{17}COOH$

$C_{28}H_{52}O_3$ M 436.717

(Z)-form
Isol. from seed oil of *Cuspidaria pterocarpa*.
Me ester: Cryst. (MeOH). Mp 50-51.5°.

Smith, C.R., *Lipids*, 1966, **1**, 268 (*isol, synth, pmr*)

8-Oxo-9,11-octadecadiynoic acid O-00334

[64144-72-5]

$H_3C(CH_2)_5C{\equiv}CC{\equiv}CCO(CH_2)_6COOH$

$C_{18}H_{26}O_3$ M 290.402

Isol. from Isano oil (*Ongokea gore*).

Miller, R.W. *et al*, *Phytochemistry*, 1977, **16**, 947.

2-Oxooctadecanoic acid, 8CI O-00335

2-Oxostearic acid

[4468-17-1]

$H_3C(CH_2)_{15}COCOOH$

$C_{18}H_{34}O_3$ M 298.465

Mp 74.3-74.6°.

Me ester: Cryst. (MeOH). Mp 54.5-55°.

Bergström, S. *et al*, *Acta Chem. Scand.*, 1952, **6**, 1157 (*synth*)
Ryhage, R., *Ark. Kemi*, 1960, **15**, 545 (*ms*)
Fischmeister, I., *Ark. Kemi*, 1961, **18**, 101; 1967, **26**, 453 (*ir*)
Kitagawa, I. *et al*, *J. Am. Oil Chem. Soc.*, 1962, **39**, 217 (*ir*)
Tulloch, A.P., *J. Am. Oil Chem. Soc.*, 1964, **41**, 873 (*glc*)

3-Oxooctadecanoic acid O-00336

3-Oxostearic acid. 3-Ketostearic acid

[16694-29-4]

$H_3C(CH_2)_{14}COCH_2COOH$

$C_{18}H_{34}O_3$ M 298.465

Mp 102-103°.

Me ester: [14531-34-1].
$C_{19}H_{36}O_3$ M 312.492
Mp 40.1°.

Et ester: [62625-23-4].
$C_{20}H_{38}O_3$ M 326.518
Cryst. (EtOH). Mp 37-38°. Alc. $FeCl_3$ → wine-red col.

Helferich, B. *et al*, *Ber.*, 1923, **56**, 2088.
Tulloch, A.P., *J. Am. Oil Chem. Soc.*, 1964, **41**, 833 (*glc*)
Fischmeister, I., *Ark. Kemi*, 1967, **26**, 453 (*ir*)
Morris, L.J. *et al*, *J. Chromatogr.*, 1968, **33**, 471.
Tulloch, A.P., *Can. J. Chem.*, 1977, **55**, 1135 (*cmr*)

4-Oxooctadecanoic acid O-00337

4-Oxostearic acid. 4-Ketostearic acid

[16694-30-7]

$H_3C(CH_2)_{13}COCH_2CH_2COOH$

$C_{18}H_{34}O_3$ M 298.465

Occurs in Po-Yoak oil. Mp 97°.

▷ WI4570000.

Semicarbazone: Mp 119°, Mp 137°.

Oxime:
$C_{18}H_{35}NO_3$ M 313.479
Mp 85°.

Steger, A. *et al*, *CA*, 1947, **41**, 1472 (*isol*)
Tulloch, A.P., *J. Am. Oil Chem. Soc.*, 1964, **41**, 833 (*glc*)
Fischmeister, I., *Ark. Kemi*, 1967, **26**, 453 (*ir*)
Morris, L.J. *et al*, *J. Chromatogr.*, 1968, **33**, 471.
Swern, D. *et al*, *Cancer Res.*, 1970, **30**, 1037.
Tulloch, A.P., *Can. J. Chem.*, 1977, **55**, 1135 (*cmr*)
Lewis, R.J., *Sax's Dangerous Properties of Industrial Materials*, 8th Ed., Van Nostrand-Reinhold, 1992, KGK150.

5-Oxooctadecanoic acid O-00338

5-Oxostearic acid. 5-Ketostearic acid

[16694-31-8]

$H_3C(CH_2)_{12}CO(CH_2)_3COOH$

$C_{18}H_{34}O_3$ M 298.465

Trace constit. of milk fat.

Me ester: [2380-20-3].
$C_{19}H_{36}O_3$ M 312.492
Mp 49.5°.

Semicarbazone: Mp 197°.

Dodecyl ester: Dodecyl 5-oxooctadecanoate. Actinomycin B1†. Actinomycin J2
$C_{30}H_{58}O_3$ M 466.786
Isol. from cultures of *Actinomyces flavus*. Cryst. Mp 59-61°, Mp 81.5°.

Brown, W.B. *et al*, *Biochem. J.*, 1935, **29**, 631.
Hirata, Y. *et al*, *Bull. Chem. Soc. Jpn.*, 1949, **22**, 121 (*Dodecyl ester*)
Tulloch, A.P., *J. Am. Oil Chem. Soc.*, 1964, **41**, 833 (*glc*)
Weihrauch, J.L. *et al*, *Lipids*, 1974, **9**, 883 (*isol, glc, ms*)
Tulloch, A.P., *Can. J. Chem.*, 1977, **55**, 1135 (*cmr*)

6-Oxooctadecanoic acid O-00339

Lactarinic acid. 6-Oxostearic acid. 6-Ketostearic acid

[502-71-6]

$H_3C(CH_2)_{11}CO(CH_2)_4COOH$

$C_{18}H_{34}O_3$ M 298.465

Constit. of *Lactarius rufus*. Cryst. (EtOH). Mp 87°.

Et ester:
$C_{20}H_{38}O_3$ M 326.518
Mp 41°.

Oxime:
$C_{18}H_{35}NO_3$ M 313.479
Cryst. (pet. ether). Mp 59-61°.

Amide:
$C_{18}H_{35}NO_2$ M 297.480
Cryst. (EtOH). Mp 104°.

Robinson, G.M. *et al*, *J. Chem. Soc.*, 1925, **127**, 175 (*synth*)
Schneider, A.K. *et al*, *J. Biol. Chem.*, 1942, **142**, 345 (*synth*)
Tulloch, A.P., *J. Am. Oil Chem. Soc.*, 1964, **41**, 833 (*glc*)
Kuranova, I.L. *et al*, *Zh. Org. Khim.*, 1973, **9**, 260.
Tulloch, A.P., *Can. J. Chem.*, 1977, **55**, 1135 (*cmr*)
Tulloch, A.P., *Lipids*, 1977, **12**, 92 (*synth*)

7-Oxooctadecanoic acid O-00340

7-Oxostearic acid. 7-Ketostearic acid

[16694-32-9]

$H_3C(CH_2)_{10}CO(CH_2)_5COOH$

$C_{18}H_{34}O_3$ M 298.465

Constit. of Taliawood. Cryst. (EtOH). Mp 75°.

Tulloch, A.P., *J. Am. Oil Chem. Soc.*, 1964, **41**, 833 (*glc*)
Kuranova, I.L. *et al*, *Zh. Org. Khim.*, 1973, **9**, 260.
Abe, Z. *et al*, *CA*, 1976, **85**, 74986c (*isol*)
Tulloch, A.P., *Can. J. Chem.*, 1977, **55**, 1135 (*cmr*)
Tulloch, A.P., *Lipids*, 1977, **12**, 92 (*synth*)

8-Oxooctadecanoic acid, 9CI O-00341

8-Oxostearic acid

[16694-33-0]

$H_3C(CH_2)_9CO(CH_2)_6COOH$

$C_{18}H_{34}O_3$ M 298.465

Cryst. ($CHCl_3$/pet. ether). Mp 84.2-85.2°.

Me ester: Mp 46.4-46.9°.

Bergström, S. *et al*, *Acta Chem. Scand.*, 1952, **6**, 1157 (*synth*)
Crombie, L. *et al*, *J. Chem. Soc.*, 1958, 4435 (*synth*)
Ryhage, R., *Ark. Kemi*, 1960, **15**, 545 (*ms*)
Fischmeister, I., *Ark. Kemi*, 1961, **18**, 101; 1967, **26**, 453 (*ir*)
Breusch, F.L. *et al*, *Fette, Seifen, Anstrichm.*, 1963, **68**, 995 (*synth*)
Tulloch, A.P., *J. Am. Oil Chem. Soc.*, 1964, **41**, 833 (*glc*)

9-Oxooctadecanoic acid — O-00342

9-Oxostearic acid. 9-Ketostearic acid

[4114-74-3]

$H_3C(CH_2)_8CO(CH_2)_7COOH$

$C_{18}H_{34}O_3$ M 298.465

Trace constit. of milk fat. Cryst. (EtOH). Mp 83°.

Amide:

$C_{18}H_{35}NO_2$ M 297.480

Cryst. (C_6H_6 or MeOH). Mp 79°.

Behrend, O., *Ber.*, 1896, **29**, 806 (*synth*)
Robinson, G.M. *et al*, *J. Chem. Soc.*, 1926, 2204 (*synth*)
Tulloch, A.P., *J. Am. Oil Chem. Soc.*, 1964, **41**, 833 (*glc*)
Kenney, H.E. *et al*, *J. Am. Oil Chem. Soc.*, 1971, **48**, 765 (*synth*)
Weihrauch, J.L. *et al*, *Lipids*, 1974, **9**, 883 (*isol, glc, ms*)
Tulloch, A.P., *Can. J. Chem.*, 1977, **55**, 1135 (*cmr*)

10-Oxooctadecanoic acid — O-00343

10-Oxostearic acid. 10-Ketostearic acid

[4158-12-7]

$H_3C(CH_2)_7CO(CH_2)_8COOH$

$C_{18}H_{34}O_3$ M 298.465

Trace constit. of milk fat. Cryst. Mp 76°.

Amide:

$C_{18}H_{35}NO_2$ M 297.480

Cryst. (C_6H_6/pet. ether). Mp 80°.

Baruch, J., *Ber.*, 1894, **27**, 1742 (*synth*)
Robinson, G.M. *et al*, *J. Chem. Soc.*, 1926, 2204 (*synth*)
Tulloch, A.P., *J. Am. Oil Chem. Soc.*, 1964, **41**, 833 (*glc*)
Kenney, H.E. *et al*, *J. Am. Oil Chem. Soc.*, 1971, **48**, 765 (*synth*)
Weihrauch, J.L. *et al*, *Lipids*, 1974, **9**, 883 (*isol, glc, ms*)
Zhdanov, R.I. *et al*, *Bioorg. Khim.*, 1976, **2**, 395.
Tulloch, A.P., *Can. J. Chem.*, 1977, **55**, 1135 (*cmr*)

11-Oxooctadecanoic acid — O-00344

11-Oxostearic acid. 11-Ketostearic acid

[2388-83-2]

$H_3C(CH_2)_6CO(CH_2)_9COOH$

$C_{18}H_{34}O_3$ M 298.465

Trace constit. of milk fat. Cryst. (EtOH/AcOH). Mp 65°.

Tulloch, A.P., *J. Am. Oil Chem. Soc.*, 1964, **41**, 833 (*glc*)
Weihrauch, J.L. *et al*, *Lipids*, 1974, **9**, 883 (*isol, glc, ms*)
Tulloch, A.P., *Can. J. Chem.*, 1977, **55**, 1135 (*cmr*)

12-Oxooctadecanoic acid — O-00345

12-Oxostearic acid. Lycaonic acid

[925-44-0]

$H_3C(CH_2)_5CO(CH_2)_{10}COOH$

$C_{18}H_{34}O_3$ M 298.465

Mp 81.5°.

Semicarbazone: Mp 122°.

Pyrrolidide: [83182-47-2]. ***Lycaonic acid pyrrolidide.*** *1-(1,12-Dioxooctadecyl)pyrrolidine*

$C_{22}H_{41}NO_2$ M 351.571

Alkaloid from the roots of *Achillea* spp.(Compositae). Mp 44°.

Piperidide: [83182-45-0]. ***Lycaonic acid piperidide.*** *1-(1,12-Dioxooctadecyl)piperidine*

$C_{23}H_{43}NO_2$ M 365.598

Alkaloid from the roots of *A. lycaonica* (Compositae). Mp 41°.

Belopol'skii, M.P. *et al*, *CA*, 1937, **31**, 7397 (*synth*)
Tulloch, A.P., *Can. J. Chem.*, 1977, **55**, 1135 (*cmr*)
Tulloch, A.P. *et al*, *Lipids*, 1977, **12**, 92 (*synth*)
Greger, H. *et al*, *Phytochemistry*, 1982, **21**, 1071 (*isol, ir, pmr, struct, pyrrolidide, piperidide*)
Bohlmann, F. *et al*, *Tetrahedron*, 1983, **39**, 123 (*synth*)

13-Oxooctadecanoic acid, 9CI — O-00346

13-Oxostearic acid

[2389-06-2]

$H_3C(CH_2)_4CO(CH_2)_{11}COOH$

$C_{18}H_{34}O_3$ M 298.465

Mp 82.1-82.3°.

Me ester: Mp 47.3-47.7°.

Bergström, S. *et al*, *Acta Chem. Scand.*, 1952, **6**, 1127 (*synth*)
Ryhage, R., *Ark. Kemi*, 1960, **15**, 545 (*ms*)
Fischmeister, I., *Ark. Kemi*, 1961, **18**, 101; 1967, **26**, 453 (*ir*)
Tulloch, A.P., *J. Am. Oil Chem. Soc.*, 1964, **41**, 833 (*synth, glc*)
Breusch, F.L. *et al*, *Fette, Seifen, Anstrichm.*, 1965, **67**, 4 (*synth*)

14-Oxooctadecanoic acid, 8CI — O-00347

14-Oxostearic acid

[16694-36-3]

$H_3C(CH_2)_3CO(CH_2)_{12}COOH$

$C_{18}H_{34}O_3$ M 298.465

Mp 81.7-81.9°.

Me ester: Mp 48.9-49.0°.

Bergström, S. *et al*, *Acta Chem. Scand.*, 1952, **6**, 1157 (*synth*)
Fischmeister, I., *Ark. Kemi*, 1961, **18**, 101; 1967, **26**, 453 (*ir*)
Tulloch, A.T., *J. Am. Oil Chem. Soc.*, 1964, **41**, 833 (*synth, glc*)

15-Oxooctadecanoic acid, 8CI — O-00348

15-Oxostearic acid

[16694-37-4]

$H_3CCH_2CH_2CO(CH_2)_{13}COOH$

$C_{18}H_{34}O_3$ M 298.465

Mp 83.0-83.2°.

Me ester: Mp 53.5-53.7°.

Bergström, S. *et al*, *Acta Chem. Scand.*, 1952, **6**, 1157 (*synth*)
Crossley, A.T. *et al*, *Can. J. Chem.*, 1955, **33**, 1426 (*synth*)
Ryhage, R., *Ark. Kemi*, 1960, **15**, 545 (*ms*)
Fischmeister, I., *Ark. Kemi*, 1961, **18**, 101; 1967, **26**, 453 (*ir*)
Tulloch, A.P., *J. Am. Oil Chem. Soc.*, 1964, **41**, 833 (*synth, glc*)

16-Oxooctadecanoic acid, 9CI — O-00349

16-Oxostearic acid

[13126-28-8]

$H_3CCH_2CO(CH_2)_{14}COOH$

$C_{18}H_{34}O_3$ M 298.465

Mp 93.1-93.5°.

Me ester: Mp 54.3-54.5°.

Bergström, S. *et al*, *Acta Chem. Scand.*, 1952, **6**, 1157 (*synth*)
Ryhage, R., *Ark. Kemi*, 1960, **15**, 545 (*ms*)
Tulloch, A.P., *J. Am. Oil Chem. Soc.*, 1964, **41**, 833 (*synth, glc*)
Fischmeister, I., *Ark. Kemi*, 1967, **26**, 453 (*ir*)

17-Oxooctadecanoic acid, 9CI O-00350

17-Oxostearic acid

[16694-38-5]

$H_3CCO(CH_2)_{15}COOH$

$C_{18}H_{34}O_3$ M 298.465

Mp 87.4-87.7°.

Me ester: Mp 56.1-56.5°.

Bergström, S. *et al*, *Acta Chem. Scand.*, 1952, **6**, 1157 (*synth*)
Ryhage, R., *Ark. Kemi*, 1960, **15**, 545 (*ms*)
Fischmeister, I., *Ark. Kemi*, 1961, **18**, 101; 1967, **26**, 453 (*ir*)
Tulloch, A.P., *J. Am. Oil Chem. Soc.*, 1964, **41**, 833 (*glc*)

4-Oxo-9,11,13,15-octadecatetraenoic acid O-00351

$H_3CCH_2(CH{=}CH)_4(CH_2)_4COCH_2CH_2COOH$

$C_{18}H_{26}O_3$ M 290.402

(9Z,11E,13E,15Z)-form [17699-21-7]

Obt. from *Chrysobalanus icaco* seed oil.

Gunstone, F.D. *et al*, *Chem. Phys. Lipids*, 1967, **1**, 349.

12-Oxo-9-octadecenoic acid O-00352

[75922-01-9]

$H_3C(CH_2)_5COCH_2CH{=}CH(CH_2)_7COOH$

$C_{18}H_{32}O_3$ M 296.449

(E)-form

Cryst. (EtOH/pet. ether). Mp 72-72.5°.

Me ester: Cryst. (pet. ether at −30°). Mp 24-25°.

(Z)-form [5455-97-0]

Cryst. (pet. ether). Mp 40-41°.

Me ester: [3047-65-2].

$C_{19}H_{34}O_3$ M 310.476

Solid.

Et ester:

$C_{20}H_{36}O_3$ M 324.503

Liq. Mp −7° to −6°.

Nichols, J. *et al*, *J. Am. Chem. Soc.*, 1958, **30**, 5705 (*synth*)
Frankel, E.N. *et al*, *J. Chem. Soc., Perkin Trans. 1*, 1984, 2233 (*synth, ir, pmr*)

12-Oxo-10-octadecenoic acid O-00353

$H_3C(CH_2)_5COCH{=}CH(CH_2)_8COOH$

$C_{18}H_{32}O_3$ M 296.449

(E)-form

Cryst. (EtOH aq.). Mp 50-50.5°.

Me ester: [21308-79-2].

$C_{19}H_{34}O_3$ M 310.476

Oil. Mp 12-13°.

▷ RG4100000.

(Z)-form [57818-50-5]

Solid.

Nichols, J. *et al*, *J. Am. Chem. Soc.*, 1958, **30**, 5705.
Chakrabarti, P. *et al*, *Biochemistry*, 1975, **14**, 5021 (*synth, ir, pmr*)
Frankel, E. *et al*, *J. Chem. Soc., Perkin Trans. 1*, 1984, 2233 (*synth, uv, ir, pmr, ms*)
Lewis, R.J., *Sax's Dangerous Properties of Industrial Materials*, 8th Ed., Van Nostrand-Reinhold, 1992, MNF250.

12-Oxo-9-octadecynoic acid, 9CI O-00354

[68396-17-8]

$H_3C(CH_2)_5COCH_2C{\equiv}C(CH_2)_7COOH$

$C_{18}H_{30}O_3$ M 294.433

Cryst. (pet. ether). Mp 62-63°.

Nichols, J. *et al*, *J. Am. Chem. Soc.*, 1958, **30**, 5705 (*synth*)
Frankel, E.N. *et al*, *J. Chem. Soc., Perkin Trans. 1*, 1984, 2233 (*synth, ir, pmr*)

3-Oxooctanoic acid, 9CI O-00355

Caproylacetic acid. 3-Oxocaprylic acid

[13283-91-5]

$H_3C(CH_2)_4COCH_2COOH$

$C_8H_{14}O_3$ M 158.197

Leaflets. Insol. H_2O. Mp 75-76° dec.

Et ester: [10488-95-6].

$C_{10}H_{18}O_3$ M 186.250

$Bp_{0.5}$ 73-74°.

Amide:

$C_8H_{15}NO_2$ M 157.212

Mp 100°.

Nitrile:

$C_8H_{13}NO$ M 139.197

Bp_{14} 127-128°.

Schecter, M.S. *et al*, *J. Am. Chem. Soc.*, 1949, **71**, 3165 (*synth*)
Stenhagen, E. *et al*, *Ark. Kemi*, 1951, **3**, 381; *CA*, **46**, 7050 (*synth*)
Rathke, M.W. *et al*, *Tetrahedron Lett.*, 1971, 2953 (*synth, ester*)
Oikawa, Y. *et al*, *J. Org. Chem.*, 1978, **43**, 2087 (*synth, ester*)

6-Oxooctanoic acid, 9CI, 8CI O-00356

6-Ketocaprylic acid. 5-Propionylvaleric acid

[4233-57-2]

$H_3CCH_2CO(CH_2)_4COOH$

$C_8H_{14}O_3$ M 158.197

Cryst. (C_6H_6/pet. ether). Mp 52.5°. Bp_9 160-161°.

2,4-Dinitrophenylhydrazone: Mp 136.5-137.5°.

Semicarbazone: Mp 186-187°.

Me ester: [2955-61-5].

$C_9H_{16}O_3$ M 172.224

Bp_{14} 125°.

Me ester, 2,4-dinitrophenylhydrazone: Mp 55.5-57.0°.

Lamant, M. *et al*, *Bull. Soc. Chim. Fr.*, 1965, 3041 (*synth*)
Collman, J.P. *et al*, *J. Am. Chem. Soc.*, 1972, **94**, 1788 (*synth*)
Muramoto, Y. *et al*, *Agric. Biol. Chem.*, 1974, **38**, 2197 (*synth, ester*)
Odinokov, V.N. *et al*, *Zh. Org. Khim.*, 1980, **16**, 524 (*synth, pmr, ms*)
White, R.H., *J. Am. Chem. Soc.*, 1980, **102**, 6605 (*biosynth*)

7-Oxooctanoic acid, 9CI O-00357

[14112-98-2]

$H_3CCO(CH_2)_5COOH$

$C_8H_{14}O_3$ M 158.197

Mp 29°. $Bp_{1.5}$ 144-145°.

2,4-Dinitrophenylhydrazone: Cryst. (EtOH aq.). Mp 108-112°.

Et ester: [36651-36-2].

$C_{10}H_{18}O_3$ M 186.250

Bp_2 120°.

Et ester, 2,4-dinitrophenylhydrazone: [16112-02-0].

Cryst. (Et_2O/EtOH). Mp 67°.

Bestmann, H.J. *et al, Justus Liebigs Ann. Chem.*, 1966, **699**, 33 (*synth*)
Manecke, G. *et al, Chem. Ber.*, 1967, **100**, 3413 (*synth*)
Birch, A.J. *et al, J. Chem. Soc., Perkin Trans. 1*, 1990, 1423 (*synth, ir, pmr*)

5-Oxo-7-octenoic acid O-00358

[87086-55-3]

$H_2C{=}CHCH_2CO(CH_2)_3COOH$

$C_8H_{12}O_3$ M 156.181
Oil. Bp_2 120° (bulb).

Utaka, M. *et al, J. Org. Chem.*, 1986, **51**, 935 (*synth, ir, pmr, cmr*)

2-Oxopentadecanoic acid O-00359

$H_3C(CH_2)_{12}COCOOH$

$C_{15}H_{28}O_3$ M 256.384
Mp 67-68°.
Oxime:
$C_{15}H_{29}NO_3$ M 271.399
Mp 88-89°.

Asano, M. *et al, CA*, 1951, **45**, 1518.

3-Oxopentadecanoic acid O-00360

$H_3C(CH_2)_{11}COCH_2COOH$

$C_{15}H_{28}O_3$ M 256.384
Mp 90-95°.
Et ester:
$C_{17}H_{32}O_3$ M 284.438
$Bp_{0.5}$ 137-142°.

Stenhagen, E., *Ark. Kemi*, 1951, **3**, 381.
Borodina, G.M., *CA*, 1955, **49**, 4518 (*synth*)

11-Oxopentadecanoic acid O-00361

$H_3C(CH_2)_3CO(CH_2)_9COOH$

$C_{15}H_{28}O_3$ M 256.384
Cryst. (EtOAc). Mp 70-71°.

Davies, L.A. *et al, J. Am. Chem. Soc.*, 1928, **50**, 1754 (*synth*)

5-Oxopentanoic acid, 9CI O-00362

4-Formylbutanoic acid. Glutaric semialdehyde
[5746-02-1]

$OHCCH_2CH_2CH_2COOH$

$C_5H_8O_3$ M 116.116
Me ester: [6026-86-4].
$C_6H_{10}O_3$ M 130.143
Synthon. Liq. Bp_{10} 90-98°.
Nitrile: [3350-74-1]. *4-Cyanobutanal*
C_5H_7NO M 97.116
Liq. Bp_2 83-84°.
Nitrile, 2,4-dinitrophenylhydrazone: Mp 117-118°.

Brown, H.C. *et al, J. Am. Chem. Soc.*, 1969, **91**, 4606 (*nitrile, synth*)
Huckstep, M. *et al, Synthesis*, 1982, 881.
Gannett, P.M. *et al, J. Org. Chem.*, 1988, **53**, 1064 (*ester, synth, ir, pmr, cmr, ms*)
Bergeron, R.J. *et al, J. Org. Chem.*, 1988, **53**, 3131 (*nitrile, pmr, ir*)

3-Oxo-4-pentenoic acid, 9CI O-00363

$H_2C{=}CHCOCH_2COOH$

$C_5H_6O_3$ M 114.101
Me ester: [37734-05-7]. *Methyl 3-oxo-4-pentenoate. Vinyl carboxymethyl ketone. Methyl acryloylacetate*
$C_6H_8O_3$ M 128.127
Annelating reagent. Bp_{19} 65°. Polymerises on storage.
Et ester: [22418-80-0]. *Ethyl 3-oxo-4-pentenoate. Ethyl acryloylacetate. Nazarov's reagent*
$C_7H_{10}O_3$ M 142.154
Annelating reagent. Bp_{18} 76-78°.
tert-*Butyl ester:*
$C_9H_{14}O_3$ M 170.208
Used for annelation reactions. Oil. Bp_5 80-82°.

Hohenlohe-Oehringen, K., *Monatsh. Chem.*, 1962, **93**, 576 (*use*)
Wenkert, E. *et al, J. Am. Chem. Soc.*, 1964, **86**, 2038 (*synth, ir*)
Stork, G. *et al, Tetrahedron Lett.*, 1972, 2755 (*synth*)
Trost, B.M. *et al, J. Org. Chem.*, 1974, **39**, 2648 (*synth, ir, pmr*)
Trost, B.M. *et al, J. Am. Chem. Soc.*, 1975, **97**, 7152 (*use*)
Fieser and Fieser's Reagents for Organic Synthesis, Wiley, 1977, **6**, 662 (*use*)
Ohta, S. *et al, Synthesis*, 1983, 715 (*deriv*)
Zibuck, R. *et al, J. Org. Chem.*, 1989, **54**, 4717 (*esters, synth, bibl*)
Kuehne, M.E. *et al, J. Org. Chem.*, 1991, **56**, 2701 (*synth, uv, ir, pmr, ms*)

4-Oxo-5-(2-pentenyl)-2-cyclopentene-1-octanoic acid, 9CI O-00364

12-Oxo-10,15-phytodienoic acid
[67204-66-4]

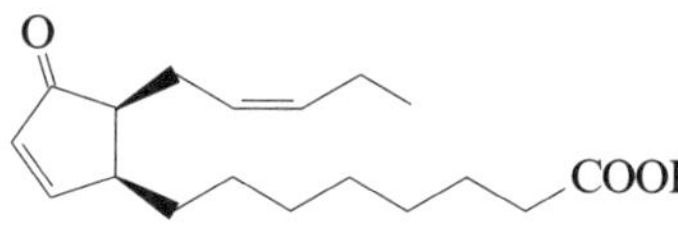

$C_{18}H_{28}O_3$ M 292.417
Isol. from Japanese mosses. $[\alpha]_D^{25}$ +104° (c, 9.5 in $CHCl_3$).
Me ester: [117179-81-4].
Oil.

[70630-16-9, 71606-07-0, 72458-40-3, 72541-23-2, 85551-10-6, 98673-77-9, 104371-93-9, 111956-15-1, 115510-01-5, 115510-02-6, 117179-94-9, 134002-16-7]

Vick, B.A. *et al, Lipids*, 1979, **14**, 734 (*synth, ir, pmr, ms*)
Sakai, K. *et al, Tetrahedron Lett.*, 1985, **26**, 2089 (*isol*)
Baertschi, S.W. *et al, Biochemistry*, 1988, **27**, 18 (*abs config, cd, biosynth, hplc*)
Brash, A.R. *et al, Proc. Natl. Acad. Sci. U.S.A.*, 1988, **85**, 3382 (*biosynth, uv, cd, pmr, O-18 nmr*)
Crombie, L. *et al, J. Chem. Soc., Perkin Trans. 1*, 1991, 581 (*synth, pmr, cmr, ms*)

6-Oxoprostaglandin E_1 O-00365

11,15-Dihydroxy-6,9-dioxoprost-13-en-1-oic acid
[67786-53-2]

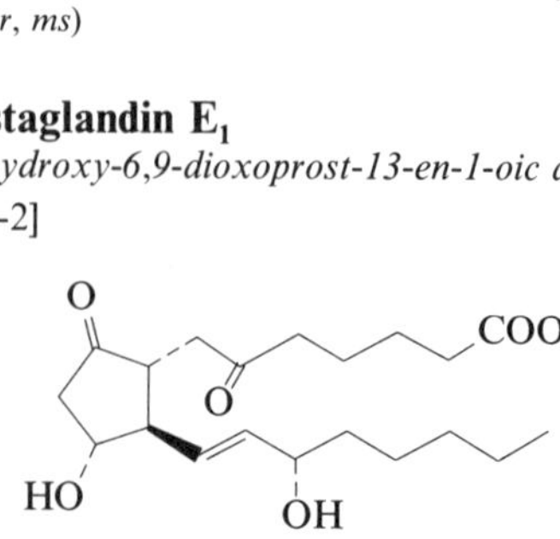

$C_{20}H_{32}O_6$ M 368.469
Powerful vasoactive metab. of Prostacyclin, P-00144. Inhibits platelet aggregation. Cryst. (hexane/Et_2O). Mp 65°. $[\alpha]_D^{20}$ −50.0° (c, 1.55 in MeOH).
Me ester: [67786-49-6].

Cryst. (hexane/Et_2O). Mp 44.0-44.5°. $[\alpha]_D^{21}$ −48.5° (c, 0.71 in MeOH).

Tanaka, T. *et al*, *Tetrahedron Lett.*, 1984, **25**, 4947 (*synth*)
Hoult, J.R.S. *et al*, *Trends Pharmacol. Sci.*, 1986, **7**, 297.
Toru, T. *et al*, *J. Am. Chem. Soc.*, 1988, **110**, 4815 (*Me ester, synth, pmr, ir*)

9-Oxoprosta-5,10,14-trien-1-oic acid, 9CI O-00366

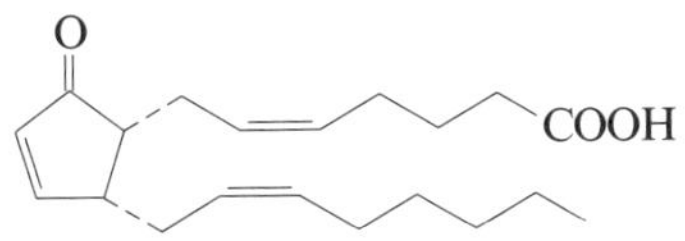

$C_{20}H_{30}O_3$ M 318.455

(5Z,12α,14Z)-form [117248-65-4] ***Preclavulone A***
Isol. from corals *Clavularia viridis*, *Pseudoplexaura porosa* and other coral spp.
Me ester: [98964-95-5].
$[\alpha]_D^{20}$ −131.8° (c, 1.14 in THF).

Corey, E.J. *et al*, *Tetrahedron Lett.*, 1985, **26**, 4171; 1987, **28**, 4247; 1988, **29**, 995 (*isol, struct, biosynth, synth, pmr, cmr, ir, ms*)
Corey, E.J. *et al*, *J. Am. Chem. Soc.*, 1987, **109**, 289 (*biosynth*)

15-Oxo-18-tetracosenoic acid O-00367

$H_3C(CH_2)_4CH{=}CHCH_2CH_2CO(CH_2)_{13}COOH$

$C_{24}H_{44}O_3$ M 380.610

(Z)-form
Isol. from seed oil of *Cuspidaria pterocarpa*.
Me ester:
$C_{25}H_{46}O_3$ M 394.637
Cryst. (MeOH aq.). Mp 40-40.5°.

Smith, C.R., *Lipids*, 1966, **1**, 268 (*isol*)

3-Oxotetradecanoic acid O-00368

3-Oxomyristic acid

$H_3C(CH_2)_{10}COCH_2COOH$

$C_{14}H_{26}O_3$ M 242.358
Me ester: [22348-97-6].
$C_{15}H_{28}O_3$ M 256.384
Mp 28.7°. $Bp_{0.9}$ 137-139°.
Et ester:
$C_{16}H_{30}O_3$ M 270.411
Bp_{10} 173-175°.

Asahina, Y. *et al*, *Chem. Ber.*, 1937, **70**, 200 (*synth*)

4-Oxotetradecanoic acid O-00369

4-Oxomyristic acid

$H_3C(CH_2)_9COCH_2CH_2COOH$

$C_{14}H_{26}O_3$ M 242.358
Plates (pet. ether). Mp 87°.

Robinson, G.M. *et al*, *J. Chem. Soc.*, 1926, 2204.
Robinson, G.M. *et al*, *J. Chem. Soc.*, 1930, 745.

6-Oxotetradecanoic acid O-00370

6-Ketomyristic acid
[1619-89-2]

$H_3C(CH_2)_7CO(CH_2)_4COOH$

$C_{14}H_{26}O_3$ M 242.358
Minor constit. of shellac and of milk fat. Cryst. (pet. ether). Mp 70-71°.
Semicarbazone: Mp 129-131°.

Keskin, H., *CA*, 1954, **48**, 10618 (*synth*)
Christie, W.W. *et al*, *J. Chem. Soc.*, 1963, 5768 (*isol*)
Christie, W.W. *et al*, *J. Chem. Soc.*, 1964, 5833 (*isol*)
Wadia, M.S. *et al*, *Tetrahedron*, 1969, **25**, 3841 (*ms*)
Weihrauch, J.L. *et al*, *Lipids*, 1974, **9**, 883 (*ms*)

13-Oxotetradecanoic acid O-00371

13-Oxomyristic acid

$H_3CCO(CH_2)_{11}COOH$

$C_{14}H_{26}O_3$ M 242.358
Mp 75°.
Et ester:
$C_{16}H_{30}O_3$ M 270.411
Mp 36°. $Bp_{1.0}$ 164-166°.
Semicarbazone: Cryst. (EtOH). Mp 105-106°.

Ruzicka, L. *et al*, *Helv. Chim. Acta*, 1927, **10**, 693 (*synth*)
Grimshaw, D.C. *et al*, *J. Chem. Soc.*, 1940, 68 (*synth*)
Haug, P. *et al*, *J. Chem. Soc.*, *Chem. Commun.*, 1967, 1130 (*ms*)
Moore, B.D. *et al*, *Aust. J. Chem.*, 1976, **29**, 1365 (*synth*)

2-Oxoundecanoic acid O-00372

[34634-92-9]

$H_3C(CH_2)_8COCOOH$

$C_{11}H_{20}O_3$ M 200.277
Mp 54°.
Phenylhydrazone: Mp 94°.
Me ester, oxime (E-)*:* [17709-08-9].
$C_{12}H_{23}NO_3$ M 229.319
Needles (C_6H_6). Mp 63.5-64°.
Me ester, 2,4-dinitrophenylhydrazone: [18281-53-3].
Yellow needles (MeOH). Mp 136.5-137.5°.
Et ester: [56275-58-2].
$C_{13}H_{24}O_3$ M 228.331
$Bp_{1.3}$ 110-111°.
Et ester, oxime:
$C_{13}H_{25}NO_3$ M 243.345
Mp 66-67°.

Bowman, R.E. *et al*, *J. Chem. Soc.*, 1952, 3945 (*synth*)
Reinheckel, H., *Monatsh. Chem.*, 1967, **98**, 1217 (*deriv*)
Anatol, J. *et al*, *Synthesis*, 1971, 538 (*synth*)
Anatol, J. *et al*, *Bull. Soc. Chim. Fr.*, 1972, 189 (*synth*)
Lapkin, I.I. *et al*, *Zh. Org. Khim.*, 1976, 14; *CA*, **88**, 190037 (*synth*)

3-Oxoundecanoic acid O-00373

[4435-85-2]

$H_3C(CH_2)_7COCH_2COOH$

$C_{11}H_{20}O_3$ M 200.277
Mp 55°.
2,4-Dinitrophenylhydrazone: Mp 86°.
Oxime:
$C_{11}H_{21}NO_3$ M 215.292
Mp 85-86°.

v. Adickes, F. *et al*, *Justus Liebigs Ann. Chem.*, 1943, **555**, 41 (*synth*)
Angelo, B., *C. R. Hebd. Seances Acad. Sci.*, 1973, **276**, 293 (*synth*)

4-Oxoundecanoic acid O-00374

[22847-06-9]

$H_3C(CH_2)_6COCH_2CH_2COOH$

$C_{11}H_{20}O_3$ M 200.277
Mp 78-79°.

Me ester:
$C_{12}H_{22}O_3$ M 214.304
Bp_1 97-98°.
Et ester: [22769-72-8].
$C_{13}H_{24}O_3$ M 228.331
Bp_5 138-139°, Bp_2 135-140°.
Semicarbazone: [22847-07-0].
Mp 129-130°.

Stenhagen, E., *Ark. Kemi*, 1945, **A20**, 19 (*synth*)
Breusch, F.L. *et al*, *CA*, 1946, **40**, 5400 (*synth*)
Ponomorev, A.A. *et al*, *CA*, 1969, **70**, 114539 (*synth*)
Takeda, A. *et al*, *Bull. Chem. Soc. Jpn.*, 1971, **44**, 1342 (*synth*)
Molina, P. *et al*, *CA*, 1974, **81**, 169065 (*synth*)

5-Oxoundecanoic acid O-00375

[16424-28-5]

$H_3C(CH_2)_5CO(CH_2)_3COOH$

$C_{11}H_{20}O_3$ M 200.277
Mp 62°.
Semicarbazone: Mp 132.5°.

Franke, A. *et al*, *Monatsh. Chem.*, 1936, **69**, 167 (*synth*)
Lukeš, R. *et al*, *CA*, 1936, **30**, 5990 (*synth*)
Ijima, A. *et al*, *Chem. Pharm. Bull.*, 1971, **19**, 1053 (*synth*)
Ijima, A. *et al*, *Chem. Pharm. Bull.*, 1973, **21**, 215.
Fresneda, P.M. *et al*, *CA*, 1979, **90**, 203449 (*synth*)

9-Oxoundecanoic acid O-00376

$H_3CCH_2CO(CH_2)_7COOH$

$C_{11}H_{20}O_3$ M 200.277
Cryst. (pet. ether). Mp 56° (43.5°).
Et ester:
$C_{13}H_{24}O_3$ M 228.331
Yellow oil. Bp_{21} 186°.
Semicarbazone: Mp 161°.

Myddleton, W.W. *et al*, *J. Am. Chem. Soc.*, 1927, **49**, 2258 (*synth*)
Abraham, E.P. *et al*, *J. Chem. Soc.*, 1937, 948 (*synth*)
Wieland, H. *et al*, *Justus Liebigs Ann. Chem.*, 1939, **540**, 103 (*synth*)
Lam, C.M. *et al*, *Chem. Phys. Lipids*, 1976, **16**, 181 (*synth*)

10-Oxoundecanoic acid O-00377

[676-00-6]

$H_3CCO(CH_2)_8COOH$

$C_{11}H_{20}O_3$ M 200.277
Needles (pet. ether). Mp 59.5°. Bp_1 166-167°.
Et ester: [36651-38-4].
$C_{13}H_{24}O_3$ M 228.331
Bp_{12} 169-170°.
Oxime:
$C_{11}H_{21}NO_3$ M 215.292
Cryst. (EtOH). Mp 68-69°.
Semicarbazone: Cryst. (EtOH). Mp 136.5°.

Myddleton, W.W. *et al*, *J. Am. Chem. Soc.*, 1927, **49**, 2258 (*synth*)
Richter, W.J. *et al*, *Helv. Chim. Acta*, 1972, **55**, 242 (*ms*)
Kannan, R. *et al*, *Fette, Seifen, Anstrichm.*, 1974, **76**, 344 (*nmr*)
Rogers, H.R. *et al*, *J. Org. Chem.*, 1975, **40**, 3577 (*synth*)
Lam, C.H. *et al*, *J. Chromatogr.*, 1976, **116**, 425 (*chromatog, synth*)
Dias, J.R. *et al*, *Org. Mass Spectrom.*, 1977, **6**, 385 (*ms*)

2,2′-Oxybisethanamine, 9CI O-00378

2,2′-Oxybisethylamine, 8CI. Di(2-aminoethyl) ether. 2,2′-Diaminodiethyl ether. Bis(2-aminoethyl)ether
[2752-17-2]

$H_2NCH_2CH_2OCH_2CH_2NH_2$

$C_4H_{12}N_2O$ M 104.152
Bp 184°.
▷ Reported synth. via the diazide resulted in explosion.
B,2HCl: [60792-79-2].
Mp 226-227°.
N,N′-*Di-Me:* [3033-62-3].
$C_6H_{16}N_2O$ M 132.205
Bp 180-182°.
▷ KR9460000.
N,N′-*Di-Me, picrate:* Mp 87°.
N,N′-*Di-Et:*
$C_8H_{20}N_2O$ M 160.259
Bp 204°.
N,N,N′,N′-*Tetra-Me:*
$C_8H_{20}N_2O$ M 160.259
Bp_{15} 79-81°.
N,N,N′,N′-*Tetra-Me; B,2HCl:* Mp 163°.
N,N,N′,N′-*Tetra-Me, picrate:* Mp 140°.
N,N,N′,N′-*Tetra-Et:* [3030-43-1].
$C_{12}H_{28}N_2O$ M 216.366
Bp_{15} 121-123°.
N,N,N′,N′-*Tetra-Et; B,2HCl:* Mp 214°.

Gabriel, S., *Ber.*, 1905, **38**, 3411 (*synth*)
Boon, W.R., *J. Chem. Soc.*, 1949, 1378 (*derivs*)
Fakstorp, J. *et al*, *Acta Chem. Scand.*, 1953, **7**, 134 (*derivs*)
Alvarez-Builla, J. *et al*, *Tetrahedron*, 1990, **46**, 967 (*derivs*)
Peacock, R.D., *Chem. Br.*, 1992, 327 (*haz*)
Lewis, R.J., *Sax's Dangerous Properties of Industrial Materials*, 8th Ed., Van Nostrand-Reinhold, 1992, BJH750.

3,3′-Oxybis[1,2-propanediol], 9CI O-00379

α,α′-Diglycerol. Di(2,3-dihydroxypropyl) ether. Bis(2,3-dihydroxypropyl) ether. 2,2′,3,3′-Tetrahydroxydipropyl ether. 4-Oxa-1,2,6,7-heptanetetrol
[627-82-7]

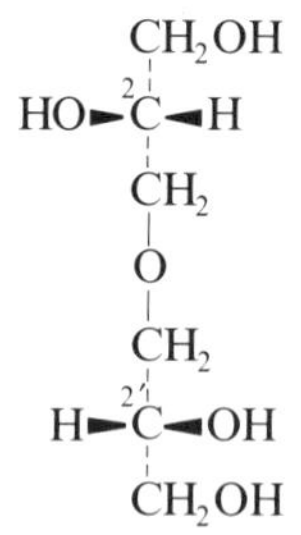

(2*R**,2′*R**)-*form*

$C_6H_{14}O_5$ M 166.174

(2R*,2′R*)-form
(−)-*form*
$Bp_{0.1}$ 175-176°. $[\alpha]_D^{20}$ −8.5°.

(2RS,2′RS)-form [18654-77-8]
(±)-*form*
d_4^{20} 1.281. $Bp_{0.1}$ 175-176°. n_D^{20} 1.4880.
Tetra-Ac: [57348-99-9].
$C_{14}H_{22}O_9$ M 334.322
Bp_2 179°.

(2RS,2′SR)-form
meso-*form*
Liq. d_4^{20} 1.281. n_D^{20} 1.4880.

Wright, H.J. *et al*, *J. Am. Chem. Soc.*, 1946, **68**, 446 (*synth*)

Istin, M., *CA*, 1947, **41**, 2392 (*synth*)
Okamoto, K. *et al*, *CA*, 1955, **49**, 14668 (*synth*)
Buchnea, D. *et al*, *Chem. Phys. Lipids*, 1967, **1**, 177, 247 (*synth*)

P

1-Palmitylthio-2-palmitoylamino-1,2-dideoxy-*sn*-glycero-3-phosphorylcholine **P-00001**

*4-Hydroxy-*N,N,N*-trimethyl-7-[(1-oxohexadecyl)amino]-3,5-dioxa-9-thia-4-phosphapentacosan-1-aminium hydroxide, inner salt, 4-oxide. Thioetheramide-PC*

[116457-99-9]

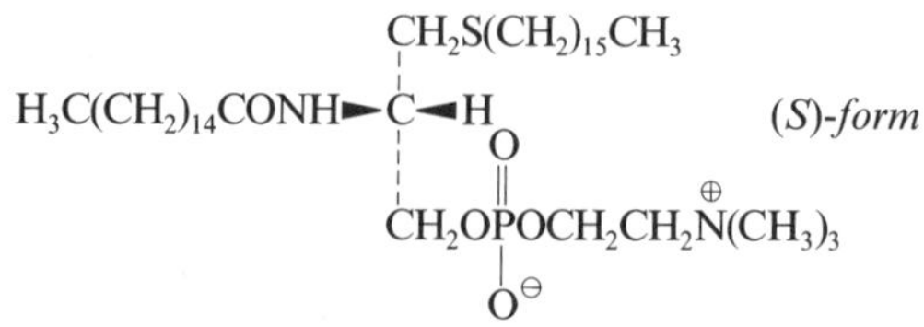

$C_{40}H_{83}N_2O_5PS$ M 735.146

Inhibitor of snake venom. Phospholipase A_2 inhibitor. Solid + $3H_2O$. $[\alpha]_D^{23}$ +9.62° (c, 1.32 in MeOH). Hygroscopic.

Bhatia, S.K. *et al*, *Tetrahedron Lett.*, 1988, **29**, 31 (*synth, pmr*)
Yu, L. *et al*, *J. Biol. Chem.*, 1990, **265**, 2657 (*biochem*)
Kan, C.C. *et al*, *Biochim. Biophys. Acta*, 1991, **1066**, 95 (*biochem*)

Panalicin **P-00002**

[133594-25-9]

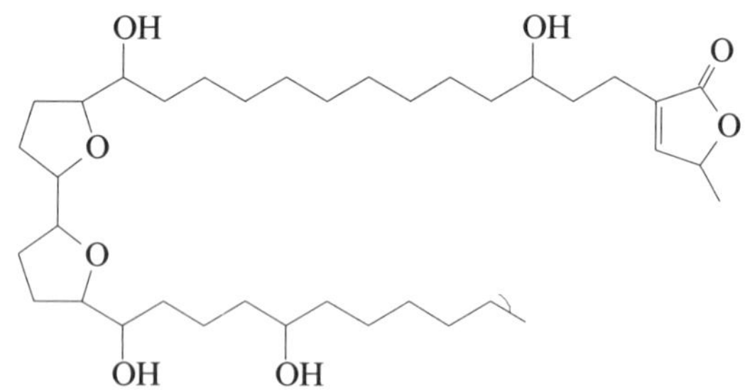

$C_{37}H_{66}O_8$ M 638.924

Constit. of *Uvaria narum*. Waxy solid.

Hisham, A. *et al*, *Phytochemistry*, 1991, **30**, 545 (*isol, pmr, cmr*)

Panaxydol **P-00003**

9,10-Epoxy-1-heptadecene-4,6-diyn-3-ol. 8-(3-Heptyloxiranyl)-1-octene-4,6-diyn-3-ol, 9CI

[72800-72-7]

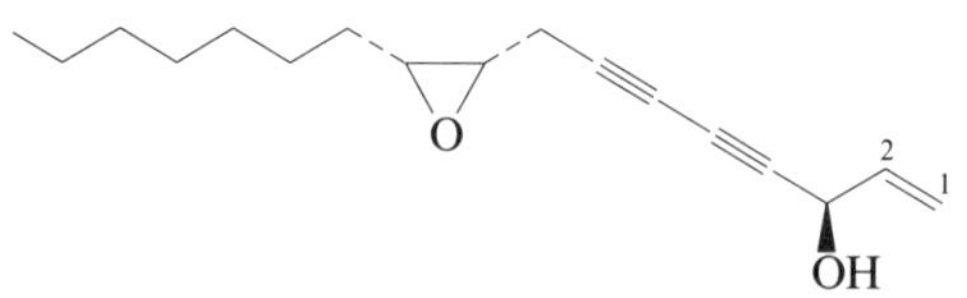

$C_{17}H_{24}O_2$ M 260.375

Isol. from *Panax ginseng*. Also from *P. quinquefolium*. Cytotoxic agent, inhibits growth of leukaemia cells in tissue culture. $[\alpha]_D$ −19.5° (c, 0.6 in MeOH).

1,2-Dihydro, 1-chloro, 2-hydroxy: [114687-51-3]. *1-Chloro-9,10-epoxy-4,6-heptadecadiyne-2,3-diol.* ***Chloropanaxydiol***
$C_{17}H_{25}ClO_3$ M 312.835
Isol. from the dried callus of *P. ginseng*. Cytotoxic agent. $[\alpha]_D$ −37.2° (c, 0.2 in MeOH).

3-Ketone: [126146-63-2]. *8-(3-Heptyloxiranyl)-1-octene-4,6-diyn-3-one, 9CI. 9,10-Epoxy-1-heptadecene-4,6-diyn-3-one.* ***Ginsenoyne E.*** *3-Oxopanaxydol. PQ 3*
$C_{17}H_{22}O_2$ M 258.360
Isol. from *P. quinquefolium* and *P. ginseng*. Shows cytotoxic activity. Pale yellow oil. $[\alpha]_D$ −36.9° (c, 0.68 in MeOH), $[\alpha]_D$ −82.9° (c, 0.8 in $CHCl_3$).

[134731-68-3, 139035-31-7, 139757-60-1]

Poplawski, J. *et al*, *Phytochemistry*, 1980, **19**, 1539 (*isol, struct, synth*)
Fujimoto, Y. *et al*, *Chem. Pharm. Bull.*, 1988, **36**, 4206; 1991, **39**, 521 (*isol, pmr, cmr, struct*)
Chung, C.B. *et al*, *Bull. Korean Chem. Soc.*, 1991, **12**, 122 (*abs config*)
Hirakawa, K. *et al*, *Phytochemistry*, 1991, **30**, 3327 (*isol, ir, cmr, uv, ms, pmr*)

12,14-Pentacosadiynoic acid **P-00004**

[101216-59-5]

$$H_3C(CH_2)_9C{\equiv}CC{\equiv}C(CH_2)_{10}COOH$$

$C_{25}H_{42}O_2$ M 374.606

Solid. Mp 57-59°.

Singh, A. *et al*, *Synth. Commun.*, 1986, **16**, 847 (*synth*)
Xu, Z. *et al*, *J. Org. Chem.*, 1991, **56**, 7183 (*synth, pmr*)

Pentacosanal **P-00005**

[58196-28-4]

$$H_3C(CH_2)_{23}CHO$$

$C_{25}H_{50}O$ M 366.669

Mp 72-76°.

4-Nitrophenylhydrazone: Cryst. (AcOH or MeOH). Mp 104-105°.

Semicarbazone: Cryst. (EtOAc). Mp 115-115.5°.

Reindel, F. *et al*, *Justus Liebigs Ann. Chem.*, 1940, **544**, 116 (*synth*)

Pentacosane **P-00006**

[629-99-2]

$$H_3C(CH_2)_{23}CH_3$$

$C_{25}H_{52}$ M 352.686

Constit. of some natural waxes. Copulation releaser pheromone of whitemarked tussock moth *Orgyia leucostigma*. Mp 53.3°. Bp_{15} 259°.

Matlack, M.B., *J. Org. Chem.*, 1940, **5**, 504 (*isol*)
Rodd's Chem. Carbon Compd. (2nd edn.), 2nd Ed., Vol 1A, 1964, 367 (*bibl*)
Karrer, W. *et al*, *Konstitution und Vorkommen der Organischen Pflanzenstoffe*, 2nd Ed., Birkhäuser Verlag, Basel, 1972-1985, no. 12 (*occur*)
Grant, G.G. *et al*, *J. Chem. Ecol.*, 1987, **13**, 345.

Pentacosanoic acid **P-00007**

Hyenic acid

[506-38-7]

$$H_3C(CH_2)_{23}COOH$$

$C_{25}H_{50}O_2$ M 382.669

Isol. from plant sources e.g. *Agave sisalana*, *Citrus bergamia*, *Shorea maranti*. Cryst. (Me_2CO). Mp 84-85°.

Me ester:
$C_{26}H_{52}O_2$ M 396.696
Cryst. (Me_2CO). Mp 61-62°.
Et ester:
$C_{27}H_{54}O_2$ M 410.722
Cryst. (Me_2CO). Mp 58-59°. $Bp_{0.5}$ 216-217°.
Nitrile:
$C_{25}H_{49}N$ M 363.669
Cryst. (Me_2CO). Mp 58-59°.

Razafindrazaka, J. *et al*, *Bull. Soc. Chim. Fr.*, 1963, 1633 (*isol*)
Schildknecht, H. *et al*, *Fette, Seifen, Anstrichm.*, 1964, **66**, 176 (*synth*)
Sundt, E. *et al*, *Helv. Chim. Acta*, 1964, **47**, 408 (*isol*)
Tember, G.A. *et al*, *CA*, 1974, **80**, 47410 (*synth*)

1-Pentacosanol P-00008

[26040-98-2]

$H_3C(CH_2)_{23}CH_2OH$

$C_{25}H_{52}O$ M 368.685
Isol. from pollen of *Cedrus atlantica* and wood of *Vitex divaricata*. Cryst. (pet. ether or Me_2CO). Mp 78.5-79.5°.

Kessel, F., *Ber.*, 1878, **11**, 2112 (*synth*)
Cocker, W. *et al*, *J. Chem. Soc.*, 1962, 5194 (*isol*)
Schildknecht, H. *et al*, *Fette, Seifen, Anstrichm.*, 1964, **66**, 176 (*synth*)

9-Pentacosanone, 8CI P-00009

[31469-36-0]

$H_3C(CH_2)_{15}CO(CH_2)_7CH_3$

$C_{25}H_{50}O$ M 366.669
Cryst. (C_6H_6). Mp 66°.
Semicarbazone: Mp 39-41°.

Brigl, P., *Hoppe Seyler's Z. Physiol. Chem.*, 1915, **95**, 178 (*synth*)
Akal, B.I., *CA*, 1970, **74**, 99408 (*synth*)

1,7,13-Pentacosatriene P-00010

$H_3C(CH_2)_{10}CH{=}CH(CH_2)_4CH{=}CH(CH_2)_4CH{=}CH_2$

$C_{25}H_{46}$ M 346.638
(Z,Z)-form [63623-55-2]
Found in the cuticle of the male stable fly. All four isomers have been synthesised.

Uebel, E.C. *et al*, *J. Chem. Ecol.*, 1977, **3**, 269.
Sonnet, P.E., *J. Org. Chem.*, 1980, **45**, 154 (*synth*)

9-Pentacosene P-00011

[69188-47-2]

$H_3C(CH_2)_7CH{=}CH(CH_2)_{14}CH_3$

$C_{25}H_{50}$ M 350.670
Component of cuticular lipids of hardback flour beetle *Tenebrio molitor* and *T. obscurus*, of cephalic secretions of anthrophorid bees *Melissodes desponsa* and female *M. denticulata* and of adult black carpet beetle *Attagenus megatoma*. Attractant for housefly.
(Z)-form [51865-00-0]
Pheromone of female housefly *Fannia canicularis*. Chemical trigger released by *Vespa crabro* (hornet) pupae for brood warming.

Sonnet, P. *et al*, *J. Org. Chem.*, 1976, **41**, 3279 (*ir*)
Lockey, K.H., *Insect Biochem.*, 1978, **8**, 237.
Veith, H.J. *et al*, *Naturwissenschaften*, 1978, **65**, 263.
Baker, J., *Insect Biochem.*, 1979, **9**, 335.

11-Pentacosene P-00012

$H_3C(CH_2)_{12}CH{=}CH(CH_2)_9CH_3$

$C_{25}H_{50}$ M 350.670
(Z)-form [56927-04-9]
Constit. of the beetle *Rhagonycha fulva* cuticular lipids.

Jacob, J., *Hoppe Seyler's Z. Physiol. Chem.*, 1978, **359**, 653 (*isol*)

16-Pentacosenoic acid P-00013

$H_3C(CH_2)_7CH{=}CH(CH_2)_{14}COOH$

$C_{25}H_{48}O_2$ M 380.653
(Z)-form [140163-39-9]
Constit. of the sponge *Mycale laevis*.

Carballeira, N.M. *et al*, *J. Nat. Prod. (Lloydia)*, 1992, **55**, 333.

17-Pentacosenoic acid P-00014

$H_3C(CH_2)_6CH{=}CH(CH_2)_{15}COOH$

$C_{25}H_{48}O_2$ M 380.653
(Z)-form [134394-72-2]
Constit. of the lichens *Cladonia* sp. and *Parmelia* sp. and from the sponge *Geodia gibberosa*.

Carballeira, N.M. *et al*, *Lipids*, 1991, **26**, 324.
Dembitskii, V.M. *et al*, *Phytochemistry*, 1991, **30**, 4015; 1992, **31**, 841 (*isol*)

18-Pentacosenoic acid P-00015

$H_3C(CH_2)_5CH{=}CH(CH_2)_{16}COOH$

$C_{25}H_{48}O_2$ M 380.653
(Z)-form [138282-15-2]
Constit. of the sponges *Euspongilla lacustris* and *Mycale laevis*.

Dembitskii, V.M. *et al*, *Comp. Biochem. Physiol., B: Comp. Biochem.*, 1991, **100**, 185 (*isol*)
Carballeira, N.M. *et al*, *J. Nat. Prod. (Lloydia)*, 1992, **55**, 333.

19-Pentacosenoic acid P-00016

$H_3C(CH_2)_4CH{=}CH(CH_2)_{17}COOH$

$C_{25}H_{48}O_2$ M 380.653
(Z)-form [140163-40-2]
Constit. of the sponge *Amphimedon compressa*.
Aldehyde: [140163-48-0]. ***19-Pentacosenal***
$C_{25}H_{48}O$ M 364.654
Constit. of *A. compressa*.

Carballeira, N.M. *et al*, *J. Nat. Prod. (Lloydia)*, 1992, **55**, 333.

2,4-Pentadecadienal P-00017

$H_3C(CH_2)_9CH{=}CHCH{=}CHCHO$

$C_{15}H_{26}O$ M 222.370
(2E,4Z)-form [85769-29-5] ***β-Triticene***
Antifungal constit. of *Triticum aestivum*. Oil.

Spendley, P.J. *et al*, *Phytochemistry*, 1982, **21**, 2403.

5,10-Pentadecadienal P-00018

$H_3C(CH_2)_3CH{=}CH(CH_2)_3CH{=}CH(CH_2)_3CHO$

$C_{15}H_{26}O$ M 222.370
(E,E)-form [64275-42-9]
$Bp_{0.05}$ 103°.

Oxime: [64275-48-5].
$C_{15}H_{27}NO$ M 237.384
Mp 44-45°.
Semicarbazone: Mp 69-70°.
(*5E,10Z*)-*form* [64275-54-3]
$Bp_{0.01}$ 120-130°.
Semicarbazone: Mp 45-55°.
(*Z,Z*)-*form* [64275-49-6]
$Bp_{0.03}$ 98°.
(*5Z,10E*)-*form* [64275-58-7]
$Bp_{0.01}$ 130°.

Ohloff, G. *et al*, *Helv. Chim. Acta*, 1977, **60**, 1161 (*synth, pmr, cmr*)

2,9-Pentadecadiene-4,6-diyne-1,8-diol P-00019

$H_3C(CH_2)_4CH{=}CHCH(OH)C{\equiv}CC{\equiv}CCH{=}CHCH_2OH$

$C_{15}H_{20}O_2$ M 232.322
(*Z,Z*)-*form*
1-Ac: [32768-87-9].
$C_{17}H_{22}O_3$ M 274.359
Constit. of *Azorella trifurcata*. Oil.

Bohlmann, F. *et al*, *Chem. Ber.*, 1971, **104**, 1322 (*isol, uv, ms, ir*)

8,14-Pentadecadiene-4,6-diyne-3,10-diol, 9CI P-00020

$H_3CCH_2CH(OH)C{\equiv}CC{\equiv}CCH{=}CHCH(OH)(CH_2)_3CH{=}CH_2$

$C_{15}H_{20}O_2$ M 232.322
(*E*)-*form*
Di-Ac: [52657-04-2]. *6,13-Diacetoxy-1,7-pentadecadiene-9,11-diyne*
$C_{19}H_{24}O_4$ M 316.396
Isol. from *Thaminophyllum multiflorum*, *Peyrousea umbellata* and *Cotula coronopifolia*. Oil.

Bohlmann, F. *et al*, *Chem. Ber.*, 1966, **99**, 2828; 1974, **107**, 1044 (*isol, struct, ir, uv, ms*)

2,9-Pentadecadiene-4,6-diyn-1-ol, 8CI P-00021

$H_3C(CH_2)_4CH{=}CHCH_2C{\equiv}CC{\equiv}CCH{=}CHCH_2OH$

$C_{15}H_{20}O$ M 216.322
(*Z,Z*)-*form* [41682-34-2]
Constit. of *Bupleurum ranunculoides*, *Hydrocotyle asiatica*, etc. Oil.
Ac: [135118-57-9]. ***Saikodiyne C***
$C_{17}H_{22}O_2$ M 258.360
Constit. of *B. falcatum*.

[41682-28-4]

Bohlmann, F. *et al*, *Chem. Ber.*, 1971, **104**, 2030 (*isol, ms, uv, ir*)
Schulte, K.E. *et al*, *Arch. Pharm. (Weinheim, Ger.)*, 1973, **306**, 197 (*isol*)
Choi, B.K. *et al*, *CA*, 1975, **87**, 71409k.
Morita, M. *et al*, *Phytochemistry*, 1991, **30**, 1543 (*isol, struct, ms, ir, pmr, cmr*)

7,13-Pentadecadiene-9,11-diyn-6-ol P-00022

$H_3C(CH_2)_4CH(OH)CH{=}CHC{\equiv}CC{\equiv}CCH{=}CHCH_3$

$C_{15}H_{20}O$ M 216.322
(*E,E*)-*form* [20311-64-2]
Constit. of *Oenanthe crocata* and *O. aquatica*. Oil.
$Bp_{0.001}$ 100°.

Bohlmann, F. *et al*, *Chem. Ber.*, 1968, **101**, 1163 (*isol*;r1971, **104**, 1362 (*synth, uv, ir, nmr, ms*)
Vincieri, F.F. *et al*, *Planta Med.*, 1985, 107 (*isol, uv, ms, ir*)

7,13-Pentadecadiene-9,11-diyn-4-one, 8CI P-00023

$H_3CCH{=}CHC{\equiv}CC{\equiv}CCH{=}CHCH_2CH_2COCH_2CH_2CH_3$

$C_{15}H_{18}O$ M 214.307
(*E,E*)-*form* [32768-92-6]
Constit. of *Oenanthe aquatica*. Oil.

Bohlmann, F. *et al*, *Chem. Ber.*, 1971, **104**, 1322 (*isol, uv, ir, nmr, ms*)
Vincieri, F.F. *et al*, *Planta Med.*, 1985, 107 (*isol, uv, ms, ir*)

8,10-Pentadecadiene-2,4,6-triyne P-00024

$H_3CC{\equiv}CC{\equiv}CC{\equiv}CCH{=}CHCH{=}CH(CH_2)_3CH_3$

$C_{15}H_{16}$ M 196.291
(*E,E*)-*form* [62885-79-4]
Constit. of *Jungia spectabilis*. Oil.

Bohlmann, F. *et al*, *Phytochemistry*, 1977, **16**, 239 (*uv, nmr, ms*)

5,7-Pentadecadiene-9,11,13-triyn-1-ol, 9CI P-00025

[13894-90-1]

$H_3CC{\equiv}CC{\equiv}CC{\equiv}CCH{=}CHCH{=}CH(CH_2)_3CH_2OH$

$C_{15}H_{16}O$ M 212.291
Cryst. (pet. ether). Mp 46°. λ_{max} 348, 325, 305, 288, 270, 260 nm.
Ac: [125003-04-5].
$C_{17}H_{18}O_2$ M 254.328
Isol. from leaves and stems of *Leuzea centauroides*.

Bohlmann, F. *et al*, *Chem. Ber.*, 1966, **99**, 3544 (*synth, uv*)
Christensen, L.P. *et al*, *Phytochemistry*, 1989, **28**, 2697.

10,12-Pentadecadienoic acid P-00026

$H_3CCH_2CH{=}CHCH{=}CH(CH_2)_8COOH$

$C_{15}H_{26}O_2$ M 238.369
(*10E,12Z*)-*form*
Me ester: [63024-90-8].
$C_{16}H_{28}O_2$ M 252.396
$Bp_{0.03}$ 112-113°.

Bestmann, H.J. *et al*, *Justus Liebigs Ann. Chem.*, 1981, 2117 (*synth, pmr*)

5,10-Pentadecadien-1-ol P-00027

$H_3C(CH_2)_3CH{=}CH(CH_2)_3CH{=}CH(CH_2)_3CH_2OH$

$C_{15}H_{28}O$ M 224.386
All four sterioisomers have been prepared; all are oils with insect pheromone props. and can be used as flavouring and perfume chemicals.

Ohloff, G. *et al*, *Helv. Chim. Acta*, 1977, **60**, 1161.

8,10-Pentadecadien-1-ol P-00028

$H_3C(CH_2)_3CH{=}CHCH{=}CH(CH_2)_6CH_2OH$

$C_{15}H_{28}O$ M 224.386
(*8E,10Z*)-*form* [80625-55-4]
$Bp_{0.05}$ 105-110°.
Ac: [80625-68-9].
$C_{17}H_{30}O_2$ M 266.423

$Bp_{0.01}$ 105-110°.

Bestmann, H.J. *et al*, *Justus Liebigs Ann. Chem.*, 1981, 2117 (*synth, pmr*)

9,11-Pentadecadien-1-ol P-00029

$H_3CCH_2CH_2CH{=}CHCH{=}CH(CH_2)_7CH_2OH$

$C_{15}H_{28}O$ M 224.386

(9E,11Z)-form [63025-03-6]

$Bp_{0.01}$ 110-120°.

Ac: [63025-08-1].

$C_{17}H_{30}O_2$ M 266.423

$Bp_{0.01}$ 95-105°.

Bestmann, H.J. *et al*, *Justus Liebigs Ann. Chem.*, 1981, 2117 (*synth, pmr*)

10,12-Pentadecadien-1-ol P-00030

$H_3CCH_2CH{=}CHCH{=}CH(CH_2)_8CH_2OH$

$C_{15}H_{28}O$ M 224.386

$Bp_{0.01}$ 108-117°.

(10E,12Z)-form [63024-96-4]

Ac: [63025-05-8].

$C_{17}H_{30}O_2$ M 266.423

$Bp_{0.01}$ 123°.

Bestmann, H.J. *et al*, *Justus Liebigs Ann. Chem.*, 1981, 2117 (*synth, pmr*)

11,13-Pentadecadien-1-ol P-00031

$H_3CCH{=}CHCH{=}CH(CH_2)_9CH_2OH$

$C_{15}H_{28}O$ M 224.386

(11E,13Z)-form [80625-61-2]

$Bp_{0.01}$ 110-115°.

Ac: [80625-73-6].

$C_{17}H_{30}O_2$ M 266.423

$Bp_{0.01}$ 110-115°.

Bestmann, H.J. *et al*, *Justus Liebigs Ann. Chem.*, 1981, 2117 (*synth, pmr*)

6,9-Pentadecadien-2-one P-00032

$H_3C(CH_2)_4CH{=}CHCH_2CH{=}CH(CH_2)_3COCH_3$

$C_{15}H_{26}O$ M 222.370

(6Z,9Z)-form [78405-86-4]

Constit. of *Chimonanthus praecox*, *Lindera benzoin* and *Neolitsea sericea*. Oil.

Ueyama, Y. *et al*, *Flavour Fragrance J.*, 1990, **5**, 85 (*occur*)
Anderson, J.E. *et al*, *J. Nat. Prod. (Lloydia)*, 1992, **55**, 71 (*isol, pmr, cmr*)

8,11-Pentadecadien-2-one P-00033

[115006-48-9]

$H_3CCH_2CH_2CH{=}CHCH_2CH{=}CH(CH_2)_5COCH_3$

$C_{15}H_{26}O$ M 222.370

(8Z,11Z)-form [116752-10-4]

Isol. from the roots of *Echinacea* spp.

Immunostimulant.

Bauer, R. *et al*, *Planta Med.*, 1988, **54**, 426; 1991, **57**, 447.

8,13-Pentadecadien-2-one P-00034

[115006-47-8]

$H_3CCH{=}CH(CH_2)_3CH{=}CH(CH_2)_5COCH_3$

$C_{15}H_{26}O$ M 222.370

(8Z,13Z)-form [126948-20-7]

Isol. from the roots of *Echinacea* spp.

Immunostimulant.

Bauer, R. *et al*, *Arzneim.-Forsch.*, 1988, **38**, 276.

Pentadecafluorooctanal, 9CI P-00035

Perfluorooctanal

[335-60-4]

$FC_3(CF_2)_6CHO$

$C_8HF_{15}O$ M 398.071

Used to derivatise primary amines for gc analysis. Liq. Bp 122°. n_D^{20} 1.2913.

Pierce, O.R. *et al*, *J. Am. Chem. Soc.*, 1954, **76**, 300 (*synth*)
Moffat, A.C. *et al*, *Anal. Lett.*, 1970, **3**, 205 (*use*)
Lang, R.W., *Helv. Chim. Acta*, 1988, **71**, 369 (*synth*)

Pentadecanal P-00036

[2765-11-9]

$H_3C(CH_2)_{13}CHO$

$C_{15}H_{30}O$ M 226.401

Isol. from essential oil of *Cinnamomum micranthum* and from lemon oil (*Citrus limon*). Mp 24-25°. Bp_{25} 185°, $Bp_{0.2}$ 103-106°. Readily forms a polymer Mp 69-70°.

2,4-Dinitrophenylhydrazone: Cryst. (EtOH). Mp 107-108°.

Oxime:

$C_{15}H_{31}NO$ M 241.416

Cryst. (EtOH aq.). Mp 86°.

Semicarbazone: Mp 108-109°.

Lauer, W.M. *et al*, *J. Am. Chem. Soc.*, 1941, **63**, 1153 (*synth*)
Fujita, Y. *et al*, *CA*, 1944, **41**, 3509; 1963, **58**, 10506 (*isol*)
Ikeda, R.M. *et al*, *J. Agric. Food Chem.*, 1962, **10**, 98 (*isol*)
Org. Synth., 1971, **51**, 39 (*synth*)

Pentadecane P-00037

[629-62-9]

$H_3C(CH_2)_{13}CH_3$

$C_{15}H_{32}$ M 212.418

Found in some plant oils and waxes. Mp 10°. Bp 270°, Bp_{10} 136°.

▷ RZ1800000.

Vogel, A.I., *J. Chem. Soc.*, 1946, 137 (*synth*)
Bergmann, G. *et al*, *Angew. Chem., Int. Ed. Engl.*, 1972, **11**, 1032 (*cmr*)
Lewis, R.J., *Sax's Dangerous Properties of Industrial Materials*, 8th Ed., Van Nostrand-Reinhold, 1992, PAY750.

2,4-Pentadecanedione P-00038

Lauroylacetone

[53759-23-2]

$H_3C(CH_2)_{10}COCH_2COCH_3$

$C_{15}H_{28}O_2$ M 240.385

Isol. from the root oil of *Ruta graveolens*. Solid. Mp 31°. Bp_3 148-150°, $Bp_{0.2}$ 195-205°.

Swamer, F.W. *et al*, *J. Am. Chem. Soc.*, 1950, **72**, 1352 (*synth*)

Tattje, D.H.E. *et al*, *CA*, 1975, **82**, 13963j (*isol*)
Kerr, J.M. *et al*, *J. Chem. Soc., Perkin Trans. 1*, 1990, 887 (*synth, uv, ir, pmr*)

Pentadecanoic acid — P-00039

[1002-84-2]

$H_3C(CH_2)_{13}COOH$

$C_{15}H_{30}O_2$ M 242.401

Isol. from some plants, eg. bark of *Zanthoxylum carolinianum*. Occurs in lichens. Cryst. (Me_2CO aq.). Mp 53°. Bp_{100} 257°, $Bp_{0.05}$ 138-139°.

▷ RZ1925000.

Me ester: [7132-64-1].
$C_{16}H_{32}O_2$ M 256.428
Cryst. (EtOH aq.). Mp 18.5°.

Et ester: [4114-00-5].
$C_{17}H_{34}O_2$ M 270.454
Cryst. (EtOH aq.). Mp 14°.

Amide: [3843-51-4].
$C_{15}H_{31}NO$ M 241.416
Cryst. (EtOH aq.). Mp 103-104°.

Wynberg, H. *et al*, *J. Am. Chem. Soc.*, 1956, **78**, 1958 (*synth*)
Corey, E.J. *et al*, *J. Am. Chem. Soc.*, 1968, **90**, 5615 (*synth*)
Lewis, R.J., *Sax's Dangerous Properties of Industrial Materials*, 8th Ed., Van Nostrand-Reinhold, 1992, PAZ000.

1-Pentadecanol — P-00040

[629-76-5]

$H_3C(CH_2)_{13}CH_2OH$

$C_{15}H_{32}O$ M 228.417

Isol. from oil of *Cinnamomum micranthum* and wax of *Agave sisalana*. Cryst. Mp 45-46°. $Bp_{0.2}$ 112-114°.

Formyl:
$C_{16}H_{32}O_2$ M 256.428
Cryst. Mp 13.7°. Bp_{30} 201°.

Ruhoff, J.R. *et al*, *J. Am. Chem. Soc.*, 1933, **55**, 3825 (*synth*)
Fujita, Y. *et al*, *CA*, 1944, **41**, 3509 (*isol*)
Dauben, W.G., *J. Am. Chem. Soc.*, 1948, **70**, 1376 (*synth*)
Razafindrazaka, J. *et al*, *Bull. Soc. Chim. Fr.*, 1963, 1633 (*isol*)

3-Pentadecanol, 9CI — P-00041

[53346-71-7]

$H_3C(CH_2)_{11}CH(OH)CH_2CH_3$

$C_{15}H_{32}O$ M 228.417

(−)-*form*
Needles (EtOH). Mp 45°. Bp_{14} 168°. $[\alpha]_D^{25}$ −5.46° (EtOH).
Hydrogen phthalate: Cryst. (pet. ether). Mp 46-47°. $[\alpha]_D^{20}$ −16.98° (EtOH).

(±)-*form*
Mp 38.2-38.8°. Bp_{12} 163°.
Hydrogen phthalate: Cryst. (pet. ether). Mp 54-55°.

Pickard, R.H. *et al*, *J. Chem. Soc.*, 1913, **103**, 1951 (*synth, resoln*)
Breusch, F.L. *et al*, *Chem. Ber.*, 1953, **86**, 669 (*synth*)

8-Pentadecanol, 9CI — P-00042

[1653-35-6]

$H_3C(CH_2)_6CH(OH)(CH_2)_6CH_3$

$C_{15}H_{32}O$ M 228.417

Waxy plates (EtOH aq.). Mp 52-52.6° (49.5-50°). Bp_1 153°, Bp_1 124-124.5°.

Tuot, M. *et al*, *Bull. Soc. Chim. Fr.*, 1947, 1086 (*synth*)
Breusch, F.L. *et al*, *Chem. Ber.*, 1953, **86**, 678 (*synth*)
Shinokai, K. *et al*, *Yakugaku Zasshi*, 1963, **12**, 516 (*synth*)

14-Pentadecanolide — P-00043

15-Methyloxacyclopentadecan-2-one, 9CI

[32539-85-8]

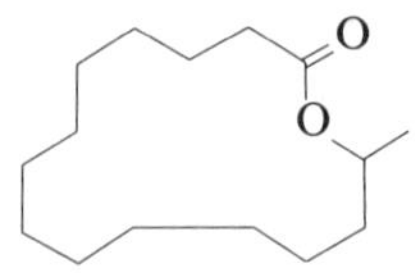

$C_{15}H_{28}O_2$ M 240.385

(−)-*form* [69297-55-8]
Constit. of Galbanum absolute. $[\alpha]_D^{20}$ −20.2° (c, 0.593 in $CHCl_3$).

Becker, J. *et al*, *Helv. Chim. Acta*, 1971, **54**, 2889 (*synth*)
Kaiser, R. *et al*, *Helv. Chim. Acta*, 1978, **61**, 2671.

2-Pentadecanone, 9CI — P-00044

[2345-28-0]

$H_3C(CH_2)_{12}COCH_3$

$C_{15}H_{30}O$ M 226.401

Isol. from hop (*Humulus lupulus*), coconut (*Cocos nucifera*) and other oils. Constit. of male mandibular gland secretions of the Hymenoptera *Philanthus basilaris* and *P. bicinctus*. Mp 39°. Bp 294°, $Bp_{0.7}$ 104-106°.

Semicarbazone: Mp 124-125°.
Oxime: Mp 63-65°.

Krafft, F., *Ber.*, 1879, **12**, 1668.
Baumgarten, P., *Ber.*, 1943, **76**, 213.
Weichet, J. *et al*, *Chem. Listy*, 1957, **51**, 127.
Rieche, A. *et al*, *Z. Chem.*, 1964, **4**, 177.
Czech. Pat., 131 766, (1969); *CA*, **72**, 132025.
McDaniel, C.A. *et al*, *J. Chem. Ecol.*, 1987, **13**, 227.

3-Pentadecanone, 9CI — P-00045

[18787-66-1]

$H_3C(CH_2)_{11}COCH_2CH_3$

$C_{15}H_{30}O$ M 226.401

Cryst. Mp 38°. Bp_{20} 174°.

Semicarbazone: Cryst. (AcOH aq.). Mp 90.5°.

Pickard, R.H. *et al*, *J. Chem. Soc.*, 1913, **103**, 1923.
Breusch, F.L. *et al*, *Chem. Ber.*, 1953, **86**, 684.

8-Pentadecanone, 9CI — P-00046

Diheptyl ketone

[818-23-5]

$H_3C(CH_2)_6CO(CH_2)_6CH_3$

$C_{15}H_{30}O$ M 226.401

Cryst. (EtOH). Mp 43-44°. Bp 178°.

▷ RZ2000000.

Oxime: Mp 19.5-20°.
Semicarbazone: Mp 40-42°, Mp 46-46.7°.

Maihle, A., *C. R. Hebd. Seances Acad. Sci.*, 1913, **157**, 219.
Deffet, L., *Bull. Soc. Chim. Belg.*, 1931, **40**, 385.
Briese, R.R. *et al*, *J. Am. Chem. Soc.*, 1933, **55**, 1697.
Pekrov, A.D. *et al*, *CA*, 1950, **44**, 1887.
Nightingale, D.V. *et al*, *J. Org. Chem.*, 1961, **26**, 2656.
Schoellner, R. *et al*, *J. Prakt. Chem.*, 1965, **27**, 140.
Lethbridge, A. *et al*, *J. Chem. Soc., Perkin Trans. 1*, 1975, 231.

2,4,8,10-Pentadecatetraen-6-yne P-00047

$H_3C(CH_2)_3CH{=}CHCH{=}CHC{\equiv}CCH{=}CHCH{=}CHCH_3$

$C_{15}H_{20}$ M 200.323

(all-E)-form [56644-61-2]

Constit. of *Pittosporum tobira*. Oil.

Bohlmann, F. *et al*, *Chem. Ber.*, 1975, **108**, 2541 (*isol, uv, ir, ms, nmr*)

3,6,9,12-Pentadecatetraen-1-yne P-00048

Laurencenyne

$H_3CCH_2(CH{=}CHCH_2)_4C{\equiv}CH$

$C_{15}H_{20}$ M 200.323

(all-Z)-form [81371-19-9]

Metab. of *Laurencia okamurai*. Liq.

(3E,6Z,9Z,12Z)-form [81066-46-8]

trans-*Laurencenyne*

Isol. from *L. okamurai*. Possible precursor of a wide range of C_{15}-halogenated cyclic ethers found in marine red algae. Liq.

Kigoshi, H. *et al*, *Tetrahedron*, 1986, **47**, 3781 (*struct, synth, bibl*)

1,4,8,10-Pentadecatetraen-6-yn-3-ol P-00049

$H_3C(CH_2)_3CH{=}CHCH{=}CHC{\equiv}CCH{=}CHCH(OH)CH{=}CH_2$

$C_{15}H_{20}O$ M 216.322

(all-E)-form [56644-50-9]

Constit. of *Pittosporum undulatum*. Oil.

Bohlmann, F. *et al*, *Chem. Ber.*, 1975, **108**, 2541 (*isol, ir, ms, uv*)

1,4,8,10-Pentadecatetraen-6-yn-3-one P-00050

$H_3C(CH_2)_3CH{=}CHCH{=}CHC{\equiv}CCH{=}CHCOCH{=}CH_2$

$C_{15}H_{18}O$ M 214.307

(all-E)-form [56644-49-6]

Constit. of *Pittosporum undulatum*. Yellow oil.

Bohlmann, F. *et al*, *Chem. Ber.*, 1975, **108**, 2541 (*isol, ir, uv, ms*)

1,8,10-Pentadecatriene P-00051

$H_3C(CH_2)_3CH{=}CHCH{=}CH(CH_2)_5CH{=}CH_2$

$C_{15}H_{26}$ M 206.370

(8E,10Z)-form [80625-32-7]

$Bp_{0.05}$ 77-82°.

Bestmann, H.J. *et al*, *Justus Liebigs Ann. Chem.*, 1981, 2117.

1,11,13-Pentadecatriene P-00052

$H_3CCH{=}CHCH{=}CH(CH_2)_8CH{=}CH_2$

$C_{15}H_{26}$ M 206.370

(11E,13Z)-form [80625-34-9]

$Bp_{0.8}$ 130°.

Bestmann, H.J. *et al*, *Justus Liebigs Ann. Chem.*, 1981, 2117 (*synth, pmr*)

2,8,10-Pentadecatriene-4,6-diyn-1-al, 8CI P-00053

$H_3C(CH_2)_3CH{=}CHCH{=}CHC{\equiv}CC{\equiv}CCH{=}CHCHO$

$C_{15}H_{16}O$ M 212.291

(all-E)-form [34443-66-8]

Constit. of the roots of *Bupleurum ranunculoides*. Yellow oil.

Bohlmann, F. *et al*, *Chem. Ber.*, 1971, **104**, 2030.

2,8,10-Pentadecatriene-4,6-diyne, 8CI P-00054

$H_3C(CH_2)_3CH{=}CHCH{=}CHC{\equiv}CC{\equiv}CCH{=}CHCH_3$

$C_{15}H_{18}$ M 198.307

(all-E)-form [19812-23-8]

Constit. of *Oenanthe crocata*, *Pittosporum buchanani* and *Jungia spectabilis*. Oil. $Bp_{0.1}$ 80°.

Bohlmann, F. *et al*, *Chem. Ber.*, 1968, **101**, 1163, 1889 (*isol*;h1971, **104**, 1362 (*synth*)

Bohlmann, F. *et al*, *Phytochemistry*, 1977, **16**, 239 (*isol*)

5,7,13-Pentadecatriene-9,11-diyne-1,3-diol, 9CI P-00055

$H_3CCH{=}CHC{\equiv}CC{\equiv}CCH{=}CHCH{=}CHCH_2CH(OH)CH_2CH_2OH$

$C_{15}H_{18}O_2$ M 230.306

Cryst. ($CHCl_3$/pet. ether). Mp 63°.

Di-Ac: ***Centaur*** X_2

$C_{19}H_{22}O_4$ M 314.380

Isol. from *Centaurea macrocephala*. Cryst. (Et_2O). Mp 54-56°. $[\alpha]_D^{20}$ −5° (MeOH).

Hellstroem, B. *et al*, *Acta Chem. Scand.*, 1952, **6**, 1024 (*isol, uv*)

Bohlmann, F. *et al*, *Chem. Ber.*, 1958, **91**, 1631, 1642 (*isol, struct, synth*)

8,10,14-Pentadecatriene-4,6-diyne-3,12-diol, 9CI P-00056

[52657-03-1]

$H_3CCH_2CH(OH)C{\equiv}CC{\equiv}CCH{=}CHCH{=}CHCH(OH)CH_2CH{=}CH_2$

$C_{15}H_{18}O_2$ M 230.306

(E,E)-form

Constit. of *Peyrousea* and *Thaminophyllum* spp. Oil.

Bohlmann, F. *et al*, *Chem. Ber.*, 1974, **107**, 1044 (*isol, uv, ir, ms*)

1,8,10-Pentadecatriene-4,6-diyn-3-ol, 8CI P-00057

$H_3C(CH_2)_3CH{=}CHCH{=}CHC{\equiv}CC{\equiv}CCH(OH)CH{=}CH_2$

$C_{15}H_{18}O$ M 214.307

(E,E)-form [19833-08-0]

Constit. of *Pittosporum buchanani*. Oil.

Bohlmann, F. *et al*, *Chem. Ber.*, 1968, **101**, 1889 (*isol*; 1971, **104**, 1362 (*synth, nmr*)

2,8,10-Pentadecatriene-4,6-diyn-1-ol, 8CI P-00058

$H_3C(CH_2)_3CH{=}CHCH{=}CHC{\equiv}CC{\equiv}CCH{=}CHCH_2OH$

$C_{15}H_{18}O$ M 214.307

(all-E)-form [32768-93-7]

Constit. of *Bupleurum ranunculoides*.

Bohlmann, F. *et al*, *Chem. Ber.*, 1971, **104**, 2030.

5,7,13-Pentadecatriene-9,11-diyn-4-ol, 9CI P-00059

$$\begin{array}{c} H_3CCH_2CH_2CH(OH) \\ CH{=}CHCH{=}CHC{\equiv}CC{\equiv}CCH{=}CHCH_3 \end{array}$$

$C_{15}H_{18}O$ M 214.307

(all-E)-form [20311-66-4]

Constit. of *Oenanthe crocata* and *O. aquatica*. Cryst. (pet. ether) or oil. Mp 34°.

(5E,7Z,13E)-form [97915-40-7]

Isol. from *O. aquatica*. Oil.

Bohlmann, F. *et al*, *Chem. Ber.*, 1968, **101**, 1163; 1971, **104**, 1362 (*synth*)
Vincieri, F.F. *et al*, *Planta Med.*, 1985, 107 (*isol, struct, uv, pmr, ms, ir*)
Vincieri, F.F. *et al*, *Chem. Ber.*, 1986, **119**, 2843 (*uv*)

1,8,10-Pentadecatriene-4,6-diyn-3-one, 8CI P-00060

$$H_3C(CH_2)_3CH{=}CHCH{=}CHC{\equiv}CC{\equiv}CCOCH{=}CH_2$$

$C_{15}H_{16}O$ M 212.291

(E,E)-form [19812-25-0]

Constit. of *Pittosporum buchanani*. Unstable oil.

Bohlmann, F. *et al*, *Chem. Ber.*, 1968, **101**, 1889; 1971, **104**, 1362 (*synth*)

5,7,13-Pentadecatriene-9,11-diyn-4-one, 9CI P-00061

$$H_3CCH{=}CHC{\equiv}CC{\equiv}CCH{=}CHCH{=}CHCOCH_2CH_2CH_3$$

$C_{15}H_{16}O$ M 212.291

(E,E,E)-form [20311-67-5]

Isol. from fruits of *Oenanthe aquatica*. Pale-yellow needles. Mp 107.5°. λ_{max} 354, 337, 319, 229, 273, 261, 210 nm.

(5E,7Z,13E)-form [97915-37-2]

Isol. from fruits of *O. aquatica*. λ_{max} 351, 336, 318 (300), 272, 260 nm (hexane).

Bohlmann, F. *et al*, *Chem. Ber.*, 1968, **101**, 1163 (*synth, pmr, uv, ir*)
Vincieri, F.F. *et al*, *Planta Med.*, 1985, 107 (*isol, uv, ms*)

6,9,12-Pentadecatrien-2-one P-00062

$$H_3CCH_2CH{=}CHCH_2CH{=}CHCH_2CH{=}CH(CH_2)_3COCH_3$$

$C_{15}H_{24}O$ M 220.354

(6Z,9Z,12Z)-form [139328-79-3]

Constit. of the ripe berries of *Lindera benzoin*. Oil.

Anderson, J.E. *et al*, *J. Nat. Prod. (Lloydia)*, 1992, **55**, 71 (*isol, pmr*)

8,11,13-Pentadecatrien-2-one P-00063

$$H_3CCH{=}CHCH{=}CHCH_2CH{=}CH(CH_2)_5COCH_3$$

$C_{15}H_{24}O$ M 220.354

(8E,11Z,13Z)-form [118853-86-4]

Isol. from the roots of *Echinacea* spp.

(8Z,11E,13Z)-form [118853-87-5]

Isol. from the roots of *E.* spp.

Bauer, R. *et al*, *Planta Med.*, 1988, **54**, 426; 1989, **55**, 367; 1991, **57**, 447 (*isol*)

4-(8,11,13-Pentadecatrienyl)-1,2-benzenediol, 9CI P-00064

4-(8,11,13-Pentadecatrienyl)catechol

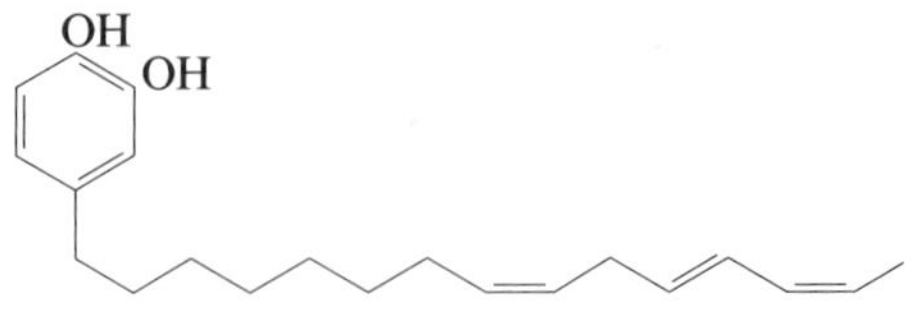

$C_{21}H_{30}O_2$ M 314.467

(8Z,11E,13Z)-form [89838-24-4]

Present in *Rhus vernicifera* (Japanese lac).

Du, Y. *et al*, *J. Chromatogr.*, 1984, **284**, 463; **295**, 179 (*occur, pmr*)

5-(8,11,14-Pentadecatrienyl)-1,3-benzenediol, 9CI P-00065

5-(8,11,14-Pentadecadienyl)resorcinol

[111047-32-6]

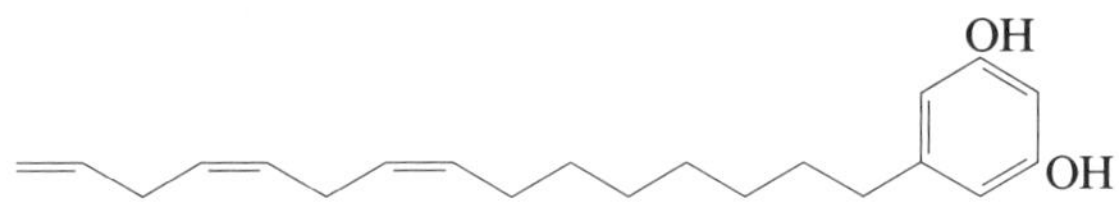

$C_{21}H_{30}O_2$ M 314.467

(Z,Z)-form [79473-24-8]

Constit. of *Anacardium occidentale*.

14,15-Dihydro: [79473-25-9]. *5-(8,11-Pentadecadienyl)-1,3-benzenediol. 5-(8,11-Pentadecadienyl)resorcinol*

$C_{21}H_{32}O_2$ M 316.483

Isol. from *A. occidentale*.

11,12,14,15-Tetrahydro: [22910-86-7]. *5-(8-Pentadecenyl)-1,3-benzenediol, 9CI. 5-(8-Pentadecenyl)resorcinol.* ***Bilobol.*** *Trifurcatol A_2*

$C_{21}H_{34}O_2$ M 318.498

Isol. from *Ginkgo biloba* fruits, *Hakea trifurcata* and several members of the Proteaceae. Cryst. (pentane). Mp 36-37°.

Cirigottis, K.A. *et al*, *Aust. J. Chem.*, 1974, **27**, 345 (*Bilobol*)
Baylis, C.J. *et al*, *J. Chem. Soc., Perkin Trans. 1*, 1981, 132 (*isol, pmr, synth*)
Tyman, J.H.P. *et al*, *J. Chromatogr.*, 1981, **213**, 287 (*isol*)
Scannell, R.T. *et al*, *J. Am. Chem. Soc.*, 1988, **110**, 3650 (*isol*)
Bruce, I.E. *et al*, *J. Liq. Chromatogr.*, 1990, **13**, 2103 (*isol*)

2,4,8-Pentadecatrien-6-yne, 9CI P-00066

$$H_3C(CH_2)_5CH{=}CHC{\equiv}CCH{=}CHCH{=}CHCH_3$$

$C_{15}H_{22}$ M 202.339

(all-E)-form

Constit. of *Pittosporum tobira*. Oil.

Bohlmann, F. *et al*, *Chem. Ber.*, 1975, **108**, 2541.

2,4,9-Pentadecatrien-6-yne, 9CI P-00067

$$H_3C(CH_2)_4CH{=}CHCH_2C{\equiv}CCH{=}CHCH{=}CHCH_3$$

$C_{15}H_{22}$ M 202.339

(2E,4E,9Z)-form [56644-62-3]

Constit. of *Pittosporum tobira*. Oil.

Bohlmann, F. *et al*, *Chem. Ber.*, 1975, **108**, 2541.

3,6,9-Pentadecatrien-1-yne P-00068

$H_3C(CH_2)_4CH{=}CHCH_2CH{=}CHCH_2CH{=}CHC{\equiv}CH$

$C_{15}H_{22}$ M 202.339

(all-Z)-form [81344-95-8] ***Neolaurencenyne***
Isol. from *Laurencia okamurai.*

(3E,6Z,9Z)-form
trans-*Neolaurencenyne*
Constit. of *L. okamurai.* Possible precursor of a wide range of C_{15}-halogenated cyclic ethers found in marine red algae. Liq.

Kigoshi, H. *et al*, *Tetrahedron*, 1986, **42**, 3781 (*struct, synth, bibl*)

3,9,12-Pentadecatrien-1-yne-6,7-diol P-00069
Laurediol

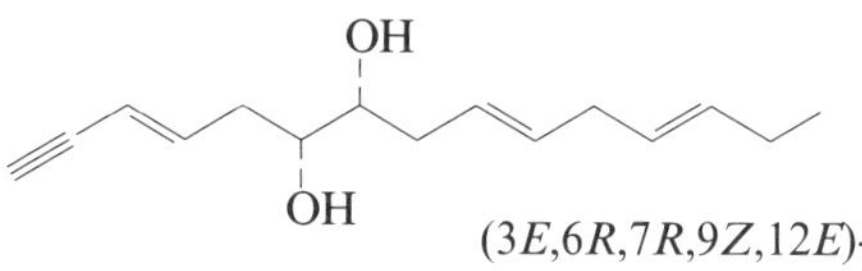

(3*E*,6*R*,7*R*,9*Z*,12*E*)-*form*

$C_{15}H_{22}O_2$ M 234.338

(3E,6R,7R,9Z,12E)-form [37866-22-1]
Constit. of *Laurencia nipponica* (Yamada oil). Unstable cryst. Mp 26-28°. $[\alpha]_D$ +27.2°.

(3Z,6S,7S,9Z,12E)-form [37866-25-4]
Constit. of *L. nipponica.* Unstable oil. $[\alpha]_D$ −19.6°.

Kurosawa, E. *et al*, *Tetrahedron Lett.*, 1972, 2121 (*synth*)

2,5,8-Pentadecatriyn-1-ol P-00070
[19960-61-3]

$H_3C(CH_2)_5C{\equiv}CCH_2C{\equiv}CCH_2C{\equiv}CCH_2OH$

$C_{15}H_{20}O$ M 216.322
Mp 34-37°. $Bp_{0.01}$ 147-150°.

Beerthius, R.K. *et al*, *Recl. Trav. Chim. Pays-Bas* (*J. R. Neth. Chem. Soc.*), 1968, **87**, 461 (*synth, ir, nmr*)

10-Pentadecenal P-00071
[60671-80-9]

$H_3C(CH_2)_3CH{=}CH(CH_2)_8CHO$

$C_{15}H_{28}O$ M 224.386

Kubo, I. *et al*, *Tetrahedron*, 1987, **43**, 2653 (*synth, ir, pmr*)

1-Pentadecene, 9CI P-00072
[13360-61-7]

$H_3C(CH_2)_{12}CH{=}CH_2$

$C_{15}H_{30}$ M 210.402
Occurs in beef, oakmoss oleoresin and *Othonna* spp. Also a component of the defence secretion of the beetles *Adelum percatum* and *A. postulosum.* Sex pheromone of the confused flour beetle *Tribolium confusum.* Mp −2.8°. Bp 268.2°, Bp_{10} 133-137°.

Schmidt, A.W. *et al*, *Ber.*, 1941, **74**, 1313.
Vinnits'kii, O.M. *et al*, *Neftekhimiya*, 1969, **9**, 767 (*synth*)
Keville, R. *et al*, *J. Insect Physiol.*, 1975, **21**, 81.

7-Pentadecene, 9CI P-00073
[15430-98-5]

$H_3C(CH_2)_6CH{=}CH(CH_2)_5CH_3$

$C_{15}H_{30}$ M 210.402

(E)-form [16416-37-8]
Occurs in Dufour's gland of ant and in corn oil. Bp_7 121.5-122°, Bp_{20} 140°.

Asinger, F. *et al*, *Ber.*, 1943, **76**, 585 (*synth*)
Villieras, J., *Bull. Soc. Chim. Fr.*, 1967, 1511 (*synth, ir*)
Kocienski, P.J. *et al*, *J. Chem. Soc., Perkin Trans. 1*, 1978, 829 (*ir, ms, pmr, synth*)

13-Pentadecene-9,11-diyn-6-one, 9CI P-00074

$H_3CCH{=}CHC{\equiv}CC{\equiv}CCH_2CH_2CO(CH_2)_3CH_2CH_3$

$C_{15}H_{20}O$ M 216.322

(E)-form [97915-38-3]
Isol. from fruits of *Oenanthe aquatica.* Oil. λ_{max} 282, 266, 252, 239 nm (hexane).

Vincieri, F.F. *et al*, *Planta Med.*, 1985, 107 (*isol, uv, ms, ir*)

6-Pentadecenoic acid P-00075

$H_3C(CH_2)_7CH{=}CH(CH_2)_4COOH$

$C_{15}H_{28}O_2$ M 240.385
$Bp_{0.01}$ 135°.

[45210-00-2]

Scholz, D., *Justus Liebigs Ann. Chem.*, 1984, 264 (*synth, ir, pmr*)

1-Pentadecen-3-ol P-00076

$H_3C(CH_2)_{11}CH(OH)CH{=}CH_2$

$C_{15}H_{30}O$ M 226.401
Mp 27-28°. Bp_6 150-152°, $Bp_{0.05}$ 90°.

Delahy, R. *et al*, *Bull. Soc. Chim. Fr.*, 1937, **4**, 738.
Weichert, R. *et al*, *Chem. Listy*, 1957, **51**, 127 (*synth*)

9-Pentadecen-1-ol, 9CI P-00077

$H_3C(CH_2)_4CH{=}CH(CH_2)_7CH_2OH$

$C_{15}H_{30}O$ M 226.401

(Z)-form [56218-94-1]
Inhibitor of the sex pheromone of *Spodoptera litura* $Bp_{1.9}$ 141-142° (≥95% (*Z*)-form), $Bp_{0.01}$ 108-111°.

Ac: [35835-77-9]. *9-Pentadecenyl acetate*
$C_{17}H_{32}O_2$ M 268.439
Inhibitor of the sex pheromones of *S. litura* and the summer fruit moth, *Adoxophyes orana.* Attractant for *Acrobasis rufilimbalis.* $Bp_{0.01}$ 118°.

[64437-40-7, 64437-41-8]

Voerman, S. *et al*, *Environ. Entomol.*, 1974, **3**, 701 (*activity*)
Bestmann, H.J. *et al*, *Chem. Ber.*, 1975, **108**, 3582 (*synth*)
Voerman, S. *et al*, *J. Chem. Ecol.*, 1975, **1**, 423 (*activity*)
Ando, T. *et al*, *Agric. Biol. Chem.*, 1977, **41**, 1485 (*synth, activity*)
Horiike, M. *et al*, *Agric. Biol. Chem.*, 1980, **44**, 257 (*synth, pmr*)
Hirano, C. *et al*, *Appl. Entomol. Zool.*, 1980, **15**, 502 (*activity*)
Bestmann, H.J. *et al*, *Z. Naturforsch., C*, 1990, **45**, 698 (*props*)

11-Pentadecen-1-ol P-00078

$H_3CCH_2CH_2CH{=}CH(CH_2)_9CH_2OH$

$C_{15}H_{30}O$ M 226.401

(*E*)-*form* [69222-14-6]

$Bp_{0.02}$ 112-113°. n_D^{26} 1.4542.

Svirskaya, P.I. *et al*, *J. Chem. Eng. Data*, 1979, **24**, 152 (*synth, ir, pmr*)

12-Pentadecen-1-ol, 9CI P-00079

$H_3CCH_2CH{=}CH(CH_2)_{10}CH_2OH$

$C_{15}H_{30}O$ M 226.401

(*Z*)-*form*

Ac: [70711-45-4]. *12-Pentadecenyl acetate*

$C_{17}H_{32}O_2$ M 268.439

Minor component of the sex pheromone of small ermine moth, *Yponomeuta* spp., and of European oak leaf roller *Tortrix viridana.*

[69222-15-7, 73304-17-3]

Svirskaya, P.I. *et al*, *J. Chem. Eng. Data*, 1979, **24**, 152 (*synth*)
Arn, H. *et al*, *Z. Naturforsch.*, *C*, 1979, **34**, 1281 (*isol*)
Van der Pers, J.N.C. *et al*, *Entomol. Exp. Appl.*, 1982, **31**, 255 (*activity*)
Pan, Y. *et al*, *Gaodeng Xuexiao Huaxue Xuebao*, 1990, **11**, 900; *CA*, **114**, 101398p (*synth, acetate*)

13-Pentadecen-1-ol P-00080

15-Hydroxy-2-pentadecene

$HOCH_2(CH_2)_{11}CH{=}CHCH_3$

$C_{15}H_{30}O$ M 226.401

Insect attractant. Cryst. (pet. ether). Mp 40-40.5°. Bp_8 170-172°.

Phenylurethane: Cryst. (EtOH). Mp 68-68.4°.

(*E*)-*form*

Ac: [39639-20-8].

Insect attractant.

Chuit, P. *et al*, *Helv. Chim. Acta*, 1927, **10**, 130.
Schwartz, H., *Synthesis*, 1972, 567.

14-Pentadecen-1-ol, 8CI P-00081

15-Hydroxy-1-pentadecene

[16346-16-0]

$HOCH_2(CH_2)_{12}CH{=}CH_2$

$C_{15}H_{30}O$ M 226.401

Cryst. (pet. ether). Mp 32-33°. Bp_{10} 170-172°, $Bp_{0.01}$ 116-120°.

Phenylurethane: Cryst. (EtOH). Mp 60.5-61°.

Chuit, P. *et al*, *Helv. Chim. Acta*, 1927, **10**, 113 (*synth*)
Luning, B. *et al*, *Acta Chem. Scand.*, 1967, **21**, 829 (*synth*)

3-(8-Pentadecenyl)phenol, 9CI P-00082

[52302-99-5]

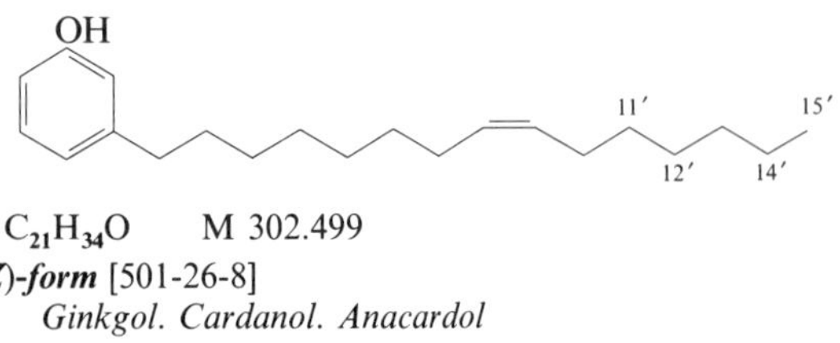

$C_{21}H_{34}O$ M 302.499

(*Z*)-*form* [501-26-8]

Ginkgol. Cardanol. Anacardol

Major constit. of lipids of *Anacardium occidentale* (cashew nut). Isol. from fruit of *Ginkgo biloba*. Oil. Bp_4 221-223°.

Me ether: 1-(8-Pentadecenyl)-3-methoxybenzene

$C_{22}H_{36}O$ M 316.526

Oil. Bp_7 224-228°.

11′,12′-Didehydro (Z-)*:* [51546-63-5]. *3-(8Z,11Z-Pentadecadienyl)phenol.* ***Cardanoldiene***

$C_{21}H_{32}O$ M 300.483

Minor component of lipids of *A. occidentale*. Oil.

11′,12′,14′,15′-Tetradehydro (Z,Z-)*:* [79353-39-2]. *3-(8,11,14-Pentadecatrienyl)phenol*

$C_{21}H_{30}O$ M 298.467

Constit. of lipids of *A. occidentale*.

Dihydro: see *3-Pentadecylphenol, P-00086*

[31910-56-2, 63502-71-6]

Loev, B. *et al*, *J. Am. Chem. Soc.*, 1958, **80**, 643 (*struct*)
Tyman, J.H.P. *et al*, *J. Chromatogr.*, 1967, **27**, 287 (*isol*)
Tyman, J.H.P. *et al*, *Chem. Ind.* (*London*), 1973, 40 (*synth*)
Tyman, J.H.P., *Chem. Soc. Rev.*, 1979, **8**, 499 (*rev*)
Caplin, J. *et al*, *J. Chem. Res.*, *Synop.*, 1982, 34 (*synth*)

Pentadecylamine P-00083

1-Aminopentadecane. 1-Pentadecanamine, 9CI

[2570-26-5]

$H_3C(CH_2)_{13}CH_2NH_2$

$C_{15}H_{33}N$ M 227.432

Cryst. Mp 36.5° (33.5°). Bp 298-301°, Bp_1 120.5°. pK_a 10.61 (25°).

▷ RZ2120000.

B,HCl: [1838-05-7].

Cryst. (EtOH). Mp 199°.

▷ RZ2134000.

N-*Ac:* Cryst. (pet. ether). Mp 72°.

Naegeli, C. *et al*, *Helv. Chim. Acta*, 1929, **12**, 227 (*synth*)
Ralston, A.W. *et al*, *Ind. Eng. Chem.*, 1940, **32**, 1093.
Murr, B.L. *et al*, *J. Am. Chem. Soc.*, 1955, **77**, 1684 (*synth*)
Krivschich, S.Y., *Toksikol. Nov. Prom. Khim. Veshchestv*, 1968, **10**, 124; *CA*, **71**, 47807 (*tox*)
Lewis, R.J., *Sax's Dangerous Properties of Industrial Materials*, 8th Ed., Van Nostrand-Reinhold, 1992, PBA000.

3-Pentadecyl-1,2-benzenediol, 9CI P-00084

3-Pentadecylresorcinol. 3-Pentadecylpyrocatechol, 8CI. 1,2-Dihydroxy-3-pentadecylbenzene. 3-Pentadecylcatechol. Hydrourushiol

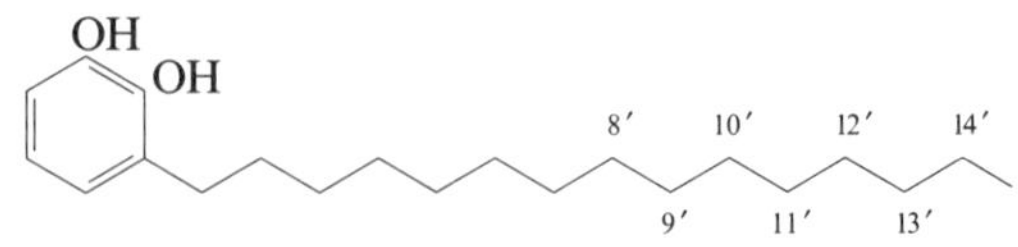

$C_{21}H_{36}O_2$ M 320.514

Minor component of Urushiol, the vesicant principle of *Rhus toxicodendron* (poison ivy) and Japanese lac (*R. vernicifera*), also isol. from other *R.* spp., *Semecarpus vernicrifera* and other plants. Irritant. Needles (pet. ether). Mp 59-60°.

▷ Vesicant.

Di-Ac: [73881-19-3].

$C_{25}H_{40}O_4$ M 404.589

Cryst. (MeOH). Mp 50-51°.

1-Me ether: [16825-58-4]. *2-Methoxy-6-pentadecylphenol*

$C_{22}H_{38}O_2$ M 334.541

Hair-like needles. Mp 43-43.5°.

Di-Me ether: [7461-75-8]. *1,2-Dimethoxy-3-pentadecylbenzene, 9CI*

$C_{23}H_{40}O_2$ M 348.568
Cryst. (EtOH). Mp 36-37°.
10',11'-Didehydro: [83532-37-0]. *3-(10-Pentadecenyl)-1,2-benzenediol.* ***Renghol***
$C_{21}H_{34}O_2$ M 318.498
Isol. from *S. heterophylla* and *Lithraea caustica.* $Bp_{0.001}$ 170-172°.
▷ Vesicant.
8',9'-Didehydro (Z-)*:* [2764-91-2]. *3-(8-Pentadecenyl)-1,2-benzenediol. 3-(8-Pentadecenyl)pyrocatechol.* ***Bhilawanol A.*** *Urushiol monoene*
$C_{21}H_{34}O_2$ M 318.498
Major constit. of Urushiol from *R. toxicodendron* (poison ivy) and *R. vernicifera*, isol. from the Indian bhilawan nut *S. anacardium*, and *S.* spp. Shows antibacterial props. Pale yellow oil. n_D^{25} 1.5115.
▷ Vesicant.
8',9',11',12'-Tetradehydro (Z,Z-)*:* [492-91-1]. *3-(8,11-Pentadecadienyl)-1,2-benzenediol*
$C_{21}H_{32}O_2$ M 316.483
Major constit. of Urushiol from *R. toxicodendron* and *R. vernicifera.* Other old names for Urushiol are Urushin, Urushic acid, Toxicodendrin and Bhilawanol.
8',9',11',12',13',14'-Hexadehydro: [21104-16-5]. *3-(8,11,13-Pentadecatrienyl)-1,2-benzenediol. 3-(8,11,13-Pentadecatrienyl)pyrocatechol*
$C_{21}H_{30}O_2$ M 314.467
Component of Urushiol from *R. vernicifera*, also from *Toxicodendron vernix* (poison sumac). At least three isomers, (8*Z*,11*E*,13*Z*), (8*Z*,11*E*,13*E*) and (8*Z*,11*Z*,13*Z*) have been reported.
▷ Causes dermititis.
8',9',11',12',14',15'-Hexadehydro (Z,Z,Z-)*:* [2790-58-1]. *3-(8,11,14-Pentadecatrienyl)-1,2-benzenediol*
$C_{21}H_{30}O_2$ M 314.467
Component of Urushiol from *R. toxicodendron, R. vernicifera* and *T. radicans.* Oil. n_D^{25} 1.5250.
▷ Vesicant.

[83532-38-1, 92687-22-4]

Backer, H.J. *et al, Recl. Trav. Chim. Pays-Bas* (*J. R. Neth. Chem. Soc.*), 1938, **57**, 225 (*Renghol*)
Dawson, C.R. *et al, J. Am. Chem. Soc.*, 1946, **68**, 534 (*synth*)
Markiewitz, K.H. *et al, J. Org. Chem.*, 1965, **30**, 1610.
Govindachari, T.R. *et al, Indian J. Chem.*, 1971, **9**, 1044 (*isol*)
Dawson, C.R. *et al, J. Med. Chem.*, 1971, **14**, 729 (*synth*)
Tyman, J.H.P. *et al, Chem. Ind.* (*London*), 1974, 526 (*synth*)
Tyman, J.H.P., *Chem. Soc. Rev.*, 1979, **8**, 499 (*rev*)
Halim, H. *et al, J. Chem. Soc., Perkin Trans. 1*, 1980, 2331 (*synth*)
Yamauchi, Y. *et al, J. Chromatogr.*, 1982, **243**, 71 (*isol, Renghol*)
El Sohly, M.A. *et al, J. Nat. Prod.* (*Lloydia*), 1982, **45**, 532 (*isol, pmr, cmr, deriv*)
Wenkert, E. *et al, J. Am. Chem. Soc.*, 1983, **105**, 2021 (*synth, spectra*)
Adawadkar, P.D. *et al, Phytochemistry*, 1983, **22**, 1280 (*isol*)
Du, Y. *et al, J. Chromatogr.*, 1984, **284**, 463; **295**, 79 (*occur, pmr*)
Gambaro, V. *et al, Planta Med.*, 1986, 20 (*Renghol*)
Sargent, M.V. *et al, J. Chem. Soc., Perkin Trans. 1*, 1990, 1429 (*synth, pmr, cmr, ms*)
Mikayoshi, T. *et al, Nippon Kagaku Kaishi*, 1990, 1271; *CA*, **114**, 101237k (*synth, config*)
Bestmann, H.J. *et al, Synlett*, 1990, 555 (*synth*)
Mikayashi, T. *et al, Synthesis*, 1990, 407 (*synth*)
Sax, N.I., *Dangerous Properties of Industrial Materials*, 5th Ed., Van Nostrand-Reinhold, 1979, 889.

2-Pentadecylcyclopropanecarboxylic acid, 9CI P-00085

2,3-Methyleneoctadecanoic acid
[40225-67-0]

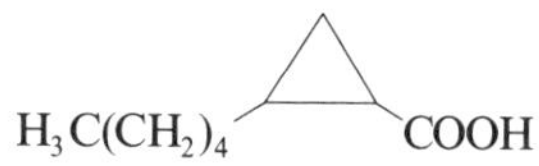

$C_{19}H_{36}O_2$ M 296.492
Both *cis-* and *trans-*forms have been synthesised.

Christie, W.W. *et al, Chem. Phys. Lipids*, 1968, **2**, 196 (*synth, glc, ir, nmr*)
Christie, W.W., *Top. Lipid Chem.*, 1970, **1**, 4 (*rev*)
Gunstone, F.D. *et al, Chem. Phys. Lipids*, 1973, **10**, 303 (*synth, glc, nmr*)

3-Pentadecylphenol P-00086

1-Hydroxy-3-pentadecylbenzene. Cyclogallipharol. Hydroginkgol
[501-24-6]

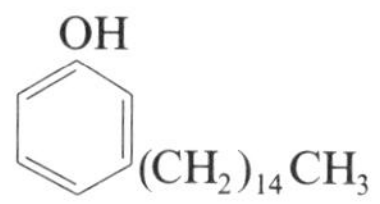

$C_{21}H_{36}O$ M 304.515
Minor constit. of lipids of *Anacardium occidentale* (cashew nut). Cryst. (pet. ether). Mp 54.5°. $Bp_{1.5}$ 197°.
Ac: [52122-74-4].
$C_{23}H_{38}O_2$ M 346.552
Mp 42°.
Benzoyl:
$C_{28}H_{40}O_2$ M 408.623
Mp 54°.
Me ether: 1-Methoxy-3-pentadecylbenzene
$C_{22}H_{38}O$ M 318.542
Mp 29-30°. Bp_{15} 240-244°, $Bp_{1.5}$ 175-177°.

Prelog, V. *et al, Helv. Chim. Acta*, 1947, **30**, 675 (*synth*)
Kaufmann, H.P. *et al, Fette, Seifen, Anstrichm.*, 1967, **69**, 437 (*isol*)
Tyman, J.H.P., *Chem. Soc. Rev.*, 1979, **8**, 499 (*rev*)
Sax, N.I., *Dangerous Properties of Industrial Materials*, 5th Ed., Van Nostrand-Reinhold, 1979, 889.

2-Pentadecynoic acid P-00087

$$H_3C(CH_2)_{11}C{\equiv}CCOOH$$

$C_{15}H_{26}O_2$ M 238.369
Cryst. (Me_2CO at −20°). Mp 42-43.7°.

Grimmer, G. *et al, Justus Liebigs Ann. Chem.*, 1965, **685**, 154 (*synth*)
Kripprath, W.G. *et al, Lipids*, 1966, **1**, 81 (*synth*)

14-Pentadecynoic acid P-00088

$$HC{\equiv}C(CH_2)_{12}COOH$$

$C_{15}H_{26}O_2$ M 238.369
Cryst. Mp 68-70°.

Christie, W.W. *et al, Chem. Phys. Lipids*, 1967, **1**, 407 (*synth*)

14-Pentadecyn-1-ol P-00089

[18202-13-6]

$HC{\equiv}C(CH_2)_{12}CH_2OH$

$C_{15}H_{28}O$ M 224.386

Cryst. Mp 35-36°.

Christie, W.W. *et al*, *Chem. Phys. Lipids*, 1967, **1**, 407 (*synth*)

8,16,26,34,36-Pentahydroxy-2,6,10,14,18,24,28,32-hentetracontaneoctone, 9CI P-00090

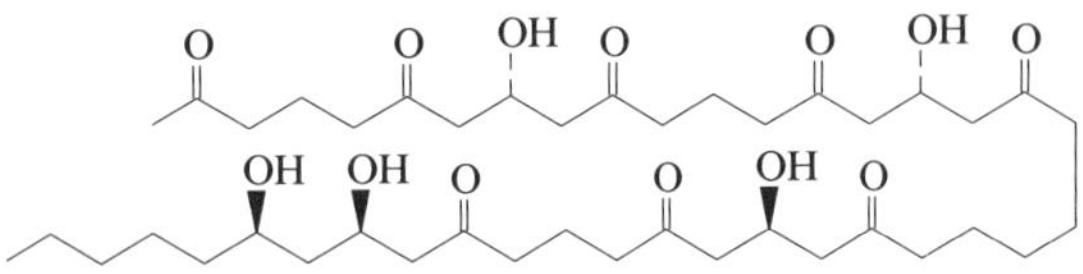

$C_{41}H_{68}O_{13}$ M 768.980

(8S,16R,26S,34R,36R)-form [72200-77-2] ***HMT Band 1 toxin***

Isol. from *Helminthosporium maydis*, race T. Host-specific pathotoxin. Powder. Mp 132-133°. $[\alpha]_D^{25}$ +12° (c, 0.12 in MeOH). Related to PM-toxins.

Kono, Y. *et al*, *Bioorg. Chem.*, 1981, **10**, 206 (*isol, struct, props*)
Kono, Y. *et al*, *Agric. Biol. Chem.*, 1985, **49**, 559 (*abs config*)

6,14,24,32,34-Pentahydroxy-4,8,12,16,22,26,30-nonatriacontaneheptone, 9CI P-00091

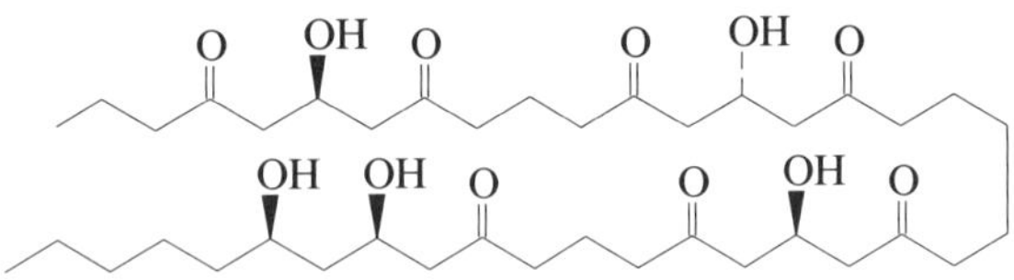

$C_{39}H_{66}O_{12}$ M 726.943

(6R,14R,24S,32R,34R)-form [79065-99-9] ***HMT Band 3 toxin***

Isol. from *Helmithosporium maydis* race T. Host-specific pathotoxin. Powder. Mp 134-135°. $[\alpha]_D^{25}$ +15.9° (c, 0.15 in MeOH).

Kono, Y. *et al*, *Bioorg. Chem.*, 1981, **10**, 206 (*isol, struct, props*)
Kono, Y. *et al*, *Agric. Biol. Chem.*, 1985, **49**, 559 (*abs config*)

6,26-Pentatriacontadien-2-one, 9CI P-00092

$H_3C(CH_2)_7CH{=}CH(CH_2)_{18}CH{=}CH(CH_2)_3COCH_3$

$C_{35}H_{66}O$ M 502.906

(Z,Z)-form [133530-17-3]

Isol. from the snake *Boiga irregularis*.

Murata, Y. *et al*, *J. Nat. Prod. (Lloydia)*, 1991, **54**, 233 (*isol, pmr, cmr*)

8,26-Pentatriacontadien-2-one, 9CI P-00093

$H_3C(CH_2)_7CH{=}CH(CH_2)_{16}CH{=}CH(CH_2)_5COCH_3$

$C_{35}H_{66}O$ M 502.906

(Z,Z)-form [133530-18-4]

Isol. from the snake *Boiga irregularis*.

Murata, Y. *et al*, *J. Nat. Prod. (Lloydia)*, 1991, **54**, 233 (*isol*)

9,26-Pentatriacontadien-18-one, 8CI P-00094

[1540-84-7]

$H_3C(CH_2)_7CH{=}CH(CH_2)_7CO(CH_2)_7CH{=}CH(CH_2)_7CH_3$

$C_{35}H_{66}O$ M 502.906

(Z,Z)-form

Oleone

Pale-yellow cryst. (EtOH). Mp 58.5° (28-29°).

Oxime: Mp 31°.

Breuer, G. *et al*, *Monatsh. Chem.*, 1935, **67**, 42.
U.S. Pat., 2 410 096, (1946); *CA*, **41**, 1234 (*synth*)
Hatt, H.H. *et al*, *Aust. J. Chem.*, 1955, **8**, 506 (*synth*)
Chakravati, K.K. *et al*, *J. Chem. Soc.*, 1965, 3475 (*synth*)

Pentatriacontane, 9CI P-00095

[630-07-9]

$H_3C(CH_2)_{33}CH_3$

$C_{35}H_{72}$ M 492.954

Occurs in many biological systems, including *Hibiscus* flowers, *Piper betel* leaves and Japanese beetles. Spar. sol. hot Et_2O. Mp 75°. Bp 490°, Bp_{10} 311°.

Clemensen, E., *Ber.*, 1913, **46**, 1837 (*synth*)
Deshpande, S.M. *et al*, *Curr. Sci.*, 1970, **39**, 372 (*isol*)
Makira, R.G., *Vestn. L'vov. Politekh. Inst.*, 1971, 20; *CA*, **78**, 57888k (*synth*)

3-Pentatriacontanone P-00096

[79097-22-6]

$H_3C(CH_2)_{31}COCH_2CH_3$

$C_{35}H_{70}O$ M 506.937

Constit. of the leaves and stems of *Hyoscamus muticus*. Cryst. (Me_2CO). Mp 86-87°.

Goswami, A. *et al*, *Phytochemistry*, 1981, **20**, 1315 (*isol*)

4-Pentatriacontanone, 9CI P-00097

[92122-81-1]

$H_3C(CH_2)_{30}COCH_2CH_2CH_3$

$C_{35}H_{70}O$ M 506.937

Constit. of the leaves and stems of *Duboisia myoporoides*. Cryst. (Me_2CO). Mp 74°.

Shukla, Y.N. *et al*, *Phytochemistry*, 1984, **23**, 799 (*isol, ir, pmr, ms*)

5-Pentatriacontanone P-00098

[70342-19-7]

$H_3C(CH_2)_{29}CO(CH_2)_3CH_3$

$C_{35}H_{70}O$ M 506.937

Constit. of the stem bark of *Ficus bengalensis*.

Subramanian, P.M. *et al*, *Pol. J. Pharmacol. Pharm.*, 1978, **30**, 559 (*isol*)

18-Pentatriacontanone, 9CI P-00099

Stearone. Diheptadecyl ketone

$H_3C(CH_2)_{16}CO(CH_2)_{16}CH_3$

$C_{35}H_{70}O$ M 506.937

Cryst. (pet. ether). Mp 88.4°.

Oxime: Mp 62-63°.

Easterfield, T.I. *et al*, *J. Chem. Soc.*, 1911, **99**, 2298 (*synth*)

Org. Synth., Coll. Vol., 4, 1963, 854 (*synth*)
Polonsky, J. *et al*, *Bull. Soc. Chim. Fr.*, 1954, 504 (*synth*)
Holman, R.T. *et al*, *Anal. Chem.*, 1956, **28**, 1533 (*ir*)

1-Pentatriaconten-17-ol, 9CI — P-00100

[133956-30-6]

$H_3C(CH_2)_{17}CH(OH)(CH_2)_{14}CH{=}CH_2$

$C_{35}H_{70}O$ M 506.937
Constit. of the leaves of *Adenocalymma alliaceum*. Cryst. (Me_2CO). Mp 87-88°.

[133956-32-8]

Misra, T.N. *et al*, *Phytochemistry*, 1991, **30**, 541 (*isol*)

2-Pentylcyclopropaneundecanoic acid, 9CI — P-00101

12,13-Methyleneoctadecanoic acid

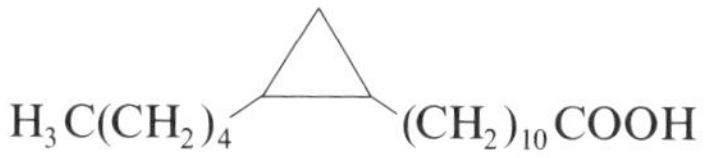

$C_{19}H_{36}O_2$ M 296.492
Both *cis-* and *trans*-forms have been synthesised as Me esters.

Christie, W.W. *et al*, *Chem. Phys. Lipids*, 1968, **2**, 196 (*synth, glc, ir, nmr*)
Christie, W.W., *Top. Lipid Chem.*, 1970, **1**, 4 (*rev*)
Gunstone, F.D. *et al*, *Chem. Phys. Lipids*, 1973, **10**, 303 (*synth, glc, nmr*)

Performic acid — P-00102

Methaneperoxoic acid, 9CI
[107-32-4]

HCOOOH

CH_2O_3 M 62.025
Oxidising agent. Liq. Misc. H_2O, EtOH, Et_2O; mod. sol. $CHCl_3$, C_6H_6. pK_a 7.1 (19.5°). Obt.in 90% purity.
▷ Explodes on heating in contact with metals or metallic oxides. Highly irritant.

Greenspan, F.P., *J. Am. Chem. Soc.*, 1946, **68**, 907 (*synth*)
Bunton, C.A. *et al*, *Chem. Ind. (London)*, 1954, 191 (*synth*)
Fieser and Fieser's Reagents for Organic Synthesis, Wiley, 1974, **4**, 373.
Bretherick, L., *Handbook of Reactive Chemical Hazards*, 2nd Ed., Butterworths, London and Boston, 1979, 299.
Lewis, R.J., *Sax's Dangerous Properties of Industrial Materials*, 8th Ed., Van Nostrand-Reinhold, 1992, PCM500.

Peroxydodecanoic acid — P-00103

Perlauric acid. Peroxylauric acid

$H_3C(CH_2)_{10}COOOH$

$C_{12}H_{24}O_3$ M 216.320
Used in the assignment of the position of ethylenic bonds by ms. Oxidising agent. Mp 52°.

Swern, D. *et al*, *J. Am. Chem. Soc.*, 1955, **77**, 4037 (*synth*)
Swern, D. *et al*, *J. Org. Chem.*, 1962, **27**, 1336 (*synth*)
Aplin, R.T. *et al*, *J. Chem. Soc., Chem. Commun.*, 1967, 858 (*use*)
Fieser and Fieser's Reagents for Organic Synthesis, Wiley, 1972, **3**, 220.
Nedelec, J.Y. *et al*, *Synthesis*, 1976, 821 (*synth*)

Peroxyhexadecanoic acid — P-00104

Perpalmitic acid. Peroxypalmitic acid

$H_3C(CH_2)_{14}COOOH$

$C_{16}H_{32}O_3$ M 272.427
Mp 61-61.5°.

Swern, D. *et al*, *J. Am. Chem. Soc.*, 1955, **77**, 4037 (*synth*)
Swern, D. *et al*, *J. Org. Chem.*, 1962, **27**, 1336 (*synth*)

Peroxyoctadecanoic acid — P-00105

Perstearic acid. Peroxystearic acid

$H_3C(CH_2)_{16}COOOH$

$C_{18}H_{36}O_3$ M 300.481
Mp 65-65.5°.

Swern, D. *et al*, *J. Org. Chem.*, 1962, **27**, 1336 (*synth*)
Markley, K.S., *Fatty Acids*, 2nd. Ed., Interscience, N.Y., 1968, 3671.

Peroxytetradecanoic acid — P-00106

Permyristic acid. Peroxymyristic acid

$H_3C(CH_2)_{12}COOOH$

$C_{14}H_{28}O_3$ M 244.373
Mp 56°.

Swern, D. *et al*, *J. Am. Chem. Soc.*, 1955, **77**, 4037 (*synth*)

Petrosynol — P-00107

4,15,26-Triacontatriene-1,12,18,29-tetrayne-3,14,17,28-tetrol
[111554-19-9]

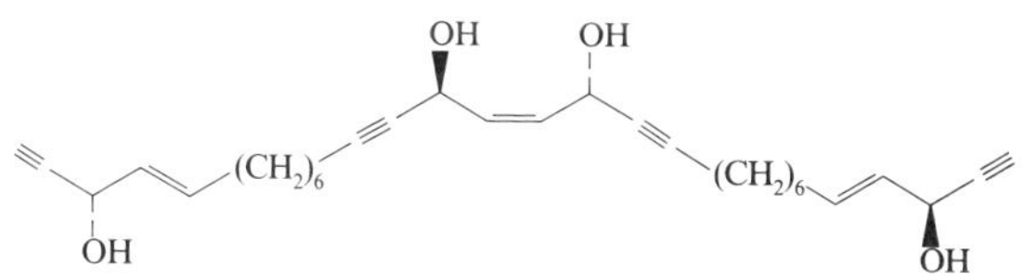

Absolute configuration

$C_{30}H_{40}O_4$ M 464.644
Isol. from marine sponge *Petrosia* sp. Exhibits antifungal props. Oil. $[\alpha]_D^{23}$ +107° (c, 0.37 in $CHCl_3$).
Tetraketone: [111397-55-8]. *4,15,26-Triacontatriene-1,12,18,29-tetrayne-3,14,17,28-tetrone, 9CI.* ***Petrosynone***
$C_{30}H_{32}O_4$ M 456.580
From *P.* sp. Active against *Bacillus subtilis*. Yellow oil.

Fusetani, N. *et al*, *Tetrahedron Lett.*, 1987, **28**, 4313 (*isol, config*)

Phanerosporic acid — P-00108

[124709-28-0]

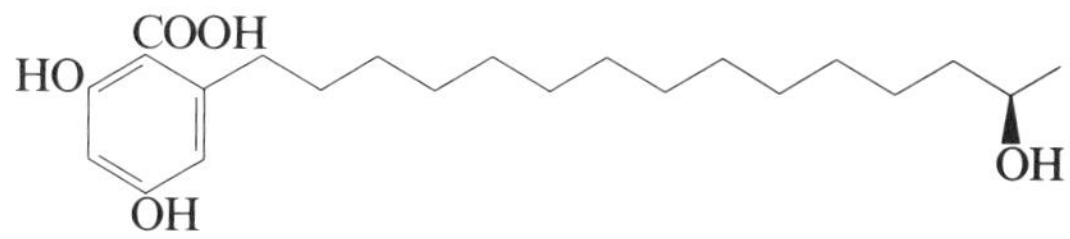

$C_{22}H_{36}O_5$ M 380.523
Metab. of *Phanerochaete chrysosporium*. Cryst. Mp 168°. $[\alpha]_D$ −6.1° (c, 0.5 in MeOH).

Arnone, A. *et al*, *Phytochemistry*, 1989, **28**, 2803 (*isol, pmr, cmr*)

10-Phenyldecanoic acid **P-00109**

Benzenedecanoic acid, 9CI. 10-Phenylcapric acid

[18017-73-7]

$Ph(CH_2)_9COOH$

$C_{16}H_{24}O_2$ M 248.364

Waxy solid. Mp 41°. Bp_{18} 228-230°.

Et ester: [58380-63-5].

$C_{18}H_{28}O_2$ M 276.418

Bp_{20} 220-224°.

Papa, D. *et al*, *J. Am. Chem. Soc.*, 1947, **69**, 3018 (*synth*)
Huisgen, R. *et al*, *Justus Liebigs Ann. Chem.*, 1954, **586**, 52 (*synth*)

12-Phenyl-1-dodecanol **P-00110**

[88336-80-5]

$Ph(CH_2)_{11}CH_2OH$

$C_{18}H_{30}O$ M 262.434

Solid (pet. ether). Mp 39-40°.

Goodman, M.M. *et al*, *J. Med. Chem.*, 1984, **27**, 390 (*synth, pmr, ms*)

16-Phenylhexadecanoic acid **P-00111**

$Ph(CH_2)_{15}COOH$

$C_{22}H_{36}O_2$ M 332.525

Cryst. (MeOH). Mp 76-77°.

Huisgen, R. *et al*, *Chem. Ber.*, 1960, **93**, 2693.
Protiva, J. *et al*, *Collect. Czech. Chem. Commun.*, 1986, **51**, 872 (*synth, pmr, ms*)

3-Phenylhexanoic acid **P-00112**

β-Propylbenzenepropanoic acid, 9CI. β-Propylhydrocinnamic acid, 8CI. 3-Phenylcaproic acid

[5703-52-6]

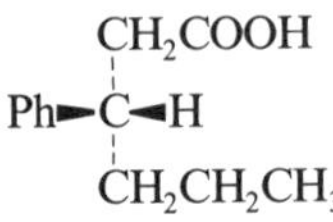

(*S*)-*form*

$C_{12}H_{16}O_2$ M 192.257

(*S*)-*form* [71845-01-7]

Bp_4 152°. $[\alpha]_D^{24}$ +20.4° (neat).

Me ester:

$C_{13}H_{18}O_2$ M 206.284

Bp_{12} 139-142°. $[\alpha]_D^{24}$ +13.4° (neat).

(±)-*form*

Cryst. (MeOH aq.). Mp 34-35°. Bp_2 122-124°.

West, B.D. *et al*, *J. Am. Chem. Soc.*, 1961, **83**, 2676.
Bott, K., *Chem. Ber.*, 1967, **100**, 2791.
Kuchar, M. *et al*, *Collect. Czech. Chem. Commun.*, 1976, **41**, 633.
Hata, T. *et al*, *Tetrahedron Lett.*, 1979, 2047.

4-Phenylhexanoic acid **P-00113**

γ-Ethylbenzenebutanoic acid, 9CI. 4-Phenylcaproic acid

[5354-04-1]

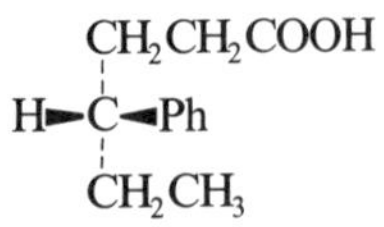

(*R*)-*form*

$C_{12}H_{16}O_2$ M 192.257

(*R*)-*form* [16557-61-2]

Bp_2 122°. $[\alpha]_D$ −9.5° (c, 0.7 in EtOH).

Et ester:

$C_{14}H_{20}O_2$ M 220.311

Bp_1 121°. $[\alpha]_D^{25}$ −1.86° (neat).

(±)-*form*

Mp 104.5-105.5°. Bp_{22} 185°, $Bp_{0.12}$ 115-117°.

Chloride:

$C_{12}H_{15}ClO$ M 210.703

Bp_{18} 138°.

Nitrile:

$C_{12}H_{15}N$ M 173.257

Bp_{19} 148-150°.

Levene, P.A. *et al*, *J. Biol. Chem.*, 1935, **110**, 329 (*synth*)
Herzog, H.L. *et al*, *J. Org. Chem.*, 1950, **15**, 370 (*synth*)
Bavin, P.M.G., *J. Med. Chem.*, 1966, **9**, 52 (*synth*)
Brienne, M.J. *et al*, *Bull. Soc. Chim. Fr.*, 1967, 613 (*synth*)
Menicagli, R. *et al*, *Chem. Ind. (London)*, 1971, 1490 (*abs config*)
Reinheckel, H. *et al*, *J. Prakt. Chem.*, 1974, **316**, 215; 1975, **317**, 273 (*synth*)

6-Phenylhexanoic acid, 8CI **P-00114**

6-Phenylcaproic acid

[5581-75-9]

$Ph(CH_2)_5COOH$

$C_{12}H_{16}O_2$ M 192.257

Cryst. or liq. Mp 22-24°. Bp_{11} 186-188°.

Et ester: [52692-51-0].

$C_{14}H_{20}O_2$ M 220.311

Bp_{12} 162-164°.

Chloride: [21389-46-8].

$C_{12}H_{15}ClO$ M 210.703

Bp_{11} 151-152°.

Amide: [31274-14-3].

$C_{12}H_{17}NO$ M 191.272

Mp 90°.

Nitrile: [17777-31-0].

$C_{12}H_{15}N$ M 173.257

Bp_{13} 161-163°.

v. Braun, J., *Ber.*, 1911, **44**, 2867 (*synth*)
Staudinger, H. *et al*, *Ber.*, 1923, **56**, 711 (*synth*)
Huisgen, R. *et al*, *Justus Liebigs Ann. Chem.*, 1954, **586**, 52 (*synth*)

9-Phenylnonanoic acid **P-00115**

Benzenenonanoic acid, 9CI

[16269-06-0]

$Ph(CH_2)_8COOH$

$C_{15}H_{22}O_2$ M 234.338

Plates (pet. ether). Mp 35-37°.

Jefferson, A. *et al*, *Aust. J. Chem.*, 1988, **41**, 19 (*synth, pmr*)

2-Phenyloctadecanoic acid **P-00116**

2-Phenylstearic acid

$H_3C(CH_2)_{15}CHPhCOOH$

$C_{24}H_{40}O_2$ M 360.579

Cryst. (pet. ether). Mp 69°. Bp_{18} 264-265°.

Amide:

$C_{24}H_{41}NO$ M 359.594

Mp 109°.

Buu-Hoï, Ng.Ph. *et al*, *Bull. Soc. Chim. Fr.*, 1943, **10**, 477 (*synth*)

10-Phenyloctadecanoic acid, 8CI P-00117

10-Phenylstearic acid

[1938-17-6]

$H_3C(CH_2)_7CHPh(CH_2)_8COOH$

$C_{24}H_{40}O_2$ M 360.579

(±)-form

Cryst. (EtOH). Mp 70°. $Bp_{0.09}$ 199-205°.

4-Bromophenacyl ester: Mp 71-72°.

4-Phenylanilide: Mp 90°.

Harmon, J. *et al*, *J. Am. Chem. Soc.*, 1932, **54**, 2515 (*synth, derivs*)
Costin, D. *et al*, *Bull. Soc. Chim. Fr.*, 1961, 218.

18-Phenyloctadecanoic acid, 8CI P-00118

18-Phenylstearic acid

[19740-00-2]

$Ph(CH_2)_{17}COOH$

$C_{24}H_{40}O_2$ M 360.579

Cryst. (pet. ether). Mp 81.7-82.6° (76-77°).

Hase, J. *et al*, *CA*, 1949, **50**, 11977 (*synth*)
Ishizawa, A. *et al*, *CA*, 1968, **69**, 76716 (*synth*)

8-Phenyloctanoic acid, 8CI P-00119

8-Phenylcaprylic acid

[26547-51-3]

$Ph(CH_2)_7COOH$

$C_{14}H_{20}O_2$ M 220.311

Cryst. (pentane). Mp 27-28°. Bp_{14} 209-210°.

Amide: Mp 93°.

Borsche, W., *Ber.*, 1919, **52**, 2084 (*synth*)
Hase, J. *et al*, *Pharm. Bull.*, 1954, **2**, 368; *CA*, **50**, 11977 (*synth*)
Huisgen, R. *et al*, *Chem. Ber.*, 1957, **90**, 1946 (*synth*)
Ansell, M.F. *et al*, *J. Chem. Soc.*, 1958, 3956 (*synth*)
Maria, H.J., *Chem. Phys.*, 1970, **52**, 3399.

8-Phenyl-6-octenoic acid P-00120

[89730-29-0]

$PhCH_2CH{=}CH(CH_2)_4COOH$

$C_{14}H_{18}O_2$ M 218.295

$Bp_{0.5}$ 165°.

Scholz, D., *Justus Liebigs Ann. Chem.*, 1984, 264 (*synth, ir, pmr*)

11-Phenylundecanoic acid P-00121

Benzeneundecanoic acid, 9CI

[3343-24-6]

$Ph(CH_2)_{10}COOH$

$C_{17}H_{26}O_2$ M 262.391

Spars (pet. ether). Mp 44-45°.

Jefferson, A. *et al*, *Aust. J. Chem.*, 1988, **41**, 19 (*synth, pmr*)

Phosphatidylcholine P-00122

Lecithin. Lecithol. Vitellin. Kelecin. Granulestin. 1,2-Diacylglycero-3-phosphorylcholine

[8002-43-5]

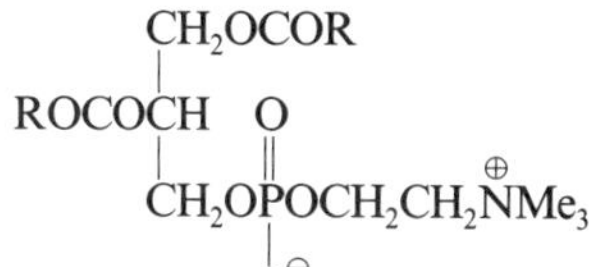

R = saturated or unsaturated fatty acids

First identified in egg yolk. Present in microorganisms and throughout the animal and vegetable kingdoms. Constit. of biol. membranes, involved in permeability, oxidative phosphorylation, phagocytosis and chemical and electrical excitation. Identified in animal tissues and organs especially the brain, nervous system, liver, heart, lungs, kidneys, blood and sperm. The degree of unsaturation, chain length and position of the fatty acid results in distinct biol. props.

Kirk-Othmer Encycl. Chem. Technol., 3rd Ed., Wiley, N.Y., 1978-1984, **14**, 250 (*rev, props, use, isol, activity*)

Phthiocerol A P-00123

3-Methoxy-4-methyl-9,11-dotriacontanediol. 3-Methoxy-4-methyl-9,11-tetratriacontanediol. Phthiocerol. Phtyoglycol

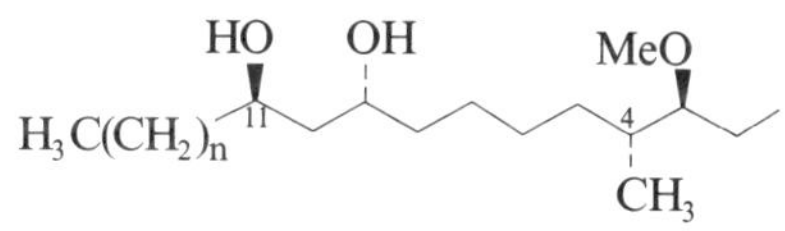

n = 20, 22
Absolute configuration

Isol. as a mixt. of $C_{34}H_{70}O_3$ and $C_{36}H_{74}O_3$ compds. Constit. of tubercle bacillus lipids. Cryst. (pet. ether). Mp 73-73-5°. $[\alpha]^{20}_{436}$ −9.5° ($CHCl_3$).

Minnikin, D.E. *et al*, *J. Chem. Soc. C*, 1966, 2107 (*isol, ms*)
Maskens, K. *et al*, *J. Chem. Soc. C*, 1966, 2113 (*stereochem*)
Welby-Gieusse, M. *et al*, *Tetrahedron*, 1970, **26**, 2875 (*stereochem*)
Maskens, K. *et al*, *J. Chem. Soc., Perkin Trans. 1*, 1973, 1117 (*stereochem*)

Phthiocerol B P-00124

2-Methoxy-3-methyl-8,10-hentriacontanediol. 2-Methoxy-3-methyl-8,10-tritriacontanediol

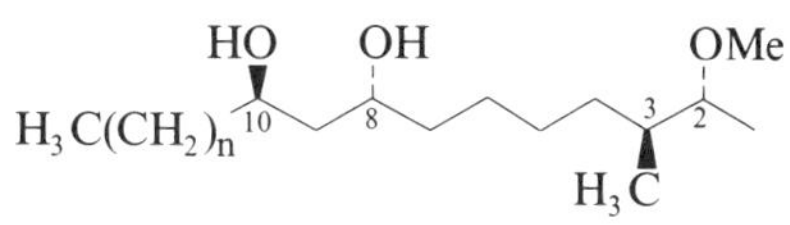

n = 20, 22
Absolute configuration

Isol. as a mixt. of $C_{33}H_{68}O_3$ and $C_{35}H_{72}O_3$ compds.

(2R,3S,8R,10R)-form

Constit. of tubercle bacillus lipids. Cryst. (pet. ether). Mp 71-71.5°. $[\alpha]^{20}_{D}$ −8.2° ($CHCl_3$), $[\alpha]^{20}_{436}$ −16.4° ($CHCl_3$).

Ethylidene deriv.: Cryst. (EtOH). Mp 35.5°.

Maskens, K. *et al*, *J. Chem. Soc. C*, 1966, 2113 (*stereochem*)
Minnikin, D.E. *et al*, *J. Chem. Soc. C*, 1966, 2107 (*isol, ms*)
Welby-Gieusse, M. *et al*, *Tetrahedron*, 1970, **26**, 2875 (*stereochem*)
Maskens, K. *et al*, *J. Chem. Soc., Perkin Trans. 1*, 1973, 1117 (*stereochem*)

Phthiodiolone A P-00125

9,11-Dihydroxy-4-methyl-3-dotriacontanone. 9,11-Dihydroxy-4-methyl-3-tetratriacontanone

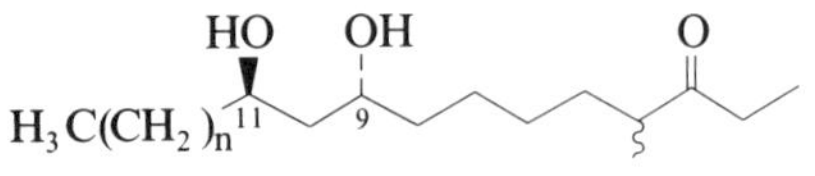

n = 20, 22
Absolute configuration

Isol. as a mixt. of $C_{33}H_{66}O_3$ and $C_{35}H_{70}O_3$ compds.

(9R,11R)-form
Constit. of tubercle bacillus lipids. Cryst. (pet. ether). Mp 76.5°. $[\alpha]_D^{20}$ −2.7,20,365° −7.35 ($CHCl_3$).
Ethylidene deriv.: Cryst. (EtOH). Mp 42.5-43°.
Di-Ac, oxime: [16518-72-2].
Cryst. (MeOH). Mp 38°. $[\alpha]_D^{20}$ −8.1° ($CHCl_3$).

Minnikin, D.E. *et al*, *J. Chem. Soc. C*, 1967, 803 (*isol, ms, pmr*)
Welby-Gieusse, M. *et al*, *Tetrahedron*, 1970, **26**, 2875 (*stereochem*)

Phytane P-00126

2,6,10,14-Tetramethylhexadecane
[638-36-8]

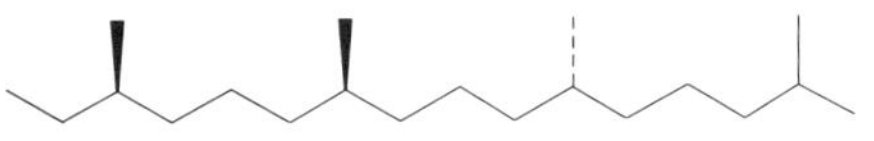

$C_{20}H_{42}$ M 282.552

(6R,10R,14R)-form [18651-07-5]
Found in human liver, meteorites, oil shale and other sediments. Oil. Bp_{15} 179.5°.

Kates, M. *et al*, *Biochemistry*, 1967, **6**, 3329.

1-Phytanoic acid P-00127

3,7,11,15-Tetramethylhexadecanoic acid. ***Phytanic acid***

(3*R*,7*R*,11*R*)-*form*

$C_{20}H_{40}O_2$ M 312.535

(3R,7R,11R)-form [18654-64-3]
Isol. from the liver of patients with Refsum's disease and constit. of oil shale and butter. Oil. $Bp_{7.5}$ 221°. $[\alpha]_{500}^{20}$ −3.8° (c, 4 in MeOH/$CHCl_3$).

(3S,7R,11R)-form [14721-66-5]
Constit. of oil shale. Oil.

Kates, M. *et al*, *Biochemistry*, 1967, **6**, 3329 (*struct*)
Cox, R.E. *et al*, *J. Chem. Soc., Chem. Commun.*, 1970, 1639 (*isol, struct*)
Lough, A.K., *Prog. Chem. Fats Other Lipids*, Part 1, 1973, **14** (*rev*)

2,6,10,14-Phytatetraen-1-ol P-00128

3,7,11,15-Tetramethyl-2,6,10,14-hexadecatetraen-1-ol

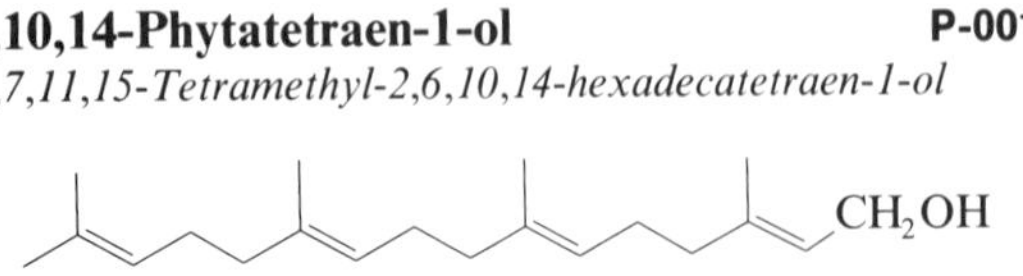

$C_{20}H_{34}O$ M 290.488

(2E,6E,10E)-form [24034-73-9] ***Geranylgeraniol***
Widespread in plants, e.g. found in linseed oil. Constit. of thoracic gland secretions of male *Xylocopa* spp. Oil. $Bp_{0.35}$ 145°. n_D^{30} 1.4932.

O-(*2,3-Dihydroxypropyl*)(R-)*:* [64604-13-3]. ***1-O-Geranylgeranylglycerol***
$C_{23}H_{40}O_3$ M 364.567
Constit. of *Dilophus fasciola*. Viscous liq. $[\alpha]_D$ −2.1° (c, 1.5 in $CHCl_3$).

Aldehyde: [13920-12-2]. *2,6,10,14-Phytatetraen-1-al. 3,7,11,15-Tetramethyl-2,6,10,14-hexadecatetraen-1-al. Geranylgeranial*
$C_{20}H_{32}O$ M 288.472
Oil. $Bp_{0.5}$ 150°.
Aldehyde, semicarbazone: Mp 118°.
Aldehyde,14ξ,15ξ-Epoxide: [108613-62-3]. *13-(3,3-Dimethyloxiranyl)-3,7,11-trimethyl-2,6,10-tridecatrienal, 9CI.* ***14,15-Epoxy-2,6,10-phytatrien-1-al***
$C_{20}H_{32}O_2$ M 304.472
Isol. from *Cystophora moniliformis*. Oil. $[\alpha]_D$ −17° (c, 0.98 in $CHCl_3$).

(2Z,6E,10E)-form
Aldehyde,14ξ,15ξ-epoxide: [115075-54-2].
Isol. from *C. moniliformis*. Oil. $[\alpha]_D$ −22° (c, 0.4 in $CHCl_3$).

Nagasampagi, B.A. *et al*, *Tetrahedron Lett.*, 1967, 189 (*pmr*)
Buggy, M.J. *et al*, *Phytochemistry*, 1974, **13**, 125 (*biosynth*)
Khan, N. *et al*, *J. Chem. Soc., Perkin Trans. 1*, 1975, 1457 (*aldehyde*)
Amico, V. *et al*, *Experientia*, 1977, **33**, 989 (*deriv*)
Altman, L.J. *et al*, *J. Am. Chem. Soc.*, 1978, **100**, 6174 (*aldehyde*)
Coates, R.M. *et al*, *J. Org. Chem.*, 1978, **43**, 4915 (*cmr*)
Ahlqvist, L. *et al*, *Prog. Chem. Fats Other Lipids*, 1978, **16**, 231 (*rev*)
Van Altena, I.A. *et al*, *Aust. J. Chem.*, 1988, **41**, 49 (*aldehyde epoxides*)
Anderson, J.F. *et al*, *J. Chem. Ecol.*, 1988, **14**, 1153.

2-Phyten-1-ol P-00129

3,7,11,15-Tetramethyl-2-hexadecen-1-ol. ***Phytol***

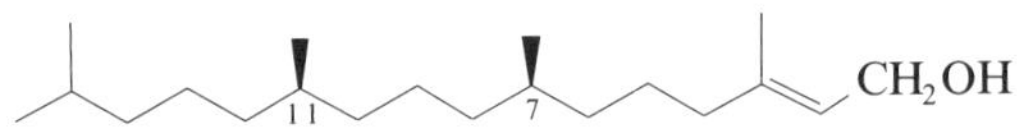

(2*E*,7*R*,11*R*)-*form*

$C_{20}H_{40}O$ M 296.535

(2E,7R,11R)-form [150-86-7]
Constit. of nettles and other plants. Oil. $Bp_{0.02}$ 132°. $[\alpha]_D^{18}$ +0.2°.
▷ TJ3490000.

1-Aldehyde: [13754-69-3]. *2-Phyten-1-al. 3,7,11,15-Tetramethyl-2-hexadecenal.* ***Phytal***
$C_{20}H_{38}O$ M 294.520
Constit. of *Tetragonia tetragonoides*. Oil.

(2Z,7R,11R)-form [5492-30-8]
Constit. of *Gracilaria andersoniana*. Oil.
Hexadecanoyl: Palmitoylphytol
$C_{36}H_{70}O_2$ M 534.948
Constit. of *Pentatropis spiralis*. Gum. $[\alpha]_D$ −14.28° (c, 0.018 in $CHCl_3$). Abs. config. not clear.

Burrell, J.W.K. *et al*, *J. Chem. Soc. C*, 1966, 2144 (*synth, abs config*)
Shilleter, D.N. *et al*, *Phytochemistry*, 1970, **9**, 153 (*biosynth*)
Goodman, R.A. *et al*, *J. Am. Chem. Soc.*, 1973, **95**, 7553 (*cmr*)
Sims, J.J. *et al*, *Phytochemistry*, 1976, **15**, 1076 (*isol*)
Schmid, M. *et al*, *Helv. Chim. Acta*, 1982, **65**, 684 (*synth*)
Aoki, T. *et al*, *Phytochemistry*, 1982, **21**, 1361 (*isol*)
Gramatica, P. *et al*, *Tetrahedron*, 1987, **43**, 4481 (*synth*)
Rasool, N. *et al*, *Phytochemistry*, 1991, **30**, 1331 (*deriv*)
Lewis, R.J., *Sax's Dangerous Properties of Industrial Materials*, 8th Ed., Van Nostrand-Reinhold, 1992, PIB600.

Pinane thromboxane A_2 P-00130

7-[3-(3-Hydroxy-1-octenyl)-6,6-dimethylbicyclo[3.1.1]hept-2-yl]-5-heptenoic acid. PTA$_2$
[71111-01-8]

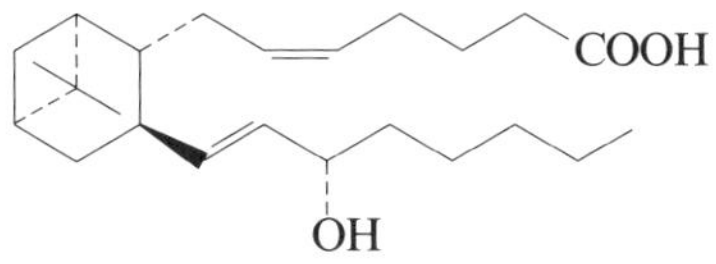

$C_{24}H_{40}O_3$ M 376.578
Antagonises coronary artery contraction and platelet aggregation. Inhibits thromboxane synthase and protects the ischaemic myocardium. Ir ν 3350br (OH), 1705 (COOH) cm^{-1}. Indefinitely stable at 25° in soln. or in the pure liq. state.
Me ester: [71111-05-2].
$C_{25}H_{42}O_3$ M 390.605
$[\alpha]_D^{22}$ +26.40° (MeOH). Ir ν 3400 (OH), 1742 (COOMe) cm^{-1}.

[71154-82-0, 71154-83-1]

Nicolaou, K.C. *et al*, *Proc. Natl. Acad. Sci. U.S.A.*, 1979, **76**, 2566 (*synth, biochem, ir, pmr, cmr*)
Schroer, K. *et al*, *Am. J. Physiol.*, 1980, **238**, H87 (*biochem*)
Wilson, N.H. *et al*, *J. Med. Chem.*, 1982, **25**, 495 (*synth, ir, pmr, ms*)
Burke, S.E. *et al*, *Br. J. Pharmacol.*, 1983, **78**, 287 (*biochem*)

Plakinic acid A P-00131

[87803-05-2]

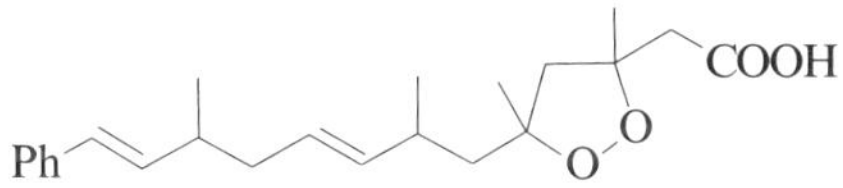

$C_{23}H_{32}O_4$ M 372.503
Isol. from a caribbean sponge from a genus of the family Plakinidae. Antifungal.
Me ester: $[\alpha]_D^{21}$ −57.8°.

Phillipson, D.W. *et al*, *J. Am. Chem. Soc.*, 1983, **105**, 7735 (*isol, ms, pmr, ir, uv*)

Plakinic acid C P-00132

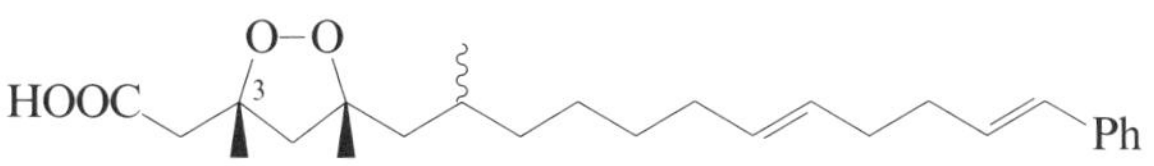

$C_{26}H_{38}O_4$ M 414.584
Constit. of a *Plakortis* sponge. $[\alpha]_D^{27}$ +31.71° (c, 0.098 in $CHCl_3$) (as Me ester).
3-Epimer: ***Epiplakinic acid C***
$C_{26}H_{38}O_4$ M 414.584
Constit. of a *P.* sponge. $[\alpha]_D^{26}$ −22.66° (c, 0.11 in $CHCl_3$) (as Me ester).

[136707-69-2, 136780-75-1]

Davidson, B.S. *et al*, *J. Org. Chem.*, 1991, **56**, 6722 (*isol, pmr, cmr*)

Plakinic acid D P-00133

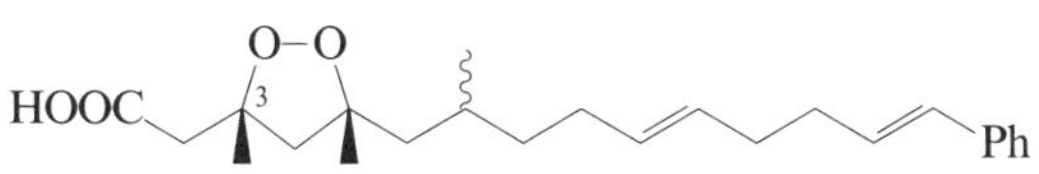

$C_{24}H_{34}O_4$ M 386.530
Constit. of a *Plakortis* sponge. $[\alpha]_D^{27}$ +40.53° (c, 0.12 in $CHCl_3$) (as Me ester).
3-Epimer: ***Epiplakinic acid D***
$C_{24}H_{34}O_4$ M 386.530
Constit. of a *P.* sponge. $[\alpha]_D^{26}$ −16.73° (c, 0.13 in $CHCl_3$) (as Me ester).

[136707-70-5, 136780-74-0]

Davidson, B.S. *et al*, *J. Org. Chem.*, 1991, **56**, 6722 (*isol, pmr, cmr*)

Plakinidone P-00134

[137319-23-4]

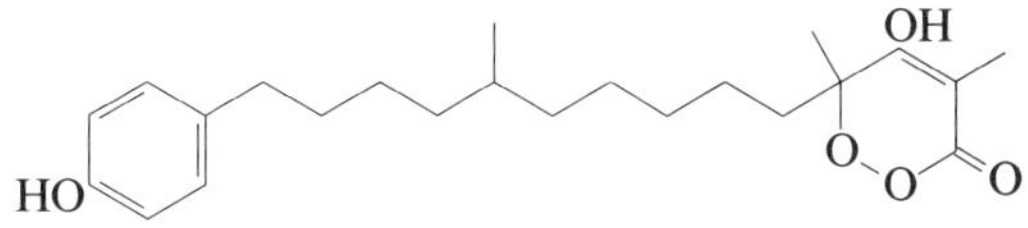

$C_{23}H_{34}O_5$ M 390.519
Metab. of *Plakortis angulospiculatus*. Oil. $[\alpha]_D$ +7.9° (c, 0.61 in MeOH).

Kushlan, D.M. *et al*, *J. Nat. Prod.* (*Lloydia*), 1991, **54**, 1451 (*isol, pmr, cmr*)

Plakorin P-00135

[124264-01-3]

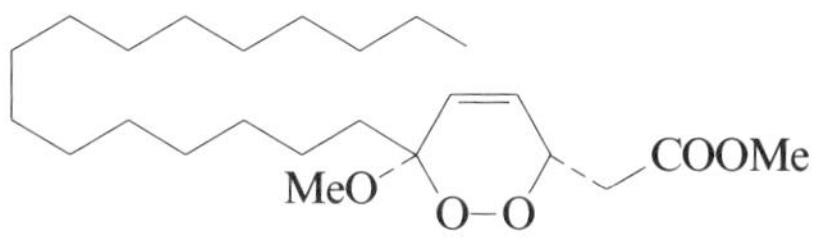

$C_{24}H_{44}O_5$ M 412.609
Metab. of a *Plakortis* sp. Ca^{2+}-ATPase activator. Oil. $[\alpha]_D^{27}$ +44.3° (c, 0.2 in $CHCl_3$).

Murayama, T. *et al*, *Experientia*, 1989, **45**, 898 (*isol, pmr, cmr*)

Platelet activating factor P-00136

*1-O-Alkyl-2-acetyl-*sn*-glyceryl-3-phosphocholine. Antihypertensive polar renomedullary lipid. PAF. APRL. Acetyl glyceryl ether phosphorylcholine. AGEPC*

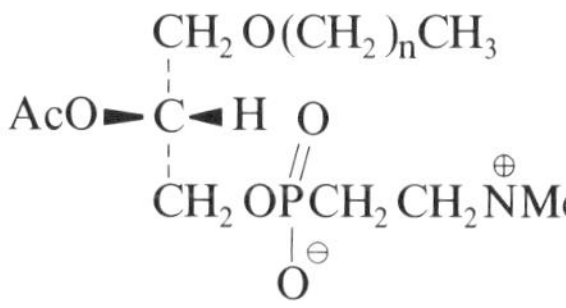

n of natural material not unequivocally detd., *n* = 15 (see 1-*O*-Hexadecyl-*sn*-glycero-3-phosphocholine, H-00280) or 17 (see 1-*O*-Octadecyl-*sn*-glycero-3-phosphocholine, O-00207) are the two major forms. Isol. from renal medulla and basophils. Chemical mediator, exerts powerful antihypertensive action, causes aggregation of blood platelets. Activates polymorphonuclear leukocytes, monocytes and macrophages. Stimulates glycogenolysis in perfused liver and uterine contractions.

Beneviste, J. *et al*, *J. Exp. Med.*, 1972, **136**, 1356 (*biosynth*)
Beneviste, J. *et al*, *Nature* (*London*), 1974, **249**, 581; 1977, **269**, 170 (*isol, struct*)
Blank, M.L. *et al*, *Biochem. Biophys. Res. Commun.*, 1979, **90**, 1194 (*struct*)
Beneviste, J. *et al*, *C. R. Hebd. Seances Acad. Sci. Ser. D*, 1979, **289**, 1037 (*struct*)
Demopoulos, C.A. *et al*, *J. Biol. Chem.*, 1979, **254**, 9355 (*struct*)
Braquet, P. *et al*, *Pharmacol. Rev.*, 1987, **39**, 97 (*rev*)
Hanahan, D.J. *et al*, *Prog. Lipids Res.*, 1987, **26**, 1 (*rev*)
Stark, A. *et al*, *Adv. Lipid Res.*, 1989, **3**, 219 (*biochem*)
Cimetiere, B. *et al*, *Bull. Soc. Chim. Fr.*, 1991, 926 (*synth*)
Braquet, P. *et al*, *Drugs*, 1991, **42**, 9, 176 (*biochem*)

Podopyrone P-00137

*2-Methoxy-3,5-dimethyl-6-undecyl-4*H*-pyran-4-one*
[106894-14-8]

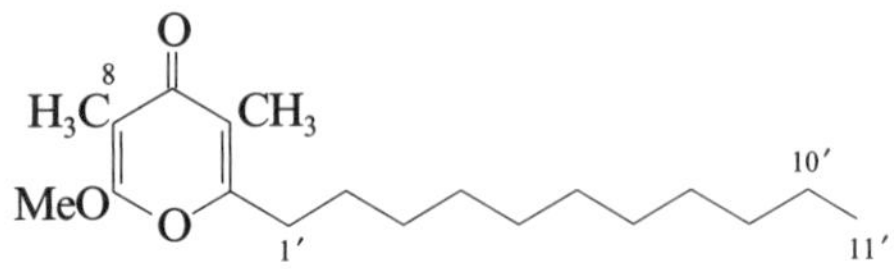

$C_{19}H_{32}O_3$ M 308.460
Constit. of *Podolepsis hieracioides*. Oil.

1′-Oxo: [126667-05-8]. ***1′-Oxopodopyrone***
$C_{19}H_{30}O_4$ M 322.444
Constit. of *P. rugata*. Oil.

10′-Oxo: [126622-71-7]. ***10′-Oxopodopyrone***
$C_{19}H_{30}O_4$ M 322.444
Constit. of *P. rugata*. Oil.

Zdero, C. *et al, Phytochemistry*, 1987, **26**, 187.
Jaensch, M. *et al, Phytochemistry*, 1989, **28**, 3497 (*isol, pmr, cmr, derivs*)

Prodigiosin P-00138

*4-Methoxy-5-[(5-methyl-4-pentyl-2*H*-pyrrol-2-ylidene) methyl]-2,2′-bi-1*H*-pyrrole, 9CI*
[82-89-3]

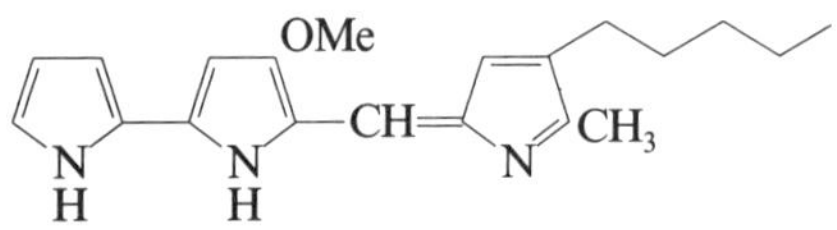

$C_{20}H_{25}N_3O$ M 323.437
Pyrrole antibiotic. Pigment from *Serratia*, *Streptomyces*, *Beneckea gazogenes* and *Cladophora* spp. Shows antileukaemic and antifungal props. Shows antimalarial activity but too toxic for clinical use. Lustrous dark-red square pyrimids. Mp 151-152°. Sinters at 70-80°.

▷ LD_{50} 18 mg/kg (i.p. mice). DW2977000.

B,HClO₄: Deep-red cryst. (EtOH). Mp 228°.
B,HCl: Magenta cryst. (C_6H_6/pet. ether). Mp 148.5-150° dec.

Wrede, F. *et al, Ber.*, 1929, **62**, 2678.
Rapoport, H. *et al, J. Am. Chem. Soc.*, 1962, **84**, 635 (*synth*)
Williams, R.P. *et al, Antibiotics (N.Y.)*, 1967, **2**, 410, 449 (*revs*)
Jackson, A.H. *et al, Tetrahedron*, 1967, **23**, 603 (*ms*)
Hearn, W.R. *et al, J. Org. Chem.*, 1970, **35**, 142 (*synth*)
Lim, D.V. *et al, J. Bacteriol.*, 1977, **129**, 124 (*biosynth*)
Sveshnikova, M.A. *et al, Antibiotiki (Moscow)*, 1983, **28**, 723 (*isol*)
Boger, D.L. *et al, J. Org. Chem.*, 1988, **53**, 1405 (*synth, biosynth*)
Wasserman, H.H. *et al, Tetrahedron Lett.*, 1989, **30**, 1725 (*synth*)

1,2,3-Propanetrithiol P-00139

[4756-13-2]

$HSCH_2CH(SH)CH_2SH$

$C_3H_8S_3$ M 140.294
d_4^{25} 1.252. Bp_{11} 111°, Bp_2 80°.

▷ TZ9500000.

1,2,3-Tri-Ac:
$C_9H_{14}O_3S_3$ M 266.406
$Bp_{0.5}$ 140°.

1-S-Me: [85365-95-3]. *3-(Methylthio)-1,2-propanedithiol, 9CI*
$C_4H_{10}S_3$ M 154.321
$Bp_{0.5}$ 59-60°.

2-S-Me: [85365-94-2]. *2-(Methylthio)-1,3-propanedithiol, 9CI*
$C_4H_{10}S_3$ M 154.321

1,2-Di-Me: [85365-96-4]. *2,3-Bis(methylthio)-1-propanethiol, 9CI*
$C_5H_{12}S_3$ M 168.348
Oil.

1,3-Di-Me: [85365-97-5]. *1,3-Bis(methylthio)-2-propanethiol, 9CI*
$C_5H_{12}S_3$ M 168.348
Oil.

1,2,3-Tri-Me: [69631-98-7]. *1,2,3-Tris(methylthio)propane, 9CI*
$C_6H_{14}S_3$ M 182.375
Bp_{15} 147°.

Stocken, L.A., *J. Chem. Soc.*, 1947, 592 (*synth*)
Doering von E., W. *et al, J. Am. Chem. Soc.*, 1955, **77**, 509 (*synth*)
Nielsen, L.E. *et al, Experientia*, 1984, **40**, 186 (*pharmacol, deriv*)

Propanoic acid, 9CI P-00140

Propionic acid. Metacetonic acid
[79-09-4]

H_3CCH_2COOH

$C_3H_6O_2$ M 74.079
Volatile component of human vaginal secretions. Food preservative used esp. in bread and grain mold inhibitors. Also used in cellulose propionate manuf. Liq. with acrid odour. Misc. H_2O. d_{15}^{15} 0.999. Mp −21.5°. Bp 141.1°, Bp_{10} 39.7°. Simple esters used as flavouring agents.

▷ Irritant causes burns, TLV 30. UE5950000.

NH₄ salt: [17496-08-1].
V. hygroscopic cryst. Mp 45°.

K salt: [327-62-8].
Plates + $1H_2O$ (EtOH aq.). Mp >300°.

Me ester: [554-12-1]. *Methyl propionate*
$C_4H_8O_2$ M 88.106
d_4^0 0.939. Bp 79.7°. n_D^{19} 1.3770.
▷ Highly flammable, fl. p. −2°. UF5970000.

Et ester: [105-37-3]. *Ethyl propionate*
$C_5H_{10}O_2$ M 102.133
Water scavenger for alkoxide-catalysed condensations. Liq. with fruity odour. Mod. sol. H_2O. d_4^{20} 0.891. Fp −73°. Bp 99°. n_D^{20} 1.3844.
▷ Flammable, fl. p. 12°. UF3675000.

Propyl ester: [106-36-5]. *Propyl propionate*
$C_6H_{12}O_2$ M 116.160
d^0 0.893. Bp 122°.
▷ UF7100000.

Isopropyl ester: [637-78-5]. *Isopropyl propionate*
$C_6H_{12}O_2$ M 116.160
d^0 0.893. Bp_{750} 109-111°.
▷ UF5140000.

Butyl ester: [590-01-2]. *Butyl propionate*
$C_7H_{14}O_2$ M 130.186
d_0^0 0.893. Bp 145.5°.
▷ UE8245000.

1-Methylpropyl ester: [591-34-4]. *1-Methylpropyl propanoate.* sec-*Butyl propionate*
$C_7H_{14}O_2$ M 130.186
d_4^{20} 0.866. Bp 132°. n_D^{25} 1.3938.

2-Methylpropyl ester: [540-42-1]. *2-Methylpropyl propanoate. Isobutyl propionate*
$C_7H_{14}O_2$ M 130.186
d_4^0 0.888. Bp 137°.
▷ UF4930000.

Pentyl ester: [624-54-4]. *Pentyl propanoate. Amyl propionate*
$C_8H_{16}O_2$ M 144.213
d_4^{15} 0.876. Fp −73.1°. Bp 168.7°. n_D^{15} 1.4096.
Ph ester: [637-27-4]. *Phenyl propionate*
$C_9H_{10}O_2$ M 150.177
Prisms. Mp 20°. Bp 211°.
Benzyl ester: [122-63-4]. *Benzyl propionate*
$C_{10}H_{12}O_2$ M 164.204
Bp 222°.
Chloride: [79-03-8].
C_3H_5ClO M 92.525
Bp 80°.
▷ Highly irritant, causes burns. Highly flammable, flash pt. 12°, reacts violently with H_2O. UG6657000.
Bromide: [598-22-1].
C_3H_5BrO M 136.976
d^{16} 1.521. Bp 103-103.6°. n_D^{16} 1.4578.
Anhydride: [123-62-6].
$C_6H_{10}O_3$ M 130.143
Bp 168.4-168.8°, Bp_{18} 67.5°.
▷ Flammable. UF9100000.
Amide: Propanamide. Propionamide
C_3H_7NO M 73.094
Plates (C_6H_6). Sol. H_2O. Bp 222.2°. Steam-volatile.
Amide, N-Ac: [19264-34-7].
$C_5H_9NO_2$ M 115.132
Mp 86°. Bp 230-240°.
Dimethylamide: [758-96-3].
$C_5H_{11}NO$ M 101.148
Bp 165-178°.
▷ UE3810000.
Anilide:
$C_9H_{11}NO$ M 149.192
Yellow cryst. $Bp_{0.3}$ 73-75°.
Hydrazide: [5818-15-5].
$C_3H_8N_2O$ M 88.109
Sol. H_2O, spar. sol. Et_2O. Mp 40°. Bp_{16} 130°.
Nitrile: [107-12-0]. *Propionitrile. Ethyl cyanide. Cyanoethane*
C_3H_5N M 55.079
Liq. with ethereal odour. Part. misc. H_2O. Fp −103.5°. Bp 97°.
▷ V. highly toxic by inhalation and skin absorption. Highly flammable, flash pt. 16°. UF9625000.
Iodide: [598-40-3]. *Propanoyl iodide*
C_3H_5IO M 183.976
Yellowish liq. Bp 127-128°, Bp_{30} 75°, Bp_{50} 52-53°.

Hardy, D.V.N. *et al, J. Chem. Soc.*, 1936, 358, 364 (*synth*)
Kirk-Othmer Encycl. Chem. Technol., 2nd Ed., Wiley, N.Y., 1963-1971, **16**, 554 (*rev*)
Crispin, P.D. *et al, Aust. J. Chem.*, 1967, **20**, 2589 (*amide*)
Fieser and Fieser's Reagents for Organic Synthesis, Wiley, 1967, **1**, 385 (*ester*)
Ger. Pat., 2 324 765, (1974); *CA*, **80**, 70359z (*ester*)
Michael, R.P. *et al, Science (Washington, D.C.)*, 1974, 1217.
Williamson, K.L. *et al, J. Magn. Reson.*, 1978, **30**, 367 (*cmr*)
Keinan, E. *et al, J. Org. Chem.*, 1990, **55**, 3922 (*iodide*)
Bretherick, L., *Handbook of Reactive Chemical Hazards*, 2nd Ed., Butterworths, London and Boston, 1979, 461, 462, 1183.
Lewis, R.J., *Sax's Dangerous Properties of Industrial Materials*, 8th Ed., Van Nostrand-Reinhold, 1992, BSJ500, EPB500, MOT000, PMU750, PMV000, PMV250, PMV500, PMV750, PMW500, PNU000.

1-Propene, 9CI P-00141

Propylene
[115-07-1]

$H_3CCH{=}CH_2$

C_3H_6 M 42.080
Used in polymerised form as polypropylene plastic. Raw material in manuf. of acetone, isopropyl compds., propylene oxide, acrylonitrile, 2-ethylhexanol and phenol. Important industrial chemical, 9th in order of volume for USA in 1990 (production 11.06 million tons/year). Gas. Mod. sol. EtOH. Mp −185.2°. Bp −47.8°. pK_a 53 (MeCN). Crit. temp. 92.1°. Liquifies under 7-8 atm. pressure. Absorbed quantitatively by Hg(*II*) solns.
▷ Extremely flammable, Fl. p. −108°. Forms an ozonide which is explosive at ambient temp.. UC6740000.

Sherwood, P.W., *Ind. Chem.*, 1960, 542 (*rev*)
Andreas, F. *et al, Chem. Tech. (Leipzig)*, 1968, **20**, 151 (*rev*)
Tokue, I. *et al, J. Mol. Struct.*, 1973, **17**, 207 (*struct*)
Breslow, R. *et al, J. Am. Chem. Soc.*, 1977, **99**, 7745 (*acidity*)
Hatch, L.F. *et al, Hydrocarbon Process.*, 1978, **57**, 149 (*use*)
Kirk-Othmer Encycl. Chem. Technol., 3rd Ed., Wiley, N.Y., 1978-1984, **19**, 228 (*rev*)
Wunder, F.G. *et al, Angew. Chem., Int. Ed. Engl.*, 1980, **19**, 126 (*synth*)
Bretherick, L., *Handbook of Reactive Chemical Hazards*, 2nd Ed., Butterworths, London and Boston, 1979, 436, 442.
Hazards in the Chemical Laboratory, (Bretherick, L., Ed.), 3rd Ed., Royal Society of Chemistry, London, 1981, 459, 460.
Lewis, R.J., *Sax's Dangerous Properties of Industrial Materials*, 8th Ed., Van Nostrand-Reinhold, 1992, PMO500.
Gibson, G.G., *Ethel Browning's Toxicity and Metabolism of Industrial Solvents*, (Snyder, R., Ed.), 2nd Ed., Elsevier, 1987, **1**, 354 (*rev, tox*)

2-Propylcyclopropanetridecanoic acid, 9CI P-00142

14,15-Methyleneoctadecanoic acid

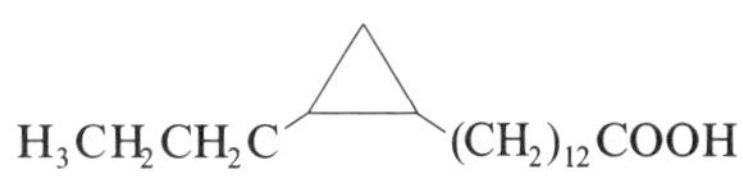

$C_{19}H_{36}O_2$ M 296.492

Christie, W.W. *et al, Chem. Phys. Lipids*, 1968, **2**, 196 (*synth, glc, ir, nmr*)
Christie, W.W., *Top. Lipid Chem.*, 1970, **1**, 4 (*rev*)
Gunstone, F.D. *et al, Chem. Phys. Lipids*, 1973, **10**, 303 (*synth, glc, nmr*)

10-Propyl-5,9-tridecadien-1-ol, 9CI P-00143

[18545-39-6]

$(H_3CCH_2CH_2)_2C{=}CHCH_2CH_2CH{=}CH(CH_2)_3CH_2OH$

$C_{16}H_{30}O$ M 238.412
(E)-form [10297-71-9]
Oil. $Bp_{0.4}$ 110-120°.
Ac: [10297-61-7]. *13-Acetoxy-4-propyl-4,8-tridecadiene.* ***Propylure***
$C_{18}H_{32}O_2$ M 280.450
Sex attractant of the female pink bollworm moth *Pectinophora gossypiella*. Liq. $Bp_{0.5}$ 110-120°.
(Z)-form [50744-64-4]
Oil.
Ac: [19889-82-8].
Oil.

Jones, W.A. *et al, Science (Washington, D.C.)*, 1966, **152**, 1516 (*struct, synth*)
Pattenden, G., *J. Chem. Soc. C*, 1968, 2385 (*ir, pmr, synth*)
Kosanyi, J. *et al, Tetrahedron Lett.*, 1973, 3459 (*synth*)
Nagarkatti, J.P. *et al, J. Indian Chem. Soc.*, 1976, **53**, 666 (*synth*)

Prostacyclin **P-00144**

6,9-Epoxy-11,15-dihydroxy-5,13-prostadienoic acid. ***Epoprostenol,* INN.** *Prostaglandin I_2. Prostaglandin X. PGI_2. PGX. U 53217*

[35121-78-9]

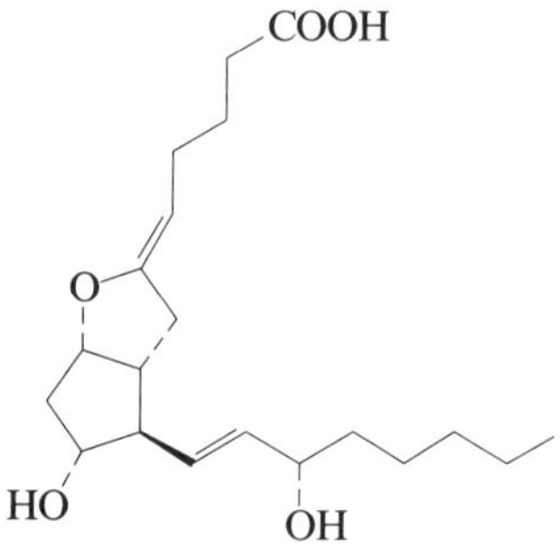

$C_{20}H_{32}O_5$ M 352.470

Inhibits blood platelet aggregation and is a vasodilator. Stereoisomers also known.

Na salt: [61849-14-7]. *Prostacyclin sodium.* ***Epoprostenol sodium,* BAN, USAN.** *Cyclo-Prostin. Flolan. U 53217A*
Hygroscopic. Mp 166-168° (block), 116-124° (capillary). $[\alpha]_D$ +88° (c, 0.808 in $CHCl_3$).

Me ester: [61799-74-4].
Mp 30-33°. $[\alpha]_D$ +78° (c, 0.882 in $CHCl_3$).

[65844-28-2]

Pace-Asciale, C. *et al*, *Biochemistry*, 1971, **10**, 3657 (*ir, pmr, ms*)
Johnson, R.A. *et al*, *Prostaglandins*, 1976, **12**, 915 (*struct*)
Corey, E.J. *et al*, *J. Am. Chem. Soc.*, 1977, **99**, 2006 (*synth, ir, pmr, ms*)
Corey, E.J. *et al*, *Tetrahedron Lett.*, 1977, 3529 (*synth, pmr*)
Bartmann, W. *et al*, *Angew. Chem., Int. Ed. Engl.*, 1982, **21**, 751 (*rev*)
Moncado, S. *et al*, *Br. J. Pharmacol.*, 1982, **76**, 3 (*rev*)
Salmon, J.A. *et al*, *Methods Enzymol.*, 1982, **86**, 91, 436 (*revs, anal*)
Martindale, The Extra Pharmacopoeia, 28th/29th Ed., Pharmaceutical Press, London, 1982/1989, 8084, 8085.
Baxter, A.D. *et al*, *J. Chem. Soc., Perkin Trans. 1*, 1984, 675 (*synth*)
Moncado, S. *et al*, *Adv. Exp. Med. Biol.*, 1985, **192**, 337 (*rev*)
Beierbeck, H. *et al*, *Can. J. Chem.*, 1985, **63**, 1143 (*pmr, cmr, conformn*)
Moncado, S. *et al*, *J. Pharmacol.*, 1985, **16**, 71 (*rev*)
Willis, L. *et al*, *Prog. Lipids Res.*, 1986, **25**, 645 (*rev, props*)
Suzuki, M. *et al*, *J. Am. Chem. Soc.*, 1988, **11**, 4718 (*Me ester, synth, ir, pmr, cmr*)

Prostaglandin F-M **P-00145**

2-(2-Carboxyethyl)-3,5-dihydroxy-ε-oxocyclopentaneoctanoic acid. 5,7-Dihydroxy-11-oxotetranorprostane-1,16-dioic acid

[69779-96-0]

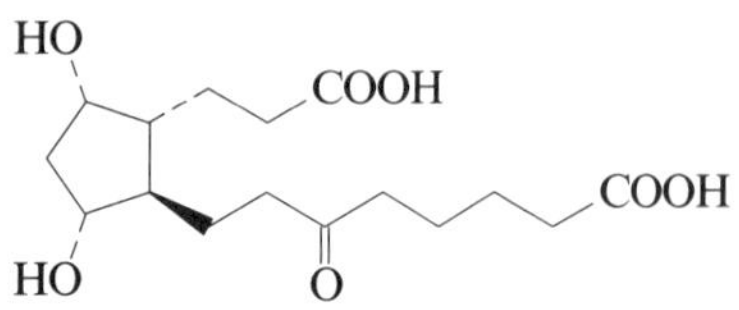

$C_{16}H_{26}O_7$ M 330.377

Urinary metab. of $PGF_{2\alpha}$ (see 9,11,15-Trihydroxyprosta-5,13-dienoic acid, T-00297).

5-Ketone: [24769-56-0]. *Prostaglandin E-M. 2-(2-Carboxyethyl)-5-hydroxy-ε,3-dioxo-cyclopentaneoctanoic acid*
$C_{16}H_{24}O_7$ M 328.361

PGE_2 (see 11,15-Dihydroxy-9-oxo-5,13-prostadienoic acid, D-00225) human urinary metab.

[23109-94-6, 31935-05-4, 55029-37-3]

Granstroem, E. *et al*, *J. Am. Chem. Soc.*, 1969, **91**, 3398 (*isol*)
Boot, J.R. *et al*, *Prostaglandins*, 1974, **8**, 439 (*synth*)
Brash, A.R. *et al*, *Biochim. Biophys. Acta*, 1979, **572**, 371 (*biosynth*)
Granstroem, E. *et al*, *Biochim. Biophys. Acta*, 1982, **713**, 46 (*biosynth*)
Rosenkranz, B. *et al*, *Biochim. Biophys. Acta*, 1983, **750**, 231 (*biosynth*)

Prostaglandin G_2 **P-00146**

9,11-Epidioxy-15-hydroperoxy-5,13-prostadienoic acid, 9CI. PGG_2

[51982-36-6]

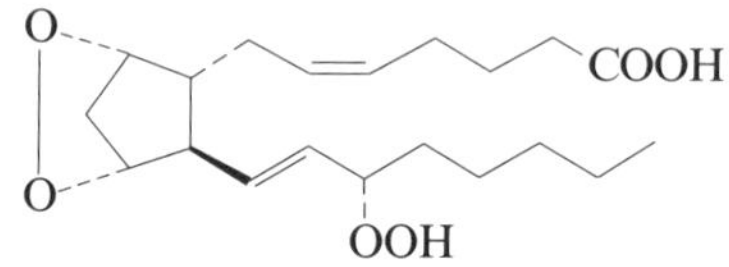

$C_{20}H_{32}O_6$ M 368.469

Intermed. in metab. of dietary arachidonic acid (see 5,8,11,14-Eicosatetraenoic acid, E-00034). Induces *in vitro* contraction of rabbit aorta and blood platelet aggregation. More active than PGE_2.

Nugteren, D.H. *et al*, *Biochim. Biophys. Acta*, 1973, **326**, 448 (*isol*)
Hamberg, H. *et al*, *Proc. Natl. Acad. Sci. U.S.A.*, 1973, **70**, 899.
Hamberg, M. *et al*, *Proc. Natl. Acad. Sci. U.S.A.*, 1974, **71**, 345, 3400 (*isol, struct*)
Ubatuba, F.B. *et al*, *Prostaglandins*, 1977, **13**, 1055 (*synth*)
Graff, G., *Methods Enzymol.*, 1982, **86**, 376 (*purifn, hplc*)
Ullrich, V. *et al*, *Biochem. Pharmacol.*, 1987, **36**, 851 (*purifn, hplc*)

Prostaglandin J_2 **P-00147**

15-Hydroxy-11-oxo-5,9,13-prostatrienoic acid, 9CI. PGJ_2

[60203-57-8]

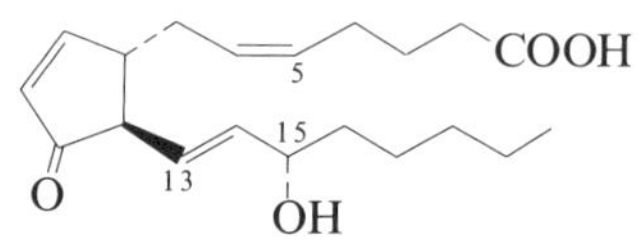

$C_{20}H_{30}O_4$ M 334.455

Intermediate in metab. of PGD_2. Inhibits leukaemia cell growth, shows cellular differentiation inducing antimetastatic and antiviral activity. Antineoplastic. Viscous pale yellow oil.

Fukushina, M. *et al*, *Biochem. Biophys. Res. Commun.*, 1982, **109**, 626.
Bundy, G.L. *et al*, *J. Med. Chem.*, 1983, **26**, 790 (*synth, ir, pmr, uv, ms*)
Mahmud, I. *et al*, *Prostaglandins, Leukotrienes Med.*, 1984, **16**, 131 (*props, pmr, ms, tlc*)
Fukushina, M., *Eicosanoids*, 1990, **3**, 189 (*rev, props*)

Prostanoic acid, 9CI **P-00148**

2-Octylcyclopentaneheptanoic acid, 8CI

[25151-81-9]

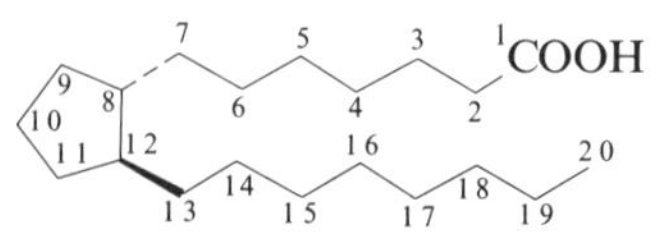

(8S,12S)-form

$C_{20}H_{38}O_2$ M 310.519

Prostanoid numbering shown.

(8S,12S)-form
(+)-trans-*form*
Parent compd. for prostaglandin nomenclature. Mp 25°. $[\alpha]_D^{20}$ +49.3° (EtOH). Configs. of derivs. according to the (*RS*)-system vary erratically according to the subn. pattern.

(8RS,12RS)-form [58986-83-7]
(±)-trans-*form*
p-*Toluidide:* Mp 85.5-86.5°.

Hamon, A. *et al*, *Tetrahedron Lett.*, 1975, 4481 (*synth*)
Sakai, K. *et al*, *Prostaglandins*, 1976, **12**, 399 (*synth, pmr, ir*)

Protopraesorediosic acid **P-00149**

3-Carboxytetrahydro-4-methylene-5-oxo-2-furanpentadecanoic acid, 9CI

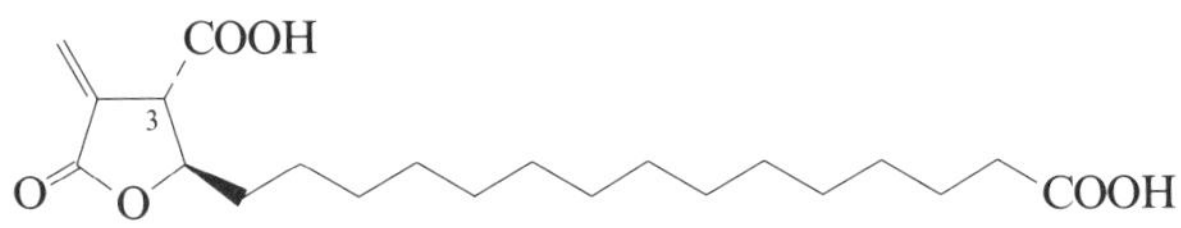

$C_{21}H_{34}O_6$ M 382.496

(2R,3S)-form [130342-69-7]
Isol. from the lichen *Parmotrema praesorediosum*. Cryst. ($CHCl_3$/pet. ether). Mp 123-124°. $[\alpha]_D^{25}$ +9.1° (MeOH).

Δ³-Isomer: [130342-70-0]. ***Praesorediosic acid.*** *3-Carboxy-2,5-dihydro-4-methyl-5-oxo-2-furanpentadecanoic acid, 9CI*
Isol. from *P. praesorediosum*. Cryst. ($CHCl_3$/pet. ether). Mp 139-140.5°. $[\alpha]_D^{25}$ +22.7° (MeOH).

Feeya, D. *et al*, *Aust. J. Chem.*, 1990, **43**, 1297 (*isol, pmr, cd, struct*)

14-Proxyloctadecanoic acid **P-00150**

2-Butyl-2-(12-carboxydodecyl)-5,5-dimethyl-1-pyrrolidinyloxy, 9CI
[66978-86-7]

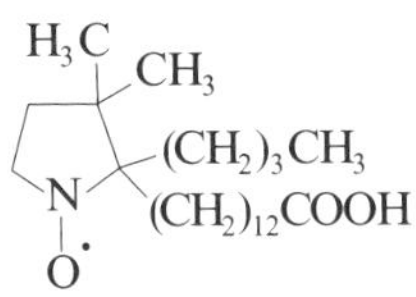

$C_{23}H_{44}NO_3$ M 382.605
Lipid spin label.

Trimethylammoniummethanesulfonate:
$C_{27}H_{57}N_2O_4S$ M 505.824
Low-melting waxy solid.

Keana, J.F.W. *et al*, *Chem. Phys. Lipids*, 1982, **31**, 339 (*synth*)

Psammaplysin A **P-00151**

[85819-66-5]

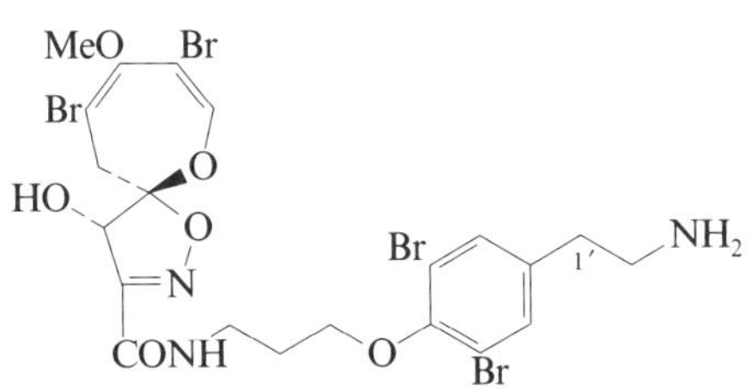

$C_{21}H_{23}Br_4N_3O_6$ M 733.045

Isol. from Red Sea sponge *Psammaplysilla purpurea*. Shows antibiotic props.; active against gram-positive bacteria. Foam. $[\alpha]_D^{22}$ −65.2° (c, 0.52 in MeOH). λ_{max} 228, 256, 272, 276 nm (MeOH).

Di-Ac: [95739-54-1].
Needles (MeOH). Mp 140°.

1′-Hydroxy: [85819-67-6]. ***Psammaplysin B***
$C_{21}H_{23}Br_4N_3O_7$ M 749.045
Isol. from *P. purpurea*. Active against gram-positive bacteria. Foam. $[\alpha]_D^{25}$ −60.2° (c, 0.632 in MeOH).

Rotem, M. *et al*, *Tetrahedron*, 1983, **39**, 667 (*isol, uv, ir, pmr, cmr, ms, struct*)
Roll, D.M. *et al*, *J. Am. Chem. Soc.*, 1985, **107**, 2916 (*uv, ir, pmr, cmr, ms, struct*)

Pseudomonic acid C **P-00152**

[71980-98-8]

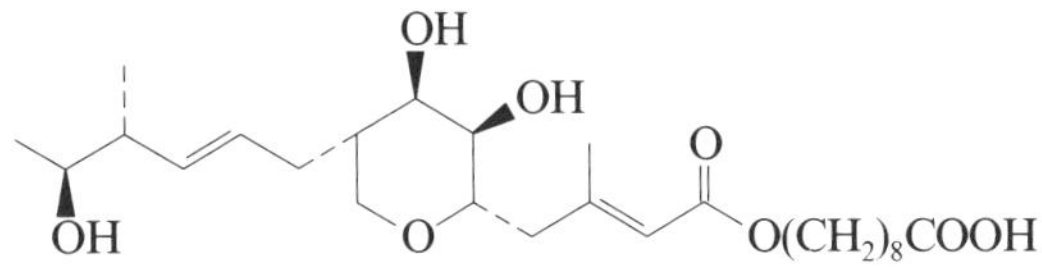

$C_{26}H_{44}O_8$ M 484.629
Metab. of *Pseudomonas fluorescens*. Active against mycoplasma infections. Oil or cryst. (EtOAc).

Me ester: [72042-22-9].
Cryst. Mp 47-49°.

[85617-79-4]

Schoenberger, B. *et al*, *Helv. Chim. Acta*, 1982, **65**, 2333 (*synth, bibl*)
Clayton, J.P. *et al*, *J. Chem. Soc., Perkin Trans. 1*, 1982, 2827 (*cryst struct*)
Beau, J.M. *et al*, *J. Am. Chem. Soc.*, 1983, **105**, 621 (*synth*)
Snider, B.B. *et al*, *J. Org. Chem.*, 1983, **48**, 3003 (*synth*)
Fleet, G.W.J. *et al*, *Tetrahedron Lett.*, 1983, **24**, 3661 (*synth*)
Keck, G.E. *et al*, *J. Org. Chem.*, 1984, **49**, 1462; 1985, **50**, 4317; 1989, **54**, 5845 (*synth, bibl*)
Barrish, J.C. *et al*, *J. Org. Chem.*, 1988, **53**, 4282 (*synth*)

Punaglandin 1 **P-00153**

5,6,7-Tris(acetyloxy)-10-chloro-12-hydroxy-9-oxoprosta-10,14,17-trien-1-oic acid methyl ester, 9CI
[96055-63-9]

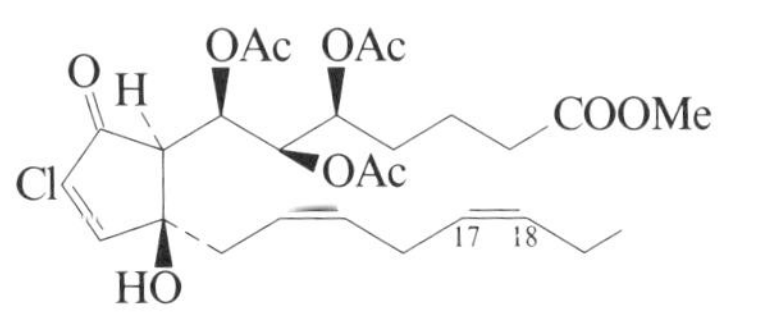

$C_{27}H_{37}ClO_{10}$ M 557.036
Isol. from Japanese octocoral *Telesto riisei*. Possesses potent antitumour props.

17,18-Dihydro: [96055-64-0]. ***Punaglandin 2***
$C_{27}H_{39}ClO_{10}$ M 559.052
From *T. riisei*. Possesses potent antitumour props.

Baker, B.J. *et al*, *J. Am. Chem. Soc.*, 1985, **107**, 2976 (*isol, struct*)
Sasai, H. *et al*, *Tetrahedron Lett.*, 1987, **28**, 333.

Punaglandin 3 **P-00154**

5,6-Bis(acetyloxy)-10-chloro-12-hydroxy-9-oxoprosta-7,10,14,17-tetraen-1-oic acid methyl ester,9CI

[96055-65-1]

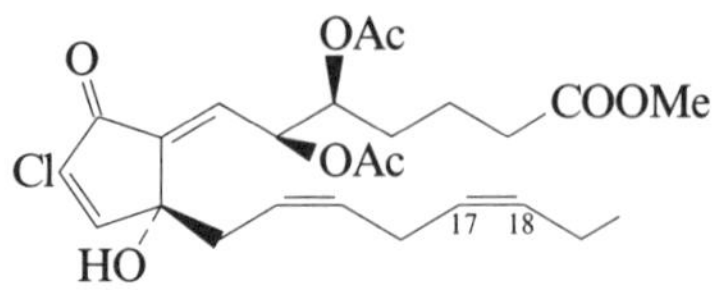

$C_{25}H_{33}ClO_8$ M 496.984

Isol. from Japanese octocoral *Telesto riisei*. Possesses potent antitumour props. Oil. The 12-config. in Punaglandins 3 and 4 is the opposite of that originally assigned and the opposite of that in Punaglandin 1, P-00153.

17,18-Dihydro: [96055-66-2]. ***Punaglandin 4***

$C_{25}H_{35}ClO_8$ M 498.999

From *T. riisei*. Possesses potent antitumour props. Oil. $[\alpha]_D^{24}$ +72.3° (c, 0.52 in $CHCl_3$) (synthetic). n_D^{24} 1.4967.

17,18-Dihydro, 14E-isomer: (E)-*Punaglandin 4*

Isol. from *T. riisei*. Oil.

Baker, B.J. *et al*, *J. Am. Chem. Soc.*, 1985, **107**, 2976 (*isol*)

Nagaoka, H. *et al*, *J. Am. Chem. Soc.*, 1986, **108**, 5019, 5021 (*synth, struct*)

Sasai, H. *et al*, *Tetrahedron Lett.*, 1987, **28**, 333 (*synth*)

Suzuki, M. *et al*, *J. Org. Chem.*, 1988, **53**, 286 (*synth, uv, pmr, ms, abs config, bibl*)

Mori, K. *et al*, *Tetrahedron*, 1988, **44**, 333 (*synth, ir, pmr, ms, cd*)

R

Retinal **R-00001**

Vitamin A_1 aldehyde. Retinene$_1$. Retinaldehyde
[116-31-4]

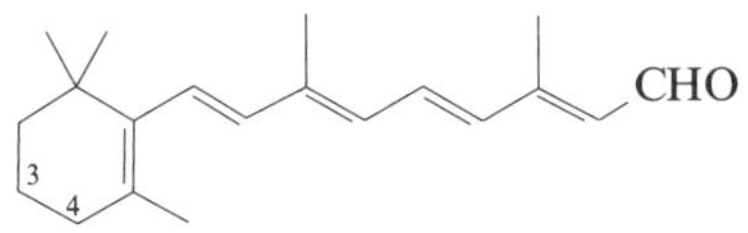

$C_{20}H_{28}O$ M 284.441
Forms part of the Rhodopsin complex. Cryst. Mp 64-65°.
▷ VH6407000.

Semicarbazone: Yellow cryst. ($CHCl_3/Et_2O$). Mp 199-201°.
3,4-Didehydro: [472-87-7]. ***Vitamin A_2 aldehyde.*** *Retinene$_2$. Dehydroretinal. Didehydroretinal*
$C_{20}H_{26}O$ M 282.425
Constit. of liver oil. Orange-red cryst. Mp 77-78°.

Ball, S. *et al*, *Biochem. J.*, 1948, **42**, 516 (*synth*)
Arens, J.F. *et al*, *Recl. Trav. Chim. Pays-Bas* (*J. R. Neth. Chem. Soc.*), 1949, **68**, 604 (*synth*)
Sebrell, W.H. *et al*, *The Vitamins*, Academic Press, Vol. 1, 1967 (*rev*)
Patel, D.J. *et al*, *Nature (London)*, 1969, **221**, 825 (*pmr*)
Hamanaka, T. *et al*, *Acta Crystallogr., Sect. B*, 1972, **28**, 214 (*cryst struct*)
Gilardi, R.G. *et al*, *Acta Crystallogr., Sect. B*, 1972, **28**, 2605 (*cryst struct*)
Yasaka, W.J. *et al*, *J. Nutr. Sci. Vitaminol.*, 1973, **19**, 369 (*isol*)
Honig, B. *et al*, *Annu. Rev. Biophys. Bioeng.*, 1974, **3**, 151 (*rev*)
Englert, G., *Helv. Chim. Acta*, 1975, **58**, 2367 (*cmr*)
Shriver, J. *et al*, *J. Am. Chem. Soc.*, 1976, **98**, 2407 (*cmr*)
Simmons, C.J. *et al*, *Acta Crystallogr., Sect. B*, 1981, **37**, 2197 (*cryst struct*)
Asato, A.E. *et al*, *J. Am. Chem. Soc.*, 1983, **105**, 2923 (*synth*)
Zhu, Y. *et al*, *Tetrahedron*, 1992, **48**, 10061 (*ir*)

Retinol, 9CI, INN **R-00002**

3,7-Dimethyl-9-(2,6,6-trimethyl-1-cyclohexen-1-yl)-2,4,6,8-nonatetraen-1-ol. 15-Apo-β-caroten-15-ol. Vitamin A_1. Afaxin. Axerol. Gadol. Prepalin. Testavol. Vitamin A. Axerophthol
[68-26-8]

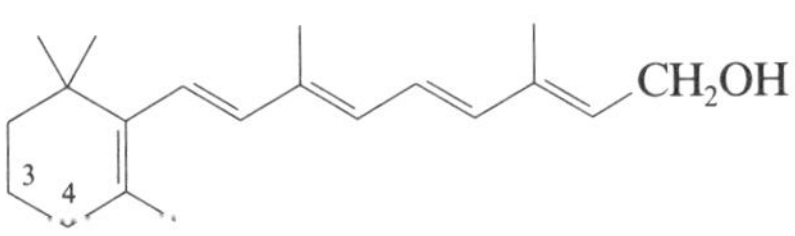

$C_{20}H_{30}O$ M 286.456
Constit. of many fish-liver oils, milk, egg-yolk, etc. Yellow cryst. Mp 63-64°. $Bp_{0.000001}$ 137-138°.
▷ Acute and chronic toxic effects in humans (large doses). Exp. and human teratogen. Exp. reprod. effects. LD_{50} (rat, orl) 2000 mg/kg. VH6750000.

Ac: [127-47-9]. *Retinol acetate. Vitamin A acetate. Other proprietary names*
$C_{22}H_{32}O_2$ M 328.494
Cryst. (MeOH). Mp 57-58°.
▷ LD_{50} (mus, orl) 4100 mg/kg. Exp. reprod. and teratogenic effects. VH6825000.

Propanoyl: [7069-42-3]. *Retinol propanoate. Axerofluid. Axerophthol propionate. Vitamin A propionate*
$C_{23}H_{34}O_2$ M 342.520
n_D^{25} 1.6042. Fluid to −10°.
Hexadecanoyl: [79-81-2]. *Retinol palmitate. Vitamin A palmitate. Arovit. Other synonyms*
$C_{36}H_{60}O_2$ M 524.869
Found in fish liver oils. Mp 28-29°.
▷ LD_{50} (mus, orl) 4760 mg/kg. Exp. reprod. and teratogenic effects. VH6860000.
3,4-Didehydro: [79-80-1]. ***3,4-Didehydroretinol.*** *!Vitamin A_2. Dehydroretinol. 3-Dehydroretinol*
$C_{20}H_{28}O$ M 284.441
Constit. of fresh-water fish oils. Golden-yellow oil. Mixt. of stereoisomers. Has 40% biological activity of Vit. A_1.
15-Aldehyde: see *Retinal, R-00001*

Isler, O. *et al*, *Adv. Org. Chem.*, 1963, **4**, 115 (*synth, rev*)
Fr. Pat., 1 452 791, (1966); *CA*, **66**, 105072r (*synth, propanoate*)
Isler, O. *et al*, *Chim. Ind. (Milan)*, 1967, **49**, 1317 (*rev*)
Sebrell, W.H. *et al*, *The Vitamins*, Academic Press, Vol. 1, 1967 (*rev*)
Westermann, F.H., *Int. J. Vitam. Nutr. Res.*, 1972, **42**, 104 (*metab, palmitate*)
Oberhänsli, W.E. *et al*, *Acta Crystallogr., Sect. B*, 1974, **30**, 161 (*cryst struct*)
Mukaiyama, T. *et al*, *Chem. Lett.*, 1975, 1201 (*synth*)
Englert, G., *Helv. Chim. Acta*, 1975, **58**, 2367 (*cmr*)
Baumann, C., *Method. Chim.*, 1978, **11**, 200 (*rev, Vit. A_2*)
Cardillo, G. *et al*, *J. Chem. Soc., Perkin Trans. 1*, 1979, 1729 (*synth*)
Mehta, R.R. *et al*, *J. Chem. Soc., Perkin Trans. 1*, 1982, **12**, 2921 (*synth, acetate*)
Martindale, The Extra Pharmacopoeia, 28th/29th Ed., Pharmaceutical Press, London, 1982/1989, 7821, 7822, 7823.
Pitt, G.A., *Proc. Nutr. Soc.*, 1983, **42**, 43 (*struct, activity, rev*)
Pure Appl. Chem., 1983, **55**, 722 (*nomencl*)
Williams, J.B. *et al*, *J. Lipid Res.*, 1984, **25**, 638 (*metab, acetate*)
Goodman, D.S. *et al*, *Retinoids*, 1984, **2**, 1, 177 (*metab, biosynth, rev*)
Liu, R.S.H. *et al*, *Tetrahedron*, 1984, **40**, 1931 (*rev*)
Harbison, G.S. *et al*, *J. Am. Chem. Soc.*, 1985, **107**, 4809 (*cmr, acetate*)
Ajima, A. *et al*, *Biotechnol. Lett.*, 1986, **8**, 547 (*synth, hexadecanoate*)
Otera, J. *et al*, *J. Org. Chem.*, 1986, **51**, 3834 (*synth*)
Sundaresan, P., *J. Sci. Ind. Res.*, 1987, **46**, 105 (*rev, metab*)
Vaezi, H.F. *et al*, *Org. Prep. Proced. Int.*, 1987, **19**, 187 (*synth, Vit. A_2*)
Negwer, M., *Organic-Chemical Drugs and their Synonyms*, 6th Ed., Akademie-Verlag, Berlin, 1987, 5941, 6719, 8103.
Hosoda, A. *et al*, *Tetrahedron Lett.*, 1987, **28**, 65 (*synth*)
Vogt, P. *et al*, *Tetrahedron Lett.*, 1991, **32**, 4115 (*synth*)
Soukup, M. *et al*, *Tetrahedron Lett.*, 1991, **32**, 4117 (*synth*)
Martindale, The Extra Pharmacopoeia, 30th edn., Pharmaceutical Press, London, 1993, 7823.
Lewis, R.J., *Sax's Dangerous Properties of Industrial Materials*, 8th Ed., Van Nostrand-Reinhold, 1992, VSK600, VSK900, VSP000.

S

Samin

S-00001

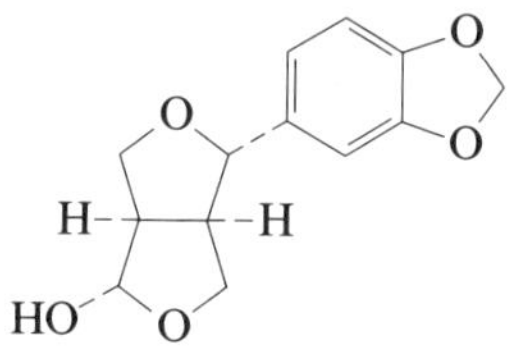

$C_{13}H_{14}O_5$ M 250.251

Isol. from sesame seed oil. Mp 106°. $[\alpha]_D$ +103°. Config. not certain.

O-(*3,4-Methylenedioxyphenyl*): [526-07-8]. ***Sesamolin***

$C_{20}H_{18}O_7$ M 370.358

Constit. of sesame oil. Plates (EtOH). Mp 94°. $[\alpha]_D$ +220° (c, 1.13 in $CHCl_3$).

Boesken, J. *et al*, *Recl. Trav. Chim. Pays-Bas* (*J. R. Neth. Chem. Soc.*), 1936, **55**, 815.
Carnmalm, B. *et al*, *Acta Chem. Scand.*, 1955, **9**, 1111.
Erdtman, H. *et al*, *Chem. Ind.* (*London*), 1955, 567.
Beroza, M., *J. Am. Chem. Soc.*, 1955, **77**, 3332.
Haslam, E. *et al*, *J. Chem. Soc.*, 1955, 827.
Haslam, E., *J. Chem. Soc. C*, 1970, 2332 (*abs config, pmr, ms*)
Fukuda, Y. *et al*, *Heterocycles*, 1986, **24**, 926.
Bradley, H.M. *et al*, *J. Chem. Soc., Chem. Commun.*, 1991, 1641 (*synth*)
Bradley, H.M. *et al*, *Synlett*, 1992, 479 (*synth*)

9,10-Secocholesta-5,7,10(19)-trien-3-ol, 9CI

S-00002

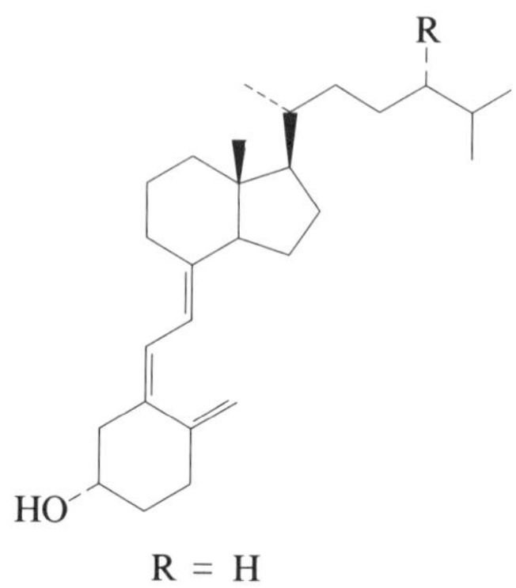

R = H

$C_{27}H_{44}O$ M 384.644

$[\alpha]_D^{20}$ +84.8° (c, 1.6 in Me_2CO).

(3S,5Z,7E)-form [67-97-0] ***Vitamin D$_3$***. *!Cholecalciferol*, BAN. *Colecalciferol*, INN. *Calciol. Delsterol. Provitina. Ricketon. Vigorsan*

Constit. of fish-liver oil, esp. tuna oil. Antirachitic vitamin with approx. the same potency as vitamin D_2 in humans. Cryst. (Me_2CO). Mp 87-88°. A lower-melting form, Mp 63-4° (from MeOH) was also reported. Deuterated and tritiated compds.have been synth.

▷ Excessive intake can cause hypercalcaemia and other effects. LD_{50} (rat, orl) 42 mg/kg. Exp. teratogen. VS2900000.

Ac: [22350-49-8].

$C_{29}H_{46}O_2$ M 426.681

Mp 131-132°. $[\alpha]_D^{18}$ +142° ($CHCl_3$).

Benzoyl:

$C_{34}H_{48}O_2$ M 488.752

Mp 63-64°. $[\alpha]_D$ +39° ($CHCl_3$).

3,5-Dinitrobenzoyl: Yellow needles. Mp 131°.

3-Sulfate: Amorph. solid (as Na salt). Mp 113-120° (as Na salt).

[2200-64-8, 3844-73-3, 7489-19-2, 10597-47-4, 13083-02-8, 22350-41-0, 57651-82-8, 57651-83-9, 72409-70-2, 72506-00-4, 74512-98-4, 74512-99-5, 77286-64-7, 80666-48-4, 81126-11-6, 81126-44-5, 98744-46-8, 98818-17-8, 118584-54-6, 122958-39-8]

Inhoffen, M. *et al*, *Angew. Chem.*, 1960, **72**, 875 (*rev*)
Haussler, M.S. *et al*, *J. Biol. Chem.*, 1972, **247**, 2328 (*rev*)
Borle, A.B., *Annu. Rev. Physiol.*, 1974, **36**, 361 (*rev, metab*)
Wing, R.M. *et al*, *J. Am. Chem. Soc.*, 1975, **97**, 4980 (*pmr, conformn*)
Okamura, W.H. *et al*, *Tetrahedron Lett.*, 1976, 4807 (*ms*)
Lythgoe, B. *et al*, *Tetrahedron Lett.*, 1977, 3685 (*synth*)
Fraser, D.R., *Br. Med. Bull.*, 1981, **37**, 37 (*rev, metab*)
Pure Appl. Chem., 1982, **54**, 1511 (*nomencl*)
Martindale, The Extra Pharmacopoeia, 28th/29th Ed., Pharmaceutical Press, London, 1982/1989, 7884.
Koshy, K.T. *et al*, *Anal. Profiles Drug Subst.*, 1984, **13**, 655 (*rev*)
Wilson, S.R. *et al*, *Tetrahedron Lett.*, 1984, **25**, 3147 (*synth*)
Nemoto, H. *et al*, *Heterocycles*, 1985, **23**, 567; *J. Org. Chem.*, 1986, **51**, 5311 (*synth*)
Dauben, W.G. *et al*, *J. Org. Chem.*, 1985, **50**, 2007 (*synth, fluoro, deriv*)
Coldwell, R.D. *et al*, *Steroids*, 1990, **55**, 418 (*rev*)
Martindale, The Extra Pharmacopoeia, 30th edn., Pharmaceutical Press, London, 1993, 7888.
Lewis, R.J., *Sax's Dangerous Properties of Industrial Materials*, 8th Ed., Van Nostrand-Reinhold, 1992, CMC750, HJV000.

Sesamin

S-00003

*5,5′-(Tetrahydro-1*H*,3*H*-furo[3,4-*c*]furan-1,4-diyl)bis-1,3-benzodioxole,9CI. 2,6-Bis(3,4-methylenedioxyphenyl)-3,7-dioxabicyclo[3.3.0]octane. 3,4:3′,4′-Bis(methylenedioxy)-7,9′:7′,9-diepoxylignan. Pseudocubebin. Fagarol*

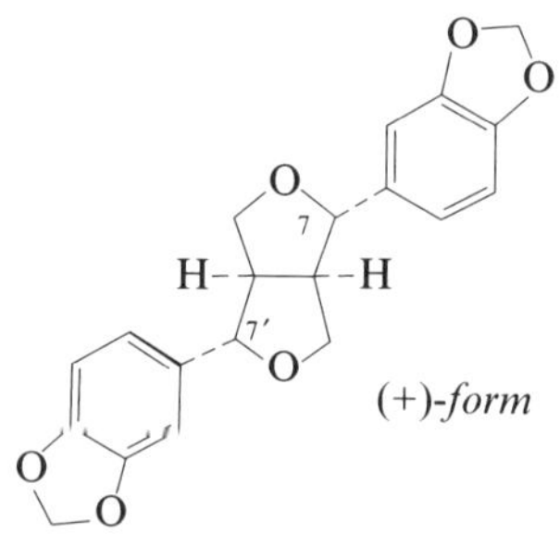

(+)-*form*

$C_{20}H_{18}O_6$ M 354.359

Lignan numbering shown.

(+)-***form*** [607-80-7]

Isol. from *Sideritis canariensis*, the guggula resin from *Commiphora mukul*, *Xanthoxylum* spp. and *Talauma hodgsoni*. Constit. of sesame oil. Isol. from pyrethrum flowers (*Chrysanthemum cinerariaefolium*). Present in the Rutaceae, Piperaceae, Scrophulariaceae, Compositae and Ginkgoaceae. Synergist for pyrethrum insecticides. Needles (EtOH). Mp 123-124°. $[\alpha]_D^{20}$ +78.4° ($CHCl_3$).

7-Epimer: [133-03-9]. (+)-***Asarinin***. *Isosesamin. Episesamin. Xanthoxylin S*

Constit. of *Acronychia muelleri*, also formed by isomerisation of (+)-Sesamin. Cryst. (EtOH or Me_2CO). Mp 121-122°. $[\alpha]_D^{18}$ +121.7° ($CHCl_3$).

7,7′-Diepimer: [551-30-4]. *Diasesamin. Epiasarinin*

$C_{20}H_{18}O_6$ M 354.359

Obt. by isom. of Asarinin. Mp 168-171°. $[\alpha]_D^{26}$ +385°.

(−)-***form*** [13079-95-3]

Isol. from the leaves of *Magnolia mutabilis* and from the bark of *X. tingoassuiba*. Also in *Acanthopanax sessiliflorus*. Formed by heating (−)-Asarinin with EtOH/HCl. Mp 122-124°. $[\alpha]_D^{17}$ −68° (c, 0.5 in $CHCl_3$).

7-Epimer: [133-04-0]. (−)-***Asarinin***

Isol. from *Asarum* and *Xanthoxylum* spp. Cryst. (EtOH). Mp 121°. $[\alpha]_D^{18}$ −122° (c, 3.54 in $CHCl_3$). Acid isom. → (−)-Sesamin.

(±)-***form***

Constit. of *X. senegalense* and *Fargaria xanthoxyloides*. Mp 129-130°.

7-Epimer: Mp 134-135°.

Davenport, J.B. *et al*, *Aust. J. Chem.*, 1954, **1**, 384 (*isol*)
Beroza, M. *et al*, *J. Am. Chem. Soc.*, 1956, **78**, 1242, 5082 (*struct, ms, synth*)
Freudenberg, K. *et al*, *Chem. Ber.*, 1961, **94**, 851 (*config*)
Takahashi, K. *et al*, *Chem. Pharm. Bull.*, 1966, **14**, 641 (*pmr, config*)
Pelter, A., *J. Chem. Soc. C*, 1967, 1376 (*ms*)
Karrer, W. *et al*, *Konstitution und Vorkommen der Organischen Pflanzenstoffe*, 2nd Ed., Birkhäuser Verlag, Basel, 1972-1985, no. 1154 (*occur*)
Krajniak, R.E. *et al*, *Aust. J. Chem.*, 1973, **26**, 687 (*isol*)
Abe, F. *et al*, *Chem. Pharm. Bull.*, 1974, **22**, 2650 (*isol*)
Talapatra, B. *et al*, *Phytochemistry*, 1975, **14**, 589 (*isol*)
Pelter, A. *et al*, *Tetrahedron*, 1976, **32**, 2783 (*struct, nmr*)
Anjaneyulu, A.S.R. *et al*, *Tetrahedron*, 1977, **33**, 133 (*pmr*)
Pelter, A. *et al*, *Tetrahedron Lett.*, 1977, 4137; 1978, 1509 (*cmr*)
Jaensch, M. *et al*, *Phytochemistry*, 1989, **28**, 3497 (*isol, pmr*)
Baures, P.W. *et al*, *Acta Crystallogr., Sect. C*, 1992, **48**, 574 (*cryst struct*)
Stevens, D.R. *et al*, *J. Chem. Soc., Perkin Trans. 1*, 1992, 185 (*synth*)

Soyacerebroside I — S-00004

1-O-β-D-Glucopyranosyl-N-(2-hydroxyhexadecanoyl)-4,8-sphingadiene

[114297-20-0]

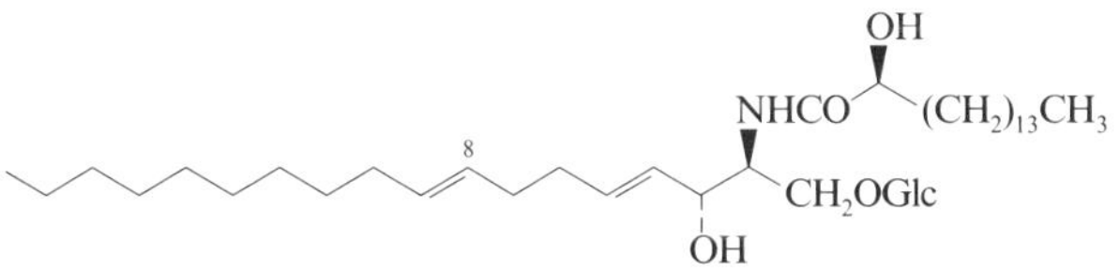

$C_{40}H_{75}NO_9$ M 714.034

Cerebroside isol. from *Tetragonia tetragonoides* (Tetragoniae Herba) and *Glycine max*. Antiulcerogenic principle. Hygroscopic granules or amorph. solid. Mp 184-186°. $[\alpha]_D^{21}$ +10.5° (c, 0.30 in $MeOH/CHCl_3$, 3:2).

8Z-Isomer: [115074-93-6]. ***Soyacerebroside II***

Isol. from *T. tetragonoides* and *G. max*. Antiulcerogenic principle. Hygroscopic granules or amorph. solid. Mp 192-194° (183°). $[\alpha]_D^{24}$ +13.4° (c, 0.43 in $MeOH/CHCl_3$ 3:2)(+4.6°).

Mori, K. *et al*, *Justus Liebigs Ann. Chem.*, 1988, 807 (*abs config, ir, pmr, cmr, synth, bibl*)
Shibuya, H. *et al*, *Chem. Pharm. Bull.*, 1990, **38**, 2933 (*isol, struct*)

Sphingofungin E — S-00005

[145401-47-4]

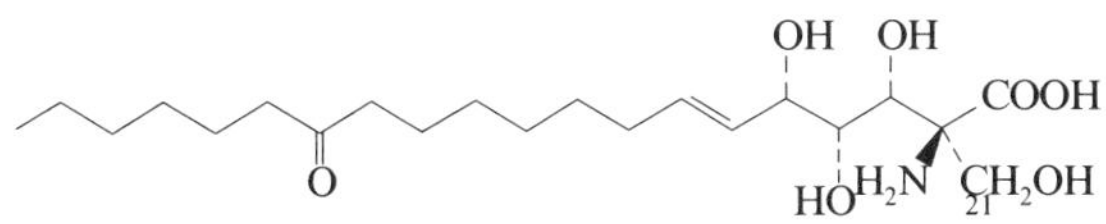

$C_{21}H_{39}NO_7$ M 417.542

Config. not confirmed. Prod. by *Paecilomyces variotii*. Serinepalmitoyl transferase inhibitor. Antifungal agent.

21-Deoxy: [145401-48-5]. ***Sphingofungin F***

$C_{21}H_{39}NO_6$ M 401.542

Prod. by *P. variotii*. Serinepalmitoyl transferase inhibitor. Antifungal agent.

Horn, W.S. *et al*, *J. Antibiot.*, 1992, **45**, 1692 (*isol, pmr, cmr, struct*)

Sphingomyelins — S-00006

A class of sphingolipids with the general structure indicated. Hydrolysis furnishes sphing-4-enine (or other long-chain amine), a fatty acid, phosphoric acid and choline (or ethanolamine).

Rodd's Chem. Carbon Compd. (2nd edn.), 2nd Ed., Vol 1E, 1976, 425.

Spiculisporic acid — S-00007

2-Carboxy-α-decyltetrahydro-5-oxo-2-furanacetic acid, 9CI. 4,5-Dicarboxy-4-pentadecanolide

[469-77-2]

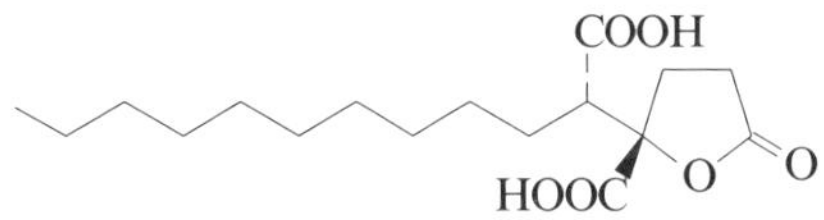

Absolute configuration

$C_{17}H_{28}O_6$ M 328.405

Produced by *Penicillium spiculisporum* and other spp. Possesses surfactant props. Mp 145°. $[\alpha]_{461}^{5}$ −14.76° (EtOH).

Birkinshaw, J.H. *et al*, *Biochem. J.*, 1934, **28**, 828 (*isol*)
Oxford, E. *et al*, *Biochem. J.*, 1934, **28**, 1321 (*isol*)
Gatenbeck, S. *et al*, *CA*, 1967, **66**, 44470v (*biosynth*)
Tabuchi, T. *et al*, *J. Ferment. Technol.*, 1977, **55**, 37 (*isol*)
Brandänge, S. *et al*, *Acta Chem. Scand., Ser. B*, 1984, **38**, 837 (*abs config, synth*)

Squalene — S-00008

2,6,10,15,19,23-Hexamethyl-2,6,10,14,18,22-tetracosahexaene. Spinacene

[111-02-4]

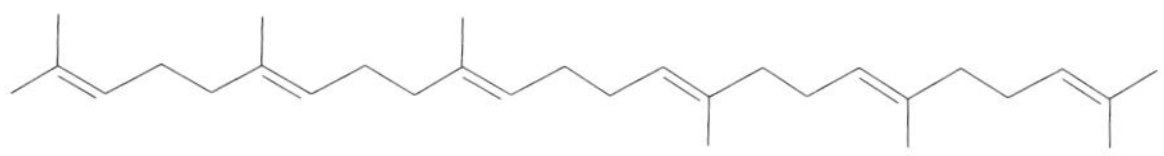

$C_{30}H_{50}$ M 410.725

(***All-E)-form***

Found in fish liver oils, yeast lipids and many vegetable oils, e.g. palm oil, cottonseed oil, rapeseed oil. Volatile component of scent material from *Saguinus oedipus* (cotton-top tamarin monkey) and *S. fuscicollis* (saddle-back tamarin monkey). Oil. Mp −4.8° to −5.2°. Bp_{25} 284-285°.

Johnson, W.S. *et al*, *J. Am. Chem. Soc.*, 1970, **92**, 741 (*synth*)
Bhalerao, U.T. *et al*, *J. Am. Chem. Soc.*, 1971, **93**, 531 (*synth*)
Biellmann, J.F. *et al*, *Tetrahedron*, 1971, **27**, 5861 (*synth*)

Karrer, W. *et al*, *Konstitution und Vorkommen der Organischen Pflanzenstoffe*, 2nd Ed., Birkhäuser Verlag, Basel, 1972-1985, no. 34 (*occur*)
Grieco, P.A. *et al*, *J. Org. Chem.*, 1974, **39**, 2135 (*synth*)
Suga, T. *et al*, *Phytochemistry*, 1975, **14**, 2411 (*biosynth*)
v. Dommelen, M.E. *et al*, *Recl. Trav. Chim. Pays-Bas* (*J. R. Neth. Chem. Soc.*), 1975, **94**, 182 (*cmr*)
Johnson, W.S. *et al*, *Angew. Chem., Int. Ed. Engl.*, 1976, **15**, 9 (*rev*)
Ernst, J. *et al*, *Angew. Chem., Int. Ed. Engl.*, 1976, **15**, 778 (*cryst struct*)
Ernst, J. *et al*, *Justus Liebigs Ann. Chem.*, 1979, 1635 (*cryst struct*)
Belcher, A. *et al*, *J. Chem. Ecol.*, 1988, **14**, 1367.
Nes, W.D. *et al*, *Phytochemistry*, 1988, **27**, 628 (*biosynth*)
Nordstrom, K.M. *et al*, *J. Chem. Ecol.*, 1989, **15**, 629.

Sterculic acid **S-00009**

9,10-Methyleneoctadec-9-enoic acid. 2-Octyl-1-cyclopropen-1-octanoic acid

[738-87-4]

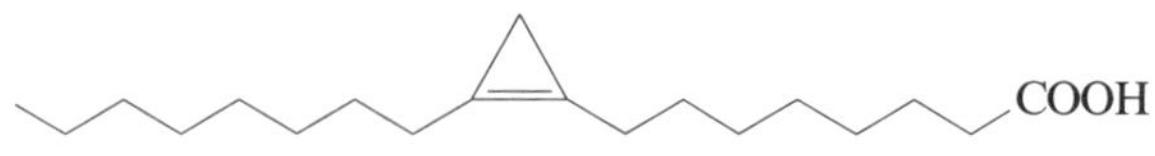

$C_{19}H_{34}O_2$ M 294.476

Constit. of some seed oils of Malvaceae, Sterculiaceae, Tiliaceae and Bombaceae. Blocks the action of Δ^9-desaturase. Mp 18.2-18.3°.

Nunn, J.R., *J. Chem. Soc.*, 1952, 313 (*isol, struct*)
Faure, P.K. *et al*, *J. Chem. Soc.*, 1956, 1818 (*isol, struct*)
Rinehart, K.L. *et al*, *J. Am. Chem. Soc.*, 1958, **80**, 503 (*nmr*)
Gensler, W.J. *et al*, *J. Am. Chem. Soc.*, 1970, **92**, 2472 (*synth*)
Gensler, W.J. *et al*, *Chem. Phys. Lipids*, 1971, **6**, 280 (*synth*)
Williams, J.L. *et al*, *J. Org. Chem.*, 1971, **36**, 3064 (*synth*)

Sterculynic acid **S-00010**

8,9-Methyleneoctadec-8-en-17-ynoic acid

[18325-59-2]

$C_{19}H_{30}O_2$ M 290.445

Constit. of the seed oil of *Sterculia alata*. Oil.

Me ester: [18459-83-1].

$C_{20}H_{32}O_2$ M 304.472

Oil.

Jevans, A.W. *et al*, *Tetrahedron Lett.*, 1968, 2167 (*isol*)

Stigmasta-5,24(28)-dien-3-ol, 9CI **S-00011**

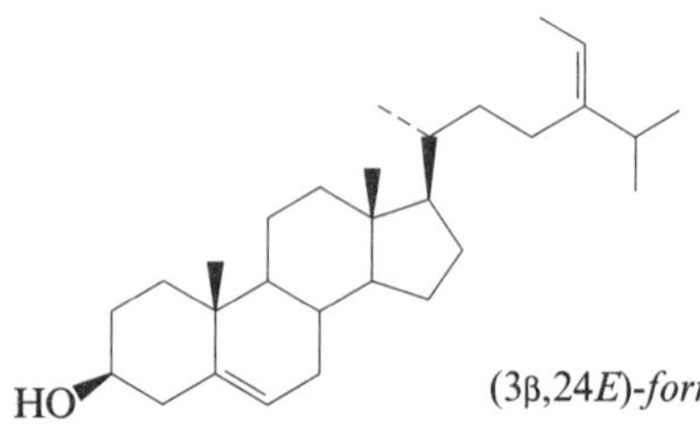

(3β,24*E*)-*form*

$C_{29}H_{48}O$ M 412.698

3β-form [18472-36-1] ***Δ⁵-Avenasterol***

Isol. from yeast, algal and fern lipids. Platelets (MeOH). Mp 137°. $[\alpha]_D^{26}$ −37.6° ($CHCl_3$). Originally considered to have the 5,11-diene struct.

Ac: [58581-06-9].

$C_{31}H_{50}O_2$ M 454.735

Plates (EtOH). Mp 134°. $[\alpha]_D^{26}$ −39.7° ($CHCl_3$).

3-O-β-D-Galactopyranoside: [121250-36-0].

$C_{35}H_{58}O_6$ M 574.840

Constit. of the seeds of *Sesbania sesban*.

(3β,24E)-form [17605-67-3] ***Fucosterol***

Characteristic sterol of seaweeds; isol. from *Fucus vesiculosus*. Needles (MeOH). Mp 124°. $[\alpha]_D^{20}$ −38.4° ($CHCl_3$).

Ac: [51297-12-2].

Cryst. (MeOH). Mp 120-121°. $[\alpha]_D$ −45°.

3-O-α-L-Rhamnopyranoside: [77539-34-5].

$C_{35}H_{58}O_5$ M 558.840

Constit. of *Cleome viscosa*. Cryst. ($CHCl_3$/MeOH). Mp 60-62°.

(3β,20R,24Z)-form [481-14-1] ***Isofucosterol***

Present in oats. Plates (MeOH). Mp 137° (130-131°). $[\alpha]_D^{26}$ −37.6° (MeOH).

Ac: [6035-62-7].

Plates (EtOH). Mp 134°. $[\alpha]_D^{26}$ −39.7° (EtOH).

(3β,20S,24E)-form [481-15-2] ***Sargasterol.*** *20S-Fucosterol*

Isol. from *Sargassum ringgoldianum*. Cryst. (EtOH). $[\alpha]_D$ −47.5° ($CHCl_3$).

Ac: Mp 138-139°. $[\alpha]_D$ −52.9° ($CHCl_3$).

[6793-19-7, 20584-62-7, 83997-22-2]

Heilbron, I. *et al*, *J. Chem. Soc.*, 1934, 1572 (*isol*)
Idler, D.R. *et al*, *J. Am. Chem. Soc.*, 1953, **75**, 1712 (*isol*)
Hayazu, R. *et al*, *Pharm. Bull.*, 1957, **5**, 452 (*synth*)
Tsuda, K. *et al*, *J. Am. Chem. Soc.*, 1958, **80**, 921 (*Sargasterol*)
Dusza, J.P., *J. Org. Chem.*, 1960, **25**, 93 (*synth, Isofucosterol*)
Goad, L.J. *et al*, *Nature (London)*, 1966, **210**, 1322 (*biosynth*)
Nes, W.R. *et al*, *Steroids*, 1966, **8**, 655 (*pmr*)
Audier, R. *et al*, *Tetrahedron Lett.*, 1966, 4341 (*ms, deriv*)
Bates, R.B. *et al*, *Tetrahedron Lett.*, 1968, 6163 (*struct*)
Brooks, C.J.W. *et al*, *Steroids*, 1972, **20**, 487 (*ms, struct*)
Yu, P.L.C. *et al*, *J. Nat. Prod. (Lloydia)*, 1974, **37**, 593 (*Sargasterol*)
Rubinstein, I. *et al*, *Phytochemistry*, 1976, **15**, 195 (*pmr*)
Bu'Lock, J.D. *et al*, *Phytochemistry*, 1976, **15**, 1249 (*biosynth*)
Massey, I.J. *et al*, *J. Org. Chem.*, 1979, **44**, 2448 (*ms, synth*)
Nicotra, F. *et al*, *Gazz. Chim. Ital.*, 1980, **110**, 579.
McInnes, A.G. *et al*, *Org. Magn. Reson.*, 1980, **13**, 302 (*cmr*)
Srivastava, S.K., *Phytochemistry*, 1980, **19**, 2510 (*isol*)
Kohli, D.V. *et al*, *Fitoterapia*, 1988, **59**, 478 (*galactoside*)
Seo, S. *et al*, *J. Chem. Soc., Perkin Trans. 1*, 1990, 105 (*biosynth*)

Stigmasta-7,22-dien-3-ol **S-00012**

$C_{29}H_{48}O$ M 412.698

(3β,5α,22E,24R)-form [481-17-4] ***Chondrillasterol.*** *Poriferasta-7,22-dien-3-ol*

Constit. of *Chondrilla nucula* and the green alga *Scenedesmus obliquus*. Cryst. (Me_2CO/MeOH). Mp 174.5-175.5°. $[\alpha]_D^{20}$ −2° ($CHCl_3$).

3-O-[β-D-Glucopyranosyl-(1→4)-β-D-glucopyranoside]:

$C_{41}H_{68}O_{11}$ M 736.982

Constit. of *Amaranthus spinosus*. Cryst. (MeOH/$CHCl_3$). Mp 290°. $[\alpha]_D^{28}$ −40.1° (c, 1 in Py).

3-O-[β-D-Glucopyranosyl-(1→2)-β-D-glucopyranosyl-(1→2)-β-D-glucopyranoside]:

$C_{47}H_{78}O_{16}$ M 899.124

Constit. of *A. spinosus*. Cryst. (MeOH/$CHCl_3$). Mp 340° dec. $[\alpha]_D^{28}$ −51.1° (c, 1. in Py).

(3β,22E,24S)-form [481-18-5] ***α-Spinasterol.*** *Bessisterol*

Constit. of *Spinacea oleracea* and *Cucumis sativus*, alfalfa and senega root. Cryst. (EtOH/C_6H_6). Mp 171-173°. $[\alpha]_D^{22}$ −2.5° ($CHCl_3$).

O-β-D-Glucopyranoside: [1745-36-4].

Constit. of *Maesa chisia* and *Pithecellobium dulce*. Cryst. Mp 292-294°. $[\alpha]_D$ −29° (c, 1 in Py).

Nigam, S.K. *et al*, *Indian J. Chem.*, 1967, **5**, 395 (*isol, glucoside*)
Thompson, M.J. *et al*, *Phytochemistry*, 1972, **11**, 1781 (*pmr*)

Kircher, H.W. *et al*, *J. Org. Chem.*, 1973, **38**, 2259 (*synth*)
Armarego, W.L.F. *et al*, *Phytochemistry*, 1973, **12**, 2181 (*biosynth*)
Sucrow, W. *et al*, *Phytochemistry*, 1976, **15**, 1533 (*config*)
Banerji, N., *J. Indian Chem. Soc.*, 1980, **57**, 417, 1043 (*isol*)
Anastasia, M. *et al*, *J. Chem. Soc., Perkin Trans. 1*, 1981, 2561 (*synth*)

Stigmast-5-en-3-ol S-00013

24-Ethylcholest-5-en-3-ol

$C_{29}H_{50}O$ M 414.713

(3β,24R)-form [83-46-5] ***β-Sitosterol.*** *Nimbosterol. Cupreol. Quebrachol. Rhamnol. Cinchol. Verosterol. Slanutosterol. Raphanisterol. Papaveristerol. Sitosterol. Angelicin†. 22,23-Dihydrostigmasterol*
Widely distributed in plants. The commonest sterol of higher plants. Hypolipidnemic agent. Cryst. (MeOH). Mp 136-137°. $[\alpha]_D^{22}$ −35° ($CHCl_3$). Early isolates usually contd. *ca.* 5-30% campesterol.

▷ Exp. roprod. effects. WJ2600000.

Ac: β-Sitosterol acetate
$C_{31}H_{52}O_2$ M 456.751
Constit. of pods of *Acacia leucophloea* and roots of *Robinia pseudoacacia*. Also present in *Alysicarpus longifolius*. Cryst. (MeOH). Mp 134°. $[\alpha]_D^{22}$ −38.5°.

O-*β-D-Xyluronofuranoside:* [133538-67-7].
$C_{34}H_{56}O_6$ M 560.813
Constit. of *Bauhinia uruguayensis*.

O-*α-D-Xyluronofuranoside:* [108195-79-5].
$C_{34}H_{56}O_6$ M 560.813
Constit. of *B. candicans* and *B. uruguayensis*.

O-*β-D-Glucopyranoside:* [474-58-8]. ***Daucosterol.*** *Eleutheroside A. Alexandrin. Daucosterin. Sitogluside, INN. Sterolin. Nimbosterin. AW 10. BSSG. EU-4906. Ipuranol*
$C_{35}H_{60}O_6$ M 576.855
Constit. of many plants. Drug used to treat prostatic hypertrophy. Cryst. (Py/EtOH). Mp 283-286° (270°) dec. $[\alpha]_D^{26}$ −41.5° (c, 0.4 in Py).

β-D-Galactoside:
$C_{35}H_{60}O_6$ M 576.855
Constit. of *Hibiscus sabdariffa*. Cryst. ($CHCl_3$). Mp 275-277°. $[\alpha]_D$ −63° (c, 0.52 in Py).

O-*α-D-Riburonofuranoside:* [96087-22-8].
$C_{34}H_{56}O_6$ M 560.813
Constit. of *B. candicans* and *B. uruguayensis*. Formerly used as a hypolipidaemic agent.

3-O-(6-O-Oleoyl-β-D-glucopyranoside): ***Sitoindoside II***
$C_{53}H_{92}O_7$ M 841.306
Constit. of *Musa paradisiaca*. Shows antiulcerogenic activity. Amorph. powder. $[\alpha]_D$ −10.2° ($CHCl_3$).

3-O-(6-O-Hexadecanoyl-β-D-glucopyranoside): [18749-71-8]. ***Sitoindoside I***
$C_{51}H_{90}O_7$ M 815.268
Constit. of fruits of *M. paradisiaca* and fern *Woodwardia orientalis*. Shows antiulcerogenic activity. Amorph. powder. Mp 163-165°. $[\alpha]_D$ −18.8° ($CHCl_3$).

3-O-[α-Rhamnopyranosyl-(1→5)-α-L-arabinofuranoside]: [74131-14-9].
$C_{40}H_{68}O_9$ M 692.972
From *Lindbergia indica*. Cryst. (MeOH). Mp 112°.

3-O-[α-D-Rhamnopyranosyl-(1→4)-β-D-glucopyranoside]: [74131-13-8].
$C_{41}H_{70}O_{10}$ M 722.998
From *L. indica*. Cryst. (MeOH). Mp 98°.

2-Deoxy-β-D-glucopyranoside:
$C_{35}H_{60}O_5$ M 560.856
Isol. from *Pteris inaequalis* var. *aequata*. Mp 216-220°. $[\alpha]_D$ −33.0° ($CHCl_3$).

3-O-β-D-Glucuronide:
$C_{35}H_{58}O_7$ M 590.839
Constit. of *Helichrysum arenarium*. Mp 280-290°. $[\alpha]_D^{20}$ −30.2° (Py).

Hexadecanoyl: [2308-85-2]. *β-Sitosterol palmitate*
$C_{45}H_{80}O_2$ M 653.126
Constit. of many plants. Cryst. Mp 89°. $[\alpha]_D$ −16.5° (c, 1.8 in $CHCl_3$).

Eicosanoyl: [59015-74-6]. *β-Sitosterol arachidate*
$C_{49}H_{88}O_2$ M 709.233
Constit. of *Angelica orotangelica*. Cryst. (EtOH). Mp 75.5°. $[\alpha]_D^{22}$ −11° (c, 1.15 in $CHCl_3$).

7,8-Didehydro: [521-04-0]. *Stigmasta-5,7-dien-3β-ol.* ***7-Dehydro-β-sitosterol***
$C_{29}H_{48}O$ M 412.698
Constit. of sponges. Cryst. (EtOH). Mp 148-150°. $[\alpha]_D^{21}$ −116° ($CHCl_3$).

3-O-[2-(4-Chlorophenoxy)-2-methylpropanoyl]: [55902-94-8]. ***Sitofibrate*, INN.** *Atherol. Longeril. Stigmasterol clofibrate*
$C_{39}H_{59}ClO_3$ M 611.346
Hypolipidaemic agent.

3-Sulfate: Mp 151-152° (as Na salt).

3-O-β-D-Xylopyranoside: [93772-33-9]. *Sitosterol 3-β-D-xyloside*
$C_{34}H_{58}O_5$ M 546.829
Isol. from the aerial parts of *Bauhinia candicans*.

Docosanoyl: [22554-56-9]. *β-Sitosterol behenate*
$C_{51}H_{92}O_2$ M 737.287
Constit. of *Cassia javanica* and *Prunus cerasoides*.

(3β,24S)-form [83-47-6] ***γ-Sitosterol.*** *Clionasterol. Poriferast-5-en-3β-ol. β-Dihydroeucosterol*
Constit. of *Xiphogorgia spongia*, *Sphicospongia vesparia* and *Cliona celata*. Cryst. (EtOH). Mp 147-148°. $[\alpha]_D$ −47.7° ($CHCl_3$).

Ac: Mp 140-141°. $[\alpha]_D^{26}$ −48.3° ($CHCl_3$).

β-D-Glucopyranoside: [18777-86-1].
Constit. of *Nicotiana tabacum*. Cryst. (EtOH). Mp 215-235°.

[53657-29-7]

Vrkoč, J., *Collect. Czech. Chem. Commun.*, 1959, **24**, 3938; 1962, **27**, 1345 (*β-Sitosterol glucuronide*)
Slomp, G. *et al*, *J. Am. Chem. Soc.*, 1962, **84**, 204 (*pmr*)
Neilsen, B.E. *et al*, *Acta Chem. Scand.*, 1963, **13**, 1161 (*derivs*)
Steele, J.A. *et al*, *J. Org. Chem.*, 1963, **28**, 571 (*isol, purifn*)
Khaletskii, A.M. *et al*, *Zh. Obshch. Khim.*, 1963, **33**, 1104 (*sulfate*)
Kiribuchi, T. *et al*, *Agric. Biol. Chem.*, 1967, **31**, 1244 (*synth, tlc, Sitogluside*)
Ikan, R. *et al*, *J. Org. Chem.*, 1971, **36**, 3944 (*synth*)
Karrer, W. *et al*, *Konstitution und Vorkommen der Organischen Pflanzenstoffe*, 2nd Ed., Birkhäuser Verlag, Basel, 1972-1985, no. 2071 (*occur*)
Naqui, S.H. *et al*, *Steroids*, 1973, **22**, 285 (*ms*)
Ger. Pat., 2 422 317, (1974); *CA*, **82**, 58013r (*Sitofibrate*)
Aizawa, K. *et al*, *Org. Mass Spectrom.*, 1974, **9**, 470 (*ms*)
Osman, A.M. *et al*, *Aust. J. Chem.*, 1975, **28**, 217 (*isol*)
Sucrow, W. *et al*, *Chem. Ber.*, 1975, **108**, 3721 (*synth*)
Murakami, T. *et al*, *Chem. Pharm. Bull.*, 1975, **23**, 1634 (*2-Deoxyglucoside*)
Lenton, J.R. *et al*, *Phytochemistry*, 1975, **14**, 1523 (*biosynth*)
Rubinstein, I. *et al*, *Phytochemistry*, 1976, **15**, 195 (*pmr*)
Sliwowski, J.K. *et al*, *J. Am. Chem. Soc.*, 1977, **99**, 4479 (*biosynth*)
Fujimoto, Y. *et al*, *J. Org. Chem.*, 1979, **44**, 1011 (*synth*)
Tiwari, K.P. *et al*, *Phytochemistry*, 1979, **18**, 2044 (*glycosides*)
Matsumoto, T. *et al*, *Phytochemistry*, 1983, **22**, 2622 (*isol*)
Iribarren, A.M. *et al*, *Phytochemistry*, 1984, **23**, 2087; 1985, **24**, 360; 1987, **26**, 857 (*glycosides*)
Seo, S. *et al*, *J. Chem. Soc., Chem. Commun.*, 1986, 1139 (*biosynth*)
Chaudhuri, K. *et al*, *J. Nat. Prod. (Lloydia)*, 1987, **50**, 1183 (*docosanate*)
Negwer, M., *Organic-Chemical Drugs and their Synonyms*, 6th Ed., Akademie-Verlag, Berlin, 1987, 7834.

Khan, M.S.Y. *et al*, *Indian Drugs Pharm. Ind.*, 1990, **28**, 97; *CA*, **114**, 118494q (*acetate*)
Nes, W.D. *et al*, *Phytochemistry*, 1992, **31**, 805 (*biosynth*)
Seo, S. *et al*, *Phytochemistry*, 1992, **31**, 3029 (*biosynth*)
Martindale, The Extra Pharmacopoeia, 30th edn., Pharmaceutical Press, London, 1993, 1359, 13237.

Stigmast-7-en-3-ol — S-00014

$C_{29}H_{50}O$ M 414.713

(3β,5α,24R)-form [521-03-9] ***Schottenol.*** *22-Dihydrochondrillasterol*
Constit. of *Cucumis sativus*. Also from *Lophocereus schottii*, *Bupleurum falcatum*, *Bryonia dioica* and others. Cryst. (MeOH). Mp 151-151.5°. $[\alpha]_D^{30}$ +9.1° (c, 0.95 in $CHCl_3$). Prob. indentical with *β*-Spinasterol.

3-O-β-D-Glucopyranoside: [61376-86-1].
$C_{35}H_{60}O_6$ M 576.855
Constit. of *Citrullus lanatus* var. *citroides*, *Styrax officinalis* and *Cucumis sativus*.

Takeda, K. *et al*, *CA*, 1959, **53**, 9284.
Terauchi, H. *et al*, *Chem. Pharm. Bull.*, 1970, **18**, 213 (*isol, struct*)
Lin, H.-K. *et al*, *Phytochemistry*, 1972, **11**, 2319 (*isol*)
Kircher, H.W. *et al*, *J. Org. Chem.*, 1973, **38**, 2259 (*synth*)
Rubinstein, I. *et al*, *Phytochemistry*, 1976, **15**, 195 (*pmr*)
Ulubelen, A. *et al*, *Planta Med.*, 1976, **30**, 221 (*isol*)
Siefert, K. *et al*, *Pharmazie*, 1977, **32**, 125 (*isol*)
Seo, S. *et al*, *J. Chem. Soc., Chem. Commun.*, 1987, 1876 (*biosynth*)
Uomori, A. *et al*, *Phytochemistry*, 1992, **31**, 3861 (*biosynth*)

Streptindole — S-00015

*β-1*H*-Indol-3-yl-1*H*-indole-3-ethanol acetate* (*ester*), *9CI.*
2,2-Di(3-indolyl)ethyl acetate
[88321-08-8]

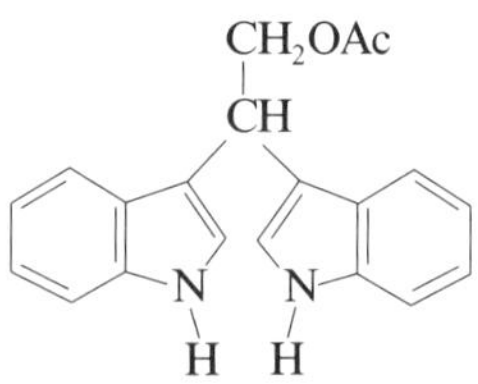

$C_{20}H_{18}N_2O_2$ M 318.374
Metab. from intestinal bacteria (*Streptococcus faecium* IB 37). Genotoxic agent.

Osawa, T. *et al*, *Tetrahedron Lett.*, 1983, **24**, 4719 (*uv, ir, pmr, ms, struct*)
Hogan, I.T. *et al*, *Synthesis*, 1984, 872 (*synth, uv, ir, pmr*)
Singh, H. *et al*, *Tetrahedron*, 1988, **44**, 5897 (*synth, ir, pmr*)

2-Sulfodocosanoic acid — S-00016

$H_3C(CH_2)_{19}CH(SO_3H)COOH$

$C_{22}H_{44}O_5S$ M 420.653
Surfactant.

Di-Na salt: [37108-68-2].
Used for parting materials for plastics/rubbers.

Org. Synth., Coll. Vol., 4, 1963, 862.

2-Sulfododecanoic acid — S-00017

2-Sulfolauric acid
[3054-88-4]

$H_3C(CH_2)_9CH(SO_3H)COOH$

$C_{12}H_{24}O_5S$ M 280.385
Mp 86.5-88°.

Disodium salt: [38841-48-4].
Antistatic agent for polymers.

Org. Synth., Coll. Vol., 4, 1963, 862 (*synth*)
Tember, G.A. *et al*, *CA*, 1974, **80**, 89067 (*ir*)

2-Sulfohexadecanoic acid — S-00018

2-Sulfopalmitic acid
[1782-10-1]

$H_3C(CH_2)_{13}CH(SO_3H)COOH$

$C_{16}H_{32}O_5S$ M 336.492
Cryst. (Me_2CO). Mp 90-91°.

1-Butyl ester, Na salt: [6061-34-3].
Surfactant.

Org. Synth., Coll. Vol., 4, 1963, 862.
Tember, G.A. *et al*, *CA*, 1974, **80**, 89067 (*ir*)
Volkov, Yu. *et al*, *CA*, 1974, **81**, 13783 (*synth*)
Gladkii, F.F. *et al*, *CA*, 1974, **81**, 151453 (*synth*)
Volkov, Yu. *et al*, *CA*, 1975, **82**, 60544 (*deriv*)

2-Sulfooctadecanoic acid — S-00019

2-Sulfostearic acid
[1115-17-9]

$H_3C(CH_2)_{15}CH(SO_3H)COOH$

$C_{18}H_{36}O_5S$ M 364.545
Surfactant. Mp 96-97°. A wide range of carboxylic acid esters-sulphonic acid salts has been prepared.

Org. Synth., Coll. Vol., 4, 1963, 862 (*synth*)
Tember, G.A. *et al*, *CA*, 1974, **80**, 89067 (*ir*)
Volkov, Yu. *et al*, *CA*, 1974, **81**, 79659u (*synth*)
Gladkii, F.F. *et al*, *CA*, 1974, **81**, 151453 (*synth*)
Ger. Pat., 2 455 891, (1975); *CA*, **83**, 78623p (*salts*)

2-Sulfotetradecanoic acid — S-00020

2-Sulfomyristic acid

$H_3C(CH_2)_{11}CH(SO_3H)COOH$

$C_{14}H_{28}O_5S$ M 308.438
Surfactant. Mp 85-86.5°.

Org. Synth., Coll. Vol., 4, 1963, 862 (*synth*)
Tember, G.A. *et al*, *CA*, 1974, **80**, 89067 (*ir*)

Sylvaticin — S-00021

[92280-14-3]

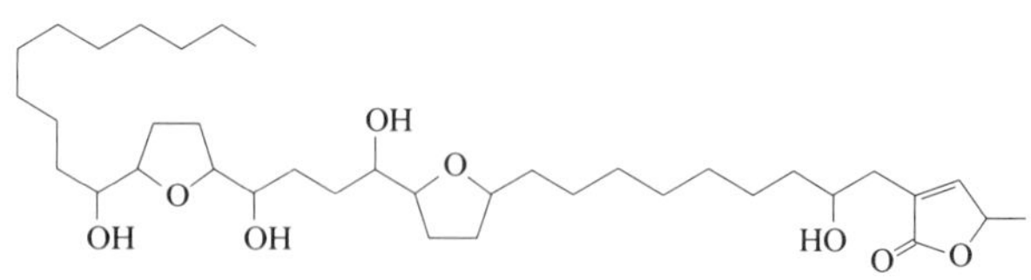

$C_{37}H_{66}O_8$ M 638.924
Constit. of *Rollinia sylvatica*. Wax. Mp 48-50°. $[\alpha]_D^{23}$ +5.9° (c, 0.524 in $CHCl_3$).

Mikolajczak, K.J. *et al*, *Experientia*, 1990, **46**, 324 (*isol, pmr, cmr*)

T

Taurocholic acid T-00001

N-(3α,7α,12α-Trihydroxy-5β-cholan-24-oyl)taurine. Cholyltaurine

[81-24-3]

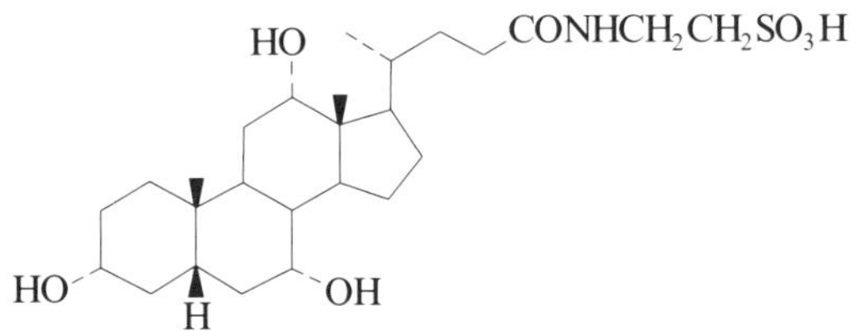

$C_{26}H_{45}NO_7S$ M 515.710

Constit. of bile. Choleretic. Hepatoprotectant. Amorph. powder. Mp 125° dec. $[\alpha]_D^{20}$ +38.8° (EtOH aq.). pK_a 1.40. Very hygroscopic.

▷ WX0350000.

Na salt: [145-42-6].

Lipase accelerator. Cryst. (EtOH/EtOAc). Mp 182-184° (anhyd.), Mp 235° (+ $2H_2O$).

▷ WX0400000.

Begemann, F., *Z. Klin. Chem. Klin. Biochem.*, 1972, **10**, 29 (*isol*)
Tserng, K.-Y., *J. Lipid Res.*, 1977, **18**, 404 (*synth*)
Shaw, R. *et al*, *Lipids*, 1978, **13**, 971 (*ms*)
Lewis, R.J., *Sax's Dangerous Properties of Industrial Materials*, 8th Ed., Van Nostrand-Reinhold, 1992, SKB500, TAH250.

9,10,12,13-Tetrabromooctadecanoic acid T-00002

Linoleic acid tetrabromide. 9,10,12,13-Tetrabromostearic acid

[1794-89-4]

$H_3C(CH_2)_4(CHBr)_2CH_2(CHBr)_2(CH_2)_7COOH$

$C_{18}H_{32}Br_4O_2$ M 600.066

(9RS,10RS,12RS,13RS)-(±)-form [56846-03-8]

From bromination of linoleic acid. Cryst. (Et_2O/pet. ether). Mp 114.7-115.2°.

Me ester:

$C_{19}H_{34}Br_4O_2$ M 614.092

Cryst. Mp 63° (50-56°).

Et ester:

$C_{20}H_{36}Br_4O_2$ M 628.119

Cryst. Mp 63° (58-58.5°).

(9RS,10RS,12SR,13SR)-(±)-form [56846-04-9]

From bromination of linoleic acid. Oil.

Org. Synth., 1942, **22**, 76 (*synth*)
Olson, E.S., *J. Am. Oil Chem. Soc.*, 1975, **52**, 319 (*stereochem, nmr*)

5,9-Tetracosadienoic acid T-00003

[59708-80-4]

$H_3C(CH_2)_{13}CH{=}CHCH_2CH_2CH{=}CH(CH_2)_3COOH$

$C_{24}H_{44}O_2$ M 364.610

Present in the lipids of the marine sponge *Microciona prolifera*.

Morales, R.W. *et al*, *Biochim. Biophys. Acta*, 1976, **431**, 206.

6,9,12,15,18,21-Tetracosahexaynoic acid, 8CI T-00004

[31814-88-7]

$H_3CCH_2(C{\equiv}CCH_2)_6(CH_2)_3COOH$

$C_{24}H_{24}O_2$ M 344.452

Mp 112.5-113.5°.

Kunau, W.H. *et al*, *Hoppe Seyler's Z. Physiol. Chem.*, 1971, **352**, 542 (*synth*)
Sprecher, H., *Prog. Chem. Fats Other Lipids*, 1978, **15**, 219.

Tetracosanal, 9CI, 8CI T-00005

[57866-08-7]

$H_3C(CH_2)_{22}CHO$

$C_{24}H_{48}O$ M 352.643

Isol. from various fruit waxes, oil shales and marine sediments. Cryst. (Et_2O). Mp 57.5-62°.

Oxime:

$C_{24}H_{49}NO$ M 367.657

Cryst. (C_6H_6/Et_2O or C_6H_6/Me_2CO). Mp 112-113.5°.

Cason, J. *et al*, *J. Org. Chem.*, 1953, **18**, 850 (*synth, oxime*)
Radler, F. *et al*, *Aust. J. Chem.*, 1965, **18**, 1059 (*isol, chromatog*)
Stephanoll, E., *Naturwissenschaften*, 1989, **76**, 464 (*isol, ms*)

Tetracosane T-00006

Lignocerane

[646-31-1]

$H_3C(CH_2)_{22}CH_3$

$C_{24}H_{50}$ M 338.659

Present in coal tar. Also isol. from plant sources eg. rose and orange oils and *Coriolus sanguineus*. Copulation releaser pheromone in female whitemarked tussock moth *Orgyia leucostigma*. Cryst. (Et_2O). Mp 54° (51°). Bp_{10} 231°.

Rodd's Chem. Carbon Compd. (2nd edn.), 2nd Ed., Vol 1A, 1964, 367.
Cambie, R.C. *et al*, *J. Chem. Soc. C*, 1966, 72 (*isol*)
Hunter, G.L.K. *et al*, *Phytochemistry*, 1966, **5**, 807 (*isol*)
Grant, G.G. *et al*, *J. Chem. Ecol.*, 1987, **13**, 345.

Tetracosanedioic acid T-00007

Docosane-1,22-dicarboxylic acid

[2450-31-9]

$HOOC(CH_2)_{22}COOH$

$C_{24}H_{46}O_4$ M 398.625

Present in Douglas fir bark. Cryst. Mp 126.9-127.1°.

Di-Me ester:

$C_{26}H_{50}O_4$ M 426.679

Cryst. Mp 75-75.2°.

Di-Et ester:

$C_{28}H_{54}O_4$ M 454.732

Cryst. Mp 65.9-66.1°.

Siina, S., *CA*, 1939, **33**, 7278.
Natrajan, A. *et al*, *J. Am. Chem. Soc.*, 1987, **109**, 7477 (*deriv, synth, ir, pmr, cmr*)

1,24-Tetracosanediol T-00008

[22513-82-2]

$HOCH_2(CH_2)_{22}CH_2OH$

$C_{24}H_{50}O_2$ M 370.658

Present in carnauba wax in an acylated form. Also from apple (*Pyrus malus*). Cryst. (C_6H_6/Me_2CO). Mp 108.4-108.6°.

Di-Ac:

$C_{28}H_{54}O_4$ M 454.732

Mp 73.6-73.7°. Bp_1 243°.

Murray, K.E., *Aust. J. Chem.*, 1955, **8**, 432.
Mazliak, P., *Phytochemistry*, 1962, **1**, 79 (*isol*)

Tetracosanoic acid T-00009

Lignoceric acid

[557-59-5]

$H_3C(CH_2)_{22}COOH$

$C_{24}H_{48}O_2$ M 368.642

Occurs in some waxes and less commonly in some triacylglycerols. Fairly widespread in plant oils. Cryst. (AcOH). Mp 87.5-88°. Polymorphous.

Me ester: [2442-49-1].

$C_{25}H_{50}O_2$ M 382.669

Mp 59.5-60°.

Et ester: [24634-95-5].

$C_{26}H_{52}O_2$ M 396.696

Mp 56-57°. $Bp_{0.24}$ 198-199°.

Tetracosyl ester: [1001-43-0]. *Tetracosanoyl tetracosanoate. Lignoceryl lignocerate*

$C_{48}H_{96}O_2$ M 705.285

Isol. from aerial parts of *Flemingia strobilifera*. Wax. Mp 82-84°.

Anhydride:

$C_{48}H_{94}O_3$ M 719.269

Mp 86-86.3°.

Nitrile: Tricosyl cyanide

$C_{24}H_{47}N$ M 349.642

Mp 61-62°.

Levene, P.A. *et al*, *J. Biol. Chem.*, 1924, **59**, 905.
Francis, F. *et al*, *Proc. R. Soc. London, A*, 1930, **28**, 214, 242.
Bleyberg, W. *et al*, *Chem. Ber.*, 1931, **64**, 2504.
Lundquist, M., *Ark. Kemi*, 1970, **32**, 27 (*polymorphism*)
Karrer, W. *et al*, *Konstitution und Vorkommen der Organischen Pflanzenstoffe*, 2nd Ed., Birkhäuser Verlag, Basel, 1972-1985, no. 710 (*occur*)
Marosi, L., *Justus Liebigs Ann. Chem.*, 1973, 584 (*esters*)
Bhatt, S., *Indian J. Chem.*, 1975, **13**, 1105 (*tetracosanoate*)
Vergoten, G. *et al*, *C. R. Hebd. Seances Acad. Sci.*, 1977, **284**, 21 (*raman*)

1-Tetracosanol T-00010

Lignoceryl alcohol

[506-51-4]

$H_3C(CH_2)_{22}CH_2OH$

$C_{24}H_{50}O$ M 354.658

Isol. from various plant waxes. Mp 77-79°. $Bp_{0.4}$ 210°.

Ac: [822-29-7].

$C_{26}H_{52}O_2$ M 396.696

Mp 58-59°.

Bleyberg, W. *et al*, *Ber.*, 1931, **64**, 2504.
Pollard, A. *et al*, *Biochem. J.*, 1931, **25**, 2111 (*isol*)
Mathis, C. *et al*, *Phytochemistry*, 1964, **3**, 379 (*isol*)
Dominguez, X.A., *CA*, 1970, **73**, 22153.
Knoop, E. *et al*, *Naturwissenschaften*, 1975, **62**, 37 (*cryst struct*)
Precht, D., *Fette, Seifen, Anstrichm.*, 1976, **78**, 145 (*cryst struct*)

2-(Tetracosanoylamino)-1,3-octadecanediol T-00011

N-[*2-Hydroxy-1-(hydroxymethyl)heptadecyl*]*tetracosanamide*

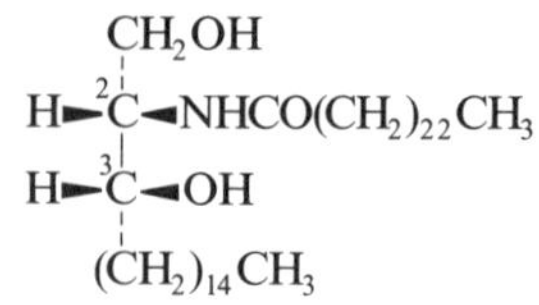

$C_{42}H_{85}NO_3$ M 652.138

(2RS,3SR)-form

(±)-erythro-*form*. N-*Tetracosanoylsphinganine*

Mp 104-105°.

Ong, D.E. *et al*, *J. Lipid Res.*, 1972, **13**, 819 (*synth*)

2-(Tetracosanoylamino)-1,3,4-octadecanetriol T-00012

N-[*2,3-Dihydroxy-1-(hydroxymethylheptadecyl*)]-*2-hydroxytetracosanamide*, *8CI*

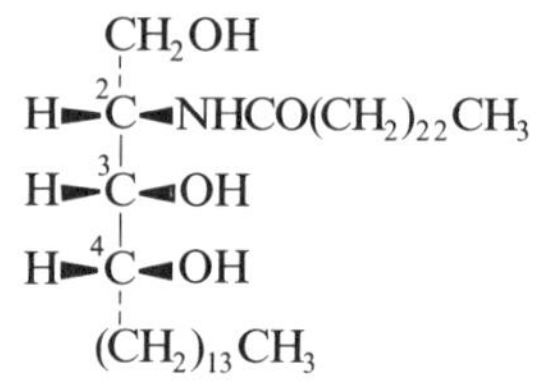

$C_{42}H_{85}NO_4$ M 668.138

(2S,3S,4R)-form [34437-74-6]

D-ribo-*form*. N-*Tetracosanoylphytosphingosine*

Karlsson, K.A., *J. Lipid Res.*, 1971, **12**, 466 (*tlc*)
Ohnishi, M. *et al*, *Agric. Biol. Chem.*, 1976, **40**, 1419 (*occur*)
Pascher, I., *Biochim. Biophys. Acta*, 1976, **455**, 423 (*cryst struct*)
Dahlen, B. *et al*, *Acta Crystallogr., Sect. B*, 1977, **28**, 2396 (*cryst struct*)

6,9,12,15,18-Tetracosapentaynoic acid, 8CI T-00013

[31814-87-6]

$H_3C(CH_2)_4(C{\equiv}CCH_2)_5(CH_2)_3COOH$

$C_{24}H_{28}O_2$ M 348.484

Mp 82.5-83.5°.

Kunau, W.H. *et al*, *Hoppe Seyler's Z. Physiol. Chem.*, 1971, **352**, 542 (*synth*)
Sprecher, H., *Prog. Chem. Fats Other Lipids*, 1978, **15**, 219.

15-Tetracosenoic acid T-00014

$H_3C(CH_2)_7CH{=}CH(CH_2)_{13}COOH$

$C_{24}H_{46}O_2$ M 366.626

(E)-form [14490-79-0]

Cryst. (EtOH). Mp 66-67° (61°).

Me ester:

$C_{25}H_{48}O_2$ M 380.653

Cryst. (Me_2CO). Mp 36.5-37.5°. $Bp_{0.00001}$ 180-190°.

Amide:

$C_{24}H_{47}NO$ M 365.641

Mp 97-98°.

(Z)-form [506-37-6]

Nervonic acid. Selacholeic acid

Present in cerebrosides and in fish oils. Also in seed oils of *Lunaria biennis*, *Brassica napus* and *Cathamus tinctorius*. Cryst. (EtOH). Mp 40.5-41°, Mp 44-45°.

Me ester: [2733-88-2].

Mp 14-15°. $Bp_{0.02}$ 165° approx.
Amide: Cryst. (EtOH). Mp 86.5-87.5°.

Hale, J.B. *et al*, *J. Am. Chem. Soc.*, 1930, **52**, 4536 (*synth*)
Bounds, D.G. *et al*, *J. Chem. Soc.*, 1954, 448 (*synth*)
Batchelor, J.G. *et al*, *J. Am. Chem. Soc.*, 1973, **95**, 6358 (*cmr*)
Lecerf, J. *et al*, *J. Physiol. (Paris)*, 1975, **70**, 493.
Morales, R.W. *et al*, *Biochim. Biophys. Acta*, 1976, **431**, 206 (*isol, biosynth*)
Murad, S. *et al*, *J. Biol. Chem.*, 1976, **251**, 5237.
Bohannon, M.B. *et al*, *Lipids*, 1976, **11**, 157 (*isol*)

5,8-Tetradecadienal T-00015

$H_3C(CH_2)_4CH{=}CHCH_2CH{=}CH(CH_2)_3CHO$

$C_{14}H_{24}O$ M 208.343

(Z,Z)-form [22644-15-1]
$Bp_{0.15}$ 94-95°.
2,4-Dinitrophenylhydrazone: Mp 54-55°.
Di-Et acetal: [22644-16-2].
$C_{18}H_{34}O_2$ M 282.465
$Bp_{0.1}$ 138-140°.

Ward, J.P. *et al*, *Recl. Trav. Chim. Pays-Bas (J. R. Neth. Chem. Soc.)*, 1969, **88**, 177.

8,10-Tetradecadienal, 9CI T-00016

$H_3CCH_2CH_2CH{=}CHCH{=}CH(CH_2)_6CHO$

$C_{14}H_{24}O$ M 208.343

(8E,10E)-form [111876-38-1]
Sex attractant for several *Acrocercops* and *Phyllonorycter* moth spp.

[100547-44-2]

Reed, D.W. *et al*, *J. Chem. Ecol.*, 1985, **11**, 1645.
Ando, T. *et al*, *Agric. Biol. Chem.*, 1987, **51**, 2691; 1988, **52**, 1415 (*synth, props, ms*)

9,11-Tetradecadienal T-00017

$H_3CCH_2CH{=}CHCH{=}CH(CH_2)_7CHO$

$C_{14}H_{24}O$ M 208.343

(9Z,11E)-form [123314-24-9]
Synergistic minor component of the sex pheromone of the Carob moth, *Ectomyelois ceratoniae*. Major component with (9,11-Tetradecadien-1-ol, T-00035, 9*Z*,11*E*-form, acetate) of the sex pheromone of *Stenoma cecropia*.

[75589-42-3, 75589-43-4, 100496-44-4, 136027-82-2]

Baker, T.C. *et al*, *Tetrahedron Lett.*, 1989, **30**, 2901 (*isol, activity*)
Ramiandrasoa, F. *et al*, *Synth. Commun.*, 1990, **20**, 333 (*isol, synth, pmr, cmr, ir, ms*)
Zagatti, P. *et al*, *J. Chem. Ecol.*, 1991, **17**, 1399 (*activity*)
Baker, T.C. *et al*, *J. Chem. Ecol.*, 1991, **17**, 1973 (*isol, synth, glc, ms*)

9,12-Tetradecadienal, 9CI T-00018

$H_3CCH{=}CHCH_2CH{=}CH(CH_2)_7CHO$

$C_{14}H_{24}O$ M 208.343

(9Z,12E)-form [87092-33-9]
Key component of sex pheromone of the European sunflower moth, *Homoeosoma nebulellum*; component of the sex pheromone of the dried-fruit moth, *Vitula edmandsae*.

Struble, D.L. *et al*, *J. Chem. Ecol.*, 1983, **9**, 785 (*isol, activity*)
Renou, M. *et al*, *J. Chem. Ecol.*, 1988, **14**, 1187 (*isol, activity*)
Zagatti, P. *et al*, *J. Chem. Ecol.*, 1991, **17**, 1399 (*glc, ms, activity*)

11,13-Tetradecadienal, 9CI T-00019

$H_2C{=}CHCH{=}CH(CH_2)_9CHO$

$C_{14}H_{24}O$ M 208.343

(E)-form [111876-44-9]
Component of sex pheromone of several *Acleris* moth species, including *A. affinitana*.

Ando, T. *et al*, *Agric. Biol. Chem.*, 1987, **51**, 2691; 1988, **52**, 1415 (*synth, activity*)

1,13-Tetradecadiene, 9CI T-00020

[21964-49-8]

$H_2C{=}CH(CH_2)_{10}CH{=}CH_2$

$C_{14}H_{26}$ M 194.359

Casteignau, G. *et al*, *Bull. Soc. Chim. Fr.*, 1968, 3893 (*synth, glc*)

1,8-Tetradecadiene-11,13-diyne T-00021

6,13-Tetradecadiene-1,3-diyne

$HC{\equiv}CC{\equiv}CCH_2CH{=}CH(CH_2)_5CH{=}CH_2$

$C_{14}H_{18}$ M 186.296

(Z)-form [37142-95-3]
Constit. of *Eriocephalus africanus*. Oil.
8,9-Epoxide: 8,9-Epoxy-1-tetradecene-11,13-diyne. 2-(6-Heptenyl)-3-(2,4-pentadiynyl)oxirane. PQ5
Constit. of *Panax quinquefolium*. Oil. $[\alpha]_D$ −90.0° (c, 1 in $CHCl_3$).

Bohlmann, F. *et al*, *Chem. Ber.*, 1972, **105**, 1783 (*isol, pmr*)
Fujimoto, Y. *et al*, *Phytochemistry*, 1992, **31**, 3499 (*isol, pmr, cmr*)

6,12-Tetradecadiene-8,10-diyne-1,3-diol T-00022

$H_3CCH{=}CHC{\equiv}CC{\equiv}CCH{=}CHCH_2CH_2CH(OH)CH_2CH_2OH$

$C_{14}H_{18}O_2$ M 218.295

(E,E)-form [120166-71-4]
Pale-yellow amorph. powder. $[\alpha]_D$ +0.73° (c, 1.1 in MeOH).
1-Ac: [131802-62-5]. *1-Acetoxy-6,12-tetradecadiene-8,10-diyn-3-ol*
$C_{16}H_{20}O_3$ M 260.332
Isol. from the rhizomes of *Atractylis koreana*. Oil. λ_{max} 312, 292.5, 276, 260.5, 247 nm (CH_2Cl_2).
3-Ac: [131802-63-6]. *3-Acetoxy-6,12-tetradecadiene-8,10-diyn-1-ol*
$C_{16}H_{20}O_3$ M 260.332
Isol. from the rhizomes of *A. koreana*. Oil.
Di-Ac: [89913-46-2].
$C_{18}H_{22}O_4$ M 302.369
Isol. from Atractylodes Rhizome (*A. japonica*) and *Coreopsis capillacea*. Oil. $[\alpha]_D$ 0° (c, 1.7 in MeOH).

Kano, Y. *et al*, *Chem. Pharm. Bull.*, 1989, **37**, 193 (*isol, uv, pmr, struct*)
Pachaly, P. *et al*, *Planta Med.*, 1990, **56**, 469 (*isol, uv, ir, ms, pmr, cmr*)

6,13-Tetradecadiene-8,10-diyne-1,5,12-triol T-00023

[40111-22-6]

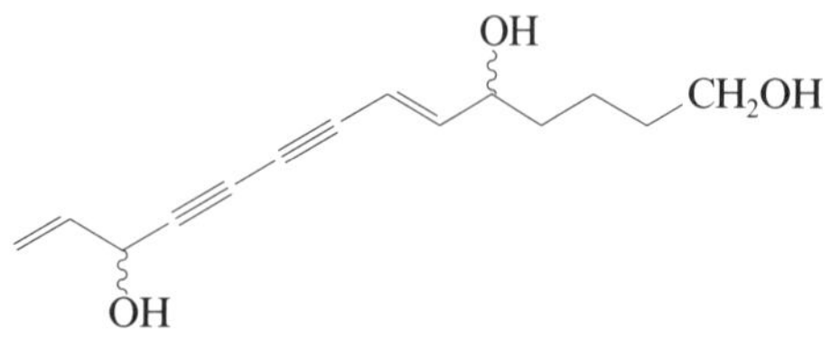

$C_{14}H_{18}O_3$ M 234.294

(E)-form

Constit. of *Campanula* spp.

Badanyan, S.O. *et al*, *J. Chem. Soc., Perkin Trans. 1*, 1973, 145 (*synth, uv, ir, pmr, ms*)

10,12-Tetradecadiene-4,6-diynoic acid, 9CI T-00024

$H_3CCH{=}CHCH{=}CHCH_2CH_2C{\equiv}CC{\equiv}CCH_2CH_2COOH$

$C_{14}H_{16}O_2$ M 216.279

(10E,12E)-form

2-Methylpropylamide: [125085-13-4]. N-*(2-Methylpropyl)-10,12-tetradecadiene-4,6-diynamide*

$C_{18}H_{25}NO$ M 271.402

Isol. from *Achillea magnifica.*

Ulubelen, A. *et al*, *Planta Med.*, 1989, **55**, 395 (*isol*)

4,6-Tetradecadiene-8,10,12-triyne-1,3-diol, 9CI T-00025

[3311-37-3]

$H_3CC{\equiv}CC{\equiv}CC{\equiv}CCH{=}CHCH{=}CHCH(OH)CH_2CH_2OH$

$C_{14}H_{14}O_2$ M 214.263

(E, E)-form [1117-69-7]

Isol. from *Dahlia* sp., *Centaurea ruthenica*, *Matricaria oreades* and *Echinops sphaerocephalus*. Cryst. (CCl_4). Mp 69.5°.

3-Ac: [14686-21-6]. *3-Acetoxy-4,6-tetradecadiene-8,10,12-triyn-1-ol*

$C_{16}H_{16}O_3$ M 256.301

Isol. from leaves of *M. caucasica* and *Coreopsis parvifolia*. Oil.

Di-Ac: [26130-92-7]. *1,3-Diacetoxy-4,6-tetradecadiene-8,10,12-triyne*

$C_{18}H_{18}O_4$ M 298.338

Isol. from roots of *Centaurea scabiosa*, *C. ruthenica*, *Matricaria oreades* and *Dahlia* sp. Cryst. (CCl_4/pet. ether). Mp 60°. $[\alpha]_D^{20}$ +211° (c, 1.6 in Et_2O).

(4E,6Z)-form [71867-30-6]

Isol. from the aerial parts of *Centaurea ruthenica.*

Di-Ac: [71866-94-9].

Isol. from aerial parts of *Centaurea ruthenica.*

[3311-35-1]

Bohlmann, F. *et al*, *Chem. Ber.*, 1964, **97**, 1193; 1965, **98**, 155, 2605; 1967, **100**, 611 (*isol, synth, ir, uv, struct*)
Lam, J. *et al*, *Phytochemistry*, 1975, **14**, 1621 (*isol*)
Andersen, A.B. *et al*, *Phytochemistry*, 1977, **16**, 1829 (*isol*)
Jente, R. *et al*, *Phytochemistry*, 1979, **18**, 829 (*isol, synth, pmr*)

4,6-Tetradecadiene-8,10,12-triyn-1-ol T-00026

[2271-25-2]

$H_3C(C{\equiv}C)_3CH{=}CHCH{=}CHCH_2CH_2CH_2OH$

$C_{14}H_{14}O$ M 198.264

(E,E)-form

Isol. from *Chrysanthemum atratum*, *Santolina procumbens* and *Anthemis* spp. Cryst. (Et_2O/pet. ether). Mp 112° (104°).

Ac: [6581-87-9]. *1-Acetoxy-4,6-tetradecadiene-8,10,12-triyne*

$C_{16}H_{16}O_2$ M 240.301

Isol. from roots of *Tanacetum vulgare* and *Rudbeckia fulgida*, from *Anthemis*, *Artemisia* and *Chrysanthemum* spp. Cryst. (pet. ether). Mp 36-37° (31.5°).

4,5-Epoxide: [17098-67-8]. *4,5-Epoxy-6-tetradecene-8,10,12-triyn-1-ol*

$C_{14}H_{14}O_2$ M 214.263

Isol. from *Chrysanthemum serotinum* and *Calea reticulata*. Cryst. Mp 74-76°.

4,5-Epoxide, Ac:

$C_{16}H_{16}O_3$ M 256.301

Isol. from *Calea reticulata* and *Chrysanthemum serotinum*. Cryst. Mp 60-62° (45°).

[7089-15-8]

Bohlmann, F. *et al*, *Chem. Ber.*, 1960, **93**, 1937; 1962, **95**, 1733; 1964, **97**, 1179; 1965, **98**, 1411; 1966, **99**, 1830, 2096 (*isol, uv, pmr, struct*)
Bohlmann, F. *et al*, *Phytochemistry*, 1982, **21**, 1793 (*isol, epoxide*)
Herz, W. *et al*, *Phytochemistry*, 1985, **24**, 89 (*isol*)

2,4-Tetradecadienoic acid T-00027

$H_3C(CH_2)_8CH{=}CHCH{=}CHCOOH$

$C_{14}H_{24}O_2$ M 224.342

(2E,4E)-form [24738-49-6]

Cryst. (pet. ether at low temp.). Mp 57-58°.

Me ester: [24738-47-4].

$C_{15}H_{26}O_2$ M 238.369

$Bp_{0.3}$ 117-118°.

2-Methylpropylamide: [13891-73-1]. ***2,4-Tetradecadienoic acid isobutylamide***

$C_{18}H_{33}NO$ M 279.465

Alkaloid from *Chrysanthemum frutescens*, *Anacyclus pyrethrum* and *Leucocyclus formosus* (Compositae) (in admixture with homologues). Cryst. (pet. ether). Mp 88°.

Piperidide: [52657-12-2]. *1-(1-Oxo-2,4-tetradecadienyl) piperidine, 9CI.* ***1-(2,4-Tetradecadienoyl)piperidine.*** *2,4-Tetradecadienoic acid piperidide*

$C_{19}H_{33}NO$ M 291.476

Alkaloid from the roots of *Otanthus maritimus* (Compositae). Oil. $Bp_{0.1}$ 150°.

Bohlmann, F. *et al*, *Chem. Ber.*, 1967, **100**, 104; 1974, **107**, 1038 (*isobutylamide, piperidide*)
Burden, R.S. *et al*, *J. Chem. Soc. C*, 1969, 2477 (*synth*)
Greger, H. *et al*, *Phytochemistry*, 1981, **20**, 2579 (*isobutylamide*)

3,4-Tetradecadienoic acid T-00028

$H_3C(CH_2)_8CH{=}C{=}CHCH_2COOH$

$C_{14}H_{24}O_2$ M 224.342

Me ester: [81981-10-4].

$C_{15}H_{26}O_2$ M 238.369

$Bp_{0.1}$ 130°.

Tsuboi, S. *et al*, *J. Org. Chem.*, 1982, **47**, 4478 (*synth, ir, pmr*)

3,5-Tetradecadienoic acid T-00029

$H_3C(CH_2)_7CH{=}CHCH{=}CHCH_2COOH$

$C_{14}H_{24}O_2$ M 224.342

(3E,5Z)-form [23400-52-4]

Megatomic acid

Sex attractant of the black carpet beetle *Attagenus megatoma*.

(Z,Z)-form [25091-12-7]

Sex pheromone of *A. elongatulus*.

Silverstein, R.M., *J. Chem. Educ.*, 1968, **45**, 794 (*ir, ms, pmr*)
Rodin, J.O. *et al*, *J. Org. Chem.*, 1970, **35**, 3152 (*synth, ir, uv, pmr*)
Fukui, H. *et al*, *J. Chem. Ecol.*, 1977, **3**, 539 (*isol*)
Abrams, S.R., *Can. J. Chem.*, 1982, **60**, 1238 (*synth, pmr*)
Tsuboi, S. *et al*, *Bull. Chem. Soc. Jpn.*, 1983, **56**, 3521 (*synth*)

5,8-Tetradecadienoic acid T-00030

[10390-16-6]

$H_3C(CH_2)_4CH{=}CHCH_2CH{=}CH(CH_2)_3COOH$

$C_{14}H_{24}O_2$ M 224.342

$Bp_{0.05}$ 110-120°.

(Z,Z)-form [39039-37-7] ***Goshuyic acid***

Isol. from the fruit of *Evodia rutaecarpa*. Minor constit. of rape oil.

Me ester: [18829-89-5].

$C_{15}H_{26}O_2$ M 238.369

Constit. of pear volatiles. n_D^{20} 1.4585, λ_{max} 230 nm.

Et ester: [28369-25-7].

$C_{16}H_{28}O_2$ M 252.396

Constit. of pear volatiles.

Sprecher, H., *Lipids*, 1968, **3**, 14 (*synth, metab, uv, ms*)
Creveling, R.K. *et al*, *J. Agric. Food Chem.*, 1970, **18**, 19 (*isol*)
Kurono, G. *et al*, *Chem. Pharm. Bull.*, 1972, **20**, 559 (*isol, ir, pmr*)
Bus, J. *et al*, *Chem. Phys. Lipids*, 1976, **17**, 501; 1977, **18**, 130 (*cmr*)
Sebedio, J.L. *et al*, *J. Am. Oil Chem. Soc.*, 1979, **56**, 15 (*isol*)

10,12-Tetradecadienoic acid T-00031

$H_3CCH{=}CHCH{=}CH(CH_2)_8COOH$

$C_{14}H_{24}O_2$ M 224.342

(10E,12Z)-form

Me ester: [80625-48-5].

$C_{15}H_{26}O_2$ M 238.369

$Bp_{0.01}$ 95-100°.

Bestmann, H.J. *et al*, *Justus Liebigs Ann. Chem.*, 1981, 2117 (*synth, pmr*)

3,5-Tetradecadien-1-ol T-00032

$H_3C(CH_2)_7CH{=}CHCH{=}CHCH_2CH_2OH$

$C_{14}H_{26}O$ M 210.359

(3Z,5E)-form [102488-79-9]

Oil.

Ac: [61360-84-7]. *3,5-Tetradecadien-1-yl acetate. 1-Acetoxy-3,5-tetradecadiene*

$C_{16}H_{28}O_2$ M 252.396

Constit. of the sex pheromone of the carpenter worm moth *Prionoxystus robiniae*. (3*Z*,5*Z*) and (3*E*,5*E*) acetates respectively inhibit and synergise attractive effect of (3*Z*,5*E*) acetate. Oil. $Bp_{0.01}$ 94°.

(3Z,5Z)-form [79532-13-1]

Ac: [60754-63-4].

$Bp_{0.35}$ 118-119°.

[61360-85-8, 61360-86-9, 115478-89-2, 115478-90-5]

Doolittle, R.E. *et al*, *J. Chem. Ecol.*, 1976, **2**, 399; 1986, **12**, 619 (*acetate*)
Rossi, R. *et al*, *Gazz. Chim. Ital.*, 1981, **111**, 173 (*synth*)
Rossi, R. *et al*, *Tetrahedron*, 1982, **38**, 631 (*acetate*)
Doolittle, R.E. *et al*, *Chem. Anal. (N.Y.)*, 1987, **91**, 84 (*ms*)
Huang, Y.Z. *et al*, *J. Org. Chem.*, 1987, **52**, 3558 (*synth, acetate*)
Zhang, Y. *et al*, *Anal. Chim. Acta*, 1990, **233**, 223 (*cmr*)
Vig, O. *et al*, *Indian J. Chem., Sect. B*, 1990, **29**, 606 (*acetate, pmr*)

7,10-Tetradecadien-1-ol T-00033

$H_3CCH_2CH_2CH{=}CHCH_2CH{=}CH(CH_2)_5CH_2OH$

$C_{14}H_{26}O$ M 210.359

(E,E)-form

$Bp_{0.01}$ 100-105°.

Ac:

$C_{16}H_{28}O_2$ M 252.396

$Bp_{0.01}$ 100-105°.

Bestmann, H.J. *et al*, *Justus Liebigs Ann. Chem.*, 1982, 536 (*synth, pmr*)

8,10-Tetradecadien-1-ol T-00034

$H_3CCH_2CH_2CH{=}CHCH{=}CH(CH_2)_6CH_2OH$

$C_{14}H_{26}O$ M 210.359

(8E,10Z)-form

$Bp_{0.05}$ 100-105°.

Ac: [80625-67-8].

$C_{16}H_{28}O_2$ M 252.396

$Bp_{0.01}$ 100-105°.

Bestmann, H.J. *et al*, *Justus Liebigs Ann. Chem.*, 1981, 2117 (*synth, pmr*)

9,11-Tetradecadien-1-ol T-00035

$H_3CCH_2CH{=}CHCH{=}CH(CH_2)_7CH_2OH$

$C_{14}H_{26}O$ M 210.359

(E,E)-form [72553-58-3]

$Bp_{0.01}$ 100-105°.

Ac: [54664-98-1].

$C_{16}H_{28}O_2$ M 252.396

$Bp_{0.01}$ 95-100°.

(9E,11Z)-form [65726-40-1]

$Bp_{0.01}$ 108-115°.

Ac: [50767-79-8].

Pheromone of cotton Armyworm. $Bp_{0.01}$ 110-111°.

(9Z,11E)-form [63025-02-5]

$Bp_{0.01}$ 100-105°.

Ac: [30562-09-5].

Major sex pheromone of the Egyptian cotton leafworm *Spodoptera littoralis*. $Bp_{0.2}$ 147-148°, $Bp_{0.003}$ 85-86°.

(Z,Z)-form [72553-59-4]

$Bp_{0.01}$ 95-100°.

Ac: [54664-97-0].

Sex pheromone component of *S. littoralis*. $Bp_{0.02}$ 90-95°.

Tamaki, Y. *et al*, *Appl. Entomol. Zool.*, 1973, **8**, 200 (*isol*)
Hall, D.R. *et al*, *Chem. Ind. (London)*, 1975, 216 (*synth, ir, uv, ms, pmr*)
Goto, G. *et al*, *Chem. Lett.*, 1975, 103 (*synth, uv, ir, pmr, ms*)
Decodts, G. *et al*, *Synthesis*, 1979, 510 (*synth, ms, ir, pmr*)
Bestmann, H.J. *et al*, *Justus Liebigs Ann. Chem.*, 1981, 2117 (*synth, pmr*)
Rossi, R. *et al*, *Tetrahedron*, 1982, **38**, 639 (*cmr*)
Bloch, R. *et al*, *Tetrahedron Lett.*, 1983, **24**, 1247 (*synth*)
Cuvigny, T. *et al*, *Tetrahedron Lett.*, 1983, **24**, 4319 (*synth*)

Yadav, J.S. *et al*, *Synth. Commun.*, 1989, **19**, 125 (*synth, ir, pmr, bibl*)
Martinez, T. *et al*, *J. Biol. Chem.*, 1990, **265**, 1381 (*biosynth*)
Moiseenkov, A.M. *et al*, *J. Chem. Soc., Perkin Trans. 1*, 1991, 2639 (*synth, pmr*)

9,12-Tetradecadien-1-ol T-00036

$H_3CCH{=}CHCH_2CH{=}CH(CH_2)_7CH_2OH$

$C_{14}H_{26}O$ M 210.359

(9Z,12E)-form [51937-00-9]
$Bp_{0.05}$ 120°.
Ac: [31654-77-0].
$C_{16}H_{28}O_2$ M 252.396
Female sex pheromone from numerous species of moth (*Plodia interpunctella, Cadra cautella, Ephestia cautella, Myelois cribrella*). $Bp_{0.1}$ 126-128°.

Brady, U.E. *et al*, *Science (Washington, D.C.)*, 1971, **171**, 802 (*isol*)
Bestmann, H.J. *et al*, *Tetrahedron Lett.*, 1974, 779 (*synth*)
Rossi, R. *et al*, *Tetrahedron*, 1981, **37**, 2617 (*synth, pmr, cmr, ms*)
Bestmann, H.J. *et al*, *Justus Liebigs Ann. Chem.*, 1982, 536 (*synth*)
Attygalle, A.B. *et al*, *J. Chem. Ecol.*, 1988, **14**, 485.

10,12-Tetradecadien-1-ol T-00037

$H_3CCH{=}CHCH{=}CH(CH_2)_8CH_2OH$

$C_{14}H_{26}O$ M 210.359

(10E,12Z)-form
$Bp_{0.01}$ 95-100°.
Ac: [69775-63-9].
$C_{16}H_{28}O_2$ M 252.396
$Bp_{0.01}$ 95-100°.

(10E,12E)-form
$Bp_{0.05}$ 95-100°.
Ac: [69775-61-7].
$Bp_{0.01}$ 100-105°.

Bestmann, H.J. *et al*, *Justus Liebigs Ann. Chem.*, 1981, 1227 (*synth, pmr*)

11,13-Tetradecadien-1-ol T-00038

$H_2C{=}CHCH{=}CH(CH_2)_9CH_2OH$

$C_{14}H_{26}O$ M 210.359

(E)-form [80625-44-1]
$Bp_{0.01}$ 100-105°.
Ac: [80625-43-0].
$C_{16}H_{28}O_2$ M 252.396
$Bp_{0.01}$ 90-95°.

Bestmann, H.J. *et al*, *Justus Liebigs Ann. Chem.*, 1981, 2117 (*synth, pmr*)

6,8-Tetradecadiynedioic acid T-00039

[33128-28-8]

$HOOC(CH_2)_4(C{\equiv}C)_2(CH_2)_4COOH$

$C_{14}H_{18}O_4$ M 250.294
Cryst. (Me_2CO/pet. ether). Mp 175°.

Morris, S.G., *J. Am. Oil Chem. Soc.*, 1971, **48**, 376 (*synth*)

2,4,6,8,10,12-Tetradecahexaene-1,14-dioic acid T-00040

$HOOCCH{=}CHCH{=}CHCH{=}CH$
$CH{=}CHCH{=}CHCH{=}CHCOOH$

$C_{14}H_{14}O_4$ M 246.262

(all-E)-form [505-53-3] ***Corticrocin***
Pigment of the fungus *Corticium croceum*. Also from paprika (*Capsicum annuum*) and from *Encephalartos* spp. Orange-red needles. Insol. most solvs. Mp 270° subl. dec. (*ca.* 300°).
Di-Me ester: [77186-00-6].
$C_{16}H_{18}O_4$ M 274.316
Orange cryst. ($CHCl_3$ or AcOH). Mp 230-232°.
Di-Et ester: [1441-61-8].
$C_{18}H_{22}O_4$ M 302.369
Mp 204-205°.

Erdtman, H., *Acta Chem. Scand.*, 1948, **2**, 209 (*isol, struct*)
Shaw, B.L. *et al*, *J. Chem. Soc.*, 1954, 3217 (*synth*)
Kovalev, B.G. *et al*, *CA*, 1963, **58**, 11209f (*synth*)
Rives-Arnau, V., *Macromolecules*, 1985, **18**, 2088.

Tetradecanal T-00041

Myristic aldehyde. Myristaldehyde. Tetradecylaldehyde
[124-25-4]

$H_3C(CH_2)_{12}CHO$

$C_{14}H_{28}O$ M 212.375
Isol. from *Ocotea usambarensis*, lemon oil etc. Common constituent of Lepidopteran sex pheromones. Mp 30° (23°). Trimerises to a white solid Mp 65°.
▷ XB7900000.
Oxime:
$C_{14}H_{29}NO$ M 227.389
Cryst. (MeOH). Mp 82-83°.
Semicarbazone: Cryst. (MeOH). Mp 106.5°.

Stephen, H., *J. Chem. Soc.*, 1925, **127**, 1874 (*synth*)
Lewis, R.J., *Sax's Dangerous Properties of Industrial Materials*, 8th Ed., Van Nostrand-Reinhold, 1992, TBX500.

Tetradecane T-00042

[629-59-4]

$H_3C(CH_2)_{12}CH_3$

$C_{14}H_{30}$ M 198.391
Isol. from wood of *Manilkara bidentata*. Oil. Mp 5.9°. Bp 253.7°, Bp_{10} 121.9°.
▷ XB8000000.

Krafft, F., *Ber.*, 1882, **15**, 1687; 1886, **19**, 2218 (*synth*)
Cocker, W. *et al*, *J. Chem. Soc.*, 1963, 677 (*isol*)
Rodd's Chem. Carbon Compd. (2nd edn.), 2nd Ed., 1964, **1**, 357.
Lewis, R.J., *Sax's Dangerous Properties of Industrial Materials*, 8th Ed., Van Nostrand-Reinhold, 1992, TBX750.

Tetradecanedioic acid T-00043

Dodecane-1,12-dicarboxylic acid
[821-38-5]

$HOOC(CH_2)_{12}COOH$

$C_{14}H_{26}O_4$ M 258.357
Mp 124-125°.
Di-Me ester: [5024-21-5].
$C_{16}H_{30}O_4$ M 286.411
Mp 58-60° (44°).
Di-Et ester: [19812-63-6].
$C_{18}H_{34}O_4$ M 314.464
Mp 27°.

Hunig, S. *et al*, *Chem. Ber.*, 1959, **92**, 652 (*synth*)
Ried, W. *et al*, *Justus Liebigs Ann. Chem.*, 1964, **678**, 113 (*synth*)
Saotome, K. *et al*, *Bull. Chem. Soc. Jpn.*, 1966, **39**, 480 (*synth*)

1,14-Tetradecanediol T-00044

[19812-64-7]

$HOCH_2(CH_2)_{12}CH_2OH$

$C_{14}H_{30}O_2$ M 230.390

Mp 84-85°.

Ansell, M.F. *et al*, *J. Chem. Soc.*, 1957, 3302 (*synth*)

1-Tetradecanesulfonic acid T-00045

[7314-37-6]

$H_3C(CH_2)_{13}SO_3H$

$C_{14}H_{30}O_3S$ M 278.455

Mp 65.5°.

Monohydrate: Mp 55-56°.

Benzylaniline salt: Mp 82.7°.

Noller, C.R. *et al*, *J. Am. Chem. Soc.*, 1935, **55**, 1090 (*synth*)

Tetradecanoic acid, 9CI T-00046

Myristic acid, 8CI

[544-63-8]

$H_3C(CH_2)_{12}COOH$

$C_{14}H_{28}O_2$ M 228.374

Occurs widely in vegetable glycerides. Major component of the lipids of nutmeg (*Myristica moschata*). Cryst. (pet. ether). Mp 54°. Bp_{100} 250.5°, Bp_{16} 199°.

▷ QH4375000.

Me ester: [124-10-7].
$C_{15}H_{30}O_2$ M 242.401
Cryst. (Me_2CO). Mp 18.5°. Bp_7 155-157°.

Et ester: [124-06-1].
$C_{16}H_{32}O_2$ M 256.428
Cryst. (Me_2CO). Mp 12.3°. Bp 295°, Bp_4 139°.

Chloride: [112-64-1].
$C_{14}H_{27}ClO$ M 246.819
Bp_{16} 174°.

Anhydride: [626-29-9].
$C_{28}H_{54}O_3$ M 438.733
Mp 53°.

Amide: [638-58-4].
$C_{14}H_{29}NO$ M 227.389
Mp 105-107°.

Anilide: [622-56-0].
$C_{20}H_{33}NO$ M 303.487
Mp 84°. Bp_{10} 113°.

Nitrile: [629-63-0].
$C_{14}H_{27}N$ M 209.374
Mp 19°. Bp_{100} 226.5°, Bp_{13} 169°.

Ames, D.E. *et al*, *J. Chem. Soc.*, 1950, 174 (*synth*)
Rodd's Chem. Carbon Compd. (2nd edn.), 2nd Ed., Vol 1, 1965, 125, 1371.
Marosi, L. *et al*, *Justus Liebigs Ann. Chem.*, 1973, 584 (*synth, struct*)
Klein, W.J., *Recl. Trav. Chim. Pays-Bas (J. R. Neth. Chem. Soc.)*, 1975, **94**, 48 (*synth*)
Lewis, R.J., *Sax's Dangerous Properties of Industrial Materials*, 8th Ed., Van Nostrand-Reinhold, 1992, ENL850, MSA250.

1-Tetradecanol T-00047

Tetradecyl alcohol. Myristyl alcohol. 1-Hydroxytetradecane

[112-72-1]

$H_3C(CH_2)_{12}CH_2OH$

$C_{14}H_{30}O$ M 214.390

Constit. of natural waxes. Cryst. (EtOH aq.). Mp 39-39.5°. Bp_{20} 170-173°, Bp_{10} 160°.

▷ XB8655000.

Formyl: [5451-63-8].
$C_{15}H_{30}O_2$ M 242.401
Bp_{17} 166°.

Ac: [638-59-5].
$C_{16}H_{32}O_2$ M 256.428
Sex pheromone of European corn borer *Ostrinia nubitalis* and numerous other lepidoptera. Mp 14°.

Watanabe, A., *Bull. Chem. Soc. Jpn.*, 1961, **34**, 1728 (*synth*)
Adams, K.R. *et al*, *Nature (London)*, 1969, **223**, 943 (*isol*)
Precht, D., *Fette, Seifen, Anstrichm.*, 1976, **78**, 145, 189 (*struct*)
Klun, J.A. *et al*, *J. Chem. Educ.*, 1977, **3**, 447.
Rossi, R. *et al*, *Tetrahedron*, 1982, **38**, 639 (*cmr*)
Lewis, R.J., *Sax's Dangerous Properties of Industrial Materials*, 8th Ed., Van Nostrand-Reinhold, 1992, TBY250.

3-Tetradecanol T-00048

$H_3C(CH_2)_{10}CH(OH)CH_2CH_3$

$C_{14}H_{30}O$ M 214.390

(+)-*form* [3760-98-3]
Mp 32-38°. $Bp_{0.05}$ 111°. $[\alpha]_D^{25}$ +6.7° (EtOH).

(−)-*form* [1653-32-3]
Cryst. (MeOH). Mp 38°. Bp_{15} 160°, Bp_{10} 146°.
Acid phthalate: Mp 33°. $[\alpha]_D^{20}$ −17.43° (EtOH).

(±)-*form*
Mp 31.2-32°. Bp_{25} 173°.
Acid phthalate: Mp 58-60°.

Coke, J.L. *et al*, *J. Org. Chem.*, 1965, **30**, 3420; 1976, **41**, 3516.

13-Tetradecanolide T-00049

[27198-63-6]

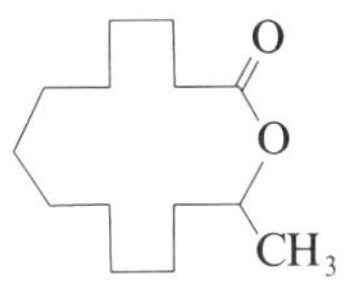

$C_{14}H_{26}O_2$ M 226.358

(−)-*form*
Constit. of *Galbanum* absolute. $[\alpha]_D^{20}$ −33.8° (c, 1.035 in $CHCl_3$).

(±)-*form*
Bp 05° 91.

Kaiser, R. *et al*, *Helv. Chim. Acta*, 1978, **61**, 2671 (*synth, isol, pmr, ms*)
Sims, R.J. *et al*, *Tetrahedron Lett.*, 1983, **24**, 253 (*synth*)

2-Tetradecanone T-00050

[2345-27-9]

$H_3C(CH_2)_{11}COCH_3$

$C_{14}H_{28}O$ M 212.375

Constit. of hop oil. Cryst. (EtOH aq.). Mp 33-34°. Bp_{100} 205-206°, Bp_{13} 134°.

Semicarbazone: Mp 125-126°.

Rodd's Chem. Carbon Compd. (2nd edn.), 2nd Ed., Vol 1, 1965, 52.
Allen, J.E. *et al*, *Lipids*, 1971, **6**, 448 (*synth*)
Bagnell, L. *et al*, *Aust. J. Chem.*, 1974, **27**, 1381 (*synth*)

3-Tetradecanone T-00051

[629-23-2]

$H_3C(CH_2)_{10}COCH_2CH_3$

$C_{14}H_{28}O$ M 212.375

Isol. from *Schedorhinotermes* spp. Cryst. (MeOH). Mp 34°. Bp_{16} 152°, Bp_{10} 148°.

Oxime:

$C_{14}H_{29}NO$ M 227.389

Cryst. (MeOH). Mp 40°.

Semicarbazone: Mp 89°.

2,4-Dinitrophenylhydrazone: Mp 45-47°.

Bertsch, H. *et al, Fette, Seifen, Anstrichm.*, 1962, **64**, 881 (*synth*)
Allen, J.E. *et al, Lipids*, 1971, **16**, 448 (*synth*)
Prestwich, G.D. *et al, Tetrahedron Lett.*, 1975, 4701 (*isol*)

5-Tetradecanone T-00052

[31857-89-3]

$H_3C(CH_2)_8CO(CH_2)_3CH_3$

$C_{14}H_{28}O$ M 212.375

Mp 23°. Bp_{16} 145-146°, $Bp_{0.7}$ 85-88°.

Carrington, R.A.G. *et al, J. Chem. Soc.*, 1957, 1701 (*synth*)
Anderson, M.W. *et al, J. Chem. Soc., Perkin Trans. 1*, 1986, 1995 (*synth, ir, pmr*)

2-(Tetradecanoylamino)-1,3-octadecanediol T-00053

2-Hydroxy-1-(hydroxymethyl)heptadecyltetradecanamide. N-*Tetradecanoylsphinganine*

$H_3C(CH_2)_{14}CH(OH)CH(CH_2OH)NHCO(CH_2)_{12}CH_3$

$C_{32}H_{65}NO_3$ M 511.870

Gaskell, S.J. *et al, J. Chromatogr.*, 1976, **126**, 591 (*glc, ms*)

5,7,9,11,13-Tetradecapentaenoic acid T-00054

[30345-96-1]

$H_2C{=}CH(CH{=}CH)_4(CH_2)_3COOH$

$C_{14}H_{18}O_2$ M 218.295

Constit. of aje, the body fat of *Llaveia axia*. Unstable.

Cason, J. *et al, J. Org. Chem.*, 1971, **36**, 2621.

3-(1,3,5,7,9-Tetradecapentaenyloxy)-1,2-propanediol, 9CI T-00055

1-O-(1,3,5,7,9-Tetradecapentaenyl)glycerol. Fecapentaene-14

[84000-58-8]

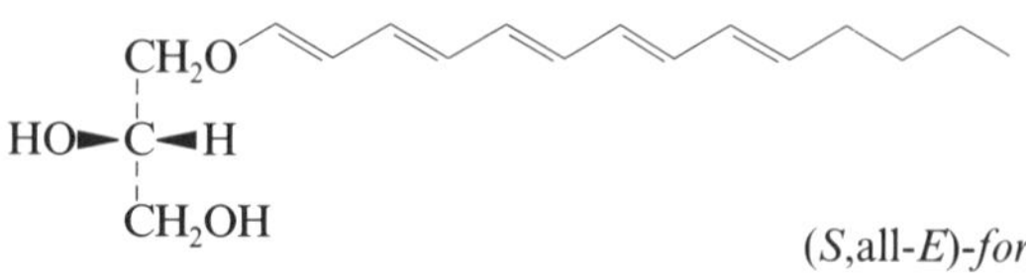

(*S*,all-*E*)-*form*

$C_{17}H_{26}O_3$ M 278.391

See also refs. under 3-(1,3,5,7,9-Dodecapentaenyloxy)-1,2-propanediol, D-00389.

(S,all-E)-form

Isol. from human faeces. Mutagen with implications for intestinal cancer. Beige plates. Mp 188-190° (sealed tube under Ar). $[\alpha]_D^{20}$ +6.5° (c, 0.46 in DMSO). Labile.

▷ Highly mutagenic.

((±)-all-E)-form

Mp 183-190° (sealed tube under Ar).

▷ Highly mutagenic.

[91379-15-6]

Gupta, I. *et al, Science (Washington, D.C.)*, 1984, **225**, 521.
Baptista, J. *et al, Angew. Chem., Int. Ed. Engl.*, 1987, **26**, 1186 (*config*)
Pfaendler, R. *et al, Justus Liebigs Ann. Chem.*, 1988, 449 (*synth, ir, pmr, cmr, ms*)

2,4,6,12-Tetradecatetraene-8,10-diynoic acid T-00056

$H_3CCH{=}CHC{\equiv}CC{\equiv}C(CH{=}CH)_3COOH$

$C_{14}H_{12}O_2$ M 212.248

(all-E)-form

Isol. from the underground parts of *Achillea wilhelmsii*.

Piperidide: [83182-49-4]. ***2,4,6,12-Tetradecatetraene-8,10-diynoic acid piperidide.*** *1-(1-Oxo-2,4,6,12-tetradecatetraene-8,10-diynyl)piperidine, 9CI*

$C_{19}H_{21}NO$ M 279.381

Alkamide from the roots of *A. grandifolia* (Compositae). Unstable yellow foam.

(2E,4E,6E,12Z)-form

Me ester: [14109-34-3].

$C_{15}H_{14}O_2$ M 226.274

Isol. from *Sanvitalia procumbens*. Cryst. (Et_2O/pet. ether). Mp 102°. λ_{max} 368, 348, 281, 275, 252 nm.

2-Methylpropylamide: [113235-98-6]. *2,4,6,12-Tetradecatetraene-8,10-diynoic acid isobutylamide*

$C_{18}H_{21}NO$ M 267.370

Isol. from underground parts of *A. wilhelmsii*. Oil. λ_{max} 368, 348, 333sh, 279, 264 nm (Et_2O). Contains 30% of the 6*Z*-isomer.

3-Methylbutylamide: [113235-95-3]. *2,4,6,12-Tetradecatetraene-8,10-diynoic acid isopentylamide*

$C_{19}H_{23}NO$ M 281.397

Isol. from underground parts of *A. wilhelmsii*. Cryst. Mp 136-139°. λ_{max} 368, 347, 333sh, 279, 217 nm (Et_2O).

(2E,4E,6Z,12Z)-form

2-Methylpropylamide: [113235-97-5].

Isol. from underground parts of *A. wilhelmsii*. Oil. λ_{max} 366sh, 348, 335sh, 277, 264 nm (Et_2O). Obtained as a mixture containing 6*E*-isomer.

3-Methylbutylamide: [113235-94-2].

Isol. from underground parts of *A. wilhelmsii*. Oil. λ_{max} 365sh, 347, 332sh, 277, 265 nm (Et_2O).

Bohlmann, F. *et al, Chem. Ber.*, 1966, **99**, 3194 (*isol, struct, uv, ir, pmr*)
Greger, H. *et al, Phytochemistry*, 1982, **21**, 1071 (*piperidide*)
Greger, H. *et al, J. Nat. Prod. (Lloydia)*, 1987, **50**, 1100 (*amides, isol, uv, ir, ms, pmr*)

1,8,10-Tetradecatriene T-00057

$H_3CCH_2CH_2CH{=}CHCH{=}CH(CH_2)_5CH{=}CH_2$

$C_{14}H_{24}$ M 192.344

(8E,10Z)-form [80625-31-6]

$Bp_{0.1}$ 74-77°.

Bestmann, H.J. *et al, Justus Liebigs Ann. Chem.*, 1981, 2117 (*synth, pmr*)

2,8,10-Tetradecatriene-4,6-diyne-1,14-diol T-00058

[29576-60-1]

$HOCH_2CH_2CH_2CH{=}CHCH{=}CH$
$C{\equiv}CC{\equiv}CCH{=}CHCH_2OH$

$C_{14}H_{16}O_2$ M 216.279

(2E,8E,10E)-form

Constit. of *Coreopsis verticillata* roots and *Bidens aurea*. Cryst. (Et_2O/pet. ether). Mp 132°.

Di-Ac: [29576-62-3].

$C_{18}H_{20}O_4$ M 300.354

Cryst. (pet. ether). Mp 41-42°.

Bohlmann, F. *et al, Chem. Ber.*, 1970, **103**, 2095; 1975, **108**, 440 (*isol, uv, ir, pmr, ms*)

4,6,12-Tetradecatriene-8,10-diyne-1,3-diol T-00059

[23414-62-2]

H$_3$CCH=CHC≡CC≡CCH=CHCH=CHCH(OH)CH$_2$CH$_2$OH

$C_{14}H_{16}O_2$ M 216.279

(4E,6E,12E)-form [59950-61-7]

Constit. of *Dahlia* spp. and from *Carthamus coeruleus*. Cryst. (CH_2Cl_2/pet. ether). Mp 110-112° (96-97°). $[\alpha]_D^{20}$ −19° (c, 1.1 in EtOH).

1-Ac: 1-Acetoxy-4,6,12-tetradecatriene-8,10-diyn-3-ol

$C_{16}H_{18}O_3$ M 258.316

Isol. from roots of *C. coeruleus* and from *Crupina vulgaris*. Oil. λ_{max} 249 (ε 34 800), 265 (29 700), 278 (16 200), 295 (29 100), 344 (39 800), 336 nm (29 000) (Et_2O).

3-Ac: 3-Acetoxy-4,6,12-tetradecatriene-8,10-diyn-1-ol

$C_{16}H_{18}O_3$ M 258.316

Isol. from *Carthamus coeruleus* and *Centaurea diluta*. Cryst. (Et_2O/pet. ether). Mp 61°. $[\alpha]_{546}$ +31.2° (c, 1.06 in Et_2O).

Di-Ac: [29576-66-7]. *1,3-Diacetoxy-4,6,12-tetradecatriene-8,10-diyne*

$C_{18}H_{20}O_4$ M 300.354

Constit. of *D.* spp., *Carthamus coeruleus*, *C. tinctorius*, *Centaurea diluta* and *Atractylodes japonica*. Mp 36°, Mp 58°. $[\alpha]_{546}^{24}$ +6.4° (c, 3.65 in Et_2O), $[\alpha]_D$ +0.4° (c, 1.0 in MeOH). Isol. as pale yellow oil by Kano *et al.*

[23414-61-1, 103994-34-9]

Bohlmann, F. *et al, Chem. Ber.*, 1966, **99**, 3433, 3544 (*isol, uv, ir, pmr*)

Bedford, C.T. *et al, J. Chem. Soc., Perkin Trans. 1*, 1976, 735 (*isol, uv, pmr, ms*)

Kano, Y. *et al, Chem. Pharm. Bull.*, 1989, **37**, 193 (*isol, ir, uv, pmr, ms*)

4,6,12-Tetradecatriene-8,10-diynoic acid T-00060

H$_3$CCH=CHC≡CC≡CCH=CHCH=CHCH$_2$CH$_2$COOH

$C_{14}H_{14}O_2$ M 214.263

(4E,6E,12Z)-form

Me ester: [101622-93-9].

$C_{15}H_{16}O_2$ M 228.290

Isol. from the aerial parts of *Mutisia retrorsa*. Oil. λ_{max} 336, 312, 296 nm (Et_2O).

Zdero, C. *et al, Phytochemistry*, 1986, **25**, 509 (*isol, ms, uv, ir, pmr*)

4,6,8-Tetradecatriene-10,12-diyn-3-ol T-00061

H$_3$CC≡CC≡CC≡CCH=CHCH=CHCH(OH)CH$_2$CH$_3$

$C_{14}H_{16}O$ M 200.280

Constit. of *Brickellia laciniata* roots. Yellow oil. Isol. as a mixt. of isomers of undetermined stereochemistry.

Bohlmann, F. *et al, Phytochemistry*, 1978, **17**, 763 (*isol, uv, ms*)

4,6,12-Tetradecatriene-8,10-diyn-1-ol, 9CI T-00062

[13081-23-7]

H$_3$CCH=CHC≡CC≡CCH=CHCH=CHCH$_2$CH$_2$CH$_2$OH

$C_{14}H_{16}O$ M 200.280

(E,E,E)-form [17090-99-2]

Isol. from *Dahlia merckii* and roots of *Anthemis saguramica*. Platelets (Et_2O/pet. ether). Mp 110-111°.

Ac: [1540-89-2].

$C_{16}H_{18}O_2$ M 242.317

Isol. from *A. carpatica*, *D. merckii*, *Cotula coronopifolia* and other plants. Cryst. (pet. ether). Mp 31-33°.

4,5-Epoxide, Ac: [34444-55-8]. ***1-Acetoxy-4,5-epoxy-6,12-tetradecadiene-8,10-diyne***

$C_{16}H_{18}O_3$ M 258.316

Isol. from *D. scapigera*. Pale yellow oil.

(4E,6Z,12E)-form [131802-64-7]

Isol. from rhizomes of *Atractylis koreana*.

[13081-22-6]

Bohlmann, F. *et al, Justus Liebigs Ann. Chem.*, 1963, **668**, 51 (*isol, deriv*)

Bohlmann, F. *et al, Chem. Ber.*, 1965, **98**, 872; 1966, **99**, 1642, 3544 (*isol uv, ir, pmr, struct, synth*)

Sörensen, J.S. *et al, Acta Chem. Scand.*, 1966, **20**, 992 (*isol, synth*)

Aplin, R.T. *et al, J. Chem. Soc., Chem. Commun.*, 1967, 140 (*ms*)

Lam, J. *et al, Phytochemistry*, 1971, **10**, 1877 (*isol, struct*)

Bedford, C.T., *J. Chem. Soc., Perkin Trans. 1*, 1976, 735 (*isol, synth*)

Pachaly, P. *et al, Planta Med.*, 1990, **56**, 469 (*isol, struct, pmr, cmr*)

2,4,5-Tetradecatrienoic acid T-00063

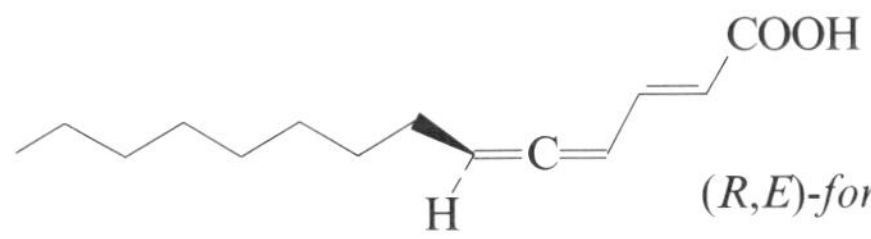

(R,E)-form

$C_{14}H_{22}O_2$ M 222.327

(R,E)-form

Me ester: [28066-21-9].

$C_{15}H_{24}O_2$ M 236.353

Sex attractant of the male dried bean beetle *Acanthoscelides obtectus*. Light yellow oil. $[\alpha]_D^{23}$ −162° (c, 0.95 in hexane).

(S,E)-form

Me ester: [65451-10-7].

Light yellow oil. $[\alpha]_D$ +160° (c, 0.75 in hexane).

Horler, D.F., *J. Chem. Soc. C*, 1970, 859 (*isol, synth*)

Kocienski, P.J. *et al, J. Org. Chem.*, 1977, **42**, 353 (*synth*)

Pirkle, W.H. *et al, J. Org. Chem.*, 1978, **43**, 2091 (*synth*)

Mori, K. *et al, Tetrahedron*, 1981, **37**, 1343 (*synth*)

Westmijze, H. *et al, Tetrahedron Lett.*, 1982, **23**, 2797 (*synth*)

Le Doussal, B. *et al, Tetrahedron*, 1983, **39**, 2185 (*synth*)

Fujisawa, T. *et al, Tetrahedron Lett.*, 1984, **25**, 4007 (*synth*)

Lang, R.W. *et al, Helv. Chim. Acta*, 1985, **68**, 2249 (*synth*)

2,5,8-Tetradecatriyn-1-ol T-00064

H$_3$C(CH$_2$)$_4$C≡CCH$_2$C≡CCH$_2$C≡CCH$_2$OH

$C_{14}H_{18}O$ M 202.296

Cryst. (pet. ether). Mp 28-30°. $Bp_{0.08}$ 135-142°.

Degenerates rapidly to give a red-brown polymer.

Osbond, M. *et al, J. Chem. Soc.*, 1961, 2779 (*synth, bibl*)

2-Tetradecenal T-00065

$H_3C(CH_2)_{10}CH{=}CHCHO$

$C_{14}H_{26}O$ M 210.359

(E)-form [51534-36-2]

Liq. $Bp_{0.1}$ 84-86°.

2,4-Dinitrophenylhydrazone: Orange needles. Mp 126-127°.

Walles, P.C. *et al*, *Aust. J. Chem.*, 1959, **12**, 173 (*synth*)
Chang, Y.-H. *et al*, *Synthesis*, 1990, 1033 (*synth, ir, pmr, ms*)

5-Tetradecenal, 9CI T-00066

$H_3C(CH_2)_7CH{=}CH(CH_2)_3CHO$

$C_{14}H_{26}O$ M 210.359

(Z)-form [63851-42-3]

Component of the sex pheromones of a large number of noctuid moth species. Principal component of the tergal gland of the Staphylinid beetle, *Aleochara curtula*. Liq.

Underhill, E.W. *et al*, *Environ. Entomol.*, 1977, **6**, 333 (*isol, activity*)
Peschke, K., *J. Chem. Ecol.*, 1982, **8**, 773; 1983, **9**, 13 (*isol, activity, ms, ir, glc*)
Latli, B. *et al*, *J. Chem. Ecol.*, 1991, **17**, 1745 (*synth, pmr, cmr, ir*)

7-Tetradecenal T-00067

[65128-96-3]

$H_3C(CH_2)_5CH{=}CH(CH_2)_5CHO$

$C_{14}H_{26}O$ M 210.359

(E)-form

Liq.

(Z)-form

Pheromone of citrus flower moth *Prays citri* and olive moth *P. oleae*. Liq. $Bp_{0.05}$ 87-89°. n_D^{20} 1.4499.

Ohloff, G. *et al*, *Helv. Chim. Acta*, 1977, **60**, 1161 (*synth*)
Sternlicht, M. *et al*, *Phytoparasitica*, 1978, **6**, 101.
Campion, D.G. *et al*, *Experientia*, 1979, **35**, 1146 (*isol*)
Becker, D. *et al*, *Tetrahedron*, 1982, **38**, 1689 (*synth, cmr*)
Brown, H.C. *et al*, *J. Org. Chem.*, 1986, **51**, 4518 (*synth, ir, pmr, cmr*)

9-Tetradecenal T-00068

$H_3C(CH_2)_3CH{=}CH(CH_2)_7CHO$

$C_{14}H_{26}O$ M 210.359

(E)-form [71377-13-4]

Sex pheromone of Lepidoptera.

(Z)-form [53939-27-8]

Pheromone of *Heliothis virescens*. Can inhibit behaviour of other pheromones. $Bp_{0.6}$ 89°.

Bestmann, H.J. *et al*, *Chem. Ber.*, 1975, **108**, 3582 (*synth*)
Renov, M., *C. R. Hebd. Seances Acad. Sci. Ser. D*, 1979, **288**, 1559.
Leadbetter, G., *J. Chem. Ecol.*, 1979, **5**, 101 (*synth, bibl*)

11-Tetradecenal, 9CI T-00069

[70893-80-0]

$H_3CCH_2CH{=}CH(CH_2)_9CHO$

$C_{14}H_{26}O$ M 210.359

(E)-form [35746-21-5]

Principal component of the sex pheromones of the eastern spruce budworm *Choristoneura furniferana*, and the western spruce budworm *C. occidentalis*. Component of the sex pheromone of the blueberry leaftier moth *Croesia curvalana*, synergist for the sex pheromone of *Cheimophila salicella*. Liq. $Bp_{0.025}$ 67-68°. n_D^{21} 1.4483 (>98%).

(Z)-form [35237-64-0]

Specificity-determining component of the sex pheromone of the oblique banded leafroller, *Choristoneura rosaceana*. $Bp_{0.02}$ 115-125°.

Weatherston, J. *et al*, *Can. Entomol.*, 1971, **103**, 1741 (*isol, activity*)
Brown, H.C. *et al*, *J. Org. Chem.*, 1986, **51**, 5282 (*synth, pmr, cmr, ir, glc, ms*)
Wolf, W.A., *J. Chem. Ecol.*, 1987, **13**, 1019 (*isol, activity*)
Vakenti, J.M. *et al*, *J. Chem. Ecol.*, 1988, **14**, 605 (*isol, activity*)
Sweeney, J.D. *et al*, *J. Chem. Ecol.*, 1990, **16**, 1517, 1531 (*activity*)
Singh, K.N. *et al*, *Indian J. Chem., Sect. B*, 1991, **30**, 867 (*synth, pmr, ir*)
Moiseenkov, A.M. *et al*, *J. Chem. Soc., Perkin Trans. 1*, 1991, 2639 (*synth, pmr*)

1-Tetradecene T-00070

[1120-36-1]

$H_3C(CH_2)_{11}CH{=}CH_2$

$C_{14}H_{28}$ M 196.375

Isol. from *Tribolium* spp. Mp −12°. Bp 251°, Bp_{15} 124.5-125°.

Dibromide: [6064-45-5]. *1,2-Dibromotetradecane*

$C_{14}H_{28}Br_2$ M 356.183

Mp 0°. $Bp_{0.22}$ 140-156°.

Sadtler Standard C-13 NMR Spectra, 601.
Sadtler Standard Infrared Spectra, 4536.
Sadtler Standard NMR Spectra, 6751 (*pmr*)
Dupont, G. *et al*, *C. R. Hebd. Seances Acad. Sci.*, 1958, **246**, 128 (*synth*)
Suzuki, T. *et al*, *Agric. Biol. Chem.*, 1975, **39**, 2207 (*isol*)

7-Tetradecene, 9CI, 8CI T-00071

[10374-74-0]

$H_3C(CH_2)_5CH{=}CH(CH_2)_5CH_3$

$C_{14}H_{28}$ M 196.375

(E)-form [41446-63-3]

Liq. Bp 250.2°.

(Z)-form [41446-60-0]

Liq. Bp 249.4°, $Bp_{0.8}$ 75-76°. n_D^{20} 1.4399.

Sadtler Standard Infrared Spectra, 40362 (*ir*)
Sadtler Standard NMR Spectra, 9909 (*pmr*)
Janak, J., *Anal. Chem.*, 1973, **45**, 293 (*chromatog*)
Brown, H.C. *et al*, *J. Org. Chem.*, 1982, **47**, 3806, 3808; 1986, **26**, 5270; 1988, **53**, 239 (*synth, pmr*)
Julia, M. *et al*, *Tetrahedron Lett.*, 1982, **23**, 2457 (*synth*)
Polyakova, A.A. *et al*, *Zh. Org. Khim.*, 1987, **23**, 1164 (*ms*)
Collazo, L.R. *et al*, *J. Org. Chem.*, 1993, **58**, 43 (*synth, ir, pmr, cmr*)

6-Tetradecene-1,3-diyne-5,8-diol, 9CI T-00072

$H_3C(CH_2)_5CH(OH)CH{=}CHCH(OH)C{\equiv}CC{\equiv}CH$

$C_{14}H_{20}O_2$ M 220.311

(Z)-form [77769-23-4]

Isol. from tomato plants. Phytoalexin. λ_{max} 229, 241, 254, 284 nm (MeOH).

De Wit, P.J.G. *et al*, *Physiol. Plant Pathol.*, 1981, **18**, 143 (*isol, struct, ir, uv, pmr, ms*)
Elgersma, D.M. *et al*, *Physiol. Mol. Plant Pathol.*, 1989, **34**, 545 (*isol*)

13-Tetradecene-1,3-diyne-6,7-diol T-00073

Panaxyne

[122855-49-6]

$H_2C{=}CH(CH_2)_5CH(OH)CH(OH)CH_2C{\equiv}CC{\equiv}CH$

$C_{14}H_{20}O_2$ M 220.311

Constit. of the roots of *Panax ginseng*. Cytotoxic.

Kim, S.I. *et al*, *Saengyak Hakhoechi*, 1989, **20**, 71; *CA*, **111**, 224891 (*isol*)

6-Tetradecene-8,10-diyne-1,5,12-triol T-00074

$H_3CCH_2CH(OH)C{\equiv}CC{\equiv}CCH{=}CHCHOH(CH_2)_3CH_2OH$

$C_{14}H_{20}O_3$ M 236.310

(E)-form [40111-21-5]

Constit. of *Campanula pyramidalis*.

Tri-Ac: [40111-32-8].

$Bp_{0.01}$ 170-180°.

Badanyan, Sh. *et al*, *J. Chem. Soc., Perkin Trans. 1*, 1973, 145 (*isol, synth*)

12-Tetradecene-8,10-diyne-1,4,5-triol T-00075

[87440-72-0]

$H_3CCH{=}CHC{\equiv}CC{\equiv}CCH_2CH_2CH(OH)CH(OH)CH_2CH_2CH_2OH$

$C_{14}H_{20}O_3$ M 236.310

Constit. of *Hyoseris lucida*. Cryst. ($CHCl_3$/EtOAc). Mp 110-112°.

[87440-73-1, 113137-50-1]

El-Masry, S. *et al*, *Phytochemistry*, 1983, **22**, 592.

13-Tetradecene-2,4-diyn-1-ol, 9CI T-00076

[125906-53-8]

$H_2C{=}CH(CH_2)_7C{\equiv}CC{\equiv}CCH_2OH$

$C_{14}H_{20}O$ M 204.311

Isol. from the hard corals, *Montipora* sp., *M. mollis* and *Pectinia lactuca*. Exhibits ichthyotoxicity and inhibits growth of some bacteria and fungi.

Higa, T. *et al*, *Chem. Lett.*, 1990, 145 (*isol, struct*)

5-Tetradecene-8,10,12-triynedioic acid T-00077

$HOOCC{\equiv}CC{\equiv}CC{\equiv}CCH_2CH{=}CH(CH_2)_3COOH$

$C_{14}H_{12}O_4$ M 244.246

(Z)-form

Metabolite of *Poria sinuosa*. Mp *ca.* 120° dec.

Di-Me ester: Oil. Mp −10°.

Cambie, R.C. *et al*, *J. Chem. Soc.*, 1963, 2056.

6-Tetradecene-8,10,12-triyn-3-ol T-00078

$H_3C(C{\equiv}C)_3CH{=}CHCH_2CH_2CH(OH)CH_2CH_3$

$C_{14}H_{16}O$ M 200.280

(−)-*(E)-form*

Isol. from *Anacyclus pyrethrum* and *Anthemis saguramica*. Cryst. (pet. ether). $[\alpha]_D^{22}$ −56° (c, 0.25 in Et_2O).

(+)-*E-form*

Ac: [126381-88-2]. *Artemisia acetate*

$C_{16}H_{18}O_2$ M 242.317

Isol. from *Artemisia carruthii*. Oil. $[\alpha]_D^{20}$ +17° (c, 0.4 in $CHCl_3$).

[126829-74-1]

Bohlmann, F. *et al*, *Chem. Ber.*, 1965, **98**, 1411; 1966, **99**, 1642 (*isol, uv, ir, pmr, ord, struct*)

Wallnofer, B. *et al*, *Phytochemistry*, 1989, **28**, 2687 (*isol, ir, uv, pmr, cd, ms*)

Zdero, C. *et al*, *Phytochemistry*, 1990, **29**, 189 (*isol*)

6-Tetradecene-8,10,12-triyn-3-one T-00079

Artemisia ketone†

[2739-59-5]

$H_3CC{\equiv}CC{\equiv}CC{\equiv}CCH{=}CHCH_2CH_2COCH_2CH_3$

$C_{14}H_{14}O$ M 198.264

The use of 'Artemisia ketone' for this compd. should be discontinued.

(E)-form [37064-08-7]

Widely distributed in the Compositae, e.g. *Anthemis, Artemisia, Chrysanthemum, Anacyclus, Matricaria* spp. (*E*-config. not always specifically demonstrated). Needles (pet. ether). Mp 57.5°.

Bohlmann, F. *et al*, *Chem. Ber.*, 1955, **88**, 361.

Bohlmann, F. *et al*, *Chem. Ber.*, 1969, **102**, 1698 (*biosynth*)

Jente, R. *et al*, *Chem. Ber.*, 1972, **105**, 1694.

Bohlmann, F. *et al*, *Chem. Ber.*, 1974, **107**, 1074 (*synth*)

2-Tetradecenoic acid T-00080

[39525-69-4]

$H_3C(CH_2)_{10}CH{=}CHCOOH$

$C_{14}H_{26}O_2$ M 226.358

(E)-form [14811-80-4]

Mp 36.8-37.2° (50-53°). Bp_8 186°.

Me ester:

$C_{15}H_{28}O_2$ M 240.385

Bp_{15} 167-168°.

Et ester: [38112-60-6].

$C_{16}H_{30}O_2$ M 254.412

$Bp_{0.02}$ 106-110°, Bp_{11} 165-167°.

U.S. Pat., 2 399 601, (1945); *CA*, **40**, 4180.

Ogibin, Yu.N. *et al*, *J. Org. Chem. USSR (Engl. Transl.)*, 1967, **3**, 257 (*synth*)

Falk, F. *et al*, *Fette, Seifen, Anstrichm.*, 1968, **70**, 709.

Jurasek, A. *et al*, *CA*, 1975, **82**, 169954 (*synth*)

3-Tetradecenoic acid T-00081

$H_3C(CH_2)_9CH{=}CHCH_2COOH$

$C_{14}H_{26}O_2$ M 226.358

(E)-form [75730-21-1]

Isol. from seed oil of *Grindelia oxylepis*.

[62472-83-7, 81236-42-2]

Kleiman, R. *et al*, *Lipids*, 1966, **1**, 301 (*isol*)

Gunstone, F.D. *et al*, *Chem. Phys. Lipids*, 1977, **18**, 115 (*cmr*)

Mizugaki, M. *et al*, *Chem. Pharm. Bull.*, 1980, **28**, 2347 (*nmr*)

Ogura, K. *et al*, *Tetrahedron Lett.*, 1981, **22**, 4499 (*synth*)

4-Tetradecenoic acid T-00082

Tsuzuic acid. Tsuduic acid

[544-65-0]

$H_3C(CH_2)_8CH{=}CHCH_2CH_2COOH$

$C_{14}H_{26}O_2$ M 226.358

(Z)-form [7089-44-3]
Occurs in seed oil of *Lindera obtusiloba* (tohaku oil), *Tetradenia glauca* ("tsuzu") and *Litsea glauca*. Mp 18-18.5°. Bp_{13} 185-188°.
Me ester: [37443-02-0].
$C_{15}H_{28}O_2$ M 240.385
Bp_{15} 158-160°.
Et ester: [54977-80-9].
$C_{16}H_{30}O_2$ M 254.412
Bp_{15} 168-170°.

Komori, S. *et al*, *Bull. Chem. Soc. Jpn.*, 1937, **12**, 433.
Hopkins, C.Y. *et al*, *Lipids*, 1966, **1**, 118.
Furukawa, K. *et al*, *CA*, 1973, **79**, 41844 (*synth, spectra*)
Gunstone, F.D. *et al*, *Chem. Phys. Lipids*, 1977, **18**, 115 (*cmr*)

5-Tetradecenoic acid T-00083

Physeteric acid
[544-66-1]

$H_3C(CH_2)_7CH{=}CH(CH_2)_3COOH$

$C_{14}H_{26}O_2$ M 226.358
(Z)-form [5684-70-8]
Occurs in sperm whale oil. Also in lipids of *Euglena gracilis* and *Streptococcus* sp. d_4^{15} 0.908. Mp 20°. Bp_{15} 192-197°. n_D^{15} 1.4571.

Tsujimoto, M., *Kogyo Kagaku Zasshi*, 1926, **29**, 102; *CA*, **23**, 1109.
Toyama, Y. *et al*, *Nippon Kagaku Kaishi*, 1935, **56**, 1050; *CA*, **30**, 315.
Korn, E.D., *J. Lipid Res.*, 1964, **5**, 352 (*isol*)
Gunstone, F.D. *et al*, *Chem. Phys. Lipids*, 1977, **18**, 115 (*cmr*)

7-Tetradecenoic acid T-00084

$H_3C(CH_2)_5CH{=}CH(CH_2)_5COOH$

$C_{14}H_{26}O_2$ M 226.358
Isol. from various microbial lipids.
(Z)-form [2430-95-7]
Constit. of lipids of *Euglena gracilis*.
2-Methylpropyl ester: [83005-39-4].
$C_{18}H_{34}O_2$ M 282.465
Pheromone of the Western grapeleaf skeletoniser *Harrisina brillians*.

Hofmann, K. *et al*, *J. Biol. Chem.*, 1959, **234**, 1672 (*synth*)
Scheuerbrandt, G. *et al*, *J. Biol. Chem.*, 1962, **237**, 2064 (*isol*)
Korn, E.D. *et al*, *J. Lipid Res.*, 1964, **5**, 352 (*isol*)
Myerson, J. *et al*, *Tetrahedron Lett.*, 1982, **23**, 2757 (*isol, synth*)
Zakharkin, L.I. *et al*, *Zh. Org. Khim.*, 1982, **18**, 1623 (*synth*)

8-Tetradecenoic acid T-00085

$H_3C(CH_2)_4CH{=}CH(CH_2)_6COOH$

$C_{14}H_{26}O_2$ M 226.358
(Z)-form [62472-85-9]
Oil. $Bp_{0.1}$ 160°. n_D^{17} 1.4569.

Baker, C.D. *et al*, *J. Chem. Soc.*, 1963, 489.
Bus, J. *et al*, *Chem. Phys. Lipids*, 1976, **17**, 501; 1977, **18**, 130 (*cmr*)
Gunstone, F.D. *et al*, *Chem. Phys. Lipids*, 1977, **18**, 115 (*cmr*)
Huhtasaari, M. *et al*, *Acta Chem. Scand., Ser. B*, 1983, **37**, 537 (*synth, ir, pmr, ms*)

9-Tetradecenoic acid T-00086

$H_3C(CH_2)_3CH{=}CH(CH_2)_7COOH$

$C_{14}H_{26}O_2$ M 226.358
(E)-form [50286-30-1]
Cryst. (pet. ether). Mp 18-18.5°.
Et ester:
$C_{16}H_{30}O_2$ M 254.412
$Bp_{0.4}$ 112°.
(Z)-form [544-64-9]
Myristoleic acid
Occurs in natural fats, eg. Cottonseed oil, *Lophira elata*, *Acacia cyclops*. Mp −4.5°. $Bp_{0.6}$ 144°, $Bp_{0.03}$ 133°.
Me ester: [56219-06-8].
$C_{15}H_{28}O_2$ M 240.385
$Bp_{3\text{-}5}$ 125-130°.
Et ester: [24880-50-0].
$C_{16}H_{30}O_2$ M 254.412

Boughton, B.W. *et al*, *J. Chem. Soc.*, 1952, 671 (*synth*)
Calam, D.H., *Nature (London)*, 1969, **221**, 856 (*isol*)
Bestmann, H.J. *et al*, *Chem. Ber.*, 1975, **108**, 3582.
Wawra, H., *Z. Naturforsch., C*, 1978, **33**, 28 (*struct*)

3-Tetradecen-1-ol T-00087

$H_3C(CH_2)_9CH{=}CHCH_2CH_2OH$

$C_{14}H_{28}O$ M 212.375
(E)-form [68900-86-7]
$Bp_{0.05}$ 105°.
Ac:
$C_{16}H_{30}O_2$ M 254.412
Sex attractant for white barred groundling moth *Recurvaria leucatella* and lesser bud moth *R. nanella*.
(Z)-form [68892-27-3]
Oil. $Bp_{0.05}$ 105°.

Doolittle, R.E., *Org. Prep. Proced. Int.*, 1981, **13**, 179 (*synth*)
Carvalho, J.F. *et al*, *J. Org. Chem.*, 1984, **49**, 1251 (*synth, ir, pmr*)
Toth, M. *et al*, *J. Chem. Ecol.*, 1992, **18**, 1093 (*isol*)

5-Tetradecen-1-ol T-00088

$H_3C(CH_2)_7CH{=}CH(CH_2)_3CH_2OH$

$C_{14}H_{28}O$ M 212.375
(E)-form [62936-14-5]
Shows pheromone activity.
(Z)-form [40642-42-0]
Ac: [35153-13-0]. *1-Acetoxy-5-tetradecene*
$C_{16}H_{30}O_2$ M 254.412
Constit. of *Hibiscus abelmoschus* oil. Sex pheromone for brownheaded leafroller *Ctenopseustis obliquana*.
Butanoyl:
$C_{18}H_{34}O_2$ M 282.465
Constit. of cephalic secretions from stingless bee *Scaptotrigona postica*.

Maurer, B., *Helv. Chim. Acta*, 1977, **60**, 1155.
Engels, E. *et al*, *J. Chem. Ecol.*, 1987, **13**, 371.
Foster, S.P. *et al*, *J. Chem. Ecol.*, 1987, **13**, 623.

6-Tetradecen-1-ol, 9CI T-00089

[39650-09-4]

$H_3C(CH_2)_6CH{=}CH(CH_2)_4CH_2OH$

$C_{14}H_{28}O$ M 212.375
(E)-form [68760-62-3]
Liq. $Bp_{1.0}$ 91° (92% isomeric purity).

Ac: [39650-10-7]. *6-Tetradecenyl acetate*
$C_{16}H_{30}O_2$ M 254.412
Component of the sex pheromones of the apple ermine moth, *Yponomeuta malinellus*. Synthetic sex attractant for the small ermine moth, *Y. rorellus*. Liq. $Bp_{0.05}$ 104-105°, Bp_3 160-165°. n_D^{25} 1.4446.

(Z)-form [68760-63-4]
Ac: [39650-11-8].
Bp_3 170-175°. n_D^{20} 1.4478.

Schwartz, M. *et al, Synthesis*, 1972, 567 (*synth*)
Bently, R.K. *et al, J. Chem. Soc., Perkin Trans. 1*, 1973, 140 (*synth, pmr, ir, glc*)
Horiike, M. *et al, Agric. Biol. Chem.*, 1978, **42**, 1963 (*synth, pmr, ir*)
McDonough, L.M. *et al, J. Chem. Ecol.*, 1990, **16**, 477 (*activity*)
Loefstedt, C. *et al, Physiol. Entomol.*, 1990, **15**, 47 (*activity*)
Gu, Y. *et al, Rapid Commun. Mass Spectrom.*, 1991, **5**, 622 (*ms*)

7-Tetradecen-1-ol — T-00090

$H_3C(CH_2)_5CH{=}CH(CH_2)_5CH_2OH$

$C_{14}H_{28}O$ M 212.375
Pheromone of male lepidoptera.

(E)-form [37011-95-3]
$Bp_{0.01}$ 100-105°.
Ac: [28540-79-6].
$C_{16}H_{30}O_2$ M 254.412
$Bp_{0.1}$ 109-110°.

(Z)-form [40642-43-1]
$Bp_{0.01}$ 94-96°.
Ac: [16974-10-0].
Sex attractant of *Amathes c-nigrum*. $Bp_{0.01}$ 102-104°.

Voaden, D.J. *et al, J. Med. Chem.*, 1972, **15**, 619 (*synth*)
Kovaleva, A.S. *et al, J. Org. Chem. USSR (Engl. Transl.)*, 1972, **8**, 2542; 1974, **10**, 700 (*synth*)
Bestmann, H.J. *et al, Chem. Ber.*, 1979, **112**, 1923 (*synth, pmr, ms*)
Bestmann, H.J. *et al, Tetrahedron Lett.*, 1979, 497 (*isol, synth*)
Brown, H.C. *et al, J. Org. Chem.*, 1988, **53**, 246 (*synth, ir, pmr, cmr*)

8-Tetradecen-1-ol, 9CI — T-00091

[129189-59-9]

$H_3C(CH_2)_4CH{=}CH(CH_2)_6CH_2OH$

$C_{14}H_{28}O$ M 212.375

(E)-form [64437-34-9]
Bp_1 115-118°. n_D^{27} 1.4538.
Formyl: [129731-27-7].
$C_{15}H_{28}O_2$ M 240.385
Potent pheromone mimic for the yellow peach moth *Dichocrocis punctiferalis*. $Bp_{0.65}$ 110-111°. $n_D^{27.5}$ 1.4440.

(Z)-form [64470-32-2]
Sex pheromone. $Bp_{0.6}$ 109-110°. n_D^{27} 1.4505 (⩽ 95% (*Z*)-form).
Ac: [35835-80-4]. *8-Tetradecen-1-yl acetate*
$C_{16}H_{30}O_2$ M 254.412
Component of the sex pheromone of several lepidoptera species incl. *Planofortix excessansa, Ctenopseustis obliquana* and *Spilonata ocellana*. $Bp_{0.4}$ 108-112°. n_D^{23} 1.4453 (⩽ 95% (*Z*)-form).
Formyl: [129731-28-8].
Mimic for *D. punctiferalis*. $Bp_{0.6}$ 106-107°. n_D^{23} 1.4411.

[56218-64-5]

Horiike, M. *et al, Agric. Biol. Chem.*, 1980, **44**, 257 (*synth, pmr, ir, ms*)

Young, H. *et al, Z. Naturforsch., C*, 1985, **40**, 262 (*isol*)
Mori, K. *et al, Justus Liebigs Ann. Chem.*, 1990, 1257 (*synth, pmr*)
Horiike, M. *et al, Org. Mass Spectrom.*, 1990, **25**, 329 (*ms*)

9-Tetradecen-1-ol — T-00092

$H_3C(CH_2)_3CH{=}CH(CH_2)_7CH_2OH$

$C_{14}H_{28}O$ M 212.375

(E)-form [52957-16-1]
The alcohol and its formate, acetate and propionate show pheromone activity.
Ac:
$C_{16}H_{30}O_2$ M 254.412
Sex pheromone for *Ephestia cautella* and *Spodoptera sunia*.

(Z)-form [35153-15-2]
The alcohol and its formate, acetate and propionate show pheromone activity. $Bp_{0.01}$ 101-105°.
Formyl: [56218-79-2].
$C_{15}H_{28}O_2$ M 240.385
$Bp_{0.01}$ 97-100°.
Ac: [16725-53-4].
$Bp_{0.01}$ 102-104°.
Propanoyl: [56218-86-1].
$C_{17}H_{32}O_2$ M 268.439
$Bp_{0.04}$ 108-110°.

Tamaki, Y. *et al, Appl. Entomol. Zool.*, 1971, **6**, 139.
Bestmann, H.J. *et al, Chem. Ber.*, 1975, **108**, 3582 (*synth*)
Leznoff, C.C. *et al, Can. J. Chem.*, 1977, **55**, 1143 (*synth*)
Rossi, R. *et al, Tetrahedron*, 1982, **38**, 639 (*cmr*)
Michelot, D., *Synthesis*, 1983, 130 (*synth*)
Barrer, P.M. *et al, J. Chem. Ecol.*, 1987, **13**, 639.
Bestmann, H.J. *et al, J. Chem. Ecol.*, 1988, **14**, 683.
Mitra, R.B. *et al, Synthesis*, 1989, 694 (*synth, ir, pmr*)

10-Tetradecen-1-ol — T-00093

$H_3CCH_2CH_2CH{=}CH(CH_2)_8CH_2OH$

$C_{14}H_{28}O$ M 212.375
Pheromone of lepidoptera. Oil.
Ac: Pheromone of lepidoptera. Oil. n_D^{25} 1.4492.

(Z)-form [57393-02-9]
Ac: [35153-16-3].
Major sex attractant of oak leaf roller *Archips semiferanus*.

Hendry, L.B. *et al, J. Chem. Ecol.*, 1975, **1**, 317 (*synth*)
Sonnet, P.E. *et al, J. Org. Chem.*, 1976, **41**, 3279.
Leznoff, C.C. *et al, Can. J. Chem.*, 1977, **55**, 1143 (*synth, ir, pmr, ms*)
Fyles, T.M. *et al, Can. J. Chem.*, 1978, **56**, 1031 (*synth*)

11-Tetradecen-1-ol — T-00094

$H_3CCH_2CH{=}CH(CH_2)_9CH_2OH$

$C_{14}H_{28}O$ M 212.375

(E)-form
Liq. $Bp_{0.03}$ 93-94°.
Ac:
$C_{16}H_{30}O_2$ M 254.412
Sex pheromone for European corn borer *Ostrinia nubitalis* and redbanded leafroller moth *Argyrotaenia velutinana*. Liq. $Bp_{0.04}$ 85-86°.

(Z)-form [34010-15-6]
The alcohol and some of its esters show pheromone activity. Oil.

Roelofs, W.L. *et al, Nature (London)*, 1970, **226**, 1172.

Tamaki, Y. *et al*, *Appl. Entomol. Zool.*, 1971, **6**, 139.
Leznoff, C.C. *et al*, *Can. J. Chem.*, 1975, **55**, 1143 (*synth*)
Brown, H.C. *et al*, *J. Org. Chem.*, 1986, **51**, 5282 (*synth, ir, pmr, cmr*)
Glover, T.J. *et al*, *J. Chem. Ecol.*, 1987, **13**, 143.
Jurenka, R.A. *et al*, *Insect Biochem.*, 1991, **21**, 81.
Moiseenkov, A.M. *et al*, *J. Chem. Soc., Perkin Trans. 1*, 1991, 2639 (*synth, pmr*)

12-Tetradecen-1-ol, 9CI — T-00095

[129189-60-2]

$H_3CCH{=}CH(CH_2)_{10}CH_2OH$

$C_{14}H_{28}O$ M 212.375

(E)-form [70711-49-8]

Minor component of the sex pheromone of the apple ermine moth, *Yponomeuta malinellus*.

Ac: [35153-21-0]. *12-Tetradecen-1-yl acetate. 14-Acetoxy-2-tetradecene*

$C_{16}H_{30}O_2$ M 254.412

Component of the sex pheromones of the Asian corn borer *Ostrinia furnacalis* and European corn borer, *O. nubilalis*, and a synthetic sex attractant for the small ermine moth, *Yponomeuta rorellus*.

(Z)-form [70711-48-7]

Ac: [35153-20-9].

Synthetic sex attractant for the small ermine moth *Y. rorellus*.

Ando, T. *et al*, *Agric. Biol. Chem.*, 1980, **44**, 2643 (*synth, glc, ms*)
Kang, S. *et al*, *Bull. Korean Chem. Soc.*, 1985, **6**, 15; *CA*, **102**, 182820x (*synth, activity*)
Koumaglo, K., *Tetrahedron Lett.*, 1986, **27**, 883 (*synth*)
Chan, T.H. *et al*, *Tetrahedron Lett.*, 1986, **27**, 883 (*synth*)
Yeh, S.F. *et al*, *J. Chem. Ecol.*, 1989, **15**, 497 (*activity*)
McDonough, L.M. *et al*, *J. Chem. Ecol.*, 1990, **16**, 477 (*activity*)
Horiike, M. *et al*, *Org. Mass Spectrom.*, 1990, **25**, 329 (*ms*)
Loefstedt, C. *et al*, *Physiol. Entomol.*, 1990, **15**, 47 (*activity*)

5-Tetradecen-13-olide — T-00096

14-Methyloxacyclotetradec-6-en-2-one, 9CI

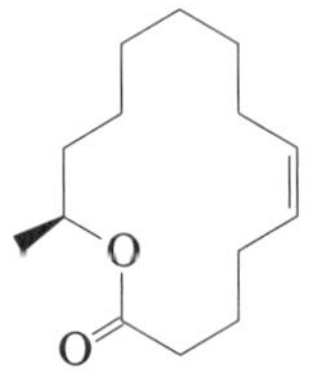

(S,Z)-form

$C_{14}H_{24}O_2$ M 224.342

(R,Z)-form [87420-70-0]

Synthetic. $[\alpha]_D^{35}$ −46.4° (c, 2.19 in $CHCl_3$).

(S,Z)-form [87420-69-7]

Synergist of aggregation pheromone of *Cryptolestes pusillus*. Oil. $[\alpha]_D^{23}$ +54.4° (c, 1.275 in $CHCl_3$), $[\alpha]_D^{32}$ +49.6° (c, 4.62 in $CHCl_3$).

(Z)-form [78418-66-3] ***Phoracantholide O***

Constit. of the metasternal secretion of *Phoracantha synonyma*. $Bp_{0.01}$ 92°.

[77761-59-2]

Moore, B.P. *et al*, *Aust. J. Chem.*, 1976, **29**, 1365 (*isol*)
Gerlach, H. *et al*, *Helv. Chim. Acta*, 1980, **63**, 2312 (*synth*)
Millar, J.G. *et al*, *J. Org. Chem.*, 1983, **48**, 4404 (*synth, ir, pmr, ms*)
Millar, J.G. *et al*, *J. Chem. Ecol.*, 1985, **11**, 1053 (*synth*)
Sakai, T. *et al*, *Agric. Biol. Chem.*, 1986, **50**, 1621 (*synth*)
Naoshima, Y. *et al*, *Chem. Lett.*, 1989, 1023 (*synth, isol*)

5-Tetradecen-7-yne — T-00097

$H_3C(CH_2)_3CH{=}CHC{\equiv}C(CH_2)_5CH_3$

$C_{14}H_{24}$ M 192.344

(Z)-form [94427-82-4]

Oil. Bp_1 86-87°. Spectroscopic props. of *E*-form also given.

[73349-17-4]

Miyaura, N. *et al*, *J. Am. Chem. Soc.*, 1985, **107**, 972 (*synth, ir, pmr, ms*)

9-Tetradecen-11-yn-1-ol — T-00098

$H_3CCH_2C{\equiv}CCH{=}CH(CH_2)_7CH_2OH$

$C_{14}H_{24}O$ M 208.343

(Z)-form

Ac: [54665-02-0].

$C_{16}H_{26}O_2$ M 250.380

$Bp_{0.05}$ 95-100°.

Bestmann, H.J. *et al*, *Justus Liebigs Ann. Chem.*, 1981, 2117 (*synth, pmr*)

2-Tetradecyl-1,4-benzenediol, 9CI — T-00099

Tetradecylquinol. Myristylquinol. Tetradecylhydroquinone. 2,5-Dihydroxytetradecylbenzene

[50677-17-3]

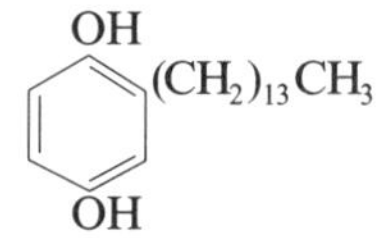

$C_{20}H_{34}O_2$ M 306.487

Cryst. (pet. ether). Mp 110°.

Di-Me ether:

$C_{22}H_{38}O_2$ M 334.541

$Bp_{0.5}$ 165°.

Di-Et ether:

$C_{24}H_{42}O_2$ M 362.595

$Bp_{0.1}$ 183°.

Quinone: Tetradecylbenzoquinone

Cryst. (EtOH/Et_2O). Mp 77.5°.

Cook, A.H. *et al*, *J. Chem. Soc.*, 1942, 659 (*synth*)

2-Tetradecylcyclopropaneacetic acid, 9CI — T-00100

3,4-Methyleneoctadecanoic acid

$C_{19}H_{36}O_2$ M 296.492

Both *cis*- and *trans*-forms have been synthesised as Me esters.

Christie, W.W. *et al*, *Chem. Phys. Lipids*, 1968, **2**, 196 (*synth, glc, ir, nmr*)
Christie, W.W., *Top. Lipid Chem.*, 1970, **1**, 4 (*rev*)
Gunstone, F.D., *Chem. Phys. Lipids*, 1973, **10**, 303 (*synth, glc, nmr*)

1-Tetradecylcyclopropanecarboxylic acid — T-00101

[70858-10-5]

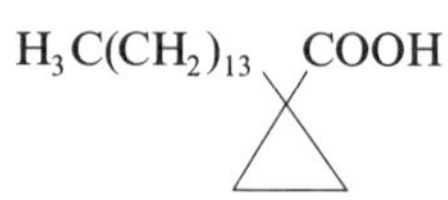

$C_{18}H_{34}O_2$ M 282.465
Cryst. (MeOH). Mp 53-55°.

Ho, W. *et al*, *J. Med. Chem.*, 1986, **29**, 2184 (*synth*)

3-Tetradecylphenol T-00102

[90866-51-6]

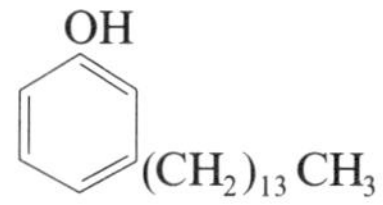

$C_{20}H_{34}O$ M 290.488
Cryst. Mp 38-40°.

Wissner, A. *et al*, *J. Med. Chem.*, 1984, **27**, 1174 (*synth, ir, pmr*)

4-Tetradecylphenol T-00103

[25401-89-2]

$C_{20}H_{34}O$ M 290.488
Mp 70-71°.

Wissner, A. *et al*, *J. Med. Chem.*, 1984, **27**, 1174 (*synth, ir, pmr*)

3-Tetradecynoic acid, 9CI T-00104

[55182-76-8]

$H_3C(CH_2)_9C{\equiv}CCH_2COOH$

$C_{14}H_{24}O_2$ M 224.342
Mp 58-60°.

Gilman, N.W. *et al*, *Chem. Phys. Lipids*, 1974, **13**, 239 (*synth*)
Gunstone, F.D. *et al*, *Chem. Phys. Lipids*, 1976, **17**, 1 (*cmr*)
Sprecher, H., *Prog. Chem. Fats Other Lipids*, 1977, **15**, 219.

4-Tetradecynoic acid, 9CI T-00105

[55182-78-0]

$H_3C(CH_2)_8C{\equiv}CCH_2CH_2COOH$

$C_{14}H_{24}O_2$ M 224.342
Mp 59-60°.

Gilman, N.W. *et al*, *Chem. Phys. Lipids*, 1974, **13**, 239 (*synth*)
Gunstone, F.D. *et al*, *Chem. Phys. Lipids*, 1976, **17**, 1 (*cmr*)
Sprecher, H., *Prog. Chem. Fats Other Lipids*, 1977, **15**, 239.

5-Tetradecynoic acid, 9CI T-00106

[55182-81-5]

$H_3C(CH_2)_7C{\equiv}C(CH_2)_3COOH$

$C_{14}H_{24}O_2$ M 224.342
Cryst. (MeOH aq.). Mp 34-36°.

Baker, B.W. *et al*, *J. Chem. Soc.*, 1955, 2218 (*synth*)
Williams, J.C. *et al*, *Chem. Phys. Lipids*, 1972, **9**, 320 (*pmr*)
Gilman, N.W. *et al*, *Chem. Phys. Lipids*, 1974, **13**, 239 (*synth*)
Gunstone, F.D. *et al*, *Chem. Phys. Lipids*, 1976, **17**, 1 (*cmr*)
Sprecher, H., *Prog. Chem. Fats Other Lipids*, 1977, **15**, 219.

6-Tetradecynoic acid, 9CI T-00107

[55182-84-8]

$H_3C(CH_2)_6C{\equiv}C(CH_2)_4COOH$

$C_{14}H_{24}O_2$ M 224.342
$Bp_{0.3}$ 139-144°.

Gilman, N.W. *et al*, *Chem. Phys. Lipids*, 1974, **13**, 239 (*synth*)
Gunstone, F.D. *et al*, *Chem. Phys. Lipids*, 1976, **17**, 1 (*cmr*)
Sprecher, H., *Prog. Chem. Fats Other Lipids*, 1977, **15**, 219 (*props*)

7-Tetradecynoic acid, 9CI T-00108

[55182-86-0]

$H_3C(CH_2)_5C{\equiv}C(CH_2)_5COOH$

$C_{14}H_{24}O_2$ M 224.342
Mp 29.5-30°. $Bp_{0.4}$ 135-139°.

Gilman, N.W. *et al*, *Chem. Phys. Lipids*, 1974, **13**, 239 (*synth*)
Gunstone, F.D. *et al*, *Chem. Phys. Lipids*, 1976, **17**, 1 (*cmr*)
Sprecher, H., *Prog. Chem. Fats Other Lipids*, 1977, **15**, 219.

8-Tetradecynoic acid, 9CI T-00109

[55182-90-6]

$H_3C(CH_2)_4C{\equiv}C(CH_2)_6COOH$

$C_{14}H_{24}O_2$ M 224.342
Mp 22-22.5°. $Bp_{0.15}$ 150-153°.

Baker, C.D. *et al*, *J. Chem. Soc.*, 1963, 489 (*synth*)
Gilman, N.W. *et al*, *Chem. Phys. Lipids*, 1974, **17**, 239 (*synth*)
Gunstone, F.D. *et al*, *Chem. Phys. Lipids*, 1976, **17**, 1 (*cmr*)
Sprecher, H., *Prog. Chem. Fats Other Lipids*, 1977, **15**, 219 (*rev*)
Huhtasaari, M. *et al*, *Acta Chem. Scand., Ser. B*, 1983, **37**, 537 (*synth, pmr, ir, ms*)

9-Tetradecynoic acid, 9CI T-00110

[55182-92-8]

$H_3C(CH_2)_3C{\equiv}C(CH_2)_7COOH$

$C_{14}H_{24}O_2$ M 224.342
Mp 35-38° (31.5-32°). $Bp_{0.15}$ 135-138°.

Ames, D.E. *et al*, *J. Chem. Soc.*, 1963, 775 (*synth*)
Gilman, N.W. *et al*, *Chem. Phys. Lipids*, 1974, **13**, 239 (*synth*)
Sprecher, H., *Prog. Chem. Fats Other Lipids*, 1977, **15**, 219.
Coppola, G.M., *J. Heterocycl. Chem.*, 1985, **22**, 491 (*synth*)

10-Tetradecynoic acid, 9CI T-00111

[54373-84-1]

$H_3CCH_2CH_2C{\equiv}C(CH_2)_8COOH$

$C_{14}H_{24}O_2$ M 224.342
$Bp_{0.2}$ 135°.

Gilman, N.W. *et al*, *Chem. Phys. Lipids*, 1974, **13**, 239 (*synth*)
Gilman, N.W. *et al*, *Synth. Commun.*, 1974, **4**, 199 (*synth*)
Sprecher, H., *Prog. Chem. Fats Other Lipids*, 1977, **15**, 219.

11-Tetradecynoic acid, 9CI T-00112

[55182-96-2]

$H_3CCH_2C{\equiv}C(CH_2)_9COOH$

$C_{14}H_{24}O_2$ M 224.342
Mp 47-49°.

Gilman, N.W. *et al*, *Chem. Phys. Lipids*, 1974, **13**, 239 (*synth*)
Sprecher, H., *Prog. Chem. Fats Other Lipids*, 1977, **15**, 219.

12-Tetradecynoic acid, 9CI T-00113

[55182-97-3]

$H_3CC{\equiv}C(CH_2)_{10}COOH$

$C_{14}H_{24}O_2$ M 224.342
Mp 55-57°.

Gilman, N.W. *et al*, *Chem. Phys. Lipids*, 1974, **13**, 239 (*synth*)
Sprecher, H., *Prog. Chem. Fats Other Lipids*, 1977, **15**, 219.

13-Tetradecynoic acid T-00114

$HC{\equiv}C(CH_2)_{11}COOH$

$C_{14}H_{24}O_2$ M 224.342
Cryst. Mp 53-55°.

Christie, W.W. *et al*, *Chem. Phys. Lipids*, 1967, **1**, 407 (*synth*)
Hébert, N. *et al*, *J. Org. Chem.*, 1992, **57**, 1777 (*synth, pmr, cmr*)

2-Tetradecyn-1-ol T-00115

[51309-22-9]

$H_3C(CH_2)_{10}C{\equiv}CCH_2OH$

$C_{14}H_{26}O$ M 210.359
Cryst. (EtOH aq.). Mp 40-42°.

Ames, D.E., *J. Chem. Soc.*, 1963, 5889 (*synth*)

7-Tetradecyn-1-ol T-00116

[37011-94-2]

$H_3C(CH_2)_5C{\equiv}C(CH_2)_5CH_2OH$

$C_{14}H_{26}O$ M 210.359
$Bp_{0.01}$ 117-118°.

Voaden, D.J. *et al*, *J. Med. Chem.*, 1972, **15**, 619 (*synth*)
Brown, H.C. *et al*, *J. Org. Chem.*, 1986, **51**, 4518 (*synth, ir, pmr, cmr*)

8-Tetradecyn-1-ol T-00117

[89216-52-4]

$H_3C(CH_2)_4C{\equiv}C(CH_2)_6CH_2OH$

$C_{14}H_{26}O$ M 210.359
Liq. $Bp_{0.03}$ 120°.

Huhtasaari, M. *et al*, *Acta Chem. Scand., Ser. B*, 1983, **37**, 537 (*synth, pmr, ir, ms*)

13-Tetradecyn-1-ol T-00118

[18202-12-5]

$HC{\equiv}C(CH_2)_{11}CH_2OH$

$C_{14}H_{26}O$ M 210.359
Cryst. Mp 34-36°. $Bp_{0.5}$ 112-115°.

Christie, W.W. *et al*, *Chem. Phys. Lipids*, 1967, **1**, 407 (*synth*)
Hébert, N. *et al*, *J. Org. Chem.*, 1992, **57**, 1777 (*synth, pmr, cmr*)

3,3,4,4-Tetrafluorocyclobutene, 9CI T-00119

[2714-38-7]

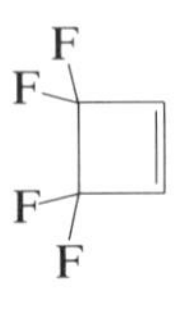

$C_4H_2F_4$ M 126.053
Liq. d_4^{20} 1.358. Bp 54-56°. n_D^{25} 1.3086.

Anderson, J.L. *et al*, *J. Am. Chem. Soc.*, 1961, **83**, 382 (*synth*)
Park, J.D. *et al*, *J. Org. Chem.*, 1963, **28**, 1008 (*synth*)
Harris, R.K. *et al*, *J. Magn. Reson.*, 1969, **1**, 362 (*F nmr*)

Tetrahectane T-00120

[84451-54-7]

$H_3C(CH_2)_{102}CH_3$

$C_{104}H_{210}$ M 1460.803
Mp 115°.

Paynter, O.I. *et al*, *J. Chem. Soc., Chem. Commun.*, 1982, 1165 (*synth*)

5-[Tetrahydro-4,6-dihydroxy-2-(3-hydroxy-1-octenyl)-2*H*-pyran-3-yl]-3-pentenoic acid T-00121

Dinorthromboxane B_2
[63250-09-9]

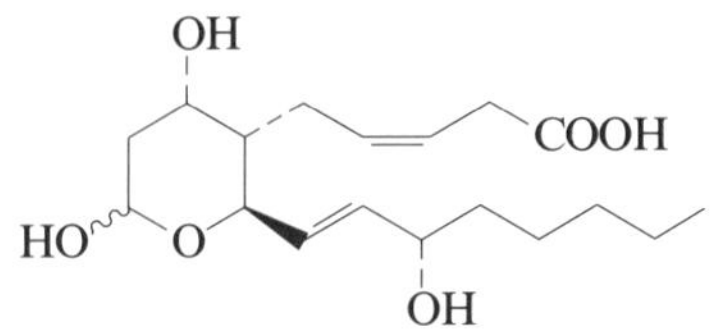

$C_{18}H_{30}O_6$ M 342.431
Major human urinary thromboxane metab.
[63635-82-5, 103190-15-4]

Robers, L.J. *et al*, *J. Biol. Chem.*, 1977, **252**, 7415; 1981, **256**, 8384 (*biosynth*)
Nelson, N.A. *et al*, *Prostaglandins*, 1978, **16**, 85 (*synth, Me ester*)
Patrignani, P. *et al*, *Biochim. Biophys. Acta*, 1989, **992**, 71 (*biosynth*)

Tetrahydro-2,5-furandiacetic acid, 9CI T-00122

3,6-Epoxyoctanedioic acid

(2RS,5RS)-form

$C_8H_{12}O_5$ M 188.180

(2RS,5RS)-form [92520-46-2]
(±)-trans-*form*
Component of human urine (excretion ~3-5 mg/day), poss. in an enantiomeric form. Higher homologues occur in trace amounts in urine.

(2RS,5SR)-form [51250-33-0]
cis-*form*
Minor consit. of urine (~10% of *trans*-isomer). Cryst. (cyclohexane/AcOH). Mp 92-96°.
Di-Me ester: [92520-54-2].
$C_{10}H_{16}O_5$ M 216.233
$Bp_{0.01}$ 211°. Cont. 4% *trans*-isomer.
[92520-55-3]

Rau, O. *et al*, *Justus Liebigs Ann. Chem.*, 1984, 1504, 1513 (*isol, synth, pmr, ms*)

Tetrahydro-5-(1-hydroxyheptyl)heptanoic acid, 9CI T-00123

6,9-Epoxy-10-hydroxyhexadecanoic acid

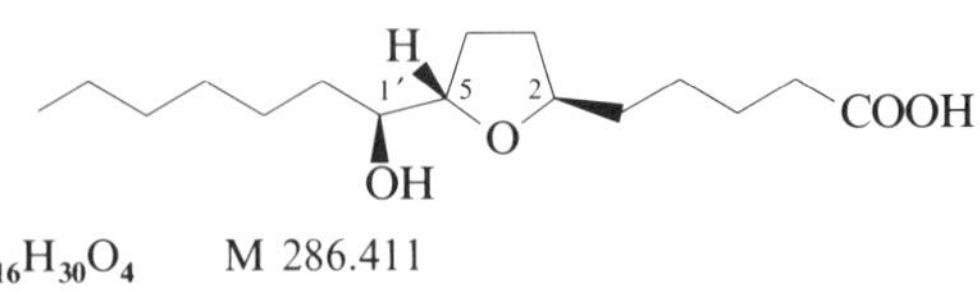

$C_{16}H_{30}O_4$ M 286.411

(2R,5S,1′S)-form [34499-28-0]
Constit. of wool fat. Mp 43.5°. $[\alpha]_D^{18}$ +10.8° (neat).
Me ester: $[\alpha]_D$ +9.8° (neat).

Ito, S. *et al*, *Tetrahedron Lett.*, 1971, 4011 (*isol, struct*)

5-[Tetrahydro-4-hydroxy-2-(3-hydroxy-1-octenyl)-6-oxo-2*H*-pyran-3-yl]-3-pentenoic acid T-00124

2,3-Dinor-11-dehydrothromboxane B_2

[79250-60-5]

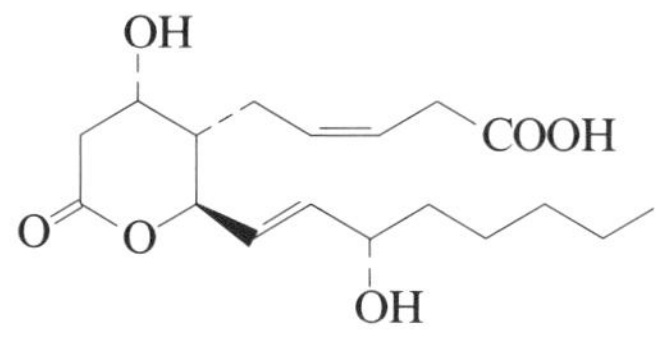

$C_{18}H_{28}O_6$ M 340.416

Urinary metab. of thromboxane B_2.

Roberts, L.J. *et al*, *J. Biol. Chem.*, 1981, **256**, 8384 (*biosynth*)

Tetrahydro-6-(3-hydroxy-4,7-tridecadienyl)-2*H*-pyran-2-one, 9CI T-00125

5,8-Dihydroxy-9,12-octadecadienoic acid δ lactone. Psi Aα

[129926-99-4]

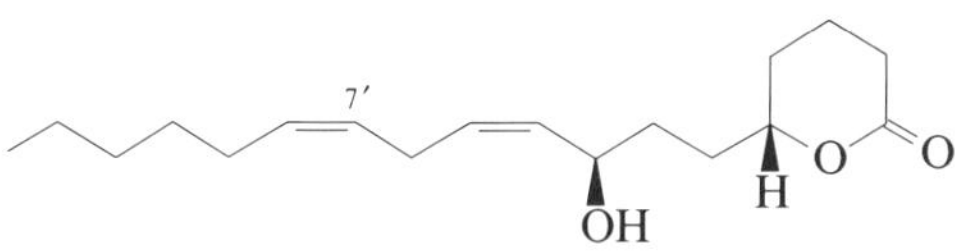

$C_{18}H_{30}O_3$ M 294.433

Metab. of *Aspergillus nidulans*. Sporogenic agent. $[\alpha]_D^{30}$ +68.3° (c, 0.0041 in MeCN).

7′,8′-Dihydro: [129927-00-0]. *Tetrahydro-6-(3-hydroxy-4-tridecenyl)-*2H*-pyran-2-one, 9CI.* ***Psi Aβ***

$C_{18}H_{32}O_3$ M 296.449

Metab. of *A. nidulans*. Sporogenic agent. $[\alpha]_D^{30}$ +63.7° (c, 0.0042 in MeCN).

Champe, S. *et al*, *J. Bacteriol.*, 1989, **171**, 3982 (*isol*)
Mazur, P. *et al*, *Tetrahedron Lett.*, 1990, **31**, 3837 (*struct, abs config, pmr, cmr*)
Mazur, P. *et al*, *J. Org. Chem.*, 1992, **57**, 1047 (*synth*)

Tetrahydro-6-(1-hydroxyundecyl)-2*H*-pyran-2-one, 9CI T-00126

6-Hydroxy-5-hexadecanolide

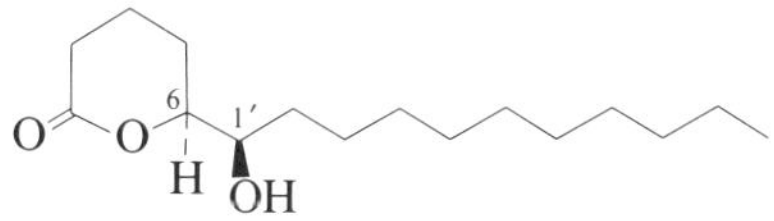

(6*S*,1′*R*)-*form*

$C_{16}H_{30}O_3$ M 270.411

(6S,1′R)-form [89102-36-3]

Cryst. (hexane). Mp 66.5-68°. $[\alpha]_D^{20}$ +12.7° (c, 0.96 in $CHCl_3$).

Ac: [85551-39-9].

$C_{18}H_{32}O_4$ M 312.448

$[\alpha]_D^{21.5}$ +38.8° (c, 1.2 in $CHCl_3$).

(6R,1′S)-form [89102-35-2]

Mp 67-68°. $[\alpha]_D^{20}$ −12.5° (c, 0.54 in $CHCl_3$).

Ac: [85551-42-4].

$C_{18}H_{32}O_4$ M 312.448

Major component of the oviposition attractant pheromone of the mosquito *Culex pipiens fatigans*. $[\alpha]_D^{21.5}$ −38.5° (c, 0.51 in $CHCl_3$).

Laurence, B.R. *et al*, *J. Chem. Soc., Chem. Commun.*, 1982, 59 (*isol, synth*)
Mori, K. *et al*, *Tetrahedron*, 1983, **39**, 3267 (*synth*)
Hwang, Y.-H. *et al*, *J. Chem. Ecol.*, 1987, **13**, 245.
Wang, Z.M. *et al*, *Tetrahedron*, 1990, **46**, 1191 (*synth*)
Ramaswamy, S. *et al*, *Tetrahedron*, 1991, **47**, 1145 (*synth, bibl*)

7,7′,8,8′-Tetrahydrolycopene T-00127

7,7′,8,8′-Tetrahydro-ψ,ψ-carotene. 2,6,10,14,19,23,27,31-Octamethyl-2,6,10,12,14,16,18,20,22,26,30-dotriacontaundecaene. ξ-Carotene

[13587-06-9]

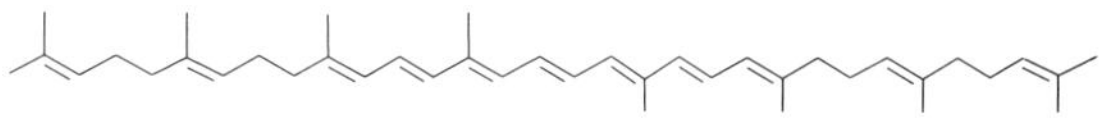

$C_{40}H_{60}$ M 540.914

Constit. of carrot oil and many other natural products. Cryst. ($CHCl_3$/MeOH). Mp 42-46°. λ_{max} 360, 380, 401, 425 nm.

(9Z,9′Z)-form [72746-33-9]

Isol. from *Lycopersicon esculentum* var. 'Tangella'. Viscous orange oil. λ_{max} 284, 294, 361, 379, 401, 427 nm (hexane).

Davis, J.B. *et al*, *J. Chem. Soc. C*, 1966, 2154 (*pmr, uv, ir, bibl*)
Davies, B.H. *et al*, *Phytochemistry*, 1974, **13**, 1209 (*ms*)
Clough, J.M. *et al*, *J. Chem. Soc., Perkin Trans. 1*, 1983, 3011 (*cmr*)
Frecknell, E.A. *et al*, *Phytochemistry*, 1984, **23**, 1707 (*isol, occur*)
Straub, O. *et al*, *Key to Carotenoids*, 2nd Ed., Birkhauser Verlag, Basel and Boston, 1987, 38 (*bibl*)

Tetrahydro-2-(1,7-nonadiene-3,5-diynyl)-2*H*-pyran-3-ol, 9CI T-00128

1,5-Epoxy-6,12-tetradecadiene-8,10-diyn-4-ol

[55661-24-0]

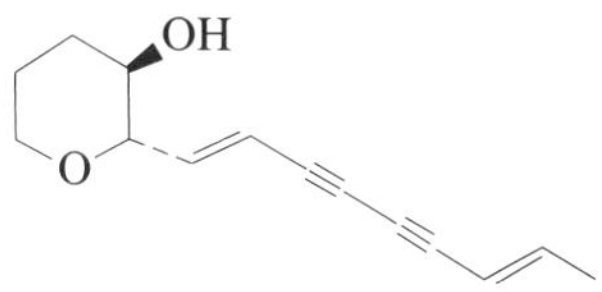

$C_{14}H_{16}O_2$ M 216.279

Constit. of the plant *Dahlia scapigera* and many *Bidens* spp. Mp 64-65°.

Ac: [59981-61-2].

$C_{16}H_{18}O_3$ M 258.316

Constit. of *D. scapigera* and *Matricaria suffructicosa*. Mp 75.5-76.5°. $[\alpha]_D^{22}$ +17° (c, 0.5 in CCl_4).

[55297-88-6]

Chin, C. *et al*, *J. Chem. Soc. C*, 1970, 314 (*isol, ir, uv, pmr, ms*)
Lam, J. *et al*, *Phytochemistry*, 1971, **10**, 1877 (*isol*)
Bohlmann, F. *et al*, *Chem. Ber.*, 1975, **108**, 437 (*isol, struct*)

Tetrahydro-2-nonyl-3-pyranol T-00129

2-Nonyl-3-tetrahydropyranol

[26130-98-3]

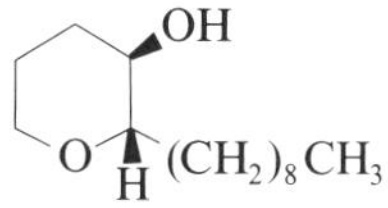

Absolute configuration

$C_{14}H_{28}O_2$ M 228.374

(2S,3R)-form

$Bp_{0.2}$ 140°. $[\alpha]_D$ −42° (c, 0.5 in CCl_4).

Ac: [26130-97-2].

$C_{16}H_{30}O_3$ M 270.411

$Bp_{0.3}$ 120°. $[\alpha]_D$ −44° (c, 0.8 in CCl_4).

Chin, C. *et al*, *J. Chem. Soc. C*, 1970, 314 (*synth*)

Tetrahydro-6-pentyl-2H-pyran-2-one, 9CI T-00130

δ-Decalactone. 5-Decanolide. 5-Hydroxydecanoic acid lactone

[705-86-2]

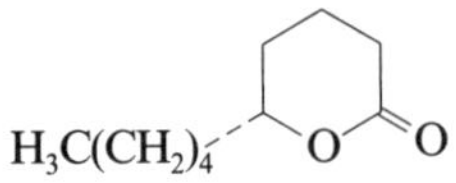

(R)-form

$C_{10}H_{18}O_2$ M 170.251

Produced by *Sporobolomyces odorus* and *Cladosporium suaveolens* cultures. Present in coconut oil, butter oil, apricots, peaches, cheese, cane sugar, pork fat, margarine, lavender oil, and other substances. Fragrance raw material. Shows antibacterial activity against *Streptococcus faecalis*.

▷ UQ1355000.

(R)-form [2825-91-4]

Liq. Bp_5 160-170°. $[\alpha]_D$ +48.2° (c, 3.69 in MeOH), $[\alpha]_D$ +60.2° (c, 1.74 in THF).

(S)-form [59285-67-5]

Liq. $[\alpha]_D$ −60.2°.

(±)-form [35221-79-5]

Liq. Mp −27°. $Bp_{0.2}$ 100-105°, Bp_3 121-122°. n_D^{25} 1.4550 (1.4675).

Meiyer, T.M., *Recl. Trav. Chim. Pays-Bas (J. R. Neth. Chem. Soc.)*, 1940, **59**, 191 (*synth*)
Crombie, L., *J. Chem. Soc.*, 1955, 1007 (*synth*)
Allen, R.R., *Chem. Ind. (London)*, 1965, 1560 (*isol, ir*)
Parliemerit, T.H. *et al*, *Chem. Ind. (London)*, 1966, 1845 (*synth, ir*)
Lardelli, G. *et al*, *Recl. Trav. Chim. Pays-Bas (J. R. Neth. Chem. Soc.)*, 1967, **86**, 481 (*synth*)
Kohver, O., *Tetrahedron*, 1970, **26**, 2391 (*abs config, ord, cd*)
Ijima, A. *et al*, *Chem. Pharm. Bull.*, 1971, **19**, 1053; 1972, **20**, 197; 1973, **21**, 215 (*synth*)
Tahara, S. *et al*, *Agric. Biol. Chem.*, 1975, **39**, 281 (*isol*)
Pyysalo, H. *et al*, *Finn. Chem. Lett.*, 1975, 129, 136 (*synth, cmr, ms*)
Opdyke, D.L.J., *Food Cosmet. Toxicol.*, 1976, **14**, 739 (*rev, tox*)
Kaiser, R. *et al*, *Tetrahedron Lett.*, 1976, 1659 (*isol, struct*)
Bacardit, R. *et al*, *J. Chem. Ecol.*, 1983, **9**, 703 (*synth, props*)
Griese, B. *et al*, *Chem. Ber.*, 1984, **117**, 859 (*synth*)
Francotte, E. *et al*, *Helv. Chim. Acta*, 1987, **70**, 1569 (*resoln*)
Utaka, M. *et al*, *J. Org. Chem.*, 1987, **52**, 4363 (*synth*)
Gessner, M. *et al*, *Z. Naturforsch., C*, 1987, **42**, 1159 (*synth*)
Mosandl, A. *et al*, *Z. Lebensm.-Unters. -Forsch.*, 1988, **187**, 40 (*synth, props*)
Inokouchi, T. *et al*, *J. Org. Chem.*, 1990, **55**, 462 (*synth*)
Fronza, G. *et al*, *J. Chem. Soc., Perkin Trans. 1*, 1991, 2977 (*synth*)
Cardillo, R. *et al*, *J. Org. Chem.*, 1991, **56**, 5237 (*synth*)
Albrecht, W. *et al*, *J. Org. Chem.*, 1992, **57**, 1954 (*synth*)
Lewis, R.J., *Sax's Dangerous Properties of Industrial Materials*, 8th Ed., Van Nostrand-Reinhold, 1992, DAF200.

Tetrahydro-6-undecyl-2H-pyran-2-one, 9CI T-00131

5-Hexadecanolide. δ-Hexadecanolactone

[7370-44-7]

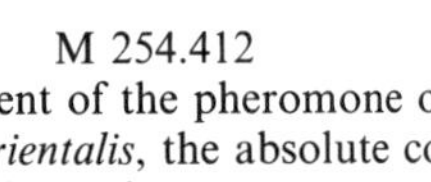

(R)-form

$C_{16}H_{30}O_2$ M 254.412

A component of the pheromone of the oriental hornet *Vespa orientalis*, the absolute config. of the natural product is not known.

(R)-form [59812-96-3]

Mp 40-41°. $[\alpha]_D^{20}$ +39.97° (c, 1 in THF).

(S)-form [59812-97-4]

Mp 40-41°. $[\alpha]_D^{20}$ −39.2° (c, 1 in THF).

Servi, S., *Tetrahedron Lett.*, 1983, **24**, 2023 (*synth, abs config, bibl*)
Kikukawa, T. *et al*, *Chem. Lett.*, 1984, 1935 (*synth*)
Larcheveque, M. *et al*, *Tetrahedron*, 1984, **40**, 1061 (*synth*)
Mori, K. *et al*, *Tetrahedron*, 1985, **41**, 547 (*synth*)
Taber, D.F. *et al*, *J. Am. Chem. Soc.*, 1987, **109**, 7488 (*synth, pmr, cmr, ir, ms*)

9,10,12,13-Tetrahydroxydocosanoic acid T-00132

Ventosic acid

$H_3C(CH_2)_8CH(OH)CH(OH)CH_2CH(OH)CH(OH)(CH_2)_7COOH$

$C_{22}H_{44}O_6$ M 404.586

Struct. uncertain. Isol. from various lichens, *Parmelia* spp. *Haematomma ventosum*, *Cetraria* spp. etc. Mp 183-185°.

Tetra-Ac: Mp 54°.

Tetrakis-4-phenylazobenzoyl: Mp 153-157°.

Solberg, Y.J., *Acta Chem. Scand.*, 1957, **11**, 1477; 1960, **14**, 2152.

9,10,12,13-Tetrahydroxyoctadecanoic acid T-00133

9,10,12,13-Tetrahydroxystearic acid. Sativic acid. Sativinic acid

[541-82-2]

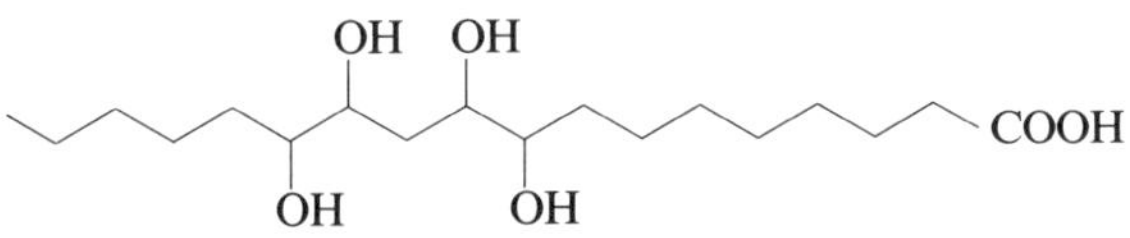

$C_{18}H_{36}O_6$ M 348.479

Found in plant oils as autoxidn. artifact of stearic acid. All stereoisomers prepd.

[17673-37-9, 17673-38-0, 17673-39-1, 17673-41-5, 17673-42-6, 17673-43-7, 17673-44-8, 17673-45-9, 17673-46-0, 17673-47-1, 17673-48-2, 18312-29-3, 18312-30-6, 18432-31-0]

Bharucha, K.E. *et al*, *J. Chem. Soc.*, 1956, 1611 (*synth*)
Christie, W.W. *et al*, *Lipids*, 1967, **2**, 286 (*stereochem*)
Khuddus, M.A. *et al*, *J. Am. Oil Chem. Soc.*, 1973, **50**, 524 (*synth*)

9,11,15,19-Tetrahydroxy-5,13-prostadienoic acid T-00134

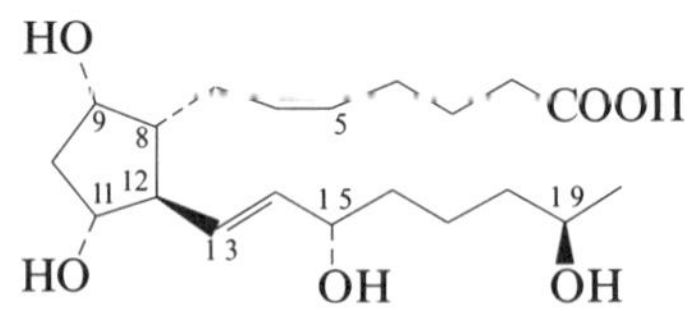

(5Z,9S,11R,13E,15S,19R)-form

$C_{20}H_{34}O_6$ M 370.485

(5Z,9S,11R,13E,15S,19R)-form [56932-60-6]

19R-Hydroxy-PGF$_{2\alpha}$

Occurs in human semen. Bronchodilator.

(5Z,8R,9S,11R,13E,15S,19R)-form [70287-57-9]

Occurs in human semen.

Taylor, P.L. *et al*, *FEBS Lett.*, 1975, **57**, 22.
U.K. Pat., 1 501 864, (1978); *CA*, **89**, 127756n (*synth*)
Taylor, P.L. *et al*, *Prostaglandins*, 1979, **17**, 259 (*ms*)

9,11,15,19-Tetrahydroxy-13-prostenoic acid T-00135

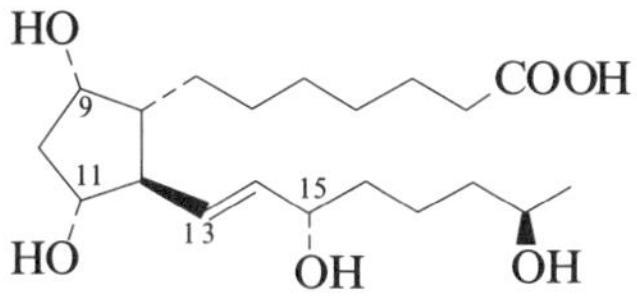

(9*S*,11*R*,13*E*,15*S*)-*form*

$C_{20}H_{36}O_6$ M 372.501

(9S,11R,13E,15S)-form [56932-59-3]

19-Hydroxy-PGF$_{1\alpha}$

Occurs in semen.

Taylor, P.L. *et al*, *FEBS Lett.*, 1975, **57**, 22.
Sih, J.C., *Prostaglandins*, 1977, **13**, 831 (*synth*)
Taylor, P.L. *et al*, *Prostaglandins*, 1979, **17**, 259.

2,4,6,8-Tetramethyldecanoic acid T-00136

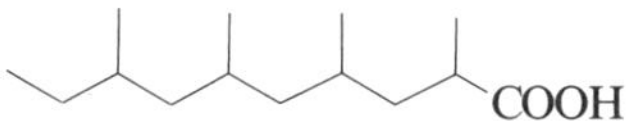

$C_{14}H_{28}O_2$ M 228.374

A main constit. of the preen gland wax of the common goose (*Anser anser*), also present in small amounts in the preen gland wax of other spp. $[\alpha]_D^{23}$ −16.4°, $[\alpha]_D^{23}$ −26.6°.

Me ester: Present in preen gland waxes of ruddy shelduck (*Tadorna ferruginea*) and common shelduck (*T. tadorna*). $[\alpha]_D^{23}$ −32°.

Murray, K.E., *Aust. J. Chem.*, 1962, **15**, 510 (*isol*)
Noble, R.E. *et al*, *Nature (London)*, 1963, **199**, 600 (*biosynth*)
Odham, G., *Ark. Kemi*, 1965, **23**, 431; 1966, **25**, 543; 1967, **27**, 295.
Smith, C.R., *Top. Lipid Chem.*, 1970, **1**, 307 (*rev, bibl*)
Scherrer, F. *et al*, *J. Chem. Soc., Chem. Commun.*, 1976, 127 (*ms*)

2,4,6,8-Tetramethyldotriacontanoic acid T-00137

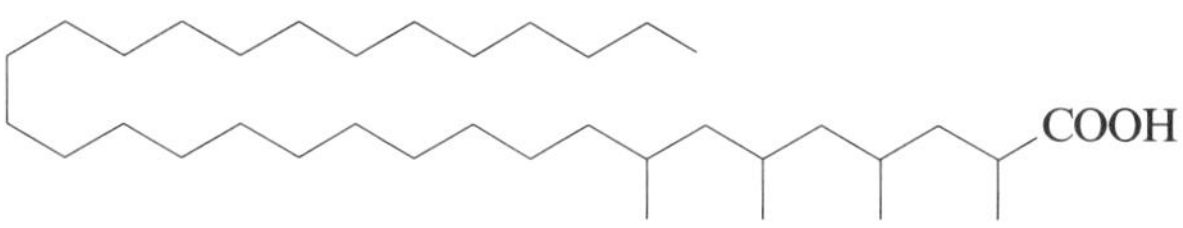

$C_{36}H_{72}O_2$ M 536.964

Component of Mycoceranic acid from tubercle bacilli.

Polgar, N. *et al*, *Chem. Ind. (London)*, 1961, 1958 (*rev*)
Polgar, N., *Top. Lipid Chem.*, 1971, **2**, 207 (*rev*)

2,4,6,8-Tetramethyltetratriacontanoic acid T-00138

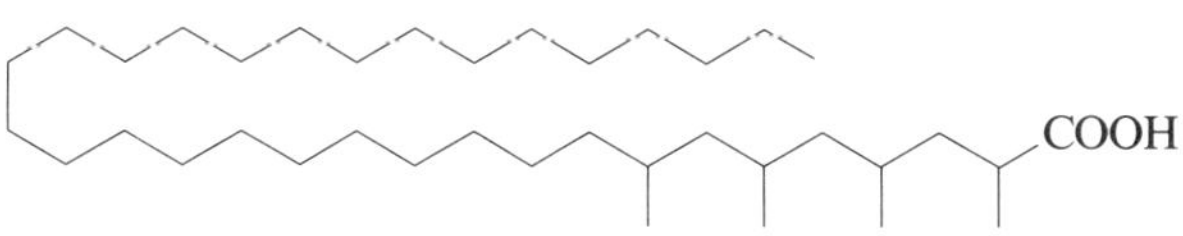

$C_{38}H_{76}O_2$ M 565.017

Constit. of Mycocerosic acid, a laevorotatory acid obt. from the lipids of tubercle bacilli.

Asselineau, C. *et al*, *Acta Chem. Scand.*, 1959, **13**, 822.
Polgar, N., *Top. Lipid Chem.*, 1971, **2**, 207 (*rev, bibl*)

2,4,6,8-Tetramethyltriacontanoic acid T-00139

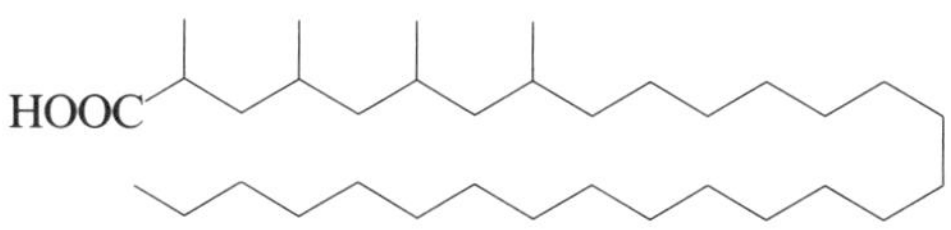

$C_{34}H_{68}O_2$ M 508.910

Component of Mycoceranic acid from tubercle bacilli.

Polgar, N. *et al*, *Chem. Ind. (London)*, 1961, 1958 (*rev*)
Polgar, N., *Top. Lipid Chem.*, 1971, **2**, 207 (*rev*)

3,4,7,11-Tetramethyl-6,10-tridecadienal T-00140

Faranal

[65395-77-9]

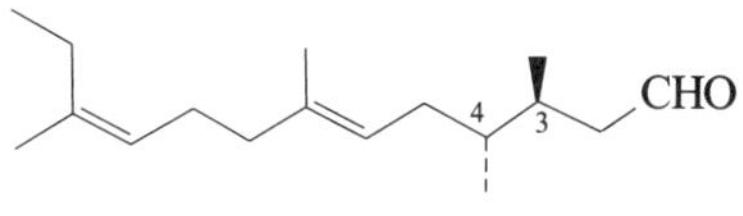

(3*S*,4*R*)-*form*

$C_{17}H_{30}O$ M 250.423

(3S,4R,6E,10Z)-form

Trail pheromone of *Monomorium pharaonis*. Oil. $[\alpha]_D^{23}$ +16.2° (c, 0.5 in hexane).

(3R,4S,6E,10Z)-form

Oil. $[\alpha]_D^{23}$ −16.4° (c, 0.22 in hexane).

Ritter, F.J. *et al*, *Tetrahedron Lett.*, 1977, 2617 (*isol*)
Knight, D.W. *et al*, *J. Chem. Soc., Perkin Trans. 1*, 1983, 955 (*synth, bibl*)
Baker, R. *et al*, *J. Chem. Soc., Perkin Trans. 1*, 1983, 1387 (*synth*)
Poppe, L. *et al*, *Tetrahedron Lett.*, 1986, **27**, 5769 (*synth*)

4,6,10,12-Tetramethyl-2,4-tridecadien-7-one T-00141

Matsuone

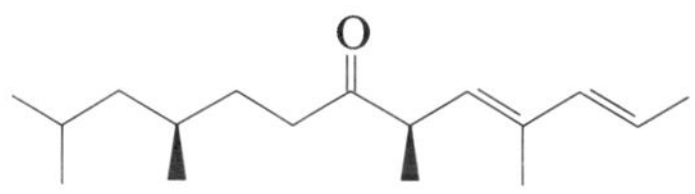

$C_{17}H_{30}O$ M 250.423

(2E,4E,6R*,10R*)-form

Constit. of the red pine scale *Matsucoccus resinosae* and other *M.* spp. Sex pheromone. Relative config. known.

[121981-51-9, 133697-99-1, 133698-00-7, 133698-01-8, 133698-02-9]

Lanier, G.N. *et al*, *J. Chem. Ecol.*, 1989, **15**, 1645; 1991, **17**, 89 (*isol, props*)
Cywin, C.L. *et al*, *J. Org. Chem.*, 1991, **56**, 2953 (*synth, abs config*)

2,4,6,8-Tetramethylundecanoic acid, 9CI T-00142

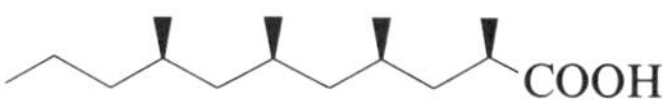

$C_{15}H_{30}O_2$ M 242.401

(2R,4R,6R,8R)-form [10553-03-4]

The octadecanyl and other long chain ester derivatives are main constits. of the preen gland wax of the graylag goose *Anser anser*; also present in small amounts in the preen gland wax of other species. $[\alpha]_D^{23}$ −15.8°.

Me ester: [2490-55-3].

Present in preen gland waxes of ruddy shelduck (*Tadorna ferruginea*) and common shelduck (*T. tadorna*). $[\alpha]_D^{21}$ −24° (c, 0.253 in $CHCl_3$). n_D^{22} 1.4317.

Murray, K.E., *Aust. J. Chem.*, 1962, **15**, 510 (*isol*)
Noble, R.E. *et al*, *Nature (London)*, 1963, **199**, 600 (*biosynth*)
Odham, G., *Ark. Kemi*, 1964, **21**, 379; 1965, **23**, 431; 1966, **25**, 543; 1967, **27**, 295 (*isol*)
Smith, C.R., *Top. Lipid Chem.*, 1970, **1**, 307 (*rev, bibl*)
Mori, K. *et al*, *Justus Liebigs Ann. Chem.*, 1987, 555 (*synth, abs config, ir, pmr, ms*)

Tetranonacontane, 9CI T-00143

[1574-32-9]

$H_3C(CH_2)_{92}CH_3$

$C_{94}H_{190}$ M 1320.535

Mp 114.07-114.47°. Viscosity 33.95 Cp at 125°.

Reinhard, R.R. *et al*, *J. Org. Chem.*, 1965, **30**, 1450 (*synth*)

Tetrapalmitoylphosphatidylglycerol T-00144

Tetrahexadecanoylphosphatidylglycerol. Bis(1,2-dihexadecanoylglycero)-3-phosphate

[84905-98-6]

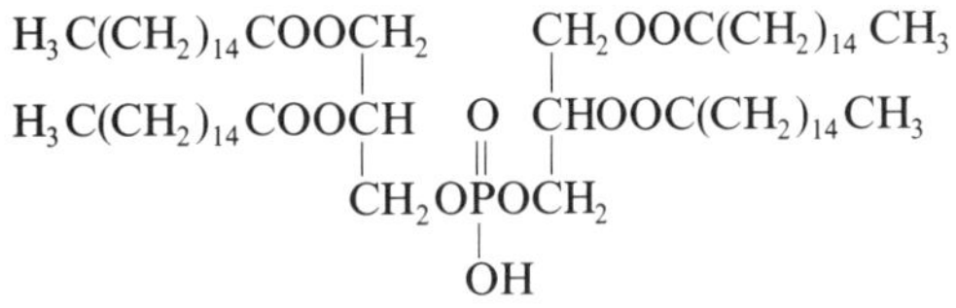

$C_{70}H_{135}O_{12}P$ M 1199.803

Cryst. (C_6H_6/Me_2CO). Mp 62-63°. $[\alpha]_D^{23}$ +6.7° (c, 4 in C_6H_6).

Baer, E., *J. Biol. Chem.*, 1952, **198**, 853.
Baer, E., *Prog. Chem. Fats Other Lipids*, 1963, **6**, 33.
Dang, Q.Q. *et al*, *Chem. Phys. Lipids*, 1983, **33**, 33 (*synth, nmr*)

Tetrastearoylphosphatidylglycerol T-00145

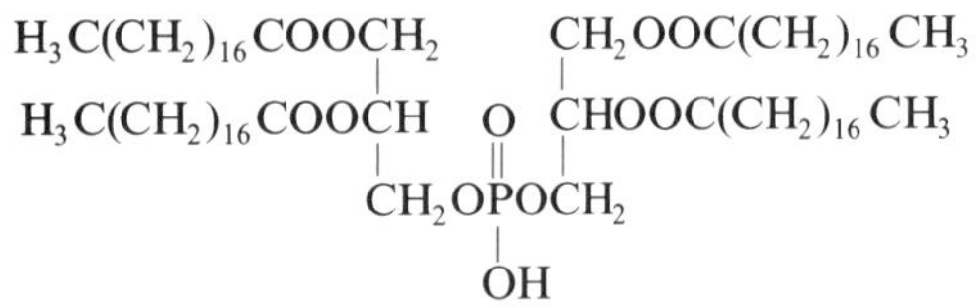

$C_{78}H_{151}O_{12}P$ M 1312.017

Mp 69.5-70.5°. $[\alpha]_D^{24}$ +6.2° (c, 4 in C_6H_6).

Baer, E., *J. Biol. Chem.*, 1952, **198**, 153.
Baer, E. *et al*, *Arch. Biochem. Biophys.*, 1958, **78**, 294.
Baer, E., *Prog. Chem. Fats Other Lipids*, 1963, **6**, 33.

Tetratriacontane T-00146

[14167-59-0]

$H_3C(CH_2)_{32}CH_3$

$C_{34}H_{70}$ M 478.927

Constit. of mineral oils and plant waxes. Cryst. (Et_2O). Mp 72.7°. Bp_3 285.4°.

Mazee, W.M., *Recl. Trav. Chim. Pays-Bas* (*J. R. Neth. Chem. Soc.*), 1948, **67**, 197 (*synth*)

1-Tetratriacontanol, 9CI, 8CI T-00147

Sapiol

[28484-70-0]

$H_3C(CH_2)_{32}CH_2OH$

$C_{34}H_{70}O$ M 494.926

Isol. from candelilla, cotton waxes and leaves of *Sapium indicum*. Mp 92°. Bp_1 200-250° subl.

[103215-66-3]

Chibnall, A.C. *et al*, *Biochem. J.*, 1934, **28**, 2189 (*isol*)
Mazliak, P. *et al*, *C. R. Hebd. Seances Acad. Sci.*, 1960, **251**, 2393 (*isol*)
Japan. Pat., 80 129 292, (1980); *CA*, **94**, 120844r (*synth*)
Tasumin, M. *et al*, *Spectrochim. Acta*, 1974, **20**, 629 (*ir*)
Wanatabe, A. *et al*, *Bull. Chem. Soc. Jpn.*, 1961, **34**, 1728 (*synth*)

8-(2-Thienyl)-3,5-octadien-7-yn-1-ol T-00148

$C_{12}H_{12}OS$ M 204.292

(3E,5E)-form [55309-63-2]

Constit. of *Chrysanthemum macrotum*. Oil.

Ac: [55309-62-1].

$C_{14}H_{14}O_2S$ M 246.329

Constit. of *C. macrotum*. Oil.

Phenylazobenzoyl: Yellow cryst. (pet. ether). Mp 129°.

Bohlmann, F. *et al*, *Chem. Ber.*, 1975, **108**, 739 (*isol, uv, pmr, ms*)

Thromboxane B_1 T-00149

9α,11,15S-Trihydroxythrombox-13E-en-1-oic acid

[64626-32-0]

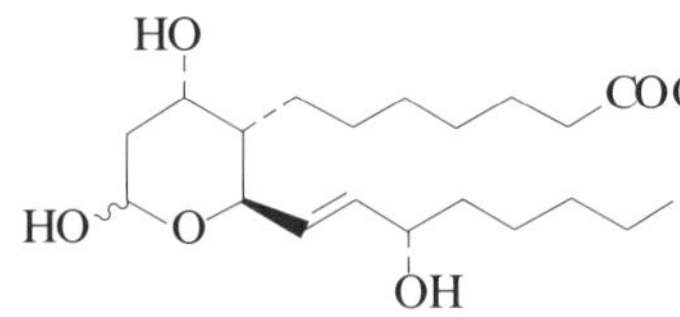

$C_{20}H_{36}O_6$ M 372.501

Ger. Pat., 2 715 883, (1977); *CA*, **88**, 190215m (*synth*)

Thromboxane B_2 T-00150

7-[Tetrahydro-4,6-dihydroxy-2-(3-hydroxy-1-octenyl)-2H-pyran-3-yl]-5-heptenoic acid, 9CI. 9α,11,15S-Trihydroxythromboxa-5Z,13E-dien-1-oic acid. TXB_2

[54397-85-2]

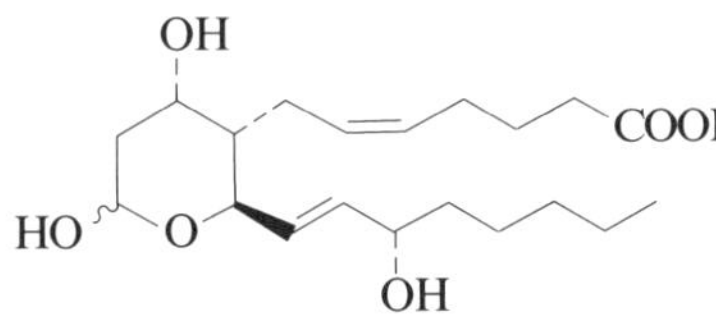

$C_{20}H_{34}O_6$ M 370.485

(+)-form

One of the major prods. of the biosynth. system which converts arachidonic acid to prostaglandins. Plates ($EtOAc/Et_2O$/pet. ether). Mp 95-96°. $[\alpha]_D^{25}$ +57.4° (c, 0.26 in EtOAc).

11-Dehydro: [67910-12-7]. *9,15-Dihydroxy-11-oxothromboxa-5,13-dien-1-oic acid. 11-Dehydrothromboxane B_2. 11-Dehydro-TXB_2*

Thromboxane B_2 metab.

11-Dehydro, Me ester: [67719-11-3].

$C_{21}H_{34}O_6$ M 382.496

13,14-Dihydro, 15-Oxo: [62107-18-0]. *9,11-Dihydroxy-15-oxothrombox-5-en-1-oic acid. 15-Dehydro-13,14-dihydrothromboxane B_2*

$C_{20}H_{34}O_6$ M 370.485

Thromboxane B_2 metab.

(±)-form

Plates ($EtOAc/Et_2O$/pet. ether). Mp 92-94° (89-90°).

Hamberg, M. *et al*, *Proc. Natl. Acad. Sci. U.S.A.*, 1974, **71**, 3400 (*struct, ms*)
Nelson, N.A. *et al*, *Tetrahedron Lett.*, 1976, 3275 (*synth*)
Kelly, R.C. *et al*, *Tetrahedron Lett.*, 1976, 3279 (*synth*)
Schneider, W.P. *et al*, *Tetrahedron Lett.*, 1976, 3283 (*synth, ms*)
Hanessian, S. *et al*, *Can. J. Chem.*, 1977, **55**, 562 (*synth*)

Corey, E.J. *et al*, *Tetrahedron Lett.*, 1977, 785 (*synth*)
Nelson, N.A. *et al*, *Prostaglandins*, 1978, **16**, 85 (*synth, ir, pmr, ms*)
Hernandez, O., *Tetrahedron Lett.*, 1978, 219 (*synth*)
Langs, D.A. *et al*, *Nature (London)*, 1979, **281**, 237 (*cryst struct, conformn*)
Roberts, L.J. *et al*, *J. Biol. Chem.*, 1981, **256**, 8384 (*biosynth*)
Robinson, C. *et al*, *Biochim. Biophys. Acta*, 1982, **712**, 315; 1983, **754**, 190 (*biosynth*)
Watanabe, K. *et al*, *Chem. Pharm. Bull.*, 1989, **37**, 140 (*detn*)

α-Tocopherol T-00151

3,4-Dihydro-2,5,7,8-tetramethyl-2-(4,8,12-trimethyltridecyl)-2H-1-benzopyran-6-ol, 9CI. 5,7,8-Trimethyltocol. Vitamin E. Antisterility vitamin. Ephynal. Syntopherol. Profecundin

[59-02-9]

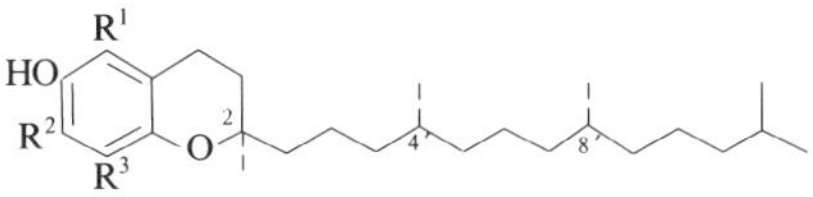

$R^1 = R^2 = R^3 = CH_3$

$C_{29}H_{50}O_2$ M 430.713

Constit. of many vegetable oils. Nutritional factor. Has been used to promote fertility. Viscous oil, cryst. (MeOH at −35°). Mp 2.5-3.5°. $Bp_{0.000001}$ 140°. $[\alpha]_D^{25}$ +0.65° (EtOH). Thermostable.

▷ DJ2900000.

Ac: [58-95-7]. *Alfacol. Fertilvit. Tocophrin.* ***α-Tocopherol acetate.*** *Vitamin E acetate. Numerous proprietary names*
$C_{31}H_{52}O_3$ M 472.750
Vitamin E supplement. Oil. Mp 26.5-27.5°. $Bp_{0.02}$ 205° (bath). $[\alpha]_D^{25}$ +3.03° (c, 5 in EtOH).
▷ GP8588000.

p-*Phenylazobenzoyl:* Cryst. Mp 62-64°. $[\alpha]_{600}$ +6.99° (c, 3.33 in $CHCl_3$).

3-Carboxypropanoyl: [4345-03-3]. ***α-Tocopherol succinate.*** *Numerous proprietary names*
$C_{33}H_{54}O_5$ M 530.787
Needles (pet. ether). Mp 76-77°.

3-Pyridinecarbonyl: [16676-75-8]. ***α-Tocopherol nicotinate.*** *Vitamin E nicotinate. Numerous proprietary names*
$C_{35}H_{53}NO_3$ M 535.809
Vasodilator. Yellow oil, solid. Mp *ca.* 38°.

Ester with Polyethylene glycol succinate: [30999-06-5]. ***Tocophersolan*, USAN.** *Tocofersolan, INN*
Vitamin E supplement. Antioxidant.

Baxter, J.G. *et al*, *J. Am. Chem. Soc.*, 1943, **65**, 918 (*succinate*)
Praill, P.F. *et al*, *J. Chem. Soc.*, 1959, 3100 (*synth*)
Robeson, C.D. *et al*, *J. Am. Chem. Soc.*, 1962, **84**, 3196 (*isol*)
Meyer, H. *et al*, *Helv. Chim. Acta*, 1963, **46**, 650, 963 (*synth, abs config*)
Fr. Pat., M6320, (1968); *CA*, **74**, 79582 (*nicotinate*)
Nakamura, A. *et al*, *Chem. Pharm. Bull.*, 1971, **19**, 2318 (*synth*)
Scheppele, S.E. *et al*, *Lipids*, 1972, **7**, 297 (*ms*)
Rudy, B.C. *et al*, *Anal. Profiles Drug Subst.*, 1974, **3**, 111 (*bibl*)
Gatlin, L.A. *et al*, *Drug Intell. Clin. Pharm.*, 1975, **9**, 655 (*Tocophersolan, rev*)
Scott, J.W. *et al*, *Helv. Chim. Acta*, 1976, **59**, 290 (*synth*)
Cohen, N. *et al*, *J. Org. Chem.*, 1976, **41**, 3505 (*synth*)
Schultz, G. *et al*, *Phytochemistry*, 1976, **15**, 1383 (*biosynth*)
Matsuo, M. *et al*, *Tetrahedron*, 1976, **32**, 229 (*pmr*)
Nakamura, T. *et al*, *J. Nutr. Sci. Vitaminol.*, 1978, **24**, 459; 1979, **25**, 159 (*pharmacol, esters*)
Chan, K.-K. *et al*, *J. Org. Chem.*, 1978, **43**, 3435 (*synth*)
Cohen, N. *et al*, *J. Am. Chem. Soc.*, 1979, **101**, 6710 (*synth*)
Akkerman, J.M. *et al*, *J. Chem. Soc., Perkin Trans. 1*, 1979, 2124 (*synth*)
Urano, S. *et al*, *Chem. Pharm. Bull.*, 1980, **28**, 1992 (*cmr*)
Olson, G.L. *et al*, *J. Org. Chem.*, 1980, **45**, 803 (*synth*)
Cohen, N. *et al*, *Helv. Chim. Acta*, 1981, **64**, 1158 (*synth, isom*)
Martindale, The Extra Pharmacopoeia, 28th/29th Ed., Pharmaceutical Press, London, 1982/1989, 7893, 7894, 7895, 78.
Suzuki, N. *et al*, *J. Nutr. Sci. Vitaminol.*, 1983, **29**, 93 (*nicotinate, metab*)
Ballester-Rodes, M. *et al*, *Synth. Commun.*, 1984, **14**, 515 (*nicotinate, synth*)
Burton, G.W. *et al*, *Acc. Chem. Res.*, 1986, **19**, 194 (*rev, props*)
Inoue, S. *et al*, *J. Org. Chem.*, 1987, **52**, 5495 (*synth*)
Negwer, M., *Organic-Chemical Drugs and their Synonyms*, 6th Ed., Akademie-Verlag, Berlin, 1987, 7835, 7936, 8009.
Furuya, T. *et al*, *Phytochemistry*, 1987, **26**, 2741 (*biosynth*)
Hübscher, M. *et al*, *Helv. Chim. Acta*, 1990, **73**, 782, 1068 (*bibl, hplc*)
Cohen, N. *et al*, *J. Org. Chem.*, 1992, **57**, 5783 (*synth, bibl*)
Lewis, R.J., *Sax's Dangerous Properties of Industrial Materials*, 8th Ed., Van Nostrand-Reinhold, 1992, TGJ050, TGJ060, VSZ450.

β-Tocopherol T-00152

3,4-Dihydro-2,5,8-trimethyl-2-(4,8,12-trimethyltridecyl)-2H-1-benzopyran-6-ol, 9CI. 2,5,8-Trimethyl-2-(4,8,12-trimethyltridecyl)chroman-6-ol, 8CI. 5,8-Dimethyltocol. Cumotocopherol. Neotocopherol. p-*Xylotocopherol*

[148-03-8]

As α-Tocopherol, T-00151 with

$R^1 = R^3 = CH_3, R^2 = H$

$C_{28}H_{48}O_2$ M 416.686

Isol. from vegetable sources, e.g. spinach chloroplasts (*Spinacea oleracea*), soybean oil, corn oil. Antioxidant. Pale-yellow viscous oil. $Bp_{0.1}$ 200-210°. $[\alpha]_D^{20}$ +6.37°. Thermostable; resistant to acids and alkalis.

3,5-Dinitrobenzoyl: Mp 86-87°.

Ac: Pale-yellow oil. $Bp_{0.3}$ 215-220°.

[16662-70-7, 16698-35-4]

Karrer, P. *et al*, *Helv. Chim. Acta*, 1938, **21**, 1234; 1939, **22**, 260 (*synth, struct*)
Mayer, H. *et al*, *Helv. Chim. Acta*, 1963, **46**, 963 (*abs config*)
Pearson, C.K. *et al*, *Br. J. Nutr.*, 1970, **24**, 581 (*metab*)
Nakamura, A. *et al*, *Chem. Pharm. Bull.*, 1971, **19**, 2318 (*synth*)
Scheppele, S.E. *et al*, *Lipids*, 1972, **7**, 297 (*ms*)
Matsuo, M. *et al*, *Tetrahedron*, 1976, **32**, 229 (*cmr*)
Tangney, C.C. *et al*, *J. Chromatogr.*, 1979, **172**, 513 (*purifn*)
Burton, G.W. *et al*, *J. Am. Chem. Soc.*, 1981, **103**, 6472 (*props*)
Pure Appl. Chem., 1982, **54**, 1507 (*nomencl*)
Urano, S. *et al*, *Heterocycles*, 1985, **23**, 2793 (*synth*)

δ-Tocopherol T-00153

3,4-Dihydro-2,8-dimethyl-2-(4,8,12-trimethyltridecyl)-2H-1-benzopyran-6-ol, 9CI. 8-Methyltocol

[119-13-1]

As α-Tocopherol, T-00151 with

$R^1 = R^2 = H, R^3 = CH_3$

$C_{27}H_{46}O_2$ M 402.659

Widespread occurrence in vegetable oils. Pale-yellow oil. $Bp_{0.001}$ 150° (bath).

p-*Phenylazobenzoyl:* Orange-yellow leaflets. Mp 41-42°. $Bp_{0.001}$ 250°.

Stein, M.H. *et al*, *J. Am. Chem. Soc.*, 1947, **69**, 869 (*isol*)
Green, J. *et al*, *J. Chem. Soc.*, 1959, 3374 (*synth*)
Matsuo, M. *et al*, *Tetrahedron*, 1976, **32**, 229 (*cmr*)
Tangney, C.C. *et al*, *J. Chromatogr.*, 1979, **172**, 513 (*purifn*)

γ-Tocopherol T-00154

3,4-Dihydro-2,7,8-trimethyl-2-(4,8,12-trimethyltridecyl)-2H-1-benzopyran-6-ol, 9CI. 7,8-Dimethyltocol. Vitamin E

[7616-22-0]

As α-Tocopherol, T-00151 with

$R^1 = H, R^2 = R^3 = CH_3$

$C_{28}H_{48}O_2$ M 416.686
Occurs in many nut and other vegetable oils. Viscous oil. Mp −3° to −2°. $[\alpha]^{25}_{546}$ +2.2° (EtOH), $[\alpha]^{25}_{546}$ −2.4° (C_6H_6).
p-*Nitrophenylurethane:* Mp 119-121°.

Karrer, P. *et al*, *Helv. Chim. Acta*, 1939, **22**, 260 (*struct*)
Tsukida, K. *et al*, *J. Vitaminol.*, 1972, **18**, 24 (*pmr*)
Scheppele, S.E. *et al*, *Lipids*, 1972, **7**, 297 (*ms*)
Matsuo, M. *et al*, *Tetrahedron*, 1976, **32**, 229 (*cmr*)
Tangney, C.C. *et al*, *J. Chromatogr.*, 1979, **172**, 513 (*purifn*)

η-Tocopherol T-00155

7-Methyltocol
[91-86-1]

As α-Tocopherol, T-00151 with

$R^1 = R^3 = H, R^2 = CH_3$

$C_{27}H_{46}O_2$ M 402.659
Constit. of rice. Oil.
4-Phenylazobenzoyl: Cryst. (2-propanol). Mp 55-56°.

Marcinkiewicz, S. *et al*, *J. Chem. Soc.*, 1959, 3377.

ζ_2-Tocopherol T-00156

5,7-Dimethyltocol
[17976-95-3]

As α-Tocopherol, T-00151 with

$R^1 = R^2 = CH_3, R^3 = H$

$C_{28}H_{48}O_2$ M 416.686
Constit. of rice. Oil. Mp −4°.
4-Phenylazobenzoyl: Cryst. (2-propanol). Mp 61°.

McHale, D. *et al*, *J. Chem. Soc.*, 1958, 1600 (*synth*)
Green, J. *et al*, *J. Chem. Soc.*, 1959, 3362 (*synth*)

α-Tocotrienol T-00157

*3,4-Dihydro-2,5,7,8-tetramethyl-2-(4,8,12-trimethyl-3,7,11-tridecatrienyl)-2*H*-1-benzopyran-6-ol, 9CI. ζ_1-Tocopherol. 5,7,8-Trimethyltocotrienol*
[1721-51-3]

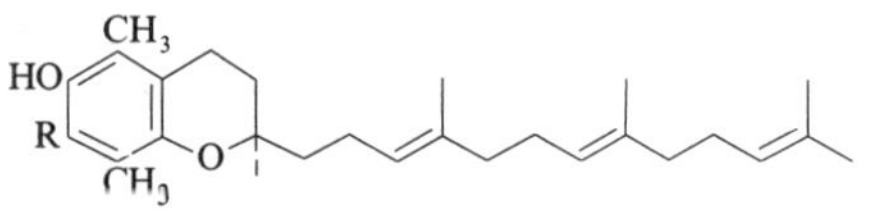

R = CH_3
(−)-*form*

$C_{29}H_{44}O_2$ M 424.665
The name α-Tocotrienol is IUPAC preferred (*Eur. J. Biochem.*, 1974, **46**, 217).
(−)-*form* [58864-81-6]
Occurs in wheat, barley, rye and palm oil. Mp 30-31°. $[\alpha]^{25}_D$ −5.7° (c, 1.01 in $CHCl_3$).
p-*Phenylazobenzoyl:* Cryst. (2-propanol). Mp 70-71°.
(±)-*form*
Cryst. Mp 30-31°.
p-*Phenylazobenzoyl:* Cryst. Mp 84-85.5°.

Green, J. *et al*, *Chem. Ind. (London)*, 1960, 73 (*struct*)
Mayer, H. *et al*, *Helv. Chim. Acta*, 1963, **46**, 963 (*abs config*)
Schudel, P. *et al*, *Helv. Chim. Acta*, 1963, **46**, 2517 (*synth*)
Scott, J.W. *et al*, *Helv. Chim. Acta*, 1976, **59**, 290 (*synth*)

β-Tocotrienol T-00158

*3,4-Dihydro-2,5,8-trimethyl-2-(4,8,12-trimethyl-3,7,11-tridecatrienyl)-4*H*-1-benzopyran-6-ol, 9CI. ε-Tocopherol*
[490-23-3]

As α-Tocotrienol, T-00157 with

R = H

$C_{28}H_{42}O_2$ M 410.639
Isol. from wheat bran oil. Yellow oil. $Bp_{0.00005}$ 140° (bath). Hydrogenation gives β-Tocopherol, T-00152.
p-*Phenylazobenzoyl:* Cryst. (2-propanol). Mp 70-71° (49-52°).

Green, J. *et al*, *J. Chem. Soc.*, 1959, 3362 (*isol, struct*)
Green, J. *et al*, *Chem. Ind. (London)*, 1960, 73 (*struct, isol*)
Schudel, P. *et al*, *Helv. Chim. Acta*, 1963, **46**, 2517 (*synth*)
McHale, D. *et al*, *J. Chem. Soc.*, 1963, 784 (*struct*)

Triacontanal, 9CI T-00159

[22725-63-9]

$H_3C(CH_2)_{28}CHO$

$C_{30}H_{60}O$ M 436.803
Isol. from berries of *Vitis vinifera*, from *Brassica oleracea* leaves and from apple cuticle wax.

Radler, F. *et al*, *Aust. J. Chem.*, 1965, **18**, 1059.
Schmid, H.H.O. *et al*, *Hoppe Seyler's Z. Physiol. Chem.*, 1969, **350**, 462 (*isol*)
Japan. Pat., 59 95 236, (1982); *CA*, **101**, 170706f (*synth*)

Triacontane T-00160

[638-68-6]

$H_3C(CH_2)_{28}CH_3$

$C_{30}H_{62}$ M 422.820
Constit. of mineral oils and plant waxes. Cryst. (Et_2O/Me_2CO). Mp 65.9°. Bp 449.7°, Bp_3 258.5°.

Mazee, W.M., *Recl. Trav. Chim. Pays-Bas* (*J. R. Neth. Chem. Soc.*), 1948, **67**, 197 (*synth*)
Karrer, W. *et al*, *Konstitution und Vorkommen der Organischen Pflanzenstoffe*, 2nd Ed., Birkhäuser Verlag, Basel, 1972-1985, no. 17 (*occur*)
Piesczek, W. *et al*, *Acta Crystallogr., Sect. B*, 1974, **30**, 1278 (*cryst struct*)

Triacontanedioic acid T-00161

Octacosane-1,28-dicarboxylic acid. Equisetolic acid
[6708-53-8]

$HOOC(CH_2)_{28}COOH$

$C_{30}H_{58}O_4$ M 482.786
Constit. of *Equisetum telmateja* and *E. arvense* spores. Mp 126-127°.
K salt: Mp 119°.
Di-Me ester:
$C_{32}H_{62}O_4$ M 510.839
Mp 83-84°.

Adams, K.R. *et al*, *Phytochemistry*, 1971, **10**, 1885 (*isol, synth, ms*)

10,11-Triacontanediol T-00162

Aeneadiol
[52945-88-7]

$HC_3(CH_2)_{18}CH(OH)CH(OH)(CH_2)_8CH_3$

$C_{30}H_{62}O_2$ M 454.819
Constit. of *Argemone* spp. flowers. Cryst. (MeOH). Mp 115-116°.
Di-Ac:
$C_{34}H_{66}O_4$ M 538.893
Mp 34-38°.

Domínguez, X.A. *et al*, *J. Org. Chem.*, 1965, **30**, 2049 (*isol*)
McCullough, T., *J. Chem. Educ.*, 1974, **51**, 228 (*isol*)

Triacontanoic acid T-00163

Melissic acid
[506-50-3]

$H_3C(CH_2)_{28}COOH$

$C_{30}H_{60}O_2$ M 452.803
Found in some plant waxes, eg. cotton, carnauba, various leaf waxes. Main constit. of acid fraction of *Vitex divaricata* wood. Cryst. (C_6H_6/Me_2CO). Mp 93.5-94°.
Me ester: [629-83-4].
$C_{31}H_{62}O_2$ M 466.830
Cryst. (pet. ether). Mp 71.5°.
Et ester: [7505-12-6].
$C_{32}H_{64}O_2$ M 480.856
Cryst. (EtOH). Mp 70.5°.
Pentyl ester: [116044-06-5].
$C_{35}H_{70}O_2$ M 522.937
Constit. of the leaves of *Ilex aquifolium*.
Hexyl ester: [108706-11-2].
$C_{36}H_{72}O_2$ M 536.964
Constit. of the waxes of *Abies* sp., *Ilex* sp. and *Picea* sp.
Tetradecyl ester: [104932-41-4].
$C_{44}H_{88}O_2$ M 649.178
Constit. of *Abutilon pakistanicum* and the green alga *Chlorella kessleri*.
Hentriacontyl ester: [135729-36-1].
$C_{61}H_{122}O_2$ M 887.634
Constit. of the leaves and stem of *Artemisia annua*.

Robinson, G.M., *J. Chem. Soc.*, 1934, 1543 (*synth*)
Cocker, W. *et al*, *J. Chem. Soc.*, 1962, 5194 (*isol*)
Karrer, W. *et al*, *Konstitution und Vorkommen der Organischen Pflanzenstoffe*, 2nd Ed., Birkhäuser Verlag, Basel, 1972-1985, no. 713 (*occur*)

1-Triacontanol T-00164

Melissyl alcohol. Myricyl alcohol
[593-50-0]

$H_3C(CH_2)_{28}CH_2OH$

$C_{30}H_{62}O$ M 438.819
Older isolates consisted of a mixt. of 1-triacontanol and 1-hentriancontanol formerly referred to as myricyl alcohol. Since they were separately distinguished in 1963 the name myricyl alcohol is used for 1-triacontanol. Constit. of many plant waxes. Plant growth stimulant. Cryst. (C_6H_6). Mp 86.5°.
Ac: [41755-58-2].
$C_{32}H_{64}O_2$ M 480.856
Cryst. (pet. ether). Mp 69°.

Robinson, G.M., *J. Chem. Soc.*, 1934, 1543 (*synth*)
Razafindrazaka, J. *et al*, *Bull. Soc. Chim. Fr.*, 1963, 1633 (*glc, purifn, nomencl*)
Karrer, W. *et al*, *Konstitution und Vorkommen der Organischen Pflanzenstoffe*, 2nd Ed., Birkhäuser Verlag, Basel, 1972-1985, no. 100 (*bibl*)
Ries, S.K. *et al*, *Science (Washington, D.C.)*, 1977, **195**, 1339 (*pharmacol*)
Maruyama, K. *et al*, *J. Org. Chem.*, 1980, **45**, 737 (*synth*)
Rama Rao, A.V. *et al*, *Tetrahedron*, 1981, **37**, 227 (*synth*)
Qian, C. *et al*, *CA*, 1982, **96**, 51743 (*synth*)
Gibson, T., *Tetrahedron Lett.*, 1982, **23**, 157 (*synth*)
Villemin, D., *Tetrahedron Lett.*, 1983, **24**, 2855 (*synth*)
Bhalerao, U.T. *et al*, *Tetrahedron Lett.*, 1984, **25**, 5439 (*synth*)
Kirchner, G. *et al*, *Justus Liebigs Ann. Chem.*, 1985, 214 (*synth*)
Iyer, R.R. *et al*, *Chem. Ind. (London)*, 1986, 398 (*synth*)
Rao, B.V.S.K. *et al*, *Org. Prep. Proced. Int.*, 1992, **24**, 67 (*synth*)

5-Triacontanol T-00165

$H_3C(CH_2)_{24}CH(OH)(CH_2)_3CH_3$

$C_{30}H_{62}O$ M 438.819
Ac: [146356-95-8]. *5-Acetoxytriacontane*
$C_{32}H_{64}O_2$ M 480.856
Constit. of *Leucas aspera*. Cryst. (hexane). Mp 90-91°.

Misra, T.N. *et al*, *Phytochemistry*, 1993, **32**, 199.

21-Triacontenoic acid T-00166

$H_3C(CH_2)_7CH{=}CH(CH_2)_{19}COOH$

$C_{30}H_{58}O_2$ M 450.787
(Z)-form [67329-09-3]
Lumequic acid
Isol. from seed oil of *Ximenia americana* and other *X.* spp. Plates. Mp 60.8-61.2°.

Lightelm, S.P. *et al*, *J. Sci. Food Agric.*, 1954, **5**, 281.

Trichlorodecylsilane, 9CI T-00167

1-Trichlorosilyldecane
[13829-21-5]

$H_3C(CH_2)_9SiCl_3$

$C_{10}H_{21}Cl_3Si$ M 275.720
Liq. d_4^{20} 1.05. Bp_5 133-137°.

Whitmore, F.C. *et al*, *J. Am. Chem. Soc.*, 1946, **68**, 475 (*synth*)
Nesmeyanov, A.N. *et al*, *Tetrahedron*, 1957, **1**, 248 (*synth*)

Trichlorododecylsilane, 9CI T-00168

Dodecyltrichlorosilane. 1-Trichlorosilyldodecane
[4484-72-4]

$H_3C(CH_2)_{11}SiCl_3$

$C_{12}H_{25}Cl_3Si$ M 303.773
Synth. from $Me(CH_2)_{11}MgBr$ and $SiCl_4$. Commercially available waterproofing agent. Moisture-sensitive liq. d 1.02. Bp 293-294°, Bp_3 152-153°. n_D^{20} 1.454.
▷ VV3940000.

Rosenberg, H. *et al*, *J. Org. Chem.*, 1960, **25**, 243 (*synth*)
Lewis, R.J., *Sax's Dangerous Properties of Industrial Materials*, 8th Ed., Van Nostrand-Reinhold, 1992, DYA800.

1,1,1-Trichloro-4-hydroxy-11-methyl-3,5,7,9-dodecatetraen-2-one, 9CI T-00169

Neocarzirin B
[125002-00-8]

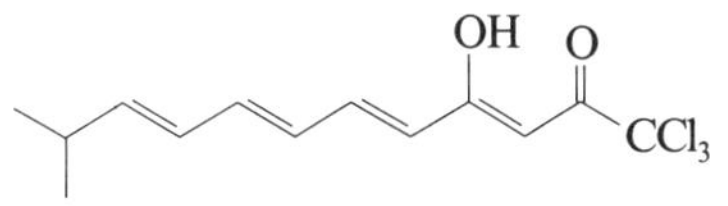

$C_{13}H_{15}Cl_3O_2$ M 309.618
Prod. by *Streptomyces carzinostaticus*. Antitumour agent.
1-Dechloro: [124958-30-1]. *1,1-Dichloro-4-hydroxy-11-methyl-3,5,7,9-dodecatetraen-2-one, 9CI.* ***Neocarzirin C***
$C_{13}H_{16}Cl_2O_2$ M 275.174
Prod. by *S. carzinostaticus*. Antitumour agent.

Japan. Pat., 89 224 341, (1989); *CA*, **112**, 75322 (*isol*)

1,1,1-Trichloro-4-hydroxy-11-methyl-3,5,7,9-tridecatetraen-2-one T-00170

Neocarzirin A

[124958-29-8]

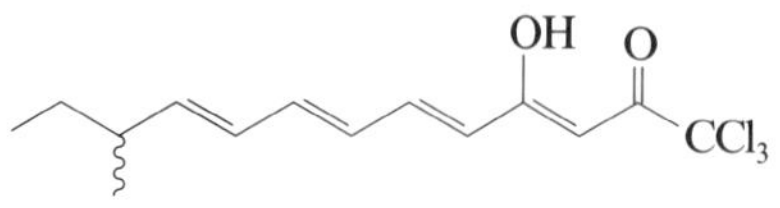

$C_{14}H_{17}Cl_3O_2$ M 323.645

Prod. by *Streptomyces carzinostaticus*. Antitumour agent.

Japan. Pat., 89 224 341, (1989); *CA*, **112**, 75322 (*isol*)

Trichlorooctadecylsilane, 9CI T-00171

1-Trichlorosilyloctadecane

[112-04-9]

$H_3C(CH_2)_{17}SiCl_3$

$C_{18}H_{37}Cl_3Si$ M 387.934

Packing material for glc columns. Liq. d^4 0.95. Bp_{13} 160°.

▷ VV4680000.

Barry, A.J. *et al*, *J. Am. Chem. Soc.*, 1947, **69**, 2916 (*synth*)
Lewis, R.J., *Sax's Dangerous Properties of Industrial Materials*, 8th Ed., Van Nostrand-Reinhold, 1992, OBI000.

14,20-Tricosadiene-3,5,10,12,22-pentayne-1,2-diol, 9CI T-00172

Siphonodiol

[90934-31-9]

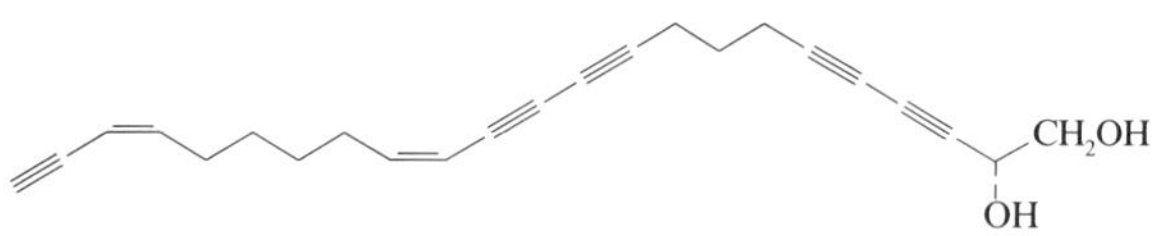

$C_{23}H_{24}O_2$ M 332.441

Metab. of sponge *Siphonochalina truncata*. Shows antibacterial activity. H,K-ATPase inhibitor. Cryst. Mp 31-32°. $[\alpha]_D$ −6.7° (c, 0.5 in MeOH).

14,15-Dihydro: 20-Tricosene-3,5,10,12,22-pentayne-1,2-diol. ***14,15-Dihydrosiphonodiol***

$C_{23}H_{26}O_2$ M 334.457

From *S. truncata*. H,K-ATPase inhibitor.

12,13,14,15-Tetrahydro: 12,20-Tricosadiene-3,5,10,22-tetrayne-1,2-diol. ***12,13,14,15-Tetrahydrosiphonodiol***

$C_{23}H_{28}O_2$ M 336.473

From *S. truncata*. H,K-ATPase inhibitor.

Tada, H. *et al*, *Chem. Lett.*, 1984, 779.
Fusetani, N. *et al*, *Tetrahedron Lett.*, 1987, **28**, 4311 (*isol, struct*)

Tricosane T-00173

[638-67-5]

$H_3C(CH_2)_{21}CH_3$

$C_{23}H_{48}$ M 324.632

Constit. of mineral oils and plant waxes. Cryst. (Et_2O/Me_2CO). Mp 47.4°. Bp 380.2°, Bp_3 199.5°.

Mazee, W.M., *Recl. Trav. Chim. Pays-Bas* (*J. R. Neth. Chem. Soc.*), 1948, **67**, 197 (*synth*)
Hunter, G.L.K. *et al*, *Phytochemistry*, 1966, **5**, 807 (*isol*)
Carnduff, J. *et al*, *Phytochemistry*, 1966, **5**, 1029 (*isol*)

Tricosanedioic acid T-00174

Heneicosane-1,21-dicarboxylic acid

$HOOC(CH_2)_{21}COOH$

$C_{23}H_{44}O_4$ M 384.598

Constit. of some plant waxes. Cryst. ($CHCl_3$). Mp 127.5°.

Di-Me ester: [42235-77-8].

$C_{25}H_{48}O_4$ M 412.652

Mp 70.5-71.4°.

Di-Et ester: [42235-78-9].

$C_{27}H_{52}O_4$ M 440.705

Mp 60.8-61.3°.

Marosi, L. *et al*, *Justus Liebigs Ann. Chem.*, 1973, 584.

Tricosanoic acid T-00175

[2433-96-7]

$H_3C(CH_2)_{21}COOH$

$C_{23}H_{46}O_2$ M 354.615

Constit. of some plant waxes and oils, e.g. *Ulex europaeus*, *Shorea maranti*, *Pinus contorta* and *Citrus bergamia*. Cryst. (C_6H_6). Mp 79.1°.

Me ester: [2433-97-8].

$C_{24}H_{48}O_2$ M 368.642

Cryst. (Me_2CO). Mp 53.4°.

Et ester: [18281-07-7].

$C_{25}H_{50}O_2$ M 382.669

Cryst. (Me_2CO). Mp 51.2°.

Nitrile:

$C_{23}H_{45}N$ M 335.615

Cryst. (Me_2CO). Mp 53.5-54.5°.

Francis, F. *et al*, *J. Am. Chem. Soc.*, 1939, **61**, 577.
Cocker, W. *et al*, *Perfum. Essent. Oil Rec.*, 1963, **54**, 235; 1964, **55**, 442 (*isol*)
Rowe, J.W. *et al*, *J. Org. Chem.*, 1964, **29**, 1554 (*isol*)
Aleby, S. *et al*, *Ark. Kemi*, 1969, **31**, 267 (*ir, polymorphism*)
Marosi, L. *et al*, *Justus Liebigs Ann. Chem.*, 1973, 584 (*cryst struct*)

1-Tricosanol T-00176

Tricosyl alcohol

[3133-01-5]

$H_3C(CH_2)_{21}CH_2OH$

$C_{23}H_{48}O$ M 340.632

Constit. of conifer waxes and of *Dryopteris filix-mas* wax. Cryst. (Me_2CO). Mp 73.5-74.5°. $Bp_{0.7}$ 191-193°.

Ac:

$C_{25}H_{50}O_2$ M 382.669

Mp 52-52.5°.

Levene, P.A. *et al*, *J. Biol. Chem.*, 1924, **59**, 905 (*synth*)
Thomas, T.L. *et al*, *Can. J. Chem.*, 1962, **40**, 1302 (*isol, synth, ir*)

12-Tricosanol T-00177

[24897-74-3]

$H_3C(CH_2)_{10}CH(OH)(CH_2)_{10}CH_3$

$C_{23}H_{48}O$ M 340.632

Constit. of *Platanus orientalis* fruits and of leaf wax of *Artemisia vulgaris*. Mp 75.5°.

Kipping, S., *J. Chem. Soc.*, 1890, **57**, 980 (*synth*)
Fujita, A. *et al*, *CA*, 1954, **48**, 3252 (*isol*)
Streibl, V.M. *et al*, *Fette, Seifen, Anstrichm.*, 1972, **74**, 566 (*synth*)
Ubik, K. *et al*, *Collect. Czech. Chem. Commun.*, 1975, **40**, 1718 (*ms*)
Dhar, D.N. *et al*, *Planta Med.*, 6th Ed., 1976, **29**, 91 (*isol*)

12-Tricosanone T-00178

Laurone. Diundecyl ketone. 12-Oxotricosane
[540-09-0]

$H_3C(CH_2)_{10}CO(CH_2)_{10}CH_3$

$C_{23}H_{46}O$ M 338.616
Mp 69°.
▷ YD0805000.
Oxime:
$C_{23}H_{47}NO$ M 353.630
Mp 39-40°.

Org. Synth., 1951, **31**, 68 (*synth*)
Malta, V. *et al*, *Gazz. Chim. Ital.*, 1974, **104**, 921 (*cryst struct*)
Malta, V. *et al*, *CA*, 1976, **85**, 152158a (*cryst struct*)
Lewis, R.J., *Sax's Dangerous Properties of Industrial Materials*, 8th Ed., Van Nostrand-Reinhold, 1992, TJF250.

17-Tricosenal T-00179

$H_3C(CH_2)_4CH{=}CH(CH_2)_{15}CHO$

$C_{23}H_{44}O$ M 336.600
(Z)-form [140163-47-9]
Constit. of the sponge *Amphimedon compressa.*

Carballeira, N.M. *et al*, *J. Nat. Prod.* (*Lloydia*), 1992, **55**, 333.

9-Tricosene T-00180

[52078-48-5]

$H_3C(CH_2)_7CH{=}CH(CH_2)_{12}CH_3$

$C_{23}H_{46}$ M 322.616
(E)-form [35857-62-6]
$Bp_{0.35}$ 164-165°. n_D^{25} 1.4529.
(Z)-form [27519-02-4]
Muscalure
Sex attractant of the housefly. Oil. $Bp_{0.15}$ 170-172°, $Bp_{0.2}$ 160-161°. n_D^{25} 1.4524.

Gribble, G.W. *et al*, *J. Chem. Soc., Chem. Commun.*, 1973, 735 (*synth*)
Küpper, F.-W. *et al*, *Chem.-Ztg.*, 1975, **99**, 464; *Z. Naturforsch., B*, 1976, **31**, 1256 (*synth*)
Rossi, R., *Chim. Ind. (Milan)*, 1975, **57**, 242 (*synth, pmr*)
Abe, K. *et al*, *Bull. Chem. Soc. Jpn.*, 1977, **50**, 2792 (*synth, ir, pmr*)
Shani, A., *J. Chem. Ecol.*, 1979, **5**, 557 (*synth*)
Cormier, R.A. *et al*, *J. Chem. Educ.*, 1979, **56**, 345 (*synth*)
Naoshima, Y. *et al*, *Agric. Biol. Chem.*, 1981, **45**, 1723 (*synth*)
Brown, H.C. *et al*, *J. Org. Chem.*, 1982, **47**, 3806 (*synth*)
Odinokov, V.N. *et al*, *Tetrahedron Lett.*, 1982, **23**, 1371 (*synth*)
Yadav, A.K. *et al*, *Helv. Chim. Acta*, 1984, **67**, 1698 (*synth*)
Koumaglo, K. *et al*, *Tetrahedron Lett.*, 1984, **25**, 717 (*synth*)
Wenkert, E. *et al*, *J. Org. Chem.*, 1985, **50**, 719 (*synth, pmr*)
Subramanian, G.B.V. *et al*, *Tetrahedron*, 1986, **42**, 3967 (*synth, ir, pmr*)
Kelkar, S.V. *et al*, *Synth. Commun.*, 1989, **19**, 1369 (*synth*)

16-Tricosene-2,4-diyn-1-ol T-00181

[63987-86-0]

$H_3C(CH_2)_5CH{=}CH(CH_2)_{10}C{\equiv}CC{\equiv}CCH_2OH$

$C_{23}H_{38}O$ M 330.553
(Z)-form
Renierin 2
Constit. of *Reniera fulva*. Mp 35°.

Cimino, G. *et al*, *Tetrahedron Lett.*, 1977, 1325 (*isol, struct*)

16-Tricosenoic acid T-00182

$H_3C(CH_2)_5CH{=}CH(CH_2)_{14}COOH$

$C_{23}H_{44}O_2$ M 352.599
(Z)-form [140163-36-6]
Constit. of the sponges *Amphimedon compressa* and *Mycale laevis.*

Carballeira, N.M. *et al*, *J. Nat. Prod.* (*Lloydia*), 1992, **55**, 333.

17-Tricosenoic acid T-00183

$H_3C(CH_2)_4CH{=}CH(CH_2)_{15}COOH$

$C_{23}H_{44}O_2$ M 352.599
(Z)-form [140163-37-7]
Constit. of the sponge *Amphimedon compressa.*

Carballeira, N.M. *et al*, *J. Nat. Prod.* (*Lloydia*), 1992, **55**, 333.

18-Tricosenoic acid T-00184

$H_3C(CH_2)_3CH{=}CH(CH_2)_{16}COOH$

$C_{23}H_{44}O_2$ M 352.599
(Z)-form [140163-38-8]
Constit. of the sponge *Amphimedon compressa.*

Carballeira, N.M. *et al*, *J. Nat. Prod.* (*Lloydia*), 1992, **55**, 333.

4,7-Tridecadienal T-00185

[13552-95-9]

$H_3C(CH_2)_4CH{=}CHCH_2CH{=}CHCH_2CH_2CHO$

$C_{13}H_{22}O$ M 194.316
(Z,Z)-form
$Bp_{0.15}$ 82-83°.
2,4-Dinitrophenylhydrazone: [22644-13-9].
Mp 59°.

Ward, J.P. *et al*, *Recl. Trav. Chim. Pays-Bas* (*J. R. Neth. Chem. Soc.*), 1969, **88**, 177 (*synth*)

1,2-Tridecadiene T-00186

[52629-63-7]

$H_3C(CH_2)_9CH{=}C{=}CH_2$

$C_{13}H_{24}$ M 180.333
$Bp_{0.1}$ 63-64°.

Chan, T.H. *et al*, *J. Org. Chem.*, 1978, **43**, 1526 (*synth, ir, pmr*)
Crandall, J.K. *et al*, *J. Org. Chem.*, 1991, **56**, 1153 (*synth, ir, pmr, cmr*)

5,7-Tridecadiene T-00187

$H_3C(CH_2)_4CH{=}CHCH{=}CH(CH_2)_3CH_3$

$C_{13}H_{24}$ M 180.333

Baba, S. *et al*, *J. Am. Chem. Soc.*, 1976, **98**, 6729 (*synth*)
Okukado, N. *et al*, *Tetrahedron Lett.*, 1978, 1027 (*synth*)
Negishi, E. *et al*, *J. Am. Chem. Soc.*, 1987, **109**, 2393 (*synth, ir, pmr*)

5,11-Tridecadiene-7,9-diyne-1,3-diol T-00188

Atractylodiol

$H_3CCH{=}CHC{\equiv}CC{\equiv}CCH{=}CHCH_2CH(OH)CH_2CH_2OH$

$C_{13}H_{16}O_2$ M 204.268
Di-Ac: [55078-26-7]. ***Diacetylatractylodiol***

$C_{17}H_{20}O_4$ M 288.343
Constit. of *Atractylodes japonica*. $[\alpha]_D$ $-4.8°$ ($CHCl_3$).

Yosioka, I. *et al, Chem. Pharm. Bull.*, 1974, **22**, 1943 (*isol, struct, uv, pmr*)

5,11-Tridecadiene-7,9-diyne-3,4-diol T-00189

3-Hydroxyaethusanol A

$H_3CCH{=}CHC{\equiv}CC{\equiv}CCH{=}CHCH(OH)CH(OH)CH_2CH_3$

$C_{13}H_{16}O_2$ M 204.268
Constit. of *Aethusa cynapium*.

Bohlmann, F. *et al, Chem. Ber.*, 1964, **97**, 2598 (*isol, uv, ir*)

5,11-Tridecadiene-7,9-diyn-4-ol, 9CI T-00190

Aethusanol A. Ethusanol A

[462-67-9]

$H_3CCH_2CH_2CH(OH)CH{=}CHC{\equiv}CC{\equiv}CCH{=}CHCH_3$

$C_{13}H_{16}O$ M 188.269
Isol. from *Aethusa cynapium* and *Pimpinella magna*. Mp $< -20°$. $[\alpha]_D^{25.5}$ $+14.4°$ (c, 4.25 in MeOH).

Ac:
$C_{15}H_{18}O_2$ M 230.306
Isol. from *A. cynapium*.

Bohlmann, F. *et al, Chem. Ber.*, 1960, **93**, 981; 1964, **97**, 2598; 1972, **105**, 355 (*isol*)
Schulte, K.E. *et al, Arch. Pharm.* (*Weinheim, Ger.*), 1970, **303**, 912 (*isol*)

1,3-Tridecadiene-5,7,9,11-tetrayne, 9CI T-00191

[18668-88-7]

$H_3CC{\equiv}CC{\equiv}CC{\equiv}CC{\equiv}CCH{=}CHCH{=}CH_2$

$C_{13}H_8$ M 164.206

(E)-form [3760-28-9]
Trace constituent of roots of *Rudbeckia amplexicaulis, R. bicolor, Flourensia heterolepsis, Scaevola lobelia, Helichrysum vestitum* and *Arnica chamissonis*. Unstable cryst. (pet. ether). λ_{max} 273, 289.5, 311, 331, 355.5, 382 nm (Et_2O).

1,2-Epoxide: [104017-81-4]. *2-(1-Undecen-3,5,7,9-tetraynyl) oxirane. 1,2-Epoxy-3-tridecene-5,7,9,11-tetrayne*
$C_{13}H_8O$ M 180.206
Isol. from *Carthamus, Centaurea* and *Silybum* spp. and from roots of *Arctium lappa*. Cryst. Mp 95.5-97°. $[\alpha]_D$ $+75.1°$ ($CHCl_3$).

(Z)-form [124604-45-1]
Isol. from yellow starthistle (*Centaurea solstitialis*) and safflower (*Carthamus tinctorius*).

Bohlmann, F. *et al, Chem. Ber.*, 1961, **94**, 3179; 1964, **97**, 809, 2125; 1965, **98**, 3081; 1966, **99**, 3201, 3433, 3544; 1968, **101**, 2056; 1975, **108**, 433 (*isol, uv, ir, synth, struct*)
Schulte, K.E. *et al, J. Nat. Prod.* (*Lloydia*), 1969, **32**, 360 (*isol, struct*)
Bohlmann, F. *et al, Phytochemistry*, 1979, **18**, 1189; 1980, **19**, 331 (*isol*)
Washino, T. *et al, Nippon Nogei Kagaku Kaishi*, 1986, **60**, 377; *CA*, **105**, 111980a (*isol*)
Binder, R.G. *et al, J. Agric. Food Chem.*, 1990, **38**, 764, 1245 (*isol*)

1,5-Tridecadiene-7,9,11-triyne-3,4-diol T-00192

$H_3CC{\equiv}CC{\equiv}CC{\equiv}CCH{=}CH(OH)CH(OH)CH{=}CH_2$

$C_{13}H_{12}O_2$ M 200.237

(Z)-form [27267-72-7]
Constit. of *Pentzia grandiflora*. Cryst. (pet. ether). Mp 70°.

Bohlmann, F. *et al, Chem. Ber.*, 1970, **103**, 1879 (*isol*)

3,11-Tridecadiene-5,7,9-triyne-1,2-diol, 9CI T-00193

[65398-35-8]

$H_3CCH{=}CHC{\equiv}CC{\equiv}CC{\equiv}CCH{=}CHCH(OH)CH_2OH$

$C_{13}H_{12}O_2$ M 200.237

(R,E,E)-form [27978-14-9] ***Safynol***
Isol. from diseased *Carthamus tinctorius, Bidens* spp. and *Centaurea* spp. Antifungal, antibiotic phytoalexin. Cryst. (Et_2O/pet. ether or $CHCl_3$/hexane). Mp 122.5-123° (112°). $[\alpha]_D^{20}$ $-17°$ (MeOH). An opt. rotn. of $[\alpha]_D^{23}$ $+38°$ (MeOH) is given without explanation by Bohlmann for a prod. from saponification of the natural monoacetate. Stable in dark at $-18°$, resinifies in air.

1-Ac: [71867-02-2]. *1-Acetoxy-3,11-tridecadiene-5,7,9-triyn-2-ol*
$C_{15}H_{14}O_3$ M 242.274
Isol. from aerial parts of *Centaurea ruthenica*. Oil.

2-Ac: 2-Acetoxy-3,11-tridecadiene-5,7,9-triyn-1-ol
$C_{15}H_{14}O_3$ M 242.274
Isol. from *Carthamus* and *Centaurea* spp. Cryst. (Et_2O/pet. ether). Mp 86°. $[\alpha]_D$ $+39°$ ($CHCl_3$).

Di-Ac: 12,13-Diacetoxy-2,10-tridecadiene-4,6,8-triyne
$C_{17}H_{16}O_4$ M 284.311
Isol. from roots of *Centaurea ruthenica*. Cryst. (Et_2O or Et_2O/pet. ether). Mp 64° (80°). $[\alpha]_D^{20}$ $+102.2°$ ($CHCl_3$).

2-Me ether: [71867-03-3]. *2-Methoxy-3,11-tridecadiene-5,7,9-triyn-1-ol*
$C_{14}H_{14}O_2$ M 214.263
Isol. from aerial parts of *C. ruthenica*. Oil.

(S,E,E)-form [65207-93-4]
$[\alpha]_D^{23}$ $+30.8°$ (MeOH).

(3E,11Z)-form

1-Ac: [71867-04-4].
Isol. from aerial parts of *C. ruthenica*. Oil.

2-Me ether: [71867-05-5].
Isol. from aerial parts of *C. ruthenica*. Oil.

Bohlmann, F. *et al, Chem. Ber.*, 1958, **91**, 1642; 1959, **92**, 1319; 1961, **94**, 3179; 1964, **97**, 520 (*isol, struct, biosynth*)
Thomas, C.A. *et al, Phytopathology*, 1970, **60**, 261 (*isol*)
Allen, E.H. *et al, Phytochemistry*, 1971, **10**, 1579 (*isol*)
Nakada, H. *et al, Agric. Biol. Chem.*, 1977, **41**, 1761 (*synth, abs config*)
Jente, R. *et al, Phytochemistry*, 1979, **18**, 829 (*isol, struct, ms, uv, pmr*)
MacRae, W.D. *et al, Experientia*, 1980, **36**, 1096 (*pharmacol*)

3,5-Tridecadienoic acid T-00194

$H_3C(CH_2)_6CH{=}CHCH{=}CHCH_2COOH$

$C_{13}H_{22}O_2$ M 210.316

(3E,5E)-form

Me ester: $Bp_{0.04}$ 97-100°.
Me ester, maleic anhydride adduct: Cryst. (hexane). Mp 94-96°.

(3E,5Z)-form

Me ester: $Bp_{0.17}$ 93-94°.

Celmer, W.D. *et al, J. Am. Chem. Soc.*, 1953, **75**, 3430 (*synth*)

10,12-Tridecadienoic acid **T-00195**

$H_2C{=}CHCH{=}CH(CH_2)_8COOH$

$C_{13}H_{22}O_2$ M 210.316

(E)-form

Me ester: [80625-41-8].

$C_{14}H_{24}O_2$ M 224.342

$Bp_{0.01}$ 90-95°.

Bestmann, H.J. *et al*, *Justus Liebigs Ann. Chem.*, 1981, 2117 (*synth, pmr*)

4,7-Tridecadien-1-ol **T-00196**

$H_3C(CH_2)_4CH{=}CHCH_2CH{=}CHCH_2CH_2CH_2OH$

$C_{13}H_{24}O$ M 196.332

(4Z,7E)-form

Ac: [61389-12-6].

$C_{15}H_{26}O_2$ M 238.369

Sex pheromone in male potato tuberworm moth *Phthorimaea operculella*.

(4E,7Z)-form [57981-61-0]

Shows pheromone activity in *P. operculella*.

Ac: [57981-60-9].

Sex attractant for leaf miner moth *Lithocolletis corylifoiella*; sex pheromone of *P. operculella*. $Bp_{0.005}$ 93°.

Bacon, O.G. *et al*, *J. Econ. Entomol.*, 1976, **69**, 569.
Alexakis, A., *Tetrahedron Lett.*, 1978, 2027 (*synth*)
Fujisawa, T. *et al*, *Tetrahedron Lett.*, 1982, **23**, 3583 (*synth*)
Nishiyama, H. *et al*, *Tetrahedron Lett.*, 1984, **25**, 223 (*synth*)
Nonoshita, K. *et al*, *J. Am. Chem. Soc.*, 1990, **112**, 316 (*synth, pmr, ir, bibl*)
Moiseenkov, A.M. *et al*, *J. Chem. Soc., Perkin Trans. 1*, 1991, 2639 (*synth, Ac, pmr*)

8,10-Tridecadien-1-ol **T-00197**

$H_3CCH_2CH{=}CHCH{=}CH(CH_2)_6CH_2OH$

$C_{13}H_{24}O$ M 196.332

(8E,10Z)-form

Ac: [65603-26-1].

$C_{15}H_{26}O_2$ M 238.369

$Bp_{0.01}$ 95-100°.

Bestmann, H.J. *et al*, *Justus Liebigs Ann. Chem.*, 1981, 2117 (*synth, pmr*)

9,11-Tridecadien-1-ol **T-00198**

$H_3CCH{=}CHCH{=}CH(CH_2)_7CH_2OH$

$C_{13}H_{24}O$ M 196.332

(9E,11Z)-form [80625-57-6]

$Bp_{0.05}$ 90-95°.

Ac: [80625-70-3].

$C_{15}H_{26}O_2$ M 238.369

$Bp_{0.01}$ 86-90°.

Bestmann, H.J. *et al*, *Justus Liebigs Ann. Chem.*, 1981, 2117 (*synth, pmr*)

10,12-Tridecadien-1-ol **T-00199**

$H_2C{=}CHCH{=}CH(CH_2)_8COOH$

$C_{13}H_{24}O$ M 196.332

(E)-form [69737-74-2]

$Bp_{0.1}$ 85-90°.

Ac: [80625-42-9].

$C_{15}H_{26}O_2$ M 238.369

$Bp_{0.01}$ 90-95°.

Bestmann, H.J. *et al*, *Justus Liebigs Ann. Chem.*, 1981, 2117 (*synth, pmr*)

2-(1,12-Tridecadienyl)furan **T-00200**

Avocadienofuran

[34227-08-2]

$CH{=}CH(CH_2)_9CH{=}CH_2$

$C_{17}H_{26}O$ M 246.392

Component of Avocatin X, a constit. of the seeds of *Persea gratissima*.

Alves, H.M. *et al*, *Ann. Acad. Bras. Ciénc.*, 1970, **42S**, 45 (*isol, struct*)

2-(2,12-Tridecadienyl)furan **T-00201**

Isoavocadienofuran

[34227-09-3]

$CH_2CH{=}CH(CH_2)_8CH{=}CH_2$

$C_{17}H_{26}O$ M 246.392

Component of Avocatin X, a constit. of the seeds of *Persea gratissima*.

Alves, H.M. *et al*, *Ann. Acad. Bras. Ciénc.*, 1970, **42S**, 45 (*isol, struct*)

2,5-Tridecadiyn-1-ol **T-00202**

[13488-13-6]

$H_3C(CH_2)_6C{\equiv}CCH_2C{\equiv}CCH_2OH$

$C_{13}H_{20}O$ M 192.300

Oil. $Bp_{0.01}$ 90-94°.

Struijk, C.B. *et al*, *Recl. Trav. Chim. Pays-Bas* (*J. R. Neth. Chem. Soc.*), 1966, **85**, 1233 (*synth*)

Tridecanal **T-00203**

Tridecyl aldehyde

[10486-19-8]

$H_3C(CH_2)_{11}CHO$

$C_{13}H_{26}O$ M 198.348

Volatile flavour component of coriander leaf. Mp 14°. Bp_{10} 126-128°.

Oxime:

$C_{13}H_{27}NO$ M 213.362

Cryst. (EtOH aq.). Mp 80.5°.

Semicarbazone: Cryst. (EtOH). Mp 113-114°.

Stoll, M., *Helv. Chim. Acta*, 1947, **30**, 122 (*synth*)
Bott, K., *Fette, Seifen, Anstrichm.*, 1974, **76**, 443 (*synth*)
Kemp, T.R. *et al*, *J. Agric. Food Chem.*, 1974, **22**, 717 (*isol*)
Kummer, R. *et al*, *CA*, 1974, **80**, 14569f (*isol*)
Mac Leod, A.J. *et al*, *J. Sci. Food Agric.*, 1976, **27**, 721 (*isol*)

Tridecane **T-00204**

[629-50-5]

$H_3C(CH_2)_{11}CH_3$

$C_{13}H_{28}$ M 184.364

Minor constit. of wood of *Manilkara bidentata*, also isol. from lime oil. Fp −5.5°. Bp 234°, Bp_{15} 114°.

▷ YD3025000.

Vogel, A.I., *J. Chem. Soc.*, 1946, 13 (*synth*)
Kováts, E., *Helv. Chim. Acta*, 1963, **46**, 2705 (*isol*)
Cocker, W. *et al*, *J. Chem. Soc.*, 1963, 677 (*isol*)
Ivanov, C. *et al*, *CA*, 1972, **77**, 111474s (*isol, struct*)
Critchley, B.R. *et al*, *Bull. Entomol. Res.*, 1973, **63**, 1 (*tox*)
Clark, R.G. *et al*, *J. Sci. Food Agric.*, 1976, **27**, 713 (*glc, ms*)
Tonelli, A.E., *Macromolecules*, 1976, **9**, 863.
Lewis, R.J., *Sax's Dangerous Properties of Industrial Materials*, 8th Ed., Van Nostrand-Reinhold, 1992, TJH500.

Tridecanedioic acid T-00205

Brassylic acid

$HOOC(CH_2)_{11}COOH$

$C_{13}H_{24}O_4$ M 244.330

Cryst. (EtOH aq. or EtOAc). Mp 114° (111°).

Me ester: [3927-59-1].
Mp 58°.

Di-Me ester: [1472-87-3].
Mp 36°. Bp 326-328°, Bp 194°.

Di-Et ester: [15423-05-9].
Mp 20°. Bp_{12} 204°.

Et ester, chloride: Bp_5 185-188°.

Et ester, amide: Mp 91-92°.

Diamide: Mp 177°.

Dinitrile: 1,11-Dicyanoundecane
Bp_{16} 210-215°.

Cyclic ethylene ester: Ethylene brassylate
Used in perfumery.

Holde, D. *et al*, *Ber.*, 1923, **56**, 2052 (*synth*)
Potter, M.D. *et al*, *J. Chem. Soc.*, 1951, 3513 (*synth*)
Stetter, H. *et al*, *Chem. Ber.*, 1952, **85**, 290 (*synth*)
Armstead, D.E.F., *Sch. Sci. Rev.*, 1973, **55**, 416.
Marosi, L. *et al*, *Justus Liebigs Ann. Chem.*, 1973, 584.

1,2-Tridecanediol T-00206

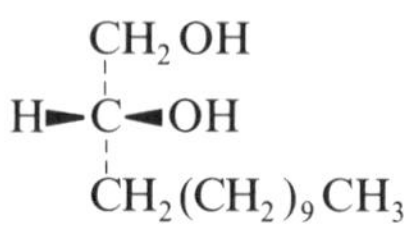

(*R*)-*form*

$C_{13}H_{28}O_2$ M 216.363

(*R*)-*form* [96883-16-8]
Plates (Et_2O/pet. ether). Mp 67-68°. $[\alpha]_D^{21}$ +10.1° (c, 1.06 in MeOH).

(*S*)-*form* [96883-15-7]
Plates (Et_2O/pet. ether). Mp 67-68°. $[\alpha]_D^{21}$ −9.63° (c, 0.122 in MeOH).

Mori, K. *et al*, *Tetrahedron*, 1985, **41**, 547 (*synth, ir, pmr*)

1,5-Tridecanediol T-00207

[39516-30-8]

$H_3C(CH_2)_7CH(OH)CH_2CH_2CH_2CH_2OH$

$C_{13}H_{28}O_2$ M 216.363

(±)-*form*
Cryst. (pet. ether). Mp 51-52°.

Massy-Westropp, R.A. *et al*, *Aust. J. Chem.*, 1984, **37**, 1303 (*synth, ir, pmr*)

1,6-Tridecanediol T-00208

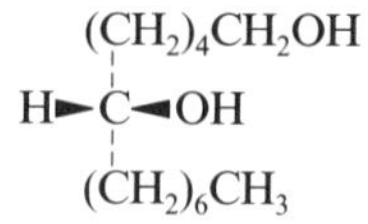

(*R*)-*form*

$C_{13}H_{28}O_2$ M 216.363

(*R*)-*form* [130409-61-9]
Cryst. (hexane). Mp 49-50°. $[\alpha]_D^{24}$ −1.6° (2.13 in $CHCl_3$).

[130433-66-8]

Matsumoto, K. *et al*, *J. Am. Chem. Soc.*, 1990, **112**, 9614 (*synth, pmr, ir, ms*)

1,13-Tridecanediol, 9CI T-00209

[13362-52-2]

$HOCH_2(CH_2)_{11}CH_2OH$

$C_{13}H_{28}O_2$ M 216.363

Mp 81°.

Di-Ph ether: [141620-05-5]. *1,1′-[1,13-Tridecanediylbis(oxy)] bisbenzene, 9CI. 1,13-Diphenoxytridecane*
$C_{25}H_{36}O_2$ M 368.558
Isol. from the leaves of *Arachis hypogaea* infected with *Puccinia arachidis*. Phytoalexin.

Guyer, A. *et al*, *Helv. Chim. Acta*, 1955, **38**, 976 (*synth*)
Gold'farb, Y.L. *et al*, *J. Gen. Chem. USSR* (*Engl. Transl.*), 1955, **25**, 1321 (*synth*)
Bianchetti, G., *Farmaco, Ed. Sci.*, 1957, **12**, 441; *CA*, **51**, 17914 (*synth*)
Colonge, J. *et al*, *Bull. Soc. Chim. Fr.*, 1959, 1248 (*synth*)
Rao, P.V.S. *et al*, *Oleagineux*, 1991, **46**, 501; *CA*, **116**, 252152 (*di-Ph ether*)

5,8-Tridecanediol T-00210

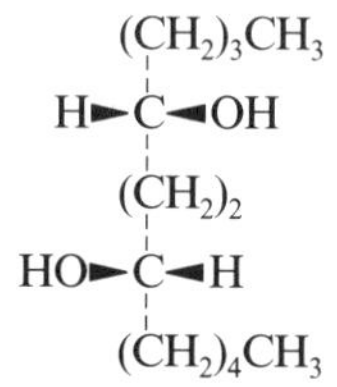

$C_{13}H_{28}O_2$ M 216.363

(*5S,8S*)-*form* [131618-46-7]
Needles (hexane). Mp 78°. $[\alpha]_D^{26}$ +4.0° (c, 0.99 in $CHCl_3$).

[131618-54-7]

Machinaga, N. *et al*, *J. Org. Chem.*, 1991, **56**, 1386 (*synth, pmr, cmr, ms*)

1-Tridecanesulfonic acid T-00211

[31169-62-7]

$H_3C(CH_2)_{12}SO_3H$

$C_{13}H_{28}O_3S$ M 264.428

Surfactant. Cryst. Mp 58°. Hygroscopic.

Na salt: [5802-89-1].
Kraft point 35.5°.

Hydrate: Cryst. Mp 48-49°.

Noller, C.R. *et al*, *J. Am. Chem. Soc.*, 1933, **55**, 1090 (*synth*)
Wilcox, L.A. *et al*, *J. Am. Chem. Soc.*, 1953, **75**, 5761 (*cryst struct*)

Tridecanoic acid T-00212

[638-53-9]

$H_3C(CH_2)_{11}COOH$

$C_{13}H_{26}O_2$ M 214.347

Constit. of *Erythrina crista-galli* trunk wood and bark. Cryst. (Me_2CO). Mp 44.5-45.5°. Bp_{24} 199-200°.

▷ YD3850000.

Zn salt: Needles (3-methylbutanol). Mp 128°.

Me ester: [1731-88-0].

$C_{14}H_{28}O_2$ M 228.374

Mp 6.5°. Bp_1 90-95°.

Et ester: [28267-29-0].

$C_{15}H_{30}O_2$ M 242.401

Bp_{60} 197-198°, Bp_5 163-165°.

Amide: [34778-57-9].

$C_{13}H_{27}NO$ M 213.362

Mp 100°.

Nitrile: [629-60-7].

$C_{13}H_{25}N$ M 195.347

Mp 9.7°. Bp 275°.

▷ YD3500000.

Org. Synth., 1936, **16**, 35.
Karlsson, K. *et al*, *Acta Chem. Scand.*, 1972, **26**, 2837 (*isol*)
Quraishi, M.S., *Can. Entomol.*, 1972, **104**, 1505 (*tox*)
Peng, A.C., *Lipids*, 1974, **9**, 299 (*isol*)
Imamura, H. *et al*, *Gigu Daigaku Nogakubu Kenkyu Hokoku*, 1981, 77; *CA*, **96**, 196524y (*occur*)
Lewis, R.J., *Sax's Dangerous Properties of Industrial Materials*, 8th Ed., Van Nostrand-Reinhold, 1992, TJH750, TJI250.

1-Tridecanol T-00213

Tridecyl alcohol

[112-70-9]

$H_3C(CH_2)_{11}CH_2OH$

$C_{13}H_{28}O$ M 200.364

Rare component of lepidopteran pheromones e.g. summer fruit tortrix moth *Adoxophyes orana fasciata*. Possesses oral, dermal and inhalation toxicity. Mp 32.5-33.5°. Bp_{15} 155-156°, $Bp_{0.5}$ 117°.

▷ YD4200000.

Propanoyl:

$C_{16}H_{32}O_2$ M 256.428

Mp −0.4°. Bp_{30} 195°.

4-Nitrobenzoyl: Mp 37.4°.

Levene, P.A. *et al*, *J. Biol. Chem.*, 1915, **20**, 528.
Armstrong, M.D. *et al*, *J. Am. Chem. Soc.*, 1943, **65**, 2252 (*synth*)
Scala, R.A. *et al*, *Am. Ind. Hyg. Assoc. J.*, 1973, **34**, 493 (*tox*)
Lee, A.G. *et al*, *Proc. R. Soc. London, B*, 1976, **193**, 253 (*cmr*)
Noguchi, H. *et al*, *Nippon Oyo Dobutsu Konchu Gakkaishi*, 1985, **29**, 278.
Lewis, R.J., *Sax's Dangerous Properties of Industrial Materials*, 8th Ed., Van Nostrand-Reinhold, 1992, TJI750.

2-Tridecanol T-00214

[1653-31-2]

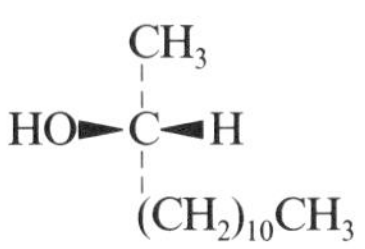

(*R*)-*form*

$C_{13}H_{28}O$ M 200.364

(R)-form [59812-98-5]

$Bp_{0.5}$ 106°. $[\alpha]_D^{25}$ −6.0° (c, 2.015 in EtOH).

(S)-form

Mp 30°. Bp 265°, Bp_{11} 156-157°. $[\alpha]_D$ +8.7° (C_6H_6).

Ac: [41015-28-5].

$C_{15}H_{30}O_2$ M 242.401

Male produced aggregation pheromone of *Drosophila mulleri*. Bp_{20} 162°. $[\alpha]_D^{20}$ +4.63°.

Hydrogen phthalate: Mp 26°. $[\alpha]_D$ +36° ($CHCl_3$), $[\alpha]_D$ +41° (EtOH).

(±)-form [112711-14-5]

Bp_{11} 151°.

Hydrogen phthalate: Mp 58-59°.

Pickard, R.H. *et al*, *J. Chem. Soc.*, 1911, **99**, 58 (*synth*)
Gunstone, F.D. *et al*, *Chem. Phys. Lipids*, 1973, **10**, 73 (*synth*)
Coke, J.L. *et al*, *J. Org. Chem.*, 1976, **41**, 3516 (*synth*)
Bartlett, R.J. *et al*, *J. Chem. Ecol.*, 1989, **15**, 399.
Gopalan, A.S., *Tetrahedron Lett.*, 1990, **31**, 5575 (*synth, deriv*)
Enders, D. *et al*, *Justus Liebigs Ann. Chem.*, 1991, 1241 (*synth*)

3-Tridecanol T-00215

3-Hydroxytridecane

[10289-68-6]

$H_3C(CH_2)_9CH(OH)CH_2CH_3$

$C_{13}H_{28}O$ M 200.364

(+)-form

Mp 32°. Bp_{12} 139°. $[\alpha]_D^{20}$ +12.4°.

Hydrogen phthalate: Mp 35-35.5°. $[\alpha]_D^{20}$ +17.6° (EtOH).

(−)-form [54831-34-4]

Mp 32°. Bp_{15} 140°. $[\alpha]_D^{20}$ −6.7° (EtOH).

Hydrogen phthalate: Mp 35-35.5°. $[\alpha]_D^{20}$ −17.7° (EtOH).

(±)-form

Mp 14.5°. Bp_{20} 148°.

Hydrogen phthalate: Mp 46-47°.

Pickard, R.H. *et al*, *J. Chem. Soc.*, 1913, **103**, 1948 (*synth*)
Yosioka, I. *et al*, *Chem. Pharm. Bull.*, 1974, **22**, 1943 (*isol*)
Krupcik, J. *et al*, *Petrochemia*, 1975, **15**, 137.

7-Tridecanol T-00216

7-Hydroxytridecane

[927-45-7]

$H_3C(CH_2)_5CHOH(CH_2)_5CH_3$

$C_{13}H_{28}O$ M 200.364

Mp 41-42°.

α-Naphthylurethane: Mp 51°.

Kipping, F.S., *J. Chem. Soc.*, 1890, **57**, 536 (*synth*)
Krupcik, J. *et al*, *Petrochemia*, 1975, **15**, 137.
Hughes, R.J. *et al*, *Tetrahedron Lett.*, 1976, 87 (*synth*)

12-Tridecanolide T-00217

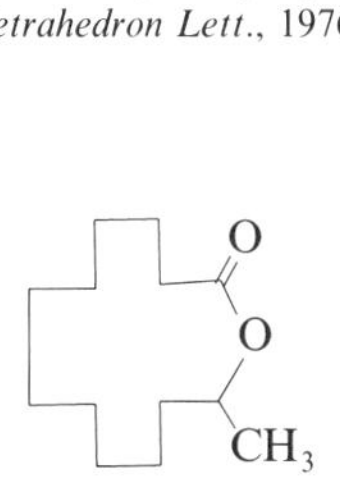

$C_{13}H_{24}O_2$ M 212.331

(−)-form

Constit. of Galbanum absolute. $[\alpha]_D^{20}$ −20.8° (c, 0.125 in $CHCl_3$).

Kaiser, R. *et al*, *Helv. Chim. Acta*, 1978, **61**, 2671.

2-Tridecanone T-00218

Methyl undecyl ketone

[593-08-8]

$H_3C(CH_2)_{10}COCH_3$

$C_{13}H_{26}O$ M 198.348

Isol. from plant oils e.g. *Schizandra nigra*, coconut oil. Ant alarm pheromone. Constit. of male mandibular gland secretions of *Philanthus basilaris* and *P. bicinctus*. Used as a flavouring essence. Mp 29°. Bp 260-265°, Bp_{16} 160°.

Oxime:

$C_{13}H_{27}NO$ M 213.362

Insol. pet. ether. Mp 56-57°.

Semicarbazone: Mp 126°.

Pickard, R.H. *et al, J. Chem. Soc.*, 1911, **55**, 45 (*synth*)
Morgan, G.T. *et al, J. Soc. Chem. Ind., London*, 1925, **44**, 108r.
Allen, R.R., *Chem. Ind. (London)*, 1965, 1560 (*isol*)
Regnier, F.E. *et al, J. Insect Physiol.*, 1968, **43**, 411.
Hajek, M. *et al, Tetrahedron Lett.*, 1974, **36**, 3193 (*synth*)
McDaniel, C.A. *et al, J. Chem. Ecol.*, 1987, **13**, 227.

3-Tridecanone T-00219

Ethyl decyl ketone

[1534-26-5]

$H_3C(CH_2)_9COCH_2CH_3$

$C_{13}H_{26}O$ M 198.348

Mp 25°. Bp_{17} 140°.

Semicarbazone: Mp 90°.

Pickard, R.H. *et al, J. Chem. Soc.*, 1913, **103**, 1948 (*synth*)

7-Tridecanone T-00220

Dihexyl ketone. Enanthone. Oenanthone

[462-18-0]

$H_3C(CH_2)_5CO(CH_2)_5CH_3$

$C_{13}H_{26}O$ M 198.348

Mp 33°. Bp_{766} 255°.

4-Nitrophenylhydrazone: Chocolate-red needles. Mp 97°.

Pickard, R.H. *et al, J. Chem. Soc.*, 1912, **101**, 629 (*synth*)
Badger, G.M. *et al, J. Chem. Soc.*, 1954, 4162 (*synth*)
Gottfried, S. *et al, Synthesis*, 1974, **2**, 117 (*synth*)
Roland, J., *Acta Pharmacol. Toxicol.*, 1975, **37**, 56 (*tox*)

3,5,7,9,11-Tridecapentayne-1,2-diol T-00221

[1771-89-7]

$H_3CC{\equiv}CC{\equiv}CC{\equiv}CC{\equiv}CC{\equiv}CCH(OH)CH_2OH$

$C_{13}H_8O_2$ M 196.205

Isol. from above-ground parts of *Bidens leucantha* and *Centaurea dealbata*. Unstable light-yellow cryst. (Et_2O/pet. ether).

Di-Ac: [3420-89-1].

$C_{17}H_{12}O_4$ M 280.279

Isol. from *Helichrysum adenocarpum*. Yellowish cryst. (pet. ether). Mp 90° dec. $[\alpha]_D^{25}$ +52.5° (c, 0.19 in Et_2O).

Di-Me ether: [124902-02-9]. *12,13-Dimethoxy-2,4,6,8,10-tridecapentayne*

$C_{15}H_{12}O_2$ M 224.259

Isol. from roots of *Pterocaulon allopecuroides*. V. unstable liq. λ_{max} 237.5, 250, 264 nm (Et_2O).

Bohlmann, F. *et al, Chem. Ber.*, 1965, **98**, 1228; 1967, **100**, 1209; 1967, **102**, 1034 (*isol, uv, ir, ord, struct*)
Bohlmann, F. *et al, Phytochemistry*, 1980, **19**, 873 (*isol*)
Magalhaes, A.F. *et al, Phytochemistry*, 1989, **28**, 2497 (*isol, struct, uv, pmr*)
Christensen, L.P. *et al, Phytochemistry*, 1991, **30**, 3289 (*isol*)

1,3,5,11-Tridecatetraene-7,9-diyne T-00222

$H_2C{=}CHCH{=}CHCH{=}CHC{\equiv}CC{\equiv}CCH{=}CHCH_3$

$C_{13}H_{12}$ M 168.238

(E,E,E)-form

Isol. from various Compositae, e.g. *Coreopsis* spp., *Bidens* spp., *Carthamus coeruleus*, *Centaurea pullata*. Yellowish cryst. (pet. ether at −60°). Mp 71-72°. V. unstable.

Sörensen, J.S. *et al, Acta Chem. Scand.*, 1954, **8**, 1741; 1961, **15**, 1884 (*isol, struct*)
Bohlmann, F. *et al, Chem. Ber.*, 1955, **88**, 1330; 1964, **97**, 1193, 2583; 1966, **99**, 1223, 3433, 3544 (*isol, synth*)

3,5,7,8-Tridecatetraene-10,12-diynoic acid, 9CI T-00223

Mycomycin

[544-51-4]

$HC{\equiv}CC{\equiv}CCH{=}C{=}CHCH{=}CHCH{=}CHCH_2COOH$

$C_{13}H_{10}O_2$ M 198.221

▷ YD4587000.

(3E,5Z)-form

Isol. from *Nocardia acidophilus*. Antibiotic. Highly unstable solid. Mp 75° (explodes). $[\alpha]_D^{25}$ −130° (c, 0.4 in EtOH).

▷ Explodes on melting.

Me ester: Cryst. (hexane). Mp 44° dec. $[\alpha]_D^{25}$ −120° (MeOH).

Celmer, W.D. *et al, J. Am. Chem. Soc.*, 1952, **74**, 1870, 2245 (*isol, uv, ir, struct*)
Bohlmann, F. *et al, Angew. Chem.*, 1964, **76**, 611 (*synth*)
Ahmed, M. *et al, J. Chem. Res., Synop.*, 1977, 125 (*biosynth*)

2,4,6,8-Tridecatetrayne T-00224

$H_3CC{\equiv}CC{\equiv}CC{\equiv}CC{\equiv}C(CH_2)_3CH_3$

$C_{13}H_{12}$ M 168.238

Metabolite of *Fistulina hepatica*. Mp 25-26°.

Jones, E.R.H. *et al, J. Chem. Soc. C*, 1966, 139 (*isol, synth*)

2,4,7-Tridecatrienal, 9CI T-00225

$H_3C(CH_2)_4CH{=}CHCH_2CH{=}CHCH{=}CHCHO$

$C_{13}H_{20}O$ M 192.300

(2E,4Z,7Z)-form [13552-96-0]

Component of cooked chicken flavour. Also from *n*-6 polyene acyl chains.

Harkes, P.D. *et al, J. Am. Oil Chem. Soc.*, 1974, **51**, 356.

1,8,10-Tridecatriene T-00226

$H_3CCH_2CH{=}CHCH{=}CH(CH_2)_5CH{=}CH_2$

$C_{13}H_{22}$ M 178.317

(8E,10Z)-form [80625-30-5]

$Bp_{0.01}$ 65-67°.

Bestmann, H.J. *et al, Justus Liebigs Ann. Chem.*, 1981, 2117 (*synth, pmr*)

1,3,5-Tridecatriene-7,9-diyne, 9CI T-00227

$H_3CCH_2CH_2C{\equiv}CC{\equiv}CCH{=}CHCH{=}CHCH{=}CH_2$

$C_{13}H_{14}$ M 170.254

(E,E,E)-form

Isol. from *Bidens aurea* and *Dahlia* spp.

Bohlmann, F. *et al*, *Chem. Ber.*, 1975, **108**, 440 (*isol*)
Bedford, C.T. *et al*, *J. Chem. Soc., Perkin Trans. 1*, 1976, 735 (*isol*)

1,5,11-Tridecatriene-7,9-diyne-3,4-diol T-00228

[59950-60-6]

$H_3CCH{=}CHC{\equiv}CC{\equiv}CCH{=}CHCH(OH)CH(OH)CH{=}CH_2$

$C_{13}H_{14}O_2$ M 202.252

(E,E)-form

Constit. of *Dahlia* spp. Mp 83-84°. $[\alpha]_D^{25}$ +28° (c, 0.64 in EtOH).

Di-Ac: [23414-60-0].
$C_{17}H_{18}O_4$ M 286.327
Mp 52-53°. $[\alpha]_D^{23}$ +147° (c, 1.01 in EtOH).

Bedford, C.T. *et al*, *J. Chem. Soc., Perkin Trans. 1*, 1976, 735 (*isol, synth*)

3,5,7-Tridecatriene-9,11-diyn-1-ol, 9CI T-00229

Chamomilla alcohol

$H_3CC{\equiv}CC{\equiv}CCH{=}CHCH{=}CHCH{=}CHCH_2CH_2OH$

$C_{13}H_{14}O$ M 186.253

(all-E)-form

Isol. from roots of *Matricaria chamomilla* and *M. discoidea*. Cryst. (Et_2O/pet. ether). Mp 86.5°.

Ac: [506-16-1]. *1-Acetoxy-3,5,7-tridecatriene-9,11-diyne*. ***Chamomilla ester***
$C_{15}H_{16}O_2$ M 228.290
Isol. from *M.* sp. Cryst. (pet. ether). Mp 68.5°.

(3E,5Z,7E)-form

Ac: From *M.* sp.

Christensen, P.K., *CA*, 1960, **54**, 6531 (*isol*)
Bohlmann, F. *et al*, *Chem. Ber.*, 1960, **93**, 1931; 1970, **103**, 2860 (*isol, synth, ir, uv, struct, biosynth*)

1,3,5-Tridecatriene-7,9,11-triyne T-00230

[2739-64-2]

$H_3CC{\equiv}CC{\equiv}CC{\equiv}CCH{=}CHCH{=}CHCH{=}CH_2$

$C_{13}H_{10}$ M 166.222

Found in many Compositae spp. without detn. of geom. isom. (prob. (*E, E*)), e.g. *Achillea ptarmica*, *Anthemis* spp., *Anacyclus* spp., *Flaveria repanda*, *Matricaria oreades*, and *Maruta fuscata*.

(E,E)-form [6581-77-7]

Constit. of *Achillea ptarmica*. Mp 88-93°.

3,4:5,6-Diepoxide: 3,4:5,6-Diepoxy-1-tridecene-7,9,11-triyne. ***Annuadiepoxide***
$C_{13}H_{10}O_2$ M 198.221
Isol. from *Artemisia annua*. Unstable cryst.

(3E,5Z)-form [126381-91-7]

Isol. from *Artemisia vulgaris*.

Bohlmann, F. *et al*, *Chem. Ber.*, 1962, **95**, 1320; 1973, **106**, 1328 (*isol, uv, ir, pmr*)
David, D. *et al*, *Phytochemistry*, 1974, **13**, 455 (*isol, struct*)
Kenichi, I. *et al*, *Agric. Biol. Chem.*, 1975, **39**, 1103 (*isol*)
Wallnoefer, B. *et al*, *Phytochemistry*, 1989, **28**, 2687 (*isol, pmr*)
Manns, D. *et al*, *J. Nat. Prod.* (*Lloydia*), 1992, **55**, 29 (*Annuadiepoxide*)

1,3,11-Tridecatriene-5,7,9-triyne T-00231

[18668-89-8]

$H_3CCH{=}CH(C{\equiv}C)_3CH{=}CHCH{=}CH_2$

$C_{13}H_{10}$ M 166.222

Constit. of *Carthamus tinctorius*, *Bidens ferulaefolia*, *Centaurea* spp. and *Dahlia* spp. Oil. Partial stereochem. known for some isolates.

1,2-Epoxide: [2780-71-4]. *(1,9-Undecadiene-3,5,7-triynyl) oxirane, 9CI. 12,13-Epoxy-2,10-undecadiene-4,6,8-triyne*
$C_{13}H_{10}O$ M 182.221
Isol. from several Compositae spp. Cryst. (pet. ether). Mp 76.5°. $[\alpha]_D^{20}$ +63° ($CHCl_3$). Has (3*E*,7*E*)-config., abs. config. not detd.

[50739-51-0, 58998-11-1, 61434-49-9, 124604-44-0]

Sorensen, N.A. *et al*, *Acta Chem. Scand.*, 1958, **12**, 756; 1959, **13**, 2101 (*isol, synth*)
Bohlmann, F. *et al*, *Chem. Ber.*, 1958, **91**, 1631, 1642; 1961, **94**, 3179; 1964, **97**, 1193, 2583; 1966, **99**, 3433 (*isol, synth, epoxide*)
Chin, C. *et al*, *J. Chem. Soc. C*, 1970, 314 (*isol*)
Lam, J., *Planta Med.*, 1973, **24**, 107 (*isol*)
Kenichi, I. *et al*, *Agric. Biol. Chem.*, 1975, **39**, 1103 (*isol*)
Kogiso, S. *et al*, *Agric. Biol. Chem.*, 1976, **40**, 2085 (*isol, struct, uv, ir, ms, pmr activity*)
Bedford, C.T. *et al*, *J. Chem. Soc., Perkin Trans. 1*, 1976, 735 (*isol, synth*)
Binder, R.G. *et al*, *J. Agric. Food Chem.*, 1990, **38**, 764, 1053, 1245 (*isol*)

2,10,12-Tridecatriene-4,6,8-triyn-1-ol T-00232

[20252-40-8]

$H_2C{=}CHCH{=}CH(C{\equiv}C)_3CH{=}CHCH_2OH$

$C_{13}H_{10}O$ M 182.221

(E,E)-form [26130-90-5]

Isol. from *Cosmos sulphureus*, *C. hybridus*, *Coreopsis tinctoria*, *Dahlia coccinea*, *D. scapigera* and others. Cryst. (Et_2O/pet. ether). Mp 100° dec.

Ac: [26130-88-1]. *13-Acetoxy-1,3,11-tridecatriene-5,7,9-triyne*
$C_{15}H_{12}O_2$ M 224.259
Isol. from *Cosmos sulphureus*, *Bidens andicola*, *D. coccinea* and *D. scapigera*. Needles. Mp 31°.

Aldehyde: [132237-71-9]. *2,10,12-Tridecatriene-4,6,8-triyn-1-al*
$C_{13}H_8O$ M 180.206
Constit. of the root of *Carlina vulgare*, *Cosmos sulphureus* and of *C. hybridus*. Cryst. (pet. ether). Mp 92° dec. Also descr. as yellow oil.

(E,Z)-form [15400-70-1]

Isol. from *Carlina* sp. Yellowish oil.

(Z,Z)-form [15400-66-5]

Isol. from *Carlina acaulis* and *C. vulgaris*. Yellowish oil.

[20252-38-4, 20252-39-5]

Sorensen, J.S. *et al*, *Acta Chem. Scand.*, 1954, **8**, 1769 (*isol*)
Bohlmann, F. *et al*, *Chem. Ber.*, 1956, **89**, 21; 1964, **97**, 2583; 1967, **100**, 1507 (*isol, uv, ir, pmr, struct, synth*)
Lam, J. *et al*, *Phytochemistry*, 1968, **7**, 269; 1971, **10**, 2227 (*isol*)
Chin, C. *et al*, *J. Chem. Soc. C*, 1970, 314 (*isol*)
Clinton, C. *et al*, *J. Chem. Soc. C*, 1970, 314 (*isol, struct*)
Bedford, C.T. *et al*, *J. Chem. Soc., Perkin Trans. 1*, 1976, 735 (*isol*)
Bohlmann, F. *et al*, *Phytochemistry*, 1983, **22**, 1281 (*acetate*)
Norton, R.A. *et al*, *J. Plant Physiol.*, 1985, **120**, 273.
Christensen, L.P. *et al*, *Phytochemistry*, 1990, **29**, 3153 (*isol*)

4,7,10-Tridecatrien-1-ol T-00233

$H_3CCH_2CH{=}CHCH_2CH{=}CHCH_2CH{=}CHCH_2CH_2CH_2OH$

$C_{13}H_{22}O$ M 194.316

(4E,7Z,10Z)-form

Ac: [61810-56-8].

$C_{15}H_{24}O_2$ M 236.353

Sex pheromone of potato tuberworm moth *Phthorimaea operculella.*

(4Z,7E,10Z)-form

Ac: [67467-16-7].

Sex pheromone of *P. operculella.*

Persoons, C.J. *et al*, *Entomol. Exp. Appl.*, 1976, **20**, 289.
Voerman, S. *et al*, *Potato Res.*, 1977, **20**, 123 (*bibl*)
Voerman, S. *et al*, *J. Chem. Ecol.*, 1978, **4**, 531 (*synth*)
Japan. Pat., 78 37 612, (1978); *CA*, **89**, 108177.
Toth, M. *et al*, *J. Chem. Ecol.*, 1984, **10**, 271.

2,5,8-Tridecatriyn-1-ol T-00234

[19960-60-2]

$H_3C(CH_2)_3C{\equiv}CCH_2C{\equiv}CCH_2C{\equiv}CCH_2OH$

$C_{13}H_{16}O$ M 188.269

Mp 23-25°. $Bp_{0.001}$ 118-123°.

Beerthuis, R.K. *et al*, *Recl. Trav. Chim. Pays-Bas* (*J. R. Neth. Chem. Soc.*), 1968, **87**, 461 (*synth, ir, nmr*)

2-Tridecenal T-00235

[7774-82-5]

$H_3C(CH_2)_9CH{=}CHCHO$

$C_{13}H_{24}O$ M 196.332

(*E*)-*form* [7069-41-2]

Liq. $Bp_{0.16}$ 80°.

2,4-Dinitrophenylhydrazone: Cryst. Mp 117°.

(*Z*)-*form* [71277-06-0]

Liq. $Bp_{0.3}$ 90-92°.

Di-Et acetal: [13043-66-8]. *1,1-Diethoxy-2-tridecene*

$C_{17}H_{34}O_2$ M 270.454

$Bp_{0.2}$ 111°.

Hirai, K. *et al*, *Tetrahedron Lett.*, 1972, 2743 (*synth, pmr*)
Bestmann, H.J. *et al*, *Chem. Ber.*, 1982, **115**, 161 (*synth*)
Chang, Y.-H. *et al*, *Synthesis*, 1990, 1033 (*synth, pmr, ir, ms*)

1-Tridecene T-00236

1-Tridecylene

[2437-56-1]

$H_3C(CH_2)_{10}CH{=}CH_2$

$C_{13}H_{26}$ M 182.348

Fp −13°. Bp_{10} 102-103°.

▷ YD4595000.

Kozacik, A.P. *et al*, *J. Am. Chem. Soc.*, 1938, **60**, 2436 (*synth*)
Hoebold, W. *et al*, *J. Prakt. Chem.*, 1975, **317**, 1054 (*cmr*)
Schmidt, A.W. *et al*, *Chem. Ber.*, 1977, **74**, 1313 (*synth*)

2-Tridecene-4,7-diynal, 9CI T-00237

$H_3C(CH_2)_4C{\equiv}CCH_2C{\equiv}CCH{=}CHCHO$

$C_{13}H_{16}O$ M 188.269

(*E*)-*form* [74448-18-3]

Liq.

Rosenberger, M. *et al*, *J. Am. Chem. Soc.*, 1983, **105**, 3656 (*synth, pmr*)

1-Tridecene-6,12-diyne T-00238

[74585-60-7]

$HC{\equiv}C(CH_2)_4C{\equiv}C(CH_2)_3CH{=}CH_2$

$C_{13}H_{18}$ M 174.285

Oil. $Bp_{0.01}$ 50°.

Sternberg, E.D. *et al*, *J. Org. Chem.*, 1984, **49**, 1564 (*synth, ir, pmr, ms*)

7-Tridecene-10,12-diynoic acid, 9CI T-00239

$HC{\equiv}CC{\equiv}CCH_2CH{=}CH(CH_2)_5COOH$

$C_{13}H_{16}O_2$ M 204.268

(*Z*)-*form*

2-Methylpropylamide: N-*Isobutyl-7-tridecene-10,12-diynamide*

$C_{17}H_{25}NO$ M 259.391

Isol. from flower heads of *Spilanthes acmella.* Oil.

Nakatani, N. *et al*, *Biosci., Biotechnol., Biochem.*, 1992, **56**, 759 (*isol, ms, ir, cmr, pmr*)

3-Tridecene-5,7,9-triyne-1,2-diol T-00240

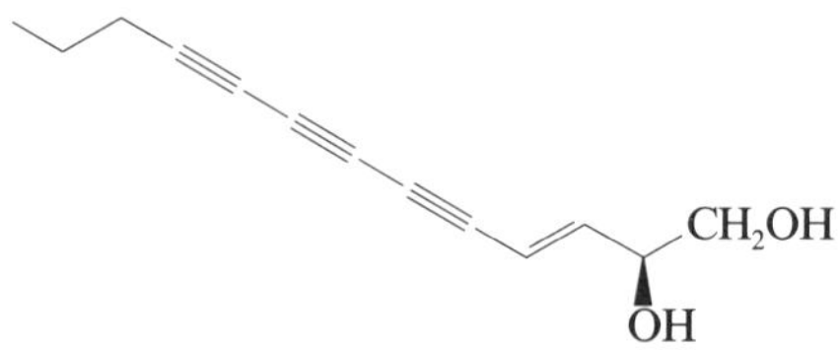

(*S,E*)-*form*

$C_{13}H_{14}O_2$ M 202.252

(*S*)-(*E*)-*form* [54542-14-2]

Isol. from cultures of *Fistulina pallida.* Cryst. (Et_2O/pet. ether). Mp 98-98.5°. $[\alpha]_D^{20}$ −9.85° (c, 0.5 in EtOH).

(*S*)-(*Z*)-*form* [54541-81-0]

Needles (Et_2O/pet. ether). Mp 73-74°. $[\alpha]_D^{20}$ −38° (c, 0.253 in EtOH).

Ahmed, M. *et al*, *J. Chem. Soc., Perkin Trans. 1*, 1974, 1981 (*isol, synth*)

2-Tridecene-4,6,8-triyn-1-ol T-00241

$H_3C(CH_2)_3C{\equiv}CC{\equiv}CC{\equiv}CCH{=}CHCH_2OH$

$C_{13}H_{14}O$ M 186.253

(*E*)-*form*

Metabolite of *Fistulina hepatica.* Mp 50-52°.

Jones, E.R.H. *et al*, *J. Chem. Soc. C*, 1966, 139 (*isol, synth*)

2-Tridecenoic acid T-00242

[6969-16-0]

$H_3C(CH_2)_9CH{=}CHCOOH$

$C_{13}H_{24}O_2$ M 212.331

(*E*)-*form* [32466-55-0]

Chloride:

$C_{13}H_{23}ClO$ M 230.777

Bp_3 131°.

Amide:

$C_{13}H_{25}NO$ M 211.347

Mp 116°.

Zaar, B., *CA*, 1930, **24**, 2108 (*synth*)
Sayeed, Q.M., *Can. Entomol.*, 1972, **105**, 1505.

7-Tridecenoic acid T-00243

[73292-41-8]

$H_3C(CH_2)_4CH{=}CH(CH_2)_5COOH$

$C_{13}H_{24}O_2$ M 212.331

$Bp_{0.5}$ 145°.

Scholz, D., *Justus Liebigs Ann. Chem.*, 1984, 264 (*synth, ir, pmr*)

12-Tridecenoic acid T-00244

$H_2C{=}CH(CH_2)_{10}COOH$

$C_{13}H_{24}O_2$ M 212.331

Cryst. (EtOH aq.). Mp 38-39°. Bp_{20} 192°, Bp_3 162°.

Me ester: [29780-00-5].

$C_{14}H_{26}O_2$ M 226.358

Bp_3 133°.

Et ester:

$C_{15}H_{28}O_2$ M 240.385

Bp_8 150°.

Chuit, P. *et al*, *Helv. Chim. Acta*, 1927, **10**, 118 (*synth*)
Tomecko, C.G. *et al*, *J. Am. Chem. Soc.*, 1927, **49**, 522 (*synth*)

4-Tridecen-1-ol, 9CI T-00245

$H_3C(CH_2)_7CH{=}CHCH_2CH_2CH_2OH$

$C_{13}H_{26}O$ M 198.348

(E)-form [75568-02-4]

Liq. $Bp_{0.3}$ 96-99°, $Bp_{0.1}$ 106-107°. n_D^{21} 1.4550.

Ac: [72269-48-8]. *1-Acetoxy-4-tridecene*

$C_{15}H_{28}O_2$ M 240.385

Sole component of the sex pheromone of the tomato pinworm moth, *Keiferia lycopersicella*. Commercially available. $Bp_{0.01}$ 80-82°, Bp_1 107-108°. n_D^{21} 1.4445.

(Z)-form [75568-01-3]

Pre-orbital gland secretion of the small neotragine antelope, *Raphicerus melanotis* (grysbok). Liq. Bp_1 120°.

Ac: [65954-19-0].

Synthetic sex attractant when used as a minor constituent of (E/Z)-acetate mixture for *K. lycopersicella*. Commercially available.

[122705-57-1]

Sonnet, P.E., *Synthesis*, 1980, 828 (*synth*)
Wenkert, E. *et al*, *J. Org. Chem.*, 1985, **50**, 719 (*isol, synth, pmr, cmr, ir*)
Patterson, J.W., *Synthesis*, 1985, 337 (*synth*)
Yadav, J.S. *et al*, *Tetrahedron*, 1988, **44**, 7243 (*synth, pmr, ir*)
Kelkar, S.V. *et al*, *Indian J. Chem., Sect. B*, 1989, **28**, 980 (*synth*)
Moiseenkov, A.M. *et al*, *J. Chem. Soc., Perkin Trans. 1*, 1991, 2639 (*synth, pmr*)
Bikulova, L.M. *et al*, *Khim. Prir. Soedin.*, 1991, 444 (*synth, ir, pmr*)

6-Tridecen-2-ol T-00246

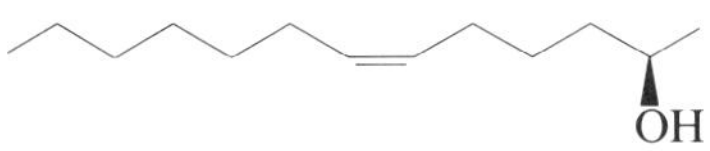

$C_{13}H_{26}O$ M 198.348

(R,Z)-form

Constit. of the thoracic gland secretions of *Acanthaclisis occitanica* and *Synclysis baetica*.

Bergstrom, G. *et al*, *J. Chem. Ecol.*, 1992, **18**, 1177 (*isol*)

8-Tridecen-1-ol, 9CI T-00247

$H_3C(CH_2)_3CH{=}CH(CH_2)_6CH_2OH$

$C_{13}H_{26}O$ M 198.348

(E)-form [64437-29-2]

Liq. Bp_1 110-111°. n_D^{23} 1.4539.

Ac: [64437-30-5].

Liq. $Bp_{0.015}$ 97-98°. n_D^{23} 1.4440.

(Z)-form [64502-56-3]

Liq. Bp_{10} 145-147°, $Bp_{0.02}$ 90-91° (>95% *Z*). n_D^{23} 1.4539.

Ac: [60198-15-4].

$C_{15}H_{28}O_2$ M 240.385

Major component of the sex pheromones of several Lepidoptera species, including the tomato looper *Plusia chalcites* and the sugarcane internode borer, *Chilo sacchariphagus indicus*.

Ando, T. *et al*, *Agric. Biol. Chem.*, 1977, **41**, 1485.
Horiike, M. *et al*, *Agric. Biol. Chem.*, 1980, **44**, 257 (*synth, pmr*)
Nesbitt, B.F. *et al*, *J. Chem. Ecol.*, 1986, **12**, 1377 (*isol, biochem*)
Dunkelblum, E. *et al*, *J. Chem. Ecol.*, 1987, **13**, 991 (*isol*)
Trehan, I.R. *et al*, *Indian J. Chem., Sect. B*, 1991, **30**, 563 (*synth, pmr, ir*)
Moiseenkov, A.M. *et al*, *J. Chem. Soc., Perkin Trans. 1*, 1991, 2639 (*synth, pmr, cmr*)

10-Tridecen-1-ol, 9CI T-00248

$H_3CCH_2CH{=}CH(CH_2)_8CH_2OH$

$C_{13}H_{26}O$ M 198.348

(Z)-form [64437-31-6]

Oil. $Bp_{0.35}$ 101-106° (>95% (*Z*)-form).

Ac: [64437-24-7]. *10-Tridecenyl acetate*

$C_{15}H_{28}O_2$ M 240.385

Minor component of the sex pheromone of *Phyllonorycter watanebei*, *Cosmopterix victor*, European grape berry moth *Eupoecilia ambiguella*; shows synergistic effect with principal component 9-Dodecen-1-ol, D-00420 (acetate, *Z* form).

[64437-32-7, 64437-33-8]

Ando, T. *et al*, *Agric. Biol. Chem.*, 1977, **41**, 1485 (*synth, glc*)
Svirskaya, P.I. *et al*, *J. Chem. Eng. Data*, 1979, **24**, 152 (*synth*)
Horiike, M. *et al*, *Agric. Biol. Chem.*, 1980, **44**, 257 (*synth, pmr, ms, glc, activity*)
Rauscher, S. *et al*, *J. Chem. Ecol.*, 1984, **10**, 253 (*activity*)
Willemse, L.P.M. *et al*, *J. Appl. Entomol.*, 1987, **103**, 508 (*activity*)
Streinz, L. *et al*, *Collect. Czech. Chem. Commun.*, 1990, **55**, 1555 (*synth*)

11-Tridecen-1-ol T-00249

$H_3CCH{=}CH(CH_2)_9CH_2OH$

$C_{13}H_{26}O$ M 198.348

Bp_{11} 150-152°.

(Z)-form [34010-24-7]

Ac: [33951-95-0].

Insect attractant for male red-banded leaf roller moths *Argyrotaenia velutinana*. $Bp_{0.05}$ 101°.

Chuit, P. *et al*, *Helv. Chim. Acta*, 1927, **10**, 113 (*synth*)
Bestmann, H.J. *et al*, *Chem. Ber.*, 1978, **111**, 248 (*synth, pmr, ir*)

12-Tridecen-2-one T-00250

[60437-21-0]

$H_2C{=}CH(CH_2)_9COCH_3$

$C_{13}H_{24}O$ M 196.332

Isol. from bark of *Litsea elliptica*. Bp_{30} 150°.

Arbain, D. *et al*, *Aust. J. Chem.*, 1990, **43**, 1949 (*isol, ir, pmr, cmr*)

5-(8-Tridecenyl)-1,3-benzenediol, 9CI T-00251

5-(8-Tridecenyl)resorcinol

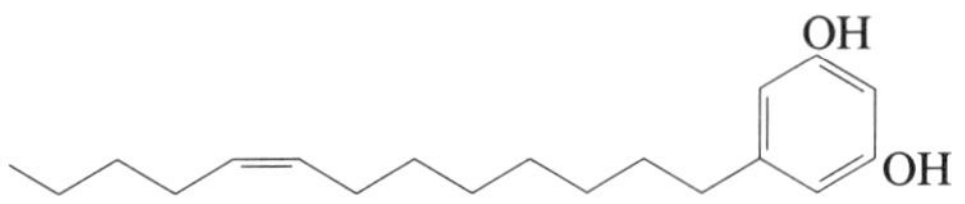

$C_{19}H_{30}O_2$ M 290.445

(*Z*)-*form* [62897-10-3] ***Ardisinol II.*** *Roupelliol*

Isol. from *Rapanea laetevirens* seed, *Ardisia japonica* and *Protea roupelliae*. Shows antitubercular activity. Yellow oil.

▷ Skin irritant.

Di-Me ether: [125187-38-4]. *1,3-Dimethoxy-5-(8-tridecenyl)-benzene, 9CI*

$C_{21}H_{34}O_2$ M 318.498

Isol. from heartwood of *P. roupelliae*. Pale yellow oil.

8′,9′-Dihydro: see *5-Tridecyl-1,3-benzenediol, T-00254*

Madriwal, R.V. *et al*, *Lipids*, 1977, **12**, 402 (*isol*)
Hu, H. *et al*, *CA*, 1980, **92**, 124878m (*isol*)
Bullock, L.M. *et al*, *S. Afr. J. Chem.*, 1989, **42**, 76 (*isol, pmr, ms*)

2-(12-Tridecenyl)furan T-00252

Avocadenofuran

[25346-24-1]

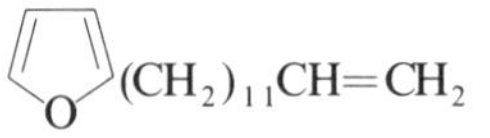

$C_{17}H_{28}O$ M 248.408

Constit. of Avocado pear.

Kashman, Y. *et al*, *Tetrahedron*, 1969, **25**, 4617 (*isol*)
Alves, H.M. *et al*, *Ann. Acad. Bras. Ciênc.*, 1970, **42S**, 45.

Tridecylamine T-00253

1-Aminotridecane. 1-Tridecanamine, 9CI

[2869-34-3]

$H_3C(CH_2)_{11}CH_2NH_2$

$C_{13}H_{29}N$ M 199.379

Mp 27°. Bp 276°, Bp_4 121°. pK_a 10.63 (25°).

B,HCl: Mp 192-193°.

Benzoyl: Mp 70.6°.

Ralston, A.W. *et al*, *Ind. Eng. Chem.*, 1940, **32**, 1093.
Suida, H. *et al*, *Ber.*, 1942, **75**, 991.
Murr, B.L. *et al*, *J. Am. Chem. Soc.*, 1955, **77**, 1684 (*synth*)
Goto, R. *et al*, *Nippon Kagaku Zasshi*, 1966, **87**, 1220; *CA*, **66**, 75599.

5-Tridecyl-1,3-benzenediol, 9CI T-00254

5-Tridecylresorcinol. ***Grevillol.*** *Trifurcatol A_1*

[5259-01-8]

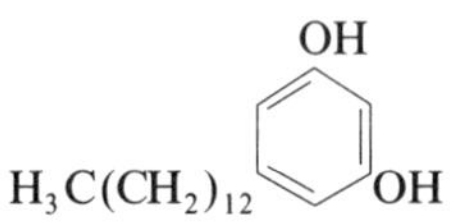

$C_{19}H_{32}O_2$ M 292.461

Constit. of the woods of *Grevillea* spp. and *Opisthiolepis heterophylla*. Needles (C_6H_6). Mp 82-83°.

Di-Ac: [5259-02-9].

$C_{23}H_{36}O_4$ M 376.535

Needles (EtOH). Mp 48°.

Mono-Me ether: [66284-51-3]. *3-Methoxy-5-tridecylphenol*

$C_{20}H_{34}O_2$ M 306.487

Constit. of *Hakea saligna*. Syrup.

Di-Me ether: [5259-04-1]. *1,3-Dimethoxy-5-tridecylbenzene*

$C_{21}H_{36}O_2$ M 320.514

Constit. of *H. saligna*. Mp 109-110°.

*2′*R*,8′*S*-Dihydroxy:* [129277-44-7]. ***5-(2,8-Dihydroxytridecyl)-1,3-benzenediol***

$C_{19}H_{32}O_4$ M 324.459

Constit. of *Ononis natrix* ssp. *hispanica*.

*2′*R*,8′*S*-Dihydroxy, 3-Me ether:* [129277-45-8]. ***5-(2,8-Dihydroxytridecyl)-3-methoxyphenol***

$C_{20}H_{34}O_4$ M 338.486

Constit. of *O. natrix* ssp. *hispanica*.

*2′*R*-Acetoxy:* [129277-43-6]. ***5-(2-Acetoxytridecyl)-1,3-benzenediol***

$C_{21}H_{34}O_4$ M 350.497

Constit. of *O. natrix* ssp. *hispanica*.

*2′*R*-Hydroxy, 3-Me ether:* ***5-(2-Hydroxytridecyl)-3-methoxyphenol***

$C_{20}H_{34}O_3$ M 322.487

Constit. of *O. natrix* ssp. *hispanica*. Cryst. (CH_2Cl_2/hexane). Mp 70°. $[\alpha]_D$ −0.6° (c, 1.2 in $CHCl_3$).

*2′*R*-Acetoxy, 3-Me ether:* ***5-(2-Acetoxytridecyl)-3-methoxyphenol***

$C_{22}H_{36}O_4$ M 364.524

Constit. of *O.* ssp. *hispanica*. Oil. $[\alpha]_D$ +0.7° (c, 1.24 in $CHCl_3$).

*2′*R*-Acetoxy, 12′*S*-hydroxy:* [133361-21-4]. ***5-(2-Acetoxy-12-hydroxytridecyl)-1,3-benzenediol***

$C_{21}H_{34}O_5$ M 366.497

Constit. of *O. viscosa*. Oil. $[\alpha]^{20}_D$ +6.9° (c, 0.63 in $CHCl_3$).

*8′-Oxo, 2′*R*-hydroxy:* [129277-46-9]. ***13-(2,4-Dihydroxyphenyl)-12-hydroxy-6-tridecanone.*** *5-(2-Hydroxy-8-oxotridecyl)-1,3-benzenediol*

$C_{19}H_{30}O_4$ M 322.444

Constit. of *O. natrix* ssp. *hispanica*.

Ritchie, E. *et al*, *Aust. J. Chem.*, 1965, **18**, 2015 (*isol, struct, synth*)
Cirigottis, K.A. *et al*, *Aust. J. Chem.*, 1974, **27**, 345 (*isol*)
Cleaver, L. *et al*, *Aust. J. Chem.*, 1976, **29**, 1989 (*synth*)
Manju, M. *et al*, *Indian J. Chem., Sect. B*, 1977, **15**, 1090 (*isol, derivs*)
Feliciano, A.S. *et al*, *Phytochemistry*, 1983, **22**, 2031 (*derivs*)
Barrero, A.F. *et al*, *Phytochemistry*, 1990, **29**, 1967; 1991, **30**, 641 (*derivs*)

2-Tridecylcyclopropanepropanoic acid, 9CI T-00255

4,5-Methyleneoctadecanoic acid

$C_{19}H_{36}O_2$ M 296.492

Both *cis*- and *trans*-forms have been synthesised as Me esters.

Christie, W.W. *et al*, *Chem. Phys. Lipids*, 1968, **2**, 196 (*synth, glc, ir, nmr*)
Christie, W.W., *Top. Lipid Chem.*, 1970, **1**, 4 (*rev*)
Gunstone, F.D. *et al*, *Chem. Phys. Lipids*, 1973, **10**, 303 (*synth, glc, nmr*)

3-Tridecynoic acid, 9CI T-00256

[54373-83-0]

$H_3C(CH_2)_8C\equiv CCH_2COOH$

$C_{13}H_{22}O_2$ M 210.316

Mp 54.5-56.5°.

Gilman, N.W. *et al*, *Chem. Phys. Lipids*, 1974, **13**, 239 (*synth*)
Holland, B.C. *et al*, *Synth. Commun.*, 1974, **4**, 203 (*synth*)
Gunstone, F.D. *et al*, *Chem. Phys. Lipids*, 1976, **17**, 1 (*cmr*)
Specher, H., *Prog. Chem. Fats Other Lipids*, 1977, **15**, 279.

4-Tridecynoic acid, 9CI T-00257

[25593-88-8]

$H_3C(CH_2)_7C\equiv CCH_2CH_2COOH$

$C_{13}H_{22}O_2$ M 210.316

Mp 53-55°.

Williams, J.C. *et al*, *Chem. Phys. Lipids*, 1972, **9**, 320 (*pmr*)
Gilman, N.W. *et al*, *Chem. Phys. Lipids*, 1974, **13**, 239 (*synth*)
Sprecher, H., *Prog. Chem. Fats Other Lipids*, 1977, **15**, 219.

5-Tridecynoic acid, 9CI T-00258

[55182-80-4]

$H_3C(CH_2)_6C\equiv C(CH_2)_3COOH$

$C_{13}H_{22}O_2$ M 210.316

$Bp_{0.35}$ 123-125°.

Gilman, N.W. *et al*, *Chem. Phys. Lipids*, 1974, **13**, 239 (*synth*)
Gunstone, F.D. *et al*, *Chem. Phys. Lipids*, 1976, **17**, 1 (*cmr*)
Sprecher, H., *Prog. Chem. Fats Other Lipids*, 1977, **15**, 219.

6-Tridecynoic acid, 9CI T-00259

[39727-86-1]

$H_3C(CH_2)_5C\equiv C(CH_2)_4COOH$

$C_{13}H_{22}O_2$ M 210.316

Mp 7.5-8.5°. Bp_1 149-151°, $Bp_{0.2}$ 135-145°.

Anilide: Mp 41-41.5°.

Taylor, W.R. *et al*, *J. Am. Chem. Soc.*, 1950, **72**, 4263 (*synth*)
Gilman, N.W. *et al*, *Chem. Phys. Lipids*, 1974, **13**, 239 (*synth*)
Sprecher, H., *Prog. Chem. Fats Other Lipids*, 1977, **15**, 219.

7-Tridecynoic acid, 9CI T-00260

[55182-85-9]

$H_3C(CH_2)_4C\equiv C(CH_2)_5COOH$

$C_{13}H_{22}O_2$ M 210.316

Mp 22-24°. Bp_1 152-153°, $Bp_{0.15}$ 115°.

Anilide: Mp 51.5-52°.

Taylor, W.R. *et al*, *J. Am. Chem. Soc.*, 1950, **72**, 4263 (*synth*)
Gilman, N.W. *et al*, *Chem. Phys. Lipids*, 1974, **13**, 239 (*synth*)
Gunstone, F.D. *et al*, *Chem. Phys. Lipids*, 1976, **17**, 1 (*cmr*)
Sprecher, H., *Prog. Chem. Fats Other Lipids*, 1977, **15**, 219.

8-Tridecynoic acid, 9CI T-00261

[55182-89-3]

$H_3C(CH_2)_3C\equiv C(CH_2)_6COOH$

$C_{13}H_{22}O_2$ M 210.316

$Bp_{1.2}$ 130-134°.

Gilman, N.W. *et al*, *Chem. Phys. Lipids*, 1974, **13**, 239 (*synth*)
Sprecher, H., *Prog. Chem. Fats Other Lipids*, 1977, **15**, 219.

9-Tridecynoic acid, 9CI T-00262

[55182-91-7]

$H_3CCH_2CH_2C\equiv C(CH_2)_7COOH$

$C_{13}H_{22}O_2$ M 210.316

Mp 29-31°.

Grimmer, G. *et al*, *Chem. Ber.*, 1963, **96**, 3370 (*synth*)
Gilman, N.W. *et al*, *Chem. Phys. Lipids*, 1974, **13**, 239 (*synth*)
Gunstone, F.D. *et al*, *Chem. Phys. Lipids*, 1976, **17**, 1 (*cmr*)
Sprecher, H., *Prog. Chem. Fats Other Lipids*, 1977, **15**, 219.

10-Tridecynoic acid, 9CI T-00263

[55182-93-9]

$H_3CCH_2C\equiv C(CH_2)_8COOH$

$C_{13}H_{22}O_2$ M 210.316

Mp 34-36°.

Gilman, N.W. *et al*, *Chem. Phys. Lipids*, 1974, **13**, 239 (*synth*)
Gunstone, F.D. *et al*, *Chem. Phys. Lipids*, 1976, **17**, 1 (*cmr*)
Sprecher, H., *Prog. Chem. Fats Other Lipids*, 1977, **15**, 219.

11-Tridecynoic acid, 9CI T-00264

[55182-95-1]

$H_3CC\equiv C(CH_2)_9COOH$

$C_{13}H_{22}O_2$ M 210.316

Mp 61-63°.

Gilman, N.W. *et al*, *Chem. Phys. Lipids*, 1974, **13**, 239 (*synth*)
Sprecher, H., *Prog. Chem. Fats Other Lipids*, 1977, **15**, 219.

12-Tridecynoic acid T-00265

[14502-46-6]

$HC\equiv C(CH_2)_{10}COOH$

$C_{13}H_{22}O_2$ M 210.316

Cryst. (pet. ether). Mp 54-55°.

Christie, W.W. *et al*, *Chem. Phys. Lipids*, 1967, **1**, 407 (*synth*)

2-Tridecyn-1-ol, 9CI T-00266

[51887-25-3]

$H_3C(CH_2)_9C\equiv CCH_2OH$

$C_{13}H_{24}O$ M 196.332

Mp 37-38°.

Christie, W.W. *et al*, *Chem. Phys. Lipids*, 1967, **1**, 407 (*synth*)
Ermilova, E.V., *Zh. Org. Khim.*, 1975, **11**, 520 (*synth*)

12-Tridecyn-1-ol T-00267

[18202-11-4]

$HC\equiv C(CH_2)_{10}CH_2OH$

$C_{13}H_{24}O$ M 196.332

Mp 30-32°. $Bp_{0.3}$ 95-98°.

Christie, W.W. *et al*, *Chem. Phys. Lipids*, 1967, **1**, 407 (*synth*)
Knapp, F.F. *et al*, *J. Med. Chem.*, 1984, **27**, 57 (*synth*, *pmr*)

2-(12-Tridecynyl)furan T-00268

Avocadynofuran

[24708-33-6]

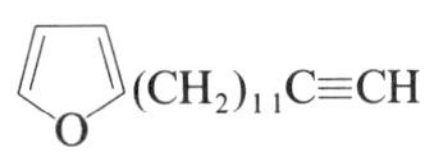

$C_{17}H_{26}O$ M 246.392

Constit. of Avocado pear.

Kashman, Y. *et al*, *Tetrahedron*, 1969, **25**, 4617 (*isol*)
Alves, H.M. *et al*, *Ann. Acad. Bras. Ciénc.*, 1970, **42S**, 45.

1,3,5-Tridodecanoylbenzene T-00269

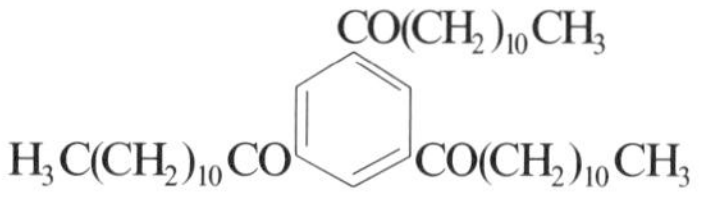

$C_{42}H_{72}O_3$ M 625.029

Constit. of root bark of *Cochlospermum planchonii*. Cryst. Mp 34-35°.

Addae-Mensah, I. *et al*, *Justus Liebigs Ann. Chem.*, 1985, 1284.

Trifurcatol B_1 T-00270

5,5'-(6-Tetradecene-1,14-diyl)bis[1,3-benzenediol], *9CI*. *1,14-Bis(3,5-dihydroxyphenyl)-6-tetradecene*

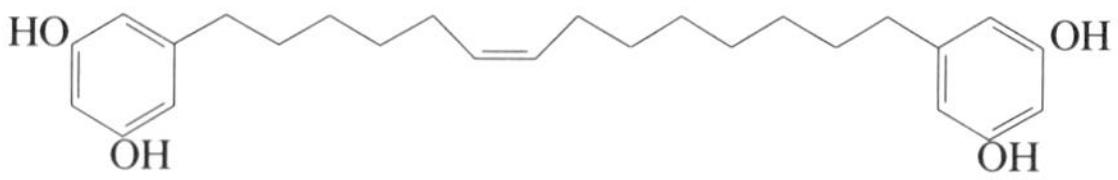

$C_{26}H_{36}O_4$ M 412.568

(Z)-form [52483-24-6]

Isol. from *Opisthiolepis heterophylla* and *Hakea trifurcata*. Mediates DNA strand scission.

Dihydro: [26050-60-2]. *5,5'-(1,4-Tetradecanediyl)bis[1,3-benzenediol]*, *9CI*. ***Bisnorstriatol***

$C_{26}H_{38}O_4$ M 414.584

Isol. from *Grevillea robusta*. Cryst. (C_6H_6 or C_6H_6/EtOAc). Mp 97-99°, Mp 140-141°.

[60796-67-0, 60796-69-2, 117940-50-8]

Cirigottis, K.A. *et al*, *Aust. J. Chem.*, 1974, **27**, 345 (*isol*)
Varma, R.S. *et al*, *Phytochemistry*, 1976, **15**, 1418 (*Bisnorstriatol*)
Scannell, R.T. *et al*, *J. Am. Chem. Soc.*, 1988, **110**, 3650 (*isol*)
Barr, J.P. *et al*, *J. Org. Chem.*, 1989, **54**, 494 (*struct*)

3,6,7-Trihydroxycholan-24-oic acid T-00271

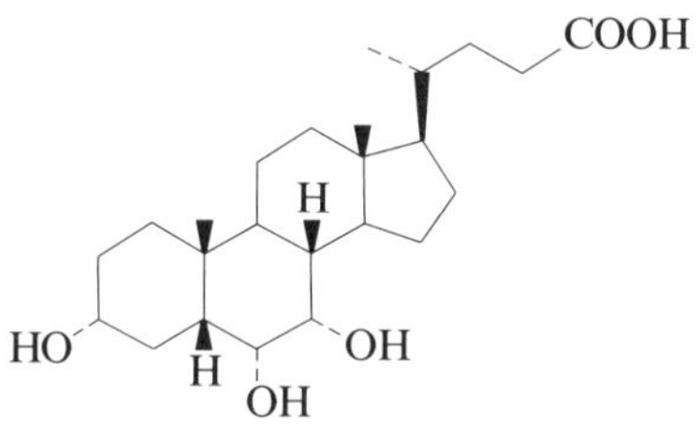

$C_{24}H_{40}O_5$ M 408.577

(3α,5α,6α,7α)-form [29311-00-0]

Allohyocholic acid. Allo-γ-muricholic acid

Cryst. (MeOH aq.). Mp 272-273°. $[\alpha]_D^{25}$ +30° (c, 0.97 in MeOH).

Me ester: [34186-28-2].

$C_{25}H_{42}O_5$ M 422.604

Cryst. (EtOAc/heptane). Mp 210-211°. $[\alpha]_D^{25}$ +46° (c, 0.57 in MeOH).

(3α,5α,6β,7α)-form [34186-25-9]

Allo-α-murocholic acid

Cryst. (MeOH aq.). Mp 261-262°. $[\alpha]_D^{25}$ 0° (MeOH).

Me ester: [34186-26-0].

Cryst. (EtOAc/heptane). Mp 149-150°. $[\alpha]_D^{25}$ +3° (c, 0.93 in MeOH).

(3α,5α,6β,7β)-form

Allo-β-muricholic acid

Cryst. (EtOAc/MeOH). Mp 256-257°. $[\alpha]_D^{25}$ +49.5° (c, 0.417 in MeOH).

Me ester: [28071-14-9].

Cryst. (Et_2O/hexane). Mp 142-144°. $[\alpha]_D^{23}$ +47.9° (c, 0.624 in EtOH).

(3α,5β,6α,7α)-form [547-75-1]

Hyocholic acid. γ-Muricholic acid

Isol. from hog bile. Cryst. (Me_2CO aq.). Mp 188-189°. $[\alpha]_D^{21}$ +5.5° (c, 1.7 in EtOH).

Me ester: [10190-22-4].

Mp 83-85°.

Me ester, tri-Ac:

$C_{31}H_{48}O_8$ M 548.715

Plates (Et_2O/hexane). Mp 188-190°. $[\alpha]_D^{20}$ +20.95° (c, 1.22 in dioxan).

(3α,5α,6α,7β)-form

Allo-ω-muricholic acid

Me ester: [34186-29-3].

Isol. from rat bile. Cryst. (Me_2CO/heptane). Mp 192-193°. $[\alpha]_D^{25}$ +19.9° (c, 0.7 in MeOH).

(3α,5β,6α,7β)-form [6830-03-1]

ω-Muricholic acid

Isol. from rat bile. Mp 150-153° and 184-188° (double Mp). $[\alpha]_D^{25}$ +36° (c, 0.38 in MeOH).

(3α,5β,6β,7α)-form [2393-58-0]

α-Muricholic acid

Isol. from rat bile. Needles (MeOH). Mp 199-200°. $[\alpha]_D^{25}$ +35° (c, 0.514 in MeOH).

(3α,5β,6β,7β)-form [2393-59-1]

β-Muricholic acid

Isol. from rat bile. Needles (MeOH). Mp 226-228°. $[\alpha]_D^{26}$ +64° (c, 0.528 in MeOH).

Ziegler, P. *et al*, *Can. J. Chem.*, 1956, **34**, 523, 1528 (*synth, struct*)
Hsia, S.L. *et al*, *J. Biol. Chem.*, 1957, **225**, 811; **226**, 667; 1958, **230**, 573, 581, 589, 593; **233**, 1337; 1960, **235**, 1963.
Kagan, H.B. *et al*, *Bull. Soc. Chim. Fr.*, 1960, 871 (*synth*)
Lehmann, G. *et al*, *Biochem. Prep.*, 1966, **11**, 122 (*isol, struct*)
Hsia, S.L., *Bile Acids*, 1971, **1**, 95 (*rev*)
Kelsey, M.I. *et al*, *Steroids*, 1971, **18**, 261.
Shaw, R. *et al*, *J. Chromatogr.*, 1980, **202**, 347 (*hplc*)

3,7,12-Trihydroxycholan-24-oic acid T-00272

$C_{24}H_{40}O_5$ M 408.577

(3α,5α,7α,12α)-form [2464-18-8] ***Allocholic acid***

Isol. from many animal sources. Bile acid in penguin (*Pinguinus impennis*) and sturgeon (*Acipenser sturio*). Cryst. (Me_2CO aq.). Mp 250-251°. $[\alpha]_D^{25}$ +27.8° (c, 0.75 in MeOH).

Me ester:

$C_{25}H_{42}O_5$ M 422.604

Cryst. (Me_2CO/hexane). Mp 225-226°. $[\alpha]_D^{25}$ +26.7° (c, 0.84 in MeOH).

(3α,5β,7α,12α)-form [81-25-4] ***Cholic acid.*** *Cholalic acid. Cholagit. Colalin. Hypocholate. NSC 6135*

Isol. from bile and animal excretions. Widespread primary bile acid found in many spp. Choleretic, laxative and antihypertensive agent. Cryst. + 1 H_2O (H_2O or Et_2O aq.). Cryst. + 1EtOH (EtOH). Mp 197° (anhyd.). $[\alpha]_D^{20}$ +37° (EtOH). pK_{a1} 4.98 (20°).

▷ LD_{50} (mus, orl) 4950 mg/kg. FZ9350000.

Me ester: Exists in two cryst. forms (EtOH). Mp 155°, Mp 162°. $[\alpha]_D^{20}$ +25°.

3,7-Di-Ac:

$C_{28}H_{44}O_7$ M 492.651

Mp 261-262°. $[\alpha]_D^{17}$ +49.8° (EtOH).

7,12-Di-Ac:

$C_{28}H_{44}O_7$ M 492.651

Mp 204-205°. $[\alpha]_D^{28}$ +71° (MeOH).
3,7,12-Tri-Ac, Me ester: Mp 89-91°. $[\alpha]_D^{27}$ +77° (MeOH).
Taurine conjugate: see *Taurocholic acid, T-00001*
(3α,5α,7β,12β)-form
Cryst. (MeOH). Mp 256-257°. $[\alpha]_D$ +50° (EtOH).
(3α,5β,7α,12β)-form [71883-64-2]
Cryst. (EtOH). Mp 200-201.5°. $[\alpha]_D$ +30.5° (EtOH).
(3α,5β,7β,12α)-form [28050-56-8]
Cryst. (EtOAc/hexane). Mp 127-129°. $[\alpha]_D$ +62° (dioxan).
(3α,5β,7β,12β)-form [81938-67-2]
Cryst. (EtOH aq.). Mp 191.5-193.0°.
(3β,5β,7α,12α)-form [3338-16-7]
Cryst. (EtOH). Mp 196.5-197.0°. $[\alpha]_D$ +30° ($CHCl_3$).
(3β,5β,7α,12β)-form [71883-66-4]
Cryst. (EtOH). Mp 225-226°. $[\alpha]_D$ +23.4° (EtOH).
(3β,5β,7β,12α)-form [10322-18-6]
Cryst. (EtOAc). Mp 164.5-165.5°.
(3β,5β,7β,12β)-form [81873-90-7]
Cryst. (EtOAc/MeOH). Mp 164.5-167.0°.

Cortese, F., *J. Am. Chem. Soc.*, 1937, **59**, 2532 (*synth*)
Ziegler, P., *Can. J. Chem.*, 1956, **34**, 1528 (*synth*)
Anderson, I.G. *et al*, *Biochem. J.*, 1957, **67**, 323; 1962, **85**, 236 (*isol, struct*)
Danielsson, H. *et al*, *J. Biol. Chem.*, 1962, **237**, 2657 (*3β,5β,7α,12α-form*)
Eneroch, P. *et al*, *J. Lipid Res.*, 1966, **7**, 524 (*ms*)
Mitra, M.N. *et al*, *J. Org. Chem.*, 1968, **33**, 175 (*synth*)
Leibfritz, D. *et al*, *J. Am. Chem. Soc.*, 1973, **95**, 4996 (*cmr*)
Percy-Robb, I.W. *et al*, *Scott. Med. J.*, 1973, **18**, 166 (*biosynth, rev*)
Chang, F.C., *J. Org. Chem.*, 1979, **44**, 4567 (*synth*)
Iida, T. *et al*, *J. Org. Chem.*, 1982, **47**, 2972 (*synth*)
Kirk, D.N. *et al*, *J. Chem. Soc., Perkin Trans. 2*, 1990, 1567 (*pmr*)
Martindale, The Extra Pharmacopoeia, 30th edn., Pharmaceutical Press, London, 1993, 998.
Lewis, R.J., *Sax's Dangerous Properties of Industrial Materials*, 8th Ed., Van Nostrand-Reinhold, 1992, CME750.

9,11,15-Trihydroxy-10,10-dimethyl-5,13-prostadienoic acid — T-00273

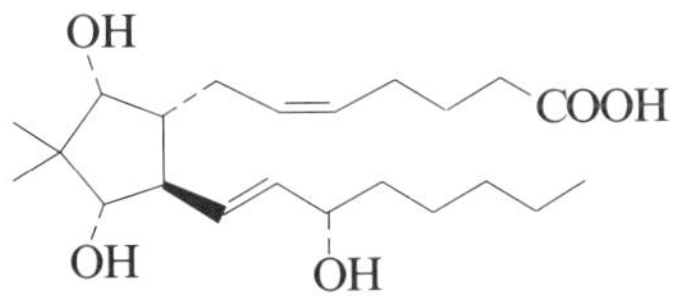

$C_{22}H_{38}O_5$ M 382.539
Prostaglandin F analogue.

Plantema, O.G. *et al*, *Recl. Trav. Chim. Pays-Bas* (*J. R. Neth. Chem. Soc.*), 1982, **102**, 268 (*synth, spectra*)

9,11,15-Trihydroxy-10,10-dimethylprost-13-enoic acid, 9CI — T-00274

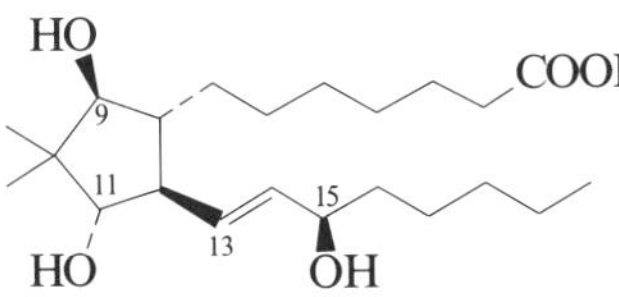

(9S,11S,13E,15R)-form

$C_{22}H_{40}O_5$ M 384.555
Prostaglandin F analogue.
(9S,11S,13E,15R)-form [57613-37-3]
Mp 90-91°.
(9S,11S,13E,15S)-form [57613-36-2]
10,10-Dimethyl-PGF$_{1\alpha}$
Mp 71-72°.

Plantema, O.G. *et al*, *Tetrahedron Lett.*, 1975, 2945 (*synth*)
Plantema, O.G. *et al*, *Recl. Trav. Chim. Pays-Bas* (*J. R. Neth. Chem. Soc.*), 1982, **102**, 268 (*synth, spectra*)

5,6,15-Trihydroxy-7,9,11,13,17-eicosapentaenoic acid, 9CI — T-00275

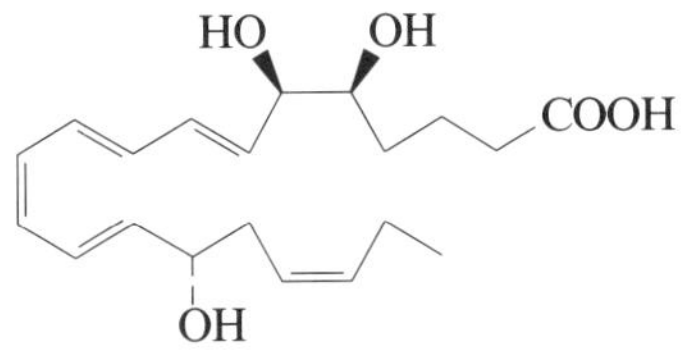

$C_{20}H_{30}O_5$ M 350.454
(5S,6R,7E,9E,11Z,13E,15S,17Z)-form [110657-98-2] ***Lipoxin A$_5$***. *LXA$_5$*
Isol. from incubation mixt. of porcine leucocytes and 15-hydroperoxyicosapentaenoic acid. Prob. stereochem. by analogy.
Me ester: $[\alpha]_D^{23}$ +11.2° (c, 1.76 in $CHCl_3$).

Wong, P.Y.-K. *et al*, *Biochem. Biophys. Res. Commun.*, 1985, **126**, 763 (*isol*)
Nicolaou, K.C. *et al*, *Angew. Chem., Int. Ed. Engl.*, 1987, **26**, 1019 (*synth, ir, pmr, uv*)

5,14,15-Trihydroxy-6,8,10,12,17-eicosapentaenoic acid, 9CI — T-00276

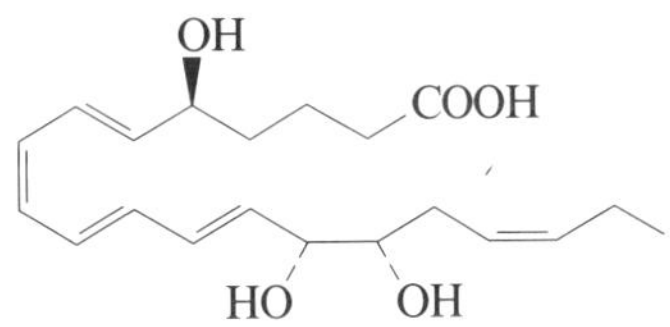

$C_{20}H_{30}O_5$ M 350.454
(5S,6E,8Z,10E,12E,14R,15S,17Z)-form [110657-99-3] ***Lipoxin B$_5$***. *LXB$_5$*
Isol. from incubation of porcine leucocytes with 15-hydroperoxyeicosapentaenoic acid. Prob. stereochem. by analogy.
Me ester: $[\alpha]_D^{23}$ +14° (c, 0.570 in MeOH).

Wong, P.Y.-K. *et al*, *Biochem. Biophys. Res. Commun.*, 1985, **126**, 763 (*isol*)
Nicolaou, K.C. *et al*, *Angew. Chem., Int. Ed. Engl.*, 1987, **26**, 1019 (*synth, ir, pmr, uv*)

5,6,15-Trihydroxy-7,9,11,13-eicosatetraenoic acid — T-00277

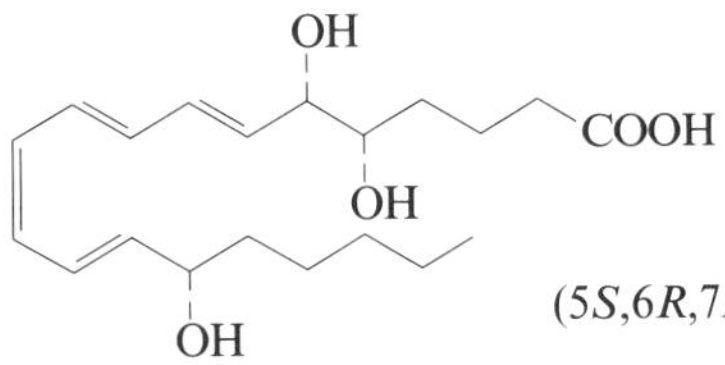

(5S,6R,7E,9E,11Z,13E,15S)-form

$C_{20}H_{32}O_5$ M 352.470
(5S,6R,7E,9E,11Z,13E,15S)-form [89663-86-5] ***Lipoxin A***. *Lipoxin A$_4$*. *LXA$_4$*
Arachidonic acid metab. isol. from human leucocytes. Stimulates elastase and superoxide ion release from neutrophils. Other isomers have been synthesised.
Me ester: [97643-35-1].
$C_{21}H_{34}O_5$ M 366.497

$[\alpha]_D^{23}$ +38.2° (c, 0.86 in MeOH). Ir 3450, 1730 cm^{-1}. Uv 309, 296, 294 (MeOH).

(5S,6S,7E,9E,11Z,13E,15S)-form [94292-80-5]
6-epi-*Lipoxin A_4*

(5S,6R,7E,9E,11E,13E,15S)-form [94292-83-8]
11-trans-*Lipoxin A_4*. *All*-trans-*form*

11,12-Didehydro: [117694-92-5]. *5,6,15-Trihydroxy-7,9,13-eicosatrien-11-ynoic acid. (11,12)-Acetylenic lipoxin A_4*
$C_{20}H_{30}O_5$ M 350.454
Intermed. in lipoxin synth.

[94297-87-7, 95341-55-2]

Serhan, C.N. *et al, Biochem. Biophys. Res. Commun.*, 1984, **118**, 943 (*isol, ms*)
Serhan, C.N. *et al, Proc. Natl. Acad. Sci., India, Sect. A*, 1984, **81**, 5335 (*isol, biochem*)
Adams, J. *et al, J. Am. Chem. Soc.*, 1985, **107**, 464 (*synth, pmr, uv, ms*)
Nicolaou, K.C. *et al, J. Am. Chem. Soc.*, 1985, **107**, 7517 (*synth*)
Corey, E.J. *et al, Tetrahedron Lett.*, 1985, **26**, 281; 1989, **30**, 4181 (*synth, pmr, uv, ir*)
Nicolaou, K.C. *et al, J. Org. Chem.*, 1989, **54**, 5527 (*synth*)
Gravier-Pelletier, C. *et al, Tetrahedron Lett.*, 1991, **32**, 1165 (*synth*)

5,14,15-Trihydroxy-6,8,10,12-eicosatetraenoic acid, 9CI — T-00278

[92950-25-9]

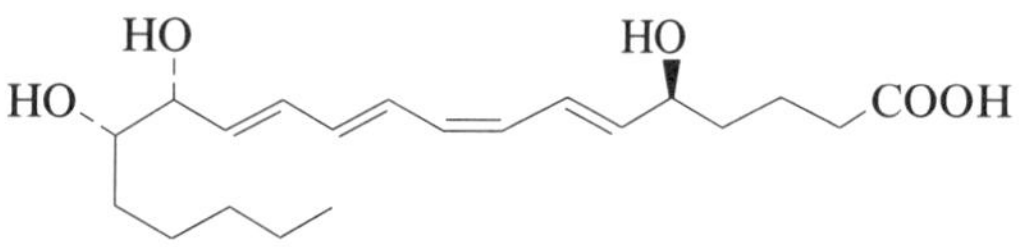

(5*S*,6*E*,8*Z*,10*E*,12*E*,14*R*,15*S*)-*form*

$C_{20}H_{32}O_5$ M 352.470

(5S,6E,8Z,10E,12E,14R,15S)-form [106062-83-3]
Lipoxin B. LX-B. Lipoxin B_4
Formed in human leukocytes from arachidonic acid.
Major biol. isomer.

8,9-Didehydro: [106062-85-5]. *5,14,15-Trihydroxy-6,10,12-eicosatrien-8-ynoic acid. 8,9-Dehydrolipoxin B*
$C_{20}H_{30}O_5$ M 350.454
Lipoxin intermed.

8,9-Didehydro, Me ester: [106062-80-0].
$C_{21}H_{32}O_5$ M 364.481
Oil. $[\alpha]_D^{20}$ −21.72° (c, 1.0 in $CHCl_3$).

(5S,6E,8E,10E,12E,14R,15S)-form
8-trans-*Lipoxin B. 8*-trans-*LXB. 8*-trans-*Lipoxin B_4*
Minor biologically derived isomer.

(5S,6E,8E,10E,12E,14S,15S)-form
(*14*S)-*8*-trans-*Lipoxin B.* (*14*S)-*8*-trans-*LXB*
Minor biologically derived isomer.

[102906-22-9, 123285-64-3]

Serhan, C.N. *et al, Proc. Natl. Acad. Sci. U.S.A.*, 1984, **81**, 5335; 1986, **83**, 1983 (*biosynth, isol, ms, uv, stereochem*)
Fitzsimmonos, B.J. *et al, Tetrahedron Lett.*, 1985, **26**, 3939 (*biosynth, struct*)
Nicolaou, K.C. *et al, Synthesis*, 1986, 453 (*synth, ir, uv, pmr*)
Morris, J. *et al, Tetrahedron Lett.*, 1986, **27**, 803 (*synth*)
Nicolaou, K.C. *et al, J. Org. Chem.*, 1989, **54**, 5527 (*synth, ir, uv, pmr, ms*)

10,11,12-Trihydroxyeicosa-5,8,14-trienoic acid — T-00279

Trioxilin B_3
[94161-08-7]

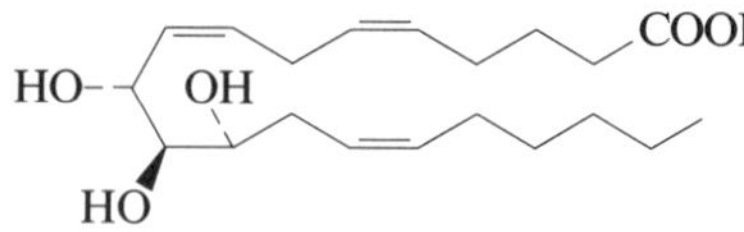

(5*Z*,8*Z*,10*R*,11*R*,12*R*,14*Z*)-*form*

$C_{20}H_{34}O_5$ M 354.486
Metab. of arachidonic acid in mammalian blood platelets.

(5Z,8Z,10R,11R,12R,14Z)-form
Me ester: [119119-49-2].
$[\alpha]_D^{20}$ −16.1° (c, 3.2 in $CHCl_3$).

(5Z,8Z,10S,11R,12R,14Z)-form
$[\alpha]_D$ +30.2° (c, 1.8 in Me_2CO).
Me ester: [119238-83-4].
$[\alpha]_D^{22}$ +29.2° (c, 2.0 in Me_2CO).

[85589-25-9]

Sun, L.M. *et al, Tetrahedron Lett.*, 1988, **29**, 4237 (*synth, pmr, hplc*)
Wu, W.-L. *et al, J. Chem. Soc., Perkin Trans. 1*, 1992, 2705 (*synth, pmr, ms*)
Yadav, J.S. *et al, Tetrahedron Lett.*, 1992, **33**, 135 (*synth, pmr*)
Wu, W.-L. *et al, J. Org. Chem.*, 1993, **58**, 2760 (*synth*)

2,15,16-Trihydroxyhexadecanoic acid — T-00280

2,15,16-Trihydroxypalmitic acid

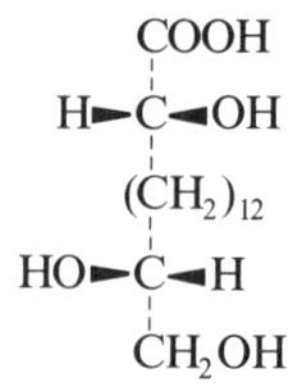

$C_{16}H_{32}O_5$ M 304.426

(2R,15R)-form
Ustilic acid B
Mp 140-141°. $[\alpha]_D$ −10° (c, 1.0 in MeOH).

Me ester: Methyl ustilate B
Cryst. (MeOH). Mp 74-75°. $[\alpha]_D$ −8.5° (c, 1.0 in MeOH).

Hydrazide: Ustilohydrazide B
Cryst. (EtOH). Mp 164-165°.

Lemieux, R.U., *Can. J. Chem.*, 1953, **31**, 396, 1054 (*struct, ir, config*)
U.S. Pat., 2 809 205, (1957); *CA*, **52**, 1212.
Canadian Pat., 600 121, (1960); *CA*, **54**, 21634 (*manuf*)

7,8,16-Trihydroxyhexadecanoic acid — T-00281

Isoaleuritic acid

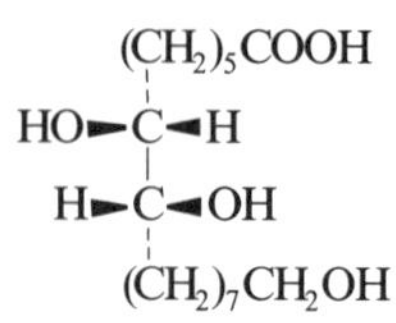

(7*RS*,8*RS*)-*form*

$C_{16}H_{32}O_5$ M 304.426

(7RS,8RS)-form [60390-03-6]
(±)-threo-*form*
Cryst. (EtOAc). Mp 94-95°.

(7RS,8SR)-form [60390-05-8]

(±)-erythro-*form*
Cryst. (EtOAc). Mp 121°.

Sanz, V. *et al*, *J. Chem. Soc., Perkin Trans. 1*, 1982, 1837 (*synth, ir, pmr*)

9,10,16-Trihydroxyhexadecanoic acid T-00282

9,10,16-Trihydroxypalmitic acid
[533-87-9]

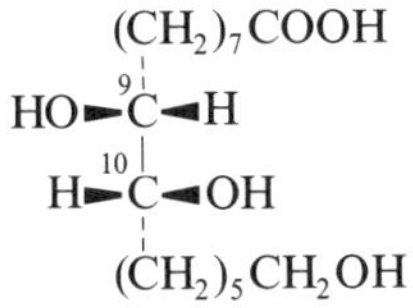

$C_{16}H_{32}O_5$ M 304.426
▷ML9850000.
(8R*,9R*)-form [33447-77-7]
(+)-threo-*form*
Mp 102.5-103.5°. $[\alpha]_D$ +24.6°.
Me ester:
$C_{17}H_{34}O_5$ M 318.453
Mp 71-72°. $[\alpha]_D$ +23.45°.
(8S*,9S*)-form [33447-75-5]
(−)-threo-*form*
Mp 103-104°. $[\alpha]_D$ −24.45°.
Me ester: Mp 72-73°. $[\alpha]_D$ −23.45°.
(8RS,9RS)-form [18299-27-9]
(±)-threo-*form*. *Aleuritic acid*
Constit. of shellac resin. Mp 102°.
Me ester: Mp 63-64°, Mp 73°. $Bp_{0.1}$ 235°.
Tri-Ac:
$C_{22}H_{38}O_8$ M 430.537
$Bp_{0.15}$ 232-234°.

Ames, D.E. *et al*, *J. Chem. Soc. C*, 1968, 268 (*synth*)
Eswaran, S.V. *et al*, *Indian J. Chem.*, 1971, **9**, 196 (*pmr*)
McGhie, J.F. *et al*, *Chem. Ind. (London)*, 1971, 1074 (*resoln*)
Prasad, S.M. *et al*, *Indian J. Phys.*, 1975, **49**, 72 (*struct*)
Singh, A.N. *et al*, *Tetrahedron*, 1978, **34**, 595 (*synth*)
Lewis, R.J., *Sax's Dangerous Properties of Industrial Materials*, 8th Ed., Van Nostrand-Reinhold, 1992, TKP000.

1,3,8-Trihydroxy-7-methyl-2-methylene-4-nonanone T-00283

Ketalin
[75692-33-0]

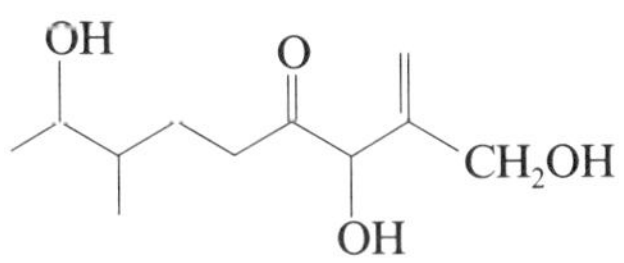

$C_{11}H_{20}O_4$ M 216.277
Metab. of *Streptomyces* Tü 1668. Ionophore.

König, W.A., *Justus Liebigs Ann. Chem.*, 1980, 1384 (*isol, spectra*)
Drautz, H. *et al*, *CA*, 1982, **97**, 37455.

9,11,15-Trihydroxy-11-methyl-prosta-5,13-dienoic acid T-00284

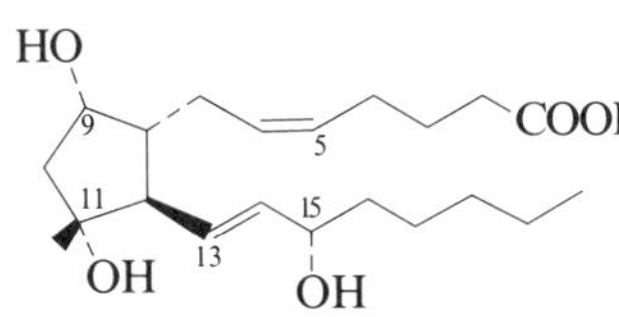

$C_{21}H_{36}O_5$ M 368.512
(5Z,9S,11R,13E,15S)-form
Oil.

Guzman, A. *et al*, *Prostaglandins*, 1974, **8**, 85 (*synth*)
Lin, C.H., *Chem. Ind. (London)*, 1976, 994 (*pmr, ir, ms*)

9,11,15-Trihydroxy-12-methyl-prosta-5,13-dienoic acid T-00285

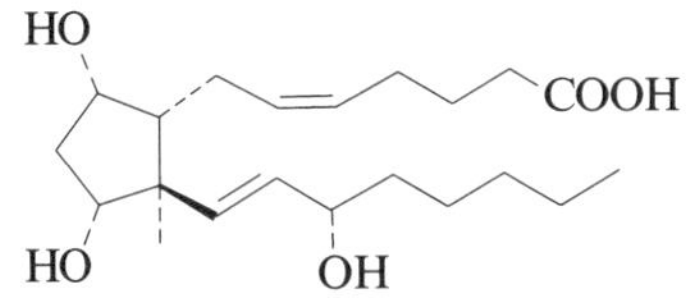

$C_{21}H_{36}O_5$ M 368.512
(5Z,9RS,11SR,13E,15RS)-form [57261-09-3]
(±)-*12-Methyl-PGF$_{2\alpha}$*
Mp 57-59°.
Me ester: [63557-49-3].
Mp 51-53°.

Grieco, P.A. *et al*, *J. Am. Chem. Soc.*, 1977, **99**, 4111 (*synth, ir, pmr*)

9,12,13-Trihydroxy-10,15-octadecadienoic acid, 9CI T-00286

[51146-90-8]

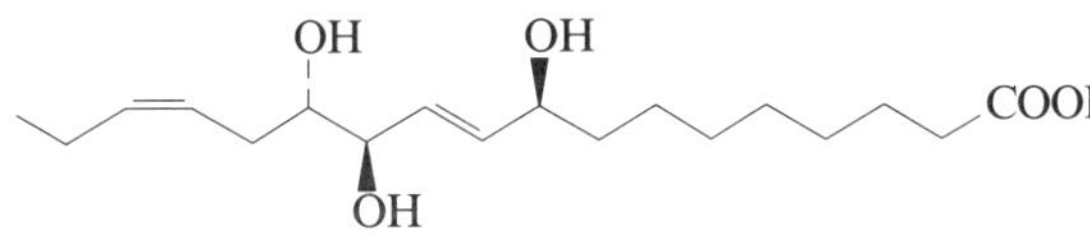

(9*S*,10*E*,12*R*,13*S*,15*Z*)-*form*

$C_{18}H_{32}O_5$ M 328.448
(9R,10E,12R,13R,15Z)-form [95783-97-4] ***Fulgidic acid***
Isol. from *Rudbeckia fulgida*. No phys. props. reported.
(9R,10E,12R,13S,15Z)-form [95341-35-8]
Isol. from roots of *Bryonia alba*.
(9R,10E,12S,13S,15Z)-form [95341-38-1]
From *B. alba*.
Me ester: [114488-84-5].
Oil. $[\alpha]_D^{25}$ −10.9° (c, 1.01 in $CHCl_3$).
(9S,10E,12R,13S,15Z)-form [74923-93-6] ***Malyngic acid***
Isol. from *Lyngbya majuscula*; the roots of *B. alba* and *Wedelia pinetorum*. Waxy solid. Mp 48.5-51°. $[\alpha]_D$ +7.7° (MeOH).
(9S,10E,12S,13S,15Z)-form [95341-44-9]
Isol. from *B. alba*, *Helianthus heterophyllus* and rice infected with blast disease. Confers activity against rice blast disease. Unnatural stereoisomers also synthesised.
Me ester: [101219-65-2].
Oil. $[\alpha]_D^{25}$ −16.83° (c, 1.83 in $CHCl_3$).
15,16-Dihydro: [97134-11-7]. *9*S*,12*S*,13*S*-Trihydroxy-10*E*-octadecenoic acid*
$C_{18}H_{34}O_5$ M 330.464
Isol. from a rice cultivar. Confers resistance to rice blast disease.
15,16-Dihydro, Me ester: [93380-89-3].
$[\alpha]_D^{25}$ −7.03° (c, 1.28 in $CHCl_3$). Unnatural stereoisomers also synthesised.

Panosyan, A.G. *et al*, *Khim. Prir. Soedin.*, 1980, 825 (*isol*)
Cardellina, J.H. *et al*, *Tetrahedron*, 1980, **36**, 993 (*isol*)
Panosyan, A.G. *et al*, *Bioorg. Khim.*, 1985, **11**, 126 (*isol, struct*)

Herz, W. *et al*, *Phytochemistry*, 1985, **24**, 89; 1986, **25**, 1481, 1913 (*isol*)
Kato, T. *et al*, *Tetrahedron Lett.*, 1985, **26**, 2357 (*isol*)
Suemune, H. *et al*, *Chem. Pharm. Bull.*, 1988, **36**, 3632 (*synth, ir, pmr, ms, esters*)
Gurjar, M.K. *et al*, *Tetrahedron Lett.*, 1990, **31**, 1783 (*synth*)

11,12,13-Trihydroxy-9,15-octadecadienoic acid T-00287

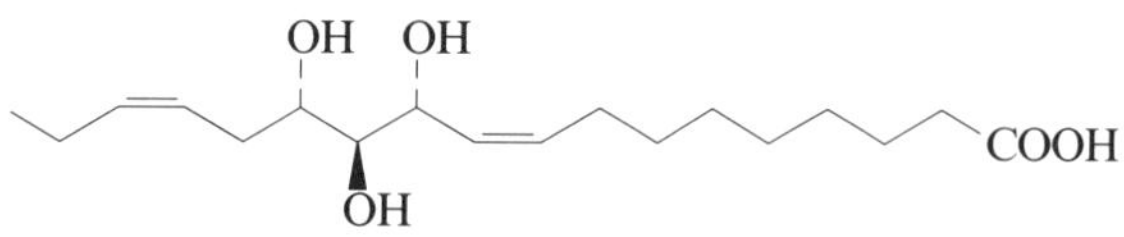

(9*Z*,11*R*,12*S*,13*S*,15*Z*)-*form*

$C_{18}H_{32}O_5$ M 328.448

Metab. of plants suffering from rice blast disease. Phytoalexin. Stereochem. of nat. prod. not detd.

(9Z,11R,12S,13S,15Z)-form [102735-55-7]

$[\alpha]_D$ −16.6° (c, 1.4 in $CHCl_3$).

Me ester: [130980-39-1].

$[\alpha]_D$ −16° (c, 0.17 in $CHCl_3$).

[101219-67-4, 102693-71-0]

Kato, T. *et al*, *Chem. Lett.*, 1986, 577 (*isol*)
Yadav, J.-S. *et al*, *Tetrahedron Lett.*, 1990, **31**, 4349 (*synth*)
Wu, W.-L. *et al*, *Tetrahedron Lett.*, 1992, **33**, 3887 (*synth*)

9,10,12-Trihydroxyoctadecanoic acid T-00288

9,10,12-Trihydroxystearic acid

[25027-95-6]

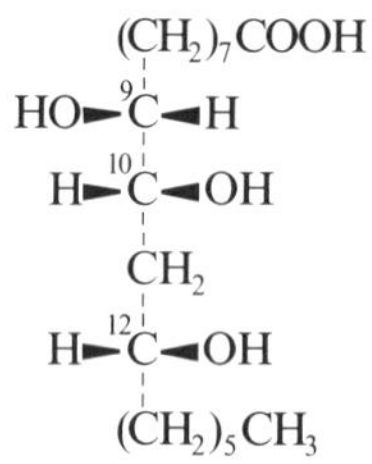

(9*R*,10*R*,12*R*)-*form*

$C_{18}H_{36}O_5$ M 332.479

(9R,10R,12R)-form [17673-32-4]

Cryst. ($CHCl_3$). Mp 85-86°.

(9R,10S,12R)-form [17673-35-7]

Cryst. (EtOH). Mp 136-137°.

(9S,10R,12R)-form [17673-34-6]

Cryst. (EtOH). Mp 110-112°.

(9S,10S,12R)-form [35017-67-5]

Cryst. ($CHCl_3$). Mp 109-111°.

Abbot, G.G. *et al*, *Chem. Phys. Lipids*, 1971, **7**, 279.
Harris, L.J. *et al*, *Lipids*, 1972, **7**, 372.

9,10,13-Trihydroxyoctadecanoic acid T-00289

9,10,13-Trihydroxystearic acid

[50439-74-2]

$H_3C(CH_2)_4CH(OH)CH_2CH_2CH(OH)CH(OH)(CH_2)_7COOH$

$C_{18}H_{36}O_5$ M 332.479

Constit. of *Phaseolus vulgaris* roots. Hatching stimulant of nematode eggs. Mp 135-137°.

Me ester: [50439-75-3].

$C_{19}H_{38}O_5$ M 346.506

Mp 117.5-119.5°.

Takasugi, M. *et al*, *Chem. Lett.*, 1973, 445 (*isol, pmr, ms*)

9,10,18-Trihydroxyoctadecanoic acid T-00290

[496-86-6]

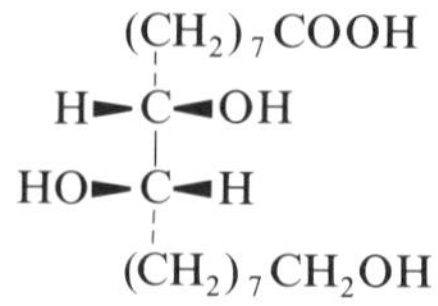

(9*S*,10*S*)-*form*

$C_{18}H_{36}O_5$ M 332.479

(9S,10S)-form [17705-68-9] ***Phloionolic acid.*** *Floionolic acid*

Constit. of cork and other plants, eg. *Olea europaea*, *Agave americana*, *Chamaepeuce afra*, *Ligustrum japonicum*, apple wax and also cutins. Cryst. (MeOH). Mp 104-105°. $[\alpha]_{589}$ +22.75°. pK_{a1} 4.95. The (9*R*,10*R*)-(−) form has also been prepared.

Me ester: [36693-05-7].

$C_{19}H_{38}O_5$ M 346.506

Mp 79-80.5°. $[\alpha]_D$ +21.52°.

(9RS,10RS)-form [583-86-8]

(±)-threo-*form*

Cryst. (EtOAc). Mp 104°.

(9RS,10SR)-form [17673-81-3]

erythro-*form*. *Isophloionolic acid*

Cryst. (EtOAc). Mp 132-133°. Opt. inactive (*meso*-).

Duhamel, L., *Bull. Soc. Chim. Fr.*, 1965, 399 (*struct*)
Ames, D.E. *et al*, *J. Chem. Soc. C*, 1967, 1556 (*synth*)
Eglinton, G. *et al*, *Phytochemistry*, 1968, **7**, 313 (*ms*)
McGhie, J.F. *et al*, *Chem. Ind. (London)*, 1972, 463 (*stereochem*)

9,12,13-Trihydroxyoctadecanoic acid T-00291

9,12,13-Trihydroxystearic acid

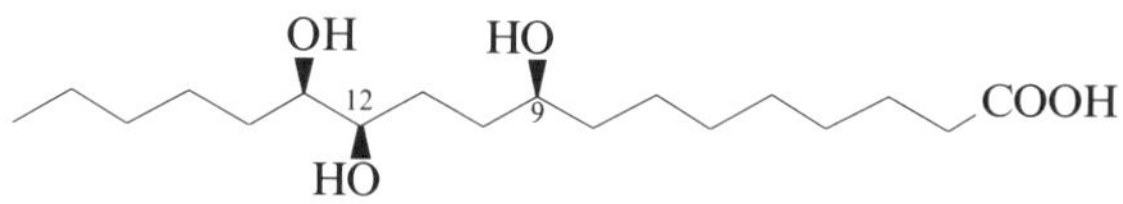

(9*S*,12*R*,13*R*)-*form*

$C_{18}H_{36}O_5$ M 332.479

(9S,12R,13R)-form [35017-70-0]

Mp 87-89°.

(9S,12R,13S)-form [35017-68-6]

Mp 101-103°.

(9S,12S,13R)-form [35017-69-7]

Mp 148-150°.

Abbot, G.G. *et al*, *Chem. Phys. Lipids*, 1971, **7**, 279 (*synth, glc*)

5,8,12-Trihydroxy-9-octadecenoic acid T-00292

$H_3C(CH_2)_5CH(OH)CH_2CH{=}CHCH(OH)CH_2CH_2CH(OH)(CH_2)_3COOH$

$C_{18}H_{34}O_5$ M 330.464

Constit. of wheat bran.

Albro, P.W. *et al*, *Phytochemistry*, 1971, **10**, 631 (*isol, pmr, ms*)

9,10,18-Trihydroxy-12-octadecenoic acid T-00293

[50632-65-0]

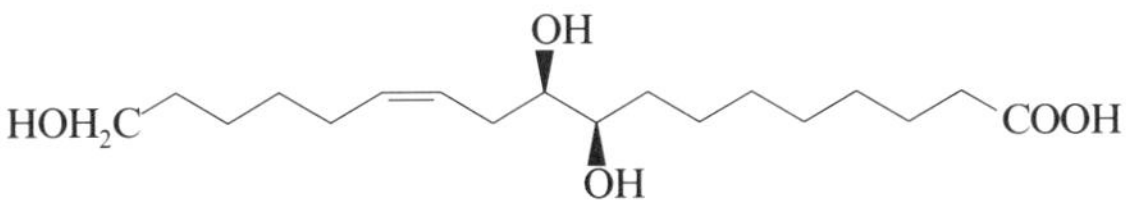

$C_{18}H_{34}O_5$ M 330.464

(9R*,10R*,12Z)-form [17705-69-0]

(+)-threo-trans-*form*

Constit. of *Chamaepeuce afra* seed oil.

Mikolajczak, K.L. *et al*, *Lipids*, 1967, **2**, 261 (*isol*)
Holloway, P.J. *et al*, *Phytochemistry*, 1973, **12**, 1721 (*isol*)

9,11,15-Trihydroxy-6-oxoprost-13-enoic acid, 9CI T-00294

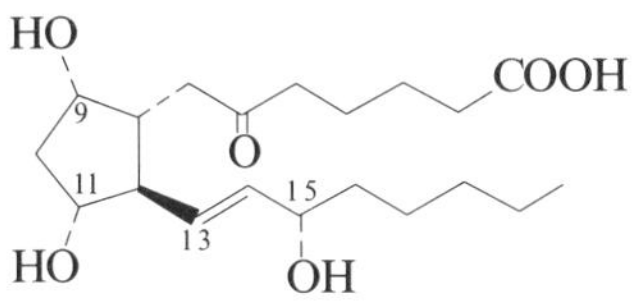

$C_{20}H_{34}O_6$ M 370.485

(9S,11R,13E,15S)-form [58962-34-8]

6-Oxoprostaglandin $F_{1\alpha}$. *6-Oxo-PGF*$_{1\alpha}$

Metab. of Prostacyclin. Cryst. Mp 75-78°.

Me ester: [63557-55-1].

Mp 68-74°.

Johnson, R.A. *et al*, *J. Am. Chem. Soc.*, 1977, **99**, 4182; 1978, **100**, 7690 (*ir*, *pmr*, *ms*)
Tanaka, T. *et al*, *Tetrahedron Lett.*, 1984, **25**, 4947 (*synth*)

11,15,19-Trihydroxy-9-oxoprost-13-enoic acid T-00295

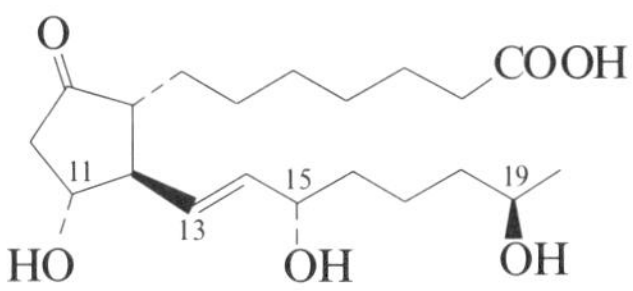

(11*R*,13*E*,15*S*,19*R*)-*form*

$C_{20}H_{34}O_6$ M 370.485

(11R,13E,15S,19R)-form [68295-73-8]

*19*R-*Hydroxy-PGE*$_1$

Occurs in human semen. Mp 62-64°.

(11R,13E,15S,19S)-form

*19*S-*Hydroxy-PGE*$_1$

Mp 89.5-91°.

Jonsson, H.T. *et al*, *Science* (*Washington, D.C.*), 1975, **187**, 1093 (*ms*)
Sih, J.H., *Prostaglandins*, 1977, **13**, 831 (*synth*)
Leuthy, C. *et al*, *J. Am. Chem. Soc.*, 1978, **100**, 6211 (*synth*, *ir*, *pmr*, *ms*)

9,11,15-Trihydroxyprosta-4,13-dienoic acid, 9CI T-00296

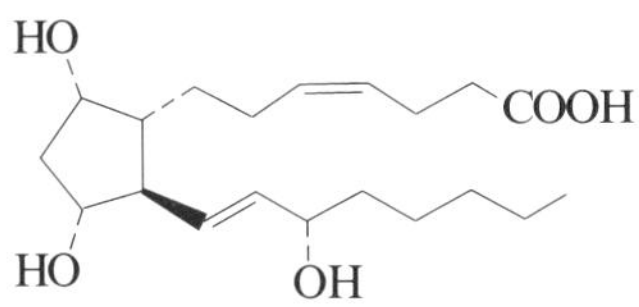

$C_{20}H_{34}O_5$ M 354.486

(4Z,9S,11R,13Z,15S)-form [67148-63-4]

Ger. Pat., 2 736 782, (1978); *CA*, **89**, 108338t (*synth*)
U.S. Pat., 4 099 014, (1978); *CA*, **90**, 71824r (*synth*)

9,11,15-Trihydroxyprosta-5,13-dienoic acid, 9CI T-00297

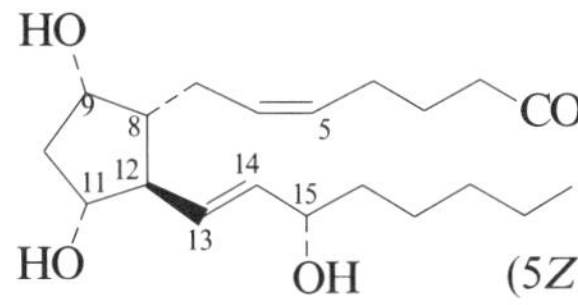

(5*Z*,8*R*,9*S*,11*R*,12*S*,13*E*,15*S*)-*form*

$C_{20}H_{34}O_5$ M 354.486

(5Z,8R,9S,11R,12S,13E,15S)-form [551-11-1]

(*9α,11α,15α*)-(*5Z,13*E)-*form*. *Prostaglandin* $F_{2\alpha}$. *PGF*$_{2\alpha}$. ***Dinoprost, BAN, INN, JAN.*** *Other proprietary names*

Common naturally occurring prostaglandin.

Abortifacient, oxytocic and smooth muscle stimulant.

Oil or solid. Mp 25-35°. $[\alpha]_D^{25}$ +23.5° (c, 1 in THF).

▷ UK8020000.

2-Amino-2-hydroxymethyl-1,3-propanediol salt (1:1): [38562-01-5]. *Dinoprost trometamol*, *BAN*. *Dinoprost tromethamine*, *USAN*. *Lutalyse*. *Prostin F2 Alpha*

Veterinary smooth muscle activator. Cryst.

▷ UK8025000.

16,16-Dimethyl: [43157-24-0]. *16,16-Dimethylprostaglandin* $F_{2\alpha}$

$C_{22}H_{38}O_5$ M 382.539

Shows anti-fertility activity. Inhibits gastric secretion.

(5Z,8RS,9SR,11RS,12SR,13E,15SR)-form

(±)-*PGF*$_{2\alpha}$

Oil which cryst. with difficulty from Et_2O. Mp 30-35°.

(5Z,8R,9R,11R,12S,13E,15S)-form [4510-16-1]

(*9β,11α,15α*)-(*5Z,13*E)-*form*. *Prostaglandin* $F_{2\beta}$. *PGF*$_{2\beta}$. *9*-epi-*PGF*$_{2\alpha}$. *9β,11α-Prostaglandin* F_2

Powerful bronchodilator. Cryst. Mp 94-95°. $[\alpha]_{365}^{25}$ +12.7° (MeOH).

▷ MJ9658000.

2-Amino-2-hydroxymethyl-1,3-propanediol salt(1:1): [89847-02-9].

Bronchodilator. Cryst. Props. similar to PGF_2.

Me ester: Mp 84-85.5°. $[\alpha]_{365}^{25}$ +10.1° (MeOH).

(5Z,8RS,9RS,11RS,12SR,13E,15SR)-form

(±)-*PGF*$_{2\beta}$

Prisms (EtOAc). Mp 96.5-97° (90-92°).

Me ester: Cryst. Mp 85-87°.

(5E,8R,9R,11R,12S,13E,15S)-form [36150-02-4]

Cryst. Mp 68-69°. $[\alpha]_D$ −8° (EtOH).

(5E,8R,9S,11R,12S,13E,15S)-form [36150-01-3]

(*9α,11α,15α*)-(5E,*13*E)-*form*. (5E)-*PGF*$_{2\alpha}$

Cryst. Mp 94.8-95.8°. $[\alpha]_D$ +9° (EtOH).

(5Z,8R,9S,11S,12S,13E,15S)-form [38432-87-0]

11-epi-*PGF*$_{2\alpha}$

Cryst. (EtOAc/hexane or MeCN). Mp 112-113.5°, Mp 120-122°. $[\alpha]_D^{25}$ +80.6° (c, 1.0 in THF), +89.6 (MeOH).

(5Z,8RS,9SR,11RS,12SR,13E,15RS)-form [39762-37-3]

(±)-*15*-epi-*PGF*$_{2\alpha}$

Oil.

Corey, E.J. *et al*, *J. Am. Chem. Soc.*, 1969, **91**, 5675; *Tetrahedron Lett.*, 1970, 307, 311; 1977, 3529; *J. Chem. Soc., Chem. Commun.*, 1975, 658 (*synth*)
Pike, J.E. *et al*, *J. Org. Chem.*, 1969, **34**, 3552 (*synth*, *pmr*, *ir*, *ms*)
Rabinowitz, I. *et al*, *Nature* (*London*), *New Biol.*, 1971, **233**, 88 (*conformn*)
Bundy, G.L. *et al*, *J. Am. Chem. Soc.*, 1972, **94**, 2124 (*synth*)
Floyd, D.M. *et al*, *Tetrahedron Lett.*, 1972, 3269 (*synth*)

Schneider, W.P. *et al*, *J. Org. Chem.*, 1973, **38**, 397 (*synth*)
Roseman, T.J. *et al*, *J. Pharm. Sci.*, 1973, **62**, 1680 (*props*)
Magerlein, J. *et al*, *Prostaglandins*, 1973, **4**, 143 (*synth, biochem*)
Lukacs, G. *et al*, *Tetrahedron Lett.*, 1973, 515 (*cmr*)
Sun, F.F. *et al*, *Biochim. Biophys. Acta*, 1974, **348**, 249 (*metab*)
Leovey, E.M.K. *et al*, *J. Am. Chem. Soc.*, 1975, **97**, 4148 (*conformn, cd*)
Corey, E.J. *et al*, *J. Chem. Soc., Chem. Commun.*, 1975, 658 (*synth*)
Ernest, I., *Angew. Chem., Int. Ed. Engl.*, 1976, **15**, 207 (*rev*)
Goyings, L.S. *et al*, *Am. J. Vet. Res.*, 1977, **38**, 1445; 1978, **39**, 1415 (*pharmacol, toxicol*)
Martinez, E. *et al*, *J. Org. Chem.*, 1977, **42**, 1087 (*synth*)
Andersen, N.H. *et al*, *Prostaglandins*, 1977, **14**, 61 (*synth*)
Stork, G. *et al*, *J. Am. Chem. Soc.*, 1978, **100**, 8272 (*synth*)
Lee, T.V. *et al*, *J. Chem. Soc., Perkin Trans. 1*, 1978, 1176 (*synth*)
Kondo, K. *et al*, *Tetrahedron Lett.*, 1978, 3927 (*synth*)
Nicolaou, K.C. *et al*, *J. Chem. Soc., Chem. Commun.*, 1979, 83 (*synth*)
Newton, R.F. *et al*, *J. Chem. Soc., Perkin Trans. 1*, 1979, 2789 (*synth, ir, pmr*)
de Titta, G.T. *et al*, *Acta Crystallogr., Sect. B*, 1980, **36**, 638 (*cryst struct, conformn*)
Andersen, N.H. *et al*, *Prostaglandins*, 1981, **22**, 809, 841 (*synth, biochem*)
Stehle, R.G., *Methods Enzymol.*, 1982, **86**, 436 (*rev*)
Martindale, The Extra Pharmacopoeia, 28th/29th Ed., Pharmaceutical Press, London, 1982/1989, 8081, 2931.
Bito, L.Z. *et al*, *Exp. Eye Res.*, 1984, **38**, 181 (*biochem*)
Kozikowski, A.P. *et al*, *J. Org. Chem.*, 1984, **49**, 2301 (*synth*)
Stier, C.T. *et al*, *J. Pharmacol. Exp. Ther.*, 1987, **243**, 487 (*biochem*)
Negwer, M., *Organic-Chemical Drugs and their Synonyms*, 6th Ed., Akademie-Verlag, Berlin, 1987, 5991.
Suzuki, M. *et al*, *J. Am. Chem. Soc.*, 1988, **110**, 4718 (*Me ester, synth, ir, pmr, cmr*)
Lewis, R.J., *Sax's Dangerous Properties of Industrial Materials*, 8th Ed., Van Nostrand-Reinhold, 1992, POC500, POC750.

9,11,15-Trihydroxyprosta-5,13,17-trienoic acid, 9CI T-00298

7-[3,5-Dihydroxy-2-(3-hydroxy-1,5-octadienyl)cyclopentyl]-5-heptenoic acid, 8CI

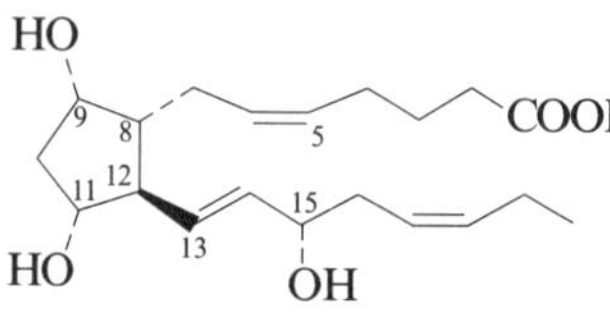

$C_{20}H_{32}O_5$ M 352.470

(5Z,9S,11R,13E,15S,17Z)-form [745-64-2]

(9α,11α,15α)-(5Z,13E,17Z)-form. Prostaglandin $F_{3α}$. $PGF_{3α}$

Biosynth. from eicosapentaenoic acid. Viscous oil. $[\alpha]_D^{26}$ +29.6° (c, 0.54 in THF).

Corey, E.J. *et al*, *J. Am. Chem. Soc.*, 1971, **93**, 1490 (*synth*)

9,11,15-Trihydroxyprost-13-enoic acid T-00299

3,5-Dihydroxy-2-(3-hydroxy-1-octenyl)cyclopentaneheptanoic acid, 8CI

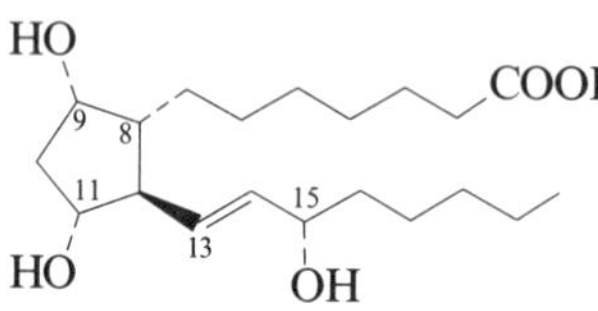

$C_{20}H_{36}O_5$ M 356.501

(8R,9S,11R,13E,15S)-form [745-62-0]

$PGF_{1α}$. Prostaglandin $F_{1α}$. Prostaglandin F_1

Constit. of sheep prostate gland. Stimulates rabbit duodenum *in vitro*. Cryst. Mp 102-103°.

▷GY4569700.

Me ester: [13227-94-6].

$C_{21}H_{38}O_5$ M 370.528

Oil. $[\alpha]_D^{21}$ +29.3° (c, 0.30 in MeOH).

13,14-Dihydro: [20592-20-5]. *9,11,15-Trihydroxyprostanoic acid*

Acts on depressor and pressor receptor systems. Cryst. (EtOAc. hexane). Mp 66-68°. $[\alpha]_D$ +41° ($CHCl_3$).

6,15-Dioxo, 13,14-dihydro: [63983-53-9]. *9,11-Dihydroxy-6,15-dioxoprostanoic acid*

$C_{20}H_{34}O_6$ M 370.485

Human urinary metab. of prostacyclin. Viscous oil.

(8S,9S,11R,13E,15S)-form [42607-55-6]

8-epi-$PGF_{1α}$

Occurs in semen.

(8RS,9SR,11RS,13E,15SR)-form

(±)-$PGF_{1α}$

Amorph. Mp 81°.

Me ester: Mp 75-77°.

(8R,9R,11R,13E,15S)-form [10164-73-5]

Prostaglandin $F_{1β}$. $PGF_{1β}$

Mp 128°.

(8RS,9RS,11RS,13E,15SR)-form [20348-60-1]

(±)-$PGF_{1β}$

Mp 116.4-116.8°.

Borgström, S. *et al*, *Acta Chem. Scand.*, 1960, **14**, 1693.
Eglinton, G. *et al*, *Nature (London)*, 1963, **200**, 960 (*isol*)
Corey, E.J. *et al*, *J. Am. Chem. Soc.*, 1968, **90**, 3247 (*synth*)
Miyano, M. *et al*, *J. Org. Chem.*, 1972, **37**, 1810 (*synth*)
Lincoln, F.W. *et al*, *J. Org. Chem.*, 1973, **38**, 951 (*synth, ir*)
v. Hulle, F. *et al*, *Tetrahedron Lett.*, 1973, 2213 (*synth, ir, pmr, ms*)
Jones, R.L., *Adv. Prostaglandin, Thromboxane, Leukotriene Res.*, 1976, **1**, 221 (*biochem*)
de Clercq, P. *et al*, *J. Org. Chem.*, 1977, **42**, 3140 (*pmr*)
Desiderio, D.M., *GC-MS News*, 1979, **7**, 32 (*ms*)
Taylor, P., *Prostaglandins*, 1979, **17**, 259 (*occur*)
Foerstermann, U. *et al*, *Biochim. Biophys. Acta*, 1982, **712**, 684 (*biochem*)
Brash, A.R. *et al*, *J. Pharmacol. Exp. Ther.*, 1983, **226**, 78 (*biochem*)
Cainelli, G. *et al*, *J. Chem. Soc., Perkin Trans. 1*, 1985, 203 (*synth*)
Suzuki, M. *et al*, *J. Am. Chem. Soc.*, 1988, **110**, 4718 (*Me ester, synth, ir, pmr, cmr*)
Lewis, R.J., *Sax's Dangerous Properties of Industrial Materials*, 8th Ed., Van Nostrand-Reinhold, 1992, POC400.

1,11,15-Trihydroxy-13-prosten-9-one T-00300

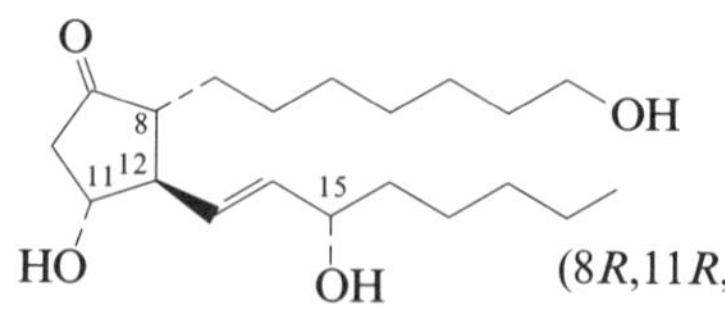

(8*R*,11*R*,12*R*,13*E*,15*S*)-*form*

$C_{20}H_{36}O_4$ M 340.502

(8R,11R,12R,13E,15S)-form [21562-57-2]

Prostaglandin E_1 alcohol. TR 4161

Non-irritant prostaglandin bronchodilator. Solid. Mp 96-98°.

Marino, J.P. *et al*, *J. Org. Chem.*, 1987, **52**, 4898 (*synth, ir, cmr, pmr, ms*)

3,7,11-Trimethyl-1,3,6,10-dodecatetraene T-00301

α-Farnesene. *Sesquicitronellene*

[502-61-4]

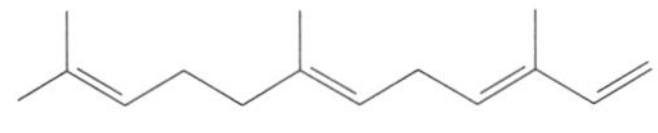

(3*E*,6*E*)-*form*

$C_{15}H_{24}$ M 204.355

(3E,6E)-form [21499-64-9]

Constit. of the natural coating of apples and pears and other fruit. Attractant for the codling moth (*Laspeyresia pomonella*). Oil.

(3Z,6E)-form [26560-14-5]

Constit. of the oil of *Perilla frutscens*. Main constit. of Dufour's gland of the ant *Pheidole pallidula* and trail recruitment pheromone of the imported fire ant *Solenopsis invicta*. Oil.

Murray, K.E., *Aust. J. Chem.*, 1969, **22**, 197 (*isol, synth*)
Brieger, G. *et al*, *J. Org. Chem.*, 1969, **34**, 3789 (*synth*)
Anet, E.F.L.J. *et al*, *Aust. J. Chem.*, 1970, **23**, 2101 (*synth*)
Vig, O.P. *et al*, *J. Indian Chem. Soc.*, 1970, **47**, 851 (*synth*)
Sutherland, O.R.W. *et al*, *J. Insect Physiol.*, 1973, **19**, 723 (*isol*)
Detrain, C. *et al*, *Experientia*, 1987, **43**, 345.
Van der Meer, R.K. *et al*, *J. Chem. Ecol.*, 1988, **14**, 825 (*isol*)

3,7,11-Trimethyl-2,7,10-dodecatrienal T-00302

[115396-97-9]

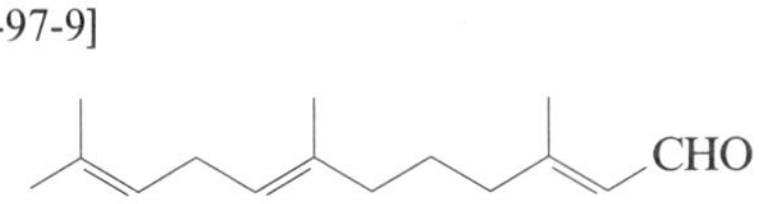

$C_{15}H_{24}O$ M 220.354

Constit. of the thoracic gland secretions of *Xylocopa* spp. (bees).

Andersen, J.F. *et al*, *J. Chem. Ecol.*, 1988, **14**, 1153.

3,7,11-Trimethyl-2,6,10-dodecatrienoic acid T-00303

Farnesenic acid. Farnesic acid. Farnesylic acid. Farnesolic acid

[7548-13-2]

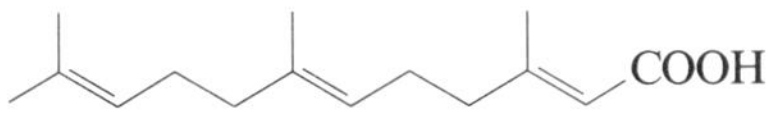

$C_{15}H_{24}O_2$ M 236.353

All isomers (*E,E; E,Z; Z,E;* and *Z,Z*) are known. Oil. Bp_{16} 202-206°.

Me ester: [10485-70-8].

$C_{16}H_{26}O_2$ M 250.380

Juvenile hormone. Oil. Bp_{10} 177-185°.

2,3-Dihydroxypropyl ester:

$C_{18}H_{30}O_4$ M 310.433

Constit. of *Archidoris odhneri*. Oil.

2-Acetoxy-3-hydroxypropyl ester:

$C_{20}H_{32}O_5$ M 352.470

Constit. of *A. odhneri*. Oil.

3-Acetoxy-2-hydroxypropyl ester:

$C_{20}H_{32}O_5$ M 352.470

Constit. of *A. odhneri*. Oil.

Me ester, 10,11-epoxide: [22963-93-5]. *Juvenile hormone III*

$C_{16}H_{26}O_3$ M 266.380

Oil. $Bp_{0.08}$ 125-126°.

Ohki, M. *et al*, *Agric. Biol. Chem.*, 1974, **38**, 175 (*synth*)
Crombie, L. *et al*, *J. Chem. Soc., Perkin Trans. 1*, 1975, 913 (*cmr*)
Katzenellenbogen, J.A. *et al*, *J. Am. Chem. Soc.*, 1976, **98**, 4925 (*synth*)
Pitzele, B.S. *et al*, *Tetrahedron*, 1976, **32**, 1347 (*synth*)
Anderson, R.J. *et al*, *Tetrahedron Lett.*, 1980, 797 (*isol, deriv*)
Kuhnz, W. *et al*, *Org. Magn. Reson.*, 1981, **16**, 138 (*cmr*)
Mori, K. *et al*, *Tetrahedron*, 1987, **43**, 4097 (*synth*)

3,7,11-Trimethyl-2,6,10-dodecatrien-1-ol T-00304

Farnesol

[4602-84-0]

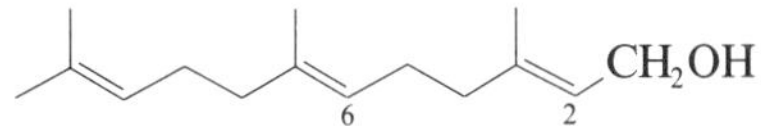

(2*E*,6*E*)-*form*

$C_{15}H_{26}O$ M 222.370

Used in perfumery industry.

▷ LD_{50} (rat, orl) 6000 mg/kg. JR4979000.

(2E,6E)-form [106-28-5]

Widespread. The main farnesol isomer, e.g. >90% of the farnesol component of the oil of *Hibiscus abelmoschus*. Nasonov pheromone of the honey bee *Apis mellifera*; constit. of temporal gland secretions of the African elephant (*Loxodonta africana*). Bp_3 137°.

Aldehyde: [502-67-0]. *3,7,11-Trimethyl-2,6,10-dodecatrienal.* ***Farnesal***

$C_{15}H_{24}O$ M 220.354

Isol. from thoracic glands of male carpenter bees (*Xylocopa* spp.). Shows juvenile hormone activity. Oil. Bp_{14} 172-174°.

Aldehyde, semicarbazone: Mp 133-135°.

Hexanoyl: Farnesyl hexanoate

Constit. of Dufour's gland secretion of *Andrena* spp.

(2Z,6E)-form [3790-71-4]

Minor constit. of various essential oils including petitgrain oil. Bp_3 135°.

(2E,6Z)-form [3879-60-5]

Bp_3 132°.

(2Z,6Z)-form [16106-95-9]

Bp_3 127°.

Bates, R.B. *et al*, *J. Org. Chem.*, 1963, **28**, 1086 (*isol, struct*)
Burrell, J.W.K. *et al*, *J. Chem. Soc. C*, 1966, 2144 (*synth*)
Corey, E.J. *et al*, *J. Am. Chem. Soc.*, 1967, **89**, 4245 (*synth*)
Hill, H.C. *et al*, *J. Chem. Soc. C*, 1968, 93 (*ms*)
Richards, J.B. *et al*, *Biochem. J.*, 1972, **128**, 1345 (*biosynth*)
Jacob, G. *et al*, *Phytochemistry*, 1972, **11**, 1683 (*biosynth*)
Nilles, G. *et al*, *J. Agric. Food Chem.*, 1973, **21**, 342 (*synth*)
Morris, W.W., *J. Assoc. Off. Anal. Chem.*, 1973, **56**, 1037 (*ir*)
Evans, R.H. *et al*, *J. Chem. Soc., Chem. Commun.*, 1973, 465 (*biosynth*)
Tengo, J. *et al*, *J. Chem. Ecol.*, 1975, **1**, 253 (*hexanoyl, isol*)
Crombie, L. *et al*, *J. Chem. Soc., Perkin Trans. 1*, 1975, 913 (*cmr*)
Pauling, H. *et al*, *Helv. Chim. Acta*, 1976, **59**, 1233 (*synth*)
Cardillo, C.R. *et al*, *J. Chem. Soc., Chem. Commun.*, 1976, 190 (*synth*)
Pickett, J.A. *et al*, *J. Chem. Ecol.*, 1980, **6**, 425 (*isol*)
Wheller, J.W. *et al*, *J. Chem. Ecol.*, 1982, **8**, 821 (*isol*)
Andersen, J.F. *et al*, *J. Chem. Ecol.*, 1988, **14**, 1153 (*isol*)
Yoshioka, T. *et al*, *Phytochemistry*, 1990, **29**, 3469 (*biosynth*)
Lewis, R.J., *Sax's Dangerous Properties of Industrial Materials*, 8th Ed., Van Nostrand-Reinhold, 1992, FAB800.

2,4,6-Trimethylheptacosanoic acid T-00305

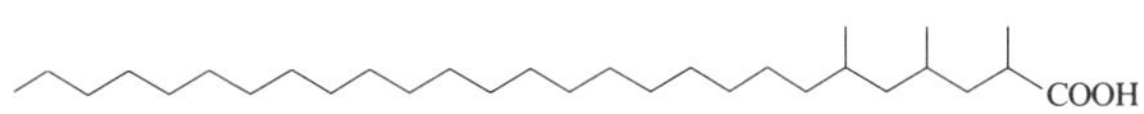

$C_{30}H_{60}O_2$ M 452.803

Component of Mycoceranic acid from tubercle bacilli.

Polgar, N. *et al*, *Chem. Ind. (London)*, 1961, 1958 (*rev*)
Polgar, N., *Top. Lipid Chem.*, 1971, **2**, 207 (*rev*)

2,4,6-Trimethylhexacosanoic acid T-00306

Mycoceranic acid. Mycocerosic acid. Phthianoic acid

[5691-97-4]

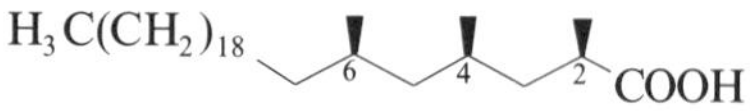

(2*R*,4*R*,6*R*)-*form*

$C_{29}H_{58}O_2$ M 438.776

(2R,4R,6R)-form

Constit. of the tubercle bacillus. Cryst. (Me_2CO). Mp 45.5-45.8°. $[\alpha]_D^{26}$ −6.1° (c, 2.2 in $CHCl_3$).

(2R,4S,6R)-form [27829-52-3]

Cryst. (Me_2CO). Mp 32°, Mp 52-53°. $[\alpha]_D^{20}$ −16.0° (c, 1.0 in $CHCl_3$).

Polgar, N. *et al*, *J. Chem. Soc.*, 1963, 3081, 3085 (*struct, synth*)
Cason, J. *et al*, *Tetrahedron*, 1964, **20**, 91 (*isol*)
Odham, G. *et al*, *Ark. Kemi*, 1970, **31**, 533 (*synth, ms*)
Polgar, N., *Top. Lipid Chem.*, 1971, **2**, 207 (*rev*)

2,4,6-Trimethylnonacosanoic acid T-00307

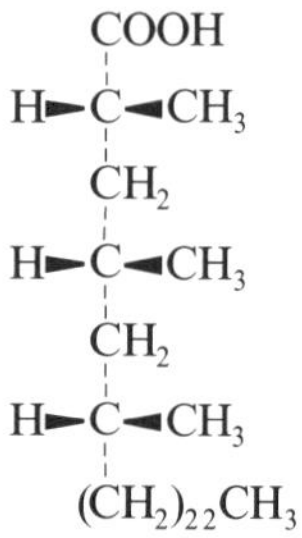

$C_{32}H_{64}O_2$ M 480.856

(2R,4R,6R)-form

Me ester: Mp 37.9-38.2°. $[\alpha]_D^{20}$ −7.2° (c, 5.5 in $CHCl_3$).

Asselineau, J. *et al*, *Acta Chem. Scand.*, 1957, **11**, 196; 1959, **13**, 822 (*struct, ms, synth*)
Ahlquist, L. *et al*, *Bull. Soc. Chim. Fr.*, 1960, 1316 (*synth*)
Cason, J. *et al*, *Tetrahedron*, 1964, **20**, 91 (*isol*)
Polgar, N., *Top. Lipid Chem.*, 1971, **2**, 207 (*rev*)

2,4,6-Trimethylnonanoic acid T-00308

[3005-07-0]

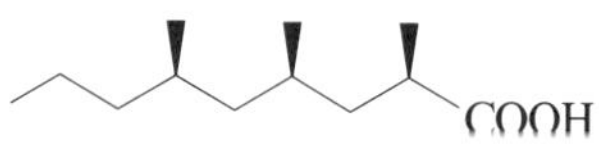

(2*R*,4*R*,6*R*)-*form*

$C_{12}H_{24}O_2$ M 200.320

(2R,4R,6R)-form

Present in the uropygial gland waxes of several birds.

Me ester: [2490-57-5].

Constit. of the preen gland waxes of the ruddy shelduck (*Tadorna ferruginea*) and common shelduck (*T. tadorna*). $[\alpha]_D^{21}$ −22.9°.

(2S,4R,6R)-form

Me ester: [18450-80-1].
$[\alpha]_D^{21}$ +9°.

(2S,4S,6R)-form

Me ester: [18450-81-2].
$[\alpha]_D^{24}$ +1.7°.

(2R,4S,6R)-form

Me ester: [18450-82-3].
$[\alpha]_D^{24}$ −35.2°.

Odham, G., *Ark. Kemi*, 1966, **25**, 543; 1967, **27**, 231, 251 (*isol, synth*)
Jacob, J. *et al*, *Z. Naturforsch., B*, 1970, **25**, 1438; 1972, **27**, 695 (*isol*)
Polgar, N., *Top. Lipid Chem.*, 1971, **2**, 207 (*rev, bibl*)

2,4,6-Trimethylpentacosanoic acid T-00309

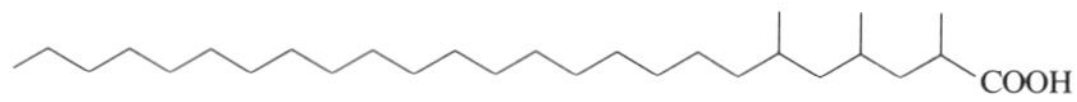

$C_{28}H_{56}O_2$ M 424.749

Component of Mycoceranic acid from tubercle bacilli.

Polgar, N. *et al*, *Chem. Ind.* (*London*), 1961, 1958 (*rev*)
Polgar, N., *Top. Lipid Chem.*, 1971, **2**, 207 (*rev, bibl*)

6,10,14-Trimethyl-2-pentadecanol, 9CI T-00310

1,2-Dinor-3-phytanol

[69729-17-5]

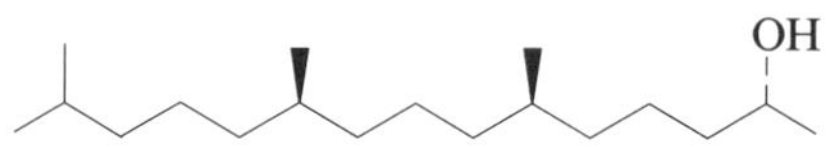

(2*R*,6*R*,10*R*)-*form*

$C_{18}H_{38}O$ M 270.498

(2R,6R,10R)-form [87247-27-6]

Major component of the sex pheromone of the female rice moth *Corcyra cephalonica*. Liq. $[\alpha]_D^{18}$ −64° (c, 1.07 in pentane) (>99.5% e.e.).

Ac: [104000-15-9].
$C_{20}H_{40}O_2$ M 312.535
Liq.

(2R,6S,10S)-form [87247-32-3]

Liq. $[\alpha]_D^{21}$ −7.8° (c, 0.99 in pentane) (>99.5% e.e.).

(2S,6R,10R)-form [87247-28-7]

Possible minor component of the sex pheromone of the female rice moth *Corcyra cephalonica*. Liq. $[\alpha]_D^{18}$ +8.0° (c, 1.05 in pentane) (98.2% e.e.).

Ac: [104068-79-3].
Liq.

(2S,6S,10S)-form [87247-31-2]

Liq. $[\alpha]_D^{21}$ +6.6° (c, 1.08 in pentane) (>99.5% e.e.).

[87247-05-0, 87247-09-4, 87247-10-7, 87247-29-8, 87247-30-1, 87247-33-4, 87247-34-5, 131656-34-3]

Rowland, S.J., *J. Chromatogr. Sci.*, 1983, **21**, 298 (*deriv, glc*)
Tori, M. *et al*, *Bull. Chem. Soc. Jpn.*, 1985, **58**, 2669 (*deriv, synth, pmr, ir, ms*)
Hall, D.R. *et al*, *J. Chem. Ecol.*, 1987, **13**, 1575 (*isol*)
Naoshima, Y. *et al*, *J. Chem. Soc., Perkin Trans. 1*, 1991, 549 (*synth, pmr, ir, ms*)
Mori, K. *et al*, *Justus Liebigs Ann. Chem.*, 1991, 259 (*synth, pmr, cmr, ir, glc, activity*)

2,4,6-Trimethyl-2,15-tetracosadienoic acid T-00311

Mycolipodienic acid

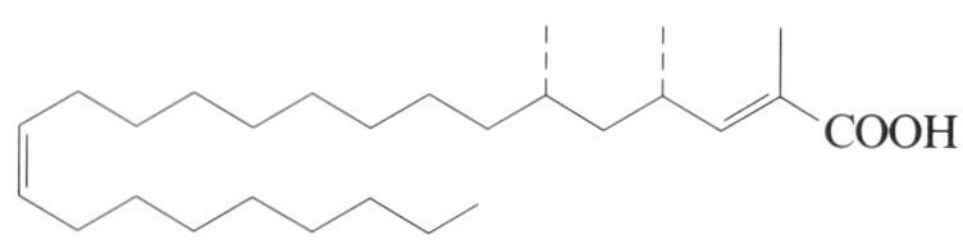

$C_{27}H_{50}O_2$ M 406.691

This is the main constit. (71%) of the mycolipodienic acid mixt. which contains other homologues up to C_{29}. Constit. of tubercle bacilli lipids.

Me ester:
$C_{28}H_{52}O_2$ M 420.718

Oil. $[\alpha]_D^{20}$ +12.2° (CCl_4).

Polgar, N. *et al*, *J. Chem. Soc. C*, 1969, 23 (*isol, ms, pmr*)
Polgar, N., *Top. Lipid Chem.*, 1971, **2**, 207 (*rev*)

2,4,6-Trimethyl-2-tetracosenoic acid T-00312

[504-13-2]

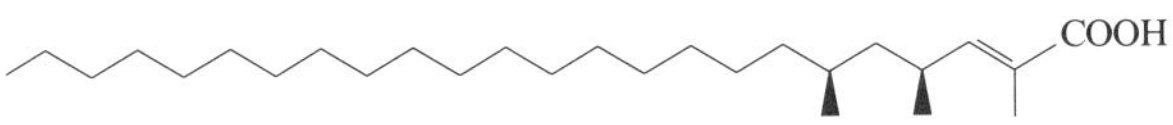

$C_{27}H_{52}O_2$ M 408.707

(2E,4S,6S)-form

Mycolipenic acid. C_{27}-Phthienoic acid

Constit. of lipids of tubercle bacilli. Cryst. (Me_2CO at −8°). Mp 28°. $[\alpha]_D^{21}$ +19.3° (c, 3.0 in $CHCl_3$). Racemic and opt. active stereoisomers have been synthesised.

Me ester:

$C_{28}H_{54}O_2$ M 422.733

$[\alpha]_D^{20}$ +16.4° (c, 4.5 in $CHCl_3$).

Quinine salt: Needles (Me_2CO at −8°). Mp 87-88°.

Asselineau, C. *et al*, *Acta Chem. Scand.*, 1956, **10**, 478, 1035 (*synth*)
Ahlquist, L. *et al*, *Ark. Kemi*, 1958, **13**, 543 (*synth*)
Millin, D.J. *et al*, *J. Chem. Soc.*, 1958, 1902 (*synth, uv, ir*)
Ryhage, R. *et al*, *Ark. Kemi*, 1961, **18**, 179 (*ms*)
Cason, J. *et al*, *Tetrahedron*, 1962, **18**, 437 (*struct*)
Tilak, B.D. *et al*, *Indian J. Chem.*, 1969, **7**, 1175 (*synth, bibl*)
Polgar, N., *Top. Lipid Chem.*, 1971, **2**, 207 (*rev, bibl*)

4,6,8-Trimethyl-7,9-undecadien-5-ol T-00313

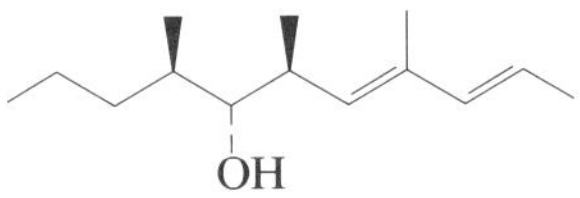

(4*R*,5*R*,6*S*,7*E*,9*E*)-*form*

$C_{14}H_{26}O$ M 210.359

(4R,5R,6S,7E,9E)-form [137492-55-8]

Female-specific constit. of the woodroach *Cryptocereus punctulatus* (rel. config. only of the nat. prod. currently known). Oil. Bp_3 78-80°. $[\alpha]_D^{20}$ −34.7° (c, 1.04 in hexane).

(4S,5S,6R,7E,9E)-form [137492-72-9]

Oil. Bp_3 79-82°. $[\alpha]_D^{21}$ +34.3° (c, 1.04 in hexane).

[129851-26-9, 129938-82-5, 129938-83-6, 129938-84-7, 129939-00-0, 129939-01-1, 129939-02-2, 129939-03-3]

Mori, K. *et al*, *Justus Liebigs Ann. Chem.*, 1990, 1249; 1992, 87 (*struct, synth*)

2,6,10-Trimethylundecanoic acid T-00314

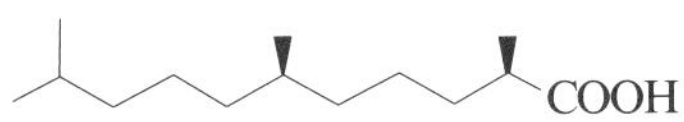

$C_{14}H_{28}O_2$ M 228.374

(2R,6R)-form [31653-13-1]

Constit. of Californian petroleum.

Cason, J. *et al*, *Tetrahedron*, 1965, **21**, 471 (*isol*)
Cox, R.E. *et al*, *J. Chem. Soc., Chem. Commun.*, 1970, 1639 (*struct*)

4,6,8-Trimethyl-2-undecanol, 9CI T-00315

[83474-31-1]

$C_{14}H_{30}O$ M 214.390

(1R,3R,5R,7R)-form

Formyl: [83540-84-5]. ***Lardolure***

$C_{15}H_{30}O_2$ M 242.401

Aggregation pheromone of the Acarid mite, *Lardoglyphus konoi*.

Kiswahara, Y. *et al*, *Agric. Biol. Chem.*, 1982, **46**, 2283 (*isol, struct*)
Mori, K. *et al*, *Tetrahedron*, 1986, **42**, 5539, 5545 (*synth, abs config*)

Trioctylamine, 8CI T-00316

N,N-*Dioctyl-1-octanamine, 9CI. Tricaprylamine*

[1116-76-3]

$[H_3C(CH_2)_7]_3N$

$C_{24}H_{51}N$ M 353.674

Used as 10% soln. in CCl_4 for extraction separation of Ce, Cr, Co, Ge, (as ion-associates with the reagent). Viscous oily liq. Sol. common org. solvs.; insol. H_2O. d 0.809. Bp 365-367°, $Bp_{0.7}$ 164-168°. n_D^{20} 1.4485.

▷ RG8225000.

B,HBr: [14846-47-0].

Mp 252-253°.

Borrows, E.T. *et al*, *J. Chem. Soc.*, 1947, 197 (*synth*)
Goszczynski, S. *et al*, *CA*, 1964, **61**, 2956 (*synth*)
Green, H., *Talanta*, 1964, **11**, 1561; 1973, **20**, 139 (*use*)
Ioffe, E.S. *et al*, *Zavod. Lab.*, 1967, **33**, 1502 (*detn, Co*)
Andrianov, A.M. *et al*, *Zh. Anal. Khim.*, 1969, **24**, 117 (*detn, Ge*)
Adam, J. *et al*, *Talanta*, 1971, **18**, 91 (*detn, Cr*)
Eggert, H. *et al*, *J. Am. Chem. Soc.*, 1973, **95**, 3710 (*cmr*)
Kliger, G.A. *et al*, *Neftekhimiya*, 1973, **13**, 265 (*synth*)
Bui, T.K. *et al*, *J. Org. Chem.*, 1981, **46**, 1759 (*synth*)
Mu, G. *et al*, *CA*, 1984, **101**, 26612 (*synth*)
Lewis, R.J., *Sax's Dangerous Properties of Industrial Materials*, 8th Ed., Van Nostrand-Reinhold, 1992, DVL000.

3,7,12-Trioxo-5β-cholan-24-oic acid T-00317

Dehydrocholic acid, BAN, INN, USAN. Decholin. Dehydrocholin. Atrocholin. Dilabil. Hykolex. Ketochol

[81-23-2]

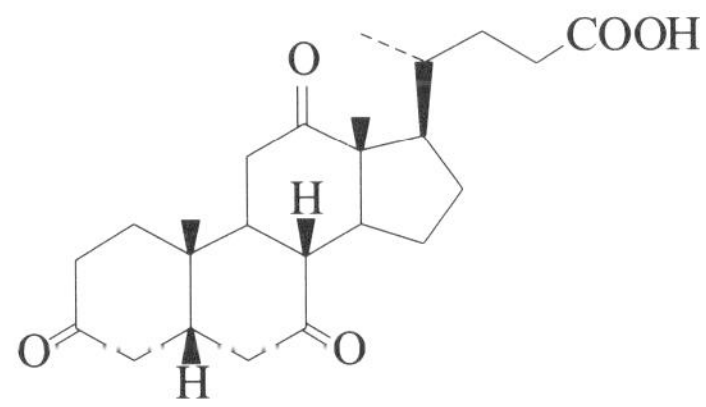

$C_{24}H_{34}O_5$ M 402.530

Oxidn. prod. of cholic acid. Choleretic agent, laxative, diagnostic aid. Cryst. (Me_2CO). Mp 237°. $[\alpha]_D^{20}$ +26° (c, 1.4 in EtOH).

▷ FZ2300000.

Na salt: [145-41-5]. *Dehydrocholate sodium, USAN*

▷ FZ2500000.

Et ester: [52718-49-7]. ***Ethyl dehydrocholate.*** *Coledrin-Compresse. Deidrosan. Eucolil. Fluxobil. Lisacol*

$C_{26}H_{38}O_5$ M 430.583

Choloretic agent. Mp 218-220°.

Diester with 3,3-bis(4-hydroxyphenyl)-2-oxoindole: [54063-34-2]. ***Cofisatin,*** **INN**

$C_{68}H_{79}NO_{11}$ M 1086.372

Laxative. Mp 235-236°.

Glycine amide: [3415-45-0]. *Dehydrocholylglycine. Glycodehydrocholic acid*

$C_{26}H_{37}NO_6$ M 459.581

Cryst. (EtOH). Mp 246-248° dec.

Fr. Pat., 1 161 345, (1958); *CA*, **54**, 18552 (*Cofisatin*)
Shimiza, T. *et al*, *Hoppe Seyler's Z. Physiol. Chem.*, 1932, **213**, 136 (*synth*)
Takeda, K. *et al*, *Yakugaku Zasshi*, 1952, **72**, 1213.
Fransson, B. *et al*, *Acta Pharm. Suec.*, 1975, **12**, 107 (*synth*)
Martindale, The Extra Pharmacopoeia, 28th/29th Ed., Pharmaceutical Press, London, 1982/1989, 3716.
Negwer, M., *Organic-Chemical Drugs and their Synonyms*, 6th Ed., Akademie-Verlag, Berlin, 1987, 7197 (*synonyms*)
Lewis, R.J., *Sax's Dangerous Properties of Industrial Materials*, 8th Ed., Van Nostrand-Reinhold, 1992, DAL000, SGD500.

22-Tritetracontanone T-00318

Behenone

[591-71-9]

$H_3C(CH_2)_{20}CO(CH_2)_{20}CH_3$

$C_{43}H_{86}O$ M 619.152
Mp 96.1-96.3°.

Grün, A. *et al*, *Angew. Chem.*, 1926, **39**, 423.
Becker, H.J. *et al*, *Recl. Trav. Chim. Pays-Bas* (*J. R. Neth. Chem. Soc.*), 1940, **59**, 933.

12,14-Tritriacontanedione, 9CI T-00319

[63142-47-2]

$H_3C(CH_2)_{18}COCH_2CO(CH_2)_{10}CH_3$

$C_{33}H_{64}O_2$ M 492.867
Constit. of leaf wax of *Festuca* spp. and many other plants.

Horn, D.H.S. *et al*, *Chem. Ind.* (*London*), 1962, 2036 (*isol, chromatog*)
Tulloch, A.P. *et al*, *Phytochemistry*, 1979, **18**, 267; 1980, **19**, 837; 1982, **21**, 661 (*isol*)

14,16-Tritriacontanedione, 9CI T-00320

[71149-65-0]

$H_3C(CH_2)_{16}COCH_2CO(CH_2)_{12}CH_3$

$C_{33}H_{64}O_2$ M 492.867
Constit. of leaf wax of *Andropogon* spp.

Tulloch, A.P. *et al*, *Phytochemistry*, 1979, **18**, 267 (*isol*)

16,18-Tritriacontanedione T-00321

[24514-86-1]

$H_3C(CH_2)_{14}COCH_2CO(CH_2)_{14}CH_3$

$C_{33}H_{64}O_2$ M 492.867
Constit. of plant waxes of *Eucalyptus* spp. Mp 67.5-68°. Forms Cu complex.

Semicarbazone: Mp 125-126°.

Horn, D.H.S. *et al*, *Chem. Ind.* (*London*), 1962, 2036 (*isol*)
Yoshikazu, Y. *et al*, *Phytochemistry*, 1976, **15**, 1180 (*isol*)

3-Tritriacontanone T-00322

[79097-23-7]

$H_3C(CH_2)_{29}COCH_2CH_3$

$C_{33}H_{66}O$ M 478.884
Constit. of *Hyoscyamus muticus* and *Solanum torvum*. Cryst. (Me_2CO/MeOH). Mp 83°.

Goswami, A. *et al*, *Phytochemistry*, 1981, **20**, 1315 (*isol*)
Mahmood, U. *et al*, *Phytochemistry*, 1983, **22**, 167 (*isol*)

Trivernolin T-00323

Glycerol tris(12,13-epoxy-9-octadecanoate). Glycerol trivernolate. Glycerol triepoxyoleate

[20449-88-1]

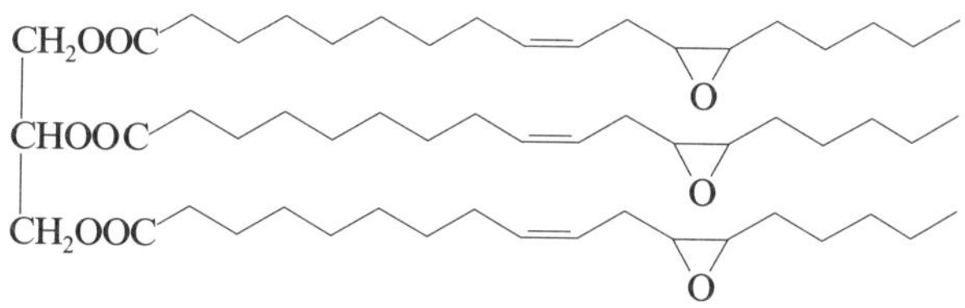

$C_{57}H_{98}O_9$ M 927.396
Triglyceride from seeds of *Vernonia anthelmintica*. Mp 25°. $[\alpha]_D^{25}$ +2.01° (hexane).

Krewson, C.F. *et al*, *J. Am. Oil Chem. Soc.*, 1962, **39**, 334 (*isol*)

U

Umbraculumin A — U-00001

Glycerol 2-(3-methyl-2-butenoate) 1-(2,4,11-tridecatrienoate)
[116944-99-1]

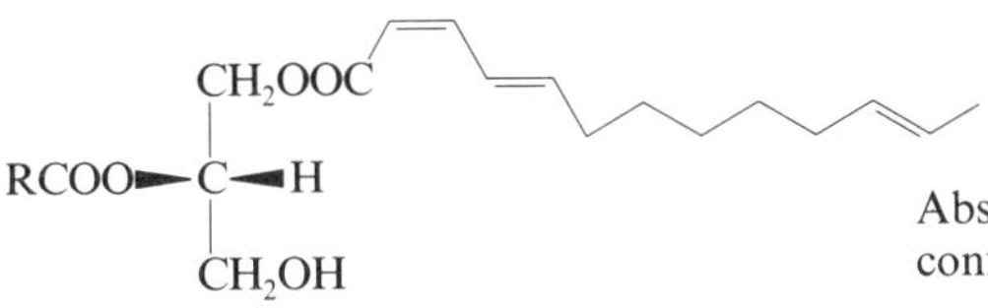

Absolute configuration

R = $-CH{=}C(CH_3)_2$

$C_{21}H_{32}O_5$ M 364.481
Constit. of mollusc *Umbraculum mediterraneum*.
Ichthyotoxin. $[\alpha]_D$ −24.3° (c, 0.8 in $CHCl_3$).

De Madeiros, E.F. *et al*, *J. Chem. Soc., Perkin Trans. 1*, 1991, 2725 (*synth, abs config*)

Umbraculumin B — U-00002

[123444-54-2]

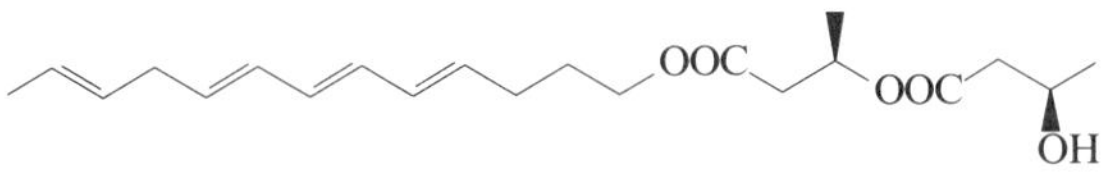

$C_{21}H_{32}O_5$ M 364.481
Isol. from the skin extracts of the mollusc *Umbraculum mediterraneum*.

Cimino, G. *et al*, *Tetrahedron Lett.*, 1989, **30**, 1147 (*isol, struct*)

Umbraculumin C — U-00003

Glycerol 2-(3-methylthio-2-propenoate) 1-(2,4,11-tridecatrienoate)
[116963-50-9]

As Umbraculumin A, U-00001 with

R = $-CH{=}CHSMe$ (*E*−)

$C_{20}H_{30}O_5S$ M 382.520
Constit. of mollusc *Umbraculum mediterraneum*.
Ichthyotoxin. $[\alpha]_D$ +7° (c, 0.5 in $CHCl_3$).

Cimino, G. *et al*, *Tetrahedron Lett.*, 1988, **29**, 3613 (*struct*)
De Madeiros, E.F. *et al*, *J. Chem. Soc., Perkin Trans. 1*, 1991, 2725 (*synth, abs config*)

2,5-Undecadienal, 9CI — U-00004

$H_3C(CH_2)_4CH{=}CHCH_2CH{=}CHCHO$

$C_{11}H_{18}O$ M 166.263
(2E,5Z)-form [13553-02-1]
Component of cooked chicken flavour. Also from *n*-6 polyene acyl chains.

Harkes, P.D. *et al*, *J. Am. Oil Chem. Soc.*, 1974, **51**, 356.

1,9-Undecadiene-5,7-diyne, 9CI — U-00005

$H_2C{=}CHCH_2CH_2C{\equiv}CC{\equiv}CCH{=}CHCH_3$

$C_{11}H_{12}$ M 144.216
(E)-form [69904-73-0]
Constit. of *Cineraria* spp. Oil.

Bohlmann, F. *et al*, *Phytochemistry*, 1978, **17**, 1629.

5,6-Undecadiene-8,10-diyne-1,3-diol — U-00006

$HC{\equiv}CC{\equiv}CCH{=}C{=}CHCH_2CH(OH)CH_2CH_2OH$

$C_{11}H_{12}O_2$ M 176.215
Metab. of *Flammula sapinea*. Oil.
Isopropylidene deriv.:
$C_{14}H_{16}O_2$ M 216.279
$[\alpha]_D^{21}$ −210° (c, 0.053 in EtOH).

Bew, R.E. *et al*, *J. Chem. Soc. C*, 1966, 129.

5,6-Undecadiene-8,10-diyne-1,4-diol — U-00007

$HC{\equiv}CC{\equiv}CCH{=}C{=}CHCH(OH)CH_2CH_2CH_2OH$

$C_{11}H_{12}O_2$ M 176.215
Metabolite of *Odontia bicolor*. Oil.

Bew, R.E. *et al*, *J. Chem. Soc. C*, 1966, 129 (*uv, ir, isol*)

2,8-Undecadiene-4,6-diynoic acid — U-00008

$H_3CCH_2CH{=}CHC{\equiv}CC{\equiv}CCH{=}CHCOOH$

$C_{11}H_{10}O_2$ M 174.199
(2E,8E)-form
Mp 143-145°.
Me ester:
$C_{12}H_{12}O_2$ M 188.226
Cryst. (hexane at −70°). Mp <10°.
(2E,8Z)-form
Mp 85-88°.
Me ester: Cryst. (hexane at −70°). Mp <10°.

Thaller, V. *et al*, *J. Chem. Soc., Perkin Trans. 1*, 1972, 552 (*synth*)

2,8-Undecadiene-4,6-diyn-1-ol — U-00009

$H_3CCH_2CH{=}CHC{\equiv}CC{\equiv}CCH{=}CHCH_2OH$

$C_{11}H_{12}O$ M 160.215
(2E,8E)-form [35042-77-4]
Cryst. (Et_2O/hexane). Mp 48.5°.
(2E,8Z)-form [35042-78-5]
Cryst. (Et_2O/hexane). Mp *ca.* 10°.

Thaller, V. *et al*, *J. Chem. Soc., Perkin Trans. 1*, 1972, 552 (*synth*)

5,6-Undecadiene-8,10-diyn-1-ol, 9CI — U-00010

$HC{\equiv}CC{\equiv}CCH{=}C{=}CH(CH_2)_3CH_2OH$

$C_{11}H_{12}O$ M 160.215
(−)-form [41793-46-8]
Metab. of *Odontia bicolor*, *Flammula sapinea* and Fungus B285. $[\alpha]_D^{21}$ −160° (c, 0.017 in EtOH).

Bew, R.E. *et al*, *J. Chem. Soc. C*, 1966, 129 (*isol, struct, ir, uv*)

Landor, P.D. *et al*, *Tetrahedron Lett.*, 1973, 1019 (*synth*)
Landor, P.D. *et al*, *J. Chem. Soc., Perkin Trans. 1*, 1975, 1628 (*synth, uv, ir*)

1,9-Undecadiene-3,5,7-triyne, 9CI **U-00011**

$H_2C{=}CHC{\equiv}CC{\equiv}CC{\equiv}CCH{=}CHCH_3$

$C_{11}H_8$ M 140.184
Constit. of *Cineraria* spp. Unstable oil.
[69904-71-8, 69904-72-9]

Bohlmann, F. *et al*, *Phytochemistry*, 1978, **17**, 1629 (*isol*)

2,4-Undecadienoic acid **U-00012**

$H_3C(CH_2)_5CH{=}CHCH{=}CHCOOH$

$C_{11}H_{18}O_2$ M 182.262
(2E,4E)-form
2-Methylpropylamide: [74267-82-6]. ***2,4-Undecadienoic acid isobutylamide.*** N-*(2-Methylpropyl)-2,4-undecadienamide*
$C_{15}H_{27}NO$ M 237.384
Alkaloid from *Leucocyclus formosus* (Compositae). Yellow cryst. Obt. only in admixture with homologues.

Greger, H. *et al*, *Phytochemistry*, 1981, **20**, 2579 (*isol, ms, struct*)
Plobeck, N.A. *et al*, *J. Org. Chem.*, 1991, **56**, 4508 (*synth, cmr*)

7,9-Undecadienoic acid **U-00013**

$H_3CCH{=}CHCH{=}CH(CH_2)_5COOH$

$C_{11}H_{18}O_2$ M 182.262
(7E,9Z)-form
Et ester: [80625-82-7].
$C_{13}H_{22}O_2$ M 210.316
$Bp_{0.01}$ 75°.

Bestmann, H.J. *et al*, *Justus Liebigs Ann. Chem.*, 1981, 2117 (*synth, pmr*)

1,5-Undecadien-3-ol **U-00014**

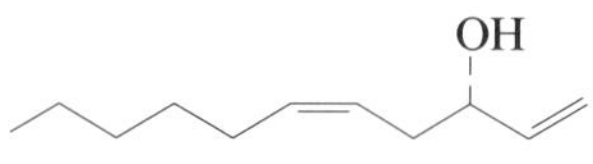

$C_{11}H_{20}O$ M 168.278
(3S,5Z)-form [81703-01-7]
Dictyoprolenol
Ac: [72538-16-0]. ***Dictyoprolene***
$C_{13}H_{22}O_2$ M 210.316
Odoriferous constit. of brown alga *Dictyopteris prolifera*. Liq. $[\alpha]_D^{27}$ +13° (c, 1.30 in $CHCl_3$).

Yamada, K. *et al*, *Tetrahedron*, 1986, **42**, 3775 (*struct, synth*)

7,9-Undecadien-1-ol **U-00015**

$H_3CCH{=}CHCH{=}CH(CH_2)_5CH_2OH$

$C_{11}H_{20}O$ M 168.278
$Bp_{0.1}$ 82-85°.
[80625-50-9, 80625-52-1]

Bestmann, H.J. *et al*, *Justus Liebigs Ann. Chem.*, 1981, 2117 (*synth, pmr*)

4,10-Undecadiynal **U-00016**
[87226-60-6]

$HC{\equiv}C(CH_2)_4C{\equiv}CCH_2CH_2CHO$

$C_{11}H_{14}O$ M 162.231
Liq. $Bp_{0.15}$ 97-99°.

Duñach, E. *et al*, *J. Am. Chem. Soc.*, 1985, **107**, 1664 (*synth, ir, pmr, ms*)

2,4-Undecadiyne **U-00017**
[90320-90-4]

$H_3C(CH_2)_5C{\equiv}CC{\equiv}CCH_3$

$C_{11}H_{16}$ M 148.247
Liq. $Bp_{0.1}$ 41-43°.

Negishi, E. *et al*, *J. Org. Chem.*, 1984, **49**, 2629 (*synth, ir, pmr*)

2,5-Undecadiyn-1-ol **U-00018**
[35378-79-1]

$H_3C(CH_2)_4C{\equiv}CCH_2C{\equiv}CCH_2OH$

$C_{11}H_{16}O$ M 164.247
Oil. Mp 1.2-1.5°. $Bp_{0.03}$ 86-91°.

Ege, S.N., *J. Am. Chem. Soc.*, 1961, **83**, 3081 (*synth*)
Osbond, J.M. *et al*, *J. Chem. Soc.*, 1961, 2779 (*synth*)

Undecanal **U-00019**
Undecyl aldehyde. Undecylic aldehyde
[112-44-7]

$H_3C(CH_2)_9CHO$

$C_{11}H_{22}O$ M 170.294
Found in many essential oils, e.g. *Citrus* spp., *Cunila lythrifolia* and *Fortunella marginata*. Perfumery and flavouring ingredient. d_4^{23} 0.825. Mp −4°. Bp_{18} 116-117°. Readily polymerises. n_D^{23} 1.4322.
▷ Mod. irritant. YQ1500000.
Oxime:
$C_{11}H_{23}NO$ M 185.309
Cryst. (MeOH). Mp 72°.
Semicarbazone: Cryst. (MeOH). Mp 103°.
2,4-Dinitrophenylhydrazone: Yellow cryst. (EtOH). Mp 104°.

Blaise, E.E., *Bull. Soc. Chim. Fr.*, 1904, **31**, 492 (*synth*)
Allen, C.F.H. *et al*, *J. Org. Chem.*, 1941, **6**, 596 (*synth*)
Davies, R.R. *et al*, *J. Soc. Chem. Ind., London*, 1943, **62**, 128 (*synth*)
Org. Synth., 1971, **51**, 39 (*synth*)
Lewis, R.J., *Sax's Dangerous Properties of Industrial Materials*, 8th Ed., Van Nostrand-Reinhold, 1992, UJJ000.

1-Undecanamine, 9CI **U-00020**
1-Aminoundecane. Undecylamine
[7307-55-3]

$H_3C(CH_2)_9CH_2NH_2$

$C_{11}H_{25}N$ M 171.325
Mp 15-16°. Bp_{727} 231.2°, Bp_{25} 129.5-130.5°. pK_a 10.63 (25°).
B,HCl: Mp 190.2°.
N-*Ac:* Cryst. (pet. ether). Mp 48°. Bp_{12} 192°.
N-*Benzoyl:* Cryst. (C_6H_6/pet. ether). Mp 60°.

Murr, L.B. *et al*, *J. Am. Chem. Soc.*, 1955, **77**, 1684 (*deriv*)

Rusin, V.Y., *CA*, 1960, **54**, 7931 (*tox*)
Gohlke, R.S. *et al*, *Anal. Chem.*, 1962, **34**, 1281 (*ms*)
Yoshio, N. *et al*, *Kogyo Kagaku Zasshi*, 1969, **72**, 696 (*synth*)

2-Undecanamine, 9CI — U-00021

2-Aminoundecane
[13205-56-6]

$H_3C(CH_2)_8CH(NH_2)CH_3$

$C_{11}H_{25}N$ M 171.325
(±)-*form*
Bp_{741} 231°, Bp_{26} 113°.
B,HCl: Cryst. (pet. ether). Mp 84°.
Ac: Cryst. (EtOH aq.). Mp 58°.

Thoms, H. *et al*, *Ber.*, 1903, **36**, 2554.

Undecane — U-00022

$H_3C(CH_2)_9CH_3$

$C_{11}H_{24}$ M 156.311
Constit. of natural petroleum. Also from turpentines (*Pinus* spp.) and other plant sources. Fp −25.6°. Bp 196°, Bp_2 61°.

Clemmensen, E., *Ber.*, 1913, **46**, 1837 (*synth*)
Hess, V.K. *et al*, *Justus Liebigs Ann. Chem.*, 1925, **441**, 151 (*synth*)
Fujimoto, T. *et al*, *Bull. Chem. Soc. Jpn.*, 1974, **47**, 1930 (*synth, glc*)
U.S. Pat., 4 070 410, (1979); *CA*, **89**, 8726 (*manuf*)

Undecanedioic acid — U-00023

[1852-04-6]

$HOOC(CH_2)_9COOH$

$C_{11}H_{20}O_4$ M 216.277
Isol. from leaves and stems of *Anthyllis sericea*. Cryst. ($CHCl_3/C_6H_6$ or H_2O). Mp 111-112.5°.
Mono-Me ester:
$C_{12}H_{22}O_4$ M 230.303
Mp 44-46°. Bp_2 165-168°.
Di-Me ester: [4567-98-0].
$C_{13}H_{24}O_4$ M 244.330
Mp 17°. Bp_2 123-124.5°.
Mono-Et ester:
$C_{13}H_{24}O_4$ M 244.330
Mp 43-44°. Bp_3 176-178°.
Di-Et ester: [22543-29-9].
$C_{15}H_{28}O_4$ M 272.384
Bp_6 155-158°, Bp_2 146°.
Dichloride:
$C_{11}H_{18}Cl_2O_2$ M 253.167
Bp_{22} 191-192°.

Walker, J. *et al*, *J. Chem. Soc.*, 1901, **79**, 1191 (*synth*)
v.Braun, J. *et al*, *Ber.*, 1912, **45**, 1930, 1975 (*synth*)
Buu-Hoi, Ng.Ph. *et al*, *C. R. Hebd. Seances Acad. Sci.*, 1955, **240**, 442 (*synth*)
Org. Synth., Coll. Vol., 4, 1963, 510, 635 (*synth*)
Holmes, J.L. *et al*, *Org. Mass Spectrom.*, 1970, **3**, 1505 (*ms*)
Crowell, E.P. *et al*, *J. Chromatogr. Sci.*, 1971, **9**, 296 (*chromatog*)
Marco, J.A. *et al*, *Phytochemistry*, 1978, **17**, 1438 (*isol*)

1,3-Undecanediol — U-00024

[6071-32-5]

$H_3C(CH_2)_7CH(OH)CH_2CH_2OH$

$C_{11}H_{24}O_2$ M 188.309
(±)-*form*
Cryst. (pet. ether). Mp 37.5-38.5°.
Isopropylidene deriv.:
$C_{14}H_{28}O_2$ M 228.374
Oil.

Bew, R.E. *et al*, *J. Chem. Soc. C*, 1966, 129 (*synth, ir*)
U.S. Pat., 3 846 326, (1974); *CA*, **82**, 60554 (*use*)

1,11-Undecanediol, 9CI — U-00025

[765-04-8]

$HOCH_2(CH_2)_9CH_2OH$

$C_{11}H_{24}O_2$ M 188.309
Cryst. (C_6H_6). Mp 61° (57-59°).
Bisphenylurethane: Mp 132-133°.
Di-Ph ether: [141620-04-4]. *1,1′-[1,11-Undecanediylbis(oxy)]bisbenzene, 9CI. 1,11-Diphenoxyundecane*
$C_{23}H_{32}O_2$ M 340.505
Isol. from the leaves of *Arachis hypogaea* infected with *Puccinia arachidis*. Phytoalexin.

Colonge, J. *et al*, *Bull. Soc. Chim. Fr.*, 1959, 1248 (*synth*)
Dulou, R. *et al*, *Bull. Soc. Chim. Fr.*, 1959, 1362 (*synth*)
Tolyanova, S.G. *et al*, *Zh. Obshch. Khim.*, 1963, **34**, 565 (*synth*)
Reinheckel, H., *CA*, 1965, **62**, 8990 (*synth*)
Rao, P.V.S. *et al*, *Oleagineux*, 1991, **46**, 501; *CA*, **116**, 252152 (*di-Ph ether*)

1-Undecanesulfonic acid — U-00026

Hendecanesulfonic acid

$H_3C(CH_2)_{10}SO_3H$

$C_{11}H_{24}O_3S$ M 236.375
Surfactant. Cryst. Mp 49°. Hygroscopic.
Na salt: [5838-34-6].
Kraft point 30°.
Hydrate: Mp 40-41°.

Noller, C.R. *et al*, *J. Am. Chem. Soc.*, 1933, **55**, 1090 (*synth*)
Wilcox, L.A. *et al*, *J. Am. Chem. Soc.*, 1953, **75**, 5761 (*cryst struct*)

Undecanoic acid — U-00027

[112-37-8]

$H_3C(CH_2)_9COOH$

$C_{11}H_{22}O_2$ M 186.294
Cryst. (AcOH aq.). Mp 28.2-28.6°. Bp_{100} 212.5°, Bp_{15} 164°.
▷ YQ2275000.
Me ester: [1731-86-8].
$C_{12}H_{24}O_2$ M 200.320
Used in perfumery. $Bp_{9\text{-}10}$ 123°, Bp_1 70-72°.
Et ester: [627-90-7].
$C_{13}H_{26}O_2$ M 214.347
Used in perfumery. Bp_{20} 140°.
Anhydride:
$C_{22}H_{42}O_3$ M 354.572
Cryst. (Me_2CO). Mp 35°. $Bp_{15\text{-}20}$ 210-245°.
Amide:
$C_{11}H_{23}NO$ M 185.309
Cryst. (EtOH). Mp 79.5-80°.
Nitrile:
$C_{11}H_{21}N$ M 167.294
Bp 253-254°.

Levene, P.A. *et al*, *J. Biol. Chem.*, 1914, **18**, 464 (*synth*)
Cason, J. *et al*, *J. Org. Chem.*, 1950, **15**, 139 (*synth*)
Hayashi, S. *et al*, *Bull. Inst. Chem. Res., Kyoto Univ.*, 1968, **46**, 213; *CA*, **70**, 110173 (*ir*)
Gunstone, F.D. *et al*, *Chem. Phys. Lipids*, 1976, **17**, 1 (*cmr*)

Vergoten, G. *et al*, *C. R. Hebd. Seances Acad. Sci. Ser. B*, 1977, **284**, 21 (*raman*)
Lewis, R.J., *Sax's Dangerous Properties of Industrial Materials*, 8th Ed., Van Nostrand-Reinhold, 1992, UKA000.

1-Undecanol U-00028

[112-42-5]

$H_3C(CH_2)_9CH_2OH$

$C_{11}H_{24}O$ M 172.310

Isol. from hop oil (*Humulus lupulus*). Acetate possibly a minor component of the sex pheromone of the female fruit tree leaf roller *Archips argyrospilus*. Perfumery ingredient. d_4^{23} 0.833. Fp 19°. Bp_{15} 131°, Bp_6 123-125°. n_D^{23} 1.4392.

▷ YQ3155000.

4-Nitrobenzoyl: Mp 29.2°.

Org. Synth., 1930, **10**, 63 (*synth*)
Dickinson, G.H. *et al*, *J. Am. Chem. Soc.*, 1937, **59**, 1094.
Jahnsen, V.J., *Nature (London)*, 1962, **196**, 474 (*isol*)
Smentouski, F.J. *et al*, *J. Am. Oil Chem. Soc.*, 1972, **49**, 48 (*nmr*)
Morris, W.W. *et al*, *J. Assoc. Off. Anal. Chem.*, 1973, **56**, 1037 (*ir*)
Roelofs, W. *et al*, *Environ. Entomol.*, 1974, **3**, 747.
Levina, O.V. *et al*, *CA*, 1976, **85**, 93918y (*derivs*)
Lewis, R.J., *Sax's Dangerous Properties of Industrial Materials*, 8th Ed., Van Nostrand-Reinhold, 1992, UNA000.

2-Undecanol U-00029

$H_3C(CH_2)_8CH(OH)CH_3$

$C_{11}H_{24}O$ M 172.310

▷ Mod. toxic.

(+)-form

Isol. from cocoa butter. Mp 12°. Bp_{20} 128°. $[\alpha]_D^{20}$ +8.11° (EtOH), $[\alpha]_D^{20}$ +10.29° (C_6H_6).

Ac: [14936-67-5].

$C_{13}H_{26}O_2$ M 214.347

Bp_{15} 84°.

Benzoyl:

$C_{18}H_{28}O_2$ M 276.418

$Bp_{3.5}$ 159°. $[\alpha]_D^{20}$ +27.55°.

Hydrogen phthalate: Mp 31-32°. $[\alpha]_D$ +37.3° ($CHCl_3$), $[\alpha]_D$ +44.3° (EtOH).

(−)-form

Isol. from leaf oil of *Litsea odorifera* and a few other sources. Bp 231-233°.

(±)-form

Bp 228-229°, Bp_{12} 119°.

Power, F.B. *et al*, *J. Chem. Soc.*, 1902, **101**, 1593 (*isol*)
v. Romburgh, P., *C. R. Hebd. Seances Acad. Sci.*, 1911, 1863 (*isol*)
Pickard, R.H. *et al*, *J. Chem. Soc.*, 1914, **105**, 850 (*synth*)
Keil, W., *Hoppe Seyler's Z. Physiol. Chem.*, 1942, **274**, 175.
VanderMoss, J.H. *et al*, *Spectrochim. Acta, Part A*, 1974, **30**, 2005 (*ir*)
Sax, N.I., *Dangerous Properties of Industrial Materials*, 5th Ed., Van Nostrand-Reinhold, 1979, 1077.

3-Undecanol U-00030

[6929-08-4]

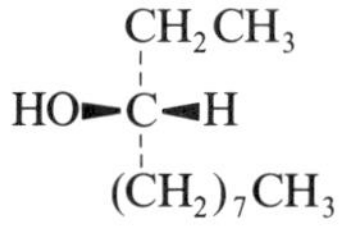

(R)-form

$C_{11}H_{24}O$ M 172.310

(R)-form

Mp 17°. $[\alpha]_D^{20}$ −6.22° (EtOH).

(±)-form

Bp 229°.

Pickard, R.H. *et al*, *J. Chem. Soc.*, 1913, **103**, 1946 (*isol, synth*)
Levene, P.A. *et al*, *J. Org. Chem.*, 1936, **1**, 76 (*abs config*)
Singliar, M., *Petrochemia*, 1971, **11**, 99 (*chromatog*)

5-Undecanol U-00031

$H_3C(CH_2)_5CH(OH)(CH_2)_3CH_3$

$C_{11}H_{24}O$ M 172.310

(±)-form

Fp −3.5°. Bp 229-230.5°.

Byrtschenko, M.S., *Chem. Zentralbl.*, 1910, **2**, 1744 (*synth*)
Obe, Y. *et al*, *CA*, 1972, **77**, 48546 (*synth*)

6-Undecanol U-00032

[23708-56-7]

$H_3C(CH_2)_4CHOH(CH_2)_4CH_3$

$C_{11}H_{24}O$ M 172.310

(±)-form

Mp 16°. Bp_{754} 235°, Bp_{12} 111°.

Hess, V.K. *et al*, *Justus Liebigs Ann. Chem.*, 1925, **441**, 150 (*synth*)
Dreger, E.E. *et al*, *Ind. Eng. Chem.*, 1944, **36**, 610 (*synth*)

11-Undecanolide U-00033

Oxacyclododecan-2-one, 9CI

[1725-03-7]

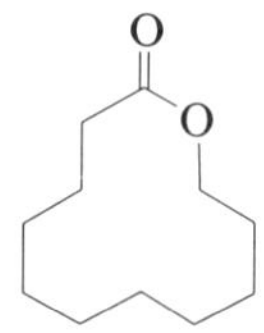

$C_{11}H_{20}O_2$ M 184.278

Bp_{10-11} 116°, Bp_1 75-78°.

Corey, E.J. *et al*, *J. Am. Chem. Soc.*, 1974, **96**, 5614 (*synth*)
Corey, E.J. *et al*, *Tetrahedron Lett.*, 1976, 3405 (*synth*)
Mukaiyama, T. *et al*, *Chem. Lett.*, 1977, 441 (*synth*)
Narasaka, K. *et al*, *Chem. Lett.*, 1977, 763 (*synth*)
Inanaga, J. *et al*, *Bull. Chem. Soc. Jpn.*, 1979, **52**, 1989 (*synth*)
Kruizinga, W.M. *et al*, *J. Chem. Soc., Chem. Commun.*, 1979, 286 (*synth*)

2-Undecanone U-00034

Methyl nonyl ketone. Enodyl. Luparone

[112-12-9]

$H_3C(CH_2)_8COCH_3$

$C_{11}H_{22}O$ M 170.294

Found in palm kernel oil and soya bean oil. Important constit. of rue oil (*Ruta graveolens*) and found in many other essential oils. Constit. of alarm pheromone of weaver ant *Oecophylla longinoda*. Used in perfumery and flavourings. Fp 15°. Bp 228°, Bp_{12} 105-106°.

▷ Mod. toxic orally. YQ2820000.

Oxime: [2158-28-3].

$C_{11}H_{23}NO$ M 185.309

Mp 44-45°.

Semicarbazone: [820-35-9].

Cryst. (EtOH). Mp 122-123°.

2,4-Dinitrophenylhydrazone: [2121-90-6].

Orange-yellow cryst. Mp 63°.

Haller, A. *et al*, *C. R. Hebd. Seances Acad. Sci.*, 1910, **151**, 698 (*isol*)
Dhont, J.H. *et al*, *Analyst (London)*, 1969, **94**, 1090 (*chromatog*)
Karrer, W. *et al*, *Konstitution und Vorkommen der Organischen Pflanzenstoffe*, 2nd Ed., Birkhäuser Verlag, Basel, 1972-1985, no. 425 (*occur*)
Kuleza, J. *et al*, *CA*, 1973, **78**, 136186 (*synth*)
Opdyke, D.L.J., *Food Cosmet. Toxicol.*, 1975, **13**, 869 (*rev*)
Bradshaw, J.W.S. *et al*, *Nature (London)*, 1975, **258**, 230 (*biochem*)
Cainelli, G. *et al*, *J. Am. Chem. Soc.*, 1976, **98**, 6737 (*synth*)
Anderson, M.W. *et al*, *J. Chem. Soc., Perkin Trans. 1*, 1986, 1995 (*synth, ir, pmr*)
Lewis, R.J., *Sax's Dangerous Properties of Industrial Materials*, 8th Ed., Van Nostrand-Reinhold, 1992, UKS000.

3-Undecanone U-00035

[2216-87-7]

$H_3C(CH_2)_7COCH_2CH_3$

$C_{11}H_{22}O$ M 170.294

Bp 227°, Bp_{11} 104-106°.

Semicarbazone: Cryst. (EtOH). Mp 90°.

Pickard, R.H. *et al*, *J. Chem. Soc.*, 1913, **103**, 1936, 1946 (*synth*)
Collman, J.P. *et al*, *J. Am. Chem. Soc.*, 1972, **94**, 1788 (*synth*)
Mekhliev, L.F. *et al*, *CA*, 1977, **86**, 43139 (*synth*)

4-Undecanone U-00036

[14476-37-0]

$H_3C(CH_2)_6CO(CH_2)_2CH_3$

$C_{11}H_{22}O$ M 170.294

Oil with characteristic fruity odour. Mp 4-5°. Bp_{13} 106-107°.

Oxime:

$C_{11}H_{23}NO$ M 185.309

Bp_{13} 141°.

Semicarbazone: Cryst. (EtOH). Mp 54-56°.

v. Braun, J. *et al*, *Chem. Ber.*, 1929, **62**, 2882 (*synth*)
Manfred, K. *et al*, *Org. Mass Spectrom.*, 1968, **1**, 617 (*ms*)
Gol'dfarb, Y.L. *et al*, *Khim. Geterotsikl. Soedin.*, 1971, **7**, 1633; *CA*, **76**, 139896 (*synth*)
Mekhtiev, L.F. *et al*, *CA*, 1977, **86**, 43139 (*synth*)

5-Undecanone U-00037

[33083-83-9]

$H_3C(CH_2)_5CO(CH_2)_3CH_3$

$C_{11}H_{22}O$ M 170.294

Mp 2°. Bp_{12} 105-106°, Bp_5 82-83°.

Oxime:

$C_{11}H_{23}NO$ M 185.309

Bp_{14} 143°.

Semicarbazone: Cryst. (MeOH). Mp 75°.

v. Braun, J. *et al*, *Chem. Ber.*, 1929, **62**, 2884 (*synth*)
Yamamoto, Y. *et al*, *Bull. Chem. Soc. Jpn.*, 1975, **48**, 3682 (*nmr, synth*)

6-Undecanone U-00038

[927-49-1]

$H_3C(CH_2)_4CO(CH_2)_4CH_3$

$C_{11}H_{22}O$ M 170.294

Mp 14-15°. Bp 226°.

▷ YQ2828000.

Oxime: [32504-26-0].

$C_{11}H_{23}NO$ M 185.309

Has bactericidal activity.

Sabatier, E.E. *et al*, *C. R. Hebd. Seances Acad. Sci.*, 1914, **158**, 832 (*synth*)
Lippmaa, E. *et al*, *CA*, 1968, **69**, 111930 (*cmr*)
Di Paolo, T., *J. Pharm. Sci.*, 1978, **67**, 566 (*pharmacol, tox*)
Lewis, R.J., *Sax's Dangerous Properties of Industrial Materials*, 8th Ed., Van Nostrand-Reinhold, 1992, ULA000.

1,3,5,8-Undecatetraene U-00039

$H_3CCH_2CH{=}CHCH_2CH{=}CHCH{=}CHCH{=}CH_2$

$C_{11}H_{16}$ M 148.247

(3E,5Z,8Z)-form [29837-19-2] ***Finaverrene***

Constit. of the essential oil of algae of the genera *Dictyopteris* and *Spermatochnus*. Smelling principle of the gametes of the brown algae *S. paradoxus*. Liq. n_D^{14} 1.5285.

(3Z,5E,8Z)-form

n_D^{14} 1.5304.

Pettus, J.A. *et al*, *J. Chem. Soc., Chem. Commun.*, 1970, 1093 (*isol, struct*)
Mueller, D.G. *et al*, *Naturwissenschaften*, 1981, **68**, 478 (*isol*)
Janicke, L. *et al*, *Angew. Chem., Int. Ed. Engl.*, 1982, **21**, 643 (*rev*)
Marner, F.-J. *et al*, *Justus Liebigs Ann. Chem.*, 1982, 579 (*synth, spectra*)

1,3,5-Undecatriene U-00040

Cystophorene

$H_3C(CH_2)_4CH{=}CHCH{=}CHCH{=}CH_2$

$C_{11}H_{18}$ M 150.263

Occurs in Galbanum extract (*Ferula galbaniflua*) and in algae.

(3E,5E)-form [19883-29-5]

Bp_{10} 100-105°.

(3E,5Z)-form [51447-08-6] ***Galbanolene***

Semiochemical of the Hawaiian seaweed *Dictyopteris* sp. Bp_{10} 100-105°.

(3Z,5Z)-form [19883-26-2]

Oil. Bp_6 100°. Thermally unstable.

Chretien-Bessiere, Y. *et al*, *Bull. Soc. Chim. Fr.*, 1967, 97 (*isol*)
Teisseire, P. *et al*, *Recherches*, 1967, **16**, 5 (*synth*)
Moore, R.E. *et al*, *J. Org. Chem.*, 1974, **39**, 2201 (*isol*)
Naef, F. *et al*, *Helv. Chim. Acta*, 1975, **58**, 1016 (*synth*)
Giraudi, E. *et al*, *Tetrahedron Lett.*, 1983, **24**, 489 (*synth, pmr*)
Hayashi, T. *et al*, *Tetrahedron Lett.*, 1983, **24**, 2665 (*synth*)
Block, E. *et al*, *J. Am. Chem. Soc.*, 1986, **108**, 4568 (*synth, bibl*)
Ratovelomanana, V. *et al*, *Bull. Soc. Chim. Fr.*, 1987, 174 (*synth, ir, cmr, ms, bibl*)
Andreini, B.P. *et al*, *Tetrahedron*, 1987, **43**, 4591 (*synth, pmr, ir, ms*)
Tsuboi, S. *et al*, *Bull. Chem. Soc. Jpn.*, 1990, **63**, 1888 (*synth*)
Tellier, F. *et al*, *Tetrahedron*, 1991, **47**, 7767 (*synth*)

1,3,9-Undecatriene-5,7-diyne, 9CI U-00041

$H_2C{=}CHCH{=}CHC{\equiv}CC{\equiv}CCH{=}CHCH_3$

$C_{11}H_{10}$ M 142.200

Constit. of *Cineraria* spp. Oil. λ_{max} 264, 277, 294, 312, 334 nm (Et_2O).

[69904-74-1, 69904-75-2, 132998-79-9]

Bohlmann, F. *et al*, *Phytochemistry*, 1978, **17**, 1629; 1982, **21**, 2531 (*isol, struct, pmr, uv, ir*)
Gonser, P. *et al*, *Phytochemistry*, 1990, **29**, 3940 (*isol*)

3,9,10-Undecatriene-5,7-diynoic acid U-00042

$H_2C{=}C{=}CHC{\equiv}CC{\equiv}CCH{=}CHCH_2COOH$

$C_{11}H_8O_2$ M 172.183

(+)-(Z)-form

Drosophilin D

Metab. of *Drosophila subatrata*. Plates (hexane). Mp 21-28°.

Jones, E.R.H. *et al*, *J. Chem. Soc.*, 1960, 2257 (*isol*)

1,5,8-Undecatrien-3-ol U-00043

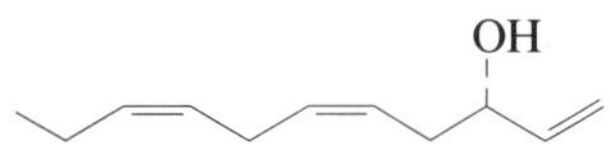

$C_{11}H_{18}O$ M 166.263

(3S,5Z,8Z)-form

Ac: [73276-42-3]. ***Neodictyoprolene***

$C_{13}H_{20}O_2$ M 208.300

Odoriferous constit. of brown alga *Dictyopteris prolifera*. Liq. $[\alpha]_D^{22}$ +20° (c, 1.00 in $CHCl_3$).

Yamada, K. *et al*, *Tetrahedron*, 1986, **42**, 3775 (*struct, synth*)

5,7,9-Undecatriyn-1-ol U-00044

$H_3C(C{\equiv}C)_3(CH_2)_3CH_2OH$

$C_{11}H_{12}O$ M 160.215

Plates (hexane). Mp 38-40°.

Bew, R.E. *et al*, *J. Chem. Soc. C*, 1966, 129 (*isol, struct*)

4-Undecenal U-00045

$H_3C(CH_2)_5CH{=}CHCH_2CH_2CHO$

$C_{11}H_{20}O$ M 168.278

(Z)-form [68820-32-6]

Liq. Bp_3 78°.

Wenkert, E. *et al*, *J. Org. Chem.*, 1985, **50**, 719 (*synth, ir, pmr, cmr*)
Davis, T.L. *et al*, *Synthesis*, 1989, 936 (*synth, ir*)

10-Undecenal, 9CI U-00046

[112-45-8]

$H_2C{=}CH(CH_2)_8CHO$

$C_{11}H_{20}O$ M 168.278

Perfumery ingredient. Bp_{10} 101-103°.

▷ YQ2835000.

2,4-Dinitrophenylhydrazone: Mp 91°.

Weyand, F. *et al*, *Angew. Chem.*, 1953, **65**, 525 (*synth*)
Nigam, S.S. *et al*, *J. Chem. Soc.*, 1957, 3320 (*synth*)
Gringas, B.A. *et al*, *Can. J. Chem.*, 1960, **38**, 712 (*ir*)
Hideg, K. *et al*, *J. Chem. Soc., Perkin Trans. 1*, 1986, 1431 (*synth*)
Lewis, R.J., *Sax's Dangerous Properties of Industrial Materials*, 8th Ed., Van Nostrand-Reinhold, 1992, ULJ000.

10-Undecen-1-amine U-00047

11-Amino-1-undecene

$H_2C{=}CH(CH_2)_8CH_2NH_2$

$C_{11}H_{23}N$ M 169.309

Bp 238-240°, Bp_{16} 123°.

Benzoyl: Plates (C_6H_6). Mp 41-42°.

Krafft, E. *et al*, *Ber.*, 1900, **33**, 3581.

1-Undecene U-00048

[821-95-4]

$H_3C(CH_2)_8CH{=}CH_2$

$C_{11}H_{22}$ M 154.295

Oil from flower stalk of *Petasites japonicus*. Defensive secretion of tenebrionid beetles. d_4^{20} 0.763. Fp −49.1°. Bp 192.7°, Bp_{10} 73°. n_D^{20} 1.4284.

Asinger, F. *et al*, *Chem. Ber.*, 1964, **97**, 1555.
Kikuchi, M., *Yakugaku Zasshi*, 1973, **93**, 123 (*isol*)
Bongini, A. *et al*, *J. Organomet. Chem.*, C4, 1974, **72** (*synth*)
Tschinkel, W.R. *et al*, *J. Insect Physiol.*, 1975, **21**, 753 (*isol*)

2-Undecene U-00049

[2244-02-2]

$H_3C(CH_2)_7CH{=}CHCH_3$

$C_{11}H_{22}$ M 154.295

(E)-form [693-61-8]

Fp −48.3°. Bp 195°, Bp_{10} 75.5°.

(Z)-form [821-96-5]

Fp −66.5°. Bp 196°, Bp_{10} 75-78°.

Asinger, F. *et al*, *Chem. Ber.*, 1964, **97**, 1555.
Naef, F. *et al*, *Helv. Chim. Acta*, 1971, **54**, 1939 (*synth*)
Sonnet, P.E. *et al*, *J. Am. Oil Chem. Soc.*, 1974, **51**, 371 (*synth*)
Hara, S. *et al*, *Tetrahedron Lett.*, 1975, 1545 (*synth*)
Org. Synth., 1976, **55**, 103 (*synth*)
Neumann, H. *et al*, *Tetrahedron Lett.*, 1976, 4839 (*synth*)
Savoia, D. *et al*, *J. Chem. Soc., Perkin Trans. 1*, 1977, 123 (*synth*)
Normant, I.F. *et al*, *Bull. Soc. Chim. Fr.*, II, 1979, 309 (*synth*)

3-Undecene, 9CI, 8CI U-00050

[60669-40-1]

$H_3C(CH_2)_6CH{=}CHCH_2CH_3$

$C_{11}H_{22}$ M 154.295

(E)-form [1002-68-2]

Occurs in shale oil. Oil. Fp −62.1°. Bp_{10} 73.9°.

(Z)-form [821-97-6]

Occurs in shale oil. Oil. Fp −69.6°. Bp_{10} 73.1°. λ_{max} 236, 247, 261, 288, 307, 328 and 352 nm (Et_2O).

Asinger, F. *et al*, *Chem. Ber.*, 1964, **97**, 1555 (*synth*)
Rang, S. *et al*, *CA*, 1977, **87**, 67504 (*ms*)

4-Undecene, 9CI, 8CI U-00051

[20810-21-3]

$H_3CCH_2CH_2CH{=}CH(CH_2)_5CH_3$

$C_{11}H_{22}$ M 154.295

(E)-form [693-62-9]

Found in beef. Oil. Fp −63.7°. Bp_{10} 73.3°.

Asinger, F. *et al*, *Chem. Ber.*, 1964, **97**, 1555 (*synth*)
Fuqua, S.M. *et al*, *J. Org. Chem.*, 1965, **30**, 2543.
Rienaecker, R. *et al*, *Justus Liebigs Ann. Chem.*, 1970, **737**, 182 (*synth*)
Rang, S. *et al*, *CA*, 1977, **87**, 67504 (*ms*)
Van Horn, D.E. *et al*, *J. Am. Chem. Soc.*, 1978, **100**, 2252 (*synth*)

5-Undecene U-00052

[4941-53-1]

$H_3C(CH_2)_4CH{=}CH(CH_2)_3CH_3$

$C_{11}H_{22}$ M 154.295

(E)-form [764-97-6]

Fp −61.1°. Bp 193°, Bp_{10} 73.2°.

(Z)-form

Fp −106.5°. Bp 192°, Bp_{10} 72°.

Asinger, F. *et al*, *Chem. Ber.*, 1964, **97**, 1555 (*synth*)
Sybilska, D. *et al*, *Chem. Anal.* (*Warsaw*), 1972, **17**, 1031 (*isol*)
Elvelt, A. *et al*, *CA*, 1975, **82**, 21908k.

2-Undecenedioic acid, 9CI, 8CI U-00053

1-Nonene-1,9-dicarboxylic acid

[28598-92-7]

$HOOC(CH_2)_7CH{=}CHCOOH$

$C_{11}H_{18}O_4$ M 214.261

Short needles (CCl_4). Mp 102-103°.

Dichloride:

$C_{11}H_{16}Cl_2O_2$ M 251.152

Liq. Bp_2 184°.

Diamide:

$C_{11}H_{20}N_2O_2$ M 212.291

Cryst. (butyl acetate). Mp 160-161°.

Bergmann, F., *J. Am. Chem. Soc.*, 1940, **62**, 3255 (*synth, derivs*)
English, J. Jr., *J. Am. Chem. Soc.*, 1941, **63**, 941 (*synth*)
Kennedy, J. *et al*, *Proc. Chem. Soc., London*, 1964, 148 (*ester*)
Doležal, S., *Collect. Czech. Chem. Commun.*, 1970, **35**, 1932 (*synth*)

1-Undecene-5,10-diyne U-00054

[94732-60-2]

$HC{\equiv}C(CH_2)_3C{\equiv}CCH_2CH_2CH{=}CH_2$

$C_{11}H_{14}$ M 146.232

Liq. Bp_{20} 95-97°.

Duñach, E. *et al*, *J. Am. Chem. Soc.*, 1985, **107**, 1664 (*synth, ir, pmr, ms*)

2-Undecene-4,6-diyne-1,11-dioic acid U-00055

$HOOC(CH_2)_3C{\equiv}CC{\equiv}CCH{=}CHCOOH$

$C_{11}H_{10}O_4$ M 206.198

(E)-form

Pale yellow cryst. (Et_2O/hexane). Mp 157-159°.

Di-Me ester:

$C_{13}H_{14}O_4$ M 234.251

Metab. of *Drosophila subatrata*. Plates (hexane). Mp 15-16°.

Jones, E.R.H. *et al*, *J. Chem. Soc.*, 1960, 2257 (*isol*)

9-Undecene-5,7-diyn-3-ol U-00056

$H_3CCH{=}CHC{\equiv}CC{\equiv}CCH_2CH(OH)CH_2CH_3$

$C_{11}H_{14}O$ M 162.231

(E)-form [69904-79-6]

Constit. of *Cineraria* spp. Oil. λ_{max} 239, 253, 269 and 284 nm (Et_2O).

Bohlmann, F. *et al*, *Phytochemistry*, 1978, **17**, 1629 (*isol*)

1-Undecene-3,5,7,9-tetrayne, 9CI U-00057

[69904-77-4]

$H_3CC{\equiv}CC{\equiv}CC{\equiv}CC{\equiv}CCH{=}CH_2$

$C_{11}H_6$ M 138.168

Isol. from aerial parts of *Cineraria fruticulorum*. Oil. λ_{max} 373, 346, 323, 303, 286, 265, 251, 240 nm (Et_2O).

Bohlmann, F. *et al*, *Phytochemistry*, 1978, **17**, 1629; 1982, **21**, 2531 (*isol, pmr, uv*)

1-Undecene-5,7,9-triyne U-00058

[31333-25-2]

$H_3CC{\equiv}CC{\equiv}CC{\equiv}CCH_2CH_2CH{=}CH_2$

$C_{11}H_{10}$ M 142.200

Constit. of *Psilothonna tagetes*. Oil. $Bp_{0.1}$ 70°.

Bohlmann, F. *et al*, *Chem. Ber.*, 1971, **104**, 954 (*isol, synth*)

3-Undecene-5,7,10-triynoic acid, 9CI, 8CI U-00059

$HC{\equiv}CCH_2C{\equiv}CC{\equiv}CCH{=}CHCH_2COOH$

$C_{11}H_8O_2$ M 172.183

(Z)-form [506-00-3]

Drosophilin C

Isol. from *Drosophila subatrata* and *D. vestita*. Shows antibiotic props. Unstable needles (hexane). Mp 115-120° dec.

Me ester:

$C_{12}H_{10}O_2$ M 186.210

Needles. Mp 27-31°.

Jones, E.R.H. *et al*, *J. Chem. Soc.*, 1960, 2257 (*isol, uv, ir, struct*)
Ahmed, M. *et al*, *J. Chem. Res., Synop.*, 1977, 125 (*biosynth*)

4-Undecene-6,8,10-triynoic acid U-00060

Nemotin A

[505-99-7]

$HC{\equiv}CC{\equiv}CC{\equiv}CCH{=}CHCH_2CH_2COOH$

$C_{11}H_8O_2$ M 172.183

Needles. Mp 50-60° dec. Photosensitive.

Bu'Lock, J.D. *et al*, *J. Chem. Soc.*, 1956, 3767 (*struct, uv*)
Bohlmann, F. *et al*, *Chem. Ber.*, 1961, **94**, 948 (*synth*)

2-Undecenoic acid U-00061

[4189-02-0]

$H_3C(CH_2)_7CH{=}CHCOOH$

$C_{11}H_{20}O_2$ M 184.278

(E)-form

Amide:

$C_{11}H_{21}NO$ M 183.293

Cryst. (C_6H_6). Mp 114-115°.

Nitrile:

$C_{11}H_{19}N$ M 165.278

Bp_{10} 127.9-128.1°.

(Z)-form

Amide: Cryst. (C_6H_6). Mp 76-77°.

Nitrile: Bp_{10} 119.7-119.9°.

[5332-51-4]

Stempel, G.H. *et al*, *J. Am. Chem. Soc.*, 1942, **64**, 470 (*synth*)
Gershon, H. *et al*, *Antimicrob. Agents Chemother.*, 1973, **4**, 435.
Heslinga, L. *et al*, *Recl. Trav. Chim. Pays-Bas* (*J. R. Neth. Chem. Soc.*), 1973, **92**, 287 (*synth*)
Shimoji, K. *et al*, *J. Am. Chem. Soc.*, 1974, **96**, 1620 (*synth*)
Davies, J.E.D. *et al*, *Chem. Phys. Lipids*, 1975, **15**, 157 (*ir*)
Larock, R.C., *J. Org. Chem.*, 1975, **40**, 3237 (*deriv, synth, ir, pmr*)
Lam, C.H. *et al*, *Chem. Phys. Lipids*, 1976, **16**, 181 (*synth*)
Larock, R.C. *et al*, *J. Org. Chem.*, 1984, **49**, 3411 (*synth, pmr*)

3-Undecenoic acid U-00062

$H_3C(CH_2)_6CH{=}CHCH_2COOH$

$C_{11}H_{20}O_2$ M 184.278

(E)-form [56770-26-4]

Shows antimicrobial activity in combination with formaldehyde. Both (*E*)- and (*Z*)-forms have been synthesised.

Lie Ken Jie, M.S.F. *et al*, *J. Chromatogr.*, 1974, **97**, 165 (*synth, glc*)
Davies, J.E.D. *et al*, *Chem. Phys. Lipids*, 1975, **15**, 157 (*ir*)
Bus, J.D. *et al*, *Chem. Phys. Lipids*, 1976, **17**, 501 (*cmr*)

4-Undecenoic acid U-00063

$H_3C(CH_2)_5CH{=}CHCH_2CH_2COOH$

$C_{11}H_{20}O_2$ M 184.278

Kiyoshi, F. *et al*, *CA*, 1972, **76**, 126325x; 1973, **79**, 125767t; 1975, **82**, 155384u (*synth*)
Lie Ken Jie, M.S.F. *et al*, *J. Chromatogr.*, 1974, **97**, 165 (*synth, glc*)
Davies, J.E.D. *et al*, *Chem. Phys. Lipids*, 1975, **15**, 157 (*ir, raman*)
Bus, J.D. *et al*, *Chem. Phys. Lipids*, 1976, **17**, 501 (*cmr*)

5-Undecenoic acid U-00064

[56579-27-2]

$H_3C(CH_2)_4CH{=}CH(CH_2)_3COOH$

$C_{11}H_{20}O_2$ M 184.278

Lie Ken Jie, M.S.F. *et al*, *J. Chromatogr.*, 1974, **97**, 165 (*synth, glc*)
Davies, J.E.D. *et al*, *Chem. Phys. Lipids*, 1975, **15**, 157 (*ir, raman*)
Bus, J.D. *et al*, *Chem. Phys. Lipids*, 1976, **17**, 501 (*cmr*)
Gunstone, F.D. *et al*, *Chem. Phys. Lipids*, 1977, **18**, 115 (*cmr*)

6-Undecenoic acid U-00065

[65423-22-5]

$H_3C(CH_2)_3CH{=}CH(CH_2)_4COOH$

$C_{11}H_{20}O_2$ M 184.278

(Z)-form

Amide:

$C_{11}H_{21}NO$ M 183.293

Mp 56-57°.

Taylor, W.R. *et al*, *J. Am. Chem. Soc.*, 1950, **72**, 4263 (*synth*)
Lie Ken Jie, M.S.F. *et al*, *J. Chromatogr.*, 1974, **97**, 165 (*synth, glc*)
Lam, C.H. *et al*, *Chem. Phys. Lipids*, 1976, **16**, 181 (*synth*)
Kabara, J.J. *et al*, *Lipids*, 1977, **12**, 753.

7-Undecenoic acid U-00066

$H_3CCH_2CH_2CH{=}CH(CH_2)_5COOH$

$C_{11}H_{20}O_2$ M 184.278

Kovaleva, A.S. *et al*, *CA*, 1974, **81**, 37206v (*synth*)
Lie Ken Jie, M.S.F. *et al*, *J. Chromatogr.*, 1974, **97**, 165 (*synth, glc*)
Davies, J.E.D. *et al*, *Chem. Phys. Lipids*, 1975, **15**, 157 (*ir, raman*)
Bus, J.D. *et al*, *Chem. Phys. Lipids*, 1976, **17**, 501 (*cmr*)

8-Undecenoic acid U-00067

[62472-78-0]

$H_3CCH_2CH{=}CH(CH_2)_6COOH$

$C_{11}H_{20}O_2$ M 184.278

Lie Ken Jie, M.S.F. *et al*, *J. Chromatogr.*, 1974, **97**, 165 (*synth, glc*)
Davies, J.E.D. *et al*, *Chem. Phys. Lipids*, 1975, **15**, 157 (*ir, raman*)
Bus, J.D. *et al*, *Chem. Phys. Lipids*, 1976, **17**, 501 (*cmr*)
Gunstone, F.D. *et al*, *Chem. Phys. Lipids*, 1977, **18**, 115 (*cmr*)

9-Undecenoic acid U-00068

[593-25-9]

$H_3CCH_2CH{=}CH(CH_2)_6COOH$

$C_{11}H_{20}O_2$ M 184.278

(E)-form [37973-84-5]

Mp 17-18°. $Bp_{0.7}$ 121-123°.

Et ester:

$C_{13}H_{24}O_2$ M 212.331

$Bp_{0.5}$ 87-89°.

(Z)-form

Mp 1-2°. $Bp_{0.8}$ 118-120°.

Et ester: $Bp_{0.6}$ 88-90°.

Ames, D.E. *et al*, *J. Chem. Soc.*, 1952, 677 (*synth*)
Davies, J.E.D. *et al*, *Chem. Phys. Lipids*, 1975, **15**, 157 (*ir*)
Lam, C.H. *et al*, *Chem. Phys. Lipids*, 1976, **16**, 181 (*synth*)
Kabara, J.J. *et al*, *Lipids*, 1977, **12**, 753.

10-Undecenoic acid U-00069

Undecylenic acid, USAN

[112-38-9]

$H_2C{=}CH(CH_2)_8COOH$

$C_{11}H_{20}O_2$ M 184.278

Metab. of *Rhodotorula glutinis* var. *lusitanica.* Also from leaf oils of *Thujopsis dolabrata* and *Juniperus chinensis.* Formed on pyrol. of ricinoleic acid. Antifungal agent. Simple esters are perfumery and flavour ingredients. Mp 24.5°. Bp 275°, Bp_{15} 165°. Component of Cruex, Desenex, Fulvidex and Ting.

▷ YQ2975000.

Zn salt: Zinc undecylenate, USAN

Mp 116-121°.

Me ester: [111-81-9].

$C_{12}H_{22}O_2$ M 198.305

Fp −27.5°. Bp 248°, Bp_{10} 124°.

Et ester: [692-86-4].

$C_{13}H_{24}O_2$ M 212.331

Bp 263.5-265.5°, Bp_{10} 131.5°

▷ YQ2978400.

Krafft, F. *et al*, *Ber.*, 1884, **17**, 2985 (*synth*)
Osman, S.M. *et al*, *Fette, Seifen, Anstrichm.*, 1975, **77**, 473.
McGrattan, C.J. *et al*, *J. Phycol.*, 1976, **12**, 129.
Gunstone, F.D. *et al*, *Chem. Phys. Lipids*, 1977, **18**, 115 (*cmr*)
Tamao, K. *et al*, *J. Am. Chem. Soc.*, 1978, **100**, 2268 (*synth*)
Hunting, A.L.L., *Cosmet. Toiletries*, 1981, **96**, 29 (*rev, uses*)
Dalavoy, V.S. *et al*, *J. Sci. Ind. Res.*, 1981, **40**, 520 (*rev*)
Martindale, The Extra Pharmacopoeia, 28th/29th Ed., Pharmaceutical Press, London, 1982/1989, 3012, 3014.
Brauner, A. *et al*, *Org. Mass Spectrom.*, 1985, **20**, 578 (*ms*)
Negwer, M., *Organic-Chemical Drugs and their Synonyms*, 6th Ed., Akademie-Verlag, Berlin, 1987, 2165.
Lewis, R.J., *Sax's Dangerous Properties of Industrial Materials*, 8th Ed., Van Nostrand-Reinhold, 1992, ULS000.

1-Undecen-3-ol U-00070

[35329-42-1]

$H_3C(CH_2)_7CH(OH)CH{=}CH_2$

$C_{11}H_{22}O$ M 170.294

Bp_{10} 112-114°.

Phenylurethane: Cryst. (pet. ether). Mp 44°.

Willems, J.F., *Bull. Soc. Chim. Belg.*, 1955, **64**, 409 (*synth*)
Nii, H. *et al*, *Nippon Kagaku Zasshi*, 1971, **92**, 1214 (*synth*)

2-Undecen-4-ol U-00071

[22381-86-8]

$H_3C(CH_2)_6CH(OH)CH{=}CHCH_3$

$C_{11}H_{22}O$ M 170.294
Bp_{13} 114-115°.

Normant, H., *C. R. Hebd. Seances Acad. Sci.*, 1955, **240**, 314 (*synth*)
Kraft, M. *et al*, *Org. Mass Spectrom.*, 1968, **1**, 617 (*ms*)

4-Undecen-1-ol U-00072

$H_3C(CH_2)_5CH{=}CHCH_2CH_2CH_2OH$

$C_{11}H_{22}O$ M 170.294
(Z)-form [21676-07-3]
Oil.

Wenkert, E. *et al*, *J. Org. Chem.*, 1985, **50**, 719 (*synth, ir, pmr, cmr*)
Davis, T.L. *et al*, *Synthesis*, 1989, 936 (*synth, ir, pmr*)

6-Undecen-2-ol U-00073

Nostrenol

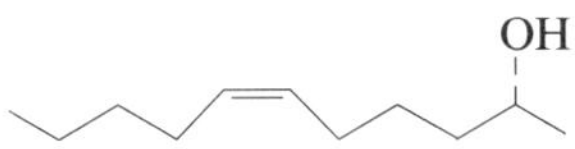

$C_{11}H_{22}O$ M 170.294
(2R,6Z)-form [120926-44-5]
Constit. of the thoracic gland of *Euroleon nostras* and *Grocus bore*.

[67233-89-0, 67233-90-3, 120876-12-2, 120926-45-6]

Baeckstroem, P. *et al*, *J. Chem. Ecol.*, 1989, **15**, 61 (*isol, synth*)

7-Undecen-1-ol, 9CI U-00074

$H_3CCH_2CH_2CH{=}CH(CH_2)_5CH_2OH$

$C_{11}H_{22}O$ M 170.294
(E)-form [52957-13-8]
$Bp_{0.02}$ 75°. n_D^{26} 1.4502.
Ac: [52957-18-3].
(Z)-form [52957-09-2]
$Bp_{0.3}$ 84-86° (94% Z-form).
Ac: [35835-83-7]. *Undec-7-en-1-yl acetate*
$C_{13}H_{24}O_2$ M 212.331
Possible sex attractant for the almond moth, *Cadra cautella*.

Takahashi, S. *et al*, *Bochu Kagaku*, 1971, **36**, 24; *CA*, **76**, 82170m (*activity*)
Ando, T. *et al*, *Agric. Biol. Chem.*, 1977, **41**, 1485 (*synth, activity*)
Svirskaya, P.I. *et al*, *J. Chem. Eng. Data*, 1979, **24**, 152 (*synth, pmr, ir*)
Horiike, M. *et al*, *Agric. Biol. Chem.*, 1980, **44**, 2229 (*synth, pmr, ir*)
Kuwahara, Y. *et al*, *Appl. Entomol. Zool.*, 1985, **20**, 354 (*synth, activity*)
Crisp, G.T. *et al*, *Aust. J. Chem.*, 1988, **41**, 935 (*synth, pmr, cmr, ms, glc*)

9-Undecen-1-ol U-00075

[112-46-9]

$HOCH_2(CH_2)_7CH{=}CHCH_3$

$C_{11}H_{22}O$ M 170.294
(Z)-form [58257-52-6]
Noctuid sex pheromone.
Ac: [56218-82-7].
$C_{13}H_{24}O_2$ M 212.331
Bp_7 127-128°.

Chuit, P. *et al*, *Helv. Chim. Acta*, 1926, **9**, 1085 (*synth*)
Mamoru, S. *et al*, *Nippon Yakurigaku Zasshi*, 1966, **62**, 241.
Bestmann, H.J. *et al*, *Chem. Ber.*, 1975, **108**, 3582 (*synth*)
Priesner, E. *et al*, *Z. Naturforsch., C*, 1975, **30**, 283.

10-Undecen-1-ol U-00076

[112-43-6]

$H_2C{=}CH(CH_2)_8CH_2OH$

$C_{11}H_{22}O$ M 170.294
Pheromone. Liq. Fp −2°. Bp_{14} 133°, Bp_6 123-125°.
▷ YQ2985000.
Ac:
$C_{13}H_{24}O_2$ M 212.331
Bp_7 125-127°.

Bestmann, H.J. *et al*, *Tetrahedron Lett.*, 1977, 121 (*synth*)
Miyaura, N. *et al*, *Tetrahedron*, 1982, **39**, 3271 (*synth*)
Lewis, R.J., *Sax's Dangerous Properties of Industrial Materials*, 8th Ed., Van Nostrand-Reinhold, 1992, UMA000.

10-Undecen-2-ol U-00077

$H_2C{=}CH(CH_2)_7CH(OH)CH_3$

$C_{11}H_{22}O$ M 170.294
Constit. of essential oil of *Litsea odorifera*. Oil. Bp 233°. $[\alpha]_D$ −51.1°.

v. Romburgh, P., *CA*, 1912, **6**, 2977.

1-Undecen-3-one U-00078

[42832-47-3]

$H_3C(CH_2)_7COCH{=}CH_2$

$C_{11}H_{20}O$ M 168.278
Minor constit. of *Dictyopteris plagiogramma*. Unstable oil.

Roller, P. *et al*, *J. Chem. Soc., Chem. Commun.*, 1971, 503 (*isol*)
Stetter, H. *et al*, *Chem. Ber.*, 1978, **111**, 431 (*synth*)

5-Undecen-2-one, 9CI U-00079

[40657-56-5]

$H_3C(CH_2)_4CH{=}CHCH_2CH_2COCH_3$

$C_{11}H_{20}O$ M 168.278
(E)-form [58761-29-8]
Bp_5 91-95°, $Bp_{0.001}$ 39-39.5°.
(Z)-form [21944-96-7]
Sex pheromone from *Damaliscus dorcas dorcas* (bontebok). Bp_5 87-89°.

Teisseire, P. *et al*, *Recherches*, 1956, 30; *CA*, **51**, 2565 (*synth*)
South African Pat., 6 706 415, (1968); *CA*, **70**, 88108.
Burger, V. *et al*, *Z. Naturforsch., C*, 1976, **31**, 21 (*synth*)
Mori, K. *et al*, *Agric. Biol. Chem.*, 1977, **41**, 2295 (*synth*)
Yamashita, M. *et al*, *Bull. Chem. Soc. Jpn.*, 1988, **61**, 3368 (*synth, pmr, bibl*)

5-Undecen-1-yn-3-ol, 9CI U-00080

(R,Z)-form

$C_{11}H_{18}O$ M 166.263

(R,Z)-form [111137-93-0]

Intermed. in the synth. of Leukotriene B_4. Oil. $[\alpha]_D^{20}$ +34° (c, 0.008 in CCl_4) (ee 96%).

Ac: [129030-91-7].

$C_{13}H_{20}O_2$ M 208.300

Oil. $[\alpha]_D^{20}$ +36.3° (0.01 in CCl_4).

(±)-***(Z)-form*** [129030-85-9]

Oil.

Avignon-Tropis, M. *et al*, *J. Org. Chem.*, 1992, **57**, 651 (*synth, ir, pmr*)
Treilhou, M. *et al*, *J. Org. Chem.*, 1992, **57**, 3203 (*synth, ir, pmr*)
Chemin, D. *et al*, *Tetrahedron*, 1992, **48**, 1943.

2-Undecylcyclopropanepentanoic acid, 9CI U-00081

6,7-Methyleneoctadecanoic acid

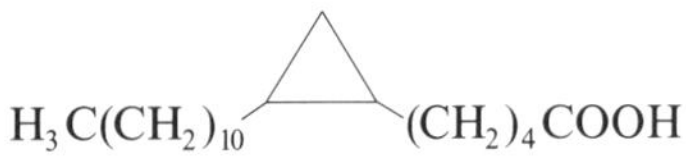

$C_{19}H_{36}O_2$ M 296.492

Both *cis-* and *trans-* forms have been synthesised as Me esters.

Christie, W.W. *et al*, *Chem. Phys. Lipids*, 1968, **2**, 196 (*synth, glc, ir, nmr*)
Christie, W.W., *Top. Lipid Chem.*, 1970, **1**, 4 (*rev*)
Gunstone, F.D. *et al*, *Chem. Phys. Lipids*, 1973, **10**, 303 (*synth, glc, nmr*)

1-Undecyne U-00082

[2243-98-3]

$H_3C(CH_2)_8C{\equiv}CH$

$C_{11}H_{20}$ M 152.279

Mp −17.7°. Bp 194.9°, Bp_{13} 78°.

Asinger, F. *et al*, *Chem. Ber.*, 1964, **97**, 1555 (*synth*)
Villieras, J. *et al*, *Synthesis*, 1975, **7**, 458 (*synth*)
Orav, A. *et al*, *J. Chromatogr.*, 1976, **119**, 451 (*glc*)
Lehtveer, M. *et al*, *CA*, 1978, **89**, 146016 (*ir*)

2-Undecyne U-00083

[60212-29-5]

$H_3C(CH_2)_7C{\equiv}CCH_3$

$C_{11}H_{20}$ M 152.279

Mp −30.1°. Bp 204.2°, Bp_{10} 84.2°.

Asinger, F. *et al*, *Chem. Ber.*, 1964, **97**, 1555 (*synth*)
Orav, A. *et al*, *J. Chromatogr.*, 1976, **119**, 451 (*glc*)
Lehtveer, M. *et al*, *CA*, 1978, **89**, 146016 (*ir*)

5-Undecyne U-00084

[2294-72-6]

$H_3C(CH_2)_3C{\equiv}C(CH_2)_4CH_3$

$C_{11}H_{20}$ M 152.279

Mp −74.1°. Bp 198.1°, Bp_{10} 78.3°.

Asinger, F. *et al*, *Chem. Ber.*, 1964, **97**, 1555 (*synth*)
Brattesani, D.N. *et al*, *Synth. Commun.*, 1973, **3**, 245 (*synth*)
Orav, A. *et al*, *J. Chromatogr.*, 1976, **119**, 451 (*glc*)
Lehtveer, M. *et al*, *CA*, 1978, **89**, 146016 (*ir*)

2-Undecynoic acid U-00085

[54299-08-0]

$H_3C(CH_2)_7C{\equiv}CCOOH$

$C_{11}H_{18}O_2$ M 182.262

$Bp_{0.03}$ 103°.

Me ester: [10522-18-6].

$C_{12}H_{20}O_2$ M 196.289

Liq. $Bp_{0.2}$ 116-120°.

Et ester:

$C_{13}H_{22}O_2$ M 210.316

Bp_{10} 128°.

S-*Benzylisothiouronium salt:* Mp 174-175°.

Barker, S.A. *et al*, *Tetrahedron*, 1962, **18**, 177 (*synth*)
Lie Ken Jie, M.S.F. *et al*, *J. Chromatogr.*, 1974, **97**, 165 (*synth*)
Davies, J.E.D. *et al*, *Chem. Phys. Lipids*, 1975, **15**, 157 (*ir, uv*)
Bus, J.D. *et al*, *Chem. Phys. Lipids*, 1976, **17**, 501 (*cmr*)

3-Undecynoic acid, 9CI U-00086

[54299-10-4]

$H_3C(CH_2)_6C{\equiv}CCH_2COOH$

$C_{11}H_{18}O_2$ M 182.262

Mp 41-42.5°.

Gilman, N.W. *et al*, *Chem. Phys. Lipids*, 1974, **13**, 239 (*synth*)
Lie Ken Jie, M.S.F. *et al*, *J. Chromatogr.*, 1974, **97**, 165 (*synth*)
Davies, J.E.D. *et al*, *Chem. Phys. Lipids*, 1975, **15**, 157 (*ir, raman*)
Bus, J.D. *et al*, *Chem. Phys. Lipids*, 1976, **17**, 501 (*cmr*)

4-Undecynoic acid, 9CI U-00087

[54299-11-5]

$H_3C(CH_2)_5C{\equiv}CCH_2CH_2COOH$

$C_{11}H_{18}O_2$ M 182.262

Mp 38.5-40.5°.

Gilman, N.W. *et al*, *Chem. Phys. Lipids*, 1974, **13**, 239 (*synth*)
Lie Ken Jie, M.S.F. *et al*, *J. Chromatogr.*, 1974, **97**, 165 (*synth*)
Davies, J.E.D. *et al*, *Chem. Phys. Lipids*, 1975, **15**, 157 (*ir, raman*)
Gunstone, F.D. *et al*, *Chem. Phys. Lipids*, 1976, **17**, 1 (*cmr*)
Bus, J.D. *et al*, *Chem. Phys. Lipids*, 1976, **17**, 501 (*cmr*)

5-Undecynoic acid, 9CI U-00088

[822-02-6]

$H_3C(CH_2)_4C{\equiv}C(CH_2)_3COOH$

$C_{11}H_{18}O_2$ M 182.262

$Bp_{0.15}$ 118-122°.

Gilman, N.W. *et al*, *Chem. Phys. Lipids*, 1974, **13**, 239 (*synth*)
Lie Ken Jie, M.S.F. *et al*, *J. Chromatogr.*, 1974, **97**, 165 (*synth*)
Davies, J.E.D. *et al*, *Chem. Phys. Lipids*, 1975, **15**, 157 (*ir, raman*)
Gunstone, F.D. *et al*, *Chem. Phys. Lipids*, 1976, **17**, 1 (*cmr*)
Bus, J.D. *et al*, *Chem. Phys. Lipids*, 1976, **17**, 501 (*cmr*)

6-Undecynoic acid U-00089

[55182-83-7]

$H_3C(CH_2)_3C{\equiv}C(CH_2)_4COOH$

$C_{11}H_{18}O_2$ M 182.262

$Bp_{0.17}$ 124-125°.

Me ester: [54299-00-2].

$C_{12}H_{20}O_2$ M 196.289

$Bp_{0.7}$ 97-97.5°.

Lie Ken Jie, M.S.F. *et al*, *J. Chromatogr.*, 1974, **97**, 165 (*synth*)
Davies, J.E.D. *et al*, *Chem. Phys. Lipids*, 1975, **15**, 157 (*ir, uv*)
Gunstone, F.D. *et al*, *Chem. Phys. Lipids*, 1976, **17**, 1 (*cmr*)
Bus, J.D. *et al*, *Chem. Phys. Lipids*, 1976, **17**, 501 (*cmr*)

7-Undecynoic acid, 9CI U-00090

[54373-86-3]

$H_3CCH_2CH_2C{\equiv}C(CH_2)_5COOH$

$C_{11}H_{18}O_2$ M 182.262

$Bp_{0.5}$ 127-135°.

Me ester: [54299-01-3].

$Bp_{0.8}$ 98-99°.

Gilman, N.W. *et al*, *Chem. Phys. Lipids*, 1974, **97**, 239 (*synth*)
Gilman, N.W. *et al*, *Synth. Commun.*, 1974, **4**, 199 (*synth*)
Lie Ken Jie, M.S.F. *et al*, *J. Chromatogr.*, 1974, **97**, 165 (*synth*)
Davies, J.E.D. *et al*, *Chem. Phys. Lipids*, 1975, **15**, 157 (*ir, raman*)
Bus, J.D. *et al*, *Chem. Phys. Lipids*, 1976, **17**, 501 (*cmr*)

8-Undecynoic acid, 9CI U-00091

[55182-87-1]

$H_3CCH_2C{\equiv}C(CH_2)_6COOH$

$C_{11}H_{18}O_2$ M 182.262

$Bp_{0.25}$ 120-123°.

Me ester: [54299-02-4].

Bp_1 104-105°.

Gilman, N.W. *et al*, *Chem. Phys. Lipids*, 1974, **97**, 239 (*synth*)
Lie Ken Jie, M.F.S. *et al*, *J. Chromatogr.*, 1974, **97**, 165 (*synth*)
Davies, J.E.D. *et al*, *Chem. Phys. Lipids*, 1975, **15**, 157 (*ir, raman*)
Gunstone, F.D. *et al*, *Chem. Phys. Lipids*, 1976, **17**, 1 (*cmr*)
Bus, J.D. *et al*, *Chem. Phys. Lipids*, 1976, **17**, 501 (*cmr*)

9-Undecynoic acid U-00092

[22202-65-9]

$H_3CC{\equiv}C(CH_2)_7COOH$

$C_{11}H_{18}O_2$ M 182.262

Cryst. (EtOH aq.). Mp 61°, Mp 58°. $Bp_{0.05}$ 120-130°.

Me ester: [18937-76-3].

$C_{12}H_{20}O_2$ M 196.289

Bp_1 93-94°.

Et ester:

$C_{13}H_{22}O_2$ M 210.316

Bp_{49} 197°.

Lie Ken Jie, M.S.F. *et al*, *J. Chromatogr.*, 1974, **97**, 165 (*synth*)
Davies, J.E.D. *et al*, *Chem. Phys. Lipids*, 1975, **15**, 157 (*ir, uv*)
Bus, J.D. *et al*, *Chem. Phys. Lipids*, 1976, **17**, 501 (*cmr*)

10-Undecynoic acid U-00093

[2777-65-3]

$HC{\equiv}C(CH_2)_8COOH$

$C_{11}H_{18}O_2$ M 182.262

Metabolite of *Rhodotorula glutinis*. Cryst. (pet. ether). Mp 43°. Bp 116-120°.

Me ester: [2777-66-4].

$C_{12}H_{20}O_2$ M 196.289

Liq. Bp_{15} 121°.

Et ester:

$C_{13}H_{22}O_2$ M 210.316

$Bp_{6.5}$ 134.5°.

Hydrazide:

$C_{11}H_{20}N_2O$ M 196.292

Cryst. (EtOH). Mp 84-85°.

Vogel, J., *J. Chem. Soc.*, 1948, 677.
Org. Synth., Coll. Vol., 4, 1963, 969 (*synth*)
Lie Ken Jie, M.S.F. *et al*, *J. Chromatogr.*, 1974, **97**, 165 (*chromatog*)
Davies, J.E.D. *et al*, *Chem. Phys. Lipids*, 1975, **15**, 157 (*ir, raman*)
Bus, J.D. *et al*, *Chem. Phys. Lipids*, 1976, **17**, 501 (*cmr*)
Vinczer, P. *et al*, *Org. Prep. Proced. Int.*, 1989, **21**, 232 (***synth***, *pmr*)

2-Undecyn-1-ol, 9CI U-00094

[34683-71-1]

$H_3C(CH_2)_7C{\equiv}CCH_2OH$

$C_{11}H_{20}O$ M 168.278

$Bp_{0.03}$ 97-103°.

Christie, W.W. *et al*, *Chem. Phys. Lipids*, 1967, **1**, 407 (*synth*)
Castuma, J.C., *An. Asoc. Quim. Argent.*, 1972, **60**, 55 (*synth*)
Ermilova, E.V., *Zh. Org. Khim.*, 1975, **11**, 520 (*synth*)

3-Undecyn-2-ol U-00095

[124126-36-9]

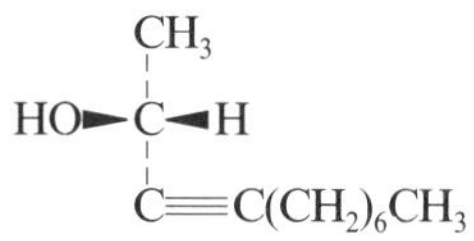

$C_{11}H_{20}O$ M 168.278

***(R)*-form** [124126-37-0]

$[\alpha]_D$ +18.7° (c, 2.37 in $CHCl_3$) (e.e. 92%).

(±)-form [131101-36-5]

Liq.

Marshall, J.A. *et al*, *J. Org. Chem.*, 1991, **56**, 960, 4913 (*synth, ir, pmr, ms*)

6-Undecyn-1-ol U-00096

[69222-10-2]

$H_3C(CH_2)_3C{\equiv}C(CH_2)_4CH_2OH$

$C_{11}H_{20}O$ M 168.278

Liq. Bp_2 108-110°, $Bp_{0.03}$ 90-92°.

Svirskaya, P.I. *et al*, *J. Chem. Eng. Data*, 1979, **24**, 152 (*synth*)
Brown, H.C. *et al*, *J. Org. Chem.*, 1986, **51**, 4518 (*synth, ir, pmr, cmr*)

10-Undecyn-1-ol U-00097

[2774-84-7]

$HC{\equiv}C(CH_2)_8CH_2OH$

$C_{11}H_{20}O$ M 168.278

Viscous oil. Bp_{16} 138°.

Miyaura, N. *et al*, *Tetrahedron*, 1983, **39**, 3271 (*synth, ir, pmr*)

1-Undecyn-5-one U-00098

[72971-55-2]

$H_3C(CH_2)_5COCH_2CH_2C{\equiv}CH$

$C_{11}H_{18}O$ M 166.263

Oil. Bp_3 98-100°, $Bp_{0.6}$ 71°.

Subramaniam, C.S. *et al*, *J. Chem. Soc., Perkin Trans. 1*, 1979, 2346 (*synth, ir*)
Haruta, J. *et al*, *J. Org. Chem.*, 1990, **55**, 4853 (*synth, ir, pmr*)

12-Ursen-3-ol U-00099

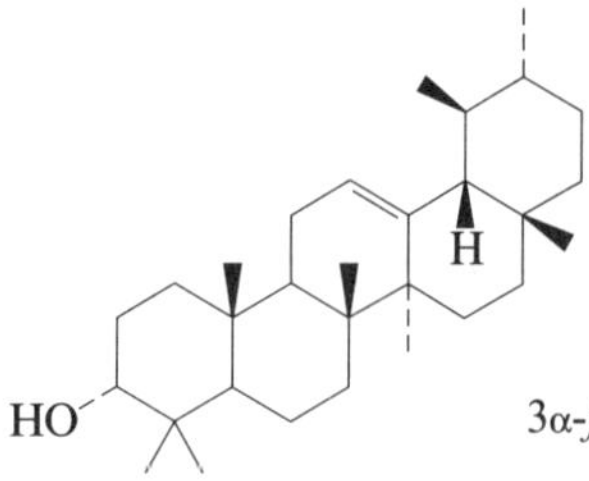

3α-*form*

$C_{30}H_{50}O$ M 426.724

3α-form [5937-48-4] ***Epi-α-amyrin***

Isol. from leaves of *Rosmarinus officinalis* and *Dacryodes edulis*. Mp 110-111°, Mp 138°.

3β-form [638-95-9] ***α-Amyrin***

Constit. of many spp. including *Artemisia vulgaris* and *Lactuca denticulata*. Very widely distributed. Cryst. (EtOH). Mp 186°. $[\alpha]_D^{17}$ +83.5° ($CHCl_3$).

Ac: [863-76-3].

$C_{32}H_{52}O_2$ M 468.762

Constit. of many spp. including *A. vulgaris*. Cryst. (pet. ether). Mp 225-226°. $[\alpha]_D^{18}$ +77° (C_6H_6).

O-[*α-L-Rhamnopyranosyl-(1→5)-β-D-xylofuranoside*]: [72782-80-0].

$C_{41}H_{68}O_9$ M 704.983

Constit. of the stem bark of *Wisteria sinensis*. Light brown solid. Mp 169-171°.

Octanoyl: [22000-04-0]. *α-Amyrin caprylate*

$C_{38}H_{64}O_2$ M 552.922

Isol. from *Mimusops manilkara*. Silky needles. Mp 155-160°. $[\alpha]_D^{34}$ +76.5° (c, 1 in $CHCl_3$).

Dodecanoyl: [55803-93-5].

$C_{42}H_{72}O_2$ M 609.030

Isol. from *Spilanthes acmella*.

Tetradecanoyl:

$C_{44}H_{76}O_2$ M 637.083

From *S. acmella*.

Hexadecanoyl: [22255-10-3]. *α-Amyrin palmitate*

$C_{46}H_{80}O_2$ M 665.137

Isol. from *Viburnum* spp., *Santalum album*, *Sambucus* spp. and others. Prisms or needles (EtOH/Et_2O). Mp 73-74°. $[\alpha]_D^{23}$ +52.5° (c, 1.0 in C_6H_6).

Octadecanoyl: [63195-78-8].

$C_{48}H_{84}O_2$ M 693.190

Isol. from *Saussurea lappa*.

Cinnamoyl: [13161-35-8]. *α-Amyrin cinnamate*

$C_{39}H_{56}O_2$ M 556.870

Isol. from *Hoya*, *Marsdenia*, *Mimusops*, *Himatanthus* etc. spp. Mp 179°. $[\alpha]_D^{34}$ +79° (68°).

O-(*3-Methylbutanoyl*): [93722-35-1].

$C_{35}H_{58}O_2$ M 510.842

Isol. from latex of *Calotropis gigantea*.

Me ether: [14021-28-4]. ***3β-Methoxy-12-ursene***

$C_{31}H_{52}O$ M 440.751

Constit. of *Cortaderia toetoe* (*Arundo conspicua*) and other spp. Cryst. Mp 221-223°.

3-Ketone: [638-96-0]. *12-Ursen-3-one*. ***α-Amyrone.*** *α-Amyrenone*

$C_{30}H_{48}O$ M 424.709

Found in Deertongue leaf and *Inertia hypoleuca*, *Quercus*, *Sambucus nigra* and others. Cryst. (MeOH). Mp 125-126°. $[\alpha]_D$ +106.5° (c, 1.6 in $CHCl_3$).

Shannon, J.S., *Aust. J. Chem.*, 1963, **16**, 683 (*ms*)
Witz, P. *et al*, *Bull. Soc. Chim. Fr.*, 1963, 1101 (*cd, ketone*)
Laurie, W. *et al*, *Phytochemistry*, 1964, **3**, 267 (*isol, ketone*)
Brieskorn, C.H. *et al*, *Arch. Pharm.* (*Weinheim, Ger.*), 1966, **299**, 663 (*isol, Epiamyrin*)
Bryce, T.A. *et al*, *Phytochemistry*, 1967, **6**, 727 (*isol*)
Misra, G. *et al*, *Phytochemistry*, 1968, **7**, 2173; 1969, **8**, 249 (*cinnamate, caprylate*)
Inoue, T. *et al*, *Chem. Pharm. Bull.*, 1969, **17**, 124 (*palmitate*)
Wahlberg, I. *et al*, *Acta Chem. Scand.*, 1972, **26**, 1383 (*isol*)
v. Tamelen, E.E. *et al*, *J. Am. Chem. Soc.*, 1972, **94**, 8229 (*synth*)
Lou-Cam, C.A., *Phytochemistry*, 1973, **12**, 475 (*isol*)
Knight, S.A. *et al*, *Org. Magn. Reson.*, 1974, **6**, 603 (*cmr*)
Romeo, G. *et al*, *Org. Magn. Reson.*, 1977, **9**, 29 (*pmr*)
Tiwari, K.P. *et al*, *Indian J. Chem., Sect. B*, 1979, **18**, 389 (*glycoside*)
Thakur, S. *et al*, *Phytochemistry*, 1984, **23**, 2085 (*3-methylbutanoyl*)

V

Vitamin K_1 V-00001

2-Methyl-3-(3,7,11,15-tetramethyl-2-hexadecenyl)-1,4-naphthalenedione, 9CI. 2-Methyl-3-phytyl-1,4-naphthoquinone. ***Phytomenadione,*** **BAN, INN.** *Phytonadione, USAN. Phylloquinone. Numerous proprietary names*

[84-80-0]

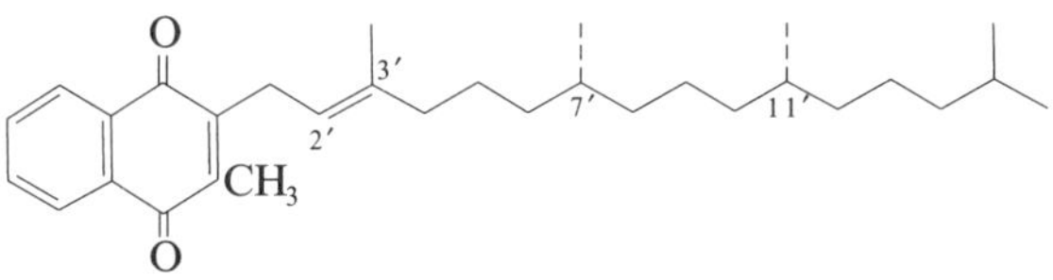

$C_{31}H_{46}O_2$ M 450.703

Fat sol. dietary factor essential for blood coagulation, widely distributed in green leaves and vegetables, esp. cabbage and spinach. Anticoagulant. Prothrombogenic vitamin. Pale-yellow oil, yellow cryst. (Me_2CO at −20°). Mp −4°. All 4 stereoisomers have been synth.

▷ Local effects in humans following intramuscular or subcataneous administration. LD_{50} (mus, orl) 25000 mg/kg. QJ5800000.

[16033-41-3, 132487-93-5, 132487-94-6, 132487-95-7]

Fieser, L.F., *J. Am. Chem. Soc.*, 1939, **61**, 3467.
Doisy, E.A. *et al*, *Chem. Rev.*, 1941, **28**, 477 (*rev, bibl*)
Jacobsen, B.K. *et al*, *Biochim. Biophys. Acta*, 1960, **40**, 211 (*isol, purifn*)
Mayer, H. *et al*, *Helv. Chim. Acta*, 1964, **47**, 221 (*abs config*)
Jackman, L.M. *et al*, *Helv. Chim. Acta*, 1965, **48**, 1332 (*config*)
Barkhan, P., *Proc. R. Soc. Med.*, 1977, **70**, 93 (*rev, metab*)
Chenard, B.L. *et al*, *J. Org. Chem.*, 1980, **45**, 378 (*synth*)
Naruta, Y., *J. Org. Chem.*, 1980, **45**, 4097 (*synth*)
Doetz, K.H. *et al*, *J. Organomet. Chem.*, 1981, **209**, C4 (*synth*)
Schmid, M. *et al*, *Helv. Chim. Acta*, 1982, **65**, 684 (*synth*)
Martindale, The Extra Pharmacopoeia, 28th/29th Ed., Pharmaceutical Press, London, 1982/1989, 7902, 7903.
Bentley, R. *et al*, *J. Nat. Prod. (Lloydia)*, 1983, **46**, 44 (*biosynth*)
Masaki, Y. *et al*, *Chem. Pharm. Bull.*, 1984, **32**, 3959 (*synth*)
Negwer, M., *Organic-Chemical Drugs and their Synonyms*, 6th Ed., Akademie-Verlag, Berlin, 1987, 7926.
Hassan, M.M.A. *et al*, *Anal. Profiles Drug Subst.*, 1988, **17**, 449 (*rev*)
Schmid, R. *et al*, *HCA*, 1990, **73**, 1276 (*synth, ord hplc bibl*)
Schmid, R. *et al*, *Helv. Chim. Acta*, 1990, **73**, 1276 (*synth, ord, hplc, bibl*)
Martindale, The Extra Pharmacopoeia, 30th edn., Pharmaceutical Press, London, 1993, 7902, 7903.
Lewis, R.J., *Sax's Dangerous Properties of Industrial Materials*, 8th Ed., Van Nostrand-Reinhold, 1992, VTA000.

Vitamin K_2 V-00002

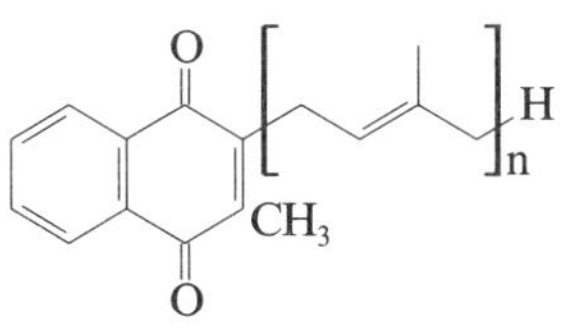

Fat sol. dietary factor effective in controlling blood coagulation. Widely distributed in green leaves and vegetables. Consists of a mixt. of homologues. The figure in brackets after the name indicates the number of carbon atoms in the sidechain. The name vitamin K_2 is sometimes applied to vitamin K_2(30). Various partially hydrogenated derivatives Dihydromenaquinones, Tetrahydromenaquinones, Hexahydromenaquinone 9 and Octahydromenaquinone 9 with partly saturated side-chains occur, e.g. in cultures of *Oerskovia turbata, Breisbacterium lipopolyticum* and *Actinomadura madurae.*

Vitamin K₂(30) [84-81-1]

Menaquinone-6. Menaquinone K_6. Farnoquinone. β-Phylloquinone

$C_{41}H_{56}O_2$ M 580.892

Minor constit. of vitamin K_2 complex. Yellow cryst. (Me_2CO/EtOH or pet. ether). Mp 50°.

Vitamin K_2(35) [2124-57-4]

Menaquinone-7. Menaquinone K_7

$C_{46}H_{64}O_2$ M 649.010

Constit. of chestnut leaves. Cryst. (Me_2CO/EtOH or MeOH/$CHCl_3$). Mp 54°.

Vitamin K_2(40) [523-38-6]

Menaquinone-8. Menaquinone K_8

$C_{51}H_{72}O_2$ M 717.129

Present in *E. coli, Proteus vulgaris* and *Chromatium vinosum.*

Vitamin K_2(45) [523-39-7]

Menaquinone-9. Menaquinone K_9

$C_{56}H_{80}O_2$ M 785.247

Constit. of lipids of *Mycobacterium tuberculosis.* Cryst. Mp 58-59°.

Vitamin K_2(50) [523-40-0]

Menaquinone-10. Menaquinone K_{10}

$C_{61}H_{88}O_2$ M 853.365

Cryst. Mp 62°.

Isler, O. *et al*, *Helv. Chim. Acta*, 1958, **41**, 786.
Isler, O. *et al*, *Vitam. Horm. (N.Y.)*, 1959, **17**, 53 (*rev*)
Noll, H. *et al*, *J. Biol. Chem.*, 1960, **235**, 2207; *Helv. Chim. Acta*, 1960, **43**, 433 (*isol*)
Osnitskaya, L.K. *et al*, *Nature (London)*, 1964, **204**, 80.
Sato, K. *et al*, *J. Chem. Soc., Perkin Trans. 1*, 1973, 2289 (*synth*)
Snyder, C.D. *et al*, *J. Am. Chem. Soc.*, 1974, **96**, 8046 (*synth*)
Yamada, Y. *et al*, *Biochim. Biophys. Acta*, 1977, **486**, 195 (*isol, derivs*)
Yamada, Y. *et al*, *Agric. Biol. Chem.*, 1980, **44**, 1693; 1982, **46**, 1105 (*isol, derivs*)

W

Wyerone acid W-00001

3-[5-(1-Oxo-4-hepten-2-ynyl)-2-furanyl]-2-propenoic acid, 9CI. 4,7-Epoxy-8-oxo-2,4,6,11-tetradecatetraen-9-ynoic acid

[54954-14-2]

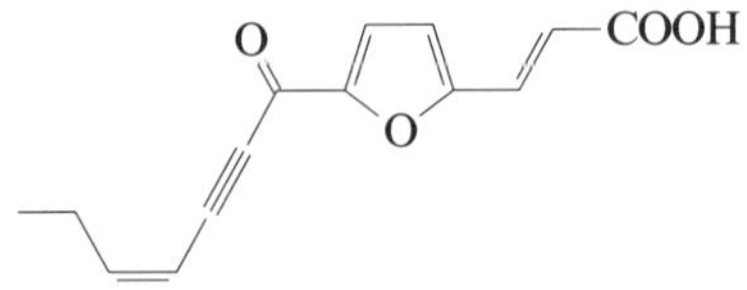

$C_{14}H_{12}O_4$ M 244.246

(2E,11Z)-form

Isol. from broad beans (*Vicia faba*) infected with *Botrytis* spp.

Me ester: [20079-30-5]. ***Wyerone***

$C_{15}H_{14}O_4$ M 258.273

Constit. of *V. faba*. Fungitoxic. Cryst. (hexane or cyclohexane). Mp 63.5-64°.

11,12-Dihydro: [70711-57-8]. ***Dihydrowyerone acid***

$C_{14}H_{14}O_4$ M 246.262

Constit. of *Vicia faba*. Phytoalexin.

11,12-Dihydro, Me ester: ***Dihydrowyerone***

$C_{15}H_{16}O_4$ M 260.289

Minor constit. of *V. faba*. Small prisms. Mp 79-80°.

8-Alcohol, Me ester: [20450-52-6]. ***Wyerol.*** *Methyl 3-[5-(1-hydroxy-4-hepten-2-ynyl)-2-furanyl]-2-propenoate*

$C_{15}H_{16}O_4$ M 260.289

Isol. from *V. faba*. Phytoalexin. Needles (hexane). Mp 55-56°.

8-Alcohol, 11,12-dihydro, Me ester: [70711-58-9]. ***Dlhydrowyerol***

$C_{15}H_{18}O_4$ M 262.305

Constit. of *V. faba*. Phytoalexin.

11,12-Epoxide, Me ester: [60375-16-8]. ***Wyerone epoxide.*** *3-[5-[3-(3-ethyloxiranyl)-1-oxo-2-propynyl]-2-furanyl]-2-propenoic acid methyl ester, 9CI*

$C_{15}H_{14}O_5$ M 274.273

Constit. of *V. faba*. Cryst. (Et_2O/hexane). Mp 74-76°.

Fawcett, C.H. *et al*, *J. Chem. Soc. C*, 1968, 2455 (*isol, struct, alcohol*)

Letcher, R.M. *et al*, *Phytochemistry*, 1970, **9**, 249 (*isol*)

Mansfield, J.W. *et al*, *Nature (London)*, 1974, **252**, 316 (*isol*)

Hargreaves, J.A. *et al*, *Ann. Appl. Biol.*, 1975, **81**, 271 (*isol, epoxide*)

Hearn, M.T.W., *Aust. J. Chem.*, 1976, **29**, 107 (*cmr*)

Hargreaves, J.A. *et al*, *Phytochemistry*, 1976, **15**, 651, 1119 (*isol, epoxide*)

Knight, D.W. *et al*, *J. Chem. Soc., Perkin Trans. 1*, 1982, 623 (*synth, alcohol*)

COMPOUND NAME INDEX

COMPOUND NAME INDEX

FORMULA INDEX

FORMULA INDEX

CHEMICAL ABSTRACTS SERVICE
REGISTRY NUMBER INDEX

CHEMICAL ABSTRACTS SERVICE REGISTRY NUMBER INDEX